LANDOLT-BÖRNSTEIN

Numerical Data and Functional Relationships in Science and Technology

New Series

Editor in Chief: K.-H. Hellwege

Group III: Crystal and Solid State Physics

Volume 16

Revised, Updated and Extended Edition of Volumes III/3 and III/9

Ferroelectrics and Related Substances

Subvolume b: **Non-oxides**

T. Mitsui · E. Nakamura

K. Gesi · T. Ikeda · Y. Makita · M. Marutake · S. Nomura
E. Sawaguchi · T. Shigenari · Y. Shiozaki · I. Tatsuzaki · K. Toyoda
and
M. Adachi · J. Harada · T. Shiosaki · K. Wakino · T. Yamada

Editors: K.-H. Hellwege and A. M. Hellwege

Springer-Verlag Berlin · Heidelberg · New York 1982

LANDOLT-BÖRNSTEIN

Zahlenwerte und Funktionen aus Naturwissenschaften und Technik

Neue Serie

Gesamtherausgabe: K.-H. Hellwege

Gruppe III: Kristall- und Festkörperphysik

Band 16

Neubearbeitung und Erweiterung der Bände III/3 und III/9

Ferroelektrika und verwandte Substanzen

Teilband b: **Nicht-Oxide**

T. Mitsui · E. Nakamura

K. Gesi · T. Ikeda · Y. Makita · M. Marutake · S. Nomura
E. Sawaguchi · T. Shigenari · Y. Shiozaki · I. Tatsuzaki · K. Toyoda
und
M. Adachi · J. Harada · T. Shiosaki · K. Wakino · T. Yamada

Herausgeber: K.-H. Hellwege und A. M. Hellwege

Springer-Verlag Berlin · Heidelberg · New York 1982

CIP-Kurztitelaufnahme der Deutschen Bibliothek: *Zahlenwerte und Funktionen aus Naturwissenschaften und Technik*
Landolt-Börnstein. — Berlin, Heidelberg, New York: Springer. Parallelt.: Numerical data and functional relationships
in science and technology.
NE: Landolt Hans [Begr.]; PT N.S./Gesamthrsg.: K.-H. Hellwege. N.S., Gruppe 3, Kristall- und Festkörperphysik. N.S., Gruppe 3, Bd. 16.
Ferroelektrika und verwandte Substanzen/Hrsg.: K.-H. Hellwege u. A.M. Hellwege. N.S., Gruppe 3, Bd. 16 Teilbd. b. Nicht-Oxide/
T. Mitsui... — 1982

ISBN 3-540-10484-4 (Berlin, Heidelberg, New York)
ISBN 0-387-10484-4 (New York, Heidelberg, Berlin)

NE: Hellwege, Karl-Heinz [Hrsg.]; Mitsui, Toshio [Mitverf.]

Printed in Germany

Typesetting, printing and bookbinding: Brühlsche Universitätsdruckerei, 6300 Giessen
2163/3020—543210

Contributors — Mitarbeiter

Adachi, Masatoshi, Kyoto University, Kyoto, Japan
Pyroelectricity, index of substances

Gesi, Kazuo, Japan Atomic Energy Research Institute, Tokai-mura, Ibaraki-ken, Japan
Dielectric properties, inelastic neutron scattering

Harada, Jimpei, Nagoya University, Nagoya, Japan
Structure of oxide crystals, inelastic neutron scattering

Ikeda, Takuro, Tohoku University, Sendai, Japan
Solid solutions, electromechanical and elastic properties

Makita, Yasuharu, Science University of Tokyo, Shinjuku-ku, Tokyo, Japan
Thermal properties, thermal expansion

Marutake, Masakazu, University of Electro-Communications, Chofu, Tokyo, Japan
Electromechanical and elastic properties

Mitsui, Toshio, Osaka University, Toyonada, Osaka, Japan
Introduction, systematization

Nakamura, Eiji, Hiroshima University, Hiroshima, Japan
Dielectric dispersion, dielectric constants in infrared region, paraelectric resonance

Nomura, Shōichiro, Tokyo Institute of Technology, Meguro-ku, Tokyo, Japan
Perovskite-type oxides of complex composition, magnetic properties, Mössbauer study

Sawaguchi, Etsuro, Hokkaido University, Sapporo, Japan
Electrical transport phenomena (semiconductive and photoconductive properties, etc.)

Shigenari, Takeshi, University of Electro-Communications, Chofu, Tokyo, Japan
Raman scattering, infrared spectroscopy

Shiosaki, Tadashi, Kyoto University, Kyoto, Japan
Thin film, acoustic surface wave

Shiozaki, Yōichi, Hokkaido University, Sapporo, Japan
Structures of crystals, diffuse X-ray scattering, thermal expansion, references

Tatsuzaki, Itaru, Hokkaido University, Sapporo, Japan
Ultrasonic properties, Brillouin scattering, Rayleigh scattering, properties studied by means of NMR, NQR, ESR and ENDOR

Toyoda, Kōichi, Shizuoka University, Hamamatsu, Japan
Crystal growth, optical properties in ultraviolet region, domain structure, dynamical properties of domains, surface layer, radiation damage (including ESR), references

Wakino, Kikuo, Murata Manufacturing Co., Ltd, Nagaokakyo, Kyoto, Japan
Pyroelectricity

Yamada, Tomoaki, The Electrical Communication Laboratories, Musashino, Tokyo, Japan
Linear and nonlinear optical properties, electromechanical properties

Vorwort

Die wachsende Zahl der Veröffentlichungen mit wichtigen Daten über Ferroelektrika ließ eine vollständig neue und erweiterte Daten-Zusammenstellung sinnvoll erscheinen.

Dieser neue Band, III/16 enthält überprüfte, auf den neuesten Stand gebrachte und erweiterte Informationen über ferroelektrische, antiferroelektrische und eng mit diesen verwandte Substanzen, und zwar so vollständig wie möglich und so ausführlich wie nötig. Alle zuverlässigen Daten über reine Verbindungen und Mischkristalle aus der Zeit zwischen 1921 und 1980 wurden kritisch ausgewertet und aufgenommen. In den Figuren und Tabellen werden das dielektrische und ferroelektrische Verhalten, aber auch alle anderen charakteristischen Eigenschaften dieser Substanzen dargestellt.

Anordnung und Abgrenzung des Materials sind dieselben wie in den früheren Bänden III/3 und III/9. Diese bleiben eine nützliche Quelle für die Verfolgung der historischen Entwicklung und für zusätzliche ältere Daten.

Alle Werte dieses Bandes sind in SI-Einheiten angegeben. Eine Tabelle der Umrechnungsfaktoren zwischen traditionellen und SI-Einheiten wurde der Einleitung angefügt.

Aus technischen Gründen wurde der neue Band in zwei Teilbände aufgeteilt:
III/16a, schon 1980 erschienen, behandelt die Oxide;
III/16b, hiermit vorgelegt, enthält die „Nicht-Oxide", d. h. die restlichen anorganischen Substanzen, organische Kristalle, und einige flüssige Kristalle, Polymere, sowie Kristalle, deren ferroelektrisches Verhalten unsicher ist.

Wir sind überzeugt, daß diese neue Datensammlung von großem Wert für mehr als ein Gebiet der Festkörperforschung und -technologie sein wird.

Unser besonderer Dank gilt den Autoren für ihre hervorragend organisierte, schnelle, gründliche und kompetente Arbeit, sowie für ihre enge Mitwirkung während der herausgeberischen Bearbeitung der Manuskripte.

Zu danken ist auch wieder der Landolt-Börnstein-Redaktion, besonders Herrn Dr. W. Polzin für die redaktionelle Betreuung dieses Teilbandes, sowie Frau B. Ade und Frau R. Schattschneider für die zuverlässige Überprüfung der Manuskripte und der Korrekturfahnen.

Dem Springer-Verlag danken wir für seine verständnisvolle Geduld und die gewohnte Sorgfalt bei der Buchherstellung.

Wie alle anderen Landolt-Börnstein-Bände wird auch dieser Teilband ohne Zuschüsse von anderer Seite publiziert.

Darmstadt, November 1981 **Die Herausgeber**

Preface

The growing number of publications on ferroelectrics with their increasing amount of valuable data requires the publication of a complete new compilation.

This new volume, III/16, contains revised, updated and extended information on ferroelectric, antiferroelectric and closely related substances. The compilation is as complete as possible and as detailed as reasonable. All reliable data on both pure compounds and solid solutions published between 1921 and 1980 are critically evaluated and included. As well as the dielectric and ferroelectric behaviour, all other properties relevant to the characterization of these substances are presented in tables and figures.

The arrangement and scope of information is the same as in the previous volumes III/3 and III/9. These will still be useful to follow the lines of historical development, and as sources of additional data.

All values in this volume are given in SI units. A table of conversion factors between the traditional and the SI units is added to the introduction.

For technical reasons, the new volume has been divided into two subvolumes:
III/16a: already published in 1980, deals with the oxides, and
III/16b: presented herewith, covers the non-oxides, i.e. the inorganic crystals other than oxides, the organic crystals, some liquid crystals, polymers, and crystals of uncertain ferroelectric behaviour.

We are convinced that this new compilation will be of great value for various fields of solid state research and technology.

We wish to thank all authors for their excellently organized, quick, thorough and competent work and for their fine cooperation during the editorial discussion of the manuscripts.

Thanks are also due to the editorial staff of Landolt-Börnstein, especially to Dr. W. Polzin who was in charge of the editorial preparation of this subvolume and to Frau B. Ade and Frau R. Schattschneider for their careful checking of the manuscripts and galleys.

We are also grateful to Springer-Verlag for their patient care and understanding help in the final preparation.

This subvolume, as all volumes of Landolt-Börnstein, has been published without any support from other sources.

Darmstadt, November 1981 **The Editors**

Table of contents

Ferroelectrics and related substances

(Following the title of each chapter, the name of the author responsible for the manuscript preparation of the chapter is given)

Subvolume b: Non-oxides

Table of contents

Subvolume a: Oxides

Ferroelectrics and related substances

I Introduction

IA General remarks

This volume III/16 consisting of two parts a and b is a revised, updated and extended edition of Landolt-Börnstein, New Series, Group III, Volume 3 "Ferro- and antiferroelectric substances" [69m1] and its supplement, Volume 9 [75m2]. It was designed for the use independent of [69m1] and [75m2]*). Materials treated in this volume are ferroelectric and antiferroelectric substances (including solid solutions) as well as substances closely related to them. This subvolume b is substantially for non-oxides but Chapter 43 (Miscellaneous crystals and polymers) includes some oxides. The data included are those which give information on the ferroelectric or antiferroelectric character of the substances and are important in connection with it. Figures and tables are presented only when they seemed to be reliable. In some cases, references are given without presenting numerical data for user's convenience. The compilation of data has been made on the basis of a bibliography prepared by Toyoda [79t1], which includes about 20,000 references to literature on ferroelectrics and related materials from 1921, when Valasek found the ferroelectric effect in Rochelle salt, till about June, 1978. More recent data, published up to early 1980 have been taken into account.

The definition of ferroelectrics and antiferroelectrics adopted in this volume is given in Section IB. Remarks on some fundamental concepts and quantities are given in IC, including some basic relations between various quantities. Tables concerning symbols and units are given in ID. A detailed survey of contents is made in IE. Textbooks, review articles and proceedings on ferroelectrics are listed in Section IF.

How to find data

For location of data, it is recommended to look first into Table IE-2 "Two-dimensional survey of contents" or Table V "Index of substances." They list all materials which appear in this subvolume III/16b. (In Table V the substances of subvolume a are also included.) Table IE-2 indicates where the required data can be found under special properties and individual substances and is useful for the reader acquainted with ferroelectrics. In case the reader is not familiar with ferroelectrics, Table V (p. 766) will be easier to handle since the substances are arranged alphabetically.

IB Definition of ferroelectrics and antiferroelectrics

A ferroelectric crystal is defined as a crystal which belongs to the pyroelectric family (i.e. shows a spontaneous electric polarization), and whose direction of spontaneous polarization can be reversed by an electric field. An antiferroelectric crystal is defined as a crystal whose structure can be considered as being composed of two sublattices polarizing spontaneously in antiparallel directions at least in one projection**), and whose ferroelectric phase can be induced by applying an electric field. Experimentally the reversal of the spontaneous polarization in ferroelectrics is observed as a single hysteresis loop (Fig. IB-1), and the induced phase transition in antiferroelectrics as a double hysteresis loop (Fig. IB-2), when a low-frequency ac field of a suitable strength is applied.

The spontaneous polarization in ferroelectrics and the sublattice polarizations in antiferroelectrics are analogous to their magnetic counterparts. As described above, however, these polarizations are a necessary, but not a sufficient condition for ferroelectricity or antiferroelectricity. In other words, ferroelectricity and antiferroelectricity are concepts based not only upon crystal structure, but also upon the dielectric behavior of the

*) These previous volumes may be useful to find historically important data and to see how the studies advanced in the last 60 years, and will remain in the list of available volumes.

**) For instance, the sublattice polarizations of $PbZrO_3$ in the antiferroelectric phase are antiparallel only in the *c*-projection (cf. Fig. 405 of subvolume a).

crystal. It is a common dielectric character of ferroelectrics and antiferroelectrics that in a certain temperature range the dielectric polarization is observed as a two-valued function of the electric field strength as shown in Figs. IB-1 and IB-2.

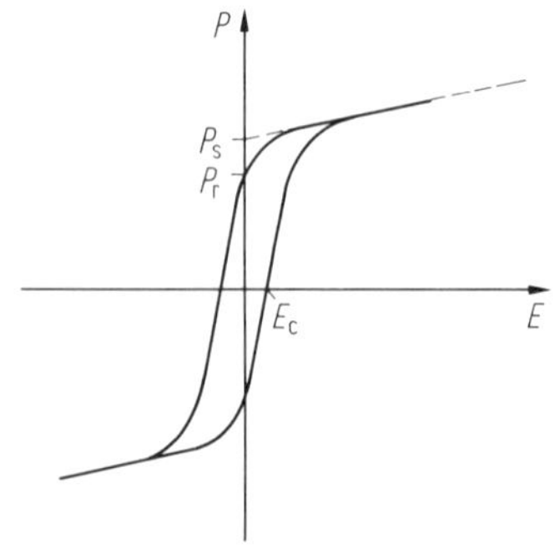

Fig. IB-1. Ferroelectric hysteresis loop. P_s: spontaneous polarization (see remarks on P_s in IC). P_r: remanent polarization. E_c: coercive field.

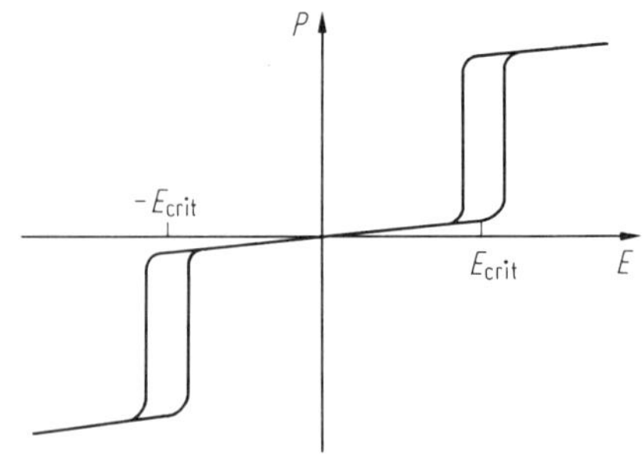

Fig. IB-2. Antiferroelectric hysteresis loop. E_{crit}: critical field.

In general, the spontaneous polarization in ferroelectrics or the sublattice polarization in antiferroelectrics can be regarded as a structural perturbation on a nonpolarized crystal. Such a non-perturbed crystal structure is sometimes called a paraelectric structure. By definition, the paraelectric structure can be realized by making the ferroelectric polarization or the sublattice polarization equal to zero. The phase having a paraelectric structure is called a paraelectric phase. Usually ferroelectrics or antiferroelectrics exhibit the paraelectric phase at high temperatures.

Some structural perturbation on a paraelectric state results in a state which neither has a spontaneous polarization nor fits in the definition of antiferroelectrics. Such a crystal state is called nonpolar in this volume.

There are several concepts which have been discussed recently in connection with ferro- or antiferroelectricity: ferroelasticity [69A5, 74N14, 74N15]; improper ferroelectricity [74d1, 7413, 76i1]; structural phase transition [71s1, 74s2]; incommensurate structure [75M11, 76M26, 77r1]. These terms seem, however, to be used with somewhat different implications by different authors and thus they will not be used in this volume. Instead, data will be given simply on anomalous ways of behavior of special quantities (e.g. elastic stiffness or frequency of a mode of lattice vibration, etc.).

I C Remarks on some fundamental concepts and quantities

Data at room temperatures

Throughout this volume, data in tables and figures should be understood as room temperature data if not otherwise stated.

Presentation of experimental errors

In the tables of this volume, experimental errors are given in parentheses referring to the last decimal places. For example, 1.356 (12) stands for 1.356 ± 0.012 and 127.3 (21) for 127.3 ± 2.1.

Molecular weight

The molecular weight M is given following the chemical formula at the beginning of each section. It was calculated with the atomic weights (the ratio of the mass per mole of the elements to 1/12 of the mass of one mole of nuclide ^{12}C) recommended by IUPAC in 1977.

The following remarks are made according to the order of the subsections of Table IE-1. The number in parentheses corresponds to that of the relevant subsection.

History of research (1)

Brief description is made on the history of research of the substance in each subsection 1a. Figures IC-1 and -2 give a survey on the history of the discovery of ferroelectrics.

Names of phases (1)

Phases are numbered as I, II,… beginning with the high-temperature phase.

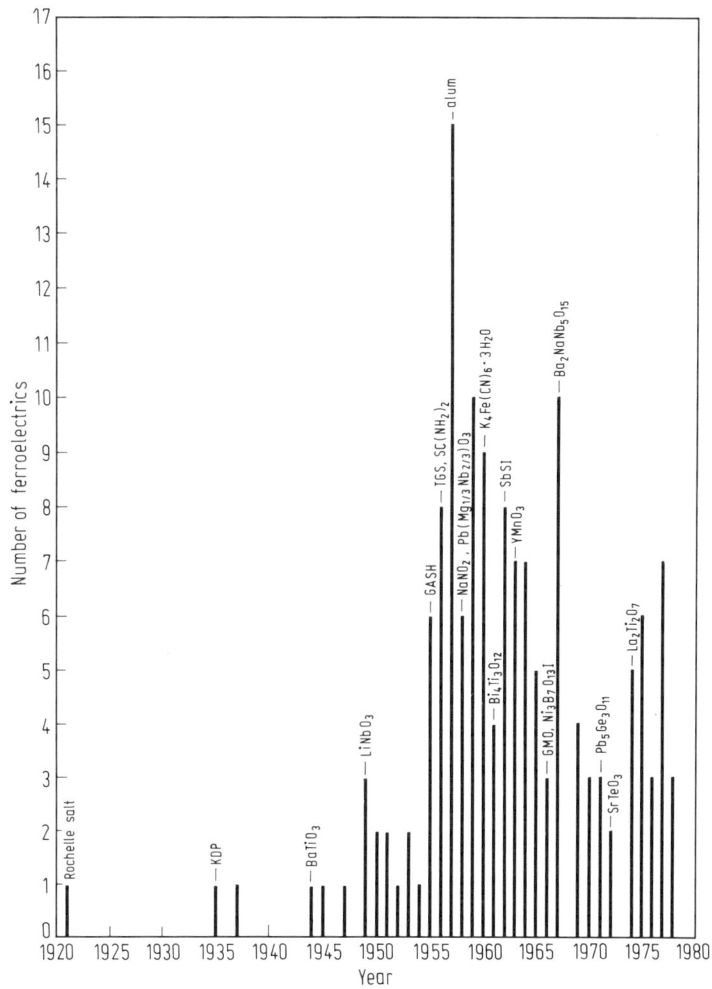

Fig. IC-1. Number of ferroelectrics discovered in each year. Only pure compounds are included. Representative ferroelectrics are indicated at the year of discovery. Prepared by M. Adachi using the information given in subsection 1a of each substance.

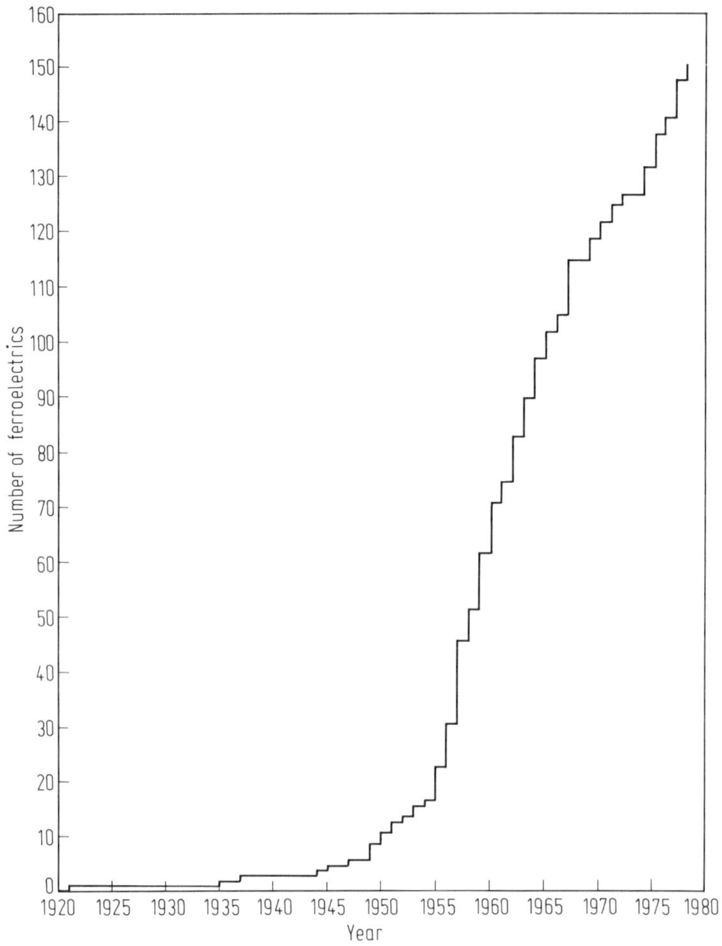

Fig. IC-2. Total number of ferroelectrics known at the end of each year. Only pure compounds are included. Prepared by M. Adachi and F. Kuwahara using the information given in subsection 1a of each substance.

Crystal structure (3)

Most crystallographic nomenclatures are the same as in the International Tables for Crystallography [6511]. Data expressed in kX units have been transformed to Å units according to the conversion factor 1 kX = 1.00202 Å [48 B 2]. For some crystals, different systems of crystallographic *a*, *b*, *c* axes are used in literatures. The *a*, *b*, *c* axes adopted in this volume are those which seem to be most widely used.

The notation [uvw] should be understood as a direction referred to the *a*, *b*, *c* axes in the paraelectric phase unless otherwise stated. The *X*, *Y*, *Z* axes (the rectangular coordinates to describe vector and tensor quantities) are parallel to the *a*, *b*, *c* axes, respectively, when the *a*, *b*, *c* axes are perpendicular to each other. In the monoclinic system (taking *b* parallel to the unique axis) $Y \| b$, $Z \| c$, and *X* forms a right-handed rectangular system along with the *Y* and *Z* axes.

In crystal structure analysis, the atomic scattering factor is corrected for the thermal vibration by the temperature factor. Various expressions have been proposed for it. The most frequently used ones are as follows.
Temperature factor

$$=\exp\{-(B_{11}h^2+B_{22}k^2+B_{33}l^2+B_{12}hk+B_{23}kl+B_{31}lh)\} \quad \text{(a)}$$

$$=\exp\{-(b_{11}h^2+b_{22}k^2+b_{33}l^2+2b_{12}hk+2b_{23}kl+2b_{31}lh)\} \quad \text{(b)}$$

$$=\exp\{-\tfrac{1}{4}(a^{*2}\beta_{11}h^2+b^{*2}\beta_{22}k^2+c^{*2}\beta_{33}l^2+2a^*b^*\beta_{12}hk+2b^*c^*\beta_{23}kl+2c^*a^*\beta_{31}lh)\} \quad \text{(c)}$$

$$=\exp\{-2\pi^2(a^{*2}U_{11}h^2+b^{*2}U_{22}k^2+c^{*2}U_{33}l^2+2a^*b^*U_{12}hk+2b^*c^*U_{23}kl+2c^*a^*U_{31}lh)\}, \quad \text{(d)}$$

where B_{ij}, b_{ij}, β_{ij}, and U_{ij} are called temperature parameters.

The isotropic temperature parameter (sometimes called Debye parameter) B is defined by:
Temperature factor

$$=\exp(-B\sin^2\theta/\lambda^2)\,. \qquad \text{(e)}$$

In a few publications the mean square displacement of the atom, $\overline{u^2}=B/(8\pi^2)$, is listed instead of B.

Dielectric and electrocaloric properties (5)

For methods of measuring dielectric constants at various frequencies readers may consult Field and Westphal's article in [54v1]. The isothermal dielectric permittivity ε_{ij} and pyroelectric (or electrocaloric) coefficient p_i are defined by the equations*):

$$\Delta D_i=\sum_j \varepsilon_{ij}E_j+p_i\Delta T,$$

$$\Delta S=\sum_j p_jE_j+(\varrho c_p/T)\Delta T,$$

where ΔD_i, ΔT, and ΔS are changes of the component of electric displacement, temperature and entropy, respectively, and E_j, ϱ, c_p and T are component of electric field strength, density, specific heat capacity and the absolute temperature, respectively. With $\Delta S=0$ in the above equations we have the expression for the adiabatic dielectric permittivity ε^S_{ij}:

$$\varepsilon^S_{ij}=\varepsilon_{ij}-(T/\varrho c_p)p_ip_j\,.$$

The dielectric permittivities measured with the alternating current should be regarded as ε^S_{ij}.

Nonlinear properties of the polarization $\boldsymbol{P}$ and the strain $\mathbf{S}$ against the electric field $\boldsymbol{E}$ and the stress $\mathbf{T}$ are characteristic of ferro- and antiferroelectrics. At constant temperature and stress, the electric field strength E is expressed by the Taylor series expansion as

$$E=(1/\chi_p)P+\xi P^3+\zeta P^5\,.$$

Therefore, values of ξ and ζ are given as measures of the dielectric nonlinearity. They should be understood as the values at atmospheric pressure unless otherwise stated. Values of χ_p are not given in this volume, but, in the paraelectric phase, they can be calculated with the relation

$$\chi_p=\varepsilon_0(\varkappa-1)\,,$$

where $\varkappa$ is the dielectric constant. In general, $\varkappa$ obeys the Curie-Weiss law above the Curie point:

$$\varkappa=\varkappa_\infty+C/(T-\Theta_p)\,,$$

where $\varkappa_\infty$ is a temperature-independent term and sometimes small enough to be neglected compared with the term $C/(T-\Theta_p)$.

Usually the spontaneous polarization P_s is determined by the extrapolation of the linear section of the hysteresis loop from high values of E to the P axis as shown in Fig. IB-1. But a few research workers determine P_s by putting it equal to the remanent polarization P_r. In general, however, the difference between P_s obtained by the extrapolation and P_r is within the experimental error.

Electromechanical and elastic properties (7, 8)

Latin subscripts i and j stand for 1, 2, 3, and Greek subscripts λ, μ, and ν for 1···6. The correspondence between i, j and λ (or μ, ν) is as usual: 1,1↔1; 2,2↔2; 3,3↔3; 2,3↔4; 3,1↔5; 1,2↔6.

In this volume the sign of the component T_{ij} of the stress tensor is defined as positive when it produces a positive strain S_{ij}. The strain component S_λ is related to the component S_{ij} of strain tensor by

$$S_\lambda=S_{ij} \quad \text{for} \quad i=j$$

$$S_\lambda=2S_{ij} \quad \text{for} \quad i\neq j\,.$$

Taking the electrostriction into account, $\mathbf{S}$ is given by**)

$$S_\lambda=\sum_\mu s^P_{\lambda\mu}T_\mu+\sum_i f_{i\lambda}P_i+\sum_{i,j} Q_{\lambda ij}P_iP_j\,.$$

---

*) Compare Landolt-Börnstein, Vol. III/11, p. 473ff.
**) Compare Landolt-Börnstein, Vol. III/11, pp. 289 and 296.

The electrostrictive constants $Q_{\lambda\mu}$ are defined by

$Q_{\lambda\mu}=Q_{\lambda ij}$ for $i=j$

$Q_{\lambda\mu}=2Q_{\lambda ij}$ for $i\neq j$.

Other constants related to electromechanical and elastic properties are defined in the same manner and expressed by the same symbols as in 2.1 of [66b1]*). Relations associated with electromechanical and elastic constants are summarized in Table IC-1. Definitions and symbols of quantities related to electrooptic and piezooptic effects are the same as in 3.1 of [66b1]**). Relations associated with electrooptic and piezooptic constants are summarized in Table IC-2. For detailed discussions on interaction between vector and tensor quantities readers are referred to [57n1] and [66m1]. For general reviews on electromechanical properties readers may consult [46c1], [50m1], and [64m2]. For piezoelectric ceramics, see [71j1].

Acoustic surface waves (8)

The acoustic surface waves were treated first by Lord Rayleigh in 1885. They propagate through the thin surface region with smaller velocities than bulk elastic waves. Recently engineers have been interested in these properties of the surface waves for constructing signal-processing devices. Basic data on the surface waves are presented in subsection 8a. Readers may consult [70w1], [70f1] and [78o1] for basic properties of the surface waves and recent development in application, and refer to [68c1] and [72f1] for analytic treatment of various devices.

Optical properties (9)

For general reviews on crystal optics readers may consult [61r1].

To describe nonlinear optical properties related to second harmonic generation (SHG), a nonlinear susceptibility tensor is used, whose components d_{ijk} are defined by^{+})

$$P_i(2\nu)=\sum_{j,k}\varepsilon_0 d_{ijk}E_j(\nu)E_k(\nu) \qquad \text{(f)}$$

in the SI (MKSA) unit system, and by

$$P_i(2\nu)=\sum_{j,k}d_{ijk}E_j(\nu)E_k(\nu) \qquad \text{(g)}$$

in the cgs-esu system. Here $P_i(2\nu)$ is a component of the second harmonic polarization of frequency 2ν generated by the electric field of light of frequency ν, $E(\nu)$, and ε_0 is the dielectric permittivity of vacuum. The SHG susceptibility $d_{i\lambda}$ is defined by $d_{i\lambda}=d_{ijk}$ for all combinations of j and k. (For the convention on the contraction of these suffixes, see the preceeding note on electromechanical and elastic properties.) Kleinman showed that the SHG susceptibility tensor may exhibit greater symmetry than the piezoelectric tensors [62K12]. The relations deduced by him (the Kleinman conditions) are as follows:

$$d_{12}=d_{26},\ d_{13}=d_{35},\ d_{14}=d_{25}=d_{36},\ d_{15}=d_{31},\ d_{16}=d_{21},\ d_{23}=d_{34},\ d_{24}=d_{32}\,.$$

According to Eqs. (f) and (g), $d_{i\lambda}$ (MKSA) has the unit of V^{-1} m and is converted to $d_{i\lambda}$ (esu) by

$$d_{i\lambda}\,(\text{esu})=\frac{3}{4\pi}\cdot 10^4\cdot d_{i\lambda}\,(\text{MKSA})\,. \qquad \text{(h)}$$

In most cases, $d_{i\lambda}$ are given by relative values referring to d_{36}^{KDP}, d_{36}^{ADP} or d_{11}^{quartz}, which are d_{36} of KH_2PO_4, d_{36} of $NH_4H_2PO_4$ and d_{11} of quartz, respectively. Their values proposed by various authors are as follows, in units of $10^{-12}\,V^{-1}$ m:

	[71s2]	[72L5]	[79k1]	λ [μm]
d_{36}^{KDP}	0.473 (28)	0.44 (2)	0.63 (6)	1.06
d_{36}^{ADP}	0.57 (7)	0.53 (2)	0.76 (12)	1.064
d_{11}^{quartz}	0.364 (32)	0.335 (16)	0.50 (8)	1.064

*) See also Landolt-Börnstein, Vol. III/11, sections 1–3.
**) See also Landolt-Börnstein, Vol. III/11, section 5.
$^{+}$) See also Landolt-Börnstein, Vol. III/11, p. 674 ff.

It seems worthwhile giving the following comments to avoid any misconception.

(i) The components $P_i(2\nu)$ and $E_j(\nu)$ stand for amplitudes of the polarization and the electric field, and not the effective values of them.

(ii) Conversion from Eq. (f) to Eq. (g) is not made by the conventional substitution: $\varepsilon_0 \rightarrow 1/4\pi$. This is the reason why the factor $(3/4\pi)\cdot 10^4$ appears in Eq. (h), whereas usually the conversion factor for the quantity given in V^{-1} m is $3\cdot 10^4$ as shown in Table ID-3.

(iii) Although the symbol $d_{i\lambda}$ of SHG susceptibility is the same as that of the piezoelectric strain constant, the relations between the $d_{i\lambda}$ and the tensor component d_{ijk} are different in the two cases: $d_{i\lambda}=d_{ijk}$ for all i, j, k in case of SHG whereas $d_{i\lambda}=d_{ijk}$ for j=k and $d_{i\lambda}=2d_{ijk}$ for $j \neq k$ in case of the piezoelectric strain constant.

Properties studied by light scattering (10)

Scattered light gives information on elementary excitations or optical inhomogeneities in materials when they are irradiated by monochromatic light (usually laser beam). We shall use notations ν_i, ν_s, $\boldsymbol{q}_i$, $\boldsymbol{q}_s$ for ν (frequency) and $\boldsymbol{q}$ (wave number vector) of the incident and scattered lights, respectively, and $\Delta\nu=\nu_i-\nu_s$ and $\Delta\boldsymbol{q}=\boldsymbol{q}_i-\boldsymbol{q}_s$. Naturally the unit of $\Delta\nu$ is Hz in the SI (MKSA) unit system but in many publications the frequency shift is expressed by the wave number $(\Delta\nu/c)$ in vacuum in units of cm^{-1} (kayser). In this volume $\Delta\nu$ is given in both units in most cases.

Data concerning Raman, Brillouin and Rayleigh scatterings are presented in the subsection 10 except for elastic constants which are given in the subsection 8a. In Raman scattering, the frequency shift $\Delta\nu$ is usually caused by molecular vibrations or optical lattice vibrations. Then $\Delta\nu$ is in general larger than $1.5\cdot 10^{11}$ Hz (5 cm^{-1}). The light scattering caused by acoustic phonons is called Brillouin scattering, for which $\Delta\nu$ is around $3\cdot 10^8 \cdots 1.5\cdot 10^{11}$ Hz ($0.01\cdots 5$ cm^{-1}). There are elastic or quasielastic ($\Delta\nu \approx 0$) scattering processes which are caused by inhomogeneous distribution of the refractive index in a crystal. They are called Rayleigh scattering.

The scattering geometry is expressed by a notation such as $Z(X, Y)X$ for light scattering by a crystal. The notation stands for $\boldsymbol{q}_i(\boldsymbol{e}_i, \boldsymbol{e}_s)\boldsymbol{q}_s$ where $\boldsymbol{e}_i$ and $\boldsymbol{e}_s$ are polarization vectors of the incident and scattered lights, respectively. Thus, the first and last letters of $Z(X, Y)X$ indicate that propagation directions of the incident and scattered lights are parallel to the Z and X axes, respectively, and the two letters in parentheses indicate that polarization directions of the incident and scattered lights are parallel to the X and Y axes, respectively. Therefore, $Z(X, Y)X$ corresponds to the experimental set-up illustrated in Fig. IC-3a. In case of isotropic materials (or in case the detailed specification of the scattering geometry is difficult or unnecessary), the geometry is expressed by two letters such as HV. They stand for directions of $\boldsymbol{e}_i$ and $\boldsymbol{e}_s$ referring to the plane determined by $\boldsymbol{q}_i$ and $\boldsymbol{q}_s$. For instance, HV means that $\boldsymbol{e}_i$ is horizontal (parallel) to the plane and $\boldsymbol{e}_s$ is vertical (perpendicular) to the plane, as illustrated in Fig. IC-3b. Readers may consult [76b1] or [78h1] for details of light scattering by crystals, and [74s1] in the relation with ferroelectrics.

Modes of lattice vibrations are classified according to the irreducible representations of the 32 crystal point groups. Table IC-3 gives symbols for them together with information on their Raman and infrared activities.

When a crystal has several infrared active modes with the same symmetry, they are usually numbered by the order of increasing frequency. In this volume, the number is given as a subscript. For instance, $E(TO_n)$ [or $E(LO_n)$] is used to express the nth-lowest-frequency transverse (or longitudinal) optic mode with the E symmetry.

Readers are referred to [57k2] or [72b2] for the mathematical theory of symmetry, and to [64h1] for the Raman and infrared activities.

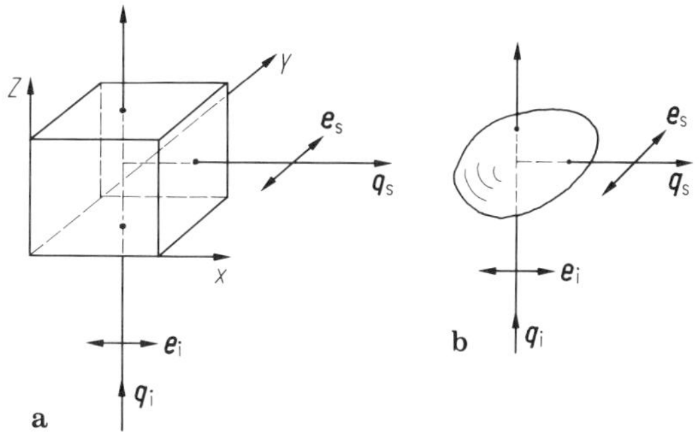

Fig. IC-3. Symbols on scattering geometries. (a) $Z(X, Y)X$. (b) HV.

Magnetic properties (12)

There are two ways to define the magnetic counterpart of the dielectric polarization $\boldsymbol{P}=\boldsymbol{D}-\varepsilon_0\boldsymbol{E}$: the magnetic polarization $\boldsymbol{J}=\boldsymbol{B}-\mu_0\boldsymbol{H}$ and the magnetization $\boldsymbol{M}=\boldsymbol{B}/\mu_0-\boldsymbol{H}$, with $\boldsymbol{J}=\mu_0\boldsymbol{M}$. In this volume magnetic properties are based upon $\boldsymbol{M}$ in accordance with other Landolt-Börnstein volumes [78n1]. Their values can be transformed into those based upon $\boldsymbol{J}$ by multiplying by μ_0.

The specific magnetization σ is defined by M/ϱ with the unit of $\mathrm{A\,m^2\,kg^{-1}}$, where ϱ is the density. The molar magnetization σ_{m} is defined by $M/$(mol/volume).

The magnetic susceptibility χ_{magn} is M/H. The specific susceptibility (or mass magnetic susceptibility) $\chi_{\mathrm{magn}\,\varrho}$ is $M/(H\varrho)$. The molar susceptibility $\chi_{\mathrm{magn\,m}}$ is $M/(H$(mol/volume)).

Magnetic resonance and Mössbauer effect (13)

Various spin Hamiltonians used in the analysis of experimental data on ESR are listed in Table IC-4. The number assigned to each Hamiltonian in this table is used when the respective Hamiltonian is referred to in the following chapters. An explanation of the spin Hamiltonian (8) is given in [64R2]. Interpretation of other Hamiltonians can be found in [53b1, 55b1, 60l1, 64h2] or [70a1].

For general information on the Mössbauer studies, readers are referred to [63f2] and [68g1]. Chapters 4 and 5 of [73b1] are devoted to applications of Mössbauer effect to studies of lattice dynamics and ferroelectrics.

Diffraction related with secondary structures (14)

Subsection 14 presents experimental data on temperature dependence of Bragg reflection, satellite and diffuse scatterings, and inelastic neutron scattering. They all are related with imperfections (including lattice vibrations) in the crystal. Readers may consult [62w1], [64g1] or [67j1] for details of diffuse X-ray scattering. Several conventional symbols are used to indicate special points or lines in the Brillouin zone for discussing the inelastic neutron scattering. Figure IC-4 shows them for the simple cubic lattice. Readers may consult [72b2] or [57k2] for other crystal lattices. Fundamentals of thermal neutron scattering are described e.g. in [65e1], [66b2], and [77m1].

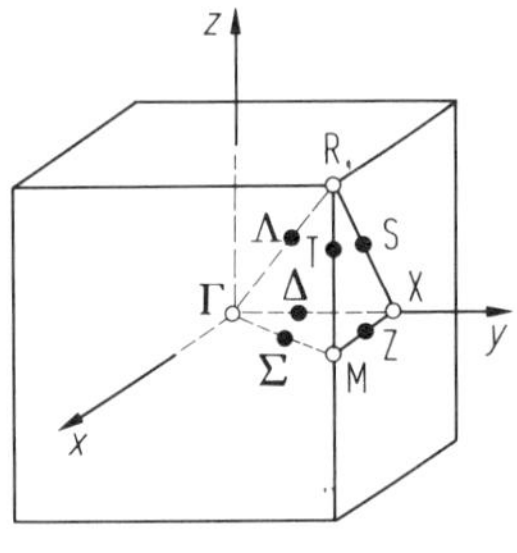

Fig. IC-4. Conventional symbols of special points in reciprocal space (q-space, $q=2\pi/\lambda$) for the simple cubic lattice. Γ: $\pi/a\{0,0,0\}$, X: $\pi/a\{1,0,0\}$, M: $\pi/a\{1,1,0\}$, R: $\pi/a\{1,1,1\}$. The symbols S, T, Z, Δ, Λ, Σ represent the positions along the lines on which the filled circles sit.

Domains (15)

Usually a ferroelectric crystal is not polarized uniformly in one direction, but is composed of many twin components which are called domains. The spontaneous polarization, P_{s}, is uniform within the domain. The boundary between two domains is called a wall. Some ferroelectrics (e.g. TGS) have only two possible (antiparallel) orientations of P_{s}, so that the domain walls separate antiparallel domains and are therefore called 180° walls. Other ferroelectrics (e.g. $BaTiO_3$) have more than two possible orientations of P_{s} and therefore a more complicated domain structure may occur. The domain wall which separates two domains having P_{s} at right angles to each other is called a 90° wall. Occasionally a crystal consists of only one domain and is called a single domain crystal. Usually, however, crystals are multi-domain crystals. A relatively strong electric field (e.g. $100\,\mathrm{kV\,m^{-1}}$) may change a multi-domain crystal into a single domain crystal, or may reverse the spontaneous polarization P_{s} of a single domain crystal. Such dynamical processes are called switching and are composed of domain wall motion and nucleation of new domains. The switching current is defined as an electric current which flows into the crystal condenser during the switching. Quantities which characterize the switching current are shown in Fig. IC-5.

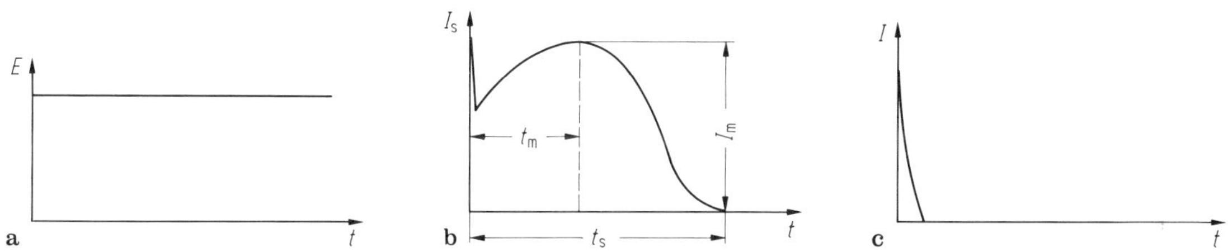

Fig. IC-5. Switching of ferroelectric crystal [67f1]. E: Electric field strength, I: electric current, I_s: switching current, t: time. (a) constant field applied at $t=0$. (b) switching current I_s obtained with an electric field antiparallel to the original P_s of the crystal. t_s: switching time, t_m: time for maximum switching current. I_m: maximum switching current. (c) Current I obtained with an electric field parallel to P_s.

Table IC-1. Electromechanical and elastic constants: relations and definitions.

(1) Piezoelectric equations

$$\begin{cases} D_i = \Sigma\, \varepsilon^{\mathbf{T}}_{ij} E_j + \Sigma\, d_{i\mu} T_\mu \\ S_\lambda = \Sigma\, d_{j\lambda} E_j + \Sigma\, s^{\boldsymbol{E}}_{\lambda\mu} T_\mu \end{cases} \qquad \begin{cases} E_i = \Sigma\, \beta^{\mathbf{T}}_{ij} D_j - \Sigma\, g_{i\mu} T_\mu \\ S_\lambda = \Sigma\, g_{j\lambda} D_j + \Sigma\, s^{\boldsymbol{D}}_{\lambda\mu} T_\mu \end{cases}$$

$$\begin{cases} D_i = \Sigma\, \varepsilon^{\mathbf{S}}_{ij} E_j + \Sigma\, e_{i\mu} S_\mu \\ T_\lambda = -\Sigma\, e_{j\lambda} E_j + \Sigma\, c^{\boldsymbol{E}}_{\lambda\mu} S_\mu \end{cases} \qquad \begin{cases} E_i = \Sigma\, \beta^{\mathbf{S}}_{ij} D_j - \Sigma\, h_{i\mu} S_\mu \\ T_\lambda = -\Sigma\, h_{j\lambda} D_j + \Sigma\, c^{\boldsymbol{D}}_{\lambda\mu} S_\mu \end{cases}$$

(2) Relations between various constants

$$\Sigma\, s^{+}_{\lambda\nu} c^{+}_{\mu\nu} = \delta_{\lambda\mu}\ (^{+} = \boldsymbol{E} \text{ or } \boldsymbol{D}) \qquad \Sigma\, \beta^{*}_{ik} \varepsilon^{*}_{ik} = \delta_{ij}\ (^{*} = \mathbf{S} \text{ or } \mathbf{T})$$

$$e_{i\lambda} = \Sigma\, d_{i\mu} c^{\boldsymbol{E}}_{\mu\lambda} \qquad d_{i\lambda} = \Sigma\, e_{i\mu} s^{\boldsymbol{E}}_{\mu\lambda}$$

$$h_{i\lambda} = \Sigma\, g_{i\mu} c^{\boldsymbol{D}}_{\mu\lambda} \qquad g_{i\lambda} = \Sigma\, h_{i\mu} s^{\boldsymbol{D}}_{\mu\lambda}$$

$$e_{i\lambda} = \Sigma\, \varepsilon^{\mathbf{S}}_{ij} h_{j\lambda} \qquad d_{i\lambda} = \Sigma\, \varepsilon^{\mathbf{T}}_{ij} g_{j\mu}$$

$$h_{i\lambda} = \Sigma\, \beta^{\mathbf{S}}_{ij} e_{j\lambda} \qquad g_{i\lambda} = \Sigma\, \beta^{\mathbf{T}}_{ij} d_{j\lambda}$$

$$c^{\boldsymbol{D}}_{\lambda\mu} - c^{\boldsymbol{E}}_{\lambda\mu} = \Sigma\, h_{j\lambda} e_{j\mu} \qquad s^{\boldsymbol{D}}_{\lambda\mu} - s^{\boldsymbol{E}}_{\lambda\mu} = -\Sigma\, g_{j\lambda} d_{j\mu}$$

$$\beta^{\mathbf{S}}_{ij} - \beta^{\mathbf{T}}_{ij} = \Sigma\, h_{i\mu} g_{j\mu} = \Sigma\, h_{j\mu} g_{i\mu} \qquad \varepsilon^{\mathbf{S}}_{ij} - \varepsilon^{\mathbf{T}}_{ij} = -\Sigma\, e_{i\mu} d_{j\mu} = -\Sigma\, e_{j\mu} d_{i\mu}$$

$$\delta_{\lambda\mu} = \begin{cases} 1\ (\lambda = \mu) \\ 0\ (\lambda \neq \mu) \end{cases} \qquad \delta_{ij} = \begin{cases} 1\ (i = j) \\ 0\ (i \neq j) \end{cases}$$

(3) Definitions of various electromechanical coupling factors.
The coupling factor, $k_{i\lambda}$ is defined by

$$k^2_{i\lambda} = \frac{d^2_{i\lambda}}{\varepsilon^{\mathbf{T}}_{ii}\, s^{\boldsymbol{E}}_{\lambda\lambda}}.$$

The following two coupling factors are used as well:

a) planar coupling factor:

$$k^2_p = \frac{2 d^2_{31}}{\varepsilon^{\mathbf{T}}_{33}(s^{\boldsymbol{E}}_{11} + s^{\boldsymbol{E}}_{12})}$$

$$= \left(\frac{2}{1-\sigma^{\boldsymbol{E}}}\right) k^2_{31} \qquad (\sigma^{\boldsymbol{E}}\text{: Poisson ratio})$$

for a thin disc of piezoelectric ceramics which is poled perpendicular to it.

b) thickness coupling factor: $k^2_t = \dfrac{e^2_{33}}{\varepsilon^{\mathbf{S}}_{33} c^{\boldsymbol{D}}_{33}}$ for a thin plate the thickness of which is along Z.

(4) Definitions of electrostrictive constants.

$$S_\lambda = \Sigma\, Q_{\lambda ij} P_i P_j$$

and $Q_{\lambda\mu} = \begin{cases} Q_{\lambda ij}, \text{ for } i = j \\ 2Q_{\lambda ij}, \text{ for } i \neq j \end{cases}$

Table I C-2. Electrooptic and piezooptic constants: relations and definitions.

(1) Index ellipsoid

$$a_1x^2+a_2y^2+a_3z^2+2a_4yz+2a_5zx+2a_6xy=1\,.$$

(2) Piezooptic and electrooptic equations

$$\Delta a_\lambda=\Sigma\,\Pi^{\boldsymbol{E}}_{\lambda v}T_v+\Sigma\,r^{\mathbf{T}}_{\lambda i}E_i+\Sigma\,L^{\mathbf{T}}_{\lambda ij}E_iE_j$$

$$\Delta a_\lambda=\Sigma\,p^{\boldsymbol{E}}_{\lambda v}S_v+\Sigma\,r^{\mathbf{S}}_{\lambda i}E_i+\Sigma\,L^{\mathbf{S}}_{\lambda ij}E_iE_j$$

$$\Delta a_\lambda=\Sigma\,\Pi^{\boldsymbol{P}}_{\lambda v}T_v+\Sigma\,\varrho^{\mathbf{T}}_{\lambda i}P_i+\Sigma\,M^{\mathbf{T}}_{\lambda ij}P_iP_j$$

$$\Delta a_\lambda=\Sigma\,p^{\boldsymbol{P}}_{\lambda v}S_v+\Sigma\,\varrho^{\mathbf{S}}_{\lambda i}P_i+\Sigma\,M^{\mathbf{S}}_{\lambda ij}P_iP_j$$

and

$$L^*_{\lambda\mu}=\begin{cases}L^*_{\lambda ij}\ (i=j)\\ 2L^*_{\lambda ij}\ (i\neq j)\end{cases}\qquad M^*_{\lambda\mu}=\begin{cases}M^*_{\lambda ij}\ (i=j)\\ 2M^*_{\lambda ij}\ (i\neq j)\end{cases}\qquad (*=\mathbf{S}\text{ or }\mathbf{T})\,.$$

Table IC-3. Point groups, irreducible representations, and Raman and infrared activities.

Crystal system	Point group *)	Point group **)	Irreducible representations: Raman active modes***)	Irreducible representations: Raman inactive modes***)
Triclinic	1	C_1	A(X, Y, Z)	
	$\bar{1}$	C_i	A_g	$A_u(X, Y, Z)$
Monoclinic	2	C_2	A(Y), B(X, Z)	
	m	C_s	A′(X, Z), A″(Y)	
	2/m	C_{2h}	A_g, B_g	$A_u(Y)$, $B_u(X, Z)$
Orthorhombic	222	D_2	A, $B_1(Z)$, $B_2(Y)$, $B_3(X)$	
	2mm	C_{2v}	$A_1(Z)$, A_2, $B_1(X)$, $B_2(Y)$	
	mmm	D_{2h}	A_g, B_{1g}, B_{2g}, B_{3g}	A_u, $B_{1u}(Z)$, $B_{2u}(Y)$, $B_{3u}(X)$
Tetragonal	4	C_4	A(Z), B, E(X, Y)	
	$\bar{4}$	S_4	A, B(Z), E(X, Y)	
	4/m	C_{4h}	A_g, B_g, E_g	$A_u(Z)$, B_u, $E_u(X,Y)$
	4mm	C_{4v}	$A_1(Z)$, B_1, B_2, E(X, Y)	A_2
	422	D_4	A_1, B_1, B_2, E(X, Y)	$A_2(Z)$
	$\bar{4}$2m	D_{2d}	A_1, B_1, $B_2(Z)$, E(X, Y)	A_2
	4/mmm	D_{4h}	A_{1g}, B_{1g}, B_{2g}, E_g	A_{2g}, A_{1u}, $A_{2u}(Z)$, B_{1u}, B_{2u}, $E_u(X, Y)$
Trigonal	3	C_3	A(Z), E(X, Y)	
	$\bar{3}$	C_{3i}	A_g, E_g	$A_u(Z)$, $E_u(X, Y)$
	32	D_3	A_1, E(X, Y)	$A_2(Z)$
	3m	C_{3v}	$A_1(Z)$, E(X, Y)	A_2
	$\bar{3}$m	D_{3d}	A_{1g}, E_g	A_{1u}, $A_{2u}(Z)$, $E_u(X, Y)$
Hexagonal	6	C_6	A(Z), $E_1(X, Y)$, E_2	B
	$\bar{6}$	C_{3h}	A′, E′(X, Y), E″	A″(Z)
	6/m	C_{6h}	A_g, E_{1g}, E_{2g}	B_g, $A_u(Z)$, $E_{1u}(X, Y)$, E_{2u}
	622	D_6	A_1, $E_1(X, Y)$, E_2	$A_2(Z)$, B_1, B_2
	6mm	C_{6v}	$A_1(Z)$, $E_1(X, Y)$, E_2	A_2, B_1, B_2
	$\bar{6}$m2	D_{3h}	A_1', E′(X, Y), E″	A_2', A_1'', $A_2''(Z)$
	6/mmm	D_{6h}	A_{1g}, E_{1g}, E_{2g}	A_{2g}, B_{1g}, B_{2g}, A_{1u}, $A_{2u}(Z)$, B_{1u}, B_{2u}, $E_{1u}(X, Y)$, E_{2u}
Cubic	23	T	A, E, F(X, Y, Z)	
	m3	T_h	A_g, E_g, F_g	A_u, E_u, $F_u(X, Y, Z)$
	432	O	A_1, E, F_2	A_2, $F_1(X, Y, Z)$
	$\bar{4}$3m	T_d	A_1, E, $F_2(X, Y, Z)$	A_2, F_1
	m3m	O_h	A_{1g}, E_g, F_{2g}	A_{2g}, F_{1g}, A_{1u}, A_{2u}, E_u, $F_{1u}(X, Y, Z)$, F_{2u}

For footnotes see next page

Table IC-4. Spin Hamiltonians.

$$\mathscr{H} = \mu_B(g_x H_x S_x + g_y H_y S_y + g_z H_z S_z) \tag{1}$$
$$\mathscr{H} = g_\perp \mu_B(H_x S_x + H_y S_y) + g_{||}\mu_B H_z S_z \tag{2}$$
$$\mathscr{H} = g_\perp \mu_B(H_x S_x + H_y S_y) + g_{||}\mu_B H_z S_z + {}^{n}A_\perp(I_x S_x + I_y S_y) + {}^{n}A_{||} I_z S_z \tag{3}$$
$$\mathscr{H} = g_\perp \mu_B(H_x S_x + H_y S_y) + g_{||}\mu_B H_z S_z + D[S_z^2 - (1/3)\, S(S+1)] \tag{4}$$
$$\mathscr{H} = \mu_B(g_x H_x S_x + g_y H_y S_y + g_z H_z S_z) + D[S_z^2 - (1/3)\, S(S+1)] + E(S_x^2 - S_y^2) + {}^{n}A\boldsymbol{I} \cdot \boldsymbol{S} \tag{5}$$
$$\mathscr{H} = \mu_B(g_x H_x S_x + g_y H_y S_y + g_z H_z S_z) + D[S_z^2 - (1/3)\, S(S+1)] + E(S_x^2 - S_y^2) + \boldsymbol{S} \cdot {}^{n}\mathbf{A} \cdot \boldsymbol{I} \tag{6}$$
$$\mathscr{H} = g\mu_B(H_x S_x + H_y S_y + H_z S_z) + D[S_z^2 - (1/3)\, S(S+1)] + E(S_x^2 - S_y^2) + (F/180)\, [35\, S^4 - 30\, S(S+1)\, S_z^2 + 25\, S_z^2 - 6\, S(S+1) + 3\, S^2(S+1)^2] + (a/6)\, [S_\xi^4 + S_\eta^4 + S_\zeta^4 - (1/5)\, S(S+1)\, (3S^2 + 3S - 1)] + {}^{n}A\, \boldsymbol{I} \cdot \boldsymbol{S} \tag{7}$$
$$\mathscr{H} = g\mu_B(H_x S_x + H_y S_y + H_z S_z) + b_{20} Y_{2,0} + b_{40}[Y_{4,0} + ((70)^{1/2}/4)\,(Y_{4,4} + Y_{4,-4})] + b_{60}[Y_{6,0} - ((14)^{1/2}/2)\,(Y_{6,4} + Y_{6,-4})] + {}^{n}A\, \boldsymbol{I} \cdot \boldsymbol{S} \tag{8}$$
$$\mathscr{H} = g\mu_B(H_x S_x + H_y S_y + H_z S_y) + (b_{40}/60)\,(O_4^0 + 5O_4^4) + (b_{60}/1260)\,(O_6^0 - 21O_6^4) + {}^{n}A\, \boldsymbol{I} \cdot \boldsymbol{S} \tag{9}$$
$$\mathscr{H} = \mu_B(g_x H_x S_x + g_y H_y S_y + g_z H_z S_z) + \boldsymbol{S} \cdot {}^{n}\mathbf{A} \cdot \boldsymbol{I} + Q_z[I_z^2 - (1/3)\, I(I+1)] + \sum_\lambda \boldsymbol{S} \cdot {}^{n}\mathbf{A}_\lambda \cdot \boldsymbol{I}_\lambda \tag{10}$$
$$\mathscr{H} = \mu_B \boldsymbol{H} \cdot \mathbf{g} \cdot \boldsymbol{S} + \sum_\lambda \boldsymbol{S} \cdot {}^{n}\mathbf{A}_\lambda \cdot \boldsymbol{I}_\lambda \tag{11}$$

The summations in (10) and (11) extend over all nuclei of the ligand which are contributing to the HFS. Q_z: constants of the quadrupole interaction in the spin Hamiltonian.

I D Symbols and units

Most symbols adopted in this volume are those which are recommended by S.U.N. Commission of IUPAP [78i1] or used in the International Tables for X-Ray Crystallography [65I1]. The international system of units (Le Système International (SI) d'Unités) is used throughout this volume. At present, however, the SI units are not commonly used in a few fields (e.g. magnetism), or the traditional use prefers other units for special quantities [e.g. kayser (cm^{-1}) in optics]. In such cases, the quantities are given in both SI units and traditional units in most figures and tables in this volume. An exception is Å instead of 0.1 nm. Crystallographic data are presented in Å without conversion to nm.

In Table ID-1 symbols used frequently in this volume are listed along with their units. In Table ID-2 symbols used as subscripts and superscripts are listed. Table ID-3 gives conversion factors from the SI (MKSA) units to the electrostatic and the electromagnetic units for important quantities.

Table I D-1. Symbols and units.

Symbols are arranged first in the order of the Latin alphabet and then of the Greek alphabet. Symbols such as ΔH or $\delta' H$ are arranged according to the letters following Δ or δ' (e.g. ΔH is to be found in the place of H, not Δ).

a) Letters and signs

a, b, c, ⋯	Wyckoff notations for atomic positions, see [65I1]
(hkl)	Miller indices
{hkl}	equivalent (hkl)'s in cubic or other systems
[uvw]	direction parallel to $\boldsymbol{r} = u\boldsymbol{a} + v\boldsymbol{b} + w\boldsymbol{c}$; u, v, w: integers (referred to the a, b, c axes of the paraelectric phase unless special remarks are made.)
⟨uvw⟩	equivalent [uvw]'s in cubic or other systems
‖	parallel to
⊥	perpendicular to

Footnotes of Table IC-3

*) International symbol.

**) Schoenflies symbol.

***) Infrared active modes are indicated by adding parenthesized X, Y, Z. For instance, A(X, Y, Z) and A(Z) are infrared active whereas A_g and B_{1g} are not. The letters X, Y, Z mean that the mode is infrared active to the incident radiation polarized along the orthogonal X, Y, Z axes (cf. IC), respectively.

b) Abbreviations

space group symbol	international symbol-Schoenflies symbol
I, II, III, ···	names of phases
A	antiferroelectric
(A)	possibility of being antiferroelectric
A, B, C, ···	symbols for atom in various compositions
ac	alternating current
cub	cubic
DTA	differential thermal analysis
dc	direct current
ESR	electron spin resonance
F	ferroelectric
(F)	possibility of being ferroelectric
FS	fine structure
hex	hexagonal
HFS	hyperfine structure
HT	high temperature
Im	imaginary part
IR	infrared
liq	liquid
LA	longitudinal acoustic
LO	longitudinal optic
LT	low temperature
mon	monoclinic
NMR	nuclear magnetic resonance
NQR	nuclear quadrupole resonance
orth	orthorhombic
P	paraelectric
(P)	possibility of being paraelectric
pol	polarization
R	Raman
rh	rhombohedral
RT	room temperature
SHG	second harmonic generation
ss	solid solution
tetr	tetragonal
tri	trigonal
TA	transverse acoustic
TO	transverse optic

c) Symbols and units

Symbols	Units	Definitions
A	$\mathrm{kg\,kg^{-1}}$, $\mathrm{kg\,m^{-3}}$, wt%	solubility
A, B	$\mathrm{m^{-1}}$	HFS parameters
$^{n}A_{ij}$	$\mathrm{A\,m^{-1}}$	hyperfine splitting tensor, n: nuclear mass number
$^{n}A_{i}$	$\mathrm{A\,m^{-1}}$	principal value of hyperfine splitting tensor, n: nuclear mass number
$^{n}A_{\parallel}, {}^{n}A_{\perp}$	$\mathrm{A\,m^{-1}}$	hyperfine splitting parameters in axial symmetry. n: nuclear mass number
a	$\mathrm{m^{-1}}$	FS parameter
$\boldsymbol{a}, \boldsymbol{b}, \boldsymbol{c}$	Å	unit cell vector
a, b, c	Å	length of unit cell edges
$\boldsymbol{a}^*, \boldsymbol{b}^*, \boldsymbol{c}^*$	$\mathrm{Å^{-1}}$	unit cell vector in reciprocal space
$\boldsymbol{B}$	$\mathrm{T(\equiv Wb\,m^{-2} \equiv V\,s\,m^{-2})}$	magnetic induction, flux density
B	$\mathrm{Å^2}$	isotropic temperature parameter [cf. Eq. (e) in Introduction]
B_{ij}	numeral	anisotropic temperature parameter [cf. Eq. (a) in Introduction]

Symbols	Units	Definitions
b	m^{-1}	FS parameter
b_{ij}	numeral	anisotropic temperature parameter [cf. Eq. (b) in Introduction]
c	$m\,s^{-1}$	light velocity in vacuum ($c=2.99792458(1)\cdot 10^{8}$ m s^{-1})
C	K	Curie-Weiss constant with regard to $\varkappa$: $\varkappa=\varkappa_{\infty}+C/(T-\Theta_{p})$
C_p	$J\,K^{-1}\,mol^{-1}$	molar heat capacity at constant pressure
c_p	$J\,K^{-1}\,kg^{-1}$	specific heat capacity at constant pressure
$c_{\lambda\mu}$	$N\,m^{-2}$	elastic stiffness
D	m^{-1}	FS parameter
D	numeral	dissipation factor
$\boldsymbol{D}$	$C\,m^{-2}$	electric displacement
d	m	thickness of specimen
d_{hkl}	Å	interplanar spacing of planes (hkl)
$d_{i\lambda}$	$C\,N^{-1}$	piezoelectric strain constant
$d_{i\lambda}$	$m\,V^{-1}$	susceptibility for second harmonic generation (usually given in relative values, see section IC)
E	$N\,m^{-2}$	Young's modulus
$\boldsymbol{E}$	$V\,m^{-1}$	electric field strength
ΔE, ΔE_{quad}	$m\,s^{-1}$.	quadrupole splitting in Mössbauer spectrum
E_{bias}	$V\,m^{-1}$	bias field
E_c	$V\,m^{-1}$	coercive field
E_{crit}	$V\,m^{-1}$	critical field
E_G	eV	energy separation, band gap
e	C	charge of electron ($\|e\|=1.6021892\,(46)\cdot 10^{-19}$ C)
$\boldsymbol{e}$	m^{-1}	polarization vector of phonon
$e_{i\lambda}$	$C\,m^{-2}$	piezoelectric stress constant
e^2qQ/h	Hz	nuclear quadrupole coupling constant
$eQ\phi_{ij}/h$	Hz	component of nuclear quadrupole coupling tensor
F	m^{-1}	FS parameter
$F.M._{R_I}$	$C\,m\,J^{-1}$	$F.M._{R_I}=p_i\varrho^{-1}c_p^{-1}$, figure of merit for current responsivity
$F.M._{R_V}$	$C\,m\,J^{-1}$	$F.M._{R_V}=p_i\varrho^{-1}c_p^{-1}\varkappa^{-1}$, figure of merit for voltage responsivity
$F.M._N$	$C\,m\,J^{-1}$	$F.M._N=p_i\varrho^{-1}c_p^{-1}\varkappa^{-1/2}$, figure of merit for material noise limited signal-to-noise ratio
$F.M._{D*}$	$C\,m^{3/2}\,J^{-1}\,\Omega^{1/2}$	$F.M._{D*}=p_i\varrho^{-1}c_p^{-1}\sigma^{-1/2}$, figure of merit for detectivity
f	Hz	frequency
f_A	Hz	antiresonance frequency
$f_{i\lambda}$	$m^2\,C^{-1}$	piezoelectric strain constant
f_R	Hz	resonance frequency
Δf_R	Hz	f_A-f_R
$f_R\cdot r$	Hz m	radial frequency constant
g_i	numeral	principal value of g-tensor
g_{ij}	numeral	component of gyration tensor for optical activity
g_{ij}	numeral	component of g-tensor
$g_{i\lambda}$	$m^2\,C^{-1}$	piezoelectric strain constant
$\boldsymbol{H}$	$A\,m^{-1}$	magnetic field strength
H_c	$A\,m^{-1}$	coercive field strength, coercive force
$\mathscr{H}$	$J\,m^{-3}$	spin Hamiltonian
ΔH	$A\,m^{-1}$	magnetic resonance half width
$\Delta' H$	$A\,m^{-1}$	separation between maximum and minimum of the derivative of the magnetic resonance curve
$\langle \Delta H^2\rangle$	$A^2\,m^{-2}$.	second moment of magnetic resonance curve
$\delta' H$	$A\,m^{-1}$	magnetic resonance line separation
h	J s	Planck constant ($h=6.626176\,(36)\cdot 10^{-34}$ J Hz^{-1})
$h_{i\lambda}$	$N\,C^{-1}$	piezoelectric stress constant
I	A	electric current
I	$J\,s^{-1}$	scattering intensity
I	numeral	nuclear spin quantum number

Symbols	Units	Definitions
I_{hkl}	$J\,s^{-1}$	integrated intensity of Bragg reflection (hkl)
I_s	A	switching current
$\boldsymbol{J}$	$V\,s\,m^{-2}$	magnetic polarization, $\boldsymbol{J}=\boldsymbol{B}-\mu_0\boldsymbol{H}$
j	$A\,m^{-2}$	electric current density
k	$J\,K^{-1}$	Boltzmann constant ($k=1.380662\,(44)\cdot 10^{-23}\,J\,K^{-1}$)
$\boldsymbol{k}$	m^{-1}	wave number vector
$k_{i\lambda}$	numeral	electromechanical coupling factor
k_p	numeral	planar coupling factor
k_t	numeral	thickness coupling factor
$L_{\lambda\mu}$	$m^2\,V^{-2}$	quadratic electrooptic constant for $\boldsymbol{E}$
l	m	length
$\Delta l/l$	$m\,m^{-1}$	elongation per unit length
l_{ij}	m	coherence length
M	numeral	molecular weight
$\boldsymbol{M}$	$A\,m^{-1}$	magnetization, $\boldsymbol{M}=\boldsymbol{B}/\mu_0-\boldsymbol{H}$
M_s	$A\,m^{-1}$	spontaneous magnetization
$M_{\lambda\mu}$	$m^4\,C^{-2}$	quadratic electrooptic constant for $\boldsymbol{P}$
m_0	kg	electron mass ($m_0=9.109534\,(47)\cdot 10^{-31}$ kg)
m^*	m_0	effective mass of charge carrier
$N_{\lambda\mu}$	Hz m	frequency constant
N/N_0	numeral	counting rate in Mössbauer spectrum
n	numeral	refractive index
n_a, n_b, n_c	numeral	refractive indices for $\boldsymbol{E}$ of light $\parallel a, b, c$
n_e	numeral	refractive index for extraordinary light
n_o	numeral	refractive index for ordinary light
$n_\alpha, n_\beta, n_\gamma$	numeral	principal refractive indices ($n_\alpha < n_\beta < n_\gamma$)
Δn	numeral	birefringence
$\boldsymbol{P}$	$C\,m^{-2}$	dielectric polarization
P_r	$C\,m^{-2}$	remanent polarization
P_s	$C\,m^{-2}$	spontaneous polarization
p	Pa	hydrostatic pressure ($Pa=N\,m^{-2}$)
p_{eff}	μ_B	effective magnetic moment of atom or ion
p_i	$C\,K^{-1}\,m^{-2}$	pyroelectric coefficient: $p_i=\partial P_{si}/\partial T$
$p_{\lambda\mu}$	$m\,m^{-1}$	piezooptic constant for $\mathbf{S}$
Q	numeral	quality factor
$Q_E=1/\tan\delta$	numeral	electrical quality factor
ΔQ_m	$J\,mol^{-1}$	transition heat per mole
$Q_{\lambda\mu}$	$m^4\,C^{-2}$	electrostrictive constant for $\boldsymbol{P}$
q	m^{-1}	wave number ($2\pi/\lambda$)
$\boldsymbol{q}$	m^{-1}	wave number vector
R	Ω	resistance
R	numeral	reflectivity
R	numeral, %	discrepancy index (or reliability factor)
R_H	$m^3\,C^{-1}$	Hall constant
R_M	numeral	counting rate
r_c	$m\,V^{-1}$	$r_c=r_{33}-(n_o/n_e)^3 r_{13}$
$r_{\lambda i}$	$m\,V^{-1}$	electrooptic constant for $\boldsymbol{E}$
S	numeral	spin quantum number of atom or ion
S	$V\,K^{-1}$	Seebeck coefficient
S	$J\,K^{-1}\,m^{-3}$	entropy per unit volume
S_m	$J\,K^{-1}\,mol^{-1}$	entropy per mole
ΔS_m	$J\,K^{-1}\,mol^{-1}$	transition entropy per mole
$\mathbf{S}$	$m\,m^{-1}$	strain tensor
S_{ij}	$m\,m^{-1}$	component of strain tensor
S_λ	$m\,m^{-1}$	component of strain tensor: $S_\lambda=S_{ij}$ for i=j, $S_\lambda=2S_{ij}$ for $i\neq j$
s	numeral	electron spin quantum number

Symbols	Units	Definitions
$s_{\lambda\mu}$	$m^2\ N^{-1}$	elastic compliance
T	K, °C	temperature
T_{melt}	K, °C	melting point
$\mathbf{T}$	$N\ m^{-2}$	stress tensor
T_{ij}	$N\ m^{-2}$	component of stress tensor
T_λ	$N\ m^{-2}$	component of stress tensor, $T_\lambda = T_{ij}$
T_1	sec, s	spin-lattice relaxation time
$T_{1\varrho}$	sec, s	spin lattice relaxation time in rotating frame
T_2	sec, s	spin-spin relaxation time
t	sec, s	time
t_s	sec, s	switching time
t_m	sec, s	time for maximum switching current
ΔU	eV, $J\ mol^{-1}$	activation energy
U_{ij}	$Å^2$	anisotropic temperature parameter [cf. Eq. (d) in Introduction]
V	V	voltage
V	° ′	$\frac{1}{2}$ (optical axial angle)
V	$Å^3$	volume of unit cell
V_m	$Å^3$	volume for formula unit
V_π	V	half-wave voltage
v	$m\ s^{-1}$	sound velocity
v	$m\ s^{-1}$	velocity of absorber in Mössbauer effect
v_m	$m^3\ mol^{-1}$	molar volume
X, Y, Z	Å, m	orthogonal coordinate system (When b is the unique axis, $Y \parallel b$, $Z \parallel c$ and the X axis forms a rectangular coordinate system together with the Y and Z axes.)
x, y	numeral	molar fraction
x, y, z	numeral	fractional coordinates of atoms in the unit cell
x, y, z	Å, m	principal axes of tensor ellipsoid
Z	numeral	number of formula units per cell
α	m^{-1}, $dB\ m^{-1}$, neper m^{-1}	acoustic absorption coefficient
α	m^{-1}	optical absorption coefficient
α	° ′	rhombohedral angle
α_{ij}	K^{-1}	linear thermal expansion coefficient
α, β, γ	° ′	interaxial angles: $b \wedge c$, $c \wedge a$, $a \wedge c$
$\alpha^*, \beta^*, \gamma^*$	° ′	interaxial angles: $b^* \wedge c^*$, $c^* \wedge a^*$, $a^* \wedge b^*$
β_{ij}	$Å^2$	anisotropic temperature parameter [cf. Eq. (c) in Introduction]
Γ	Hz	full width at half maximum of spectral line
Γ	Hz	damping parameter
Γ	$m\ s^{-1}$	line width of Mössbauer absorption line (half width at half maximum)
γ	$m\ A^{-1}\ s^{-1}$	gyromagnetic ratio
γ	K^{-1}	cubic thermal expansion coefficient
δ	$m\ s^{-1}$	isomer shift
δ	rad	dielectric loss angle
$\tan\delta$	numeral	dielectric loss tangent: $\tan\delta = \varkappa''/\varkappa'$
ε	$F\ m^{-1}$	permittivity, $D = \varepsilon E$
ε_{ij}	$F\ m^{-1}$	components of dielectric permittivity tensor
ε_0	$F\ m^{-1}$	permittivity of vacuum ($\varepsilon_0 = 8.85418782(5) \cdot 10^{-12}\ F\ m^{-1}$)
ζ	$V\ C^{-5}\ m^9$	coefficient of power series expansion of electric field strength: $E = (1/\chi_p)P + \xi P^3 + \zeta P^5$
ζ	numeral	reduced wave vector coordinate
η	numeral	asymmetry parameter
Θ	K, °C	transition temperature
Θ_{II-I}	K, °C	transition temperature between the phases II and I
Θ_c	K, °C	superconducting transition temperature

Symbols	Units	Definitions
Θ_f	K, °C	ferroelectric transition temperature
Θ_a	K, °C	antiferroelectric transition temperature
Θ_p	K, °C	paraelectric Curie temperature in the Curie-Weiss law
$\Theta_{f\,magn}$	K, °C	ferromagnetic transition temperature
Θ_N	K, °C	antiferromagnetic Néel temperature
$\Theta_{p\,magn}$	K, °C	paramagnetic Curie temperature in the Curie-Weiss law
Θ_ϕ	°C	phase match temperature
θ_B	rad, ° ′	Bragg angle (scattering angle $=2\,\theta_B$)
$\varkappa$	numeral	dielectric constant: $\varkappa=\varepsilon/\varepsilon_0$
$\varkappa_0$	numeral	static dielectric constant
$\varkappa_\infty$	numeral	temperature-independent term in Curie-Weiss law: $\varkappa=\varkappa_\infty+C/(T-\Theta_p)$
$\varkappa', \varkappa''$	numeral	real and imaginary parts of complex dielectric constant: $\varkappa=\varkappa'+i\varkappa''$
$\varkappa_{ij}$	numeral	component of dielectric constant tensor
$\varkappa_{[uvw]}$	numeral	dielectric constant for $\boldsymbol{E}\parallel[uvw]$
$\varkappa_{(hkl)}$	numeral	dielectric constant of crystal cut parallel to (hkl)
$\varkappa_a, \varkappa_b, \varkappa_c$	numeral	dielectric constant along the *a*, *b*, *c* axes
λ	m	wave length
λ	$J\,s^{-1}\,m^{-1}\,K^{-1}$	thermal conductivity
$1/\lambda$	m^{-1}	wave number per unit length
λ, μ, ν	numeral	direction cosines
μ_0	Hm^{-1}	permeability of vacuum ($\mu_0=4\pi\cdot10^{-7}\,Hm^{-1}$)
μ_B	$J\,T^{-1}$	Bohr magneton ($\mu_B=9.274078\,(36)\cdot10^{-24}\,J\,T^{-1}$)
μ_n	$J\,T^{-1}$	nuclear magneton ($\mu_n=5.050824\,(20)\cdot10^{-27}\,J\,T^{-1}$)
μ_H	$m^2\,V^{-1}\,s^{-1}$	Hall mobility
ν	Hz	frequency (mainly used in optical properties)
$\Delta\nu$	Hz	splitting of the resonance frequencies
$\Delta\nu$	Hz	frequency shift
ν_H	Hz	frequency of resonance field
ν_L	Hz	Larmor frequency
ν_c	Hz	correlation frequency
$\delta\nu$	Hz	half width at half maximum
$\delta\nu$	Hz	doublet separation of NQR spectrum
ξ	$V\,C^{-3}\,m^5$	coefficient of power series expansion of electric field strength: $E=(1/\chi_p)\,P+\xi P^3+\zeta P^5$
ξ, η, ζ	numeral	fractional coordinates in the reciprocal unit cell
$\Pi_{\lambda\mu}$	$m^2\,N^{-1}$	piezooptic constant for **T**
$\pi_{i\lambda}$	$m^2\,N^{-1}$	piezoresistive constant
ϱ	$kg\,m^{-3}$	density
ϱ_a	$kg\,m^{-3}$	apparent density
ϱ_X	$kg\,m^{-3}$	X-ray density
ϱ_e	$Å^{-3}$	electron density (number of electrons per $Å^3$)
ϱ	Ω m	resistivity
ϱ	$°\,m^{-1}$	specific rotatory power
$\varrho_{\lambda i}$	$m^2\,C^{-1}$	electrooptic constant for $\boldsymbol{P}$
σ	$\Omega^{-1}\,m^{-1}$	conductivity
σ	$\Omega^{-1}, (\Omega\square)^{-1}$	surface conductivity
σ	numeral	standard deviation in fractional coordinates of atomic position
σ	$Å^2$	standard deviation in temperature parameter
σ	$A\,m^2\,kg^{-1}$	magnetic moment per unit mass, specific magnetization
σ_m	$A\,m^2\,mol^{-1}$	magnetic moment per mole
τ	sec, s	dielectric relaxation time
τ_c	sec, s	correlation time
ϕ	° ′	rotation angle
ϕ_o	° ′	rotation angle of optical indicatrix
ϕ_{ij}	$V\,m^{-2}$	components of the electric field gradient tensor

Symbols	Units	Definitions
χ	$F\,m^{-1}$	electric susceptibility
χ_p	$F\,m^{-1}$	paraelectric susceptibility
$\chi_{magn\,m}$	$m^3\,mol^{-1}$	molar magnetic susceptibility
$\chi_{magn\,\varrho}$	$m^3\,kg^{-1}$	mass magnetic susceptibility, specific susceptibility
χ_{me}	$s\,m^{-1}$	magnetoelectric susceptibility

Table I D-2. Subscripts and superscripts arranged alphabetically.

Inferior		Inferior	
a	antiferroelectric	amagn	antiferromagnetic
a, *b*, *c*	parallel to the *a*, *b*, *c* axes	anis	anisotropic
A, B, a, b, c, …	sublattice A, B, a, b, c, …	brk	breakdown
d	diffuse	calc	calculated
e	extraordinary light	crit	critical
f	ferroelectric	cub	cubic
ϱ	per gram (only for magnetic quantities)	det	determined
		dia	diamagnetic
HT	high temperature	eff	effective
i, j, k	numerals (1···3)	exp	experimental
λ, μ, ν	numerals (1···6)	ext	external
LT	low temperature	fmagn	ferromagnetic
m	per mole; per molecule; maximum (in a few cases)	hex	hexagonal
		int	internal
n	nuclear, per nucleus	loc	local
n	normalized	magn	magnetic
o	ordinary light	max	maximum
p	paraelectric	me	magnetoelectric
p	at pressure *p*	mech	mechanical
R	resonance	melt	melting
r	remanent	min	minimum
s	spontaneous; switching	mon	monoclinic
v	at volume *v*	obs	observed
[uvw]	parallel to the [uvw] direction	orth	orthorhombic
(hkl)	perpendicular to the (hkl) plane	pmagn	paramagnetic
‖	parallel	prk	perovskite
⊥	perpendicular	ps	pseudo
		pyro	pyroelectric
		quad	quadrupole
		res	resonance
		rh	rhombohedral
		sol	solid
		tetr	tetragonal
		th	theoretical
		thr	threshold

Superior	
0	at 0 K
numeral (left)	mass number
D	at electric displacement ***D***
E	at field strength ***E***
P	at polarization ***P***
S	at entropy *S* (adiabatic)
S	at strain **S**
T	at temperature *T*
T	at stress **T**

Table I D-3. Conversion factors from the SI system to the cgs-esu and the cgs-emu systems.
(Replace the value of column 3 by the value of column 4 or column 5).
Arrangement according to the order of the properties of Table I E-1.

Quantities	Symbols	SI	cgs-esu (non-rationalized)	cgs-emu
Length	l	m	10^2 cm	10^2 cm
Mass	m	kg	10^3 g	10^3 g
Time	t	s	1 s	1 s
Density	ϱ	$kg\,m^{-3}$	$10^{-3}\,g\,cm^{-3}$	$10^{-3}\,g\,cm^{-3}$
Electric field strength	$\boldsymbol{E}$	$V\,m^{-1}$	$10^{-4}/3$ esu	10^6 emu
Electric displacement	$\boldsymbol{D}$	$C\,m^{-2}$	$12\pi \cdot 10^5$ esu	$4\pi \cdot 10^{-5}$ emu
Dielectric polarization	$\boldsymbol{P}$	$C\,m^{-2}$	$3 \cdot 10^5$ esu	10^{-5} emu
Piezooptic constant	$p_{\lambda\mu}$	Dimensionless	1 (dimensionless)	1 (dimensionless)
Dielectric constant	$\varkappa$	Dimensionless	1 (dimensionless)	1 (dimensionless)
Electric susceptibility	χ	$F\,m^{-1}$	$9 \cdot 10^9$ (dimensionless)	10^{-11} emu
Pyroelectric coefficient	p_i	$C\,m^{-2}\,K^{-1}$	$3 \cdot 10^5$ esu K^{-1}	10^{-5} emu K^{-1}
Electric current density	$\boldsymbol{j}$	$A\,m^{-2}$	$3 \cdot 10^5$ esu	10^{-5} emu
Resistivity	ϱ	$\Omega\,m$	$10^{-9}/9$ esu	10^{11} emu
Conductivity	σ	$\Omega^{-1}\,m^{-1}$	$9 \cdot 10^9$ esu	10^{-11} emu
Molar heat capacity at constant pressure	C_p	$J\,K^{-1}\,mol^{-1}$	$10^7\,erg\,K^{-1}\,mol^{-1}$	$10^7\,erg\,K^{-1}\,mol^{-1}$
Transition heat per mol	ΔQ_m	$J\,mol^{-1}$	$10^7\,erg\,mol^{-1}$	$10^7\,erg\,mol^{-1}$
Transition entropy per mol	ΔS_m	$J\,K^{-1}\,mol^{-1}$	$10^7\,erg\,K^{-1}\,mol^{-1}$	$10^7\,erg\,K^{-1}\,mol^{-1}$
Thermal conductivity	λ	$J\,m^{-1}\,s^{-1}\,K^{-1}$	$10^5\,erg\,cm^{-1}\,s^{-1}\,K^{-1}$	$10^5\,erg\,cm^{-1}\,s^{-1}\,K^{-1}$
Stress tensor	$\mathbf{T}$	$N\,m^{-2}$	$10\,dyn\,cm^{-2}$	$10\,dyn\,cm^{-2}$
Hydrostatic pressure	p	$Pa\,(=N\,m^{-2})$	$10\,dyn\,cm^{-2}$	$10\,dyn\,cm^{-2}$
Strain tensor	$\mathbf{S}$	$m\,m^{-1}$	$1\,cm\,cm^{-1}$	$1\,cm\,cm^{-1}$
Elastic stiffness	$c_{\lambda\mu}$	$N\,m^{-2}$	$10\,dyn\,cm^{-2}$	$10\,dyn\,cm^{-2}$
Elastic compliance	$s_{\lambda\mu}$	$m^2\,N^{-1}$	$10^{-1}\,cm^2\,dyn^{-1}$	$10^{-1}\,cm^2\,dyn^{-1}$
Piezoelectric strain constant	$d_{i\lambda}$	$C\,N^{-1}$	$3 \cdot 10^4$ esu	10^{-6} emu
Piezoelectric stress constant	$e_{i\lambda}$	$C\,m^{-2}$	$3 \cdot 10^5$ esu	10^{-5} emu
Piezoelectric strain constant	$g_{i\lambda}$	$m^2\,C^{-1}$	$10^{-5}/3$ esu	10^5 emu
Piezoelectric stress constant	$h_{i\lambda}$	$N\,C^{-1}$	$10^{-4}/3$ esu	10^6 emu
Electrostrictive constant (for $\boldsymbol{P}$)	$Q_{\lambda\mu}$	$m^4\,C^{-2}$	$10^{-10}/9$ esu	10^{10} emu
Electrooptic constant (for $\boldsymbol{P}$)	$\varrho_{i\lambda}$	$m^2\,C^{-1}$	$10^{-5}/3$ esu	10^5 emu
Electrooptic constant (for $\boldsymbol{E}$)	$r_{i\lambda}$	$m\,V^{-1}$	$3 \cdot 10^4$ esu	10^{-6} emu
Piezooptic constant (for $\mathbf{T}$)	$\Pi_{\lambda\mu}$	$m^2\,N^{-1}$	$10^{-1}\,cm^2\,dyn^{-1}$	$10^{-1}\,cm^2\,dyn^{-1}$
Quadratic electrooptic constant (for $\boldsymbol{E}$)	$L_{\lambda\mu}$	$m^2\,V^{-2}$	$9 \cdot 10^8$ esu	10^{-12} emu
Quadratic electrooptic constant (for $\boldsymbol{P}$)	$M_{\lambda\mu}$	$m^4\,C^{-2}$	$10^{-10}/9$ esu	10^{10} emu
Hall constant	R_H	$m^3\,C^{-1}$	$10^{-3}/3$ esu	10^7 emu
Magnetic induction (flux density)	$\boldsymbol{B}$	$T\,(\equiv Wb\,m^{-2} \equiv V\,s\,m^{-2})$	$10^{-6}/3$ esu	10^4 G
Magnetic field	$\boldsymbol{H}$	$A\,m^{-1}$	$12\pi \cdot 10^7$ esu	$4\pi \cdot 10^{-3}$ Oe
Magnetic polarization ($\boldsymbol{B} = \mu_o \boldsymbol{H} + \boldsymbol{J}$)	$\boldsymbol{J}$	$T\,(\equiv Wb\,m^{-2} \equiv V\,s\,m^{-2})$	$10^{-6}/(12\pi)$ esu	$10^4/(4\pi)$ G
Magnetization ($\boldsymbol{B} = \mu_o(\boldsymbol{H} + \boldsymbol{M})$)	$\boldsymbol{M}$	$A\,m^{-1}$	$3 \cdot 10^7$ esu	10^{-3} G
Magnetic moment per unit mass, specific magnetization	σ	$A\,m^2\,kg^{-1}$	$3 \cdot 10^{10}\,esu \cdot cm^3\,g^{-1}$	$1\,emu\,g^{-1}$
Magnetic moment per mol	σ_m	$A\,m^2\,mol^{-1}$	$3 \cdot 10^{13}\,esu \cdot cm^3\,mol^{-1}$	$10^3\,emu\,mol^{-1}$
Mass magnetic susceptibility, specific susceptibility	$\chi_{magn\,\varrho}$	$m^3\,kg^{-1}$	$10^3/(4\pi)\,esu\,cm^3\,g^{-1}$	$10^3/(4\pi)\,emu\,g^{-1}$
Molar magnetic susceptibility	$\chi_{magn\,m}$	$m^3\,mol^{-1}$	$10^6/(4\pi)\,esu\,cm^3\,mol^{-1}$	$10^6/(4\pi)\,emu\,mol^{-1}$
Magnetic susceptibility	χ_{magn}	Dimensionless	$1/(4\pi)$ dimensionless	$1/(4\pi)$ dimensionless
Magnetic resonance half width	ΔH	$A\,m^{-1}$	$12\pi \cdot 10^7$ esu	$4\pi \cdot 10^{-3}$ Oe

Table I D-3 (continued)

Quantities	Symbols	SI	cgs-esu (non-rationalized)	cgs-emu
Second moment of magnetic resonance curve	$\langle \Delta H^2 \rangle$	$A^2\,m^{-2}$	$144\,\pi^2 \cdot 10^{14}$ esu	$16\,\pi^2 \cdot 10^{-6}\,Oe^2$
Gyromagnetic ratio	γ	$m\,A^{-1}\,s^{-1}$	$10^{-7}/(12\,\pi)$ esu	$10^3/(4\,\pi)\,Oe^{-1}\,s^{-1}$
Isomer shift	δ	$m\,s^{-1}$	$10^2\,cm\,s^{-1}$	$10^2\,cm\,s^{-1}$
Magnetoelectric susceptibility	χ_{me}	$s\,m^{-1}$	$10^{-2}/(4\,\pi)\,s\,cm^{-1}$	$10^{-2}/(4\,\pi)\,s\,cm^{-1}$

I E Survey of contents

Each chapter 14···44 of II as listed in the table of contents corresponds to one family consisting of similar substances. In addition to ferro- and antiferroelectrics, the chapter includes data on crystals closely related to them as an end material of solid solutions, etc. The last two chapters for miscellanea contain ferroelectrics reported recently, and crystals called ferro- or antiferroelectrics on the basis of definitions different from those adopted in this volume (see IB) as well as high-dielectric constant substances, etc. These two chapters include both oxides and non-oxides.

Each section in the chapters is devoted to describing properties of one substance (pure compound or solid solution). Data on crystals in which H is replaced by D are presented in the same section as non-deuterated crystals. The information given in each section is surveyed at the beginning of the section according to the order of subsections shown in Table I E-1 and then tables are presented in the same order.

A detailed two-dimensional table of contents is given in Table I E-2 which gives the substances along the ordinate and the properties along the abscissa and indicates on which page the description of a particular property of an individual substance can be found.

Figures concerning each substance are given in III. The pages on which the figures can be found are indicated on the top of each page of II. Conversely, the page on which the tables can be found are indicated on the top of each page of III.

In IV, all references for I, II, and III are listed. An alphabetical index of substances is given in V for all compounds of subvolumes a and b.

Table I E-1. Arrangement of data in each section.

Subsection	Information
1	History and fundamental quantities.
a	History (discoverer, year of discovery).
b	Fundamental quantities (phases, state (F, A, P), crystal system, space group of each phase, transition temperatures, direction of spontaneous polarization, melting point, density, transparency and color, cleavage plane, deliquescence and efflorescence, phase diagram for solid solution).
2	Material preparation and crystal growth.
a	Method, solubility in fluxes or solvents.
b	Crystal forms, *a*, *b*, *c* axes, *X*, *Y*, *Z* axes.
3	Crystal structure.
a	Unit cell parameters.
b	Crystal structure (*Z*, table of positional and temperature parameters, interatomic distances and bond angles, figure of crystal structures, structural change associated with phase transitions).
4	Lattice distortions (thermal expansion, lattice deformation associated with spontaneous polarization).
5	Dielectric properties.
a	Dielectric constants ($\varkappa$ vs. T, Curie-Weiss law constants, $\varkappa$ vs. p, $\varkappa$ vs. two- or one-dimensional pressure, $\varkappa$ vs. frequency, phase diagram in regard to p and E_{bias}).
b	Nonlinear dielectric properties (effect of E_{bias} on $\varkappa$, values of ξ and ζ).
c	Spontaneous polarization and coercive field (or critical field for antiferroelectrics).
d	Pyroelectric and electrocaloric effect.

Subsection	Information
6	Thermal properties.
a	Heat capacity, transition heat, transition entropy.
b	Thermal conductivity.
7	Electromechanical properties.
a	Piezoelectricity.
b	Electrostriction.
c	Nonlinear electromechanical properties.
8	Elastic properties.
a	Elastic compliances and stiffnesses (including data on acoustic surface wave).
b	Nonlinear elastic properties.
9	Optical properties.
a	Refractive indices, birefringence, reflection, absorption (infrared region, visible region, ultra-violet region).
b	Electrooptic effect.
c	Piezooptic effect (photoelastic effect).
d	Optical activity (rotatory power).
e	Nonlinear optical properties.
10	Properties studied by light scattering.
a	Raman scattering.
b	Brillouin scattering and Rayleigh scattering (Elastic constants are given in 8a).
11	Electrical conduction (conductivity, breakdown strength, thermoelectric effect, photoconductivity and photoemission, superconductivity, band structure).
12	Magnetic properties (magnetic susceptibility, spontaneous magnetization, magnetic structure, magnetoelectric effect).
13	Properties studied by magnetic resonance and Mössbauer effect.
a	NMR.
b	ESR and ENDOR.
c	Mössbauer effect.
14	Diffraction phenomena related with secondary structures.
a	Bragg reflections due to structural modulations.
b	Diffuse or inelastic scattering.
15	Domains.
a	Domain structure.
b	Effects of electric field and mechanical stress.
16	Miscellanea (thin layer, surface layer, radiation damage, plasticity, dislocation, etchant, point defects, twin structure, stripe pattern, paraelectric resonance).

Table I E-2. Two-dimensional survey of contents.

This table indicates the pages where the required data for special properties and individual substances can be found. Detailed items of the property columns 1···16 on the abscissa can be found in Table I E-1.

This table gives also the ferro- and antiferroelectric transition temperatures Θ_f and Θ_a; they were calculated by means of $T=(t+273.15)$ K, $T=(t+273.2)$ K or $T=(t+273)$ K when the original value was given by t [°C] with two, one or no decimal places, respectively.

Table I E-2. Two-dimensional survey of contents

Nr.	Substance	ferroelectric or antiferroelectric	Θ_f or Θ_a [K]	1 history, fundamental quantities	2 material preparation, crystal growth	3 crystal structure	4 thermal expansion, lattice distortion	5 dielectric properties	6 heat capacity thermal conductivity	7 electromechanical properties	8 elastic properties	9 optical properties	10 Raman, Brillouin, Rayleigh scatterings	11 electrical conduction	12 magnetic properties	13 NMR, ESR, Mössbauer study	14 secondary structure	15 domain switching	16 miscellaneous properties
14	**SbSI family**																		
14A	*Pure compounds*																		
1	BiSCl			35	35	35							35	35	35				
2	BiSeCl			35	35	35								35					
3	SbSBr	F	93	36	36	36		36	36			36	36	36	36	36			
4	BiSBr	F	103	38	38	38		38					38	38					
5	SbSeBr			38	38	39								39		39			
6	BiSeBr			39	39	39								39	39				
7	SbSI	F	292 ~ 295	39	40	40	40	40	41	41	41	41	41	42	42	42	42	42	42
8	BiSI	F	113	44	44	44		44	44					44	44	44			
9	SbSeI			45	45	45		45	45			45		45	45	45			
10	BiSeI			45	46	46								46	46				
11	SbTeI			46	46	46						46				46			
14B	*Solid solutions*																		
1	SbSBr-SbSI	F		46	46	46							46						
2	SbSI-SbOI	F		46	46														
3	SbSI-AsSI	F		46	46														
4	SbSI-BiSI	F		47	47			47					47						
5	SbSI-SbSeI	F		47		47		47				47							
15	**HCl family**																		
15A	*Pure compounds*																		
1	HCl	F	98.36	47	47	47	47	48	48		48	48	48			48	48		
2	HBr	F	89.75	49	49	49	49	49	49		49	50	50			50	50		
3	HI			51		51		51	51							51			

Nr.	Substance	ferroelectric or antiferroelectric	Θ_f or Θ_a [K]	1 history, fundamental quantities	2 material preparation, crystal growth	3 crystal structure	4 thermal expansion, lattice distortion	5 dielectric properties	6 heat capacity thermal conductivity	7 electromechanical properties	8 elastic properties	9 optical properties	10 Raman, Brillouin, Rayleigh scatterings	11 electrical conduction	12 magnetic properties	13 NMR, ESR, Mössbauer study	14 secondary structure	15 domain switching	16 miscellaneous properties
16	**$NaNO_2$ family**																		
16A	*Pure compounds*																		
1	$NaNO_2$	F	437.1	52	52	52	52	52	52	53	53	53	53	53		53	53	53	54
2	$AgNa(NO_2)_2$	F	311.2	57	57	57	58	58	58	58	58		58			58			
16B	*Solid solution*																		
1	$NaNO_2$-$AgNO_2$	F		58															
17	**KNO_3 family**																		
17A	*Pure compounds*																		
1	$NaNO_3$	(F)		59	59	59	59	59			59		59						
2	KNO_3	F	398	59	60	60	60	60	60		60	60	60	60		60	60	60	60
3	$RbNO_3$	(A)	492	64	64	64	64	64	64			64	64	64			64		
4	$CsNO_3$			65	65	65		66	66	66		66	66	66					
17B	*Solid solutions*																		
1	KNO_3-NH_4NO_3	F		66		66		66											
2	KNO_3-$RbNO_3$	F		66		66	66	66											
3	KNO_3-KNO_2	F		67		67													
4	KNO_3-KI	F						67											
5	$RbNO_3$-$CsNO_3$	(A)		67		67	67	67											
18	**$PbHPO_4$ family**																		
18A	*Pure compounds*																		
1	$PbHPO_4$	F	310	67	67	68	68	68				68	68						
2	$PbHAsO_4$	F	313.7	68	68	68		68				69	69						

				1	2	3	4	5	6	7	8	9	10	11	12	13	14	15	16
19	**KDP(KH_2PO_4) family**																		
19A	*Pure compounds*																		
1	KH_2PO_4	F	123 [213]	69	69	70	70	70	71	71	71	71	72	72		72	72	73	73
2	RbH_2PO_4	F	147 [218]	83		84		84	84	84	84	84	84	84		84			84
3	CsH_2PO_4	F	151.3 ~ 153.5 [264.5]	86	86	86		87	87							87	87		
4	TlH_2PO_4			88	88	88		88	88				88						
5	$NH_4H_2PO_4$	(A)	148 [242]	88	88	88	89	89	89	89	89	89	89	90		90	90		90
6	KH_2AsO_4	F	97 [161 ~ 159]	96	96	96	96	96	96	96	96	96	97	97		97		97	
7	RbH_2AsO_4	F	110 [173]	99		99	99	99		99	99	99	99			99			99
8	CsH_2AsO_4	F	143 [212]	100	100	100	100	100	100		100	100	101			101	101		
9	$NH_4H_2AsO_4$	(A)	216 [304]	101	102	102	102	102	102	102	102	102	102	102		102			
19B	*Solid solutions*																		
1	KH_2PO_4-$NH_4H_2PO_4$	F		104		104		104											104
2	TlH_2PO_4-$NH_4H_2PO_4$	(A)		105				105		105	105								
3	TlH_2AsO_4-$NH_4H_2AsO_4$	(A)		105		105	105	105											
20	**$(NH_4)_2SO_4$ family**																		
20A	*Pure compounds*																		
1	$(NH_4)_2SO_4$	F	223.7	105	105	105	105	105	106	106	106	106	106	106		106	106		
2	K_2SeO_4	F	93	113	113	113	113	113	114		114		114			114	114		
3	Tl_2MoO_4			115	116	116			116			116							
4	Tl_2WO_4			116	116	116			116			116							
5	$(NH_4)_2BeF_4$	F	176	116	116	116	116	117	117		117	117	117			117	117		117
6	K_2ZnCl_4	F	403	123	123	123		123								123	123		
7	Rb_2ZnCl_4	F	192	123	123	123		123			123		124			124	124		
8	$[N(CH_3)_4]_2CoCl_4$	F	266.1	124	124	124		124											
9	$[N(CH_3)_4]_2ZnCl_4$	F	279	124	124	124		125			125					125	125		
10	Rb_2ZnBr_4	F	187	125	125	125		125				125	125				125		
20B	*Solid solutions*																		
1	$(NH_4)_2SO_4$-$(NH_4)_2BeF_4$	F		125												125			
2	$(NH_4)_2SO_4$-K_2SO_4	F		126	126	126		126								126			

Nr.	Substance	ferroelectric or antiferroelectric	Θ_f or Θ_a [K]	1 history, fundamental quantities	2 material preparation, crystal growth	3 crystal structure	4 thermal expansion, lattice distortion	5 dielectric properties	6 heat capacity thermal conductivity	7 electromechanical properties	8 elastic properties	9 optical properties	10 Raman, Brillouin, Rayleigh scatterings	11 electrical conduction	12 magnetic properties	13 NMR, ESR, Mössbauer study	14 secondary structure	15 domain switching	16 miscellaneous properties
3	$(NH_4)_2SO_4$-Rb_2SO_4	F		127	127	127		127								127			
4	$(NH_4)_2SO_4$-Cs_2SO_4	F		127	127	127		127								127			
21	**NH_4HSO_4 family**																		
21A	*Pure compounds*																		
1	$RbHSO_4$	F	263.8	127	127	127		128	128		128	128				128			
2	NH_4HSO_4	F	270	131	131	132		132	132			132	132			132		132	
3	$RbHSeO_4$	F	446	134	134	134	134	134				135						135	
4	NH_4HSeO_4	F	247	135	136	136		136											
22	**$(NH_4)_3H(SO_4)_2$ family**																		
22A	*Pure compounds*																		
1	$(NH_4)_3H(SO_4)_2$	F	pressure induced	136	136	136	136	136	137			137							
2	$(NH_4)_3H(SeO_4)_2$	F	181	137	137		137	137											
23	**NH_4LiSO_4 family**																		
23A	*Pure compounds*																		
1	$KLiSO_4$			138	138	138		138				138							
2	$RbLiSO_4$	F	461	138	138	138		138						139					
3	NH_4LiSO_4	F	459.7	139	139	139	139	139			139		139			139		139	
24	**Langbeinite-type family**																		
24A	*Pure compounds*																		
1	$K_2Mg_2(SO_4)_3$					141		141											

				1	2	3	4	5	6	7	8	9	10	11	12	13	14	15	16
2	$Rb_2Mg_2(SO_4)_3$					141		141											
3	$K_2Ca_2(SO_4)_3$			142								142							
4	$Rb_2Ca_2(SO_4)_3$			142		142		142											
5	$(NH_4)_2Ca_2(SO_4)_3$			142		142		142											
6	$K_2Mn_2(SO_4)_3$			142	142	142	142	143			143	143							
7	$K_2Fe_2(SO_4)_3$			143		143		143											
8	$K_2Co_2(SO_4)_3$			143	143	143		143	143										
9	$K_2Zn_2(SO_4)_3$	F	136	143	143	143		143											
10	$K_2Cd_2(SO_4)_3$			144	144	144	144	144			144								
11	$Rb_2Cd_2(SO_4)_3$	F	129	147	147	147	147	147	148									148	
12	$Tl_2Cd_2(SO_4)_3$	F	130 ~ 128	148	148	148		148	148		148								
13	$(NH_4)_2Cd_2(SO_4)_3$	F	95 ~ 89	148	148	149	149	149	149	149	149	149	149			149			
24B	*Solid solutions*																		
1	$Tl_2Cd_2(SO_4)_3$-$(NH_4)_2Cd_2(SO_4)_3$	F		149	149	149		149			149								
2	$(NH_4)_2Mn_2(SO_4)_3$ -$(NH_4)_2Cd_2(SO_4)_3$					149													
25	**Lecontite ($NaNH_4SO_4 \cdot 2H_2O$) family**																		
25A	*Pure compounds*																		
1	$NaNH_4SO_4 \cdot 2H_2O$	F	101	151	151	151	151	152	152			152	152			152			
2	$NaNH_4SeO_4 \cdot 2H_2O$	F	180	153	154	154		154	154	154	154	154	154			154			
26	**Alum ($NH_4Fe(SO_4)_2 \cdot 12H_2O$) family**																		
26A	*Pure compounds*																		
1	$NH_4V(SO_4)_2 \cdot 12H_2O$	F	116	157	157	157		157											
2	$NH_4Fe(SO_4)_2 \cdot 12H_2O$	F	88	157	157	157		158				158				158			
3	$NH_4In(SO_4)_2 \cdot 12H_2O$	F	127	158	159	159		159											
4	$CH_3NH_3Al(SO_4)_2 \cdot 12H_2O$	F	177	159	159	159	159	159	159		159	160				160			
5	$CH_3NH_3V(SO_4)_2 \cdot 12H_2O$	F	164	161	161	161		161											
6	$CH_3NH_3Cr(SO_4)_2 \cdot 12H_2O$	F	164	161	162	162		162	162							162			
7	$CH_3NH_3Fe(SO_4)_2 \cdot 12H_2O$	F	169	162	162	162		162				162							

Nr.	Substance	ferroelectric or antiferroelectric	Θ_f or Θ_a [K]	1 history, fundamental quantities	2 material preparation, crystal growth	3 crystal structure	4 thermal expansion, latice distortion	5 dielectric properties	6 heat capacity thermal conductivity	7 electromechanical properties	8 elastic properties	9 optical properties	10 Raman, Brillouin, Rayleigh scatterings	11 electrical conduction	12 magnetic properties	13 NMR, ESR, Mössbauer study	14 secondary structure	15 domain switching	16 miscellaneous properties
8	$CH_3NH_3Ga(SO_4)_2 \cdot 12H_2O$	F	171	163	163			163				163				163			
9	$CH_3NH_3In(SO_4)_2 \cdot 12H_2O$	F	164	163	163	163		163											
10	$CO(NH_2)_2HCr(SO_4)_2 \cdot 12H_2O$	(F)		164															
11	$CH_3NH_3Al(SeO_4)_2 \cdot 12H_2O$	F	216	164	164			164				164				164			
27	**GASH($C(NH_2)_3Al(SO_4)_2 \cdot 6H_2O$) family**																		
27A	*Pure compounds*																		
1	$C(NH_2)_3Al(SO_4)_2 \cdot 6H_2O$	F	none	164	165	165	165	165		165	165	165		165		165		165	
2	$C(NH_2)_3V(SO_4)_2 \cdot 6H_2O$	F	none	169		169		169											
3	$C(NH_2)_3Cr(SO_4)_2 \cdot 6H_2O$	F	none	169		169		169								169			
4	$C(NH_2)_3Ga(SO_4)_2 \cdot 6H_2O$	F	none	171	171	171	171	171			171	171				171			
5	$C(NH_2)_3Al(SeO_4)_2 \cdot 6H_2O$	F	none	172	172	172	173	173			173	173				173			
6	$C(NH_2)_3Cr(SeO_4)_2 \cdot 6H_2O$	F	none	173		173		173											
7	$C(NH_2)_3Ga(SeO_4)_2 \cdot 6H_2O$	F	none	173	174	174	174	174			174	174				174			
28	**$LiH_3(SeO_3)_2$ family**																		
28A	*Pure compounds*																		
1	$LiH_3(SeO_3)_2$	F	none	174	174	175	175	175		175	175	175	175	175		175			
2	$NaH_3(SeO_3)_2$	F	194	178	178	178	178	179	179	179		179	179	179		179	179		
3	$KH_3(SeO_3)_2$			184	185	185	185	185	185		185	185	185			185			
4	$RbH_3(SeO_3)_2$	F	153	187	187	187	187	187			187	187				188	188		
5	$CsH_3(SeO_3)_2$	(A)	145	190	190	190		190	190			190							
6	$NH_4H_3(SeO_3)_2$			194	194	194		194				194							
29	**Colemanite ($Ca_2B_6O_{11} \cdot 5H_2O$)**																		
29A	*Pure compound*																		
1	$Ca_2B_6O_{11} \cdot 5H_2O$	F	266	196	196	196		196								196			

				1	2	3	4	5	6	7	8	9	10	11	12	13	14	15	16
30	**$K_4Fe(CN)_6 \cdot 3H_2O$ family**																		
30A	*Pure compounds*																		
1	$K_4Mn(CN)_6 \cdot 3H_2O$	F	232.8	199				199											
2	$K_4Fe(CN)_6 \cdot 3H_2O$	F	248.7 [255]	199	200	200		200	200		200	200	200	200		200	200	200	
3	$K_4Ru(CN)_6 \cdot 3H_2O$	F	258.7 [265.9]	202		202		202				202				202			
4	$K_4Os(CN)_6 \cdot 3H_2O$	F	270.8 [275]	202		202		203											
30B	*Solid solutions*																		
1	$K_4Fe(CN)_6 \cdot 3H_2O$-$Rb_4Fe(CN)_6 \cdot 3H_2O$	F		203				203											
2	$K_4Fe(CN)_6 \cdot 3H_2O$-$Tl_4Fe(CN)_6 \cdot 3H_2O$	F		203				203											
3	$K_4Fe(CN)_6 \cdot 3H_2O$-$(NH_4)_4Fe(CN)_6 \cdot 3H_2O$	F		203				203											
31	**$SC(NH_2)_2$**																		
31A	*Pure compound*																		
1	$SC(NH_2)_2$	F	169 [185]	203	204	204	204	204	205		205	205	205	205		205	205		
32	**$N(CH_3)_4HgCl_3$ family**																		
32A	*Pure compounds*																		
1	$N(CH_3)_4 \cdot HgCl_3$	F	none	208	209	209		209					209					209	
2	$N(CH_3)_4 \cdot HgBr_3$	F	none	209	209	210		210											
3	$P(CH_3)_4 \cdot HgBr_3$	F	none	210	211			211											
4	$N(CH_3)_4 \cdot HgI_3$	F	none	211	211	211		211				211							
33	**$Cu(HCOO)_2 \cdot 4H_2O$**																		
33A	*Pure compound*																		
1	$Cu(HCOO)_2 \cdot 4H_2O$	A	235.5	211	212	212	212	212	212		212	212	212	212	212	212			

Nr.	Substance	ferroelectric or antiferroelectric	Θ_f or Θ_a [K]	1 history, fundamental quantities	2 material preparation, crystal growth	3 crystal structure	4 thermal expansion, lattice distortion	5 dielectric properties	6 heat capacity thermal conductivity	7 electromechanical properties	8 elastic properties	9 optical properties	10 Raman, Brillouin, Rayleigh scatterings	11 electrical conduction	12 magnetic properties	13 NMR, ESR, Mössbauer study	14 secondary structure	15 domain switching	16 miscellaneous properties
34	**DSP($Ca_2Sr(CH_3CH_2COO)_6$) family**																		
34A	*Pure compounds*																		
1	$Ca_2Sr(CH_3CH_2COO)_6$	F	282.6	217	217	217	217	217	217	217	217	217	218			218		218	
2	$Ca_2Ba(CH_3CH_2COO)_6$	F	pressure induced	219	220	220	220	220	220		220					220			
3	$Ca_2Pb(CH_3CH_2COO)_6$	F	333	220	220	220	220	220	220	220	221	221	221			221	221		
34B	*Solid solution*																		
1	$Ca_2Sr(CH_3CH_2COO)_6$ -$Ca_2Pb(CH_3CH_2COO)_6$	F		221	221		221												
35	**$CH_2ClCOONH_4$ family**																		
35A	*Pure compounds*																		
1	$CH_2ClCOONH_4$	F	123	221	221	221		221											
2	$(CH_2ClCOO)_2H \cdot NH_4$	F	120.4 [131.7]	222	222	222		222	222							222			
36	**TGS ($(NH_2CH_2COOH)_3 \cdot H_2SO_4$) family**																		
36A	*Pure compounds*																		
1	$(NH_2CH_2COOH)_3 \cdot H_2SO_4$	F	322.57	223	223	223	224	224	224	224	224	224	225	225		225	225	225	225
2	$(NH_2CH_2COOH)_3 \cdot H_2SeO_4$	F	295	236	236	236	236	236	237	237	237	237	237	237		237		237	
3	$(NH_2CH_2COOH)_3 \cdot H_2BeF_4$	F	348	241	241	241	241	241	242		242	242		242		242		242	
4	Triglycine tellurate			244															
36B	*Solid solution*																		
1	$(NH_2CH_2COOH)_3 \cdot H_2SO_4$ -$(NH_2CH_2COOH)_3 \cdot H_2SeO_4$	F		244				244	244			244		244					244

				1	2	3	4	5	6	7	8	9	10	11	12	13	14	15	16
37	**$NH_2CH_2COOH \cdot AgNO_3$**																		
37A	*Pure compound*																		
1	$NH_2CH_2COOH \cdot AgNO_3$	F	218	245	245	246	246	246		246						246			246
38	**$(NH_2CH_2COOH)_2 \cdot HNO_3$**																		
38A	*Pure compound*																		
1	$(NH_2CH_2COOH)_2 \cdot HNO_3$	F	206	248	248	248		248	248	248	248	248				248			
39	**$(NH_2CH_2COOH)_2 \cdot MnCl_2 \cdot 2H_2O$**																		
39A	*Pure compound*																		
1	$(NH_2CH_2COOH)_2 \cdot MnCl_2 \cdot 2H_2O$	F	none	250	251	251		251											
40	**$(CH_3NHCH_2COOH)_3 \cdot CaCl_2$**																		
40A	*Pure compound*																		
1	$(CH_3NHCH_2COOH)_3 \cdot CaCl_2$	F	127	251	251	251		251	251		251					251			
41	**Rochelle salt ($NaKC_4H_4O_6 \cdot 4H_2O$) family**																		
41A	*Pure compounds*																		
1	$NaKC_4H_4O_6 \cdot 4H_2O$	F	255, 297	253	253	254	254	254	255	255	255	255	255	255		255	255	255	256
2	$NaNH_4C_4H_4O_6 \cdot 4H_2O$	F	109	260	260	260		260		260	260	260				260	260		
3	$NaRbC_4H_4O_6 \cdot 4H_2O$			262		262		262											
41B	*Solid solution*																		
1	$NaKC_4H_4O_6 \cdot 4H_2O$ -$NaNH_4C_4H_4O_6 \cdot 4H_2O$	F		262	262			262		262	262	262				262			

Nr.	Substance	ferroelectric or antiferroelectric	Θ_f or Θ_a [K]	1 history, fundamental quantities	2 material preparation, crystal growth	3 crystal structure	4 thermal expansion, lattice distortion	5 dielectric properties	6 heat capacity thermal conductivity	7 electromechanical properties	8 elastic properties	9 optical properties	10 Raman, Brillouin, Rayleigh scatterings	11 electrical conduction	12 magnetic properties	13 NMR, ESR, Mössbauer study	14 secondary structure	15 domain switching	16 miscellaneous properties
42	**$LiNH_4C_4H_4O_6 \cdot H_2O$ family**																		
42A	*Pure compounds*																		
1	$LiNH_4C_4H_4O_6 \cdot H_2O$	F	106	263		263		263	263	263	263					264			
2	$LiTlC_4H_4O_6 \cdot H_2O$	F	10	266	266	266		266		266	266					266			
43	**Miscellaneous crystals and polymers**																		
1	TiO_2			269															
2	KIO_3			269	270	270	270	270		270		270	270	270		270			
3	$KMnF_3$			272															
4	NH_4IO_3	F	355	272	272	272	272	272	272	272	272	272	273					273	
5	$AgNbO_3$ group																		
5-i	$AgNbO_3$			274		274	274	274											
5-ii	$AgTaO_3$			274		274	274												
6	Stibiotantalite group																		
6-i	Stibiotantalite, $Sb(Ta\text{-}Nb)O_4$	(F)	673	274		274		275		275	275	275						275	
6-ii	$SbNbO_4$	(F)	676 ~ 687	275	275	275		276		276		276		276					
6-iii	$SbTaO_4$	(F)	668	276	276	276		276				276							
6-iv	Bismutotantalite, $Bi(Ta\text{-}Nb)O_4$					276													
6-v	$BiNbO_4$	(F)	633	277		277		277											
6-vi	$BiTaO_4$					277													
7	$DyVO_4$			277				277			277								
8	Cr_2BeO_4	(F)	28	277		277		278							278				
9	Sb_5O_7I	F	438	278	278	278	278	278			278	278	278						
10	$Rb_3MoO_3F_3$	F	538	280	280	280	281	281											
11	Pb_4SiO_6	(A)	~ 118	281			281	281											
12	$Ca_3(VO_4)_2$	(F)	1380	282	282	282	282	282		282	282	282							
13	$BaMnF_4$ group																		

				1	2	3	4	5	6	7	8	9	10	11	12	13	14	15	16
13-i	$BaMgF_4$	F	none	283	283	283	283	283		283	283	283				283			
13-ii	$BaMnF_4$	F	none	284	284	284		284		285	285	285	285	285	285	285			
13-iii	$BaFeF_4$	(F)	none	287	287	287		287						287	287				
13-iv	$BaCoF_4$	F	none	287	287	287		287							287	287			
13-v	$BaNiF_4$	F	none	288	288	288		288							288	288			
13-vi	$BaZnF_4$	F	none	288	288	288		288				288							
14	$PrCl_3$	(A)		288															
15	Sb_2S_3	F	420 ~ 490	289	289	289		289			289	289		289		289		289	
16	$Sn_2P_2S_6$	F	339	290	290	290		290				290	290	290					
17	$Li(N_2H_5)SO_4$ group																		
17-i	$Li(N_2H_5)SO_4$			291	291	291	291	291	291			291	291	291		291	291		
17-ii	$Li(N_2H_5)BeF_4$			293	293	293		293				293		293					
18	$[C(NH_2)_3]_2UO_2(SO_4)_2 \cdot 3H_2O$	(F)	none	293	294			294										294	
19	$(NH_2CH_2COOH)_2 \cdot H_2SO_4 \cdot H_2O$			294															
20	$H_2(UO_2)_2(AsO_4)_2 \cdot 8H_2O$	(F)	253	294		294		294											
21	$NaTh_2(PO_4)_3$ group																		
21-i	$NaTh_2(PO_4)_3$	F	none	294	294		294	294		294		294							
21-ii	$AgTh_2(PO_4)_3$	F	none	295	295	295		295											
21-iii	$NaU_2(PO_4)_3$	F	none	295	295	295		295											
22	$Ag_2H_3IO_6$ group																		
22-i	$Ag_2H_3IO_6$			295	295	296		296	296			296				296	296		
22-ii	$(NH_4)_2H_3IO_6$			296	296	296		296	296			296	297			297			
23	$C_4H_2O_4$			297		297		297	297			297	297			297	297		
24	$SnCl_2 \cdot 2H_2O$			299	299	300	300	300	300				300			300			
25	$NH_4PF_6 \cdot NH_4F$	A	172	300		300		300							300				
26	$H_2NCONHNH_2 \cdot HCl$	(F)	294	301	301	301		301				301	301				301		
27	$C_9H_{18}NO$ (tanane)	(F)	286.58	301	302	302	302	302	302			302							302
28	H_2O			302				302											
29	Alkali halides with additives	(F)		302															
30	SnTe group																		
30-i	GeTe	(F)	673	303	303	303	303	303	303			303	303	303		303			
30-ii	SnTe			303	304	304	304	304	304		304	304	304	304		304	304		
30-iii	GeTe-SnTe	(F)		305	305	305	305		305							305			
31	Polyvinylidene fluoride (PVF_2)	(F)		305	305	305	305	306		306	306	306		306		306			

Nr.	Substance	ferroelectric or antiferroelectric	Θ_f or Θ_a [K]	1 history, fundamental quantities	2 material preparation, crystal growth	3 crystal structure	4 thermal expansion, lattice distortion	5 dielectric properties	6 heat capacity thermal conductivity	7 electromechanical properties	8 elastic properties	9 optical properties	10 Raman, Brillouin, Rayleigh scatterings	11 electrical conduction	12 magnetic properties	13 NMR, ESR, Mössbauer study	14 secondary structure	15 domain switching	16 miscellaneous properties
44	**Liquid crystals and related liquids**																		
1	$C_{14}H_{14}N_2O_3$ (p-azoxyanisole)			306															
2	$C_{16}H_{18}N_2O_3$ (p-azoxyphenetole)			307															
3	$C_{11}H_{14}O_3$ (p-butoxybenzoic acid)			307															
4	$C_{10}H_{10}O_3$ (p-methoxycinnamic acid)			307															
5	$C_{12}H_{10}N_2$ (azobenzene)			307															
6	$C_{12}H_{10}N_2O$ (azoxybenzene)			307															
7	$C_{31}H_{43}O_3N$ (DOBAMBC)	F	368	307				308		308		308							
8	$C_{33}H_{47}O_3N$ (DDOBAMBC)	F	360	308				308											
9	$C_{35}H_{51}O_3N$ (TDOBAMBC)	F	357	308				308											

I F Publications on ferroelectrics

Research papers

Research activities concerning ferroelectrics are increasing steadily. The situation is illustrated in Fig. I F-1.

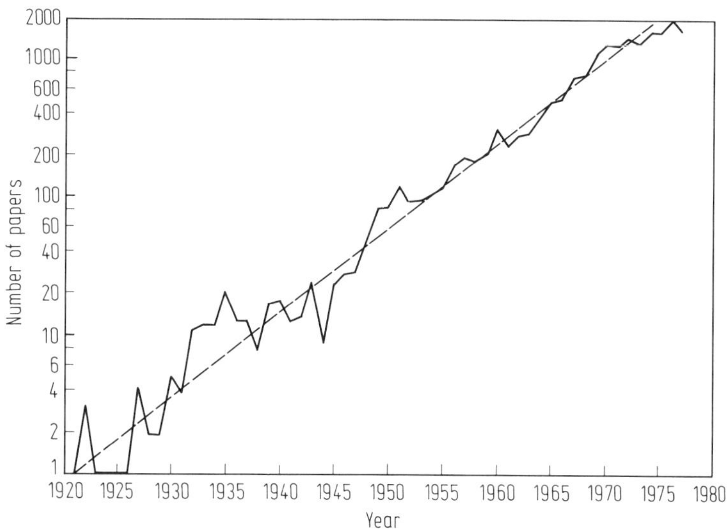

Fig. I F-1. Number of research papers on ferroelectric and related substances published within each year. Prepared by M. Adachi on the basis of [79t1]. The dashed line correponds to $N = \exp\{0.14(Y\text{-}1921)\}$, where N is the number of papers and Y is the year.

Textbooks, monographs, and review articles

Recently many books and review articles have been published on general properties of ferroelectrics: [54d1], [56f1], [56s1], [57k1], [57m1], [62j1], [64m1], [67b3], [67e1], [67f1], [68s2], [68t1], [68z1], [69m2], [70g4], [70n1], [70s2], [71k1], [71s3], [72c2], [73v1], [73z1], [74a1], [74b2], [75n2], [76b2], [77l1], [77s1].

Discussions are given on special ferroelectrics in the following books and review articles.

Perovskite-type oxides: [69g1], [71p1]; crystal chemistry [71f1]; acoustic properties [70b1]; magnetic properties [78n1].
$BaTiO_3$: [55t1]; semiconductive properties [71h1], [76d1], [77p1], [77t2].
PLZT: [74l1], [75m1].
$LiNbO_3$ and $LiTaO_3$: [75k1], [78r1].
Pyrochlore-type oxides: [76s2].
Boracite-type oxides: [74n1].
SbSI: [71k2].
Hydrogen-bonded crystals: [67b2], [76b3], [76s1], [76s3]; ice [76w2].
KDP: [70t2].
Organic crystals: [78k1].
Rochelle salt: [33k1], [49m1].
PVF_2: [76w1].
Liquid crystals: [77m2].

Discussions are given on various topics concerning ferroelectricity in the following books and review articles. The number in parentheses indicates the number of relevant subsection in Table IE-1.

History (1): [71m2], [71v1].
Symmetry (1): [71z1].
Crystal growth (2): [76n1].
Ceramic technology (2): [69o1], [76b2].
Crystal structure (3): [72p1], [73m2].
Pyroelectricity (5): [74l2], [79c1]; devices [70p1], [74g1], [77p2], [77s2], [78l1].
Effect of pressure (5): [69s1], [78s1].
Piezoelectricity (7): [46c1], [50m1], [64b1], [71b2], [79c1]; ceramics [71b2], [71j1], [73t1], [74e1]; devices [69h1], [69t1].
Acoustic and ultrasonic properties (8): [70b1], [70g1], [73r1].
Acoustic surface waves (8): [70f1], [70w1], [78o1].
Optical properties (9): [79c1]; infrared spectroscopy [71p1]; nonlinear optics [72k1], [72k2], [72w1], [72w2], [74v1].
Electrooptic effect (9): [72w1], [72w2]; devices [74k1], [74m1]. Elastooptic effect (9): [72p2].
Holographic storage (9): [75v1], [76v1], [77s3].
Light scattering (10): [70d1], [70f2], [75n1].
Electrical conductivity (11): [69g2].
Magnetic properties (12): [78n1].
Magneto-electric effect (12): [68s1], [70o1], [70s1].
NMR (13): [67b2], [74b1], [76b3].
Mössbauer effect (13): [73b2].
Lattice dynamics and soft modes (14): [60c1], [61c1], [69c1], [72a2], [73r1], [74b2], [74s1], [74s2], [74v1], [77d1].
Incommensurate structure (14): [76a1].
Dynamics of order-disorder-type ferroelectrics: [72b1], [73m3].
Critical phenomena: [65b1], [70g1], [73m3], [76t1].
Materials research: [75w1].

Proceedings

Recently many international or semi-international meetings have been held concerning ferroelectricity, and the following proceedings have been published.

The International Meeting on Ferroelectricity: [66d1], [70t1], [74c1], [78s2].
The European Meeting on Ferroelectricity: [70m1], [72g1], [76m1].
The Symposium on Application of Ferroelectrics: [69s2], [72c1], [76l1].
The NATO Advanced Study Institute: [71s1], [74r1], [75r1], [77r1].
The International Conference on Light Scattering in Solids: [69w1], [71b1], [76b1].
Others: [67w1], [75b1], [75f1], [76m2], [77a1], [77g1], [77t1], [78b1].

 Figs. p. 309

II Data

Inorganic crystals other than oxides

14 SbSI family

14A Pure compounds

Nr. 14A-1 BiSCl, Bismuth sulfide chloride

($M = 276.49$)

1a	No ferroelectric transition was found between −180° and −100 °C.	64N3
b	crystal system: orthorhombic; space group: Pnam-D_{2h}^{16}	50D2
	$\varrho = 6.0_0 \cdot 10^3$ kg m^{-3}, ($\varrho_X = 6.0_4 \cdot 10^3$ kg m^{-3}). Color: red (needles).	50D2 60N1
2a	Synthesis of BiSCl and preparation of single crystalline needles: similar method as described in subsection 2a of Nr. 14A-3.	50D2, 60N1, 64N3
b	Crystal form, axes: Fig. 1.	
3a	Unit cell parameters: $a = 7.7_0$ Å, $b = 9.8_7$ Å, $c = 4.0_0$ Å at RT.	50D2
b	$Z = 4$. All atoms are at 4c positions (Pnam-D_{2h}^{16}), (cf. subsection 3 in Nr. 14A-7): Table 1.	50D2
10a	Raman scattering: Fig. 2.	
11	Electrical conduction and photoconduction: see	60N1, 64N3
12	Magnetic susceptibility: Fig. 3.	

Table 1. BiSCl. Fractional coordinate of atoms [50D2].

	x	y	z
Bi	0.14_0	0.13_8	0.25
S	0.7_7	0.0_4	0.25
Cl	0.5_0	0.7_9	0.25

Nr. 14A-2 BiSeCl, Bismuth selenide chloride

($M = 323.39$)

1b	crystal system: orthorhombic; space group: Pmmm-D_{2h}^{1}	50D3
	$\varrho = 7.0_3 \cdot 10^3$ kg m^{-3} ($\varrho_X = 7.0_5 \cdot 10^3$ kg m^{-3}). Color: dark red (needles).	50D3 60N1
2a	Synthesis and preparation, as described in subsection 2a of Nr. 14A-4.	
3a	Unit cell parameter: $a = 12.3_7$ Å, $b = 18.1_0$ Å, $c = 4.0_8$ Å at RT.	50D3
b	$Z = 12$.	50D3
11	Photoconductive with high dark current at RT.	60N1

Nr. 14A-3 SbSBr, Antimony sulfide bromide
($M = 233.71$)

1a	Ferroelectric transition of SbSBr at −180 °C was found by Nitsche et al. in 1964.	64N3
b	phase: II / I; state: F[a]) / P[a]); crystal system: — / orthorhombic[b]); space group: — / Pnam-D_{2h}^{16} [b]); Θ: −180 °C [a])	[a]) 64N3 [b]) 50D2
	$\varrho = 4.8_6 \cdot 10^3$ kg m^{-3} (observed), ($\varrho_X = 4.9_4 \cdot 10^3$ kg m^{-3}) at RT. Color: orange (needles). Cleavage: crystals cleave along the needle axis (*c*-axis).	50D2 60N1
2a	Synthesis: $Sb_2S_3 + SbBr_3 \rightarrow 3SbSBr$. Mixtures of Sb, S and Br, or of 2Sb, 3S and $SbBr_3$, or of Sb_2S_3 and $SbBr_3$ were reacted in sealed, evacuated glass ampoules. *c*-axis oriented polycrystals: densely packed single-crystalline needles whose axes are all parallel to each other were obtained by using Bridgman furnace. Single crystals: needle crystals were obtained by slowly cooling the molten compound. Hydrothermal method: see	50D2 60N1, 64N3 64N3 60N1 69P7, 70P7, 71K6
3a	Unit cell parameters: $a = 8.2_0$ Å, $b = 9.7_0$ Å, $c = 3.9_5$ Å at RT.	50D2
b	$Z = 4$ All atoms are at 4c positions of Pnam-D_{2h}^{16} in phase I: Table 2; Fig. 4.	50D2 58M5
5a	Dielectric constant: Fig. 5. Curie-Weiss constant: $C = 1.2 \cdot 10^5$ K, $\Theta_p = 80$ K. Effect of E_{bias} on Θ_f: $d\Theta_f/dE_{bias} \approx 1.1 \cdot 10^{-5}$ KV^{-1} m.	68P2
b	$\xi = -44 \cdot 10^8$ Jm^5 C^{-4}, $\zeta = 350 \cdot 10^9$ Jm^9 C^{-6}.	68P2
c	$P_s \approx 7.5 \cdot 10^{-2}$ Cm^{-2}.	68P2
6a	Transition heat and transition entropy at $\Theta_{II \rightarrow I}$ (= 91 K): $\Delta Q_m = 15.07$ J mol^{-1}, $\Delta S_m = 0.17$ J K^{-1} mol^{-1}.	68P2
9b	$n_a = 2.64$, $n_b = 3.13$, $n_c = 4.00$ at $\lambda = 570$ nm, see also	68O4
10a	Raman scattering: Fig. 6.	
11	Conductivity: order of 10^{-7} Ω^{-1} m^{-1} at RT. Photosensitivity: order of 10^{-8} Ω^{-1} m^2 W^{-1} (maximum at $\lambda \approx 550$ nm). Band gap estimated from light absorption: $E_G \approx 2.2$ eV. $dE_G/dT \approx -9 \cdot 10^{-4}$ eV deg^{-1}.	68P2 60N1 68P2
12	Magnetic susceptibility: Fig. 3.	
13a	NQR frequencies: Fig. 7; Table 3.	
c	Mössbauer spectrum: Fig. 8, Table 4.	

Table 2. SbSBr. Fractional coordinate of atoms [58M5].

	x	*y*	*z*
Sb	0.12_0	0.14_5	0.25
S	0.8_{25}	0.08_0	0.25
Br	0.5_{15}	0.8_{35}	0.25

Table 3. SbSBr, SbSI. NQR frequencies of ^{121}Sb and ^{123}Sb, asymmetry parameter η, and quadrupole coupling constant [69P8]. Subscripts, 121 and 123, on (e^2qQ/h) indicate that the quantities are for ^{121}Sb and ^{123}Sb, respectively.

Compound	T °C	ν MHz	Nucleus	Transition	η	$\left(\frac{e^2qQ}{h}\right)_{123}$ MHz	$\left(\frac{e^2qQ}{h}\right)_{121}$ MHz	$\left(\frac{e^2qQ}{h}\right)_{123} / \left(\frac{e^2qQ}{h}\right)_{121}$
SbSI	+60	ν_1 27.78	^{123}Sb	$1/2-3/2$	0.096 (1)	376.4 (5)	295.3 (5)	1.274
	+60	ν_2 44.75	^{121}Sb	$1/2-3/2$				
	+60	ν_3 53.496	^{123}Sb	$3/2-5/2$				
	≈32	ν_4 80.77	^{123}Sb	$5/2-7/2$				
	≈32	ν_5 88.59	^{121}Sb	$3/2-5/2$				
	−195	ν_1 36.966	^{123}Sb	$1/2-3/2$	0.311 (2)	398.1 (5)	312.4 (5)	1.274
	−195	ν_2 51.552	^{121}Sb	$1/2-3/2$				
	−195	ν_3 54.507	^{123}Sb	$3/2-5/2$				
	−195	ν_4 84.55	^{123}Sb	$5/2-7/2$				
	−195	ν_5 92.10	^{121}Sb	$3/2-5/2$				
SbSBr	+30	ν_1 31.548	^{123}Sb	$1/2-3/2$	0.120 (1)	420.2 (5)	329.7 (5)	1.274
	+30	ν_2 50.231	^{121}Sb	$1/2-3/2$				
	+30	ν_3 59.554	^{123}Sb	$3/2-5/2$				
	+30	ν_4 62.375	^{81}Br	$1/2-3/2$	—	—	—	—
	−195	ν_1 32.509	^{123}Sb	$1/2-3/2$	0.132 (1)	428.9 (5)	336.4 (5)	1.275
	−195	ν_2 51.407	^{121}Sb	$1/2-3/2$				
	−195	ν_4 60.695	^{123}Sb	$3/2-5/2$				
	−195	ν_4 62.350	^{81}Br	$1/2-3/2$	—	—	—	—

Table 4. SbSBr, SbSI, SbSeI, SbTeI. Mössbauer parameters [75D7]. δ: isomer shift, eQV_{zz}: quadrupole coupling constant, Γ: line width, η: asymmetry parameter, A: maximum absorption effect. The values are with respect to $Ba^{121}SnO_3$ ($Ca^{121}SnO_3$ and $^{121}SnO_2$ for data quoted from [71K6] and [72A6], respectively). Probable experimental errors in present data are ± 0.1 mm/s in δ, ± 0.5 mm/s in eQV_{zz}, and ± 0.02 mm/s in Γ.

Compound	T [K]	δ [mm/s]	eQV_{zz} [mm/s]	Γ [mm/s]	η	A [%]	Ref.
SbSBr	4.2	−14.9	10.4	2.9	0.14	4.9	75D11
SbSBr	77	—	11.2	—	0.139	—	72S35
SbSBr	77	−14.6	—	5.2	—	—	71K6
SbSBr	77	−15.5	—	4.5	—	—	72A6
SbSeBr	4.2	−14.5	10.6	4.0	0	5.0	75D11
SbSeBr	77	−14.6	—	5.8	—	—	71K6
SbSI	4.2	−14.8	12.3	3.0	0.31	9.5	75D11
SbSI	77	−15.1	10.3	3.0	0.31	4.6	75D11
SbSI	77	—	10.41	—	0.31	—	72S35
SbSI	77	−15.2	—	4.3	—	—	71K6
SbSI	77	−14.2	—	4.0	—	—	72A6
SbSeI	4.2	−14.9	6.4	3.1	~1	5.7	75D11
SbSeI	77	−17.4	—	4.1	—	—	72A6
SbTeI	4.2	−14.0	−12.0	3.0	0.2	7.5	75D11
SbTeI	77	−14.7	—	3.5	—	—	72A6

Nr. 14A-4 BiSBr, Bismuth sulfide bromide
($M = 320.94$)

1a	Ferroelectric transition of BiSBr at −170 °C was found by Nitsche et al. in 1964.	64N3
b	phase: II / I; state: F[a]) / P[a]); crystal system: orthorhombic[b]) (I); space group: Pnam-D_{2h}^{16} [b]) (I); Θ: −170 °C [a])	[a]) 64N3 [b]) 50D2
	$\varrho = 6.5_7 \cdot 10^3$ kg m^{-3} ($\varrho_X = 6.8_3 \cdot 10^3$ kg m^{-3}).	50D2
	Color: dark red (needles).	60N1
2a	Synthesis: $Bi_2S_3 + BiBr_3 \rightarrow 3BiSBr$ (400 °C).	50D2
	Mixtures of Bi, S, and Br, or of 2Bi, 3S, and $BiBr_3$, or of Bi_2S_3 and $BiBr_3$ were reacted in sealed, evacuated glass ampoules (500 ··· 600 °C).	60N1, 64N3
	c-axis oriented polycrystals: BiSBr needles whose axes are all parallel to each other were obtained by means of Bridgman furnace.	64N3
	Single crystals: needle crystals were obtained by slowly cooling the molten compound.	60N1
	See also	69P7
3a	Unit cell parameters: $a = 8.0_2$ Å, $b = 9.7_0$ Å, $c = 4.0_1$ Å at RT.	50D2
b	$Z = 4$.	50D2
	All atoms are at 4c positions of Pnam-D_{2h}^{16} in phase I: Table 5.	
5a	Curie-Weiss constant: $C = 1.5 \cdot 10^5$ K, $\Theta_p = 99$ K, $\Theta_f = 108$ K.	68P2
b	$\xi = -3 \cdot 10^8$ Jm^{5}C^{-4}, $\zeta = 29 \cdot 10^9$ Jm^{9}C^{-6}.	68P2
c	$P_s = 8.5 \cdot 10^{-2}$ Cm^{-2}.	68P2
10a	Raman scattering: Fig. 9.	
11	Photoconductive properties, see	60N1, 68P2

Phase table (1b):

phase	II	I
state	F[a])	P[a])
crystal system		orthorhombic[b])
space group		Pnam-D_{2h}^{16} [b])
Θ	−170 °C [a])	

Table 5. BiSBr. Fractional coordinate of atoms [50D2].

	x	y	z
Bi	0.13_8	0.13_5	0.25
S	0.8_0	0.0_4	0.25
Br	0.4_8	0.8_2	0.25

Nr. 14A-5 SbSeBr, Antimony selenide bromide
($M = 280.61$)

1a	No ferroelectric transition was found in the temperature range between −180° and 70 °C.	64N3
b	crystal system: orthorhombic; space group: Pnam-D_{2h}^{16}	50D3
	$\varrho = 5.4_6 \cdot 10^3$ kg m^{-3} (observed), ($\varrho_X = 5.5_7 \cdot 10^3$ kg m^{-3}).	50D3
	Color: dark red (needles).	60N1
2a	Synthesis, preparation of ingots and single crystal growth: similar method as described in subsection 2a of Nr. 14A-3.	60N1, 64N3
	Vapor transport method.	68H2
	Hydrothermal method. continued	69P7, 70P7

Nr. 14A-5 SbSeBr, continued

3a	Unit cell parameters: $a=8.3_0$ Å, $b=10.2_0$ Å, $c=3.9_5$ Å at RT.	50D3
b	$Z=4$. All atoms are at 4c positions of Pnam-D_{2h}^{16}.	50D3
11	Photoconductive properties: see	60N1
13c	Mössbauer parameters: Table 4.	

Nr. 14A-6 BiSeBr, Bismuth selenide bromide
($M=367.84$)

1a	No ferroelectric transition was found between $-180°$ and -100 °C.	64N3
b	crystal system: orthorhombic space group: Pnam-D_{2h}^{16}	50D3
	$\varrho=6.9_2\cdot10^3$ kg m^{-3} ($\varrho_X=6.9_4\cdot10^3$ kg m^{-3}). Color: black (needles).	50D3 60N1
2a	Synthesis and preparation of single crystalline needles: similar method as described in subsection 2a of Nr. 14A-4. See also	60N1, 64N3 69P7, 68H2
3a	Unit cell parameters: $a=8.1_8$ Å, $b=10.4_7$ Å, $c=4.1_1$ Å at RT.	50D3
b	$Z=4$. All atoms are at 4c positions of Pnam-D_{2h}^{16}: Table 6.	50D3
11	Photoconductive with high dark current at RT.	60N1, 68P2
12	Magnetic susceptibility: Fig. 3.	

Table 6. BiSeBr. Fractional coordinate of atoms [50D3].

	x	y	z
Bi	0.13_5	0.13_0	0.25
Se	0.8_0	0.0_4	0.25
Br	0.4_7	0.8_2	0.25

Nr. 14A-7 SbSI, Antimony sulfide iodide
($M=280.71$)

1a Ferroelectricity in SbSI was discovered by Fatuzzo et al. in 1962. SbSI is the first example of ferroelectric substance showing large photoconductivity. — 62F3

b

phase	II	I
state	F [a]	p [a]
crystal system	orthorhombic [c]	orthorhombic [b, c]
space group	Pna2_1 − C_{2v}^9 [c]	Pnam-D_{2h}^{16} [b, c]
Θ	about 20, (22 [a]), 19 [d]) °C	

[a]) 62F3
[b]) 50D2
[c]) 64A3
[d]) 64M5

$P_s \parallel [001]$. — 62F3
$T_{melt} \approx 400$ °C. — 64M5
$\varrho=5.1_9\cdot10^3$ kg m^{-3} (observed), ($\varrho_X=5.2\cdot10^3$ kg m^{-3}). — 50D2
Color: dark red, transparent at RT.
Cleavage: fibrous, along [001] direction. — 60N1

continued

Nr. 14A-7 SbSI, continued

2a	Preparation: $SbI_3 + Sb_2S_3 \rightarrow 3SbSI$ (reaction in sealed, evacuated ampoule, 500°···600 °C).	60N1
	Crystal growth:	
	vapor transport, chemical transport method;	62K9, 67K4, 71N7
	flux method;	64M5, 70N2
	zone melting;	71N7
	Bridgman-Stockbarger method;	64N3, 69M2, 69R5, 71N7, 71S37
	hydrothermal method;	69R1, 69P7, 70P7
	high gas pressure synthesis	72B13
	Phase diagram SbI_3–Sb_2S_3: Fig. 10, see also	70P11
b	Crystal form: Fig. 11.	
3a	Unit cell parameters: a=8.522 (3), b=10.130 (4), c=4.088 (2) at 27 °C. Additional data: Table 7; temperature dependence, Fig. 12.	74I14
b	Z=4 in phases I and II.	50D2
	Crystal structure: space group Pnam-D_{2h}^{16} in phase I, and Pna2_1-C_{2v}^9 in phase II. Atomic arrangement: Fig. 13; positional parameters at 1.3 °C, 27 °C, and 48 °C: Tables 8, 9; for additional data at 5° and 35 °C, see	67K4
	Interatomic distances in phases I and II: Table 10; Figs. 14, 15;	
	atomic shifts in phase II relative to phase I, see	76I7
	Neutron diffraction method, see	66I7, 76I10
4	For lattice distortion, see Fig. 12	
	Effects of light illumination: Fig. 16.	
5a	Dielectric constant:	
	$\varkappa_c = C/(T-\Theta_p)$, $T > \Theta_f$, where $C = 2.33 \cdot 10^5$ K, Θ_p=16.0 °C, Θ_f=19.2(2) °C, see also	64M5, 68P2, 69T4
	low frequency range, Figs. 17, 18; see also	78I4, 74I7
	$\varkappa$ at microwave frequencies, Figs. 19, 20;	
	$\varkappa$ at wide frequency range, see	71B4
	effects of laser light illumination on $\varkappa$, see	71B4
	$\varkappa$ of thin films, and thickness dependence of $\varkappa$, see	73Y3
	$\varkappa$ of pressed powder, and particle size dependence of $\varkappa$, see	77M11
	effects of hydrostatic pressure on $\varkappa$ and on Θ_f, Figs. 21, 22, additional information, see	75S6, 65G6, 68S4, 69G4, 68G4
	tricritical point at 235 K, $p \approx 1.4 \cdot 10^8$ Pa;	78S2
	effect of uniaxial stress on Θ_f, Fig. 23;	
	bias field along the c axis, $d\Theta_f/dE = 2.66 \cdot 10^{-2}$ °C kV^{-1} m.	69T4
	See also	68P2

continued

Nr. 14A-7 SbSI, continued

5b	$\xi = -2.6 \cdot 10^8$ J m^5 C^{-4}, $\zeta = 5.4 \cdot 10^9$ J m^9 C^{-6},	68P2
	see also	68K2, 69T4
c	Spontaneous polarization: Figs. 24, 25; effects of hydrostatic pressure: Fig. 26;	
	jump of P_s at Θ_f, $\Delta P_s \approx 18 \cdot 10^{-2}$ C m^{-2},	68K2
	data for polycrystalline sample, see	69M2, 69P1
d	Effects of poling and of light illumination upon the pyroelectric effect, see	66I4, 68M4
	Pyroelectric detector sensitivity, $\approx 2.6 \cdot 10^{-3}$ C m^{-2} K^{-1} at 300 K.	70p1
6a	Heat capacity: Fig. 27, additional data, transition heat and transition entropy information, see	65M8, 68P2, 72T2
b	Thermal conductivity: Fig. 28.	
7a	Piezoelectric constant: Fig. 29; see also	64B4, 72H2, 72S9
	Electromechanical coupling factor: Fig. 30; see also	64B4
b	Electrostrictive constant Q_{33}: Fig. 31; see also	72S9
	$\|Q_{13}\| < 10^{-2}$ m^4 C^{-2}, $Q_{23} = -0.05$ m^4 C^{-2}, $Q_{33} = 0.19$ m^4 C^{-2}.	72H2
8a	Elastic compliance and stiffness: Fig. 32; Table 11; see also	64B4, 72H2
	Sound velocity and attenuation: Figs. 33, 34, 35; for additional information, see	70Z1, 75Z3, 70B4
b	Second acoustic harmonic generation and Grüneisen coefficient, see	71S10
9a	SbSI is negative biaxial. Optic angle $2V$ (the acute angle between the optic axes) is approximately 83°.	67J1
	Refractive index at $\lambda = 633$ nm: $n_a = 2.87$, $n_b = 3.63$, $n_c = 4.55$ at 22 °C; $n_a = 2.87$, $n_b = 3.57$, $n_c = 4.44$ at 12 °C.	70S5
	Refractive indices: dependence on λ, Figs. 36, 37 temperature dependence, Figs. 38, 39, 40 effects of electric bias, Figs. 41, 42	
	Reflectivity spectra, and dielectric constant obtained by applying Kramers-Kronig method: Figs. 43···47; see also	69R3, 69P5, 70S24, 70P4, 71A7
	Absorption and transmission: Figs. 48, 49, 50. See also band gap E_G data in subsection 11 below. Electroreflectance, see subsection 11 below.	
b	Electrooptic constant, see	70O3
e	Quadratic electrooptic constant, see	77M9
	Second harmonic generation; information on the optical nonlinear susceptibilities and effects of electric field, see	70H2
10a	Raman scattering: intensity vs. frequency, Figs. 51, 52; effects of pressures Figs. 53, 54, 55, 56; see also	70H6, 70C10, 71A7, 70P10, 71S30
	continued	

Nr. 14A-7 SbSI, continued

10b	Brillouin scattering, see	70S5
11	Conductivity: Fig. 57. Conductivity σ_{dark} of as grown crystal at RT is usually of the order of $10^{-6}\,\Omega^{-1}\mathrm{m}^{-1}$ along the *c*-axis, see also	 64S5, 65S7, 71k2, 69N4, 72A16, 75A1
	Activation energy and information on trapping center, see	66N4, 66A3, 73I5
	Photoconductivity at RT shows a maximum at about $\lambda = 600$ nm.	60N1, 66H9, 73F11, 70B7, 70F7, 71N9
	Information on photoluminescence,	75A4
	Absorption edge: Fig. 58. See also subsection 9 above. Amount of shift ΔE_G of the absorption edge depends on P_s. $\Delta E_G = \beta P_s^2 + \gamma P_s^4$ with $\beta = 1.38$ eV m^{4}C^{-2}, $\gamma = -7.23$ eV m^{8}C^{-4}.	 72I6
	Effect of E_{bias} on the absorption edge: Figs. 59, 60; see also	63H2
	Effect of hydrostatic pressure on the absorption edge: Fig. 61.	
	Life time of photo-excited charge carriers, see	66N4
	Excitation of trapped carriers, thermally stimulated current behavior.	70T12, 73I5
	Piezoresistance data, see	70Z2
	Thermo-electric power: Fig. 62	
	Electroreflectance $\Delta R/R$: Fig. 63; see also	78Z1
	Electronic band structure: Figs. 64, 65; for the band structure and the density of state, see also	71N11, 74F2
12	Magnetic susceptibility: Figs. 3, 66	
13a	NMR and NQR: Fig. 67; Table 3 see also	66K7, 69K10, 70P8, 72K14
c	Mössbauer effect: Fig. 68; Table 4; see also	72A6, 75D7, 73T6
14	Diffraction phenomena: Fig. 69. For diffuse X-ray scattering, see	77L7
	Inelastic neutron scattering, see	77P4
15	Domains, see	70K12, 70K13, 71B6
	Domain switching: Fig. 70; see also	69B12, 69K12, 69R6
16	Dilatation caused by photo-illumination: Fig. 16; see also	70G10
	Acoustoelectric effect for ultrasonic waves, see	69K14
	Characteristic patterns observed with a microscope: Figs. 71, 72; see also	62K9 69T8
	Thin film and surface layer, see	70M3 72H1, 78B8

Table 7. SbSI. Unit cell parameters from different sources (in Å) [64A3, 74I14, 76I7].

T [°C]	a	b	c	Ref.
1.3	8.52_5	10.13_7	4.09_7	76I7
6	8.52_3	10.12_4	4.11_1	64A3
27	8.522 (3)	10.130 (4)	4.088 (2)	74I14
35	8.52_7	10.13_8	4.08_9	64A3
48	8.533 (3)	10.169 (4)	4.090 (2)	74I14

Table 8. SbSI. Fractional coordinates and temperature parameters at 1.3 °C [76I7]. U_{ij} (in Å^2) is defined by Eq. (d) in Introduction.

	Sb	S	I
x	0.12022 (8)	0.8450 (2)	0.50854 (7)
y	0.12335 (6)	0.0472 (2)	0.82839 (5)
z	0.27519 (34)	0.2547 (13)	0.25
U_{11}	0.0146 (2)	0.0104 (7)	0.0196 (2)
U_{22}	0.0184 (2)	0.0123 (7)	0.0149 (2)
U_{33}	0.0254 (3)	0.0139 (8)	0.0161 (2)
U_{12}	−0.0026 (2)	0.0009 (5)	−0.0029 (2)
U_{13}	−0.0003 (5)	0.0014 (16)	−0.0008 (5)
U_{23}	0.0011 (5)	−0.0017 (17)	−0.0009 (4)

Table 9. SbSI. Fractional coordinates and temperature parameters at 27° and 48 °C [74I14]. b_{ij} is defined by Eq. (b) in Introduction.

	27 °C			48 °C		
	Sb	S	I	Sb	S	I
x	0.1198 (1)	0.8453 (3)	0.5085 (1)	0.1197 (1)	0.8447 (3)	0.5086 (1)
y	0.1232 (1)	0.0468 (2)	0.8281 (1)	0.1232 (1)	0.0469 (2)	0.8281 (1)
z	0.25	0.25	0.25	0.25	0.25	0.25
b_{11}	0.00477 (7)	0.00381 (20)	0.00652 (7)	0.00528 (7)	0.00455 (23)	0.00702 (8)
b_{22}	0.00407 (5)	0.00320 (15)	0.00324 (4)	0.00452 (6)	0.00307 (15)	0.00358 (4)
b_{33}	0.03562 (39)	0.01921 (96)	0.01892 (23)	0.03777 (43)	0.02034 (103)	0.02127 (26)
b_{12}	−0.00076 (5)	−0.00008 (1)	−0.00069 (5)	−0.00086 (5)	0.00020 (15)	−0.00080 (5)

Table 10. SbSI. Interatomic distances at 35 °C and 5 °C [67K4].

Interatomic distances Å	Paraelectric phase	Ferroelectric phase
Sb (III)−S (IV)	2.764 (21)	2.605 (27) 2.835 (28)
Sb (III)−S (III)	2.492 (26)	2.475 (17)
Sb (III)−I (I)	3.098 (5)	2.987 (11) 3.247 (12)
Sb (III)−I (IV)	3.813 (6)	3.714 (10) 3.926 (10)
S (III)−S (IV)	3.551 (21)	3.516 (19)
S (IV)−I (I)	4.084 (19)	4.071 (17)
S (IV)−I (II)	3.713 (16)	3.691 (23) 3.741 (23)
S (IV)−I (IV)	3.620 (19)	3.631 (18)
S (IV)−I (III′)	3.815 (16)	3.812 (23) 3.860 (22)
I (I)−I (IV)	4.546 (7)	4.548 (6)
I (I)−I (II′)	4.052 (5)	4.043 (4)

Table 11. SbSI. Elastic constant $c^{D}_{\lambda\mu}$ measured by Brillouin scattering [70S5].

$c^{D}_{\lambda\mu}$ 10^{10} N m^{-2}	Paraelectric phase ≈22 °C	Ferroelectric phase ≈12 °C
c_{11}	3.09 (9)	3.06 (9)
c_{22}	3.27 (4)	3.14 (4)
c_{33}	4.95 (6)	5.18 (6)
c_{44}	2.21 (3)	2.24 (3)
c_{55}	0.92 (3)	0.99 (3)
c_{66}	0.60 (2)	0.59 (2)
c_{12}	0.96 (9)	0.85 (9)
c_{13}	0.93 (30)	0.97 (30)
c_{23}	1.58 (12)	1.44 (12)

Nr. 14A-8 BiSI, Bismuth sulfide iodide
($M=367.94$)

1a	Ferroelectric transition of BiSI at −160 °C was found by Nitsche et al. in 1964.	64N3
b	phase: II / I; state: F[a]) / P[a]); crystal system: — / orthorhombic[b]); space group: — / Pnam-D_{2h}^{16} [b]); Θ: −160 °C[a])	[a]) 64N3 [b]) 50D2
	$\varrho=6.60\cdot10^3$ kg m^{-3} ($\varrho_X=6.75\cdot10^3$ kg m^{-3}).	50D2, 73H7
	Color: black (needles).	60N1
2a	Synthesis: similar method as described in subsection 2a of Nr. 14A-7.	60N1, 64N3, 65H9, 70P6
3a	Unit cell parameters: $a=8.519$ (5) Å, $b=10.177$ (8) Å, $c=4.172$ (6) Å.	73H7
b	$Z=4$. Crystal structure in phase I: all atoms are at 4c positions of Pnam-D_{2h}^{16}. Table 12; see also	50D2
5a	Curie-Weiss constant: $C=1.9\cdot10^5$ K, $\Theta_p=88$ K.	68P2
b	$\xi=-89\cdot10^8$ Jm^{5}C^{-4}, $\zeta=950\cdot10^9$ Jm^{9}C^{-6}	
c	$P_s=7\cdot10^{-2}$ C m^{-2}.	
6a	Transition heat: $\Delta Q_m=8.8$ J mol^{-1}, transition entropy: $\Delta S_m=0.08$ J K^{-1}mol^{-1}.	68P2
11	Conducting properties, and photoconductivity: see	60N1, 65H9, 65S6
12	Magnetic susceptibility: Fig. 3.	
13a	NQR data: Table 13; see also	69K10

phase	II	I
state	F[a])	P[a])
crystal system		orthorhombic[b])
space group		Pnam-D_{2h}^{16} [b])
Θ	−160 °C[a])	

Table 12. BiSI. Structure of phase I [73H7]. Fractional coordinates and temperature factors at RT. B (in Å^2) is defined by Eq. (e) in Introduction.

	Bi	S	I
x	0.1318 (2)	0.8403 (8)	0.5077 (3)
y	0.1230 (2)	0.0500 (8)	0.8307 (3)
z	0.25	0.25	0.25
B	1.88 (2)	1.46 (10)	1.83 (3)

Table 13. BiSI. Nuclear quadrupole resonance frequencies *f*, asymmetry parameters η, and e^2qQ/h of ^{209}Bi [75P12].

T [°C]	frequency [MHz]	^{209}Bi transition	η	e^2qQ/h [MHz]
30	19.609 (9)	1/2–3/2	0.1830 (3)	372.77 (1)
	29.992 (2)	3/2–5/2		
	46.299 (2)	5/2–7/2		
	61.959 (2)	7/2–9/2		
–196	19.611 (4)	1/2–3/2	0.1750 (2)	378.91 (1)
	30.558 (2)	3/2–5/2		
	47.090 (2)	5/2–7/2		
	62.996 (3)	7/2–9/2		

Nr. 14A-9 SbSeI Antimony selenide iodide
(M = 327.61)

1a	No ferroelectric transition was found in the temperature range between –180 °C and 70 °C.	64N3
b	Crystal system: orthorhombic; space group: Pnam-D_{2h}^{16}	
	$\varrho = 5.7_7 \cdot 10^3$ kg m^{-3}, ($\varrho_X = 5.8_8 \cdot 10^3$ kg m^{-3}).	50D3
	Color: black (needles).	60N1
2a	Synthesis, single crystal growth: similar method as described in subsection 2a of Nr. 14A-7.	60N1, 64N3, 66H9, 69P7, 70P7
3a	Unit cell parameters: $a = 8.6_5$ Å, $b = 10.3_8$ Å, $c = 4.1_2$ Å at RT.	50D3
b	$Z = 4$, all atoms are at 4c positions of Pnam-D_{2h}^{16}.	50D3
5d	Pyroelectric effect: see	67G10
6a	Heat capacity: see	68P2
9a	Reflectivity: see	66H9, 70R7
11	Electrical conductivity, photoconduction: see	60N1, 66H9, 68P2
12	Magnetic susceptibility: Fig. 3	
13c	Mössbauer spectra: Fig. 68; Table 4.	

Nr. 14A-10 BiSeI, Bismuth selenide iodide
(M = 414.84)

1a	No ferroelectric transition was found between –180° and –100 °C.	64N3
b	crystal system: orthorhombic; space group: Pnam-D_{2h}^{16}	50D3
	$\varrho = 7.0_5 \cdot 10^3$ kg m^{-3} ($\varrho_X = 7.1_6 \cdot 10^3$ kg m^{-3}).	50D3
	Color: black (needles). continued	60N1

Nr. 14A-10 BiSeI, continued

2a	Synthesis and crystal growth: as described in subsection 2a of Nr. 14A-7.	60 N 1, 64 N 3, 70 H 12
3a	Unit cell parameters: $a=8.7_1$ Å, $b=10.5_4$ Å, $c=4.1_9$ Å at RT.	50 D 3
b	$Z=4$, all atoms are at 4c positions (Pnam-D_{2h}^{16}).	50 D 3
11	Photoconduction, energy band gap ($E_G \approx 0.39$ eV), see also	60 N 1 70 H 12, 72 B 28
12	Magnetic susceptibility: Figs. 3, 66.	

Nr. 14A-11 SbTeI, Antimony telluride iodide
($M=376.25$)

1b	Crystal system: orthorhombic. $\varrho_X=5.96 \cdot 10^3$ kg m^{-3}. Color: black (needles).	51 D 1 60 N 1
2a	Crystal growth, see	60 N 1, 66 H 9, 70 P 7
3a	Unit cell parameters: $a=9.1_8$ Å, $b=10.8$ Å, $c=4.2_3$ Å at RT.	51 D 1
b	$Z=4$.	51 D 1
9a	Reflectivity, see	66 H 9
13c	Mössbauer spectra: Fig. 68; Table 4.	

14B Solid solutions

Nr. 14B-1 SbSBr-SbSI

1b	The solid solution $(SbSBr)_x \cdot (SbSI)_{1-x}$ shows a ferroelectric transition. The ferroelectric transition temperature Θ_f falls with increasing x: Figs. 73, 74.	64 N 3
2a	Synthesis and crystal growth: similar method as described in subsection 2a of Nr. 14A-7.	64 N 3
3a	Unit cell parameters at RT (orthorhombic): Fig. 75.	
10a	Raman scattering: Fig. 76.	

Nr. 14B-2 SbSI-SbOI

1b	$SbO_{0.2}S_{0.8}I$ shows a ferroelectric transition: $\Theta_f=65$ °C. Data on Θ_f as a function of sample location in polycrystalline ingots were given for $SbO_{0.05}S_{0.95}I$ and $SbO_{0.2}S_{0.8}I$.	64 N 3
2a	Partial replacement of S by O in SbSI is possible under conservation of the SbSI structure (see Nr. 14A-7), although the isostructural compound SbOI is not known to exist.	64 N 3

Nr. 14B-3 SbSI-AsSI

1b	$As_{0.1}Sb_{0.9}SI$ shows a ferroelectric transition: $\Theta_f=0$ °C.	64 N 3
2a	Partial replacement of Sb by As in SbSI is possible under conservation of the SbSI structure (see Nr. 14A-7), although the isostructural pure compound AsSI is not known to exist.	64 N 3

Nr. 14B-4 SbSI-BiSI

1b	Ferroelectric transition temperature: Fig. 77; see also	76T6
2a	Synthesis: similar to those described in subsection 2a of Nr. 14A-7.	64N3
5a	Dielectric properties, see	74I18
10a	Raman scattering: Fig. 78; see also	76T6, 78T5

Nr. 14B-5 SbSI-SbSeI

1b	The ferroelectric transition temperature falls as the SbSeI content increases: Fig. 74.	
3a	Unit cell parameters: Fig. 79.	
5a	Dielectric constant: Fig. 80.	
9a	Optical data: see	73P14

15 HCl family

15A Pure compounds

Nr. 15A-1 HCl, Hydrogen chloride

($M = 36.46$)

1a Ferroelectricity in HCl was discovered by Hoshino et al. in 1967. — 67H11

b

phase	II	I*
state	F	P
crystal system	orthorhombic	cubic
space group	$Bb2_1m$-C_{2v}^{12} [a]	$Fm3m$-O_h^5 [b]
Θ	98.36 [c] K	

[a] 67S2, [b] 67S3, [c] 28G1

$\Theta_{II-I} = 105.03$ K for deuterated compound. — 47C1

$T_{melt} = 158.91$ K [a] (158.44 K for deuterated compound [b]). — [a] 28G1, [b] 47C1

*) It has been reported by Niimura et al. [72N8] that there exists an extra phase transition at 120 K. However, this has not been confirmed as yet by other investigators.

2a Crystal growth: cooling method from liquid phase. — 67H11

3a Unit cell parameters: — 67S2

Compound	Phase	T	a	b	c
		K	Å		
HCl	I	118.5	5.482 (1)		
	II	92.4	5.082 (1)	5.410 (1)	5.826 (1)
DCl	I	118.5	5.475 (1)		
	II	92.4	5.0685 (1)	5.399 (1)	5.828 (1)

b $Z = 4$ in phases I and II. — 67S2
Crystal structure: Fig. 81, Tables 14, 15; angular displacements due to molecular vibration, Fig. 82.

4 Thermal expansion of DCl in phase I: Fig. 83; see also — 75P14

continued

Nr. 15A-1 HCl, continued

5a	Dielectric constant: Fig. 84. Dielectric dispersion: see Phase diagram with regard to p: Fig. 85.	 54S7, 55H1 57C1 67G12
c	Spontaneous polarization: Fig. 86.	
6a	Heat capacity: Figs. 87, 88 Transition heat and transition entropy of the II–I transition:	
	Compound | ΔQ_m J mol^{-1} | ΔS_m J K^{-1} mol^{-1} HCl [a]) | 1187 | 12.1 DCl [b]) | 1338 | 12.75	 [a]) 28G1 [b]) 47C1
8a	Elastic stiffnesses of DCl obtained from phonon dispersion curves measured by neutron inelastic scattering: $c_{11}=6.29$ (23) $\cdot 10^9$ N m^{-2}, $c_{12}=3.95$ (22) $\cdot 10^9$ N m^{-2}, $c_{44}=2.47$ (12) $\cdot 10^9$ N m^{-2}, $T=109$ K (in phase I). Volume compressibility: $\beta=3.67\cdot10^{-10}$ Pa^{-1} at 130 K.	 75P14 62S15
9a	Far infrared absorption: Fig. 89; see also	55H4, 62A2, 71C2
10a	Raman scattering: lattice mode frequencies, Table 16; see also	65B9
13a	NMR: Figs. 90···92. NQR: Figs. 93, 94.	
14b	Phonon dispersion: Fig. 95. Inelastic scattering of cold neutrons: see Incoherent neutron scattering by proton motions in phase I: see	 65B9 78F4

Table 14. DCl. Positional and temperature parameters. B, β_{ij} are defined by Eqs. (e), (c) in Introduction.

77.4 K [67S2]. $R=2.9\%$

	x	y	z	B [Å^2]
D	0.081 (2)	0.170 (4)	0	2.7 (5)
Cl	0.25	0	0	1.71 (1)

111.5 K [67S3]. $R=1.9\%$

	$x(=y)$	z	$\beta_{11}(=\beta_{22})$ [Å^2]	β_{33} [Å^2]	β_{12} [Å^2]	$\beta_{13}(=\beta_{23})$
D (1/12 atom in 48-fold position)	0.151 (5)	0	7.5 (15)	15.7 (48)	−3.65 (180)	0
Cl (in 4-fold position)	0	0	3.55 (30)	3.55 (30)	0	0

Table 15. DCl. Bond lengths and bond angles.

T=77.4 K [67S2]	
D–Cl	1.25 (2) Å
Cl–Cl	3.688 (1) Å
Cl–Cl–Cl	93°31 (2)′
T=111.5 K [67S3]	
D–Cl	1.17 (4) Å

Table 16. HCl. Lattice mode frequencies in [cm^{-1}].

Raman				IR
[69I3] T=77 K	[70W2] 77 K	[72S33] 79 K	[77V3] 74 K	[63A1]
—	50?	—	49	—
55 (A_2)	59 (A_2)	58 (A_2)	58	—
84 (A_1)	87 (A_1)	84 (A_1)	85	86
—	113 ($2A_2$)	110 (B_1)	110	109
133 (A_2+A_1)	138?	135 (A_2)	134 (A_1)	—
218 (B_2)	221 (B_2)	217 (A_1)	219 (A_2)	217
290 (A_1)	—	—	—	296
—	—	333 (B_2)	—	—
—	408?	399 (B_1)	403 (B_1)	—
—	—	—	—	496

Nr. 15A-2 HBr, Hydrogen bromide
(M=80.91)

1a	Ferroelectricity in HBr was discovered by Hoshino et al. in 1967.	67H11
b	see table below	a) 71S23, b) 68S5, c) 70S15, d) 47C1

phase	IV	III	II	I
state	F			
crystal system	ortho-rhombic	ortho-rhombic	cubic a)	cubic
space group	$Bb2_1m - C_{2v}^{12}$ b)	$Bb\,cm - D_{2h}^{18}$ b)		$Fm\,3m - O_h^5$ c)
Θ	89.75 d)	113.62 d)	116.86 d)	K

	In deuterated compound, phase II does not appear; phase diagram of HBr–DBr system, Fig. 96.	
	Θ_{III-I}=120.26 K, Θ_{IV-III}=93.5 K in deuterated compound.	47C1
	T_{melt}=186.28 K (185.62 K in deuterated compound).	47C1
2a	Crystal growth: cooling method from liquid phase.	67H11
3a	Unit cell parameters: Table 17.	
b	Z=4 in phases I a), III b), IV b). Crystal structures: Table 18.	a) 70S15, b) 68S5
4	Thermal expansion: Table 19.	
5a	Dielectric constant: Fig. 97. Dielectric dispersion: see	53B2, 54S7, 55H1, 57C1, 67G12
c	Spontaneous polarization: Fig. 98.	
6a	Heat capacity: Fig. 99. Transition heat and transition entropy: Table 20.	
8a	Volume compressibility: $\beta=3.72\cdot10^{-10}$ Pa^{-1} at 150 K. continued	62S15

Nr. 15A-2 HBr, continued

9a	Infrared absorption: Fig. 100; see also	71C2
10a	Raman scattering: Figs. 101, 102; see also	64A2, 70I4
13a	NMR: Figs. 90, 92, 103, 104 NQR: Fig. 105; see also	68N4
14b	Inelastic scattering of cold neutrons: see	65B9

Table 17. HBr, DBr. Unit cell parameters.

Compound	Phase	T K	Unit cell parameters Å	Ref.
HBr	I	130	a = 5.815 (15)	71S23
	II	115	a = 5.785 (20)	
	III	100	a = 5.555 (25) b = 5.648 (20) c = 6.086 (20)	
	IV	85	a = 5.462 (3) b = 5.643 (3) c = 6.111 (3)	
DBr	III	107	a = 5.559 (5) b = 5.649 (5) c = 6.095 (5)	68S5
	IV	84	a = 5.44 (2) b = 5.614 (5) c = 6.120 (5)	

Table 18. DBr. Fractional coordinates and temperature parameters [68S5]. B is defined by Eq. (e) in Introduction.

Atom	x	y	z	B [Å²]	x	y	z	B [Å²]
	T = 84 K (phase IV)				T = 107 K (phase III)			
D	0.075 (2)	0.179 (3)	0	5.0 (5)	−0.172 (6)	0.177 (4)	0	8.4 (8)
Br	0.25	0	0	2.1 (4)	0	0	0	4.4 (8)

Table 19. HBr. Linear thermal expansion coefficients [71S23].

Phase	T K	α_{ij} 10^{-6} K^{-1}
I	130	α_{11} = 225 (15)
II	115	α_{11} = 225 (40)
III	100	α_{11} = 430 (20) α_{22} = 370 (20) α_{33} = −60 (30)

Table 20. HBr, DBr. Transition heat and transition entropy.

Compound	Transition	ΔQ_m J mol^{-1}	ΔS_m J K mol^{-1}	Ref.
HBr	I–II–III	691	5.98	24E1
	III–IV	472	5.27	
DBr	I–III	1267	10.5	47C1
	III–IV	709	7.6	

Figs. p. 329, 333

Nr. 15A-3 HI, Hydrogen iodide
($M = 127.91$)

		Ref.
1a	A peak in dielectric constant has been observed at the III–II transition. However, ferroelectricity in phase III has not been recognized so far.	57C1
b	see table below	a) 70S15 b) 47C1
	$\Theta_{II-I} = 128.28$ K, $\Theta_{III-II} = 77.3$ K for deuterated compound.	47C1
	$T_{melt} = 221.31$ K (221.23 K for deuterated compound).	47C1
3a	Unit cell parameters: see table below	70S15
b	$Z = 4$ in phases I, II, III.	70S15
5a	Dielectric constant: Fig. 106. Dielectric dispersion: see	55H1, 57C1, 67G12
6a	Heat capacity: Fig. 107. Transition heat and transition entropy: Table 21.	
13a	NMR: Figs. 90, 108.	

1b:

phase	III	II	I
crystal system		orthorhombic	cubic
space group		Bbcm–D_{2h}^{18} a)	Fm3m–O_h^5 a)
Θ		70.1 b)	125.68 b)

3a:

Phase	T	a	b	c
	K	Å		
I	130	6.290		
II	100	6.049	6.127	6.566

Table 21. HI, DI. Transition heat and transition entropy per mole.

Compound	Transition	ΔQ_m J mol^{-1}	ΔS_m J K mol^{-1}	Ref.
HI	II–I	526	4.18	24E1
	III–II	295	4.22	
DI	II–I	1615	12.6	47C1
	III–II	734	9.49	

16 $NaNO_2$ family

16A Pure compounds

Nr. 16A-1 $NaNO_2$, Sodium nitrite
(M=69.00)

1a Ferroelectricity in $NaNO_2$ was discovered by Sawada et al. in 1958. — 58S2

b

phase	IV	III	II	I
state	F	F	A*	P
crystal system		orthorhombic		orthorhombic
space group		Im2m-C_{2v}^{20} [a]		Immm-D_{2h}^{25} [b]
Θ	−95 [c]	163.9 [d]	165.2 [d]	°C

[a] 31Z1
[b] 43S1
[c] 69G6
[d] 75E2

*) A long period structural modulation was found in phase II; see sections 3b and 14a.

$P_s \parallel [010]$. — 58S2
T_{melt}=281.0 °C. — 63C6
$\varrho = 2.157 \cdot 10^3\ kg\ m^{-3}$ at RT. — 31Z2
Transparent, light yellowish. Deliquescent.
Cleavage plane: {101}.

2a Crystal growth: evaporation or cooling of aqueous solution; solubility in H_2O, Table 22; cooling from melt. — 61N4

b Crystal form, Fig. 109.

3a Unit cell parameters: a=3.560 (10) Å, b=5.563 (5) Å, c=5.384 (5) Å at 26 °C; unit cell parameters at different temperatures, Table 27. — 61K3

b Z=2 in phases III and I.
Crystal structures of phases III and I: Figs. 110, 111; Tables 23···26; see also — 54T1, 55C2, 61K3, 62K6

Structure of phase II: The net polarization of the (100) plane is sinusoidally modulated along the [100] direction with a period of $8a \cdots 11a$. — 63Y2, 67H10, 78B11

4 Thermal expansion: Figs. 112···114; see also — 61H6
Thermal expansion in the vicinity of the III–II–I transitions, Fig. 115; see also — 65S1, 65M5, 69G6

5a Dielectric constant: temperature dependence, Fig. 116; temperature dependence in the vicinity of the III–II–I transitions, Figs. 117, 118; deviation from Curie-Weiss law, Fig. 119.
Dielectric dispersion: Figs. 120, 121.
Phase diagram with regard to p: Fig. 122.
$[d\Theta_{III-II}/dp]_{p=0} = 4.9 \cdot 10^{-8}$ K Pa^{-1}, $[d\Theta_{II-I}/dp]_{p=0} = 5.6 \cdot 10^{-8}$ K Pa^{-1}. — 65G3
Effect of bias field: Fig. 123; phase diagram with regard to E_{bias}, Fig. 124.

b Non-linear dielectric property:
$\xi = -1 \cdot 10^{11}\ V\ m^5\ C^{-3}$, $\zeta = 4.8 \cdot 10^{13}\ V\ m^9\ C^{-5}$. — 61N3

c Spontaneous polarization: Fig. 125; spontaneous polarization at high pressure, Fig. 126.

d Pyroelectricity: see — 61S2

6a Heat capacity: Figs. 127, 128; critical behavior of the molar heat capacity anomaly, Fig. 129.
Transition heat and transition entropy for the III→II→I transitions: ΔQ_m=2215 J mol^{-1}, ΔS_m=5.27 J K^{-1} mol^{-1}. — 65S1

b Thermal conductivity: Figs. 130, 131, 132.

7a	Piezoelectricity: Figs. 133, 134.	
b	Electrostriction: Figs. 135, 136.	
8a	Elastic compliance and stiffness: Figs. 137, 138; Table 28; temperature dependence of elastic stiffness coefficients obtained from ultrasonic measurements, Fig. 139; elastic constants obtained from Brillouin scattering, see	70H8, 74S18, 74S17
	Linear compressibility: Fractional variation of unit cell parameters with p; $-\Delta a/a=\gamma_{11}p+\gamma_{12}p^2$, $-\Delta b/b=\gamma_{21}p+\gamma_{22}p^2$, $-\Delta c/c=\gamma_{31}p+\gamma_{32}p^2$, with $\gamma_{11}=2.63\,(10)\cdot10^{-2}\,\text{GPa}^{-1}$, $\gamma_{12}=-2.1\,(7)\cdot10^{-3}\,\text{GPa}^{-2}$, $\gamma_{21}=1.22\,(6)\cdot10^{-2}\,\text{GPa}^{-1}$, $\gamma_{22}=-9\,(3)\cdot10^{-4}\,\text{GPa}^{-2}$, $\gamma_{31}=7.3\,(7)\cdot10^{-3}\,\text{GPa}^{-1}$, $\gamma_{32}=-1.1\,(4)\cdot10^{-3}\,\text{GPa}^{-2}$, $p<2.7$ GPa.	79H6
	Ultrasonic attenuation: Figs. 140···144.	
9a	Refractive indices: Fig. 145; infrared refractive indices: Table 29. Far-infrared absorption: Figs. 146, 147.	
b	Electrooptic effect: Fig. 148. Electrooptic constants: $r_{42}=-3.0\,(2)\cdot10^{-12}\,\text{m V}^{-1}$, $r_{51}=-1.9\,(2)\cdot10^{-12}\,\text{m V}^{-1}$ for $\lambda=546$ nm.	69J2
c	Piezooptic effect: Table 30; see also	72H6
d	Optical rotatory power: Fig. 149.	
e	Nonlinear optical property: Fig. 150; see also	70V3, 70Y3, 72I3
10a	Raman scattering: Figs. 151···153; Table 31.	
b	Brillouin scattering: see	70H8, 74S18, 74S17
11	Electrical conduction: Figs. 154, 155; Table 32; see also Electrical breakdown: Fig. 156.	67T1
13a	NMR: Fig. 157, 158; see also	74S21, 75K5, 76P13
	NQR: Figs. 159···165; effect of high pressure, see	74S15
b	ESR: zero field parameters of NO_2, $D=0.444\,(5)\cdot10^2\,\text{m}^{-1}$, $E=-0.045\,(5)\cdot10^2\,\text{m}^{-1}$, $g=2.002\,(3)$,	73D4
	see also	76D7
	ESR of Mn^{2+} doped crystal, see	72U3, 78J2
	ESR of irradiated crystals, see	61Z1, 64T3, 69L7, 69G12
14a	Bragg reflections due to structural modulation: Figs. 166···168. Diffuse and critical scattering: Figs. 169···173. Neutron inelastic scattering: Fig. 174.	
15a	Domains: Domain structure, Fig. 175; domains were observed by etching or by X-ray diffraction topograph; see continued	61N4,

Nr. 16A-1 $NaNO_2$, continued

15a		66S26, 67S5, 71S36
b	Switching: Fig. 176; see also	65H4
	switching process was observed by time-resolved X-ray diffraction.	77I10
16	Twin structure: Fig. 177.	
	Plastisity: see	66T1
	Etch pits: see	67S5
	Whiskers: grown in [010] from aqueous solution.	67U1
	Radiation effect: see	64G2, 79J2
	Surface layer: see	76M3
	Positron annihilation: see	79K5

Table 22. $NaNO_2$. Solubility A in H_2O [62k1].

A [wt%]	41.63	43.7	45.07	52.00	54.07	56.95	61.5
T [°C]	0	15	22	56.8	64.7	79.5	99.9

Table 23. $NaNO_2$. Fractional coordinates, temperature parameters, occupancy factors [72K7]. The second column under 150 °C gives the coordinates from the refinement including the rotational motion. The second column under 225 °C includes results refined with anisotropic extinction coefficients. b_{ij} is defined by Eq. (b) in Introduction. f_1: fraction in the obverse position in phase III, f_2: fraction in the reverse position, f_3: fraction in dynamical rotation in two-or four-fold wells. $f_1+f_2+f_3=1$. Ordinary order-disorder model corresponds to the case of $f_3=0$

T [°C]	150		185	225	
Fractional coordinates					
Na(y)	0.5386 (21)	0.5428 (22)	0.05390 (20)	0.5352 (35)	0.5337 (43)
N(y)	0.0689 (12)	0.0767 (23)	0.0719 (4)	0.0699 (5)	0.0704 (5)
O(y)	−0.0489 (10)	−0.0492 (18)	−0.0435 (10)	−0.0465 (10)	−0.0468 (8)
O(z)	0.1951 (11)	0.1951	0.1920 (4)	0.1910 (4)	0.1912 (4)
Temperature parameters					
Na b_{11}	0.055 (9)	0.057 (8)	0.069 (4)	0.084 (4)	0.082 (4)
b_{22}	0.014 (4)	0.012 (4)	0.031 (6)	0.032 (8)	0.036 (9)
b_{33}	0.027 (4)	0.025 (3)	0.028 (1)	0.033 (2)	0.032 (2)
N b_{11}	0.078 (6)	0.058 (7)	0.117 (2)	0.131 (3)	0.130 (4)
b_{22}	0.029 (2)	0.027 (2)	0.032 (1)	0.032 (1)	0.032 (1)
b_{33}	0.025 (2)	0.031 (3)	0.030 (1)	0.036 (1)	0.033 (1)
O b_{11}	0.082 (5)	0.083 (6)	0.116 (3)	0.136 (3)	0.131 (3)
b_{22}	0.032 (3)	0.034 (3)	0.045 (3)	0.044 (4)	0.044 (3)
b_{33}	0.020 (2)	0.028 (3)	0.031 (1)	0.0354 (9)	0.0352 (8)
b_{23}	0.003 (2)	0.004 (2)	0.004 (1)	0.007 (1)	0.007 (1)
Occupancies					
f_1	0.875 (4)	0.79 (2)			
f_2	0.125	0.05 (2)			
f_3		0.16 (2)			

Table 24. $NaNO_2$. Interatomic distances, bond angles, principal values of root-mean-square amplitude of thermal vibrations [72K7]. μ_{xx}, μ_{yy}, μ_{zz}: root-mean-square amplitudes refer to principal axes in the x, y, z-direction, respectively. μ_y, μ_z, angle: refer to the principal axes closest to the y and z-directions, and the angle between μ_y and the b-axis for the oxygen atom motion. Both refinements at 150 °C without and with rotation, in that order, are given. The 225 °C results are from the refinement with anisotropic extinction parameters.

T[°C]		150	150	185	225
Bond distances and angles					
N–O	[Å]	1.242 (8)	1.267 (6)	1.217 (5)	1.214 (6)
Na–O	[Å]	2.499 (9)	2.505 (9)	2.509 (6)	2.510 (8)
Na–O	[Å]	2.557 (17)	2.533 (9)	2.579 (13)	2.597 (19)
Na–N	[Å]	2.655 (20)	2.635 (25)	2.648 (13)	2.642 (24)
O–N–O	[°]	115.0 (10)	112.0 (10)	114.9 (6)	113.2 (6)
Thermal ellipsoids					
Nitrogen					
μ_{xx}	[Å]	0.230 (9)	0.197 (6)	0.282 (3)	0.300 (4)
μ_{yy}	[Å]	0.215 (8)	0.209 (5)	0.227 (3)	0.232 (4)
μ_{zz}	[Å]	0.193 (8)	0.193 (6)	0.210 (3)	0.218 (4)
Oxygen					
μ_{xx}	[Å]	0.235 (7)	0.236 (5)	0.282 (3)	0.300 (4)
μ_y	[Å]	0.228 (11)	0.239 (11)	0.273 (9)	0.276 (10)
Angle	[°]	10 (7)	23 (8)	10 (3)	21 (5)
μ_z	[Å]	0.168 (9)	0.212 (4)	0.208 (3)	0.214 (5)
Sodium					
μ_{xx}	[Å]	0.19 (2)	0.197 (8)	0.217 (4)	0.238 (7)
μ_{yy}	[Å]	0.15 (2)	0.137 (48)	0.229 (2)	0.243 (28)
μ_{zz}	[Å]	0.20 (2)	0.193 (7)	0.203 (4)	0.212 (6)

Table 25. $NaNO_2$. Fractional coordinates, temperature parameters, occupancy factors [75K10]. The second column under 162 °C includes rotational motion about c. Definitions of b_{ij} f_i: see caption of Table 23.

T[°C]		RT	145	150	162	162
Nitrogen	y	0.1188 (4)	0.0727 (11)	0.0723 (6)	0.0764 (6)	0.0783 (12)
Oxygen	y	0	−0.0435	−0.0457	−0.0408 (15)	−0.0416 (15)
	z	0.1953 (4)	0.1949 (9)	0.1937 (6)	0.1931 (6)	0.1930 (5)
Sodium	y	0.5830 (11)	0.5346 (23)	0.5353 (13)	0.5439 (12)	0.5448 (14)
Nitrogen	b_{11}	0.0501 (20)	0.0784 (44)	0.0912 (33)	0.0935 (38)	0.0861 (47)
	b_{22}	0.0081 (9)	0.0140 (14)	0.0165 (11)	0.0216 (11)	0.0204 (12)
	b_{33}	0.0109 (11)	0.0191 (23)	0.0217 (16)	0.0237 (13)	0.0244 (11)
Oxygen	b_{11}	0.0562 (19)	0.0900 (44)	0.0939 (33)	0.1009 (36)	0.1032 (32)
	b_{22}	0.0135 (11)	0.0292 (29)	0.0288 (18)	0.0337 (17)	0.0352 (24)
	b_{33}	0.0113 (7)	0.0163 (16)	0.0210 (12)	0.0231 (12)	0.0240 (10)
	b_{23}	0.0011 (10)	0.0044 (22)	0.0006 (13)	0.0040 (13)	0.0041 (14)
Sodium	b_{11}	0.0300 (38)	0.0594 (91)	0.0599 (61)	0.0689 (58)	0.0695 (44)
	b_{22}	0.0130 (17)	0.0231 (48)	0.0220 (22)	0.0170 (28)	0.0154 (39)
	b_{33}	0.0139 (19)	0.0174 (36)	0.0247 (24)	0.0253 (23)	0.0253 (17)
f_1				0.883	0.768 (5)	0.710 (19)
f_2				0.117	0.232 (5)	0.220 (19)
f_3						0.070 (33)

Table 26. $NaNO_2$. Interatomic distances, bond angles, principal values of root-mean-square amplitudes of thermal vibrations [75 K 10]. μ_{xx}, μ_{yy}, μ_{zz}: root-mean-square amplitues of motion along the principal axes of the thermal ellipsoids in the x, y, and z-directions. μ_y, μ_z: refer to motion along principal axes close to y and z. Angle: angle between the major principal axis of oxygen in the (100) plane and the y-axis.

T[°C]		RT	145	150	162
N–O	[Å]	1.244 (4)	1.237 (5)	1.236 (4)	1.229 (4)
O–N–O	[°]	115.6 (3)	115.8 (6)	114.7 (4)	114.8 (4)
Na–O	[Å]	2.554 (8)	2.605 (15)	2.587 (29)	2.566 (9)
Na–N	[Å]	2.590 (10)	2.611 (17)	2.617 (32)	2.642 (10)
Root mean square amplitudes of motion					
Nitrogen	μ_{xx} [Å]	0.180 (4)	0.230 (7)	0.203 (9)	0.251 (5)
	μ_{yy} [Å]	0.126 (6)	0.167 (10)	0.193 (10)	0.187 (5)
	μ_{zz} [Å]	0.113 (6)	0.150 (8)	0.193 (10)	0.186 (5)
Oxygen	μ_{xx} [Å]	0.190 (3)	0.247 (6)	0.253 (5)	0.261 (5)
	μ_y [Å]	0.148 (6)	0.218 (11)	0.219 (7)	0.237 (6)
	μ_z [Å]	0.127 (5)	0.154 (8)	0.180 (5)	0.179 (5)
	Angle [°]	17.2 (14)	5.1 (80)	0.6 (70)	15.1 (46)
Sodium	μ_{xx} [Å]	0.143 (9)	0.200 (15)	0.203 (9)	0.216 (9)
	μ_{yy} [Å]	0.143 (10)	0.194 (20)	0.193 (10)	0.192 (9)
	μ_{zz} [Å]	0.139 (9)	0.160 (16)	0.193 (10)	0.166 (14)

Table 27. $NaNO_2$. Unit cell parameters [72 K 7].

T[°C]	150	185	225
a [Å]	3.642 (19)	3.671 (12)	3.695 (17)
b [Å]	5.653 (20)	5.670 (8)	5.701 (30)
c [Å]	5.375 (16)	5.341 (18)	5.298 (22)

Table 29. $NaNO_2$. n_b, n_c vs. λ [74 C 6].

λ µm	n_c	n_b
1.014	1.631_3	1.405
1.3673	1.621_4	1.401_8
1.5295	1.616_0	1.401_0
1.7109	1.613_6	1.401_0
2.25	1.610_2	1.399_7
3.4	1.593_3	1.398_0
4.4	1.540_0	1.395_0
5.4	1.495_0	1.390_7
6.	1.462_6	1.388_0

Table 28. $NaNO_2$. $c_{\lambda\mu}$, $s_{\lambda\mu}$ [70 H 8]. T=21 °C.

$c_{\lambda\mu}[10^9\ N\ m^{-2}]$								
c_{11}	c_{22}	c_{33}	c_{44}	c_{55}	c_{66}	c_{12}	c_{13}	c_{23}
30.1 (2)	55.8 (2)	63.4 (3)	12.0 (2)	10.0 (2)	4.8 (2)	12.3 (30)	16.4 (30)	17.0 (30)
$s_{\lambda\mu}[10^{-11}\ m^2\ N^{-1}]$								
s_{11}	s_{22}	s_{33}	s_{44}	s_{55}	s_{66}	s_{12}	s_{13}	s_{23}
4.06	2.05	1.91	8.33	10	20.8	−0.63	−0.80	−0.39

Table 30. $NaNO_2$. Piezooptic constants measured by Brillouin scattering [75 S 21].

$\|p_{11}\|=0.44$	$\|p_{12}\|=0.37$	$\|p_{13}\|=0.36$
$\|p_{21}\|=0.39$	$\|p_{22}\|=0.33$	$\|p_{23}\|=0.27$
$\|p_{31}\|=0.18$	$\|p_{32}\|=0.19$	$\|p_{33}\|=0.15$

Table 31. $NaNO_2$. ν_0/c, γ_{TO}/c at RT [72H3]. ν_0: frequency of phonon mode, γ_{TO}: damping constant.

	ν_0/c [cm^{-1}] (IR)				ν_0/c [cm^{-1}] (Raman)						γ_{TO}/c [cm^{-1}]	Dielectric transition strengths[a]) S	Assignment
	[68A6]		[68B20]		[67C11]		[68T8]		[72H3]				
	TO	LO	TO	LO	TO	LO	TO	LO	TO	LO			
A_1	194	269	186.8									2.085 [b])	
	826	829	825.3		825		830		827.5	829	1.5	0.0715 [b])	symmet. stretch
	1323	1336	1320.5		1327		1323		1326	1328	7.5	0.0041 [b])	symmet. bend
A_2					119		117		119.5		10	0.000	libration
B_1	157	163	150.9		153		158		154	165	8	1.519	libration
	188	250	180.7		177		191		184.5	236	15	1.420	translation
	1235	1368	1227.3		1280		1230		1225	1356	14	0.608	asymmetric stretch
B_2	149	193	145.7						149.5	201	16	2.085	translation
	223	261	231.7		220		223		227.5	254	25	0.178	libration

a) Calculation was made with Eq. (5) of [71H12], using the values of [72H3].
b) Calculation uses the three LO and TO frequencies of the external mode (194···TO) from [68A6].

Table 32. $NaNO_2$. ΔU. $\varrho = \varrho_0 \exp(\Delta U/kT)$.

Temperature range	ΔU [eV] in			
	a-direction	b-direction		c-direction
$T < 155$ °C	1.2_5	1.5_5	1.05	1.8_5
165 °C $< T <$ 178 °C			1.39	
$T > 178$ °C	1.2_5	1.3	1.23	1.1_5
Ref.	62A4	62A4	67T1	62A4

Nr. 16A-2 $AgNa(NO_2)_2$, Silver sodium nitrite
($M = 222.87$)

1a Ferroelectricity in $AgNa(NO_2)_2$ was discovered by Gesi in 1969. — 69G7

b

phase	II	I
state	F	P
crystal system	orthorhombic	orthorhombic
space Group	Fd2d-C_{2v}^{19} a)	Fddd-D_{2h}^{24} a)
Θ_f	38 b)	°C

a) 72I5
b) 70G3

P_s ‖[010]. — 69G7
$\varrho = 3.26 \cdot 10^3$ kg m^{-3} at RT. — 53C1
Transparent, yellowish. Cleavage plane: {101}.

2a Crystal growth: evaporation of aqueous solution; solubility of $AgNO_2$–$NaNO_2$–H_2O, Fig. 178.

b Crystal form: Fig. 179.

3a Unit cell parameters: — 74I9

phase	T	a	b	c
	°C	Å		
II	20	7.89_1	10.70_8	10.84_6
I	50	8.05_4	10.77_3	10.76_0

continued

Nr. 16A-2 $AgNa(NO_2)_2$, continued

3b	$Z=8$ in phases II and I. Crystal structure: Figs. 180, 181; Tables 33, 34.	74I9
4	Thermal expansion: Figs. 182, 183.	
5a	Dielectric constant: Fig. 184. Curie-Weiss behavior, $\varkappa_0=\varkappa_\infty+C/(T-\Theta_p)$, $T>\Theta_{II-I}$, where $C=4.63\cdot10^3$ K, $\Theta_p=36.89$ °C, $\varkappa_\infty=6$. Dielectric relaxation: Figs. 185⋯187. Effect of hydrostatic pressure: $[d\Theta_f/dp]_{p=0}=9.4\,(2)\cdot10^{-8}$ K Pa^{-1}, $[d\Theta_p/dp]_{p=0}=9.3\,(2)\cdot10^{-8}$ K Pa^{-1}; see also Effect of hydrostatic pressure on dielectric relaxation: Fig. 188.	70G4 73G2 70B5
b	Nonlinear dielectric property: $\xi=-1.45\,(25)\cdot10^{10}$ V m^5C^{-3}, $\zeta=7.4\,(7)\cdot10^{12}$ V m^9C^{-5}.	77H9
c	Spontaneous polarization: Fig. 189.	
6a	Specific heat capacity: Figs. 190, 191; transition heat and transition entropy, $\Delta Q_m=3.41$ kJ mol^{-1}, $\Delta S_m=12.2$ J K^{-1} mol^{-1}.	70G5
7a	Piezoelectricity: Figs. 192, 193.	
8a	Elastic compliance: Fig. 194.	
10a	Raman scattering: Figs. 195α, 195β, 195γ.	
13a	NMR: Figs. 196, 197; see also NQR: Fig. 198	74G11

Table 33. $AgNa(NO_2)_2$. Structure of phase II [74I9]. Fractional coordinates and anisotropic temperature parameters. $T=20$ °C. R-factor: 0.0484. B_{ij} is defined by Eq. (a) in Introduction.

Atom	x	y	z	B_{11}	B_{22}	B_{33}	B_{12}	B_{13}	B_{23}
Ag	0.0	0.0000 (3)	0.0	0.0185 (2)	0.00325 (4)	0.00638 (7)	0.0	−0.0014 (5)	0.0
Na	0.0	0.509 (1)	0.0	0.011 (1)	0.0060 (3)	0.0036 (3)	0.0	−0.002 (2)	0.0
N(1)	0.0	0.274 (1)	0.0	0.010 (1)	0.0016 (5)	0.0043 (8)	0.0	−0.000 (2)	0.0
O(1)	−0.045 (1)	0.206 (1)	0.0907 (6)	0.018 (2)	0.0032 (4)	0.0038 (5)	−0.000 (1)	0.005 (1)	−0.000 (1)
N(2)	0.0	0.793 (1)	0.0	0.010 (2)	0.0074 (7)	0.0043 (8)	0.0	−0.002 (3)	0.0
O(2)	0.041 (1)	0.738 (1)	0.0931 (7)	0.017 (2)	0.0086 (9)	0.0053 (7)	−0.000 (2)	−0.005 (2)	0.003 (1)

Table 34. $AgNa(NO_2)_2$. Structure of phase I [74I9]. Fractional coordinates and anisotropic temperature parameters. $T=50$ °C. The split atom model was used for the analysis. N(α), O(α) refer to 70% fraction of atoms and N(β), O(β) to 30% fraction. U_{ij} (in 10^{-2} Å^2) is defined by Eq. (d) in Introduction.

Atoms	x	y	z	U_{11}	U_{22}	U_{33}	U_{12}	U_{13}	U_{23}
Ag	0.0	0.0	0.0	7.22 (6)	2.87 (3)	5.11 (5)	0.0	0.0	0.0
Na	0.0	0.50	0.0	4.03 (21)	4.87 (22)	2.77 (17)	0.0	0.0	0.0
N(α)	0.0	0.2709 (8)	0.0	4.25 (41)	3.47 (38)	2.75 (32)	0.0	−1.03 (43)	0.0
N(β)	0.0	0.1974 (26)	0.0	3.53 (100)	6.03 (144)	6.29 (140)	0.0	−0.23 (144)	0.0
O(α)	−0.0480 (8)	0.2055 (5)	0.0901 (5)	6.60 (41)	3.60 (24)	3.19 (25)	0.30 (25)	1.67 (26)	0.18 (20)
O(β)	−0.0543 (20)	0.2575 (16)	0.0897 (13)	8.31 (114)	4.47 (67)	5.97 (82)	−1.15 (82)	3.25 (83)	−2.29 (76)

16B Solid solution

Nr. 16B-1 $NaNO_2$-$AgNO_2$

1b	Phase transitions in $NaNO_2$-$AgNO_2$ system: Fig. 199.	

 Figs. p. 358 ff.

17 KNO_3 family

17A Pure compounds

Nr. 17A-1 $NaNO_3$, Sodium nitrate
($M = 84.99$)

1a	Ferroelectricity in high pressure phase III in $NaNO_3$ was reported by Barnett et al. in 1969.	69B1
b	(see table below)	a) 69B1 b) 37b1 c) 64M3

phase	II	I	III*)
state			F a)
crystal system	trigonal	trigonal	trigonal
space group	$R\bar{3}c$-D_{3d}^6 b)		$R3c$-C_{3v}^6 a)
Θ	274.2 c) °C		

*) Phase III is stable for $p \gtrsim 4.3 \cdot 10^9$ Pa at RT: see Fig. 202.

$T_{melt} = 308$ °C. — 65O5

$\varrho = 2.260 \cdot 10^3$ kg m^{-3} at 25 °C. — 56S7

Transparent, colorless. Cleavage plane: {112}.

2a	Crystal growth: cooling from melt.	69B1
3a	Unit cell parameters: $a = 6.326$ Å, $\alpha = 47°17'$ at 24 °C.	55M4
b	$Z = 2$ in phases III a), II, I b).	a) 69B1 b) 55M4
4	Lattice distortion at high pressures: Fig. 200.	
5a	Dielectric constant: pressure dependence, Fig. 201. Phase diagram with regard to p: Fig. 202.	
c	Spontaneous polarization: Fig. 203.	
8a	Linear compressibilities: fractional variation of unit cell parameters with p: $-\Delta a/a = \gamma_{11}p + \gamma_{12}p^2$, $-\Delta c/c = \gamma_{31}p + \gamma_{32}p^2$, with $\gamma_{11} = 7.18(6) \cdot 10^{-3}$ GPa^{-1}, $\gamma_{12} = -3.7(4) \cdot 10^{-4}$ GPa^{-2}, $\gamma_{31} = 2.5(1) \cdot 10^{-2}$ GPa^{-1}, $\gamma_{32} = -2.4 \cdot 10^{-3}$ GPa^{-2} $p < 2.6$ GPa.	79H6
10a	Raman scattering: Fig. 204.	

Nr. 17A-2 KNO_3, Potassium nitrate
($M = 101.10$)

1a	Ferroelectricity in KNO_3 was discovered by Sawada et al. in 1958.	58S1
b	(see table below)	a) 31E1 b) 39B1 c) 47T1

phase	II	III	I
state		F	P
crystal system	orthorhombic	trigonal	trigonal
space group	$Pnma$–D_{2h}^{16} a)	$R3m$–C_{3v}^5 b)	$R\bar{3}m$–D_{3d}^6 c)

Transition temperatures: — 61S3

on heating	phase	II		I
	Θ	130 °C		
on cooling	phase	II	III	I
	Θ	115	125	°C

continued

Nr. 17A-2 KNO_3, continued

1b Phase III is metastable at atmospheric pressure and appears only on cooling. See also Fig. 217.
$P_s \| [111]$. 61 S 3
$T_{melt} = 333$ °C. 60 S 4
$\varrho = 2.109 \cdot 10^3$ kg m^{-3} at RT. 31 E 1
Transparent, colorless.

2a Crystal growth: cooling or evaporation of aqueous solution; solubility in H_2O, Table 35.

b Crystal form: Fig. 205.

3a Unit cell parameters at RT: $a = 5.4119$ Å, $b = 9.1567$ Å, $c = 6.4213$ Å at 25 °C; unit cell parameters in various phases. Table 36. 73 N 11

b

phase	I	II	III	IV*
Z	1 [a]	4 [b]	1 [c]	4 [d]

[a]) 47 T 1
[b]) 31 E 1
[c]) 39 B 1
[d]) 69 W 2

* High pressure phase; see Fig. 217.

Crystal structures: Figs. 206···210; Tables 37α, β, 38, 39.

4 Temperature dependence of unit cell parameters, Figs. 211, 212;
thermal expansion coefficients, Table 40; thermal expansion below room temperature, see 68 F 1

5a Dielectric constants: Figs. 213···215;
Curie-Weiss behavior, Fig. 216; $\varkappa_c = C/(T - \Theta_p)$, $T > \Theta_{III-I}$ with $C = 5.62 \cdot 10^3$ K, $\Theta_p = 308$ K, where $\varkappa_c$ is the dielectric constant along the trigonal [111] direction. 61 S 3
Phase diagram with regard to p: Fig. 217.
Dielectric dispersion: see 67 C 3

b Non-linear dielectric property: $\xi = -7.75 \cdot 10^{11}$ V m^{5}C^{-3}, $\zeta = 6.48 \cdot 10^{13}$ V m^{9}C^{-5}. 71 M 5

c Spontaneous polarization and coercive field: Figs. 218, 219;
spontaneous polarization at high pressure, Fig. 220.

6a Heat capacity: Figs. 221, 222.
Transition heat, 30 K 2

Transition	II→I	I→III	III→II
ΔQ_m[J mol^{-1}]	4427	2328	2098

b Thermal conductivity: Fig. 223.

8a Elastic stiffnesses: Table 41.

9a Refractive indices: Figs. 224, 225, Table 42.
Far-infrared spectra: Fig. 226; see also 79 H 5

10a Raman scattering: Fig. 227.

11 Electrical conduction: Fig. 228; see also 67 F 8

13b ESR of irradiated crystal: see 62 Z 2, 62 C 2, 64 L 3, 67 H 8

c Mössbauer effect: Fig. 229.

14b Diffuse X-ray scattering: Fig. 230.

15b Domain switching: Figs. 231, 232.

16 Radiation damage: see 67 G 5

Table 35. KNO_3. Solubility A in H_2O [62k1]

A [wt%]	T [°C]	A [wt%]	T [°C]
12.2	0	45.5	50
17.5	10	51.5	60
24.0	20	57.5	70
27.5	25	62.5	80
31.2	30	67.0	90
38.0	40	71.0	100

Table 36. KNO_3. Unit cell parameters in various phases.

Phase	T °C	Unit cell parameters Å	Ref.
II	100	$a=5.4283$, $b=9.1849$, $c=6.5034$	73N11
III	91*)	$a=5.487$ (1), $c=9.156$ (3)**)	76N7
I	151	$a=5.425$ (1), $c=9.836$ (4)**)	
IV***)		$a=5.58$, $b=7.52$, $c=6.58$	69W2

*) A quenched state. The temperature region in which phase III is obtained depends on previous treatment, and that in this study was lower than those generally reported (cf. subsection 1b).
**) For the hexagonal cells.
***) High pressure phase. p, T: not given. Unit cell parameters are given for $P2_1nb-C_{2v}^9$ cell. See also [62D1].

Table 37α. KNO_3. Structure of phases I and III [76N7]. Fractional coordinates and anisotropic temperature parameters. b_{ij} is defined by Eq. (b) in Introduction. R-factor: 0.18 for phase III, 0.13 for phase I.

Phase		x	y	z	b_{11}	b_{22}	b_{33}	b_{12}	b_{13}	b_{23}
III (91 °C)*)	K	0	0	0	0.038 (6)	b_{11}	0.030 (5)	$b_{11}/2$	0	0
	N	0	0	0.405 (6)	0.09 (3)	b_{11}	0.00 (1)	$b_{11}/2$	0	0
	O	0.131 (2)	$\bar{x}$	0.434 (3)	0.05 (2)	b_{11}	0.007 (4)	0.03 (2)	−0.001 (3)	$-b_{13}$
I (151 °C)	K	0	0	0	0.109 (7)	b_{11}	0.029 (3)	$b_{11}/2$	0	0
	N	0	0	$\frac{1}{2}$	0.03 (2)	b_{11}	0.06 (1)	$b_{11}/2$	0	0
	O	0.130 (2)	$\bar{x}$	0.475 (5)	0.05 (2)	b_{11}	0.10 (1)	0.04 (2)	0.008 (8)	$-b_{13}$

*) A quenched state. See also footnotes of Table 36.

Table 37β. KNO_3. Structure of phases I and III [76N7]. Interatomic distances and bond angles.

Phase	III (91 °C)*)	I (151 °C)
Interatomic distances [Å]		
K···O	2.87 (2)	2.68 (4)
	2.91 (1)	3.07 (2)
N–O	1.28 (2)	1.25 (2)
Interatomic angles [°]		
O–N–O	116 (2)	116 (1)

*) A quenched state. See also footnotes of Table 36.

Table 38. KNO_3. Structure of phase II [73N11]. Fractional coordinates and anisotropic temperature parameters. b_{ij}: defined by Eq. (b) in Introduction. R-factor: 0.058 at 25 °C, 0.141 at 100 °C.

T[°C]		x	y	z	b_{11}	b_{22}	b_{33}	b_{12}	b_{13}	b_{23}
25 (1)	K	0.2500	0.4166 (1)	0.7568 (2)	0.0184 (5)	0.0061 (1)	0.0180 (4)	0	0	0.0005 (2)
	N	0.2500	0.7548 (1)	−0.0848 (1)	0.0184 (2)	0.0071 (1)	0.0111 (2)	0	0	0.0001 (1)
	O (1)	0.2500	0.8902 (1)	−0.0893 (2)	0.0273 (4)	0.0066 (1)	0.0249 (3)	0	0	−0.0024 (1)
	O (2)	0.4492 (1)	0.6866 (1)	−0.0849 (1)	0.0173 (3)	0.0097 (1)	0.0236 (2)	0.0019 (1)	−0.0027 (1)	0.0016 (1)
100 (1)	K	0.2500	0.4155 (3)	0.7646 (8)	0.0158 (5)	0.0101 (3)	0.0238 (16)	0	0	−0.0089 (6)
	N	0.2500	0.7539 (1)	−0.0829 (2)	0.0256 (4)	0.0083 (1)	0.0181 (4)	0	0	0.0018 (2)
	O (1)	0.2500	0.8890 (2)	−0.0900 (6)	0.0425 (9)	0.0090 (3)	0.0258 (10)	0	0	0.0011 (5)
	O (2)	0.4442 (2)	0.6849 (1)	−0.0835 (5)	0.0247 (4)	0.0116 (2)	0.0332 (7)	0.0017 (3)	−0.0048 (5)	−0.0007 (3)

Table 39. KNO_3. Structure of phase II [73N11]. Interatomic distances and bond angles. See also Fig. 209.

			25 °C	100 °C
Uncorrected interatomic distances [Å]				
	Metal-oxygen	K_4–O_3 (1)	2.845 (4)	2.926 (13)
		K_4–O_1 (2)	2.885 (3)	2.864 (8)
		K_4–O_4 (1)	2.925 (1)	2.951 (5)
		K_4–O'_2 (2)	2.828 (2)	2.815 (6)
		K_4–O_4 (2′)	2.884 (3)	2.938 (7)
	Within the NO_3 group	N–O (1)	1.241 (2)	1.241 (5)
		N–O (2), N–O (2′)	1.246 (1)	1.229 (3)
		O (1)–O (2)	2.155 (2)	2.150 (4)
		O (2)–O (2′)	2.157 (1)	2.106 (2)
Corrections to be added [Å]				
	(i) Assuming rigid-body motion of the NO_3 group	N–O (1)	0.017 (6)	0.014 (6)
		N–O (2), N–O (2′)	0.012 (4)	0.016 (4)
		O (1)–O (2)	0.031 (11)	0.035 (10)
		O (2)–O (2′)	0.027 (10)	0.040 (10)
	(ii) Assuming "riding" motion of the O on N	N–O (1)	0.016 (1)	0.018 (1)
		N–O (2), N–O (2′)	0.014 (1)	0.018 (1)
Uncorrected bond angles [°]				
	O (2)–N–O (1)		120.06 (8)	121.02 (17)
	O (2)–N–O (2′)		119.85 (15)	117.90 (34)

Table 40. KNO_3. Linear thermal expansion coefficients. $\alpha_{[111]}$: thermal expansion coefficient along the trigonal [111] direction.

phase	T [°C]	α_{ii} [10^{-4} K^{-1}]			Ref.
		α_{11}	α_{22}	α_{33}	
II	−180···20	1.36	0.20	0.16	63G4
III	120	$\alpha_{[111]}=3.33$			65K4
I	125	$\alpha_{[111]}=3.15$			65K4

Table 41. KNO_3. Elastic stiffnesses [71M4]. $T=20$ °C.

Crystal	c_{11}	c_{22}	c_{33}	c_{44}	c_{55}	c_{66}	c_{12}	c_{13}	c_{23}
	$\cdot 10^{10}$ N m^{-2}								
	3.58 (4)	3.00 (3)	2.04 (2)	0.67 (1)	0.54 (1)	0.83 (1)	1.34 (15)	1.16 (15)	0.92 (15)

Table 42. KNO_3. Refractive indices and optical axial angle [62k2].

λ nm	n_α	n_β	n_γ	$2V$	Ref.
253.0	1.367	*)	1.698	*)	33K1
260.0	1.349	*)	1.671	*)	
270.0	1.361	*)	1.644	*)	
280.0	1.358	*)	1.624	*)	
290.0	1.356	*)	1.608	*)	
300.0	1.3543	*)	1.5450	*)	
310.0	1.3526	*)	1.5861	*)	
320.0	1.3514	*)	1.5782	*)	
330.0	1.3499	*)	1.5703	*)	
350.0	1.3479	*)	1.5582	*)	
366.3	1.3460	*)	1.5502	*)	
396.9	1.3435	*)	1.5384	*)	
435.8	1.3410	*)	1.5279	*)	
589.3	1.3346	*)	1.5056	*)	
686.8	1.3328	*)	1.4988	*)	
405	1.3384	1.5299	1.5304	*)	30m1
436	1.3368	1.5224	1.5229	*)	
486	1.3347	1.5139	1.5144	*)	
546	1.3330	1.5072	1.5077	*)	
656	1.3310	1.4998	1.5002	*)	
706	*)	1.4975	1.4979	*)	
768	*)	1.4951	1.4955	*)	
527	1.3365	1.5124	1.5135	*)	1860S1
589	1.3346	1.5056	1.5064	−7°12′**)	
687	1.3328	1.4988	1.4994	*)	

*) No value given. **) Calculated [44d1].

Nr. 17A-3 $RbNO_3$, Rubidium nitrate
($M = 147.47$)

1a Antiferroelectricity in phase III of $RbNO_3$ was suggested by Dantsiger et al. [a]) in 1963 from dielectric measurements. Pyroelectricity in phase IV was found by Bury et al. [b]) in 1969. — [a]) 63D1; [b]) 69B19

b

phase	IV	III	II	I
state		(A) [a])		
crystal system	trigonal	cubic	trigonal	cubic
space group	$P3_1$-C_3^2 or $P3_2$-C_3^3 [b])	$Pa3$-T_h^6 [c])	$R\bar{3}m$-D_{3d}^5 [d])	$Fm3m$-O_h^5 [d])
Θ [e])	164	219	291	°C

[a]) 63D1; [b]) 69B19; [c]) 51K1; [d]) 80A1; [e]) 56P3

$T_{melt} = 314$ °C. — 56P3
$\varrho = 3.11 \cdot 10^3$ kg m^{-3}. — 62B9
Transparent, colorless.

2a Crystal growth: evaporation of aqueous solution; solubility in H_2O, Table 43; growth from melt.

3a Unit cell parameters in each phase:

phase	T[K]	p[GPa]	unit cell parameters
IV	302	0	$a = 10.49$ Å, $c = 7.44$ Å [a])*)
III	463	0	$a = 4.3718$ (2) Å [b])
II	503	0	$a = 4.8048$ (4) Å, $\alpha = 70.491$ (5)° [b])
I	568	0	$a = 7.3150$ (22) Å [b])
V**)	RT	2.8	$a = 5.667$ Å, $b = 5.181$ Å, $c = 4.798$ Å [c])

[a]) 68S1; [b]) 80A1; [c]) 79K4

*) For the hexagonal cell.
**) High pressure phase; see Fig. 240.

b

phase	IV	III	II	I
Z	9*)	1	1	4

68S1

*) For the hexagonal cell.

Crystal structures: Figs. 233, 234; Tables 44α, 44β, 44γ, 44δ; structure of high pressure phase, see 79K4

4 Thermal expansion: Figs. 235···237.

5a Dielectric constant: Figs. 238, 239. Curie-Weiss behavior in phase II, $\varkappa_a = C/(T - \Theta_P)$ with $C = 5.3 \cdot 10^3$ K, $\Theta_P = 201.3$ °C. — 63D1
$\varkappa_a$: dielectric constant perpendicular to the trigonal axis, $f = 1$ MHz.
Phase diagram with regard to p: Fig. 240; see also 77F7

b Effect of bias field: Fig. 241.

d Pyroelectricity: Fig. 242.

6a Specific heat capacity: Fig. 243.

9a Infrared absorption: see 79F4

10a Raman scattering: see 73B9, 74B1, 79O5

11 Electric conduction: Fig. 244.

14b X-ray diffuse scattering: Figs. 245, 246.

 Fig. p. 368

Table 43. $RbNO_3$. Solubility A in H_2O [62k1].

A [wt%]	T [°C]	A [wt%]	T [°C]
16.3	0	66.7	60
24.8	10	71.5	70
34.6	20	75.6	80
44.8	30	78.9	90
53.9	40	81.9	100
60.9	50		

Table 44α. $RbNO_3$. Structure of phase III [80A1]. Fractional coordinates of atoms and isotropic temperature parameters. T=463 K.

	x	y	z	B [Å^2]
Rb	$\frac{1}{2}$	$\frac{1}{2}$	$\frac{1}{2}$	7.7 (2)
N	0.065 (5)	0	0	8.6 (8)
O (1)	0.350	0	0	8.4 (14)
O (2)	0	0.244 (6)	−0.100 (7)	

Table 44β. $RbNO_3$. Structure of phase II [80A1]. Fractional coordinates of atoms and overall isotropic temperature parameter B (overall). T=503 K.

	x	y	z
Rb	$\frac{1}{2}$	$\frac{1}{2}$	$\frac{1}{2}$
N	−0.0318 (7)	−0.0318	−0.0318
O	−0.0726 (7)	−0.0726	0.2463 (7)

B (overall) = 8.0 (1) Å^2

Table 44γ. $RbNO_3$. Structure of phase I [80A1]. Fractional coordinates of atoms and overall isotropic temperature parameter B (overall). T=568 K.

	x	y	z
Rb	$\frac{1}{2}$	$\frac{1}{2}$	$\frac{1}{2}$
N	−0.0373 (16)	−0.0373	−0.0373
O	0.0659 (21)	0.0659	0.1562 (20)

B (overall) = 10.8 (6) Å^2

Table 44δ. $RbNO_3$. Interatomic distances and bond angles [80A1]. Numbers in brackets mark the bonds involved.

Phase	III		II		I	
N—O (1)	1.25 Å	[1]	1.24 (2) Å	[3]	1.77 (5) Å	[1]
N—O (2)	1.10 (8) Å	[2]			1.17 (4) Å	[2]
O (1)—N—O (2)	104 (2)°	[2]	102 (2)°	[3]	101 (4)°	[2]
O (2)—N—O (2)	120°	[1]			158 (4)°	[1]
O (1)—O (2)	1.92 (4) Å	[2]	2.15 (4) Å	[3]	2.30 (4) Å	[3]
O (2)—O (2)	2.13 (6) Å	[1]				
Rb—O (1)	3.16 (1) Å	[4]	3.06 (2) Å	[6]	2.61 (2) Å	[6]
Rb—O (2)	3.01 (3) Å	[2]				

Nr. 17A-4 $CsNO_3$, Cesium nitrate
(M = 194.91)

		Ref.
1a	Pyroelectricity in $CsNO_3$ was reported by Bury et al. in 1969.	69B19
b	phase: II / I; crystal system: trigonal / cubic; space group: $P3_1$-C_3^2 or $P3_2$-C_3^3 a) / $Pa3$-T_h^6 b); Θ: 154 c) °C	a) 69B19; b) 53K1; c) 63S10
	T_{melt} = 404 °C.	65O5
	ϱ_X = 3.635 · 10^3 kg m^{-3}.	60S12
	Transparent, colorless.	
2a	Crystal growth: cooling from melt, evaporation of aqueous solution.	63S10
3a	Unit cell parameters: a = 10.950 Å, c = 7.716 Å in phase II for the hexagonal cell at RT.	60S12
	a = 8.980 Å in phase I.	53K1
	a = 5.837 Å, b = 5.143 Å, c = 4.978 Å in phase III (pressure induced phase; see Fig. 248). T = RT, p = 4.5 GPa. continued	79K4

Nr. 17A-4 $CsNO_3$, continued

3b	$Z=9$ in phase II (for the hexagonal unit cell). $Z=8$ in phase I. Structure of high pressure phase: see	60S12 53K1 79K4
5a	Dielectric constant: Fig. 247; see also	63S10, 72F2
	Phase diagram with regard to p: Fig. 248.	
c	Pyroelectricity: Fig. 249.	
6a	Heat capacity: Fig. 250; transition heat, $\Delta Q_m = 3.3\,(2)$ kJ mol^{-1}.	79B1
b	Thermal conductivity: Fig. 251.	
7a	Piezoelectricity: Fig. 252.	
9a	Infrared absorption: see	79F4
d	Optical rotatory power: Fig. 253.	
10a	Raman scattering: see	73B9, 74B1, 79O5
11	Electrical conduction: Fig. 254; see also	63S10, 75M5

17B Solid solutions

Nr. 17B-1 KNO_3-NH_4NO_3

1b	Phase diagram: Fig. 255.	
3b	Crystal structure: see	75H8
5c	Spontaneous polarization and coercive field: Figs. 256, 257.	

Nr. 17B-2 KNO_3-$RbNO_3$

1b	Phase diagram: Fig. 258.	
3a	Unit cell parameters: Table 45.	
4	Thermal expansion: Fig. 237.	
5a	Dielectric constant: Fig. 259.	
c	Spontaneous polarization and coercive field: Figs. 260, 261.	

Table 45. $K_xRb_{1-x}NO_3$. Unit cell parameters and unit cell volume for the hexagonal lattice [69S2]. T=RT.

x	a_{hex} Å	c_{hex} Å	V Å^3
0.02	10.480	7.443	707.92
0.04	10.480	7.444	708.02
0.05	10.483	7.446	708.61

Nr. 17B-3 KNO_3-KNO_2

1b	Phase diagram: Fig. 262.	
3a	Unit cell parameters: Fig. 263.	

Nr. 17B-4 KNO_3-KI

5c	Spontaneous polarization: Fig. 264.	

Nr. 17B-5 $RbNO_3$-$CsNO_3$

1b	Phase diagram: Fig. 265.	
3a	Unit cell parameters: Table 46.	
4	Thermal expansion: see Fig. 237.	
5a	Dielectric constant: Fig. 266.	

Table 46. $Rb_{1-x}Cs_xNO_3$. Unit cell parameters and unit cell volume for the hexagonal lattice at RT [69S2].

x	a_{hex} Å	c_{hex} Å	V Å^3
0	10.490	7.450	710.24
0.05	10.513	7.458	713.81
0.10	10.534	7.484	719.15
0.20	10.580	7.504	727.41
0.25	10.608	7.521	732.92
0.30	10.619	7.518	734.15
0.50	10.710	7.602	755.10

18 $PbHPO_4$ family

18A Pure compounds

Nr. 18A-1 $PbHPO_4$, Lead hydrogen phosphate (Lead monetite)
(M = 303.2)

1a	Ferroelectricity in $PbHPO_4$ was discovered by Negran et al. in 1974.	74N5
b	(see table below)	a) 76A13; b) 74N5
	For deuterated compound: Θ_{II-I} = 179 °C.	74N5
	P_s makes an angle 22 (5)° to the [100] direction in the (010) plane, see Fig. 268.	74N5
	Transparent. Colorless.	
2a	Crystal growth: evaporation from aqueous solution a), or growth from gel b).	a) 74N5; b) 76B11
b	Crystal form: Fig. 267.	

phase	II	I
state	F	P
crystal system	monoclinic	monoclinic
space group	Pc-C_s^2 a)	P2/c-C_{2h}^4 b)
Θ	37 b) °C	

continued

Nr. 18A-1 $PbHPO_4$, continued

3a	Unit cell parameters at RT: $a=4.688\,(4)$ Å, $b=6.649\,(5)$ Å, $c=5.781\,(5)$ Å, $\beta=97.11\,(2)^\circ$ (for deuterated compound, $a=4.6855\,(5)$ Å, $b=6.6911\,(7)$ Å, $c=5.7867\,(8)$ Å, $\beta=97.10\,(1)^\circ$).	74N5
b	$Z=2$ in phases II and I. Crystal structure: Fig. 268; disorder in hydrogens, see	74N5 78N4
4	Linear thermal expansion coefficients: $\alpha_a=1.12\cdot10^{-5}\,K^{-1}$, $\alpha_b=1.34\cdot10^{-5}\,K^{-1}$, $\alpha_c=2.1\cdot10^{-5}\,K^{-1}$. $T>\Theta_f$.	78H15
5a	Dielectric constant: Figs. 269α, 269β, 269γ. Curie constant: $C=2.5\cdot10^3$ K ($1.6\cdot10^3$ K for deuterated compound). Effect of hydrostatic pressure: $d\Theta_{II-I}/dp=-14.0\cdot10^{-8}\,K\,Pa^{-1}$.	 74N5 78Y5
b	Nonlinear dielectric property: $\xi=7.7\cdot10^{12}\,V\,m^5\,C^{-3}$ ($1.3\cdot10^{14}\,V\,m^5\,C^{-3}$ for deuterated compound), Figs. 270α, 270β.	74N5
c	Spontaneous polarization: Figs. 271, 272.	
d	Performance parameters of pyroelectric detector: see	77N7
9a	Optical data: $\Delta n_{\gamma\beta}=0.0725\,(30)$, $\Delta n_{\beta\alpha}=0.008$, $2V=38\,(2)^\circ$. $\lambda=0.633$ μm. Figs. 273α, 273β. Infrared and far-infrared absorption: see	77K12 76A13, 76B7, 78K20
10a	Raman scattering: see	76B7, 77L2
b	Brillouin scattering: see	78L6

Nr. 18A-2 $PbHAsO_4$, Lead hydrogen arsenate (Schultenite)
($M=347.1$)

1a	Ferroelectricity in $PbHAsO_4$ was first reported by Březina et al. in 1975.	75B19
b	(see table below)	[a]) 76A13 [b]) 75B19

phase	II	I
state	F	P
crystal system	monoclinic	monoclinic
space group	Pc-C_s^2 [a])	
Θ	40.5 [b]) °C	

	For deuterated compound: $\Theta_{II-I}=162$ °C. P_s lies on (010) making an angle ~7° to [100]. Transparent. Colorless.	75B19 75B19
2a	Crystal growth: growth from gel.	75B19
3a	Unit cell parameters: $a=4.842$ Å, $b=6.743$ Å, $c=5.821$ Å, $\beta=95.47^\circ$ at RT.	76A13
b	$Z=2$. Crystal structure: isomorphous with $PbHPO_4$.	76A13 75B19
5a	Dielectric constant: Fig. 274. Effect of hydrostatic pressure: $d\Theta_{II-I}/dp=-12.5\cdot10^{-8}\,K\,Pa^{-1}$.	 78Y5
c	Spontaneous polarization: Fig. 275.	

 Figs. p. 374

9a	Optical data: $n_\alpha=1.8903$, $n_\beta=1.9097$, $n_\gamma=1.9765$, $\Delta n_{\beta\alpha}=0.00194$, $\Delta n_{\gamma\beta}=0.0194$, $2V=58.23°$. $\lambda=0.589$ μm.	27S1
	Infrared and far-infrared absorption: see	76A13, 78K20
10a	Raman scattering: see	76B7, 77L2

19 KDP (KH_2PO_4) family

19A Pure compounds

Data on deuterated compounds are given in bracket as [D:].

Nr. 19A-1 KH_2PO_4, Potassium dihydrogen phosphate (KDP)

($M=136.09$; [D: 138.10])

1a Ferroelectricity of KH_2PO_4 was first reported by Busch and Scherrer in 1935. — 35B2

b — 35B2

phase	IV	III	II	I
state	F	P		
crystal system	orthorhombic	tetragonal	monoclinic [c, d]	
space group	Fdd2-C_{2v}^{19}	$I\bar{4}2d$-D_{2d}^{12} ($F\bar{4}d2$*)	$P2_1$-C_2^2 or $P2_1/m$-C_{2h}^2 [b]	
Θ	123 [D: 213 K]	450 [a]) [D: 383 K] [c])	~506 K [a])	

[a]) 70R3, [b]) 75I6, [c]) 72G14, [d]) 68B8

Phase I was reported to exist between 506 K and T_{melt} for KH_2PO_4. See the phase diagram Fig. 276.

Slightly different values for Θ_{III-II} were reported: 448 K [c, d]) and 453 K [e]) for KH_2PO_4, and 375 K [e]) for KD_2PO_4. Phase II is metastable at RT, and can be brought to RT by cooling the crystal from above Θ_{III-II}. Highly deuterated KDP crystallizes in the monoclinic form at RT: Fig. 277. — [e]) 67O5

Very highly deuterated crystals, $KH_{2(1-x)}D_{2x}PO_4$ with $x \geqq 0.98$ were reported to be stable or at least metastable in the monoclinic phase from RT to 6 K [f]) when cooled very slowly. — [f]) 76K6

Transition from phase III to IV is first order [g]) with a thermal hysteresis of 0.83 K [h]); [D: 1.01 K [i]]. — [g]) 71O1, [h]) 71K11, [i]) 72S23

$\Theta_f=\Theta_{IV-III}$ of $KH_{2(1-x)}D_{2x}PO_4$: Fig. 291 in 5a, $\Theta_f=(121.7+107\,x)$ K with the maximum of $x=0.985$ and $\Theta_f=228$ K. — 74L9

Θ_{III-II} of $KH_{2(1-x)}D_{2x}PO_4$: Fig. 278.

$P_s \| [001]$ in the phase IV. — 35B2

$T_{melt}=260\,(1)$ °C. — 70R3

$\varrho=2.338\cdot10^3$ kg m^{-3}; [D: $2.345\cdot10^3$ kg m^{-3}]. — 66b1

Transparent, colorless.

* For the relation between the two space groups, see illustration in 3.

2a	Crystal growth: evaporation or cooling method from aqueous solution, see	74J2, 77J1
	Industrial growth of deuterated KDP crystals: see	75L8
	Also refer to the method for $NH_4H_2PO_4$.	50W1
	Solubility in H_2O: Table 47; see also	77S36
	Crystal growth in gel: see	71B24, 78J4
	Effect of deuteration degree of a solution on that of a crystal: see	77S21
b	Crystal form: Fig. 279.	38B3

continued

Nr. 19A-1 KH_2PO_4, continued

3a	Unit cell parameters: Tables 48···51.	
b	Crystal structure $Z=4$ in a $I\bar{4}2d$ cell, $Z=8$ in a $F\bar{4}d2$ cell [D: $Z=8$ in a monoclinic cell], Figs. 280, 281, 282.	30W1, 47U1
	Fractional coordinates of atoms: Tables 52···55; see also	72N3, 74T10
	Interatomic distances: Tables 56, 57.	
	Structure related to hydrogen bonds, see also	74E3, 74N7, 76K6
	Thermal parameters: Table 58; see also	74N1
4	Temperature dependence of lattice constants: Fig. 283.	
	Effect of E_{bias}: Figs. 284, 285.	
	Mean square displacement: Fig. 286.	
	Thermal expansion: Table 59; Fig. 287; see also	70K18
	Spontaneous shear strain measured by neutron diffraction: see	74S13, 76Z2, 78B2
	Effect of p: Fig. 288; see also	72M18
5a	Dielectric constant $\varkappa_c$ vs. T: Figs. 289···291. For $\varkappa_c$ below 6 K, see	76H9
	Curie-Weiss law: Figs. 292, 293; see also	73S1
	$\varkappa_c = C/(T-\Theta_p)$ for $T>\Theta_f$	
	C and Θ_p: Table 60 in 5b and Fig. 294; see also	74Z5, 72S28, 73S1, 63S8
	$\varkappa_a$ vs. T: Figs. 295, 296.	
	$\varkappa_c$ measured by a thermal noise method agrees with that obtained by the usual bridge method in phase III.	77T15
	$\varkappa_c$ vs. T for monoclinic DKDP (=deuterated KDP): Fig. 297.	
	$\varkappa$ vs. p: Fig. 298.	
	$\varkappa$ and $\tan\delta$ at 9.2 GHz: Figs. 299···303.	
	$\varkappa$ at $f=5···80$ MHz: see	77T11
	$\varkappa$ vs. f for DKDP: Figs. 304···307.	
	$\varkappa$ and $\tan\delta$ for DKDP at 620 MHz: Figs. 308, 309.	
	$\varkappa$ at 139 GHz: Figs. 310···313; see also	76P8
	Increase of $\varkappa'_c$ below 0.1 Hz was reported in	76P2
	For $\varkappa$ at far-infrared and infrared region, see 9a and 10a.	
	Dielectric dispersion at $f=1.8\cdot10^{11}···3.6\cdot10^{11}$ Hz (6···12 cm^{-1}) was studied in	78K17
	Phase diagram: Figs. 276, 314, 315; Table 61.	
	Effect of E_{bias} on Θ_f: Fig. 316.	
	$d\Theta_f/dE_{bias}=1.92\cdot10^{-6}$ K m V^{-1} [a]) [D: 1.25 (5) 10^{-6} K m V^{-1} [b])].	[a]) 71K11, [b]) 71G5
	$T_{crit}=212.6$ (15) K, $E_{crit}=8.3$ (6) $\cdot 10^5$ V m^{-1} [b]). (T_{crit}: temperature above which the double hysteresis loop vanishes. E_{crit}: see Fig. IB-2 in Introduction).	
b	Nonlinear dielectric properties: Table 60; see also	74Z5
	$\xi=-1.54$ (8) $\cdot 10^{10}$ V m^5 C^{-3}, $\zeta=3.9$ (3) $\cdot 10^{13}$ V m^9 C^{-5} [a]).	[a]) 77O1
	For $KH_{2(1-x)}D_{2x}PO_4$ ($x\approx0.85$), $\xi=-4.5\cdot10^{10}$ V m^5 C^{-3}, $\zeta=1.8\cdot10^{13}$ V m^9 C^{-5}, $\Theta_f=211.73$ K, $\Theta_p=210.80$ K [b]).	[b]) 72S23
	Effect of p on ξ, ζ, and C: Table 62, Fig. 317.	
	Tricritical point is reported to exist at $p=2.3$ (3) $\cdot 10^8$ Pa.	78W3, 78B2
	Effect of E_{bias} on $\varkappa_c$: Fig. 318.	

5c P_s and E_c vs. T: Figs. 319···323.
Effect of E_{bias} on P_s: Figs. 324···327.

d Electrocaloric effect: Fig. 328.

6a Heat capacity: Figs. 329···333.
Effect of E_{bias}: Fig. 334.
c_p at low temperature, see — 76L2
Transition heat ΔQ_m and transition entropy ΔS_m:

ΔQ_m [J mol^{-1}]	ΔS_m [J K^{-1} mol^{-1}]		Ref.
238 [D: 418]	1.97 [D: 1.97]	a)	a) 42B1
364	3.09	b)	b) 44S3
	4.01 [D: 2.89]	c)	c) 68S21

See also Fig. 335.

b Thermal conductivity: Figs. 336···338.

7a Piezoelectricity: Table 63; Figs. 339, 340.

8a Elastic compliance and stiffness: Table 64; Figs. 341···347.
Elastic stiffness obtained from Brillouin scattering: Fig. 348, c^E_{ij} at RT in the unit of 10^{10} Nm^{-2}.

c^E_{11}	c^E_{33}	c^E_{44}	c^E_{66}	c^E_{12}	c^E_{13}	Ref.
7.17 (3)	5.68 (3)	1.27 (1)	0.64 (2)	−0.63 (9)	1.60 (5)	71B23

Effect of p: Figs. 349, 350.
Effect of p on strain: see — 78B2
Ultrasonic attenuation: Fig. 351; see also — 70S12

9a Refractive indices: Tables 65, 66; Figs. 352, 353; see also — 76V5
Birefringence: Figs. 354···358.
Transmission in the vacuum ultraviolet region: Fig. 359.
Infrared absorption: Figs. 360···362.
Far-infrared spectra: Figs. 363···368; see also — 70S22, 77B9, 78R4

b Electrooptic effect: Tables 67, 68, 69; Figs. 369···373; see also — 44Z1, 69A2, 77S30
Electrooptic effect in the ultraviolet region near the absorption edge: see — 76O5
Pockels cell for a short light pulse: see — 78J3

c Piezooptic effect (photoelastic effect): Table 70; Figs. 374, 375; see also — 77M2
Effect of p on refractive indices: Fig. 376.
p^E_{66} obtained from Brillouin scattering: Fig. 377.

d gyration tensor: $g_{11} = 2.54 \cdot 10^{-4}$ for $\lambda = 462$ nm and $2.28 \cdot 10^{-4}$ for $\lambda = 506$ nm. — 78K10
Verdet constant $V = 204\,(11)$ K^{-1} m^{-1} for $\lambda = 632.8$ nm. — 76K15

e Nonlinear susceptibility for SHG: — 72L5
KH_2PO_4: $d_{36} = 0.44 \cdot 10^{-12}$ mV^{-1} for $\lambda = 1.06$ μm.
$|d_{14}/d^{KDP}_{36}| = 1.00$ (6) or 0.86 (5) for $\lambda = 694.3$ nm and 1.03 (5) for $\lambda = 1.06$ μm.
KD_2PO_4: $|d_{36}/d^{KDP}_{36}| = 1.05$ (5) for $\lambda = 694.3$ nm and 1.06 (5) for $\lambda = 1.06$ μm.
$|d_{14}/d^{KDP}_{36}| = 0.97$ (8) or 0.85 (10) for $\lambda = 694.3$ nm and 1.05 (5) for $\lambda = 1.06$ μm. — 69h1
See also Figs. 378, 379. — 77N4
Third order non-linear susceptibilities for the difference frequency generation: see — 77E1
Third and fourth harmonic generation of Nd-glass laser at $\lambda = 1.06$ μm: see — 77A11
Two photon absorption of DKDP at $\lambda = 266.1$ nm: see — 77R6

continued

Nr. 19A-1 KH_2PO_4, continued

10a	Raman scattering:	
	Spectrum at RT: Fig. 380; see also Fig. 524.	
	Temperature dependence of B_2 modes: Fig. 381, 382, 383; see also	73H3, 73R5
	Effect of deuteration: Fig. 384; see also	76B8, 76P4, 77S22
	Coupled mode analysis: Fig. 385; see also	72S17, 76M9, 77K4
	Spectra of E modes: Fig. 386; see also	70W7
	Effect of p: Figs. 387, 388.	
	Polariton and near forward scattering: see	78T6, 77A5
	Spectra in the monoclinic phase (phase II): see	75S19
	Spectra of highly deuterated KDP in the monoclinic phase at RT: see	78R4
b	Brillouin scattering: c_{ij}^E obtained from Brillouin scattering, see 8a.	
	For the details of sound velocity and c_{ij}^E, see	74B19
	p_{66}^E obtained from Brillouin scattering, see 9c, see also	74A8
	Brillouin and Rayleigh scattering spectra: Figs. 389, 390, 391; see also	77L1
	Rayleigh scattering in the vicinity of Θ_f: Fig. 392; see also	78C4
	Rayleigh scattering at $T \approx \Theta_{III-II}$: see	73S10
11	Electrical conductivity: Figs. 393, 394, 395; see also	50m1, 62S4, 66H5
	Conductivity of Co^{2+}-doped KDP: see	77S20
	Anomalies in σ are reported to occur at 180 °C in KH_2PO_4 and at 102 °C in KD_2PO_4, see also	67O5, 72G14
	Electric breakdown: see	46D1
	Activation energy: Table 71; see also	66H5
13a	NMR of proton or deuteron: Tables 72, 73; Figs. 396, 397, 398, see also	74B22, 73G10, 62S4, 67B9, 71B18
	NMR of ^{39}K: Table 74; Figs. 399, 400; see also	69T6
	NMR of ^{31}P: Table 75; Figs. 401, 402, 403; see also	74T5, 74B21
	NQR of ^{17}O: Fig. 404.	
b	ESR: Tables 76, 77, 78.	
	ESR of X-ray irradiated crystals; see	73H4, 78H13, 75V11, 77D1
	ESR of γ-ray irradiated crystals; see	73T4, 75T13, 78M6
	Slow ferroelectric cluster dynamics was studied by ESR of ^{54}Cr in KH_2PO_4.	76M24
	ENDOR: see	77D1
c	Mössbauer effect: Fig. 405.	
14b	Inelastic neutron scattering: Figs. 406, 407; see also Fig. 451, and	63W2

15a	Domain structure: domain walls are perpendicular to the *a* axis of the tetragonal phase. Observation by polarized light: see Observation by dew method: Fig. 408.	53M2
b	Switching characteristics: Figs. 409, 410; see also	76B2
16	Surface layer: In the vicinity of Θ_f, a layer of low dielectric constant ($\varkappa \approx 21$) with the thickness of about $6.63 \cdot 10^{-7}$ m is reported to exist.	76M4
	Whiskers are grown along [100] (rarely along [101]) from aqueous solution.	67U1
	Internal friction measured at about 1 Hz: see	78G12
	Strain distribution and domain switching studied by γ-ray diffractometry: see	76B2
	Dislocations and interdomain interactions: see	72B15
	X-ray topographic studies of dislocations: see	76B3

Table 47. KH_2PO_4(KDP). Solubility *A* in water [62k1].

A [wt%]	*T* [°C]	*A* [wt%]	*T* [°C]	*A* [wt%]	*T* [°C]
12.88	0	20.04	25	29.00	50
14.00	5	21.90	30	33.40	60
15.50	10	23.65	35	37.05	70
16.87	15	25.10	40	41.30	80
18.45	20	26.90	45	45.50	90

Table 48. KH_2PO_4(KDP) family. Unit cell parameters at RT.

crystals	*a* [Å]	*c* [Å]	*c*/*a*	*T* [°C]	Ref.
KH_2PO_4	7.4529 (2)	6.9751 (6)	0.9359 (1)	25	67C5
	7.452 (2)	6.959	0.9338	20	47U1
KD_2PO_4	7.4697 (3)	6.9766 (5)	0.9340 (1)	25	67C5
	7.471_4	6.956_1	0.9310	20	47U1, 39U1*)
KH_2AsO_4	7.6295 (6)	7.1605 (8)	0.9385 (2)	25	67C5
	7.6300 (4)	7.1630 (3)	0.9388 (2)	27, 25	65D4
	7.625_6 (10)	7.164_8 (10)	0.9396 (2)	25	50D1
KD_2AsO_4	7.6410 (10)	7.1636 (9)	0.9375 (2)	25	67C5
	7.641_5	7.165	0.9376	25	50D1
RbH_2PO_4	7.608 (8)	7.296 (8)	0.9590 (21)		64H3
RbH_2AsO_4	7.7933 (5)	7.4671 (5)	0.9581 (1)	25	67C5
RbD_2AsO_4	7.8063 (7)	7.4674 (10)	0.9566 (2)	25	67C5
CsH_2AsO_4	7.9852 (4)	7.8928 (3)	0.9884 (1)	25	67C5
$NH_4H_2PO_4$	7.4991 (4)	7.5493 (12)	1.0067 (2)	20	63D2
$ND_4D_2PO_4$	7.5193 (9)	7.5400 (19)	1.0028 (4)	20	63D2, 42U1*)
$NH_4H_2AsO_4$	7.6944 (4)	7.720 (2)	1.0033 (3)	25	67C3
	7.694 (8)	7.718 (8)	1.0031 (21)		64H3
	7.699 (5)	7.729 (5)	1.0039 (13)		58D2

*) Recalculated in [67C5].

Table 49. KH_2PO_4(KDP) family. Unit cell parameters at Θ_f [67C5].

Crystals	Θ_f or Θ_{IV-III} °C	Unit cell parameters [Å] at Θ_f or Θ_{IV-III}		c/a at Θ_f or Θ_{IV-III}	c/a at 25 °C
		a	c		
KH_2PO_4	−150	7.4256 (5)	6.9296 (9)	0.9332 (2)	0.9359 (1)
KD_2PO_4	− 51	7.4586 (7)	6.9559 (8)	0.9326 (2)	0.9340 (1)
KH_2AsO_4	−178	7.6034 (9)	7.1006 (11)	0.9339 (3)	0.9385 (2)
KD_2AsO_4	−112	7.6268 (14)	7.1195 (17)	0.9335 (4)	0.9375 (2)
	−114				
RbH_2AsO_4	−163	7.7727 (8)	7.4019 (10)	0.9523 (3)	0.9581 (1)
RbD_2AsO_4	−100	7.7950 (14)	7.4216 (19)	0.9521 (4)	0.9566 (2)
RbH_2PO_4	−126	7.586 (9)	7.250 (9)	0.9556 (23)	0.9590 (21)
CsH_2AsO_4	−130	7.9703 (14)	7.8329 (13)	0.9828 (4)	0.9884 (1)
$NH_4H_2PO_4$	−126	7.4710 (8)	7.5374 (18)	1.0089 (4)	1.0065 (2)
$NH_4H_2AsO_4$	− 57	7.6790 (12)	7.715 (2)	1.0047 (4)	1.0033 (3)

Table 50. $KH_{2(1-x)}D_{2x}PO_4$. Unit cell parameters in the monoclinic phase (phase II).

	KDP 468 K in $P2_1$ or $P2_1/m$ cell	DKDP 403 K in $P2_1$ cell	DKDP (x ≈0.9) RT in $P2_1$ cell	DKDP (x ≧0.98) 77 K in $P2_1$ cell
a [Å]	7.47	7.471	7.45 (1)	7.439 (1)
b [Å]	14.49	14.711	14.71 (2)	14.626 (2)
c [Å]	7.43	7.167	7.14 (1)	7.090 (1)
β [deg]	92.2	92.34	92.31 (1)	92.28 (1)
Ref.	75I6	74T10	75T10	76K6

Table 52. KH_2PO_4 (KDP). Fractional coordinates of atoms in the paraelectric phase III.

		RT [53B1] (Small cell, I$\bar{4}$2d)	RT [53B1] (Large cell, F$\bar{4}$d2)	132 K [55B1] (Large cell, F$\bar{4}$d2)
K	x	0	0	0
	y	0	0	0
	z	$\frac{1}{2}$	$\frac{1}{2}$	$\frac{1}{2}$
P	x	0	0	0
	y	0	0	0
	z	0	0	0
O	x	0.1487 (3)	0.0328 (4)	0.0331 (3)
	y	0.0828 (3)	0.1158 (40)	0.1161 (3)
	z	0.1262 (6)	0.1261 (60)	0.1270 (3)
H	x	0.1476 (14)	{0.961 0.937}	{0.960 (2) 0.937 (2)}
	y	{0.226 (25) 0.274 (25)}	{0.187 0.211}	{0.187 (2) 0.210 (2)}
	z	0.125 (3)	0.125	{0.123 (2) 0.127 (2)}

Table 51. KD_2PO_4(DKDP). Unit cell parameters in a face centered cell [74N1].

T	293 K	232 K	211 K
a [Å]	10.557 (1)	10.549 (4)	10.602 (2)
b [Å]	10.557 (1)	10.549 (4)	10.498 (2)
c [Å]	6.975 (1)	6.961 (1)	6.961 (1)

Table 53. KH_2PO_4(KDP). Fractional coordinates of atoms in the ferroelectric phase IV [55B1].

		77 K (Orthorhombic cell, Fdd2)	
K	x	0	
	y	0	
	z	$\frac{1}{2}$+0.016 (2)	
P	x	0	
	y	0	
	z	0	
O		(OH)	(O)
	x	0.0339 (3)	0.1164 (3)
	y	0.1158 (3)	0.9661 (3)
	z	0.139 (2)	0.883 (2)
H	x	0.963 (1)	
	y	0.187 (1)	
	z	0.138 (4)	

Table 54. KD_2PO_4(DKDP). Fractional coordinates of atoms in phase III (T=293, 232 K) and in phase IV (T=212 K) [74N1].

Atom		293 K	232 K	211 K	
P	x	0	0	0	
	y	0	0	0	
	z	0	0	0	
K	x	0	0	0	
	y	0	0	0	
	z	0.5	0.5	0.5178 (1)	
O	x	0.0338 (1)	0.0342 (1)	0.0343 (1)	0.1156 (1)
	y	0.1151 (1)	0.1152 (1)	0.1147 (1)	−0.0350 (1)
	z	0.1259 (2)	0.1265 (2)	0.1370 (3)	−0.1147 (3)

Table 55. KD_2PO_4(DKDP). Fractional coordinates of atoms at RT in the monoclinic phase [75T10].

Atom	x	y	z	Atom	x	y	z
K1	0.2312 (5)	0.1150 (1)	0.3664 (6)	O9	0.0565 (4)	0.4450 (3)	0.5316 (6)
K2	0.7184 (5)	0.1046 (3)	0.6593 (6)	O10	0.2062 (3)	0.2913 (3)	0.5308 (4)
K3	0.6253 (5)	0.4076 (3)	0.2086 (7)	O11	0.9813 (3)	0.3339 (3)	0.2747 (4)
K4	0.1436 (5)	0.3780 (3)	0.9266 (7)	O12	0.2795 (3)	0.4129 (3)	0.2939 (4)
P1	0.7374 (3)	0.1359 (3)	0.1766 (4)	O13	0.7234 (3)	0.4347 (3)	0.5833 (4)
P2	0.2274 (3)	0.1267 (3)	0.8682 (4)	O14	0.4861 (3)	0.4167 (3)	0.8220 (4)
P3	0.1408 (3)	0.3697 (2)	0.4069 (4)	O15	0.7807 (3)	0.3301 (3)	−0.1523 (4)
P4	0.6306 (3)	0.3726 (3)	0.7131 (4)	O16	0.5517 (4)	0.2879 (3)	0.6018 (5)
O1	0.8762 (3)	0.1696 (3)	0.3187 (4)	D1	0.5646 (4)	0.0163 (3)	0.2349 (5)
O2	0.6453 (3)	0.2123 (3)	0.0672 (5)	D2	0.9586 (5)	0.0664 (3)	0.0210 (6)
O3	0.5880 (3)	0.0814 (2)	0.2783 (4)	D3	0.4387 (5)	0.2060 (3)	0.0203 (6)
O4	0.8179 (3)	0.0658 (3)	0.0409 (5)	D4	0.1073 (4)	0.2220 (3)	0.6665 (5)
O5	0.1408 (3)	0.0556 (3)	−0.0125 (4)	D5	0.9221 (4)	0.4362 (3)	0.5519 (6)
O6	0.3667 (3)	0.0932 (3)	0.7429 (4)	D6	0.9397 (4)	0.2691 (3)	0.2984 (5)
O7	0.0669 (3)	0.1750 (3)	0.7551 (4)	D7	0.7358 (4)	0.2802 (3)	−0.0699 (5)
O8	0.3024 (5)	0.2054 (3)	0.0002 (5)	D8	0.4216 (5)	0.2897 (3)	0.5790 (6)

Table 56, see p. 77.

Table 57. KD_2PO_4(DKDP). Bond lengths and angles related to PO_4 tetrahedron and O–D–O bonding [74N1].

	293 K	232 K	211 K	
Angle between pairs of bonds related by symmetry axis along c	110°34′ (6)	110°28′ (9)	105°38′ (10)	116°05′ (12)
Lengths of P–O bond [Å]	1.541 (1)	1.544 (1)	1.578 (2)	1.509 (2)
Angle between O–D–O bonds and x–y plane	17′ (2)	28′ (3)	17′ (4)	
Length of O–D–O bond [Å]	2.519 (1)	2.521 (2)	2.530 (2)	

Table 58. KH_2PO_4 and KD_2PO_4. Fractional coordinates and temperature factors at 20 °C [73N4]. U_{ij} is defined by Eq. (d) in Introduction. Small cell $I\bar{4}2d$. Values for KD_2PO_4 are given in brackets.

Atom	x	y	z	U_{11} [Å²]	U_{22} [Å²]	U_{33} [Å²]	U_{12} [Å²]	U_{23} [Å²]	U_{31} [Å²]
K	0	0	0.5	0.0213 (3)	0.0213 (3)	0.0141 (2)	0	0	0
	[0]	[0]	[0.5]	[0.0215 (2)]	[0.0215 (2)]	[0.0146 (1)]	[0]	[0]	[0]
P	0	0	0	0.0122 (3)	0.0122 (3)	0.0155 (2)	0	0	0
	[0]	[0]	[0]	[0.0123 (2)]	[0.0123 (2)]	[0.0172 (2)]	[0]	[0]	[0]
O	0.1486 (1)	0.0825 (1)	0.1262 (2)	0.0154 (3)	0.0157 (3)	0.0199 (3)	0.0020 (3)	−0.0042 (3)	−0.0066 (3)
	[0.1489 (1)]	[0.0813 (1)]	[0.1259 (2)]	[0.0152 (2)]	[0.0161 (2)]	[0.0201 (2)]	[0.0022 (2)]	[−0.0043 (2)]	[−0.0060 (2)]

Table 59. KH_2PO_4 family. Thermal expansion [67C5].

Crystals	Sample length mm	Total contraction, 25 °C to Θ_f or Θ_{IV-III} $\Delta l/l$ [10^{-4}]	Expansion coefficient α [10^{-6} K^{-1}] Θ_f or Θ_{IV-III} to 25 °C	Expansion coefficient α [10^{-6} K^{-1}] −50 °C to +50 °C	Change of length at Θ_f or Θ_{IV-III} (on cooling) $\Delta L/L$ [10^{-4}]	Change in volume at Θ_f or Θ_{IV-III} $\Delta V/V$ [10^{-4}]
Along *a* axis						
KH_2PO_4	7.2	37.8	22.0	24.9	+ 5.5	+17.5
KD_2PO_4	4.9	15.1	19.4	20.1	+ 0.3	+10.5
KH_2AsO_4	5.2	34.2	17.5	24.2	+ 4.3	+15.1
KD_2AsO_4	6.5	18.3	12.9	15.1	+ 0.4, −1.9	+ 9.5···+4.9
RbH_2AsO_4	2.8	26.4	14.3	16.9	+ 7.5	+20.1
RbD_2AsO_4	2.4	14.5	11.2	14.3	+ 1.4	+ 9.8
$NH_4H_2PO_4$	7.3	40.8	27.2	32.0	+56	+42
$NH_4H_2AsO_4$	1.5	21.4	25.1	27.4	+56	+47.5
Along *c* axis						
KH_2PO_4	4.5	67.4	39.2	44.0	+ 6.5	
KD_2PO_4	3.9	29.6	39.5	40.7	+ 9.9	
KH_2AsO_4	5.8	83.6	43.6	47.1	+ 6.5	
KD_2AsO_4	5.0	61.5	43.3	45.6	+ 8.7	
RbH_2AsO_4	3.5	87.3	46.7	49.5	+ 5.1	
RbD_2AsO_4	2.5	61.4	47.3	49.5	+ 7.0	
$NH_4H_2PO_4$	6.1	16.2	10.7	4.2	−70	
$NH_4H_2AsO_4$	3.1	6.5	7.6	5.8	−64.5	

Table 56. KH_2PO_4(KDP). Interatomic distances in Å at different temperatures [55B1]. O and O_H are two sorts of hydrogen atoms at 77 K: O_H with the nearest hydrogen atom 1.04 Å away, O with the nearest hydrogen atom at 1.44 Å. The primed and unprimed atoms are in an opposite z direction relative to the central atom of the PO_4 tetrahedron.

Bond	RT	132 K	77 K obs	77 K mean
$P–O_H$ } $P–O$	1.538 (5)	1.538 (4)	1.583 (20) } 1.508 (20)	1.545
$O_H–O_H$ } $O–O$	2.528 (7)	2.528 (9)	2.521 (6) } 2.549 (6)	2.535
$O–O'_H$ } $O_H–O'$	2.503 (8)	2.503 (5)	2.519 (20) } 2.512 (20)	2.515
O–H–O	2.487 (5)	2.475 (4)	2.486 (4)	
$O_H–H$	1.07 (1)	1.07 (1)	1.050 (14)	
O–H	1.42 (1)	1.41 (1)	1.430 (14)	
$K–O_H$ } $K–O$	2.888 (6)	2.870 (3)	2.896 (30) } 2.833 (30)	2.864
$K–O'_H$ } $K–O'$	2.819 (4)	2.802 (5)	2.812 (10) } 2.777 (10)	2.795

Table 60. KH_2PO_4(KDP) family. Parameters in the formula $E=\frac{1}{\varepsilon_0 C}(T-\Theta_p)P+\xi P^3+\zeta P^5+\nu P^n$. [77C1]. The content of deuterium in DKDP is about 70%.

Crystals	Θ_p K	C 10^3K	ξ 10^{10} V m^5 C^{-3}	ζ 10^{13} V m^9 C^{-5}	n	ν
KH_2PO_4	122.76	2.9	− 2.57	5.95	11	$4.2 \cdot 10^{21}$ V m^{21} C^{-11}
KD_2PO_4	207.16	3.9	− 4.94	2.14	17	$9.9 \cdot 10^{27}$ V m^{33} C^{-17}
RbH_2PO_4	147.66	3.7	− 3.24	8.38	19	$1.94 \cdot 10^{32}$ V m^{37} C^{-19}
CsH_2AsO_4	146.15	2.9	− 9.18	6.13	15	$5.81 \cdot 10^{29}$ V m^{29} C^{-15}
KH_2AsO_4	96	2.49	−17.0	6.54	29	$3.60 \cdot 10^{44}$ V m^{57} C^{-29}

Table 61. KD_2PO_4, RbH_2PO_4, CsH_2PO_4. Transition temperatures as a function of pressure [78R3].

Phase boundary	Least-squares fit p in 10^8 Pa	Standard deviation K
RbH_2PO_4 II/I	$\Theta[°C]=281+5.125p-0.05715p^2$	1.7
RbH_2PO_4 V/I	$\Theta[°C]=350+7.143(p-16.4)-0.0919(p-16.4)^2$	1.9
RbH_2PO_4 V/Liquid	$\Theta[°C]=402+3.909(p-24.4)-0.0270(p-24.4)^2$	2.0
RbH_2PO_4 I/Liquid	$\Theta[°C]=292.1+9.175p-0.1955p^2$	1.9
CsH_2PO_4 II/I	$\Theta[°C]=230+2.80p$	3.0
CsH_2PO_4 V/I	$\Theta[°C]=262+8.716(p-11)-0.1107(p-11)^2$	2.9
CsH_2PO_4 V/VI	$\Theta[°C]=384.9+2.016(p-29.9)$	2.5
CsH_2PO_4 VI/I	$\Theta[°C]=384.9+6.998(p-29.9)-0.0828(p-29.9)^2$	2.0
CsH_2PO_4 VI/Liquid	$\Theta[°C]=457-0.0411(p-42)$	0.7
CsH_2PO_4 I/Liquid	$\Theta[°C]=345.7+8.340p-0.1405p^2$	4.1
KD_2PO_4 I/Liquid	$\Theta[°C]=250+7.334p-0.1081p^2$	1.9
KD_2PO_4 VI/Liquid	$\Theta[°C]=368.5+4.883(p-25)-0.0724(p-25)^2$	1.4

Table 62. KH_2PO_4(KDP). C, ξ, ζ, vs. p [78W3]. Note the change of sign of ξ with the increase of pressure.

p 10^8 Pa	Sample no.	C 10^3 K	ξ 10^{10} V m^5 C^{-3}	ζ 10^{13} V m^9 C^{-5}
0.001	1	2.92 (14)	−1.90 (32)	4.3 (11)
0.0016	2	3.20 (6)	−1.20 (4)	2.3 (1)
0.001	3	3.21 (3)	−1.02 (4)	2.32 (9)
1.00	2	3.45 (8)	−0.72 (4)	2.6 (1)
2.00	3	3.22 (16)	−0.52 (5)	4.29 (10)
2.40	3 (Run 1)	3.16 (2)	−0.02 (8)	3.1 (2)
2.40	3 (Run 2)	3.31 (4)	−0.25 (29)	6.5 (12)
2.40	3 (Run 3)	3.81 (9)	+0.02 (16)	3.2 (4)
2.40	3*)	3.23 (3)	−0.02 (5)	3.2 (2)
3.00	2	3.11 (8)	+0.73 (4)	4.5 (2)

*) Weighted average.

Table 63. KH_2PO_4(KDP), KD_2PO_4(DKDP). Piezoelectric constants [66b1].

Crystals	d_{14} 10^{-12} CN^{-1}	d_{36} 10^{-12} CN^{-1}	e_{36} Cm^{-2}	g_{14} 10^{-3} m^2 C^{-1}	g_{36} 10^{-3} m^2 C^{-1}	h_{14} 10^8 NC^{-1}	h_{36} 10^8 NC^{-1}	k_{36}	T °C	Ref.
KH_2PO_4(KDP)	1.28	20.9							20	38S2
		21.7			112				27	56B2
			0.142						20	50V1
				3.3	125	0.42	8.1		0	50V1
		16.8	0.097		110		6.48		100	46M1
		18.0	0.104		112		6.63		80	
		19.7	0.117		116		6.96		60	
		21.1	0.127		116		7.05		40	
		23.2	0.142		118		7.32		20	
		25.4	0.158		119		7.50		0	
		28.6	0.180		121		7.68		−20	
		32.9	0.208		123		7.98		−40	
		39.7	0.252		125		8.19		−60	
		51.0	0.325		126		8.37		−80	
		67.3	0.427		127		8.43		−100	
		111	0.69		129		8.79		−120	
		160	0.97		130		8.91		−130	
		292	1.58		130		8.94		−140	
		488	2.33		131		9.06		−145	
		1470	4.3		132		9.21		−150	
KD_2PO_4(DKDP)		51.7							20	45B3
		53.3							27	55B3
		58.0			130			0.22	RT	63S8

Table 64. KH_2PO_4(KDP), KD_2PO_4(DKDP). Elastic constants.

Crystals	s_{11}	s_{12}	s_{13}	s_{33}	s_{44}	s_{66}	T	Ref.
	$10^{-11}\,m^2\,N^{-1}$						°C	
KH_2PO_4	1.75	−0.4	−0.75	2.0	7.9	16.6	25	46M1
(KDP)	1.53	0.2	−0.38	1.96	7.75	16.8	20	46Z1
KD_2PO_4					7.77	16.1	RT	45J1
(DKDP)						16.5		50J1
						17.1	27	56B2
						16.5		63S8
	1.575 (8)	0.21 (6)	−0.40 (3)	2.014 (8)	7.90 (2)	16.83 (4)	20	66S17
	c_{11}	c_{12}	c_{13}	c_{33}	c_{44}	c_{66}		
	$10^{10}\,N\,m^{-2}$							
KH_2PO_4	8.0	3.4	4.1	8.0	1.28	0.61	20	46M1
(KDP)	6.91	−0.600	1.22	5.56	1.290	0.600	20	46Z1
	7.14	−0.49	1.29	5.62	1.27	0.624		50P1
	7.4	1.8	2.7	6.8	1.35	0.63		45B3
	7.165	−0.627	1.494	5.640	1.248	0.621	20	64H3
	7.17 (3)	−0.63 (9)	1.60 (5)	5.68 (3)	1.27 (1)	0.64 (2)		71B23
KD_2PO_4	7.04	0.46				0.607	26	50J1
(DKDP)	6.93	−0.78	1.22	5.45	1.265	0.594	20	66S17

Table 65. KH_2PO_4(KDP). Refractive indices [64Z1]. $T = 24.8$ °C.

λ nm	Index in air		Absolute index	
	n_o	n_e	n_o	n_e
200	1.621996	1.563315	1.622630	1.563913
300	1.545084	1.497691	1.545570	1.498153
400	1.524035	1.479814	1.524481	1.480244
500	1.514498	1.472068	1.514928	1.472486
600	1.508851	1.467856	1.509274	1.468267
700	1.504817	1.465193	1.505235	1.465601
800	1.501508	1.463303	1.501924	1.463708
900	1.498514	1.461830	1.498930	1.462234
1000	1.495628	1.460590	1.496044	1.460993
1100	1.492730	1.459481	1.493147	1.459884
1200	1.489751	1.458443	1.490169	1.458845
1300	1.486645	1.457436	1.487064	1.457838
1400	1,483381	1.456437	1.483803	1.456838
1500	1.479938	1.455427	1.480363	1.455829
1600	1.476302	1.454395	1.476729	1.454797
1700	1.472459	1.453333	1.472890	1.453735
1800	1.468400	1.452234	1.468834	1.452636
1900	1.464118	1.451093	1.464555	1.451495
2000	1.459603	1.449906	1.460044	1.450308

Table 66. KD_2PO_4(DKDP). Refractive indices with respect to air [66P3]. Accuracy: ±0.0001.

λ nm	Index at 298 K		Increase at 228 K	
	n_o	n_e	n_o	n_e
690.7	1.5022	1.4639	0.0019	0.0014
623.4	1.5044	1.4656	0.0020	0.0015
577.9	1.5063	1.4670	0.0019	0.0014
546.1	1.5079	1.4683	0.0020	0.0015
491.6	1.5111	1.4710	0.0020	0.0014
435.8	1.5155	1.4747	0.0020	0.0015
407.8	1.5185	1.4772	0.0020	0.0014
404.7	1.5189	1.4776	0.0020	0.0014

Table 67. KH_2PO_4(KDP) and KD_2PO_4(DKDP). Electrooptic constants.

Crystals	r^T_{41}	r^S_{41}	r^T_{63}	r^S_{63}	ϱ^T_{63}	T °C	λ nm	Ref.
	10^{-12} m V^{-1}				10^{-3} m^2 C^{-1}			
KH_2PO_4(KDP)	8.6		+10.5			22	556	50C1
			−11.6			20	546	44Z1
	8.8		10.3			RT	546	64O2
			12.0	3.5				63M5
			9.7					53C2
			10.0				546.1	61N2
					−62.7			44Z1, 50B3
			9.7					64V2
				8.3			400···725	66V2
				8.8			633	67R5
			9.4	8.15			633	67O1
	8.5			9.3			633	76F10
	−8.28	−8.25	10.22	8.89		21	589	78V1
KD_2PO_4 (deuterated KDP)			−23.3			RT		59J1
			−26.4			RT	546	63S8
			−20			20	546	44Z1
					−41.7			44Z1, 50B3
83···92% deuterated KDP	8.8							64O2

Table 68. KH_2PO_4 family. ϱ_{63} vs. T [71V1]. ϱ_{63}: electrooptic constants for $\boldsymbol{P}$. $\lambda = 535$ nm.

Crystals	ϱ_{63} $\cdot 10^{-2}$ m^2 C^{-1}							
	20 °C	0 °C	−20 °C	−40 °C	−60 °C	−80 °C	−100 °C	−120 °C
KH_2PO_4(KDP)	5.7	6.3	6.6	6.9	6.9	6.9	6.9	6.9
KD_2PO_4(DKDP)	5.7	6.0	6.0	6.0				
RbH_2PO_4(RDP)	5.7	6.0	6.3	6.6	6.6	6.6		
KH_2AsO_4(KDA)	6.9	7.2	7.5	7.8	7.8	7.8	7.8	7.8
RbH_2AsO_4(RDA)	6.6	6.9	7.5	7.8	8.1	8.4	8.4	8.4
CsH_2AsO_4(CDA)	7.5	8.1	8.4	8.7	8.7	8.7	8.7	8.7

Table 69. $KH_{2(1-x)}D_{2x}PO_4$. $r_{i\lambda}$ vs. x [73V2].
$r_{i\lambda}$: electrooptic constants.

x	r_{63} 10^{-12} m V^{-1}	r_{41} 10^{-12} m V^{-1}
0	9.9 (2)	8.6 (2)
0.18	10.1 (3)	
0.30	11.7 (3)	
0.40	13.4 (4)	
0.50	12.8 (4)	9.3 (3)
0.70	16.0 (6)	
0.86	19.8 (7)	10.7 (3)
0.90	22.7 (12)	10.9 (3)
0.96	23.3 (12)	
0.98	25.8 (2)	10.7 (3)
0.998	26.4 (7)	

Table 71. KH_2PO_4(DKP), KD_2PO_4(DKDP). Activation energy ΔU associated with the electric conductivity along the c axis [67O5]. Estimated from the conductivity vs. T curves, $\sigma \propto \exp(-\Delta U/kT)$.

Crystals	Temperature range	ΔU [eV]
KH_2PO_4	$T < \approx 180$ °C	0.543 (5)
	$T > \approx 180$ °C	0.78 (7)
KD_2PO_4	$T < \approx 102$ °C	0.53 (1)
	$T > \approx 102$ °C	0.82 (4)

Table 72. KH_2PO_4(KDP). $\langle \Delta\nu^2 \rangle^{1/2}$ of proton resonance [62B5].

	$\langle \Delta\nu^2 \rangle^{1/2}$ (exp)	$\langle \Delta\nu^2 \rangle^{1/2}$ (calc)
H along c	5.2 kHz	4.6 kHz
H along a	5.6	5.3

Table 70. KH_2PO_4(KDP). Piezooptic constants.

$\pi_{11}-\pi_{12}$	π_{11}	π_{12}	π_{13}	π_{31}	π_{33}	π_{44}	π_{66}	λ nm	Ref.
10^{-12} m^2 N^{-1}									
> -0.3, < 0							-11.25 (π^E_{66})	560	50W2
							-8.6 (π^E_{66})		59J1
	4.15	4.08	1.12	3.60	0.44	-1.47	-10.26	589	75V4
3*)							10*)		76S6
$p_{11}-p_{12}$	p_{11}	p_{12}	p_{13}	p_{31}	p_{33}	p_{44}	p_{66}		
dimensionless									
> -0.7, < 0							-6.85 (p^E_{66})	560	50W2
	0.251*)	0.249*)	0.246*)	0.225*)	0.221*)		0.058*) (p^E_{66})	633	67D2
	0.25*)	0.259*)		0.219*)				633	73M6
	0.287	0.282	0.174	0.241	0.122	-0.019	-0.064	589	75V4
	0.254*)	0.230*)	0.233*)	0.221*)	0.212*)			633	77A7

*) Absolute value.

Table 73. KD_2PO_4(DKDP), KD_2AsO_4(DKDA), CsD_2AsO_4(DCDA). Principal values, eigenvectors and angles between the X-bond deuteron quadrupole coupling tensor eigensystem and the $X(a)$, $Y(b)$, $Z(c)$ crystal axes system [71B20].

		$(eQ/h)\phi_{zz}$[kHz]			$(eQ/h)\phi_{yy}$[kHz]			$(eQ/h)\phi_{xx}$[kHz]		
KD_2PO_4										
	$T = 30$ °C		119.7			-62.6			-56.9	
		± 1	0	0	0	0	± 1	0	± 1	0
		0°	90°	90°	90°	90°	0°	90°	0°	90°
	$T = -100$ °C		127.5			-71.1			-56.4	
		± 1	0	0	0	± 0.58	0.81	0	0.81	± 0.58
		0°	90°	90°	90°	55°	$\pm 35°$	90°	$\pm 35°$	55°

continued

Table 73, continued

	$(eQ/h)\phi_{zz}$[kHz]			$(eQ/h)\phi_{yy}$[kHz]			$(eQ/h)\phi_{xx}$[kHz]		
KD_2AsO_4									
T=30 °C		118.5			−62.0			−55.9	
	±0.99	0	±0.04	±0.04	0	±0.99	0	±1	0
	2°	90°	88°	88°	90°	2°	90°	0°	90°
T=−123 °C		125.1			−65.4			−54.6	
	±0.99	0.02	∓0.030	±0.04	±0.55	0.82	0	0.83	±0.55
	2°	88°	88°	88°	56°	±34°	90°	±34°	56°
CsD_2AsO_4									
T=30 °C		121.5			−64.0			−57.5	
	±1	0	0	0	0	±1	0	±1	0
	0°	90°	90°	90°	90°	0°	90°	0°	90°
T=−68 °C		126.7			70.0			−54.0	
	±1	0	0	0	∓0.54	0.84	0	0.84	∓0.54
	0°	90°	90°	90°	57°	±33°	90°	±33°	57°

Table 74. KH_2PO_4(KDP). ^{39}K electric field gradient tensor components at 77 K [68T7]. There are two non-equivalent sites A and B for K in a unit cell (see illustration in section 3b). X, Y, Z axes are the tetragonal a, b, c axes in the paraelectric phase.

	$eQ\phi_{zz}/12\,h$ kHz	$eQ\phi_{yy}/12\,h$ kHz	$eQ\phi_{xx}/12\,h$ kHz	$eQ\phi_{xy}/12\,h$ kHz
Site A	123.3 (7)	−19.3 (12)	−104.0 (18)	7.1 (33)
Site B	123.3 (7)	−104.0 (12)	− 19.3 (12)	7.1 (33)

Table 75. KH_2PO_4(KDP), KD_2PO_4(DKDP), $NH_4H_2PO_4$(ADP). Principal values of ^{31}P chemical shift tensor. $\Delta\sigma=\sigma_{\|}-\sigma_{\perp}$, $\bar{\sigma}=\frac{1}{3}(\sigma_{\|}+2\sigma_{\perp})$. $\sigma_{\|}$, $\sigma_{\perp}$: components parallel and perpendicular to the symmetry axis (c-axis).

	$\sigma_{\|}$	$\sigma_{\perp}$	$\Delta\sigma$	$\bar{\sigma}$	T	Ref.
KH_2PO_4	− 1.1	−16.5	15.4	−11.2	77K	74T5
KH_2PO_4	12.8 (2)	−14.7 (2)	27.5	− 5.5	RT	
KD_2PO_4	12.8 (2)	−13.1 (2)	25.9	− 4.5	RT	76J2
$NH_4H_2PO_4$	20.0 (2)	−12.5 (2)	32.5	− 1.5	RT	

Table 76. KH_2PO_4(KDP):VO^{2+}. Spin Hamiltonian parameters for VO^{2+} in KH_2PO_4 at RT [78K12]. A: in the unit of $10^{-2}\,m^{-1}$. θ, ϕ: direction of the symmetriy axis (V–O bond) relative to the tetragonal axis. Three distinct hyperfine patterns of different intensity were observed corresponding to three chemically distinct sites I, II, III.

Parameters	Site I	Site II	Site III
$g_{\|}$	1.924 (1)	1.929 (2)	1.933 (2)
$g_{\perp}$	1.975 (2)	1.982 (2)	1.985 (2)
$A_{\|}$	182.0 (5)	177.1 (10)	180.8 (10)
$A_{\perp}$	70.9 (10)°	68.1 (10)°	68.9 (20)°
θ	18.5 (10)°	90 (1)°	34 (10)°
ϕ	20 (2)°	20 (2)°	36 (10)°
Intensity	15	2	1

Table 77. KH_2PO_4(KDP): Fe^{3+}. Principal values and direction cosines of fine structure D tensor [76T17]. D_{ii}: in the unit of $10^3\,Am^{-1}$.

Principal values		Direction cosines		
		l	m	n
D_{xx}	41.9 (4)	0.628	0.739	0.243
D_{yy}	−38.3 (4)	0.761	−0.649	0.003
D_{zz}	− 3.6 (4)	−0.160	−0.183	0.970

Fig. p. 404

Table 78 α. KH_2PO_4(KDP). ESR data [67S4, 69O3].

Paramagnetic center	Site	S	$\mathcal{H}$	ν [GHz]	T [K]	g-factor	HFS $\lvert^{n}A\rvert$ [10^{-2} m^{-1}]
Mn^{2+} [67S4]	*)	5/2	(6)	9.5	RT	$g_{x_1}=0.63$ (1)	$\lvert^{55}A_{z_1}\rvert=95$
						$g_{y_1}=0.98$	
						$g_{z_1}=2.2$	
						$g_{x_2}=0.70$	$\lvert^{55}A_{z_2}\rvert=75$
						$g_{y_2}=1.07$	
						$g_{z_2}=3.5$	
						$g_{x_3}=0.54$	$\lvert^{55}A_{x_3}\rvert=70$ (10)
						$g_{y_3}=1.01$	$\lvert^{55}A_{y_3}\rvert=70$ (10)
						$g_{z_3}=2.28$	$\lvert^{55}A_{z_3}\rvert=110$ (5)
Cu^{2+} [69O3]	**)	1/2		≈9	109	$g_{\parallel}=2.364$	$A_{\parallel}=142$
						$g_{\perp}=2.080$	$A_{\perp}=19$
						$g_{\parallel}=2.340$	$A_{\parallel}=145$
						$g_{\perp}=2.072$	$A_{\perp}=20$

*) Mn^{2+} ions are shown to be located on three nonequivalent sites, with C_2 symmetry.
**) Two different K^+ sites accompanied by vacancies.

Table 78 β. Direction cosines of principal axes of **g** and **A** tensors relative to the crystallographic axes a, b, c.

	x_1	y_1	z_1	x_2	y_2	z_2	x_3	y_3	z_3
a	0.951	0.309	0	0.766	0.643	0	0.985	0.174	0
b	0.309	0.951	0	0.643	0.766	0	0.174	0.985	0
c	0	0	1	0	0	1	0	0	1

Nr. 19A-2 RbH_2PO_4, Rubidium dihydrogen phosphate (RDP)
($M=182.45$; [D: 184.47])

1a Ferroelectricity of RbH_2PO_4 was first reported by Baertschi et al. in 1945. — 45B2

b — 45B2

phase	IV	III	II	I
state	F	P		
crystal system	orthorhombic	tetragonal	monoclinic a, b)	
space group	Fdd2-C_{2v}^{19}	$I\bar{4}2d$-D_{2d}^{12}	$P2_1$-C_2^2	
Θ	−126 °C [D: −55 °C]	86 °C a, b) 86…111 °C d)	280 °C c)	

a) 69B10
b) 72G14
c) 78R3
d) 78M7

Phase I was reported to exist between 175 °C and T_{melt} b). See the phase diagram Fig. 411. Phase II is metastable at RT, and can be brought to RT by cooling the crystal from above Θ_{III-II}.

Different values of Θ_f were reported ranging from 137 K e) to 147.6 K f). — e) 68L4, f) 71G4

Θ_f was reported as 140.4 (2) K in — 73P3

$P_s \parallel [001]$ in phase IV. — 45B2

Transparent, colorless.

continued

Nr. 19A-2 RbH_2PO_4, continued

3a	Unit cell parameters: $a=7.622\,(7)$ Å, $c=7.315\,(9)$ Å at RT. Lattice constants at RT: Table 48, and at Θ_f: Table 49.	78A5
b	Crystal structure: $Z=4$ in a $I\bar{4}2d$ cell. Fractional coordinates and temperature parameters: Table 79. Interatomic distances: Table 80.	45B2
5a	Dielectric constant: $\varkappa$ vs. T: Fig. 412, 413. $\varkappa$ vs. T near Θ_{III-II}: Fig. 414. $\varkappa_c=C/(T-\Theta_p)$ for $T>\Theta_p$, where $C=3.430\,(15)\cdot10^3$ K and $\Theta_p=141.4$ (2) K. $\varkappa_c$ vs. T under p: Fig. 415; see also $\varkappa_c$, C vs. p: Figs. 416, 417. $\varkappa_c=C^*/(p-p_0)$; $C^*=1530\,(5)\cdot10^8$ Pa and $p_0=0.24\cdot10^8$ Pa at 140 K. $\varkappa$ and $\tan\delta$ for several frequencies: see Phase diagram: Fig. 411. Θ_f vs. p: Fig. 418; see also Table 61 in 19A-1.	73P3 70F6 73P3 65Z1 70F6
b	Nonlinear dielectric properties: Table 60 in 19A-1; see also	74Z5
c	P_s vs. T: Fig. 419. Effect of E_{bias} on P_s: Fig. 420.	
6a	Heat capacity: Fig. 421. Differential scanning calorimetric study of the III→II transition: see	78M7
7a	Piezoelectricity: Fig. 422. $d_{14}=3.0\cdot10^{-12}$ CN^{-1}, $d_{36}=26.7\cdot10^{-12}$ CN^{-1} at 25 °C, see also	69A3 49J1
8a	Elastic compliance and stiffness: Table 81; Figs. 423, 424, c_{66} obtained from Brillouin scattering: Fig. 425. Effect of p on c_{ij}: Fig. 426. Ultrasonic velocity: Fig. 427.	
9a	Refractive indices: Table 82; Fig. 428; see also	76V5
b	Electrooptic effect: Tables 83 and 68 in 19A-1; Figs. 369, 371; see also	69A2, 77S30
c	Piezooptic effect (photoelastic effect): $p_{11}=0.273$, $p_{12}=0.240$, $p_{13}=0.218$, $p_{31}=0.210$, $p_{33}=0.208$ for $\lambda=632.8$ nm [a]) and $p_{66}=0.032$ [b]). p_{66} obtained from Brillouin scattering: Fig. 429 and Fig. 374.	[a]) 77A7 [b]) 69V5
e	Nonlinear susceptibility for SHG: $d_{14}=0.39\,(4)\cdot10^{-12}$ m V^{-1} $d_{36}=0.41\,(5)\cdot10^{-12}$ m V^{-1} for $\lambda=1.06$ μm; see also Induced birefringence: Fig. 358.	74K9 77N4
10a	Raman scattering: Figs. 430, 431; see also Fig. 524 and Effect of p; Fig. 432. Coupled mode analysis: see	73H3 73P3, 76S8
b	Photoelastic constant obtained by Brillouin scattering, see Fig. 429. Elastic stiffness obtained by Brillouin scattering, see Fig. 425.	
11	Conductivity change in regard to the high temperature phase transition III–II: see	78S8
13a	NMR: Table 84; Figs. 433···437. NMR study of NH_4^+ doped RbH_2PO_4: see	74B9
16	Internal friction measured at about 1 Hz, see	78G12

Table 79. RbH_2PO_4(RDP). Fractional coordinates and temperature parameters at RT [78A5]. b_{ij} is defined by Eq. (b) in Introduction.

	x	y	z	$b_{11}\cdot10^4$	$b_{22}\cdot10^4$	$b_{33}\cdot10^4$	$b_{12}\cdot10^4$	$b_{13}\cdot10^4$	$b_{23}\cdot10^4$
Centered-hydrogen model									
O	0.1422 (1)	0.0866 (2)	0.1200 (1)	71 (3)	73 (5)	89 (4)	8 (1)	–26 (2)	–19 (2)
H	0.1387 (3)	0.25	0.125	89 (5)	257 (18)	129 (7)	0	0	49 (7)
Rb	0.0	0.0	0.5	79 (4)	79 (4)	58 (5)	0	0	0
P	0.0	0.0	0.0	51 (6)	51 (6)	70 (8)	0	0	0
Disordered-hydrogen model									
O	0.1421 (2)	0.0863 (2)	0.1199 (1)	71 (3)	72 (5)	89 (3)	6 (1)	–23 (2)	–22 (2)
H	0.1382 (6)	0.223	0.125	94 (5)	77 (17)	137 (7)	12 (8)	–49 (14)	46 (6)
Rb	0.0	0.0	0.5	79 (4)	79 (4)	58 (5)	0	0	0
P	0.0	0.0	0.0	51 (5)	51 (5)	71 (8)	0	0	0

Table 80. RbH_2PO_4(RDP). Interatomic distances and bond angles at RT [78A5]. The primed and unprimed atoms are in an opposite z direction relative to the central atom of the PO_4 tetrahedron.

Distances [Å]		Angles [deg]	
O–H	1.043 (2)	O–H···O	177.17 (15)
O–P	1.541 (1)	O–P–O′	108.91 (6)
O(H)···O	2.497 (2)	O–P–O	110.61 (6)
O···O′	2.508 (2)	P–O–H	115.25 (17)
O···O	2.534 (2)	P–O(H)···O	116.32 (7)

Table 81. RbH_2PO_4(RDP). Elastic compliances $s_{\lambda\mu}$ and stiffnesses $c_{\lambda\mu}$.

c_{11}	c_{12}	c_{13}	c_{33}	c_{44}	c_{66}	T	Ref.
10^9 N m^{-2}						°C	
67	2				4	10	50J1
66.97	–5.49	14.92	52.96	10.20	3.58	20	64H3
s_{11}	s_{12}	s_{13}	s_{33}	s_{44}	s_{66}		
10^{-12} m^2 N^{-1}							
18.5	2.2	–1.3	22.9	87.6	285.8	25	69A3

Table 82. $Rb(H_{1-x}D_x)_2PO_4$. Refractive indices [71V5].

x	500 nm		550 nm		600 nm		650 nm	
	n_0	n_e	n_0	n_e	n_0	n_e	n_0	n_e
0	1.5125	1.4813	1.5093	1.4804	1.5067	1.4784	1.5046	1.4767
0.30	1.5114	1.4808	1.5082	1.4799	1.5057	1.4778	1.5037	1.4762
0.50	1.5109	1.4805	1.5079	1.4796	1.5054	1.4776	1.5033	1.4760
0.60	1.5105	1.4803	1.5075	1.4794	1.5050	1.4774	1.5030	1.4759
0.88	1.5086	1.4792	1.5056	1.4782	1.5031	1.4763	1.5012	1.4750

Table 83. $Rb(H_{1-x}D_x)_2PO_4$. Electro-optic constants [71 V 5]. $\lambda = 633$ nm.

x	r_{41}	r_{63}
	10^{-12} m V^{-1}	
0	7.7 (3)	12.5 (7)
0.196	8.0 (3)	13.7 (7)
0.30		16.0 (7)
0.42	8.3 (3)	16.7 (7)
0.50	8.5 (3)	17.3 (7)
0.60		18 (1)
0.65		18.5 (10)
0.79		23 (1)
0.88		28 (2)

Table 84. RbH_2PO_4(RDP). ^{87}Rb quadrupole coupling tensor components in the tetragonal $X(a)$, $Y(b)$, $Z(c)$ coordinates system. [69 B 9]. A and B are two physically non-equivalent Rb sites in the unit cell. Primed and unprimed symbols refer to oppositely polarized domains.

T		$eQ\phi_{zz}/h$	$eQ\phi_{yy}/h$	$eQ\phi_{xx}/h$	$eQ\phi_{xy}/h$
K		MHz			
108	A	5.76 (6)	−5.34 (6)	−0.42 (6)	0.21 (3)
	B	5.76 (6)	−0.42 (6)	−5.34 (6)	0.21 (3)
	B′	5.76 (6)	−0.42 (6)	−5.34 (6)	−0.21 (3)
	A′	5.76 (6)	−5.34 (6)	−0.42 (6)	−0.21 (3)
370	A=B	6.90 (6)	−3.45 (6)	−3.45 (6)	0

Nr. 19A-3 CsH_2PO_4, Cesium dihydrogen phosphate (CDP)

($M = 229.89$; [D: 231.91])

1a Ferroelectricity of CsH_2PO_4 was first reported by Seidl in 1950. 50 S 2

b

phase	IV	III	II	I
state	F	P		
crystal system		monoclinic		
space group		$P2_1/m - C^2_{2h}$ [a]) or $B2_1/m - C^2_{2h}$ [b])		
Θ		$\Theta_f = -121.7$ °C [a]) −119.5 °C [c]) [D: −8.5 °C [d])]	~149 °C [c])	~230 °C [c])

[a]) 76 U 2
[b]) 78 N 2
[c]) 78 R 3
[d]) 78 G 7

For phase I and II, see the phase diagram Fig. 438. Antiferroelectric phase (phase VII) was reported to exist under hydrostatic pressure, see Figs. 444, 445.
$P_s \| [010]$. 76 U 2
$T_{melt} = 345$ °C [c]).
$\varrho = 3.20 \cdot 10^3$ kg m^{-3} [b]).

2a Crystal growth: evaporation method. 77 R 4

b Crystal form: Fig. 439.

3a Unit cell parameters and Z at RT:

	a [Å]	b [Å]	c [Å]	β	Z	
$P2_1/m$ cell	7.90065 (8)	6.36890 (9)	4.87254 (12)	107.742 (9)°	2	76 U 2
$B2_1/m$ cell	4.8725 (1)	6.3689 (1)	15.0499 (8)	90.22 (1)°	4	78 N 2

b Crystal structure: Figs. 440, 441.
Fractional coordinates: Table 85, 86.
Structure related to hydrogen bonds: Table 87; see also 76 U 2

		Ref.
5a	Dielectric constant $\varkappa$ vs. T: Figs. 442, 443. Curie-Weiss law: $\varkappa_b = C/(T-\Theta_p)$ for $T > \Theta_f$.	

	C K	Θ_p °C	Θ_f °C	Ref.
CsH_2PO_4	$4.51 \cdot 10^4$	−122.4	−121.7	76U2
	$4.3 \cdot 10^4$	−121.4	−119.5	75L5
CsD_2PO_4	$3.2 \cdot 10^4$	− 16.4	− 5.5	75L5

	$\varkappa$ at f=100 Hz, 10 kHz, see	56L1
	Phase diagram: Figs. 438, 444 [D: Fig. 445], see also Table 61 in 19A-1.	
	Effect of p on $\varkappa_b$: Fig. 446. $d\Theta_f/dE_{bias} = 5.8 \cdot 10^{-2}$ °C kV^{-1}m.	75L5
c	P_s vs. T: Figs. 447, 448.	
6a	Heat capacity: differential calorimetry for the transitions III→II and II→I: see	78M7
13a	NMR: Fig. 449.	
14b	Inelastic neutron scattering: Figs. 450, 451.	

Table 85. CsH_2PO_4(CDP). Fractional coordinates and the isotropic temperature parameter B at RT [76U2]. B is defined by Eq. (e) in Introduction.

Atom	Wyckoff notation	x	y	z	B [Å^2]
Cs	e	0.2659 (3)	0.2500	0.0357 (5)	1.04 (3)
P	e	0.2342 (7)	0.7500	0.5308 (12)	0.74 (14)
O1	e	0.3935 (22)	0.7500	0.3900 (35)	1.33 (23)
O2	c	0.3256 (30)	0.7500	0.8413 (45)	3.86 (35)
O3	f	0.1232 (22)	0.5524 (29)	0.4155 (36)	2.99 (28)

Table 86. CsH_2PO_4(CDP). Fractional coordinates in a pseudo-orthorhombic cell at RT [78N2].

Atom	Site symmetry	x	y	z
Cs	m	0.0965 (7)	0.25	0.1331 (2)
P	m	0.5899 (6)	0.75	0.1187 (2)
O1	m	0.8091 (6)	0.75	0.1955 (2)
O2	m	0.3152 (6)	0.75	0.1613 (2)
O3	1	0.6454 (6)	0.5530 (5)	0.0633 (2)
H1	m	0.495 (1)	0.25	0.3226 (3)
H2	1	0.973 (5)	0.500 (6)	0.487 (1)

Table 87. CsH_2PO_4(CDP). Interatomic distances and angles related with O–H···O bonds at RT [78N2]. H 1 is a special position on the mirror plane. H 2 is a special position at the centre of inversion ($\bar{1}$). O 3 is a general position. O 3 and O 3′ are related by the centre of inversion at the H 2 site (0, 1/2, 1/2); H 2 and H 2′ are symmetrically displaced from (0, 1/2, 1/2) along the O 3–O 3′ line.

O1−H1···O2		O3−H2−H2′−O3′	
Distances [Å]:			
O1−H1	0.995 (7)	O3−H2	1.00 (2)
H1···O2	1.530 (7)	H2···O3′	1.46 (2)
O1···O2	2.521 (7)	O3···O3′	2.46 (1)
		H2−H2′	0.48 (4)
Angles [°]:			
O1−H1···O2	173.2 (5)	O3−H2···O3′	174 (4)

Nr. 19A-4 TlH_2PO_4, Thallium dihydrogen phosphate
($M=301.36$; [D: 303.37])

1a Phase transition of TlH_2PO_4 was first clearly shown by P. V. Huong et al. in 1976.*) — 76H10

b

phase	II	I
state		P
crystal system		monoclinic
space group		C2/m-C_{2h}^3
Θ	229.76 (5) K [a]) [D: 350 K] [b])	

[a]) 77M14
[b]) 76V4

Space group of TlD_2PO_4 at RT: Pnma-D_{2h}^{16} (orthorhombic). — 76H10
$\varrho=4.752\cdot10^3$ kg m^{-3}. — 72O1

*) Measurement of the dielectric constant was reported in [47M2]. It does not seem to have been detailed enough to detect the phase transition.

2a	Crystal growth: slow evaporation [c]) or cooling of aqueous solution.	[c]) 76H10
3a	Unit cell parameters at RT: $a=14.34$ (1) Å, $b=4.51$ (1) Å, $c=6.52$ (1) Å, $\beta=91°30$ (10)′.	72O1
b	$Z=4$ ($Z=2$ in a primitive cell); [D: $Z=8$].	72O1
5a	Dielectric constants: Fig. 452. Effect of p: Fig. 453.	
6a	Heat capacity: Fig. 454. Anomalous entropy: Fig. 455.	
10a	Raman scattering: Fig. 456; see also	76V4

Nr. 19A-5 $NH_4H_2PO_4$, Ammonium dihydrogen phosphate (ADP)
($M=115.03$; [D: 117.04])

1a Anomalous dielectric behavior of $NH_4H_2PO_4$ was first reported by Busch in 1937. — 37B2

b — 66D2

phase	II	I
state	(A)*)	P
crystal system	orthorhombic	tetragonal
space group	$P2_12_12_1$-D_2^4	$I\bar{4}2d$-D_{2d}^{12}
Θ	−125 °C [D: −31 °C]	

Other phases III and IV were reported to exist under hydrostatic pressure, see Fig. 457.
$T_{melt}\approx190$ °C. — 69C2
$\varrho=1.803\cdot10^3$ kg m^{-3}. — 66b1
Transparent, colorless.

*) Antiferroelectric model in phase II was proposed by Nagamiya — 52N1

2a	Crystal growth: evaporation or cooling from aqueous solution. Solubility in H_2O: Fig. 458; see also Mass production apparatus for growing large crystals, see	73K12 58O1 50W1
b	Crystal form: see subsection 2b in 19A-1.	
3a	Unit cell parameters at RT: $a=b=7.4997$ (4) Å, $c=7.5494$ (12) Å.	63D2

3a	For DADP (= deuterated ADP), $a=7.502$ Å, $c=7.520\,(2)$ Å at 295 K and $a=7.507$ Å, $b=7.529\,(2)$ Å, $c=7.445\,(2)$ Å at 77 K. see also Tables 48, 49 in 19A-1.	73H10
b	Crystal structure: $Z=4$ in $I\bar{4}2d$ cell. Fig. 459. Fractional coordinates of atoms: Tables 88, 89; see also Interatomic distances and bond angle: Table 90. Root mean square of thermal displacements: Table 91.	73K12 58T3, 55K2, 48F1, 48U1, 73H10
4	Lattice distortion: Fig. 460; see also Thermal expansion: Figs. 461, 462; see also Table 59 in 19A-1, subsection 4. Effect of p: Fig. 463.	67C5
5a	Dielectric constant: Figs. 464, 465, 466. Curie-Weiss law: $\varkappa_c=\varkappa_\infty+C/(T-\Theta_p)$, $T>\Theta_{II-I}$, where $C=2.67\cdot10^3$ K, $\Theta_p=-14$ K, $\varkappa_\infty=7.0$. $\varkappa'$ and $\tan\delta$ at 9.2 GHz: Figs. 467, 468. $\varkappa$ for 10 kHz···36 GHz: Table 92. Phase diagram: Fig. 457. Θ_{II-I} vs. p: Fig. 469. Effect of p on $\varkappa_a$: Fig. 470.	46M1
6a	Heat capacity: Figs. 471, 472, 473. Transition heat ΔQ_m and transition entropy ΔS_m at Θ_{II-I}	
	ΔQ_m [J mol^{-1}] | ΔS_m [J K^{-1} mol^{-1}]	
	644 [a]) | 4.4 [a])	[a]) 44S5
	690 [D: 727] [b]) | 4.6 [D: 3.05] [b])	[b]) 70A5
7a	Piezoelectricity: Table 93; Figs. 474, 475.	
8a	Elastic compliance and stiffness: Table 94; Figs. 476···480. Phonon damping coefficient measured by the phonon echo method: see For the discrepancy between c_{44} and c_{55}: see	 77B10 61J1
9a	Refractive indices: Table 95; Figs. 481, 482. [D: $n_o=1.521$ at RT for $\lambda=550$ nm]. Birefringence vs. T: Fig. 483. $\varkappa''$ obtained from far-infrared spectra: Figs. 484, 485. Transmission in vacuum ultraviolet: Fig. 359 of 19A-1.	69A2
b	Electrooptic effect: Tables 96, 97; Figs. 486··491. See also Fig. 369 of 19A-1. Electrooptic effect in ultraviolet region near the absorption edge: see	 76O5
c	Piezooptic effect: Tables 98, 99; see also Figs. 374, 375 of 19A-1. Effect of p on refractive indices: Fig. 492.	
d	Verdet constant $V=234\,(11)$ K m^{-1} for $\lambda=632.8$ nm.	76K15
e	Nonlinear susceptibility for SHG: $\|d_{36}\|=0.53\cdot10^{-12}$ m V^{-1} $\|d_{36}/d_{36}^{KDP}\|=1.21\,(5)$ at RT for $\lambda=1.064$ µm $\|d_{14}/d_{14}^{KDP}\|=1.02\,(5)$ or $1.07\,(5)$ at RT for $\lambda=0.6943$ µm, and $1.17\,(5)$ at RT for $\lambda=1.0582$ µm. [D: $\|d_{36}/d_{36}^{KDP}\|=1.10$ for $\lambda=0.6943$ µm]. Second harmonic generation near $\lambda=250$ nm: see Third and fourth harmonic generation of Nd-glass laser at $\lambda=1.06$ µm; see Two photon absorption: see	 72L5 69b1 76J1, 77N4 77A11 77R6
10a	Raman scattering: Figs. 493, 494; see also Fig. 524 of 19A-6.	

continued

Nr. 19A-5 $NH_4H_2PO_4$, continued

11	Electrical conductivity: Figs. 495, 496, 497.	
		conductivity at RT	Activation energy ΔU	
	$NH_4H_2PO_4$	$2.9 \cdot 10^{-8}\ \Omega^{-1}\ m^{-1}$	53.2 kJ mol^{-1} (0.55 eV)	
	$ND_4D_2PO_4$	$1.3 \cdot 10^{-8}\ \Omega^{-1}\ m^{-1}$	54.4 kJ mol^{-1} (0.56 eV)	76P10

	For additional conductivity data: see	73H5, 50m1
	Conductivity of a ADP crystal doped with $(HSO_4)^-$ ion, see	74K14
	Conductivity of Co^{2+} doped ADP crystals: see	77S20
	Breakdown field: Figs. 498, 499; see also	50m1
13a	NMR: Tables 100, 101, 102.	
	T_1 vs. $1/T$: Fig. 500; see also	75K8, 74G10
	$\langle\Delta\omega^2\rangle$ vs. θ: Fig. 501. Chemical shift of ^{3c}P, see Table 75 in 19A-1. Double resonance study of ^{14}N of ADP, see	69H4
b	ESR: parameters of ADP: Cr^{3+} at 215 K: Table 103.	
	ESR of γ-ray irradiated crystal ADP: $(AsO_4)^{4-}$, see	71S33, 71S35, 72K6, 75O6
	ESR of irradiated crystal ADP doped with K_2SeO_4, see	73K9, 74K11
c	Mössbauer effect: Fig. 502. No anomaly is observed in the temperature dependence of isomer shift and normalized resonance intensity at Θ_{II-I}. However, a sudden increase in the linewidth near Θ_{II-I} is observed, consistent with the expected increase in electric field gradient created by the onset of antiferroelectric ordering.	72S4
14b	Inelastic neutron scattering of DADP: Fig. 503.	
16	Domain structures:	
	Etchant for revealing etch pits: a saturated solution of potassium acid phthalate.	66D2
	X-ray topographic study of twin boundaries and dislocations: see	78O3

Table 88. $NH_4H_2PO_4$(ADP), $NH_4H_2AsO_4$(ADA). Fractional coordinates and temperature parameters for the P, As, N, and O atoms at RT [73K12]. b_{ij} is defined by Eq. (b) in Introduction. b_{ij}'s are in the unit of 10^{-4}. Isotropic temperature parameters [defined in Eq. (e) of Introduction] for the H atoms are listed under b_{11}. See Fig. 459 for H_N and H_O.

	Site symmetry	x	y	z	b_{11}	b_{22}	b_{33}	b_{12}	b_{13}	b_{23}
$NH_4H_2PO_4$(ADP)										
P	$\bar{4}$	0	0	0	55 (1)	55 (1)	84 (1)	0	0	0
N	$\bar{4}$	0	0	$\frac{1}{2}$	110 (2)	110 (2)	88 (3)	0	0	0
O	1	0.0843 (1)	0.1466 (1)	0.1151 (1)	75 (2)	83 (2)	118 (2)	13 (1)	−22 (2)	−37 (2)
H_N	1	−0.002 (5)	0.089 (3)	0.563 (2)	2.8 (5) Å^2					
H_O	2	$\frac{1}{4}$	0.150 (6)	$\frac{1}{8}$	5 (1) Å^2					
$NH_4H_2AsO_4$(ADA)										
As	$\bar{4}$	0	0	0	57 (1)	57 (1)	89 (1)	0	0	0
N	$\bar{4}$	0	0	$\frac{1}{2}$	112 (2)	112 (2)	89 (3)	0	0	0
O	1	0.0869 (2)	0.1578 (2)	0.1233 (2)	78 (2)	88 (2)	145 (2)	19 (1)	−28 (2)	−48 (2)
H_N	1	0.014 (5)	0.120 (3)	0.567 (3)	2.4 (6) Å^2					
H_O	2	$\frac{1}{4}$	0.199 (9)	$\frac{1}{8}$	4.6 (2) Å^2					

Table 89. $ND_4D_2PO_4$(DADP). Fractional coordinates of deuterium in O–D···O [73H10]. $T=295$ K and 77.4 K.

	$T=295$ K				$T=77.4$ K			
	x	y	z	Occupation	x	y	z	Occupation
D(O, 1)	0.277 (3)	0.144 (1)	0.140 (4)	$\frac{1}{2}$	0.275 (3)	0.144 (3)	0.134 (3)	1
D(O, 2)	0.223	0.144	0.110	$\frac{1}{2}$				0
D(O, 3)	−0.144	0.277	−0.140	$\frac{1}{2}$	−0.146 (3)	0.277 (3)	−0.145 (3)	1
D(O, 4)	−0.144	0.223	−0.110	$\frac{1}{2}$				0

Table 90. $NH_4H_2PO_4$(ADP), $NH_4H_2AsO_4$(ADA). Interatomic distances and angles at RT [73K12]. O(u) and O(l) represent, respectively, the upper and the lower oxygen atoms of a given tetrahedral group as viewed along the c axis. O(f) and O(s) represent, respectively, the oxygen atoms of the flat and steep disphenoids around the N atoms.

		$NH_4H_2PO_4$(ADP)	$NH_4H_2AsO_4$(ADA)
4×	X–O(X=P or As)	1.537 (1) Å	1.682 (1) Å
4×	O(u)–X–O(u)	111.17 (4)°	111.08 (7)°
2×	O(u)–X–O(l)	108.63 (4)	108.67 (6)
2×	O(u)–O(u)	2.536 (1) Å	2.774 (2) Å
4×	O(u)–O(l)	2.497 (1)	2.733 (2)
	O–H···O	2.490 (1)	2.512 (2)
4×	N–H···O(f)	2.909 (1)	2.889 (1)
4×	N–H···O(s)	3.170 (1)	3.220 (2)
4×	O(f)–N···O(f)	97.04 (2)°	96.58 (4)°
2×	O(f)–N···O(f)	139.01 (2)	140.42 (5)
2×	O(s)–N···O(s)	47.16 (2)	50.92 (4)
4×	O(s)–N···O(s)	147.14 (2)	144.60 (4)

Table 91. $NH_4H_2PO_4$(ADP), $NH_4H_2AsO_4$(ADA). Root mean square thermal displacements along principal axes of the thermal ellipsoid and their orientation relative to the crystallographic axes at RT [73K12].

	Axis	Displacement	[100]	[010]	[001]
$NH_4H_2PO_4$(ADP)					
P	1	0.126 (1) Å	0°	90°	90°
	2	0.126	90	0	90
	3	0.156 (1)	90	90	0
N	1	0.156 (3)	90	90	0
	2	0.177 (2)	0	90	90
	3	0.177	90	0	90
O	1	0.130 (1)	86 (7)	33 (3)	57 (2)
	2	0.138 (1)	161 (2)	77 (7)	103 (4)
	3	0.208 (1)	108 (1)	120 (1)	36 (1)
$NH_4H_2AsO_4$(ADA)					
As	1	0.131 (1) Å	0°	90°	90°
	2	0.131	90	0	90
	3	0.164 (1)	90	90	0
N	1	0.164 (3)	90	90	0
	2	0.183 (2)	0	90	90
	3	0.183	90	0	90
O	1	0.135 (2)	106 (7)	29 (2)	67 (3)
	2	0.143 (2)	155 (5)	97 (7)	114 (3)
	3	0.235 (2)	108 (1)	108 (1)	34 (1)

Table 93. $NH_4H_2PO_4$(ADP), $ND_4D_2PO_4$(DADP). Piezoelectric constants [66 b 1].

Crystals	d_{14}	d_{36}	e_{14}	e_{36}	g_{14}	g_{36}	h_{14}	h_{36}	k_{36}	T	Ref.
	10^{-12} C N^{-1}		C m^{-2}		10^{-3} m^2 C^{-1}		10^8 N C^{-1}			°C	
$NH_4H_2PO_4$ (ADP)	1.51	−45.6								20	38 S 2
		48								27	48 J 1
	1.3	48								RT	61 J 1
			0.015	0.319	3.1	375	0.27	26.3		0	50 V 1
			0.015	−0.294						20	57 B 3
									0.32		62 B 4
		43.2		0.254		354		22.4		100	46 M 1
		44.2		0.261		348		22.6		80	
		45.3		0.271		348		23.0		60	
		47.3		0.287		353		23.6		40	
		49.3		0.301		356		24.3		20	
		51.7		0.318		360		25.1		0	
		54		0.333		354		25.3		−20	
		57		0.355		363		26.2		−40	
		60		0.377		366		26.8		−60	
		66		0.413		375		28.1		−80	
		69		0.43		384		28.8		−100	
		81		0.50		390		30.0		−110	
		87		0.51		396		30.6		−120	
		90		0.52		396		30.9		−122	
$ND_4D_2PO_4$ (DADP)	10	75								20	62 G 1

Table 94. $NH_4H_2PO_4$(ADP), $ND_4D_2PO_4$(DADP). Elastic constants [66 b 1].

Crystals	s_{11}	s_{12}	s_{13}	s_{33}	s_{44}	s_{66}	T	Ref.
	10^{-11} m^2 N^{-1}						°C	
$NH_4H_2PO_4$					11.6	16.6	20	45 J 1
	1.75	0.75	−1.15	4.35	11.5	16.4	25	46 M 1
	2.00	0.17	−1.29	4.57	11.70	18.5	20	46 Z 1
	1.81	0.19	−1.18	4.35	11.53	16.46	20	57 B 3
$ND_4D_2PO_4$	1.9	0.2	−1.15	4.4	11.0	16.4	20	52 M 1
	c_{11}	c_{12}	c_{13}	c_{33}	c_{44}	c_{66}		
	10^{10} N m^{-2}							
$NH_4H_2PO_4$	7.57	−2.43	1.30	2.96	0.87	0.61	25	46 M 1
	6.17	0.72	1.94	3.28	0.85	0.592	20	46 Z 1
	6.89	0.40	1.89	3.35	0.856	0.595		50 P 1
	6.76	0.59	2.00	3.38	0.87	0.61	20	57 B 3
	6.80			3.42	0.862	0.602		62 A 1
	6.877	0.406	2.038	3.402	0.862	0.601	20	64 H 3
$ND_4D_2PO_4$	6.28	0.395	1.74	3.18	0.909	0.610	RT	52 M 1

Table 92. $NH_4H_2PO_4$(ADP). Dielectric constants [61R1].

f	$\varkappa_a$	$\varkappa_c$	T [°C]
10 kHz	57.6 (5)	14.0 (2)	18.5
1 MHz	57.2 (6)		21.5
1 MHz		14.0 (2)	21
36 GHz	57.1 (6)		21.5
36 GHz		14.0 (3)	21

Table 95. $NH_4H_2PO_4$(ADP). n vs. λ [64Z1]. T=24.8 °C.

λ [μm]	Index in air		Absolute index	
	n_o	n_e	n_o	n_e
0.2000	1.648335	1.587012	1.649083	1.587632
0.3000	1.563478	1.512318	1.563951	1.512787
0.4000	1.540308	1.492136	1.540785	1.492571
0.5000	1.529792	1.483315	1.530276	1.483737
0.6000	1.523539	1.478412	1.524024	1.478828
0.7000	1.519047	1.475202	1.519528	1.475614
0.8000	1.515340	1.472818	1.515813	1.473227
0.9000	1.511969	1.470859	1.512433	1.471268
1.0000	1.508705	1.469123	1.509156	1.469530
1.1000	1.505418	1.467494	1.505853	1.467901
1.2000	1.502029	1.465904	1.502447	1.466311
1.3000	1.498490	1.464311	1.498888	1.464718
1.4000	1.494766	1.462687	1.495142	1.463094
1.5000	1.490834	1.461012	1.491187	1.461419
1.6000	1.486676	1.459272	1.487004	1.459679
1.7000	1.482279	1.457458	1.482580	1.457865
1.8000	1.477631	1.455562	1.477903	1.455970
1.9000	1.472724	1.453578	1.472965	1.453986
2.0000	1.467548	1.451501	1.467756	1.451910

Table 96. $NH_4H_2PO_4$(ADP). Electrooptic constants.

r^T_{41}	r^S_{41}	r^T_{63}	r^S_{63}	T	λ	Ref.
10^{-12} m V^{-1}				°C	nm	
20.8		8.47		22	556	66b1
24.5				RT	546	
		8.73		RT	408	
		8.66			436	
		8.62			468	
		8.59			492	
		8.56			546	
		8.54			579	
		8.49			623	
		8.48			500	
		8.4			500	
24.42					405	69b1
24.27					436	
24.26					492	
23.76					546	
23.32					578	
23.67					611.8	
23.41					632.8	
23.38					730.6	
22.65					1080	
−22.18	−22.03	6.96	11.43	21	589	76V1

Table 97. $NH_4H_2PO_4$(ADP). r_{63}, ϱ_{63}, vs. T [51C1, 66b1]. λ=546 nm.

T [°C]	r^T_{63}	r^S_{63}	ϱ^T_{63}	ϱ^S_{63}
	10^{-12} m V^{-1}		10^{-3} m^2 C^{-1}	
−120	−15.7	−8.33	−74.7	−50.7
−110	−15.0	−7.97	−77.4	−51.3
−100	−14.4	−7.60	−78.6	−51.0
−90	−13.7	−7.27	−79.2	−50.7
−80	−13.1	−7.03	−79.2	−50.7
−70	−12.5	−6.67	−78.0	−49.2
−60	−11.9	−6.40	−76.8	−48.6
−50	−11.3	−6.10	−75.0	−46.8
−40	−10.8	−5.80	−73.2	−45.9
−30	−10.3	−5.60	−71.4	−44.7
−20	−9.87	−5.37	−70.2	−43.5
−10	−9.50	−5.23	−69.0	−43.2
0	−9.13	−5.03	−67.8	−42.3
10	−8.80	−4.87	−66.6	−41.7
20	−8.57	−4.77	−66.0	−41.4
30	−8.33	−4.63	−65.7	−40.8

Table 98. $NH_4H_2PO_4$(ADP). Piezooptic constants.

$\Pi_{11}-\Pi_{12}$	Π_{11}	Π_{12}	Π_{13}	Π_{31}	Π_{33}	Π_{44}	Π_{66}	T °C	λ nm	Ref.
$10^{-12}\,m^2\,N^{-1}$										
							−18.0			59S5
							−18.15		560	48W1
							−12.5			50W3
0.7	8.6	7.9		12.3		−5.8	−12.2			54C1
							−13.6			59J1
0.78						5.78*)	19.6*)			55D3
1.04							−18.15			50W2
			−37.3		35.7			RT		61A1
	4.40	3.71	0.34	2.02	2.65	−6.70	−15.25	21	589	73N7
$\frac{1}{2}(p_{11}-p_{12})$	p_{11}	p_{12}	p_{13}	p_{31}	p_{33}	p_{44}	p_{66}			
dimensionless										
0.052							−0.1107		560	50W2
							−0.0744			54C1
							−0.076			50W3
	−0.11	−0.15	−0.93	0.20	−0.71			RT		61A1
	0.302*)	0.246*)	0.236*)	0.195*)	0.263*)		0.075*) (p^E_{66})		633	67D2
	0.319	0.277	0.169	0.197	0.167	−0.058	−0.091	21	589	73N7
	0.292*)	0.243*)		0.185*)				22	633	73M6
	0.296*)	0.243*)	0.208*)	0.188*)	0.228*)				633	77A7

*) Absolute value.

Table 99. $NH_4H_2PO_4$(ADP). Π_{66}, p_{66} vs. T [51C1, 66b1]. $\lambda = 546$ nm.

T	Π^E_{66}	Π^D_{66}	Π^P_{66}	p^E_{66}	p^D_{66}	p^P_{66}
°C	$10^{-12}\,m^2\,N^{-1}$			10^{-2}		
−120	−14.0	−7.8	−7.5	−8.5	−6.0	−5.8
−110	−14.4	−8.5	−8.3	−8.9	−6.5	−6.4
−100	−14.9	−9.4	−9.1	−9.2	−7.0	−6.9
−90	−15.0	−9.8	−9.6	−9.4	−7.3	−7.2
−80	−14.8	−10.0	−9.6	−9.3	−7.4	−7.3
−70	−14.6	−10.1	−9.7	−9.3	−7.4	−7.3
−60	−14.4	−10.0	−9.7	−9.1	−7.4	−7.3
−50	−14.1	−10.0	−9.7	−8.9	−7.3	−7.2
−40	−13.9	−10.0	−9.7	−8.8	−7.2	−7.1
−30	−13.7	−9.9	−9.7	−8.6	−7.1	−7.0
−20	−13.5	−9.7	−9.6	−8.4	−7.0	−6.9
−10	−13.0	−9.6	−9.5	−8.1	−6.8	−6.7
0	−12.9	−9.6	−9.4	−8.0	−6.7	−6.7
10	−12.8	−9.6	−9.4	−7.9	−6.6	−6.6
20	−12.6	−9.6	−9.4	−7.7	−6.5	−6.5
30	−12.6	−9.6	−9.5	−7.7	−6.6	−6.5

Table 100. $ND_4D_2PO_4$(DADP) and $NH_4H_2PO_4$(ADP). NMR data of acid D, ammonium D and ^{14}N. η: asymmetry parameter: F. g.: field gradient tensor. q_I, q_{II}, q_{III} are defined as the principal components of the F. g. tensor with $|q_I| > |q_{II}| > |q_{III}|$.

T [°C]	Crystals	Nuclei	e^2qQ/h [kHz]	η	Remarks	Ref.
RT	$NH_4H_2PO_4$	^{14}N	24.6	0	F. g. ∥ c	65C4
RT	$ND_4D_2PO_4$	^{14}N	27.3			65C4
RT		D (acid)	119.6 (8)	0.053 (8)	q_I in a-axis, q_{II} in c-axis, q_{III} in a-axis directions	64C2
RT		D (ammonium)	3.4	0	F. g. ∥ c	65C4
−40		D (ammonium)	6.0	0.89	F. g.*)	65C4

*) F. g. component

F. g. component	a (or b)	b (or a)	c
Z	0.226	0.413	0.882
Y	0.681	0.580	−0.447
X	−0.696	0.702	−0.150

And three other sets of F. g. direction cosines which differ from the one listed by the sign of either a, or b, or c axis component.

Table 101. $ND_4D_2PO_4$(DADP). Principal components of deuteron quadrupole coupling tensors ($eQ\,\phi_{ii}/h$) and their orientations in parenthesis. θ and ϕ are directions of principal axes of quadrupole-coupling tensor relative to the crystal axes in a I$\bar{4}$2d cell.

T °C	Type	$\|eQ\phi_{zz}/h\|$ (θ_z, ϕ_z) kHz	$\|eQ\phi_{yy}/h\|$ (θ_y, ϕ_y)	$\|eQ\phi_{xx}/h\|$ (θ_x, ϕ_x)	Ref.
25	x	118 (90°, −2°)	62 (2°, −96°)	56 (92°, −92°)	68G2
	y	120 (90°, −93°)	61 (6°, −179°)	59 (84°, −3°)	
−65		131 (90°, 0°)	72 (25°, 90°)	59 (65°, −90°)	
25		120 (90°, 0°)	63 (0°, −)	57 (90°, 90°)	
−70		128.2 (15)	71.0 (15)	57.2 (15)	74B11
		$\eta = 0.108$ (10)			

Table 102. $ND_4D_2PO_4$(DADP). Direction cosines of the largest principal values of the 16 physically nonequivalent O−D···O deuteron electric field gradient tensors below Θ_{II-I} [74B11].

O−D···O bonds along the a direction

	Domain A				Domain B			
a	0.996	0.996	0.996	0.996	0.996	0.996	0.996	0.996
b	0.083	−0.083	0.083	−0.083	0.028	−0.028	0.028	−0.028
c	0.028	0.028	−0.028	−0.028	0.083	0.083	−0.083	−0.083

O−D···O bonds along the b direction

	Domain A				Domain B			
a	0.083	−0.083	0.083	−0.083	0.028	−0.028	0.028	−0.028
b	0.996	0.996	0.996	0.996	0.996	0.996	0.996	0.996
c	0.028	0.028	−0.028	−0.028	0.083	0.083	−0.083	−0.083

Nr. 19A-6 KH_2AsO_4, Potassium dihydrogen arsenate (KDA)
(M=180.03; [D:182.05])

1a Ferroelectricity of KH_2AsO_4 was first reported by Busch in 1937. — 37B2

b — 37B2

phase	II	I
state	F	P
crystal system	orthorhombic	tetragonal
space group	$Fdd2-C_{2v}^{19}$	$I\bar{4}2d-D_{2d}^{12}$
Θ	−176 °C [D: −112···−114 °C]	

[D: −112···−114 °C] — 67C5

$P_s \| [001]$
$\varrho = 2.870 \cdot 10^3\ kg\ m^{-3}$. — 66b1
Transparent, colorless.

2a Crystal growth: solution method (solvent H_2O) with addition of small amounts of alkali ($pH \approx 5.5$). — 64H3

3a Unit cell parameters: Table 104; see also Tables 48, 49 in 19A-1, 3a.
a=7.626 (1) Å, c=7.158 (1) Å at 20 °C; — 73N3
[D: a=7.641 (5) Å, c=7.165 Å at 25 °C]. — 67C5
Substitution of D for H causes expansion along a. — 50D1

b Crystal structure: Z=4, in $I\bar{4}2d$ cell.
Fractional coordinates of atoms: Table 105.

4 Thermal expansion: Figs. 504, 505; see also Table 59 in 19A-1.

5a Dielectric constant: Figs. 506, 507.
Curie-Weiss law:
$\varkappa_c = C/(T-\Theta_p)$ $T > \Theta_f$, where $C = 2.70 \cdot 10^3$ K, $\Theta_p = 93.5$ K; see also — 73G8
$\varkappa'$ and $\tan\delta$ at ν=9.2 GHz: Figs. 508, 509. — 74C4
Low frequency dielectric dispersion: see — 73A4
Phase diagram: Fig. 510. Effect of p on $\varkappa_c$: Fig. 511; Table 106.
Effect of p on Θ_f: Fig. 512. $d\Theta_f/dp = -3.3 \cdot 10^{-8}\ K\ Pa^{-1}$. — 70F6

b Nonlinear dielectric properties: $\xi = -1.94 \cdot 10^{11}\ Vm^5C^{-3}$, $\zeta = 9.5 \cdot 10^{13}\ Vm^9C^{-5}$. — 73G8
See also Table 60 in 19A-1, 5a, and — 74C4

c P_s vs. T: Fig. 513. Effect of E_{bias} on P_s: Fig. 514.

6a Heat capacity: Fig. 515.
Transition heat ΔQ_m and transition entropy ΔS_m

ΔQ_m [J mol^{-1}]	ΔS_m [J K^{-1} mol^{-1}]	
351	3.64	42B1
351	3.76	44S4

b Thermal conductivity: Fig. 516; see also Fig. 336 in 19A1, 6b.

7a Piezoelectricity: Fig. 517; Table 107.

8a Elastic constants: Table 108; Figs. 518, 519.
Elastic constants obtained from Brillouin scattering: Fig. 520.

9a Refractive indices: Table 109.
Infrared reflectivity: see — 78S8

b Electrooptic effect: Table 110; Figs. 521, 522; see also Table 68 in 9b of 19A-1.

9c	Piezooptic effect: Fig. 523. Piezooptic constant: $p_{66}=0.020$.	 69V5
e	Nonlinear optical susceptibility: $d_{14}/d_{36}^{KDP}=0.86\,(10)$ for $\lambda=0.6943$ µm, 1.16 (5) for $\lambda=1.0582$ µm, $d_{36}/d_{36}^{KDP}=1.0\,(1)$ for $\lambda=0.6943$ µm, 1.10 (6) for $\lambda=1.0582$ µm; see also	 69b1 77N4
10a	Raman scattering: Figs. 524, 525, 526. Effect of p on Raman spectrum: see	 77L5
b	Brillouin scattering: Fig. 527; see also Fig. 521 in 9b.	
11	Electrical conductivity: Fig. 528.	73F1

	σ_c at RT	Activation energy
KH_2AsO_4	$\approx 10^{-8}\,\Omega^{-1}\,m^{-1}$	0.66 eV
KD_2AsO_4	$\approx 10^{-6}\,\Omega^{-1}\,m^{-1}$	0.64 eV

	σ_c depends much on the impurity of specimen.	
13a	NMR: Figs. 529···534; see also e^2qQ/h of ^{75}As vs. T: Fig. 535; see also Table 73 in 19A-1, 13a. Double resonance of ^{39}K: see NQR: Figs. 536···538. Effect of E_{bias} on NQR and level crossing, see	73B6 74H5 68G6
b	ESR: hyperfine splitting, Fig. 539 ESR of Cr^{5+} in KH_2AsO_4: Fig. 540; see also Slow ferroelectric cluster dynamics above Θ_f was studied by ESR; see also ESR of $(AsO_4)^{4-}$ and ENDOR: see ESR of γ-ray and X-ray irradiated crystals: see	78M8 76M24 77G1 72D2 76G3 72D2, 66H3, 72D2, 75G1, 67B11, 75C3, 76G2
15a	Domain structure: observation by polarized light, see	76K11

Table 103. $NH_4H_2PO_4$(ADP): Cr^{3+}. Spin Hamiltonian parameters and direction cosines of the principal Z axis with respect to the a, b, and c axis at 215 K [78D7]. $B_2^0=D/3$, $B_2^2=E$, in units of $10^{-2}\,m^{-1}$. Three solutions are shown. See also [70K7].

	a	b	c
B_2^0	−385	49	336
B_2^2	−287	−721	434
$g_{xx}=g_{yy}=g_{zz}$	1.97	1.97	1.97
l*)	0.540 (57.3)	−0.744 (138.1)	0.003 (89.8)
m*)	0.810 (35.9)	−0.586 (125.9)	−0.276 (106.2)
n*)	−0.227 (103.1)	0.321 (71.2)	0.961 (16.1)

*) l, m, and n are the direction cosines of the principal Z-axis with respect to the a-, b-, and c-axis, respectively (in parentheses the corresponding angles in degrees of arc).

Table 104. KH_2AsO_4(KDA). Lattice constants at different temperatures [65D4].

T [°C]	a [Å]	c [Å]	c/a
27	7.6300	7.1630	0.9388
29	7.6299	7.1637	0.9389
54	7.6344	7.1699	0.9392
55	7.6340	7.1703	0.9393
90	7.6410	7.1811	0.9398
117	7.6453	7.1921	0.9407
129	7.6492	7.1976	0.9410
134	7.6502	7.1994	0.9411
147	7.6524	7.2049	0.9415

Table 106. XH_2AsO_4, XD_2AsO_4 (X=K, Rb, Cs). Θ_f, $d\Theta_f/dp$ [75L9].

	Θ_f[°K]	$d\Theta_f/dp$ [10^{-8} K Pa^{-1}]
KH_2AsO_4	96.0 (7)	−2.63 (12)
RbH_2AsO_4	110.3 (2)	−4.59 (23)
CsH_2AsO_4	148.3 (1)	−5.63 (18)
KD_2AsO_4	159.7 (1)	−1.60 (8)
RbD_2AsO_4	169.8 (1)	−2.70 (14)
CsD_2AsO_4	203.3 (1)	−4.04 (11)

Table 105. KH_2AsO_4(KDA). Fractional coordinates in the I42d unit all at 20 °C [73N3]. Temperature parameter U_{ij} is defined by Eq. (d) in Introduction.

Atom	x	y	z	U_{11}[Å²]	U_{22}[Å²]	U_{33}[Å²]	U_{12}[Å²]	U_{23}[Å²]	U_{31}[Å²]
K	0	0	0.5	0.0224 (6)	0.0224 (6)	0.0178 (5)	0	0	0
As	0	0	0	0.0131 (4)	0.0131 (4)	0.0164 (3)	0	0	0
O	0.1610 (5)	0.0867 (5)	0.1336 (7)	0.0180 (9)	0.0203 (11)	0.0281 (16)	0.0034 (9)	−0.0049 (12)	−0.0068 (12)

Table 107. KH_2AsO_4(KDA). Piezoelectric constants [66b1].

d_{14}	d_{36}	e_{14}	e_{36}	g_{14}	g_{36}	h_{14}	h_{36}	T	Ref.
10^{-12} C N^{-1}		C m^{-2}		10^{-3} m^2 C^{-1}		10^8 N C^{-1}		°C	
26.6	22.4							0	49N1
25	22	0.29	0.14	85.8	101	9.9	6.9	25	50V1

Table 108. KH_2AsO_4(KDA). Elastic constants [66b1].

s_{11}	s_{12}	s_{13}	s_{33}	s_{44}	s_{66}	T	Ref.
10^{-12} m^2 N^{-1}						°C	
19	2	1	27	86	152	25	48O1
c_{11}	c_{12}	c_{13}	c_{33}	c_{44}	c_{66}		
10^9 N m^{-2}							
53	−6	−2	37	12	7	25	48O1
64.82	0.77	13.58	48.24	10.75	6.63	20	64H3

Table 109. KH_2AsO_4(KDA). n vs. λ [69b1].

λ nm	n_e	n_o
535	1.5252	1.5762
589	1.5179	1.5674
656	1.5146	1.5632

Table 110. KH_2AsO_4(KDA). Electrooptic constants [66b1].

r_{41}	r_{63}^T	ϱ_{63}^T	T	λ	Ref.
10^{-12} m V^{-1}		10^{-3} m^2 C^{-1}	°C	nm	
	−13			546	63W1
	−11.0		RT	546	64S7
12.5	10.9		RT	546	64O2
		−79.5			50B3

Nr. 19A-7 RbH_2AsO_4, Rubidium dihydrogen arsenate (RDA)
(M=226.40; [D: 228.42])

1a	Ferroelectricity of RbH_2AsO_4 was first reported by Matthias et al. in 1947.	47M1
b	phase: II / I; state: F / P; crystal system: — / tetragonal; Θ: −163 °C [D: −100 °C]; $P_s \parallel [001]$	47M1
3a	Unit cell parameters at 25 °C: a=7.7933 (5) Å, c=7.4671 (5) Å [D: a=7.8063 (7) Å, c=7.4674 (7) Å]; see also Tables 48, 49 in 19A-1, 3a.	67C5
4	Thermal expansion: Fig. 541; see also Table 59 in 19A-1, 4.	
5a, b	Curie-Weiss law and nonlinear dielectric properties. $E=\frac{1}{\varepsilon_0 C}(T-\Theta_p)P+\xi P^3+\zeta P^5$, where $C=3.1\cdot10^3$ K, $\xi=-1.12\cdot10^{11}$ V m^5 C^{-3}, $\zeta=6.4\cdot10^{13}$ V m^9 C^{-5}. Fig. 542; see also Effect of p on $\varkappa_c$: Fig. 543; see also Table 106 in 19A-6, 5a and	74K2 75M4 75L9
7a	Piezoelectricity: $d_{14}=9.9\cdot10^{-12}$ CN^{-1}; [D: $14.6\cdot10^{-12}$ CN^{-1}]; $d_{36}=28.5\cdot10^{-12}$ CN^{-1}; [D: $45.8\cdot10^{-12}$ CN^{-1}] at 25 °C.	69A1
8a	Elastic compliance: Table 111.	
9a	Refractive indices: n_o=1.559, n_e=1.520 at RT for λ=546 nm; [D: n_o=1.557 at RT for λ=550 nm].	64O2 69A2
b	Electrooptic effect: Table 112; see also Figs. 521, 522 and Table 68.	
c	Piezooptic effect: p_{12}=0.239, p_{13}=0.200, p_{31}=0.205, p_{33}=0.182, for λ=632.8 nm. p_{66}=0.023. See also Fig. 523.	77A7 69V5
e	Nonlinear optical susceptibility: $d_{36}=0.39\,(4)\cdot10^{-12}$ m V^{-1} [a]), $0.44\cdot10^{-12}$ m V^{-1} [b]); λ=694 nm. See also	[a]) 74K7 [b]) 78S1 77N4
10a	Raman scattering: Fig. 544; see also Effect of p; Fig. 545.	74L10
13a	NMR, NQR: Figs. 535, 536, 537 of 19A-6, 13a; see also	73B6, 69B9
b	ESR data for irradiated crystal: see	72D1
16	Internal friction measured at about 1 Hz, see	78G12

Table 1b (phase data):

phase	II	I
state	F	P
crystal system		tetragonal
Θ	−163 °C [D: −100 °C]	

Table 111. RbH_2AsO_4(RDA), RbD_2AsO_4(DRDA). Elastic compliance $s_{\lambda\mu}$ at 25 °C [69A1].

Crystals	s_{11}	s_{12}	s_{13}	s_{33}	s_{44}	s_{66}	T
	$\cdot10^{-12}$ m^2 N^{-1}						°C
RbH_2AsO_4	22.8	8.5	−1.8	25.4	96.0	232.0	25
RbD_2AsO_4	24.7	10.1	−4.4	27.1	105.5	246.0	

Table 112. RbH_2AsO_4(RDA), RbD_2AsO_4(DRDA). Electrooptic constants.

	r^T_{63} 10^{-12} m V^{-1}	T °C	λ nm	Ref.
RDA	13.0 (2)	RT	546	64O2
	14.8	RT	550	69A2
DRDA	21.4	RT	550	69A2

Nr. 19A-8 CsH_2AsO_4, Cesium dihydrogen arsenate (CDA)
(M=273.84; [D: 275.85])

1a	Ferroelectricity of CsH_2AsO_4 was first reported by Frazer et al. in 1953.	53F1
b	phase: III / II / I; state: F / P / –; crystal system: – / tetragonal / (monoclinic) [a]; Θ: III–II −130 °C [D: −61 °C]; II–I 160 (5) °C (on heating) [a], 140 (5) °C (on cooling) [a]	53F1 [a]) 77G14
	Θ_f of $CsH_{2(1-x)}D_{2x}AsO_4$=(140.0+101x) K. $P_s \| [001]$ $\varrho=3.644 \cdot 10^3$ kg m^{-3} for $Cs(H_{0.14}D_{0.86})_2AsO_4$. Transparent, colorless.	75L8 53S6 77L6 53S6
2a	Crystal growth: see	75L8, 77L6
3a	Unit cell parameters: Tables 48, 49 in 19A-1, 3a. $CsH_{2(1-x)}D_{2x}AsO_4$ with x=0.95: a=7.99 (1) Å, c=7.88 (1) Å at Θ_f+5 K (Θ_f=203 K).	 76H5
4	Thermal expansion: Table 59 in 19A-1, 4. Spontaneous shear was observed by neutron diffraction.	 76M15, 76M16
5a	Dielectric constant: Figs. 546, 547, 548. Curie-Weiss law, see b below. $\varkappa_c$ near Θ_{II-I}: Fig. 549. Effect of p on $\varkappa_c$: Fig. 550. Effect of p on Θ_f: Fig. 551. See also Table 106 in 19A-6, 5a. $\varkappa$ at f=5⋯80 MHz: see	 77T11
b	Nonlinear dielectric properties: $E=\frac{1}{\varepsilon_0 C}(T-\Theta_p)P+\xi P^3+\zeta P^5$, where $C=2.92 \cdot 10^3$ K, $\xi=9.18 \cdot 10^{10}$ V m^5 C^{-3}, $\zeta=6.13 \cdot 10^{13}$ V m^9 C^{-5}. Θ_f=146.8 K, $\Theta_f-\Theta_p$=0.65 K, those parameters are valid for $P \leqq 5 \cdot 10^{-2}$ Cm^{-2}.	 75C4
c	P_s vs. T and effect of E_{bias}: Fig. 552.	
6a	Heat capacity: Figs. 553, 554.	
8a	Elastic stiffness obtained from ultrasonic measurements: Fig. 555. c^E_{66} obtained from Brillouin scattering: see Fig. 557 in 10a, and	 74A8
9a	Refractive indices: Fig. 556. n_o=1.567, n_e=1.546 for λ=632.8 nm; [D: n_o=1.567 at RT for λ=550 nm].	 77A7 69A2

9b	Electrooptic effect: Fig. 521, 522 of 19A-6, 9b. $r^T_{63} = 18.6 \cdot 10^{-12}$ m V^{-1} [D: $36.6 \cdot 10^{-12}$ m V^{-1}] at RT for $\lambda = 550$ nm. See also Table 68 in 19A-1, 9b.	69A2
c	Piezooptic effect: Fig. 523 of 19A-6, 9c. $p_{11} = 0.267$, $p_{12} = 0.225$, $p_{13} = 0.200$, $p_{31} = 0.195$, $p_{33} = 0.227$, for $\lambda = 632.8$ nm. $p_{66} = 0.025$.	77A7 69V5
e	Nonlinear optical susceptibilities: d_{36}(CDA) = d_{36}(DCDA = deuterated CDA) = 0.40 (5) $\cdot 10^{-12}$ m V^{-1} for $\lambda = 1.064$ μm; see also	74K6 77N4, 77S7
10a	Raman scattering: Fig. 557. Coupled mode analysis; Fig. 558; see also Effect of p: see	71K2 77L5
b	Brillouin scattering: Figs. 559, 560. c^E_{66} measured by Brillouin scattering: see Fig. 558 in 10a and Brillouin and Rayleigh scattering spectra: Fig. 561.	74A8
13a	NMR of proton in CsH_2AsO_4: Fig. 562. NMR of deuteron in CsD_2AsO_4: Figs. 563, 564. NMR of ^{133}Cs: Table 113; see also NQR of ^{75}As: Figs. 535, 536 of 19A-6, 13a; see also	72B11 73B6
b	ESR of irradiated crystals: see	72D1
14	Neutron inelastic scattering: Spontaneous shear and the splitting of the Bragg spot was observed at $T < \Theta_f$, giving the shear angle $\approx 1.75°$. A peak corresponding to Σ_2 acoustic phonon was observed in See also	76M15 76M15 74D7

Table 113. CsH_2AsO_4. ^{133}Cs quadrupole coupling tensor components [kHz] in the crystal field fixed $X(a)$, $Y(b)$, $Z(c)$ system [69G13]. A and B are two physically non-equivalent Cs sites.

T		$eQ\phi_{zz}/h$	$eQ\phi_{yy}/h$	$eQ\phi_{xx}/h$
K		kHz		
128	A	359 (10)	−359 (10)	0
	B	359 (10)	0	−359 (10)
300		359 (10)	−179.5 (50)	−179.5 (50)

Nr. 19A-9 $NH_4H_2AsO_4$, Ammonium dihydrogen arsenate (ADA)

($M = 158.97$; [D: 160.99])

1a	Anomalous dielectric behavior of $NH_4H_2AsO_4$ was first reported by Busch in 1937.	37B2
b	(see table below)	37B2

phase	II	I
state	(A)*)	P
crystal system	orthorhombic	tetragonal
space group	$P2_12_12_1$-D_2^4	$I\bar{4}2d$-D_{2d}^{12}
Θ	−57 °C [a]) [D: +31 °C] [a])	

[a]) 53F1

Θ_{II-I} of DADA with 98% deuteration is 302.3 K. — 78B6

$\varrho = 2.311 \cdot 10^3$ kg m^{-3}. — 66b1

*) See 19A-5, 1b.

continued

Nr. 19A-9 $NH_4H_2AsO_4$, continued

2a	Crystal growth: evaporation or cooling from aqueous solution with addition of small amounts of alkali (pH ≈5.5). Table 114. Crystal growth of $(NH_4H_2)_{1-x}(ND_4D_2)_xAsO_4$: see	78B6
3a	Unit cell parameters: see Tables 48, 49 in 19A-1, 3a. $a=2.6998$ Å, $c=7.7518$ Å.	73K12
b	Crystal structure: Positional and thermal parameters: Table 115. Bond distances and angles: Tables 116, 117. See also Tables 88, 90 in 19A-5.	
4	Thermal expansion: Table 59 in 19A-1, 4. Mean square displacement: see Table 91 in 19A-5.	
5a	Dielectric constants: Figs. 565···568.	
6a	Heat capacity: Fig. 569. Transition heat ΔQ_m and transition entropy ΔS_m: $\Delta Q_m = 990$ J mol^{-1}, $\Delta S_m = 4.51$ J K^{-1} mol^{-1}.	44S1
7a	Piezoelectricity: Fig. 570; see also	66b1
8a	Elastic compliances and stiffnesses: Fig. 571; Table 118.	
9a	Refractive indices:	

λ[nm]	486	589	656
n_o	1.5859	1.5766	1.5721
n_e	1.5296	1.5217	1.5186

(9a ref.: 1874T1)

b	Electrooptic constant: $r_{63}^T = 9.2 \cdot 10^{-12}$ m V^{-1} at RT for $\lambda = 550$ nm; see also Fig. 521 in 19A-6, 9b.	69A2
c	Piezooptic constant: $\pi_{66}^E = -8.3 \cdot 10^{-12}$ m^2 N^{-1}.	66b1
e	Second harmonic generation of a dye-laser by ADA: see	77F4
10a	Raman scattering: see Fig. 523 of 19A-6, 10a.	
11	Electrical conductivity: Fig. 572.	
13a	NMR of proton in ADA: Fig. 573; see also NMR of ^{75}As: Table 119; Figs. 574, 575; see also NQR of ^{75}As: Table 120; Fig. 576; see also Figs. 536, 537	73G11, 50N1 76A1
b	ESR of $(AsO_4)^{4-}$ in a γ-ray irradiated crystal, see ESR of Cr^{5+}: see ENDOR of irradiated crystals: see	72D1, 76C3 78S8 74G2, 74G3

Table 114. $NH_4H_2AsO_4$(ADA). Solubility A in water [62k1].

A[wt%]	T[°C]
25.23	0
32.74	20
38.96	40
45.37	60
51.75	80
55.04	90

Table 115. $NH_4H_2AsO_4$(ADA). Fractional coordinates and anisotropic temperature parameters [78K8]. b_{ij} is defined by Eq. (b) in Introduction.

	x	y	z	$b_{ij}\cdot 10^4$					
				b_{11}	b_{22}	b_{33}	b_{12}	b_{13}	b_{14}
Oxygen	0.0878 (3)	0.1576 (4)	0.1233 (4)	75 (7)	80 (5)	142 (6)	19 (4)	−30 (5)	−47 (4)
Arsenic	0.0	0.0	0.0	47 (5)	47 (5)	91 (8)			
Hydrogen (O)	0.2581 (115)	0.1536 (12)	0.1234 (52)	279 (45)	107 (14)	129 (12)	−15 (73)	37 (17)	91 (37)
Nitrogen	0.0	0.0	0.5	109 (4)	109 (4)	92 (6)			
Hydrogen (N)	0.0154 (10)	0.1065 (10)	0.5720 (6)	185 (11)	274 (17)	247 (17)	28 (15)	−39 (12)	−128 (13)

Table 116. $NH_4H_2AsO_4$(ADA). Interatomic distances, bond angles and principal root mean square amplitudes, U_i in phase I [78K8].

Arsenate		Ammonium	
As–O	1.684 Å	N–H	0.997 (6) 1.05 (8) Å
O–As–O	111.2 (2)°	H–N–H	112.3 (8)°
O–As–O	108.6 (1)°	H–N–H	108.1 (4)°
Arsenic		H···O	1.947 (9) Å
$U_{\parallel c}$	0.166 (7) Å	N–H–O	157.2 (6)°
$U_{\perp c}$	0.119 (7) Å	H–H	1.614 (8) Å
Oxygen		H–H	1.656 (15) Å
U_1	0.127 (6) Å	Nitrogen	
U_2	0.139 (7) Å	$U_{\perp c}$	0.181 (4) Å
U_3	0.233 (4) Å	$U_{\parallel c}$	0.166 (6) Å
		NH_4 Hydrogen	
		U_1	0.197 (11) Å
		U_2	0.230 (8) Å
		U_3	0.347 (8) Å

Table 117. $NH_4H_2AsO_4$(ADA). Interatomic distances and angles of O–H···O bond and principal root mean square amplitudes U_i of the hydrogen [78K8].

Distances		Principal amplitudes		Angle from O–H···O bond
O–H	1.19 (9) Å	U_1	0.08 (6) Å	98 (10)°
H···O	1.31 (9) Å	U_2	0.25 (3) Å	102 (40)°
O–H···O	177 (2)°	U_3	0.29 (2) Å	166 (29)°

Table 118. $NH_4H_2AsO_4$(ADA). Elastic constants [48J1, 64H3].

T	s_{44}	s_{66}	c_{11}	c_{12}	c_{13}	c_{33}	c_{44}	c_{66}	Ref.
°C	$10^{-12}\,m^2\,N^{-1}$		$10^9\,N\,m^{-2}$						
	136	156							48J1
20			62.20	8.64	18.38	29.56	6.69	6.20	64H3

Table 119. $ND_4D_2AsO_4$(DADA). Principal values (in kHz) and direction cosines with respect to *a*, *b*, *c* axes of the ammonium ion deuteron quadrupole coupling tensors. [77 B 13].

$T = 293\ K > \Theta_{II-I}$			
	$\frac{3}{2h} eQ\phi_{zz} = -5.36$	$\frac{3}{2h} eQ\phi_{yy} = 2.68$	$\frac{3}{2h} eQ\phi_{xx} = 2.68$
a	0	0	1
b	0	1	0
c	1	0	0

$T = 184.8\ K < \Theta_{II-I}$												
	$\frac{3}{2h} eQ\phi_{zz} = -12.2$				$\frac{3}{2h} eQ\phi_{yy} = 11.4$				$\frac{3}{2h} 2h\ eQ\phi_{xx} = 0.8$			
a	0.213	0.432	−0.213	−0.432	0.714	0.543	−0.714	−0.543	0.667	−0.720	−0.667	0.720
b	0.432	−0.213	−0.432	0.213	0.543	−0.714	−0.543	0.714	−0.720	−0.667	0.720	0.667
c	−0.876	−0.876	−0.876	−0.876	0.441	0.441	0.441	−0.441	−0.193	−0.193	−0.193	−0.193

Table 120. $NH_4H_2AsO_4$(ADA). Eigenvalues and direction cosines of the principal axes of the ^{75}As electric field gradient tensors with respect to *a*, *b*, *c* axes in a domain with the antipolarization axis along [100]. 3, 4, 3′, 4′: type of $H_2AsO_4^-$ configuration. $T = 180$ K.

	Eigenvalues [MHz]								
	59.2			0.0			−59.2		
3	−0.11,	0.70,	0.70	0.97,	0.23,	−0.07	−0.21,	0.68,	−0.71
4	0.11,	−0.70,	0.70	−0.97,	−0.23,	−0.07	0.21,	−0.68,	−0.71
3′	0.21,	0.68,	0.70	0.97,	−0.23,	−0.07	0.11,	0.70,	−0.70
4′	−0.21,	−0.68,	0.70	−0.97,	0.23,	−0.07	−0.11,	−0.70,	−0.70

19B Solid Solutions

Nr. 19B-1 KH_2PO_4-$NH_4H_2PO_4$(KDP-ADP)

1a	Dielectric and X-ray investigations were done by Nitta et al. in 1951.	51 N 1
3a	Unit cell parameters: Table 121.	
5a	Dielectric constants: Fig. 577.	
16	$(NH_4)_xK_{1-x}H_2PO_4$ (x = 0.07) thin film optical waveguides were grown epitaxially on KDP substrates.	72 R 1

Table 121. $K_{1-x}(NH_4)_xH_2PO_4$. Unit cell parameters [51 N 1]. $T = 15 \cdots 25$ °C.

x	a_x	c_x	$(a_x - a_0)/a_0$	$(c_x - c_0)/c_0$
0	7.449	6.973	0	0
1/5	7.47	7.15⋯7.25	0.0035	0.033
2/3	7.482	7.375	0.0044	0.058
4/5	7.48	7.39	0.0048	0.060
1	7.492	7.522	0.0058	0.079

Nr. 19B-2 TlH_2PO_4-$NH_4H_2PO_4$(TlH_2PO_4-ADP)

1a	This system was studied by Matthias et al. in 1947.	47M1
b	ϱ: see Fig. 580 in 5a.	
5a	Dielectric constants: Figs. 578, 579. Θ_{II-I} vs. x: Fig. 580.	
7a	Piezoelectric constants: Fig. 581.	
8a	Elastic compliance and stiffness: Fig. 582, 583.	

Nr. 19B-3 TlH_2AsO_4-$NH_4H_2AsO_4$(TlH_2AsO_4-ADA)

1a	The system was first studied by LeMontagner in 1957.	57L2
3a	Unit cell parameters: Fig. 584.	
4	Thermal expansion: Fig. 585.	
5a	Dielectric constants: Fig. 586. Θ_{II-I} vs. x: Fig. 587; see also	65L1, 61G2

20 $(NH_4)_2SO_4$ family

20A Pure compounds

Nr. 20A-1 $(NH_4)_2SO_4$, Ammonium sulfate
(M = 132.13)

1a	Ferroelectric activity in $(NH_4)_2SO_4$ was discovered by Matthias and Remeika in 1956.	56M2
b	see table below	a) 56M2 b) 66S13

phase	II [a]	I [a]
state	F [a]	P [a]
crystal system	orthorhombic [b]	orthorhombic [b]
space group	$Pna2_1$-C_{2v}^9 [b]	Pnam-D_{2h}^{16} [b]
Θ	−49.5 [a] °C	

	$P_s \parallel [001]$. $\varrho = 1.55 \cdot 10^3$ kg m^{-3} at RT Transparent, colorless.	58H3
2a	Crystal growth: cooling method from aqueous solution.	62k1
b	Solubility in water.	62k1
3a	Unit cell parameters: $a = 7.782$ Å, $b = 10.636$ Å, $c = 5.993$ Å at RT $a = 7.837$ (7) Å, $b = 10.01$ (1) Å, $c = 5.967$ (6) Å at 180 K	 66S13
b	$Z = 4$ in phase I and II. Crystal structure: Tables 122···130; Figs. 588, 589.	66S13
4	Thermal expansion: Figs. 590, 591.	
5a	Dielectric constants: Figs. 592, 593; see also $T > \Theta_f$, $\varkappa_c - \varkappa_\infty = C/(T - \Theta_p)$, $\varkappa_\infty = 9.4$ (2), $C = 29$ (2) K. $\Theta_p = 214.8$ (4) K, $\Theta_f = 223$ K. (See also Fig. 594.) $d\Theta_f/dp = -4.1 \cdot 10^{-8}$ KPa^{-1} (up to $6 \cdot 10^8$ Pa); for $[(NH_4)_2SO_4]_{0.9}[NH_4BeF_3]_{0.1}$ $d\Theta_f/dp = -5.2 \cdot 10^{-8}$ KPa^{-1} (up to $6 \cdot 10^8$ Pa); see also	73I3 76M13 72T17
b	Effect of E_{bias} on $\varkappa_c$: Figs. 595, 596. continued	

Nr. 20A-1 $(NH_4)_2SO_4$, continued

5c	Spontaneous polarization: Fig. 597; see also	58H3
6a	Heat capacity: Figs. 598, 599. $\Delta Q_m = 3892$ J mol^{-1}, $\Delta S_m = 17.6$ J K^{-1} mol^{-1}. For ΔS_m: see	 58H3 45S1
7a	Electromechanical coupling: Fig. 600.	
b	Electrostriction: Fig. 601.	
8a	Elastic compliance: Fig. 602. Elastic stiffness: Table 131; Fig. 603.	
9a	Refractive indices: Table 132; Figs. 604, 605. Infrared reflection spectra: see	 75Z6
b	Electrooptic effect: Figs. 606, 607.	
c	Piezooptic constant: Tables 133, 134.	
e	Nonlinear optical susceptibility: Fig. 608; Table 135.	
10a	Raman scattering: Figs. 609···612. Raman scattering from the internal modes associated with $(SO_4)^{2-}$ and $(NH_4)^+$ ions: see	 73J2
b	Brillouin scattering: Fig. 603.	
11	Electrical conductivity: Figs. 613, 614. Effects of heat treatment on σ, effects of surface layer on σ, mobility $\mu = 1.6 \cdot 10^{-5} \exp(-0.49/kT)$ m^2 V^{-1}s^{-1}: see $\sigma_b = \sigma_0 \exp(-\Delta U/kT)$, $\Delta U = 0.76$ eV, $\sigma_0 = 2.9 \cdot 10^3\ \Omega^{-1}$ m^{-1}	 74K14 63S4
13a	NMR: Figs. 615, 616, 617. Pressure effect on the second moment of ^{1}H: see ^{14}N quadrupole coupling constants and asymmetry parameters in phase I: $(e^2qQ/h)_I = 117$ kHz, $(e^2qQ/h)_{II} = 91$ kHz, $\eta_I = 0.64$, $\eta_{II} = 0.67$ (Subscripts I and II point to two chemically nonequivalent sets of N sites in the unit cell). Proton-enhanced nuclear induction spectra of ^{15}N: see Figs. 618, 619.	 78R1 72B10 72G6
b	ESR: Figs. 620, 621; Tables 136···140; see also ESR data of Cr^{5+}: see ESR data of $(NH_3)^+$ and $(SeO_3)^-$ radicals: see ESR data for irradiated crystal: see	77K6, 76C6 78M3 76S12 75B1
14b	X-ray diffuse scattering: Figs. 622, 623.	

Table 122. $(NH_4)_2SO_4$. Structure of phase I [66 S 13]. Fractional coordinates of the heavy atoms at RT by various authors.

		X-ray			Electron diffraction	Neutron diffraction	
Ref.		16 O 1, 28 O 1, 30 O 1	28 T 1	62 S 9[1])	63 U 1	62 S 9[1])	66 S 13
S	*x*	0.250	0.252	0.2441 (4)	0.255	0.238 (2)	0.2440 (9)
	y	0.417	0.420	0.4110 (3)	0.420	0.422 (2)	0.4192 (6)
	z	0.250	0.250	0.250	0.250		0.250
O (1)	*x*	0.056	0.053	0.060 (2)	0.046	0.055 (2)	0.0612 (6)
	y	0.417	0.411	0.394 (2)	0.412	0.395 (2)	0.3887 (60)
	z	0.250	0.250	0.250	0.250		0.250
O (2)	*x*	0.315	0.311	0.273 (2)	0.309	0.269 (3)	0.2701 (7)
	y	0.549	0.544	0.554 (2)	0.542	0.558 (3)	0.5559 (4)
	z	0.250	0.250	0.250	0.250		0.250
O (3)	*x*	0.315	0.311	0.320 (1)	0.317	0.325 (2)	0.3239 (5)
	y	0.351	0.345	0.364 (1)	0.352	0.367 (2)	0.3665 (3)
	z	0.045	0.045	0.052 (2)	0.041		0.0488 (5)
N (1)	*x*	0.685	0.680	0.687 (2)	0.690	0.688 (1)	0.6895 (3)
	y	0.417	0.413	0.403 (2)	0.405	0.402 (1)	0.4025 (2)
	z	0.250	0.250	0.250	0.250		0.250
N (2)	*x*	1.000	0.999	0.966 (2)	0.985	0.968 (1)	0.9677 (4)
	y	0.689	0.722	0.703 (2)	0.692	0.704 (1)	0.7050 (2)
	z	0.250	0.250	0.250	0.250		0.250

[1]) *hk*0 data only

Table 123. $(NH_4)_2SO_4$. Structure of phase I [66 S 13]. Fractional coordinates of H atoms at RT by various authors.

		Electron diffraction	Neutron diffraction and X-ray	Neutron diffraction
Ref.		63 U 1	62 S 9[1])	66 S 13
H (1)	*x*	0.652	0.813	0.810 (2)
	y	0.304	0.372	0.372 (1)
	z	0.250	0.250	0.250
H (2)	*x*	0.545	0.597	0.600 (2)
	y	0.435	0.341	0.343 (1)
	z	0.250	0.250	0.250
H (3)	*x*	0.758	0.683	0.676 (1)
	y	0.437	0.468	0.457 (1)
	z	0.111	0.100	0.122 (2)
H (4)	*x*	0.882	1.016	0.991 (2)
	y	0.780	0.777	0.792 (1)
	z	0.250	0.250	0.250
H (5)	*x*	1.143	1.069	1.069 (1)
	y	0.700	0.634	0.649 (1)
	z	0.250	0.250	0.250
H (6)	*x*	0.957	0.883	0.893 (1)
	y	0.639	0.686	0.682 (1)
	z	0.111	0.100	0.124 (2)

[1]) Based on average coordinates for pairs of H atoms from a twofold disorder model.

Table 124. $(NH_4)_2SO_4$. Structure of phase I and II [66S13]. Fractional coordinates above and below Θ_f. RT ≃ 298 K, LT ≃ 180 K.

	T	x	y	z
S	RT	0.244 (1)	0.419 (1)	0.250
	LT	0.243 (2)	0.421 (1)	0.250
	LT [62S9]	0.2421 (4)	0.4230 (3)	0.250
O (1)	RT	0.0612 (6)	0.3887 (6)	0.250
	LT	0.061 (1)	0.402 (1)	0.212 (4)
	LT [62S9]	0.0602 (13)	0.4024 (90)	0.2127 (16)
O (2)	RT	0.2701 (7)	0.5559 (4)	0.250
	LT	0.284 (1)	0.557 (1)	0.222 (4)
	LT [62S9]	0.2873 (15)	0.5565 (11)	0.2255 (18)
O (3)	RT	0.3239 (5)	0.3665 (3)	0.0488 (5)
	LT	0.339 (1)	0.346 (1)	0.083 (4)
	LT [62S9]	0.3375 (11)	0.3462 (8)	0.0872 (14)
O (4)	RT	0.3239 (5)	0.3665 (3)	0.4512 (5)
	LT	0.294 (1)	0.382 (1)	0.477 (4)
	LT [62S9]	0.2885 (12)	0.3849 (9)	0.4795 (15)
N (1)	RT	0.6895 (3)	0.4025 (2)	0.250
	LT	0.6808 (7)	0.3958 (5)	0.245 (4)
	LT [62S9]	0.6789 (15)	0.3966 (11)	0.2533 (18)
N (2)	RT	0.9677 (4)	0.7050 (2)	0.250
	LT	0.9796 (6)	0.6999 (5)	0.256 (4)
	LT [62S9]	0.9786 (14)	0.7007 (10)	0.2656 (16)
H (1)	RT	0.810 (2)	0.372 (1)	0.250
	LT	0.771 (3)	0.336 (2)	0.202 (5)
H (2)	RT	0.600 (2)	0.343 (1)	0.250
	LT	0.564 (3)	0.360 (2)	0.206 (5)
H (3)	RT	0.676 (1)	0.457 (1)	0.122 (2)
	LT	0.691 (2)	0.480 (1)	0.157 (5)
H (4)	RT	0.676 (1)	0.457 (1)	0.378 (2)
	LT	0.682 (2)	0.412 (2)	0.408 (5)
H (5)	RT	1.069 (1)	0.649 (1)	0.250
	LT	1.092 (2)	0.658 (2)	0.222 (5)
H (6)	RT	0.991 (2)	0.792 (1)	0.250
	LT	0.990 (3)	0.793 (2)	0.265 (8)
H (7)	RT	0.893 (1)	0.682 (1)	0.124 (2)
	LT	0.896 (3)	0.670 (2)	0.142 (5)
H (8)	RT	0.893 (1)	0.682 (1)	0.376 (2)
	LT	0.945 (3)	0.672 (2)	0.412 (5)

Table 125. $(NH_4)_2SO_4$. Structure of phase I [66S13]. Anisotropic temperature parameters at RT. The second row for each atom gives the standard deviations from the least-squares varience-covariance matrix. U_{ij} (in Å^2) is defined by Eq. (d) in Introduction.

	U_{11}	U_{22}	U_{33}	U_{12}	U_{13}	U_{23}
S	0.0243	0.0292	0.0231	0.0025	0	0
	0.0028	0.0034	0.0025	0.0025	0	0
O (1)	0.0221	0.0802	0.0632	−0.0130	0	0
	0.0021	0.0040	0.0031	0.0025	0	0
O (2)	0.0519	0.0229	0.0519	0.0008	0	0
	0.0028	0.0017	0.0025	0.0017	0	0
O (3)	0.0522	0.0476	0.0306	0.0092	0.0090	−0.0129
	0.0018	0.0017	0.0013	0.0013	0.0012	0.0010

continued

Table 125, continued

	U_{11}	U_{22}	U_{33}	U_{12}	U_{13}	U_{23}
N (1)	0.0261	0.0309	0.0284	0.0025	0	0
	0.0012	0.0011	0.0009	0.0013	0	0
N (2)	0.0322	0.0287	0.0335	−0.0013	0	0
	0.0012	0.0011	0.0013	0.0008	0	0
H (1)	0.0663	0.1129	0.1043	0.0486	0	0
	0.0068	0.0109	0.0093	0.0071	0	0
H (2)	0.0985	0.0945	0.0950	−0.0453	0	0
	0.0092	0.0097	0.0087	0.0084	0	0
H (3)	0.0758	0.1559	0.1096	−0.0050	−0.0111	0.0824
	0.0049	0.0103	0.0071	0.0059	0.0050	0.0074
H (4)	0.0556	0.0865	0.0905	0.0193	0	0
	0.0055	0.0074	0.0073	0.0054	0	0
H (5)	0.1458	0.0470	0.1421	−0.0281	0	0
	0.0147	0.0057	0.0135	0.0075	0	0
H (6)	0.1108	0.0785	0.1065	−0.0059	−0.0526	0.0039
	0.0071	0.0052	0.0066	0.0050	0.0059	0.0048

Table 126. $(NH_4)_2SO_4$. Structure of phase II [66S13]. The second row for each atom gives the standard deviations from the least-squares variance-covariance matrix. U_{ij} (in Å^2) is defined by Eq. (d) in Introduction.

	U_{11}	U_{22}	U_{33}	U_{12}	U_{13}	U_{23}
S	0.0221	0.0091	0.0307	0.0004	−0.0021	−0.0019
	0.0072	0.0063	0.0078	0.0051	0.0034	0.0024
O (1)	0.0165	0.0280	0.0265	−0.0046	−0.0054	−0.0006
	0.0037	0.0040	0.0047	0.0029	0.0033	0.0035
O (2)	0.0265	0.0080	0.0245	−0.0059	0.0033	−0.0029
	0.0044	0.0034	0.0047	0.0025	0.0033	0.0035
O (3)	0.0286	0.0137	0.0231	0.0021	0.0090	0.0013
	0.0050	0.0040	0.0038	0.0034	0.0036	0.0035
O (4)	0.0277	0.0131	0.0177	0.0004	−0.0021	0.0019
	0.0050	0.0046	0.0038	0.0034	0.0033	0.0035
N (1)	0.0293	0.0080	0.0224	−0.0046	−0.0036	0.0067
	0.0034	0.0029	0.0034	0.0021	0.0028	0.0032
N (2)	0.0246	0.0057	0.0256	0.0004	−0.0088	0.0048
	0.0037	0.0034	0.0029	0.0021	0.0031	0.0032
H (1)	0.1030	0.0365	0.0732	0.0375	0.0479	0.0045
	0.0177	0.0103	0.0166	0.0118	0.0147	0.0099
H (2)	0.0429	0.0462	0.0586	−0.0110	−0.0073	0.0071
	0.0109	0.0091	0.0128	0.0076	0.0090	0.0093
H (3)*	0.0498	0.0154	0.0310	0.0093	0.0190	0.0298
	0.0103	0.0080	0.0083	0.0059	0.0064	0.0074
H (4)	0.0342	0.0536	0.0483	0.0042	0.0102	−0.0157
	0.0103	0.0120	0.0123	0.0072	0.0081	0.0093
H (5)	0.0461	0.0708	0.0256	0.0253	0.0092	0.0241
	0.0103	0.0103	0.0088	0.0088	0.0078	0.0090
H (6)	0.0921	0.0086	0.0761	0.0067	−0.0301	−0.0225
	0.0156	0.0086	0.0117	0.0080	0.0149	0.0103
H (7)	0.0654	0.0360	0.0561	−0.0139	−0.0308	0.0170
	0.0149	0.0103	0.0110	0.0088	0.0104	0.0090
H (8)	0.0261	0.0639	0.0458	−0.0046	−0.0040	0.0109
	0.0096	0.0126	0.0128	0.0072	0.0078	0.0106

*) These parameters do not describe a real ellipsoid since $U_{23}^2 > U_{22} \cdot U_{33}$.

Table 127. $(NH_4)_2SO_4$. Structures of phases I and II [66S13]. Interatomic distances and angles in ammonium ions. RT≅298 K, LT≅180 K.

	RT	LT
	Distances in [Å]	
N (1)–H (1)	1.08 (1)	1.07 (2)
N (1)–H (2)	1.04 (2)	1.06 (2)
N (1)–H (3)	1.09 (1)	1.06 (2)
N (1)–H (4)	1.09 (1)	1.02 (3)
N (2)–H (5)	1.05 (1)	1.04 (2)
N (2)–H (6)	1.06 (2)	1.06 (2)
N (2)–H (7)	1.07 (1)	1.04 (3)
N (2)–H (8)	1.07 (1)	1.05 (2)
Average N–H	1.08 (1)	1.05 (1)
	Bond angles in [°]	
H (1)–N (1)–H (2)	118.5 (15)	109. 9 (21)
H (1)–N (1)–H (3)	107.6 (9)	111.6 (16)
H (1)–N (1)–H (4)	107.6 (9)	111.3 (20)
H (2)–N (1)–H (3)	108.8 (9)	106.1 (16)
H (2)–N (1)–H (4)	108.8 (9)	107.4 (17)
H (3)–N (1)–H (4)	104.7 (9)	110.4 (19)
H (5)–N (2)–H (6)	116.2 (14)	112.2 (18)
H (5)–N (2)–H (7)	108.4 (8)	107.6 (20)
H (5)–N (2)–H (8)	108.4 (8)	107.1 (16)
H (6)–N (2)–H (7)	111.2 (8)	114.1 (18)
H (6)–N (2)–H (8)	111.2 (8)	104.7 (24)
H (7)–N (2)–H (8)	100.2 (8)	111.0 (17)

Table 128. $(NH_4)_2SO_4$. Structures of phases I and II [66S13]. Interatomic distances and angles in sulfate ion. RT≅298 K, LT≅180 K.

	RT	LT
	Distances in [Å]	
S–O (1)	1.49 (1)	1.47 (1)
S–O (2)	1.49 (1)	1.50 (1)
S–O (3)	1.49 (1)	1.48 (1)
S–O (4)	1.49 (1)	1.47 (1)
	Bond angles in [°]	
O (1)–S–O (2)	110.8 (5)	108.8 (10)
O (1)–S–O (3)	109.1 (4)	108.9 (12)
O (1)–S–O (4)	109.1 (4)	111.6 (14)
O (2)–S–O (3)	108.7 (4)	109.4 (13)
O (2)–S–O (4)	108.7 (4)	108.5 (12)
O (3)–S–O (4)	110.5 (4)	109.6 (10)

Table 129. $(NH_4)_2SO_4$. Structures of phases I and II [66S13] O–H distances and O···H–N bond angles. RT≅298 K, LT≅180 K.

	RT	LT
	Distances in [Å]	
O (1)–H (1)	1.97	2.38
O (1)–H (2)	2.48	2.78
O (1)–H (7)	2.39	2.70
O (1)–H (8)	2.39	1.96
O (2)–H (5)	1.85	1.85
O (2)–H (3)	2.27	2.63
O (2)–H (4)	2.27	1.92
O (2)–H (6)	2.36	2.28
O (3)–H (8)	2.05	2.46
O (3)–H (6)	2.43	2.39
O (3)–H (2)	2.48	1.91
O (3)–H (4)	2.14	2.78
O (4)–H (7)	2.05	1.87
O (4)–H (6)	2.43	2.59
O (4)–H (2)	2.48	2.68
O (4)–H (3)	2.14	1.82
	Bond angles in [°]	
O (1)···H (1)–N (1)	155.7	119.2
O (1)···H (2)–N (1)	139.3	—
O (1)···H (7)–N (2)	135.4	—
O (1)···H (8)–N (2)	135.4	164.3
O (2)···H (5)–N (2)	174.9	165.5
O (2)···H (3)–N (1)	136.0	—
O (2)···H (4)–N (1)	136.0	172.7
O (2)···H (6)–N (2)	122.2	129.0
O (3)···H (8)–N (2)	159.7	130.4
O (3)···H (6)–N (2)	115.7	108.9
O (3)···H (2)–N (1)	124.9	160.2
O (3)···H (4)–N (1)	155.0	—
O (4)···H (7)–N (2)	159.7	168.2
O (4)···H (6)–N (2)	115.7	111.9
O (4)···H (2)–N (1)	124.9	—
O (4)···H (3)–N (1)	155.0	174.3

Table 131. $(NH_4)_2SO_4$. Elastic stiffness [78Y7]. c_{ij}: in units of 10^9 N m^{-2}. Brillouin scattering.

c_{ij}	72L9	65H5	78Y7
c_{11}	35.2 (3)	36.1	35.6 (5)
c_{22}	29.7 (3)	29.8	28.1 (5)
c_{33}	36.0 (3)	35.3	36.3 (5)
c_{44}	9.5 (3)	10.3	9.3 (5)
c_{55}	7.0 (3)	7.2	7.5 (5)
c_{66}	10.3 (3)	9.7	9.9 (5)
c_{12}	14.1 (2)	16.5	
c_{23}	17.3 (2)	14.6	
c_{13}	15.7 (2)	15.8	

Table 130. $(NH_4)_2SO_4$. Structures of phases I and II [66S13]. Root mean square thermal displacements (in Å) in directions of principal axes of the thermal ellipsoid. RT≈298 K, LT≈180 K.

Atom	T	Axis 1	Axis 2	Axis 3
S	RT	0.152 (9)	0.152 (8)	0.175 (10)
	LT	0.09 (3)	0.15 (3)	0.18 (2)
O (1)	RT	0.139 (8)	0.251 (6)	0.288 (7)
	LT	0.11 (2)	0.17 (1)	0.17 (1)
O (2)	RT	0.150 (6)	0.228 (5)	0.228 (6)
	LT	0.08 (2)	0.15 (1)	0.18 (1)
O (3)	RT	0.138 (4)	0.228 (4)	0.244 (4)
	LT	0.12 (2)	0.13 (1)	0.19 (1)
O (4)	RT	0.138 (4)	0.228 (4)	0.244 (4)
	LT	0.10 (2)	0.14 (2)	0.17 (1)
N (1)	RT	0.158 (4)	0.168 (3)	0.180 (4)
	LT	0.07 (3)	0.15 (1)	0.18 (1)
N (2)	RT	0.169 (4)	0.181 (4)	0.183 (3)
	LT	0.07 (3)	0.13 (1)	0.19 (1)
H (1)	RT	0.19 (1)	0.32 (1)	0.38 (2)
	LT	0.12 (4)	0.23 (3)	0.38 (4)
H (2)	RT	0.23 (1)	0.31 (1)	0.38 (2)
	LT	0.18 (3)	0.22 (2)	0.26 (3)
H (3)	RT	0.21 (1)	0.28 (1)	0.47 (1)
	LT		0.18 (2)	0.27 (2)
H (4)	RT	0.21 (1)	0.28 (1)	0.47 (1)
	LT	0.16 (3)	0.21 (2)	0.26 (2)
H (5)	RT	0.21 (1)	0.30 (1)	0.31 (1)
	LT	0.12 (3)	0.18 (3)	0.31 (2)
H (6)	RT	0.20 (1)	0.38 (2)	0.39 (2)
	LT	0.04 (11)	0.24 (3)	0.34 (3)
H (7)	RT	0.24 (1)	0.28 (1)	0.40 (1)
	LT	0.16 (3)	0.18 (3)	0.32 (2)
H (8)	RT	0.24 (1)	0.28 (1)	0.40 (1)
	LT	0.16 (3)	0.20 (3)	0.26 (3)

Table 132. $(NH_4)_2SO_4$. Refractive indices and optical axial angle [03T1].

	RT				T=80 °C				
λ [nm]	n_α	n_β	n_γ	$2V$	λ [nm]	n_α	n_β	n_γ	$2V$
431	1.5318	1.5340	1.5445	—	431	1.5287	1.5315	1.5406	—
486	1.5270	1.5291	1.5394	+52° 00′	486	1.5239	1.5266	1.5355	—
535	1.5237	1.5288	1.5359	+52° 07′	535	1.5206	1.5233	1.5320	—
589	1.5209	1.5230	1.5330	+52° 12′	589	1.5178	1.5205	1.5291	—
656	1.5182	1.5204	1.5302	+52° 17′	656	1.5151	1.5179	1.5263	—
671	1.5177	1.5199	1.5297	+52° 18′	671	1.5146	1.5174	1.5258	—

Table 133. $(NH_4)_2SO_4$. Piezooptic constants at RT [72L9]. λ=633 nm.

$\|p_{12}\|=0.270$	$\|p_{21}\|=0.230$	$\|p_{13}\|=0.260$	$\|p_{31}\|=0.234$
$\|p_{23}\|=0.254$	$\|p_{32}\|=0.130$	$\|p_{11}-p_{21}\|=0.130$	$\|p_{11}-p_{31}\|=0.120$
$p_{22}\cong p_{32}\cong p_{12}$	$p_{33}\cong p_{13}\cong p_{23}$	$p_{44}=+0.015$	$\|p_{55}\|\leqq 0.0015$
$p_{66}\cong 0.012$			

Table 134. $(NH_4)_2SO_4$. Piezooptic constants [73M6]. $T = 22$ °C, $\lambda = 632.8$ nm.

Crystal	p_{11}	p_{12}	p_{31}
$(NH_4)_2SO_4$	0.261 (11)	0.191 (11)	0.197 (11)

Table 135. $(NH_4)_2SO_4$. Nonlinear optical susceptibilities [77S41]. $T = -100$ °C. l_c: coherence length. A Q-switched Nd–YAG laser was used.

d_{ij}	Relative to $d_{11}(SiO_2)$	l_c µm	Relative sign
d_{31}	0.56 (6)	10.0 (5)	$d_{31} \cdot d_{33} < 0$
d_{32}	0.59 (6)	11 (1)	$d_{32} \cdot d_{33} < 0$
d_{33}	1.3 (1)	18 (1)	

Table 136. $(NH_4)_2SO_4$. ESR of Cu^{2+}. [77M10]. Spin Hamiltonian parameters and direction cosines of the principal axes (x, y, z), relative to the crystalline axes (a, b, c). See also [68C2].

Direction cosines:

	x	y	z
a	0.2940	0.1101	0.9241
b	0.3007	0.9272	0.1435
c	0.9463	0.1732	0.9255

Spin Hamiltonian parameters:

T K	g_x	g_y	g_z	A_x $\cdot 10^{-2}$ m^{-1}	A_y	A_z
300	2.2890 (20)	2.2821 (20)	2.0446 (20)	79.3 (9)	56.27 (90)	189.09 (80)
77	2.2519 (20)	2.2779 (20)	2.0253 (20)	85.03 (9)	57.74 (90)	168.57 (90)

Table 137. $(NH_4)_2SO_4$. ESR of V–O. [75P1]. Spin Hamiltonian parameters at 25 °C for site I. (The VO^{2+} occupy four type of sites in the lattice. Only the spectra corresponding to V–O sites I···III have been analyzed. See also Figs. 588, 589.) Direction cosines relating the crystal axes (a, b, c) to the principal axes of **g** and **A** tensors. 1′, 2′, 3′: principal axes of **g** tensor. 1, 2, 3: principal axes of **A** tensor. **A**: in units of 10^{-2} m^{-1}. See also [77K6].

Spin Hamiltonian parameters:

$\lvert g_{1'}\rvert = 1.976$	$\lvert A_1\rvert = 68.2$
$\lvert g_{2'}\rvert = 1.982$	$\lvert A_2\rvert = 73.7$
$\lvert g_{3'}\rvert = 1.930$	$\lvert A_3\rvert = 180.2$

Direction cosines:

For **g**	a	b	c
1′	0.976	−0.212	−0.046
2′	0.172	0.887	−0.428
3′	0.132	0.410	0.902

For **A**	a	b	c
1	0.813	−0.570	0.117
2	0.561	0.714	−0.419
3	0.156	0.406	0.900

Table 140. $(NH_4)_2SO_4$. ESR of $(SeO_3)^-$. [77O5]. Spin Hamiltonian parameters of $(SeO_3)^-$ radical.

T K	$g_{\parallel}$	$g_{\perp}$	$A_{\parallel}$	$A_{\perp}$
246	2.006	2.027	541.8	389.6
206	2.006	2.027	549.2	389.6

Table 138. $(NH_4)_2SO_4$. ESR of V–O [75P1]. Spin Hamiltonian parameters and direction cosines at 25 °C for site II. For explanations, see Table 137.

Spin Hamiltonian parameters:

$\|g_{1'}\|=1.981$	$\|A_1\|=67.4$
$\|g_{2'}\|=1.934$	$\|A_2\|=178.9$
$\|g_{3'}\|=1.979$	$\|A_3\|=72.5$

Direction cosines:

For **g**	*a*	*b*	*c*
1′	0.797	−0.384	0.466
2′	0.338	0.923	0.182
3′	−0.500	0.012	0.866

For **A**	*a*	*b*	*c*
1	0.807	−0.141	−0.574
2	0.300	0.934	0.192
3	0.509	−0.327	0.796

Table 139. $(NH_4)_2SO_4$. ESR of V–O [75P1]. Spin Hamiltonian parameters and direction cosines at 25 °C for site III. For explanations, see Table 137.

Spin Hamiltonian parameters:

$\|g_{1'}\|=1.984$	$\|A_1\|=69.3$
$\|g_{2'}\|=1.978$	$\|A_2\|=70.5$
$\|g_{3'}\|=1.928$	$\|A_3\|=180.2$

Direction cosines:

For **g** or **A**	*a*	*b*	*c*
1′ or 1	0.000	1.000	0.000
2′ or 2	−0.647	0.000	0.763
3′ or 3	0.763	0.000	0.647

Nr. 20A-2 K_2SeO_4, Potassium selenate
($M=221.15$)

1a	Ferroelectric activity in K_2SeO_4 was discovered by Aiki et al. in 1969.					69A4
b	phase	IV	III	II	I	
	state	F [a])			P	
	crystal system	orthorhombic	orthorhombic	orthorhombic	hexagonal	[a]) 69A4 [b]) 76S7 [c]) 70K2
	space group	$Pna2_1$-C_{2v}^{9} [d])		Pnam-D_{2h}^{16} [c])	$P6_3/mmc$-D_{6h}^{4} [b])	
	Θ	93	129.5	745.0 K		[d]) 74O1
	$P_s\|\|[001]$ $\varrho_x=3.032\cdot10^3$/kg m^{-3}, $\varrho=3.052\cdot10^3$ kg m^{-3} in phase II [c]). Transparent.					70A3
2a	Crystal growth: evaporation method from aqueous solution at 50 °C.					70A3
3a	Unit cell parameters: $a=7.661$ (4) Å, $b=10.466$ (8) Å, $c=6.003$ (3) Å in phase II.					70K2
b	$Z=2$ in phase I, $Z=4$ in phase II. In phase III, modulated structure appears. [a, b]) Crystal structure: Tables 141⋯144; Figs. 624, 625.					70K2 [a]) 72S19 [b]) 77I2
4	Thermal expansion: Figs. 626α, 626β.					
5a	Dielectric constant: Figs. 627α, 627β, 627γ, 628α, 628β.					
b	Effect of E_{bias} on Θ_f: Fig. 629.					
c	Spontaneous polarization and coercive field: Figs. 630α, 630β. continued					

Nr. 20A-2 K_2SeO_4, continued

6a	Heat capacity anomaly: see	70A3
8a	Elastic compliance: Figs. 631α, 631β.	
10a	Raman scattering: Figs. 632, 633, 634; see also	79M2
b	Brillouin scattering: Fig. 635.	
13b	ESR: Fig. 636; Table 145. For γ-ray irradiated spectrum, see	79J1
14a	Bragg reflections due to structural modulations: Figs. 637, 638α, 638β.	
b	X-ray diffuse scattering: Figs. 639···641. Neutron inelastic scattering: Fig. 642.	

Table 141. K_2SeO_4. Structure of phase II [70K2]. Fractional coordinates.

	x	y	z
Se	0.2242 (2)	0.4200 (1)	$\frac{1}{4}$
K (1)	0.1705 (5)	0.0843 (3)	$\frac{1}{4}$
K (2)	−0.0057 (4)	−0.2905 (3)	$\frac{1}{4}$
O (1)	0.2931 (10)	0.3471 (6)	0.0271 (10)
O (2)	0.3024 (16)	−0.4356 (8)	$\frac{1}{4}$
O (3)	0.0126 (21)	0.4251 (10)	$\frac{1}{4}$

Table 142. K_2SeO_4. Structure of phase II [70K2]. U_{ij} is defined by Eq. (d) in Introduction.

	U_{11} [Å^2]	U_{22} [Å^2]	U_{33} [Å^2]	U_{12} [Å^2]	U_{13} [Å^2]	U_{23} [Å^2]
Se	0.0125 (11)	0.0160 (10)	0.0172 (10)	−0.0006 (4)	0	0
K (1)	0.0164 (19)	0.0295 (17)	0.0283 (16)	0.0032 (11)	0	0
K (2)	0.0135 (15)	0.0198 (15)	0.0264 (16)	0.0002 (11)	0	0
O (1)	0.0474 (43)	0.0417 (36)	0.0239 (35)	0.0132 (31)	0.0060 (31)	−0.0107 (31)
O (2)	0.0373 (70)	0.0213 (49)	0.0465 (60)	−0.0090 (41)	0	0
O (3)	0.0389 (83)	0.0583 (75)	0.0584 (72)	−0.0088 (52)	0	0

Table 143. K_2SeO_4. Structure of phase II [70K2]. Interatomic distances and bond angles for the SeO_4 group. Corrections were made for thermal vibrations. O (1′) is mirror-symmetric related mate of O (1).

Distance	Å	Angle	deg
Se-O (1)	1.649 (6)		
Se-O (1′)	1.649 (6)		
Se-O (2)	1.644 (9)		
Se-O (3)	1.648 (16)		
O (1) ···O (1′)	2.676 (9)	O (1) -Se-O (1′)	110.50 (37)
O (1) ···O (2)	2.640 (10)	O (1) -Se-O (2)	108.43 (33)
O (1) ···O (3)	2.660 (15)	O (1) -Se-O (3)	109.84 (34)
O (1′)···O (2)	2.640 (10)	O (1′)-Se-O (2)	108.43 (33)
O (1′)···O (3)	2.660 (15)	O (1′)-Se-O (3)	109.84 (34)
O (2) ···O (3)	2.656 (19)	O (2) -Se-O (3)	109.76 (59)

Table 144. K_2SeO_4. Structure of phase II [70K2]. Potassium oxygen interatomic distances. The coordinates x, y, z of O (1), O (2), O (3) are given in Table 141. Coordinates of O [I], O [II], etc. are given in the third column of this table, and numbers of these oxygen atoms can be identified in Figs. 624 and 625. The fourth column gives interatomic distances between K (1) and O [I], etc.

K (1) (x, y, z)		Coordinates of O [I],···	K (1)-O [I],···[Å]
O [I]	O (1)	(x, y, z)	3.20
O [II]	O (1)	(x, y, $\frac{1}{2}-z$)	3.20
O [III]	O (1)	($-\frac{1}{2}+x$, $\frac{1}{2}-y$, $\frac{1}{2}-z$)	3.26
O [IV]	O (1)	($-\frac{1}{2}+x$, $\frac{1}{2}-y$, z)	3.26
O [V]	O (1)	($\frac{1}{2}-x$, $-\frac{1}{2}+y$, $\frac{1}{2}+z$)	3.00
O [VI]	O (1)	($\frac{1}{2}-x$, $-\frac{1}{2}+y$, $-z$)	3.00
O [VII]	O (2)	($-\frac{1}{2}+x$, $-\frac{1}{2}-y$, $\frac{1}{2}-z$)	3.23
O [VIII]	O (2)	($\frac{1}{2}-x$, $\frac{1}{2}+y$, $\frac{1}{2}+z$)	3.02
O [IX]	O (2)	($\frac{1}{2}-x$, $\frac{1}{2}+y$, $-z$)	3.02
O [X]	O (3)	($\frac{1}{2}+x$, $\frac{1}{2}-y$, $\frac{1}{2}-z$)	2.64
O [XI]	O (3)	(x, y, z)	3.76
K (2) ($x, 1+y, z$)		**Coordinates of O [1],…**	**K (2)-O [XII]…**
O [XII]	O (1)	($-x$, $-y$, $-z$)	2.83
O [XIII]	O (1)	($-x$, $-y$, $\frac{1}{2}+z$)	2.83
O [V′]	O (1)	($\frac{1}{2}-x$, $\frac{1}{2}+y$, $\frac{1}{2}+z$)	2.73
O [VI′]	O (1)	($\frac{1}{2}-x$, $\frac{1}{2}+y$, $-z$)	2.73
O [XIV]	O (2)	(x, $1+y$, z)	2.80
O [VII′]	O (2)	($-\frac{1}{2}+x$, $\frac{1}{2}-y$, $\frac{1}{2}-z$)	2.79
O [XI]	O (3)	(x, y, z)	2.98
O [XV]	O (3)	($-x$, $-y$, $-z$)	3.32
O [XVI]	O (3)	($-x$, $-y$, $\frac{1}{2}+z$)	3.32

Table 145. K_2SeO_4. ESR data of Cu^{2+} [72K13].

Paramagnetic center	Cu^{2+}
S	$\frac{1}{2}$
$\mathscr{H}$	(3)
ν [GHz]	≈9
T [K]	300
g-factor	$g_x=2.0347$, $g_y=2.1646$, $g_z=2.3905$
HFS [10^{-2} m^{-1}]	$A_x=52.20$, $A_y=18.79$, $A_z=114.67$

Nr. 20A-3 Tl_2MoO_4, Thallium molibdate
($M=568.68$)

1a Pyroelectric effects in Tl_2MoO_4 were observed below 38 °C by Sleight et al. in 1975. 75S25

b

phase	IV	III	II	I
crystal system	orthorhombic			
space group	$Pna2_1$-C_{2v}^9 [a])			
Θ	38 [a])	400 [b, *])	503 [b]) °C	

[a]) 75S25
[b]) 72G1

$T_{melt}=645$ °C [a])
Fragile plate.

*) Sleight et al. could find no evidence for a transition near 400 °C. They found transitions at 37 and 499 °C.

continued

Nr. 20A-3 Tl_2MoO_4, continued

2a	Crystal growth: Precipitation from $Tl_2CO_3 + Tl_2MoO_4 \cdot 2H_2O$ salt.	75S25
3a	Unit cell parameters: $a = 8.028$ (6) Å, $b = 11.08$ (2) Å, $c = 6.190$ (8) Å at 25 °C. See also	 75S25 72G1
6a	Differential scanning calorimetry measurement reveals a lambda hape at 499 °C.	75S25
9e	Second harmonic generation effect. See	75S25

Nr. 20A-4 Tl_2WO_4, Thallium tungstate.
($M = 656.69$)

1a Center of symmetry disappears in Tl_2WO_4 below 11 °C, according to Sleight et al. — 75S25

b

phase	IV	III	II	I
crystal system		hexagonal		
Θ	11	37	562 °C	

75S25

$T_{melt} = 609$ °C
Above 11 °C: optically uniaxial.
Below 11 °C: optically biaxial.
Fragile plate.

2a	Crystal growth: Precipitation from $Tl_2CO_3 + Na_2WO_4 \cdot 2H_2O$ salt.	75S25
3a	Unit cell parameters: $a = 6.281$ (2) Å, $c = 8.115$ (8) Å at 25 °C	75S25
6a	Differential scanning calorimetry measurements revealed that the transition at 562 °C has a distinct lambda shape, but the others do not.	75S25
9e	Second harmonic generation effect is observed in phase VI.	75S25

Nr. 20A-5 $(NH_4)_2BeF_4$, Ammonium fluoroberyllate
($M = 121.08$)

1a Ferroelectric activity in $(NH_4)_2BeF_4$ was discovered by Pepinsky and Jona in 1957. — 57P1

b

phase	III	II	I
state	F [b, c]		P
crystal system	orthorhombic	orthorhombic	orthorhombic
space group	$Pn2_1$-C_{2v}^{9} [b, c, d]		Pnam-D_{2h}^{16} [a, b, c]
Θ	−97 [e]	−91 °C [f]	

[a]) 44M1
[b]) 74M3
[c]) 77O3
[d]) 77T2
[e]) 77I3
[f]) 73S20

$P_s \| [010]$
Transparent, colorless.

2a	Crystal growth: evaporation or cooling method from aqueous solution.	58H3
3a	Unit cell parameters: $a = 7.646$ (1) Å, $b = 10.430$ (1) Å, $c = 5.918$ (1) Å at 20 (1) °C $a = 15.1048$ (11) Å, $b = 10.4821$ (6) Å, $c = 5.9101$ (3) Å at −140 °C.	77O3
b	$Z = 4$ in phase I. $Z = 8$ in phase III. Tables 146···157; Figs. 643···648.	79O1
4	Thermal expansion: Figs. 649, 650, 651.	

5a	Dielectric constants: Figs. 652, 653. Effect of hydrostatic pressure: Figs. 654, 655.	
c	Spontaneous polarization: Fig. 656. Coercive field: Fig. 657.	
6a	Heat capacity: Figs. 658, 659; Table 158.	
8a	Elastic stiffness: Table 159.	
9a	Birefringence: Figs. 660, 661.	
10a	Raman scattering: Fig. 662.	
b	Brillouin scattering: Figs. 663, 664; see also	77H12
13a	NMR: Figs. 615, 665···668; Table 160.	
b	ESR data for irradiated crystal: see	78M4
14a	Bragg reflections due to structural modulations: Figs. 669···671.	
b	X-ray diffuse scattering: Figs. 672···675.	
16	Surface layer effect on the dielectric constant: see	78R2

Table 146, see p. 118.

Table 147. $(NH_4)_2BeF_4$. Structure of phase I [81O1]. Interatomic distances and angles in BeF_4 ion at 20 (1) °C. Corrected: corrected for thermal vibration.

	Bond lengths [Å]		Bond angles [°]	
	uncorrected	corrected		
Be–F (1)	1.534 (3)	1.550 (3)	F (1)–Be–F (2)	110.83 (20)
Be–F (2)	1.536 (3)	1.548 (3)	F (1)–Be–F (3)	109.15 (13)
Be–F (3)	1.536 (2)	1.548 (2)	F (2)–Be–F (3)	108.37 (13)
average	1.538	1.549	F (3)–Be–F (3)	110.99 (18)
			average	109.84

Table 148. $(NH_4)_2BeF_4$. Structure of phase I [81O1]. Interatomic distances and angles in ammonium ions at 20 (1) °C.

Bond lengths [Å]		Bond angles [°]	
N (1)–H (1)	0.84 (4)	H (1)–N (1)–H (2)	108.8 (47)
N (1)–H (2)	0.81 (5)	H (1)–N (1)–H (3)	115.0 (25)
N (1)–H (3)	0.91 (4)	H (2)–N (1)–H (3)	113.6 (35)
average	0.87	H (3)–N (1)–H (3)	90.2 (38)
		average	106.9
N (2)–H (4)	0.74 (4)	H (4)–N (2)–H (5)	117.3 (44)
N (2)–H (5)	0.74 (5)	H (4)–N (2)–H (6)	110.6 (25)
N (2)–H (6)	0.86 (2)	H (5)–N (2)–H (6)	103.6 (32)
average	0.80	H (6)–N (2)–H (6)	110.8 (32)
		average	110.6

Table 146. $(NH_4)_2BeF_4$. Structure of phase I [78 O 5]. Fractional coordinates and temperature parameters at 20 (1) °C. The anisotropic parameters b_{ij} are for nonhydrogen atoms and the isotropic B are for hydrogen atoms. b_{ij} and B are defined by Eqs. (b) and (e) in Introduction.

	x	y	z	b_{11} [$\cdot 10^{-4}$] B [Å^2]	b_{22} [$\cdot 10^{-4}$]	b_{33} [$\cdot 10^{-4}$]	b_{12} [$\cdot 10^{-4}$]	b_{13} [$\cdot 10^{-4}$]	b_{23} [$\cdot 10^{-4}$]
N (1)	0.1799 (3)	0.0969 (2)	0.2500	80 (3)	62 (2)	167 (6)	– 1 (2)	0	0
N (2)	0.4692 (2)	0.8051 (2)	0.2500	78 (3)	43 (2)	137 (5)	4 (1)	0	0
Be	0.2475 (4)	0.4179 (3)	0.2500	72 (4)	44 (2)	121 (7)	5 (3)	0	0
F (1)	0.0497 (2)	0.3932 (2)	0.2500	74 (2)	140 (2)	327 (6)	–25 (2)	0	0
F (2)	0.2864 (2)	0.5624 (1)	0.2500	173 (3)	45 (1)	335 (6)	–14 (1)	0	0
F (3)	0.3278 (2)	0.3589 (1)	0.0361 (2)	146 (2)	112 (1)	179 (3)	26 (1)	14 (2)	62 (1)
H (1)	0.281 (5)	0.129 (4)	0.250	3.0 (9)					
H (2)	0.109 (8)	0.155 (5)	0.250	5.5 (13)					
H (3)	0.160 (4)	0.037 (3)	0.141 (6)	5.6 (9)					
H (4)	0.494 (6)	0.736 (4)	0.250	3.6 (9)					
H (5)	0.541 (6)	0.853 (5)	0.250	4.5 (11)					
H (6)	0.410 (5)	0.824 (3)	0.130 (7)	6.8 (10)					

Table 151. $(NH_4)_2BeF_4$. Structure of phase I [79 O 1]. Fractional coordinates. Anisotropic temperature parameters for nonhydrogen atoms and isotropic temperature parameters for hydrogen atoms at –84 °C. b_{ij} and B are defined in Eqs. (b) and (e) in Introduction.

Atom	x	y	z	b_{11} [$\cdot 10^{-4}$], B [Å^2]	b_{22} [$\cdot 10^{-4}$]	b_{33} [$\cdot 10^{-4}$]	b_{12} [$\cdot 10^{-4}$]	b_{13} [$\cdot 10^{-4}$]	b_{23} [$\cdot 10^{-4}$]
N (1)	0.1862 (3)	0.0993 (3)	0.2500	62 (4)	37 (2)	98 (8)	4 (3)	0	00
N (2)	0.4598 (3)	0.8037 (3)	0.2500	62 (4)	31 (2)	101 (8)	8 (3)	0	00
Be	0.2510 (5)	0.4187 (4)	0.2500	52 (5)	27 (3)	69 (10)	2 (3)	0	00
F (1)	0.0540 (3)	0.3812 (2)	0.2500	57 (3)	88 (3)	287 (9)	– 19 (2)	0	00
F (2)	0.2726 (3)	0.5643 (2)	0.2500	123 (4)	31 (2)	306 (9)	– 7 (2)	0	00
F (3)	0.3362 (2)	0.3648 (2)	0.0347 (2)	100 (2)	89 (2)	116 (5)	29 (3)	–5 (3)	–45 (3)
H (1)	0.306 (7)	0.127 (5)	0.250	2.4 (1.3)					
H (2)	0.117 (7)	0.168 (5)	0.250	1.6 (1.1)	Bonded to N (1)				
H (3)	0.169 (4)	0.048 (3)	0.141 (6)	1.8 (0.8)					
H (4)	0.480 (7)	0.727 (5)	0.250	2.5 (1.2)					
H (5)	0.556 (7)	0.849 (5)	0.250	2.5 (1.3)	Bonded to N (2)				
H (6)	0.392 (6)	0.824 (4)	0.131 (7)	2.9 (0.9)					

Table 149. $(NH_4)_2BeF_4$, $(NH_4)_2SO_4$, K_2SeO_4. Structure at RT [81 O 1]. Root mean square (r.m.s.) displacements and root mean square amplitudes by thermal motions of BeF_4, SO_4, SeO_4 ions.

	r.m.s. displacement in Å along principal axis			r.m.s. amplitude in degree around principal axis		
	~*a* axis	~*b* axis	*c* axis	~*a* axis	~*b* axis	*c* axis
BeF_4	0.12_4	0.16_6	0.14_7	7.5_3	6.5_4	7.8_9
SO_4	0.15_2	0.16_3	0.15_4	6.3_2	7.4_3	7.8_7
SeO_4	0.12_2	0.13_4	0.13_0	5.2_6	7.3_3	6.5_8

Table 150. $(NH_4)_2SO_4$, $(NH_4)_2BeF_4$, K_2SeO_4. Structures at RT [79 O 1]. The r.m.s. (root mean square) displacements along the principal axes of N atoms in $(NH_4)_2SO_4$ and $(NH_4)_2BeF_4$, and of K atoms in K_2SeO_4. The orientation of the principal axis is given in parentheses in degrees such as (θ_a, θ_b, θ_c). For example, θ_a is the angle between its principal axis and the *a* axis.

	axis-1	axis-2	axis-3
$(NH_4)_2SO_4$			
N (1)	0.180 (67, 23, 90)	0.168 (90, 90, 0)	0.158 (23, 67, 90)
N (2)	0.183 (90, 90, 0)	0.181 (18, 72, 90)	0.167 (72, 18, 90)
$(NH_4)_2BeF_4$			
N (1)	0.185 (87, 3, 90)	0.172 (90, 90, 0)	0.154 (3, 87, 90)
N (2)	0.158 (49, 41, 90)	0.156 (90, 90, 0)	0.147 (41, 49, 90)
K_2SeO_4			
K (1)	0.174 (77, 13, 90)	0.168 (90, 90, 0)	0.125 (13, 77, 90)
K (2)	0.162 (90, 90, 0)	0.141 (88, 2, 90)	0.116 (2, 88, 90)

Table 152. $(NH_4)_2BeF_4$. Structure of phase I [79 O 1]. Root mean square (r.m.s.) displacements and amplitudes of BeF_4 by thermal motion.

	r.m.s. displacements along principal axes		r.m.s. amplitudes around principal axes	
	T=RT	*T*= −84 °C	*T*=RT	*T*= −84 °C
~*a* axis	0.12_4 [Å]	0.10_7 [Å]	7.5_3 [°]	7.8_8 [°]
~*b* axis	0.16_6	0.13_4	6.5_4	5.0_2
c axis	0.14_7	0.12_5	7.8_9	6.2_8

Table 153, see p. 120.

Table 154. $(NH_4)_2BeF_4$. Structure [79 O 1]. Correspondence of the constituent atoms between phases I and III.

phase I	phase III	phase I	phase III
N (1)	N (11), N (12)	H (1)	H (11), H (12)
N (2)	N (21), N (22)	H (2)	H (21), H (22)
Be	Be (1), Be (2)	H (3)	H (31), H (32), H (33), H (34)
F (1)	F (11), F (12)	H (4)	H (41), H (42)
F (2)	F (21), F (22)	H (5)	H (51), H (52)
F (3)	F (31), F (32), F (33), F (34)	H (6)	H (61), H (62), H (63), H (64).

Table 153. $(NH_4)_2BeF_4$. Structure of phase III [79O1]. Fractional coordinates. Anisotropic temperature parameters for non-hydrogen atoms and isotropic temperature parameters for hydrogen atoms at −140 °C. b_{ij} and B are defined in Eqs. (b) and (e) in Introduction. See Fig. 645 about the choice of the unit cell and Table 154 about the correspondence to between the atoms in phases I and III.

Atom	x	y	z	b_{11} [$\cdot 10^{-4}$], B [Å^2]	b_{22} [$\cdot 10^{-4}$]	b_{33} [$\cdot 10^{-4}$]	b_{12} [$\cdot 10^{-4}$]	b_{13} [$\cdot 10^{-4}$]	b_{23} [$\cdot 10^{-4}$]
N (11)	0.0962 (2)	0.0999 (3)	0.2658 (6)	15 (2)	23 (3)	83 (9)	9 (2)	− 6 (3)	− 4 (4)
N (21)	0.2294 (2)	0.8020 (4)	0.2584 (6)	16 (2)	50 (4)	73 (9)	6 (2)	− 4 (3)	− 9 (5)
Be (1)	0.1289 (3)	0.4182 (5)	0.2580 (8)	9 (2)	29 (4)	48 (12)	− 8 (3)	− 1 (4)	− 2 (7)
F (11)	0.0279 (2)	0.3778 (3)	0.2264 (4)	14 (1)	50 (3)	135 (7)	− 6 (1)	− 0 (2)	− 4 (4)
F (21)	0.1326 (2)	0.5649 (2)	0.2954 (4)	18 (1)	19 (2)	108 (6)	3 (1)	− 4 (2)	− 9 (3)
F (31)	0.1743 (2)	0.3841 (2)	0.0351 (4)	19 (1)	31 (2)	70 (6)	7 (1)	16 (2)	− 2 (3)
F (33)	0.1632 (2)	0.3503 (2)	0.4648 (4)	12 (1)	43 (2)	78 (6)	− 0 (1)	− 7 (2)	17 (3)
N (12)	0.3426 (2)	0.3968 (4)	0.2528 (5)	6 (3)	41 (3)	51 (8)	8 (2)	− 0 (3)	7 (5)
N (22)	0.4755 (2)	0.6979 (3)	0.2391 (5)	5 (1)	2 (2)	53 (8)	3 (1)	− 5 (3)	− 3 (4)
Be (2)	0.3738 (3)	0.0813 (5)	0.2570 (8)	11 (2)	8 (4)	80 (13)	−11 (3)	4 (4)	−12 (6)
F (12)	0.2795 (2)	0.1271 (3)	0.2836 (4)	7 (1)	46 (2)	130 (7)	4 (1)	6 (2)	1 (4)
F (22)	0.3816 (2)	0.9345 (2)	0.2198 (4)	18 (1)	19 (2)	104 (6)	0 (1)	4 (2)	3 (3)
F (32)	0.4278 (1)	0.1153 (2)	0.4771 (4)	13 (1)	40 (2)	60 (6)	− 4 (1)	2 (2)	− 8 (3)
F (34)	0.4176 (2)	0.1466 (2)	0.0467 (4)	11 (1)	33 (2)	75 (6)	− 5 (1)	−10 (2)	15 (3)
H (11)	0.153 (4)	0.119 (7)	0.287 (11)	2.3 (1.7)	Bonded to N (11)				
H (21)	0.047 (5)	0.161 (7)	0.303 (13)	3.1 (1.8)					
H (31)	0.096 (4)	0.072 (7)	0.101 (12)	2.5 (1.7)					
H (33)	0.086 (4)	0.031 (6)	0.369 (11)	1.8 (1.6)					
H (41)	0.242 (4)	0.722 (7)	0.241 (11)	2.5 (1.8)	Bonded to N (21)				
H (51)	0.286 (4)	0.848 (6)	0.194 (11)	1.7 (1.5)					
H (61)	0.178 (4)	0.818 (7)	0.137 (12)	2.8 (1.7)					
H (63)	0.212 (4)	0.829 (7)	0.383 (11)	2.2 (1.6)					
H (12)	0.406 (4)	0.382 (6)	0.273 (11)	1.6 (1.6)	Bonded to N (12)				
H (22)	0.309 (5)	0.328 (8)	0.244 (13)	2.8 (1.8)					
H (32)	0.338 (4)	0.421 (7)	0.107 (11)	2.0 (1.6)					
H (34)	0.327 (4)	0.468 (7)	0.332 (13)	2.9 (1.8)					
H (42)	0.481 (4)	0.776 (7)	0.250 (11)	2.3 (1.8)	Bonded to N (22)				
H (52)	0.525 (5)	0.637 (8)	0.229 (14)	3.6 (2.1)					
H (62)	0.437 (4)	0.682 (7)	0.176 (12)	2.1 (1.9)					
H (64)	0.459 (4)	0.680 (6)	0.382 (11)	1.3 (1.6)					

Table 155. $(NH_4)_2BeF_4$. Structure of phase III [79O1]. Interatomic distances and angles of ammonium ions at −140 °C.

Distance [Å]		Angle [°]	
N (11)–H (11)	0.89 (6)	H (11)–N (11)–H (21)	122.5 (62)
–H (21)	1.01 (7)	H (11)– –H (31)	101.9 (55)
–H (31)	1.01 (7)	H (11)– –H (33)	103.9 (56)
–H (33)	0.96 (7)	H (21)– –H (31)	112.6 (57)
⟨N (11)–H⟩	0.97	H (21)– –H (33)	172.8 (56)
		H (31)– –H (33)	113.2 (58)
N (21)–H (41)	0.87 (7)	H (41)–N (21)–H (51)	103.5 (56)
–H (51)	1.04 (6)	H (41)– –H (61)	102.6 (60)
–H (61)	1.07 (7)	H (41)– –H (63)	120.3 (66)
–H (63)	0.84 (7)	H (51)– –H (61)	105.9 (52)
⟨N (21)–H⟩	0.96	H (51)– –H (63)	115.4 (59)
		H (61)– –H (63)	107.7 (58)
N (12)–H (12)	0.97 (6)	H (12)–N (12)–H (22)	116.3 (61)
–H (22)	0.89 (8)	H (12)– –H (32)	104.1 (52)
–H (32)	0.90 (7)	H (12)– –H (34)	108.1 (53)
–H (34)	0.91 (8)	H (22)– –H (32)	97.0 (64)
⟨N (12)–H⟩	0.92	H (22)– –H (34)	123.8 (63)
		H (32)– –H (34)	104.1 (65)
N (22)–H (42)	0.83 (7)	H (42)–N (22)–H (52)	124.2 (64)
–H (52)	0.98 (8)	H (42)– –H (62)	111.0 (72)
–H (62)	0.71 (7)	H (42)– –H (64)	99.2 (60)
–H (64)	0.90 (6)	H (52)– –H (62)	116.1 (72)
⟨N (22)–H⟩	0.86	H (52)– –H (64)	97.6 (61)
		H (64)– –H (64)	103.1 (69)

Table 156. $(NH_4)_2BeF_4$. Structure of phase III [79O1]. Interatomic distances and angles related with hydrogen bonds at −140 °C.

X Y	X−Y	H−Y	∢X−H···Y
N (11)−H (11)···F (12)	2.785 Å	1.92 Å	166°
N (11)−H (31)···F (21)	2.822	1.84	162
N (11)−H (33)···F (31)	2.783	1.83	171
N (21)−H (51)···F (22)	2.690	1.72	153
N (21)−H (61)···F (33)	2.816	1.75	174
N (12)−H (12)···F (11)	2.809	1.85	170
N (12)−H (32)···F (22)	2.848	1.99	158
N (12)−H (34)···F (32)	2.809	1.91	168
N (22)−H (52)···F (21)	2.760	1.80	165
N (22)−H (62)···F (34)	2.796	2.09	173

Table 157, see p. 122.

Table 158. $(NH_4)_2BeF_4$. Transition heat and transition entropy per mol.

Transition heat ΔQ_m $J \cdot mol^{-1}$	Transition entropy ΔS_m $J \cdot K^{-1} \cdot mol^{-1}$	Ref.
1300	7.9	58H3
1590	9.5	58H3
836 (84)		73S20

Table 157. $(NH_4)_2BeF_4$. Structure [79 O 1]. Root mean square thermal displacements in Å in direction of principal axes. The orientations of principal axes relative to the crystallographic axes are given in parentheses in degree such as $(\theta_a, \theta_b, \theta_c)$. For example, θ_a is the angle between its principal axis and the a axis.

Temperature Atom (i)	RT	−84 °C	−140 °C	
			Atom (i1)	Atom (i2)
N (1) or	0.154 (3, 87, 90)	0.133 (16, 74, 90)	0.142 (32, 63, 74)	0.081 (12, 79, 85)
N (11), N (12)	0.185 (87, 3, 90)	0.146 (74, 16, 90)	0.103 (62, 28, 88)	0.153 (80, 11, 86)
	0.172 (90, 90, 0)	0.131 (90, 90, 0)	0.119 (77, 81, 16)	0.094 (85, 87, 6)
N (2) or	0.147 (41, 49, 90)	0.139 (40, 50, 90)	0.136 (17, 77, 79)	0.077 (22, 79, 71)
N (21), N (22)	0.158 (49, 41, 90)	0.127 (50, 40, 90)	0.170 (76, 14, 86)	0.027 (78, 12, 89)
	0.156 (90, 90, 0)	0.134 (90, 90, 0)	0.112 (80, 82, 12)	0.099 (72, 85, 19)
Be or	0.143 (27, 63, 90)	0.124 (33, 57, 90)	0.097 (31, 66, 71)	0.126 (53, 69, 45)
Be (1), Be (2)	0.159 (63, 27, 90)	0.121 (57, 33, 90)	0.133 (64, 26, 90)	0.048 (66, 25, 83)
	0.147 (90, 90, 0)	0.111 (90, 90, 0)	0.091 (73, 82, 19)	0.113 (47, 77, 46)

Table 159. $(NH_4)_2BeF_4$. Elastic stiffness in 10^{10} N m^{-2} [76 A 5]. Brillouin scattering. $\lambda = 632.8$ nm.

T [K]	c_{11}	c_{22}	c_{33}	c_{44}	c_{55}	c_{66}	c_{12}	c_{13}	c_{23}
290	3.333	2.377	3.182	0.856	0.672	0.867	1.32	1.60	1.23
280	3.376	2.365	3.215	0.865	0.672	0.876	1.30	1.61	1.22
270	3.412	2.353	3.251	0.875	0.677	0.887	1.29	1.62	1.21
260	3.449	2.340	3.280	0.884	0.661	0.896	1.27	1.63	1.20
250	3.490	2.328	3.309	0.896	0.652	0.909	1.26	1.64	1.19
240	3.527	2.315	3.332	0.907	0.642	0.924	1.25	1.65	1.19
230	3.576	2.317	3.346	0.919	0.631	0.939	1.26	1.66	1.18
220	3.649	2.307	3.353	0.932	0.615	0.958	1.30	1.66	1.17
210	3.740	2.280	3.360	0.948	0.598	0.974	1.33	1.68	1.15
200	3.856	2.250	3.367	0.967	0.581	0.989	1.35	1.70	1.12
190	3.948	2.237	3.377	0.989	0.559	1.002	1.37	1.71	1.12
185	3.967	2.246	3.389	0.989	0.549	1.006	1.38	1.72	1.12
183	3.964	2.246	3.394	1.003	0.547	1.000	1.39	1.71	1.12
182	3.961	2.245	3.394	1.003	0.546	0.999	1.38	1.71	1.12

Table 160. $(ND_4)_2BeF_4$. Quadrupole coupling tensor components of deuteron [69 A 8]. θ: z axis of the electric field gradient tensor rotates gradually in the ac plane with decreasing temperature. The angle of rotation of the z axis is measured from the a axis.

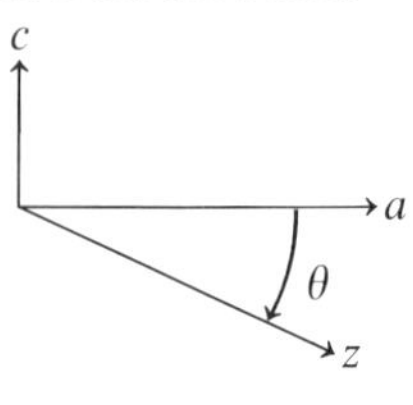

T	$eQ\phi_{xx}/h$	$eQ\phi_{yy}/h$	$eQ\phi_{zz}/h$	θ
°C	kHz			deg
+ 20	±2.7	±1.0	±3.7	0
− 80	±3.6	±0.9	±4.5	10
−110	±5.3	±0.1	±5.4	10
−150	±5.7	±0.2	±5.9	−15

Nr. 20A-6 K_2ZnCl_4, Potassium tetrachlorozincate (II).
($M = 285.39$)

1a	Ferroelectricity in K_2ZnCl_4 was discovered by Gesi in 1978.	78G4
b	(see phase table below)	[a]) 45K1 [b]) 79G1

phase	III	II	I
state	F		P
crystal system	orthorhombic	orthorhombic	orthorhombic[a])
space group	P2cm-C_{2v}^4		Pmcn-D_{2h}^{16}
Θ	130[b])	280 °C [b])	

	$P_s \parallel [100]$ Cleavage plane {010}	78G4
2a	Crystal growth: evaporation from aqueous solution.	78G4
3a	Unit cell parameters: $a = 26.70$ Å, $b = 12.26$ Å, $c = 7.28$ Å at RT. $b \simeq \sqrt{3}a$ in phase I, c: pseudohexagonal axis	45K1
b	$Z = 12$ in phase III.	45K1
5a	Dielectric constant: Fig. 676α.	
c	Spontaneous polarization: Fig. 676β.	
13a	NQR: Fig. 677.	
14a	Bragg reflections due to structural modulation: Figs. 678α, 678β.	

Nr. 20A-7 Rb_2ZnCl_4, Rubidium tetrachlorozincate (II)
($M = 378.13$)

1a	Ferroelectricity in Rb_2ZnCl_4 was found by Sawada et al. in 1977.	77S10
b	(see phase table below)	

phase	III	II	I
state	F		P
crystal system	orthorhombic	orthorhombic	orthorhombic
space group			Pmcn-D_{2h}^{16}
Θ	−81	29 °C	

	$P_s \parallel [100]$ Transparent, colorless. New phase transition at −201 °C was reported.[a])	[a]) 80F1
2a	Crystal growth: evaporation method from aqueous solution.	77S10
3a	Unit cell parameters: $a = 7.282$ Å, $b = 12.726$ Å, $c = 9.257$ Å at RT neglecting structure modulation (cf. 14a below). $b \simeq \sqrt{3}a$, c: pseudohexagonal axis	78H11
b	$Z = 4$	79W1
5a	Dielectric constant: Figs. 679α, 679β, 679γ.	
c	Spontaneous polarization and coercive field: Fig. 680.	
8a	Sound velocity and attenuation: Figs. 680α, 680β. continued	

Nr. 20A-7 Rb_2ZnCl_4, continued

10a Raman scattering: Figs. 681α, 681β. — 80F1
New phase transition temperature $\Theta_{IV-III} \simeq 72$ K is suggested from Raman scattering data. See — 80F1

13a NMR: Figs. 682α, 682β.
NQR: Fig. 682γ; see — 79M3

14a Bragg reflection due to structural modulation: see Fig. 683. — 70J1

Nr. 20A-8 $\{N(CH_3)_4\}_2CoCl_4$, Tetramethylammonium tetrachlorocobaltate

($M = 348.956$)

1a Ferroelectric activity in $\{N(CH_3)_4\}_2CoCl_4$ was discovered by Sawada et al. in 1978. — 78S5

b

	VII	VI	V	IV	III	II	I
state					F		
crystal system	ortho-rhombic		mono-clinic	ortho-rhombic	ortho-rhombic	ortho-rhombic	ortho-rhombic
space group							Pmcn-D_{2h}^{16}

Θ −151 −81 3.0 4.6 7.1 20° °C

$P_s \parallel [100]$

2a Crystal growth: evaporation method from aqueous solution.

3a Unit cell parameters: $a = 9.001$ (1) Å, $b = 15.539$ (2) Å, $c = 12.276$ (1) Å at RT. — 67W6
($b \simeq \sqrt{3}a$, c: pseudohexagonal axis)

b $Z = 4$ in phase I — 59M6

5a Dielectric constant: Figs. 684α, 684β, 684γ.

c Spontaneous polarization and coercive field: Fig. 685.

Nr. 20A-9 $\{N(CH_3)_4\}_2ZnCl_4$, Tetramethylammonium tetrachlorozincate

($M = 355.48$)

1a Ferroelectric activity in $\{N(CH_3)_4\}_2ZnCl_4$ was discovered by Sawada et al. in 1978. — 78S6

b

phase	VI	V	IV	III	II	I
state				F		P
crystal system	ortho-rhombic		ortho-rhombic	ortho-rhombic	ortho-rhombic	ortho-rhombic
space group						Pmcn-D_{2h}^{16}

Θ −112 −92 3.3 6.0 20 °C

$P_s \parallel [100]$
Transparent, colorless.

2a Crystal growth: evaporation method from aqueous solution. — 78S6

3a Unit cell parameters: $a = 8.946$ Å, $b = 15.515$ Å, $c = 12.268$ Å at RT. — 78S6
$b \simeq \sqrt{3}a$, c: pseudohexagonal axis.

b $Z = 4$ at RT. — 59M6

5a	Dielectric constant: Fig. 686α. Phase diagram in regard to *p*: Fig. 686β.	
c	Spontaneous polarization and coercive field: Fig. 686γ.	
8a	Sound velocity: Fig. 687α. Sound attenuation: Fig. 687β.	
13a	NMR: see	79B2
14a	Integrated intensities and satellite reflections due to structural modulation: Fig. 688. Neutron scattering: see	80G2

Nr. 20A-10 Rb_2ZnBr_4, Rubidium tetrabromozincate (II)
($M = 555.93$)

1a	Ferroelectricity in Rb_2ZnBr_4 was discovered by Sawada et al. in 1977.	77S11
b	phase: III | II | I state: F | | P crystal system: orthorhombic[a]) | orthorhombic[a]) | orthorhombic space group: | | Pmcn-D_{2h}^{16} Θ: −86 [b]) (III–II); 73 °C [a]) (II–I)	[a]) 79S1 [b]) 77S11
	$P_s \parallel [100]$ Colorless	77S11
2a	Crystal growth: evaporation from aqueous solution or fusion and cooling of RbBr:$ZnBr_2$ = 2:1 mixture.	77S11
3a	Unit cell parameters: $a = 7.656$ Å, $b = 13.343$ Å, $c = 9.708$ Å at RT. (average structure) $b \simeq \sqrt{3}a$, c: pseudohexagonal axis.	78D4
b	$Z = 4$.	80B2
5a	Dielectric constant: 689α; see also	78D3
c	Spontaneous polarization and coercive field: Fig. 689β; see also	78D3
9a	Far infrared transmission: New phase transition at 50 K is suggested.	80R1
10a	Raman scattering: Fig. 690.	
14a	Bragg reflections due to structural modulations: see Figs. 691α, 691β; see also	72J1 78D4
b	Neutron diffuse scattering: Fig. 691γ.	

20B Solid solutions

Nr. 20B-1 $(NH_4)_2SO_4$-$(NH_4)_2BeF_4$

1b	Transition temperature: Table 161. Phase diagram: Fig. 692.	
13a	NMR: Fig. 693.	

Table 161. $(NH_4)_2(SO_4)_x(BeF_4)_{1-x}$. Transition temperatures, Θ, [62M5].

x	Θ [K] from measurements of		
	$\varkappa$	C_p	T_1
1.00	223	223	230
0.95	≈220	219	225
0.78	≈203	180	189
0	176	—	178
Ref.	58H3	53G2	

Nr. 20B-2 $(NH_4)_2SO_4$-K_2SO_4

1b	Phase diagram: Fig. 694.	
2b	Concentration in mixed crystals: Fig. 695.	
3a	$[(NH_4)_{1-x}K_x]_2SO_4$: unit cell parameters at 20 °C; $a=7.738$ (1) Å, $b=10.373$ (2) Å, $c=5.905$ (1) Å for $x=0.31_4$; $a=7.613$ (6) Å, $b=10.131$ (2) Å, $c=5.8107$ (5) Å for $x=0.70_0$. See also Fig. 696.	
b	Crystal structure: Table 162.	
5a	Dielectric constants: Fig. 697. $[(NH_4)_{0.7}K_{0.3}]_2SO_4$: $\varkappa_c=\varkappa_\infty+C/(T-\Theta_p)$, $T\gtrsim\Theta_f$ with $C=35$ K, $\varkappa_\infty=9.5$, $\Theta_p=-66$ °C.	75S8
c	Spontaneous polarization: Fig. 698.	
13a	NMR: Fig. 699.	

Table 162. $[(NH_4)_{1-x}K_x]_2SO_4$. $x=0.31_4$, $x=0.70_0$. Structure [77S27]. Fractional coordinate ($\cdot 10^{-4}$) and temperature factors ($\cdot 10^{-4}$) at 20 °C. b_{ij} is defined by Eq. (b) in Introduction.

molar fraction $=0.31_4$

	x	y	z	b_{11}	b_{22}	b_{33}	b_{12}	b_{13}	b_{23}
N, K(1)	1831 (4)	945 (3)	2500	87 (3)	48 (2)	145 (6)	5 (2)	0	0
N, K(2)	4778 (2)	7936 (2)	2500	76 (2)	50 (1)	161 (4)	− 1 (1)	0	0
S	2398 (1)	4192 (1)	2500	58 (1)	37 (1)	106 (2)	− 3 (1)	0	0
0(1)	537 (5)	3989 (5)	2500	67 (5)	125 (6)	332 (16)	−32 (4)	0	0
0(2)	2793 (6)	5574 (3)	2500	139 (7)	39 (2)	250 (12)	−15 (3)	0	0
0(3)	3143 (4)	3603 (3)	464 (5)	163 (5)	80 (3)	168 (7)	19 (3)	26 (5)	−28 (4)

molar fraction $=0.70_0$

	x	y	z	b_{11}	b_{22}	b_{33}	b_{12}	b_{13}	b_{23}
N, K(1)	1775 (1)	908 (1)	2500	74 (1)	41 (1)	106 (2)	0 (1)	0	0
N, K(2)	4850 (1)	7955 (1)	2500	60 (1)	38 (0)	134 (1)	− 1 (0)	0	0
S	2355 (1)	4193 (1)	2500	53 (1)	25 (0)	79 (1)	− 1 (0)	0	0
0(1)	433 (3)	4098 (3)	2500	59 (3)	84 (3)	263 (9)	−18 (2)	0	0
0(2)	2902 (4)	5592 (2)	2500	115 (4)	32 (1)	172 (6)	−14 (2)	0	0
0(3)	3055 (3)	3547 (2)	421 (3)	141 (3)	56 (1)	118 (3)	16 (2)	25 (3)	−22 (2)

Nr. 20B-3 $(NH_4)_2SO_4$-Rb_2SO_4

1b	Phase diagram: Fig. 694.	
2b	Concentration in mixed crystals: Fig. 695.	
3	Unit cell parameters: Fig. 696.	
5a	Dielectric constants: Figs. 700, 701.	
c	Spontaneous polarization: Fig. 702.	
13a	NMR: Fig. 703.	

Nr. 20B-4 $(NH_4)_2SO_4$-Cs_2SO_4

1b	Phase diagram: Fig. 694.	
2b	Concentration in mixed crystal: Fig. 695.	
3	Unit cell parameters: Fig. 696.	
5a	Dielectric constant: Figs. 704, 705.	
c	Spontaneous polarization: Fig. 706.	
13a	NMR: Fig. 707.	

21 NH_4HSO_4 family

21A Pure compounds

Nr. 21A-1 $RbHSO_4$, Rubidium bisulfate
($M = 182.53$)

1a	Ferroelectricity in $RbHSP_4$ was discovered by Pepinsky et al. in 1960.	60P1
b		60P1

phase	II	I
state	F	P
crystal system	monoclinic	monoclinic
space group	Pc-C_s^2	$P2_1/c$-C_{2h}^5
Θ	-9.4 °C [a])	

[a]) 75C5

Θ_{II-I} of deuterated crystal was reported as -22 °C. (74K4)
$P_s \| [001]$
$\varrho_x = 1.734 \cdot 10^3$ kg m^{-3} (75A13)
Transparent, colorless. Cleavage plane: (001) (60P1)
Before 1960, the crystal was believed to belong to the orthorhombic system. In this pseudo-orthorhombic lattice, the space group is $B2_1/a$ and the volume of the pseudo-unit cell is twice that of the true unit cell. Pepinsky et al. reported the ferroelectric transition temperature to be about -15 °C, but more recent studies showed that it is around -10 °C. The value -9.4 °C was obtained by Chihara from a heat capacity measurement.

2a	Crystal growth: slow cooling or slow evaporation from aqueous solution containing equimolar ratio of Rb_2SO_4 and H_2SO_4.	62M9
3a	Unit cell parameters: $a = 14.354\,(14)$ Å, $b = 4.618\,(7)$ Å, $c = 14.808\,(13)$ Å, $\beta = 120.92\,(16)°$ at RT. In the pseudo-orthorhombic lattice, the crystallographic axes designated as a', b', and c' lie along the [201], [010], and [001] directions, respectively.	75A13
	$a' = 24.61_2$ Å, $b' = 4.62_2$ Å, $c' = 14.80_7$ Å, $\beta = 90°3'$ at RT. continued	60P1

Nr. 21A-1 $RbHSO_4$, continued

b	$Z=8$ in phase I. Crystal structure: Figs. 708α, 708β; Tables 163···167.	70A9
5a	Dielectric constant: Figs. 709α, 709β. $\varkappa_c = C/(T-\Theta_p)$: $T \geqq \Theta_f$, $C=144$ K, $\Theta_p=-8$ °C. See also Effect of hydrostatic pressure: Figs. 710α, 710β. $d\Theta/dp = 12.0\,(1) \cdot 10^{-8}$ K Pa^{-1}. For deuterated crystal: $d\Theta/dp = 12.7\,(2) \cdot 10^{-8}$ K Pa^{-1}	 77K1 78N1 75G7
b	Nonlinear dielectric properties: $\xi = 7.2 \cdot 10^{14}$ V C^{-3} m^5, $\zeta = 4.2 \cdot 10^{19}$ V C^{-5} m^9.	77K1
c	Spontaneous polarization and coercive field: Figs. 711α, 711β, 711γ.	
6a	Heat capacity: Fig. 712. $\Delta S = 0.91\,(8)$ J K^{-1} mol^{-1}. See also	 76A4 76F7
8a	Sound velocity and attenuation: Figs. 713α, 713β.	
9a	Refractive index: $n_\beta = 1.473$ for $\lambda = 589$ nm. Optical axial angle: $2V = 55°52'$ Infrared spectra, see	 31w1 72M14
c	Piezooptic constant: Fig. 714.	
13a	NMR: Fig. 715; Tables 168, 169, 170.	

Table 163. $RbHSO_4$. Structure of phase I [75A13]. Fractional coordinates obtained by neutron diffraction and X-ray diffraction. $T=23$ °C. See also Fig. 708α.

	Neutron diffraction	X-ray diffraction		Neutron diffraction	X-ray diffraction
Rb1 *x*	0.1231 (3)	0.1233 (2)	O14 *x*	0.1050 (6)	0.1031 (20)
y	0.1414 (12)	0.1405 (4)	*y*	0.4912 (17)	0.4893 (20)
z	0.4158 (3)	0.4160 (2)	*z*	0.1194 (7)	0.1185 (20)
Rb2 *x*	0.3748 (3)	0.3749 (2)	O21 *x*	0.4665 (4)	0.4662 (20)
y	0.7017 (12)	0.7018 (4)	*y*	0.6903 (15)	0.6914 (33)
z	0.3361 (3)	0.3360 (2)	*z*	0.0737 (4)	0.0709 (18)
S1 *x*	0.1257 (8)	0.1246 (5)	O22 *x*	0.4102 (5)	0.4064 (11)
y	0.2113 (27)	0.2171 (8)	*y*	0.9944 (17)	0.9963 (26)
z	0.1743 (8)	0.1741 (5)	*z*	0.1676 (5)	0.1652 (10)
S2 *x*	0.3728 (7)	0.3715 (5)	O23 *x*	0.2851 (4)	0.2848 (13)
y	0.7549 (29)	0.7614 (9)	*y*	0.8826 (16)	0.8838 (27)
z	0.0795 (7)	0.0787 (5)	*z*	−0.0166 (3)	−0.0171 (11)
O11 *x*	0.0322 (4)	0.0318 (10)	O24 *x*	0.3356 (4)	0.3392 (11)
y	0.1199 (17)	0.1096 (30)	*y*	0.5229 (16)	0.5205 (25)
z	0.1769 (4)	0.1790 (10)	*z*	0.1175 (4)	0.1167 (9)
O12 *x*	0.1410 (6)	0.1370 (10)	H1 *x*	0.1312 (9)	—
y	0.0164 (25)	0.0160 (30)	*y*	0.8081 (32)	—
z	0.0986 (6)	0.0955 (10)	*z*	0.1098 (8)	—
O13 *x*	0.2221 (4)	0.2223 (20)	H2 *x*	0.3778 (8)	—
y	0.2061 (21)	0.2032 (40)	*y*	0.1956 (25)	—
z	0.2735 (5)	0.2775 (20)	*z*	0.1388 (8)	—

Table 164. $RbHSO_4$. Structure of phase I [75A13]. Anisotropic temperature factors (· 10^4 Å^2). X-ray results are on the first line and neutron results on the second for each atom. U_{ij} is defined by Eq. (d) in Introduction.

	U_{11}	U_{22}	U_{33}	U_{12}	U_{13}	U_{23}
Rb1	207 (12)	243 (9)	261 (12)	6 (7)	114 (9)	– 13 (7)
	269 (17)	375 (30)	249 (20)	– 20 (25)	122 (15)	16 (26)
Rb2	235 (13)	297 (8)	250 (12)	9 (8)	122 (9)	11 (8)
	306 (19)	451 (38)	291 (21)	28 (24)	180 (15)	21 (24)
S1	256 (34)	143 (17)	205 (29)	– 2 (18)	134 (25)	– 20 (17)
	309 (43)	470 (75)	285 (49)	– 66 (60)	136 (36)	13 (62)
S2	178 (31)	175 (18)	202 (28)	21 (16)	88 (24)	– 5 (16)
	244 (35)	408 (62)	282 (48)	154 (51)	128 (34)	– 26 (56)
O11	228 (92)	455 (77)	335 (97)	– 65 (69)	156 (73)	– 42 (71)
	339 (25)	494 (41)	437 (29)	– 109 (32)	211 (22)	– 66 (35)
O12	488 (109)	350 (70)	239 (83)	– 166 (67)	249 (75)	– 139 (59)
	725 (48)	605 (72)	493 (41)	– 156 (49)	457 (35)	– 155 (47)
O13	404 (141)	856 (128)	409 (134)	– 117 (96)	169 (106)	– 147 (94)
	354 (25)	983 (71)	438 (31)	– 95 (38)	138 (22)	– 98 (42)
O14	1202 (195)	129 (62)	928 (161)	47 (83)	760 (149)	217 (78)
	1006 (58)	254 (40)	1101 (64)	156 (38)	721 (48)	179 (47)
O21	306 (116)	530 (91)	577 (136)	153 (75)	279 (96)	52 (83)
	313 (24)	487 (49)	405 (28)	95 (27)	199 (22)	6 (31)
O22	391 (98)	221 (59)	195 (73)	57 (60)	85 (67)	– 11 (54)
	433 (29)	308 (39)	297 (29)	25 (32)	118 (23)	– 78 (33)
O23	236 (91)	339 (68)	108 (70)	73 (59)	– 27 (62)	26 (54)
	340 (23)	404 (37)	269 (22)	9 (29)	117 (18)	35 (34)
O24	303 (87)	222 (57)	137 (69)	– 89 (54)	74 (62)	– 36 (48)
	420 (28)	404 (43)	365 (28)	– 62 (31)	208 (23)	37 (35)
H1	—	—	—	—	—	—
	480 (59)	532 (91)	397 (61)	19 (65)	195 (47)	– 26 (67)
H2	—	—	—	—	—	—
	491 (52)	344 (67)	390 (58)	80 (52)	220 (46)	71 (56)

Table 165. $RbHSO_4$. Structure of phase I [75A13] Interatomic distances (in Å) and angles (deg) in sulfate ions. Translations are indicated in parantheses after the atom label.

Atoms	Neutron	X-ray
S1 –O11	1.43 (1)	1.46 (2)
–O12	1.54 (2)	1.57 (2)
–O13	1.41 (1)	1.45 (2)
–O14	1.48 (2)	1.45 (2)
S2 –O21	1.42 (1)	1.46 (3)
–O22	1.58 (1)	1.55 (2)
–O23	1.45 (1)	1.44 (2)
–O24	1.44 (2)	1.43 (2)
O11–O12	2.42 (1)	2.43 (3)
–O13	2.37 (1)	2.39 (3)
–O14	2.38 (1)	2.42 (3)
O12–O13	2.39 (1)	2.47 (3)
–O14	2.31 (1)	2.30 (2)
O13–O14	2.41 (1)	2.46 (3)
O21–O22	2.39 (1)	2.43 (3)
–O23	2.40 (1)	2.40 (3)
–O24	2.41 (1)	2.38 (2)

Atoms	Neutron	X-ray
O22–O23	2.43 (1)	2.40 (2)
–O24	2.37 (1)	2.36 (2)
O23–O24	2.40 (1)	2.40 (2)
O12–O14 (*b*)	2.53 (1)	2.54 (2)
O22–O24 (*b*)	2.62 (1)	2.57 (2)
O11–S1–O12	110 (1)	107 (1)
O11–S1–O13	114 (1)	110 (2)
O11–S1–O14	110 (1)	113 (2)
O12–S1–O13	109 (1)	110 (2)
O12–S1–O14	100 (1)	100 (1)
O13–S1–O14	114 (1)	117 (1)
O12–S2–O22	105 (1)	108 (1)
O21–S2–O23	113 (1)	112 (1)
O21–S2–O24	115 (1)	111 (1)
O22–S2–O23	106 (1)	107 (2)
O22–S2–O24	104 (1)	105 (1)
O23–S2–O24	112 (1)	114 (1)

Table 166. $RbHSO_4$. Structure of phase I [75A13]. Distances (in Å) between rubidium and oxygen at 23 °C. Atoms are those in the same asymmetric zone unless otherwise indicated. A double prime denotes atoms obtained by a twofold screw operation, while a single prime denotes a glide operation. Translations are indicated in parantheses after the atom label.

Atoms	Neutron	X-ray	Atoms	Neutron	X-ray
Rb1–O11	3.07 (1)	3.05 (2)	Rb2–O21′(*a*)	2.99 (1)	3.00 (2)
–O11′	2.93 (1)	2.90 (2)	–O21′(*a*–*b*)	3.07 (1)	3.07 (2)
–O11′(–*b*)	3.09 (1)	3.12 (2)	–O21″(*b*)	3.10 (1)	3.06 (3)
–O12″	3.03 (1)	3.02 (2)	–O22	3.10 (1)	3.10 (2)
–O13	3.10 (1)	3.05 (3)	–O22′(*a*–*b*)	3.26 (1)	3.30 (2)
–O14′(–*b*)	3.12 (1)	3.10 (3)	–O23″(*b*)	3.08 (1)	3.07 (2)
–O14″	3.22 (1)	3.22 (2)	–O24	3.10 (1)	3.13 (2)
–O23″	3.14 (1)	3.14 (1)	–O13	3.00 (1)	2.98 (2)
–O23″(*b*)	2.98 (1)	2.97 (2)	–O13(*b*)	2.97 (1)	2.99 (2)
–O24	3.07 (1)	3.08 (2)			

Table 167. $RbHSO_4$. Structure of phase I [75A13]. Interatomic distances (in Å) and angles (in deg) involving hydrogens obtained by neutron diffraction. $T=23$ °C. Atoms are those in the same asymmetric unit unless otherwise indicated. A double prime denotes atoms obtained by a two fold operation, while a single prime denotes a glide operation. Translations are indicated in parantheses after the atom label.

O12–H1	0.99 (1)	O22–H2	1.04 (1)
O12–O14(*b*)	2.53 (1)	O22–O24(*b*)	2.62 (1)
H1···O14(*b*)	1.54 (2)	H2···O24(*b*)	1.59 (1)
∠ O12–H1···O14(*b*)	174 (1)	∠ O22–H2···O24(*b*)	169 (1)
∠ S1–O12–H1	111 (1)	∠ S2–O22–H2	114 (1)
H1–Rb1′(–*a*)	3.50 (1)	H2–Rb1″(–*c*)	3.52 (1)
–Rb1″(–*c*)	3.56 (1)	–Rb2(–*a*)	3.73 (1)
–Rb1″(*b*–*c*)	3.80 (1)	–Rb2	3.76 (1)
–Rb2	3.41 (1)	–Rb2′	3.38 (1)

Table 168. $RbDSO_4$. $|eQ\phi_{ii}/h|$ of ^{87}Rb [74K3]. Principal values of the nuclear quadrupole coupling tensor and their direction cosines with respect to the crystallographic axes a', b', c'^*. c'^* is perpendicular to the $a'b'$ plane. A, B and A_1, A_2, B_1, B_2 correspond to different ^{87}Rb sites above and below Θ_f, respectively.

T °C	$\lvert eQ\phi_{ii}/h\rvert$ kHz		Direction cosine			$\lvert eQ\phi_{ii}/h\rvert$ kHz		Direction cosine		
			a'	b'	c'^*			a'	b'	c'^*
18	A	4006	0.890	0.492	0.154	B	3652	0.913	0.216	0.354
		3038	0.453	−0.795	−0.404		3272	0.005	−0.854	0.521
		968	0.051	−0.429	0.902		380	−0.407	0.474	0.781
−50	A_1	6323	0.879	0.473	0.054	B_1	4118	0.995	−0.031	0.093
		4687	0.458	−0.807	−0.372		4472	−0.073	−0.866	0.494
		1636	0.132	−0.352	0.927		354	−0.065	0.499	0.864
	A_2	2676	0.859	0.384	0.344	B_2	3795	0.836	0.374	0.401
		2583	0.485	−0.826	−0.285		2037	0.432	−0.900	0.061
		93	−0.174	−0.412	0.895		1758	−0.338	0.224	0.914

Table 169. $RbDSO_4$. $eQ\phi_{ii}/h$ of ^2D [74 K 4]. Principal values of the nuclear quadrupole coupling tensor and their direction cosines with respect to the crystallographic axes a', b', c'^*. c'^* is perpendicular to the $a'b'$ plane. $T = 20$ °C. See also caption of Table 170.

	$eQ\phi_{ii}/h$ kHz	Direction cosine a'	b'	c'^*		$eQ\phi_{ii}/h$ kHz	Direction cosine a'	b'	c'^*
D_1	−171.5	−0.106	±0.970	0.218	D_2	−167.6	0.395	±0.908	0.143
	96.3	±0.989	0.081	±0.120		92.3	±0.802	−0.264	∓0.535
	75.2	−0.099	∓0.228	0.968		75.3	0.448	∓0.326	0.832

Table 170. $RbDSO_4$. $eQ\phi_{ii}/h$ of ^2D [75 K 6]. Principal values of the nuclear quadrupole coupling tensor and their direction cosines with respect to the crystallographic axes a', b', c'^*. c'^* is perpendicular to the $a'b'$ plane. $T = -120$ °C. Two crystallographically inequivalent sites in the paraelectric (P) phase result in four sites in the ferroelectric phase. D_1 in the P-phase splits into D_1 and D'_1, D_2 into D_2 and D'_2.

	$eQ\phi_{ii}/h$ kHz	Direction cosine a'	b'	c'^*		$eQ\phi_{ii}/h$ kHz	Direction cosine a'	b'	c'^*
D_1	−192.1	−0.002	±0.982	0.177	D_2	−166.4	0.382	±0.911	0.157
	110.5	±0.999	−0.001	±0.015		92.8	±0.789	−0.233	∓0.568
	81.6	−0.015	∓0.177	0.984		73.6	0.481	∓0.341	0.808
D'_1	−165.2	−0.307	±0.933	0.190	D'_2	−168.2	0.456	±0.882	0.119
	92.3	±0.886	0.207	±0.416		94.3	±0.703	−0.275	∓0.656
	72.9	−0.348	∓0.296	0.889		73.9	0.545	∓0.383	0.745

Nr. 21A-2 NH_4HSO_4, Ammonium bisulfate
($M = 115.10$)

1a Ferroelectricity in NH_4HSO_4 was discovered by Pepinsky et al. in 1958. — 58 P 2

b — 58 P 2

phase	III	II	I
state		F	P
crystal system	triclinic	monoclinic	monoclinic
space group	P1-C_1^1	Pc-C_s^2	$P2_1$/c-C_{2h}^5
Θ	−119	−3	°C

$P_s \| [001]$ in phase II.
$T_{melt} = 147$ °C [a] — [a]) 70 O 2
Transparent, colorless, deliquescent.
Cleavage plane: (001).
Before 1958, the crystal was believed to belong to the orthorhombic system. In this pseudo-orthorhombic lattice, the crystallographic axes designated as a', b', and c' lie along the [201], [010], and [001] directions, respectively, in the above-mentioned crystal system. The volume of the pseudo-unit cell is twice that of the true unit cell, its space group is $B2_1/a$, Ba, and B1 in phase I, II, and III, respectively. — 71 N 8

2a Crystal growth: slow cooling of $(NH_4)_2SO_4$ dissolved in concentrated H_2SO_4, or slow hydrolysis of chloroacetate-amide in dilute H_2SO_4. The latter gave better crystals. — 58 P 2

continued

Nr. 21A-2 NH_4HSO_4, continued

3a	Unit cell parameters: $a=14.51_1$ Å, $b=4.54_0$ Å, $c=14.90_2$ Å, $\beta=120°\,18'$ at RT. $Z=8$.	58P2
	$a'=24.66\,(2)$ Å, $b'=4.60\,(1)$ Å, $c'=14.82\,(2)$ Å, $\beta'=89.87\,(1)°$ at RT. $Z=16$.	71N8
	$a'=24.48\,(4)$ Å, $b'=4.58\,(2)$ Å, $c'=14.75\,(3)$ Å, $\beta'=90.00\,(5)°$ at $-70\,(5)$ °C. $Z=16$. Tables 171α, 171β, 171γ; Figs. 716α, 716β, 716γ.	72N4
b	$Z=16$ in phases I and II. (pseudo-orthorhombic system) Bragg reflection: Fig. 717.	71N8, 72N4
5a	Dielectric constants: Figs. 718α, 718β. [For deuterated crystals: $T>\Theta_{\text{II-I}}$; $\varkappa_c=C/(T-\Theta_p)$, $C=250\,(10)$ K, $\Theta_p=\Theta_{\text{II-I}}=-10.9$ °C, $\Theta_{\text{III-II}}=-115\,(1)$ °C.] Figs. 719α, 719β.	70K4
	Effect of hydrostatic pressure: Figs. 720α, 720β. $d\Theta_{\text{II-I}}/dp=13.8\cdot10^{-8}$ K Pa^{-1}, $d\Theta_{\text{III-II}}/dp=75.5\cdot10^{-8}$ K Pa^{-1}	68P4
c	Spontaneous polarization: Fig. 721. Coercive field: Fig. 722.	
6a	Heat capacity: Fig. 723. $\Delta Q_m=502$[a]), 602[b]) J mol^{-1} for transition II−I. $=1420$[a]) J mol^{-1} for transition III−II. $\Delta S_m=2.1$[a]), 2.2[b]), 1.66[c]) J K^{-1} mol^{-1} for transition II−I, 8.8[a]), 6.7[c]) J K^{-1} mol^{-1} for transition III−II.	[a]) 58P2 [b]) 63S14 [c]) 76A4
9a	Infrared spectra: Fig. 724; see also Refractive indices and optical angle: $n_\alpha=1.463$, $n_\beta=1.473$, $n_\gamma=1.510$, $2V=60°$ for $\lambda=589$ nm.	71S12
10a	Raman spectra: Fig. 725.	
b	Brillouin spectra: Fig. 726.	
13a	NMR: Figs. 727α, 727β. T_1 of deuteron: see	72T16
15a	Domain structure	
b	Effects of mechanical stress: see	74K19

Table 171α. NH_4HSO_4. Structure of phase I [71N8]. Pseudo-orthorhombic cell. Fractional coordinates and temperature factors b_{ij}. b_{ij} is defined by Eq. (b) in Introduction.

	x	y	z
S (1)	0.06208 (2)	0.71422 (9)	0.11079 (3)
S (2)	0.31343 (1)	0.76186 (9)	0.10707 (2)
O (1)	0.01521 (5)	0.61791 (36)	0.15926 (9)
O (2)	0.10986 (6)	0.70472 (51)	0.16386 (11)
O (3)	0.05250 (9)	0.98660 (39)	0.06552 (15)
O (4)	0.07093 (8)	0.51118 (40)	0.02896 (11)
O (5)	0.26636 (5)	0.68751 (34)	0.15988 (9)
O (6)	0.35675 (5)	0.88429 (30)	0.15996 (8)
O (7)	0.29432 (6)	0.99568 (32)	0.03814 (10)
O (8)	0.33160 (5)	0.52234 (27)	0.05042 (9)
N (1)	0.06223 (5)	0.63909 (38)	0.35245 (10)
N (2)	0.31258 (5)	0.79758 (38)	0.34988 (10)
H (9)	0.0691 (140)	0.3614 (890)	0.0502 (240)
H (10)	0.3057 (100)	0.1299 (650)	0.0449 (190)

continued

Table 171 α, continued

	b_{11}	b_{22}	b_{33}	b_{23}	b_{31}	b_{12}
S (1)	0.00107 (1)	0.02160 (19)	0.00355 (2)	−0.00098 (5)	0.00031 (1)	−0.00054 (3)
S (2)	0.00084 (1)	0.02100 (16)	0.00237 (1)	0.00048 (5)	−0.00008 (1)	−0.00051 (3)
O (1)	0.00134 (2)	0.05505 (94)	0.00445 (7)	−0.00231 (23)	0.00079 (3)	−0.00255 (12)
O (2)	0.00137 (3)	0.1112 (170)	0.00582 (9)	−0.00586 (34)	−0.00049 (4)	−0.00134 (17)
O (3)	0.00411 (6)	0.03212 (87)	0.00981 (14)	0.00605 (30)	0.00238 (7)	0.00353 (17)
O (4)	0.00300 (4)	0.03886 (87)	0.00350 (7)	−0.00233 (22)	0.00124 (4)	−0.00190 (16)
O (5)	0.00116 (2)	0.05329 (91)	0.00372 (6)	0.00044 (20)	0.00057 (3)	−0.00204 (11)
O (6)	0.00121 (2)	0.03728 (69)	0.00316 (6)	0.00064 (17)	−0.00062 (3)	−0.00190 (10)
O (7)	0.00162 (3)	0.02373 (60)	0.00403 (7)	0.00185 (18)	−0.00089 (3)	−0.00061 (10)
O (8)	0.00144 (2)	0.02413 (59)	0.00385 (6)	−0.00102 (16)	0.00009 (3)	0.00061 (9)
N (1)	0.00097 (2)	0.03649 (77)	0.00347 (7)	−0.00007 (20)	−0.00005 (3)	−0.00024 (11)
N (2)	0.00118 (2)	0.04439 (96)	0.00282 (6)	0.00007 (21)	0.00007 (3)	−0.00065 (13)
H (9)	B=3.5 (10) Å² [a])					
H (10)	B=3.0 (10) Å² [a])					

[a]) Isotropic temperature factors for the hydrogen atoms of the HSO_4^- groups. B is defined by Eq. (e) in Introduction.

Table 171β. NH_4HSO_4. Structure of phase I. [71 N 8]. Interatomic distances and bond angles. See Figs. 716α, β about the method of labellings symmetry related atoms.

Bond lengths	[Å]	Bond lengths	[Å]
S (1)−O (1)	1.430 (2)	S (2)−O (5)	1.440 (2)
O (2)	1.419 (2)	O (6)	1.441 (2)
O (3)	1.441 (3)	O (7)	1.557 (2)
O (4)	1.546 (3)	O (8)	1.455 (2)
N (1)−O (6b)	3.134 (5)	N (2)−O (5b)	2.965 (4)
O (6b″)	2.972 (4)	O (5b″)	3.070 (5)
O (1h′)	3.070 (5)	O (2d)	3.006 (5)
[O (1h′)]″	2.921 (4)	O (2d″)	2.990 (5)
O (4i)	3.079 (4)	O (7k)	3.131 (4)
O (3i)	3.219 (5)	O (8k)	3.122 (4)
O (1)	3.093 (4)	O (5)	3.082 (4)
O (2)	3.044 (4)	O (6)	3.042 (4)
O (8b)	3.081 (3)	O (7b″)	3.254 (3)
O (3h′)	3.156 (3)		
O (3)−H (9″)	1.790 (40)	O (7)−H (10″)	0.690 (30)
H (9″)−O (4″)	0.760 (40)	H (10″)−O (8″)	1.920 (30)
O (3)−O (4″)	2.514 (6)	O (7)−O (8″)	2.598 (5)

Bond angles	[deg]	Bond angles	[deg]
O (1)−S (1)−O (2)	112.5 (1)	O (5)−S (2)−O (6)	113.3 (1)
O (1)−S (1)−O (3)	111.7 (1)	O (5)−S (2)−O (7)	106.0 (1)
O (1)−S (1)−O (4)	108.6 (1)	O (5)−S (2)−O (8)	112.4 (1)
O (2)−S (1)−O (3)	115.0 (1)	O (6)−S (2)−O (7)	108.2 (1)
O (2)−S (1)−O (4)	107.5 (1)	O (6)−S (2)−O (8)	112.5 (1)
O (3)−S (1)−O (4)	100.5 (1)	O (7)−S (2)−O (8)	103.8 (1)
O (3)···H (9″)−O (4″)	160 (4)	O (7)−H (10″)···O (8″)	172 (6)

Table 171γ. NH_4HSO_4. Structure of phase II [72N4]. Pseudo-orthorhombic cell. Coordinates and temperature factors of X, Z: coordinates along a', b'^*, respectively. B is defined by Eq. (e) in Introduction. These parameters correspond to the atoms of the subcells labelled as a, b, c and d in Fig. 716β. a' and b' are pseudo-orthorhombic lattice parameters. b'^* is perpendicular to the a', c' plane.

Atoms	X [Å]	Z [Å]	B [Å^2]	Atoms	X [Å]	Z [Å]	B [Å^2]
S (1)	1.52	1.61	0.9 (1)	S (1′)	10.73 (6)	5.74 (7)	1.5 (2)
O (1)	0.46 (2)	2.44 (2)	2.0 (4)	O (1′)	11.98 (2)	5.16 (2)	2.2 (4)
O (2)	2.70 (2)	2.40 (2)	1.7 (4)	O (2′)	9.59 (2)	4.96 (2)	1.8 (4)
O (3)	1.11 (2)	0.76 (2)	1.8 (4)	O (3′)	10.76 (3)	6.25 (3)	3.1 (5)
O (4)	1.98 (2)	0.56 (2)	1.5 (3)	O (4′)	10.75 (2)	7.08 (2)	1.4 (3)
N (1)	1.50 (2)	5.23 (2)	0.6 (3)	N (1′)	10.71 (3)	2.24 (3)	2.3 (6)
S (2)	7.700 (5)	1.620 (6)	0.8 (1)	S (2′)	4.60 (6)	5.82 (6)	0.7 (1)
O (5)	6.53 (2)	2.45 (2)	1.4 (3)	O (5′)	5.77 (1)	5.09 (2)	0.2 (3)
O (6)	8.76 (2)	2.33 (2)	2.0 (4)	O (6′)	3.57 (1)	4.96 (2)	1.0 (4)
O (7)	7.23 (2)	0.52 (2)	1.8 (4)	O (7′)	5.14 (2)	6.78 (2)	2.0 (4)
O (8)	8.20 (2)	0.82 (2)	0.9 (3)	O (8′)	4.16 (2)	6.68 (2)	2.1 (4)
N (2)	7.65 (2)	5.18 (2)	0.2 (3)	N (2′)	4.61 (3)	2.29 (3)	2.1 (6)

Nr. 21A-3 $RbHSeO_4$, Rubidium biselenate
(M = 229.43)

1a Ferroelectricity in $RbHSeO_4$ below 97.6 °C was found by Czapla et al.[a]) in 1979. In addition, Suzuki et al.[b]) found a new ferroelectric phase below 173 °C in 1979. — [a]) 79C2; [b]) 79S6

b

phase	III	II	I
state	F	F	P
crystal system	triclinic	monoclinic	
space group	C_1^1-P1		
Θ	97.6[a])	173 °C[b])	

[a]) 79C2; [b]) 79S6

$P_s \parallel [010]$[c]) — [c]) 79P3

$\varrho_X = 3.38 \cdot 10^3$ kg m^{-3} — 78W8

The crystal in phase III can be looked upon pseudo-orthorhombic. Their unit cell vectors are denoted as $\boldsymbol{a}'$, $\boldsymbol{b}'$, and $\boldsymbol{c}'$. — 79P32

2a Crystal growth: evaporation from aqueous solution. — 79P2

b Crystal form: Fig. 728.

3a Unit cell parameters: $a = 19.359(3)$ Å, $b = 4.619(3)$ Å, $c = 7.572(4)$ Å.
$\alpha = 90.60(5)°$, $\beta = 89.80(5)°$, $\gamma = 90.73(5)°$ at RT. — 78W8
$a = 19.464$ Å, $b = 4.624$ Å, $c = 7.599$ Å, $\gamma = 90.66°$ at 150 °C. — 79S6
The unit cell vectors of the pseudo-orthorhombic cell in phase III are given by $\boldsymbol{a}' = \boldsymbol{b}$, $\boldsymbol{b}' = \boldsymbol{c}$, and $\boldsymbol{c}' = \boldsymbol{a} - \boldsymbol{b}$.

b $Z = 6$ in phase III. — 78W8
Crystal structure: Fig. 729; Table 172.

4 Lattice distortions: Figs. 730α, 730β; Table 173.

5a Dielectric constant: Figs. 731α, β, γ, δ.
Curie-Weiss law constants: $C = 3.6 \cdot 10^5$ K, $\Theta_p = 317$ K in phase II. — 79P3

b Nonlinear dielectric properties: $\xi = -1.63 \cdot 10^{10}$ V C^{-3} m^5; $\zeta = 3.37 \cdot 10^3$ V C^{-5} m^9. — 80P1

c Spontaneous polarization and coercive field: Figs. 732α, β.

d Pyroelectric coefficient: Fig. 732β.

9a	Far infrared reflectivity and dielectric constant: Figs. 733 α, β, γ.	
15a	Domain structure: see	79P3, 80C1

Table 172. $RbHSeO_4$. Structure of phase II [78W8]. Fractional coordinates of atoms in the unit cell.

	x	y	z
Rb (1)	0.0840 (2)	0.7207 (8)	0.1585 (5)
Rb (2)	0.2495 (2)	0.3045 (10)	0.7583 (5)
Rb (3)	0.4151 (2)	0.8914 (8)	0.1601 (5)
Se (1)	0.0830*)	0.8400*)	0.6700*)
Se (2)	0.2512 (2)	0.2992 (9)	0.2469 (4)
Se (3)	0.4175 (2)	0.7744 (6)	0.6701 (4)
O (1)	0.1476 (15)	0.7367 (57)	0.7924 (30)
O (2)	0.0213 (12)	0.9862 (54)	0.7806 (35)
O (3)	0.1218 (11)	0.1006 (47)	0.5328 (31)
O (4)	0.0589 (13)	0.5838 (43)	0.5347 (30)
O (5)	0.3171 (11)	0.4325 (47)	0.1262 (27)
O (6)	0.1895 (11)	0.1906 (47)	0.1325 (28)
O (7)	0.2818 (11)	0.0391 (42)	0.3773 (27)
O (8)	0.2217 (10)	0.5622 (45)	0.3857 (26)
O (9)	0.3842 (13)	0.5224 (46)	0.5326 (31)
O (10)	0.3555 (11)	0.8768 (50)	0.7942 (28)
O (11)	0.4482 (13)	0.0300 (47)	0.5388 (30)
O (12)	0.4848 (14)	0.6406 (60)	0.7734 (39)

*) Fixed parameters.

Table 173. $RbHSeO_4$. Unit cell parameters at selected temperatures [79P2].

T	a	b	c	α	β	γ
K	Å			deg		
298.78	19.35927	4.61883	7.57104	90.648	89.792	90.720
358.48	19.42285	4.62636	7.59484	90.563	89.820	90.699
370.61	19.43649	4.62783	7,59866	89.927	89.955	90.689
381.88	19.44872	4.62932	7.60458	89.995	89.989	90.695
395.64	19.46457	4.63110	7.61094	89.988	89.998	90.693

Nr. 21A-4 NH_4HSeO_4, Ammonium biselenate

($M = 162.00$)

1a Ferroelectricity in NH_4HSeO_4 below −22 °C was discovered by Czapla et al. in 1979. In addition, phase transitions at −180 °C and 13 °C were discovered by Gesi in 1980. 79C2

b

phase	IV	III	II	I	
state		F		P	
crystal system		triclinic[a])	monoclinic[b])		a) 80G1
space group		C_1^1-P1[c])	C_2^3-C2[a, b])		b) 79C2
Θ	−180[a])	−26[a])	13[a]) °C		c) 79K3

The crystal in phase III can be looked upon pseudo-orthorhombic. Their unit cell vectors are denoted as ***a***′, ***b***′, and ***c***′. Phase II, III, IV are pyroelectric and polar along the *a* axis. P_s||[100] in the phase III. 79K3

continued

Table 21A-4 NH_4HSeO_4, continued

2a	Crystal growth: slow evaporation of aqueous solution.	79C2
b	Crystal form: Fig. 734.	
3a	Unit cell parameters: a=4.601 Å, b=7.547 Å, c=19.258 Å, $\alpha=\gamma=90°$, $\beta=90.90°$ at RT (pseudo-orthorhombic setting).	79C2
	a=4.562 Å, b=7.486 Å, c=19.050 Å, $\alpha=\gamma=90°$, $\beta=91.02°$ at −50 °C.	79K3
	The unit cell vectors of the pseudo-orthorhombic cell in phase III are given by $\boldsymbol{a}'=\boldsymbol{a}$, $\boldsymbol{b}'=\boldsymbol{b}$, $\boldsymbol{c}'=\boldsymbol{c}-\boldsymbol{a}$.	
5a	Dielectric constants: Figs. 735 α, β. Curie-Weiss law constant: $C=5.453\cdot10^3$; $\Theta_p=248$ K.	79C2
b	Nonlinear dielectric properties: $\xi=-2.64\cdot10^{12}$ V C^{-3} m^5, $\zeta=5.3\cdot10^{16}$ V C^{-5} m^9.	80P1
c	Spontaneous polarization and coercive field: Figs. 736 α, β.	
d	Pyroelectric charge: Fig. 737. Pyroelectric coefficient: Fig. 738.	

22 $(NH_4)_3H(SO_4)_2$ family

22A Pure compounds

Nr. 22A-1 $(NH_4)_3H(SO_4)_2$, Triammonium hydrogen disulfate
(M=247.24)

1a Ferroelectricity in pressure-induced phases of $(NH_4)_3H(SO_4)_2$ was discovered by Gesi in 1976. — 76G8

b

phase	VII*)	V	IV	III	II	I	VI
state	F						F
crystal system					mono-clinic	trigonal	
space group					A2/a-C_{2h}^6 a)		

Θ: −211 b) −140 c) −136 c) −8 c) 140 c) °C

a) 78S15
b) 80G3
c) 76G6, 77G4

The phase VI is stable at high pressures ($p>5\cdot10^8$ Pa); see Fig. 751 α. By deuteration, it is stabilized at atmospheric pressure; see Fig. 751 β.
Phase diagram of $(NH_4)_3H(SO_4)_2$–$(ND_4)_3D(SO_4)_2$: Fig. 739.
P_s: Perpendicular to (001). — 77G4
$\varrho=1.831\cdot10^3$ kg m^{-3}. — 04G1
Transparent, colorless. Deliquescent.

*) Phase VII is identical with pressure-induced ferroelectric phase; see Figs. 751 α, 739.

2a	Crystal growth: Evaporation of aqueous solution.	76G6
b	Crystal form: Fig. 740.	
3a	Unit cell parameters: a=10.153(3) Å, b=5.854(2) Å, c=15.410(6) Å, β=101.76(2)° at RT.	78S15
b	Z=4 in phase II.	78S15
	Structure of phase II: Figs. 741, 742; Tables 174, 175.	
4	Thermal expansion: Figs. 743 α, 743 β.	
5a	Dielectric constant: Figs. 744···748.	
	Dielectric constants of deuterated compounds: Figs. 749, 750.	
	Phase diagrams with regard to p for normal and deuterated compounds: Figs. 751 α, 751 β.	

 Figs. p. 493 ff.

c	Spontaneous polarization in high pressure phases: Fig. 752. Spontaneous polarization of deuterated compounds: Fig. 753.	
6a	Heat capacity: Figs. 754, 755. Transition heat and transition entropy: Table 176.	
9a	Infrared absorption: see	73A2

Table 174. $(NH_4)_3H(SO_4)_2$. Structure of phase II [78S15]. Fractional coordinates of non-hydrogen atoms. T = RT.

	x	y	z
N (1)	0.7500	0.2692 (4)	0
N (2)	0.6528 (2)	0.7240 (3)	0.1988 (1)
S	0.4615 (1)	0.2181 (1)	0.1142 (1)
O (1)	0.3983 (2)	0.0268 (3)	0.1495 (1)
O (2)	0.4427 (2)	0.1845 (3)	0.0148 (1)
O (3)	0.6054 (1)	0.2225 (3)	0.1504 (1)
O (4)	0.3985 (2)	0.4330 (3)	0.1284 (1)

Table 175. $(NH_4)_3H(SO_4)_2$. Structure of phase II [78S15]. Fractional coordinates of hydrogen atoms. T = RT.

	x	y	z	Bonded to
H (1)	0.311 (3)	0.655 (7)	0.022 (3)	N (1)
H (2)	0.226 (4)	0.797 (7)	0.046 (3)	N (1)
H (3)	0.723 (4)	0.708 (7)	0.172 (3)	N (2)
H (4)	0.660 (5)	0.720 (8)	0.258 (4)	N (2)
H (5)	0.597 (5)	0.626 (9)	0.150 (3)	N (2)
H (6)	0.617 (5)	0.860 (9)	0.163 (3)	N (2)
H (7)	0	0	0	O (2)

Table 176. $(NH_4)_3H(SO_4)_2$. Transition heat and transition entropy per mol [79S7].

Phase transition	ΔQ_m J mol^{-1}	ΔS_m J K^{-1} mol^{-1}
I–II	4290 (340)	10.5 (8)
II–III	2120 (80)	8.9 (3)
III–IV	265 (16)	2.1 (1)
IV–V	301 (16)	2.2 (1)

Nr. 22A-2 $(NH_4)_3H(SeO_4)_2$, Triammonium hydrogen diselenate
(M = 341.04)

1a	Ferroelectricity in $(NH_4)_3H(SeO_4)_2$ was discovered by Gesi in 1977.	77G3
b	(see phase table below)	a) 79O4 b) 77G5
	$P_s \parallel [111]$. Transparent, colorless.	77G5
2a	Crystal growth: evaporation from aqueous solution.	77G5
4	Thermal expansion: Fig. 756.	
5a	Dielectric constant: Figs. 757, 758. Curie constant and paraelectric Curie temperature of deuterated compound (D-content: 80%): $C = 3.7 \cdot 10^3$ K, $\Theta_p = -63$ °C. Phase diagram with regard to p: Fig. 759.	 79O4
c	Spontaneous polarization: Figs. 760, 761.	

phase	V		IV		III		II		I	
state	F									
crystal system							trigonal a)			
Θ		−92 b)		2 b)		29 a)		55 a)		°C

23 NH_4LiSO_4 family

23A Pure compounds

Nr. 23A-1 $KLiSO_4$, Lithium potassium sulfate
($M = 142.10$)

1a	Dielectric anomaly associated with a phase transition was observed in $KLiSO_4$ and the possibility of ferroelectricity was discussed by Ando in 1962.	62A3
b	(see table below)	[a]) 62A3 [b]) 25B1
	Pyroelectric axis ∥ [001]. Transparent.	25B1
2a	Crystal growth: slow evaporation of an aqueous solution of $2KHSO_4$ and Li_2CO_3.	62A3
3a	Unit cell parameters: $a = 5.13$ Å, $c = 8.60$ Å at RT.	25B1
b	$Z = 2$ in phase II. Projection of crystal structure: Figs. 762, 763.	25B1
5a	Dielectric constants: Fig. 764.	
9a	Refractive index: $n_o = 1.4738$ for $\lambda = 546.1$ nm. Birefringence: $\Delta n = 0.0006$ for $\lambda = 546.1$ nm.	31W1 64S9
b	Electrooptic constant: $\|r_{13} - r_{33}\| = 1.60(16) \cdot 10^{-12}$ m V^{-1} for $\lambda = 546$ nm.	64S9
d	Optical rotary power: $= 41\ (2) \cdot 10^2$ °m^{-1} for $\lambda = 546.1$ nm.	64S9
e	Nonlinear optical susceptibilities: $d_{31}/d_{36}^{KDP} = 0.8$, $d_{33}/d_{36}^{KDP} = 1.5$, at RT for $\lambda = 694.3$ nm.	68S17

Table 1b:

phase	II [a])	I [a])
state		P [a])
crystal system	hexagonal [b])	
space group	$P6_3$-C_6^6 [b])	
Θ	435 [a]) K	

Nr. 23A-2 $RbLiSO_4$, Rubidium lithium sulfate
($M = 188.47$)

1a	Ferroelectric activity in $RbLiSO_4$ was discovered between 166 and 188 °C by Shiroishi et al. in 1976.	76S26
b	(see table below)	76S26
	P_s ∥ [100] Colorless, $T_{melt} = 740$ °C.	76S26
2a	Crystal growth: fusion method or water solution method from stoichiometric mixtures of Rb_2SO_4 and $Li_2SO_4 \cdot H_2O$.	76S26
b	Crystal forms: pseudohexagonal	76S26
3a	$a = 5.288$ Å*), $b = 9.105$ Å*), $c = 8.731$ Å, $\gamma = 89°53'$ *) Here the a and b axes are defined following [76S26]. The axes are defined differently in [69H1]: $a \to -b$, $b \to a$.	69H1
5a	Dielectric constant: Fig. 765.	

Table 1b:

phase	V	IV	III	II	I
state		F			
crystal system	triclinic or monoclinic	monoclinic	orthorhombic	orthorhombic	orthorhombic
Θ	166	188	202	204 °C	

5c	Spontaneous polarization: $1.0 \cdot 10^{-3}\,Cm^{-2}$ at 168 °C Coercive field: $2.9 \cdot 10^2\,kV \cdot m^{-1}$ at 168 °C	76S26 76S26
11	Specific resistivity: $\sim 1 \cdot 10^8\ \Omega m$ along the *a* axis at 205 °C.	76S26

Nr. 23A-3 NH_4LiSO_4, Ammonium lithium sulfate
($M = 121.04$)

1a	Ferroelectric activity in NH_4LiSO_4 was discovered by Mitsui et al. in 1975.	75M16
b	phase: III / II / I state: P / F / P crystal system: — / orthorhombic / — space group: — / $P2_1cn$ [a])-C_{2v}^9 / — Θ: 10 [b]) (III–II); 186.5 [c]) °C (II–I) $P_s \parallel [100]$ [c]) $\varrho_X = 1.896 \cdot 10^3\,kg\,m^{-3}$ at RT [a]) Colorless.	[a]) 69D3 [b]) 58P4 [c]) 77S24
2a	Crystal growth: evaporation method from an aqueous solution at about 30 °C.	77S24
3a	Unit cell parameters: $a = 5.280$ (2) Å, $b = 9.140$ (7) Å, $c = 8.786$ (6) Å, at RT.	69D3
b	Crystal structure: Tables 177, 178; Fig. 766. $Z = 4$ in phase II	69D3
4	Thermal expansion: Figs. 767, 768.	
5a	Dielectric constant: Fig. 769; see also Effect of hydrostatic pressure: Fig. 770. Phase diagram with regard to *p*: Fig. 771. See also	74Y5, 75A6, 75H6 77S24 77A6
c	Spontaneous polarization: Fig. 772.	
8a	Elastic stiffness: Table 179.	
10a	Raman scattering: Figs. 773, 774.	
13a	NMR: Figs. 775, 776. NMR under high pressure: see	 77A8
15a	Domain structure	75H6
b	Effects of mechanical stress: see	75H6

Table 177. NH_4LiSO_4. Structure of phase II [69D3]. Fractional coordinates and temperature parameters at RT. Root-mean-square amplitude of thermal vibrations for oxygens are also shown. b_{ij} is defined by Eq. (b), *B* is defined by Eq. (e) in Introduction. φ's are the angles between the ellipsoid axes and the unit cell axes.

	x	*y*	*z*	*B*
NH_4	0.4912 (25)	0.2137 (7)	0.4999 (7)	1.8 (1) $Å^2$
Li	0.0084 (73)	0.4117 (14)	0.3234 (15)	1.5 (2)
S	0*)	0.0836 (2)	0.2030 (2)	0.90 (4)
O (1)	0.0003 (33)	0.0961 (10)	0.0384 (7)	
O (2)	0.3323 (15)	0.4631 (7)	0.2496 (9)	
O (3)	0.2585 (15)	0.0537 (7)	0.2565 (9)	
O (4)	0.9059 (16)	0.2191 (7)	0.2705 (11)	

continued

Table 177, continued

	b_{11}	b_{22}	b_{33}	$2b_{12}$	$2b_{13}$	$2b_{23}$
O (1)	0.028 (3)	0.030 (2)	0.003 (1)	−0.004 (4)	0.001 (3)	0.003 (1)
O (2)	0.016 (3)	0.006 (1)	0.011 (1)	−0.005 (1)	0.006 (2)	−0.004 (1)
O (3)	0.009 (2)	0.007 (1)	0.009 (1)	0.002 (1)	−0.004 (1)	−0.004 (1)
O (4)	0.019 (3)	0.004 (1)	0.021 (2)	0.001 (1)	0.002 (2)	−0.004 (1)

	Ellipsoid axis	r.m.s. amplitude	φa	φb	φc
O (1)	1	0.20 (1) Å	9 (12)°	86 (6)°	82 (13)°
	2	0.36 (1)	95 (7)	8 (4)	84 (2)
	3	0.09 (2)	97 (13)	97 (2)	10 (9)
O (2)	1	0.11 (2)	44 (21)	46 (20)	88 (20)
	2	0.13 (2)	122 (24)	59 (24)	48 (4)
	3	0.25 (1)	64 (4)	120 (4)	42 (4)
O (3)	1	0.10 (2)	23 (15)	87 (19)	67 (13)
	2	0.13 (2)	105 (21)	35 (6)	60 (11)
	3	0.23 (1)	107 (5)	125 (5)	40 (5)
O (4)	1	0.17 (1)	16 (10)	74 (10)	90 (4)
	2	0.10 (2)	106 (10)	21 (8)	77 (3)
	3	0.29 (1)	86 (4)	103 (3)	13 (3)

*) Fixed, to define x as origin.

Table 178. NH_4LiSO_4. Structure of phase II [69D3]. Interatomic distances and angles at RT. See also Fig. 766.

Bonds	Uncorrected length	Corrected*) length	Bonds	Angle
S–O (1)	1.450 (7) Å	1.496 Å	O (1)–S–O (2)	110.0 (7)°
–O (2)	1.474 (8)	1.493	O (1)–S–O (3)	109.4 (8)
–O (3)	1.469 (8)	1.490	O (1)–S–O (4)	109.8 (6)
–O (4)	1.460 (7)	1.489	O (2)–S–O (3)	109.2 (5)
Li–O (1)	1.891 (105)		O (2)–S–O (4)	108.3 (5)
–O (2)	1.889 (36)		O (3)–S–O (4)	110.1 (5)
–O (3)	1.979 (28)		O (1)–Li–O (2)	111.9 (20)
–O (4)	1.899 (18)		O (1)–Li–O (3)	111.3 (20)
NH_4–O (1)	3.139 (19)		O (1)–Li–O (4)	101.7 (13)
–O (1′)	3.219 (19)		O (2)–Li–O (3)	108.6 (19)
–O (2)	3.277 (10)		O (2)–Li–O (4)	113.9 (20)
–O (2′)	2.851 (11)		O (3)–Li–O (4)	109.4 (13)
–O (3)	2.867 (11)		S–O (1)–Li	173.2 (17)
–O (3′)	3.333 (11)		S–O (2)–Li	129.3 (7)
–O (4)	2.976 (14)		S–O (3)–Li	128.9 (7)
–O (4′)	3.290 (13)		S–O (4)–Li	142.0 (13)

*) Assuming oxygen atoms "riding" on sulfur atoms.

Table 179. NH_4LiSO_4. Elastic stiffness [75A6]. Ultrasonic method. f = 10 MHz. In units of $\cdot 10^{10}$ Nm^{-2}.

T [°C]	c_{11}	c_{22}	c_{23}	c_{44}	c_{55}	c_{66}	c_{12}	c_{13}	c_{23}
32.0	4.150	4.399	4.958	1.538	0.987	1.558	1.97	2.61	2.28
27.0	4.159	4.410	4.973	1.540	0.988	1.573	1.97	2.61	2.28
22.0	4.165	4.422	4.984	1.543	0.990	1.586	1.97	2.62	2.29
17.0	4.170	4.434	4.996	1.545	0.993	1.598	1.97	2.62	2.30
14.0	4.169	4.439	4.999	1.546	0.993	1.602	1.97	2.62	2.30
12.0	4.161	4.431	5.001	1.547	0.994	1.600	1.97	2.61	2.30

24 Langbeinite-type family

24A Pure compounds

Nr. 24A-1 $K_2Mg_2(SO_4)_3$ Potassium magnesium sulfate
($M = 414.98$)

3a	Unit cell parameters: see Table 180.	
b	The langbeinite is a mineral name of $K_2Mg_2(SO_4)_3$. Crystals belonging to this family are isomorphous to this crystal. Space group: $P2_13$-T^4. Crystal structure: Fig. 777.	65w1
5a	Dielectric constant: Fig. 778.	

Table 180. Langbeinite family, $M_2^{1+}M_2^{2+}(SO_4)_3$. Unit cell parameters [Å] [65w1] and phase transition temperatures [°C]. (The number in parentheses indicates the section number.)

M^{1+} \ M^{2+}	Ni	Mg	Zn	Co	Fe	Mn	Cd	Ca
Cs								10.724 b)
Tl				10.033 b)	10.108	10.229 b)	10.385 −145, −153, −181 (24A-12)	
Rb	9.930 b)	10.005 b) (24A-2)		10.026 b)	10.098	10.218 b)	10.382 −144, −170 (24A-11)	10.570 −90 a) (24A-4)
NH_4	9.904 c)	9.979 c)		9.997 b)	10.068	10.192 c)	10.350 −178 (24A-13)	10.536 −115 a) (24A-5)
K	9.838 b)	9.920 b) (24A-1)	9.925 −135 −188 (24A-9)	9.929 −165 −265 (24A-8)	10.007 −128 (24A-7)	10.114 −72 (24A-6)	10.280 b) d) (24A-10)	−187 (24A-3)

a) Dielectric measurement was made for polycrystalline sample [77H13, 77H14].
b) Dielectric anomalies were not found for polycrystalline sample in the temperature range between room temperature and liquid nitrogen temperature [77H13, 77H14].
c) Phase transitions were not found for crystals in the same temperature range as in b [77H13, 77H14].
d) Orthorhombic at RT [78A1].

Nr. 24A-2 $Rb_2Mg_2(SO_4)_3$ Rubidium magnesium sulfate
($M = 507.72$)

3a	Unit cell parameters: Table 180.	
5a	Dielectric constant: Fig. 778.	

Nr. 24A-3 $K_2Ca_2(SO_4)_3$ Potassium calcium sulfate
($M = 446.53$)

1a	Dielectric anomaly was found by Yamada et al. in 1980.	80Y1
b	phase: II | I; state: nonpolar | P; Θ: −187 °C	80Y1
9a	Birefringence: Fig. 779α.	

phase	II	I
state	nonpolar	P
Θ	−187 °C	

Nr. 24A-4 $Rb_2Ca_2(SO_4)_3$ Rubidium calcium sulfate
($M = 539.27$)

1a	Dielectric anomaly was found by Hikita et al. in 1977.	77H14
b	phase: II | I; Θ: −90 °C	77H14
3a	Unit cell parameters: Table 180.	
5a	Dielectric constant: Fig. 779β.	

phase	II	I
Θ	−90 °C	

Nr. 24A-5 $(NH_4)_2Ca_2(SO_4)_3$ Ammonium calcium sulfate
($M = 404.40$)

1a	Dielectric anomaly was found by Hikita et al. in 1977.	77H14
b	phase: II | I; Θ: −115 °C	77H14
3a	Unit cell parameters: Table 180.	
5a	Dielectric constant: Fig. 779β.	

phase	II	I
Θ	−115 °C	

Nr. 24A-6 $K_2Mn_2(SO_4)_3$, Potassium manganese sulfate
($M = 476.25$)

1a	A phase transition was found by Hikita et al. in 1977.	77H13
b	phase (see table below)	78H10
2a	Crystal growth: evaporation from aqueous solution at about 80 °C.	77H13
b	Crystal habit: Fig. 780.	77H13
3a	Unit cell parameters: $a = 10.114$ Å at RT, Table 180. $a = 10.082$ Å, $b = 10.103$ Å, $c = 10.039$ Å at −85 °C	78Y1
4	Thermal distortion: Fig. 781.	

phase	II	I
state	nonpolar	P
crystal system	orthorhombic	cubic
space group	$P2_12_12_1$-D_2^4	$P2_13$-T^4
Θ	−72 °C	

Figs. p. 501 ff.

5a	Dielectric constant: Figs. 782, 783.	
8a	Elastic compliance: Fig. 782.	
9a	Birefringence: Fig. 784.	
b	Electrooptic constant: $r_{41}=2.0(2)\cdot 10^{-12}$ m V^{-1} for $\lambda=453$ nm.	68E3
c	Piezooptic constant: $p_{11}-p_{12}\simeq p_{11}-p_{13}=-0.020(5)$ at RT.	78Y1

Nr. 24A-7 $K_2Fe_2(SO_4)_3$ Potassium iron sulfate
($M=478.06$)

1a	Dielectric anomaly was found in polycrystalline sample by Yamada et al. in 1980.	80Y1
b		

phase	II	I
Θ	−128 °C	

3a	Unit cell parameters: Table 180.	
5a	Dielectric constant: Fig. 785.	

Nr. 24A-8 $K_2Co_2(SO_4)_3$ Potassium cobalt sulfate
($M=484.24$)

1a	A dielectric anomaly was found by Hikita et al. in 1977.	77H14
b		80Y1

phase	III	II	I
state		F	P
Θ	−265	−165 °C	

2a	Crystal growth: Bridgman method	80Y1
3a	Unit cell parameters: Table 180.	
5a	Dielectric constant: Fig. 786α.	
6a	Heat capacity: Fig. 786β.	

Nr. 24A-9 $K_2Zn_2(SO_4)_3$ Potassium zinc sulfate
($M=497.13$)

1a	Dielectric anomalies were found for polycrystalline samples by Hikita et al. in 1977.	77H13
b		79H2

phase	III	II	I
state	nonpolar	F	P
Θ	−188	−135 °C	

2a	Crystal growth: Czochralski method at about 480 °C	79H2
3a	Unit cell parameters: Table 180.	
5a	Dielectric constant: Fig. 787α.	
c	Spontaneous polarization: Fig. 787β.	

Nr. 24A-10 $K_2Cd_2(SO_4)_3$, Potassium cadmium sulfate
(M = 591.19)

1a	Dielectric anomaly was not found below RT. A phase transition was found above RT by Abrahams et al. in 1978.	77 H 13 78 A 1
b	(see phase table below)	78 A 1 a) 77 A 2
2a	Crystal growth: Bridgman and Czochralski techniques.	77 N 2
b	Phase diagram of K_2SO_4-$CdSO_4$ system: Fig. 788.	
3a	Unit cell parameters: a = 10.2082 (2) Å, b = 10.2837 (3) Å, c = 10.1661 (1) Å at 298.2 K.	77 A 2
b	Crystal structure: Figs. 789, 790, 791. Structural parameters: Tables 181, 182.	
4	Thermal expansion: Fig. 792.	
5a	Dielectric constant: Fig. 779.	
8b	Reorientation from *abc* to *bca* observed under the applied stress of 5 MPa normal to (010).	77 A 2

phase	II	I
state	nonpolar	P
crystal system	orthorhombic	cubic
space group	$P2_12_12_1$-D_2^4 a)	$P2_13$-T^4
Θ	432 K	

Table 181. $K_2Cd_2(SO_4)_3$. Atomic coordinates [78 A 1]. Five rows given for each atom correspond to the coordinates at 298, 351, 390.5, 417.5 K (phase II) and 443.5 K (phase I).

Atom	x	y	z	Atom	x	y	z
K (1)	0.17012 (9)	0.31356 (10)	0.69223 (10)	O (4)	0.2110 (3)	0.4345 (3)	0.1572 (3)
	0.16941 (20)	0.31319 (21)	0.69278 (24)		0.2104 (7)	0.4344 (7)	0.1567 (7)
	0.17104 (12)	0.31240 (13)	0.69221 (12)		0.2158 (4)	0.4333 (4)	0.1548 (4)
	0.17315 (21)	0.31260 (19)	0.69123 (23)		0.2206 (9)	0.4318 (7)	0.1542 (8)
	0.18522 (37)	0.31478	0.68522		0.2594	0.4178	0.1431
K (2)	0.03815 (10)	0.05562 (12)	0.06101 (10)	O (5)	0.3054 (3)	0.9308 (3)	0.3384 (3)
	0.03830 (24)	0.05571 (27)	0.06070 (26)		0.3042 (8)	0.9315 (7)	0.3391 (8)
	0.03993 (15)	0.05457 (16)	0.06013 (12)		0.3028 (4)	0.9343 (5)	0.3382 (4)
	0.04203 (25)	0.05334 (24)	0.05926 (23)		0.2981 (7)	0.9418 (8)	0.3349 (6)
	0.05282 (39)	0.05282	0.05282		0.2786 (4)	0.9616 (7)	0.3165 (4)
Cd (1)	0.30681 (2)	0.34105 (3)	0.33136 (3)	O (6)	0.3909 (3)	0.1427 (3)	0.2881 (3)
	0.30791 (5)	0.34086 (5)	0.33162 (5)		0.3912 (9)	0.1435 (7)	0.2879 (9)
	0.30957 (3)	0.34058 (3)	0.33177 (3)		0.3937 (6)	0.1445 (5)	0.2843 (5)
	0.31304 (6)	0.33944 (4)	0.33198 (6)		0.3968 (9)	0.1460 (6)	0.2796 (7)
	0.32987 (9)	0.32987	0.32987		0.4178 (6)	0.1431 (4)	0.2594 (7)
Cd (2)	0.39831 (2)	0.09039 (3)	0.92580 (3)	O (7)	0.5176 (3)	0.9550 (3)	0.2517 (4)
	0.39920 (5)	0.08996 (5)	0.92526 (5)		0.5214 (8)	0.9532 (8)	0.2526 (11)
	0.40030 (3)	0.08974 (3)	0.92451 (3)		0.5168 (4)	0.9531 (5)	0.2529 (6)
	0.40262 (6)	0.08920 (5)	0.92298 (5)		0.5129 (8)	0.9499 (8)	0.2476 (9)
	0.41305 (9)	0.08695	0.91305		0.4904	0.9303	0.2423
S (1)	0.21410 (8)	0.37222 (8)	0.02556 (9)	O (8)	0.3284 (4)	0.9881 (4)	0.1161 (3)
	0.21523 (15)	0.37238 (15)	0.02501 (18)		0.3265 (8)	0.9908 (8)	0.1132 (7)
	0.21600 (8)	0.37225 (9)	0.02394 (9)		0.3278 (5)	0.9923 (5)	0.1128 (4)
	0.21810 (17)	0.37313 (15)	0.02209 (18)		0.3245 (8)	0.9941 (8)	0.1122 (6)
	0.22678 (8)	0.37504 (8)	0.01178 (9)		0.3216	0.0053	0.0966

continued

Table 181, continued

Atom	x	y	z	Atom	x	y	z
S (2)	0.38451 (8)	0.00655 (9)	0.24697 (9)	O (9)	0.0561 (3)	0.0940 (3)	0.3170 (3)
	0.38351 (15)	0.00769 (15)	0.24599 (17)		0.0559 (8)	0.0915 (7)	0.3186 (9)
	0.38339 (8)	0.00793 (9)	0.24467 (9)		0.0540 (6)	0.0951 (5)	0.3184 (5)
	0.38141 (19)	0.00895 (15)	0.24202 (17)		0.0441 (9)	0.0939 (7)	0.3187 (9)
	0.37504	0.01178	0.22678		0.0053	0.0966	0.3216
S (3)	0.00036 (8)	0.21604 (8)	0.36900 (8)	O (10)	0.9031 (3)	0.1885 (4)	0.4702 (3)
	0.00051 (15)	0.21700 (15)	0.36951 (17)		0.9052 (11)	0.1896 (12)	0.4721 (9)
	0.00129 (9)	0.21723 (9)	0.36943 (9)		0.9059 (5)	0.1903 (7)	0.4731 (5)
	0.00030 (17)	0.21873 (14)	0.37001 (19)		0.9089 (9)	0.1979 (12)	0.4736 (9)
	0.01178	0.22678	0.37504		0.9303 (6)	0.2423 (8)	0.4904 (5)
O (1)	0.3056 (3)	0.2628 (3)	0.0336 (3)	O (11)	0.9353 (3)	0.2817 (4)	0.2570 (4)
	0.3037 (10)	0.2626 (8)	0.0294 (10)		0.9339 (9)	0.2828 (9)	0.2603 (9)
	0.3076 (5)	0.2638 (5)	0.0272 (5)		0.9376 (6)	0.2868 (6)	0.2629 (5)
	0.3067 (10)	0.2645 (7)	0.0189 (9)		0.9417 (10)	0.2928 (9)	0.2654 (9)
	0.3165	0.2786	0.9616		0.9616	0.3165	0.2786
O (2)	0.0824 (3)	0.3259 (3)	0.9880 (3)	O (12)	0.1063 (3)	0.2980 (4)	0.4217 (3)
	0.0829 (7)	0.3286 (8)	0.9885 (9)		0.1074 (6)	0.2966 (7)	0.4240 (9)
	0.0832 (4)	0.3274 (5)	0.9899 (5)		0.1091 (4)	0.2955 (5)	0.4235 (5)
	0.0848 (7)	0.3261 (7)	0.9909 (8)		0.1155 (9)	0.2896 (10)	0.4230 (8)
	0.0966 (4)	0.3216 (5)	0.0053 (6)		0.1431	0.2594	0.4178
O (3)	0.2593 (3)	0.4634 (3)	0.9226 (3)				
	0.2607 (7)	0.4640 (7)	0.9266 (8)				
	0.2599 (5)	0.4644 (5)	0.9250 (5)				
	0.2582 (10)	0.4659 (8)	0.9223 (7)				
	0.2423	0.4904	0.9303				

Table 182. $K_2Cd_2(SO_4)_3$. Temperature factors at the same temperatures as those in Table 181 [78A1]. B_{ij} is defined by Eq. (a) in Introduction. The values are in 10^{-5} units.

Atom	B_{11}	B_{22}	B_{33}	B_{12}	B_{13}	B_{23}
K (1)	396 (7)	408 (7)	411 (7)	−98 (11)	−66 (11)	32 (11)
	471 (14)	454 (13)	468 (17)	−183 (18)	−72 (20)	20 (20)
	473 (8)	519 (8)	594 (9)	−133 (11)	−72 (11)	33 (12)
	500 (20)	594 (16)	636 (23)	−30 (12)	−36 (14)	4 (13)
	775 (23)	775	775	−80 (46)	−80	−80
K (2)	476 (8)	603 (8)	387 (7)	−87 (13)	94 (12)	−80 (13)
	515 (15)	722 (17)	408 (17)	−192 (23)	89 (23)	−96 (24)
	596 (9)	781 (11)	465 (8)	−140 (14)	83 (12)	−91 (13)
	677 (25)	805 (20)	515 (26)	−67 (15)	94 (16)	−2 (15)
	822 (26)	822	822	21 (50)	21	21
Cd (1)	214 (2)	240 (2)	247 (2)	19 (3)	−40 (3)	17 (3)
	252 (3)	259 (3)	245 (3)	17 (4)	−44 (5)	20 (5)
	281 (2)	295 (2)	335 (2)	4 (3)	−50 (3)	6 (3)
	388 (7)	377 (4)	450 (7)	−12 (3)	−39 (4)	−10 (3)
	527 (5)	527	527	−154 (11)	−154	−154
Cd (2)	257 (2)	266 (2)	246 (2)	3 (3)	−52 (3)	−23 (3)
	292 (3)	301 (3)	234 (3)	2 (5)	−62 (5)	−23 (5)
	308 (2)	337 (2)	320 (2)	7 (3)	−61 (3)	−29 (3)
	401 (7)	400 (5)	391 (7)	0 (3)	−38 (4)	−25 (3)
	458 (4)	458	458	−40 (10)	−40	−40

continued

Table 182, continued

Atom	B_{11}	B_{22}	B_{33}	B_{12}	B_{13}	B_{23}
S (1)	180 (6)	194 (6)	229 (6)	50 (9)	8 (9)	−9 (9)
	249 (10)	220 (9)	150 (11)	43 (13)	−26 (14)	−28 (14)
	227 (6)	248 (6)	256 (6)	73 (8)	24 (7)	−34 (8)
	228 (17)	273 (12)	225 (18)	35 (8)	−32 (10)	−24 (9)
	377 (6)	358 (6)	315 (6)	167 (10)	10 (9)	50 (9)
S (2)	213 (6)	217 (6)	181 (6)	−34 (9)	−28 (9)	1 (10)
	272 (10)	218 (9)	178 (11)	−1 (14)	−39 (15)	33 (15)
	267 (6)	254 (6)	257 (5)	−24 (8)	−82 (8)	15 (9)
	312 (18)	219 (12)	300 (21)	−9 (9)	−100 (11)	17 (9)
	358	315	377	50	167	10
S (3)	187 (6)	211 (6)	210 (6)	−22 (9)	−49 (9)	−65 (9)
	207 (10)	209 (9)	206 (11)	−26 (13)	−39 (14)	−93 (14)
	256 (6)	248 (6)	250 (5)	−15 (8)	−60 (8)	−88 (8)
	242 (17)	238 (11)	261 (19)	−8 (8)	−28 (11)	−40 (9)
	315	377	358	10	50	167
O (1)	419 (23)	453 (25)	503 (27)	−511 (38)	246 (43)	−88 (43)
	937 (66)	494 (48)	409 (58)	−703 (78)	281 (97)	5 (76)
	644 (30)	438 (26)	731 (33)	−593 (40)	189 (51)	−21 (46)
	989 (93)	350 (55)	1198 (102)	−310 (52)	496 (77)	−198 (50)
	839	924	2346	817	493	−1181
O (2)	327 (23)	445 (25)	479 (26)	211 (38)	−21 (40)	380 (41)
	344 (42)	563 (48)	473 (57)	269 (67)	145 (68)	340 (76)
	294 (25)	604 (29)	727 (31)	159 (39)	−2 (38)	562 (45)
	313 (63)	728 (68)	1117 (100)	128 (49)	−49 (58)	445 (64)
	438 (32)	1245 (42)	2132 (64)	−514 (62)	569 (82)	−1632 (86)
O (3)	536 (27)	337 (25)	463 (26)	−53 (41)	128 (45)	381 (41)
	381 (43)	405 (43)	575 (53)	−11 (58)	269 (71)	587 (65)
	577 (31)	501 (28)	623 (27)	−8 (43)	217 (45)	657 (41)
	1285 (110)	774 (74)	287 (85)	48 (62)	163 (74)	282 (55)
	2261	1208	1981	432	684	2334
O (4)	408 (23)	420 (25)	274 (22)	122 (39)	−6 (38)	87 (38)
	514 (45)	487 (45)	214 (46)	177 (66)	202 (64)	21 (67)
	625 (30)	541 (29)	325 (25)	283 (42)	246 (35)	205 (38)
	1138 (102)	403 (61)	913 (101)	172 (54)	431 (73)	190 (53)
	2127	1356	513	−371	−1043	−275
O (5)	309 (22)	584 (28)	307 (22)	14 (40)	−258 (39)	−224 (39)
	498 (47)	475 (44)	355 (48)	−49 (69)	−256 (76)	−248 (67)
	337 (25)	819 (34)	455 (25)	−155 (43)	−179 (36)	−366 (44)
	492 (70)	1138 (79)	30 (63)	−336 (51)	−146 (43)	−114 (47)
	924 (37)	2346 (73)	839 (39)	−1181 (81)	817 (54)	493 (83)
O (6)	541 (27)	325 (24)	478 (27)	209 (41)	−81 (46)	−274 (42)
	681 (57)	265 (42)	777 (68)	387 (68)	−190 (94)	−590 (70)
	740 (37)	311 (28)	858 (36)	48 (47)	−308 (55)	−503 (45)
	920 (88)	320 (50)	568 (71)	1 (46)	−71 (58)	−57 (37)
	1356 (59)	513 (34)	2127 (72)	−275 (73)	−371 (132)	−1043 (76)
O (7)	196 (21)	529 (25)	686 (28)	333 (37)	278 (42)	96 (52)
	463 (53)	519 (51)	613 (66)	184 (76)	349 (88)	89 (92)
	288 (27)	645 (31)	869 (37)	362 (41)	276 (45)	97 (59)
	366 (76)	856 (67)	1072 (92)	274 (55)	269 (59)	9 (66)
	1208	1981	2261	2334	432	684

continued

Table 182, continued

Atom	B_{11}	B_{22}	B_{33}	B_{12}	B_{13}	B_{23}
O (8)	672 (30)	473 (27)	301 (25)	−479 (45)	177 (43)	53 (42)
	725 (53)	681 (53)	124 (45)	−674 (76)	264 (68)	−170 (69)
	756 (33)	702 (34)	242 (23)	−537 (50)	175 (38)	−125 (41)
	628 (72)	954 (68)	129 (66)	−425 (58)	139 (49)	42 (43)
	1245	2132	438	−1632	−514	569
O (9)	492 (24)	329 (23)	455 (26)	341 (39)	67 (42)	−139 (40)
	545 (49)	273 (44)	628 (63)	276 (63)	−5 (84)	−211 (74)
	947 (40)	457 (31)	567 (32)	532 (50)	−7 (55)	−246 (46)
	1311 (108)	263 (52)	1160 (104)	245 (54)	124 (82)	−231 (52)
	2132	438	1245	569	−1632	−514
O (10)	538 (27)	626 (30)	450 (28)	−472 (47)	−402 (45)	−80 (48)
	840 (69)	1037 (87)	253 (57)	−419 (117)	−666 (84)	−90(101)
	576 (31)	1169 (51)	464 (29)	−610 (58)	−499 (41)	−61 (59)
	707 (90)	1909 (144)	451 (89)	−58 (98)	−381 (68)	−123 (89)
	1981 (61)	2261 (92)	1208 (45)	684 (146)	2334 (71)	432(113)
O (11)	347 (24)	643 (30)	526 (27)	64 (44)	188 (45)	−483 (48)
	637 (60)	701 (56)	490 (56)	12 (83)	286 (83)	−813 (75)
	685 (36)	760 (34)	672 (30)	159 (52)	369 (46)	−868 (42)
	942 (107)	1097 (96)	827 (105)	217 (72)	366 (77)	−538 (72)
	2346	839	924	493	−1181	817
O (12)	324 (23)	592 (28)	491 (28)	408 (41)	−92 (45)	190 (47)
	168 (37)	589 (48)	666 (59)	362 (59)	−170 (71)	246 (82)
	401 (26)	743 (32)	653 (30)	611 (40)	−14 (43)	306 (51)
	529 (77)	1134 (101)	829 (102)	359 (64)	−198 (64)	235 (65)
	513	2127	1356	−1043	−275	−371

Nr. 24A-11 $Rb_2Cd_2(SO_4)_3$, Rubidium cadmium sulfate

($M=683.93$)

1a	Ferroelectricity was found by Hikita et al. in 1976.	76H7
b	phase: III, II, I state: F, F, P crystal system: triclinic, monoclinic, cubic space group: P1-C_1^1, $P2_1$-C_2^2, $P2_13$-T^4 Θ (III/II): −170; (II/I): −144 °C	77Y3
2a	Crystal growth: evaporation method.	
b	Solubility in water: Fig. 793.	
3a	Unit cell parameters: $a=10.3482$ (5) Å, $b=10.3430$ (5) Å, $c=10.3475$ (5) Å, $\beta=90.040$ (6)° at −150 °C (phase II) $a=10.3392$ (5) Å, $b=10.3330$ (5) Å, $c=10.3382$ (5) Å, $\alpha=89.988$ (6)°, $\beta=90.038$ (6)°, $\gamma=89.979$ (6)° at −190 °C (phase III).	77Y3
4	Lattice distortion: Fig. 794.	
5a	Dielectric constant: Fig. 795.	
c	Spontaneous polarization: Fig. 796.	

phase	III	II	I
state	F	F	P
crystal system	triclinic	monoclinic	cubic
space group	P1-C_1^1	$P2_1$-C_2^2	$P2_13$-T^4
Θ	−170	−144 °C	

continued

Nr. 24A-11 $Rb_2Cd_2(SO_4)_3$, continued

6a	Heat capacity: Fig. 785.	
15a	Domain structures: see	79Y2

Nr. 24A-12 $Tl_2Cd_2(SO_4)_3$, Thallium cadmium sulfate
($M = 921.73$)

1a	Dielectric anomalies and ferroelectricity were found by Březina et al. in 1972.	72B16

b

phase	IV	III	II	I	
state	nonpolar	F	F	P	
crystal system	orthorhombic	triclinic	monoclinic	cubic	
space group	$P2_12_12_1$-D_2^4	P1-C_1^1	$P2_1$-C_2^2	$P2_13$-T^4	75I2
Θ	−175	−146	−143 °C		72B16
	−181	−153	−145		75I2

2a	Crystal growth: evaporation method. Phase diagram of Tl_2SO_4–$CdSO_4$–H_2O system: Fig. 797.	72B16
b	Solubility: Fig. 793.	
3	Unit cell parameters: Table 180.	
5a	Dielectric constant: Fig. 798.	
c	Spontaneous polarization: Fig. 799.	
6a	Heat capacity: Figs. 785, 800. Transition entropy (in J mol^{-1} K^{-1}): ΔS_m(IV–III) = 5.73, ΔS_m(III–II) = 1.71, ΔS_m(II–I) = 2.01.	75F7
8a	Elastic compliance: Figs. 801, 802. Elastic constants determined by Brillouin scattering measurement (in 10^{10} N m^{-2}): $c_{11} = 7.80$, $c_{12} = 3.06$, $c_{44} = 1.95$.	75S1

Nr. 24A-13 $(NH_4)_2Cd(SO_4)_3$, Ammonium cadmium sulfate
($M = 549.07$)

1a	Ferroelectricity was discovered by Jona and Pepinsky in 1956.	56J1

b

phase	II	I	
state	F[a])	P[a])	[a]) 56J1
crystal system	monoclinic	cubic[a])	
space group	$P2_1$-C_2^2 [c])	$P2_13$-T^4 [b])	[b]) 57E1
Θ_f	−184 °C[a])		[c]) 73G9
	−178[b])		
	−181[c])		

$\varrho = 3.28 \cdot 10^3$ kg m^{-3} [a])
$P_s \parallel [010]$ [c])

2a	Crystal growth: evaporation method.	62j1
b	Phase diagram of $(NH_4)_2SO_4$–$CdSO_4$–H_2O system: Fig. 803. Crystal habit: Fig. 804.	

3a	Unit cell parameters: Table 180.	
b	$Z=4$. Crystal structure: Tables 183, 184; Fig. 805.	57E1
4	Thermal expansion: Fig. 806.	
5a	Dielectric constant: Figs. 807, 808. Pressure effect on the ferroelectric transition temperature: $(\mathrm{d}\Theta_f/\mathrm{d}p)_{p\to 0}=3.35\cdot 10^{-8}$ K Pa^{-1}. Fig. 809.	72G10
c	Spontaneous polarization: Fig. 810. Coercive field: Fig. 811.	
6a	Transition heat and entropy: Table 185. Heat capacity: Fig. 812; see also	73F8
7b	Electrostrictive constant: Fig. 813.	
8a	Elastic compliance: Fig. 814. Elastic constant determined by Brillouin scattering measurement (in 10^{10} N m^{-2}): $c_{11}=7.66$, $c_{12}=2.95$, $c_{44}=2.04$.	75S1
9a	Birefringence: Fig. 815; see also	73G9
b	Electrooptic effect: Fig. 816; see also	64B10
c	Piezooptic constant: $\lvert p_{11}\rvert=0.175$ (5), $\lvert p_{21}\rvert$ or $\lvert p_{31}\rvert=0.166$ (5), $\lvert p_{55}\rvert=0.025$ for $\lambda=633$ nm determined by Dixon-Cohen method.	74K23
d	Optical activity: Figs. 817, 818.	
10a	Raman scattering: Fig. 819; see also	76F2
13a	NMR: Fig. 820.	
b	Electron paramagnetic resonance: Table 186.	

Tables 183, 185, 186, see p. 150; Table 184, see p. 151.

24B Solid solutions

Nr. 24B-1 $Tl_2Cd_2(SO_4)_3$-$(NH_4)_2Cd_2(SO_4)_3$

1b	Transition temperature: Fig. 821.	
2b	Distribution curve for water solution: Fig. 822.	
3a	Unit cell parameter: Fig. 822.	
5a	Dielectric constant: Fig. 823.	
8a	Elastic compliance: Figs. 824, 825.	

Nr. 24B-2 $(NH_4)_2Mn_2(SO_4)_3$-$(NH_4)_2Cd_2(SO_4)_3$

3a	$(NH_4)_2(Cd_{1-x}Mn_x)_2(SO_4)_3$, x = 0.580 (25): a = 10.262 (8) Å.	75H7
b	Space group: $P2_13$-T^4 $Z=4$ Accommodation of Mn in Cd sites: site 1: 0.82 (11) Mn; site 2: 0.38 (11) Mn.	75H7

Table 183. $(NH_4)_2Cd_2(SO_4)_3$. Atomic coordinates and temperature parameters at RT [75H7]. U_{ij} (in Å^2) is defined by Eq. (d) in Introduction.

Atom	Position type	x	y	z	U_{11}	U_{22}	U_{33}	U_{12}	U_{13}	U_{23}
Cd (1)	4b	0.5806 (1)	0.5806 (1)	0.5806 (1)	0.013 (1)	0.013 (1)	0.013 (1)	−0.0008 (8)	0.0008 (8)	0.0008 (8)
Cd (2)	4b	0.8368 (1)	0.8386 (1)	0.8368 (1)	0.013 (1)	0.013 (1)	0.013 (1)	−0.0012 (9)	0.0012 (9)	0.0012 (9)
N (1)	4b	0.065 (2)	0.065 (2)	0.065 (2)	0.022 (8)	0.022 (8)	0.022 (8)	−0.0010 (10)	0.0010 (10)	0.0010 (10)
N (2)	4b	0.301 (2)	0.301 (2)	0.301 (2)	0.028 (10)	0.028 (10)	0.028 (10)	−0.0010 (10)	0.0010 (10)	0.0010 (10)
S	12a	0.6253 (2)	0.4745 (2)	0.2603 (2)	0.0111 (9)	0.010 (1)	0.009 (1)	0.0022 (7)	−0.0004 (8)	−0.0009 (8)
O (1)	12a	0.6588 (7)	0.5002 (7)	0.3959 (6)	0.040 (3)	0.046 (4)	0.022 (3)	−0.001 (4)	−0.001 (3)	−0.015 (3)
O (2)	12a	0.7406 (7)	0.4963 (7)	0.1834 (7)	0.026 (3)	0.047 (3)	0.032 (4)	0.005 (3)	0.022 (3)	0.008 (3)
O (3)	12a	0.5788 (6)	0.3423 (6)	0.2451 (6)	0.033 (3)	0.014 (3)	0.047 (4)	−0.005 (3)	−0.020 (3)	0.004 (3)
O (4)	12a	0.5257 (6)	0.5622 (6)	0.2152 (6)	0.024 (3)	0.030 (4)	0.047 (4)	0.010 (3)	−0.012 (3)	−0.001 (3)

Table 185. $(NH_4)_2Cd_2(SO_4)_3$. Transition heat and entropy per mol [62S14].

ΔQ_m J mol^{-1}	ΔS_m J K^{-1} mol^{-1}	Ref.
907	9.91	73F8
903 (8)	9.87	62S14

Table 186. $(NH_4)_2Cd_2(SO_4)_3$. Spin Hamiltonian parameters of Mn^{2+} [75H7]. A, D and a–F: in $10^{-2}\,m^{-1}$.

Parameter	Site I	Site II
$g_{\|\|}$	1.992 (1)	1.992 (1)
$\|A\|$	90 (3)	90 (3)
D	−159 (5)	−151 (5)
$(a–F)$	4.0 (1)	4.5 (1)

Table 184. $(NH_4)_2Cd_2(SO_4)_3$. Interatomic distances and bond angles at RT [75H7].

Atom or group	Bond	Distance Å	Bonds	Angle deg
SO_4 group	S–O (1)	1.472 (7)	O (1)–S–O (2)	107.5 (4)
	–O (2)	1.454 (8)	–O (3)	110.5 (4)
	–O (3)	1.460 (7)	–O (4)	111.3 (4)
	–O (4)	1.452 (7)	O (2)–S–O (3)	110.9 (4)
			–O (4)	108.1 (4)
			O (3)–S–O (4)	108.5 (4)
NH_4(I) group and environment	N (1)–O (1) 11 × 3	2.97 (1)		
	–O (2) 4 × 3	3.22 (1)		
	–O (2) 6 × 3	3.37 (1)		
	–O (3) 4 × 3	3.36 (1)		
	–O (4) 4 × 3	3.22 (1)		
NH_4(II) group and environment	N (2)–O (2) 6 × 3	3.18 (1)		
	–O (3) × 3	2.97 (1)		
	–O (4) 6 × 3	3.06 (1)		
Cd (1) environment	Cd (1)–O (1) × 3	2.239 (7)	O (1)–Cd (1)–O (1) 2	97.3 (3)
	–O (2) 5 × 3	2.281 (8)	–O (2) 5	80.5 (3)
			–O (2) 10	88.6 (3)
			–O (2) 9	174.0 (3)
			O (2) 5–Cd (1)–O (2) 9	93.9 (3)
Cd (2) environment	Cd (2)–O (3) 5 × 3	2.261 (7)	O (3) 5–Cd (2)–O (3) 9	97.7 (2)
	–O (4) 4 × 3	2.283 (6)	–O (4) 4	89.4 (2)
			–O (4) 8	86.9 (2)
			–O (4) 12	170.9 (2)
			O (4) 4–Cd (2)–O (4) 8	85.3 (2)

25 $NaNH_4SO_4 \cdot 2H_2O$ (lecontite) family

25A Pure compounds

Nr. 25A-1 $NaNH_4SO_4 \cdot 2H_2O$, Sodium ammonium sulfate dihydrate (lecontite) ($M = 173.12$)

1a Ferroelectric activity in $NaNH_4SO_4 \cdot 2H_2O$ was discovered by Pepinsky et al. in 1959. — 59P2

b — 65M4

phase	II	I
state	F	P
crystal system		orthorhombic
space group		$P2_12_12_1$-D_2^4
Θ	101 K	

$P_s \parallel [001]$.
Dehydration point is 338 K. — 67M1
$\varrho = 1.745 \cdot 10^3$ kg m^{-3}. — 67C7
Transparent, colorless.
Efflorescent.

2a Crystal growth: evaporation or cooling method from an aqueous solution of the constituents with equi-molar ratio of Na_2SO_4 and $(NH_4)_2SO_4$ or with excess $(NH_4)_2SO_4$. — 65M4

3a Unit cell parameters: $a = 8.216$ Å, $b = 12.854$ Å, $c = 6.232$ Å at RT. — 67C7

b $Z = 4$ in phase I. — 65M4
Crystal structure: Table 187, 188, 189; Figs. 826, 827.

4 Lattice distortions: Figs. 828, 829. continued

Nr. 25A-1 $NaNH_4SO_4 \cdot 2\,H_2O$, continued

5a	Dielectric constants: Figs. 830, 831.	
	$C=300$ K.	65M4
	$d\Theta/dE=2.9 \cdot 10^{-4}$ K m kV^{-1}.	78O6
	A dielectric anomaly was reported to occur at 92 K below the ferroelectric transition temperature.	65M4
c	Spontaneous polarization: Fig. 832.	
6a	Heat capacity: transition heat and transition entropy: $\Delta Q_m=326$ J mol^{-1}, $\Delta S_m=3.30$ J K^{-1} mol^{-1}. See also Fig. 833.	78O6
9a	Refractive indices: $n_a=1.4534$, $n_b=1.4554$, $n_c=1.4386$ for $\lambda=632.8$ nm, $T=296$ K. Fig. 834.	75A11
b	Electrooptic constants: $r_{41}=0.37 \cdot 10^{-12}$ m V^{-1}, $r_{52}=0.11 \cdot 10^{-12}$ m V^{-1}, $r_{63}=1.3 \cdot 10^{-12}$ m V^{-1} for $\lambda=632.8$ nm, $T=296$ K.	75A11
10a	Raman scattering: Fig. 835.	
13a	NMR: Table 190; Figs. 836, 837; see also	66E2

Table 187. $NaNH_4SO_4 \cdot 2H_2O$ (lecontite). Crystal structure of phase I [67C7]. Fractional coordinates at RT.

Atom	x	y	z
S	0.0842 (2)	0.1277 (1)	0.3733 (6)
Na	0.2652 (4)	0.4847 (2)	0.9133 (8)
O (1)	0.0690 (9)	0.2001 (6)	0.1890 (28)
O (2)	0.0617 (12)	0.1894 (3)	0.5740 (27)
O (3)	0.2450 (10)	0.0807 (5)	0.3719 (24)
O (4)	0.0392 (10)	0.5455 (5)	0.1380 (20)
O (5)	0.2115 (7)	0.6350 (4)	0.6949 (16)
O (6)	0.0793 (6)	0.4039 (4)	0.6625 (19)
N	0.1719 (8)	0.8541 (5)	0.3678 (24)
H (1)	0.910	0.145	0.700
H (2)	0.710	0.190	0.780
H (3)	0.570	0.175	0.370
H (4)	0.480	0.090	0.360
H (5)	0.395	0.200	0.855
H (6)	0.410	0.060	0.860
H (7)	0.230	0.160	0.990
H (8)	0.240	0.160	0.780

Table 188. $NaNH_4SO_4 \cdot 2H_2O$ (lecontite). Crystal structure of phase I [67C7]. Temperature factors at RT. B_{ij} is defined by Eq. (a) in Introduction.

Atom	B_{11}	B_{22}	B_{33}	B_{12}	B_{13}	B_{23}
S	0.0048 (1)	0.0026 (0)	0.0069 (16)	−0.0010 (3)	0.0004 (10)	0.0001 (6)
Na	0.0091 (2)	0.0041 (1)	0.0000 (24)	−0.0017 (7)	−0.0005 (19)	0.0009 (12)
O (1)	0.0146 (5)	0.0048 (4)	0.0245 (78)	−0.0013 (22)	−0.0011 (63)	0.0127 (45)
O (2)	0.0171 (9)	0.0055 (1)	0.0191 (66)	−0.0080 (16)	0.0132 (63)	−0.0074 (24)
O (3)	0.0051 (1)	0.0055 (3)	0.0386 (71)	0.0007 (15)	−0.0073 (41)	0.0007 (36)
O (4)	0.0066 (4)	0.0043 (3)	0.0108 (51)	0.0038 (16)	0.0017 (42)	0.0035 (30)
O (5)	0.0103 (6)	0.0038 (2)	0.0041 (41)	0.0016 (14)	0.0019 (39)	−0.0051 (20)
O (6)	0.0073 (3)	0.0048 (2)	0.0164 (50)	−0.0005 (14)	−0.0024 (36)	−0.0042 (24)
N	0.0108 (3)	0.0030 (2)	0.0198 (62)	−0.0011 (15)	−0.0165 (38)	0.0010 (30)

Fig. p. 511

Table 189. $NaNH_4SO_4 \cdot 2H_2O$ (lecontite). Crystal structure of phase I [67C7]. Interatomic distances and angles at RT. See also Fig. 826.

Distances	Å	Distances	Å	Angles	deg
Na–O (4)	2.453(10)	S–O (1)	1.483(14)	O (4″)–NH_4–O (3′)	46.8 (2)
Na–O (5)	2.404(8)	S–O (2)	1.492(14)	O (4″)–NH_4–O (3)	85.6 (4)
Na–O (6)	2.468(9)	S–O (3)	1.452(8)	O (4″)–NH_4–O (3″)	84.3 (4)
Na–O (4′)	2.382(11)	S–O (4)	1.466(7)	O (4″)–NH_4–O (1)	127.3 (5)
Na–O (5′)	2.341(9)			O (4″)–NH_4–O (2)	125.7 (5)
Na–O (6′)	2.419(9)	Angles	deg	O (4″)–NH_4–O (1′)	99.7 (3)
		O (1)–S–O (2)	107.8 (6)	O (3′)–NH_4–O (3)	72.8 (7)
Angles	deg	O (1)–S–O (3)	109.4 (6)	O (3′)–NH_4–O (3″)	73.5 (7)
O (4)–Na–O (5)	85.9 (3)	O (1)–S–O (4)	110.9 (6)	O (3′)–NH_4–O (1)	94.5 (4)
O (4)–Na–O (6)	81.3 (3)	O (2)–S–O (3)	109.8 (7)	O (3′)–NH_4–O (2)	92.4 (3)
O (4)–Na–O (4′)	166.7 (7)	O (2)–S–O (4)	109.8 (6)	O (3′)–NH_4–O (1′)	146.1 (4)
O (4)–Na–O (5′)	81.0 (3)	O (3)–S–O (4)	109.1 (4)	O (3)–NH_4–O (3″)	141.7 (5)
O (4)–Na–O (6′)	91.6 (3)			O (3)–NH_4–O (1)	44.2 (3)
O (5)–Na–O (6)	89.1 (3)	Distances	Å	O (3)–NH_4–O (2)	119.3 (4)
O (5)–Na–O (4′)	81.2 (3)	NH_4–O (4″)	3.009(9)	O (3)–NH_4–O (1′)	116.1 (6)
O (5)–Na–O (5′)	165.3 (6)	NH_4–O (3′)	2.973(9)	O (3″)–NH_4–O (1)	122.7 (4)
O (5)–Na–O (6′)	82.2 (3)	NH_4–O (3)	3.322(20)	O (3″)–NH_4–O (2)	45.4 (4)
O (6)–Na–O (4′)	101.4 (3)	NH_4–O (3″)	3.274(20)	O (3″)–NH_4–O (1′)	102.0 (5)
O (6)–Na–O (5′)	82.4 (3)	NH_4–O (1)	3.004(16)	O (1)–NH_4–O (2)	80.9 (4)
O (6)–Na–O (6′)	169.1 (4)	NH_4–O (2)	2.908(16)	O (1)–NH_4–O (1′)	114.7 (7)
O (4′)–Na–O (5′)	112.2 (3)	NH_4–O (1′)	2.821(10)	O (2)–NH_4–O (1′)	108.3 (5)
O (4′)–Na–O (6′)	83.7 (3)				
O (5′)–Na–O (6′)	104.7 (3)				

Table 190. $NaND_4SO_4 \cdot 2D_2O$ (deuterated lecontite). Nuclear quadrupole coupling tensors $|eQ\phi_{ii}/h|$ of deuterons and their orientations [69G2]. $\theta_{x,y,z}$ and $\phi_{x,y,z}$ are spherical coordinates representing the orientations of three principal components (x, y, z) of nuclear quadrupole coupling tensors with respect to the crystallographic c and a axes, respectively.

T K	Type	$\|eQ\phi_{zz}/h\|(\theta_z, \phi_z)$ kHz	$\|eQ\phi_{yy}/h\|(\theta_y, \phi_y)$	$\|eQ\phi_{xx}/h\|(\theta_x, \phi_x)$
298	β–D_2O	125 (149°, −131°)	112 (103°, −17°)	13 (63°, −100°)
97	ND_4^+ equatorial deuterons	58 (90°, 53°)	29 (90°, −37°)	29 (0°, ···)
	ND_4^+ axial deuterons	176 (90°, 53°)	88 (90°, −37°)	88 (0°, ···)

Nr. 25A-2 $NaNH_4SeO_4 \cdot 2H_2O$, Sodium ammonium selenate dihydrate
(M = 220.02)

1a Ferroelectric activity in $NaNH_4SeO_4 \cdot 2H_2O$ was discovered by Aleksandrov et al. in 1969. 69A7

b

phase	II	I
state	F[a]	P[a]
crystal system	monoclinic[c]	orthorhombic[b]
space group	$P2_1$-C_2^2 [c]	$P2_12_12_1$-D_2^4 [b]
Θ	180 (1) K[b]	

[a] 69A7
[b] 71A8
[c] 76K18

Transparent, colorless. 71A8

continued

Nr. 25A-2 $NaNH_4SeO_4 \cdot 2\,H_2O$, continued

2a	Crystal growth: cooling or evaporation method from aqueous solution.	71A8
	Unit cell parameters:	
3a	$a=8.405$ Å, $b=13.137$ Å, $c=6.287$ Å at RT.	73K18
	$a=8.401$ Å, $b=13.091$ Å, $c=6.205$ Å, $\gamma=91°04'$ at 148 K.	76K18
b	$Z=4$ in phases I and II	73K18
	Crystal structure at RT: Figs. 838, 839; Tables 191···193.	
	Crystal structure at 148 K: Tables 194, 195.	
	For atomic positions of hydrogen atoms, see also	78N5
5a	Dielectric constants: Fig. 840.	
	Curie-Weiss constant $C=496$ K.	71Z1
	$d\Theta/dp = -4.1\,(2) \cdot 10^{-8}\,K \cdot Pa^{-1}$ ($p \lessapprox 4 \cdot 10^8$ Pa).	73G4
c	Spontaneous polarization and coercive field:	
	Fig. 841; see also	71A8
d	Pyroelectric effect: see	75Z5
6a	Heat capacity: Fig. 842.	
	Transition heat: $\Delta Q_m=669\,(21)\,J\,mol^{-1}$.	72A7
7a	Piezoelectricity: Fig. 843.	
	$d_{36}=10 \cdot 10^{-12}\,CN^{-1}$, $d_{25}=0.9 \cdot 10^{-12}\,CN^{-1}$, $d_{14}=0.42 \cdot 10^{-12}\,C\,N^{-1}$ at RT.	71Z1
8a	Elastic compliance and stiffness: Figs. 844···847.	
	$s_{66}=2.4 \cdot 10^{-12}\,m^2N^{-1}$, $s_{55}=2.56 \cdot 10^{-12}\,m^2N^{-1}$, $s_{44}=1.66 \cdot 10^{-12}\,m^2N^{-1}$ at RT.	71Z1
b	For nonlinear properties of the deuterated crystal, see	78S18
9a	Refractive indices: $n_a=1.4902$, $n_b=1.4943$, $n_c=1.4745$ for $\lambda=632.8$nm, $T=296$ K.	75A11
	Birefringence: Fig. 848.	
b	Electrooptic effect: Figs. 849, 850.	
	Electrooptic constants: $r_{41}=0.33 \cdot 10^{-12}\,mV^{-1}$, $r_{52}=0.07 \cdot 10^{-12}\,mV^{-1}$,	75A11
	$r_{63}=2.0 \cdot 10^{-12}\,mV^{-1}$ for $\lambda=6328$ nm, $T=296$ K.	
e	Nonlinear optical properties: Fig. 851.	
10a	Raman scattering: Fig. 852.	
13a	NMR: Table 196, 197; Fig. 853; see also	71Y6, 77A8

Table 192. $NaNH_4SeO_4 \cdot 2H_2O$. Crystal structure of phase I [73K18]. Anisotropic temperature parameters for non-hydrogen atoms at RT. The temperature parameters are defined by Eq. (a) in Introduction.

Atoms	B_{11}	B_{22}	B_{33}	B_{12}	B_{13}	B_{23}
Se	0.009	0.0036	0.017	−0.0014	−0.0002	0.0011
Na	0.009	0.0039	0.011	−0.0008	0.0019	0.0006
O 1	0.018	0.005	0.025	−0.001	0.008	−0.003
O 2	0.020	0.006	0.024	−0.005	−0.013	0.015
O 3	0.011	0.007	0.056	−0.004	−0.005	−0.005
O 4	0.121	0.006	0.015	0.001	0.000	0.000
O 5	0.129	0.005	0.014	0.002	−0.004	0.003
O 6	0.111	0.005	0.023	0.001	0.000	−0.002
N	0.015	0.005	0.021	0.000	0.002	0.004

Table 191. $NaNH_4SeO_4 \cdot 2H_2O$. Crystal structure of phase I [73K18]. Fractional coordinates and temperature parameters at RT. The temperature parameters are defined by Eq. (e) in Introduction.

Atoms	*x*	*y*	*z*	*B* Å^2
Se	0.0935	0.1284	0.3828	0.11
Na	0.2606	0.4894	0.9049	0.14
O 1	0.0783	0.2064	0.1873	1.75
O 2	0.0747	0.1907	0.6096	1.76
O 3	0.2642	0.0744	0.3730	1.98
O 4	0.0415	0.5410	0.1364	0.93
O 5	0.2050	0.6321	0.6749	0.53
O 6	0.0886	0.4042	0.6532	1.11
N	0.1596	0.8558	0.3772	0.91
H 1	0.382	0.200	0.867	
H 2	0.397	0.059	0.871	
H 3	0.217	0.160	1.024	
H 4	0.226	0.160	0.792	
H 5	0.027	0.922	0.858	
H 6	0.094	0.330	0.642	
H 7	0.281	0.688	0.679	
H 8	0.109	0.651	0.750	

Table 193. $NaNH_4SeO_4 \cdot 2H_2O$. Crystal structure of phase I [73K18]. Interatomic distances and angles at RT. See also Fig. 838.

Na–O 5′	2.23 Å	Se–O 1	1.629 Å	∠O 1′NH$_4$O 2	117°
Na–O 4	2.37	Se–O 2	1.643	O 1 NH$_4$O 4′	66
Na–O 5	2.41	Se–O 3	1.591	O 1 NH$_4$O 1	115
Na–O 6	2.43	Se–O 4	1.636	O 1′NH$_4$O 3′	104
Na–O 4′	2.44			O 1′ NH$_4$O 3	91
Na–O 6′	2.46	∠O 1 SeO 2	110°09′	O 1′ NH$_4$O 3″	98
		O 1 SeO 3	110°00′	O 2 NH$_4$O 4′	115
∠O 5′NaO 4	101°	O 1 SeO 4	108°53′	O 2 NH$_4$O 1	78
O 5′NaO 5	157	O 2 SeO 3	109°59′	O 2 NH$_4$O 3′	73
O 5′NaO 6	103	O 2 SeO 4	110°49′	O 2 NH$_4$O 3	115
O 5′NaO 4′	99	O 3 SeO 4	108°10′	O 2 NH$_4$O 3″	53
O 5′NaO 6′	83			O 4′NH$_4$O 1	118
O 4 NaO 5	81	NH$_4$–O 1′	2.80 Å	O 4 NH$_4$O 3′	95
O 4 NaO 6	83	NH$_4$–O 2	2.87	O 4 NH$_4$O 3	97
O 4 NaO 4′	170	NH$_4$–O 4′	2.99	O 4′NH$_4$O 3″	93
O 4 NaO 6′	101	NH$_4$–O 1	3.03	O 1 NH$_4$O 3′	98
O 5 NaO 6	82	NH$_4$–O 3′	3.04	O 1 NH$_4$O 3	49
O 5 NaO 4	90	NH$_4$–O 3	3.29	O 1 NH$_4$O 3″	122
O 5 NaO 6′	90	NH$_4$–O 3″	3.38	O 3′ NH$_4$O 3	100
O 6 NaO 4′	86			O 3′ NH$_4$O 3″	63
O 6 NaO 6′	172			O 3 NH$_4$O 3″	153
O 4′ NaO 6′	81				

Table 194. $NaNH_4SeO_4 \cdot 2H_2O$. Crystal structure of phase II [76K18]. Fractional coordinates and isotropic temperature parameters at 148 K. The temperature parameters are defined by Eq. (e) in Introduction.

Atom	x	y	z	B [Å^2]
Se (I)	0.34510 (13)	0.12812 (08)	0.38443 (59)	0.47
Se (II)	0.83987 (13)	0.37056 (08)	0.60884 (59)	0.47
Na (I)	0.5139 (06)	0.4913 (04)	0.8988 (11)	0.60
Na (II)	0.0089 (06)	0.0111 (04)	0.1002 (12)	0.63
O 1	0.3428 (11)	0.1835 (09)	0.6188 (22)	1.60
O 2	0.3159 (11)	0.2108 (08)	0.1931 (20)	1.31
O 3	0.2063 (09)	0.0369 (06)	0.3677 (19)	1.09
O 4	0.5157 (09)	0.0753 (08)	0.3291 (22)	2.49
O 5	0.8452 (11)	0.3002 (08)	0.8208 (21)	1.58
O 6	0.7997 (11)	0.3000 (09)	0.3942 (26)	2.27
O 7	0.7052 (08)	0.4582 (06)	0.6235 (18)	0.87
O 8	1.0129 (09)	0.4224 (07)	0.5659 (22)	1.64
O 9	0.4589 (10)	0.6313 (07)	0.6617 (15)	0.94
O 10	0.0504 (11)	0.1343 (07)	0.8258 (19)	1.07
O 11	0.1600 (08)	0.9062 (07)	0.8575 (20)	1.07
O 12	0.3324 (09)	0.4013 (07)	0.6532 (19)	1.63
NH_4(I)	0.0924 (11)	0.3582 (09)	0.1237 (22)	1.35
NH_4(II)	0.3956 (11)	0.8563 (09)	0.3938 (22)	1.12

Table 195. $NaNH_4SeO_4 \cdot 2H_2O$. Crystal structure of Phase II [76K18]. Interatomic distances and angles at 148 K. Primed atoms are connected with basis atoms by symmetry operations.

Distance [Å]		Angle [deg]		Distance [Å]		Angle [deg]	
Se (I) –O 1	1.625	O 1 Se (I) O 2	110°38′	Na (I) –O 9′	2.30	NH_4(I) –O 2	2.75
–O 2	1.628	O 1 Se (I) O 3	111°19′	–O 7	2.39	–O 5	2.89
–O 3	1.656	O 1 Se (I) O 4	113°09′	–O 9	2.40	–O 7′	2.92
–O 4	1.639	O 2 Se (I) O 3	108°43′	–O 7′	2.41	–O 8	2.95
		O 2 Se (I) O 4	105°45′	–O 12	2.44	–O 8′	3.04
		O 3 Se (I) O 4	107°00′	–O 12′	2.46	–O 6	3.07
Se (II)–O 5	1.607	O 5 Se (II) O 6	110°15′	Na (II)–O 3	2.37	NH_4(II)–O 1′	2.84
–O 6	1.655	O 5 Se (II) O 7	112°38′	–O 10	2.37	–O 3	2.88
–O 7	1.629	O 5 Se (II) O 8	109°58′	–O 3′	2.39	–O 5′	2.89
–O 8	1.614	O 6 Se (II) O 7	107°05′	–O 11′	2.41	–O 4′	2.93
		O 6 Se (II) O 8	106°28′	–O 10′	2.41	–O 4	3.05
		O 7 Se (II) O 8	110°12′	–O 11	2.41	–O 2′	3.19

Table 196. $NaNH_4SeO_4 \cdot 2H_2O$. Principal components of quadrupole coupling tensor $|eQ\phi_{ii}/h|$ of ^{23}Na and their orientations (shown in parentheses) [71A8]. Angles α, β, γ are angles between each principal component of nuclear quadrupole coupling tensors and the crystallographic a, b, c axes, respectively.

	$\|eQ\phi_{xx}/h\|$ kHz	(α, β, γ) deg	$\|eQ\phi_{yy}/h\|$ kHz	(α, β, γ) deg	$\|eQ\phi_{zz}/h\|$ kHz	(α, β, γ) deg	T °C
	190	(–17, –89, 73)	930	(90, –8, –97)	1120	(73, –97, 162)	–15
Na (I)	232	(–13, –90, 77)	752	(88, –8, –99)	984	(77, –99, 164)	–120
Na (II)	228	(–37, –96, 54)	980	(98, –172, 95)	1208	(54, –95, 144)	–120

Table 197. $NaND_4SeO_4 \cdot 2D_2O$. Deuteron quadrupole coupling constants $|eQ\phi_{ii}/h|$ and their orientations (shown in parentheses) [71A8]. Angles α, β, γ are angles between each principal component of nuclear quadrupole coupling tensors and the crystallographic a, b, c axes, respectively.

	$\|eQ\phi_{xx}/h\|$ kHz (α, β, γ) deg	$\|eQ\phi_{yy}/h\|$ kHz (α, β, γ) deg	$\|eQ\phi_{zz}/h\|$ kHz (α, β, γ) deg	T °C
ND_4	0.4 (90, 0, −90)	3.8 (−155, 90, −65)	4.2 (−65, 90, 25)	−15
ND_4(I)	0.0 (2) (···)	4.3 (−53, −135, 68)	4.3 (53, −45, −68)	−120
ND_4(II)	2.1 (99, −32, −121)	5.7 (139, 76, −52)	7.8 (51, −58, −54)	−120

26 Alum ($NH_4Fe(SO_4)_2 \cdot 12H_2O$) family

26A Pure compounds

Nr. 26A-1 $NH_4V(SO_4)_2 \cdot 12H_2O$, Ammonium vanadium sulfate dodecahydrate (M=477.28)

1a	Ferroelectric activity in $NH_4V(SO_4)_2 \cdot 12H_2O$ was discovered by Pepinsky et al. in 1957.	57P5
b	phase: II / I; state: F / P; crystal system: — / cubic; Θ: −157 °C $\varrho = 1.680 \cdot 10^3$ kg m^{-3} at RT. Transparent.	57P5
2a	Crystal growth: evaporation or cooling method from aqueous solution.	57P5
3a	Unit cell parameter: a=12.355 Å at 12 °C.	57P5
b	Z=4 in phase I.	65w1
5c	Spontaneous polarization and coercive field: $P_s = 1.0 \cdot 10^{-2}$ C m^{-2} at about 2 °C below Θ. $E_c = 12 \cdot 10^5$ V m^{-1} at about 2 °C below Θ (f=60 Hz).	57P5

Phase table (1b):

phase	II	I
state	F	P
crystal system		cubic
Θ	−157 °C	

Nr. 26A-2 $NH_4Fe(SO_4)_2 \cdot 12H_2O$, Ammonium iron sulfate dodecahydrate (M=482.18)

1a	Ferroelectric activity in $NH_4Fe(SO_4)_2 \cdot 12H_2O$ was discovered by Jona et al. in 1957.	57J5
b	phase: II[a]) / I[a]); state: F[a]) / P[a]); crystal system: — / cubic[b]); Θ: −185 °C[c]) For the deuterated crystal, Θ = −185 °C[c]). $\varrho = 1.713 \cdot 10^3$ kg m^{-3}. ($\varrho = 1.812 \cdot 10^3$ kg m^{-3} for the deuterated crystal). Transparent, colorless.	[a]) 57J5 [b]) 65w1 [c]) 57J5 62j1
2a	Crystal growth: evaporation or cooling method from aqueous solution. Solubility: Table 198.	57J2
3a	Unit cell parameter: a=12.318 Å at 25 °C (a=12.317 Å at 22 °C for the deuterated crystal). Z=4 in phase I. continued	65w1

Phase table (1b):

phase	II[a])	I[a])
state	F[a])	P[a])
crystal system		cubic[b])
Θ	−185 °C[c])	

Nr. 26A-2 $NH_4Fe(SO_4)_2$, continued

5a	Dielectric constants: Fig. 854; Table 199.	
b	Non-linear dielectric properties: Table 200.	
c	Spontaneous polarization and coercive field: $P_s=0.4\cdot10^{-2}$ C m^{-2} and $E_c=33\cdot10^5$ V m^{-1} at about 2 °C below Θ ($f=60$ Hz). ($P_s=0.4\cdot10^{-2}$ C m^{-2} and $E_c=30\cdot10^5$ Vm^{-1} at about 2 °C below Θ for the deuterated compound).	57J5
9a	Refractive indices: λ: 486, 589, 656 nm n: 1.4929, 1.4848, 1.4815 Ref. [31 w 1], cited in Landolt-Börnstein, 6th ed., II/8.	31w1
d	Optical activity: Table 203.	
13c	Mössbauer effect: Figs. 855···858.	
	The Mössbauer absorption spectra are observed for samples irradiated with protons of 1~5 MeV at varying doses.	75K18
	The Mössbauer spectra of partly magnetized crystals show a broadening and a shift of the hyperfine structure lines, reflecting the paramagnetic spin fluctuations.	73W4
	For the hyperfine-structure spectra, see also	65O1

Table 198. $NH_4Fe(SO_4)_2\cdot12H_2O$. Solubility, A, in water [62k1].

A [wt%]	11.1	16.7	20.0	24.8	28.5	34.2
T [°C]	0	15	25	40	50	70

Table 200. $NH_4Fe(SO_4)_2\cdot12H_2O$. ξ and ζ [57J5]. ξ, ζ: expansion coefficients of the free energy.

Method	From the thermodynamical relations	From the double hysteresis loop experiment
ξ	$-6.3\cdot10^{14}$ VC^{-3}m^5	$-2.7\cdot10^{14}$ VC^{-3} m^5
ζ	$3.9\cdot10^{19}$ VC^{-5} m^9	$1.9\cdot10^{19}$ VC^{-5} m^9

Table 199. $NH_4Fe(SO_4)_2\cdot12H_2O$. T_m, a, b vs. λ. T_m: T at which the maximum loss occurs. a, b: $\ln\tau=a+b/T$. τ: dielectric relaxation time [52G1].

	T_m[K]			a	b 10^3 K
	$\lambda=3.0\cdot10^{-2}$ m	$\lambda=1.18\cdot10^{-2}$ m	$\lambda=0.85\cdot10^{-2}$ m		
$CH_3NH_3Al(SO_4)_2\cdot12H_2O$	262	>300	>300	−30	1.3
$CH_3NH_3Cr(SO_4)_2\cdot12H_2O$	237	>300	>300	−30	1.2
$NH_4Fe(SO_4)_2\cdot12H_2O$	177	203	213	−32	1.3

Nr. 26A-3 $NH_4In(SO_4)_2\cdot12H_2O$, Ammonium indium sulfate dodecahydrate
($M=541.16$)

1a	Ferroelectric activity in $NH_4In(SO_4)_2\cdot12H_2O$ was discovered by Pepinsky et al. in 1957.	57P5
b	phase: II, I state: F, P crystal system: –, cubic Θ: −146 °C $\varrho=1.872\cdot10^3$ kg m^{-3} at RT. Transparent.	57P5

2a	Crystal growth: evaporation or cooling method from aqueous solution.	57P5
3a	Unit cell parameter: $a=12.427$ Å at 10 °C.	57P5
b	$Z=4$ in phase I.	65w1
5c	Spontaneous polarization and coercive field: $P_s=1.2\cdot10^2$ Cm^{-2} at about 2 °C below Θ. $E_c=10\cdot10^5$ Vm^{-1} at about 2 °C below Θ ($f=60$ Hz).	57P5

Nr. 26A-4 $CH_3NH_3Al(SO_4)_2\cdot12H_2O$, Methylammonium aluminum sulfate dodecahydrate (MASD)
($M=467.34$)

1a	Ferroelectric activity in $CH_3NH_3Al(SO_4)_2\cdot12H_2O$ was discovered by Pepinsky et al. in 1956.	56P2
b	(see table below)	a) 56P2 b) 57J2

phase	II[a]	I[a]
state	F[a]	P[a]
crystal system	monoclinic[b]	cubic[b]
space group	$P2_1$-C_2^2 [b]	$P2_13$-T^4 [b]
Θ	−96 °C[b]	

	Θ is insensitive to deuteration and remained −96 °C.[b]	
	Possibility of other space groups was proposed: $Pa3-T_h^6$ for phase I and $Pca2_1-C_{2v}^5$ for phase II.	64F5
	$P_s \parallel [100]$.	
	$\varrho=1.585\cdot10^3$ kg m^{-3} at 23 °C.	57J2
	Transparent.	
2a	Crystal growth: evaporation or cooling method from aqueous solution.	57J2
b	Crystal form: Fig. 859.	
3a	Unit cell parameter: $a=12.502$ (1) Å at 23 °C. See also Table 202.	57J2
b	$Z=4$ in phase I. Crystal structure: Table 201; Figs. 860, 861, 862.	57O1
4	Unit cell parameters: $a=12.462$ Å at −82 °C. $a=12.47_2$ Å, $b=12.32_7$ Å, $c=12.28_2$ Å, $\beta=90°40'$ at −165 °C.	57J2 57J2
5a	Dielectric constants: Figs. 863, 864. Dielectric relaxation: Figs. 865, 866. $[d\Theta/dp]_{p=0}=2.5(1)\cdot10^{-8}$ K Pa^{-1} ($p\lesssim7\cdot10^8$ Pa^{-1}, p: hydrostatic pressure).	73G4
b	Non-linear dielectric properties: expansion coefficients of the free energy. $\xi=-7.2\cdot10^{13}$ Vm^{5}C^{-3}, $\zeta=5.4\cdot10^{17}$ Vm^{9}C^{-5}.	57J2
c	Spontaneous polarization: Fig. 867. $P_s=1.0\cdot10^{-2}$ C m^{-2} at about 2 °C below Θ. Also for the deuterated crystal, $P_s=1.0\cdot10^{-2}$ C m^{-2}. Coercive field: $E_c=6\cdot10^5$ Vm^{-1} at about 2 °C below Θ ($f=60$ Hz). Also for the deuterated crystal, $E_c=6\cdot10^5$ Vm^{-1}.	57J2
6a	Heat capacity: Fig. 868. Transition entropy, transition heat: $\Delta Q_m=1676$ J mol^{-1}, $\Delta S_m=9.70$ J mol^{-1} K^{-1}.	68A4
8a	Sound velocity and elastic stiffness: Figs. 869, 870. continued	

Nr. 26A-4 $CH_3NH_3Al(SO_4)_2 \cdot 12H_2O$, continued

9a	Birefringence: Fig. 871. Refractive index: $n=1.458$ (2) at RT. $n=1.465$ (2) at RT for the deuterated crystal.	74Z2
b	Electrooptic effect: Fig. 872.	
c	Piezooptic constants: $p_{11}=0.177$ (11), $p_{12}=0.288$ (10), $p_{31}=0.245$ (10) at 22 °C, for $\lambda=632.8$ nm.	73M6
d	Optical activity: Table 203.	
13a	NMR: Fig. 873; see also	72V3
b	ESR: Table 204. ESR for γ-ray irradiated crystal: see	66K6

Table 201. $CH_3NH_3Al(SO_4)_2 \cdot 12H_2O$ (MASD). Crystal structure of phase I [57O1]. Fractional coordinates and isotropic temperature parameters are defined by Eq. (e) in Introduction. See also Fig. 862.

		x	y	z	B [Å^2]
4 Al	(a)	0.5000	0.5000	0.0000	1.00
8 S	(c)	0.3329	0.3329	0.3329	1.12
8 $O_{S(1)}$	(c)	0.2639	0.2639	0.2639	3.12
24 $O_{S(2)}$	(d)	0.2808	0.3491	0.4372	2.71
24 O (1)	(d)	0.5029	0.5024	0.1507	1.98
24 O (2)	(d)	0.2288	0.3331	0.0540	2.69
$4CH_3NH_3^+$	(d)	0.4449	0.5264	0.5348	4.44

Table 202. $CH_3NH_3Al(SO_4)_2 \cdot 12H_2O$. Lattice constants a, b, c, and volume expansion coefficient γ at various temperatures [69K8].

T K	a Å	b Å	c Å	γ 10^{-6} K^{-1}
291	12.242	—	—	35
221	12.232	—	—	33
199	12.229	—	—	—
155	12.250	12.214	12.200	−133
130	12.248	12.236	12.220	− 69
115	12.246	12.240	12.230	− 31
78	12.242	—	—	—

Table 203. Alum family. Electrogyration [76W2]. ϱ: optical rotatory power, g_{123}: axial third rank electrogyration tensor. $\Delta\varrho$ refers to 1 m crystal length and 1 V m^{-1} electric field, where the directions of wave number vector and electric field are parallel to [110] and [001], respectively.

Crystals	$\Delta\varrho$ [$10^{-7} \cdot V^{-1}$]	g_{123} [10^{-16} mV^{-1}]	T [°C]
$NH_4Fe(SO_4)_2 \cdot 12H_2O$	10.7	48.6	10
$CH_3NH_3Al(SO_4)_2 \cdot 12H_2O$	−15.2	−67.4	20
$CH_3NH_3Ga(SO_4)_2 \cdot 12H_2O$	−13.4	−59.9	20
$CH_3NH_3Fe(SO_4)_2 \cdot 12H_2O$	− 6.95	−31.4	10
$CH_3NH_3Al(SeO_4)_2 \cdot 12H_2O$	−33.2	−150.0	20

Table 204. $CH_3NH_3Al(SO_4)_2 \cdot 12H_2O$ (MASD). ESR data [55B2, 56B1, 54B1, 61B4].

Paramagnetic center	Site	S	$\mathscr{H}$	ν GHz	T K	g-factor	FS D, E, a, F [10^{-2} m^{-1}]	Ref.
Cr^{3+}	Al^{3+}	3/2	(5)	9; 3	90	1.975 (10)	D=959 (20) E= 90 (10)	55B2, 56B1 [1])
					20	1.977 (3)	D=−958 (4) E=− 92 (8)	
Fe^{3+}	Al^{3+}	5/2	(7)	9	90		D=1880 (140) a= 100 (40)	54B1 [2])
					90		D=+1893 (2) a≅+130 F≅− 40	61B4 [3])

[1]) Zero field splitting δ of the used sample (dilution is 100) is measured directly: $\delta=1790(30)\cdot10^{-2}\,\mathrm{m}^{-1}$; $(2D^2+E^2)^{1/2}=1770(20)\cdot10^{-2}\,\mathrm{m}^{-1}$.

[2]) Single crystal (dilution is 200).

[3]) Polycrystal (dilution is 100). At 90 K the zero field separation is measured directly: the separation of the $S_z=\pm3/2$ states from the $\pm1/2$ is equivalent to 12227 (5) MHz, and that of the $\pm5/2$ from the $\pm3/2$ to 22043 (10) MHz.

Nr. 26A-5 $CH_3NH_3V(SO_4)_2 \cdot 12H_2O$, Methylammonium vanadium sulfate dodecahydrate
(M=491.30)

1a	Ferroelectric activity in $CH_3NH_3V(SO_4)_2 \cdot 12H_2O$ was discovered by Pepinsky et al. in 1957.	57P5
b	phase: II / I; state: F / P; crystal system: — / cubic; Θ: −116 °C $\varrho=1.633\cdot10^3$ kg m^{-3} at RT. Transparent.	57P5
2a	Crystal growth: evaporation or cooling method from aqueous solution.	57P5
3a	Unit cell parameter: a=12.594 Å at 10 °C.	57P5
5c	Spontaneous polarization and coercive field: $P_s=0.9\cdot10^{-2}$ Cm^{-2} at about 2 °C below Θ. $E_c=6\cdot10^5$ V m^{-1} at about 2 °C below Θ (f=60 Hz).	57P5

Table for 1b:

phase	II	I
state	F	P
crystal system		cubic
Θ	−116 °C	

Nr. 26A-6 $CH_3NH_3Cr(SO_4)_2 \cdot 12H_2O$, Methylammonium chrome sulfate dodecahydrate
(M=492.36)

1a	Ferroelectric activity in $CH_3NH_3Cr(SO_4)_2 \cdot 12H_2O$ was discovered by Pepinsky et al. in 1957.	57P5
b	(see table below)	a) 57P5 b) 65w1 c) 68L5 d) 70L6
	$\varrho=1.650\cdot10^3$ kg m^{-3}. Transparent.	57P5

phase	II [a])	I [a])
state	F [a])	P [a])
crystal system	orthorhombic [d])	cubic [b])
space group	$Pca2_1$-C_{2v}^5 [d])	$Pa3$-T_h^6 [c])
Θ	−109 °C [a])	

continued

Nr. 26A-6 $CH_3NH_3Cr(SO_4)_2 \cdot 12H_2O$, continued

2a	Crystal growth: evaporation or cooling method from aqueous solution.	57P5
3a	Unit cell parameter: $a = 12.559$ Å at 20 °C.	57P5
b	$Z = 4$ in phase I.	65w1
5a	Dielectric relaxation: Figs. 874, 875, 876.	
c	Spontaneous polarization and coercive field: $P_s = 1.0 \cdot 10^{-2}$ Cm^{-2} at about 2 °C below Θ. $E_c = 6 \cdot 10^5$ Vm^{-1} at about 2 °C below Θ ($f = 60$ Hz).	57P5
6a	Heat capacity: Fig. 877. Transition entropy: $\Delta S_m = 9.70$ J mol^{-1} K^{-1}.	70B18
13b	ESR: Table 205.	

Table 205. $CH_3NH_3Cr(SO_4)_2 \cdot 12H_2O$. [50B1, 50B4, 49B1, 56B1] ESR data.

Paramagnetic center	S	$\mathcal{H}$	ν GHZ	T[°K]	g-factor	FS D, E [10^{-2} m^{-1}]	Ref.
Cr^{3+} [a])	3/2	(5)	9; 3	20	1.976 (7)	$D = -871$ (7) $E = -92$ (8)	50B1 50B4
				90	1.975 (10)	$D = 870$ (20) $E = 90$ (10)	49B1
				290	1.98	$D = 825$ (30)	56B1

[a]) Direction cosines of rhombic axes with respect to tetragonal axes a, b, c, are.

	a	b	c
x	−0.35 (5)	−0.35 (5)	+0.87 (3)
y	+0.71 (4)	−0.71 (4)	0.00 (5)
z	+0.61 (4)	+0.61 (4)	+0.50 (4)

Nr. 26A-7 $CH_3NH_3Fe(SO_4)_2 \cdot 12\,H_2O$, Methylammonium iron sulfate dodecahydrate ($M = 496.21$)

1a	Ferroelectric activity in $CH_3NH_3Fe(SO_4)_2 \cdot 12H_2O$ was discovered by Pepinsky et al. in 1957.	57P5
b	see table below; $\varrho = 1.647 \cdot 10^3$ kg m^{-3} at RT. Transparent.	57P5
2a	Crystal growth: evaporation or cooling method from aqueous solution.	57P5
3a	Unit cell parameter: $a = 12.60$ Å at 12 °C.	57P5
5c	Spontaneous polarization and coercive field: $P_s = 1.3 \cdot 10^{-2}$ C m^{-2} at about 2 °C below Θ. $E_c = 6 \cdot 10^5$ V m^{-1} at about 2 °C below Θ ($f = 60$ Hz).	57P5
9d	Optical activity: Table 203.	

1b:

phase	II	I
state	F	P
crystal system		cubic
Θ	−104 °C	

$\varrho = 1.647 \cdot 10^3$ kg m^{-3} at RT.
Transparent.

Nr. 26A-8 $CH_3NH_3Ga(SO_4)_2 \cdot 12H_2O$, Methylammonium gallium sulfate dodecahydrate
($M = 510.08$)

1a	Ferroelectric activity in $CH_3NH_3Ga(SO_4)_2 \cdot 12H_2O$ was discovered by Dunne in 1959.	59D4
b	phase: II, I; state: F, P; crystal system: –, cubic[a]; Θ: $-102\,°C$ Transparent.	59D4 [a]) 62j1
2a	Crystal growth: evaporation or cooling method from aqueous solution.	59D4
5a	Dielectric constant: Fig. 878. The Curie-Weiss constant $C = 360$ K $P_s = 4.10^{-3}$ Cm^{-2} just below Θ.	78Z3
9d	Optical activity: Table 203.	
13b	ESR: Table 206.	

Phase table (1b):

phase	II	I
state	F	P
crystal system		cubic[a])
Θ	$-102\,°C$	

Table 206. $CH_3NH_3Ga(SO_4)_2 \cdot 12H_2O{:}Cr^{3+}$, $CH_3NH_3Al(SeO_4)_2 \cdot 12H_2O{:}Cr^{3+}$. Spin Hamiltonian parameters of Cr^{3+} at room temperature. [76N8].

Alum	$g_\parallel$	$g_\perp$	$D[\cdot 10^2\ m^{-1}]$
$CH_3NH_3Ga(SO_4)_2 \cdot 12H_2O$	1.97	1.96	0.082
$CH_3NH_3Al(SeO_4)_2 \cdot 12H_2O$	1.975	1.96	0.072

Nr. 26A-9 $CH_3NH_3In(SO_4)_2 \cdot 12H_2O$, Methylammonium indium sulfate dodecahydrate
($M = 555.18$)

1a	Ferroelectric activity in $CH_3NH_3IN(SO_4)_2 \cdot 12H_2O$ was discovered by Pepinsky et al. in 1957.	57P5
b	phase: II, I; state: F, P; crystal system: –, cubic; Θ: $-109\,°C$ $\varrho = 1.805 \cdot 10^3$ kg m^{-3} at RT. Transparent.	57P5
2a	Crystal growth: evaporation or cooling method from aqueous solution.	57P5
3a	Unit cell parameter: $a = 12.688$ Å at 0 °C.	57P5
5c	Spontaneous polarization and coercive field: $P_s = 1.2 \cdot 10^{-2}$ Cm^{-2} at about 2 °C below Θ. $E_c = 9 \cdot 10^5$ Vm^{-1} at about 2 °C below Θ ($f = 60$ Hz).	57P5

Phase table (1b):

phase	II	I
state	F	P
crystal system		cubic
Θ	$-109\,°C$	

Nr. 26A-10 $CO(NH_2)_2HCr(SO_4)_2 \cdot 12H_2O$, Urea chrome alum
($M = 521.36$)

1a	Ferroelectric behavior was reported by Pepinsky et al. in 1958.	58P3

Nr. 26A-11 $CH_3NH_3Al(SeO_4)_2 \cdot 12H_2O$, Methylammonium aluminum selenate dodecahydrate
($M = 561.14$)

1a Ferroelectric activity in $CH_3NH_3Al(SeO_4)_2 \cdot 12H_2O$ was discovered by Pepinsky et al. in 1957. 57P5

b 57P5

phase	II	I
state	F	P
crystal system		cubic
Θ	−57 °C	

$\varrho = 1.820 \cdot 10^3$ kg m^{-3} at RT.
Transparent.

2a Crystal growth: evaporation or cooling method from aqueous solution. 57P5

5a $\varkappa_o - \varkappa_\infty = C/(T - \Theta_p)$, $T > \Theta$, $C = 185$ K, $\varkappa_\infty = 8.0$, $\Theta_p = 214.5$ K, $\Theta = 217.4$ K. 75A7
Dielectric constant: Fig. 879.
Pressure effect: Fig. 880α.
Dielectric relaxation: Figs. 880β, 880γ; for dielectric relaxation see also 75A7

c Spontaneous polarization and coercive field:
$P_s = 1.2 \cdot 10^{-2}$ Cm^{-2} at about 2 °C below Θ. 57P5
$E_c = 9 \cdot 10^5$ Vm^{-2} at about 2 °C below Θ ($f = 60$ Hz). Fig. 881.

9d Optical activity: Table 203.

13b ESR: Table 206.

27 GASH ($C(NH_2)_3Al(SO_4)_2 \cdot 6H_2O$) family

27A Pure compounds

Nr. 27A-1 $C(NH_2)_3Al(SO_4)_2 \cdot 6H_2O$, Guanidine aluminum sulfate hexahydrate (GASH)
($M = 387.27$)

1a Ferroelectricity in GASH was first reported by Holden et al. in 1955. 55H2

b

state	F
crystal system	trigonal*)
space group	P31m-C_{3v}^2 [a])
Θ	none**)

[a]) 56W1

$P_s \parallel [0001]$.
$\varrho_X = 1.804 \cdot 10^3$ kg m^{-3}. 62G2
Transparent, colorless.
Cleavage plane: (0001) (perfect), $\{11\bar{2}0\}$ (less perfect). 56W1

*) In the following descriptions, the hexagonal lattice frame will be used instead of the trigonal one.
**) Transition to the paraelectric phase does not exist up to the decomposition temperature of about 200 °C.

2a	Crystal growth: evaporation of aqueous solution. Solubility in H_2O: Table 207.	55H2
b	Crystal form: Fig. 882.	
3a	Unit cell parameters: $a=11.738$ (2) Å, $c=8.951$ (2) Å at RT.	67S7
b	$Z=3$. Crystal structure: Fig. 883; Tables 208···211.	62G2
4	Linear thermal expansion coefficients: $\alpha_a=10.05\cdot10^{-6}$ K^{-1}, $\alpha_c=92.80\cdot10^{-6}$ K^{-1}, 0 °C < T < 80 °C.	59H3
5a	Dielectric constants: $\varkappa_a=5$, $\varkappa_c=6$ at RT. The dielectric constants are practically temperature independent up to 100 °C.	56H1 56H1
c	Spontaneous polarization and coercive field: Figs. 884, 885. Effect of hydrostatic pressure on P_s: Fig. 886. Spontaneous polarization of deuterated compound: $P_s=3.5\cdot10^{-3}$ C m^{-2} at RT.	55H2
7a	Piezoelectricity: Fig. 887.	
8a	Elastic compliances and stiffnesses: Table 212.	
9a	Refractive indices: Table 213.	
e	Nonlinear optical susceptibilities: $d_{22}/d_{11}^{\text{quartz}}=-0.31$ (5), $d_{31}/d_{11}^{\text{quartz}}=+0.025$ (5), $d_{33}/d_{11}^{\text{quartz}}=0.059$ (10) for $\lambda=1.06$ µm. See also	74M13 62S2
11	Electrical conduction: Fig. 888.	
13a	NMR: Fig. 889. NQR: Tables 214, 215.	
b	ESR: Table 216; see also ESR of deuterated compound: Table 217.	74Z1, 74M4, 75M14, 77M18, 77M17
15b	Domain switching: Figs. 890, 891.	

Table 207. $C(NH_2)_3Al(SO_4)_2\cdot6H_2O$ (GASH). Solubility in H_2O [56R3]. A: kg solute/m^3 solvent.

T [°C]	A [kg m^{-3}]	T [°C]	A [kg m^{-3}]
22.7	592	75.6	901.45
26.1	637.7	79	940
28.2	686.2	81.7	990.8
42.9	728.9	86.	1044
64.45	812.5	88.2	1088.5
70.2	864.2		

Table 208, see p. 166.

Table 209. $C(NH_2)_3Al(SO_4)_2\cdot6H_2O$ (GASH). Fractional coordinates of hydrogens [67S7].

	x	y	z
H (1) on N (1)	0.9245	0.1113	0.5092
H (2) on N (2)	0.5729	0.1380	0.5611
H (3) on N (2)	0.7361	0.2309	0.5436
H (4) on O (8)	0.8714	0.0653	0.8611
H (5) on O (7)	0.9287	0.1291	0.1275
H (6) on O (9)	0.7380	0.2061	0.9032
H (7) on O (9)	0.5950	0.1343	0.9000
H (8) on O (10)	0.7957	0.2716	0.2069
H (9) on O (10)	0.8657	0.4078	0.2051

Table 208. $C(NH_2)_3Al(SO_4)_2 \cdot 6H_2O$ (GASH). Fractional coordinates (10^{-4}) and temperature parameters [67S7]. β_{ij} (in Å^2) is defined by Eq. (c) in Introduction.

	x	y	z	β_{11}	β_{22}	β_{33}	β_{12}	β_{12}	β_{23}
N (2)	5595 (5)	3430 (6)	5606 (8)	2.99 (20)	5.59 (28)	3.70 (21)	2.85 (19)	−0.35 (16)	−0.71 (17)
O (5)	8805 (3)	2241 (3)	8005 (6)	1.05 (11)	1.12 (12)	3.91 (16)	0.28 (10)	−0.01 (10)	−0.11 (11)
O (6)	7904 (4)	1161 (4)	2449 (7)	1.63 (13)	1.53 (12)	4.13 (18)	0.60 (11)	0.99 (13)	0.66 (12)
O (9)	5308 (3)	3274 (3)	9437 (6)	1.78 (13)	1.26 (12)	3.01 (15)	0.49 (9)	−0.66 (11)	0.01 (9)
O (10)	5373 (4)	1976 (4)	1773 (7)	2.08 (13)	2.37 (14)	3.37 (16)	1.35 (12)	0.79 (12)	0.95 (13)
S (1)	3486 (1)	0	7607 (5)	1.00 (4)	1.02 (4)	2.16 (5)	0.51	0.04 (4)	0
S (2)	6831 (1)	0	3167 (5)	1.06 (4)	1.14 (4)	2.04 (5)	0.57	0.03 (4)	0
N (1)	1119 (4)	0	5150 (9)	1.56 (18)	1.91 (18)	3.69 (26)	0.95	−0.43 (17)	0
O (1)	4592 (5)	0	8457 (8)	1.52 (14)	1.42 (14)	4.15 (26)	0.71	−1.01 (14)	0
O (2)	5584 (4)	0	2912 (7)	1.54 (14)	1.96 (15)	3.15 (22)	0.98	−0.65 (14)	0
O (3)	3749 (5)	0	6011 (7)	3.49 (20)	3.62 (21)	2.63 (21)	1.81	0.33 (17)	0
O (4)	7073 (4)	0	4756 (7)	2.35 (16)	3.51 (18)	2.65 (19)	1.76	0.14 (14)	0
O (7)	1324 (4)	0	1165 (7)	1.15 (13)	0.84 (15)	2.94 (19)	0.42	−0.66 (13)	0
O (8)	8680 (4)	0	8824 (7)	1.63 (11)	1.75 (11)	2.60 (20)	0.88	−0.69 (12)	0
Al (2)	1/3	2/3	0612 (4)	0.91 (4)	0.91	1.70 (9)	0.46	0	0
C (2)	1/3	2/3	5586 (11)	2.28 (45)	2.28	1.95 (27)	1.14	0	0
C (1)	0	0	5142 (12)	1.12 (21)	1.12	1.88 (35)	0.56	0	0
Al (1)	0	0	0	1.19 (7)	1.19	1.57 (11)	0.60	0	0

Table 210. $C(NH_2)_3Al(SO_4)_2 \cdot 6H_2O$ (GASH). Interatomic distances (in Å) [67S7].

Al (1)−O (7)	1.871 (6)
Al (1)−O (8)	1.873 (4)
Al (2)−O (9)	1.882 (5)
Al (2)−O (10)	1.872 (6)
C (1)−N (1)	1.313 (3)
C (2)−N (2)	1.319 (7)
S (1)−O (1)	1.505 (6)
S (1)−O (3)	1.462 (8)
S (1)−O (5)	1.477 (3)
S (2)−O (2)	1.481 (5)
S (2)−O (4)	1.451 (7)
S (2)−O (6)	1.463 (5)

Table 214. $C(NH_2)_3Al(SO_4)_2 \cdot 6H_2O$ (GASH) family. Quadrupole coupling constants and asymmetry parameters of ^{14}N. T=77 K [73O3].

Compound	e^2qQ/h kHz	η
$C(NH_2)_3Al(SO_4)_2 \cdot 6H_2O$ (GASH)	3575.4	0.39475
$C(NH_2)_3Cr(SO_4)_2 \cdot 6H_2O$ (GCrSH)	3559.4	0.39866
$C(NH_2)_3Al(SeO_4)_2 \cdot 6H_2O$ (GASeH)	3556.8	0.40643
$C(NH_2)_3Ga(SO_4)_2 \cdot 6H_2O$ (GGaSH)	3558.1	0.39918

Table 211. $C(NH_2)_3Al(SO_4)_2 \cdot 6H_2O$ (GASH). Bond angles (in [°])[67S7]. The primed number indicates that the atom is related with the unprimed atom by the symmetry operations.

O (7)−Al (1)−O (7′)	91.5 (2)
O (7)−Al (1)−O (8)	88.3 (2)
O (8)−Al (1)−O (8′)	92.0 (2)
O (9)−Al (2)−O (9′)	91.8 (2)
O (9)−Al (2)−O (10)	88.1 (2)
O (10)−Al (2)−O (10′)	92.3 (2)
O (1)−S (1)−O (3)	108.2 (4)
O (1)−S (1)−O (5)	108.8 (3)
O (3)−S (1)−O (5)	110.2 (3)
O (5)−S (1)−O (5′)	110.7 (3)
O (2)−S (2)−O (4)	110.2 (4)
O (2)−S (2)−O (6)	108.8 (3)
O (4)−S (2)−O (6)	110.7 (3)
O (6)−S (2)−O (6′)	108.8 (3)
H (1)−N (1)−H (1′)	120 (5)
H (2)−N (2)−H (3)	133 (6)
H (4)−O (8)−H (4′)	119 (4)
H (5)−O (7)−H (5′)	123 (6)
H (6)−O (9)−H (7)	109 (5)
H (8)−O (10)−H (9)	118 (6)

Table 213. $C(NH_2)_3Al(SO_4)_2 \cdot 6H_2O$ (GASH) family. Refractive indices [59H2]. λ=588.9 nm.

Compound	n_o	n_e
$C(NH_2)_3Al(SO_4)_2 \cdot 6H_2O$	1.5305	1.4416
$C(NH_2)_3Ga(SO_4)_2 \cdot 6H_2O$	1.5374	1.4513
$C(NH_2)_3Al(SeO_4)_2 \cdot 6H_2O$	1.5635	1.4713
$C(NH_2)_3Ga(SeO_4)_2 \cdot 6H_2O$	1.5709	1.4810

Table 212. $C(NH_2)_3Al(SO_4)_2 \cdot 6H_2O$ (GASH) family. $c_{\lambda\mu}$ and $s_{\lambda\mu}$ [59H2].

Compound	c_{11}	c_{33}	c_{12}	c_{13}	c_{44}	c_{66} $[=\frac{1}{2}(c_{11}-c_{12})]$	c_{14}
	10^{10} N m^{-2}						
$C(NH_2)_3Al(SO_4)_2 \cdot 6H_2O$ (GASH)	4.475	1.438	2.835	1.345	0.92	0.820	0.295
$C(NH_2)_3Ga(SO_4)_2 \cdot 6H_2O$ (GGaSH)	4.378	1.446	2.853	1.413	0.92	0.763	0.305
$C(NH_2)_3Al(SeO_4)_2 \cdot 6H_2O$ (GASeH)	4.293	1.361	2.688	1.380	0.815	0.803	0.316
$C(NH_2)_3Ga(SeO_4)_2 \cdot 6H_2O$ (GGaSeH)	4.040	1.353	2.54	1.428	0.81	0.750	0.304
	s_{11}	s_{33}	s_{12}	s_{13}	s_{44}	s_{66} $[=2(s_{11}-s_{12})]$	s_{11}
	10^{-11} m^2 N^{-1}						
$C(NH_2)_3Al(SO_4)_2 \cdot 6H_2O$ (GASH)	4.489	10.604	−2.403	−1.951	12.287	13.785	−2.210
$C(NH_2)_3Ga(SO_4)_2 \cdot 6H_2O$ (GGaSH)	4.896	11.189	−2.658	−2.186	12.530	15.018	−2.504
$C(NH_2)_3Al(SeO_4)_2 \cdot 6H_2O$ (GASeH)	4.870	12.265	−2.478	−2.425	14.479	14.696	−2.849
$C(NH_2)_3Ga(SeO_4)_2 \cdot 6H_2O$ (GGaSeH)	5.334	13.64	−2.529	−2.960	14.561	15.726	−2.951

Table 215. $C(NH_2)_3Al(SO_4)_2 \cdot 6H_2O$ (GASH) family. e^2qQ/h and $d(e^2qQ/h)/dT$ for ^{27}Al and ^{69}Ga [61B7].

Compound	Site[b])	e^2qQ/h[c]) kHz, 23 °C	$d(e^2qQ/h)dT$[c]) kHz·K^{-1}
$C(NH_2)_3Al(SO_4)_2 \cdot 6H_2O$ (GASH)	(1)	± 23	0.575
	(2)	∓ 30	0.411
$C(NH_2)_3Al(SeO_4)_2 \cdot 6H_2O$ (GASeH)	(1)	± 300	0.525
	(2)	± 200	0.228
$C(NH_2)_3Ga(SO_4)_2 \cdot 6H_2O$ (GGaSH)	(1)	∓ 485 (∓ 253)[c])	1.96 (1.02)
	(2)	∓ 635 (∓ 331)	1.51 (0.787)
$C(NH_2)_3Ga(SeO_4)_2 \cdot 6H_2O$ (GGaSeH)	(1)	± 78 (± 41)	2.29 (1.19)
	(2)	∓ 197 (∓ 103)	1.13 (0.589)
$C(ND_2)_3Al(SO_4)_2 \cdot 6D_2O$ [a])	(1)	± 23	0.575
	(2)	∓ 30	0.411

[a]) By comparing the proton resonance in this sample with that obtained from GASH, this compound is 85% deuterated.

[b]) Of the three Al or Ga ions per cell, two are equivalent to each other and called site (2). The other one is called site (1).

[c]) The values of e^2qQ/h and its temperature derivative for ^{71}Ga are multiplied by 0.521 and given in parentheses in this column.

Table 216. $C(NH_2)_3Al(SO_4)_2 \cdot 6H_2O$ (GASH). ESR data of Cr^{3+}, Fe^{3+}, and Mn^{2+}.

Para-magnetic center	Site	S	$\mathscr{H}$	ν GHz	T K	g-factor[a, b]	FS[a, b] D, a, F [10^{-2} m^{-1}]	HFS A [10^{-2} m^{-1}]	Ref.
Cr^{3+}	Al^{3+}	3/2	(5)	9.3	290	I: 1.980 (3)	I: D=750 (2)		57B5
						II: 1.977 (1)	II: D=590 (2)		
					373	I: 1.980 (5)	I: D=610 (4)		
						II: 1.980 (2)	II: D=488 (4)		
				9	35	1.975 (5)	I: D=1090 (5)		58D1
							II: D=850 (30)		
					77	1.975 (5)	I: D=1050 (30)		
							II: D=822 (10)		
					195	1.975 (5)	I: D=882 (10)		
							II: D=696 (6)		
					295	1.975 (5)	I: D=730 (5)		
							II: D=576 (5)		
				9	109		I: D=1070		61B7
							II: D=821		
					297	1.978 (7)	I: D=752		
							II: D=593		
Fe^{3+}	Al^{3+}	5/2	(7)	9.6, 23.4	77	I: 2.003	I: D=−1950.6 a=593.0 F=1639		62B8
						II: 2.003	II: D=−1890.2 a=539.6 F=1520		
					293	I: 2.003	I: D=−1850.8 a=219.7 F=1053		
						II: 2.003	II: D=−1760.1 a=205.9 F=1005		
Mn^{2+}	Al^{3+}	5/2	(7)	9	RT	2.004	D=371, 344 $(a-F)\approx 10$	−87.5	69M7

a) Two sets of spectra, a total of six lines, were observed.
b) Spectrum II has twice the intensity of I.

Table 217. $C(ND_2)_3Al(SO_4)_2 \cdot 6D_2O$. ESR data of Cr^{3+} [61B7].

Para-magnetic center	Site	S	$\mathscr{H}$	ν [GHz]	T [K]	g-factor	FS[a]) D [10^{-2} m^{-1}]
Cr^{+3}	Al^{3+}	3/2	(5)	9	77	—	I: D=1140
							II: D= 885
					293	1.977	I: D= 774
							II: D= 610

a) Spectrum II has twice the intensity of I.

Nr. 27A-2 $C(NH_2)_3V(SO_4)_2 \cdot 6H_2O$, Guanidine vanadium sulfate hexahydrate (GVSH)
($M = 411.23$)

1a	Ferroelectricity in GVSH was first reported by Remeika et al. in 1956.	56R2
b	state: F; crystal system: trigonal*); space group: P31m-C_{3v}^2; Θ: none*) $P_s \parallel [0001]$. *) See footnotes in section 1b of Nr. 27A-1.	56R2
3a	Unit cell parameters: $a = 11.809$ Å, $c = 9.026$ Å at RT.	61B7
b	$Z = 3$. Crystal structure: isomorphous with other GASH family compounds; see Figs. 883, 892, 893.	61B7
5c	Spontaneous polarization and coercive field: $P_s = 3.8 \cdot 10^{-3}$ C m^{-2}, $E_c = 6 \cdot 10^5$ V m^{-1} at RT.	56R2

Nr. 27A-3 $C(NH_2)_3Cr(SO_4)_2 \cdot 6H_2O$, Guanidine chromium sulfate hexahydrate (GCrSH)
($M = 412.28$)

1a	Ferroelectricity in GCrSH was first reported by Holden et al. in 1955.	55H2
b	state: F; crystal system: trigonal*); space group: P31m-C_{3v}^2; Θ: none*) $P_s \parallel [0001]$. *) See footnotes in section 1b of Nr. 27A-1.	55H2
3a	Unit cell parameters: $a = 11.754$ (3) Å, $c = 9.041$ (3) Å at RT.	67S7
b	$Z = 3$. Crystal structure: Tables 218···221. Structure is isomorphous with other GASH family compounds; see Figs. 883, 892, 893.	67S7
5c	Spontaneous polarization: $P_s = 3.7 \cdot 10^{-3}$ C m^{-2} at RT.	56H1
13a	NQR: Table 214.	

Table 219. $C(NH_2)_3Cr(SO_4)_2 \cdot 6H_2O$ (GCrSH).
Fractional coordinates of hydrogens [67S7].

	x	y	z
H (1) on N (1)	0.9202	0.1090	0.5194
H (2) on N (2)	0.5846	0.1407	0.5473
H (3) on N (2)	0.7349	0.2288	0.5602
H (4) on O (8)	0.8626	0.0772	0.8528
H (5) on O (7)	0.9239	0.1263	0.1302
H (6) on O (9)	0.7377	0.2028	0.8949
H (7) on O (9)	0.6066	0.1246	0.9060
H (8) on O (10)	0.8280	0.2958	0.1934
H (9) on O (10)	0.8664	0.4068	0.2095

Table 218, see p. 170.

Table 218. $C(NH_2)_3Cr(SO_4)_2 \cdot 6H_2O$ (GCrSH). Fractional coordinates ($\cdot 10^{-4}$) and Temperature parameters [67S7]. β_{ij} (in Å^2) is defined by Eq. (c) in Introduction.

	x	y	z	β_{11}	β_{22}	β_{33}	β_{12}	β_{13}	β_{23}
N (2)	5606 (7)	3431 (7)	5608 (9)	3.59 (27)	5.36 (34)	3.77 (24)	2.73 (24)	−0.41 (23)	0.26 (23)
O (5)	8801 (4)	2251 (4)	7956 (7)	1.03 (15)	1.39 (14)	4.56 (21)	0.36 (14)	0.25 (14)	0.13 (14)
O (6)	7922 (5)	1174 (4)	2522 (7)	2.47 (20)	1.51 (18)	4.19 (20)	0.80 (17)	0.57 (14)	0.58 (15)
O (9)	5239 (5)	3269 (4)	9402 (6)	2.52 (19)	1.82 (18)	3.24 (18)	0.83 (14)	−1.22 (15)	−0.13 (16)
O (10)	5336 (5)	1928 (5)	1821 (7)	2.34 (20)	2.64 (26)	4.18 (22)	1.74 (18)	0.99 (16)	1.24 (17)
S (1)	3498 (2)	0	7560 (4)	1.14 (5)	1.23 (5)	2.38 (6)	0.61	0.00 (5)	0
S (2)	6838 (2)	0	3223 (4)	1.23 (5)	1.25 (5)	2.28 (6)	0.62	−0.02 (5)	0
N (1)	1132 (5)	0	5175 (10)	1.16 (19)	1.79 (21)	4.21 (35)	0.90	0.00 (21)	0
O (1)	4577 (6)	0	8419 (8)	2.06 (22)	1.34 (18)	4.08 (30)	0.67	−1.23 (19)	0
O (2)	5605 (5)	0	2966 (7)	1.79 (30)	2.15 (19)	2.90 (23)	1.08	−0.38 (17)	0
O (3)	3765 (6)	0	5989 (8)	3.28 (24)	3.88 (27)	2.88 (25)	1.94	−0.13 (21)	0
O (4)	7080 (5)	0	4795 (7)	2.78 (20)	3.02 (22)	2.37 (22)	1.51	−0.41 (18)	0
O (7)	1378 (6)	0	1169 (8)	1.76 (19)	1.81 (19)	2.23 (26)	0.91	−0.73 (18)	0
O (8)	8622 (5)	0	8768 (8)	1.88 (19)	2.22 (21)	3.18 (27)	1.11	−1.15 (18)	0
Cr (2)	1/3	2/3	0634 (3)	0.96 (3)	0.96	1.46 (6)	0.48	0	0
C (2)	1/3	2/3	5584 (12)	2.24 (21)	2.24	2.20 (31)	1.12	0	0
C (1)	0	0	5166 (15)	1.40 (28)	1.40	1.84 (43)	0.70	0	0
Cr (1)	0	0	0	1.04 (5)	1.04	1.44 (7)	0.52	0	0

Table 220. $C(NH_2)_3Cr(SO_4)_2 \cdot 6H_2O$ (GCrSH). Interatomic distances in [Å] [67S7].

Cr (1)–O (7)	1.934 (5)
Cr (1)–O (8)	1.966 (7)
Cr (2)–O (9)	1.983 (6)
Cr (2)–O (10)	1.934 (6)
C (1)–N (1)	1.331 (6)
C (2)–N (2)	1.308 (8)
S (1)–O (1)	1.487 (7)
S (1)–O (3)	1.454 (8)
S (1)–O (5)	1.483 (6)
S (2)–O (2)	1.486 (7)
S (2)–O (4)	1.449 (7)
S (2)–O (6)	1.472 (7)

Table 221. $C(NH_2)_3Cr(SO_4)_2 \cdot 6H_2O$ (GCrSH). Bond angles in [°] [67S7]. The primed number indicates that the atom is related with the non-primed atom by the symmetry operations.

O (7)–Cr (1)–O (7′)	91.3 (3)
O (7)–Cr (1)–O (8)	88.0 (3)
O (8)–Cr (1)–O (8′)	93.0 (3)
O (9)–Cr (2)–O (9′)	91.6 (2)
O (9)–Cr (2)–O (10)	88.4 (2)
O (10)–Cr (2)–O (10′)	92.2 (2)
O (1)–S (1)–O (3)	109.0 (4)
O (1)–S (1)–O (5)	108.2 (3)
O (3)–S (1)–O (5)	110.3 (3)
O (5)–S (1)–O (5′)	110.9 (3)
O (2)–S (2)–O (4)	110.4 (4)
O (2)–S (2)–O (6)	108.7 (3)
O (4)–S (2)–O (6)	110.1 (3)
O (6)–S (2)–O (6′)	108.5 (4)
H (1)–N (1)–H (1′)	125 (6)
H (2)–N (2)–H (3)	124 (6)
H (4)–O (8)–H (4′)	115 (5)
H (5)–O (7)–H (5′)	133 (6)
H (6)–O (9)–H (7)	104 (5)
H (8)–O (10)–H (9)	104 (6)

Nr. 27A-4 $C(NH_2)_3Ga(SO_4)_2 \cdot 6H_2O$, Guanidine gallium sulfate hexahydrate (GGaSH)
($M=430.00$)

1a	Ferroelectricity in GGaSH was first reported by Holden et al. in 1955.	55H2
b	state: F; crystal system: trigonal*); space group: P31m-C_{3v}^2; Θ: none*) $P_s \parallel [0001]$. *) See footnotes in section 1b of Nr. 27A-1.	55H2
2a	Crystal growth: evaporation of aqueous solution.	56H1
3a	Unit cell parameters: $a=11.82$ (3) Å, $c=9.13$ (3) Å at RT.	59G1
b	$Z=3$. Crystal structure: Figs. 892, 893; Tables 222, 223.	59G1
4	Linear thermal expansion coefficients: $\alpha_a=14.17 \cdot 10^{-6}$ K^{-1}, $\alpha_c=83.85 \cdot 10^{-6}$ K^{-1}, 0 °C<T<80 °C.	59H3
5c	Spontaneous polarization and coercive field: Figs. 894, 895.	
8a	Elastic compliances and stiffnesses: Table 212.	
9a	Refractive indices: Table 213.	
e	Nonlinear optical properties: see	62S2
13a	NQR: Tables 214, 215.	
b	ESR: Table 224.	

Table 222. $C(NH_2)_3Ga(SO_4)_2 \cdot 6H_2O$ (GGaSH). Fractional coordinates [59G1].

Atom	Position	x	y	z
Ga^{3+} (1)	1a	0	0	0.000
Ga^{3+} (2)	2b	$\frac{1}{3}$	$\frac{2}{3}$	0.059
S (1)	3c	0.351	0	−0.250
S (2)	3c	−0.318	0	0.306
O (1)*)	3c	0.455	0	−0.156
O (2)*)	3c	−0.433	0	0.238
O (3)**)	3c	0.386	0	−0.409
O (4)**)	3c	−0.328	0	0.470
O (5)***)	6d	0.342	0.120	−0.220
O (6)***)	6d	−0.314	−0.120	0.258
O (7)	3c	0.140	0	0.130
O (8)	3c	−0.144	0	−0.135
O (9)	6d	−0.477	0.333	−0.066
O (10)	6d	0.476	−0.336	0.174
C (1)	1a	0	0	0.50
C (2)	2b	$\frac{1}{3}$	$\frac{2}{3}$	0.52
N (1)	3c	0.113	0	0.50
N (2)	6d	0.553	0.333	0.52
H (1)	6d	0.198	0.085	0.50
H (2)	6d	0.553	0.418	0.52
H (3)	6d	0.469	0.248	0.52

*) The [SO_4] "basal" O atoms in the symmetry planes.
**) The [SO_4] "peak" O atoms in the symmetry planes.
***) The [SO_4] "basal" O atoms related by the symmetry planes.

Table 223. $C(NH_2)_3Ga(SO_4)_2 \cdot 6H_2O$ (GGaSH). Interatomic distances [59G1].

Distances	Å	Distances	Å
$Ga^{3+}-H_2O$ distances:		$H_2O-[SO_4]$ oxygen distances:	
Ga (1)–O (7)	2.04	O (7)–O (6)	2.54
Ga (1)–O (8)	2.10	O (8)–O (5)	2.61
Ga (2)–O (9)	2.05	O (10)–O (6)	2.63
Ga (2)–O (10)	2.04	O (9)–O (1)	2.51
H_2O-H_2O distances:		O (10)–O (2)	2.59
O (7)–O (7)	2.87	O (9)–O (5)	2.64
O (8)–O (8)	2.95	N–O distances:	
O (7)–O (8)	2.94	N (1)–O (2)	3.33
O (9)–O (9)	2.94	N (1)–O (4)	3.42
O (10)–O (10)	2.92	N (2)–O (3)	3.33
O (9)–O (10)	2.77	N (2)–O (4)	2.67
		Guanidinium hydrogen-$[SO_4]$ oxygen distances:	
		H (1)–O (3)	2.98
		H (1)–O (4)	2.50
		H (2)–O (3)	4.35
		H (2)–O (4)	2.36
		H (3)–O (3)	2.67
		H (3)–O (4)	2.34

Table 224. $C(NH_2)_3Ga(SO_4)_2 \cdot 6H_2O$ (GGaSH). ESR data of Cr^{3+} [61B7].

Paramagnetic center	Site	S	$\mathscr{H}$	ν [GHz]	T [K]	g-factor	FS[a]) $D\,[10^{-2}\,m^{-1}]$
Cr^{3+}	Ga^{3+}	3/2	(5)	9	77	1.978	I: $D=866$
							II: $D=664$
					297	1.976	I: $D=576$
							II: $D=456$

[a]) Spectrum II has twice the intensity of I.

Nr. 27A-5 $C(NH_2)_3Al(SeO_4)_2 \cdot 6H_2O$, Guanidine aluminum selenate hexahydrate (GASeH)
($M=481.07$)

1a	Ferroelectricity in GASeH was first reported by Holden et al. in 1955.	55H2
b	state: F; crystal system: trigonal*); space group: P31m-C_{3v}^2; Θ: none*) $P_s \parallel [0001]$. *) See footnotes in section 1b of Nr. 27A-1.	55H2
2a	Crystal growth: evaporation of aqueous solution.	56H1
3a	Unit cell parameters: $a=11.989$ Å, $c=9.062$ Å at RT.	61S13
b	$Z=3$. Crystal structure: isomorphous with other GASH family compounds; see Figs. 883, 892, 893.	61S13

4	Linear thermal expansion coefficients: $\alpha_a = 10.16 \cdot 10^{-6}\ K^{-1}$, $\alpha_c = 83.89 \cdot 10^{-6}\ K^{-1}$, $0\ °C < T < 80\ °C$.	59H3
5c	Spontaneous polarization: Fig. 896.	
8a	Elastic compliances and stiffnesses: Table 212.	
9a	Refractive indices: Table 213.	
e	Nonlinear optical properties: see	62S2
13a	NQR: Tables 214, 215. ESR: Table 225.	

Table 225. $C(NH_2)_3Al(SeO_4)_2 \cdot 6H_2O$ (GASeH). ESR data of Cr^{3+} [61B7].

Paramagnetic center	Site	S	$\mathscr{H}$	ν[GHz]	T[K]	g-factor	FS[a]) $D[10^{-2}\ m^{-1}]$
Cr^{3+}	Al^{3+}	3/2	(5)	9	77	1.976	I: D=1850 II: D=1310
					297	1.977	I: D=1430 II: D=1110

[a]) Spectrum II has twice the intensity of I.

Nr. 27A-6 $C(NH_2)_3Cr(SeO_4)_2 \cdot 6H_2O$, Guanidine chromium selenate hexahydrate (GCrSeH) (M = 506.08)

1a	Ferroelectricity in GCrSeH was first reported by Holden et al. in 1955.	55H2
b	state: F; crystal system: trigonal*); space group: P31m-C_{3v}^2; Θ: none*) P_s∥[0001]. *) See footnotes in section 1b of Nr. 27A-1.	55H2
3b	Crystal structure: isomorphous with other GASH family compounds; see Figs. 883, 892, 893.	
5c	Spontaneous polarization: $P_s = 4.7 \cdot 10^{-3}\ C\ m^{-2}$ at RT.	56H1

Nr. 27A-7 $C(NH_2)_3Ga(SeO_4)_2 \cdot 6H_2O$, Guanidine gallium selenate hexahydrate (GGaSeH) (M = 523.80)

1a	Ferroelectricity in GGaSeH was first reported by Holden et al. in 1955.	55H2
b	state: F; crystal system: trigonal*); space group: P31m-C_{3v}^2; Θ: none*) P_s ∥ [0001]. *) See footnotes in section 1b of Nr. 27A-1.	55H2

continued

Nr. 27A-7 $C(NH_2)_3Ga(SeO_4)_2 \cdot 6H_2O$, continued

2a	Crystal growth: evaporation of aqueous solution.	
3a	Unit cell parameters: $a = 12.020$ Å, $c = 9.144$ Å at RT.	61S13
b	$Z = 3$.	61S13
	Crystal structure: isomorphous with other GASH family compounds; see Figs. 883, 892, 893.	
4	Linear thermal expansion coefficients: $\alpha_a = 12.07 \cdot 10^{-6}\ K^{-1}$, $\alpha_c = 71.16 \cdot 10^{-6}\ K^{-1}$, $0\ °C < T < 80\ °C$.	59H3
5c	Spontaneous polarization: Fig. 897.	
8a	Elastic compliances and stiffnesses: Table 212.	
9a	Refractive indices: Table 213.	
e	Nonlinear optical properties: see	62S2
13a	NQR: Table 214, 215. ESR: Table 226.	

Table 226. $C(NH_2)_3Ga(SeO_4)_2 \cdot 6H_2O$ (GGaSeH). ESR data of Cr^{3+} [61B7].

Paramagnetic center	Site	S	$\mathscr{H}$	ν[GHz]	T[K]	g-factor	FS [a]) $D[10^{-2}\ m^{-1}]$
Cr^{3+}	Ga^{3+}	3/2	(5)	9	77	1.975	I: D = — II: D = 1190
					297	1.977	I: D = — II: D = 1000

[a]) Spectrum II has twice the intensity of I.

28 $LiH_3(SeO_3)_2$ family

28A Pure Compounds

Nr. 28A-1 $LiH_3(SeO_3)_2$, Lithium trihydrogen selenite ($M = 263.88$)

1a	Ferroelectric activity in $LiH_3(SeO_3)_2$ was discovered by Pepinsky et al. in 1959.	59P3
b	state: F; crystal system: monoclinic; space group: Pn-C_s^2; Θ *): —	60V1
	P_s has a major component along the direction perpendicular to (001) plane. $T_{melt} = 110\ °C$. $\varrho = 3.391 \cdot 10^3\ kg\ m^{-3}$. Transparent.	72T5
	*) Ferroelectric transition temperature does not exist under the atmospheric pressure below the melting point, but has been observed under high pressures. See subsection 5a.	
2a	Crystal growth: slow cooling or evaporation of aqueous solution of lithium hydroxide or lithium carbonate and selenious acid in stoichiometric proportions.	
b	Crystal form: Fig. 898.	

3a	Unit cell parameters: $a=6.2554$ (3) Å, $b=7.8823$ (5) Å, $c=5.4339$ (5) Å, $\beta=105.325$ (5)°.	72T5
b	$Z=2$ Rectangular coordinates X_c, Y_c, Z_c are defined as $X_c \parallel a$, $Y_c \parallel b$, $Z_c \perp$ (X_c and Y_c). Crystal structure: Tables 227…230; Figs. 899…901. The unit cell parameters given in 3a were used in determination of the crystal structure. See also	59P3 60V1, 69S15, 79C3
4	Volume change with hydrostatic pressure: Fig. 902.	
5a	Dielectric constants: Figs. 903…907. Curie-Weiss constant, $C=4.1 \cdot 10^4$ K. $(d\Theta_f/dp)_{p=0}=-6.0 \cdot 10^{-8}$ KPa^{-1}. For the deuterated crystal, $C=2.9 \cdot 10^4$ K, $(d\Theta_f/dp)_{p=0}=-5.9 \cdot 10^{-8}$ KPa^{-1}.	 68S2
b	ξ, ζ: non-linear coefficients. $E=\frac{1}{\chi_p}P+\xi P^3+\zeta P^5$ $\xi=9.88 \cdot 10^9$ V m^5 C^{-3}, $\zeta=2.17 \cdot 10^{11}$ V m^9 C^{-5}.	68S2
c	Spontaneous polarization: $P_s=15 \cdot 10^{-2}$ C m^{-2} at RT. Coercive field: Fig. 908.	68S2
7a	Piezoelectricity: Table 231.	
8a	Elastic compliances: Table 232.	
9a	Refractive indices: Figs. 909…911. Optical indicatrix: Fig. 912.	
d	Optical activity: Fig. 913.	
10a	Raman scattering: Fig. 914.	
11	Electrical conduction: Fig. 915.	
13a	NMR: Tables 233…236; Fig. 916. Dipolar spin-lattice relaxation time, see	 71B18
b	ESR in γ-ray irradiated crystals and for VO^{2+} [a]), see	76E1, 65I6, 79J3[a])

Table 227. $LiH_3(SeO_3)_2$. Crystal structure [72T5]. Fractional coordinates at RT. For assignment of atomic numbers see Figs. 899, 900.

Atoms	x	y	z
	10^{-5}		
Se (1)	23300	39507 (23)	08800
Se (2)	76690 (20)	10074 (21)	91095 (26)
O (1)	19252 (40)	22747 (32)	−13341 (47)
O (2)	79816 (38)	27366 (31)	110145 (46)
O (3)	30176 (43)	29114 (33)	36189 (49)
O (4)	70944 (42)	20193 (35)	61072 (49)
O (5)	47888 (42)	47641 (32)	05452 (46)
O (6)	51917 (40)	02447 (29)	91695 (44)
Li	00241 (150)	24866 (146)	48066 (160)
H (1)	32880 (84)	14685 (64)	−09938 (82)
H (2)	61997 (106)	38177 (71)	07074 (98)
H (3)	55051 (60)	23084 (52)	52793 (75)

Table 228. $LiH_3(SeO_3)_2$. Crystal structure [72T5]. Anisotropic temperature parameters ($\cdot 10^5$). b_{ij} is defined by Eq. (b) in Introduction. For assignment of atomic numbers, see Figs. 899, 900.

Atoms	b_{11}	b_{22}	b_{33}	b_{12}	b_{13}	b_{23}
	10^{-5}					
Se (1)	562 (30)	479 (21)	793 (34)	083 (19)	118 (26)	033 (21)
Se (2)	478 (30)	418 (19)	915 (36)	− 018 (19)	070 (25)	027 (21)
O (1)	917 (43)	682 (30)	1015 (51)	− 014 (35)	060 (40)	−195 (38)
O (2)	777 (44)	602 (29)	1221 (54)	013 (34)	065 (40)	−206 (36)
O (3)	747 (42)	823 (35)	1017 (47)	062 (32)	146 (39)	260 (36)
O (4)	797 (46)	924 (38)	1093 (54)	− 053 (31)	161 (42)	266 (37)
O (5)	774 (44)	684 (35)	1646 (70)	− 119 (34)	373 (48)	117 (38)
O (6)	756 (43)	555 (31)	1621 (67)	− 131 (31)	392 (47)	− 17 (37)
Li	1207 (112)	921 (68)	1245 (138)	33 (62)	355 (96)	28 (78)
H (1)	1903 (104)	910 (55)	1833 (108)	− 153 (58)	350 (83)	−139 (62)
H (2)	2771 (150)	1408 (88)	2187 (126)	−1046 (99)	732 (111)	−187 (86)
H (3)	1097 (85)	1270 (59)	1967 (114)	− 025 (53)	249 (73)	129 (65)

Table 229. $LiH_3(SeO_3)_2$. Crystal structure [72T5]. Interatomic distances and angles at RT. For assignment of atomic numbers, see Figs. 899, 900.

Distances	Å	Angles	deg	Angles	deg
Se(1)–O(1)	1.760 (3)	O(1)–Se(1)–O(3)	101.52 (15)	O(1)–Li–O(3)	90.22 (35)
–O(3)	1.653 (3)	O(1)–Se(1)–O(5)	100.29 (12)	O(1)–Li–O(4)	88.03 (33)
–O(5)	1.720 (3)	O(3)–Se(1)–O(5)	104.64 (12)	O(1)–Li–O(5)	86.84 (37)
–O(2)	2.903 (2)	Se(1)–O(1)–H(1)	111.97 (31)	O(1)–Li–O(6)	92.36 (41)
–O(4)	3.184 (3)	Se(1)–O(5)–H(2)	116.31 (35)	O(2)–Li–O(3)	92.43 (34)
–O(5)	3.086 (3)	Se(1)–O(3)···H(3)	128.23 (21)	O(2)–Li–O(4)	89.34 (35)
–Li	3.097 (9)			O(2)–Li–O(5)	91.90 (41)
		O(2)–Se(2)–O(4)	99.43 (15)	O(2)–Li–O(6)	88.83 (38)
Se(2)–O(2)	1.691 (3)	O(2)–Se(2)–O(6)	103.64 (13)	O(3)–Li–O(5)	90.39 (41)
–O(4)	1.765 (3)	O(4)–Se(2)–O(6)	102.55 (13)	O(3)–Li–O(6)	91.36 (38)
–O(6)	1.670 (3)	Se(2)–O(6)···H(1)	119.21 (24)	O(4)–Li–O(5)	90.27 (38)
–O(1)	2.913 (3)	Se(2)–O(2)···H(2)	117.74 (28)	O(4)–Li–O(6)	87.95 (40)
–O(3)	3.113 (3)	Se(2)–O(4)–H(3)	116.92 (30)		
–O(6)	2.955 (3)				
Li–O(1)	2.126 (9)				
–O(2)	2.127 (9)				
–O(3)	2.162 (9)				
–O(4)	2.163 (9)				
–O(5)	2.216 (12)				
–O(6)	2.187 (12)				

Table 230. $LiH_3(SeO_3)_2$. Crystal structure [72T5]. Hydrogen bond distances and angles at RT. For assignment of atomic numbers, see Figs. 899, 900.

X H Y	Distances			Angles X–H···Y deg
	X–H Å	H···Y Å	X···Y Å	
O(1)–H(1)···O(6)	1.040 (6)	1.517 (6)	2.552 (3)	173.25 (43)
O(5)–H(2)···O(2)	1.141 (7)	1.378 (7)	2.518 (3)	176.56 (54)
O(4)–H(3)···O(3)	1.003 (4)	1.649 (5)	2.646 (4)	172.27 (40)

Figs. p. 526

Table 231. $LiH_3(SeO_3)_2$. $d_{i\lambda}$, $g_{i\lambda}$ at 25 °C [63B2].

d_{11}	d_{12}	d_{13}	d_{15}	d_{24}	d_{26}	d_{31}	d_{32}	d_{33}	d_{35}
10^{-12} C N^{-1}									
23.2	−22.3	−12.1	−46.6	−12.8	−14.6	−18.4	5.5	19.9	53.2

g_{11}	g_{12}	g_{13}	g_{15}	g_{24}	g_{26}	g_{31}	g_{32}	g_{33}	g_{35}
m^2 C^{-1}									
0.078	−0.107	−0.0051	−0.0935	−0.113	−0.129	−0.0302	−0.0391	0.0734	0.152

Table 232. $LiH_3(SeO_3)_2$. $s_{\lambda\mu}$ at 25 °C [63B2].

s_{11}	s_{22}	s_{33}	$s_{44}+2s_{23}$	$s_{55}+2s_{13}$	$s_{66}+2s_{12}$	$s_{35}+s_{15}$
10^{-11} m^2 N^{-1}						
2.94	3.41	8.67	23.6	10.8	20.7	7.5

Table 233. $LiD_3(SeO_3)_2$. e^2qQ/h, η of ^{7}Li and the direction cosines of the principal axes of the field-gradient tensors at RT. [69S15]. For X_c, Y_c, and Z_c axes, see Fig. 898. Signs ± in the column for the Y_c correspond to the two equivalent sets of direction of the principal axes.

$e^2Qq/h = 33.9\,(3)$ kHz, $\eta = 0.593\,(13)$

	X_c	Y_c	Z_c
ϕ_{xx}	−0.6273 (61)	±0.0506 (133)	0.7771 (54)
ϕ_{yy}	−0.2874 (105)	±0.9124 (15)	−0.2914 (100)
ϕ_{zz}	0.7238 (42)	±0.4062 (38)	0.5578 (59)

Table 235. $LiD_3(SeO_3)_2$. Deuteron e^2qQ/h and η at RT [69S15]. Three kinds of spectra, D_a, D_b, D_c, were observed. For assignment of atomic numbers, see Fig. 901.

	e^2Qq/h [kHz]	η	Assignment
D_a	107.8 (10)	0.172 (9)	O (6)–$D_\alpha\cdots$O′ (1)
D_b	127.9 (10)	0.154 (6)	O (2)–$D_\beta\cdots$O′ (5)
D_c	181.1 (5)	0.101 (4)	O (4)$\cdots D_\gamma$–O′ (3)

Table 234. $LiD_3(SeO_3)_2$. Direction cosines of the principal axes of the electric field-gradient tensors at the positions of deuterons at RT [69S15]. For X_c, Y_c, and Z_c axes, see Fig. 898. For positions D_α, D_β, D_γ, see Fig. 901. D_a, D_b, D_c are three kinds of observed spectra.

		X_c	Y_c	Z_c
D_a	ϕ_{xx}	0.6552 (14)	0.7516 (7)	0.0763 (108)
(D_α)	ϕ_{yy}	−0.1246 (72)	0.0079 (81)	0.9922 (8)
	ϕ_{zz}	0.7451 (7)	−0.6596 (8)	0.0988 (12)
D_b	ϕ_{xx}	0.3827 (139)	0.6631 (122)	0.6433 (209)
(D_β)	ϕ_{yy}	−0.5349 (106)	−0.4087 (185)	0.7395 (179)
	ϕ_{zz}	0.7533 (6)	−0.6271 (10)	0.1982 (14)
D_c	ϕ_{xx}	−0.4379 (30)	−0.4978 (210)	0.7486 (123)
(D_γ)	ϕ_{yy}	−0.0606 (133)	0.8471 (116)	0.5279 (171)
	ϕ_{zz}	0.8970 (6)	−0.1858 (36)	0.4012 (10)

Table 236. $LiH_3(SeO_3)_2$. e^2qQ/h, η and direction cosines for principal components of the electric field gradient tensor for ^{7}Li at hydrostatic pressures 10^5 Pa, $3.3 \cdot 10^8$ Pa, $5.9 \cdot 10^8$ Pa at RT. [75S17]. x, y, z: Principal coordinates system. $X \| a$, $Y \| b$, $Z \| c'$. The $\pm$ signs correspond to the two EFG tensors.

P	Quadrupole coupling constant and η	Direction cosines X	Y	Z
$1 \cdot 10^5$ Pa	$\frac{e^2Qq}{h} = 35.64\,(100)$ kHz	ϕ_{xx} −0.6674	±0.0897	+0.7393
		ϕ_{yy} ∓0.2513	+0.9074	∓0.3369
	$\eta = 0.487\,(25)$	ϕ_{zz} +0.7010	±0.4106	+0.5830
$3.3 \cdot 10^8$ Pa	$\frac{e^2Qq}{h} = 35.23\,(100)$ kHz	ϕ_{xx} −0.7110	±0.2062	+0.6722
		ϕ_{yy} ∓0.1677	+0.8787	∓0.4469
	$\eta = 0.361\,(25)$	ϕ_{zz} +0.6829	±0.4304	+0.5902
$5.9 \cdot 10^8$ Pa	$\frac{e^2Qq}{h} = 34.63\,(100)$ kHz	ϕ_{xx} −0.7465	±0.3484	+0.5668
		ϕ_{yy} ∓0.0486	+0.8211	∓0.5687
	$\eta = 0.263\,(25)$	ϕ_{zz} +0.6636	±0.4521	+0.5960

Nr. 28A-2 $NaH_3(SeO_3)_2$, Sodium trihydrogen selenite
($M = 279.93$)

1a Ferroelectric activity of $NaH_3(SeO_3)_2$ was discovered by Pepinsky et al. in 1959. — 59P3

b

phase	III[b])	II[a])	I[a])
state	F[b])	F[a])	P[a])
crystal system		triclinic[c])	monoclinic[c])
space group		P1-C_1^1 [c])	$P2_1/a$-C_{2h}^5 [c])
Θ	−162 °C*)		−79 °C[a])

[a]) 59P3; [b]) 64M1; [c]) 59P4

For the deuterated crystal, $\Theta_{II-I} = -3.5$ °C[d]). — [d]) 65B6
The spontaneous polarization of $NaH_3(SeO_3)_2$ has a large component along the a axis in phases II and III. — 65B6, 67S15
$\varrho = 3.20\,(1) \cdot 10^{-3}$ kg m^{-3}. — 68V2
Phase diagram of $Na(H_{1-x}D_x)_3(SeO_3)_2$: Fig. 917; see also — 72S22
Transparent, colourless.

*) II−III phase transition has thermal hysteresis of about 10 °C.
Phase III is electrically uniaxial [67S15].

2a Crystal growth: slow cooling or evaporation of aqueous solution of sodium hydroxide or sodium carbonate and selenious acid in stoichiometric proportions. — 67C2

b Crystal form: Fig. 918.

3a Unit cell parameters: $a = 11.77$ Å, $b = 4.84$ Å, $c = 5.80$ Å, $\beta = 118.5°$ at RT. — 59P4
For the deuterated crystal: $a = 11.7$ Å, $b = 4.9$ Å, $c = 5.8$ Å, $\beta = 118.5°$ at RT. — 65B6
Kaplan et al. selected another axial system: $a' = 10.3450\,(4)$ Å, $b' = 4.8440\,(2)$ Å, $c' = 5.7866\,(2)$ Å, $\beta' = 91.133\,(4)°$ at RT. $\boldsymbol{a}' = \boldsymbol{a} + \boldsymbol{c}$, $\boldsymbol{b}' = -\boldsymbol{b}$, $\boldsymbol{c}' = -\boldsymbol{c}$. — 70K3
Vijayan et al. selected another axial system: $\boldsymbol{a}'' = \boldsymbol{a} + \boldsymbol{c}$, $\boldsymbol{b}'' = \boldsymbol{b}$, $\boldsymbol{c}'' = \boldsymbol{c}$, $\beta'' = 180° - \beta'$. — 68V2
Unit cell parameters: Table 237.

b $Z = 2$ in phase I. — 70K3
Crystal structure: Figs. 919···921; Tables 238···241.

4 Lattice distortions: Figs. 922···924.

5a	Dielectric constant: Figs. 925···929; see also Pressure effect: Table 242. $d\Theta_{I-II}/dp = -1.24\,(7) \cdot 10^{-8}$ K Pa^{-1}. For $NaD_3(SeO_3)_2$, $d\Theta/dp = -3.3 \cdot 10^{-8}$ K Pa^{-1}.	78L7 74S31 66S5
b	Nonlinear dielectric properties: Fig. 930.	
	$E = \frac{1}{\varepsilon_0 C}(T - \Theta_p)\,P + \xi P^3 + \zeta P^5$	70B10
	$C = 1.6 \cdot 10^3$ K, $\xi = -1.9 \cdot 10^{10}$ $Vm^5 C^{-3}$, $\zeta = 9 \cdot 10^{13}$ $Vm^9 C^{-5}$.	70B10
c	Spontaneous polarization: Figs. 931, 932.	
6a	Heat capacity: Fig. 933. Transition heat, transition entropy:	64M1

	III–II	II–I	
ΔQ_m	≃196	794	J mol^{-1}
ΔS_m	≃ 2.1	4.1	J K^{-1} mol^{-1}

7a	Piezoelectricity: Fig. 934, Table 243.	
9a	Reflection: Fig. 935. Infrared absorption: Fig. 936.	
d	Optical activity: Fig. 937.	
10a	Raman scattering: Fig. 938···940; Table 244.	
b	Brillouin scattering: Fig. 941.	
11	For both crystals of $NaH_3(SeO_3)_2$ and $NaD_3(SeO_3)_2$: Conductivity: $\sigma \approx 10^{-7}$ $\Omega^{-1} m^{-1}$ at RT. Activation energy: $\Delta U = 0.74 \sim 0.83$ eV.	 67K5 67K5
13a	NMR: Figs. 942···946; Tables 245···250. For ^{23}Na NMR, see See for T_1 data Proton spin lattice relaxation time for γ-ray irradiated specimens: see In the ferroelectric phase of $NaD_3(SeO_3)_2$, there are two inequivalent Na sites.	 77A8 76M1 71L1 78K3
b	ESR: Fig. 947; Table 251.	
14	Bragg reflection: Figs. 948···953.	

Table 237. $NaH_3(SeO_3)_2$. Unit cell parameters of phases I and II [73M14].

	Phase I (at 298 K)	Phase II (at 183 K)
a'	10.33_8 Å	20.32_0 Å
b'	4.84_0 Å	9.59_2 Å
c'	5.77_5 Å	5.75_9 Å
α'	90°00′	89°53′
β'	91°10′	91°07′
γ'	90°00′	90°20′

Table 238. $NaH_3(SeO_3)_2$. Crystal structure of phase I [70K3]. Fractional coordinates at RT. Parameters listed for $H_d(1)$ are obtained on the assumption that the H(1) atom is disordered. The coordinates refer to the a', b' c' axial system.

	x	y	z
Na	0	0	0
Se	0.1691 (2)	0.3861 (6)	0.5060 (4)
	0.1691	0.3850	0.5068
O (1)	0.4473 (4)	0.6904 (10)	0.1257 (6)
	0.4487	0.6948	0.1264
O (2)	0.6443 (3)	0.2072 (9)	0.2877 (6)
	0.6410	0.2015	0.2975
O (3)	0.3071 (3)	0.2067 (9)	0.4509 (6)
	0.3111	0.1980	0.4365
H (1)	0	0	1/2
$H_d(1)$	0.0180 (33)	0.0205 (199)	0.4851 (92)
H (2)	0.1721 (7)	0.4919 (21)	0.9152 (15)

Table 239. $NaH_3(SeO_3)_2$. Crystal structure of phase I [70K 3]. Anisotropic temperature parameters. B_{ij} is defined by Eq. (a) in Introduction, referring to the a', b', c' axial system. For explanation of $H_d(1)$, see caption of Table 238.

	B_{11}	B_{22}	B_{33}	B_{12}	B_{13}	B_{23}
	10^{-4}					
Na	16 (7)	137 (4)	69 (2)	− 20 (16)	0 (11)	18 (31)
Se	38 (2)	147 (12)	93 (6)	− 7 (5)	− 2 (3)	7 (10)
O (1)	44 (3)	220 (22)	49 (9)	56 (8)	−18 (5)	− 13 (16)
O (2)	40 (3)	129 (19)	45 (8)	28 (8)	10 (5)	13 (14)
O (3)	30 (3)	149 (18)	70 (10)	17 (7)	12 (5)	38 (15)
H (1)	90 (12)	497 (77)	189 (31)	116 (31)	−45 (20)	− 47 (54)
H_d (1)	9 (33)	425 (272)	149 (103)	25 (100)	24 (46)	9 (152)
H (2)	66 (7)	634 (65)	206 (23)	90 (20)	15 (12)	144 (42)

Table 240. $NaH_3(SeO_3)_2$. Crystal structure of phase I [70K 3]. Interatomic distances and angles. For explanation of $H_d(1)$, see caption of Table 238. Atom with primed number is related to basis atom by an inversion center.

Distances	[Å]	Angles	[deg]
(a) involving oxygen atoms about the sodium ion			
Na−O (1)	2.407 (4)	O (1)−Na−O (2)	84.2 and 95.8 (2)
Na−O (2)	2.413 (4)	O (1)−Na−O (3)	87.0 and 93.0 (2)
Na−O (3)	2.462 (4)	O (2)−Na−O (3)	83.7 and 96.3 (2)
(b) involving selenite group			
Distances	[Å]	Angles	[deg]
Se−O (1)	1.700 (5)	O (1)−Se−O (2)	99.3 (3)
Se−O (2)	1.716 (5)	O (1)−Se−O (3)	102.7 (3)
Se−O (3)	1.707 (5)	O (2)−Se−O (3)	100.6 (3)
(c) involving hydrogen bonds			
Distances	[Å]	Angles	[deg]
O (1)−H (1)	1.301 (5)	O (1)−H (1)−O (1)	180
O (1)−O (1′)	2.602 (9)	O (2)−H (2)−O (3)	176.1 (9)
O (2)−H (2)	1.24 (1)		
O (3)−H (2)	1.31 (2)		
O (2)−O (3)	2.556 (6)		
O (1)−H_d(1)	1.11 (8)		
O (1′)−H_d(1)	1.50 (8)	O (1)−H_d(1)−O (1′)	170 (4)

Table 242. $XH_3(SeO_3)_2$, $XD_3(SeO_3)_2$, X = Na, K, Rb, Cs. Θ, $d\Theta/dp$ [71S20].

Material	$NaH_3(SeO_3)_2$	$NaD_3(SeO_3)_2$	$KH_3(SeO_3)_2$	$KD_3(SeO_3)_2$	$RbH_3(SeO_3)_2$	$CsH_3(SeO_3)_2$
$\frac{d\Theta}{dp}$ [deg Pa^{-1}]	−1.1 (2)	−1.7 (1)	− 5.3 (4)	− 5.4 (2)	7.1 (6)	0.7 (1)
Θ [°C]	−78.8	−5.6	−58.8	−19.2	−113	−124

Table 241. $NaHD_2(SeO_3)_2$ and $NaH_2D(SeO_3)_2$. Crystal structures of phase I. Fractional coordinates and temperature parameters. The temperature parameters are defined by Eq. (e) in Introduction. Compound I = $NaHD_2(SeO_3)_2$ [74N12], II = $NaH_2D(SeO_3)_2$ [75N6]. The coordinates refer to the a', b', c' axial system.[a])

Atom	Compound	x	y	z	B, $Å^2$	Multiplicity of the position
Na	I	0	0	0	1.04	2
	II	0	0	0	1.16	2
Se	I	0.1692	0.3841	0.5056	1.45	4
	II	0.1690	0.3850	0.5061	1.38	4
O 1	I	0.4466	0.6902	0.1253	1.54	4
	II	0.4471	0.6910	0.1256	1.63	4
O 2	I	0.6440	0.2078	0.2876	1.16	4
	II	0.6443	0.2075	0.2876	1.29	4
O 3	I	0.3077	0.2098	0.4509	1.25	4
	II	0.3071	0.2063	0.4510	1.18	4
H 1	I	0	0	$^1/_2$	2.65	1.65
	II	0	0	$^1/_2$	2.74	1.76
H 2	I	0.1734	0.4995	0.9144	2.96	4.36
	II	0.1751	0.5000	0.9122	2.69	4.26

[a]) The rate of substitution of H for D at site I is different from the rate at site II.

Table 243. $NaH_3(SeO_3)_2$. Piezoelectric constants at -80 °C in $10^{-11}CN^{-1}$ [75S14]. Primed quantities are coefficients in the coordinate system rotated by 45° around the y (d'_{23}) and x (d'_{12}) axis. x: direction perpendicular to c' and b' axes; $y = b'$ axis.

$d_{11} = 1.6$	$d_{21} = -0.3$
$d_{12} = 3.4$	$d_{22} = 5.5$
$d_{13} = -0.6$	$d_{23} = 0.9$
$d'_{12} = -0.6$	$d'_{23} = 1.0$
$d_{14} = -4.0$	$d_{25} = 1.7$

Table 244. $Na(D_xH_{1-x})_3(SeO_3)_2$. $\Delta\nu$ vs. x [77F2]. Frequencies of lattice modes below 300 cm^{-1} at RT obtained from Raman scattering spectra. Parameter x: concentration of deuterium. 1 cm^{-1} is equal to $3 \cdot 10^{10}$ Hz.

Symmetry type	Deuterium concentration in crystal, x				
	0	0.18	0.22	0.62	0.96
	cm^{-1}	cm^{-1}	cm^{-1}	cm^{-1}	cm^{-1}
B_g	57	57	57	56	57
A_g	70	70	69	69	70
B_g	92	92	91	91	92
B_g	112	112	111	111	112
A_g	113	113	112	112	112
A_g	125	125	125	124	123
A_g	133	132	131	132	133
B_g	156	149	147	145	143
B_g	159	158	155	154	154
A_g	206	—	—	—	212
B_g	252	≈250	≈250	≈250	250
A_g	278	—	—	—	264

Table 245. $NaD_3(SeO_3)_2$. Deuteron e^2qQ/h, η and relative intensity [69 S 16]. In phase I two pairs of DMR lines, D_a and D_b were observed. Below Θ_f, D_a splits into 2 lines, D_b into 4 lines.

	e^2Qq/h [kHz]	η	Relative intensity
298 K			
D_a	145.5 (4)	0.156 (4)	1
D_b	129.7 (2)	0.135 (3)	2
134 K			
D_{a1}	179.5 (4)	0.130 (3)	1
D_{a2}	141.4 (3)	0.172 (3)	1
D_{b1}	167.1 (3)	0.165 (3)	1
D_{b2}	121.2 (2)	0.216 (3)	1
D_{b3}	142.6 (2)	0.148 (2)	1
D_{b4}	111.3 (2)	0.248 (2)	1

Table 246. $NaD_3(SeO_3)_2$. Direction cosines of the principal axes of the deuteron quadrupole coupling tensors at the deuteron positions at 298 K and 134 K [69 S 16]. $X_c \parallel a'$ axis; $Y_c \parallel b'$ axis; Z_c: perpendicular to the $a'b'$ plane (compare also Fig. 898).

		X_c	Y_c	Z_c
298 K				
D_a	ϕ_{xx}	0.5863 (87)	0.3359 (77)	0.7372 (37)
	ϕ_{yy}	0.7246 (70)	−0.6243 (39)	−0.2918 (91)
	ϕ_{zz}	0.3623 (13)	0.7053 (10)	−0.6094 (14)
D_b	ϕ_{xx}	−0.5595 (86)	−0.4099 (50)	0.7204 (40)
	ϕ_{yy}	0.8130 (59)	−0.4406 (43)	0.3807 (78)
	ϕ_{zz}	0.1613 (12)	0.7987 (7)	0.5797 (10)
134 K				
D_{a1}	ϕ_{xx}	0.6092 (87)	0.3147 (77)	0.7279 (41)
	ϕ_{yy}	0.7122 (73)	−0.6209 (37)	−0.3276 (91)
	ϕ_{zz}	0.3488 (11)	0.7180 (8)	−0.6024 (11)
D_{a2}	ϕ_{xx}	0.6232 (67)	0.3361 (59)	0.7062 (33)
	ϕ_{yy}	0.7148 (58)	−0.6112 (30)	−0.3400 (69)
	ϕ_{zz}	0.3173 (12)	0.7166 (8)	−0.6211 (12)
D_{b1}	ϕ_{xx}	−0.3009 (72)	−0.4962 (31)	0.8144 (11)
	ϕ_{yy}	0.9252 (24)	−0.3588 (41)	0.1232 (63)
	ϕ_{zz}	0.2311 (10)	0.7906 (7)	0.5671 (10)
D_{b2}	ϕ_{xx}	−0.3119 (58)	−0.5263 (28)	0.7911 (10)
	ϕ_{yy}	0.9163 (20)	−0.3867 (36)	0.1041 (50)
	ϕ_{zz}	0.2512 (10)	0.7573 (8)	0.6028 (11)
D_{b3}	ϕ_{xx}	−0.8351 (43)	−0.2755 (39)	0.4760 (54)
	ϕ_{yy}	0.5485 (66)	−0.4827 (20)	0.6828 (40)
	ϕ_{zz}	0.0417 (10)	0.8313 (5)	0.5542 (8)
D_{b4}	ϕ_{xx}	−0.3679 (41)	−0.5156 (21)	0.7739 (9)
	ϕ_{yy}	0.9082 (17)	−0.3778 (25)	0.1801 (35)
	ϕ_{zz}	0.1996 (9)	0.7691 (6)	0.6072 (9)

Table 247. $NaD_3(SeO_3)_2$. e^2qQ/h, η of ^{23}Na and the direction cosines of the principal axes of the quadrupole coupling tensor [69 S 16]. In the column of Y_c, the signs correspond to the two equivalent sets of direction cosines of the principal axes. The unit cell refers to the a', b', c' axial system.

T [K]		e^2Qq/h [MHz]		η	
298	Na	1.19 (2)		0.09 (2)	
256	Na (1)	1.24 (1)		0.61 (2)	
	Na (2)	1.22 (1)		0.13 (1)	
			X_c	Y_c	Z_c
298	Na	ϕ_{xx}	0.270 (21)	∓0.182 (24)	0.945 (39)
		ϕ_{yy}	0.936 (35)	∓0.179 (25)	−0.302 (122)
		ϕ_{zz}	0.224 (7)	±0.967 (2)	0.122 (9)
256	Na (1)	ϕ_{xx}	−0.221 (16)	±0.073 (9)	0.973 (4)
		ϕ_{yy}	0.940 (4)	∓0.251 (6)	0.233 (17)
		ϕ_{zz}	0.261 (6)	±0.965 (2)	−0.013 (8)
	Na (2)	ϕ_{xx}	0.963 (11)	∓0.236 (7)	0.127 (73)
		ϕ_{yy}	−0.143 (71)	∓0.049 (18)	0.989 (9)
		ϕ_{zz}	0.227 (6)	±0.971 (1)	0.081 (6)

Table 248. $NaD_3(SeO_3)_2$. Quadrupole coupling tensors and direction cosines with respect to the crystalline a', b', and c^* axes [78 K 3]. $\boldsymbol{c}^* = \boldsymbol{a}' \times \boldsymbol{b}'$.

	$eQ\phi_{ii}/h$ [kHz]	Direction cosine with respect to		
		a'	b'	c^*
11.8 °C Na	1211.9	−0.1915	0.9780	−0.0828
	677.3	0.9790	0.1844	−0.0867
	534.6	0.0695	0.0976	0.9928
−79.5 °C Na (1)	1307.9	−0.2715	0.9622	0.0196
	1139.3	0.9369	0.2596	0.2341
	168.6	−0.2213	−0.0819	0.9720
Na (2)	1281.3	−0.2354	0.9707	−0.0477
	738.5	0.1691	0.0892	0.9816
	542.8	0.9571	0.2230	−0.1851

Table 250. $NaH_3(SeO_3)_2$. e^2qQ/h and η of ^{23}Na under high pressure [76 A 7]. $p = 7 \cdot 10^8$ Pa, $T = -90$ °C.

	$\frac{e^2qQ}{h}$ kHz	η	Phase
Na	1198	0.08	I
NaI	1311	0.13	IV
NaII	1340	0.04	IV
NaI	1299	0.33	II
NaII	1280	0.26	II
NaIII	1245	0.22	II
NaIV	1280	0.27	II
NaI	1321	0.20	III
NaII	1342	0.68	III

Table 249. $NaD_3(SeO_3)_2$. e^2qQ/h, η of ^{23}Na and direction cosines of the largest principal axes with respect to the crystallographic a', b', and c^* axes [77 K 5]. $\boldsymbol{c}^* = \boldsymbol{a}' \times \boldsymbol{b}'$.

T [°C]	Site	e^2qQ/h kHz	η	Direction cosines with respect to		
				a'	b'	c^*
20	I, II	1207 (25)	+0.16 (5)	0.182	±0.981	0.068
−75	I	1309 (40)	+0.74 (8)	0.267	±0.960	−0.078
−75	II	1309 (40)	−0.16 (8)	0.254	±0.966	0.042

Table 251. $NaH_3(SeO_3)_2$. Frequencies and assignments of vibrational modes at −196 °C [65K 5]. $10^2\ m^{-1}$ is equal to $3 \cdot 10^{10}$ Hz.

Frequency of absorption maxima $[10^2\ m^{-1}]$	Assignment	Frequency of absorption maxima $[10^2\ m^{-1}]$	Assignment
2800 v.b.	O–H···O (st.)	770, 755, 730	$\nu_3''HSeO_3^-$
2500, 2460, 2400, 2300 v.b.	O–H···O (st.)	705, 694, 690	$\nu_3'H_2SeO_3$
1302, 1290, 1270	O–H···O in-plane bend	647, 633, 622	$\nu_3''H_2SeO_3$ and $\nu_1HSeO_3^-$
1245, 1235, 1225	O–H···O in-plane bend	478, 464, 438	$\nu_2HSeO_3^-$
1005, 990, 975	O–H···O out-of-plane bend	430, 407, 382	$\nu_2H_2SeO_3$
		382, 375, 358	$\nu_4'H_2SeO_3$ $\nu_4''H_2SeO_3$
950, 917, 907	O–H···O out-of-plane bend		
867, 857, 848	$\nu_3'HSeO_3^-$	350, 348, 334	$\nu_4'HSeO_3^-$
		334, 320, 317	$\nu_4''HSeO_3^-$
827, 812, 804	$\nu_1H_2SeO_3$		

v.b. = very broad; st. = stretching.

Nr. 28A-3 $KH_3(SeO_3)_2$, Potassium trihydrogen selenite
($M = 296.04$)

1a Elastic anomaly in $KH_3(SeO_3)_2$ associated with the phase transition was found by Ivanov et al. in 1975. 75I7

b

phase	II	I
state		P [a])
crystal system	monoclinic [c])	orthorhombic [b])
space group	$P2_1/b\text{-}C_{2h}^5$ [c])	$Pbcn\text{-}D_{2h}^{14}$ [b])
Θ	−61.6 °C [a])	

[a]) 67S14
[b]) 71G8
[c]) 77M5

For $KD_3(SeO_3)_2$, $\Theta = +23.5$ °C. 69Y1

$T_{melt} \approx 70$ °C. 67S14

$\varrho = 3.01 \cdot 10^3\ kg\ m^{-3}$, $\varrho_X = 3.08 \cdot 10^3\ kg\ m^{-3}$. 71G8

2a	Crystal growth: cooling method from a saturated aqueous solution.	67S14
b	Crystal form: Fig. 954.	
3a	Unit cell parameters: $a=16.150$ (2) Å, $b=6.25$ (1) Å, $c=6.298$ Å at RT. Sometimes another axial system is adopted, which are $\boldsymbol{a}'=-\boldsymbol{b}$, $\boldsymbol{b}'=\boldsymbol{a}$, $\boldsymbol{c}'=\boldsymbol{c}$. For notation of a', b', c', see Fig. 954.	71G8
b	$Z=4$. Crystal structure: Figs. 955···957; Tables 252α, 252β, 253α, 253β, 254; see also For positions of hydrogen atoms in phase I, see	71G8 71G8 72P5
4	Thermal expansion: Figs. 958···961; see also	78P4
5a	Dielectric constants: Figs. 962, 963. For deuterated crystals, see Fig. 964.	
6a	Heat capacity: Fig. 965. Transition heat, transition entropy: $\Delta Q_m=439$ (22) J · mol^{-1}, $\Delta S_m=2.47$ (13) J K^{-1} mol^{-1}.	77M6
8a	Elastic compliance and stiffness: Figs. 966···969; see also	75I7
9a	Infrared absorption and optical indicatrix: Figs. 970, 971; see also	78I5
10a	Raman scattering: Fig. 972.	
b	Brillouin scattering and Rayleigh scattering: Figs. 973···975. Rayleigh and Brillouin scattering spectra for a slightly deuterated crystal. Isotope effects on the central peak.	 78T3 78T2
13a	NMR: Table 255; Figs. 976, 977; see also	78G11
b	ESR: Tables 256, 257.	

Table 252α. $KH_3(SeO_3)_2$. Crystal structure of phase I [69H2]. Fractional coordinates. T=RT. The crystallographic axes referred to are a, b, c.

Atoms	x	y	z
Se	0.15153 (4)	0.18806 (9)	0.21329 (9)
K	0.50000 (0)	0.18848 (32)	0.25000 (0)
O (1)	0.11114 (26)	0.38798 (67)	0.07151 (66)
O (2)	0.06727 (26)	0.11189 (70)	0.35833 (68)
O (3)	0.20691 (29)	0.32281 (75)	0.40645 (75)
H (1)	0.1830 (500)	0.4415 (1350)	0.4735 (1225)
H (2)	0.0135 (1000)	0.1164 (2600)	0.2700 (2700)

Table 253β. $KH_3(SeO_3)_2$. Crystal structure of phase I [69H2]. Bond angles.

Angles	deg
O (1)–Se–O (2)	100.61 (21)
O (1)–Se–O (3)	102.41 (21)
O (2)–Se–O (3)	99.82 (22)
Se–O (3)–H	111.96
Se–O (3)–H	119.34

Table 252β, see p. 186.

Table 253α. $KH_3(SeO_3)_2$. Crystal structure of phase I [69H2]. Temperature parameters. The temperature parameters are defined by Eq. (d) in Introduction. The crystallographic axes referred to are a, b, c.

Atoms	U_{11}	U_{22}	U_{33}	U_{12}	U_{13}	U_{23}
	10^{-4} Å^2					
Se	161 (3)	243 (3)	239 (3)	23 (2)	10 (2)	8 (2)
K	256 (10)	284 (10)	246 (9)	0 (0)	48 (7)	0 (0)
O (1)	245 (24)	299 (23)	268 (21)	24 (19)	−77 (18)	65 (17)
O (2)	225 (22)	390 (25)	235 (20)	−43 (19)	19 (17)	34 (18)
O (3)	243 (23)	429 (27)	366 (25)	16 (21)	−130 (20)	−118 (22)

Table 252β. $KD_3(SeO_3)_2$. Structure of phase II [79I1]. Fractional coordinates by neutron diffraction method. The crystallographic axes referred to are *a*, *b*, *c*. $T = 80$ K.

	x	*y*	*z*
Se (1)	0.1498 (2)	0.1914 (7)	0.2149 (8)
Se (2)	0.3483 (2)	0.3187 (7)	0.7033 (7)
K	0.5005 (5)	0.1906 (19)	0.2321 (44)
O (1)	0.1096 (3)	0.3860 (10)	0.0685 (11)
O (2)	0.0694 (3)	0.0939 (10)	0.3615 (10)
O (3)	0.2110 (3)	0.3107 (9)	0.4093 (9)
O (4)	0.3891 (3)	0.1025 (9)	0.5733 (10)
O (5)	0.4341 (3)	0.3906 (11)	0.8615 (11)
O (6)	0.2949 (3)	0.1891 (12)	0.9058 (12)
D (1)	0.1763 (3)	0.4327 (13)	0.4690 (12)
D (2)	0.3312 (4)	0.0609 (11)	0.9667 (11)
D (3)	0.4871 (4)	0.3754 (12)	0.7373 (21)

Table 254. $KD_3(SeO_3)_2$. Crystal structure of phase II [79I1]. Bond lengths (in Å) and angles. $T = 80$ K. Primed atoms are related to basis atoms by symmetry operation.

$DSeO_3$					
Se 1–O 1	1.669 (8)	O 1–O 2	2.695 (8)	O 1–Se 1–O 2	105.7 (4)°
Se 1–O 2	1.712 (7)	O 1–O 3	2.744 (8)	O 1–Se 1–O 3	108.0 (3)°
Se 1–O 3	1.723 (7)	O 2–O 3	2.670 (7)	O 2–Se 1–O 3	102.0 (3)°
De_2SeO_3					
Se 2–O 4	1.687 (7)	O 4–O 5	2.613 (9)	O 4–Se 2–O 5	98.7 (3)°
Se 2–O 5	1.757 (7)	O 4–O 6	2.627 (9)	O 4–Se 2–O 6	99.7 (4)°
Se 2–O 6	1.749 (8)	O 5–O 6	2.594 (8)	O 5–Se 2–O 6	95.4 (4)°
Hydrogen Bond					
D 1–O 3	1.007 (9)	D 2–O 6	1.068 (9)	D 3–O 5	1.160 (11)
D 1–O 4′	1.619 (9)	D 2–O 1′	1.594 (9)	D 3–O 2′	1.482 (9)
O 3–O 4′	2.612 (7)	O 6–O 1′	2.659 (9)	O 5–O 2′	2.600 (8)

Table 255. $KD_3(SeO_3)_2$. The principal values and direction cosines of electric field gradient tensors of deuterons. Direction cosines at −53 °C are referred to the orthorhombic axes. [78K2]. See also [78G11].

	$eQ\phi_{ii}/h$ kHz	Direction cosine with respect to		
		a	*b*	*c*
at 40.0 °C				
D (1)	151.4	0.5527	0.7349	−0.3930
	87.4	−0.2987	0.6149	0.7298
	64.0	−0.7780	0.2860	−0.5594
D (2)	129.0	0.8298	0.0	0.5580
	71.4	0.0	1.0000	0.0
	57.6	0.5580	0.0	−0.8298
at −53.0 °C				
D (1)	175.8	0.5503	0.7427	−0.3815
	100.5	−0.2719	0.5914	0.7591
	75.3	−0.7894	0.3140	−0.5274
D′ (1)	134.9	0.5490	0.7353	−0.3974
	79.8	−0.3512	0.6344	0.6886
	55.1	−0.7585	0.2384	−0.6065
D (2)	131.2	0.8234	−0.0564	0.5647
	75.2	−0.1321	0.9487	0.2874
	56.0	0.5519	0.3112	−0.7737

 Figs. p. 544ff.

Table 256. $KH_3(SeO_3)_2$:Cr^{3+}. Spin Hamiltonian parameters at RT [78Z2]. A, B: two kinds of paramagnetic centers.

	g_x	g_y	g_z	D	E
				$\cdot 10^2$ m^{-1}	
A	1.984	1.985	1.978	0.218	0.037
B	1.967	1.966	1.968	0.141	0.024

Table 257. $KH_3(SeO_3)_2$: VO^{2+}. g and A tensors and the direction cosines of the principal axes with respect to the a, b, and c axes. [78J1].

	Direction cosines with respect to		
	a	b	c
g tensor			
1.925 (1)	0.9683	0.0044	−0.2496
1.992 (2)	−0.0666	−0.9590	−0.2832
1.998 (2)	0.2406	−0.2832	0.9284
A tensor ($\cdot 10^{-2}$ m^{-1})			
174.8 (10)	0.9961	0.0878	0.0010
71.3 (20)	−0.0775	0.8739	0.4800
65.4 (20)	0.0412	−0.4782	0.8773

Nr. 28A-4 $RbH_3(SeO_3)_2$, Rubidium trihydrogen selenite
(M = 342.41)

1a	Ferroelectric activity in $RbH_3(SeO_3)_2$ was discovered by Shuvalov et al. in 1969.	69S7
b	phase: II*) / I; state: F / P; crystal system: monoclinic[a]) / orthorhombic[a]); space group: $P2_1$-C_2^2 [c]) / $P2_12_12_1$-D_2^4 [b]); Θ: −120 °C	a) 69S7 b) 73T1 c) 78G13 78G14
	$P_s \parallel [010]$. T_{melt} = 57.1 °C. $\varrho = 3.42 \cdot 10^{-3}$ kg m^{-3}. Transparent, colorless.	69S7 69S7 73T1
	*) An intermediate phase between the paraelectric and ferroelectric phases was reported to be existing by Gladkii et al. [78G14].	
2a	Crystal growth: solution method (solvent H_2O), cooling from temperatures lower than 10 °C in a dark room.	69S7
b	Crystal form: Fig. 978.	
3a	Unit cell parameters: a = 5.9192 (2) Å, b = 17.9506 (5) Å, c = 6.2519 (3) Å at RT.	73T1
b	Z = 4 in phase I. Crystal structure: Tables 258, 259α, 259β; Fig. 979; see also	73T1 77N8, 73T1
4	Thermal expansion: Fig. 980.	
5a	Dielectric constant: Fig. 981. $d\Theta/dp = (5.8\ (2))$ K $\cdot$ Pa^{-1} [a]). See also [b])	a) 73G6 b) 71S20
c	Spontaneous polarization and coercive field: Figs. 982α, 982β, 982γ.	
8a	Elastic compliance and stiffness: Fig. 983.	
9a	Refractive indices: n_a = 1.667, n_b = 1.627, n_c = 1.634 at RT, λ = 589.0 nm. Fig. 984. Birefringence: Fig. 985. Absorption: Fig. 986.	69S7
b	Electrooptic effect: Fig. 987. continued	

Nr. 28A-4 $RbH_3(SeO_3)_2$, continued

13a	NMR: Figs. 988α, 988β, 988γ, 988δ; Tables 260α, 260β; see also	74S20, 79V2
b	ESR: see	75S16
14a	Bragg reflections: Fig. 989.	
b	Inelastic scattering: Figs. 990, 991.	

Table 258. $RbH_3(SeO_3)_2$. Crystal structure of phase I [77T7]. Fractional coordinates (· 10^5) at 25 °C.

Atom	x	y	z
Se (1)	21603 (16)	39522 (5)	29592 (16)
Se (2)	73074 (17)	24230 (5)	16852 (15)
O (1)	19248 (30)	42042 (9)	02792 (26)
O (2)	50871 (27)	22527 (10)	32176 (38)
O (3)	49199 (22)	42598 (8)	33834 (29)
O (4)	77270 (27)	33469 (8)	18980 (28)
O (5)	07888 (26)	46154 (8)	42234 (31)
O (6)	95431 (29)	20776 (10)	31866 (40)
Rb	26060 (20)	08290 (6)	23376 (21)
H (1)	27392 (52)	46900 (16)	−00433 (50)
H (2)	97911 (55)	23704 (18)	45659 (73)
H (3)	60320 (52)	38474 (18)	28169 (51)

Table 259α. $RbH_3(SeO_3)_2$. Crystal structure of phase I [77T7]. Anisotropic temperature parameters and the root-mean-square components R_i(· 10^3 Å) of thermal displacement along the ellipsoids axes. T=25 °C. b_{ij} is defined by Eq. (b) in Introduction.

Atom	b_{11}	b_{22}	b_{33}	b_{12}	b_{13}	b_{23}	R_1	R_2	R_3
Se (1)	1311 (24)	0091 (2)	1059 (22)	−0014 (5)	−0027 (17)	−0017 (5)	121 (2)	145 (2)	153 (1)
Se (2)	1443 (24)	0130 (3)	0947 (21)	−0015 (6)	0096 (19)	−0037 (5)	132 (2)	148 (1)	163 (1)
O (1)	2447 (45)	0192 (4)	1169 (35)	−0259 (12)	−0399 (33)	0076 (9)	144 (2)	149 (2)	234 (2)
O (2)	1605 (39)	0228 (5)	2448 (53)	−0209 (11)	0703 (40)	−0163 (13)	139 (2)	181 (2)	250 (2)
O (3)	1314 (34)	0148 (4)	1891 (42)	−0004 (10)	−0275 (30)	−0066 (10)	145 (2)	155 (2)	200 (2)
O (4)	1701 (38)	0129 (3)	1965 (41)	0018 (9)	0462 (35)	0025 (9)	145 (2)	158 (2)	211 (2)
O (5)	1427 (34)	0163 (4)	2208 (45)	−0019 (9)	0351 (34)	−0238 (11)	135 (2)	157 (2)	229 (2)
O (6)	1778 (41)	0180 (4)	3036 (62)	0219 (11)	−0772 (45)	−0134 (15)	135 (2)	183 (3)	264 (3)
Rb	1535 (29)	0165 (3)	1522 (30)	0028 (8)	0071 (25)	0043 (6)	159 (2)	164 (2)	179 (2)
H (1)	2281 (73)	0227 (7)	2010 (68)	−0059 (21)	−0115 (62)	0081 (18)	183 (3)	196 (4)	214 (4)
H (2)	2295 (78)	0251 (9)	3559 (130)	0062 (23)	−0675 (88)	0121 (30)	180 (4)	210 (4)	276 (5)
H (3)	2227 (72)	0272 (9)	1941 (74)	−0135 (22)	−0360 (62)	0113 (22)	179 (4)	188 (4)	234 (4)

Table 260β. $RbD_3(SeO_3)_2$. Principal values and the direction cosines of EFG tensor of deuterons at 14 °C. [79S8]. See also Figs. 988γ, 988δ.

	$eQ\phi_{ii}/h$	Direction cosine with respect to		
	kHz	a	b	c
D (1)	− 72.25	0.3106	0.0220	0.9503
	− 88.83	0.8194	−0.5129	−0.2560
	161.08	0.4817	0.8582	−0.1773
D (2)	− 65.02	0.7649	0.5992	−0.2364
	− 84.58	−0.6299	0.6192	−0.4689
	149.60	−0.1346	0.5076	0.8510
D (3)	− 48.46	0.7553	−0.6066	−0.2481
	− 71.19	−0.0302	−0.4105	0.9114
	119.65	0.6547	0.6809	0.3284

Table 259β. $RbH_3(SeO_3)_2$. Crystal structure of phase I [77T7]. Interatomic distances [Å] and angles [deg]. $T=25$ °C. Atoms outside the asymmetric unit carry a subscript which indicates how the atomic parameters can be derived from those of the asymmetric unit. The first three digits code the lattice translation (the asymmetric unit is in cell 555), and the fourth digit specifies the symmetry operation: $1=x, y, z$; $2=1/2-x, -y, 1/2+z$; $3=1/2+x, 1/2-y, -z$; $4=x, 1/2+y, 1/2-z$.

Covalent bonds			
O–Se–O	O–Se	Se–O	∢O–Se–O
O (1)–Se (1)–O (3)	1.741 (2)	1.745 (2)	98.0 (1)
O (1)–Se (1)–O (5)	1.741 (2)	1.644 (2)	103.6 (1)
O (3)–Se (1)–O (5)	1.745 (2)	1.644 (2)	99.2 (1)
O (2)–Se (2)–O (4)	1.655 (2)	1.682 (2)	104.7 (1)
O (2)–Se (2)–O (6)	1.655 (2)	1.737 (2)	103.1 (1)
O (4)–Se (2)–O (6)	1.682 (2)	1.737 (2)	101.3 (1)

Ionic bonds	
Rb–O $(1)_{5553}$	3.036 (2)
–O (2)	2.998 (2)
–O $(3)_{4563}$	3.116 (2)
–O $(3)_{6454}$	3.207 (2)
–O $(4)_{4553}$	3.034 (2)
–O $(5)_{5563}$	2.968 (2)
–O $(5)_{5454}$	3.120 (2)
–O $(6)_{4551}$	2.931 (2)

Hydrogen bonds			
O–H···O	O–H	H···O	O···O
O (1)–H (1)···O $(5)_{5642}$	1.017 (3)	1.589 (3)	2.599 (2)
O (6)–H (2)···O $(2)_{5563}$	1.020 (5)	1.552 (5)	2.570 (3)
O (3)–H (3)···O (4)	1.052 (3)	1.464 (3)	2.512 (2)
Z–O–H···O–W	∢Z–O–H	∢O–H···O	∢H···O–W
Se (1)–O (1)–H (1)···O (5)–Se (1)	112.1 (2)	172.1 (3)	115.9 (2)
Se (2)–O (6)–H (2)···O (2)–Se (2)	112.5 (2)	174.4 (4)	121.7 (2)
Se (1)–O (3)–H (3)···O (4)–Se (2)	108.9 (2)	173.0 (3)	122.3 (2)

Table 260α. $RbH_3(SeO_3)_2$. Direction cosines and the principal values of the EFG tensor of ^{87}Rb at 13 °C and −155.5 °C [79T5]. See also Figs. 988α, 988β. Four inequivalent Rb sites A_1, A_2, B_1, B_2 were observed in the ferroelectric phase.

T °C		$\lvert eQ\phi_{ii}/h\rvert$	Direction cosine with respect to		
		kHz	a	b	c
13.0		5130	±0.299	±0.359	0.884
		3832	±0.789	0.615	∓0.017
		1298	0.538	0.702	∓0.467
−155.5	A_1	6129	−0.349	−0.455	0.819
		5566	−0.775	0.631	0.020
		563	0.526	0.628	0.573
	A_2	5151	−0.275	−0.379	0.884
		2968	0.119	0.898	0.423
		2183	0.954	−0.221	0.202
	B_1	5404	0.239	0.417	0.877
		4412	−0.811	0.583	−0.056
		992	0.535	0.697	−0.048
	B_2	5975	0.393	0.410	0.823
		3605	−0.589	0.800	−0.118
		2370	0.707	0.438	−0.555

Nr. 28A-5 $CsH_3(SeO_3)_2$, Cesium trihydrogen selenite
($M=389.85$)

1a	Dielectric anomaly associated with a phase transition was observed and possibility of antiferroelectricity was first reported by Makita et al. in 1962.	62M3 65M2
b	see table below	a) 62M3 b) 65M2

phase	II[a])	I[a])
state	(A)[a])	P[a])
crystal system	triclinic[b])	triclinic[b])
space group	$P\bar{1}$-C_i^1 [b]) *)	$P\bar{1}$-C_i^1 [b])
Θ	−128 °C[b])	

Isotope effect: Fig. 992.
Transparent, colorless.
Deliquescent.

*) Phase II has superstructure: length of the c axis is twice as big as that in phase I.

2a	Crystal growth: slow evaporation from aqueous solution of CsOH and H_2SeO_3 in stoichiometric proportion of one mole of CsOH and two moles of H_2SeO_3.	65M2
3a	Unit cell parameters: $a=9.32_6$ Å, $b=6.52_5$ Å, $c=5.83_5$ Å, $\alpha=91.5°$, $\beta=105.1°$, $\gamma=91.1°$ at RT. Phase II has superlattice structure with $c_{II}\simeq 2c_I$ (c_I: c in phase I).	65M2
b	Crystal structure: Tables 261···266; Figs. 993, 994, 995.	
5a	Dielectric constant: Figs. 996, 997. $\varkappa_{[100]}=C/(T-\Theta_p)$, where $C=1.14\cdot10^3$ K, $\Theta_p=65$ K.	65M2
b	Bias effect on Θ: Fig. 998.	
6a	Heat capacity: Fig. 999. Transition entropy: $\Delta S_m=4.51$ J mol^{-1} K^{-1} ($\Delta S_m=5.56$ J mol^{-1} K^{-1} for $CsD_3(SeO_3)_2$).	79O3
9a	Birefringence: Fig. 1000. Infrared absorption: Fig. 1001; Table 267.	

Table 261. $CsH_3(SeO_3)_2$. Crystal structure of phase I at RT [72S5]. Fractional coordinates and the isotropic temperature parameters. The temperature parameters are defined by Eq. (e) in Introduction.

Atoms	x	y	z	B [Å^2]
Cs	0.161	0.760	0.110	1.7
Se (1)	0.214	0.255	0.589	2.0
Se (2)	0.528	0.250	0.189	2.1
O (1)	0.154	0.260	0.284	2.4
O (2)	0.132	0.029	0.633	2.8
O (3)	0.096	0.430	0.668	3.3
O (4)	0.483	0.014	0.273	3.5
O (5)	0.594	0.379	0.449	1.9
O (6)	0.362	0.350	0.082	2.3
H (1)	0.249	0.295	0.211	4.1
H (2)	0.029	0.004	0.534	2.1
H (3)	0.033	0.475	0.541	1.3
H (4)	0.491	−0.038	0.452	1.7
H (5)	0.520	0.481	0.459	2.4

Table 262. $CsH_3(SeO_3)_2$. Crystal structure of phase I [74T4]. Fractional coordinates at 25 °C ($\cdot 10^5$ for heavy atoms and $\cdot 10^3$ for hydrogens). The lattice parameters used are $a = 9.3474$ (5) Å, $b = 6.5398$ (4) Å, $c = 5.8498$ (3) Å, $\alpha = 91.443$ (6)°, $\beta = 105.336$ (6)°, $\gamma = 91.629$ (6)°.

Atom	x	y	z
Cs	83899 (2)	23929 (3)	89065 (3)
Se (1)	21376 (3)	25642 (4)	59253 (5)
Se (2)	52820 (3)	25001 (4)	18867 (5)
O (1)	15274 (28)	26035 (40)	28574 (44)
O (2)	13294 (30)	02932 (37)	63418 (49)
O (3)	09940 (28)	42592 (42)	66694 (41)
O (4)	48506 (33)	01557 (36)	27616 (45)
O (5)	59499 (24)	37950 (33)	45356 (37)
O (6)	35998 (23)	34866 (34)	08362 (35)
H (1)	195 (9)	256 (13)	246 (14)
H (2)	951 (8)	−022 (13)	438 (14)
H (3)	044 (9)	452 (13)	579 (14)
H (4)	485 (10)	006 (14)	380 (12)
H (5)	545 (7)	447 (10)	482 (12)
H (6)	283 (9)	305 (11)	191 (13)

Table 263. $CsH_3(SeO_3)_2$. Crystal structure of phase I [74T4].
A. Anisotropic temperature parameters for heavy atoms ($\cdot 10^5$) and root-mean-square components R_i ($\cdot 10^3$ Å) of thermal displacement along ellipsoid axes. Temperature parameters are defined by Eq. (b) in Introduction.
B. Isotropic temperature parameters for hydrogens. The temperature parameters are defined by Eq. (e) in Introduction.

A	b_{11}	b_{22}	b_{33}	b_{12}	b_{13}	b_{23}
Cs	0698 (3)	1376 (5)	1797 (6)	0078 (2)	0248 (3)	0043 (4)
Se (1)	0505 (3)	1186 (6)	1470 (8)	0017 (3)	0115 (4)	0058 (5)
Se (2)	0694 (4)	1250 (6)	1416 (8)	−0008 (3)	0376 (4)	−0129 (5)
O (1)	0681 (29)	1984 (63)	1489 (66)	0025 (32)	0183 (37)	−0083 (47)
O (2)	1059 (35)	1653 (57)	3033 (91)	−0470 (36)	0040 (45)	0675 (56)
O (3)	0940 (31)	2435 (67)	1628 (69)	0779 (37)	0205 (37)	−0072 (54)
O (4)	1526 (40)	1110 (49)	1972 (79)	−0232 (33)	0342 (45)	0029 (50)
O (5)	0761 (27)	1386 (50)	1747 (64)	0059 (29)	0030 (32)	−0422 (44)
O (6)	0741 (26)	2082 (57)	1485 (61)	0218 (30)	0089 (31)	0105 (47)

A	R_1	R_2	R_3	B	B [Å^2]
Cs	166 (1)	171 (1)	177 (1)	H (1)	1.2 (20)
Se (1)	140 (1)	160 (1)	164 (1)	H (2)	2.9 (17)
Se (2)	140 (1)	163 (1)	175 (1)	H (3)	2.9 (18)
O (1)	152 (3)	172 (4)	208 (3)	H (4)	3.0 (20)
O (2)	150 (3)	195 (3)	272 (3)	H (5)	1.5 (14)
O (3)	148 (4)	161 (3)	265 (3)	H (6)	3.0 (15)
O (4)	150 (3)	178 (4)	259 (3)		
O (5)	141 (3)	174 (3)	210 (3)		
O (6)	148 (3)	182 (3)	216 (3)		

Table 264. $CsH_3(SeO_3)_2$. Crystal structure of phase I [74T4]. Interatomic distances (in Å) and angles (in deg). Primed atoms are related to basis atoms by symmetry operations.

Covalent bonds				Hydrogen bonds		
O−Se−O	O−Se	Se−O	<O−Se−O	O−H···O	O−H	O···O
O (1)−Se (1)−O (2)	1.735 (3)	1.699 (2)	98.9 (1)	O (2)−H (2)···O (2)	0.78 (7)	2.577 (5)
O (1)−Se (1)−O (3)	1.735 (3)	1.688 (2)	100.2 (1)	O (3)−H (3)···O (3)	0.65 (7)	2.546 (5)
O (2)−Se (1)−O (3)	1.699 (2)	1.688 (2)	101.8 (1)	O (4)−H (4)···O (4)	0.61 (7)	2.573 (5)
O (4)−Se (2)−O (5)	1.700 (2)	1.703 (2)	101.8 (1)	O (5)−H (5)···O (5)	0.70 (6)	2.561 (4)
O (4)−Se (2)−O (6)	1.700 (2)	1.683 (2)	102.4 (1)	O (1)−H (1)···O (6)	0.51 (7)	2.582 (3)
O (5)−Se (2)−O (6)	1.703 (2)	1.683 (2)	102.6 (1)	O (6)−H (6)···O (1)	1.11 (7)	—

Short contacts around the cesium ion

Cs−O (5)	3.120 (2)	Cs−H (6)	3.22 (7)
Cs−O (1)	3.215 (2)	Cs−H (3)	3.28 (9)
Cs−O (3)	3.268 (3)	Cs−H (1)	3.31 (8)
Cs−O (3′)	3.275 (3)		
Cs−O (2)	3.281 (3)		
Cs−O (4)	3.313 (3)		
Cs−O (6)	3.340 (2)		

Table 266. $CsH_3(SeO_3)_2$. Crystal structure of phase I [78C1]. Interatomic distances (in Å) and angles (in deg) at RT. A prime indicates an atom outside the asymmetric unit.

Ionic bonds

Cs−O (1)	3.223 (1)	Cs−O (4)	3.308 (2)
Cs−O (2)	3.281 (2)	Cs−O (5)	3.123 (2)
Cs−O (3)	3.270 (2)	Cs−O (6)	3.342 (2)
Cs−O (3)	3.278 (2)	Cs−H (1)	3.217 (4)
		Cs−H (6)	3.229 (9)

Covalent bonds

O−Se−O	O−Se	Se−O	∠O−Se−O
O (1)−Se (1)−O (2)	1.735 (1)	1.702 (1)	99.06 (7)
O (1)−Se (1)−O (3)	1.735 (1)	1.689 (1)	100.22 (6)
O (2)−Se (1)−O (3)	1.702 (1)	1.689 (1)	101.92 (7)
O (4)−Se (2)−O (5)	1.703 (1)	1.705 (1)	101.79 (7)
O (4)−Se (2)−O (6)	1.703 (1)	1.683 (1)	102.33 (7)
O (5)−Se (2)−O (6)	1.705 (1)	1.683 (1)	102.52 (6)

Hydrogen bonds

O−H···O	O−H	H···O	O···O
O (1)−H (1)···O (6)	1.000 (4)	1.581 (4)	2.577 (1)
O (6)−H (6)···O (1)	1.028 (9)	1.558 (9)	2.577 (1)
O (2)−H (2)···O (2′)	1.013 (5)	1.563 (5)	2.576 (3)
O (3)−H (3)···O (3′)	1.000 (6)	1.543 (5)	2.539 (2)
O (4)−H (4)···O (4′)	1.011 (6)	1.558 (6)	2.566 (3)
O (5)−H (5)···O (5′)	1.040 (11)	1.511 (11)	2.529 (2)

Se−O−H···O−Se	∠Se−O−H	∠O−H···O	∠H···O−Se
Se (1)−O (1)−H (1)···O (6)−Se (2)	111.2 (2)	174.0 (4)	116.4 (2)
Se (1)−O (2)−H (2)···O (2′)−Se (1′)	112.0 (4)	177.7 (7)	114.3 (3)
Se (1)−O (3)−H (3)···O (3′)−Se (1′)	114.7 (3)	174.0 (5)	120.2 (2)
Se (2)−O (4)−H (4)···O (4′)−Se (2′)	112.0 (5)	173.8 (8)	118.1 (3)
Se (2)−O (5)−H (5)···O (5′)−Se (2′)	110.7 (6)	175.7 (10)	113.7 (5)
Se (2)−O (6)−H (6)···O (1)−Se (1)	113.1 (5)	170.8 (9)	111.7 (4)

Table 265. $CsH_3(SeO_3)_2$. Crystal structure of phase I [78C1]. Fractional coordinates at RT. $\alpha[H(1)]=0.68$, $\alpha[H(6)]=0.32$, α: occupation number.

	x	y	z
Cs	83884 (12)	23298 (16)	89073 (20)
Se (1)	21390 (7)	25648 (10)	59247 (12)
Se (2)	52845 (7)	24999 (10)	18862 (12)
O (1)	15393 (11)	26063 (16)	28549 (18)
O (2)	13300 (15)	2888 (18)	63388 (25)
O (3)	9928 (14)	42605 (19)	66614 (21)
O (4)	48467 (15)	1545 (15)	27668 (26)
O (5)	59459 (12)	38031 (16)	45347 (18)
O (6)	36030 (12)	34863 (17)	8263 (18)
H (1)	2386 (4)	2904 (6)	2152 (7)
H (2)	286 (5)	95 (11)	5266 (12)
H (3)	252 (5)	4826 (10)	5284 (11)
H (4)	4991 (9)	130 (13)	4541 (10)
H (5)	5138 (14)	4728 (13)	3890 (20)
H (6)	2851 (10)	3040 (14)	1735 (18)

Table 267. $CsH_3(SeO_3)_2$. Frequencies and assignments of vibrational modes [66K4]. Frequencies[1]) in $10^2\ m^{-1}$.

phase I (−120 °C)	phase II (−175 °C)	Assignment	phase I (−120 °C)	phase II (−175 °C)	Assignment
310	275 308 323	ν_4', ν_4'' (O_2*)SeO)	850 vw	847 873	ν_1 (O_2*)SeO)
322	323 326 335?	ν_4' O*) SeO_2^-	970	967 988 992? 1025	O−H···O out-of-plane bend
360 vw[1])	358 371	ν_4'' O*) SeO_2^-	1220 1270 1300	1225 1275 1325 1330?	O−H···O in-plane bend
395	400 417	ν_2 (O_2^*) SeO)	1600 2250 vb[1]) 2380 2600 2720 vb 2950 vb 3400	1600 2260 vb 2390 2630 2700 3240 3400	O−H···O stretch combinations[2]) and overtones
417	417 437	ν_2 (O*) SeO_2^-)			
630 vw	630 vw (background)	ν_3' (O_2^*) SeO)			
650	649 663	ν_1 (O*) SeO_2^-)			
685 vw	675 693	ν_3'' (O_2^*) SeO)			
755	755 767	ν_3 (O*) SeO_2^-)			
825	817 823	ν_3'' (O*) SeO_2^-)			

*) Phase II has superlattice structure; length of the c axis is twice as big as that in phase I.

[1]) vw: very weak, vb: very broad.

[2]) Observed in the spectra of the polycrystalline film deposited on a BaF_2 substrate. In general, the frequencies observed in the spectra of the polycrystalline film are different from the corresponding frequencies in the mulled spectra.

Nr. 28A-6 $NH_4H_3(SeO_3)_2$, Ammonium trihydrogen selenite
(M=274.98)

1a	A single crystal of $NH_4H_3(SeO_3)_2$ was grown by Shuvalov et al. in 1969.	69S7
b	Space group: $P2_12_12_1$-D_2^4. T_{melt}=40 °C. ϱ=2.75 · 10^3 kg m^{-3}. Transparent, colorless.	72T4
2a	Crystal growth: solution method (solvent H_2O), cooling from temperatures lower than 10 °C in a dark room.	69S7
b	Crystal form: Fig. 1002.	
3a	Unit cell parameters: a=5.837 Å, b=17.718 Å, c=6.307 Å at RT. a=5.81 (1) Å, b=17.56 (3) Å, c=6.25 (1) Å at −150 °C.	 72G11 76G13
b	Z=4. Crystal structure: Tables 268···271; Fig. 1003. For neutron diffraction data, see	75T3 74T3
5a	Dielectric constants: Fig. 1004.	
9a	Refractive indices: n_a=1.684, n_b=1.651, n_c=1.652 for λ=589 nm.	69S7
b	Electrooptic constants: r_{41}^T=1.8 (4) · 10^{-12} m V^{-1}, r_{52}^T=1.5 (5) · 10^{-12} m V^{-1} for λ=633 nm at 20 °C..	 70I8
d	Optical activity: Fig. 1005	

Table 268. $NH_4H_3(SeO_3)_2$. Crystal structure [72G11]. Fractional coordinates and temperature parameters at RT. The temperature parameters are defined by Eq. (e) in Introduction.

Atom	x	y	z	B[Å^2]
Se 1	0.459	0.3565	0.0549	0.14
Se 2	0.493	0.01055	0.0770	0.32
O 1	0.328	0.2906	0.196	0.77
O 2	0.416	0.3253	−0.202	0.91
O 3	0.736	0.3243	0.0911	0.90
O 4	0.467	−0.08428	0.0569	2.01
O 5	0.306	0.04532	−0.102	0.89
O 6	0.752	0.02873	−0.0434	1.15
N	0.977	0.3333	0.539	0.85

Table 269. $NH_4H_3(SeO_3)_2$. Structure [72G11]. Interatomic distances at RT (in Å). For primed atoms, see Fig. 1003. Primed atoms are related to basis atoms by symmetry operations.

Se 1–O 1	1.65	Se 2–O 4	1.69	O 5–O 6	2.61
Se 1–O 2	1.73	Se 2–O 5	1.69	O 3–O 4	2.55
Se 1–O 3	1.73	Se 2–O 6	1.72	O 1–O 2	2.62
O 1–O 3	2.54	O 5–O 6	2.65	N–O 5	2.85
O 1–O 2	2.63	O 5–O 4	2.67	N–O 4	2.95
O 2–O 3	2.63	O 4–O 6	2.68	N–O 6	2.96
				N–O 1	2.99
				N–O 2	3.04
				N–O 1′	3.07
				N–O 3	3.16
				N–O 3′	3.27

Table 270. $NH_4H_3(SeO_3)_2$. Crystal structure [76G13]. Fractional coordinates and temperature parameters at −150 °C. The temperature parameters are defined by Eq. (a) in Introduction.

	x	y	z	B_{11}	B_{22}	B_{33}	B_{12}	B_{13}	B_{23}
Se 1	0.46252 (10)	0.35712 (4)	0.05835 (12)	0.00553 (16)	0.00048 (2)	0.00558 (15)	0.00000 (9)	−0.00065 (34)	−0.00002 (10)
Se 2	0.49036 (11)	0.01015 (3)	0.07872 (11)	0.00674 (16)	0.00063 (2)	0.00538 (14)	0.00006 (10)	0.00032 (35)	0.00011 (9)
O 3	0.74187 (77)	0.32223 (28)	0.09228 (90)	0.00822 (115)	0.00085 (14)	0.01011 (146)	0.00041 (73)	0.00068 (251)	0.00094 (85)
O 2	0.42677 (83)	0.32864 (26)	−0.20782 (88)	0.01049 (140)	0.00095 (15)	0.00776 (123)	0.00083 (79)	−0.00193 (239)	0.00060 (74)
O 6	0.75305 (80)	0.02733 (28)	−0.04817 (109)	0.00827 (118)	0.00115 (16)	0.001173 (164)	−0.00080 (70)	−0.00173 (297)	0.00006 (94)
O 5	0.30059 (81)	0.04587 (29)	−0.09715 (102)	0.01030 (124)	0.00086 (15)	0.01101 (175)	0.00160 (73)	−0.00356 (268)	0.00070 (87)
O 4	0.46324 (80)	0.08378 (25)	0.05146 (91)	0.01060 (122)	0.00067 (12)	0.01093 (126)	0.00011 (71)	−0.00114 (311)	−0.00069 (74)
O 1	0.32492 (81)	0.29006 (28)	0.19301 (90)	0.00830 (115)	0.00101 (14)	0.00822 (127)	0.00083 (75)	−0.00596 (242)	0.00225 (83)
N	0.99015 (109)	0.33234 (31)	0.52843 (99)	0.01208 (166)	0.00077 (16)	0.00630 (143)	−0.00147 (91)	−0.00199 (330)	−0.00041 (74)
H 1	0.010	0.788	0.250						
H 2	0.250	0.495	0.330						
H 3	0.320	0.647	0.530						

Table 271. $NH_4H_3(SeO_3)_2$. Crystal structure [76G13]. Interatomic distances at −150 °C (in Å).

	−150 °C		−150 °C		−150 °C
Se 1–O 1	1.653 (4)	O 3···H–O′ 6	2.580 (4)	Se 1–Se′ 1	4.911 (3)
				Se 1–Se′ 2	3.807 (2)
Se 1–O 2	1.747 (5)	O 3–H···O 4	2.538 (5)	Se 1–Se″ 2	3.857 (1)
Se 1–O 3	1.750 (3)	O 1···H–O 2	2.611 (3)	Se 1–Se‴ 2	4.010 (2)
O 1–O 3	2.566 (4)	O 1–H 1	1.79	Se 1–Se⁗ 2	4.249 (3)
O 1–O 2	2.662 (5)	O 2–H 1	0.84	Se 1–O′ 1	3.664 (2)
O 2–O 3	2.624 (4)	O 5–H 2	1.71	Se 1–O′ 2	4.236 (1)
Se 2–O 4	1.669 (4)	O 6–H 2	0.85	Se 1–O 3	3.633 (1)
Se 2–O 5	1.680 (3)	O 3–H 3	1.04	Se 1–O′ 5	2.767 (3)
Se 2–O 6	1.747 (3)	O 4–H 3	1.56	Se 1–O 5	3.070 (4)
				Se 1–O″ 1	3.416 (2)
O 5–O 6	2.668 (6)	N–O 5	2.829 (6)		
O 5–O 4	2.640 (4)	N–O 4	2.916 (6)		
O 4–O 6	2.656 (5)	N–O 6	2.958 (5)		
		N–O 1	3.010 (4)		
		N–O 2	3.018 (5)		
		N–O′ 1	3.026 (6)		
		N–O 3	3.089 (5)		
		N–O′ 3	3.160 (4)		

29 Colemanite ($Ca_2B_6O_{11} \cdot 5H_2O$)

29A Pure compound

Nr. 29A-1 $Ca_2B_6O_{11} \cdot 5H_2O$, Colemanite
Another chemical formula $CaB_3O_4(OH)_3 \cdot H_2O$ was proposed from the crystal-structural view point [58C1]. Number of atoms in this formula is the same as in $1/2(Ca_2B_6O_{11} \cdot 5H_2O)$
(M=411.09)

1a	Ferroelectric activity in $Ca_2B_6O_{11} \cdot 5H_2O$ was discovered by Goldsmith in 1956.	56G1
b	phase: II [a]), I [a]); state: F [a]), P [a]); crystal system: monoclinic [b]), monoclinic [c]); space group: $p2_1$-C_2^2 [b]), $P2_1/a$-C_{2h}^5 [c]); Θ: ≈ −7 °C [d])	a) 56G1 b) 58H2 c) 58C1 d) 59W3
	$P_s \| [010]$.	58C1
	$\varrho = 2.42 \cdot 10^3$ kg m^{-3} at RT. Transparent, colorless. Cleavage plane ∥ (010).	
2a	Crystal growth: Small crystals are obtained by the dehydration of higher order hydrates of some calcium borates. Colemanite of high quality, however, has not yet been produced artificially. All data for the physical properties of colemanite described here are concerned with natural minerals.	60W5
3a	Unit cell parameters: a=8.743 (4) Å, b=11.264 (2) Å, c=6.102 (3) Å, β=110°7 (5)′ at RT.	58C1
b	Z=2 in phase I [a]) and in phase II [b]) [Z=4 if the molecular unit is expressed as $CaB_3O_4(OH)_3 \cdot H_2O$]. Crystal structure: Tables 272···275; Figs. 1006···1008.	a) 58C1 b) 65H1
5a	Dielectric constants: Figs. 1009···1012. $\Theta_p = -7$ °C, $C = 5 \cdot 10^2$ K.	59W3
b	$\xi = 7.6 \cdot 10^{13}$ V m^5 C^{-3}, $\zeta = 9.2 \cdot 10^{17}$ V m^9 C^{-5}.	59W3
c	Spontaneous polarization: Fig. 1013.	
13a	NMR: Figs. 1014···1017.	

Phase table (1b):

phase	II [a])	I [a])
state	F [a])	P [a])
crystal system	monoclinic [b])	monoclinic [c])
space group	$p2_1$-C_2^2 [b])	$P2_1/a$-C_{2h}^5 [c])
Θ	≈ −7 °C [d])	

Table 272. $Ca_2B_6O_{11} \cdot 5H_2O$ (colemanite). Crystal structure of phase I [66H1]. T=23 °C. Fractional coordinates, and isotropic temperature parameters obtained by neutron diffraction. For A, B, C, D sites, see Figs. 1007, 1008. The temperature parameters are defined by Eq. (e) in Introduction.

Atoms	x	y	z	B
	10^{-4}			Å^2
H (63)	775 (15)	246 (12)	6959 (26)	1.8 (2)
H (83)	−789 (11)	830 (8)	−1276 (26)	1.5 (2)
H (54) (C)	3799 (35)	−843 (31)	6539 (69)	3.0 (6)
H (54) (D)	3128 (26)	−992 (30)	6784 (49)	1.6 (4)
H (99) (A)	232 (63)	5037 (99)	9336 (99)	5.7 (11)
H (99) (B)	729 (55)	5199 (52)	7120 (91)	4.0 (8)
H (96)	2006 (15)	5204 (15)	229 (32)	3.4 (3)
Ca	1349 (8)	2884 (6)	7340 (17)	1.20 (9)
B (1)	3345 (6)	2117 (15)	3104 (11)	0.47 (7)
B (2)	2204 (6)	523 (4)	4943 (23)	0.44 (7)
B (3)	402 (5)	1709 (5)	1656 (11)	0.56 (7)

continued

Table 272, continued

Atoms	x	y	z	B
	10^{-4}			Å^2
O (1)	3424 (6)	1499 (5)	5078 (12)	0.78 (9)
O (2)	1889 (5)	2314 (4)	1411 (10)	0.43 (6)
O (3)	880 (5)	554 (5)	2701 (10)	0.43 (6)
O (4)	4808 (5)	2530 (5)	3139 (11)	0.59 (6)
O (5)	2961 (7)	− 671 (7)	5151 (27)	1.11 (9)
O (6)	1669 (7)	776 (5)	6924 (12)	0.82 (7)
O (8)	− 801 (6)	1632 (5)	− 724 (13)	0.67 (7)
O (9)	1143 (7)	4823 (6)	9047 (15)	1.52 (10)

Table 273. $Ca_2B_6O_{11} \cdot 5H_2O$ (colemanite). Crystal structure of phase II [66H1]. $T = -20\,°C$. Fractional coordinates and isotropic temperature parameters. A primed atom is related to an unprimed atom by the a glide plane in phase I. For A, B, C, D sites, see Figs. 1007, 1008. See also captions of Tables 274, 275. The temperature parameters are defined by Eq. (e) in Introduction.

Atoms	Unprimed	Primed	Unprimed	Primed	Unprimed	Primed	Unprimed	Primed
	x 10^{-4}		y 10^{-4}		z 10^{-4}		B Å^2	
H (63)	3331 (40)	8224 (34)	198 (36)	4711 (38)	6922 (93)	6989 (75)	1.7 (6)	1.2 (5)
H (83)	1606 (27)	6847 (26)	878 (29)	4180 (31)	−1365 (55)	−1204 (71)	0.7 (4)	0.9 (4)
H (54) (C)		1322 (30)		5769 (37)		6574 (62)		2.9 (5)
H (54) (D)	5645 (25)		− 937 (36)		6823 (43)		1.6 (4)	
H (99) (A)		7600 (69)		52 (81)		9312 (96)		5.0 (7)
H (99) (B)	3293 (36)		5229 (34)		7232 (96)		2.4 (6)	
H (96)	4524 (52)	9564 (33)	5212 (76)	− 143 (55)	66 (99)	466 (64)	4.3 (10)	1.8 (5)
Ca	3929 (27)	8781 (25)	2878 (18)	2110 (18)	7320 (47)	7434 (46)	0.97 (37)	1.05 (34)
B (1)	5821 (19)	876 (23)	2164 (28)	2907 (30)	3040 (55)	3030 (69)	0.23 (30)	0.37 (32)
B (2)	4704 (17)	9715 (20)	604 (26)	4585 (29)	4906 (62)	4883 (91)	0.18 (30)	0.61 (38)
B (3)	2938 (19)	7868 (17)	1784 (25)	3361 (24)	1640 (48)	1614 (40)	0.70 (30)	0.39 (25)
O (1)	5970 (20)	883 (17)	1473 (29)	3495 (24)	5054 (46)	5063 (38)	1.19 (32)	0.28 (23)
O (2)	4426 (18)	9346 (19)	2291 (30)	2681 (32)	1381 (31)	1446 (37)	0.15 (25)	0.69 (29)
O (3)	3318 (21)	8409 (18)	581 (29)	4509 (29)	2680 (48)	2704 (37)	0.93 (34)	0.24 (24)
O (4)	7302 (20)	2310 (21)	2537 (30)	2490 (32)	3165 (42)	3139 (44)	0.51 (30)	0.78 (33)
O (5)	5481 (26)	442 (20)	− 614 (37)	5708 (30)	5162 (82)	5075 (75)	1.81 (43)	0.40 (29)
O (6)	4104 (23)	9238 (21)	790 (33)	4244 (29)	6696 (46)	6984 (69)	1.50 (41)	0.59 (29)
O (8)	1705 (21)	6705 (16)	1568 (32)	3329 (27)	− 691 (49)	− 781 (32)	1.46 (33)	0.16 (21)
O (9)	3599 (18)	8698 (27)	4878 (23)	256 (29)	8974 (32)	9155 (52)	0.54 (22)	2.47 (46)

Table 274. $Ca_2B_6O_{11} \cdot 5H_2O$ (colemanite). Crystal structures of phases I and II [66H1]. $T = +23$ °C (phase I) and -20 °C (phase II). Interatomic distances and angles for the hydroxyl groups. A primed atom is related to an unprimed atom by a glide plane in phase I; to a two primed atom by the 2_1 axis in both phases; to a three primed atom by inversion in phase I. For sites C, D, see Figs. 1007, 1008.

Bond system	T [°C]	Distances [Å]			Angles [deg]		
O (6)–H (63)		O (6)···O (3)	O (6)–H (63)	H (63)···O (3)	B (2)–O (6)···O (3)	B (2)–O (6)–H (63)	O (6)–H (63)···O (3)
B (2)–O (6)–H (63)···O‴ (3)	+23	2.76 (1)	1.01 (2)	1.79 (2)	117	112	172
B (2)–O (6)–H (63)···O‴ (3)	−20	2.76 (4)	1.01 (5)	1.79 (5)	124	118	163
B (2)–O′ (6)–H′ (63)···O″ (3)	−20	2.76 (4)	1.05 (5)	1.73 (5)	113	111	173
O (8)–H (83)		O (8)···O (3)	O (8)–H (83)	H (83)···O (3)	B (3)–O (8)···O (3)	B (3)–O (8)–H (83)	O (8)–H (83)···O (3)
B (3)–O (8)–H (83)···O‴ (3)	+23	2.73 (1)	0.99 (2)	1.79 (2)	113	108	172
B (3)–O (8)–H (83)···O‴ (3)	−20	2.61 (4)	0.89 (5)	1.74 (5)	117	122	175
B (3)–O′ (8)–H′ (83)···O″ (3)	−20	2.79 (3)	1.02 (4)	1.80 (4)	108	97	165
O (5)–H (54), site C		O (5)···O (4)	O (5)–H (54)	H (54)···O (4)	B (2)–O (5)···O (4)	B (2)–O (5)–H (54)	O (5)–H (54)···O (4)
B (2)–O (5)–H (54)···O‴ (4)	+23	2.81 (1)	0.98 (4)	2.25 (4)	160	117	120
B′ (2)–O′ (5)–H′ (54)···O″ (4)	−20	2.80 (5)	1.01 (5)	2.30 (4)	161	109	111
O (5)–H (54), site D		O (5)···O (2)	O (5)–H (54)	H (54)···O (2)	B (2)–O (5)···O (2)	B (2)–O (5)–H (54)	O (5)–H (54)···O (2)
B (2)–O (5)–H (54)···O″ (2)	+23	3.06 (1)	1.04 (3)	2.21 (3)	130	108	140
B (2)–O (5)–H (54)···O″ (2)	−20	3.15 (4)	1.06 (5)	2.29 (4)	131	109	139

Table 276. $K_4Me(CN)_6 \cdot 3H_2O$, $K_4Me(CN)_6 \cdot 3D_2O$. Ferroelectric transition temperature. [60W1].

Me	H	D
	Θ_f [°C]	
Mn	−40	—
Fe	−24.5	−18.0
Ru	−14.5	− 7.3
Os	− 2.4	1.8

Tables 276···278 see chapter 30.

Table 277. $K_4Fe(CN)_6 \cdot 3H_2O$. Solubility A in water [62K2].

A wt%	T °C	A wt%	T °C
1.25	− 0.24	26.0	30
4.0	− 0.62	27.8	35
8.0	− 1.09	32.6	50
11.6	− 1.6	36.8	65
12.5	0	38.2	70
15.0	5	40.1	80
17.4	10	41.34	87.3
19.3	15	41.99	94
22.0	20	42.63	99.6
24.0	25	44.77	104.1

Table 278. $K_4Fe(CN)_6 \cdot 3H_2O$. Atomic coordinates at RT (in phase I) [64K5].

Atom	Equivalent positions	x	y	z
Fe	4e	0.000	0.179	0.250
K (1)	8f	0.188	0.850	0.156
K (2)	8f	0.095	0.646	0.444
C (1)	8f	0.190	0.179	0.314
C (2)	8f	0.067	0.179	0.055
C (3)	4e	0.000	0.052	0.250
C (4)	4e	0.000	0.303	0.250
N (1)	8f	0.301	0.179	0.353
N (2)	8f	0.095	0.179	0.937
N (3)	4e	0.000	0.003	0.250
N (4)	4e	0.000	0.354	0.250
O (1)	4e	0.000	0.467	0.750
O (2)	8f	0.250	0.500	0.000

Table 275. $Ca_2B_6O_{11} \cdot 5H_2O$ (colemanite). Crystal structures of phases I and II [66H1]. Interatomic distances and angles for water molecules at different sites. See captions of Tables 273, 274 for primed atoms. For sites A, B, see Figs. 1007, 1008.

Distances [Å]			Angles [deg]		
T:	+23 °C	−20 °C		+23 °C	−20 °C
Site:	A or A′	A		A or A′	A′
O‴(9)···O′(6)	2.77 (1)	2.68 (4)	O(9)···O‴(9)···O(6)	87	86
O‴(9)–H‴(96)	0.99 (4)	1.06 (5)	O‴(9)–H‴(99)···O(9)	158	150
H‴(96)···O′(6)	1.86 (3)	1.71 (5)	O‴(9)–H‴(96)···O′(6)	163	157
O‴(99)···O(9)	2.67 (1)	2.66 (4)	H‴(96)–O‴(9)–H‴(99)	105	104
O‴(9)–H‴(99)	0.99 (8)	1.05 (5)			
H‴(99)···O(9)	1.85 (9)	1.74 (5)			
Ca‴–H‴(99)	3.02 (8)	2.93 (7)			
Ca‴–O‴(9)	2.45 (1)	2.35 (4)			
Site:	B or B′	B		B or B′	B
O(9)···O″(6)	2.77 (1)	2.90 (4)	O″(5)···O(9)···O″(6)	122	123
O(9)–H(96)	0.99 (2)	0.96 (7)	O(9)–H(99)···O″(5)	114	118
H 96···O″(6)	1.86 (3)	2.03 (7)	O(9)–H(96)···O″(6)	163	165
O(9)–H(99)	1.23 (7)	1.11 (5)	H(96)–O(9)–H(99)	119	119
Ca···H(99)	2.66 (6)	2.70 (5)			
Ca···O(9)	2.45 (1)	2.53 (4)			

30 $K_4Fe(CN)_6 \cdot 3H_2O$ family

30A Pure compounds

Data on deuterated salts are given in bracket as [D:].

Nr. 30A-1 $K_4Mn(CN)_6 \cdot 3H_2O$. Potassium manganese cyanide trihydrate
($M = 421.48$)

1a	Ferroelectricity was found by Waku et al. in 1960.	60W2
b	Ferroelectric transition temperature: $\Theta_f = -40.4$ °C. Table 276. Color: dark violet.	60W2
5a	Dielectric constant: Fig. 1018.	

Table 276, see p. 198.

Nr. 30A-2 $K_4Fe(CN)_6 \cdot 3H_2O$. Potassium ferrocyanide trihydrate
($M = 422.39$)

1a Ferroelectricity was discovered by Waku et al. in 1959. 59W1

b

phase	II	I
state	F	P
crystal system	monoclinic	monoclinic
space group	Cc-C_s^4	C2/c-C_{2h}^6
Θ_f	−24.5 °C [D: −18.0 °C]	

Θ_f: 64K5; [D]: 60W1

$P_s \| [10\bar{1}]$ 59W1
$\varrho = 1.89 \cdot 10^3$ kg m^{-3}.
Transparent, yellow.

continued

Nr. 30A-2 $K_4Fe(CN)_6 \cdot 3H_2O$

2a	Evaporation or cooling method using aqueous solution. Solubility in water: Table 277.	59W1
b	Crystal form: Fig. 1019.	
3a	Unit cell parameters: $a=9.39_6$ Å, $b=16.8_6$ Å, $c=9.41_3$ Å, $\beta=90°3'$ at RT (phase I). $a=9.35$ Å, $b=16.7$ Å, $c=9.37$ Å, $\beta=90°6'$ at -90 °C (phase II). There is a tetragonal modification which is not ferroelectric: $a=9.41$ Å, $c=33.67$ Å, space group: $I4_1/a\text{-}C_{4h}^6$. The tetragonal form is irreversibly transformed to the monoclinic form, when cooled down to low temperatures, e.g. -55 °C.	64K5 60T3
b	Crystal structure: Table 278; Fig. 1020. Crystal structure of deuterated salt: Table 279; Fig. 1021.	
5a	Dielectric constant: Figs. 1022, 1023, 1024. Several dielectric peaks often observed at lower temperatures are due to the irreversible transformations between monoclinic and tetragonal forms. Pressure effect on the dielectric properties: Figs. 1025, 1026. $d\Theta_f/dp=2.3 \cdot 10^{-8}$ K Pa^{-1}.	66K8
c	Spontaneous polarization and coercive field: Figs. 1027, 1028. $P_s=1.4 \cdot 10^{-2}$ Cm^{-2} at -80 °C [D: $P_s=1.5 \cdot 10^{-2}$ Cm^{-2} at -40 °C].	60W2 60W1
6a	Heat capacity: Fig. 1029. Transition heat and entropy: $Q_m=2592$ J mol^{-1} [D: 3017 J mol^{-1}], $S_m=12.41$ J K^{-1} mol^{-1} [D: 14.36 J K^{-1} mol^{-1}].	75O1
8a	Sound velocity: Fig. 1030. $v_l^{[010]}=4137\,(29)$ m s^{-1}, $v_l^{[101]} \approx v_l^{[10\bar{1}]}=4108\,(15)$ m s^{-1} at 20 °C.	67S6
9a	Refractive indices and optical axial angle: Table 280. Infrared absorption: Fig. 1031.	
10a	Raman scattering: Fig. 1032; see also	77S42
11	Electrical conductivity: Fig. 1033. Activation energy: $\Delta U=0.53$ eV ($T<\Theta_f$), 0.30 eV ($T>\Theta_f$).	72A10
13a	NMR: Figs. 1034, 1035; Table 281; see also	61B1, 67O7, 72K16
b	ESR: Fig. 1036; see also	76S5
c	Mössbauer effect: Fig. 1037; see also	72K10, 75K15, 76P15
14a	Neutron scattering: Fig. 1038.	
15a	Domain structures: see	71K15

Tables 277, 278, see p. 198.

Table 279. $K_4Fe(CN)_6 \cdot 3D_2O$. Atomic coordinates and temperature factors. [70T4]. b_{ij} is defined by Eq. (b) in Introduction. Water II is associated with O (2) and O (3) in general positions. See Fig. 1021.

Atoms	x [10^{-4}]	z	y	b_{11}	b_{22}	b_{33}	b_{12}	b_{13}	b_{23}
Fe	0	2500	1765 (2)	0029 (2)	0016 (1)	0027 (2)	0	0001 (2)	0
N (1)	0	2500	−0053 (1)	0182 (6)	0012 (1)	0084 (4)	0	0047 (4)	0
N (2)	0	2500	3598 (2)	0099 (5)	0019 (5)	0067 (5)	0	0014 (4)	0
N (3)	6924 (2)	1390 (2)	1748 (2)	0038 (2)	0040 (1)	0060 (2)	−0014 (1)	−0013 (1)	0002 (2)
N (4)	6101 (2)	4394 (2)	3219 (2)	0071 (2)	0033 (1)	0044 (2)	0006 (1)	0011 (1)	−0013 (1)
C (1)	0	2500	0643 (2)	0066 (5)	0023 (1)	0062 (5)	0	0006 (5)	0
C (2)	0	2500	2915 (2)	0024 (4)	0012 (1)	0054 (4)	0	−0006 (4)	0
C (3)	8076 (2)	1806 (2)	1774 (2)	0034 (2)	0021 (1)	0039 (2)	−0005 (1)	−0002 (1)	0003 (1)
C (4)	9325 (2)	4419 (2)	1778 (2)	0040 (2)	0021 (1)	0033 (2)	0002 (1)	0003 (1)	−0008 (1)
K (1)	5963 (6)	4350 (5)	1431 (3)	0050 (5)	0034 (3)	0050 (5)	0001 (2)	−0018 (3)	−0019 (2)
K (2)	6907 (5)	1576 (6)	3574 (3)	0056 (6)	0049 (3)	0043 (5)	−0021 (3)	−0015 (4)	0007 (3)
O (1)	5391 (8)	2276 (4)	0242 (4)	0093 (8)	0016 (2)	0133 (10)	0000 (3)	0023 (6)	−0007 (4)
D (1)	5176 (13)	2275 (11)	−0272 (4)	0150 (12)	0030 (2)	0096 (12)	0037 (6)	0041 (8)	0016 (4)
D (2)	6144 (14)	1702 (13)	0283 (6)	0209 (16)	0037 (3)	0393 (26)	−0010 (6)	0046 (16)	0012 (8)
O (2)	7750 (130)	0030 (130)	0000 (70)	disordered atom – coordinates from Fourier synthesis					
D (3)	8300 (190)	0780 (190)	0060 (110)	disordered atom – coordinates from Fourier synthesis					
D (4)	8376 (12)	−0784 (16)	−0214 (6)	0168 (13)	0554 (41)	0059 (4)	0015 (7)	−0055 (22)	−0096 (12)
O (3)	7500 (130)	0190 (130)	−0050 (70)	disordered atom – coordinates from Fourier synthesis					
D (5)	8300 (190)	0690 (190)	−0230 (110)	disordered atom – coordinates from Fourier synthesis					
D (6)	7042 (25)	0926 (17)	0281 (10)	0524 (49)	0391 (32)	0095 (8)	0121 (16)	−0010 (33)	−0104 (13)
O (2)*	7604 (7)	0164 (6)	−0013 (5)	0257 (10)	0191 (9)	0037 (2)	0028 (3)	0100 (10)	−0024 (3)
D (3)*	8316 (5)	0675 (8)	−0087 (5)	0122 (6)	0337 (14)	0122 (5)	−0003 (4)	0029 (9)	0100 (7)

O (2)* is a composite of the superposed O (2) and O (3) in Water II.
D (3)* is a composite of the superposed D (3) and D (5) in Water II.

Table 280. $K_4Fe(CN)_6 \cdot 3H_2O$. Refractive indices and optical axial angle. [62K6].

λ [nm]	n_α	n_β	n_γ	2 V	Ref.
589	—	1.5772	—	−78° 10′	1895D1
486	1.582	1.587	1.594		65B2
520	1.576	1.581	1.587		
555	1.573	1.578	1.583		
589	1.570	1.575	1.580		
620	1.567	1.572	1.577		
656	1.564	1.569	1.574		

Table 281. $K_4Fe(CN)_6 \cdot 3D_2O$. e^2qQ/h, η and directions of the principal axes at −73 °C. [65T7]. A, B, C, and D are assigned to four pairs of lines. η: asymmetry parameter. The θ and ϕ_z are spherical coordinate angles when taking $X \| [101]$ and $Z \| [010]$.

		A	B	C	D
e^2qQ/h [kHz]		−126 (1)	−126 (1)	−123 (2)	−123 (2)
η		0.96 (10)	0.96 (10)	0.63 (2)	0.63 (2)
X direction	θ	37 (1)°	37 (1)°	70 (1)°	70 (1)°
	ϕ_z	78 (9)°	− 3 (6)°	50 (3)°	− 40 (3)°
Y direction	θ	127 (1)°	127 (1)°	81 (2)°	81 (2)°
	ϕ_z	88 (3)°	− 5 (3)°	144 (2)°	54 (2)°

Nr. 30A-3 $K_4Ru(CN)_6 \cdot 3H_2O$. Potassium ruthenium cyanide trihydrate
($M = 467.62$)

				Ref.
1a	Ferroelectricity of $K_4Ru(CN)_6 \cdot 3H_2O$ was discovered by Waku et al. in 1960.			60W2
b	phase	II	I	60W2
	state	F	P	
	crystal system		monoclinic	
	space group		$C2/c\text{-}C_{2h}^6$	
	Θ_F	−14.5 °C [D: −7.3 °C]		60W1
3a	Unit cell parameters: $a = 9.30$ Å, $b = 16.80$ Å, $c = 9.3$ Å, $\beta = 90° 6'$ at RT.			47P1
b	$Z = 4$			47P1
5a	Dielectric constant: Figs. 1039, 1040.			
c	Spontaneous polarization and coercive field: Fig. 1041. [D: see→] $P_s = 1.4 \cdot 10^{-2}$ Cm^{-2} at −65 °C [D: $1.5 \cdot 10^{-2}$ Cm^{-2} at −35 °C]			60W1
9a	$2V = 54° 0'$			60W2
13a	NMR: Fig. 1042; see also			69L5

Nr. 30A-4 $K_4Os(CN)_6 \cdot 3H_2O$. Potassium osmium cyanide trihydrate
($M = 556.7$)

		Ref.
1a	Ferroelectricity of $K_4Os(CN)_6 \cdot 3H_2O$ was discovered by Waku et al. in 1960.	60W2
b	$\Theta_f = -2.4$ °C [D: 1.8 °C].	60W2
3a	Unit cell parameters: $a:b:c = 0.3929:1:0.3941$, $\beta = 90° 6'$	60W1
b	$Z = 4$	60W1

5a	Dielectric constant: Figs. 1043, 1044.	
c	Spontaneous polarization and coercive field: Fig. 1045 [D: see→] $P_s = 3.5 \cdot 10^{-2}\,Cm^{-2}$ at $-45\,°C$ [D: $1.3 \cdot 10^{-2}\,C\,m^{-2}$ at $-20\,°C$].	60W1 60W2, 60W1

30B Solid solutions
Nr. 30B-1 $K_4Fe(CN)_6 \cdot 3H_2O$–$Rb_4Fe(CN)_6 \cdot 3H_2O$

1b	Ferroelectric transition temperature: Fig. 1046.	
5a	Dielectric constant: Fig. 1047.	
c	Spontaneous polarization and coercive field: see	61W1

Nr. 30B-2 $K_4Fe(CN)_6 \cdot 3H_2O$–$Tl_4Fe(CN)_6 \cdot 3H_2O$

1b	Ferroelectric transition temperature: Fig. 1048.	
5a	Dielectric constant: Fig. 1049.	

Nr. 30B-3 $K_4Fe(CN)_6 \cdot 3H_2O$–$(NH_4)_4Fe(CN)_6 \cdot 3H_2O$

1b	Ferroelectric transition temperature: Fig. 1050.	
5a	Dielectric constant: Fig. 1051.	
c	Spontaneous polarization and coercive field: see	61W1

Organic crystals

31 $SC(NH_2)_2$
31A Pure compound
Nr. 31A-1 $SC(NH_2)_2$, Thiourea
($M = 76.12$)

1a Ferroelectricity in $SC(NH_2)_2$ was discovered by Solomon in 1956. — 56S6

b

phase*)	V [a]	IV [a]	III [a]	II [a]	I [a]
state	F [b]		F [a]		P [b]
crystal system	orthorhombic [a]				orthorhombic [c, d]
space group	$Pb2_1m$-C_{2v}^2 [a]				$Pbnm$-D_{2h}^{16} [c, d]
Θ	169	176	180	202 K	
[D**]:	185	192	196	213 K]	

References: [a] 59G2; [b] 56S6; [c] 28D1; [d] 32W1

$P_s \parallel [010]$ in phases III and V. — 59G2

$\varrho = 1.406 \cdot 10^3\,kg\,m^{-3}$ at RT. — 10g1

continued

Nr. 31A-1 $SC(NH_2)_2$, continued

	A different system of the crystallographic axes (a', b', c') is adopted by several workers: $\boldsymbol{a}'=\boldsymbol{b}$, $\boldsymbol{b}'=\boldsymbol{c}$, $\boldsymbol{c}'=\boldsymbol{a}$. Transparent, colorless. Cleavage plane: (010) There is a report that the phase IV is not a single phase but contains a ferroelectric region in it; see subsection 5c. *) The phases are numbered from the high-temperature phase obeying the general rule adopted in this volume (cf. I C of Introduction). Thus the names of phases here are different from those in III/3 and III/9, where the phases were numbered from the low-temperature phase following Goldsmith and White [59G2]. **) D: data on deuterated thiourea $SC(ND_2)_2$.	 56S6, 59G2
2a	Crystal growth: evaporation method from a saturated solution in methanol	56S6
b	Crystal form: Fig. 1052.	
3a	Unit cell parameters: Table 282.	
b	$Z=4$ in phases I [a]) and V [b]).	[a]) 32W1, 28D1 [b]) 59G2
	Crystal structures of phases I and V: Tables 283, 284; Figs. 1053···1056; see also	32W1, 28D1, 58K5, 60D2, 61V1, 67T6
	Precise molecular structure, see	78M9, 78M10
	Crystal structure of phase II: Table 285. In phases II, III, and IV, crystal structure is modulated by a sinusoidal wave. The amplitude of the modulation wave are given as follows: $\alpha_j=a_j(\Theta_{II-I}-T)$ $a_S=0.0008$ $a_C=0.0010$ $a_N=0.0013$ $\beta_j=b_j(\Theta_{II-I}-T)^{\frac{1}{2}}$ $b_S=0.0067$ $b_C=0.0015$ $b_N=0.0005$ $M=8.0+0.0018(\Theta_{II-I}-T)^2$ α_j, β_j: x, y components of the amplitude of the modulation wave. Θ_{II-I}: transition temperature between phase I and phase II. M: period of the modulation wave. $M=1/\delta$.	71S17
	Temperature dependence of δ: Fig. 1057; see also	71S17, 79T2, 67F7
	See subsection 14a about satellite reflections due to the structural modulation.	
4	Lattice distortions: Fig. 1058.	
5a	Dielectric constants: Figs. 1059···1067. $\varkappa_b=C/(T-\Theta_p)$ $T>\Theta_{II-I}$, where $C=3.7\cdot10^3$ K, $\Theta_p=185$ K, $\Theta_{II-I}=202$ K $\varkappa_a$, $\varkappa_c\approx3$ ($f=1$ kHz). Their temperature variations are negligible in the temperature range from 77 K to RT. Phase diagram in regard to biasing field: Fig. 1068; Table 286. Phase diagram in regard to hydrostatic pressure: Figs. 1069···1071. Table 287.	59G2 59G2

5b	Effects of hydrostatic pressure on $\varkappa_b$: Fig. 1072. Effects of hydrostatic pressure on transition temperatures: see	69G5
c	Spontaneous polarization: Figs. 1073, 1074. Coercive field: Fig. 1075. In phases II, III, and IV, double, single-double, double-double hysteresis loops are observed. According to Futama there is a very narrow ferroelectric region at about 171 K within the phase IV. Effects of hydrostatic pressure on P_s: Figs. 1076, 1077; see also	62F8 76K12
d	Pyroelectricity: Fig. 1078.	
6a	Transition heats and transition entropies: Table 288. Heat capacity: Figs. 1079, 1080.	
8a	Velocity of ultrasonic wave: Fig. 1081. Elastic stiffness constants: Figs. 1082, 1083. Attenuation coefficient of ultrasonic wave: Figs. 1084, 1085.	
9a	Optical axial plane is normal to ***a***. Optical axial angle: (−) $2V=36°$ for white light. Reflection: Fig. 1086.	71J1
10a	Spectra of Raman scattering: Figs. 1087, 1088, 1089. Frequencies of lattice vibrations obtained by Raman spectra: Table 289; Fig. 1090. For the assignment of the lattice vibration in Raman, IR and far IR studies, see Effects of hydrostatic pressure on Raman spectra: Fig. 1091.	67T2, 70M3, 72L2, 72B1
11	Conductivity: Fig. 1092. Resistivity of a pressed powder sample can be expressed as $\varrho=\exp(\Delta U/kT)$ with the activation energy $\Delta U=1.35$ eV. Seebeck coefficient: Fig. 1093. Crystals X-ray irradiated at low temperature emit a broad spectrum of thermoluminescence with peaks centered at about 490 and 530 nm. Activation energy of this process is 0.55 (5) eV.	70Y5 74R6
13a	NMR: Figs. 1094···1096; Tables 290, 291; see also NQR: Table 292.	61E2
14a	X-ray satellite reflections due to structural modulation in phases II, III, and IV: Fig. 1097.	
b	Diffuse scattering of X-ray: Figs. 1098···1101. Diffuse scattering of neutrons: Fig. 1102. Inelastic neutron scattering: Fig. 1103. About elastic stiffness constants estimated by inelastic neutron scattering, see	75M10

Table 282. $SC(NH_2)_2$. Unit cell parameters *a*, *b*, *c* of phases I and V determined by X-ray diffraction.

Phase	T [°C]	a [Å]	b [Å]	c [Å]	Ref.
I	RT	5.520	7.655	8.537	67T6
V	−153	5.494 (5)	7.516 (7)	8.519 (10)	59G2

Table 283. $SC(NH_2)_2$ and $SC(ND_2)_2$. Structures of phases I and V determined by neutron diffraction [68E2]. Fractional coordinates and temperature factors at RT and 110 K. b_{ij} is defined by Eq. (b) in Introduction.

	x	y	z	b_{11}	b_{22}	b_{33}	b_{12}	b_{13}	b_{23}
	$[10^{-4}]$								
$SC(NH_2)_2$, phase I at RT									
S	1148 (13)	−0073 (12)	2500	185 (24)	227 (17)	100 (11)	− 1 (18)	0	0
C	−1635 (6)	0906 (4)	2500	298 (13)	148 (6)	81 (5)	49 (7)	0	0
N	−2773 (5)	1307 (3)	3829 (2)	488 (11)	328 (5)	95 (4)	192 (6)	11 (5)	− 0 (3)
H (1)	−4405 (18)	1842 (13)	3761 (9)	706 (39)	614 (26)	154 (10)	351 (27)	48 (16)	10 (12)
H (2)	−2054 (10)	0955 (8)	4870 (6)	498 (22)	353 (73)	95 (9)	71 (17)	− 1 (11)	− 0 (8)
$SC(ND_2)_2$, phase I at RT									
S	1144 (11)	−0061 (10)	2500	239 (19)	244 (14)	122 (9)	6 (14)	0	0
C	−1628 (5)	0911 (4)	2500	340 (10)	165 (5)	122 (4)	43 (6)	0	0
N	−2755 (4)	1305 (2)	3830 (2)	497 (8)	328 (4)	137 (3)	178 (5)	24 (4)	− 8 (3)
D (1)	−4386 (7)	1870 (6)	3797 (4)	668 (14)	611 (12)	168 (5)	412 (12)	34 (7)	−16 (7)
D (2)	−2046 (5)	0930 (4)	4866 (3)	503 (13)	325 (7)	116 (4)	116 (7)	− 7 (5)	− 4 (4)
$SC(NH_2)_2$, phase V at 110 K									
S (1)	3454 (19)	0062 (16)	0000	52 (30)	99 (21)	26 (12)	− 20 (19)	0	0
C (1)	0455 (7)	0733 (10)	0000	54 (13)	73 (10)	19 (5)	6 (8)	0	0
N (1)	−0756 (4)	1000	1337 (3)	94 (8)	113 (4)	36 (4)	33 (7)	5 (4)	1 (5)
H (1)	−2506 (13)	1365 (12)	1305 (10)	151 (26)	242 (25)	70 (10)	94 (17)	14 (11)	− 3 (11)
H (2)	0052 (14)	0772 (18)	2378 (8)	198 (20)	183 (17)	50 (9)	34 (18)	− 7 (12)	14 (13)
S (2)	1215 (17)	0272 (16)	5000	34 (27)	86 (17)	35 (12)	− 14 (18)	0	0
C (2)	3860 (8)	−0985 (9)	5000	61 (13)	51 (7)	49 (7)	9 (9)	0	0
N (2)	4912 (6)	−1475 (5)	3656 (3)	117 (8)	84 (4)	40 (4)	37 (5)	3 (5)	− 9 (5)
H (3)	6506 (14)	−2169 (13)	3698 (9)	189 (25)	174 (14)	85 (10)	118 (20)	15 (13)	− 9 (11)
H (4)	4271 (13)	−1028 (16)	2614 (10)	217 (23)	133 (13)	66 (12)	40 (14)	21 (12)	− 5 (11)
$SC(ND_2)_2$, phase V at 110 K									
S (1)	3412 (21)	0113 (17)	0000	133 (34)	70 (21)	75 (15)	6 (22)	0	0
C (1)	0427 (8)	0744 (9)	0000	111 (15)	70 (10)	66 (6)	14 (11)	0	0
N (1)	−0785 (5)	1000	1333 (3)	160 (9)	128 (5)	50 (4)	47 (7)	− 1 (4)	− 2 (5)
D (1)	−2569 (8)	1358 (9)	1323 (5)	173 (14)	241 (13)	105 (7)	92 (11)	15 (8)	− 5 (8)
D (2)	0013 (8)	0777 (10)	2381 (5)	215 (11)	146 (10)	81 (6)	28 (12)	10 (8)	− 6 (7)
S (2)	1217 (20)	0328 (16)	5000	123 (32)	69 (19)	64 (14)	13 (21)	0	0
C (2)	3852 (10)	−0956 (9)	5000	168 (15)	62 (8)	54 (7)	19 (11)	0	0
N (2)	4863 (6)	−1463 (6)	3659 (3)	184 (9)	94 (4)	67 (4)	36 (5)	3 (6)	1 (5)
D (3)	6441 (8)	−2191 (8)	3690 (6)	191 (13)	159 (8)	98 (6)	89 (9)	− 6 (8)	− 4 (7)
D (4)	4232 (8)	−0999 (8)	2622 (6)	241 (14)	130 (9)	79 (7)	43 (10)	1 (7)	6 (7)

Table 284. $SC(NH_2)_2$ and $SC(ND_2)_2$. Structures of phases I and V [68 E 2]. Intermolecular N–H···S distances and angles. The atoms with asterisks are related by the symmetry elements to those given in Table 283. They can be identified in Figs. 1053, 1055 where all the N–S distances in $SC(NH_2)_2$ are given. (I): the H atom involved is noncoplanar with its molecule. (II), (III): the H atom involved is planar with its molecule.

Substance and phase		Atoms			Distances [Å] N–S	Distances [Å] H···S	Angles deg
$SC(NH_2)_2$, I	(I)	N	H (2)	S*	3.394 (4)	2.397 (6)	168.9 (5)
$SC(ND_2)_2$, I		N	D (2)	S*	3.393 (3)	2.398 (3)	169.1 (3)
$SC(NH_2)_2$, V		N (1)	H (2)	S (2)	3.348 (4)	2.354 (8)	169.5 (6)
$SC(ND_2)_2$, V		N (1)	D (2)	S (2)	3.350 (5)	2.352 (5)	170.4 (5)
$SC(NH_2)_2$, V		N (2)	H (4)	S (1)	3.417 (6)	2.415 (10)	170.4 (7)
$SC(ND_2)_2$, V		N (2)	D (4)	S (1)	3.429 (6)	2.427 (7)	170.6 (5)
$SC(NH_2)_2$, I	(II)	N*	H (1)*	S*	3.526 (8)	2.767 (12)	133.7 (7)
$SC(ND_2)_2$, I		N*	D (1)*	S*	3.539 (7)	2.772 (8)	133.9 (4)
$SC(NH_2)_2$, V		N (1)*	H (1)*	S (1)*	3.580 (11)	3.038 (13)	115.2 (6)
$SC(ND_2)_2$, V		N (1)*	D (1)*	S (1)*	3.596 (12)	3.074 (13)	113.0 (4)
$SC(NH_2)_2$, V		N (2)*	H (3)*	S (2)	3.438 (10)	2.549 (13)	145.4 (7)
$SC(ND_2)_2$, V		N (2)*	D (3)*	S (2)	3.429 (12)	2.526 (11)	146.7 (5)
$SC(NH_2)_2$, I	(III)	N*	H (1)*	S	3.696 (7)	3.055 (12)	123.5 (7)
$SC(ND_2)_2$, I		N*	D (1)*	S	3.705 (6)	3.082 (8)	121.6 (3)
$SC(NH_2)_2$, V		N (1)*	H (1)*	S (1)	3.452 (10)	2.669 (12)	135.3 (7)
$SC(ND_2)_2$, V		N (1)*	D (1)*	S (1)	3.449 (11)	2.649 (12)	135.5 (5)
$SC(NH_2)_2$, V		N (2)	H (3)	S (2)*	3.876 (10)	3.360 (13)	113.2 (6)
$SC(ND_2)_2$, V		N (2)	D (3)	S (2)*	3.912 (11)	3.423 (12)	111.2 (4)

Table 285. $SC(NH_2)_2$. Parameters concerning sinusoidal structural modulation in phase II determined at −86 °C [71 S 17]. The structural modulation can be described as a transverse wave with the propagation vector parallel to the c axis. x, y: fractional coordinates of atoms in the unit cell without the structural perturbation. α, β: x and y components of the amplitude of the modulation wave. ϕ_x, ϕ_y: relative phases of the modulation wave concerning each atom.

	x	α	ϕ_x	y	β	ϕ_y
S	0.115	0.012	0.5	−0.009	0.026	0
C	−0.165	0.015	0.5	0.106	0.006	0
N	−0.292	0.019	0.5	0.131	0.002	0.5

Table 288, see p. 208.

Table 286. $SC(NH_2)_2$. Shifts of the transition temperatures per unit field strength obtained from 50 Hz hysteresis loop [62 F 8].

$d\Theta_{V-IV}/dE_{bias}$	$0.83\,(5) \cdot 10^{-5}$ K m V^{-1}
$d\Theta_{IV-III}/dE_{bias}$	$-0.5\,(1) \cdot 10^{-5}$ K m V^{-1}
$d\Theta_{III-II}/dE_{bias}$	$0.50\,(5) \cdot 10^{-5}$ K m V^{-1}

Table 287. $SC(NH_2)_2$. Θ, $d\Theta/dp$ [75 K 13].

Transition	Θ [K]	$d\Theta/dp$ [10^{-8} K Pa^{-1}]
II– I	202.3	−16.8
III– II	180	−35.2
IV–III	177.5	−32.2
V–IV	171.4	−25.5
V– II		−35.5

Table 289. $SC(NH_2)_2$. Frequencies of the lattice vibrations obtained by Raman spectra at RT. $(\partial\nu/c\partial p)_T$ is the rate of the frequency shift due to hydrostatic pressure. 1 cm^{-1} = $3 \cdot 10^{10}$ Hz.

Ref.	[67 T 2]		[72 L 2]		[76 I 4]	
Assignment	ν/c cm^{-1}	Relative intensity	ν/c cm^{-1}	Relative intensity	ν/c cm^{-1}	$(\partial\nu/c\partial p)_T$ cm^{-1} (10^8 Pa)$^{-1}$
A_g	122	3	118 (1)	1.5	116	2.29
			32 (1)	0.75		
B_{1g}	101	4	101 (1)	3	99.5	2.08
B_{2g}	63	6	62 (1)	6	60.0	2.00
B_{3g}	45	2	47 (1)	1.87	44.0	1.73

Table 288. $SC(NH_2)_2$. Transition heats and transition entropies per mol.

Transition	ΔQ_m J mol^{-1}	ΔS_m J K^{-1} mol^{-1}	Ref.
V→IV	27.6	0.163	62F8
	12.5	0.075	63N1
		0.167	63C1
IV→III→II		0.042	63C1
II→I	4.2	0.025	63N1
		0.710	63C1

Table 292. $SC(NH_2)_2$. e^2qQ/h and η of ^{14}N [64S9]. For molecules 1 and 2, see caption of Table 291.

Molecule	e^2qQ/h kHz	η %
1	3121.6	39.54
2	3099.6	39.30

Table 290. $SC(ND_2)_2$. Electric field gradient tensors and the direction cosines of their principal axes with respect to crystallographic *a*, *b*, *c* axes at 300 K [71O2]. Estimated error in angles is ±2°.

i	ϕ_{ii} [kHz]	Direction cosines	i	ϕ_{ii} [kHz]	Direction cosines
Inner deuterons			Outer deuterons		
x	−93 (3)	(0.000, 0.000, 1.000)	x	−21 (2)	(0.841, 0.541, 0.000)
y	−104 (3)	(−0.406, 0.914, 0.000)	y	−114 (2)	(−0.541, 0.841, 0.000)
z	197 (3)	(0.914, 0.406, 0.000)	z	135 (2)	(0.000, 0.000, 1.000)

Table 291. $SC(ND_2)_2$. Electric field gradient tensors and the direction cosines of their principal axes at 120 K [71O2]. At $T<169$ K (phase V) there are two crystallographically inequivalent molecules; the molecular phase of one type (molecule 1) intersects the *a* axis with an angle of 19.6° and the other (molecule 2) with 29.5°.

i	ϕ_{ii} [kHz]	Direction cosines	i	ϕ_{ii} [kHz]	Direction cosines
Inner deuterons, molecule 1			Outer deuterons, molecule 1		
x	−69	(−0.006, 0.036, 0.999)	*x*	−65	(0.829, −0.335, 0.447)
y	−93	(−0.291, 0.956, −0.036)	*y*	−91	(0.288, 0.942, 0.173)
z	162	(0.956, 0.291, −0.005)	*z*	157	(−0.479, −0.015, 0.878)
Inner deuterons, molecule 2			Outer deuterons, molecule 2		
x	−64	(0.000, 0.000, 1.000)	*z*	−62	(0.657, −0.667, 0.350)
y	−93	(−0.504, 0.861, 0.000)	*y*	−91	(0.594, 0.744, 0.305)
z	157	(0.861, 0.504, 0.000)	*x*	153	(−0.464, 0.008, 0.886)

32 $N(CH_3)_4 \cdot HgCl_3$ family

32A Pure compounds

Nr. 32A-1 $N(CH_3)_4 \cdot HgCl_3$, Tetramethylammonium-trichloro-mercurate ($M=381.09$)

1a Ferroelectricity in $N(CH_3)_4 \cdot HgCl_3$ was discovered by Fatuzzo and Nitsche in 1960. 60F4

b

state	F
crystal system	monoclinic [a])
space group	$P2_1$-C_2^2 [a])
Θ	none

[a]) 63W5

Temperature of decomposition ≈200 °C. 60F4
Colorless. 60F4

2a	Crystal growth: slow evaporation from an aqueous solution containing $[N(CH_3)_4]Cl$ and $HgCl_2$ in the molar ratio 2:1.	60F4
b	Crystal form: needle shape.	60F4
3a	Unit cell parameters: $a=8.68$ (4) Å, $b=15.75$ (8) Å, $c=7.69$ (4) Å, $\beta=93.0$ (2)° at RT.	63W5
5a	$\varkappa \approx 60$ at RT; almost independent of T.	60F4
c	$P_s \approx 1.2 \cdot 10^{-2}$ C m^{-2} at RT; only slightly T dependent. $E_c \approx 3 \cdot 10^5$ V m^{-1} at RT.	60F4 60F4
10a	Raman scattering: Table 293.	
15b	Switching: Figs. 1104, 1105.	

Table 293. $N(CH_3)_4 \cdot HgCl_3$. Frequencies and assignments of lattice vibrational modes obtained from Raman scattering spectra [75H4]. Scattering intensities are represented by vs (very strong), s (strong), and w (weak). br indicates that the mode gives a broad spectrum. In the column of assignment following abbreviations are used, ν_s: symmetric stretch, ν_{as}: asymmetric stretch, δ_s: symmetric bend, δ_{as}: asymmetric bend. Frequencies are given in the unit of cm^{-1}. 1 cm^{-1} = $3 \cdot 10^{10}$ Hz.

Frequencies cm^{-1}	Scattering intensities	Assignments	Symmetries
274	vs	ν_s (Hg-Cl)	A_1
244	w, br	ν_{as} (Hg-Cl)	E
90	s, br	δ_s (ClHgCl)	A_1
74	s	δ_{as} (ClHgCl)	E
56	s	external vibrations	
37	w		
33	w		

Table 294. $N(CH_3)_4 \cdot HgBr_3$. Crystal structure [63W5]. Fractional coordinates.

	x	y	z	e.s.d. [Å]
Hg (1)	0.165	0.000	0.025	0.009
Br (1)	0.421	−0.045	−0.023	0.022
Br (2)	−0.052	−0.101	−0.007	
Br (3)	0.148	0.127	0.223	
Hg (2)	0.278	0.046	0.528	0.009
Br (4)	0.250	−0.106	0.430	0.022
Br (5)	0.514	0.126	0.550	
Br (6)	0.089	0.100	0.720	
N (1)	0.67	−0.13	0.42	
C (1)	0.57	−0.18	0.35	
C (2)	0.59	−0.08	0.51	
C (3)	0.74	−0.08	0.30	
C (4)	0.77	−0.17	0.50	
N (2)	0.64	0.19	0.07	
C (5)	0.56	0.25	−0.01	
C (6)	0.56	0.13	0.14	
C (7)	0.73	0.23	0.19	
C (8)	0.73	0.15	−0.05	

Nr. 32A-2 $N(CH_3)_4 \cdot HgBr_3$, Tetramethylammonium-tribromo-mercurate
($M=514.45$)

1a	Ferroelectricity in $N(CH_3)_4 \cdot HgBr_3$ was discovered by Fatuzzo et al. in 1962.	62F4
b	state: F; crystal system: monoclinic [a]); space group: $P2_1$-C_2^2 [a]); Θ: none	[a]) 63W5
	Temperature of decomposition ≈ 170 °C. $\varrho \approx 2.96 \cdot 10^3$ kg m^{-3}. Color: white.	62F4 63W5 62F4
2a	Crystal growth: slow cooling from a saturated solution containing $[N(CH_3)_4]Br$ and $HgBr_2$ in 1.3 normal HBr. Solubility: Fig. 1106.	62F4
b	Crystal form: needle shape.	62F4

continued

Nr. 32A-2 $N(CH_3)_4 \cdot HgBr_3$, continued

3a	Unit cell parameters: $a = 9.05\,(2)$ Å, $b = 15.90\,(5)$ Å, $c = 7.94\,(2)$ Å, $\beta = 93.6\,(2)°$ at RT.	63W5
b	$Z = 2$. Crystal structure: Tables 294, 295, 296; Figs. 1107···1110.	63W5
5a	$\varkappa \approx 10$; almost independent of T (-40 °C $< T <$ 160 °C).	62F4
c	$P_s \approx 1 \cdot 10^{-2}$ C m^{-2}; almost independent of T (-40 °C $< T <$ 160 °C). E_c: Fig. 1111.	62F4

Table 295. $N(CH_3)_4 \cdot HgBr_3$. Crystal structure [63W5]. Interatomic distances and bond angles. See also Fig. 1109.

Hg–Br distances			Br–Hg–Br bond angles		
Atoms	Intra-anion distance [Å]	e.s.d. [Å]	Atoms	Angle [°]	e.s.d. [°]
Hg (1)–Br (1)	2.48	0.025	Br (1)–Hg (1)–Br (2)	121.8	0.8
Hg (1)–Br (2)	2.53		Br (2)–Hg (1)–Br (3)	119.1	
Hg (1)–Br (3)	2.56		Br (1)–Hg (1)–Br (3)	114.4	
Hg (2)–Br (4)	2.55		Br (4)–Hg (2)–Br (5)	125.3	
Hg (2)–Br (5)	2.49		Br (5)–Hg (2)–Br (6)	113.3	
Hg (2)–Br (6)	2.52		Br (4)–Hg (2)–Br (6)	116.9	
Hg (1)–Br (6)	2.94				
Hg (2)–Br (3)	2.92				

Table 296. $N(CH_3)_4 \cdot HgBr_3$. Crystal structure [63W5]. C–Br distances less than 4 Å. Br with coordinates related by the screw axis to those of Table 294 are dashed, and atoms with coordinates related by a translation along a, b or c to the standard molecule are indicated by the corresponding symbols in parentheses. ►

Atoms	Distance [Å]
C (1)–Br (1)	3.88
C (1)–Br (4)	3.28
C (1)–Br (5)′ ($+a-b+c$)	3.25
C (2)–Br (4)	3.14
C (2)–Br (5)	3.43
C (3)–Br (1)	3.80
C (3)–Br (2) ($+a$)	3.18
C (5)–Br (1)′ ($+a$)	3.27
C (6)–Br (1)	3.25
C (6)–Br (3)	3.78
C (6)–Br (5)	3.28
C (7)–Br (5)	3.91
C (8)–Br (5) ($-c$)	3.65

Nr. 32A-3 $P(CH_3)_4 \cdot HgBr_3$, Tetramethylphosphonium-tribromo-mercurate
($M = 531.41$)

1a	Ferroelectricity in $P(CH_3)_4 \cdot HgBr_3$ was discovered by Fatuzzo et al. in 1962.	62F4
b	state: F; Θ: none	62F4
	Temperature of decomposition ≈ 170 °C.	62F4
	Color: white.	62F4

2a	Crystal growth: vacuum evaporation of the solution containing $[P(CH_3)_4]Br$ and $HgBr_2$.	62F4
b	Crystal form: flat plate.	62F4
5a	$\varkappa \approx 10$; independent of T (-40 °C < T < 160 °C).	62F4
c	$P_s \approx 3 \cdot 10^{-2}$ C m^{-2}; independent of T (-40 °C < T < 160 °C). E_c: Fig. 1112.	62F4

Nr. 32A-4 $N(CH_3)_4 \cdot HgI_3$, Tetramethylammonium-triiodo-mercurate
(M = 655.45)

1a	Ferroelectricity in $N(CH_3)_4 \cdot HgI_3$ was discovered by Fatuzzo et al. in 1962.	62F4
b	(see table below)	a) 65P1
	Temperature of decomposition ≈ 170 °C. $\varrho = 3.15 \cdot 10^3$ kg m^{-3}. Color: pale-yellow. Cleavage plane: (100).	62F4 67G3 62F4 67G3
2a	Crystal growth: slow evaporation from an equimolar mixture of $[N(CH_3)_4]I$ and HgI_2 in cyclohexanone.	62F4
b	Crystal form: polyhedral shape.	62F4
3a	Unit cell parameters: a = 9.46 Å, b = 16.48 Å, c = 8.56 Å at RT.	67G3
b	Z = 4. Crystal structure: Table 297; Fig. 1113. For approximate positions of light atoms, see	65P1 65P1
5a	$\varkappa \approx 10$; independent of T (-40 °C < T < 100 °C).	62F4
c	$P_s = 1.2 \cdot 10^{-2}$ C m^{-2} at -40 °C; $1.7 \cdot 10^{-2}$ C m^{-2} at RT; $2.3 \cdot 10^{-2}$ C m^{-2} at 80 °C. E_c: Fig. 1114.	62F4
9a	Birefringence: Figs. 1115, 1116.	

Table for 1b:

state	F
crystal system	orthorhombic [a]
space group	$Pb2_1m$-C_{2v}^2 [a]
Θ	none

Table 297, see p. 213.

33 $Cu(HCOO)_2 \cdot 4H_2O$

33A Pure compound

Nr. 33A-1 $Cu(HCOO)_2 \cdot 4H_2O$, Cupric formate tetrahydrate
(M = 225.642)

1a	Antiferroelectric activity of $Cu(HCOO)_2 \cdot 4H_2O$ was first reported by Okada in 1965. Antiferromagnetic activity of $Cu(HCOO)_2 \cdot 4H_2O$ was reported independently by Itoh et al. in 1962 and Kobayashi et al. in 1963.	65O2 62I5 63K5
b	(see table below) For the deuterated crystal, Θ_{II-I} = 245.7 K [e].	a) 62K10 b) 65O2 c) 63K5, 62I5 d) 54K2 e) 67O2 f) 71B14 g) 69S14

Table for 1b:

phase	III [c]		II [a]		I [a]
state	A [b], A_{magn}* [c]		A [b], P_{magn} [c]		P [b], P_{magn} [c]
crystal system					monoclinic [d]
space group					C_{2h}^5-$P2_1/a$ [d]
Θ		17		235.5 K [a]	

continued

Nr. 33A-1 $Cu(HCOO)_2 \cdot 4H_2O$, continued

1b	Two different assignments were proposed for the space group of phase II (monoclinic): $P2_1$ [f]) and $P2_1/n$ [g, h]).	[h]) 73M2
	Antiferroelectric axis‖[010].	65O2
	Transparent, blue. Cleavage plane: (001).	54K2
	Efflorescent.	
	*) The antiferromagnetic phase below 17 K is accompanied by weak ferromagnetism [63K5].	
2a	Crystal growth: evaporation or cooling method from aqueous solution.	65O2
b	Crystal form: Fig. 1117.	
3a	Unit cell parameters: $a=8.18\,(2)$ Å, $b=8.15\,(2)$ Å, $c=6.35\,(2)$ Å, $\beta=101°\,5\,(20)'$ at RT	63K5
b	$Z=2$ in phase I [a]), $Z=4$ in phase II [b]).	[a]) 54K2 [b]) 72K8
	Crystal structure: positional and temperature parameters: Tables 298, 299. Bond lengths and bond angles: Tables 300, 301. Projection of crystal structure: Figs. 1118···1120.	
	An approximate structure of phase II has been determined by neutron diffraction technique on the assumption that the space group is $P2_1/n$. The crystal has a superstructure with the c axis doubled in phase II.	72K8 67T8
4	Thermal expansion: Fig. 1121.	
5a	Dielectric constant: Figs. 1122···1124. Dielectric relaxation: Figs. 1125···1127.	
	$\varkappa_b=C/(T-\Theta_p)$, where $C=3.2\cdot10^4$ K, $\Theta_p=215$ K.	65O2
b	Nonlinear dielectric properties: Fig. 1128.	
6a	Heat capacity: Figs. 1129···1132. Transition heat and transition entropy:	
	$\Delta Q_m=836.0\,(10)\ \mathrm{J\cdot mol^{-1}}$, $\Delta S_m=3.546\,(5)\ \mathrm{J\,K^{-1}\,mol^{-1}}$. For $Cu(HCOO)_2\cdot 4D_2O$: $\Delta Q_m=936.9\,(5)\ \mathrm{J\cdot mol^{-1}}$, $\Delta S_m=3.814\,(2)\ \mathrm{J\,K^{-1}\,mol^{-1}}$.	76M6
b	Thermal conductivity: Fig. 1133.	
8a	Elastic compliance and stiffness: Table 302; Figs. 1134, 1135.	
9a	Birefringence: Figs. 1136, 1137.	
10a	Raman scattering: Fig. 1138.	
11	Electrical conduction: Fig. 1139.	
	Protonic conductivity parallel to the water planes (perpendicular to (110)): $\sigma=4.5\,(13)\cdot10^{-5}\,\Omega^{-1}\,\mathrm{m^{-1}}$ at RT.	75W2
	Activation Energy: $\Delta U=54.5\ \mathrm{kJ\cdot mol^{-1}}$ ($T>\Theta_a$), $\Delta U=63.5\ \mathrm{kJ\cdot mol^{-1}}$ ($T<\Theta_a$).	75W2
	Protonic conductivity perpendicular to the water layer is small. $\sigma_c\approx10^{-6}\,\Omega^{-1}\,\mathrm{m^{-1}}$ at RT.	75W2
	Ohmic conductivity at 298 K (protonic): $\sigma_a=8.5\,(21)\cdot10^{-6}\,\Omega^{-1}\,\mathrm{m^{-1}}$; $\sigma_b=5.5\,(16)\cdot10^{-5}\,\Omega^{-1}\,\mathrm{m^{-1}}$; $\sigma_c=7\,(2)\cdot10^{-6}\,\Omega^{-1}\,\mathrm{m^{-1}}$.	77M21
	Ohmic and non-ohmic behaviors have been found; for details see	77M21
12	Magnetic properties: Figs. 1140···1143.	
	$\Theta_N=17$ K.	63K5
13a	NMR: Tables 303···306; Figs. 1144···1147.	
	Proton magnetic resonance at 4.2 K [for $Cu(HCOO)_2\cdot 4D_2O$], see also	62K11, 74M7
b	ESR: see	68S10

Table 297. $N(CH_3)_4 \cdot HgI_3$. Crystal structure [67G3]. Fractional coordinates of Hg and I. For the notations of atoms, see Fig. 1113.

Atoms	x	y	z
Hg (1)	0.154	0.000	0.000
Hg (2)	0.278	0.045	0.500
I (1)	0.427	−0.039	0.000
I (2)	−0.068	−0.106	0.000
I (3)	0.109	0.115	0.258
I (4)	0.246	−0.114	0.500
I (5)	0.503	0.131	0.500
I (6)	0.109	0.115	0.742

Table 298, see p. 214.

Table 299. $Cu(HCOO)_2 \cdot 4H_2O$. Crystal structure of phase I [75K9]. Fractional coordinates and temperature parameters. The temperature parameters are defined by Eq. (b) in Introduction. Occupancy parameters f_i for the hydrogen site H−i in the water molecules are assumed as below. $f_1=0.85$, $f_2=1.5-f_1$, $f_3=0.5$, $f_4=1.0$, $f_5=0.5$, $f_6=1-f_1$, $f_7=f_1-0.5$. For the hydrogen sites, see Figs. 1119, 1120. See also Fig. 1118. H_F: H belonging to formate.

Atoms	x	y	z	b_{11}	b_{22}	b_{33}	b_{12}	b_{13}	$b_{23} \cdot 10^4$
Copper-Cu	0	0	0	35 (2)	27 (2)	129 (4)	1 (2)	28 (2)	− 6 (2)
Formate ion									
Carbon C	0.2387 (1)	0.2695 (2)	0.0244 (2)	54 (2)	45 (2)	144 (4)	− 7 (2)	37 (2)	− 6 (2)
Oxygens OF-1	0.2024 (2)	−0.0921 (2)	−0.0794 (2)	55 (2)	40 (2)	198 (5)	14 (2)	50 (2)	19 (3)
OF-2	0.1145 (2)	0.2119 (2)	0.0896 (2)	62 (2)	52 (2)	183 (5)	−23 (2)	55 (2)	− 23 (3)
Hydrogen H_F	0.2037 (4)	0.6933 (4)	0.0826 (6)	190 (6)	127 (5)	457 (13)	70 (5)	207 (7)	106 (6)
Water molecules									
OW−1	0.4287 (6)	0.4006 (4)	0.6381 (4)	98 (5)	94 (5)	156 (8)	−10 (5)	36 (5)	− 17 (5)
OW−2	0.0890 (5)	0.3596 (4)	0.4849 (6)	118 (9)	110 (8)	202 (10)	−12 (7)	55 (8)	− 29 (8)
H−1	0.3107 (13)	0.3920 (6)	0.5847 (8)	100 (8)	142 (8)	200 (15)	−16 (11)	10 (10)	− 8 (8)
H−2	0.4810 (12)	0.2978 (14)	0.5979 (14)	135 (16)	136 (19)	281 (27)	2 (13)	20 (13)	21 (16)
H−3	0.4746 (10)	0.4739 (8)	0.5424 (11)	188 (23)	103 (21)	210 (43)	−20 (17)	41 (18)	52 (18)
H−4	0.0880 (4)	0.3257 (4)	0.3394 (8)	149 (6)	171 (7)	242 (11)	−12 (4)	53 (7)	− 29 (8)
H−5	0.0265 (12)	0.4587 (9)	0.4907 (18)	136 (39)	111 (36)	307 (28)	8 (24)	51 (25)	33 (42)
H−6	0.2154 (69)	0.3782 (35)	0.5355 (50)	86 (52)	157 (50)	288 (83)	12 (71)	− 34 (66)	− 22 (47)
H−7	0.0330 (18)	0.2753 (28)	0.5438 (25)	129 (32)	132 (35)	306 (48)	−47 (21)	89 (25)	3 (30)

Table 298. $Cu(HCOO)_2 \cdot 4H_2O$. Crystal structure of phase I [54K2, 66O2]. Fractional coordinates and temperature parameters. *B* is defined by Eq. (e) in Introduction.

[1])		[54K2] X-ray	[66O2] neutron diffraction
Cu	*x*	0	0
	y	0	0
	z	0	0
	B	1.35 Å^2	1.02 (14) Å^2
C	*x*	0.238	0.237 (1)
	y	0.270	0.269 (1)
	z	0.018	0.024 (1)
	B	1.35 Å^2	1.04 (16) Å^2
$O_F(1)$	*x*	0.206	0.205 (1)
	y	−0.092	−0.092 (1)
	z	−0.080	−0.080 (1)
	B	1.35 Å^2	1.27 (16) Å^2
$O_F(2)$	*x*	0.117	0.116 (1)
	y	0.210	0.213 (1)
	z	0.086	0.088 (1)
	B	1.35 Å^2	1.34 (1) Å^2
$O_W(1)$	*x*	0.423	0.430 (2)
	y	0.399	0.401 (2)
	z	0.647	0.641 (2)
	B	1.35 Å^2	1.94 (2) Å^2
$O_W(2)$	*x*	0.086	0.089 (2)
	y	0.349	0.357 (2)
	z	0.483	0.482 (2)
	B	1.35 Å^2	2.59 (22) Å^2

[1]) O_F: O belonging to formate. O_W: O belonging to water. Figures in parentheses correspond to those given in Fig. 1118.

Table 301. $Cu(HCOO)_2 \cdot 4H_2O$. Crystal structure of phase I [75K9]. Root-mean-square amplitudes of thermal motion of hydrogens along the principal axes and angles between the axes and O−O hydrogen bonds.

Atom	Amplitude	Angle [deg]
H−1	0.172 (9)	14 (10)
	0.206 (6)	81 (11)
	0.223 (6)	79 (12)
H−2	0.204 (15)	51
	0.213 (8)	80 (30)
	0.245 (14)	41 (13)
H−3	0.149 (25)	22 (8)
	0.225 (8)	87 (9)
	0.254 (14)	69 (8)
H−4	0.205 (5)	20 (18)
	0.218 (4)	71 (18)
	0.250 (5)	89 (6)
H−5	0.186 (28)	37 (42)
	0.208 (30)	58 (50)
	0.252 (20)	73 (25)
H−6	0.152 (50)	14 (30)
	0.224 (34)	85 (41)
	0.268 (36)	77 (24)
H−7	0.148 (29)	19 (15)
	0.231 (14)	79 (15)
	0.258 (16)	75 (15)

Table 300. $Cu(HCOO)_2 \cdot 4H_2O$. Crystal structure of phase I [75K9]. Interatomic distances and angles. "Riding correction" is given in brackets for O–H distances. See also Figs. 1119, 1120. Primed atoms are related to unprimed atoms by symmetry operations.

Bond	Distance [Å]	Bond	Bond Angle [deg]
Copper octahedron			
Cu−OF-1	1.970 (4)	OF-1−Cu−OF-2	91.8 (2)
Cu−OF-2	1.994 (4)	OF-1−Cu−OW-1	93.2 (2)
Cu−OW-1	2.401 (7)	OF-2−Cu−OW-1	86.8 (2)
Formate ion			
C−OF-1	1.250 (3)	OF-1−C−OF-2	122.5 (2)
C−OF-2	1.259 (3)	HF −C−OF-1	119.8 (2)
C−HF	1.093 (4) [1.142]	HF −C−OF-2	117.7 (2)
Water-1			
OW-1−H-1	0.96 (1) [0.98]	H-1−OW-1−H-2	107.4 (5)
OW-1−H-2	1.00 (2) [1.01]	H-1−OW-1−H-3	107.7 (6)
OW-1−H-3	0.98 (1) [1.00]	H-2−OW-1−H-3	96.9 (8)
Water-2			
OW-2−H-4	0.96 (1) [0.98]	H-4−OW-2−H-5	104.8 (9)
OW-2−H-5	0.96 (1) [0.97]	H-4−OW-2−H-6	111.8 (8)
OW-2−H-6	1.03 (6) [1.05]	H-4−OW-2−H-7	99.5 (15)
OW-2−H-7	0.94 (3) [0.96]		

Table 300, continued

Bond	Distance [Å]	Bond	Bond Angle [deg]
Hydrogen Bond Angles			
		H-1 – OW-1 - - - H-3′	110.2 (4)
		H-1 – OW-1 - - - H-7	109.1 (5)
		H-2 – OW-1 - - - H-6	105.2 (10)
		H-2 – OW-1 - - - H-3′	92.9 (6)
		H-3 – OW-1 - - - H-6	107.8 (11)
		H-3 – OW-1 - - - H-7	91.0 (6)
		H-6 - - - OW-1 - - - H-3′	110.2 (9)
		H-6 - - - OW-1 - - - H-7	106.9 (9)
		H-3′ - - - OW-1 - - - H-7	87.0 (4)
		H-4 – OW-2 - - - H-2	105.4 (4)
		H-4 – OW-2 - - - H-1	101.5 (4)
		H-4 – OW-2 - - - H-5	114.6 (5)
		H-5 – OW-2 - - - H-1	111.6 (8)
		H-5 – OW-2 - - - H-2	106.3 (7)
		H-6 – OW-2 - - - H-1	121.6 (16)
		H-6 – OW-2 - - - H-5′	111.5 (16)
		H-7 – OW-2 - - - H-5′	105.4 (7)
		H-7 – OW-2 - - - H-1	119.7 (8)
Hydrogen Bonds			
OW-2 – OF-2	2.828 (8)	OW-2 – H-4 – OF-2	165.3 (3)
H-4 – OF-2	1.886 (7)		
OW-1 – OW-1	2.801 (8)	OW-1 – H-3 – OW-1	175.8 (7)
H-3 – OW-1	1.826 (6)		
OW-2 – OW-2	2.739 (8)	OW-2 – H-5 – OW-2	177.2 (10)
H-5 – OW-2	1.779 (6)		
OW-1 – OW-2	2.782 (8)	OW-1 – H-1 – OW-2	175.9 (4)
H-1 – OW-2	1.822 (14)	OW-1 – H-6 – OW-2	175.7 (24)
H-6 – OW-1	1.751 (60)		
OW-1 – OW-2	2.767 (7)	OW-1 – H-2 – OW-2	167.9 (6)
H-2 – OW-2	1.785 (19)	OW-1 – H-7 – OW-2	174.6 (11)
H-7 – OW-1	1.827 (31)		
		H-1 - - - OW-2 - - - H-2	120.1 (4)
		H-5′ - - - OW-2 - - - H-2	104.5 (4)
		H-5′ – OW-2 - - - H-1	111.0 (4)
		H-1 - - - OW-1 - - - Cu	113.9 (4)
		H-2 - - - OW-1 - - - Cu	119.5 (4)
		H-3 - - - OW-1 - - - Cu	109.7 (6)

Table 302. $Cu(HCOO_2 \cdot 4H_2O$. Elastic stiffness at room temperature, in units of 10^{10} Nm^{-2} [73K2].

$c_{11}=4.45$	$c_{22}=4.99$	$c_{33}=2.11$
$c_{44}=0.54$	$c_{55}=0.71$	$c_{66}=2.35$
$c_{12}=3.41$	$c_{23}=1.49$	$c_{13}=1.42$
$c_{15}=-0.07$	$c_{25}=0.27$	$c_{35}=-0.29$
$c_{16}=-0.24$		

Table 303. $Cu(DCOO)_2 \cdot 4D_2O$. Direction cosines of the principal axes of the field gradient tensors at D of formate ions above (D_F) and below Θ_a [$D_F(1)$ and $D_F(2)$] [69S14]. $X_c \| a$, $Y_c \| b$, $Z_c \| c^*$ (c^* = reciprocal c axis).

		X_c	Y_c	Z_c
291 K	ϕ_{xx}	0.329 (84)	∓0.495 (76)	0.805 (14)
D_F	ϕ_{yy}	0.735 (36)	±0.669 (55)	0.111 (94)
	ϕ_{zz}	−0.593 (6)	±0.555 (5)	0.584 (7)
153 K	ϕ_{xx}	0.267 (172)	∓0.544 (144)	0.796 (41)
$D_F(1)$	ϕ_{yy}	0.756 (60)	±0.630 (123)	0.177 (183)
	ϕ_{zz}	−0.598 (5)	±0.555 (4)	0.579 (6)
$D_F(2)$	ϕ_{xx}	0.357 (192)	∓0.507 (174)	0.785 (25)
	ϕ_{yy}	0.738 (92)	±0.668 (131)	0.096 (206)
	ϕ_{zz}	−0.573 (4)	±0.545 (3)	0.612 (4)

Table 304. $Cu(DCOO)_2 \cdot 4D_2O$. Principal values and asymmetry parameter (η) of field gradient tensors at D of formate ion above and below Θ_a [69S14].

		$eQ\phi_{zz}/h$	$eQ\phi_{yy}/h$	$eQ\phi_{xx}/h$	η
		kHz			
291 K	D_F	155.3 (25)	−84.2 (29)	−71.0 (24)	0.085 (24)
253 K	D_F	157.2 (19)	−82.7 (21)	−74.5 (18)	0.053 (18)
244 K	$D_F(1)$	157.9 (17)	−83.3 (21)	−74.8 (19)	0.057 (16)
	$D_F(2)$	157.2 (23)	−82.5 (19)	−74.7 (17)	0.050 (15)
153 K	$D_F(1)$	161.4 (20)	−83.4 (21)	−77.9 (34)	0.034 (20)
	$D_F(2)$	160.4 (17)	−82.1 (20)	−78.2 (17)	0.025 (16)

Table 305. $Cu(DCOO)_2 \cdot 4D_2O$. Direction cosines of the principal axes of the field gradient tensors at D of water molecules in phase II at 153 K [69S14]. $X_c \| a$, $Y_c \| b$, $Z_c \| c^*$ (c^* = reciprocal c axis). $D_a \cdots D_h$: D at 8 different sites. For numbering of hydrogen atoms, see [66O2].

		X_c	Y_c	Z_c
D_a	ϕ_{xx}	−0.773 (66)	0.327 (67)	0.544 (133)
D (2)	ϕ_{yy}	0.395 (125)	0.424 (55)	−0.815 (89)
	ϕ_{zz}	−0.497 (6)	0.845 (3)	0.199 (9)
D_c	ϕ_{xx}	−0.243 (55)	0.874 (84)	0.421 (143)
D (1′)	ϕ_{yy}	−0.298 (41)	−0.480 (152)	0.825 (74)
	ϕ_{zz}	0.923 (3)	0.075 (15)	0.377 (6)
D_b	ϕ_{xx}	−0.054 (64)	0.038 (32)	0.998 (5)
D (5)	ϕ_{yy}	0.853 (5)	−0.519 (6)	0.066 (71)
	ϕ_{zz}	0.520 (5)	0.854 (3)	−0.005 (10)
D_e	ϕ_{xx}	0.531 (127)	0.790 (88)	0.308 (15)
D (4)	ϕ_{yy}	−0.829 (82)	0.560 (125)	−0.007 (50)
	ϕ_{zz}	−0.178 (9)	−0.252 (16)	0.951 (5)
D_d	ϕ_{xx}	0.279 (34)	0.513 (111)	−0.812 (59)
D (1)	ϕ_{yy}	0.241 (36)	−0.856 (66)	−0.458 (106)
	ϕ_{zz}	−0.930 (3)	−0.069 (15)	−0.362 (7)
D_h	ϕ_{xx}	−0.691 (74)	−0.083 (110)	−0.715 (59)
D (3)	ϕ_{yy}	−0.495 (100)	0.776 (12)	0.391 (104)
	ϕ_{zz}	0.523 (6)	0.625 (4)	−0.580 (6)
D_f	ϕ_{xx}	−0.764 (60)	0.645 (69)	0.041 (34)
D (4′)	ϕ_{yy}	0.608 (76)	0.696 (59)	0.381 (18)
	ϕ_{zz}	0.217 (13)	0.316 (19)	−0.924 (7)
D_g	ϕ_{xx}	0.073 (61)	0.539 (33)	−0.839 (26)
D (7)	ϕ_{yy}	−0.828 (6)	−0.437 (41)	−0.352 (62)
	ϕ_{zz}	−0.556 (5)	0.721 (3)	0.414 (7)

Table 306. $Cu(DCOO)_2 \cdot 4D_2O$. $|e^2qQ/h|$ and η of D in water molecules in phase II at 153 K. For assignment, see [69S14]. $D_a \cdots D_h$: D at 8 different sites.

	$\|e^2qQ/h\|$ [kHz]	η	Assignment
D_a	215.5 (37)	0.074 (20)	D (2) > $O_{I'}$
D_c	216.6 (23)	0.079 (27)	D (1′)
D_b	213.6 (36)	0.136 (21)	D (5) > O_{II}
D_e	220.6 (47)	0.091 (40)	D (4)
D_d	214.2 (23)	0.104 (28)	D (1) > O_I
D_h	214.0 (32)	0.094 (17)	D (3)
D_f	220.7 (43)	0.112 (36)	D (4′) > $O_{II'}$
D_g	215.0 (33)	0.134 (20)	D (7)

34 DSP ($Ca_2Sr(CH_3CH_2COO)_6$) family

34A Pure compounds

Nr. 34A-1 $Ca_2Sr(CH_3CH_2COO)_6$, Dicalcium strontium propionate (DSP)
($M = 606.21$)

1a	Ferroelectricity in DSP was discovered by Matthias and Remeika in 1957.	57M2
b	(see table below)	[a] 65N2 [b] 57M2 [c] 67M7 [d] 67M4 [e] 78H14

phase	III [a]	II [b]	I [b]
state		F [b]	P [b]
crystal system	monoclinic [e]	tetragonal [c]	tetragonal [d]
space group		$P4_1$-C_4^2 or $P4_3$-C_4^4 [c]	$P4_12_12$-D_4^4 or $P4_32_12$-D_4^8 [d]
Θ	104.2 [a]	282.6 [a]	K

	$P_s \parallel [001]$ Transparent, colorless.	57M2
2a	Evaporation or heating method from aqueous solution. Solubility: Fig. 1148.	65K7
b	Crystal form: Fig. 1149.	
3a	Unit cell parameters: $a = 12.53$ Å, $c = 17.26$ Å at RT.	67M4
b	$Z = 4$. Atomic positions: Tables 307, 308; Figs. 1150, 1151.	67M4
4	Lattice distortion: Fig. 1152.	
5a	Dielectric constant: Figs. 1153···1156. $\varkappa_a$ decreases from 6.6 at −178 °C to 4.3 at 32 °C with increasing temperature ($f = 326$ kHz). Effect of hydrostatic pressure: Figs. 1157···1159. Effect of CH_3COO addition: see	61K4 69N1
c	Spontaneous polarization: Figs. 1160, 1161; see also Fig. 1154.	
d	Pyroelectric coefficient: see Fig. 1160.	
6a	Heat capacity: Fig. 1162. Transition heat, transition entropy:	65N2

transition	III − II	II − I	
ΔQ_m	818	940	J mol^{-1}
ΔS_m	2.55	13.4	J K^{-1} mol^{-1}

7a	Piezoelectricity: Fig. 1163.	
b	Electrostriction: Figs. 1164, 1165.	
8a	Elastic stiffness: Fig. 1166; see also Ultrasonic attenuation: Fig. 1167. Sound wave velocity: Figs. 1168, 1169.	66H2
b	The tetragonal c-axis is rotated by 90° under appropriate external stress in phase I.	77S9
9a	Refractive indices: Tables 309, 310. Birefringence: Fig. 1170.	
b	Electrooptic constant: Fig. 1171.	
d	Optical activity: Figs. 1172, 1173.	continued

Nr. 34A-1 $Ca_2Sr(CH_3CH_2COO)_6$, continued

e	Nonlinear optical properties: Fig. 1174. Nonlinear optical susceptibilities: $\|d_{31}\|=0.19\,(2)\,d_{11}^{\text{quartz}}$, $\|d_{33}\|=0.37\,(04)\,d_{11}^{\text{quartz}}$, $d_{31}\cdot d_{33}<0$ at $-20\,°C$ for $\lambda=1.06\,\mu m$	74I10
10a	Raman scattering: Figs. 1175, 1176.	
b	Brillouin scattering: see	73S8
13a	NMR: Figs. 1177, 1178; see also	68N1, 71S18
b	ESR: Fig. 1179. ESR of Mn^{2+} doped crystal:	73R1
15a	Ferroelectric domains are observed by powder pattern method [a]) and by electron mirror microscope. [b])	[a]) 67M8 [b]) 71S29

Table 307. $Ca_2Sr(CH_3CH_2COO)_6$(DSP). Fractional coordinates of atoms in the unit cell and isotropic temperature parameters in the phase I (20 °C) [67M7]. X(ij): The jth atom X in the ith propionate radical. *B* is defined by Eq. (e) in Introduction. See Fig. 1150.

Atom	*x*	*y*	*z*	*B* [Å^2]
Sr	0.2771	0.2771	0.0000	3.4
Ca	0.2596	0.0660	0.1582	4.6
O (1)	0.184	0.086	0.037	6.9
O (2)	0.383	0.463	0.040	8.4
O (31)	0.395	0.156	0.091	5.6
O (32)	0.546	0.171	0.032	8.4
O (41)	0.196	0.236	0.168	8.7
O (42)	0.184	0.387	0.114	5.5
C (11)	0.10	0.10	0.00	7.0
C (12)	0.01	0.01	0.00	
C (21)	0.46	0.46	0.00	7.5
C (22)	0.56	0.56	0.00	
C (31)	0.493	0.152	0.086	4.5
C (32)	0.56	0.14	0.18	
C (41)	0.173	0.332	0.172	5.1
C (42)	0.13	0.38	0.24	

Table 309. $Ca_2Sr(CH_3CH_2COO)_6$ (DSP). Refractive indices [1882F1].

λ [nm]	n_0	n_e
535	1.4897	1.4987
589	1.4871	1.4956
671	1.4839	1.4917

Table 310. $Ca_2Sr(CH_3CH_2COO)_6$ (DSP). n_o vs. *T* [71K7]. n_o: refractive index for ordinary light. $\lambda=589.3$ nm.

T [°C]	n_o	*T* [°C]	n_o
70	1.4944	−40	1.5037
60	1.4952	−50	1.5045
50	1.4960	−60	1.5054
40	1.4968	−70	1.5062
30	1.4976	−80	1.5069
20	1.4984	−90	1.5077
10	1.4991	−100	1.5084
0	1.4999	−110	1.5091
−10	1.5009	−120	1.5098
−20	1.5019	−130	1.5103
−30	1.5028	−140	1.5107
		−150	1.5113

Figs. p. 586, 594

Table 308. $Ca_2Sr(CH_3CH_2COO)_6$(DSP). Fractional coordinates of atoms in the unit cell and isotropic temperature parameters in the phase II (−50 °C) [67M4]. X(ij): The jth atom X in the ith propionate radical. *B* is defined by Eq. (e) in Introduction. See Fig. 1151.

Atom	*x*	*y*	*z*	*B* [Å²]
Sr	0.2780	0.2776	0.0000	3.0
Ca (1)	0.2587	0.0670	0.1578	3.8
Ca (2)	0.0684	0.2610	−0.1597	3.6
O (11)	0.189	0.087	0.044	6.3
O (12)	0.079	0.189	−0.033	5.9
O (21)	0.384	0.470	0.041	8.3
O (22)	0.474	0.373	−0.044	6.9
O (31)	0.392	0.157	0.089	3.8
O (32)	0.562	0.171	0.031	7.3
O (41)	0.162	0.389	−0.085	3.2
O (42)	0.175	0.546	−0.032	8.4
O (51)	0.199	0.234	0.166	6.1
O (52)	0.182	0.394	0.112	5.1
O (61)	0.228	0.197	−0.161	7.5
O (62)	0.398	0.179	−0.109	4.9
C (11)	0.10	0.10	0.00	5.3
C (12)	0.02	−0.01	−0.02	6.9
C (13)	0.04	−0.10	−0.01	13.0
C (21)	0.46	0.48	0.01	4.7
C (22)	0.53	0.55	0.06	9.8
C (23)	0.52	0.63	0.08	17.0
C (31)	0.49	0.16	0.09	4.1
C (32)	0.53	0.10	0.16	6.2
C (33)	0.57	0.21	0.18	13.3
C (41)	0.15	0.50	−0.09	4.3
C (42)	0.10	0.54	−0.16	5.4
C (43)	0.15	0.55	−0.23	14.3
C (51)	0.17	0.34	0.18	4.5
C (52)	0.11	0.39	0.24	4.6
C (53)	0.18	0.36	0.30	15.3
C (61)	0.34	0.17	−0.17	6.4
C (62)	0.37	0.15	−0.25	9.5
C (63)	0.34	0.19	−0.33	17.5

Nr. 34A-2 $Ca_2Ba(CH_3CH_2COO)_6$, Dicalcium barium propionate (DBP)
(M=655.92)

1a Pressure-induced ferroelectricity in DBP was reported by Sawada et al. in 1979. — 79S2

b

phase	III [b])	II [a])	I [a])
crystal system			cubic [c])
space group			Fd3m-O_h^7 [c])
Θ	204 [b])	266.9 [a]) K	

[a]) 55S1
[b]) 68N1
[c]) 35N1

Under hydrostatic pressure of $p=0.90\cdot10^8$ Pa: — 79S2

phase	V	IV
state	F	P
crystal system	tetragonal	tetragonal
space group	$P4_1$-C_4^2 or $P4_3$-C_4^4	$P4_12_12$-D_4^4 or $P4_32_12$-D_4^8
Θ_f	−6.5 °C	

Pressure induced phases: see Fig. 1182.

$\varrho=1.44\cdot10^3$ kg m^{-3} at 27 °C. — 35B1

Transparent, colorless.

continued

Nr. 34A-2 $Ca_2Ba(CH_3CH_2COO)_6$, continued

2a	Evaporation method from aqueous solution.	35 B 1
3a	Unit cell parameter: $a = 18.3\,(1)$ Å at RT.	35 B 1
b	$Z = 8$ in phase I.	35 B 1
4	Thermal expansion: Fig. 1180.	
5a	Dielectric constant in phase IV: $\varkappa_c = \varkappa_\infty + C/(T - \Theta_p)$, $\varkappa_\infty = 5.5$, $C = 49$ K, $\Theta_p = -6.5$ °C at $0.9 \cdot 10^8$ Pa. Fig. 1181. Phase diagram in regard to p: Fig. 1182.	79 S 2
6a	Transition heat: ΔQ_m (III–II) = 7277 J mol^{-1}. Transition entropy: ΔS_m (II–I) = 27.4 J mol^{-1} K^{-1}.	65 N 2 65 N 2
8a	Sound velocity: Figs. 1183···1185.	
13a	NMR: Fig. 1186; see also	71 S 18

Nr. 34A-3 $Ca_2Pb(CH_3CH_2COO)_6$, Dicalcium lead propionate (DLP)
($M = 725.8$)

1a	Ferroelectric activity in DLP was discovered by Takashige et al. in 1975.	75 T 3

phase	III [a]	II [a]	I [a]
state	(F) [b]	F [c]	P [c]
crystal system		tetragonal [c]	tetragonal [c]
space group		$P4_1$-C_4^2 or $P4_3$-C_4^4 *)	$P4_12_12$-D_4^4 or $P4_32_12$-D_4^8 *)
Θ		−82 [a])	60 °C [a])

[a]) 65 N 2
[b]) 75 O 5
[c]) 75 T 3

	$P_s \parallel [001]$	75 T 3
	Transparent, colorless.	
	$\varrho = 1.76 \cdot 10^3$ kg m^{-3}.	59 F 4
	*) Estimated by Osaka et al. [75 O 5].	
2a	Evaporation method from aqueous solution.	72 T 3
3a	Unit cell parameters: $a = 12.50$ Å, $c = 17.26$ Å, at RT.	59 F 4
b	$Z = 4$	59 F 4
4	Thermal expansion: Figs. 1187, 1188.	
5a	Dielectric constant: Figs. 1189···1191. Phase diagram in regard to p: Fig. 1192. Thermal hysteresis of III–II transition: see Fig. 1159.	
b	Effect of E_{bias} on $\varkappa$: Fig. 1193.	
c	Spontaneous polarization and coercive field: Fig. 1194.	
d	Pyroelectric charge: Fig. 1195.	
6a	Heat capacity: Fig. 1196. Transition heat: ΔQ_m(III–II) = 4850 J mol^{-1}. Transition entropy: ΔS_m(III–II) = 24 J mol^{-1} K^{-1}.	65 N 2 65 N 2
7a	Piezoelectric constant: Fig. 1197.	

8a	Elastic compliance: Fig. 1198. Sound velocity: Fig. 1199.	
9a	Birefringence: Fig. 1200.	
d	Rotatory power: Fig. 1201.	
10a	Raman scattering: Fig. 1202α; see also Fig. 1175.	
b	Brillouin scattering: Fig. 1202β.	
13a	NMR: Figs. 1203α, 1203β.	
14b	Inelastic neutron scattering: see	75D6

34B Solid solution

Nr. 34B-1 $Ca_2Sr(CH_3CH_2COO)_6$-$Ca_2Pb(CH_3CH_2COO)_6$ (DSP-DLP)

1b	Phase diagram: Fig. 1204α.	
2a	Relation between contents of DLP in the solution and in the crystal: Fig. 1204β.	
4	Thermal expansion: see Fig. 1187.	

35 $CH_2ClCOONH_4$ family

35A Pure compounds

Nr. 35A-1 $CH_2ClCOONH_4$, Ammonium monochloroacetate
($M=111.528$)

1a	$CH_2ClCOONH_4$ crystallizes either in the monoclinic or in the orthorhombic system. Ferroelectricity in monoclinic $CH_2ClCOONH_4$ was reported by Pepinsky et al. in 1957.	57P4
b	(see table below)	57P4

phase	II	I
state	F	P
crystal system	monoclinic*)	monoclinic
space group	C2-C_2^3*)	C2/c-C_{2h}^6
Θ	−150 °C	

$P_s \| [010]$
$\varrho = 1.558_5 \cdot 10^3$ kg m^{-3} at RT.

*) Speculated by Pepinsky et al. [57P4].

2a	Evaporation method from the cold aqueous solution.	57P4
3a	Unit cell parameters: $a=8.42$ Å, $b=11.63$ Å, $c=9.82$ Å, $\beta=110°$ at RT.	57P4
b	$Z=8$.	57P4
5a	$\varkappa_b=5$ at RT, $\varkappa_b=15$ at −150 °C.	57P4
c	$P_s \approx 0.1 \cdot 10^{-2}$ C m^{-2} and $E_c \approx 1 \cdot 10^6$ V m^{-1} at −170 °C	57P4

Nr. 35A-2 $(CH_2ClCOO)_2H \cdot NH_4$, Ammonium hydrogen bis-chloroacetate
(M=206.025)

1a	Ferroelectricity in $(CH_2ClCOO)_2H \cdot NH_4$ was discovered by Ichikawa and Mitsui in 1966.	66I2
b	(see table below)	66I2

phase	II	I	
state	F	P	
crystal system	monoclnic [a])	monoclinic	[a]) 73C4
space group	C_c-C_s^4	$C2/c$-C_{2h}^6	
Θ	120.4 (3) K [a])		

Θ=131.7 (3) K for $(CH_2ClCOO)_2D \cdot ND_4$. 73C4
$P_s \parallel [102]$ 66I2
$\varrho = 1.528 \cdot 10^3$ kg m^{-3} at 18 °C. 66I2
Transparent, colorless.

2a	Evaporation method from methyl alcohol solution.	66I2
	Polycrystal is prepared by trickling concentrated aqueous ammonia into aqueous solution of monochloroacetic acid until pH 4.5 is reached.	73C4
b	Crystal form: Fig. 1205.	
3a	Unit cell parameters: a=10.521 (4) Å, b=11.576 (4) Å, c=8.387 (3) Å, β=119.48 (3)° at RT.	72I1
b	Z=4	66I2
	Crystal structure: Tables 311, 312.	
5a	$\varkappa_{[102]} = C/(T-\Theta_p)$, $T > \Theta_{II-I}$, C=44 K, Θ_p=−145 °C.	66I2
	Dielectric constant: Figs. 1206, 1207.	
b	Effect of E_{bias} on $\varkappa$: Fig. 1208.	
c	Spontaneous polarization: Fig. 1209.	
	Coercive field: Fig. 1210.	
6a	Heat capacity: Figs. 1211, 1212.	
	Transition heat and transition entropy:	76C5

	ΔQ_m [J mol^{-1}]	ΔS_m [J K mol^{-1}]
$(CH_2ClCOO)_2H \cdot NH_4$	300 (21)	2.90 (22)
$(CH_2ClCOO)_2D \cdot ND_4$	340	2.95

13a	NMR: Figs. 1213···1216; Tables 313, 314.	

Tables 312. $(CH_2ClCOO)_2H \cdot NH_4$. Structure of phase I [72I1]. Temperature parameters at RT. B_{ij} is defined by Eq. (a) in Introduction.

	B_{11}	B_{22}	B_{33}	B_{12}	B_{13}	B_{23}
	10^{-4}					
Cl	144 (1)	127 (1)	207 (2)	− 69 (2)	76 (2)	98 (2)
O (1)	140 (3)	96 (2)	243 (5)	−107 (4)	182 (6)	8 (5)
O (2)	92 (2)	69 (2)	193 (4)	− 48 (3)	121 (5)	− 4 (4)
N	80 (3)	55 (2)	291 (9)	0	126 (9)	0
C (1)	106 (3)	92 (3)	232 (7)	− 61 (5)	84 (8)	30 (7)
C (2)	89 (3)	54 (2)	208 (5)	− 21 (4)	133 (7)	−16 (5)

Figs. p. 602 ff.

Tables 311. $(CH_2ClCOO)_2H \cdot NH_4$. Structure of phase I [72I1]. Fractional coordinates at RT.

	x	y	z
	10^{-4}		
Cl	1842 (1)	4237 (1)	− 607 (1)
O (1)	1770 (2)	2548 (2)	3340 (3)
O (2)	3401 (2)	3781 (2)	3363 (3)
N	0	453 (3)	2500
C (1)	1161 (3)	3404 (3)	575 (4)
C (2)	2242 (3)	3255 (2)	2580 (4)
H (1)	2500	2500	5000
H (2)	727 (37)	− 45 (31)	2366 (47)
H (3)	351 (40)	820 (30)	3521 (43)
H (4)	906 (37)	2678 (31)	− 55 (45)
H (5)	238 (43)	3785 (32)	479 (50)

Tables 313. $(CH_2ClCOO)_2H \cdot NH_4$. ^{35}Cl, direction cosines of the principal axes of the electric field gradient tensors with respect to the orthorhombic crystal axes *a*, *b*, and *c** [73C4].

	T K		Principal axis X	Y	Z
Cl	290	$X_a =$	−0.515 (67)	0.514 (74)	0.686 (7)
		$X_b =$	0.829 (15)	0.097 (116)	0.550 (8)
		$X_{c*} =$	0.217 (124)	0.852 (26)	−0.476 (5)
Cl (1)	78	$X_a =$	−0.501 (51)	0.521 (46)	0.692 (8)
		$X_b =$	0.847 (14)	0.131 (86)	0.515 (9)
		$X_{c*} =$	0.177 (86)	0.844 (16)	−0.507 (13)
Cl (2)	78	$X_a =$	−0.489 (91)	0.500 (56)	0.714 (9)
		$X_b =$	0.836 (13)	0.038 (144)	0.546 (6)
		$X_{c*} =$	0.247 (144)	0.865 (39)	−0.438 (11)

Tables 314. $(CH_2ClCOO)_2H \cdot NH_4$. e^2qQ/h and η for ^{35}Cl [73C4].

	T K	e^2qQ/h MHz	η
Cl	290	67.416 (15)	0.064 (10)
Cl (1)	78	69.774 (13)	0.061 (9)
Cl (2)	78	70.230 (16)	0.054 (12)

36 TGS ($(NH_2CH_2COOH)_3 \cdot H_2SO_4$) family

36A Pure compounds

Nr. 36A-1 $(NH_2CH_2COOH)_3 \cdot H_2SO_4$, Triglycine sulfate (TGS) ($M = 323.27$)

1a Ferroelectricity in TGS was discovered by Matthias et al. in 1956. — 56M1

b

phase	II	I
state	F	P
crystal system	monoclinic	monoclinic
space group	$P2_1\text{-}C_2^2$ [a])	$P2_1/m\text{-}C_{2h}^2$ [a])
Θ	49.42 °C [b])	

[a]) 57W3

[b]) 66G3

$\Theta = 60$ °C for deuterated crystal. — 61S9

$P_s \parallel [010]$ — 56M1

$\varrho = 1.69 \cdot 10^3$ kg m^{-3} at RT. — 57W3

Transparent, colorless. Cleavage plane: (010).

2a Crystal growth: evaporation method and cooling method from aqueous solution. Solubility in H_2O: Fig. 1217. — 66K10, 58N1

b Crystal form: Figs. 1218···1220.

3a Unit cell parameters:
Axial system by Wood et al.: $a = 9.15$ Å, $b = 12.69$ Å, $c = 5.73$ Å, $\beta = 105°40'$ at RT. Axial system by Hoshino et al.: $a' = 9.41_7$ Å, $b' = 12.64_3$ Å, $c' = 5.73_5$ Å, $\beta' = 110°23'$ at RT. — 57W3, 59H6

$\boldsymbol{a}' = \boldsymbol{a} + \boldsymbol{c}$, $\boldsymbol{b}' = -\boldsymbol{b}$, $\boldsymbol{c}' = -\boldsymbol{c}$. See Fig. 1220.

continued

Nr. 36A-1 $(NH_2CH_2COOH)_3 \cdot H_2SO_4$, continued

3b	$Z=2$. Positional and temperature parameters: Tables 315···319. Effect of X-irradiation: see Interatomic distances and bond angles: Tables 320, 321. Crystal structure: Fig. 1221. Structure change associated with phase transition: Tables 322···324; Fig. 1222.	59H6 76F9
4	Thermal expansion: Figs. 1223···1225; see also Effect of X-irradiation: see	62S6 74S36
5a	Dielectric constants: Figs. 1226···1237. Effect of H_2PO_3F addition: see Change of Θ determined by dielectric constant measurements: Figs. 1238···1240. Change of Θ_f for two dimensional pressure **T** in the plane perpendicular to the ferroelectric *b*-axis: $d\Theta_f/d\mathbf{T}=13(1)\cdot10^{-8}$ K m² N⁻¹.	 61D1 72M17
b	Nonlinear dielectric properties: Figs. 1241···1244. $\xi=6.25\cdot10^{11}$ V m⁵ C⁻³, $\zeta=4.02\cdot10^{14}$ V m⁹ C⁻⁵. See also:	 77C3 74Z3
c	Spontaneous polarization: Figs. 1245···1247. Coercive field: Figs. 1248···1250.	
d	Pyroelectric coefficient: Figs. 1251, 1252. Electrocaloric effect: Figs. 1253···1255. Figure of merit for $(NH_2CH_2COOH)_3 \cdot H_2SO_4$ at 25 °C: $p_2=4\cdot10^{-4}$ C K⁻¹ m⁻², $F.M._{R_I}=1.6\cdot10^{-10}$ C m J⁻¹, $F.M._{R_V}=4.6\cdot10^{-12}$ C m J⁻¹, $F.M._N=2.7\cdot10^{-11}$ C m J⁻¹, $F.M._{D*}=2.1\cdot10^{-6}$ C m^{3/2} J⁻¹ Ω^{1/2}. Figure of merit for $(ND_2CD_2COOD)_3 \cdot D_2SO_4$ at 25 °C: $p_2=2.7\cdot10^{-4}$ C K⁻¹ m⁻², $F.M._{R_I}=1.1\cdot10^{-10}$ C m J⁻¹, $F.M._{R_V}=6\cdot10^{-2}$ C m J⁻¹, $F.M._N=2.5\cdot10^{-11}$ C m J⁻¹, $F.M._{D*}=2.4\cdot10^{-6}$ C m^{3/2} J⁻¹ Ω^{1/2}.	 75B7 75B7
6a	Heat capacity: Figs. 1256···1259. Transition heat, transition entropy:	
b	Thermal conductivity: Fig. 1260.	
7a	Piezoelectric constant: Figs. 1261···1263.	
b	Electrostrictive constant: Fig. 1264; Table 325.	
8a	Elastic constant: Figs. 1265···1267; Table 326. See also 10b. Acoustic surface wave: Figs. 1268, 1269. Ultrasonic wave velocity and attenuation: Figs. 1270···1277; Table 327.	
b	Third order elastic constants: Figs. 1278, 1279.	
9a	Refractive indices: Figs. 1280···1287. Birefringence: Figs. 1288···1290. Reflection in infrared region: Figs. 1291, 1292.	

Transition heat, transition entropy (6a):

ΔQ_m [J mol⁻¹]	ΔS_m [J K mol⁻¹]	
$(NH_2CH_2COOH)_3 \cdot H_2SO_4$		
627	2.01	57H2
1446	4.60	64S12
1409	4.47	70T6
614 (32)	2.01 (10)	75C1
1629		77T5
Deuterated TGS		
646 (34)	2.11 (10)	75C1

b	Electrooptic effect: Figs. 1293···1299.	
c	Piezooptic effect: Fig. 1300.	
e	Nonlinear optical susceptibility: $d_{23}/d_{36}^{KDP}=0.07$ for $\lambda=693.4$ nm. See also	67S19 62S2
10a b	Raman scattering: Fig. 1301, 1302. Brillouin scattering: Figs. 1303, 1304; Table 328. Elastic constant from Brillouin scattering: see Fig. 1267.	
11	Electrical conduction: Figs. 1305···1307; Table 329. Additional data for conductivity: see Electric breakdown strength: $2.2 \cdot 10^4$ kV/m at RT. Thermally excited light emission ($\lambda \approx 425\cdots455$ nm) has been studied ($T=20\cdots70$ °C).	 61G4 69g2 73J3
13a	NMR: Figs. 1308···1312; Tables 330···334. Anisotropy of the chemical shift tensor: see	 76B12
b	ESR and ENDOR: Figs. 1313···1316; Tables 335···340. Cr^{3+} doped crystal: Fig. 1315; see also Cu^{2+} doped crystal: see VO^{2+} doped crystal: Fig. 1315; see also X-irradiated crystal: see γ-irradiated crystal: see	 74K8, 74S33, 75W5, 75W3, 76Y5, 78W4 77B17 75F9, 76F15, 77F5 74W3, 75W3, 76K16 70K5, 69K1
c	Mössbauer effect: Table 341.	
14	Integrated intensity of Bragg reflection: Figs. 1317···1320. Diffuse scattering: Figs. 1321···1323.	
15a	Domain structure: 180° domains of lenticular shape are observed by means of etching method and powder pattern method [a]). Domains are also observed by laser scattering [b]), scanning electron microscopy [c]) and X-ray topography [d]).	 [a]) 62j1 [b]) 67S11 [c]) 73N5 [d]) 78T1
b	Domain switching: Figs. 1324···1328.	
16	Thin layer: depolarization effect is reported in thin films. Surface layer: surface layers with low dielectric constant are estimated through measurements of dielectric constant. Radiation damage: effect of ^{60}Co γ-ray on $\varkappa_b$. Effect of X-ray irradiation on lattice constant Etchant: H_2O, ethyl alcohol.	76W6 67S10 68C3 67O3 77S40

Table 315. $(NH_2CH_2COOH)_3 \cdot H_2SO_4$ (TGS). Structure of phase I [73I10]. Fractional coordinates and temperature parameters at 57 °C. U_{ij} is defined by Eq. (d) in Introduction. Crystallographic axes are referred to those of Hoshino et al., see 3a.

Atoms	x	y	z	U_{11}	U_{22}	U_{33}	U_{23}	U_{31}	U_{12}
				10^{-2} Å^2					
(Sulfate ion)									
S	1.0003 (1)	0.2500 (0)	0.2252 (2)	1.8 (0)	1.8 (0)	1.9 (0)	0.0 (0)	0.8 (0)	0.0 (0)
O (1)	0.8585 (4)	0.2538 (37)	0.0058 (7)	2.1 (2)	4.8 (7)	2.4 (2)	0.9 (8)	0.4 (1)	1.1 (8)
O (2)	0.9651 (5)	0.2396 (28)	0.4549 (7)	4.1 (2)	2.3 (12)	2.4 (2)	−0.1 (3)	1.6 (2)	0.0 (3)
O (3)	1.0864 (3)	0.1550 (2)	0.2100 (5)	3.5 (1)	2.5 (1)	4.6 (2)	0.1 (1)	1.7 (1)	0.9 (1)
(Glycine I)									
O (11)	0.6077 (5)	0.2626 (18)	1.0709 (8)	2.3 (2)	3.6 (10)	3.3 (2)	−0.6 (3)	0.8 (2)	−0.1 (2)
O (12)	0.4937 (6)	0.2568 (33)	0.6609 (10)	4.1 (3)	9.6 (11)	3.8 (3)	0.5 (8)	1.2 (2)	−2.4 (9)
C (11)	0.4885 (7)	0.2600 (33)	0.8682 (12)	2.7 (2)	1.6 (14)	3.6 (3)	−0.3 (3)	1.2 (2)	−0.3 (4)
C (12)	0.3375 (7)	0.2543 (19)	0.9096 (12)	2.5 (2)	4.4 (5)	3.8 (3)	−0.5 (7)	0.9 (2)	−2.3 (6)
N (1)	0.3576 (7)	0.2136 (6)	1.1602 (13)	2.8 (3)	6.6 (6)	4.7 (4)	0.6 (3)	2.1 (3)	1.0 (3)
(Glycine II)									
O (21)	0.7817 (3)	0.4994 (3)	0.2313 (5)	3.3 (1)	5.3 (2)	3.5 (1)	1.2 (1)	0.9 (1)	−0.3 (1)
O (22)	0.5439 (3)	0.4741 (3)	0.2186 (6)	2.7 (1)	6.2 (2)	4.9 (2)	0.9 (2)	0.7 (1)	0.9 (1)
C (21)	0.6903 (4)	0.4722 (3)	0.3254 (7)	3.1 (2)	2.8 (2)	3.1 (2)	0.1 (1)	0.7 (1)	0.1 (1)
C (22)	0.7381 (4)	0.4297 (3)	0.5879 (7)	3.1 (2)	4.4 (2)	3.1 (2)	0.7 (1)	1.1 (1)	−0.1 (2)
N (2)	0.9061 (3)	0.4282 (3)	0.7009 (5)	3.4 (2)	3.2 (2)	2.3 (1)	0.3 (1)	0.3 (1)	−0.4 (1)

Table 317. $(NH_2CH_2COOH)_3 \cdot H_2SO_4$ (TGS). Structure of phase II [76F8]. Anisotropic temperature factor and principal values of root mean square amplitudes of thermal vibration. Anisotropic temperature parameter: see Table 315. Values of U_{ij}'s are $\times 10^4$ for S and $\times 10^3$ for O, N, and C in Å^2.

Atom	U_{11}	U_{22}	U_{33}	U_{12}	U_{13}	U_{23}	Min.	Inter.	Max.
S	117 (9)	95 (9)	145 (9)	− 8 (4)	67 (4)	− 3 (5)	0.094 Å	0.100 Å	0.123 Å
O (1)	14 (1)	35 (2)	20 (1)	− 4 (2)	5 (1)	− 2 (2)	0.115	0.147	0.188
O (2)	30 (2)	27 (2)	19 (1)	− 4 (2)	16 (1)	− 1 (2)	0.107	0.158	0.184
O (3)	29 (2)	9 (2)	39 (2)	− 8 (2)	19 (2)	0 (2)	0.069	0.160	0.203
O (4)	20 (2)	16 (2)	36 (2)	6 (2)	11 (2)	1 (2)	0.108	0.152	0.189
O (11)	15 (1)	42 (2)	27 (2)	2 (2)	7 (1)	− 2 (2)	0.121	0.167	0.205
O (12)	31 (2)	56 (4)	27 (2)	9 (2)	11 (1)	− 1 (2)	0.158	0.170	0.245
C (11)	19 (2)	17 (3)	29 (2)	4 (2)	10 (1)	1 (2)	0.115	0.147	0.169
C (12)	15 (2)	26 (2)	29 (2)	0 (2)	6 (1)	− 4 (3)	0.121	0.154	0.181
N (1)	16 (2)	47 (3)	29 (2)	− 7 (2)	14 (2)	− 3 (2)	0.101	0.172	0.220
O (21)	25 (2)	38 (2)	29 (2)	− 1 (2)	10 (2)	12 (2)	0.137	0.160	0.218
O (22)	19 (2)	41 (2)	31 (2)	5 (2)	4 (2)	7 (2)	0.129	0.181	0.211
C (21)	21 (2)	15 (2)	17 (2)	1 (2)	4 (2)	4 (2)	0.110	0.137	0.156
C (22)	18 (2)	30 (3)	27 (3)	− 1 (2)	11 (2)	1 (2)	0.121	0.165	0.176
N (2)	21 (2)	14 (2)	15 (2)	− 1 (2)	5 (2)	3 (2)	0.105	0.132	0.151
O (31)	25 (2)	41 (2)	25 (2)	− 3 (2)	10 (2)	10 (2)	0.130	0.161	0.217
O (32)	18 (2)	64 (3)	39 (2)	12 (2)	9 (2)	11 (2)	0.121	0.194	0.262
C (31)	22 (2)	17 (2)	33 (2)	2 (2)	9 (2)	− 2 (2)	0.126	0.151	0.184
C (32)	22 (2)	25 (2)	24 (2)	2 (2)	10 (2)	6 (2)	0.130	0.147	0.176
N (3)	19 (2)	19 (2)	16 (2)	− 2 (2)	5 (2)	2 (2)	0.124	0.130	0.153

Fig. p. 603

Table 316. $(NH_2CH_2COOH)_3 \cdot H_2SO_4$ (TGS). Structure of phase II [76F8]. Fractional coordinates of the non-hydrogen atoms. Crystallographic axial system by Hoshino et al. is adopted, see 3a. $a' = 9.419\,(1)$ Å, $b' = 12.647\,(1)$ Å, $c' = 5.727\,(1)$ Å, $\beta' = 110.32°$ at RT. See also Fig. 1221.

Atom	x	y	z
S	0.99956 (8)	0.25	0.22648 (13)
O (1)	0.8572 (3)	0.2470 (4)	0.0072 (5)
O (2)	0.9649 (3)	0.2500 (4)	0.4565 (5)
O (3)	1.0847 (5)	0.3455 (3)	0.2091 (8)
O (4)	1.0873 (4)	0.1550 (3)	0.2131 (8)
O (11)	0.6068 (3)	0.2379 (4)	1.0798 (5)
O (12)	0.4946 (4)	0.2292 (4)	0.6677 (6)
O (11)	0.4883 (4)	0.2367 (5)	0.8727 (8)
C (12)	0.3398 (4)	0.2418 (6)	0.9164 (7)
N (1)	0.3580 (4)	0.2856 (4)	1.1643 (8)
O (21)	0.7790 (5)	0.4927 (4)	0.2232 (8)
O (22)	0.5387 (5)	0.4659 (4)	0.1974 (8)
C (21)	0.6825 (5)	0.4659 (4)	0.3118 (8)
C (22)	0.7273 (5)	0.4290 (4)	0.5764 (9)
N (2)	0.8934 (5)	0.4284 (3)	0.6989 (8)
O (31)	0.2138 (5)	0.4958 (4)	0.7611 (8)
O (32)	0.4475 (5)	0.5164 (4)	0.7527 (9)
C (31)	0.2997 (5)	0.5229 (4)	0.6560 (9)
C (32)	0.2472 (5)	0.5665 (4)	0.3979 (9)
N (3)	0.0810 (5)	0.5727 (4)	0.2949 (8)

Table 318. $(NH_2CH_2COOH)_3 \cdot H_2SO_4$ (TGS). Structure of phase II [76F8]. Fractional coordinates and isotropic temperature factors of hydrogen atoms. Crystallographic axial system by Hoshino et al. is adopted, see 3a. Isotropic temperature parameter U is given by the formula $\exp\{-8\pi^2 U (\sin\theta/\lambda)^2\}$.

Atom	x	y	z	U [Å^2]
H (11)	0.432 (7)	0.335 (6)	1.235 (12)	0.04 (2)
H (12)	0.273 (8)	0.285 (6)	1.186 (12)	0.05 (2)
H (13)	0.403 (8)	0.250 (10)	1.275 (12)	0.06 (2)
H (14)	0.272 (5)	0.286 (4)	0.798 (8)	0.01 (1)
H (15)	0.293 (5)	0.174 (4)	0.908 (8)	0.01 (1)
H (16)	0.691 (9)	0.273 (8)	1.015 (14)	0.08 (3)
H (21)	0.922 (6)	0.484 (5)	0.719 (10)	0.02 (1)
H (22)	0.911 (6)	0.394 (5)	0.846 (10)	0.02 (1)
H (23)	0.929 (6)	0.383 (5)	0.608 (11)	0.02 (2)
H (24)	0.694 (8)	0.478 (6)	0.707 (12)	0.03 (2)
H (25)	0.662 (7)	0.372 (5)	0.572 (12)	0.03 (2)
H (26)	0.511 (7)	0.489 (5)	0.054 (9)	0.02 (1)
H (31)	0.014 (14)	0.521 (10)	0.290 (22)	0.12 (4)
H (32)	0.044 (7)	0.596 (5)	0.104 (12)	0.03 (2)
H (33)	0.043 (8)	0.629 (7)	0.402 (14)	0.05 (2)
H (34)	0.286 (8)	0.533 (6)	0.322 (13)	0.04 (2)
H (35)	0.286 (6)	0.637 (5)	0.388 (10)	0.02 (1)

Table 319. $(NH_2CH_2COOH)_3 \cdot H_2SO_4$ (TGS). Structure of phase II determined by neutron diffraction [73 K 10]. Fractional coordinates and temperature parameters. B_{ij} is defined by Eq. (a) in Introduction. Inverse polarity of the structure is shown in Fig. 1221. Hoshino et al.'s crystal axes are adopted, see 3a.

				$B_{ij} \cdot 10^4$					
	x	y	z	B_{11}	B_{22}	B_{33}	B_{12}	B_{13}	B_{23}
Sulfate group									
S	0.0002 (12)	0.02500 (22)	0.2282 (19)	37 (15)	25 (10)	104 (39)	– 27 (40)	25 (40)	101 (66)
O (1)	0.8568 (6)	0.2472 (14)	0.0046 (11)	38 (8)	49 (6)	160 (22)	– 2 (24)	12 (22)	110 (39)
O (2)	0.9652 (7)	0.2468 (15)	0.4579 (11)	86 (10)	41 (6)	151 (24)	– 16 (24)	115 (25)	– 95 (38)
O (4)	0.0836 (13)	0.1497 (11)	0.2064 (26)	96 (16)	7 (8)	286 (54)	14 (17)	122 (42)	– 27 (30)
O (3)	0.0882 (12)	0.3397 (14)	0.2172 (27)	52 (13)	33 (9)	282 (54)	– 5 (18)	103 (39)	36 (32)
Glycine groups									
O (11)	0.6086 (9)	0.2548 (15)	1.0777 (12)	57 (10)	54 (7)	166 (27)	4 (25)	34 (28)	42 (42)
O (12)	0.4953 (19)	0.2653 (14)	0.6660 (15)	76 (10)	86 (14)	204 (31)	5 (20)	77 (27)	– 55 (38)
C (11)	0.4890 (6)	0.2571 (12)	0.8732 (10)	47 (8)	35 (6)	177 (23)	– 30 (15)	54 (23)	– 10 (27)
C (12)	0.3409 (7)	0.2510 (14)	0.9182 (12)	34 (8)	42 (6)	230 (28)	24 (21)	80 (26)	76 (40)
N (1)	0.3584 (8)	0.2092 (11)	1.1645 (14)	38 (9)	110 (11)	245 (31)	21 (13)	94 (26)	00 (24)
H_1	0.707 (2)	0.253 (2)	1.049 (2)	73 (18)	71 (12)	225 (42)	122 (40)	59 (46)	134 (64)
H_2	0.292 (3)	0.330 (2)	0.897 (5)	215 (44)	98 (23)	694 (15)	77 (53)	478 (140)	264 (85)
H_3	0.263 (2)	0.198 (2)	0.777 (4)	82 (24)	133 (22)	286 (77)	– 90 (37)	86 (72)	– 71 (64)
H_4	0.415 (3)	0.143 (2)	1.192 (4)	224 (42)	56 (17)	303 (75)	56 (43)	297 (83)	154 (59)
H_5	0.252 (2)	0.189 (2)	1.178 (31)	63 (22)	120 (18)	357 (70)	– 2 (29)	173 (65)	61 (50)
H_6	0.411 (2)	0.260 (4)	1.301 (4)	118 (27)	187 (32)	336 (74)	120 (60)	41 (75)	– 240 (128)
O (31)	0.2202 (11)	0.5020 (12)	0.7774 (22)	40 (13)	59 (10)	167 (43)	– 7 (17)	– 4 (40)	116 (34)
O (32)	0.4600 (11)	0.5298 (13)	0.7973 (22)	23 (12)	79 (10)	125 (39)	– 3 (17)	– 19 (32)	– 47 (29)
C (31)	0.3190 (10)	0.5302 (11)	0.6898 (17)	96 (15)	20 (6)	150 (35)	– 10 (15)	99 (36)	19 (22)
C (32)	0.2683 (13)	0.5653 (13)	0.4166 (21)	74 (14)	44 (10)	230 (49)	– 42 (22)	88 (47)	82 (31)
N (3)	0.1046 (8)	0.5673 (8)	0.3017 (15)	53 (9)	39 (8)	52 (34)	– 19 (13)	– 42 (27)	21 (23)
H_8	0.327 (3)	0.525 (3)	0.327 (5)	97 (28)	143 (30)	313 (91)	21 (41)	120 (80)	– 49 (72)
H_9	0.319 (2)	0.645 (2)	0.417 (5)	94 (27)	97 (23)	415 (105)	– 167 (44)	33 (79)	121 (68)
H_{10}	0.064 (3)	0.495 (3)	0.295 (4)	176 (35)	62 (22)	284 (88)	– 36 (47)	20 (81)	252 (69)
H_{11}	0.062 (3)	0.600 (2)	0.121 (5)	227 (42)	70 (20)	165 (109)	24 (47)	260 (115)	40 (75)
H_{12}	0.056 (3)	0.614 (2)	0.385 (5)	161 (40)	66 (21)	327 (101)	102 (40)	132 (106)	35 (65)

continued

Table 319, continued

	x	y	z	B_{11}	B_{22}	B_{33}	B_{12}	B_{13}	B_{23}
				$B_{ij} \cdot 10^4$					
O (21)	0.7848 (14)	0.4972 (13)	0.2405 (26)	100 (18)	42 (9)	280 (50)	− 15 (20)	150 (52)	19 (36)
O (22)	0.5515 (16)	0.4808 (15)	0.2480 (32)	99 (19)	91 (12)	342 (59)	51 (23)	166 (54)	67 (42)
C (21)	0.6991 (10)	0.4727 (11)	0.3404 (19)	22 (10)	40 (8)	212 (41)	17 (12)	41 (35)	11 (22)
C (22)	0.7485 (12)	0.4273 (13)	0.5982 (19)	99 (17)	41 (96)	162 (43)	34 (22)	134 (47)	− 42 (33)
N (2)	0.9192 (10)	0.4232	0.7088 (22)	98 (13)	21 (7)	264 (45)	− 35 (16)	151 (46)	− 5 (28)
H_7	0.515 (2)	0.501 (2)	1.050 (4)	53 (20)	58 (12)	387 (114)	28 (24)	28 (26)	144 (62)
H_{13}	0.716 (2)	0.482 (2)	0.731 (4)	111 (32)	113 (21)	309 (83)	52 (37)	129 (85)	−118 (67)
H_{14}	0.715 (2)	0.355 (3)	0.606 (6)	109 (31)	72 (22)	586 (128)	− 60 (39)	− 30 (94)	328 (81)
H_{15}	0.960 (3)	0.502 (3)	0.711 (5)	169 (38)	59 (23)	483 (113)	−193 (45)	112 (94)	−200 (72)
H_{16}	0.947 (3)	0.394 (2)	0.870 (7)	125 (35)	34 (18)	371 (127)	59 (34)	−272 (100)	− 54 (78)
H_{17}	0.950 (2)	0.373 (2)	0.589 (5)	99 (28)	110 (26)	240 (85)	14 (36)	262 (84)	64 (67)

Table 320. $(NH_2CH_2COOH)_3 \cdot H_2SO_4$ (TGS). Structure of phase II [76F8]. Interatomic distances (in Å) and bond angles (in deg).

Sulfate ion		Glycine I		Glycine II		Glycine III	
S–O (1)	1.486 (2)	C (11)–O (11)	1.316 (4)	C (21)–O (21)	1.232 (8)	C (31)–O (31)	1.214 (9)
S–O (2)	1.463 (3)	C (11)–O (12)	1.200 (6)	C (21)–O (22)	1.284 (6)	C (31)–O (32)	1.310 (7)
S–O (3)	1.472 (4)	C (11)–C (12)	1.506 (6)	C (21)–C (22)	1.499 (8)	C (31)–C (32)	1.492 (9)
S–O (4)	1.476 (4)	C (12)–N (1)	1.477 (7)	C (22)–N (2)	1.475 (7)	C (32)–N (3)	1.471 (7)
O (1)–S–O (2)	110.1 (2)	O (11)–C (11)–O (12)	124.6 (4)	O (21)–C (21)–O (22)	126.0 (5)	O (31)–C (31)–O (32)	124.5 (6)
O (1)–S–O (3)	108.5 (2)	O (11)–C (11)–C (12)	113.3 (4)	O (21)–C (21)–C (22)	120.9 (5)	O (31)–C (31)–C (32)	123.2 (5)
O (1)–S–O (4)	107.0 (2)	O (12)–C (11)–C (12)	122.0 (3)	O (22)–C (21)–C (22)	113.1 (6)	O (32)–C (31)–C (32)	112.3 (6)
O (2)–S–O (3)	111.0 (3)	C (11)–C (12)–N (1)	111.8 (3)	C (21)–C (22)–N (2)	111.2 (6)	C (31)–C (32)–N (3)	110.9 (6)
O (2)–S–O (4)	110.5 (3)						
O (3)–S–O (4)	109.7 (2)						

Table 322. $(NH_2CH_2COOH)_3 \cdot H_2SO_4$ (TGS). Temperature variations of lattice constants [73I10]. Hoshino et al.'s axes are adopted, see 3a. ►

	19 °C	37 °C	57 °C
a' [Å]	9.42_4	9.41_3	9.41_5
b' [Å]	12.64_3	12.64_1	12.64_3
c' [Å]	5.73_5	5.74_1	5.74_0
β' [deg]	110.3_6	110.3_5	110.3_3

Table 321, see p. 230.

Table 321. $(NH_2CH_2COOH)_3 \cdot H_2SO_4$ (TGS). Structure of phase II [76F8]. Interatomic distances and bond angles involving hydrogen atoms.

Donor Acceptor	D–A [Å]	D–H [Å]	H–A [Å]	D–H–A [deg]
O (11)–H (16)–O (1)^{I}	2.536 (4)	1.08 (9)	1.62 (8)	139 (6)
N (1)–H (11)–O (22)^{I}	2.812 (7)	0.92 (7)	1.98 (8)	149 (6)
N (1)–H (12)–O (3)^{II}	2.780 (7)	0.85 (8)	1.98 (8)	156 (7)
N (1)–H (13)–O (12)^{I}	2.811 (5)	0.78 (8)	2.13 (7)	147 (7)
C (12)–H (14)		0.94 (5)		
C (12)–H (15)		0.96 (5)		
O (22)–H (26)–O (32)^{III}	2.473 (7)	0.82 (6)	1.65 (7)	171 (7)
N (2)–H (21)–O (4)^{IV}	2.904 (6)	0.74 (8)	2.21 (8)	156 (8)
N (2)–H (22)–O (1)^{I}	2.986 (7)	0.91 (6)	2.21 (6)	143 (4)
N (2)–H (22)–O (3)^{I}	3.037 (6)	0.92 (6)	2.23 (5)	146 (5)
N (2)–H (23)–O (2)	2.850 (7)	0.92 (9)	1.96 (8)	160 (7)
C (22)–H (24)		1.10 (9)		
C (22)–H (24)		0.95 (7)		
N (3)–H (31)–O (2)^{V}	2.911 (7)	0.91 (9)	2.13 (9)	143 (8)
N (3)–H (31)–O (3)^{V}	2.918 (6)	0.91 (9)	2.41 (10)	116 (7)
N (3)–H (32)–O (4)^{VI}	2.976 (6)	1.07 (7)	1.95 (6)	159 (6)
N (3)–H (33)–O (2)^{VII}	2.769 (7)	1.08 (8)	1.74 (8)	156 (7)
C (32)–H (34)	0.78 (7)			
C (32)–H (35)	0.98 (7)			

Superscripts I···VII stand for atoms whose coordinates are: I: $x, y, 1+z$. II: $x-1, y, 1+z$. III: $x, y, z-1$. IV: $2-x, \frac{1}{2}+y, 1-z$. V: $x-1, y, z$. VI: $1-x, \frac{1}{2}+y, -z$. VII: $1-x, \frac{1}{2}+y, 1-z$. The coordinates x, y, z are given in Tables 316 and 318.

Table 323. $(NH_2CH_2COOH)_3 \cdot H_2SO_4$ (TGS). Temperature variations of bond lengths and angles in the glycine molecules [73I10]. i=1, 2, 3.

		19 °C	37 °C	57 °C
Bond lengths [Å]: O (i1)–C (i1)	Glycine I	1.306 (5)	1.309 (6)	1.307 (7)
	Glycine II	1.213 (7)	1.197 (16)	1.214 (6)
	Glycine III	1.218 (8)	1.238 (16)	1.214 (6)
O (i2)–C (i1)	Glycine I	1.198 (7)	1.209 (9)	1.208 (10)
	Glycine II	1.299 (5)	1.332 (13)	1.300 (4)
	Glycine III	1.305 (6)	1.263 (12)	1.300 (4)
C (i1)–C (i2)	Glycine I	1.527 (7)	1.518 (9)	1.522 (11)
	Glycine II	1.544 (8)	1.520 (16)	1.514 (5)
	Glycine III	1.494 (7)	1.509 (16)	1.514 (5)
C (i2)–N (i)	Glycine I	1.478 (8)	1.485 (11)	1.477 (13)
	Glycine II	1.540 (7)	1.543 (15)	1.487 (4)
	Glycine III	1.460 (6)	1.457 (14)	1.487 (4)
Bond angles [deg]: O (i1)–C (i1)–O (i2)	Glycine I	125.4 (5)	124.1 (6)	124.2 (7)
	Glycine II	125.7 (5)	125.5 (10)	125.8 (4)
	Glycine III	126.1 (5)	125.2 (11)	125.8 (4)
O (i1)–C (i1)–C (i2)	Glycine I	113.0 (4)	114.7 (6)	115.0 (6)
	Glycine II	122.7 (4)	124.1 (9)	122.1 (3)
	Glycine III	122.1 (4)	120.8 (8)	122.1 (3)
O (i2)–C (i1)–C (i2)	Glycine I	121.5 (4)	121.1 (5)	120.7 (6)
	Glycine II	111.5 (5)	110.4 (11)	112.1 (4)
	Glycine III	111.7 (5)	113.9 (11)	112.1 (4)
C (i1)–C (i2)–N (i)	Glycine I	111.5 (4)	110.1 (6)	111.0 (8)
	Glycine II	106.2 (5)	107.7 (11)	110.0 (3)
	Glycine III	113.1 (5)	111.4 (11)	110.0 (3)

Table 324. $(NH_2CH_2COOH)_3 \cdot H_2SO_4$ (TGS). Temperature variations of bond lengths and bond angles in the sulfate ion [73I10].

	19 °C	37 °C	57 °C
Bond lengths [Å]			
S–O (1)	1.488 (2)	1.488 (3)	1.485 (4)
S–O (2)	1.469 (3)	1.467 (5)	1.473 (6)
S–O (3)	1.472 (4)	1.475 (9)	1.468 (3)
S–O (4)	1.477 (4)	1.469 (8)	1.468 (3)
Bond angles [deg]			
O (1)–S–O (2)	110.2 (2)	110.0 (2)	110.3 (3)
O (1)–S–O (3)	108.4 (3)	108.1 (5)	109.2 (15)
O (1)–S–O (4)	107.5 (3)	107.0 (5)	109.2 (15)
O (2)–S–O (3)	109.5 (3)	111.6 (6)	106.3 (12)
O (2)–S–O (4)	111.0 (3)	110.1 (6)	106.3 (12)
O (3)–S–O (4)	110.2 (2)	109.8 (5)	109.8 (2)

Table 325. $(NH_2CH_2COOH)_3 \cdot H_2SO_4$ (TGS). $Q_{\lambda\mu}$ at RT. Coordinate systems: see Fig. 1219.

Q_{12}	Q_{22}	Q_{32}	Q_{44}	Q_{46}	Q_{52}	Q_{64}	Q_{66}	Q_{h2} *)	Ref.
$m^4 C^{-2}$									
1.35	2.25	−4	−3.9	0.54	−0.36	3.8	2	−0.4	63I2
1.6	2.4	−4.45			<0.1			−0.45	63S6
								−0.515	60J3

*) Volume electrostrictive constant: $Q_{h2} = Q_{12} + Q_{22} + Q_{32}$.

Table 326. Deuterated triglycine sulfate. Elastic stiffness obtained by Brillouin scattering [77V2]. c_{ij}: in units of 10^{10} N m^{-2}. Coordinate systems: see Fig. 1219.

c_{11}	c_{22}	c_{33}	c_{44}	c_{55}	c_{66}	c_{12}	c_{13}	c_{15}	c_{23}	c_{25}	c_{35}	c_{46}
5.06	3.51	2.81	1.10	1.21	0.73	2.04	2.05	—	2.40	—	—	+0.02

Table 327. $(NH_2CH_2COOH)_3 \cdot H_2SO_4$ (TGS). Change of the sound velocity Δv_l relative to the sound velocity at 50.0 °C [64O1]. Pulse method. All measurements are for longitudinal waves perpendicular to $(\bar{1}01)$.

f / T [°C]	15 MHz	45 MHz	65 MHz
	Δv_l [10 m s^{-1}]		
49.50	− 0.55	0	0
49.20	− 1.4	− 1.2	− 1.2
49.15	− 1.8	− 1.5	− 2.1
49.10	− 2.7	− 1.9	− 2.6
49.00	− 6.9	− 6.0	− 6.5
48.90	− 8.9	− 9.5	−10.30
48.80	− 9.9	−11.0	−11.1
48.00	−11.3	−11.0	−11.7

Table 328. $(NH_2CH_2COOH)_3 \cdot H_2SO_4$ (TGS), $(NH_2CH_2COOH)_3 \cdot H_2BeF_4$ (TGFB). Pockels (piezooptic) coefficients determined from Brillouin scattering intensity [73M12]. $\lambda = 633$ nm.

Crystal	p_{21}	p_{23}	p_{25}
TGFB	$20 \cdot 10^{-2}$	$20 \cdot 10^{-2}$	~0
TGS	$27 \cdot 10^{-2}$	$27 \cdot 10^{-2}$	~0

Table 329. $(NH_2CH_2COOH)_3 \cdot H_2SO_4$ (TGS). Effect of radiation damage and impurity doping on ϱ and ΔU of TGS [61 T 2]. In the X-ray irradiation, the X-ray is excited by 35 kV; in the neutron irradiation, the dose is $3 \cdot 10^{16}$ nvt. Asterisks (*) indicate that the crystals were grown from aqueous solutions containing 0.2 mol% of Fe^{3+} or Cr^{3+} ions. Activation energy ΔU is estimated from the relation $\varrho \propto \exp(\Delta U/kT)$.

Direction of E	Treatment	ϱ [Ωm]		ΔU [eV]	
		27 °C	48 °C	Ferroelectric phase	Paraelectric phase
b	Normal	$1.5 \cdot 10^{12}$	$8.5 \cdot 10^{10}$	1.5	0.7
b	X-ray 15 min	$1.9 \cdot 10^{11}$	$6.5 \cdot 10^{9}$	1.35	0.73
b	X-ray 1 h	$2 \cdot 10^{11}$	$6.5 \cdot 10^{9}$	1.39	0.84
b	X-ray 4 h	$2.1 \cdot 10^{11}$	$6.0 \cdot 10^{9}$	1.52	0.92
b	Neutron	$1.8 \cdot 10^{11}$	$6.2 \cdot 10^{9}$	1.58	0.9
b	*Fe^{3+}	$3 \cdot 10^{12}$	$1 \cdot 10^{11}$	1.47	0.9
b	*Cr^{3+}	$3.5 \cdot 10^{12}$	$7 \cdot 10^{10}$	1.68	0.9
b	*Cr^{3+} X-ray 4 h	$3.5 \cdot 10^{11}$	$8 \cdot 10^{9}$	1.6	0.9
c	Normal	$2.7 \cdot 10^{11}$	$1.2 \cdot 10^{10}$	1.3	0.75
c	X-ray 15 min	$7 \cdot 10^{9}$	$1.4 \cdot 10^{8}$	1.9	0.6
c	X-ray 1 h	$8.5 \cdot 10^{9}$	$1.5 \cdot 10^{8}$	1.86	0.67
c	X-ray 4 h	$7 \cdot 10^{9}$	$2.5 \cdot 10^{8}$	1.65	0.71
c	*Fe^{3+}	$1.3 \cdot 10^{10}$	$3.1 \cdot 10^{8}$	1.65	0.6
c	*Fe^{3+} X-ray 1 h	$8.5 \cdot 10^{9}$	$1.6 \cdot 10^{8}$	1.84	0.68

Table 330. $(ND_2CH_2COOD)_3 \cdot D_2SO_4$. Eigenvalues $3eQ\,\phi_{ij}/2h$ (in kHz) and direction cosines λ, μ, ν of the largest principal axes for the O–D- -O and ND_3 electric quadrupole coupling constant tensors [67 B 12]. G: Glycine. See Fig. 1219ff.

O–D- -O (2.55 Å) *)				O–D- -O (2.44 Å)			
T=73 °C		T=20 °C		T=73 °C		T=20 °C	
$3eQ\,\phi_{ii}/2h$	λ μ ν	$3eQ\,\phi_{ii}/2h$	λ μ ν	$3eQ\,\phi_{ii}/2h$	λ μ ν	$3eQ\,\phi_{ii}/2h$	λ μ ν
	0.861		0.860		±0.313		±0.325
195.8	0.022	197.5	0.012	118.3	0.275	127.9	0.237
	0.508		0.510		±0.909		±0.901
	−0.206		−0.122		±0.141		±0.026
−115.5	0.928	−114.6	0.975	− 70.5	0.933	− 78.2	0.955
	0.309		0.184		±0.331		±0.295
	−0.464		−0.495		±0.939		±0.945
− 80.3	−0.371	− 83.0	−0.221	− 47.8	0.232	− 49.7	0.072
	0.804		0.840		±0.253		±0.318
$(e^2qQ)/h$=131.0		$(e^2qQ)/h$=130.5		$(e^2qQ)/h$=78.8		$(e^2qQ)/h$=85.0	
η= 0.160		η= 0.180		η= 0.192		η= 0.223	

*) Since the two $R_{O\cdots O}$=2.54 Å H-bonds have nearly the same direction, only the average direction cosines of the coupling tensor for this bond are given.

continued

Table 330, continued

GI				GII, GIII		GII		GIII	
$T=73$ °C		$T=20$ °C		$T=73$ °C		$T=20$ °C			
$3eQ\phi_{ii}/2h$	λ μ ν	$3eQ\phi_{ii}/2h$	λ μ ν	$3eQ\phi_{ii}/2h$	λ μ ν	$3eQ\phi_{ii}/2h$	λ μ ν	$3eQ\phi_{ii}/2h$	λ μ ν
	±0.024		±0.039		±0.995		±0.990		±0.997
56.0	0	70.5	−0.400	77.9	−0.067	80.7	−0.135	79.2	−0.075
	±0.997		±0.916		±0.076		±0.035		±0.025
	±0.997		±0.442		±0.013		±0.075		±0.070
−28.6	0	−43.9	0.828	−44.7	−0.676	−45.7	0.729	−45.4	−0.686
	±0.0236		±0.343		±0.736		±0.680		±0.724
	0		±0.896		±0.098		±0.117		±0.037
−27.4	1	−26.5	0.392	−33.2	0.734	−35.0	0.671	−33.7	0.724
	0		0.209		±0.672		±0.732		±0.689
$(e^2qQ)/h=37.3$		$(e^2qQ)/h=47.0$		$(e^2qQ)/h=51.9$		$(e^2qQ)/h=54.8$		$(e^2qQ)/h=52.8$	
$\eta=$ 0.021		$\eta=$ 0.247		$\eta=$ 0.148		$\eta=$ 0.133		$\eta=$ 0.147	

Table 332. $(NH_2CH_2COOH)_3 \cdot H_2SO_4$ (TGS). Components of ^{14}N quadrupole coupling tensors $eQ\phi_{ij}/h$ (in kHz) expressed in the crystal fixed coordinate system $a'\sin\beta'$, b', c' [71 B 17]. Hoshino et al.'s axes are adopted. See also Fig. 1221.

	$T<\Theta_f$			$T>\Theta_f$		
	$a'\sin\beta'$	b'	c'	$a'\sin\beta'$	b'	c'
Glycine I	−747	0	280	−740	0	208
	0	−107	±350	0	−140	0
	280	±350	853	208	0	880
Glycine II	960	0	−243	960	0	−208
	0	−493	± 73	0	−480	± 52
	−243	± 73	−467	−208	± 52	−480
Glycine III	880	0	−243	960	0	−208
	0	−467	± 73	0	−480	± 52
	−243	± 73	−453	−208	± 52	−480

Tables 331, 333, see p. 234.

Table 334. $(ND_2CH_2COOD)_3 \cdot D_2SO_4$. e^2qQ/h and η of the ND_3 group deuterons [77 H 17]. The data of [67 B 12] (see Table 330) are given in parantheses.

<table>
<tr><th>Glycine</th><th colspan="2">Ferro (294 K)</th><th colspan="2">Para (345 K)</th></tr>
<tr><th></th><th>e^2qQ/h [kHz]</th><th>η</th><th>e^2qQ/h [kHz]</th><th>η</th></tr>
<tr><td>I</td><td>47.2 (47.0)</td><td>0.291 (0.247)</td><td>38.09 (37.3)</td><td>0.033 (0.021)</td></tr>
<tr><td>II</td><td>53.6 (54.8)</td><td>0.217 (0.133)</td><td rowspan="2">52.3 (51.9)</td><td rowspan="2">0.212 (0.148)</td></tr>
<tr><td>III</td><td>52.3 (52.8)</td><td>0.212 (0.147)</td></tr>
</table>

Table 331. $(ND_2CH_2COOD)_3 \cdot D_2SO_4$. ND_3^+ deuteron electric field gradient eigenvalues, $3eQ\,\phi_{ij}/2h$ (in kHz), and direction cosines λ, μ, ν for each eigenvalue [67 B 12]. $|\phi_{ZZ}| > |\phi_{YY}| > |\phi_{XX}|$; the orientational accuracy is within $\pm 1°$; $A_{1,2}$, $B_{1,2}$, and $C_{1,2}$ associate with $N_{II}D_3^+$, $N_{III}D_3^+$, and $N_ID_3^+$, respectively.

Site	T = 65 °C: $3eQ\,\phi_{ii}/2h$	λ	μ	ν	Site	T = 30 °C: $3eQ\,\phi_{ii}/2h$	λ	μ	ν
$A_1(B_1)$, $A_2(B_2)$	79.6	0.995	∓0.054	0.078	A_1, A_2	76.3	0.895	∓0.084	0.057
	−46.2	−0.097	∓0.617	0.782		−44.9	−0.102	∓0.749	0.658
	−33.4	0.015	∓0.780	−0.622		−31.4	−0.013	∓0.663	−0.751
					B_1, B_2	79.3	0.993	∓0.027	0.112
						−46.3	−0.109	∓0.625	0.773
						−33.0	0.043	∓0.782	−0.662
C	56.6	0	0	1	C_1, C_2	69.0	0.023	∓0.386	0.922
	−29.6	1	0	0		−42.7	0.441	±0.832	0.337
	−27.0	0	1	0		−26.3	−0.897	±0.397	0.189

Table 333. $(NH_2CH_2COOH)_3 \cdot H_2SO_4$ (TGS). ^{14}N and $-ND_3^+$ deuteron quadrupole coupling constants e^2qQ/h (in kHz), asymmetry parameters η and direction cosines λ, μ, ν of the largest principal axis of the electric field gradient tensors [71 B 17]. The direction cosines of the C–N bond directions, taken from the structural data on the ferroelectric phase, are listed as well.

		$(^{14}N)_{65\,°C}$	$(^{14}N)_{20\,°C}$	$(\text{-}ND_3)_{70\,°C}$	$(\text{-}ND_3)_{20\,°C}$	(-C-N-)$_{20\,°C}$
Glycine I	e^2qQ/h	906	1008	37.3	47.0	—
	η	0.691	0.586	0.021	0.247	—
	λ	± 0.125	± 0.150	± 0.024	± 0.039	±0.150
	μ	0.000	0.296	0	0.400	0.218
	ν	± 0.992	± 0.943	± 0.997	± 0.916	±0.964
Glycine II	e^2qQ/h	990	983	51.9	54.8	—
	η	0.10	0.137	0.148	0.133	—
	λ	0.990	0.985	0.995	0.990	0.999
	μ	± 0.005	± 0.009	± 0.067	± 0.135	±0.054
	ν	0.140	− 0.175	0.076	0.035	0.005
Glycine III	e^2qQ/h	990	1020	51.9	52.8	—
	η	0.10	0.130	0.148	0.147	—
	λ	0.990	0.986	0.995	0.997	0.996
	μ	± 0.05	± 0.008	± 0.068	± 0.075	±0.009
	ν	0.140	− 0.164	0.076	0.025	0.088

Table 335. $(NH_2CH_2COOH)_3 \cdot H_2SO_4$ (TGS). g and A values of Cu^{2+}-doped TGS [73K7]. A: hyper fine structure factor. $T = 77$ K. l, m, n: direction cosines of the principal axes relative to the X, Y, Z frame. $X \| a$, $Y \| b$ of Wood and Holden's cell, see Fig. 1219. A_s^N, A_s^H: Super hyperfine structure factors due to Fermi contact interaction at nitrogen nucleus and hydrogen nucleus, respectively.

Principal value		Direction cosine		
		l	m	n
Spectrum I				
g	2.4332	0.712	∓0.703	∓0.006
	2.0379	0.556	0.558	±0.617
	2.1748	−0.430	∓0.442	0.787
$A(\cdot 10^{-2}\ m^{-1})$	127.3	0.735	∓0.675	0.063
	77.7	±0.588	0.589	∓0.555
	17.7	0.338	±0.444	0.830
Spectrum II				
g	2.092	0.792	∓0.266	−0.549
	2.259	∓0.081	0.846	∓0.527
	2.027	0.605	±0.462	0.649
$A\ (\cdot 10^{-2}\ m^{-1})$	24.2	0.756	±0.332	0.564
	165.4	∓0.008	0.867	∓0.499
	84.8	−0.654	±0.373	0.658

$A_s^N \cdots 10.0 \cdot 10^{-2}\ m^{-1}$, $A_s^H \cdots 10.0 \cdot 10^{-2}\ m^{-1}$

Table 338. $(ND_2CH_2COOD)_3 \cdot D_2SO_4{:}VO^{2+}$. Spin Hamiltonian parameters of $(VO^{2+})_s$ at room temperature and direction cosines of the principal axes of g and A tensors [74F5]. $(VO^{2+})_s$: vanishing complex during annealing process. $A_\perp$, $A_\|$ in units of $10^{-2}\ m^{-1}$. Reference axes of direction cosines are $\bar{a}$, b, c axes. $\bar{a} \perp (100)$, see Fig. 1219.

Tensors		Principal values	Direction cosines
g	$g_\perp$	1.9939	0.735, ±0.394, −0.552
		1.9850	∓0.607, 0.745, ∓0.277
	$g_\|$	1.9296	0.302, ±0.539, 0.787
^{51}V	$A_\perp$	68.9	0.803, ∓0.581, 0.135
		57.1	±0.560, 0.655, ∓0.507
	$A_\|$	173.0	0.206, ±0.483, 0.851

Table 340. $(NH_2CH_2COOH)_3 \cdot H_2SO_4$ (TGS). Relaxation time τ for the exchange of protons in the radical [70K5, 72K4]. $\tau^{-1} = f_0 \exp(-E/kT)$.

Radical	E [eV]	f_0 [s^{-1}]	Ref.
$\dot{C}H_2COO^-$	0.16 (3)	$1.2 \cdot 10^{12}$	70K5
$NH_3\dot{C}HCOO^-$	0.17	$5 \cdot 10^{11}$	72K4

Table 336. $(NH_2CH_2COOH)_3 \cdot H_2SO_4{:}Cu^{2+}$. Spin Hamiltonian parameters of $^{63}Cu^{2+}$ [74S34].

$g_\|$	2.261	$A_\|^{Cu}$	$150.1 \cdot 10^{-2}\ m^{-1}$
$g_\perp$	2.054	$A_\perp^{Cu}$	$30 \cdot 10^{-2}\ m^{-1}$
$A_\|^N$	$7.4 \cdot 10^{-2}\ m^{-1}$	A^H	$3.6 \cdot 10^2\ A\,m^{-1}$
$A_\perp^N$	$11.4 \cdot 10^{-2}\ m^{-1}$		

Table 337. $(ND_2CH_2COOH)_3 \cdot D_2SO_4{:}VO^{2+}$. Spin Hamiltonian parameters of $(VO^{2+})_1$ and $(VO^{2+})_2$ at room temperature and direction cosines of the principal axes of g and A tensors [74F5]. $(VO^{2+})_1$, $(VO^{2+})_2$: observed centers after the annealing process. Reference axes of the direction cosines are $\bar{a}$, b, c axes, where $\bar{a} \perp (100)$, see Fig. 1219. $A_\perp$, $A_\|$ in units of $10^{-2}\ m^{-1}$.

Tensors		Principal values	Direction cosines
g_1	$g_{1\perp}$	1.9852	0.712, ∓0.674, 0.188
		1.9982	−0.305, ∓0.058, 0.949
	$g_{1\|}$	1.9288	±0.631, 0.736, ±0.246
g_2	$g_{2\perp}$	1.9854	0.745, ∓0.661, ∓0.096
		1.9981	−0.146, ∓0.132, 0.976
	$g_{2\|}$	1.9289	±0.648, 0.737, ±0.195
$^{51}V_1$	$A_{1\perp}$	71.2	0.611, ∓0.540, ±0.565
		73.7	∓0.407, 0.410, 0.806
	$A_{1\|}$	177.8	±0.662, 0.730, ±0.169
$^{51}V_2$	$A_{2\perp}$	70.1	−0.446, ±0.380, 0.810
		73.9	∓0.567, 0.581, ∓0.568
	$A_{2\|}$	178.2	±0.692, 0.719, ±0.044

Table 339, see p. 236.

Table 341. $(NH_2CH_2COOH)_3 \cdot H_2SO_4$: ^{57}Fe. Mössbauer data [76M20]. δ: isomer shift. ΔE_{quad}: quadrupole splitting. Γ: linewidth.

	T K	δ mm s^{-1}	E_{quad} mm s^{-1}	Γ mm s^{-1}
Iron-doped TGS	80	0.542 (3)	0.598 (6)	0.23 (1)
	180	0.493 (7)	0.706 (6)	0.26 (2)
	315	0.419 (7)	0.687 (8)	0.26 (3)
	325	0.413 (7)	0.684 (8)	0.24 (2)
	335	0.408 (9)	0.678 (9)	0.25 (2)
	345	0.394 (5)	0.718 (5)	0.260 (5)
Iron-doped TGS after heating	330	0.413 (5)	0.720 (5)	0.274 (5)
	315	0.419 (5)	0.708 (5)	0.267 (5)

Table 339. $(NH_2CH_2COOH)_3 \cdot H_2SO_4$ (TGS). ESR data for irradiated crystals: see [61 B 2, 61 O 2].

Paramagnetic center	S	$\mathscr{H}$	ν GHz	T K	g-factor	FS D, E $[10^{-2}\,m^{-1}]$	HFS $^nA, ^nA_\lambda, Q_z$ $[10^{-2}\,m^{-1}]$	Ref.
diglycine-Cr^{3+}-chelate*)	3/2	(5)	9.4	RT	I: 1.985 (50)	I: $\|D\|$ = 4252 (15)	$\|^{53}A\| < 13$[1])	66 W 4
						$\|E\|$ = 668 (10)		
					II: 1.985 (5)	II: $\|D\|$ = 3964 (15)		
						$\|E\|$ = 489 (10)		
diglycine-Cu^{2+}-chelate*)	1/2	(10)	9	77	g_x = 2.048 (5)		$^{63}A_x$ = − 37 (4)[2])	
					g_y = 2.063 (5)		$\|^{63}A_y\|$ ≦ 10	67 W 7
					g_z = 2.252 (5)		$^{63}A_z$ = − 150 (3)	65 L 2
				300	g_x = 2.052 (5)		$^{63}A_x$ = − 28 (4)[2])	66 S 22
					g_y = 2.064 (3)		$\|^{63}A_y\|$ ≦ 9	
					g_z = 2.259 (5)		$^{63}A_z$ = − 150 (3)	
				77,			$\|(^{14}A_N)_{\|\|}\|$ = 10.8 (6)[2])	
				300			$\|(^{14}A_N)_\perp\|$ = 7.7 (4)	
							$\|^1A_H\|$ ≈ 5	
							$\|Q_z\|$ ≦ 5[2])	

*) Formula:

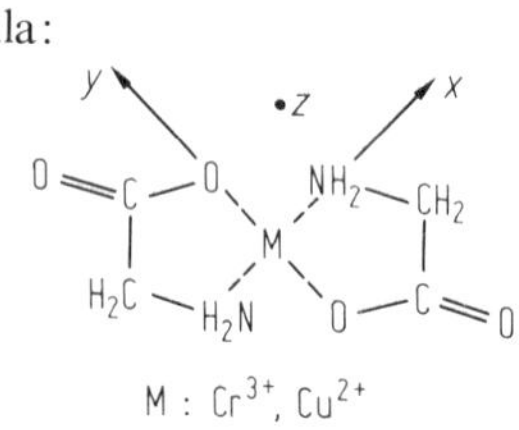

[1]) Orientation of the principal axes x, y, z relative to the a', b', c' axes, which are the crystallographic axes according to Hoshino et al. (see 1b).
I: α = 17.5 (25)°, γ = 23 (2)°
II: α = 17.5 (25)°, γ = 20 (2)°

[2]) Orientation of the principal axes x, y, z, ||, ⊥: parallel, perpendicular to x.

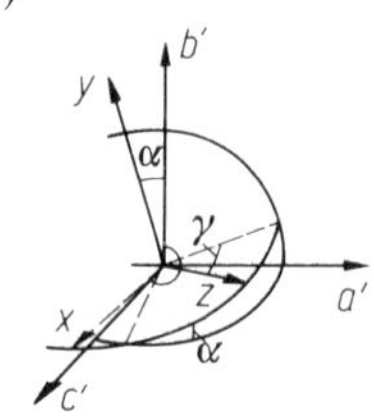

Nr. 36A-2 $(NH_2CH_2COOH)_3 \cdot H_2SeO_4$, Triglycine selenate (TGSe)
(M = 370.17)

1a	Ferroelectricity in TGSe was discovered by Matthias et al. in 1956.	56 M 1
b	see table below; P_s ∥ [010]; $\varrho = 1.845_2 \cdot 10^3$ kg m^{-3} at 20 °C. Transparent, colorless. Cleavage plane: (010).	56 M 1; a) 57 P 3; b) 56 M 1; 57 P 3
2a	Crystal growth: evaporation or slow cooling from aqueous solution (solvent H_2O). Solubility in H_2O: see Fig. 1217.	58 N 1
3a	Unit cell parameters: a = 9.5028 (6) Å, b = 12.8279 (6) Å, c = 5.8564 (4) Å, β = 110.565 (6)° at 24 °C (phase I). (For deuterated TGSe, see Table 345.)	79 Y 1
b	Crystal structure: isomorphous with TGS. Z = 2. Tables 342···351.	79 Y 1
4	Thermal expansion: Figs. 1329···1332.	
5a	Dielectric constant: $\varkappa_b = C/(T-\Theta_p)$, $T > \Theta$, where C = 4050 K, Θ_p = 22.7 °C. Figs. 1333···1335. Phase diagram: Figs. 1336···1338.	79 S 4

Table for 1b:

phase	II	I
state	F	P
crystal system	monoclinic	monoclinic
space group	$P2_1$-C_2^2 a)	$P2_1/m$-C_{2h}^2 a)
Θ	22[b]) °C	

 Figs. p. 618ff., 629ff.

5b	Nonlinear dielectric properties: $\xi = 1.12 \cdot 10^{11}$ V m^5 C^{-3}, $\zeta = 3.02 \cdot 10^{13}$ V m^9 C^{-5}, $\eta = 2.55 \cdot 10^{17}$ V m^{13} C^{-7}, $E = \chi_p^{-1} P + \xi P^3 + \zeta P^5 + \eta P^7$. Figs. 1339, 1340.	79S4
c	Spontaneous polarization: Figs. 1341, 1342.	
6a	Heat capacity: Figs. 1343, 1344.	
b	Transition entropy: $\Delta S_m = 2.38$ J K^{-1} mol^{-1}.	68S22
7a	Piezoelectric constant: Fig. 1345.	
8a	Elastic compliance: Figs. 1346, 1347. Ultrasonic wave velocity: Figs. 1348···1350. Attenuation coefficient of ultrasonic wave: Figs. 1351, 1352.	
9a	Refractive index: Figs. 1353, 1354. Birefringence: Fig. 1355. Optical axial angle: Figs. 1356, 1357. Infrared absorption: Fig. 1358; see also Figs. 1291, 1292.	
b	Electrooptic constants: see	74S23
10a	Raman scattering: Figs. 1359, 1360.	
b	Brillouin scattering: Fig. 1361.	
11	Electrical conduction: Fig. 1362.	
13a	NMR: Fig. 1363. Electric field gradient tensor: see	75A8, 76A6
b	ESR of γ-irradiated crystal: see	71S34
15	Domains: 180° domains of lenticular shape are observed by powder pattern method.	78S14

Table 342. $(NH_2CH_2COOH)_3 \cdot H_2SeO_4$ (TGSe). Structure of phase I [79Y1]. Fractional coordinates of nonhydrogen atoms at 24 °C.

	x	y	z
Se	0.9990 (1)	0.2500	0.2250 (1)
O (1)	0.8433 (4)	0.2500	−0.0158 (6)
O (2)	0.9636 (5)	0.2500	0.4788 (6)
O (3)	1.0923 (3)	0.1458 (2)	0.2036 (5)
O (11)	0.5984 (4)	0.2500	0.0577 (7)
O (12)	0.4820 (6)	0.2637 (33)	0.6559 (7)
C (11)	0.4809 (5)	0.2500	0.8594(10)
C (12)	0.3341 (5)	0.2500	0.9078(10)
N (1)	0.3385 (7)	0.2097 (7)	1.1544(12)
O (31)	0.7810 (3)	0.4975 (3)	0.2349 (5)
O (32)	0.5434 (3)	0.4763 (3)	0.2153 (6)
C (31)	0.6877 (4)	0.4700 (3)	0.3225 (7)
C (32)	0.7325 (4)	0.4306 (3)	0.5786 (7)
N (3)	0.8986 (4)	0.4294 (3)	0.6970 (5)

Table 343. $(NH_2CH_2COOH)_3 \cdot H_2SeO_4$ (TGSe). Structure of phase I [79Y1]. Anisotropic temperature parameters for nonhydrogen atoms at 24 °C. U_{ij} (in $Å^2$) is defined by Eq. (d) in Introduction and the values are multiplied by 10^2.

	U_{11}	U_{22}	U_{33}	U_{12}	U_{13}	U_{23}
Se	1.83 (1)	1.88 (2)	2.28 (2)	0	1.01 (1)	0
O (1)	2.31 (15)	5.46 (25)	2.60 (16)	0	0.70 (12)	0
O (2)	4.72 (22)	4.01 (21)	2.51 (15)	0	1.96 (15)	0
O (3)	3.95 (14)	2.29 (11)	5.27 (15)	1.05 (11)	2.44 (12)	0.43 (11)
O (11)	2.16 (16)	7.05 (19)	3.66 (19)	0	1.07 (14)	0
O (12)	3.94 (24)	8.01 (201)	3.60 (21)	−1.44 (19)	1.10 (19)	0.52 (46)
C (11)	2.42 (20)	3.67 (26)	3.50 (23)	0	1.02 (17)	0
C (12)	2.25 (20)	4.43 (31)	3.97 (26)	0	0.94 (19)	0
N (1)	2.75 (27)	8.16 (63)	4.51 (33)	1.29 (30)	2.13 (26)	0.18 (33)
O (31)	3.86 (13)	6.20 (21)	3.80 (14)	−0.48 (14)	1.20 (12)	1.33 (14)
O (32)	2.94 (13)	6.87 (23)	5.22 (17)	1.25 (15)	0.88 (12)	0.99 (16)
C (31)	3.21 (16)	2.72 (16)	4.12 (18)	−0.10 (14)	1.26 (14)	0.03 (14)
C (32)	3.70 (18)	3.53 (19)	3.84 (18)	−0.32 (15)	1.53 (15)	0.24 (15)
N (3)	4.16 (16)	3.22 (16)	2.68 (13)	−0.49 (13)	0.77 (12)	0.21 (12)

Table 344. $(NH_2CH_2COOH)_3 \cdot H_2SeO_4$ (TGSe). Structure of phase I [79Y1]. Fractional coordinates and isotropic temperature parameters of hydrogens at 24 °C. Values of U are multiplied by 10^2. Temperature correction is given by the formula $\exp(-8\pi^2 U \sin^2\theta/\lambda^2)$.

	x	y	z	U [$Å^2$]
H (1)	0.687 (11)	0.250	1.012 (18)	6.6 (33)
H (2)	0.281 (7)	0.316 (5)	0.851 (11)	6.0 (20)
H (4)	0.412 (9)	0.149 (7)	1.175 (14)	0.5 (19)
H (5)	0.256 (10)	0.183 (7)	1.200 (16)	2.0 (24)
H (6)	0.398 (10)	0.250	1.174 (16)	5.4 (28)
H (7)	0.531 (8)	0.489 (6)	0.088 (12)	0.0 (16)
H (13)	0.689 (6)	0.472 (5)	0.675 (10)	4.1 (17)
H (14)	0.699 (6)	0.365 (5)	0.588 (10)	4.1 (17)
H (15)	0.936 (6)	0.496 (4)	0.715 (10)	3.3 (15)
H (16)	0.910 (5)	0.397 (4)	0.836 (9)	2.4 (13)
H (17)	0.937 (8)	0.387 (6)	0.616 (12)	7.0 (24)

Table 345. $(ND_2CH_2COOD)_3 \cdot D_2SeO_4$. Structure of phase I [79Y1]. Fractional coordinates of nonhydrogen atoms at 37 °C. Unit cell parameters: $a = 9.5118$ (8) Å, $b = 12.8321$ (6) Å, $c = 5.8557$ (5) Å, $\beta = 110.539$ (5)°.

	x	y	z
Se	0.9987 (0)	0.2500	0.2242 (1)
O (1)	0.8436 (3)	0.2500	−0.0166 (5)
O (2)	0.9627 (4)	0.2500	0.4753 (5)
O (3)	1.0926 (2)	0.1462 (2)	0.2039 (4)
O (11)	0.5980 (3)	0.2500	0.0568 (5)
O (12)	0.4818 (5)	0.2517 (36)	0.6554 (7)
C (11)	0.4793 (4)	0.2500	0.8593 (8)
C (12)	0.3347 (4)	0.2500	0.9089 (8)
N (1)	0.3584 (5)	0.2082 (5)	1.1529 (9)
O (31)	0.7807 (2)	0.4973 (2)	0.2342 (4)
O (32)	0.5445 (2)	0.4720 (2)	0.2162 (4)
C (31)	0.6882 (3)	0.4706 (2)	0.3225 (5)
C (32)	0.7342 (3)	0.4299 (3)	0.5803 (6)
N (3)	0.8983 (3)	0.4302 (2)	0.6968 (4)

Table 346. $(ND_2CH_2COOD)_3 \cdot D_2SeO_4$. Structure of phase I [79Y1]. Anisotropic temperature parameters for nonhydrogen atoms at 37 °C. U_{ij} (in Å^2) is defined by Eq. (d) in Introduction and the values are multiplied by 10^2.

	U_{11}	U_{22}	U_{33}	U_{12}	U_{13}	U_{23}
Se	1.96 (1)	1.87 (1)	2.31 (1)	0	0.83 (1)	0
O (1)	2.36 (12)	6.12 (20)	2.53 (13)	0	0.46 (10)	0
O (2)	4.89 (17)	4.03 (16)	2.67 (13)	0	1.90 (13)	0
O (3)	4.04 (11)	2.51 (9)	5.51 (12)	1.04 (8)	2.14 (10)	0.20 (9)
O (11)	2.26 (13)	7.95 (25)	3.57 (15)	0	0.89 (11)	0
O (12)	4.60 (21)	9.72 (179)	3.76 (18)	1.59 (71)	1.35 (16)	−0.96 (55)
C (11)	2.43 (17)	3.79 (21)	4.09 (21)	0	0.75 (15)	0
C (12)	2.32 (17)	4.52 (24)	4.24 (22)	0	0.81 (16)	0
N (1)	3.01 (23)	7.90 (45)	4.02 (25)	0.84 (24)	1.75 (20)	0.63 (25)
O (31)	3.88 (11)	5.85 (15)	4.15 (11)	−0.38 (11)	1.04 (9)	1.39 (11)
O (32)	3.30 (11)	6.96 (18)	5.74 (14)	1.16 (12)	0.78 (10)	0.75 (13)
C (31)	3.48 (13)	3.05 (14)	4.18 (15)	0.43 (11)	0.92 (12)	0.09 (12)
C (32)	4.06 (16)	3.91 (16)	4.28 (16)	−0.18 (13)	1.70 (13)	0.47 (13)
N (3)	4.47 (13)	2.89 (11)	2.65 (10)	−0.36 (10)	0.52 (9)	0.25 (9)

Table 347. $(ND_2CH_2COOD)_3 \cdot D_2SeO_4$. Structure of phase I [79Y1]. Fractional coordinates and isotropic temperature parameters of hydrogens at 37 °C. Values of U's are multiplied by 10^2. Temperature correction is given by the formula $\exp(-8\pi^2 U \sin^2\theta/\lambda^2)$.

	x	y	z	U [Å^2]
D (1)	0.693 (6)	0.250	1.021 (10)	3.0 (15)
H (2)	0.275 (5)	0.312 (3)	0.841 (7)	5.1 (14)
D (4)	0.432 (9)	0.363 (7)	1.175 (14)	4.4 (25)
D (5)	0.264 (9)	0.316 (6)	1.169 (14)	4.6 (26)
D (6)	0.399 (8)	0.250	1.275 (13)	7.4 (25)
D (7)	0.516 (19)	0.490 (15)	0.065 (25)	3.8 (46)
H (13)	0.694 (4)	0.473 (3)	0.684 (6)	1.7 (09)
H (14)	0.695 (4)	0.363 (3)	0.577 (6)	2.5 (10)
D (15)	0.932 (4)	0.494 (3)	0.706 (7)	1.7 (10)
D (16)	0.924 (4)	0.409 (3)	0.836 (6)	2.6 (10)
D (17)	0.937 (5)	0.386 (4)	0.609 (6)	5.4 (14)

Table 348. $(ND_2CH_2COOD)_3 \cdot D_2SeO_4$. Structure of phase II [79Y1]. Fractional coordinates of nonhydrogen atoms at 21 °C. Unit cell parameter: $a = 9.415$ (2) Å, $b = 12.836$ (2) Å, $c = 5.839$ (1) Å, $\beta = 110.547$ (19)°.

	x	y	z
Se	0.99847 (5)	0.25000 (0)	0.22525 (7)
O (1)	0.8429 (4)	0.2492 (11)	−0.0152 (6)
O (2)	0.9632 (4)	0.2473 (11)	0.4791 (6)
O (3)	1.0893 (9)	0.1501 (7)	0.1973 (19)
O (4)	1.0951 (10)	0.3570 (6)	0.2118 (20)
O (11)	0.5986 (4)	0.2611 (12)	1.0605 (7)
O (12)	0.4821 (6)	0.2695 (9)	0.6577 (8)
C (11)	0.4809 (6)	0.2599 (16)	0.8635 (10)
C (12)	0.3351 (5)	0.2588 (16)	0.9116 (10)
N (1)	0.3597 (6)	0.2132 (5)	1.1579 (10)
O (21)	0.2204 (10)	0.5109 (8)	0.7777 (20)
O (22)	0.4588 (10)	0.5373 (7)	0.8096 (16)
C (21)	0.3191 (13)	0.5343 (11)	0.6868 (22)
C (22)	0.2827 (12)	0.5746 (12)	0.4360 (23)
N (2)	0.1132 (11)	0.5734 (7)	0.3008 (18)
O (31)	0.7834 (11)	0.5048 (8)	0.2450 (19)
O (32)	0.5504 (11)	0.4847 (8)	0.2450 (17)
C (31)	0.6965 (14)	0.4748 (10)	0.3334 (27)
C (32)	0.7474 (10)	0.4342 (8)	0.5990 (19)
N (3)	0.9125 (11)	0.4335 (9)	0.6977 (21)

Table 349. $(ND_2CH_2COOD)_3 \cdot D_2SeO_4$. Structure of phase II [79Y1]. Anisotropic temperature parameters for nonhydrogens at 21 °C. U_{ij} (in Å^2) is defined by Eq. (d) in Introduction and the values are multiplied by 10^2.

	U_{11}	U_{22}	U_{33}	U_{12}	U_{13}	U_{23}
Se	1.790 (15)	1.646 (14)	2.044 (15)	−0.231 (45)	1.155 (17)	−0.050 (40)
O (1)	2.26 (14)	5.19 (26)	2.35 (13)	−0.66 (45)	0.70 (16)	−0.28 (34)
O (2)	4.29 (19)	3.41 (24)	2.30 (13)	−0.91 (40)	2.80 (19)	−0.42 (25)
O (3)	2.65 (34)	2.88 (40)	4.28 (45)	1.31 (28)	1.43 (43)	0.24 (23)
O (4)	4.24 (47)	1.37 (30)	5.36 (50)	−0.62	4.02	−0.23
O (11)	2.05 (15)	7.04 (50)	3.43 (17)	0.85 (37)	1.24 (18)	0.39 (27)
O (12)	4.37 (24)	8.05 (80)	3.58 (20)	−1.74 (36)	1.98 (26)	0.04 (21)
C (11)	2.54 (20)	4.52 (56)	3.32 (21)	0.69 (45)	1.48 (23)	−0.21 (32)
C (12)	2.04 (18)	3.74 (40)	4.06 (23)	0.20 (47)	1.21 (23)	0.46 (36)
N (1)	1.76 (21)	6.32 (39)	3.84 (26)	0.65 (21)	0.21 (27)	0.54 (16)
O (21)	2.55 (35)	4.78 (41)	3.65 (39)	−0.43 (29)	1.07 (39)	0.44 (21)
O (22)	2.74 (27)	3.95 (31)	4.79 (42)	0.55 (22)	0.53 (37)	−0.15 (19)
C (21)	3.14 (45)	3.19 (50)	2.73 (36)	0.44 (37)	1.54 (45)	0.35 (23)
C (22)	2.71 (43)	5.41 (57)	3.05 (41)	−0.08 (39)	0.53 (45)	0.22 (25)
N (2)	5.73 (38)	1.99 (28)	2.15 (28)	−0.74 (25)	1.10 (35)	0.00 (15)
O (31)	4.11 (43)	5.41 (44)	3.15 (35)	−0.50 (33)	1.63 (41)	1.23 (21)
O (32)	3.07 (33)	7.34 (64)	4.65 (42)	1.33 (36)	1.74 (40)	0.71 (27)
C (31)	2.40 (39)	2.02 (38)	6.06 (67)	0.26 (29)	1.23 (54)	0.46 (27)
C (32)	1.86 (30)	2.84 (31)	2.89 (35)	−0.32 (24)	1.16 (37)	0.53 (17)
N (3)	2.85 (32)	3.62 (41)	2.68 (35)	−0.45 (20)	0.53 (36)	0.57 (20)

Table 350. $(ND_2CH_2COOD)_3 \cdot D_2SeO_4$. Structure of phase II [79Y1]. Fractional coordinates and isotropic temperature parameters of hydrogens at 21 °C. Temperature correction is given by the formula, $\exp(-8\pi^2 U \sin^2\theta/\lambda^2)$.

	x	y	z	U [Å^2]
D (1)	0.695 (7)	0.260 (15)	1.012 (11)	1.6 (18)
H (2)	0.274 (14)	0.313 (11)	0.812 (24)	3.8 (45)
H (3)	0.286 (15)	0.188 (12)	0.894 (25)	4.8 (45)
D (4)	0.430 (12)	0.151 (10)	1.188 (19)	5.2 (34)
D (5)	0.248 (14)	0.192 (10)	1.177 (22)	7.6 (44)
D (6)	0.417 (13)	0.222 (12)	1.364 (20)	8.2 (47)
D (7)	0.473 (16)	0.482 (12)	0.920 (22)	9.2 (52)
H (8)	0.311 (14)	0.519 (11)	0.329 (25)	1.9 (39)
H (9)	0.293 (21)	0.640 (14)	0.431 (34)	5.0 (61)
D (10)	0.075 (15)	0.496 (9)	0.289 (26)	1.1 (35)
D (11)	0.070 (19)	0.605 (13)	0.189 (32)	4.0 (51)
D (12)	0.048 (19)	0.607 (12)	0.424 (30)	3.9 (33)
H (13)	0.697 (15)	0.484 (11)	0.709 (25)	1.7 (39)
H (14)	0.709 (19)	0.358 (13)	0.358 (33)	3.4 (51)
D (15)	0.925 (18)	0.500 (11)	0.681 (31)	4.4 (52)
D (16)	0.927 (22)	0.409 (14)	0.854 (34)	6.6 (65)
D (17)	0.942 (19)	0.388 (12)	0.589 (29)	3.3 (00)

Figs. p. 602, 635 ff.

Table 351. $(NH_2CH_2COOH)_3 \cdot H_2SeO_4$ (TGSe). Unit cell parameters at 17, 27, and 39.6 °C [77O2].

	17 °C	27 °C	39.6 °C
a	9.544 Å	9.530 Å	9.540 Å
b	12.719	12.761	12.845
c	5.841	5.861	5.863
β	110.46°	110.52°	110.86°

Nr. 36A-3 $(NH_2CH_2COOH)_3 \cdot H_2BeF_4$, Triglycine fluoroberyllate (TGFB)
($M = 312.223$)

1a Ferroelectricity in TGFB was first reported by Pepinsky et al. in 1957. — 57P3

b — 57P3

phase	II	I
state	F	P
crystal system	monoclinic	monoclinic
space group	$P2_1\text{-}C_2^2$	$P2_1/m\text{-}C_{2h}^2$
Θ	75 °C	

$\Theta = 77$ °C for deuterated crystal. — 66B8

$P_s \| [010]$ — 57P3

$\varrho = 1.631_5 \cdot 10^3$ kg m^{-3} at 20 °C.

Transparent, colorless. Cleavage plane: (010).

2a Crystal growth: cooling from aqueous solution (solvent H_2O). Solubility in H_2O: see Fig. 1217. — 58N1

3a Unit cell parameters: $a = 9.378\,(1)$ Å, $b = 12.617\,(1)$ Å, $c = 5.692\,(1)$ Å, $\beta = 110.28\,(1)°$ at 25 °C. — 73W1

b Crystal structure: isomorphous with TGS, $Z = 2$. Tables 352···356. — 73W1

4 Thermal expansion: Figs. 1364···1367; see also Thermal expansion coefficients: Fig. 1368. — 72Z1

5a Dielectric constant: Fig. 1369.
Phase diagram in regard to p: Fig. 1370.

b Nonlinear dielectric properties: Figs. 1371, 1372.

$$E = \frac{1}{\varepsilon_0 C}(T - \Theta)\,P + \xi P^3 + \zeta P^5,\ C = 2.5 \cdot 10^3\ \text{K},\ \xi = 2.4 \cdot 10^{11}\ \text{V m}^5\,\text{C}^{-3},$$ — 77G15

$\zeta = 4.2 \cdot 10^{14}$ V m^9 C^{-5}, $\Theta = 72.82\,(5)$ °C.

c Spontaneous polarization and coercive field: Figs. 1373···1377.

d Pyroelectric coefficient: Fig. 1378.

continued

Nr. 36A-3 $(NH_2CH_2COOH)_3 \cdot H_2BeF_4$, continued

6a	Heat capacity: Fig. 1379.	
b	Transition heat and transition entropy: $\Delta Q_m = 1670$ J mol^{-1}, $\Delta S_m = 4.89$ J K^{-1} mol^{-1}.	57H2
8a	Ultrasonic attenuation: Fig. 1380.	
9a	Refractive indices: $n_a = 1.526$, $n_b = 1.502$, $n_c = 1.526$ for $\lambda = 633$ nm.	73M12
c	Pockels (piezooptic) coefficient: see Table 328.	
11	Electric conduction: Fig. 1381; Table 357.	
13a	NMR: Figs. 1382···1385; Table 358.	
b	ESR: Figs. 1386, 1387; Tables 359, 360.	
15a	Domains have been observed by dew method.	66S1
b	Domain switching: see	58P8

Table 352. $(NH_2CH_2COOH)_3 \cdot H_2BeF_4$ (TGFB). Fractional coordinates and isotropic temperature parameters in phase I (85 °C) [73W2]. B is defined by Eq. (e) in Introduction. Unit cell parameters: $a = 9.401\,(1)$ Å, $b = 12.627\,(1)$ Å, $c = 5.713$ Å, $\beta = 110.13\,(1)°$ at 85 °C.

	x	y	z	B [Å^2]
F (3)	0.0830 (4)	0.1503 (3)	0.1983 (8)	4.12 (8)
O (3)	0.2187 (6)	0.5002 (4)	0.7649 (10)	4.73 (11)
O (4)	0.4547 (6)	0.5275 (5)	0.7790 (10)	5.13 (12)
O (2)	0.4878 (9)	0.2655 (11)	0.6594 (16)	5.11 (20)
C (3)	0.3112 (7)	0.5285 (6)	0.6722 (12)	3.63 (13)
C (4)	0.2644 (7)	0.5710 (7)	0.4110 (13)	4.08 (13)
N (2)	0.0972 (6)	0.5727 (5)	0.2961 (10)	3.85 (11)
N (1)	0.3490 (15)	0.2144 (13)	0.1502 (26)	6.14 (34)
Be	0.9966 (11)	0.25	0.2222 (20)	2.49 (18)
F (1)	0.8456 (6)	0.25	−0.0022 (10)	3.61 (11)
F (2)	0.9676 (6)	0.25	0.4737 (11)	4.18 (12)
O (1)	0.5995 (9)	0.25	0.0615 (15)	4.94 (17)
C (1)	0.5791 (12)	0.25	0.8633 (20)	4.12 (20)
C (2)	0.3268 (12)	0.25	0.8997 (21)	4.45 (21)

Table 353. $(NH_2CH_2COOH)_3 \cdot H_2BeF_4$ (TGFB). Interatomic distances and bond angles in phase I (85 °C) [73W2].

Distances [Å]				Bond angles	
Be –F (1)	1.552 (10)	O (3)–C (3)	1.217 (10)	F (2)–Be–F (3)	110.52°
–F (2)	1.553 (15)	O (4)–C (3)	1.276 (08)	F (2)–Be–F (1)	111.27°
–F (3)	1.528 (08)	C (3)–C (4)	1.503 (10)	F (1)–Be–F (3)	106.81°
F (1)–F (2)	2.563 (08)	C (4)–N (2)	1.481 (09)	O (1)–C (1)–O (2)	120.81°
–F (3)	2.473 (06)	N (1)–F (3)	2.732 (17)	O (1)–C (1)–C (2)	117.51°
F (2)–F (3)	2.533 (08)	–O (4)	2.937 (17)	O (2)–C (1)–C (2)	120.76°
F (3)–F (4)	2.516 (10)	–O (2)	2.824 (16)	N (1)–C (2)–C (1)	108.37°
O (1)–C (1)	1.297 (12)	N (2)–F (1)	2.955 (08)	O (3)–C (3)–O (4)	125.83°
O (2)–C (1)	1.213 (16)	–F (2)	2.767 (08)	O (3)–C (3)–C (4)	121.91°
C (1)–C (2)	1.516 (18)	–F (4)	2.865 (09)	O (4)–C (3)–C (4)	112.26°
C (2)–N (1)	1.445 (19)	O (4)–O (6)	2.471 (11)	N (2)–C (4)–C (3)	110.43°
		O (1)–F (1)	2.464 (11)		

Table 354. $(NH_2CH_2COOH)_3 \cdot H_2BeF_4$ (TGFB). Fractional coordinates and isotropic temperature parameters in phase II (25 °C) [73W1]. B is defined by Eq. (e) in Introduction.

	x	y	z	B [Å²]
Be	0.9959 (9)	0.2464 (9)	0.2259 (15)	1.60 (13)
F (1)	0.8443 (4)	0.2484 (5)	0.0001 (8)	2.42 (7)
F (2)	0.9672 (5)	0.2430 (5)	0.4766 (9)	2.81 (8)
F (3)	0.0813 (5)	0.1467 (4)	0.1970 (10)	2.74 (9)
F (4)	0.0854 (5)	0.3478 (4)	0.2039 (9)	2.66 (9)
O (1)	0.6008 (6)	0.2579 (6)	0.0719 (11)	3.19 (11)
O (2)	0.4867 (7)	0.2716 (6)	0.6628 (12)	3.78 (13)
O (3)	0.2232 (7)	0.5053 (6)	0.7779 (13)	3.31 (12)
O (4)	0.4631 (7)	0.5338 (6)	0.8089 (13)	3.22 (12)
O (5)	0.7912 (8)	0.5000 (6)	0.2460 (13)	3.47 (13)
O (6)	0.5573 (8)	0.4828 (7)	0.2574 (13)	3.89 (14)
C (1)	0.4811 (9)	0.2588 (9)	0.8707 (15)	3.01 (15)
C (2)	0.3307 (9)	0.2567 (9)	0.9113 (14)	2.81 (14)
C (3)	0.3194 (9)	0.5323 (7)	0.6930 (15)	2.60 (14)
C (4)	0.2765 (10)	0.5696 (7)	0.4229 (17)	2.88 (15)
C (5)	0.6998 (9)	0.4743 (7)	0.3481 (16)	2.59 (14)
C (6)	0.7518 (10)	0.4288 (8)	0.6085 (17)	2.96 (15)
N (1)	0.3484 (9)	0.2109 (7)	0.1616 (15)	2.36 (15)
N (2)	0.1085 (7)	0.5696 (6)	0.2968 (12)	2.25 (11)
N (3)	0.9219 (7)	0.4196 (6)	0.7106 (12)	2.41 (12)

Table 355. $(NH_2CH_2COOH)_3 \cdot H_2BeF_4$ (TGFB). Interatomic distances and bond angles in phase II (25 °C) [73W1].

Distances [Å]				Bond angles	
Be –F (1)	1.558 (8)	C (4)–N (2)	1.492 (11)	F (2)–Be–F (3)	109.88°
–F (2)	1.541 (11)	O (5)–C (5)	1.232 (14)	F (2)–Be–F (4)	111.52°
–F (3)	1.529 (12)	O (6)–C (5)	1.263 (11)	F (2)–Be–F (1)	111.49°
–F (4)	1.558 (12)	C (5)–C (6)	1.507 (13)	F (1)–Be–F (3)	106.75°
F (1)–F (2)	2.561 (6)	C (6)–N (3)	1.505 (11)	F (1)–Be–F (4)	106.57°
–F (3)	2.478 (7)	N (1)–F (3)	2.705 (11)	F (3)–Be–F (4)	110.50°
–F (4)	2.498 (7)	–O (4)	2.818 (12)	O (1)–C (1)–O (2)	123.25°
F (2)–F (3)	2.513 (9)	–O (2)	2.808 (10)	O (1)–C (1)–C (2)	115.82°
–F (4)	2.562 (9)	N (2)–F (1)	2.942 (10)	O (2)–C (1)–C (2)	120.50°
F (3)–F (4)	2.536 (9)	–F (2)	2.753 (10)	N (1)–C (2)–C (1)	110.44°
O (1)–C (1)	1.302 (9)	–F (4)	2.843 (10)	O (3)–C (3)–O (4)	126.49°
O (2)–C (1)	1.213 (12)	N (3)–F (2)	2.703 (10)	O (3)–C (3)–C (4)	120.94°
C (1)–C (2)	1.505 (13)	–F (3)	2.916 (10)	O (4)–C (3)–C (4)	112.58°
C (2)–N (1)	1.493 (13)	–F (4)	2.850 (8)	N (2)–C (4)–C (3)	110.93°
O (3)–C (3)	1.209 (13)	O (1)–F (1)	2.457 (9)	O (5)–C (5)–C (6)	126.44°
O (4)–C (3)	1.282 (10)	O (4)–O (6)	2.485 (10)	O (5)–C (5)–C (6)	121.42°
C (3)–C (4)	1.525 (13)			O (6)–C (5)–C (6)	112.13°
				N (3)–C (6)–C (5)	110.69°

Table 356. $(NH_2CH_2COOH)_3 \cdot H_2BeF_4$ (TGFB). Displacements of atoms from the best fit planes of the glycine molecules (G I, II, III) in phase I [73W2] and phase II [73W1].

	Phase II (25 °C)			Phase I (85 °C)	
Atoms	G I	G II	G III	G I	G II, III
N	+0.417 Å	−0.037 Å	−0.008 Å	+0.246 Å	−0.021 Å
C^N	−0.026	0.000	−0.001	−0.048	−0.001
C	+0.038	0.000	+0.004	+0.056	+0.004
O	−0.005	0.000	−0.002	0.000	−0.001
O^N	−0.008	0.000	−0.002	−0.009	−0.002

Table 357. $(NH_2CH_2COOH)_3 \cdot H_2BeF_4$ (TGFB). Activation energy ΔU estimated from electrical conductivity. Ionic conductivity $\sigma = \sigma_0 \exp(-\Delta U/kT)$ [61 G 5].

Estimated from	Ferroelectric phase	Paraelectric phase
σ_a	0.93 eV	0.64 eV
σ_b	0.76	0.70
σ_c	0.72	0.65

Table 358. $(ND_2CD_2COOD)_3 \cdot D_2BeF_4$. e^2qQ/h [kHz], η for ^9Be and direction cosines λ, μ, ν of the principal axes of the electric field gradient tensors in the crystal fixed coordinate system $a \sin\beta$, b, c [71 B 19].

	T=361 K	T=340 K	T=340 K
e^2qQ/h	130 (3)	126 (3)	126 (3)
η	0.43 (3)	0.64 (3)	0.64 (3)
ϕ_{zz} λ	$\pm$0.370	0.363	−0.363
μ	0.000	0.152	0.152
ν	$\pm$0.929	0.920	−0.920
ϕ_{yy} λ	$\pm$0.929	0.828	−0.828
μ	0.000	0.400	0.400
ν	0.370	−0.392	0.392
ϕ_{xx} λ	0.000	−0.428	0.428
μ	1.000	0.904	0.904
ν	0.000	0.020	−0.020

Table 359. $(NH_2CH_2COOH)_3 \cdot H_2BeF_4{:}Cu^{2+}$. ESR data [72 S 26].

Paramagnetic center	S	$\mathscr{H}$	ν GHz	T	g-factor	FS 10^{-2} m^{-1}	HFS 10^{-2} m^{-1}	Comment
Cu^{2+}	1/2	(3)	≈9	−100 ≈120 °C	$g_{\parallel}=2.24$ $g_{\perp}=2.06$		$^nA_{\parallel}=132$ $^nA_{\perp}=32$	for FS due to two-ion aggregate; see Fig. 1386.
Cu^{2+}	2	(5)	9.4	160 K	$g_z=2.12$ $g_y=2.15$ $g_x=2.08$	D=1080 E=300		Temperature change; see Fig. 1387.

Table 360. $(NH_2CH_2COOH)_3 \cdot H_2BeF_4{:}Cu^{2+}$. Spin Hamiltonian parameters of copper aggregates with spin $S=2$ [72 M 1].

D $\cdot 10^2$ m^{-1}	E $\cdot 10^2$ m^{-1}	a $\cdot 10^2$ m^{-1}	b $\cdot 10^2$ m^{-1}	c $\cdot 10^2$ m^{-1}	g_x	g_y	g_z
0.108	0.030	−0.003	0	0.009	2.08	2.15	2.12

Nr. 36A-4 Triglycine tellurate

1a	Dielectric properties of triglycine tellurate were reported by Kaczmarek in 1961. Any evidence of ferroelectricity has not been found in the temperature range from 4 K to 373 K.	61 K 1

36 B Solid Solution

Nr. 36B-1 $(NH_2CH_2COOH)_3 \cdot H_2SO_4$–$(NH_2CH_2COOH)_3 \cdot H_2SeO_4$ (TGS–TGSe)

1b	Phase diagram: Fig. 1388. Properties of L-alanine doped crystals: Table 361.	
5a	Dielectric constant: Fig. 1389; Table 362. For L-alanine doped crystals: see Table 361.	
d	Pyroelectric coefficient: see Table 361; Fig. 1389, Table 362.	
6a	Heat capacity of L-alanine doped crystals: see Table 361.	
9a	Refractive index: Figs. 1390α, 1390β.	
11	Electrical conductivity: see Table 362; for L-alanine doped crystal, Table 361.	
16	Effect of X-ray irradiation: see Table 362.	

Table 361. $\{(NH_2CH_2COOH)_3 \cdot H_2SO_4\}_x \{(NH_2CH_2COOH)_3 \cdot H_2SeO_4\}_{1-x}$ with L-alanine addition. Properties at 21 °C [76B18].

% L-alanine in solution	x in solution	x in crystal	Θ_f °C	$\varrho_{d.c.}$ Ωm	$\varrho_{1.6\,kHz}$ Ωm	$\varkappa_{1.6\,kHz}$	p_2 10^{-4} C K^{-1} m^{-2}	ϱC_p 10^6 Jm^{-3} K^{-1}	ϱ 10^3 kg m^{-3}
40	1	1	49.5	$2 \cdot 10^{13}$	$1.4 \cdot 10^8$	27	3.6	2.1	1.69
20	0.75	0.92	46.8	$3 \cdot 10^{11}$	$3.1 \cdot 10^3$	34	4.0	2.2	—
20	0.60	0.89	45.6	$5 \cdot 10^{12}$	$3.3 \cdot 10^7$	36	3.8	2.0	—
20	0.35	0.70	40.6	$3 \cdot 10^{12}$	$2.1 \cdot 10^8$	41	6.1	2.0	—
20	0.15	0.38	32.1	$3 \cdot 10^{10}$	$1.4 \cdot 10^7$	65	7.1	2.0	—
20	0.05	0.16	26.3	$3 \cdot 10^{11}$	$5.6 \cdot 10^7$	122	12.2	1.9	—
40	0	0	22.5	$2.5 \cdot 10^{12}$	$7.0 \cdot 10^6$	364	28.7	1.8	1.87

Table 362. $\{(NH_2CH_2COOH)_3 \cdot H_2SO_4\}_y\{(NH_2CH_2COOH)_3 \cdot H_2SeO_4\}_{1-y}$ (TGS–TGSe). Pyroelectric coefficient, dielectric constant and electrical resistivity at 21 °C before and after X-ray irradiation field treatment for 1 h [76B18].

x	p_2 10^{-4} C K^{-1} m^{-2}		$\varkappa_{1.6\,kHz}$		ϱ 10^7 Ωm	
	Before	After	Before	After	Before	After
1.00	3.5	4.0	38	30	1.2	10
.95	3.4	4.1	26	26	1.5	4.5
.90	3.5	4.0	40	30	1.5	2.2
.80	3.0	4.9	38	38	1.3	2.1
.35	5.6	9.8	73	120	0.6	0.3
.05	14.8	17.5	220	130	0.2	0.3
0	42	10	420	1000	0.01	0.4

37 $NH_2CH_2COOH \cdot AgNO_3$

37A Pure compound

Nr. 37A-1 $NH_2CH_2COOH \cdot AgNO_3$, Glycine silver nitrate
($M = 244.940$)

1a Ferroelectricity in $NH_2CH_2COOH \cdot AgNO_3$ was discovered by Pepinsky et al. in 1957. 57P2

b 57P2

phase	II	I
state	F	P
crystal system		monoclinic
space group		$P2_1/a\text{-}C_{2h}^5$
Θ	−55 °C	

$\Theta = -43$ °C for $ND_2CH_2COOD \cdot AgNO_3$. 77G6
$P_s \| [010]$ 57P2
$T_{melt} = 145$ °C 72M16
$\varrho = 2.79 \cdot 10^3$ kg m^{-3}. 72M16
$\varrho_x = 2.806 \cdot 10^3$ kg m^{-3}. 72M16
Transparent, colorless but usually brown as a result of illumination by light. 64M4
Cleavage plane: (010). 69T1

2a Crystal growth: cooling and evaporation methods from aqueous solution. 57P2
Slow introduction of ethyl alcohol vapor into the saturated aqueous solution. 57P2
Crystal form: Fig. 1391.

continued

Nr. 37A-1 $NH_2CH_2COOH \cdot AgNO_3$, continued

3a	Unit cell parameters: $a = 5.451$ Å, $b = 19.493$ Å, $c = 5.541$ Å, $\beta = 100.20\,(25)°$ in phase I.	72M16
b	$Z = 4$ in phase I. Crystal structure: Figs. 1392···1396; Tables 363α, 363β···366.	72M16
4	Lattice distortions: Figs. 1397, 1398.	
5a	Dielectric constants: Figs. 1399, 1400, 1401. Effect of hydrostatic pressure on partially deuterated crystal: see $\varkappa_b = C/(T-\Theta_p)$, $T > \Theta_p$, $C = 446$ K, $\Theta_p = -55$ °C. Curie-Weiss constant: Curie-Weiss constant changes with hydrostatic pressure as $C = C_o(1+\beta p)$, $C_o = 860\,(40)$ K, $\beta = -0.17\,(02) \cdot 10^{-8}$ Pa^{-1} in the range $p < 1.8 \cdot 10^8$ Pa. Phase diagram in regard to p: Fig. 1402. Phase diagram for partially deuterated crystal: see	 77G6 64M4 77G6 77G6
b	Nonlinear dielectric properties: $\xi = 5.2 \cdot 10^{14}$ V m^5 C^{-3}.	64M4
c	Spontaneous polarization and coercive field: Figs. 1403, 1404.	
7b	Electrostrictive constants: $\frac{\Delta a}{aP_s^2} = Q_{12} \approx -18$ C^{-2} m^4, $\frac{\Delta b}{bP_s^2} = Q_{22} \approx 36$ C^{-2} m^4, $\frac{\Delta c}{cP_s^2} \approx 300$ C^{-2} m^4, $\frac{\Delta\beta}{\beta P_s^2} \approx 220$ C^{-2} m^4 at -100 °C.	64M4
13a	NMR: Fig. 1405.	
16	Plastically soft.	64M4

Table 363β. $NH_2CH_2COOH \cdot AgNO_3$. Structure of phase I [72M16]. Temperature parameter B_{ij} is defined by Eq. (a) in Introduction.

Ag	B_{11}	0.0275 (6)	O (3)	B_{11}	0.035 (7)	N (1)	B_{11}	0.030 (7)
	B_{22}	0.00293 (4)		B_{22}	0.0030 (5)		B_{22}	0.0021 (6)
	B_{33}	0.0314 (6)		B_{33}	0.049 (8)		B_{33}	0.046 (9)
	B_{12}	−0.0061 (3)		B_{12}	0.004 (3)		B_{12}	−0.003 (3)
	B_{13}	0.0041 (11)		B_{13}	−0.022 (12)		B_{13}	−0.010 (14)
	B_{23}	−0.0035 (3)		B_{23}	−0.006 (3)		B_{23}	0.005 (4)
O (1)	B_{11}	0.029 (6)	O (4)	B_{11}	0.043 (11)	N (2)	B_{11}	0.024 (7)
	B_{22}	0.0032 (5)		B_{22}	0.0053 (9)		B_{22}	0.0020 (5)
	B_{33}	0.032 (7)		B_{33}	0.052 (10)		B_{33}	0.052 (10)
	B_{12}	−0.003 (3)		B_{12}	−0.002 (4)		B_{12}	0.001 (3)
	B_{13}	−0.006 (11)		B_{13}	−0.001 (18)		B_{13}	−0.005 (15)
	B_{23}	−0.004 (3)		B_{23}	0.002 (5)		B_{23}	−0.002 (3)
O (2)	B_{11}	0.021 (5)	O (5)	B_{11}	0.040 (8)	C (1)	B_{11}	0.036 (9)
	B_{22}	0.0027 (5)		B_{22}	0.0021 (5)		B_{22}	0.0013 (5)
	B_{33}	0.032 (6)		B_{33}	0.054 (9)		B_{33}	0.033 (10)
	B_{12}	−0.001 (3)		B_{12}	−0.003 (4)		B_{12}	−0.004 (3)
	B_{13}	−0.004 (9)		B_{13}	−0.018 (15)		B_{13}	−0.036 (14)
	B_{23}	−0.003 (3)		B_{23}	−0.003 (4)		B_{23}	−0.001 (4)
						C (2)	B_{11}	0.028 (9)
							B_{22}	0.0023 (8)
							B_{33}	0.034 (13)
							B_{12}	−0.007 (4)
							B_{13}	−0.014 (19)
							B_{23}	−0.006 (5)

Table 363α. $NH_2CH_2COOH \cdot AgNO_3$. Structure of phase I [72M16]. Fractional coordinates.

		Fractional coordinates			Fractional coordinates
Ag	*x*	0.1450 (4)	O (5)	*x*	−0.095 (4)
	y	0.01631 (10)		*y*	0.1360 (10)
	z	−0.1855 (4)		*z*	−0.038 (4)
O (1)	*x*	0.166 (3)	N (1)	*x*	0.689 (4)
	y	0.0430 (9)		*y*	0.1589 (11)
	z	0.401 (3)		*z*	0.403 (5)
O (2)	*x*	0.412 (3)	N (2)	*x*	0.047 (4)
	y	0.0645 (9)		*y*	0.1884 (10)
	z	0.126 (3)		*z*	−0.030 (4)
O (3)	*x*	0.151 (3)	C (1)	*x*	0.355 (5)
	y	0.2074 (10)		*y*	0.0710 (12)
	z	0.180 (4)		*z*	0.333 (5)
O (4)	*x*	0.103 (5)	C (2)	*x*	0.516 (5)
	y	0.2124 (13)		*y*	0.1166 (15)
	z	−0.222 (5)		*z*	0.524 (6)

Table 364. $NH_2CH_2COOH \cdot AgNO_3$. Structure of phase I [72M16]. Interatomic distances around Ag.

	Distances [Å]
Ag–O (1) (−*x*, −*y*, −*z*)	2.22 (2)
Ag–O (2)	2.25 (2)
Ag–O (1) (*x*, *y*, *z*−1)	2.37 (2)
Ag–O (2) (−*x*+1, −*y*, −*z*)	2.85 (2)
Ag–O (5)	2.86 (2)
Ag–Ag (−*x*, −*y*, −*z*)	2.877 (6)

Table 365. $NH_2CH_2COOH \cdot AgNO_3$. Structure of phase I [72M16]. Interatomic angles around Ag.

	Angles [deg]
O (1) (−*x*, −*y*, −*z*)–Ag–O (2)	163.1 (7)
O (5)–Ag–O (2) (−*x*+1, −*y*, −*z*)	146.7 (5)
O (1) (*x*, *y*, *z*−1)–Ag–Ag (−*x*, −*y*, −*z*)	149.9 (4)
O (1) (−*x*, −*y*, −*z*)–Ag–O (5)	103.5 (6)
O (5)–Ag–O (2)	73.0 (6)
O (2)–Ag–O (2) (−*x*+1, −*y*, −*z*)	73.9 (6)
O (2) (−*x*+1, −*y*, −*z*)–Ag–O (1) (−*x*, −*y*, −*z*)	109.0 (6)
O (1) (*x*, *y*, *z*−1)–Ag–O (5)	101.5 (6)
O (5)–Ag–Ag (−*x*, −*y*, −*z*)	68.9 (5)
Ag (−*x*, −*y*, −*z*)–Ag–O (2) (−*x*+1, −*y*, −*z*)	110.9 (4)
O (2) (−*x*+1, −*y*, −*z*)–Ag–O (1) (*x*, *y*, *z*−1)	92.8 (5)
O (1) (−*x*, −*y*, −*z*)–Ag–O (1) (*x*, *y*, *z*−1)	76.0 (7)
O (1) (*x*, *y*, *z*−1)–Ag–O (2)	120.9 (6)
O (2)–Ag–Ag (−*x*, −*y*, −*z*)	84.7 (4)
Ag (−*x*, −*y*, −*z*)–Ag–O (1) (−*x*, −*y*, −*z*)	79.1 (5)
C (1)–O (2)–Ag	120.8 (15)
O (1)–C (1)–O (2)	124.9 (23)
Ag (−*x*, −*y*, −*z*)–O (1)–C (1)	130.0 (17)

Table 366. $NH_2CH_2COOH \cdot AgNO_3$. Structure of phase I [72M16]. Interatomic distances and angles in NO_3 group.

	Distances [Å]		Angles [deg]
N (2)–O (3)	1.26 (3)	O (3)–N (2)–O (4)	123 (2)
N (2)–O (4)	1.25 (4)	O (3)–N (2)–O (5)	116 (2)
N (2)–O (5)	1.28 (3)	O (4)–N (2)–O (5)	120 (2)

38 $(NH_2CH_2COOH)_2 \cdot HNO_3$

38A Pure compound

Nr. 38A-1 $(NH_2CH_2COOH)_2 \cdot HNO_3$, Diglycine nitrate ($M = 213.15$)

1a	Ferroelectricity in $(NH_2CH_2COOH)_2 \cdot HNO_3$ was discovered by Pepinsky et al. in 1958.	58P1
b	(table below)	58P1

phase	II	I
state	F	P
crystal system	monoclinic	monoclinic
space group	Pa-C_s^2	$P2_1$/a-C_{2h}^5
Θ	−67 °C	

$P_s \| [101]$. — 58P1

$\varrho_X = 1.580(5) \cdot 10^3$ kg m^{-3}. — 68S7

Transparent colorless, deliquescent.

2a	Crystal growth: cooling or evaporation of aqueous solution.	58P1
b	Crystal form: Fig. 1406.	
3a	Unit cell parameters:	68S7

phase	T °C	a Å	b	c	β °
I	22	9.459 (7)	5.172 (5)	9.225 (6)	97.19 (3)
II	−150	9.485 (8)	5.132 (5)	9.089 (7)	97.89 (4)

b	$Z = 2$ in phases I and II. Crystal structure of phase I: Fig. 1407; Tables 367, 368. Crystal structure of phase II: Figs. 1408, 1409; Tables 369, 370.	68S7
5a	Dielectric constant: $\varkappa_{[010]} = 9$, $\varkappa_{[101]} = 14$, $\varkappa_{[10\bar{1}]} = 20$ at RT.	58P1
	$\varkappa_{[101]} = \varkappa_\infty + C/(T - \Theta_p)$, $T > \Theta_f$, where $C = 565$ K. Temperature dependence of $\varkappa_{[101]}$: Fig. 1410α; Temperature dependence of $\varkappa_{ij}$ in the vicinity of Θ_f: Fig. 1410β. Dielectric dispersion: Figs. 1411α, 1411β.	63N2
	Effect of hydrostatic pressure: $[d\Theta_p/dp]_{p=0} = 4.05(2) \cdot 10^{-8}$ K Pa^{-1}.	73G1
c	Spontaneous polarization and coercive field: Figs. 1412α, 1412β.	
6a	Transition heat and transition entropy of the II−I transistion: $\Delta Q_m = 627$ J mol^{-1}, $\Delta S_m = 3.1$ J K^{-1} mol^{-1}.	58P1
7a	Piezoelectricity: Fig. 1413.	
8a	Elastic compliance: Fig. 1414.	
9a	Infrared absorption: see	66S6
13a	NMR: Figs. 1415, 1416.	
b	ESR of γ-ray irradiated crystal: see	73O1

Table 367. $(NH_2CH_2COOH)_2 \cdot HNO_3$. Structure of phase I [68S7]. Fractional coordinates and isotropic temperature parameters. $T=22$ °C. B is defined by Eq. (e) in Introduction.

	x	y	z	B [Å^2]
(Glycine A)*)				
N (1)	0.3483 (9)	−0.0344 (18)	0.1993 (9)	2.01 (15)
C (1)	0.2866 (11)	0.1243 (22)	0.3127 (11)	2.05 (18)
C (2)	0.1557 (11)	−0.0035 (22)	0.3632 (11)	1.95 (17)
O (1)	0.1076 (8)	0.1127 (17)	0.4662 (8)	2.66 (14)
O (2)	0.1042 (8)	−0.1927 (15)	0.2962 (8)	2.25 (13)
(Glycine B′)*)				
N (2)	0.3590 (9)	−0.0427 (18)	0.2126 (9)	1.88 (14)
C (3)	0.2983 (10)	0.1015 (21)	0.3291 (10)	1.68 (16)
C (4)	0.1612 (10)	−0.0189 (19)	0.3581 (10)	1.34 (15)
O (3)	0.0971 (8)	0.1046 (16)	0.4520 (8)	2.36 (13)
O (4)	0.1231 (7)	−0.2326 (15)	0.3049 (7)	1.91 (12)
(Nitrate)				
N (3)	0.0007 (16)	0.0418 (20)	0.0108 (14)	2.22 (16)
O (5)	0.0510 (10)	0.2368 (21)	0.0595 (10)	3.97 (19)
O (6)	−0.1095 (13)	−0.0482 (27)	0.0098 (13)	6.13 (28)
O (7)	0.1096 (10)	−0.0805 (20)	−0.0492 (10)	3.77 (18)

*) In phase I, a glycine ion takes two states A and B′ in equal probability. The atomic coordinates for glycine B′ are related to those of glycine B in Tables 369, 370 by a symmetry operation 2_1.

Table 368. $(NH_2CH_2COOH)_2 \cdot HNO_3$. Structure of phase I [68S7]. Fractional coordinates and isotropic temperature parameters of hydrogen atoms. $T=22$ °C. B is defined by Eq. (e) in Introduction.

	x	y	z	B [Å^2]
(Glycine A)*)				
H (1)	0.3576 (70)	0.1649 (168)	0.3960 (76)	2.87 (115)
H (2)	0.2308 (77)	0.3167 (313)	0.2649 (85)	4.36 (142)
H (3)	0.2846 (99)	−0.0926 (191)	0.1344 (65)	3.09 (117)
H (4)	0.4170 (50)	0.0714 (164)	0.1409 (66)	1.01 (92)
H (5)	0.4211 (50)	−0.1609 (138)	0.2411 (61)	0.92 (84)
H (11)	0.0082 (99)	−0.0006 (429)	0.4817 (79)	2.69 (90)
(Glycine B′)*)				
H (6)	0.3819 (70)	0.0565 (211)	0.4320 (76)	2.90 (114)
H (7)	0.2827 (55)	0.2987 (149)	0.2922 (62)	1.54 (86)
H (8)	0.2773 (113)	−0.1966 (185)	0.1721 (83)	4.42 (142)
H (9)	0.3792 (84)	0.1146 (255)	0.1251 (107)	4.18 (143)
H (10)	0.4621 (72)	−0.1331 (243)	0.2646 (88)	4.01 (130)

*) See footnote of Table 367.

Table 369. $(NH_2CH_2COOH)_2 \cdot HNO_3$. Structure of phase II [68S7]. Fractional coordinates and isotropic temperature parameters. $T = -150$ °C. B is defined by Eq. (e) in Introduction.

	x	y	z	B [Å^2]
(Glycine A)				
N (1)	0.3454 (11)	−0.0313 (21)	0.2056 (14)	0.87 (18)
C (1)	0.2893 (13)	0.1197 (27)	0.3216 (17)	1.17 (23)
C (2)	0.1517 (12)	−0.0004 (24)	0.3606 (15)	0.50 (18)
O (1)	0.1009 (9)	0.1218 (18)	0.4640 (11)	1.02 (15)
O (2)	0.1012 (10)	−0.1997 (19)	0.2976 (12)	1.31 (16)
(Glycine B)				
N (2)	0.1401 (10)	0.4613 (22)	0.7903 (13)	0.87 (18)
C (3)	0.2011 (12)	0.6126 (24)	0.6790 (15)	0.65 (20)
C (4)	0.3363 (12)	0.4816 (26)	0.6384 (16)	0.86 (21)
O (3)	0.3972 (9)	0.6073 (17)	0.5423 (11)	0.93 (15)
O (4)	0.3775 (8)	0.2780 (17)	0.7020 (10)	0.68 (14)
(Nitrate)				
N (3)	0.0110 (13)	0.0391 (22)	0.0036 (17)	1.23 (19)
O (5)	0.0407 (10)	0.2583 (22)	0.0611 (11)	1.48 (16)
O (6)	−0.1112 (11)	−0.0614 (21)	0.0048 (14)	1.83 (18)
O (7)	0.1082 (10)	−0.0823 (20)	−0.0505 (13)	1.53 (17)

Table 370. $(NH_2CH_2COOH)_2 \cdot HNO_3$. Structure of phase II [68S7]. Fractional coordinates and isotropic temperature parameters of hydrogen atoms. $T = -150$ °C. B is defined by Eq. (e) in Introduction.

	x	y	z	B [Å^2]
(Glycine A)				
H (1)	0.3699 (56)	0.1021 (142)	0.4216 (72)	2.64 (118)
H (2)	0.2591 (51)	0.2889 (115)	0.2885 (52)	1.18 (83)
H (3)	0.2655 (41)	−0.0904 (109)	0.1126 (53)	1.06 (85)
H (4)	0.4056 (37)	0.0860 (108)	0.1446 (46)	0.12 (82)
H (5)	0.4202 (49)	−0.1780 (103)	0.2411 (59)	1.56 (95)
H (11)	0.0151 (44)	0.0300 (93)	0.5029 (64)	0.72 (67)
(Glycine B)				
H (6)	0.1378 (48)	0.5848 (135)	0.5747 (55)	0.83 (78)
H (7)	0.2153 (69)	0.7968 (167)	0.7277 (80)	3.89 (145)
H (8)	0.2134 (53)	0.3376 (117)	0.8374 (63)	1.73 (90)
H (9)	0.1066 (68)	0.6023 (175)	0.8692 (84)	3.39 (126)
H (10)	0.0465 (56)	0.3696 (131)	0.7290 (61)	2.08 (99)

39 $(NH_2CH_2COOH)_2 \cdot MnCl_2 \cdot 2H_2O$

39A Pure compound

Nr. 39A-1 $(NH_2CH_2COOH)_2 \cdot MnCl_2 \cdot 2H_2O$, Diglycine manganese chloride dihydrate
($M = 312.008$)

1a	Ferroelectric activity in $(NH_2CH_2COOH)_2 \cdot MnCl_2 \cdot 2H_2O$ was discovered by Pepinsky et al. in 1958.	58P3

				Ref.
1b	state	F*)		58P3
	crystal system	monoclinic		
	space group	$P2_1$-C_2^2		
	$P_s \| [010]$. $\varrho = 1.875_0 \cdot 10^3$ kg m^{-3} at 20 °C. Transparent. *) Transition to the paraelectric phase does not exist up to the decomposition temperature.			58P3 58P3
2a	Crystal growth: evaporation or cooling method from aqueous solution.			58P3
3a	Unit cell parameters: $a = 9.96$ Å, $b = 8.53$ Å, $c = 6.86$ Å, $\beta = 107.0(5)°$ at RT.			58P3
b	$Z = 2$ in phase II.			58P3
5a	Dielectric constants: Fig. 1417. Dielectric constants at RT: $\varkappa_a = 6.6$, $\varkappa_b = 8.1$, $\varkappa_c = 7.4$.			58P3
c	Spontaneous polarization and coercive field: Figs. 1418, 1419.			

40 $(CH_3NHCH_2COOH)_3 \cdot CaCl_2$

40A Pure compound

Nr. 40A-1 $(CH_3NHCH_2COOH)_3 \cdot CaCl_2$, Tris-sarcosine calcium chloride ($M = 378.27$)

				Ref.
1a	Ferroelectric activity in $(CH_3NHCH_2COOH)_3 \cdot CaCl_2$ was discovered by Pepinsky et al. in 1962.			62P1
b	phase	II	I	65M3
	state	F	P	
	crystal system	orthorhombic	orthorhombic	
	space group	C_{2v}^9-$Pn2_1a$*)	D_{2h}^{16}-Pnma	
	Θ	127 K		
	$P_s \| [010]$. $\varrho = 1.533 \cdot 10^3$ kg m^{-3} at RT. Transparent, colorless. *) Proposed space group. [65M3].			65M3
2a	Crystal growth: evaporation or cooling method from aqueous solution.			65M3
3a	Unit cell parameters: $a = 9.156(10)$ Å, $b = 17.460(5)$ Å, $c = 10.265(5)$ Å at RT.			65M3
b	$Z = 4$ in phases I and II. Crystal structure: Tables 371···373; Figs. 1420···1424.			65M3
5a	Dielectric constants: Figs. 1425α, 1425β. $\Theta_f - \Theta_p = 0.4(1)$ K, $C = 59(3)$ K.			65M3
c	Spontaneous polarization and coercive field: Figs. 1426α, 1426β.			
6a	Heat capacity: Fig. 1427. Transition energy ΔQ_m and transition entropy ΔS_m: $\Delta Q_m = 446(80)$ Jmol^{-1}, $\Delta S_m = 4.26(80)$ J K^{-1} mol^{-1}			79M4
8a	Elastic stiffness: Fig. 1428.			
13a	NMR: Table 374; Figs. 1429, 1430, 1431.			continued

Nr. 40A-1 $(CH_3NHCH_2COOH)_3 \cdot CaCl_2$, continued

13b	ESR: Tables 375, 376; Figs. 1432···1434.	
	Parameters of Spin Hamiltonian of Mn^{2+} at RT: $g=2.005\,(1)$, $D=192\,(1)\cdot 10^{-2}\,m^{-1}$, $E=-14.2\,(5)\cdot 10^{-2}\,m^{-1}$, $A=-88\,(1)\cdot 10^{-2}\,m^{-1}$, $a=6\,(3)\cdot 10^{-2}\,m^{-1}$.	72B3
	For ESR of Mn^{2+}, see also	75W4

Table 371. $(CH_3NHCH_2COOH)_3 \cdot CaCl_2$. Crystal structure of phase I [72A17]. Fractional coordinates and temperature parameters. The temperature parameters are defined by Eq. (a) in Introduction. For molecules 1, 2, see also Fig. 1422.

	x	y	z	B_{11}	B_{22}	B_{33}	B_{12}	B_{13}	B_{23}
	10^{-4}			10^{-4}					
Ca	386 (2)	2500 (0)	2249 (1)	42 (2)	6 (1)	16 (2)	0 (0)	2 (3)	0 (0)
Cl	4912 (2)	833 (1)	7148 (1)	96 (2)	15 (1)	47 (2)	10 (2)	0 (3)	− 2 (1)
Molecule 1									
O (11)	1720 (8)	2500 (0)	4142 (6)	120 (10)	38 (3)	40 (6)	0 (0)	−84 (14)	0 (0)
O (12)	4070 (7)	2500 (0)	4718 (5)	76 (8)	27 (2)	34 (6)	0 (0)	44 (11)	0 (0)
N (11)	3511 (8)	2500 (0)	7265 (6)	84 (10)	15 (2)	5 (6)	0 (0)	7 (13)	0 (0)
C (11)	2747 (10)	2500 (0)	4942 (8)	120 (14)	8 (3)	24 (8)	0 (0)	−11 (18)	0 (0)
C (12)	2228 (9)	2500 (0)	6378 (8)	67 (11)	20 (3)	29 (8)	0 (0)	−18 (16)	0 (0)
C (13)	3003 (12)	2500 (0)	8659 (9)	141 (16)	31 (4)	30 (9)	0 (0)	11 (20)	0 (0)
Molecule 2									
O (21)	6588 (5)	1447 (2)	3713 (4)	103 (6)	18 (2)	74 (5)	36 (5)	0 (10)	25 (5)
O (22)	8936 (6)	1601 (3)	3240 (5)	145 (8)	23 (2)	77 (5)	−55 (6)	54 (11)	21 (5)
N (21)	7000 (6)	114 (3)	4937 (5)	91 (7)	10 (2)	42 (5)	− 1 (6)	15 (10)	13 (5)
C (21)	7899 (7)	1263 (3)	3732 (5)	91 (8)	10 (2)	25 (5)	−12 (7)	8 (11)	− 4 (5)
C (22)	8309 (7)	532 (3)	4464 (6)	55 (7)	13 (2)	42 (6)	9 (6)	16 (11)	10 (6)
C (23)	7380 (9)	− 576 (4)	5733 (7)	166 (12)	14 (2)	100 (9)	−12 (9)	−17 (18)	48 (8)

Table 372. $(CH_3NHCH_2COOH)_3 \cdot CaCl_2$. Crystal structure of phase I [72A17]. Fractional coordinates of hydrogen atoms.

Atom	x	y	z	Bonded to
H (11)	0.165	0.193	0.653	C (12)
H (13)	0.410	0.202	0.712	N (11)
H (15)	0.373	0.250	0.910	C (13)
H (17)	0.248	0.193	0.858	C (13)
H (21)	0.878	0.014	0.384	C (22)
H (22)	0.894	0.069	0.530	C (22)
H (23)	0.648	−0.011	0.420	N (21)
H (24)	0.655	0.037	0.546	N (22)
H (25)	0.634	−0.082	0.595	C (23)
H (26)	0.805	−0.025	0.660	C (23)
H (27)	0.808	−0.093	0.521	C (23)

Table 373. $(CH_3NHCH_2COOH)_3 \cdot CaCl_2$. Crystal structure of phase I [72A17]. Hydrogen bonds I, II, III (see Figs. 1420, 1421). C (M): carbon atom of the methyl group. C (α): carbon atom of the carboxyl group of sarcosine molecule.

H-bond	N···Cl	H···Cl	∢NHCl	∢C(α) NCl	∢C(M) NCl
I	3.182 Å	2.20 Å	164°	107.2°	99.2°
II	3.222	2.43	165	121.2	93.9
III	3.222	2.27	166	118.3	94.3

N-Cl-N angle between I and II 128.5°
I and III 106.0
II and III 69.8

Table 374. $(CH_3NHCH_2COOH)_3 \cdot CaCl_2$. Nuclear quadrupole resonance frequencies, e^2qQ/h and η of ^{14}N, ^{35}Cl and ^{37}Cl. [75B14].

T K	Nucleus	Resonance frequencies kHz			e^2qQ/h kHz	η
292	N (1)	1040,	520,	520	1040	1
	N (2), N (3)	940,	470,	470	940	1
	^{35}Cl		2160	...	...	...
	^{37}Cl		1710		...	...
77	N	980,	530,	450	1010	0.89

Table 375. $(CH_3NHCH_2COOH)_3 \cdot CaCl_2$ (TSCC): VO^{2+}. Spin-Hamiltonian parameters [78S4].

Spectrum	$g_{\parallel}$	$g_{\perp}$	A $10^2\,m^{-1}$	B $10^2\,m^{-1}$
I	1.911 (2)	1.987 (2)	174 (2)	79 (2)
II	1.930 (2)	1.991 (2)	178 (2)	72 (2)

Table 376. $(CH_3NHCH_2COOH)_3 \cdot CaCl_2$: Mn^{2+}, $(CH_3NHCH_2COOH)_3 \cdot CaBr_2$: Mn^{2+}. Spin Hamiltonian parameters of Mn^{2+} [75N3]. A, D, E and a are expressed in units of $10^{-2}\,m^{-1}$. For spin lattice relaxation time, see [78B10].

	g	A	D	E	a
$(CH_3NHCH_2COOH)_3 \cdot CaCl_2$	2.005	−88	192	14	6
$(CH_3NHCH_2COOH)_3 \cdot CaBr_2$	2.005	−84	151	21	5

41 Rochelle salt ($NaKC_4H_4O_6 \cdot 4H_2O$) family

41A Pure compounds

Nr. 41A-1 $NaKC_4H_4O_6 \cdot 4H_2O$, Sodium potassium tartrate tetrahydrate (Seignette salt, Rochelle salt, RS) ($M = 282.2209$)

1a The ferroelectricity of Rochelle salt was first reported by Valasek in 1921. 21V1

b (24V1, 37J1)

phase	III	II	I
state	P	F	P
crystal system	orthorhombic	monoclinic	orthorhombic
space group	$P2_12_12\text{-}D_2^3$	$P2_1\text{-}C_2^2$	$P2_12_12\text{-}D_2^3$
Θ	−18	24	°C

Θ_{II-I} and Θ_{III-II} are often called upper and lower Curie temperature, and here denoted by Θ_f^u and Θ_f^l, respectively. For deuterated Rochelle salt ($NaKC_4H_2D_2O_6 \cdot 4D_2O$), $\Theta_f^l = 22$ °C, $\Theta_f^u = 35$ °C. 40H1

For Θ_f of partly deuterated RS, see 40H1

$P_s \parallel [100]$.

$T_{melt} = 55$ °C, see also Fig. 1435.

$\varrho = 1.775 \cdot 10^3\,kg\,m^{-3}$, see also Fig. 1436. 39M1

For deuterated RS, $\varrho = 1.830(3) \cdot 10^3\,kg\,m^{-3}$ at RT. 40H1

Transparent, colorless.

Deliquescent and efforescent: Fig. 1435.

2a Crystal growth: cooling or evaporation method from aqueous solution.

Solubility in H_2O: Fig. 1436.

continued

Nr. 41A-1 $NaKC_4H_4O_6 \cdot 4H_2O$, continued

2b Crystal form: Figs. 1437, 1438.
Note: In the ferroelectric state (monoclinic system), the crystallographic axes *a*, *b*, *c*, usually adopted, are parallel or nearly parallel to the *a*, *b*, *c* axes of the paraelectric state, respectively; namely the diad axis is *a* instead of *b*.

3a Unit cell parameters: $a = 11.867\,(7)$ Å, $b = 14.236\,(8)$ Å, $c = 6.213\,(4)$ Å at 20 °C. — 46U1
See also Table 377.

b Crystal structure: $Z = 4$, Tables 378···380; Figs. 1439···1443. — 46U1
For the structure of phase II under dc. bias field, see — 74M14

4 Thermal expansion: Fig. 1444, 1445; see also — 35V1, 35H1
For temperature dependence of lattice constants, see Table 377.
Density change: Fig. 1446.

5a Temperature dependence of dielectric constants: Figs. 1447···1450; see also — 48K1, 61T1, 67O6

Curie-Weiss law: — 35M1

T	Θ_p	C
25 °C < T < 32 °C	23.0 °C	$2.24 \cdot 10^3$ K
−18 °C > T > −28 °C	−17.9 °C	$-1.18 \cdot 10^3$ K

See also — 48K1, 66C4
For deuterated RS, see — 39H1
Effects of hydrostatic pressure on dielectric constants: Figs. 1451, 1452. Pressure dependence of the Curie temperature: — 38B1

Compound	$(d\Theta_{II-I}/dp)_{p=0}$ 10^{-8} K Pa^{-1}	$(d\Theta_{III-II}/dp)_{p=0}$ 10^{-8} K Pa^{-1}	Ref.
$NaKC_4H_4O_6 \cdot 4H_2O$	10.93	3.84; see also	65S2
$NaKC_4H_2D_2O_6 \cdot 4D_2O$	10.2	3.3	68S3

Two phase transitions of the first kind exist at $37 \cdot 10^8$ Pa and $59 \cdot 10^8$ Pa, at 20 °C. — 72M21
Effects of two dimensional pressure perpendicular to [100]: — 72M17
$d\Theta_{II-I}/dT = -22\,(1) \cdot 10^{-8}$ K m^2 N^{-1}; $d\Theta_{III-II}/dT = 30\,(2) \cdot 10^{-8}$ K m^2 N^{-1}.
For the effects of uniaxial pressure, see — 67U3
Dielectric constants over microwave range: Figs. 1453, 1454; see also — 60J1, 67A7, 73P10

b Nonlinear dielectric properties:

$$E = \frac{1}{\varepsilon_0 C}(T - \Theta_p)\, P + \xi P^3,\quad \xi = 4.7\,(6) \cdot 10^{13}\ \mathrm{V\,m^5\,C^{-3}};$$ — 40M1

$$E_a = (\beta_{11} + \xi_{11} D_a^2 + \xi_{12} D_b^2 + \xi_{13} D_c^2)\, D_a,$$ — 68F4
$$\xi_{11} = 5.9 \cdot 10^{13}\ \mathrm{V\,m^5\,C^{-3}},\quad \xi_{12} = 3.3 \cdot 10^{13}\ \mathrm{V\,m^5\,C^{-3}},$$
$$\xi_{13} = -3.5 \cdot 10^{12}\ \mathrm{V\,m^5\,C^{-3}}.$$

c Spontaneous polarization, coercive field: Figs. 1455, 1456.
For frequency, applied field dependence of the coercive field, see — 58W1, 64A1

d Electrocaloric effect: Fig. 1457; see also — 30K1

6a	Heat capacity: Figs. 1458···1460. Transition heat ΔQ_m, transition entropy ΔS_m:	
		ΔQ_m	ΔS_m	78T4
	III – II transition	7.51 J mol^{-1}	0.029 J K^{-1} mol^{-1}	
	II – I transition	9.86 J mol^{-1}	0.034 J K^{-1} mol^{-1}	
b	Thermal conductivity: Fig. 1461.	
7a	Piezoelectricity: Table 381; Figs. 1462···1472.	
b	Electrostriction: Figs. 1473, 1474.	
c	Nonlinear electromechanical properties: Fig. 1475.	
8a	Elastic compliances, stiffnesses: resonance method: Table 382; Figs. 1476···1480. Pulse method: 1481, 1482. Brillouin scattering: Fig. 1483. Piezoelectric line in NQR: Fig. 1484.	
9a	Refractive indices: Fig. 1485. Optical axial angle: Fig. 1486. Birefringence: Fig. 1487.	
b	Electrooptic effect: Table 383; Figs. 1488···1494.	
c	Piezooptic effect: Table 384.	
d	Rotatory power: $\varrho = 1.35(15) \cdot 10^3$ °m^{-1} for $\lambda = 583.9$ nm. Gyration tensor: $g_{11} = -8.6(5) \cdot 10^{-6}$, $g_{22} = -6.7(15) \cdot 10^{-6}$, $g_{33} = 7.2(16) \cdot 10^{-6}$ at 21 °C, for $\lambda = 632.8$ nm.	30w1 75V9
10a	Raman scattering: see	57B1
b	Brillouin scattering: Fig. 1495; c_{ij} obtained from Brillouin scattering, see 8a.	
11	Electrical conductivity: Figs. 1496, 1497; see also Breakdown field: $2 \cdot 10^4$ kV m^{-1} at RT.	46C1, 36S1, 38S1, 49T1 69g2
13a	NMR: Tables 385···387; Figs. 1498···1503. For additional data, see	 68B9, 73F4
b	ESR: Table 388; Fig. 1504. For additional data, see	 61M3, 62A5, 67S29, 68S24, 74S39, 75D3, 78S3
14a	Integrated intensity of Bragg reflection: Figs. 1505···1507.	
15a	Domain structure: Fig. 1508; see also	40M2, 48K2, 50C3, 50F1, 59P1, 64A1
b	Dynamical properties: see continued	60W4, 58W1, 56A1

Nr. 41A-1 $NaKC_4H_4O_6 \cdot 4H_2O$, continued

16	Surface layer: see	76M4
	Radiation damage: see	53A1, 61M3, 61O1, 61T1, 62A5, 63B10, 63K7, 65E1, 67O3, 67S29
	Dislocations and macroscopic defects: see	61N1
	Etchant: H_2O.	61N1
	Whisker: see	67U1
	Effects of humidity: see	43K1, 47K1, 47K2, 48K1, 49K1, 55M1

Table 377. $NaKC_4H_4O_6 \cdot 4H_2O$ (RS). Unit cell parameters *a*, *b*, *c*.

	[34S1]	[41B2]	[46U1] −50 °C	[46U1] +20 °C	[46U1] +35 °C	
a	11.93 (4)	11.93	11.815	11.867 (7)	11.878	Å
b	14.32 (5)	14.30	14.203	14.236 (8)	14.246	Å
c	6.20 (2)	6.17	6.195	6.213 (4)	6.218	Å

Table 378. $NaKC_4H_4O_6 \cdot 4H_2O$ (RS). Structure of phase I (RT) [41B2]. Fractional coordinates. See also Fig. 1439.

	X=a	Y=b	Z=c		X=a	Y=b	Z=c
2K on (a)	0.00	0.00	0.05	$4H_2O$ on (7)	0.40	0.08	0.50
2K on (b)	0.00	0.50	0.15	$4H_2O$ on (8)	0.25	0.05	0.87
4Na on	0.23	0.99	0.52	$4H_2O$ on (9)	0.44	0.30	0.05
4O on (1)	0.12	0.10	0.37	$4H_2O$ on (10)	0.42	0.40	0.45
4O on (2)	0.22	0.20	0.12	4C on	0.15	0.18	0.28
4O on (3)	0.23	0.40	0.82	4C on	0.12	0.28	0.42
4O on (4)	0.06	0.37	0.85	4C on	0.17	0.27	0.65
4OH on (5)	0.16	0.36	0.32	4C on	0.15	0.35	0.80
4OH on (6)	0.29	0.24	0.63				

Table 380. $NaKC_4H_4O_6 \cdot 4H_2O$ (RS). Structure of phase I, III [57M3]. Interatomic distances between oxygens at −64 °C, 35 °C.

	Distances [Å] at −64 °C	Distances [Å] at +35 °C	Δ
O (1)−O (10)	2.55	2.56	−0.01
O (3)−O (8)	2.71	2.72	−0.01
O (4)−O (9)	2.74	2.75	−0.01
O (6)−O (7)	2.74	2.76	−0.02
O (6)−O (10)	2.79	2.83	−0.04
O (2)−O (8)	2.80	2.82	−0.02
O (9)−O (10)	2.80	2.84	−0.04
O (4)−O (7)	2.94	2.98	−0.04
O (7)−O (8)	2.98	2.97	+0.01
O (5)−O (8)	3.11	3.10	+0.01
O (2)−O (9)	3.10	3.11	−0.01

Table 379. $NaKC_4H_4O_6 \cdot 4H_2O$ (RS). Structure of phase I (35 °C), phase III (−64 °C) [57M3]. Fractional coordinates and temperature parameters. B_M, B_m: along the direction of the maximum, minimum thermal vibration. δ: angle between the direction of the maximum thermal vibration and a^* axis. Temperature factor $= e^{-B_\varphi \frac{\sin^2\theta}{\lambda^2}}$ $B_\varphi = B_M \cos^2(\varphi-\delta) + B_m \sin^2(\varphi-\delta)$. B is defined by Eq. (e) in Introduction. φ: angle between radius vector (h k 0) and the a^* axis in reciprocal space.

Atoms	Structure at −64 °C			Structure at +35 °C		
	coordinates		temperature factors [Å²]	coordinates		temperature factors [Å²]
	x	y		x	y	
K (1)	0	0	$B_M=5.45$, $B_m=1.19$, $\delta=-30°$	0	0	$B_M=7.78$, $B_m=1.49$, $\delta=-35°$
K (2)	1/2	0	$B_M=1.78$, $B_m=1.19$, $\delta=-35°$	1/2	0	$B_M=2.25$, $B_m=1.49$, $\delta=-35°$
Na	0.232	0.992	$B_M=1.14$, $B_m=0.47$, $\delta=0°$	0.232	0.992	$B_M=1.14$, $B_m=0.47$, $\delta=0°$
C (1)	0.153	0.188	$B=1.11$	0.153	0.188	$B=1.11$
C (2)	0.127	0.271	$B=1.11$	0.127	0.271	$B=1.11$
C (3)	0.175	0.267	$B=1.11$	0.175	0.267	$B=1.11$
C (4)	0.170	0.356	$B=1.11$	0.170	0.356	$B=1.11$
O (1)	0.121	0.109	$B=1.11$	0.121	0.109	$B=1.11$
O (2)	0.211	0.202	$B=1.11$	0.209	0.202	$B=1.11$
O (3)	0.232	0.407	$B_M=2.59$, $B_m=1.11$, $\delta=0°$	0.235	0.406	$B_M=2.59$, $B_m=1.11$, $\delta=0°$
O (4)	0.050	0.359	$B=1.11$	0.054	0.361	$B=1.11$
O (5)	0.154	0.356	$B=1.11$	0.154	0.356	$B=1.11$
O (6)	0.298	0.249	$B=1.11$	0.297	0.249	$B=1.11$
O (7)	0.397	0.084	$B=1.11$	0.397	0.084	$B=1.11$
O (8)	0.244	0.040	$B_M=4.79$, $B_m=1.85$, $\delta=0°$	0.248	0.040	$B_M=5.53$, $B_m=3.84$, $\delta=0°$
O (9)	0.440	0.305	$B=2.56$	0.440	0.301	$B_M=5.53$, $B_m=3.84$, $\delta=90°$
O (10)	0.426	0.395	$B=2.56$	0.427	0.396	$B=2.59$

Table 381. $NaKC_4H_4O_6 \cdot 4H_2O$ (RS), $NaKC_4H_2D_2O_6 \cdot 4D_2O$ (deuterated RS). Piezoelectric constants [66b1].

Crystal	d_{14}	d_{25}	d_{36}	e_{14}	e_{25}	e_{36}	g_{14}	g_{25}	g_{36}	h_{14}	h_{25}	h_{36}	k_{14}	T	Ref.
	10^{-12} C N^{-1}			C m^{-2}			10^{-3} m^2 C^{-1}			10^8 N C^{-1}				°C	
RS	770	+46	9.4											0	27V1
	333	−55	11.8											RT	28M1
		−56	13.1												39M2
	2300	−56	11.8											25	50m1
	512	−54	12	2.98	−0.16	0.12	180	−660	140	23	−21	14		30, −22	50V1
	187	−54	12	1.65	−0.16	0.12	180	−660	140	23	−21	14		40, −28	
	345	54	12											34	64B2
				3.7	−0.16	0.11									50m1
													0.76		62B4
deuterated		−77	13												40H1
RS										23.4				RT	40H1

Table 382. $NaKC_4H_4O_6 \cdot 4H_2O$ (RS). Elastic constants [66b1].

s_{11}	s_{22}	s_{33}	s_{12}	s_{13}	s_{23}	s^E_{44}	s^D_{44}	s^E_{55}	s^D_{55}	s^E_{66}	s^D_{66}	T	Ref.
10^{-12} m² N⁻¹												°C	
52.30	34.30	32.40	−21.80	−16.80	−13.3	96.3		337		118			39H2
52.4	35.0	33.7	−15.4	− 9.8	− 9.1		74.6	350	312	104	102	20	47H1
50.2	30.4	31.7	−11.6	−21.4	− 8.9	82.0		333		106			49S1
51.8	34.9	33.4	−15.3	−21.1	−10.3	172		360		102.7		30, −22	50m1
51.8	34.9	33.4	−15.3	−21.1	−10.3	113.4		346		102		40, −28	50V1
51.8	34.9	33.4	−15.3	−21.1	−10.3		79.6		328		101	30, −22	39M2
51.8	34.9	33.4	−15.3	−21.1	−10.3		78.7		310		101	40, −28	39M2
52.0	36.8	35.9	−16.3	−11.6	−12.2	150.2		350.3		104.2		34	64B2
c_{11}	c_{22}	c_{33}	c_{12}	c_{13}	c_{23}	c^E_{44}	c^D_{44}	c^E_{55}	c^D_{55}	c^E_{66}	c^D_{66}		
10^9 N m⁻²													
34.7	47.3	80.6	− 8.04	31.6	−34.4	(16.4)		(3.24)		(12.4)		20	27M1
26.4	31.8	39.1	18.1	22.3				(3.04)		(9.96)			39M2
25.5	38.1	37.1	14.1	11.6	14.6		13.4	2.86	3.21	9.60	9.79	20	47H1
40.6	52.0	64.0	25.6	34.6	32.0	12.2		3.0		9.5			49S1
25.8	38.0	37.5	14.0	11.2				3.14		9.97		RT	50J1
28.0	41.4	39.4	17.4	15.0	19.7	6.66		2.85		9.60		34	64B2

Table 383. $NaKC_4H_4O_6 \cdot 4H_2O$ (RS). Electrooptic constants [1894P1].

r_{41}	r_{52}	r_{63}	λ
10^{-12} m V⁻¹			nm
−2.0	−1.7	0.32	589

Table 384. $NaKC_4H_4O_6 \cdot 4H_2O$ (RS). Piezooptic constants [69N2]. $\lambda = 589$ nm. See also [1894P1].

p_{11}	0.35	Π_{11}	3.1	10^{-12} m² N⁻¹
p_{12}	0.41	Π_{12}	3.4	
p_{13}	0.42	Π_{13}	3.2	
p_{21}	0.37	Π_{21}	5.4	
p_{22}	0.28	Π_{22}	0.76	
p_{23}	0.34	Π_{23}	2.1	
p_{31}	0.36	Π_{31}	5.0	
p_{32}	0.35	Π_{32}	2.9	
p_{33}	0.36	Π_{33}	1.6	
p_{44}	−0.030	Π_{44}	−2.5	
p_{55}	0.0046	Π_{55}	1.5	
p_{66}	−0.025	Π_{66}	−2.5	

Fig. p. 653

Table 385. $NaKC_4H_4O_6 \cdot 4H_2O$ (RS), $NaKC_4H_2D_2O_6 \cdot 4D_2O$ (deuterated RS). ^{23}Na quadrupole coupling constant e^2qQ/h and asymmetry parameter η. (A), (B): monoclinic sublattice at 0 °C.

T °C	RS		Deuterated RS	
	e^2qQ/h MHz	η	e^2qQ/h MHz	η
+30	1.313 (14)	0.809 (21)	1.313 (15)	0.800 (30)
0 (A)	1.2850 (185)	0.872 (22)	1.323 (15)	0.803 (30)
0 (B)	1.407 (20)	0.675 (29)	1.419 (15)	0.639 (30)
−40			1.395 (15)	0.636 (30)
−25	1.363 (4)*)	0.737 (4)*)		
Ref.	66 M 2 *) 70 F 2		64 B 6	

Table 386. $NaKC_4H_4O_6 \cdot 4H_2O$ (RS). Direction cosines of the principal axes (XYZ) of an electric-field gradient tensor of Na spectrum 1 relative to the crystallographic axes (abc) [66 M 2]. (A), (B): monoclinic sublattice at 0 °C.

T [°C]	Crystal axis	Principal axes X	Y	Z
30	a	0.178	0.574	0.799
	b	−0.021	−0.810	0.586
	c	−0.984	0.122	0.132
0 (A)	a	0.208	0.559	0.803
	b	−0.068	−0.809	0.583
	c	−0.976	0.182	0.123
0 (B)	a	0.157	0.592	0.791
	b	−0.008	−0.801	0.599
	c	−0.988	0.090	0.126

Table 387. $NaKC_4H_4O_6 \cdot 4H_2O$ (RS). Direction cosines of the proton-proton vectors of the water molecules with respect to the crystallographic axes. See also [55 L 1, 69 E 2]. (7), (8), (9), and (10) in the 1st column are Nos. of the water molecule as shown in Fig. 1439.

	Direction cosines of p−p pairs [70 K 6]	[67 B 10]	Differences between both directions
(7)	0.52	0.61	~ 7°
	0.47	0.45	
	0.72	0.65	
(8)	0.21	0.15	~ 4°
	0.97	0.98	
	0.14	0.15	
(9)	0.88	0.88	~ 1°
	0.34	0.34	
	0.34	0.33	
(10)	0.39	0.55	~11°
	0.33	0.33	
	0.86	0.77	

Table 388. $NaKC_4H_4O_6 \cdot 4H_2O$ (RS): Cu^{2+}. ESR data [71 V 4].

Paramag. Center	Cu^{2+} ($S=1/2$)
g-factor	$g_x=2.041$ (5)
	$g_y=2.094$ (5)
	$g_z=2.326$ (2)
HFS [10^{-2} m^{-1}]	$A_x=-39$ (4)
	$A_y=-21$ (4)
	$A_z=-123$ (1)
	$Q_z \cong 10$

Nr. 41A-2 $NaNH_4C_4H_4O_6 \cdot 4H_2O$, Sodium ammonium tartrate tetrahydrate (Ammonium Rochelle salt, ARS) ($M = 261.1609$)

1a	Dielectric anomaly was observed by Jona et al. in 1953, and polarization reversal was investigated by Takagi et al. in 1958.	53 J 1 58 T 1
b	phase: II / I; state: (F) / P; crystal system: monoclinic / orthorhombic	
	space group: $P2_1\text{-}C_2^2$ / $P2_12_12\text{-}D_2^3$	58 T 1
	Θ: 109 K	58 T 1
	$P_s \parallel [010]$; P_s can be reversed only by the mechanical stress.	58 T 1
	$\varrho = 1.587 \cdot 10^3\ kg\ m^{-3}$.	33 B 1
	Transparent, colorless. Deliquescent.	
2a	Crystal growth: cooling or evaporation method from aqueous solution.	
3a	Unit cell parameters: $a = 12.15$ Å, $b = 14.40$ Å, $c = 6.18$ Å at RT.	41 B 1
b	Superlattice structure: X-ray superlattice reflections due to $a' = 2a$ appear in phase II.	71 S 6
5a	Dielectric constants: Fig. 1509.	
c	Spontaneous polarization and coercive field: $P_s = 2.1 \cdot 10^{-3}\ C\ m^{-2}$, coercive stress $T_5 \approx 5 \cdot 10^6\ N\ m^{-2}$ at 92 K. See 7c.	58 T 1
d	Pyroelectricity: Fig. 1510.	
7a	Piezoelectricity: Table 389; Figs. 1511, 1512.	
c	Nonlinear electromechanical properties: Fig. 1513.	
8a	Elastic compliances, stiffnesses: Table 390; Fig. 1514.	
9a	Refractive indices: $n_\alpha = n_c = 1.4953$, $n_\beta = n_a = 1.4985$, $n_\gamma = n_b = 1.4496$, $2V = 119°20'$ at 20 °C, for NaD line.	1894 L 1
d	Rotatory power: $\varrho = 1.55\,(15) \cdot 10^3\ °\ m^{-1}$ for $\lambda = 589.3$ nm.	30 a 1
13a	NMR: Tables 391, 392; Figs. 1515, 1516.	
b	ESR: Table 393; see also	74 S 9
14a	Bragg reflections due to structural modulation: Fig. 1517.	

Sub-table for 1b:

phase	II	I
state	(F)	P
crystal system	monoclinic	orthorhombic
space group	$P2_1\text{-}C_2^2$	$P2_12_12\text{-}D_2^3$
Θ	109 K	

Table 390. $NaNH_4C_4O_6 \cdot 4H_2O$. Elastic constants [66 b 1].

s_{11}	s_{22}	s_{33}	s_{12}	s_{13}	s_{23}	s_{44}	s_{55}	s_{66}	T	Ref.
$10^{-12}\ m^2\ N^{-1}$									°C	
53.7	38.4	37.3	− 8.7	−34.3	− 5.0	87.4	360	118		28 M 1
57.0	38.5	40	−15.5	−22	−15.5	94.5	330	115	25	50 m 1

c_{11}	c_{22}	c_{33}	c_{12}	c_{13}	c_{23}	c_{44}	c_{55}	c_{66}		Ref.
$10^9\ N\ m^{-2}$										
53.1	34.1	77.8	18.7	51.3	21.6	11.8	2.9	8.8		28 M 1

Table 389. $NaNH_4C_4H_4O_6 \cdot 4H_2O$. Piezoelectric constants [66b1].

d_{14}	d_{25}	d_{36}	e_{14}	e_{25}	e_{36}	T	Ref.
10^{-12} C N^{-1}			C m^{-2}			°C	
18.7	−49.8	9.4					28M1
±19.0	±31.7	±10.3				25	50m1
			0.212	0.137	−0.080	20	28M1

Table 391. $NaND_4C_4H_2D_2O_6 \cdot 4D_2O$. Eigenvalues (in kHz) and transformation matrices of ND_4^+ deuteron electric field gradient tensors [76E4].

		ϕ_{xx}(1)	ϕ_{yy}(2)	ϕ_{zz}(3)
297 K		− 2.5	− 2.8	5.3
(Volkoff's method)	1	90	90	0
	2	108	18	90
	3	18	72	90
297 K		− 2.5	− 2.9	5.4
(El Saffar's method)	1	90	90	0
	2	107	17	90
	3	17	73	90
224 K		− 2.7	− 3.2	5.9
173 K		− 2.8	− 3.3	6.1
122 K		Ferroelectric transition		
113 K		− 1.2	− 3.4	4.7
	1	109	36	119
	2	119	72	34
	3	35	60	74
93 K		− 1.3	− 3.6	4.9
77 K		− 0.7	− 3.6	4.3
	1	101	11	89
	2	110	94	21
	3	23	81	70

Table 392. $NaND_4C_4H_2D_2O_6 \cdot 4D_2O$. Eigenvalues (in kHz) and transformation matrices of the deuteron electric field gradient tensors for D_2O [76E4].

Site		ϕ_{xx}(1)	ϕ_{yy}(2)	ϕ_{zz}(3)
α		− 7	−104	111
	1	− 0.279	0.958	− 0.065
	2	0.900	0.286	0.329
	3	− 0.337	− 0.035	0.941
β		−35	−100	135
	1	0.074	0.956	− 0.284
	2	0.698	0.153	0.700
	3	0.712	− 0.250	− 0.656
γ		− 2	−108	110
	1	0.761	− 0.558	0.330
	2	0.312	0.764	0.564
	3	− 0.567	− 0.328	0.755
δ		− 9	−105	114
	1	− 0.219	0.168	0.961
	2	− 0.155	0.984	− 0.092
	3	0.945	0.184	0.271

Table 393. $NaNH_4C_4H_4O_6 \cdot 4H_2O$. Spin Hamiltonian parameters of Cu^{2+} at RT [72S34]. A_1, A_2, A_3: in units of 10^{-2} m^{-1}.

	g_1	g_2	g_3	A_1	A_2	A_3
ARS	2.064	2.095	2.316	58.7	28.3	128

Nr. 41A-3 $NaRbC_4H_4O_6 \cdot 4H_2O$, Sodium rubidium tartrate tetrahydrate
(M = 328.5904)

1a	Dielectric properties were studied by Jona et al. in 1953 and no anomalies were found.	53J1
b	Crystal system: orthorhombic. Space group: $P2_12_12$-D_2^3.	
3a	Unit cell parameters: a = 12.05 Å, b = 14.40 Å, c = 6.21 Å at RT.	41B2
5a	Dielectric constants (f = 10 kHz):	53J1

T[K]	4.2	88	298	333
$\varkappa_a$	6.8	7.3	9.7	11.2
$\varkappa_b$	5.8	6.0	7.6	8.4
$\varkappa_c$	5.3	5.6	8.3	13.7

41B Solid solution

Nr. 41B-1 $NaKC_4H_4O_6 \cdot 4H_2O$-$NaNH_4C_4H_4O_6 \cdot 4H_2O$

1a	This solid solution was first studied by Kurtschatow et al. in 1932.	32K1
b	Phase diagram: Fig. 1518. $NaK_{1-x}(NH_4)_xC_4H_4O_6 \cdot 4H_2O$ is classified into four groups as follows: (see the phase diagram)	58M1

group	range of x	state (phase) low T→high T		
I	0···0.03	P	F(I)	P
II	0.03···0.18	P	II*)	P
III	0.18···0.88	F(III)		P
(boundary between III and IV)	0.88···0.93	(F)**) (IV)	F(III)	P
IV	0.93···1.00	(F)**)(IV)		P

*) At the temperature corresponding to II, the broad dielectric peak appears.
**) P_s can be reversed only by the mechanical stress.

P_s||[100] in F(I), ||[100] in F(III), ||[010] in F(IV).

2a	Crystal growth: cooling or evaporation method from aqueous solution, Fig. 1519.	
5a	Dielectric constants: Figs. 1520, 1521, 1522.	
d	Pyroelectricity: Figs. 1523, 1524.	
7a	Piezoelectricity: Fig. 1525.	
8a	Elastic properties: Figs. 1526, 1527.	
9a	Refractive indices: Fig. 1528.	
13a	NMR: Table 394; Fig. 1529.	

Table 394. $NaK_{1-x}(NH_4)_xC_4H_4O_6 \cdot 4H_2O$. Minimum proton spin-lattice relaxation time $(T_1)_{min}$ [74M15].

x	Rotational axis	$(T_1)_{min}$ 10^{-3} s
0.10	Threefold	260
	Twofold	280
0.20	Threefold	160
	Twofold	180
0.40	Threefold	75
	Twofold	74
0.60	Threefold	46
	Twofold	47
0.87	Threefold	24
	Twofold	60, 60
0.91	Threefold	23
	Twofold	54, 55
1.00	Threefold	18
	Twofold	40, 40

42 $LiNH_4C_4H_4O_6 \cdot H_2O$ family

42A Pure compounds

Nr. 42A-1 $LiNH_4C_4H_4O_6 \cdot H_2O$, Lithium ammonium tartrate monohydrate (LAT) ($M = 191.07$)

1a Ferroelectricity in $LiNH_4C_4H_4O_6 \cdot H_2O$ was discovered independently by Matthias et al. and Merz in 1951. — 51M1, 51M3

b

phase	II [a, b]	I [a, b]
state	F [a, b]	P [a, b]
crystal system	monoclinic [c, d]	orthorhombic [c]
space group	P12₁1-C_2^2 [d]	P2₁2₁2-D_2^3 [c]
Θ	106 [b] K	

[a]) 51M1
[b]) 51M3
[c]) 53V1
[d]) 62j1

$P_s \| [010]$. — 51M1
Transparent, colorless.

3a Unit cell parameters: $a = 7.884$ (6) Å, $b = 14.565$ (8) Å, $c = 6.409$ (4) Å at RT. — 72H10

b Structure of phase I at RT. $Z = 4$. — 53V1, 72H10

Fractional coordinates of atoms in the unit cell: Table 395.
Temperature parameters: Table 396.
Fractional coordinates of hydrogen atoms and their temperature parameters: Table 397.
Interatomic distances and angles: Table 398.
Schematic drawing of the tartrate molecule: Fig. 1530.
The structure projected onto (001) and (100): Figs. 1531, 1532.

5a Dielectric constant: Figs. 1533, 1534.

c Spontaneous polarization and coercive field, $P_s = 0.22 \cdot 10^{-2}$ C m^{-2} at 78 K, $E_c \approx 1 \cdot 10^{-6}$ V m^{-1} at 78 K. — 51M1, 51M3

6a Heat capacity: Fig. 1535.
Transition heat and entropy: $\Delta Q_m = 175$ J mol^{-1}, $\Delta S_m = 1.96$ J K^{-1} mol^{-1}. — 73M3

7a Piezoelectric constant $d_{i\lambda}$: Fig. 1536.

8a Elastic compliance $s_{\lambda\mu}$: Figs. 1537, 1538.
Elastic constant $c_{\lambda\mu}$ obtained from Brillouin scattering: Fig. 1539.
Additional information, see — 50m1

continued

Nr. 42A-1 $LiNH_4C_4H_4O_6 \cdot H_2O$, continued

13a	NMR: ^7Li quadrupole coupling tensor, asymmetry parameter, Table 399. Deuteron quadrupole coupling tensor in deuterated crystal: Tables 400, 401. Additional information, see	72E2
b	ESR: Mn^{2+} doped, Table 402. Cr^{3+} doped, Table 403. Additional information, see	75M2

Table 395. $LiNH_4C_4H_4O_6 \cdot H_2O$ (LAT). Structure of phase I [72H10]. Fractional coordinates at RT. O(W): oxygen of the water molecule, see Fig. 1531.

	x	y	z
O (1)	0.2176 (2)	0.1260 (1)	0.1878 (2)
O (2)	0.3134 (2)	0.2113 (1)	−0.0762 (2)
O (3)	0.3736 (2)	0.4384 (1)	0.6018 (3)
O (4)	0.1218 (2)	0.3692 (1)	0.6210 (3)
O (5)	0.2355 (2)	0.3662 (1)	0.0947 (2)
O (6)	0.5046 (2)	0.3163 (1)	0.3564 (2)
O (W)	0.3265 (2)	0.0731 (1)	0.6483 (3)
N (1)	0.0000	0.0000	−0.0534 (5)
N (2)	0.0000	0.5000	−0.0842 (5)
C (1)	0.2527 (2)	0.2011 (1)	0.1027 (3)
C (2)	0.2127 (2)	0.2885 (1)	0.2255 (2)
C (3)	0.3342 (2)	0.2984 (1)	0.4147 (2)
C (4)	0.2717 (2)	0.3756 (1)	0.5560 (3)
Li	0.1266 (5)	0.0780 (2)	0.4501 (6)

Table 396. $LiNH_4C_4H_4O_6 \cdot H_2O$ (LAT). Structure of phase I [72H10]. Temperature parameters at RT. b_{ij} is defined by Eq. (b) in Introduction.

	b_{11}	b_{22}	b_{33}	b_{12}	b_{13}	b_{23}
	10^{-4}					
O (1)	89 (2)	14 (1)	105 (3)	−6 (1)	13 (2)	− 2 (1)
O (2)	129 (3)	23 (1)	86 (3)	−9 (1)	40 (2)	− 9 (1)
O (3)	61 (2)	14 (1)	182 (4)	−3 (1)	−12 (2)	−19 (1)
O (4)	67 (2)	29 (1)	147 (3)	−8 (1)	45 (2)	−27 (1)
O (5)	116 (2)	14 (1)	93 (3)	3 (1)	−28 (2)	6 (1)
O (6)	36 (1)	28 (1)	135 (3)	3 (1)	5 (2)	−11 (1)
O (W)	93 (2)	24 (1)	148 (4)	2 (1)	−13 (3)	−15 (1)
N (1)	93 (4)	26 (1)	132 (6)	6 (2)	0	0
N (2)	82 (3)	21 (1)	162 (6)	7 (2)	0	0
C (1)	49 (2)	14 (1)	72 (3)	−3 (1)	0 (2)	− 6 (1)
C (2)	46 (2)	12 (1)	72 (3)	2 (1)	− 5 (2)	− 3 (1)
C (3)	43 (2)	11 (1)	69 (3)	1 (1)	1 (2)	− 1 (1)
C (4)	54 (2)	12 (1)	78 (3)	0 (1)	− 1 (2)	− 6 (1)
Li	68 (5)	19 (1)	112 (7)	−5 (2)	6 (5)	6 (3)

Table 402. $LiNH_4C_4H_4O_6 \cdot H_2O$: Mn^{2+}. Spin Hamiltonian parameter of Mn^{2+} in the crystal [76N4]. D, E, A, a: in units of $10^{-2}\,m^{-1}$. Groups A and B refer to the two chemically inequivalent Mn^{2+} complexes, respectively.

Group	$g_{\parallel}$	$g_{\perp}$	D	E	A	a
Group A	2.008	1.95	552	−92.8	−80.8	7
Group B	2.008	1.95	565	−98.2	−80.8	6

Table 397. $LiNH_4C_4H_4O_6 \cdot H_2O$ (LAT). Structure of phase I [72H10]. Fractional coordinates of hydrogen atoms and their isotropic temperature factors at RT. Temperature factor B is defined by Eq. (e) in Introduction.

	x	y	z	B $Å^2$	Bonded to
H (1)	0.059 (4)	−0.034 (2)	−0.139 (5)	2.5 (7)	N (1)
H (2)	0.075 (4)	0.037 (2)	0.027 (6)	3.2 (9)	N (1)
H (3)	0.066 (6)	0.462 (3)	−0.032 (7)	6.6 (16)	N (2)
H (4)	0.078 (7)	0.530 (4)	−0.190 (8)	9.9 (18)	N (2)
H (5)	0.320 (5)	0.112 (3)	0.746 (6)	6.2 (10)	O (W)
H (6)	0.409 (5)	0.074 (2)	0.581 (6)	6.2 (7)	O (W)
H (7)	0.284 (4)	0.342 (2)	−0.013 (6)	4.3 (9)	O (5)
H (8)	0.552 (4)	0.259 (2)	0.346 (4)	2.4 (6)	O (6)
H (9)	0.095 (3)	0.288 (2)	0.262 (4)	1.2 (4)	C (2)
H (10)	0.327 (3)	0.238 (2)	0.488 (4)	1.2 (5)	C (3)

Table 398. $LiNH_4C_4H_4O_6 \cdot H_2O$ (LAT). Structure of phase I [72H10]. Interatomic distances and angles related to hydrogen atoms at RT.

Distance	[Å]	Angle	[deg]
N (1)–H (1)	0.90	C (2)–O (5)–H (7)	102
N (1)–H (2)	0.91	C (3)–O (6)–H (8)	105
N (2)–H (3)	0.83		
N (2)–H (4)	1.01	C (1)–C (2)–H (9)	107
O (W)–H (5)	0.87	C (3)–C (2)–H (9)	113
O (W)–H (6)	0.87	O (5)–C (2)–H (9)	104
O (5)–H (7)	0.88	C (2)–C (3)–H (10)	105
O (6)–H (8)	0.89	C (4)–C (3)–H (10)	110
C (2)–H (9)	0.99	O (6)–C (3)–H (10)	111
C (3)–H (10)	0.99	⟨e.s.d.⟩ =	1.6
⟨e.s.d.⟩ =	0.04		

Table 399. $LiNH_4C_4H_4O_6 \cdot H_2O$ (LAT). ^7Li quadrupole coupling tensor, asymmetry parameter η, and orientation of the principal axes (1, 2, 3) of the electric field gradient tensor with respect to the crystal axes (a, b, c) at 300 K [72E2].

	$eq\phi_{xx}/h$	$eq\phi_{yy}/h$	$eq\phi_{zz}/h$
Eigenvalues [kHz]	−33.0	−44.6	77.6
	a	b	c
1	0.135	0.797	−0.604
2	0.267	0.579	0.770
3	0.964	−0.172	−0.205

$e^2qQ/h = 77.6$ kHz $\eta = 0.150$

Table 400. $LiND_4C_4H_4O_6 \cdot D_2O$. Eigenvalues and transformation matrices of deuteron quadrupole coupling tensor of D_2O in $LiND_4C_4H_4O_6 \cdot D_2O$ at 330 K. (1, 2, 3) are the principal axes of the tensor and (a, b, c) are the crystal axes [72E2].

	$eq\phi_{xx}/h$	$eq\phi_{yy}/h$	$eq\phi_{zz}/h$
Eigenvalues [kHz]	−14	−102	116
	a	b	c
1	−0.678	0.702	0.217
2	−0.395	−0.597	0.699
3	0.620	0.388	0.682

$e^2qQ/h = 116$ kHz $\eta = 0.761$

Table 401. $LiND_4C_4H_4O_6 \cdot D_2O$. Eigenvalues and transformation matrices of deuteron quadrupole coupling tensor of ND_4^+ in the crystal at 111 K and 77 K. (1, 2, 3) are the principal axes of the tensor and (a, b, c) are the crystal axes [72E2].

T = 111 K	$eq\phi_{xx}/h$	$eq\phi_{yy}/h$	$eq\phi_{zz}/h$	T = 77 K	$eq\phi_{xx}/h$	$eq\phi_{yy}/h$	$eq\phi_{zz}/h$
Eigenvalues [kHz]	−2.3	−3.3	5.6	Eigenvalues [kHz]	−0.5	−3.9	4.4
	a	b	c		a	b	c
1	0.09	−0.07	0.99	1	0.69	−0.43	0.58
2	0.71	−0.69	−0.11	2	−0.56	0.18	0.81
3	0.69	0.71	−0.03	3	0.45	0.88	0.12

Table 403. $LiNH_4C_4H_4O_6 \cdot H_2O$: Cr^{3+}. The principal values of D-tensor and their direction cosines with respect to the a, b, c axes for A-sites and B-sites [75M3]. D: Fine structure parameter.

	The principal value		l	m	n
A-site	D_{zz}	$-25.59 \cdot 10^{-1}\ m^{-1}$	0.238	−0.283	0.929
	D_{yy}	$15.80 \cdot 10^{-1}\ m^{-1}$	−0.021	0.955	0.296
	D_{xx}	$9.79 \cdot 10^{-1}\ m^{-1}$	0.971	0.091	−0.221
B-site	D_{zz}	$31.69 \cdot 10^{-1}\ m^{-1}$	0.868	−0.135	0.479
	D_{yy}	$-21.81 \cdot 10^{-1}\ m^{-1}$	0.242	0.956	−0.168
	D_{xx}	$-\ 9.86 \cdot 10^{-1}\ m^{-1}$	−0.435	0.262	0.862

Nr. 42A-2 $LiTlC_4H_4O_6 \cdot H_2O$, Lithium thallium tartrate monohydrate (LTT)
($M = 377.40$)

1a	Ferroelectricity in $LiTlC_4H_4O_6 \cdot H_2O$ was discovered by Matthias et al. in 1951.	51M1
b	phase: II [a] / I [a]; state: F [a] / P [a]; crystal system: — / orthorhombic [b]; space group: — / $P2_12_12\text{-}D_2^3$ [b]; Θ: 10 (1) K [a]	[a] 51M1 [b] 62j1
	$P_s \parallel [100]$. $\varrho_X = 3.38 \cdot 10^3\ kg\ m^{-3}$. Transparent, colorless	51M1 71M3
2a	Crystals can be grown by cooling a saturated aqueous solution with seed crystals. Solubility coefficient: 0.5 g °C^{-1} per 100 g solvent in 10···50 °C range.	70B16
b	Crystal form: Fig. 1540.	
3a	Unit cell parameters: $a = 7.894$ (3) Å, $b = 14.658$ (6) Å, $c = 6.401$ (3) Å at 20 °C (phase I). Deuterated crystal (more than 80% of H is replaced by D), $a = 7.91$ (1) Å, $b = 14.69$ (6) Å, $c = 6.43$ (4) Å.	71M3 70B16
b	$Z = 4$ in phase I. Fractional coordinate of atoms in the unit cell and temperature parameters: Table 404. Interatomic distances and bond angles: Table 405. The structure projected onto (001): Fig. 1541.	55S2, 71M3
5a	Dielectric constant: Figs. 1542, 1543.	
b	Curie-Weiss law: $\varkappa_a = \varkappa_\infty + C/(T - \Theta_p)$ ($T > \approx 15$ K) with $\varkappa_\infty = 9$, $C = 1330$ K, $\Theta_p = 11$ K for free crystal (low frequency, $\varkappa_a^T$): and $\varkappa_\infty = 9$, $C = 1500$ K, $\Theta_p = -50$ K for clamped crystal (high frequency, $\varkappa_a^S$). $\varkappa_a^{-1}$ vs. T for normal and deuterated crystals: Fig. 1544. $\varkappa'_a$ and $\varkappa''_a$ at various measuring field strengths: Figs. 1545, 1546. $\varkappa_a$ vs. measuring field (1 kHz, $10^{-1} \cdots 10^4\ V\ m^{-1}$) at 4.2 K; see Effect of dc bias field on $\varkappa_a$: Fig. 1547. $\varkappa_a^{-1}$ vs. T at various hydrostatic pressures: Fig. 1548. Θ_p vs. hydrostatic pressure: $(d\Theta_p/dp)_{p=0} = 11.3 \cdot 10^{-8}\ K\ Pa^{-1}$: Fig. 1549.	72S7 71S8 70H9
c	Polarization data: Fig. 1550; see also	51M1, 70F4, 72S7
7a	Piezoelectric strain constant d_{14}: Fig. 1551. Electromechanical coupling factor: Fig. 1552.	
8a	Elastic compliances: Fig. 1553. Effects of dc field on the elastic compliances: Fig. 1554.	
13a	NMR data, see	62B11

Table 404. $LiTlC_4H_4O_6 \cdot H_2O$ (LTT). Structure of phase I [78 K 9]. (a) Fractional coordinates and (b) temperature parameters b_{ij}. The b_{ij} is defined by Eq. (b) in Introduction. See also Fig. 1541.

Atom		(a) Fractional coordinates			(b) Temperature parameters					
		X	Y	Z	b_{11}	b_{22}	b_{33}	b_{12}	b_{13}	b_{23}
Carbon	1	0.2551 (2)	0.2010 (1)	0.0998 (3)	75 (4)	14.1 (10)	139 (5)	− 3 (1)	8 (4)	− 5 (2)
	2	0.2188 (3)	0.2887 (1)	0.2215 (3)	52 (4)	17.3 (10)	141 (5)	1 (2)	− 5 (4)	− 9 (2)
	3	0.3342 (2)	0.2994 (1)	0.4124 (3)	54 (3)	16.9 (10)	116 (5)	− 2 (1)	1 (3)	− 9 (2)
	4	0.2724 (3)	0.3768 (1)	0.5549 (3)	61 (4)	16.9 (10)	142 (6)	0 (2)	3 (4)	− 9 (2)
Oxygen	1	0.2214 (4)	0.1265 (2)	0.1896 (4)	127 (5)	15.5 (11)	152 (6)	0 (2)	11 (5)	− 4 (2)
	2	0.3160 (4)	0.2102 (2)	−0.0766 (4)	170 (6)	23.8 (12)	139 (7)	−14 (2)	50 (5)	− 8 (2)
	3	0.3744 (3)	0.4375 (2)	0.6066 (4)	64 (4)	16.8 (10)	235 (8)	− 4 (2)	−39 (5)	−24 (2)
	4	0.1212 (3)	0.3709 (2)	0.6141 (5)	66 (5)	41.3 (14)	226 (7)	− 9 (2)	34 (6)	−40 (3)
	5	0.2379 (4)	0.3645 (2)	0.0878 (5)	152 (6)	15.9 (12)	166 (9)	6 (2)	−31 (5)	4 (3)
	6	0.5054 (3)	0.3159 (2)	0.3559 (4)	38 (4)	39.1 (15)	205 (8)	6 (2)	−15 (6)	−20 (3)
	W [a]	0.3253 (5)	0.0737 (2)	0.6428 (5)	136 (7)	24.2 (13)	153 (6)	15 (3)	−43 (6)	− 4 (3)
Hydrogen	W 1	0.3277 (6)	0.1221 (4)	0.7498 (8)	176 (11)	34.7 (22)	153 (12)	13 (4)	8 (9)	−14 (5)
	W 2	0.4307 (8)	0.0809 (4)	0.5751 (12)	121 (9)	65.4 (39)	354 (20)	10 (5)	−13 (13)	−35 (8)
	1	0.2665 (10)	0.3393 (4)	−0.0455 (9)	431 (19)	33.4 (24)	148 (14)	−24 (6)	52 (15)	− 2 (5)
	2	0.5583 (7)	0.2607 (5)	0.3283 (10)	94 (9)	61.7 (40)	384 (20)	11 (5)	−15 (9)	−87 (8)
	3	0.0875 (6)	0.2836 (3)	0.2761 (7)	86 (9)	50.8 (25)	201 (12)	7 (4)	13 (9)	−29 (5)
	4	0.3258 (6)	0.2366 (3)	0.4997 (8)	160 (8)	24.5 (21)	214 (12)	0 (3)	−16 (9)	2 (4)
Lithium		0.1220 (9)	0.0803 (5)	0.4437 (11)	54 (11)	46.9 (40)	166 (18)	−20 (6)	− 6 (12)	10 (8)
Thallium	1	0.0	0.0	0.9521 (5)	123 (4)	43.7 (13)	200 (7)	3 (2)	—	—
	2	0.0	0.5	0.9109 (4)	90 (4)	31.9 (11)	181 (7)	− 5 (2)	—	—

[a]) W = Water.

Table 405. $LiTlC_4H_4O_6 \cdot H_2O$ (LTT). Structure of phase I [78 K 9]. Interatomic distances (in Å) and angles (in deg). Values corrected for "riding motion" are presented in square brackets. Primes are attached to atoms to remark that the atoms are symmetry related to those without primes.

1. Tartrate ion

Atoms	Distance or angle
a) Carbon skeleton	
C 1–C 2	1.528 (6)
C 2–C 3	1.531 (6)
C 3–C 4	1.534 (5)
C 1–C 2–C 3	112.4 (2)
C 2–C 3–C 4	111.2 (3)
b) Carboxyl groups	
O 1–C 1	1.260 (6)
O 2–C 1	1.234 (6)
O 1–C 1–O 2	126.4 (3)
O 1–C 1–C 2	117.0 (4)
O 2–C 1–C 2	116.6 (3)
O 3–C 4	1.242 (5)
O 4–C 4	1.253 (7)
O 3–C 4–O 4	125.7 (3)
O 3–C 4–C 3	118.7 (3)
O 4–C 4–C 3	115.5 (2)
c) Hydroxyl groups and methyl hydrogen atoms	
C 2–O 5	1.409 (6)
O 5–H 1	0.956 [1.012] (9)
C 2–H 3	1.095 [1.118] (8)
H 1–O 5–C 2	105.3 (5)
O 5–C 2–H 3	110.4 (3)
O 5–C 2–C 1	109.3 (4)
H 3–C 2–C 1	106.3 (3)
C 3–O 6	1.418 (7)
O 6–H 2	0.925 [0.962] (10)
C 3–H 4	1.076 [1.102] (7)
H 2–O 6–C 3	109.2 (5)
O 6–C 3–H 4	109.6 (3)
O 6–C 3–C 4	109.1 (3)
H 4–C 3–C 4	107.6 (4)
O 6–C 3–C 2	112.3 (3)
O 5–C 2–C 3	109.9 (3)
H 4–C 3–C 2	106.9 (3)
H 3–C 2–C 3	108.3 (4)

2. Lithium coordination (distorted quadrilateral pyramid)

Li–O 1	1.928 (10)
Li–O 3	1.995 (11)
Li–O 3′	2.114 (12)
Li–OW	2.049 (11)
Li–O 6	2.190 (10)
O 1–Li–O 3	107.9 (4)
O 1–Li–O 3′	102.3 (4)
O 1–Li–OW	102.9 (5)
O 1–Li–O (6)	114.9 (5)
O 3–Li–O 3′	82.0 (2)
O 3–Li–O 6	76.9 (3)
O 3′–Li–OW	92.2 (3)
OW–Li–O 6	89.8 (4)
O 3′–Li–O 6	141.3 (4)
O 3–Li–OW	149.2 (4)
O 3–O 3′	2.696 (10)
O 3′–OW	2.999 (9)
OW–O 6	2.994 (12)
O 6–O 3	2.608 (9)
O 3–O 3′–OW	86.2 (4)
O 3′–OW–O 6	85.3 (3)
OW–O 6–O 3	87.9 (3)
O 6–O 3–O 3′	99.9 (3)
	$\sum = 359.3$

3. Thallium coordination

Thallium is on a two-fold axis and all oxygen atoms listed have related atoms equidistant from Tl (OJ and OJ′)

a) Thallium 1

Tl 1–O 5	2.874 (10)
Tl 1–O 1	2.963 (9)
Tl 1–O 3	3.130 (13)
Tl 1–O 6	3.338 (12)
Tl 1–OW	3.413 (12)
O 5–Tl 1–O 5′	169.8 (2)
O 1–Tl 1–O 1′	118.3 (3)
O 3–Tl 1–O 3	51.0 (3)
O 6–Tl 1–O 6	107.6 (4)

b) Thallium 2

Tl 2–O 4	2.845 (9)
Tl 2–O 1	2.943 (10)
Tl 2–O 5	2.954 (9)
Tl 2–OW	3.349 (12)
O 4–Tl 2–O 4	96.2 (4)
O 1–Tl 2–O 1	154.8 (2)
O 5–Tl 2–O 5	134.9 (3)
OW–Tl 2–OW′	62.9 (3)

Table 405, continued

OW–Tl 1–OW	109.1 (4)	O 4–Tl 2–O 1	71.7 (3)
O 5–Tl 1–O 1	92.2 (4)	O 1′	91.2 (3)
O 1	93.1 (4)	O 5	66.2 (3)
O 3	69.6 (2)	O 5′	156.2 (1)
O 3′	120.6 (1)	OW	119.6 (3)
O 6	53.1 (2)	OW′	130.2 (2)
O 6′	119.7 (2)	O 1–Tl 2–O 1	97.8 (4)
OW	105.7 (3)	O 5′	91.9 (4)
OW′	68.1 (3)	OW	71.2 (2)
O 1–Tl 1–O 3	62.7 (3)	OW′	134.1 (1)
O 3′	62.2 (2)	O 4–Tl 2–O 1	73.6 (3)
O 6	78.0 (3)	OW′	68.2 (2)
O 6′	144.5 (2)		
OW	70.0 (3)		
OW′	159.5 (1)		
O 3–Tl 1–O 6	107.5 (3)		
O 6′	139.9 (1)		
OW	132.0 (2)		
OW′	112.2 (3)		
O 6–Tl 1–OW	52.6 (2)		
OW′	85.5 (2)		

4. Hydrogen Bonding and water

Bond	Donor H	H-acceptor	Donor-acceptor	O–H–O angle
OW–HW 1···O 2	0.985 [0.997] (9)	1.705 (8)	2.687 (9)	174.7 (5)
OW–HW 2···O 4	0.943 [0.986] (11)	2.053 (10)	2.967 (11)	162.5 (5)
O 5–H 1···O 2	0.956 [1.012] (9)	1.939 (10)	2.566 (10)	121.0 (4) (internal bond)
O 6–H 2···O 4	0.925 [0.962] (10)	2.022 (12)	2.887 (13)	155.0 (6)
HW 1–OW–HW 2		102.8 (5)		

Miscellanea

43 Miscellaneous crystals (including oxides) and polymers

Nr. 43-1 TiO_2, Titanium dioxide (Rutile)
(M=79.90)

1a	Ferroelectric activity in polycrystalline TiO_2 (rutile) was reported by Nicolini in 1959. According to Jona and Shirane, however, the exact chemical composition of his sample is open to doubt. Single crystals of rutile do not show ferroelectric properties. A review on properties of rutile was made by Grant.	59N1 62j1 59g1

Nr. 43-2 KIO_3, Potassium iodate
(M=214.001)

1a Ferroelectric behavior in KIO_3 was reported by Herlach in 1961. — 61H3

b

phase	V[a])	IV[a])	III[a])	II[a])	I[a])
state		F[a])	F[a])	F[a])	P[a])
crystal system			triclinic[b])	monoclinic[c])	trigonal[d])
space group			P1-C_1^1 [b])	Cm-C_s^3 [c])	R3m-C_{3v}^5 [d])
Θ	−190	−18, −10[a, *])	70	212 °C	

[a]) 61H3
[b]) 70H10
[c]) 72C7, 73H2
[d]) 72C7, 75C8

*) −10 °C on heating, −18 °C on cooling.

continued

Nr. 43-2 KIO_3, continued

1b	$T_{melt}=560$ °C. Transparent, colorless.	70H10
2a	Crystal growth: hydrothermal method[a]), temperature-gradient method[b]), combined method of convection and evaporation[c]). Solubility in H_2O: Fig. 1555.	[a]) 61H3, 70H10 [b]) 62K4, 70M1 [c]) 74H3
3a	Unit cell parameters in phase I, II, and III: Phase I: $a=6.404$ (1) Å, $\alpha=58.33$ (1)° at 215 °C. Phase II: $a=12.797$ (2) Å, $b=12.644$ (2) Å, $c=7.705$ (1) Å, $\beta=144.63$ (1)° at 161 °C. Phase III: $a=8.923$ (5) Å, $b=8.942$ (7) Å, $c=7.709$ (6) Å, $\alpha=54.4$ (7)°, $\beta=125.3$ (3)°, $\gamma=90.6$ (1)° at 22 °C. Crystallographic axes $\boldsymbol{a}$, $\boldsymbol{b}$, $\boldsymbol{c}$ in these phases are related with the edge vectors $\boldsymbol{a}_c$, $\boldsymbol{b}_c$, $\boldsymbol{c}_c$ of the pseudocubic cell containing one molecular unit as follows: In phase I: $a=a_c+c_c$, $b=a_c+b_c$, $c=b_c+c_c$. In phase II: $a=2a_c-2b_c$, $b=2a_c+2b_c$, $c=-a_c+b_c+c_c$. In phase III: $a=-2b_c$, $b=2a_c$, $c=-a_c+b_c+c_c$.	73H2
b	$Z=2$ in phase I; $Z=8$ in phase II; $Z=4$ in phase III. Crystal structure in phase III: Table 406.	73H2
4	Lattice deformation in phase I: see	61H3
5a	Dielectric constant: Figs. 1556···1558.	
b	Effect of E_{bias} on $\varkappa$: Fig. 1559.	
c	Spontaneous polarization: Figs. 1560, 1561.	
d	Pyroelectricity: $dP_s/dT=5.85\cdot10^{-5}$ C m^{-2} K^{-1} at RT, along the ⟨111⟩ direction of the pseudocubic cell.	72C7
7a	Piezoelectricity: see	72S1
9a	Refractive indices. $n_X=1.700$, $n_Y=1.828$, $n_Z=1.832$ for $\lambda=589.3$ nm. The X axis is the polar axis. The orientations of Y and Z axes are not given in the original paper. Optical axial angle: Fig. 1562. Birefringence: Fig. 1563. Optical absorption edge: see Infrared spectra: see	 64w1 67V5 75B2
b	Electrooptic effect: Fig. 1564; see also	72S1
10a	Raman scattering: see	74S4
11	Electrical conductivity: Fig. 1565.	
13a	NQR: Table 407; see also	61H3, 75B2
c	Mössbauer effect: Figs. 1566, 1567.	

Table 406. KIO_3. Crystal structure in phase III [73H2]. Fractional coordinates.

	x	y	z
K (1)	0.023 (4)	0.004 (5)	0.020 (7)
K (2)	0.008 (6)	0.495 (6)	0.009 (7)
K (3)	0.484 (3)	−0.011 (5)	0.012 (3)
K (4)	0.494 (2)	0.495 (3)	−0.010 (2)
I (5)	0.506 (5)	0.504 (2)	0.494 (3)
I (6)	0.498 (2)	0.007 (2)	0.499 (2)
I (7)	0.010 (2)	0.492 (3)	0.510 (1)
I (8)	0.011 (2)	0.013 (1)	0.489 (1)
O (9)	0.266 (3)	0.267 (3)	0.020 (2)
O (10)	0.779 (9)	0.211 (9)	0.026 (13)
O (11)	0.733 (2)	0.741 (5)	−0.020 (4)
O (12)	0.231 (1)	0.768 (3)	0.018 (3)
O (13)	0.034 (9)	0.219 (10)	0.440 (10)
O (14)	0.001 (10)	0.730 (11)	0.475 (14)
O (15)	0.284 (10)	0.016 (10)	0.449 (10)
O (16)	0.764 (3)	0.018 (5)	0.462 (2)
O (17)	0.267 (2)	0.550 (4)	0.520 (3)
O (18)	0.757 (11)	0.532 (10)	0.554 (2)
O (19)	0.540 (3)	0.266 (6)	0.534 (4)
O (20)	0.433 (2)	0.756 (5)	0.564 (7)

Table 407. KIO_3. Nuclear quadrupole resonance frequencies ν, the quadrupole coupling constant e^2Qq_{zz}/h, and the asymmetric parameters η of ^{127}I [75B3]. n: number of lines observed in the NQR spectrum. Values for $T=298$ K are from [56L3].

T K	$\nu(\pm{}^1/_2\leftrightarrow\pm{}^3/_2)$ MHz	$\nu(\pm{}^3/_2\leftrightarrow\pm{}^5/_2)$ MHz	n	e^2Qq_{zz}/h MHz	η %
4.2	150.567		3		
	149.837				
	149.786				
20	150.548		3		
	149.827				
	149.772				
77	150.046	299.925	4	999.83	2.1
	149.943	299.475		998.45	3.3
	149.256	298.350		994.58	2.1
	149.188	298.200		994.10	1.4
116	149.422		3		
	149.106				
	148.513				
161	148.654	296.930	3	989.98	3.1
	148.190	296.019		986.92	3.1
	147.666	295.050		983.65	2.7
252	146.736		3		
	146.180				
	145.646				
298	145.38	290.71	3	969.0	
	144.91	289.85		966.2	
	144.37	288.82		962.7	
329	144.423		3		
	144.016				
	143.617				
344	142.964		1		
366	142.323		1		

Nr. 43-3 $KMnF_3$, Potassium manganese fluoride
($M=151.030$)

1a	Many studies have been done on the phase transitions in $KMnF_3$, the structural, magnetic and optical properties. Some of the phase transitions are due to the zone-boundary phonon condensation. Readers are referred to [70g3, 78n1] for general review on the properties of $KMnF_3$ and its related compounds. Readers may also consult [75m2], which contains some descriptions on the properties of $KMnF_3$.	

Nr. 43-4 NH_4IO_3, Ammonium iodate
($M=192.94$)

1a	Polarization reversal in NH_4IO_3 by poling was first reported by Oka et al. in 1976.	76O3
b	(see table below)	a) 76O3 b) 71K3 c) 75V7 d) 74S3

phase	IV	III	II	I
state		F a)		
crystal system		orthorhombic b)	orthorhombic c)	cubic c)
space group		$Pc2_1n\text{-}C_{2v}^9$ b)		
Θ	−170 d)	82 b*)	115 b)	°C

*) Values of Θ_{III-II} ranging from 82 to 95 °C were reported by different authors: see 69C6, 76O3

$P_s \parallel [010]$.
$\varrho=3.31\cdot10^3\,\mathrm{kg\,m^{-3}}$. — 69C6
Transparent, colorless.

2a	Crystal growth: evaporation of aqueous solution.	69C6
b	Crystal form: Fig. 1568.	
3a	Unit cell parameters: $a=6.4115\,(5)$ Å, $b=9.1706\,(5)$ Å, $c=6.3740\,(5)$ Å at 24 °C.	71K3
b	phase: III, II, I; Z: 4 a), 4 b), 1 b)	a) 71K3 b) 75V7
	Crystal structure of phase II: Fig. 1569; Tables 408···411	
4	Thermal expansion: temperature dependence of unit cell parameters, Fig. 1570; Table 412; linear thermal expansion coefficients, Fig. 1571.	
5a	Dielectric constant: Fig. 1572; see also	74S3
	Effect of hydrostatic pressure on Θ_{III-II}: $[d\Theta_{III-II}/dp]_{p=0}=9.5\cdot10^{-8}\,\mathrm{K\,Pa^{-1}}$.	79S3
c	Spontaneous polarization: Fig. 1573.	
d	Pyroelectric effect: see	69C6
6a	Transition heat of the III−II transition: $\Delta Q_m=3.35\,(4)\cdot10^3\,\mathrm{J\,mol^{-1}}$.	71B15
7a	Piezoelectricity: see	74S3
8a	Volume compressibility: Fig. 1574.	
9a	Refractive indices and optical axial angle: $n_a=1.777$, $n_b=1.785$, $n_c=1.826$, $2V=48.2°$ at RT.	75V7
b	Electrooptic effect: $r_{43}=4.9\cdot10^{-12}\,\mathrm{m\,V^{-1}}$, $r_{61}=1.0\cdot10^{-12}\,\mathrm{m\,V^{-1}}$, $r_{12}n_1^3-r_{32}n_3^3=59.6\cdot10^{-12}\,\mathrm{m\,V^{-1}}$, $r_{22}n_2^3-r_{32}n_3^3=116.8\cdot10^{-12}\,\mathrm{m\,V^{-1}}$.	77S3

10a	Raman scattering: Table 413.	
15	Domains: see	76O3

Table 408. NH_4IO_3. Fractional coordinates [71K3]. T=25 °C. A spherical model was taken for the NH_4-ion.

Atom	x	y	z
I	−0.01124 (8)	0	0.00895 (8)
O (1)	0.174 (2)	−0.004 (3)	0.222 (1)
O (2)	−0.240 (2)	0.033 (1)	0.156 (2)
O (3)	0.037 (2)	0.304 (1)	0.047 (2)
NH_4	0.002 (2)	0.262 (1)	0.503 (3)

Table 409. NH_4IO_3. Anisotropic temperature parameters [71K3]. T=25 °C. b_{ij} is defined by Eq. (b) in Introduction. An isotropic model was taken for the NH_4-ion.

	b_{11}	b_{22}	b_{33}	b_{12}	b_{13}	b_{23}
I	0.0071 (1)	0.00299 (4)	0.0073 (1)	0.0009 (2)	0.0020 (1)	−0.0005 (2)
O (1)	0.015 (2)	0.013 (2)	0.010 (2)	0.009 (3)	−0.006 (2)	−0.003 (3)
O (2)	0.013 (2)	0.008 (1)	0.011 (2)	0.003 (1)	0.009 (1)	−0.002 (1)
O (3)	0.022 (3)	0.005 (1)	0.015 (2)	−0.002 (1)	−0.003 (2)	−0.001 (1)
NH_4	0.012 (1)	0.006	0.012	0	0	0

Table 410. NH_4IO_3. Interatomic distances in Å [71K3]. T=25 °C.

I−O (1)	1.806 (8)	I−O (1)	2.830 (8)
I−O (2)	1.765 (8)	I−O (2)	2.778 (9)
I−O (3)	1.836 (12)	I−O (3)	2.819 (11)
O (1)−O (2)	2.71 (1)	N−O (2)	2.86 (2)
O (2)−O (3)	2.78 (2)	N−O (3)	2.94 (2)
O (1)−O (3)	2.81 (2)	N−O (3)	3.00 (2)
		N−O (1)	3.00 (2)

Table 411. NH_4IO_3. Bond angles [71K3]. T=25 °C. Primed atoms are symmetry related to corresponding unprimed atoms.

O (1)−I−O (2)	102.3°	O (1)−N−O (2)	96.2°
O (1)−I−O (3)	105.5°	O (3)−N−O (3)′	89.9°
O (2)−I−O (3)	105.6°	O (1)−N−O (3)	109.6°
		O (1)−N−O (3)′	92.8°
		O (2)−N−O (3)	129.4°
		O (2)−N−O (3)′	115.1°

Table 412. NH_4IO_3. Unit cell parameters at different temperature [75V7].

Phase	T [°C]	a [Å]	b [Å]	c [Å]	V [Å^3]
III	20	6.403 (2)	9.168 (4)	6.337 (2)	374.6 (4)
	80	6.413	9.156	6.411	376.5
II	83	6.426	9.104	6.466	378.3
I	120	4.556			

Table 413. NH_4IO_3. ν [74S3]. T=RT. ν: frequency of $\boldsymbol{q}=0$ optical modes obtained from Raman scattering.

ν [THz]	Polarization	ν [THz]	Polarization
1.60	—	8.99	B
2.40	B (TO)	9.79	A
3.10	A (TO)	11.10	A+B
3.99	A+B (TO)	21.89	A
4.80	B (TO)	22.18	B
5.25	(LO)	22.79	A+B
7.19	(LO)		

Nr. 43-5. $AgNbO_3$ group
Nr. 43-5-i. $AgNbO_3$, Silver niobate
($M = 248.773$)

1a	Dielectric anomaly in $AgNbO_3$, associated with a phase transition, was reported by Francombe et al. in 1958.	58F1
b	phase: III / II / I; crystal system: orthorhombic / tetragonal / cubic; Θ: 325, 550 °C	58F1
3a	$a = 11.212$ Å, $b = 15.660$ Å, $c = 11.101$ Å at 20 °C. The cubic cell in phase I becomes, in phase III, a pseudo-cubic monoclinic cell, whose lattice parameters a', b', c', and β are as follows: $a' = c' = 7.888$ Å, $b' = 15.660$ Å, $\beta = 90° 34'$ at 20 °C ($a = 2a' \sin \beta/2$, $b = b'$, $c = 2c' \cos \beta/2$)	58F1
4	Unit cell parameters: Fig. 1575.	
5a	Dielectric constant: Fig. 1576.	

phase	III	II	I
crystal system	orthorhombic	tetragonal	cubic
Θ	325	550 °C	

Nr. 43-5-ii. $AgTaO_3$, Silver tantalate

1a	Structure changes, associated with phase transitions, were observed in $AgTaO_3$ by Francombe et al. in 1958.	58F1
b	phase: III / II / I; crystal system: orthorhombic / tetragonal / cubic; Θ: 370, 485 °C	58F1
3a	$a = 5.577$ Å, $b = 3.914$ Å, $c = 5.542$ Å at 20 °C. The cubic cell in phase I becomes, in phase III, a pseudo-cubic monoclinic cell, whose lattice parameters a', b', c', and β are as follows: $a' = c' = 3.931$ Å, $b' = 3.914$ Å, $\beta = 90° 21'$ at 20 °C ($a = 2a' \sin \beta/2$, $b = b'$, $c = 2c' \cos \beta/2$).	
4	Unit cell parameters: Fig. 1577.	

phase	III	II	I
crystal system	orthorhombic	tetragonal	cubic
Θ	370	485 °C	

Nr. 43-6 Stibiotantalite group
Nr. 43-6-i Stibiotantalite Sb (Ta, Nb)O_4

1a	Ferroelectricity in mineral stibiotantalite was reported by Gavrilova et al. in 1970.	70G1
b	phase: II / I; state: (F) / P; crystal system: orthorhombic; space group: $Pna2_1$-C_{2v}^9; Θ: 400 °C	70G1
	Stibiotantalite is the solid solution Sb (Ta, Nb) O_4, but the composition is different sample by sample. $P_s \parallel [001]$ $\varrho = 7.28(17) \cdot 10^3$ kg m^{-3}	70G1
3a	Unit cell parameters: $a = 4.916$ Å, $b = 5.542$ Å, $c = 11.78$ Å (Based on this cell, the space group should be represented by $Pbn2_1$, referring to [76R1].) See also Table 414.	70G1

phase	II	I
state	(F)	P
crystal system	orthorhombic	
space group	$Pna2_1$-C_{2v}^9	
Θ	400 °C	

5a	Dielectric constant: Fig. 1578. $\varkappa_c = C/(T-\Theta)$, $C = 2.49\,(67) \cdot 10^5$ K, $\Theta_p = 400$ °C.	70G1
b	Nonlinear dielectric coefficient: $\xi = 9.3 \cdot 10^9$ Vm^5 C^{-3}.	70G1
c	Spontaneous polarization and coercive field: Fig. 1578.	
7a	Piezoelectricity: Fig. 1579.	
8a	Elastic constant: Fig. 1579.	
9a	Refractive index: $n = 2.4$	70G1
b	Electrooptic effect: $r_{33} = 26 \cdot 10^{-12}$ mV^{-1} and $r_1 = a_1 r_{13} + a_2 r_{23} = 13.5 \cdot 10^{-12}$ mV^{-1} for $\lambda = 633$ nm, where a_i seems to be given by $a_i = \frac{1}{2}(n_i/\bar{n})^3 \approx 1/2$, for i = 1 and 2.	70G1
15a	Domain structure: see	71N13, 72K3

Table 414. $SbNbO_4$ group. Unit cell parameters and space groups.

	Space group	Unit cell parameters [Å] a	b	c	Ref.
$SbNbO_4$		5.562	11.822	4.930	62K8
		4.929	11.797	5.559	63R2
		5.561 (3)	11.810 (3)	4.939 (2)	65S16
	$Pn2_1a$	5.5613 (4)	11.8080 (8)	4.9323 (3)	74J1
	$Pna2_1$	5.535	4.904	11.809	76R1
$SbTaO_4$		5.540	11.807	4.906	62K8
		4.911	11.814	5.535	63R2
	$Pn2_1a$	5.5372 (3)	11.8200 (5)	4.9152 (2)	74J1
Stibiotantalite		5.553	11.80	4.926	38D1
[Sb(Nb, Ta)O_4]		4.916	5.542	11.78	70G1
α-$BiNbO_4$		4.980	11.707	5.675	63R2
	$Pn2_1a$	5.6360 (3)	11.7692 (7)	4.9589 (3)	74J1
α-$BiTaO_4$		4.957	11.763	5.633	63R2
	$Pn2_1a$	5.6360 (3)	11.7692 (7)	4.9589 (3)	74J1
Bismutotantalite		5.57···5.69	17.78···17.89	4.97···5.00	55F1, 57H3

Nr. 43-6-ii. $SbNbO_4$ Antimony ortho-niobate
($M = 278.65$)

1a	Ferroelectricity in $SbNbO_4$ was reported by Lobachev et al. in 1972.	72L7
b	phase: II / I; state: (F) / P; crystal system: orthorhombic; space group: $Pna2_1$-C_{2v}^9	76R1
	Θ: 414 (1) °C	72L7
	403 (2) °C	73P13
	P_s‖[001] (the longest axis). Color: greenish-brown, (dependent on impurities).	72L7
2	Crystal growth: hydrothermal synthesis Crystal habit: see	72L7 74I16
3a	Uni cell parameters: Table 414 $a = 5.535$, $b = 4.904$, $c = 11.809$ Å continued	76R1

Nr. 43-6-ii $SbNbO_4$, continued

b	$Z=4$ Crystal structure: Table 415; Fig. 1580.	76R1
5a	Dielectric constant: Figs. 1581, 1582. $\varkappa_a = C/(T-\Theta_p)$, $T>\Theta_{II-I}$, $C=1.6\cdot10^5$ K, $\Theta_p=361$ °C	72L7
c	Spontaneous polarization and coercive field: $P_s=15\cdots18\cdot10^{-2}$ Cm^{-2}, $E_c=10\cdots11\cdot10^5$ V m^{-1} at 200 °C (for $E_{max}\to\infty$).	73P13
7a	Piezoelectricity: $d_{33}=30\cdot10^{-12}$ CN^{-1} at 220 °C	72L7
9e	Nonlinear optical susceptibility: $d_{31}/d_{36}^{KDP}=11.5$ (20) at 20 °C for Nd-glass laser line. $d_{31}/d_{31}^{BaTiO_3}=0.28$ (5), $\lambda=1.06$ μm.	73S15 74I16
11	Activation energy for conductivity: $\Delta U\approx0.56$ eV Anomalous photovoltaic effect: see	72L7 76V3

Table 415. $SbNbO_4$. Atomic coordinates, temperature parameters [76R1]. b_{ij} and B are defined by equations (b) and (e) in the Introduction, respectively. (b_{ij} in 10^{-3}).

Atom	x	y	z	b_{11}	b_{22}	b_{33}	b_{12}	b_{13}	b_{23}	B [Å^2]
Sb	0.4745 (3)	0.0463 (2)	0.2456 (9)	7.01 (3)	5.33 (6)	1.57 (1)	0.33 (1)	0.02 (1)	0.29 (1)	
Nb	0.1159 (2)	0	0	4.79 (3)	2.91 (6)	0.91 (1)	0.26 (1)	0.12 (1)	0.02 (1)	
O_1	0.173 (5)	−0.166 (1)	−0.153 (2)							0.69 (7)
O_2	0.417 (6)	−0.209 (1)	0.056 (8)							0.68 (5)
O_3	0.151 (6)	0.163 (1)	0.157 (2)							0.64 (3)
O_4	0.366 (2)	0.277 (4)	0.048 (6)							0.66 (5)

Nr. 43-6-iii. $SbTaO_4$ Antimony orthotantalate
($M=366.70$)

1a	Ferroelectricity was reported by Ivanova et al. in 1974.	
b	phase: II / I state: (F) / P crystal system: orthorhombic / — space group: $Pna2_1$-C_{2v}^9 / — Θ: 395 °C	74J1
2	Crystal growth: hydrothermal method	74I16
3a	Unit cell parameters: Table 414. $a=5.5372$, $b=4.9152$, $c=11.8200$ Å (b and c are interchanged).	74J1
5a	Dielectric constant: Fig. 1583. The ferroelectric transition temperature was reported to be 605 °C, but it was also pointed in the same literature that the SHG signal vanished above 395 °C. Therefore the Θ is assumed to be 395 °C in 1b.	74I16
9e	Nonlinear optical susceptibility: $d_{31}/d_{36}^{KDP}=10$ (2) at 20 °C for Nd-glass laser line.	73S15

43-6-iv. Bismutotantalite $Bi(Ta, Nb)O_4$

3a	Unit cell parameters: Table 414.	

Nr. 43-6-v. $BiNbO_4$ Bismuth ortho-niobate
($M=365.88$)

1a	Ferroelectricity was reported by Sych et al. and Ivanova et al. in 1975.	75S33
b		75I8

phase	II	I
state	(F)	P
crystal system	orthorhombic	
space group	$Pna2_1$-C_{2v}^9	
Θ	360 °C	

3a	Unit cell parameters: $a=5.6360$, $b=4.9589$, $c=11.7692$ Å (b and c are interchanged.) See also Table 414.	74J1
	$BiNbO_4$ is dimorphic. The low-temperature α-form is orthorhombic and shows ferroelectricity. The high-temperature β-form is triclinic and the β-to-α transition is very sluggish.	73K11 75S26
5a	Dielectric constant: Fig. 1584; see also	75I8

Nr. 43-6-vi. $BiTaO_4$ Bismuth ortho-tantalite
($M=453.93$)

3a	Unit cell parameters: Table 414; see also	75I8
b	Dimorphism is found like in $BiNbO_4$.	75S26

Nr. 43-7 $DyVO_4$, Dysprosium vanadate
($M=277.43$)

1b (70C7)

phase	III	II	I
state	A_{magn}		
crystal system		orthorhombic [a])	tetragonal [a])
space group		$Imma$-D_{2h}^{28} [a])	$I4_1/amd$-D_{4h}^{19} [a])
Θ	3.0	14 K	

[a]) 71W1

The phase transition at 14 K is a Jahn-Teller effect induced[b]), and associated with a dielectric anomaly[c]).

[b]) 70C7, 71W1
[c]) 77U3

5a	Dielectric constant: Fig. 1585.	
8a	Elastic stiffness determined by the Brillouin scattering: $c_{11}=24.2(2)\cdot10^{10}$ N m^{-2}, $c_{12}=5.0(4)\cdot10^{10}$ N m^{-2}, $\frac{1}{2}(c_{11}-c_{12})=9.6(2)\cdot10^{10}$ N m^{-2}, $c_{66}=1.50(5)\cdot10^{10}$ N m^{-2} at RT. Fig. 1586.	72S2
	Elastic stiffness determined by ultrasonic absorption: see	72M13

Nr. 43-8 Cr_2BeO_4, Chromium chrysoberyl
($M=177.000$)

1a	Magnetoferroelectricity in Cr_2BeO_4 was observed by Newnham et al. in 1978.	78N3
b		64S4 78N3

phase	II	I
state	A_{magn}, (F)	P_{magn}, (P)
crystal system	orthorhombic	orthorhombic
space group		$Pbnm$-D_{2h}^{16}
Θ	28 K	

3a	Unit cell parameters: $a=4.555$ Å, $b=9.792$ Å, $c=5.663$ Å. continued	60W3

Nr. 43-8 Cr_2BeO_4, continued

3b	Crystal structure: Parameters of atomic positions are determined by neutron diffraction.	69C5
5	A weak pyroelectric effect appeared below 28 K. Poling by an electric field changed the sign of pyroelectric coefficient suggesting reversal of the remanent polarization, which was estimated to be $10^{-4}\cdots10^{-6}$ times as small as the spontaneous polarization of normal ferroelectrics. No anomaly was observed in dielectric constant at the Néel point.	78N3
12	The magnetic structure is a cycloidal spiral having a propagation vector directed along the c axis. The spiral periodicity is about 65 Å.	69C5
	Magnetic susceptibility: see	64S4
	Antiferromagnetic resonance: see	67E1

Nr. 43-9 Sb_5O_7I, Antimony (III) oxide iodide
(M = 847.65)

1a	Sb_5O_7I crystallizes in several polytypes (see 1b below). Ferroelectric activity in Sb_5O_7I of 2MA type was first reported by Nitsche et al. in 1977.	77N6
b	Eight polytypes of Sb_5O_7I are known: Table 416. Most abundant polytype is 2MC-Sb_5O_7I (or α-Sb_5O_7I), which undergoes a phase transition at Θ = 208 °C. But it shows only an elastic anomaly [a]). Ferroelectric polytype is 2MA-Sb_5O_7I (or β-Sb_5O_7I). Data shown below concern primarily with 2MA-Sb_5O_7I.	[a]) 73K17

phase	II	I
state	F	P
crystal system	monoclinic	hexagonal
space group	Pc-C_s^2	P$\bar{6}$-C_{3h}^1
Θ	165 °C	

(77N6)

	$\varrho = 5.55 \cdot 10^3$ kg m^{-3}. Transparent, colorless.	75K21
2a	Crystal growth: vacuum sublimation growth technique, see	74K21, 77N6
b	Crystal form: flat pseudohexagonal prism	75K21
3a	Unit cell parameters: a = 6.759 (3) Å, b = 12.718 (6) Å, c = 13.405 (6) Å, β = 120.1 (1)°, see also Table 416.	77N6
b	Crystal structure: Z = 4 in phase II, Z = 2 in phase I. Figs. 1587, 1588; Tables 417, 418.	77N6
4	Fractional coordinates and thermal parameters of 2MC-Sb_5O_7I are given in	75K21
5b	Ferroelectric hysteresis loop: see	77N6
c	$P_s = 5 \cdot 10^{-3}$ C m^{-2} at RT. P_s is not reversible but reorientable by ±120° by coercive fields of 200···500 kV m^{-1}.	77N6
8b	Spontaneous strain: $2 \cdot 10^{-2}$. Coercive stress at 20 °C: $1.8 \cdot 10^6$ Nm^{-2}, see also Table 416.	77N6
9a	Refractive indices of 2MC-Sb_5O_7I: n_1 = 2.25, n_2 = 2.30 (6) for the two main vibration directions in the a–c plane. Double refraction in the a–c plane: $\Delta n = 3.8 \cdot 10^{-2}$ for 2MC-Sb_5O_7I and $3.5 \cdot 10^{-2}$ for 2MA-Sb_5O_7I at RT for λ = 632.8 nm.	77N6
e	Nonlinear susceptibilities for SHG of Nd-glass laser (λ = 1.06 μm) is about an order of magnitude below that of quartz.	77N6
10a	Raman scattering: Table 419; Figs. 1589, 1590. Raman scattering of 2MC-Sb_5O_7I: Fig. 1591; see also	75P15

Table 416. Sb_5O_7I. Polytypes of Sb_5O_7I and their properties [77N6].

	Modified Ramsdell symbol	2MC	2MA	4MC	4MA	$6MC_1$	$6MA_1$	$6MA_2$	$8MC_1$
	Former symbol	α	β	γ	η	ζ	δ	ε	$\varkappa$
Low-temp. phase	Relative abundance [%]	85	11	2	0.7	0.7	0.2	0.3	0.1
	Space group	$P2_1/c$	Pc	$P2_1/c$	Pc				
	Lattice parameters: a [Å]	6.772	6.759	6.766	6.765	6.77	6.77	6.77	6.77
	b [Å]	12.726	12.718	25.458	25.462	38.46	38.6	38.50	51.00
	c [Å]	13.392	13.405	13.405	13.402	13.40	13.40	13.40	13.40
	β [deg]	120.1	120.1	120.1	120.1	120.1	120.1	120.1	120.1
	Formula units per cell (Z)	4	4	8	8	12	12	12	16
	Optical extinction angle [deg]	13	0	0	8	6	4	0	0
	Piezoelectric, ferroelectric	–	+	–	+	–	+	+	–
	Coercive stress at 20 °C*) [10^{+6} N/m²]	1.5	1.8	0.5	0.7	0.2	0.9	0.8	
	Spontaneous strain [$\cdot 10^{-2}$]**)	8	2	6	7	7	7	7	7
	Coercive field [10^2 kV m^{-1}]	–	2	–	4–6	–	8–10	8–10	–
	Spontaneous polarization [10^{-3} C m^{-2}]	–	5	–		–			–
	Θ [°C]	208	165	184	182	189	195	175	190
	Modified Ramsdell symbol	2HC	2HA	4TC	4HA	6TC	6HA	6HA	
High-temp. phase	Space group	P6/m	$P\bar{6}$	$P\bar{3}m$	$P\bar{6}$	$P\bar{3}$	$P\bar{6}$	$P\bar{6}$	
	Lattice parameters: a [Å]	6.77	6.78	6.78	6.78	6.79	6.77	6.77	6.73
	c [Å]	12.80	12.86	25.68	25.72	38.61	38.51	38.53	51.10
	Formula units per cell (Z)	2	2	4	4	6	6	6	8

*) Stress required for convertation three orientation states into each other (mechanical switching).
**) Derived with the procedure of Aizu [70A13].

Table 417. Sb_5O_7I (polytype 2MA). Fractional coordinates in the unit of 10^{-4} [78K18]. Non-primed and primed numbers in the brackets of the first column indicate that corresponding atoms are located in the lower and upper half of an unit cell along *b*-axis, as explicitely shown in Fig. 1588 (A). Note that Fig. 1588 (A) represents a polytype 2 MC, in which only the positions of Sb (1′), Sb (2′) and Sb (3′) differ from those for polytype 2 MA.

	x	*y*	*z*
Sb (1)	−701 (7)	2345 (1)	4079 (3)
Sb (1′)	39 (7)	7548 (1)	4301 (3)
Sb (2)	3206 (7)	2460 (2)	2652 (3)
Sb (2′)	2914 (7)	7541 (2)	2352 (4)
Sb (3)	−3561 (7)	2451 (1)	700 (3)
Sb (3′)	−3859 (6)	7652 (1)	917 (3)
Sb (4)	2931 (7)	551 (1)	802 (3)
Sb (5)	3028 (7)	4520 (1)	807 (3)
Sb (4′)	−3664 (7)	5551 (1)	−800 (3)
Sb (5′)	−3581 (7)	9519 (1)	−805 (3)
I (1)	0	0	2500
I (2)	−654 (5)	4997 (2)	2504 (3)
O (1)	6396 (41)	2484 (13)	4163 (22)
O (1′)	3073 (37)	7513 (11)	824 (20)
O (2)	509 (49)	1678 (17)	432 (26)
O (3)	−15 (42)	3858 (14)	125 (30)
O (2′)	−636 (37)	6243 (12)	−42 (19)
O (3′)	−1146 (50)	8429 (17)	−36 (24)
O (4)	4378 (43)	1247 (14)	46 (22)
O (5)	3742 (44)	3367 (15)	36 (22)
O (4′)	4873 (51)	6666 (17)	−354 (26)
O (5′)	4820 (48)	8871 (15)	−140 (24)
O (6)	4744 (54)	1396 (18)	2204 (26)
O (7)	4258 (47)	3665 (15)	2252 (22)
O (6′)	4894 (56)	6316 (17)	2802 (26)
O (7′)	5495 (59)	8567 (20)	2771 (28)

Table 418, see p. 281.

Table 419. Sb_5O_7I (polytype 2MA and 2MC). Frequencies of Raman scattering lines at RT in the unit of cm^{-1}, $1\,cm^{-1} = 3 \cdot 10^{10}$ Hz [79P1]. Frequencies of strong lines which give similar structures in the Raman spectra for both polytypes are underlined.

$\nu\,[cm^{-1}]$

2MA A′	2MC A_g	2MA A″	2MC B_g
34.5	31	33	31
37	39.5	38.5	
41.5	41		40
		46	45.5
48			48.5
		53	
		58	57
		63	70
78	79	75	80
86		88	
	104		
111	110		111
$\underline{146}$	$\underline{145}$	$\underline{147}$	$\underline{145}$
149			
		152	
$\underline{156}$	$\underline{156}$		
161.5			
		165	
		$\underline{182}$	$\underline{182.5}$
		$\underline{190}$	$\underline{190}$
213	210	210	212
		232	
262			259
273.5	272		
$\underline{330.5}$	$\underline{330}$		
$\underline{394}$	$\underline{387}$		
$\underline{408}$	$\underline{415}$		
$\underline{459}$	{ $\underline{456}$		
	} $\underline{478}$		
	534		
			568
	600		
	750		

Nr. 43-10 $Rb_3MoO_3F_3$, Rubidium fluoroxymolybdate
($M = 457.34$)

1a	Ferroelectricity in $Rb_3MoO_3F_3$ was reported by Péraudeau et al. in 1977.	77P1
b	see phase table below	77P1
2a	Sample preparation: synthesized by a reaction of $3RbF + MoO_3 \rightarrow Rb_3MoO_3F_3$.	77P1
3a	Unit cell parameters: $a = 31.74$ (2) Å, $c = 45.27$ (3) Å at RT.	77P1
	Crystal structure: $(NH_4)_3FeF_6$-type, see	77P1
	The unit cell parameters in phase III are nearly five times as large as those in phase II.	77P1

1b:

phase	III	II	I
state	F	F	P
crystal system	tetragonal	tetragonal	cubic
Θ		150	265 °C

4	Thermal expansion: Fig. 1592.	
5a	Dielectric constant: Fig. 1593.	
c	Spontaneous polarization can be reversed by poling. Temperature dependence of spontaneous polarization: Fig. 1594.	77P1

43-9, continued

Table 418. Sb_5O_7I (polytype 2MA). Interatomic distances in the unit of Å [78K18]. For the meaning of primes in brackets, such as Sb (2′), see the caption of Table 417. Square brackets denote the fourth nearest neighbor.

Sb (1)–O (1)	2.03 (3)	Sb (1′)–O (1′)	2.05 (2)
O (2)	2.00 (3)	O (2′)	1.94 (2)
O (3)	1.97 (2)	O (3′)	1.92 (3)
[O (5)	2.76 (3)]	[O (7′)	3.00 (3)]
I (1)	3.821 (3)	I (1)	3.935 (3)
I (2)	3.988 (4)	I (2)	3.928 (4)
Sb (2)–O (1)	2.09 (2)	Sb (2′)–O (1′)	2.10 (3)
O (6)	1.98 (4)	O (6′)	1.94 (3)
O (7)	1.88 (3)	O (7′)	2.02 (3)
[O (2)	2.78 (3)]	[O (4′)	2.86 (3)
I (1)	3.754 (3)	I (1)	3.754 (4)
I (2)	4.092 (5)	I (2)	4.101 (5)
Sb (3)–O (1)	2.05 (3)	Sb (3′)–O (1′)	2.02 (3)
O (4)	1.96 (2)	O (4′)	1.94 (3)
O (5)	1.96 (2)	O (5′)	1.99 (2)
[O (6)	3.08 (4)]	[O (3′)	2.88 (4)]
I (1)	3.940 (2)	I (1)	3.832 (2)
I (2)	3.930 (4)	I (2)	4.002 (3)
Sb (4)–O (2)	2.04 (3)	Sb (4′)–O (2′)	1.98 (2)
O (4)	1.94 (3)	O (4′)	1.98 (3)
O (6)	1.97 (3)	O (7′)	2.04 (3)
[O (5′)	3.07 (3)]	[O (3)	3.03 (2)]
I (1)	3.756 (6)	I (1)	3.777 (4)
	4.199 (4)		4.178 (5)
I (2)	3.896 (5)	I (2)	3.902 (6)
Sb (5)–O (3)	1.97 (3)	Sb (5′)–O (3′)	2.00 (2)
O (5)	1.99 (3)	O (5′)	1.90 (4)
O (7)	2.00 (3)	O (6′)	1.94 (3)
[O (2′)	3.07 (2)]	[O (4)	3.10 (3)]
I (1)	3.889 (4)	I (1)	3.890 (4)
I (2)	4.118 (5)	I (2)	3.731 (8)
	4.170 (7)		4.188 (5)

Nr. 43-11. Pb_4SiO_6, Lead silicate
(M=952.9)

1a	Possibility of antiferroelectricity was proposed by Isupov in 1965, but no double hysteresis loop has been observed.	65I5
b	phase: II | I; state: (A) | (P); Θ: ≈155 °C	
4	Thermal expansion: Fig. 1595.	
5a	Dielectric constant: Fig. 1595. Curie-Weiss law: $C \approx 1000$ K, $\Theta_p = 277$ K, $\varkappa_\infty = 33.3$.	65I5

phase	II	I
state	(A)	(P)
Θ	≈155 °C	

Nr. 43-12 $Ca_3(VO_4)_2$, Calcium ortho-vanadate
($M = 350.12$)

1a	Pyroelectric behavior which reverses sense by poling was observed in $Ca_3(VO_4)_2$ and ferroelectric activity was mentioned by Glass et al. in 1978.	78G8

b	phase	II	I	78G8
	state	(F)	(P)	
	crystal system	trigonal [a])		[a]) 73G13
	space group	R3c-C_{3v}^6 [a])		
	Θ	1380 (10) K		

$P_s \| [001]$
$T_{melt} \approx 1700$ K
$\varrho = 3.172 \cdot 10^3$ kg m^{-3}
$\varrho_x = 3.177 \cdot 10^3$ kg m^{-3} — 78H5

2a	Crystal growth: Czochralski method.	78G8
3a	Unit cell parameters: $a = 10.81322$ (8) Å, $c = 38.03915$ (22) Å at 298 K.	78G8
b	Crystal structure: see	73G13
4	Thermal expansion: $\alpha_{11} = 12.9 \cdot 10^{-6}$ K^{-1}, $\alpha_{33} = 11.8 \cdot 10^{-6}$ K^{-1} at 293 K	78H5
5a	Dielectric constants: $\varkappa_a = 16.60$, $\varkappa_c = 10.62$ at 10 MHz, 293 K. Fig. 1596.	78H5
d	Pyroelectric coefficient: $p_3 = 0.80$ (5) C m^{-2} K^{-1}.	78G8
7a	Piezoelectric constants: $d_{22} = -1.5 \cdot 10^{-12}$ C N^{-1}, $d_{24} = 1.6 \cdot 10^{-12}$ C N^{-1}, $d_{32} = 0.3 \cdot 10^{-12}$ C N^{-1}, $d_{33} = -2.0 \cdot 10^{-12}$ C N^{-1}.	78H5
8a	Elastic constants:	78H5

c_{11}^E	c_{33}^E	c_{44}^E	c_{66}^E	c_{14}^E	c_{13}^E
$\cdot 10^{10}$ N m^{-2}					
9.801	11.976	3.167	2.626	0.448	4.887

9a	Refractive index: Table 420. Birefringence: $n_a - n_c = 0.032$ (3)	78G8
b	Electrooptic constants for $\boldsymbol{E}$: $r_{22} = -0.7 \cdot 10^{-12}$ m V^{-1}, $r_{24} = -1.1 \cdot 10^{-12}$ m V^{-1}, $r_{32} = -1.0 \cdot 10^{-12}$ m V^{-1}, $r_{33} = 1.0 \cdot 10^{-12}$ m V^{-1}.	78H5
e	Nonlinear optical susceptibilities: $\|d_{31}/d_{36}^{KDP}\| = 1.51$ (15), $\|d_{33}/d_{36}^{KDP}\| = 0.78$ (9), $\|d_{22}/d_{36}^{KDP}\| = 0.53$ (6), $\|d_{15}/d_{36}^{KDP}\| = 1.29$ (19) for $\lambda = 1.064$ µm, $d_{31}/d_{33} > 0$, $d_{22}/d_{15} > 0$.	78B4

Table 420. $Ca_3(VO_4)_2$. n_o, n_e vs. λ [78B4]. T=295 K.

Wavelength λ [μm]	n_o	n_e	Wavelength λ [μm]	n_o	n_e
0.54607	1.901551	1.877298	1.69202	1.841400	1.821359
0.57696	1.894779	1.870378	1.70727	1.841039	1.820900
0.57907	1.894000	1.870402	1.81307	1.839478	1.819644
0.69072	1.875754	1.853362	1.97009	1.837699	1.818000
1.01398	1.855658	1.834621	2.15000	1.835150	1.815581
1.12866	1.851950	1.831057	2.32542	1.833128	1.813305
1.36728	1.846631	1.825971	2.46400	1.830720	1.811755
1.52953	1.843680	1.823557	2.55000	1.830045	1.810834
1.67100	1.841616	1.821504	2.72000	1.827707	1.808410

Nr. 43-13 $BaMnF_4$ group
Nr. 43-13-i $BaMgF_4$
(M=237.63)

1a	Ferroelectric activity was found by Eibschütz et al. in 1969.	69E1
b	Phase, state, crystal system, space group, Θ: as described in subsection 1a of Nr. 43-13-ii for $BaMnF_4$.	
	$P_s \parallel [100]$.	69D2
	T_{melt}=865 (5) °C.	
	Transparent, colorless.	74R2
2a	High quality crystal can be grown by Czochralski technique using Iridium crucible.	74R2
3a	Unit cell parameters: a=5.810 (2) Å, b=14.509 (6) Å, c=4.125 (2) Å.	69K3
b	Z=4	69K3
4	Thermal expansion coefficient at RT: $\alpha_a = 11.5 \cdot 10^{-6}$ °C^{-1}, $\alpha_b = 21 \cdot 10^{-6}$ °C^{-1}, $\alpha_c = 20 \cdot 10^{-6}$ °C^{-1}	74R2
5a	Dielectric constant along the ferroelectric a axis at 100 MHz: Fig. 1597.	
	$\varkappa_a = 8$, $\varkappa_b = 14$, $\varkappa_c = 8$ at 100 MHz at RT, and $\varkappa_a \approx 21$ near the melting point (≈865 °C).	69D2
	See also	74R2
c	$P_s = 7.7\,(3) \cdot 10^{-2}$ C m^{-2}.	69E1
7a	Piezoelectric constant: Table 421.	
8a	Elastic stiffness and thermoelastic constant: Table 422.	
9a	Refractive indices: Table 423. See also	74R2
e	Nonlinear optical susceptibilities: Table 424.	
13b	ESR and ENDOR, see	80F4

Table 421. $BaMgF_4$. Piezoelectric strain constant $d^T_{i\lambda}$ and piezoelectric stress constant $e^T_{i\lambda}$ under isothermal condition at 20 °C [74R2]. Differences between these isothermal constants and the corresponding adiabatic ones are smaller than a few percent.

i, λ	11	12	13	26	35	Unit
$d^T_{i\lambda}$	8.1	2.5	−4.2	−5.3	−1.2	$\cdot 10^{-12}$ C N^{-1}
$e^T_{i\lambda}$	6.1	2.8	0.63	−2.6	−1.4	$\cdot 10^{-1}$ C m^{-2}

Table 422. $BaMgF_4$. Elastic constant $c^E_{\lambda\mu}$ at 20 °C and thermoelastic constant $T^E_{\lambda\mu}$ (= d log c^E_{ij}/dT) at 0 °C [74R2].

$\lambda\mu$	11	22	33	44	55	66	12	13	23	Unit
$c^E_{\lambda\mu}$	10.40	8.10	12.97	3.21	5.51	2.47	2.87	6.37	3.58	$\cdot 10^{10}$ N m^{-2}
$T^E_{\lambda\mu}$	−0.94	−1.37	−3.97	−2.7	−2.0	−2.7	−0.6	−2.4	−0.5	$\cdot 10^{-4}$ °C^{-1}

Table 423. $BaMgF_4$, $BaZnF_4$. Refractive indices [75B11].

	λ [μm]	$BaMgF_4$	$BaZnF_4$
n_a	1.06	1.458	1.507
n_b	1.06	1.439	1.490
n_c	1.06	1.467	1.514
n_a	0.53	1.467	1.517
n_b	0.53	1.450	1.499
n_c	0.53	1.473	1.524

Table 424. $BaMgF_4$, $BaZnF_4$. Nonlinear optical susceptibilities [75B11]. λ = 1.06 μm.

	$BaMgF_4$	$BaZnF_4$
d_{11}/d_{11}^{quartz}	0.05	0.11
d_{31}/d_{11}^{quartz}	0.07	0.025
d_{15}/d_{11}^{quartz}	0.07	0.033
d_{12}/d_{11}^{quartz}	0.13	0.25
d_{26}/d_{11}^{quartz}	0.07	...

Nr. 43-13-ii $BaMnF_4$
(M = 268.26)

1a	It was suggested by Eibschütz et al. in 1969 that $BaMnF_4$ is ferroelectric because the dielectric behavior shows very close resemblance to the ferroelectric $BaMF_4$ (M = Mg, Zn, Co, Ni).	69E1
b	state: F; crystal system: orthorhombic[a]); space group: $A2_1am$[b])-C^{12}_{2v}; Θ: none*)	[a]) 67C8 [b]) 68S8
	P_s ‖ [100]	
	T_{melt} = 755 (5) °C.	69D2
	ϱ = 4.59 (5) · 10^3 kg m^{-3}.	69K3
	ϱ_x = 4.634 · 10^3 kg m^{-3}.	
	Hardness: 3.5 on Mohs' scale.	
	Transparent to visible light up to 10 μm thickness.	68E1
	*) Ferroelectric state persists up to the melting point. A phase transition was found by Spencer et al. at $\Theta_{II-I} \approx$ 255 K. For a doubling of the primitive unit cell below Θ_{II-I} and	70S26
	the incommensurate structural nature, see	79C4
	Also, antiferromagnetic transition was found at lower temperature.	69H8
2a	Crystals can be synthesized from the melt of high purity binary components BaF_2 and MnF_2 in an HF atmosphere. Single crystals can be prepared by a horizontal pass method.	68E1
3a	Unit cell parameters: a = 5.9845 (3) Å, b = 15.098 (2) Å, c = 4.2216 (3) Å at 298 K.	69K3
b	Z = 4. Fractional coordinates of atoms in the unit cell at 298 K: Table 425. Schematic view of the structure: Figs. 1598, 1599. Temperature parameters and the thermal vibration of atoms at 298 K: Tables 426, 427. Interatomic distances and angles: Tables 428, 429. Symmetry relation nomenclature for crystal data: Table 430.	69K3
5a	Dielectric constant: Figs. 1597, 1600, 1601. See also Effect of hydrostatic pressure: Figs. 1602, 1603.	80F2
c	The coercive field is so large that polarization reversal was not found.	69E1

7, 8	Strong piezoelectric activity has been found. Elastic stiffness: $c^E_{44}=1.985\cdot10^{10}$ N m^{-2}. Piezoelectric stress constant: $e_{24}=0.360$ C m^{-2}. Electromechanical coupling factor: $k_{24}=0.255$. Transit time data for modes propagating parallel to the crystal axes: Figs. 1604···1606.	68E1
9	Crystal is biaxial negative, optic plane (100), $2V=67.2°$. The principal refractive indices for $\lambda=589.3$ nm (along b, a, and c axes, respectively): $n_\alpha=1.48$, $n_\beta=1.499$, $n_\gamma=1.505$.	68E1
10a	Raman scattering: Figs. 1607···1609; see also	74R7, 80L1
b	Brillouin scattering, see	78B12
11	Electrical conductivity along a: $\sigma_a=1.1\cdot10^{-4}\,\Omega^{-1}$ m^{-1} at RT.	69E1
12	Antiferromagnetic at low temperatures: Néel point $\Theta_N\approx25$ K. Magnetic susceptibility: Fig. 1610. Magnetization curves at various T: Fig. 1611.	69H8
	Spin-flop transition: the critical field $H_c\approx1.15\cdot10^6$ A m^{-1} at 20 K and $H_c=0.83\cdot10^6$ A m^{-1} at 1.4 K for $\boldsymbol{H}\|b$-axis.	69H8
	Transverse magnetoelectric effect; see	72A9, 80F3
13b	Antiferromagnetic resonance at 4.2 K, see	72P2

Table 425. $BaMnF_4$. Crystal structure [69K3]. Fractional coordinates at 298 K.

	x	y	z
Ba	0.4537	0.34383 (5)	½
Mn	0.0000 (4)	0.4160 (1)	0
F (1)	0.196 (2)	0.2982 (6)	0
F (2)	−0.275 (3)	0.3363 (7)	0
F (3)	0.337 (3)	0.4651 (7)	0
F (4)	0.016 (3)	0.4217 (10)	½

Table 427. $BaMnF_4$. Crystal structure [69K3]. Root-mean-square amplitude of thermal vibration of atoms along the principal axes at 298 K.

Atom	*1* [Å]	*2* [Å]	*3* [Å]
Ba	0.096 (1)	0.115 (2)	0.194 (2)
Mn	0.095 (1)	0.103 (2)	0.134 (6)
F (1)	0.090 (14)	0.171 (18)	0.177 (13)
F (2)	0.142 (18)	0.147 (13)	0.179 (16)
F (3)	0.033 (42)	0.212 (13)	0.245 (20)
F (4)	0.089 (15)	0.187 (21)	0.275 (18)

Table 426. $BaMnF_4$. Crystal structure [69K3]. Temperature factors at 298 K. Temperature factor b_{ij} is defined by Eq. (b) in Introduction.

	b_{11}	b_{22}	b_{33}	b_{12}	b_{13}	b_{23}
Ba	0.0127 (3)	0.00240 (3)	0.0103 (3)	0.00262 (9)	0	0
Mn	0.0099 (9)	0.00092 (4)	0.0099 (5)	0.00007 (13)	0	0
F (1)	0.016 (3)	0.0007 (2)	0.035 (5)	−0.0009 (7)	0	0
F (2)	0.013 (4)	0.0024 (4)	0.024 (4)	−0.0012 (9)	0	0
F (3)	0.022 (5)	0.0018 (4)	0.050 (6)	−0.0061 (11)	0	0
F (4)	0.025 (6)	0.0057 (7)	0.009 (3)	0.0038 (15)	0	0

Table 428. $BaMnF_4$. Crystal structure [69 K 3]. Interatomic distances at 298 K. Symmetry related atoms are denoted by subscripts explained in Table 430. All atoms lie on mirror planes.

Distance	[Å]	Distance	[Å]
Mn – F (2)	2.037 (15)	Mn – Mn_i	3.922 (2)
Mn – F $(3)_{iv}$	2.042 (9)	Mn – Mn_{vi}	4.222 (3)
Mn – F (4)	2.115 (1)		
Mn – F (1)	2.129 (11)	F (1) – F (3)	2.658 (12)
Mn – F (3)	2.151 (15)	F (1) – F (2)	2.874 (21)
		F (4) – F $(3)_{iv}$	2.919 (11)
Ba – F $(1)_{ii}$	2.588 (9)	F (3) – F (4)	2.929 (17)
Ba – F $(2)_v$	2.666 (10)	F (1) – F (4)	3.015 (13)
Ba – F (1)	2.704 (7)	F (2) – F (4)	3.024 (17)
Ba – F (4)	2.871 (16)	F (2) – F $(3)_{iv}$	3.073 (18)
Ba – F (3)	2.880 (10)	F (3) – F $(3)_{iv}$	3.173 (7)
Ba – F $(2)_{ii}$	3.044 (12)		
Ba – F $(4)_i$	3.559 (15)	F (1) – F $(2)_{ii}$	2.935 (10)
Ba – F $(4)_v$	3.565 (18)	F (3) – F $(2)_v$	3.028 (23)
		F (1) – F $(2)_v$	3.220 (21)
Ba – Ba_{ii}	4.630 (10)		

Table 429. $BaMnF_4$. Crystal structure [69 K 3]. Interatomic angles at 298 K. Symmetry related atoms are denoted by subscripts, see Table 430 for the symmetry nomenclature.

Angles	[deg]	Angles	[deg]
F (1) – Mn – F (3)	76.8 (5)	F $(3)_{iv}$ – Mn – F (2)	97.7 (7)
F (4) – Mn – F (3)	86.7 (5)	F (3) – Mn – F $(3)_{iv}$	98.3 (3)
F (2) – Mn – F (1)	87.2 (5)	F (4) – Mn – F $(4)_{viii}$	173.0 (9)
F (4) – Mn – F $(3)_{iv}$	89.2 (4)	Mn – F (3) – Mn_i	138.6 (9)
F (4) – Mn – F (1)	90.5 (4)	Mn – F (4) – Mn_{vi}	173.0 (9)
F (4) – Mn – F (2)	93.5 (5)		

Table 430. $BaMnF_4$. Symmetry relation nomenclature for structural data [69 K 3].

Subscript	Operation	Subscript	Operation
i	$\frac{1}{2}+x, 1-y, z$	vi	$x, y, 1+z$
ii	$\frac{1}{2}+x, \frac{1}{2}-y, \frac{1}{2}+z$	vii	$\frac{1}{2}+x, \frac{1}{2}-y, z-\frac{1}{2}$
iii	$x-\frac{1}{2}, \frac{1}{2}-y, z-\frac{1}{2}$	viii	$x, y, z-1$
iv	$x-\frac{1}{2}, 1-y, z$	ix	$x, \frac{1}{2}+y, \frac{1}{2}+z$
v	$1+x, y, z$	x	$\overline{x}, y, z$

Nr. 43-13-iii $BaFeF_4$
($M = 269.17$)

1a	It was suggested by Eibschütz et al. in 1969 that $BaFeF_4$ is ferroelectric because the dielectric behavior shows close resemblance to the ferroelectric $BaMF_4$ (M = Mg, Zn, Co, Ni).	69E1
b	Phase, state, crystal system, space group, Θ: as described in subsection 1a of Nr. 43-13-ii for $BaMnF_4$. $P_s \parallel [100]$. $T_{melt} = 720\ (5)\ °C$. $\varrho = 4.74 \cdot 10^3\ kg\ m^{-3}$. $\varrho_X = 4.726 \cdot 10^3\ kg\ m^{-3}$.	 69D2 68E1
2a	Single crystals can be prepared with a similar method as described in subsection 2a of Nr. 43-13-ii.	
3a	Unit cell parameters: $a = 5.829$ Å, $b = 14.837$ Å, $c = 4.238$ Å.	68S8
b	$Z = 4$.	
5a	Dielectric constant along the ferroelectric a axis at 100 MHz: see Fig. 1597. $\varkappa_a = 11$, $\varkappa_b = 17$, $\varkappa_c = 8$ at 100 MHz at RT, and $\varkappa_a \approx 200$ near the melting point ($\approx 720\ °C$).	 69D2
11	Conductivity along the a axis: $\sigma_a \approx 3 \cdot 10^{-5}\ \Omega^{-1}\ m^{-1}$	69E1
12	Magnetic properties and Mössbauer effect: Antiferromagnetic at low temperatures, Néel point $\Theta_N = 60\ (2)$ K; susceptibility of the polycrystalline sample $\chi_{magn\,\varrho} \approx 81.6 \cdot 10^{-6}\ m^3\ kg^{-1}$. Magnetic structure and magnetoelectric effect.	68E1 75D9

Nr. 43-13-iv $BaCoF_4$
($M = 272.26$)

1a	Ferroelectric activity was found by Eibschütz et al. in 1969.	69E1
b	Phase, state, crystal system, space group, Θ: as described in subsection 1a of Nr. 43-13-ii for $BaMnF_4$. $P_s \parallel [100]$. $T_{melt} = 855\ (5)\ °C$.	 69D2
2a	Single crystals can be prepared with a similar method as described in subsection 2a of Nr. 43-13-ii.	
3a	Unit cell parameters: $a = 5.8519\ (3)$ Å, $b = 14.628\ (2)$ Å, $c = 4.210\ (3)$ Å.	69K3
5a	Dielectric constant along the ferroelectric a axis at 100 MHz: see Fig. 1597. $\varkappa_a = 10$, $\varkappa_b = 22$, $\varkappa_c = 8$ at 100 MHz at RT, and $\varkappa_a \approx 400$ near the melting point ($\approx 855\ °C$).	 69D2
c	$P_s = 8\ (3) \cdot 10^{-2}\ C\ m^{-2}$.	69E1
12	Magnetic susceptibility: Fig. 1612.	
13	Neutron diffraction pattern: Fig. 1613. Magnetic structure: Fig. 1614. Two dimensional antiferromagnet with $\Theta_N = 69.6\ (1)$ K. Magnetic moment of Co^{2+} at 4.7 K, 3.4 (2) μ_B. Two magnetic phases, A and B. Magnetoelectric effect, see	 72E1 72E1 75D9, 72Z2

Nr. 43-13-v $BaNiF_4$
($M = 272.02$)

1a	Ferroelectric activity was found by Eibschütz et al. in 1969.	69E1
b	Phase, state, crystal system, space group. Θ: as described in subsection 1a of Nr. 43-13-ii for $BaMnF_4$. $P_s \| [100]$. $T_{melt} = 965$ (5) °C.	 69D2
2a	Crystals can be prepared with a similar method as described in subsection 2a of Nr. 43-13-ii. For BaF_2-NiF_2 phase diagram, see	67C8
3a	Unit cell parameters: $a = 5.799$ Å, $b = 14.458$ Å, $c = 4.153$ Å.	68S8
b	$Z = 4$.	68S8
5a	Dielectric constant along the ferroelectric a axis at 100 MHz: see Fig. 1597. $\varkappa_a = 8$, $\varkappa_b = 14$, $\varkappa_c = 8$ at 100 MHz at RT, and $\varkappa_a \approx 20$ near the melting point (≈ 965 °C).	 69D2
c	$P_s = 6.7$ (3) $\cdot 10^{-2}$ C m^{-2}	69E1
d	Pyroelectric behavior, see	77G11
12	Magnetic susceptibility: Fig. 1615.	
13	Magnetic structure: Fig. 1616. The magnetic moments are directed along the b axis and the ordered moment ≈ 2.0 μ_B per Ni ion. Magnetoelectric effect: see	 70C8 75D9

Nr. 43-13-vi $BaZnF_4$
($M = 278.70$)

1a	Ferroelectric activity was found by Eibschütz et al. in 1969.	69E1
b	Phase, state, crystal system, space group, Θ: as described in subsection 1a of Nr. 43-13-ii for $BaMnF_4$. $P_s \| [100]$ $T_{melt} = 745$ (5) °C.	 69D2
2a	Single crystals can be prepared with a similar method as described in subsection 2a of Nr. 43-13-ii.	
3a	Unit cell parameters: $a = 5.841$ Å, $b = 14.563$ Å, $c = 4.206$ Å	68S8
b	$Z = 4$. Crystal model seen along the c-direction: Fig. 1617. Approximate fractional coordinate of atoms at RT: All are 4(c) position (x, y, $z = 0$) of $A2_1am$ setting; Ba (0.249, 0.1477), Zn (0.282, 0.413), F (1) (0.592, 0.472), F (2) (0.472, 0.300), F (3) (0.021, 0.334), F (4) (0.311, 0.925).	68S8 68S8
5a	Dielectric constant along the a axis at 100 MHz: see Fig. 1597. $\varkappa_a = 11$, $\varkappa_b = 17$, $\varkappa_c = 10$ at 100 MHz at RT, and $\varkappa_a \approx 64$ near the melting point (≈ 745 °C). $P_s = 9.7$ (3) $\cdot 10^{-2}$ C m^{-2}.	 69D2 69E1
d	Pyroelectric behavior, see	77G11
9a	Refractive indices: Table 423.	
e	Nonlinear optical susceptibilities: Table 424.	

Nr. 43-14 $PrCl_3$, Praseodymium chloride
($M = 240.267$)

1	Dielectric anomaly associated with a phase transition in $PrCl_3$ was observed and the possibility of antiferroelectric ordering was reported by Harrison et al. in 1976.	76H2

Nr. 43-15 Sb_2S_3, Stibnite
(M=339.68)

1a	Ferroelectric activity in Sb_2S_3 was reported by Rinkyavichyus and Mikalkevichyus in 1967.	67R3
b	phase: III / II / I; state: F / F / P; crystal system: – / orthorhombic[a]) / orthorhombic*, [a]); space group: – / $Pbn2_1$-C_{2v}^{9}*, [a]) / Pbnm-D_{2h}^{16}*, [a]); Θ: 290···310, 420···490 K	74O4 [a]) 60S3 74O3
	$P_s \parallel [001]$	
	T_{melt}=548 °C.	55I1
	ϱ=4.63 (2) · 10^3 kg m^{-3}.	41d1
	Gray with metallic luster. Cleavage plane: (100)	
	*) Estimated by Orliukas and Grigas [74O4] on the basis of the structure determined by Šćavničar [60S3].	
2a	Slow sublimation of Sb_2S_3 at 600 °C.	60S3, 62K3
3a	Unit cell parameters: a=11.2285 (5) Å, b=11.3107 (9) Å, c=3.8363 (4) Å.	72B4
b	Z=4. Crystal structure: Table 431.	72B4
5a	Dielectric constant: Figs. 1618, 1619. Effect of hydrostatic pressure: see	75K1
c	Spontaneous polarization: Fig. 1620.	
8a	Absorption of ultrasonic wave: see	72G13, 74S7
9a	Infrared absorption: see	73P6
11	Electrical conductivity: see	70A10, 75O4
	Photoconductivity: see	55I1
13a	NQR of ^{121}Sb and ^{123}Sb: see	77A1
15a	Domain structure is observed by etching technique:	76G17

Phase table (1b):

phase	III	II	I
state	F	F	P
crystal system		orthorhombic[a])	orthorhombic*, [a])
space group		$Pbn2_1$-C_{2v}^{9}*, [a])	Pbnm-D_{2h}^{16}*, [a])
Θ	290···310	420···490 K	

Table 431. Sb_2S_3. Fractional coordinates and temperature parameters [72B4]. Temperature parameters are defined by Eq. (a) in Introduction.

Atom	Fractional coordinates			Temperature parameters · 10^3			
	x	y	z	B_{11}	B_{22}	B_{33}	B_{12}
Sb_I	0.3261 (1)	0.0293 (1)	1/4	2.5 (2)	3.0 (2)	14.4 (15)	0.6 (1)
Sb_{II}	0.0360 (1)	0.1495 (1)	3/4	3.6 (2)	2.7 (2)	23.0 (15)	1.3 (1)
S_I	0.8769 (3)	0.0497 (3)	1/4	2.7 (3)	3.2 (2)	18.0 (25)	0.4 (3)
S_{II}	0.4386 (3)	0.1251 (3)	3/4	2.4 (3)	3.5 (3)	17.7 (25)	0.8 (3)
S_{III}	0.1917 (3)	0.2079 (3)	1/4	2.3 (3)	3.1 (3)	26.7 (26)	0.2 (3)

Nr. 43-16 $Sn_2P_2S_6$, Tin-hypothiodiphosphate
(M=491.68)

1a	Ferroelectricity in $Sn_2P_2S_6$ was discovered by Carpentier and Nitsche in 1974.	74C3
b	phase: II / I; state: F[a]) / P[a]); crystal system: monoclinic[b]) / (monoclinic)[b]); space group: Pc-C_s^2 [b),*]) / ($P2_1/c$-C_{2h}^5)[b]); Θ: 66 (2)[a]) °C	a) 74C3, b) 74C2
	P_s‖[101] in phase II.	74C3
	T_{melt}=775 °C.	74C2
	ϱ=3.56 · 10^3 kg m^{-3}.	74C2
	Transparent, yellowish-brown.	74C2
	*) See 3b.	
2a	Crystal growth: vapor transport with iodine.	74C2
b	Crystal form: Fig. 1621.	
3a	Unit cell parameters: a=6.529 (2) Å, b=7.485 (2) Å, c=11.317 (3) Å, β=124.11 (3)° at RT (phase II).	74C2
b	Z=2.	74C2
	Crystal structure analysis of phase II was made by Dittmar and Schäfer based on the space group Pn-C_s^2. Fig. 1622; Table 432.	74D8
5a	Dielectric constant: Fig. 1623.	
c	Spontaneous polarization: Fig. 1624.	
	$P_{s[101]}$=14 · 10^{-2} C m^{-2}, $E_{c[101]}$=7.5 · 10^4 V m^{-1} at 20 °C.	74C3
	Coercive field: Fig. 1625.	
9a	Infrared spectrum: see	77G19
	Optical absorption edge: see	77G19
10a	Raman scattering: Fig. 1626.	
11	Electrical conductivity: see	77B19
	Photoconductivity: see	77G19

Sub-table (1b):

phase	II	I
state	F[a])	P[a])
crystal system	monoclinic[b])	(monoclinic)[b])
space group	Pc-C_s^2 [b),*])	($P2_1/c$-C_{2h}^5)[b])
Θ	66 (2)[a]) °C	

Table 432. $Sn_2P_2S_6$. Crystal structure in the phase II [74D8]. Fractional coordinates of atoms in the unit cell based on the space group Pn, where unit cell parameters a=9.378(5) Å, b=7.488(5) Å, c=6.513(5) Å, β=91.15(5)° are used.

Atom	x	y	z
Sn (1)	0.5270 (2)	0.3856 (2)	0.7224 (3)
Sn (2)	0.0279 (2)	0.1245 (2)	0.7870 (2)
P (1)	0.6836 (4)	0.8608 (5)	0.8108 (6)
P (2)	0.8174 (4)	0.64447 (6)	0.6908 (6)
S (1)	0.4898 (4)	0.7511 (6)	0.8491 (6)
S (2)	0.7788 (4)	0.9462 (6)	0.0750 (6)
S (3)	0.6942 (4)	0.0517 (5)	0.5921 (6)
S (4)	0.0149 (4)	0.7437 (6)	0.6497 (7)
S (5)	0.7147 (4)	0.5633 (5)	0.4288 (6)
S (6)	0.8050 (4)	0.4480 (6)	0.9025 (6)

Nr. 43-17-i $Li(N_2H_5)SO_4$, Lithium hydrazinium sulfate
($M = 136.05$)

1a	Ferroelectric activity in $Li(N_2H_5)SO_4$ was reported by Pepinsky et al. in 1958.[a]) It has, however, been suggested later that this crystal is not ferroelectric.[b])	[a]) 58P4 [b]) 64N2, 71S11, 72A11
b	crystal system: orthorhombic space group: $Pna2_1$-C^9_{2v}	74A5
	Temperature of decomposition ≈285 °C.	58P4
	$\varrho = 1.96\,(1) \cdot 10^3$ kg m^{-3}.	64B8
	Transparent, colorless.	64V1
2a	Crystal growth: evaporation from an aqueous solution of $LiCO_3$ and $(N_2H_6)SO_4$.	71S11
3a	Unit cell parameters: $a = 9.929\,(5)$ Å, $b = 8.973\,(3)$ Å, $c = 5.181\,(2)$ Å at RT.	74A5
b	$Z = 4$.	74A5
	Crystal structure: Table 433.	
	For projections of the structure, see	64B8
4	Thermal expansion: see	66D1
5a	Dielectric constant: Figs. 1627, 1628. No dielectric peak was found in the region from −196 to 140 °C.	58P4
6a	No heat capacity anomaly was found in the region from −120 to 205 °C.	58P4
9a	Refractive indices: Table 434. Birefringence: $\Delta n_{ab} = n_a - n_b = 2.350\,(9) \cdot 10^{-3}$, $\Delta n_{ca} = n_c - n_a = 3.70\,(2) \cdot 10^{-3}$, $\Delta n_{cb} = n_c - n_b = 6.02\,(3) \cdot 10^{-3}$ for $\lambda = 551$ nm at RT.	77B18
	Infrared spectra: see	66W3
10a	Raman scattering: see	65K6
11	Electrical conductivity: Figs. 1629, 1630. The apparent hysteresis loops seem to result from saturation of the ac conduction.	71S11, 72S11
	See also	64V1
13a	NMR: see	63C5, 69H9, 70S9, 71K9, 72H5, 72P1
14a	The mechanism of the previously supposed ferroelectric switching in $Li(N_2H_5)SO_4$ crystals was examined by studying the effect of the anomalous scattering of neutrons from ^6Li on the structure factors of Bijvoet pairs both before and after ferroelectric switching. The absence of any observable change confirms that $Li(N_2H_5)SO_4$ is not ferroelectric.	72A11

Table 434. $Li(N_2H_5)SO_4$. Refractive indices at RT [77B18].

λ [nm]	n_a	n_b	n_c
436	1.5458	1.5435	1.5500
546	1.5378	1.5354	1.5416
590	1.5343	1.5318	1.5380

Table 433, see p. 292.

Table 433. $Li(N_2H_5)SO_4$. Crystal structure [74A5]. Fractional coordinates and temperature parameters. U_{ij} (in Å^2) is defined by Eq. (d) in Introduction and given for non-hydrogen atoms. U (in Å^2) is related with B in Eq. (e) by $U = B/(8\pi^2)$ and given for hydrogen atoms.

	x	y	z	U or U_{11}	U_{22}	U_{33}	U_{12}	U_{13}	U_{23}
Li	0.3312 (3)	0.4333 (3)	0.2533 (12)	0.021 (1)	0.019 (1)	0.018 (1)	−0.001 (1)	−0.001 (2)	−0.002 (2)
S	0.1580 (1)	0.1285 (1)	0.2500	0.0110 (1)	0.0124 (1)	0.0117 (1)	−0.0014 (1)	−0.0009 (2)	0.0001 (2)
O (1)	0.1909 (2)	0.1052 (2)	0.5271 (3)	0.0266 (7)	0.0243 (7)	0.0127 (6)	−0.0025 (5)	−0.0053 (5)	0.0022 (5)
O (2)	0.0115 (1)	0.1540 (2)	0.2282 (5)	0.0115 (4)	0.0317 (7)	0.0281 (8)	−0.0005 (4)	−0.0013 (7)	0.0061 (8)
O (3)	0.2299 (1)	0.2593 (2)	0.1533 (4)	0.0219 (6)	0.0202 (7)	0.0242 (7)	−0.0089 (5)	0.0022 (6)	0.0039 (6)
O (4)	0.3040 (2)	0.4957 (2)	0.6023 (3)	0.0320 (7)	0.0194 (7)	0.0220 (7)	−0.0062 (6)	0.0048 (6)	−0.0087 (6)
N (1)	0.0235 (2)	0.4157 (2)	0.7488 (7)	0.0308 (8)	0.0276 (8)	0.0353 (9)	−0.0086 (6)	−0.0010 (13)	0.0035 (13)
N (2)	0.4400 (2)	0.2136 (2)	0.7381 (6)	0.0221 (6)	0.0235 (6)	0.0218 (7)	0.0016 (5)	0.0015 (10)	−0.0001 (10)
H (1)	0.103 (4)	0.392 (4)	0.776 (9)	0.016 (9)					
H (2)	0.026 (7)	0.445 (6)	0.582 (15)	0.065 (20)					
H (3)	0.358 (4)	0.192 (6)	0.680 (11)	0.040					
H (4)	0.446 (6)	0.234 (8)	0.895 (17)	0.069 (26)					
H (5)	0.462 (3)	0.282 (4)	0.616 (8)	0.008 (8)					

Table 435, see Table Nr. 43-17-ii.

Table 435. $Li(N_2H_5)BeF_4$. Crystal structure [73A7]. Fractional coordinates and temperature parameters. U_{ij} (in Å^2) is defined by Eq. (d) in Introduction and given for non-hydrogen atoms. U (in Å^2) is related with B in Eq. (e) by $U = B/(8\pi^2)$ and given for hydrogen atoms.

	x	y	z	U or U_{11}	U_{22}	U_{33}	U_{12}	U_{13}	U_{23}
Li	0.322 (1)	0.436 (1)	0.251 (2)	0.019 (2)	0.023 (3)	0.021 (3)	−0.001 (2)	−0.004 (3)	−0.002 (3)
N (1)	0.0225 (4)	0.4116 (4)	0.7455	0.033 (2)	0.028 (2)	0.034 (2)	−0.007 (1)	0.000 (2)	0.004 (2)
N (2)	0.4422 (3)	0.2228 (4)	0.742 (1)	0.027 (1)	0.026 (1)	0.026 (2)	0.002 (1)	0.001 (2)	0.004 (2)
Be	0.1584 (4)	0.1270 (5)	0.248 (1)	0.015 (1)	0.020 (2)	0.016 (2)	−0.001 (1)	−0.002 (2)	−0.001 (2)
F (1)	0.1958 (2)	0.1051 (2)	0.540 (1)	0.035 (1)	0.025 (1)	0.018 (1)	−0.003 (1)	−0.004 (1)	0.002 (1)
F (2)	0.0002 (2)	0.1433 (3)	0.231 (1)	0.016 (1)	0.040 (1)	0.031 (1)	−0.003 (1)	−0.001 (1)	0.006 (1)
F (3)	0.2249 (2)	0.2701 (2)	0.141 (1)	0.029 (1)	0.023 (1)	0.026 (1)	−0.010 (1)	0.000 (1)	0.003 (1)
F (4)	0.2983 (2)	0.4881 (2)	0.591 (1)	0.036 (1)	0.025 (1)	0.022 (1)	−0.006 (1)	0.004 (1)	−0.008 (1)
H (1)	0.097 (6)	0.389 (6)	0.803 (17)	0.049 (19)					
H (2)	0.011 (5)	0.463 (5)	0.598 (11)	0.013 (12)					
H (3)	0.362 (4)	0.192 (5)	0.690 (12)	0.018 (13)					
H (4)	0.450 (7)	0.259 (8)	0.908 (15)	0.033 (20)					
H (5)	0.472 (6)	0.296 (7)	0.614 (16)	0.025 (19)					

Nr. 43-17-ii $Li(N_2H_5)BeF_4$, Lithium hydrazinium fluoroberyllate
($M = 125.000$)

1a	Dielectric behavior similar to those of $Li(N_2H_5)SO_4$ was observed in $Li(N_2H_5)BeF_4$ by Palau and Lassabatère in 1971.	71P1
b	crystal system: orthorhombic	71T1
	space group: $Pna2_1$-C_{2v}^9	
	$\varrho = 1.87 \cdot 10^3$ kg m^{-3}.	77Z4
	Colorless.	77Z4
2a	Crystal growth: evaporation from an aqueous solution of $(N_2H_6)BeF_4$ and $LiCO_3$.	71T1
3a	Unit cell parameters: $a = 9.811$ (4) Å, $b = 8.880$ (8) Å, $c = 5.139$ (4) Å at RT.	73A7
b	$Z = 4$.	73A7
	Crystal structure: Table 435.	
	The crystal structure was also determined by neutron diffraction.	75A9
5a	Dielectric constant: Fig. 1631.	
9a	Refractive indices: Table 436.	
	Birefringence: Table 437.	
e	Nonlinear optical susceptibilities:	
	$d_{31}/d_{11}^{\text{quartz}} = 0.2$, $d_{32}/d_{11}^{\text{quartz}} = 0.5$, $d_{33}/d_{11}^{\text{quartz}} = 1$ for $\lambda = 1060$ nm at RT.	77Z4
11	Electrical conductivity: see	71P1
	For discussion on hysteresis loop, see	77P3

Table 436. $Li(N_2H_5)BeF_4$. Refractive indices at RT [77Z4].

λ [nm]	468	480	509	546	577	644	1060
n_a	1.4061	1.404	1.4023	1.4011	1.3999	1.3985	1.3962
n_b	1.4114	1.4099	1.4077	1.4064	1.4054	1.4038	1.4005
n_c	1.4101	1.4085	1.4068	1.4053	1.4042	1.4027	1.3983

Table 437. $Li(N_2H_5)BeF_4$. Birefringence at RT [77Z4].

λ [nm]	468.0	546.1	589.3	656.3
$n_a - n_c$	$-3.31 \cdot 10^{-3}$	$-3.29 \cdot 10^{-3}$	$-3.28 \cdot 10^{-3}$	$-3.28 \cdot 10^{-3}$
$n_b - n_c$	$1.9 \cdot 10^{-3}$	$0.7 \cdot 10^{-3}$	$0.9 \cdot 10^{-3}$	$1 \cdot 10^{-3}$

Nr. 43-18 $[C(NH_2)_3]_2UO_2(SO_4)_2 \cdot 3H_2O$, Guanidine uranyl sulfate trihydrate
($M = 636.35$)

1a	Ferroelectricity in $[C(NH_2)_3]_2UO_2(SO_4)_2 \cdot 3H_2O$ was reported by de Dussel et al. in 1978.	78D2
b	state: (F)	
	crystal system: monoclinic [a])	[a]) 78D2
	Θ: none *)	
	P_s: Perpendicular to (001) (or nearly so).	78D2
	Cleavage plane: (001).	78D2
	Decomposes at 60···100 °C.	78D2

*) Transition to the paraelectric phase does not exist up to the decomposing temperature.

continued

Nr. 43-18 $[C(NH_2)_3]_2UO_2(SO_4)_2 \cdot 3H_2O$, continued

2a	Crystal growth: evaporation of aqueous solution.	78D2
5c	$P_s \approx 6 \cdot 10^{-4}$ C m^{-2}.	78D2
15	Domains: see	78D2

Nr. 43-19 $(NH_2CH_2COOH)_2 \cdot H_2SO_4 \cdot H_2O$, Diglycine sulfate monohydrate
($M = 266.22$)

1a	Dielectric anomaly in $(NH_2CH_2COOH)_2 \cdot H_2SO_4 \cdot H_2O$ was observed and the possibility of antiferroelectric phase transition was mentioned by Dominquez et al. in 1972. According to Piślewski et al., however, this anomaly cannot be attributed to a phase transition, but to the dehydration process. For anhydrous diglycine sulfate $(NH_2CH_2COOH)_2 \cdot H_2SO_4$, see	72D5 75P8 72W3

Nr. 43-20 $H_2(UO_2)_2(AsO_4)_2 \cdot 8H_2O$, Hydrogen uranospinite
($M = 964.031$)

1a	Ferroelectricity of $H_2(UO_2)_2(AsO_4)_2 \cdot 8H_2O$ was reported by de Benyacar et al. in 1975.	75D1
b	(see table below)	75D1

phase	III	II	I
state	(F)	(A)	P
crystal system			tetragonal
space group			
Θ	−20 °C	≈20 °C	

3a	Unit cell parameters: see	75D1
5a	Dielectric constant: Fig. 1632.	
c	Remanent polarization is reported to be of the order of $2.5 \cdot 10^{-2}$ C m^{-2}.	75D1

Nr. 43-21-i $NaTh_2(PO_4)_3$, Sodium dithorium trisphosphate
($M = 771.980$)

1a	Ferroelectricity in $NaTh_2(PO_4)_3$ was found at RT by Topić et al. in 1969.	69T3
b	(see table below)	72T12

state	F
crystal system	monoclinic
space group	Cc-C_s^4

	P_s: direction close to the a axis.	69T3
2a	Crystal growth: B_2O_3 flux method.	69T3
4a	Thermal expansion coefficients: -130 °C $< T <$ 1050 °C. $\alpha_{[100]} = 9 \cdot 10^{-6}$ K^{-1} (±5%), $\alpha_{[010]} = 4 \cdot 10^{-6}$ K^{-1} (±10%), $\alpha_{[001]} = 8 \cdot 10^{-6}$ K^{-1} (±10%)	72T12
5a	Dielectric constants: Fig. 1633.	
c	Remanent polarization: Fig. 1634. Coercive field: Fig. 1635.	
7a	Piezoelectric coefficient: $\alpha \approx 3 \cdot 10^{-15}$ CN^{-1} for elongation parallel to the spontaneous polarization, measured 21 days after poling.	69T3
9a	Refractive indices: $n_\alpha = 1.720$ (5), $n_\beta = 1.730$ (5), $n_\gamma = 1.745$ (5) for white light at 24 °C.	74K12

Nr. 43-21-ii $AgTh_2(PO_4)_3$, Silver dithorium trisphosphate
($M = 856.858$)

1a Ferroelectricity in $AgTh_2(PO_4)_3$ was found by Topić et al. in 1970. — 70T10

b — 70T10

state	F
crystal system	monoclinic
space group	Cc-C_s^4

2a Crystal growth: the crystals were prepared by heating a mixture of stoichiometric proportions of ThO_2, $AgNO_3$, and HPO_3. — 70T10

3a Unit cell parameters: $a = 17.385$ Å, $b = 6.815$ Å, $c = 8.148$ Å, $\beta = 101.10°$, at 25 (1) °C. — 70T10
Crystal habit: Fig. 1636.

5a Spontaneous polarization and coercive field: Fig. 1637.

Nr. 43-21-iii $NaU_2(PO_4)_3$, Sodium diuranium trisphosphate
($M = 783.962$)

1a Ferroelectricity in $NaU_2(PO_4)_3$ was found by Topić and Prodić in 1969. — 69T2

b — 68M2

state	F
crystal system	monoclinic
space group	Cc-C_s^4

$\varrho = 5.41 \cdot 10^3$ kg m^{-3}, $\varrho_x = 5.435 \cdot 10^3$ kg m^{-3}
Transparent and colorless. — 67S17

2a Crystal growth: flux method (B_2O_3). — 68M2
Crystal habit: Figs. 1638, 1639.

3a Unit cell parameters: $a = 17.34$ (2) Å, $b = 6.81$ (1) Å, $c = 8.13$ (1) Å, $\beta = 101°03'(10)'$, RT. — 67S17

b $Z = 4$ — 68M2

5c Spontaneous polarization and coercive field: Fig. 1640.
$P_s = 5.3 \cdot 10^{-2}$ Cm^{-2}, $E_c = 1.3 \cdot 10^7$ V m^{-1} at RT. — 69T2

Nr. 43-22-i $Ag_2H_3IO_6$, Silver trihydrogen periodate
($M = 441.661$)

1a Dielectric anomaly associated with a phase transition in $Ag_2H_3IO_6$ was observed by Baertschi in 1943,[a]) and the possibility of antiferroelectricity was mentioned by Gränicher et al. in 1954.[b]) — [a]) 43B1, [b]) 54G2

b — [a]) 45B1, [b]) 54G2, [c]) 64P2, [d]) 56G3

phase	II [a])	I [a])
state	(A) [b])	P [b])
crystal system	triclinic	trigonal
space group	P$\bar{1}$-C_i^1 [c])	R$\bar{3}$-C_{3i}^2 [c])
Θ	205···245 [a]) K [245···285 K for $Ag_2D_3IO_6$ [d])]	

$\varrho = 5.6 \cdot 10^3$ kg m^{-3}. — 54G2
Color: yellow. — 71A12

2a Crystal growth: Small crystals were grown from a mixture of 10% solution of $AgNO_3$ in 3% HNO_3 with aqueous solution of H_5IO_6. Gel growth technique: see — 45B1, 71A12, 72A14

b Crystal form: rhombohedra. — 71A12

continued

Nr. 43-22-i $Ag_2H_3IO_6$, continued

3a	Unit cell parameters: $a=5.944\,(3)$ Å, $c=12.711\,(5)$ Å at RT (phase I, hexagonal setting). In phase II, the unit cell is doubled for both a and c: $a_{II}=2a_I$, $c_{II}=2c_I$.	54G2 54G2
b	Crystal structure analysis was made with X-rays and neutrons by Peterson.	64P2
5a	Dielectric constant: Fig. 1641.	
6a	Heat capacity: Fig. 1642. Transition heat and transition entropy: $\Delta Q_m=1496\,(84)$ J mol^{-1}, $\Delta S_m=6.69\,(42)$ J K mol^{-1}. See also	 44S2 75T8
9a	Birefringence: Fig. 1643. Reflection in infrared region: Fig. 1644.	
13a	NMR: Fig. 1645. Proton magnetic resonance: see Proton spin-lattice relaxation time: $T_1=2$s at 150 K. Proton-iodine double resonance: see	 59B2 62M5 74R5
14b	Neutron incoherent scattering: see	72T6, 73T2

Nr. 43-22-ii $(NH_4)_2H_3IO_6$, Ammonium trihydrogen periodate
($M=262.0012$)

1a	Dielectric anomaly associated with a phase transition in $(NH_4)_2H_3IO_6$ was observed by Baertschi in 1943,[a]) and the possibility of antiferroelectricity was mentioned by Busch et al. in 1953.[b])	[a]) 43B1 [b]) 53B3
b	see table below	[a]) 45B1 [b]) 53B3 [c]) 68G8 [d]) 55A1

phase	II[a])	I[a])
state	(A)[b])	P[b])
crystal system	trigonal	trigonal
space group	*)	$R\bar{3}$-C_{3i}^2 [c])
Θ	253[a]) K [266 K for 87% deuterated specimen[d])]	

*) Possibility of $R\bar{3}c$-D_{3d}^6 or R32-D_3^7 was mentioned on the basis of NMR measurements [71K7], and of R3-C_3^4 based upon infrared measurements [76R8].

2a	Crystal growth: evaporation from a solution with excess NH_3.	45B1
b	Crystal form: rhombohedra.	37H1
3a	Unit cell parameters: $a=6.928\,(2)$ Å, $c=11.159\,(5)$ Å at RT (phase I, hexagonal setting); $a=13.842\,(5)$ Å, $c=11.157\,(5)$ Å at -45 °C (phase II, hexagonal setting). In phase II, the unit cell is doubled for a: $a_{II}=2a_I$, $c_{II}=c_I$.	68G8
b	Crystal structure analysis of phase I was made by Helmholz.	37H1
5a	Dielectric constant: Fig. 1646; see also	45B1
6a	Heat capacity: Fig. 1647. Transition heat and transition entropy: $\Delta Q_m=1463$ J mol^{-1}, $\Delta S_m=5.85$ J K mol^{-1}.	 45B1
9a	Birefringence: Fig. 1648. Reflection in infrared region: Fig. 1649.	

10a	Raman scattering: Fig. 1650.	
13a	NMR: Fig. 1651, 1652. Proton magnetic resonance: see NQR of ^{127}I: Table 438, see also	 59B2 71K7

Table 438. $(NH_4)_2H_3IO_6$. Quadrupole coupling constant e^2qQ/h and symmetry parameter η of ^{127}I [70K10].

T	e^2qQ/h [MHz]	η	^{127}I sites
RT (295···300 K)	25.35 (2)	0.00	All nuclei on 3-fold axis
94 K	93.2 (1)	0.00	$^1/_4$ of all nuclei on 3-fold axis
	123 (1)	0.33	Two sets of $^3/_8$ of all nuclei

Nr. 43-23 $C_4H_2O_4$, Squaric acid
($M = 114.057$)

1a	A phase transition was found optically by Semmingsen et al. in 1974.	74S14
b	phase: II a) / I a) state: — / — crystal system: monoclinic b) / tetragonal a) space group: $P2_1/m$ b)-C^2_{2h} / — Θ: 97.7 °C a)	a) 74S14 b) 73S5
	For deuterated crystal $\Theta = 243$ °C. $\varrho = 1.90 \cdot 10^3$ kg m^{-3} at RT.	74S14 73S5
3a	Unit cell parameters: $a = 6.129$ (3) Å, $b = 5.273$ (2) Å, $c = 6.1403$ (3) Å, $\beta = 90.00$ (1)° at RT.	73S5
b	$Z = 2$ in phase II. Crystal structure: Figs. 1653, 1654; Tables 439, 440.	73S5
5a	Dielectric constant: Fig. 1655α. Hydrostatic pressure effect: Fig. 1655β, 1655γ; see also	 79Y6
6a	Heat capacity: Fig. 1656.	
9a	Twin pattern, which is observable in phase II, disapperars above Θ.	
10a	Raman scattering: Fig. 1657.	
13a	NMR: see for ^{13}C NMR spectra	79M5
14b	X-ray scattering: Fig. 1658.	

Table 439. $C_4H_2O_4$. Positional and thermal parameters with estimated standard deviations at RT [77S18]. The anisotropic thermal parameters U_{ij} (in $Å^2$) are defined by Eq. (d) in Introduction. For each atom the first and the second lines give the values obtained by neutron and X-ray diffractions, respectively.

	x	y	z	U_{11}	U_{22}	U_{33}	U_{12}	U_{13}	U_{23}
C (1)	0.19951 (8)	0.25000	0.09801 (8)	0.00990 (23)	0.01984 (33)	0.00798 (34)	0.00000	−0.00103 (21)	0.00000
	0.19979 (9)	0.25000	0.09771 (10)	0.01023 (25)	0.02306 (56)	0.00996 (25)	0.00000	−0.00050 (16)	0.00000
C (2)	0.10625 (9)	0.25000	0.30835 (8)	0.00950 (32)	0.01987 (32)	0.00924 (28)	0.00000	−0.00057 (21)	0.00000
	0.10622 (9)	0.25000	0.30805 (10)	0.00991 (24)	0.02312 (35)	0.01037 (26)	0.00000	0.00054 (17)	0.00000
C (3)	0.32073 (8)	0.25000	0.41096 (8)	0.01040 (28)	0.01990 (33)	0.00787 (33)	0.00000	−0.00068 (19)	0.00000
	0.32096 (9)	0.25000	0.41056 (10)	0.01097 (25)	0.02255 (35)	0.00967 (26)	0.00000	−0.00023 (17)	0.00000
C (4)	0.42013 (8)	0.25000	0.18800 (8)	0.00935 (32)	0.01963 (32)	0.00909 (28)	0.00000	−0.00093 (19)	0.00000
	0.42005 (10)	0.25000	0.18766 (10)	0.01012 (24)	0.02255 (35)	0.01045 (27)	0.00000	−0.00028 (17)	0.00000
O (1)	0.11455 (9)	0.25000	−0.09355 (9)	0.01436 (36)	0.03003 (45)	0.00833 (38)	0.00000	−0.00293 (24)	0.00000
	0.11404 (9)	0.25000	−0.09398 (9)	0.01469 (25)	0.03489 (38)	0.00921 (24)	0.00000	−0.00210 (16)	0.00000
O (2)	−0.08668 (10)	0.25000	0.39011 (9)	0.01026 (35)	0.03008 (45)	0.01343 (36)	0.00000	0.00173 (24)	0.00000
	−0.08710 (8)	0.25000	0.39052 (9)	0.00953 (24)	0.03488 (37)	0.01476 (27)	0.00000	0.00183 (16)	0.00000
O (3)	0.39223 (9)	0.25000	0.59730 (9)	0.01488 (36)	0.03016 (46)	0.00851 (35)	0.00000	−0.00241 (22)	0.00000
	0.39254 (9)	0.25000	0.59786 (9)	0.01552 (27)	0.03334 (37)	0.00965 (24)	0.00000	−0.00199 (17)	0.00000
O (4)	0.60516 (9)	0.25000	0.11144 (9)	0.00925 (32)	0.03008 (44)	0.01336 (35)	0.00000	0.00092 (24)	0.00000
	0.60573 (9)	0.25000	0.11098 (9)	0.00955 (25)	0.03350 (36)	0.01494 (27)	0.00000	0.00176 (16)	0.00000
H (1)	0.22983 (17)	0.25000	−0.21518 (17)	0.02826 (54)	0.03929 (64)	0.01932 (40)	0.00000	−0.00212 (40)	0.00000
	0.2303 (21)	0.25000	−0.2169 (26)	—	—	—	—	—	
H (2)	−0.20928 (21)	0.25000	0.27418 (18)	0.02075 (46)	0.03990 (62)	0.02864 (51)	0.00000	0.00183 (39)	0.00000
	−0.2155 (23)	0.25000	0.2631 (23)	—	—	—	—	—	

Table 440. $C_4H_2O_4$. Bond distances (in Å) and angles (in deg) at RT [77S18].

	neutron [77S18]	X-ray [73S5]
C(1)–O(1)	1.289 (1)	1.289 (1)
C(2)–O(2)	1.287 (1)	1.289 (1)
C(3)–O(3)	1.227 (1)	1.231 (1)
C(4)–O(4)	1.230 (1)	1.232 (1)
C(1)–C(2)	1.414 (1)	1.413 (1)
C(1)–C(4)	1.464 (1)	1.459 (1)
C(2)–C(3)	1.461 (1)	1.459 (1)
C(3)–C(4)	1.500 (1)	1.497 (1)
O(1)–H(1)	1.030 (1)	1.04 (2)
O(2)–H(2)	1.037 (1)	1.11 (2)
O(1)···O(3)	2.553 (1)	2.548 (1)
O(2)···O(4)	2.554 (1)	2.548 (1)
H(1)···O(3)	1.524 (2)	1.51 (2)
H(2)···O(4)	1.517 (2)	1.44 (1)
C(4)–C(1)–C(2)	91.66 (4)	91.7 (1)
C(4)–C(1)–O(1)	136.14 (4)	136.3 (1)
C(2)–C(1)–O(1)	132.21 (5)	132.0 (1)
C(1)–C(2)–C(3)	91.72 (4)	91.6 (1)
C(1)–C(2)–O(2)	136.85 (5)	137.1 (1)
C(3)–C(2)–O(2)	131.43 (5)	131.3 (1)
C(2)–C(3)–C(4)	88.40 (3)	88.4 (1)
C(2)–C(3)–O(3)	136.59 (5)	136.4 (1)
C(4)–C(3)–O(3)	135.01 (5)	135.2 (1)
C(3)–C(4)–C(1)	88.23 (3)	88.3 (1)
C(3)–C(4)–O(4)	136.48 (5)	136.4 (1)
C(1)–C(4)–O(4)	135.30 (5)	135.3 (1)
C(1)–O(1)–H(1)	112.65 (7)	112.7 (7)
C(2)–O(2)–H(2)	113.61 (7)	112.0 (7)
C(3)–O(3)···H(1)	118.15 (5)	117.9 (5)
C(4)–O(4)···H(2)	116.24 (5)	117.2 (5)
O(1)–H(1)···O(3)	177.43 (9)	177.9 (12)
O(2)–H(2)···O(4)	117.85 (9)	175.6 (12)

Nr. 43-24 $SnCl_2 \cdot 2H_2O$, Tin chloride dihydrate
($M = 225.63$)

1a	Phase transition in $SnCl_2 \cdot 2H_2O$ accompanied by a marked dielectric anomaly was first reported by Kiriyama and Kiriyama in 1969.	70K11
b	phase: II / I; crystal system: monoclinic / monoclinic; space group: $P2_1/c$-C_{2h}^5 [a]) / $P2_1/c$-C_{2h}^5 [a]); Θ: 217.94 [b]) K	[a]) 73K13 [b]) 74M9
	$\Theta_{II-I} = 234.94$ K in deuterated compound.	74M9
	$T_{melt} = 40$ °C.	74M9
	Cleavage plane: (100). Transparent, colorless, deliquescent, efflorescent.	
2a	Crystal growth: evaporation or cooling of aqueous solution containg HCl, or cooling from melt.	73M8
b	Crystal form: Fig. 1659. continued	

phase	II	I
crystal system	monoclinic	monoclinic
space group	$P2_1/c$-C_{2h}^5 [a])	$P2_1/c$-C_{2h}^5 [a])
Θ	217.94 [b]) K	

Nr. 43-24 $SnCl_2 \cdot 2H_2O$, continued

3a	Unit cell parameters:	73K13

phase	T	a	b	c	β
	°C	Å			°
I	24	9.326 (2)	7.254 (2)	8.974 (3)	114.91 (5)
II	−185	9.158 (2)	7.156 (3)	8.942 (3)	114.50 (5)

b	$Z=4$ in phases I and II.	73K13
	Crystal structure: see	73K13, 77K11
4	Thermal expansion: see	77K11
5a	Dielectric constant: Fig. 1660; see also	70K11
	H_2O-content dependence of dielectric anomaly: see	78T8
	Effect of hydrostatic pressure: $d\Theta_{II-I}/dp=3.9\ (1) \cdot 10^{-8}$ K Pa^{-1}.	77G8
6a	Heat capacity: Fig. 1661; critical behavior of heat capacity anomaly, Table 441; see also	73M8
	Heat capacity of $SnCl_2 \cdot H_2O$-$SnCl_2 \cdot D_2O$: see	79T3, 79T4
10a	Raman scattering: see	77W4
13a	NMR: Fig. 1662; see also	76M18
	NQR of ^{35}Cl: see	78T8

Table 441. $SnCl_2 \cdot 2H_2O$, $SnCl_2 \cdot 2D_2O$. Critical behavior of heat capacity anomaly [74M9]. $\Delta C_p = A_\pm |T-\Theta_{II-I}|^{-\alpha_\pm}/\Theta_{II-I}$, ΔC_p: anomalous part of heat capacity, + and − signs correspond to temperatures above and below the Θ_{II-I} transition point, respectively.

	$SnCl_2 \cdot 2H_2O$	$SnCl_2 \cdot 2D_2O$
Θ_{II-I}	217.94 (5) K	234.47 (5) K
α_+	0.534 (21)	0.574 (10)
α_-	0.478 (10)	0.544 (10)

Nr. 43-25 $NH_4PF_6 \cdot NH_4F$, Hexafluorphosphate salt
($M=200.039$)

1a	Antiferroelectricity in $NH_4PF_6 \cdot NH_4F$ was discovered by Vedam et al. in 1959.	59V1
b	(see table below)	[a]) 59V1 [b]) 62P2

phase	III [a])	II [a])	I [a])
state	A [a])	(A) [a])	P [a])
crystal system	orthorhombic [a])	orthorhombic [a])	tetragonal [a])
space group			C4/nmm-D_{4h}^7 [b])
Θ	−101 [a])	−45 [a]) °C	

3a	Unit cell parameters: $a=7.88$ Å, $c=10.90$ Å at RT.	62P2
	X-ray diffraction reveals superstructure along the a and b axes below −45 °C and a doubling of the c axis below −101 °C.	59V1, 62P2
5a	The dielectric constant $\varkappa_a$ shows a small anomaly at −45 °C and a pronounced anomaly at −101 °C. Fig. 1663.	62P2
12a	NMR: Fig. 1664.	

Nr. 43-26 $H_2NCONHNH_2 \cdot HCl$, Semicarbazide hydrochloride
($M = 111.531$)

1a	Dielectric anomaly in $H_2NCONHNH_2 \cdot HCl$ was observed and the possibility of ferroelectric phase transition was mentioned by Rocaries and Boldrini in 1972.	72R5
b	see table below	a) 72R5 b) 60J2, 65N3

phase	IV a)	III a)	II a)	I a)
state			(F) a)	P a)
crystal system				orthorhombic b)
space group				$P2_12_12_1$-D_2^4 b)
Θ	−230 a)	19 a)	21 a) °C	

	Chemical decomposition and polymerization begin at about 30 °C.	72R5
	$\varrho = 1.58 \cdot 10^3$ kg m^{-3}.	60J2
	Colorless.	65N3
2a	Crystal growth: slow evaporation of saturated aqueous solution.	76F1
	Solubility: see	71B22
b	Crystal form: prism elongated along [001].	65N3
3a	Unit cell parameters: $a = 7.543$ (15) Å, $b = 13.224$ (26) Å, $c = 4.668$ (15) Å at RT (phase I).	60J2
b	$Z = 4$ in phase I.	65N3
	Crystal structure: Table 442; Fig. 1665; see also	60J2
5a	Dielectric constant: Fig. 1666. $\varkappa_{(110)} = C/(T - \Theta_p)$, $T > \Theta_p$ where $C = 7.2 \cdot 10^3$ K, $\Theta_p = 21$ °C.	72R5
9a	Infrared spectra above $1.2 \cdot 10^{13}$ Hz (400 cm^{-1}): see	78S12
10a	Raman scattering: Fig. 1667.	
14b	X-ray diffuse scattering: see	72R5

Table 442. $H_2NCONHNH_2 \cdot HCl$. Crystal structure in the phase I [65N3]. Fractional coordinates.

	x	y	z
Cl	0.2543 (4)	0.1399 (2)	0.9636 (6)
O	0.7587 (9)	0.0874 (5)	0.8233 (17)
N (1)	0.9267 (15)	0.1605 (8)	0.4787 (30)
N (2)	0.6335 (11)	0.1320 (6)	0.3930 (19)
N (3)	0.4909 (13)	0.0655 (7)	0.4551 (26)
C	0.7767 (12)	0.1236 (7)	0.5787 (24)

Nr. 43-27 $C_9H_{18}NO$, Tanane (Nitroxyde tetramethyl-2,2,6,6-piperidine oxyle)
($M = 156.25$)

1a	Ferroelectricity of tanane was first reported by Bordeaux et al. in 1973.	73B8
b	see table below	73B8 a) 74B15 b) 78L1

phase	III	II	I
state		(F)	P
crystal system	monoclinic	orthorhombic	tetragonal
space group	Cm-C_s^3	Fdd2-C_{2v}^{19}	$I\bar{4}2d$-D_{2d}^{12} a)
Θ		13.43 (1) °C b)	

continued

Nr. 43-27 $C_9H_{18}NO$, continued

1b	The temperature range where the monoclinic phase III exists strongly depends on the thermal hysteresis of a specimen. Phase II is metastable against the phase III. $P_s \parallel [001]$. $T_{melt} \approx 38$ °C Color: orange in phase III, red in phase I.	74B15 73B8 74B15
2a	Crystal growth: vapor deposition method.	78L1
3a	Unit cell parameters: $a=b=15.80$ Å, $c=8.130$ Å in phase I at 22 °C, $a=22.86$ Å, $b=21.48$ Å, $c=8.01$ Å in phase II at -20 °C. $a=6.62\,(1)$ Å, $b=14.36\,(2)$ Å, $c=5.88\,(1)$ Å, $\beta=119.1\,(1)°$ in phase III at 18 °C for an orange thin plate crystal obtained by sublimation at 3···10 °C.	74B15
b	Crystal structure: $Z=8$, 16, 2 in phase I, II, and III, respectively. See also Form of a tanane molecule: Fig. 1668.	74B15 72C1
4	Lattice distortion: Fig. 1669.	
5a	Dielectric constant: Fig. 1670.	
c	$P_s \approx 5 \cdot 10^{-3}\,Cm^{-2}$. Hysteresis loop has not been observed because of the low mobility of the domain wall.	76B1 73B8
6a	Heat capacity: Fig. 1671. Transition entropy: $\Delta S_m = 6.3\cdots8.8\,J\,K^{-1}\,mol^{-1}$.	73B8
9a	Birefringence: Fig. 1672.	
b	Electrooptic effect: Fig. 1673. $r_{63}^T = 2.75\,(8) \cdot 10^{-12}\,mV^{-1}$ at 170 °C.	76B1
16	Domain structure: Switching process is studied in	74B16

Nr. 43-28 H_2O, Ice
($M=18.0152$)

1a	Anomalous dielectric behavior of the hexagonal ice (Ih) and the cubic ice (Ic) for dc fields was reported and the possibility of ferroelectricity was mentioned by Dengel et al. in 1964 for Ih and by Cubiotti and Geracitano in 1967 for Ic.	64D2 67C9
5a	Phase diagram: Fig. 1674. Dielectric properties of Ih: Figs. 1675, 1676. $\varkappa-\varkappa_\infty = C/(T-\Theta_p)$, $C=2.25 \cdot 10^4$ K, $\Theta_p=46$ for $\boldsymbol{E} \parallel c$; $C=2.37 \cdot 10^4$ K, $\Theta_p=0$ K for $\boldsymbol{E} \perp c$. Dielectric properties of phase VI (see Fig. 1674): $\varkappa-\varkappa_\infty = C/(T-\Theta_p)$, $C=3.23 \cdot 10^4$ K, $\Theta_p=47$ K at $p=9 \cdot 10^8$ Pa.	 78K5 76J4

Nr. 43-29 Alkali halides with additives

1	Dielectric anomaly of KCl with KOH addition was observed and the possibility of ferroelectricity was mentioned by Känzig et al. in 1964. Dielectric hysteresis loop was observed in KCl:Li and the possibility of ferroelectricity was mentioned by Fiory in 1969. Dielectric hysteresis loops were observed for RbCl:OH, NaBr:F and KCl:CN as well as for KCl:OH and KCl:Li at low temperatures. Heat capacity anomaly, thermal conductivity, paraelectric resonance, microwave absorption: see	64K10 69F3 71F2 70N1

Nr. 43-30-i GeTe, Germanium telluride
($M = 200.19$)

1a	GeTe is a semiconductor with a narrow band gap. In 1966, Pawley et al. suggested that GeTe is a diatomic ferroelectric.	66P1
b	phase: II / I; state: (F)[a] / (P)[a]; crystal system: rhombohedral[b][c][d] / cubic[b][c][d]; transition at 400[c,d] °C	[a] 66P1, [b] 58A1, [c] 61K5, [d] 63B5
	Polar axis ∥ [111].	66G2
	$T_{melt} \approx 725(3)$ °C.	58h1
2	Crystal growth, see	69B3, 72W4
3a	Unit cell parameters: $a = 5.988$ Å, $\alpha = 88.25°$ (rhombohedral) at 300 K.	63B5
b	$Z = 4$ (phases I and II).	66G2
	Crystal structure: phase I: face centered cubic NaCl type;	51S3, 61K5, 63B5
	phase II: distorted NaCl type.	53S2, 61K5, 63B5
	Polar structure was proposed by Goldak et al.: take the origin of coordinates as midway between the Ge and Te ions in the cubic NaCl structure (Ge and Te are on face centered lattices with origins at $(\frac{1}{4}, \frac{1}{4}, \frac{1}{4})$ and $(-\frac{1}{4}, -\frac{1}{4}, -\frac{1}{4})$, respectively). The distorted structure is obtained by displacing two f.c.c. lattices to (u, u, u) and $(-u, -u, -u)$, with $u = 0.237(2)$ and expanding the lattice along the [111] to make the interaxial angle $\alpha = 88.25°$, at 300 K.	66G2
4	Thermal expansion,	70N3, 71N14
5,6	Dielectric constant, heat capacity and phonon frequency, see	69A10, 74L4, 76B5
9	Optical properties, see	67T7, 71S27
10	Raman effect, see	70S21
11	Conduction, superconductivity, see	67T7, 64H6, 66H12, 69A10, 70L7
13c	Mössbauer effect, see	71S13

Table for 1b:

phase	II	I
state	(F)[a]	(P)[a]
crystal system	rhombohedral[b][c][d]	cubic[b][c][d]
	400[c,d] °C	

Nr. 43-30-ii SnTe, Tin telluride
($M = 246.29$)

1a	On the basis of the crystal structure and of the anomaly of the lattice vibration, Pawley et al. suggested that SnTe is "nearly ferroelectric" at low temperatures.	66P1

continued

Nr. 43-30-ii SnTe, continued

1b	state: (P)[a]; crystal system: cubic[b, c, d]	[a]) 66P1, [b]) 58A1, [c]) 61K5, [d]) 63B5
	Evidence of a structural phase transition has been reported in the temperature range, 80···160 K.	74F7, 75I9, 75K26, 77S39
	$T_{melt} = 790$ °C.	58h1
2a	Crystal growth, see	62S1
3a	Unit cell parameter: $a = 6.319$ Å (cubic) at 300 K	63B5
b	$Z = 4$ (NaCl type)	61K5, 63B5, 66G2
	At high pressure the structure is orthorhombic	64K1
4	Thermal expansion, see	70B3
5	Dielectric constant, see	66P1, 65R2
6	Heat capacity, see	77H4, 76B5
8a	Elastic constant, see	69B4
9a	Refractive index, light absorption, see	69B15, 70B8, 67T7, 71S27
10a	Raman spectrum: Fig. 1677, see also	71B29, 74B18, 77S39
11	Conduction, see	64B5, 67T7, 65B11, 65B13, 65R2, 70S1, 72A8, 75K26, 76K20
	Superconductivity	66H11, 72G12, 69A10
	Band structure	69H6, 70L8, 72S6
13c	Mössbauer effect, see	68B16, 71S13, 74K13, 77F1, 76A10
14b	Phonon frequency: Figs. 1678, 1679.	

Nr. 43-30-iii GeTe-SnTe solid solution

1a	Possibility of ferroelectricity in the system GeTe–SnTe was discussed by Pawley et al. in 1966.	66P1
b	The system GeTe–SnTe forms a complete solid solution.	58A1,63B5
	phase: II / I; state: (F)[a] / (P)[a]; crystal system: rhombohedral[b][c][d] / cubic[b][c][d]	[a]) 66P1, [b]) 58A1, [c]) 61K5, [d]) 63B5
	Polar axis ‖ [111]. Transformation temperature as a function of composition; Fig. 1680.	66G2
2a	Preparation of sample, see	58A1,61K5, 63B5,70L2, 74R4
	Phase diagram: Fig. 1681.	
3a	Unit cell parameters: Figs. 1682, 1683.	
b	$Z=4$ (phases I and II). Crystal structure: phase I: f.c.c. NaCl structure; phase II: rhombohedral, polar	51S3, 58A1,61K5, 63B5,66G2
	No ordering of Sn and Ge on their sublattice was observed.	63B5
4	Unit cell parameters at Θ_{II-I} vs. composition of alloys, see	61K5
6	Heat capacity, see	77H5
8a	Acoustic properties, see	75R4, 75S35
13c	Mössbauer effect, see	70K14, 70R8

phase	II	I
state	(F)[a]	(P)[a]
crystal system	rhombohedral[b][c][d]	cubic[b][c][d]

Nr. 43-31 $(CH_2CF_2)_n$, Polyvinylidene fluoride (PVF_2, PVDF)

1a	High piezoelectric activity of the poled and stretched PVF_2 film with mm2 symmetry was reported by Kawai.	69K2
b	Stretched films, uniaxially (UO) or biaxially (BO) oriented, show ferroelectric-like $P-E$ hysteresis loop.	78k1
	Ferroelectricity was suggested by Bergman et al.; the ferroelectric transition temperature, however, has not been found.	71B9, 72M10, 71G6
2b	Usually X and Z axis stand for the direction of stretching and poling, respectively.	78k1
3	In the solid state, the molecules form a semicrystalline polymer (crystallinity of ≈50%). There are two primary crystal structures, α phase (or Form II) and β phase (or Form I)	72H4, 76W3
4	Thermal expansion coefficients at RT:	78k1

film	α_{11}	α_{22}
	10^{-4} K^{-1}	
UO	0.13	1.45
BO	1.24*)	1.00*)

*) For BO film, α_{11} and α_{22} should be equal.

continued

Nr. 43-31 $(CH_2CF_2)_n$, continued

5a	Dielectric constants: see	73B12, 76D3, 76O2
c	Saturated polarization (apparent P_s), remanent polarization P_r, coercive field E_c:	78k1

film	apparent P_s	P_r	E_c
	10^{-2} C m^{-2}		
UO	5.83	4.47	slightly less than $1 \cdot 10^8$ V m^{-1}
BO	6.33	5.65	

d	Pyroelectricity: For UO film, $p_3 = -2.74 \cdot 10^{-5}$ C K^{-1} m^{-2}; For BO film, $p_3 = -1.48 \cdot 10^{-5}$ C K^{-1} m^{-2}; $p_3^S \approx p_3/2$; at RT. p_3^S: pyroelectric coefficient at constant strain (or primary coefficient).	78k1
	For pyroelectricity under different poling conditions, see	71N3, 71B9, 74P2
	High coefficient, $p_3 = -4.1 \cdot 10^{-5}$ C K^{-1} m^{-2}, is reported in	74D3
7a	Piezoelectricity: Table 443; Fig. 1684; see also	76O6
b	Electrostriction: see	74T2
8a	Elastic compliances, stiffnesses: Fig. 1685; see also	78k1
9a	Infrared absorption: see	77T4
e	Nonlinear optical coefficients: see	72M10
11	Electrical conductivity: see	74Y3, 76D3
13a	NMR: see	76M10

Table 443. $(CH_2CF_2)_n$ (PVF_2). Piezoelectric constants at RT [78k1].

Film	d_{31}	d_{32}	d_{33}
	10^{-12} C N^{-1}		
UO	21.4	2.3	−31.5
BO	4.34*)	4.35*)	−13.5

*) For BO film, d_{31} and d_{32} should be equal.

44 Liquid crystals and related liquids

Nr. 44-1 $C_{14}H_{14}N_2O_3$

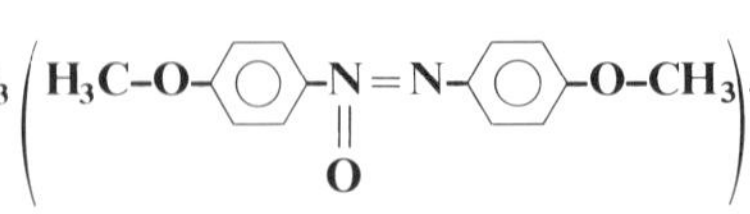

, *p*-azoxyanisole

($M = 258.276$)

1	Ferroelectric-like hysteresis loops were observed in *p*-azoxyanisole in isotropic liquid phase and nematic liquid crystal phase by Williams and Heilmeier in 1966. However, there are some doubts about the existence of the ferroelectricitv.	66W9 77I1

Nr. 44-2 $C_{16}H_{18}N_2O_3$ 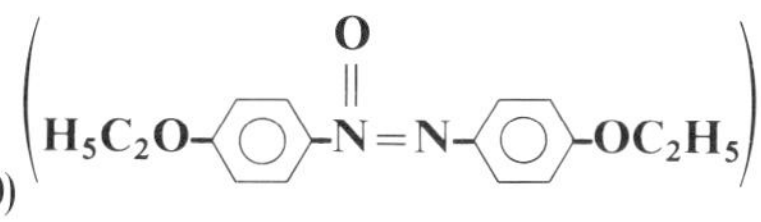, ***p*-azoxyphenetole**
($M = 286.330$)

1	Ferroelectric-like hysteresis loops were observed in *p*-azoxyphenetole in isotropic liquid phase and nematic liquid crystal phase by Kapustin and Vistin in 1965. However, there are some doubts about the existence of the ferroelectricity.	65K3 7711

Nr. 44-3 $C_{11}H_{14}O_3$ 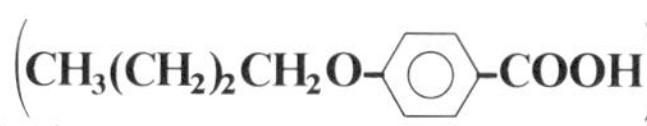, ***p*-butoxybenzoic acid**
($M = 194.230$)

1	Ferroelectric-like hysteresis loops were observed in *p*-butoxybenzoic acid in nematic liquid crystal phase by Williams and Heilmeier in 1966. However, there are some doubts about the existence of the ferroelectricity.	66W9 7711

Nr. 44-4 $C_{10}H_{10}O_3$ 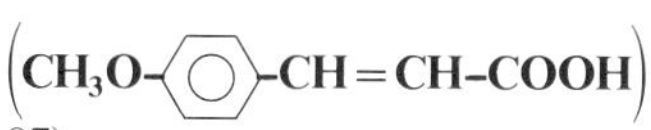, ***p*-methoxycinnamic acid**
($M = 178.187$)

1	Ferroelectric-like hysteresis loops were observed in *p*-methoxycinnamic acid in nematic liquid crystal phase by Williams and Heilmeier in 1966. However, there are some doubts about the existence of the ferroelectricity.	66W9 7711

Nr. 44-5 $C_{12}H_{10}N_2$ 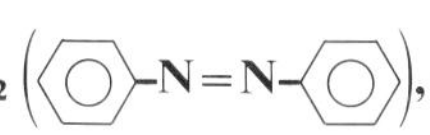, **azobenzene**
($M = 182.224$)

1	Ferroelectric-like hysteresis loops were observed in azobenzene in the liquid phase by Williams and Heilmeier in 1966. However, there are some doubts about the existence of the ferroelectricity.	66W9 7711

Nr. 44-6 $C_{12}H_{10}N_2O$ 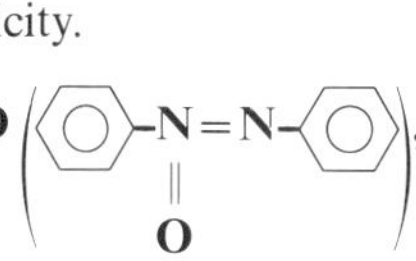, **azoxybenzene**
($M = 198.224$)

1	Ferroelectric-like hysteresis loops were observed in azoxybenzene in liquid phase by Williams and Heilmeier in 1966. However, there are some doubts about the existence of the ferroelectricity.	66W9 7711

Nr. 44-7 $C_{31}H_{43}O_3N$ 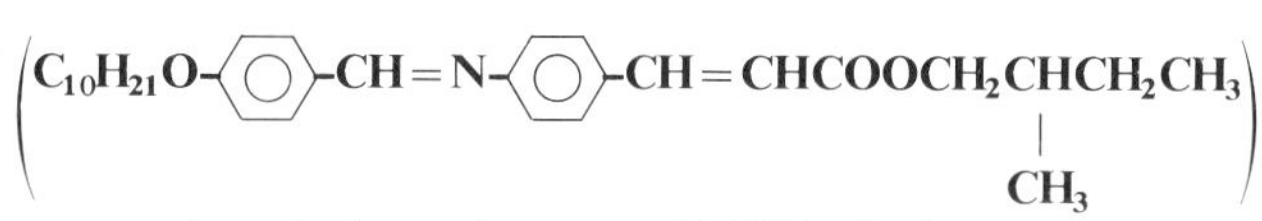
***p*-decyloxybenzylidene *p'*-amino 2 methylbutyl cinnamate (DOBAMBC)**
($M = 477.686$)

1a	Possibility of the ferroelectric activity of *L*-DOBAMBC in the smectic C phase was reported by Meyer et al. in 1975. Ferroelectric hysteresis loops were observed in the smectic C and the smectic H phases by Yoshino et al. in 1977.	75M13 77Y8

b

phase	III (smectic H)	II (smectic C)	I (smectic A)
state	F	F	P
Θ	63 (cooling)	95 °C	

77Y8, 75M13

The smectic H phase can be attained only by cooling from the smectic C phase. The smectic C phase is obtainable from crystalline solid at 76 °C by heating. (75M13)
DOBAMBC becomes an isotropic liquid above 117 °C. (75M13)

continued

Nr. 44-7 $C_{31}H_{43}O_3N$ 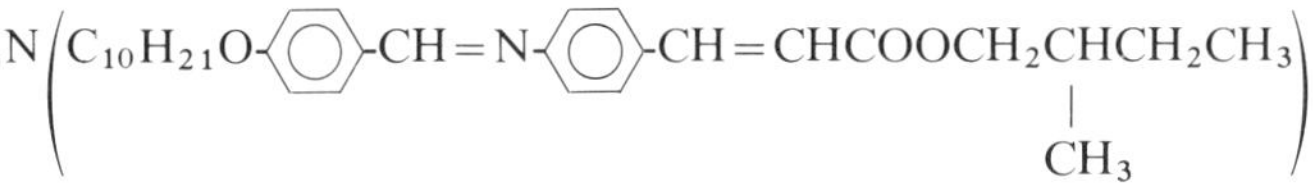 , continued

5a	Dielectric constant: Fig. 1686.	
	Frequency response: see	78H12
	The Curie-Weiss law holds in a narrow temperature range of about 1 °C above Θ_{II-I}.	78H12
c	Spontaneous polarization: Fig. 1687.	
	$P_s = A(\Theta_{II-I} - T)^\beta$, $\beta = 0.30\,(3)$ in the range $0.1 < \Theta_{II-I} - T < 20$ °C.	78H12
	Coercive field: Fig. 1688.	
d	Pyroelectricity is observed in the smectic C and the smectic H phases.	76Y4
7	Polarization is created by applying shearing force parallel to the smectic plate.	75P7
9b	Field dependence of the transmission intensity of light: see	78Y8

Nr. 44-8 $C_{33}H_{47}O_3N$ 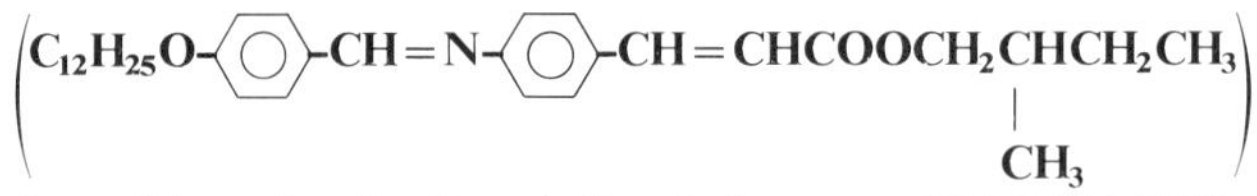

***p*-dodecyloxybenzylidene *p'*-amino 2 methylbutyl cinnamate (DDOBAMBC)**
($M = 505.739$)

1a Ferroelectric hysteresis loop was observed in the smectic C phase of *L*-DDOBAMBC by Yoshino et al. in 1977. — 77Y7

b — 77Y7

phase	II (smectic C)	I (smectic A)
state	F	P
Θ	87 °C	

5c Spontaneous polarization: $P_s \approx 3 \cdot 10^{-5}$ Cm^{-2} at 70 °C. — 77Y7
Coercive field: $E_c \approx 4 \cdot 10^2$ kV m^{-1} in the range from 60 °C to 80 °C. — 77Y7

Nr. 44-9 $C_{35}H_{51}O_3N$ 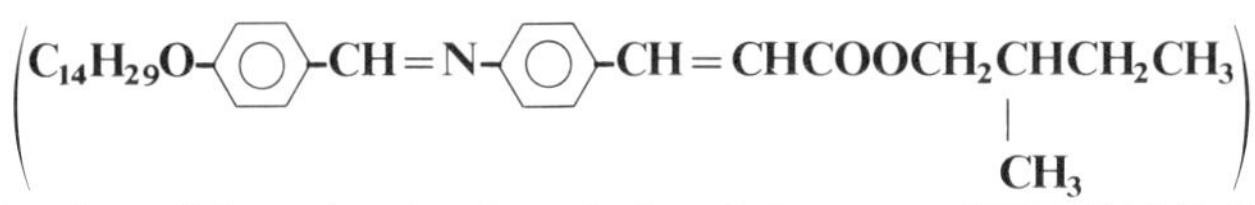

***p*-tetradecyloxybenzylidene *p'*-amino 2 methylbutyl cinnamate (TDOBAMBC)**
($M = 533.793$)

1a Ferroelectric hysteresis loop was observed in the smectic C phase of *L*-TDOBAMBC by Yoshino et al. in 1977. — 77Y7

b — 77Y7

phase	II (smectic C)	I (smectic A)
state	F	P
Θ	84 °C	

5a Temperature dependence of the dielectric constant: see — 77Y7

c Spontaneous polarization: $P_s \approx 3 \cdot 10^{-5}$ Cm^{-2} at 60 °C. — 77Y7
Coercive field: $E_c \approx 6 \cdot 10^2$ kVm^{-1} in the range from 50 °C to 70 °C. — 77Y7

III Figures

14 SbSI family

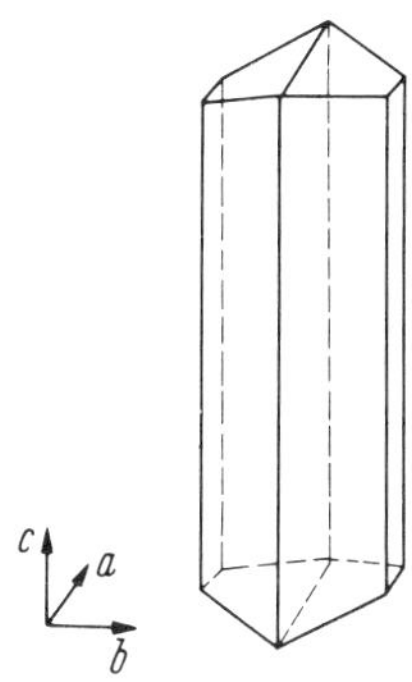

Fig. 1. BiSCl. Crystal form and axes [50 D 2].

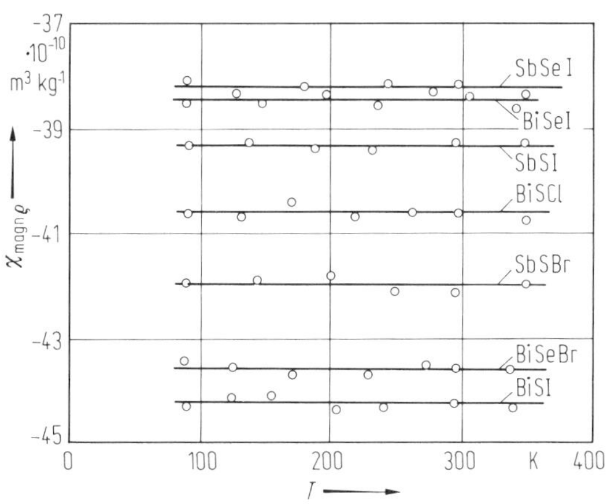

Fig. 3. BiSCl, SbSBr, BiSeBr, SbSI, BiSI, SbSeI, BiSeI. $\chi_{magn\,\varrho}$ vs. T [76 M 8].

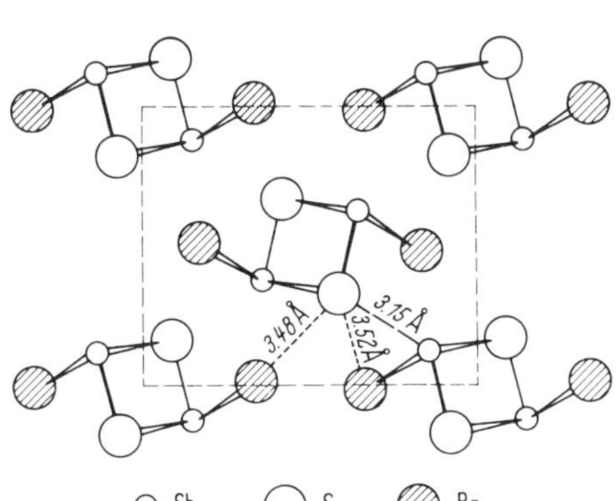

Fig. 4. SbSBr. Projection of atoms along c [58 M 5].

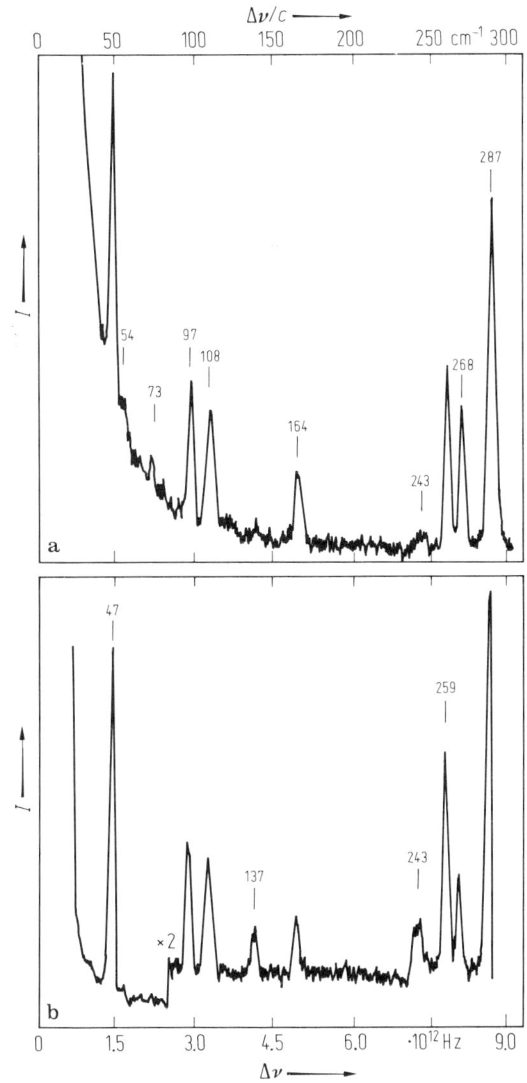

Fig. 2. BiSCl. I vs. $\Delta\nu$ at 60 K [76 F 17]. I: Raman scattering intensity. Fig. a: $X(Y,X)Y$ geometry and Fig. b: $X(Z,X)Y$. Peak frequencies are given in units of cm^{-1}. Note the change of intensity scale near $2.7\cdot10^{12}$ Hz in b).

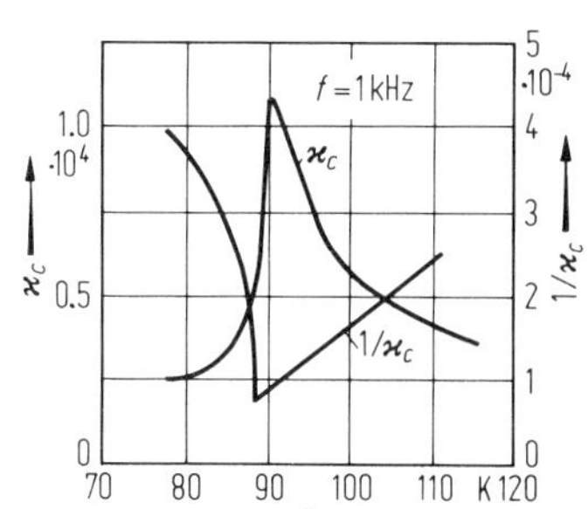

Fig. 5. SbSBr. $\varkappa_c$, $1/\varkappa_c$ vs. T [68 P 2].

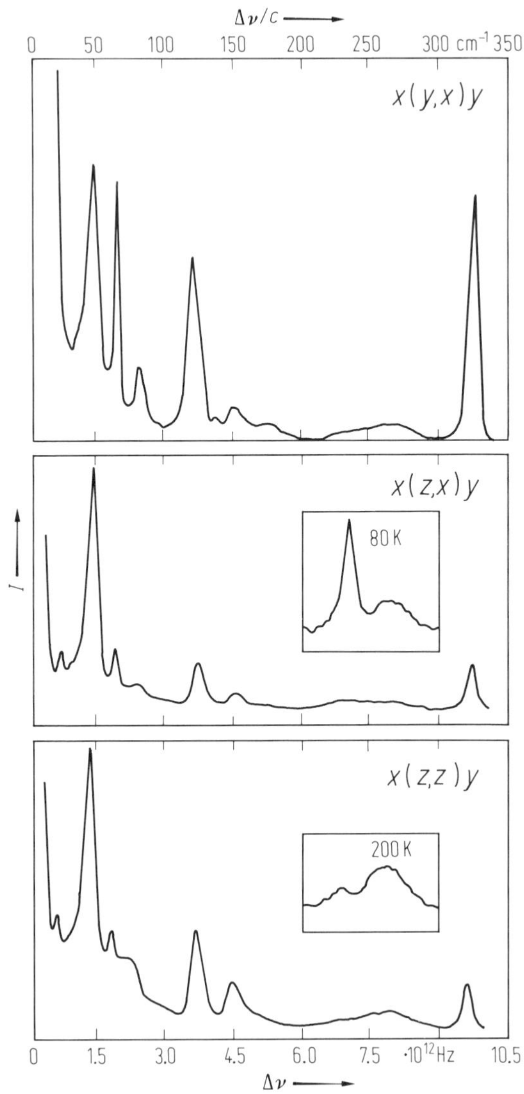

Fig. 6. SbSBr. I vs. $\Delta\nu$ at 297 K [73F13]. I: Raman scattering intensities for three different scattering geometries. $\Delta\nu$: frequency shift. The inserts show parts of the spectra at lower temperatures.

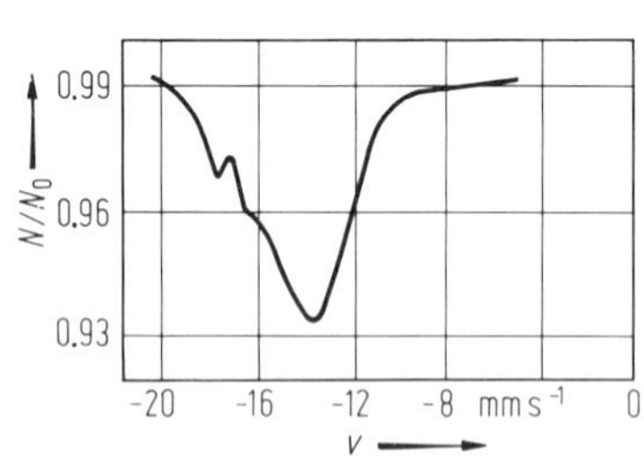

Fig. 8. SbSBr. N/N_0 vs. v [71K6]. N/N_0: counting rate in Mössbauer spectrum for ^{121}Sb nuclei, v: absorber velocity. Radioactive isotope ^{121}Sb in a $CaSnO_3$ matrix was used as source. The source and the absorber were at liquid N_2 temperature.

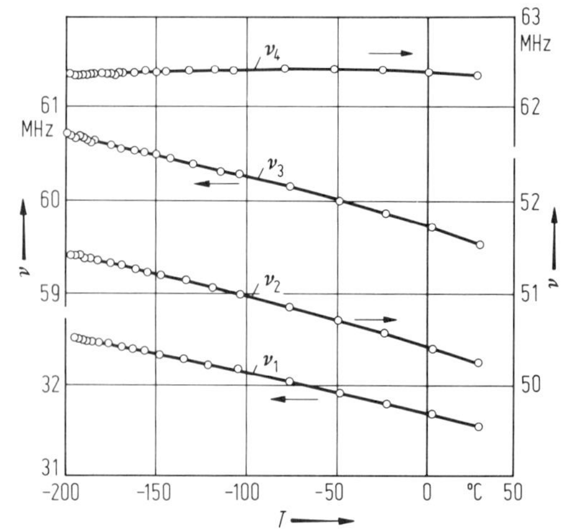

Fig. 7. SbSBr. NQR frequencies vs. T [69P8]. The ν's are the following resonance frequencies: $\nu_1(^{123}Sb_{1/2\leftrightarrow 3/2})$, $\nu_2(^{121}Sb_{1/2\leftrightarrow 3/2})$, $\nu_3(^{123}Sb_{3/2\leftrightarrow 5/2})$, and $\nu_4(^{81}Br_{1/2\leftrightarrow 3/2})$.

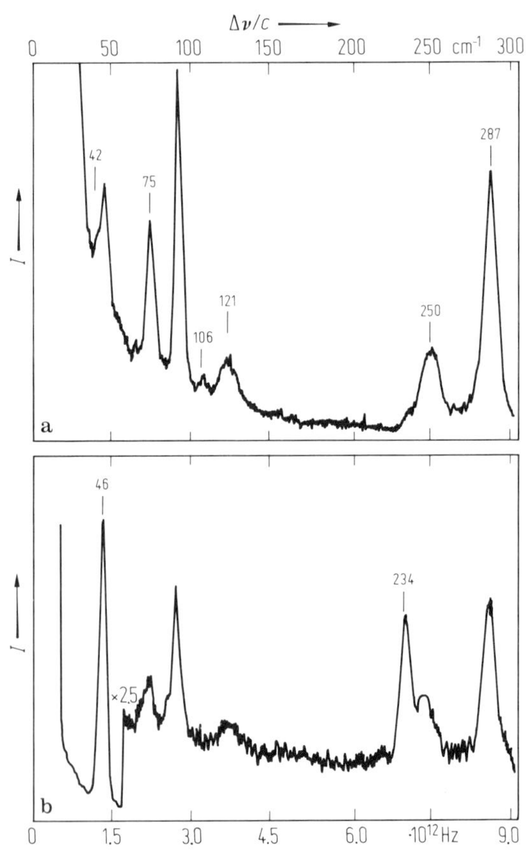

Fig. 9. BiSBr. I vs. $\Delta\nu$ at RT [76F17]. I: Raman scattering intensity. Scattering geometry is $X(Y, X)Y$ in Fig. a, $X(Z, X)Y$ in Fig. b. Peak frequencies are given in units of cm^{-1}. Note the change of intensity scale in (b) above $1.5 \cdot 10^{12}$ Hz.

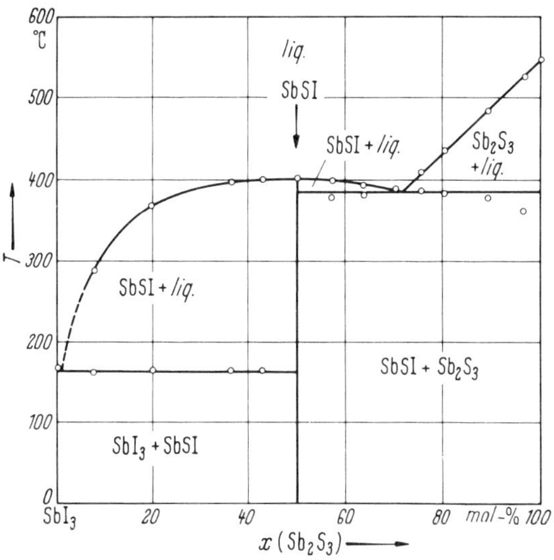

Fig. 10. SbSI. T vs. x, phase diagram of SbI_3–Sb_2S_3 system [64 M 5].

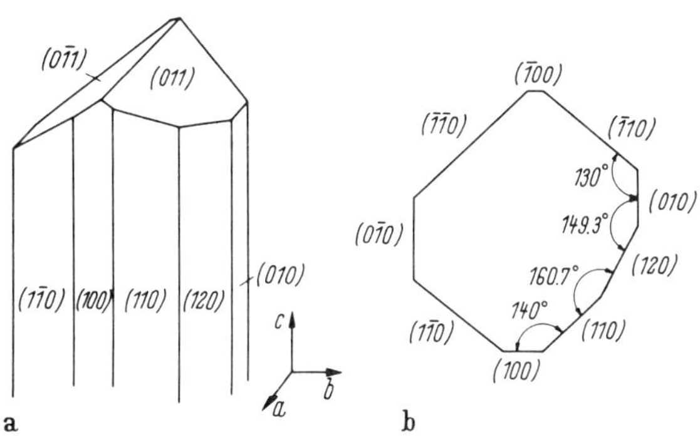

Fig. 11. SbSI. Fig. a: Crystal form; Fig. b: Cross section perpendicular to the c-axis [64 M 5], [67 K 4].

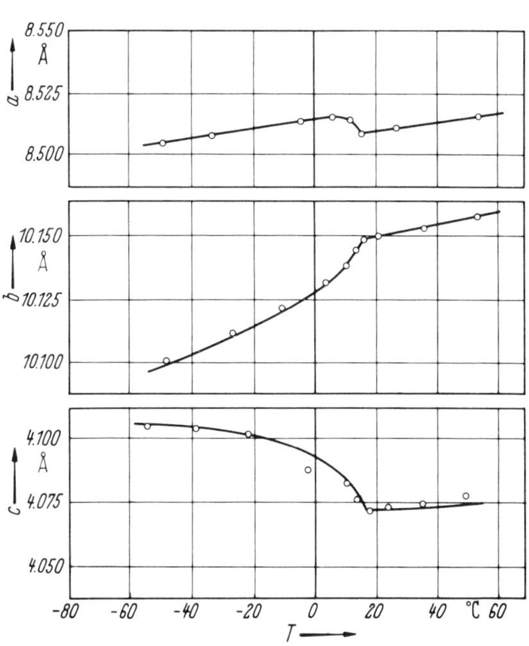

Fig. 12. SbSI. a, b, c vs. T [64 F 8]. a, b, and c are determined from (10, 0, 0), (0, 12, 0) and (0, 0, 4) reflections, respectively.

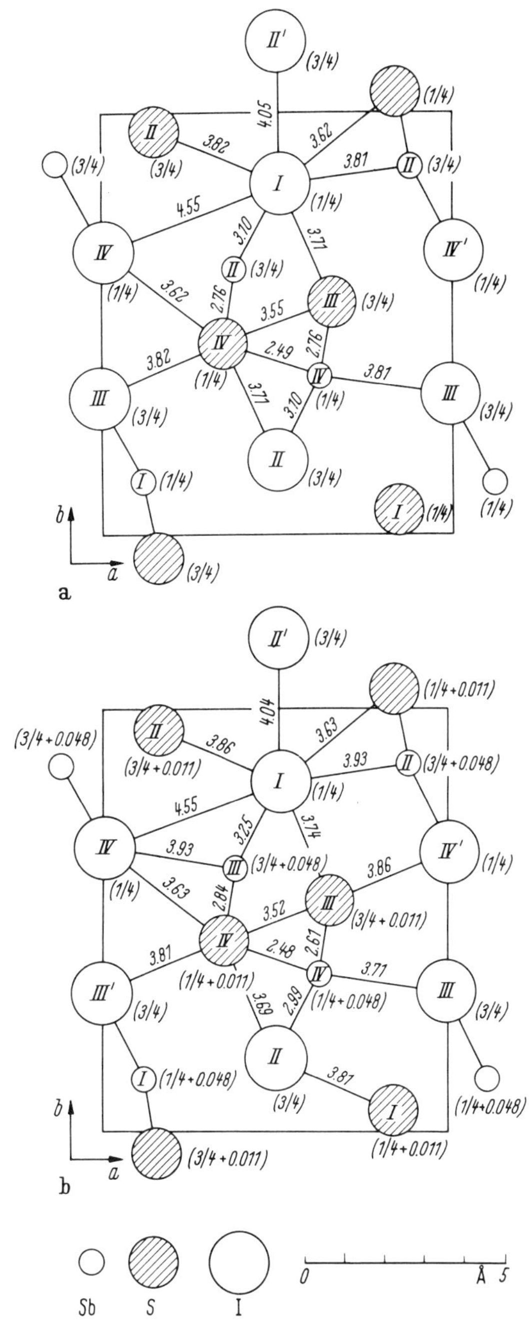

Fig. 14. SbSI. Projections of atoms along c. Interatomic distances (in Å) and z-parameters of each atom are shown. Fig. a: paraelectric phase at 35 °C. Fig. b: ferroelectric phase at 5 °C [67 K 4]. See also Table 10.

Fig. 13, see p. 312.

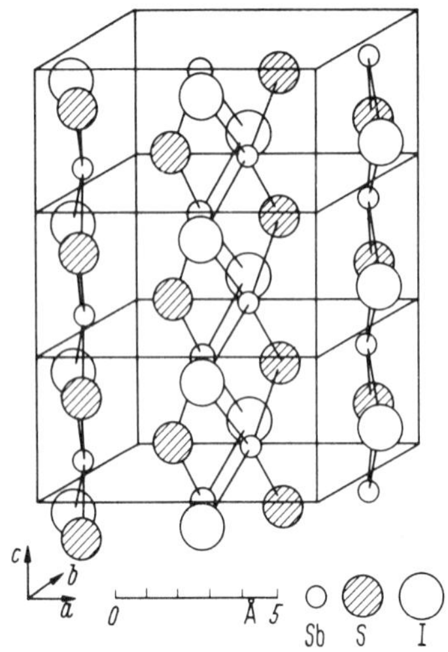

Fig. 13. SbSI. Crystal structure of paraelectric phase [50D2].

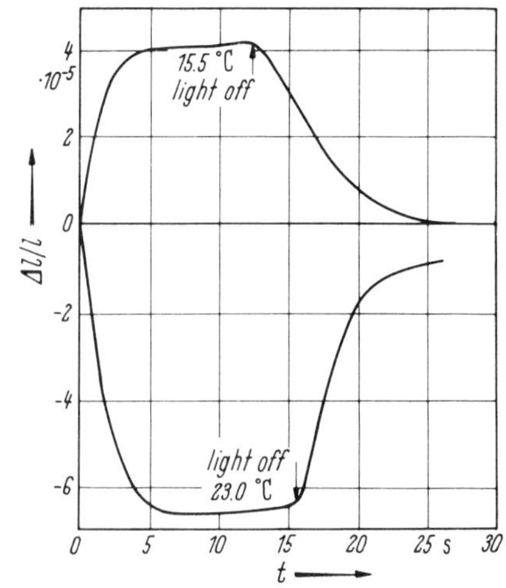

Fig. 16. SbSI. $\Delta l/l$ vs. t. Strain relaxation curve due to light illumination ($\lambda \approx 450$ nm) in the presence of constant dc field ($E = 100$ kV m^{-1}) along the c-axis [66T5].

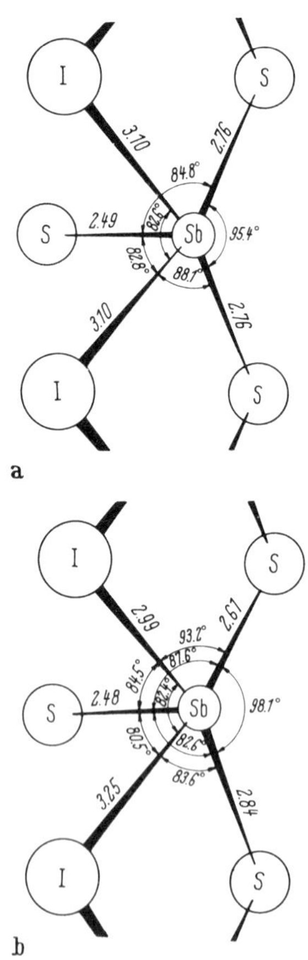

Fig. 15. SbSI. Shapes of SbSI molecule. Fig. a: paraelectric phase at 35 °C. Fig. b: ferroelectric phase at 5 °C [67K4].

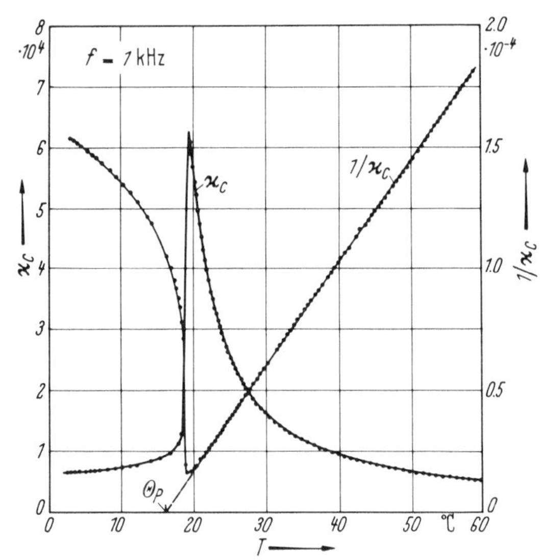

Fig. 17. SbSI. $\varkappa_c$, $1/\varkappa_c$ vs. T [64M5].

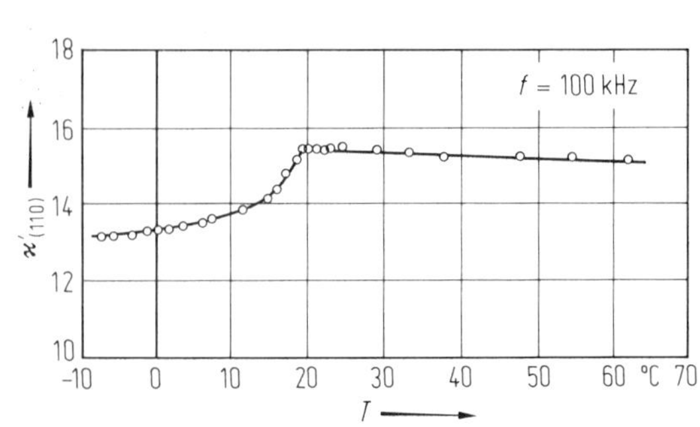

Fig. 18. SbSI. $\varkappa'_{(110)}$ vs. T [78I4]. $\varkappa'_{(110)}$: dielectric constant measured in the direction perpendicular to (110).

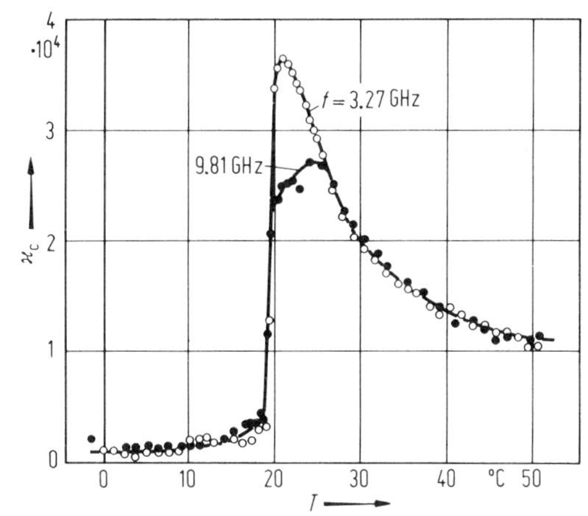

Fig. 19. SbSI. $\varkappa_c'$ vs. T at microwave frequencies [70H13]. Parameter: f.

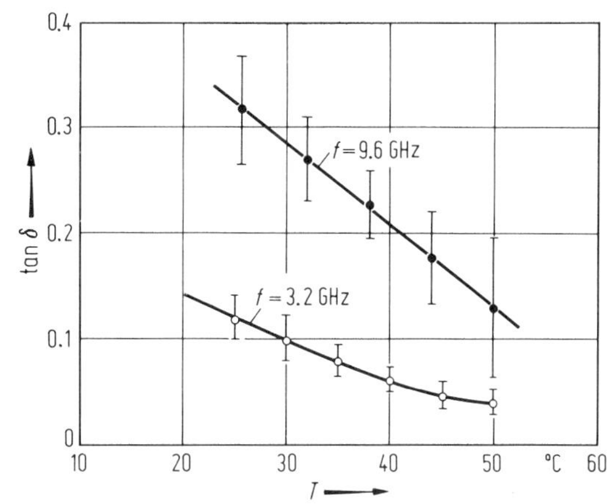

Fig. 20. SbSI. $\tan\delta$ vs. T at microwave frequencies [70H13]. Parameter: f.

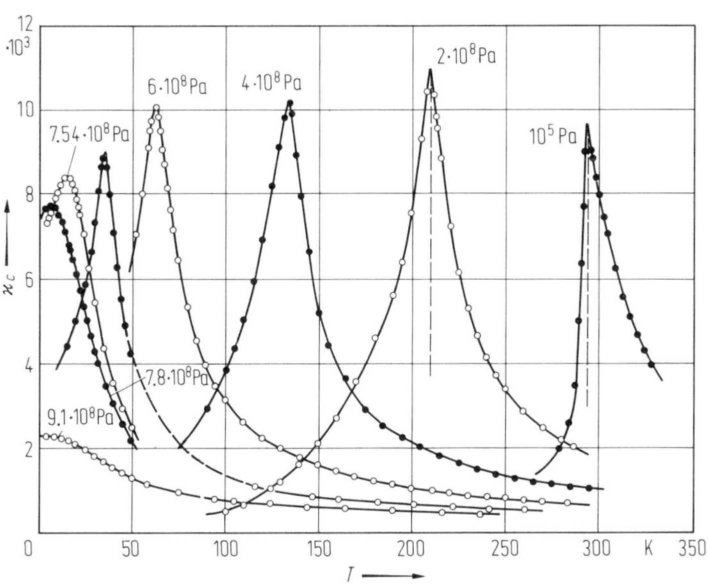

Fig. 21. SbSI. $\varkappa_c$ vs. T. [75S6]. Parameter: p, hydrostatic pressure. $\varkappa_c$ at $f = 1 \cdots 100$ kHz.

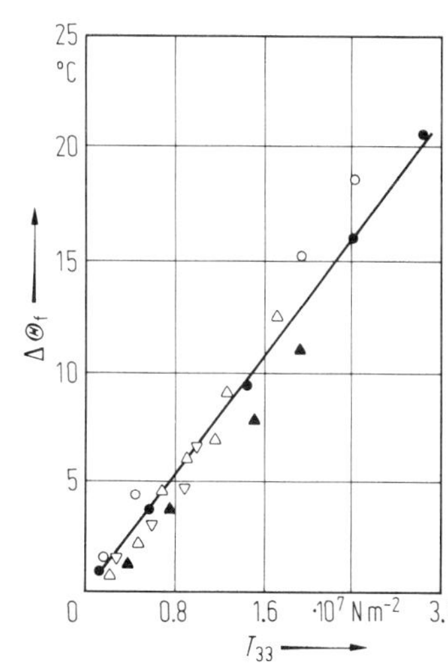

Fig. 23. SbSI. $\Delta\Theta_f$ vs. T_{33} [72V4]. $\Delta\Theta_f$: shift of the ferroelectric Curie temperature. Different symbols correspond to different specimens.

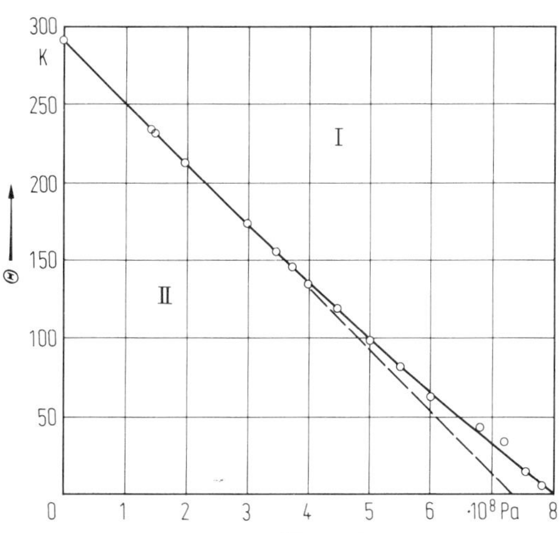

Fig. 22. SbSI. Θ vs. p [75S6].

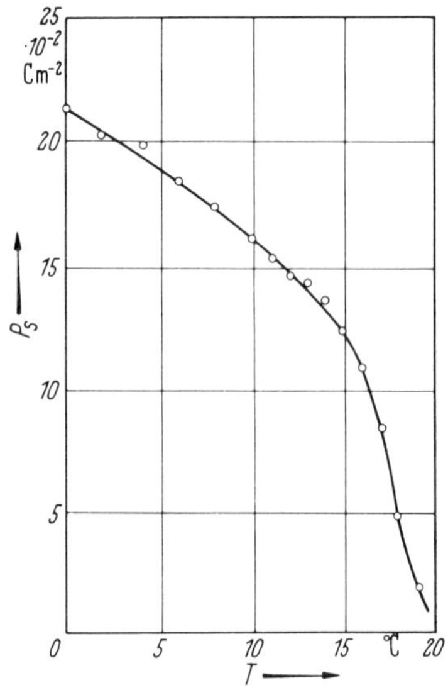

Fig. 24. SbSI. P_s vs. T estimated from D–E hysteresis loops at 50 Hz [62 F 3].

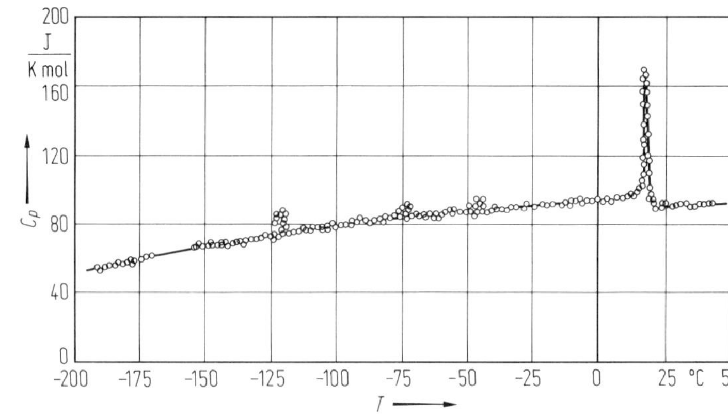

Fig. 27. SbSI. C_p vs. T [72 T 2].

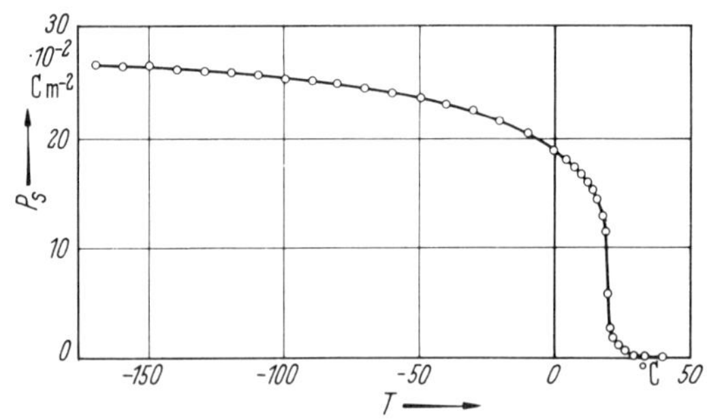

Fig. 25. SbSI. P_s vs. T estimated from the pyroelectric discharge [66 I 4].

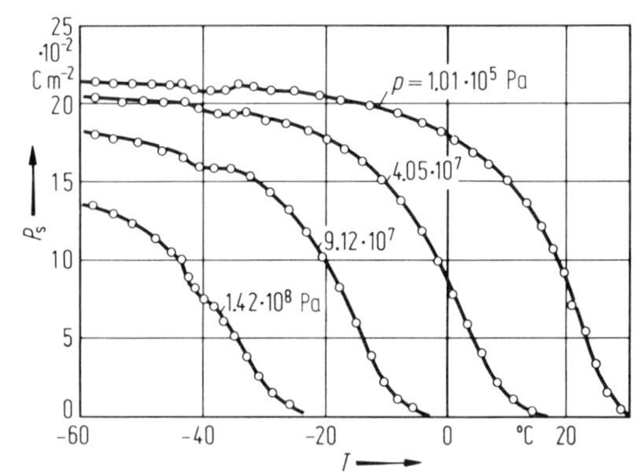

Fig. 26. SbSI. P_s vs. T [68 G 4]. Parameter: p, hydrostatic pressure.

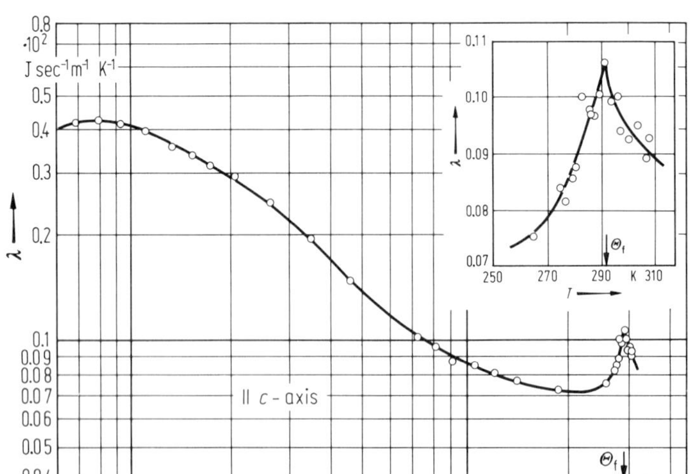

Fig. 28. SbSI. λ_c vs. T [68 S 23].

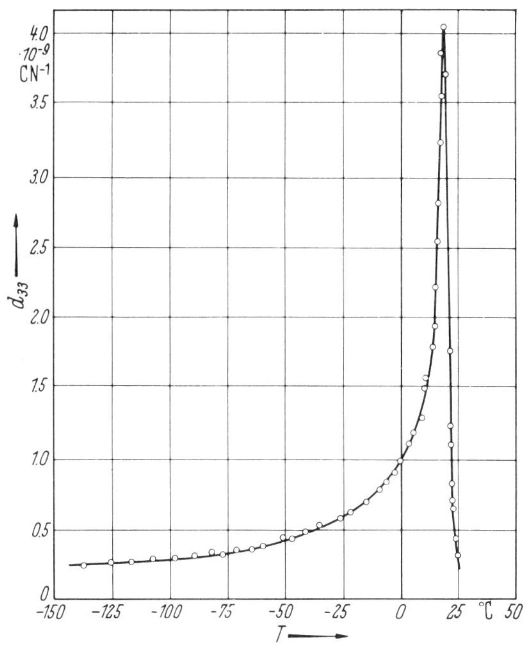

Fig. 29. SbSI. d_{33} vs. T [65H2].

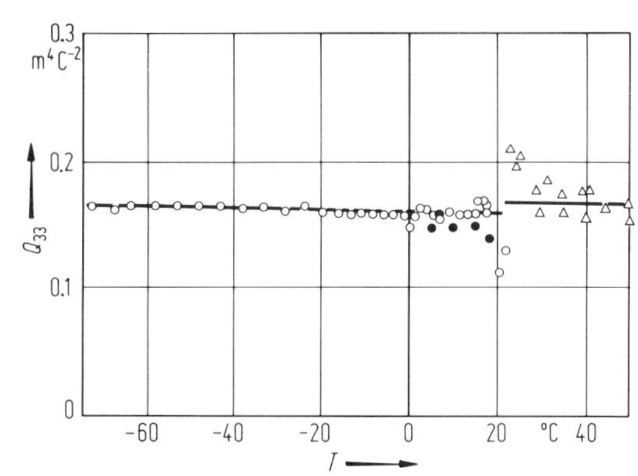

Fig. 31. SbSI. Q_{33} vs. T [72H2]. Q_{33}: electrostrictive constant, data were obtained by three different methods.

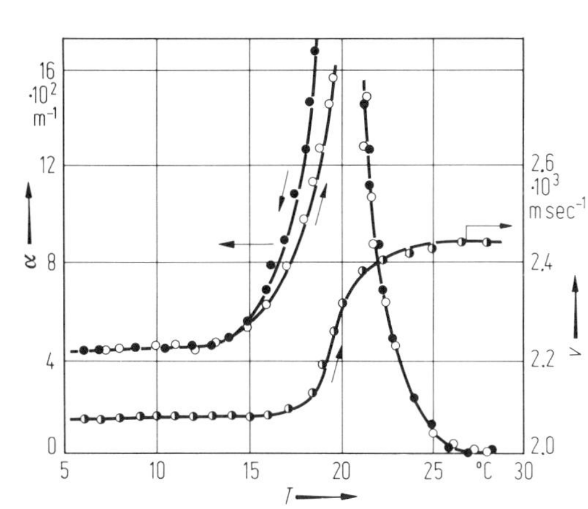

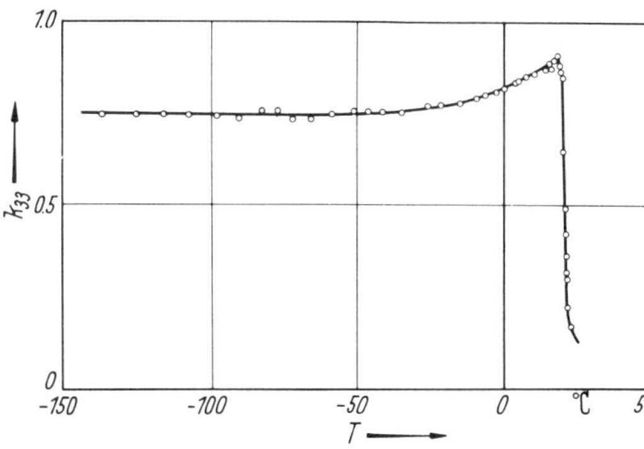

Fig. 30. SbSI. k_{33} vs. T [65H2].

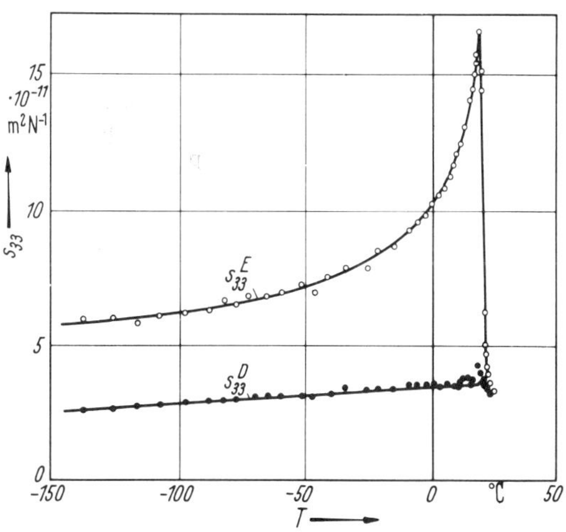

Fig. 32. SbSI. s_{33}^E, s_{33}^D vs. T [65H2].

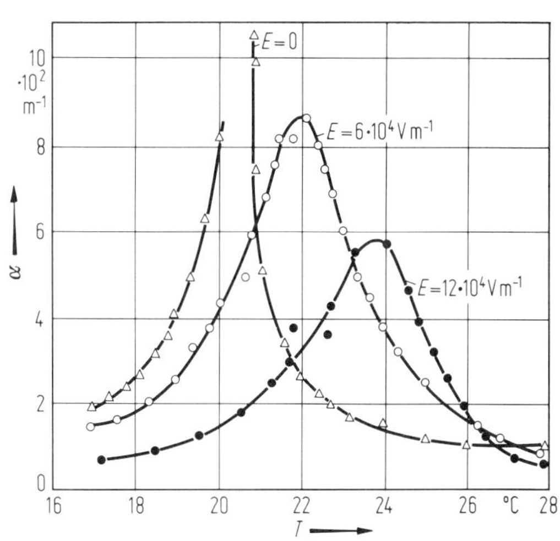

Fig. 34. SbSI. α vs. T [69S3]. Parameter: dc bias field E parallel to the c axis. α is the absorption coefficient of a longitudinal wave at 10 MHz propagating along the c axis.

◄ Fig. 33. SbSI. α, v vs. T [71S9]. α, v: absorption coefficient and velocity for 10 MHz quasi-longitudinal ultrasound wave propagating along the c axis, respectively.

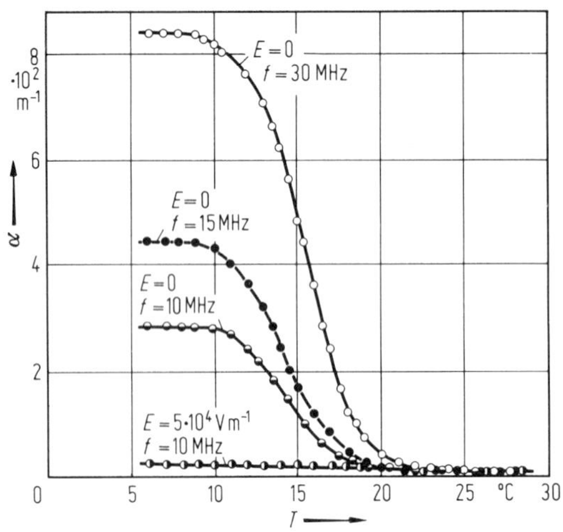

Fig. 35. SbSI. α vs. T [71S9]. Parameter: dc bias field E parallel to the c axis, and f. α is the absorption coefficient of quasi-transverse ultrasonic wave propagating perpendicular to the c axis.

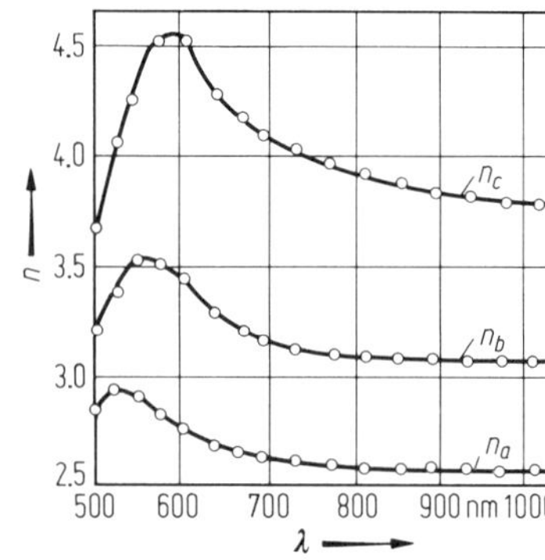

Fig. 36. SbSI. n vs. λ [68G3].

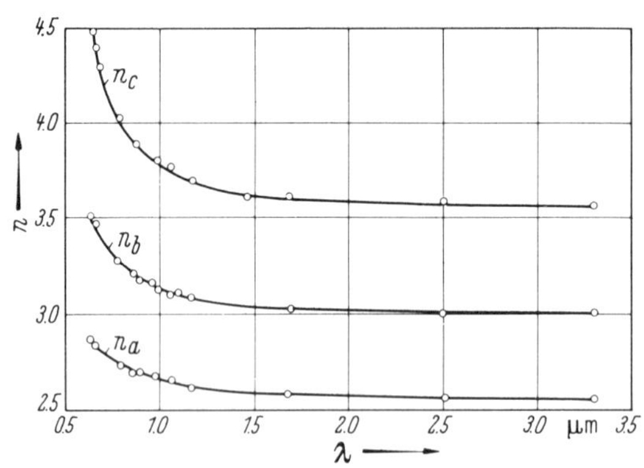

Fig. 37. SbSI. n vs. λ [67J1].

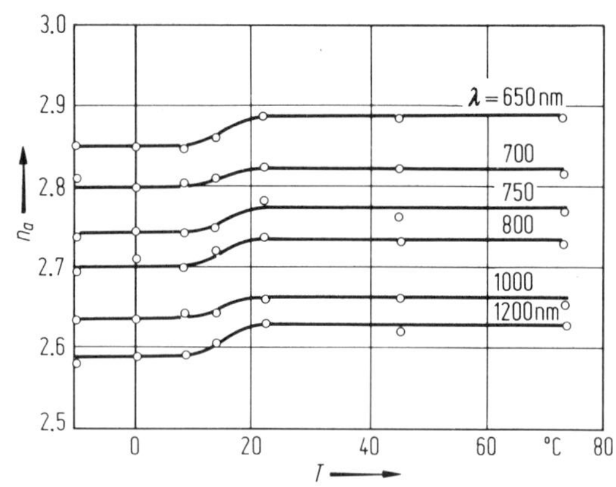

Fig. 38. SbSI. n_a vs. T [70O3]. Parameter: λ.

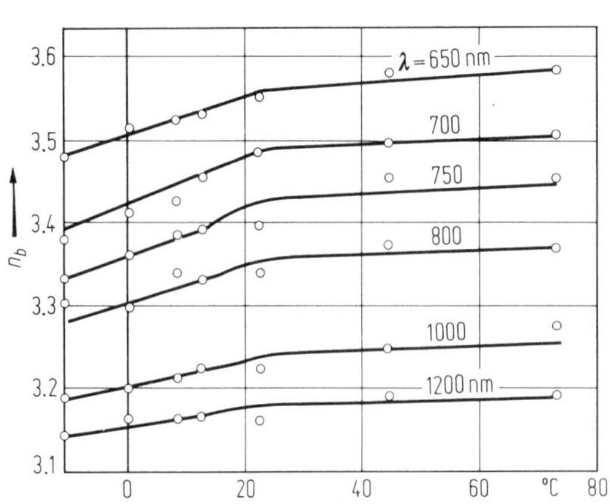

Fig. 39. SbSI. n_b vs. T [70O3]. Parameter: λ.

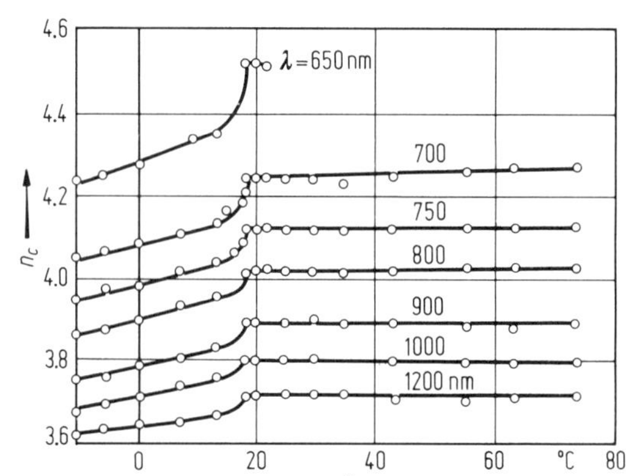

Fig. 40. SbSI. n_c vs. T [70O3]. Parameter: λ.

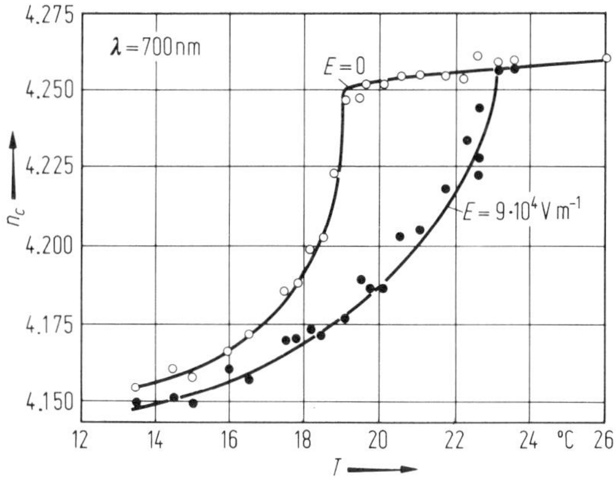

Fig. 41. SbSI. n_c vs. T [70O3]. Parameter: dc bias field.

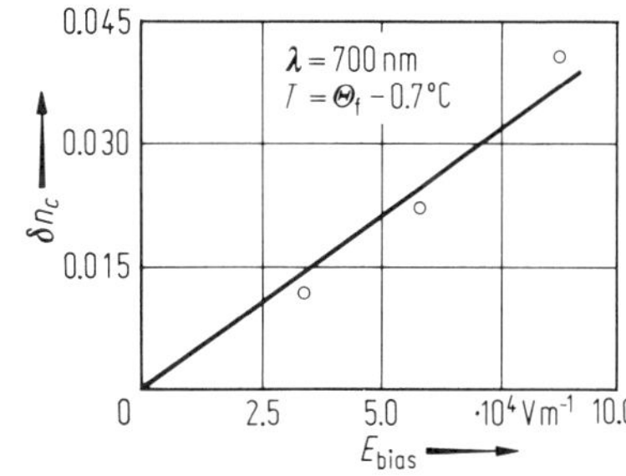

Fig. 42. SbSI. δn_c vs. E_{bias} at $\lambda = 700$ nm, at 0.7 °C below Θ_f [70O3]. $\delta n_c = n_c(E_{bias}=0) - n_c(E_{bias})$.

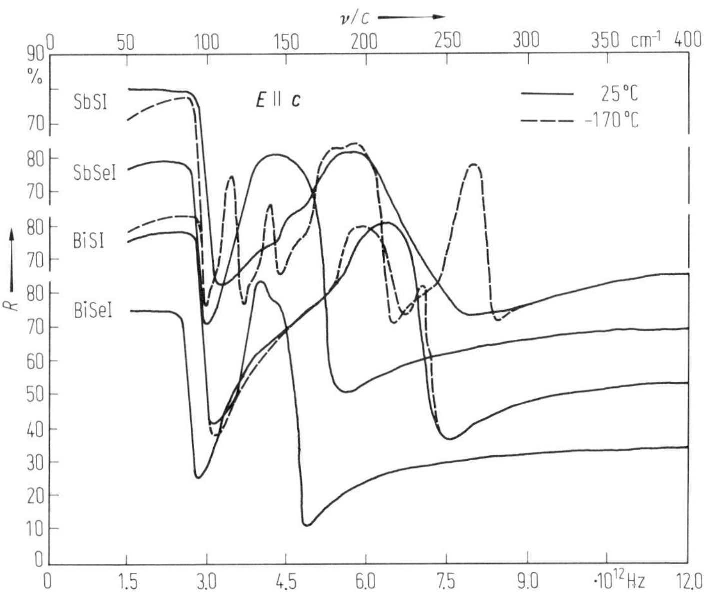

Fig. 43. SbSI, BiSI, SbSeI, BiSeI, R vs. ν at 25 °C and −170 °C [73P5]. R: reflectivity. ν: frequency of the incident infrared radiation polarized parallel to the c axis.

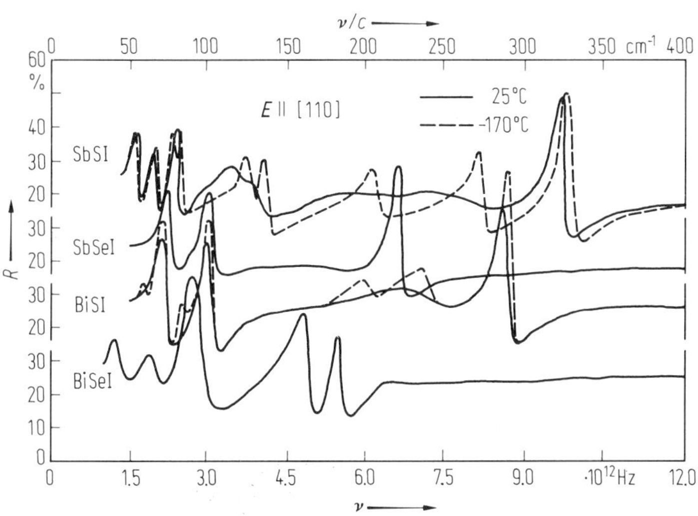

Fig. 44. SbSI, BiSI, SbSeI, BiSeI. R vs. ν at 25 °C and −170 °C [73P5]. Incident light polarizes parallel to [110].

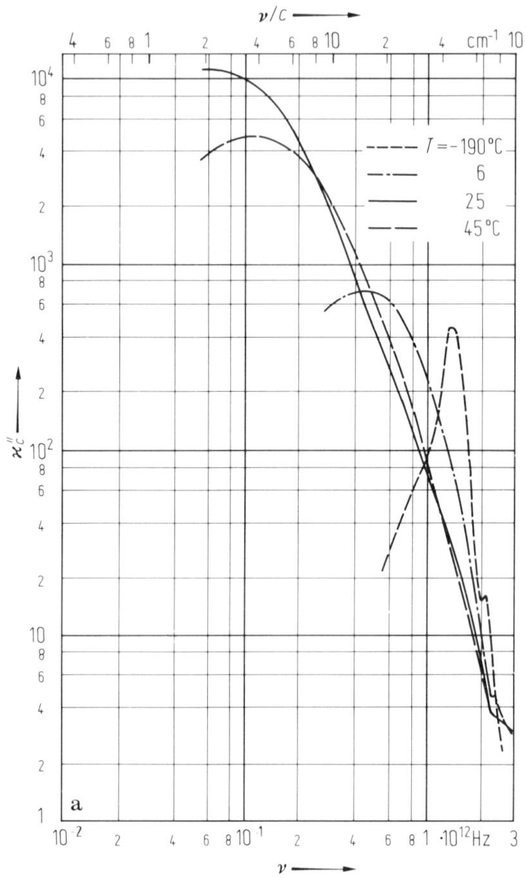

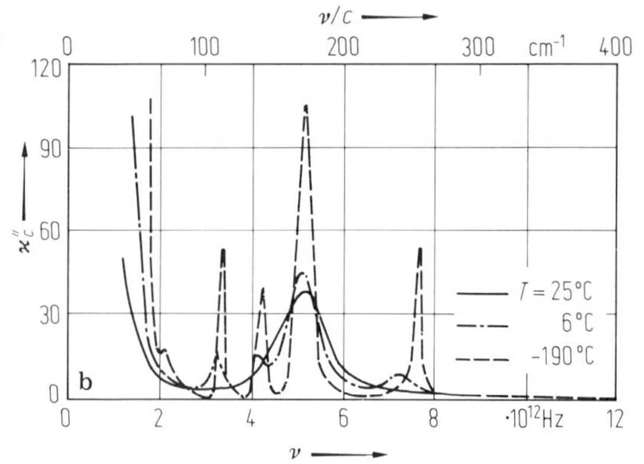

Fig. 45. SbSI. $\varkappa_c''$ vs. ν [72S31]. Parameter: T. $\varkappa_c''$ were obtained from reflectivity data using Kramers-Kronig relation.

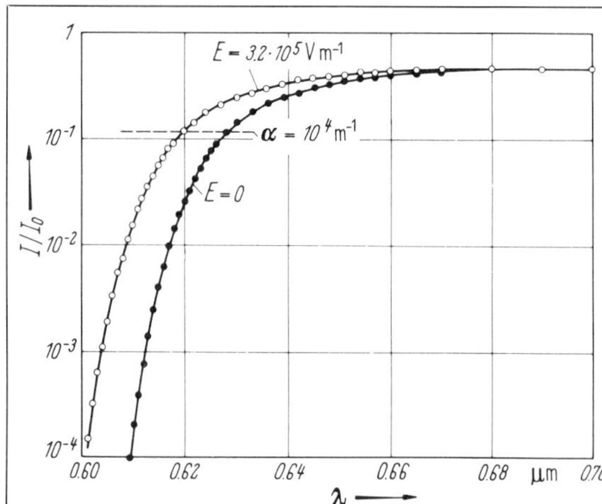

Fig. 48. SbSI. Optical transmission I/I_0 vs. λ for unpolarized light at 22 °C [62K9]. Parameter: dc bias field along the c axis.

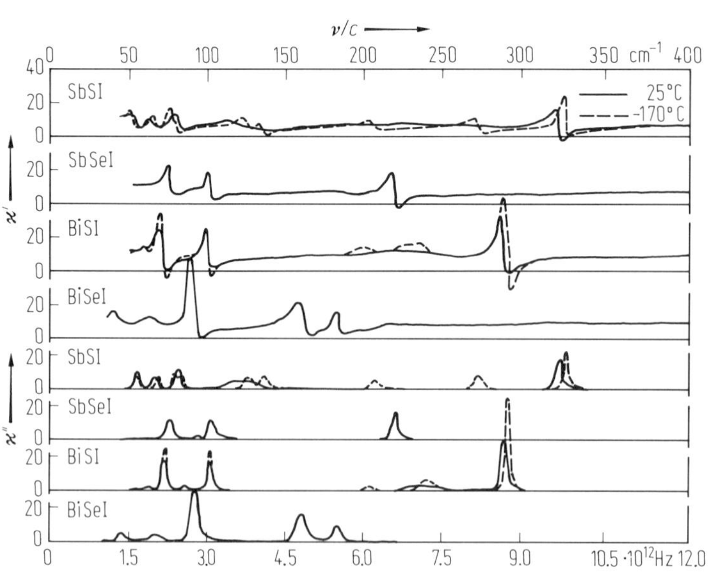

Fig. 47. SbSI, BiSI, SbSeI, BiSeI. $\varkappa'$, $\varkappa''$ vs. ν for radiation polarized parallel to (110) plane [73P5].

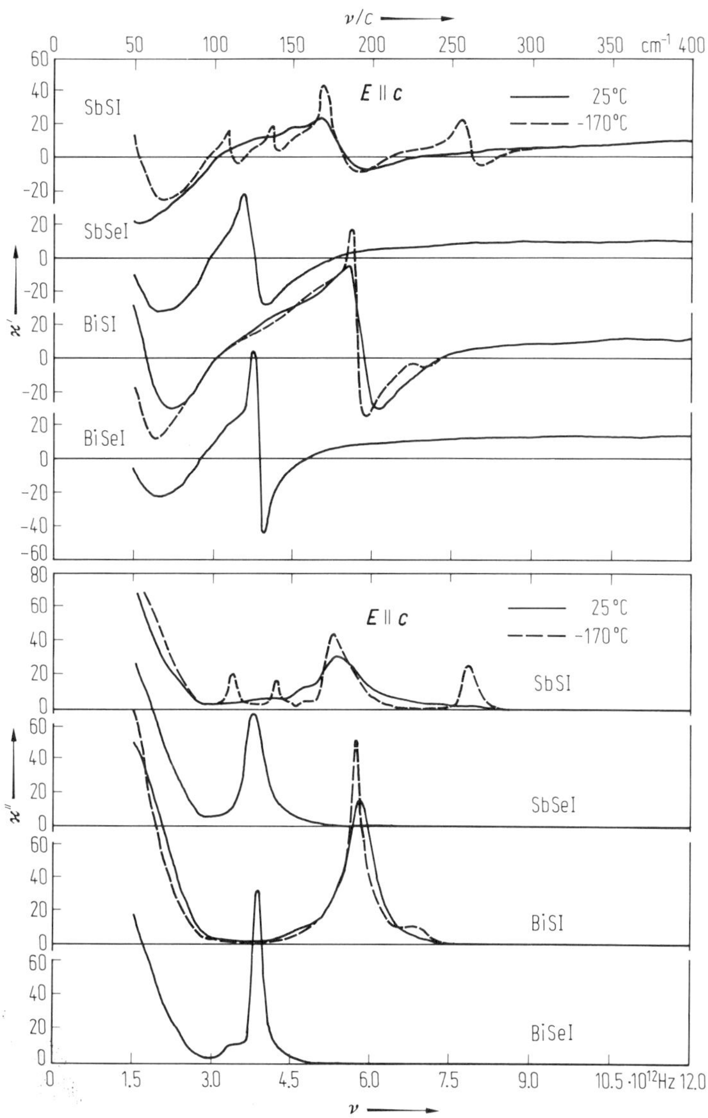

Fig. 46. SbSI, BiSI, SbSeI, BiSeI. $\varkappa'$, $\varkappa''$ vs. ν for infrared radiation polarized parallel to the c axis [73P5].

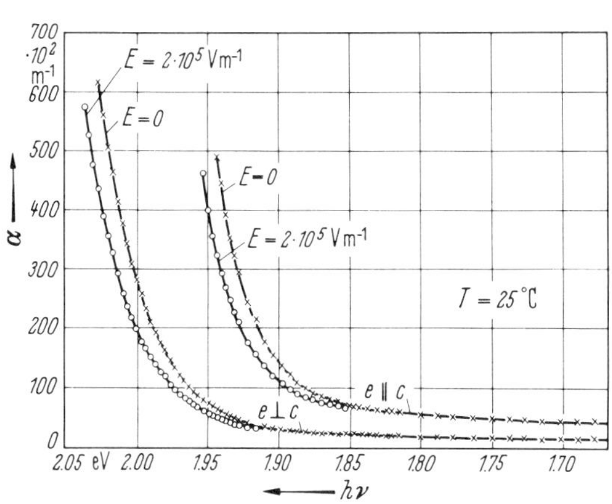

Fig. 49. SbSI. Absorption coefficient α vs. $h\nu$ at 25 °C. [63H2]. Parameter: applied dc field along c. $\boldsymbol{e}$: electric field vector of the incident polarized light.

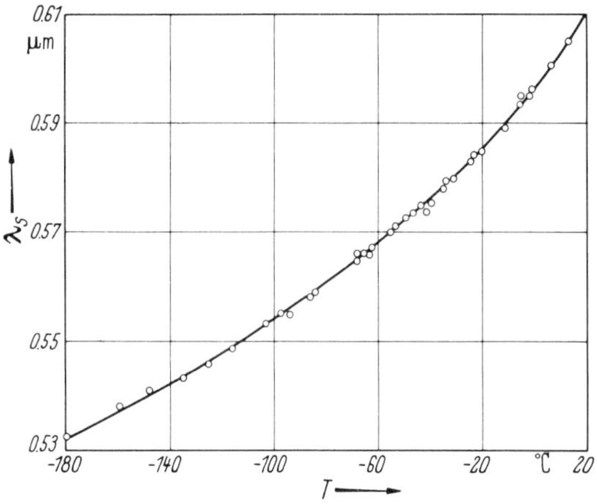

Fig. 50. SbSI. Absorption edge λ_s vs. T [62K9]. λ_s: the threshold wave length where the crystal becomes non-transparent.

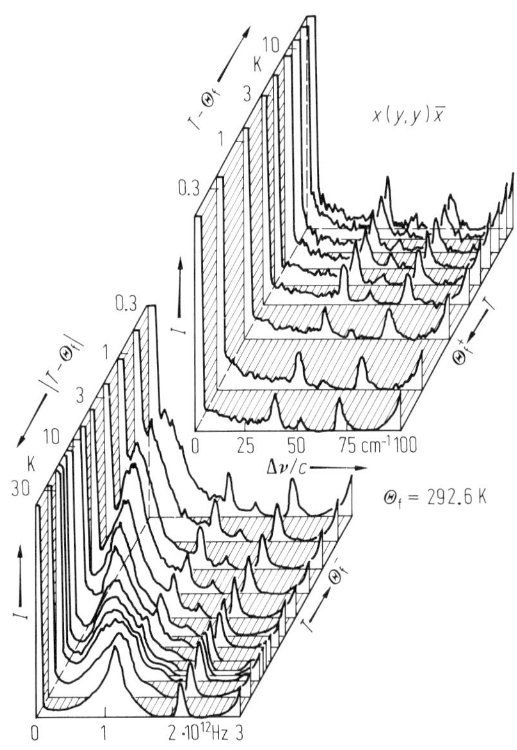

Fig. 51. SbSI. I vs. $\Delta\nu$ [75S30]. Parameter: T. I: Raman scattering intensity for the polarization geometry of $x(yy)\bar{x}$. $\Delta\nu$: frequency shift. The front part of the figure is for the ferroelectric phase and the back is for the paraelectric phase. Note that the temperature differences are plotted logarithmically.

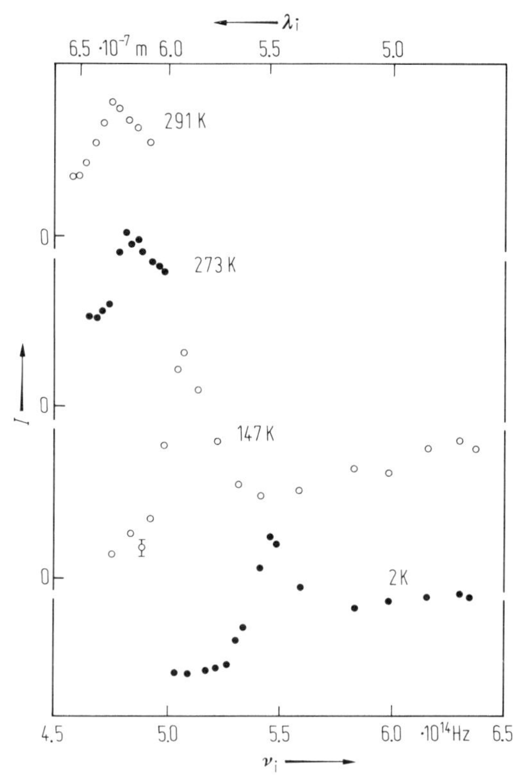

Fig. 52. SbSI. I vs. ν_i [76A11]. Parameter: T. I: Raman scattering intensity of the 112 cm^{-1} mode. ν_i: frequency of incident light. The peaks indicate that the scattering is enhanced when ν_i is near the absorption edge. Refer also to [71A7].

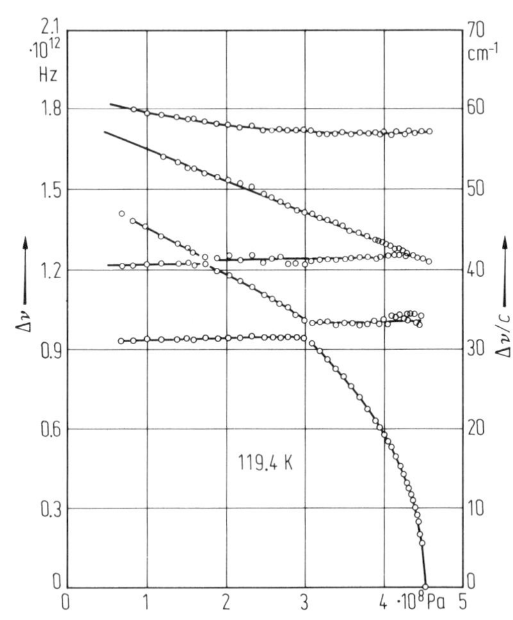

Fig. 54. SbSI. $\Delta\nu$ vs. p at 119.4 K [76P3]. p: hydrostatic pressure. $\Delta\nu$: Raman frequency shift of low frequency A_1 modes.

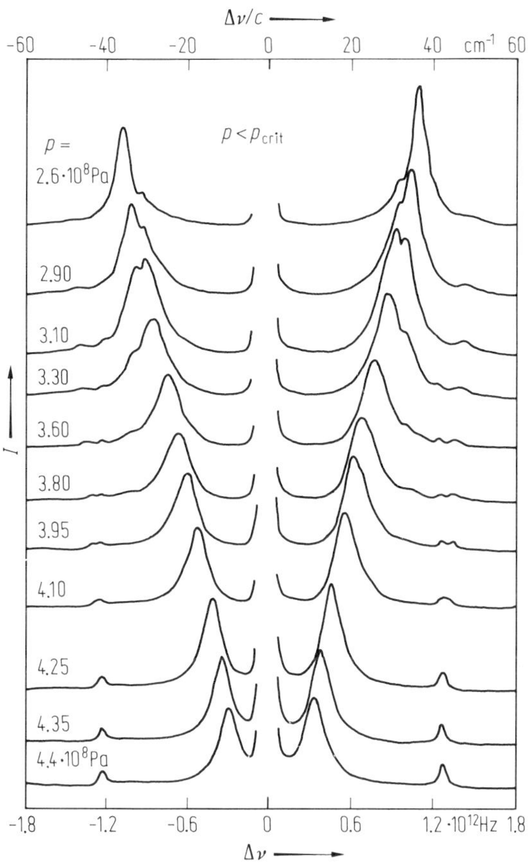

Fig. 53. SbSI. I vs. $\Delta\nu$ [76P3]. Parameter: hydrostatic pressure p. I: Raman scattering intensity at 119 K. $\Delta\nu$: frequency shift. The transition pressure at this temperature is $p_{crit} = 4.54 \cdot 10^8$ Pa. See also [75P5].

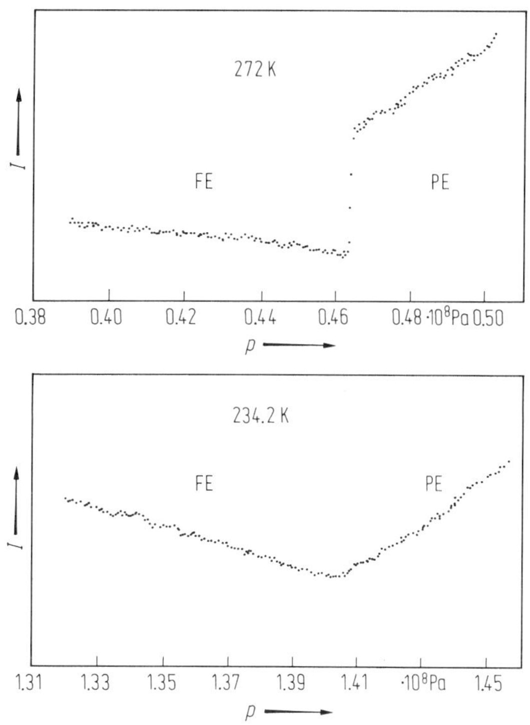

Fig. 56. SbSI. I vs. p [76P3] I: intensity of elastic light scattering obtained by measuring the transmitted light. p: hydrostatic pressure. Continuous change of I was observed for $T_{crit} \leqq 235$ K and $p_{crit} \geqq 1.40 \cdot 10^8$ Pa. Upper figure: for 272 K and lower figure: for 234.2 K. FE and PE indicate ferroelectric and paraelectric phase, respectively. See also [75P5].

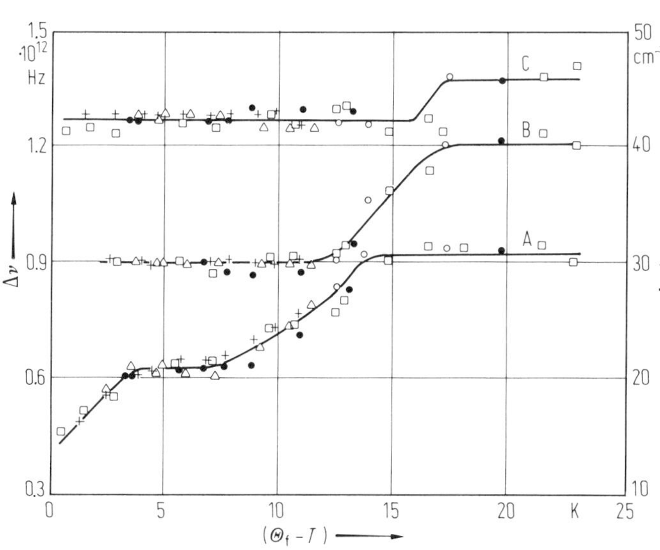

Fig. 55. SbSI. $\Delta\nu$ vs. $(\Theta_f - T)$ [72T7]. $\Delta\nu$: Raman scattering frequency shift for the three lowest frequency optical modes A, B, C. Sample temperature T was kept constant at 0 °C (closed circles), 1.5 °C (triangles) and 3 °C (crosses and open circles) and the variation of $\Theta_f - T$ was induced by hydrostatic pressure. Squares indicate the data obtained at $p = 200 \cdot 10^5$ Pa and changing the temperature.

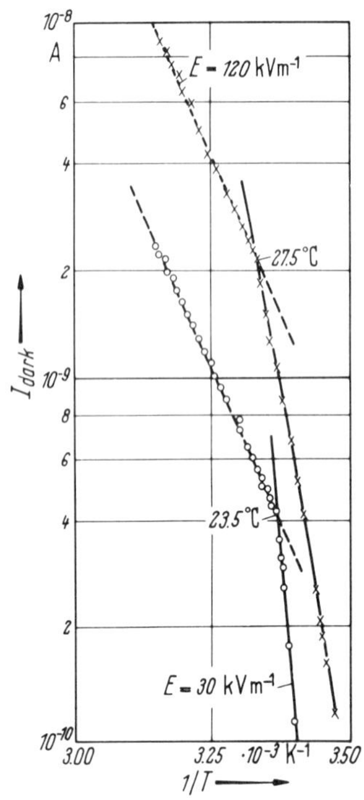

Fig. 57. SbSI. Dark current I_{dark} vs. $1/T$. [65S5]. Parameter: E, applied field along the c-axis.

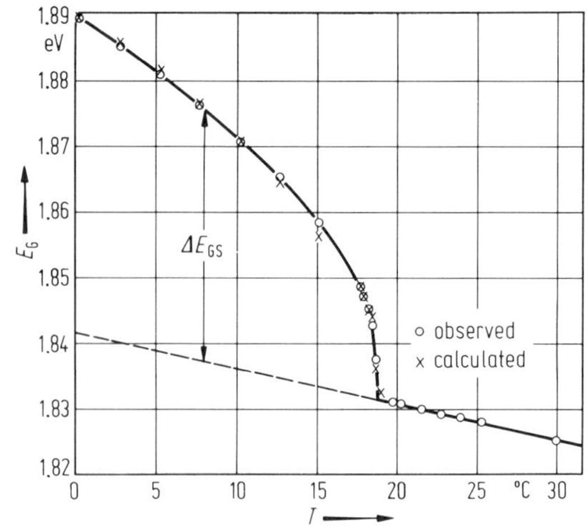

Fig. 58. SbSI. E_G vs. T for light polarized parallel to the c-axis [72I6]. E_G, absorption edge defined as the photon energy whose absorption coefficient is 10^{-4} m. The spontaneous shift ΔE_{GS} is obtained by subtracting the extrapolated values of the linear term in the paraelectric region from the observed edge energy in the ferroelectric region.

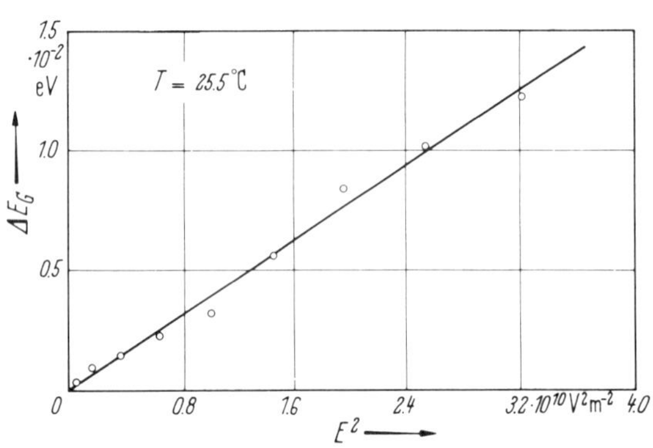

Fig. 59. SbSI. ΔE_G vs. E^2 in the paraelectric phase [63H2]. ΔE_G, increase of the absorption edge energy E_G due to a dc field applied along the c-axis.

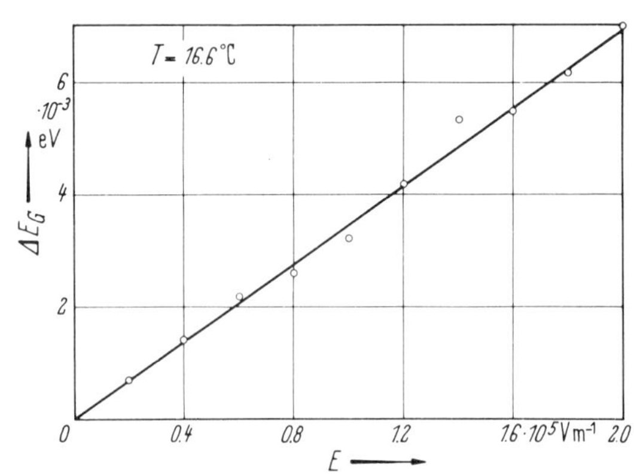

Fig. 60. SbSI. ΔE_G vs. E at the ferroelectric phase [63H2]. Cf. Fig. 59.

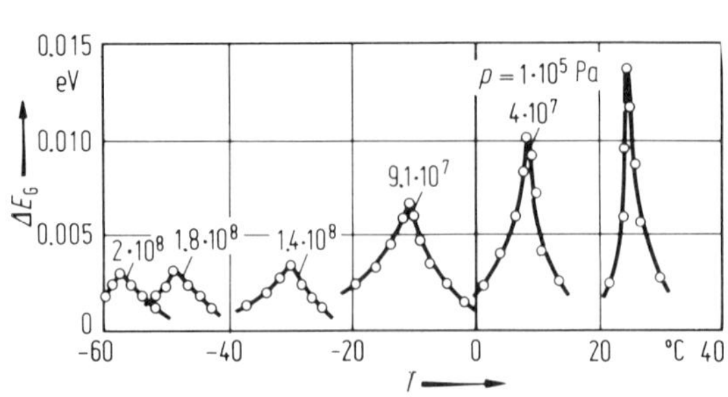

Fig. 61. SbSI. ΔE_G vs. T [69G3]. Parameter: p, hydrostatic pressure. ΔE_G, absorption edge shift induced by a constant electric field of $2 \cdot 10^5$ Vm^{-1}.

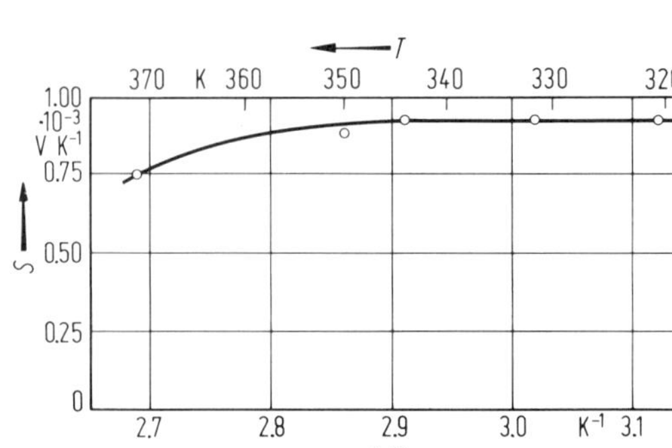

Fig. 62. SbSI. Thermoelectric power S vs. $1/T$ [70T12]. $1/T$ values have to be multiplied by 10^{-3}.

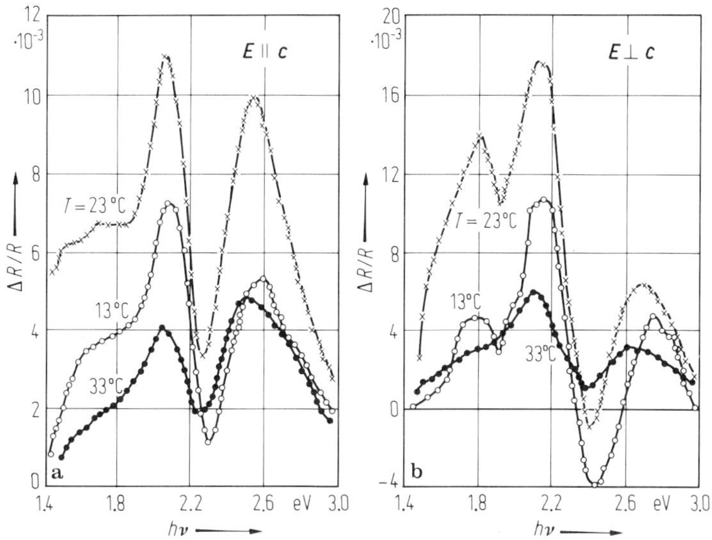

Fig. 63. SbSI. Electroreflectance ($\Delta R/R$) vs. photon energy $h\nu$ [70G8]. Parameter: T. The electric vector of incident light is, Fig. a: parallel to the c axis, Fig. b: perpendicular to the c axis.

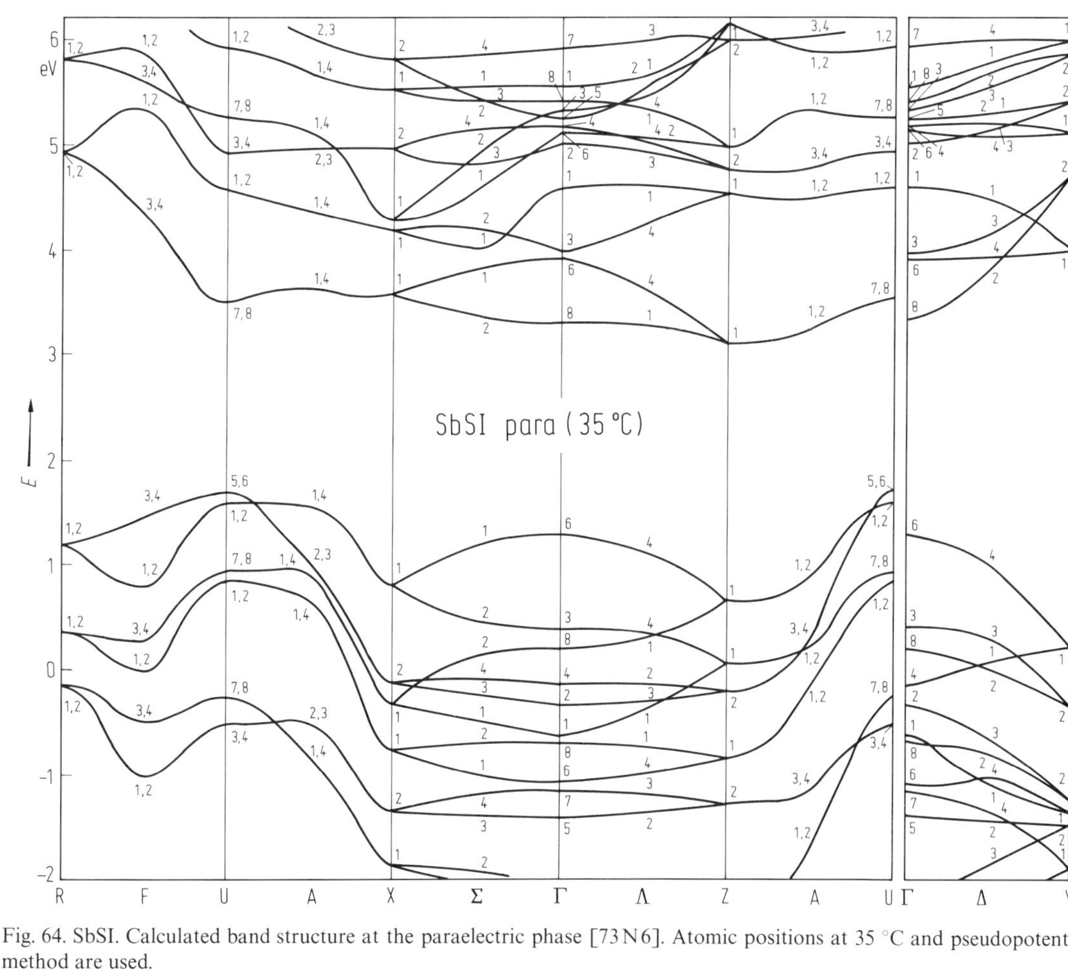

Fig. 64. SbSI. Calculated band structure at the paraelectric phase [73N6]. Atomic positions at 35 °C and pseudopotential method are used.

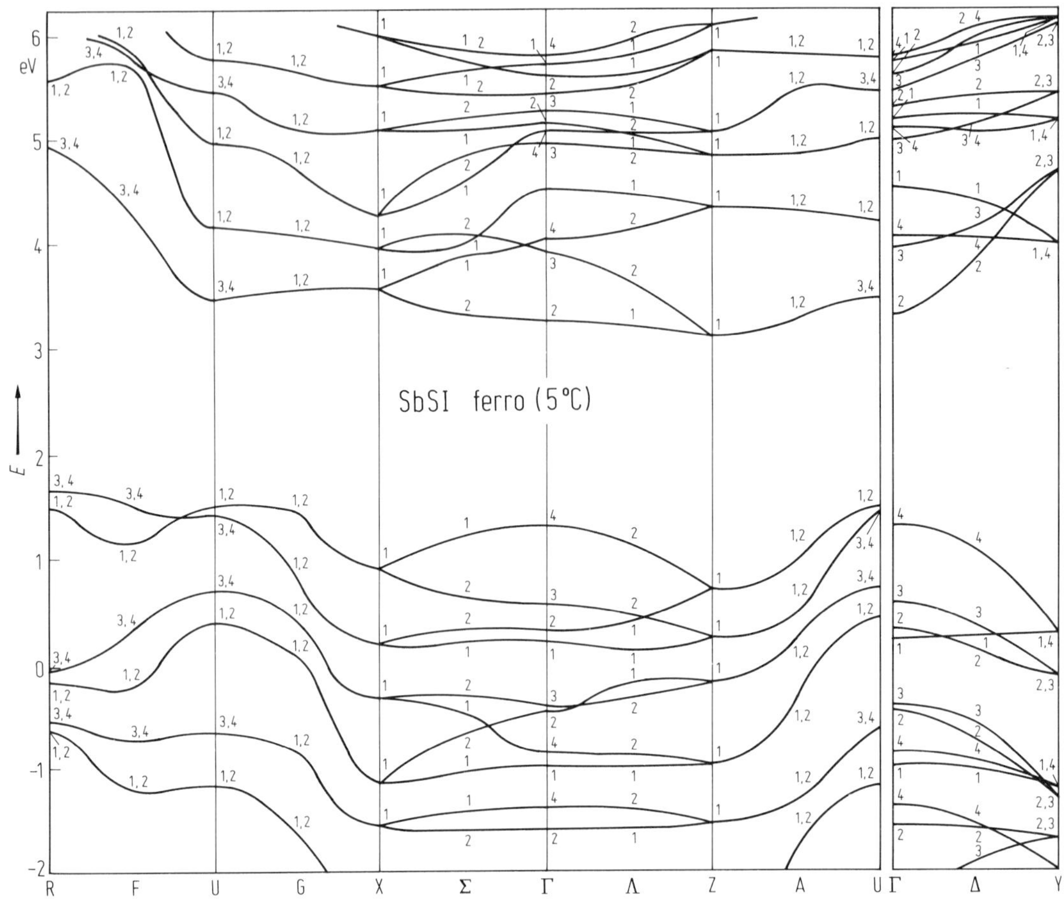

Fig. 65. SbSI. Calculated band structure at the ferroelectric phase [73N6]. Atomic positions at 5 °C and pseudopotential method are used.

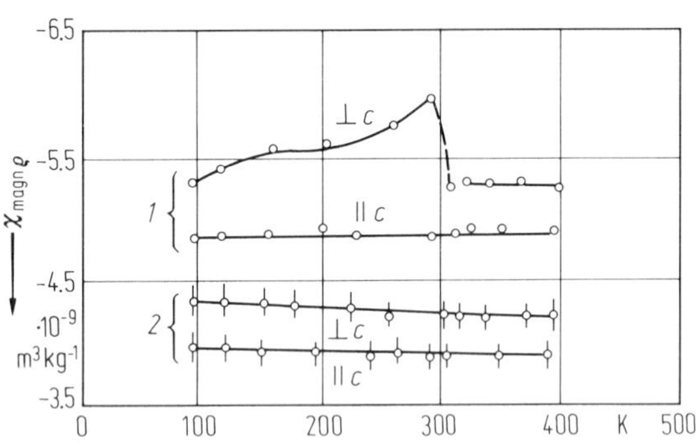

Fig. 66. SbSI, BiSeI. $\chi_{magn\,\varrho}$ vs. T [75B10]. Curves *1*: SbSI; Curves *2*: BiSeI.

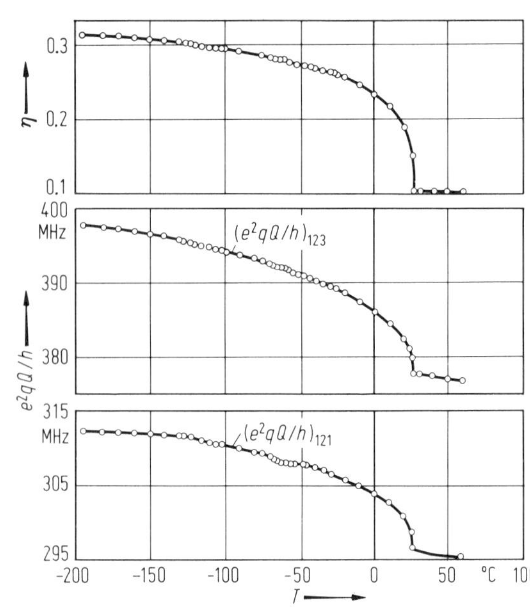

Fig. 67. SbSI. e^2qQ/h, η vs. T [70P9]. $[e^2qQ/h]_{123}$ and $[e^2qQ/h]_{121}$ are quadrupole coupling constants for ^{123}Sb and ^{121}Sb, respectively. η: asymmetry parameter.

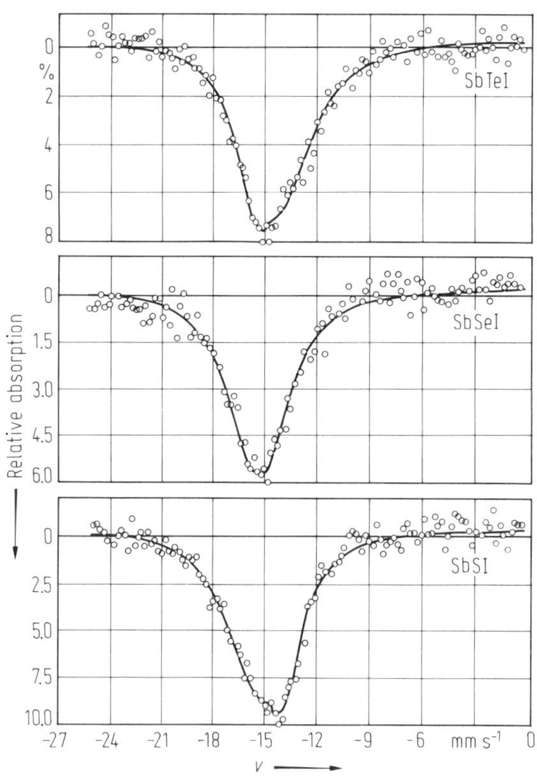

Fig. 68. SbSI, SbSeI, SbTeI. Mössbauer spectrum [75 D 7]. v: source velocity of $Ba^{121}SnO_3$, $T = 4.2$ K.

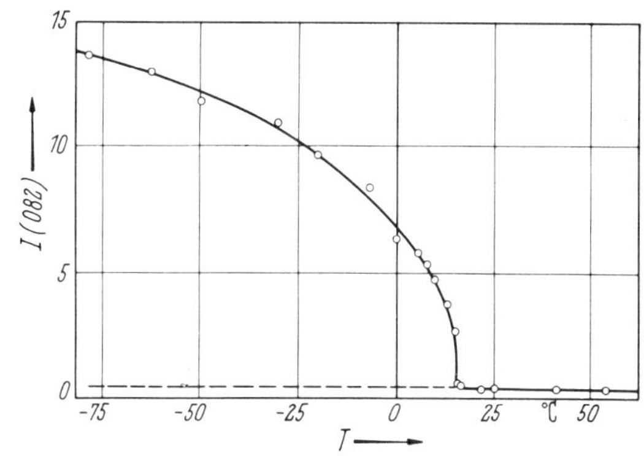

Fig. 69. SbSI. $I(082)$ vs. T [67 T 3]. $I(082)$: integrated X-ray intensity of the 082 Bragg reflections (arbitrary unit).

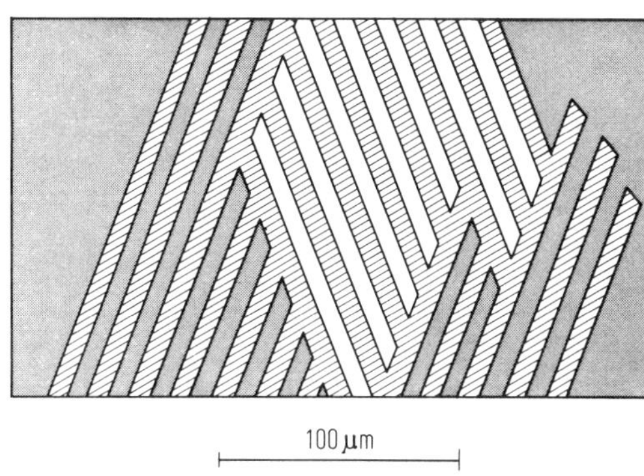

Fig. 71. SbSI. "Fish bone pattern" observed at 25 °C [70 H 2]. Sample thickness, 130 μm.

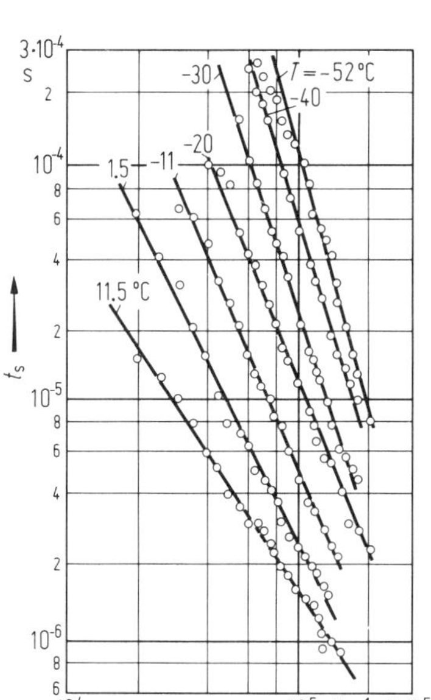

Fig. 70. SbSI. Switching time t_s vs. E [70 K 20]. Parameter: T.

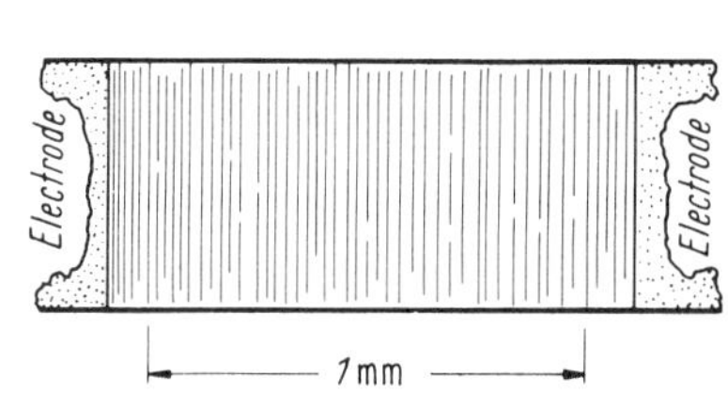

Fig. 72. SbSI. Stripe pattern observed on (110) face at 15 °C under a dc field and light illumination ($\lambda \approx 640$ nm) [66 S 9].

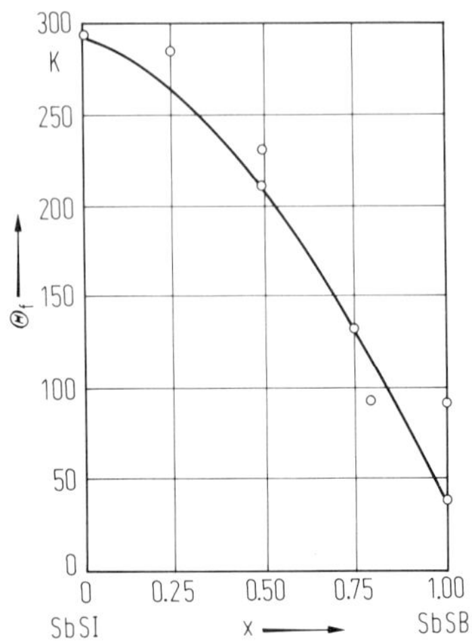

Fig. 73. SbSBr–SbSI. Θ_f vs. x [73 F 13]. x: molar fraction of SbSBr in $SbSBr_xI_{1-x}$.

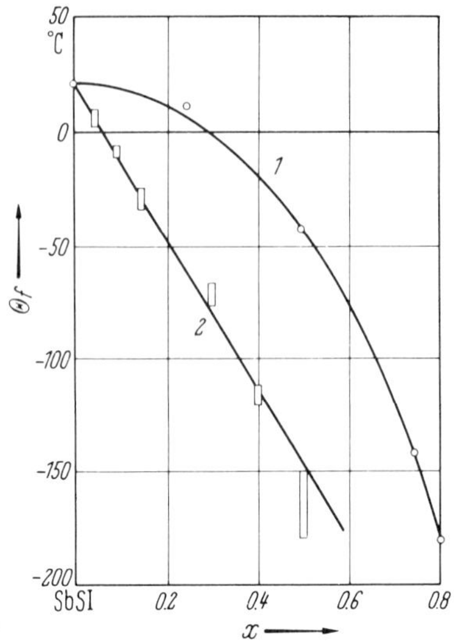

Fig. 74. SbSBr–SbSI, SbSI–SbSeI. Θ_f vs. x [64 N 3]. x: molar fraction of SbSBr in $SbSBr_xI_{1-x}$ (curve *1*) and of SbSeI in $SbSe_xS_{1-x}I$ (curve *2*).

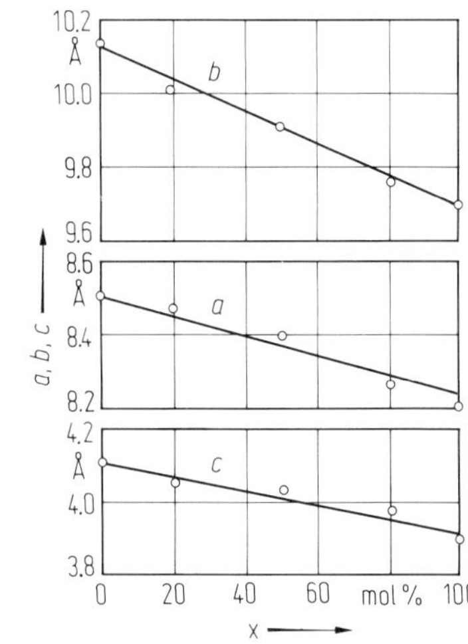

Fig. 75. SbSBr–SbSI. *a*, *b*, *c* vs. x [75 S 29]. x: mol% of SbSBr in $SbSBr_xI_{1-x}$.

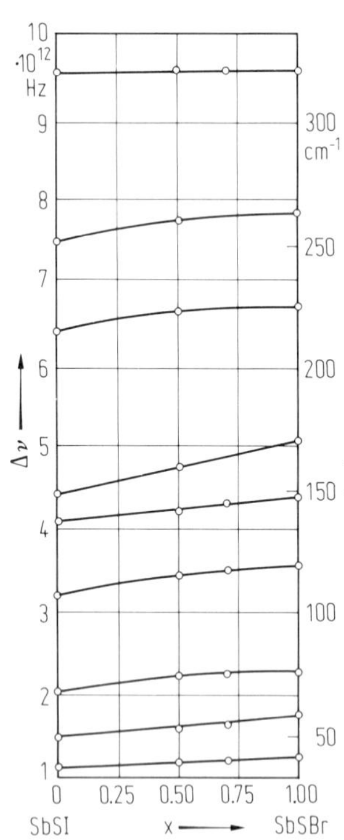

Fig. 76. SbSBr–SbSI. Raman frequency shifts $\Delta\nu$ vs. x [73 F 13]. x: molar fraction x in $SbSBr_xI_{1-x}$.

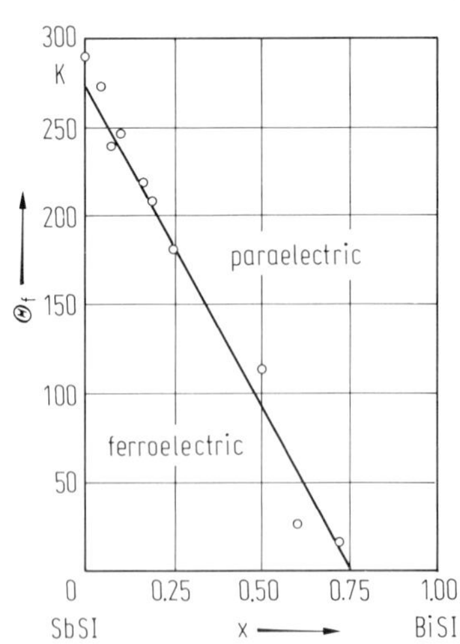

Fig. 77. SbSI–BiSI. Θ_f vs. x [78 T 5]. x: mole fraction of BiSI in $Bi_xSb_{1-x}SI$.

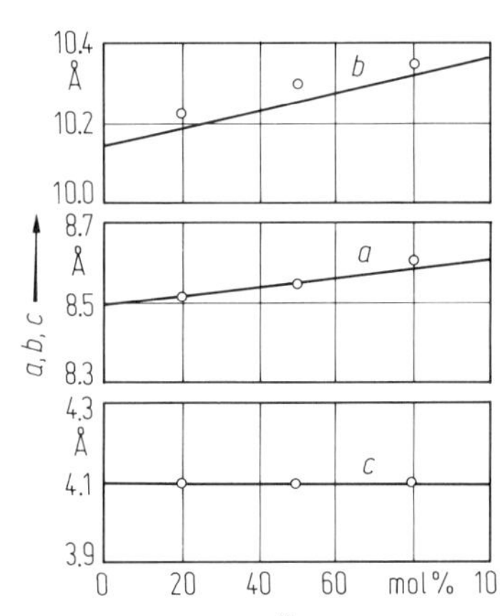

Fig. 79. SbSI–SbSeI. *a*, *b*, *c* vs. x [75 S 29]. x: mol% of SbSeI. Orthorhombic cell.

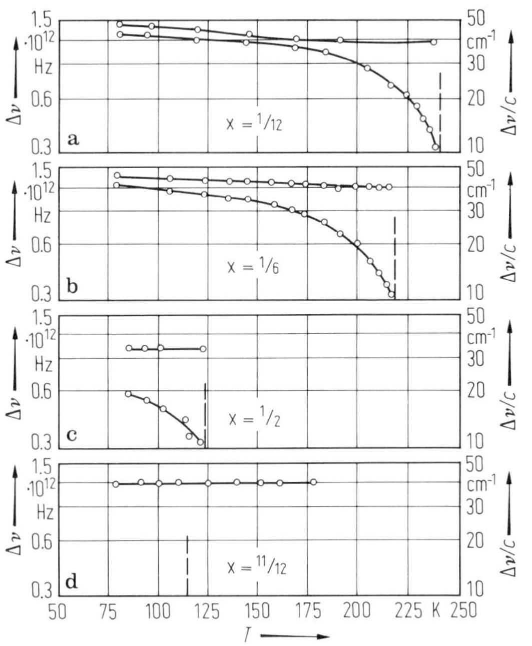

Fig. 78. SbSI–BiSI. $\Delta\nu$ vs. T [76T6]. Parameter: x, mole fraction of BiSI in $Sb_{1-x}Bi_xSI$. $\Delta\nu$: Raman frequency shift of the two lowest A_1 modes. The dashed lines indicate the position of the phase transition temperature. Note that $\Delta\nu$ is plotted logarithmically.

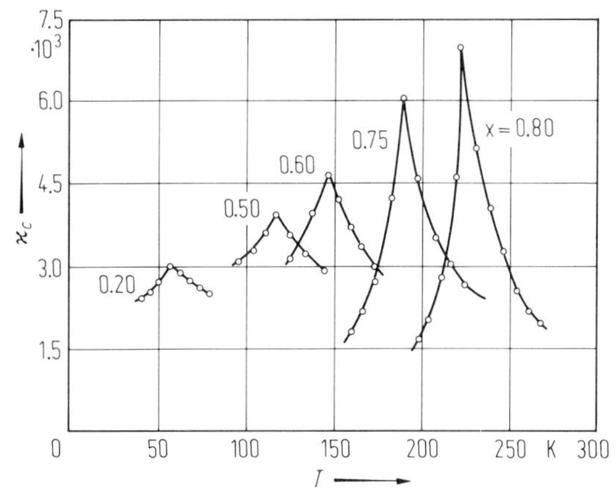

Fig. 80. SbSI–SbSeI. $\varkappa_c$ vs. T [73P14]. Parameter: x, mole fraction of SbSI in $SbS_xSe_{1-x}I$.

15 HCl family

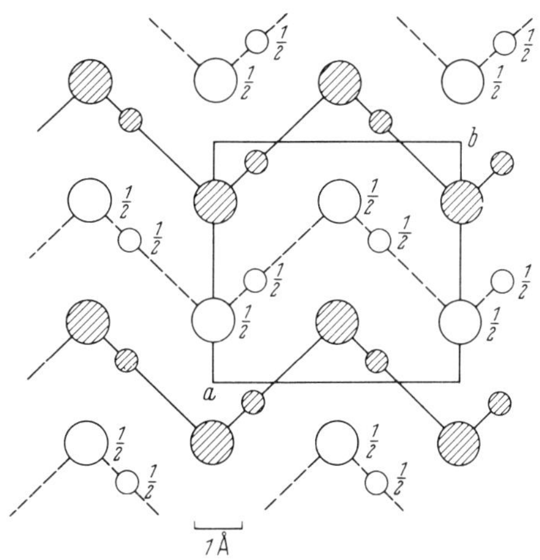

Fig. 81. DCl. Projection of the structure on (100) [67S2]. $T = 77.4$ K.

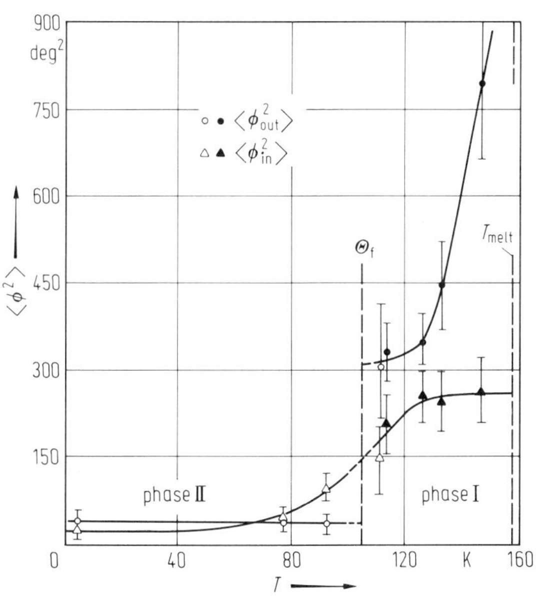

Fig. 82. DCl. $\langle\phi^2_{in}\rangle$, $\langle\phi^2_{out}\rangle$ vs. T [73N10]. $\langle\phi^2_{in}\rangle$, $\langle\phi^2_{out}\rangle$: mean square angular displacements within and out of the molecule plane, respectively. Open circles and triangles represent data given in [69S21].

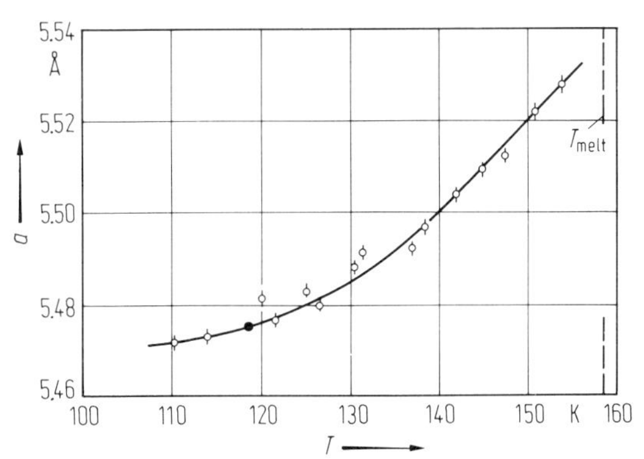

Fig. 83. DCl. a vs. T [73N10]. Data are normalized at 118.5 K so as to coincide with the value given in [69S21] shown by the black circle.

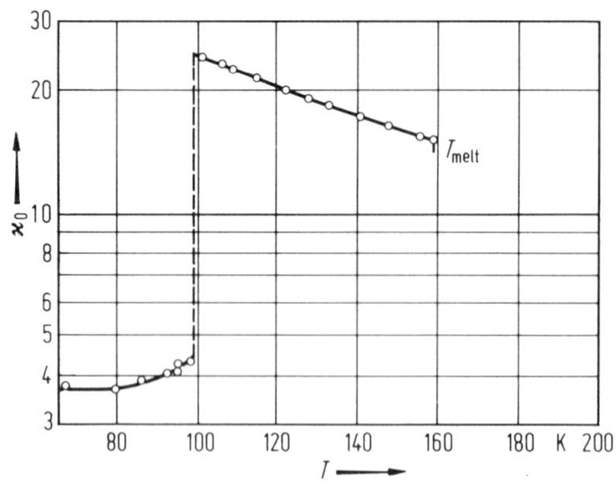

Fig. 84. HCl (polycrystalline). $\varkappa_0$ vs. T [54S7]. $\varkappa_0$: static dielectric constant.

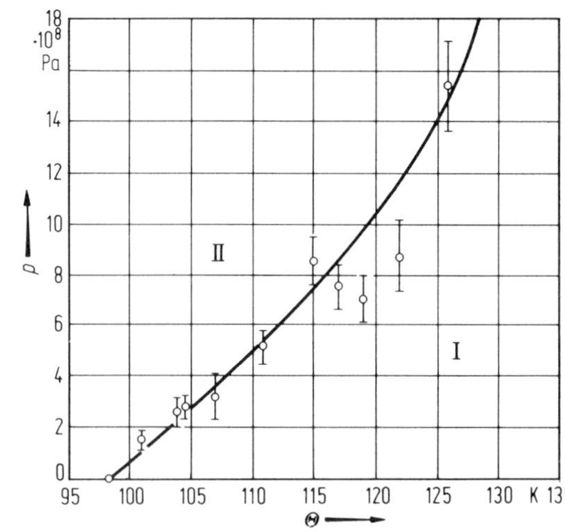

Fig. 85. HCl. Θ vs. p [62S15].

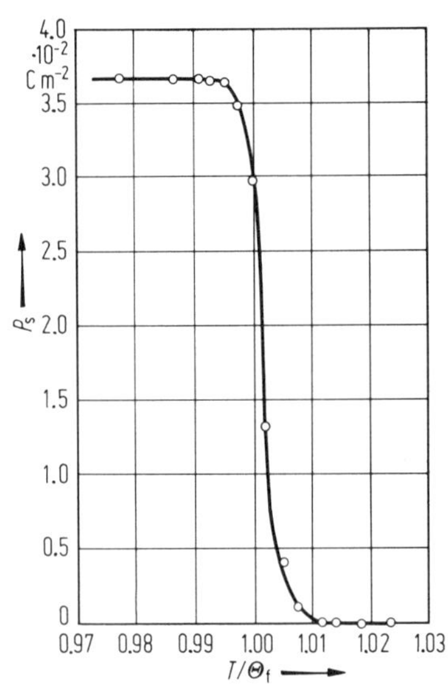

Fig. 86. HCl. P_s vs. T/Θ_f [69K6]. Obtained from pyroelectric current measurements.

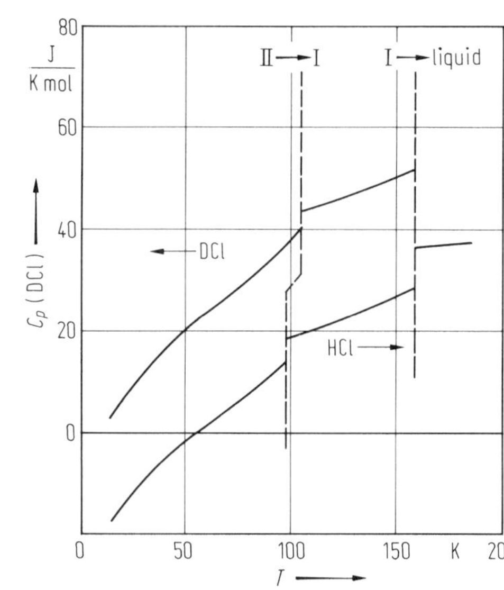

Fig. 87. HCl, DCl. C_p vs. T [47C1].

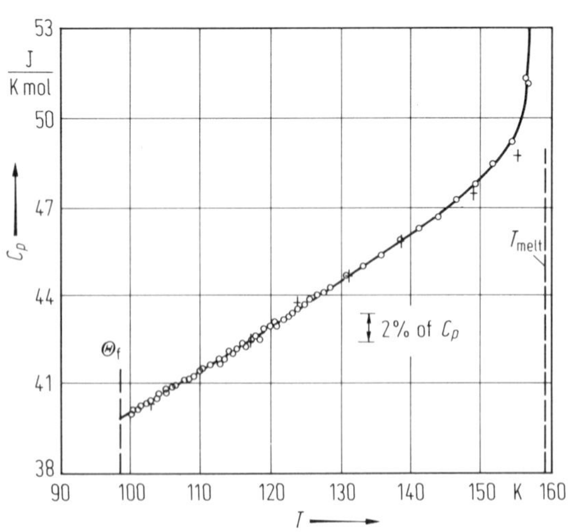

Fig. 88. HCl. C_p vs. T [76C4]. No anomaly is found at around 120 K.

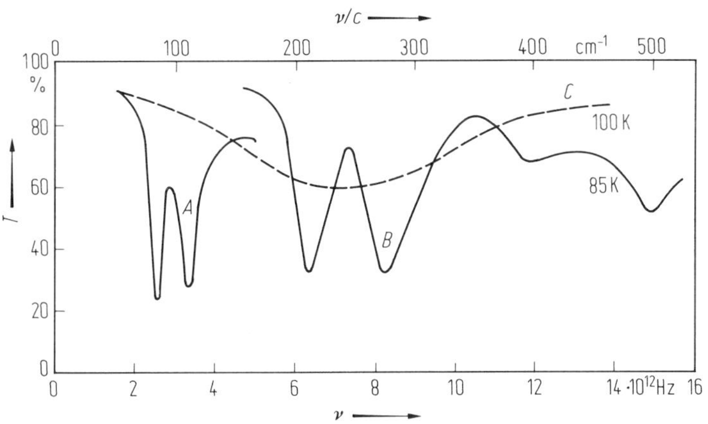

Fig. 89. HCl. T vs. ν [67A6]. T: transmittance, ν: frequency of incident infrared radiation. Specimen thickness: 20 μm for curve A, 5 μm for curve B.

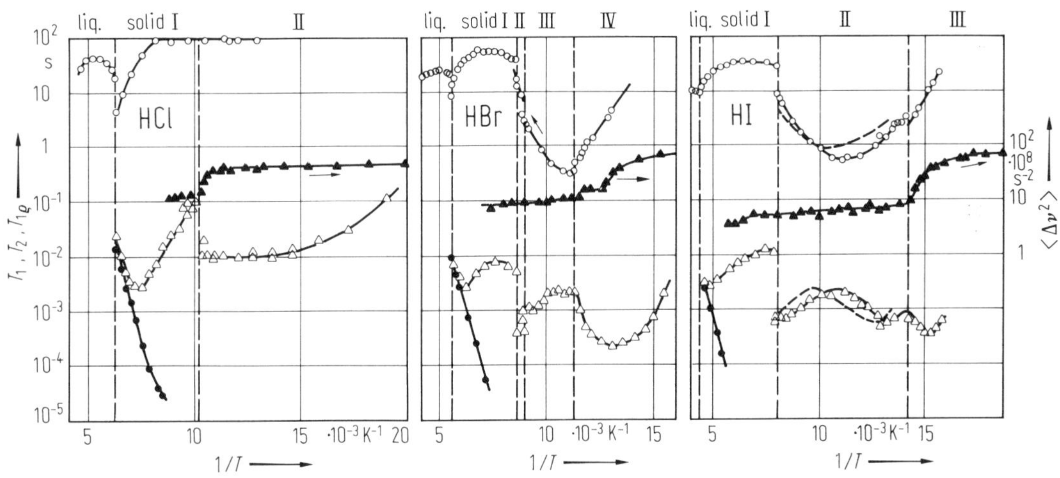

Fig. 90. HCl, HBr, HI. T_1, T_2, $T_{1\varrho}$, $\langle\Delta\nu^2\rangle$ vs. $1/T$ [68G1]. Open circles: T_1 of proton spin lattice relaxation time, full circles: spin-spin relaxation time T_2, open triangles: rotating frame relaxation time $T_{1\varrho}$, $H_1 = 1273.2\ \mathrm{A\ m^{-1}}$, full triangles: second moment $\langle\Delta\nu^2\rangle$ in [$\mathrm{s^{-2}}$]. H_1 = rf magnetic field strength.

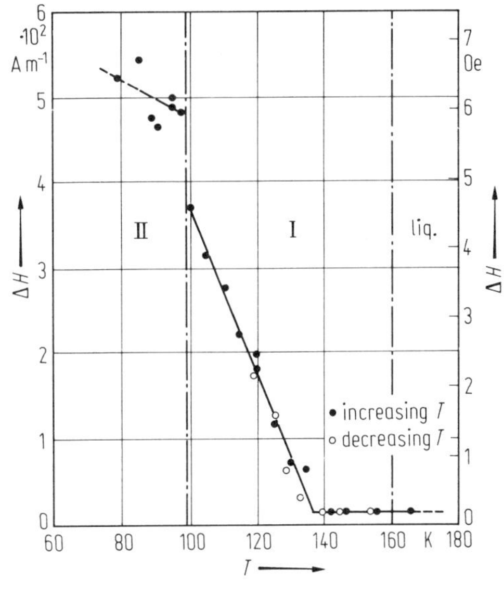

Fig. 91. HCl. ΔH vs. T [49A1]. ΔH: line width of NMR signal of proton.

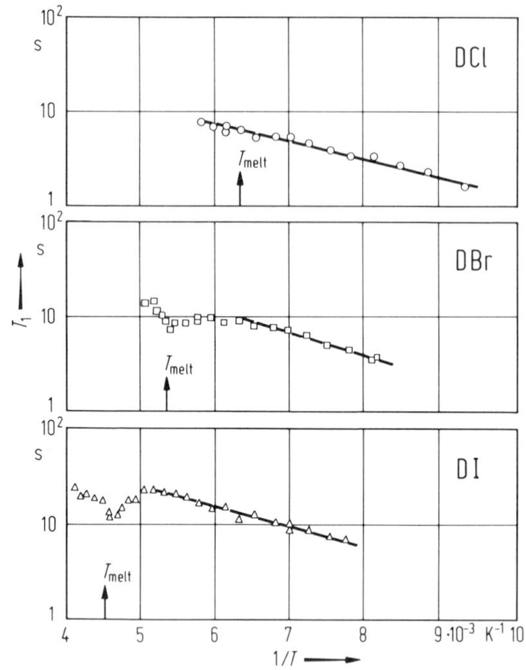

Fig. 92. DCl, DBr, DI. T_1 vs. $1/T$ [68G1]. T_1: deuteron spin lattice relaxation time.

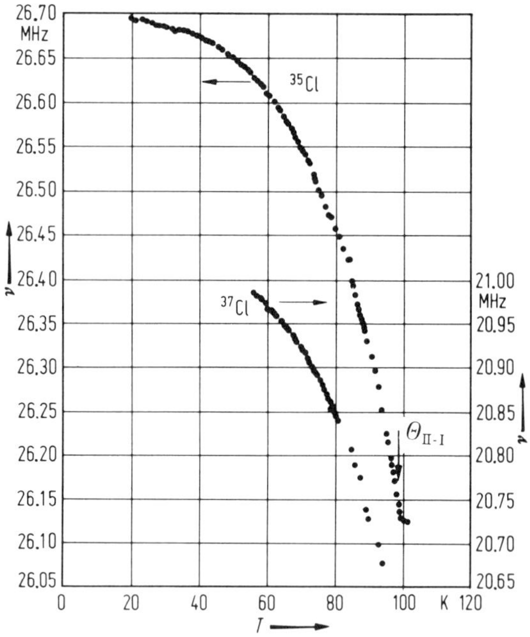

Fig. 93. HCl. ν vs. T [68O2]. ν: NQR frequency. See also [77M1].

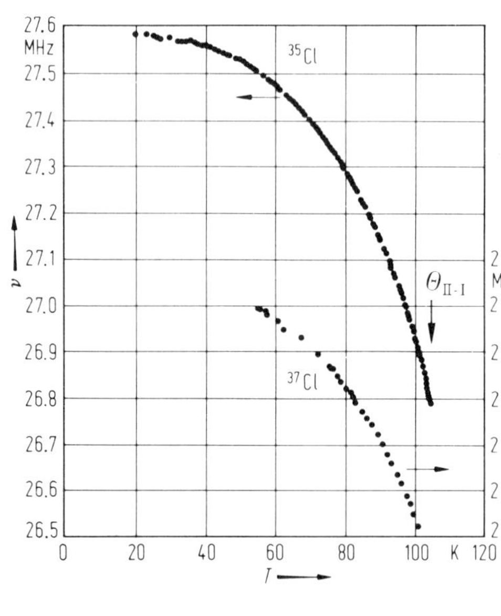

Fig. 94. DCl. ν vs. T [68O2]. ν: NQR frequency.

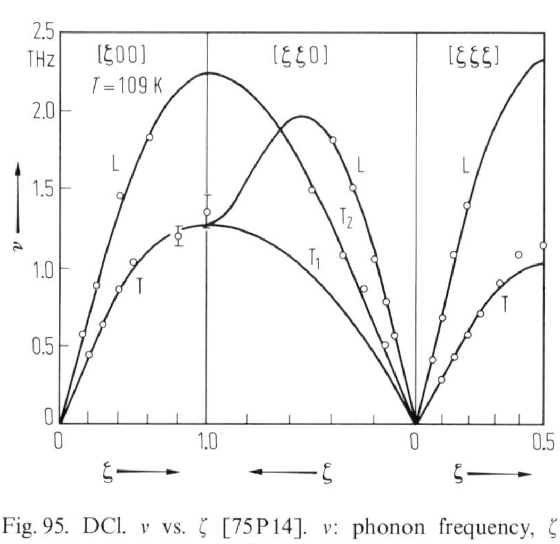

Fig. 95. DCl. ν vs. ζ [75P14]. ν: phonon frequency, ζ: reduced wave vector coordinate.

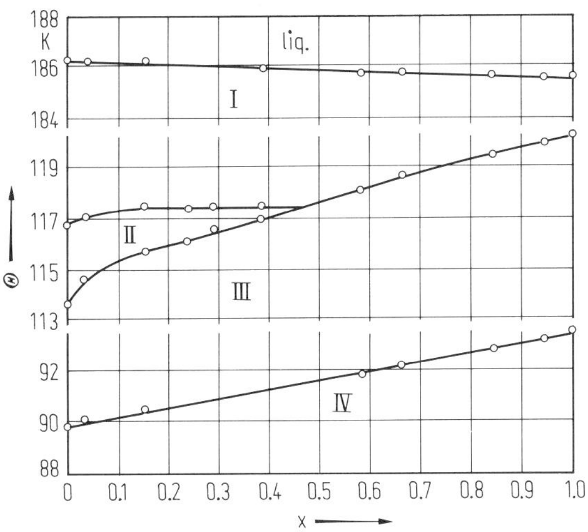

Fig. 96. $(HBr)_{1-x}(DBr)_x$. Θ vs. x [46C1].

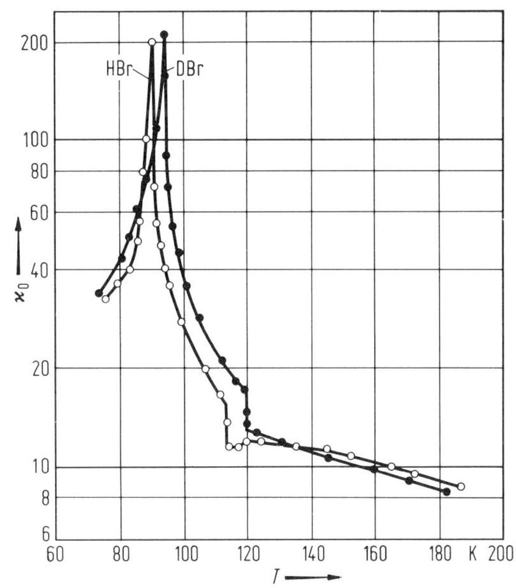

Fig. 97. HBr, DBr (Polycrystalline). $\varkappa_0$ vs. T [57C1]. $\varkappa_0$: static dielectric constant.

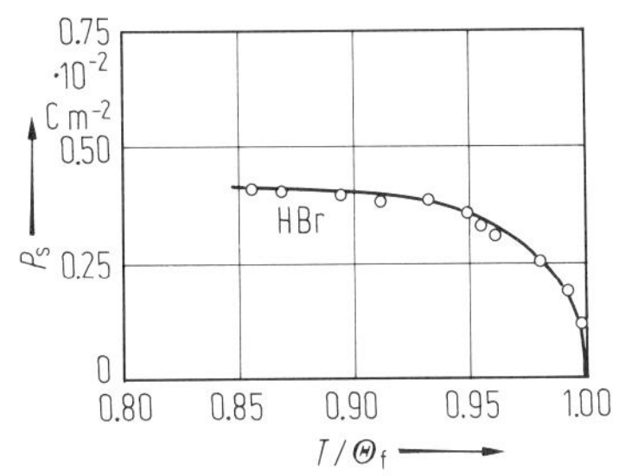

Fig. 98. HBr. P_s vs. T/Θ_f [67H11].

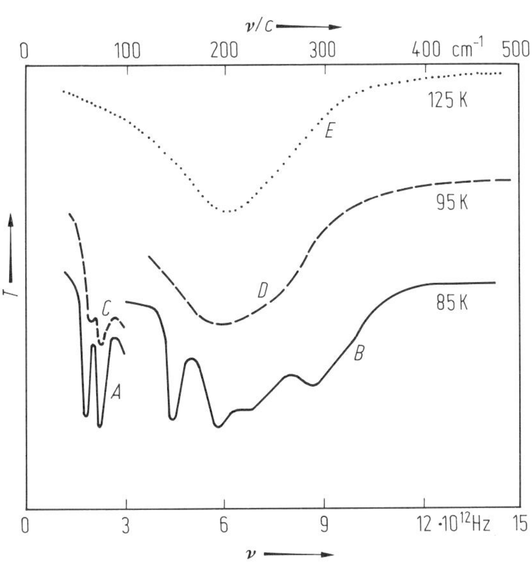

Fig. 100. DBr. T vs. ν at various temperatures [67A6]. T: transmittance, ν: frequency of incident infrared radiation. Thickness of specimen: curves A, $C \approx 20$ µm, B, D, $E \approx 10$ µm. The zero level of transmission for each curve is different.

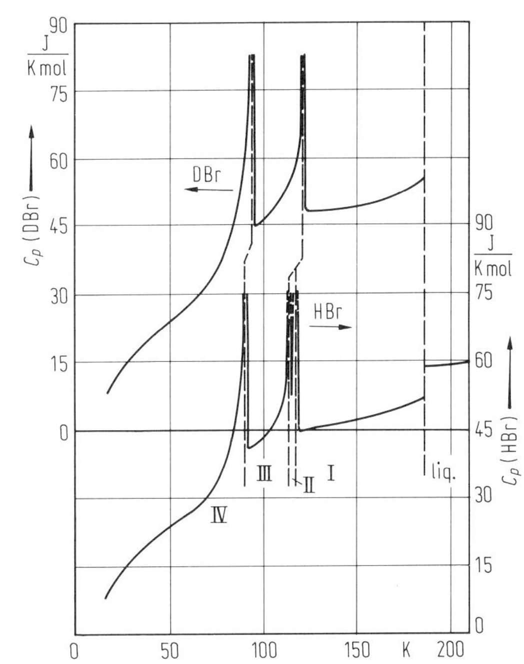

Fig. 99. HBr, DBr. C_p vs. T [47C1].

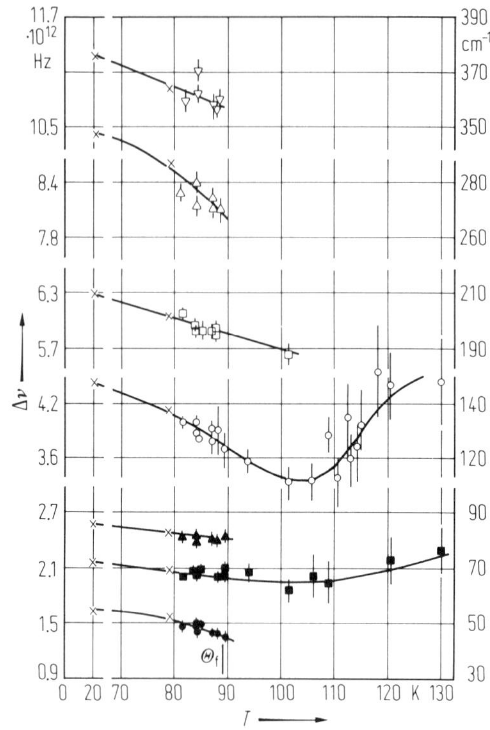

Fig. 101. HBr. Δν vs. *T* [77 V 3]. Δν: frequency shift of Raman scattering peak. Crosses: data of [72 S 33].

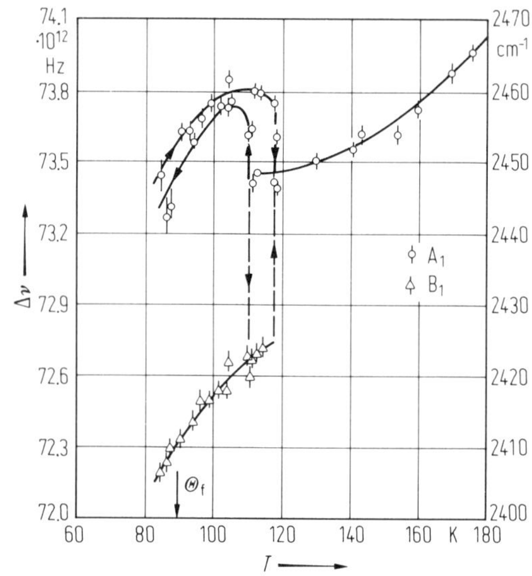

Fig. 102. HBr. Δν vs. *T* [77 V 3]. Δν: Raman frequency shift of internal mode.

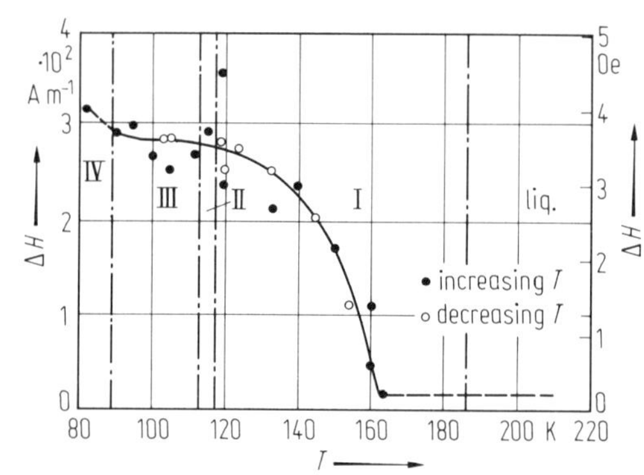

Fig. 103. HBr. Δ*H* vs. *T* [49 A 1]. Δ*H*: line width of NMR signal of proton.

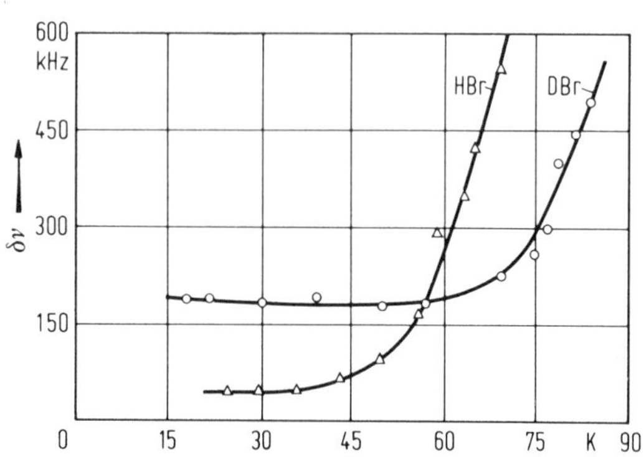

Fig. 104. HBr, DBr, δν vs. *T* [70 K 1]. δν: line width of ^{79}Br.

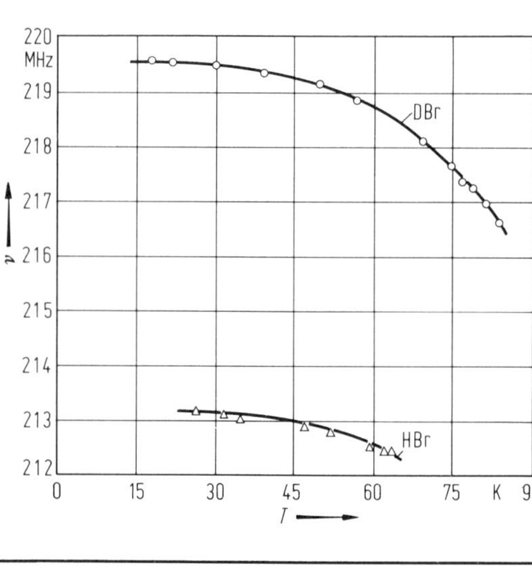

Fig. 105. HBr, DBr. ν vs. *T* [70 K 1]. ν: NQR frequency of ^{79}Br.

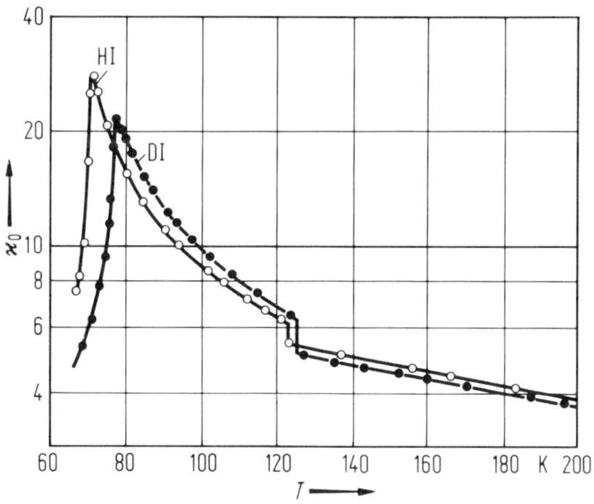

Fig. 106. HI, DI (polycrystalline). $\varkappa_0$ vs. T [57C1]. $\varkappa_0$: static dielectric constant.

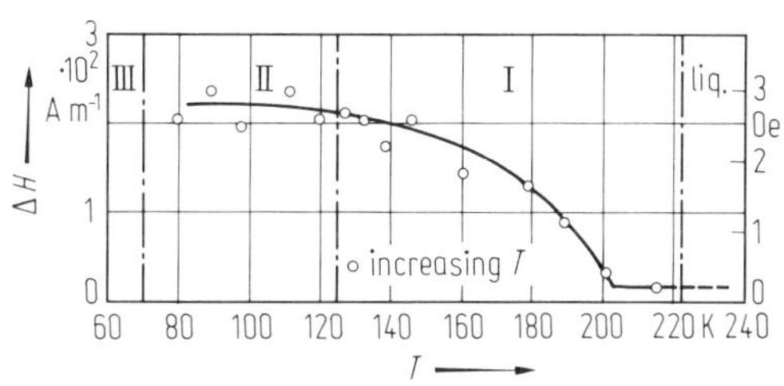

Fig. 108. HI. ΔH vs. T [49A1]. ΔH: line width of NMR signal of proton.

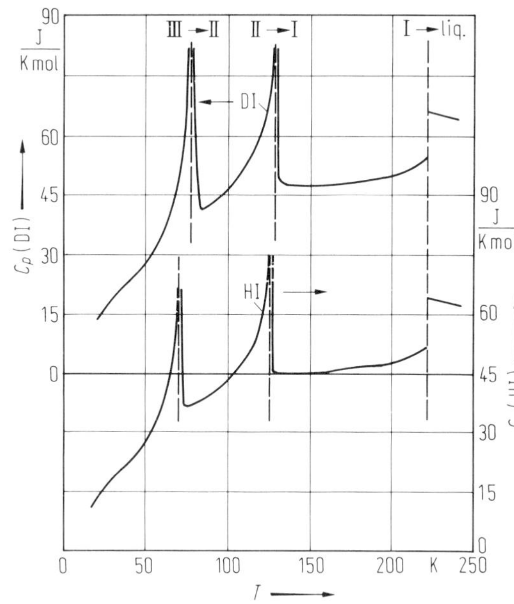

Fig. 107. HI, DI. C_p vs. T [47C1].

16 $NaNO_2$ family

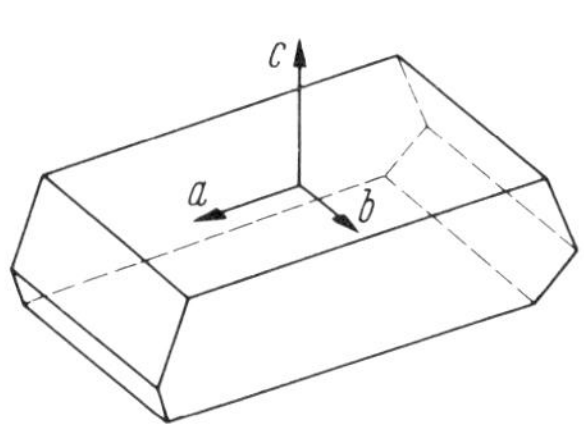

Fig. 109. $NaNO_2$. Crystal form grown from solution [61N4].

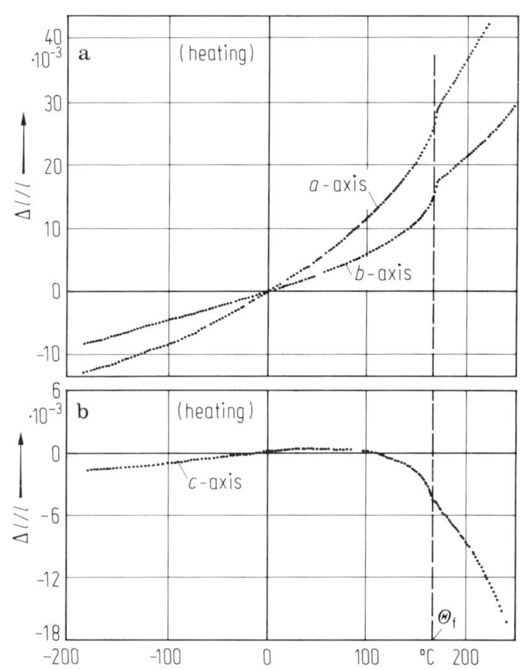

Fig. 113. $NaNO_2$. $\Delta l/l$ vs. T [75E2]. $\Delta l/l$: linear thermal expansion.

Figs. 111, 112, see p. 334.

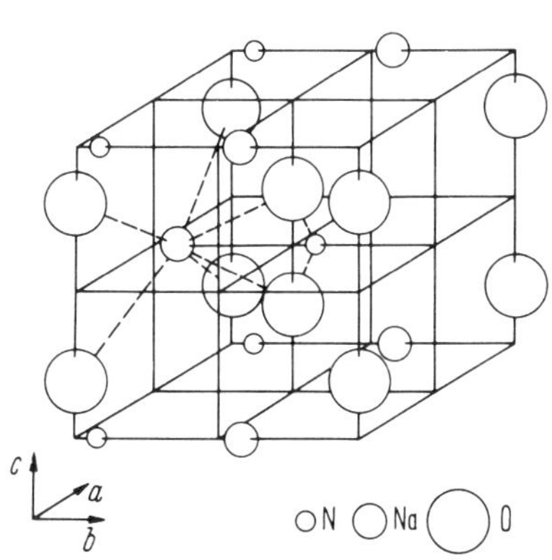

Fig. 110. $NaNO_2$. Structure of phase III [31Z1].

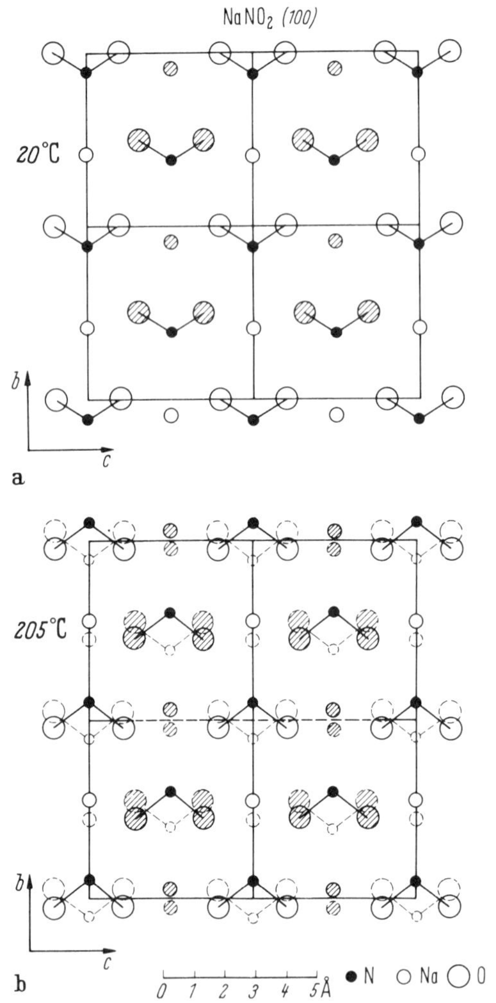

Fig. 111. $NaNO_2$. Crystal structures. Projection along the a-axis. Fig. a: phase III [64C1], Fig. b: phase I [43S1]. The two positions of the atoms in Fig. (b) are indicated by solid and dotted circles, respectively. Clear circles represent atoms placed at $a=0$ and shaded circles atoms at $a=\frac{1}{2}$.

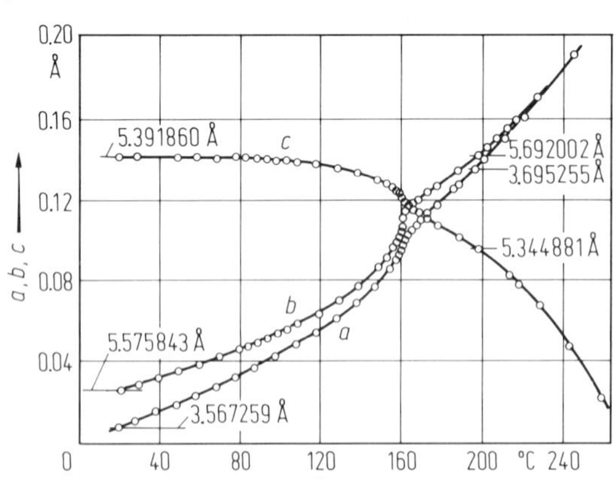

Fig. 112. $NaNO_2$. a, b, c vs. T [76K19]. Zero of ordinate scale is shifted.

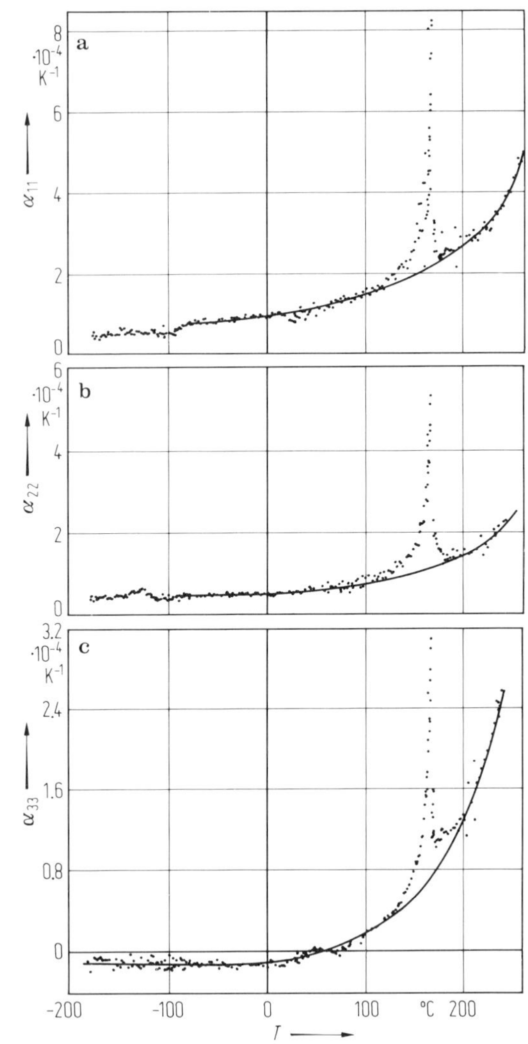

Fig. 114. $NaNO_2$. α_{11}, α_{22}, α_{33} vs. T [75E2]. α_{ii}: linear thermal expansion coefficient.

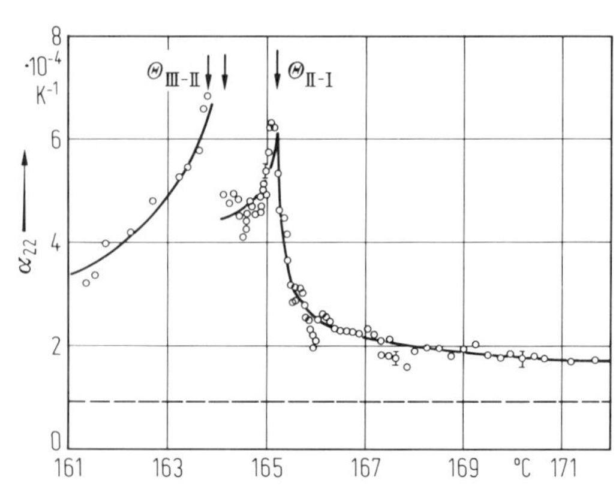

Fig. 115. $NaNO_2$. α_{22} vs. T in the vicinity of III–II–I transitions [75E2].

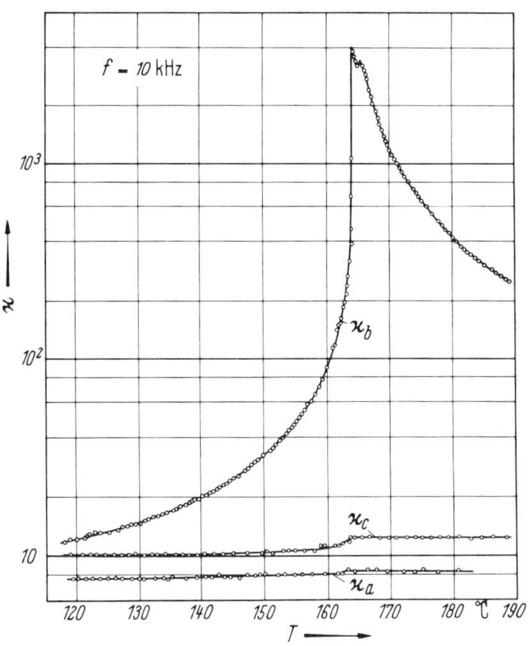

◄ Fig. 116. $NaNO_2$. $\varkappa_a$, $\varkappa_b$, $\varkappa_c$ vs. T [67G6].

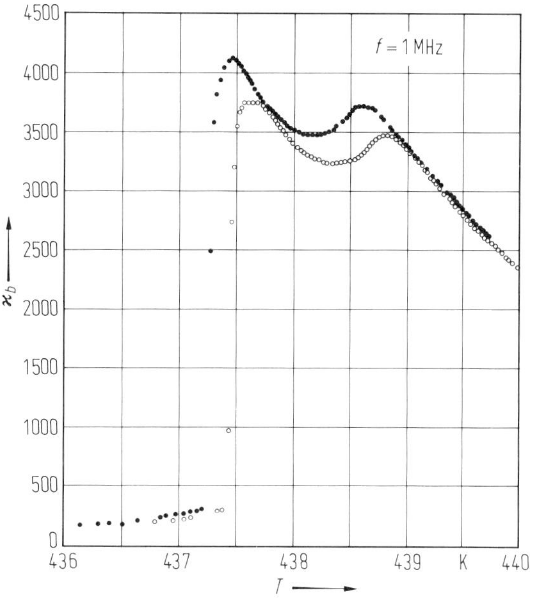

Fig. 117. $NaNO_2$. $\varkappa_b$ vs. T in the vicinity of the III–II–I transition temperatures [75H2]. Results of two specimens are shown.

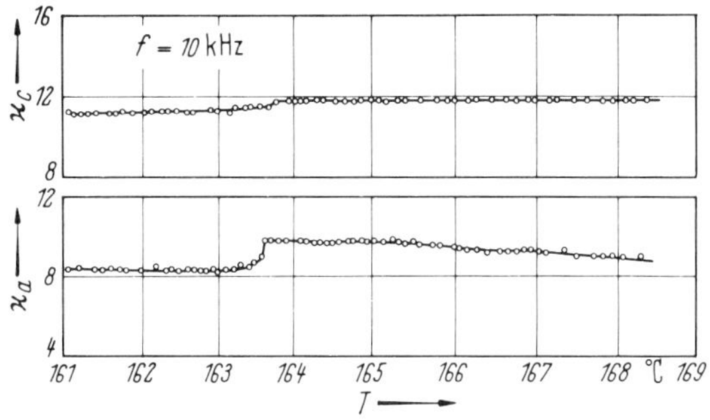

Fig. 118. $NaNO_2$. $\varkappa_a$, $\varkappa_c$ vs. T in the vicinity of the III–II–I transition temperatures [67T1].

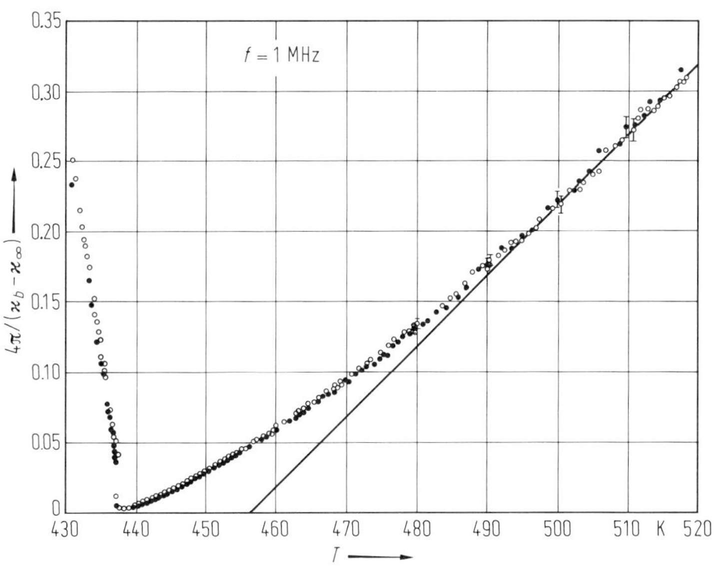

Fig. 119. $NaNO_2$. $4\pi/(\varkappa_b - \varkappa_\infty)$ vs. T [75H2]. $\varkappa_\infty$: dielectric constant at high frequency limit, $\varkappa_\infty = 6$. Open and closed circles represent results of two different specimens, respectively.

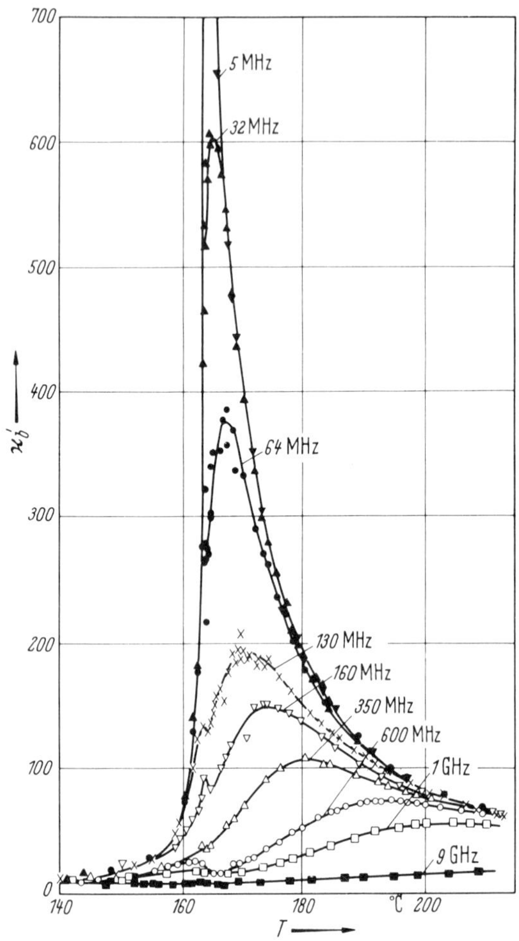

Fig. 120. $NaNO_2$. $\varkappa_b'$ vs. T [68H1]. Parameter: f.

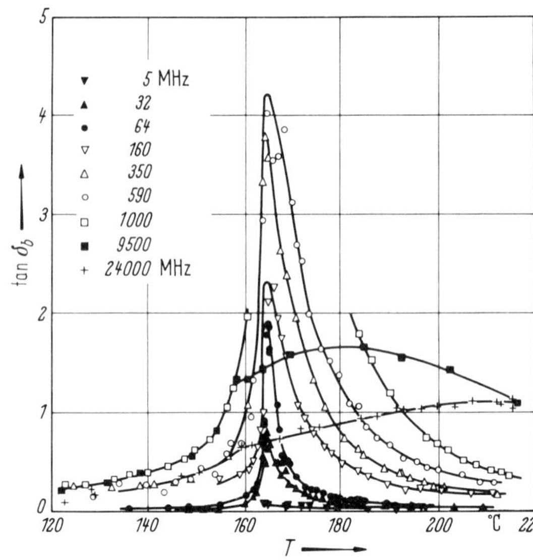

Fig. 121. $NaNO_2$. $\tan \delta_b$ vs. T [68H1]. Parameter: f.

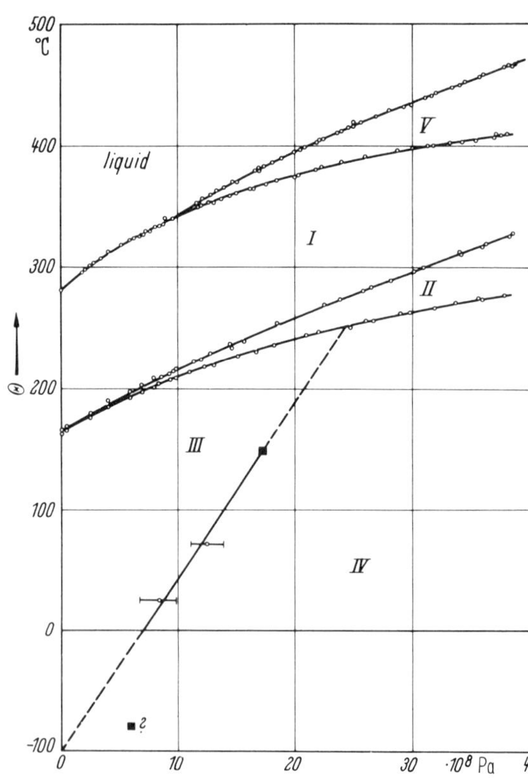

Fig. 122. $NaNO_2$. Θ vs. p [66R2]. The solid squares show results of [37B1].

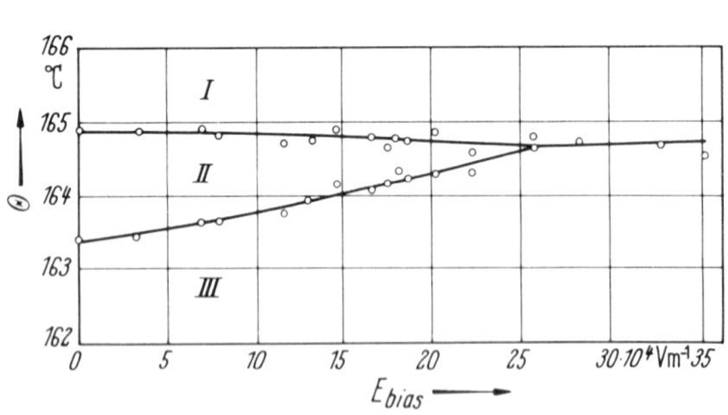

Fig. 124. $NaNO_2$. Θ vs. E_{bias} [65G2].

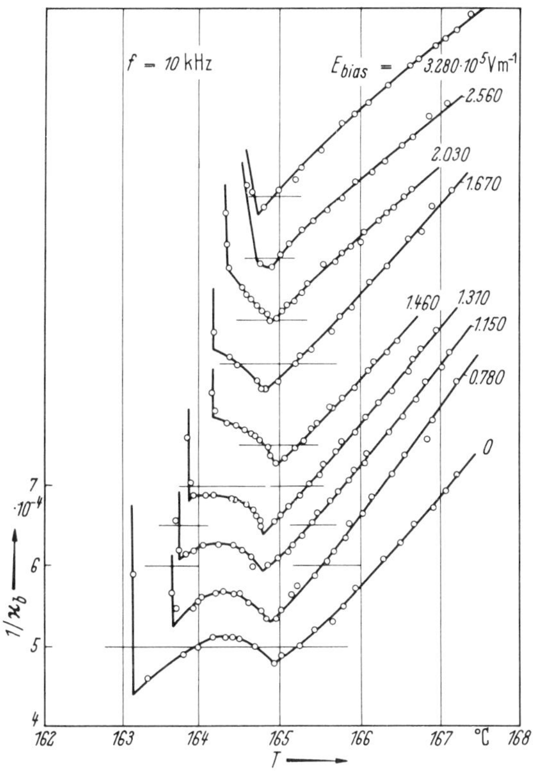

Fig. 123. $NaNO_2$. $1/\varkappa_b$ vs. T [65G2]. Parameter: E_{bias}. The line attaching to each curve indicates the level of $1/\varkappa_b = 5 \cdot 10^{-4}$.

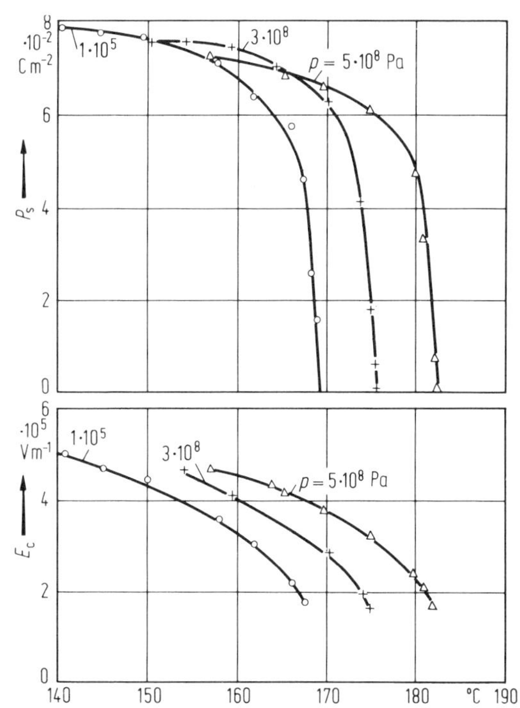

Fig. 126. $NaNO_2$. P_s, E_c vs. T [71L5]. Parameter: p.

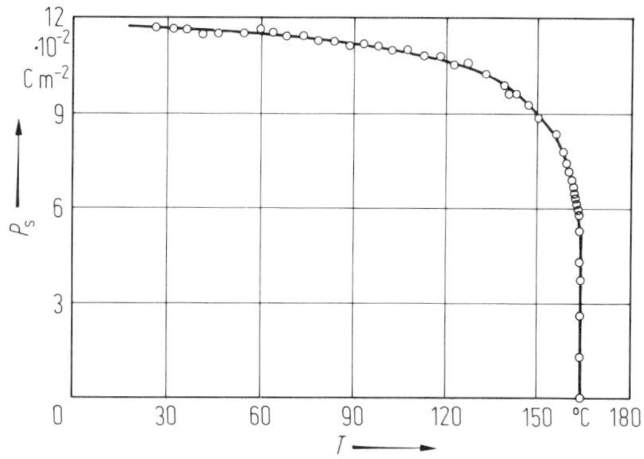

Fig. 125. $NaNO_2$. P_s vs. T [73H1]. Obtained from pyroelectric charge measurements.

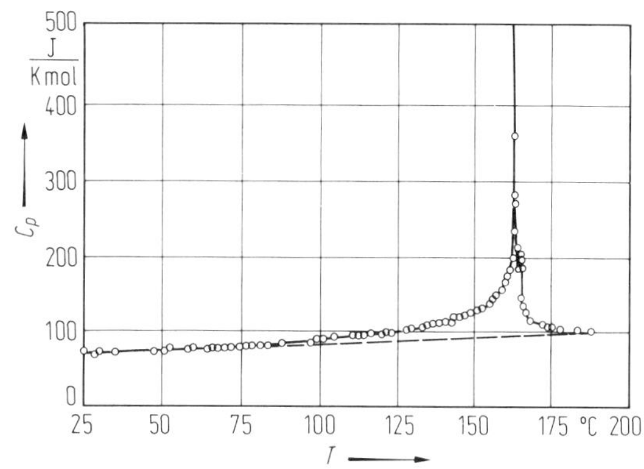

Fig. 127. $NaNO_2$. C_p vs. T [65S1].

Fig. 128, see p. 338.

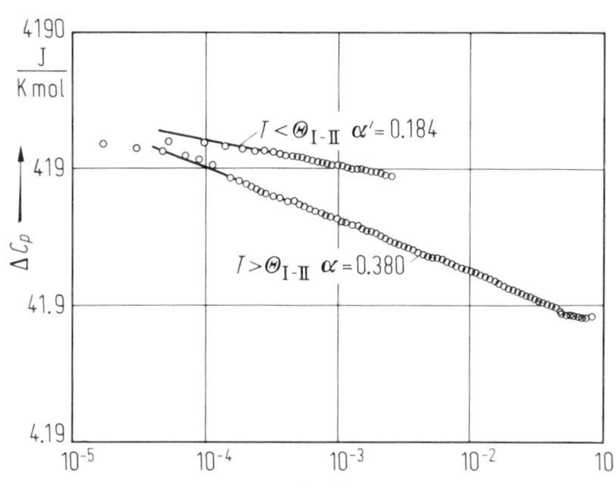

Fig. 129. $NaNO_2$. ΔC_p vs. $|\varepsilon|$ [71H5]. ΔC_p: excess molar heat capacity at constant pressure, $\varepsilon = (\Theta_{II-I} - T)/\Theta_{II-I}$, α, α': critical indices.

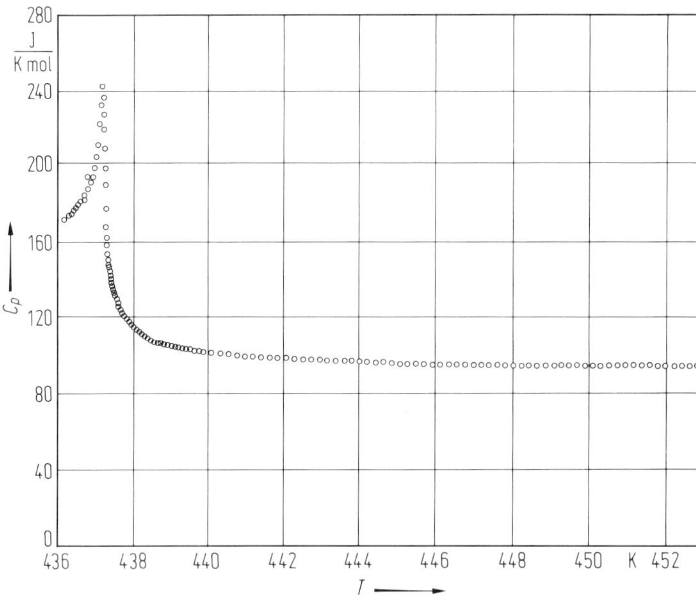

Fig. 128. $NaNO_2$. C_p vs. T [73H8]. Obtained from AC calorimeter.

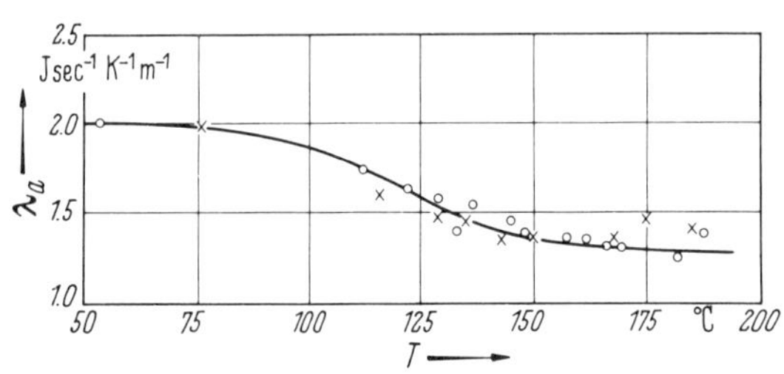

Fig. 130. $NaNO_2$. λ_a vs. T [61Y1]. λ_a: thermal conductivity along the a-axis

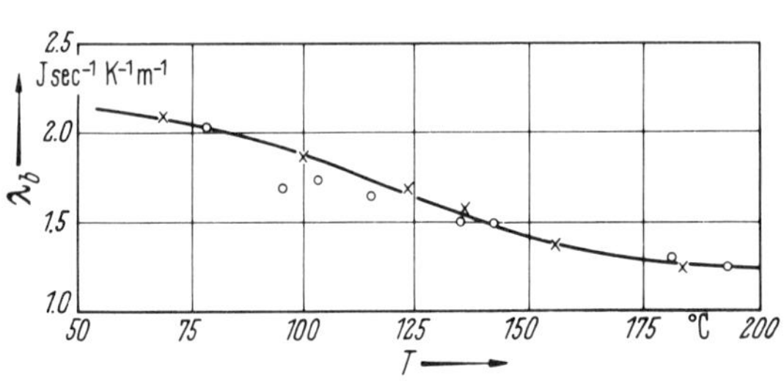

Fig. 131. $NaNO_2$. λ_b vs. T [61Y1]. λ_b: thermal conductivity along the b-axis.

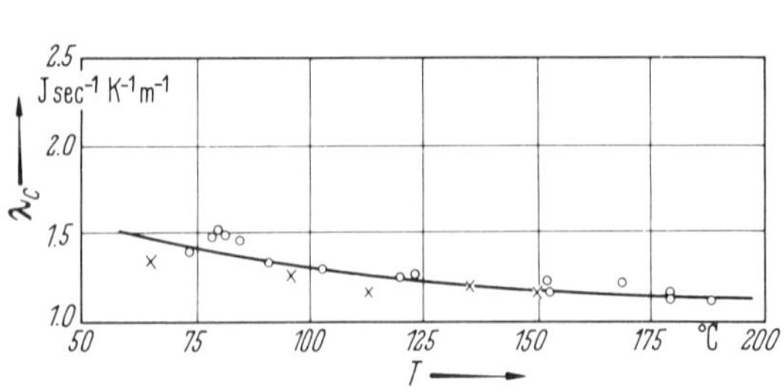

Fig. 132. $NaNO_2$. λ_c vs. T [61Y1]. λ_c: thermal conductivity along the c-axis.

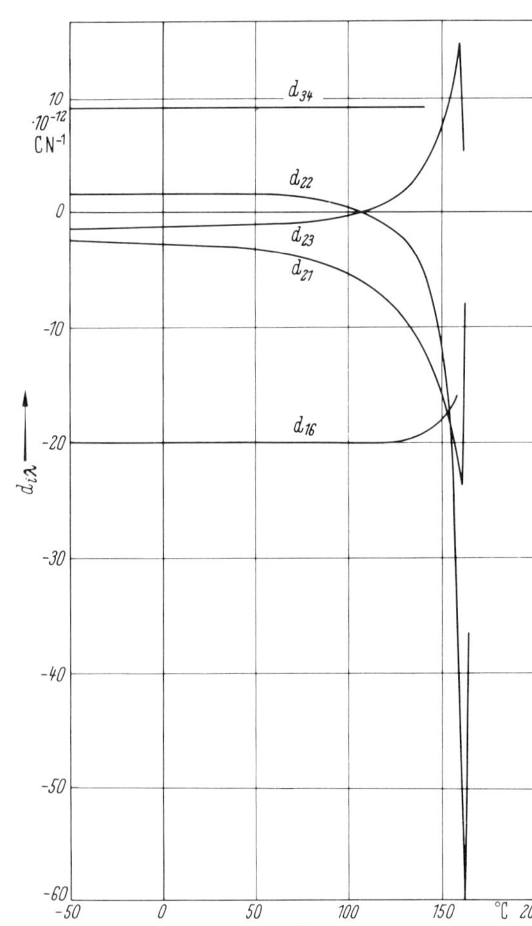

Fig. 133. $NaNO_2$. $d_{i\lambda}$ vs. T [63H1].

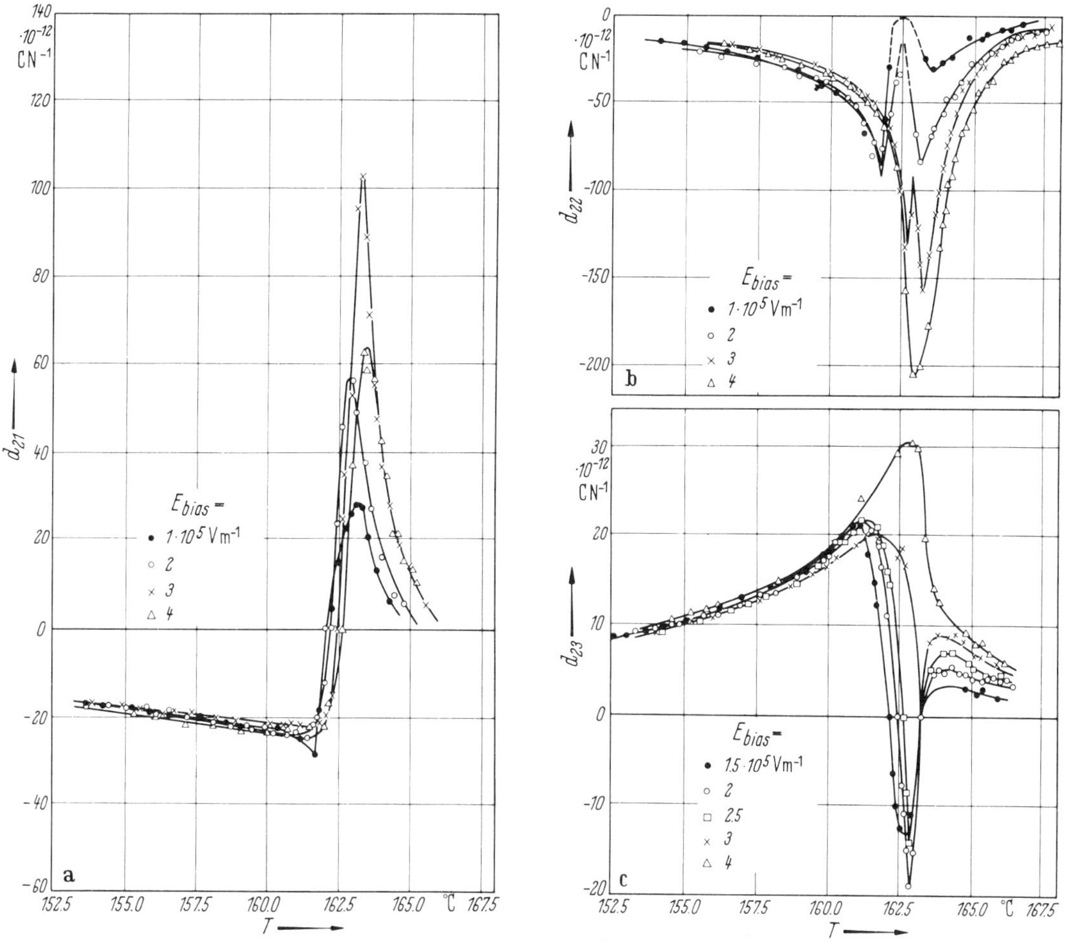

Fig. 134. $NaNO_2$. $d_{i\lambda}$ vs. T in the vicinity of the III–II–I transition temperatures [64 H 2]. Parameter: E_{bias}. Fig. a: d_{21}, b: d_{22}, c: d_{23}.

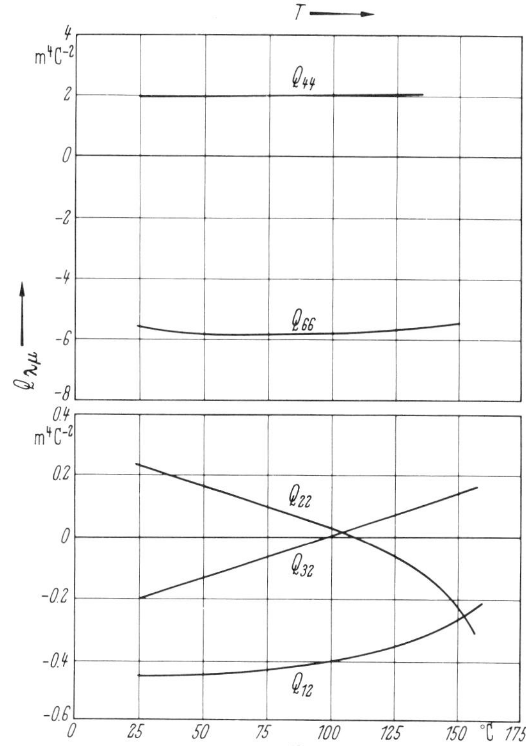

Fig. 135. $NaNO_2$. $Q_{\lambda\mu}$ vs. T [63 H 1]. Calculated from $d_{i\lambda}$, $\varkappa$ and P_s.

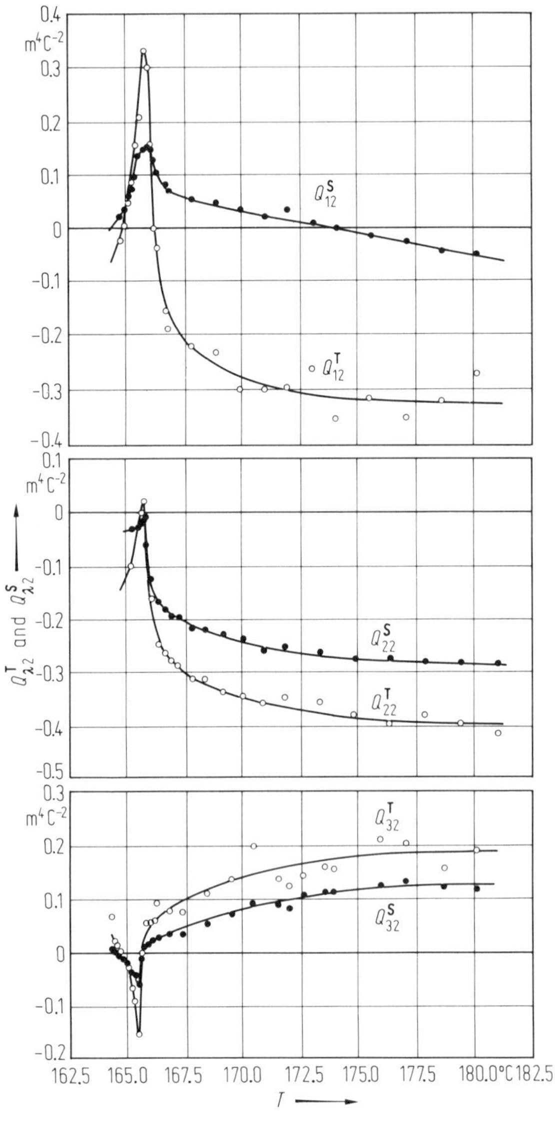

Fig. 136. $NaNO_2$. $Q^{\mathbf{T}}_{\lambda 2}$, $Q^{\mathbf{S}}_{\lambda 2}$ vs. T [76Y2]. $Q^{\mathbf{T}}_{\lambda 2}$, $Q^{\mathbf{S}}_{\lambda 2}$: isothermal and adiabatic electrostrictive constants, respectively.

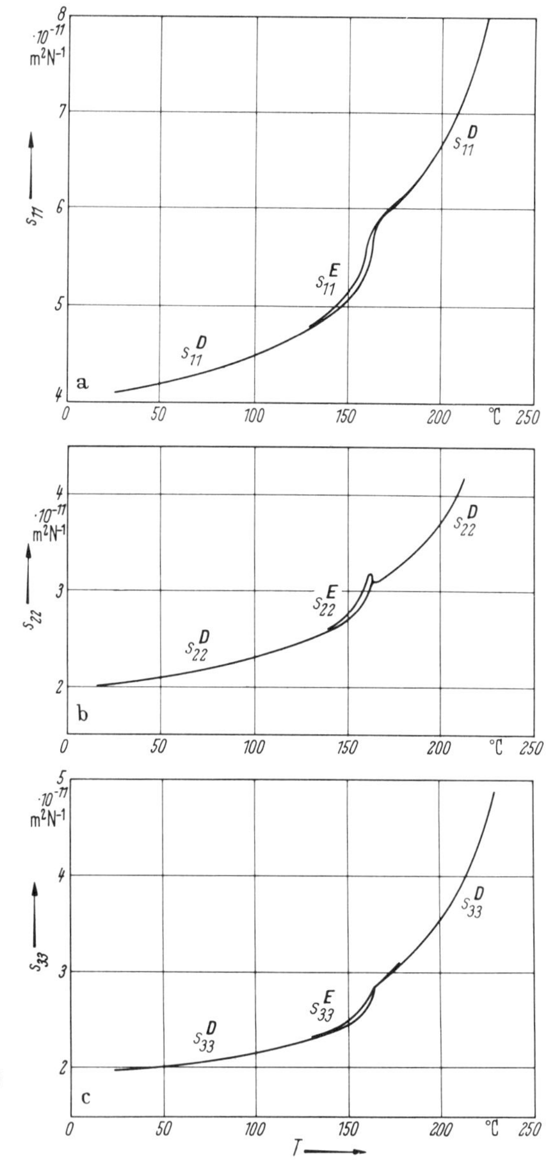

Fig. 137. $NaNO_2$. $s^{D}_{\lambda\lambda}$, $s^{E}_{\lambda\lambda}$ vs. T [63H1]. Fig. a: $\lambda = 1$, b: $\lambda = 2$, c: $\lambda = 3$.

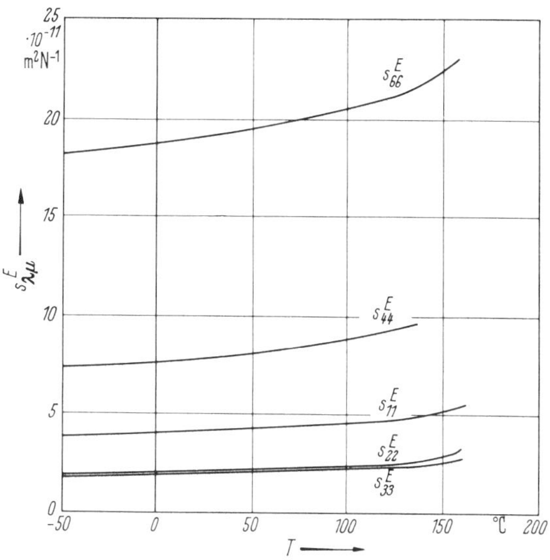

Fig. 138. $NaNO_2$. $s^E_{\lambda\mu}$ vs. T [63H1].

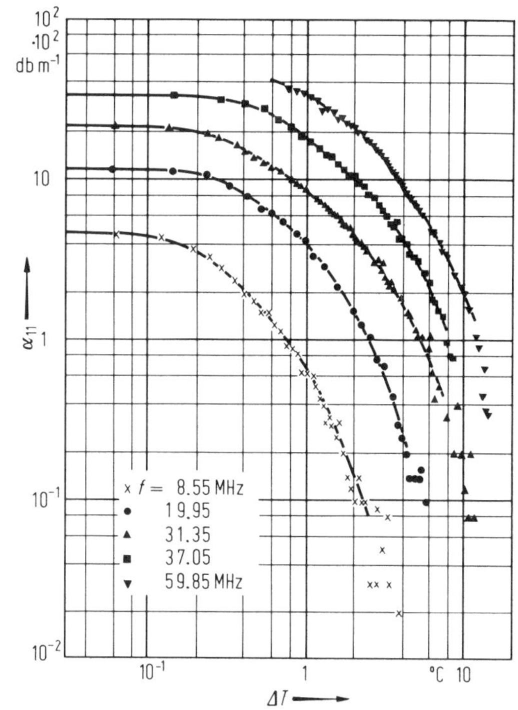

Fig. 140. $NaNO_2$. α_{11} vs. ΔT [69H5]. Parameter: f. α_{11}: attenuation coefficient of longitudinal ultrasonic waves propagating along the a-axis. $\Delta T = T - \Theta_{\text{II-I}}$.

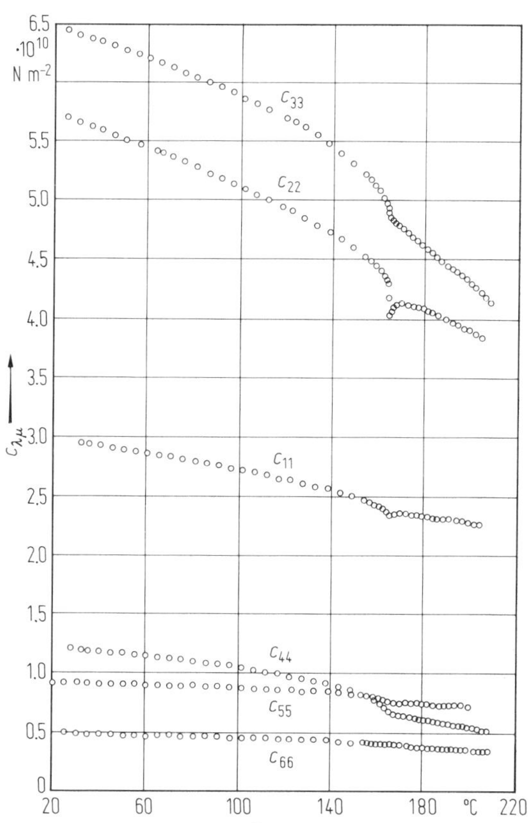

Fig. 139. $NaNO_2$. $c_{\lambda\mu}$ vs. T [78H4]. Obtained from ultrasonic measurements.

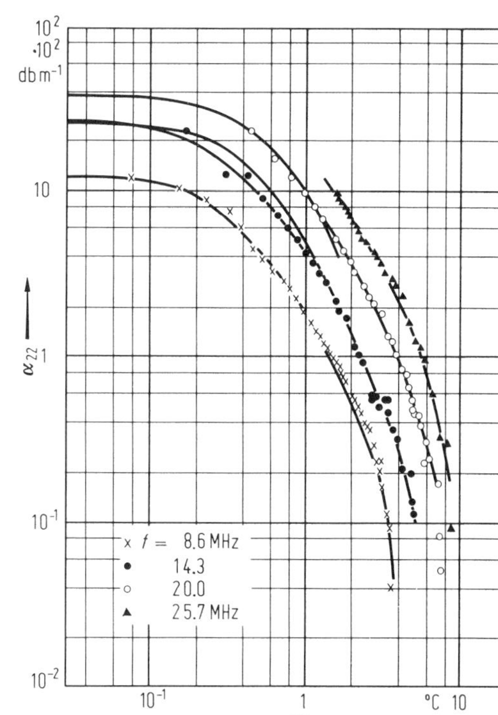

Fig. 141. $NaNO_2$. α_{22} vs. ΔT [70H7]. Parameter: f. α_{22}: attenuation coefficient of longitudinal ultrasonic waves propagating along the b-axis. $\Delta T = T - \Theta_{\text{II-I}}$.

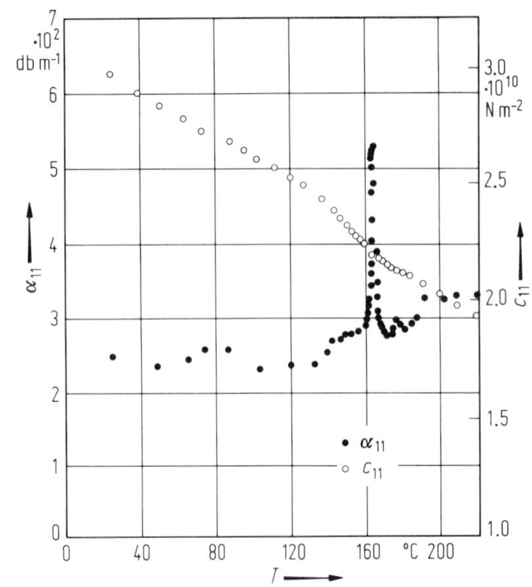

Fig. 142. $NaNO_2$. α_{11}, c_{11} vs. T [70O4]. α_{11}: attenuation coefficient of the longitudinal ultrasonic waves propagating along the a-axis. $f = 12.5$ MHz.

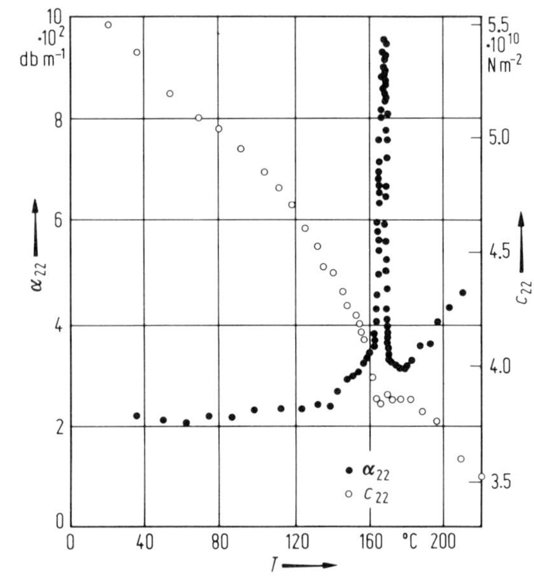

Fig. 143. $NaNO_2$. α_{22}, c_{22} vs. T [70O4]. α_{22}: attenuation coefficient of the longitudinal ultrasonic waves propagating along the b-axis. $f = 12.5$ MHz.

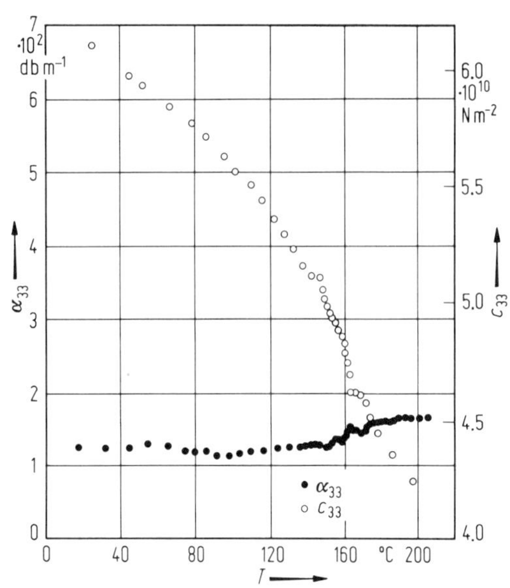

Fig. 144. $NaNO_2$. α_{33}, c_{33} vs. T [70O4]. α_{33}: attenuation coefficient of the longitudinal ultrasonic waves propagating along the c-axis. $f = 12.5$ MHz.

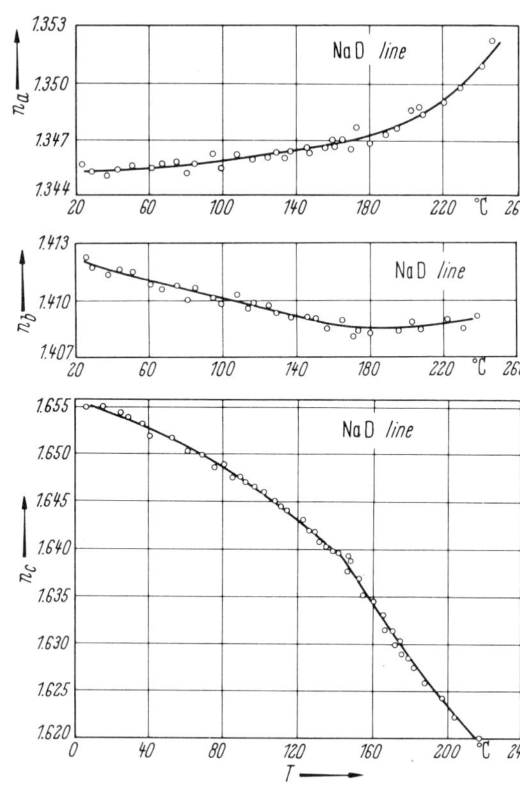

Fig. 145. $NaNO_2$. n_a, n_b, n_c vs. T [66H13].

Fig. 147. $NaNO_2$. R vs. ν [69S20]. Along the a, b, c-axis. ►
R: reflectivity. Spectral resolution is indicated by gaps. Solid curves are calculated by using classical oscillator model.

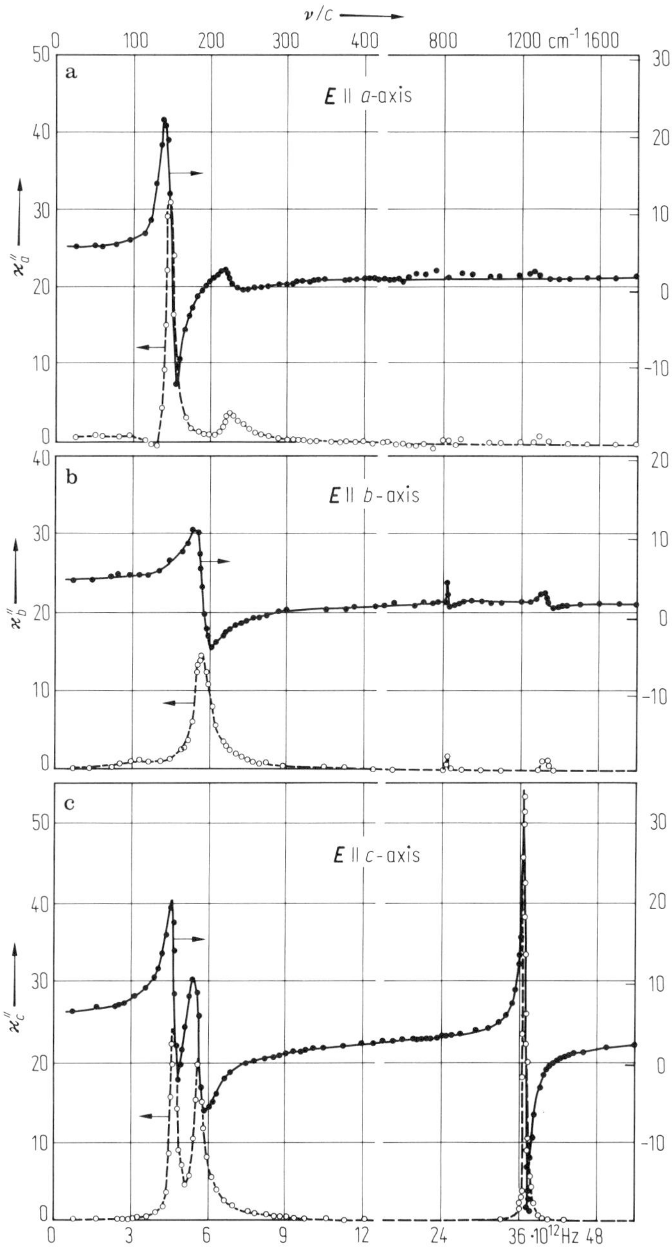

Fig. 146. $NaNO_2$. $\varkappa'$, $\varkappa''$ vs. ν [68A6]. T=RT. Polarization of light ∥ a (Fig. a), ∥ b (Fig. b), ∥ c (Fig. c). Solid lines indicate $\varkappa'$ with the right scale, broken lines indicate $\varkappa''$ with the left scale. Note change of scale at $\nu = 12 \cdot 10^{12}$ Hz.

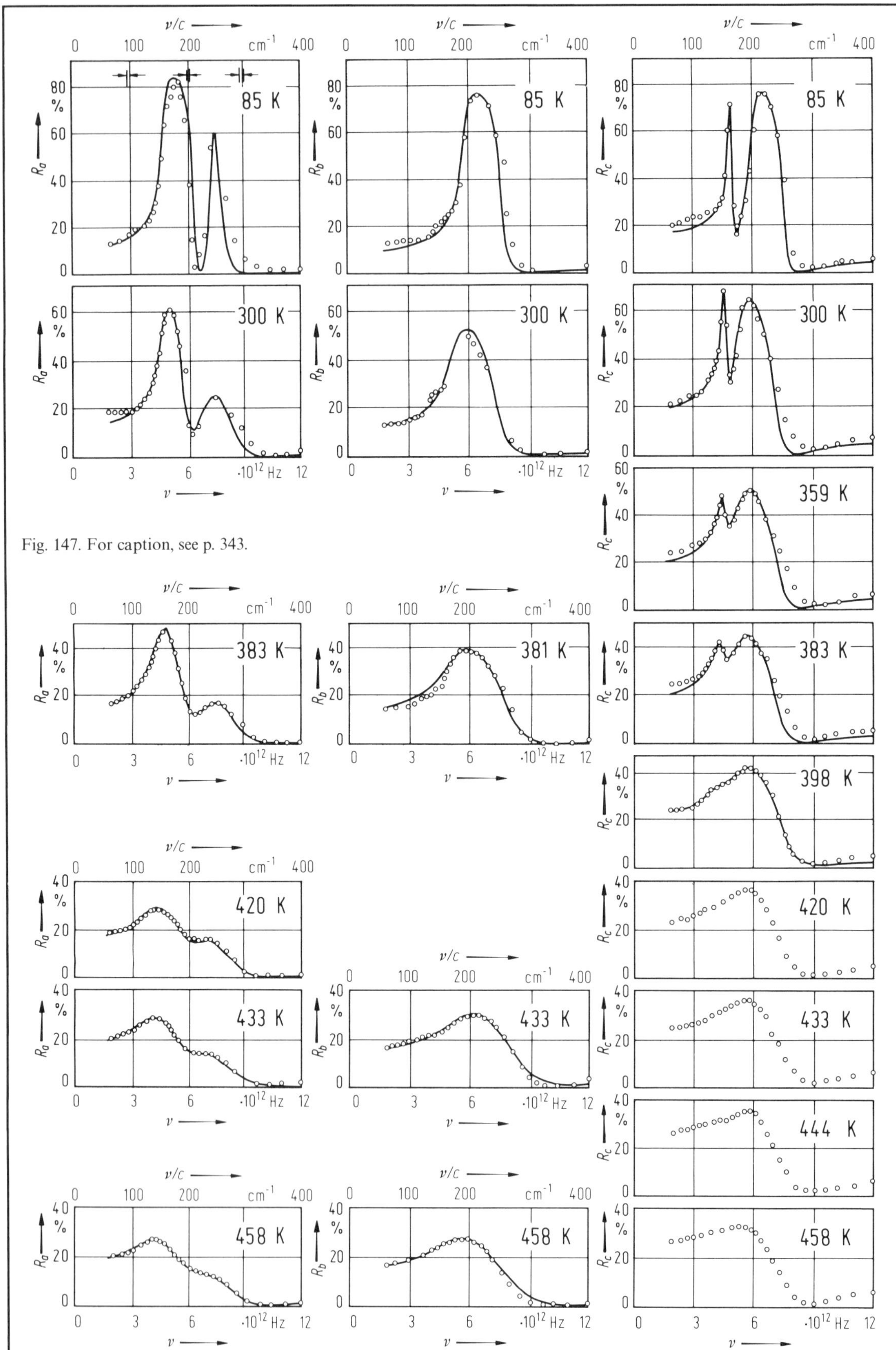

Fig. 147. For caption, see p. 343.

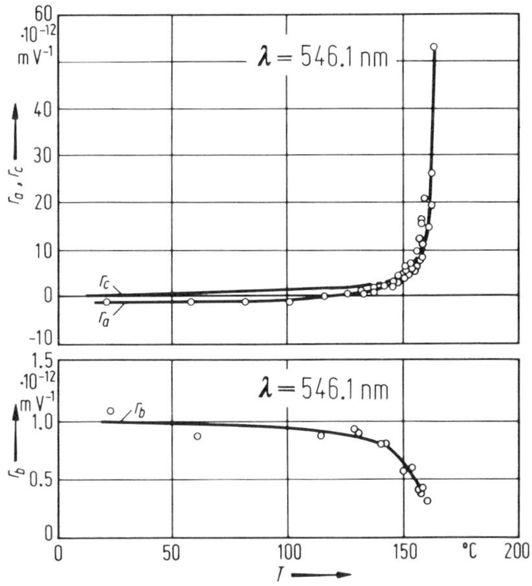

Fig. 148. $NaNO_2$. r_a, r_b, r_c vs. T [69J2]. $r_a = r_{22} - (n_3/n_2)^3 \cdot r_{32}$, $r_b = r_{32} - (n_1/n_3)^3 \cdot r_{12}$, $r_c = r_{22} - (n_1/n_2)^3 \cdot r_{12}$.

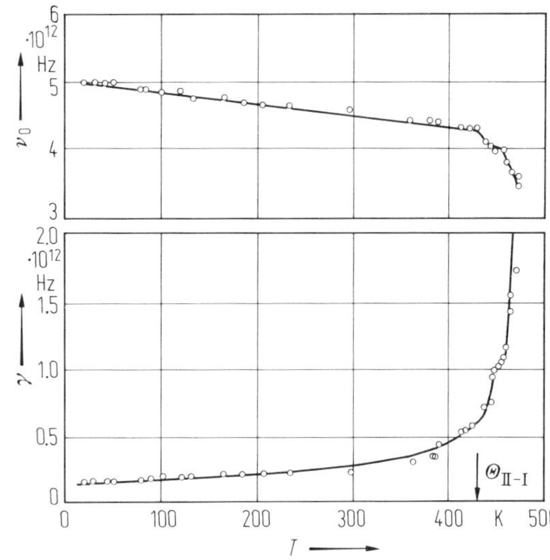

Fig. 151. $NaNO_2$. ν_0, γ vs. T [73A8]. ν_0: frequency of a B_1 phonon mode, γ: damping constant.

Fig. 152, see p. 346.

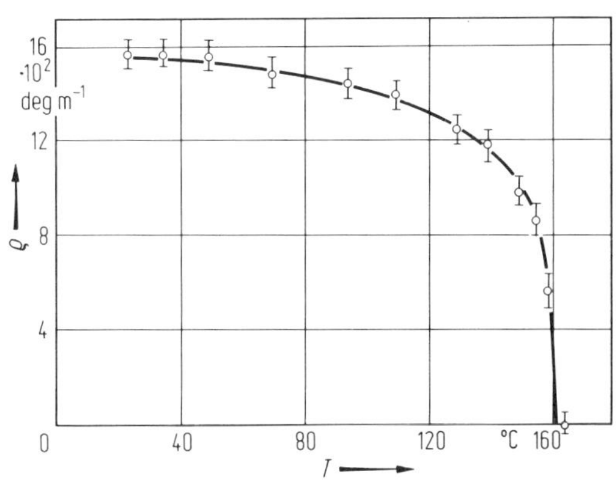

Fig. 149. $NaNO_2$. ϱ vs. T [72C3]. ϱ: optical rotatory power.

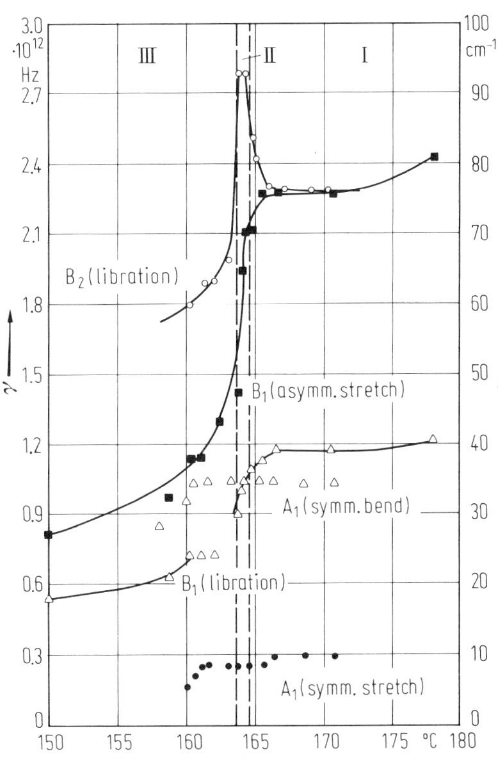

Fig. 153. $NaNO_2$. γ vs. T [76C1]. γ: full width at half maximum of Raman line. See Table 254. The dashed lines indicate range of phase II.

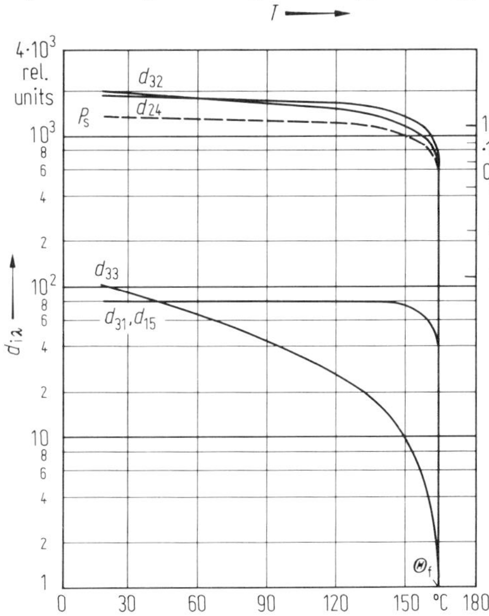

Fig. 150. $NaNO_2$. $d_{i\lambda}$ vs. T [74I6]. $d_{i\lambda}$: relative value of nonlinear optical susceptibility. P_s is also shown for comparison.

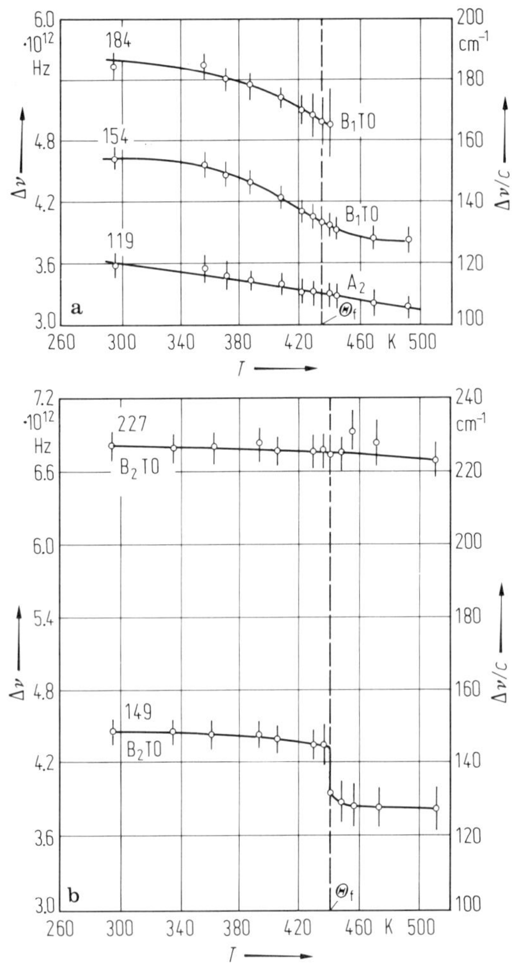

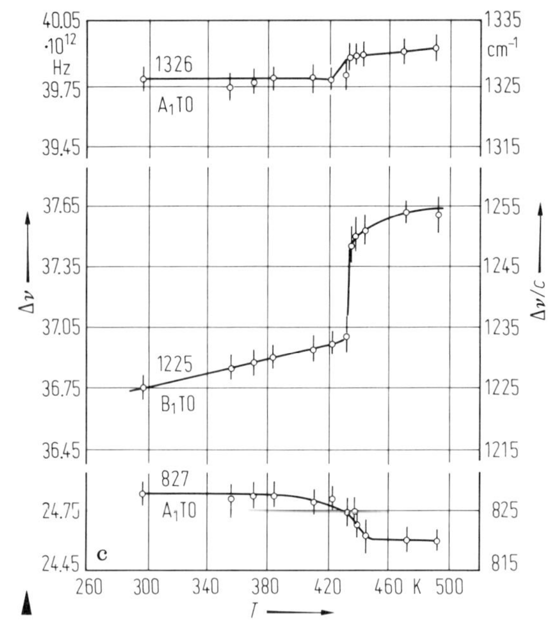

Fig. 152. $NaNO_2$. $\Delta\nu$ vs. T [76P17]. $\Delta\nu$: Raman frequency shift, Figs. a and b: for the external modes, Fig. c: for the internal modes. RT-values of the frequency shift in the figure are given in cm^{-1}. Vertical dashed lines indicate the transition temperature Θ_{III-II}.

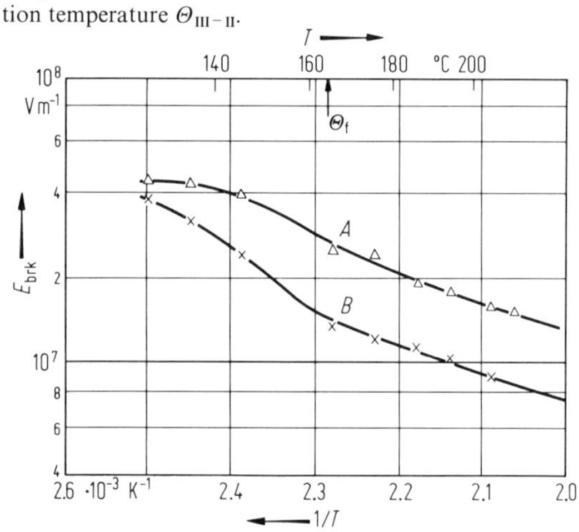

Fig. 156. $NaNO_2$. E_{brk} vs. $1/T$ [72F6]. E_{brk}: breakdown field. Curve A: for time to breakdown of 0.01 s; curve B: 0.1 s. A single flat-topped voltage pulse having a rise-time of 10^{-3} s is applied along the b-axis. $\Theta_f = \Theta_{III-II}$.

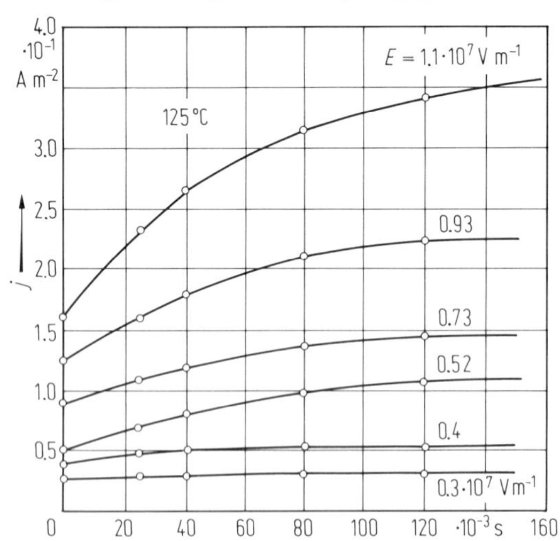

Fig. 155. $NaNO_2$. j vs. t [72F6]. Parameter: E. $T = 125\,°C$. j: current density along the b-axis, E: field strength of pulse.

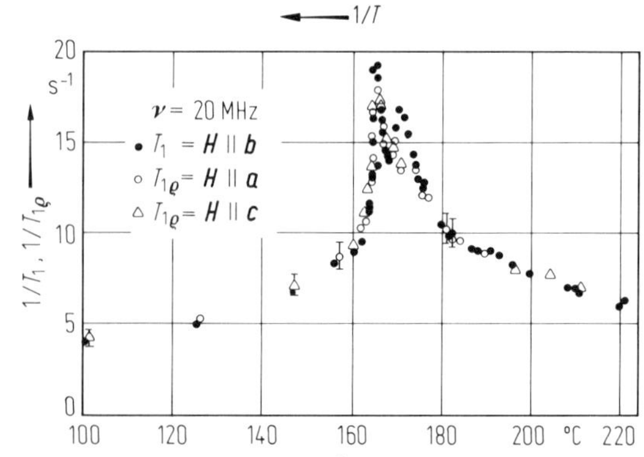

Fig. 157. $NaNO_2$. $1/T_1$, $1/T_{1\varrho}$ vs. T [75A14]. T_1: splin-lattice relaxation time for ^{23}Na, $T_{1\varrho}$: spin-lattice relaxation time for ^{23}Na in the rotating frame.

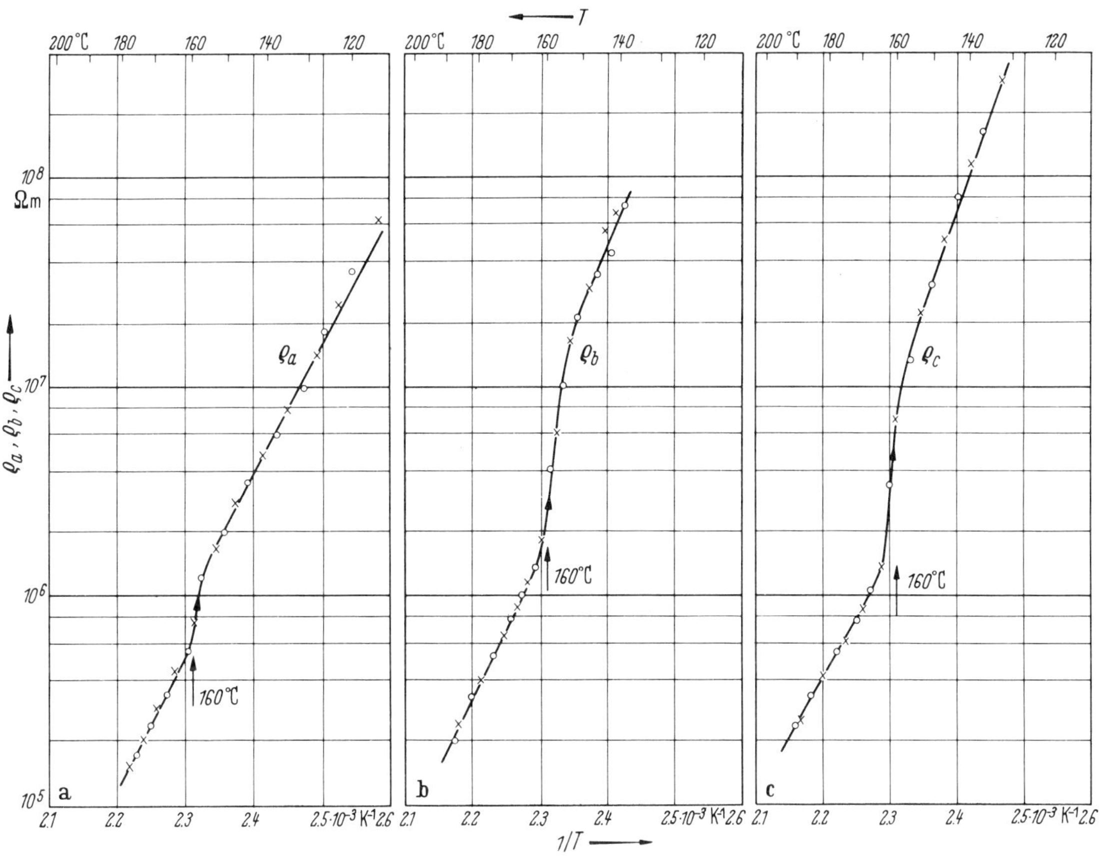

Fig. 154. $NaNO_2$. ϱ_a, ϱ_b, ϱ_c vs. $1/T$ [62A4].

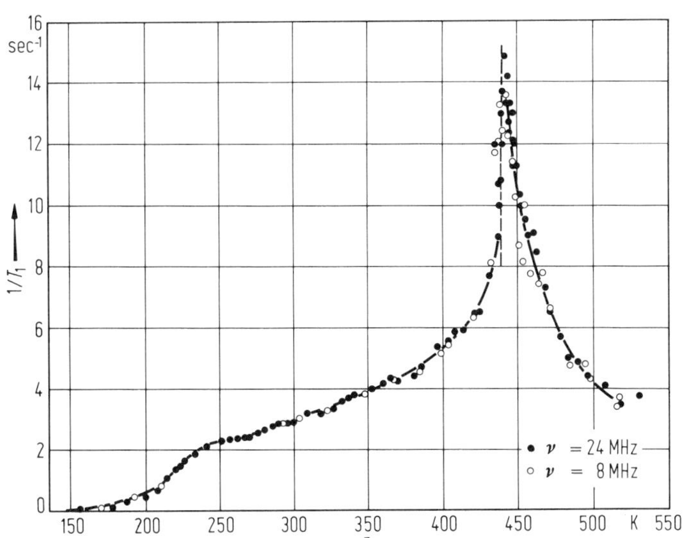

Fig. 158. $NaNO_2$. $1/T_1$ vs. T [70B13]. T_1: spin-lattice relaxation time for ^{23}Na.

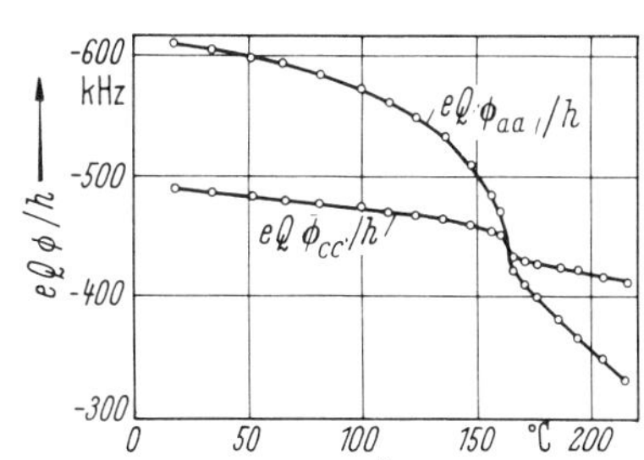

Fig. 159. $NaNO_2$. $eQ\phi_{aa}/h$, $eQ\phi_{cc}/h$ vs. T for ^{23}Na [62W1]. ϕ_{ij}: field gradient tensor components.

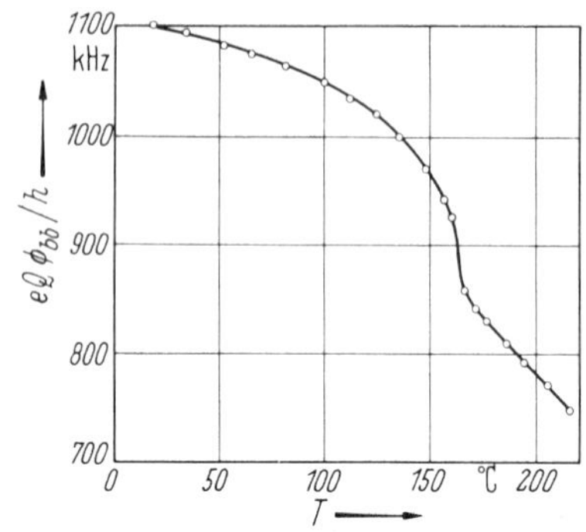

Fig. 160. $NaNO_2$. $eQ\phi_{bb}/h$ vs. T for ^{23}Na [62W1]. ϕ_{ij}: field gradient tensor components.

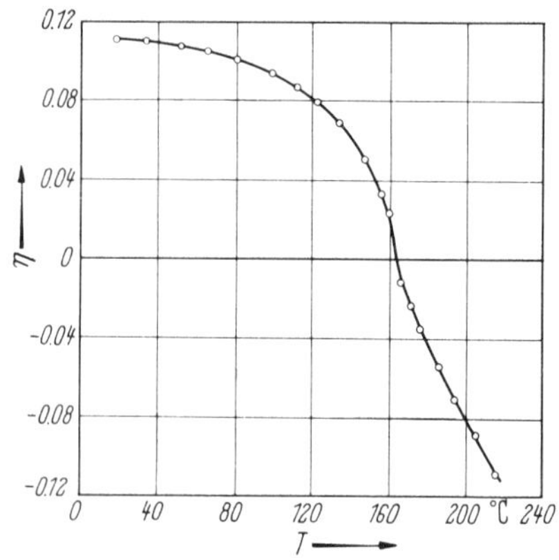

Fig. 161. $NaNO_2$. η vs. T for ^{23}Na [62W1]. $\eta=(\phi_{cc}-\phi_{aa})/\phi_{bb}$.

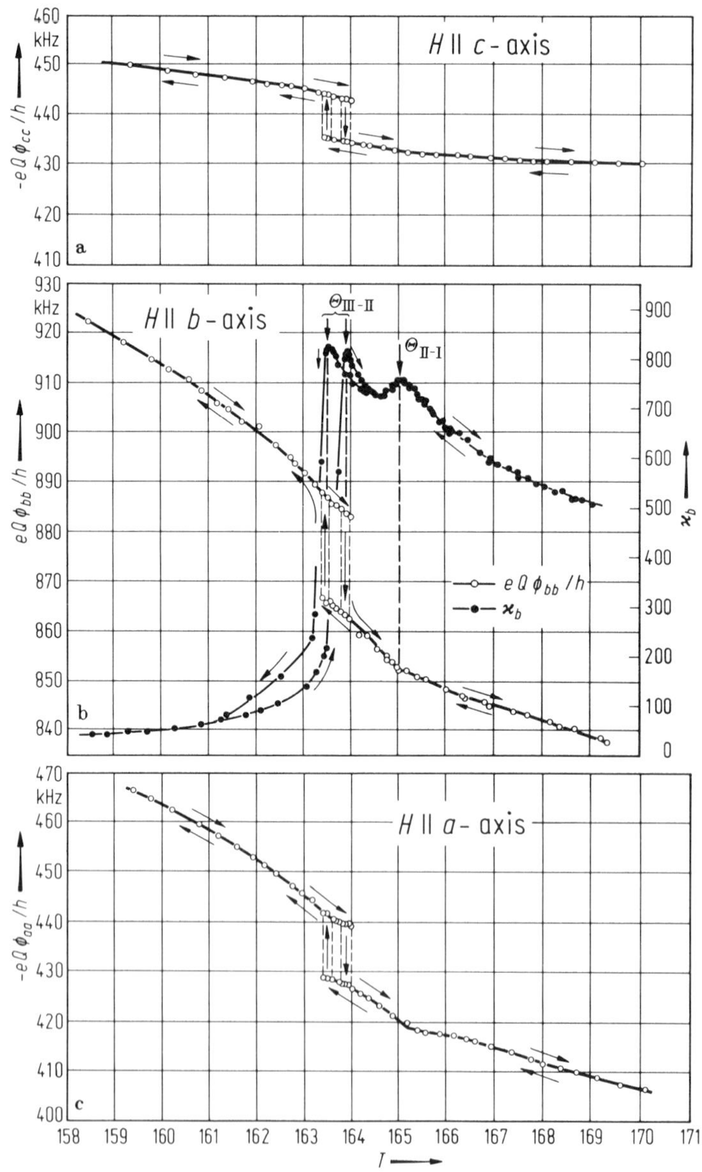

◄ Fig. 162. $NaNO_2$. $eQ\phi_{ij}/h$ vs. T [70Y1]. Fig. a: $-eQ\phi_{cc}/h$, Fig. b: $eQ\phi_{bb}/h$, Fig. c: $-eQ\phi_{aa}/h$ for ^{23}Na in the vicinity of the III–II–I transitions. In Fig. b $\varkappa_b$ vs. T is also shown $\varkappa_b$ at $f=1.5$ kHz). ϕ_{ij}: field gradient tensor components.

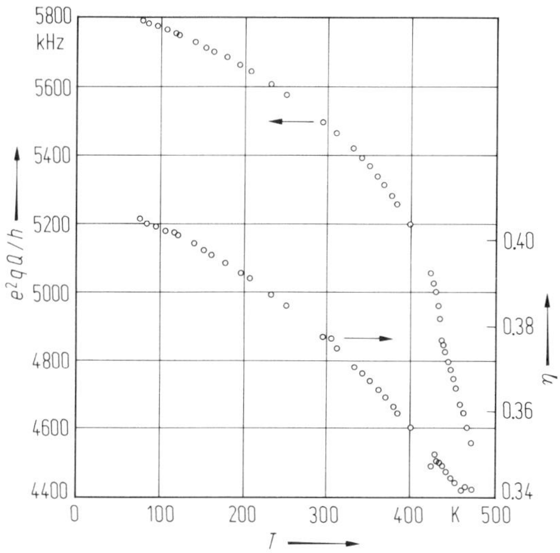

Fig. 163. $NaNO_2$. e^2qQ/h, η vs. T for ^{14}N [76P13]. η: asymmetry parameter of an electric field gradient tensor.

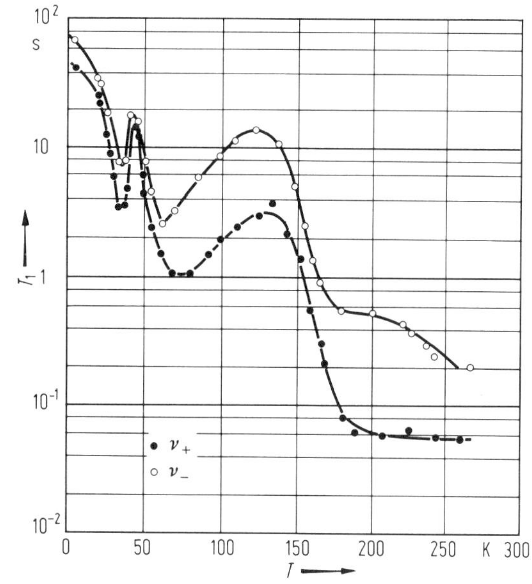

Fig. 164. $NaNO_2$. T_1 vs. T for ^{14}N [72A2]. T_1: spin-lattice relaxation time. $\nu_+ = \frac{1}{4}\ (eQ\phi_{zz}/h)\ (3+\eta)$, $\nu_- = \frac{1}{4}\ (eQ\phi_{zz}/h)\ (3-\eta)$.

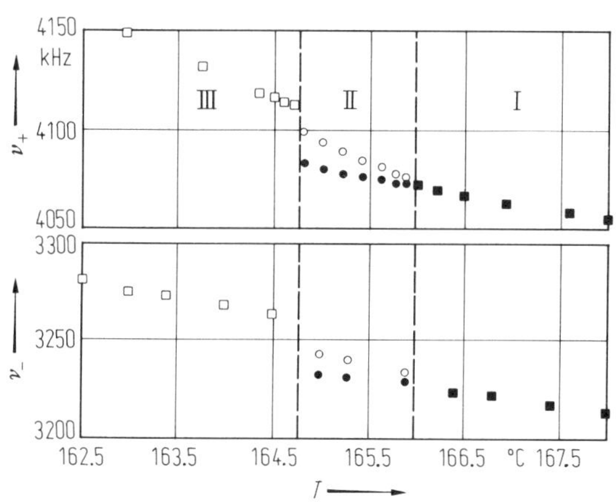

Fig. 165. $NaNO_2$. $\nu_\pm$ vs. T [77A10]. ν_+, ν_-: ^{14}N nuclear quadrupole resonance frequencies.

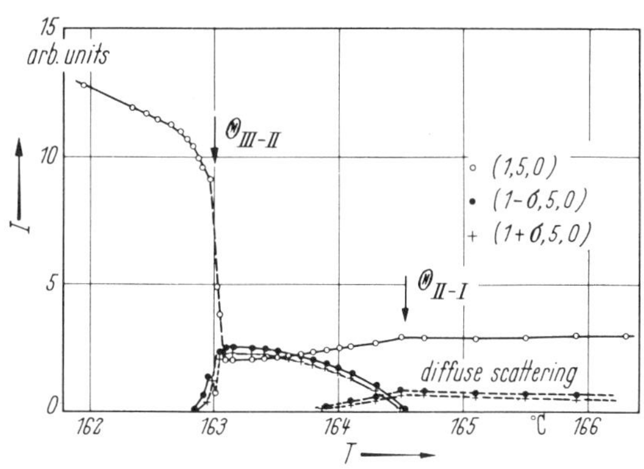

Fig. 166. $NaNO_2$. $I_{(1,5,0)}$, $I_{\text{satellite}}$ vs. T [67H10]. $I_{\text{satellite}}$: intensity of satellite reflection due to structural modulation.

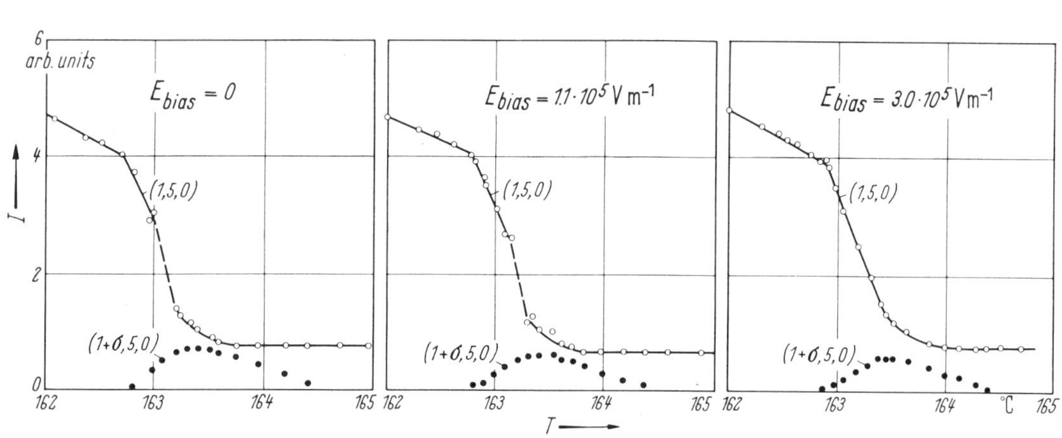

Fig. 167. $NaNO_2$, $I_{(1,5,0)}$, $I_{\text{satellite}}$ vs. T under various E_{bias} [67H10]. $I_{\text{satellite}}$: intensity of satellite reflection due to structural modulation.

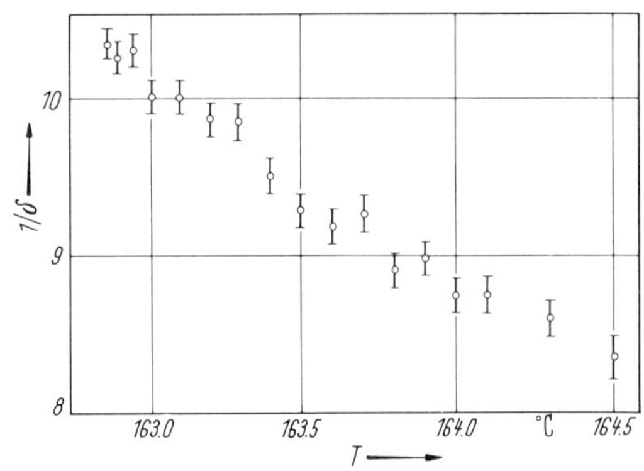

Fig. 168. $NaNO_2$. $1/\delta$ vs. T in phase II [67 H 10]. $1/\delta$: period of structural modulation in units of lattice constant a.

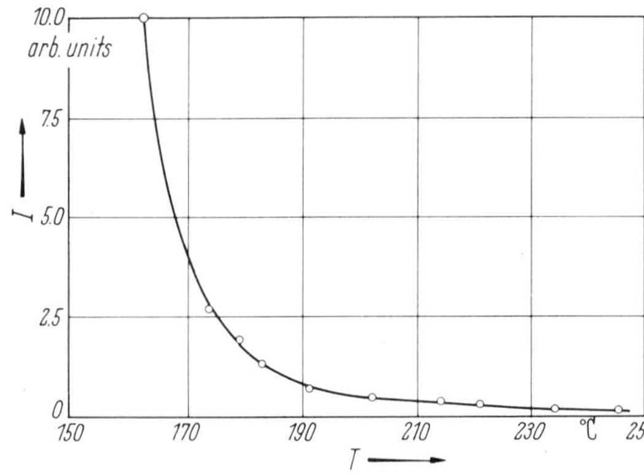

Fig. 169. $NaNO_2$. I vs. T [63 T 1]. I: intensity of the X-ray diffuse maximum near (042).

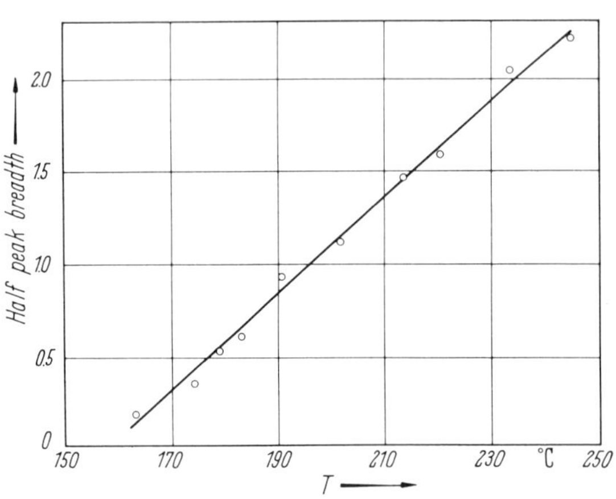

Fig. 170. $NaNO_2$. Half peak breadth of diffuse peak near (042) vs. T [63 T 1].

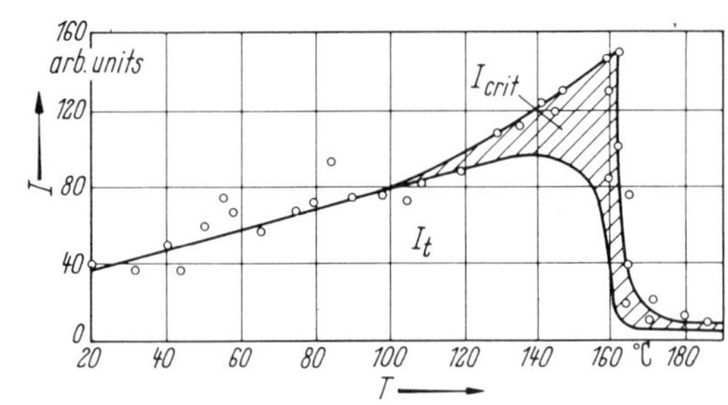

Fig. 171. $NaNO_2$. I_{crit}, I_t vs. T [61 S 6]. I_{crit}: X-ray critical scattering, I_t: thermal diffuse scattering.

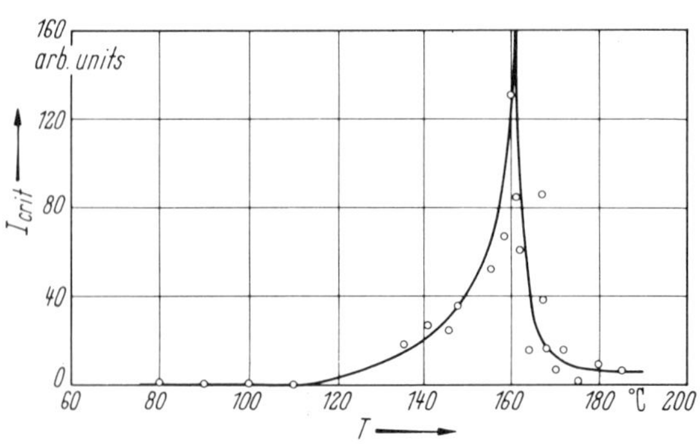

Fig. 172. $NaNO_2$. I_{crit} vs. T [61 S 6]. I_{crit}: intensity of critical scattering at $(1 \pm 0.02, 3 \pm 0.05, 0)$.

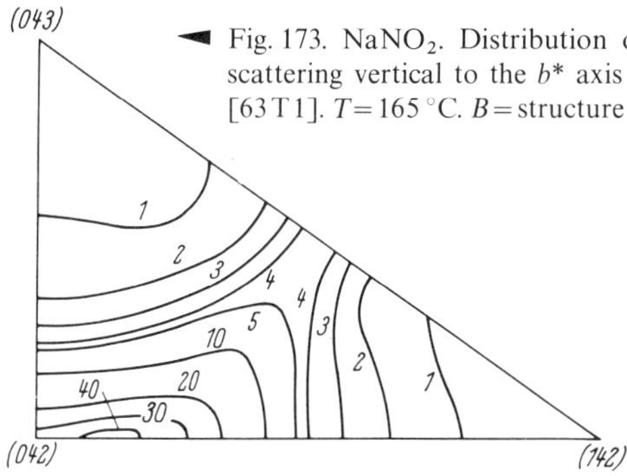

◄ Fig. 173. $NaNO_2$. Distribution of the intensity of diffuse scattering vertical to the b^* axis divided by B^2 near (042) [63T1]. $T = 165$ °C. B = structure factor (see original paper).

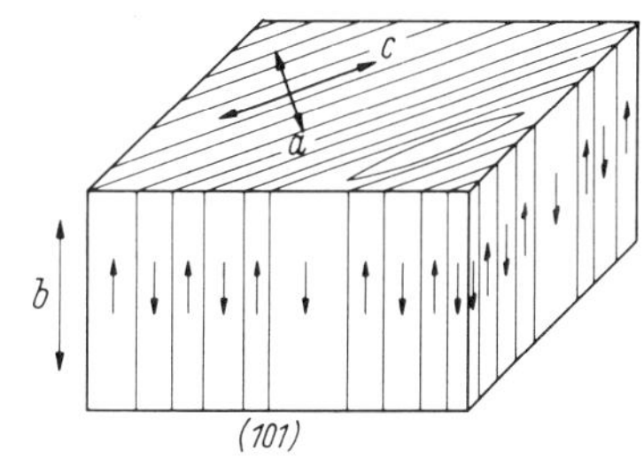

Fig. 175. $NaNO_2$. Schematic representation of domain structure [61N4].

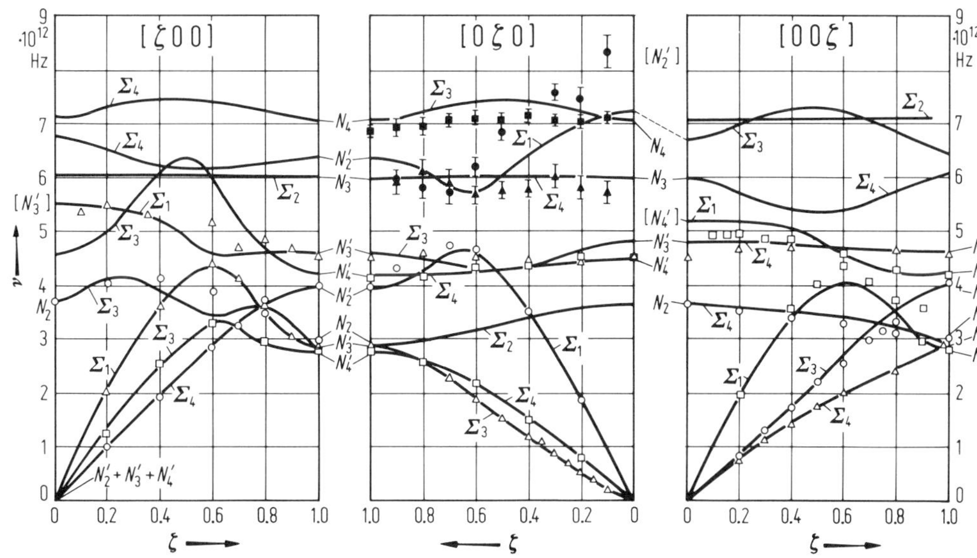

Fig. 174. $NaNO_2$. ν vs. ζ [70S3]. ν: phonon frequency, ζ: reduced wave vector coordinate. Open figures: at 23 °C, solid figures: at −163 °C.

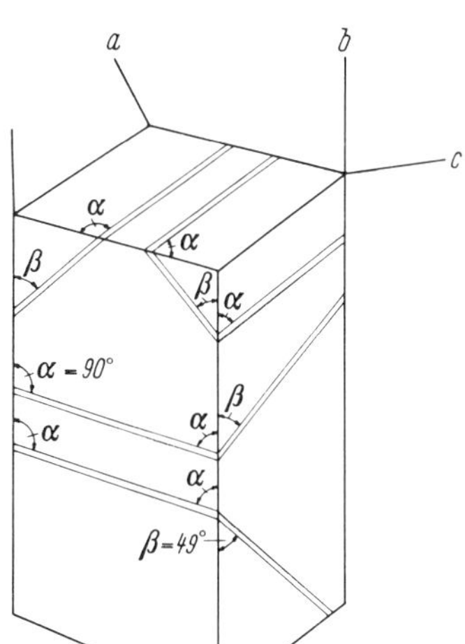

Fig. 177. $NaNO_2$. Schematic representation of mechanical twins [63N3].

Fig. 176, see p. 352.

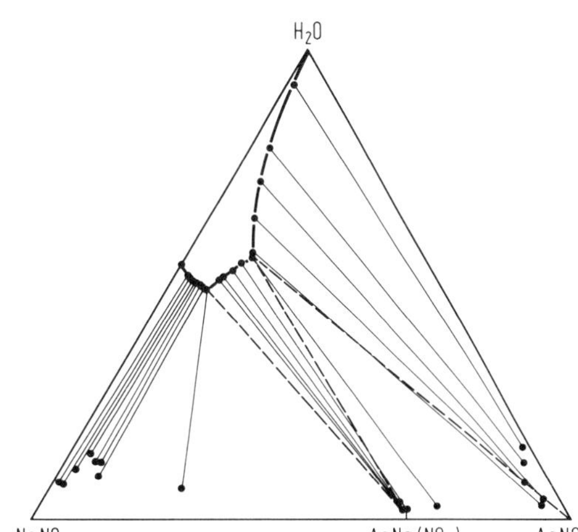

Fig. 178. $AgNa(NO_2)_2$. Solubility curve for $AgNO_2$–$NaNO_2$–H_2O system [53C1]. $T = 25$ °C. Thick line: liquidus, full circles which are joined with a thin line indicate the composition of a saturated solution and that of solid crystallized from it.

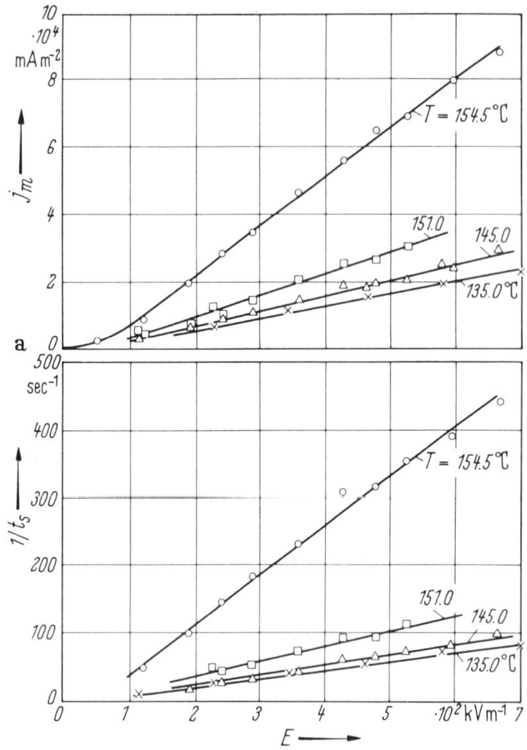

Fig. 176. $NaNO_2$. j_m, $1/t_s$ vs. E [63H3]. Parameter: T. j_m: maximum switching current density, t_s: switching time.

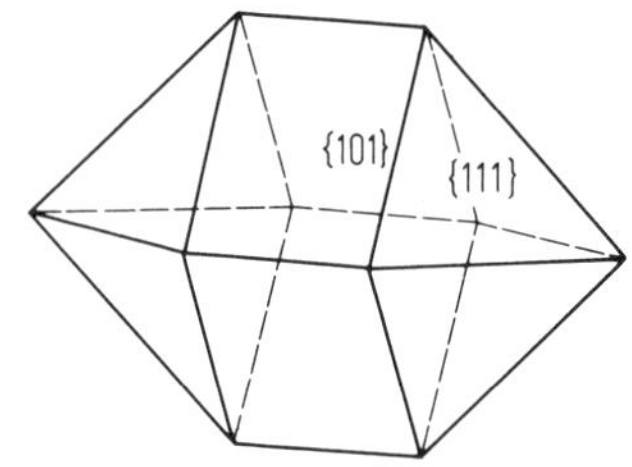

Fig. 179. $AgNa(NO_2)_2$. Crystal form [53C1].

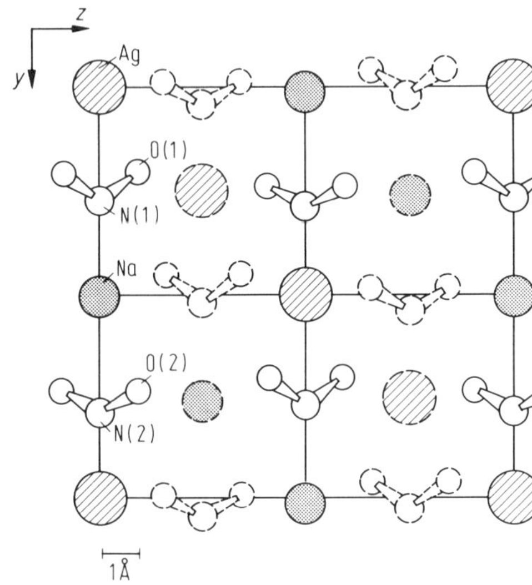

Fig. 180. $AgNa(NO_2)_2$. Structure of phase II [74I9]. $T = 20$ °C. Projection along the [100] direction. The hatched figures show the atoms by $a/4$ above (or below) the layer on which solid figures lie.

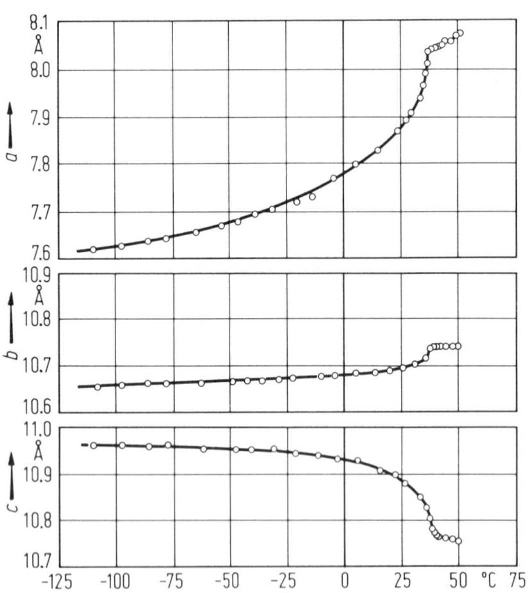

Fig. 182. $AgNa(NO_2)_2$. a, b, c vs. T [71M6].

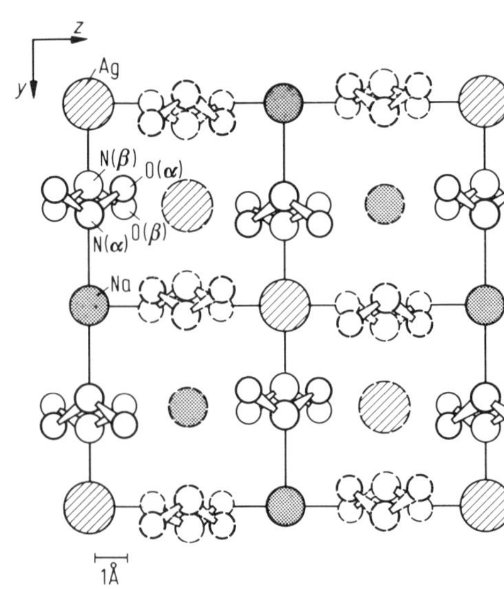

Fig. 181. $AgNa(NO_2)_2$. Structure of phase I [74I9]. $T = 50$ °C. Projection along the [100] direction. The hatched figures show the atoms by $a/4$ above (or below) the layer on which the solid figures lie. About 70% of NO_2 ions are in the positions denoted by a bold line (α) and about 30% are in that denoted by a thin line (β).

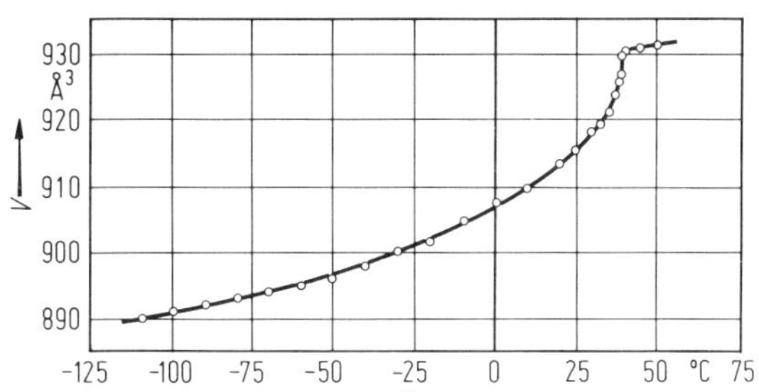

Fig. 183. $AgNa(NO_2)_2$. V vs. T [71M6]. V: unit cell volume.

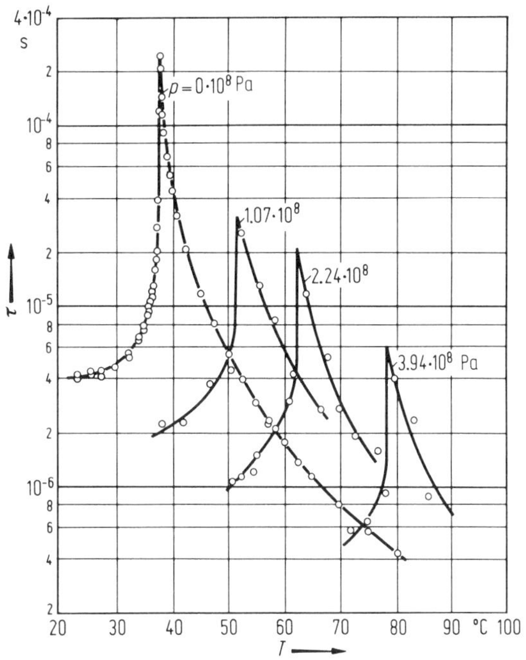

Fig. 188. $AgNa(NO_2)_2$. τ vs. T [72G5]. Parameter: p. τ: dielectric relaxation time.

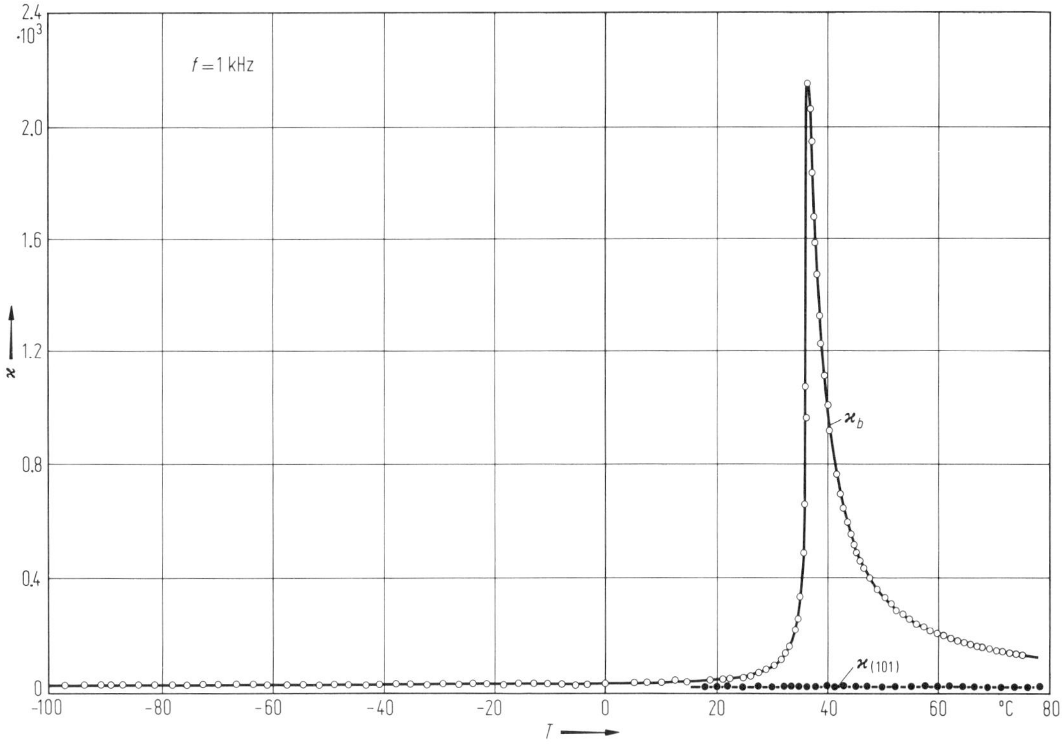

Fig. 184. $AgNa(NO_2)_2$. $\varkappa_b$, $\varkappa_{(101)}$ vs. T [70G3]. $\varkappa_{(101)}$: dielectric constant perpendicular to the (101) plane.

Figs. 185···187, see p. 354, 355.

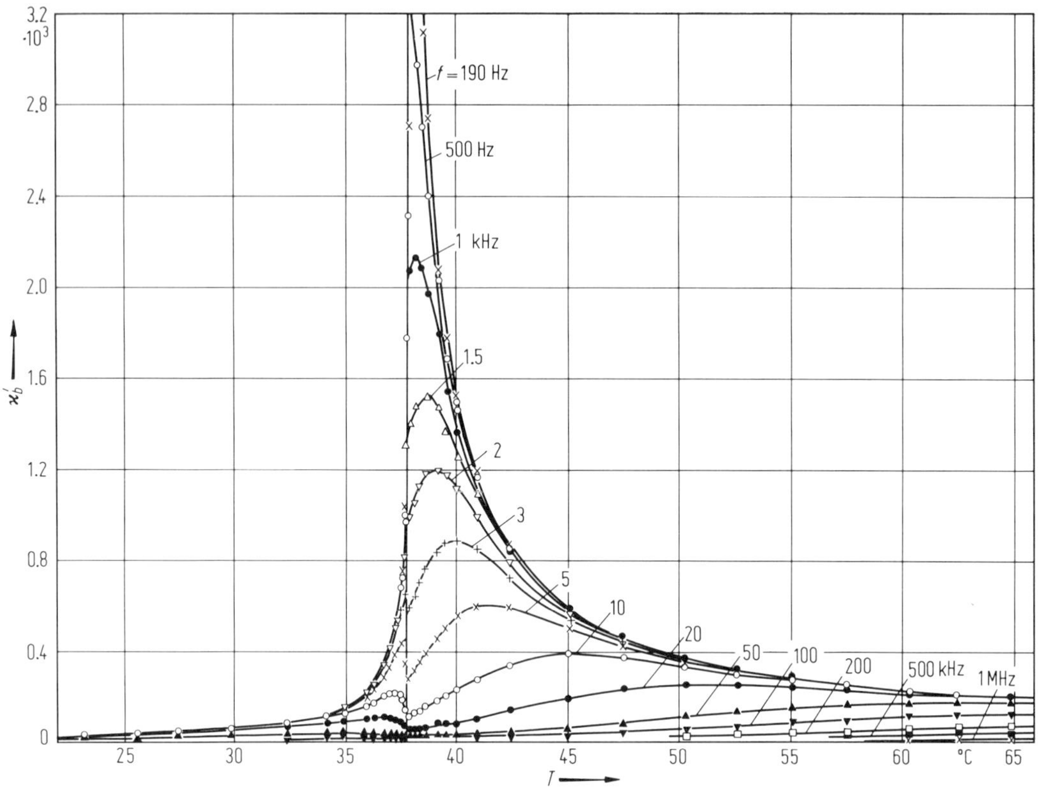

Fig. 185. $AgNa(NO_2)_2$. $\varkappa_b'$ vs. T [72G15]. Parameter: f.

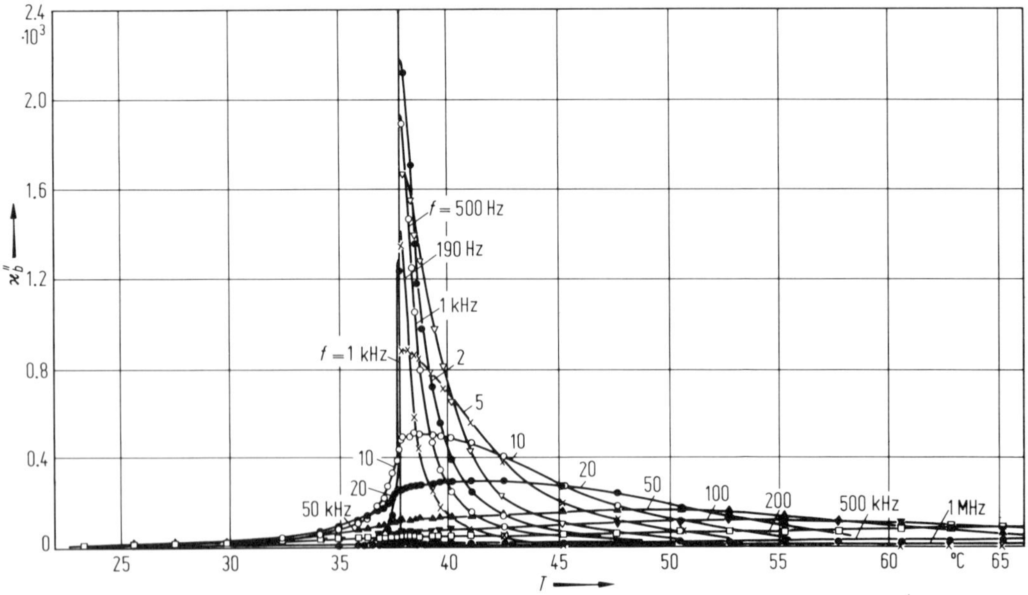

Fig. 186. $AgNa(NO_2)_2$. $\varkappa_b''$ vs. T [72G15]. Parameter: f.

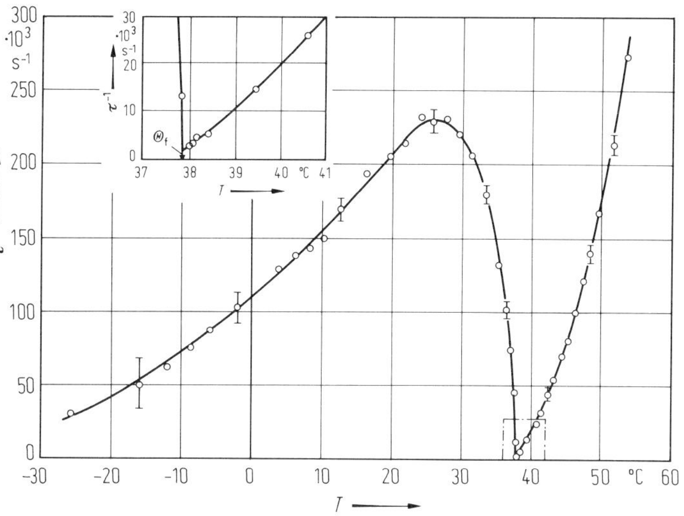

Fig. 187. $AgNa(NO_2)_2$. τ^{-1} vs. T [76G19]. τ: dielectric relaxation time.

Fig. 189. $AgNa(NO_2)_2$. P_s vs. T[72G4]. Measured by pyroelectric method.

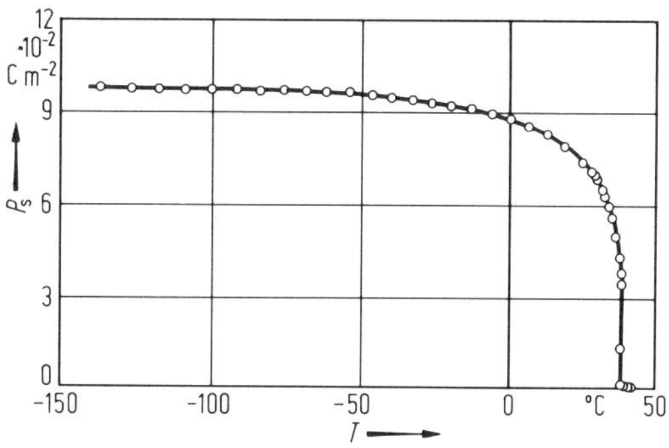

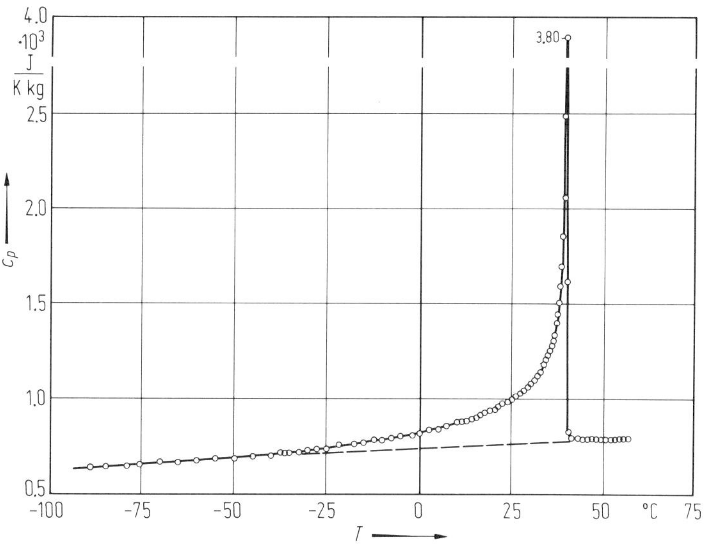

Fig. 190. $AgNa(NO_2)_2$. c_p vs. T [77H8].

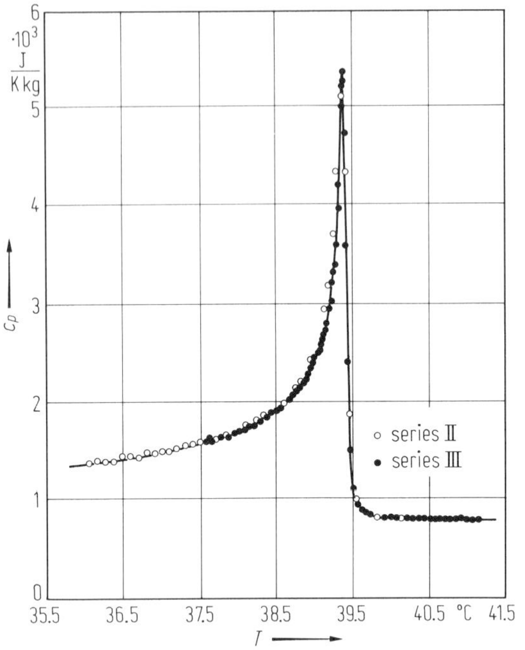

Fig. 191. $AgNa(NO_2)_2$. c_p vs. T in the vicinity of Θ_{II-I} [77H9]. Temperature increments of 0.1 K (series II) and 0.03 K (series III) were applied.

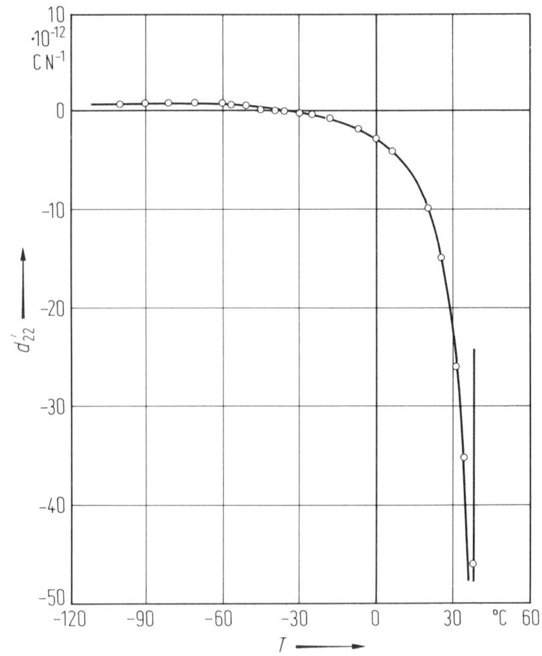

Fig. 192. $AgNa(NO_2)_2$. d'_{22} vs. T [74H2]. d'_{22}: real part of complex piezoelectric constant ($d_{22} = d'_{22} - i\,d''_{22}$).

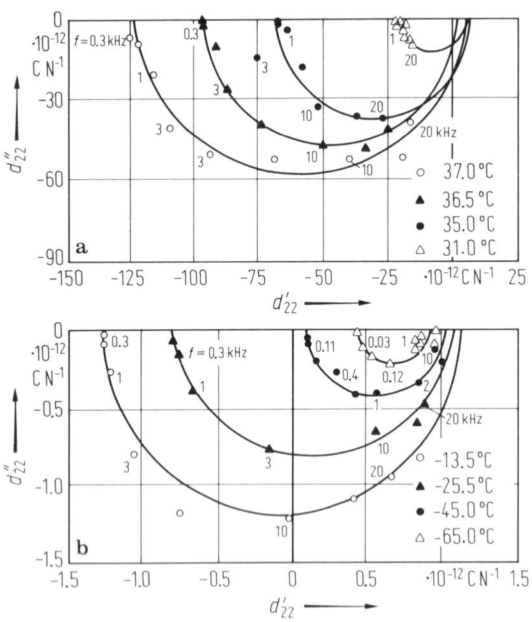

Fig. 193. $AgNa(NO_2)_2$. Cole-Cole diagrams of the complex piezoelectric constant $d_{22} = d'_{22} - i\,d''_{22}$ [74H2]. (a) Just below Θ_f, (b) below 0 °C. Frequencies (in kHz) at which the constants were measured, are also given in the figure.

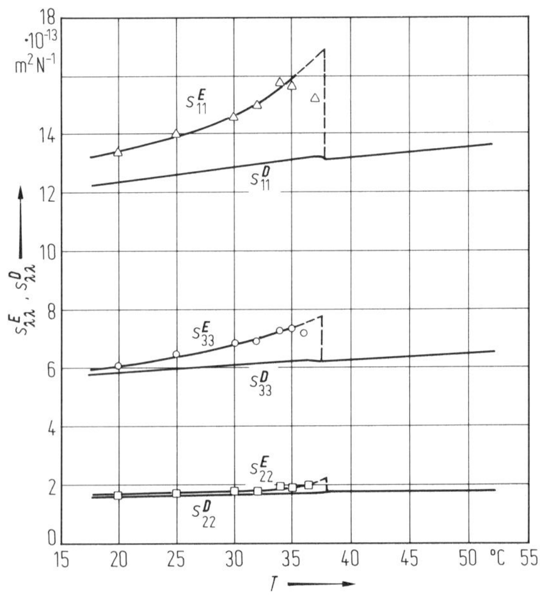

Fig. 194. $AgNa(NO_2)_2$. $s^E_{\lambda\lambda}$, $s^D_{\lambda\lambda}$ vs. T [79Y5].

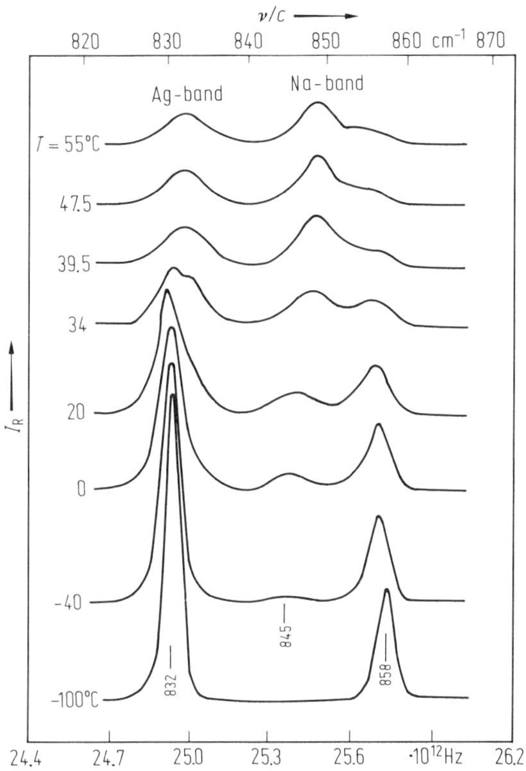

Fig. 195α. $AgNa(NO_2)_2$. Profiles of Raman scattering of ν_2-vibrational modes of NO_2^- ions at various temperatures [80H2]. I_R: see caption of Fig. 195γ. Peak frequencies are given in cm^{-1}.

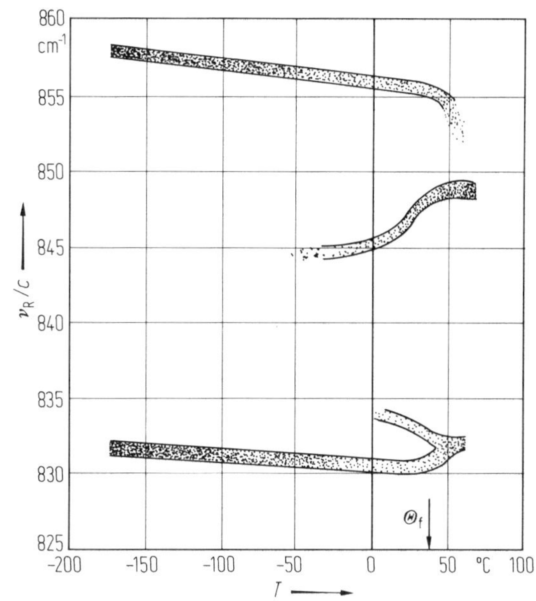

Fig. 195β. $AgNa(NO_2)_2$. ν_R/c vs. T [80H2]. ν_R/c: wave number of Raman scattering of ν_2-vibrational modes of NO_2^- ions.

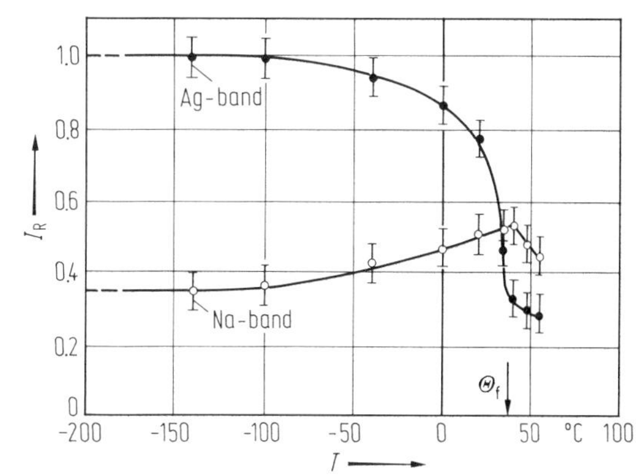

Fig. 195γ. $AgNa(NO_2)_2$. I_R vs. T [80H2]. I_R: intensity of Raman scattering of the ν_2-vibrational modes of NO_2^- ions. As to Ag- and Na-bands; see Fig. 195α.

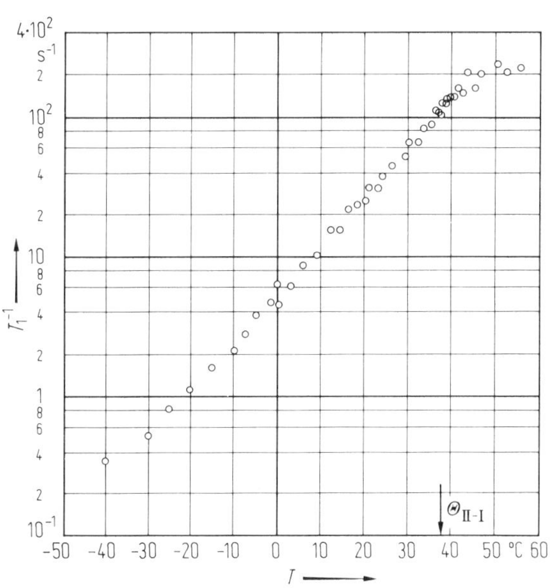

Fig. 196. $AgNa(NO_2)_2$. T_1^{-1} vs. T [75H5]. T_1: spin-lattice relaxation time of ^{23}Na.

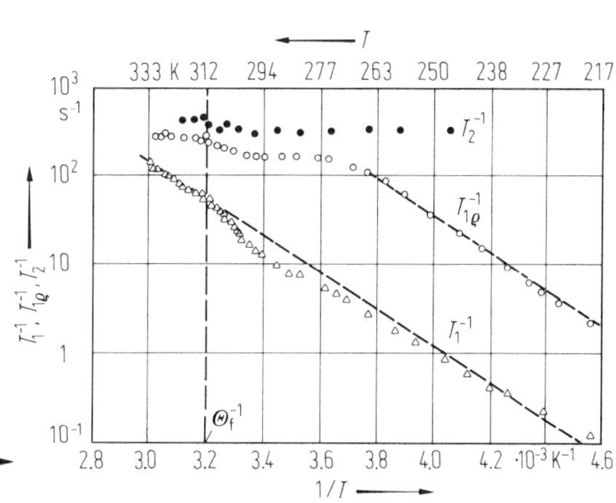

Fig. 197. $AgNa(NO_2)_2$. T_1^{-1}, $T_{1\varrho}^{-1}$, T_2^{-1} vs. T for ^{23}Na ► [78G15]. $f = 20$ MHz.

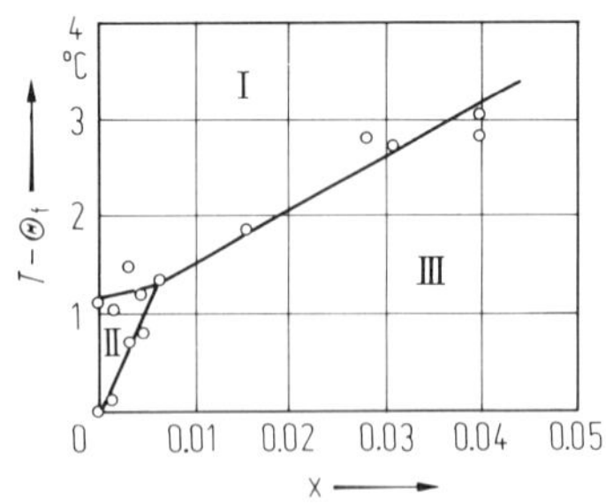

Fig. 199. $Na_{1-x}Ag_xNO_2$. Phase diagram [70G3]. Θ_f: ferroelectric transition temperature of $NaNO_2$.

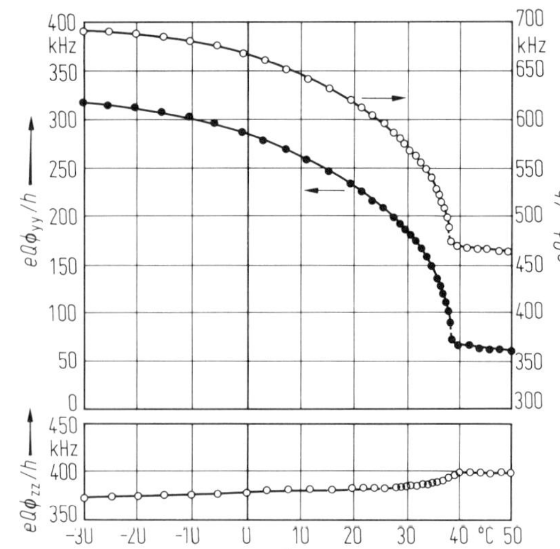

Fig. 198. $AgNa(NO_2)_2$. $eQ\phi_{ii}/h$ vs. T [78G16]. $eQ\phi_{ii}/h$: principal values of the quadrupole coupling tensor for ^{23}Na. In phase I, $x\|a$, $y\|b$, $z\|c$. In phase II, $y\|b$, x and z make angles of $\pm 4°$ with the a and c-axes, respectively.

17 KNO_3 family

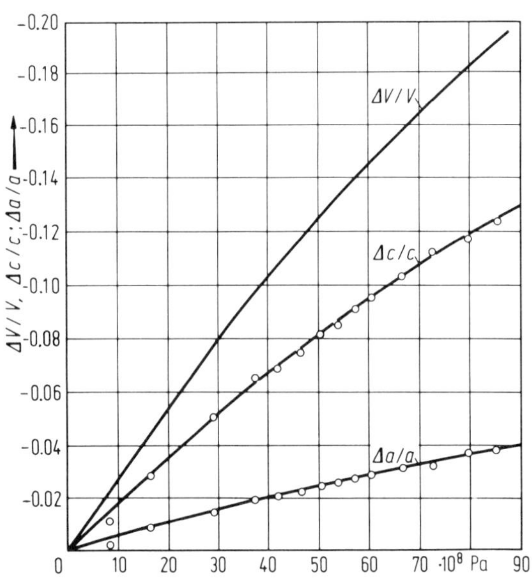

Fig. 200. $NaNO_3$. $\Delta a/a$, $\Delta c/c$, $\Delta V/V$ vs. p [69B1]. T=RT. a, c: unit cell parameters of hexagonal cell.

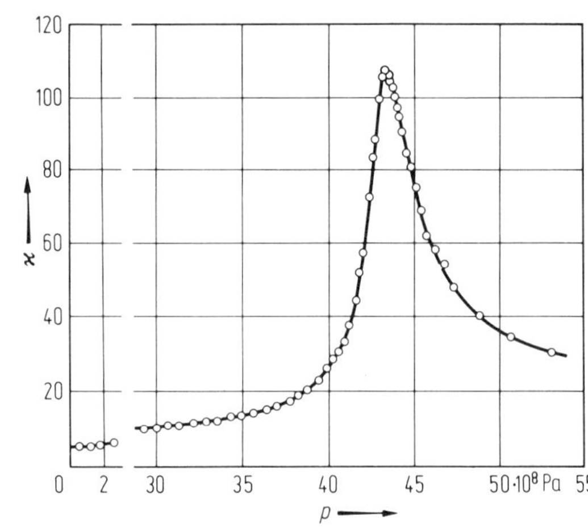

Fig. 201: $NaNO_3$ (polycrystalline). $\varkappa$ vs. p [69B1]. T=RT.

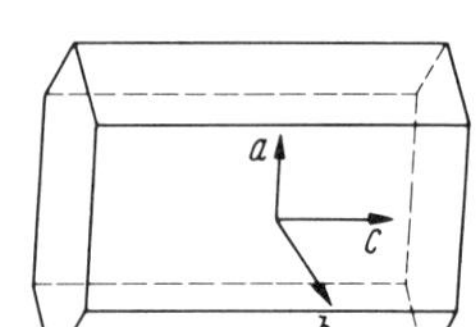

Fig. 205. KNO_3. Crystal form grown from solution [65S7]. a, b, c correspond to crystalline axes in phase II.

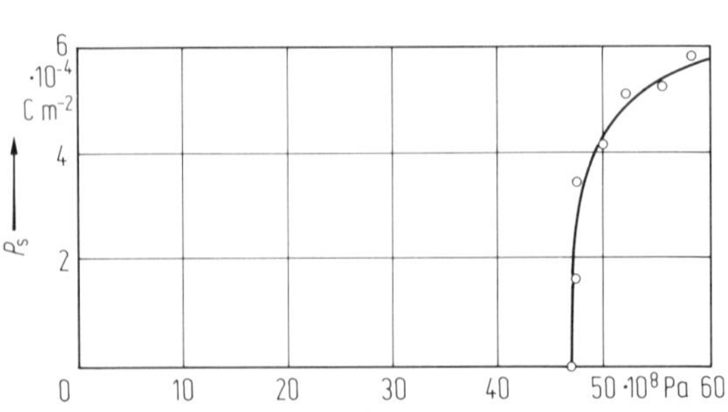

Fig. 203. $NaNO_3$ (polycrystalline). P_s vs. p [73F12]. T=RT.

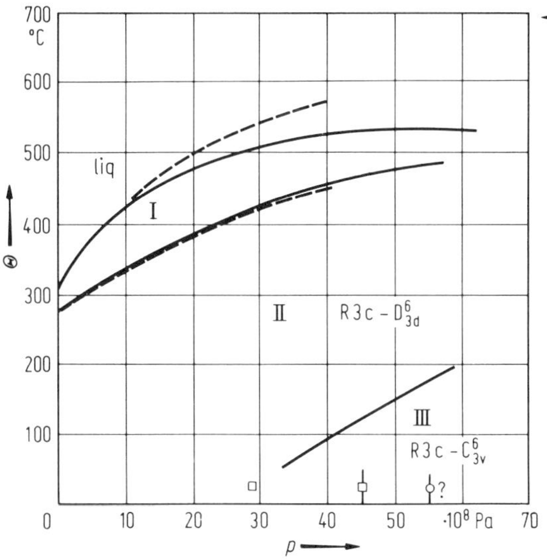

Fig. 202. $NaNO_3$. Θ vs. p. Dashed lines: [66 R 1], solid lines: [66 F 1], squares: [69 B 1], circle: [45 B 4].

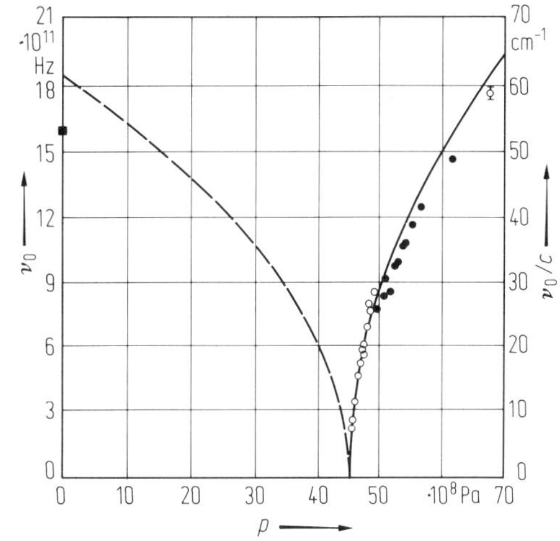

Fig. 204. $NaNO_3$. ν_0 vs. p [78 L 2]. ν_0: soft mode frequency obtained from Raman scattering spectra. T = RT. The solid square at $p = 0$ is a result of neutron scattering [74 T 11].

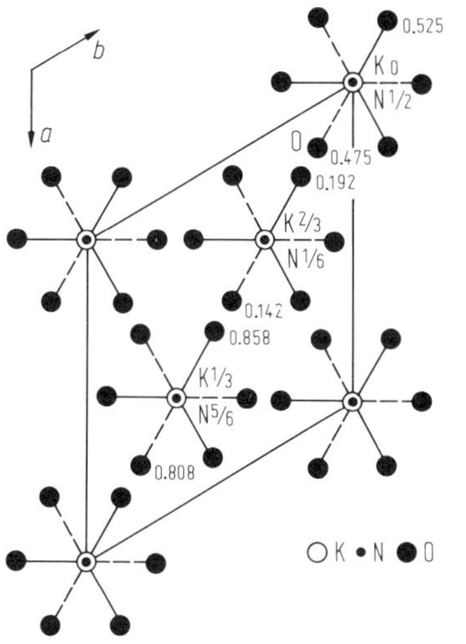

Fig. 206. KNO_3. Structure of phase I [76 N 7]. Projection along the hexagonal c-direction. NO_3 on each site is in dynamic disorder between two orientations. Numbers in figure indicate fractional coordinate z at 151 °C.

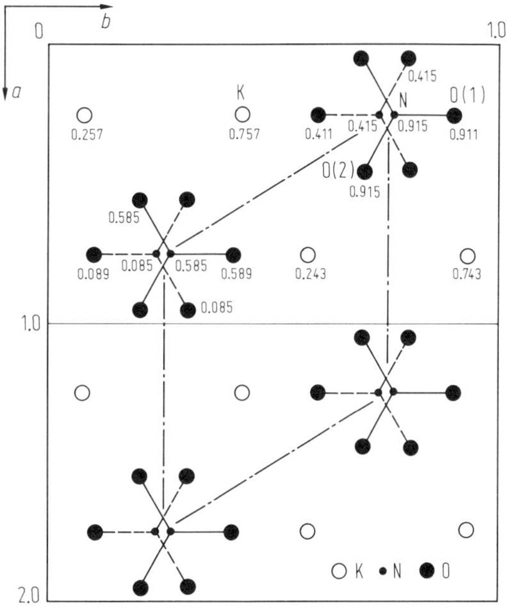

Fig. 208. KNO_3. Structure of phase II [76 N 7]. The pseudo-hexagonal cell is shown by dashed-dotted line. The figure is drawn by the reported fractional coordinates at 25 °C in [73 N 11] (see Table 38). K atoms lie on 4 sets of planes, parallel to a–b face of each unit cell.

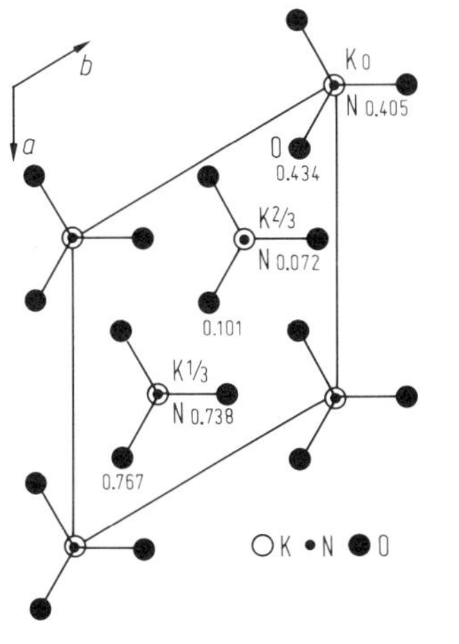

Fig. 207. KNO_3. Structure of phase III [76 N 7]. Projection along the hexagonal c-direction. Numbers in figure indicate fractional coordinate z at 91 °C.

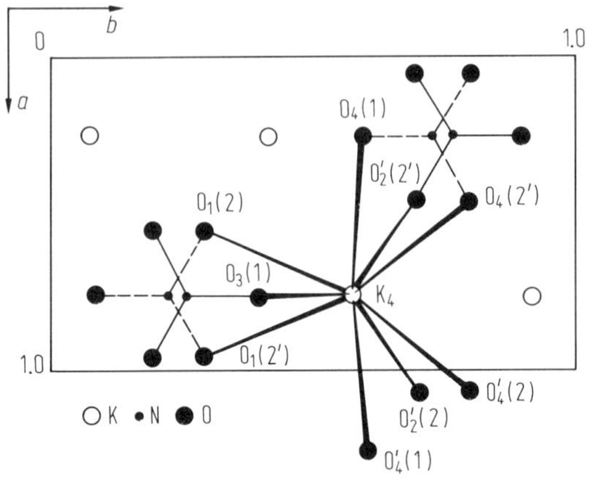

Fig. 209. KNO_3. Structure of phase II [73N11]. Projection along the c-axis.

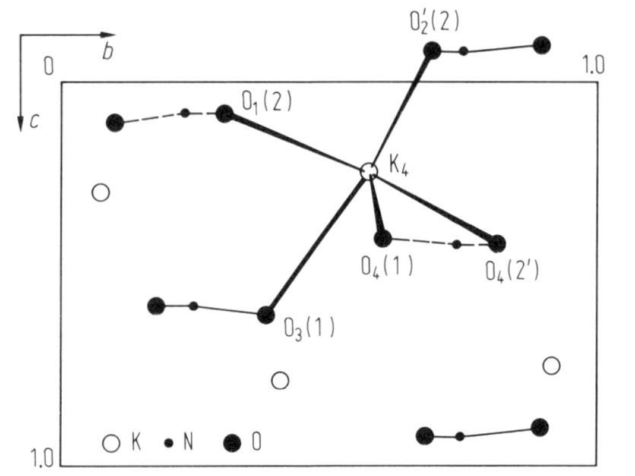

Fig. 210. KNO_3. Structure of phase II [73N11]. Projection along the a-axis.

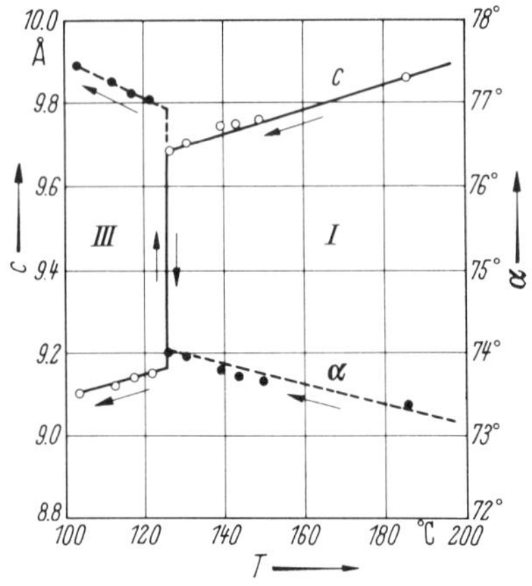

Fig. 211. KNO_3. α, c vs. T [65K4]. α: trigonal angle, c: hexagonal unit cell parameter.

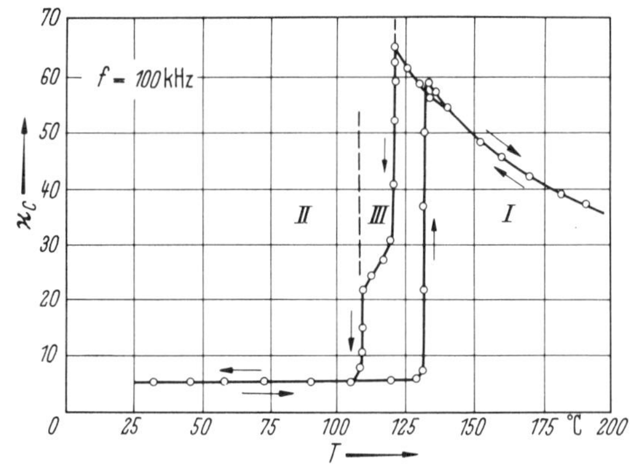

Fig. 213. KNO_3. $\varkappa_c$ vs. T [61S3].

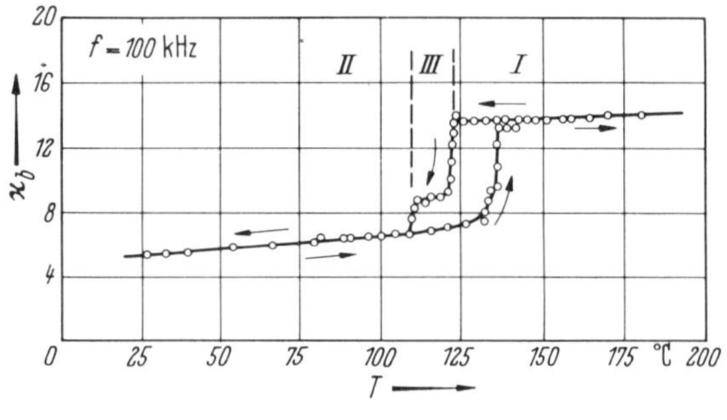

Fig. 214. KNO_3. $\varkappa_b$ vs. T [61S3].

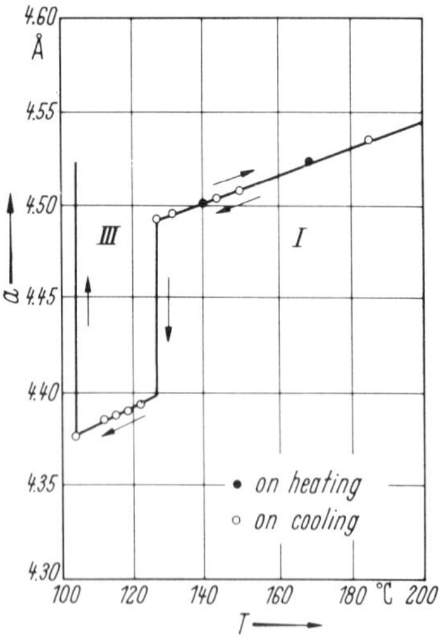

Fig. 212. KNO_3. a vs. T [65K4]. a: unit cell parameter of the trigonal cell.

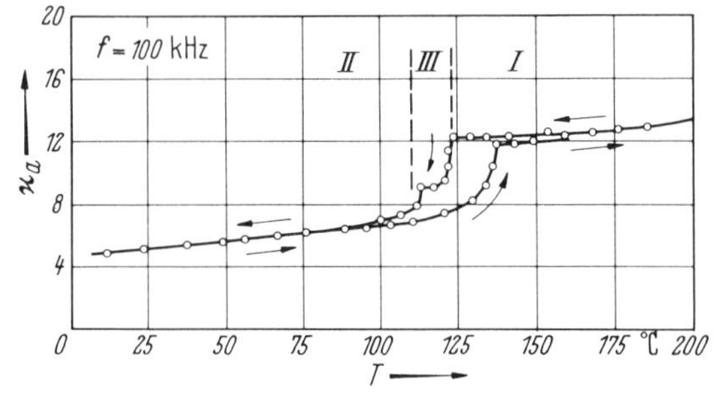

Fig. 215. KNO_3. $\varkappa_a$ vs. T [61S3].

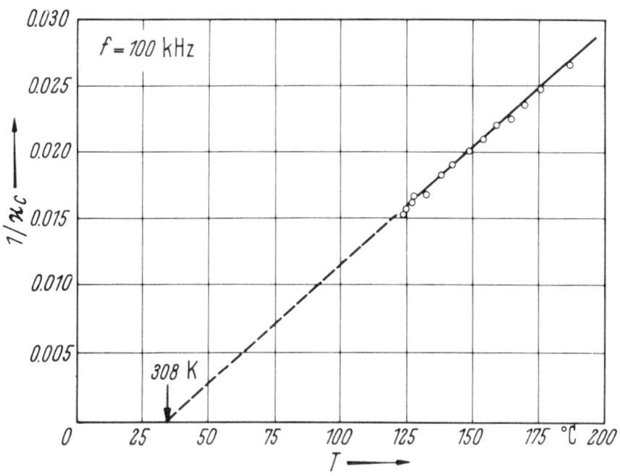

Fig. 216. KNO_3. $1/\varkappa_c$ vs. T [61S3].

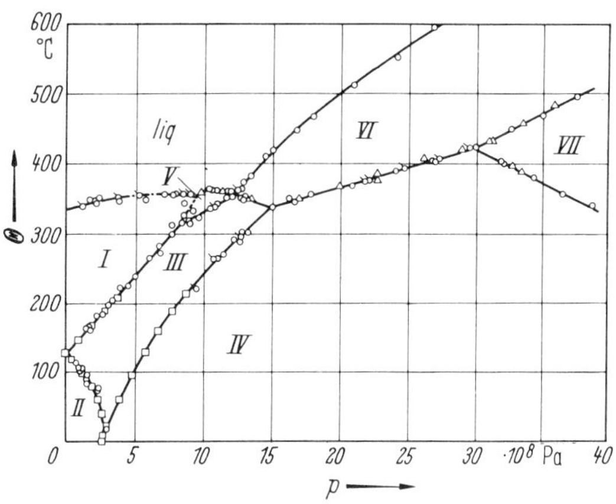

Fig. 217. KNO_3. Θ vs. p [65R1].

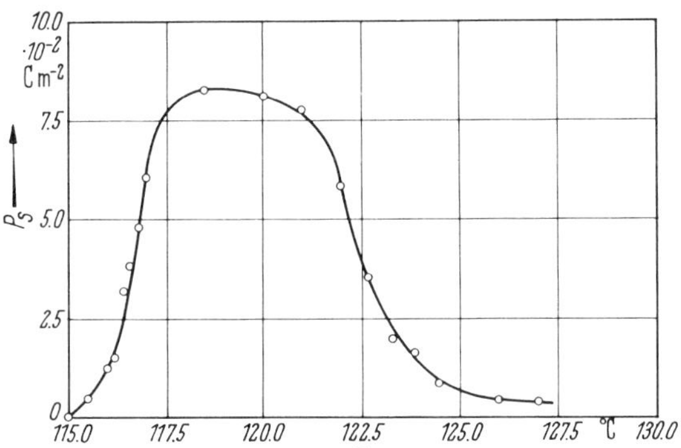

Fig. 218. KNO_3. P_s vs. T [61S3].

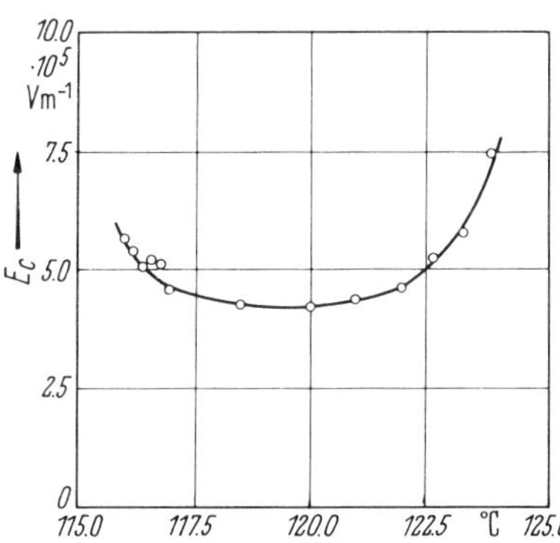

Fig. 219. KNO_3. E_c vs. T [61S3].

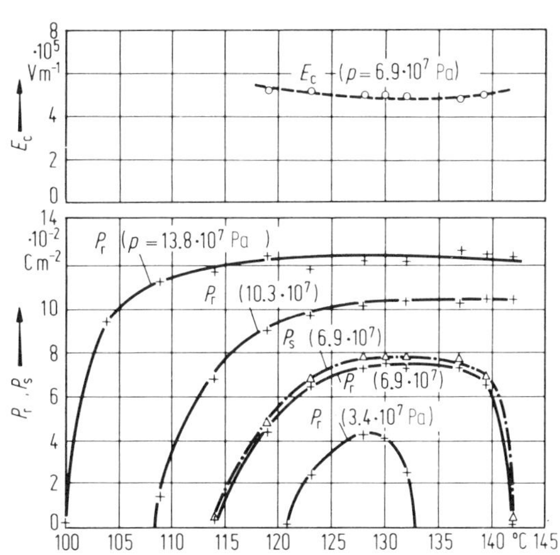

Fig. 220. KNO_3. P_r, P_s, E_c vs. T [68T2]. Parameter: p. P_r: remanent polarization. As to reproducibility of the data, see [71M5].

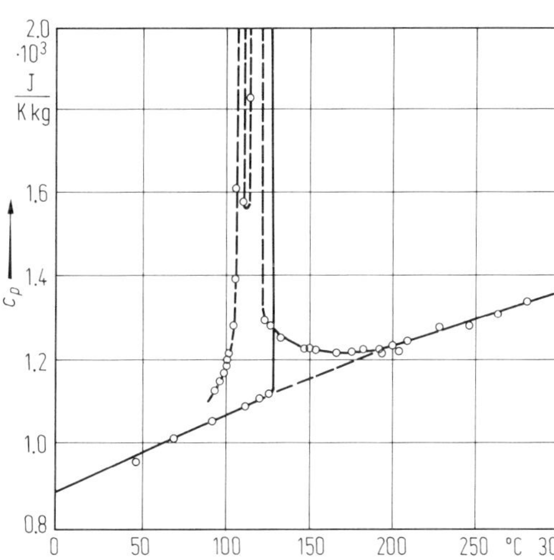

Fig. 221. KNO_3. c_p vs. T [41M1]. Full line: heating; dashed line: cooling.

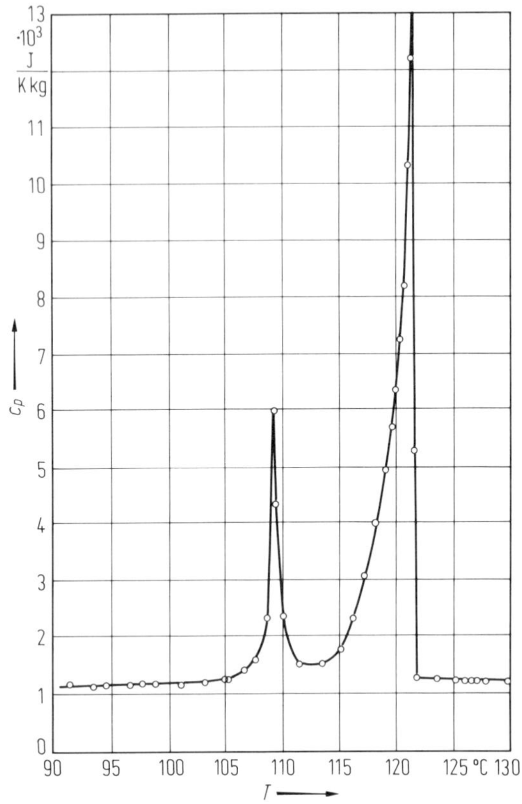

Fig. 222. KNO_3. c_p vs. T in the vicinity of transition temperatures [41 M 1].

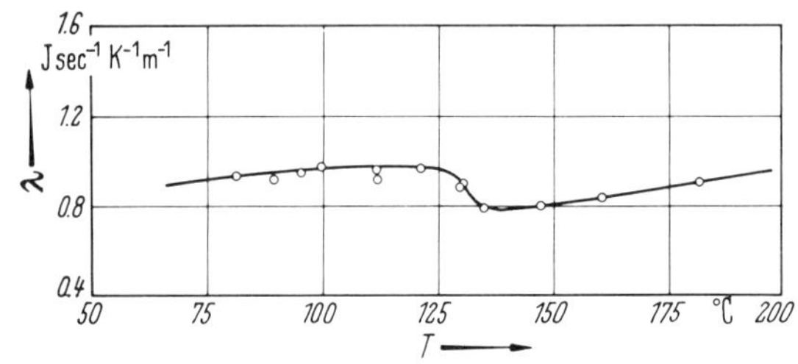

Fig. 223. KNO_3 (polycrystalline). λ vs. T [60 Y 2]. λ: thermal conductivity.

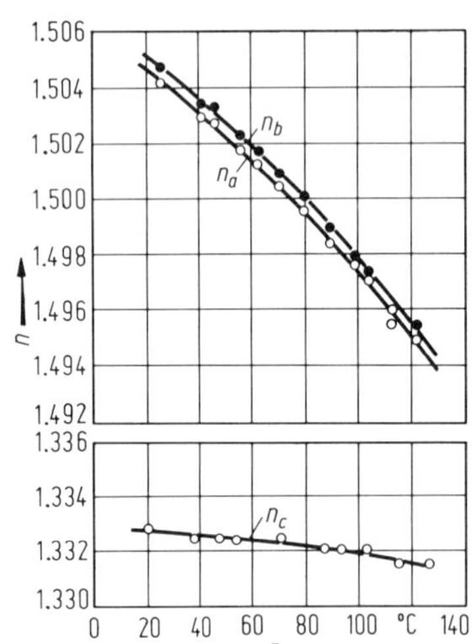

Fig. 224. KNO_3. n vs. T [68 Y 4].

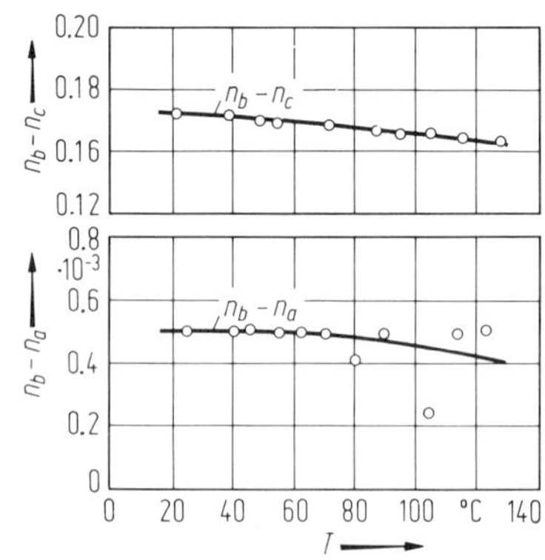

Fig. 225. KNO_3. $n_a - n_b$, $n_b - n_c$ vs. T [68 Y 4].

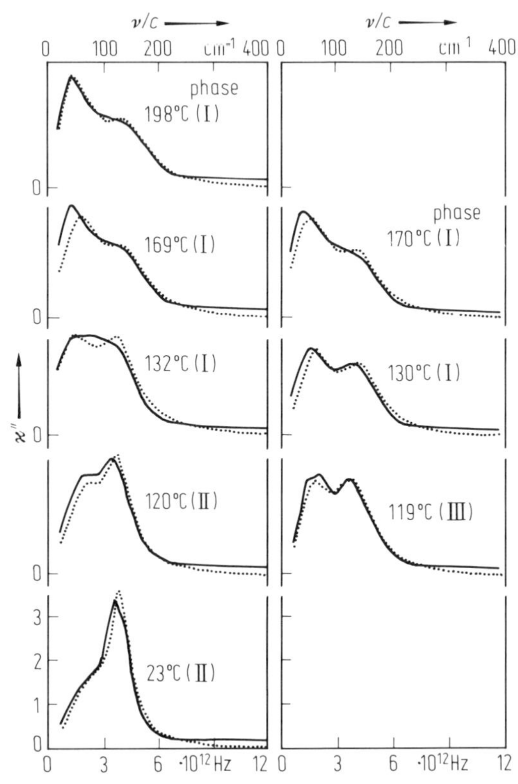

Fig. 226. KNO_3. $\varkappa''$ vs. ν [71 H 9]. Solid line is obtained from the reflectivity spectra by Kramers-Kronig analysis. Dotted line is the fit to the classical dispersion. The left and right parts of the figure correspond to results on heating and cooling, respectively.

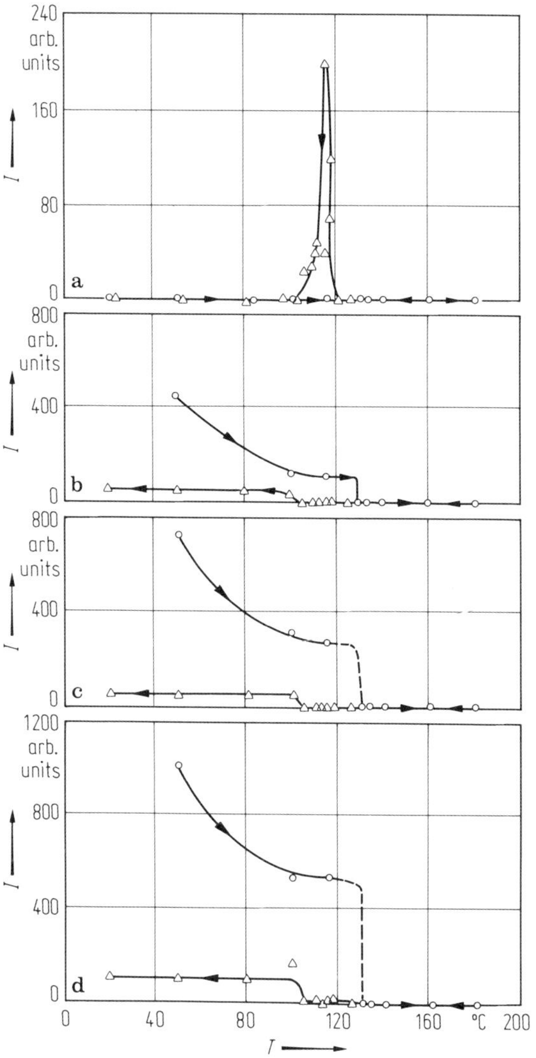

Fig. 227. KNO_3. I vs. T for several phonon modes [68 B 1]. I: Raman scattering intensity. Fig. a: line at $120\,cm^{-1}$, Fig. b: line at $83\,cm^{-1}$ performed with the geometry of $Y(Z, Y)X$, Fig. c: line at $83\,cm^{-1}$ performed with the geometry of $Y(X, Z)Y$ and Fig. d: line at $50\,cm^{-1}$. Circles: on heating; triangles: on cooling.

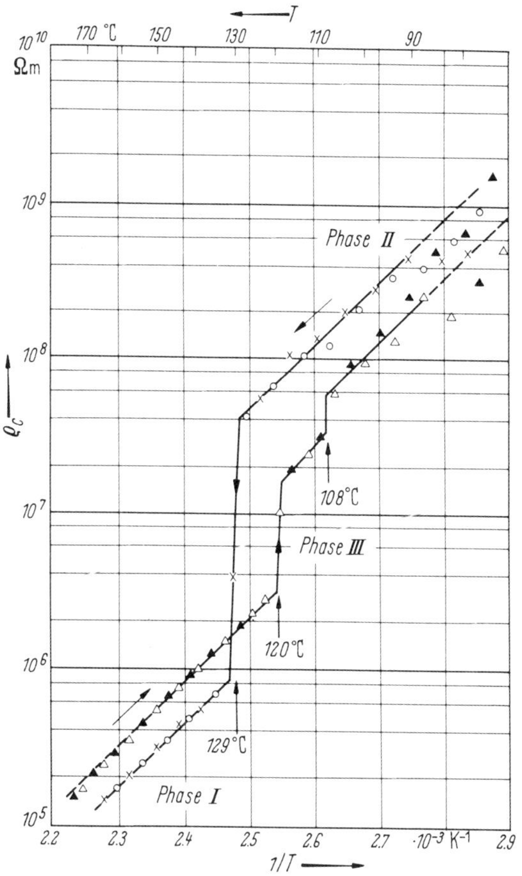

Fig. 228. KNO_3. ϱ_c vs. $1/T$ [62 A 4].

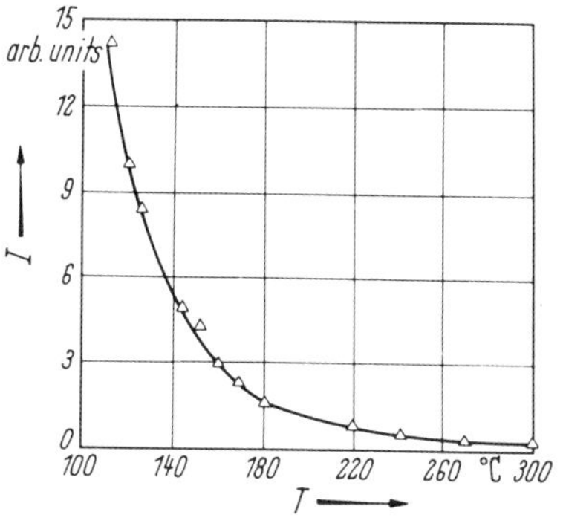

Fig. 230. KNO_3. I vs. T [62 S 7]. I: intensity of X-ray diffuse maximum in the vicinity of (111).

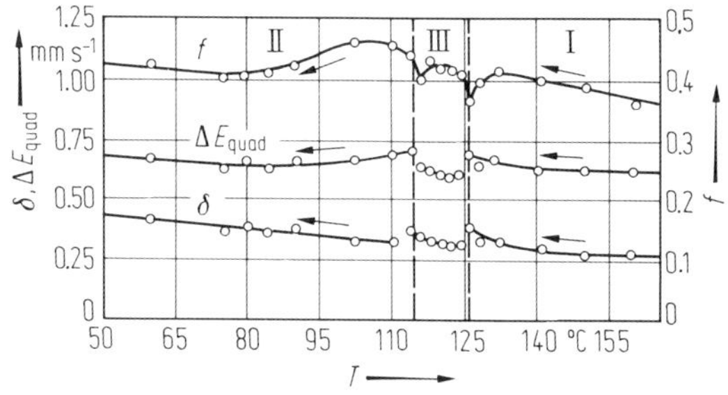

Fig. 229. KNO_3. δ, ΔE_{quad}, f vs. T [68 L 2]. δ: isomer shift, ΔE_{quad}: quadrupole splitting, f: Debye-Waller factor for ^{57}Co in KNO_3.

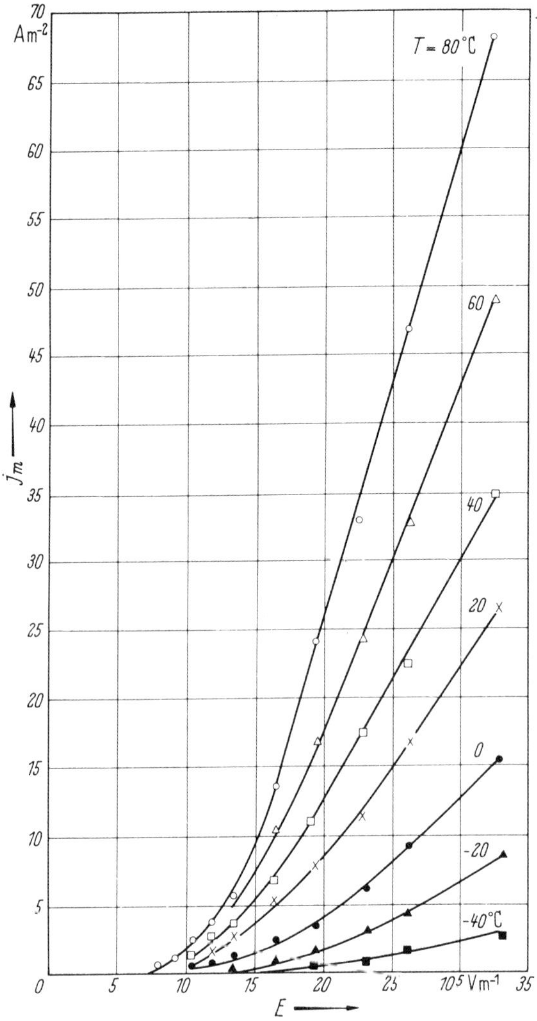

Fig. 231. KNO_3 (polycrystalline). j_m vs. E [64 D 3]. Parameter: T. j_m: maximum switching current density.

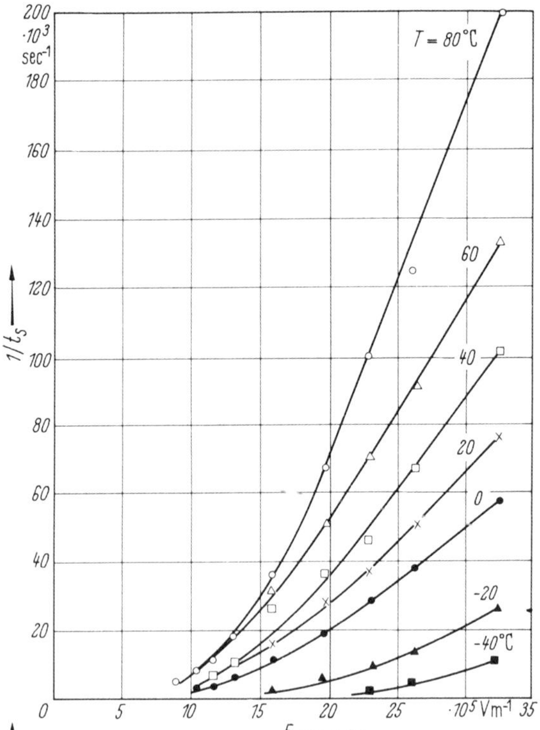

Fig. 232. KNO_3 (polycrystalline). $1/t_s$ vs. E [64 D 3]. Parameter: T. t_s: switching time.

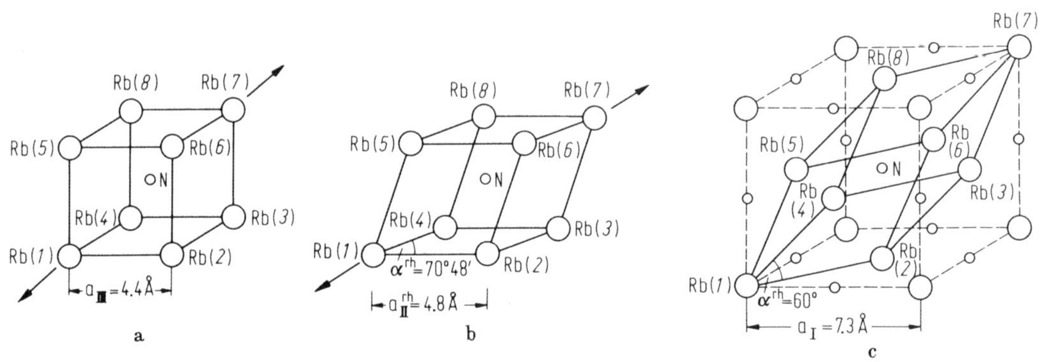

Fig. 233. $RbNO_3$. Structural relationship between Fig. a: phase III (cubic), Fig. b: phase II (trigonal) and Fig. c: phase I (cubic) [68 S 1].

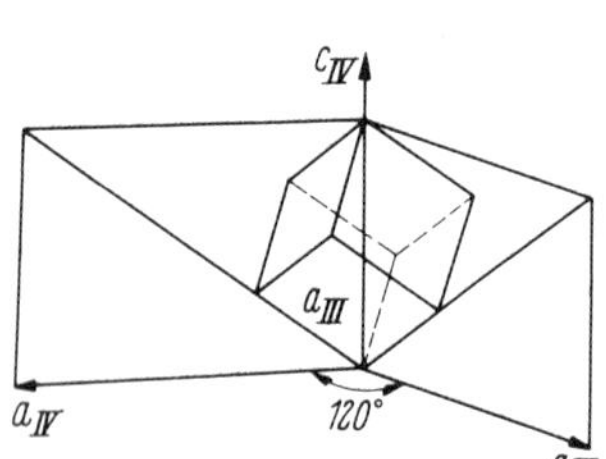

Fig. 234. $RbNO_3$. Relationship between the unit cells of phases IV and III [62 B 9].

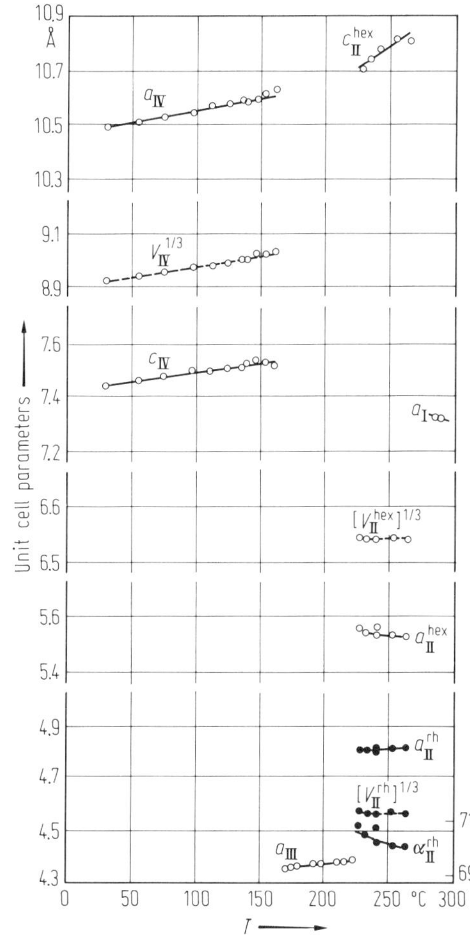

Fig. 235. $RbNO_3$. Unit cell parameters vs. T [68 S 1].

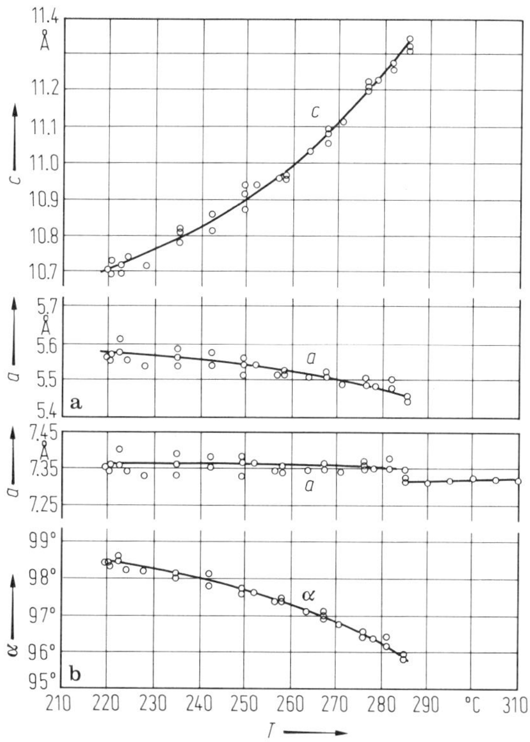

Fig. 236. $RbNO_3$. Unit cell parameters vs. T [77 Y 5]. Fig. a: Hexagonal unit cell parameters in trigonal phase II, Fig. b: cubic and distorted cubic cell parameters in cubic phase I and trigonal phase II. The relation between the rhombohedral unit cell and the distorted cubic cell is shown by the solid and dashed lines in Fig. 233.

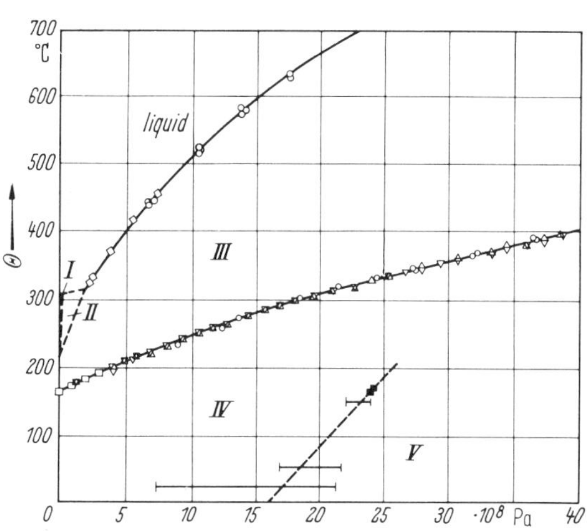

Fig. 240. $RbNO_3$. Θ vs. p [66 R 1].

Figs. 238, 239, see p. 366.

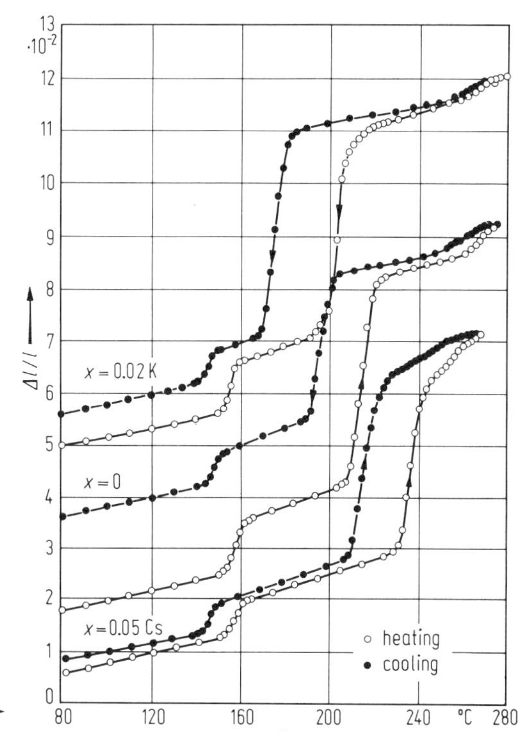

Fig. 237. $Rb_{1-x}A_xNO_3$, A = K, Cs; x = 0, x = 0.02, x = 0.05. $\Delta l/l$ vs. T [69 S 2]. Measured with an optical lever dilatometer.

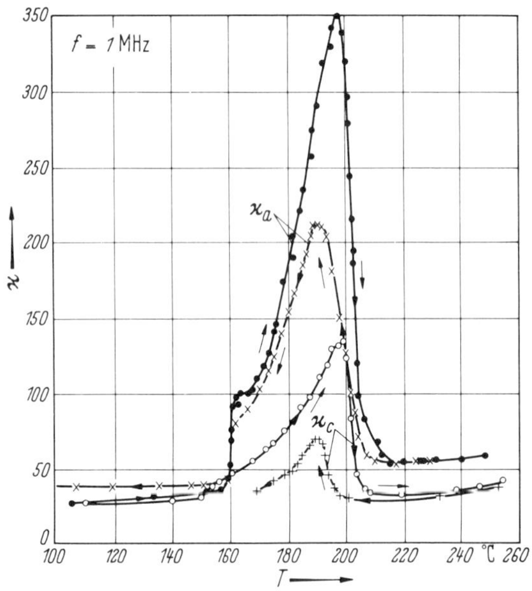

Fig. 238. $RbNO_3$. $\varkappa_c$, $\varkappa_a$ vs. T [63D1]. $\varkappa_c$, $\varkappa_a$: dielectric constants along and perpendicular to the trigonal axis, respectively.

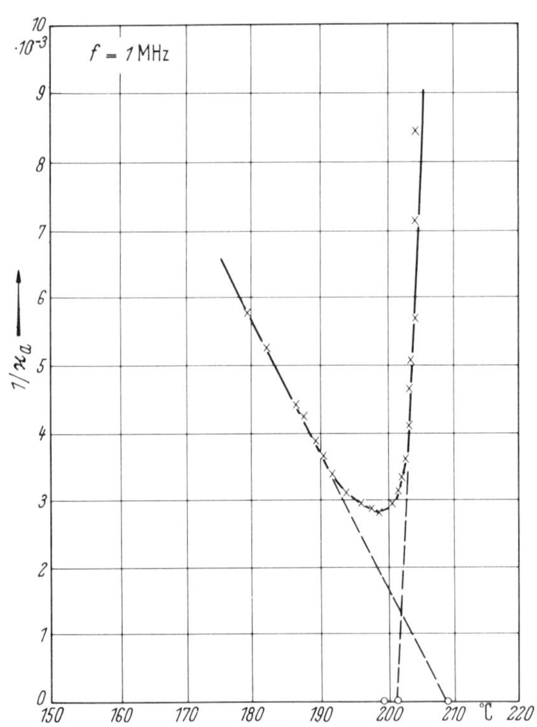

Fig. 239. $RbNO_3$. $1/\varkappa_a$ vs. T [63D1].

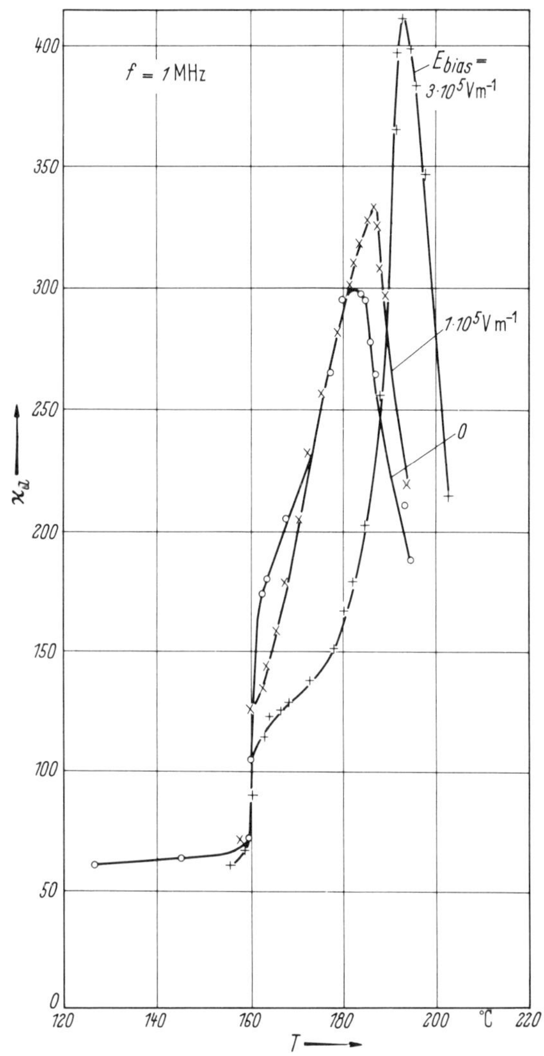

Fig. 241. $RbNO_3$. $\varkappa_a$ vs. T [63D1]. Parameter: E_{bias}.

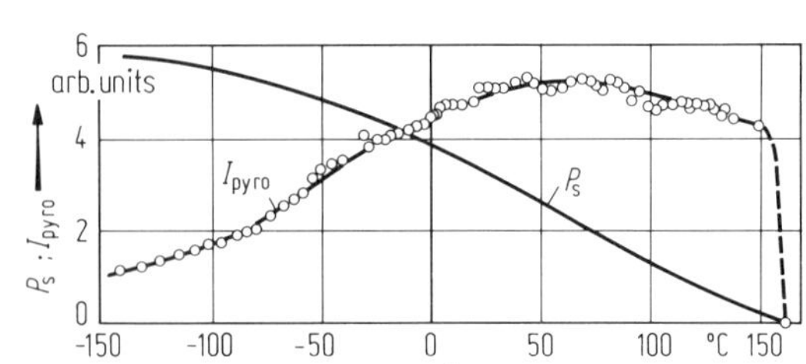

Fig. 242. $RbNO_3$. I_{pyro}, P_s vs. T [69B19]. I_{pyro}: pyroelectric current along the [111] direction. Measured by a dynamical method.

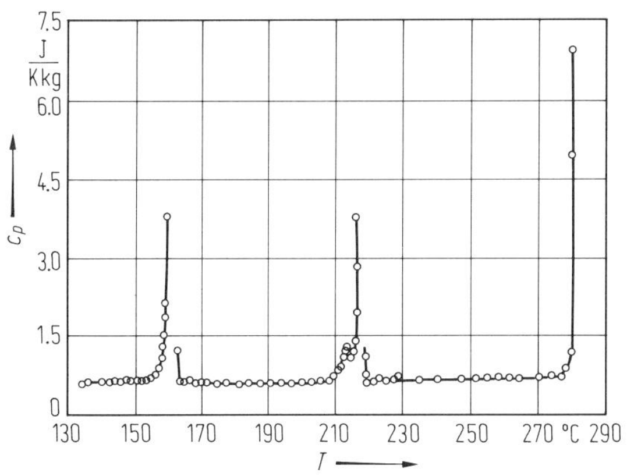

Fig. 243. $RbNO_3$. c_p vs. T [77D4].

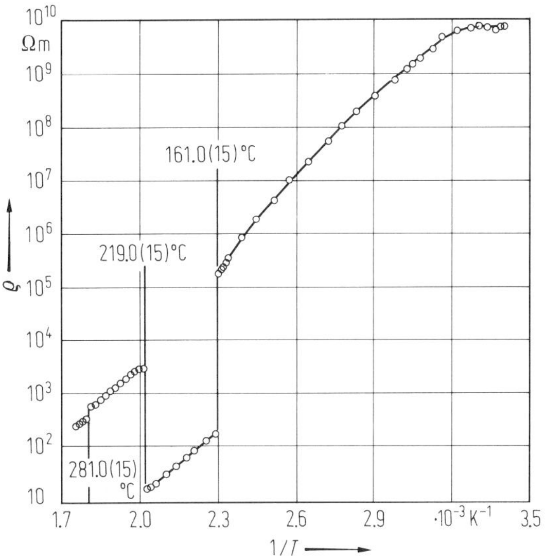

Fig. 244. $RbNO_3$. ϱ vs. $1/T$ [72F2]. Applied field: along [001] of phase IV. Measured by four-terminal method for zone-refined single crystal.

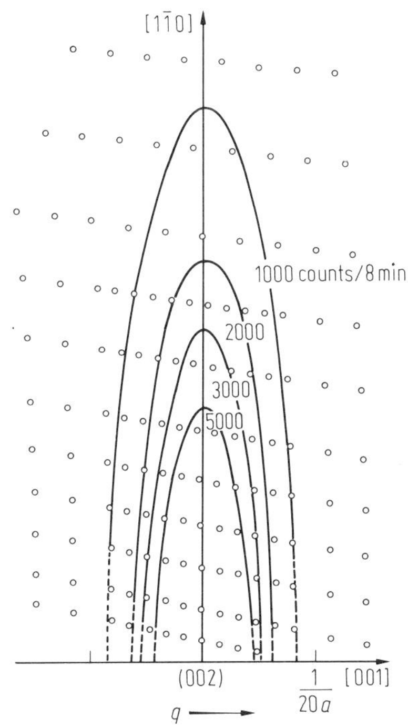

Fig. 245. $RbNO_3$. X-ray diffuse scattering [77Y5]. Intensity distribution in reciprocal space around (002) at 300 °C. The curves show equi-intensity contours after subtracting background intensity. Dots show the points where measurements were carried out.

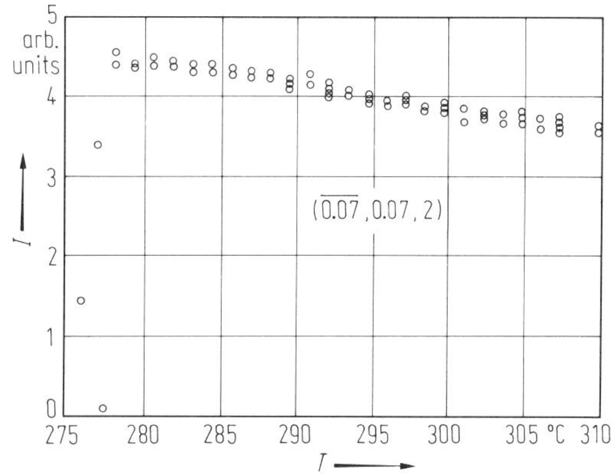

Fig. 246. $RbNO_3$. X-ray diffuse scattering [77Y5]. $I_{diffuse}$ vs. T at (−0.07, 0.07, 2). $I_{diffuse}$: X-ray diffuse scattering intensity.

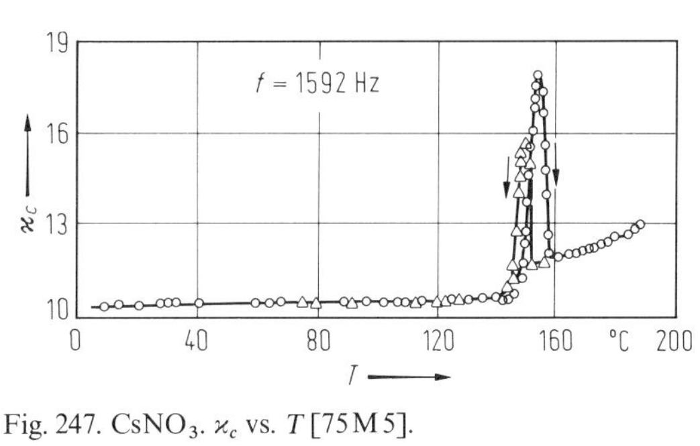

Fig. 247. $CsNO_3$. $\varkappa_c$ vs. T [75M5].

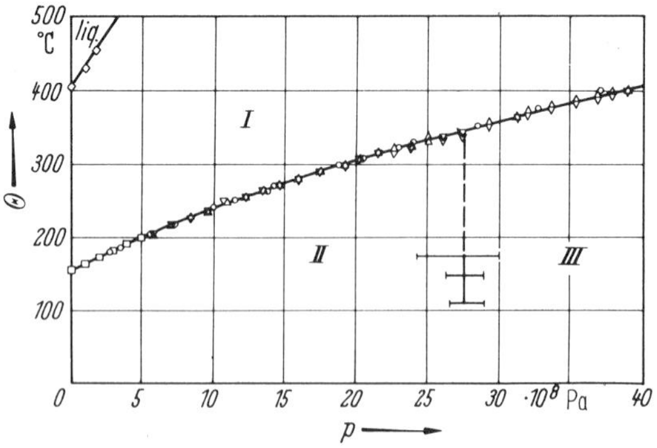

Fig. 248. $CsNO_3$. Θ vs. p [66R1].

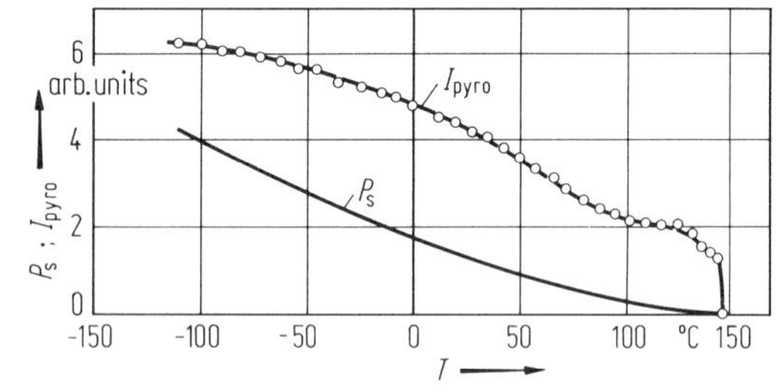

Fig. 249. $CsNO_3$. I_{pyro}, P_s vs. T [69B19]. I_{pyro}: pyroelectric current along the c-direction measured by dynamical method.

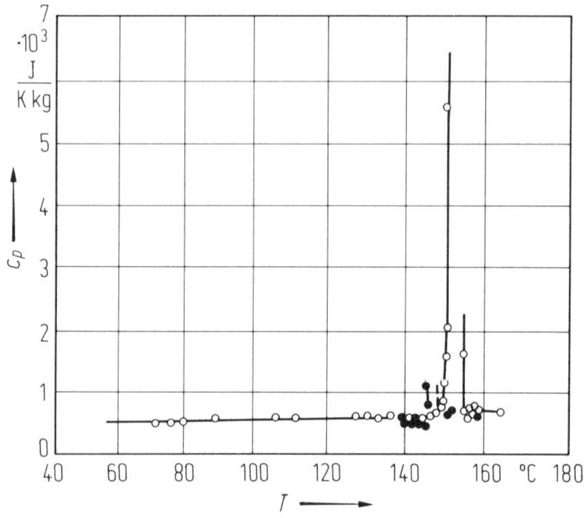

Fig. 250. $CsNO_3$. c_p vs. T [75M5]. Open circles: on heating, full circles: on cooling.

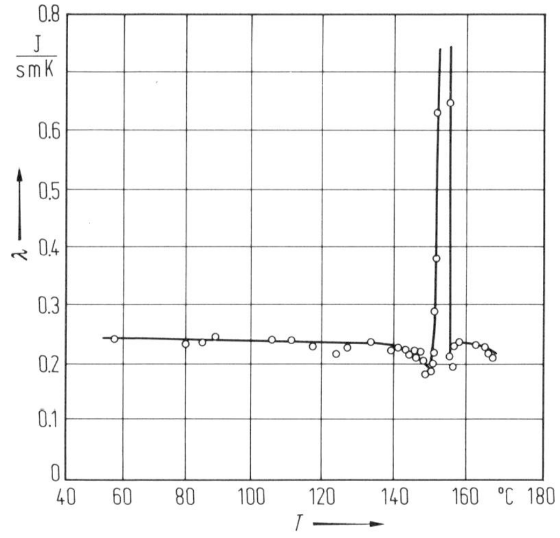

Fig. 251. $CsNO_3$. λ vs. T [75M5]. λ: thermal conductivity.

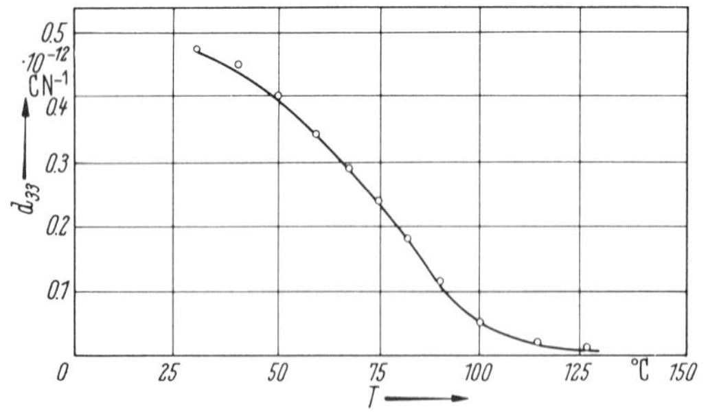

Fig. 252. $CsNO_3$. d_{33} vs. T [63S10]. Measured by a static method.

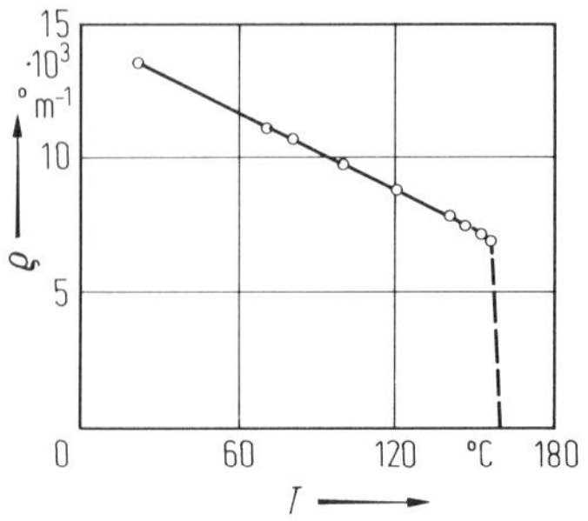

Fig. 253. $CsNO_3$. ϱ vs. T [74B23]. ϱ: optical rotatory power. $\lambda = 0.43$ μm.

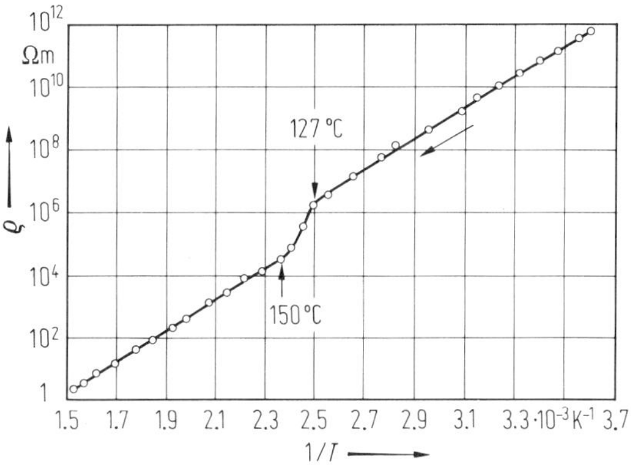

Fig. 254. $CsNO_3$. ϱ vs. $1/T$ [72 F2]. ϱ: resistivity.

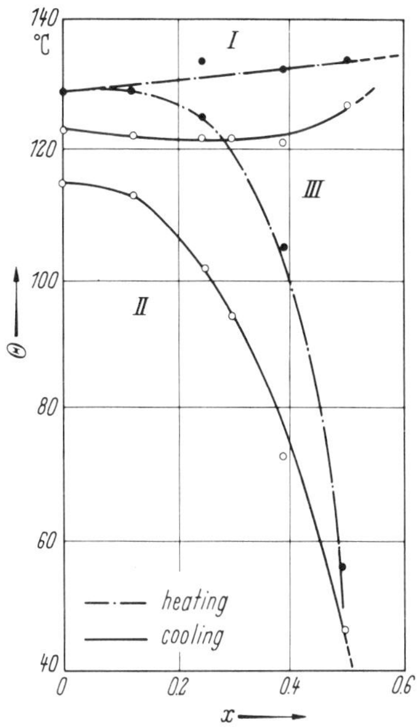

Fig. 255. $K_{1-x}(NH_4)_xNO_3$. Θ vs. x [65 Y 1].

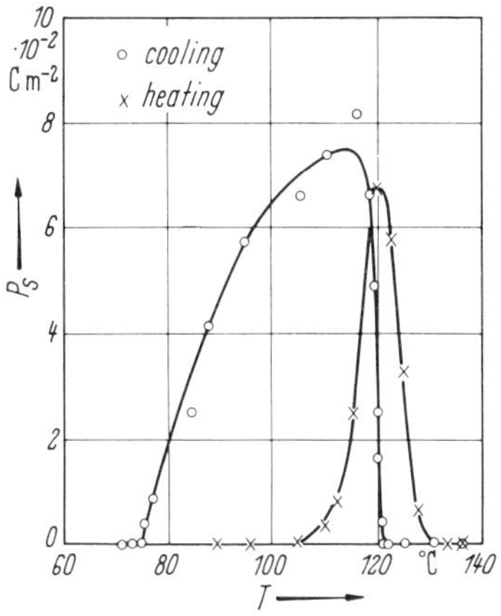

Fig. 256. $K_{0.61}(NH_4)_{0.39}NO_3$. P_s vs. T [65 Y 1].

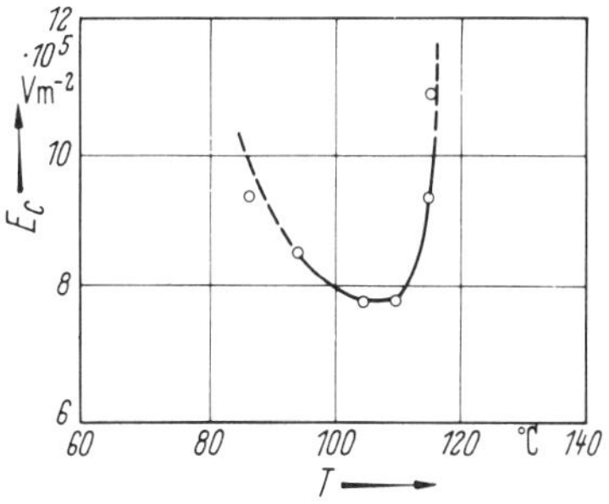

Fig. 257. $K_{0.61}(NH_4)_{0.39}NO_3$. E_c vs. T [65 Y 1].

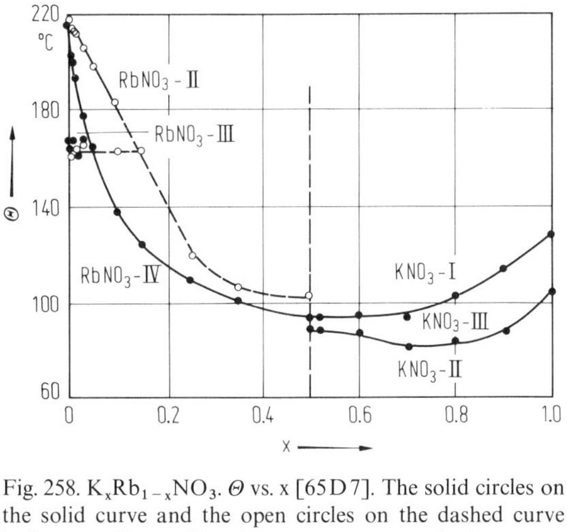

Fig. 258. $K_xRb_{1-x}NO_3$. Θ vs. x [65 D 7]. The solid circles on the solid curve and the open circles on the dashed curve indicate cooling and heating processes, respectively.

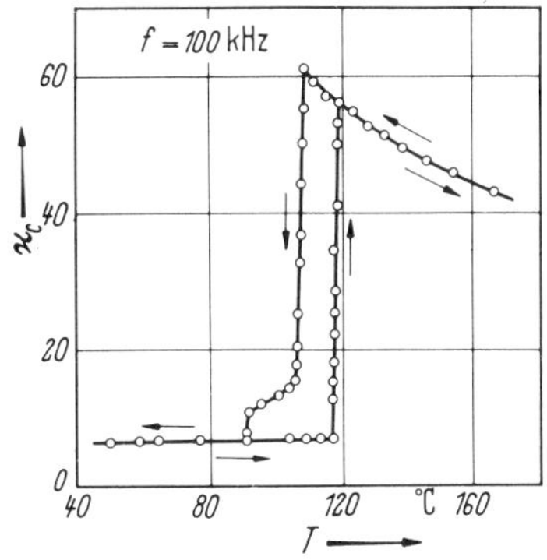

Fig. 259. $K_{0.77}Rb_{0.23}NO_3$. $\varkappa_c$ vs. T [65 K 4].

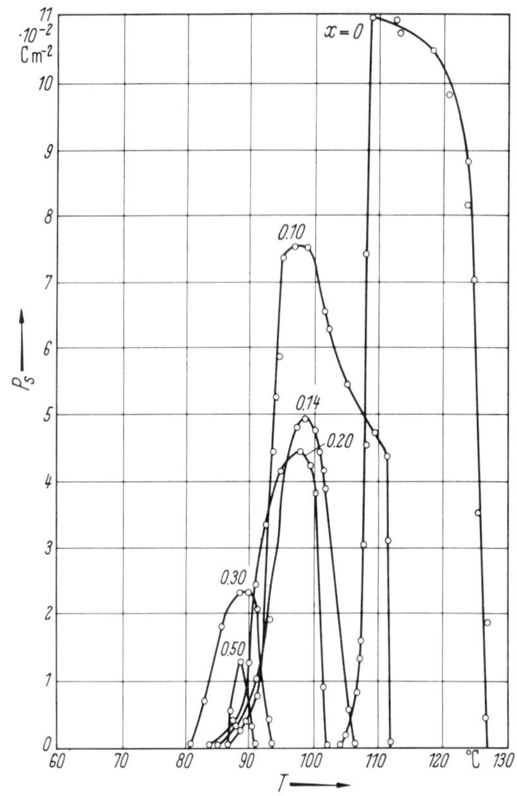

Fig. 260. $K_{1-x}Rb_xNO_3$ (polycrystalline). P_s vs. T [65D3]. Parameter: x.

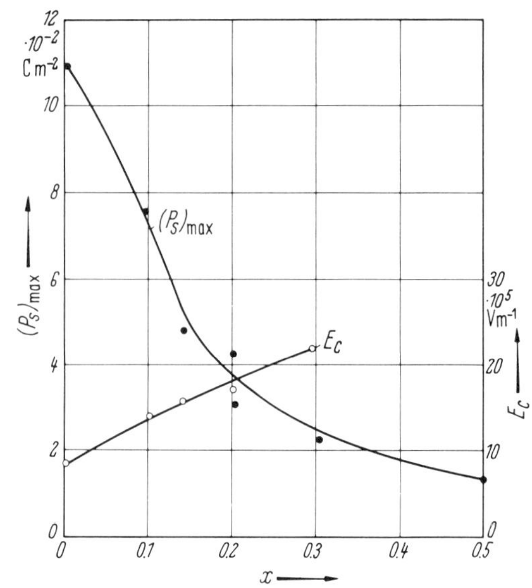

Fig. 261. $K_{1-x}Rb_xNO_3$ (polycrystalline). $(P_s)_{max}$, E_c vs. x [65D3]. $(P_s)_{max}$: maximum spontaneous polarization.

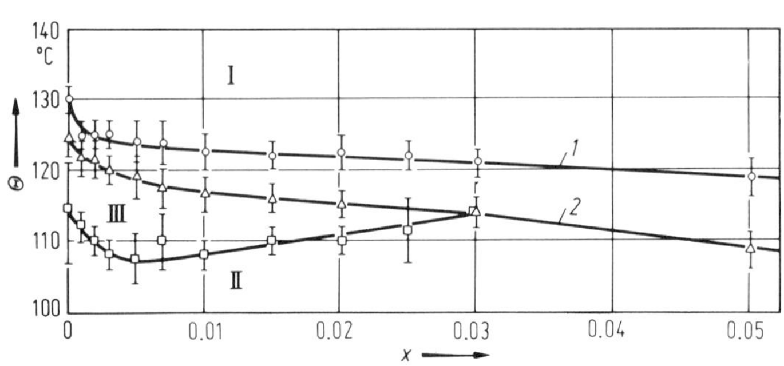

Fig. 262. $K(NO_3)_{1-x}(NO_2)_x$. Θ vs. x [70M4]. Curve *1*: on heating, curve *2*: on cooling.

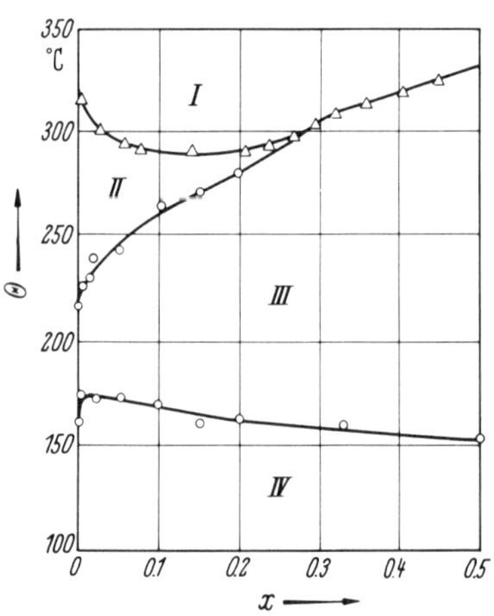

Fig. 265. $Rb_{1-x}Cs_xNO_3$. Θ vs. x [65D1].

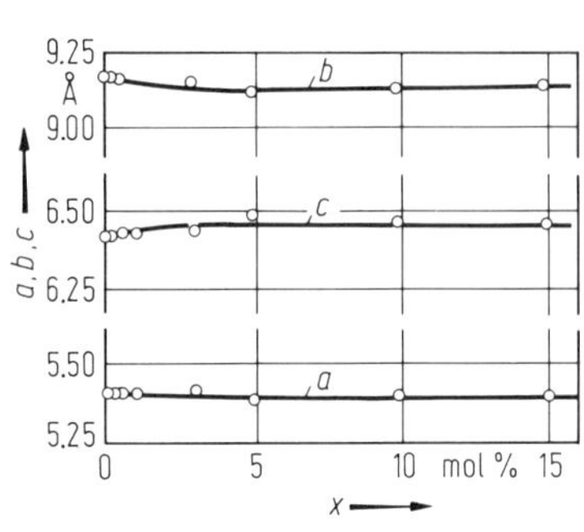

◄ Fig. 263. KNO_3–KNO_2. *a*, *b*, *c* vs. x [70M4]. x: fraction of KNO_2.

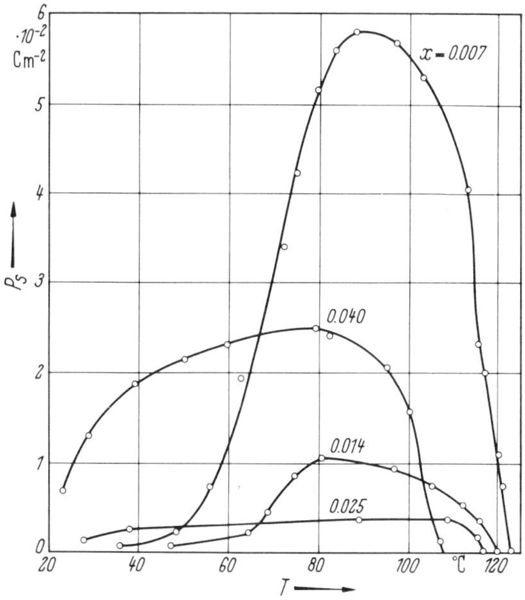

Fig. 264. $K(NO_3)_{1-x}I_x$ (polycrystalline). P_s vs. T [65D2]. Parameter: x.

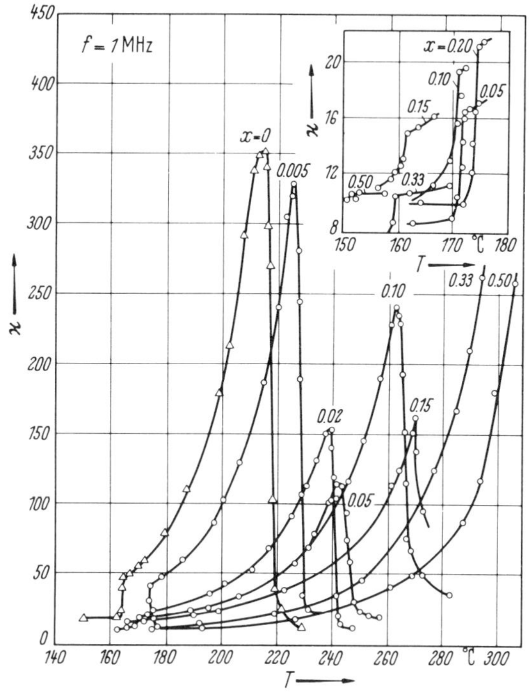

Fig. 266. $Rb_{1-x}Cs_xNO_3$ (polycrystalline). $\varkappa$ vs. T [65D1]. Parameter: x.

18 $PbHPO_4$ family

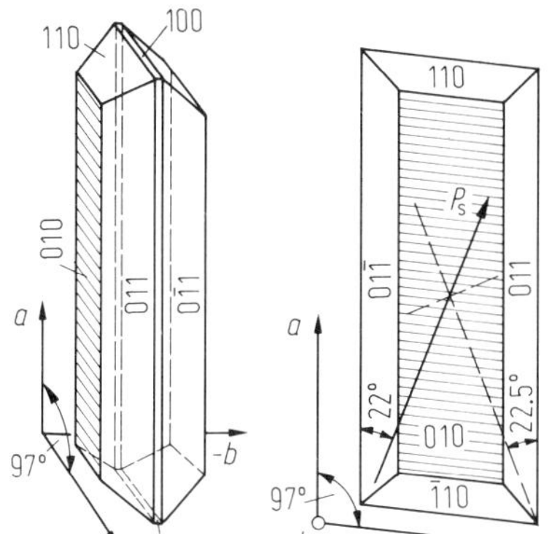

◄ Fig. 267. $PbHPO_4$. Crystal form grown from gel [76B11]. Direction of P_s is shown by the arrow. Dashed line: extinction positions under crossed nicols.

Figs. 269, 270, see p. 372.

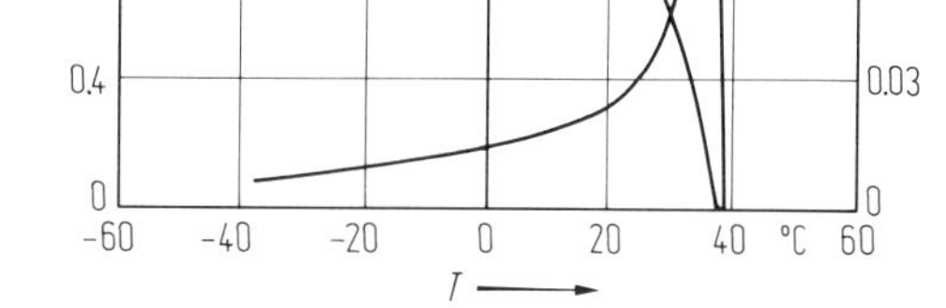

Fig. 271. $PbHPO_4$. P_s, p vs. T measured on (100) plate specimen [74N5]. $p = dP_s/dT$: pyroelectric coefficient.

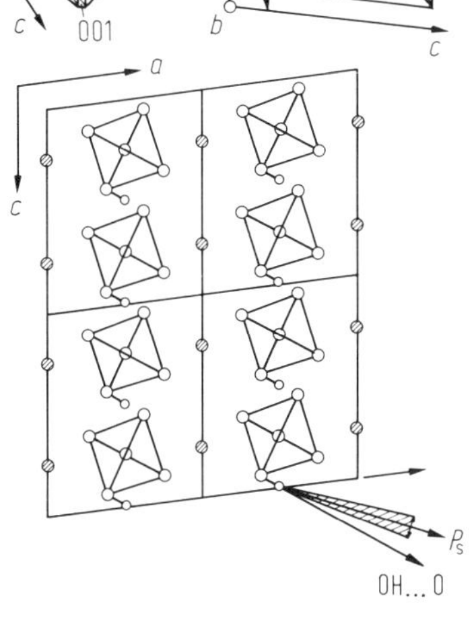

◄ Fig. 268. $PbHPO_4$. Projection of the structure along b [74N5]. Four unit cells are shown. The hatched circles and the small circles represent Pb-atoms and protons, respectively. The direction of P_s is shown.

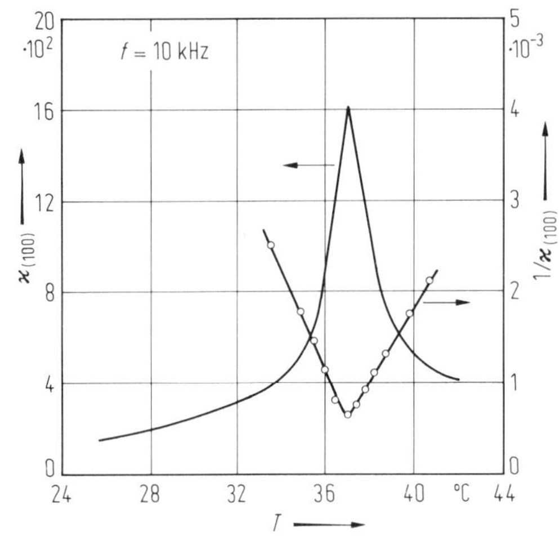

Fig. 269α. $PbHPO_4$. $\varkappa_{(100)}$, $1/\varkappa_{(100)}$ vs. T [74N5]. $\varkappa_{(100)}$: dielectric constant of (100) plate specimen.

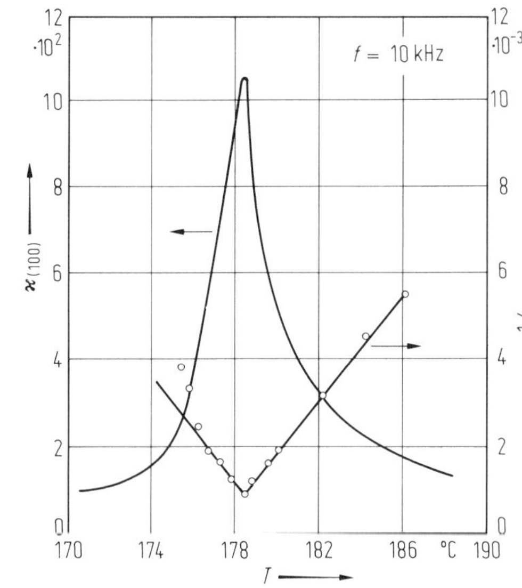

Fig. 269β. $PbDPO_4$. $\varkappa_{(100)}$, $1/\varkappa_{(100)}$ vs. T [74N5]. $\varkappa_{(100)}$: dielectric constant of (100) plate specimen. D-content is not given.

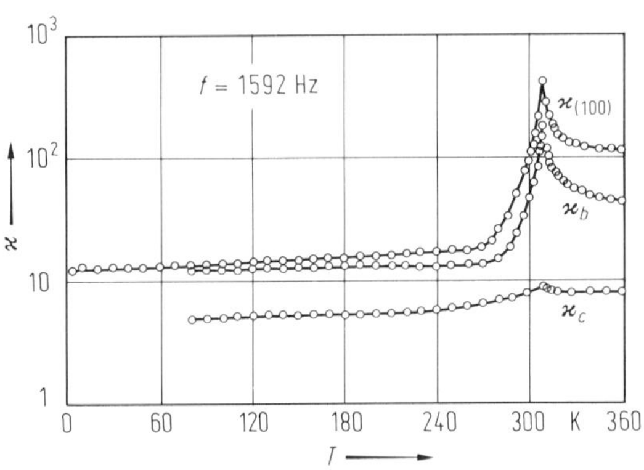

Fig. 269γ. $PbHPO_4$. $\varkappa_{(100)}$, $\varkappa_b$, $\varkappa_c$ vs. T [78S17]. $\varkappa_{(100)}$: dielectric constant perpendicular to the (100) plane.

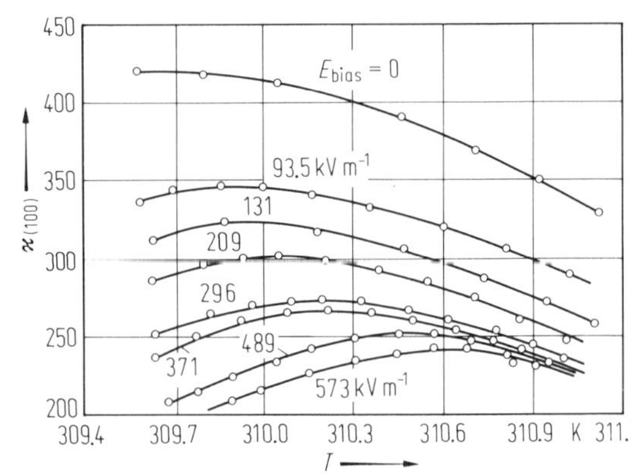

Fig. 270α. $PbHPO_4$. $\varkappa_{(100)}$ vs. T [78S17]. Parameter: E_{bias}. $\varkappa_{(100)}$: dielectric constant perpendicular to the (100) plane. $f = 1592$ Hz.

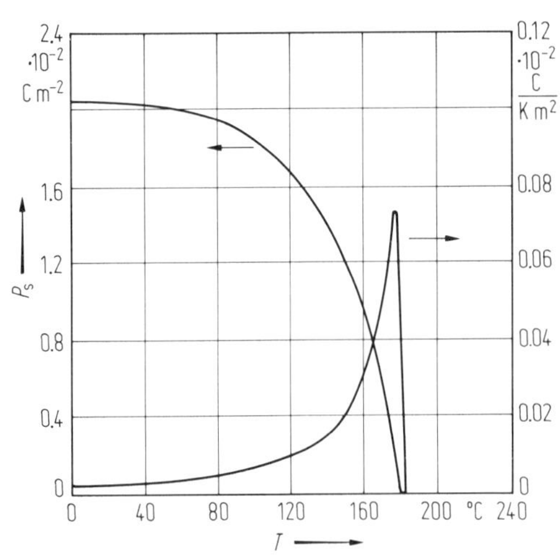

Fig. 272. $PbDPO_4$. P_s, p vs. T measured on (100) plate specimen [74N5]. $p = dP_s/dT$: pyroelectric coefficient. D-content is not given.

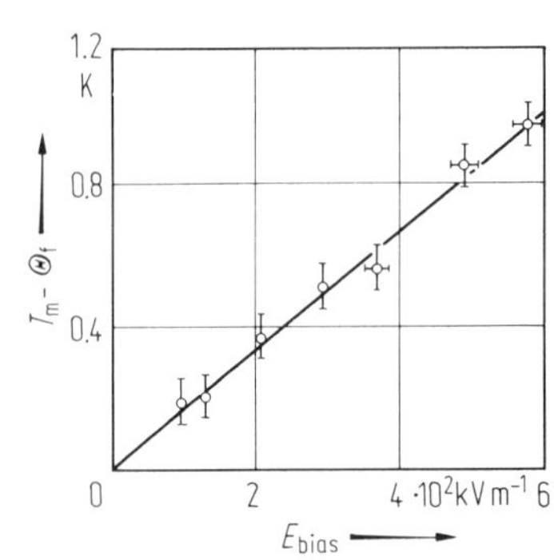

Fig. 270β. $PbHPO_4$. $T_m - \Theta_f$ vs. E_{bias} [78S17]. T_m: temperature at which $\varkappa_{(100)}$ is maximum.

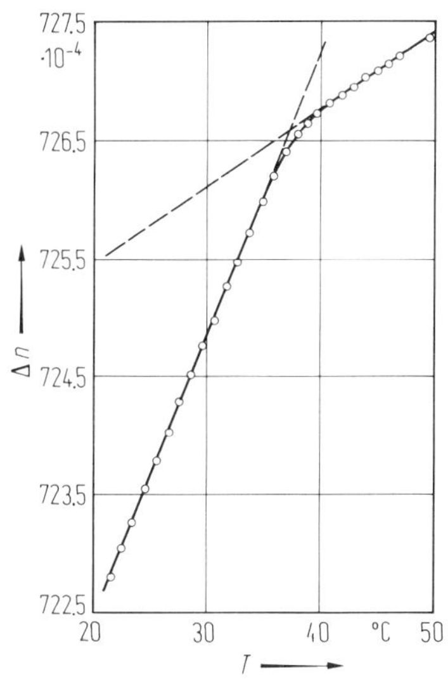

Fig. 273 α. $PbHPO_4$. Δn vs. T [77 K 12]. $\Delta n = n_\gamma - n_\beta$. $\lambda = 633$ nm.

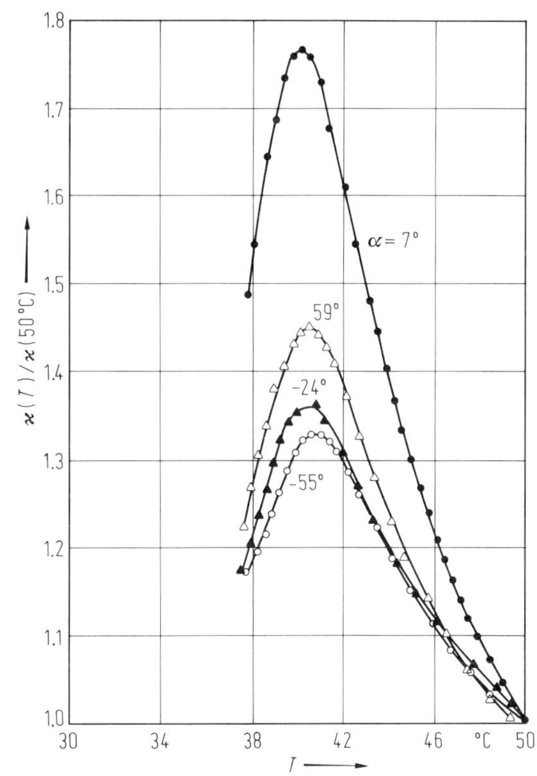

Fig. 274. $PbHAsO_4$. $\varkappa(T)/\varkappa(50\,°C)$ vs. T [75 B 19]. Parameter: α. $\varkappa$ (50 °C): dielectric constant at 50 °C, α: angle between electric field and the [100] direction. $f = 1592$ Hz.

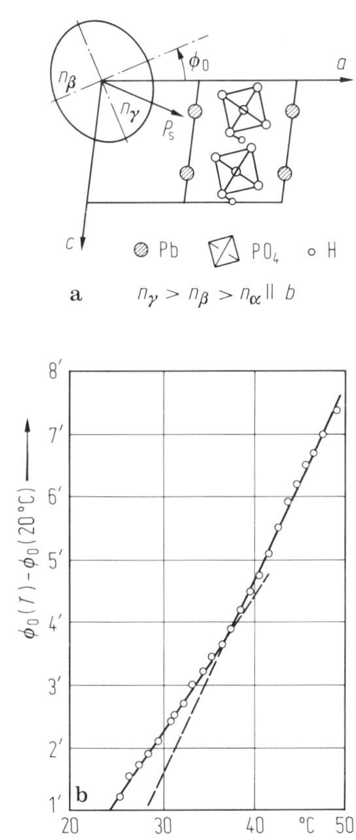

Fig. 273 β. $PbHPO_4$. $\phi_0(T) - \phi_0$ (20 °C) vs. T (b) [78 K 19]. $\phi_0(T)$: angle of indicatrix at T. (a) The orientation of indicatrix and P_s referring to the crystal structure.

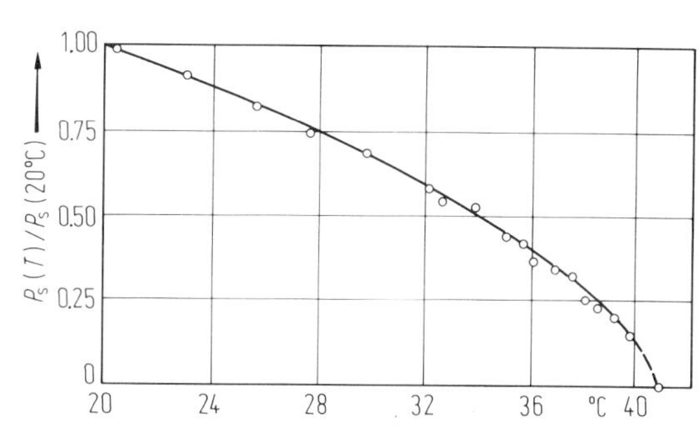

Fig. 275. $PbHAsO_4$. $P_s(T)/P_s(20\,°C)$ vs. T [75 B 19]. P_s (20 °C): spontaneous polarization at 20 °C. P_s is measured along the [100] direction.

19 KDP(KH_2PO_4) family

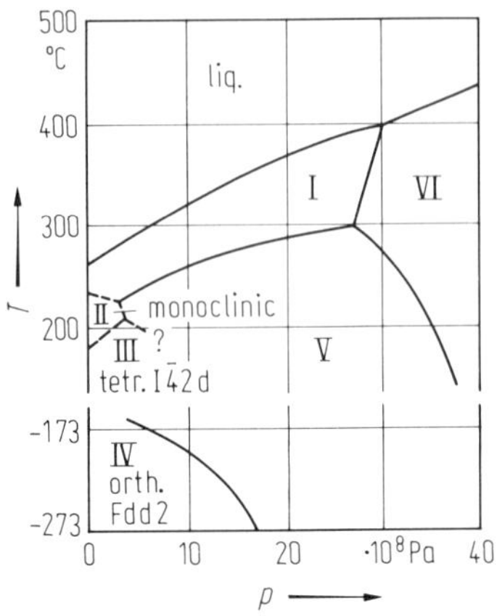

Fig. 276. KH_2PO_4(KDP). Phase diagram with regard to p [76P14].

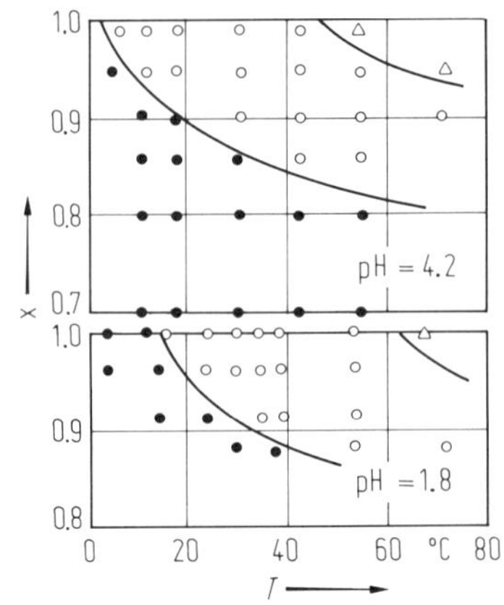

Fig. 277. $KH_{2(1-x)}D_{2x}PO_4$. Crystallization in solutions with various x and acidity. [77S36]. Full circles: tetragonal phase, triangles: monoclinic phase, open circles: mixture of both phases.

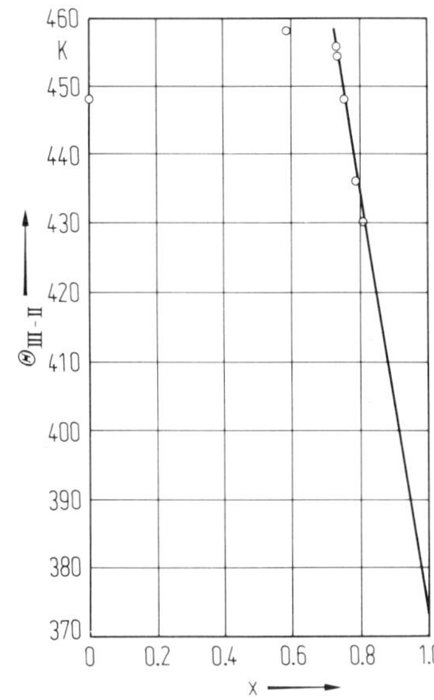

Fig. 278. $KH_{2(1-x)}D_{2x}PO_4$ Θ_{III-II} vs. x [78S8].

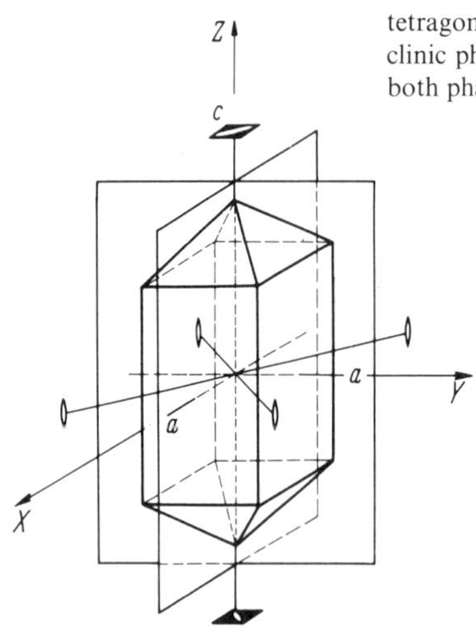

Fig. 279. KH_2PO_4(KDP). Crystal form and symmetry [38B3].

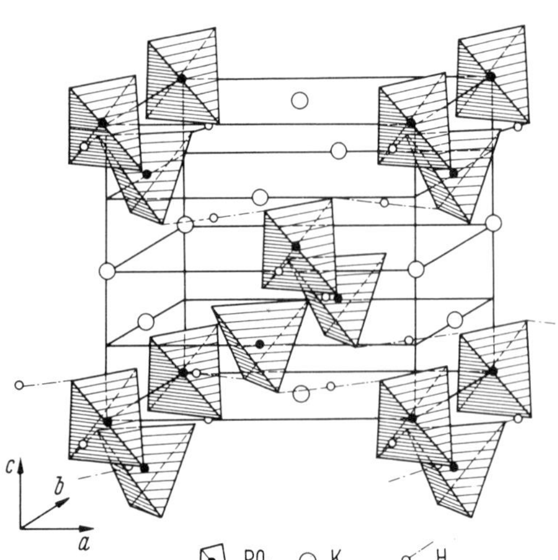

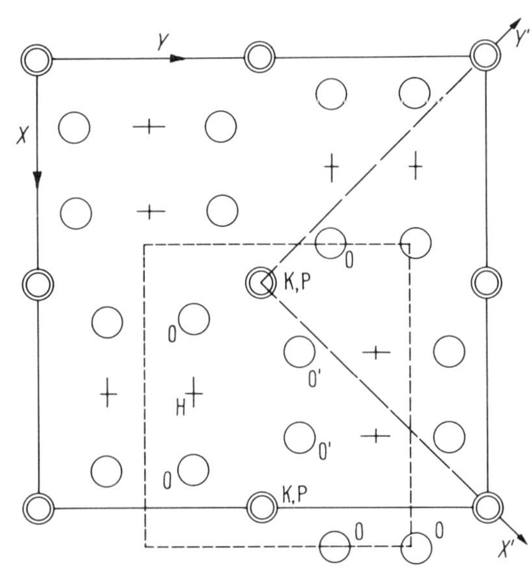

Fig. 281. KH_2PO_4(KDP). Projection on (001), representing the relation between a body centered (I$\bar{4}$2d) cell with X, Y axes and a face centered (F$\bar{4}$d2) cell with X', Y' axes. The dotted lines indicate the representative portion of the unit cell shown in Fig. 282 [55B1].

◄ Fig. 280. KH_2PO_4(KDP). Structure at RT [30W1]. Hydrogen positions are shown only schematically.

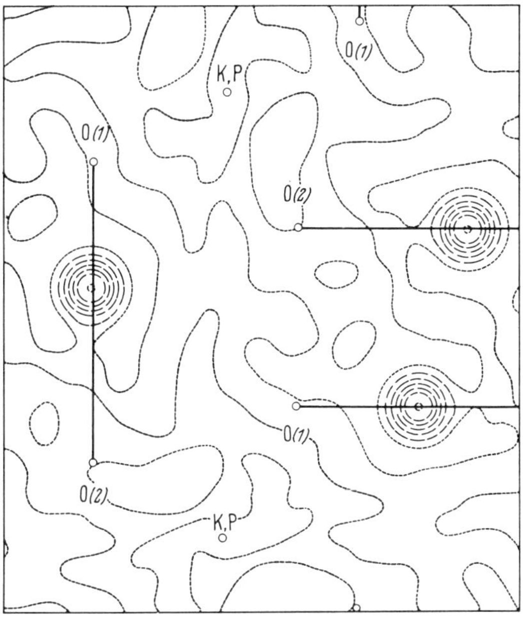

Fig. 282. KH_2PO_4(KDP). Projection of H-atoms on (001) at 77 K [55B1]. This Fourier map corresponds to the portion surrounded by the dotted lines in Fig. 281.

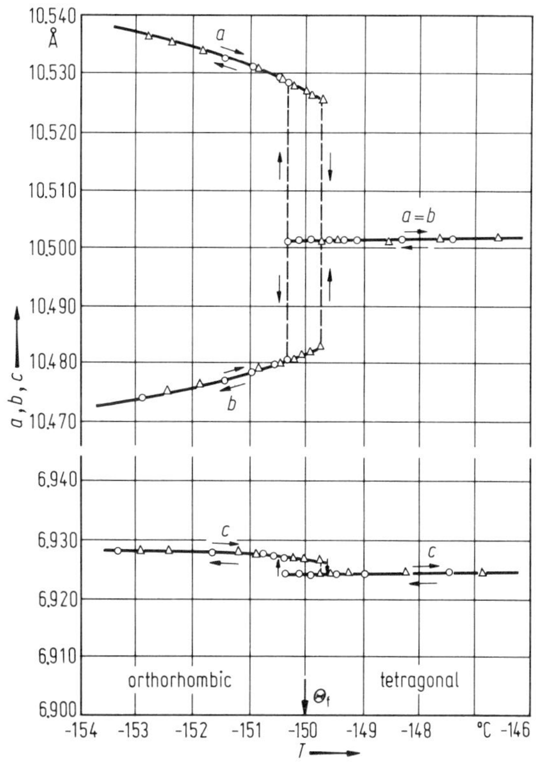

Fig. 283. KH_2PO_4(KDP). a, b, c vs. T in the vicinity of Θ_f [70K18].

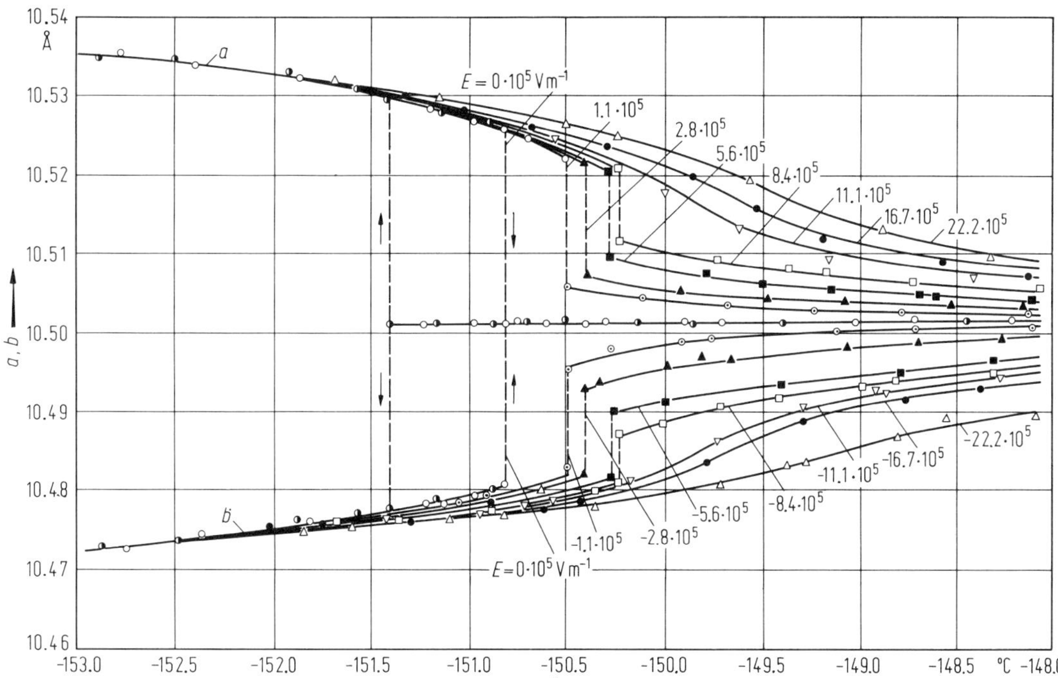

Fig. 284. KH_2PO_4(KDP). a, b vs. T under E_{bias} [71K11]. a, b: unit cell parameters.

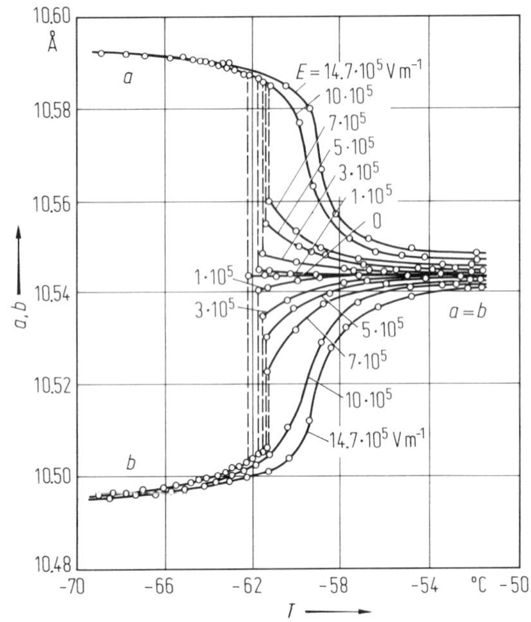

Fig. 285. KD_2PO_4(DKDP) a, b vs. T [74U3]. a, b: unit cell parameters. Parameter: E_{bias}.

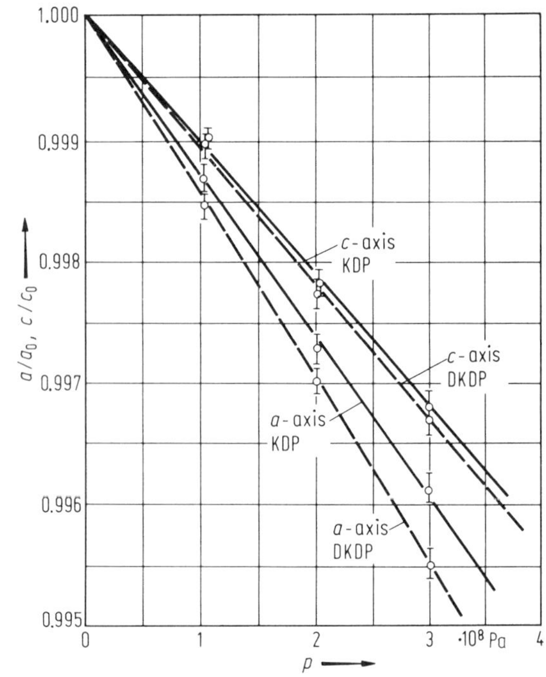

Fig. 288. KH_2PO_4(KDP), KD_2PO_4(DKDP). Reduced lattice constants a/a_0 and c/c_0 vs. p at 295 K [72M18]. a_0, b_0: unit cell parameters at the atmospheric pressure.

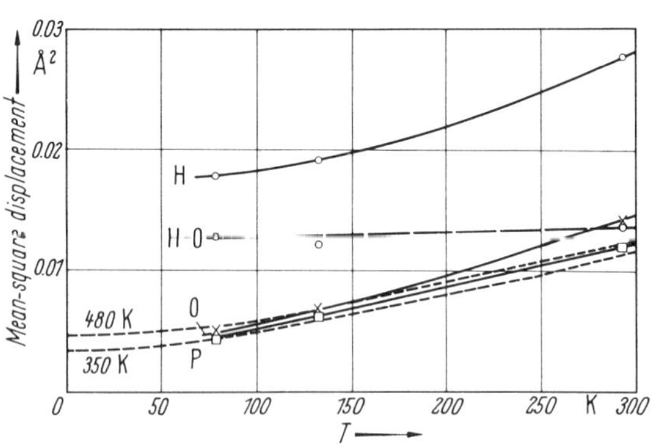

◄ Fig. 286. KH_2PO_4(KDP). Mean square displacements of x, y of H, O, P atoms vs. T [55B1]. The dotted lines show the theoretical curves for Debye temperatures of 350 K and 480 K.

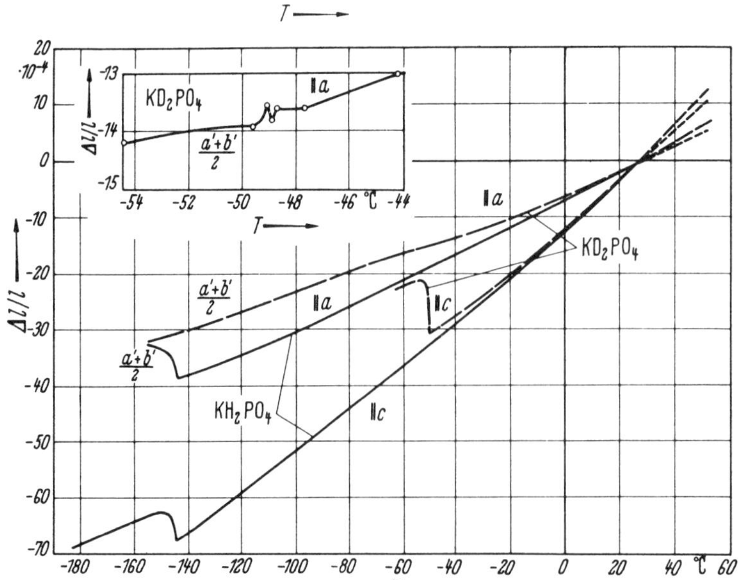

Fig. 287. KH_2PO_4(KDP), KD_2PO_4 (DKDP). Thermal expansion $\Delta l/l$ along a and c axes of the paraelectric phase III vs. T [67C5]. The inset shows an expanded plot along a near Θ_f of KD_2PO_4. $(a'+b')/2$ shows the mean values along a and b axes of the ferroelectric phase IV.

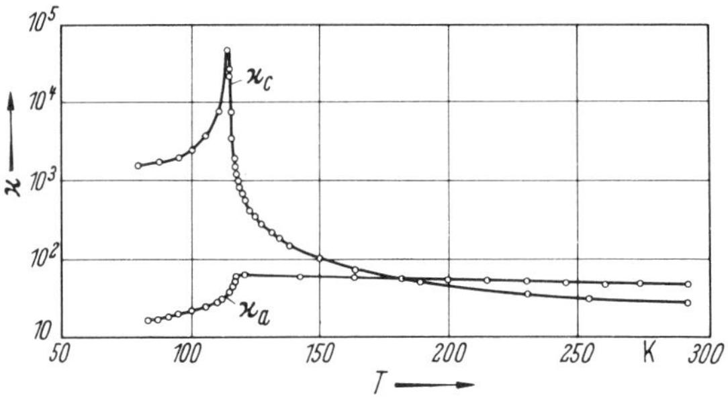

Fig. 289. KH_2PO_4(KDP). $\varkappa_a$, $\varkappa_c$ vs. T [38 B 3]. $f = 800$ Hz.

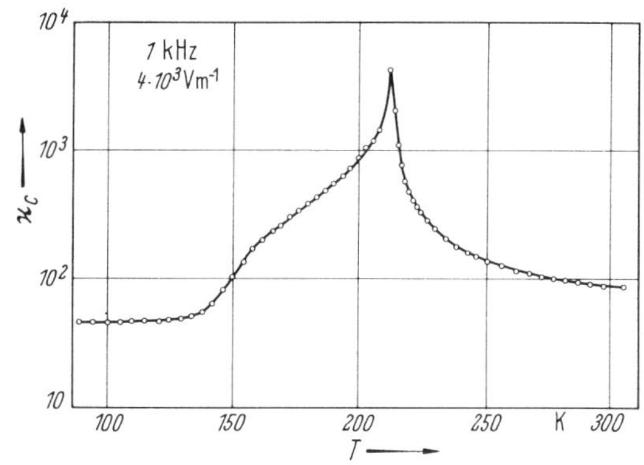

Fig. 290. Deuterated KDP. $\varkappa_c$ vs. T [41 B 1].

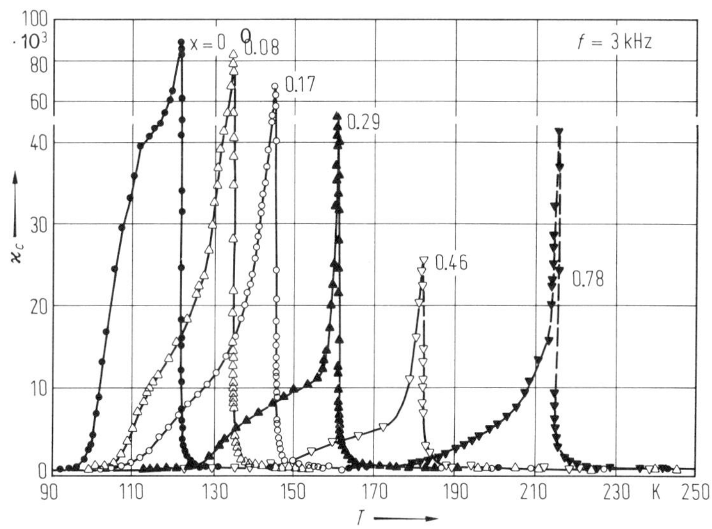

Fig. 291. $KH_{2(1-x)}D_{2x}PO_4$. $\varkappa_c$ vs. T [72 S 28]. Parameter: x.

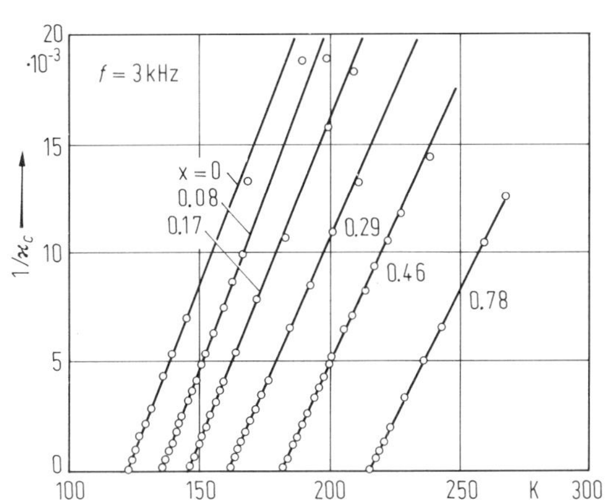

Fig. 292. $KH_{2(1-x)}D_{2x}PO_4$. $1/\varkappa_c$ vs. T [72 S 28]. Parameter: x.

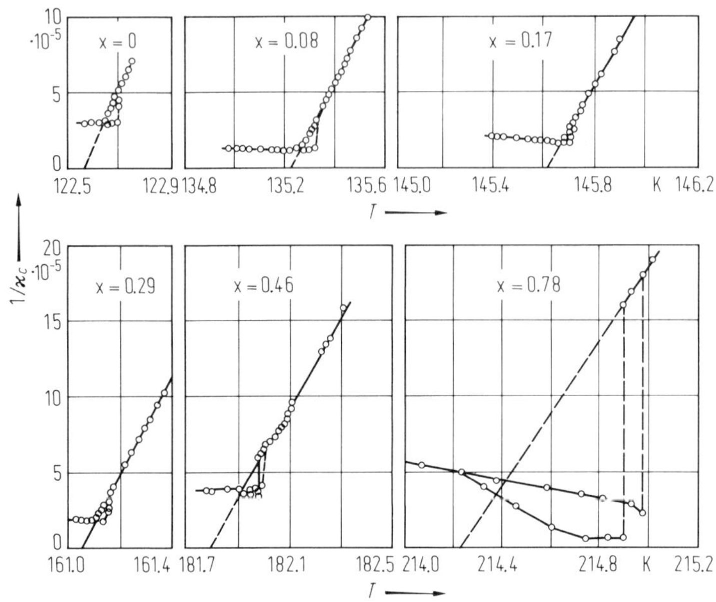

Fig. 293. $KH_{2(1-x)}D_{2x}PO_4$. $1/\varkappa_c$ vs. T on heating and cooling in the vicinity of Θ_f for various x [72S28]. $\varkappa_c$ at $f = 3$ kHz.

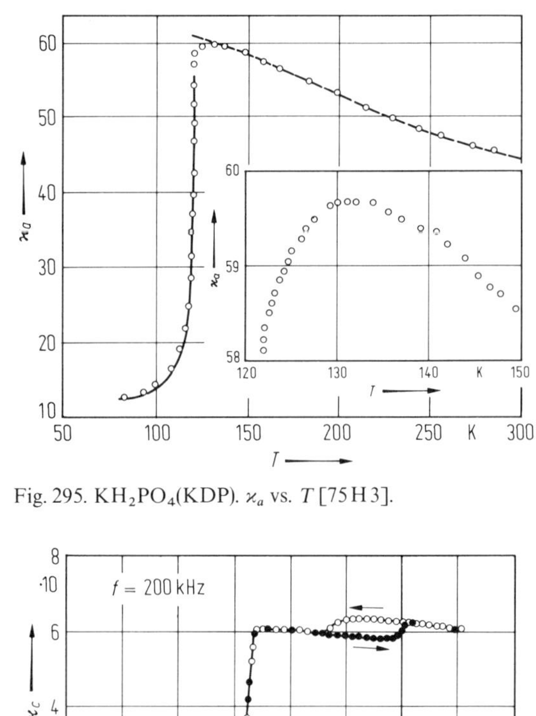

Fig. 295. KH_2PO_4(KDP). $\varkappa_a$ vs. T [75H3].

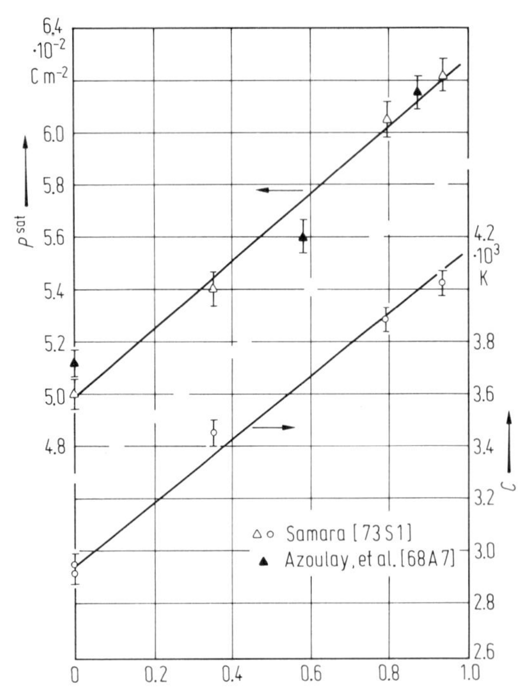

Fig. 294. $K_{2(1-x)}D_{2x}PO_4$. P^{sat}, C vs. x [73S1, 68A7]. P^{sat}: saturation polarization. C: Curie-Weiss constant.

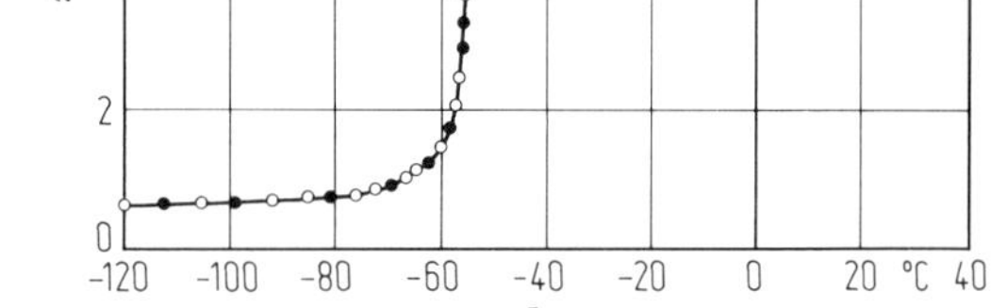

Fig. 297. KD_2PO_4(DKDP) (monoclinic). $\varkappa_c$ vs. T [75B13].

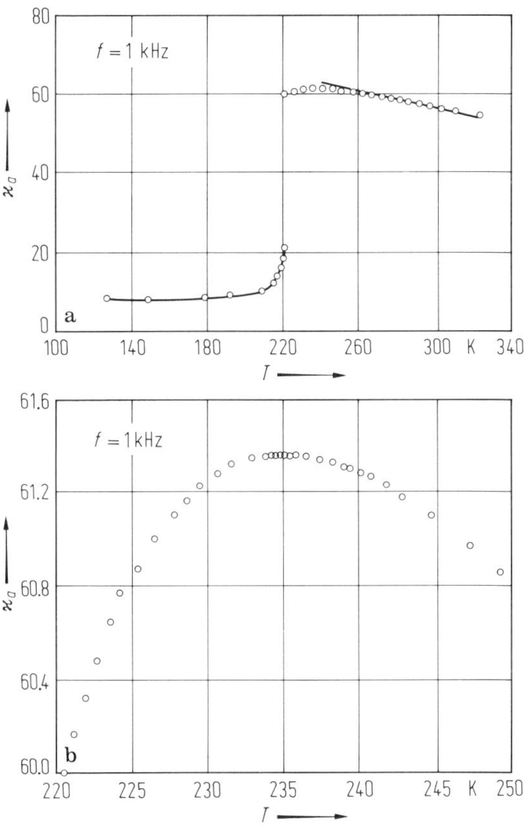

Fig. 296. KD_2PO_4(DKDP). $\varkappa_a$ vs. T [76 H 4]. Fig. a: 120···320 K. Fig. b: 220···250 K.

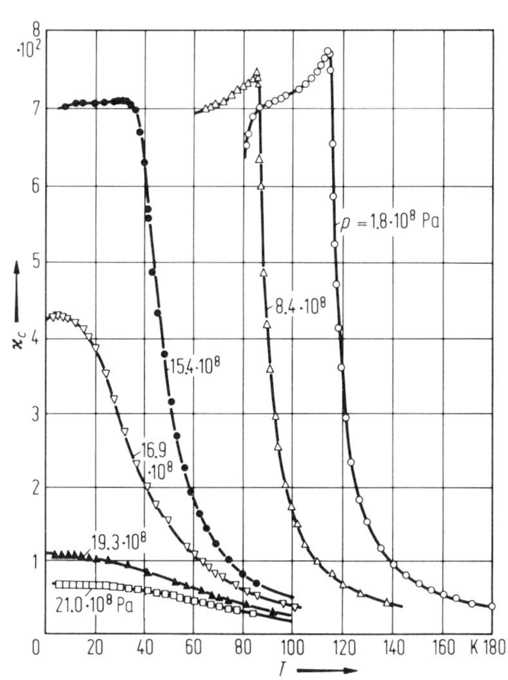

Fig. 298. KH_2PO_4(KDP). $\varkappa_c$ vs. T [71 S 4]. Parameter: p.

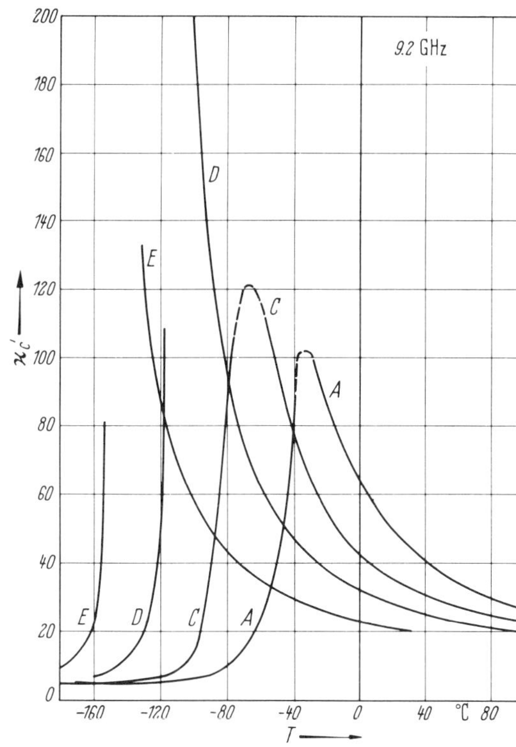

Fig. 299. $KH_{2(1-x)}D_{2x}PO_4$. $\varkappa_c'$ vs. T at 9.2 GHz [65 K 2]. Curve A: x = 0.96, C: x = 0.66, D: x = 0.35, E: x = 0.

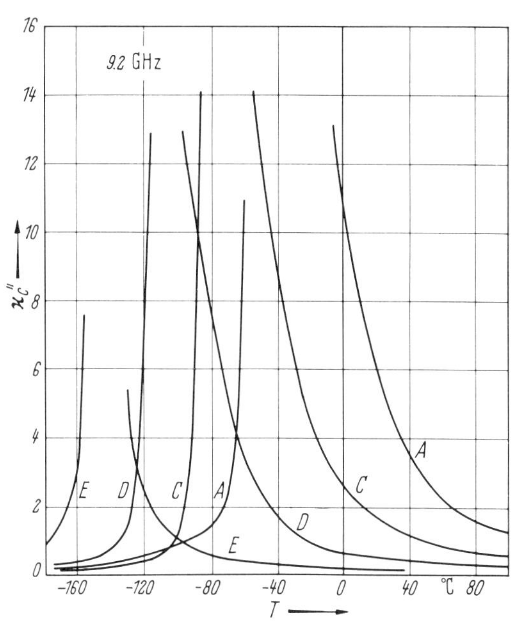

Fig. 300. $KH_{2(1-x)}D_{2x}PO_4$. $\varkappa_c''$ vs. T at 9.2 GHz [65 K 2]. Curve A: x = 0.96, C: x = 0.66, D: x = 0.35, E: x = 0.

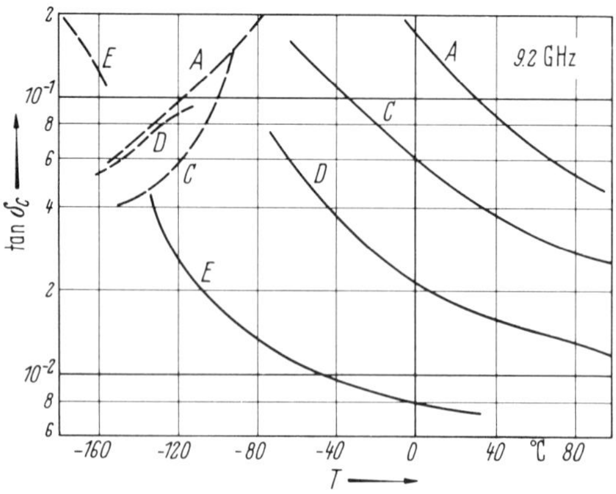

Fig. 301. $KH_{2(1-x)}D_{2x}PO_4$. $\tan\delta_c$ vs. T at 9.2 GHz [65K2]. Curve A: x = 0.96, C: x = 0.66, D: x = 0.35, E: x = 0.

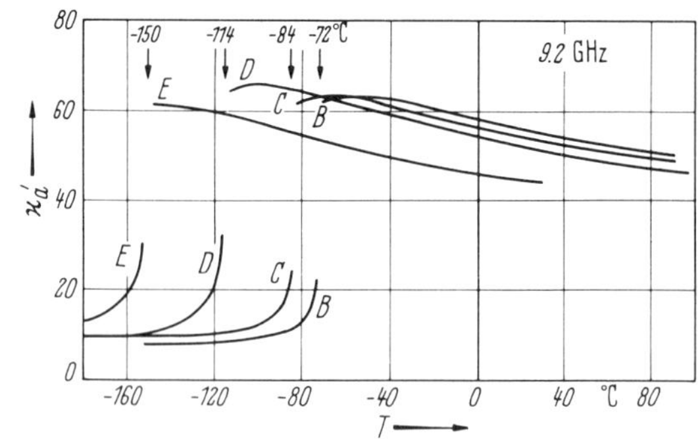

Fig. 302. $KH_{2(1-x)}D_{2x}PO_4$. $\varkappa_a'$ vs. T at 9.2 GHz [65K2]. Curve B: x = 0.68, C: x = 0.66, D: x = 0.35, E: x = 0.

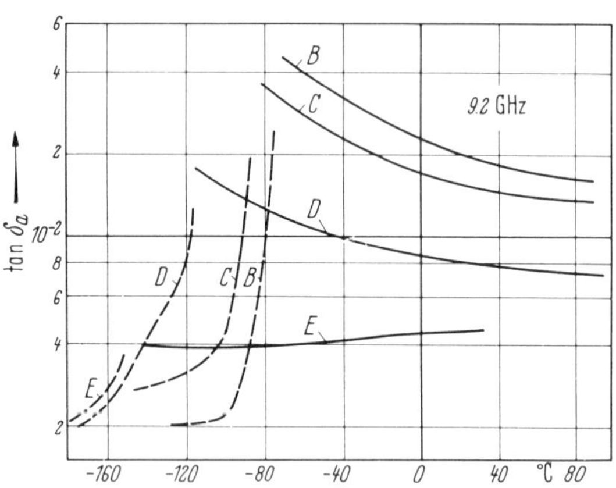

Fig. 303. $KH_{2(1-x)}D_{2x}PO_4$. $\tan\delta_a$ vs. T at 9.2 GHz [65K2]. Curve B: x = 0.68, C: x = 0.66, D: x = 0.35, E: x = 0.

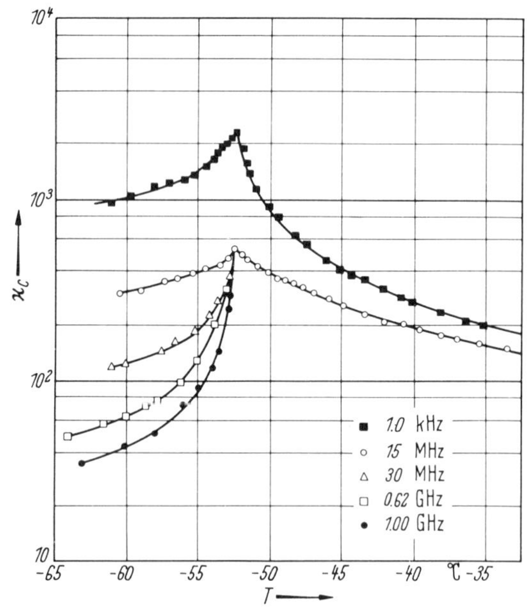

Fig. 304. KD_2PO_4(DKDP). $\varkappa_c$ vs. T [63H6]. Parameter: f.

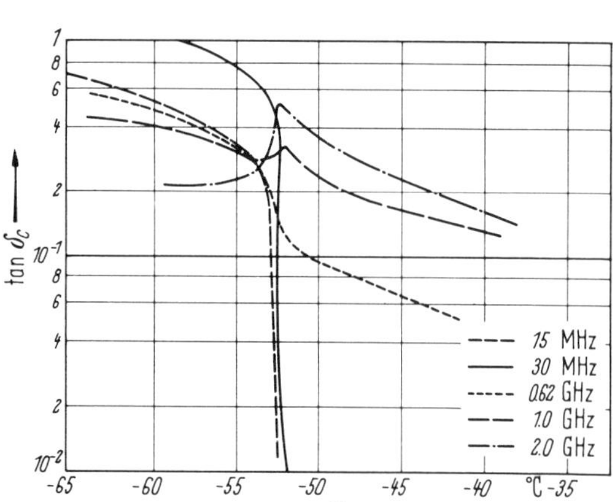

Fig. 305. KD_2PO_4(DKDP). $\tan\delta_c$ vs. T[63H6]. Parameter: f.

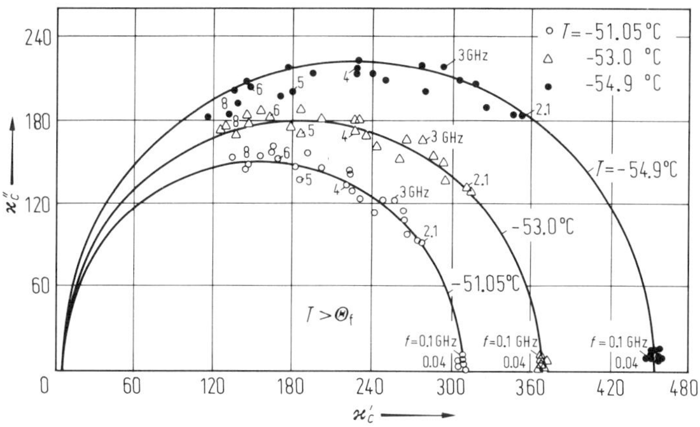

Fig. 306. KD_2PO_4(DKDP). Cole-Cole plot of the complex dielectric constant in the paraelectric phase [76L9]. $\Theta_f = -56.7$ °C.

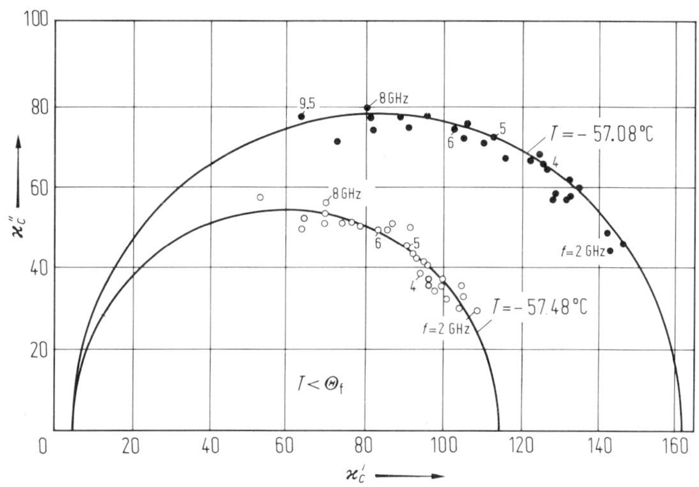

Fig. 307. KD_2PO_4(DKDP). Cole-Cole plot of the complex dielectric constant in the ferroelectric phase [76L9]. $\Theta_f = -56.7$ °C.

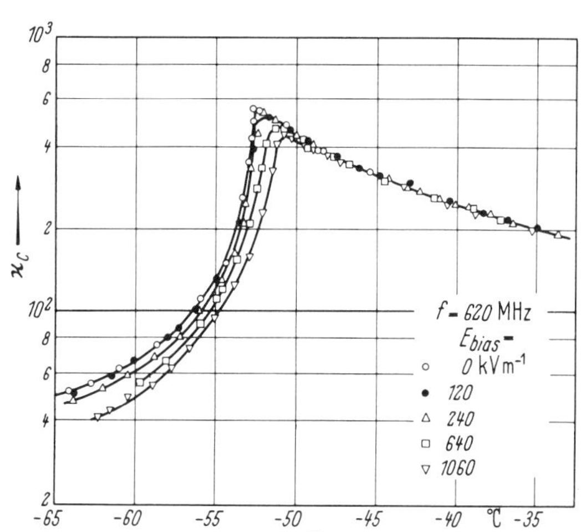

Fig. 308. KD_2PO_4(DKDP). $\varkappa_c$ vs. T at 620 MHz [63H6]. Parameter: E_{bias}.

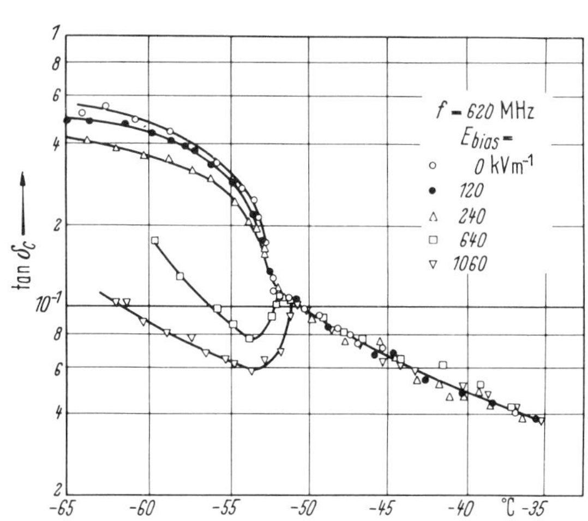

Fig. 309. KD_2PO_4(DKDP). $\tan\delta_c$ vs. T at 620 MHz [63H6]. Parameter: E_{bias}.

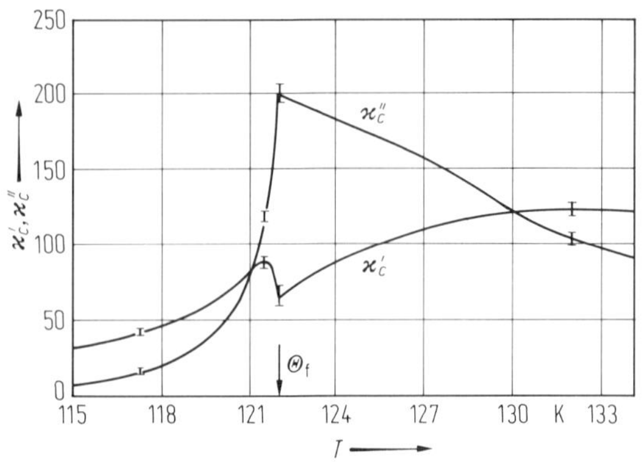

Fig. 310. KH_2PO_4(KDP). $\varkappa_c'$ and $\varkappa_c''$ vs. T at 139 GHz [75G5].

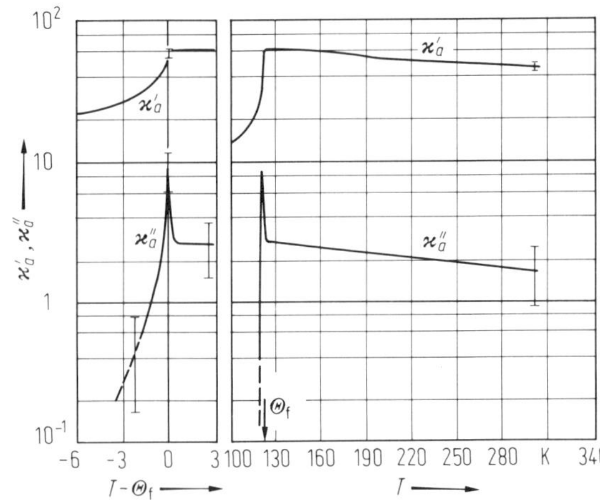

Fig. 311. KH_2PO_4(KDP). $\varkappa_a'$ and $\varkappa_a''$ vs. T at 139 GHz [75G5].

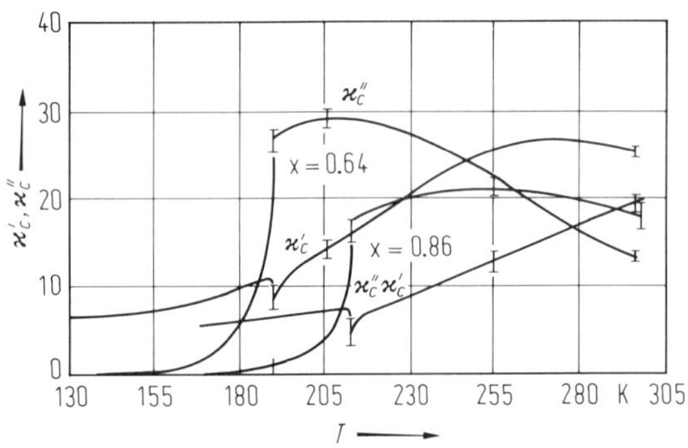

Fig. 312. $KH_{2(1-x)}D_{2x}PO_4$. $\varkappa_c'$ and $\varkappa_c''$ vs. T at 139 GHz [76G4]. Parameter: x.

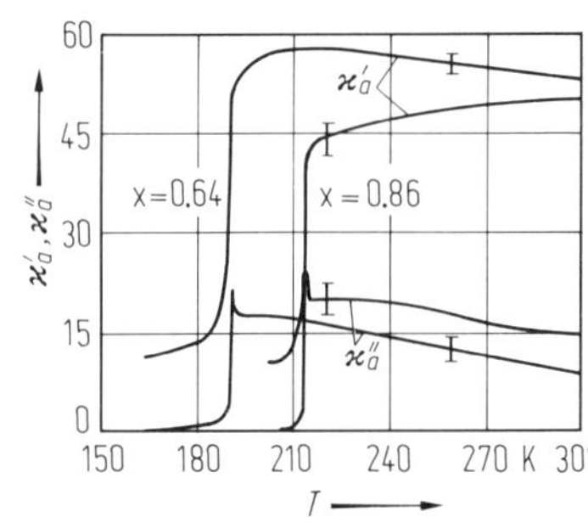

Fig. 313. $KH_{2(1-x)}D_{2x}PO_4$. $\varkappa_a'$ and $\varkappa_a''$ vs. T at 139 GHz [76G4]. Parameter: x.

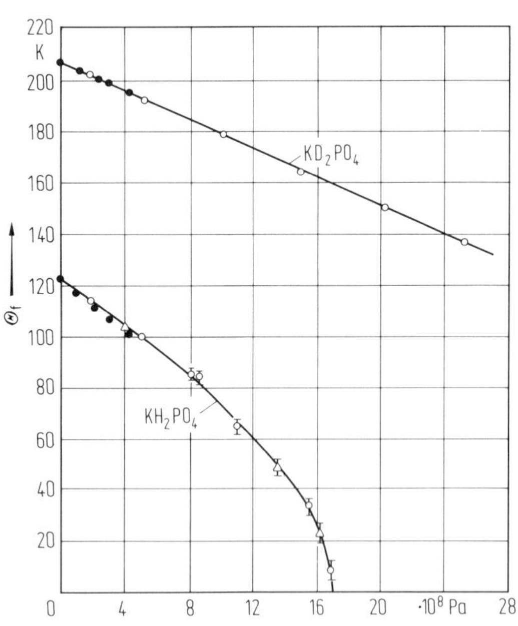

Fig. 314. KH_2PO_4(KDP) and KD_2PO_4(DKDP). Θ_f vs. p [74S6]. Θ_f: ferroelectric transition temperature. Different figures correspond to different samples and high pressure cells.

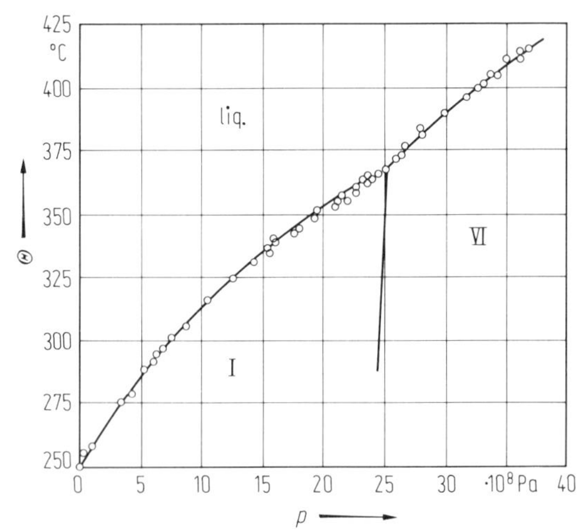

Fig. 315. KD_2PO_4(DKDP). Θ vs. p [78R3]. See also [70R3].

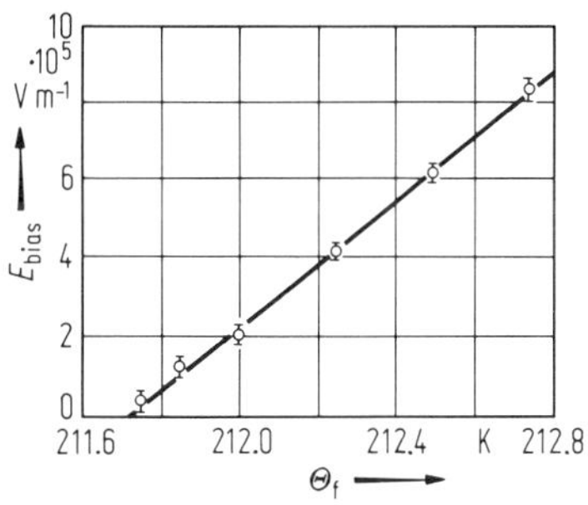

Fig. 316. KD_2PO_4(DKDP). Θ_f vs. E_{bias} [71 G 5].

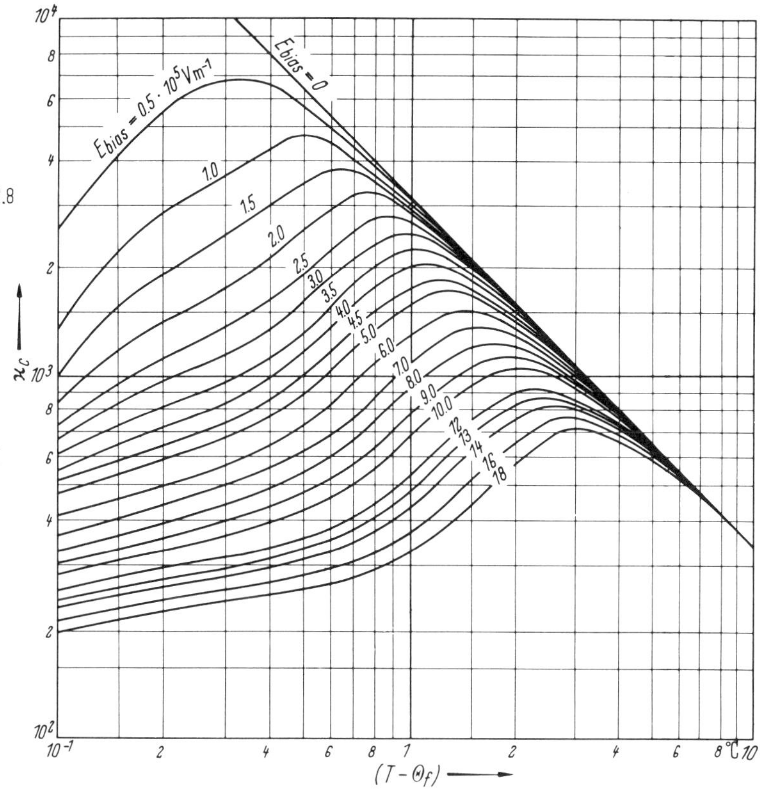

Fig. 318. KH_2PO_4(KDP). $\varkappa_c$ vs. $T-\Theta_f$ [50 B 2]. Parameter: E_{bias}. $f = 1$ kHz.

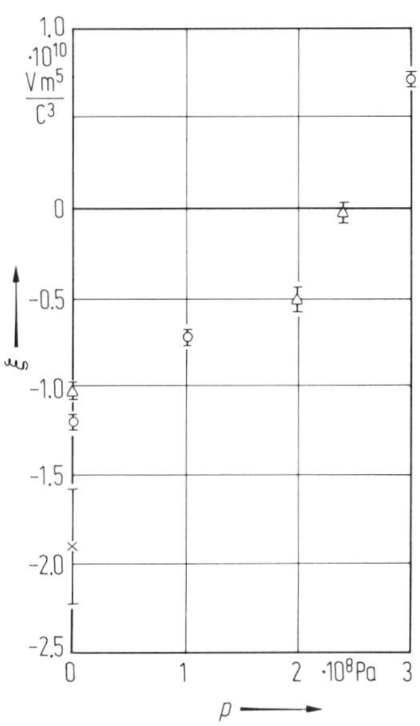

Fig. 317. KH_2PO_4(KDP). ξ vs. p [78 W 3]. Different figures correspond to different samples.

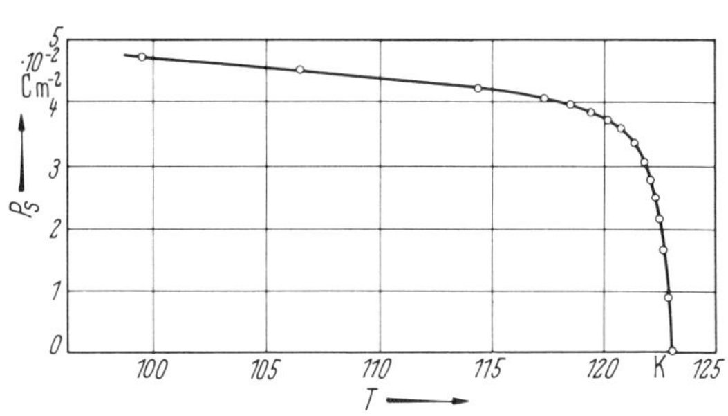

Fig. 319. KH_2PO_4(KDP). P_s vs. T [44 V 1].

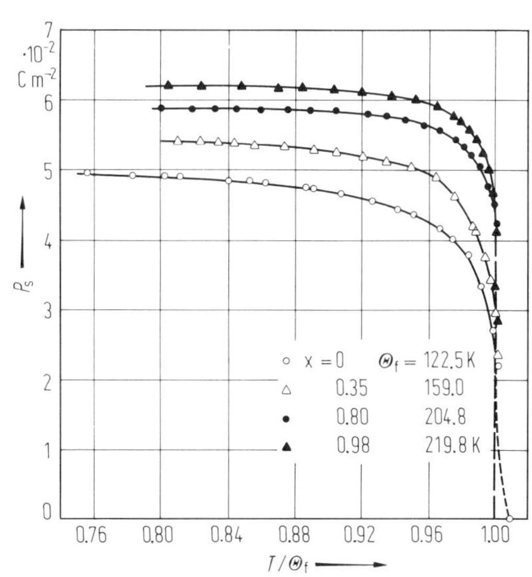

Fig. 320. $KH_{2(1-x)}D_{2x}PO_4$. P_S vs. T/Θ_f [73 S 1]. Parameter: x.

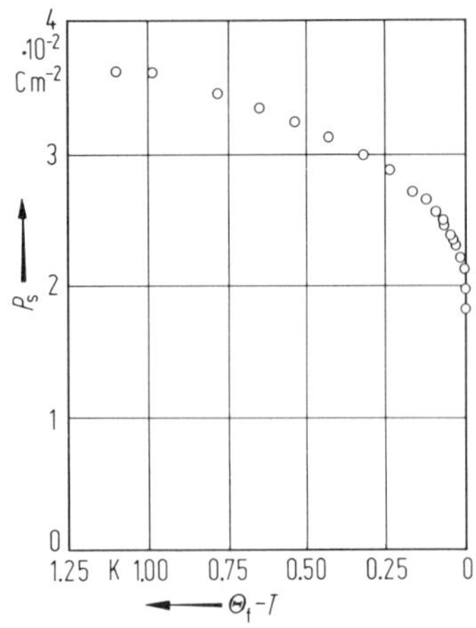

Fig. 321. KH_2PO_4(KDP). P_s vs. $(\Theta_f - T)$ [71 B 7].

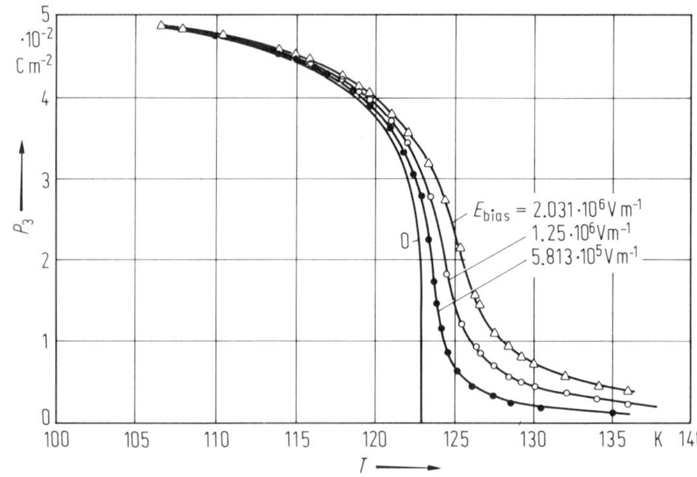

Fig. 324. KH_2PO_4(KDP). P_3 vs. T [77 C 1]. Parameter: E_{bias}. P_3: polarization along c-axis.

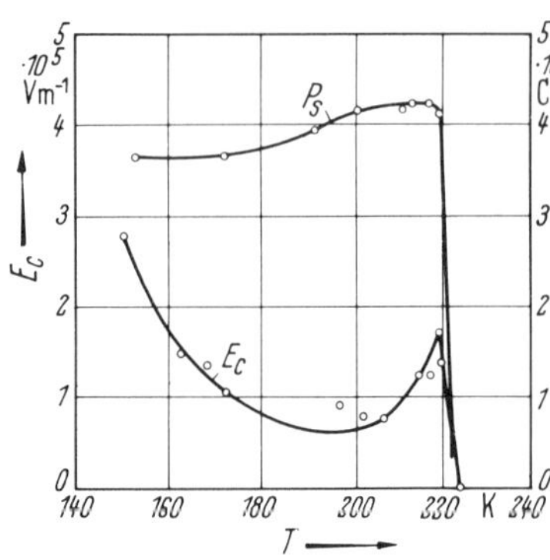

Fig. 322. Deuterated KDP. P_s and E_c vs. T [63 S 8].

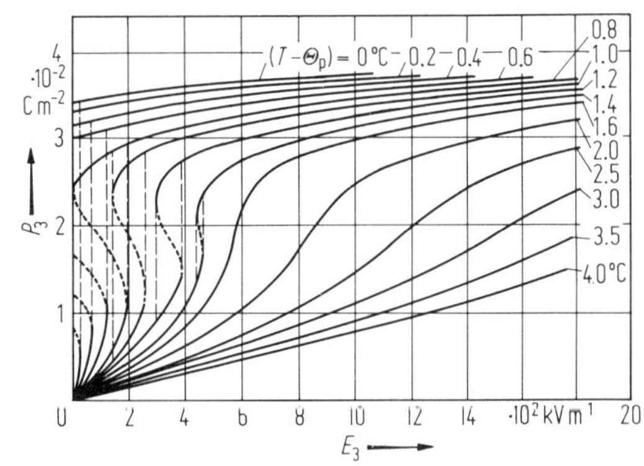

Fig. 325. KH_2PO_4(KDP). P_3 vs. E_3 [71 K 11]. Parameter: T-Θ_p. P_3: polarization along c-axis.

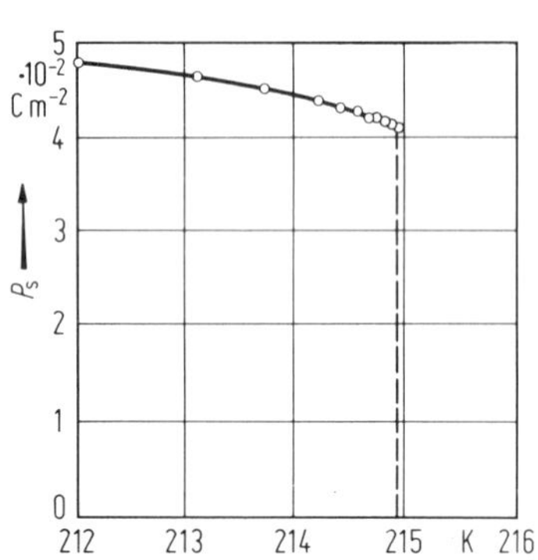

Fig. 323. KD_2PO_4(DKDP). P_s vs. T in the vicinity of Θ_f [71 S 31].

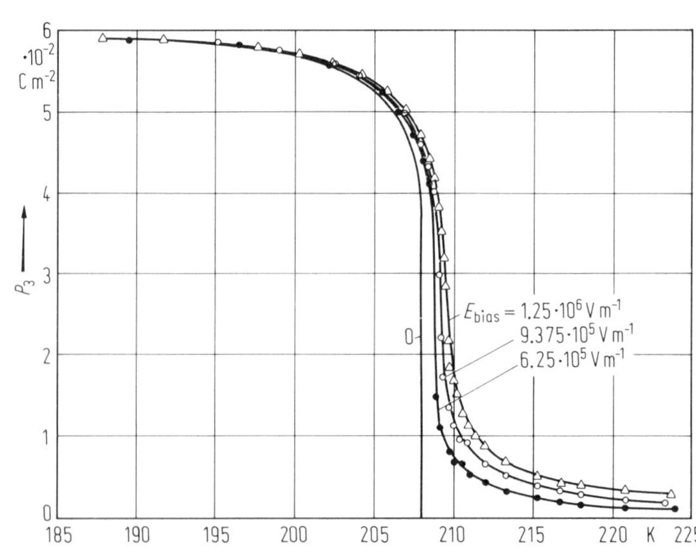

Fig. 326. KD_2PO_4(DKDP). P_3 vs. T [77 C 1]. Parameter: E_{bias}. Degree of deuteration is about 70%.

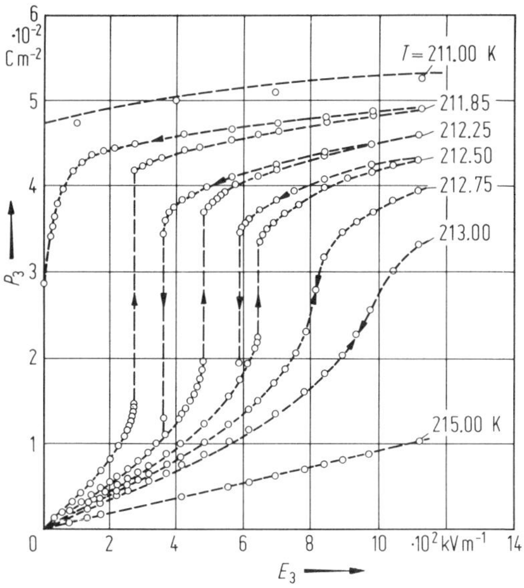

Fig. 327. KD_2PO_4(DKDP). P_3 vs. E_3 [72S23]. Parameter: T. Degree of deuteration is about 85%.

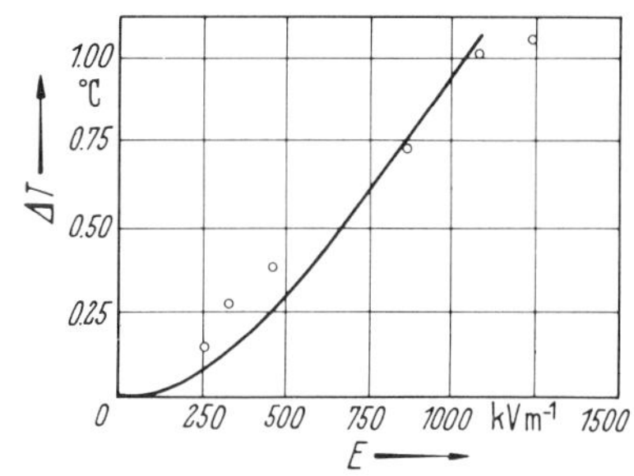

Fig. 328. KH_2PO_4(KDP). ΔT vs. E [50B2]. ΔT: electrocaloric temperature change.

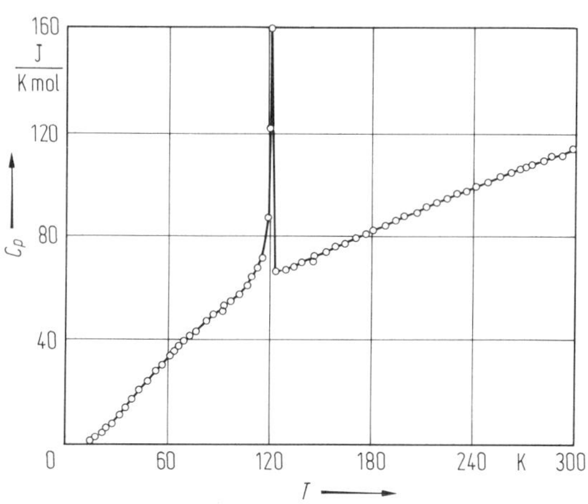

Fig. 329. KH_2PO_4(KDP). C_p vs. T [44S3]. C_p: molar heat capacity at constant pressure.

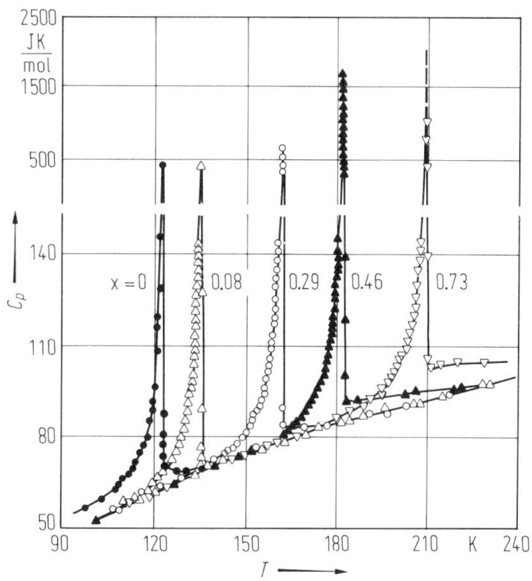

Fig. 330. $KH_{2(1-x)}D_{2x}PO_4$. C_p vs. T [72S28]. Parameter: x.

Figs. 331, 332, see p. 386.

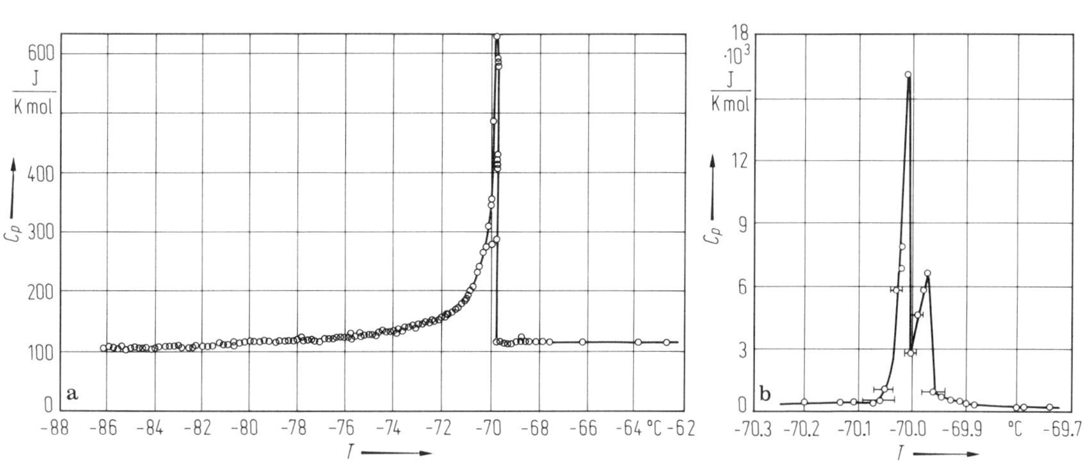

Fig. 333. KD_2PO_4(DKDP). C_p vs. T [67S26].

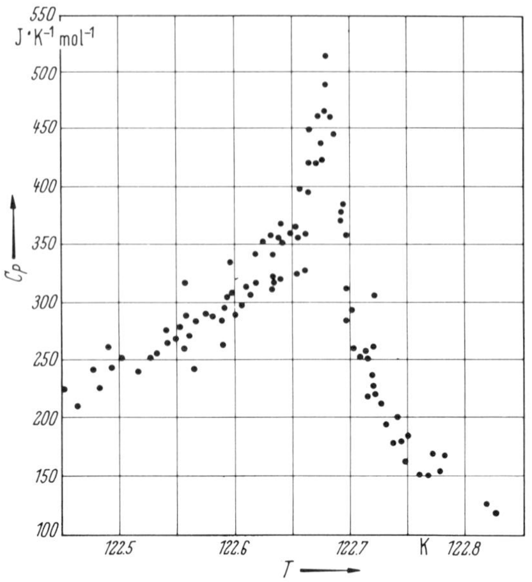

Fig. 331. KH_2PO_4(KDP). C_p vs. T in the vicinity of Θ_f [67R1]. C_p: molar heat capacity at constant pressure.

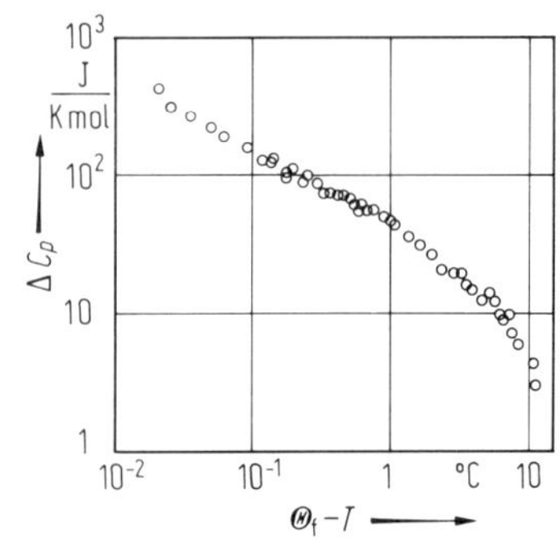

Fig. 332. KH_2PO_4(KDP). ΔC_p vs. $\Theta_f - T$ [68S21]. ΔC_p: anomalous part of C_p.

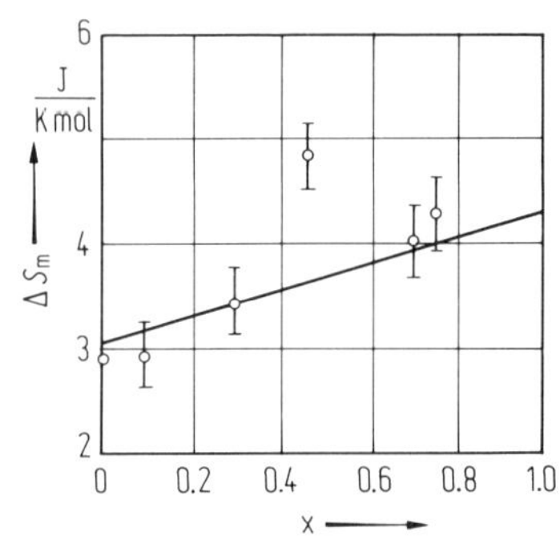

Fig. 335. $KH_{2(1-x)}D_{2x}PO_4$. ΔS_m vs. x [72S28]. ΔS_m: transition entropy.

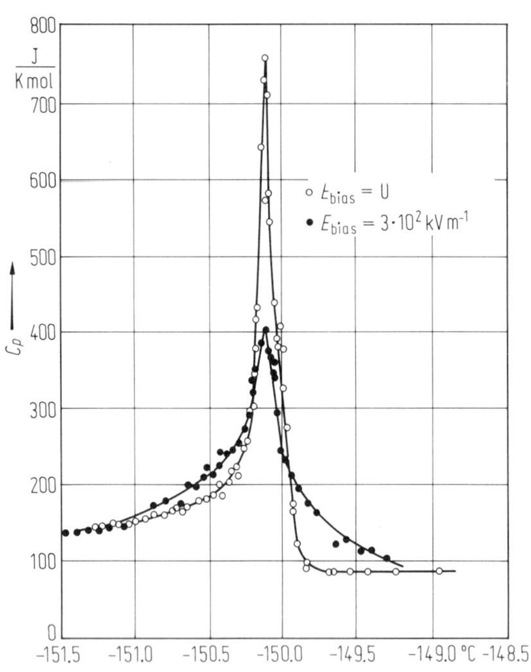

Fig. 334. KH_2PO_4(KDP). C_p vs. T under E_{bias} [68S21].

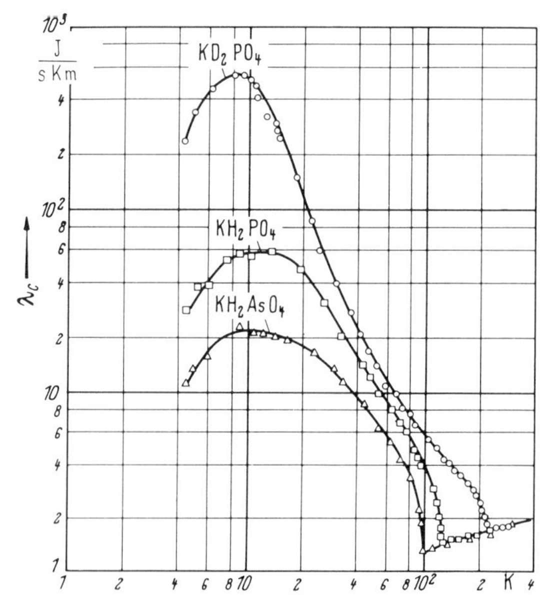

Fig. 336. KH_2PO_4(KDP), KD_2PO_4(DKDP) and KH_2AsO_4 (KDA). λ_c vs. T [67S27]. λ_c: thermal conductivity along c axis.

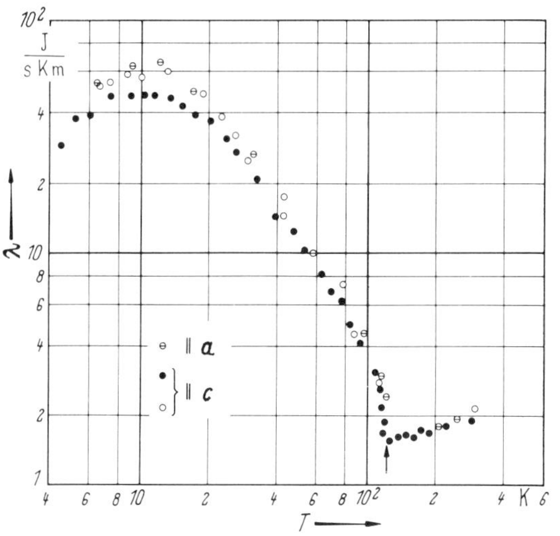

Fig. 337. KH_2PO_4(KDP). λ vs. T [67S27]. λ: thermal conductivity. Open and full circles correspond to samples of different sizes. Samples are from different crystals.

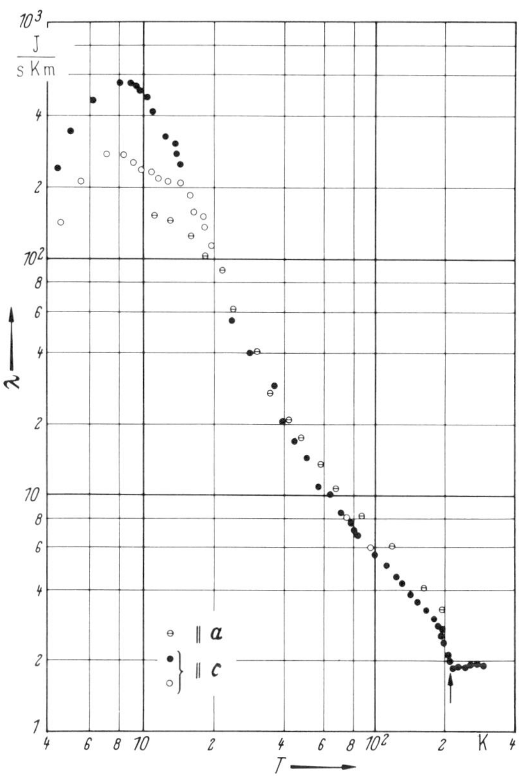

Fig. 338. KD_2PO_4(DKDP). λ vs. T [67S27]. λ: thermal conductivity. Samples are from the same crystal. See also caption of Fig. 337.

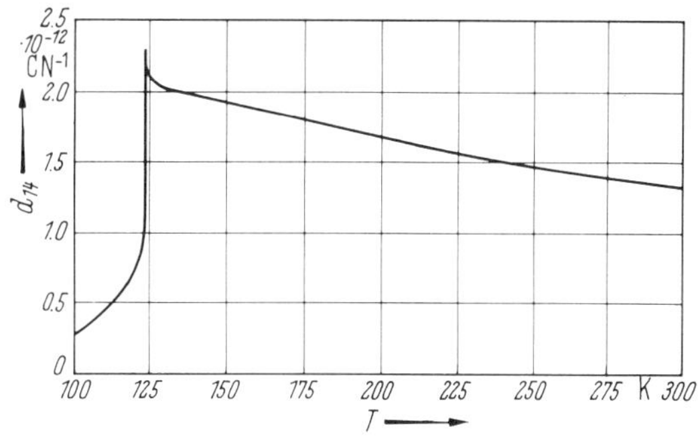

Fig. 339. KH_2PO_4(KDP). d_{14} vs. T [62j1].

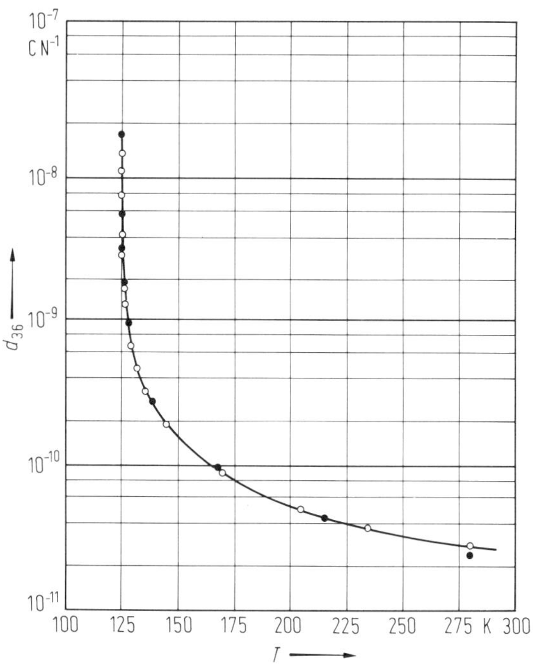

Fig. 340. KH_2PO_4(KDP). d_{36} vs. T. Measurements of the direct effect (full circles) [43B2] and of the inverse effect (open circles) [44V1].

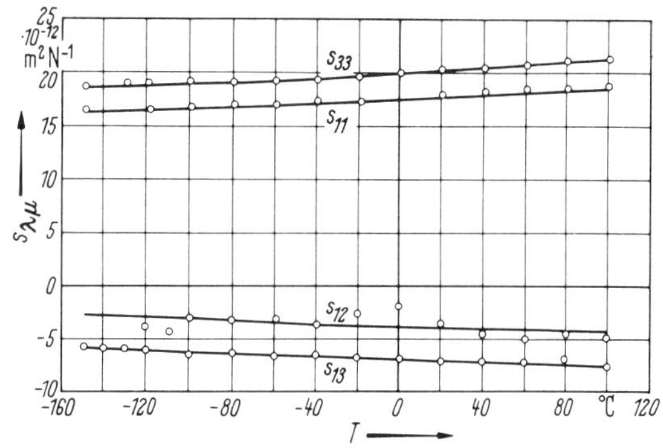

Fig. 341. KH_2PO_4(KDP). $s_{\lambda\mu}$ vs. T [46M1]. $s_{\lambda\mu}$: elastic compliance.

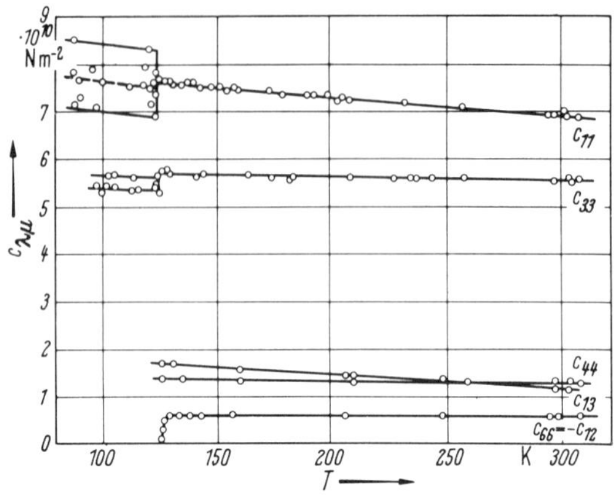

Fig. 342. KH_2PO_4(KDP). $c_{\lambda\mu}$ vs. T [46Z1]. $c_{\lambda\mu}$: elastic stiffness.

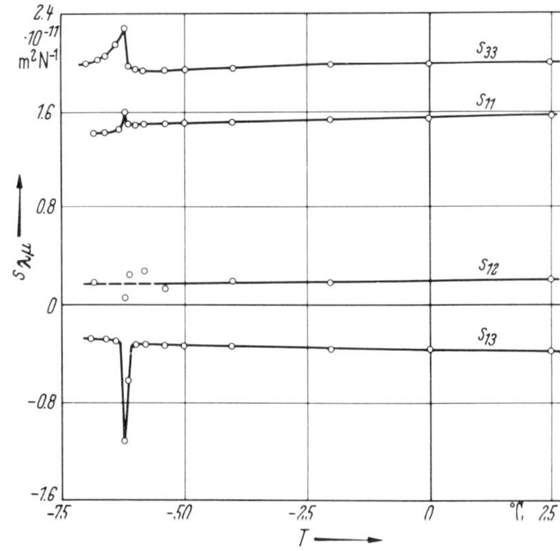

Fig. 343. KD_2PO_4(DKDP). $s_{\lambda\mu}$ vs. T [66S17]. $s_{\lambda\mu}$: elastic compliance.

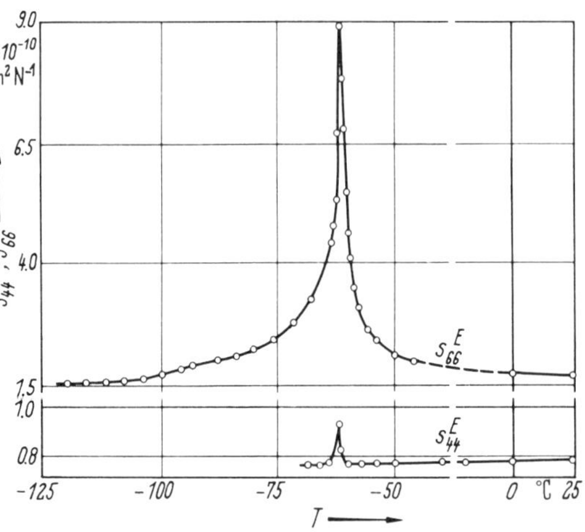

Fig. 344. KD_2PO_4(DKDP). s^E_{44} and s^E_{66} vs. T [66S17].

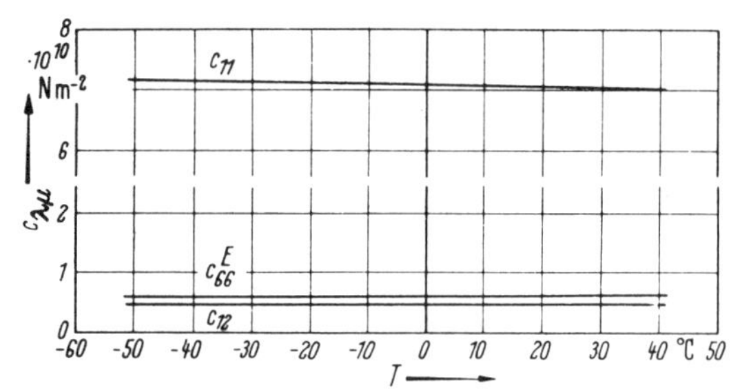

Fig. 345. KD_2PO_4(DKDP). $c_{\lambda\mu}$ vs. T [50J1]. $c_{\lambda\mu}$: elastic stiffness.

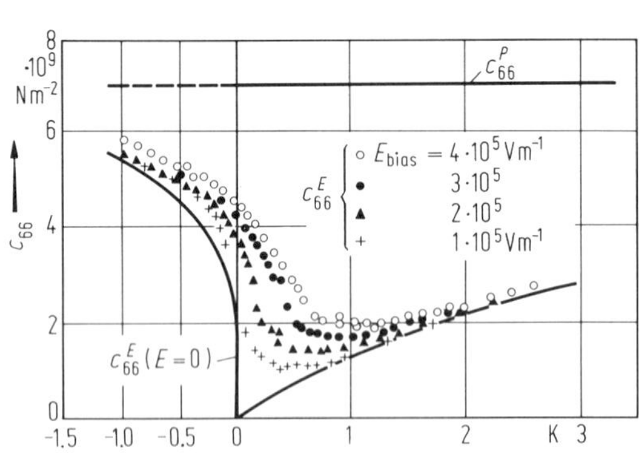

Fig. 346. KH_2PO_4(KDP). c^E_{66} and c^P_{66} vs. T-Θ_f [70L9]. Parameter: E_{bias}.

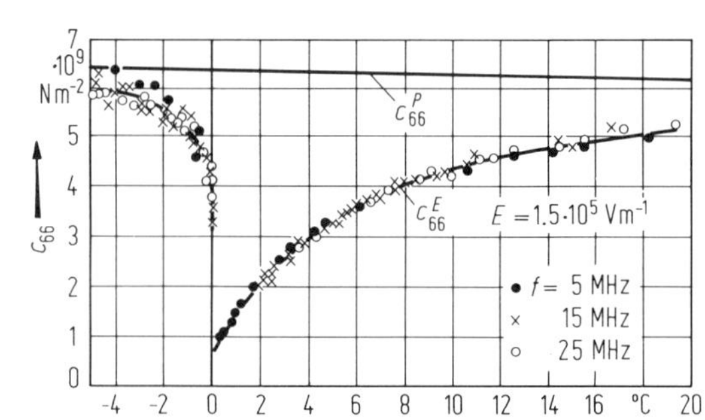

Fig. 347. KD_2PO_4(DKDP). c^E_{66} and c^P_{66} vs. T-Θ_f [70L10].

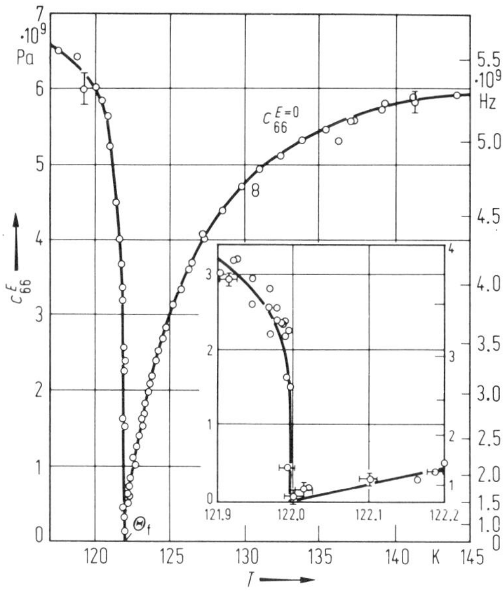

Fig. 348. KH_2PO_4(KDP). c_{66}^E vs. T, and $\Delta\nu$ vs. T [68B14]. $\Delta\nu$: Brillouin shift.

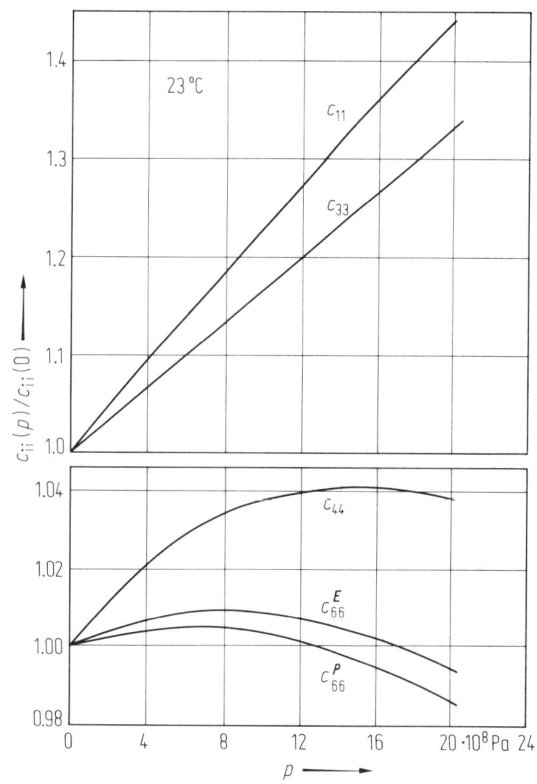

Fig. 349. KH_2PO_4(KDP). $c_{ii}(p)/c_{ii}(0)$ vs. p [76F12]. $c_{ii}(p)$: elastic stiffness under hydrostatic pressure p.

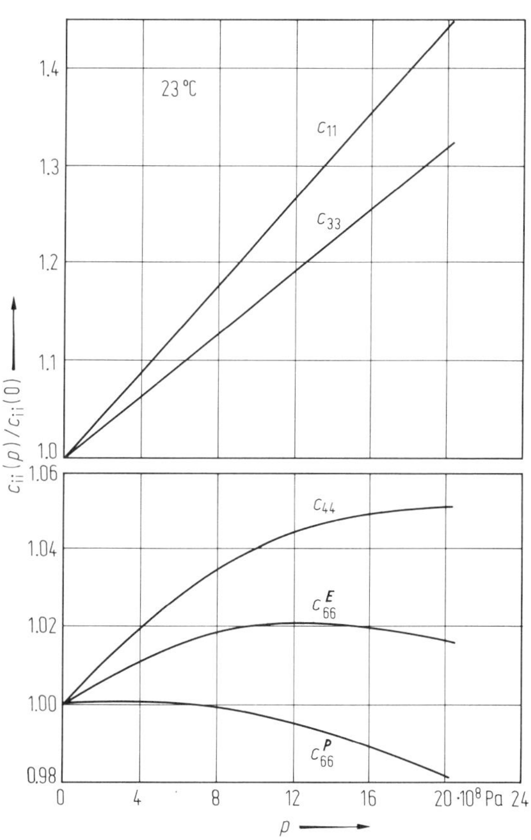

Fig. 350. KD_2PO_4(DKDP). $c_{ii}(p)/c_{ii}(0)$ vs. p [76F12]. $c_{ii}(p)$: elastic stiffness under hydrostatic pressure p.

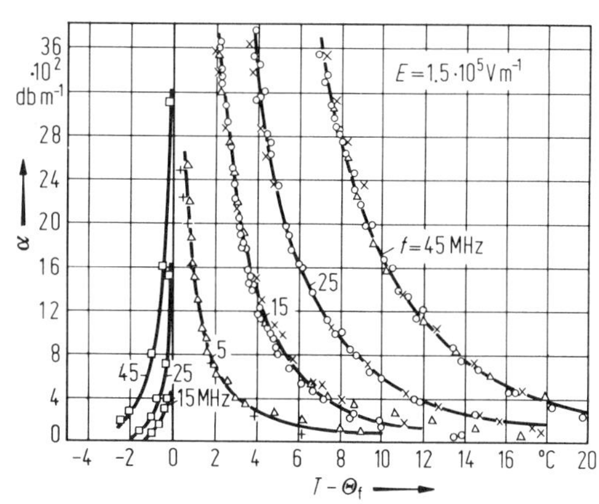

Fig. 351. KD_2PO_4(DKDP). α vs. T–Θ_f at various frequencies [70L10]. α: attenuation coefficient of transverse ultrasonic wave propagations along the *a* axis with its polarization along the *b* axis.

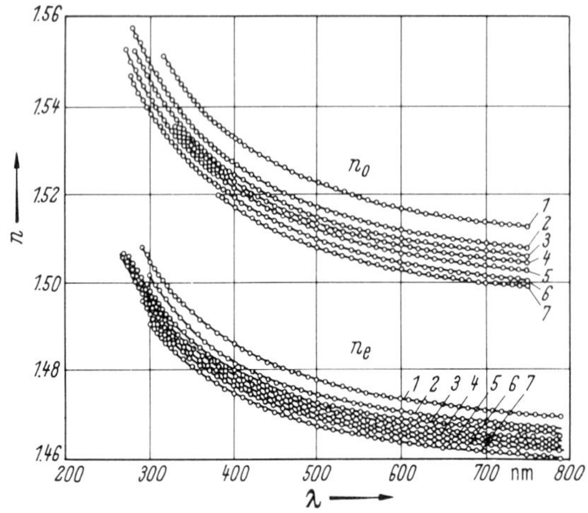

Fig. 352. KH_2PO_4(KDP). n_0 and n_e vs. λ [66 V 5]. Parameter: T[°C]. Curve *1*: −196, *2*: −50, *3*: +20, *4*: +64, *5*: +100, *6*: +150, *7*: +200.

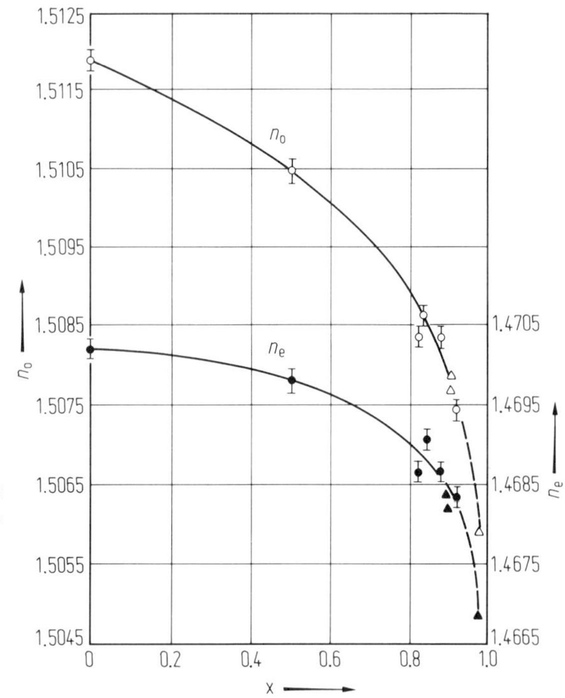

Fig. 353. $KH_{2(1-x)}D_{2x}PO_4$. n_o, n_e vs. x [75 V 8]. $\lambda = 546.1$ nm. Open and full triangles are from literature.

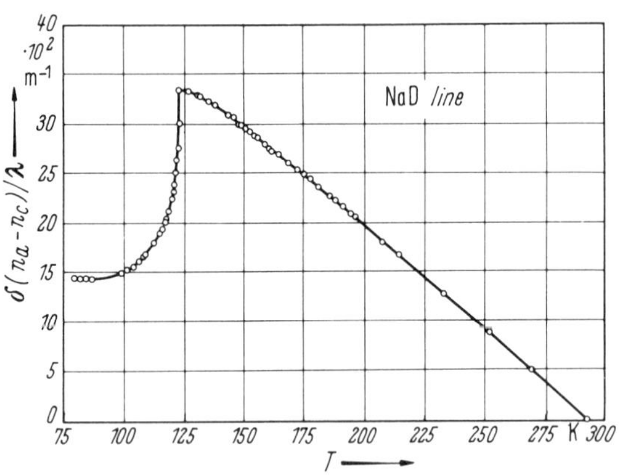

Fig. 354. KH_2PO_4(KDP). $\delta(n_a - n_c)/\lambda$ vs. T [44 Z 1].

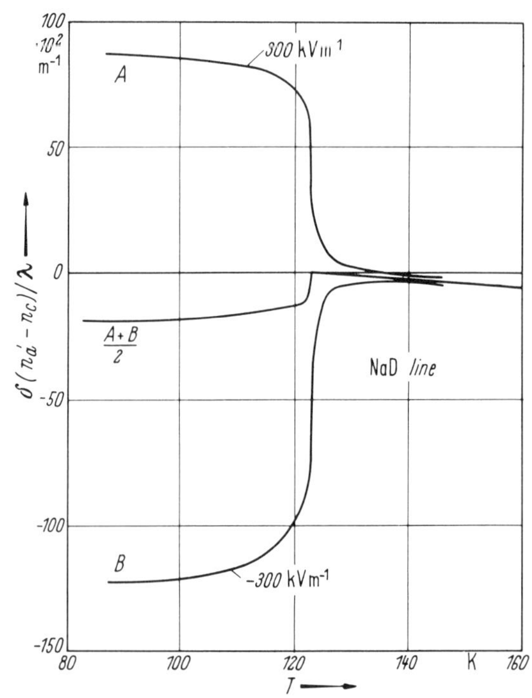

Fig. 355. KH_2PO_4(KDP). $\delta(n_a' - n_c)/\lambda$ vs. T at biasing field $E_3 = 300\ kVm^{-1}$ [44 Z 1]. n_a', n_c: refractive index along [110], [001], E_3: field $\|c$.

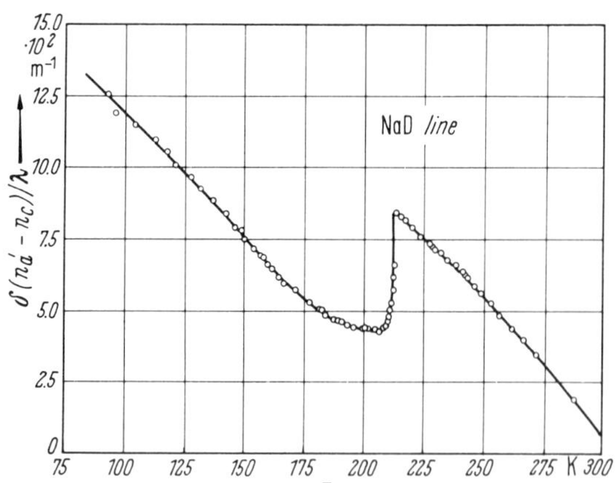

Fig. 356. KD_2PO_4(DKDP). $\delta(n_a' - n_c)/\lambda$ vs. T [44 Z 1]. n_a', n_c: refractive index along [110], [001].

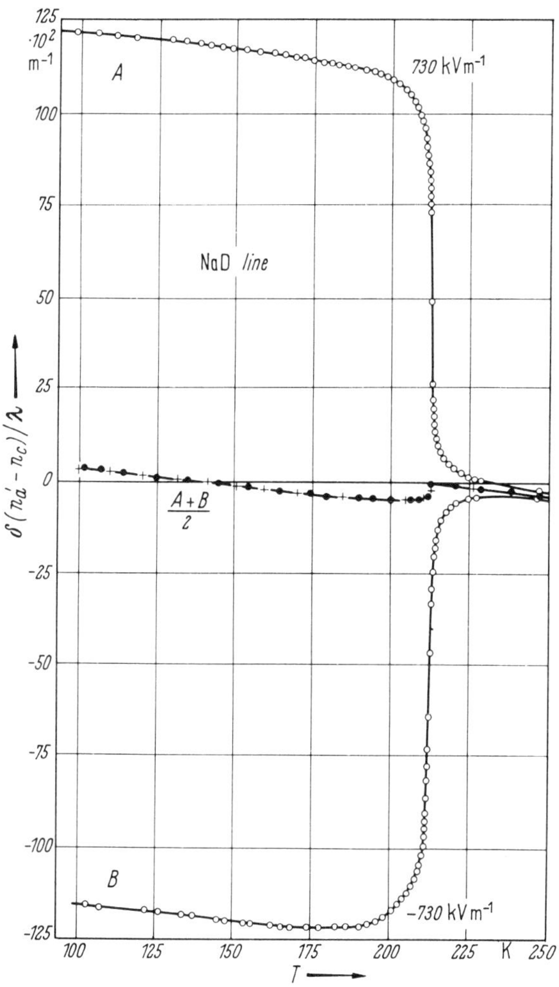

Fig. 357. KD_2PO_4(DKDP). $\delta(n_a'-n_c)/\lambda$ vs. T at biasing field $E_3=730\ kVm^{-1}$ [44Z1]. n_a', n_c: refractive index along [110], [001]. E_3: field‖c.

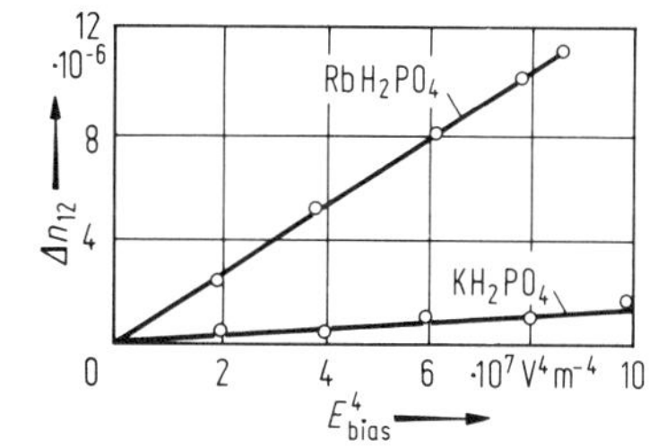

Fig. 358. KH_2PO_4(KDP), RbH_2PO_4(RDP). Induced birefringence Δn_{12} vs. E^4_{bias} [69P3]. $\Delta n_{12}=|n_1-n_2|$. $\lambda=540$ nm, $T=25\ °C$.

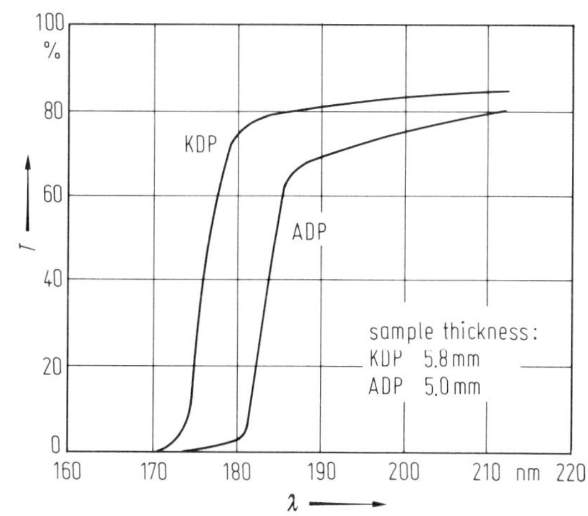

Fig. 359. KH_2PO_4(KDP), $NH_4H_2PO_4$(ADP). T vs. λ [77S31]. T: Transmission in vacuum ultraviolet region. The data include losses from surface reflection.

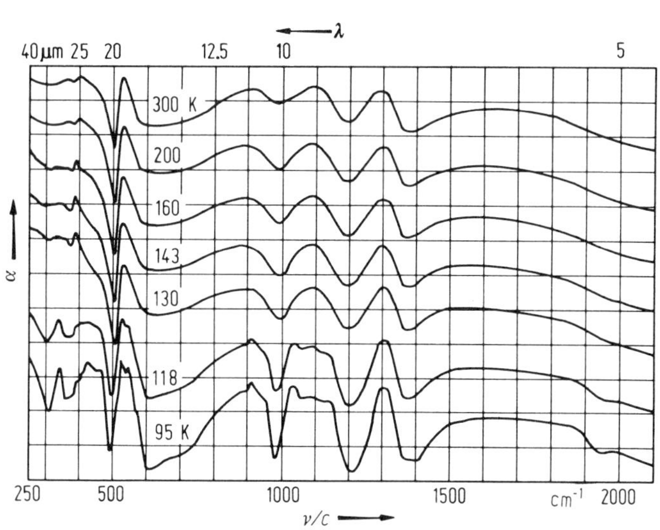

Fig. 360. KH_2PO_4(KDP). α vs. ν/c [70W5]. α: infrared absorption. Parameter: T.

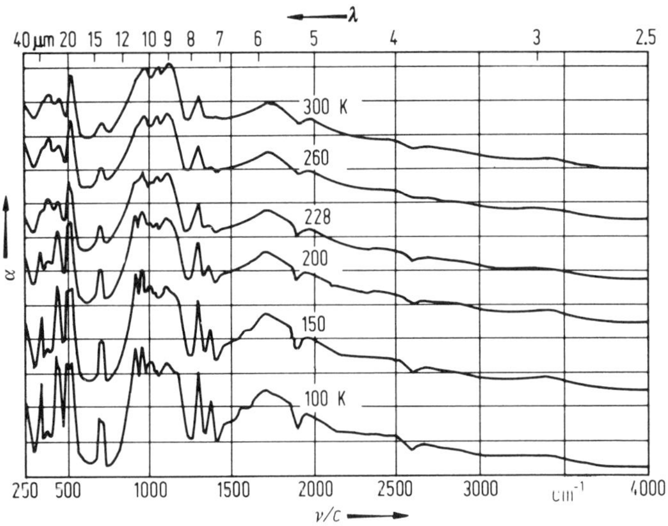

Fig. 361. $K(H_{0.1}D_{0.9})_2PO_4$. α vs. ν/c [70W5]. α: infrared absorption. Parameter: T.

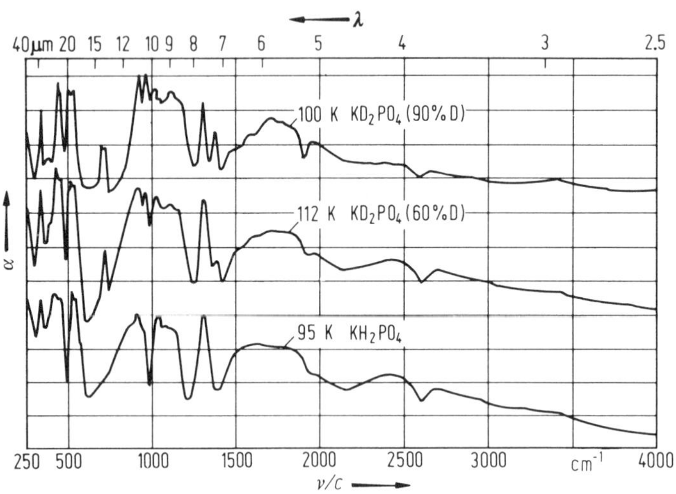

Fig. 362. $KH_{2(1-x)}D_{2x}PO_4$. α vs. ν/c in the ferroelectric states [70W5]. α: infrared absorption.

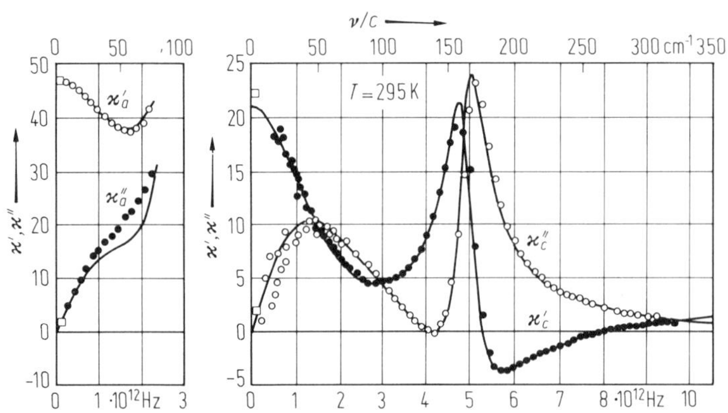

Fig. 365. KH_2PO_4(KDP). ϰ', ϰ'' vs. ν [75G5]. T = 295 K. Squares: millimeter wave points. Solid line: oscillator fit.

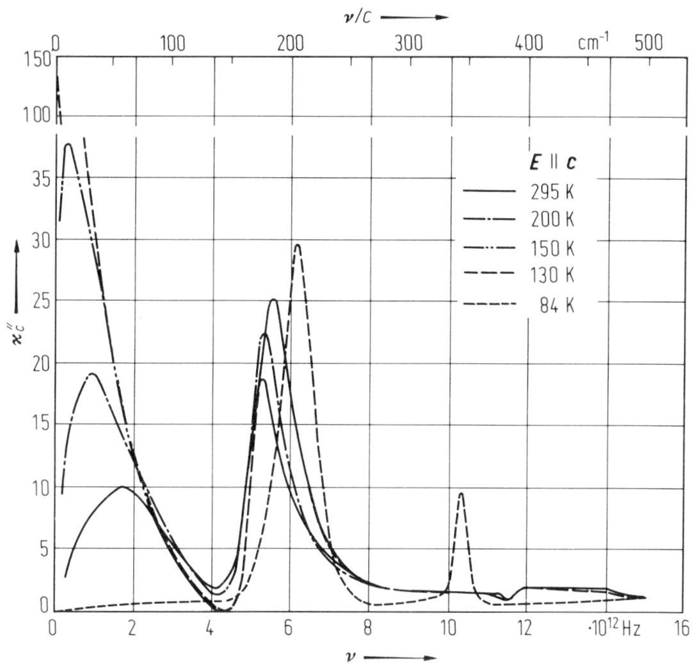

Fig. 363. KH_2PO_4(KDP). $\varkappa_c''$ vs. ν [73K8]. Parameter: *T*. The curves were obtained from reflectivity data using Kramers-Kronig relation.

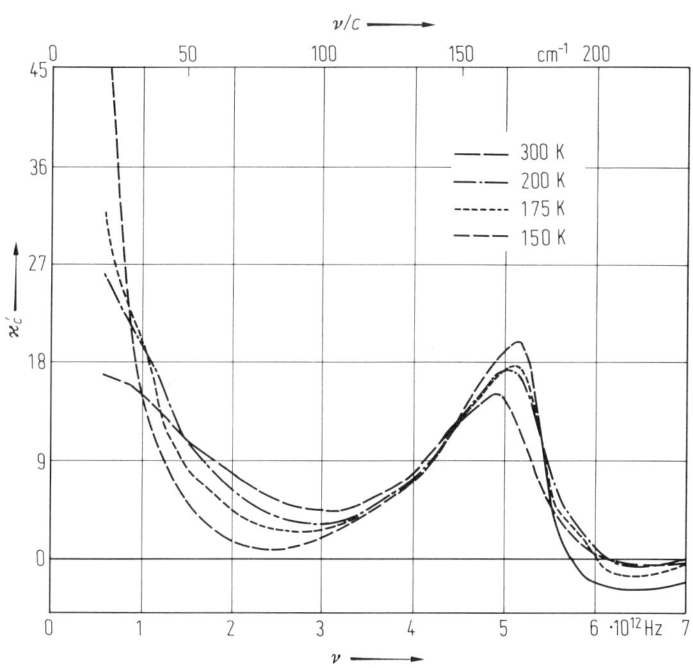

Fig. 364. KH_2PO_4(KDP). $\varkappa_c'$ vs. ν [77L4]. Parameter: *T*.

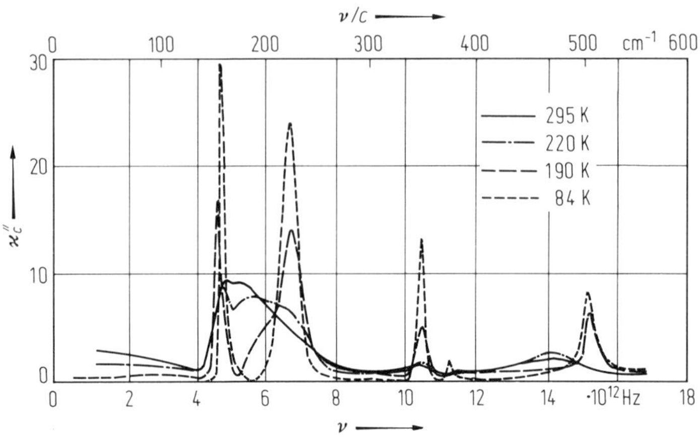

Fig. 366. KD_2PO_4(DKP). $\varkappa_c''$ vs. ν [74K10]. Parameter: T. The curves were obtained from reflectivity data using Kramers-Kronig relation.

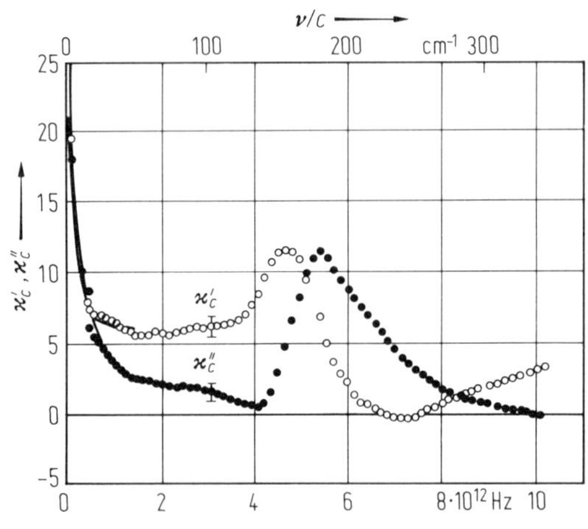

Fig. 367. $KH_{2(1-x)}D_{2x}PO_4$. $\varkappa_c'$, $\varkappa_c''$ vs. ν at 295 K [77H2]. x = 0.86.

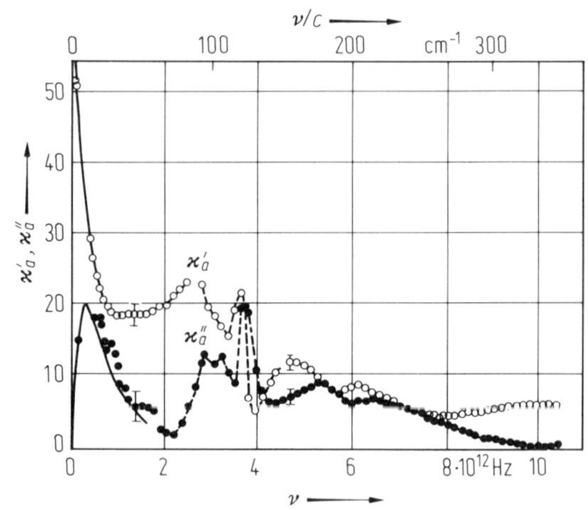

Fig. 368. $KH_{2(1-x)}D_{2x}PO_4$. $\varkappa_a'$, $\varkappa_a''$ vs. ν at 295 K [77H2]. x = 0.86.

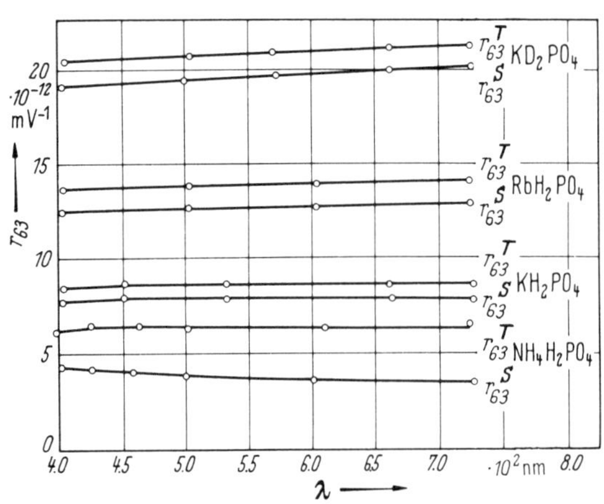

Fig. 369. KH_2PO_4(KDP), KD_2PO_4(DKDP), RbH_2PO_4 (RDP), $NH_4H_2PO_4$(ADP). r_{63} vs. λ [66V2].

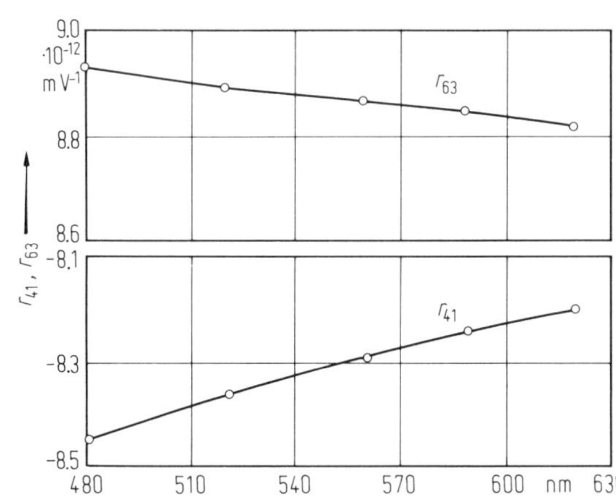

Fig. 370. KH_2PO_4(KDP). r_{41}, r_{63} vs. λ [78V1].

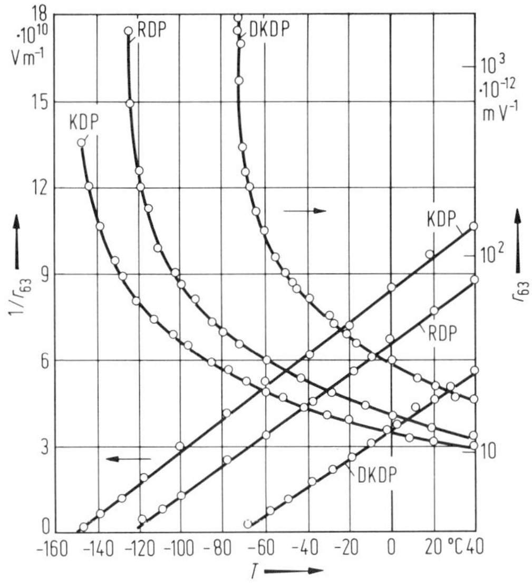

Fig. 371. KH_2PO_4(KDP), KD_2PO_4(DKDP), RbH_2PO_4 (RDP). r_{63}, $1/r_{63}$ vs. T[71V1]. r_{63}: electrooptic constant for $\boldsymbol{E}$.

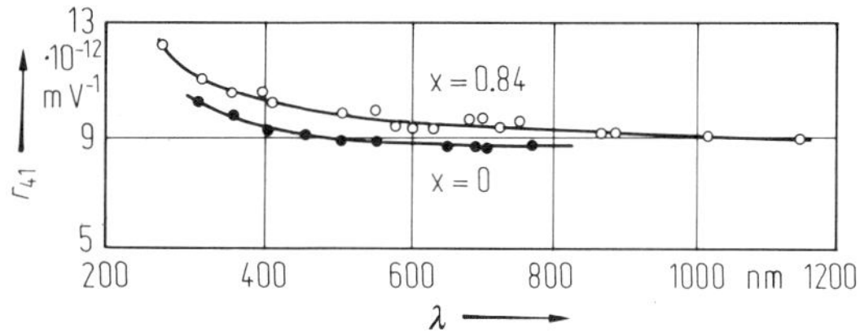

Fig. 372. $KH_{2(1-x)}D_{2x}PO_4$. r_{41} vs. λ [75V8].

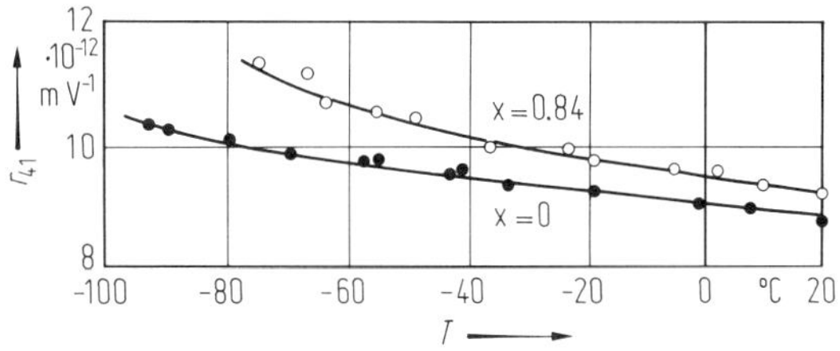

Fig. 373. $KH_{2(1-x)}D_{2x}PO_4$. r_{41} vs. T [75V8]. $\lambda = 633$ nm.

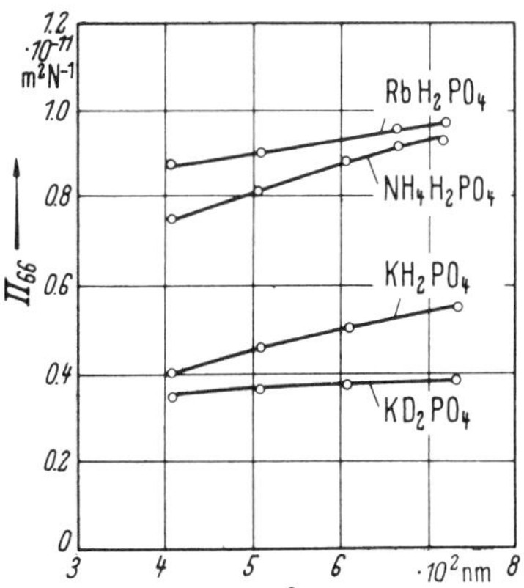

Fig. 374. KH_2PO_4(KDP), KD_2PO_4(DKDP), RbH_2PO_4 (RDP), $NH_4H_2PO_4$(ADP). Π_{66} vs. λ[66V2]. Π_{66}: piezo-optic constant.

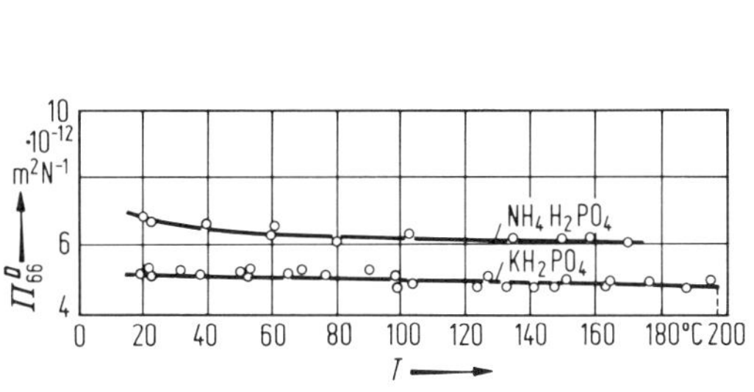

Fig. 375. KH_2PO_4(KDP), $NH_4H_2PO_4$(ADP). Π^D_{66} vs. T [71V3]. Π^D_{66}: piezooptic constant for $\mathbf{T}$ at constant $\boldsymbol{D}$. $\lambda = 450$ nm.

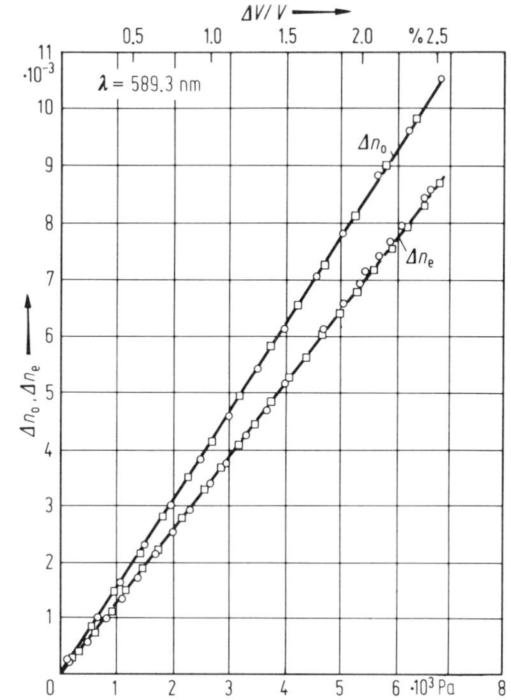

Fig. 376. KH_2PO_4(KDP). Δn_o, Δn_e vs. p [68D2]. $\Delta n_o = n_o(p) - n_o(0)$, $\Delta n_e = n_e(p) - n_e(0)$. $T = 22$ °C.

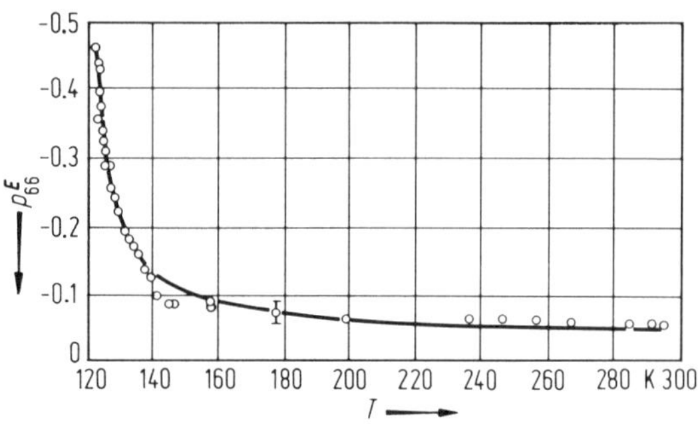

Fig. 377. KH_2PO_4(KDP). p_{66}^E vs. T [69 B 14]. p_{66}^E: piezooptic constant for **S** at constant E, obtained from Brillouin scattering spectra. $\lambda = 633$ nm.

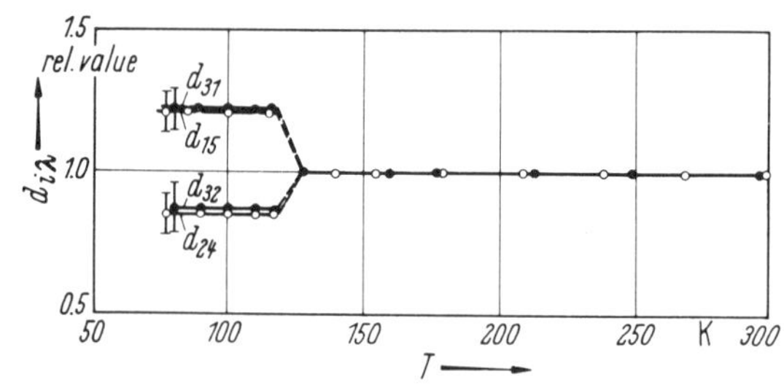

Fig. 378. KH_2PO_4(KDP). $d_{i\lambda}$ vs. T [64 V 3]. $d_{i\lambda}$: nonlinear optical susceptibility for SHG (relative value).

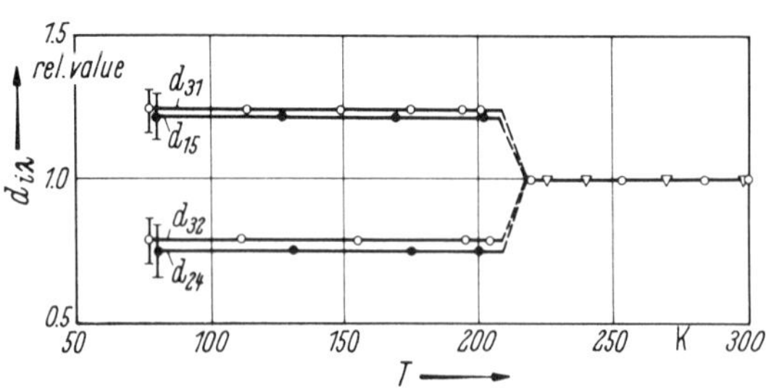

Fig. 379. KD_2PO_4(DKDP). $d_{i\lambda}$ vs. T [64 V 3]. $d_{i\lambda}$: nonlinear optical susceptibility for SHG (relative value).

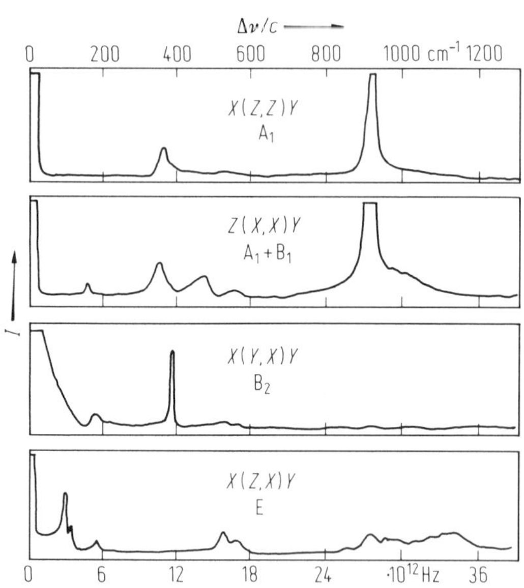

Fig. 380. KH_2PO_4(KDP). I vs. $\Delta\nu$ at RT [70 W 7]. I: Raman scattering intensity. Scattering geometries and the associated symmetry of phonons are given in the figure.

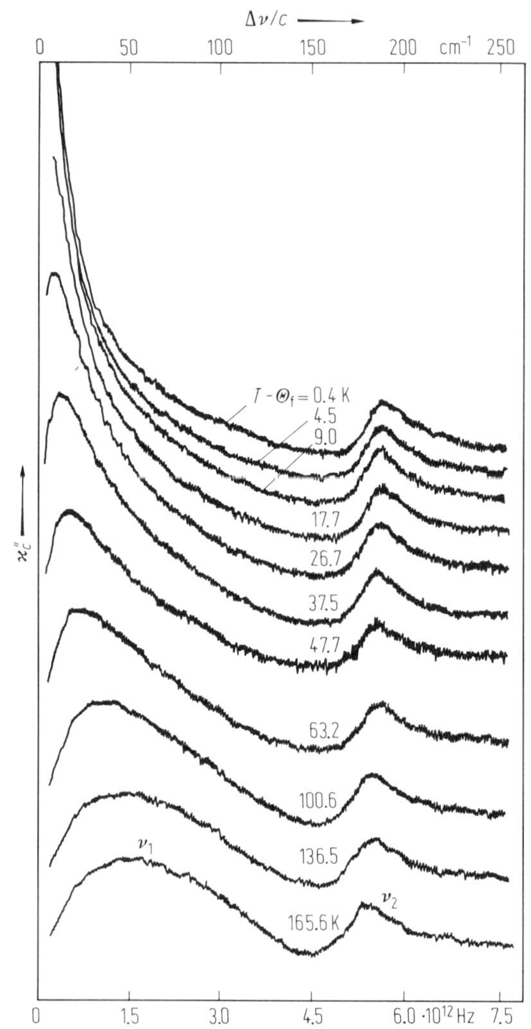

Fig. 381. KH_2PO_4(KDP). $\varkappa_c''$ vs. $\Delta\nu$ [79 T 1]. Parameter: T-Θ_f. $\Delta\nu$: frequency shift of Raman spectrum. $\varkappa_c''$ was obtained by dividing the Raman scattering intensity by the thermal factor $[1-\exp(-h\Delta\nu/kT)]^{-1}$. Scattering geometry is $X(Y, X)Y$ which gives the spectrum of B_2-symmetry phonons.

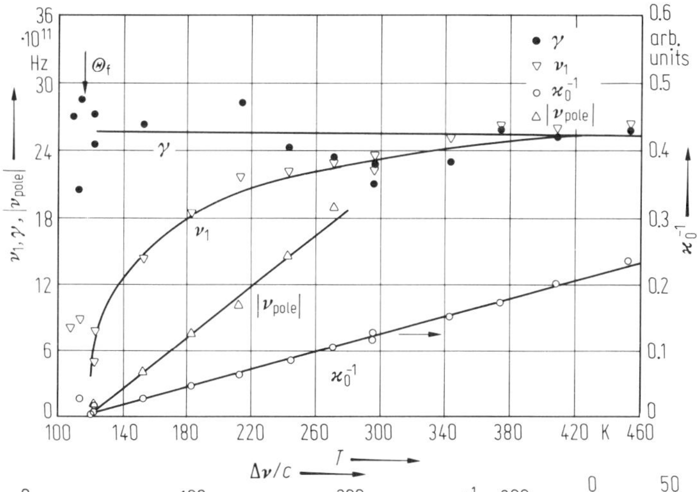

Fig. 382. KH_2PO_4(KDP) ν_1, γ, $|\nu_{pole}|$ vs. T [68 K 1]. ν_1, γ: the characteristic frequency and the damping factor of the soft B_2-mode obtained from the Raman spectra. $|\nu_{pole}|$: the distance of the pole of $\varkappa(\nu)$ from the origin.

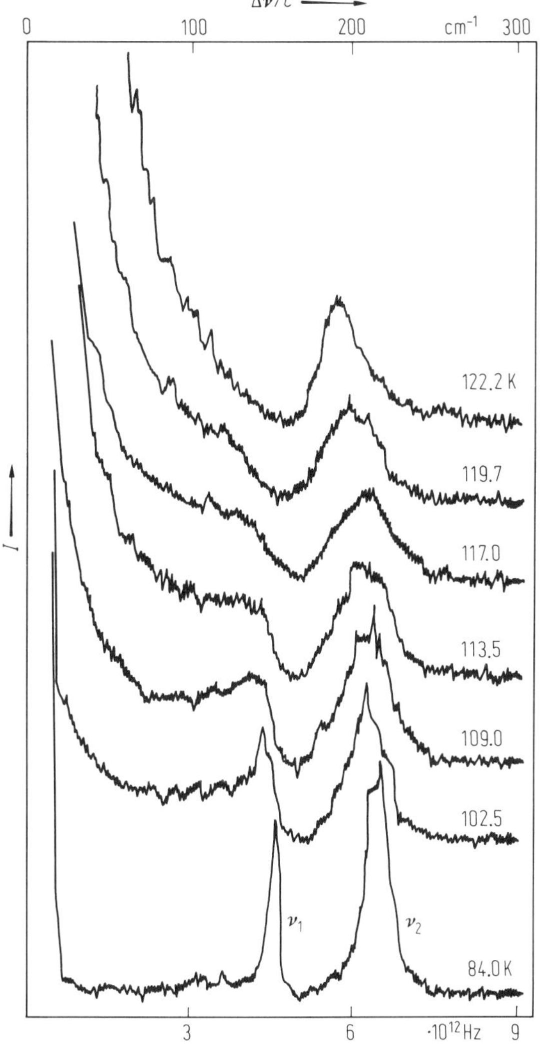

Fig. 383. KH_2PO_4(KDP). I vs. $\Delta\nu$ in the ferroelectric phase [72 S 15]. Scattering geometry is $X(Y, X)Y$ (X, Y, Z are the tetragonal axes for $T > \Theta_f$) and is same as Fig. 381. Parameter: T. See also [71 S 16, 73 B 16].

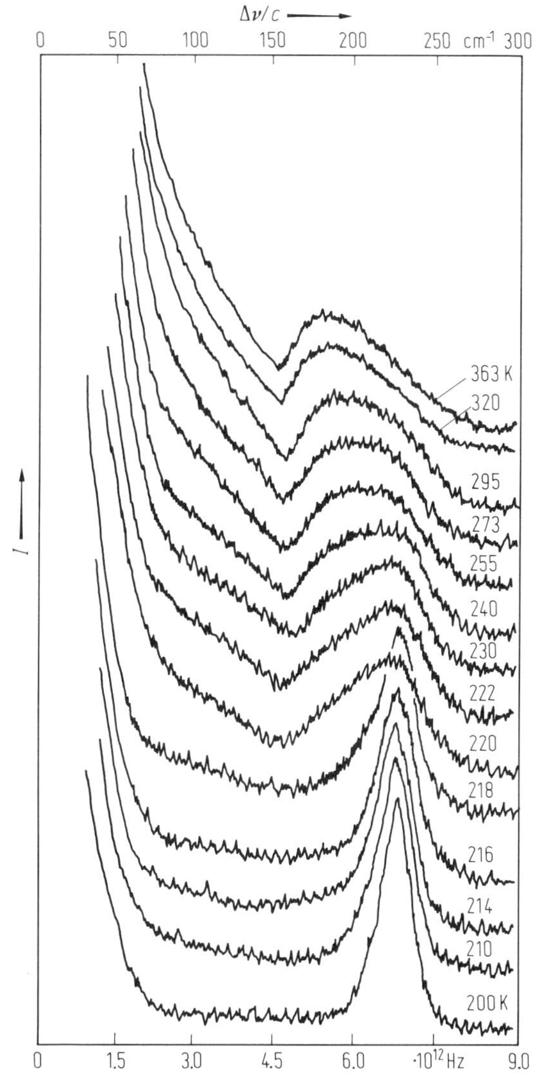

Fig. 384. KD_2PO_4(DKDP). I vs. $\Delta\nu$ [74 K 10]. I: Raman scattering intensity for the scattering geometry of $X(Y, X)Y$. Parameter: T.

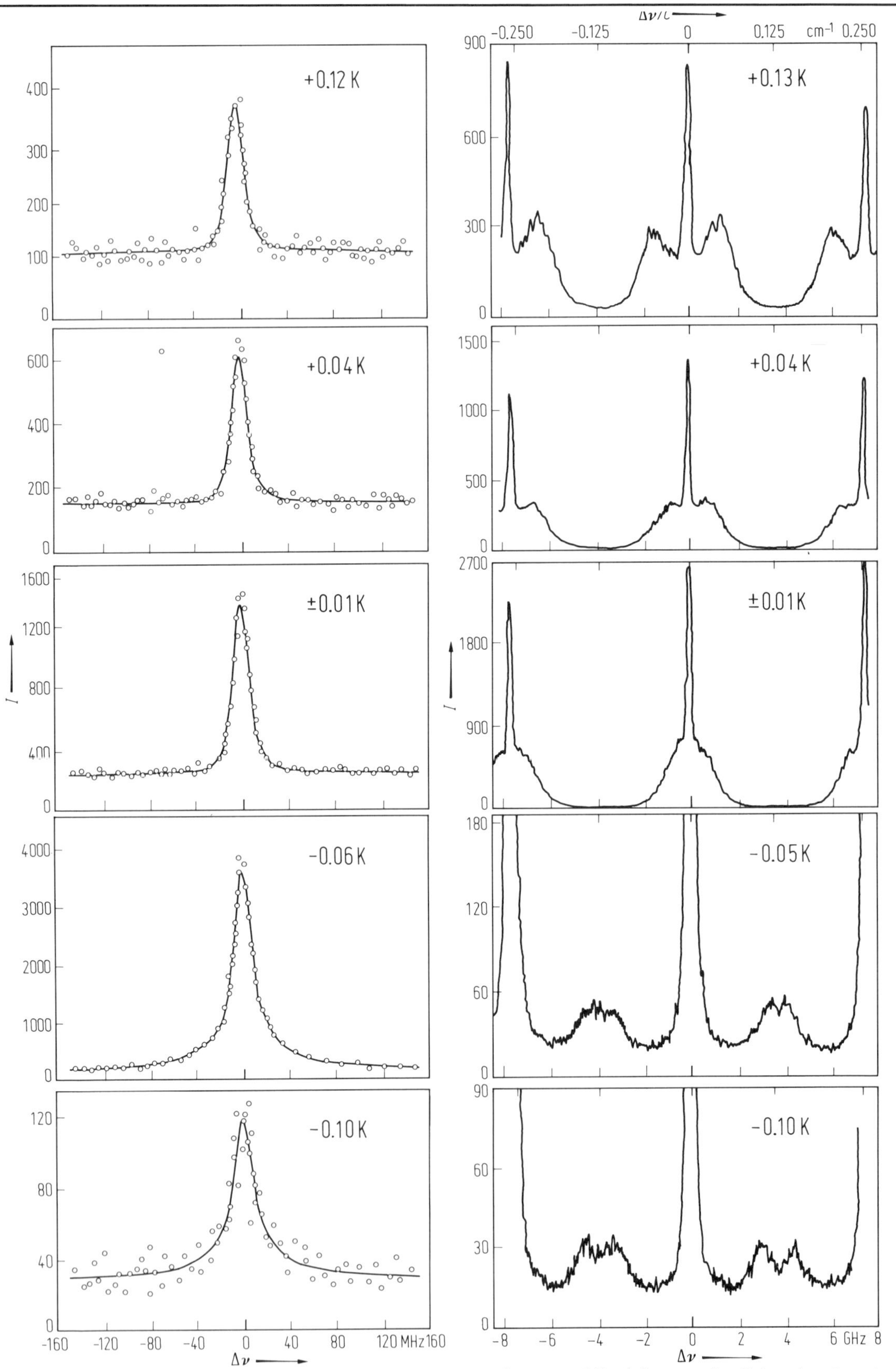

Fig. 390. KH_2PO_4(KDP). I vs. $\Delta\nu$ [77M15]. Parameter: $T-\Theta$. $\Delta\nu$: frequency shift of the scattering light. I in arb. units.

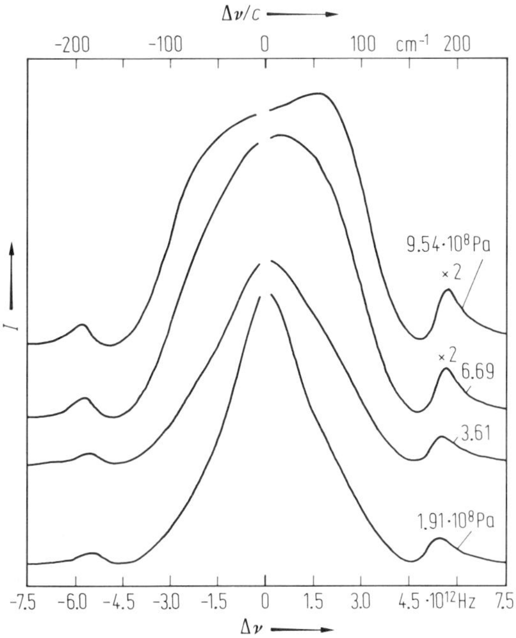

Fig. 387. KH_2PO_4(KDP). I vs. $\Delta\nu$ [75P4]. I: Raman scattering intensity for the scattering geometry of $X(Y, X)Y$. Parameter: hydrostatic pressure p. $T = 296$ K.

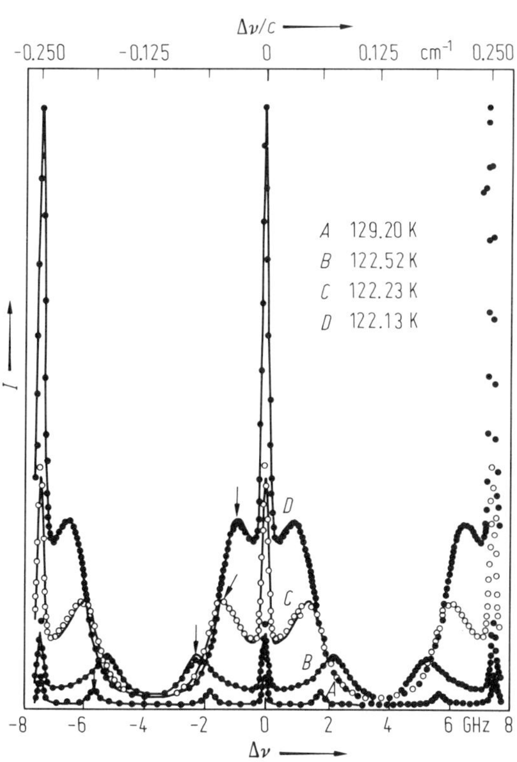

Fig. 389. KH_2PO_4(KDP). I vs. $\Delta\nu$ [74L1]. I: intensity of Brillouin spectrum and the central peak. $\Delta\nu$: frequency shift of the scattered light. Arrows indicate positions of the anti-Stokes Brillouin peaks of the order centered at zero.

Figs. 385, 386, see p. 400.

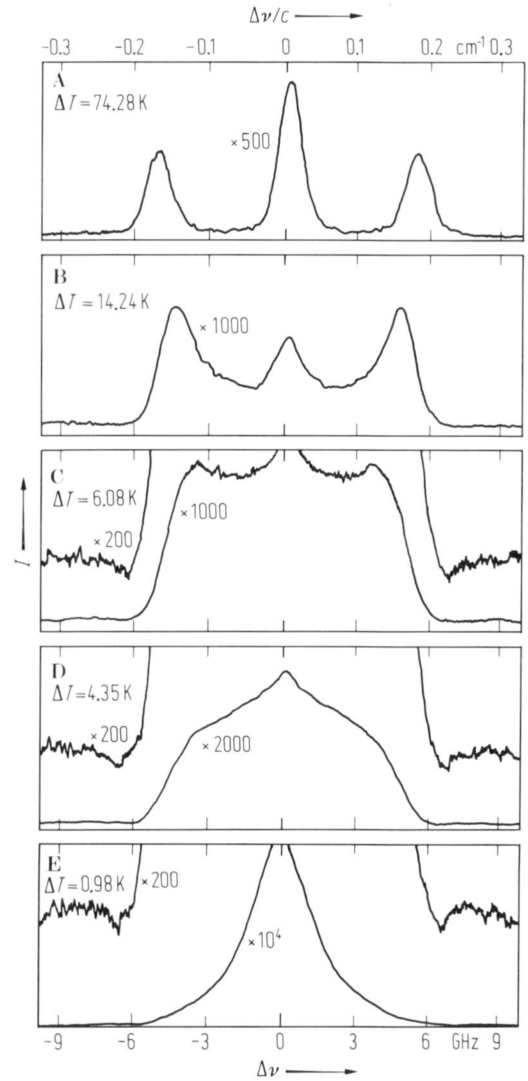

Fig. 391. KD_2PO_4(DKDP). I vs. $\Delta\nu$ [73R5]. I: Intensity of the scattered light. Scattering geometry: $XZ(-XZ, Y)-XZ$. ΔT: $T - \Theta_f$. $\Theta_f = 220.678$ K. Note the change of the vertical scale (count rate) denoted inside the figures.

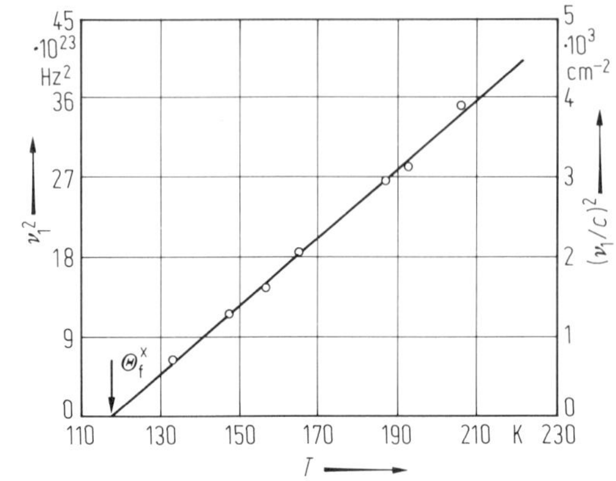

Fig. 385. KH_2PO_4(KDP). ν_1^2 vs. T [74L1]. ν_1: frequency of the lowest B_2-mode calculated from the coupled mode analysis with an imaginary coupling constant. Θ_f^x: "clamped" Curie temperature.

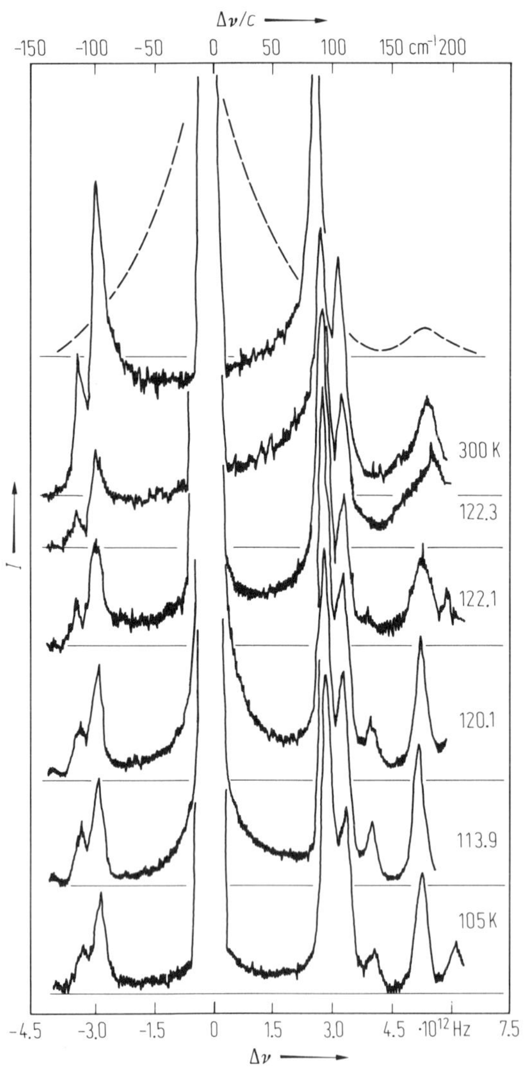

Fig. 386. KH_2PO_4(KDP). I vs. $\Delta\nu$ [75T1]. I: Raman scattering intensity of E-modes measured in the scattering geometry of $X(Y, Z)Y$. Parameter: T. The dotted line at the top is the spectrum of B_2-modes at 300 K.

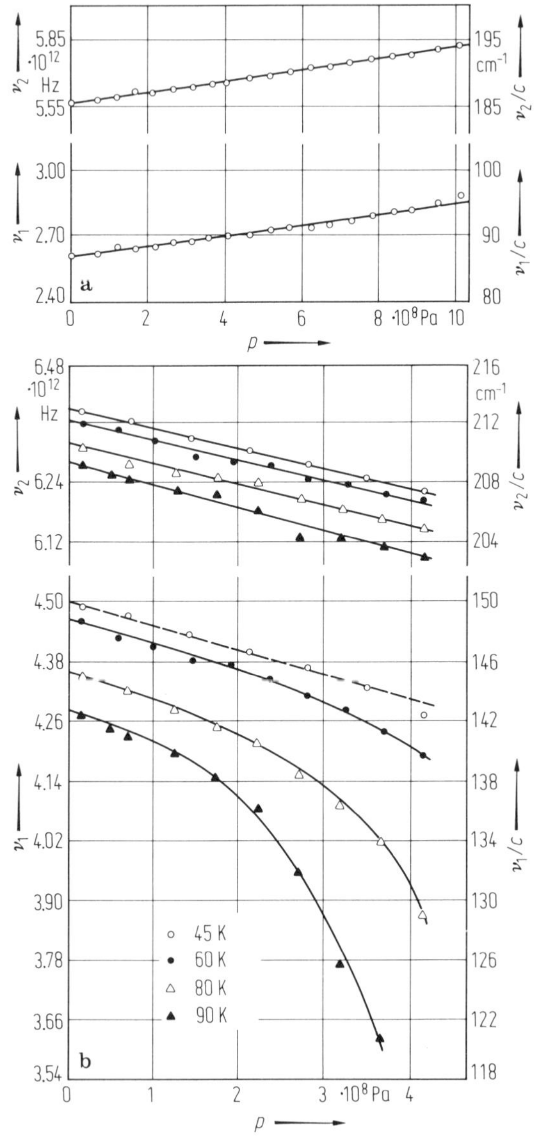

Fig. 388. KH_2PO_4(KDP). ν_1 and ν_2 vs. p [75P4]. ν_1, ν_2: frequencies of the two low-lying B_2-modes shown in Figs. 381, 383. Fig. a: $T=297$ K. Fig. b: $T<\Theta_f$.

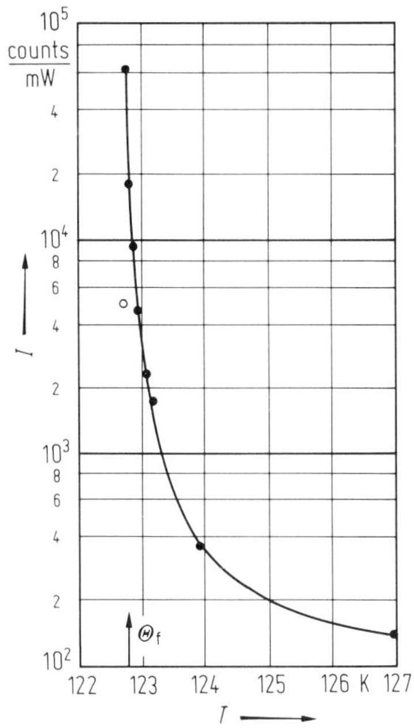

Fig. 392. KH_2PO_4(KDP). I vs. T [77D8]. I: Rayleigh scattering intensity. The open circle is a data in the ferroelectric phase.

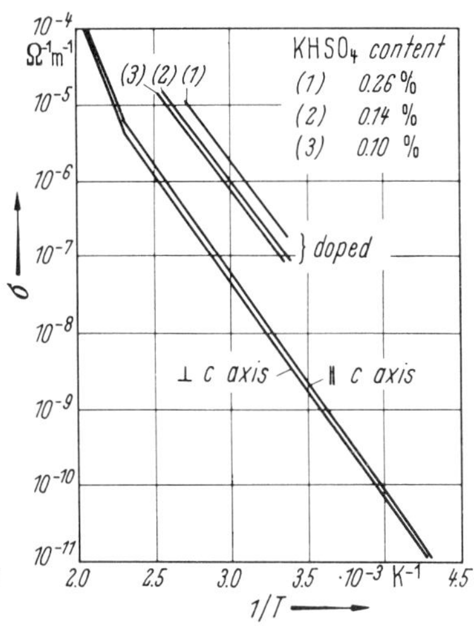

Fig. 393. KH_2PO_4(KDP). σ vs. $1/T$ for pure and doped specimens [67O4].

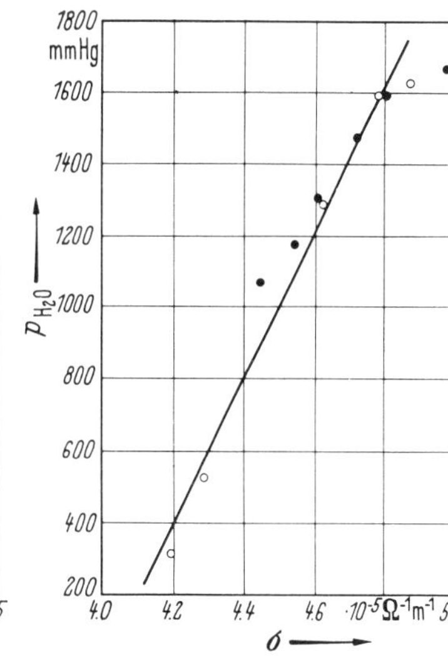

Fig. 395. KH_2PO_4(KDP). p_{H_2O} vs. σ [67O4]. p_{H_2O}: water vapor pressure. 1 mm Hg = 133.32 Pa. Open circles: increasing pressure; full circles: decreasing pressure.

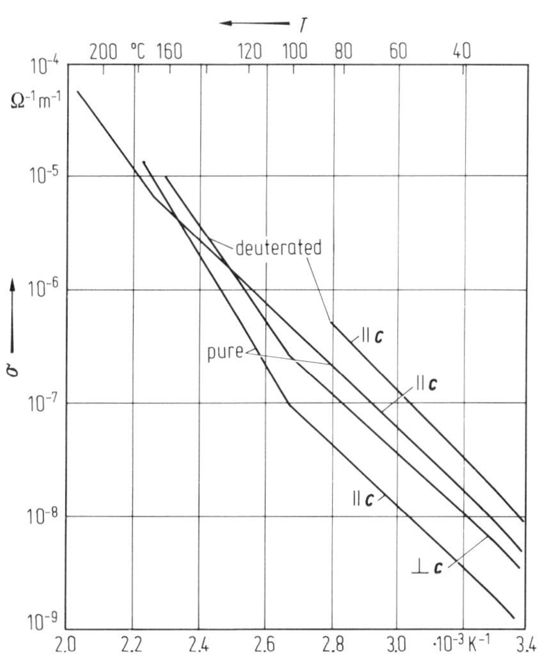

Fig. 394. KH_2PO_4(KDP), KD_2PO_4(DKDP). σ vs. $1/T$ along or perpendicular to [001] [73H5, 67O4].

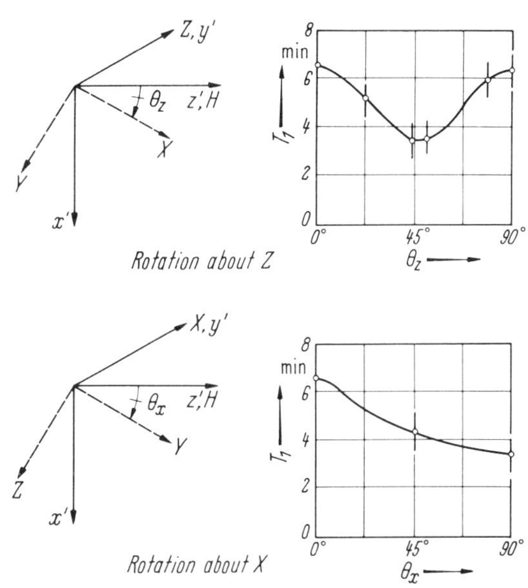

Fig. 398. KD_2PO_4(DKDP). Orientation dependence of spin-lattice relaxation time, T_1 at RT [59B1]. $f = 8$ MHz. (x', y', z'): laboratory fixed coordinates, (X, Y, Z): crystal fixed coordinates; $\boldsymbol{H} \parallel z'$.

Figs. 396, 397, see p. 402.

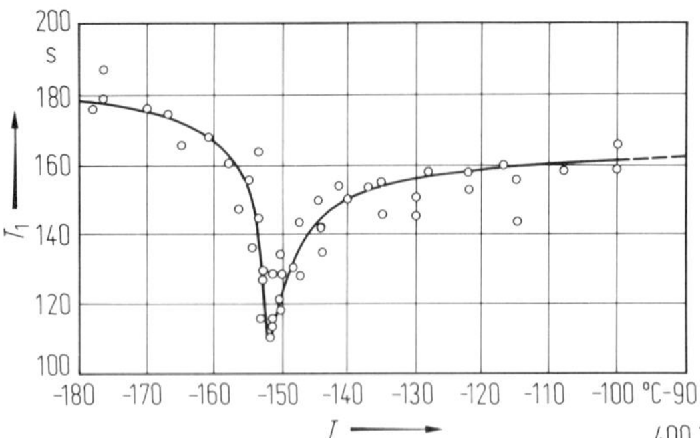

Fig. 396. KH_2PO_4(KDP). T_1 vs. T [74L2]. T_1: proton spin-lattice relaxation time. $f = 15.3$ MHz. $\boldsymbol{H} \perp c$-axis.

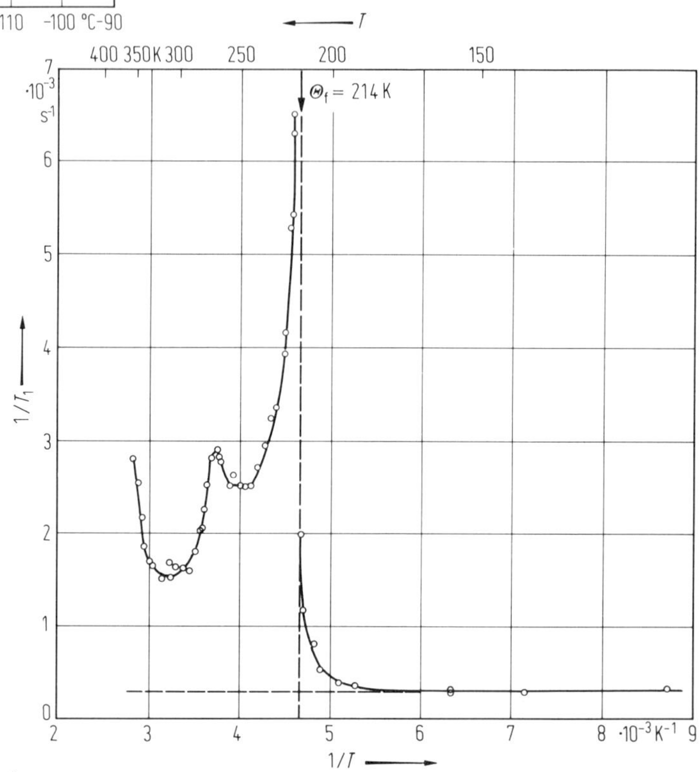

Fig. 397. KD_2PO_4(DKDP). T_1^{-1} vs. T^{-1} [76S19]. T_1: deuteron spin-lattice relaxation time. $\boldsymbol{H} \perp c$-axis, $\theta_z = 55°$. $f = 10.6$ MHz. For θ_z, see Fig. 398.

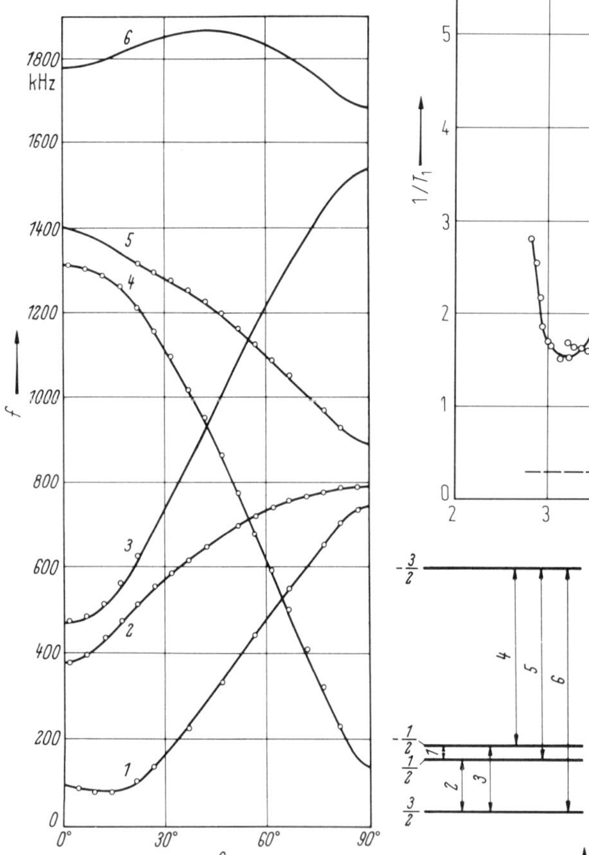

Fig. 399. KH_2PO_4(KDP). f vs. θ [67K8]. f: NMR resonance frequency of ^{39}K. θ: angle between $\boldsymbol{H}$ and the c axis in the b–c plane. Solid curves are calculated values. $e^2qQ/h = 1.68(1)$ MHz. The major principal axes of the electric field gradient tensor coincides with the c axis. Corresponding transitions are schematically shown on the right-hand side.

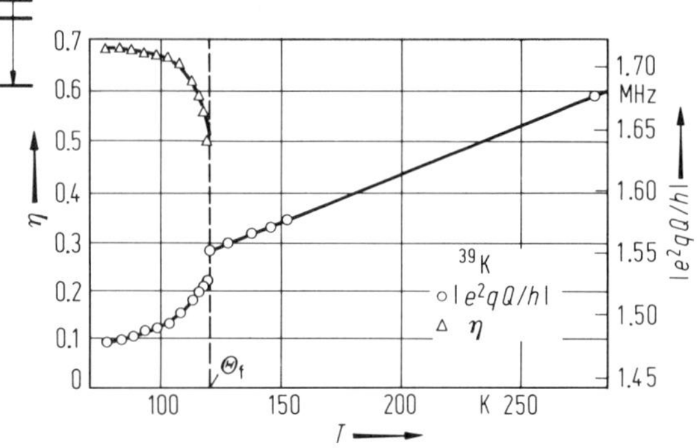

Fig. 400. KH_2PO_4(KDP). $|e^2qQ/h|$, η vs. T [70K21]. η: asymmetry parameter.

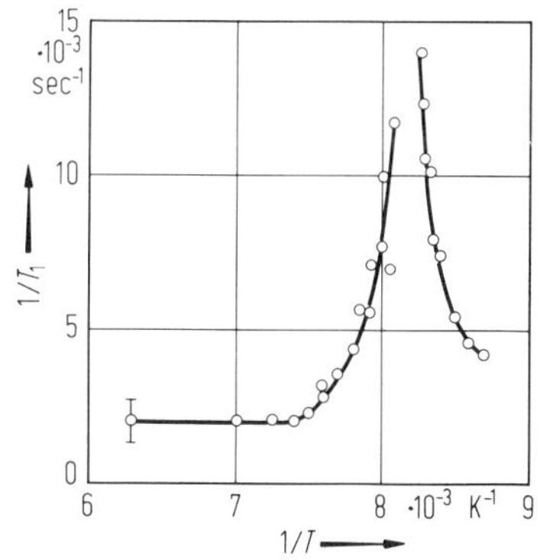

Fig. 401. KH_2PO_4(KDP). $1/T_1$ vs. $1/T$ [68 B 11]. ^{31}P spin lattice relaxation time $f=8.0$ MHz. $\boldsymbol{H}\perp c$ axis.

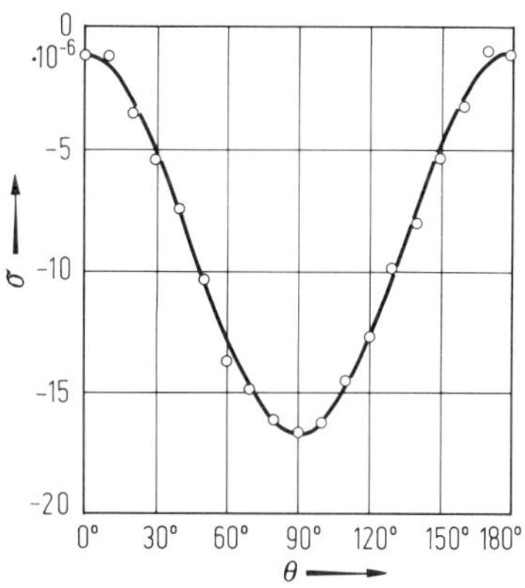

Fig. 403. KH_2PO_4(KDP). σ vs. θ [72 T 8]. σ: ^{31}P chemical shift, θ: angle between $\boldsymbol{H}$ and c-axis.

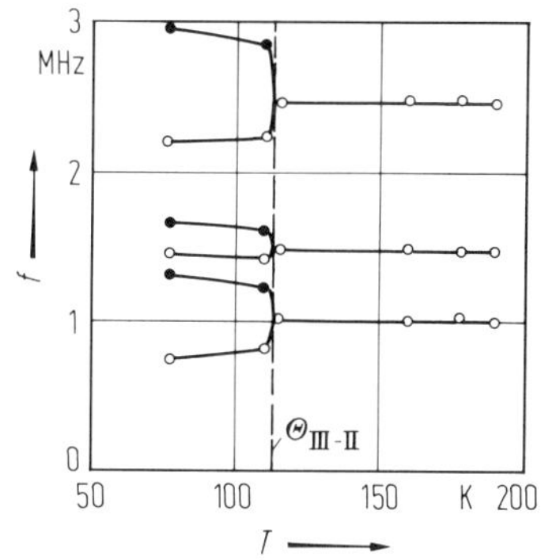

Fig. 404. KH_2PO_4(KDP). f vs. T [74 B 10]. f: ^{17}O NQR resonance frequency. $T>\Theta_f$: $e^2qQ/h=5.16$ MHz, $\eta=0.55$; $T<\Theta_f$: $(e^2qQ/h)_1=5.96$ MHz, $\eta_1=0.72$; $(e^2qQ/h)_2=4.85$ MHz, $\eta_2=0.18$. Subscripts 1 and 2 correspond to two chemically nonequivalent ^{17}O sites existing below Θ_f.

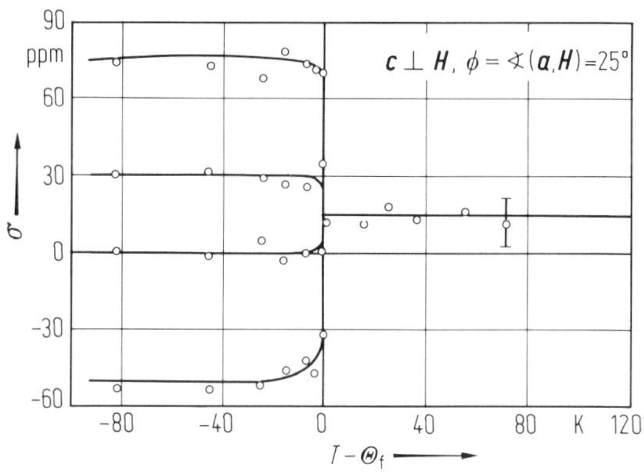

Fig. 402. KD_2PO_4(DKDP). σ vs. T-Θ_f [77 B 12]. σ: chemical shift of ^{31}P in the unit of ppm.

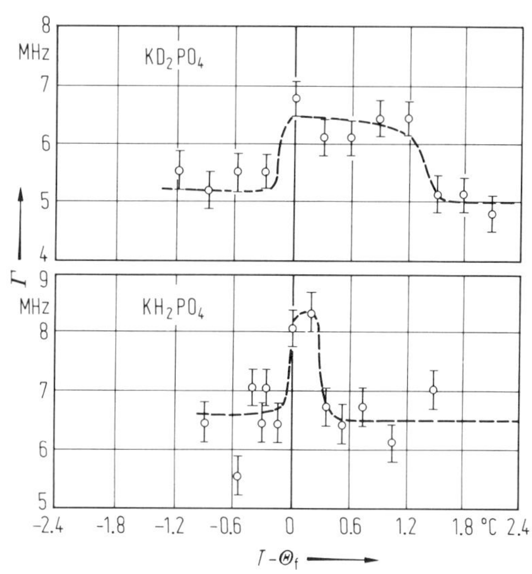

Fig. 405. KH_2PO_4(KDP), KD_2PO_4(DKDP): ^{57}Co. Γ vs. T [70 B 17]. Γ: line width of the central Mössbauer line.

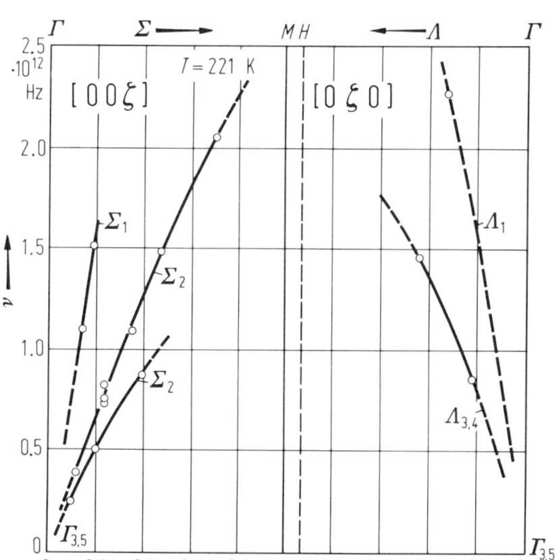

Fig. 406. KD_2PO_4(DKDP). ν vs. ζ [70 S 18]. ν: phonon frequency, ζ: reduced wave vector coordinate.

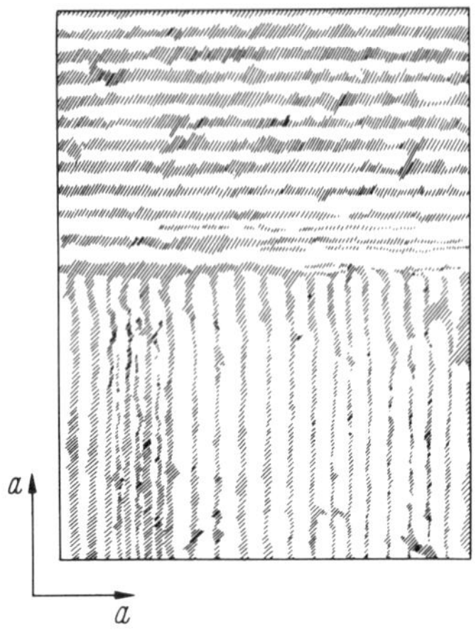

Fig. 408. KH_2PO_4(KDP). Frost pattern which shows a complicated domain structure [63 T 2].

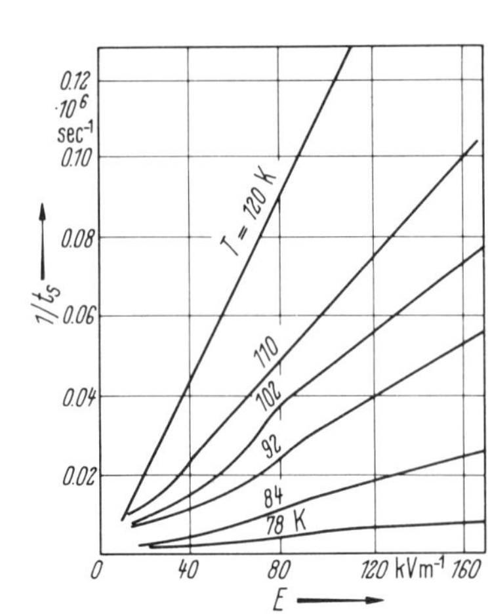

Fig. 409. KH_2PO_4(KDP) $1/t_s$ vs. E [66 G 5]. Parameter: T. t_s: switching time. E: applied field.

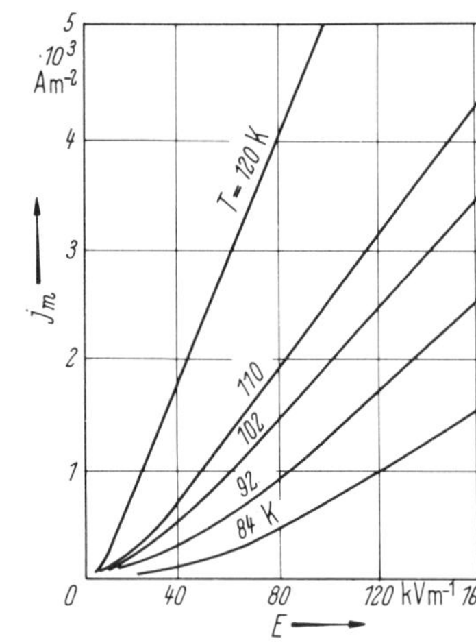

Fig. 410. KH_2PO_4(KDP). j_m vs. E [66 G 5]. Parameter: T. j_m: maximum switching current density.

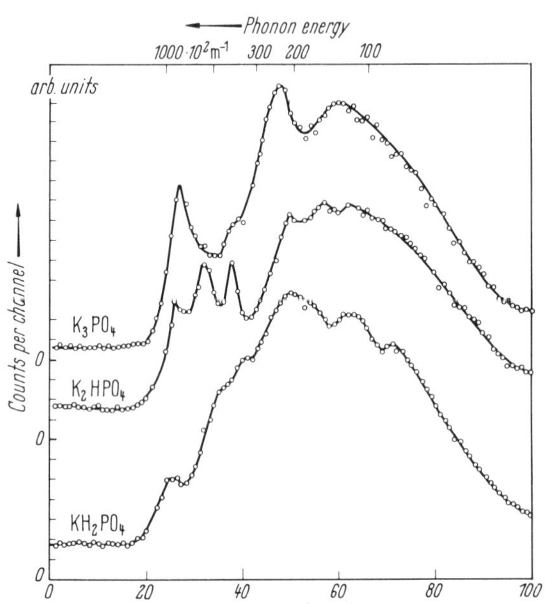

Fig. 407. KH_2PO_4(KDP) (Polycrystals). Time of flight spectra of inelastically scattered slow neutrons of 0.0005 eV at RT [59 P 4].

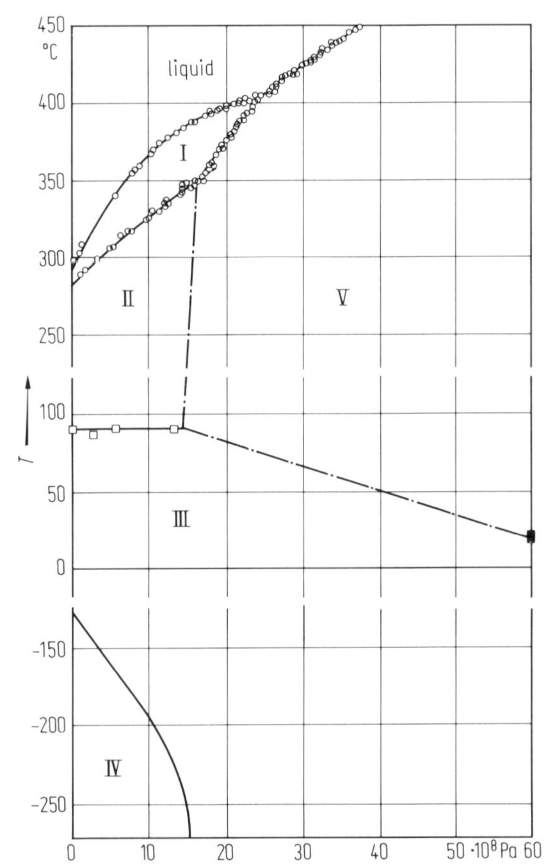

Fig. 411. RbH_2PO_4(RDP). Phase diagram [78 R 3]. Open squares: quasi-irreversible transition. Full square: [69 B 7]. The IV–III phase boundary is from [73 P 3]. Dashed-dotted line: tentative boundaries.

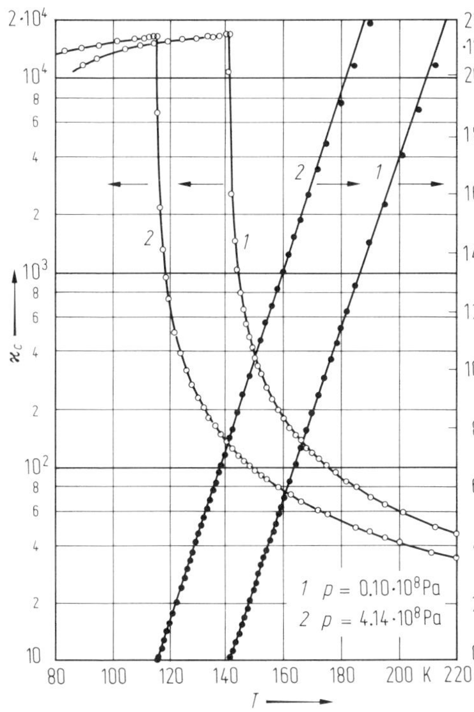

Fig. 412. RbH_2PO_4(RDP). $\varkappa_c$, $\varkappa_c^{-1}$ vs. T [73P3]. $f = 1$ kHz and 100 kHz. Parameter: p.

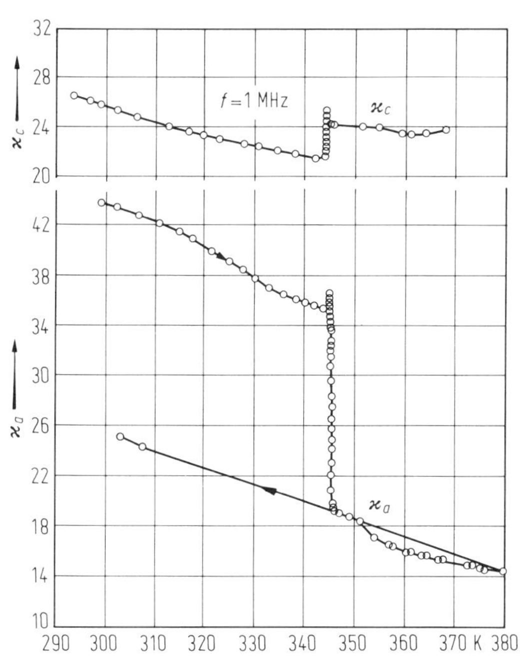

Fig. 414. RbH_2PO_4(RDP). $\varkappa_a$, $\varkappa_c$ vs. T in the neighbourhood of Θ_{III-II} [78S8].

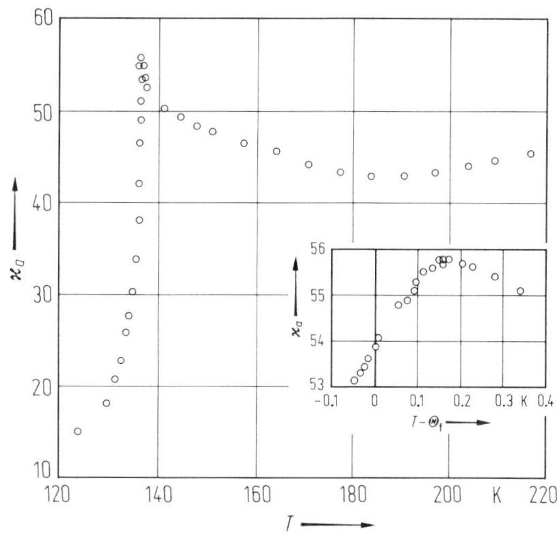

Fig. 413. RbH_2PO_4(RDP). $\varkappa_a$ vs. T [76S18]. $f = 1.596$ kHz.

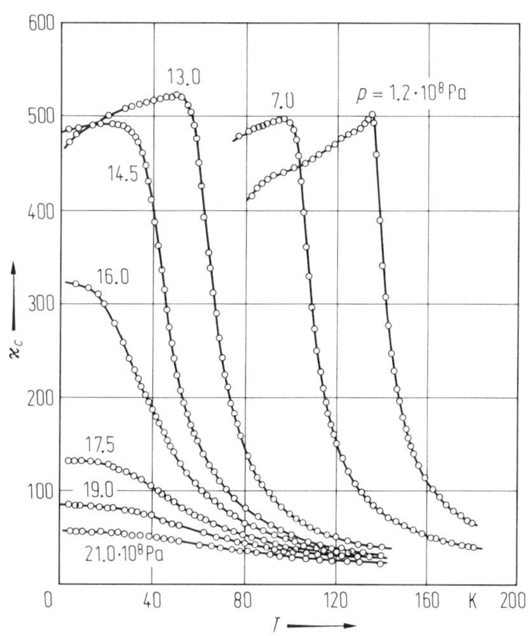

Fig. 415. RbH_2PO_4(RDP). $\varkappa_c$ vs. T [73P3]. $f = 1$ kHz and 100 kHz. Parameter: p.

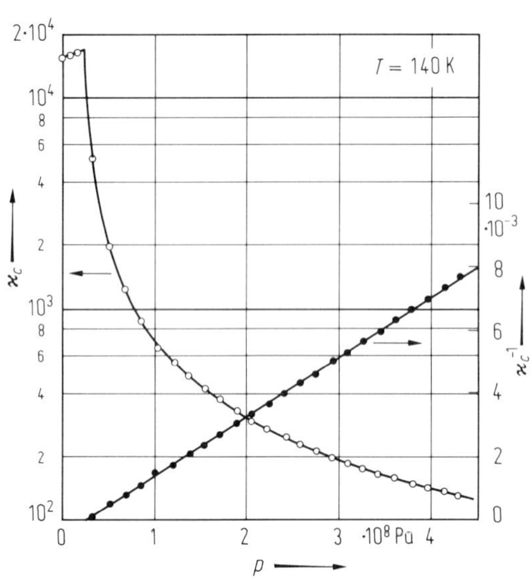

Fig. 416. RbH_2PO_4(RDP). $\varkappa_c$, $\varkappa_c^{-1}$ vs. p at 140 K [73P3]. $f = 1$ kHz and 100 kHz.

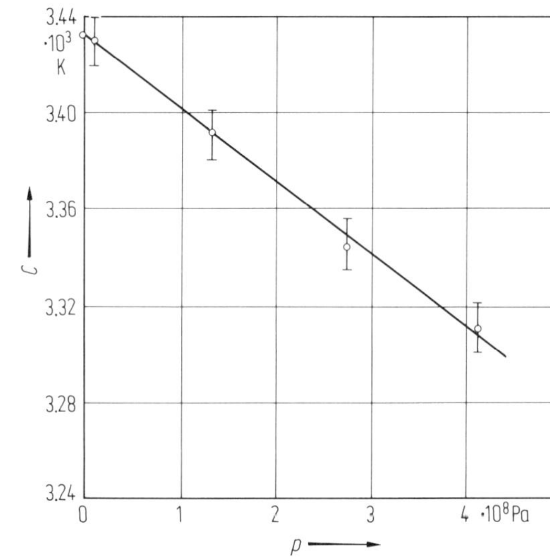

Fig. 417. RbH_2PO_4(RDP). C vs. p [73P3]. C: Curie-Weiss constant.

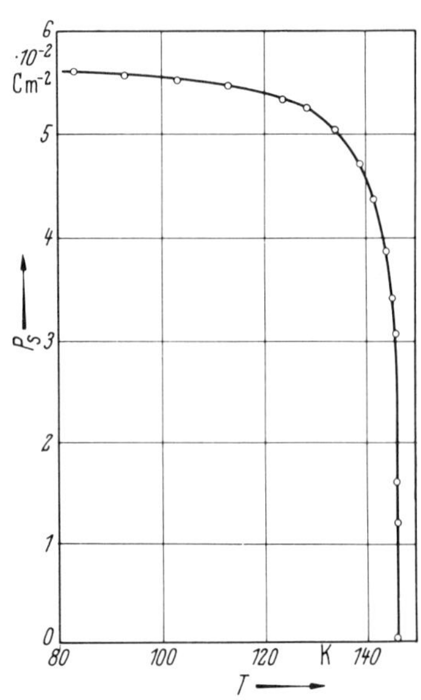

Fig. 419. RbH_2PO_4(RDP). P_s vs. T [50B3].

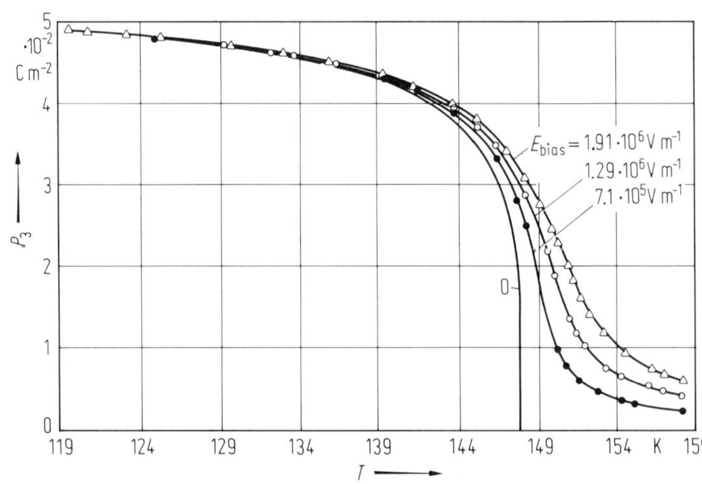

Fig. 420. RbH_2PO_4(RDP). P_3 vs. T [77C1]. Parameter: E_{bias}. P_3: polarization along c-axis.

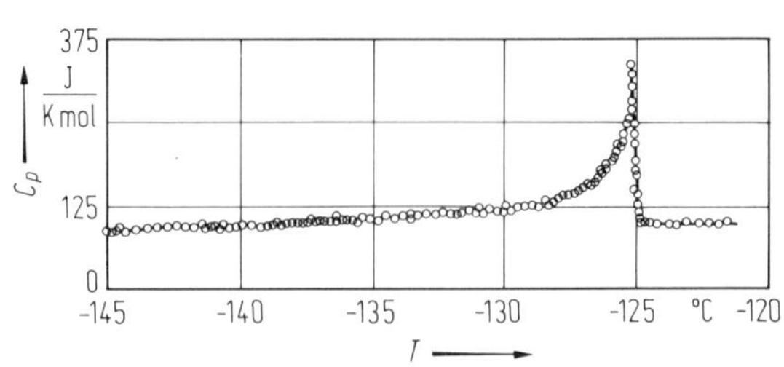

Fig. 421. RbH_2PO_4(RDP). C_p vs. T [68A3].

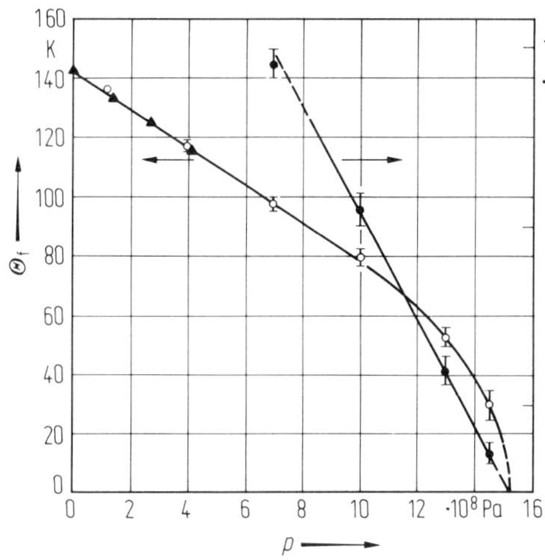

Fig. 418. RbH_2PO_4(RDP). Θ_f, Θ_f^2 vs. p [73 P 3].

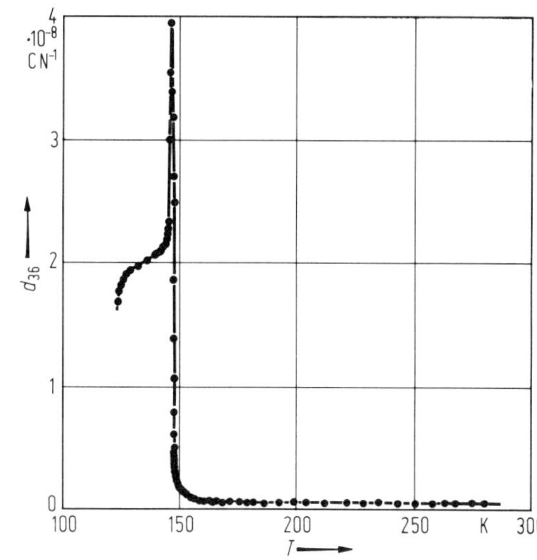

Fig. 422. RbH_2PO_4(RDP). d_{36} vs. T [71 P 4].

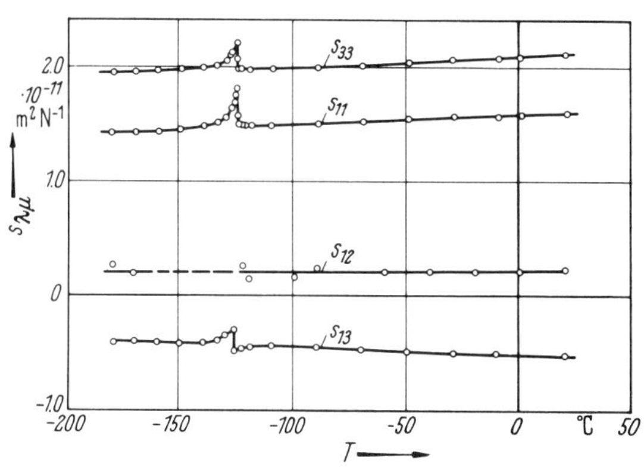

Fig. 423. RbH_2PO_4(RDP). $s_{\lambda\mu}$ vs. T [66 M 4].

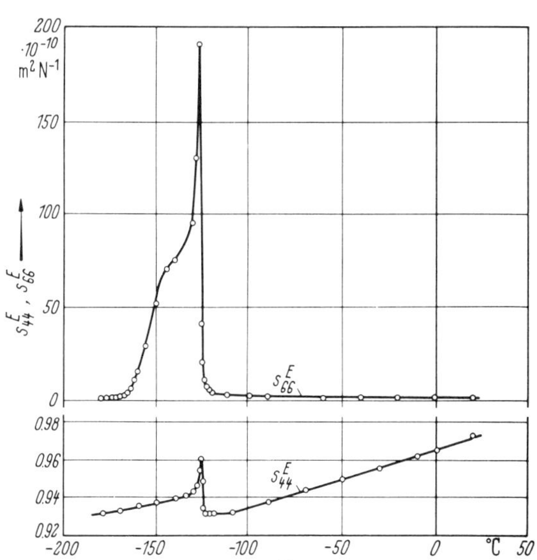

Fig. 424. RbH_2PO_4(RDP). s_{44}^E, s_{66}^E vs. T [66 M 4].

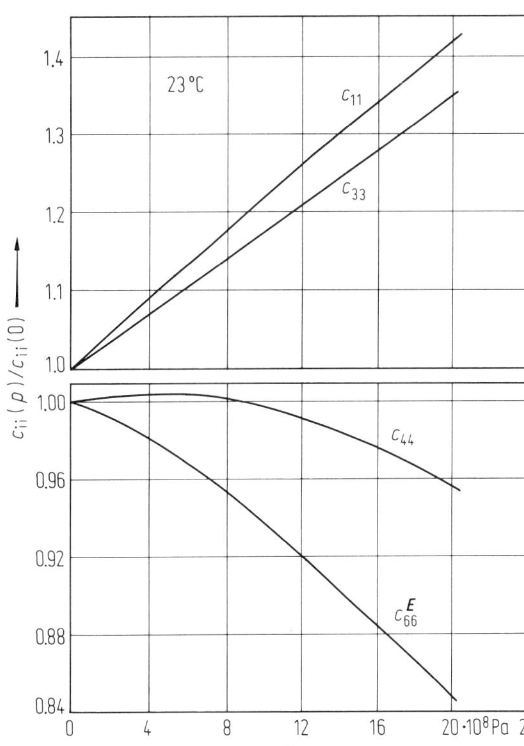

Fig. 426. RbD_2PO_4. $c_{ii}(p)/c_{ii}(0)$ vs. p [76 F 12]. $c_{ii}(p)$: elastic stiffness under hydrostatic pressure p.

Fig. 425, see p. 408.

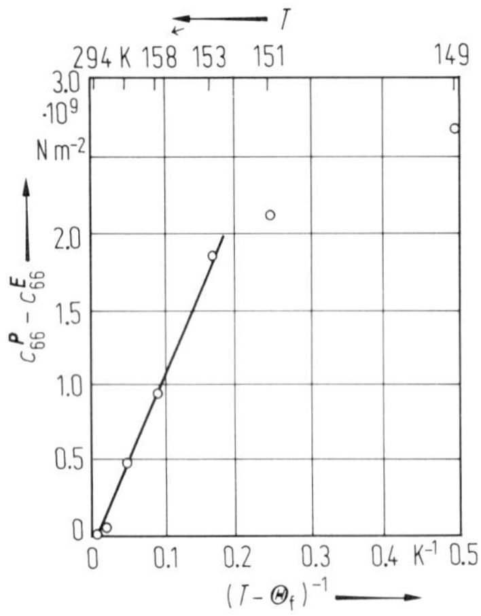

Fig. 425. RbH_2PO_4(RDP). $c^{P}_{66}-c^{E}_{66}$ vs. $(T-\Theta_f)^{-1}$ [71H6]. $c^{P}_{66}-c^{E}_{66}$: elastic stiffness obtained from Brillouin scattering.

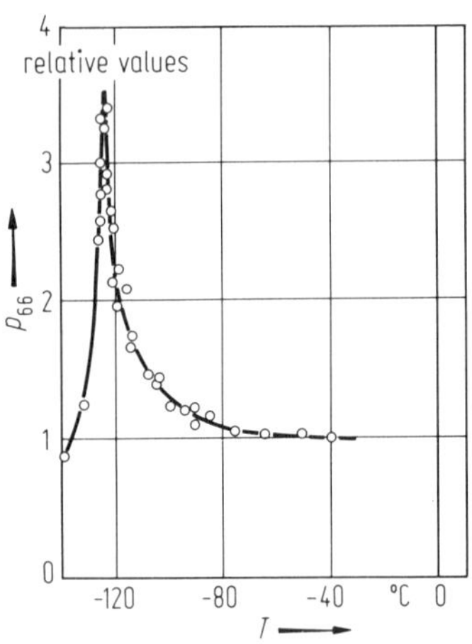

Fig. 429α. RbH_2PO_4(RDP). p^{E}_{66} vs. T [71H6]. p^{E}_{66}: piezo-optic constant obtained from Brillouin scattering intensity.

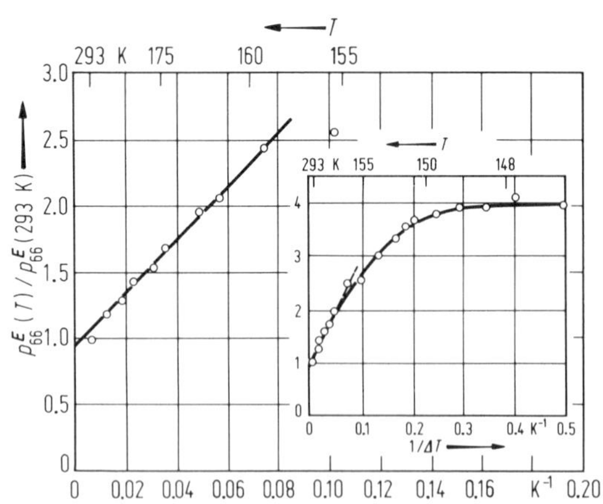

Fig. 429β. RbH_2PO_4(RDP). $p^{E}_{66}(T)/p^{E}_{66}$ (293 K) vs. $1/\Delta T$ [72H7]. p^{E}_{66}: piezooptic constant for constant $\boldsymbol{E}$ obtained from Brillouin scattering. ΔT: $T-\Theta$.

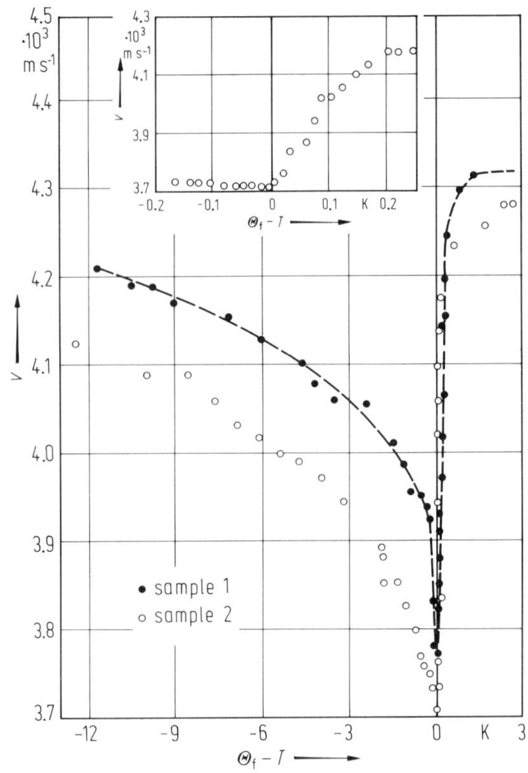

Fig. 427. RbH_2PO_4(RDP). v vs. Θ_f-T [78S9]. v: longitudinal sound velocity propagating along c axis. $f=8$ MHz.

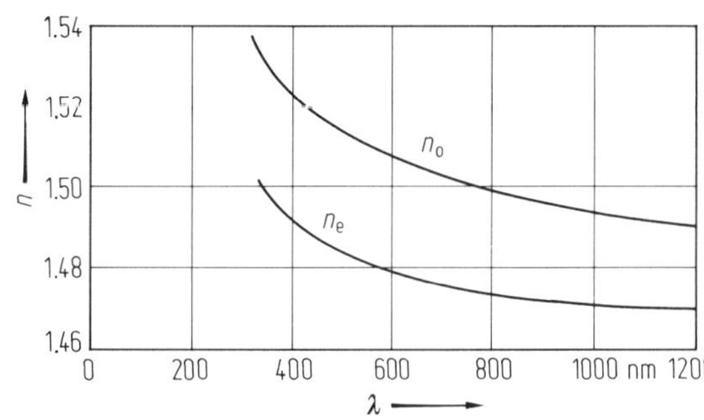

Fig. 428. RbH_2PO_4(RDP). n vs. λ at RT [70G9].

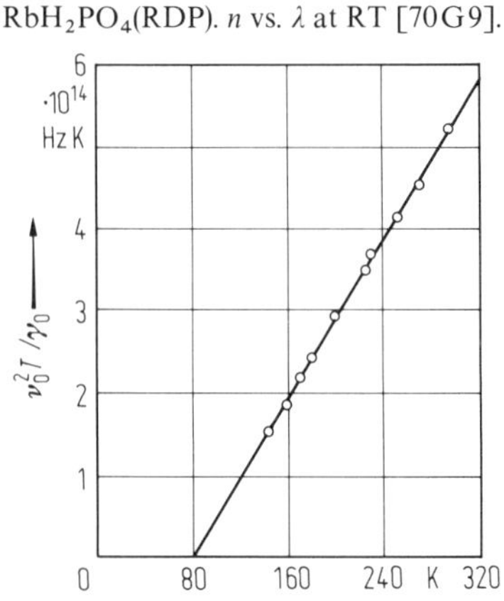

Fig. 431. RbH_2PO(RDP). $\nu_0^2 T/\gamma_0$ vs. T [73P3]. ν_0, γ_0: characteristic frequency and damping constant of the soft mode obtained from Raman spectrum.

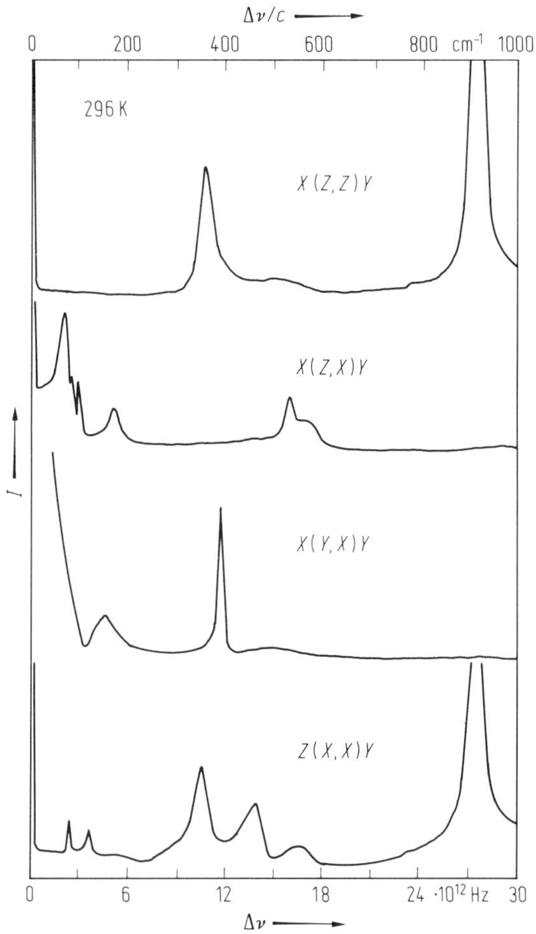

Fig. 430. RbH_2PO_4(RDP). I vs. $\Delta\nu$ [73P3]. I: Raman scattering intensity at 296 K in various scattering geometries.

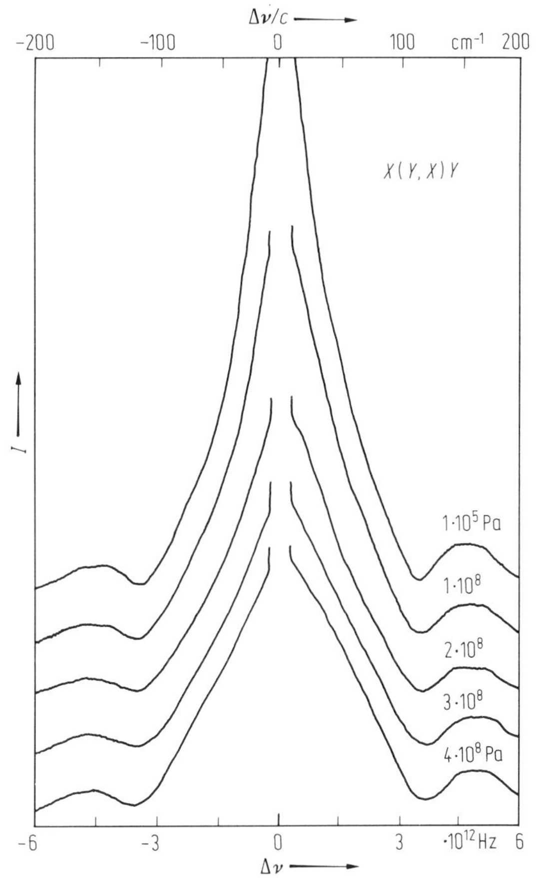

Fig. 432. RbH_2PO_4(RDP). I vs. $\Delta\nu$ [73P3]. I: Raman scattering intensity at 296 K in a scattering geometry of $X(Y, X)Y$. Parameter: p.

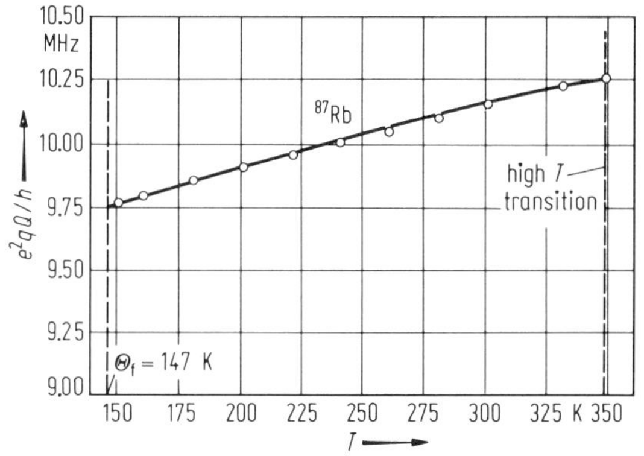

Fig. 433. RbH_2PO_4(RDP). e^2qQ/h vs. T [70B11].

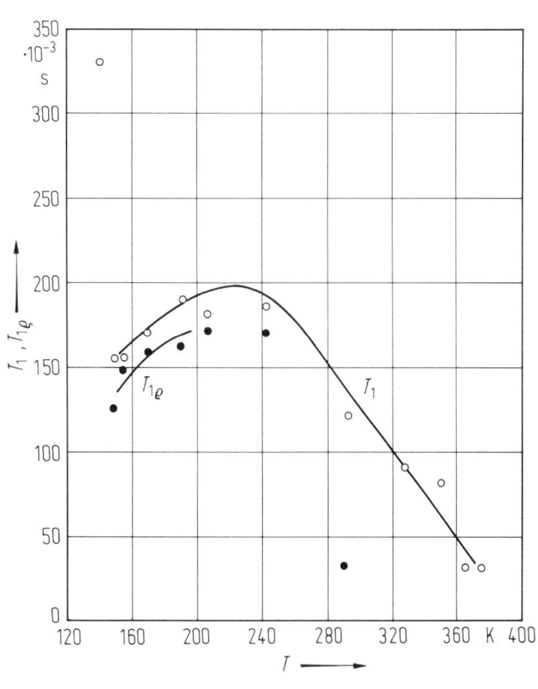

Fig. 434. RbH_2PO_4(RDP). T_1, $T_{1\varrho}$ vs. T [75B15]. Open circles: T_1 of ^{87}Rb; full circles: $T_{1\varrho}$ of ^{87}Rb, $H_1 = 3.2 \cdot 10^2$ Am^{-1}. $f = 18$ MHz. $\boldsymbol{H} \perp c$ axis.

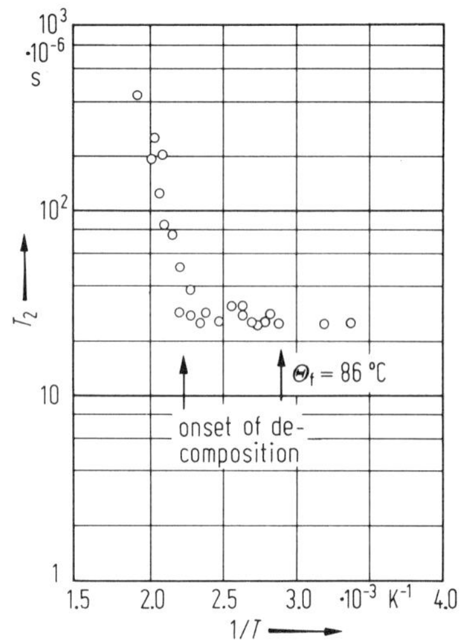

Fig. 435. RbH_2PO_4(RDP). T_2 vs. $1/T$ [69 B 10]. T_2: proton spin-spin relaxation time.

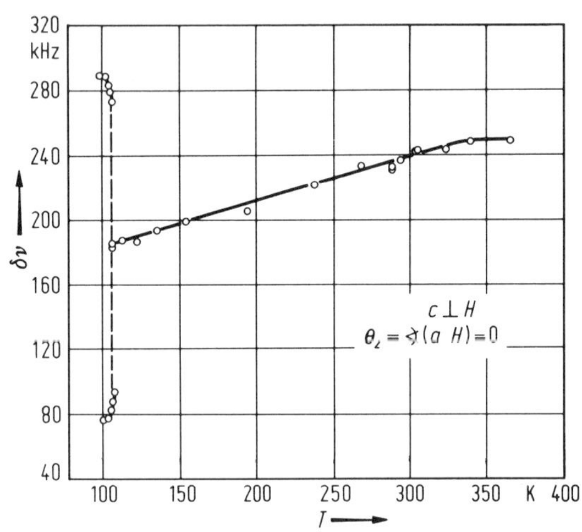

Fig. 436. RbH_2PO_4(RDP). $\delta\nu$ vs. T [69 B 10]. $\delta\nu$: second order shift of the $1/2 \rightarrow -1/2$ transition of ^{87}Rb. $\boldsymbol{H} \| a$, $\boldsymbol{H} \perp c$, $f = 9.160$ MHz.

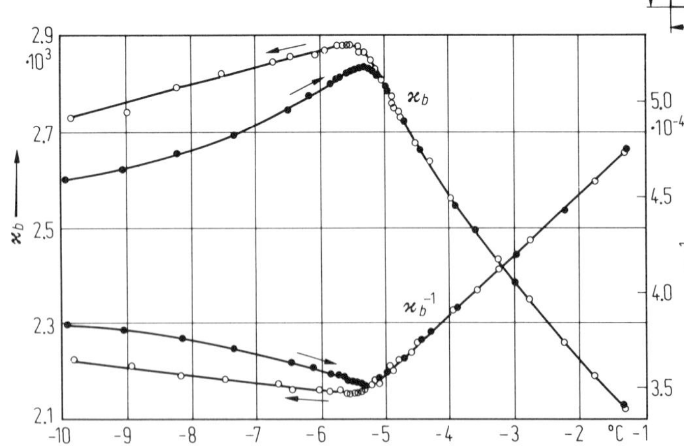

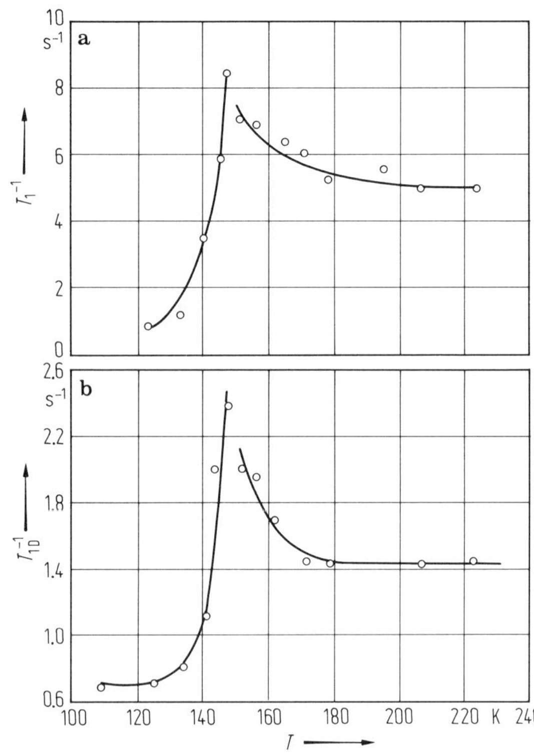

Fig. 437. RbH_2PO_4(RDP). T_1^{-1}, T_{1D}^{-1} vs. T [75 G 12]. Fig. a T_1: ^{87}Rb spin-lattice relaxation time, Fig. b T_{1D}: ^{1}H dipolar spin-lattice relaxation time.

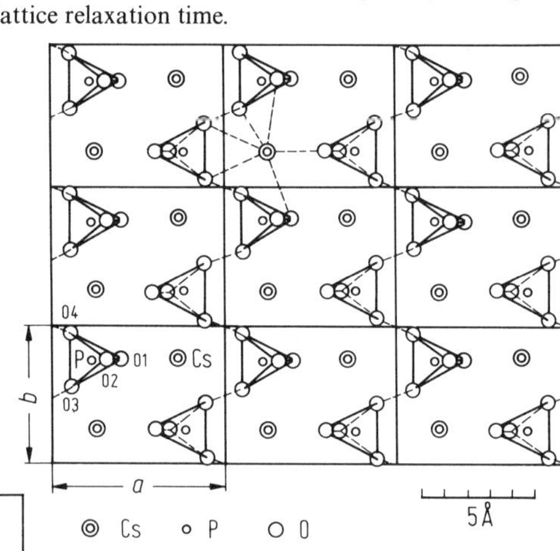

Fig. 440. CsH_2PO_4(CDP). Structure of phase III [76 U 2]. Projection on (001) plane.

◄ Fig. 443. CsD_2PO_4. $\varkappa_b$, $\varkappa_b^{-1}$ vs. T [75 L 5]. Deuteration degree is more than 90%. $f = 1$ kHz.

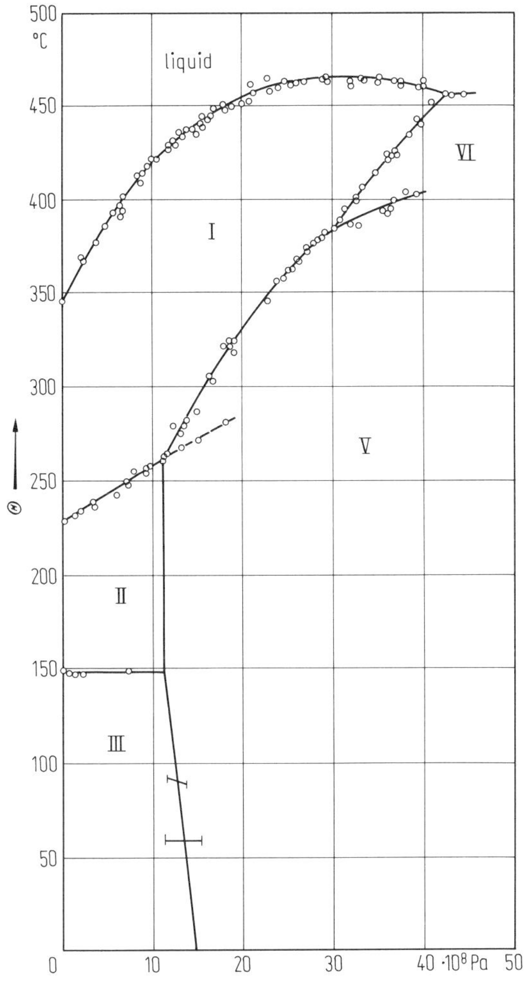

Fig. 438. CsH_2PO_4(CDP). Phase diagram [78 R 3]. Dashed line: metastable extension.

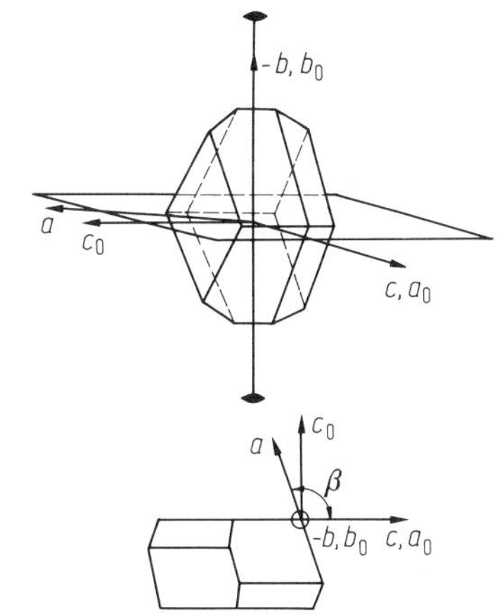

Fig. 439. CsH_2PO_4(CDP). Crystal form [76 U 2]. a, b, c: crystallographic axes in monoclinic system. a_o, b_o, c_o: those assigned as orthorhombic system in [52 F 1] and those assigned as pseudo-orthorhombic system in [78 N 2].

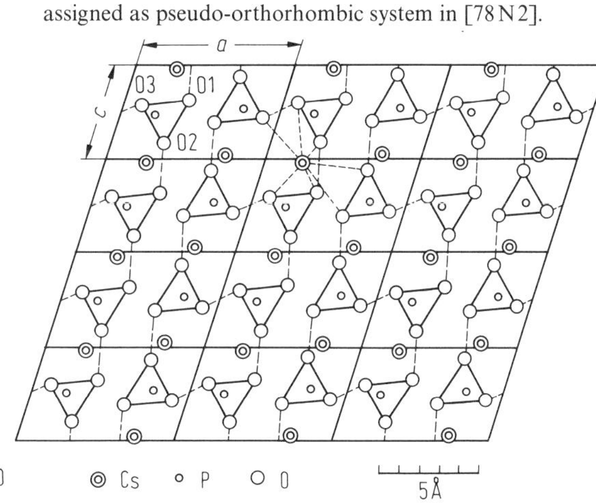

Fig. 441. CsH_2PO_4(CDP). Structure of phase III [76 U 2]. Projection on (010) plane.

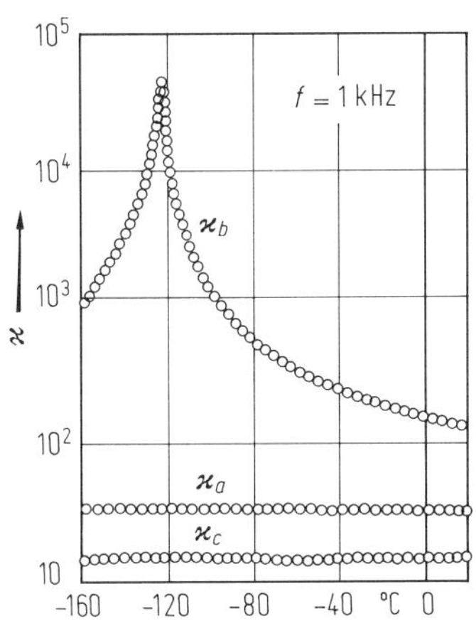

Fig. 442. CsH_2PO_4(CDP). $\varkappa_a$, $\varkappa_b$, $\varkappa_c$ vs. T [76 U 2].

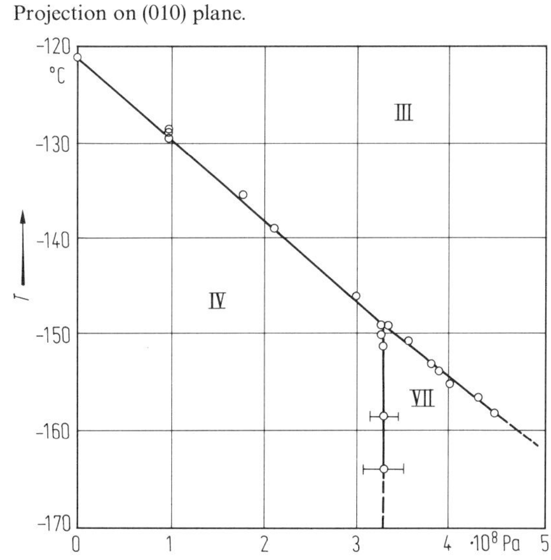

Fig. 444. CsH_2PO_4(CDP). Phase diagram [78 Y 4], see also [78 G 7].

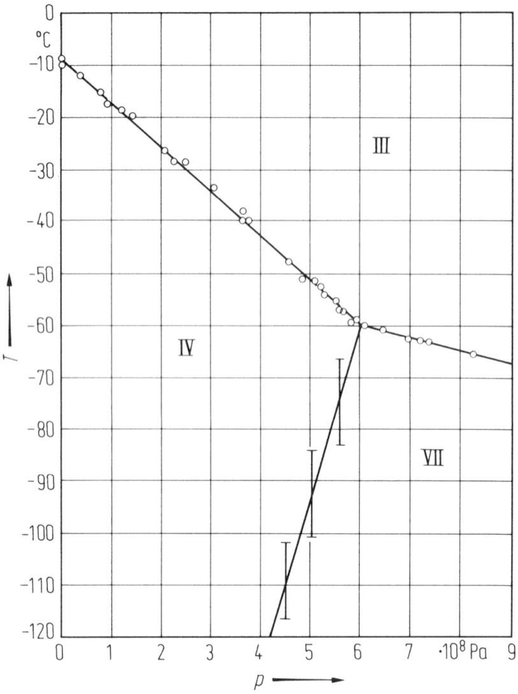

Fig. 445. CsD_2PO_4. Phase diagram [78G7]. Θ_f at $p=0$ is $-8.5\,°C$.

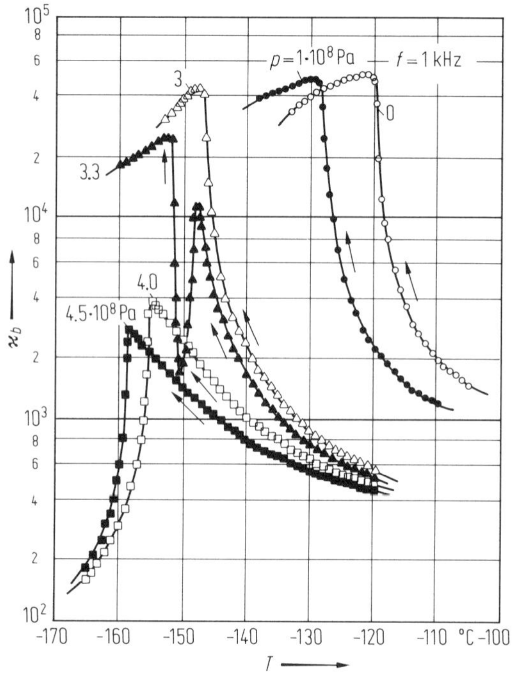

Fig. 446. CsH_2PO_4(CDP). $\varkappa_b$ vs. T [78Y4]. Parameter: p.

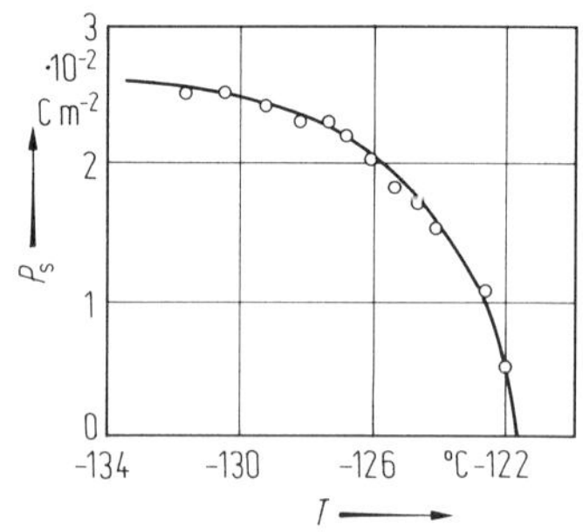

Fig. 447. CsH_2PO_4(CDP). P_s vs. T [76U2].

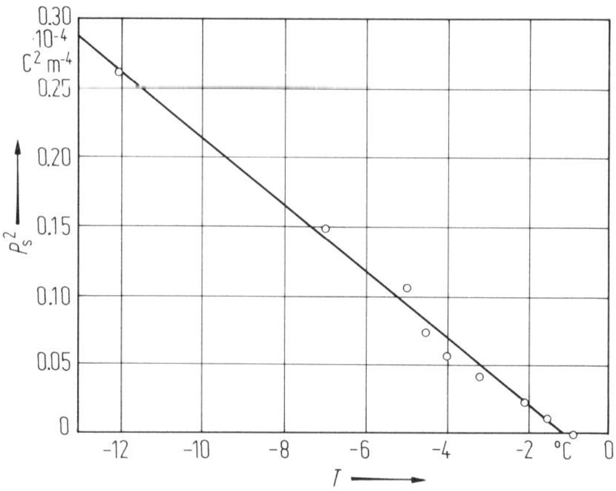

Fig. 448. CsD_2PO_4. P_s^2 vs. T [75L5].

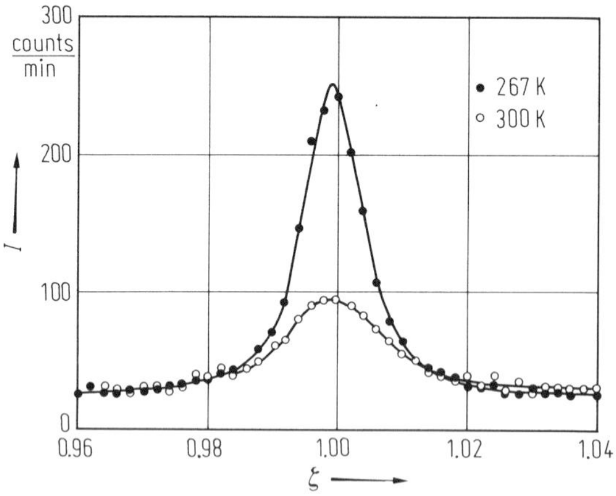

Fig. 450. CsD_2PO_4: I vs. ζ [77S17]. I: neutron scattering intensity at (2.15, ζ, 0).

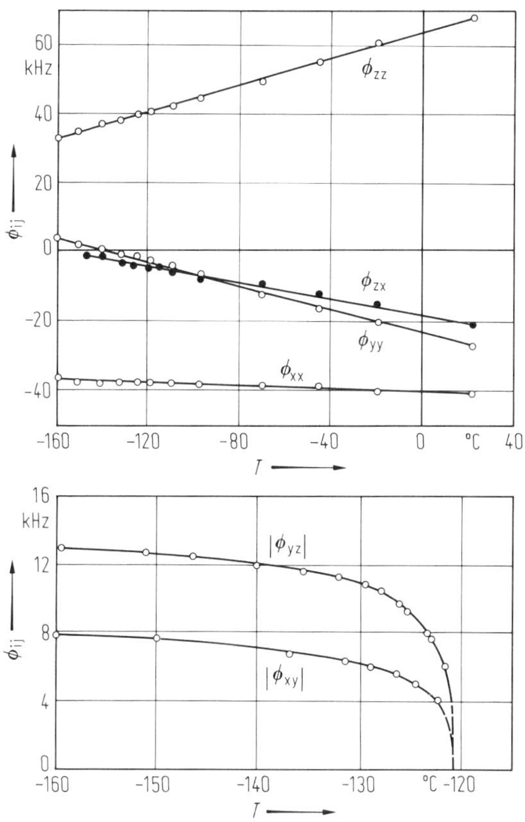

Fig. 449. CsH_2PO_4(CDP). ϕ_{ij} vs. T [77 K 2]. ϕ_{ij}: electric field gradient tensor components of ^{133}Cs. x, y, z axes are related to crystallographic a, b, c axes as follows: $x \perp b$, $z \| c$.

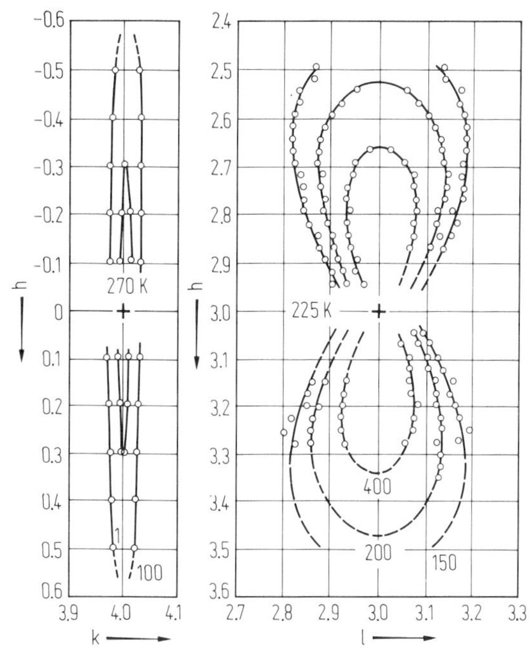

Fig. 451. CsD_2PO_4 and KD_2PO_4. Neutron scattering intensity contours [77 S 17]. (a) Contours in the (h k 0) zone near (0 4 0) for CsD_2PO_4 at $T = 270$ K. (b) Contours in the (h 0 l) zone near (3 0 3) for KD_2PO_4 at $T = 225$ K. The figures are drawn to the same scale in inverse Å. The intensity units are arbitrary. See also [70 S 18].

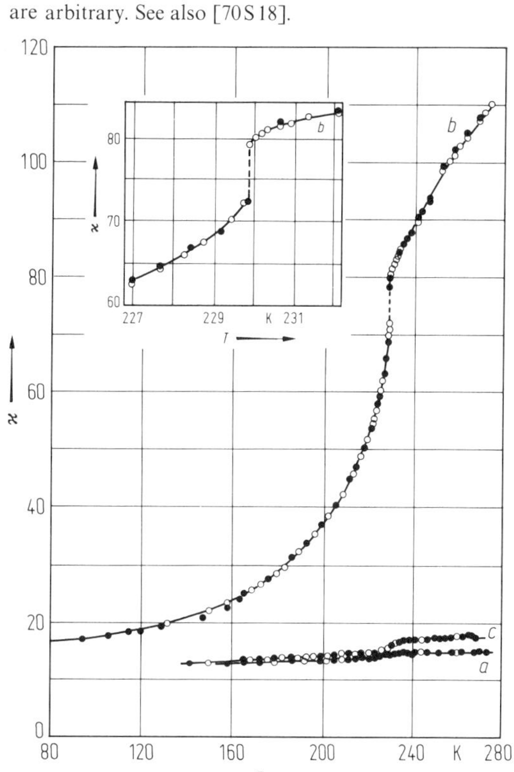

Fig. 452. TlH_2PO_4. $\varkappa$ along a, b, c axes vs. T [77 M 14]. The inset shows the behavior in the vicinity of Θ_{II-I}. $\varkappa$ at $f = 1$ kHz.

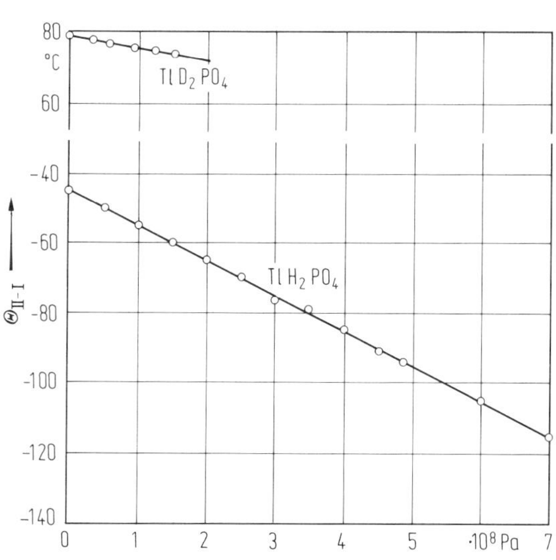

Fig. 453. TlH_2PO_4, TlD_2PO_4. Θ_{II-I} vs. p [79 Y 3].

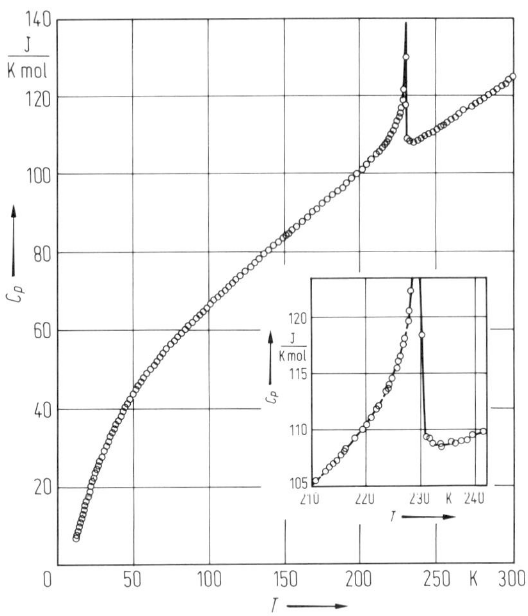

Fig. 454. TlH_2PO_4. C_p vs. T [77M14]. C_p: molar heat capacity. The inset shows the behavior in the vicinity of Θ_{II-I}.

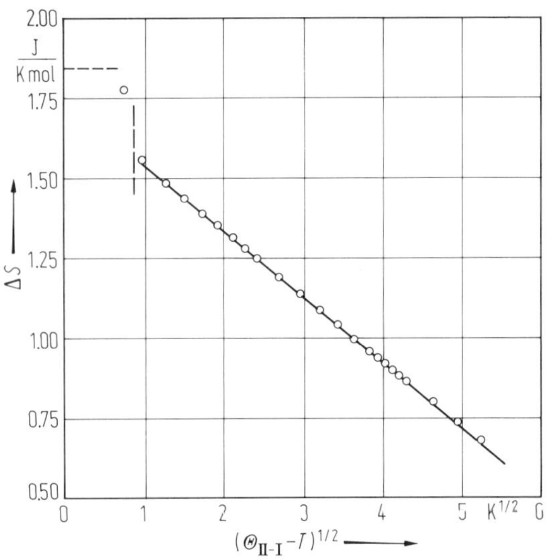

Fig. 455. TlH_2PO_4. ΔS vs. $(\Theta_{II-I}-T)^{1/2}$ [77M14]. ΔS: anomalous entropy. The horizontal broken line indicates the entropy of the phase I. The vertical broken line corresponds to the temperature of the first order transition.

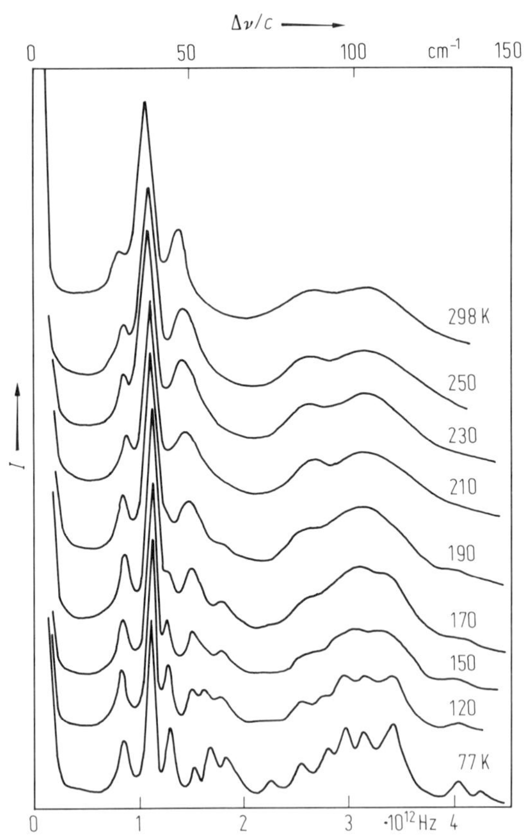

Fig. 456. TlH_2PO_4. I vs. $\Delta\nu$ [76H10]. Parameter: T. I: Raman scattering intensity of a non-oriented single crystal.

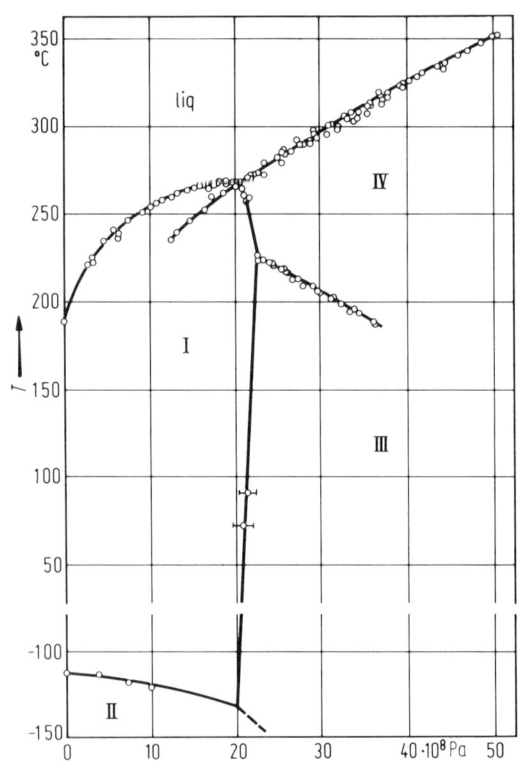

Fig. 457. $NH_4H_2PO_4$(ADP). Phase diagram [69C2].

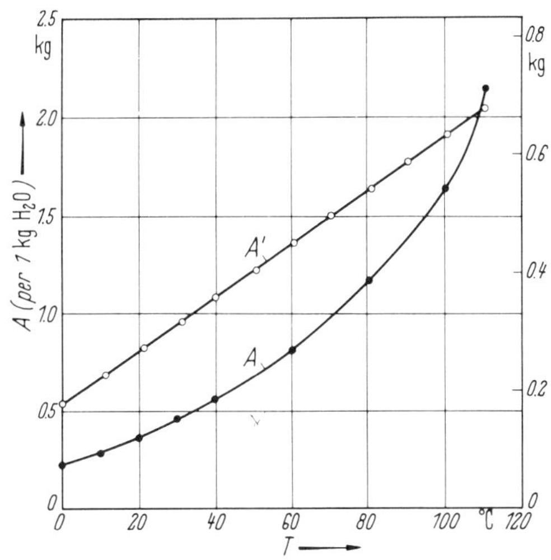

Fig. 458. $NH_4H_2PO_4$(ADP). Solubility A, A' vs. T [58 O 1]. A and A': weight of ADP soluble in 1 kg H_2O or aqueous solution, respectively.

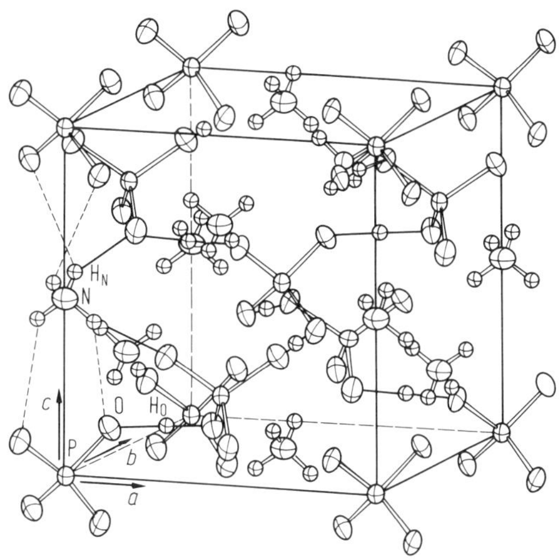

Fig. 459. $NH_4H_2PO_4$(ADP). A perspective view of the unit cell [73 K 12].

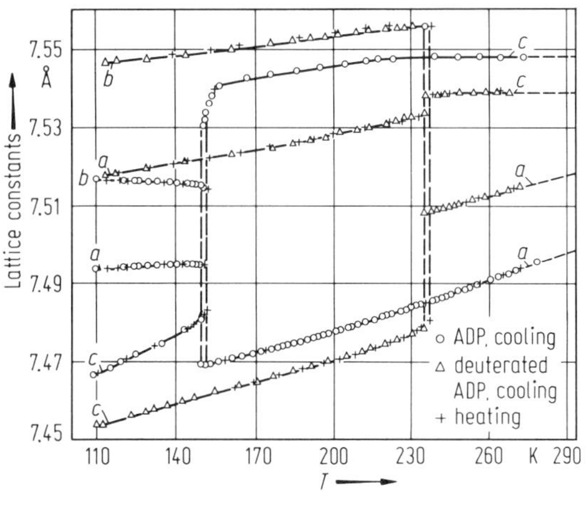

Fig. 460. $NH_4H_2PO_4$(ADP), $ND_4D_2PO_4$(DADP). a, b, c vs. T [70 B 12].

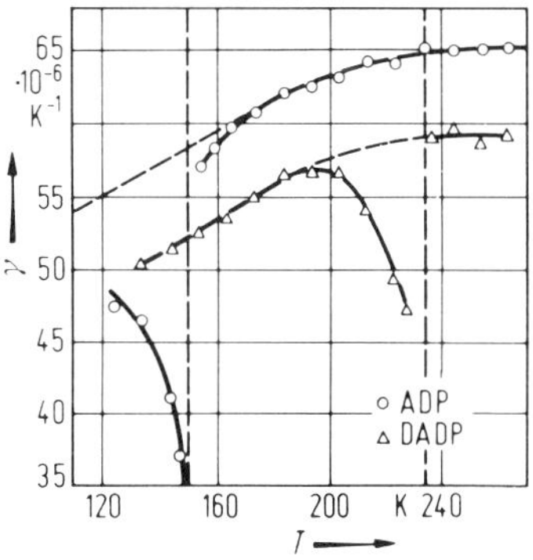

Fig. 461. $NH_4H_2PO_4$(ADP), $ND_4D_2PO_4$(DADP). γ vs. T [70 B 12]. γ: coefficient of thermal volume expansion.

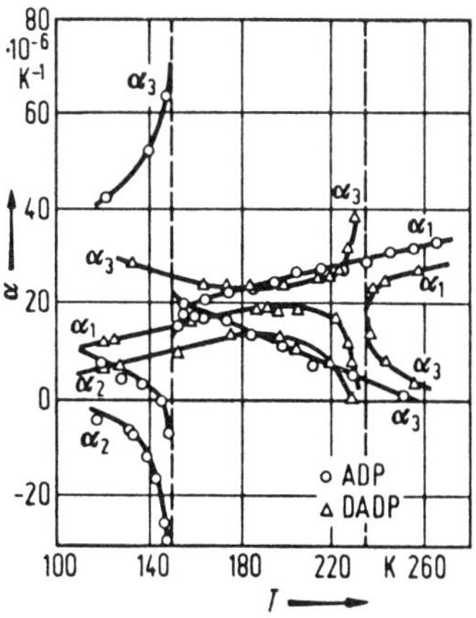

Fig. 462. $NH_4H_2PO_4$(ADP), $ND_4D_2PO_4$(DADP). α_{11}, α_{22}, and α_{33} vs. T [70 B 12]. α_{ii}: linear thermal expansion coefficients, in the Fig. given as α_i.

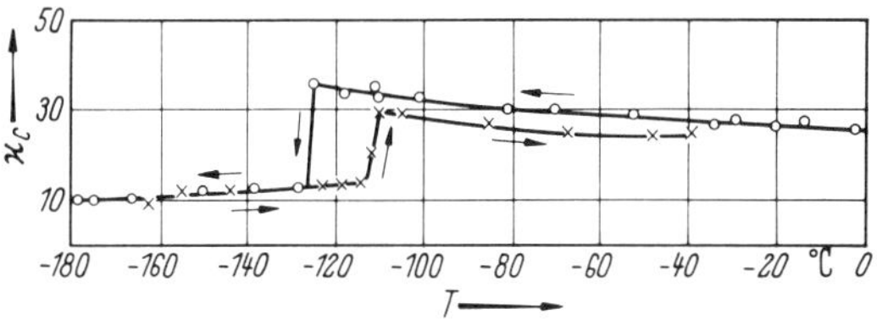

Fig. 464. $NH_4H_2PO_4$(ADP). $\varkappa_c$ vs. T [60 E 1].

Fig. 463, see p. 416.

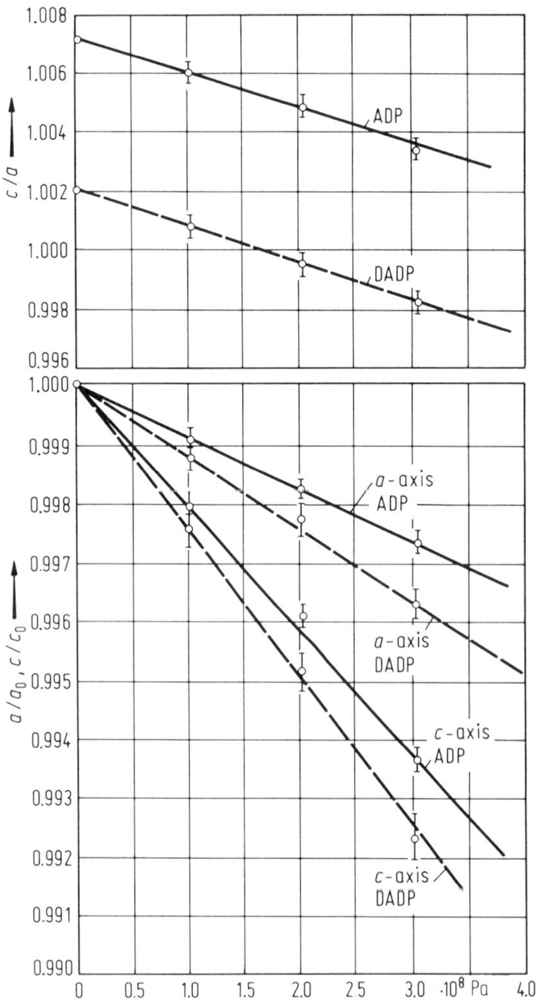

Fig. 463. $NH_4H_2PO_4$(ADP) and $ND_4D_2PO_4$(DADP). a/a_0, c/c_0, and c/a vs. p at 295 K [72M18]. a_0, c_0: lattice constants at atmospheric pressure.

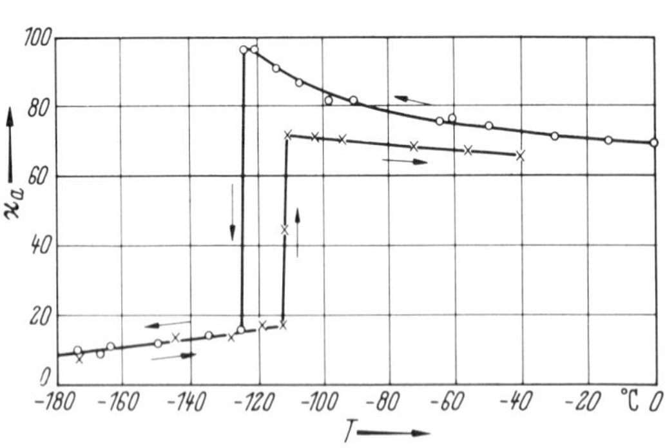

Fig. 465. $NH_4H_2PO_4$(ADP). $\varkappa_a$ vs. T [60E1]. $\varkappa_a$ at $f = 1$ kHz.

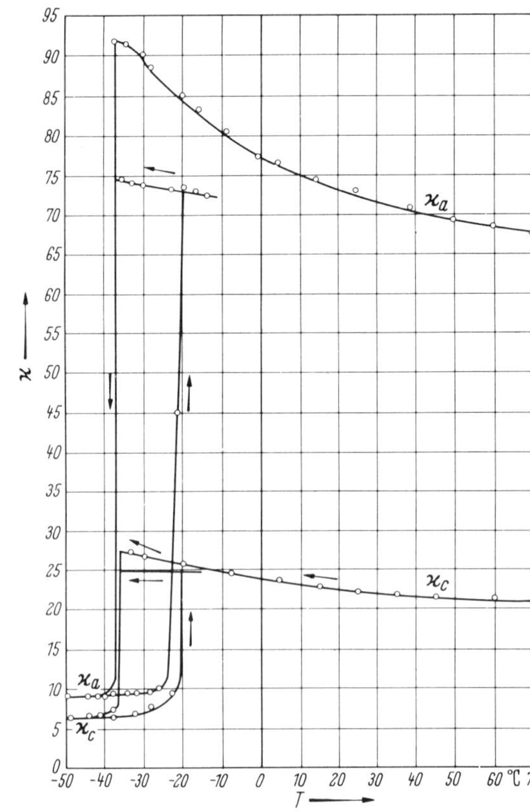

Fig. 466. Deuterated ADP. $\varkappa_a$, $\varkappa_c$ vs. T [52M1].

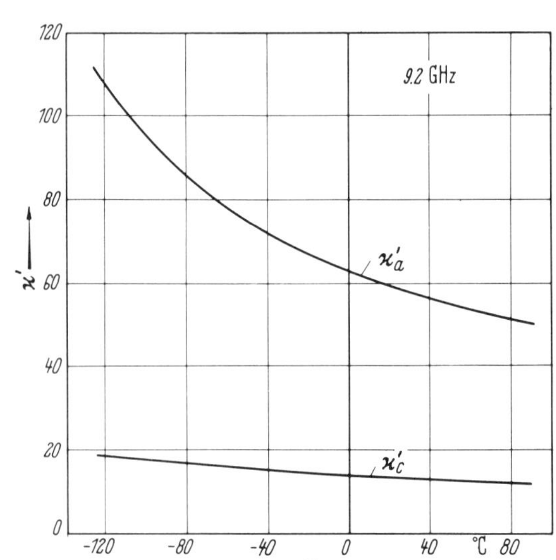

Fig. 467. $NH_4H_2PO_4$(ADP). $\varkappa'_a$ and $\varkappa'_c$ vs. T at 9.2 GHz [65K2].

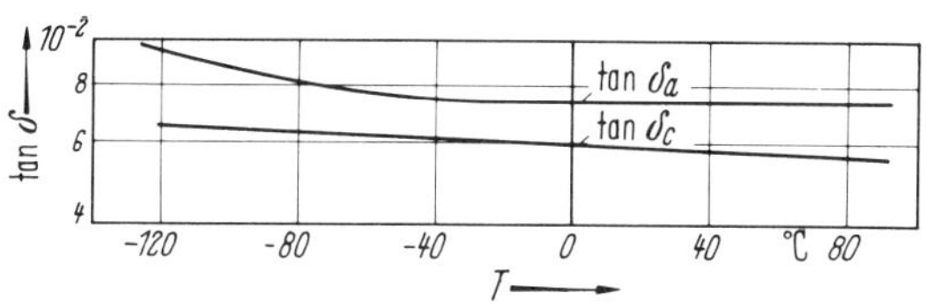

Fig. 468. $NH_4H_2PO_4$(ADP). tan δ vs. T at 9.2 GHz [65 K 2].

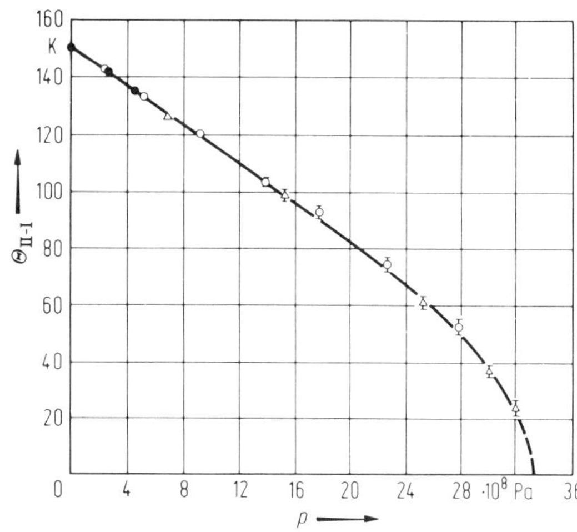

Fig. 469. $NH_4H_2PO_4$(ADP). $\Theta_{\text{II-I}}$ vs. p [71 S 4].

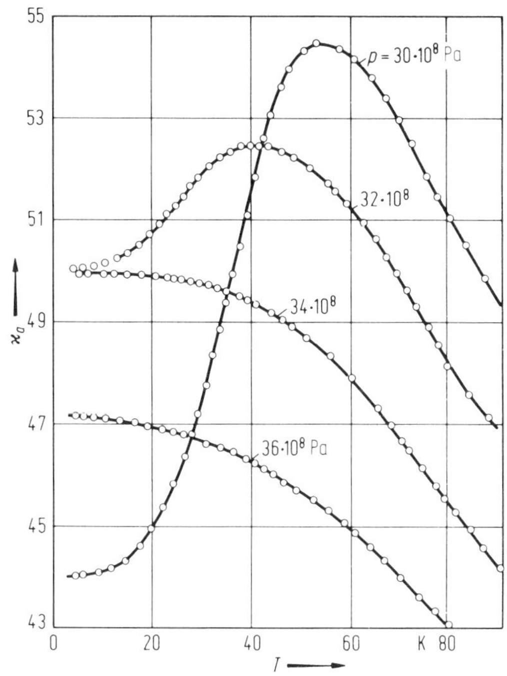

Fig. 470. $NH_4H_2PO_4$(ADP). $\varkappa_a$ vs. T [71 S 4]. Parameter: p.

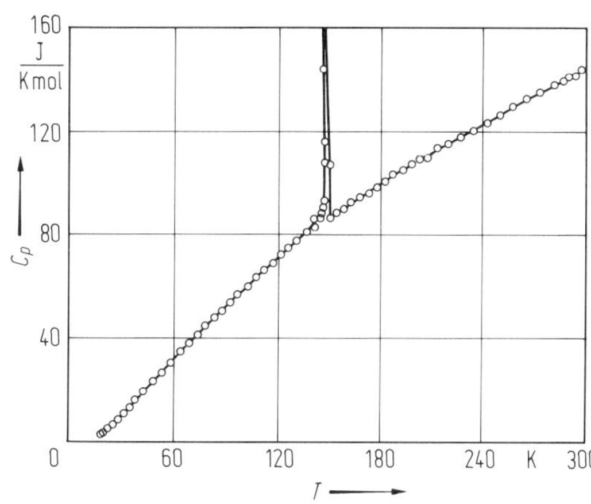

Fig. 471. $NH_4H_2PO_4$(ADP). C_p vs. T [44 S 5]. C_p: molar heat capacity at constant pressure.

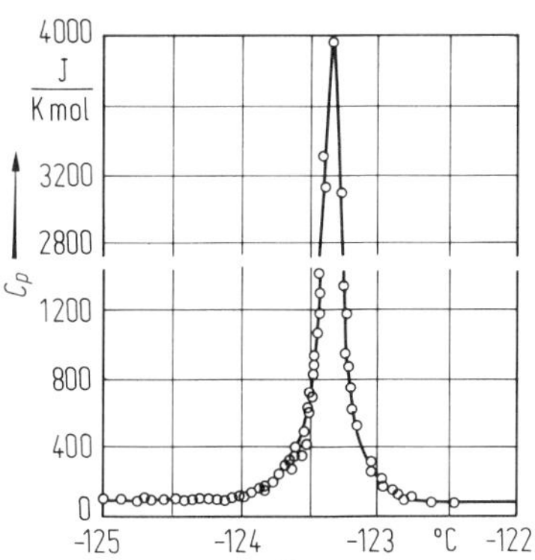

Fig. 472. $NH_4H_2PO_4$(ADP). C_p vs. T [70 A 5]. C_p: molar heat capacity in the vicinity of $\Theta_{\text{II-I}}$.

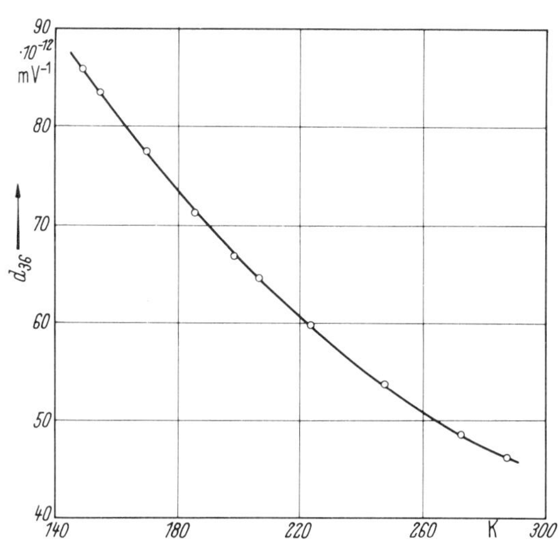

Fig. 474. $NH_4H_2PO_4$(ADP). d_{36} vs. T [47 M 1].

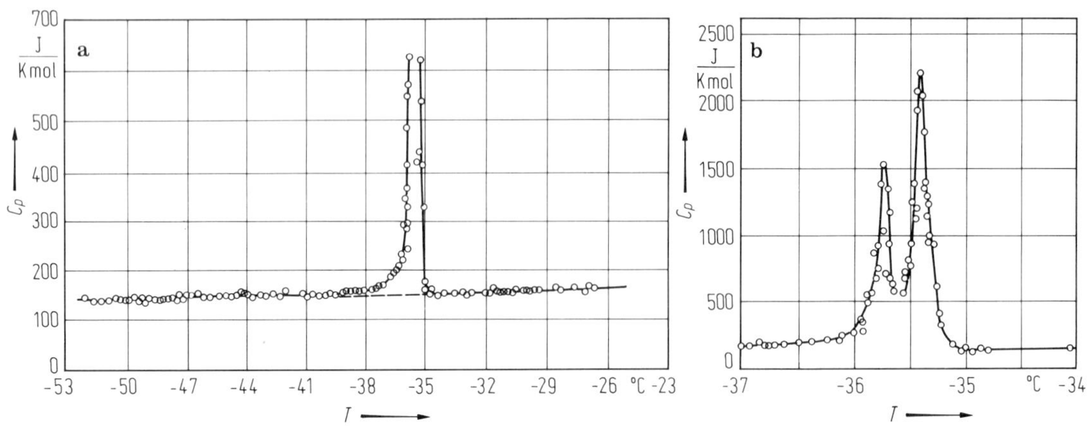

Fig. 473. $ND_4D_2PO_4$(DADP). C_p vs. T [70A5]. Fig. a: wide temperature range, Fig. b: in the vicinity of Θ_{II-I}.

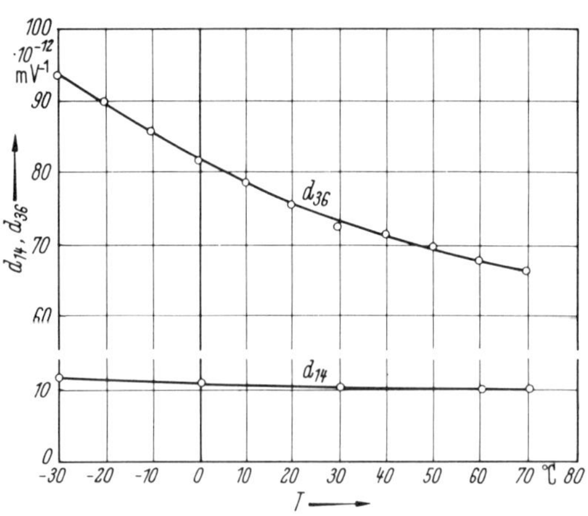

Fig. 475. $NH_4D_2PO_4$(DADP). d_{14} and d_{36} vs. T [52M1].

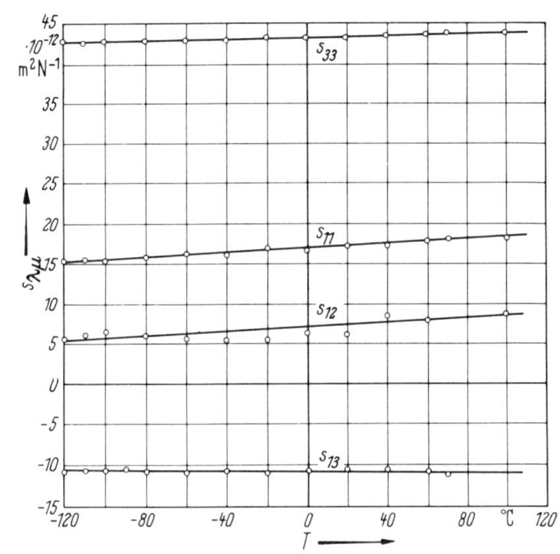

Fig. 476. $NH_4H_2PO_4$(ADP). $s_{\lambda\mu}$ vs. T [46M1].

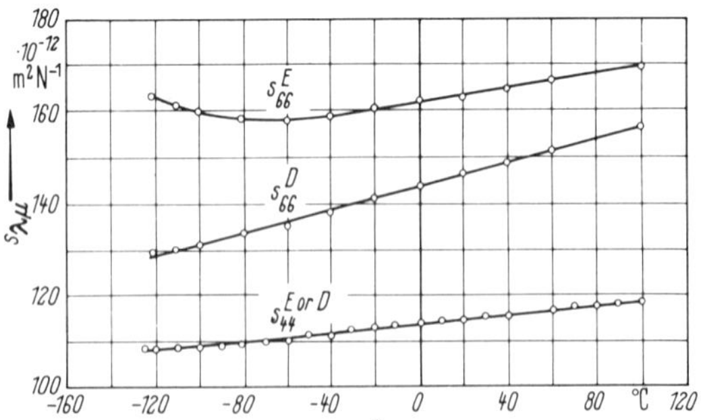

Fig. 477. $NH_4H_2PO_4$(ADP). s_{44}^{A} and s_{66}^{A} vs. T [46M1]. $\boldsymbol{A}=\boldsymbol{E}$ or $\boldsymbol{D}$.

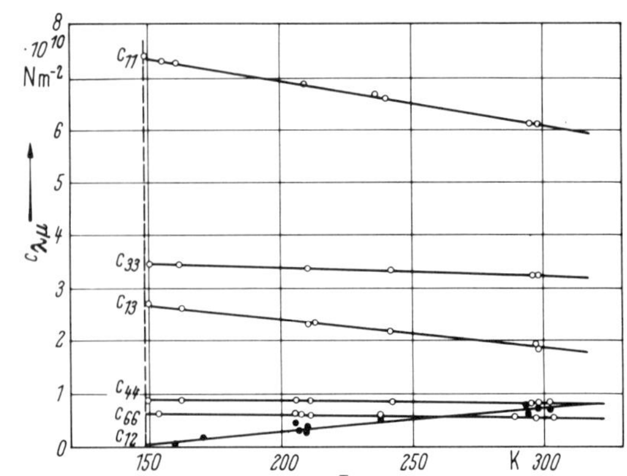

Fig. 478. $NH_4H_2PO_4$(ADP). $c_{\lambda\mu}$ vs. T [46Z1].

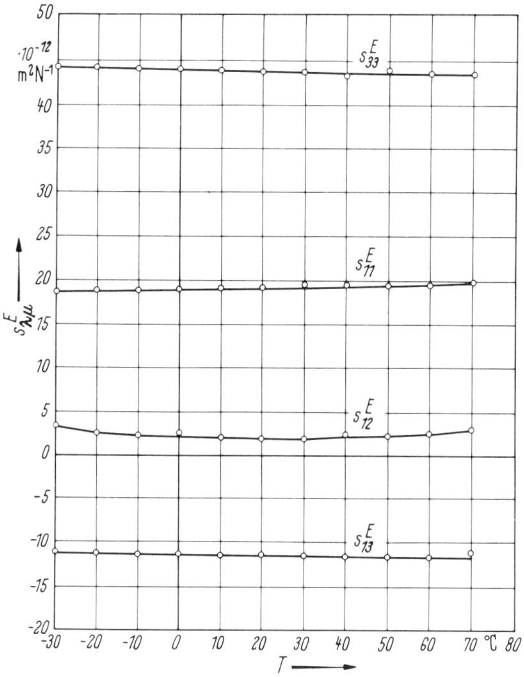

Fig. 479. $ND_4D_2PO_4$(DADP). $s_{\lambda\mu}^E$ vs. T [52 M 1].

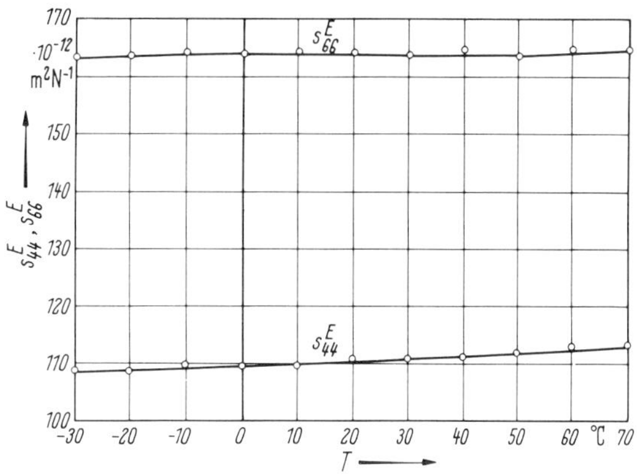

Fig. 480. $ND_4D_2PO_4$(DADP). s_{44}^E and s_{66}^E vs. T [52 M 1].

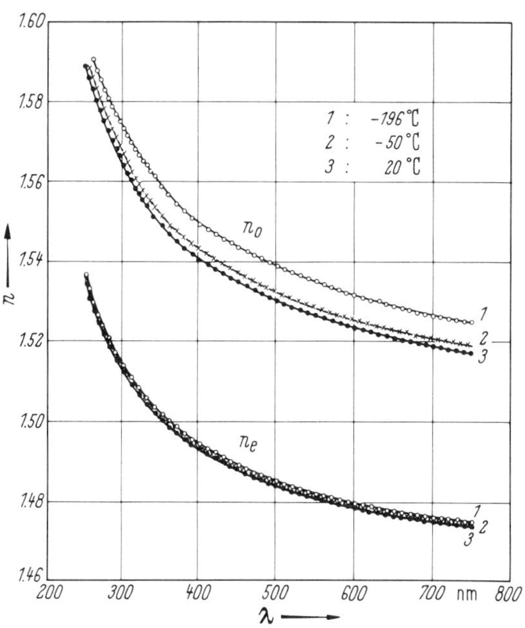

Fig. 481. $NH_4H_2PO_4$(ADP). n vs. λ [66 V 5]. Parameter: T. See also [65 V 2].

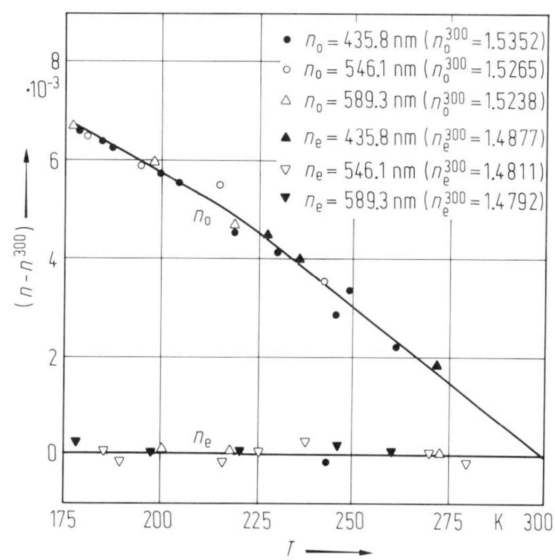

Fig. 482. $NH_4H_2PO_4$(ADP). $(n - n^{300})$ vs. T [66 Y 1]. n^{300}: refractive indices at 300 K.

Figs. 483…485, see p. 420.

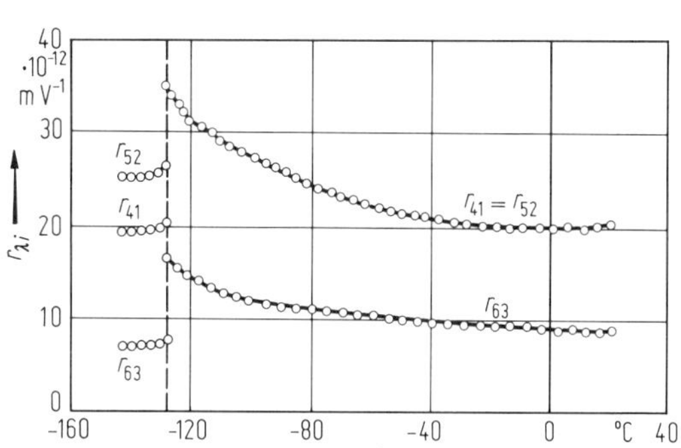

Fig. 486. $NH_4H_2PO_4$(ADP). $r_{\lambda i}$ vs. T [69 A 12]. $\lambda = 557$ nm.

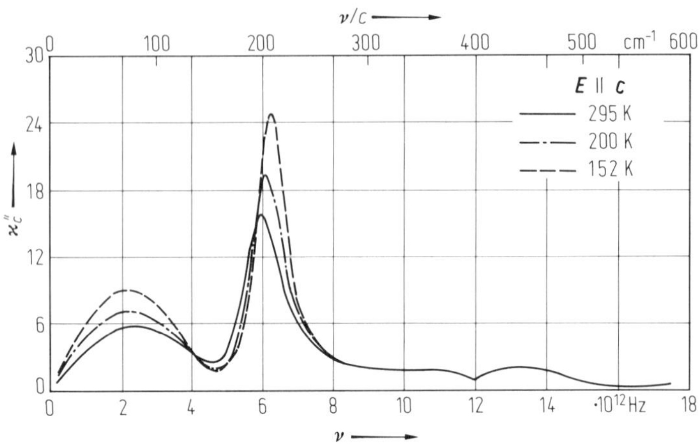

Fig. 484. $NH_4H_2PO_4$(ADP). $\varkappa_c''$ vs. ν [73 K 8]. Parameter: T. The curves were obtained from reflectivity data using Kramers-Kronig relation.

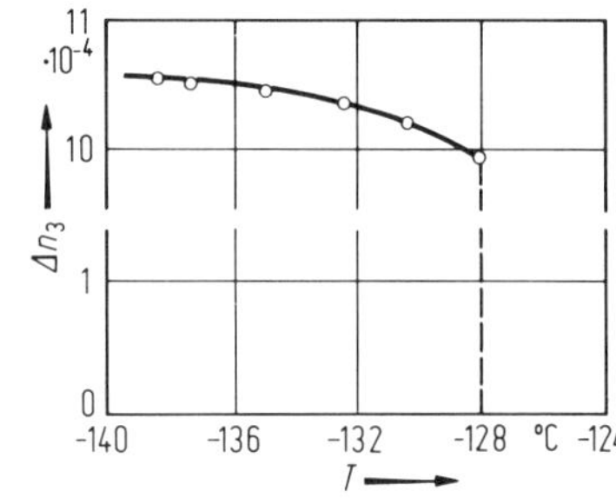

Fig. 483. $NH_4H_2PO_4$(ADP). Δn_3 vs. T [69 A 13]. Δn_3: birefringence in the (001) plane. $\lambda = 535$ nm.

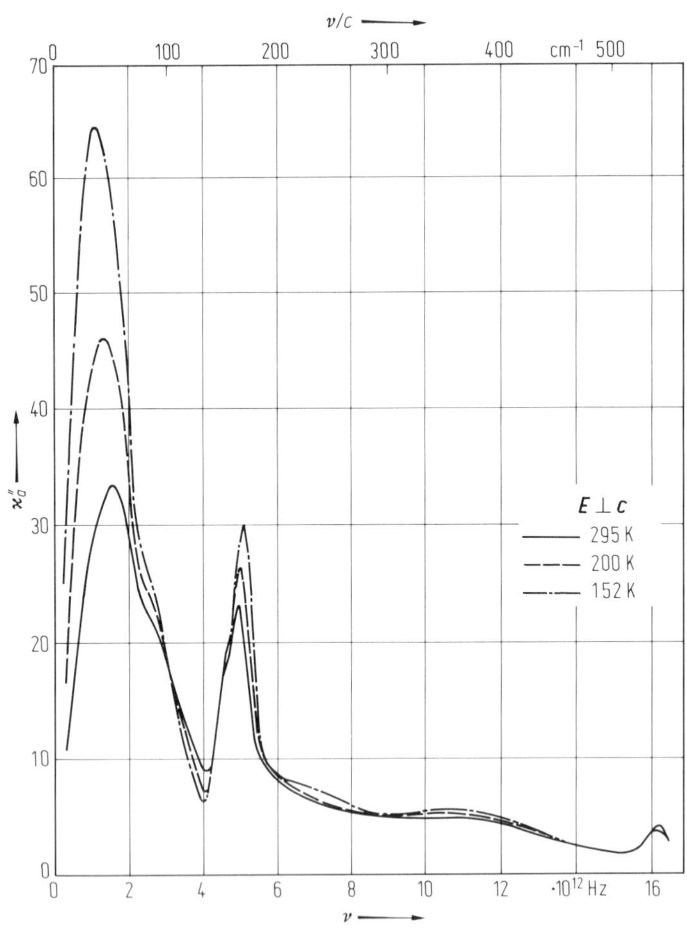

Fig. 485. $NH_4H_2PO_4$(ADP). $\varkappa_a''$ vs. ν [73 K 8]. Parameter: T. The curves were obtained from reflectivity data using Kramers-Kronig relation.

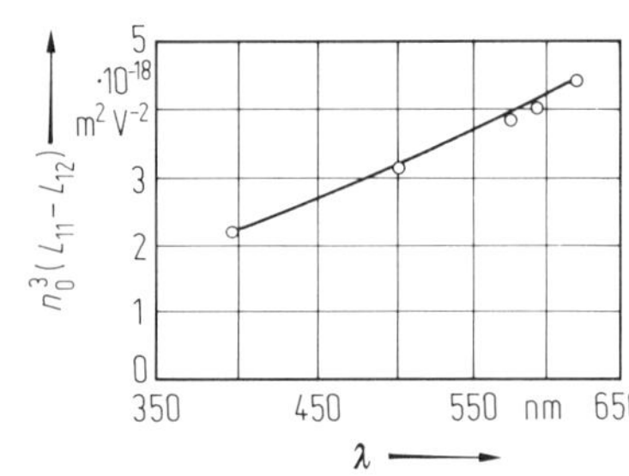

Fig. 488. $NH_4H_2PO_4$(ADP). $n_0^3 \cdot (L_{11} - L_{12})$ vs. λ at RT [69 A 13].

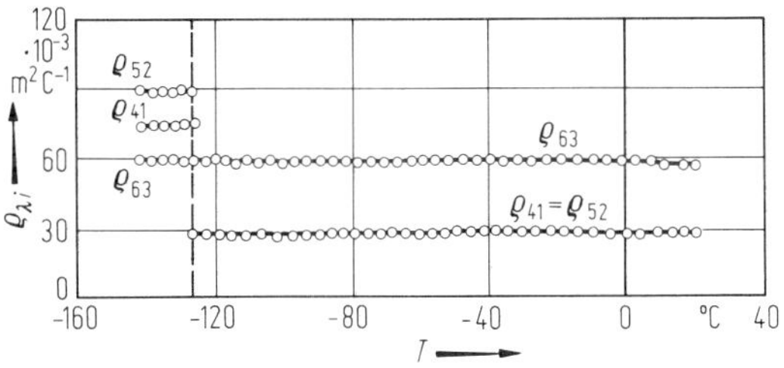

Fig. 487. $NH_4H_2PO_4$(ADP). $\varrho_{\lambda i}$ vs. T [69A12]. ϱ: electrooptic constant for $\boldsymbol{P}$. $\lambda = 577$ nm.

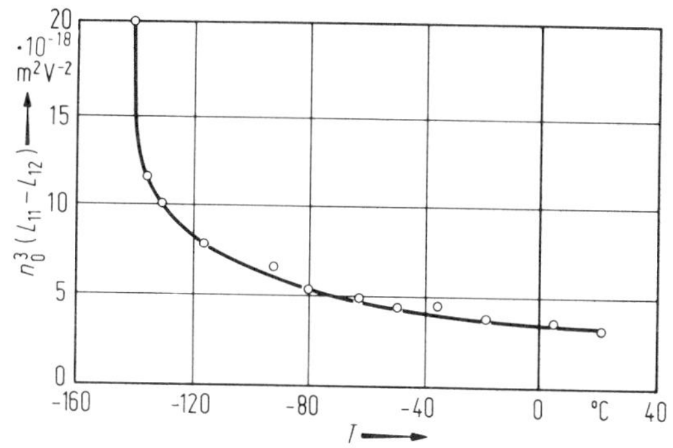

Fig. 489. $NH_4H_2PO_4$(ADP). $n_0^3(L_{11}-L_{12})$ vs. T [69A13]. $\lambda = 518$ nm. $L_{\lambda\mu}$: quadratic electrooptic constants for $\boldsymbol{E}$.

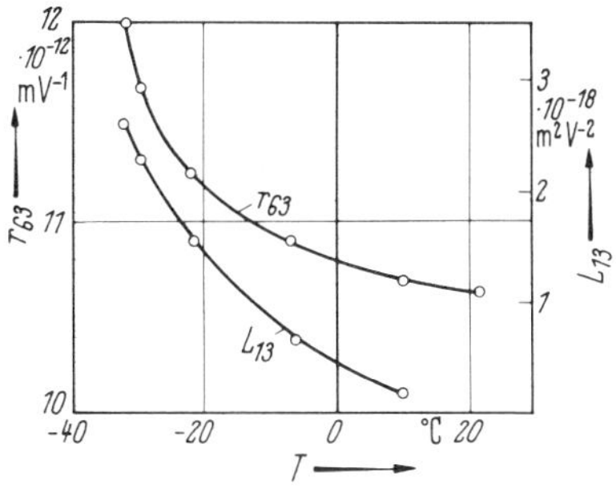

Fig. 490. $ND_4D_2PO_4$(DADP). r_{63}, L_{13} vs. T [66V3]. $\lambda = 535$ nm.

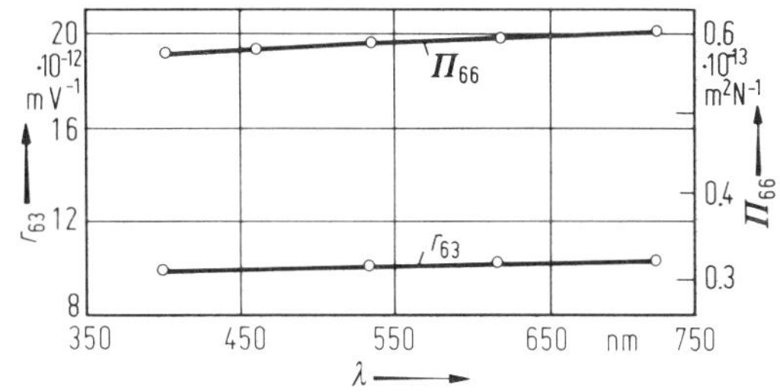

Fig. 491. $ND_4D_2PO_4$(DADP). r_{63}, Π_{66} vs. λ at RT [67V4].

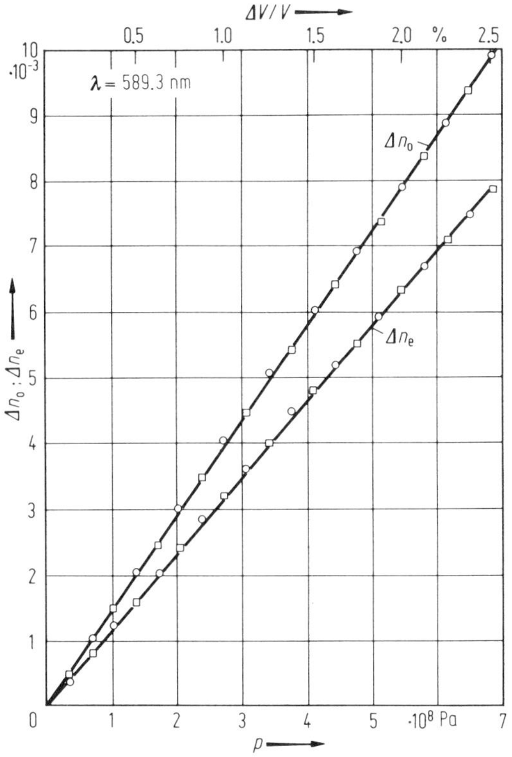

Fig. 492. $NH_4H_2PO_4$(ADP). Δn_o, Δn_e vs. p [68D2]. $\Delta n_o = n_o(p) - n_o(0)$, $\Delta n_e = n_e(p) - n_e(0)$. $n_o(0) = 1.524$, $n_e(0) = 1.479$. $T = 22$ °C.

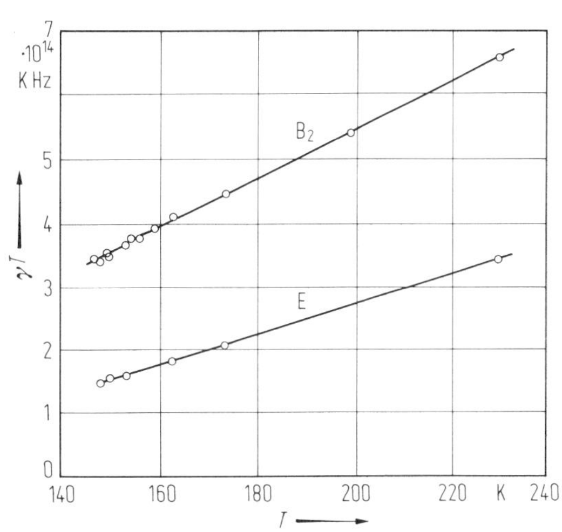

Fig. 494. $NH_4H_2PO_4$(ADP). γT vs. T [72B18]. γ: inverse of the Debye relaxational time obtained from Raman spectra of a B_2 mode and a E mode. See also [72R7].

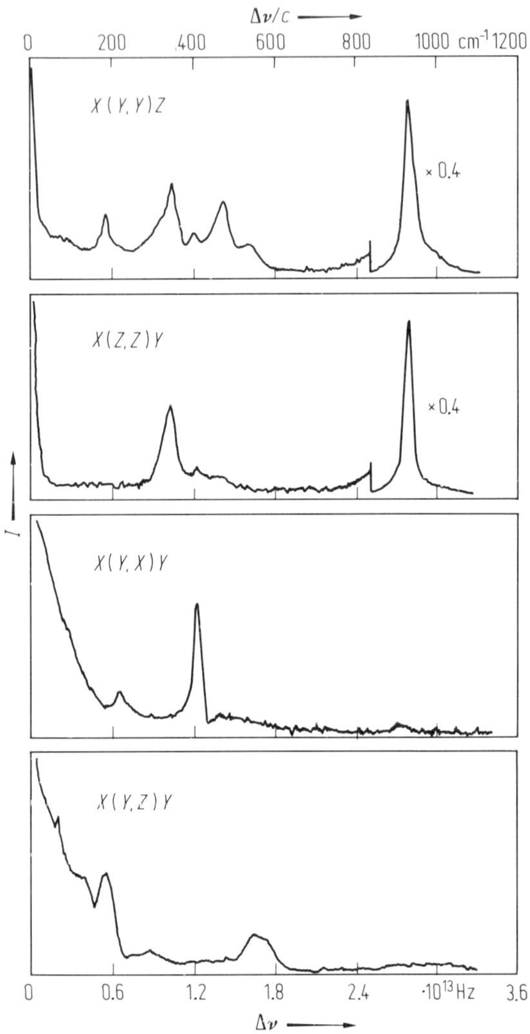

Fig. 493. $NH_4H_2PO_4$(ADP). I vs. $\Delta\nu$ at 295 K [73 K 8]. I: Raman scattering intensity in different scattering geometries.

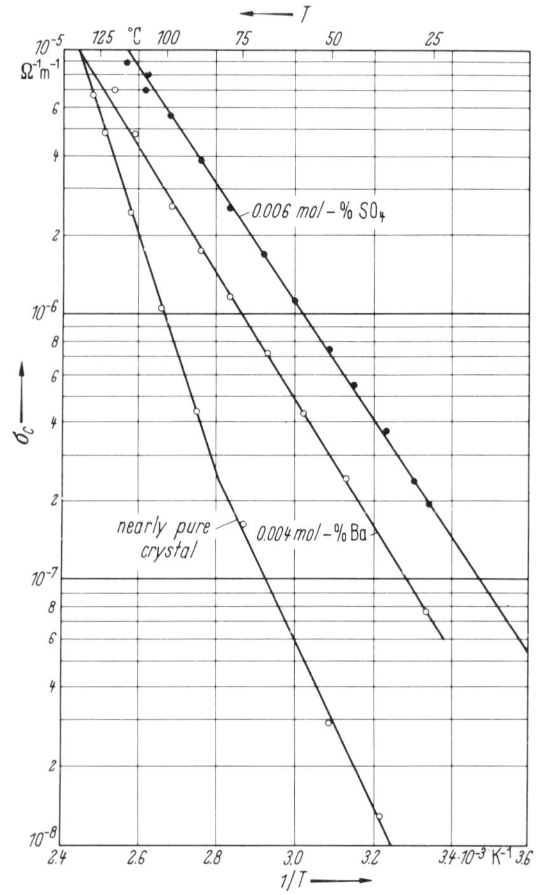

Fig. 495. $NH_4H_2PO_4$(ADP). σ_c vs. $1/T$ for crystals with and without impurities [64 M 7]. Applied field: 50···180 kVm^{-1}, specimen: 10 × 10 × 3 mm^2, electrodes: Au.

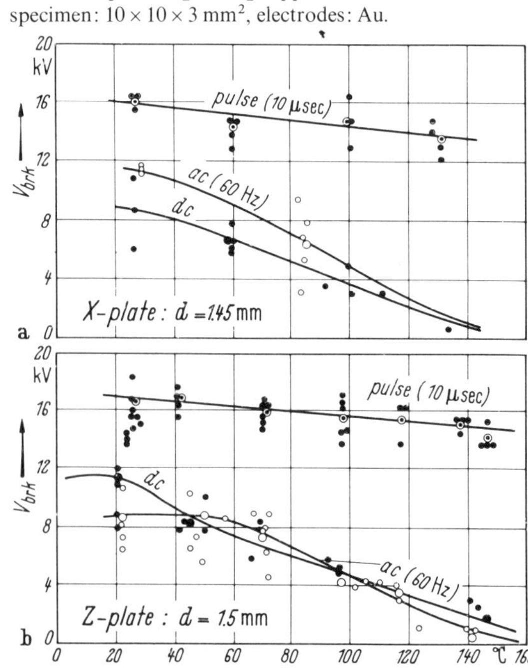

Fig. 499. $NH_4H_2PO_4$(ADP). Breakdown voltage V_{brk} vs. T [59 Y 1]. Fig. a: X-plate; Fig. b: Z-plate.

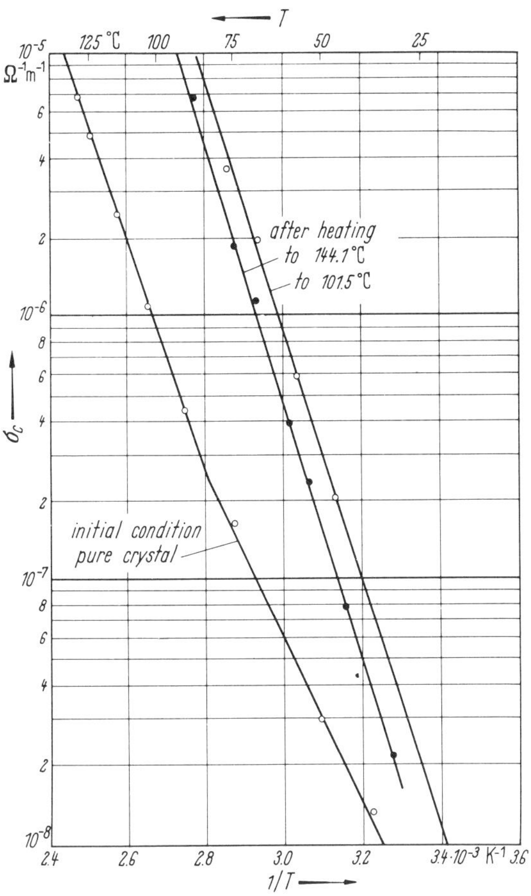

Fig. 496. $NH_4H_2PO_4$(ADP). σ_c vs. $1/T$ before and after heating for 24 hours [64M7]. Measuring condition is same as in Fig. 495.

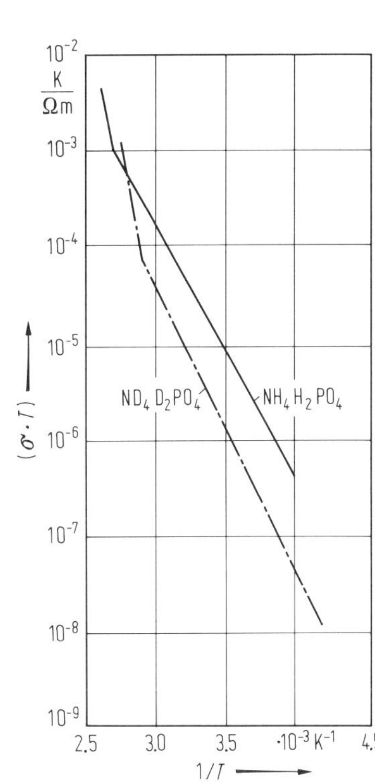

Fig. 497. $NH_4H_2PO_4$(ADP) and $ND_4D_2PO_4$(DADP). σT vs. $1/T$ [76P10].

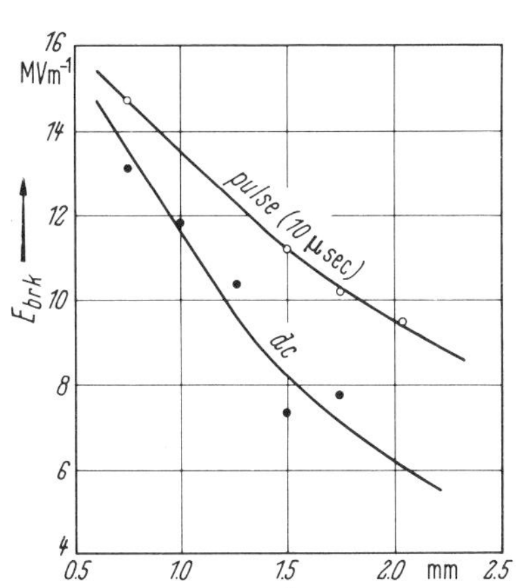

Fig. 498. $NH_4H_2PO_4$(ADP). Breakdown field E_{brk} at RT vs. thickness d when d.c. and pulse fields are applied [59Y1]. Field ∥ [001].

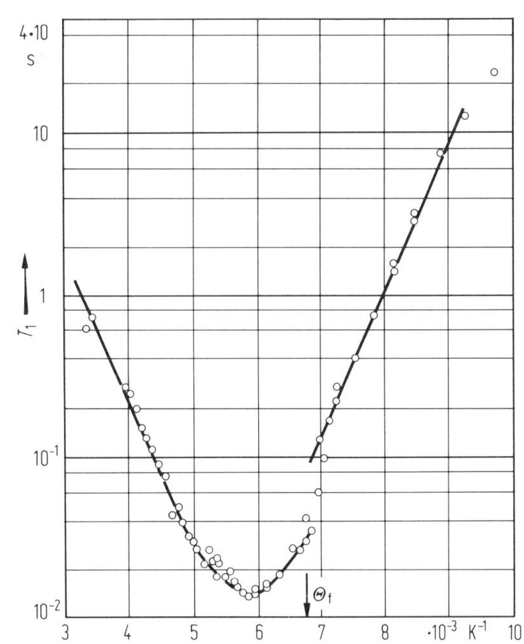

Fig. 500. $NH_4H_2PO_4$(ADP). T_1 vs. $1/T$ [68G2]. T_1: proton spin-lattice relaxation time. $f = 42$ MHz.

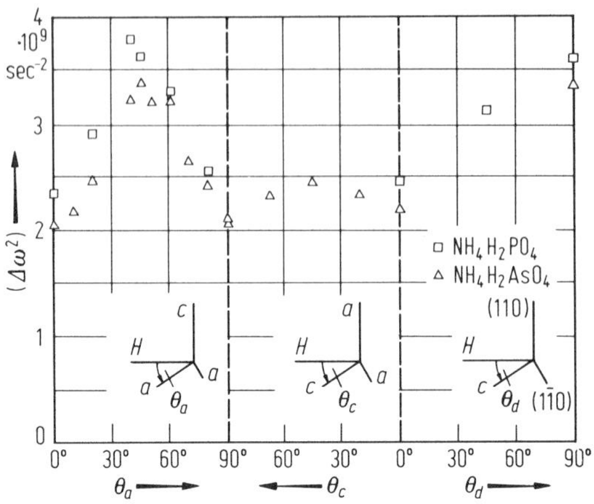

Fig. 501. $NH_4H_2PO_4$(ADP) and $NH_4H_2AsO_4$(ADA). $\langle\Delta\omega^2\rangle$ vs. θ [71A6]. $\langle\Delta\omega^2\rangle$: proton second moment in s^{-2}. θ: angle between ***H*** and the crystallographic axes as shown in the figure.

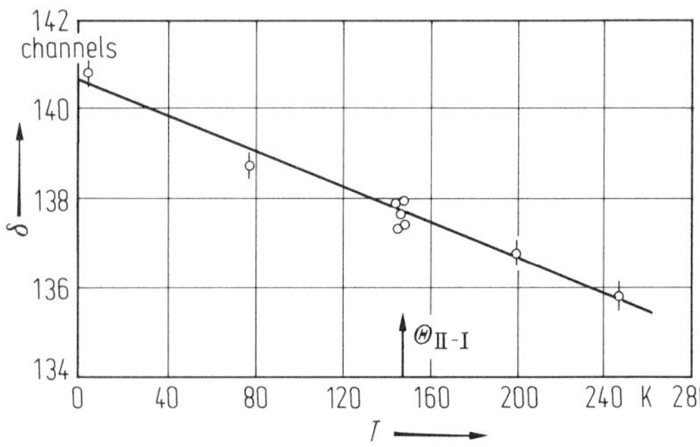

Fig. 502. $NH_4H_2PO_4$(ADP): ^{57}Fe. δ vs. T [72S4]. δ: isomer shift. 1 channel = 0.033 mm · s^{-1}.

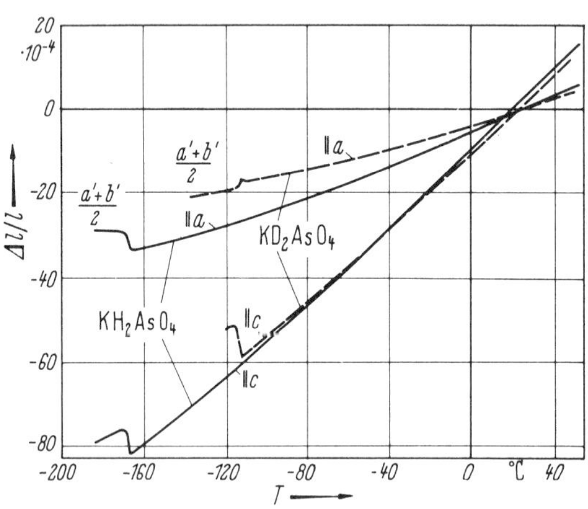

Fig. 504. KH_2AsO_4(KDA) and KD_2AsO_4(DKDA). Thermal expansion $\Delta l/l$ vs. T [67C5]. $(a'+b')/2$ shows the mean of the values along a and b axes in the phase II.

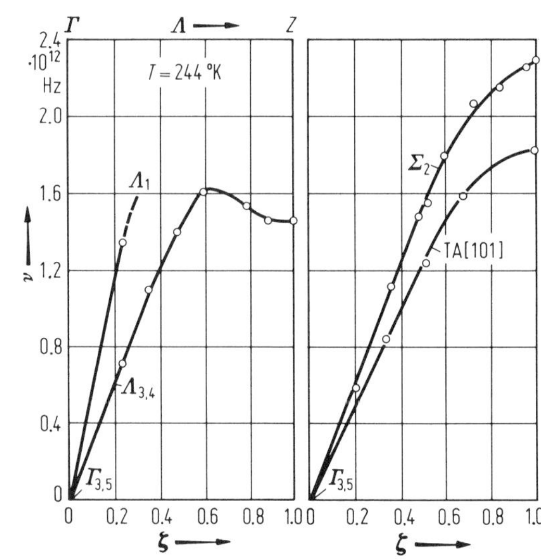

Fig. 503. $ND_4D_2PO_4$(ADP). ν vs. ζ [69M3]. ν: phonon frequency, ζ: reduced wave number coordinates. For Γ, Λ and Σ, see Introduction.

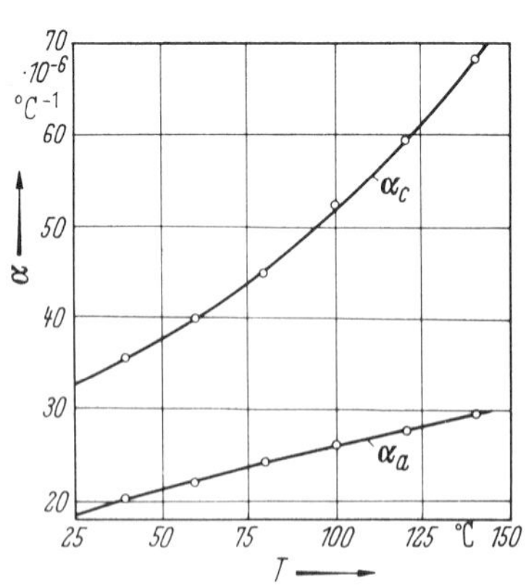

Fig. 505. KH_2AsO_4(KDA). α_a, α_c vs. T [65D4]. α_a, α_c: linear thermal expansion coefficients along the a, c axes.

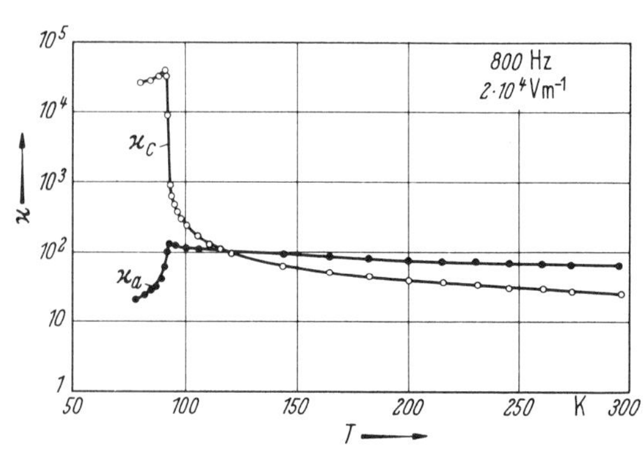

Fig. 506. KH_2AsO_4(KDA). $\varkappa_a$, $\varkappa_c$ vs. T [38B3].

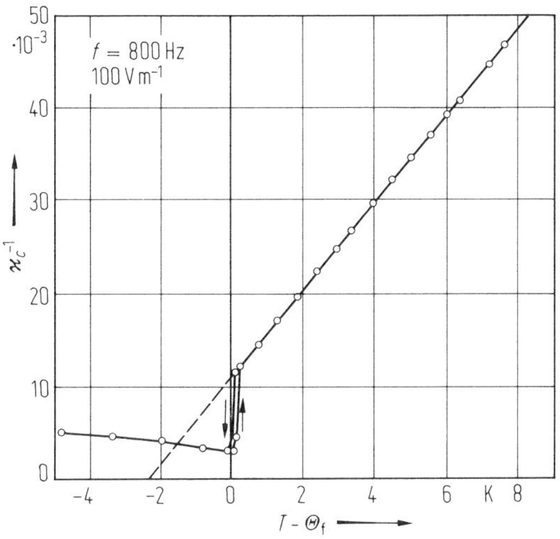

Fig. 507. KH_2AsO_4(KDA). $\varkappa_c^{-1}$ vs. $T-\Theta_f$ [73G8]. $\Theta_f=95.8$ K.

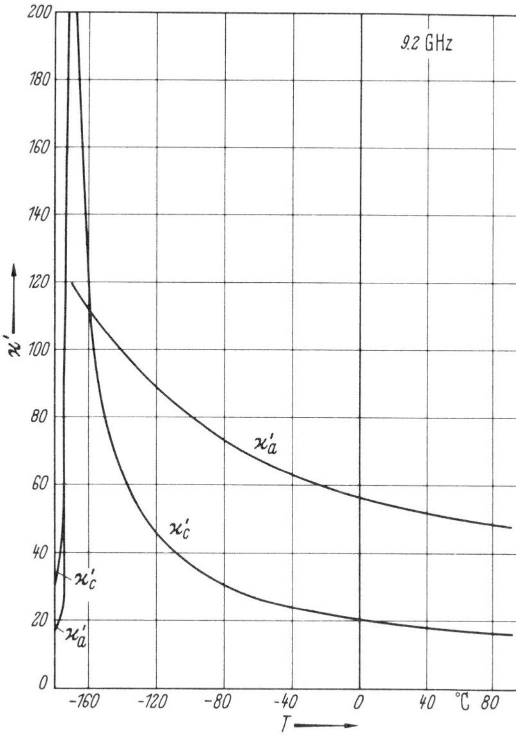

Fig. 508. KH_2AsO_4(KDA). $\varkappa'_a$ and $\varkappa'_c$ vs. T [65K2].

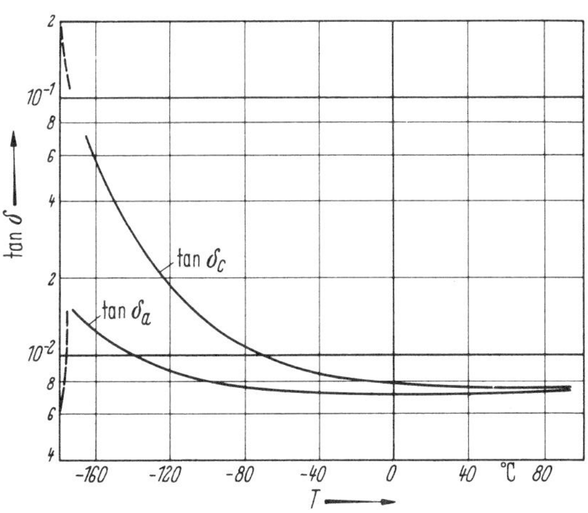

Fig. 509. KH_2AsO_4(KDA). $\tan\delta$ vs. T [65K2]. $f=9.2$ GHz. Measurements indicated by the dashed part are in doubt.

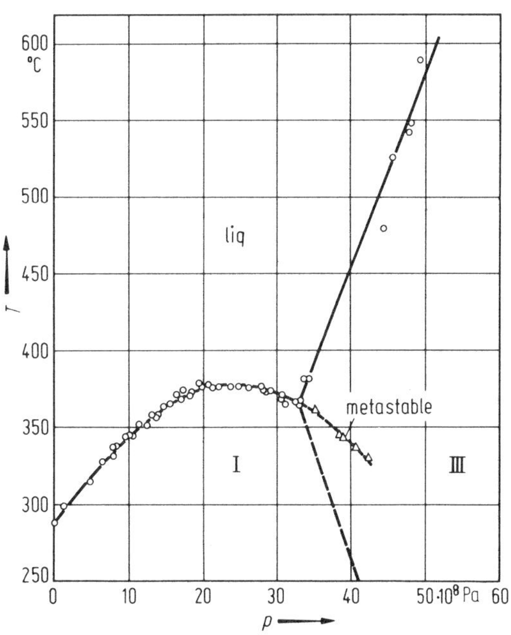

Fig. 510. KH_2AsO_4(KDA). T vs. p [69C2]. Phase diagram.

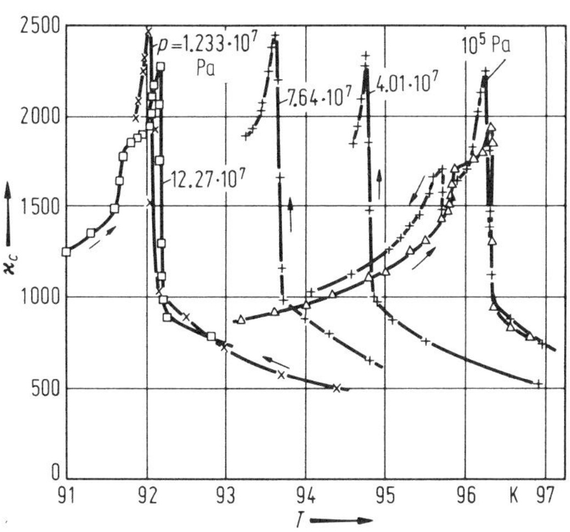

Fig. 511. KH_2AsO_4(KDA). $\varkappa_c$ vs. T [70F6]. Parameter: p. $f=800$ Hz.

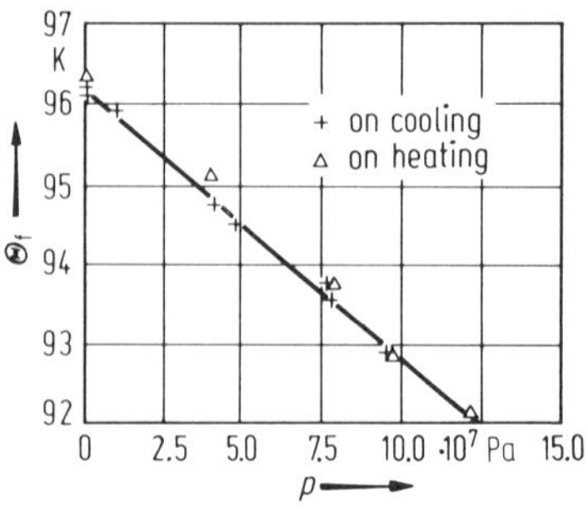

Fig. 512. KH_2AsO_4(KDA). Θ_f vs. p [70 F 6].

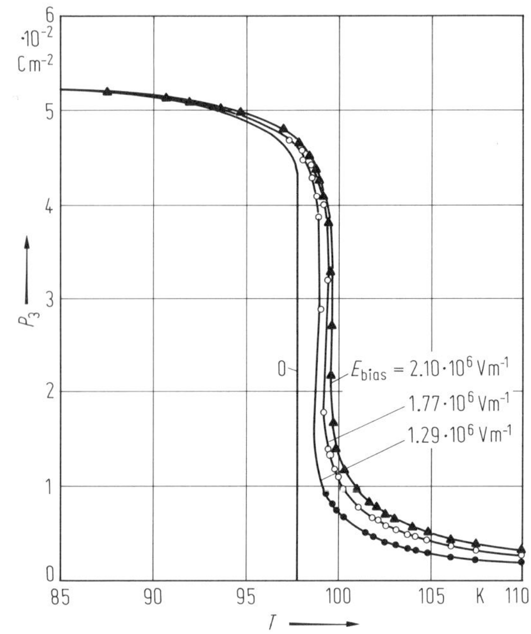

Fig. 514. KH_2AsO_4(KDA). P_3 vs. T [77 C 1]. Parameter: E_{bias}.

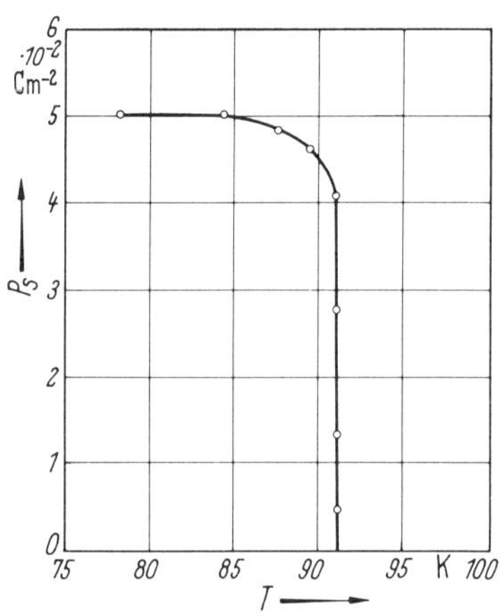

Fig. 513. KH_2AsO_4(KDA). P_s vs. T [38 B 3].

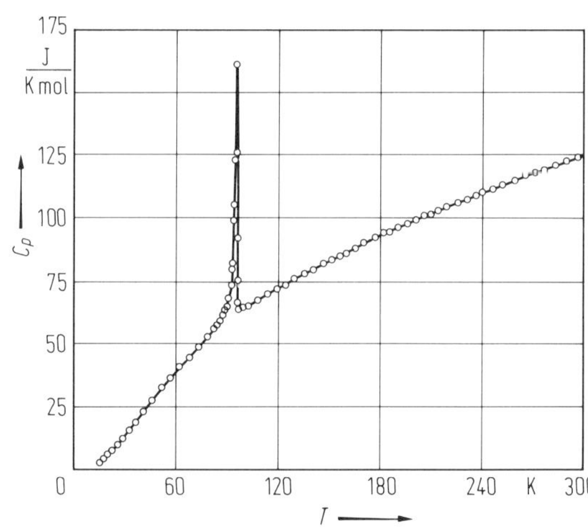

Fig. 515. KH_2AsPO_4(KDA). C_p vs. T [44 S 4]. C_p: molar heat capacity at constant pressure.

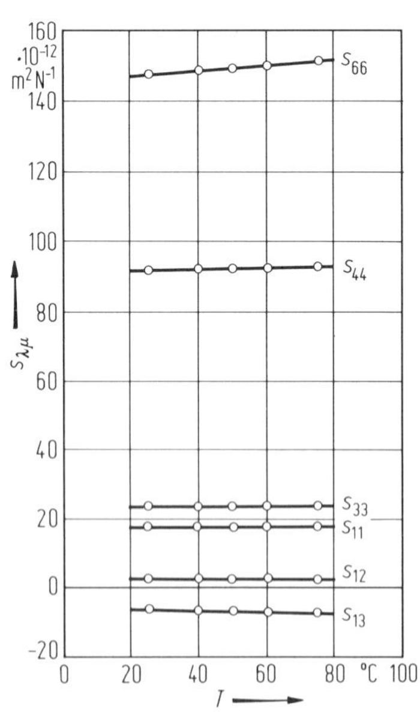

Fig. 518. KH_2AsO_4(KDA). $s_{\lambda\mu}$ vs. T [68 A 2].

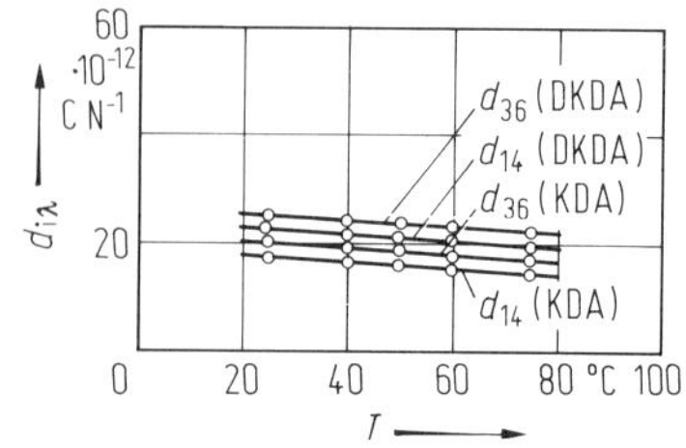

Fig. 517. KH_2AsO_4(KDA) and KD_2AsO_4(DKDA). d_{36}, d_{14} vs. T [68 A 2].

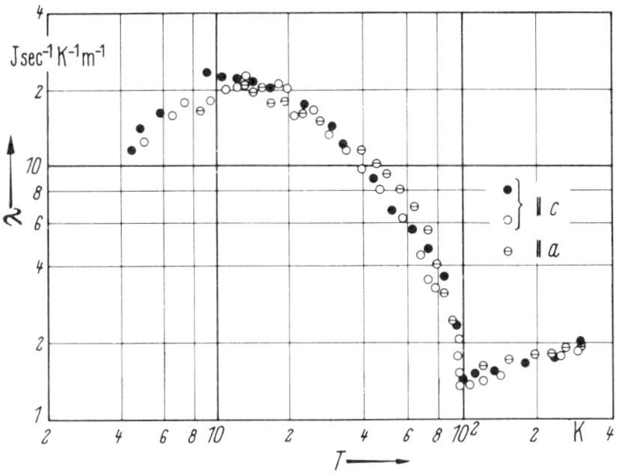

Fig. 516. KH_2AsO_4(KDA). λ vs. T [67 S 27]. λ: thermal conductivity.

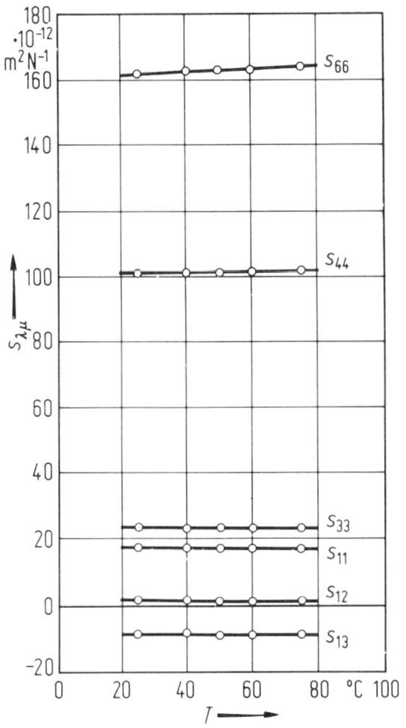

Fig. 519. KD_2AsO_4(DKDA). $s_{\lambda\mu}$ vs. T [68 A 2].

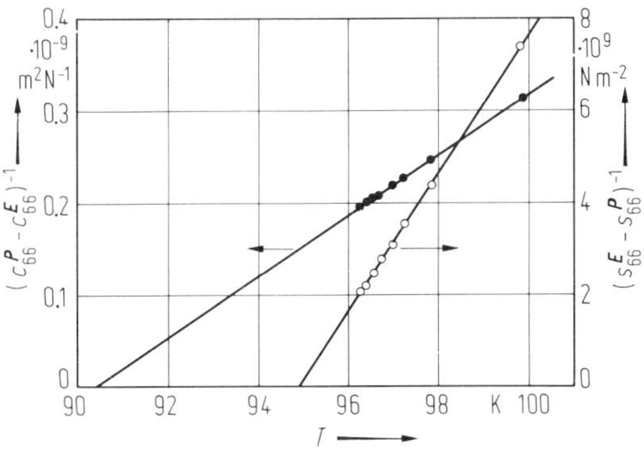

Fig. 520. KH_2AsO_4(KDA). $(c_{66}^P - c_{66}^E)^{-1}$, $(s_{66}^E - s_{66}^P)^{-1}$ vs. T [77 D 7]. Data obtained from Brillouin scattering.

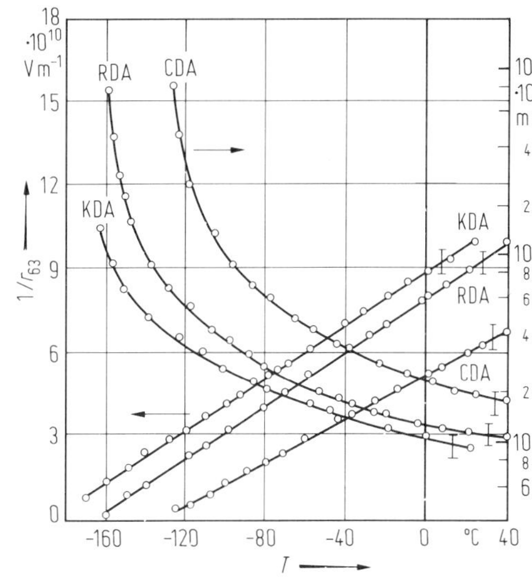

Fig. 522. KH_2AsO_4(KDA), RbH_2AsO_4(RDA), CsH_2AsO_4 (CDA). r_{63}, $1/r_{63}$ vs. T [71 V 1].

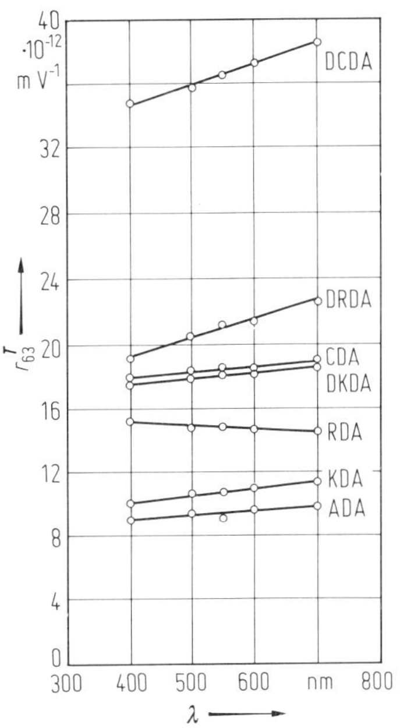

Fig. 521. KDA family. r_{63}^T vs. λ at RT [69 A 2].

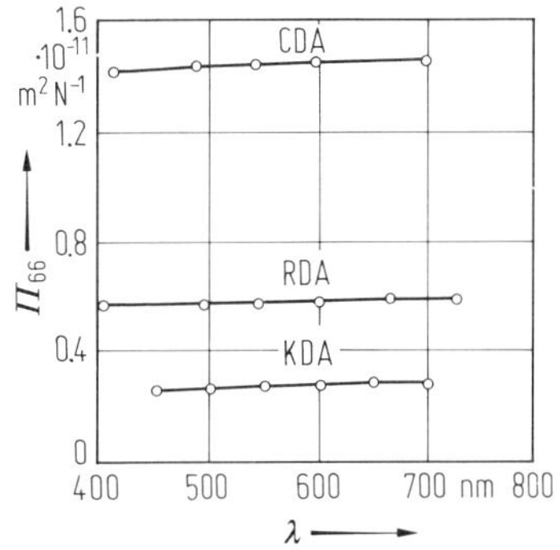

Fig. 523. KH_2AsO_4(KDA), RbH_2AsO_4(RDA), CsH_2AsO_4 (CDA). Π_{66} vs. λ [67 V 3].

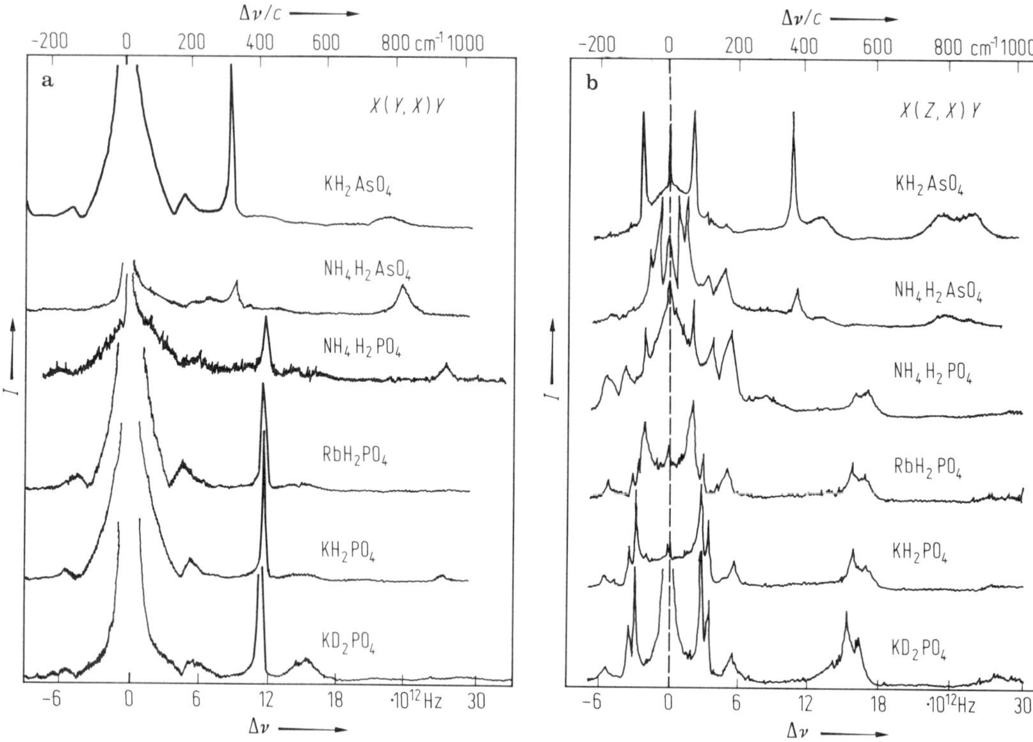

Fig. 524. KDP family. I vs. $\Delta\nu$ [71H3]. I: Raman scattering intensity. Fig. a: B_2 modes measured with the scattering geometry $X(Y, X)Y$; Fig. b: E modes measured with $X(Z, X)Y$.

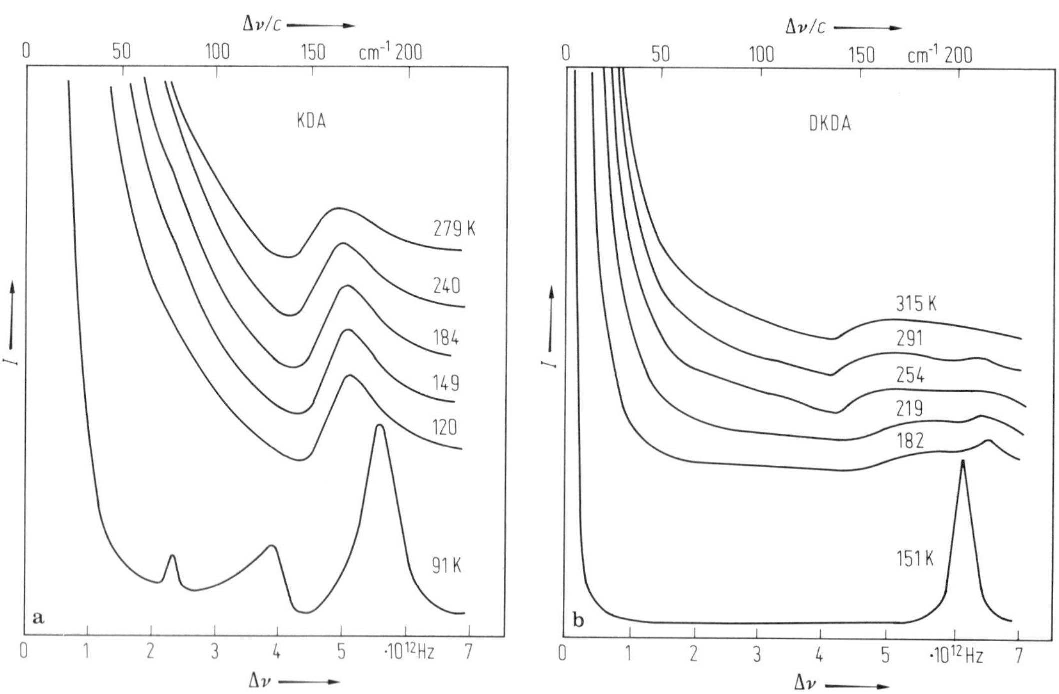

Fig. 525. KH_2AsPO_4(KDA) (Fig. a), KD_2AsPO_4(DKDA) (Fig. b). I vs. $\Delta\nu$ [74L10]. I: Raman scattering intensity of B_2 modes. Parameter: T.

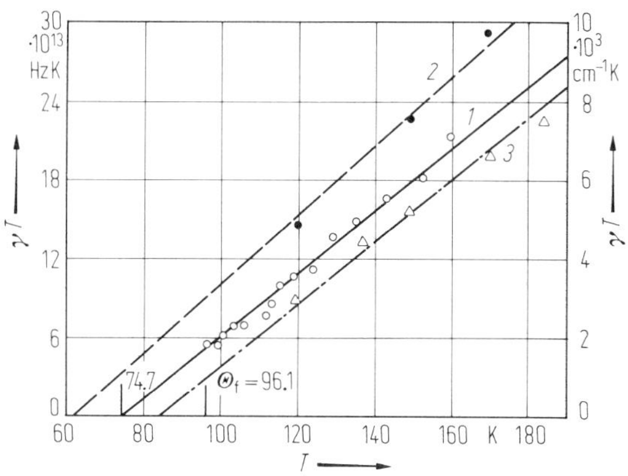

Fig. 526. KH_2AsPO_4(KDA). γT vs. T. γ: damping constant of phonons. Curve *1*: Linewidth of a B_2 mode at $1.17 \cdot 10^{13}$ Hz or 390 cm^{-1} [75L2], curves *2* and *3*: γ obtained from the coupled mode analysis of the soft mode with a real and imaginary coupling, respectively [74L10].

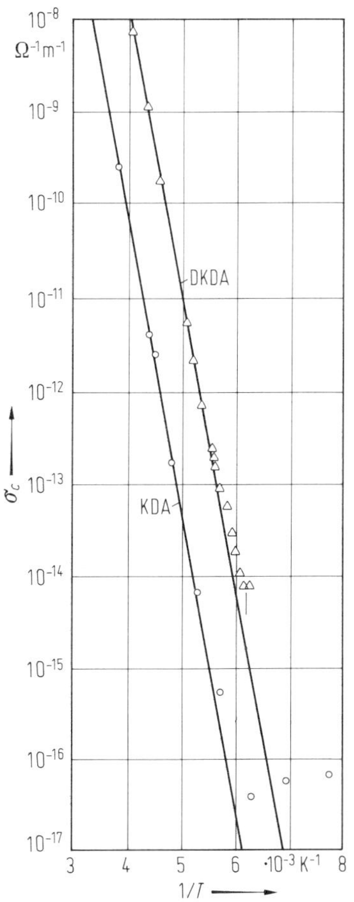

Fig. 528. KH_2AsO_4(KDA), KD_2AsO_4(DKDA). σ_c vs. $1/T$ [73F1]. Circles: KDA; triangles: DKDA.

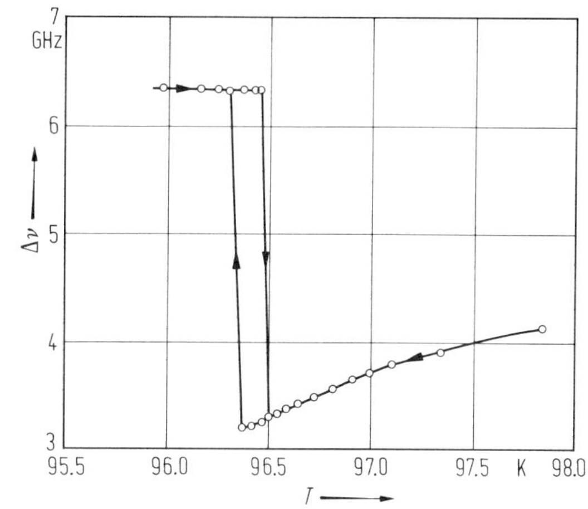

Fig. 527. KH_2AsO_4(KDA). $\Delta\nu$ vs. T [76D9]. $\Delta\nu$: Brillouin frequency shift of the XY-shear mode associated with elastic stiffness c_{66}.

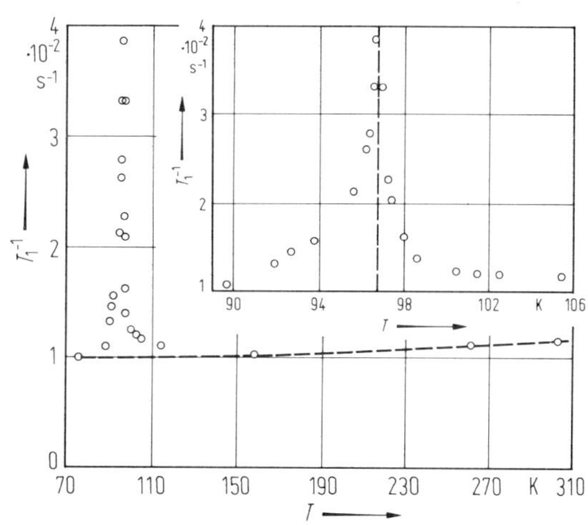

Fig. 529. KH_2AsO_4(KDA). T_1^{-1} vs. T [74B8]. T_1^{-1}: Inverse of the proton spin-lattice relaxation time. $f = 16$ MHz.

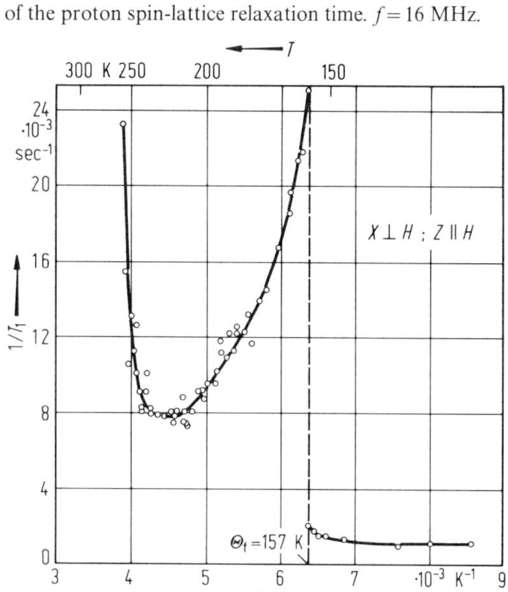

Fig. 530. KD_2AsO_4(DKDA). T_1^{-1} vs. $1/T$ [71B20]. T_1: spin-lattice relaxation time of deuteron. $\boldsymbol{H} \perp X$, $\boldsymbol{H} \parallel Z$, $f = 10.6$ MHz.

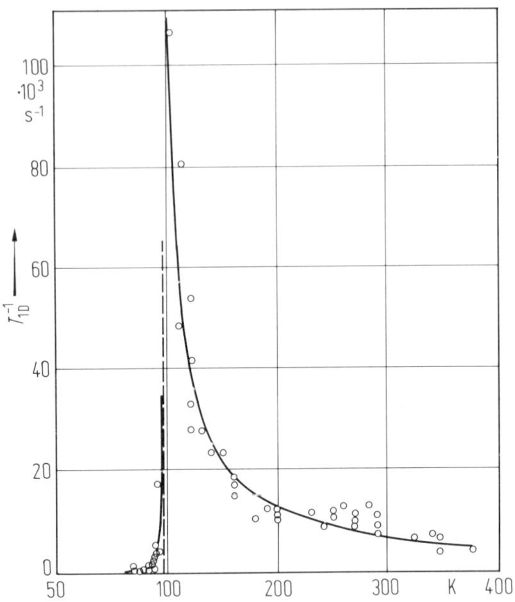

Fig. 531. KH_2AsO_4(KDA). T_{1D}^{-1} vs. T [75B16]. T_{1D}^{-1}: Inverse of ^{75}As dipolar nuclear spin-lattice relaxation time.

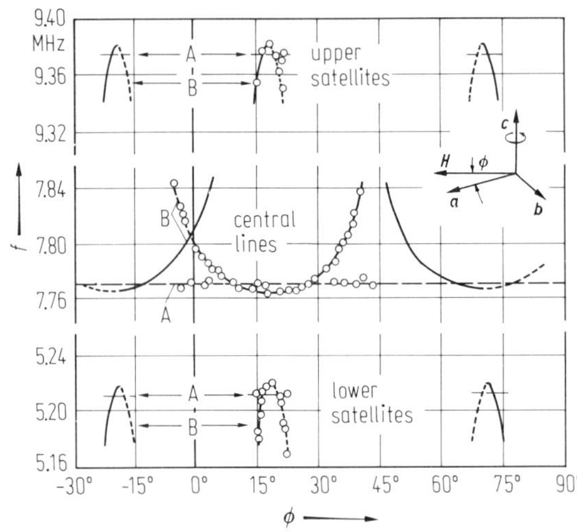

Fig. 532. KH_2AsO_4(KDA). f vs. ϕ [75A2]. f: ^{75}As NMR frequency for c axis rotation with $H=(4\pi)^{-1}\cdot 10^7$ Am^{-1} at $T=\Theta_f+1$ K, ϕ: angle of the c axis rotation shown in the figure. Dashed lines: the A spectrum, full and dotted lines: parts of the B spectrum, enhanced by a respectively negative and positive electric field along the c axis. The heavier full lines are part of the same B rotation pattern. For more detailed information about A and B lines, see original paper.

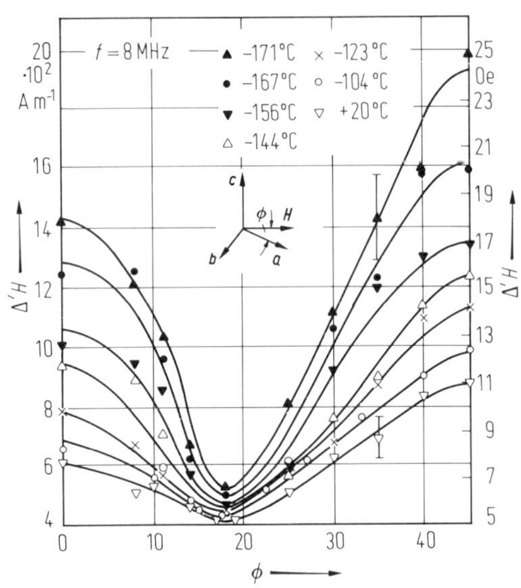

Fig. 533. KH_2AsO_4(KDA). $\Delta'H$ vs. ϕ [75T4]. $\Delta'H$: ^{75}As NMR line width. ϕ: angle between the a axis and the magnetic field in the (001) plane. See also [69B6].

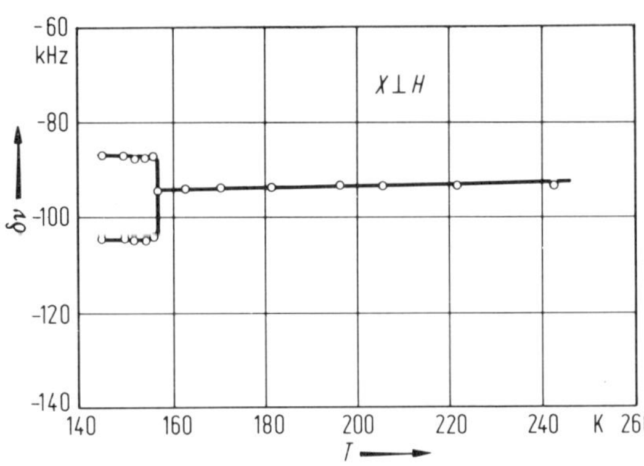

Fig. 534. KD_2AsO_4(DKDA). $\delta\nu$ vs. T [71B20]. $\delta\nu$: frequency difference between $m=1\rightarrow m=0$ and $m=0\rightarrow m=-1$ transition of deuteron. $\boldsymbol{H}\perp X$, $\theta_x=67°$. θ_x: angle between $\boldsymbol{H}$ and the crystal Y axis.

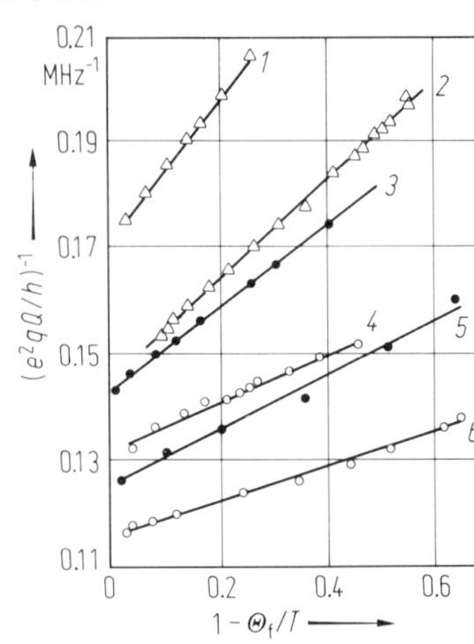

Fig. 535. KDA family. $(e^2qQ/h)^{-1}$ vs. $1-\Theta_f/T$ [76B6]. Reciprocal quadrupole coupling constant of ^{75}As. Curve *1*: CsD_2AsO_4, *2*: CsH_2AsO_4, *3*: RbD_2AsO_4, *4*: KD_2AsO_4, *5*: RbH_2AsO_4, *6*: KH_2AsO_4. See also [73B6]. ►

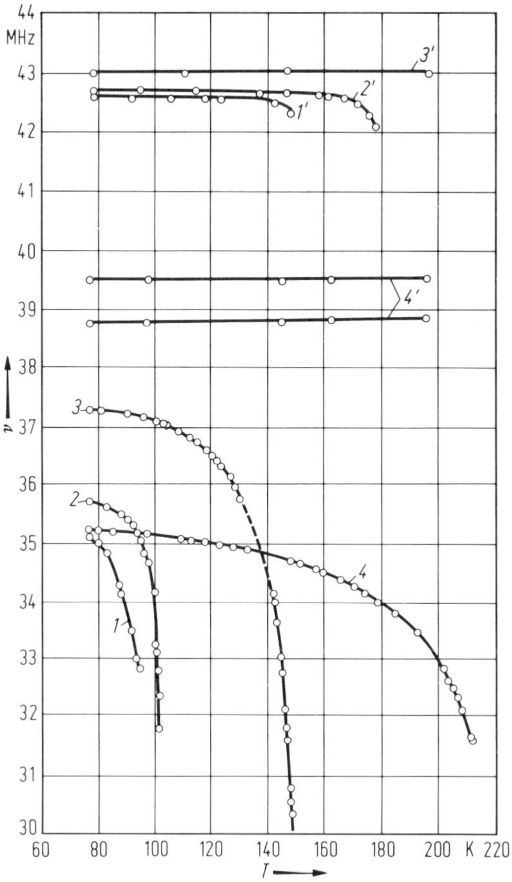

Fig. 536. KDA family. ν vs. T [69 Z 2]. Curve *1*: KH_2AsO_4, *2*: RbH_2AsO_4, *3*: CsH_2AsO_4, *4*: $NH_4H_2AsO_4$, curves *1'*…*4'*: corresponding deuterated compounds. ν: ^{75}As NQR frequency.

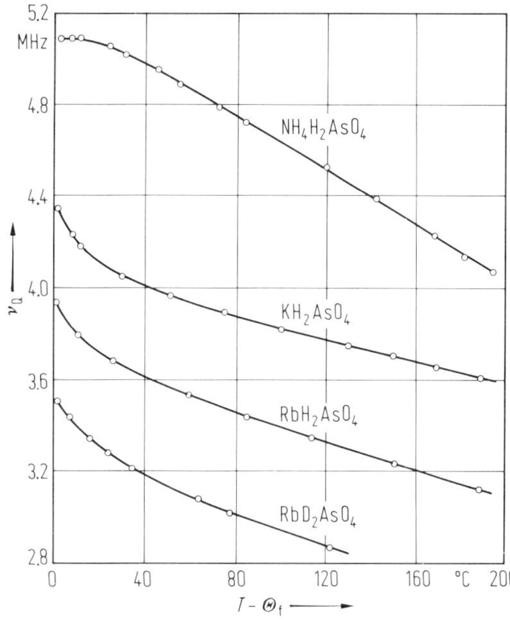

Fig. 537. KDA family. ν_Q vs. $T-\Theta_f$ in the paraelectric phase I [71 A 6]. ν_Q: ^{75}As nuclear quadrupole frequency.

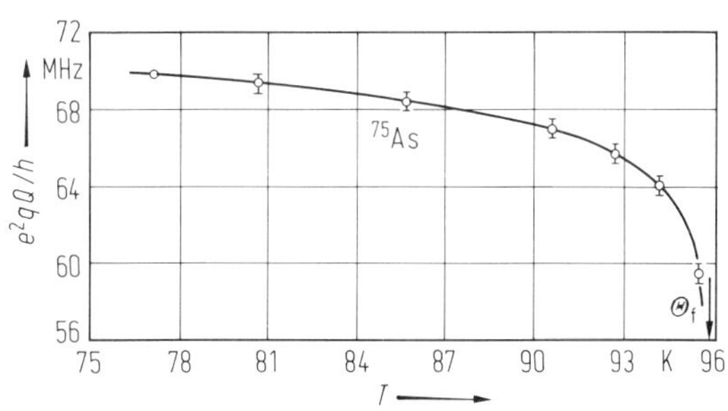

Fig. 538. KH_2AsO_4(KDA). e^2qQ/h vs. T [68 B 4]. ^{75}As NQR in phase II.

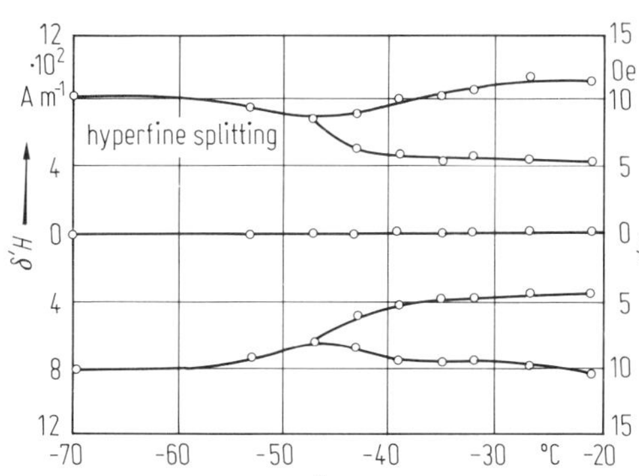

Fig. 539. KH_2AsO_4(KDA). $\delta' H$ vs. T [67 B 11]. ESR spectrum of $(AsO_4)^{4-}$ center, $\boldsymbol{H} \| c$. $\delta' H$: magnetic resonance line separation.

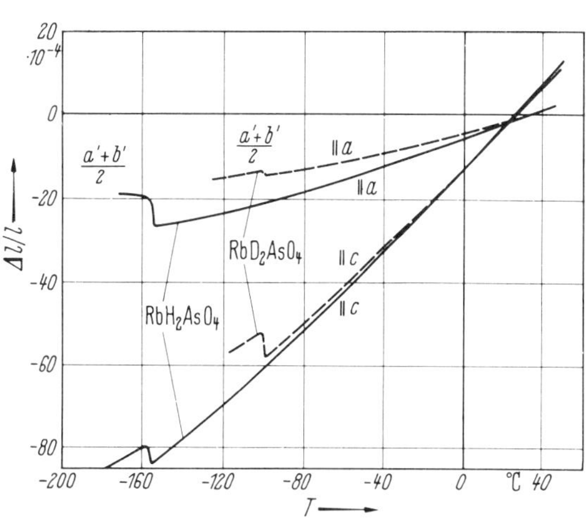

Fig. 541. RbH_2AsO_4(RDA), RbD_2AsO_4(DRDA). Thermal expansion $\Delta l/l$ vs. T [67 C 5]. $(a'+b')/2$ is the mean of the values along *a* and *b* axes in the ferroelectric state.

Fig. 540, see p. 432.

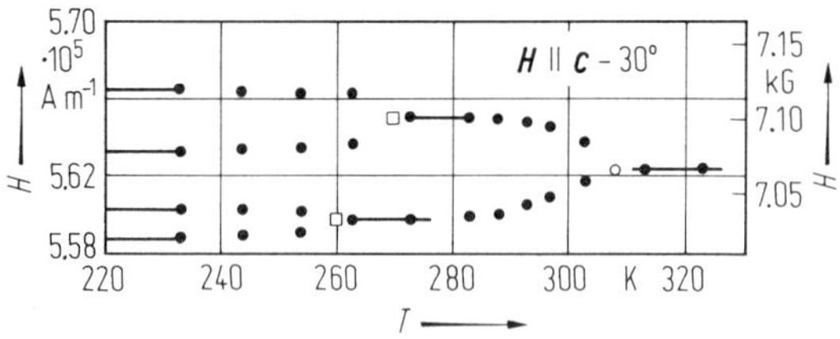

Fig. 540. KH_2AsO_4: Cr^{5+}. H vs. T [76M23]. H: Electron spin resonance magnetic field. H makes an angle of 30° with the c axis in the (010) plane. See also [76M24, 77G1].

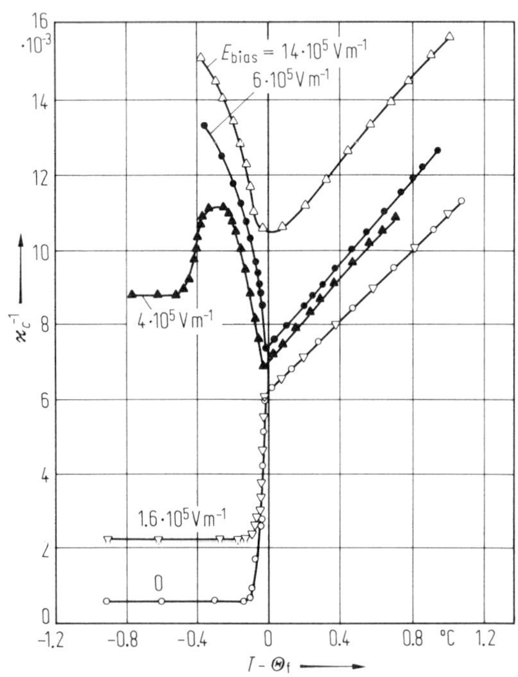

Fig. 542. RbH_2AsO_4(RDA). $\varkappa_c^{-1}$ vs. $T-\Theta_f$ [75M4]. Parameter: E_{bias}. $f=800$ Hz.

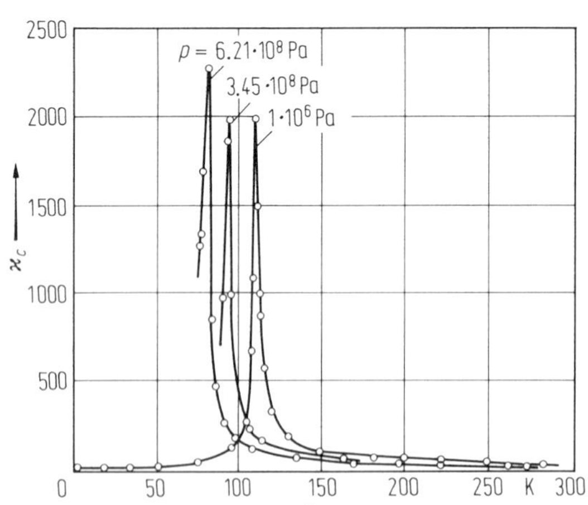

Fig. 543. RbH_2AsO_4(RDA). $\varkappa_c$ vs. T [77S34]. Parameter: p. $f=1\cdots100$ kHz.

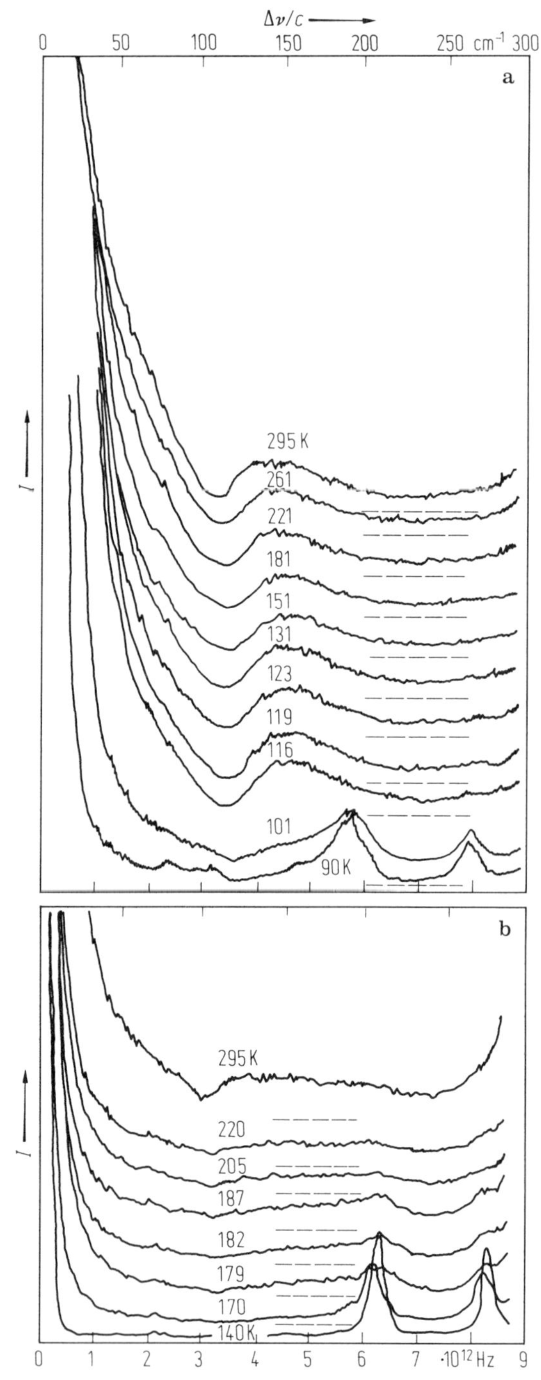

Fig. 544. RbH_2AsO_4(RDA) (Fig. a) and RbD_2AsO_4(DRDA) (Fig. b). I vs. $\Delta\nu$ [76S8]. I: Raman scattering intensity of B_2 modes measured in the scattering geometry of $Y(X, Y)X$. Parameter: T.

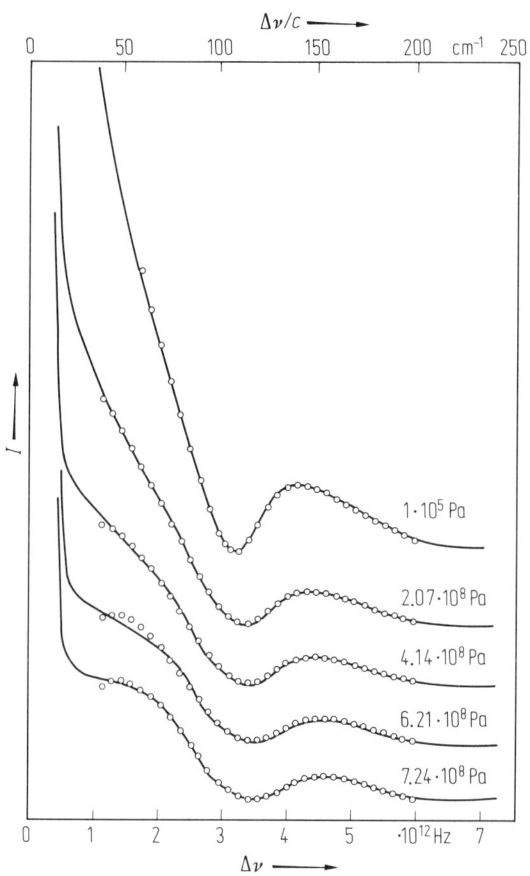

Fig. 545. RbH_2AsO_4(RDA). I vs. $\Delta\nu$ [77L5]. I: Raman scattering intensity of B_2 modes. Parameter: p. Scattering geometry: $Y(X,Y)X$.

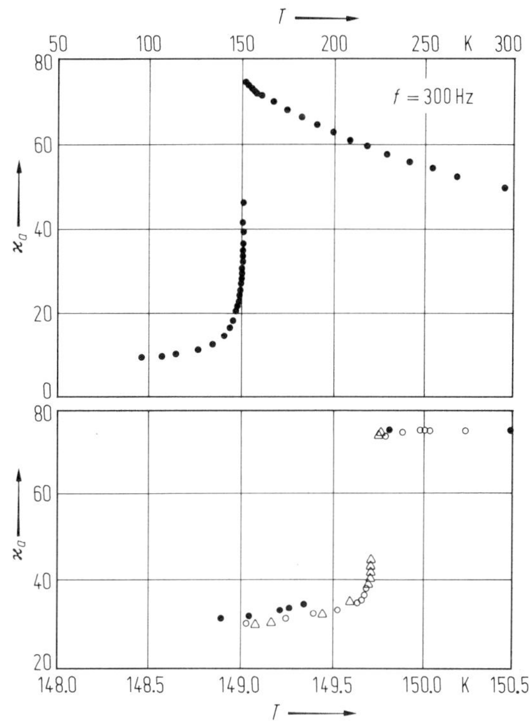

Fig. 547. CsH_2AsO_4(CDA). $\varkappa_a$ vs. T [75P11]. The lower figure shows the behavior in the vicinity of Θ_f (open and filled circles: cooling, triangles: heating).

Fig. 548, see next page.

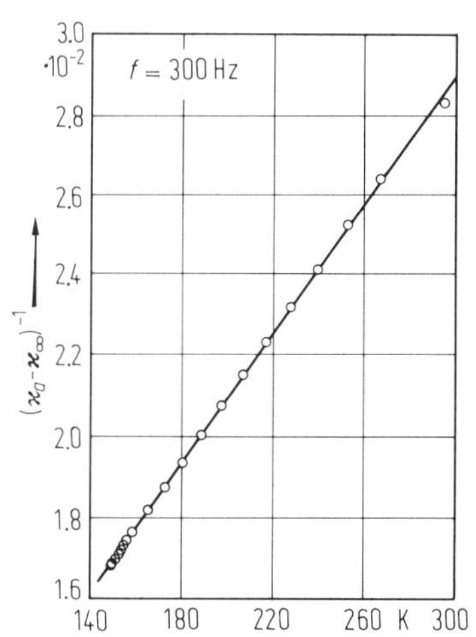

Fig. 546. CsH_2AsO_4(CDA). $(\varkappa_c - \varkappa_\infty)^{-1}$ vs. T [75P11]. $\varkappa_\infty = 15.3$.

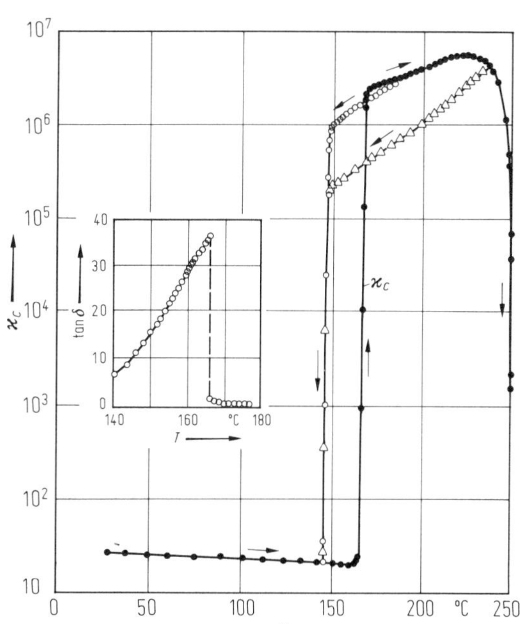

Fig. 549. CsH_2AsO_4(CDA). $\varkappa_c$, $\tan\delta$ vs. T [77G14]. $f = 1$ kHz.

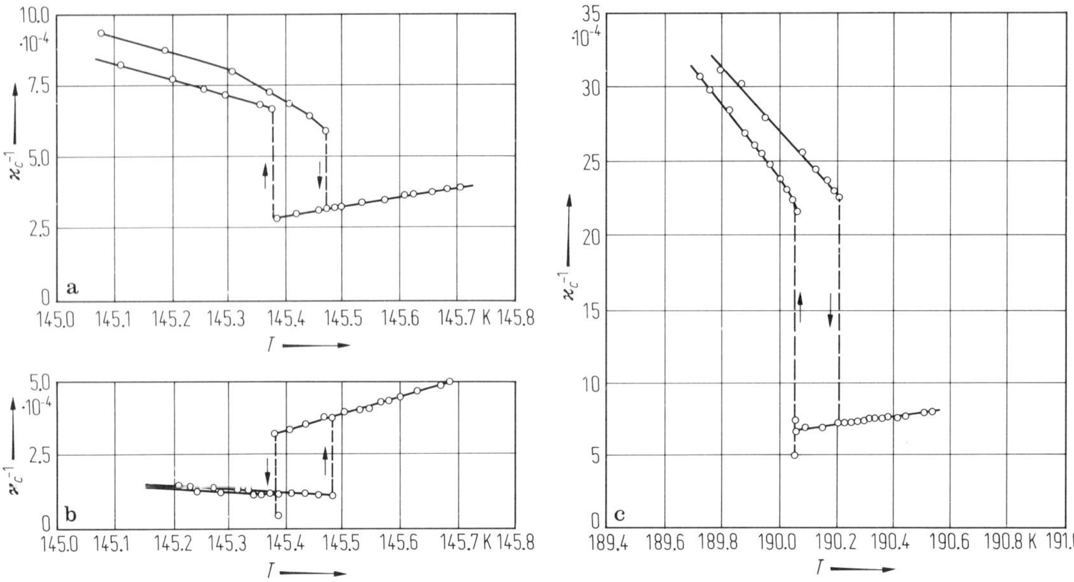

Fig. 548. CsH_2AsO_4(CDA), CsD_2AsO_4(DCDA). $\varkappa_c^{-1}$ vs. T in the vicinity of Θ_f [73S18]. Fig. a: CDA aged sample, Fig. b: CDA fresh sample, Fig. c: DCDA.

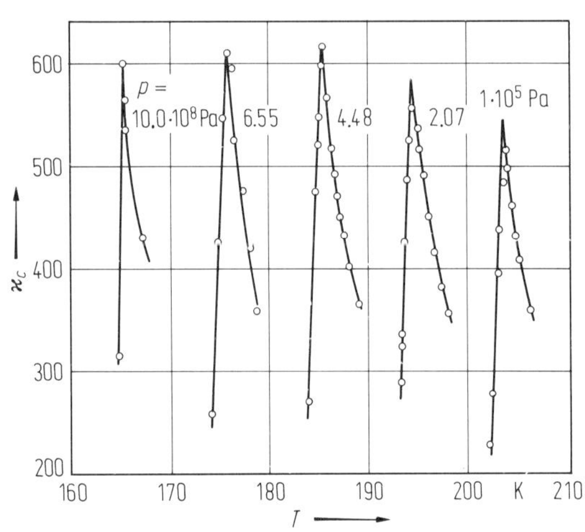

Fig. 550. CsD_2AsO_4(DCDA). $\varkappa_c$ vs. T [75L9]. Parameter: p. $f = 1 \cdots 100$ kHz.

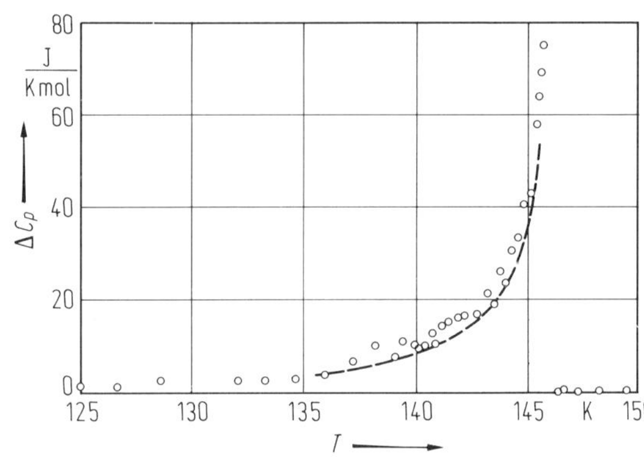

Fig. 553. CsH_2AsO_4(CDA). ΔC_p vs. T [74D6]. ΔC_p: anomalous part of the molar heat capacity at constant pressure. Dashed line is calculated.

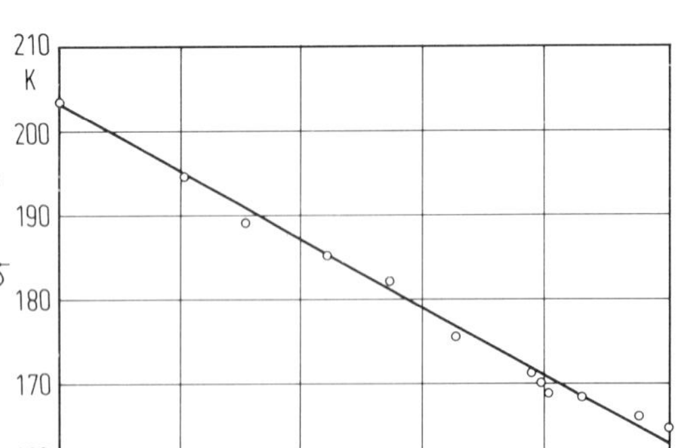

Fig. 551. CsD_2AsO_4(DCDA). Θ_f vs. p [75L9].

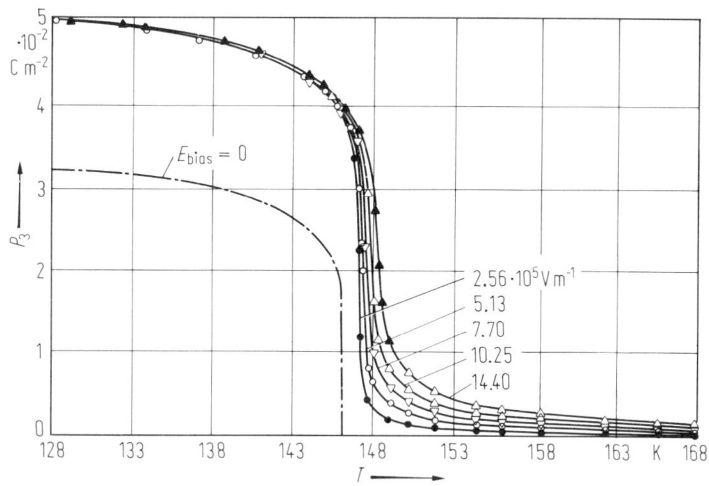

Fig. 552. CsH_2AsO_4(CDA). P_3 vs. T [75C4]. Parameter: E_{bias}.

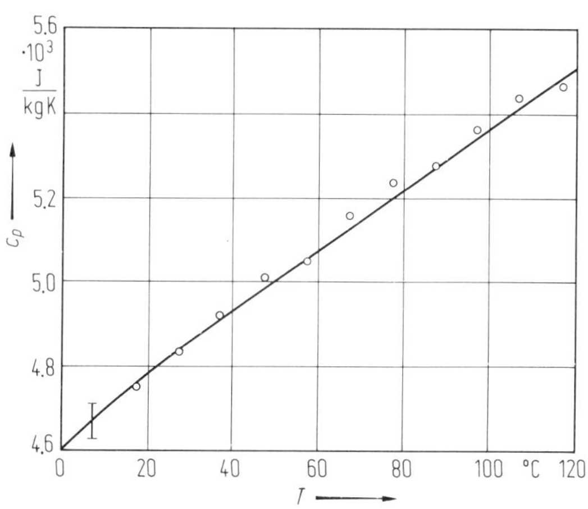

Fig. 554. CsD_2AsO_4(DCDA). c_p vs. T [75L7].

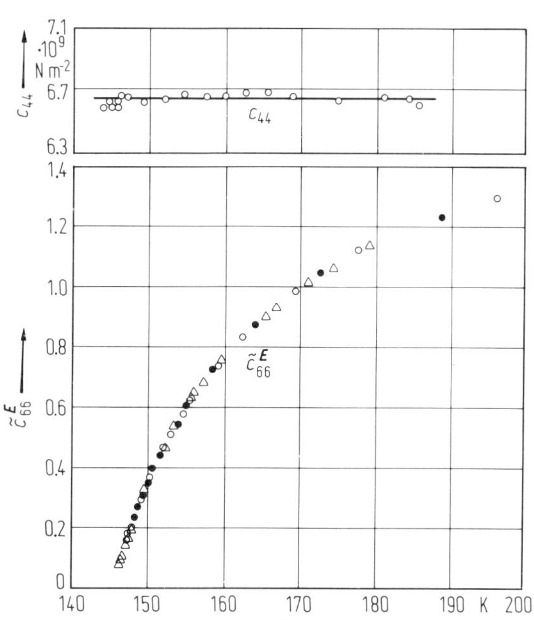

Fig. 555. CsH_2AsO_4(CDA). c_{44}, $\tilde{c}^E_{66}$ vs. T [75P11]. $\tilde{c}^E_{66} \equiv c^E_{66}(T)/c^E_{66}(170\text{ K})$. Open circles: 5 MHz; full circles: 10 MHz; triangles: 0.75 ~ 3 MHz.

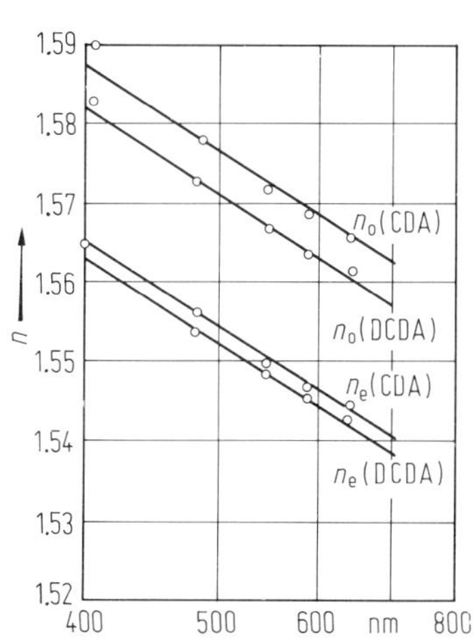

Fig. 556. CsH_2AsO_4(CDA). CsD_2AsO_4(DCDA). n vs. λ at RT [69A2].

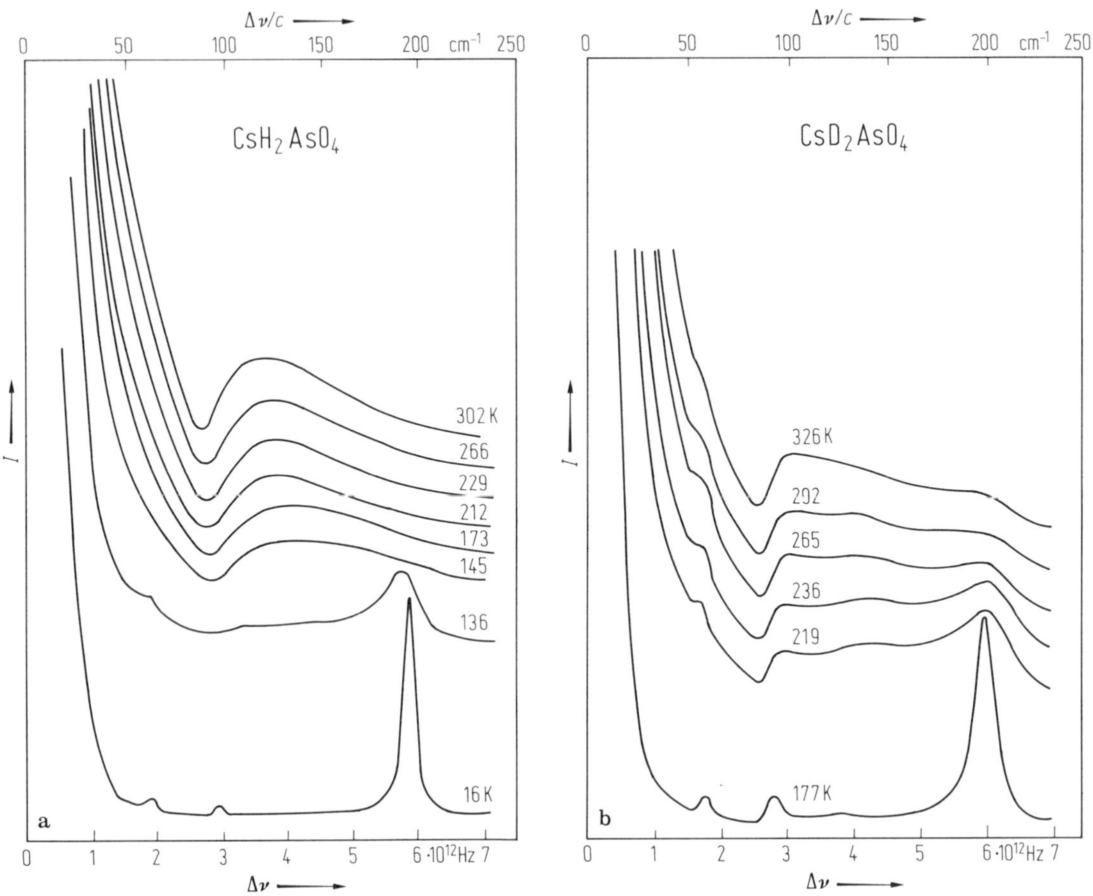

Fig. 557. CsH_2AsO_4(CDA) (Fig. a) and CsD_2AsO_4(DCDA) (Fig. b). *I* vs. $\Delta\nu$ [74L10]. *I*: Raman scattering intensity of B_2 modes measured with scattering geometry $Y(X, Y)X$. Parameter: *T*.

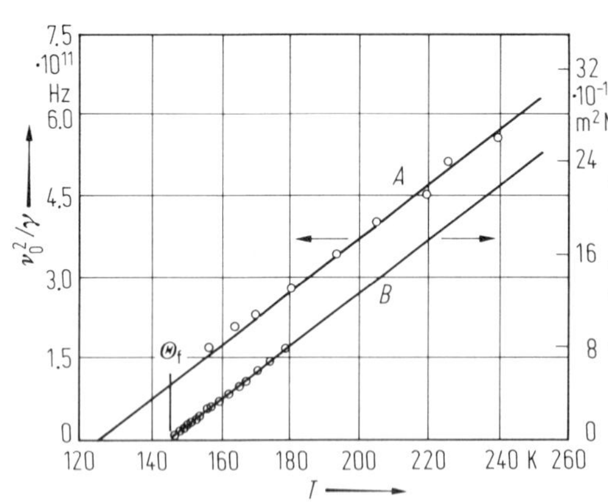

Fig. 558. CsH_2AsO_4(CDA). ν_0^2/γ vs. *T* and $(c_{66}^P - c_{66}^E)^{-1} - (c_{66}^P)^{-1}$ vs. *T* [75L1]. ν_0, γ: characteristic frequency and damping parameter of the ferroelectric soft mode deduced from Raman scattering spectra measured with a scattering geometry of $X(Y, X)Y$. (Curve *A*). Curve *B* shows elastic constants obtained from Brillouin scattering.

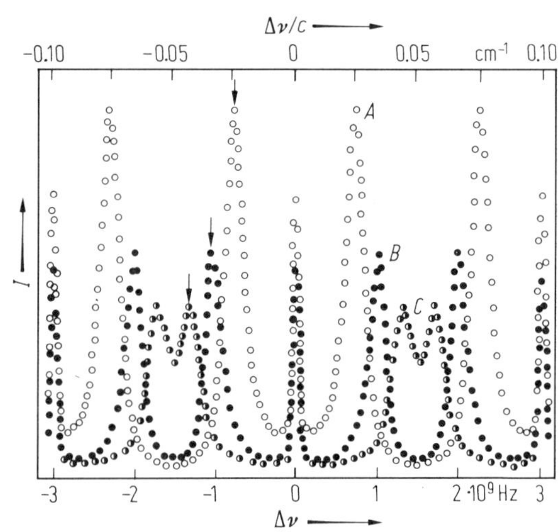

Fig. 561. CsH_2AsO_4(CDA). *I* vs. $\Delta\nu$ [75L1]. *I*: intensity of Rayleigh and Brillouin scattering, $\Delta\nu$: frequency shift. Scattering geometry is $\overline{X}\overline{Z}(X\overline{Z}, Y)X\overline{Z}$. Curve *A*: $T-\Theta_f = 0.05$ K, *B*: 0.85 K, *C*: 1.67 K. Arrows indicate the position of the anti-Stokes Brillouin peaks of the order centered at zero.

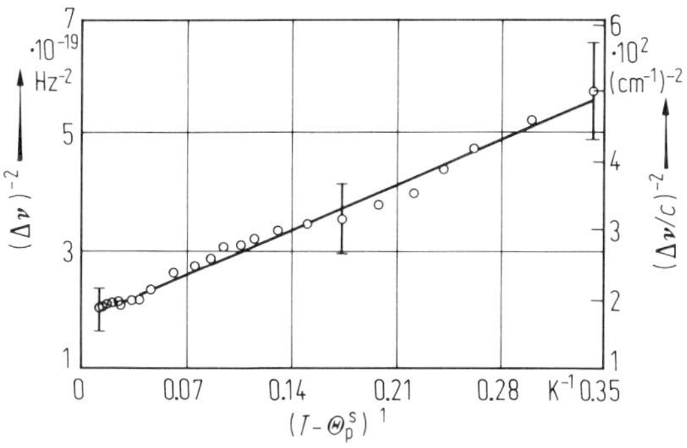

Fig. 559. CsH_2AsO_4(CDA). $(\Delta\nu)^{-2}$ vs. $|T-\Theta_{\mathrm{pl}}^{\mathrm{s}}|^{-1}$ [77A12]. $\Delta\nu$: frequency shift of Brillouin scattering, $\Theta_{\mathrm{p}}^{\mathrm{s}}$: paraelectric Curie temperature for shorted crystal.

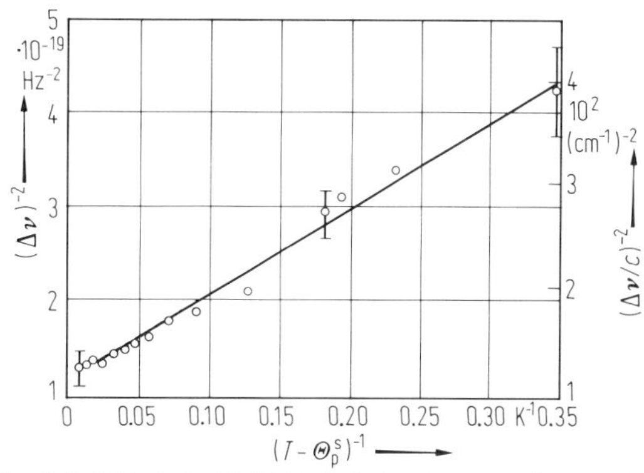

Fig. 560. CsD_2AsO_4(DCDA). $(\Delta\nu)^{-2}$ vs. $|T-\Theta_{\mathrm{pl}}^{\mathrm{s}}|^{-1}$ [77A12]. $\Delta\nu$: frequency shift of Brillouin scattering. $\Theta_{\mathrm{p}}^{\mathrm{s}}$: paraelectric Curie temperature for shorted crystal.

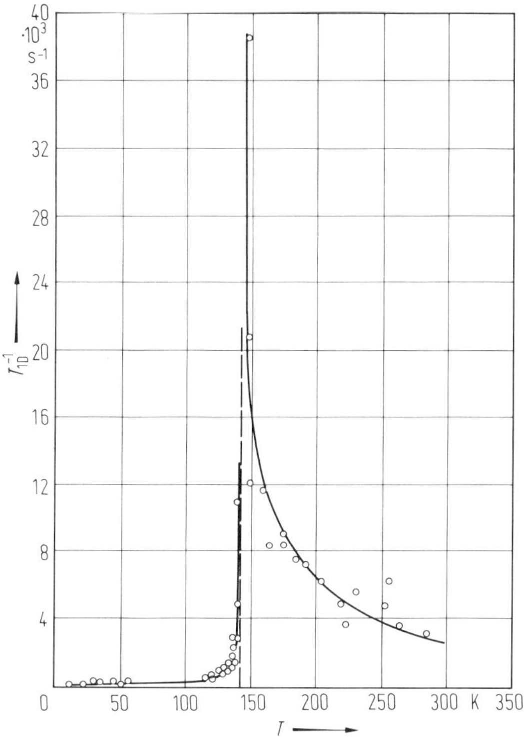

Fig. 562. CsH_2AsO_4(CDA). $T_{1\mathrm{D}}^{-1}$ vs. T [75B16]. $T_{1\mathrm{D}}^{-1}$: Inverse of ^{75}As dipolar nuclear spin-lattice relaxation time.

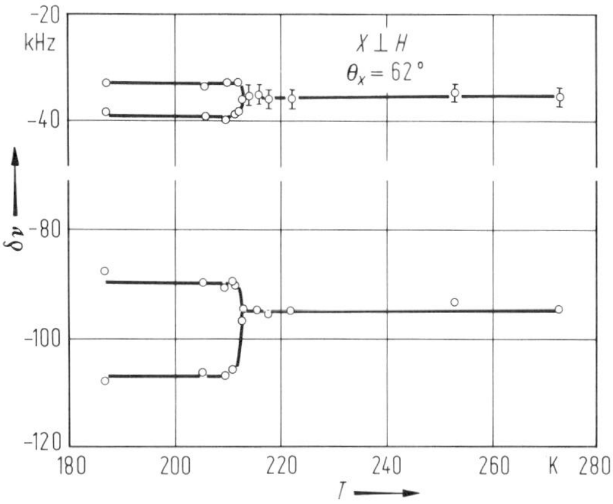

Fig. 563. CsD_2AsO_4(DCDA). $\delta\nu$ vs. T [71B20]. $\delta\nu$: frequency difference between m=1→m=0 and m=0→m=−1 transition of the deuteron, $\boldsymbol{H}\perp X$, $\theta_x=62°$, θ_x: angle between $\boldsymbol{H}$ and the crystal Y axis.

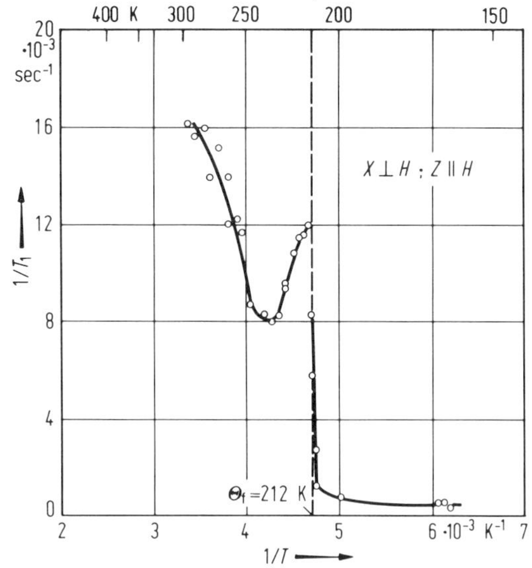

Fig. 564. CsD_2AsO_4(DCDA). $1/T_1$ vs. $1/T$ [71B20]. T_1: spin-lattice relaxation time of deuteron, $\boldsymbol{H}\perp X$, $\boldsymbol{H}\parallel Z$. $f=10.5$ MHz.

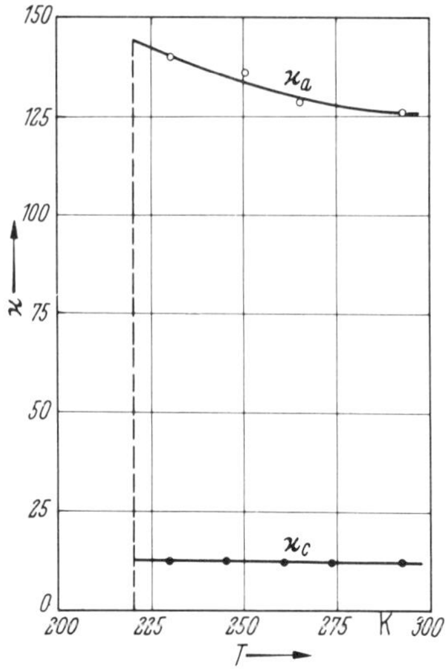

Fig. 565. $NH_4H_2AsO_4$(ADA). $\varkappa_a$, $\varkappa_c$ vs. T[38 B 3]. f = 800 Hz.

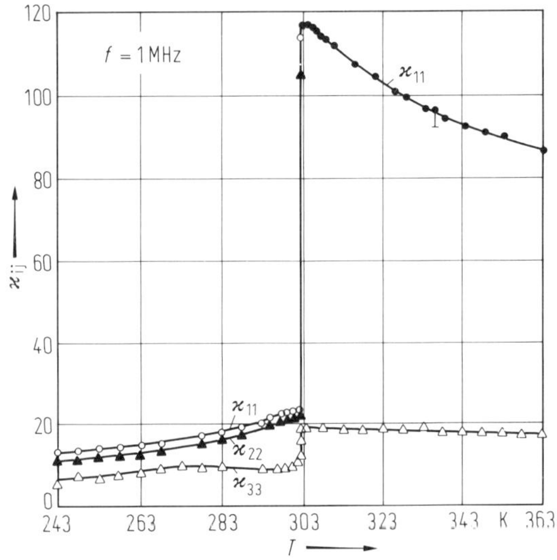

Fig. 566. $ND_4D_2AsO_4$(DADA). $\varkappa_{11}$, $\varkappa_{22}$, $\varkappa_{33}$ vs. T[78 B 6].

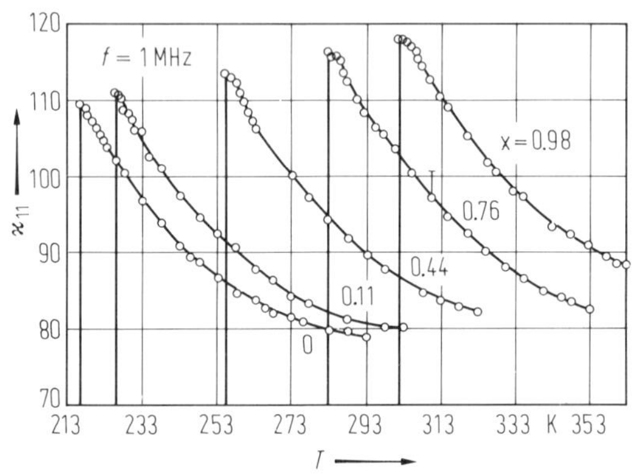

Fig. 567. $(NH_4H_2AsO_4)_{1-x}(ND_4D_2AsO_4)_x$. $\varkappa_{11}$ vs. T [78 B 6]. Parameter: x.

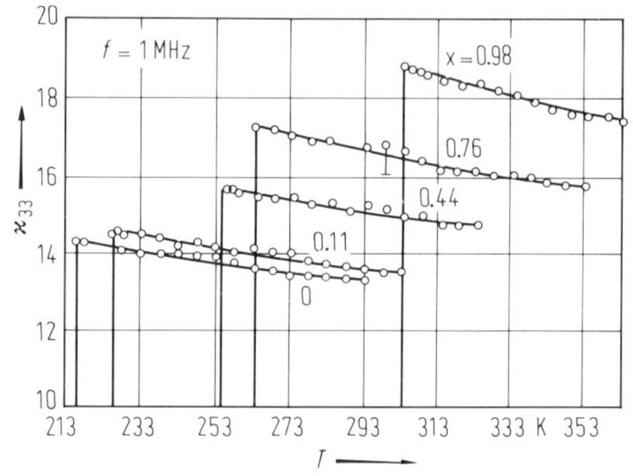

Fig. 568. $(NH_4H_2AsO_4)_{1-x}(ND_4D_2AsO_4)_x$. $\varkappa_{33}$ vs. T [78 B 6]. Parameter: x.

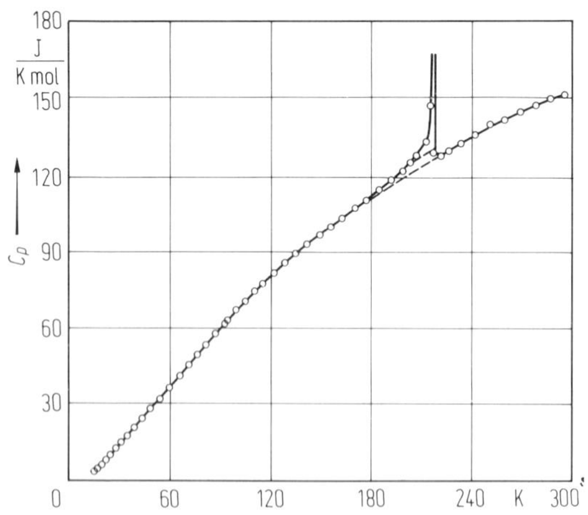

Fig. 569. $NH_4H_2AsO_4$(ADA). C_p vs. T [44 S 1]. C_p: molar heat capacity at constant pressure.

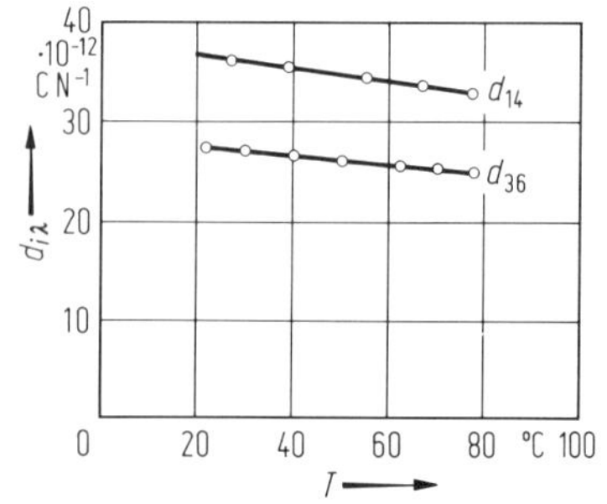

Fig. 570. $NH_4H_2AsO_4$(ADA). d_{14} and d_{36} vs. T[68 A 1].

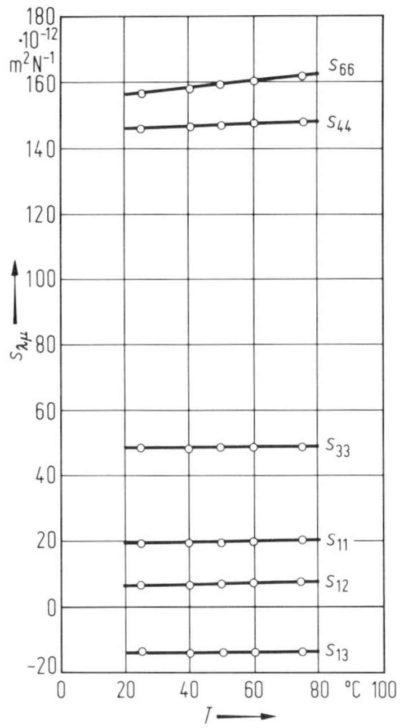

Fig. 571. $NH_4H_2AsO_4$(ADA). $s_{\lambda\mu}$ vs. T [68A1].

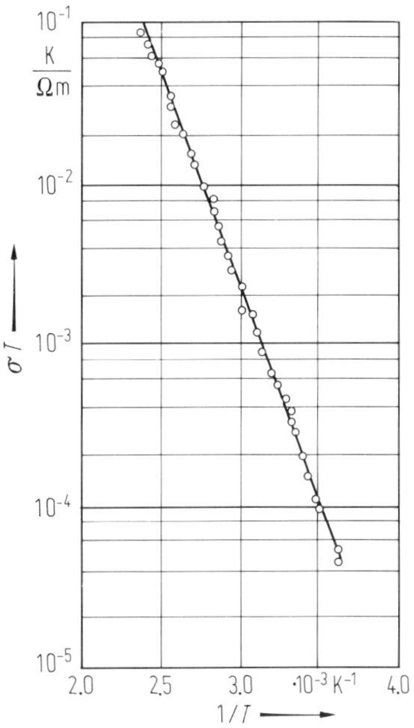

Fig. 572. $NH_4H_2AsO_4$(ADA). σT vs. $1/T$ [77P2]. σ: electrical conductivity.

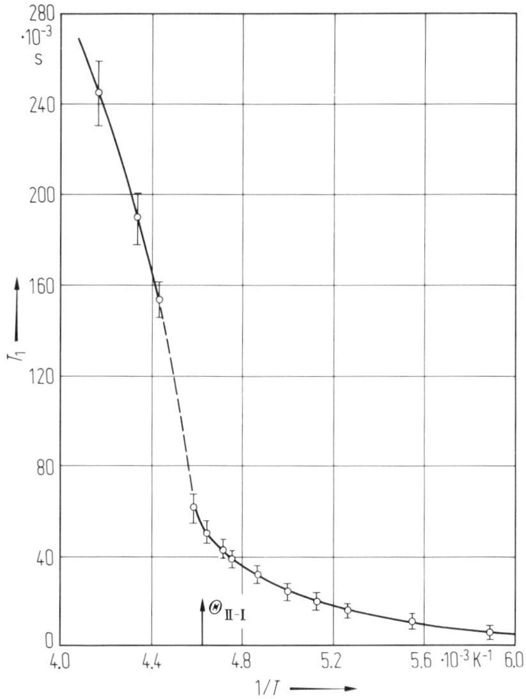

Fig. 573. $NH_4H_2AsO_4$(ADA). T_1 vs. T^{-1} [74D1]. T_1: proton spin-lattice relaxation time.

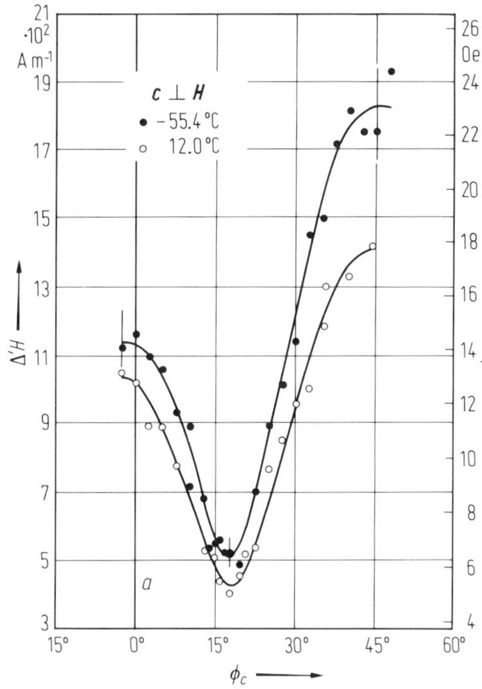

Fig. 574. $NH_4H_2AsO_4$(ADA). $\Delta'H$ vs. T [77T3]. $\Delta'H$: line width of ^{75}As NMR line. ϕ_c: angle between a axis and the magnetic field $\boldsymbol{H}$ in (001) plane.

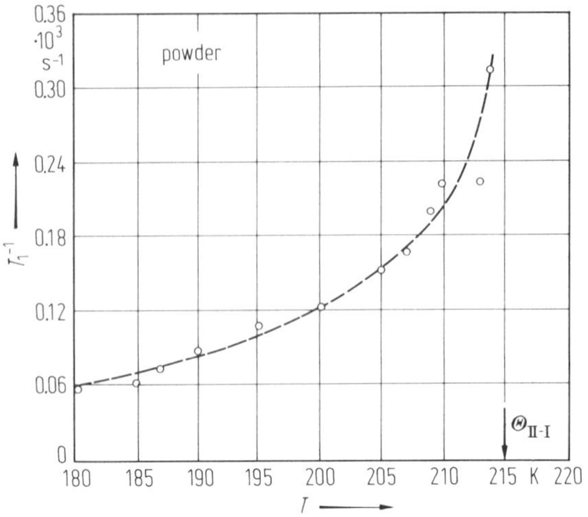

Fig. 575. $NH_4H_2AsO_4$(ADA). T_1^{-1} vs. T [75B16]. T_1^{-1}: inverse of ^{75}As nuclear spin-lattice relaxation time.

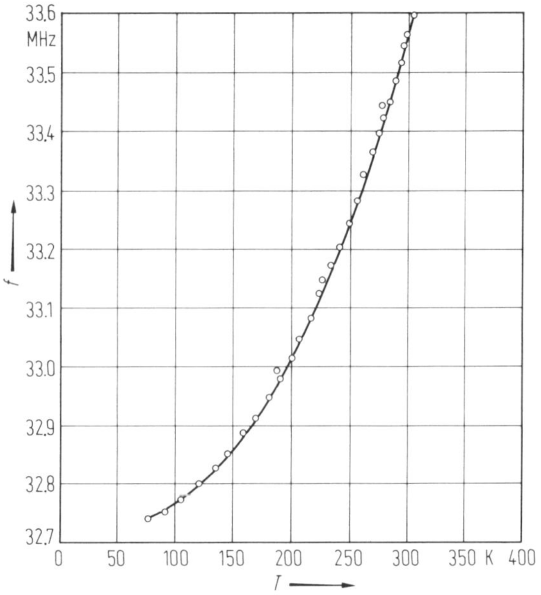

Fig. 576. $(NH_4)_2HAsO_4$. f vs. T [77M16]. f: NQR frequency of ^{75}As.

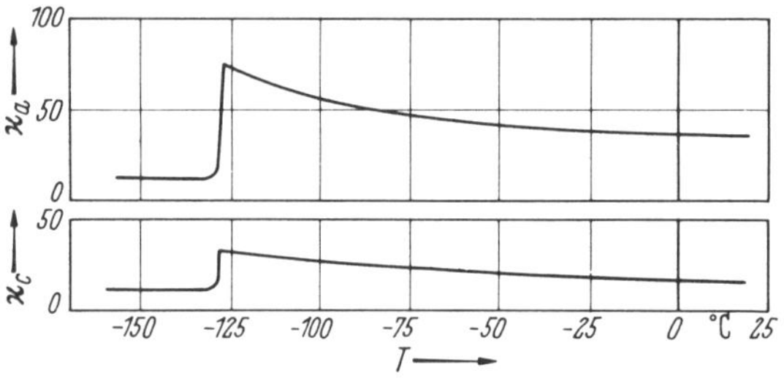

Fig. 577. $K_{0.2}(NH_4)_{0.8}H_2PO_4$. $\varkappa_a$, $\varkappa_c$ vs. T[51N1].

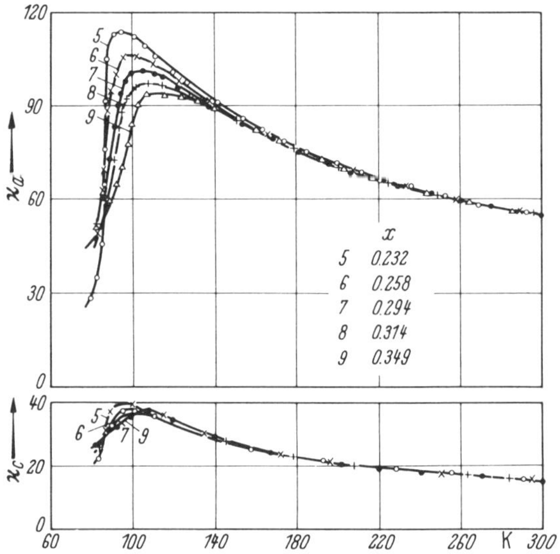

Fig. 579. $[(NH_4)_{1-x}Tl_x]H_2PO_4$. $\varkappa_a$, $\varkappa_c$ vs. T [47M1]. Parameter: x. $x>0.2$. $f=800$ Hz.

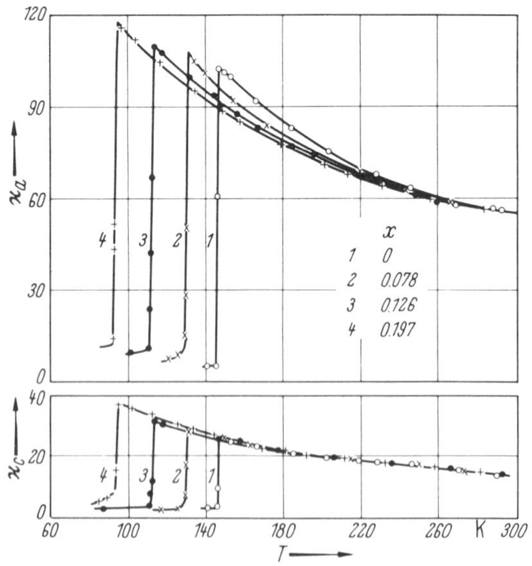

Fig. 578. $[(NH_4)_{1-x}Tl_x]H_2PO_4$. $\varkappa_a$, $\varkappa_c$ vs. T [47M1]. Parameter: x. $x<0.2$. $f=800$ Hz.

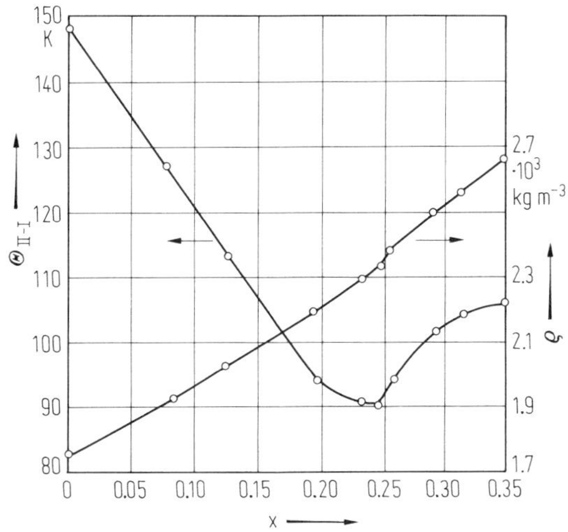

Fig. 580. $[(NH_4)_{1-x}Tl_x]H_2PO_4$. Θ_{II-I}, ϱ vs. x [47M1]. ϱ: density.

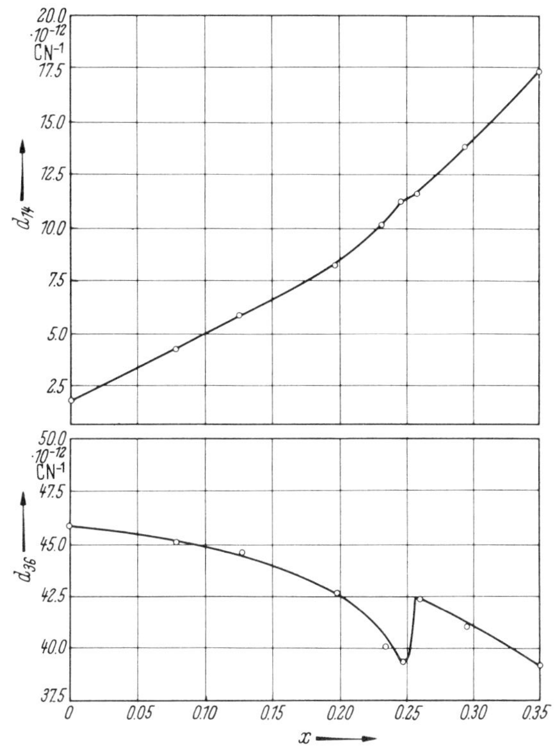

Fig. 581. $[(NH_4)_{1-x}Tl_x]H_2PO_4$. d_{14}, d_{36} vs. x at RT [47M1].

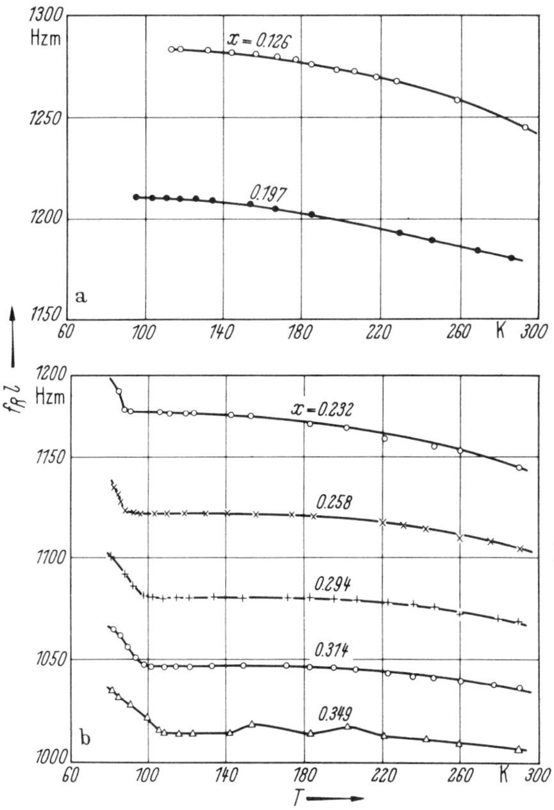

Fig. 582. $[(NH_4)_{1-x}Tl_x]H_2PO_4$. $f_R l$ vs. T [47M1]. f_R: resonance frequency. l: length of a square plate $\perp a$. Parameter: x. Fig. a: $x<0.2$, Fig. b: $x>0.2$.

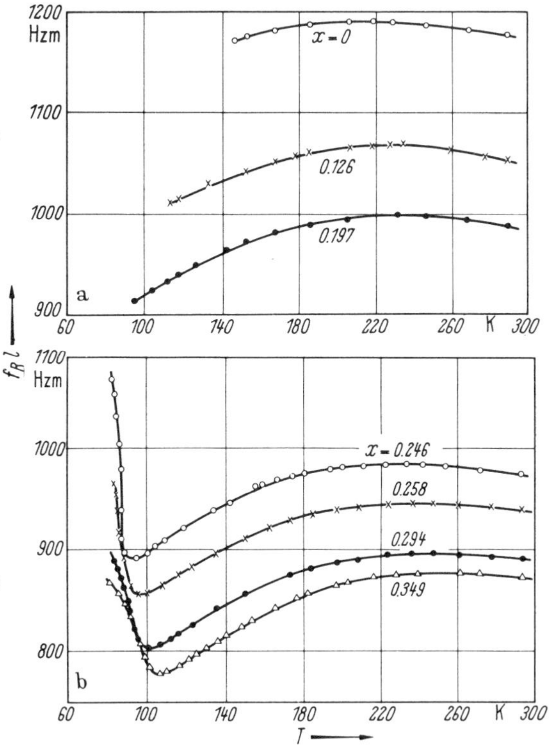

Fig. 583. $[(NH_4)_{1-x}Tl_x]H_2PO_4$. $f_R l$ vs. T [47M1]. f_R: resonance frequency l: length of a square plate $\perp c$. Parameter: x. Fig. a: $x<0.2$, Fig. b: $x>0.2$.

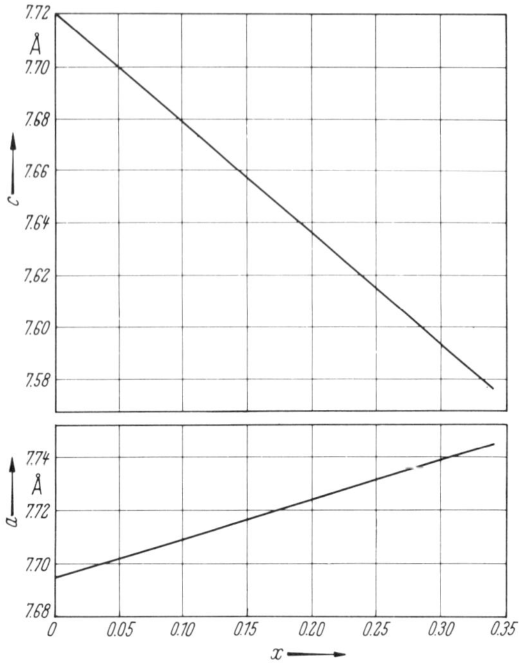

Fig. 584. $[(NH_4)_{1-x}Tl_x]\,H_2AsO_4$. Lattice constants *a* and *c* vs. x [67 C 6].

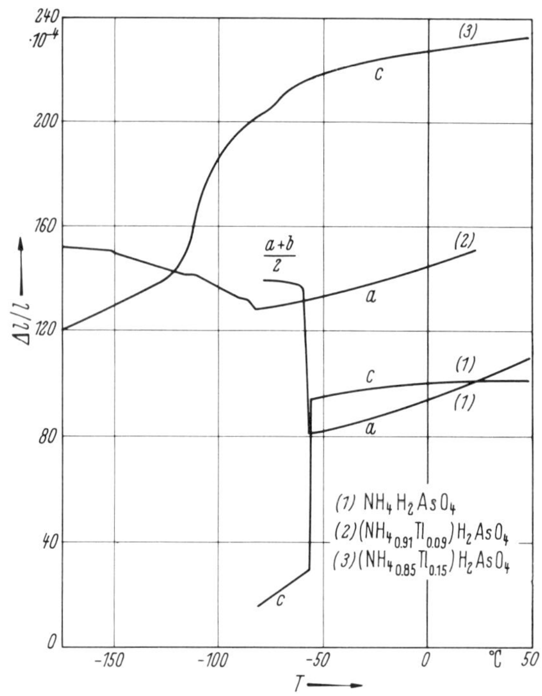

Fig. 585. $[(NH_4)_{1-x}Tl_x]\,H_2AsO_4$. Thermal expansion $\Delta l/l$ vs. *T* along *a* and *c* [67 C 6].

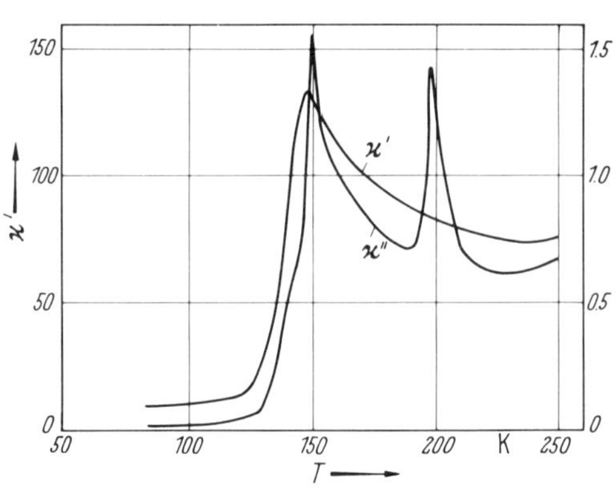

Fig. 586. $(NH_4)_4Tl_3(H_2AsO_4)_7$. $\varkappa'$, $\varkappa''$ vs. *T* [63 L 1]. $f = 1$ kHz⋯1 MHz.

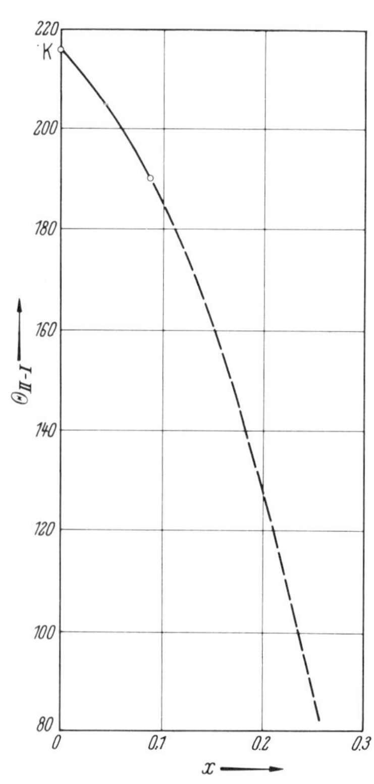

Fig. 587. $[(NH_4)_{1-x}Tl_x]\,H_2AsO_4$. Θ_{II-I} vs. x [67 C 6].

20 $(NH_4)_2SO_4$ family

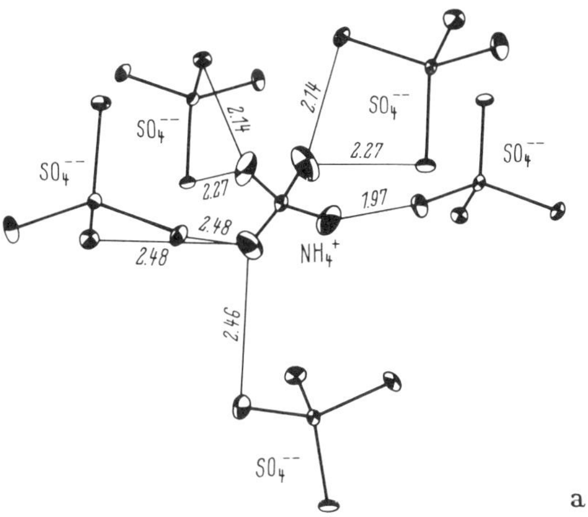

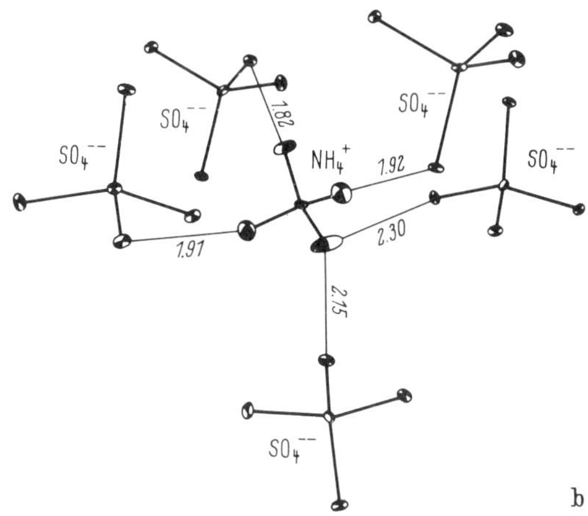

Fig. 588. $(NH_4)_2SO_4$. Structures of phases I and II [66S13]. The environment of $(NH_4)^+$ (I). Fig. a: RT, Fig. b: about 180 K. Atomic distances in Å.

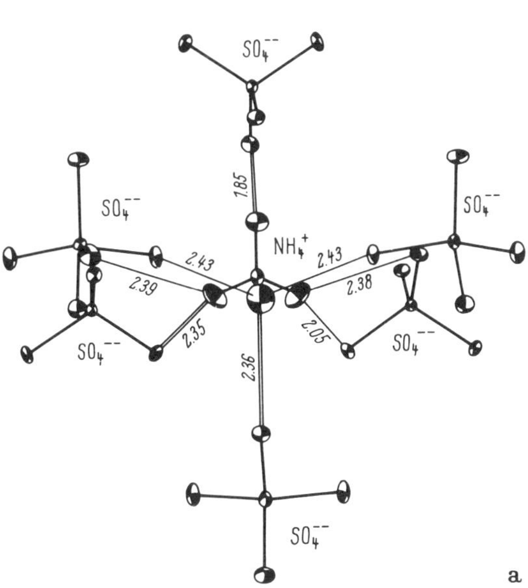

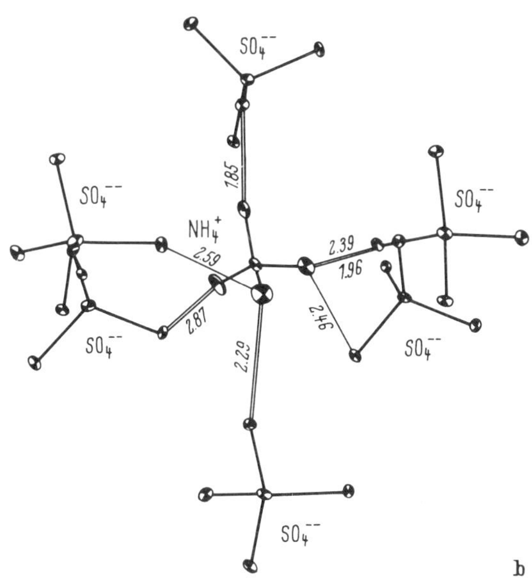

Fig. 589. $(NH_4)_2SO_4$. Structures of phases I and II [66S13]. The environment of $(NH_4)^+$ (II). Fig. a: RT, Fig. b: about 180 K. Atomic distances in Å.

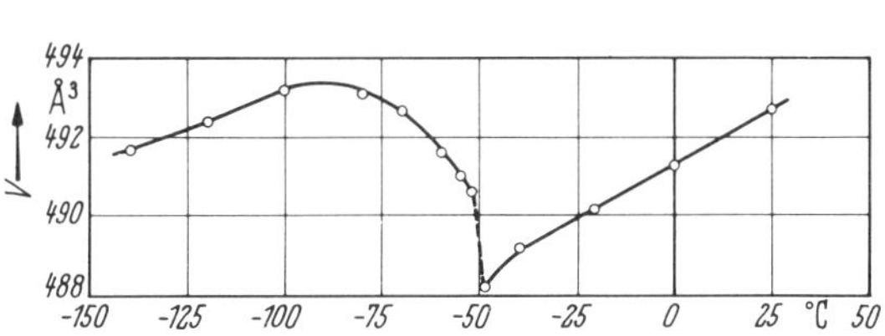

Fig. 591. $(NH_4)_2SO_4$. V vs. T [58H3]. V: volume of the unit cell.

Fig. 590, see next page.

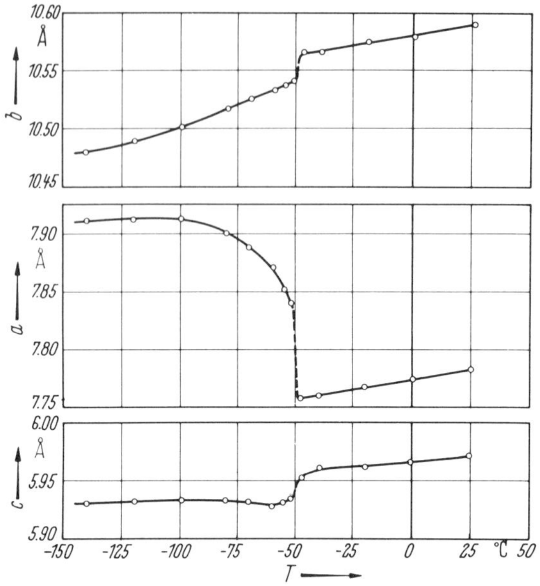

Fig. 590. $(NH_4)_2SO_4$. a, b, c vs. T [58 H 3].

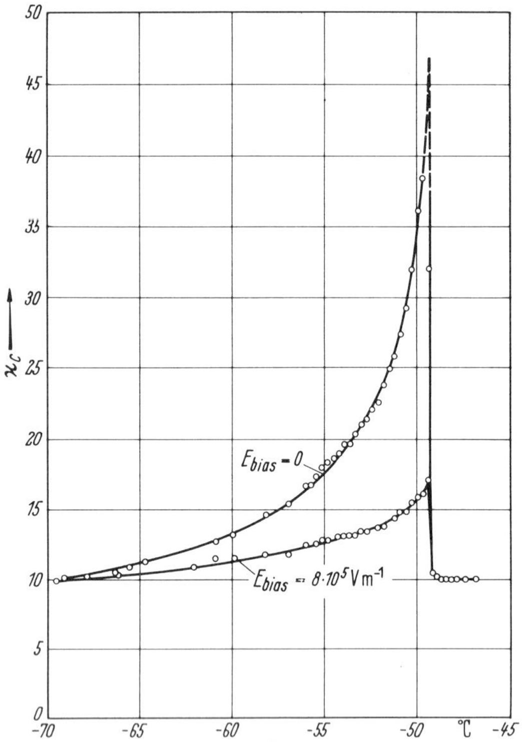

Fig. 595. $(NH_4)_2SO_4$. $\varkappa_c$ vs. T. [58 H 3]. $f = 10$ kHz. Parameter: E_{bias}.

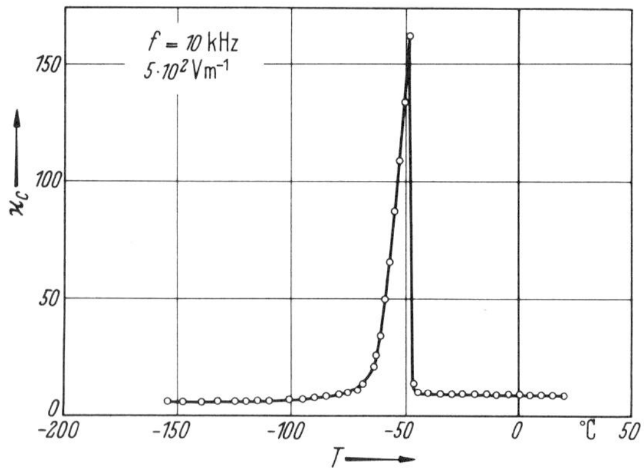

Fig. 592. $(NH_4)_2SO_4$. $\varkappa_c$ vs. T [58 H 3].

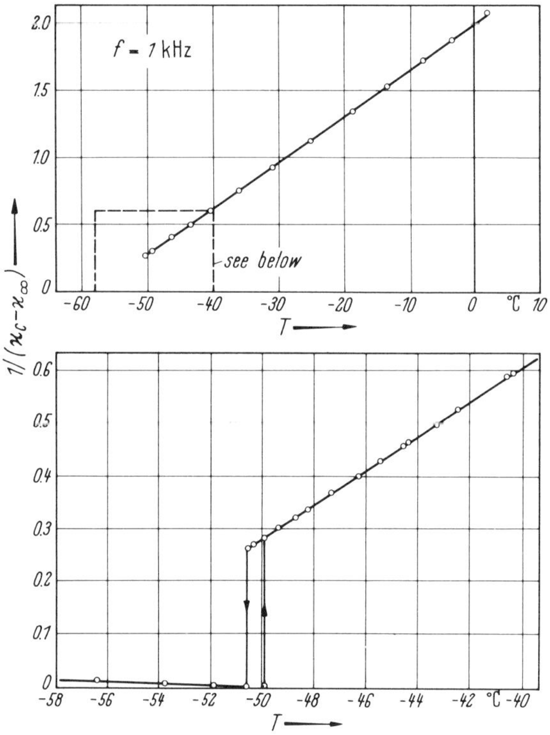

Fig. 593. $(NH_4)_2SO_4$. $1/(\varkappa_c - \varkappa_\infty)$ vs. T [65 U 3] $\varkappa_\infty = 9.4(2)$.

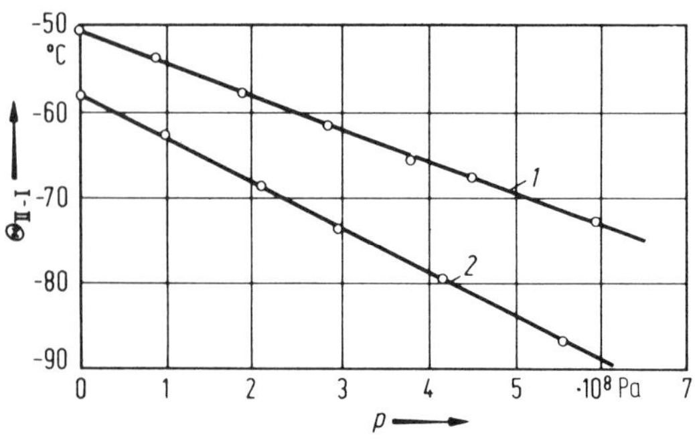

Fig. 594. $(NH_4)_2SO_4$, $[(NH_4)_2SO_4]_{0.9}$ $[NH_4BeF_3]_{0.1}$. Θ_{II-I} vs. p [72 P 4]. Curve *1*: for $(NH_4)_2SO_4$, *2*: for the solid solution.

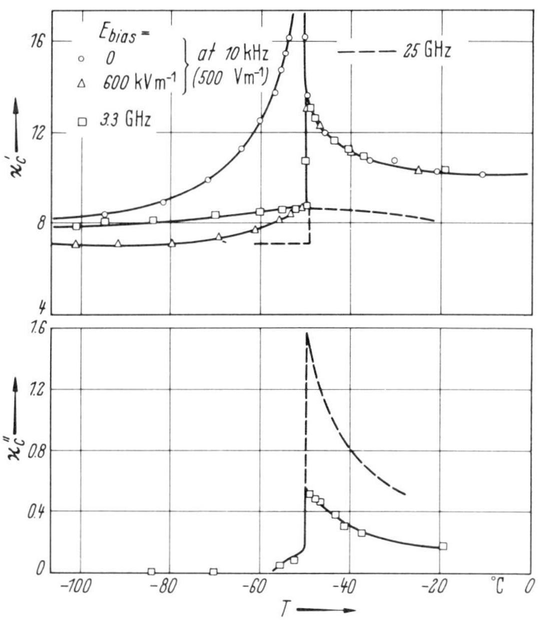

Fig. 596. $(NH_4)_2SO_4$. $\varkappa_c'$, $\varkappa_c''$ vs. T [66O1]. Dashed curve: [56C3].

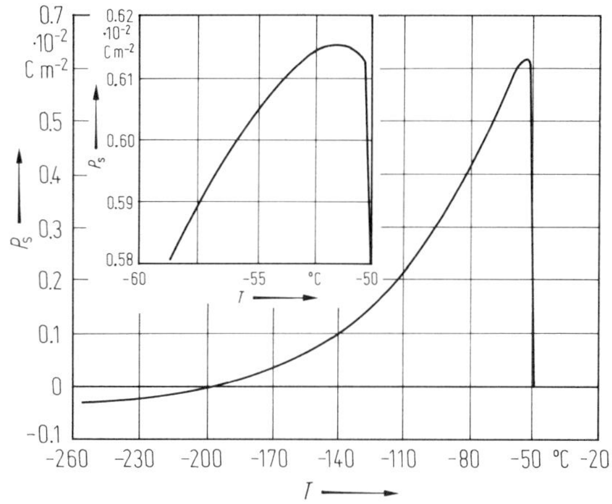

Fig. 597. $(NH_4)_2SO_4$. P_s vs. T [70U4].

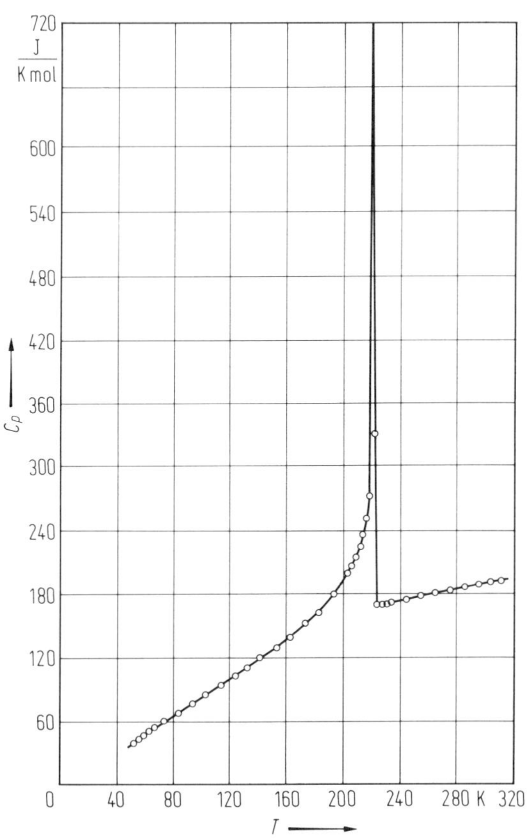

Fig. 598. $(NH_4)_2SO_4$. C_p vs. T [45S1]. C_p: molar heat capacity at constant pressure.

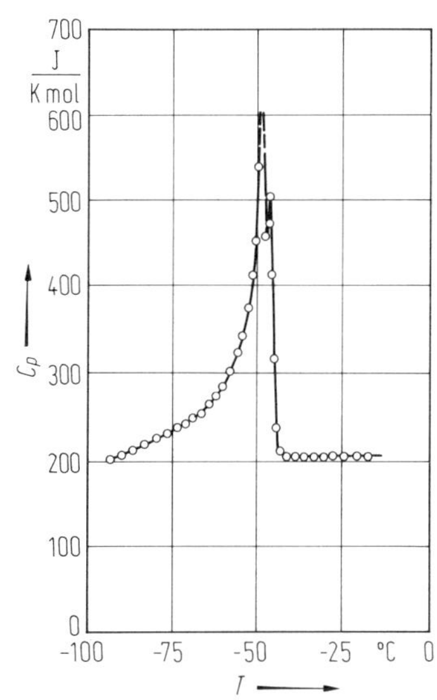

Fig. 599. $(ND_4)_2SO_4$. C_p vs. T [58H3]. C_p: molar heat capacity at constant pressure.

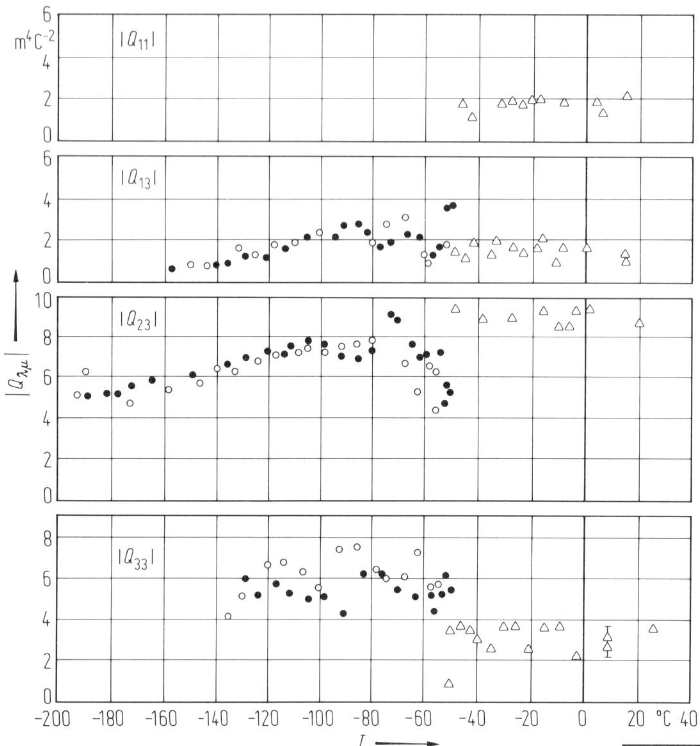

Fig. 601. $(NH_4)_2SO_4$. $|Q_{\lambda\mu}|$ vs. T [7313, 75Z2]. $Q_{\lambda\mu}$: electrostrictive constant. Values below Θ_{II-I} are obtained from piezoelectric constants assuming that P_s is independent of temperature and should be corrected by taking account of the decrease of P_s at lower temperatures [7313]. Open circles: on cooling, full circles: on heating. Those above Θ_{II-I} are determined by the measurement of electrostrictive strains at low frequencies [75Z2].

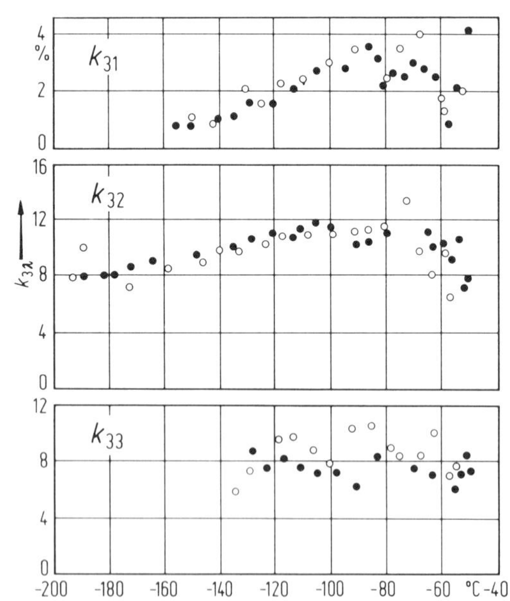

Fig. 600. $(NH_4)_2SO_4$. $k_{3\lambda}$ vs. T [7313]. $k_{3\lambda}$: electromechanical coupling coefficient. Open circles: on cooling, full circles: on heating.

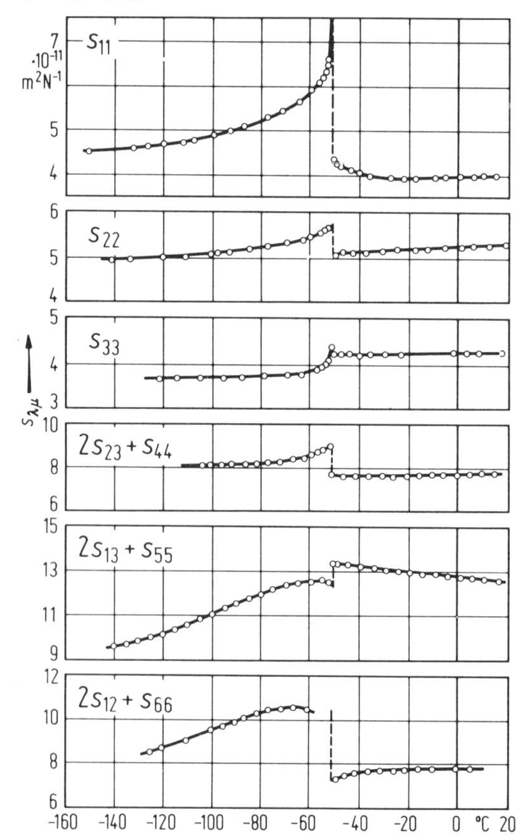

Fig. 602. $(NH_4)_2SO_4$. $s_{\lambda\mu}$ vs. T [7313]. $s_{\lambda\mu}$: elastic compliance.

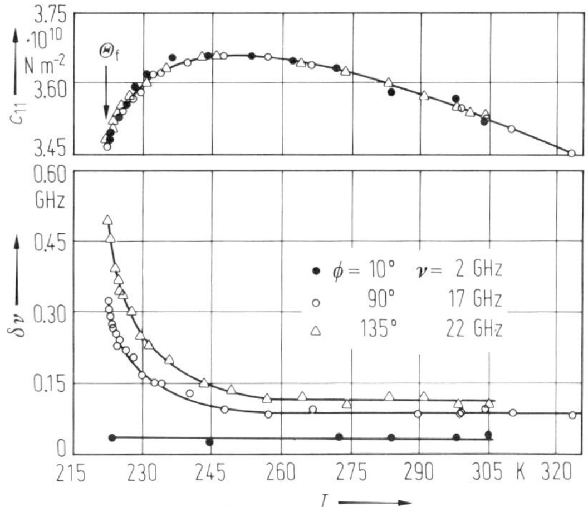

Fig. 603. $(NH_4)_2SO_4$ c_{11}, $\delta\nu$ vs. T [78U5]. c_{11}: elastic stiffness. $\delta\nu$: linewidth of phonon spectrum at different frequencies ν, i. e. different scattering angles ϕ.

Fig. 604, see next page.

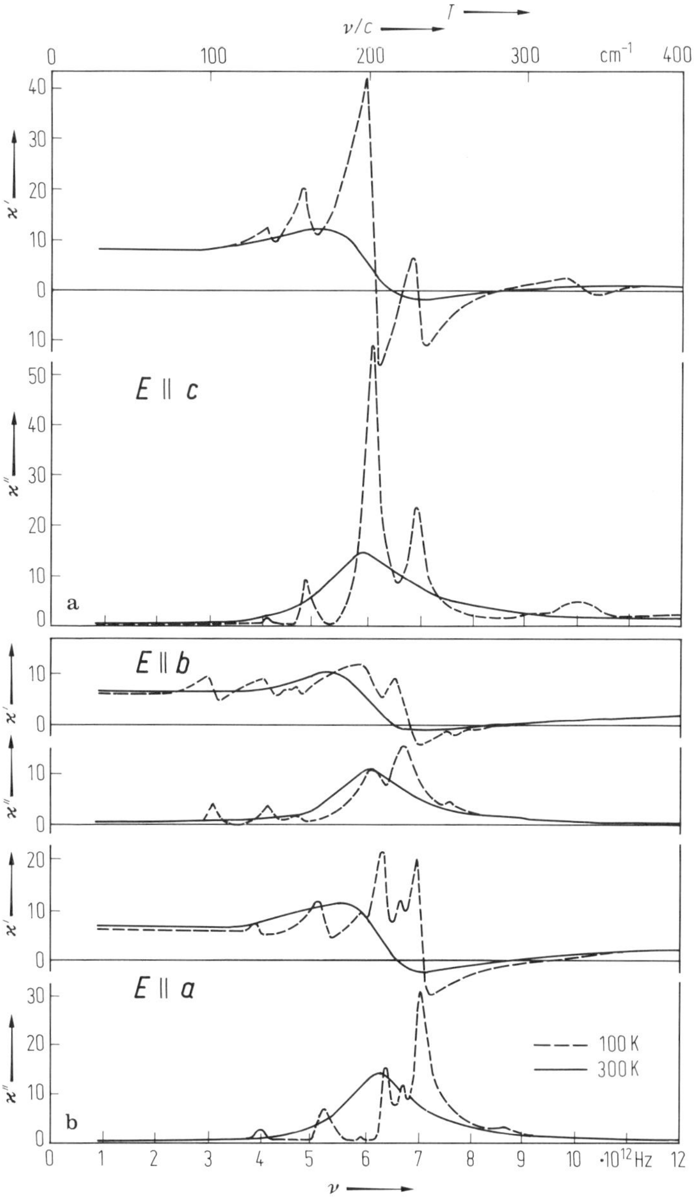

Fig. 605. $(NH_4)_2SO_4$. $\varkappa'$, $\varkappa''$ vs. ν [74P1]. ν: frequency of incident far-infrared radiation. Polarization of the radiation is parallel to the c-axis in Fig. a, and parallel to the b and the a-axis in Fig. b. $\varkappa'$ and $\varkappa''$ are obtained from the reflectivity spectra using Kramers-Kronig analysis.

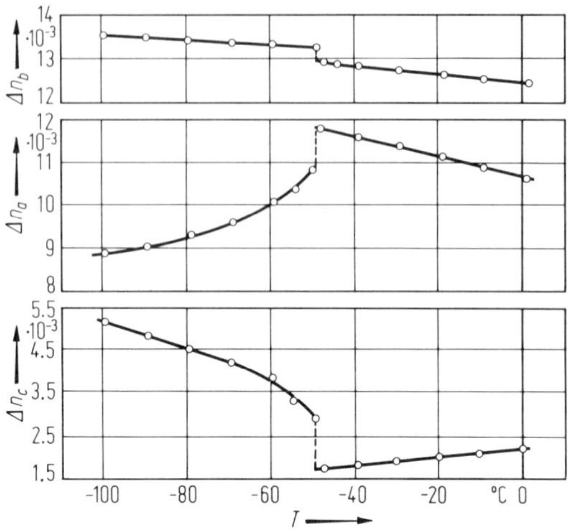

Fig. 604. $(NH_4)_2SO_4$. Δn_a, Δn_b, Δn_c vs. T [70A7]. $\Delta n_a = |n_b - n_c|$, $\Delta n_b = |n_c - n_a|$, $\Delta n_c = |n_a - n_b|$.

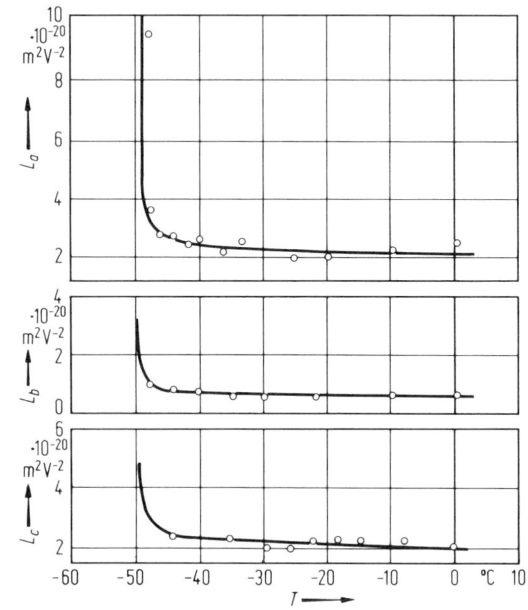

Fig. 606. $(NH_4)_2SO_4$. L_a, L_b, L_c vs. T [70A7]. $L_a = n_c^3 L_{33} - n_b^3 L_{23}$, $L_b = n_a^3 L_{13} - n_c^3 L_{33}$, $L_c = n_b^3 L_{23} - n_a^3 L_{13}$. $L_{\lambda\mu}$: quadratic electrooptic constant for E.

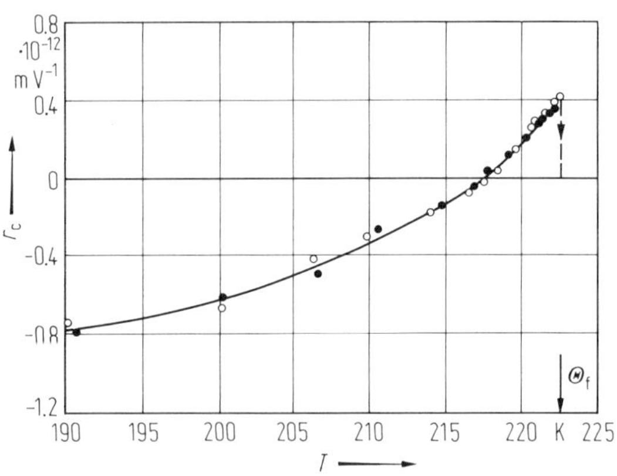

Fig. 607. $(NH_4)_2SO_4$. r_c vs. T [76F11]. $r_c = r_{13} - r_{23}$.

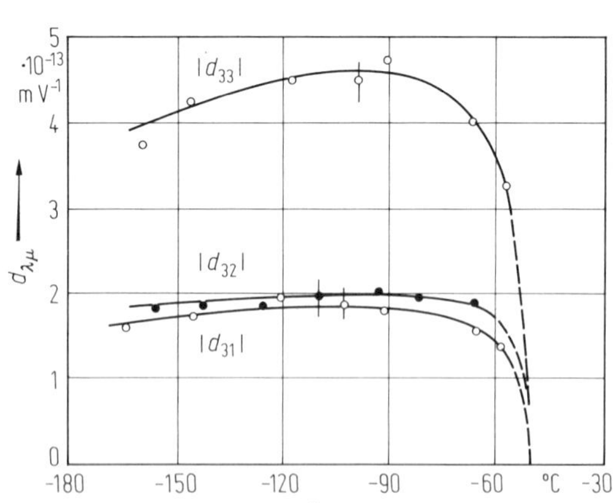

Fig. 608. $(NH_4)_2SO_4$. $d_{\lambda\mu}$ vs. T [77S41]. $d_{\lambda\mu}$: nonlinear optical susceptibility.

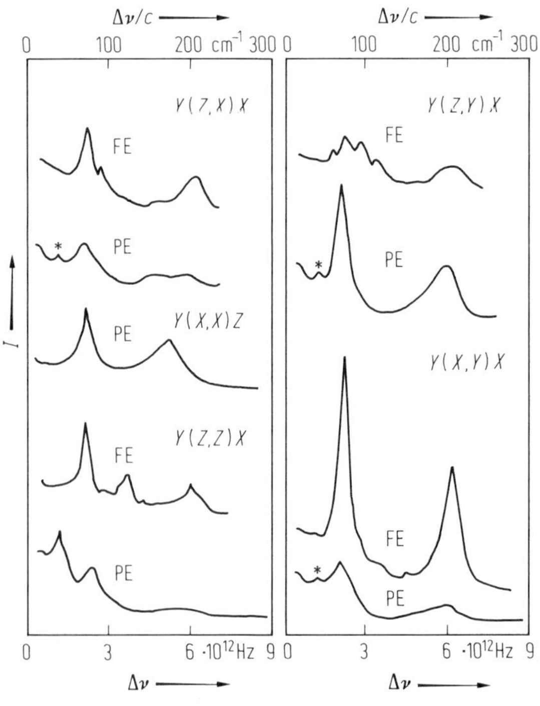

Fig. 610. $(NH_4)_2SO_4$. I vs. $\Delta\nu$ [76I3]. I: Raman scattering intensity of the low frequency region in the paraelectric phase (PE) at 240 K and ferroelectric phase (FE) at 214 K. Asterisks indicate spill-over of peaks of other symmetry.

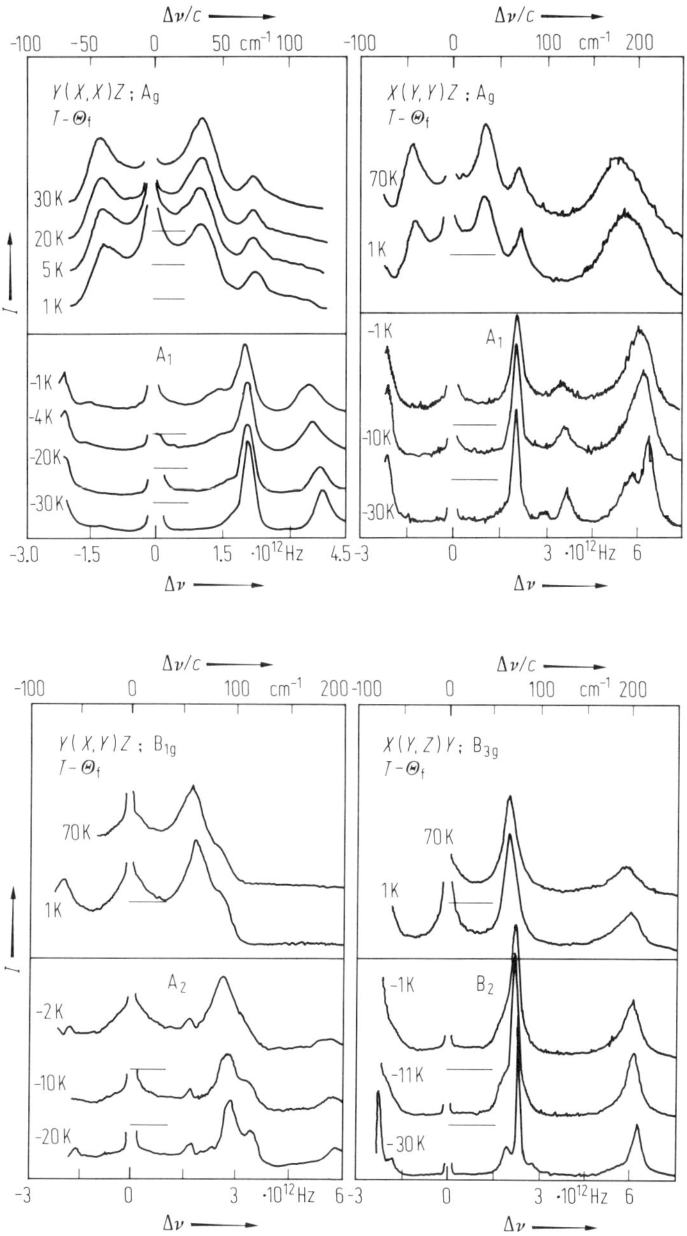

Fig. 609. $(NH_4)_2SO_4$. I vs. $\Delta\nu$ above and below Θ_f in the low frequency region [78U5]. I: Raman scattering intensity. Scattering geometries and the corresponding symmetries of phonons are given in the figure.

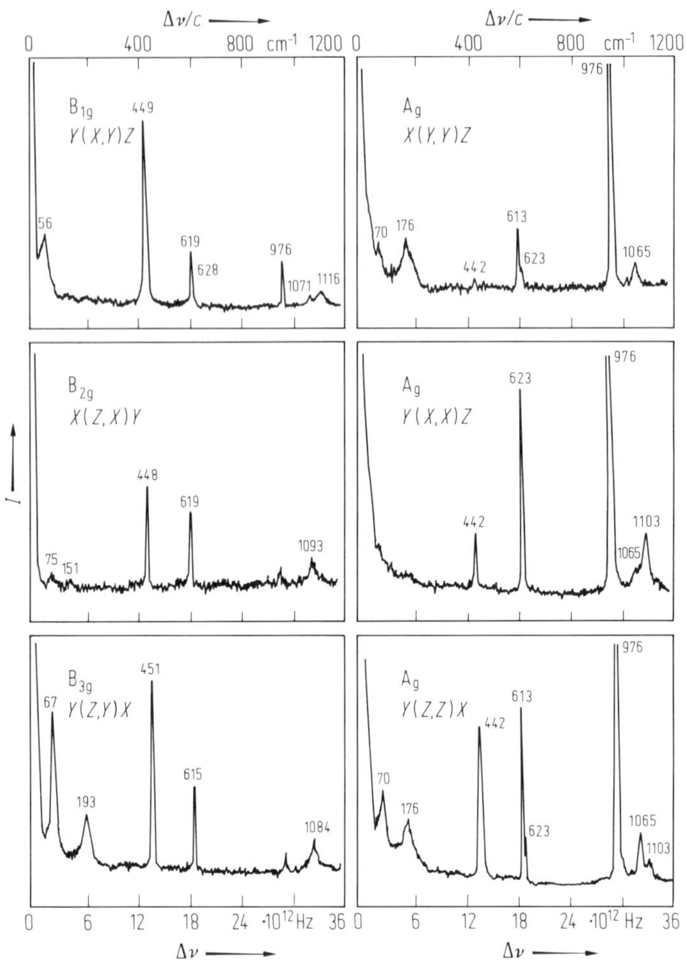

Fig. 611. $(NH_4)_2SO_4$. I vs. $\Delta\nu$ [74 U 1]. I: Raman scattering intensity at room temperature for various scattering geometries. Peak frequencies are given in the figure in units of cm^{-1}.

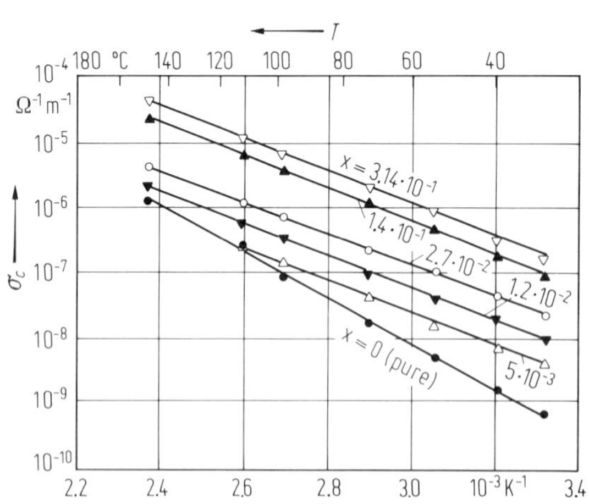

Fig. 614. $(NH_4)_2SO_4$. σ_c vs. $1/T$ for $(HSO_4)^-$ doped crystals [74 K 14]. x: contents of $(HSO_4)^-$ ions in $(NH_4)_2SO_4$ in mol%.

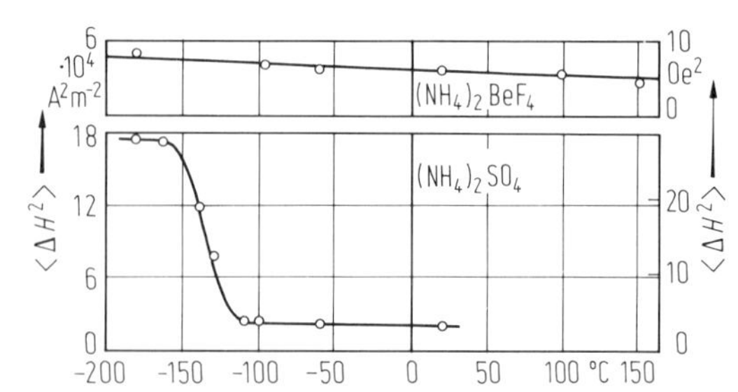

Fig. 615. $(NH_4)_2SO_4$, $(NH_4)_2BeF_4$. $\langle \Delta H^2 \rangle$ vs. T [60 B 4]. $\langle \Delta H^2 \rangle$: second moment of magnetic resonance curve of ^1H.

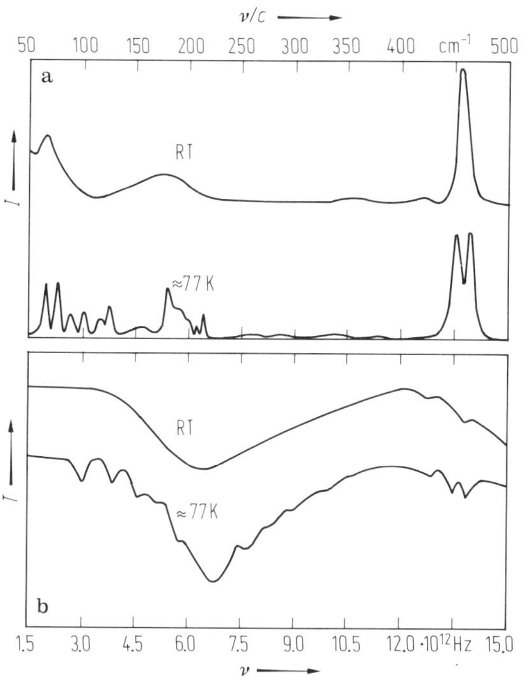

Fig. 612. $(ND_4)_2SO_4$. I, T vs. ν [72T14]. I: Raman scattering intensity (Fig. a). T: Transmittance for the infrared radiation (Fig. b). Sample: polycrystal. ν: frequency shift in (Fig. a) and frequency of the incident radiation in (Fig. b).

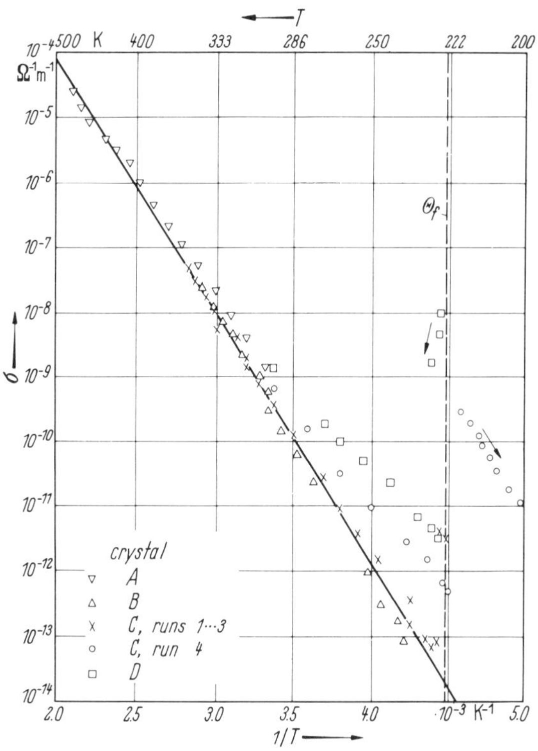

Fig. 613. $(NH_4)_2SO_4$. σ vs. T^{-1}. All points near and below Θ_f were obtained from cooling runs, except for the three isolated high-conductivity points for crystal D which were obtained just above Θ_f while warming the crystal immediately after having cooled through the transition point [63S4]. Conducting silver paint was applied on the (010) faces as electrodes: dc field of about 400 V was applied across crystals of 1…2 mm thickness. Probably the charge carriers are protons.

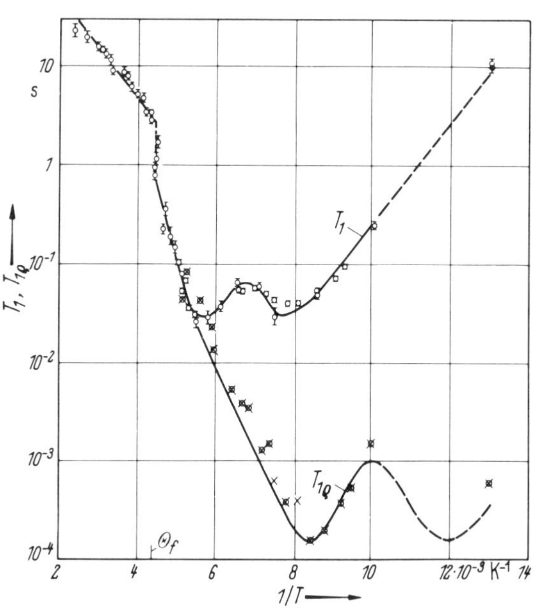

Fig. 616. $(NH_4)_2SO_4$. T_1 and $T_{1\varrho}$ vs. $1/T$ [67O6]. ○: T_1; ×: $T_{1\varrho}$ by the method of Look et al. [66L3]; ⊗: $T_{1\varrho}$ by the method of Ostroff and Waugh [66O3]. T_1: proton spin-lattice relaxation time; $T_{1\varrho}$: proton spin-lattice relaxation time along the radio frequency field.

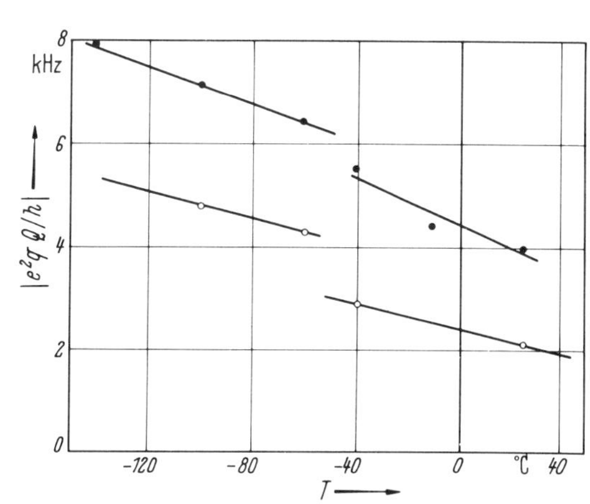

Fig. 617. $(ND_4)_2SO_4$. $|e^2qQ/h|$ vs. T [67O6]. $|e^2qQ/h|$: quadrupole coupling constant of ^2D. The different marks of data points correspond to the two inequivalent $(ND_4)^+$ ions.

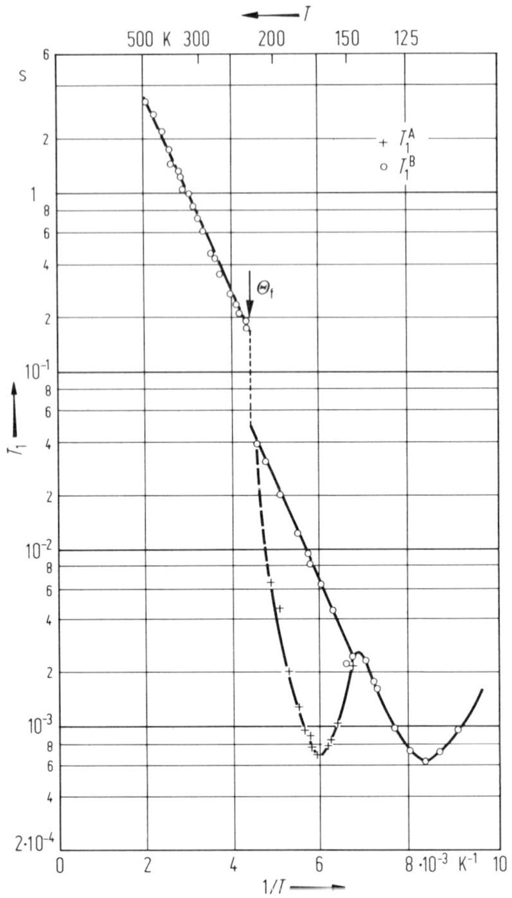

Fig. 618. $(ND_4)_2SO_4$. T_1 vs. $1/T$ [69K15]. T_1: deuteron spin-lattice relaxation time. There are two well-separated sets of T_1 around Θ_f, denoted as T_1^A and T_1^B. They suggest there are two subsystems of deuterons. $\nu = 11.5$ MHz.

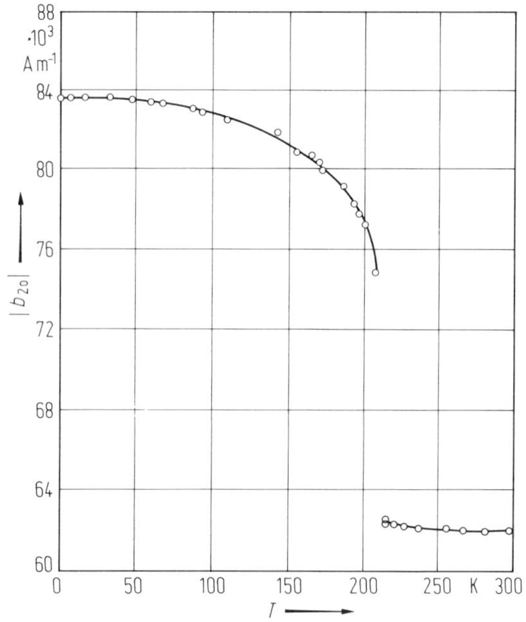

Fig. 620. $(NH_4)_2SO_4$: Mn^{2+}. $|b_{20}|$ vs. T [72A1]. ESR for Mn^{2+} doped crystal. b_{20}: FS parameter.

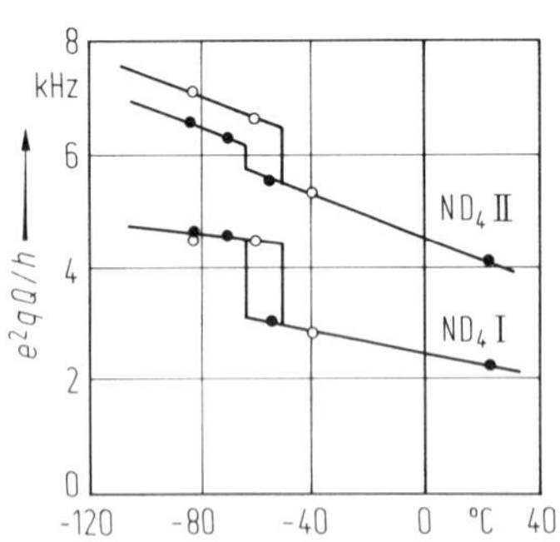

Fig. 619. $(ND_4)_2SO_4$. e^2qQ/h vs. T [75V5]. e^2qQ/h: quadrupole coupling constant of ^2D. ND_4I and ND_4II correspond to the two non-equivalent $(ND_4)^+$ ions. Open circles: at atmospheric pressure; full circles: under hydrostatic pressure of $3.5 \cdot 10^8$ Pa.

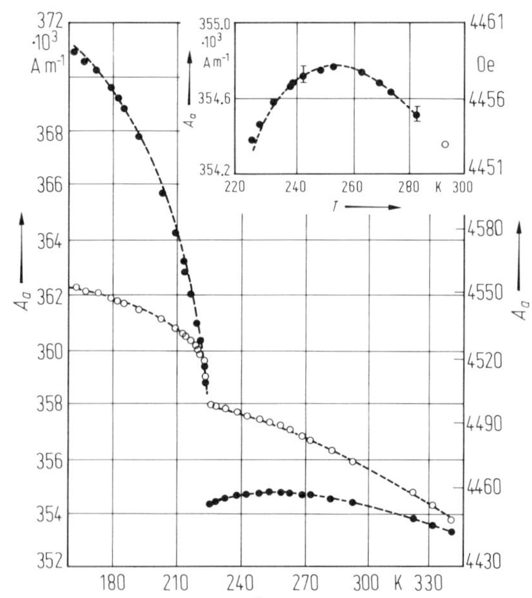

Fig. 621. $(NH_4)_2SO_4$: Cd^+. A_a vs. T [78B1]. ESR of Cd^+ impurity. A_a: Hyperfine structure parameter of Cd^+. External magnetic field is parallel to the a-axis, one of the principal axes of the **A** tensor. Full circles: Cd^+ in α-position; open circles: Cd^+ in β-position.

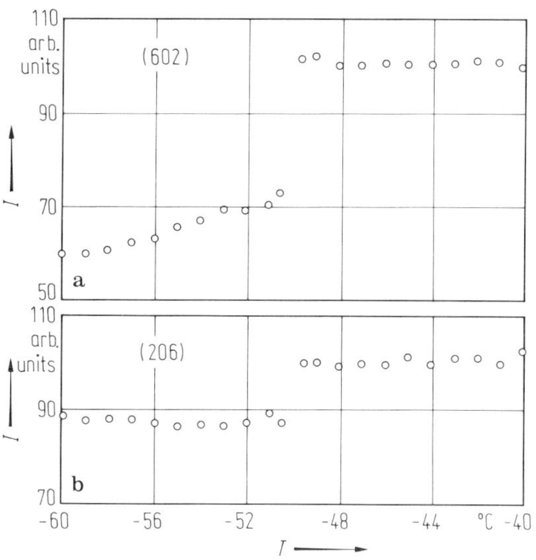

Fig. 622. $(NH_4)_2SO_4$. $I_{(6,0,2)}$, $I_{(2,0,6)}$ vs. T [77H3]. I: integrated intensity of X-ray Bragg reflection.

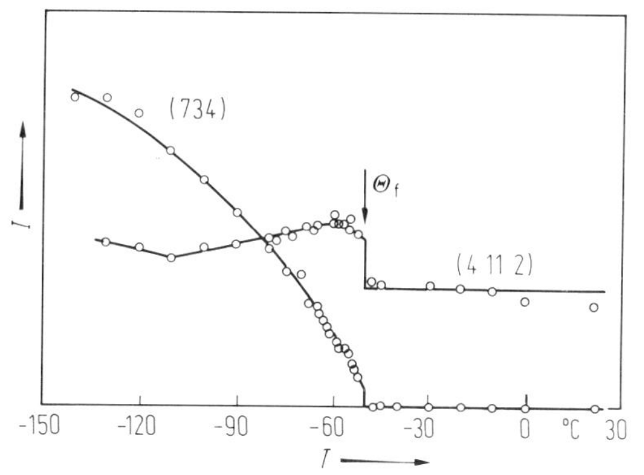

Fig. 623. $(NH_4)_2SO_4$. $I_{(7,3,4)}$ and $I_{(4,11,2)}$ vs. T [78O4]. I: integrated intensity of X-ray Bragg reflection.

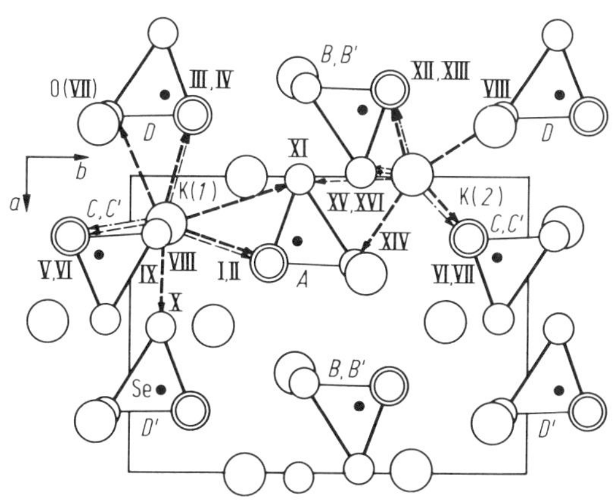

Fig. 624. K_2SeO_4. Structure of phase II [70K2]. Projection on (001). Potassium atoms are shown by large circles, oxygen atoms by smaller open circles, and selenium atoms by small full circles. For numbering of the atoms, see Tables 141, 142. Tetrahedra are labelled by capital letters. Pairs of dashed lines indicate contacts to a pair of atoms belonging to the same tetrahedron. A dashed line paired with a dash-dot line indicates that the pair of superposed atoms belong to different tetrahedra.

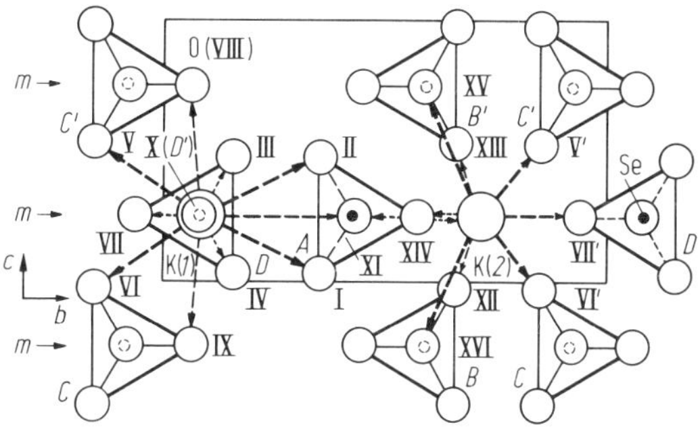

Fig. 625. K_2SeO_4. Structure of phase II [70K2]. Projection on (100). See caption of Fig. 624 for symbols and notations. m = mirror plane.

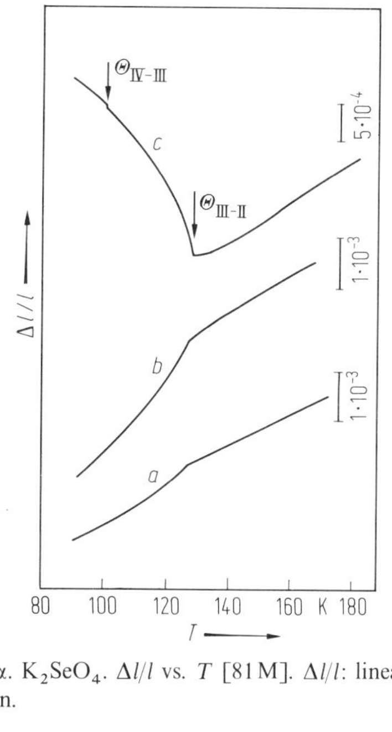

Fig. 626α. K_2SeO_4. $\Delta l/l$ vs. T [81M]. $\Delta l/l$: linear thermal expansion.

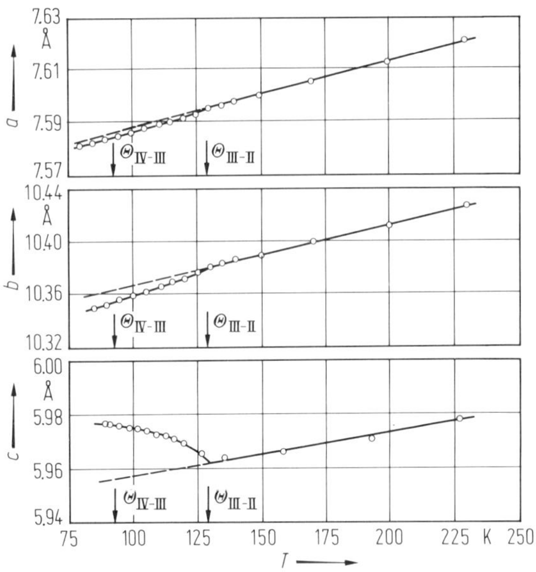

Fig. 626β. K_2SeO_4. a, b, c vs. T [81 K].

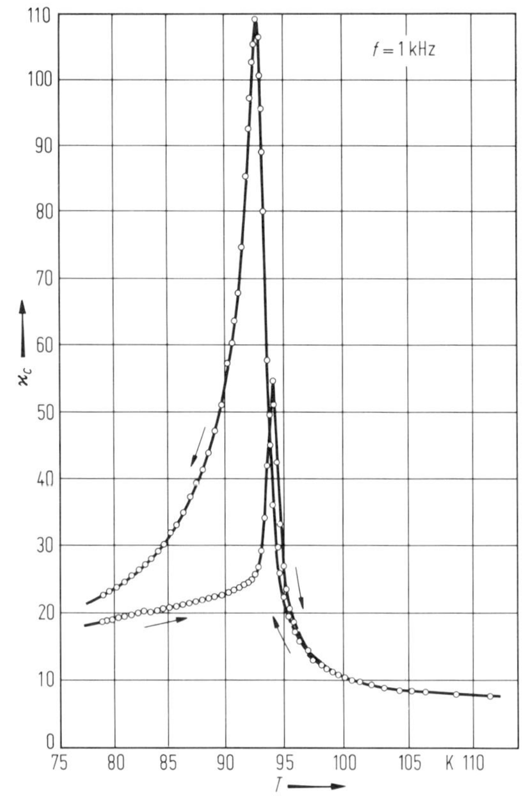

Fig. 627α. K_2SeO_4. $\varkappa_c$ vs. T [70 A 3].

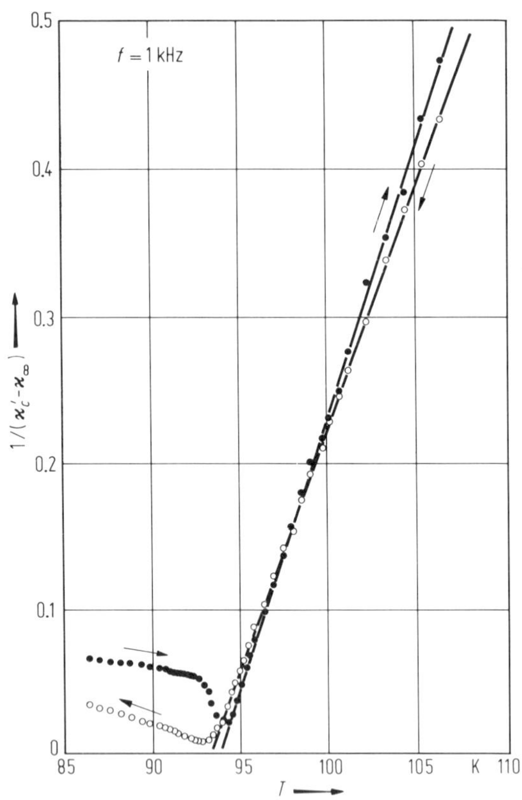

Fig. 627β. K_2SeO_4. $1/(\varkappa_c'-\varkappa_\infty)$ vs. T [70 A 3]. $\varkappa_\infty = 6.14$.

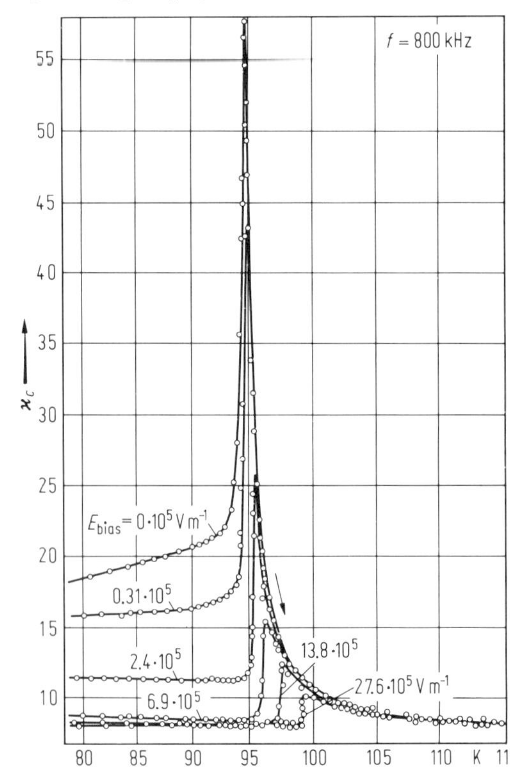

Fig. 627γ. K_2SeO_4. $\varkappa_c$ vs. T [70 A 3]. Parameter: E_{bias}.

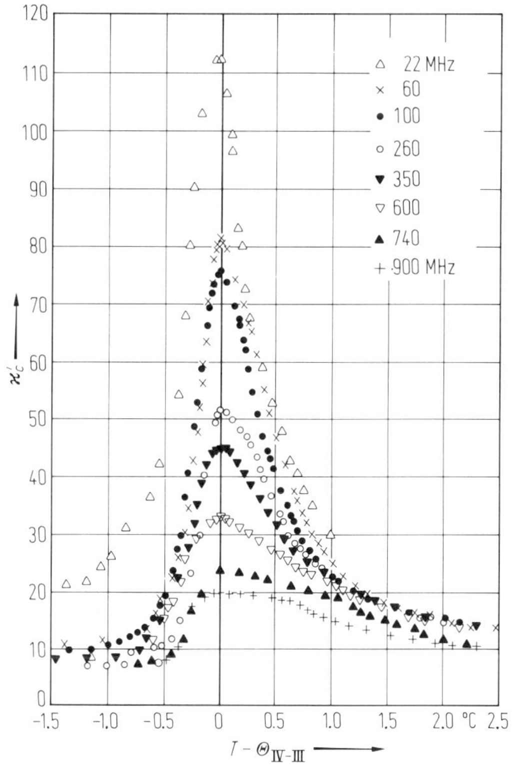

Fig. 628α. K_2SeO_4. $\varkappa_c'$ vs. $T-\Theta_{IV-III}$ [80H1]. Parameter: f.

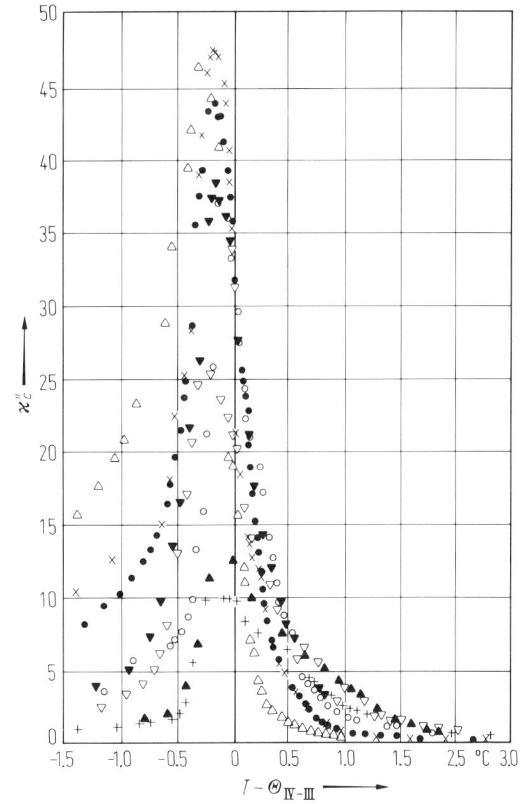

Fig. 628β. K_2SeO_4. $\varkappa_c''$ vs. $T-\Theta_{IV-III}$ [80H1]. For different symbols, see Fig. 628α.

Figs. 630, 631, see p. 456.

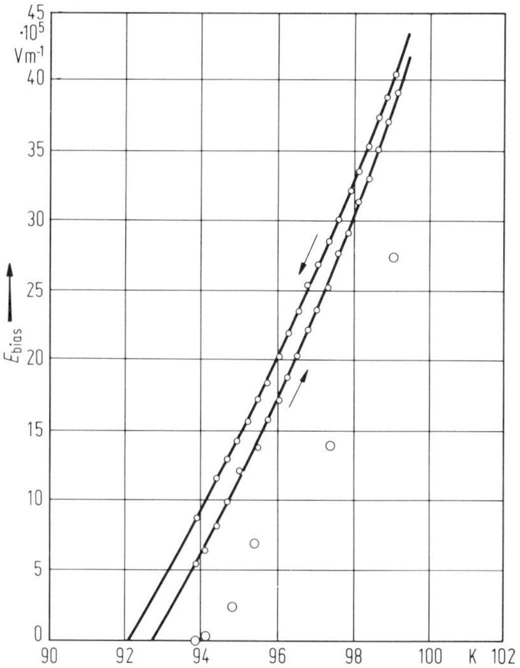

Fig. 629. K_2SeO_4. Θ_f vs. E_{bias} [70A3]. Small circles are data obtained from double hysteresis loop and large circles are those from dielectric measurements.

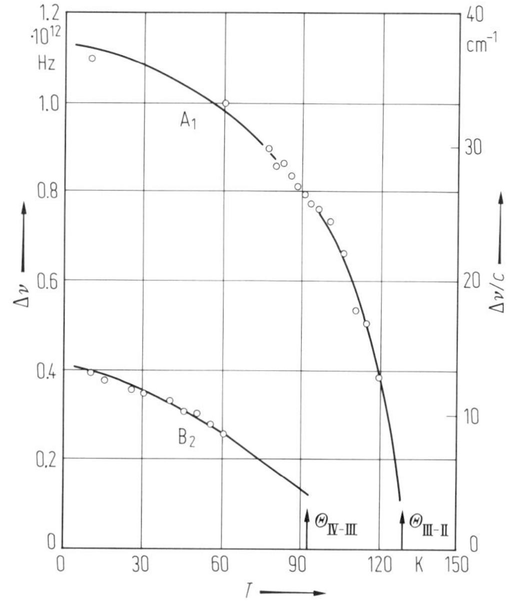

Fig. 632. K_2SeO_4. $\Delta\nu$ vs. T [77W3]. $\Delta\nu$: Raman frequency shift of the soft A_1 mode and the soft B_2 mode. The data for the A_1 mode are partly provided from [77W1].

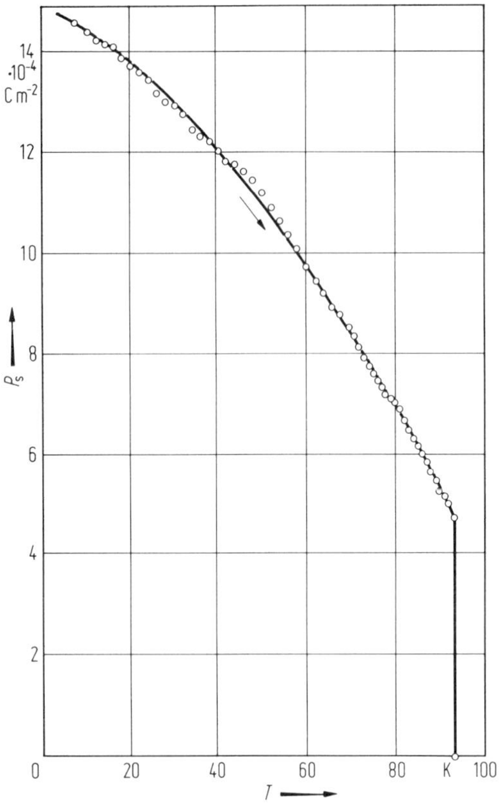

Fig. 630α. K_2SeO_4. P_s vs. T [70A3].

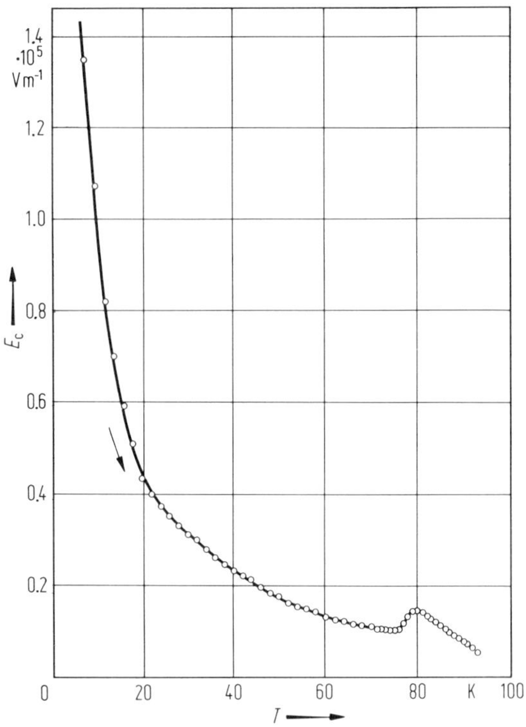

Fig. 630β. K_2SeO_4. E_c vs. T [70A3].

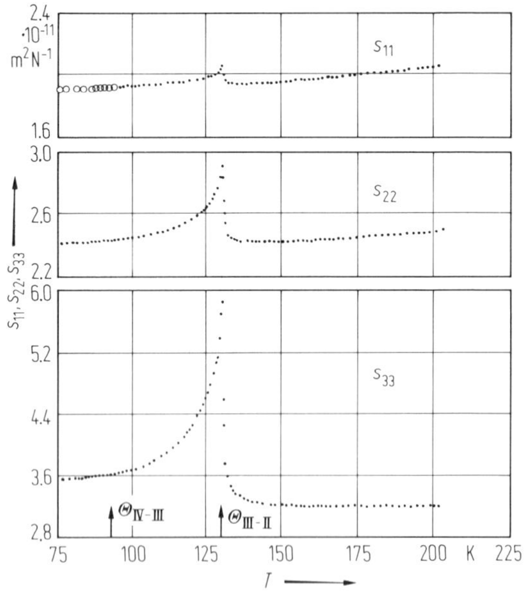

Fig. 631α. K_2SeO_4. $s_{\lambda\mu}$ vs. T [80K2].

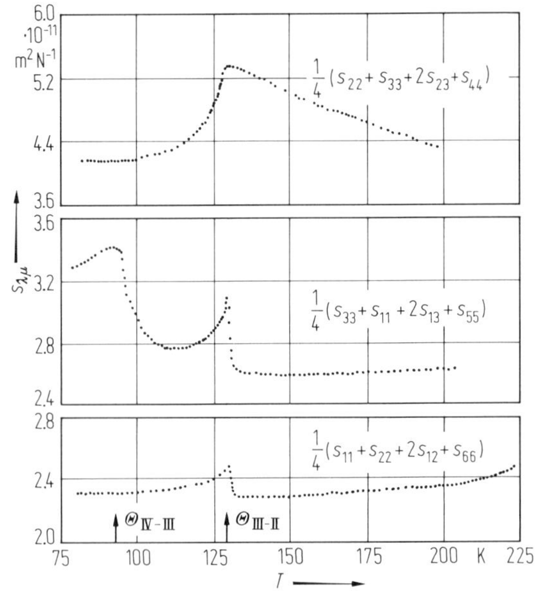

Fig. 631β. K_2SeO_4. $s_{\lambda\mu}$ vs. T [80K2].

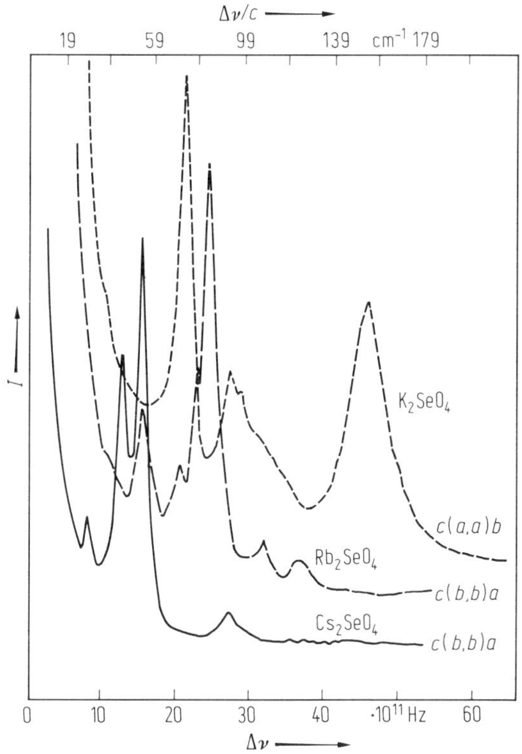

Fig. 633. K_2SeO_4, Rb_2SeO_4, and Cs_2SeO_4. I vs. $\Delta\nu$ at room temperature [76T11] I: Raman scattering intensity of low frequency A_g modes measured in the indicated scattering geometries.

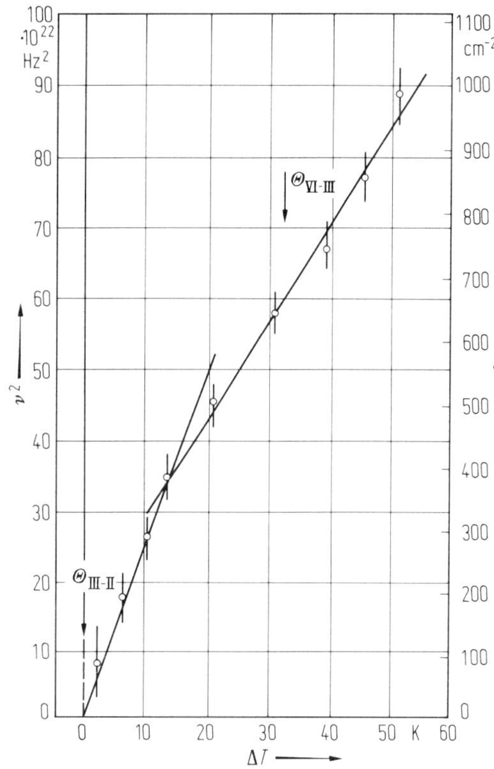

Fig. 634. K_2SeO_4. ν^2 vs. T [79F3]. $\Delta T = T_m - T$. T_m refers to the temperature at which the central peak scattering is maximum, and is approximately equal to Θ_{III-II}. ν: soft mode frequency in $c(a,a)b$ scattering geometry.

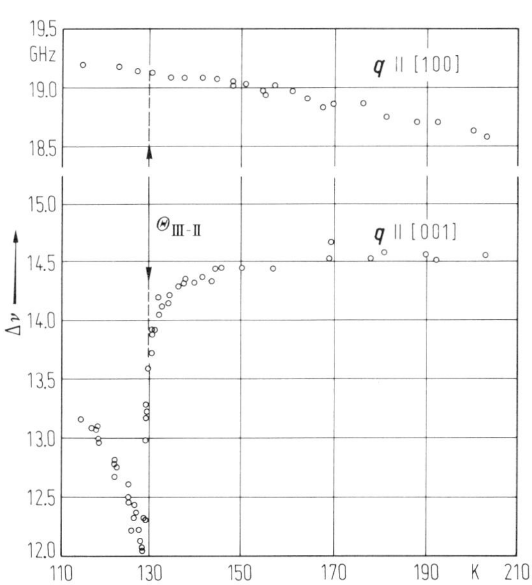

Fig. 635. K_2SeO_4. $\Delta\nu$ vs. T [79Y4]. $\Delta\nu$: Brillouin shift. $\boldsymbol{q}$: wave vector.

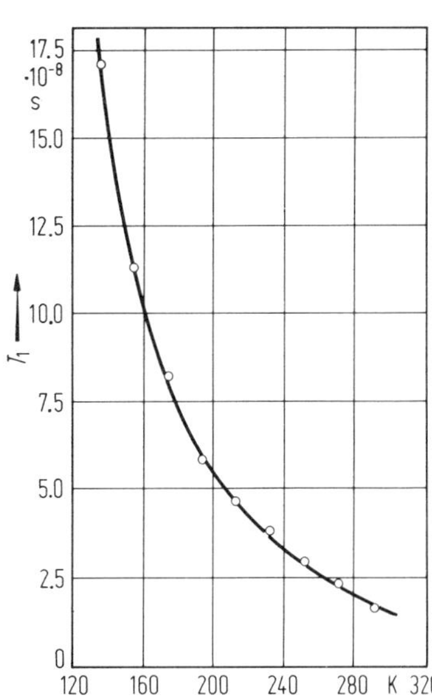

Fig. 636. K_2SeO_4. T_1 vs. T [70A2]. T_1: spin-lattice relaxation time for $(SeO_4)^-$ produced by ^{60}Co γ-rays.

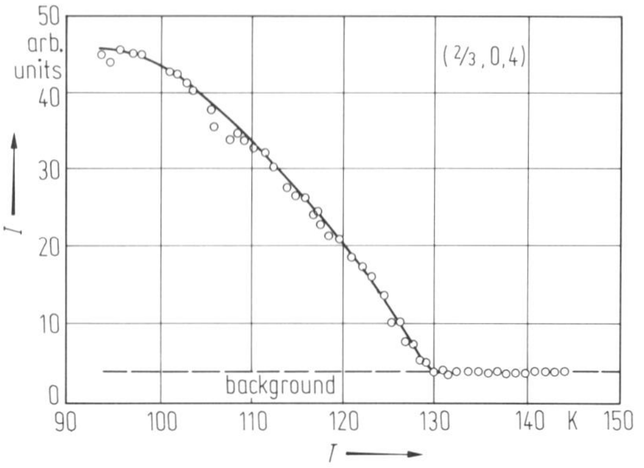

Fig. 637. K_2SeO_4. I(2/3, 0, 4) vs. T [75T9]. I: peak intensity of X-ray Bragg reflection.

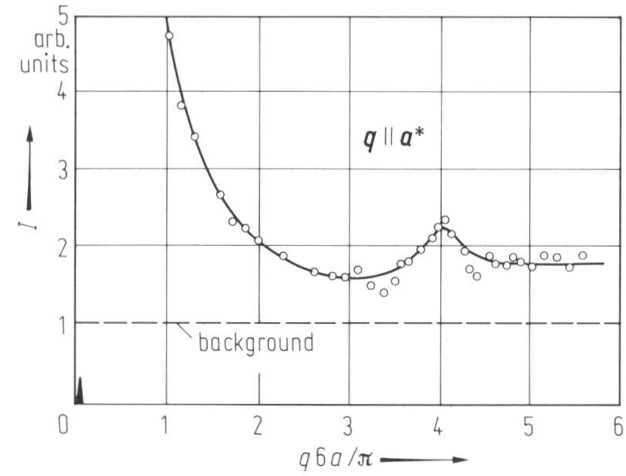

Fig. 639. K_2SeO_4. X-ray diffuse scattering around (2/3, 0, 4) in phase II [75T9]. Intensity along the line from (004) to (2/3, 0, 4). $T=\Theta_{III-II}+2$ K. q: wave number vector. Small triangle in the figure indicates the size of the resolution.

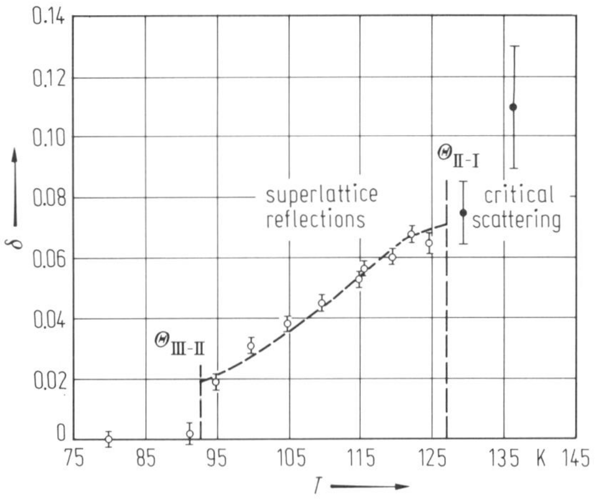

Fig. 638α. K_2SeO_4. δ vs. T [77I2]. $1/3\delta a^*$: deviation of superlattice reflection from $q=(a^*/3, 0, 0)$. $a^*=\frac{2\pi}{a}$.

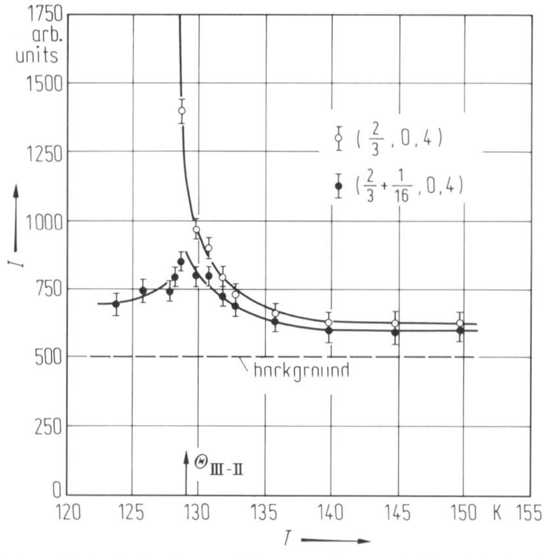

Fig. 640. K_2SeO_4. X-ray diffuse scattering in the vicinity of Θ_{III-II} [75T9]. I(2/3, 0, 4), I(2/3+1/16, 0, 4) vs. T. I: diffuse scattering intensity.

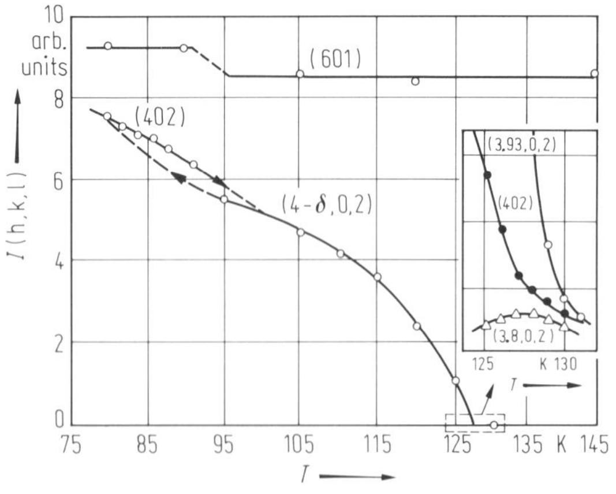

Fig. 638β. K_2SeO_4. I(h k l) vs. T [77I2]. I(h k l): intensity of neutron reflection at (h, k, l).

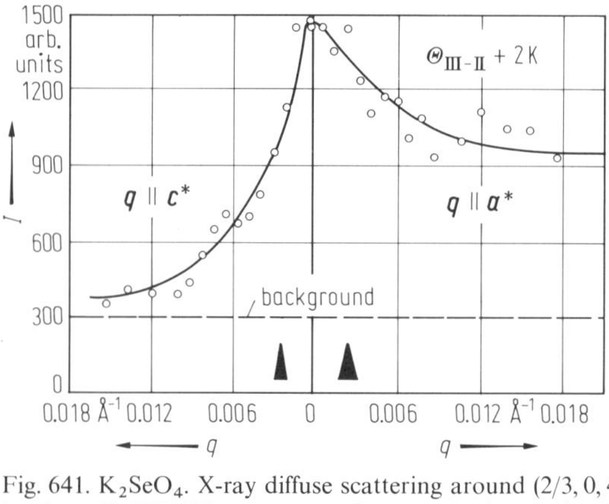

Fig. 641. K_2SeO_4. X-ray diffuse scattering around (2/3, 0, 4) in phase II. [75T9]. $T=\Theta_{III-II}+2$ K. Small triangles in the figure indicate the size of the resolution.

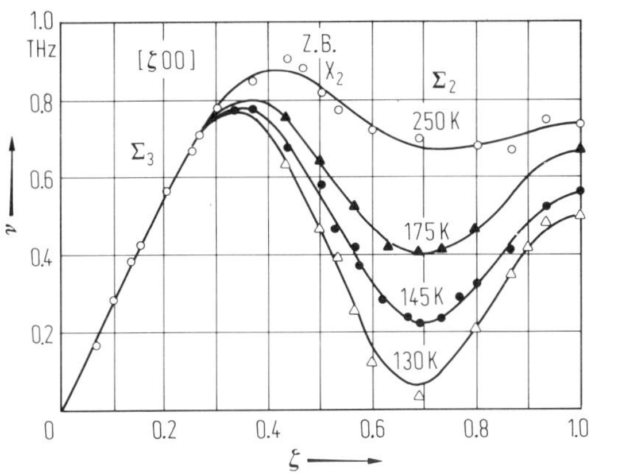

Fig. 642. K_2SeO_4. ν vs. ζ in an extended zone [77I2]. Parameter: *T*. Z. B.: zone boundary.

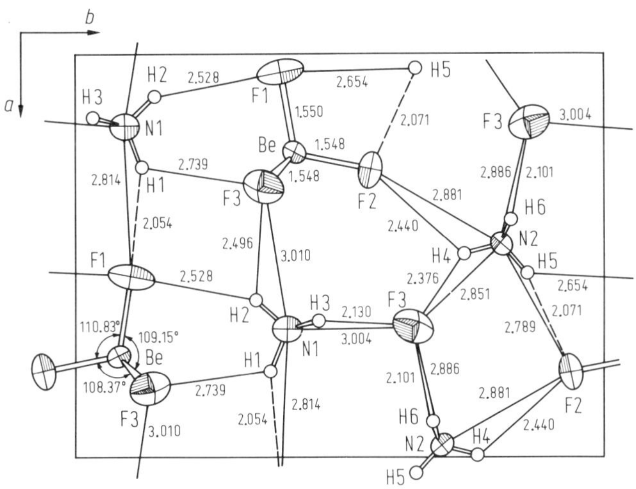

Fig. 643. $(NH_4)_2BeF_4$. Structure of phase I [79O2]. Projection on (001). Values in the figure are obtained at 20(1) °C. Dashed lines indicate proposed hydrogen bonds.

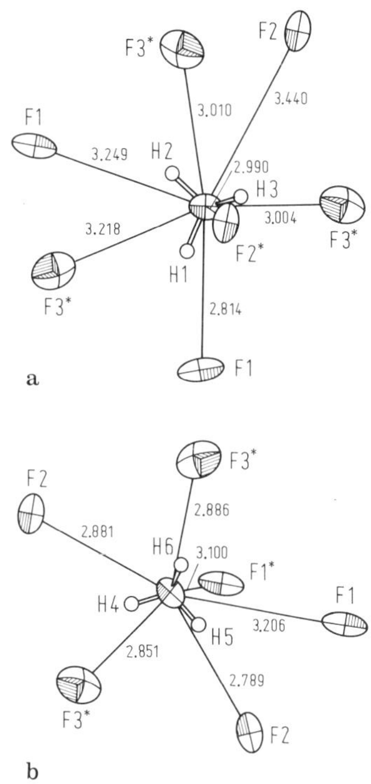

Fig. 644. $(NH_4)_2BeF_4$. Structure of phase II [79O2]. Environments of ammonium ions viewed along the *c* axis: Fig. a: around $N(1)H_4$ and Fig. b: around $N(2)H_4$. The atoms labelled by "*" are above and below the mirror plane. Interatomic distances in Å.

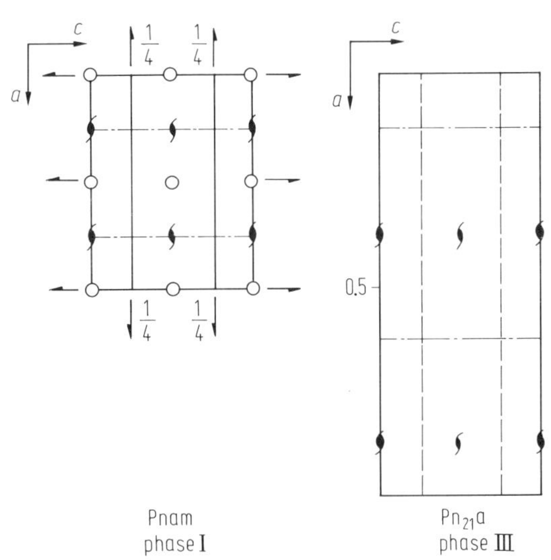

Fig. 645. $(NH_4)_2BeF_4$. Structure [79O1]. Unit cell relation between phases I and III. Solid line: mirror plane; dashed line: axial glide plane; dashed-dotted line: diagonal glide plane.

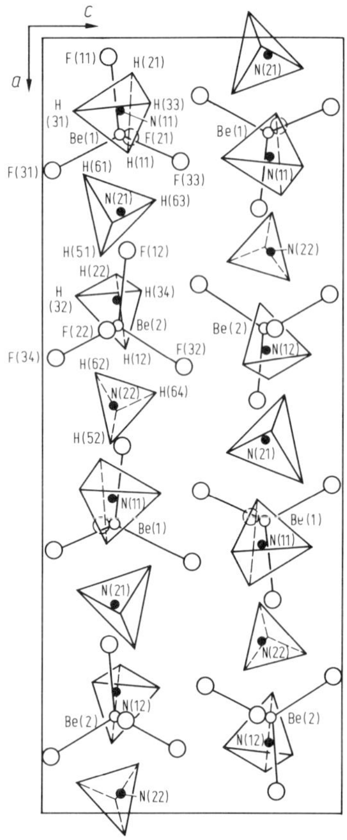

Fig. 647. $(NH_4)_2BeF_4$. Structure of phase III [79 O 1]. Projection on (010).

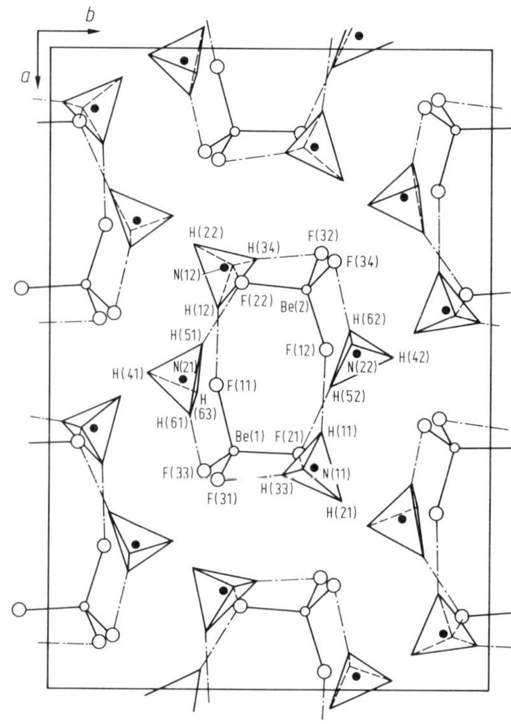

Fig. 646. $(NH_4)_2BeF_4$. Structure of phase III [79 O 1]. Projection on (001). Ammonium ions are shown by tetrahedra, where hydrogen atoms occupy the corners. The proposed hydrogen bonds are indicated by dashed-dotted lines.

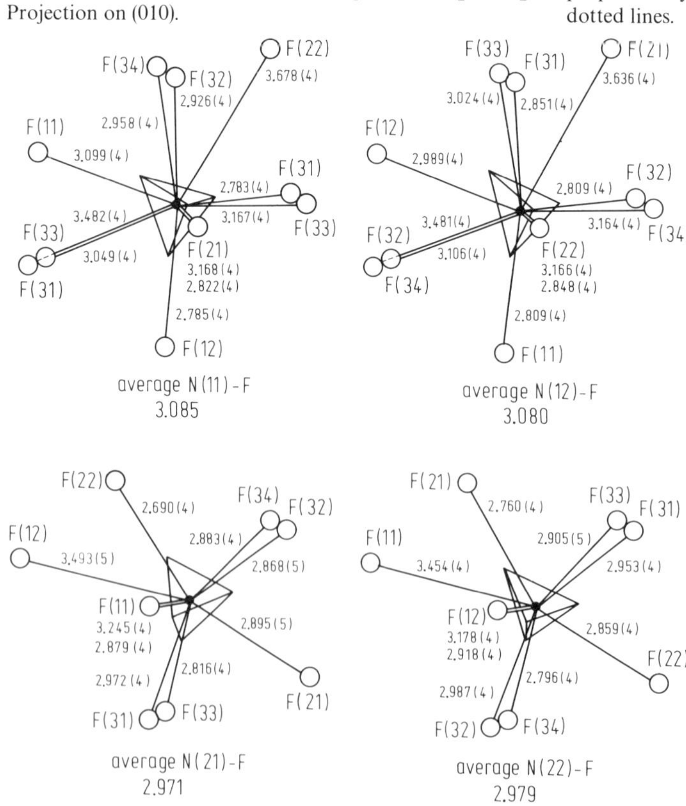

Fig. 648. $(NH_4)_2BeF_4$. Structure of phase III [79 O 1]. Environments of the ammonium ions viewed along the *c* axis. Numerals are interatomic distances at −140 °C in Å.

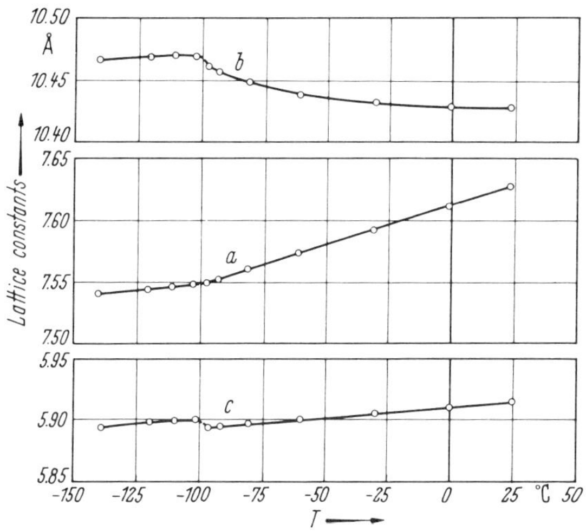

Fig. 649. $(NH_4)_2BeF_4$. Lattice constants vs. T. [58 H 3].

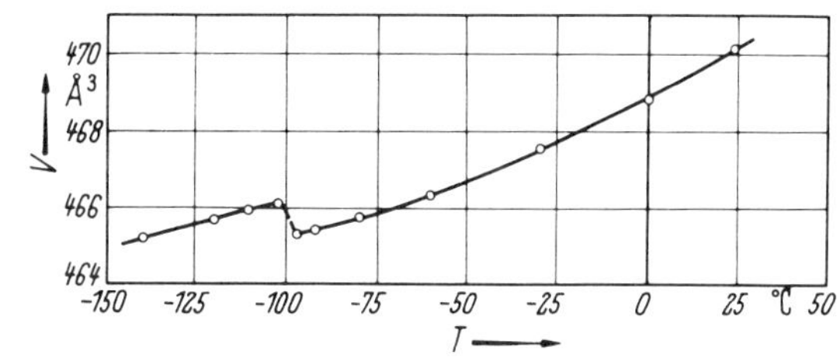

Fig. 650. $(NH_4)_2BeF_4$. Basic unit cell volume vs. T [58 H 3].

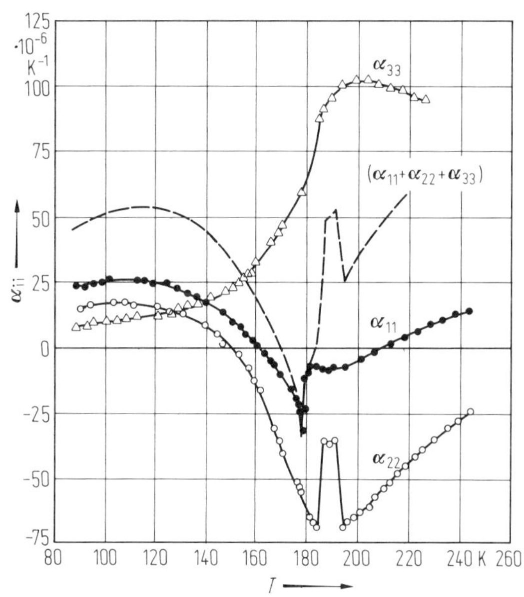

Fig. 651. $(NH_4)_2BeF_4$. α_{ii} vs. T [74 Z 4]. α_{11}, α_{22}, α_{33}: linear thermal expansion coefficients along the a, b, c, axes.

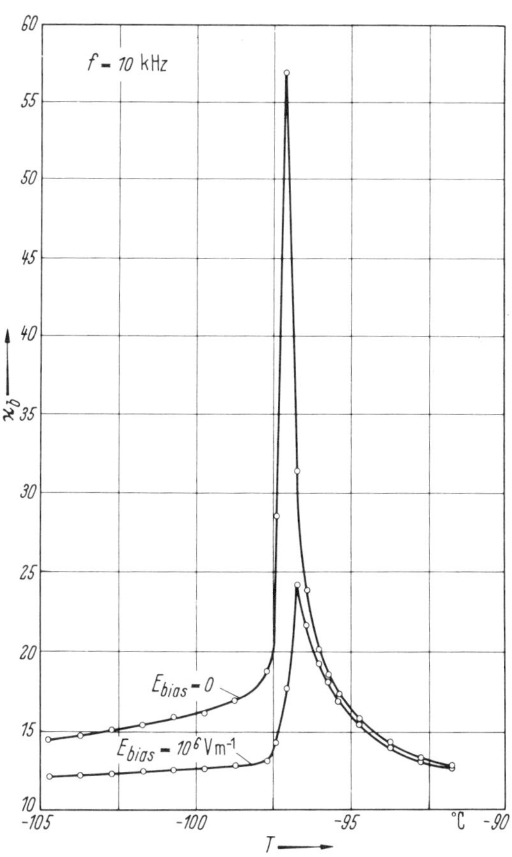

Fig. 652. $(NH_4)_2BeF_4$. $\varkappa_b$ vs. T. Parameter: E_{bias} [58 H 3].

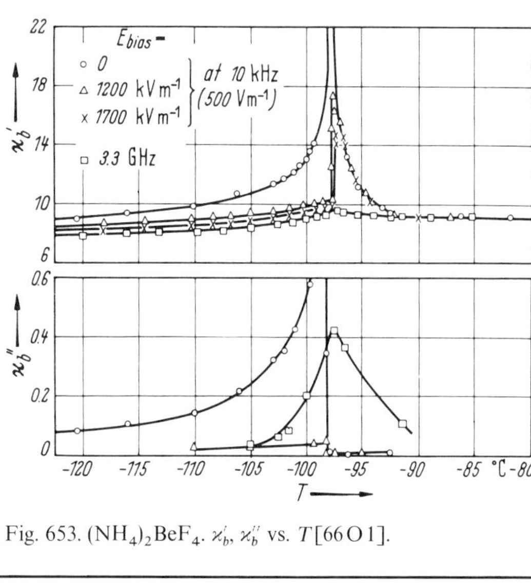

Fig. 653. $(NH_4)_2BeF_4$. $\varkappa_b'$, $\varkappa_b''$ vs. T [66 O 1].

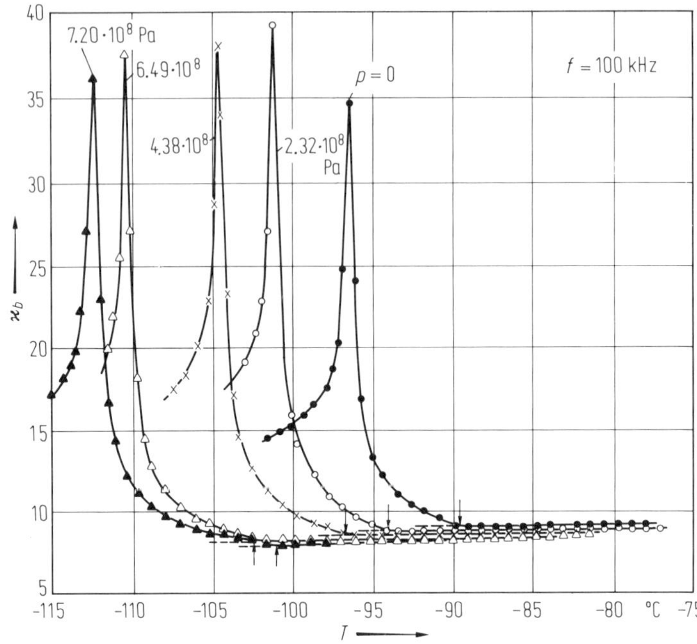

Fig. 654. $(NH_4)_2BeF_4$. $\varkappa_b$ vs. T. Parameter: p [74G5]. The vertical arrow on each curve shows the break at Θ_{II-I}.

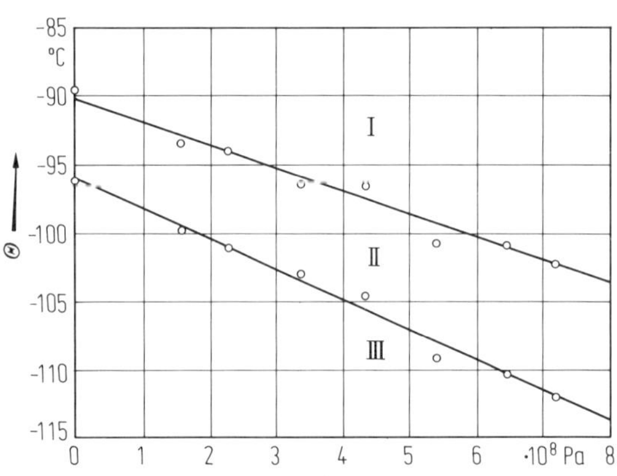

Fig. 655. $(NH_4)_2BeF_4$. Θ vs. p [74G5].

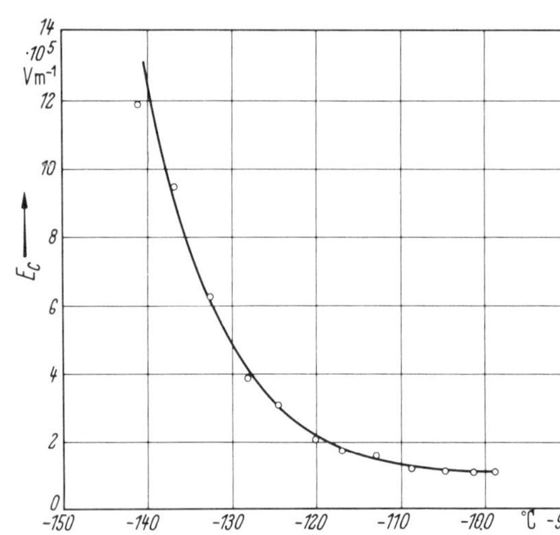

Fig. 657. $(NH_4)_2BeF_4$. E_c vs. T [58H3].

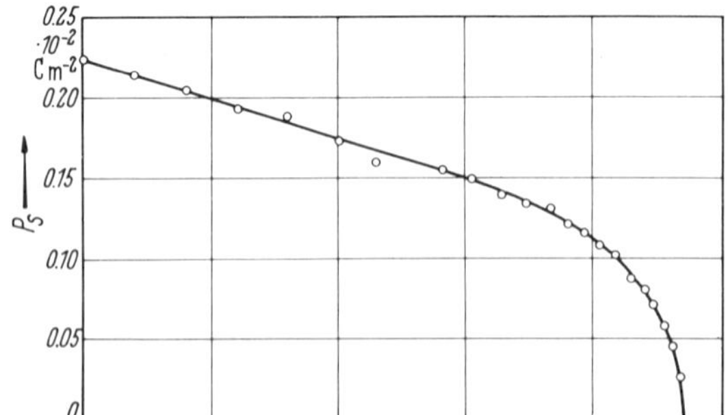

Fig. 656. $(NH_4)_2BeF_4$. P_s vs. T [58H3].

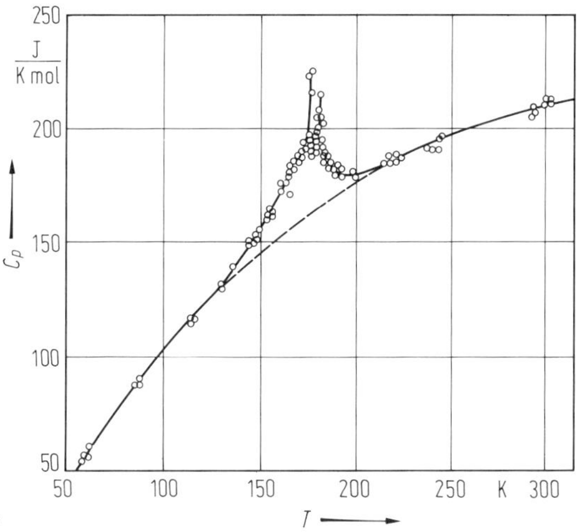

Fig. 658. $(NH_4)_2BeF_4$. C_p vs. T [73 S 20].

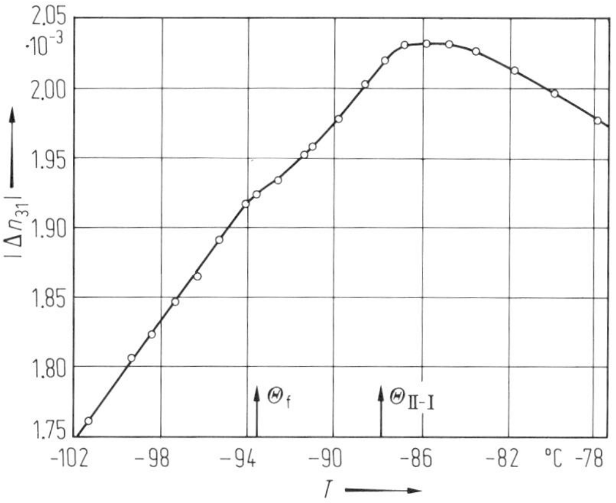

Fig. 660. $(NH_4)_2BeF_4$. $|\Delta n_{31}|$ vs. T [76 K 14].

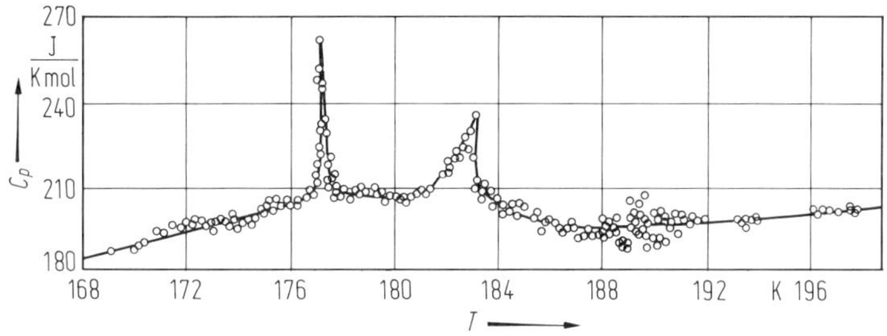

Fig. 659. $(NH_4)_2BeF_4$. C_p vs. T [73 S 20].

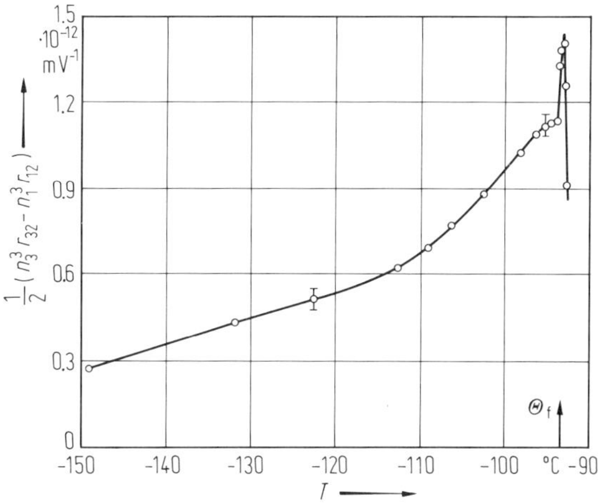

Fig. 661. $(NH_4)_2BeF_4$. $\frac{1}{2}(n_3^3 r_{32} - n_1^3 r_{12})$ vs. T [76 K 14].

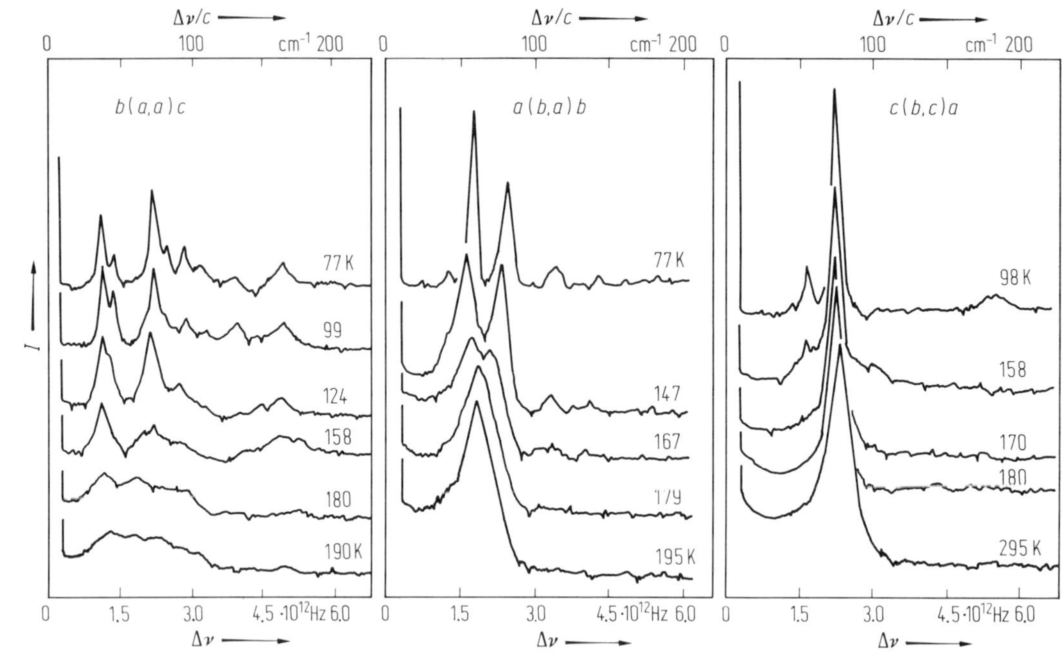

Fig. 662. $(NH_4)_2BeF_4$. I vs. $\Delta\nu$ [77 W 2]. I: Raman scattering intensity for various scattering geometries. Parameter: T.

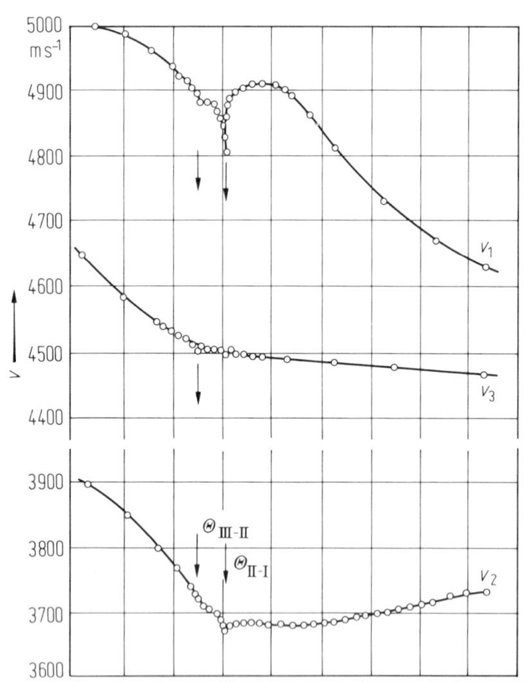

Fig. 663. $(NH_4)_2BeF_4$. v vs. T [76 A 5]. v: velocities of longitudinal sound waves. v_1: [100] direction, v_2: [010], v_3: [001]. Brillouin scattering. $\lambda = 632.81$ nm.

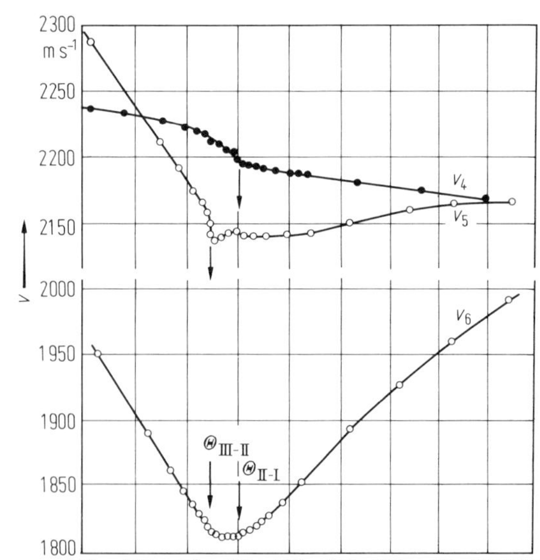

Fig. 664. $(NH_4)_2BeF_4$. v vs. T [76 A 5]. v: velocities of transverse sound waves. v_4: quasi-transverse wave propagating along the [011]. v_5: [110], v_6: [100]. Brillouin scattering. $\lambda = 632.81$ nm.

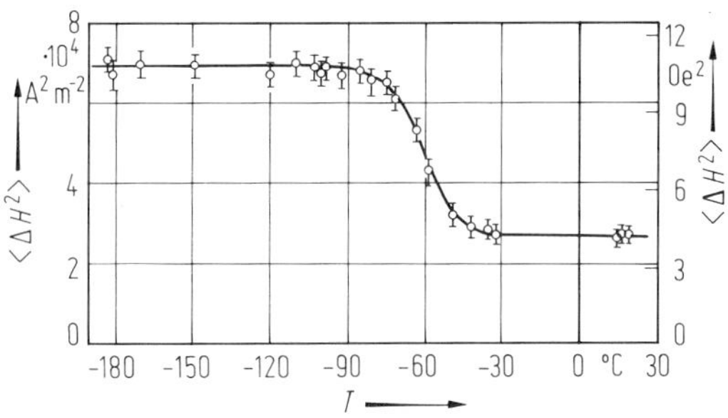

Fig. 665. $(NH_4)_2BeF_4$. $\langle \Delta H^2 \rangle$ vs. T [61 M 1]. $\langle \Delta H^2 \rangle$: second moment of magnetic resonance curve of ^{19}F.

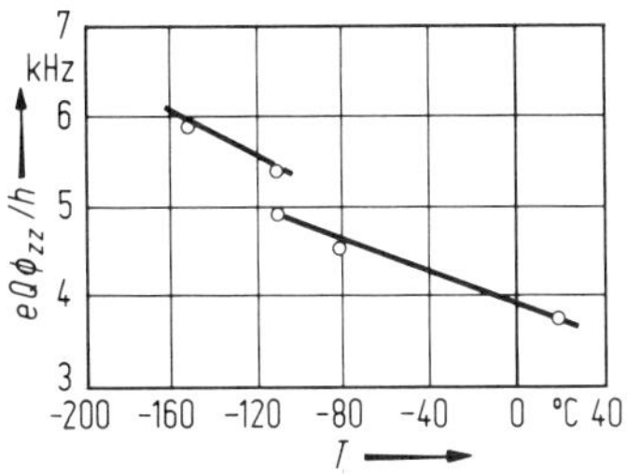

Fig. 667. $(ND_4)_2BeF_4$. $eQ\phi_{zz}/h$ vs. T [69 A 8]. $eQ\phi_{zz}$: quadrupole coupling constant of deuteron.

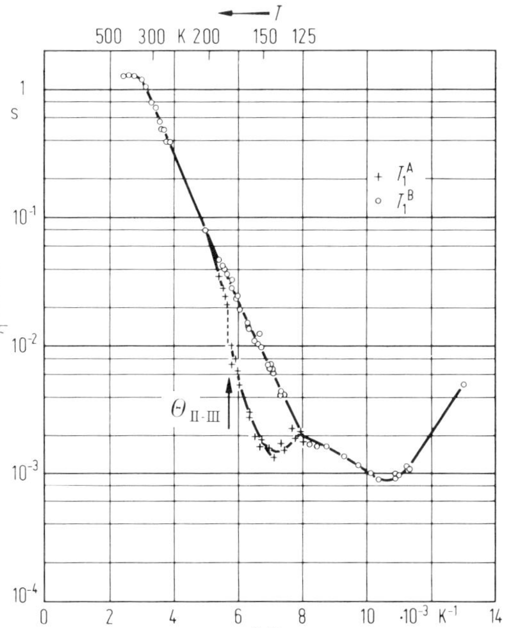

Fig. 668. $(ND_4)_2BeF_4$. T_1 vs. $1/T$ [69 K 15]. T_1: deuteron spin-lattice relaxation time. There are two well-separated sets of T_1 around Θ_{III-II}, denoted as T_1^A and T_1^B. They suggest there are two subsystems of deuterons. $\nu = 11.5$ MHz.

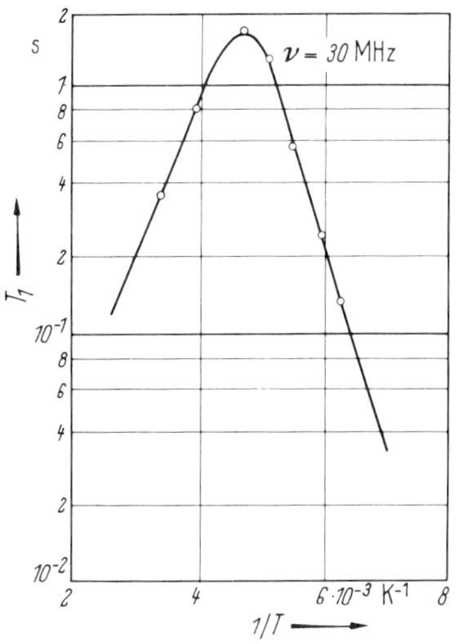

Fig. 666. $(NH_4)_2BeF_4$. T_1 vs. $1/T$ [62 M 5]. T_1: spin-lattice relaxation time of ^{19}F.

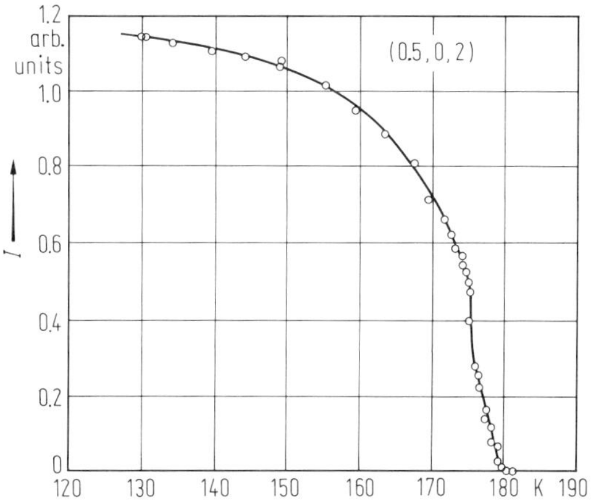

Fig. 669. $(NH_4)_2BeF_4$. $I_{(1/2,\,0,\,2)}$ vs. T [74 M 3]. I: integrated intensity of X-ray superlattice reflection.

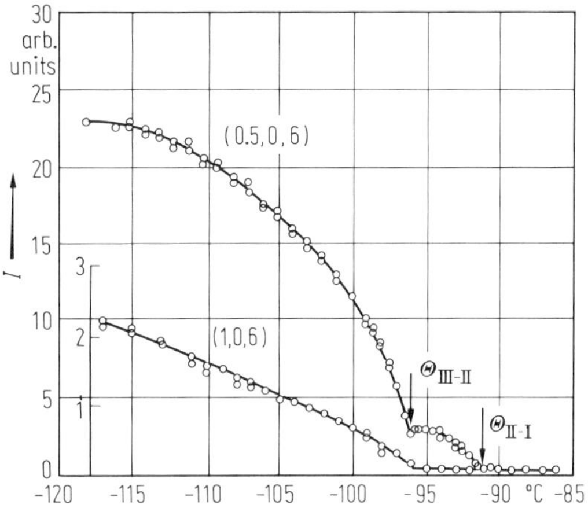

Fig. 670. $(NH_4)_2BeF_4$. $I_{(1/2,\,0,\,6)}$ and $I_{(1,\,0,\,6)}$ vs. T [77 T 2]. I: integrated intensity of X-ray Bragg reflection.

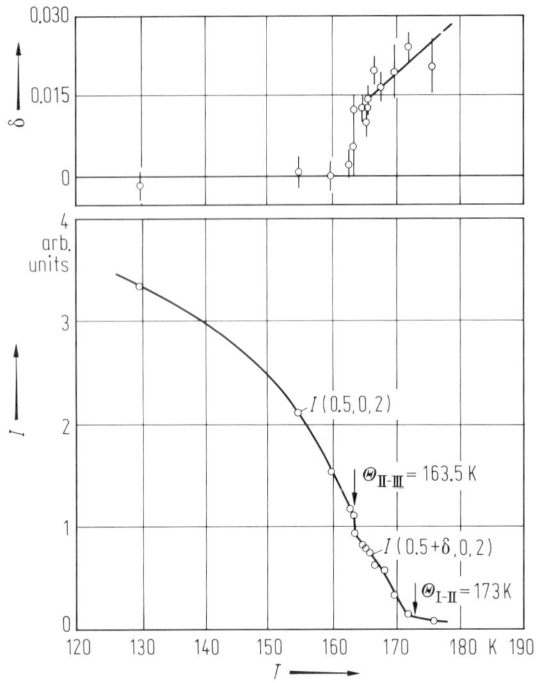

Fig. 671. $(ND_4)_2BeF_4$. I, δ vs. T [77I3]. I: integrated neutron reflection intensity at (0.5+δ, 0, 2), δ: deviation of superlattice position from (0.5, 0, 2).

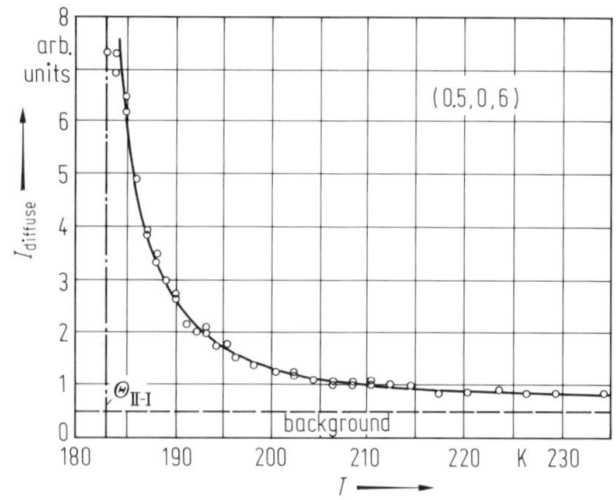

Fig. 672. $(NH_4)_2BeF_4$. $I_{diffuse}$ vs. T [77T2]. $I_{diffuse}$: X-ray diffuse scattering at (1/2, 0, 6) above Θ_{II-I}.

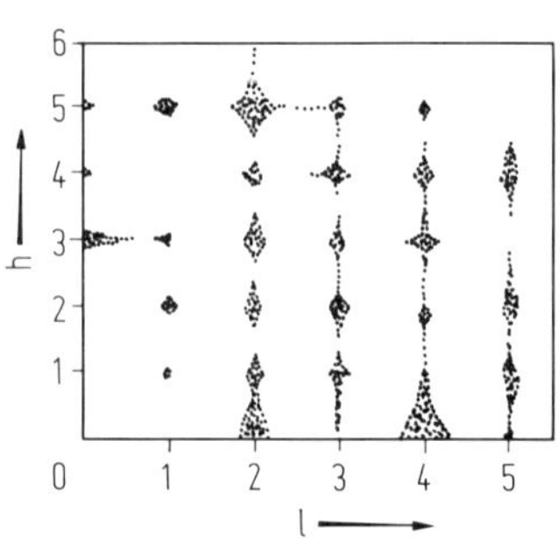

Fig. 673. $(NH_4)_2BeF_4$. X-ray diffuse scattering observed in (h, 1/2, l) [77O3].

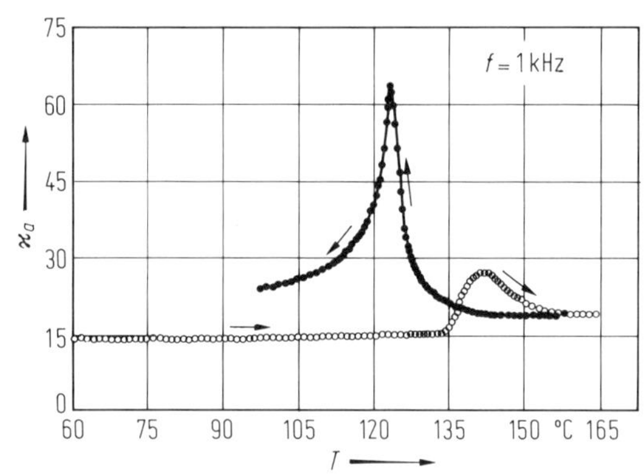

Fig. 676α. K_2ZnCl_4. $\varkappa_a$ vs. T [78G4].

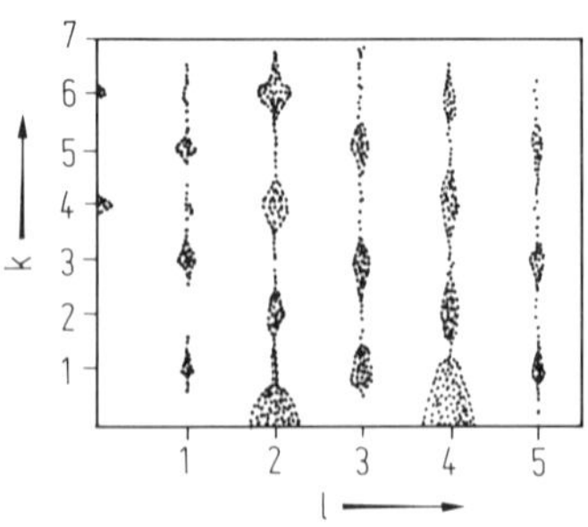

Fig. 674. $(NH_4)_2BeF_4$. X-ray diffuse scattering observed in (1/2, k, l) [77O3]. T= RT.

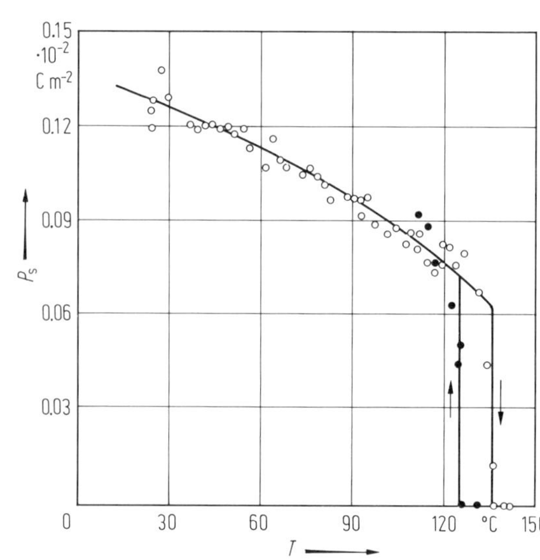

Fig. 676β. K_2ZnCl_4. P_s vs. T [78G4].

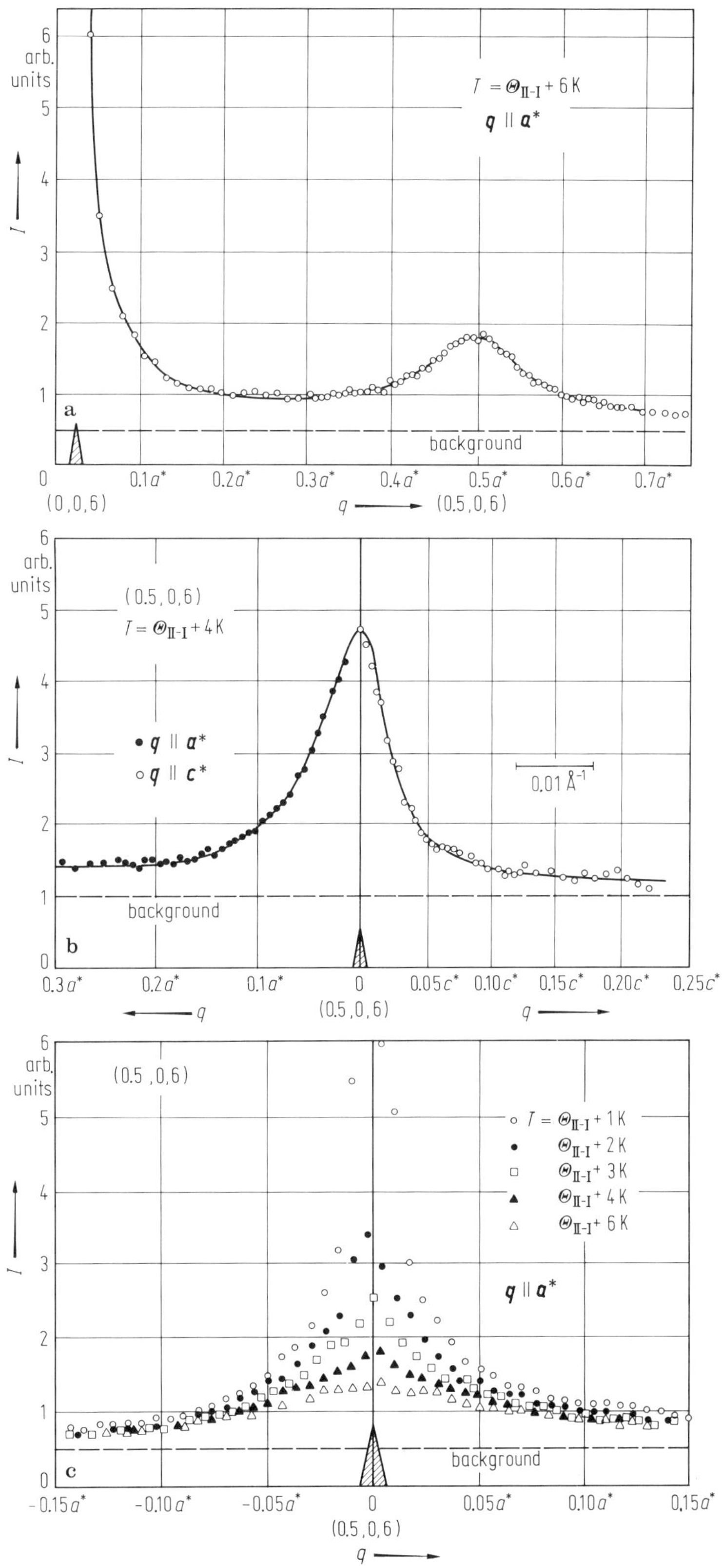

Fig. 675. $(NH_4)_2BeF_4$. X-ray diffuse scattering around (1/2, 0, 6) reciprocal lattice point in phase I [77T2]. (a) intensity along the line from (006) reciprocal lattice point to (1/2, 0, 6), $T = \Theta_{II-I} + 6$ K. (b) intensity distributions along a^* and c^* directions. $T = \Theta_{II-I} + 4$ K. (c) intensity distribution along a^* at $T = \Theta_{II-I} + 1$ K, +2 K, +3 K, +4 K, +6 K. $\Theta_{II-I} = 185$ K. – Triangles on the q-axis indicate the size of the resolution.

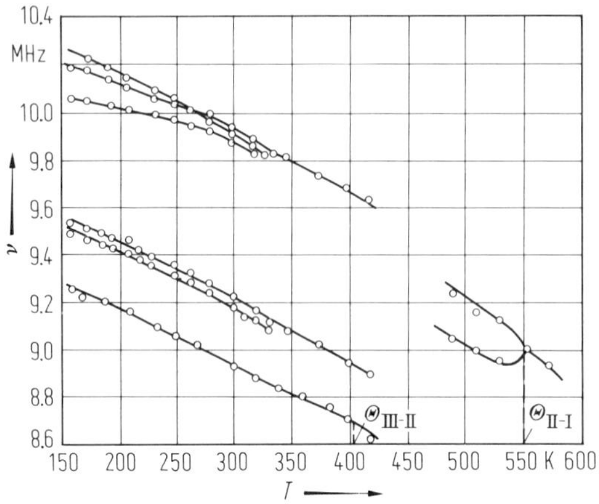

Fig. 677. K_2ZnCl_4. ν vs. T [80M3]. ν: nuclear quadrupole resonance frequency of ^{35}Cl.

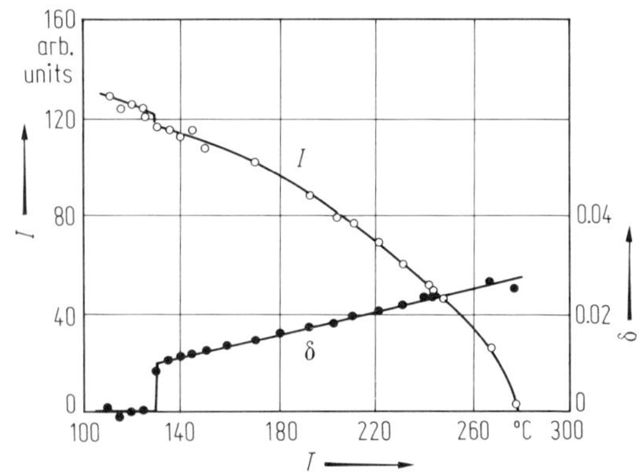

Fig. 678α. K_2ZnCl_4. I, δ vs. T [79G1]. I: integrated satellite intensity at (2, 0, 1 (1/3 − δ)), δ: parameter which measures the deviation of the period of the lattice modulation along the c-direction from $3c_0$. (c_0: unit cell parameter in phase I)

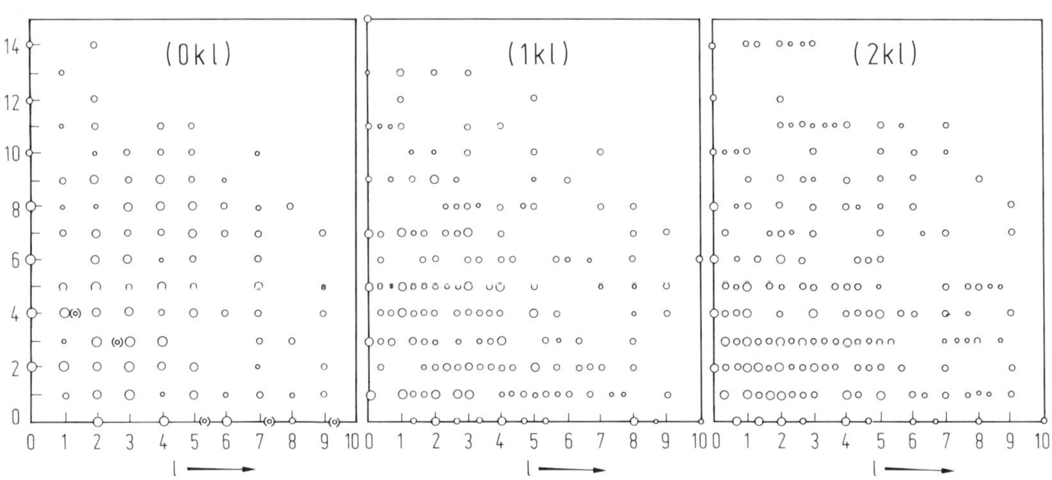

Fig. 678β. K_2ZnCl_4. Bragg intensities in (0 k l), (1 k l) and (2 k l) at RT [80I1].

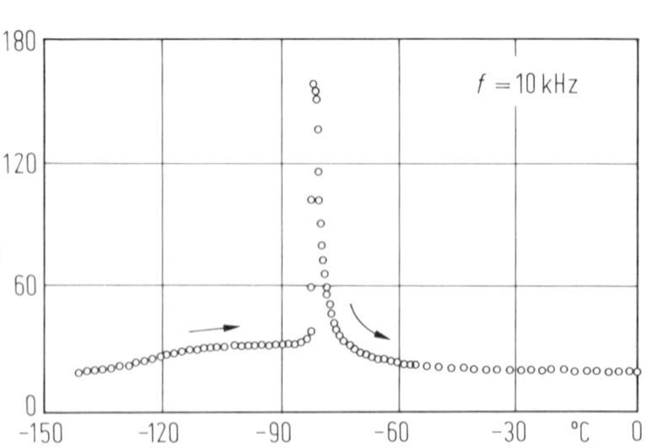

Fig. 679α. Rb_2ZnCl_4. $\varkappa_a$ vs. T [77S10].

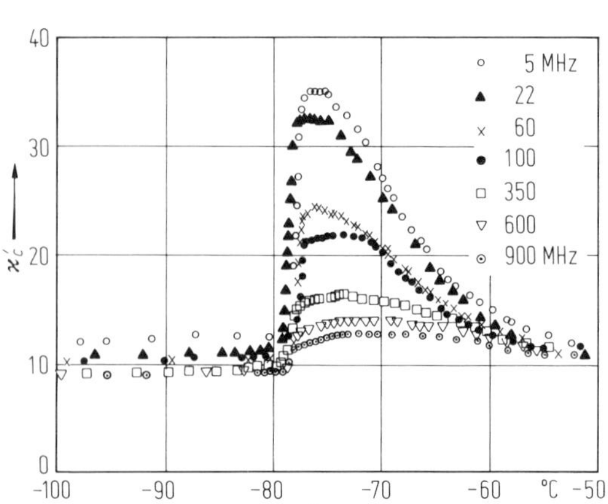

Fig. 679β. Rb_2ZnCl_4. $\varkappa_c'$ vs. T [80H1]. $\Theta_{III-II} = -76$ °C.

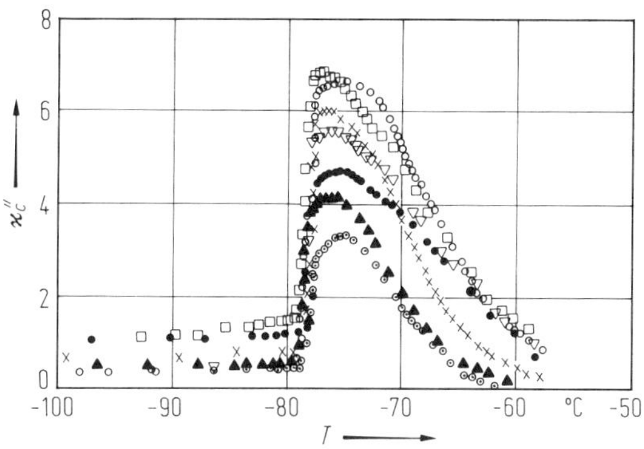

Fig. 679γ. Rb_2ZnCl_4. $\varkappa_c''$ vs. T [80 H 1]. $\Theta_{III-II} = -76$ °C. For different symbols see Fig. 679β.

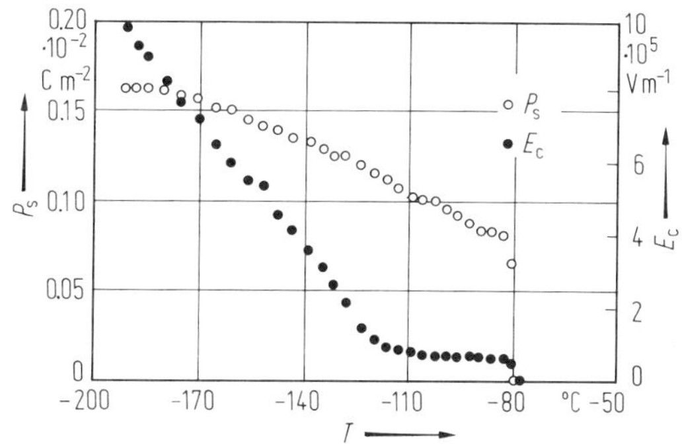

Fig. 680. Rb_2ZnCl_4. P_s, E_c vs. T [77 S 10].

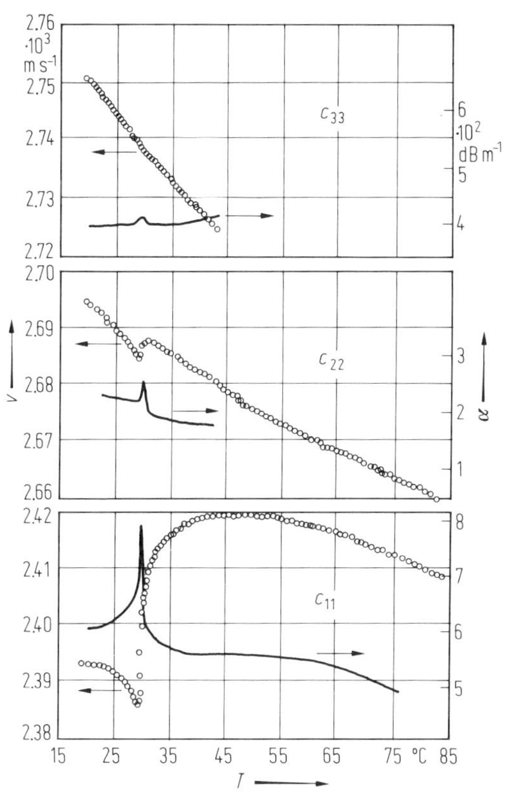

Fig. 680α. Rb_2ZnCl_4. v vs. T, α vs. T [79 H 3]. $f = 10$ MHz. v = sound velocity. α = attenuation.

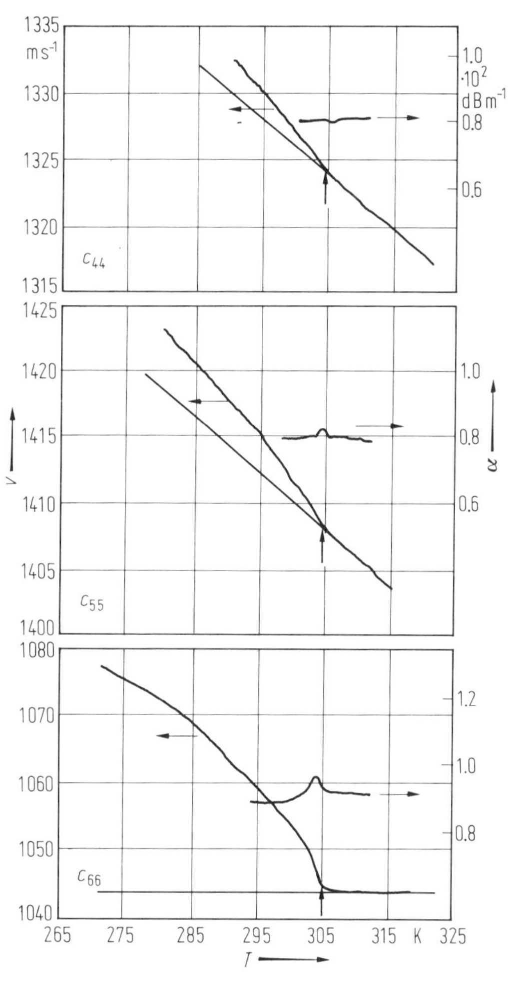

Fig. 680β. Rb_2ZnCl_4. v vs. T, α vs. T [80 M 2]. $f = 10$ MHz. Vertical arrows indicate Θ_{I-II}.

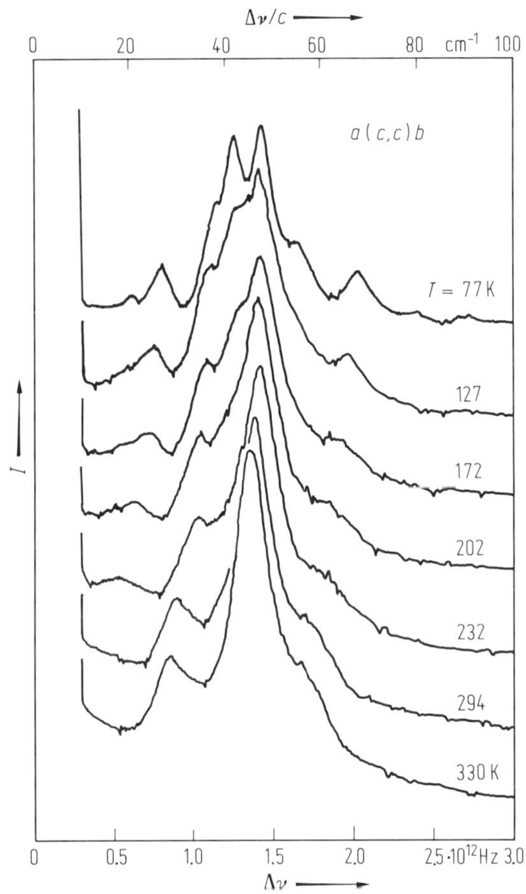

Fig. 681α. Rb_2ZnCl_4. I vs. $\Delta\nu$ [78 W 1]. I: Raman scattering intensity. $\Delta\nu$: Raman shift.

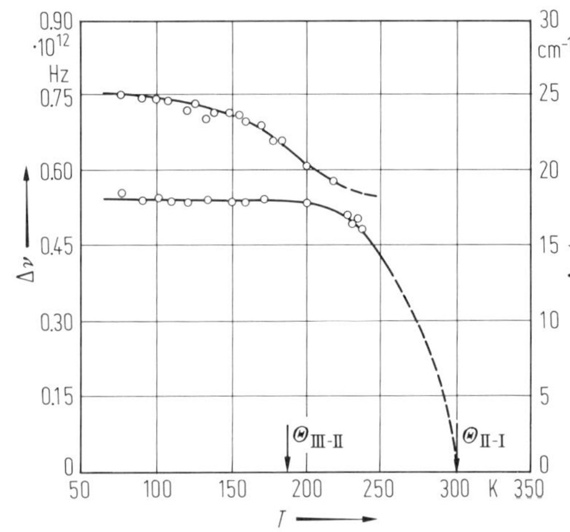

Fig. 681β. Rb_2ZnCl_4. $\Delta\nu$ vs. T [79 W 1]. $\Delta\nu$: Raman shift of the two low-lying modes.

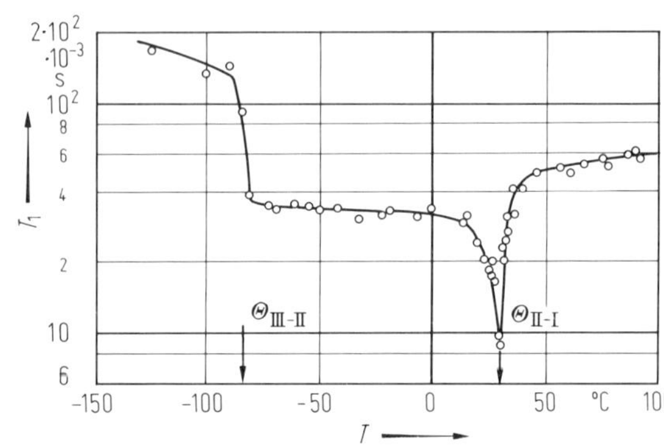

Fig. 682β. Rb_2ZnCl_4. T_1 vs. T [80 B 1]. T_1: ^{87}Rb nuclear spin-lattice relaxation time. $\nu = 88$ MHz.

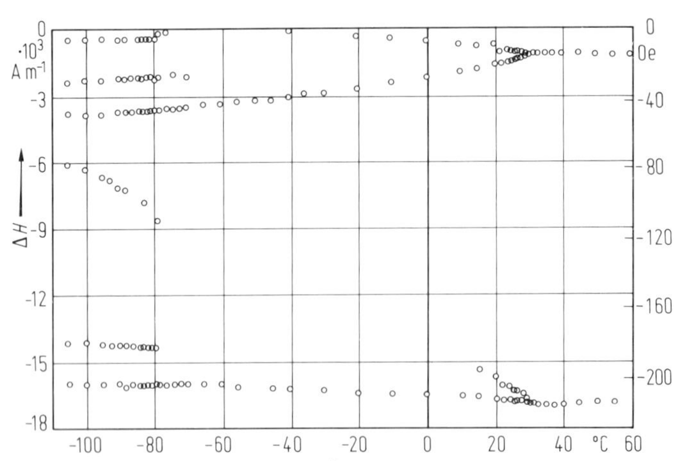

Fig. 682α. Rb_2ZnCl_4. ΔH vs. T [80 N 1]. $\Delta H = H_R - H_L$; H_R and H_L are the resonance magnetic fields of ^{87}Rb central lines perturbed and not perturbed by quadrupolar interaction, respectively. Magnetic field is parallel to the b-axis. $f = 8$ MHz.

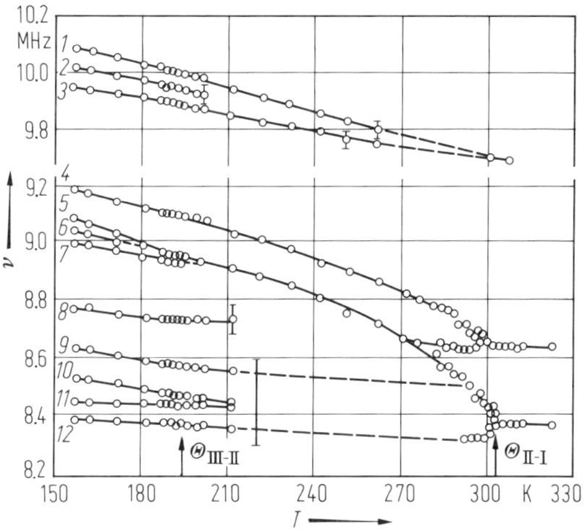

Fig. 682γ. Rb_2ZnCl_4. ν vs. T [78 M 12]. ν: nuclear quadrupole resonance frequencies of ^{35}Cl.

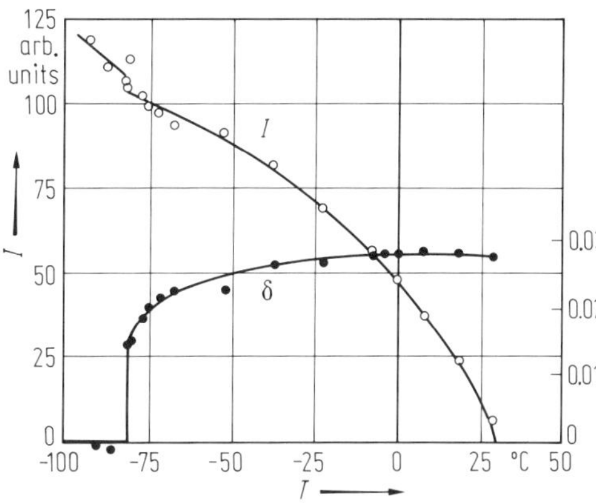

Fig. 683. Rb_2ZnCl_4. I, δ vs. T [79 G 1]. I: integrated satellite intensity at $(2, 0, 1-(1/3-\delta))$. δ: a parameter which measures the deviation of the period of the lattice modulation along the c-direction, from $3c_o$ (c_o: unit cell parameter in phase I).

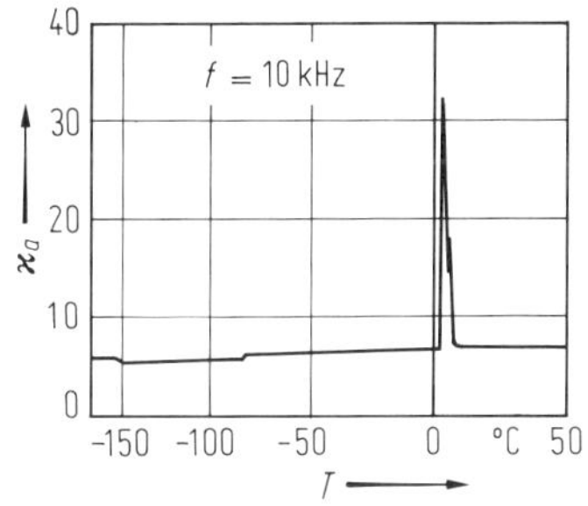

Fig. 684α. $\{N(CH_3)_4\}_2CoCl_4$. $\varkappa_a$ vs. T [78 S 5].

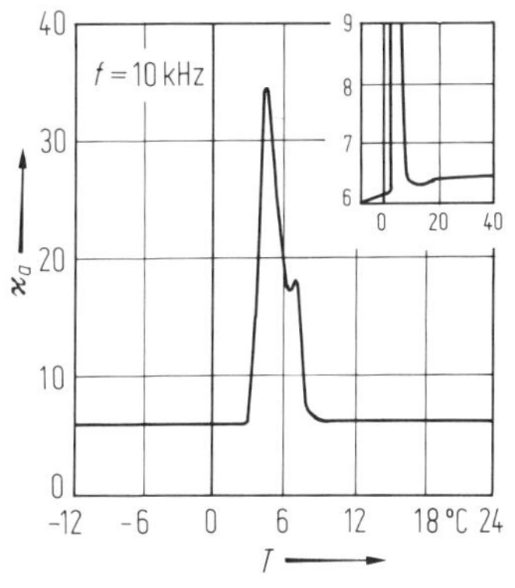

Fig. 684β. $\{N(CH_3)_4\}_2CoCl_4$. $\varkappa_a$ vs. T [78 S 5].

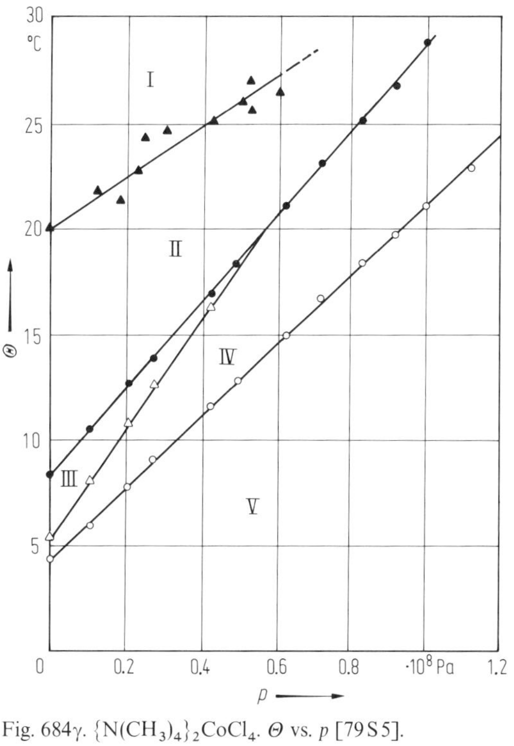

Fig. 684γ. $\{N(CH_3)_4\}_2CoCl_4$. Θ vs. p [79 S 5].

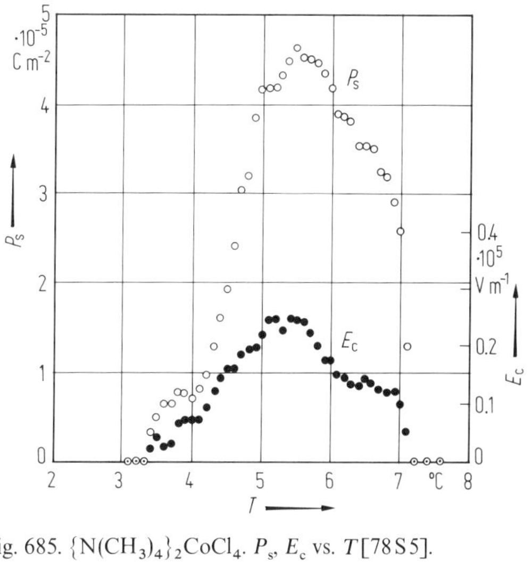

Fig. 685. $\{N(CH_3)_4\}_2CoCl_4$. P_s, E_c vs. T [78 S 5].

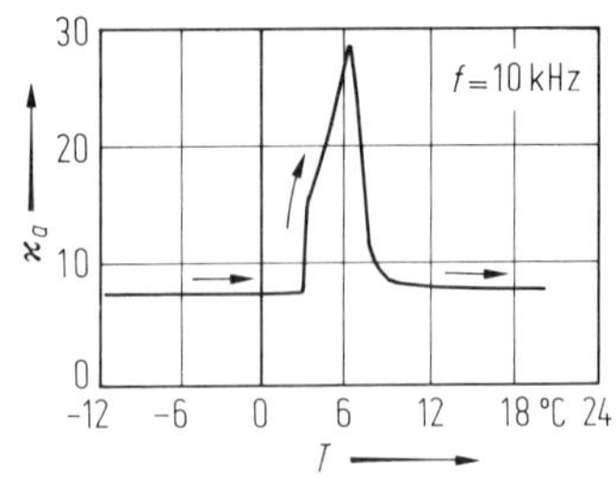

Fig. 686α. $\{N(CH_3)_4\}_2ZnCl_4$. $\varkappa_a$ vs. T [78 S6].

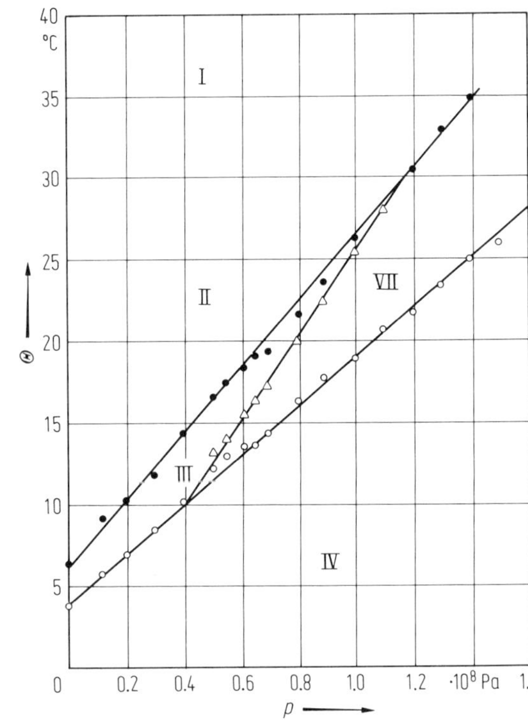

Fig. 686β. $\{N(CH_3)_4\}_2ZnCl_4$. Θ vs. p [79 S5].

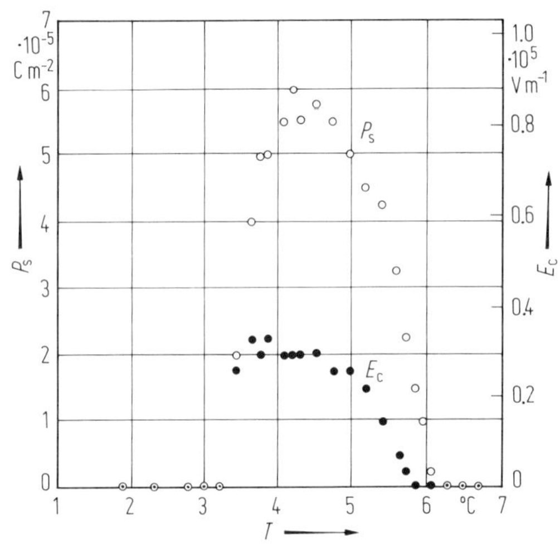

Fig. 686γ. $\{N(CH_3)_4\}_2ZnCl_4$. P_s, E_c vs. T [78 S6].

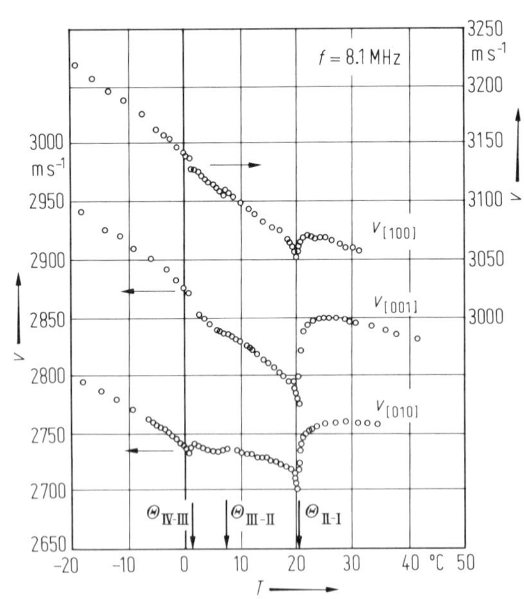

Fig. 687α. $\{N(CH_3)_4\}_2ZnCl_4$. v vs. T [79 H4]. v: sound velocity at 8.1 MHz.

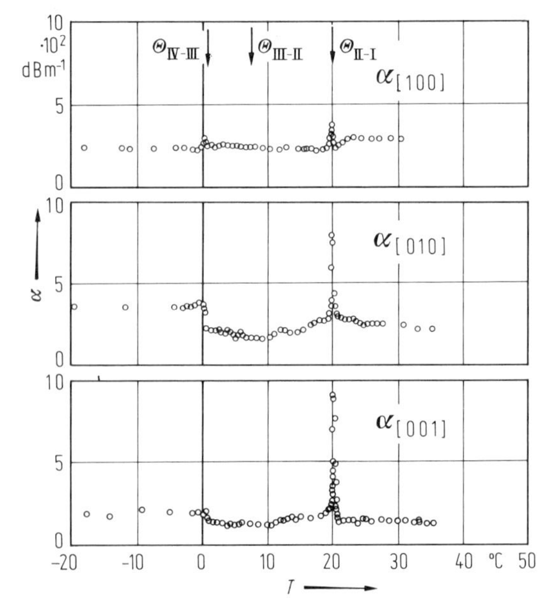

Fig. 687β. $\{N(CH_3)_4\}_2ZnCl_4$. α vs. T [79 H4]. α: sound attenuation at 8.1 MHz.

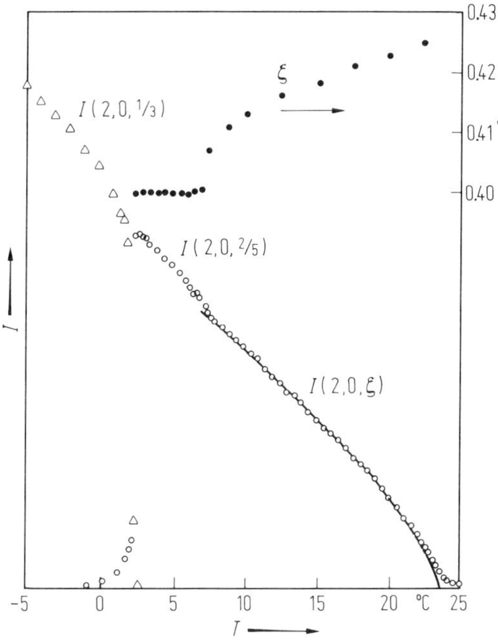

Fig. 688. $\{N(CH_3)_4\}_2ZnCl_4$. I vs. T, ζ vs. T [80M1]. I: integrated intensity of the reflections (2 0 ζ), (2 0 2/5) and (2 0 1/3) and the temperature dependence of ζ on cooling. The solid line represents the relation $I \propto (23.6 - T)^{0.74}$.

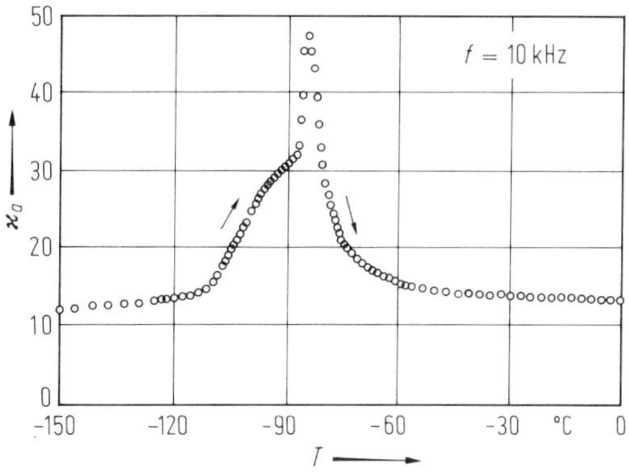

Fig. 689α. Rb_2ZnBr_4. $\varkappa_a$ vs. T [77S11].

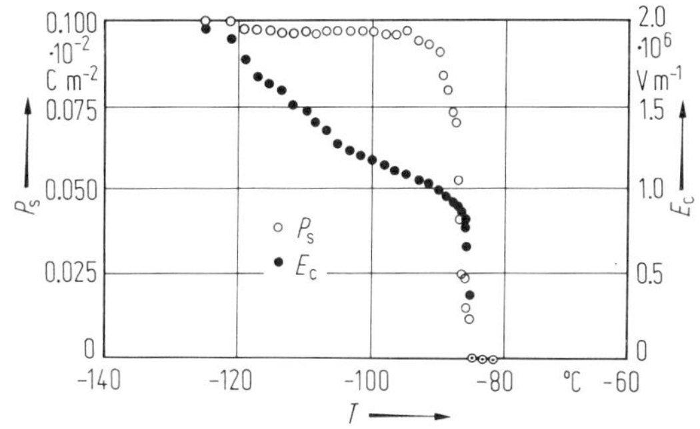

Fig. 689β. Rb_2ZnBr_4. P_s, E_c vs. T [77S11].

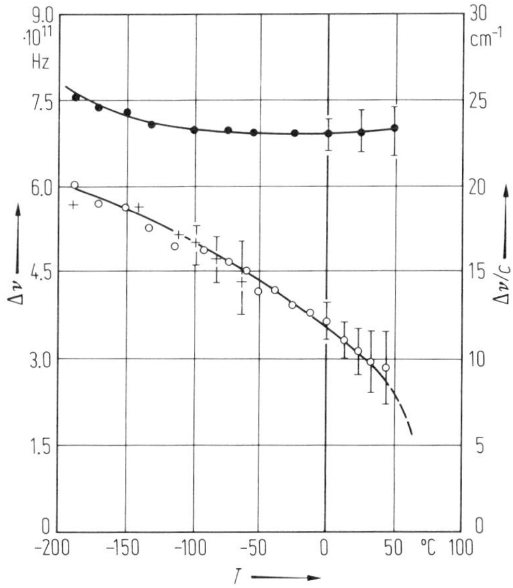

Fig. 690. Rb_2ZnBr_4. $\Delta\nu$ vs. T [80T1]. $\Delta\nu$: Raman shift. Open circles: $6 \cdot 10^{11}$ Hz (20 cm^{-1}), full circles: $7.5 \cdot 10^{11}$ Hz (25 cm^{-1}) for $b(a, a)c$. Crosses: $5.7 \cdot 10^{11}$ Hz (19 cm^{-1}) for $a(c, c)b$.

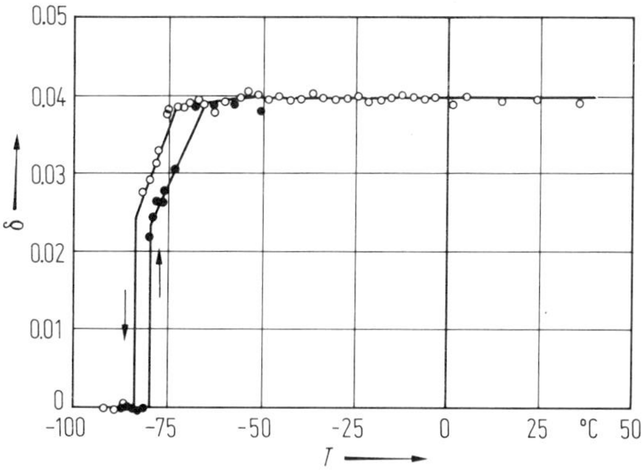

Fig. 691α. Rb_2ZnBr_4. δ vs. T [78G5]. For the meaning of δ, see Fig. 683.

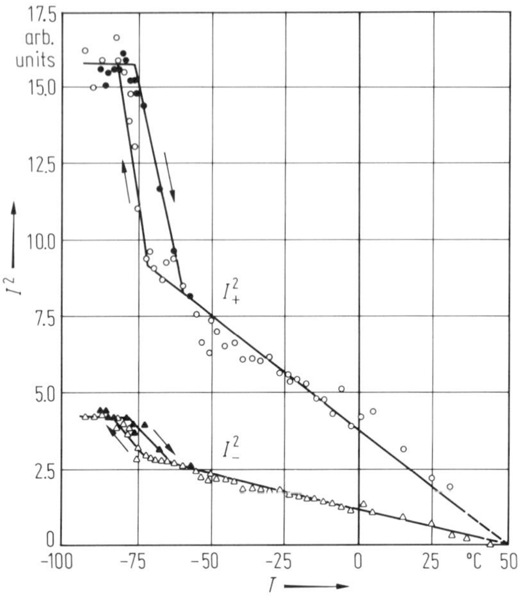

Fig. 691β. Rb_2ZnBr_4. I_+^2, I_-^2 vs. T [78G5]. I_+, I_-: integrated intensities of satellite peaks at $(2, 0, 1+(1/3-\delta))$ and at $(2, 0, 1-(1/3-\delta))$, respectively.

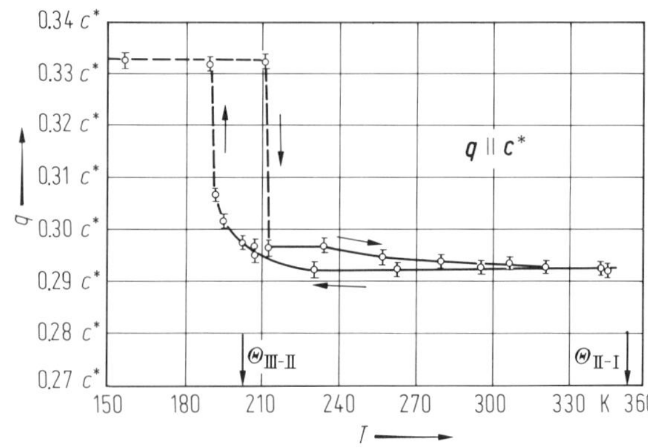

Fig. 691γ. Rb_2ZnBr_4. q vs. T [79D2]. q: modulation wave vector along the c^* axis.

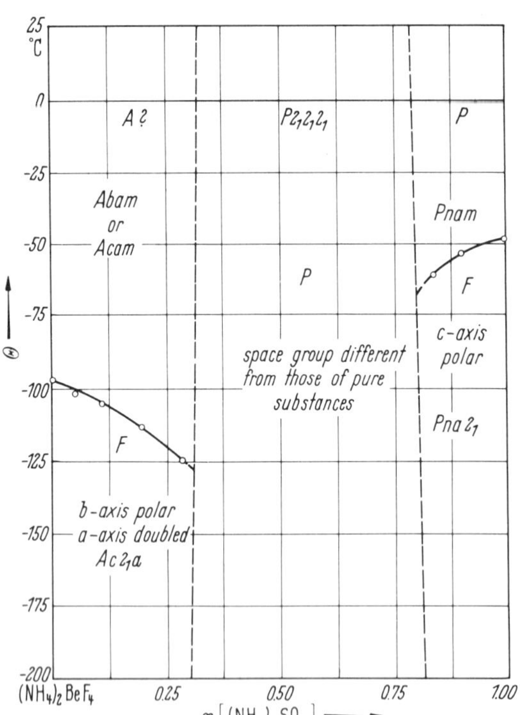

Fig. 692. $(NH_4)_2(SO_4)_x(BeF_4)_{1-x}$. Phase diagram. Θ vs. x [58H3]. Θ: transition temperature. Recently, a new phase has been found between −91 °C and −97 °C at x=0 [73S20, 74M3].

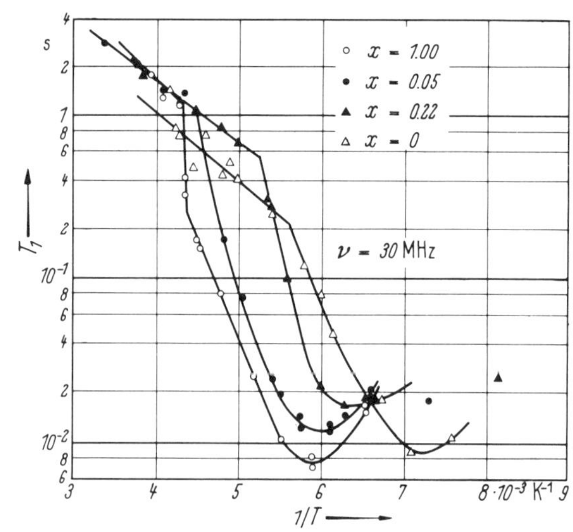

Fig. 693. $(NH_4)_2(SO_4)_x(BeF_4)_{1-x}$. T_1 vs. $1/T$ [62M5]. T_1: spin-lattice relaxation time of ^1H.

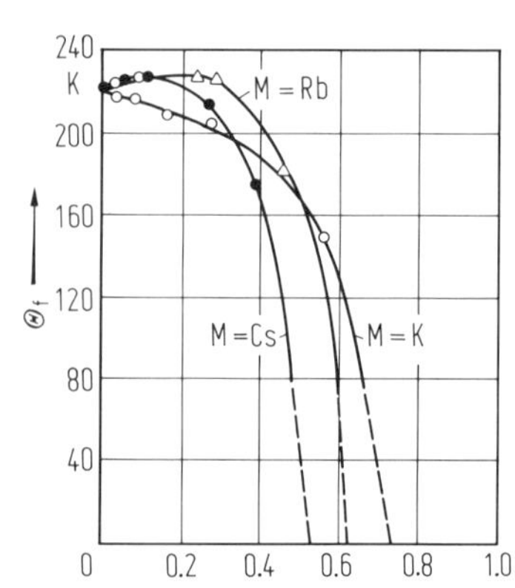

Fig. 694. $[M_x(NH_4)_{1-x}]_2SO_4$. Θ_f vs. x [78O2]. M=Rb, M=Cs: [75S8] M=K: [78O2].

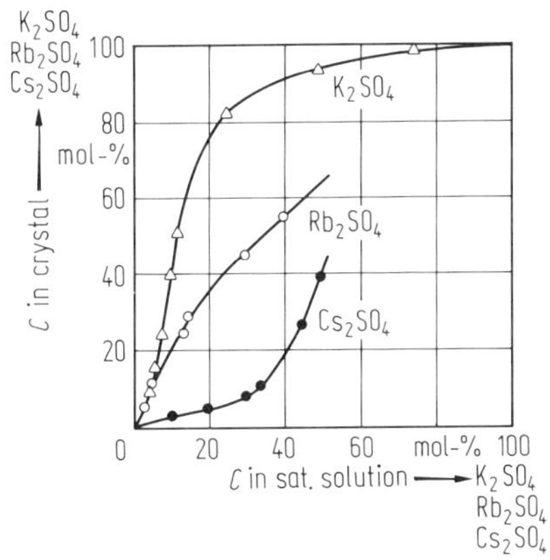

Fig. 695. $(NH_4)_2SO_4-K_2SO_4$, $(NH_4)_2SO_4-Rb_2SO_4$, $(NH_4)_2SO_4-Cs_2SO_4$. The concentrations of K_2SO_4, Rb_2SO_4, and Cs_2SO_4 in the mixed crystal systems versus those in the saturated solutions, respectively [78 O 2].

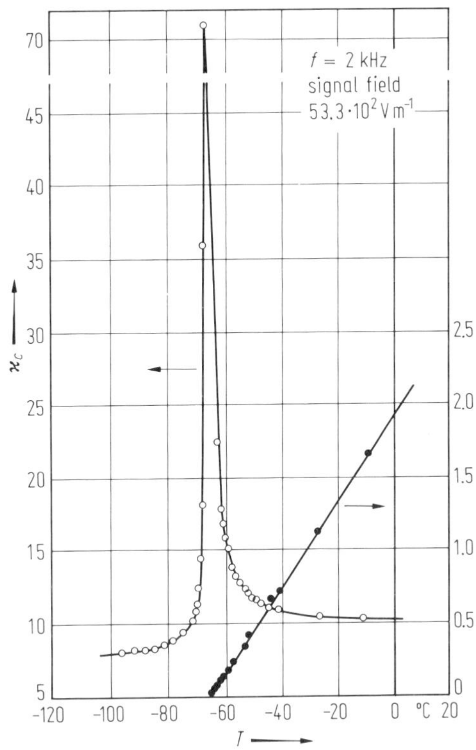

Fig. 697. $[(NH_4)_{0.7}K_{0.3}]_2SO_4$. $\varkappa_c$, $(\varkappa_c-\varkappa_\infty)^{-1}$ vs. T [75 S 8]. $\varkappa_\infty=9.5$. $\varkappa_c=\varkappa_\infty+C/(T-\Theta_p)$, $C=35$ K, $\Theta_p=-66$ °C.

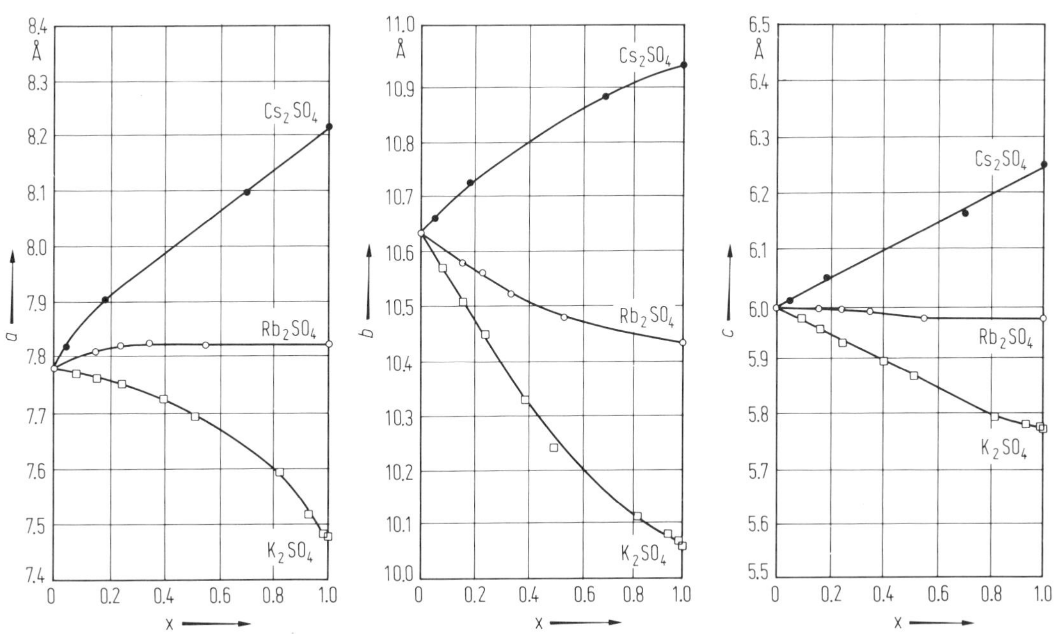

Fig. 696. $[(NH_4)_2SO_4]_{1-x}[K_2SO_4]_x$; $[(NH_4)_2SO_4]_{1-x}[Rb_2SO_4]_x$; $[(NH_4)_2SO_4]_{1-x}[Cs_2SO_4]_x$. Unit cell parameters of the mixed crystal systems at RT [78 O 2].

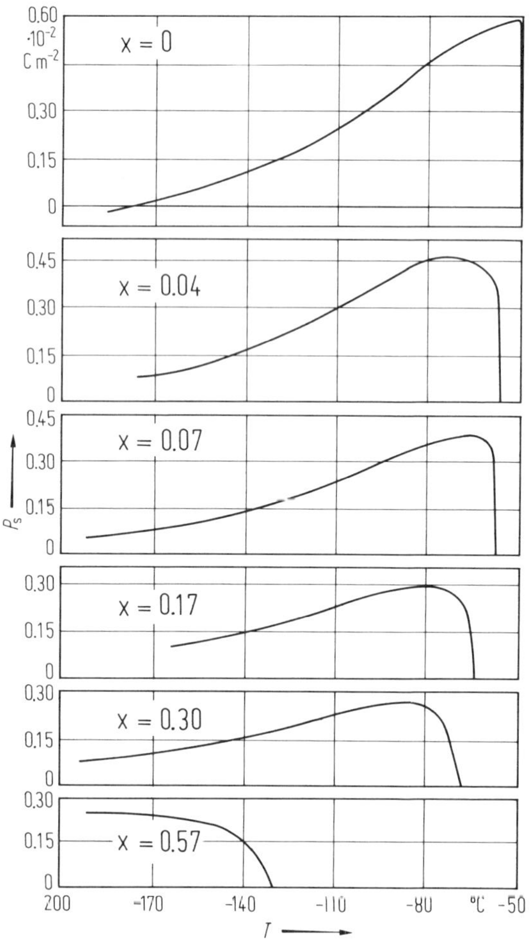

Fig. 698. $[(NH_4)_{1-x}K_x]_2SO_4$. P_s vs. T. Parameter: x [75 S 8].

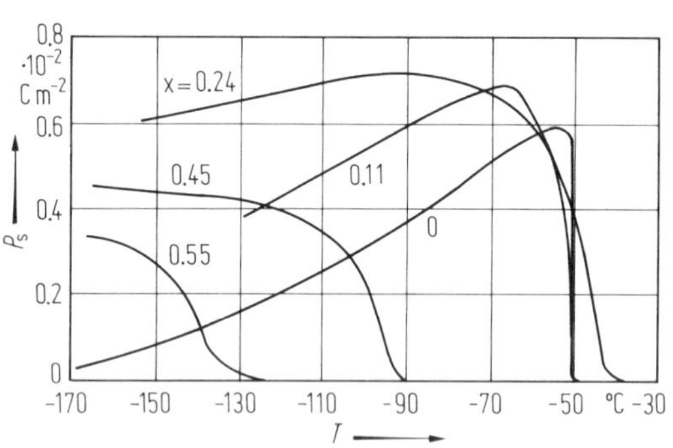

Fig. 702. $[(NH_4)_{1-x}Rb_x]_2SO_4$. P_s vs. T. Parameter: x [78 O 2].

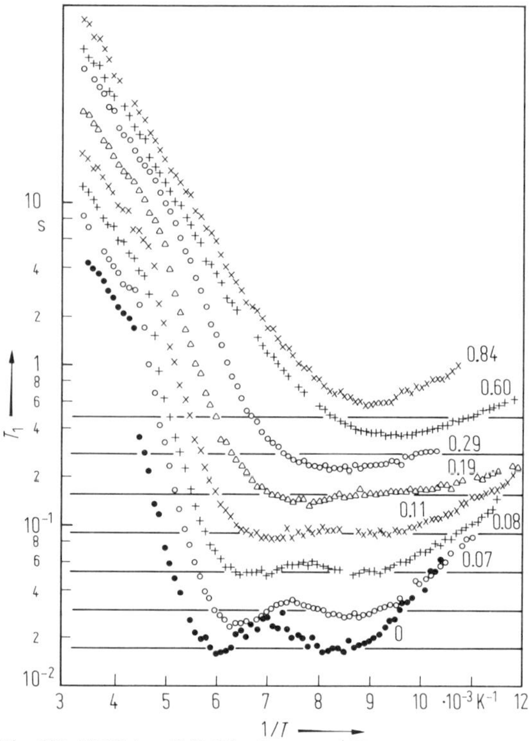

Fig. 699. $[(NH_4)_{1-x}K_x]_2SO_4$. T_1 vs. T^{-1} [75 K 7]. Parameter: x. T_1: proton spin-lattice relaxation time.

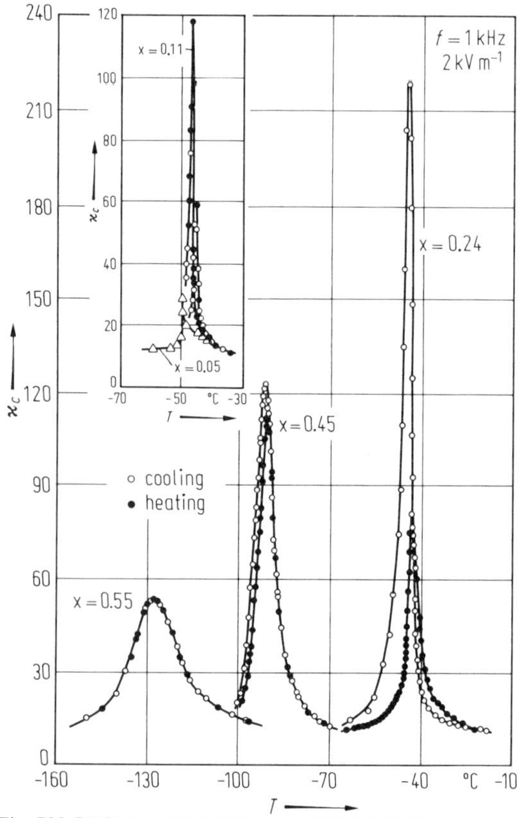

Fig. 700. $[(NH_4)_{1-x}Rb_x]_2SO_4$. $\varkappa_c$ vs. T. [78 O 2]. Parameter: x.

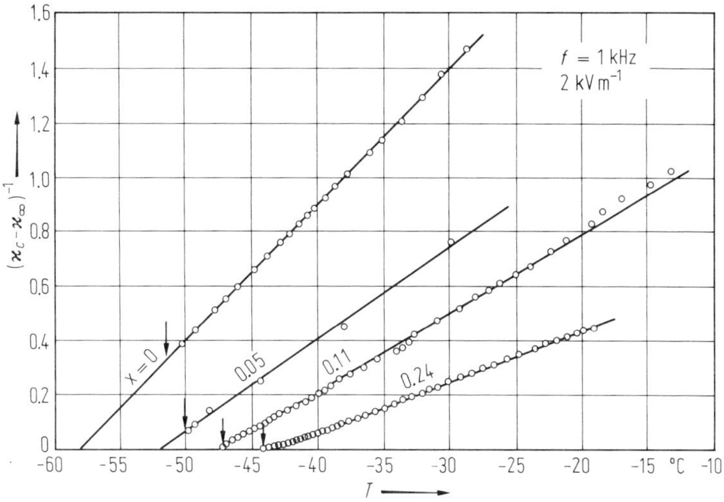

Fig. 701. $[(NH_4)_{1-x}Rb_x]_2SO_4$. $(\varkappa_c - \varkappa_\infty)^{-1}$ vs. T [78 O 2]. Vertical arrows indicate the Curie point.

x	0	0.05	0.11	0.24
$\varkappa_\infty$	8.6	13.7	9.8	10.4

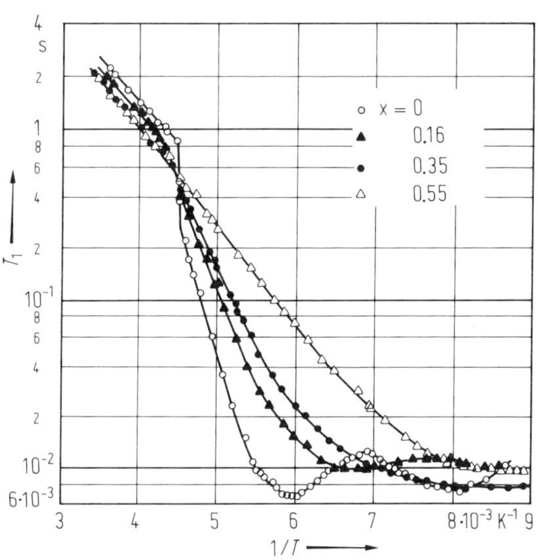

Fig. 703. $[(NH_4)_{1-x}Rb_x]_2SO_4$. T_1 vs. $1/T$ [78 O 2]. Parameter: x. T_1: proton spin-lattice relaxation time.

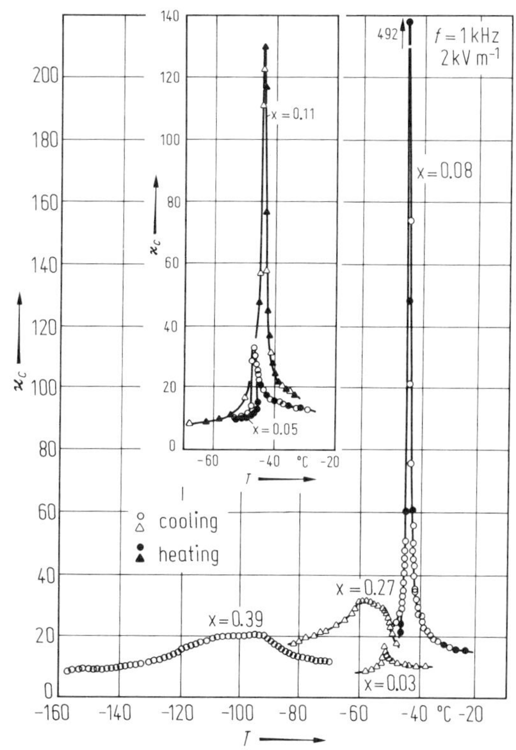

Fig. 704. $[(NH_4)_{1-x}Cs_x]_2SO_4$. $\varkappa_c$ vs. T. Parameter x [78 O 2].

Fig. 705, see p. 478.

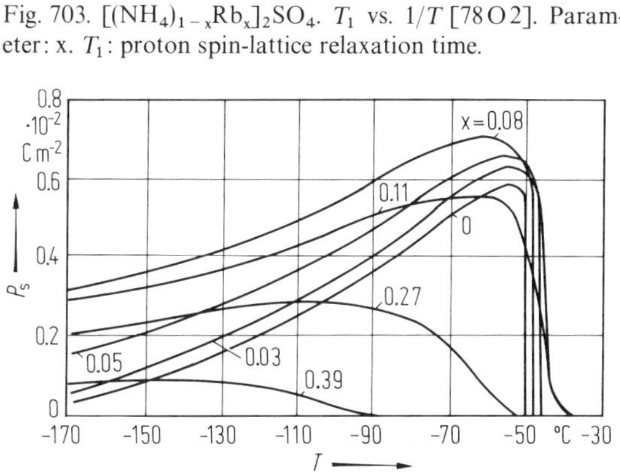

Fig. 706. $[(NH_4)_{1-x}Cs_x]_2SO_4$. P_s vs. T. Parameter: x [78 O 2].

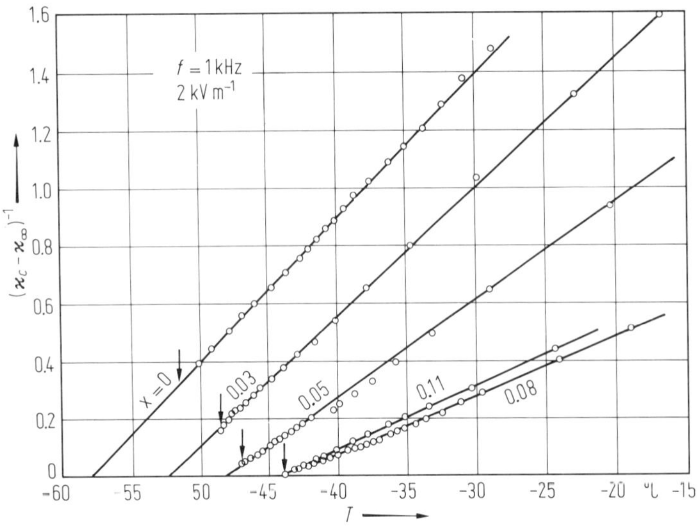

Fig. 705. $[(NH_4)_{1-x}Cs_x]_2SO_4$. $(\varkappa_c - \varkappa_\infty)^{-1}$ vs. T [78 O 2].

x	0	0.03	0.05	0.08	0.11
$\varkappa_\infty$	8.6	8.5	12.2	12.8	13.8

Vertical arrows indicate the Curie point.

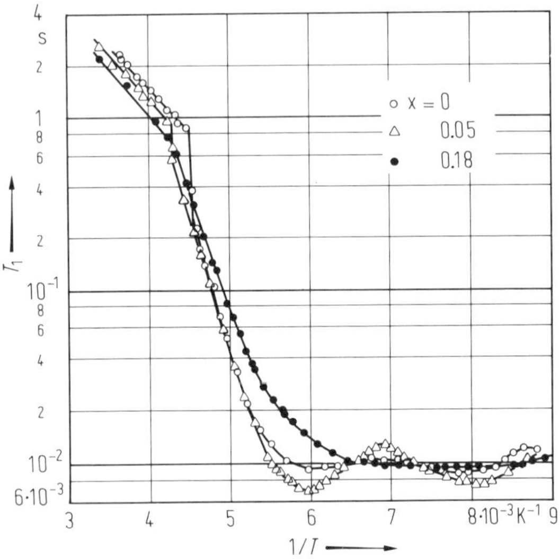

Fig. 707. $[(NH_4)_{1-x}(Cs)_x]_2SO_4$. T_1 vs. $1/T$. Parameter: x [78 O 2]. T_1: proton spin-lattice relaxation time.

21 NH_4HSO_4 family

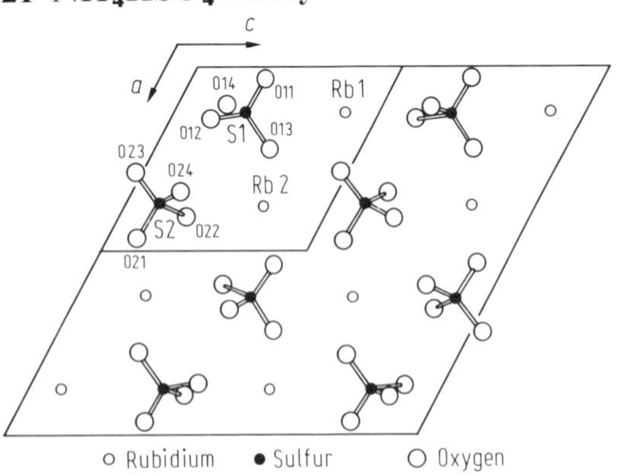

Fig. 708α. $RbHSO_4$. Structure of phase I [75 A 13]. Projection on (010).

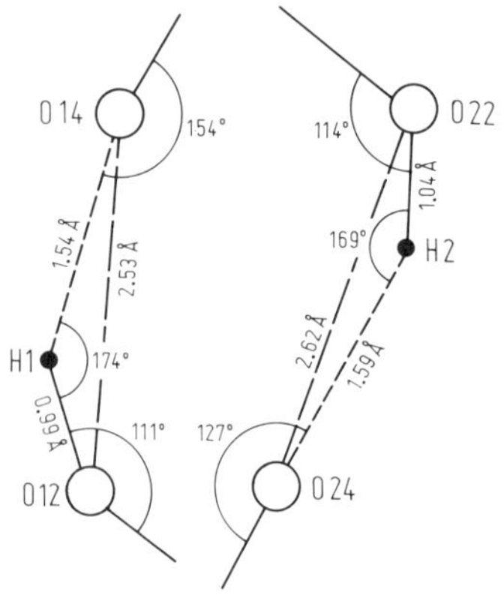

Fig. 708β. $RbHSO_4$. Structure of phase I [75 A 13]. Geometries of the two hydrogen bonds obtained by neutron diffraction at 23 °C.

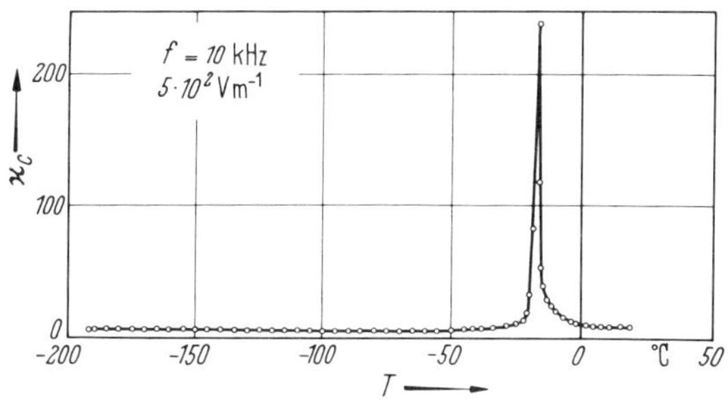

Fig. 709α. $RbHSO_4$. $\varkappa_c$ vs. T [60P1].

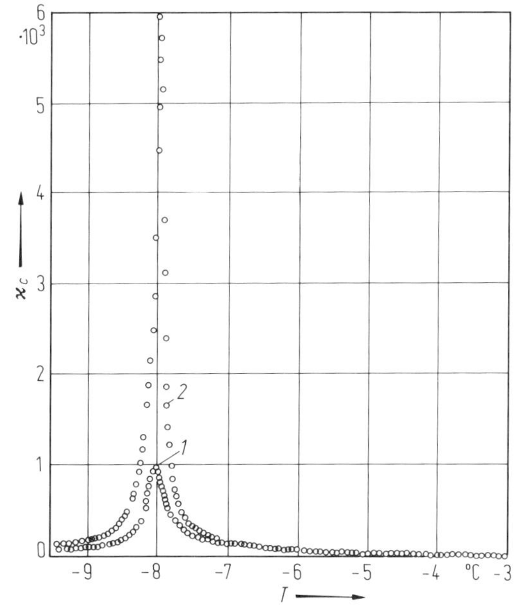

Fig. 709β. $RbHSO_4$. $\varkappa_c$ vs. T [77K1]. Curve *1*: as grown, *2*: annealed at 180 °C for 43 h. $\varkappa_c$ at $f = 10$ kHz.

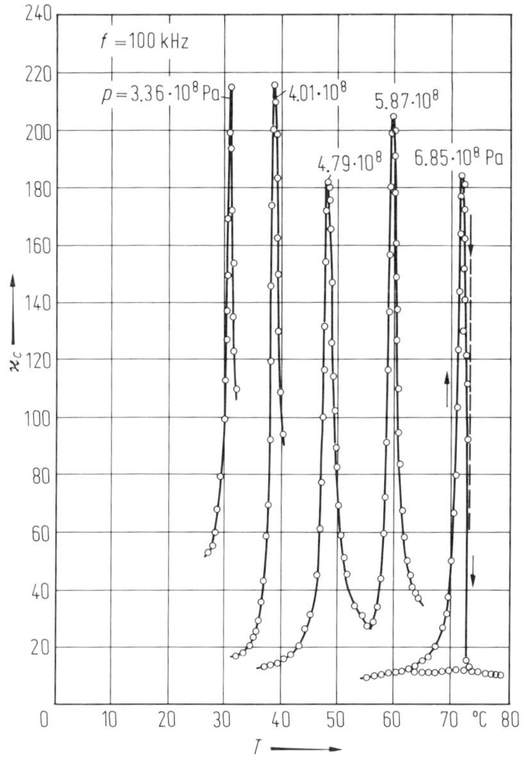

Fig. 710α. $RbHSO_4$. $\varkappa_c$ vs. T [75G7]. Parameter: p.

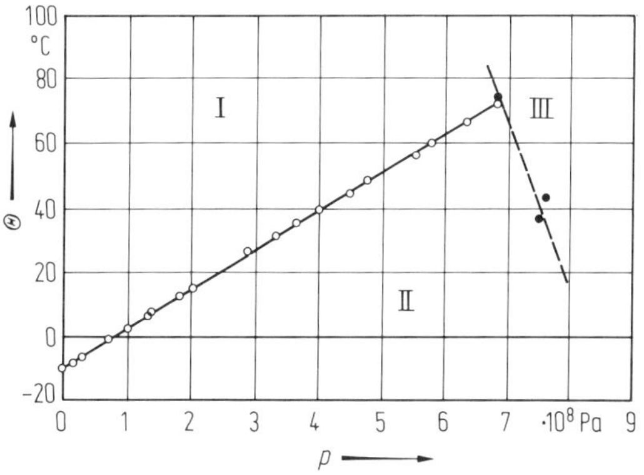

Fig. 710β. $RbHSO_4$. Θ vs. p [75G7].

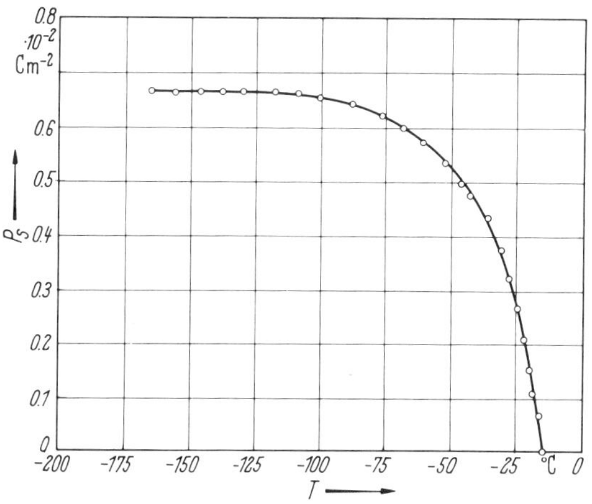

Fig. 711α. $RbHSO_4$. P_s vs. T [60P1].

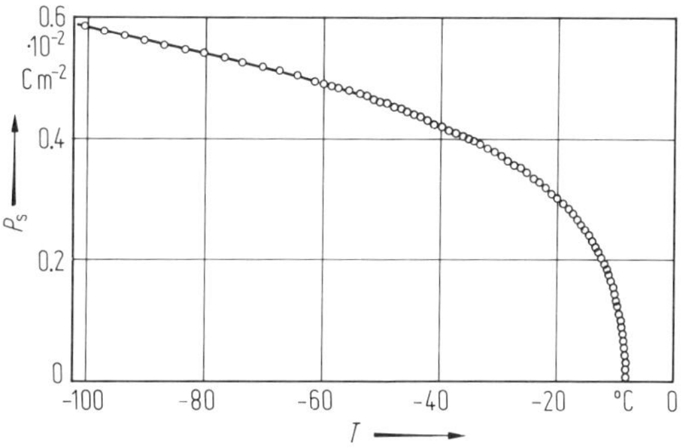

Fig. 711β. $RbHSO_4$. P_s vs. T [77K1].

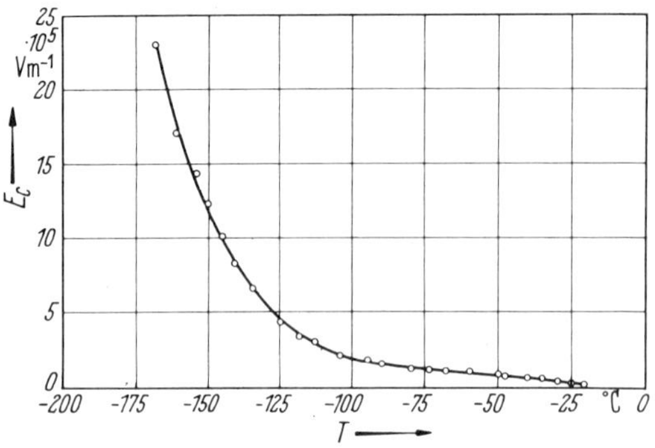

Fig. 711γ. $RbHSO_4$. E_c vs. T [60P1].

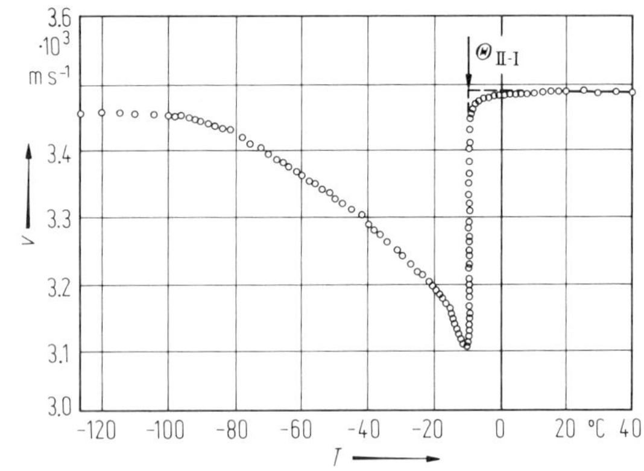

Fig. 713. $RbHSO_4$. v vs. T [78K7]. v: velocity of sound wave propagating along the b-axis at 5 MHz.

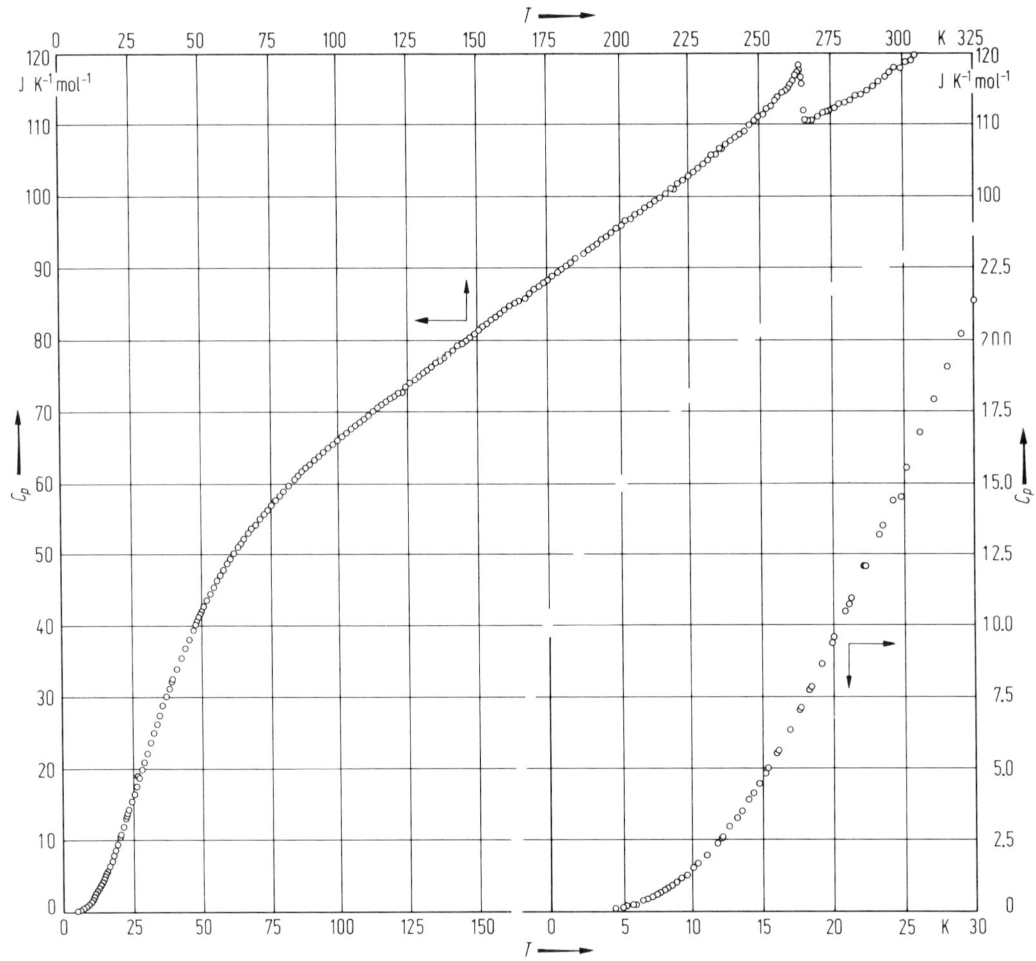

Fig. 712. $RbHSO_4$. C_p vs. T [73C3].

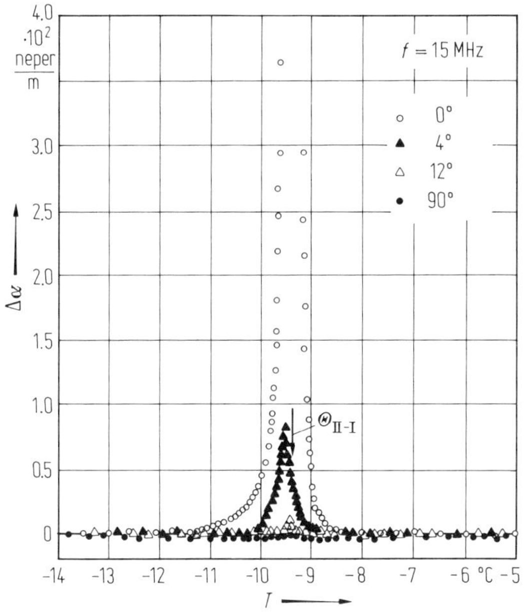

Fig. 714. $RbHSO_4$. Δα vs. T [78 K 6]. Δα: Anomalous part of the attentuation of the quasi-longitudinal sound waves propagating along the directions in the *bc* plane with angles 0°, 4°, 12° and 90° with respect to the *b*-axis.

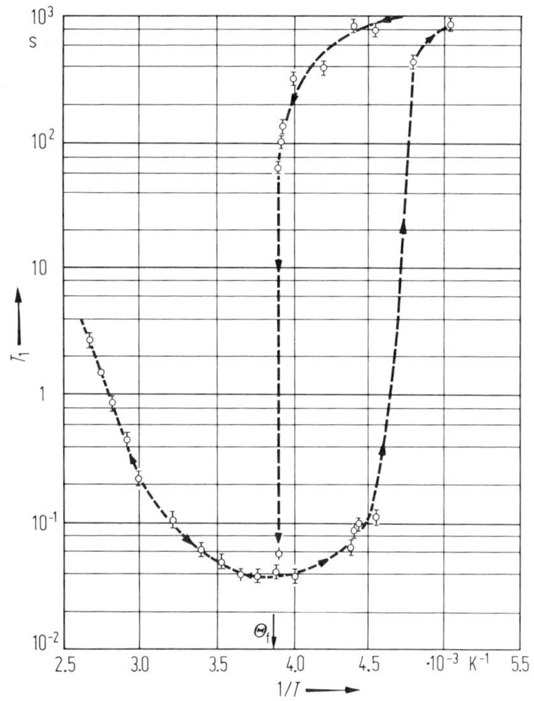

Fig. 715. $RbHSO_4$. T_1 vs. $1/T$ [69 S 10]. T_1: spin-lattice relaxation time of ^1H.

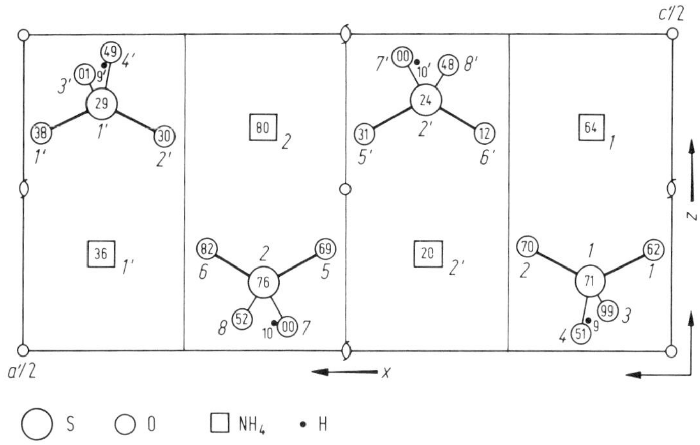

Fig. 716α. NH_4HSO_4. Structure of phase I [71 N 8]. Pseudo-orthorhombic system ($B2_1/a$) in phase I is used. A schematic projection on (010). A quarter of the pseudo-orthorhombic cell is shown, which consists of the parts labelled a, b, c and d in Fig. 716β. The atoms of the asymmetric unit are unprimed and those related by symmetry are primed. Fractional y coordinates (· 100) are inscribed in each atom, except for the hydrogen atoms, which are at heights 36, 64, 13, and 87 for hydrogen atoms 9, 9′, and 10 and 10′, respectively.

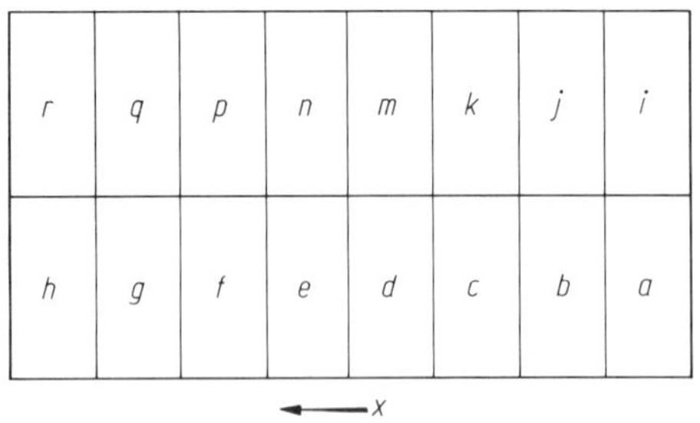

Fig. 716β. NH_4HSO_4. Structure of phase I [71N8]. The method of labelling symmetry related atoms. The full pseudo-orthorhombic cell with the cell vectors, a', b', c' is divided into sixteen "subcells" with $a'/8$, b', $c'/2$, each containing one formula unit. These cells are labelled from a to r. S(1), O(1), O(2), O(3), O(4), N(1), and H(9) are in a; S(2), O(5), O(6), O(7), O(8), N(2), and H(10) are in c. These atoms constitute the asymmetric unit. Symmetrically equivalent atoms in b are designated S(2b), O(5b), etc. Atoms in the adjacent unit cell along $\pm x$, $\pm y$, or $\pm z$ are shown by single, double and triple primes, respectively, e.g. [0(1 h')]'' is the atom symmetrically equivalent to O(1) in the h subcell of the unit cell next to the principal one along $\pm x$ and next to this second cell along $\pm y$.

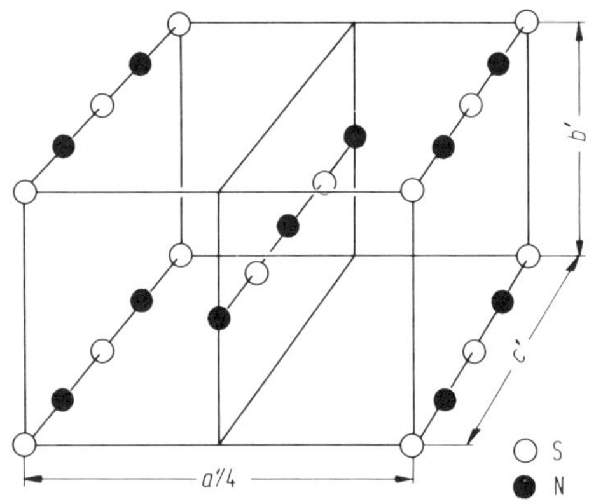

Fig. 716γ. NH_4HSO_4. Structure of phase I [71N8]. A schematic representation of the approximate disposition of the sulfur and nitrogen atoms. The origin has been moved to a sulfur atom.

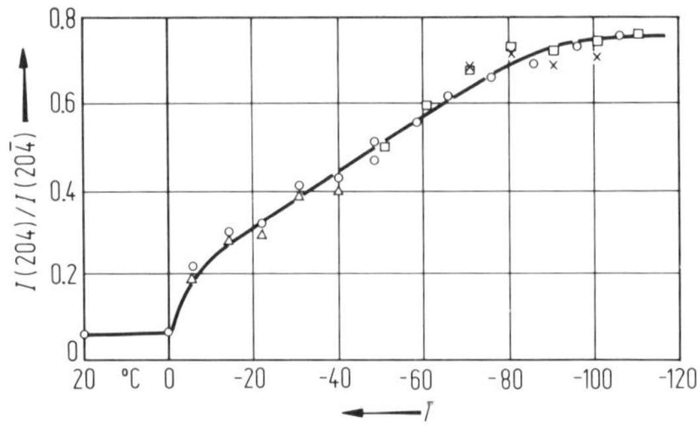

Fig. 717. NH_4HSO_4. $I_{(204)}/I_{(20\bar{4})}$ vs. T [70O2]. Intensity ratio of X-ray Bragg reflections (pseudo-orthorhombic cell).

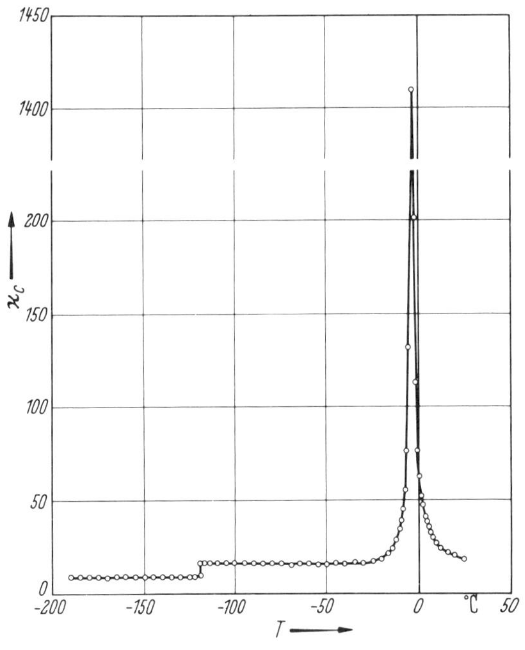

Fig. 718α. NH_4HSO_4. $\varkappa_c$ vs. T [58P2]. f = 10 kHz.

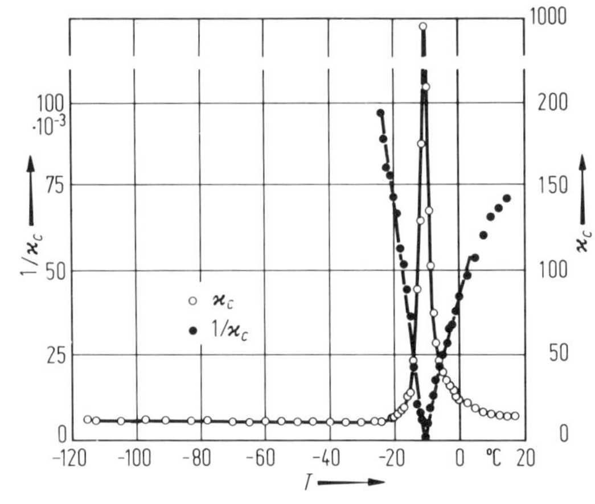

Fig. 718β. ND_4DSO_4. $\varkappa_c$, $1/\varkappa_c$ vs. T [70K4]. f = 1.5 kHz.

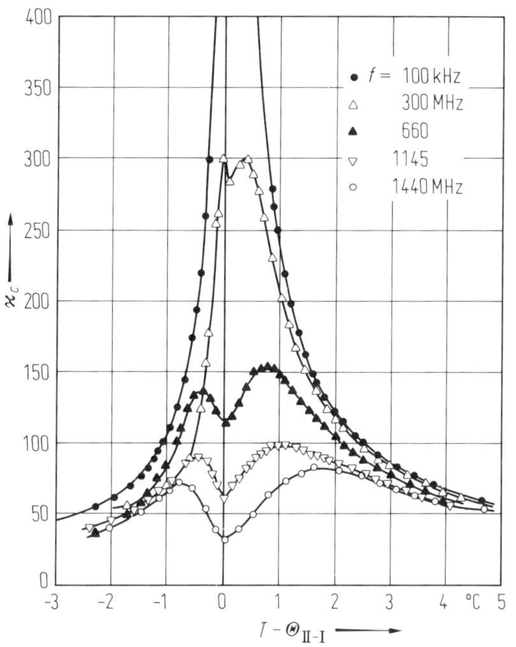

Fig. 719α. NH_4HSO_4. $\varkappa_c'$ vs. T [76T4]. Parameter: f. $\Theta_{II-I} = -2.3\,°C$.

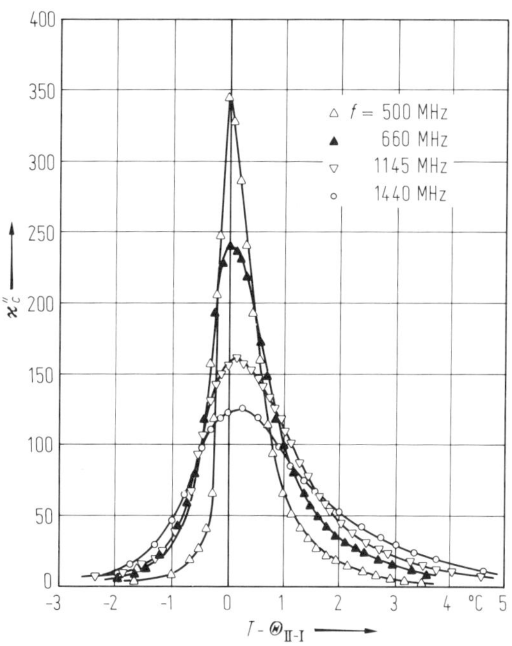

Fig. 719β. NH_4HSO_4. $\varkappa_c''$ vs. T [76T4]. Parameter: f. $\Theta_{II-I} = -2.3\,°C$.

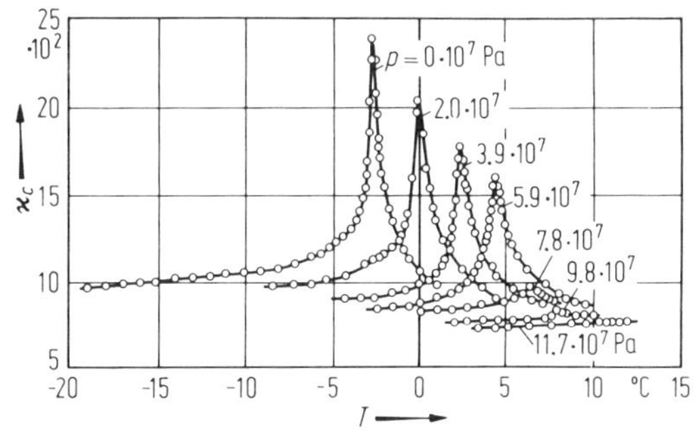

Fig. 720α. NH_4HSO_4. $\varkappa_c$ vs. T [68P4]. Parameter: p.

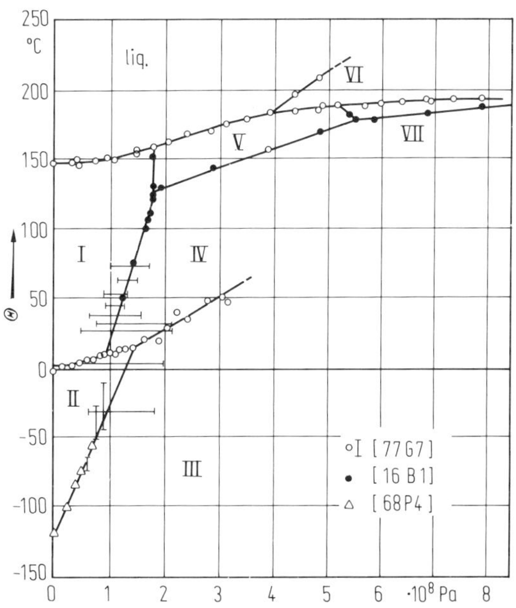

Fig. 720β. NH_4HSO_4. Θ vs. p [16B1, 68P4, 77G7].

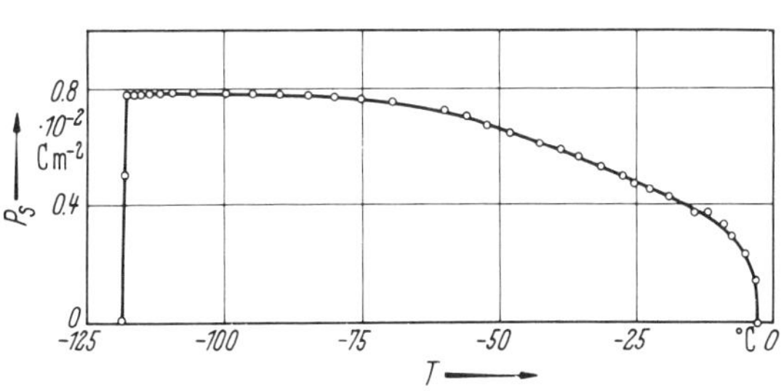

Fig. 721. NH_4HSO_4. P_s vs. T [58P2].

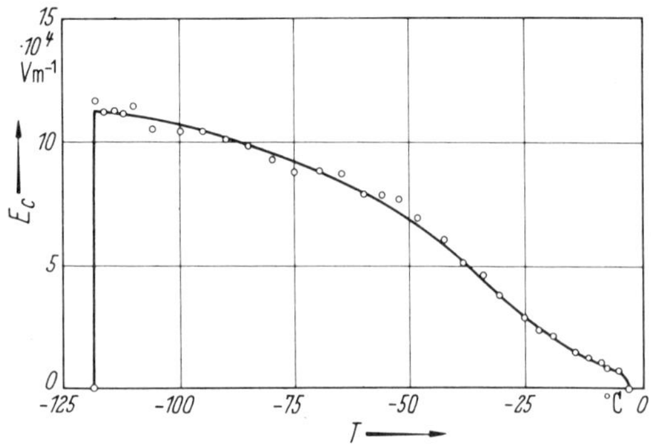

Fig. 722. NH_4HSO_4. E_c vs. T [58P2].

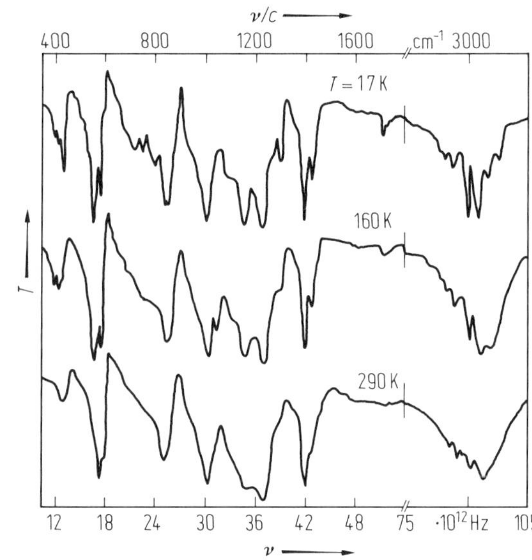

Fig. 724. NH_4HSO_4. T vs. ν [71S12]. T: Transmittance, ν: frequency of the incident infrared radiation. Parameter: temperature.

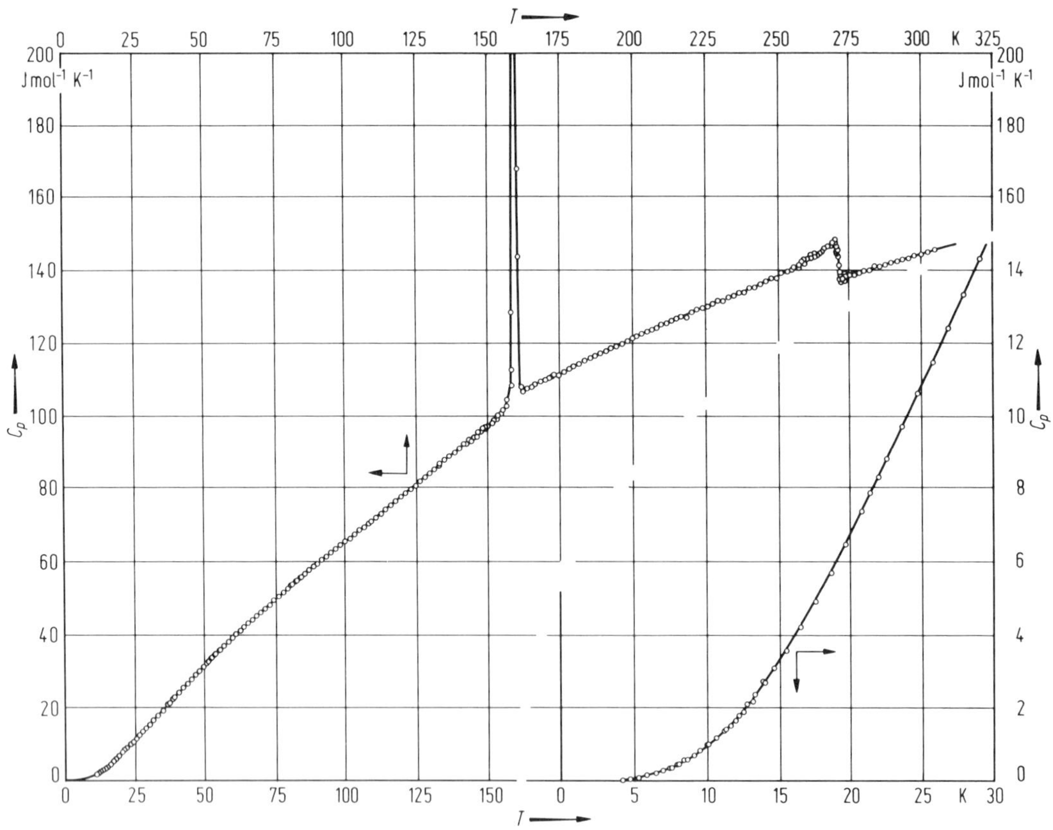

Fig. 723. NH_4HSO_4. C_p vs. T [73C3].

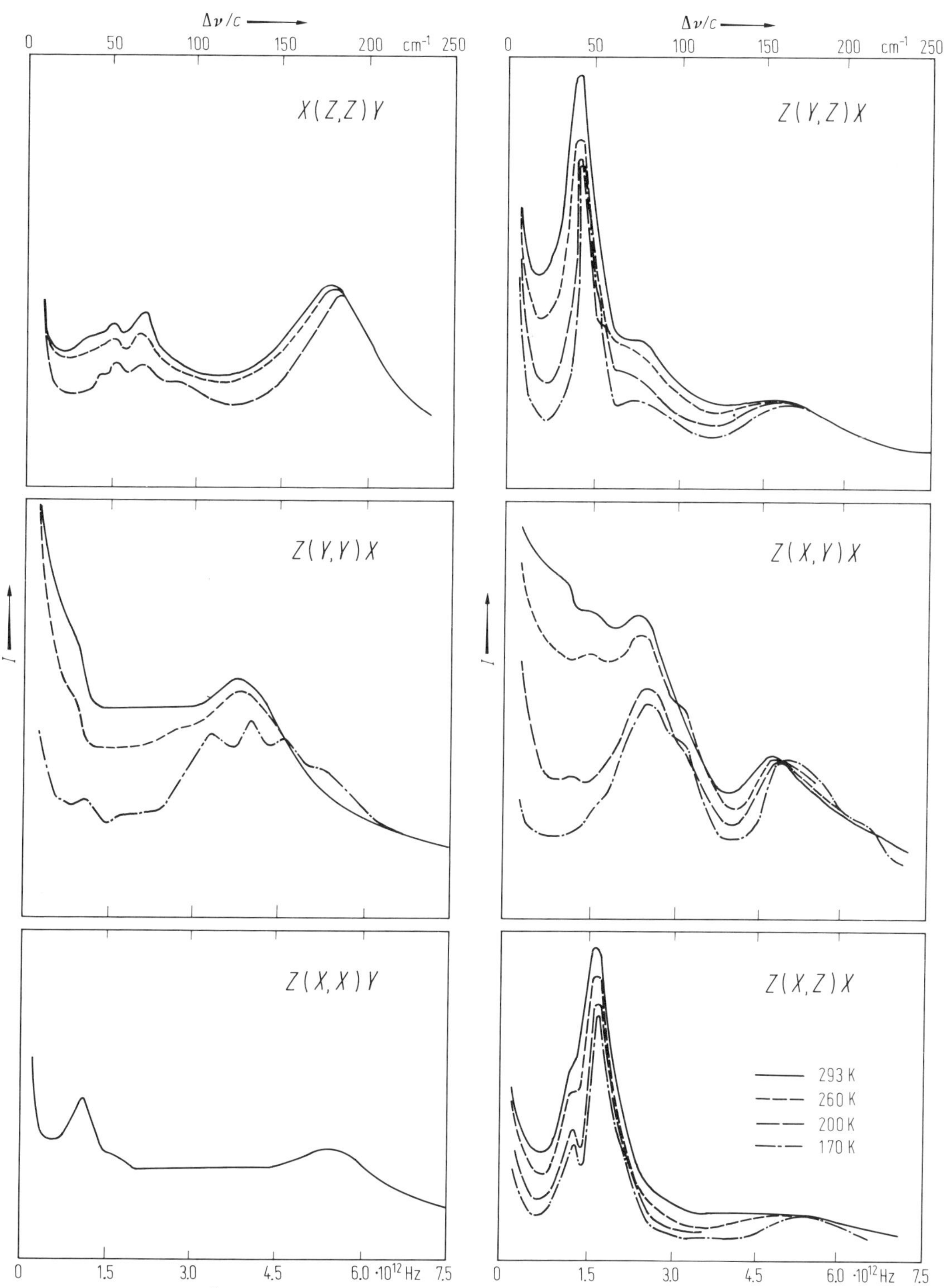

Fig. 725. NH_4HSO_4. I vs. $\Delta\nu$ [73A10]. I: Raman scattering intensity for various scattering geometries. Parameter: temperature. $X \| a'$, $Y \| b'$, $Z \| c'$ (pseudo-orthorhombic cell).

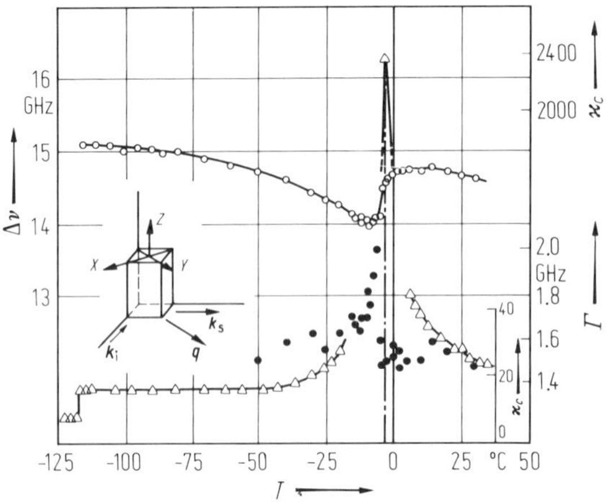

Fig. 726. NH_4HSO_4. $\Delta\nu$, Γ, $\varkappa_c$ vs. T [77H11]. Open circles: $\Delta\nu$, Brillouin shift; full circles: Γ, linewidth; triangles: $\varkappa_c$ at 50 kHz. $X \| a$, $Y \| b$, $Z \| c$ (monoclinic cell).

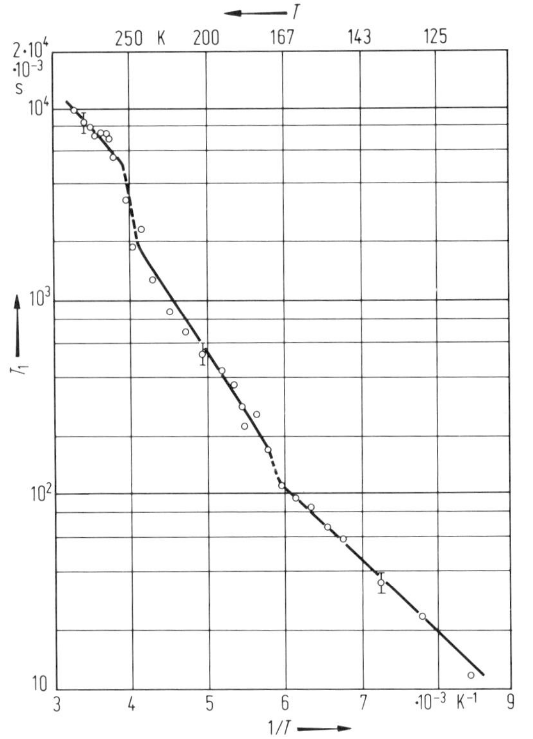

Fig. 727α. ND_4DSO_4. T_1 vs. $1/T$ [72T16]. T_1: spin-lattice relaxation time of deuteron.

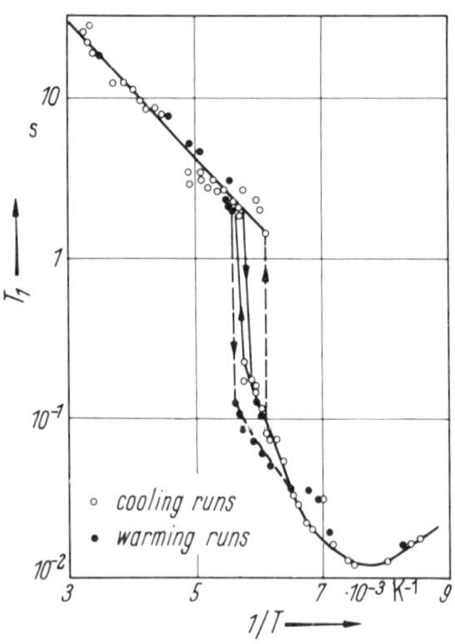

Fig. 727β. $(NH_4)HSO_4$. T_1 vs. T [62M5]. T_1: spin lattice relaxation time of proton at 30 MHz.

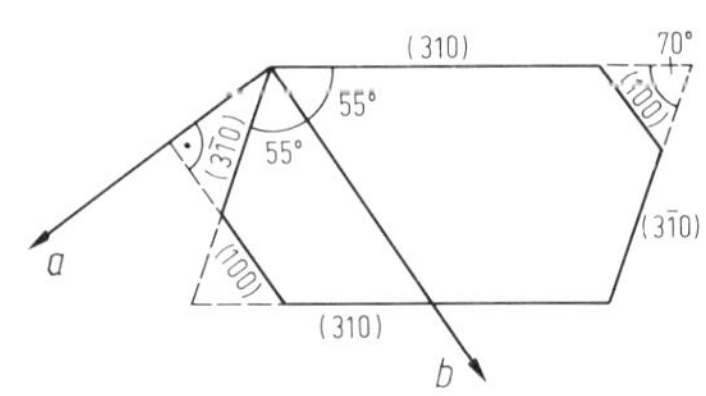

Fig. 728. $RbHSeO_4$. Crystal form [79P3].

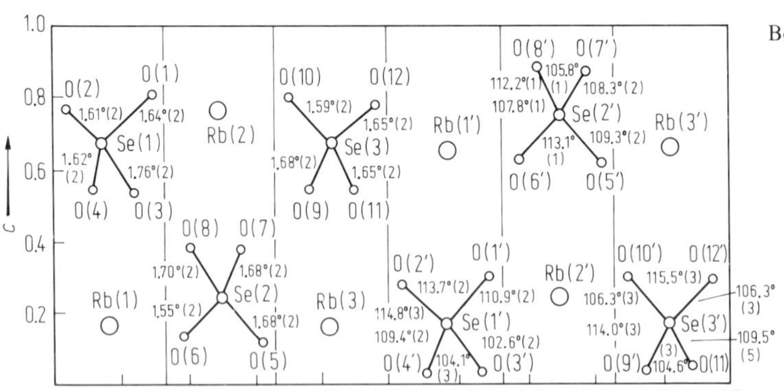

Fig. 729. $RbHSeO_4$. Structure of phase II [78W8]. Interatomic distances in Å and bond angles in degree. The atoms of the asymmetric unit are unprimed and those related by symmetry are primed.

Bond angles: ∢(O (8′), O (7′)) 105.8° (1)
∢(O (8′), O (6′)) 107.8° (1)
∢(O (8′), O (5′)) 109.3° (2)
∢(O (7′), O (5′)) 108.3° (2)
∢(O (7′), O (6′)) 112.2° (1)
∢(O (6′), O (5′)) 113.1° (1)
∢(O (2′), O (4′)) 114.8° (3)
∢(O (2′), O (3′)) 109.4° (2)
∢(O (4′), O (3′)) 104.1° (2)
∢(O (1′), O (4′)) 110.9° (2)
∢(O (1′), O (3′)) 102.6° (2)
∢(O (1′), O (2′)) 113.7° (2)
∢(O (10′), O (12′)) 115.5° (3)
∢(O (9′), O (11′) 104.6°
∢(O (9′), O (12′)) 109.5° (5)
∢(O (10′), O (11′)) 114.0° (3)
∢(O (9′), O (10′)) 106.3° (3)
∢(O (11′), O (12′)) 106.3° (3)

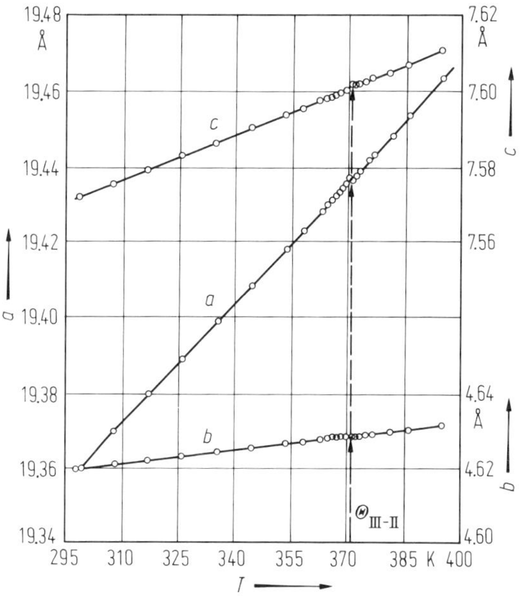

Fig. 730α. $RbHSeO_4$. a, b, c vs. T [79P2]. a, b, c: unit cell parameters.

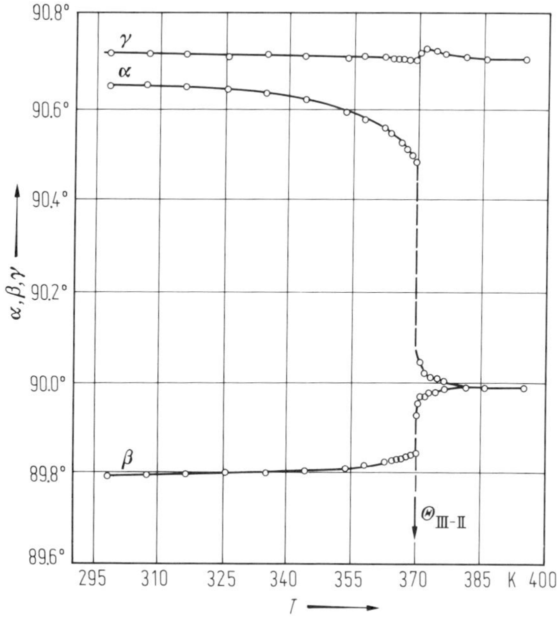

Fig. 730β. $RbHSeO_4$. α, β, γ vs. T [79P2]. α, β, γ: unit cell parameters.

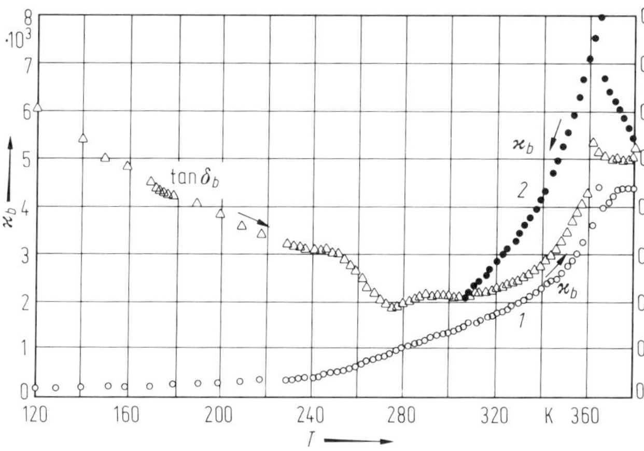

Fig. 731α. $RbHSeO_4$. $\varkappa_b$, $\tan\delta_b$ vs. T [79P3]. δ_b: dielectric loss angle along the b axis. $f = 1$ kHz. Curve *1* shows a plot for $\varkappa_b$ when the sample was first used. In each subsequent measuring cycle, curve *2* was obtained.

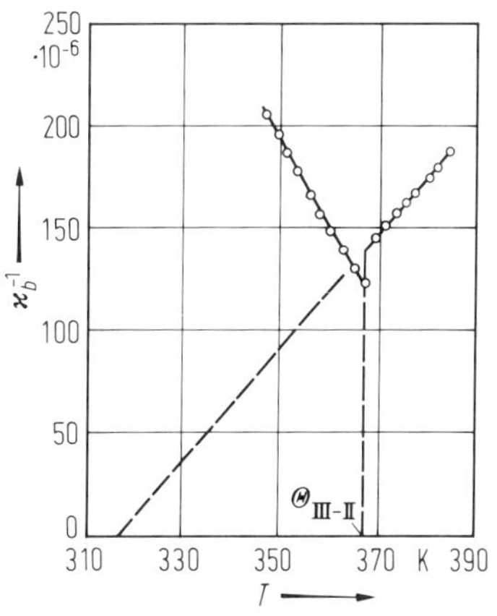

Fig. 731β. $RbHSeO_4$. κ_b^{-1} vs. T [79P3]. $f = 1$ kHz.

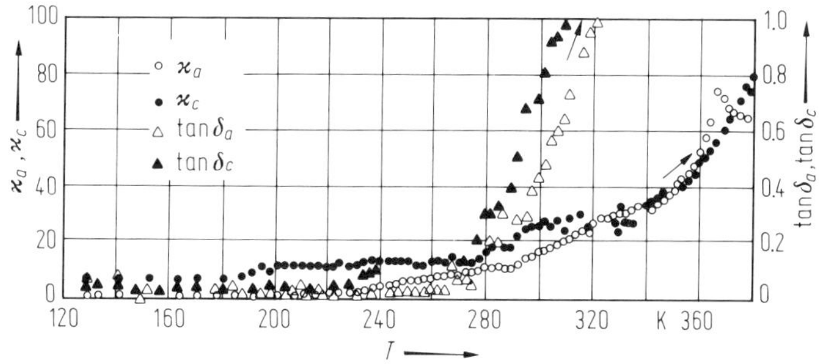

Fig. 731γ. $RbHSeO_4$. $\varkappa_a$, $\tan\delta_a$, $\varkappa_c$, $\tan\delta_c$ vs. T [79P3]. δ_a and δ_c are dielectric loss angles along the a and the c axes, respectively.

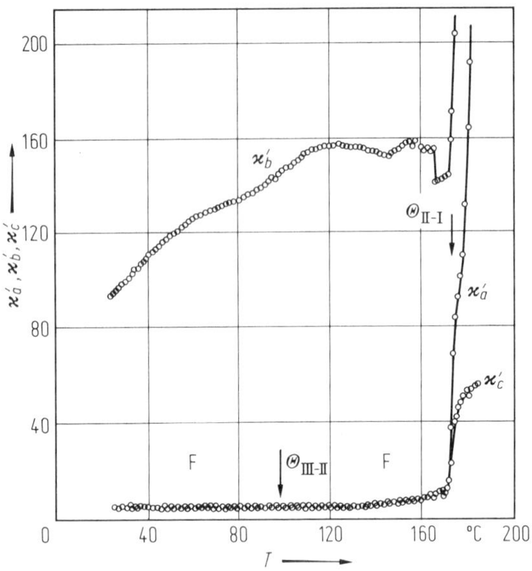

Fig. 731δ. $RbHSeO_4$. $\varkappa_a'$, $\varkappa_b'$, $\varkappa_c'$ vs. T [79S6]. $f = 1$ MHz.

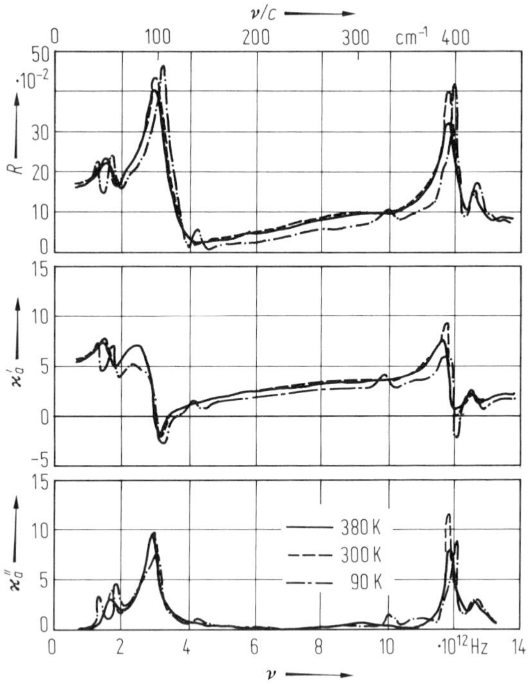

Fig. 733α. $RbHSeO_4$. R, $\varkappa_a'$, $\varkappa_a''$ vs. ν [80K1]. R: reflectivity, $\mathbf{E} \parallel a$.

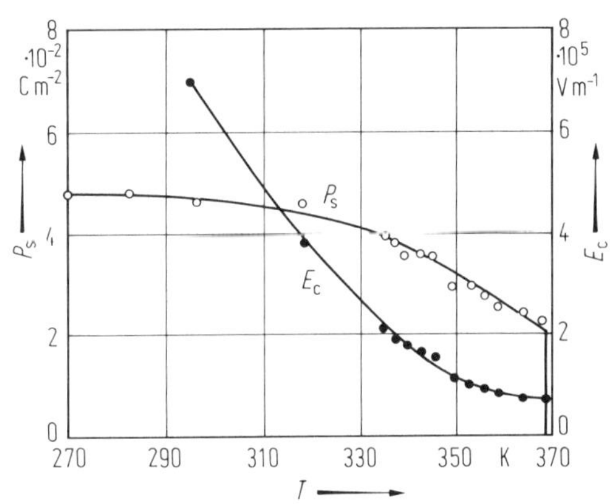

Fig. 732α. $RbHSeO_4$. P_s, E_c vs. T [79P3].

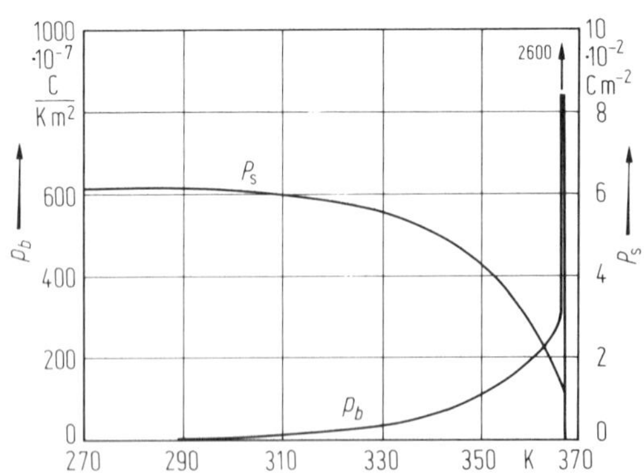

Fig. 732β. $RbHSeO_4$. p_b, P_s vs. T [79P3]. p_b: pyroelectric coefficient along b axis.

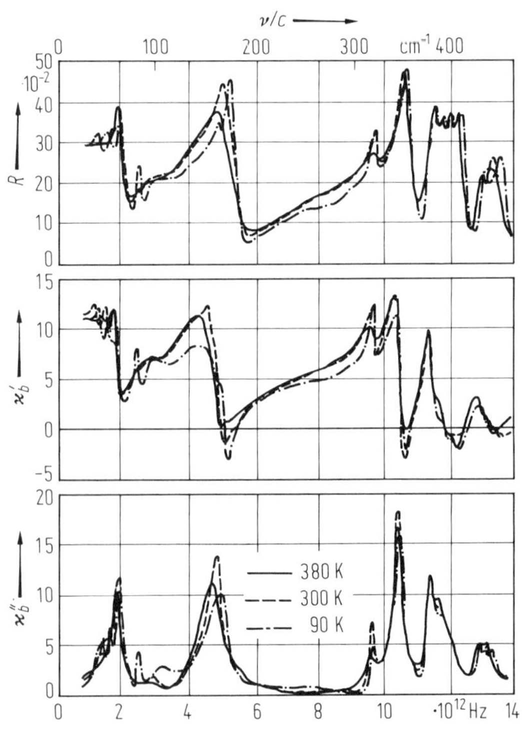

Fig. 733β. $RbHSeO_4$. R, $\varkappa_b'$, $\varkappa_b''$ vs. ν [80K1]. R: reflectivity, $\mathbf{E} \parallel b$.

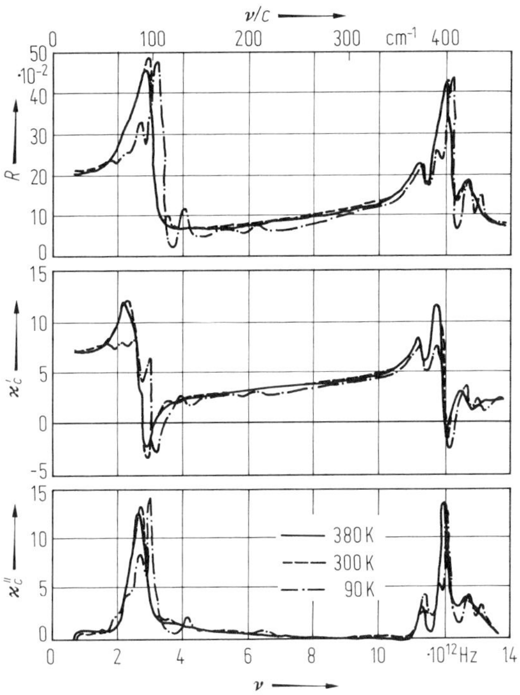

Fig. 733γ. $RbHSeO_4$. R, $\varkappa_c'$, $\varkappa_c''$ vs. ν [80 K 1]. R: reflectivity, $\boldsymbol{E} \parallel c$.

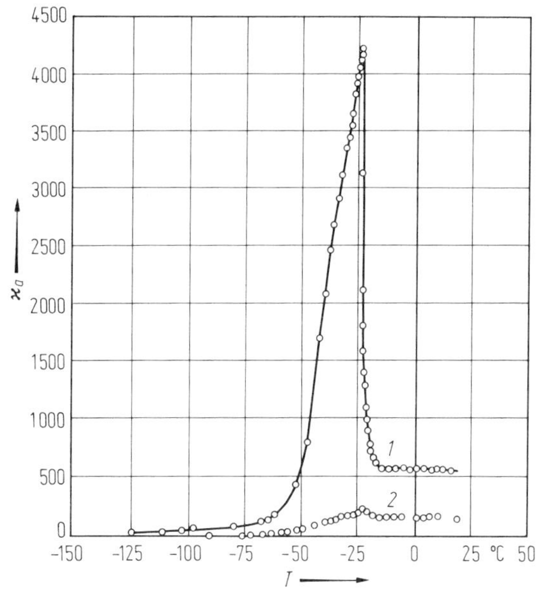

Fig. 735α. NH_4HSeO_4. $\varkappa_a$ vs. T [79 K 3]. Curve 1: $f = 800$ Hz; Curve 2: $f = 10$ MHz.

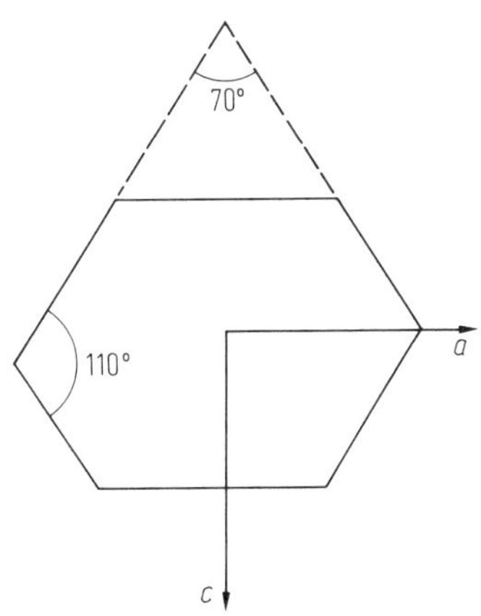

Fig. 734. NH_4HSeO_4. Crystal form [79 C 2].

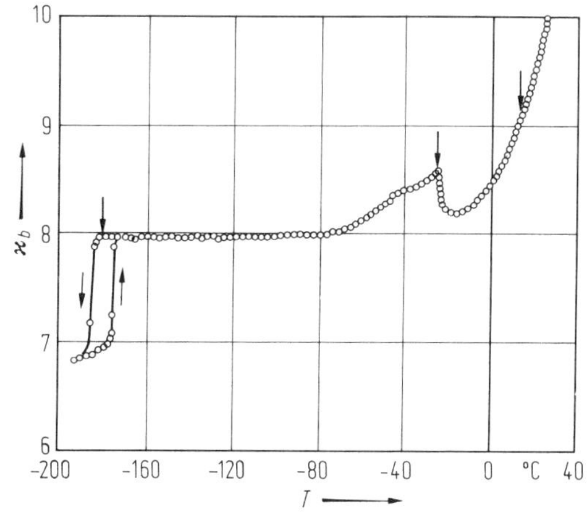

Fig. 735β. NH_4HSeO_4. $\varkappa_b$ vs. T [80 G 1]. $f = 1$ kHz. Vertical arrows point to transition temperatures.

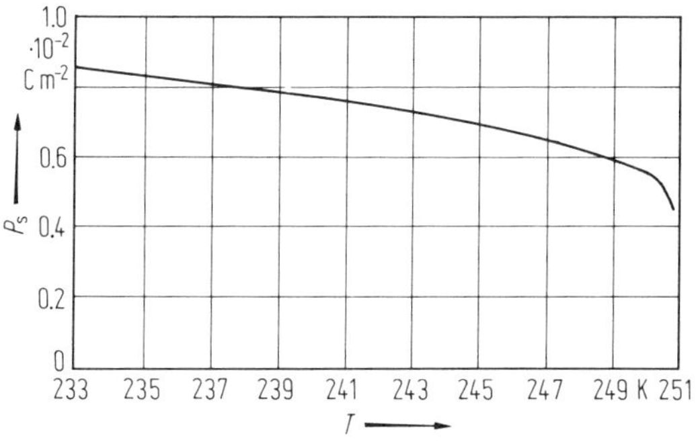

Fig. 736α. NH_4HSeO_4. P_s vs. T [80 P 1].

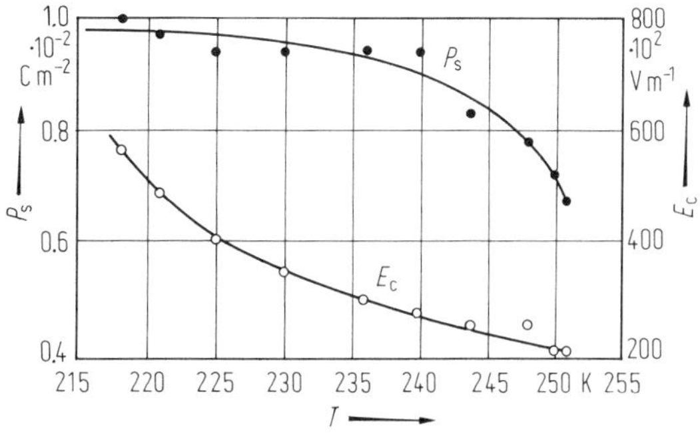

Fig. 736β. NH_4HSeO_4. P_s, E_c vs. T [79 C 2].

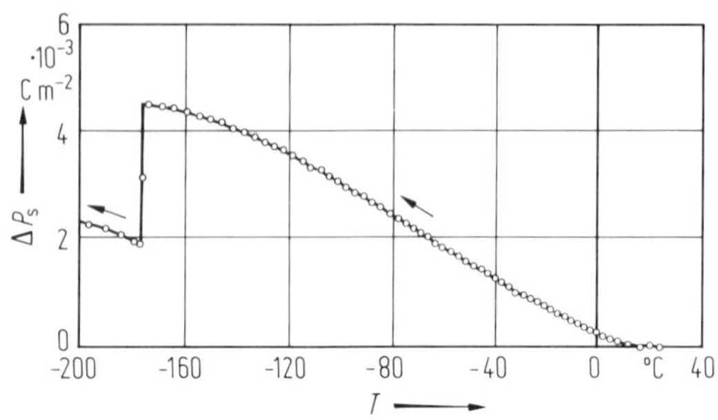

Fig. 737. NH_4HSeO_4. ΔP_s vs. T [80G1]. ΔP_s: pyroelectric charge.

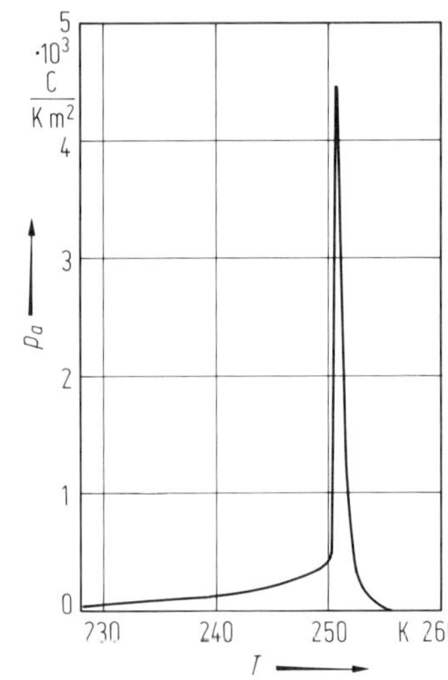

Fig. 738. NH_4HSeO_4. p_a vs. T [80P1]. p_a: pyroelectric coefficient along the a axis.

22 $(NH_4)_3H(SO_4)_2$ family

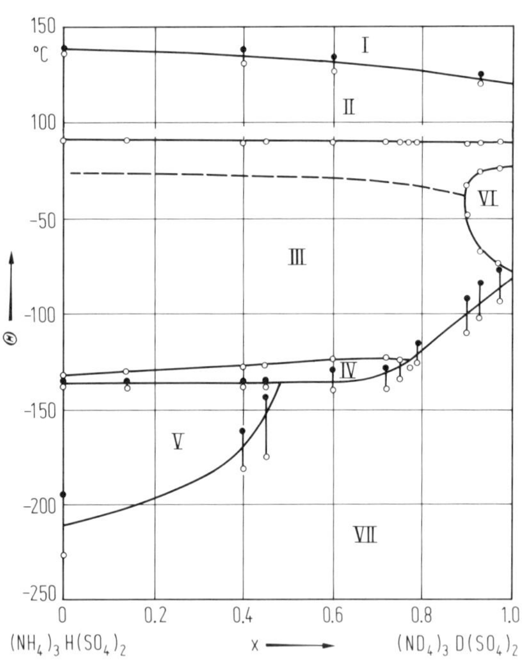

Fig. 739. $\{(NH_4)_3H(SO_4)_2\}_{1-x}$ $\{(ND_4)_3D(SO_4)_2\}_x$. Θ vs. x [80O1]. The dashed line indicates positions of broad maximum in dielectric constant. Open circles: on cooling; full circles: on heating.

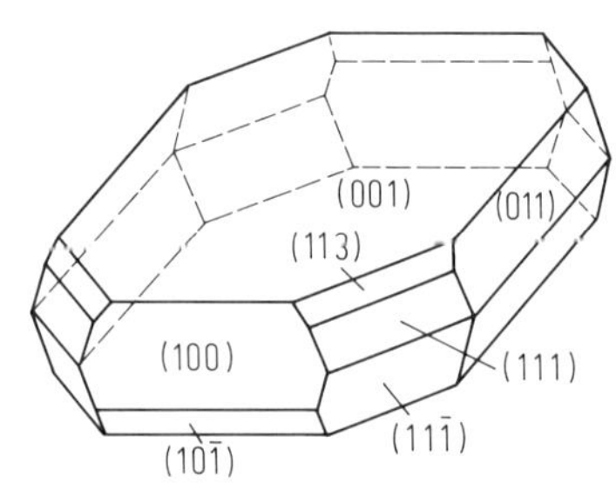

Fig. 740. $(NH_4)_3H(SO_4)_2$. Crystal form [04G1].

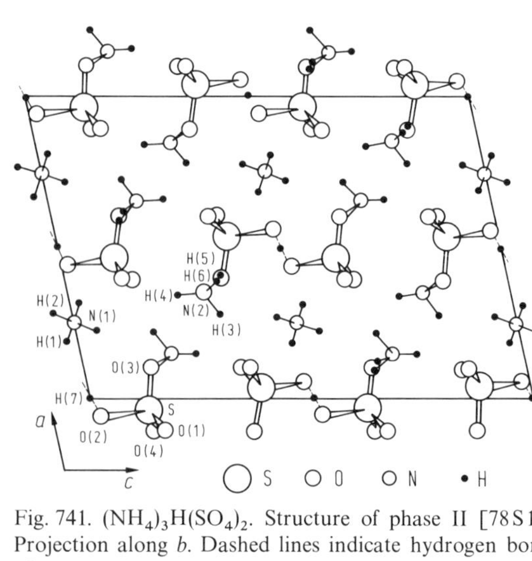

Fig. 741. $(NH_4)_3H(SO_4)_2$. Structure of phase II [78S15]. Projection along b. Dashed lines indicate hydrogen bonds of type O – H···O.

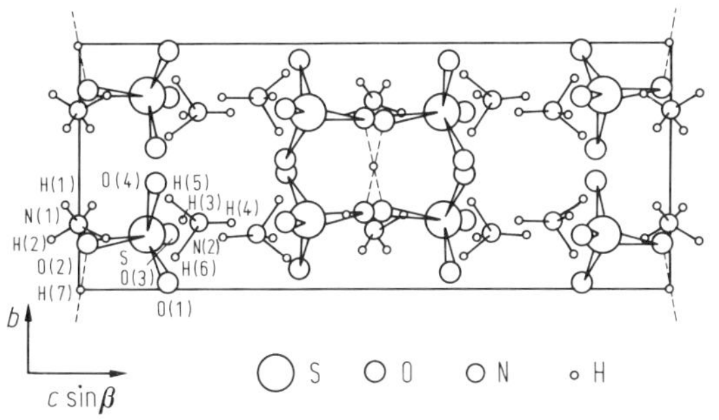

Fig. 742. $(NH_4)_3H(SO_4)_2$. Structure of phase II [78S15]. Projection along *a*. Dashed lines indicate hydrogen bonds of type O–H···O.

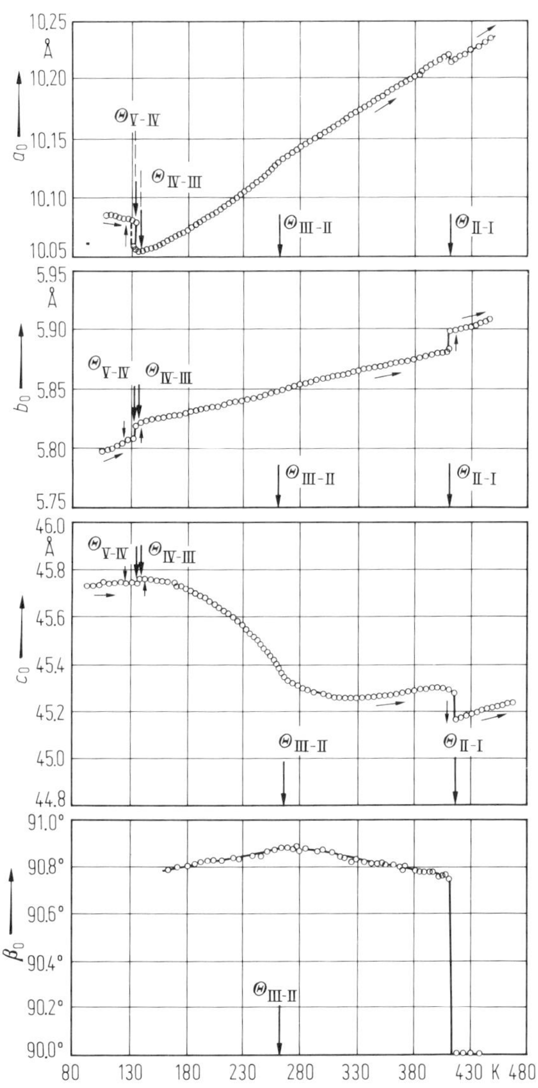

Fig. 743α. $(NH_4)_3H(SO_4)_2$. a_o, b_o, c_o, β_o vs. *T* [79S11]. a_o, b_o, c_o, β_o: lattice parameters of pseudo-orthorhombic cell. Relation between unit cell vectors $\boldsymbol{a}_o$, $\boldsymbol{b}_o$, $\boldsymbol{c}_o$ and monoclinic unit cell vectors $\boldsymbol{a}$, $\boldsymbol{b}$, $\boldsymbol{c}$: $\boldsymbol{a}_o = -\boldsymbol{a}$, $\boldsymbol{b}_o = -\boldsymbol{b}$, $\boldsymbol{c}_o = \boldsymbol{a} + 3\boldsymbol{c}$.

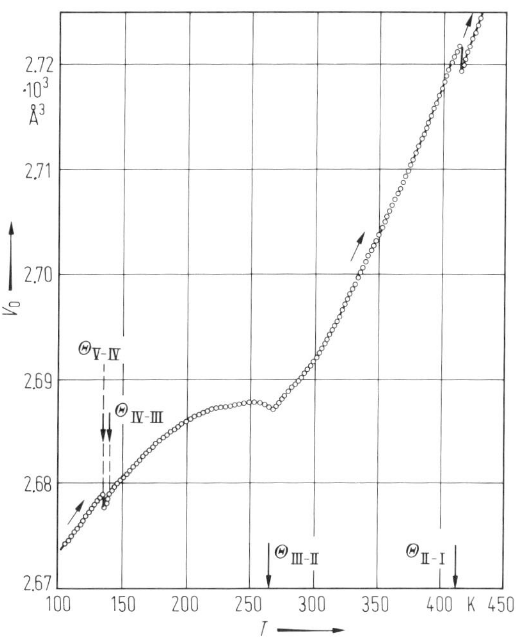

Fig. 743β. $(NH_4)_3H(SO_4)_2$. V_o vs. *T* [79S11]. V_o: volume of pseudo-orthorhombic unit cell.

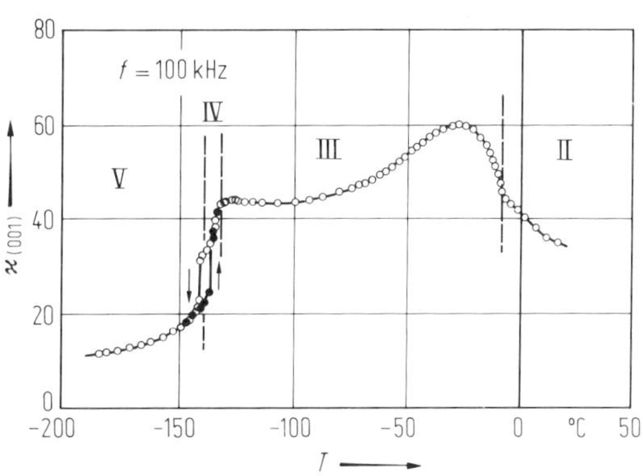

Fig. 744. $(NH_4)_3H(SO_4)_2$. $\varkappa_{(001)}$ vs. *T* [76G6]. $\varkappa_{(001)}$: dielectric constant of (001) plate specimen.

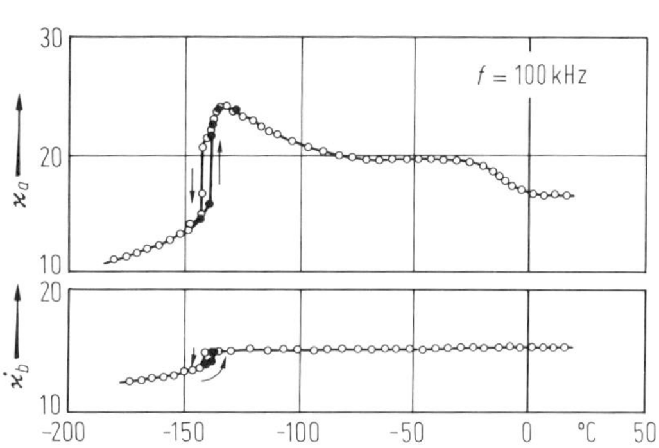

Fig. 745. $(NH_4)_3H(SO_4)_2$. $\varkappa_a$, $\varkappa_b$ vs. *T* [76G6].

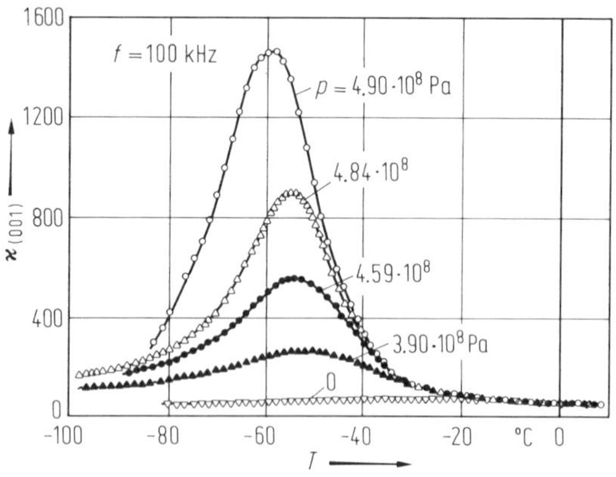

Fig. 746. $(NH_4)_3H(SO_4)_2 \cdot \varkappa_{(001)}$ vs. T[77G4]. Parameter: p. $\varkappa_{(001)}$: dielectric constant of (001) plate specimen.

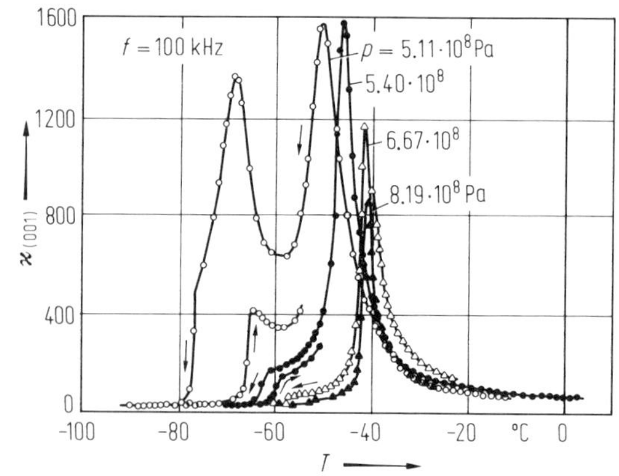

Fig. 747. $(NH_4)_3H(SO_4)_2$. $\varkappa_{(001)}$ vs. T. [77G4]. Parameter: p. $\varkappa_{(001)}$: dielectric constant of (001) plate specimen.

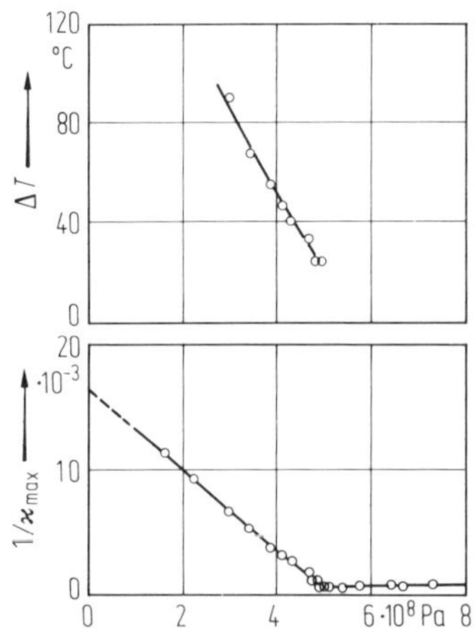

Fig. 748. $(NH_4)_3H(SO_4)_2$. ΔT, $1/\varkappa_{max}$ vs. p [77G4]. ΔT: full width at half maximum of the broad maximum in $\varkappa_{(001)}$, $\varkappa_{max}$: the maximum value of $\varkappa_{(001)}$ at $f = 100$ kHz.

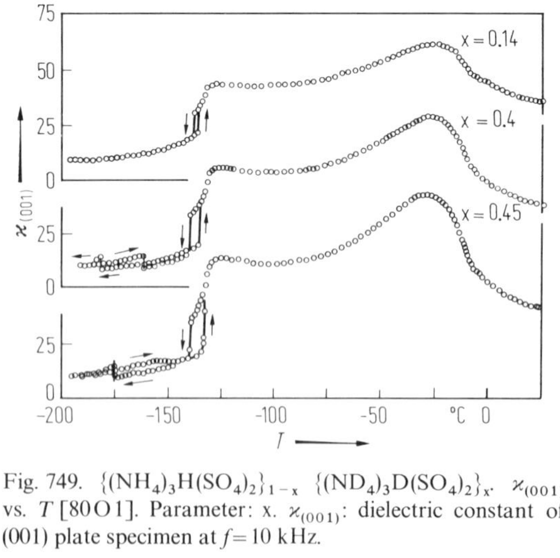

Fig. 749. $\{(NH_4)_3H(SO_4)_2\}_{1-x}$ $\{(ND_4)_3D(SO_4)_2\}_x$. $\varkappa_{(001)}$ vs. T [80O1]. Parameter: x. $\varkappa_{(001)}$: dielectric constant of (001) plate specimen at $f = 10$ kHz.

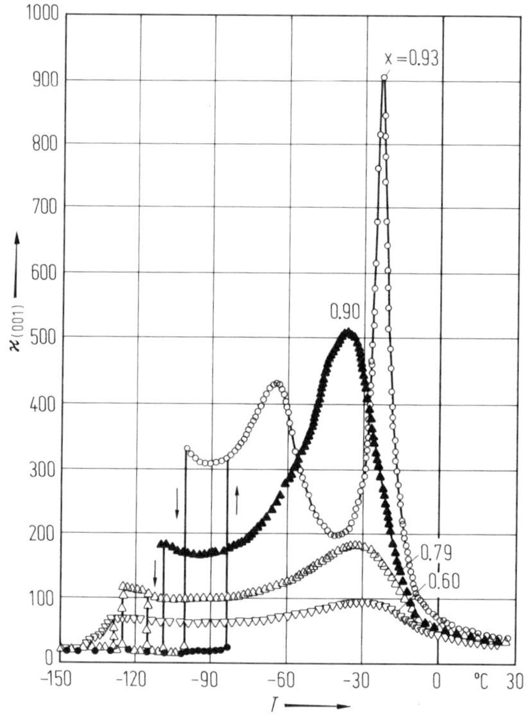

Fig. 750. $\{(NH_4)_3H(SO_4)_2\}_{1-x}$ $\{(ND_4)_3D(SO_4)_2\}_x$. $\varkappa_{(001)}$ vs. T [80O1]. Parameter: x. $\varkappa_{(001)}$: dielectric constant of (001) plate specimen at $f = 10$ kHz.

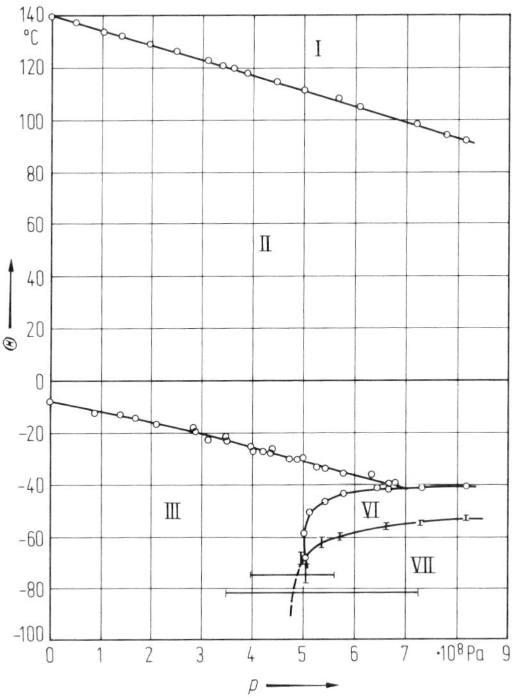

Fig. 751α. $(NH_4)_3H(SO_4)_2$. Θ vs. p [77G22].

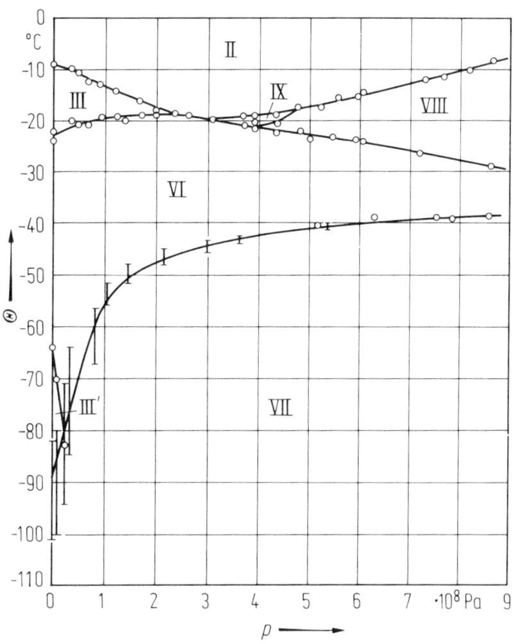

Fig. 751β. $(ND_4)_3D(SO_4)_2$. Θ vs. p [80G4]. D-content: 97%.

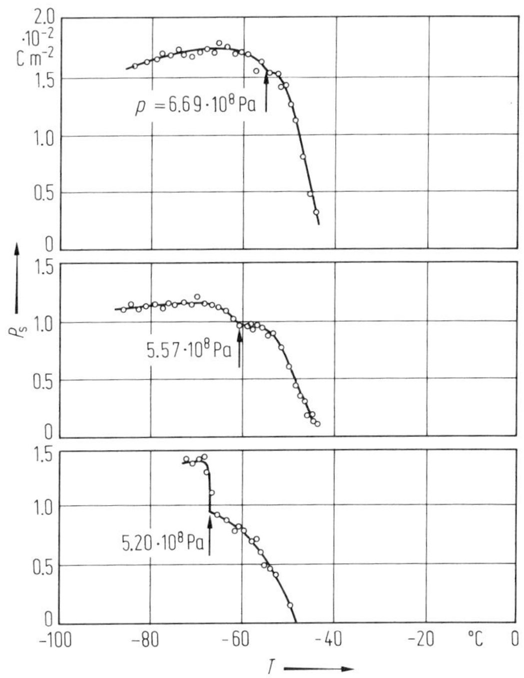

Fig. 752. $(NH_4)_3H(SO_4)_2$. P_s vs. T at various p [77G4].

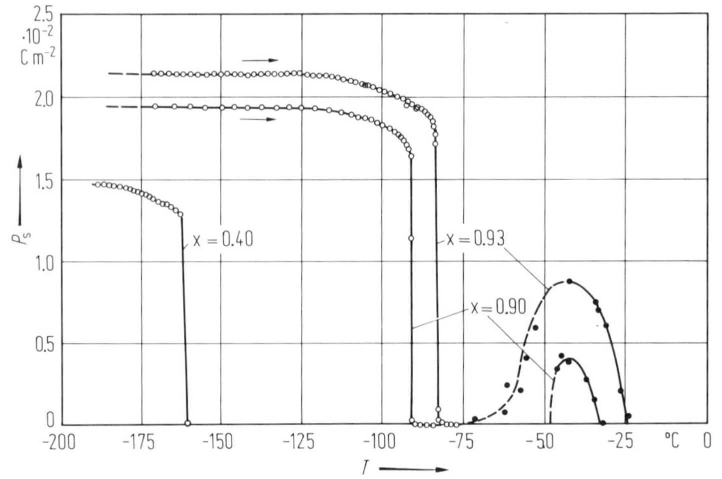

Fig. 753. $\{(NH_4)_3H(SO_4)_2\}_{1-x}$ $\{(ND_4)_3D(SO_4)_2\}_x$. P_s vs. T [80O1]. Parameter: x. Full circles: from pyroelectric measurement, open circles: from hysteresis loop measurement.

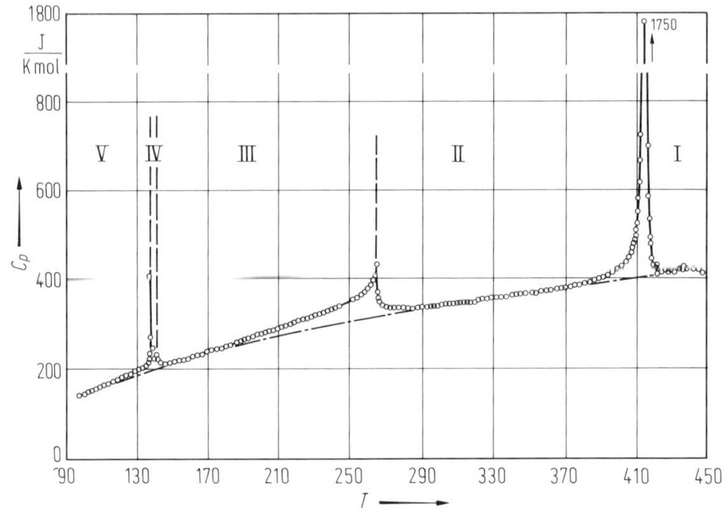

Fig. 754. $(NH_4)_3H(SO_4)_2$. C_p vs. T [79S7].

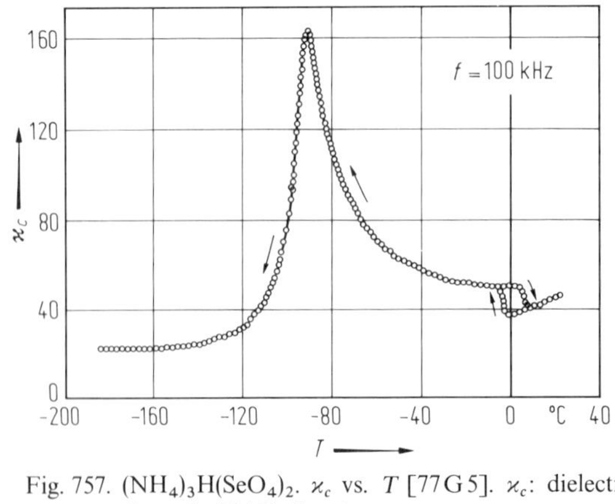

Fig. 757. $(NH_4)_3H(SeO_4)_2$. $\varkappa_c$ vs. T [77G5]. $\varkappa_c$: dielectric constant along the [111] direction.

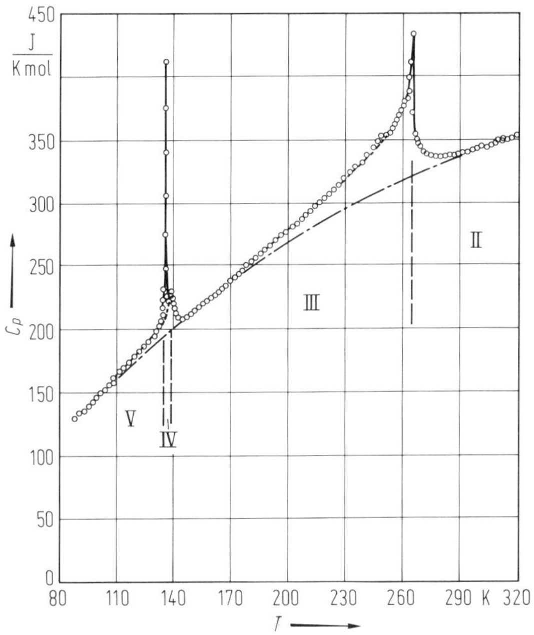

Fig. 755. $(NH_4)_3H(SO_4)_2$. C_p vs. T below RT [79S7].

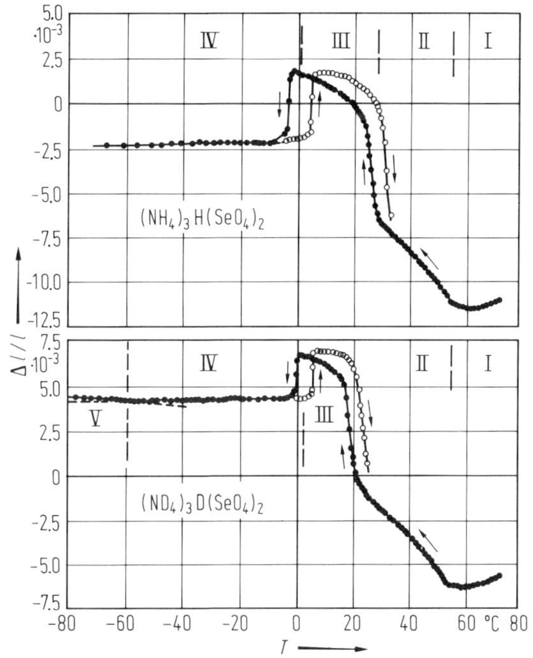

Fig. 756. $(NH_4)_3H(SeO_4)_2$, $(ND_4)_3D(SeO_4)_2$. $\Delta l/l$ along [111] [79O4]. D-content is about 80%.

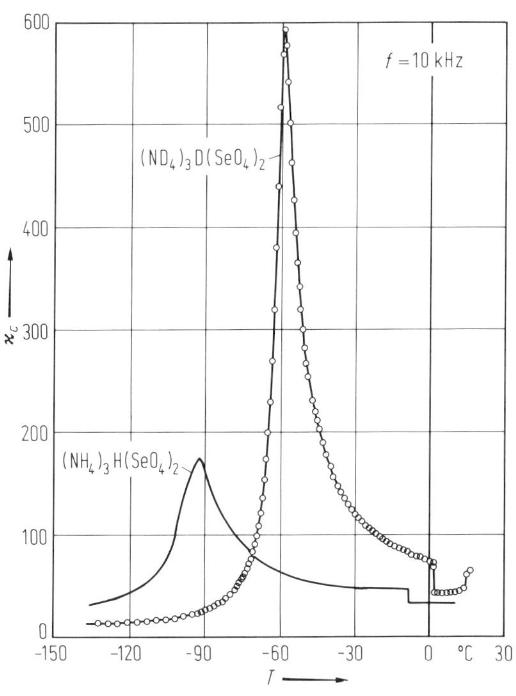

Fig. 758. $(ND_4)_3D(SeO_4)_2$ (D-content 80%). $\varkappa_c$ vs. T [79O4]. $\varkappa_c$: dielectric constant along the [111] direction. For comparision, curve for normal compound is shown.

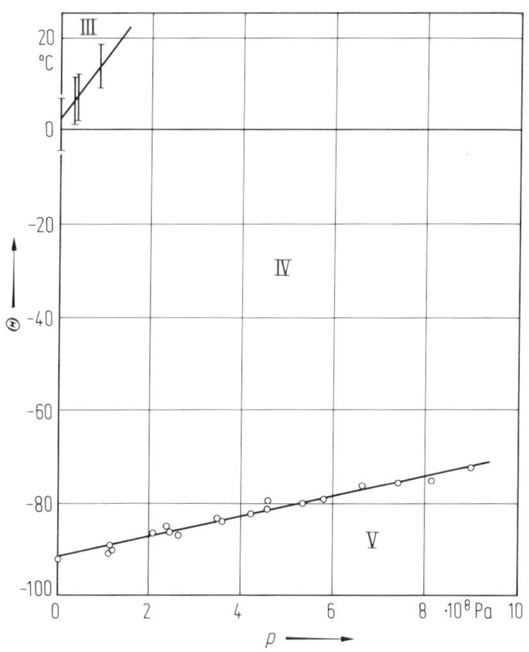

Fig. 759. $(NH_4)_3H(SeO_4)_2$. Θ vs. p below RT [77G5].

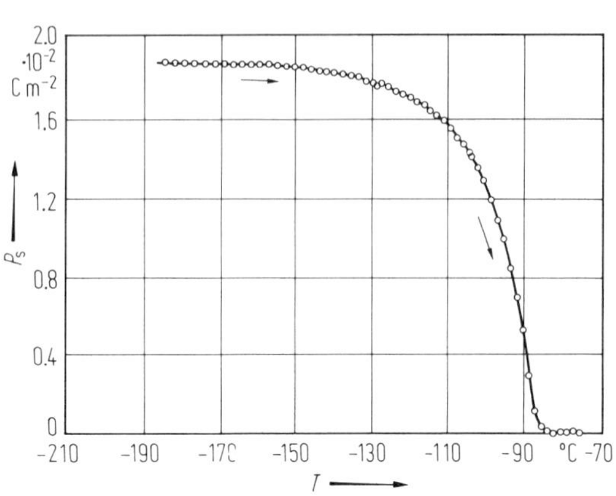

Fig. 760. $(NH_4)_3H(SeO_4)_2$. P_s vs. T [77 G 5].

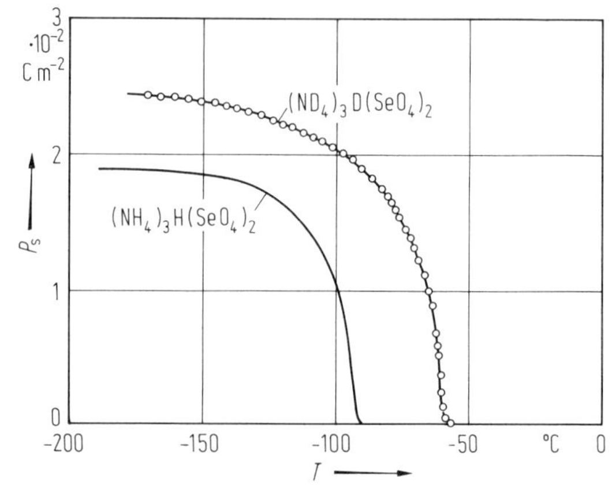

Fig. 761. $(ND_4)_3D(SeO_4)_2$ (D-content ≈80%). P_s vs. T [79 O 4]. For comparison, curve for normal compound is shown.

23 NH_4LiSO_4 family

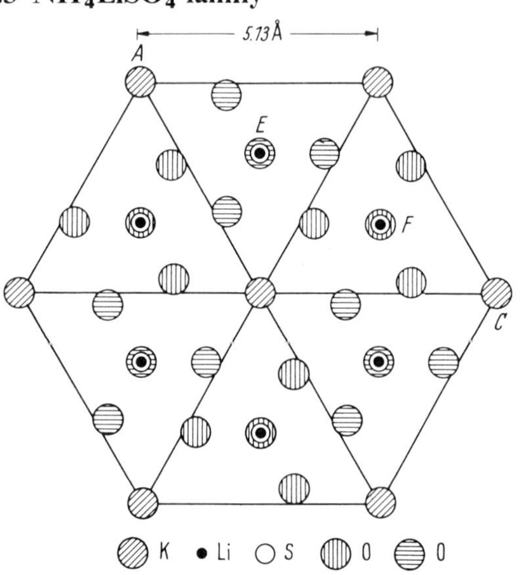

Fig. 762. $KLiSO_4$. Projection of the crystal structure along the c-axis. [25 B 1].

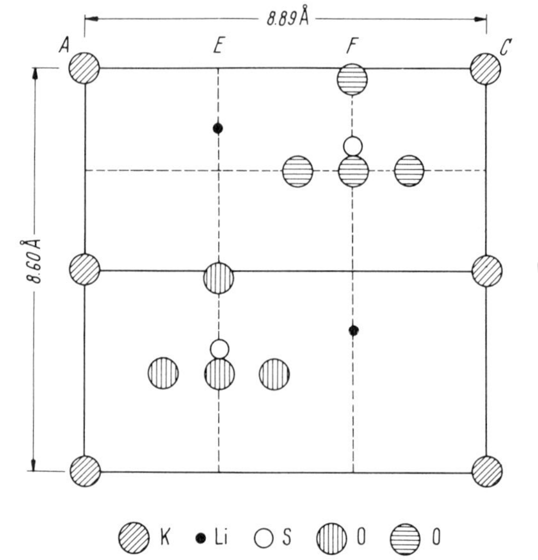

Fig. 763. $KLiSO_4$. Projection of the crystal structure along (1 0 $\bar{1}$ 0) [25 B 1]. The letters A, E, F, C refer to the same position as in Fig. 762.

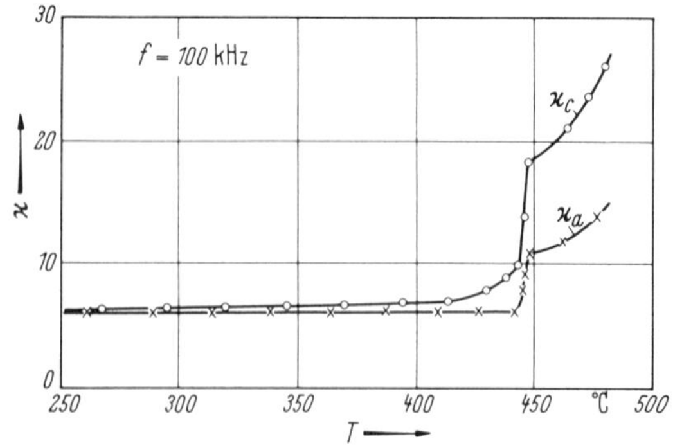

Fig. 764. $KLiSO_4$. $\varkappa_a$, $\varkappa_c$ vs. T [62 A 3].

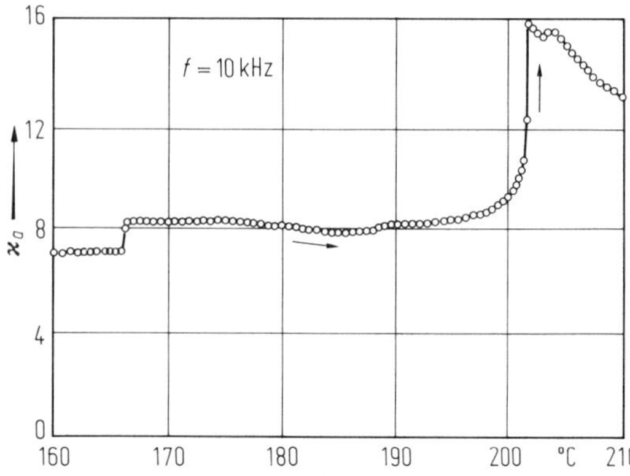

Fig. 765. $RbLiSO_4$. $\varkappa_a$ vs. T [76 S 26]. On heating.

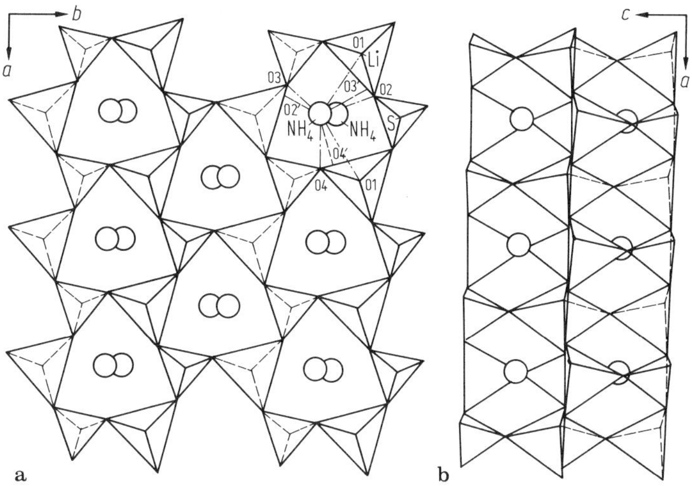

Fig. 766. NH_4LiSO_4. Structure of phase II [69D3]. Views of polyhedral linkage. Fig. a: along the *c*-axis. Only one tetrahedral layer is shown. In successive layers, which are almost exactly eclipsed in this view, a LiO_4 tetrahedron pointing up is replaced by a SO_4 tetrahedron pointing down, etc. Fractional coordinates (*z*) of labeled atoms are NH_4 0.50 and 0.00, O(1) 0.51, O(2) 0.25, O(2′) 0.75, O(3) 0.26, O(3′) 0.76, O(4) 0.27, O(4′) 0.77, Li 0.32, S 0.30. Fig. b: along the *b*-axis is showing "double chain" of tetrahedra.

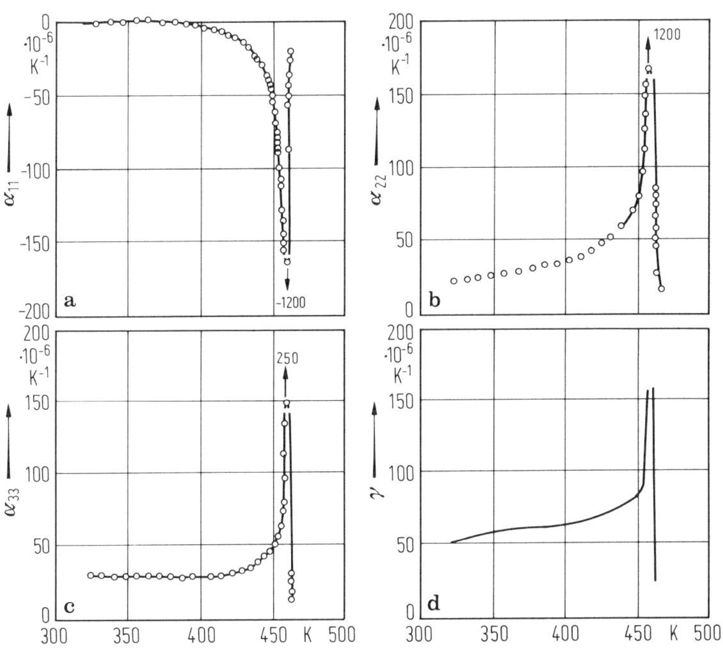

Fig. 767. NH_4LiSO_4. α_{11}, α_{22}, α_{33}, γ vs. T [77I7]. γ: cubic thermal expansion coefficient.

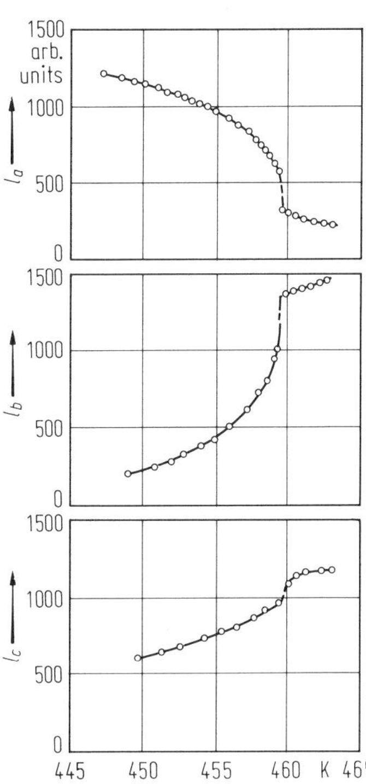

Fig. 768. NH_4LiSO_4. l_a, l_b, l_c vs. T [77I7]. l_a, l_b, l_c: linear thermal expansion along the *a*, *b*, *c* axes.

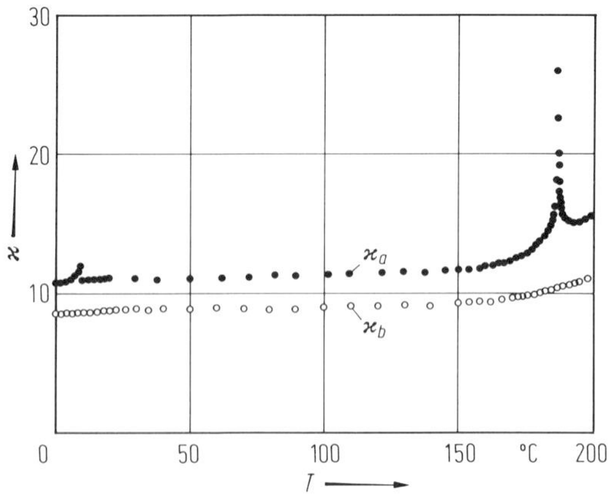

Fig. 769. NH_4LiSO_4. $\varkappa_a$, $\varkappa_b$ vs. T [75M16]. $f = 3$ kHz.

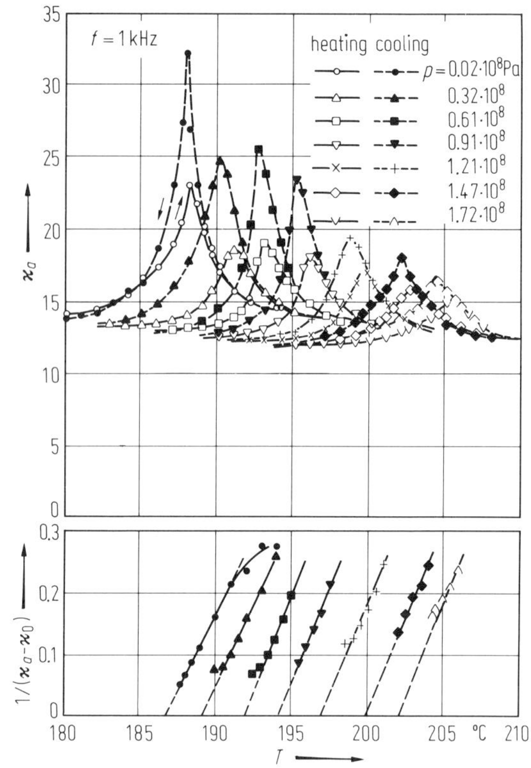

Fig. 770. NH_4LiSO_4. $\varkappa_a$, $1/(\varkappa_a - \varkappa_0)$ vs. T [77S24]. Parameter: p, $\varkappa_0 = 10.9$ ($\varkappa_0$ is the value of the flat part of $\varkappa_a$ at 50 °C).

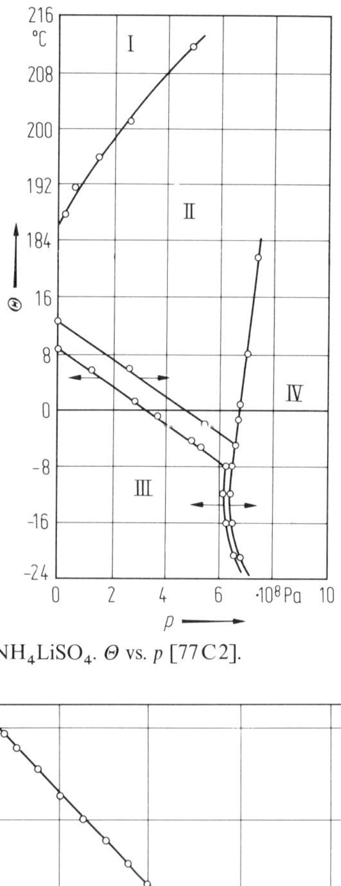

Fig. 771. NH_4LiSO_4. Θ vs. p [77C2].

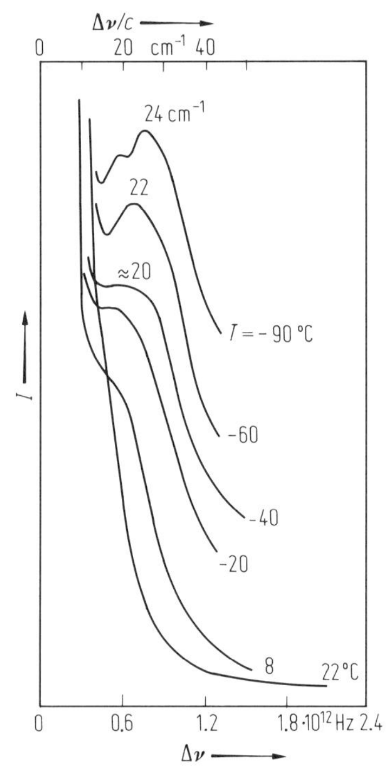

Fig. 773. NH_4LiSO_4. I vs. $\Delta\nu$ [77P6]. I: Raman scattering intensity of the lowest frequency mode in the scattering geometry of $a(b, a)b$. $\Delta\nu$: frequency shift. Parameter: temperature.

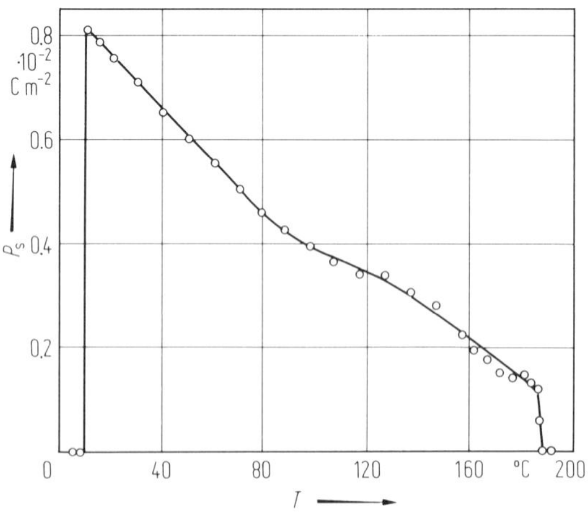

Fig. 772. NH_4LiSO_4. P_s vs. T [75M16].

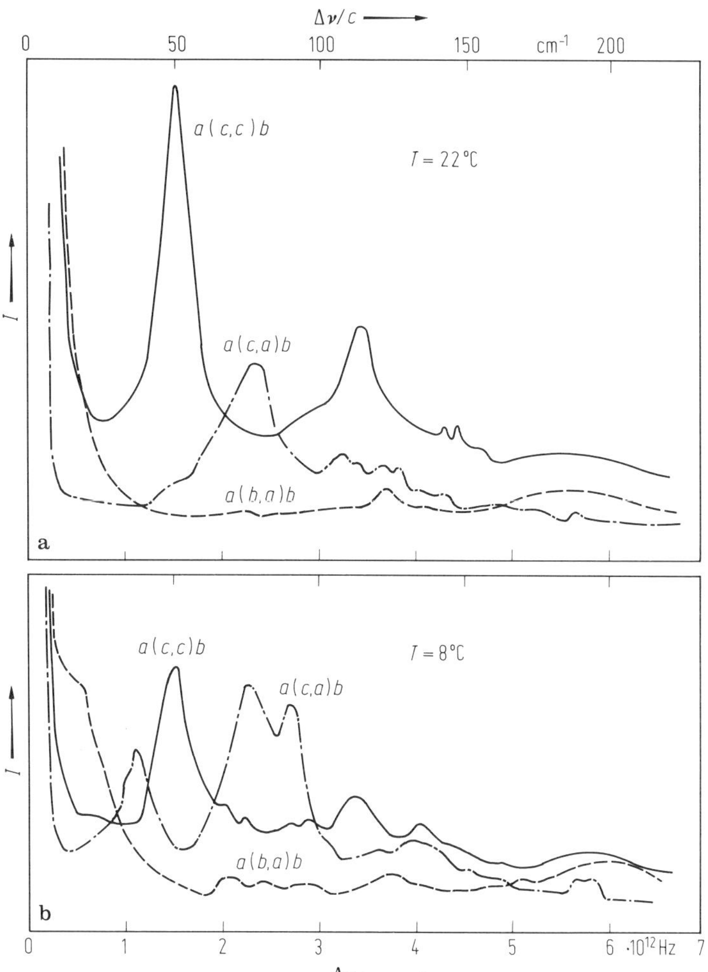

Fig. 774. NH_4LiSO_4. I vs. $\Delta\nu$ [77P6]. Fig. a: above, Fig. b: below the phase transition temperature. I: Raman scattering intensity for various scattering geometries. $\Delta\nu$: frequency shift.

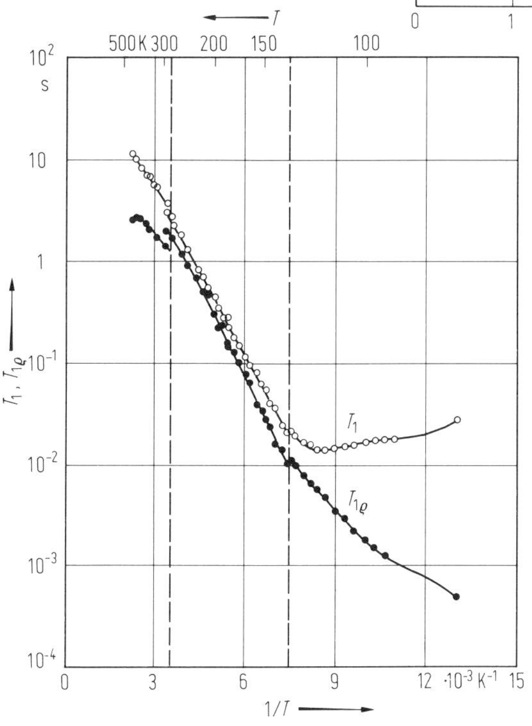

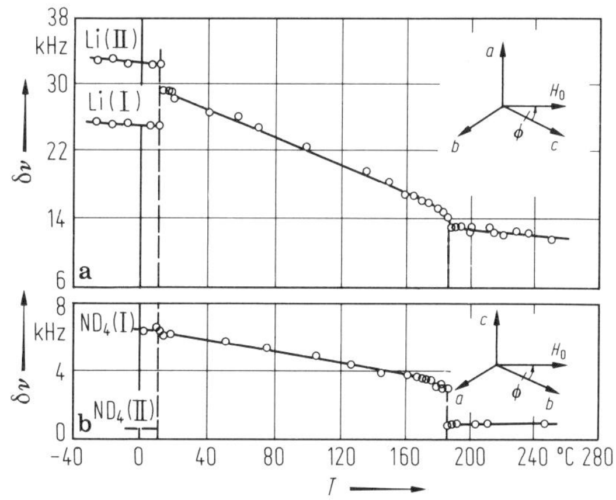

Fig. 776. ND_4LiSO_4. $\delta\nu$ vs. T [77A6]. $\delta\nu$: Quadrupole splitting. Fig. a: ^7Li, $\phi = 14°$. Fig. b: ^2D, $\phi = 45°$. See: [77Y9].

◄

Fig. 775. NH_4LiSO_4. T_1, $T_{1\varrho}$ vs. $1/T$ [78W2]. T_1: proton spin-lattice relaxation time. $T_{1\varrho}$: proton spin-lattice relaxation time in the rotary frame. $H_1 = 1.19 \cdot 10^3 Am^{-1}$.

24 Langbeinite-type family

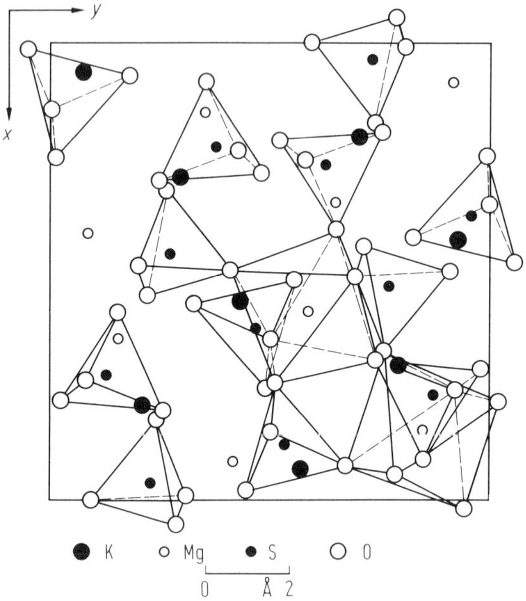

Fig. 777. $K_2Mg_2(SO_4)_3$. Crystal structure [58 V 2].

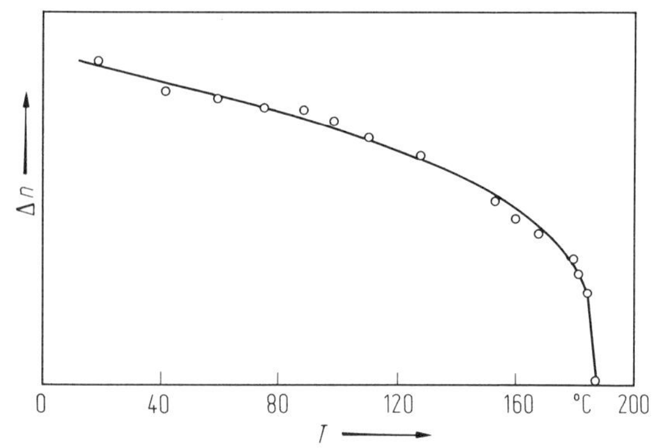

Fig. 779α. $K_2Ca_2(SO_4)_3$. Δn vs. T [80 Y 1] Δn: birefringence.

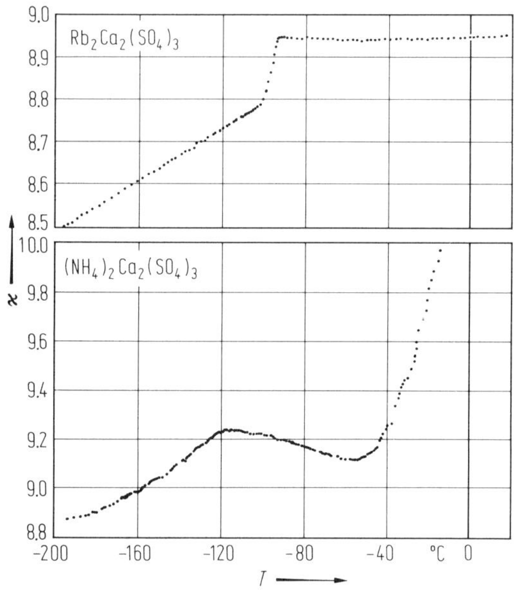

Fig. 779β. $Rb_2Ca_2(SO_4)_3$ and $(NH_4)_2Ca_2(SO_4)_3$ (polycrystals). $\varkappa$ vs. T [77 H 14] $f = 1$ kHz.

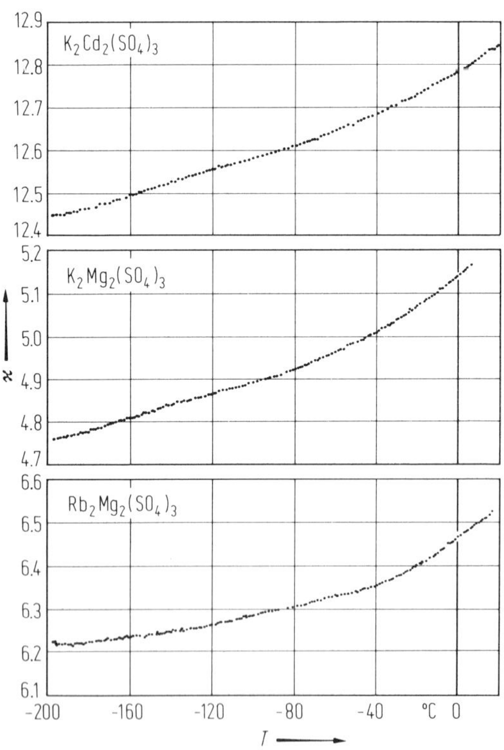

Fig. 778. $K_2Cd_2(SO_4)_3$, $K_2Mg_2(SO_4)_3$ and $Rb_2Mg_2(SO_4)_3$ (polycrystals). $\varkappa$ vs. T [77 H 13]. $f = 1$ kHz.

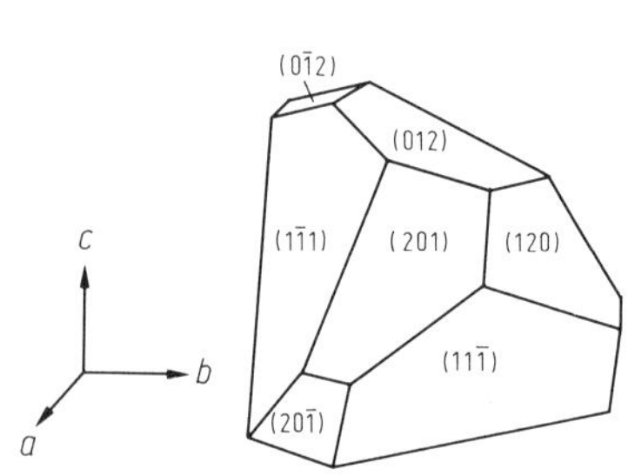

Fig. 780. $K_2Mn_2(SO_4)_3$. Crystal form of crystal grown from the aqueous solution [77 H 13].

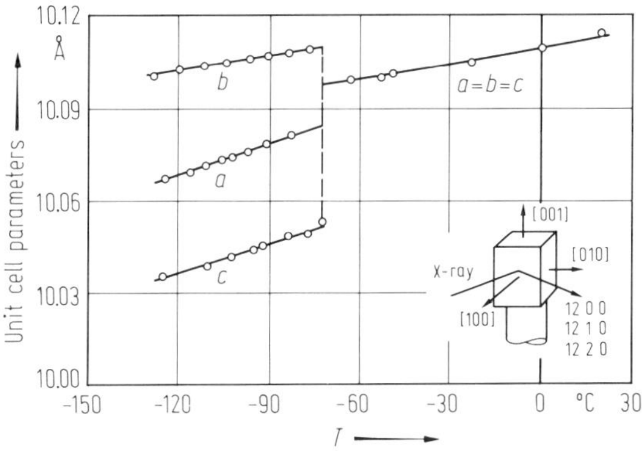

Fig. 781. $K_2Mn_2(SO_4)_3$. Unit cell parameters vs. T [78 H 10].

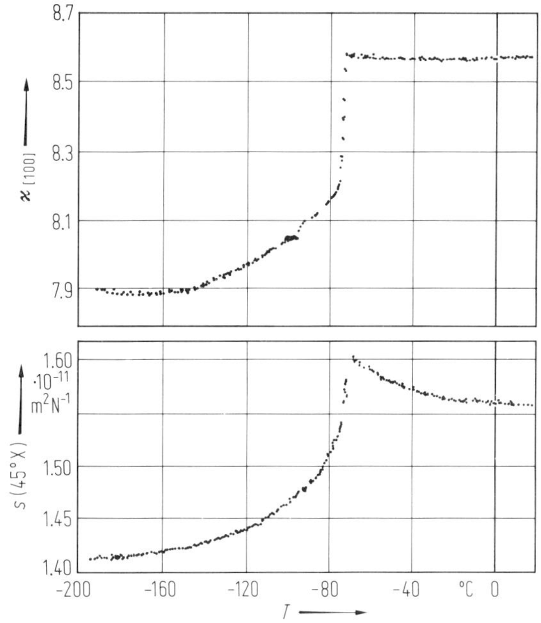

Fig. 782. $K_2Mn_2(SO_4)_3$. $\varkappa_{[100]}$, $s(45°\ X)$ vs. T [77 H 13]. $s(45°\ X)$: elastic compliance of 45° X-cut bar. $\varkappa_{[100]}$ at $f = 1$ kHz.

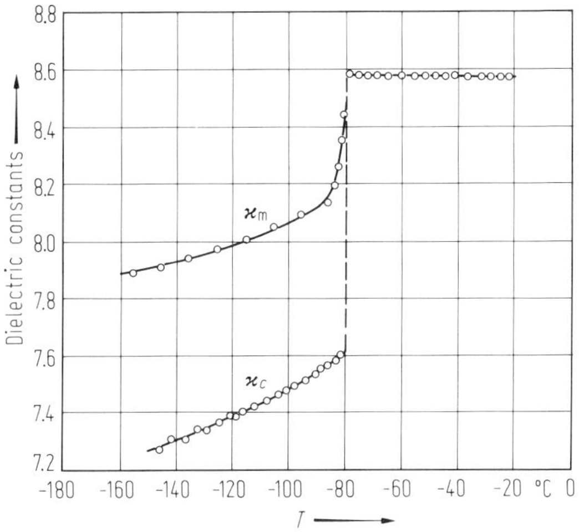

Fig. 783. $K_2Mn_2(SO_4)_3$. $\varkappa_c$, $\varkappa_m$ vs. T. [78 Y 1]. $\varkappa_m$: dielectric constant measured on multidomain crystal. $\varkappa$ at $f = 1$ kHz.

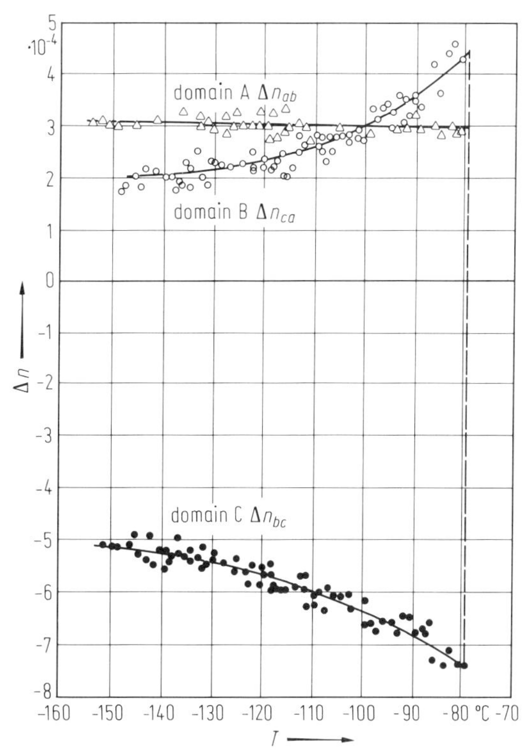

Fig. 784. $K_2Mn_2(SO_4)_3$. Δn vs. T [78 Y 1]. Three types of domains (A, B, C) were observed in the low temperature phase according to the magnitude of birefringence. $\Delta n_{ab} = n_b - n_a$, $\Delta n_{ca} = n_a - n_c$, $\Delta n_{bc} = n_c - n_b$.

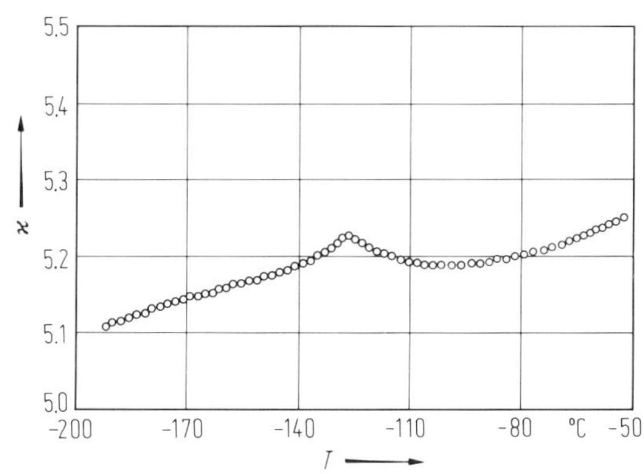

Fig. 785. $K_2Fe_2(SO_4)_3$ (polycrystal). $\varkappa$ vs. T [80 Y 1]. $f = 1$ kHz.

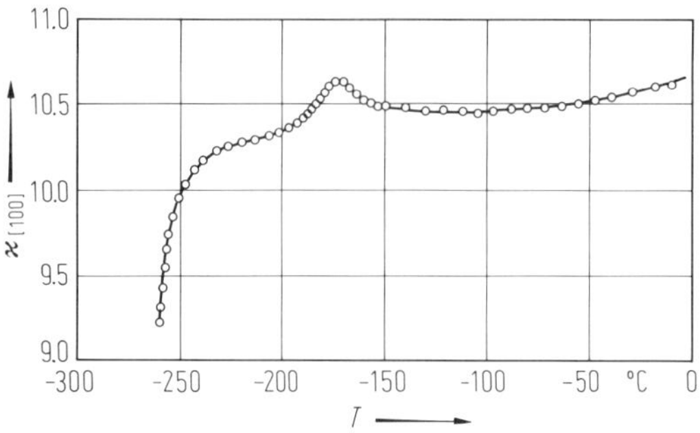

Fig. 786α. $K_2Co_2(SO_4)_3$. $\varkappa_{[100]}$ vs. T[80Y1]. f=1 kHz.

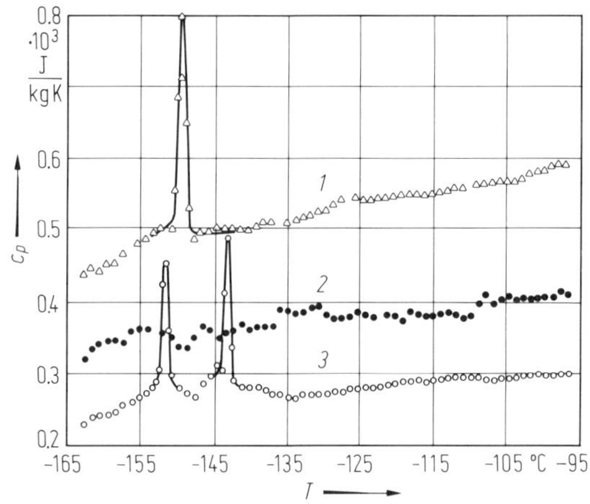

Fig. 786β. $K_2Co_2(SO_4)_3$, $Rb_2Cd_2(SO_4)_3$ and $Tl_2Cd_2(SO_4)_3$. c_p vs. T[78B9] c_p: specific heat capacity at constant pressure. Curve *1*: $K_2Co_2(SO_4)_3$ polycrystals, *2*: $Rb_2Cd_2(SO_4)_3$ single crystal, and *3*: $Tl_2Cd_2(SO_4)_3$ single crystal.

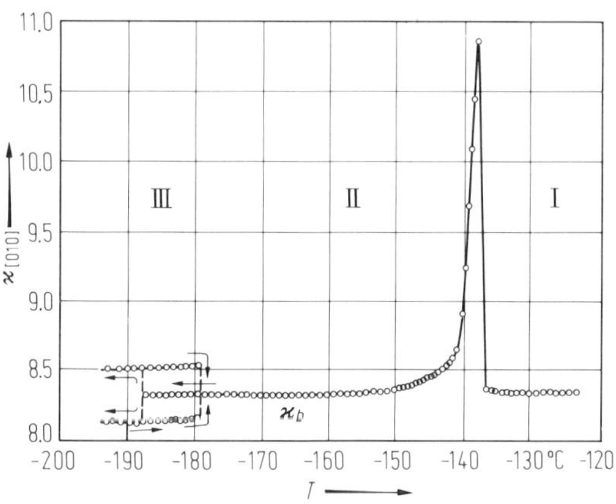

Fig. 787α. $K_2Zn_2(SO_4)_3$. $\varkappa_{[010]}$ vs. T[80Y1]. f=1 kHz. Two curves in the lowest temperature phase correspond to respective domains.

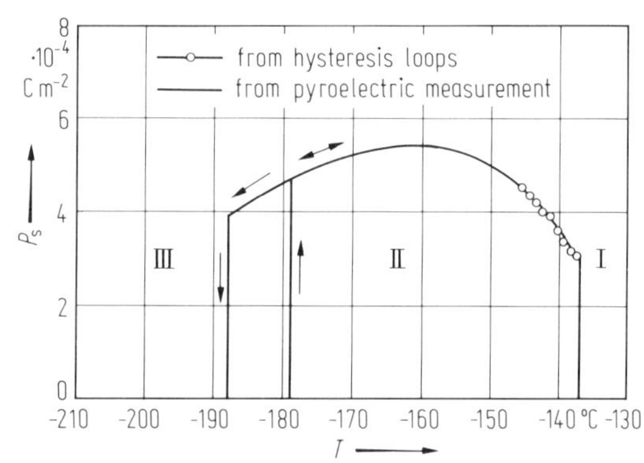

Fig. 787β. $K_2Zn_2(SO_4)_3$. P_s vs. T [80Y1]. Pyroelectric charge method.

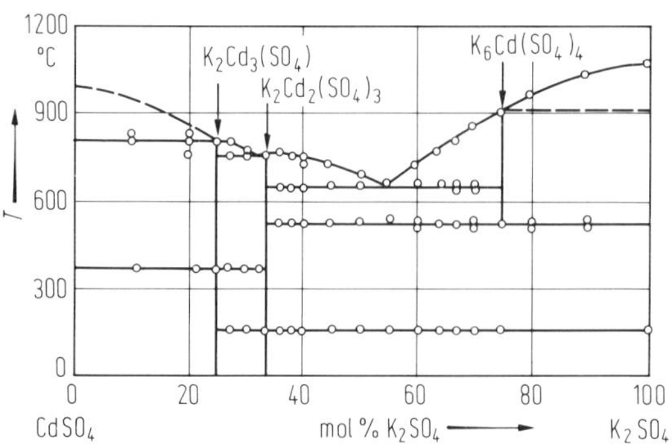

Fig. 788. $K_2Cd_2(SO_4)_3$. Phase diagram of $CdSO_4 - K_2SO_4$ system [77N2].

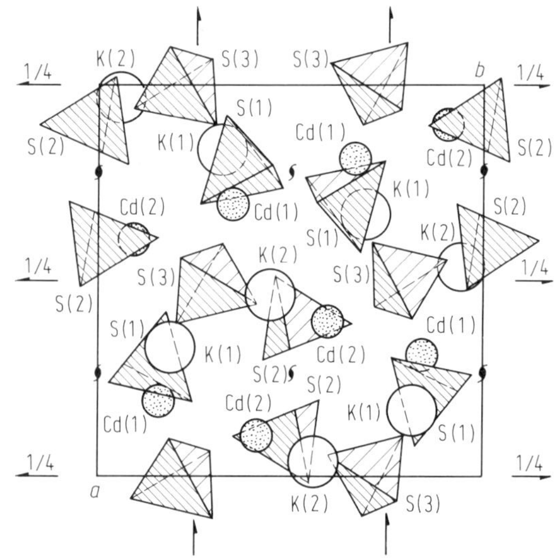

Fig. 789. $K_2Cd_2(SO_4)_3$. Crystal structure in phase II [77A2]. Projection on (001).

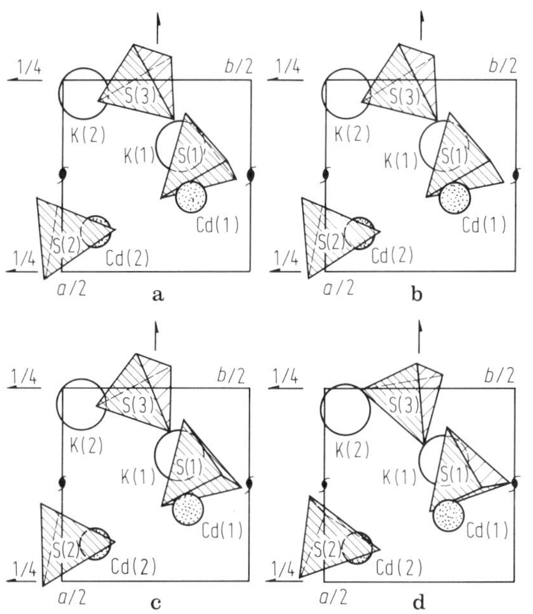

Fig. 790. $K_2Cd_2(SO_4)_3$. Crystal structure [78 A 1]. Projections of asymmetric units on (001). (a) at 298 K, (b) at 390.5 K, (c) at 417.5 K, and (d) at 443.5 K.

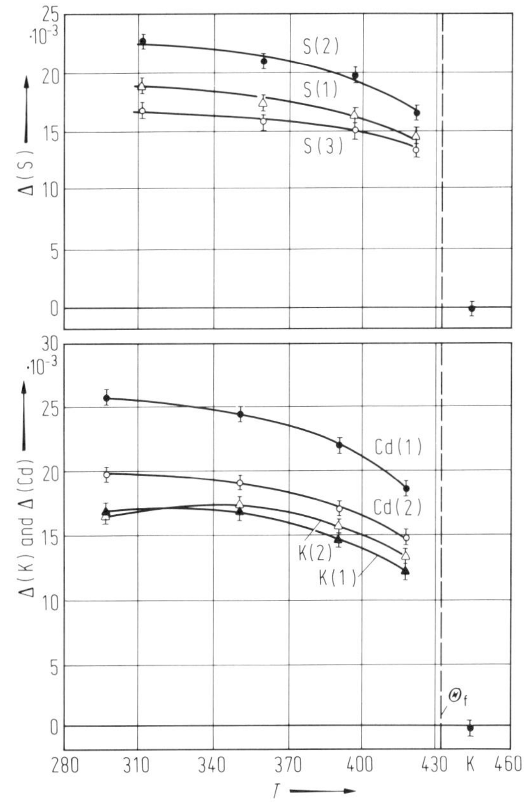

Fig. 791. $K_2Cd_2(SO_4)_3$. Δ vs. T [78 A 1]. Δ(S): displacements of sulfur atoms from the cubic positions and Δ(K) and Δ(Cd): displacements of K and Cd atoms from the [111] axis, in fractional coordinates. Average unit cell parameters are 10.219 Å at 298 K and 10.270 Å at 443.5 K.

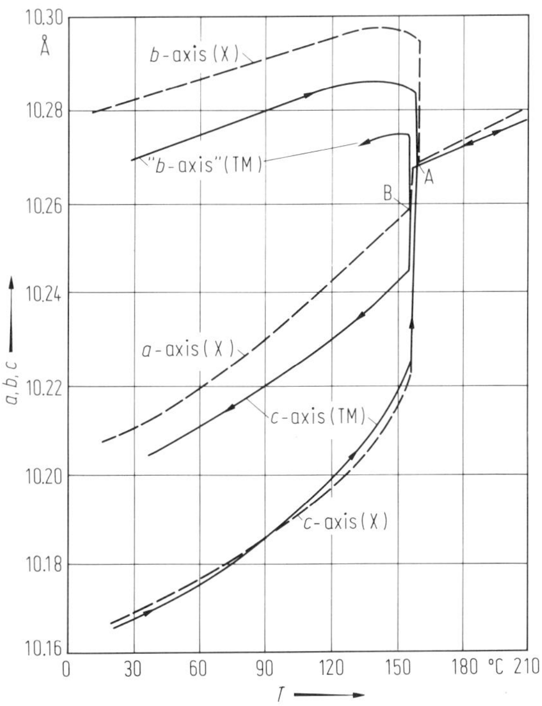

Fig. 792. $K_2Cd_2(SO_4)_3$. Unit cell parameters vs. T [77 N 2]. Full curves (TM): thermal expansion data. Broken curves (X): X-ray diffraction data after F. Lissalde et al. Superposed at point A. On cooling there was an initial contraction of 0.147% from A to B.

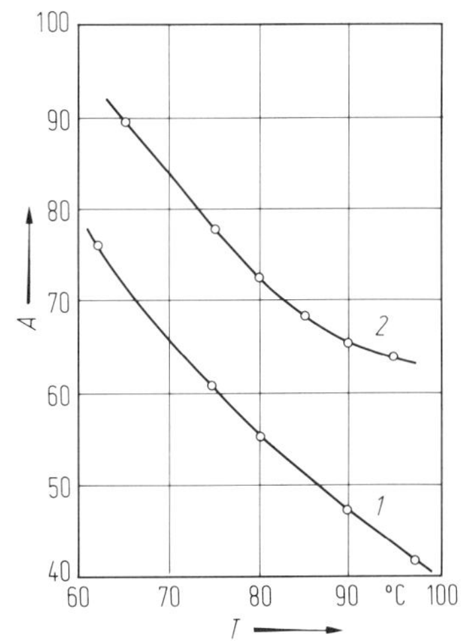

Fig. 793. $Rb_2Cd_2(SO_4)_3$ and $Tl_2Cd_2(SO_4)_3$. A vs. T. [78 B 9]. A: solubility in water (g of salt in 100 g of water). Curve 1: 50 mol% Rb_2SO_4 + 50 mol% $CdSO_4$ (pH = 1.8); Curve 2: 25 mol% Tl_2SO_4 + 75 mol% $CdSO_4$ (pH = 0.8).

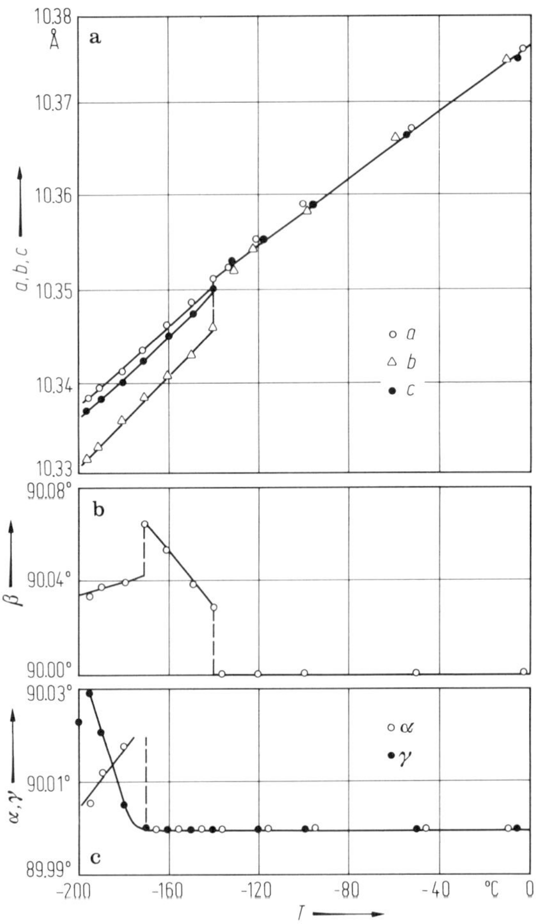

Fig. 794. $Rb_2Cd_2(SO_4)_3$. Unit cell parameters vs. T. [77 Y 3].

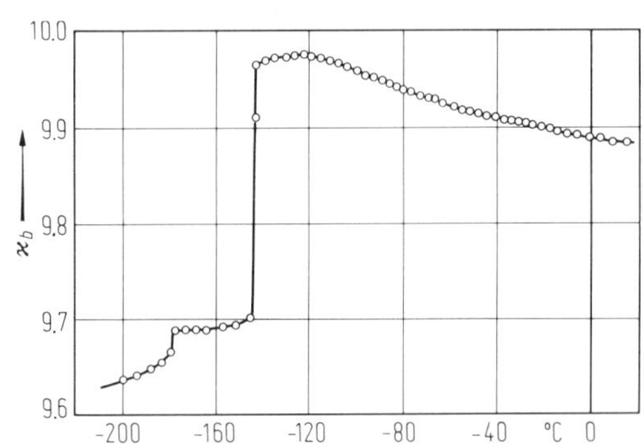

Fig. 795. $Rb_2Cd_2(SO_4)_3$. $\varkappa_b$ vs. T. [79 Y 2]. $f = 1$ kHz.

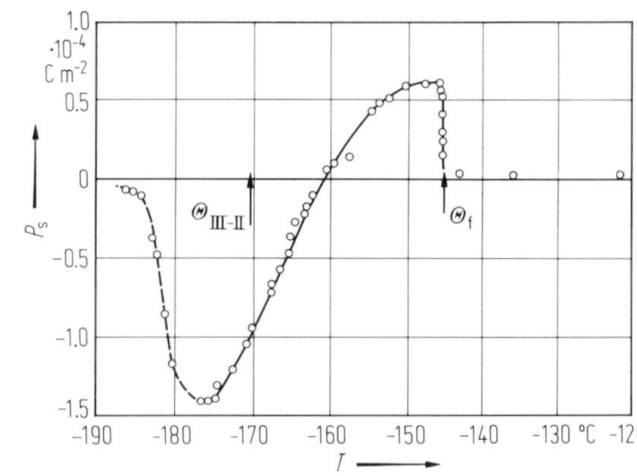

Fig. 796. $Rb_2Cd_2(SO_4)_3$. P_s vs. T. [79 Y 2]. The decrease of $|P_s|$ below -175 °C seems due to generation of polydomains.

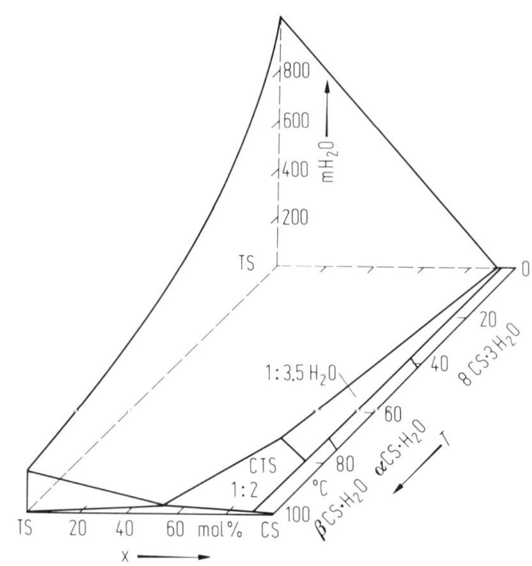

Fig. 797. $Tl_2Cd_2(SO_4)_3$(CTS). Phase diagram of $Tl_2SO_4-CdSO_4-H_2O$ system [74 B 17]. TS = Tl_2SO_4, CS = $CdSO_4$, 1:2 = $Tl_2Cd_2(SO_4)_3$, 1:3.5 H_2O = $Tl_2Cd_3(SO_4)_4 \cdot 5H_2O$. It is noticed that $Tl_2Cd_2(SO_4)_3$ are crystallized only above 80 °C. x: concentration of CS. mH_2O = mole number of H_2O which is in equilibrium with 1/b moles of the solid phase of composition given in the diagram; b equals the sum of the moles of the starting compounds TC and CS, contained in the solid phase.

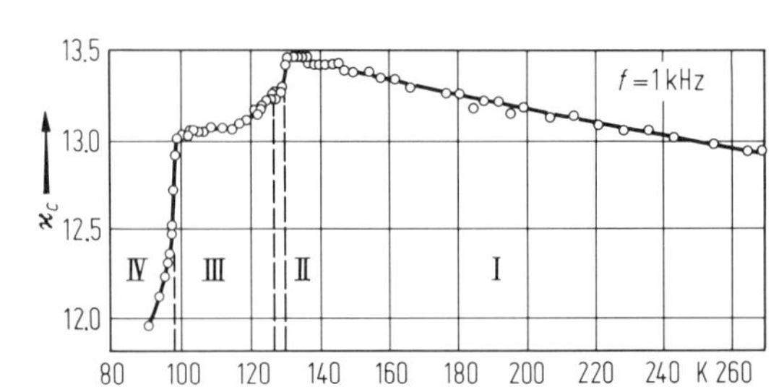

Fig. 798. $Tl_2Cd_2(SO_4)_3$. $\varkappa_c$ vs. T. [72 B 16].

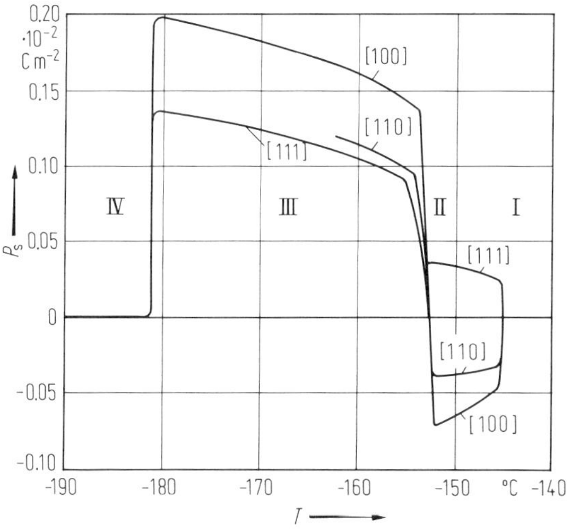

Fig. 799. $Tl_2Cd_2(SO_4)_3$. P_s vs. T. [80Y1]. Pyroelectric charge method. The zone axes indicated in the figure refer to cubic axes.

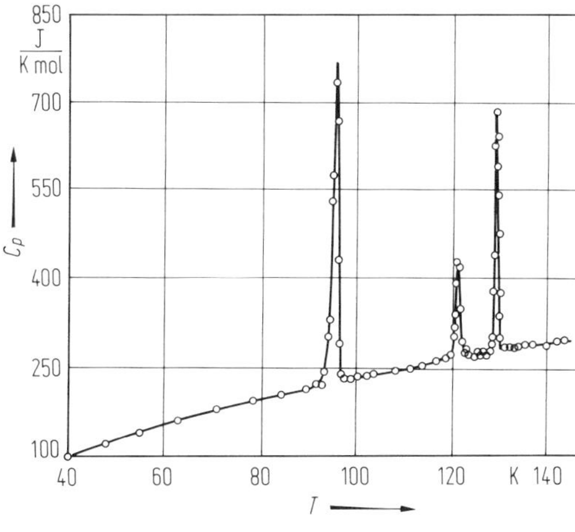

Fig. 800. $Tl_2Cd_2(SO_4)_3$. C_p vs. T. [75F7].

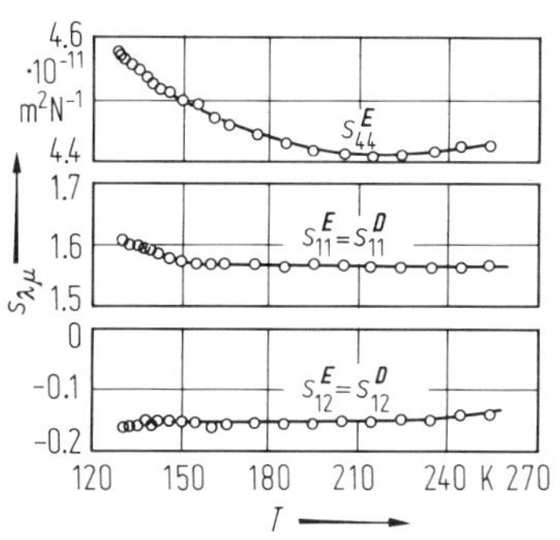

Fig. 802. $Tl_2Cd_2(SO_4)_3$. $s_{\lambda\mu}$ vs. T. ($T > \Theta_f$) [74G8].

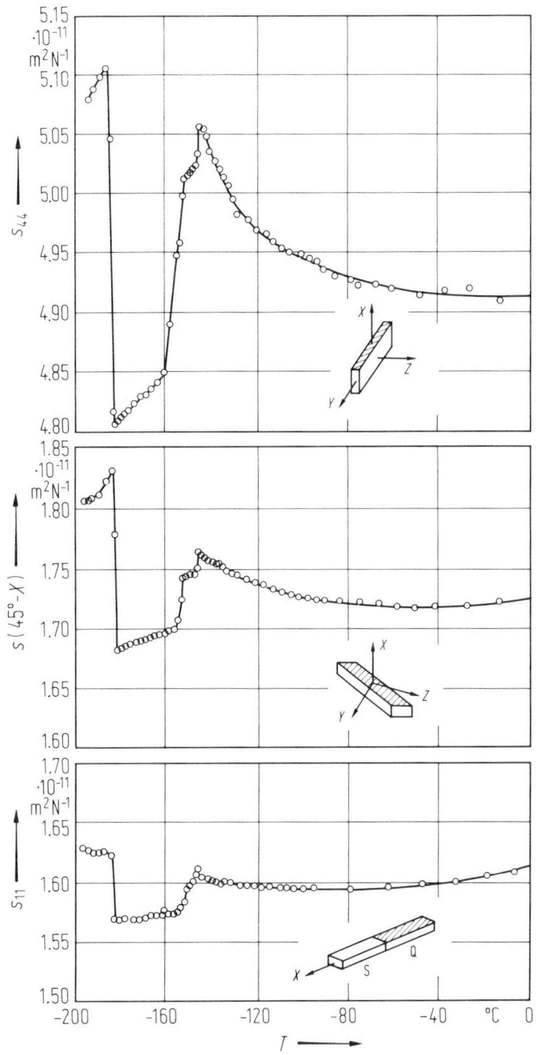

Fig. 801. $Tl_2Cd_2(SO_4)_3$. s vs. T. [75I2]. s: elastic compliance, s(45 °X): that for 45° X-cut bar. S: specimen, Q: quartz transducer.

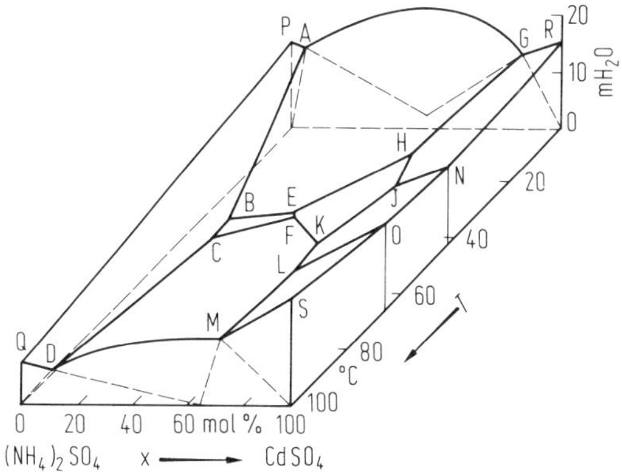

Fig. 803. $(NH_4)_2Cd_2(SO_4)_3$. Phase diagram of $(NH_4)_2SO_4-CdSO_4-H_2O$ system [32B1]. The surface CDMLKF is a crystallization surface for $(NH_4)_2Cd_2(SO_4)_3$. x: concentration of $CdSO_4$.

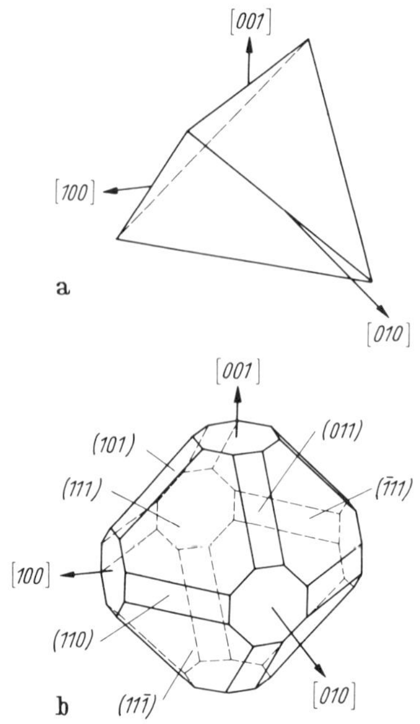

Fig. 804. $(NH_4)_2Cd_2(SO_4)_3$. Crystal forms [57 E 1]. Fig. a: Tetrahedral form; Fig. b: cubic octahedral dodecahedral form.

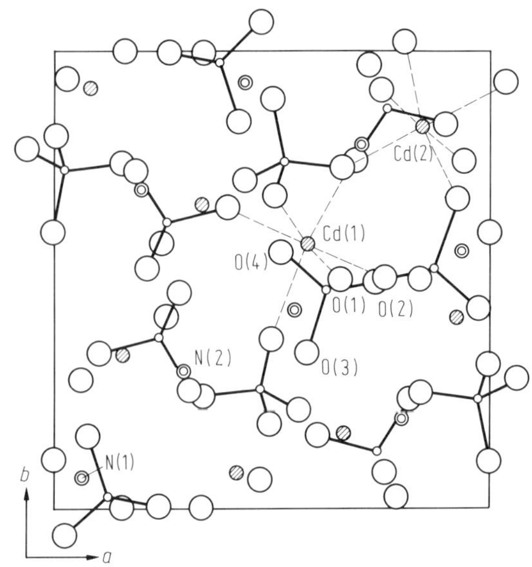

Fig. 805. $(NH_4)_2Cd_2(SO_4)_3$. Crystal structure [75 H 7]. (001) projection.

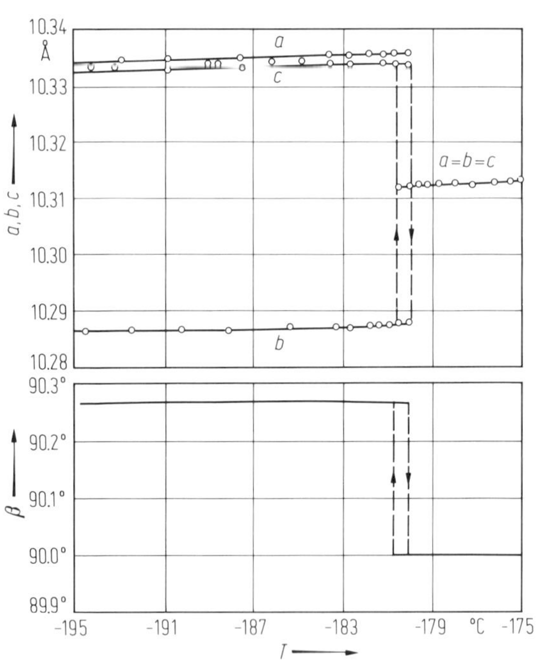

Fig. 806. $(NH_4)_2Cd_2(SO_4)_3$. Unit cell parameters vs. T [78 K 11].

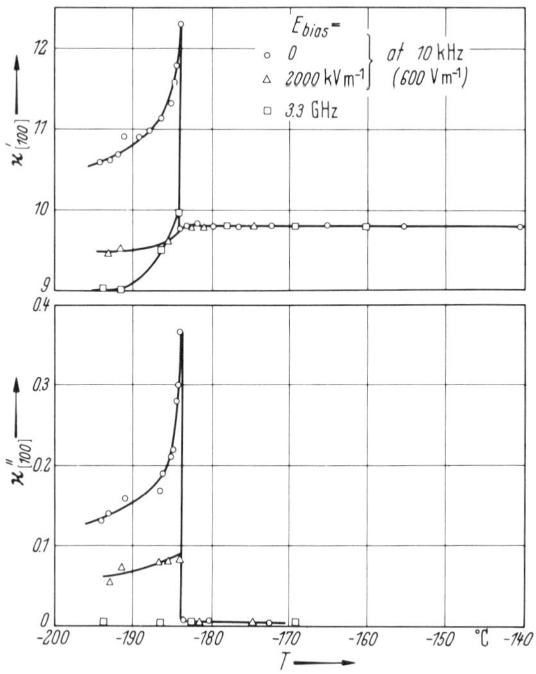

Fig. 807. $(NH_4)_2Cd_2(SO_4)_3$. $\varkappa'_{[100]}$ and $\varkappa''_{[100]}$ vs. T [66 O 1]. $\varkappa'_{[100]}$, $\varkappa''_{[100]}$: real and imaginary parts of $\varkappa_{[100]}$.

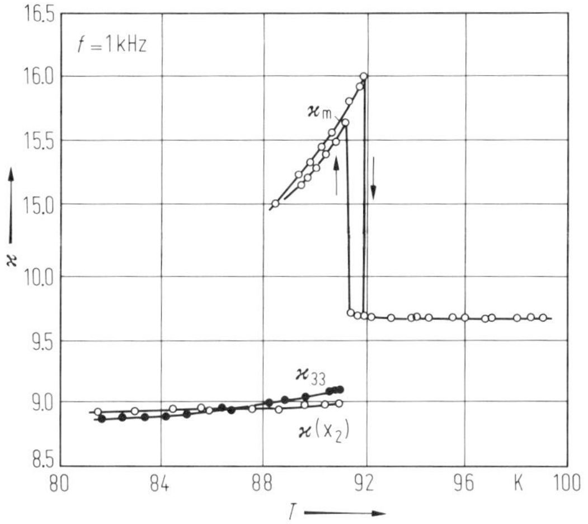

Fig. 808. $(NH_4)_2Cd_2(SO_4)_3$. $\varkappa_{33}$, $\varkappa_m$, and $\varkappa(x_2)$ vs. T [73G9]. $\varkappa_{33}$: dielectric constant along the P_s of single domain crystal, $\varkappa_m$: that of multidomain crystal and $\varkappa(x_2)$: that with field perpendicular of P_s of x_2-domain crystal.

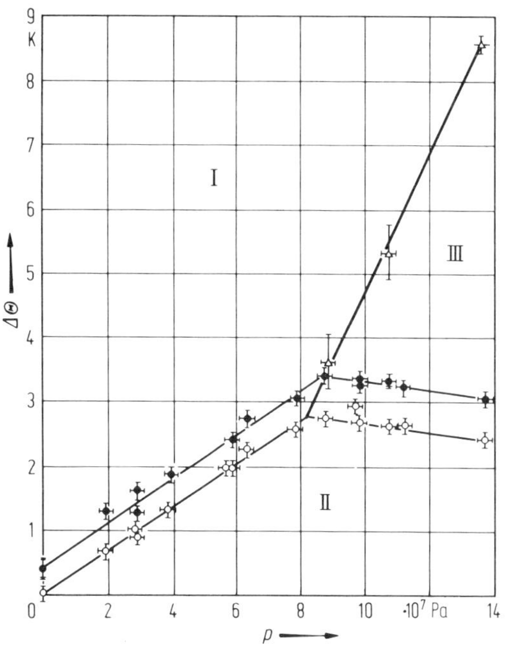

Fig. 809. $(NH_4)_2Cd_2(SO_4)_3$. $\Delta\Theta$ vs. p [72G10]. $\Delta\Theta = \Theta(p) - \Theta_f(0)$. $\Theta_f(0)$: ferroelectric transition temperature at zero pressure. Full circles: on heating; open circles: on cooling.

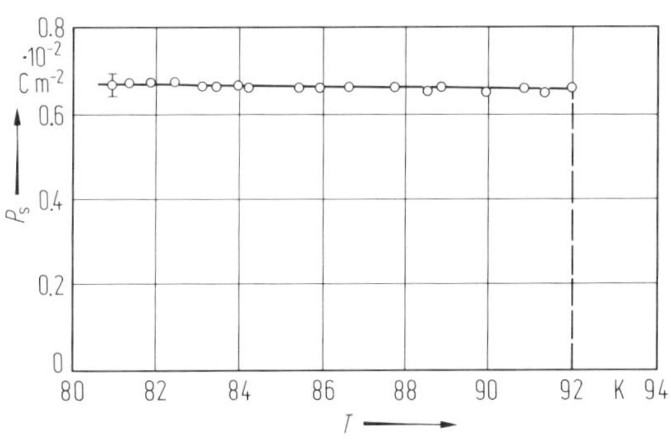

Fig. 810. $(NH_4)_2Cd_2(SO_4)_3$. P_s vs. T [73G9].

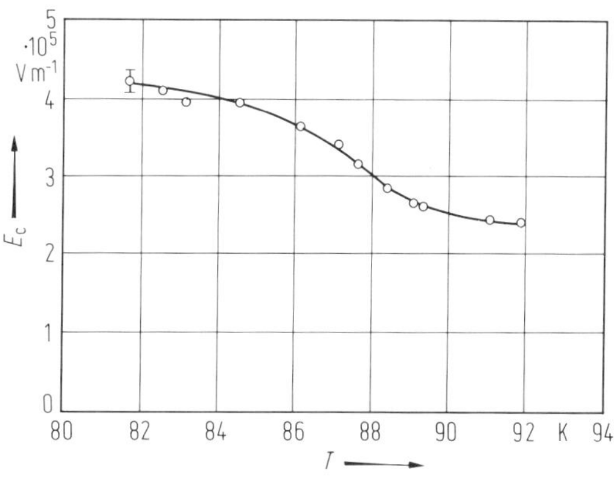

Fig. 811. $(NH_4)_2Cd_2(SO_4)_3$. E_c vs. T [73G9].

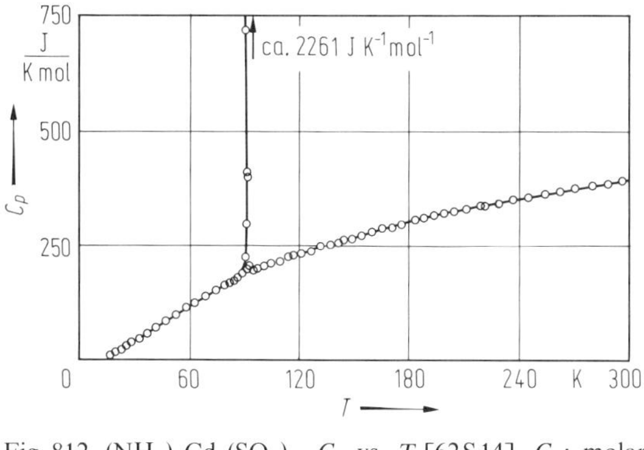

Fig. 812. $(NH_4)_2Cd_2(SO_4)_3$. C_p vs. T [62S14]. C_p: molar heat capacity at constant pressure.

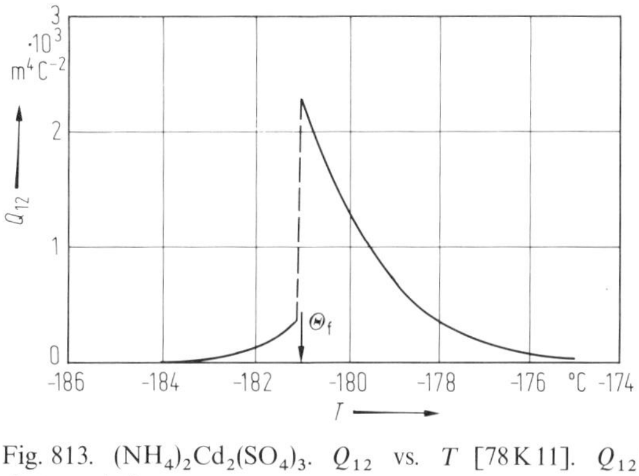

Fig. 813. $(NH_4)_2Cd_2(SO_4)_3$. Q_{12} vs. T [78K11]. Q_{12}: electrostrictive constant.

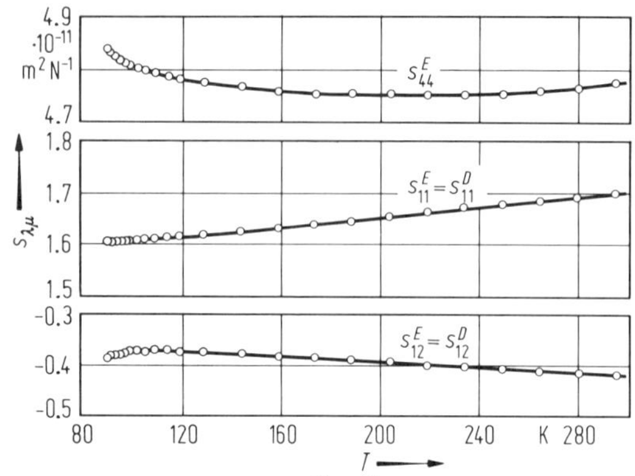

Fig. 814. $(NH_4)_2Cd_2(SO_4)_3$. $s_{\lambda\mu}^E$, $s_{\lambda\mu}^D$ vs. T [72G9].

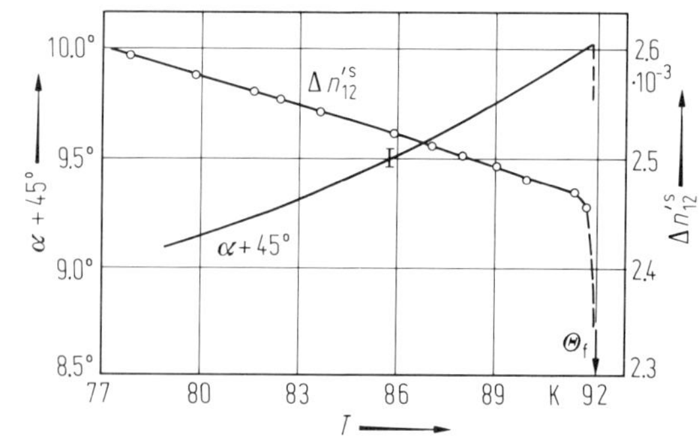

Fig. 815. $(NH_4)_2Cd_2(SO_4)_3$. $\Delta n_{12}^{\prime s}$ and $\alpha + 45°$ vs. T [74K17]. $\Delta n_{12}^{\prime s}$: spontaneous birefringence, $\alpha + 45°$: spontaneous rotation angle. $\lambda = 633$ nm.

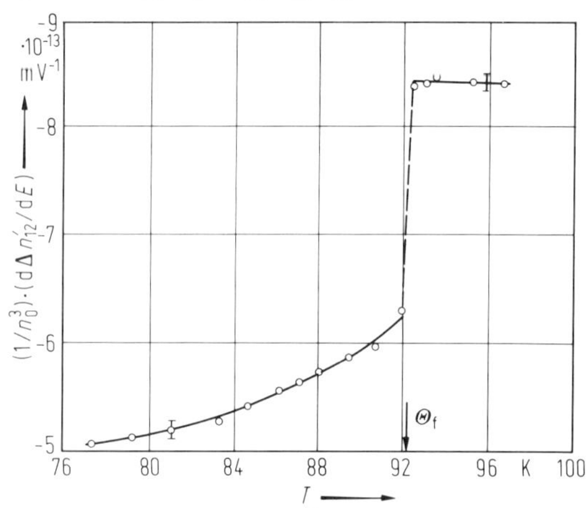

Fig. 816. $(NH_4)_2Cd_2(SO_4)_3$. $(1/n_0^3)$ $(d\Delta n_{12}'/dE)$ vs. T [74K17].

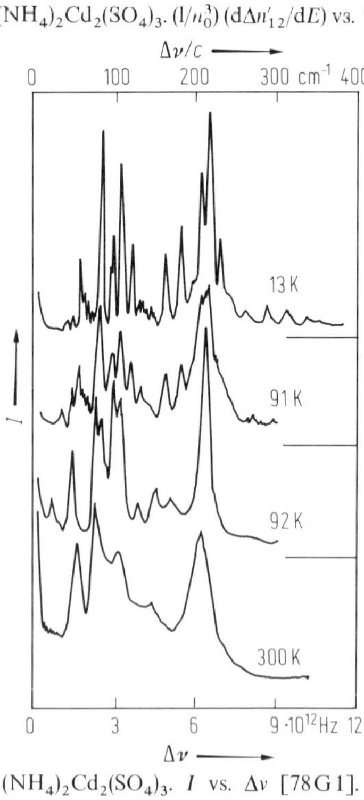

Fig. 819. $(NH_4)_2Cd_2(SO_4)_3$. I vs. $\Delta\nu$ [78G1]. I: Raman scattering intensity in the scattering geometry $Y(Z, Z)X$.

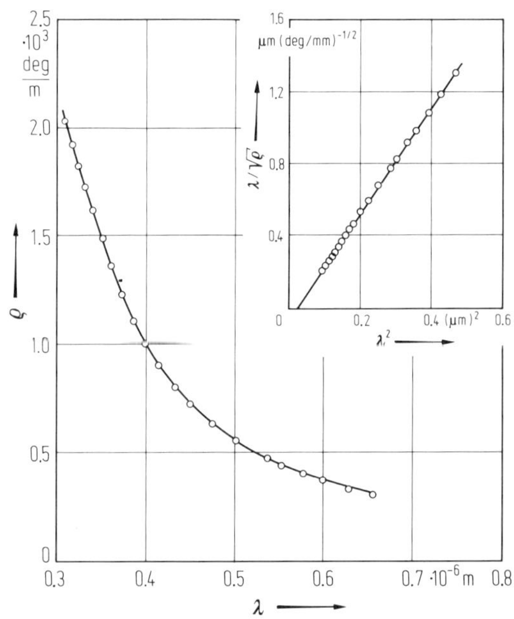

Fig. 817. $(NH_4)_2Cd_2(SO_4)_3$. ϱ vs. λ [74I15]. ϱ: optical rotatory power.

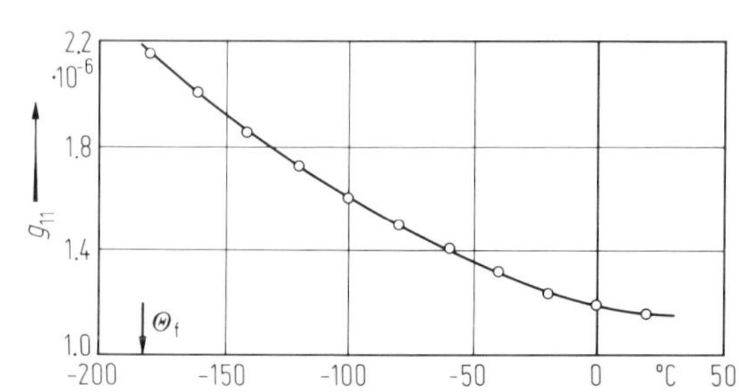

Fig. 818. $(NH_4)_2Cd_2(SO_4)_3$. g_{11} vs. T [78K11]. g_{11}: component of optical gyration tensor.

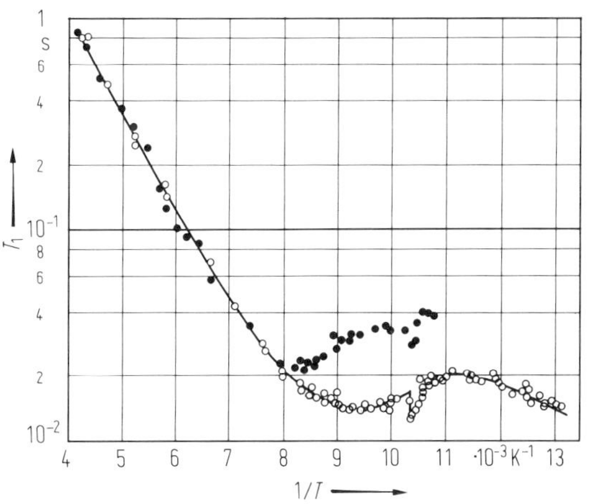

Fig. 820. $(NH_4)_2Cd_2(SO_4)_3$. T_1 vs. $1/T$ [75M8]. T_1: proton spin-lattice relaxation time. Open circles: $f = 30.8$ MHz, full circles: $f = 48.2$ MHz.

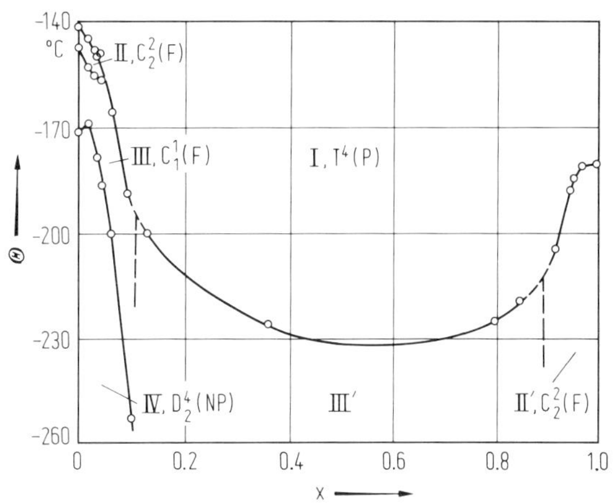

Fig. 821. $[Tl_{1-x}(NH_4)_x]Cd_2(SO_4)_3$. Θ vs. x [77I6]. I: $P2_13 - T^4$ (paraelectric), II and II′: $P2_1 - C_2^2$ (ferroelectric), III: $P1 - C_1^1$ (ferroelectric), IV: $P2_12_12_1 - D_2^4$ (nonpolar = NP) and III′: probably the same as III.

Fig. 822, see p. 510.

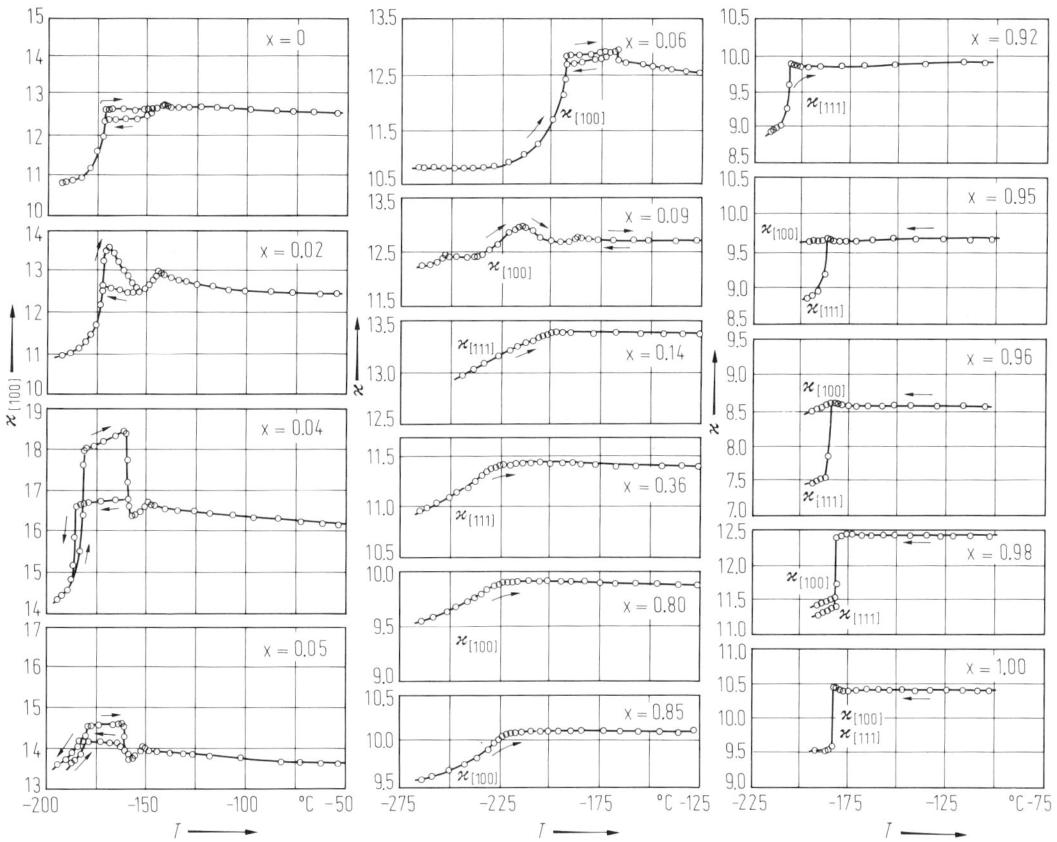

Fig. 823. $[Tl_{1-x}(NH_4)_x]_2Cd_2(SO_4)_3$. $\varkappa$ vs. T [77I6]. Parameter: x. $f = 1$ kHz.

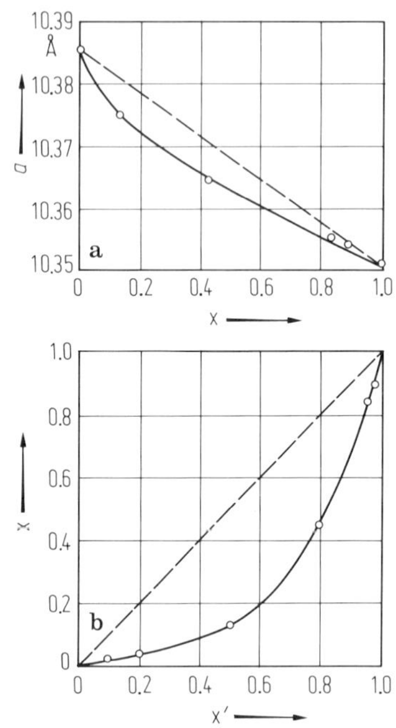

Fig. 822. $[Tl_{1-x}(NH_4)_x]_2Cd_2(SO_4)_3$. Fig. a: a vs. x and Fig. b: distribution curve [7716]. a: unit cell parameter in the cubic phase, x: composition in crystal, x': composition in solution.

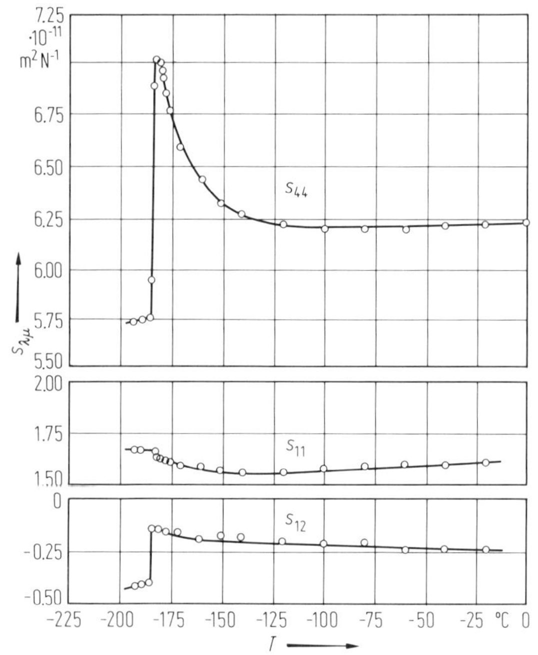

Fig. 825. $[Tl_{0.02}(NH_4)_{0.98}]_2Cd_2(SO_4)_3$. $s_{\lambda\mu}$ vs. T [7716].

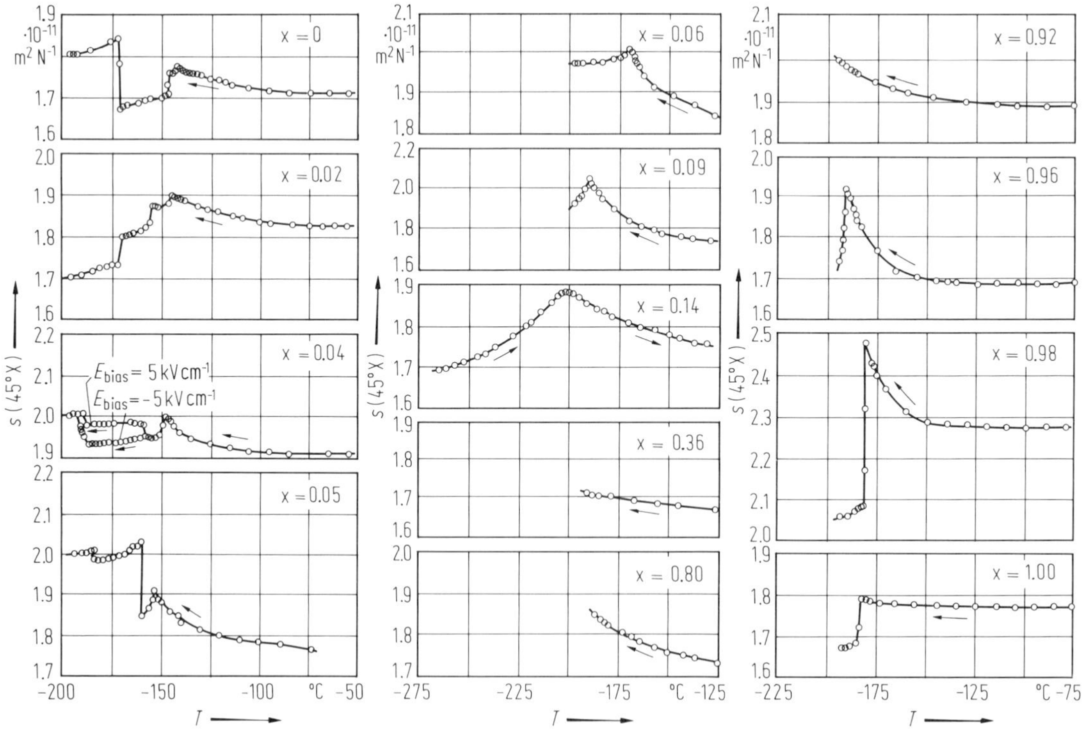

Fig. 824. $[Tl_{1-x}(NH_4)_x]_2Cd_2(SO_4)_3$. $s(45\ °X)$ vs. T [7716]. Parameter: x. $s(45\ °X)$: elastic compliance of 45 °X-cut bar.

25 $NaNH_4SO_4 \cdot 2H_2O$ (lecontite) family

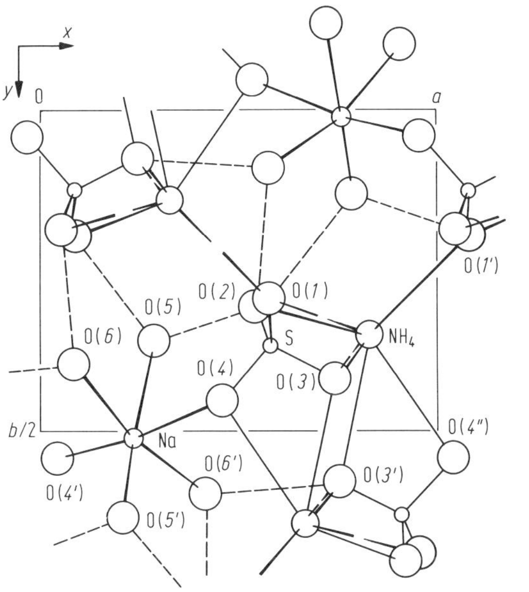

Fig. 826. $NaNH_4SO_4 \cdot 2H_2O$ (Lecontite). Crystal structure of phase I [67C7]. Projection on (001). O(3'') is probably behind O(3). See Table 189.

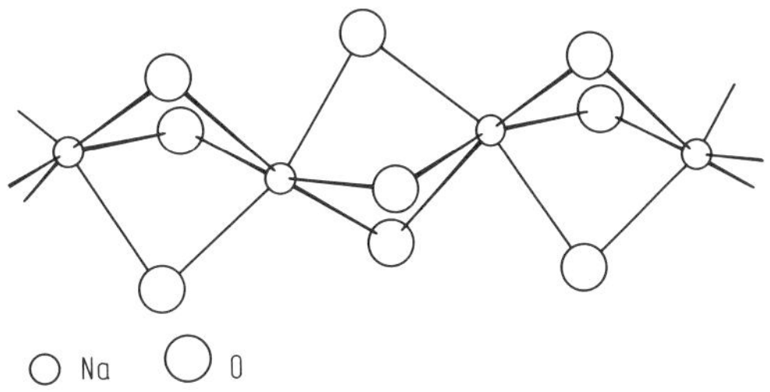

Fig. 827. $NaNH_4SO_4 \cdot 2H_2O$ (Lecontite). Crystal structure of phase I [67C7]. Chains of NaO_6 octahedra parallel to *c*.

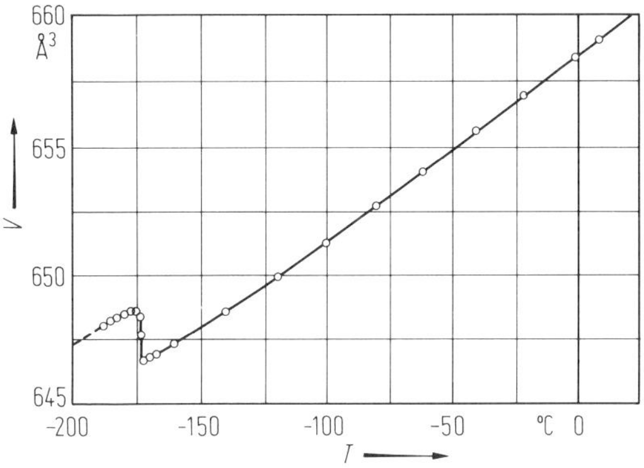

Fig. 828. $NaNH_4SO_4 \cdot 2H_2O$ (Lecontite). *V* vs. *T* [78O6]. *V*: volume of the unit cell.

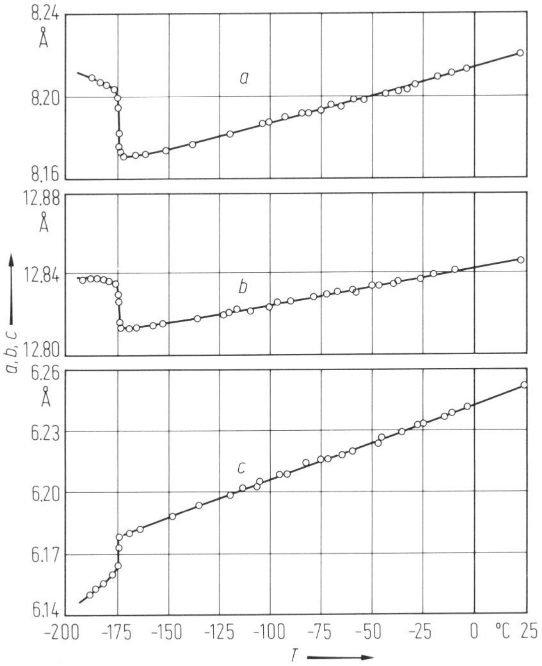

Fig. 829. $NaNH_4SO_4 \cdot 2H_2O$ (Lecontite). *a*, *b*, *c* vs. *T* [78O6]. *a*, *b*, *c*: unit cell parameters.

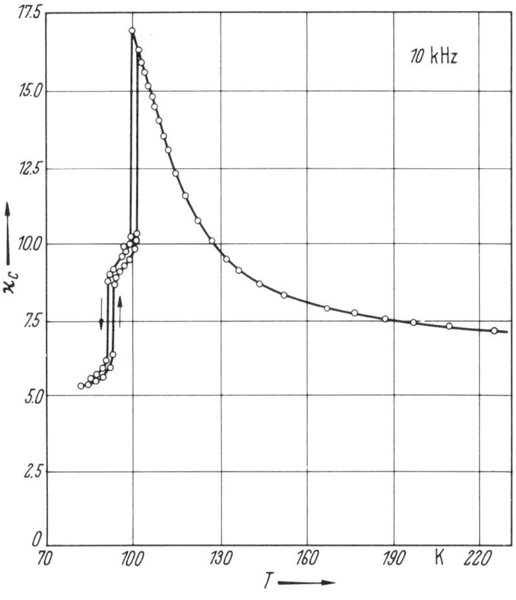

Fig. 830. $NaNH_4SO_4 \cdot 2H_2O$ (Lecontite). $\varkappa_c$ vs. *T* (f = 10 kHz) [65M4].

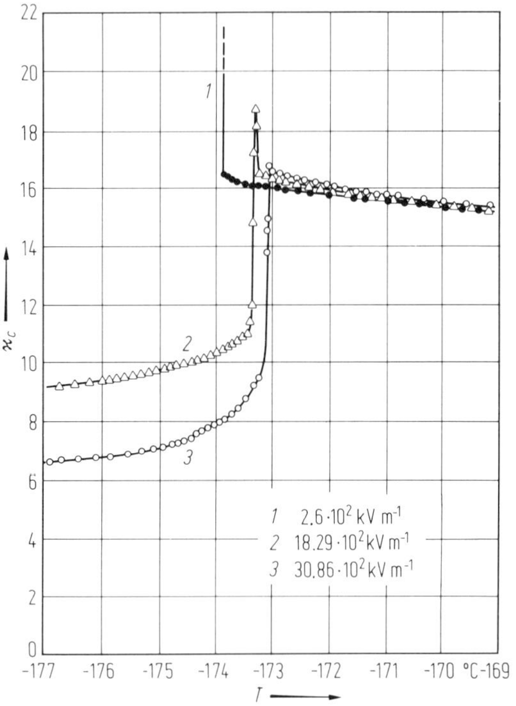

Fig. 831. $NaNH_4SO_4 \cdot 2H_2O$ (Lecontite). $\varkappa_c$ vs. T [78O6]. Parameter: E_{bias}.

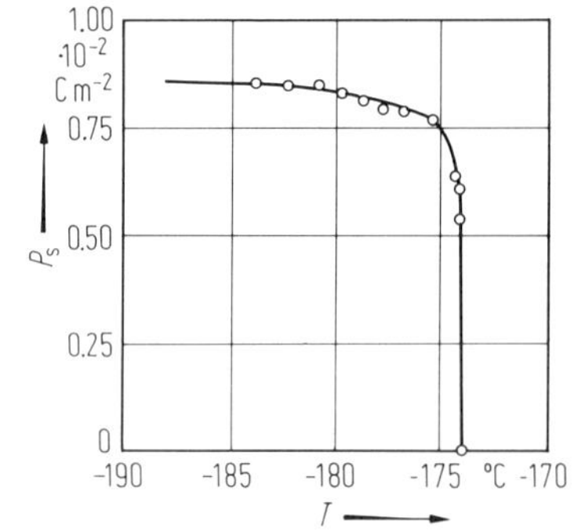

Fig. 832. $NaNH_4SO_4 \cdot 2H_2O$ (Lecontite). P_s vs. T [78O6].

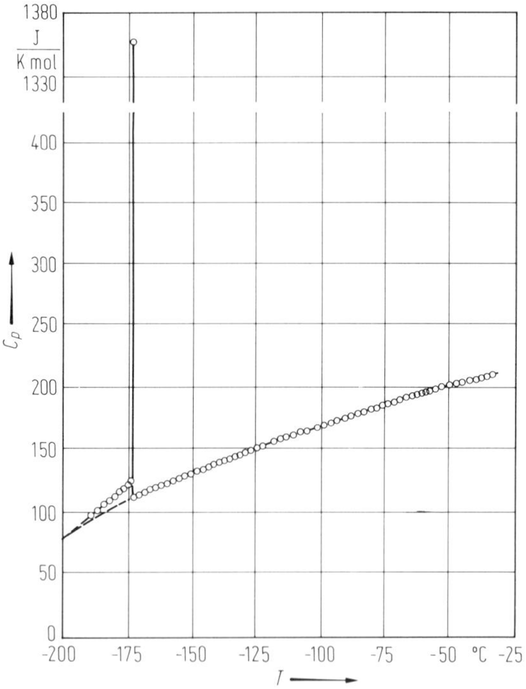

Fig. 833. $NaNH_4SO_4 \cdot 2H_2O$ (Lecontite). C_p vs. T [78O6].

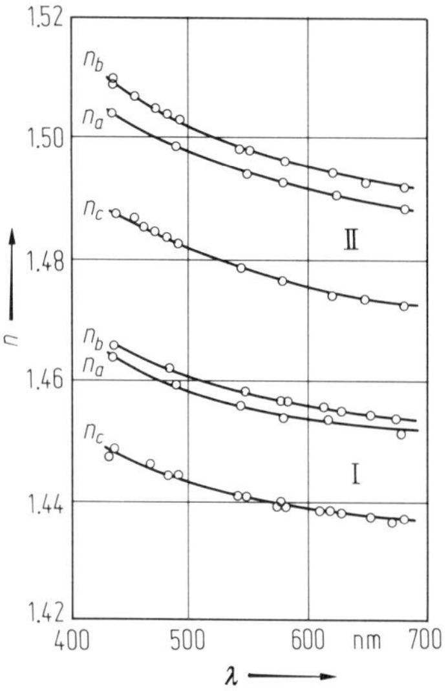

Fig. 834. $NaNH_4SO_4 \cdot 2H_2O$, $NaNH_4SeO_4 \cdot 2H_2O$. n vs. λ [75A11]. I: $NaNH_4SO_4 \cdot 2H_2O$, II: $NaNH_4SeO_4 \cdot 2H_2O$.

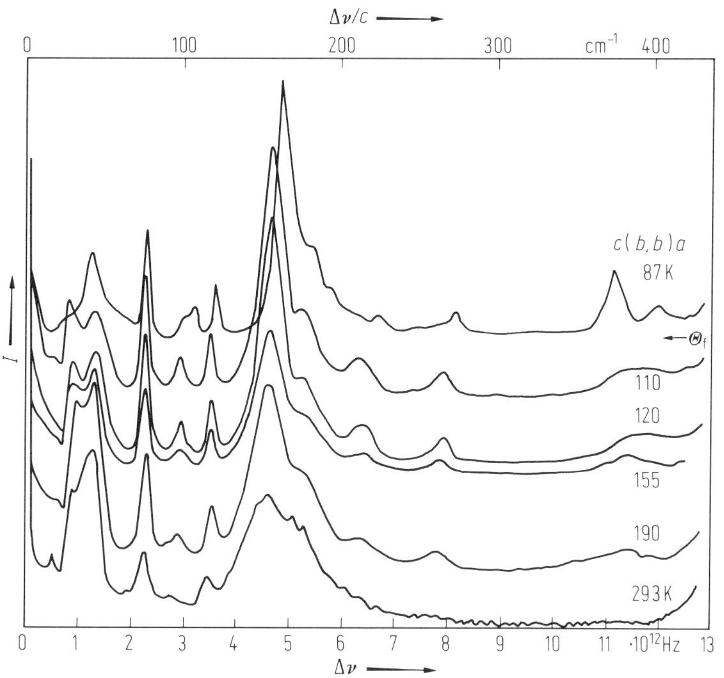

Fig. 835. $NaNH_4SO_4 \cdot 2H_2O$ (Lecontite). I vs. $\Delta\nu$ [75F3]. I: Raman scattering intensity for the scattering geometry of $c(b, b)a$. Parameter: temperature.

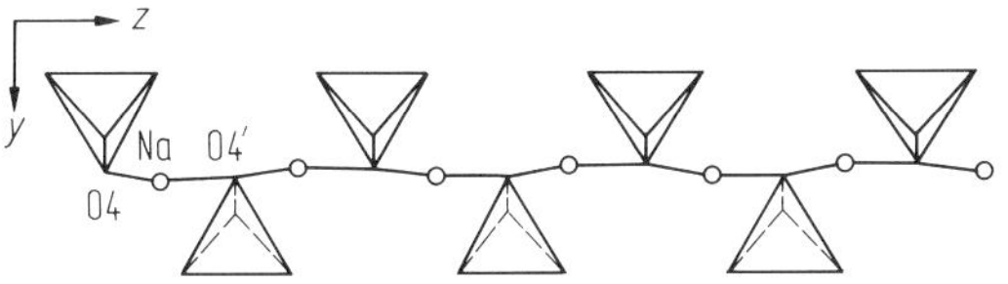

Fig. 839. $NaNH_4SeO_4 \cdot 2H_2O$. Crystal structure of phase I [73K18]. SeO_4 tetrahedra linked by Na ions viewed along [100].

Fig. 838, see p. 514.

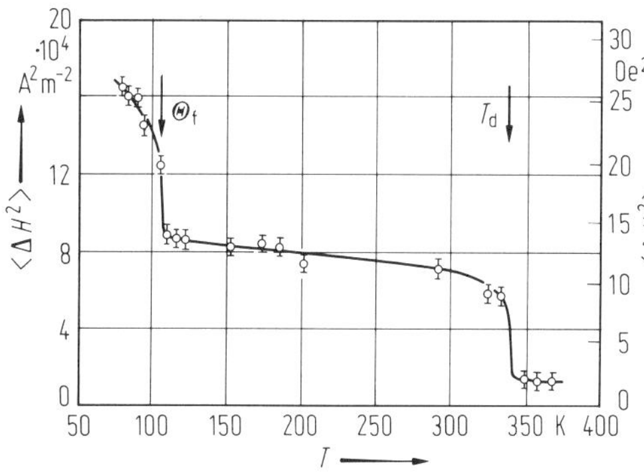

Fig. 836. $NaNH_4SO_4 \cdot 2H_2O$ (Lecontite). $\langle \Delta H^2 \rangle$ vs. T [67M1]. $\langle \Delta H^2 \rangle$: second moment of magnetic resonance curve of proton. T_d: Dehydration temperature.

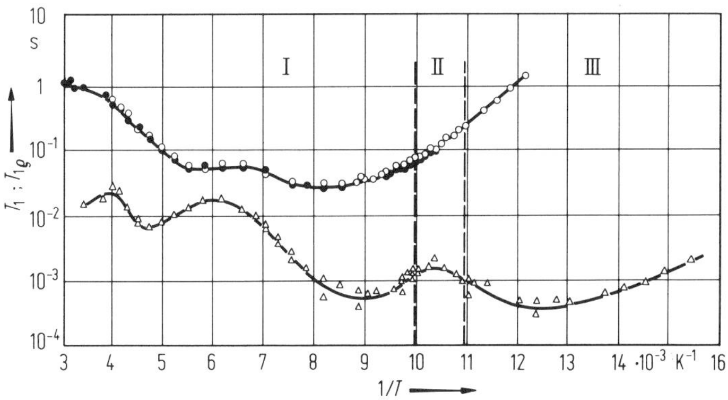

Fig. 837. $NaNH_4SO_4 \cdot 2H_2O$ (Lecontite). T_1, $T_{1\varrho}$ vs. $1/T$ [69G2]. Open circles: proton T_1 for powder; solid circles: proton T_1 for single crystal, a axis $\| H$. $f = 42$ MHz; triangles: proton $T_{1\varrho}$ for powder, $H_1 = 1.6 \cdot 10^3\ A \cdot m^{-1}$

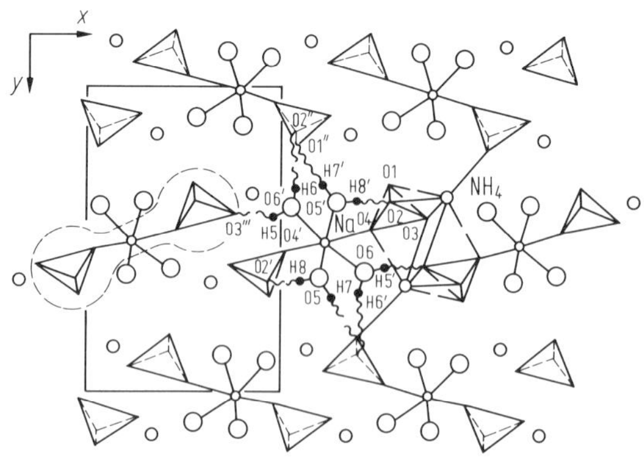

Fig. 838. $NaNH_4SeO_4 \cdot 2H_2O$. Crystal structure of phase I [73K18]. Projection on (001). SeO_4 tetrahedra are delineated. Bonds of Na and NH_4 with their surrounding oxygens are shown by straight lines and hydrogen bonds by wavy lines. Bonds shown by broken lines indicate those translationally identical along [001].

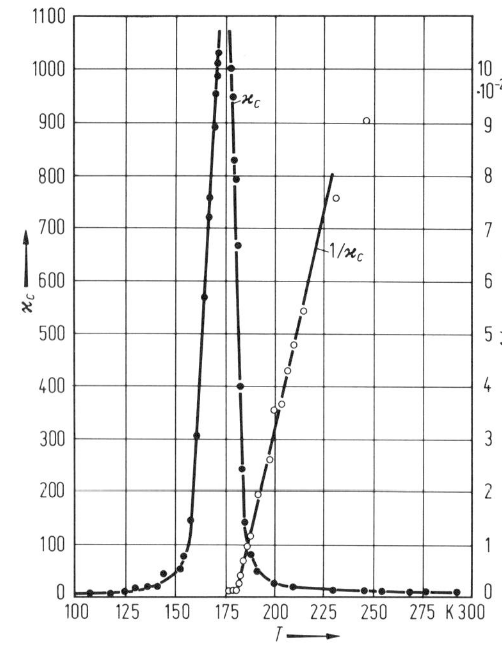

Fig. 840. $NaNH_4SeO_4 \cdot 2H_2O$. $\varkappa_c$, $1/\varkappa_c$ vs. T [71A8]. $\varkappa_c$: low frequency dielectric constant along c axis.

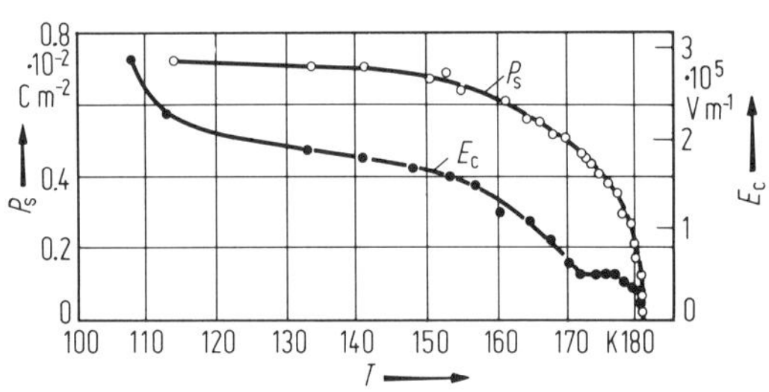

Fig. 841. $NaNH_4SeO_4 \cdot 2H_2O$. P_s, E_c vs. T [71Z1].

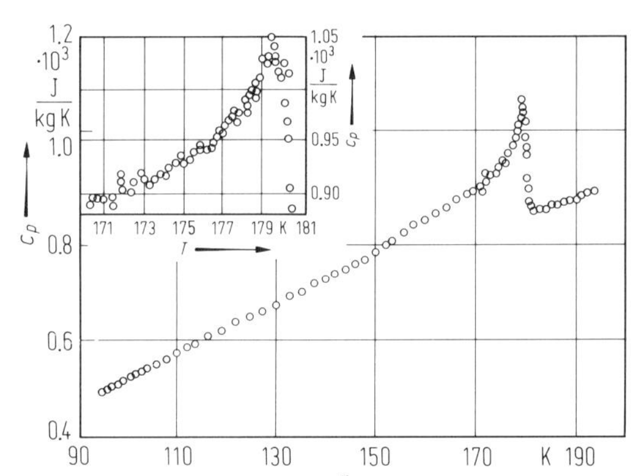

Fig. 842. $NaNH_4SeO_4 \cdot 2H_2O$. c_p vs. T [72A7]. c_p: specific heat capacity at constant pressure.

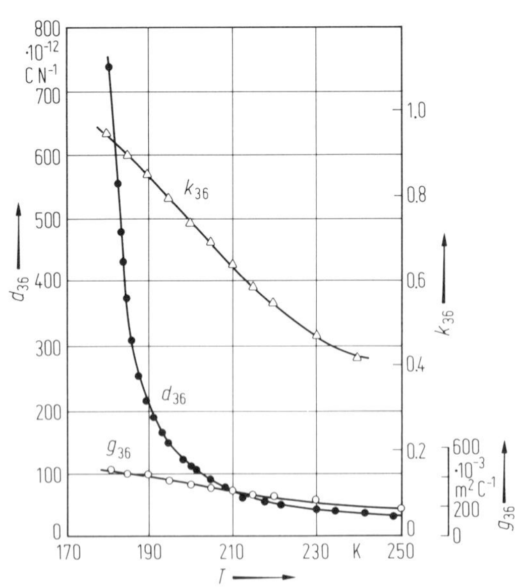

Fig. 843. $NaNH_4SeO_4 \cdot 2H_2O$. k_{36}, d_{36}, g_{36} vs. T [75Z1].

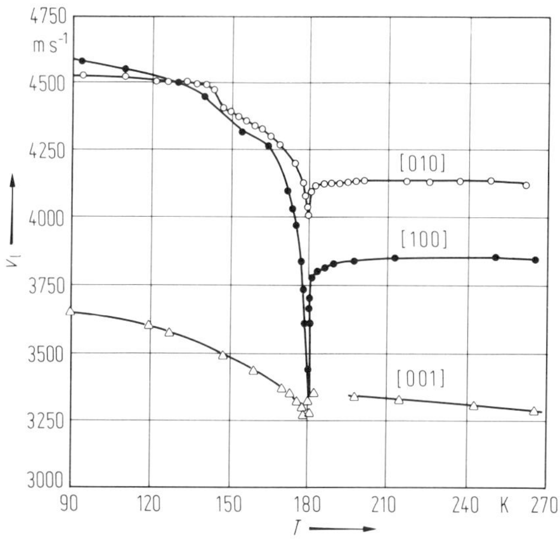

Fig. 844. $NaNH_4SeO_4 \cdot 2H_2O$. v_l vs. T [75 K 24]. v_l: velocities of the longitudinal elastic waves propagating along the [100], [010], and [001] directions.

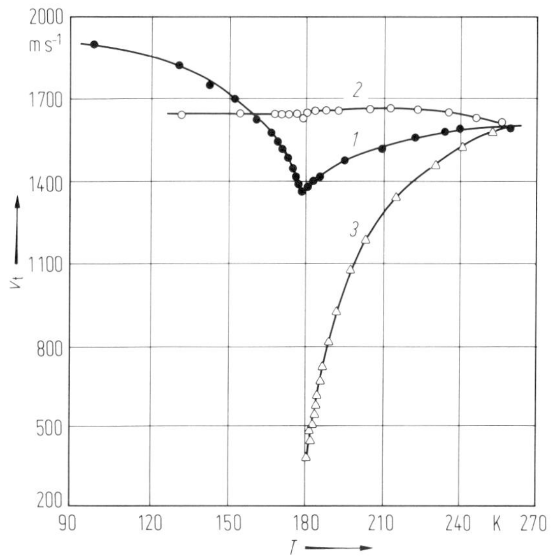

Fig. 845. $NaNH_4SeO_4 \cdot 2H_2O$. v_t vs. T [75 K 24]. v_t: sound velocities of the transverse waves with the wave number vector $\boldsymbol{q}$ and polarization vector $\boldsymbol{e}$. Curve 1: $\boldsymbol{q}\|[001]$, $\boldsymbol{e}\|[100]$; 2: $\boldsymbol{q}\|[010]$, $\boldsymbol{e}\|[001]$; 3: $\boldsymbol{q}\|[100]$, $\boldsymbol{e}\|[010]$.

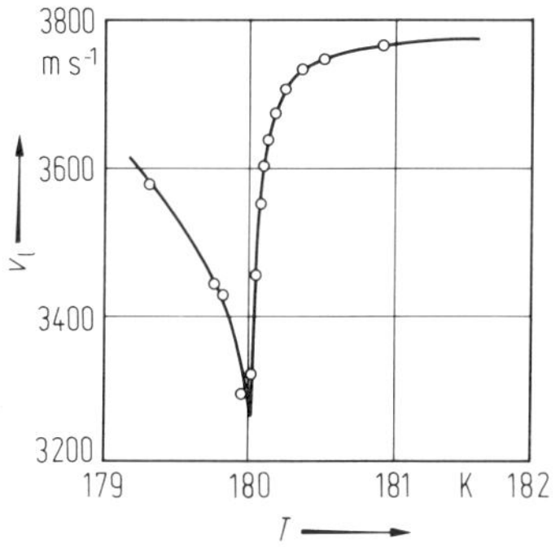

Fig. 846. $NaNH_4SeO_4 \cdot 2H_2O$. v_l vs. T [75 K 24]. v_l: sound velocity of the longitudinal wave propagating along [100] near Θ_f.

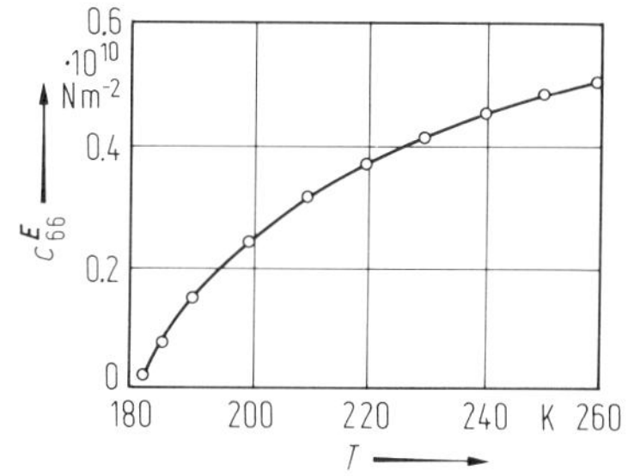

Fig. 847. $NaNH_4SeO_4 \cdot 2H_2O$. c^E_{66} vs. T [75 K 24]. c^E_{66}: elastic stiffness at constant $\boldsymbol{E}$.

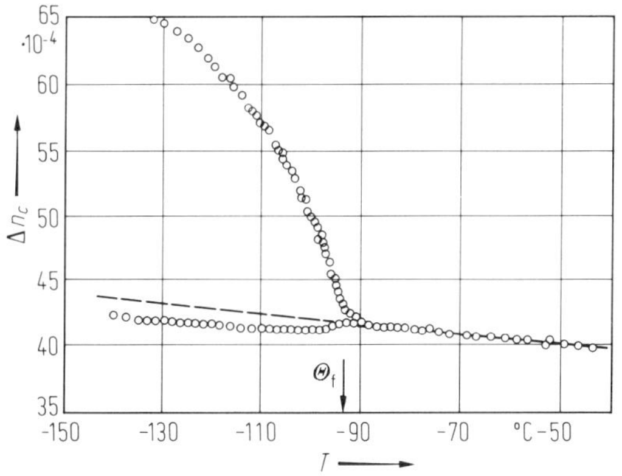

Fig. 848. $NaNH_4SeO_4 \cdot 2H_2O$. Δn_c vs. T [74 A 6]. Δn_c: birefringence. Curve 1: $E_{bias}=0$; 2: $E_{bias}=3 \cdot 10^3$ kV · m^{-1}. $\lambda=633$ nm.

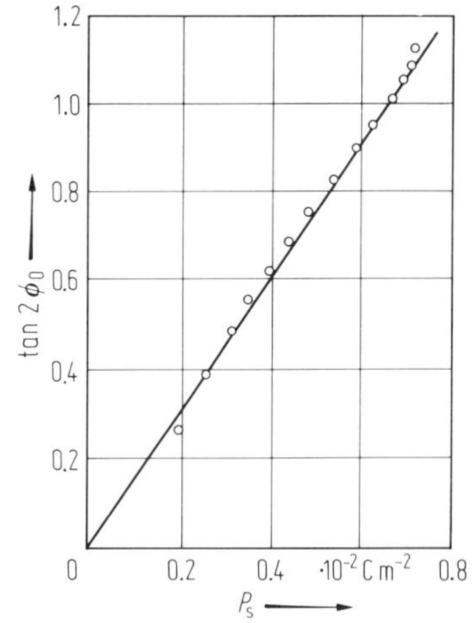

Fig. 851. $NaNH_4SeO_4 \cdot 2H_2O$. tan $2\phi_0$ vs. P_s [73 A 6]. ϕ_0: rotation angle of the optical indicatrix around the c axis.

Figs. 849, 850, see p. 516.

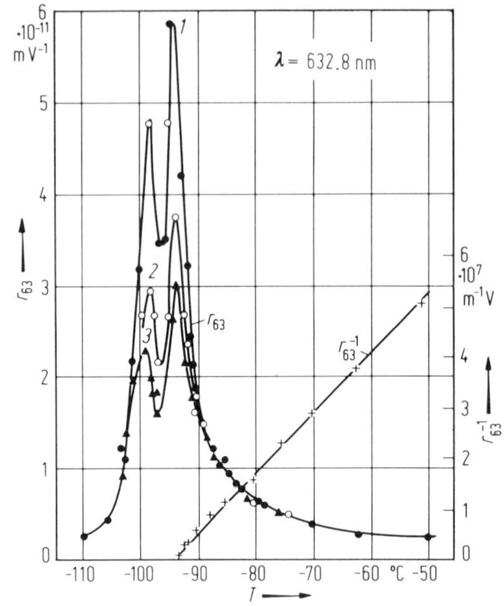

Fig. 849. $NaNH_4SeO_4 \cdot 2H_2O$. r_{63}, r_{63}^{-1} vs. T [72A12]. Curve 1: $E_{bias}=0$; 2: $E_{bias}=1\cdot10^5\,Vm^{-1}$; 3: $E_{bias}=4\cdot10^5\,Vm^{-1}$.

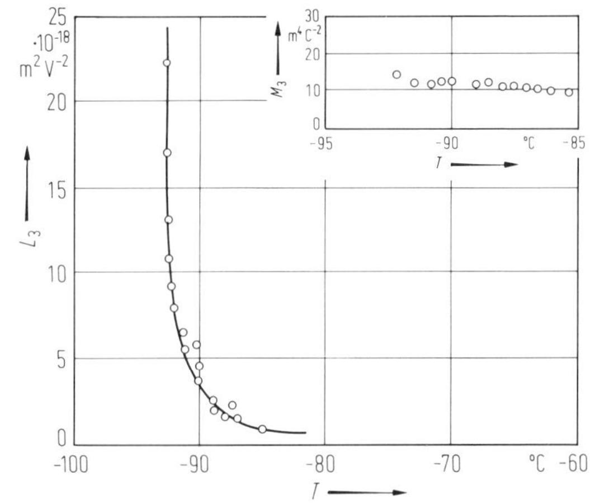

Fig. 850. $NaNH_4SeO_4 \cdot 2H_2O$. L_3, M_3 vs. T [74A6]. $L_3=n_b^3L_{23}-n_a^3L_{13}$, $M_3=n_b^3M_{23}-n_a^3M_{13}$. $L_{\lambda\mu}$: quadratic electrooptic constant for E; $M_{\lambda\mu}$: quadratic electrooptic constant for $\boldsymbol{P}$. $\lambda=633$ nm.

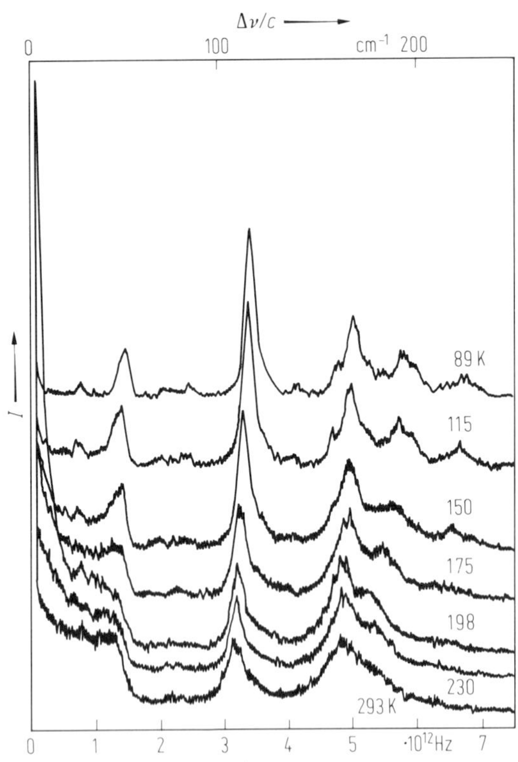

Fig. 852. $NaNH_4SeO_4 \cdot 2H_2O$. I vs. $\Delta\nu$ [75F2]. I: Raman scattering intensity in the low frequency region. Scattering geometry is $c(b,a)b$. Parameter: T.

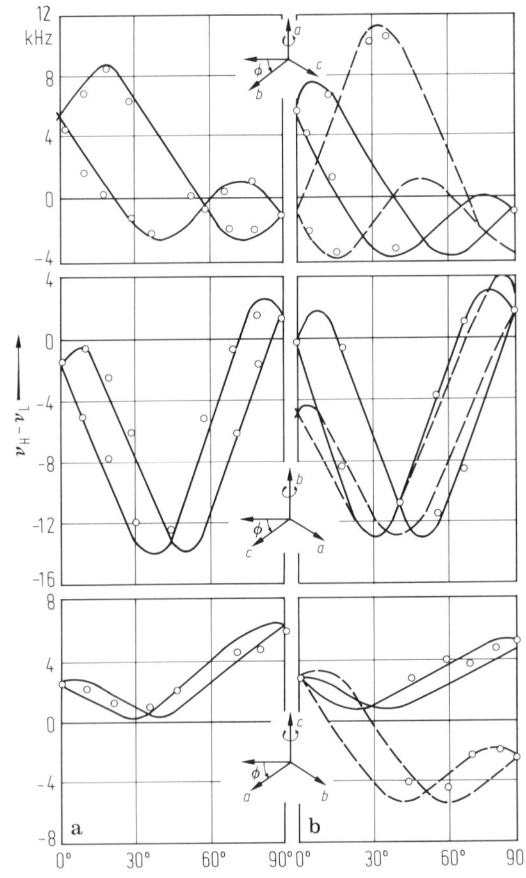

Fig. 853. $NaNH_4SeO_4 \cdot 2H_2O$. $\nu_H-\nu_L$ vs. ϕ [75S23]. $\nu_H-\nu_L$: Shift in the spectrum of ^{23}Na on rotation around the a, b and c axes. a) phase I, $T=-15\,°C$ (points for $p=9\cdot10^8$ Pa, solid lines for p=atmospheric pressure); b) phase II. (Points for $p=4.5\cdot10^8$ Pa, $T=-144\,°C$; solid and broken lines for p=atmospheric pressure, $T=-120\,°C$.) Phase II has two different Na sites.

26 Alum ($NH_4Fe(SO_4)_2 \cdot 12\,H_2O$) family

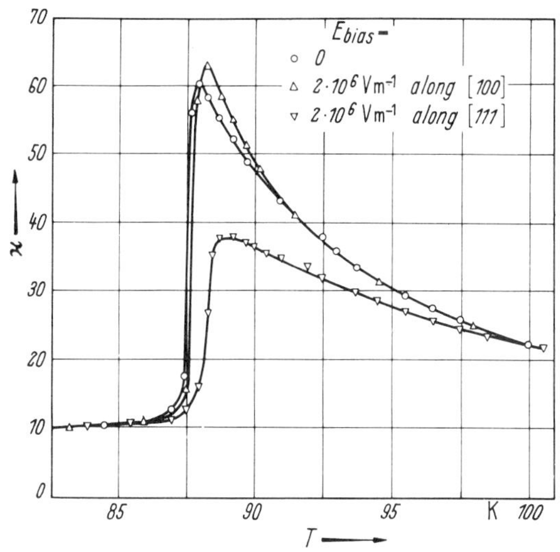

Fig. 854. $NH_4Fe(SO_4)_2 \cdot 12H_2O$. $\varkappa$ vs. T [57 J 5]. Parameter: E_{bias}. $f = 10$ kHz.

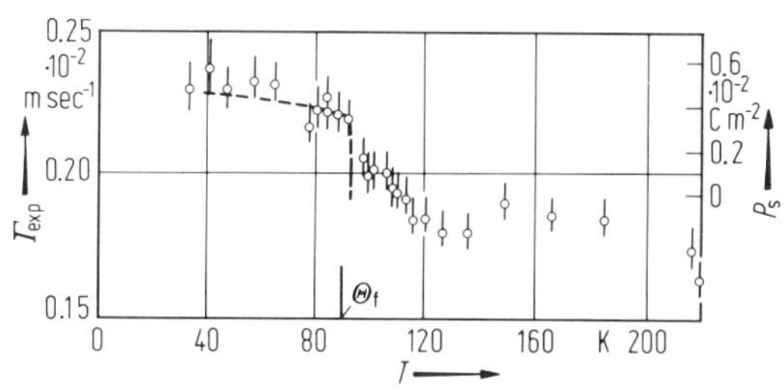

Fig. 855. $NH_4Fe(SO_4)_2 \cdot 12H_2O$. Line width of Mössbauer absorption line Γ_{exp} and P_s vs. T [71 M 11]. γ-rays are parallel to [100].

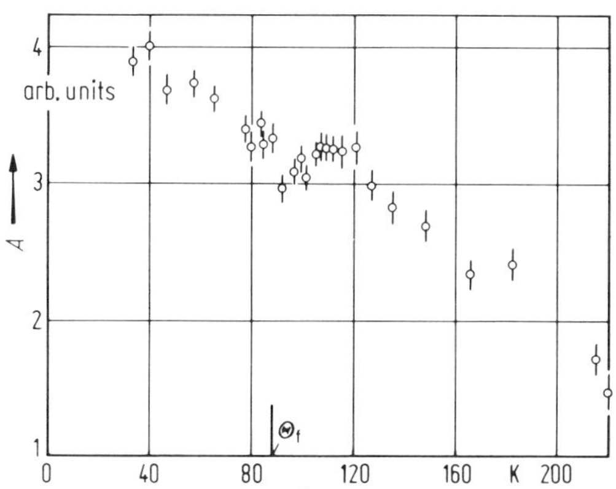

Fig. 856. $NH_4Fe(SO_4)_2 \cdot 12H_2O$. Area of Mössbauer absorption spectrum A vs. T [71 M 11]. γ-rays are parallel to [100].

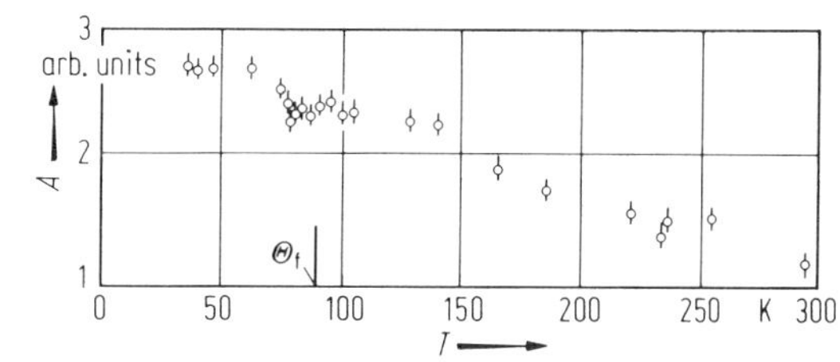

Fig. 857. $NH_4Fe(SO_4)_2 \cdot 12H_2O$. Area of Mössbauer absorption spectrum A vs. T [71 M 11]. γ-rays are parallel to [111].

Figs. 858, 859, see p. 518.

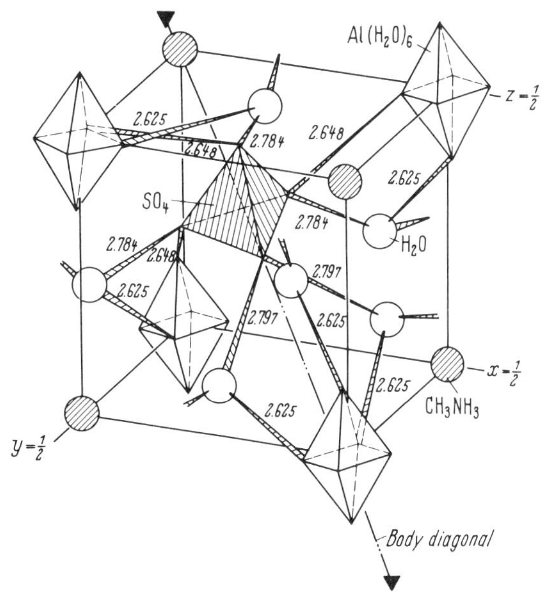

Fig. 860. $CH_3NH_3Al(SO_4)_2 \cdot 12H_2O$ (MASD). Crystal structure of phase I [57 O 1]. A sketch of one eighth of unit cell. Hydrogen bonds are shown by shaded, directed lines. Distances in Å.

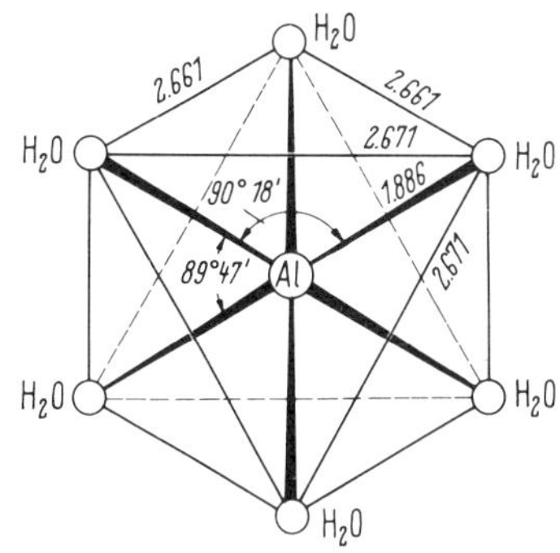

Fig. 861. $CH_3NH_3Al(SO_4)_2 \cdot 12H_2O$ (MASD). Crystal structure of phase I [57 O 1]. Shape and size of Al $(H_2O)_6$ octahedron. Distances in Å.

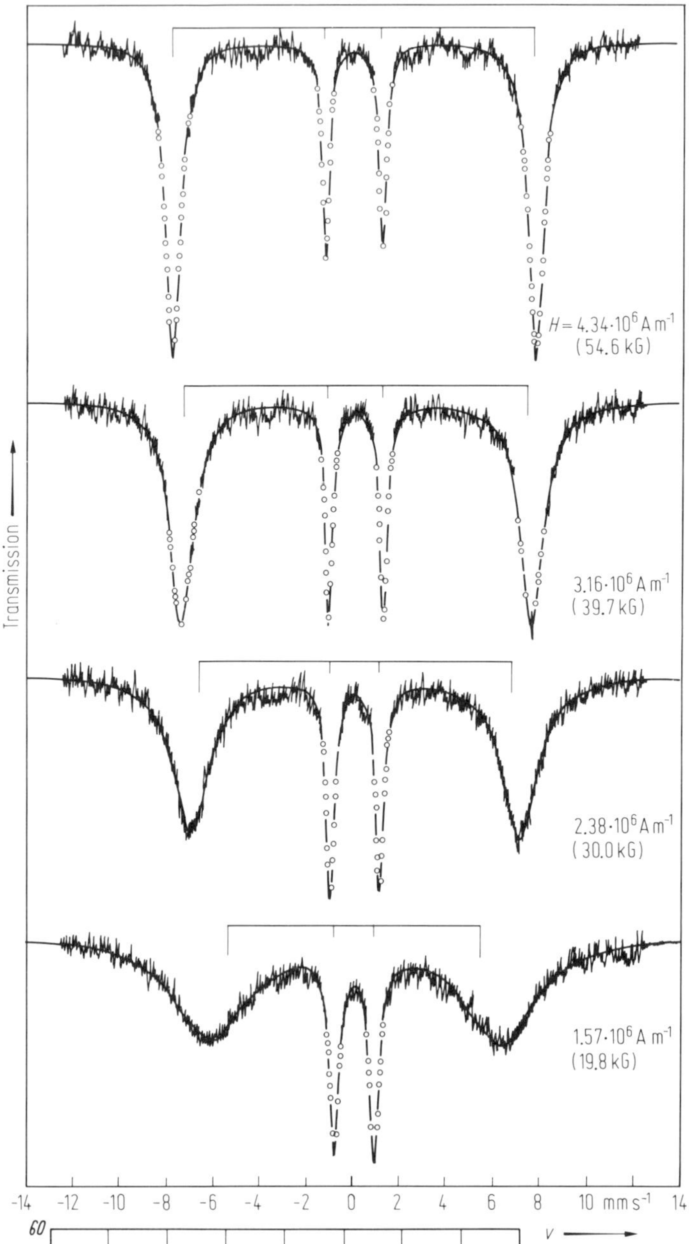

Fig. 858. $NH_4Fe(SO_4)_2 \cdot 12H_2O$. Mössbauer spectra [73W4]. Parameter: *H*. *v*: source velocity. ***H***$\|\gamma$-ray direction.

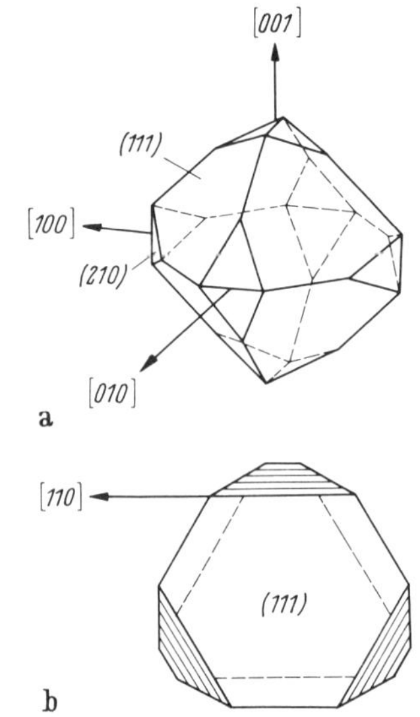

Fig. 859. $CH_3NH_3Al(SO_4)_2 \cdot 12H_2O$ (MASD). Crystal forms [56J2]. Fig. a: octahedral form and b: hexagonal plate form.

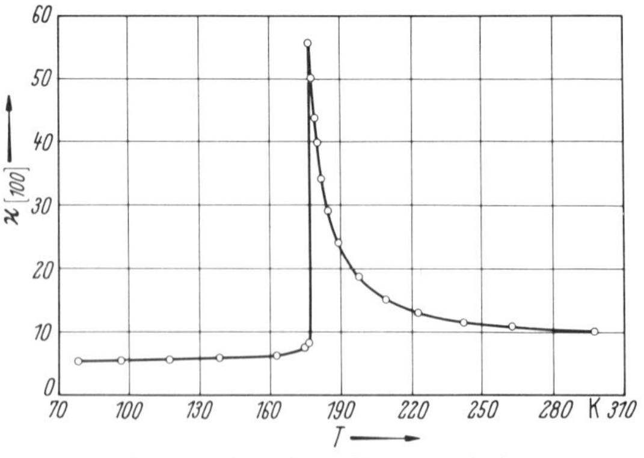

Fig. 863. $CH_3NH_3Al(SO_4)_2 \cdot 12H_2O$ (MASD). $\varkappa_{[100]}$ vs. T [57J2]. $f = 10$ kHz.

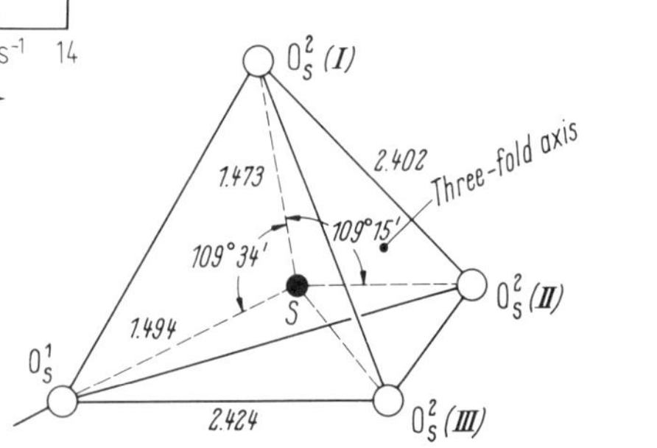

Fig. 862. $CH_3NH_3Al(SO_4)_2 \cdot 12H_2O$ (MASD). Crystal structure of phase I [57O1]. Shape and size of SO_4 tetrahedron. Distances in Å. The oxygen O_s^1 is equivalent to $O_{s(1)}$ in Table 201. The oxygen O_s^2(I), O_s^1(II) and O_s^2(III) are related to each other by the three-fold axis and equivalent to $O_{s(2)}^2$ in Table 201.

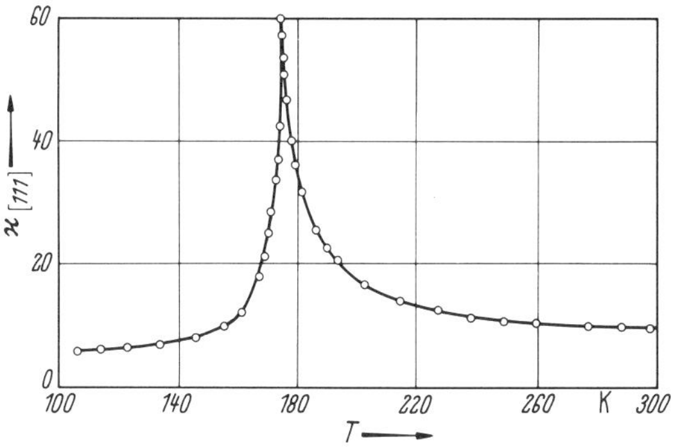

Fig. 864. $CH_3NH_3Al(SO_4)_2 \cdot 12H_2O$ (MASD). $\varkappa_{[111]}$ vs. T [56P2].

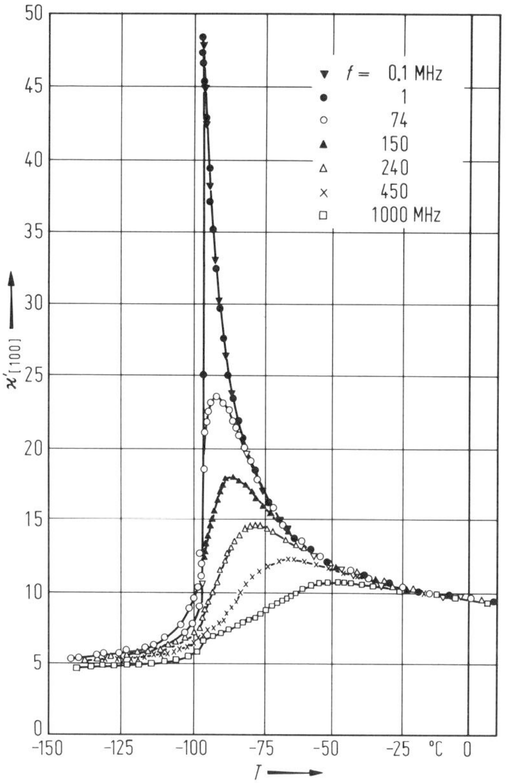

Fig. 865. $CH_3NH_3Al(SO_4)_2 \cdot 12H_2O$ (MASD). $\varkappa'_{[100]}$ vs. T [71M1]. Parameter: f.

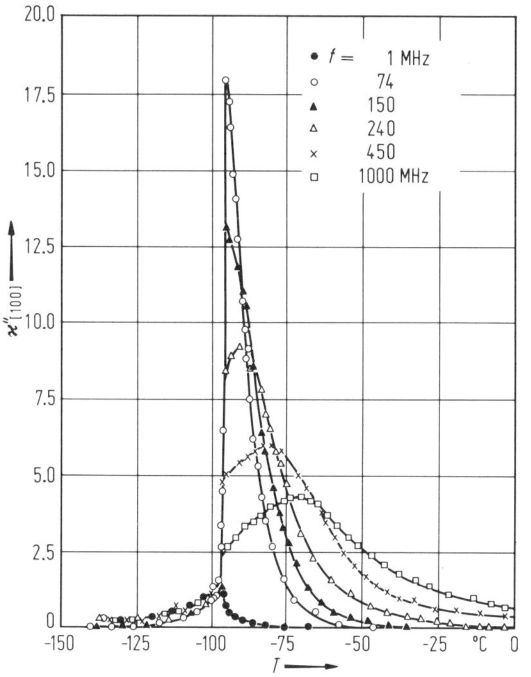

Fig. 866. $CH_3NH_3Al(SO_4)_2 \cdot 12H_2O$ (MASD). $\varkappa''_{[100]}$ vs. T [71M1]. Parameter: f.

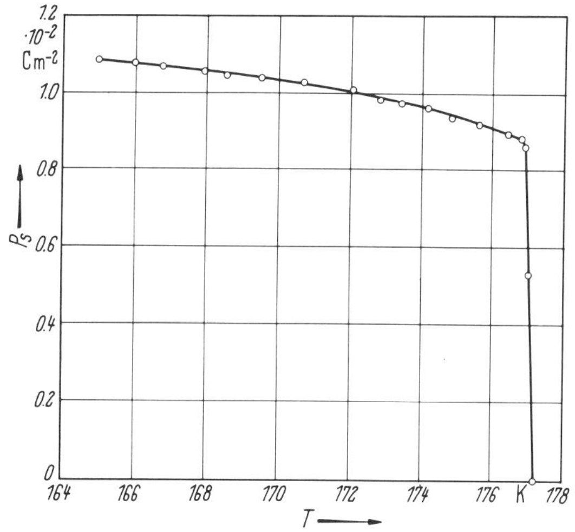

Fig. 867. $CH_3NH_3Al(SO_4)_2 \cdot 12H_2O$ (MASD). P_s vs. T [57J2].

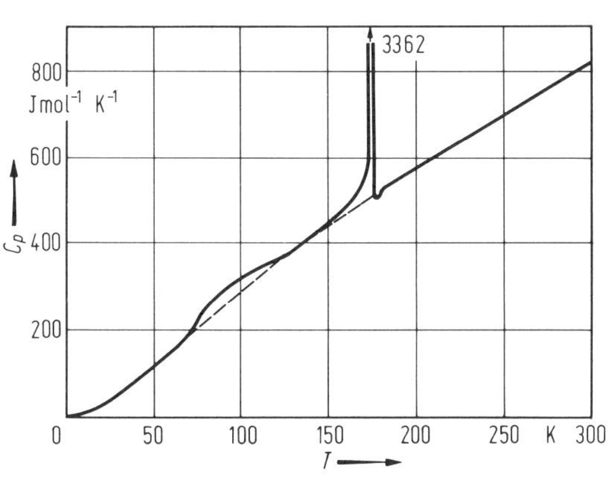

Fig. 868. $CH_3NH_3Al(SO_4)_2 \cdot 12H_2O$ (MASD). C_p vs. T [68A4]. C_p: molar heat capacity at constant pressure.

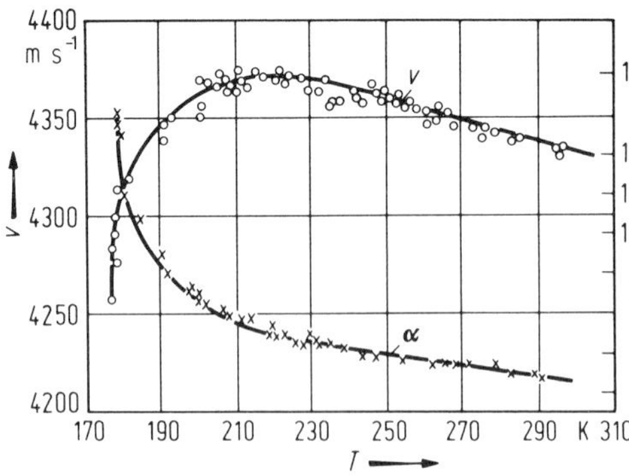

Fig. 869. $CH_3NH_3Al(SO_4)_2 \cdot 12H_2O$ (MASD). Velocity of longitudinal elastic waves in the [100] direction, v, and attenuation for the same direction (relative value), α, as functions of temperature [69A9].

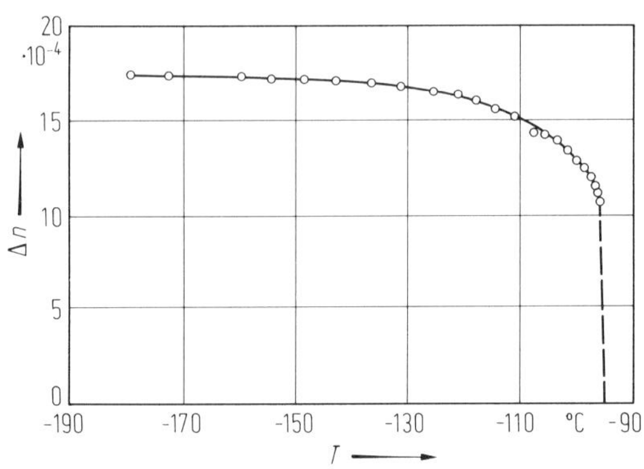

Fig. 871. $CH_3NH_3Al(SO_4)_2 \cdot 12H_2O$ (MASD). Δn vs. T [72A13]. Δn: birefringence. $\lambda = 633$ nm.

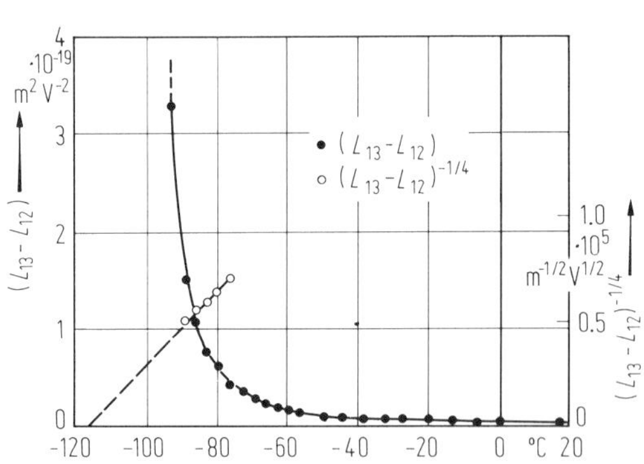

Fig. 872. $CH_3NH_3Al(SO_4)_2 \cdot 12H_2O$ (MASD). $L_{13} - L_{12}$, $(L_{13} - L_{12})^{-1/4}$ vs. T [72A13]. $L_{\lambda\mu}$: quadratic electrooptic constant. $\lambda = 633$ nm.

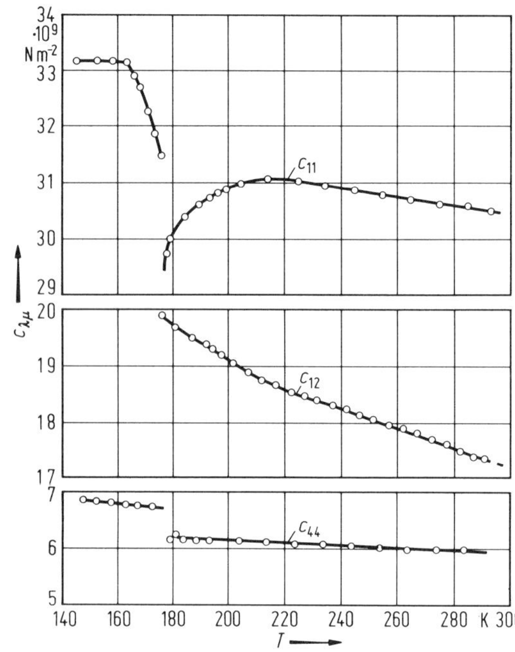

Fig. 870. $CH_3NH_3Al(SO_4)_2 \cdot 12H_2O$ (MASD). $c_{\lambda\mu}$ vs. T [69A9].

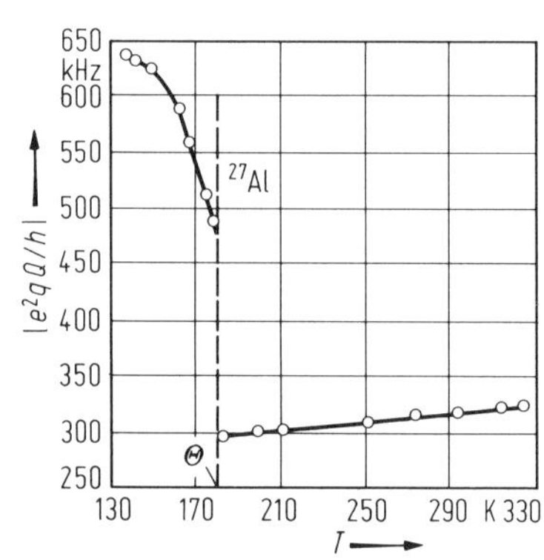

Fig. 873. $CH_3NH_3Al(SO_4)_2 \cdot 12H_2O$ (MASD). $|e^2qQ/h|$ vs. T [69V7].

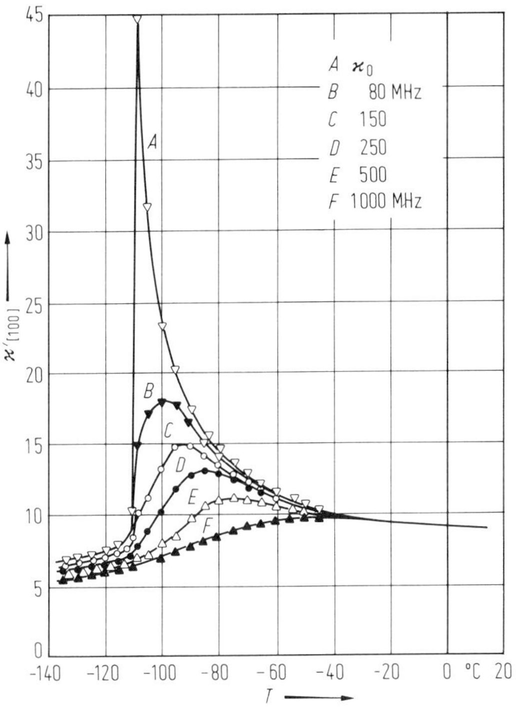

Fig. 874. $CH_3NH_3Cr(SO_4)_2 \cdot 12H_2O$. $\varkappa'_{[100]}$ vs. T. Parameter: f. [75C10].

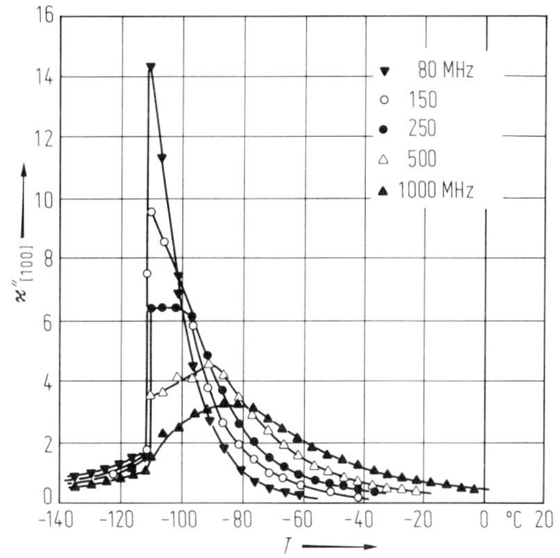

Fig. 875. $CH_3NH_3Cr(SO_4)_2 \cdot 12H_2O$. $\varkappa''_{[100]}$ vs. T. Parameter: f. [75C10].

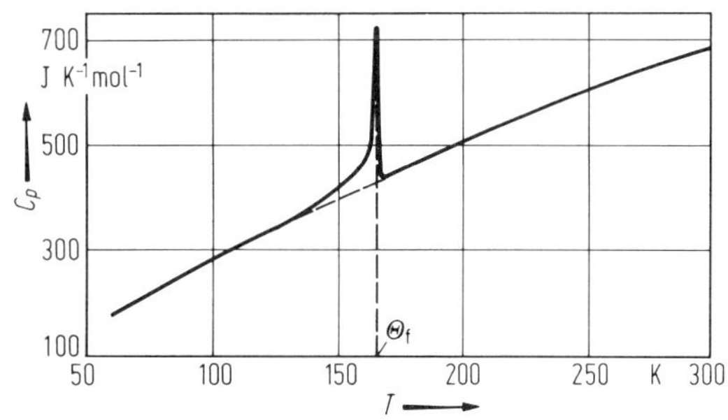

Fig. 877. $CH_3NH_3Cr(SO_4)_2 \cdot 12H_2O$. C_p vs. T [70B18]. C_p: molar heat capacity at constant pressure.

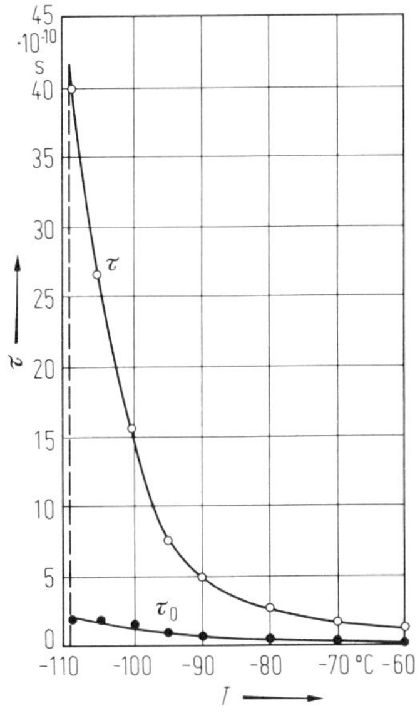

Fig. 876. $CH_3NH_3Cr(SO_4)_2 \cdot 12H_2O$. τ, τ_0 vs. T [75C10]. τ: relaxation time of the dielectric dispersion. τ_0 is obtained from the relation, $\tau = \tau_0 T (T - \Theta_f)^{-1}$.

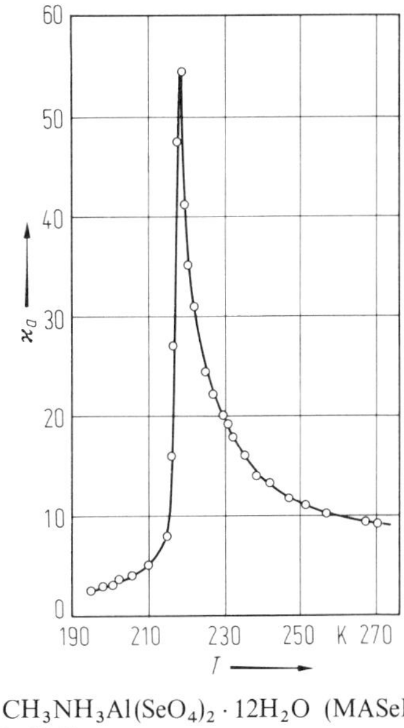

Fig. 879. $CH_3NH_3Al(SeO_4)_2 \cdot 12H_2O$ (MASeD). $\varkappa_a$ vs. T [75A7]. $f = 70$ Hz.

Fig. 878, see p. 522.

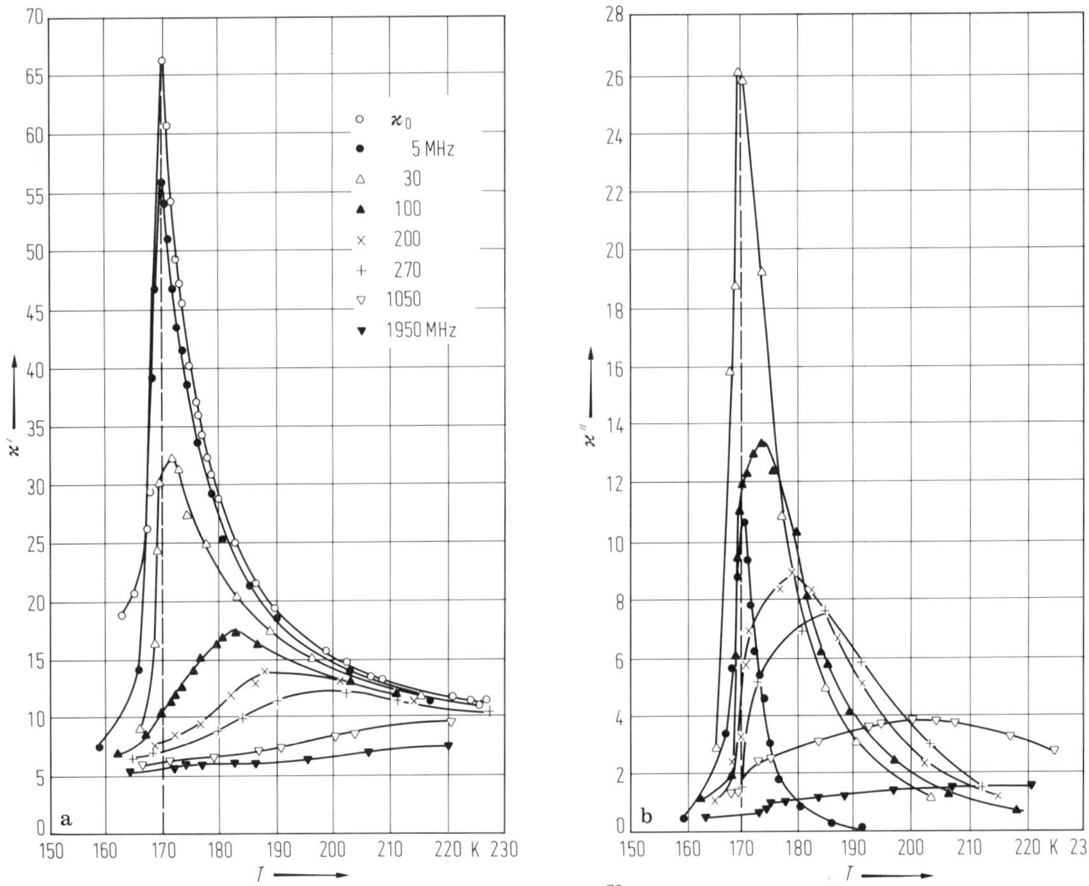

Fig. 878. $CH_3NH_3Ga(SO_4)_2 \cdot 12H_2O$. $\varkappa'$, $\varkappa''$ vs. T [79 J 4].

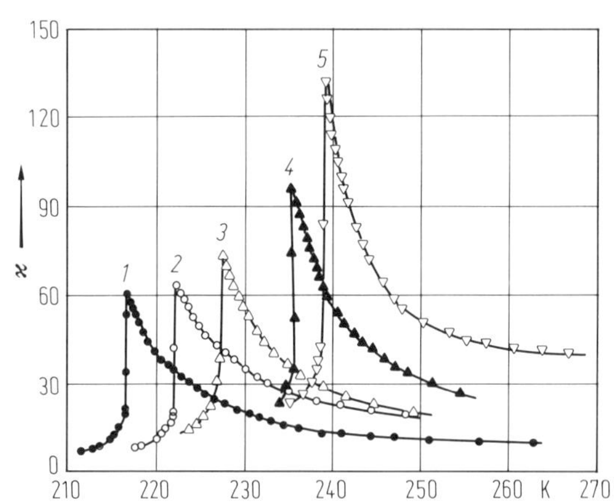

Fig. 880α. $CH_3NH_3Al(SeO_4)_2 \cdot 12H_2O$ (MASeD). $\varkappa$ vs. T. Parameter: p [78 K 22]. Curve *1*: $1.01 \cdot 10^5$ Pa (=1 atm); *2*: $1.75 \cdot 10^8$ Pa; *3*: $3.70 \cdot 10^8$ Pa; *4*: $8.25 \cdot 10^8$ Pa; *5*: $12.25 \cdot 10^8$ Pa.

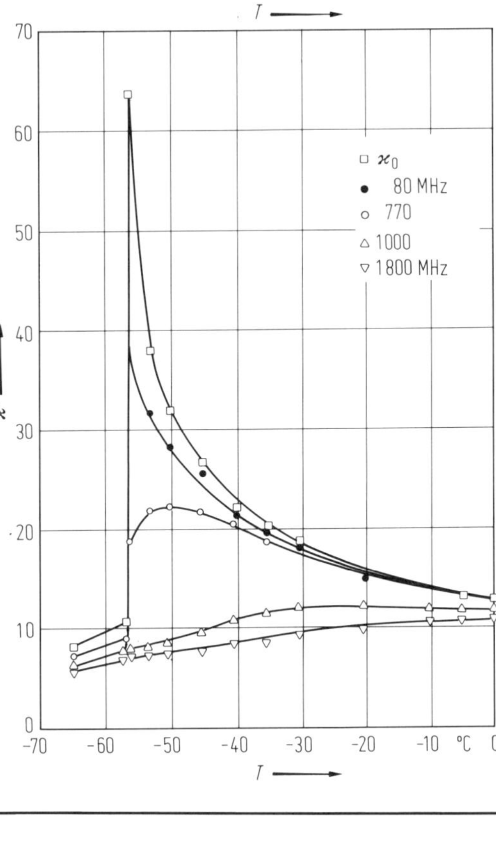

Fig. 880β. $CH_3NH_3Al(SeO_4)_2 \cdot 12H_2O$ (MASeD). $\varkappa'_{[100]}$ vs. T [77 C 6]. Parameter: f. ►

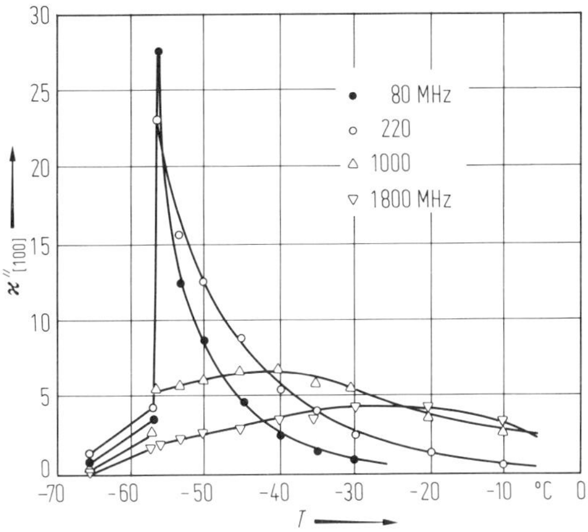

Fig. 880γ. $CH_3NH_3Al(SeO_4)_2 \cdot 12H_2O$ (MASeD). $\varkappa''_{[100]}$ vs. T[77C6]. Parameter: f.

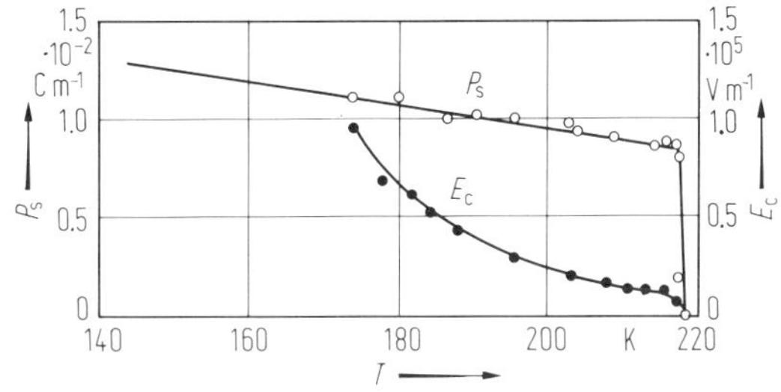

Fig. 881. $CH_3NH_3Al(SeO_4)_2 \cdot 12H_2O$ (MASeD). P_s, E_c vs. T[75A7].

27 GASH ($C(NH_2)_3Al(SO_4)_2 \cdot 6H_2O$) family

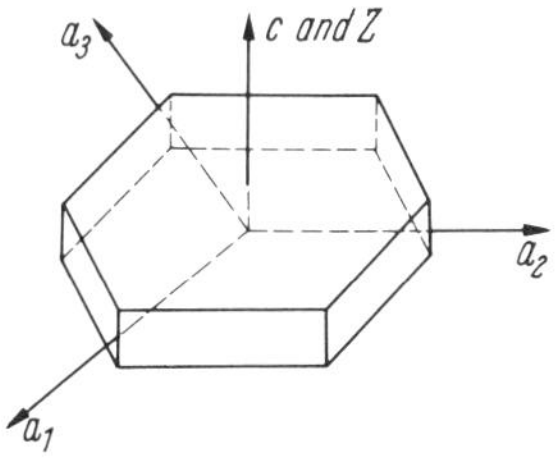

Fig. 882. $C(NH_2)_3Al(SO_4)_2 \cdot 6H_2O$ (GASH). Crystal form [61o1].

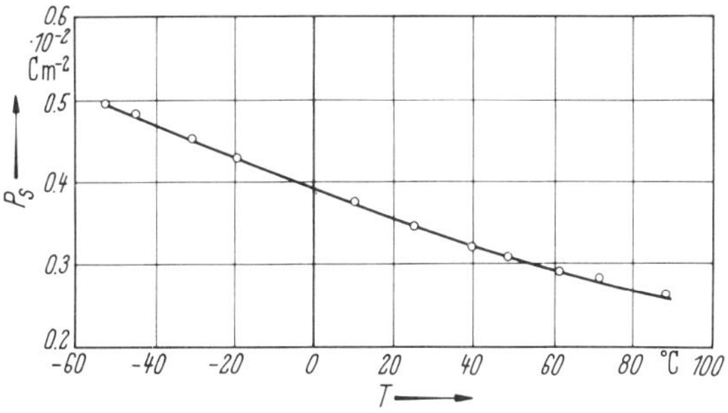

Fig. 884. $C(NH_2)_3Al(SO_4)_2 \cdot 6H_2O$ (GASH). P_s vs. T [56H1].

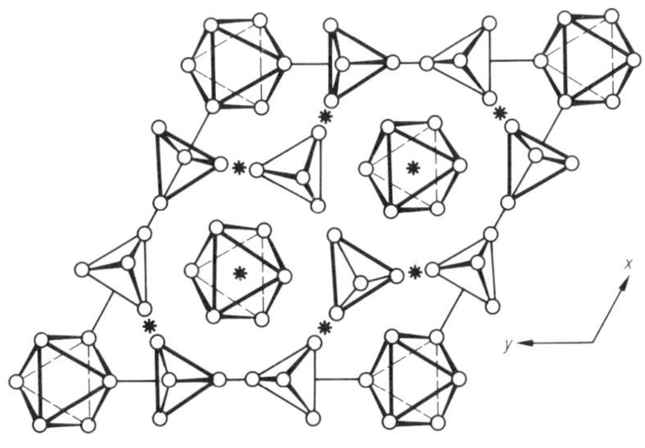

Fig. 883. $C(NH_2)_3Al(SO_4)_2 \cdot 6H_2O$ (GASH). Schematic drawing of the structure projected along the c axis. The aluminum ion in the center of each octahedron and the guanidinium ion directly below the aluminum ion have been omitted. Asterisks designate pseudocenters of inversion [67S7].

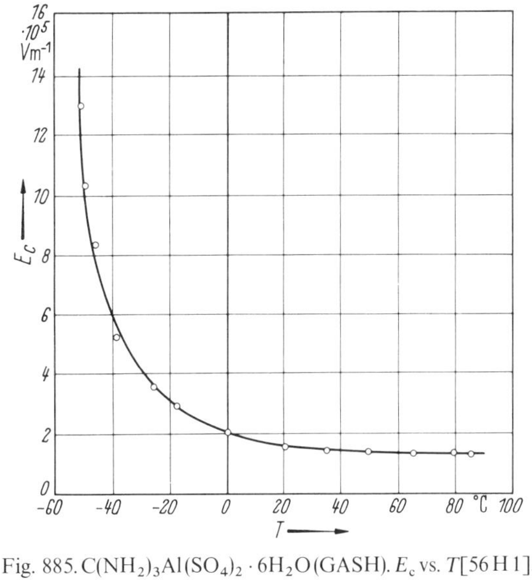

Fig. 885. $C(NH_2)_3Al(SO_4)_2 \cdot 6H_2O$ (GASH). E_c vs. T[56H1].

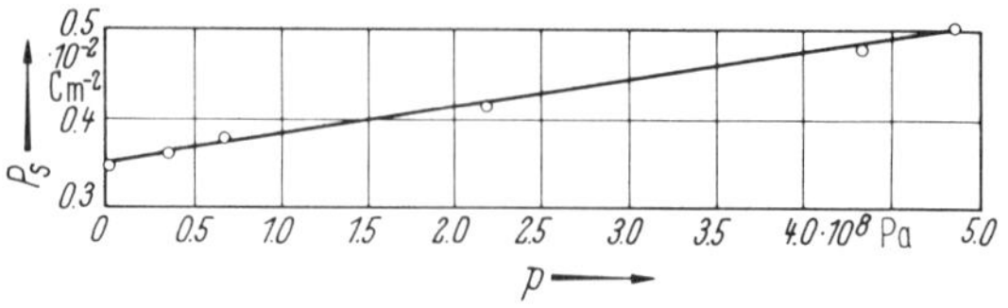

Fig. 886. $C(NH_2)_3Al(SO_4)_2 \cdot 6H_2O$ (GASH). P_s vs. p [56M3]. T=RT.

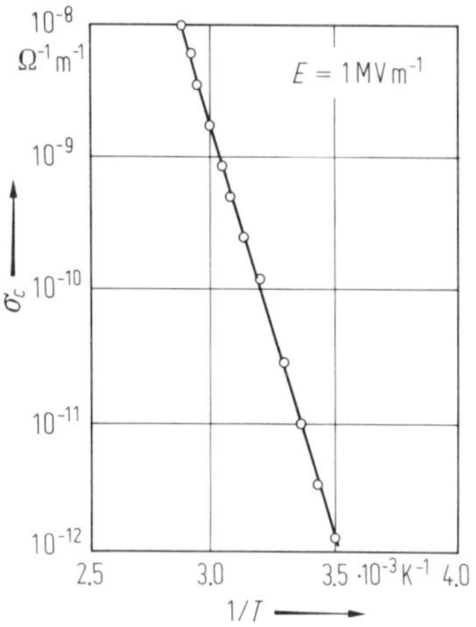

Fig. 888. $C(NH_2)_3Al(SO_4)_2 \cdot 6H_2O$ (GASH). σ_c vs. $1/T$ [62G3].

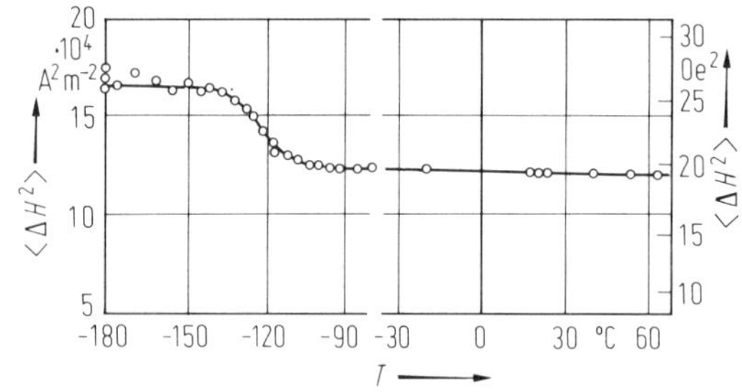

Fig. 889. $C(NH_2)_3Al(SO_4)_2 \cdot 6H_2O$ (GASH) (polycrystalline). $\langle \Delta H^2 \rangle$ vs. T [61L1]. $\langle \Delta H^2 \rangle$: second moment of the NMR signal of proton.

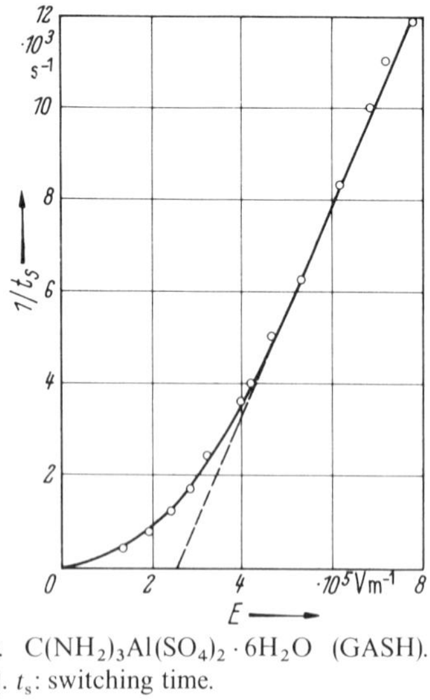

Fig. 890. $C(NH_2)_3Al(SO_4)_2 \cdot 6H_2O$ (GASH). $1/t_s$ vs. E [57W1]. t_s: switching time.

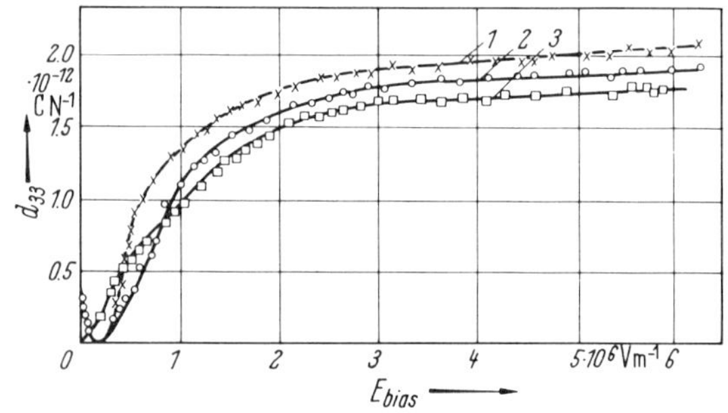

Fig. 887. $C(NH_2)_3Al(SO_4)_2 \cdot 6H_2O$ (GASH). d_{33} vs. E_{bias} [62Z3]. T=RT. Curves *1*, *2*: unipolar *Z*-cut specimens; curve *3*: specimens not unipolar. Quasi-static method.

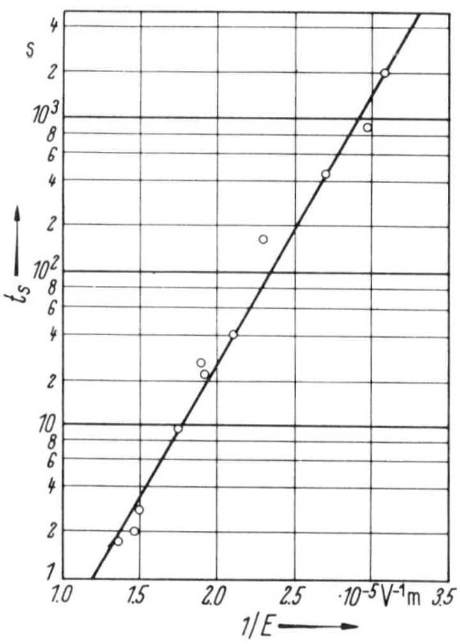

Fig. 891. $C(NH_2)_3Al(SO_4)_2 \cdot 6H_2O$ (GASH). t_s vs. $1/E$ [57W1]. t_s: switching time.

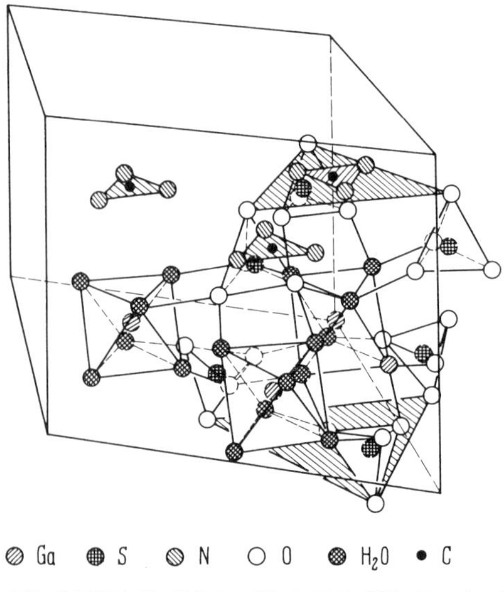

Fig. 892. $C(NH_2)_3Ga(SO_4)_2 \cdot 6H_2O$ (GGaSH). Sketch of the crystal structure [59G1].

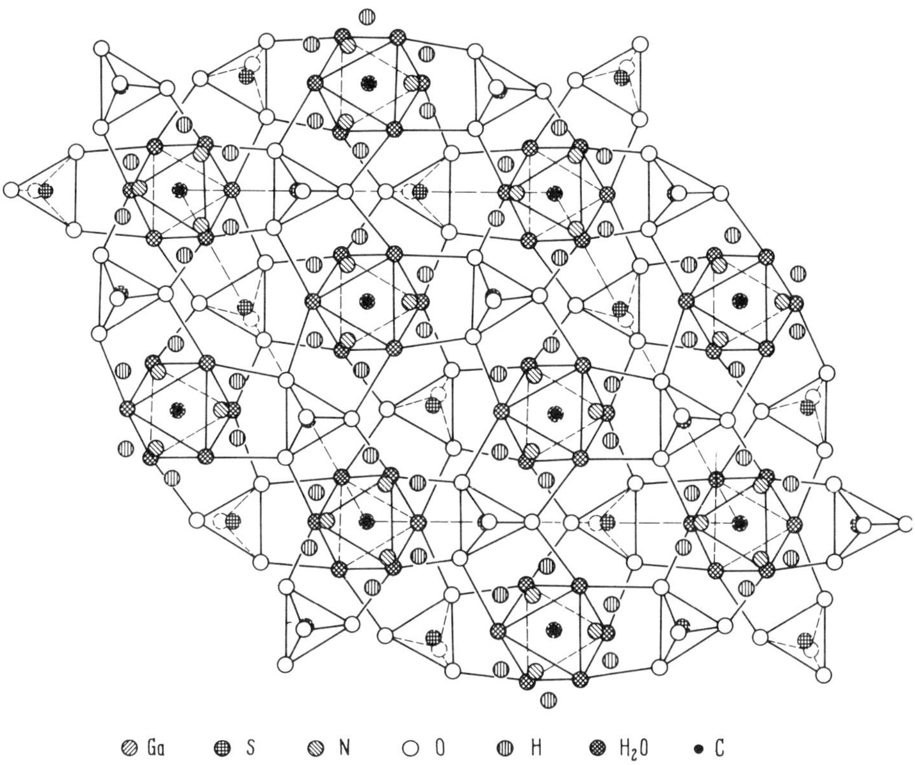

Fig. 893. $C(NH_2)_3Ga(SO_4)_2 \cdot 6H_2O$ (GGaSH). Crystal structure [59G1]. Projected along the *c* axis. Complete guanidinium ions including H atoms are shown, but none of the lines drawn are to be associated with these ions. The straight lines outline the geometrics of the water octahedra and sulfate tetrahedra, the heavy lines being associated with groupings above the plane of the paper, the light lines with those below. Tapered lines indicate hydrogen bonds.

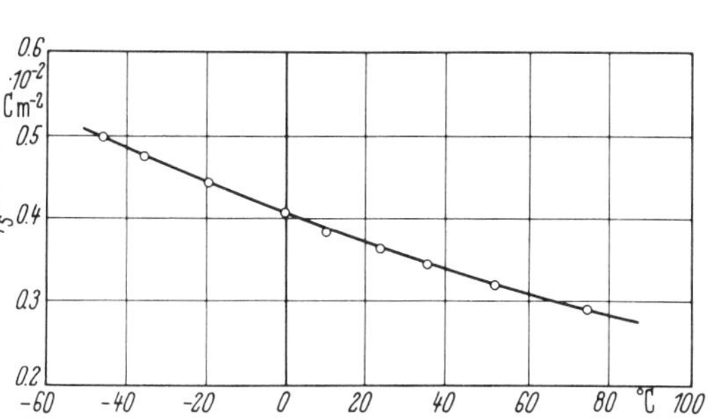

Fig. 894. $C(NH_2)_3Ga(SO_4)_2 \cdot 6H_2O$ (GGaSH). P_s vs. T [56H1].

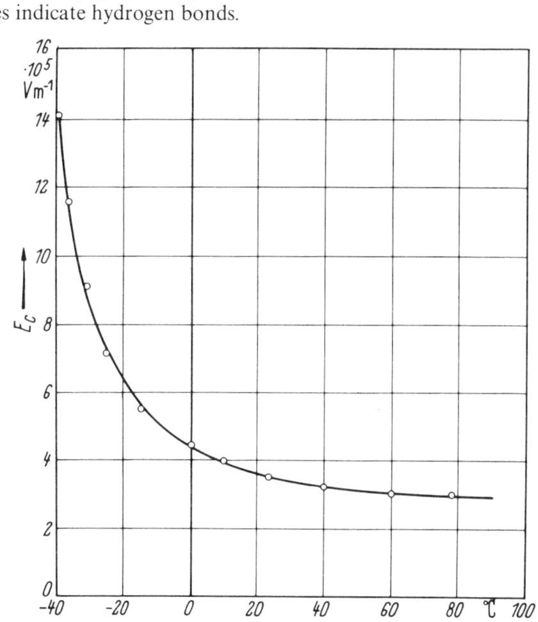

Fig. 895. $C(NH_2)_3Ga(SO_4)_2 \cdot 6H_2O$ (GGaSH). E_c vs. T [56H1].

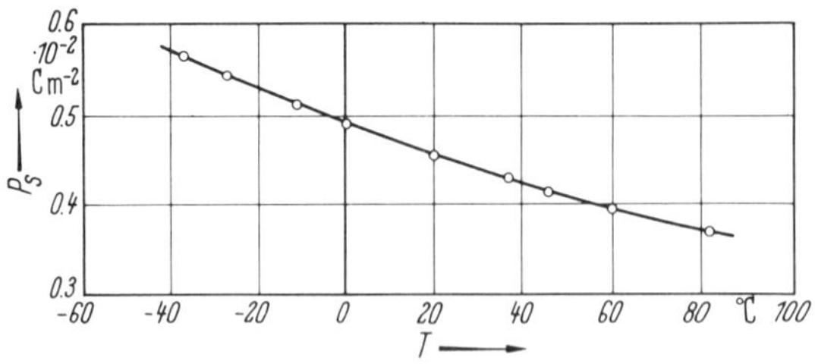

Fig. 896. $C(NH_2)_3Al(SeO_4)_2 \cdot 6H_2O$ (GASeH). P_s vs. T [56H1].

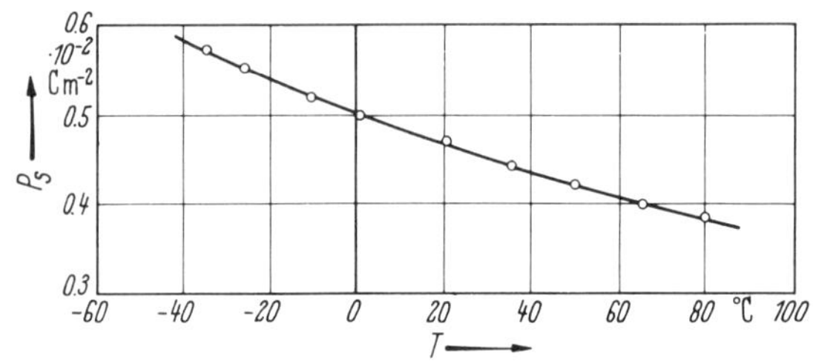

Fig. 897. $C(NH_2)_3Ga(SeO_4)_2 \cdot 6H_2O$ (GGaSeH). P_s vs. T [56H1].

28 $LiH_3(SeO_3)_2$ family

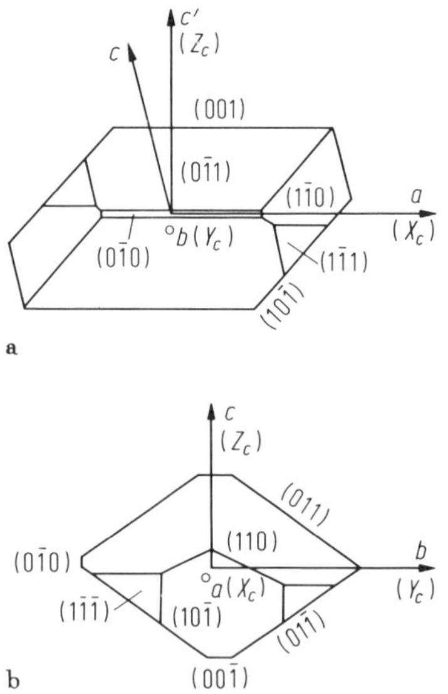

Fig. 898. $LiD_3(SeO_3)_2$. Crystal form [69S15]. In this chapter, 28A, the rectangular coordinates X_c, Y_c, Z_c are defined as $X_c \| a$, $Y_c \| b$, $Z_c \perp (X_c$ and $Y_c)$, as shown in this figure.

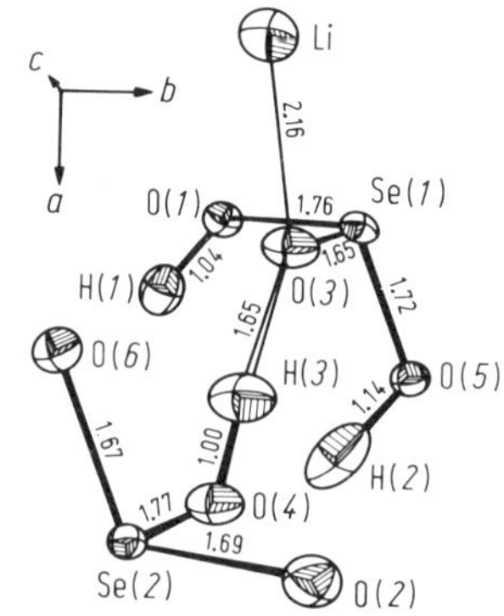

a

Fig. 899. $LiH_3(SeO_3)_2$. A perspective illustration of an asymmetric unit of RT structure [72T5]. Single thick line: covalent bonds; a pair of thin lines: hydrogen bond; single thin line: ionic bond. Distances in Å.

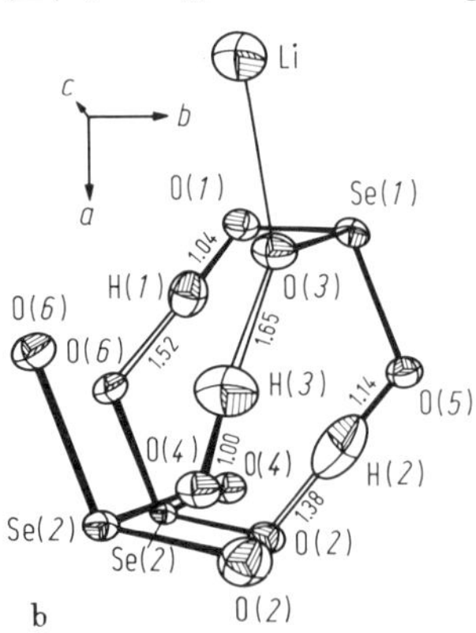

b

Fig. 900. $LiH_3(SeO_3)_2$. A perspective sketch showing the hydrogen bond scheme of RT structure [72T5]. Distances in Å.

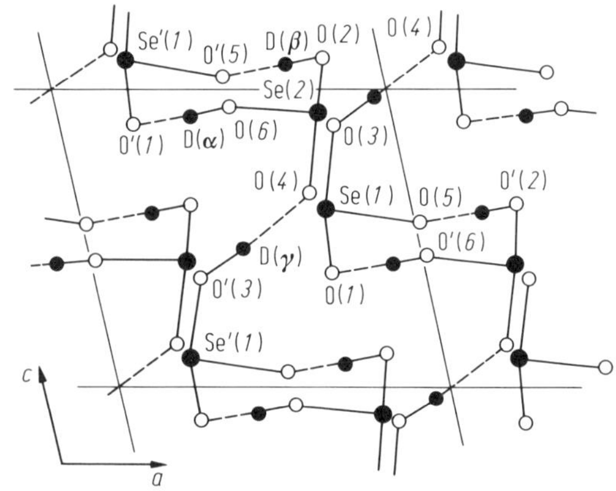

Fig. 901. $LiD_3(SeO_3)_2$. A projection of the crystal structure of $LiD_3(SeO_3)_2$ on the (010) plane [69S15]. Hydrogen bonds are indicated by dotted lines. Estimated positions of D atoms by deuteron magnetic resonance method are shown by small solid circles.

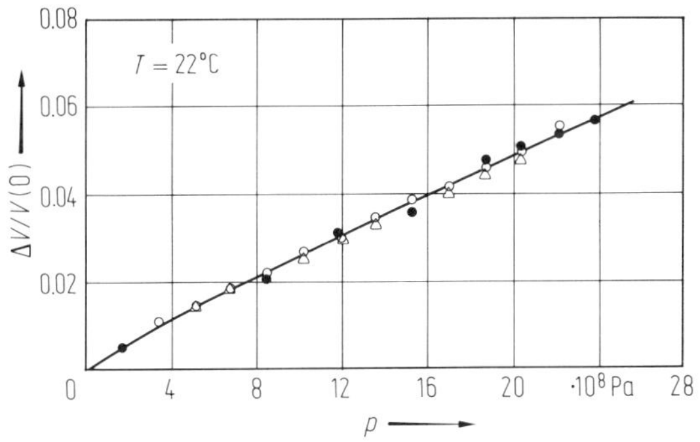

Fig. 902. $LiH_3(SeO_3)_2$. $-\Delta V/V(0)$ vs. p at $T=22\,°C$ [68 S 2]. $-\Delta V/V(0)$: relative volume change.

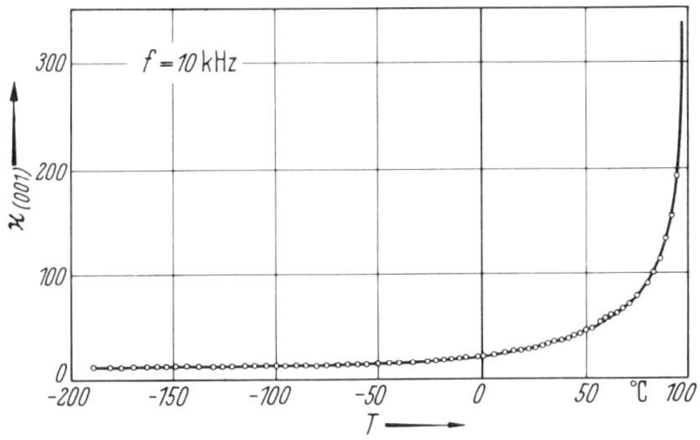

Fig. 903. $LiH_3(SeO_3)_2$. $\varkappa_{(001)}$ vs. T [59 P 3]. $\varkappa_{(001)}$: dielectric constant of (001) plate.

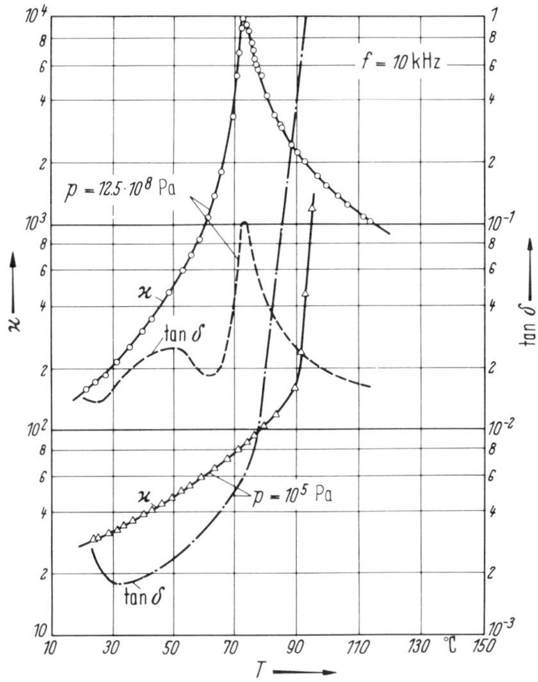

Fig. 904. $LiH_3(SeO_3)_2$. $\varkappa$, $\tan\delta$ vs. T at various p [66 S 5].

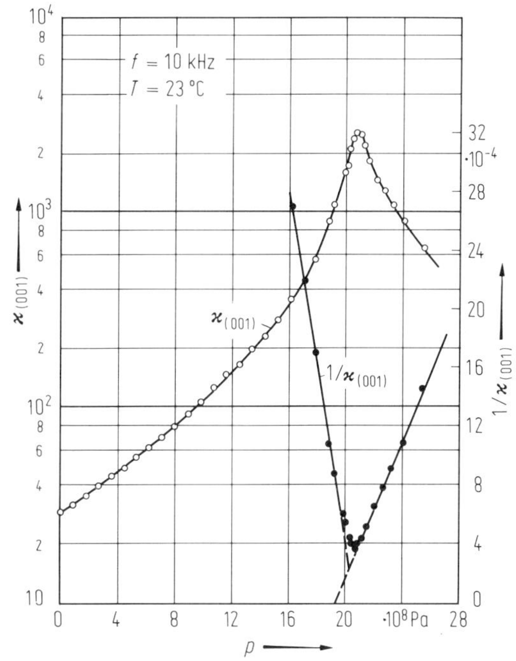

Fig. 905. $LiH_3(SeO_3)_2$. $\varkappa_{(001)}$, $1/\varkappa_{(001)}$ vs. p at $T=23\,°C$ [68 S 2].

Fig. 906, see p. 528.

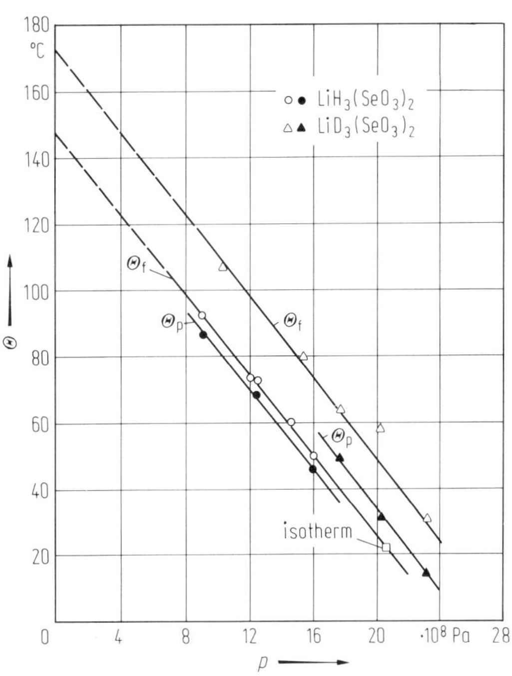

Fig. 907. $LiH_3(SeO_3)_2$, $LiD_3(SeO_3)_2$. Θ_f, Θ_p vs. p [68 S 2].

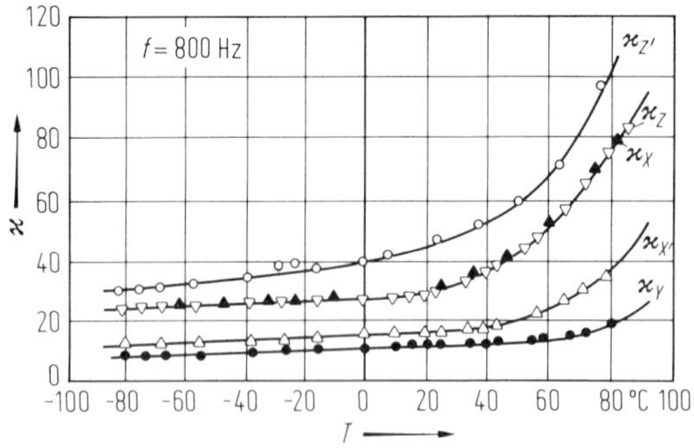

Fig. 906. $LiD_3(SeO_3)_2$. $\varkappa_X$, $\varkappa_Y$, $\varkappa_Z$, $\varkappa_{X'}$, $\varkappa_{Z'}$ vs. T [72S21]. X, Y, Z: rectangular coordinate system defined as $Y \| b$, $X \| a$, $Z \perp (001)$. X', Z': principal axes of dielectric constant tensor in the (X, Z) plane; X' makes an angle 43.7° to the X-axis.

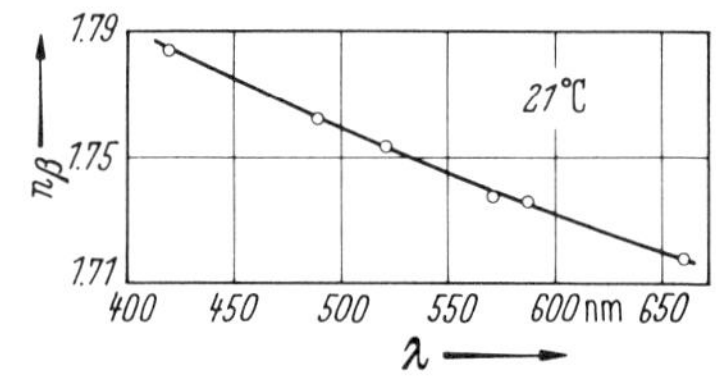

Fig. 909. $LiH_3(SeO_3)_2$. n_β vs. λ [66I6].

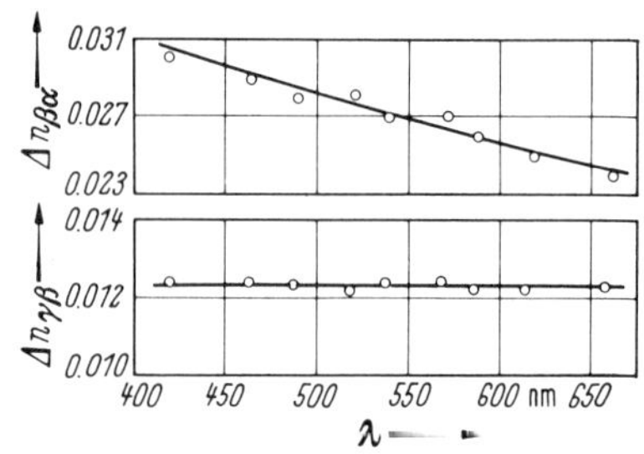

Fig. 910. $LiH_3(SeO_3)_2$. $\Delta n_{\gamma\beta}$, $\Delta n_{\beta\alpha}$ vs. λ [66I6]. $\Delta n_{\gamma\beta} = n_\gamma - n_\beta$, $\Delta n_{\beta\alpha} = n_\beta - n_\alpha$.

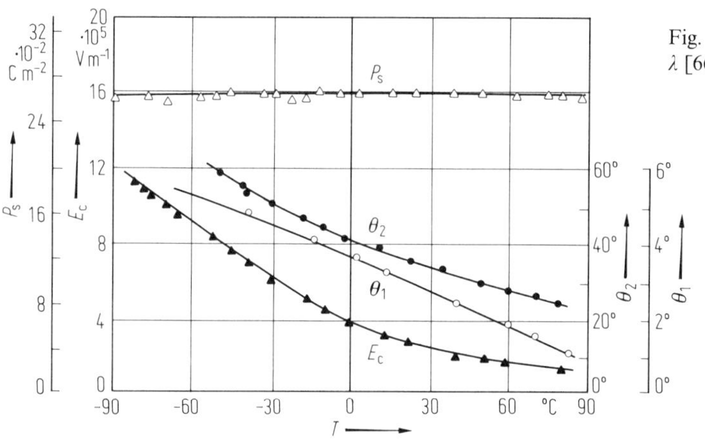

Fig. 908. $LiD_3(SeO_3)_2$. P_s, E_c, θ_1, θ_2 vs. T [72S21]. P_s, E_c: measured for (001) plate; θ_1: angle between P_s and the Z-axis; θ_2: angle between the greatest radius vector of the characteristic surface E_c^{-1} and the Z-axis.

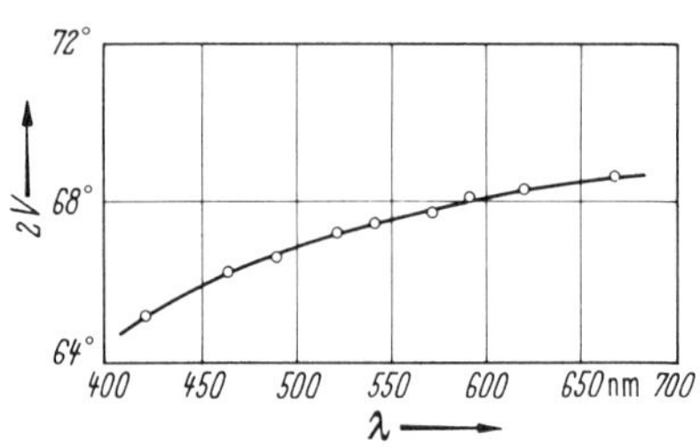

Fig. 911. $LiH_3(SeO_3)_2$. $2V$ vs. λ [66I6]. $2V$: optical axial angle.

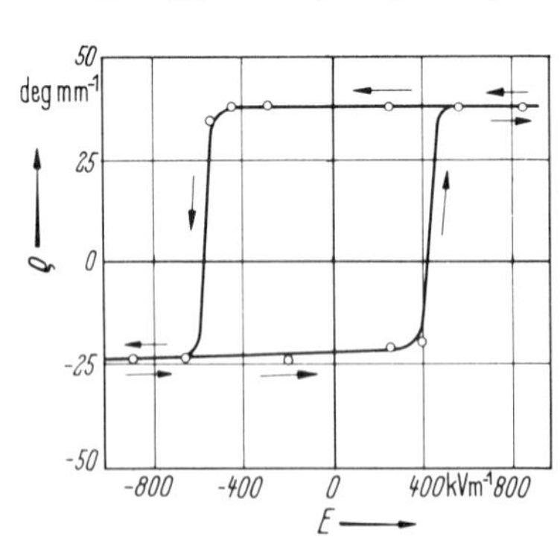

Fig. 913. $LiH_3(SeO_3)_2$. ϱ vs. E [62F9]. ϱ: optical rotatory power along the optic axis.

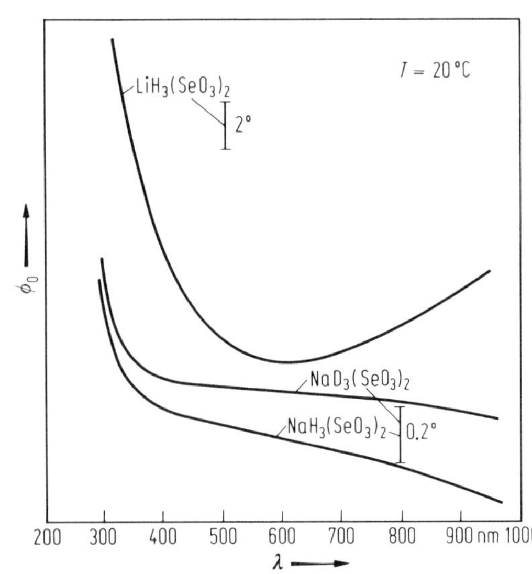

Fig. 912. $LiH_3(SeO_3)_2$, $NaH_3(SeO_3)_2$, $NaD_3(SeO_3)_2$. ϕ_0 vs. λ [70I7]. ϕ_0: rotation angle of the optical indicatrix around the b-axis.

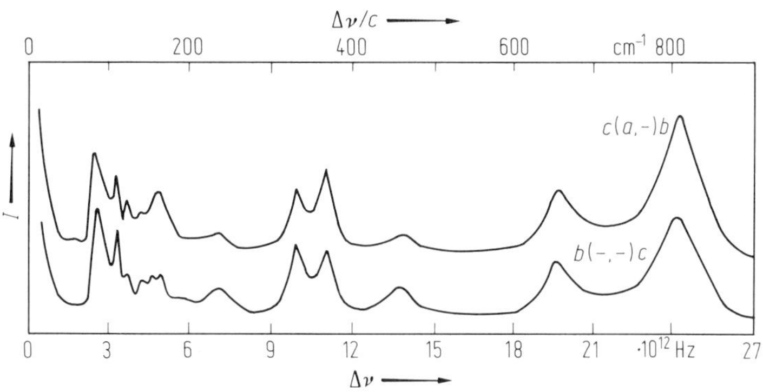

Fig. 914. $LiH_3(SeO_3)_2$. I vs. $\Delta\nu$ [73A3]. I: Raman scattering intensity at room temperature. Hyphens in the rotation of scattering geometry mean that the polarization of light is not specified.

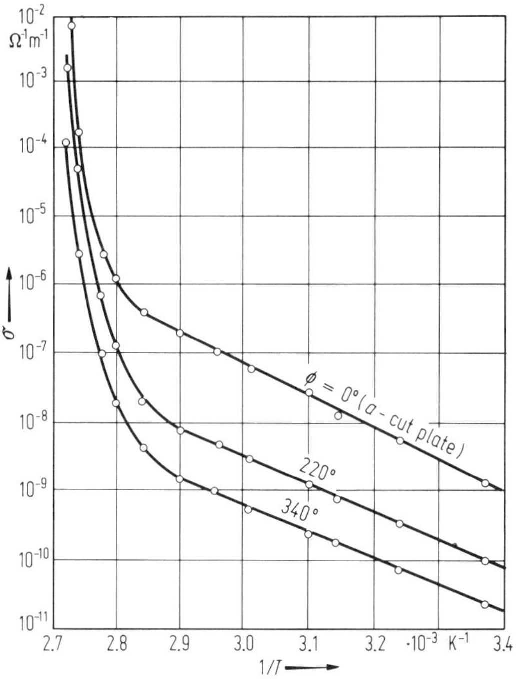

Fig. 915. $LiH_3(SeO_3)_2$. σ vs. $1/T$ for the three directions [62G3]. Parameter: ϕ. The field $\boldsymbol{E}$ applied to measure σ is perpendicular to the b axis. ϕ is the angle between $\boldsymbol{E}$ and the a axis. $E = 1$ MVm^{-1}. σ was measured by means of the two terminal method.

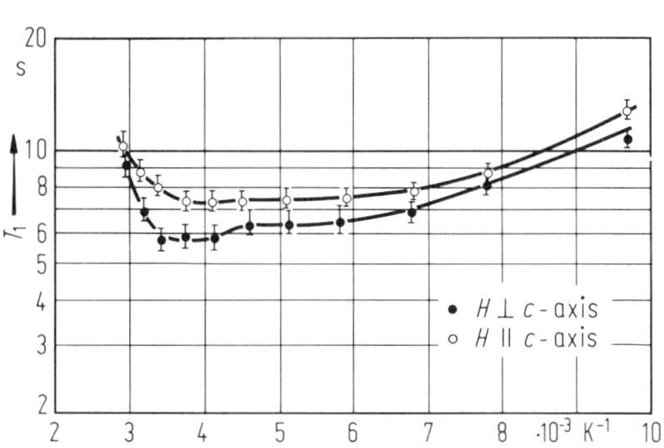

Fig. 916. $LiH_3(SeO_3)_2$. T_1 vs. $1/T$ [68S13]. T_1: proton spin-lattice relaxation time.

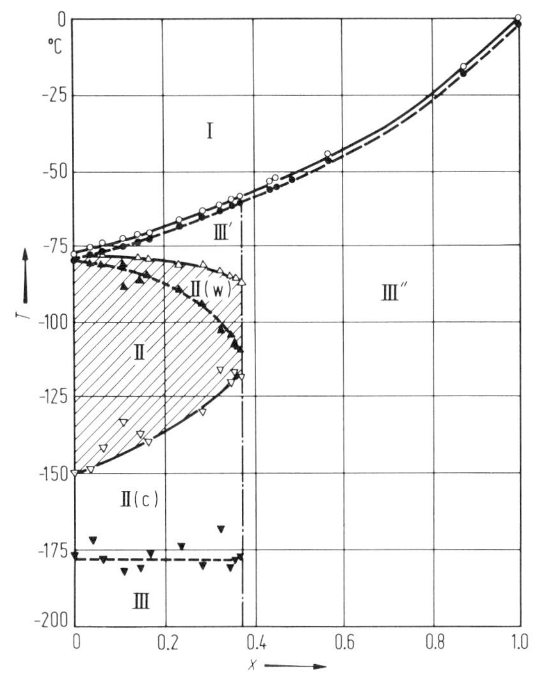

Fig. 917. $Na(H_{1-x}D_x)_3(SeO_3)_2$. Phase diagram [70M5]. Determined by anomalies in the linear thermal expansion along the b axis. Solid lines: phase boundaries observed on heating, dashed lines: on cooling; II(w) stands for phase II on warming, II(c) for phase II on cooling, etc. A similar phase diagram has been obtained by dielectric measurement [72S22]. See also [68Z1]. The hatched area shows the region of phase II(w), i.e. phase II observed on warming.

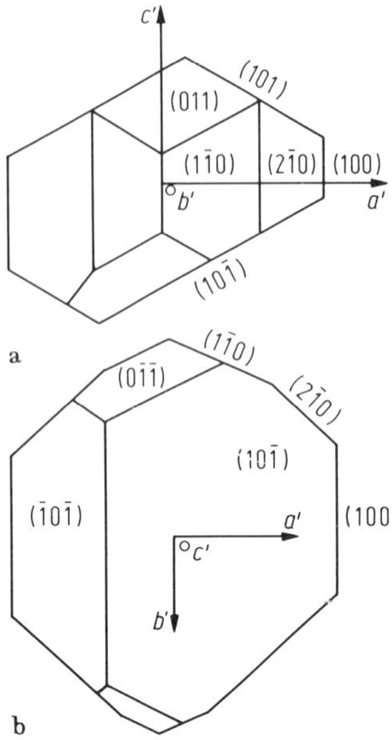

Fig. 918. $NaD_3(SeO_3)_2$. Crystal form [69 S 16]. Fig. a: viewed along the *b*' axis. Fig. b: viewed along the *c*' axis.

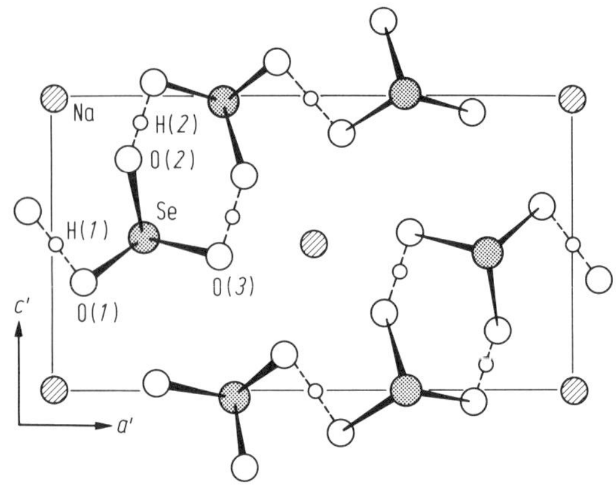

Fig. 919. $NaH_3(SeO_3)_2$. Crystal structure of phase I [70 K 3]. Viewed along the *b*′ axis.

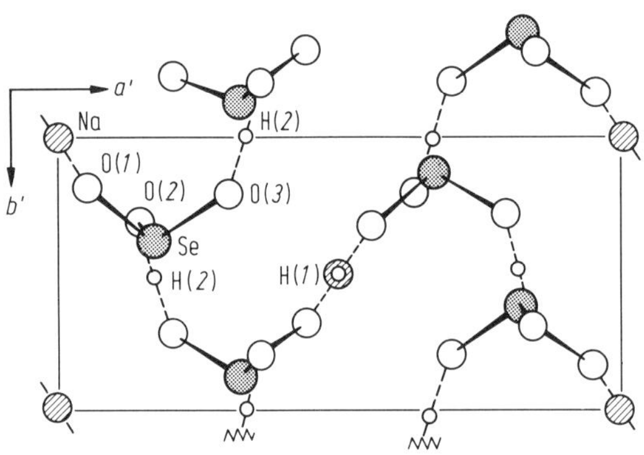

Fig. 920. $NaH_3(SeO_3)_2$. Crystal structure of phase I [70 K 3]. Viewed along the *c*′ axis.

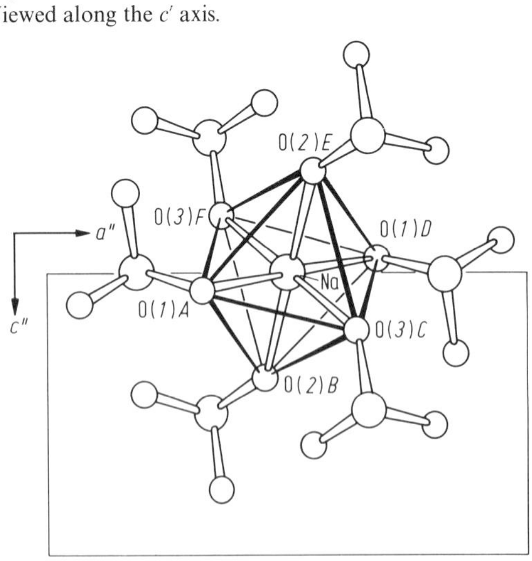

Fig. 921. $NaH_3(SeO_3)_2$. Crystal structure of phase I [68 V 2]. Environment of the sodium ion, as viewed along the *b*″ axis.

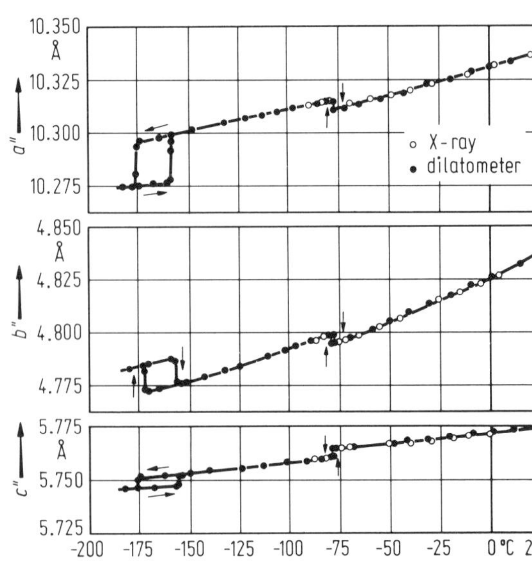

Fig. 922. $NaH_3(SeO_3)_2$. *a*″, *b*″, *c*″ vs. *T* [70 M 5].

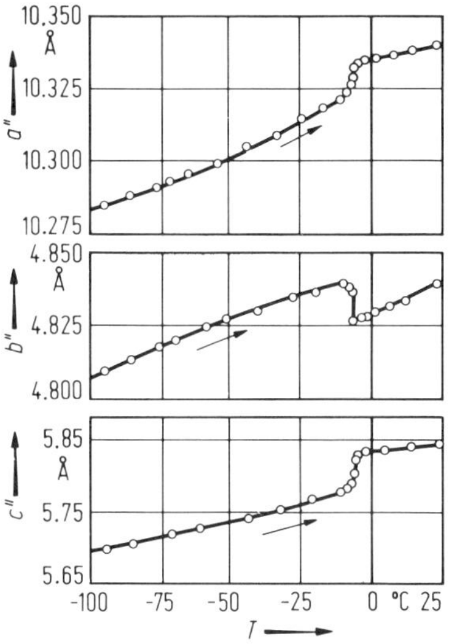

Fig. 923. $NaD_3(SeO_3)_2$. a'', b'', c'' vs. T [70 M 5].

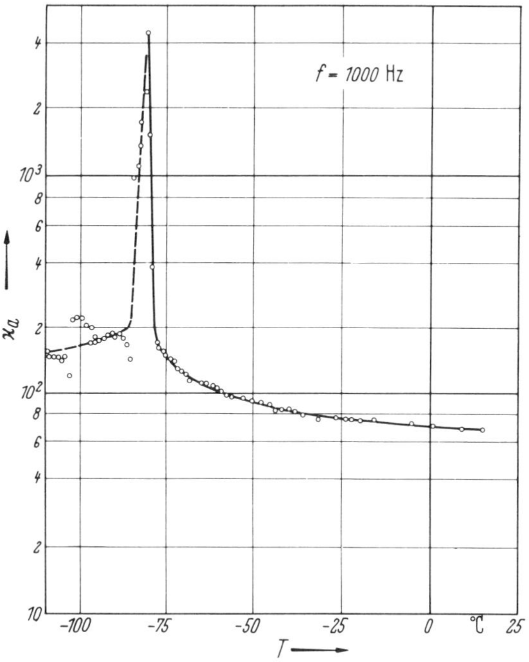

Fig. 925. $NaH_3(SeO_3)_2$. $\varkappa_a$ vs. T [65 B 6].

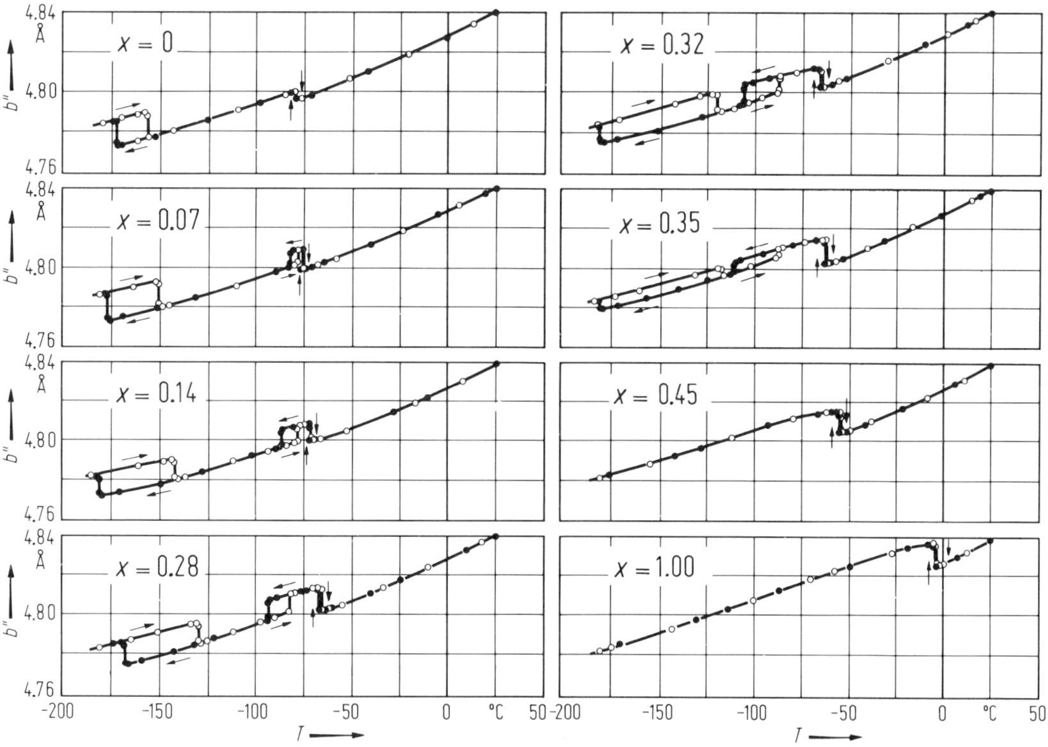

Fig. 924. $Na(H_{1-x}D_x)_3(SeO_3)_2$. b'' vs. T [70 M 5]. Parameter: x.

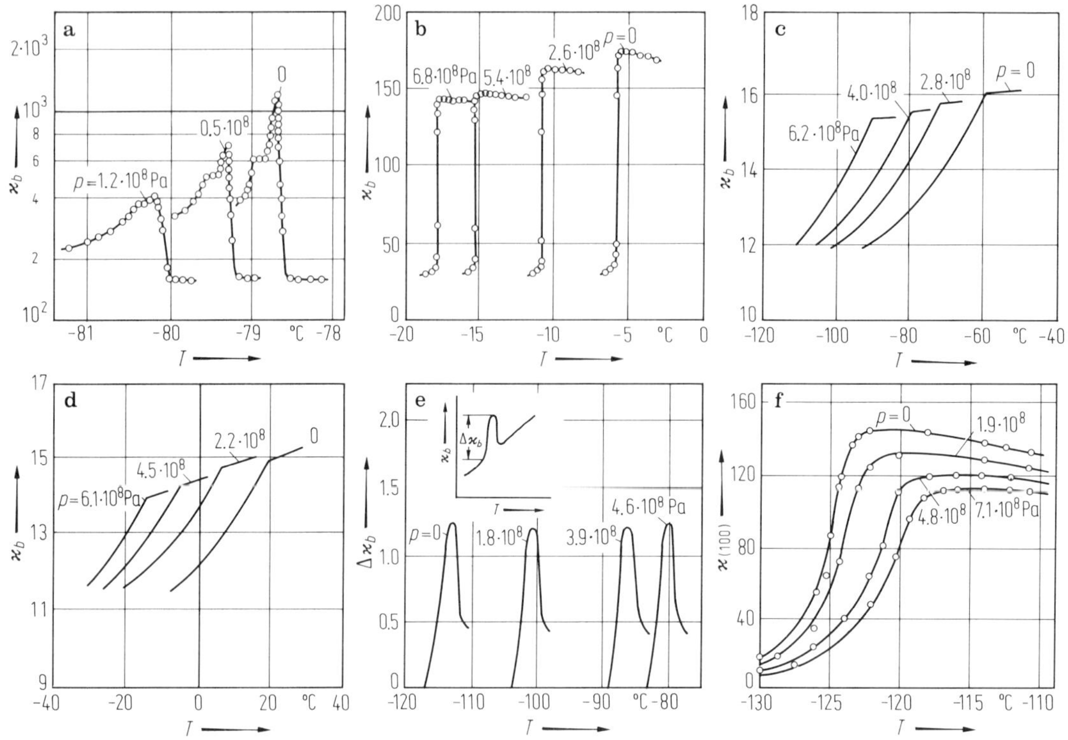

Fig. 926. $XH_3(SeO_3)_2$, $XD_3(SeO_3)_2$, X = Na, K, Rb, Cs. $\varkappa$ vs. T [71 S 20]. Parameter: p. Fig. a: $NaH_3(SeO_3)_2$; b: $NaD_3(SeO_3)_2$; c: $KH_3(SeO_3)_2$, d: $KD_3(SeO_3)_2$; e: $RbH_3(SeO_3)_2$; f: $CsH_3(SeO_3)_2$. $\varkappa_{(100)}$: Dielectric constant perpendicular to the (100) plane. $\varkappa$: low frequency dielectric constant.

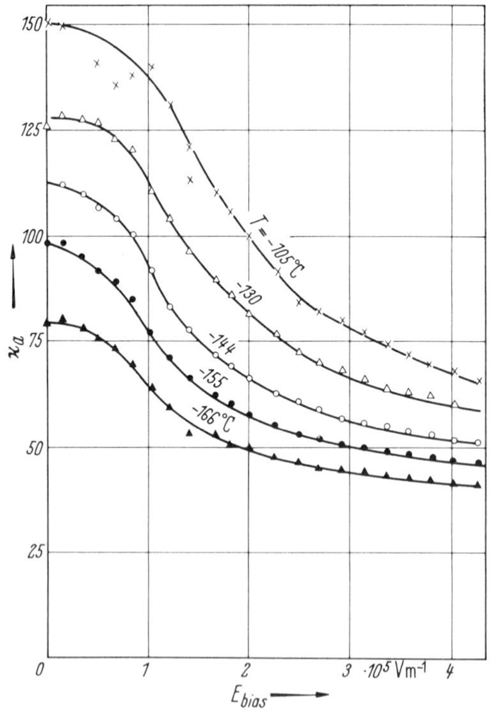

Fig. 928. $NaH_3(SeO_3)_2$. $\varkappa_a$ vs. E_{bias} [65 B 6]. f = 200 kHz. Parameter: T.

Fig. 927, see p. 533.

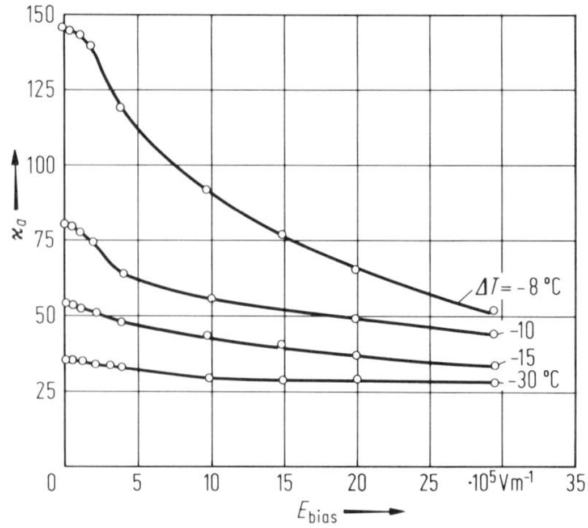

Fig. 929. $NaD_3(SeO_3)_2$. $\varkappa_a$ vs. E_{bias} [69 S 8]. (f = 800 Hz). Parameter: $\Delta T = T - \Theta_{I-II}$ [69 S 8].

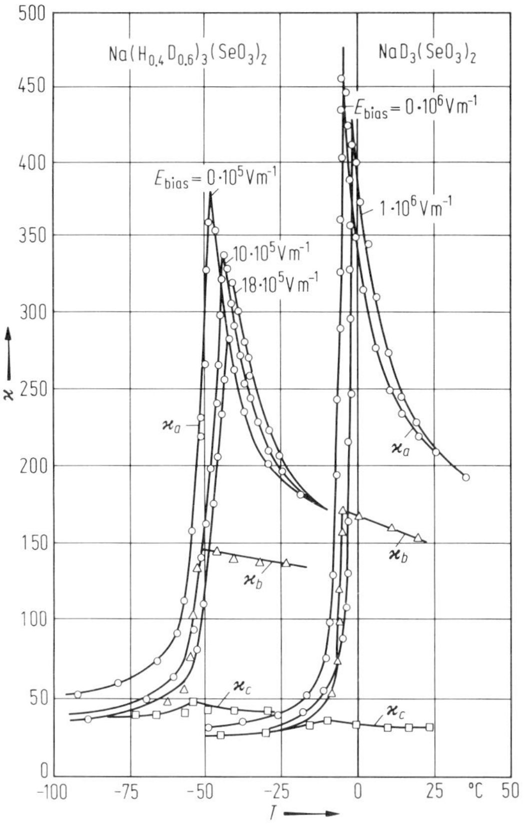

Fig. 927. $NaD_3(SeO_3)_2$, $Na(H_{0.4}D_{0.6})_3(SeO_3)_2$. $\varkappa$ vs. T [69 S 8]. $f = 800$ Hz. Parameter: E_{bias}.

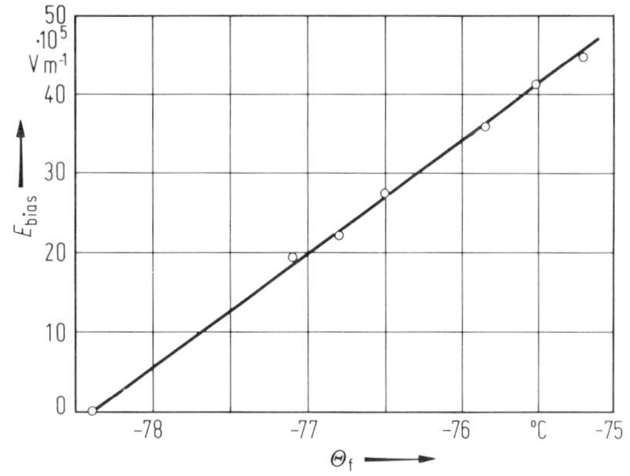

Fig. 930. $NaH_3(SeO_3)_2$. Θ_f vs. E_{bias} [70 B 10]. $E_{bias} \| a$.

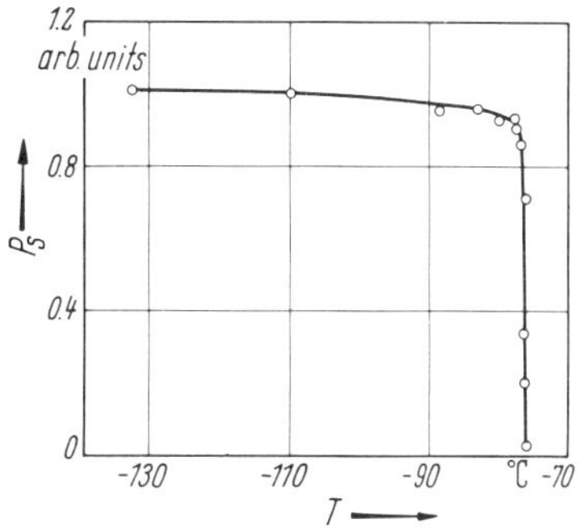

Fig. 931. $NaH_3(SeO_3)_2$. P_s vs. T [65 B 6].

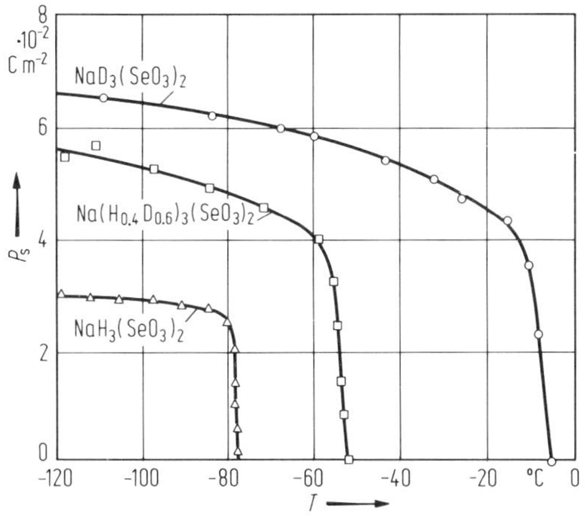

Fig. 932. $Na(H_{1-x}D_x)_3(SeO_3)_2$. P_s vs. T [69 S 8]. P_s is measured along the *a*-axis.

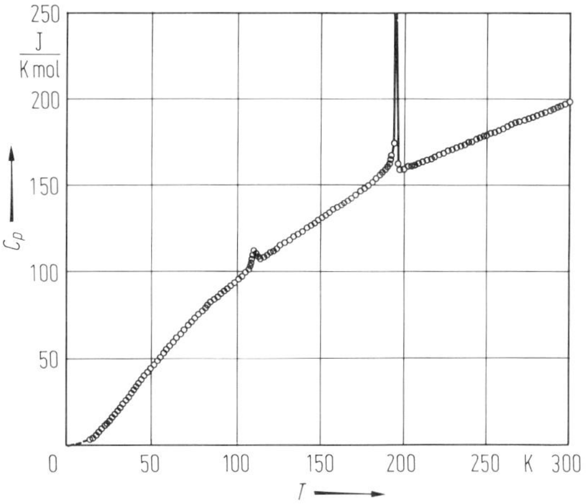

Fig. 933. $NaH_3(SeO_3)_2$. C_p vs. T [76 M 7].

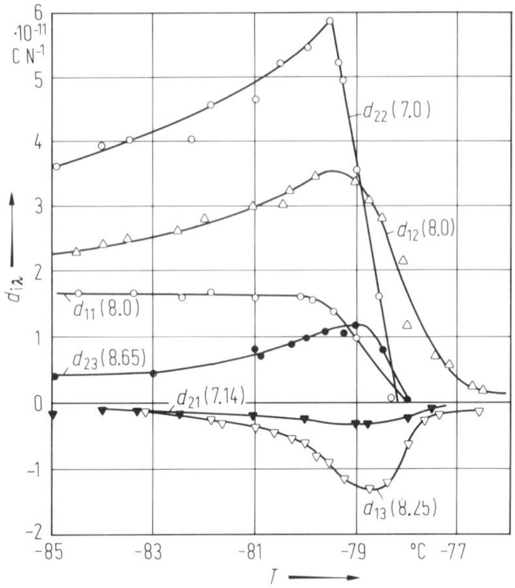

Fig. 934. $NaH_3(SeO_3)_2$. $d_{i\lambda}$ vs. T [75S14]. Figures in the parenthes indicate biasing fields in $10^5\,Vm^{-1}$.

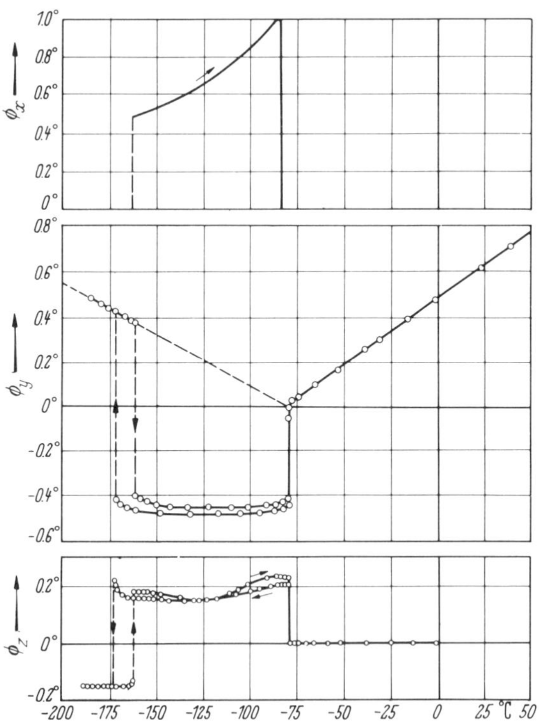

Fig. 937. $NaH_3(SeO_3)_2$. ϕ_X, ϕ_Y, ϕ_Z vs. T [67S15]. ϕ_X, ϕ_Y, ϕ_Z: rotational angles of indicatrix around X, Y, and Z.

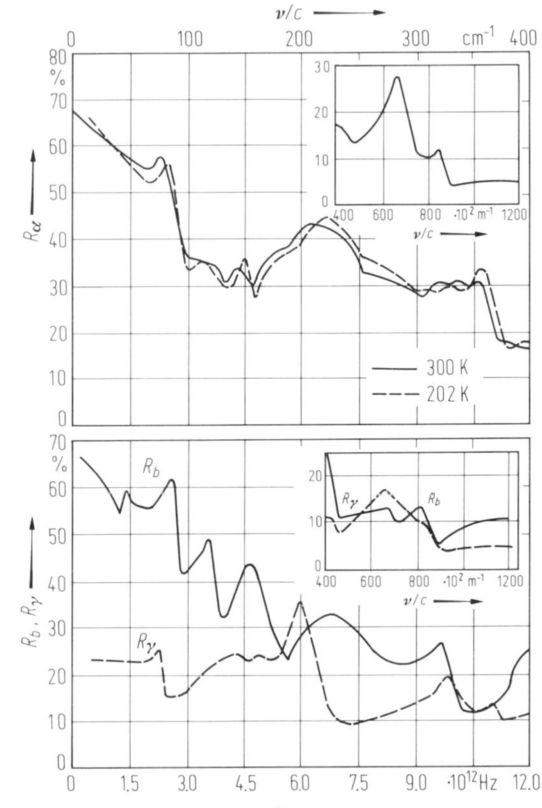

Fig. 935. $NaH_3(SeO_3)_2$. R_α, R_b and R_γ vs. ν [70S23]. R_b: reflectivity for the light with polarization parallel to the b axis. R_α, R_γ: parallel to two extinction directions in the ac plane.

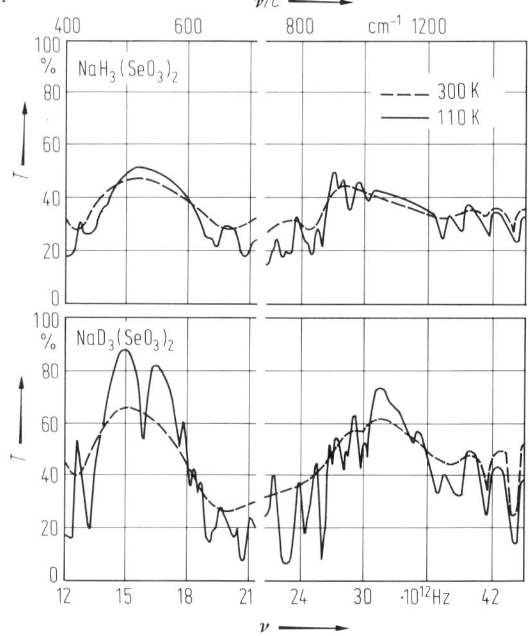

Fig. 936. $NaH_3(SeO_3)_2$, $NaD_3(SeO_3)_2$. T vs. ν [73A3]. T: transmission of the infrared radiation. The deuterium content is estimated to be more than 80%.

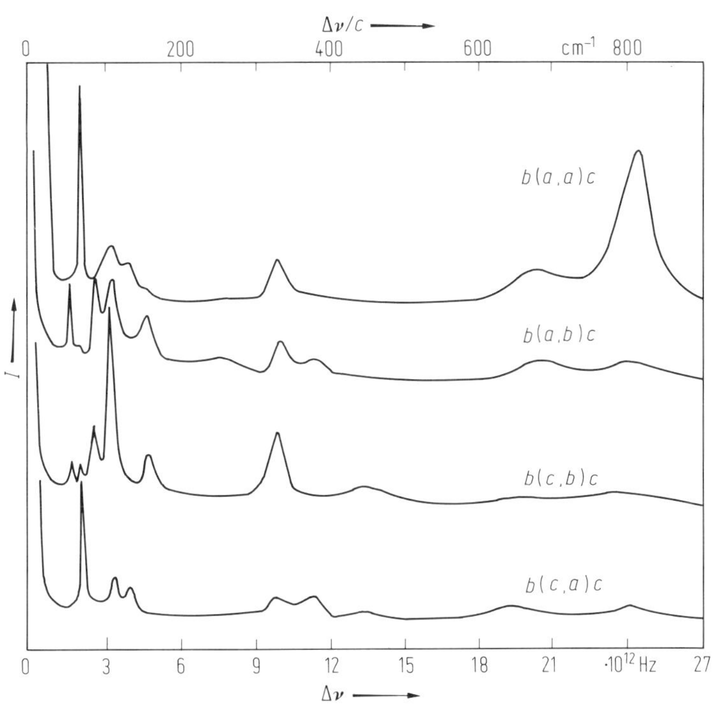

Fig. 938. $NaH_3(SeO_3)_2$. I vs. $\Delta\nu$ [73A3]. I: Raman scattering intensity at room temperature for various scattering geometries.

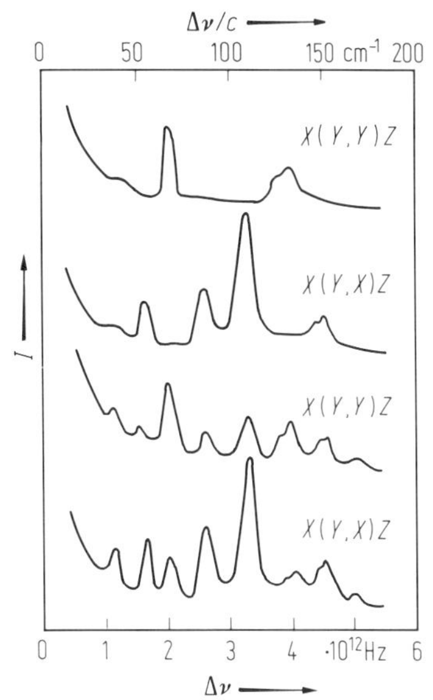

Fig. 939. $NaH_3(SeO_3)_2$. I vs. $\Delta\nu$ near the phase transition temperature [75S18]. I: Raman scattering intensity for various scattering geometries. Upper two spectra are for $T > \Theta_{II-I}$, and the lower two are for $T < \Theta_{II-I}$.

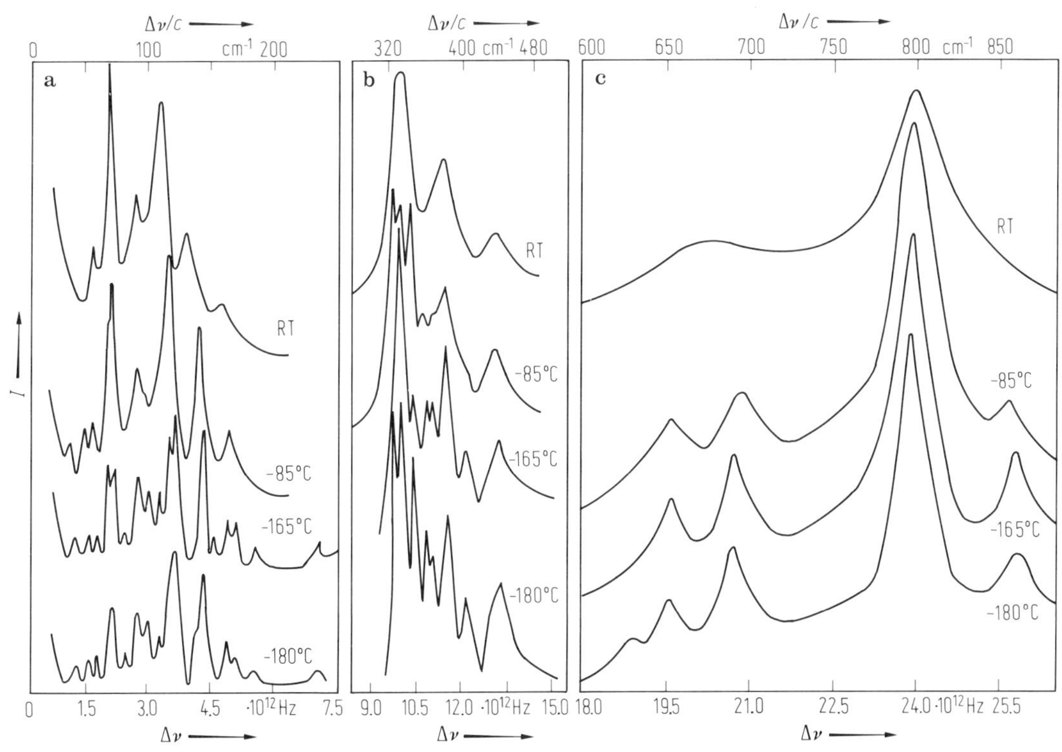

Fig. 940. $Na(H_{0.9}D_{0.1})_3(SeO_3)_2$. I vs. $\Delta\nu$ [73T3]. I: Raman scattering intensity for various temperatures. The sample was a pellet and polarization geometry was not specified.

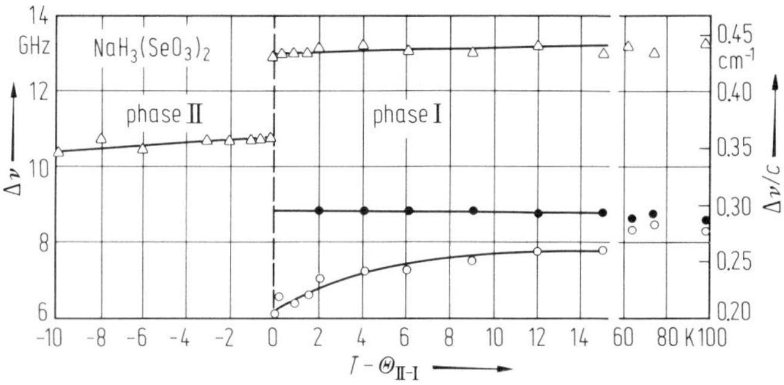

Fig. 941. $NaH_3(SeO_3)_2$. $\Delta\nu$ vs. $T-\Theta_{II-I}$ [76L1]. $\Delta\nu$: Brillouin shift. Wave vector‖[100]. Open circles: transverse mode; full circles: quasi-transverse mode, open triangles: quasi-longitudinal mode.

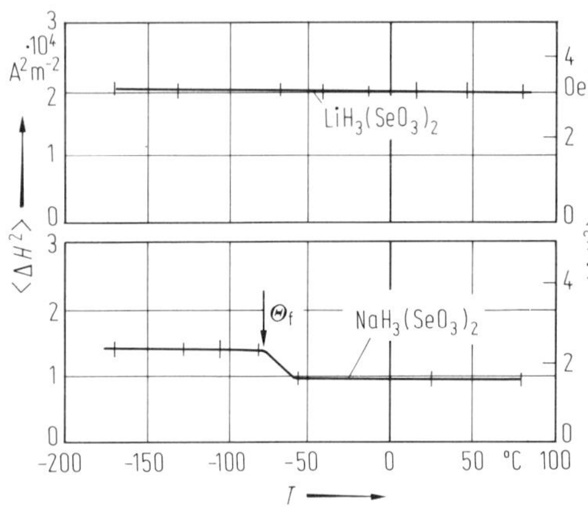

Fig. 942. $NaH_3(SeO_3)_2$, $LiH_3(SeO_3)_2$. $\langle \Delta H^2 \rangle$ vs. T [61B3]. $\langle \Delta H^2 \rangle$: second moment of proton.

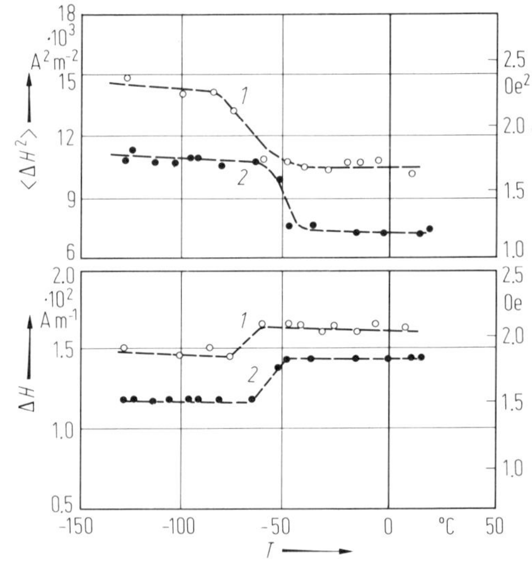

Fig. 943. $NaH_{3x}D_{3(1-x)}(SeO_3)_2$. $\langle \Delta H^2 \rangle$, ΔH vs. T [67G2]. $\langle \Delta H^2 \rangle$: second moment of proton, ΔH: half width of proton. Curve *1*: x = 1, *2*: x = 0.58.

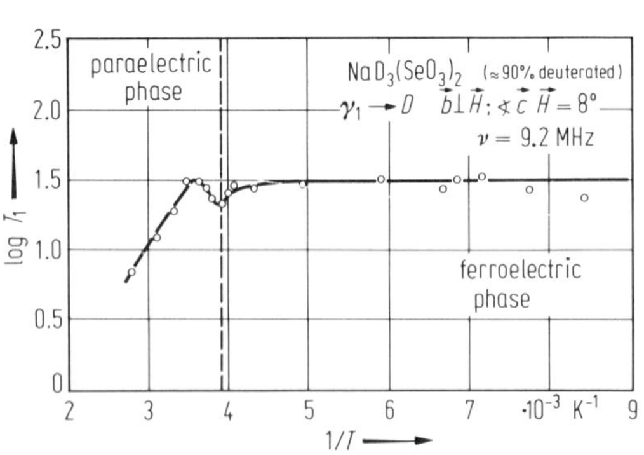

Fig. 944. $NaD_3(SeO_3)_2$. $\log T_1$ vs. $1/T$ [68B10]. T_1: deuteron spin-lattice relaxation time (in s).

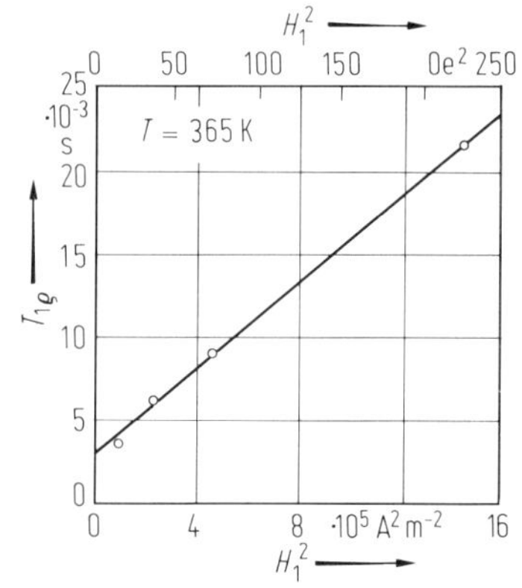

Fig. 945. $NaH_3(SeO_3)_2$. $T_{1\varrho}$ vs. H_1^2 [71B18]. $T_{1\varrho}$: roton spin-lattice relaxation time in the rotating frame; H_1: radio frequency magnetic field strength.

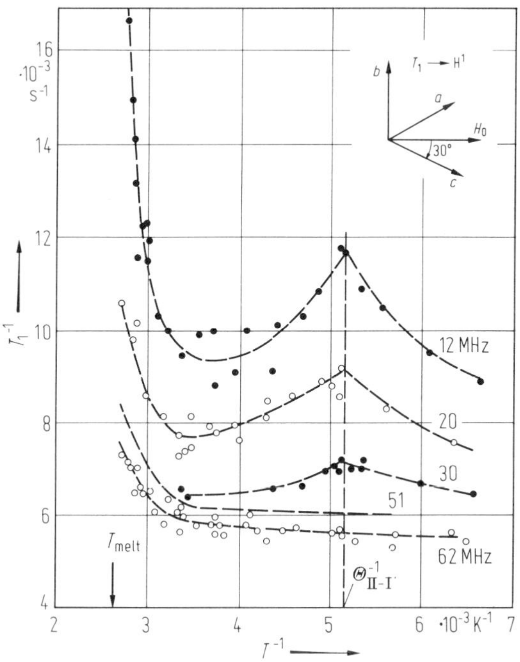

Fig. 946. $NaH_3(SeO_3)_2$. T_1^{-1} vs. T^{-1} [74 A 2]. T_1^{-1}: Inverse of the proton spin-lattice relaxation time. See also [71 K 14, 76 M 1].

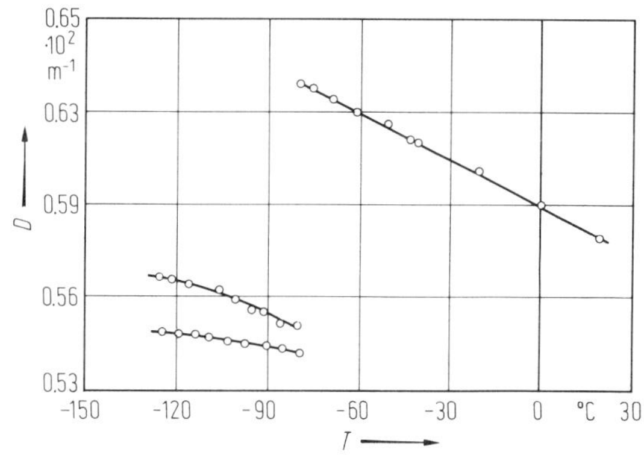

Fig. 947. $NaH_3(SeO_3)_2$. D vs. T [77 M 7]. D: Crystalline field parameter of Cr^{3+} substituted for Na^+.

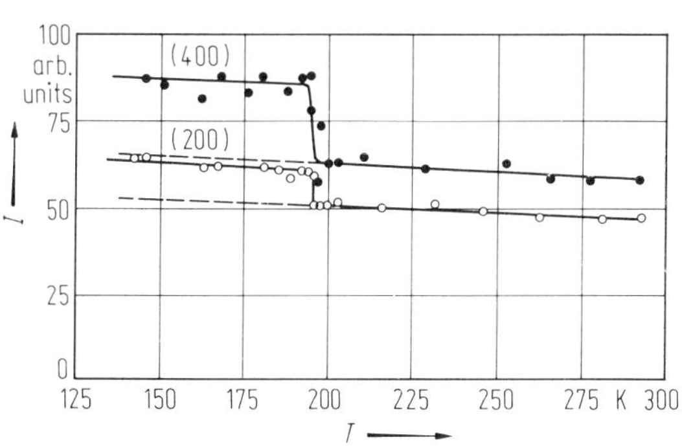

Fig. 948. $NaH_3(SeO_3)_2$. $I_{(200)}$, $I_{(400)}$ vs. T [73 M 14]. I: Integrated intensity of X-ray Bragg reflection. The coordinates refer to the a', b', c' axial system.

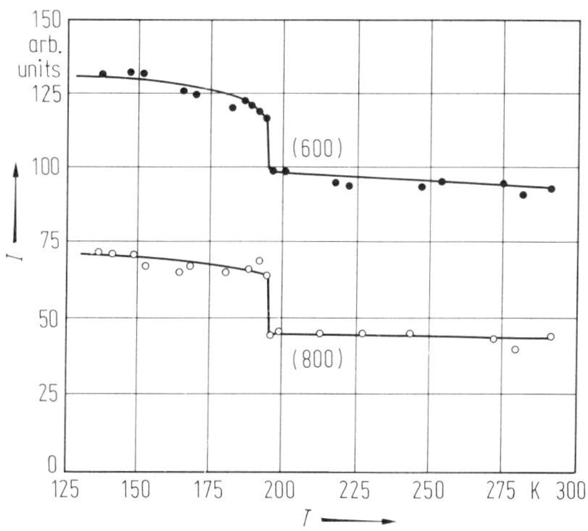

Fig. 949. $NaH_3(SeO_3)_2$. $I_{(600)}$, $I_{(800)}$ vs. T [73 M 14]. I: integrated intensity of X-ray Bragg reflection. The coordinates refer to the a', b', c' axial system.

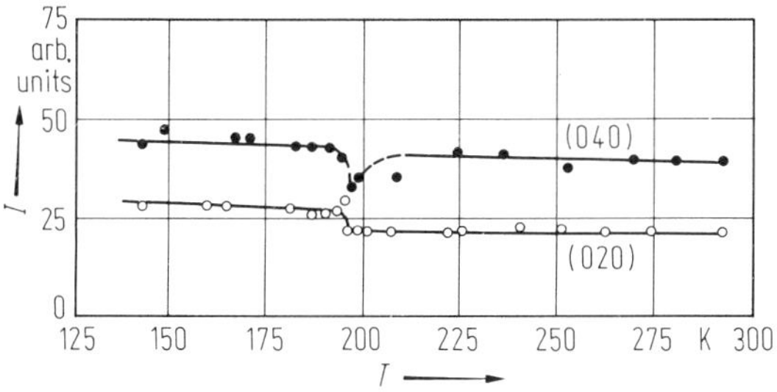

Fig. 950. $NaH_3(SeO_3)_2$. $I_{(640)}$, $I_{(020)}$ vs. T [73 M 14]. I: integrated intensity of X-ray Bragg reflection. The coordinates refer to the a', b', c' axial system.

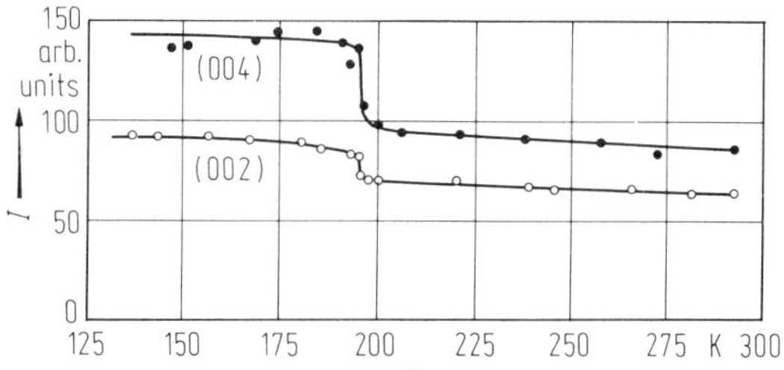

Fig. 951. $NaH_3(SeO_3)_2$. $I_{(004)}$, $I_{(002)}$ vs. T [73 M 14]. I: integrated intensity of X-ray Bragg reflection. The coordinates refer to the a', b', c' axial system.

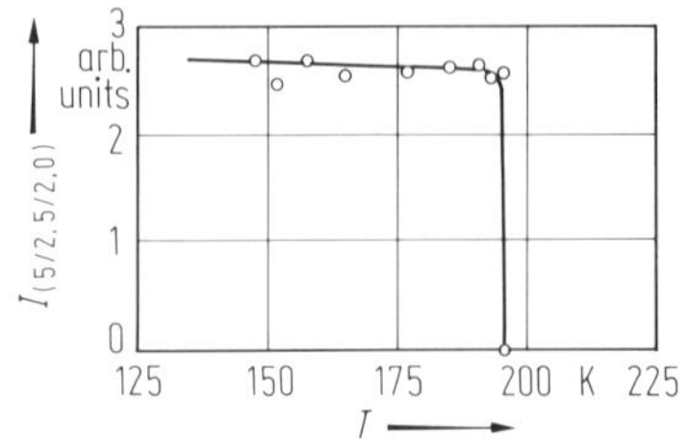

Fig. 952. $NaH_3(SeO_3)_2$. $I_{(5/2,\,5/2,\,0)}$ vs. T [73M14]. I: integrated intensity of X-ray supperlattice reflection. The coordinates refer to the a', b', c' axial system.

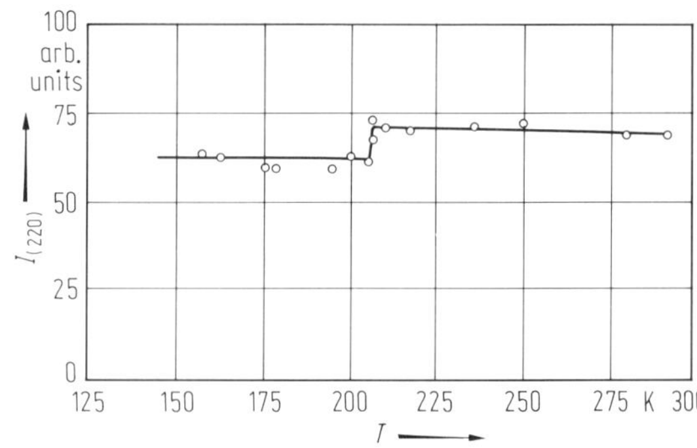

Fig. 953. $NaH_3(SeO_3)_2$. $I_{(220)}$ vs. T [73M14]. I: integrated intensity of X-ray Bragg reflection. The coordinates refer to the a', b', c' axial system.

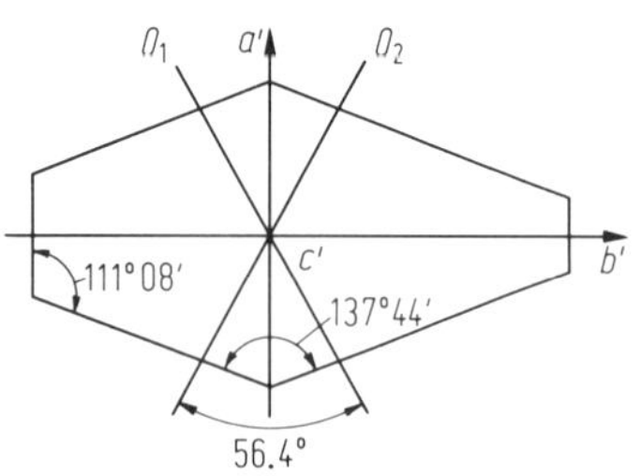

Fig. 954. $KH_3(SeO_3)_2$. Crystal form [67S14]. O_1, O_2: optical axes ($\lambda = 589.0$ nm).

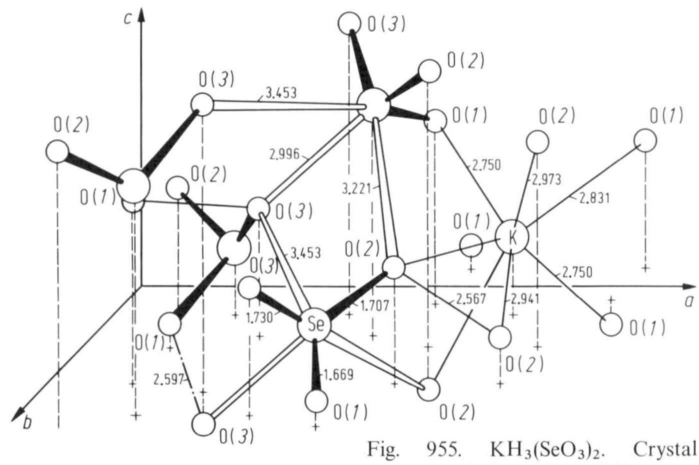

Fig. 955. $KH_3(SeO_3)_2$. Crystal structure of phase I [69H2]. Interatomic distances in Å.

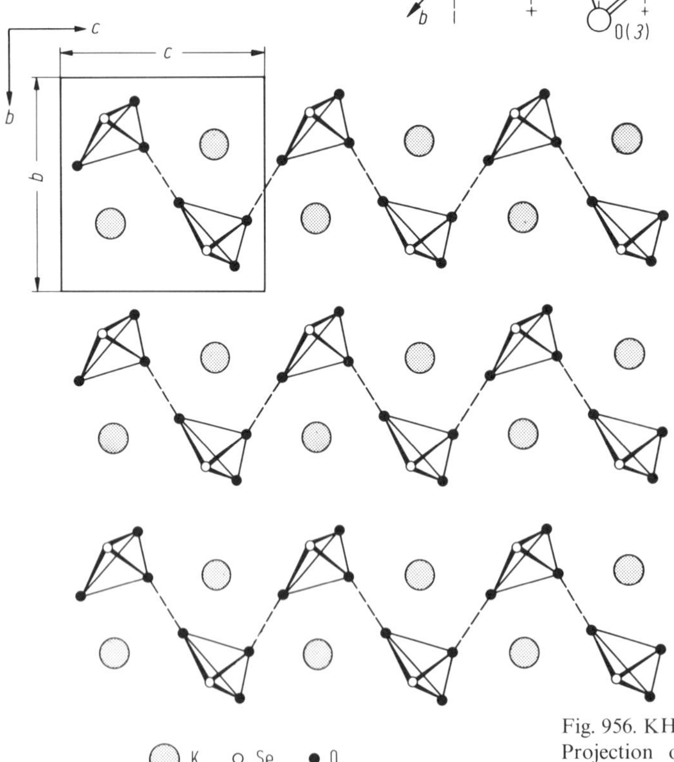

Fig. 956. $KH_3(SeO_3)_2$. Crystal structure of phase I [71G8]. Projection on (100). Dashed lines show the proposed hydrogen bonds.

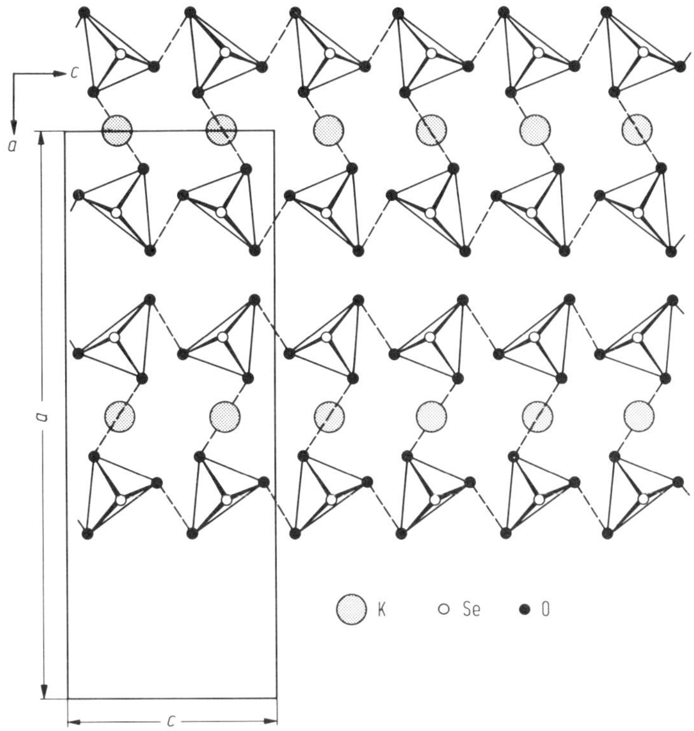

Fig. 957. $KH_3(SeO_3)_2$. Crystal structure of phase I [71G8]. Projection on (010). Dashed lines show the proposed hydrogen bonds.

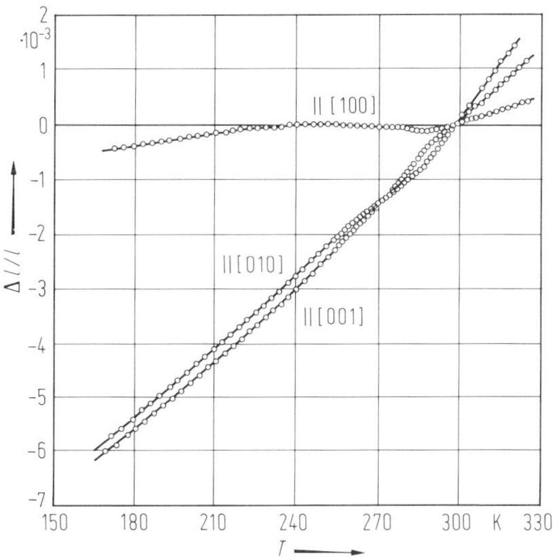

Fig. 958. $KD_3(SeO_3)_2$. $\Delta l/l$ vs. T parallel to [100], [010] and [001]. [78M2]. $\Delta l/l$: linear thermal expansion.

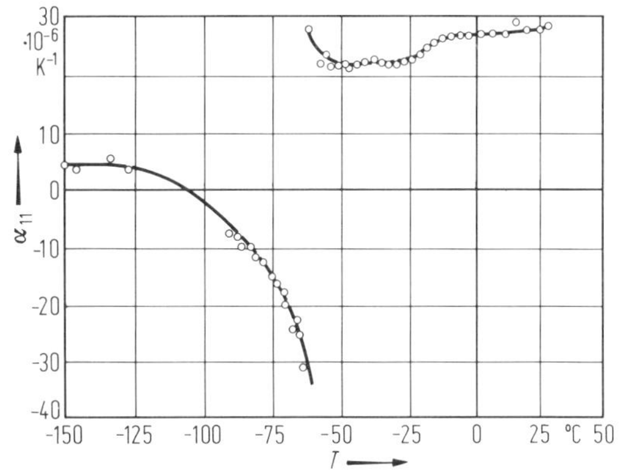

Fig. 959. $KH_3(SeO_3)_2$. α_{11} vs. T [71T2]. α_{11}: linear thermal expansion coefficient along the a axis.

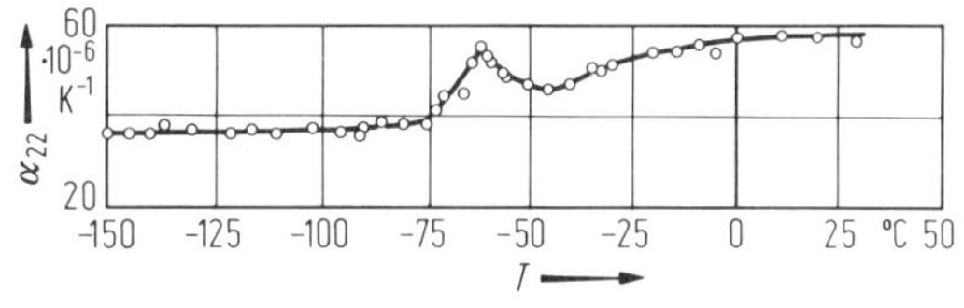

Fig. 960. $KH_3(SeO_3)_2$. α_{22} vs. T [71T2]. α_{22}: linear thermal expansion coefficient along the b axis.

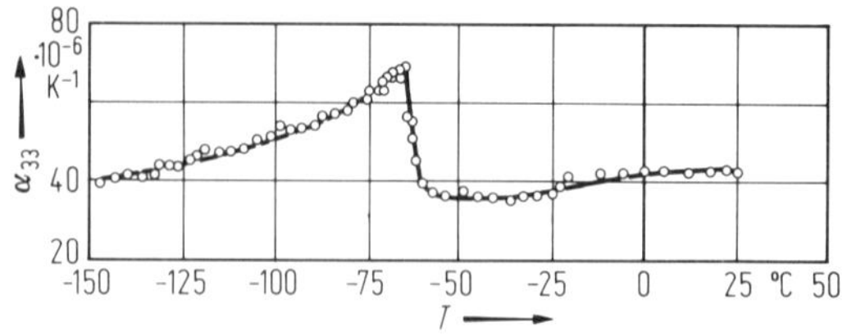

Fig. 961. $KH_3(SeO_3)_2$. α_{33} vs. T [71T2]. α_{33}: linear thermal expansion coefficient along the c axis.

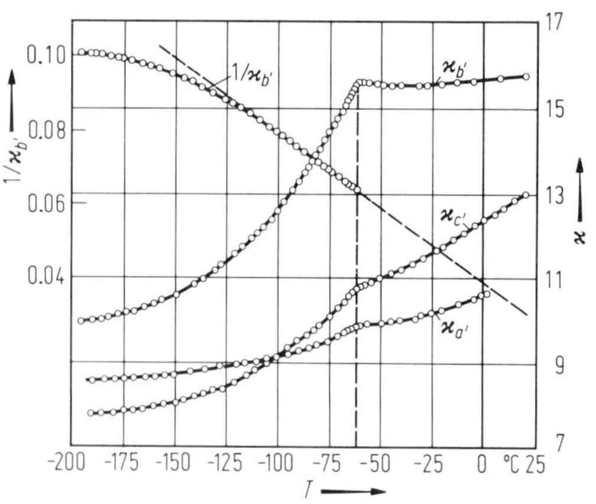

Fig. 962. $KH_3(SeO_3)_2$. $\varkappa_{a'}$, $\varkappa_{b'}$, $\varkappa_{c'}$, $1/\varkappa_{b'}$ vs. T [67S14]. $\varkappa$: low frequency dielectric constant.

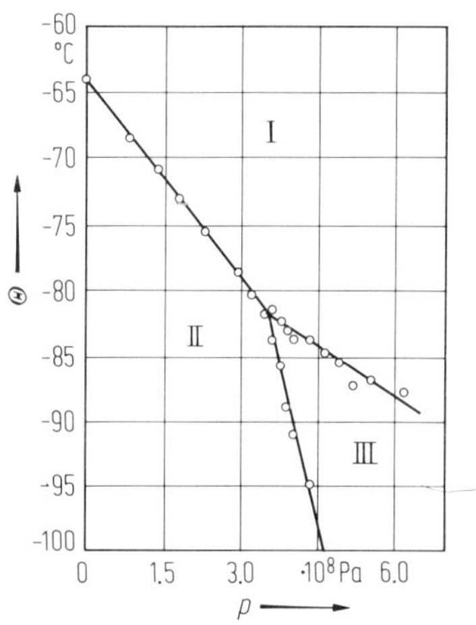

Fig. 963. $KH_3(SeO_3)_2$. Θ vs. p [73G5].

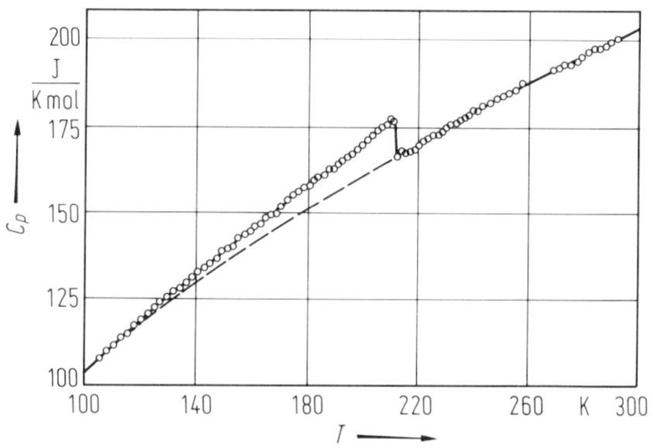

Fig. 965. $KH_3(SeO_3)_2$. C_p vs. T [77M6].

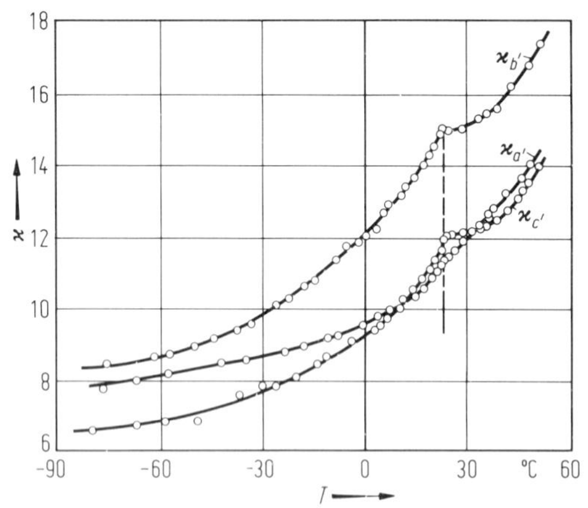

Fig. 694. $KD_3(SeO_3)_2$. $\varkappa_{a'}$, $\varkappa_{b'}$, $\varkappa_{c'}$, vs. $T(f=2\,kHz)$ [69Y1].

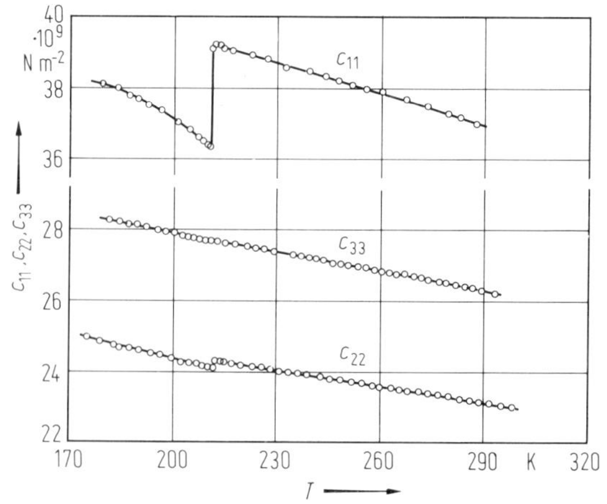

Fig. 966. $KH_3(SeO_3)_2$. c_{11}, c_{22}, c_{33} vs. T [77M5].

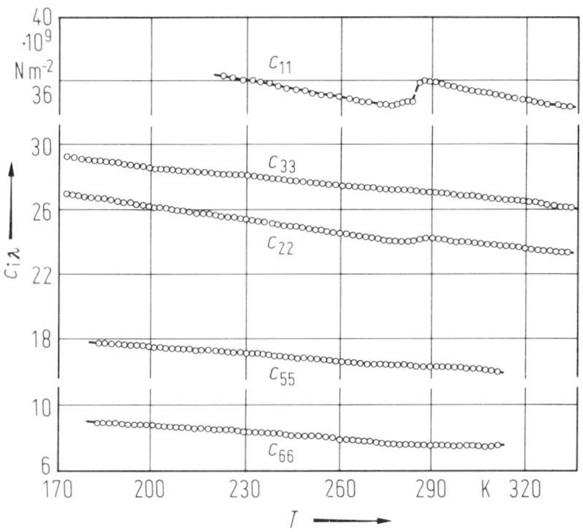

Fig. 967. $KD_3(SeO_3)_2$. $c_{i\lambda}$ vs. T [78M2]. f=10 MHz for measuring the velocity of longitudinal waves. f=5 MHz for transverse waves.

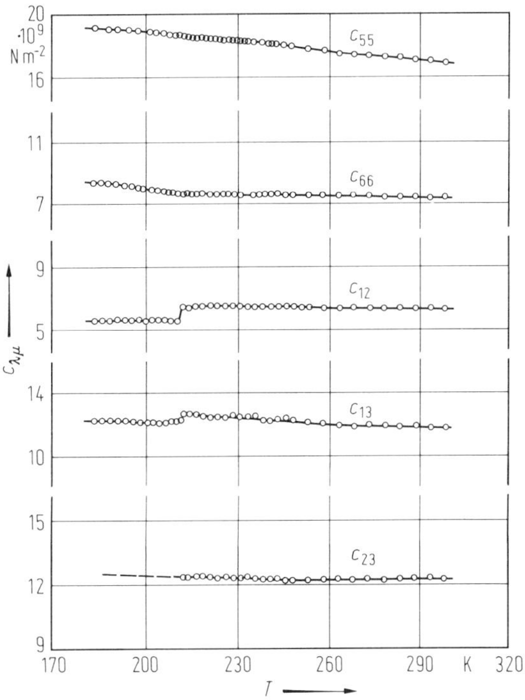

Fig. 968. $KH_3(SeO_3)_2$. c_{55}, c_{66}, c_{12}, c_{13}, c_{23} vs. T[77M5].

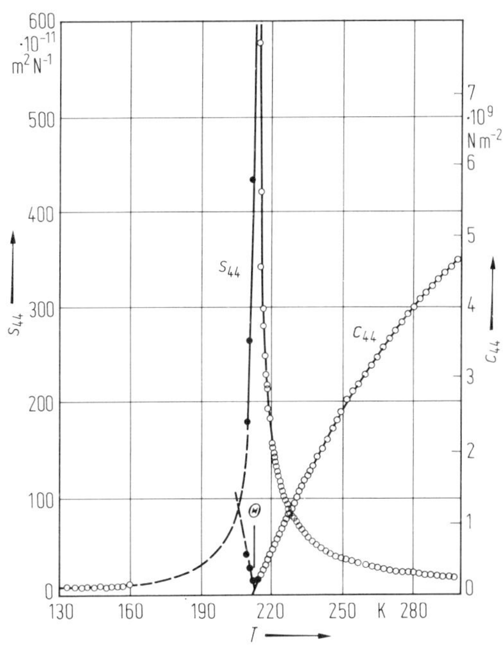

Fig. 969. $KH_3(SeO_3)_2$. s_{44}, c_{44} vs. T[77M5].

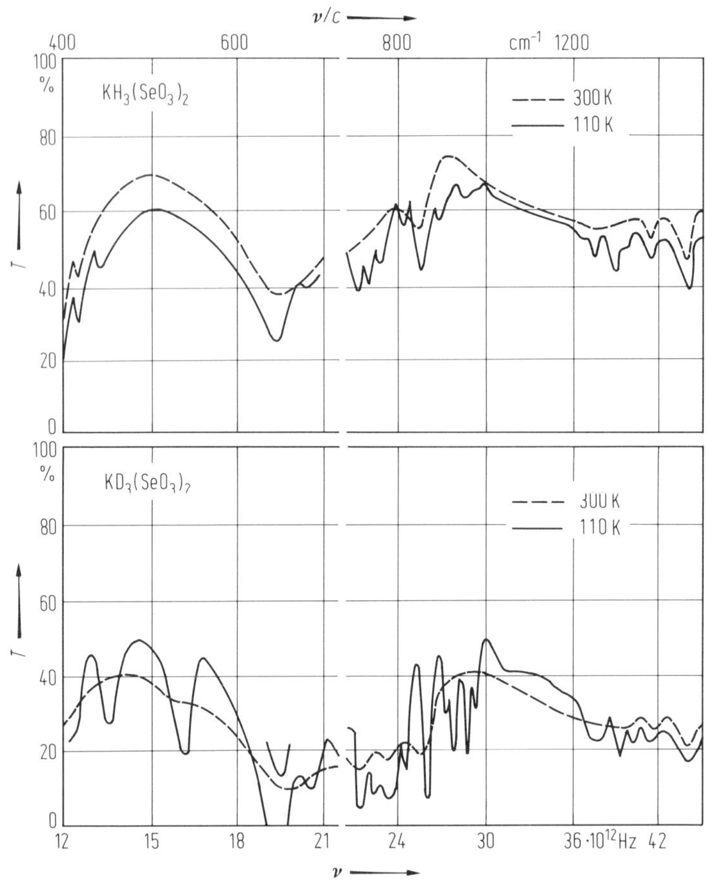

Fig. 970. $KH_3(SeO_3)_2$, $KD_3(SeO_3)_2$. T vs. ν [73A3]. T: transmission of the infrared radiation. The deuterium content is estimated to be more than 80%.

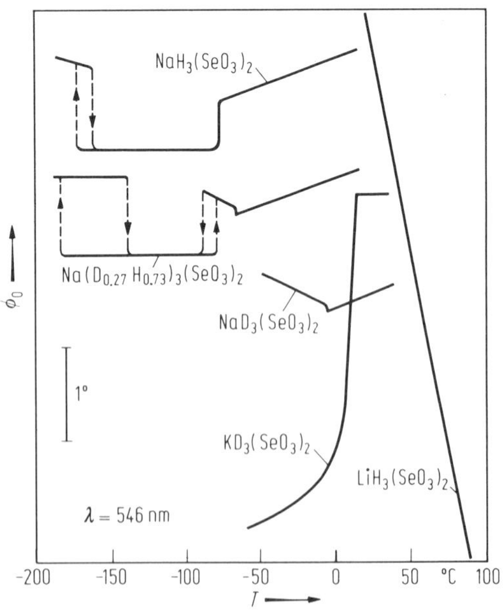

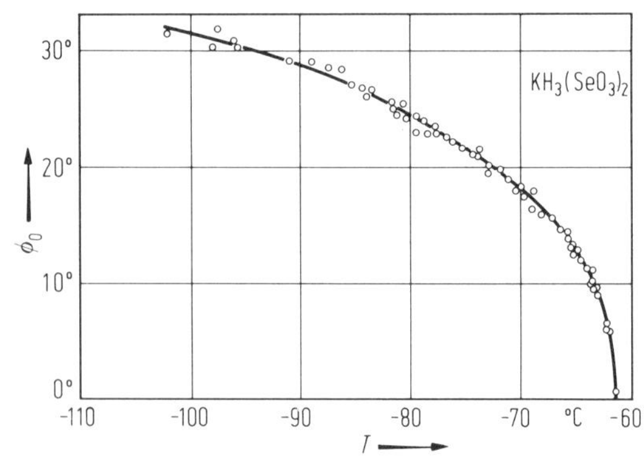

Fig. 971. $LiH_3(SeO_3)_2$, $NaH_3(SeO_3)_2$, $NaD_3(SeO_3)_2$, $Na(D_{0.27}H_{0.73})_3(SeO_3)_2$, $KH_3(SeO_3)_2$ and $KD_3(SeO_3)_2$. ϕ_0 vs. T [7017]. ϕ_0: rotation angle of the optical indicatrix around the b'-axis.

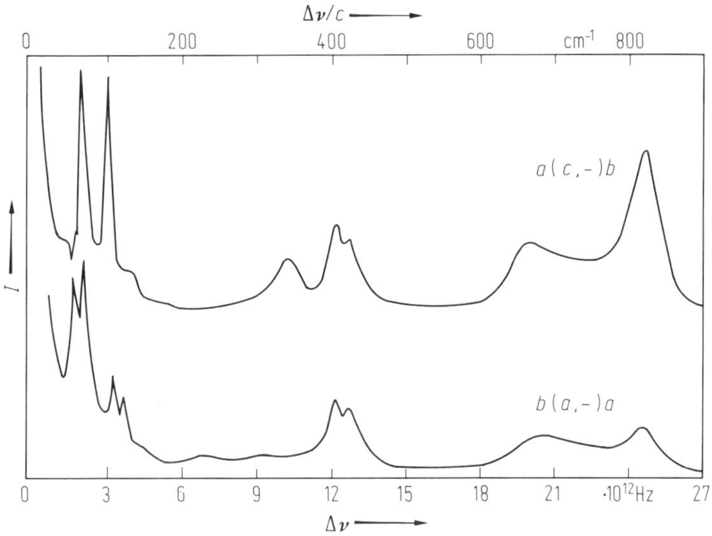

Fig. 972. $KH_3(SeO_3)_2$. I vs. $\Delta\nu$ [73A3]. I: Raman scattering intensity at RT. Hyphens in the rotation of the scattering geometry mean that the polarization of light is not specified.

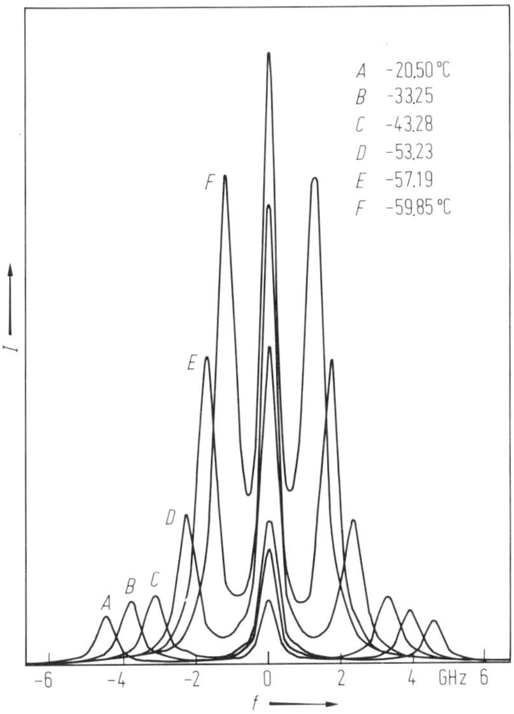

Fig. 973. $KH_3(SeO_3)_2$. I vs. f [77Y1]. I: Brillouin spectra associated with the soft acoustic mode s_{44} observed for as-grown crystal.

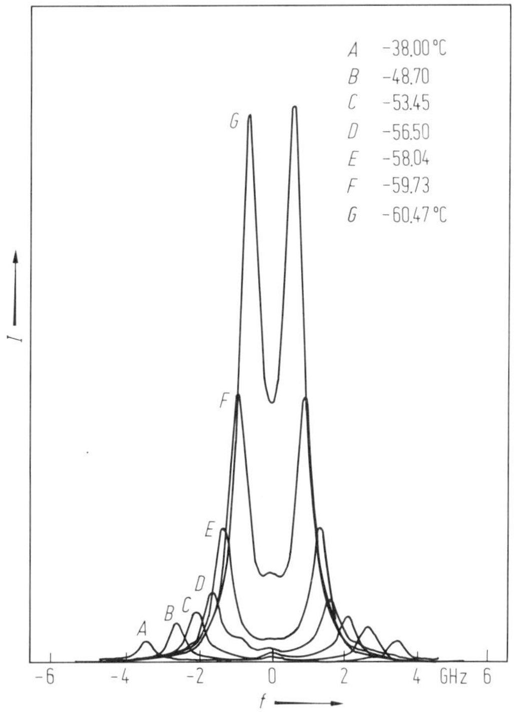

Fig. 974. $KH_3(SeO_3)_2$. I vs. f [77Y1]. I: Brillouin spectra associated with the soft acoustic s_{44} mode observed after repeated measurements.

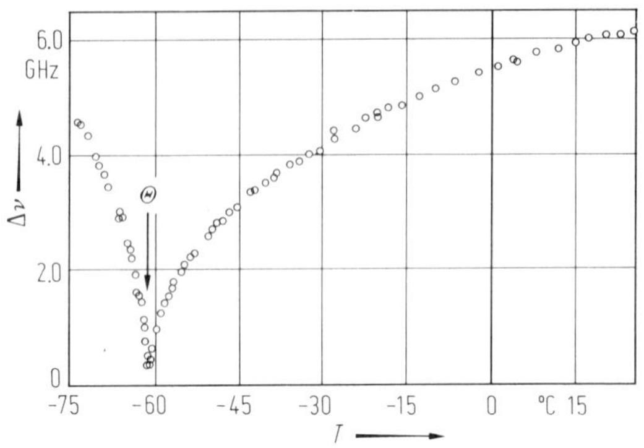

Fig. 975. $KH_3(SeO_3)_2$. $\Delta\nu$ vs. T [77Y2]. $\Delta\nu$: Brillouin shift of c_{44} mode. $\lambda = 488.0$ nm.

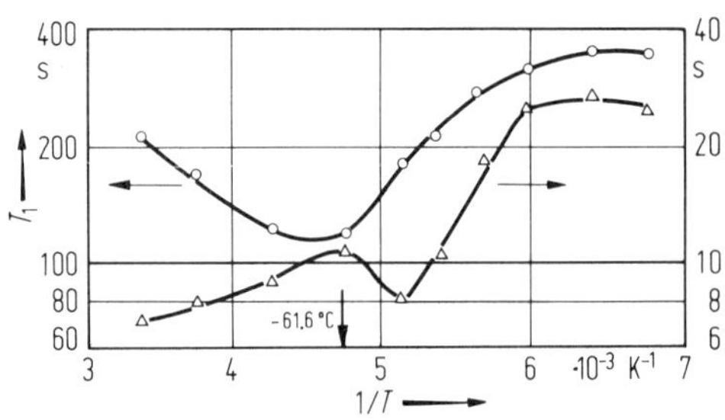

Fig. 976. $KH_3(SeO_3)_2$. T_1 vs. $1/T$ [71S22]. T_1: spin-lattice relaxation time of proton. Relaxation time was measured for single crystal. $\boldsymbol{H}\|a$. The two T_1 values are attributed to two nonequivalent sites of hydrogens.

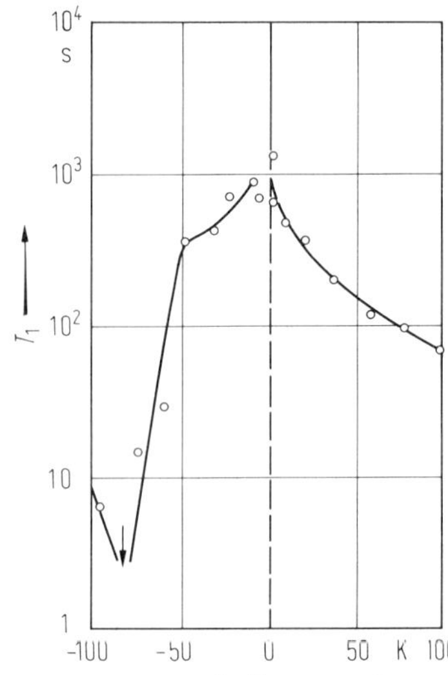

Fig. 977. $KH_3(SeO_3)_2$. T_1 vs. $T - \Theta_{II-I}$ [78S10]. T_1: proton spin-lattice relaxation time. Magnetic field $= 21.48 \cdot 10^3$ Am^{-1}.

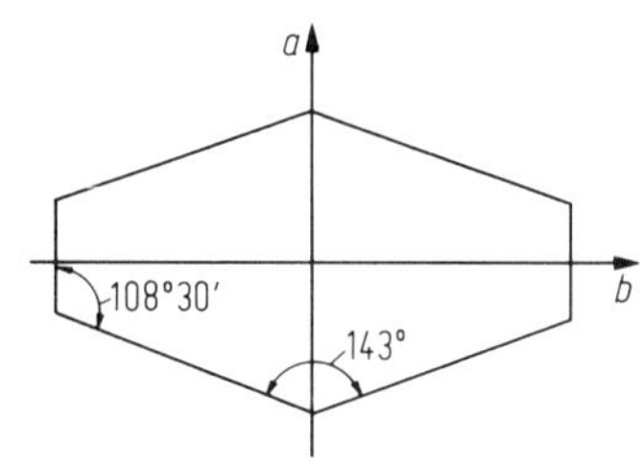

Fig. 978. $RbH_3(SeO_3)_2$. Crystal form [69S7].

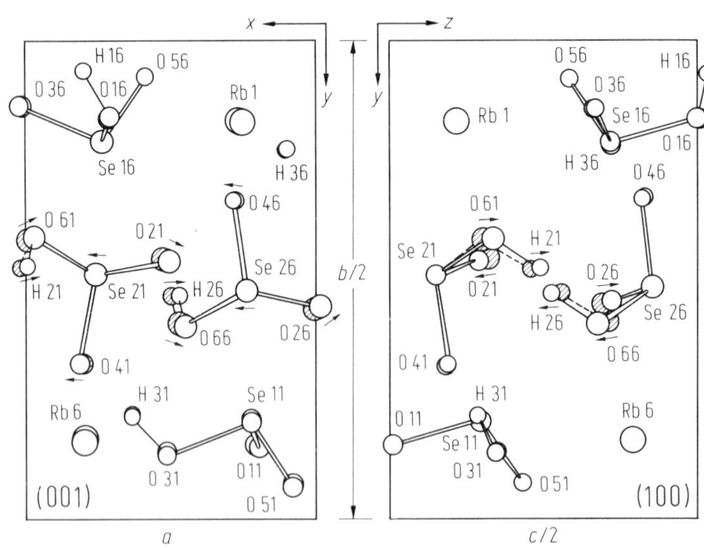

Fig. 979. $RbH_3(SeO_3)_2$. Crystal structure of phase II [78G13]. Projections on (001) and (100). Shaded atoms represent the even structure which is obtained by an assumption that the super lattice reflections are not observed. The arrows show the major displacements due to the phase transition.

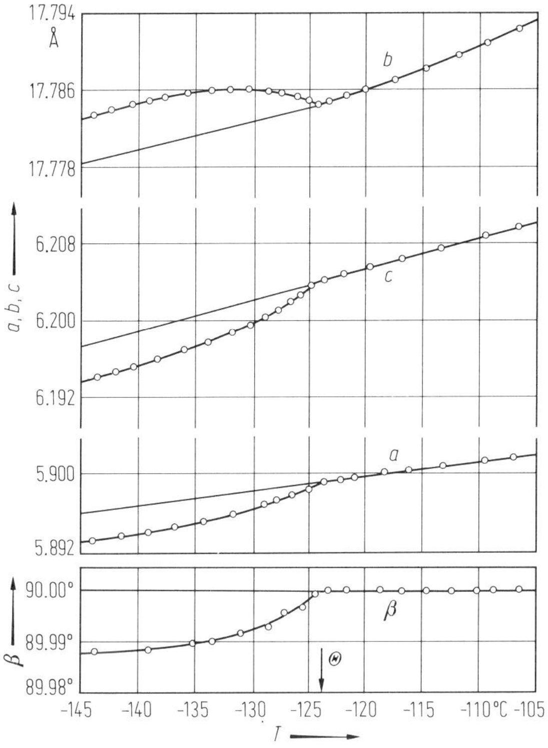

Fig. 980. $RbH_3(SeO_3)_2$. a, b, c, β vs. T[78K21].

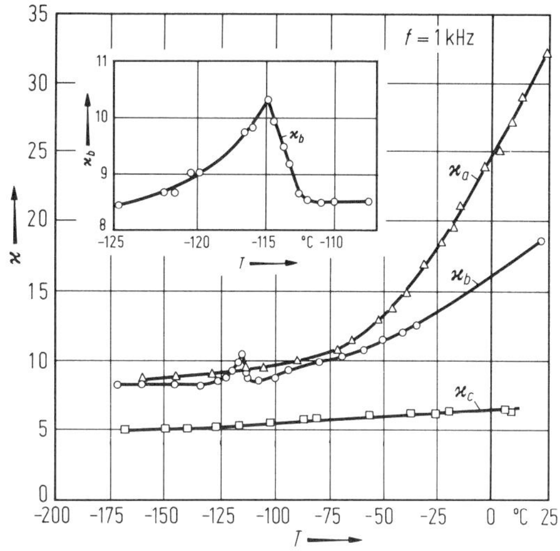

Fig. 981. $RbH_3(SeO_3)_2$. $\varkappa_a$, $\varkappa_b$, $\varkappa_c$, vs. T[69S7].

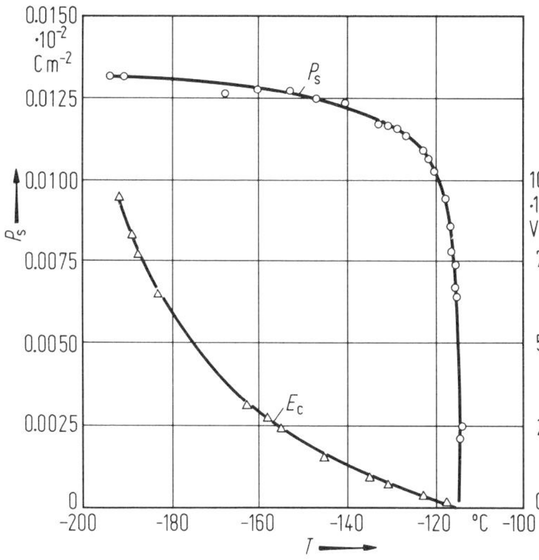

Fig. 982α. $RbH_3(SeO_3)_2$. P_s, E_c vs. T[69S7].

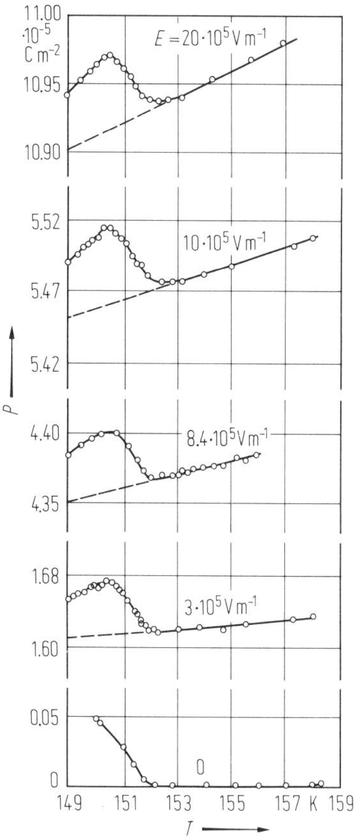

Fig. 982β. $RbD_3(SeO_3)_2$. P vs. T[76G9]. Parameter: E.

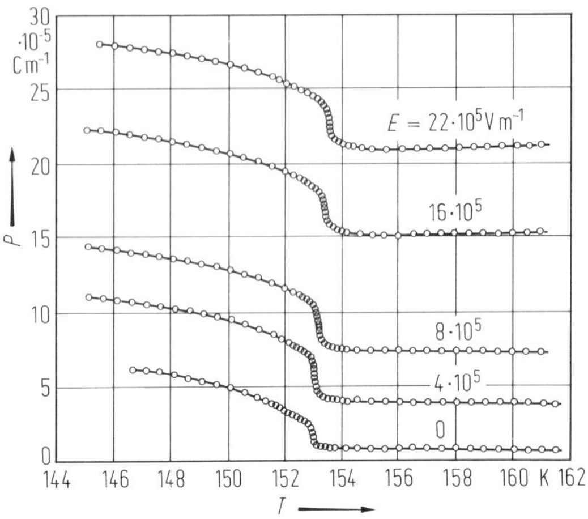

Fig. 982γ. $RbH_3(SeO_3)_2$. P vs. T [76G10]. Parameter: E.

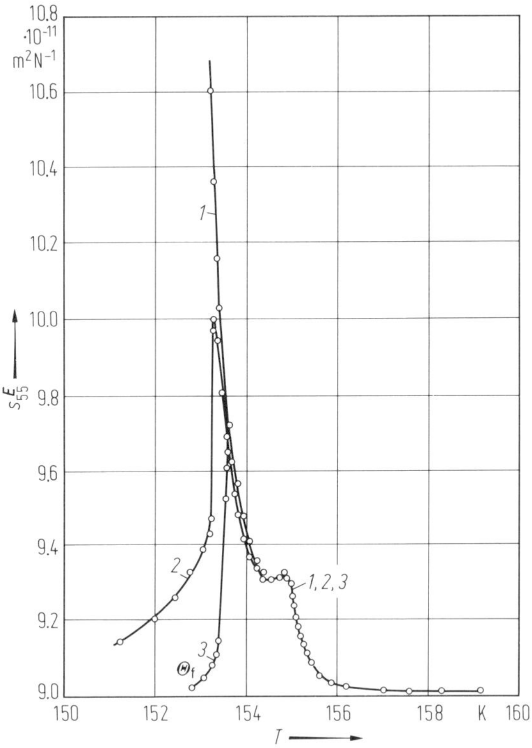

Fig. 983. $RbH_3(SeO_3)_2$. s_{55}^E vs. T [77G9]. Parameter: E_{bias}. Curve *1*: $E_{bias}=0$; *2*: $1\cdot10^3$kV m^{-1}; *3*: $2.2\cdot10^3$kV m^{-1}.

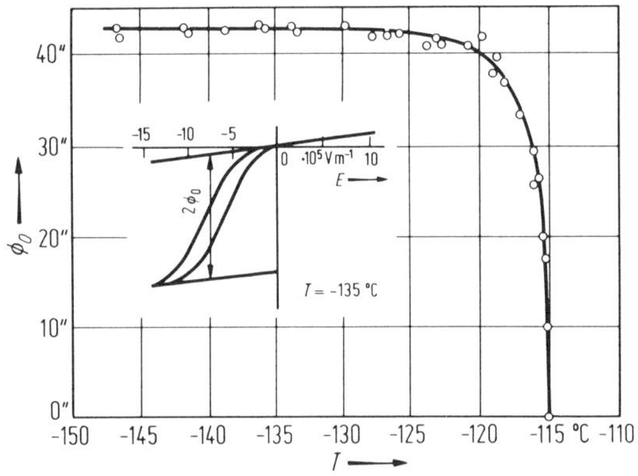

Fig. 984. $RbH_3(SeO_3)_2$. ϕ_0 vs. T, and electrooptical hysteresis loop [70I8]. ϕ_0: angle of rotation of the optical indicatrix around the b-axis.

Fig. 986, see p. 547.

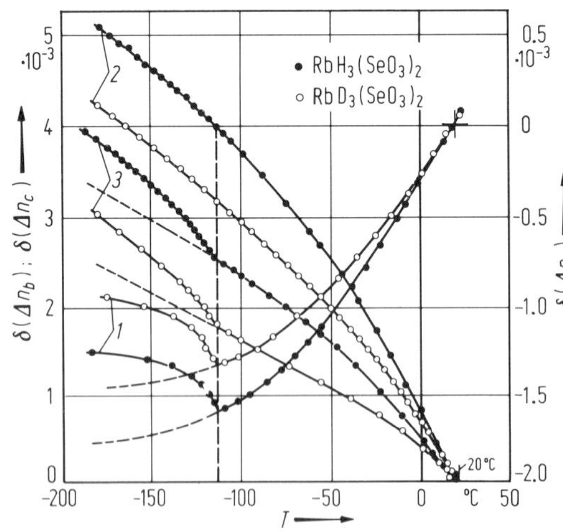

Fig. 985. $RbH_3(SeO_3)_2$, $RbD_3(SeO_3)_2$. $\delta(\Delta n)$ vs. T [70I8]. $\delta(\Delta n)=[\Delta n]_T-[\Delta n]_{20°C}$. Curves *1*: Δn_a, birefringence observed along the a-axis; *2*: Δn_b, along the b-axis; *3*: Δn_c, along the c-axis. At $\lambda=632.8$ nm. λ

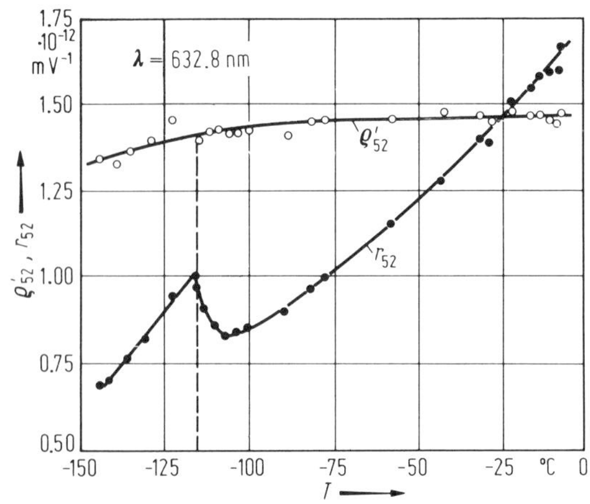

Fig. 987. $RbH_3(SeO_3)_2$. r_{52}, ϱ'_{52} vs. T [70I8]. r_{52}: electrooptic constant for E, $\varrho'_{52}=r_{52}\cdot4\pi(\varkappa_2-1)^{-1}$.

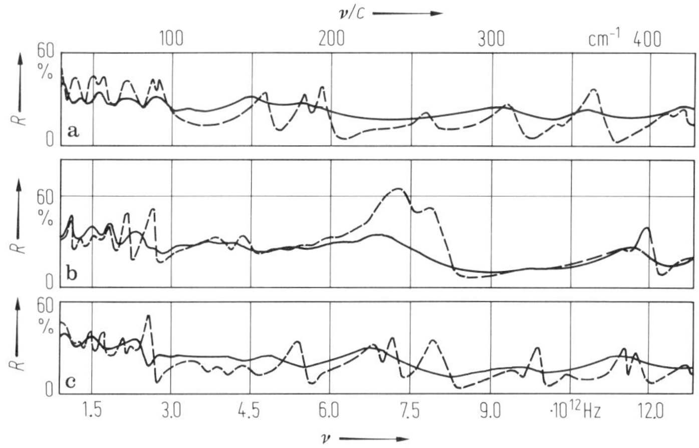

Fig. 986. $RbH_3(SeO_3)_2$. R vs. ν [73P7]. R: Reflectivity for the far infrared radiation polarized along X-axis (Fig. a), Y-axis (Fig. b), and Z-axis (Fig. c). The solid curves are at 25 °C and the dashed curves are at −170 °C.

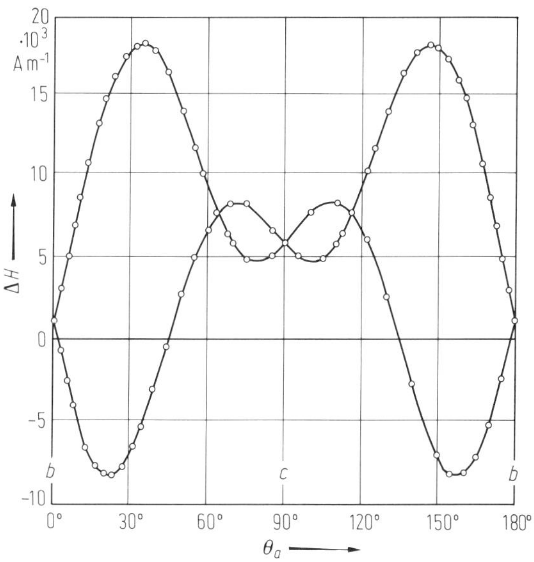

Fig. 988α. $RbH_3(SeO_3)_2$. ΔH vs. θ_a [79T5]. $\Delta H = H_R - H_L$, where H_R and H_L are the resonance field of ^{87}Rb with and without quadrupole interaction, respectively. $f = 8$ MHz, $T = 13$ °C. θ_a: angle of rotation around the a axis.

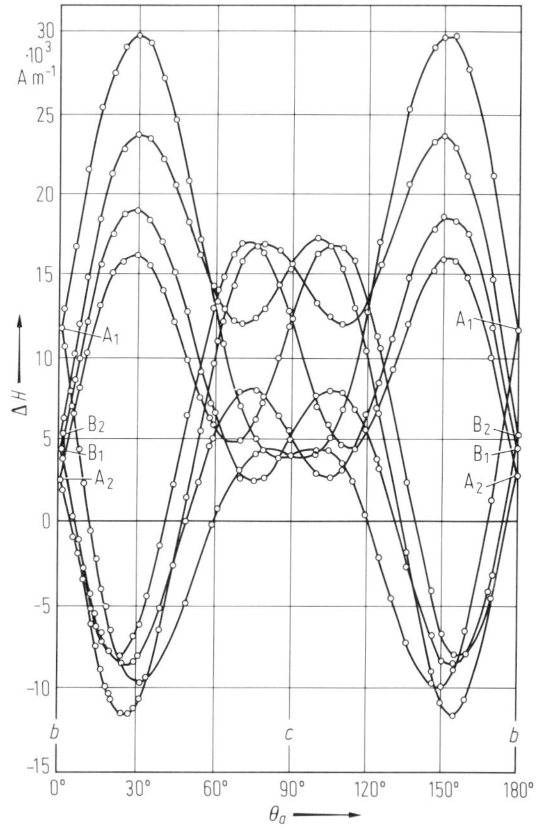

Fig. 988β. $RbH_3(SeO_3)_2$. ΔH vs. θ_a [79T5]. $\Delta H = H_R - H_L$, where H_R and H_L are the resonance field of ^{87}Rb with and without quadrupole interaction, respectively. $f = 8$ MHz, $T = -155.5$ °C. θ_a: angle of rotation around the a axis.

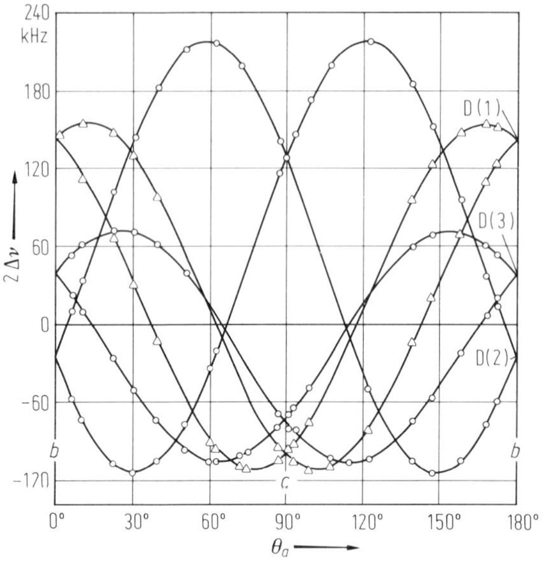

Fig. 988γ. $RbD_3(SeO_3)_2$. $2\Delta\nu$ vs. θ_a [79S8]. $2\Delta\nu=\nu(m=1\leftrightarrow 0)-\nu(m=0\leftrightarrow -1)$: frequency difference between the quadrupole-split pair lines of deuterons for the a axis rotation. $T=14\,^\circ$C. See also Fig. 988δ; Table 260β. θ_a: angle of rotation around the a axis.

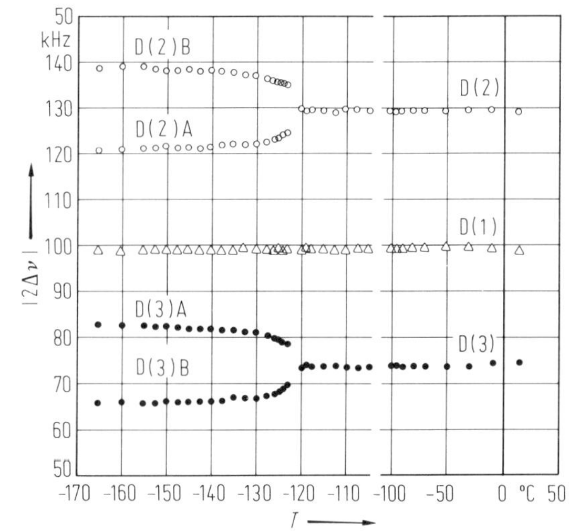

Fig. 988δ. $RbD_3(SeO_3)_2$. $|2\Delta\nu|$ vs. T [79S8]. $2\Delta\nu=\nu(m=1\leftrightarrow 0)-\nu(m=0\leftrightarrow -1)$: frequency difference between the quadrupole-split pair lines for the a axis rotation. $\boldsymbol{H}\parallel c$. See also Fig. 988γ; Table 260β.

Fig. 989, see p. 548.

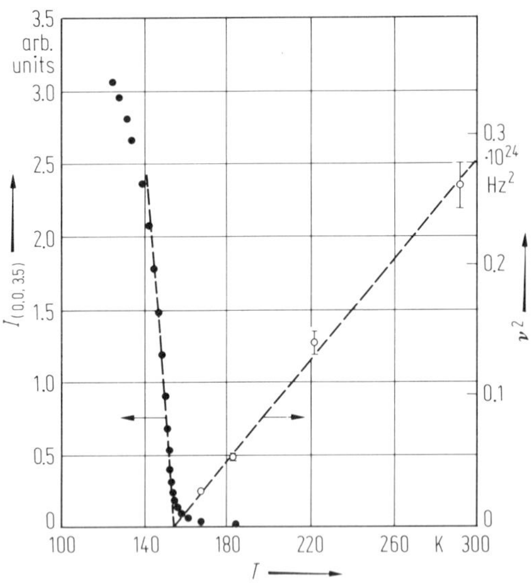

Fig. 990. $RbD_3(SeO_3)_2$. $I_{(0,0,3.5)}$, ν^2 vs. T[78G14]. $I_{(0,0,3.5)}$: intensity of superlattice neutron diffraction peak at (0, 0, 3.5); ν: soft mode frequency at the Z-point.

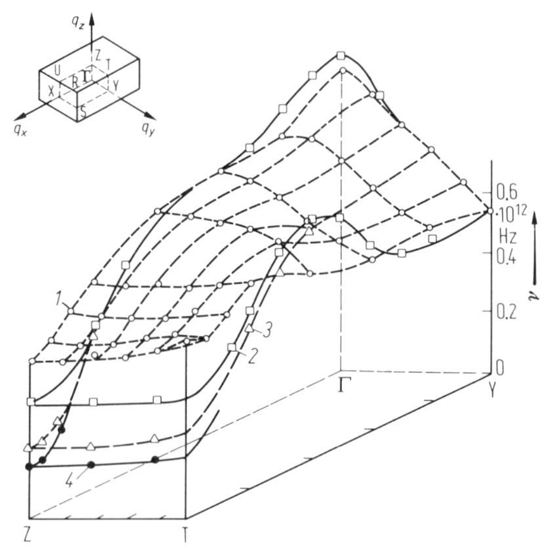

Fig. 991. $RbD_3(SeO_3)_2$. Dispersion surface in the q_y-q_z plane at various temperatures [78G14]. Insert shows the Brillouin zone.

Curve	*1*	*2*	*3*	*4*
T[K]	293	222	182	166

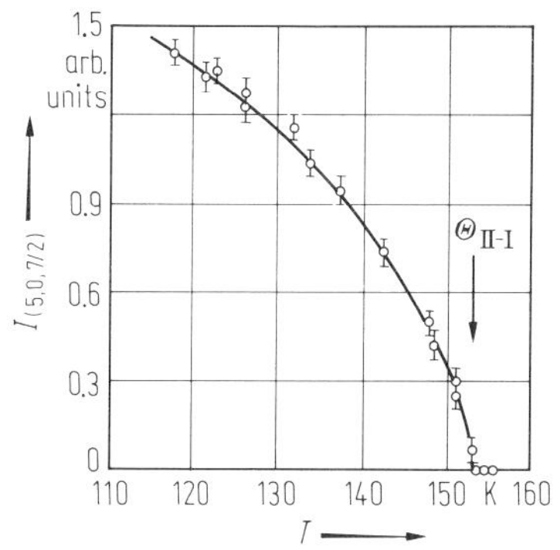

Fig. 989. $RbH_3(SeO_3)_2$. $I_{(5,0,7/2)}$ vs. T [74M2]. I: integrated intensity of X-ray diffraction.

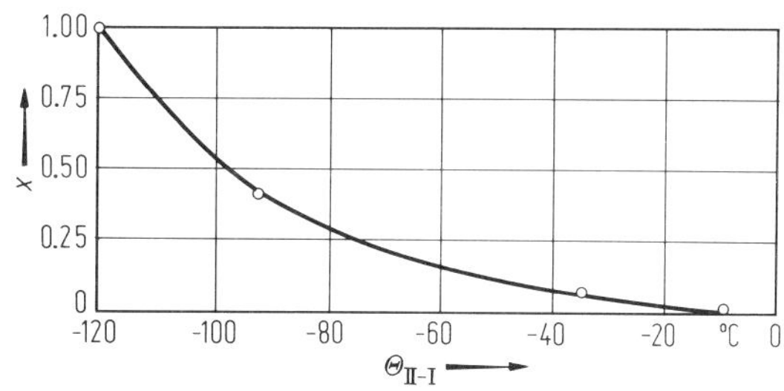

Fig. 992. $Cs(H_xD_{1-x})_3(SeO_3)_2$. x vs. Θ_{II-I} [67G9].

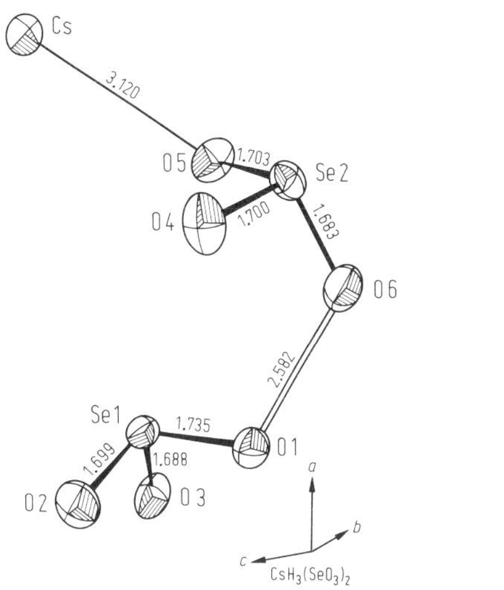

Fig. 993. $CsH_3(SeO_3)_2$. Crystal structure of phase I [74T4]. A sketch of the heavy atoms of the asymmetric unit. O- - - -O hydrogen bond is drawn as open line and ionic bond is drawn as single line. Distances in Å.

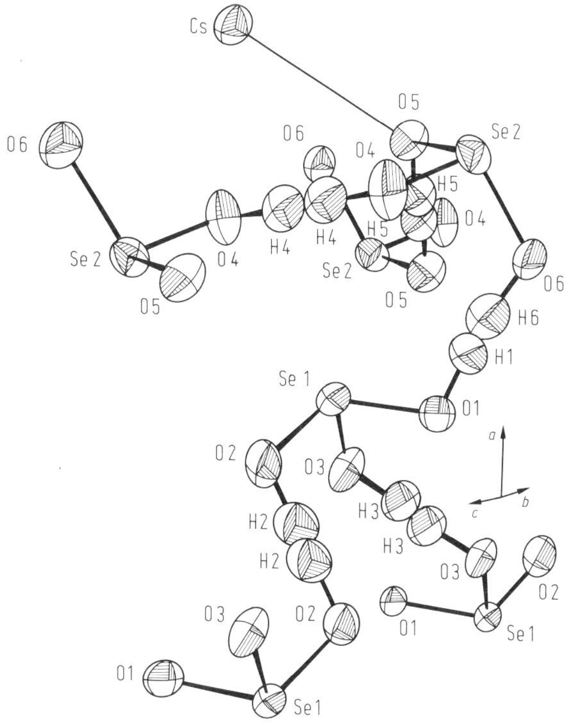

Fig. 995. $CsH_3(SeO_3)_2$. Crystal structure of phase I [78C1]. A sketch of the disordered hydrogens. T=RT.

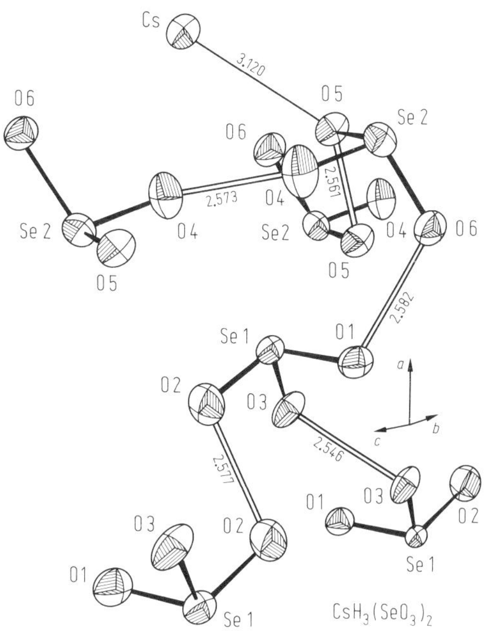

Fig. 994. $CsH_3(SeO_3)_2$. Crystal structure of phase I [74T4]. A sketch of hydrogen-bond scheme. T=RT.

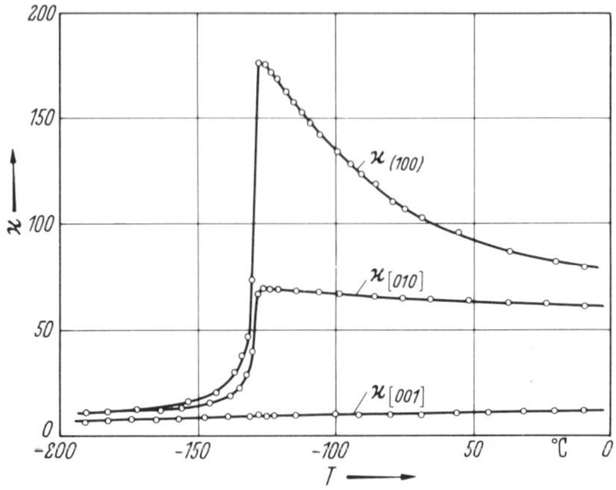

Fig. 996. $CsH_3(SeO_3)_2$. $\varkappa$ vs. T ($f=10$ kHz, $E=500$ Vm^{-1}) [65 M 2]. See Fig. 997.

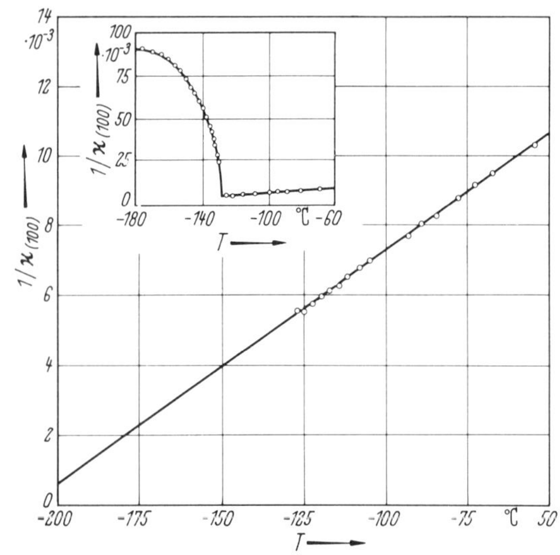

Fig. 997. $CsH_3(SeO_3)_2$. $1/\varkappa_{(100)}$ vs. T ($f=10$ kHz, $E=500$ Vm^{-1} [65 M 2]. $\varkappa_{(100)}$: dielectric constant perpendicular to the (100) plane.

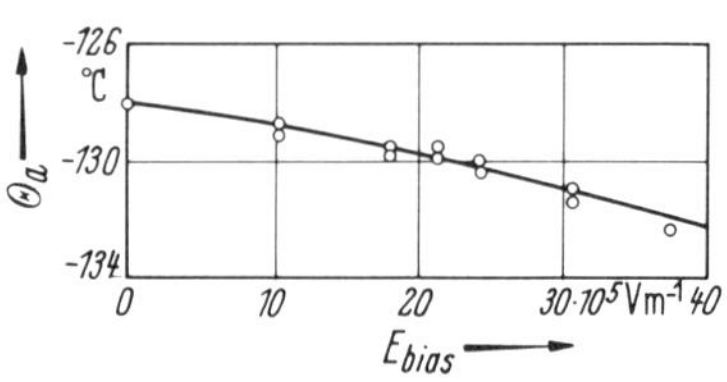

Fig. 998. $CsH_3(SeO_3)_2$. Θ_a vs. E_{bias} [65 M 2].

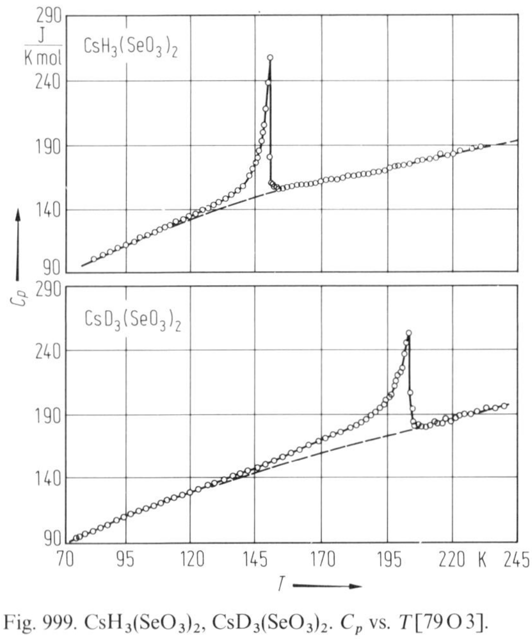

Fig. 999. $CsH_3(SeO_3)_2$, $CsD_3(SeO_3)_2$. C_p vs. T [79 O 3].

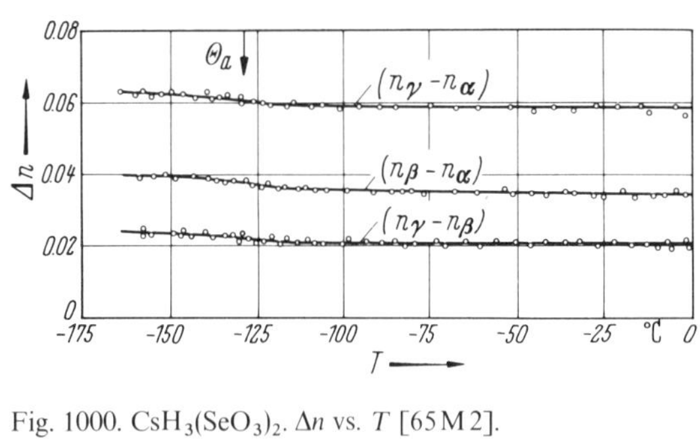

Fig. 1000. $CsH_3(SeO_3)_2$. Δn vs. T [65 M 2].

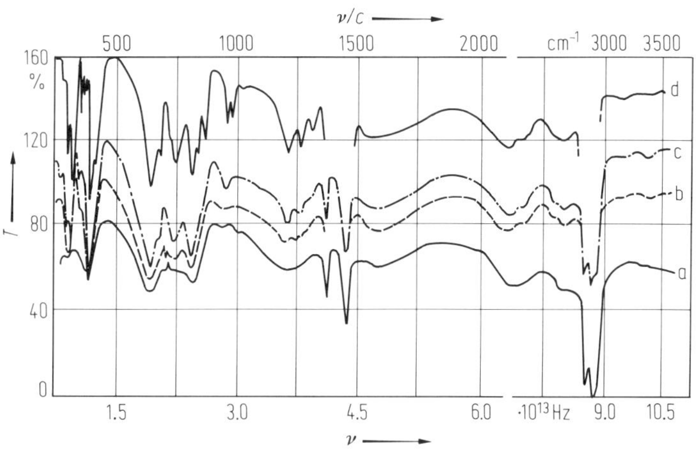

Fig. 1001. $CsH_3(SeO_3)_2$. Transmission T vs. ν [66K4]. Parameter: temperature. a: $\cong +25\,°C$, b: $\cong -120\,°C$, c: $\cong -140\,°C$, d: $\cong -175\,°C$. Note change of scale at $\nu = 6 \cdot 10^{13}$ Hz.

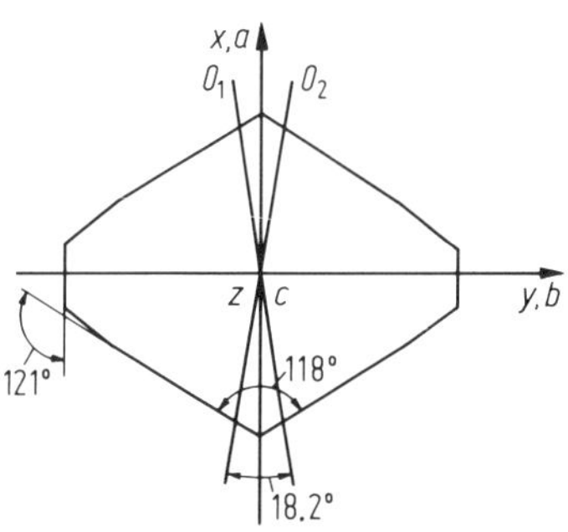

Fig. 1002. $NH_4H_3(SeO_3)_2$. Crystal form [69S7]. O_1, O_2: optical axes.

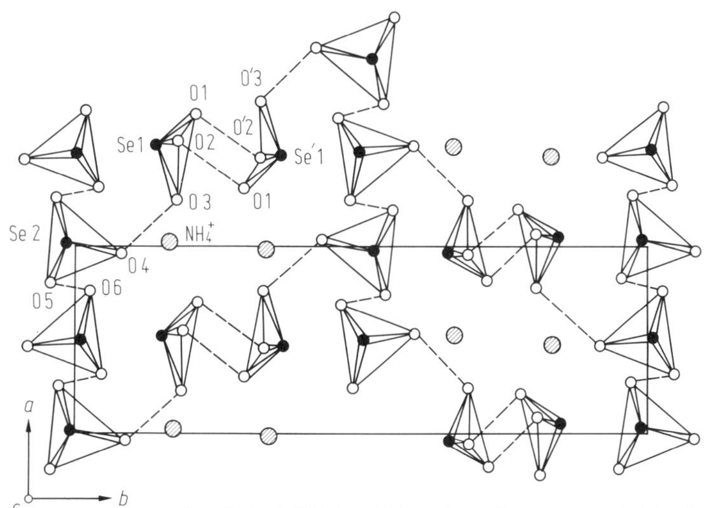

Fig. 1003. $NH_4H_3(SeO_3)_2$. Crystal structure [72G11]. Projection on (001).

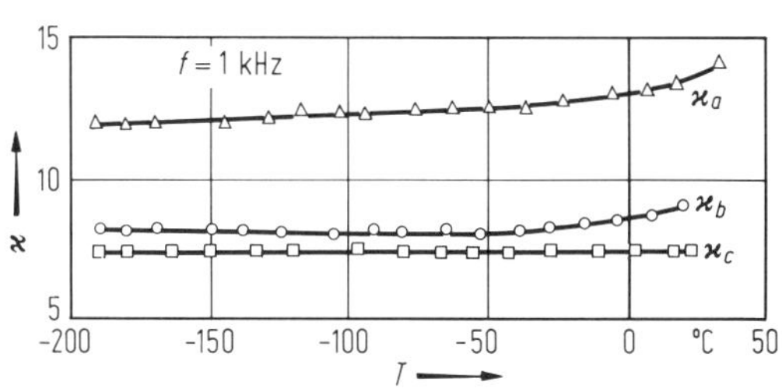

Fig. 1004. $NH_4H_3(SeO_3)_2$. $\varkappa_a$, $\varkappa_b$, $\varkappa_c$ vs. T [69S7].

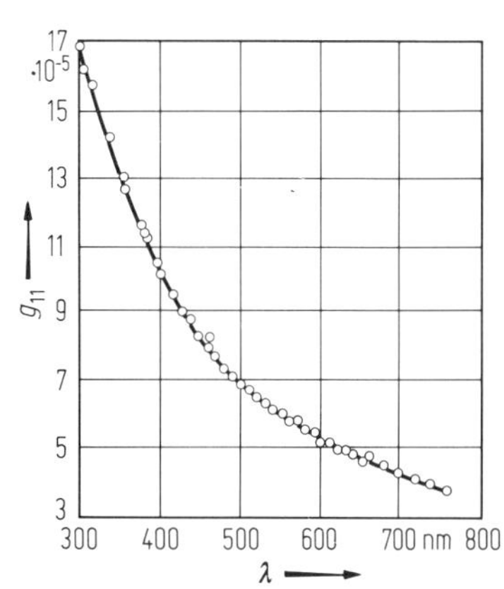

Fig. 1005. $NH_4H_3(SeO_3)_2$. g_{11} vs. λ [70I6]. g_{11}: component of gyration tensor.

29 Colemanite ($Ca_2B_6O_{11} \cdot 5H_2O$)

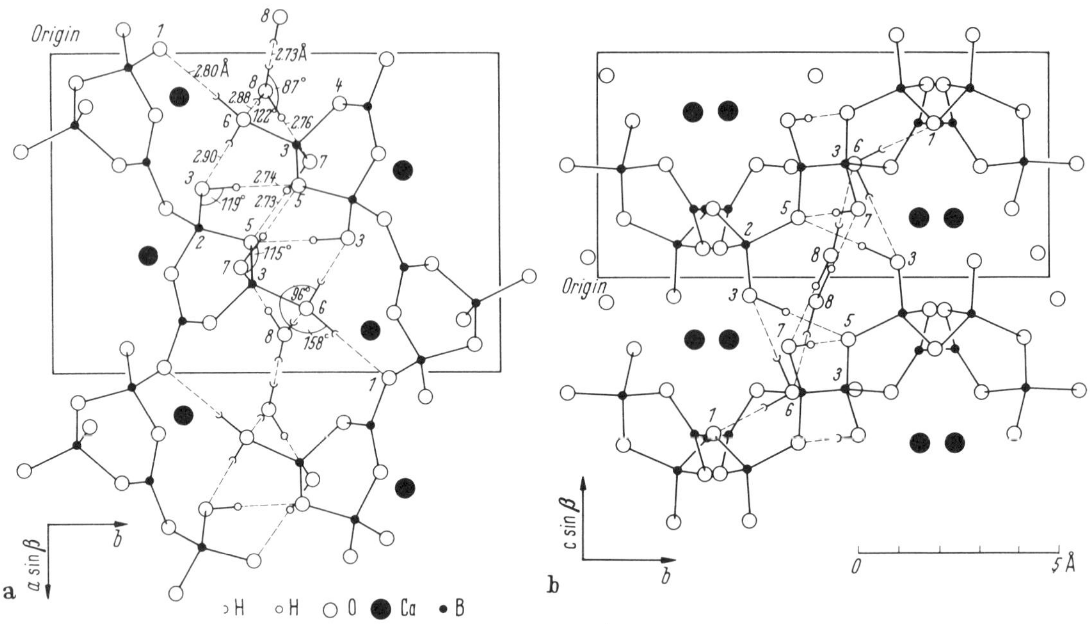

Fig. 1006. $Ca_2B_6O_{11} \cdot 5H_2O$ (colemanite). Crystal structure of phase I [58C1]. Fig. a: projection on (001); Fig. b: projection on (100). The dashed lines indicate the postulated hydrogen bonds. The smallest open circles indicate the probable positions of the hydrogen atoms, with a whole circle designating a site occupied always by a hydrogen atom, and a half-circle designating a site occupied half the time by a hydrogen atom. The numbers for oxygen and boron atoms correspond to those given in Table 273.

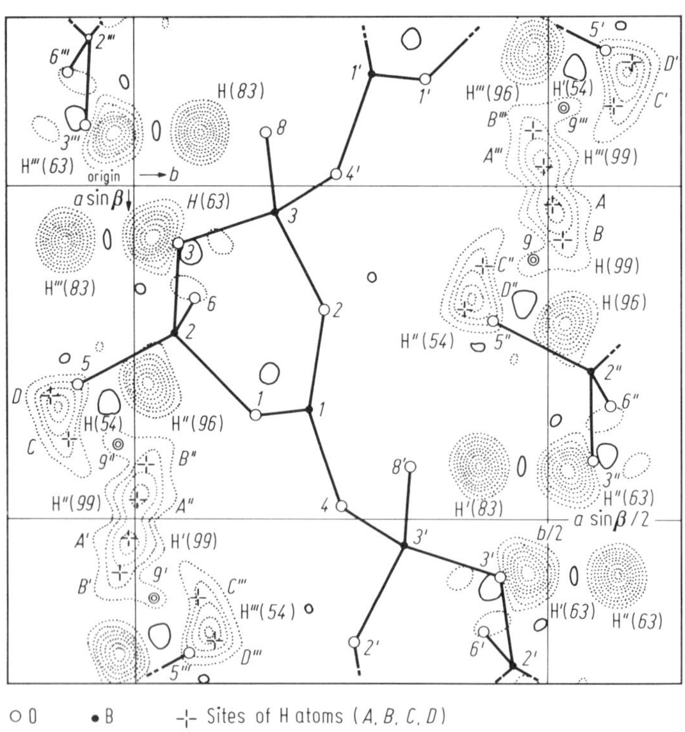

Fig. 1007. $Ca_2B_6O_{11} \cdot 5H_2O$ (colemanite). Crystal structure of phase I [66H1]. $T = 23$ °C. Fourier projection onto (001). A, B, C, D, ...: sites of H atoms. See Tables 272···275.

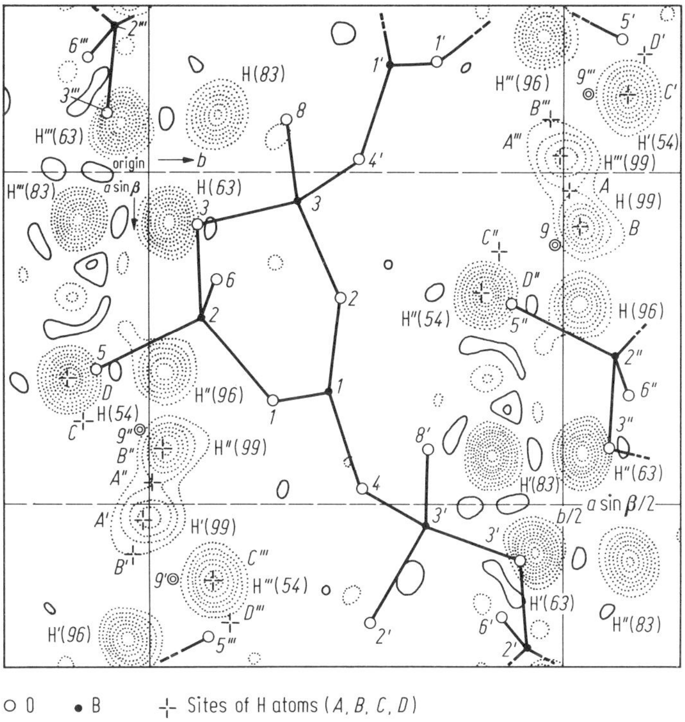

Fig. 1008. $Ca_2B_6O_{11} \cdot 5H_2O$ (colemanite). Crystal structure of phase II [66 H 1]. $T = -20$ °C. Fourier projection onto (001). A, B, C, D, ...: sites of H atoms.

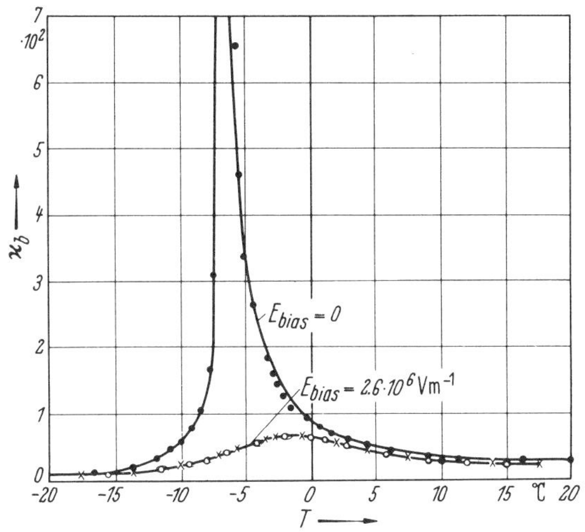

Fig. 1009. $Ca_2B_6O_{11} \cdot 5H_2O$ (colemanite). $\varkappa_b$ vs. T [59 W 3]. Parameter: E_{bias}.

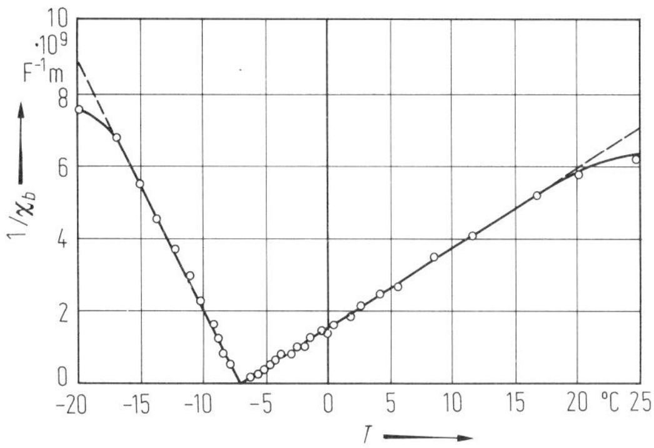

Fig. 1010. $Ca_2B_6O_{11} \cdot 5H_2O$ (colemanite). $1/\chi_b$ vs. T [59 W 3]. χ_b: dielectric susceptibility along the b-axis.

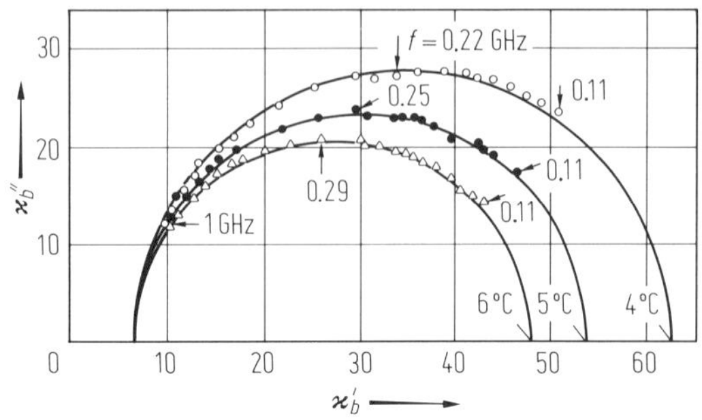

Fig. 1011. $Ca_2B_6O_{11} \cdot 5H_2O$ (colemanite). Cole-Cole plot of the complex dielectric constant above Θ_{II-I} [73 B 10]. Parameter: T. $\Theta_{II-I} \approx -3$ °C.

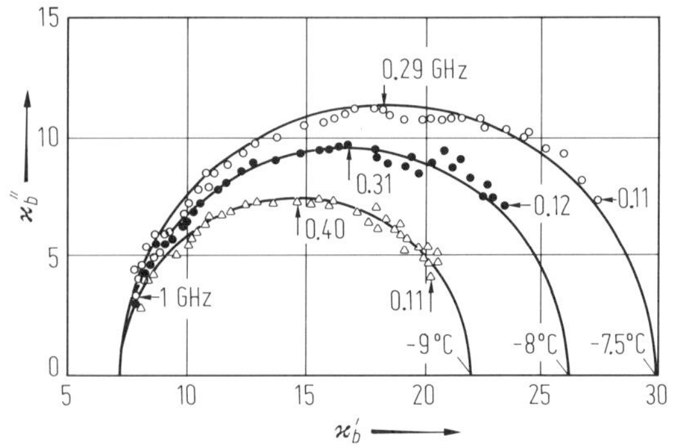

Fig. 1012. $Ca_2B_6O_{11} \cdot 5H_2O$ (colemanite). Cole-Cole plot of the complex dielectric constant below Θ_{II-I} [73 B 10]. Parameter: T.

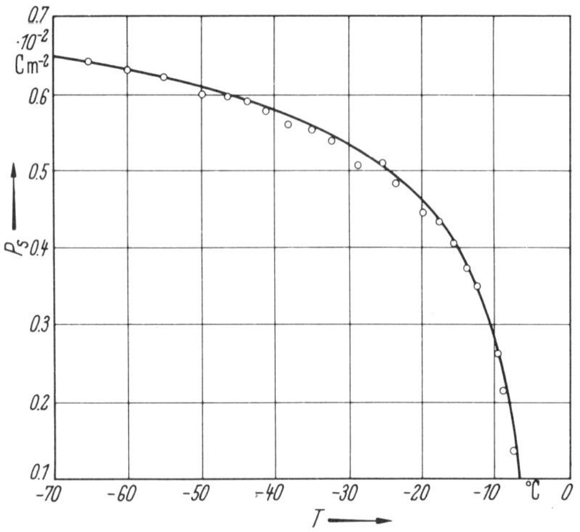

Fig. 1013. $Ca_2B_6O_{11} \cdot 5H_2O$ (colemanite). P_s vs. T [59 W 3].

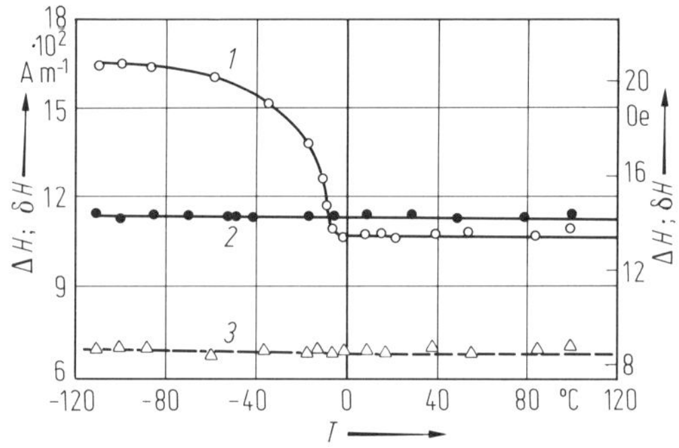

Fig. 1014. $Ca_2B_6O_{11} \cdot 5H_2O$ (colemanite). δH, ΔH vs. T [69 Z 1]. δH: maximum doublet splitting, curve *1* along one of the possible directions of the P–P (proton-proton) vectors, curve *2* along the bisector of the P–P vectors; ΔH: curve *3*, half width of the central component.

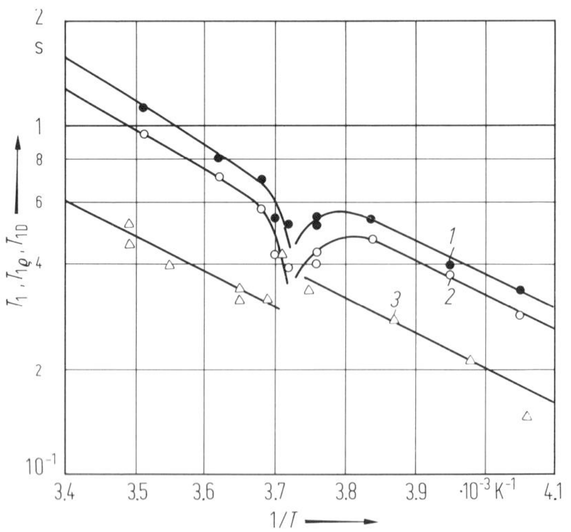

Fig. 1015. $Ca_2B_6O_{11} \cdot 5H_2O$ (colemanite). T_1, $T_{1\varrho}$, T_{1D} vs. T^{-1} [73 W 3]. Curve *1*: T_1, proton spin-lattice relaxation time: $\nu = 30$ MHz; *2*: $T_{1\varrho}$, proton spin-lattice relaxation time in the rotating frame; $H_1 = 0.795 \cdot 10^3$ A m^{-1}. *3*: T_{1D}, proton dipolar spin-lattice relaxation time.

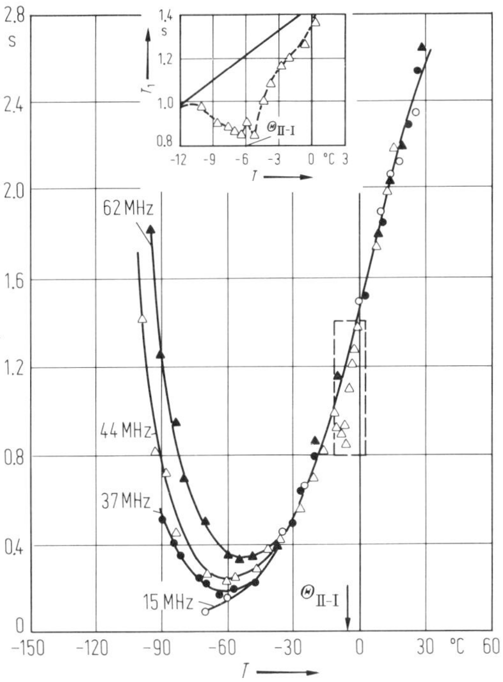

Fig. 1016. $Ca_2B_6O_{11} \cdot 5H_2O$ (colemanite). T_1 vs. T [74 B 20]. T_1: Proton spin-lattice relaxation time.

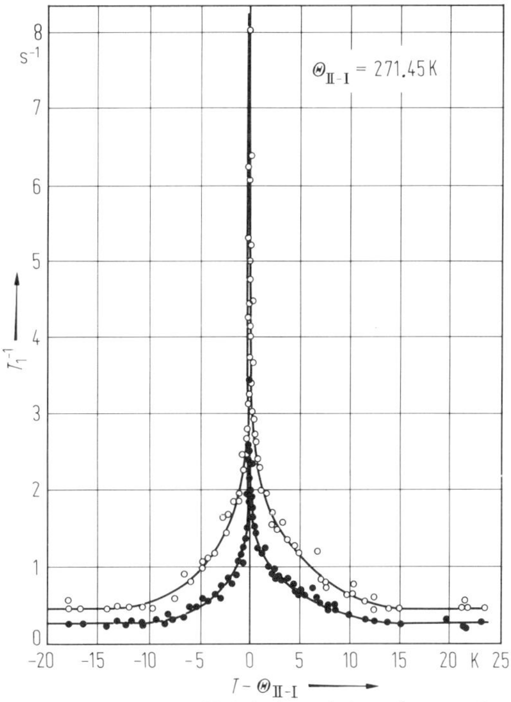

Fig. 1017. $Ca_2B_6O_{11} \cdot 5H_2O$ (colemanite). T_1^{-1} vs. $T - \Theta_{II-I}$ [76 T 8]. T_1^{-1}: Reciprocal spin-lattice relaxation time of ^{11}B. Full circles: Boron in tetrahedral sites; open circles: in triangular sites.

30 $K_4Fe(CN)_6 \cdot 3H_2O$ family

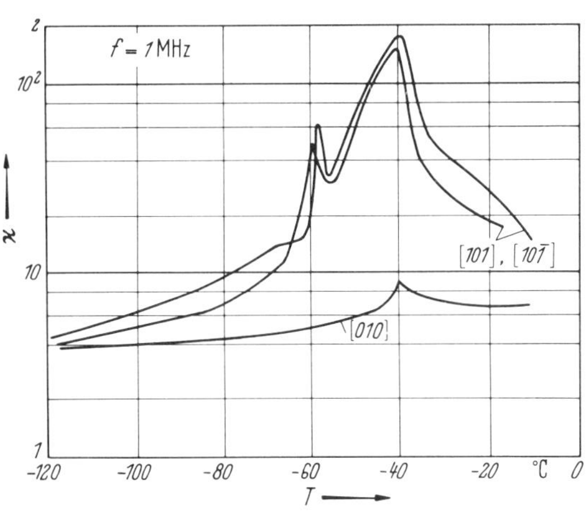

Fig. 1018. $K_4Mn(CN)_6 \cdot 3H_2O$. ϰ vs. T. [60 W 2].

Fig. 1019, see p. 556.

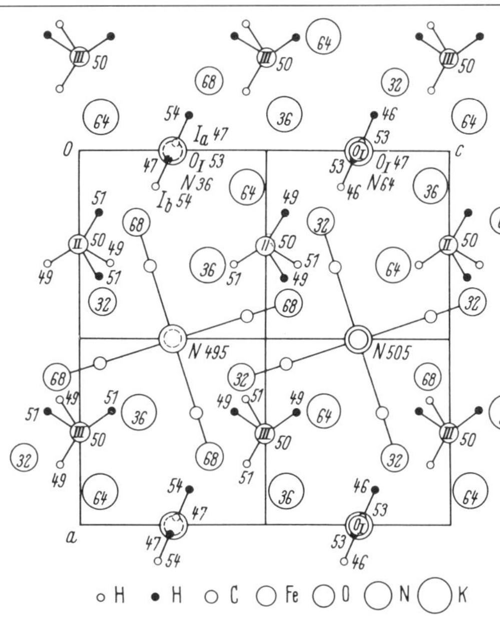

Fig. 1020. $K_4Fe(CN)_6 \cdot 3H_2O$. Crystal structure [(010) projection] in phase II [64 K 5]. The figures indicate the y parameters. Only the water layer at $b/2$ and its neighbours are included. There are two kinds of H_2O molecules in phase I; $2H_2O$ at 8(f) and H_2O at 4(e). In phase II a unit cell contains three kinds, the two at 8(f) becoming not equivalent (H_2O – I, – II, – III). H_{Ia} and H_{Ib} are two possible configurations with equal propability, provided that the positions of the heavy atoms are the same in both phases. They are indicated by open and full circles.

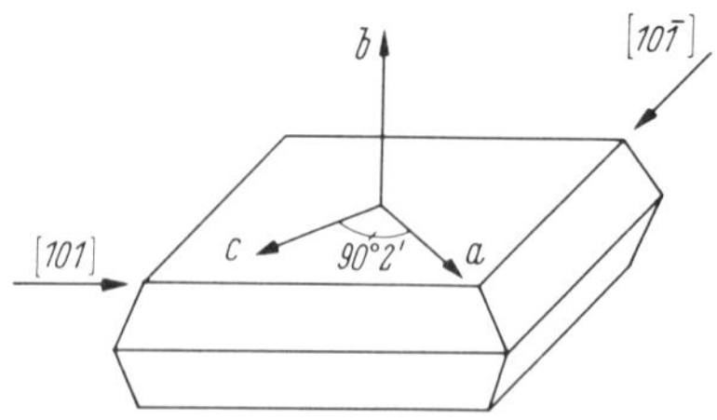

Fig. 1019. $K_4Fe(CN)_6 \cdot 3H_2O$. Crystal form [60 W 2].

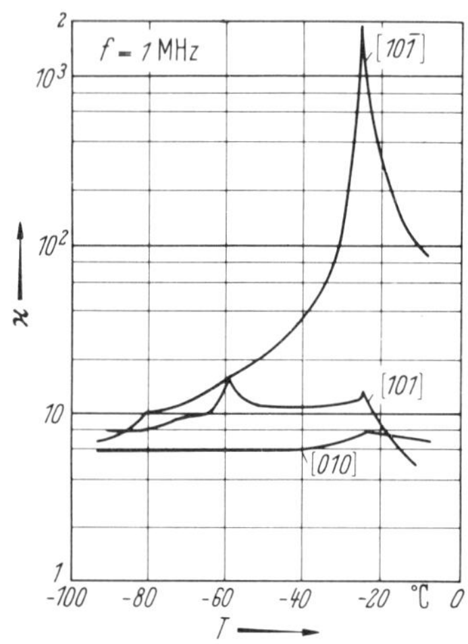

Fig. 1022. $K_4Fe(CN)_6 \cdot 3H_2O$. $\varkappa$ vs. T [60 W 2].

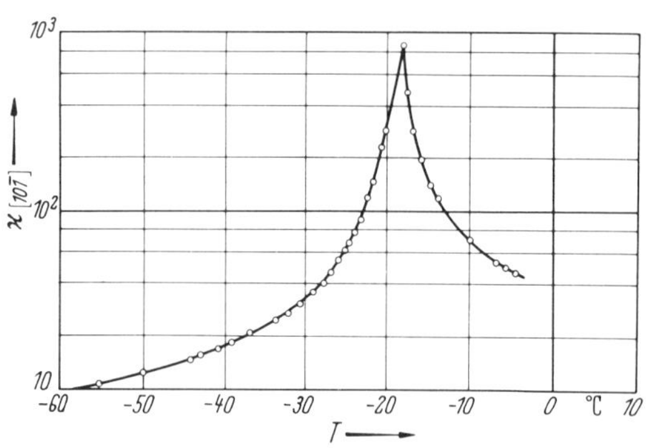

Fig. 1024. $K_4Fe(CN)_6 \cdot 3D_2O$. $\varkappa_{[10\bar{1}]}$ vs. T [60 W 1].

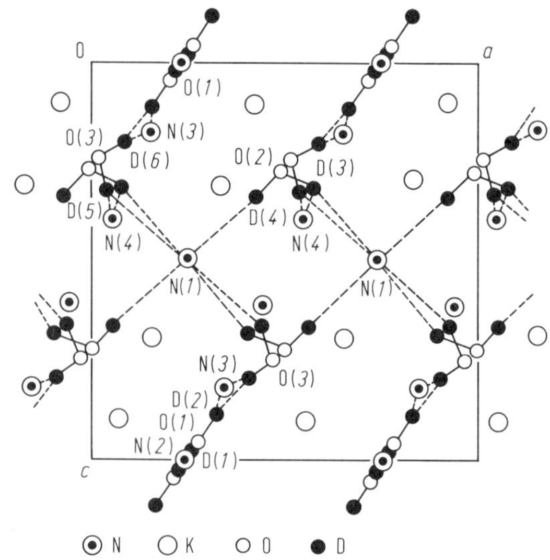

Fig. 1021. $K_4Fe(CN)_6 \cdot 3D_2O$. Crystal structure in phase I [70 T 4]. Atoms near $y = 0.5$ are projected on (010). Disordered water molecules and possible hydrogen bonds are shown.

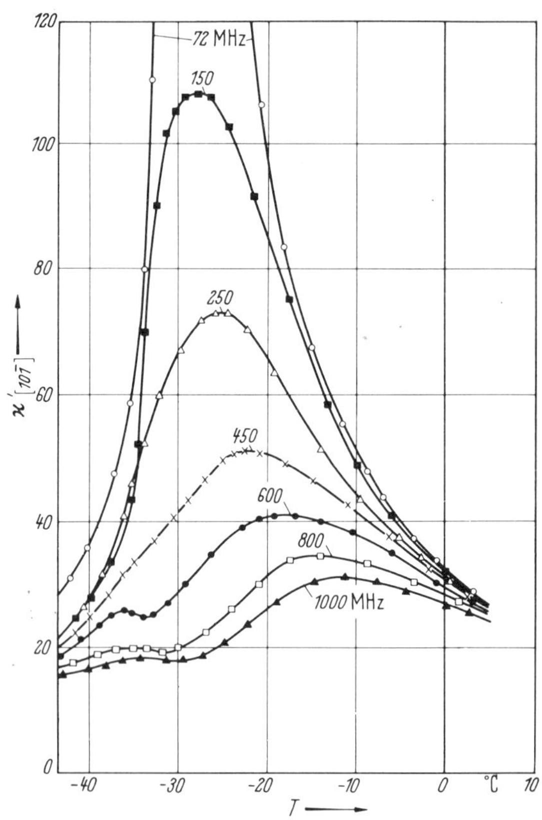

Fig. 1023. $K_4Fe(CN)_6 \cdot 3H_2O$. $\varkappa'_{[10\bar{1}]}$ vs. T [67 F 4]. Parameter: f.

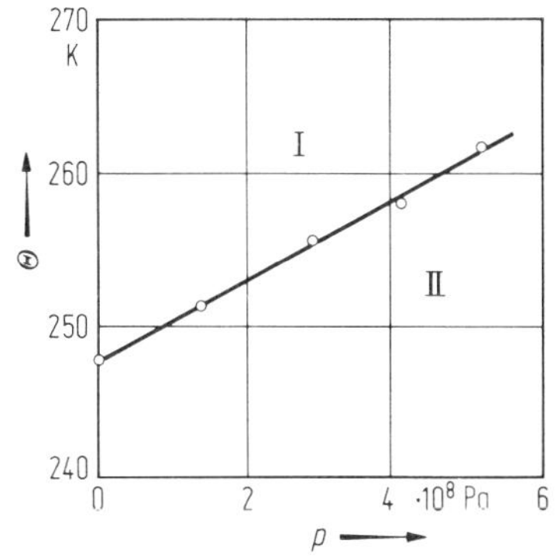

Fig. 1025. $K_4Fe(CN)_6 \cdot 3H_2O$. Θ vs. p [69 K 9]. p: hydrostatic pressure.

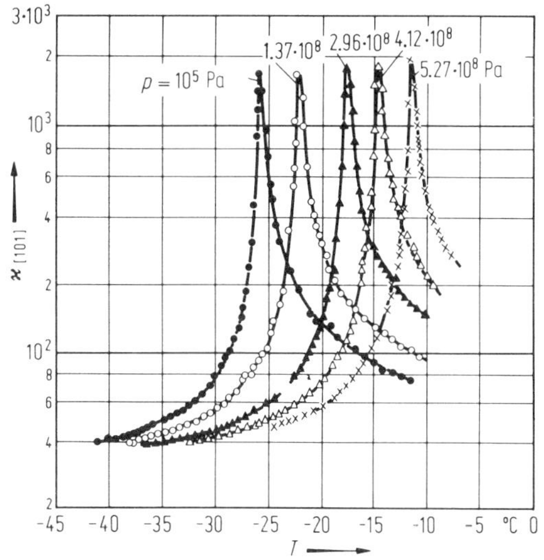

Fig. 1026. $K_4Fe(CN)_6 \cdot 3H_2O$. $\varkappa_{[10\bar{1}]}$ vs. T [69 K 9]. Parameter: p.

Fig. 1027, see p. 558.

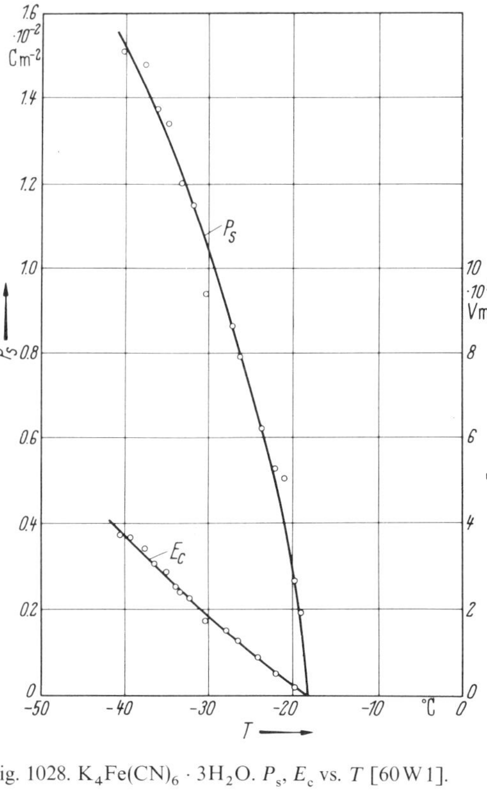

Fig. 1028. $K_4Fe(CN)_6 \cdot 3H_2O$. P_s, E_c vs. T [60 W 1].

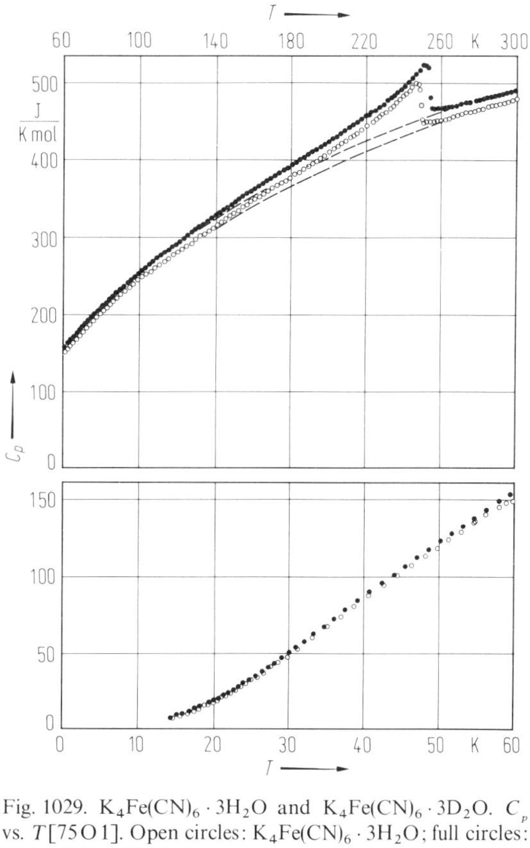

Fig. 1029. $K_4Fe(CN)_6 \cdot 3H_2O$ and $K_4Fe(CN)_6 \cdot 3D_2O$. C_p vs. T [75 O 1]. Open circles: $K_4Fe(CN)_6 \cdot 3H_2O$; full circles: $K_4Fe(CN)_6 \cdot 3D_2O$.

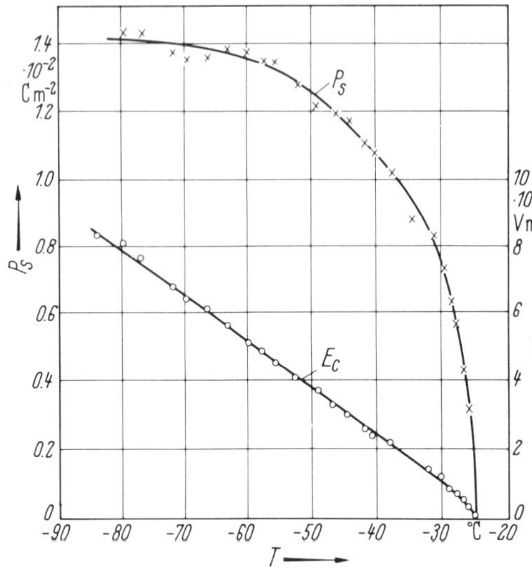

Fig. 1027. $K_4Fe(CN)_6 \cdot 3H_2O$. P_s, E_c vs. T. [60 W 2].

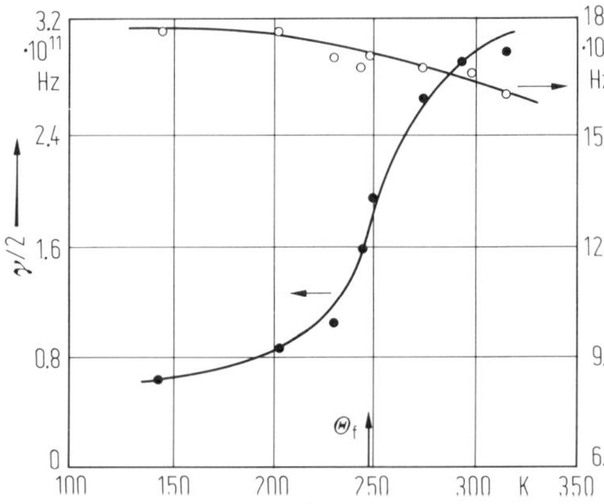

Fig. 1032. $K_4Fe(CN)_6 \cdot 3H_2O$. ν_0, $\gamma/2$ vs. T [77 S 8]. ν_0: characteristic frequency. γ: damping constant obtained from Raman scattering measurement [geometry $Y(Z, Y)Z$)].

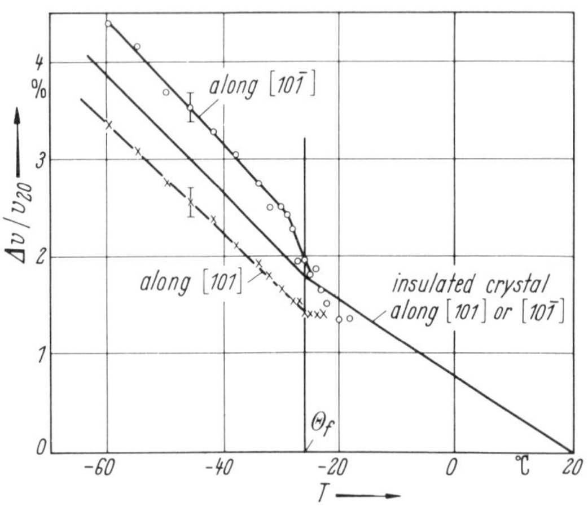

Fig. 1030. $K_4Fe(CN)_6 \cdot 3H_2O$. $\Delta v/v_{20}$ vs. T [67 S 6]. $\Delta v/v_{20}$: percent change of sound velocity from that at 20 °C.

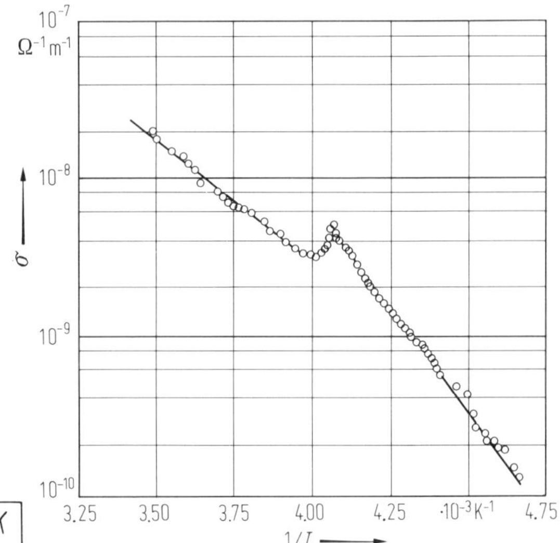

Fig. 1033. $K_4Fe(CN)_6 \cdot 3H_2O$. σ vs. $1/T$ [72 A 10]. Ferroelectric (10$\bar{1}$) section.

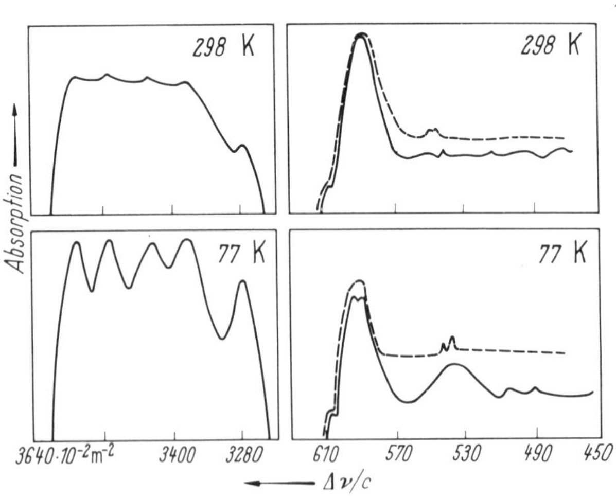

Fig. 1031. $K_4Fe(CN)_6 \cdot 3H_2O$. Infrared absorption vs. $\Delta\nu/c$ [61 B 1]. Dashed line is for $K_4Fe(CN)_6$.

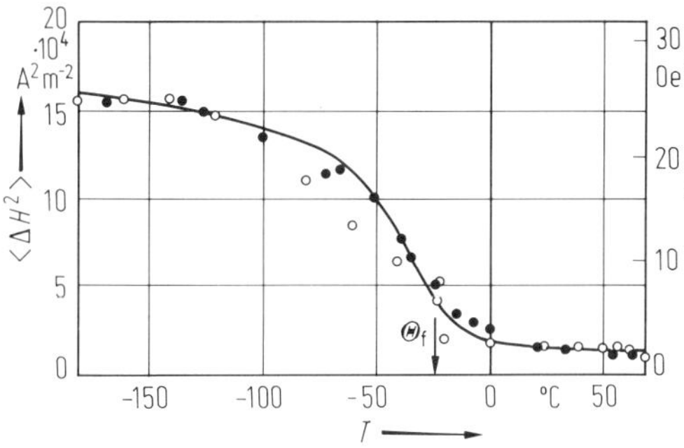

Fig. 1034. $K_4Fe(CN)_6 \cdot 3H_2O$. $\langle \Delta H^2 \rangle$ vs. T. Full circles: [64 K 5]; open circles: [61 L 2]. $\langle \Delta H^2 \rangle$: second moment of NMR of proton.

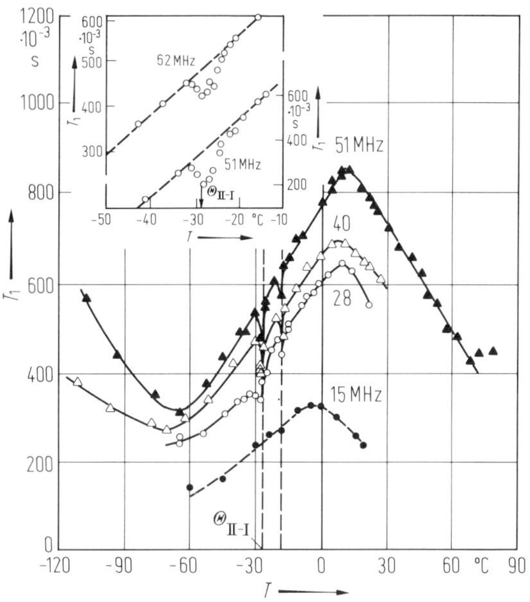

Fig. 1035. $K_4Fe(CN)_6 \cdot 3H_2O$. T_1 vs. T [74 B 20]. T_1: proton spin-lattice relaxation time.

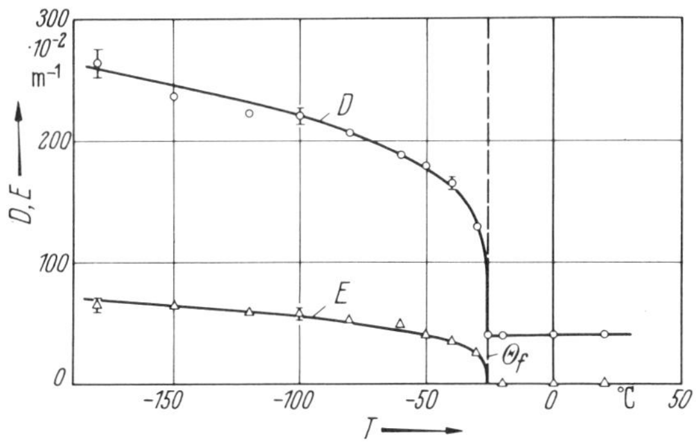

Fig. 1036. $K_4Fe(CN)_6 \cdot 3H_2O$. D, E vs. T [65 O 3]. D, E: fine structure parameters in ESR.

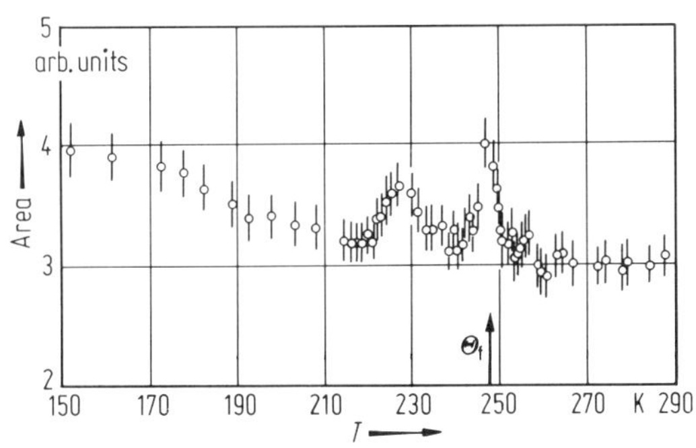

Fig. 1037. $K_4Fe(CN)_6 \cdot 3H_2O$. Area of the Mössbauer absorption line vs. T in the [10$\bar{1}$] direction [71 M 10].

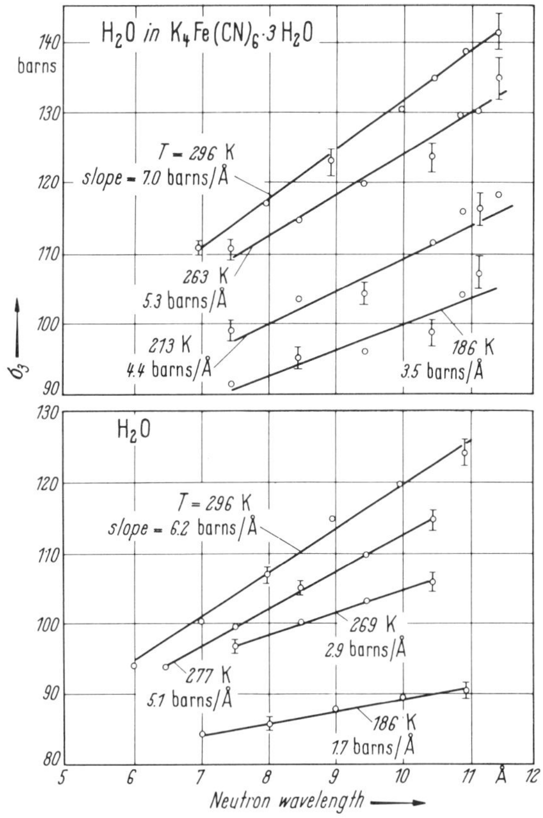

Fig. 1038. $K_4Fe(CN)_6 \cdot 3H_2O$. σ_3 vs. neutron wavelength. [66 R 7]. σ_3: total scattering cross sections per hydrogen compared with scattering in water. 1 Å = 0.1 nm.

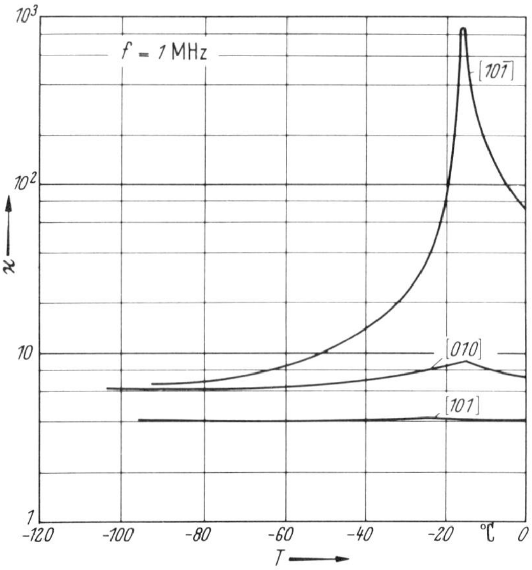

Fig. 1039. $K_4Ru(CN)_6 \cdot 3H_2O$. ϰ vs. T [60 W 2].

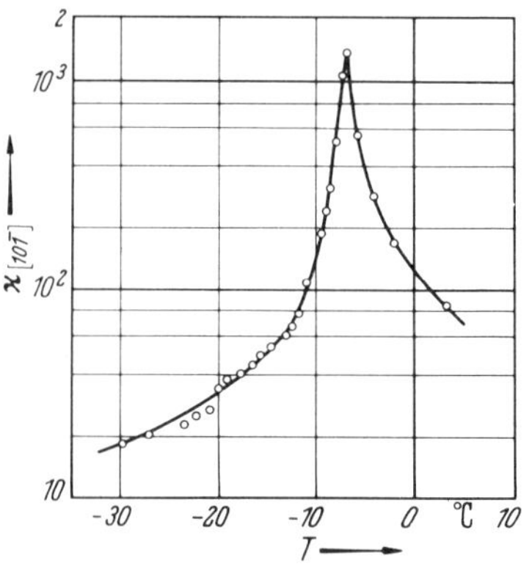

Fig. 1040. $K_4Ru(CN)_6 \cdot 3D_2O$. $ϰ_{[10\bar{1}]}$ vs. T [60 W 1].

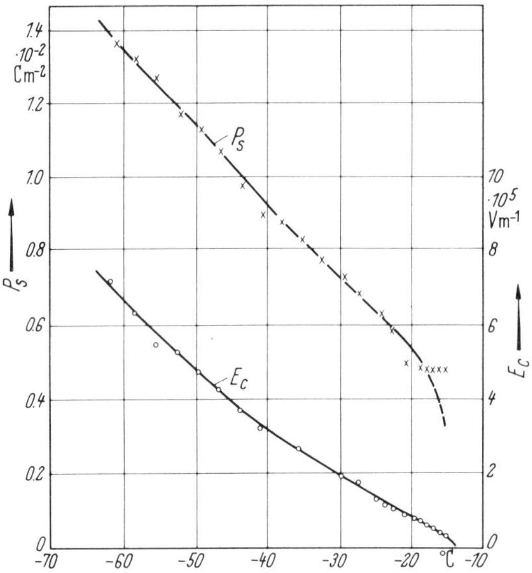

Fig. 1041. $K_4Ru(CN)_6 \cdot 3H_2O$. P_s, E_c vs. T [60 W 2].

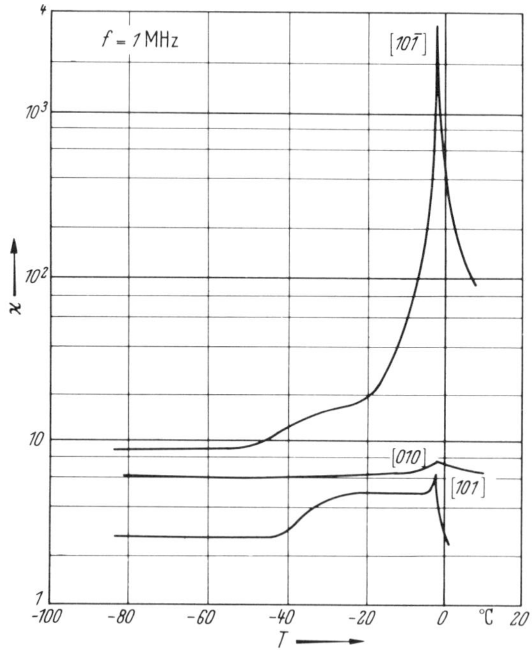

Fig. 1043. $K_4Os(CN)_6 \cdot 3H_2O$. ϰ vs. T. [60 W 2].

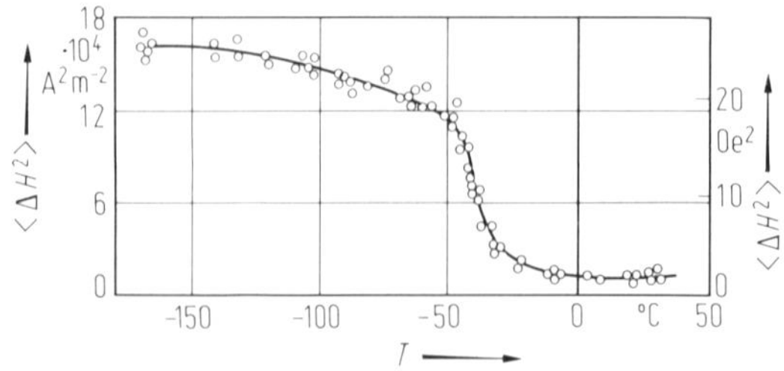

Fig. 1042. $K_4Ru(CN)_6 \cdot 3H_2O$ (powder). $\langle \Delta H^2 \rangle$ vs. T [64 K 5]. $\langle \Delta H^2 \rangle$: second moment of NMR.

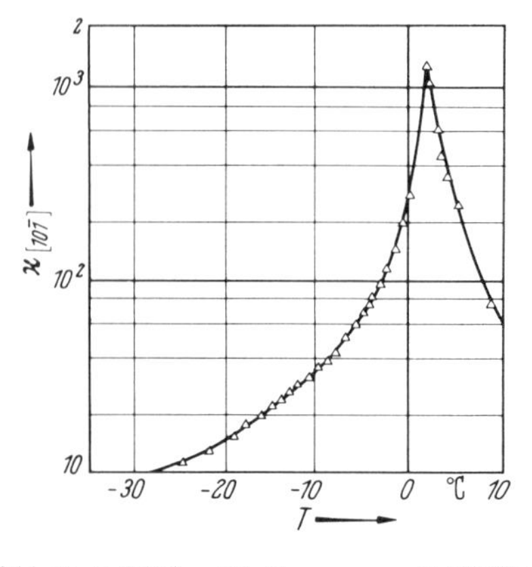

Fig. 1044. $K_4Os(CN)_6 \cdot 3D_2O$. $ϰ_{[10\bar{1}]}$ vs. T. [60 W 1].

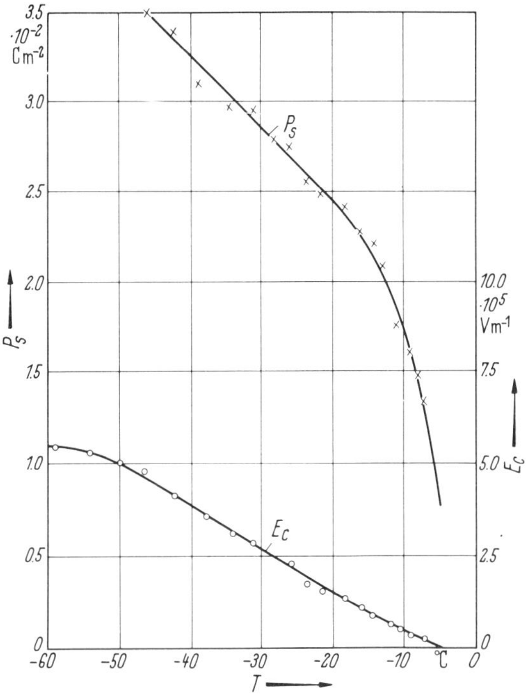

Fig. 1045. $K_4Os(CN)_6 \cdot 3H_2O$. P_s, E_c vs. T. [60W2].

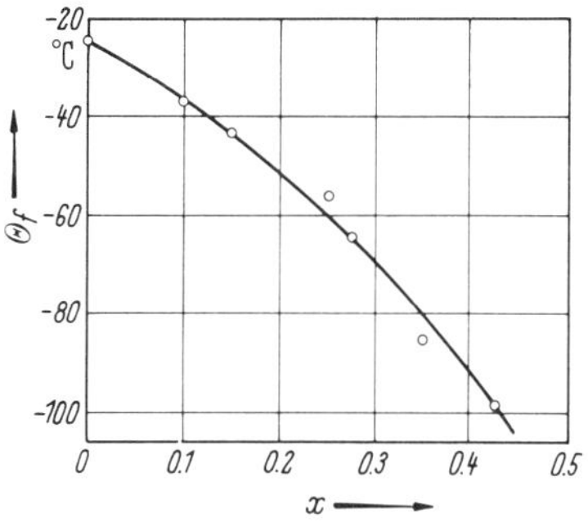

Fig. 1046. $(K_{1-x}Rb_x)_4Fe(CN)_6 \cdot 3H_2O$. Θ_f vs. x [61W1].

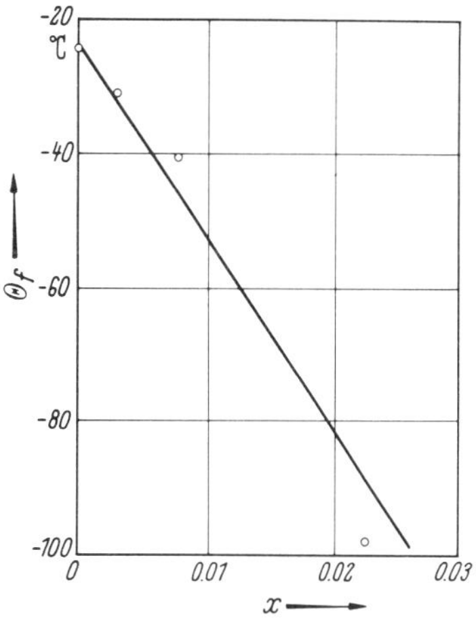

Fig. 1048. $(K_{1-x}Tl_x)_4Fe(CN)_6 \cdot 3H_2O$. Θ_f vs. x [61W1].

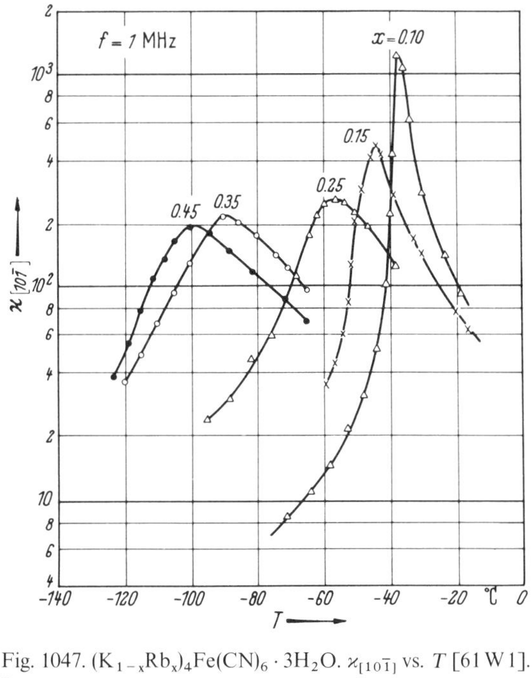

Fig. 1047. $(K_{1-x}Rb_x)_4Fe(CN)_6 \cdot 3H_2O$. $\varkappa_{[10\bar{1}]}$ vs. T [61W1]. Parameter: x.

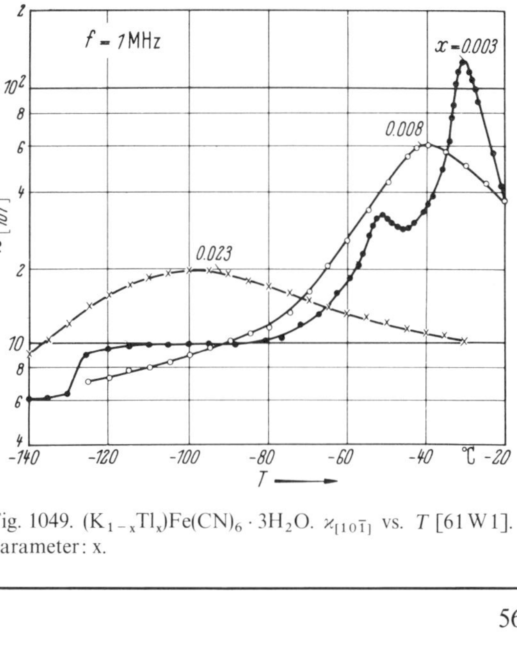

Fig. 1049. $(K_{1-x}Tl_x)Fe(CN)_6 \cdot 3H_2O$. $\varkappa_{[10\bar{1}]}$ vs. T [61W1]. Parameter: x.

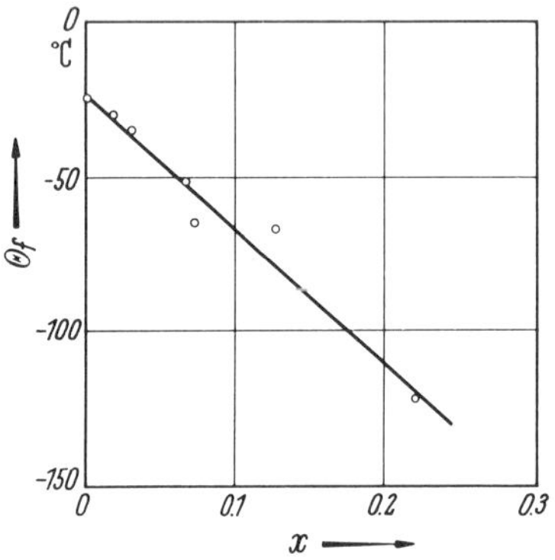

Fig. 1050. $[K_{1-x}(NH_4)_x]_4Fe(CN)_6 \cdot 3H_2O$. Θ_f vs. x [61 W 1].

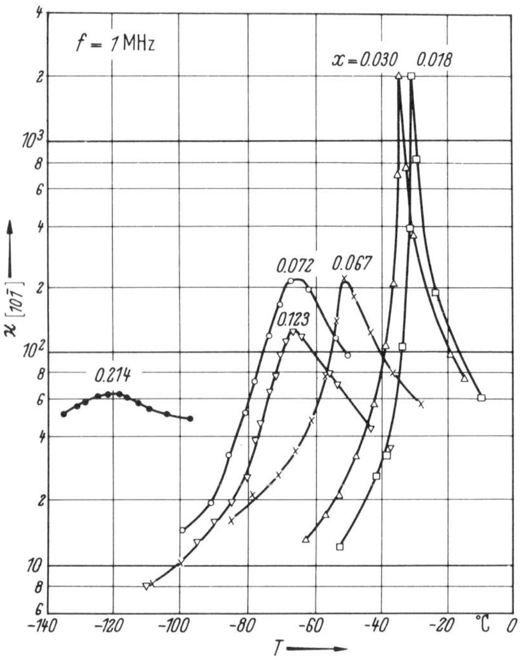

Fig. 1051. $[K_{1-x}(NH_4)_x]Fe(CN)_6 \cdot 3H_2O$. $\varkappa_{[10\bar{1}]}$ vs. T [61 W 1]. Parameter: x.

31 $SC(NH_2)_2$

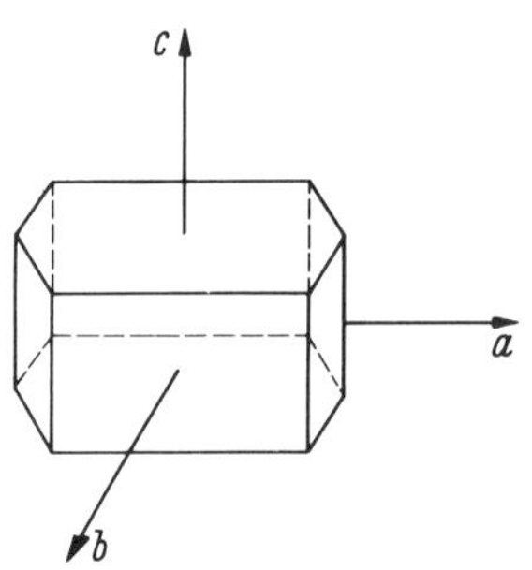

Fig. 1052. $SC(NH_2)_2$. Crystal form [59 G 2].

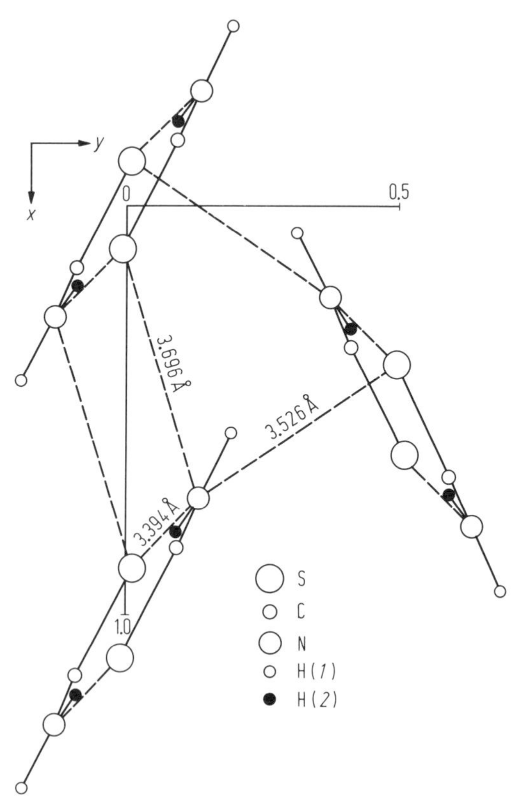

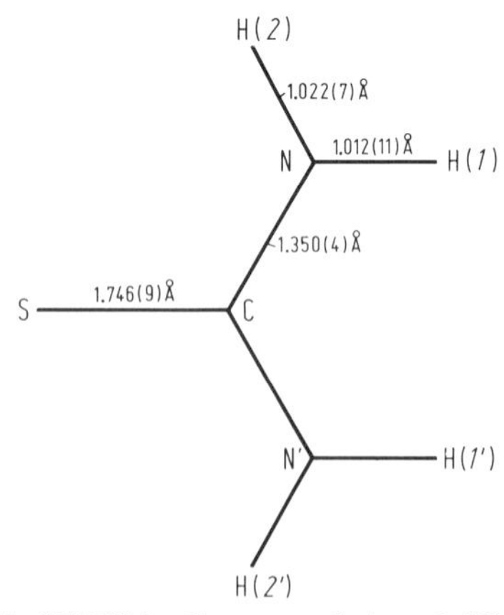

Fig. 1054. $SC(NH_2)_2$. Structure of phase I [68 E 2]. Interatomic distances in thiourea molecule at RT.

Fig. 1053. $SC(NH_2)_2$. Structure of phase I determined by neutron diffraction at RT [68 E 2]. Projection on (001).

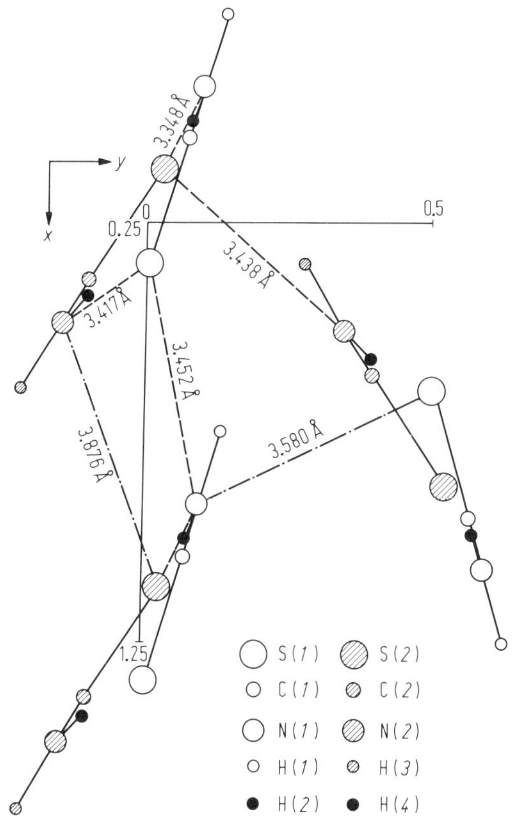

Fig. 1055. $SC(NH_2)_2$. Structure of phase V determined by neutron diffraction at 110 K [68E2]. Projection on (001).

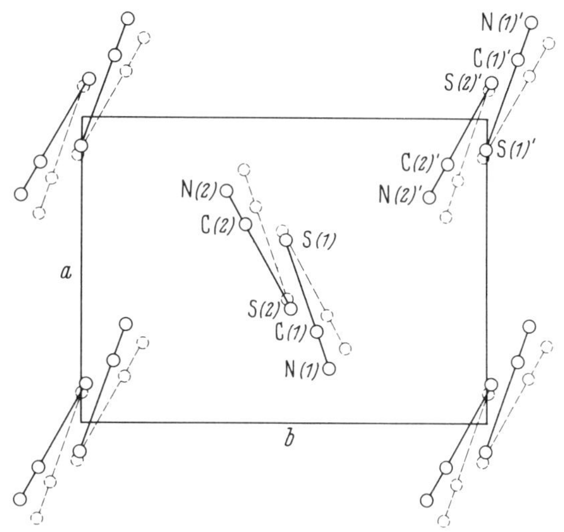

Fig. 1056. $SC(NH_2)_2$. Structure of phase V [59G2]. An illustration of the polarization reversal projected on (001). The full circles represent atomic positions for one direction of P_s; the dotted circles represent atomic positions for the opposite direction of P_s.

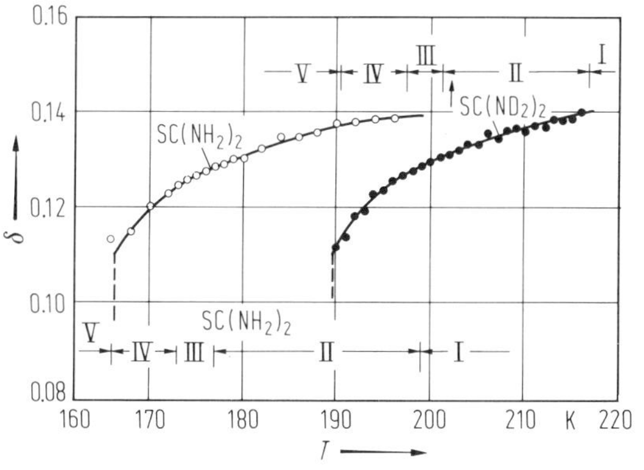

Fig. 1057. $SC(NH_2)_2$, $SC(ND_2)_2$. δ vs. T [77I4]. δ: position of satellite scattering. δ in units of $c^*(=2\pi/c)$. As for the transition temperatures the values given in subsection 1b seem to be more reliable.

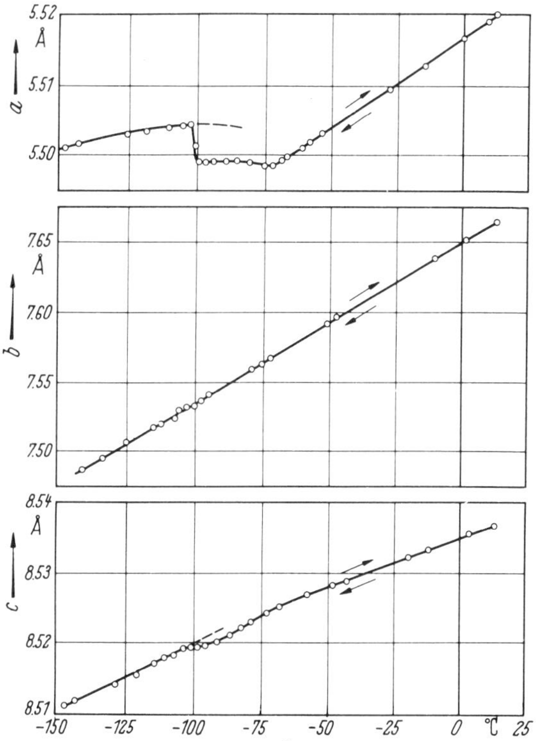

Fig. 1058. $SC(NH_2)_2$. a, b, c vs. T [62F8]. Values of a, b, c were obtained by measuring thermal elongation of specimens by means of the optical lever method and by using values of a, b, c at 20 °C determined with X-ray.

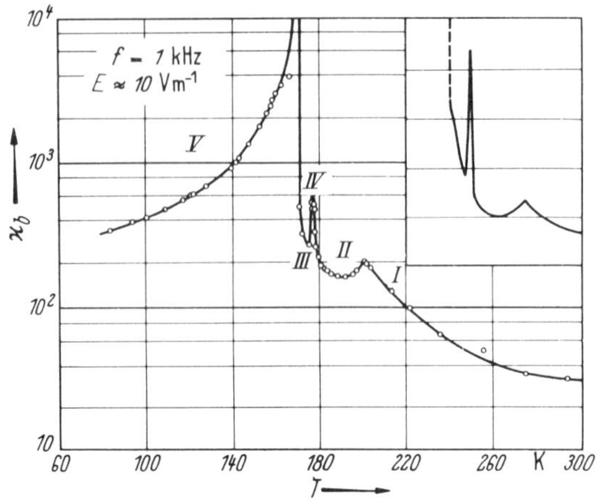

Fig. 1059. $SC(NH_2)_2$. $\varkappa_b$ vs. T [59G2].

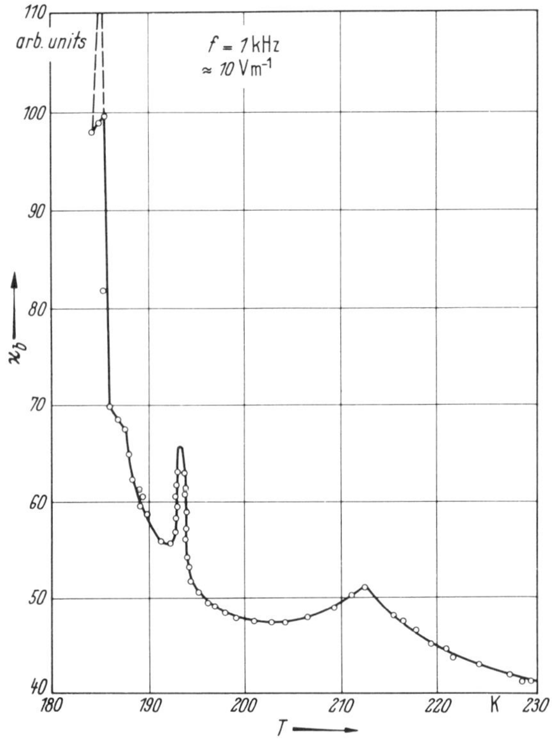

Fig. 1060. $SC(ND_2)_2$. $\varkappa_b$ vs. T [59G2].

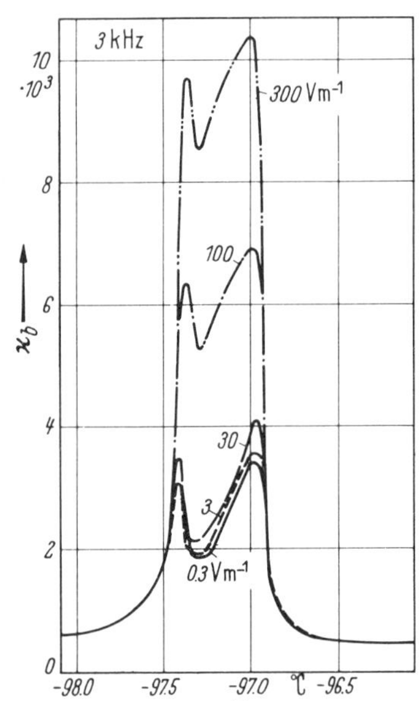

Fig. 1061. $SC(NH_2)_2$. $\varkappa_b$ vs. T near phase III [67S28]. Parameter: E, amplitude of measuring ac field.

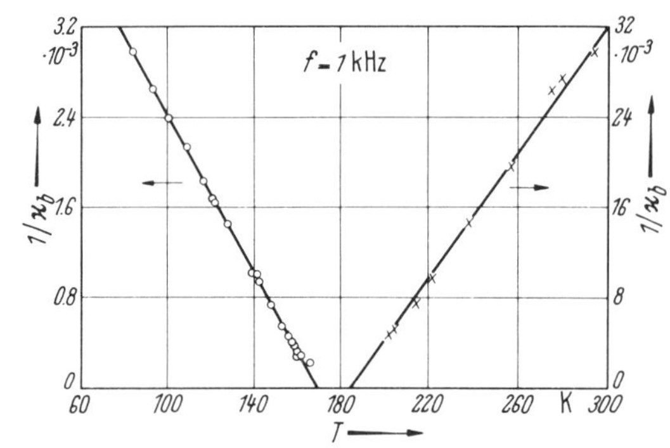

Fig. 1062. $SC(NH_2)_2$. $1/\varkappa_b$ vs. T [59G2].

Fig. 1063, see p. 565.

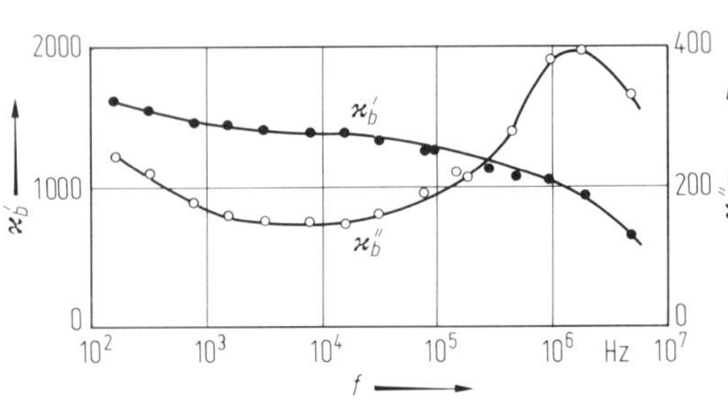

Fig. 1064. $SC(NH_2)_2$. $\varkappa_b'$ and $\varkappa_b''$ vs. f at 146 K [73M10].

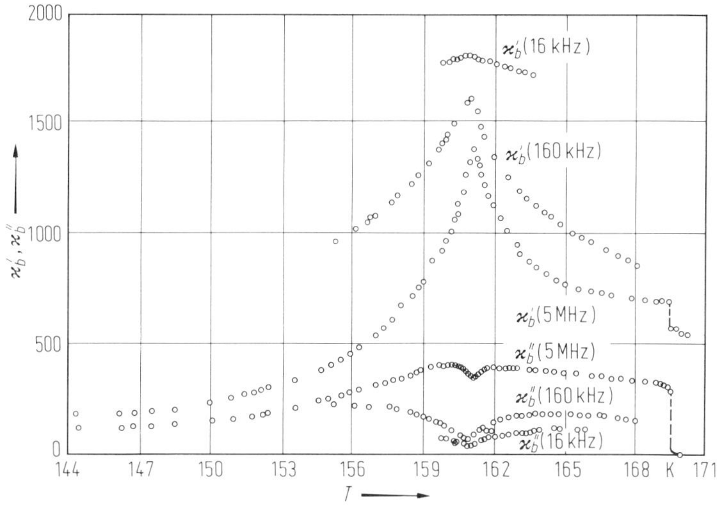

Fig. 1063. $SC(NH_2)_2$. $\varkappa_b'$ and $\varkappa_b''$ vs. T [73M10]. Parameter: f.

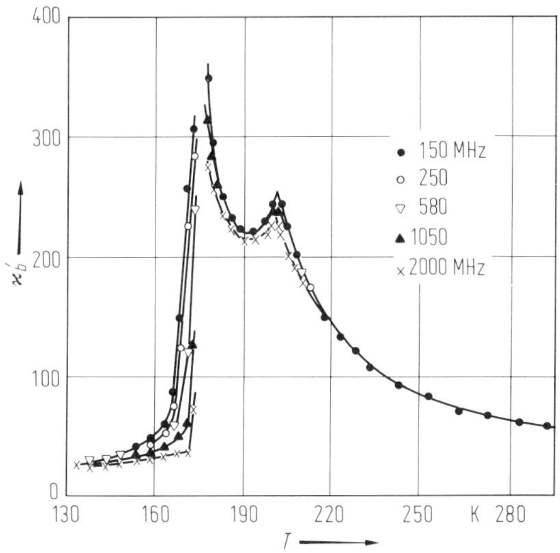

Fig. 1065. $SC(NH_2)_2$. $\varkappa_b'$ vs. T [75C9]. Parameter: f.

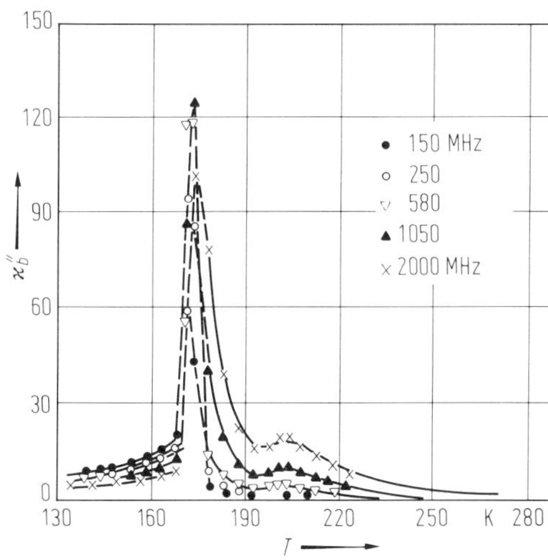

Fig. 1066. $SC(NH_2)_2$. $\varkappa_b''$ vs. T [75C9]. Parameter: f.

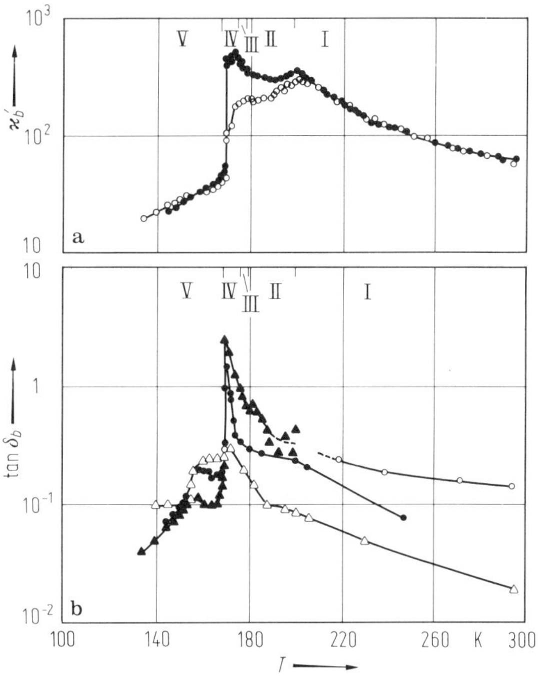

Fig. 1067. $SC(NH_2)_2$. $\varkappa_b'$ and $\tan\delta_b$ vs. T [76T13]. Parameter: f. Open triangles: 9.0; full circles: 21.0; open circles: 56.8; full triangles: 65.3 GHz. For $\varkappa_b'$ in phase I: full circles: measurements from 18.0 to 28.0 GHz, and open circles: measurements from 50.0 to 75.0 GHz.

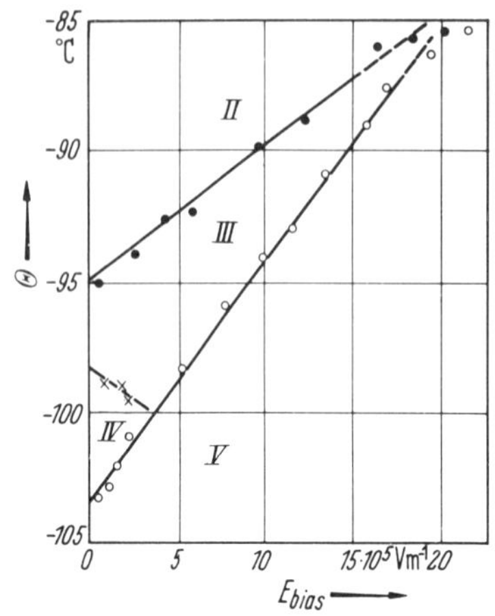

Fig. 1068. $SC(NH_2)_2$. Θ vs. E_{bias} [62F8].

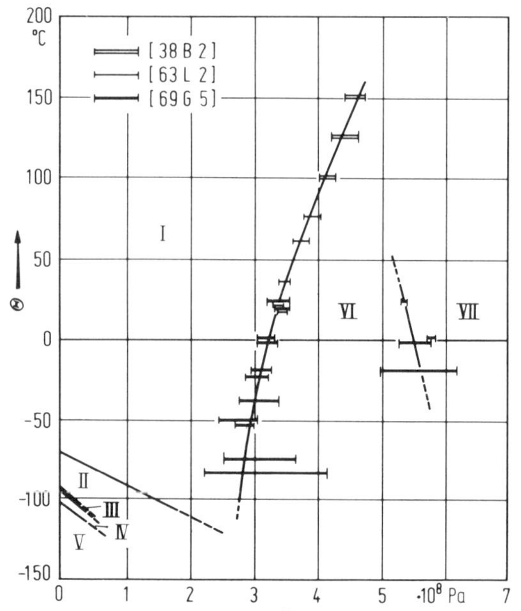

Fig. 1069. $SC(NH_2)_2$. Θ vs. p [69G5, 63L2, 38B2].

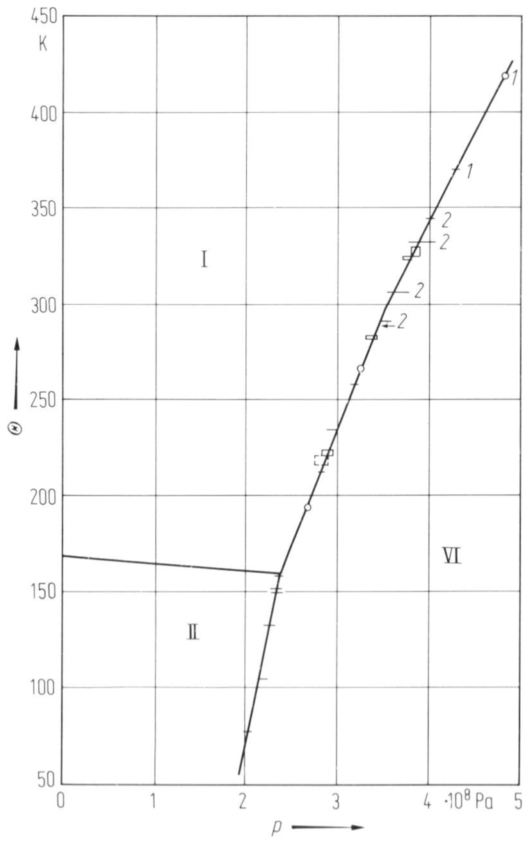

Fig. 1070. $SC(NH_2)_2$. Θ vs. p [75F4]. *1*: [38B2] and *2*: [63L2].

Fig. 1071, see p. 567

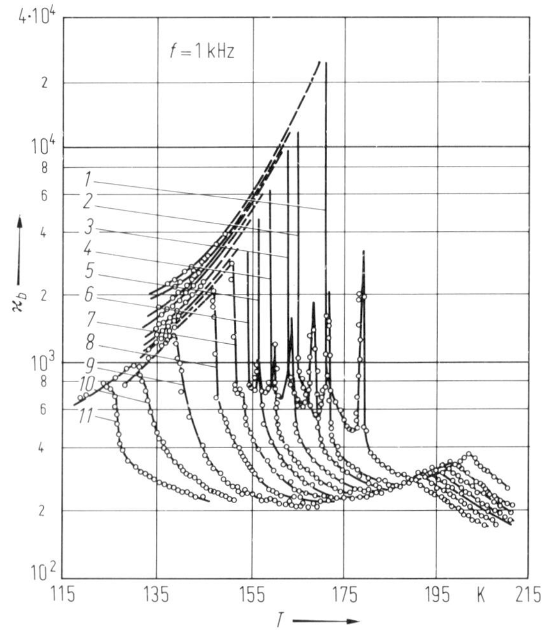

Fig. 1072. $SC(NH_2)_2$. $\varkappa_b$ vs. T [75K13]. Parameter: p. Curve *1*: atmospheric pressure, *2*: 0.22, *3*: 0.34, *4*: 0.48, *5*: 0.57, *6*: 0.67, *7*: 0.77, *8*: 0.88, *9*: 1.14, *10*: 1.30, *11*: $1.46 \cdot 10^8$ Pa.

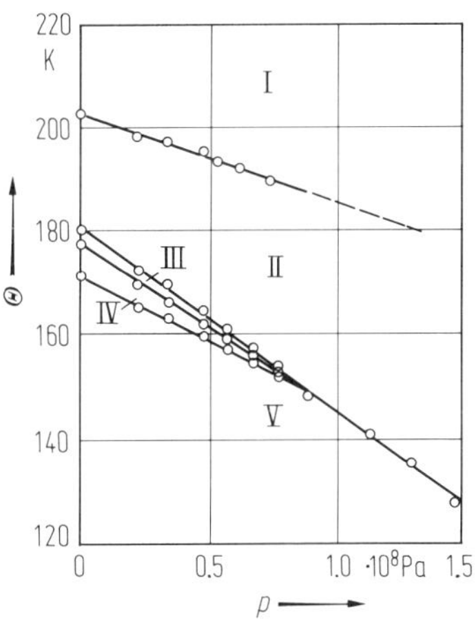

Fig. 1071. $SC(NH_2)_2$. Θ vs. p [75K13].

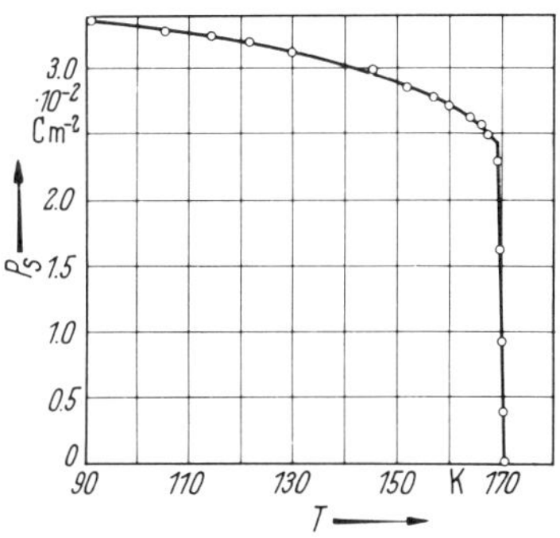

Fig. 1073. $SC(NH_2)_2$. P_s vs. T in phase V [59G2].

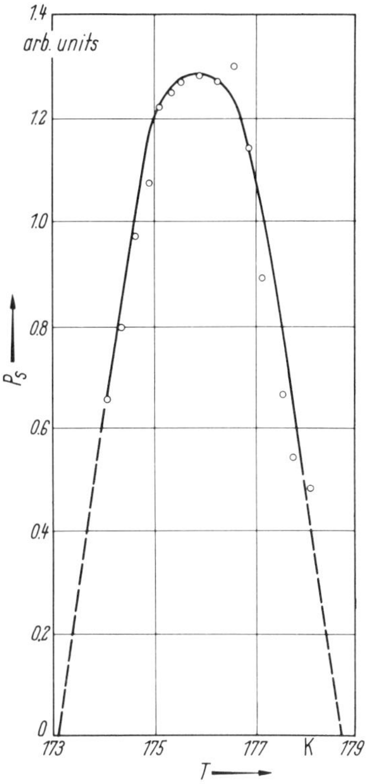

Fig. 1074. $SC(NH_2)_2$. P_s vs. T in phase III [59G2]. $E_c \simeq 10$ kV/m.

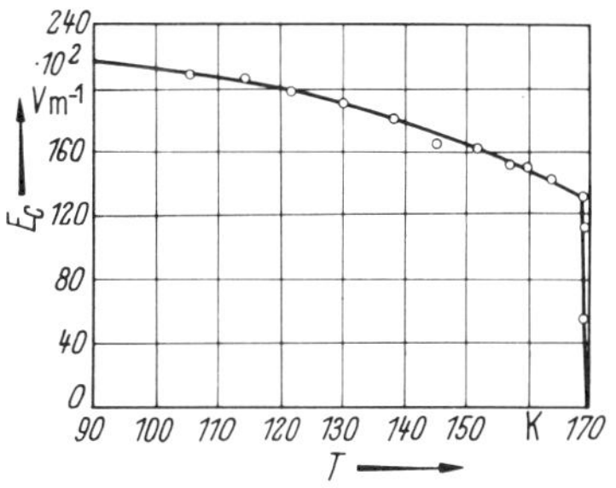

Fig. 1075. $SC(NH_2)_2$. E_c vs. T in phase V [59G2].

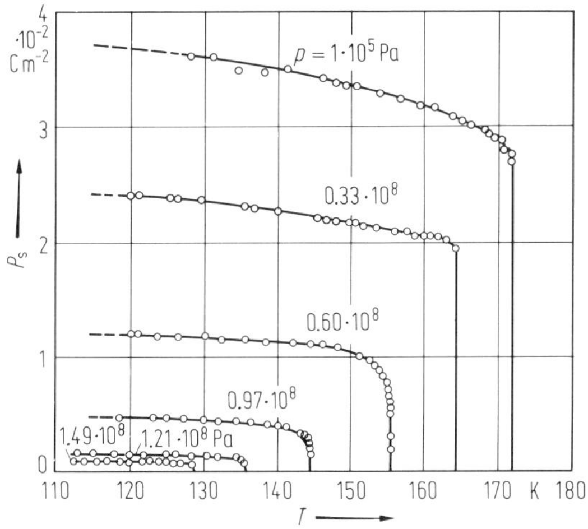

Fig. 1076. $SC(NH_2)_2$. P_s vs. T [75K13]. Parameter: p.

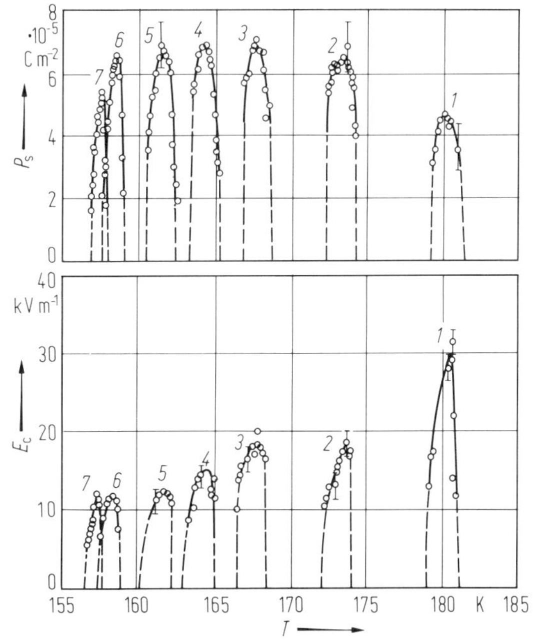

Fig. 1077. $SC(NH_2)_2$. P_s, E_c vs. T in phase *III* [75K13]. Parameter: p. Curve *1*: atmospheric pressure, *2*: 0.18, *3*: 0.37, *4*: 0.45, *5*: 0.54, *6*: 0.65, *7*: $0.70 \cdot 10^8$ Pa.

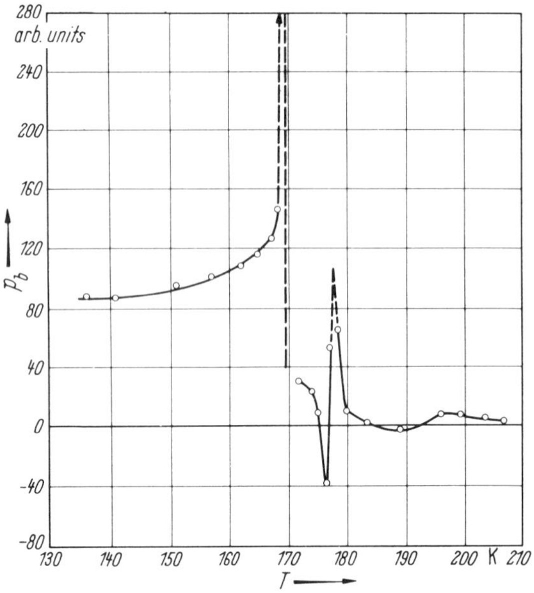

Fig. 1078. $SC(NH_2)_2$. p_b vs. T [59G2]. p_b: pyroelectric coefficient along b.

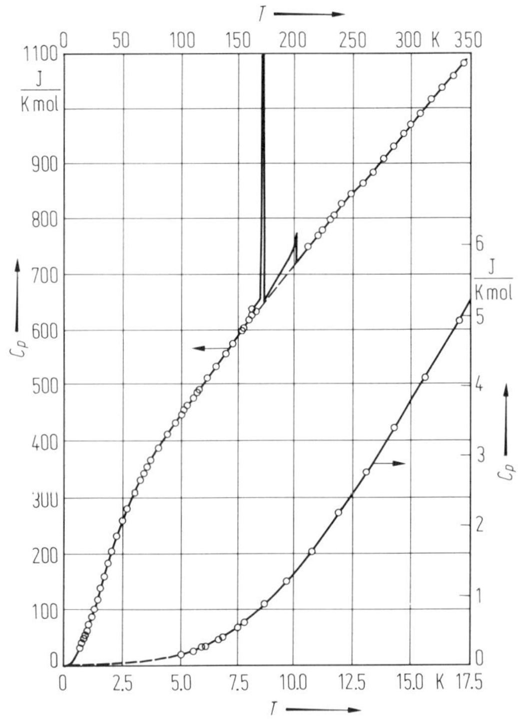

Fig. 1079. $SC(NH_2)_2$. C_p vs. T [63C1]. C_p : molar heat capacity at constant pressure. For details in the range between 160 K and 210 K, see Fig. 1080.

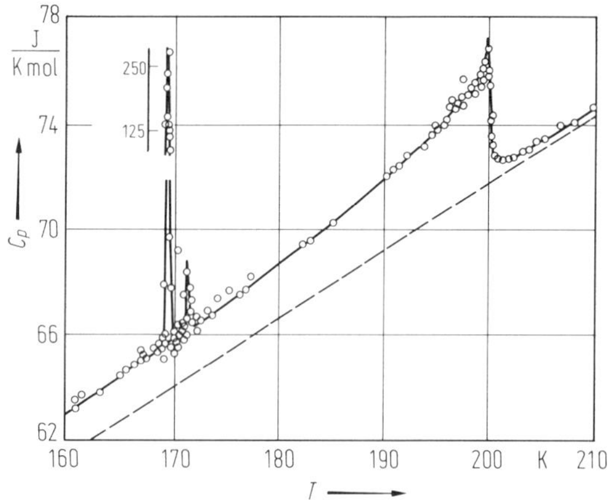

Fig. 1080. $SC(NH_2)_2$. C_p vs. T [63C1]. C_p: molar heat capacity at constant pressure. For data in wider range of T, see Fig. 1079.

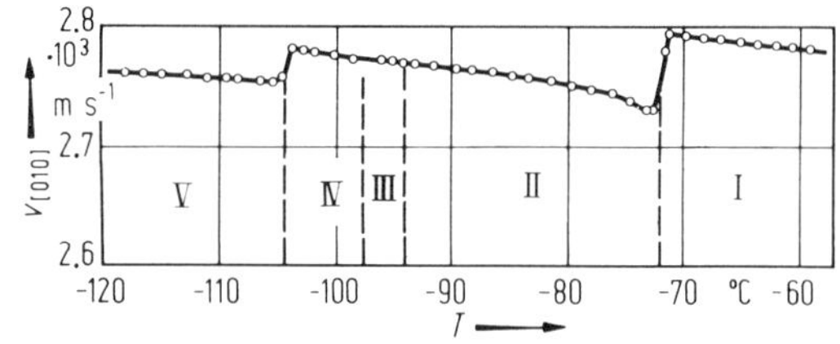

Fig. 1081. $SC(NH_2)_2$. $v_{[010]}$ vs. T [73T5]. $\nu = 7$ MHz. $v_{[010]}$: velocity of the longitudinal ultrasonic wave propagating along [010].

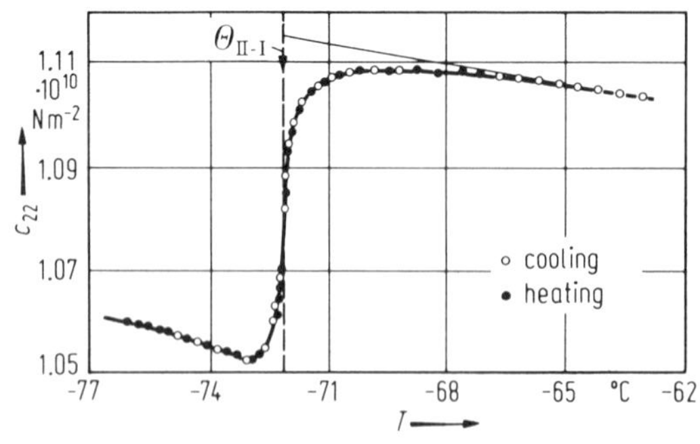

Fig. 1082. $SC(NH_2)_2$. c_{22} vs. T near $\Theta_{\text{II-I}}$ [73T5]. $\Theta_{\text{II-I}}$: transition temperature between phases I and II.

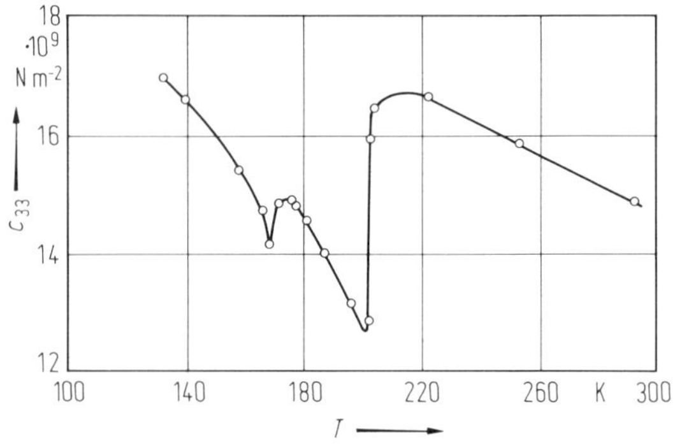

Fig. 1083. $SC(NH_2)_2$. c_{33} vs. T measured by Brillouin scattering [74B4]. $\lambda = 632.8$ nm.

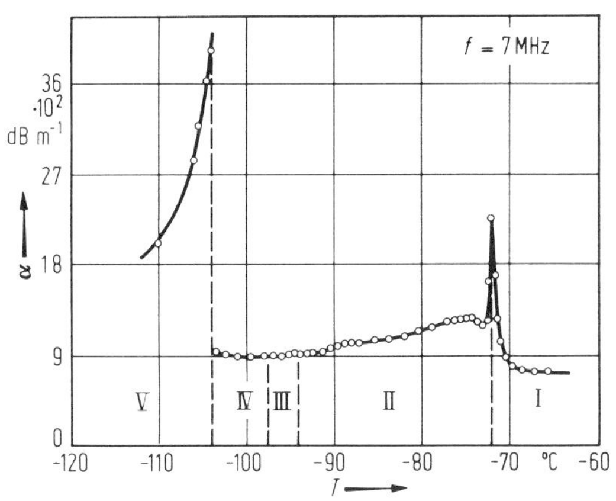

Fig. 1084. $SC(NH_2)_2$. α vs. T [73T5]. α: attenuation coefficient of the longitudinal ultrasonic wave propagating along [010].

Fig. 1086, see p. 570.

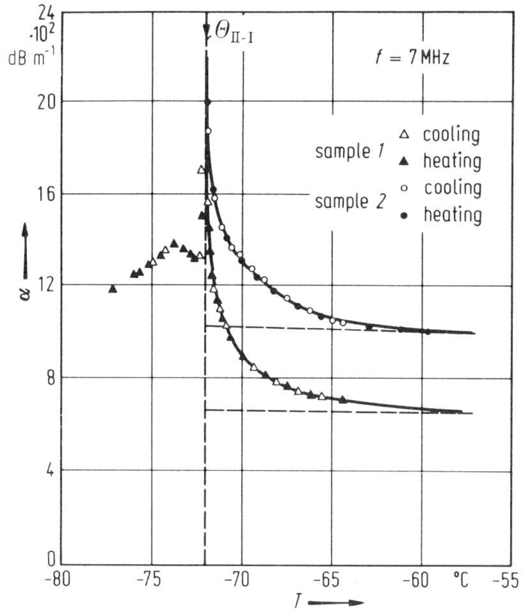

Fig. 1085. $SC(NH_2)_2$. α vs. T near Θ_{II-I} [73T5]. α: attenuation coefficient of the longitudinal ultrasonic wave propagating along [010]. Two specimens were used for this measurement. Absolute values of α differ for the two specimens, but the anomalous parts depend upon temperature in a similar manner.

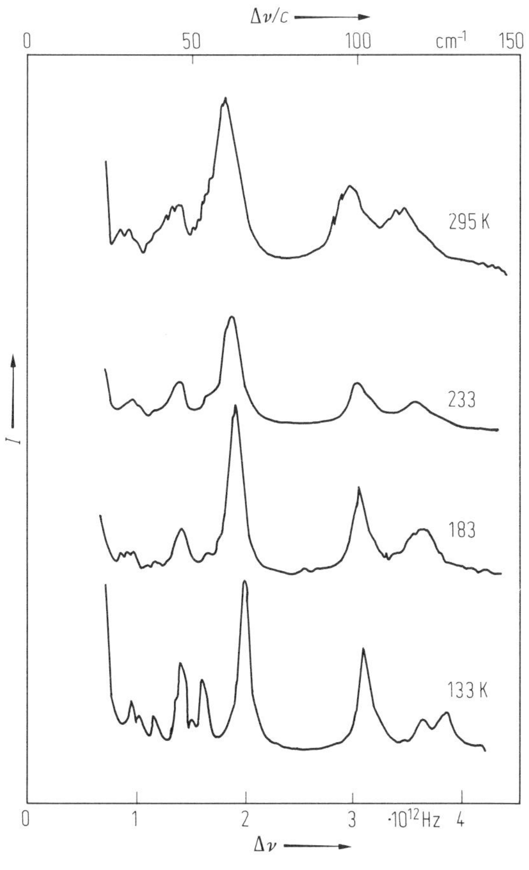

Fig. 1087. $SC(NH_2)_2$. I vs. $\Delta\nu$ [72L2]. I: Raman scattering intensity from a powder sample. Parameter: T.

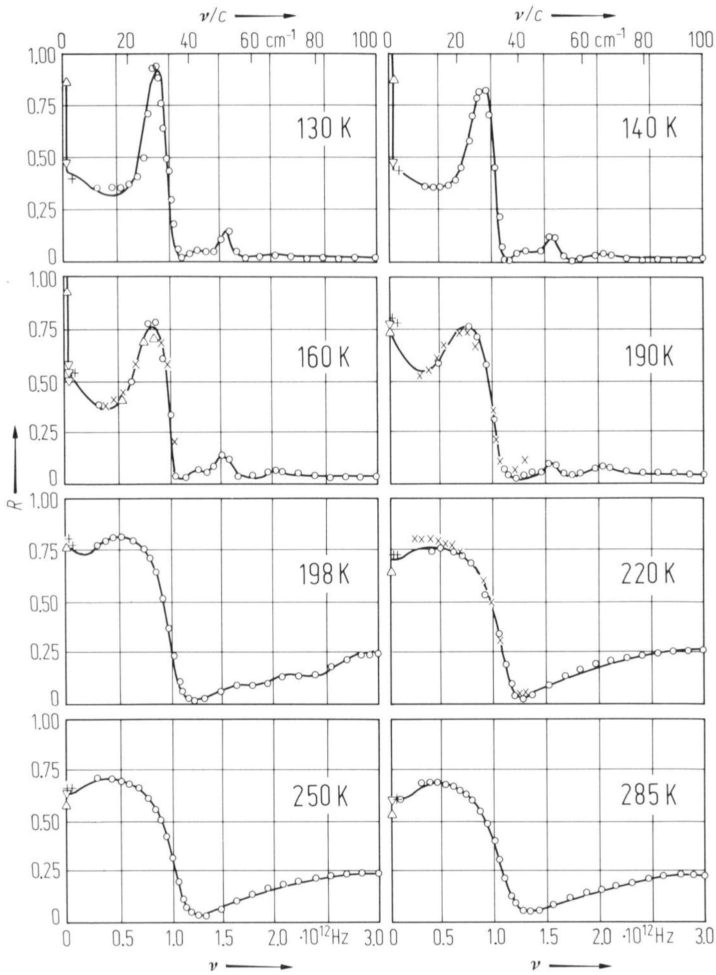

Fig. 1086. $SC(NH_2)_2$. R vs. ν along [100] at various temperatures [77K10]. ν: frequency of the infrared radiation Polarization is parallel to $\boldsymbol{a}$. Following data points are shown: △ [59G2], ▽ [75C9], + [76T13], × [73F5] and ○ [76B10]. The full curves represent the best fit reflectivity.

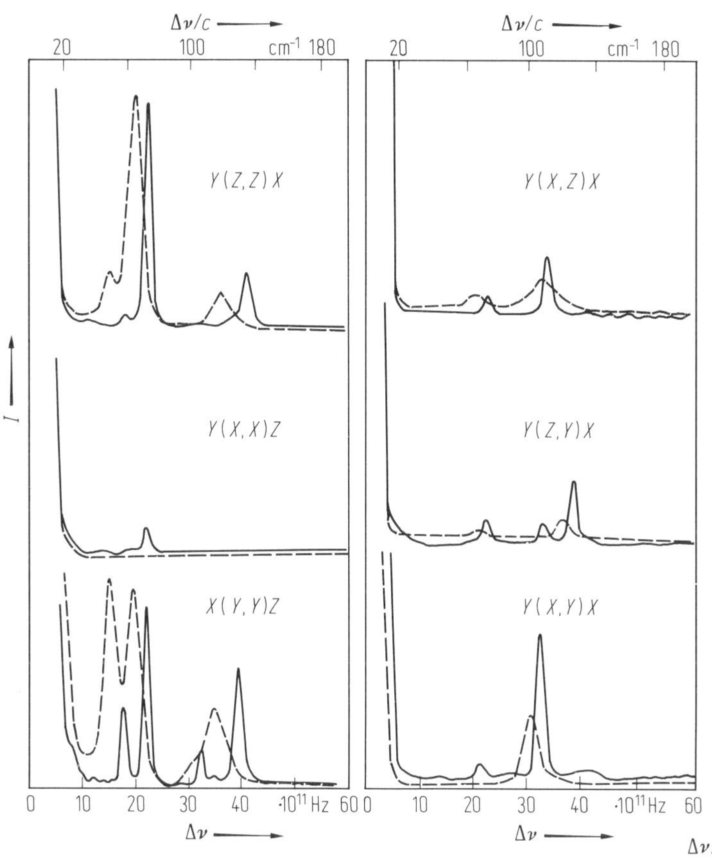

Fig. 1088. $SC(NH_2)_2$. *I* vs. Δν [72 B 1]. *I*: Raman scattering intensity for various scattering geometries. The dashed lines are data at room temperature and the solid lines are at −150 °C.

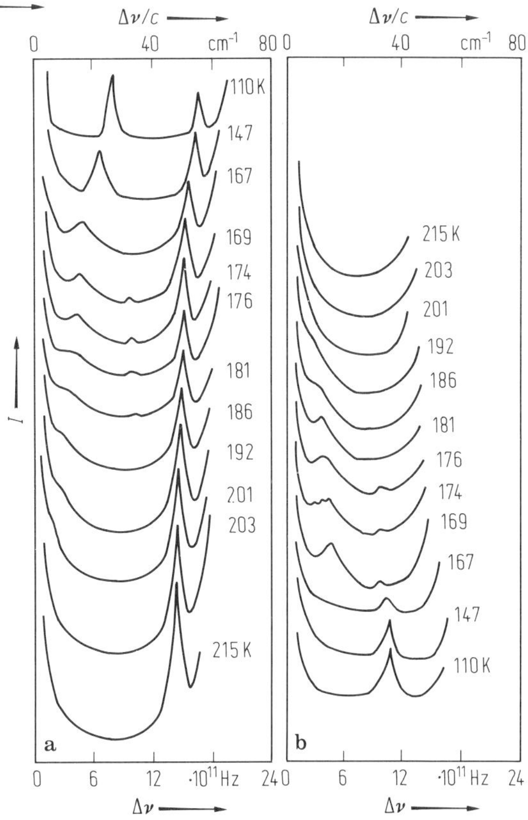

Fig. 1089. $SC(NH_2)_2$. *I* vs. Δν [75 D 4]. *I*: Raman scattering intensity at various temperatures. Spectra (a) show transverse optic modes measured in the scattering geometry of $b(c,c)\bar{b}$. Spectra (b) show longitudinal modes measured in the geometry of $a(c,c)\bar{a}$.

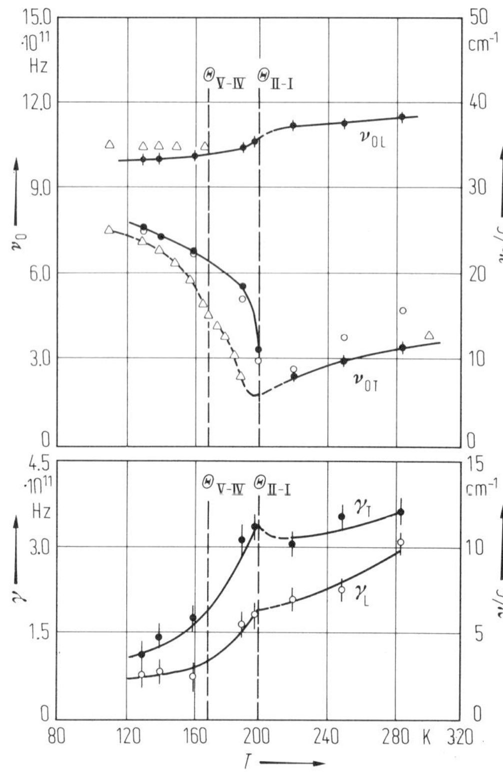

Fig. 1090. $SC(NH_2)_2$. ν_0, γ vs. T [77K10]. ν_0: soft mode frequency. γ: damping constant of the soft mode. Subscripts T and L denote transverse and longitudinal modes, respectively. Triangles are the Raman data from [75D4], others are results of the Kramers-Kronig analysis of [76B10, 73F5, 77K10].

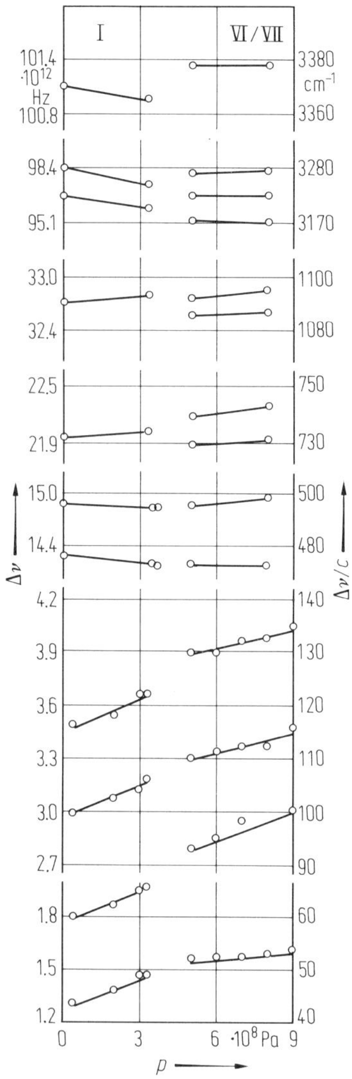

Fig. 1091. $SC(NH_2)_2$. $\Delta\nu$ vs. p [76I4]. $\Delta\nu$: Raman frequency shift at RT. p: hydrostatic pressure. cf. Figs. 1069, 1070.

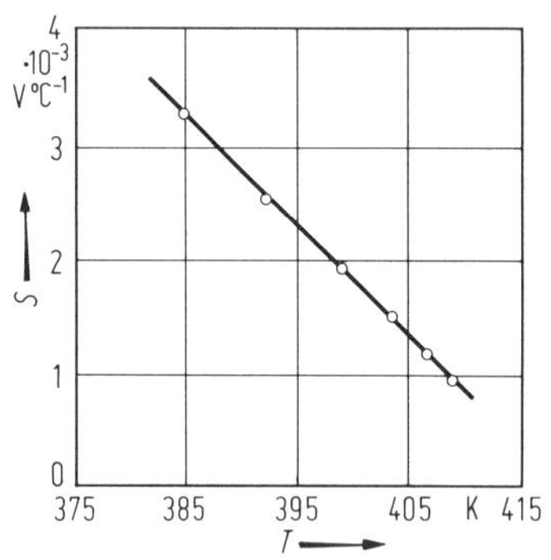

Fig. 1093. $SC(NH_2)_2$. S vs. T [70Y5]. S: Seebeck coefficient obtained from pressed powder sample.

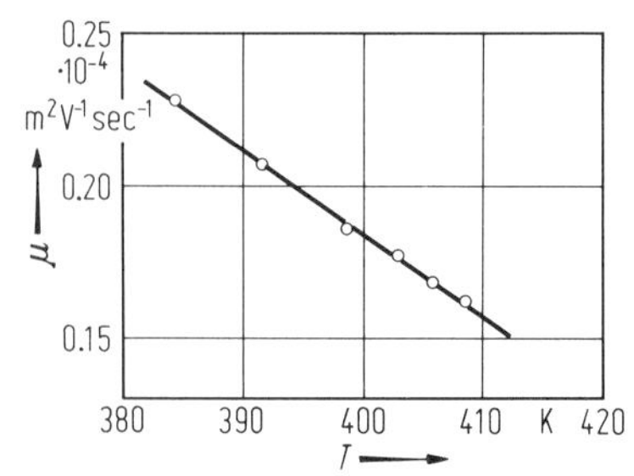

Fig. 1092. $SC(NH_2)_2$. μ vs. T [70Y5]. μ: carrier mobility.

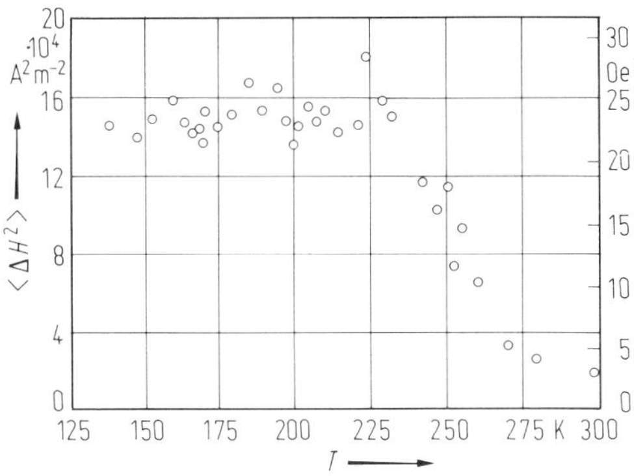

Fig. 1094. $SC(NH_2)_2$. $\langle \Delta H^2 \rangle$ of proton vs. T [61E1]. $a \wedge H = 90°$, $b \wedge H = 40°$. $\langle \Delta H^2 \rangle$: second moment obtained from a single crystal.

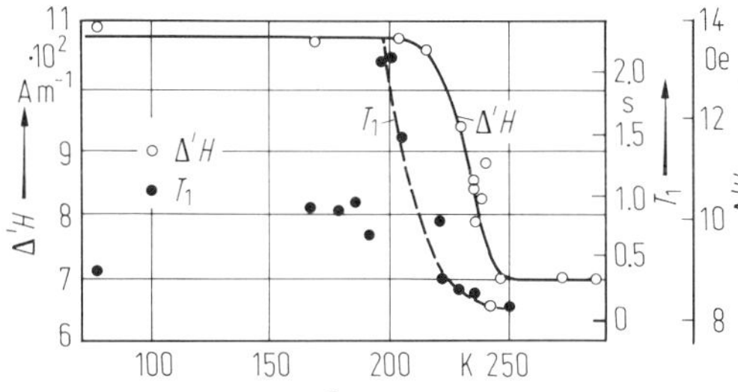

Fig. 1095. $SC(NH_2)_2$. $\Delta' H$, T_1 vs. T obtained from polycrystalline sample [61E1]. $\Delta' H$: separation between maximum and minimum of the derivative of NMR curve.

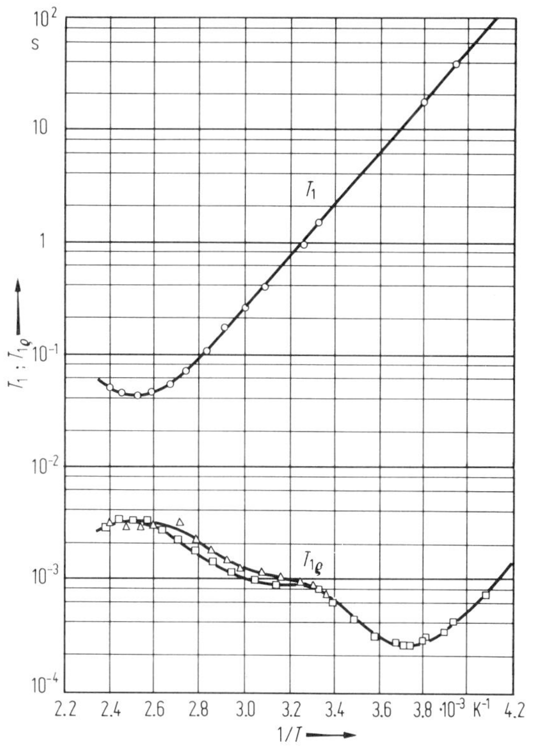

Fig. 1096. $SC(NH_2)_2$. T_1, $T_{1\varrho}$ vs. $1/T$ [71O2]. T_1: proton spin-lattice relaxation time, $T_{1\varrho}$: rotating frame relaxation time. Squares: $H_1 = 1.03 \cdot 10^3$ Am^{-1}, triangles: $H_1 = 1.51 \cdot 10^3$ Am^{-1}, where H_1 is the radio frequency magnetic field.

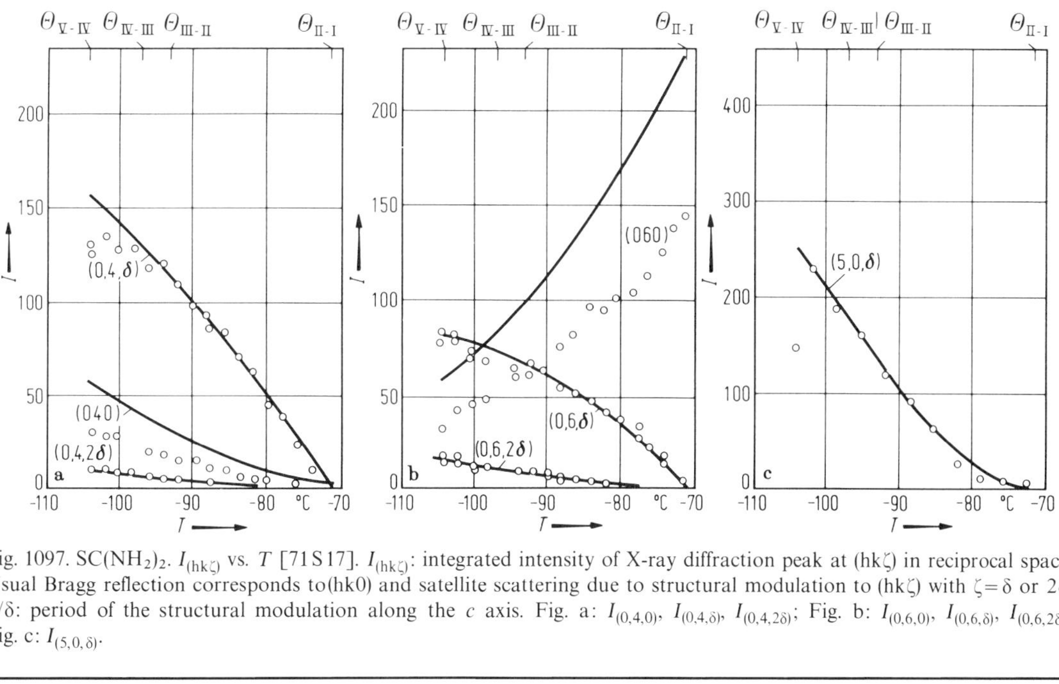

Fig. 1097. $SC(NH_2)_2$. $I_{(hk\zeta)}$ vs. T [71S17]. $I_{(hk\zeta)}$: integrated intensity of X-ray diffraction peak at (hkζ) in reciprocal space. Usual Bragg reflection corresponds to(hk0) and satellite scattering due to structural modulation to (hkζ) with $\zeta = \delta$ or 2δ. c/δ: period of the structural modulation along the c axis. Fig. a: $I_{(0,4,0)}$, $I_{(0,4,\delta)}$, $I_{(0,4,2\delta)}$; Fig. b: $I_{(0,6,0)}$, $I_{(0,6,\delta)}$, $I_{(0,6,2\delta)}$; Fig. c: $I_{(5,0,\delta)}$.

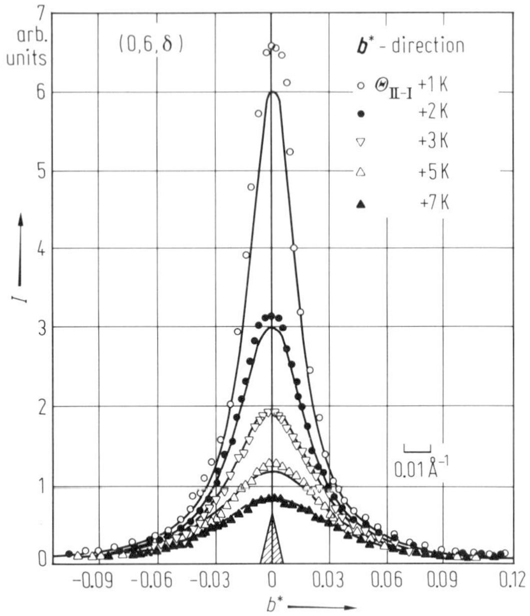

Fig. 1098. $SC(NH_2)_2$. X-ray diffuse scattering around (0,6,δ) reciprocal lattice point in phase I [79T2]. I vs. b^*. $\Theta_{II-I}=202$ K. δ: position where the first satellite scattering appears. Hatched triangle in the figure (at $b^*=0$) indicates the size of resolution.

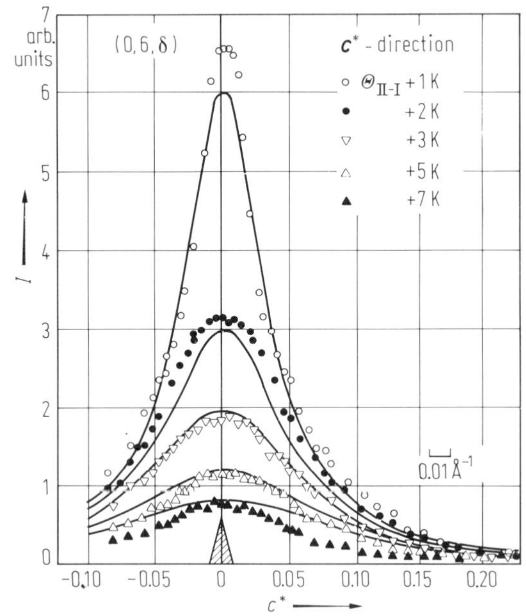

Fig. 1099. $SC(NH_2)_2$. X-ray diffuse scattering around (0,6,δ) reciprocal lattice point in phase I [79T2]. I vs. c^*. $\Theta_{II-I}=202$ K. δ: position where the first satellite scattering appears. Hatched triangle in the figure (at $c^*=0$) indicates the size of resolution.

Fig. 1100. $SC(NH_2)_2$. X-ray diffuse scattering around (0,6,0) in phase IV [79T2]. I vs. b^* and c^*. $\Theta_{V-IV}=169$ K. Hatched triangle indicates the size of resolution. ►

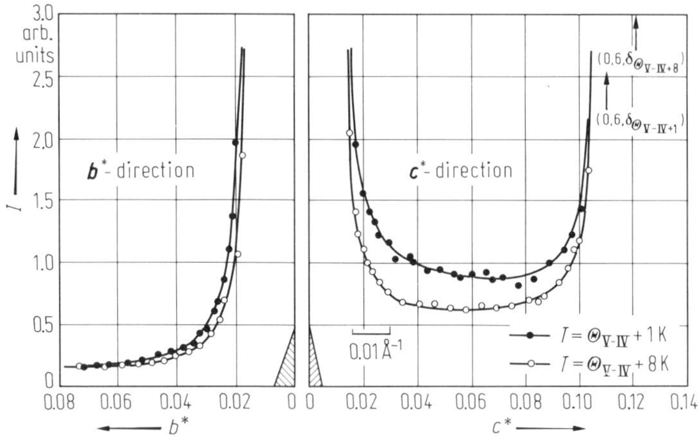

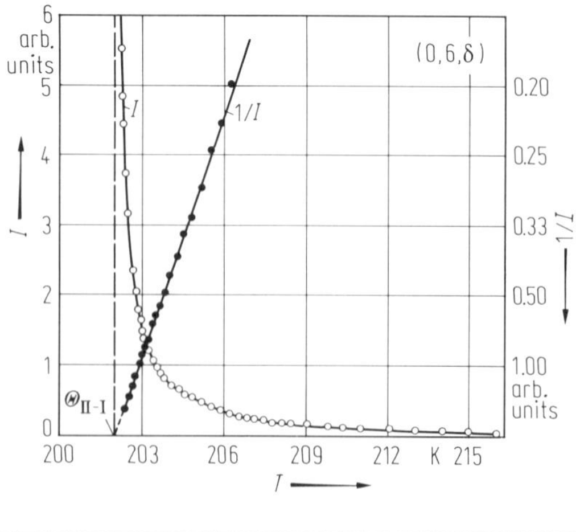

Fig. 1101. $SC(NH_2)_2$. X-ray diffuse scattering at (0,6,δ) reciprocal lattice point in phase I [79T2]. I and $1/I$ vs. T. I: intensity of X-ray diffuse scattering. $\Theta_{II-I}=202$ K. δ: position where the first satellite scattering appears.

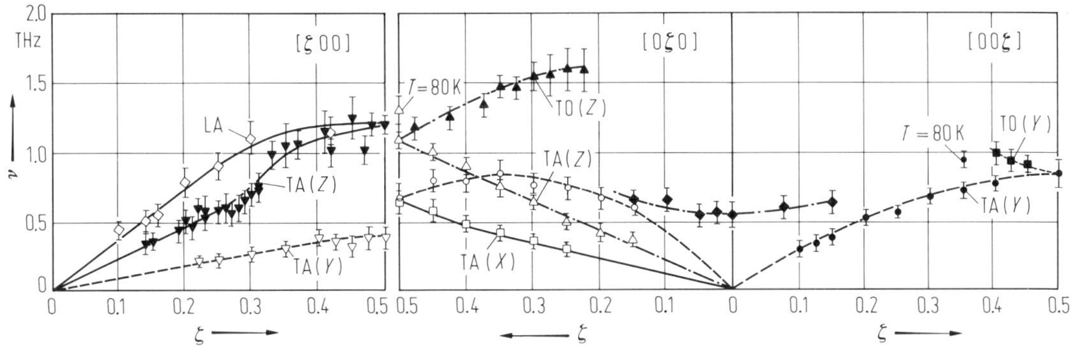

Fig. 1103. $SC(ND_2)_2$. ν vs. ζ [75M10]. ν: phonon frequency observed by neutron inelastic scattering. ζ: reduced wave vector coordinate. TA(Z), TA(Y), ...: $e\|[001]$, $e\|[010]$, ..., where e denotes polarization of phonon. Assignments for ○, ◆ are unknown.

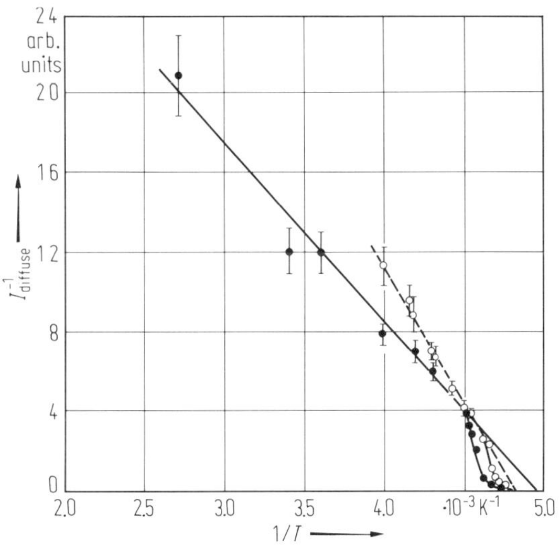

Fig. 1102. $SC(ND_2)_2$. I^{-1}_{diffuse} vs. T^{-1} [75M9]. I_{diffuse}: neutron diffuse scattering intensity at (0.4,17,5.5). Open and full circles correspond to experiments carried out on different spectrometers.

32 $N(CH_3)_4 \cdot HgCl_3$ family

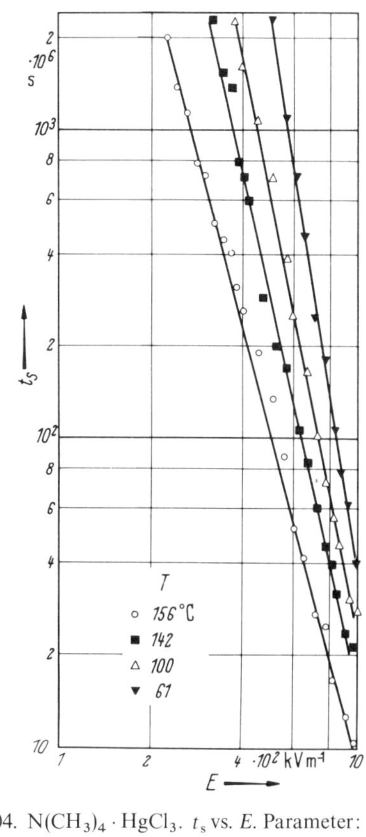

Fig. 1104. $N(CH_3)_4 \cdot HgCl_3$. t_s vs. E. Parameter: T [60F3].

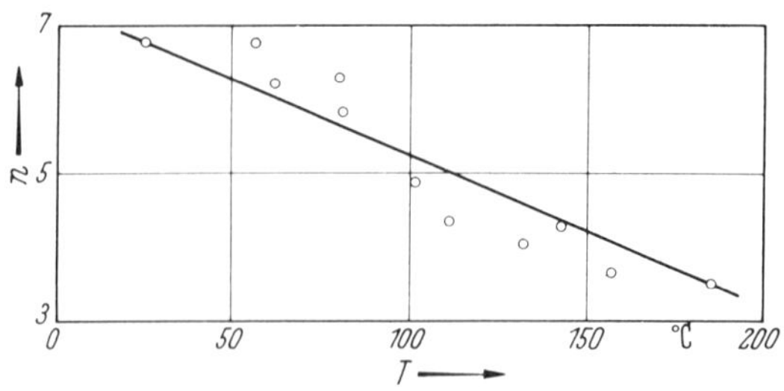

Fig. 1105. $N(CH_3)_4 \cdot HgCl_3$. n vs. T [60 F 3]. n: exponent of equation $t_s \propto E^{-n}$.

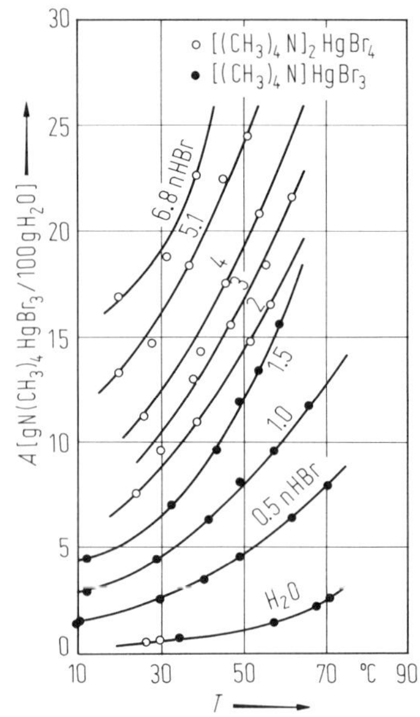

Fig. 1106. $N(CH_3)_4 \cdot HgBr_3$. A vs. T [62 F 4]. A: solubility of $[N(CH_3)_4]Br–HgBr_2$ mixture.

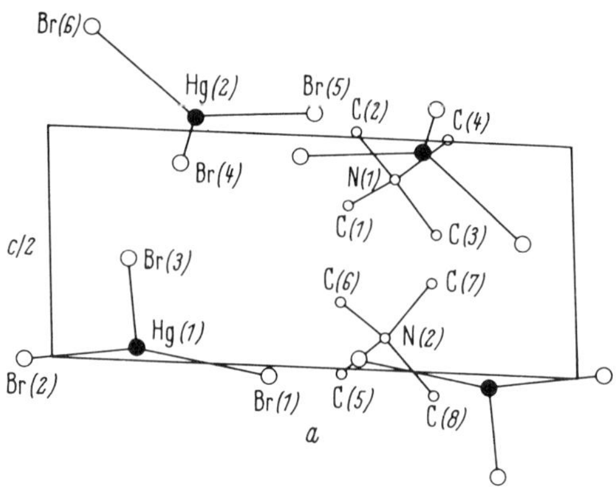

Fig. 1107. $N(CH_3)_4 \cdot HgBr_3$. Crystal structure [63 W 5]. Projection on (010).

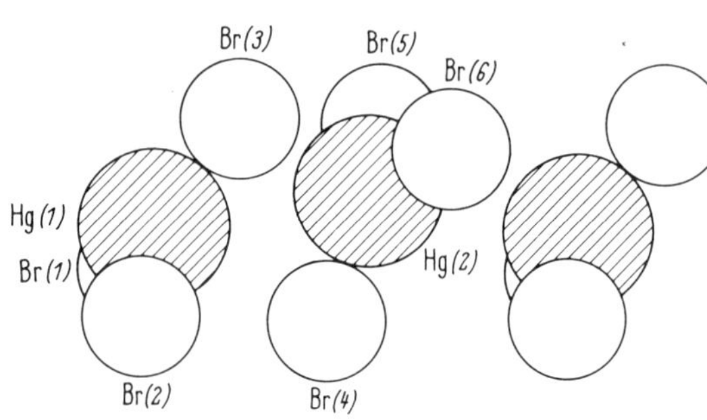

Fig. 1108. $N(CH_3)_4 \cdot HgBr_3$. Crystal structure [63 W 5]. A view of the anion chain which runs through the crystal in the c-direction.

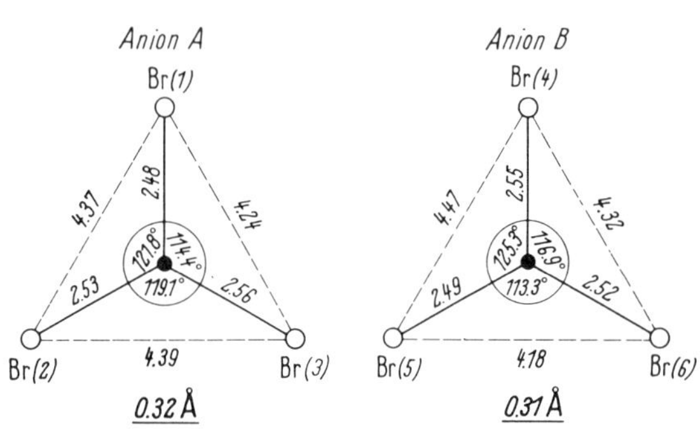

Fig. 1109. $N(CH_3)_4 \cdot HgBr_3$. Crystal structure [63 W 5]. Dimensions of the two crystallographically independent anions. Distances are in Å and angles in degrees. The underlined figures at the bottom give the distances of the Hg atoms from the planes of the Br atoms. See also Table 295.

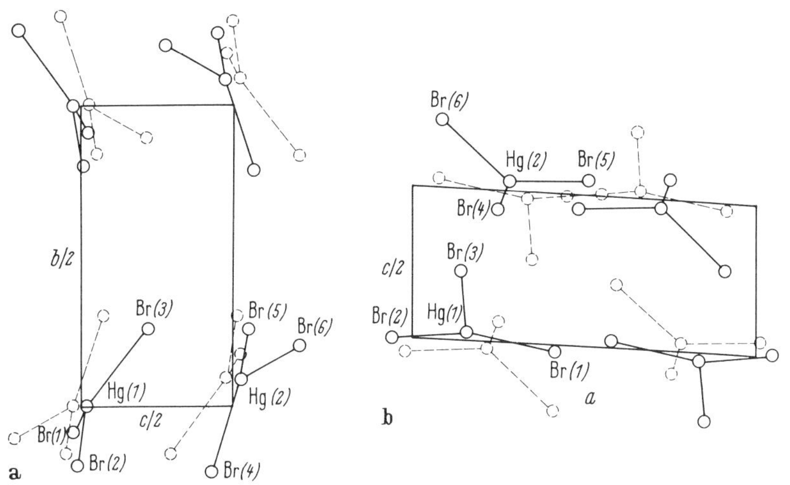

Fig. 1110. $N(CH_3)_4 \cdot HgBr_3$. Crystal structure [63W5]. An illustration of the polarization reversal. Fig. a and Fig. b are projection along a and b axes, respectively. The ions indicated by full lines correspond to one polarity of the applied field and those indicated by dashed lines to the other polarity.

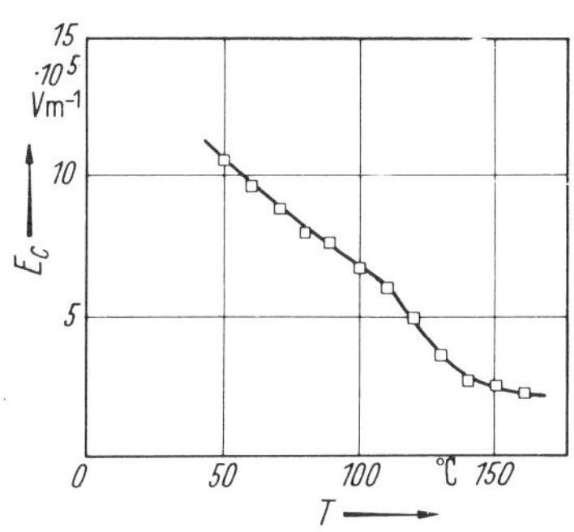

Fig. 1111. $N(CH_3)_4 \cdot HgBr_3$. E_c vs. T [62F4].

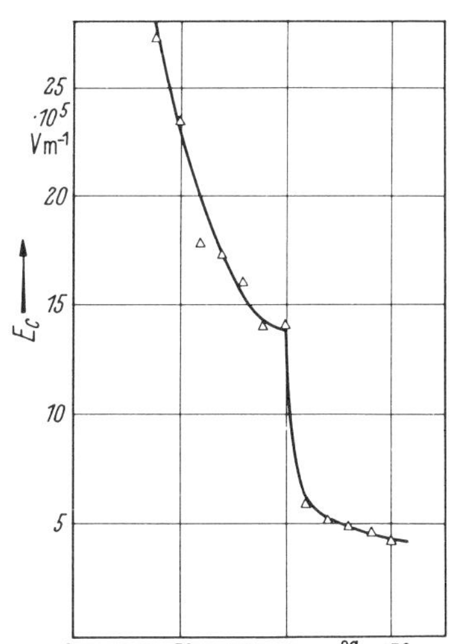

Fig. 1112. $P(CH_3)_4 \cdot HgBr_3$. E_c vs. T [62F4].

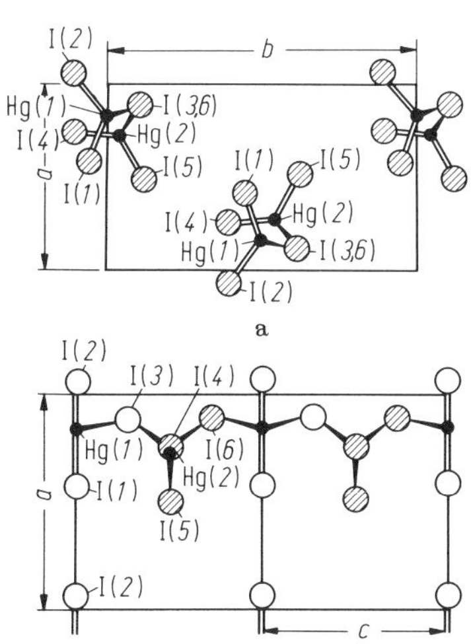

Fig. 1113. $N(CH_3)_4 \cdot HgI_3$. Crystal structure [67G3]. Fig. a: Projection on (001), Fig. b: on (010).

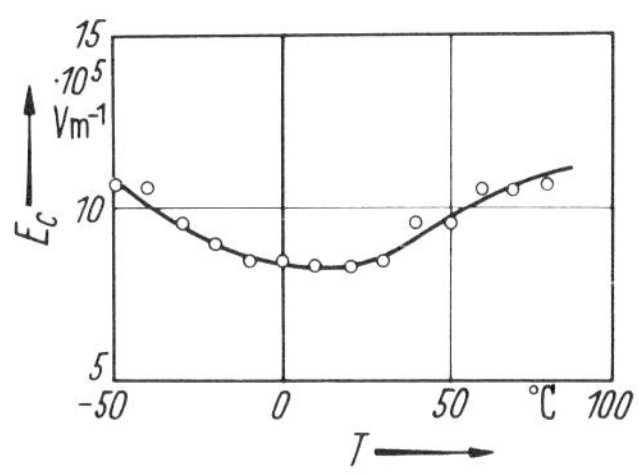

Fig. 1114. $N(CH_3)_4 \cdot HgI_3$. E_c vs. T [62F4].

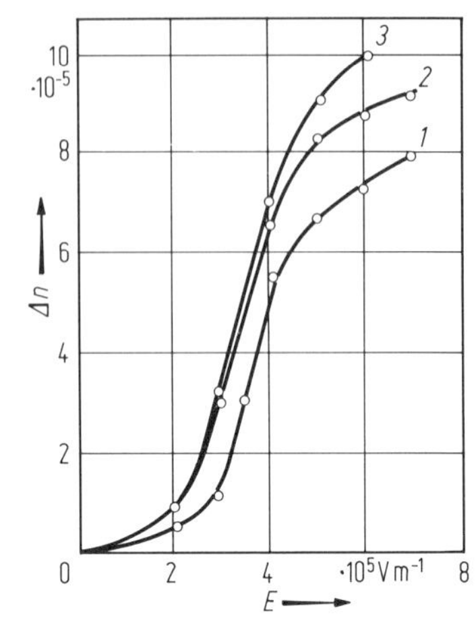

Fig. 1115. $N(CH_3)_4 \cdot HgI_3$. Δn vs. E at RT [69 S 19]. $\Delta n = \frac{1}{2}(n_1^3 r_{13} - n_3^3 r_{33})E$. Parameter: λ; curve *1*: 585 nm, *2*: 535 nm, *3*: 465 nm.

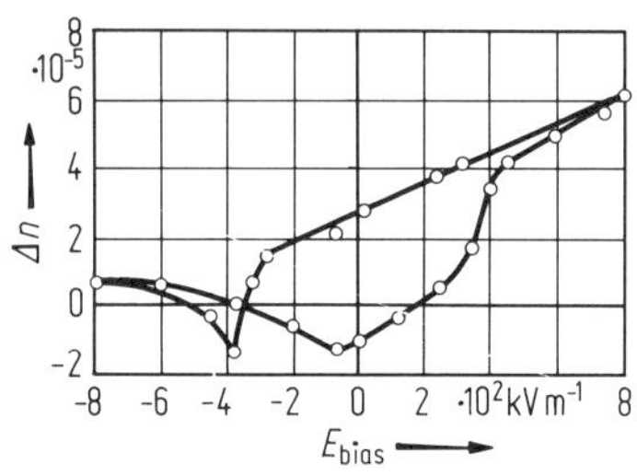

Fig. 1116. $N(CH_3)_4 \cdot HgI_3$. Δn vs. E_{bias} at RT [69 S 19]. $\Delta n = \frac{1}{2}(n_1^3 r_{13} - n_3^3 r_{33})E$. $\lambda = 535$ nm.

33 $Cu(HCOO)_2 \cdot 4H_2O$

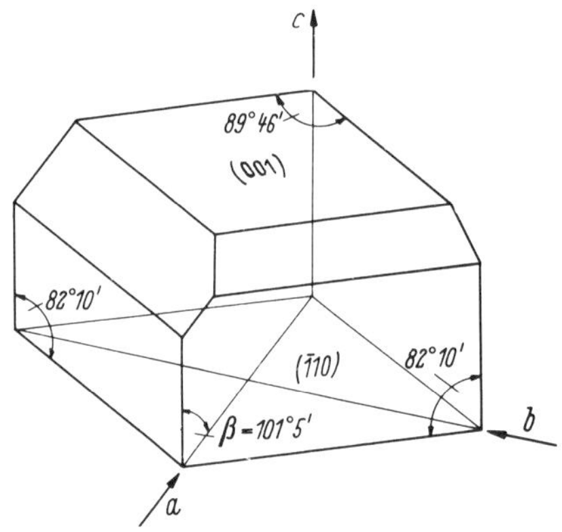

Fig. 1117. $Cu(HCOO)_2 \cdot 4H_2O$. Crystal form [66 O 2].

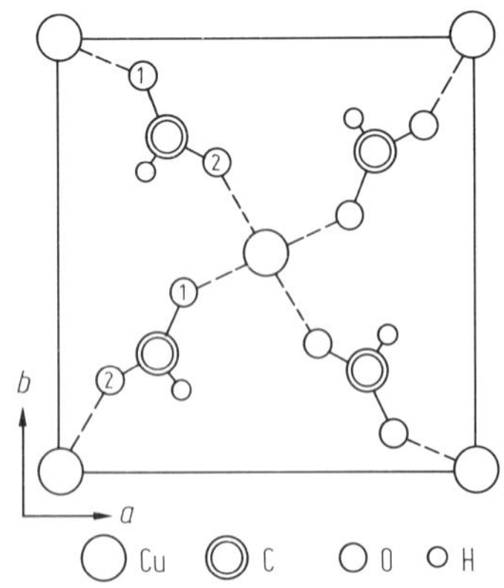

Fig. 1119. $Cu(HCOO)_2 \cdot 4H_2O$. Crystal structure of phase I [75 K 9]. Copper formate layer near $z = 0$.

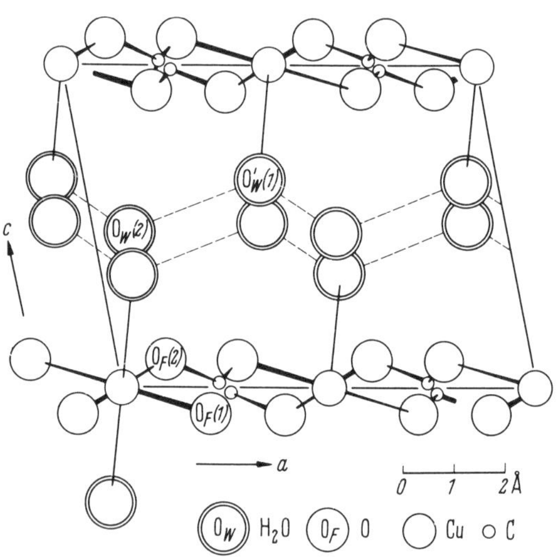

Fig. 1118. $Cu(HCOO)_2 \cdot 4H_2O$. Crystal structure of phase I [54 K 2]. Projection along *b*. Broken lines show the layer of water molecules. O_F: O belonging to formate. O_W: O belonging to water.

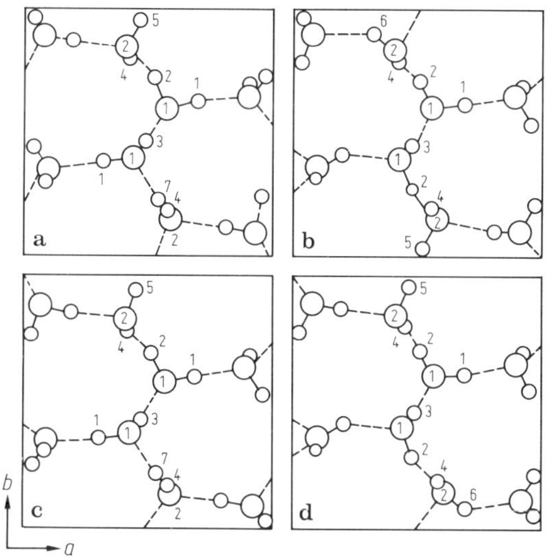

Fig. 1120. $Cu(HCOO)_2 \cdot 4H_2O$. Crystal structure of phase I [75K9]. Schematic representation of four possible fully hydrogen bonded water layers. The structures in Figs. a and b (and their centric pairs) in a 2 to 1 ratio probably make up the water layer in the paraelectric phase, although the same occupancies may be derived from Figs. c and d. Figs. a and b contain 2_1 screw axes and Figs. c and d *a* glide planes. The screw axes are retained through the transition. Large circles denote oxygen atoms. Small circles hydrogen atoms.

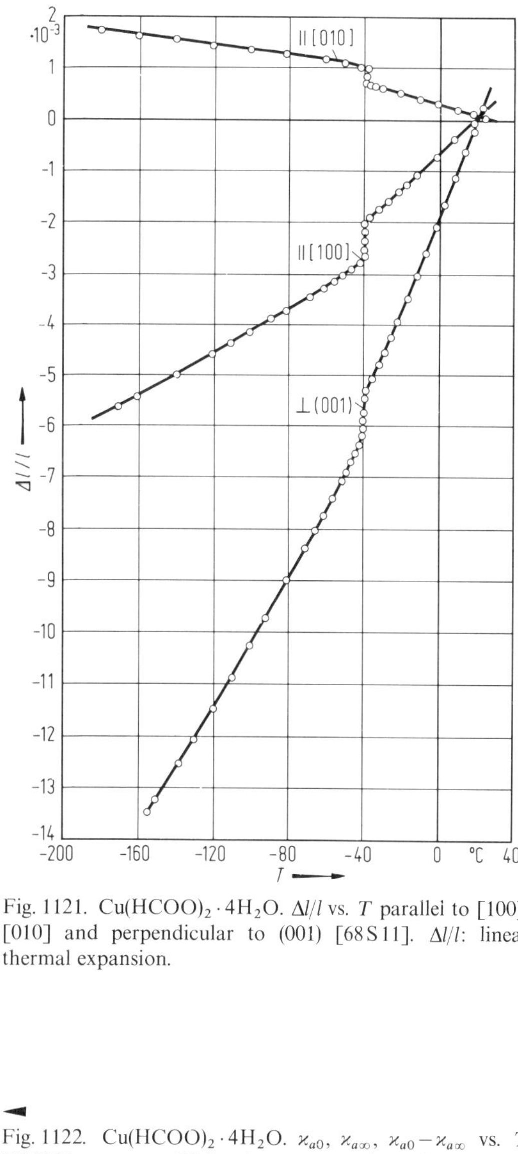

Fig. 1121. $Cu(HCOO)_2 \cdot 4H_2O$. $\Delta l/l$ vs. T parallel to [100], [010] and perpendicular to (001) [68S11]. $\Delta l/l$: linear thermal expansion.

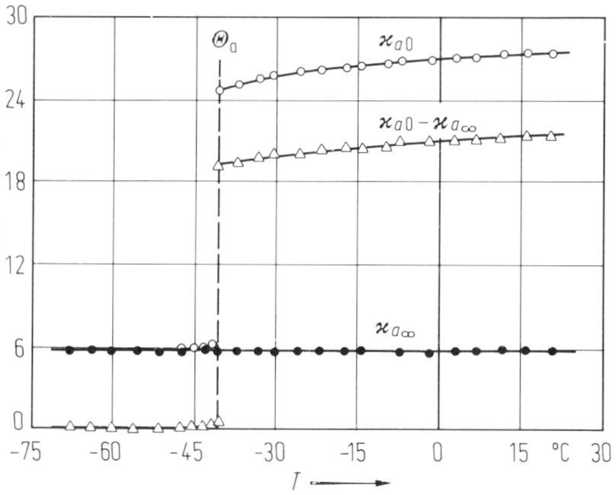

◄ Fig. 1122. $Cu(HCOO)_2 \cdot 4H_2O$. $\varkappa_{a0}$, $\varkappa_{a\infty}$, $\varkappa_{a0}-\varkappa_{a\infty}$ vs. T [78H1]. $\varkappa_{a0}$, $\varkappa_{a\infty}$: Dielectric constants along the *a*-axis at static and high-frequency limit, respectively.

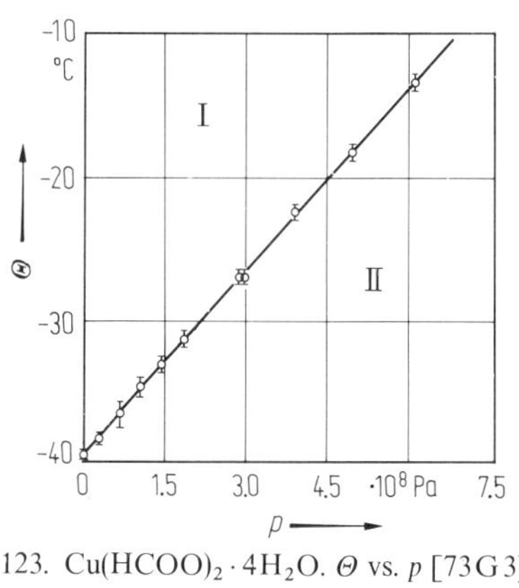

Fig. 1123. $Cu(HCOO)_2 \cdot 4H_2O$. Θ vs. p [73G3].

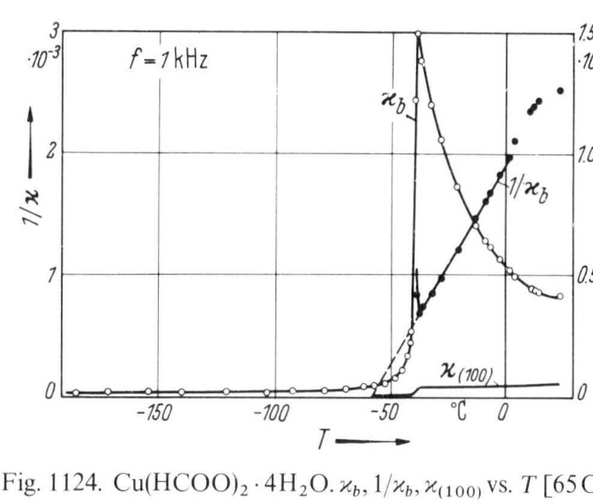

Fig. 1124. $Cu(HCOO)_2 \cdot 4H_2O$. $\varkappa_b$, $1/\varkappa_b$, $\varkappa_{(100)}$ vs. T [65O2].

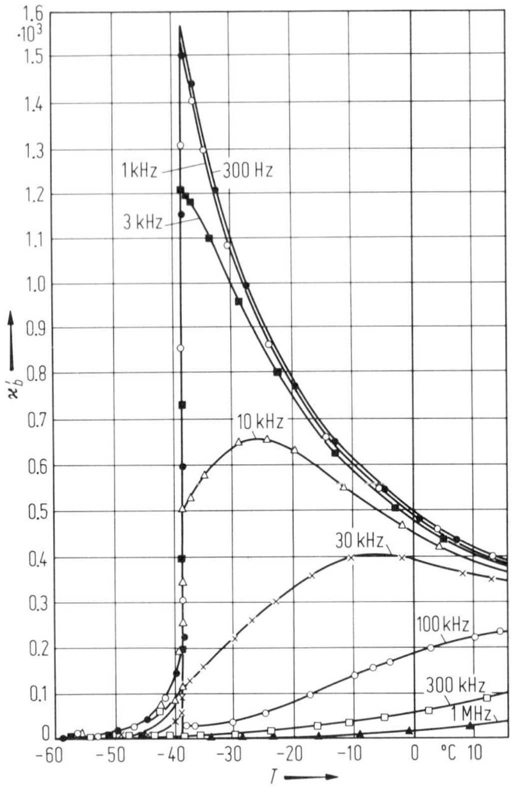

Fig. 1125. $Cu(HCOO)_2 \cdot 4H_2O$. $\varkappa_b'$ vs. T [69 M 1]. Parameter: f.

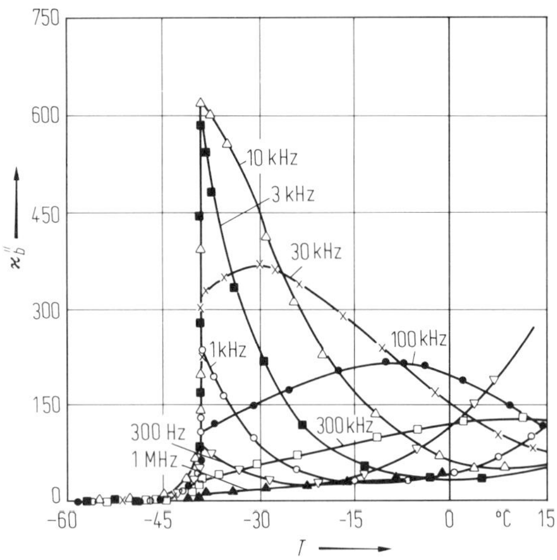

Fig. 1126. $Cu(HCOO)_2 \cdot 4H_2O$. $\varkappa_b''$ vs. T [69 M 1]. Parameter: f.

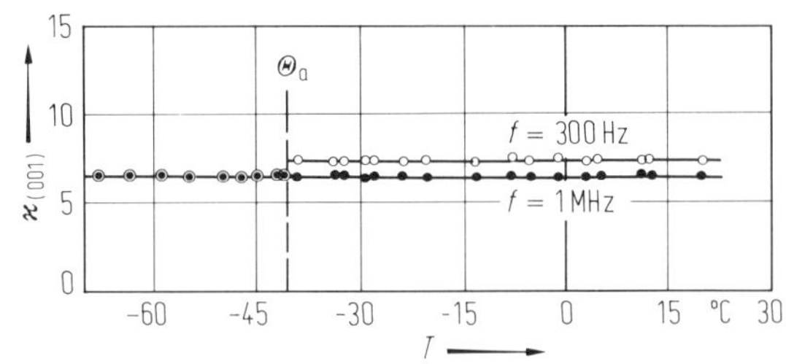

Fig. 1127. $Cu(HCOO)_2 \cdot 4H_2O$. $\varkappa_{(001)}$ vs. T [78 H 1]. Parameter: f. $\varkappa_{(001)}$: Dielectric constant for $E \perp (001)$.

For Fig. 1129, see p. 581.

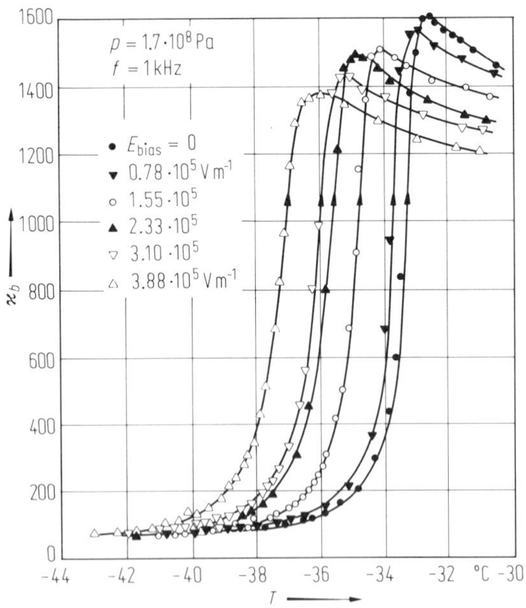

Fig. 1128. $Cu(HCOO)_2 \cdot 4H_2O$. $\varkappa_b$ vs. T [77 F 6]. Parameter: E_{bias}, $p = 1.7 \cdot 10^8$ Pa.

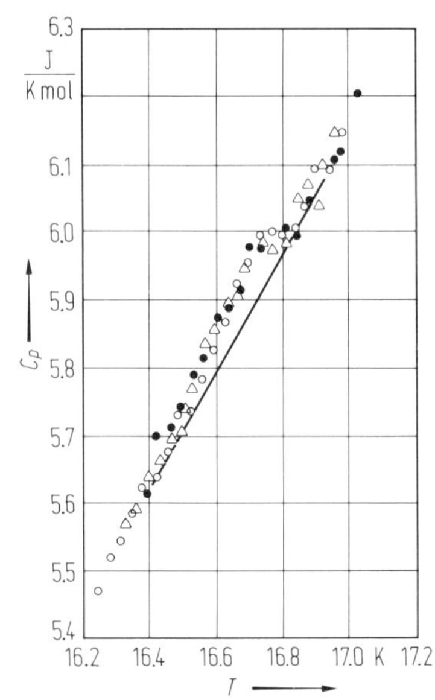

Fig. 1130. $Cu(HCOO)_2 \cdot 4H_2O$. C_p vs. T [75 Y 3]. Different symbols correspond to different measurements.

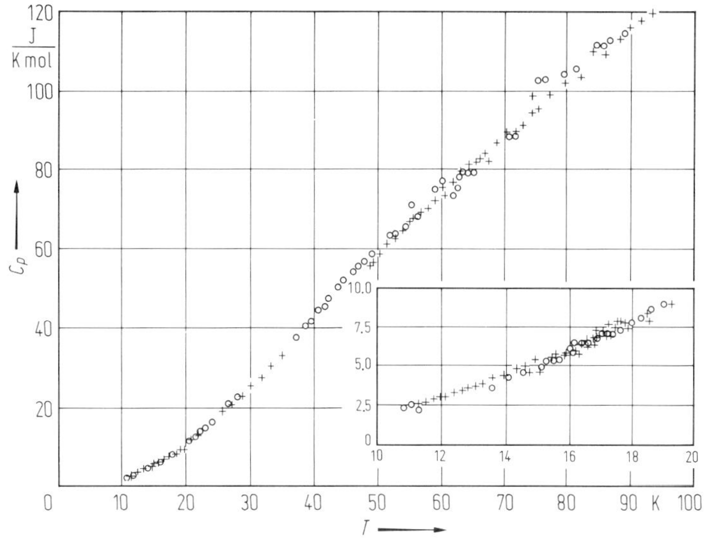

Fig. 1129. $Cu(HCOO)_2 \cdot 4H_2O$. C_p vs. T near the antiferromagnetic Néel temperature [62S11].

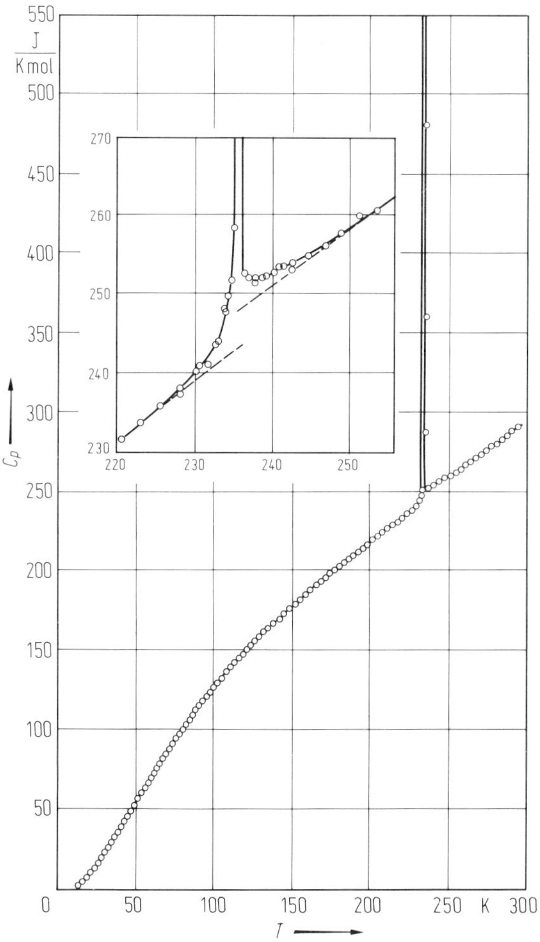

Fig. 1131. $Cu(HCOO)_2 \cdot 4H_2O$. C_p vs. T [76M6].

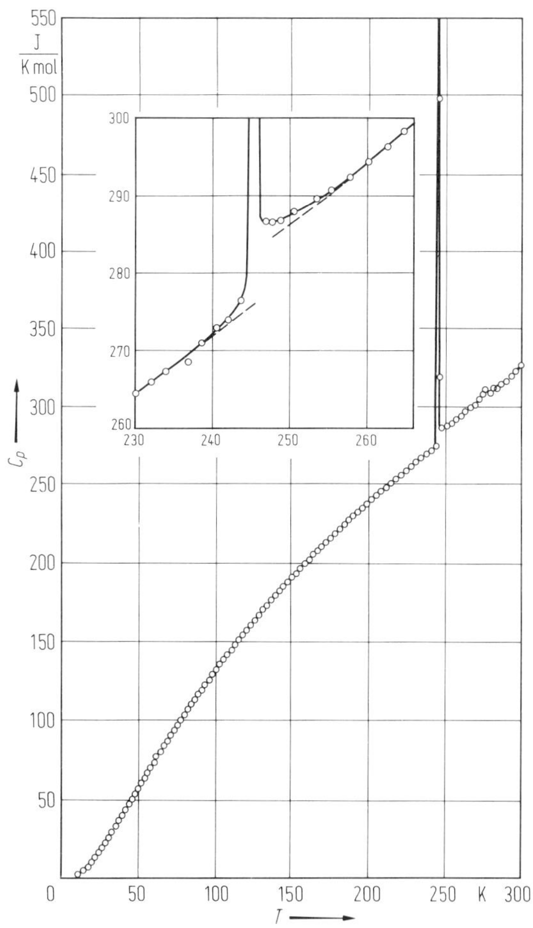

Fig. 1132. $Cu(HCOO)_2 \cdot 4D_2O$. C_p vs. T [76M6].

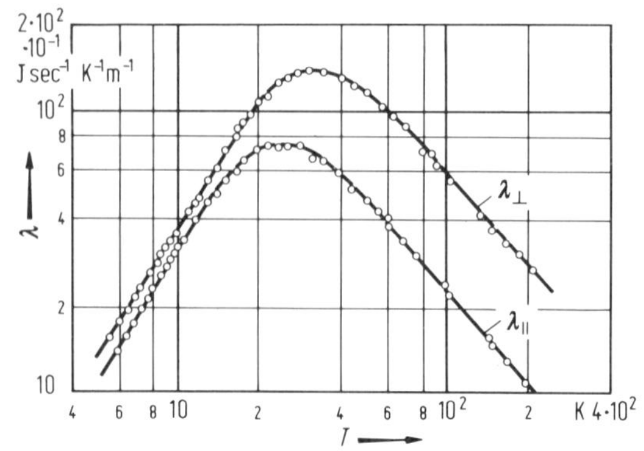

Fig. 1133. $Cu(HCOO)_2 \cdot 4H_2O$. $\lambda_\perp$, $\lambda_\parallel$ vs. T [72H11]. $\lambda_\perp$, $\lambda_\parallel$: thermal conductivity perpendicular to (001), parallel to [001], respectively.

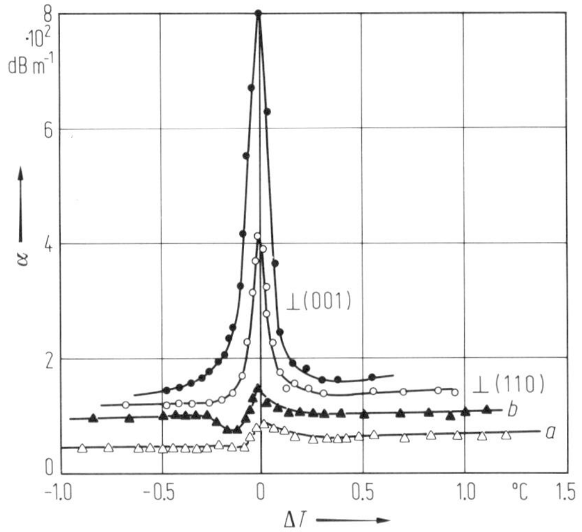

Fig. 1134. $Cu(HCOO)_2 \cdot 4H_2O$. α vs. ΔT [73K2]. α: attenuation coefficients for the longitudinal sound wave propagating along the a-axis, the b-axis, and perpendicular to the (110) habit plane and to the (001) habit plane. $\Delta T = T - \Theta_a$, $f = 30$ MHz.

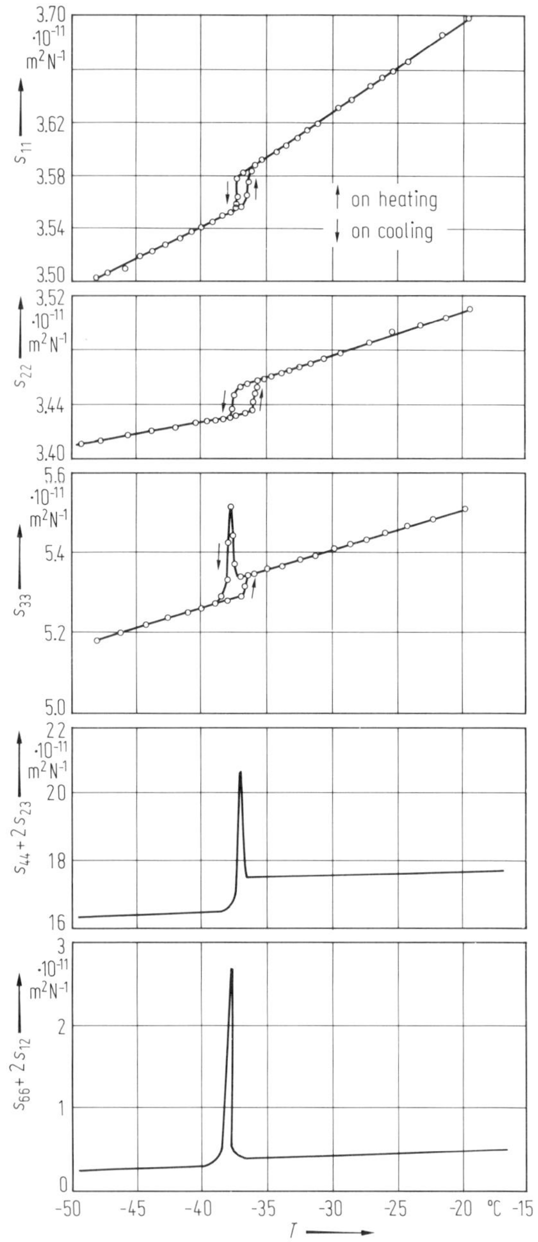

Fig. 1135. $Cu(HCOO)_2 \cdot 4H_2O$. $s_{\lambda\mu}$ vs. T [74S8].

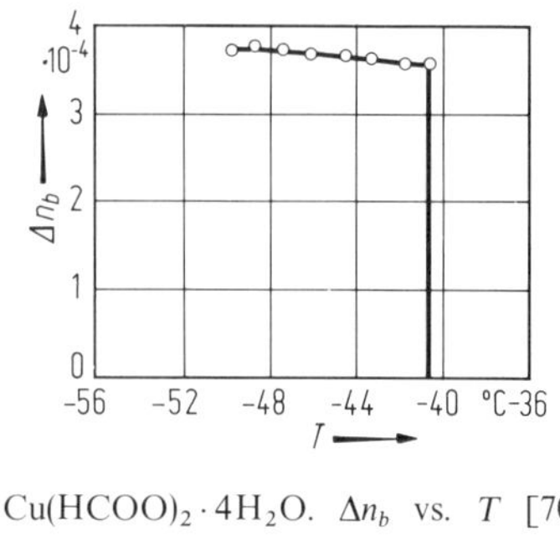

Fig. 1136. $Cu(HCOO)_2 \cdot 4H_2O$. Δn_b vs. T [70A8]. $\lambda = 632.8$ nm.

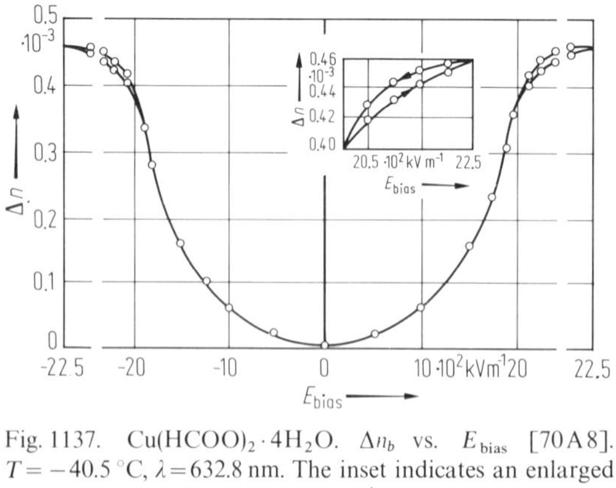

Fig. 1137. $Cu(HCOO)_2 \cdot 4H_2O$. Δn_b vs. E_{bias} [70A8]. $T = -40.5$ °C, $\lambda = 632.8$ nm. The inset indicates an enlarged version between 20 and 22.5 kV m^{-1} of E_{bias}.

Fig. 1138. $Cu(HCOO)_2 \cdot 4H_2O$. I vs. $\Delta\nu$ [73C1]. I: Raman scattering intensity for the low frequency region. Upper two spectra are in the paraelectric (250 K), and the lower two are in the antiferroelectric phase (90 K). Peak frequencies are indicated in the figure in units of cm^{-1}. D is a laser discharge line.

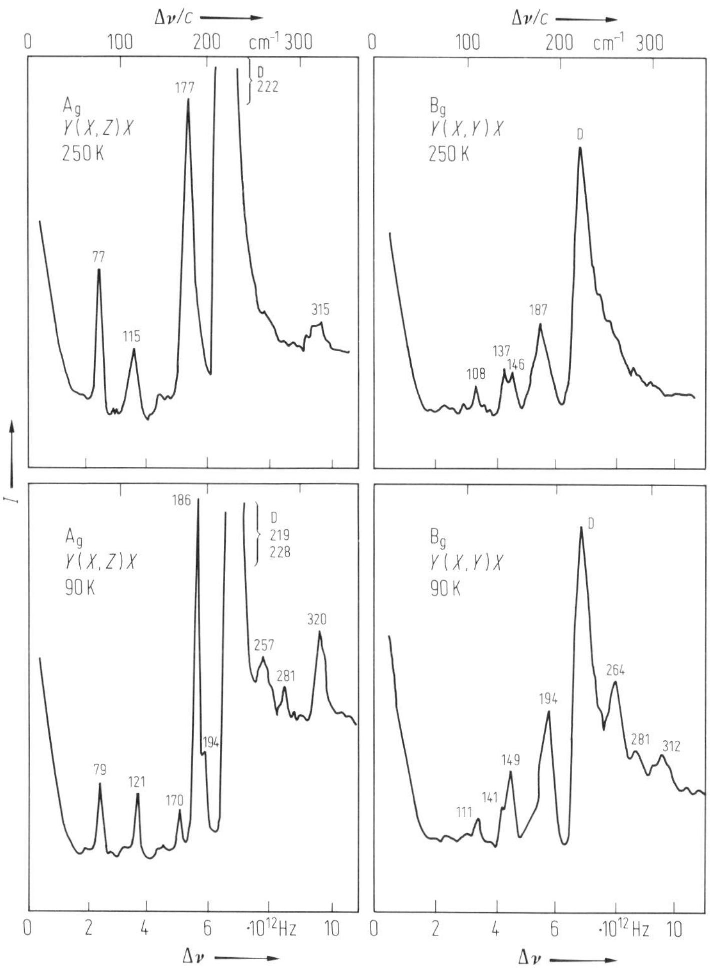

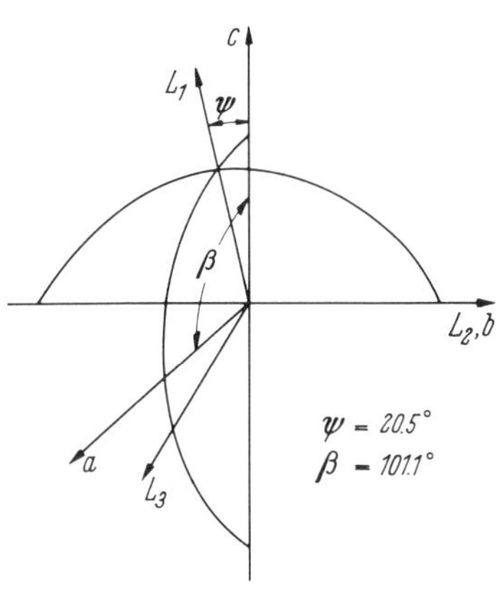

Fig. 1140. $Cu(HCOO)_2 \cdot 4H_2O$. Relation between the crystallographic axes (a, b, c) and the magnetic axes (L_1, L_2, L_3) [56S4].

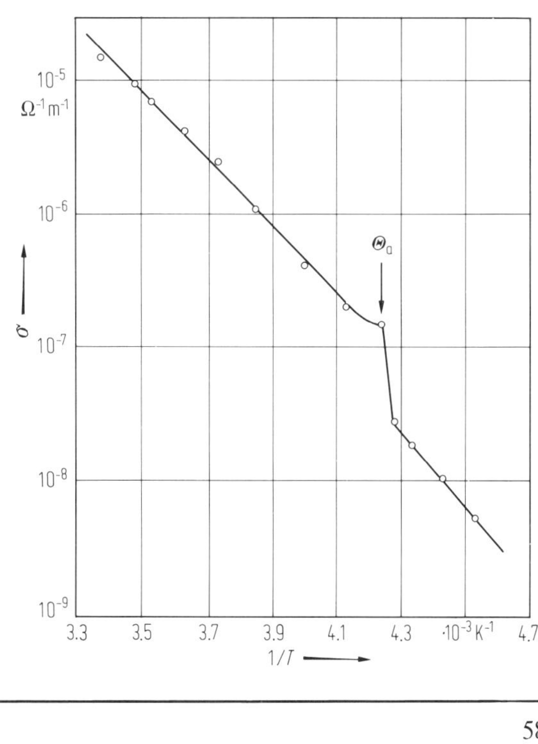

Fig. 1139. $Cu(HCOO)_2 \cdot 4H_2O$. Protonic dc conductivity σ vs. $1/T$ [75W2]. Electrodes were saturated cupric formate tetrahydrate solutions of diethylene glycol 2-methoxy ethyl ether and distilled water. σ was measured perpendicular to the naturally occuring (110) face. ►

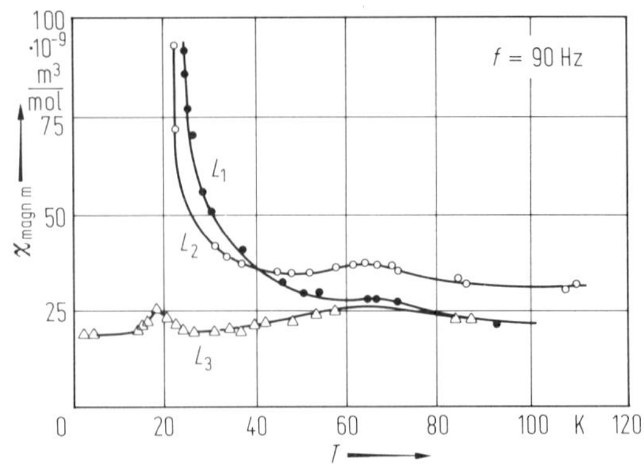

Fig. 1141. $Cu(HCOO)_2 \cdot 4H_2O$. $\chi_{magn\,m}$ vs. T [63 K 5]. $\chi_{magn\,m}$: molar magnetic susceptibilities along the L_1, L_2, and L_3 axes. See Fig. 1140.

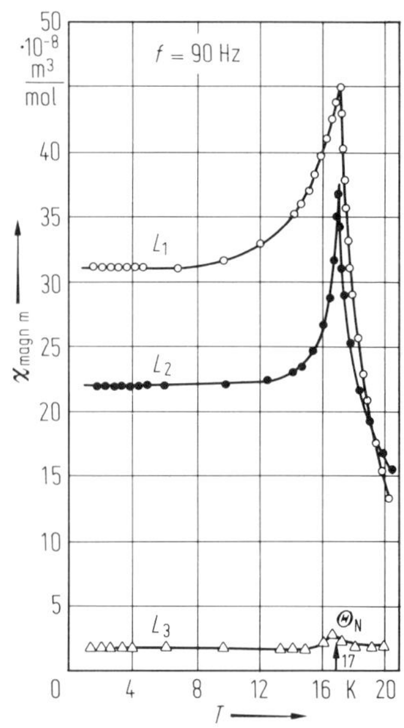

Fig. 1142. $Cu(HCOO)_2 \cdot 4H_2O$. $\chi_{magn\,m}$ vs. T [63 K 5]. $\chi_{magn\,m}$: molar magnetic susceptibilities along the L_1, L_2, and L_3 axes. Θ_N: Néel temperature. See Fig. 1140.

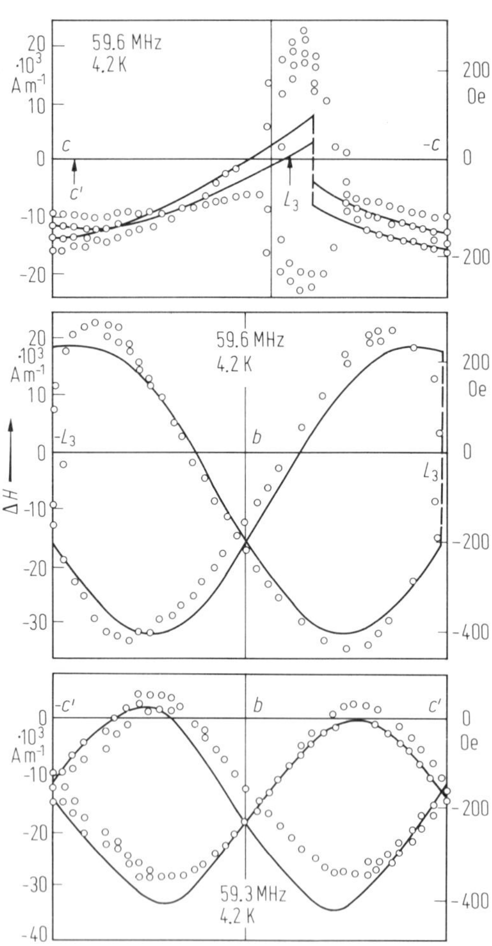

Fig. 1144. $Cu(HCOO)_2 \cdot 4D_2O$. ΔH vs. rotation angle [71 Y 2]. ΔH: shift of resonance field from free proton. Circles: experimental results; lines: calculated. c' represents the axis perpendicular to ab plane.

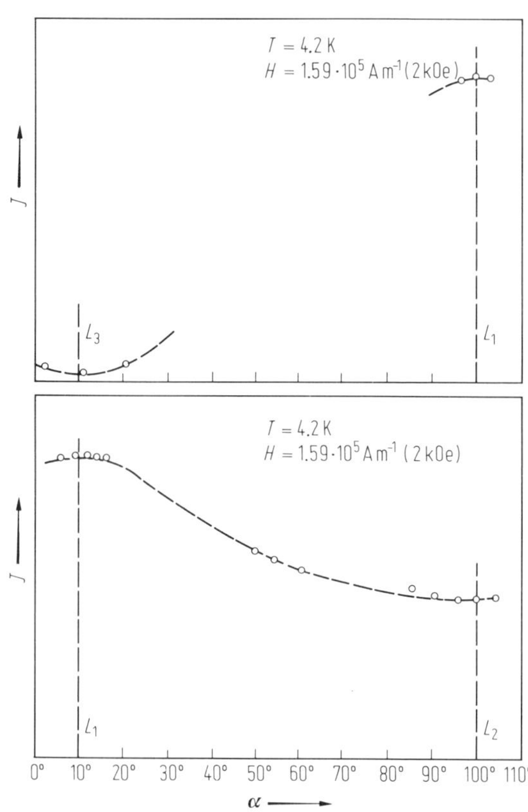

Fig. 1143. $Cu(HCOO)_2 \cdot 4H_2O$. J vs. α [63 K 5]. J: magnetic polarization, α: angle. See Fig. 1140 for L_1, L_2.

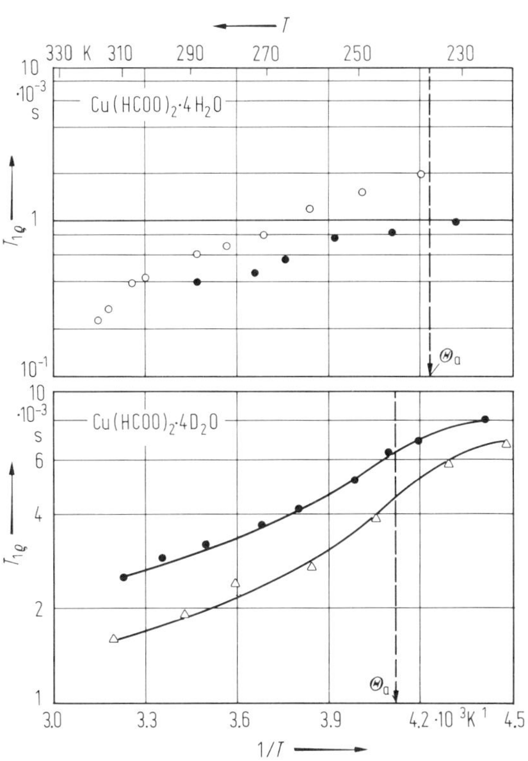

Fig. 1145. $Cu(HCOO)_2 \cdot 4H_2O$, $Cu(HCOO)_2 \cdot 4D_2O$. $T_{1\varrho}$ vs. T^{-1} [74B14]. $T_{1\varrho}$: proton spin-lattice relaxation time in the rotating frame. Open circles: $H_1 = 3.18 \cdot 10^3$ Am^{-1}; full circles: $H_1 = 1.59 \cdot 10^3$ Am^{-1}; triangles: $H_1 = 0.8 \cdot 10^3$ Am^{-1}. H_1: r.f. field intensity.

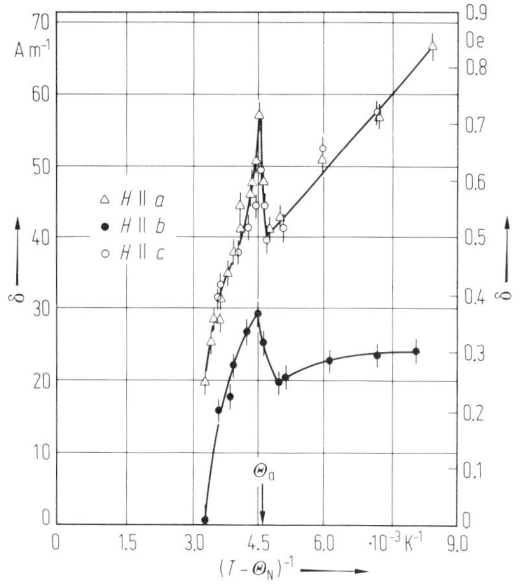

Fig. 1146. $Cu(HCOO)_2 \cdot 4H_2O$. δ vs. $(T-\Theta_N)^{-1}$ [74M7]. δ: paramagnetic shift of the formate hydrogen resonance line. $H = 278.5 \cdot 10^3$ Am^{-1}.

Fig. 1147. $Cu(HCOO)_2 \cdot 4H_2O$. $T_{1\varrho}$, T_{1D} vs. T^{-1} [74Z7]. $T_{1\varrho}$: proton spin-lattice relaxation time in the rotating frame. H_1: r.f. field intensity. Full circles: $H_1 = 3.18 \cdot 10^3$ Am^{-1}; triangles: $H_1 = 1.59 \cdot 10^3$ Am^{-1}. T_{1D}: proton spin lattice relaxation time in the dipolar frame. ►

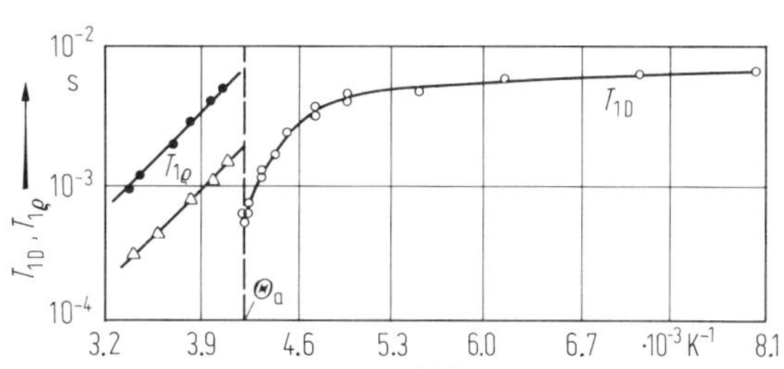

34 DSP($Ca_2Sr(CH_3CH_2COO_6)$) family

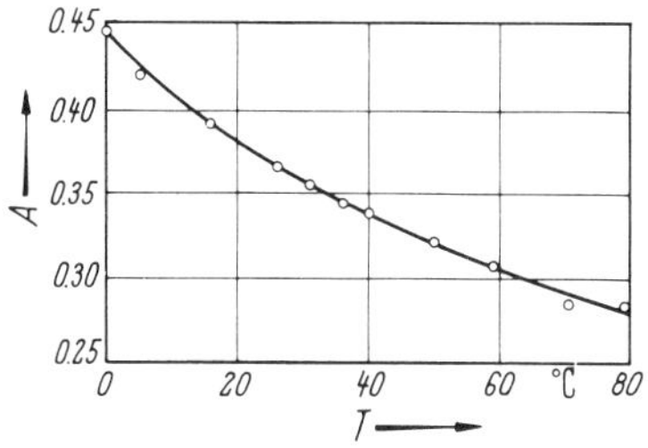

Fig. 1148. $Ca_2Sr(CH_3CH_2COO)_6$ (DSP). A vs. T [65K7]. A: solubility in H_2O (solute/solvent in kg · kg^{-1}).

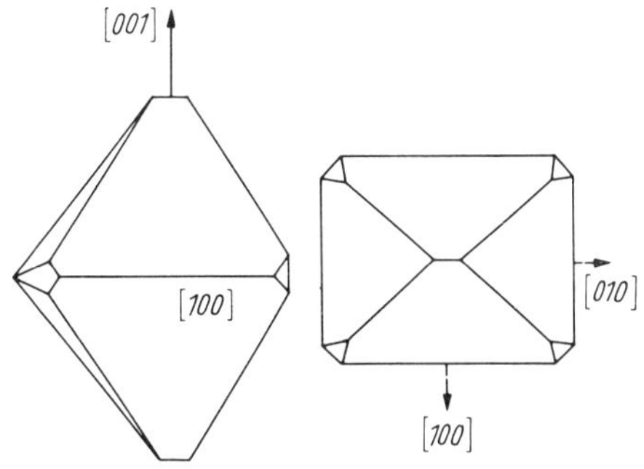

Fig. 1149. $Ca_2Sr(CH_3CH_2COO)_6$ (DSP). Crystal form [61K4].

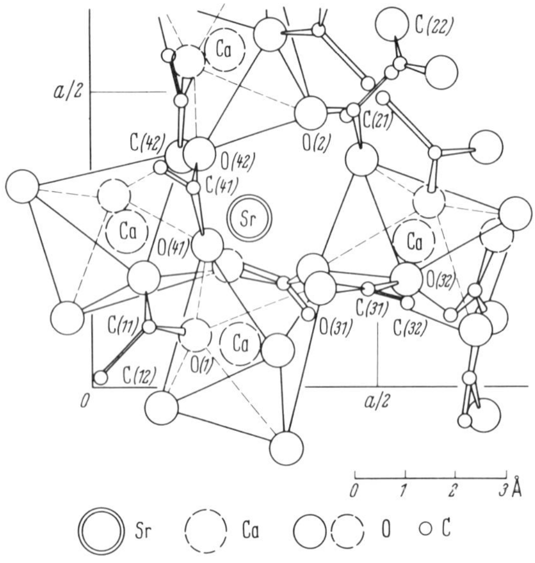

Fig. 1150. $Ca_2Sr(CH_3CH_2COO)_6$ (DSP). Structure of the phase I projected on (001) plane [67 M 4]. Numerals at each of the atoms are the same as in Table 307.

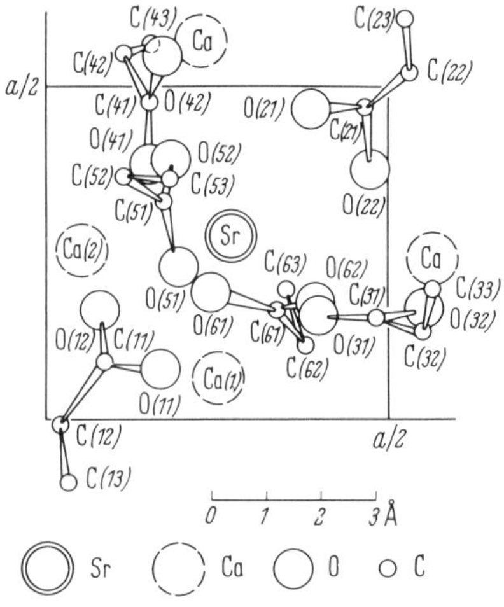

Fig. 1151. $Ca_2Sr(CH_3CH_2COO)_6$ (DSP). Structure of the phase II projected on (001) plane [67 M 7]. Numerals at each of the atoms are the same as in Table 308.

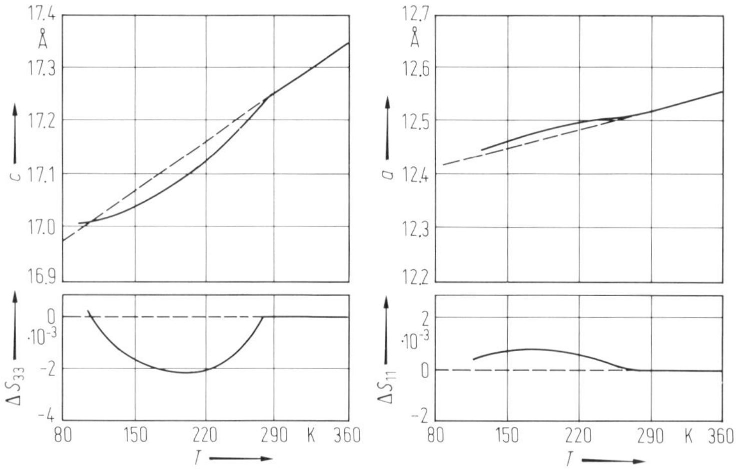

Fig. 1152. $Ca_2Sr(CH_3CH_2COO)_6$ (DSP). a, c, ΔS_{33}, ΔS_{11} vs. T [71 K 10]. ΔS_{ij}: component of spontaneous strain.

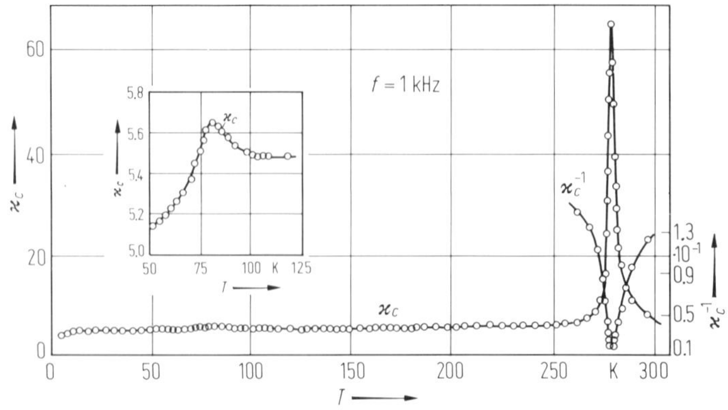

Fig. 1153. $Ca_2Sr(CH_3CH_2COO)_6$ (DSP). $\varkappa_c$ vs. T [75 V 2].

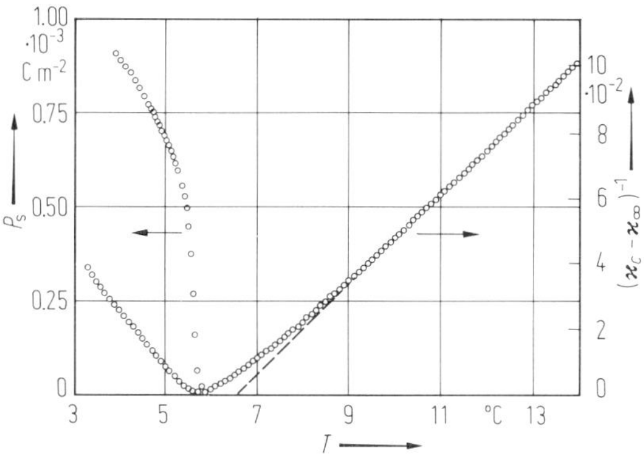

Fig. 1154. $Ca_2Sr(CH_3CH_2COO)_6$ (DSP). $(\varkappa_c - \varkappa_\infty)^{-1}$, P_s vs. T in the vicinity of Θ_f for the specimen annealed at 330 °C for 60 h [76D4]. $\varkappa_\infty = 4.0$: microwave dielectric constant.

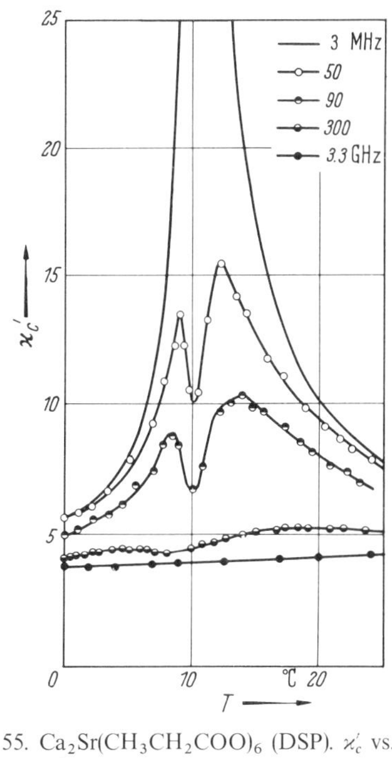

Fig. 1155. $Ca_2Sr(CH_3CH_2COO)_6$ (DSP). $\varkappa_c'$ vs. T [67N1]. Parameter: f.

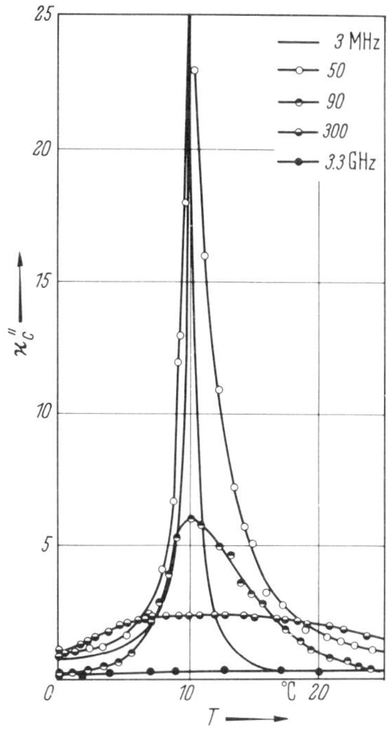

Fig. 1156. $Ca_2Sr(CH_3CH_2COO)_6$ (DSP). $\varkappa_c''$ vs. T [67N1]. Parameter: f.

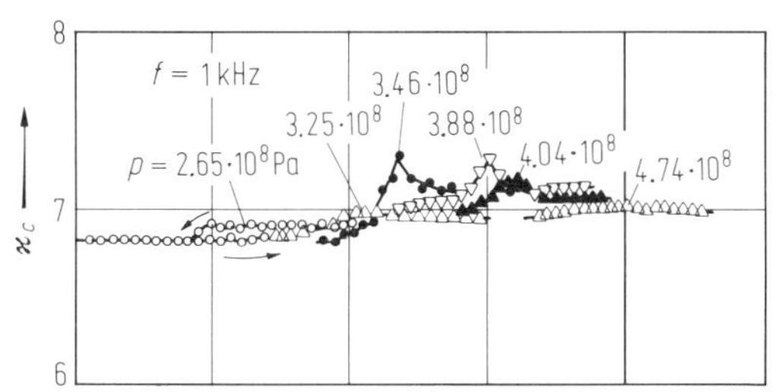

Fig. 1157. $Ca_2Sr(CH_3CH_2COO)_6$ (DSP). $\varkappa_c$ vs. T in the vicinity of II – III transition [75G9]. Parameter: p.

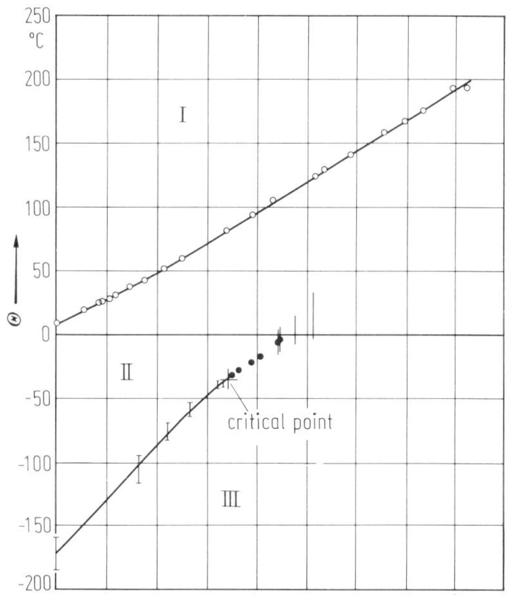

Fig. 1158. $Ca_2Sr(CH_3CH_2COO)_6$ (DSP). Θ vs. p [75G9]. Full circles indicate the maximum of dielectric constant in the supercritical region. The first order III – II phase transition terminates at the critical point.

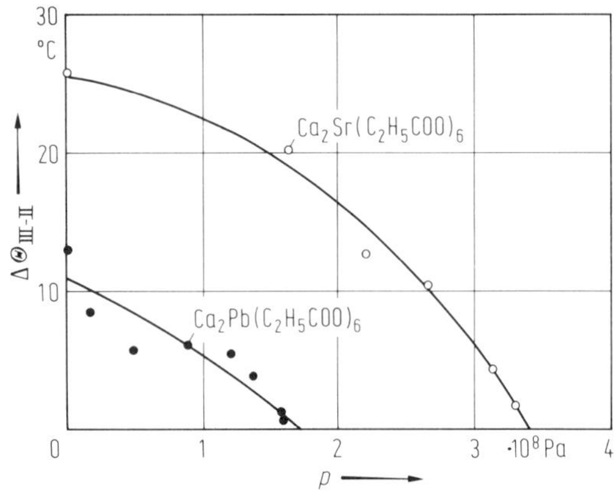

Fig. 1159. $Ca_2Sr(CH_3CH_2COO)_6$ (DSP), $Ca_2Pb(CH_3CH_2COO)_6$ (DLP). $\Delta\Theta_{III-II}$ vs. p [75G9]. $\Delta\Theta_{III-II}$: thermal hysteresis of III–II transition determined by measurements of $\varkappa_c$.

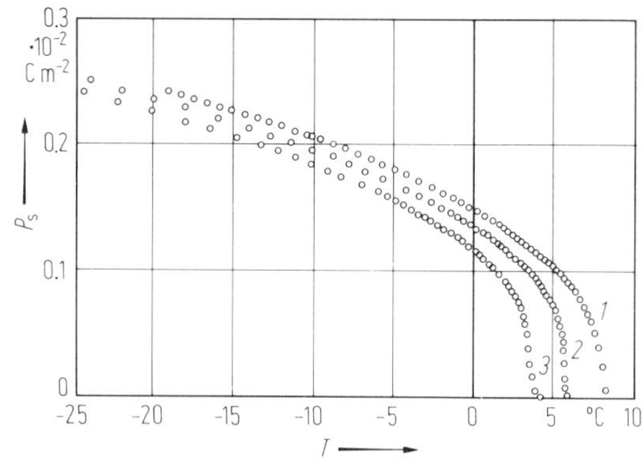

Fig. 1161. $Ca_2Sr(CH_3CH_2COO)_6$ (DSP). P_s vs. T for various specimens [76D4]. Curve *1*: unannealed crystal; *2*: annealed at 330 °C for 60 h; *3*: annealed at 330 °C for 60 h and at 390 °C for 5 h.

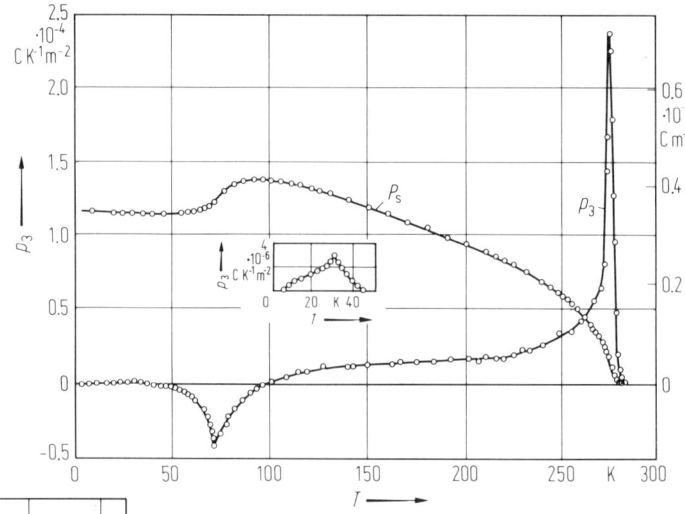

Fig. 1160. $Ca_2Sr(CH_3CH_2COO)_6$ (DSP). P_s, p_3 vs. T [75V2]. p_3: pyroelectric coefficient.

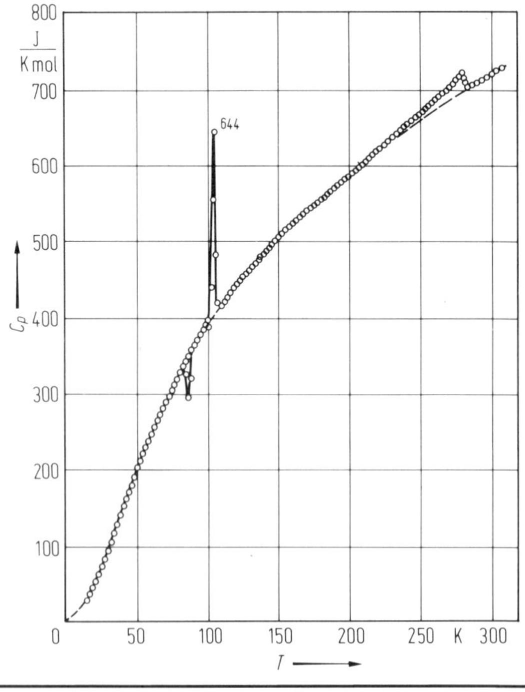

Fig. 1162. $Ca_2Sr(CH_3CH_2COO)_6$ (DSP). C_p vs. T [65N2].

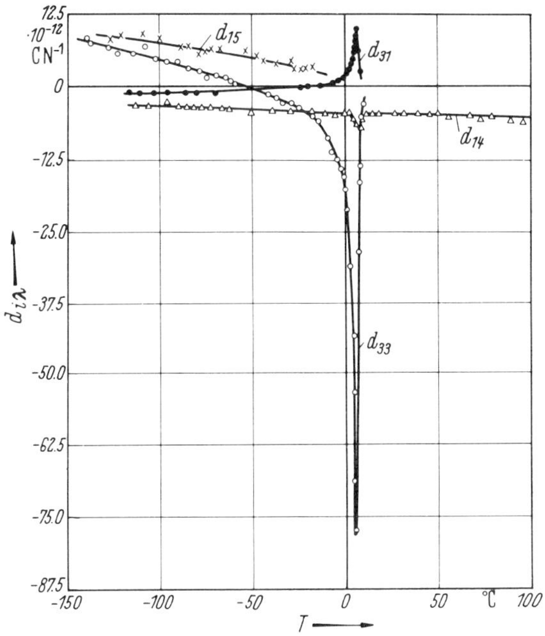

Fig. 1163. $Ca_2Sr(CH_3CH_2COO)_6$ (DSP). $d_{i\lambda}$ vs. T [66 H 2].

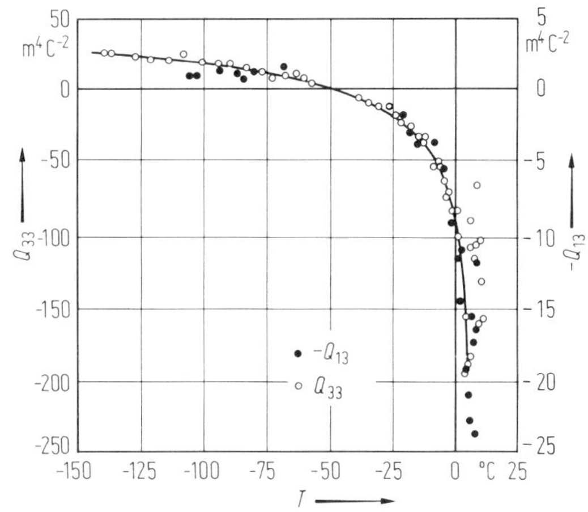

Fig. 1164. $Ca_2Sr(CH_3CH_2COO)_6$ (DSP). Q_{33}, Q_{13} vs. T [66 H 2]. Electrostrictive constants $Q_{\lambda 3}$ calculated from $Q_{\lambda 3} = d_{3\lambda}/(2\chi_{33} P_s)$.

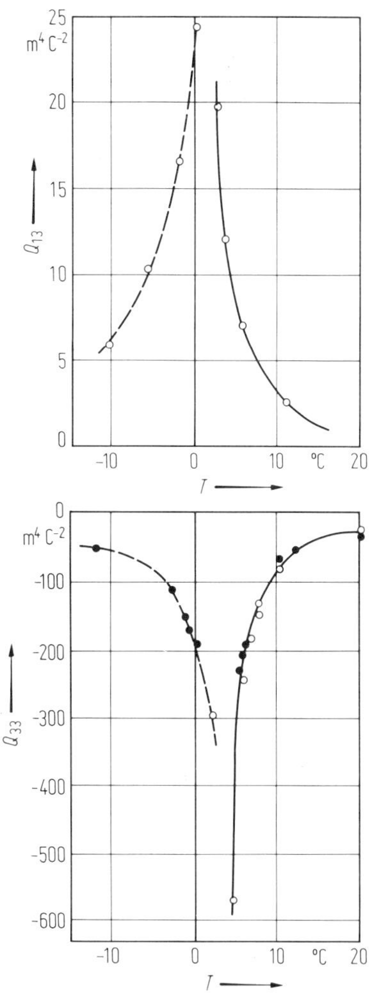

Fig. 1165. $Ca_2Sr(CH_3CH_2COO)_6$ (DSP). Q_{33}, Q_{13} vs. T [74 S 12]. Values above Θ_f were determined from $Q_{\lambda 3} = S_\lambda / D_3^2$ and those below Θ_f by $Q_{\lambda 3} = d_{3\lambda}/2P_s$ ($\lambda = 1, 3$). Dashed line: [66 H 2].

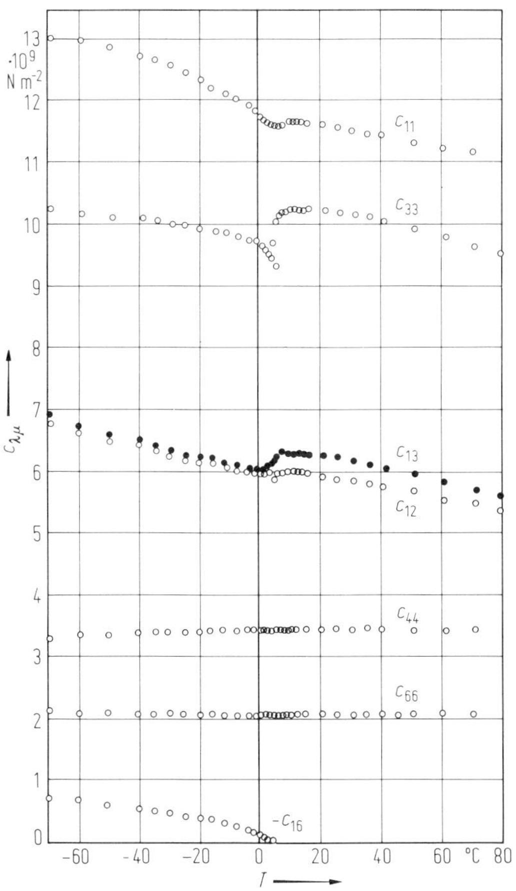

Fig. 1166. $Ca_2Sr(CH_3CH_2COO)_6$ (DSP). $c_{\lambda\mu}$ vs. T [79 K 2].

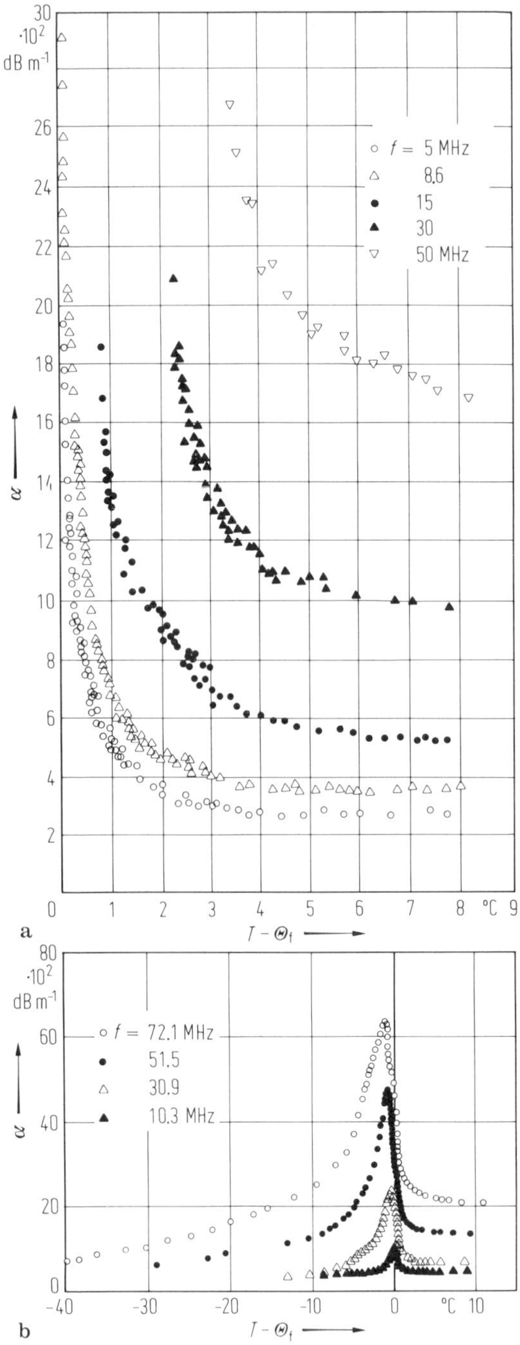

Fig. 1167. $Ca_2Sr(CH_3CH_2COO)_6$ (DSP). α vs. $T-\Theta_f$. α: attenuation coefficient of longitudinal ultrasonic wave. Fig. a: for the wave propagating along the [001] direction [71 T 3]. Fig. b: for the wave propagating along the [110] direction [79 K 1]. Parameter: f.

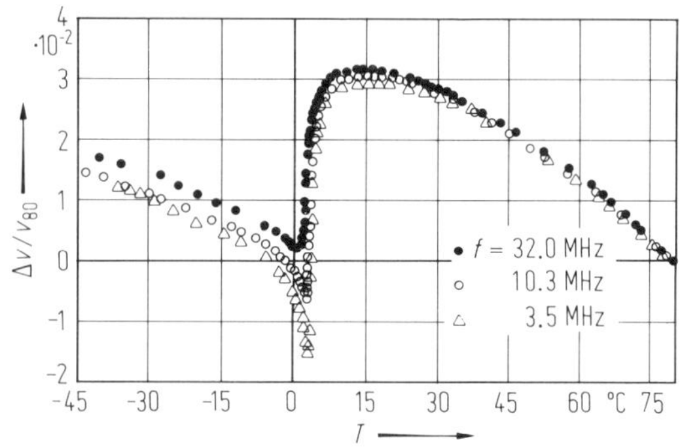

Fig. 1169. $Ca_2Sr(CH_3CH_2COO)_6$ (DSP). $\Delta v/v_{80}$ vs. T [74 K 1]. Parameter: f. $\Delta v = v - v_{80}$. v: longitudinal sound wave velocity propagating along the [001] direction. v_{80}: v at 80 °C.

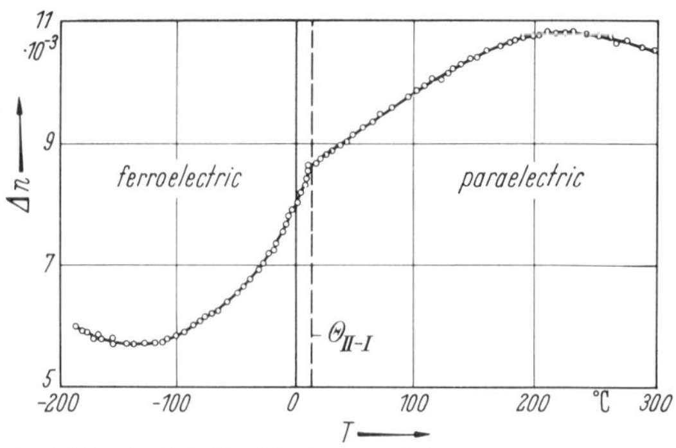

Fig. 1170. $Ca_2Sr(CH_3CH_2COO)_6$ (DSP). Δn vs. T [63 K 6].

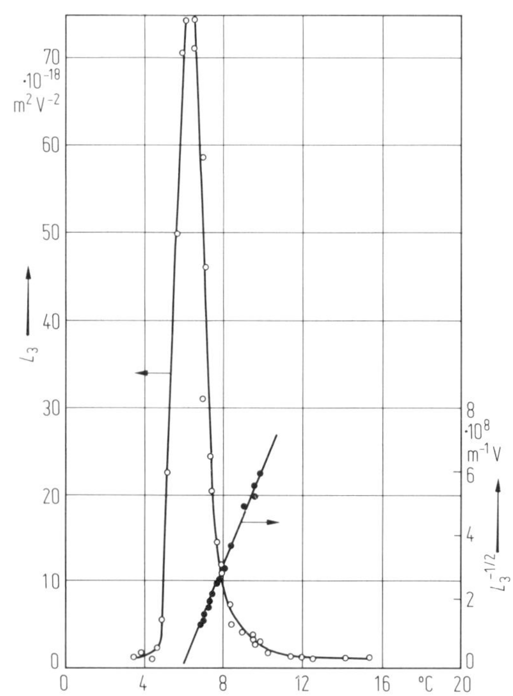

Fig. 1171. $Ca_2Sr(CH_3CH_2COO)_6$ (DSP). L_3 vs. T [75 A 12]. $L_3 = n_1^3 L_{13} - n_3^3 L_{33}$. $L_{\lambda\mu}$: quadratic electrooptic constant. $\lambda = 633$ nm.

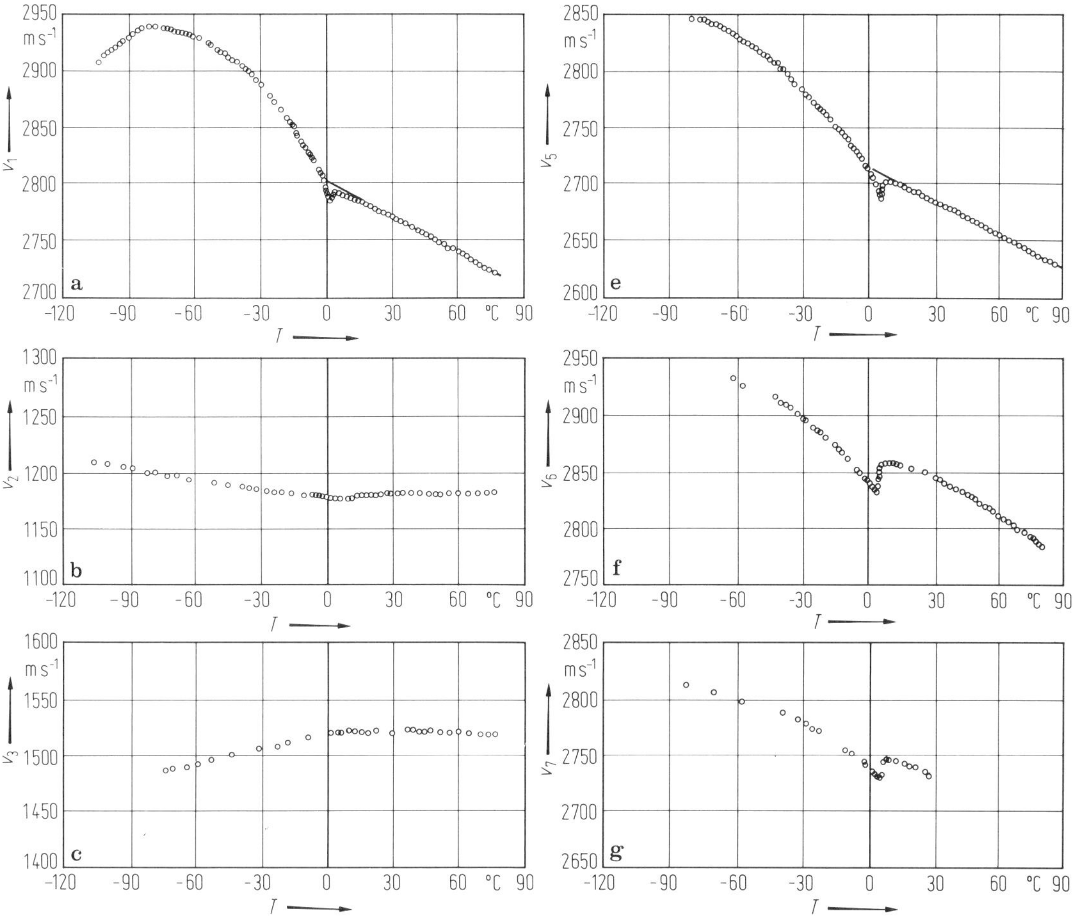

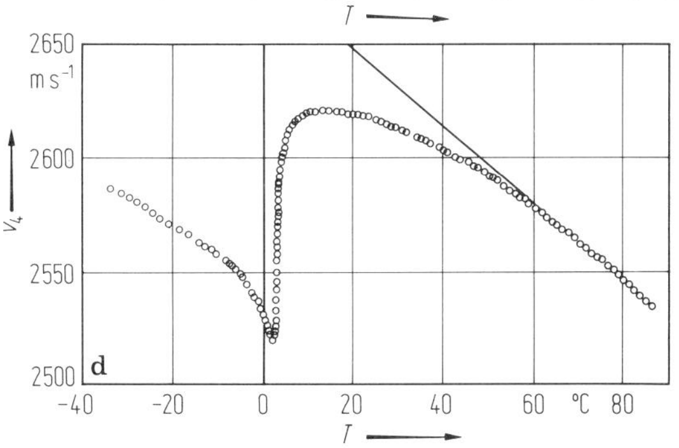

Symbol	Direction of propagation	Direction of displacement of transducer
v_1	[100]	[100]
v_2	[100]	[010]
v_3	[100]	[001]
v_4	[001]	[001]
v_5	[110]	[110]
v_6	$[l_1 0 l_3]$*	$[l_1 0 l_3]$*
v_7	$[\cos\frac{\pi}{8}, \sin\frac{\pi}{8}, 0]$	$[\cos\frac{\pi}{8}, \sin\frac{\pi}{8}, 0]$

$^*l_1 = a/\sqrt{a^2+c^2}$, $l_3 = c/\sqrt{a^2+c^2}$

Fig. 1168. $Ca_2Sr(CH_3CH_2COO)_6$ (DSP). v vs. T [79 K 2]. v: velocity of ultrasonic wave. $f = 7$ MHz.

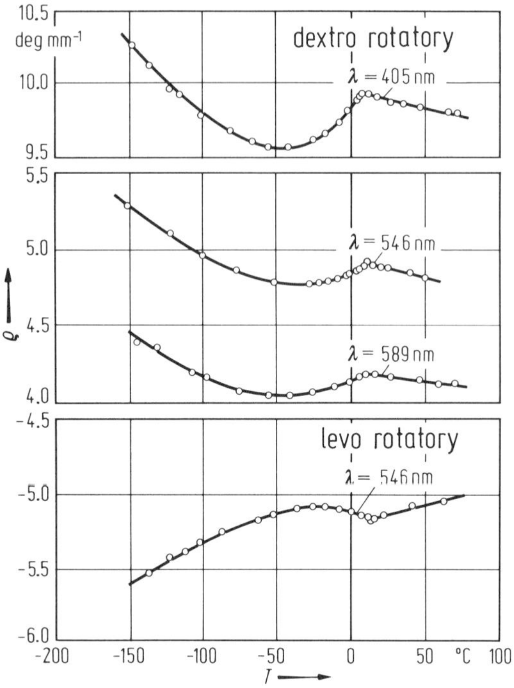

Fig. 1172. $Ca_2Sr(CH_3CH_2COO)_6$ (DSP). ϱ vs. T [71 K 10]. ϱ: optical rotatory power.

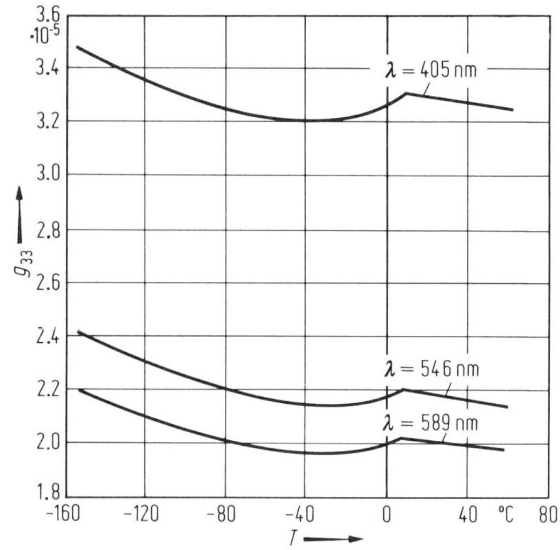

Fig. 1173. $Ca_2Sr(CH_3CH_2COO)_6$ (DSP). g_{33} vs. T [71 K 10]. g_{33}: optical gyration tensor component.

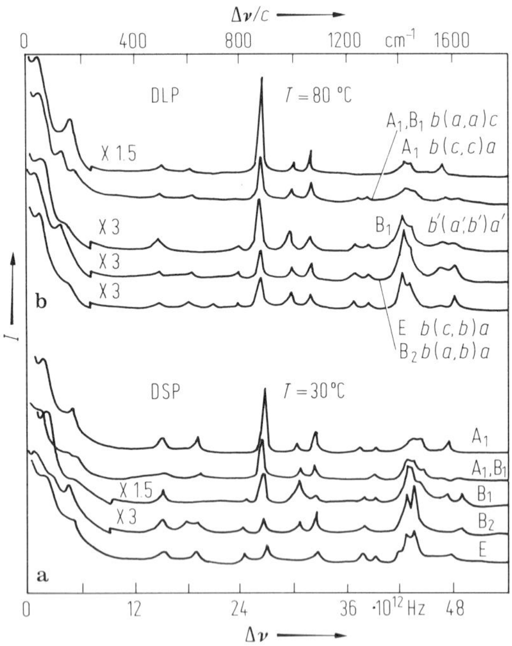

Fig. 1175. $Ca_2Sr(CH_3CH_2COO)_6$ (DSP), $Ca_2Pb(CH_3CH_2COO)_6$ (DLP). I vs. $\Delta\nu$ [76 N 2]. I: Raman scattering intensity. (a) DSP at 30 °C. (b) DLP at 80 °C. a' and b' denote the Cartesian coordinate in which a and b axes are rotated by 45° around the z-axis.

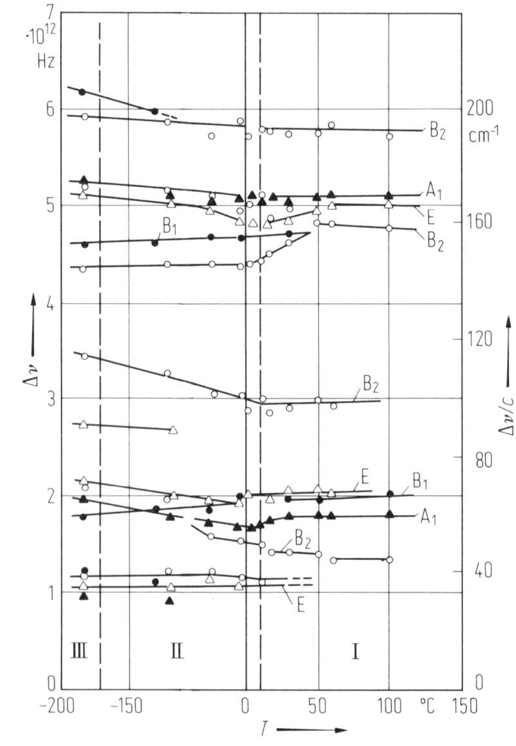

Fig. 1176. $Ca_2Sr(CH_3CH_2COO)_6$ (DSP). $\Delta\nu$ vs. T [76 N 2]. $\Delta\nu$: Raman frequency shift. Successive phases are denoted as I, II, and III.

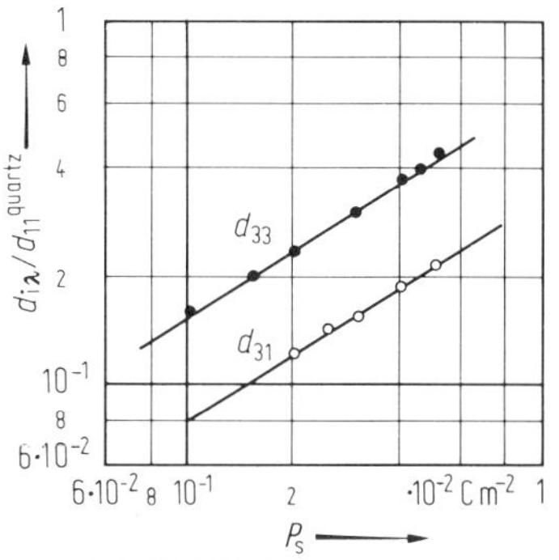

Fig. 1174. $Ca_2Sr(CH_3CH_2COO)_6$ (DSP). $d_{i\lambda}/d_{11}^{quartz}$ vs. P_s [74I10]. $d_{i\lambda}$: nonlinear optical susceptibility. $\lambda = 1.06$ μm.

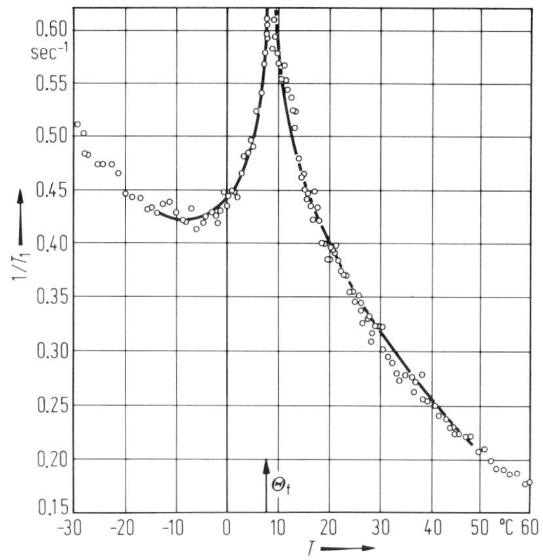

Fig. 1177. $Ca_2Sr(CH_3CH_2COO)_6$ (DSP). $1/T_1$ vs. T near Θ_f [73H11]. T_1: proton spin-lattice relaxation time.

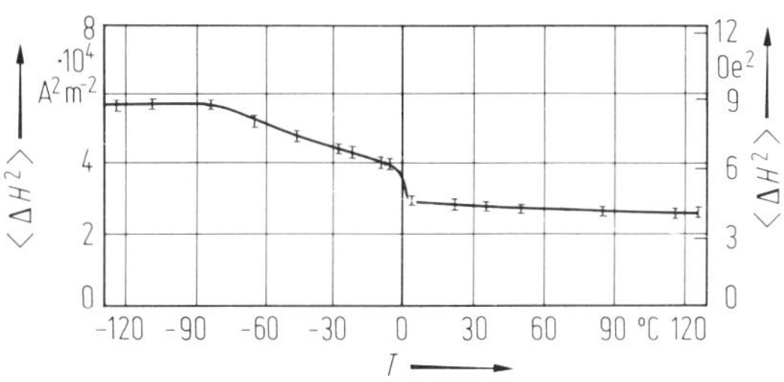

Fig. 1178. $Ca_2Sr(CH_3CH_2COO)_6$ (DSP). $\langle \Delta H^2 \rangle$ vs. T [67A4]. $\langle \Delta H^2 \rangle$: second moment of proton.

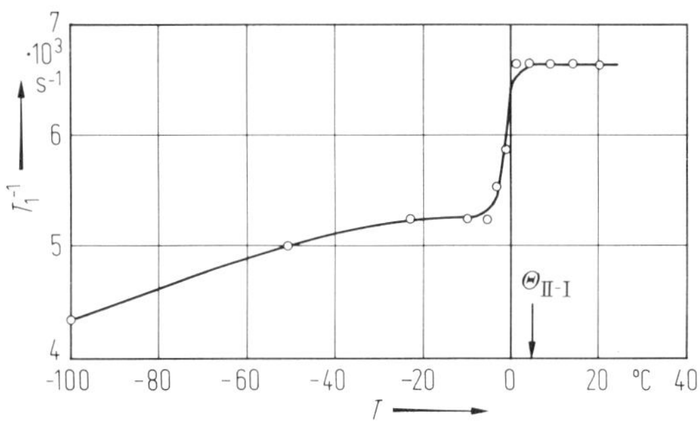

Fig. 1179. $Ca_2Sr(CH_3CH_2COO)_6$ (DSP). T_1^{-1} vs. T [74V2]. T_1: electron spin-lattice relaxation time of radical produced by X-irradiation.

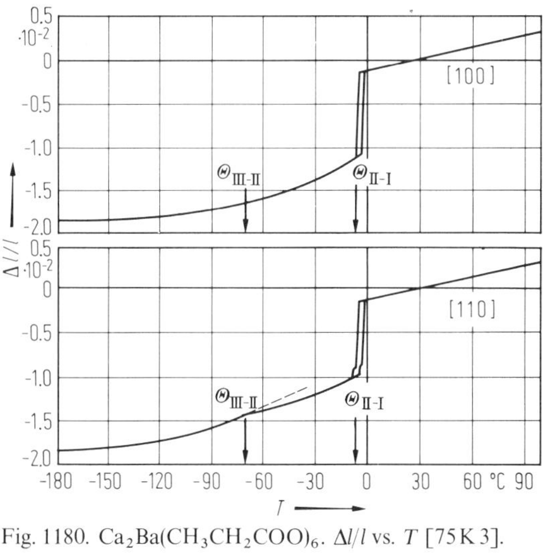

Fig. 1180. $Ca_2Ba(CH_3CH_2COO)_6$. $\Delta l/l$ vs. T [75K3].

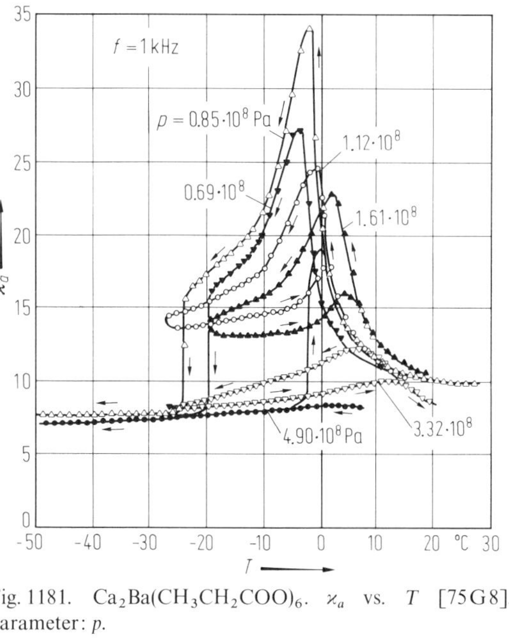

Fig. 1181. $Ca_2Ba(CH_3CH_2COO)_6$. $\varkappa_a$ vs. T [75G8]. Parameter: p.

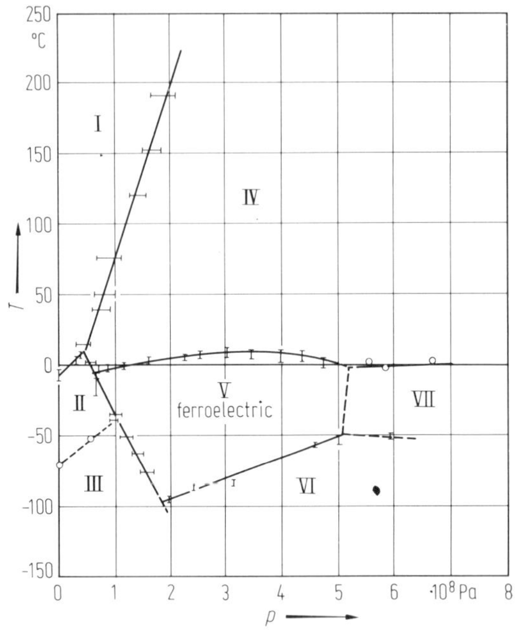

Fig. 1182. $Ca_2Ba(CH_3CH_2COO)_6$. Phase diagram [75G8].

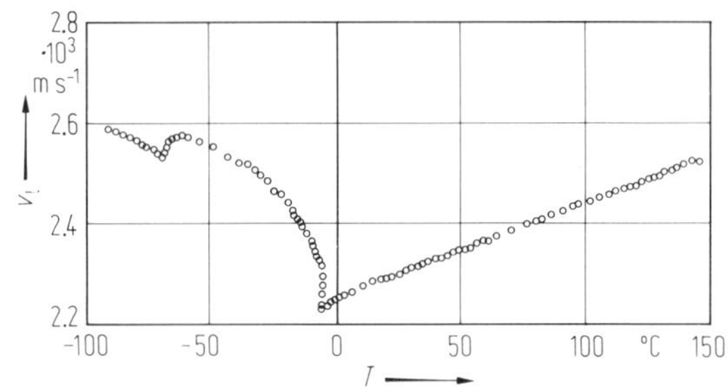

Fig. 1183. $Ca_2Ba(CH_3CH_2COO)_6$. v_l vs. T [75K3]. v_l: velocity of the longitudinal sound wave propagating along the [100] direction.

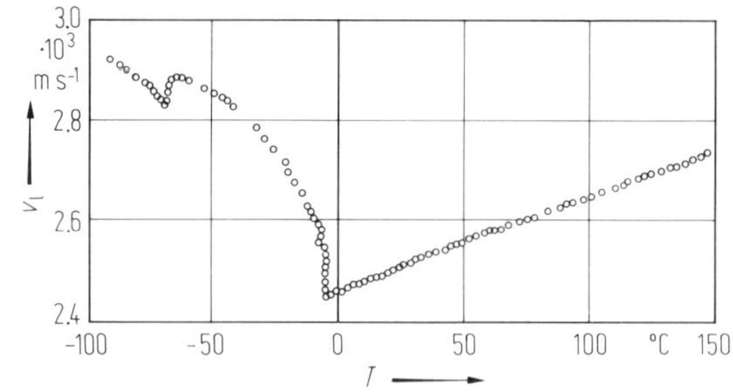

Fig. 1184. $Ca_2Ba(CH_3CH_2COO)_6$. v_l vs. T [75K3]. v_l: velocity of the longitudinal sound wave propagating along the [110] direction.

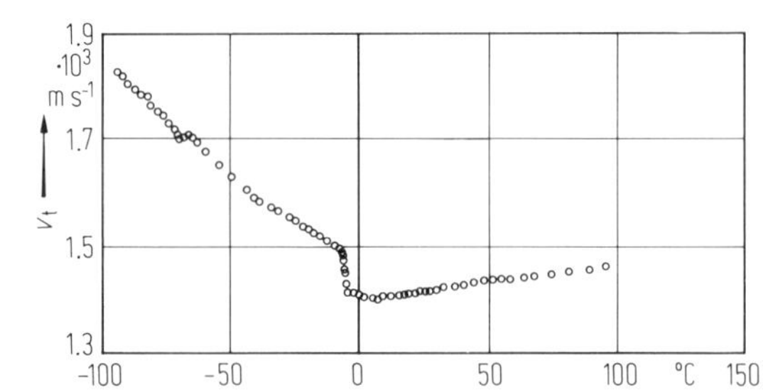

Fig. 1185. $Ca_2Ba(CH_3CH_2COO)_6$. v_t vs. T [75K3]. v_t: velocity of the transverse sound wave propagating along the [100] direction and polarized along [010].

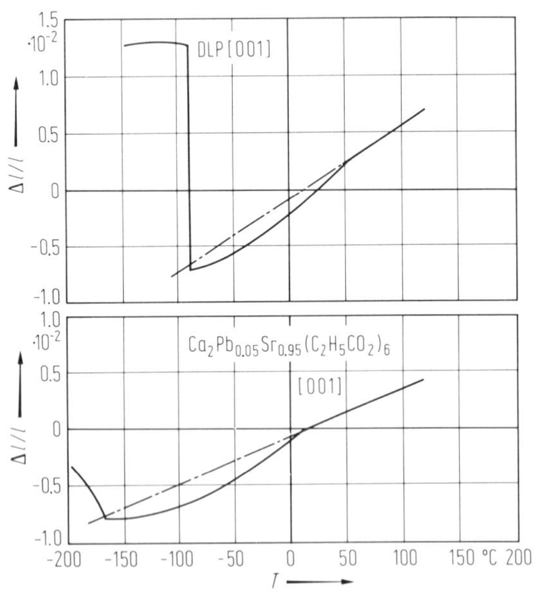

Fig. 1187. $Ca_2Pb(CH_3CH_2COO)_6$ (DLP). $Ca_2Pb_{0.05}Sr_{0.95}$ $(CH_3CH_2COO)_6$. $\Delta l/l$ vs. T along [001] direction [76N1].

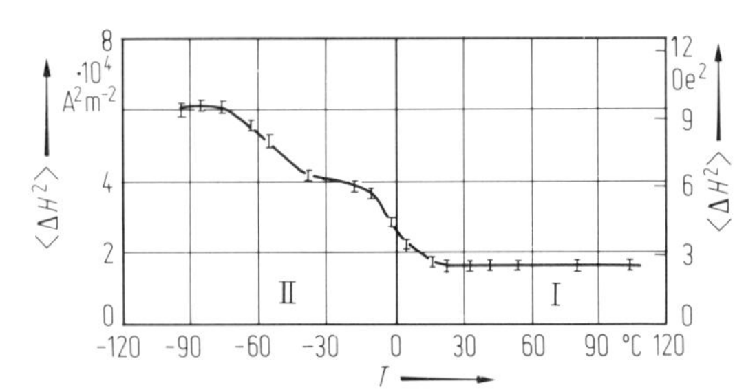

Fig. 1186. $Ca_2Ba(CH_3CH_2COO)_6$. $\langle \Delta H^2 \rangle$ vs. T [67A4]. $\langle \Delta H^2 \rangle$: second moment for proton.

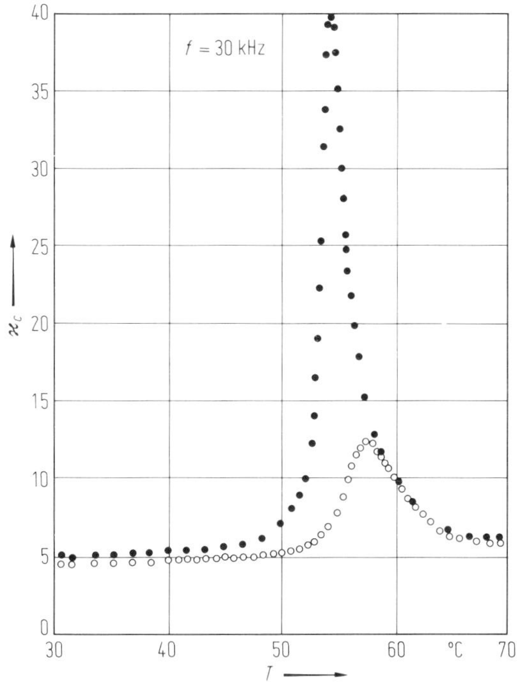

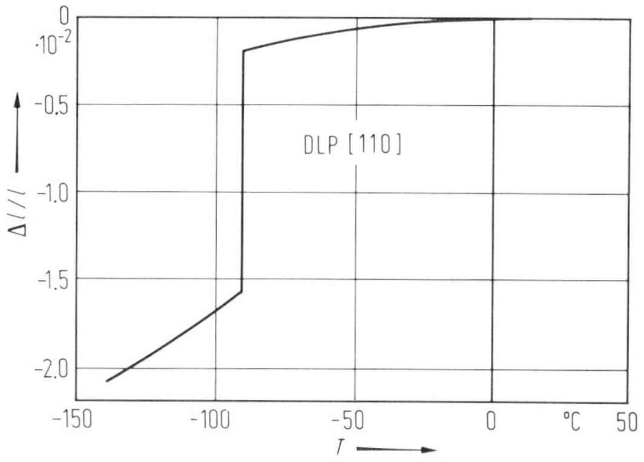

Fig. 1188. $Ca_2Pb(CH_3CH_2COO)_6$ (DLP). $\Delta l/l$ vs. T along [110] direction [76N1].

Fig. 1190. $Ca_2Pb(CH_3CH_2COO)_6$. $\varkappa_c$ vs. T [76T3]. Open circles: as-grown crystal; full circles: annealed crystal.

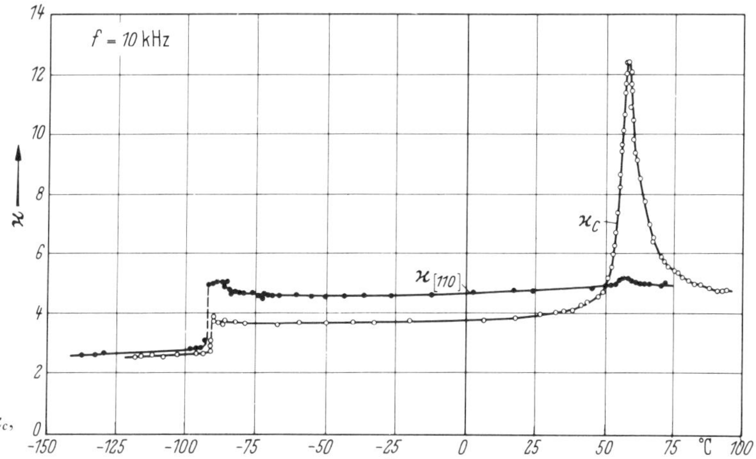

Fig. 1189. $Ca_2Pb(CH_3CH_2COO)_6$. $\varkappa_c$, $\varkappa_{[110]}$ vs. T [65N2].

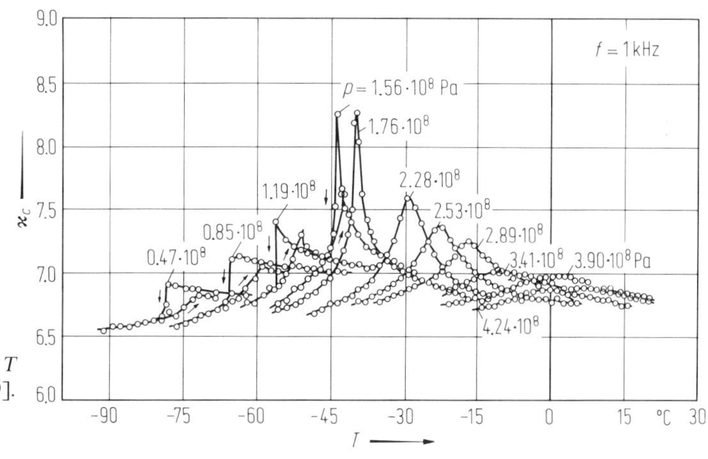

Fig. 1191. $Ca_2Pb(CH_3CH_2COO)_6$. $\varkappa_c$ vs. T in the vicinity of III–II transition [75G9]. Parameter: p (in 10^8 Pa).

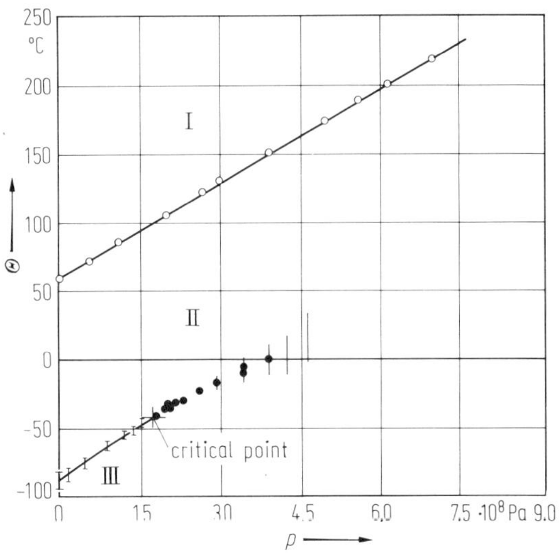

Fig. 1192. $Ca_2Pb(CH_3CH_2COO)_6$. Phase diagram [75G9].

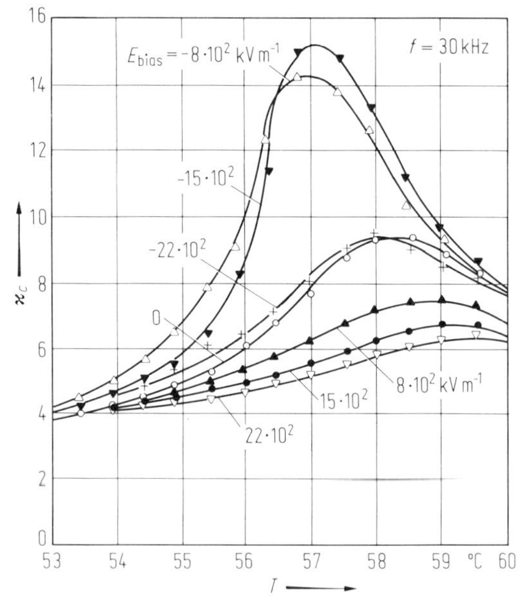

Fig. 1193. $Ca_2Pb(CH_3CH_2COO)_6$. $\varkappa_c$ vs. T of as grown crystal [76T3]. Parameter: E_{bias}.

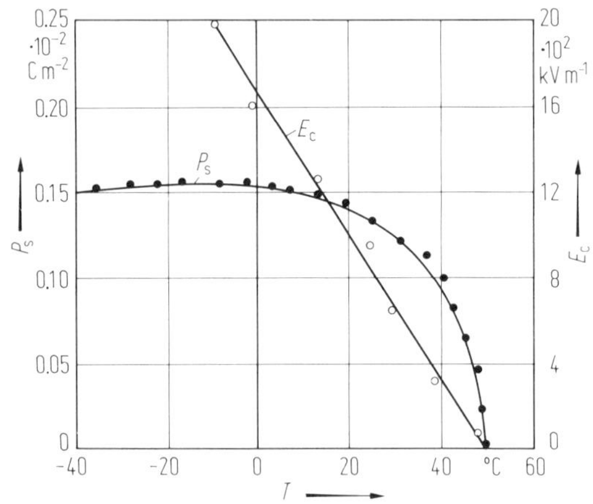

Fig. 1194. $Ca_2Pb(CH_3CH_2COO)_6$. P_s, E_c vs. T [76T3].

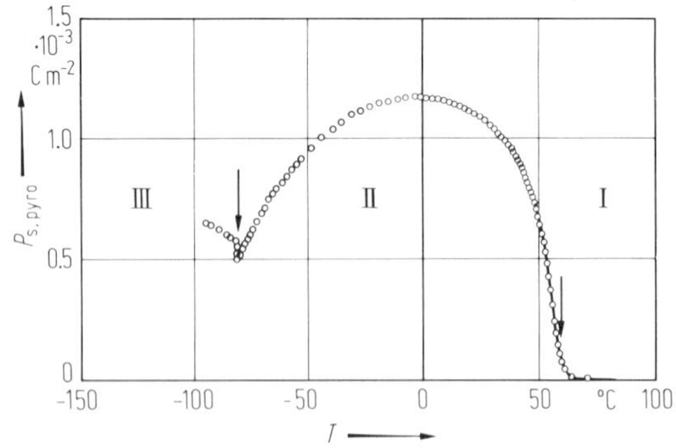

Fig. 1195. $Ca_2Pb(CH_3CH_2COO)_6$. $P_{s,\,pyro}$ vs. T [75O5]. $P_{s,\,pyro}$: spontaneous polarization obtained by pyroelectric measurement. Arrows point to transition temperatures.

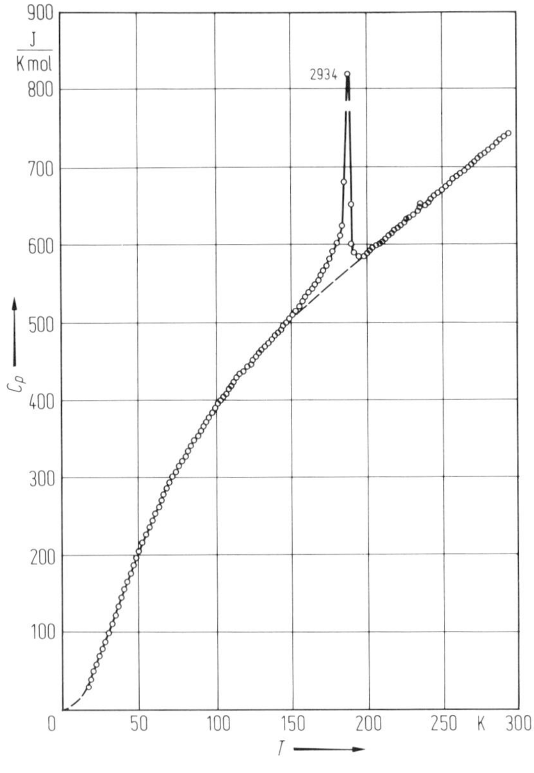

Fig. 1196. $Ca_2Pb(CH_3CH_2COO)_6$. C_p vs. T [65N2].

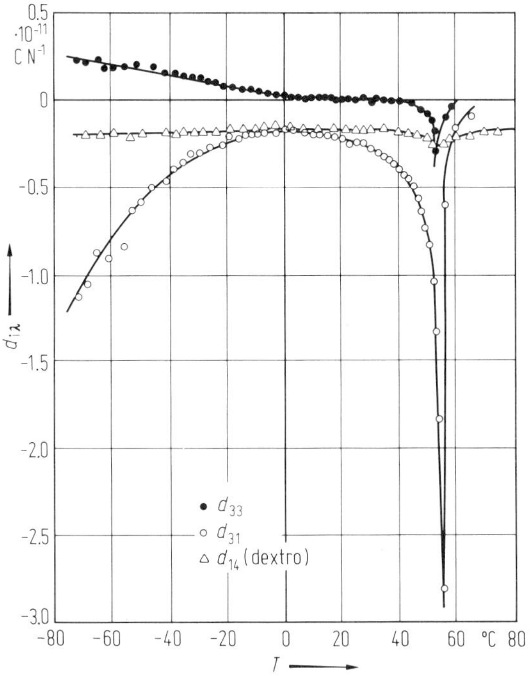

Fig. 1197. $Ca_2Pb(CH_3CH_2COO)_6$. $d_{i\lambda}$ vs. T [78T7]. Annealed specimen. Quasistatic method.

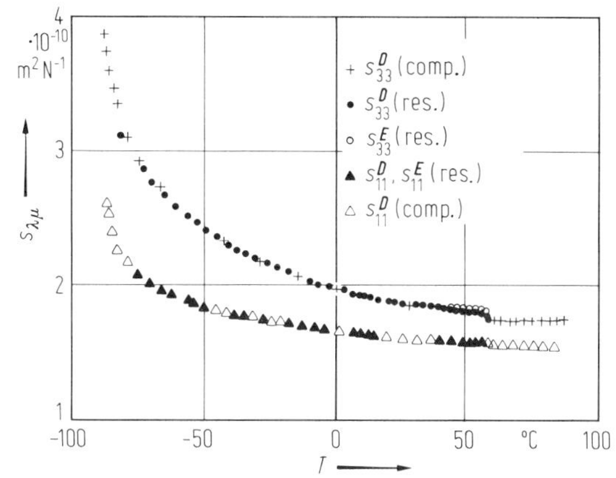

Fig. 1198. $Ca_2Pb(CH_3CH_2COO)_6$. $s_{\lambda\mu}$ vs. T [78T7]. res.: resonance method; comp.: composite bar method.

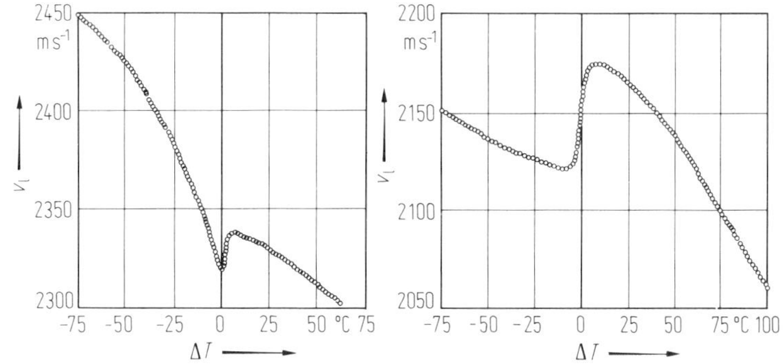

Fig. 1199. $Ca_2Pb(CH_3CH_2COO)_6$. v_l vs. ΔT [74T6]. v_l: sound velocity of the longitudinal wave. Fig. a: along the a-axis. Fig. b: along the c-axis. $\Delta T = T - \Theta_{II-I}$.

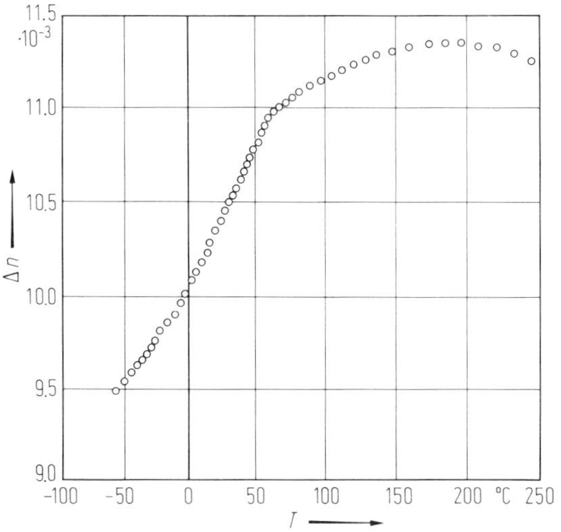

Fig. 1200. $Ca_2Pb(CH_3CH_2COO)_6$. Δn vs. T [76T3]. $\Delta n = n_c - n_a$.

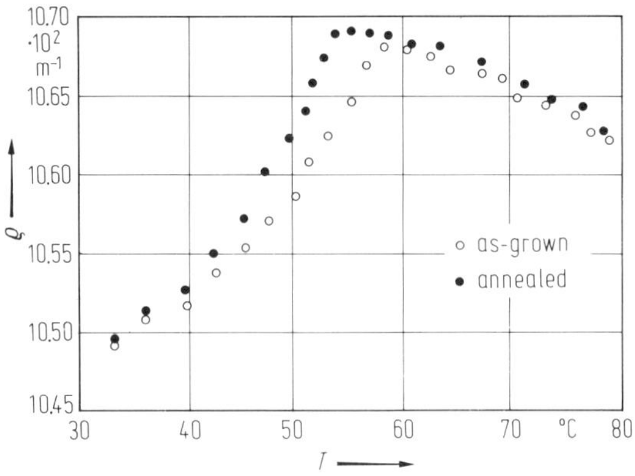

Fig. 1201. $Ca_2Pb(CH_3CH_2COO)_6$. ϱ vs. T [76 T 3]. ϱ: optical rotatory power.

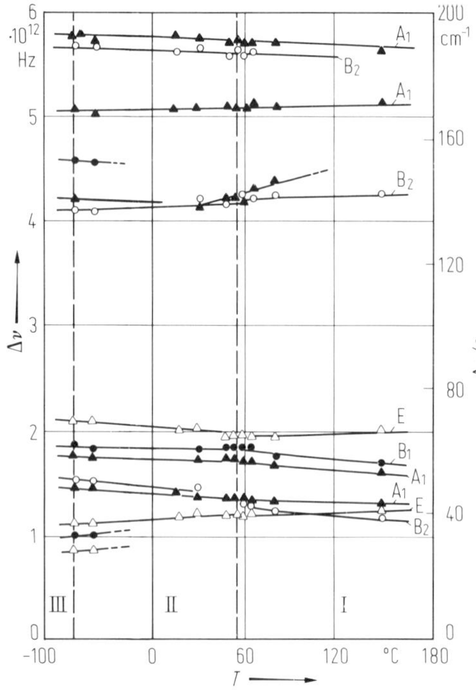

Fig. 1202 α. $Ca_2Pb(CH_3CH_2COO)_6$. Δν vs. T [76 N 2]. Δν: Raman frequency shift.

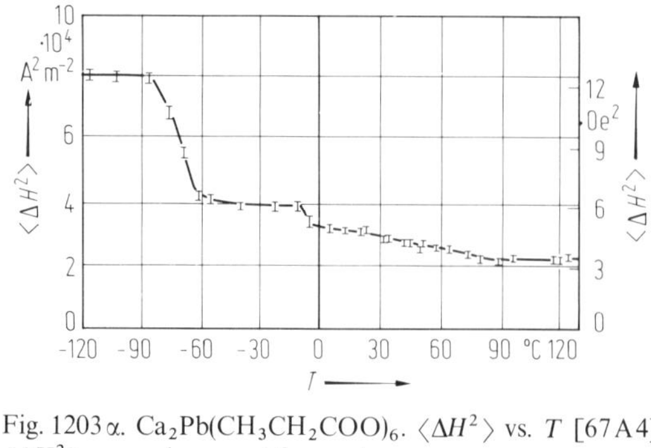

Fig. 1203 α. $Ca_2Pb(CH_3CH_2COO)_6$. $\langle \Delta H^2 \rangle$ vs. T [67 A 4]. $\langle \Delta H^2 \rangle$: second moment for proton.

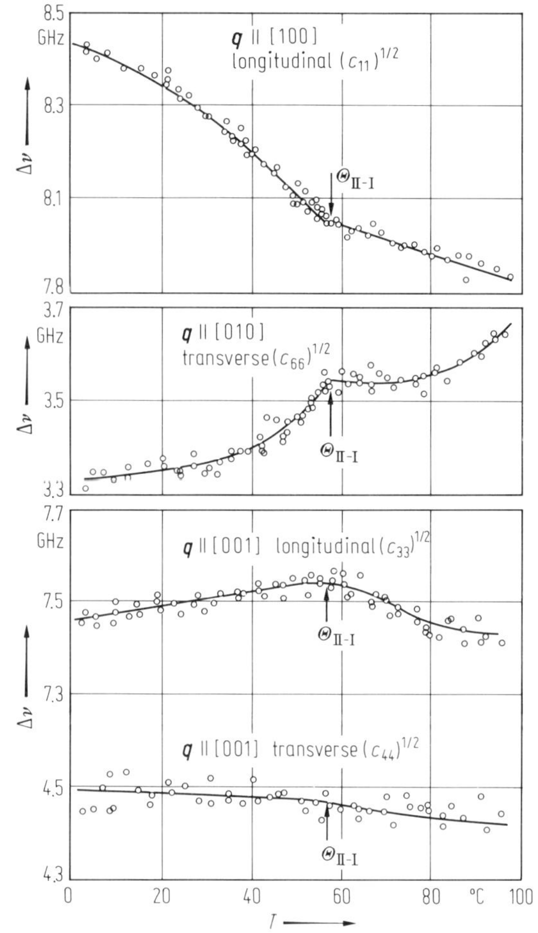

Fig. 1202 β. $Ca_2Pb(CH_3CH_2COO)_6$. Δν vs. T [74 Y 2]. Δν: frequency shift of Brillouin scattering. Scattering angle 90°. λ = 632.8 nm.

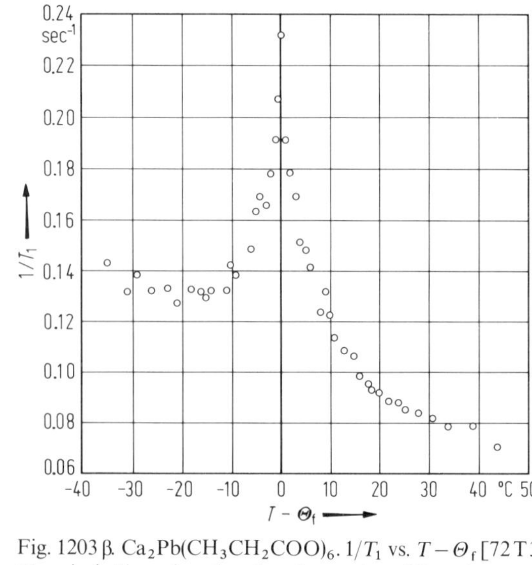

Fig. 1203 β. $Ca_2Pb(CH_3CH_2COO)_6$. $1/T_1$ vs. $T - \Theta_f$ [72 T 3]. T_1: spin-lattice relaxation time for proton. **H**∥c-axis.

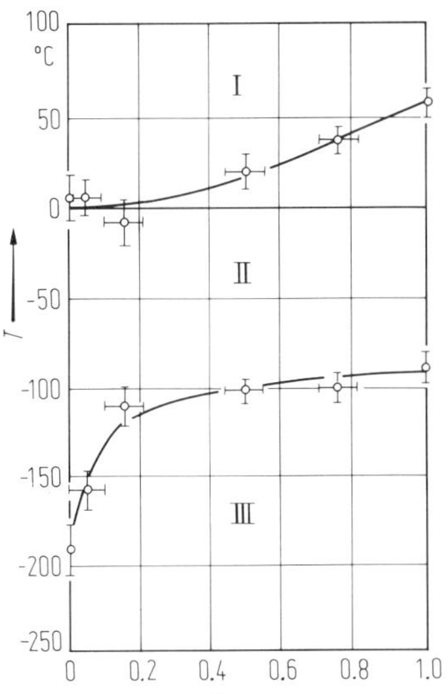

Fig. 1204α. $Ca_2Sr(CH_3CH_2COO)_6$-$Ca_2Pb(CH_3CH_2COO)_6$. Phase diagram [76N1]. x: molar fraction of $Ca_2Pb(CH_3CH_2COO)_6$.

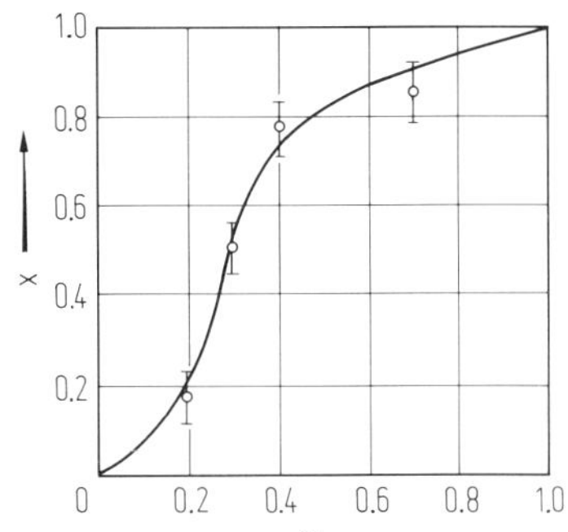

Fig. 1204β. $Ca_2Sr(CH_3CH_2COO)_6$-$Ca_2Pb(CH_3CH_2COO)_6$. x vs. y [76N1]. x: molar fraction of $Ca_2Pb(CH_3CH_2COO)_6$ in crystal. y: molar fraction of $Ca_2Pb(CH_3CH_2COO)_6$ in solution. $T = 40\,°C$.

35 $CH_2ClCOONH_4$ family

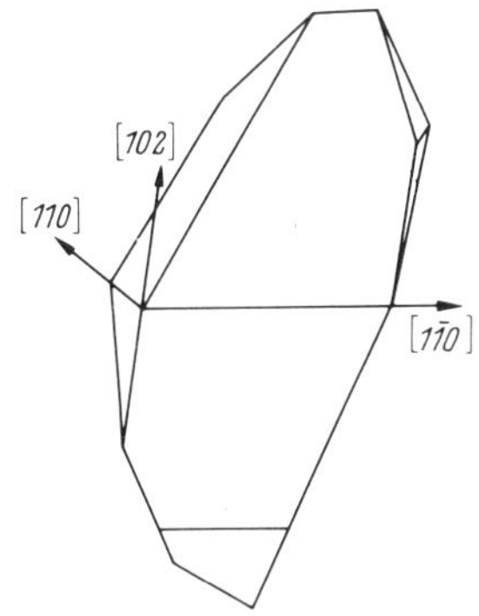

Fig. 1205. $(CH_2ClCOO)_2H \cdot NH_4$. Crystal form [66I1].

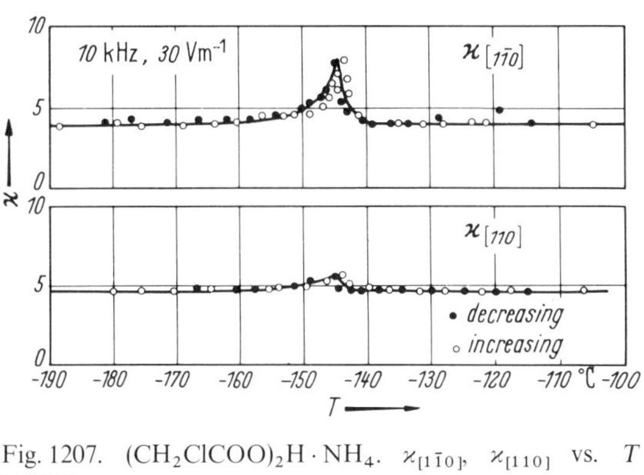

Fig. 1207. $(CH_2ClCOO)_2H \cdot NH_4$. $\varkappa_{[1\bar{1}0]}$, $\varkappa_{[110]}$ vs. T [66I2].

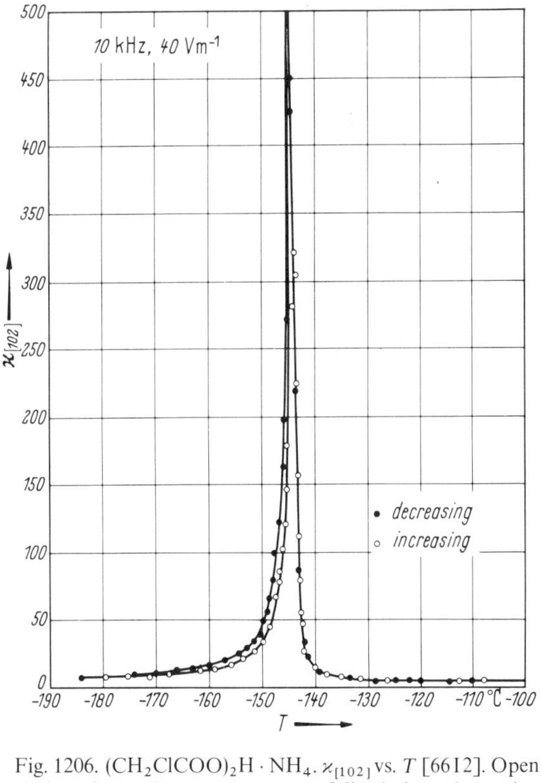

Fig. 1206. $(CH_2ClCOO)_2H \cdot NH_4$. $\varkappa_{[102]}$ vs. T [66I2]. Open circles: increasing temperature; full circles: decreasing temperature.

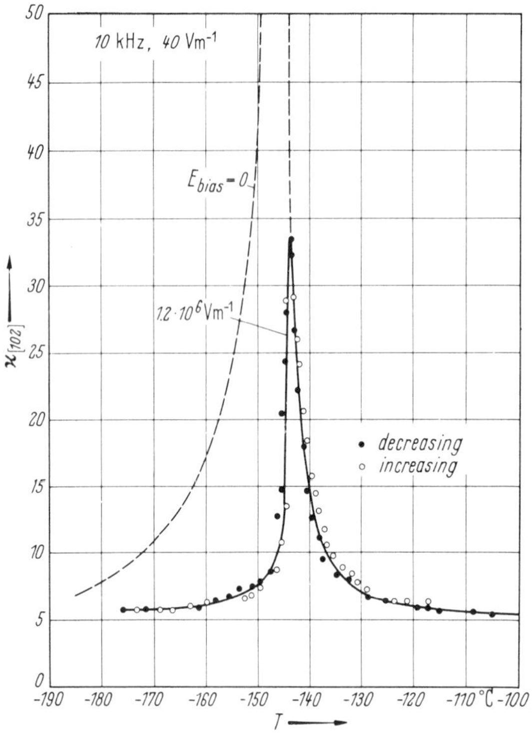

Fig. 1208. $(CH_2ClCOO)_2H \cdot NH_4$. $\varkappa_{[102]}$ vs. T with dc bias field [66I2].

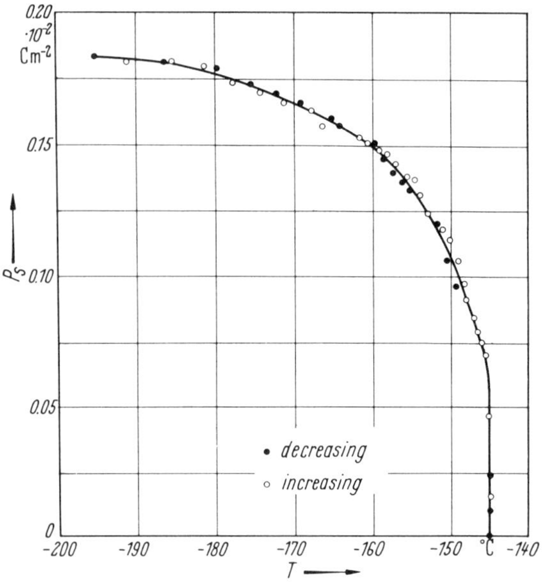

Fig. 1209. $(CH_2ClCOO)_2H \cdot NH_4$. P_s vs. T [66I2].

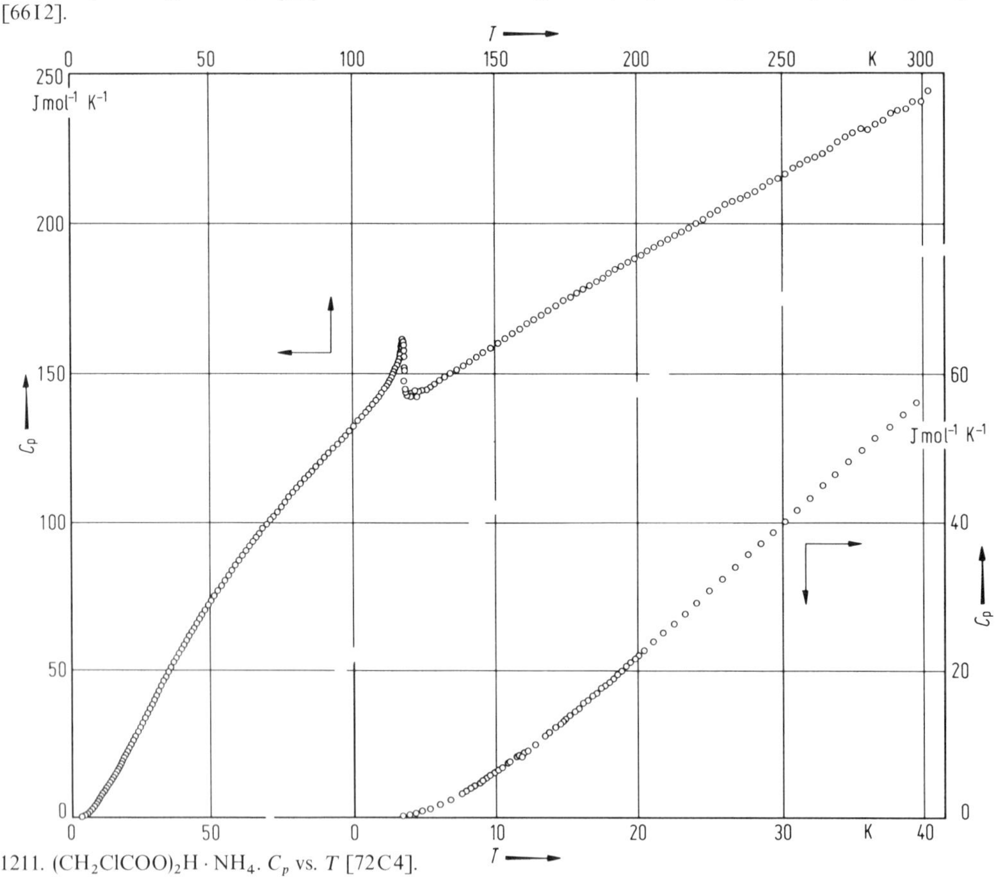

Fig. 1211. $(CH_2ClCOO)_2H \cdot NH_4$. C_p vs. T [72C4].

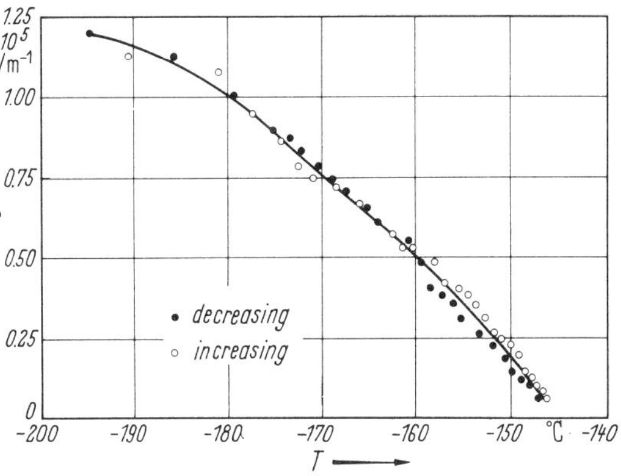

Fig. 1210. $(CH_2ClCOO)_2H \cdot NH_4$. E_c vs. T [66 I 2].

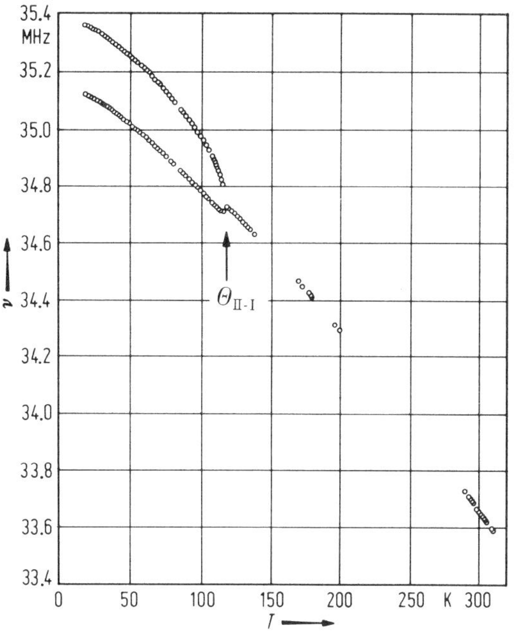

Fig. 1213. $(CH_2ClCOO)_2H \cdot NH_4$. ν vs. T [68 Y 3]. ν: ^{35}Cl nuclear quadrupole resonance frequency.

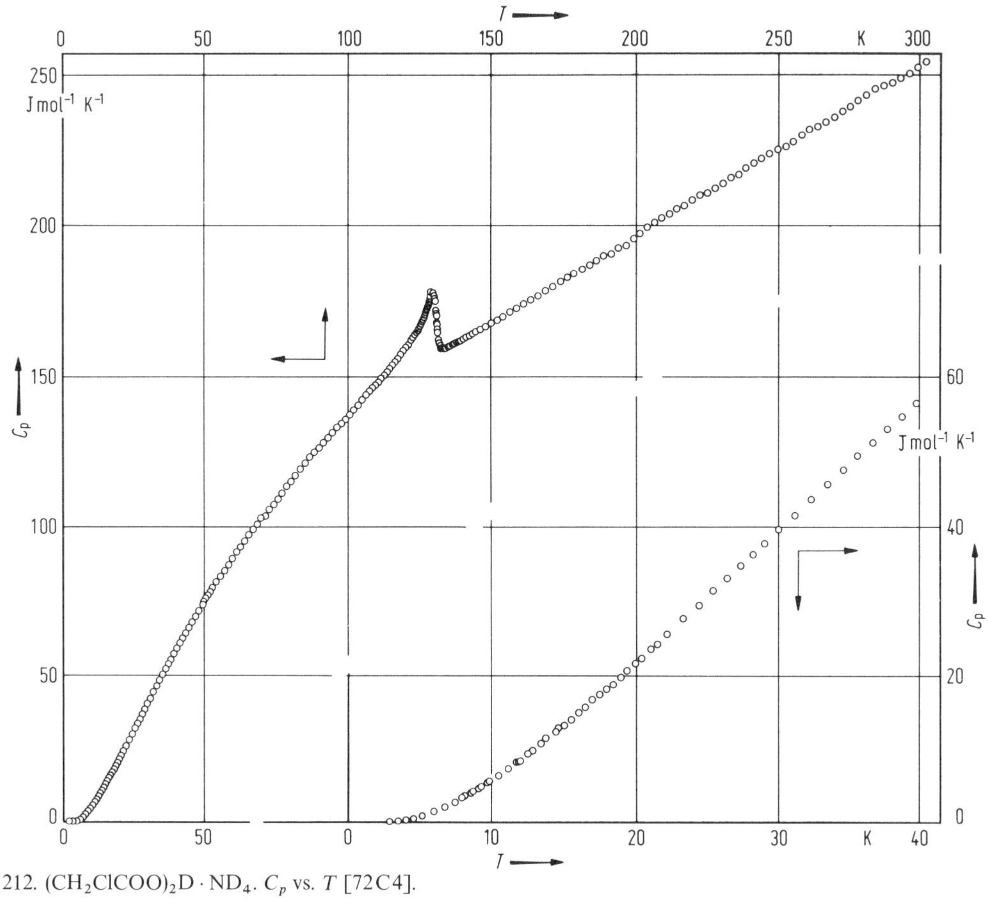

Fig. 1212. $(CH_2ClCOO)_2D \cdot ND_4$. C_p vs. T [72 C 4].

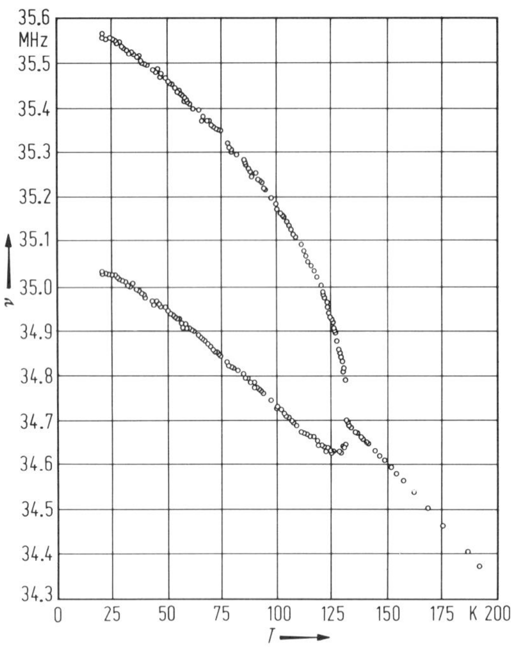

Fig. 1214. $(CH_2ClCOO)_2D \cdot ND_4$. ν vs. T [73C1]. ν: ^{35}Cl nuclear quadrupole resonance frequency. $\Theta_{II-I} = 131.7$ K.

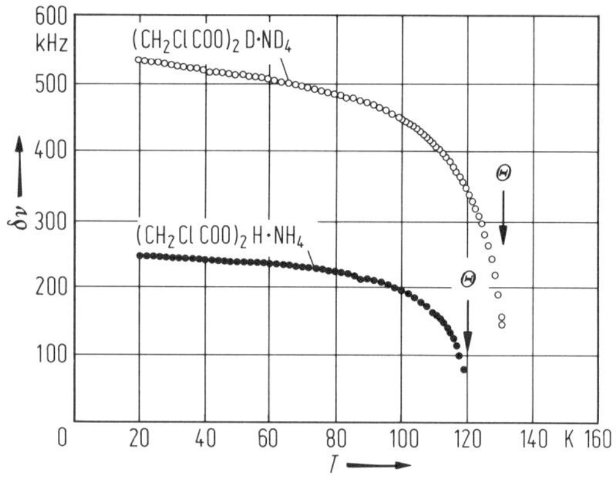

Fig. 1215. $(CH_2ClCOO)_2H \cdot NH_4$, $(CH_2ClCOO)_2D \cdot ND_4$. $\delta\nu$ vs. T [73C1]. $\delta\nu$: doublet separation of NQR spectrum for ^{35}Cl in the ferroelectric phase.

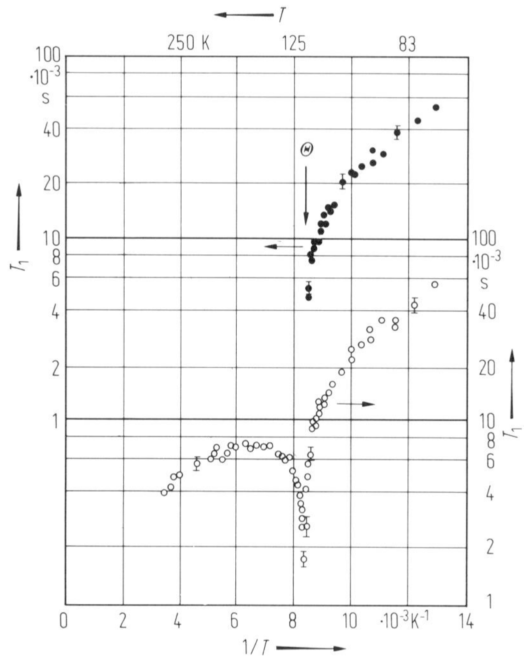

Fig. 1216. $(CH_2ClCOO)_2H \cdot NH_4$. T_1 vs. T^{-1} [74C7]. T_1: ^{35}Cl spin lattice relaxation time. Open circles: for ν_1; full circles: for ν_2. Resonance line separates in two components corresponding to ν_1 and ν_2 in the ferroelectric phase.

36 TGS($(NH_2CH_2COOH)_3H_2SO_4$) family

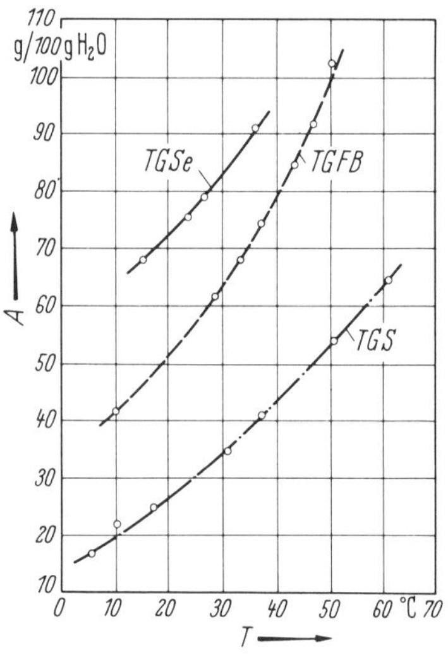

Fig. 1217. $(NH_2CH_2COOH)_3 \cdot H_2SO_4$ (TGS), $(NH_2CH_2COOH)_3 \cdot H_2SeO_4$ (TGSe), $(NH_2CH_2COOH)_3 \cdot H_2BeF_4$ (TGFB). A vs. T [58N1]. A: solubility in H_2O.

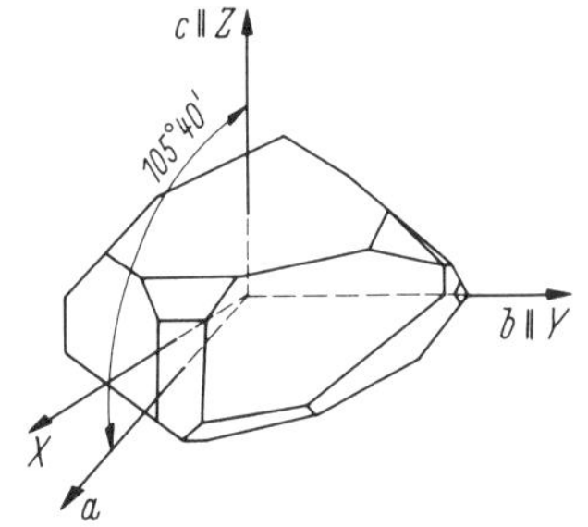

Fig. 1218. $(NH_2CH_2COOH)_3 \cdot H_2SO_4$ (TGS). Crystal form [66b1].

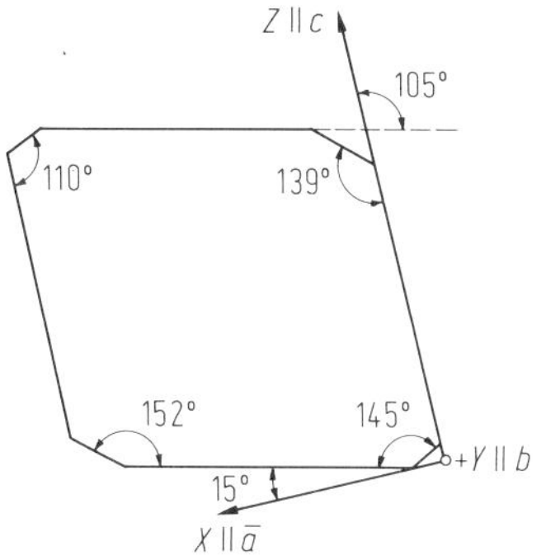

Fig. 1219. $(NH_2CH_2COOH)_3 \cdot H_2SO_4$ (TGS). Reference system of coordinate in a single crystal [59 K 2].

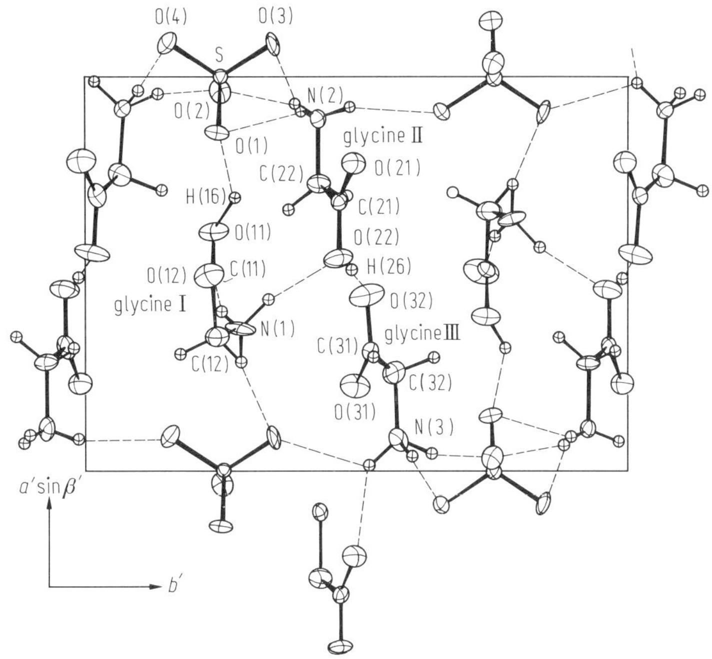

Fig. 1221. $(NH_2CH_2COOH)_3 \cdot H_2SO_4$ (TGS). Structure of phase II [76 F 8]. [001] projection. Hydrogen bonds are indicated by broken lines and thermal vibration ellipsoids are scaled to enclose 50% probability.

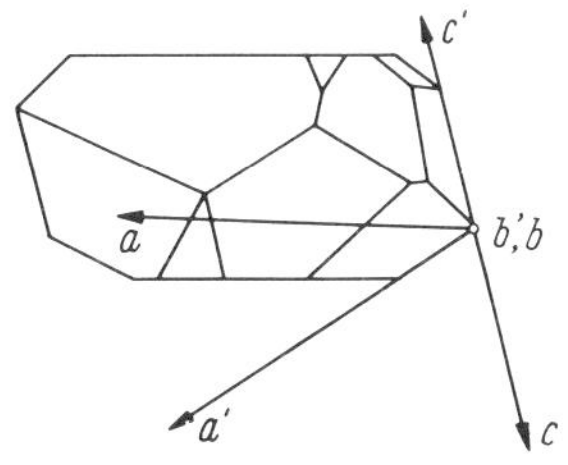

Fig. 1220. $(NH_2CH_2COOH)_3 \cdot H_2SO_4$ (TGS). Crystal and the crystallographic directions [61 K 1]. *a*, *b*, *c* and *a′*, *b′*, *c′* represent the crystallographic axes of the systems of Wood et al. [57 W 3] and of Hoshino et al. [59 H 6], respectively.

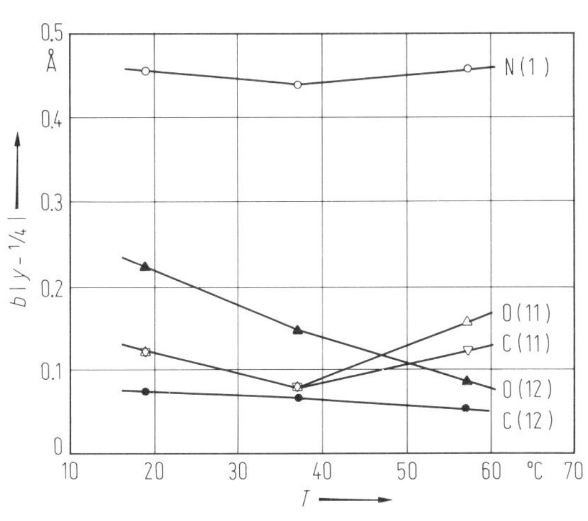

Fig. 1222. $(NH_2CH_2COOH)_3 \cdot H_2SO_4$ (TGS). $b|y-1/4|$ vs. T of the atoms in the glycine I molecules [73 I 10]. b: lattice constant.

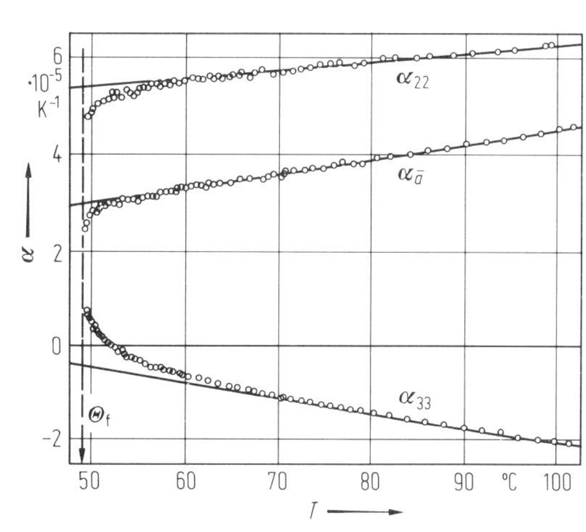

Fig. 1223. $(NH_2CH_2COOH)_3 \cdot H_2SO_4$ (TGS). $\alpha_{\bar{a}}$, α_{22}, and α_{33} vs. T [79 E 2]. $\alpha_{\bar{a}}$: thermal expansion coefficient along the $\bar{a}$-axis. The $\bar{a}$-axis is the direction perpendicular to the b' and c' axes. See Figs. 1219, 1220.

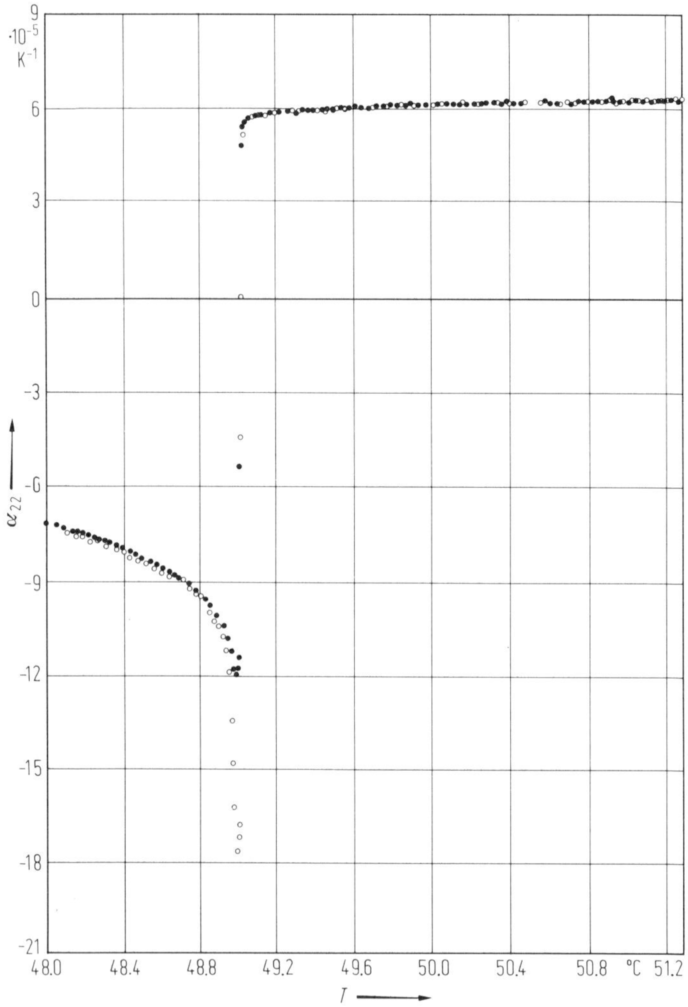

Fig. 1224. $(NH_2CH_2COOH)_3 \cdot H_2SO_4$ (TGS). α_{22} vs. T [79D1]. α_{22}: linear thermal expansion coefficient along the b-axis.

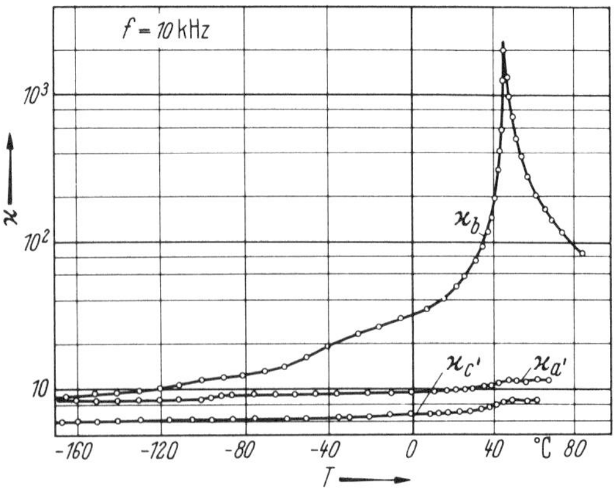

Fig. 1226. $(NH_2CH_2COOH)_3 \cdot H_2SO_4$ (TGS). $\varkappa_{a'}$, $\varkappa_{b'}$, and $\varkappa_{c'}$ vs. T [57H2].

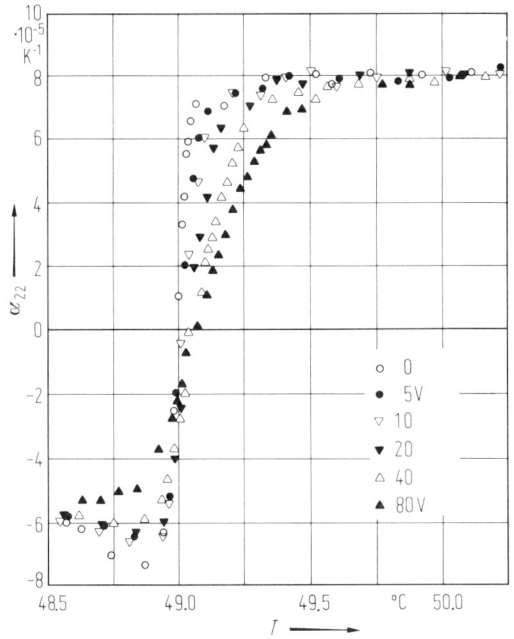

Fig. 1225. $(NH_2CH_2COOH)_3 \cdot H_2SO_4$ (TGS). α_{22} vs. T [74G7]. Parameter: applied voltage. An applied voltage of 80 V corresponds to a field of 17.8 kV m^{-1}.

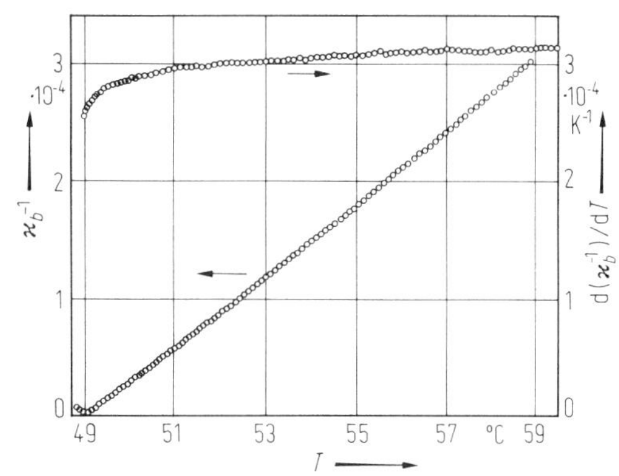

Fig. 1227. $(NH_2CH_2COOH)_3 \cdot H_2SO_4$ (TGS). $\varkappa_b^{-1}$ and $d(\varkappa_b^{-1})/dT$ vs. T in the vicinity of Θ_f ($f = 1$ kHz). [79D1].

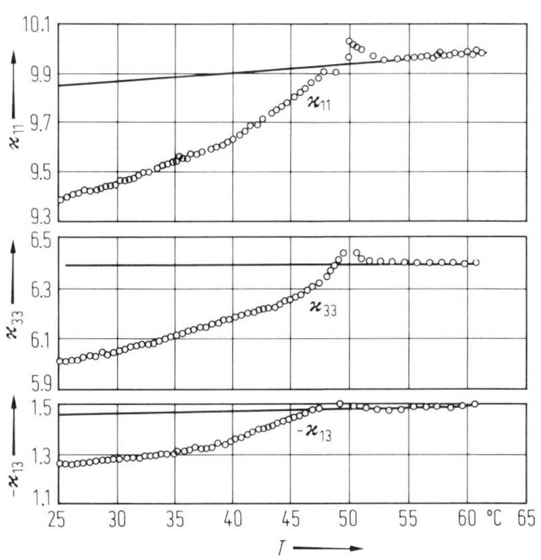

Fig. 1228. $(NH_2CH_2COOH)_3 \cdot H_2SO_4$ (TGS). $\varkappa_{11}$, $\varkappa_{33}$, and $-\varkappa_{13}$ vs. T [72B20]. $f = 1$ MHz. Coordinate system: see Fig. 1219.

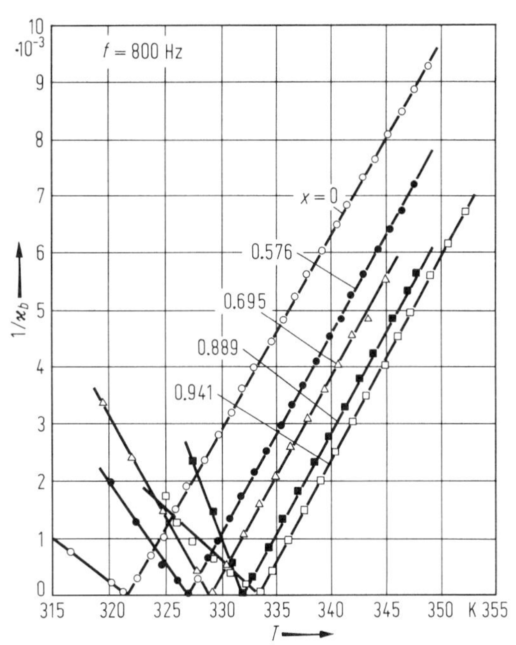

Fig. 1229. $[(NH_2CH_2COOH)_3 \cdot H_2SO_4]_{1-x} \cdot [(ND_2CH_2COOD)_3 \cdot D_2SO_4]_x$. $\varkappa_b^{-1}$ vs. T [68B13]. Parameter: x.

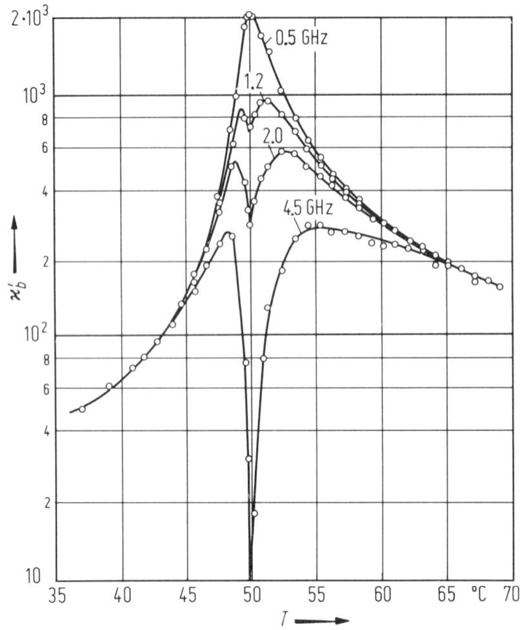

Fig. 1230. $(NH_2CH_2COOH)_3 \cdot H_2SO_4$ (TGS). $\varkappa_b'$ vs. T [69L6]. Parameter: f.

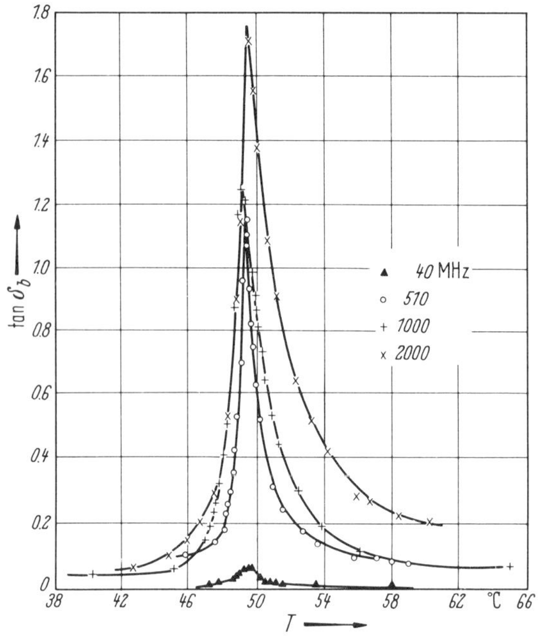

Fig. 1231. $(NH_2CH_2COOH)_3 \cdot H_2SO_4$ (TGS). $\tan\delta_b$ vs. T [60L1]. Parameter: f.

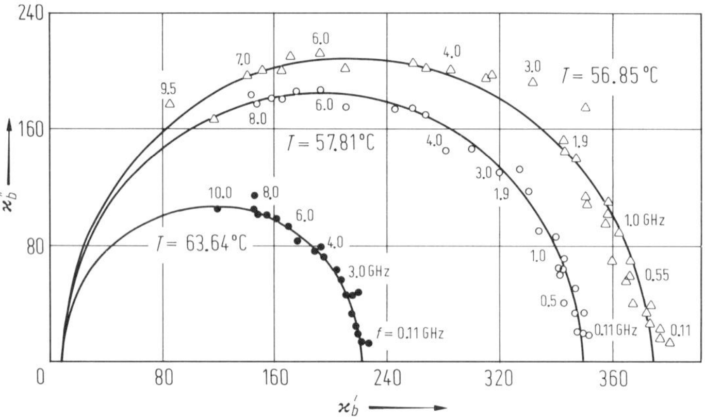

Fig. 1232. $(NH_2CH_2COOH)_3 \cdot H_2SO_4$ (TGS). Cole-Cole plot of the complex dielectric constant in the paraelectric phase [73L5]. Parameter: T.

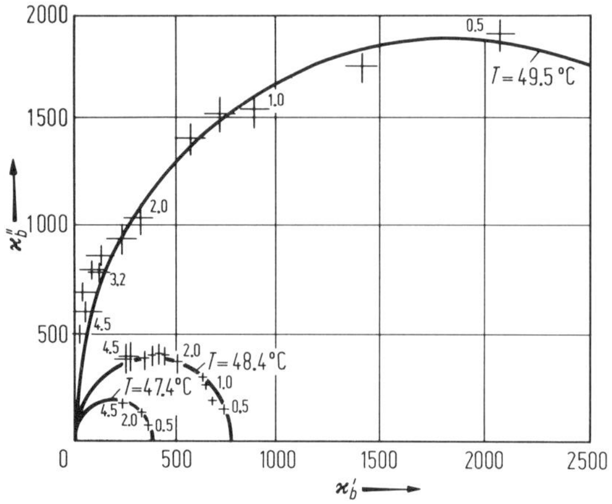

Fig. 1233. $(NH_2CH_2COOH)_3 \cdot H_2SO_4$ (TGS). Cole-Cole plot of the complex dielectric constant in the ferroelectric phase [69L6]. Parameter: T. Frequencies are in GHz.

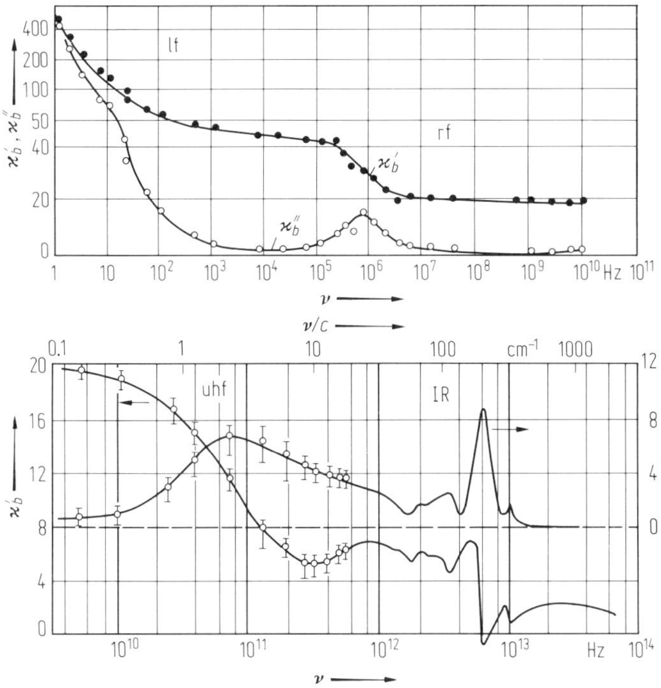

Fig. 1234. $(NH_2CH_2COOH)_3 \cdot H_2SO_4$ (TGS). $\varkappa_b'$ and $\varkappa_b''$ vs. ν at about 20 °C [73P11]. lf: low frequency; rf: radio frequency; uhf: ultra high frequency. Note change of scale of ordinate.

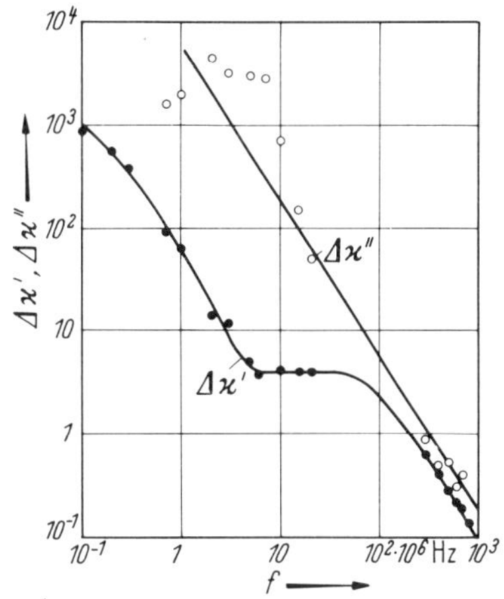

Fig. 1236. $(NH_2CH_2COOH)_3 \cdot H_2SO_4$ (TGS). $\Delta\varkappa'$ and $\Delta\varkappa''$ vs. f [62F2]. $\Delta\varkappa$: increase in dielectric constant during switching.

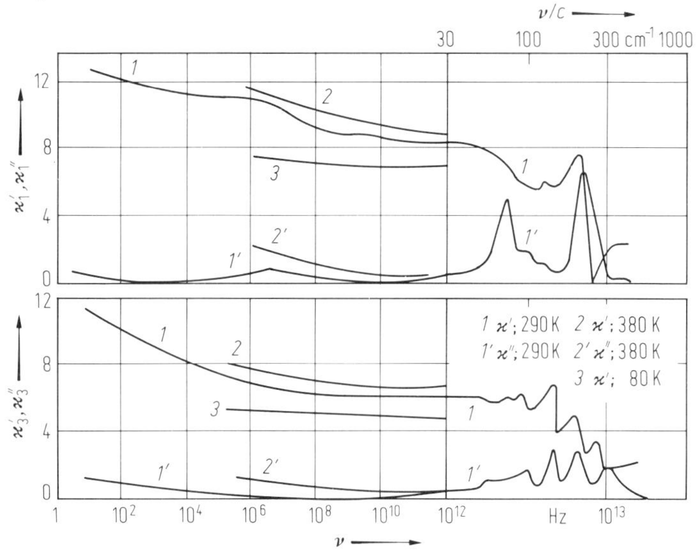

Fig. 1235. $(NH_2CH_2COOH)_3 \cdot H_2SO_4$ (TGS). $\varkappa'$ and $\varkappa''$ vs. ν [74P5]. Parameter: T. $\varkappa_1$ and $\varkappa_3$ are the principal dielectric constants along the axes perpendicular to the b-axis.

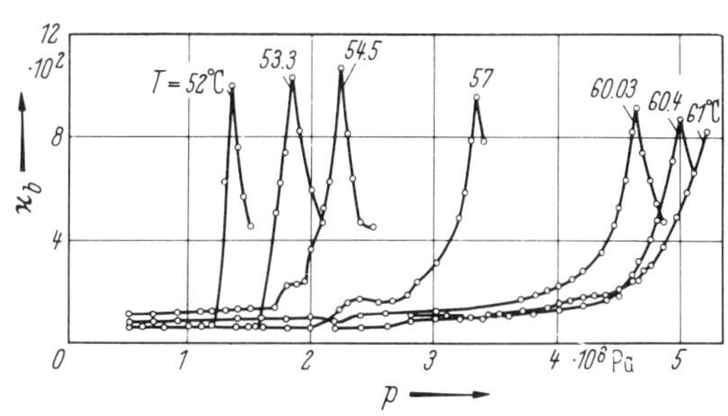

Fig. 1237. $(NH_2CH_2COOH)_3 \cdot H_2SO_4$ (TGS). $\varkappa_b$ vs. p [62L1]. Parameter: T.

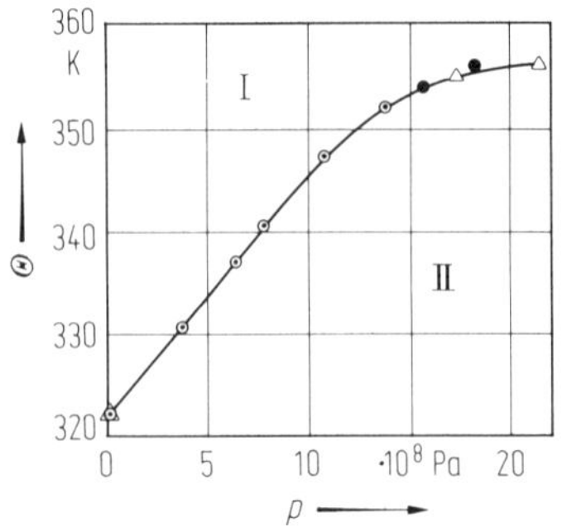

Fig. 1238. $(NH_2CH_2COOH)_3 \cdot H_2SO_4$ (TGS). Θ vs. p [74S32].

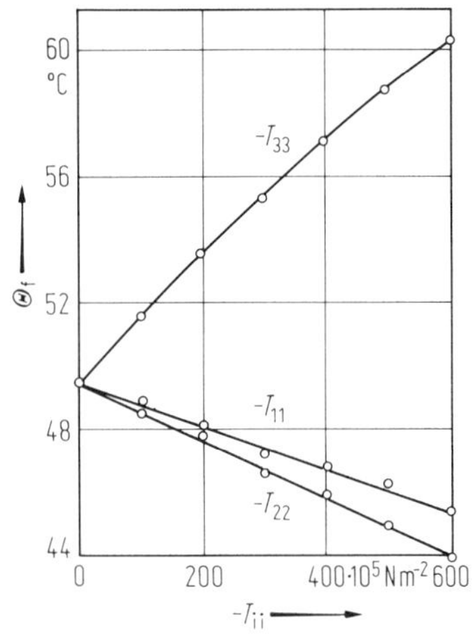

Fig. 1239. $(NH_2CH_2COOH)_3 \cdot H_2SO_4$ (TGS). Θ_f vs. $-T_{ii}$ [74I5]. Θ_f: Ferroelectric Curie temperature, $-T_{ii}$: Uniaxial stress. The orthogonal axial system of Fig. 1218 is adopted.

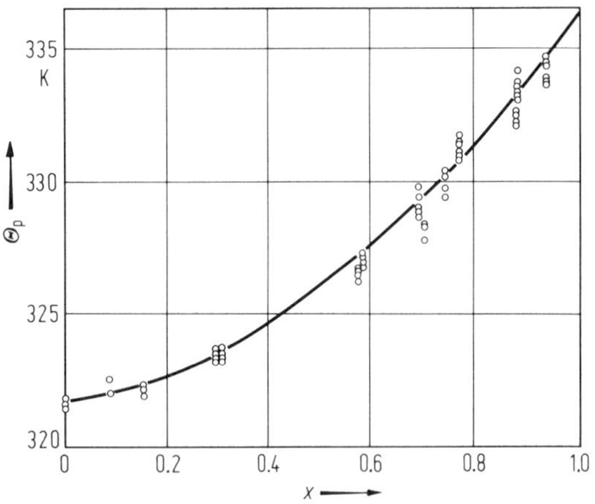

Fig. 1240. $[(NH_2CH_2COOH)_3 \cdot H_2SO_4]_{1-x} \cdot$ $[ND_2CH_2COOD)_3 \cdot D_2SO_4]_x$. Θ_p vs. x [68B13]. Θ_p: paraelectric Curie temperature.

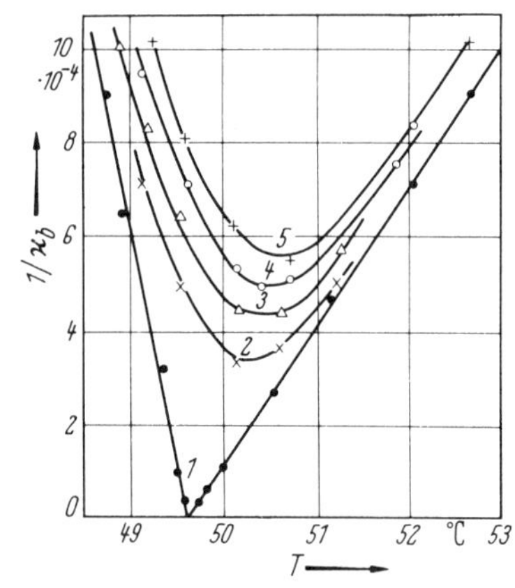

Fig. 1241. $(NH_2CH_2COOH)_3 \cdot H_2SO_4$ (TGS). $1/\varkappa_b$ vs. T [61S9]. Parameter: E_{bias}.

Curve	1	2	3	4	5	
E_{bias}	0	50	80	120	180	kV m^{-1}.

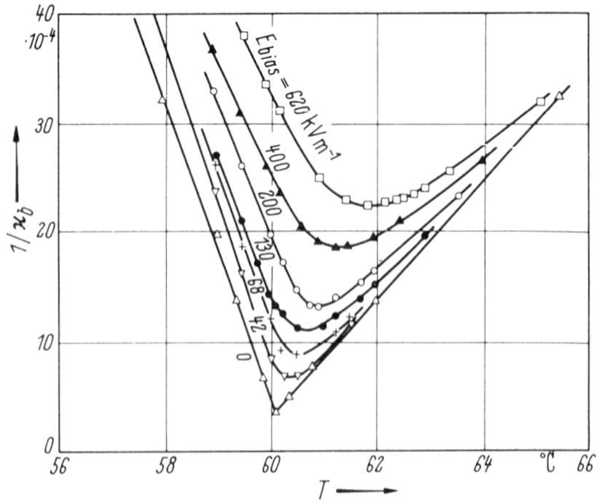

Fig. 1242. Deuterated triglycine sulfate. $\varkappa_b^{-1}$ vs. T [61S9]. Parameter: E_{bias}.

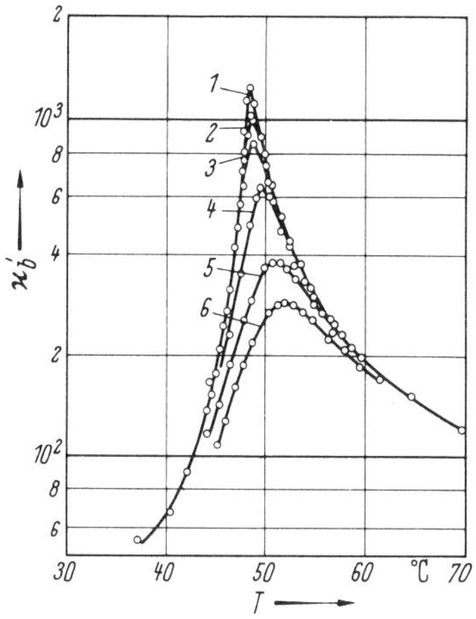

Fig. 1243. $(NH_2CH_2COOH)_3 \cdot H_2SO_4$ (TGS). $\varkappa_b'$ vs. T [63H6]. Parameter: E_{bias}. $f = 2600$ MHz.

Curve	1	2	3	4	5	6	
E_{bias}	0	60	120	240	640	1060	$kV\,m^{-1}$.

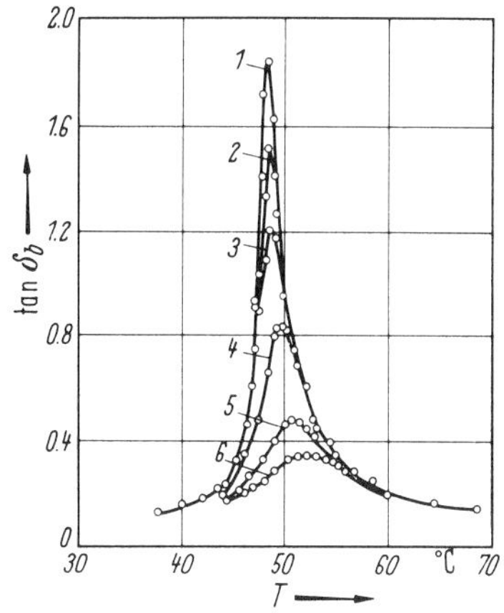

Fig. 1244. $(NH_2CH_2COOH)_3 \cdot H_2SO_4$ (TGS). $\tan\delta_b$ vs. T [63H6]. Parameter: E_{bias}. $f = 2600$ MHz. $\tan\delta_b$: $\tan\delta$ along the b-axis.

Curve	1	2	3	4	5	6	
E_{bias}	0	60	120	240	640	1060	$kV\,m^{-1}$.

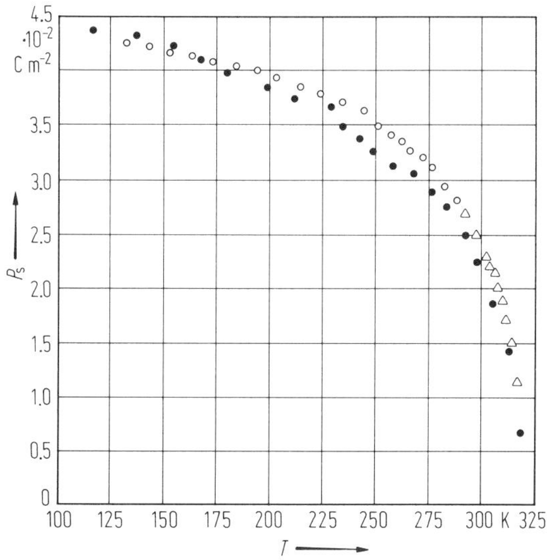

Fig. 1245. $(NH_2CH_2COOH)_3 \cdot H_2SO_4$ (TGS). P_s vs. T [64G4]. Open circles: pyroelectric [60C1]. Triangles: hysteresis loop [58T4]. Full circles: hysteresis loop [64G4].

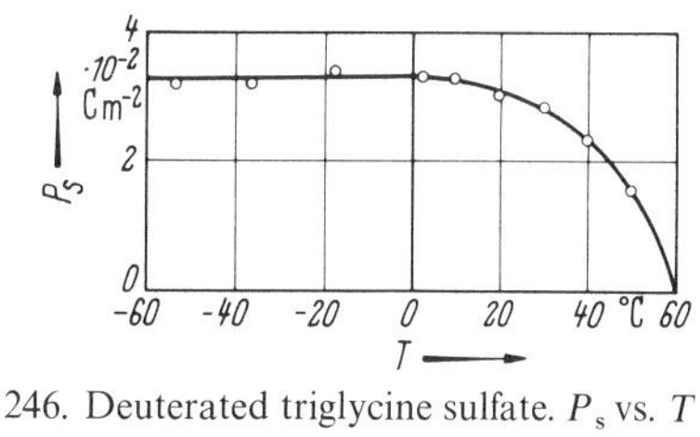

Fig. 1246. Deuterated triglycine sulfate. P_s vs. T [60K1].

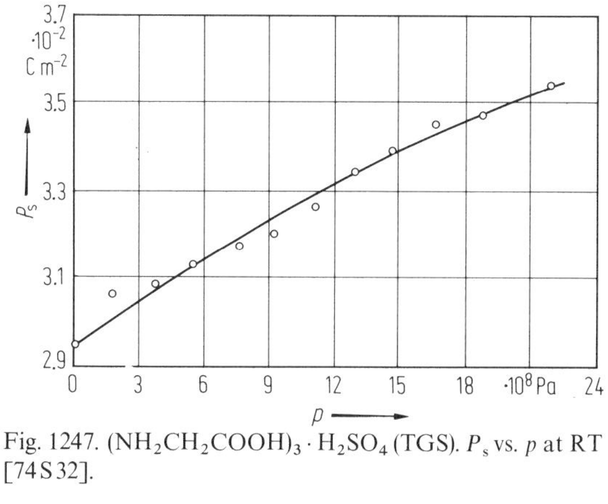

Fig. 1247. $(NH_2CH_2COOH)_3 \cdot H_2SO_4$ (TGS). P_s vs. p at RT [74S32].

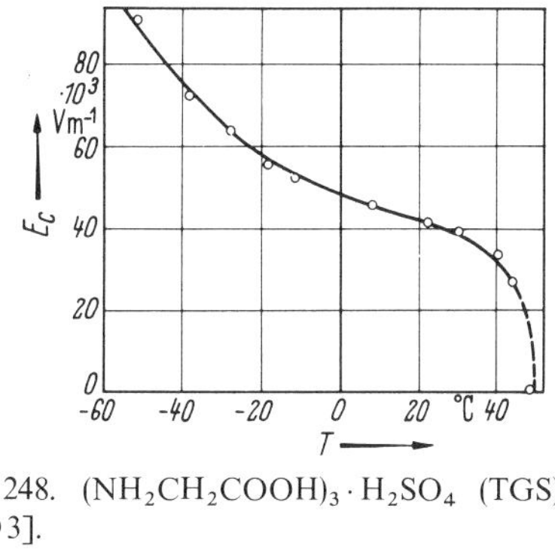

Fig. 1248. $(NH_2CH_2COOH)_3 \cdot H_2SO_4$ (TGS). E_c vs. T [58D3].

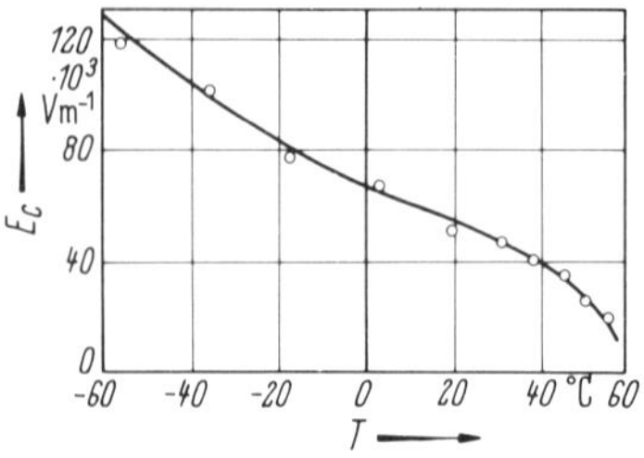

Fig. 1249. Deuterated triglycine sulfate. E_c vs. T [60K1].

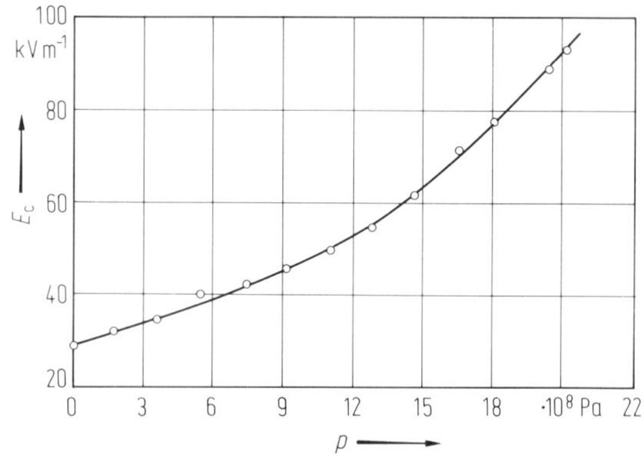

Fig. 1250. $(NH_2CH_2COOH)_3 \cdot H_2SO_4$ (TGS). E_c vs. p at RT [74S32].

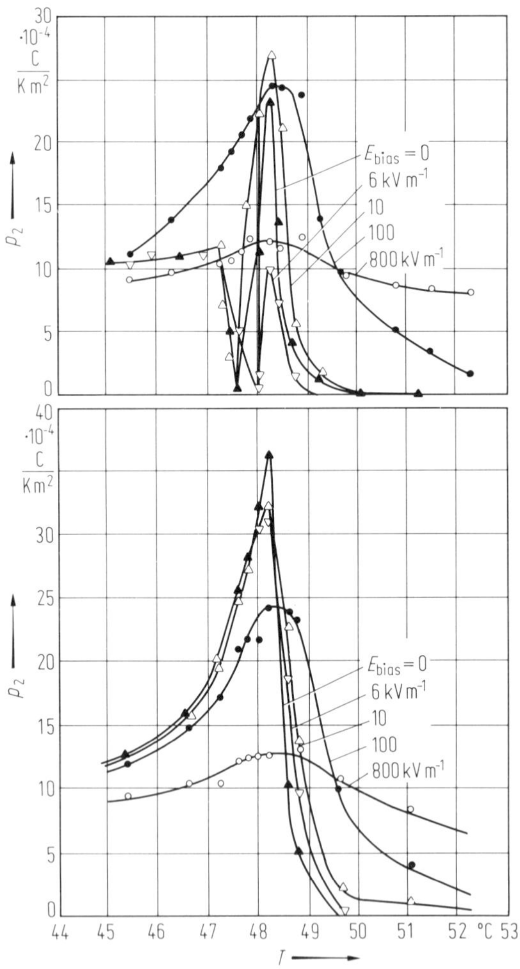

Fig. 1251. $(NH_2CH_2COOH)_3 \cdot H_2SO_4$ (TGS). p_2 vs. T [67K6]. p_2: pyroelectric coefficient along the [010] axis.

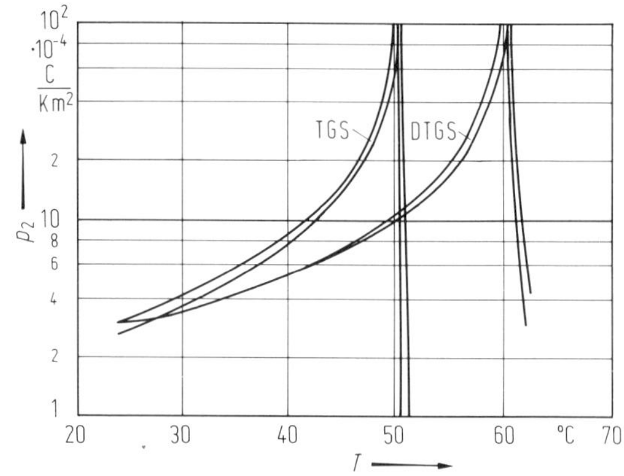

Fig. 1252. $(NH_2CH_2COOH)_3 \cdot H_2SO_4$ (TGS), deuterated triglycine sulfate (DTGS). p_2 vs. T [78F1]. p_2: pyroelectric coefficient along the [010] axis. Two curves for TGS and deuterated TGS are the extreme values obtained for several samples.

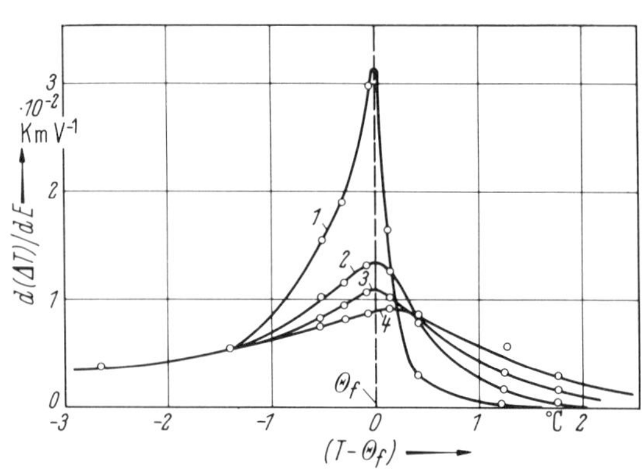

Fig. 1253. $(NH_2CH_2COOH)_3 \cdot H_2SO_4$ (TGS). $d(\Delta T)/dE$ vs. $T-\Theta_f$ [66S23]. ΔT: electrocaloric temperature change. Parameter: E. Curve *1*: $E=0$, *2*: $E=50\ kV\ m^{-1}$, *3*: $E=100\ kV\ m^{-1}$, *4*: $E=150\ kV\ m^{-1}$.

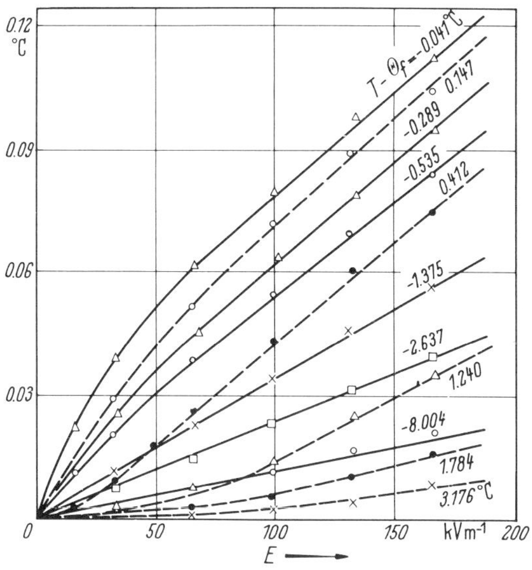

Fig. 1254. $(NH_2CH_2COOH)_3 \cdot H_2SO_4$ (TGS). ΔT vs. E [66S23]. Parameter: $T-\Theta_f$. ΔT: electrocaloric temperature change.

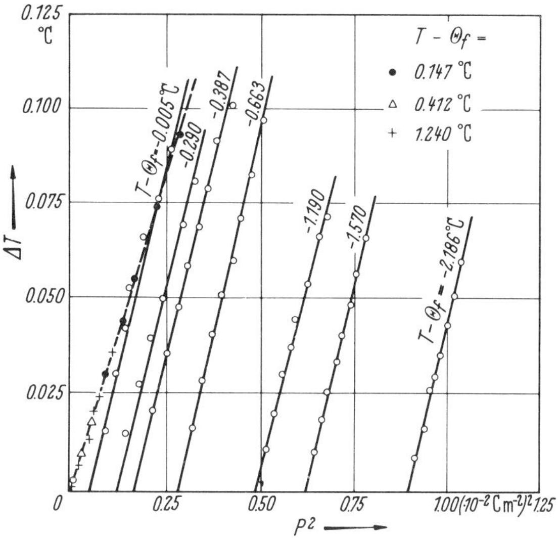

Fig. 1255. $(NH_2CH_2COOH)_3 \cdot H_2SO_4$ (TGS). ΔT vs. P^2 [66S23]. ΔT: electrocaloric temperature change.

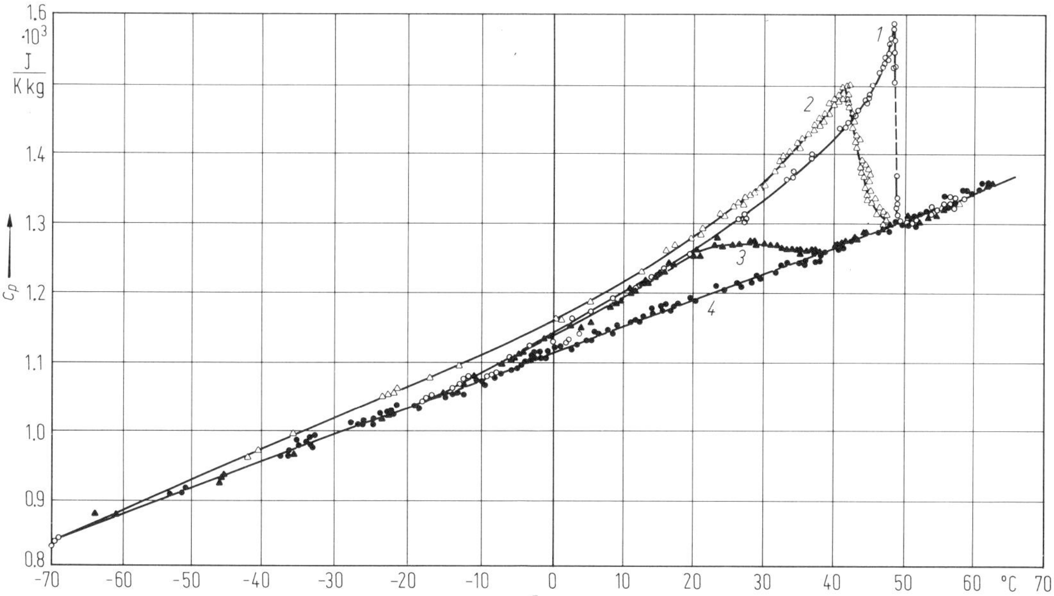

Fig. 1256. $(NH_2CH_2COOH)_3 \cdot H_2SO_4$ (TGS). c_p vs. T [77T5]. Parameter: γ-ray irradiation dose. Curve *1*: 0 mR, *2*: 3 mR, *3*: 15 mR, *4*: 30 mR. 1 R = $2.58 \cdot 10^{-4}$ C kg^{-1}.

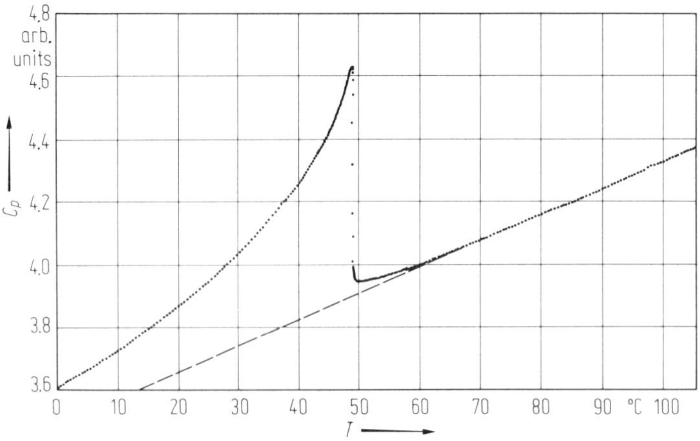

◄ Fig. 1257. $(NH_2CH_2COOH)_3 \cdot H_2SO_4$ (TGS). C_p vs. T [79E1].

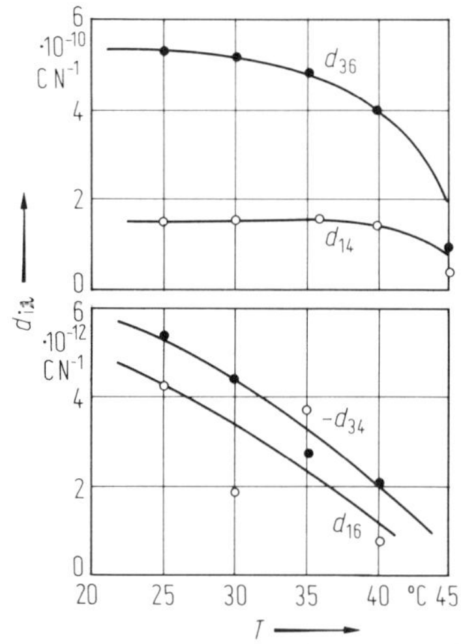

Fig. 1262. $(NH_2CH_2COOH)_3 \cdot H_2SO_4$ (TGS). $d_{i\lambda}$ ($\lambda = 4$ and 6) vs. T [62I2].

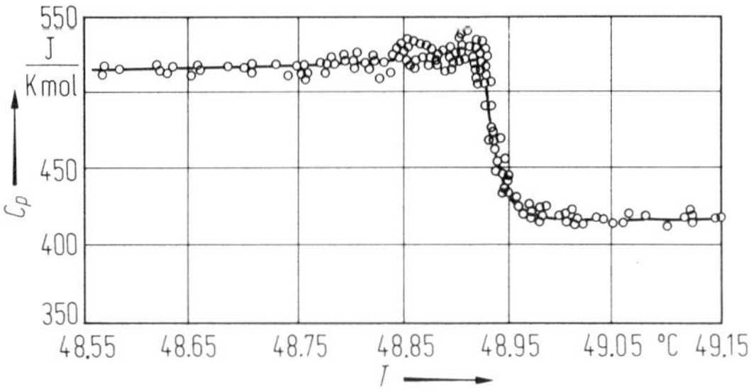

Fig. 1258. $(NH_2CH_2COOH)_3 \cdot H_2SO_4$ (TGS). C_p vs. T [70T2].

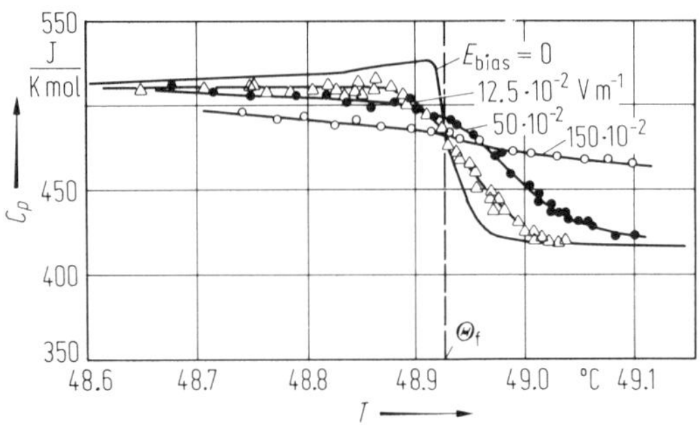

Fig. 1259. $(NH_2CH_2COOH)_3 \cdot H_2SO_4$ (TGS). C_p vs. T [70T2]. Parameter: E_{bias}.

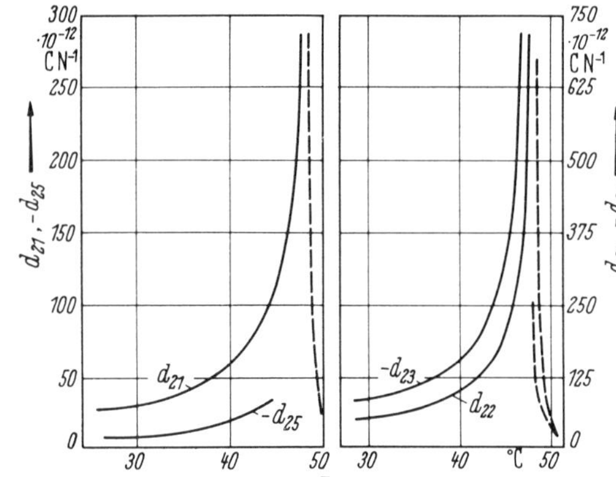

Fig. 1261. $(NH_2CH_2COOH)_3 \cdot H_2SO_4$ (TGS). $d_{i\lambda}$ vs. T [62I2]. Broken lines show apparent piezoelectricities caused by electrostriction.

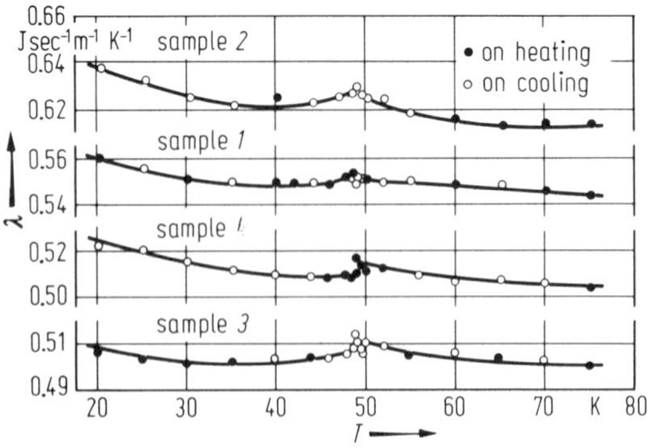

◄

Fig. 1260. $(NH_2CH_2COOH)_3 \cdot H_2SO_4$ (TGS). λ vs. T [71H7]. λ: thermal conductivity.

Sample No.	1	2	3	4
Crystallographic direction	b	$\bar{a}$	c	$\perp b$, 45° from c.

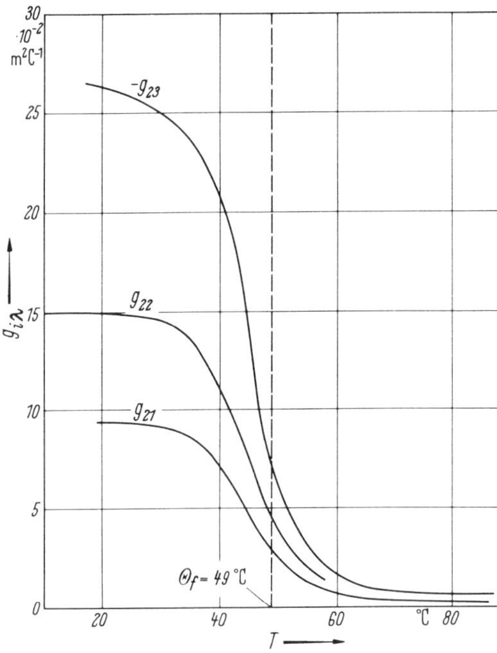

Fig. 1263. $(NH_2CH_2COOH)_3 \cdot H_2SO_4$ (TGS). $g_{i\lambda}$ vs. T [63S6].

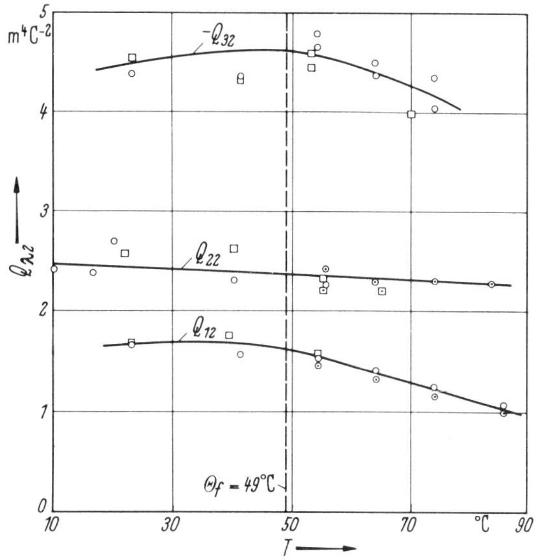

Fig. 1264. $(NH_2CH_2COOH)_3 \cdot H_2SO_4$ (TGS). $Q_{\lambda\mu}$ vs. T [63S6]. Different symbols correspond to different measurements.

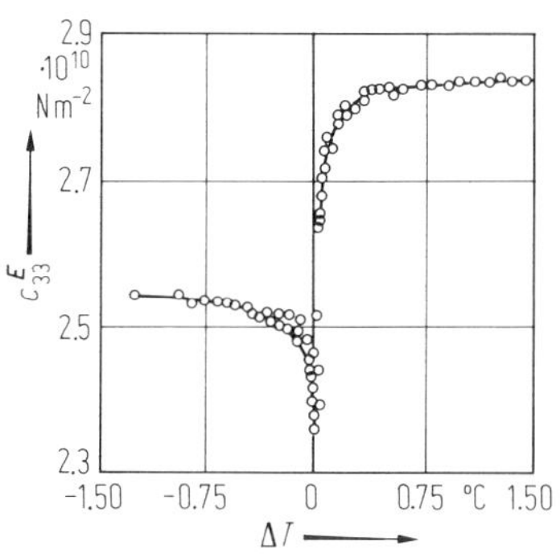

Fig. 1266. $(NH_2CH_2COOH)_3 \cdot H_2SO_4$ (TGS). $c^{\boldsymbol{E}}_{33}$ vs. ΔT [75S34]. $\Delta T = T - \Theta_f$. $f = 10$ MHz.

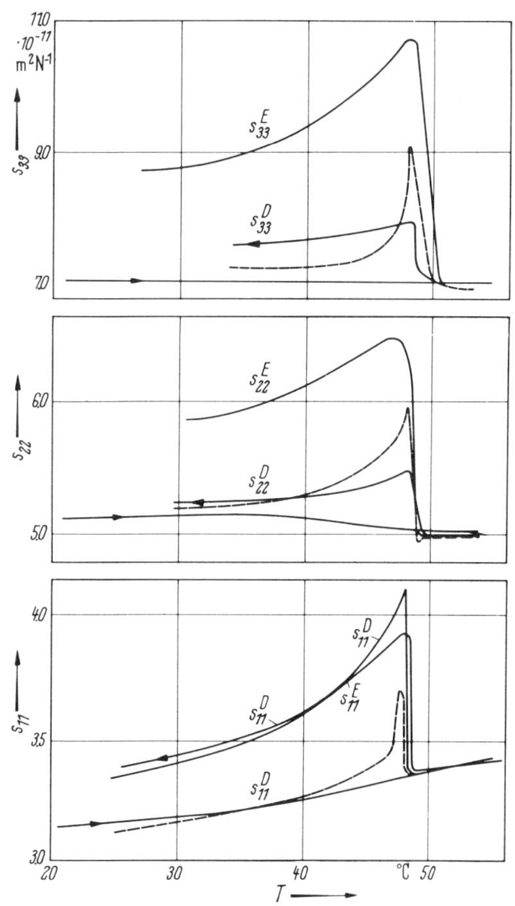

Fig. 1265. $(NH_2CH_2COOH)_3 \cdot H_2SO_4$ (TGS). $s^{\boldsymbol{D}}_{\lambda\mu}$ and $s^{\boldsymbol{E}}_{\lambda\mu}$ vs. T [62I2]. $s^{\boldsymbol{E}}$: resonance method. $s^{\boldsymbol{D}}$: calculated values from $s^{\boldsymbol{E}}$ (broken lines). The lowest flat lines show the values observed by composite-bar method (heating). The remaining curves show the values of the unpoled crystals observed by composite bar method (cooling). See further [61S8, 62M7].

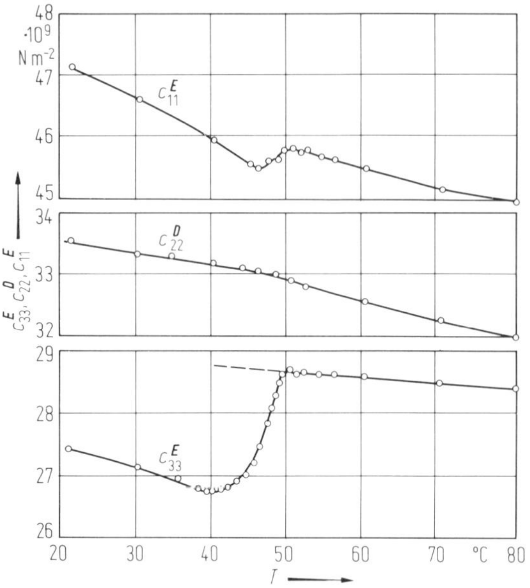

Fig. 1267. $(NH_2CH_2COOH)_3 \cdot H_2SO_4$ (TGS). c_{11}^E, c_{22}^D, c_{33}^E from Brillouin scattering vs. T [77L9].

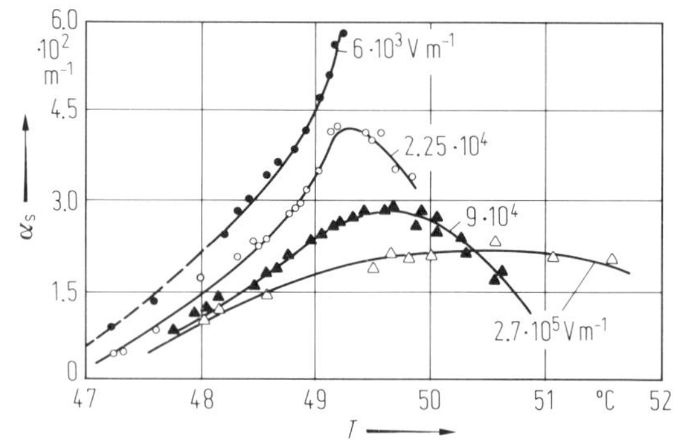

Fig. 1268. $(NH_2CH_2COOH)_3 \cdot H_2SO_4$ (TGS). α_s vs. T [76S1]. α_s: acoustic surface wave attenuation in the a'-axis on the b plane.

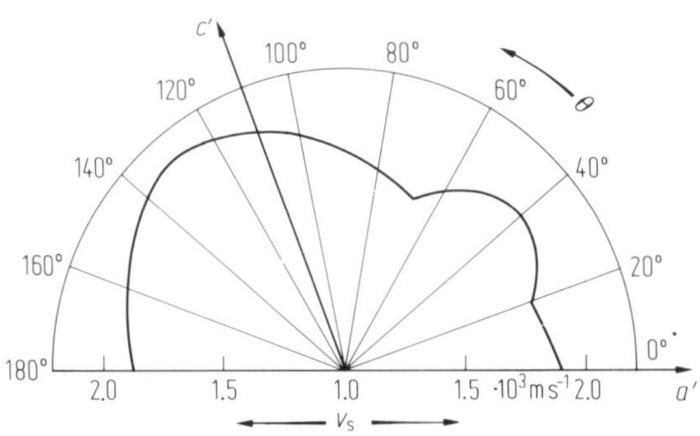

Fig. 1269. $(NH_2CH_2COOH)_3 \cdot H_2SO_4$ (TGS). v_s vs. θ [77S1]. v_s: acoustic surface wave velocity in the b plane. θ: angle between the propagation direction and the a'-axis.

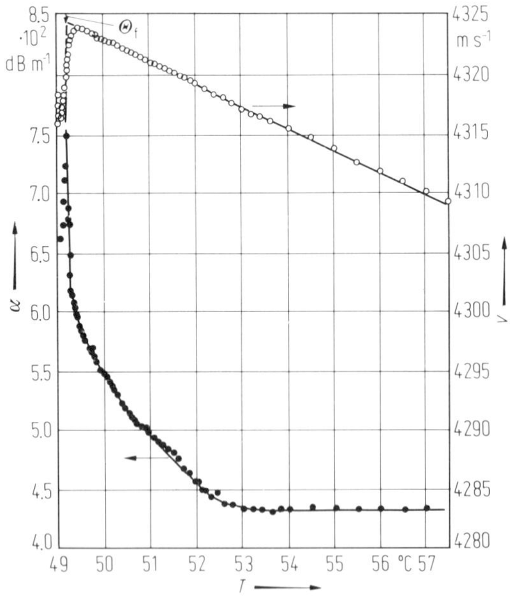

Fig. 1271. $(NH_2CH_2COOH)_3 \cdot H_2SO_4$ (TGS). α, v vs. T [77K9]. α: attenuation of longitudinal sound wave propagating along the b-axis. v: sound velocity. $f=49$ MHz.

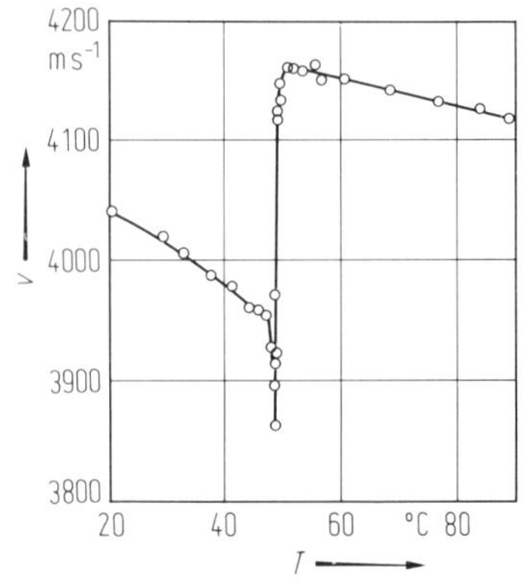

Fig. 1270. $(NH_2CH_2COOH)_3 \cdot H_2SO_4$ (TGS). v vs. T [75S34]. v: sound velocity of the quasi-longitudinal wave along the c-axis. $f=10$ MHz.

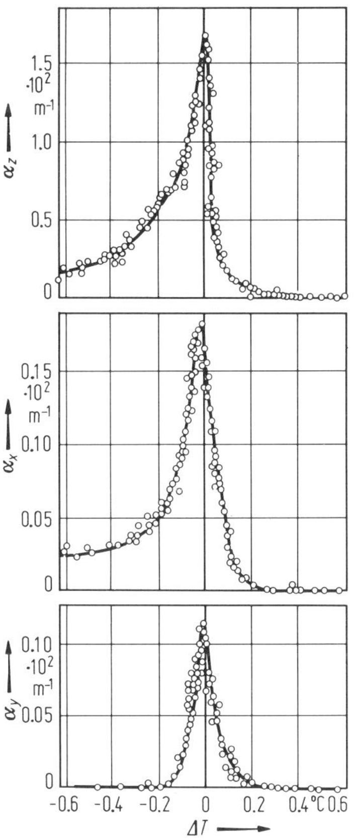

Fig. 1272. $(NH_2CH_2COOH)_3 \cdot H_2SO_4$ (TGS). α_x, α_y, α_z vs. ΔT [68 M 5]. α_j: attenuation coefficient for quasilongitudinal ultrasonic wave propagating along the $X(\bar{a})$, $Y(b)$, and $Z(c)$ axes. $\Delta T = T - \Theta_f$. See Fig. 1219.

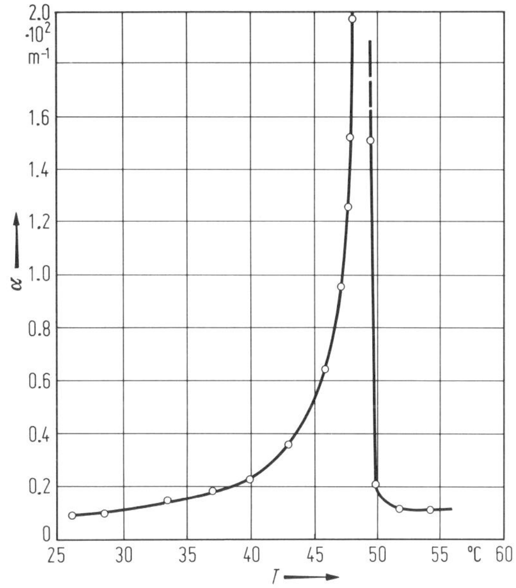

Fig. 1273. $(NH_2CH_2COOH)_3 \cdot H_2SO_4$ (TGS). α vs. T [71 B 1]. α: attenuation coefficient of quasilongitudinal waves propagating along the c-axis. $f = 58$ MHz.

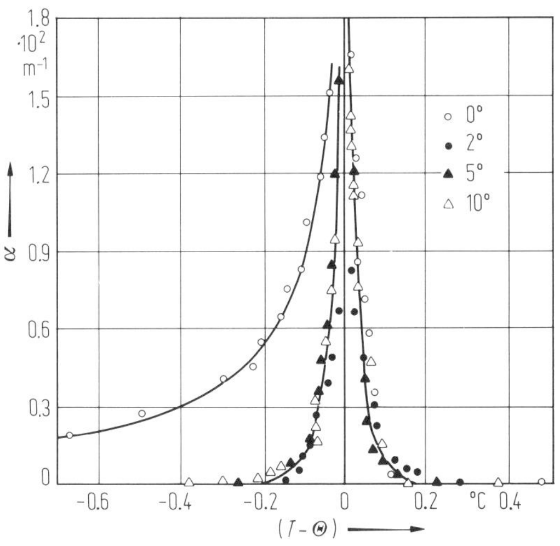

Fig. 1274. $(NH_2CH_2COOH)_3 \cdot H_2SO_4$ (TGS). α vs. $T - \Theta$ [69 M 8]. α: absorption coefficients of the quasi-longitudinal ultrasound wave propagating in the bc plane with angles 0°, 2°, and 10° to the c-axis.

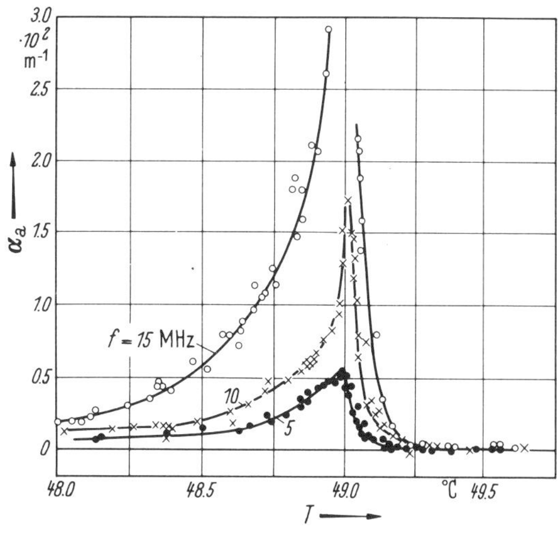

Fig. 1275. $(NH_2CH_2COOH)_3 \cdot H_2SO_4$ (TGS). α_a vs. T [65 M 6]. Parameter: f. Pulse method. Longitudinal waves along the c-axis. α_a: ultrasonic absorption coefficient.

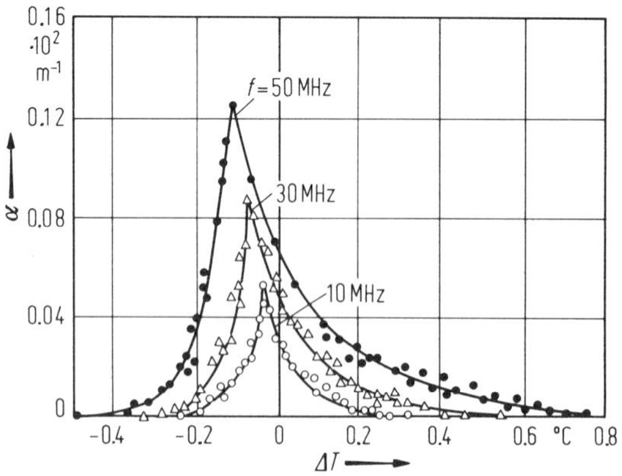

Fig. 1276. $(NH_2CH_2COOH)_3 \cdot H_2SO_4$ (TGS). α vs. ΔT at various f [70 M 7]. α: attenuation coefficient of longitudinal ultrasonic waves propagating along the b-axis. $\Delta T = T - \Theta$.

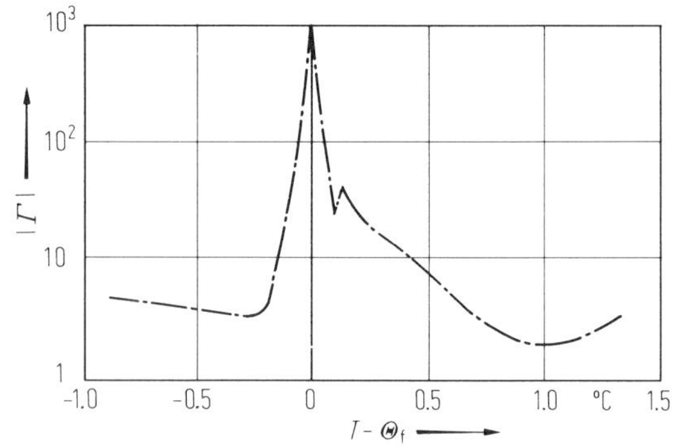

Fig. 1279. $(NH_2CH_2COOH)_3 \cdot H_2SO_4$ (TGS). $|\Gamma|$ vs. $T - \Theta_f$ [72 S 16]. $|\Gamma| = |3 + c_{333}/c_{33}|$; c_{333}: third order elastic constant. $E_b = 1.5 \cdot 10^2$ kV m^{-1}; E_b: field along ferroelectric b-axis.

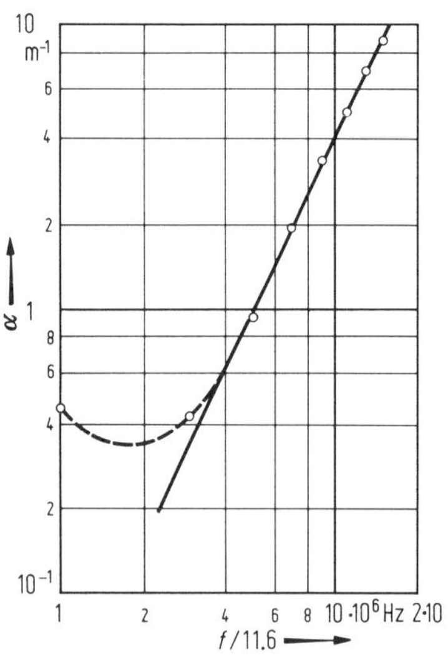

Fig. 1277. $(NH_2CH_2COOH)_3 \cdot H_2SO_4$ (TGS). α vs. f [71 B 1]. α: attenuation coefficient of quasilongitudinal ultrasonic waves propagating along the c-axis.

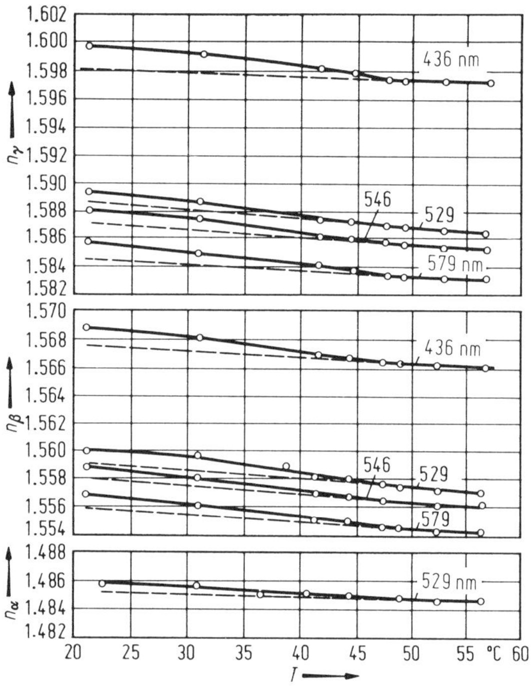

Fig. 1280. $(NH_2CH_2COOH)_3 \cdot H_2SO_4$ (TGS). n_α, n_β, n_γ vs. T [68 L 3]. Parameter: λ. $n_\alpha = n_b$.

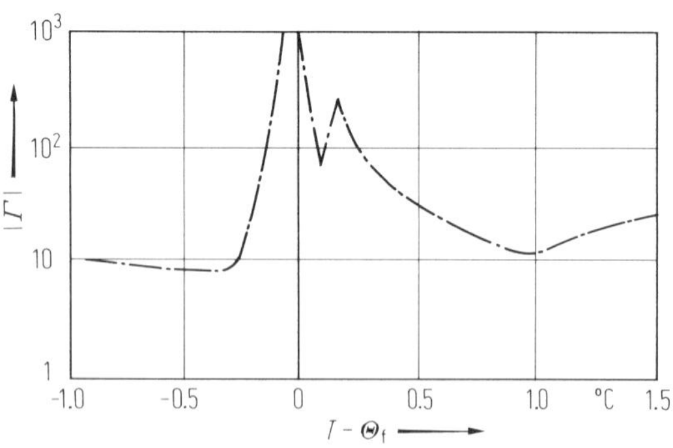

Fig. 1278. $(NH_2CH_2COOH)_3 \cdot H_2SO_4$ (TGS). $|\Gamma|$ vs. $T - \Theta_f$ [72 S 16]. $|\Gamma| = |3 + c_{333}/c_{33}|$. c_{333}: third order elastic constant. $E = 0$.

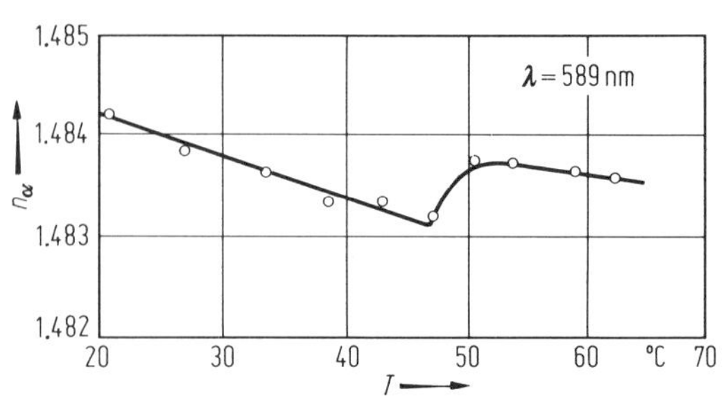

Fig. 1281. $(NH_2CH_2COOH)_3 \cdot H_2SO_4$ (TGS). n_α vs. T [68 I 2]. $\lambda = 589$ nm, $n_\alpha = n_b$.

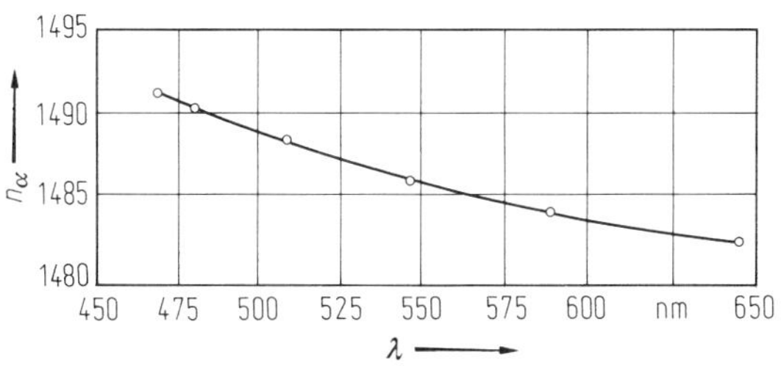

Fig. 1282. $(NH_2CH_2COOH)_3 \cdot H_2SO_4$ (TGS). n_α vs. λ at RT [68I2]. $n_\alpha = n_b$.

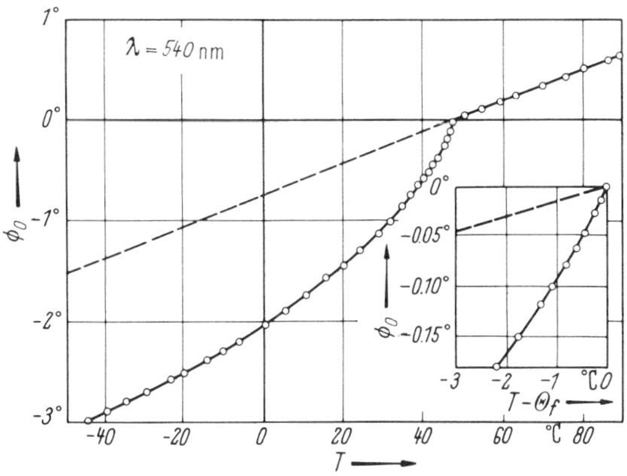

Fig. 1284. $(NH_2CH_2COOH)_3 \cdot H_2SO_4$ (TGS). ϕ_o vs. T [66I6]. ϕ_o: rotation angle of the optical indicatrix around the b-axis.

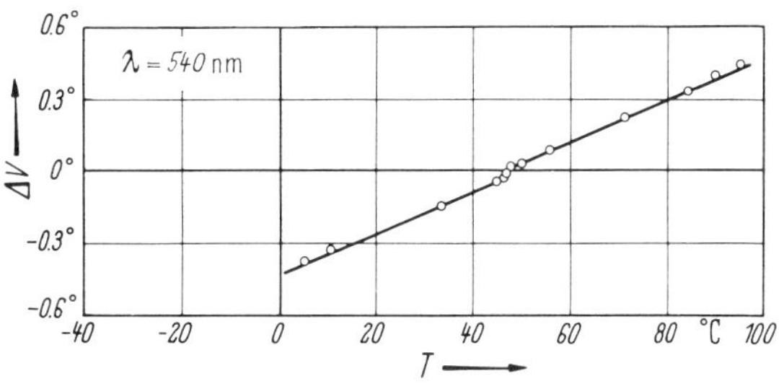

Fig. 1286. $(NH_2CH_2COOH)_3 \cdot H_2SO_4$ (TGS). ΔV vs. T [66I5]. $\Delta V = V(T) - V(\Theta)$. $2V(T)$: optical axial angle at T °C.

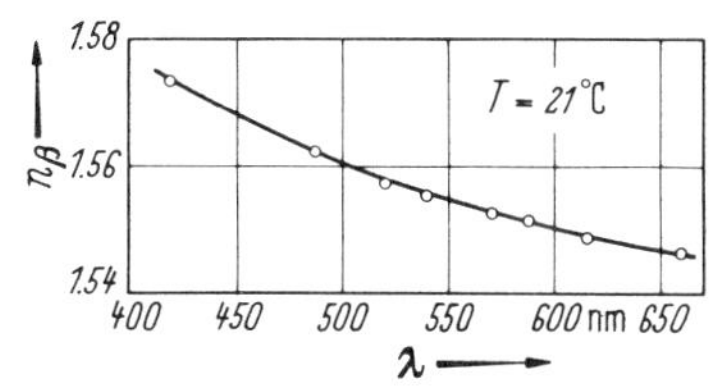

Fig. 1283. $(NH_2CH_2COOH)_3 \cdot H_2SO_4$ (TGS). n_β vs. λ [66I6].

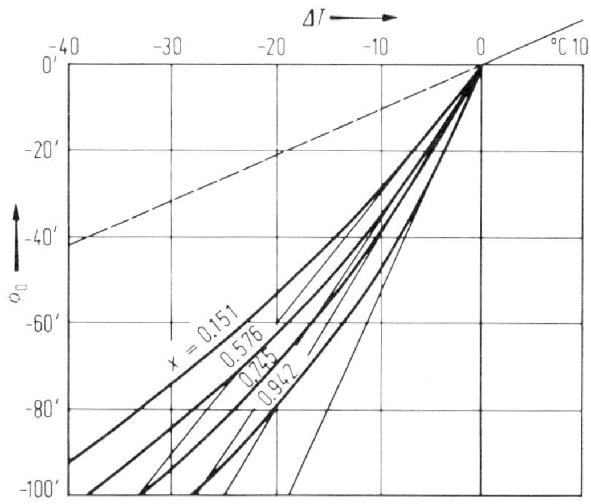

Fig. 1285. $[(NH_2CH_2COOH)_3 \cdot H_2SO_4]_{1-x}$ $[(ND_2CD_2COOD)_3 \cdot D_2SO_4]_x$. ϕ_o vs. ΔT [68B12]. ϕ_o: rotation angle of the optical indicatrix around the b-axis. $\Delta T = T - \Theta_{II-I}$.

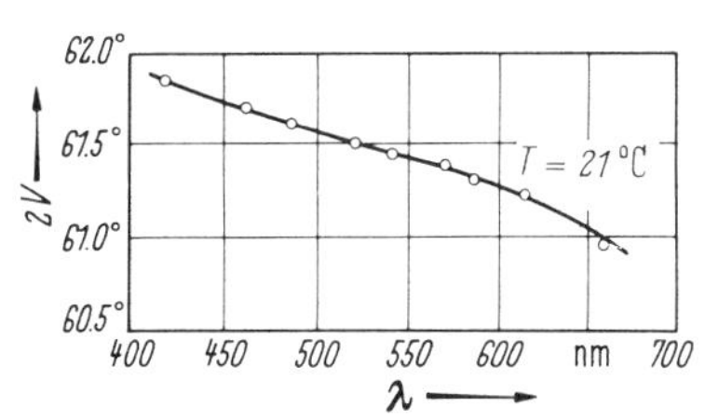

Fig. 1287. $(NH_2CH_2COOH)_3 \cdot H_2SO_4$ (TGS). $2V$ vs. λ [66I6]. $2V$: optical axial angle.

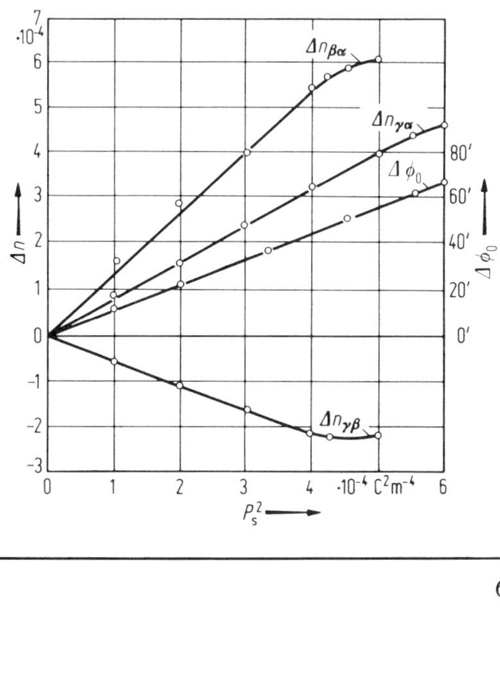

Fig. 1288. $(NH_2CH_2COOH)_3 \cdot H_2SO_4$ (TGS). Δn, $\Delta\phi_o$ vs. P_s^2 [68L3]. $\Delta n_{\lambda\mu} = n_\lambda - n_\mu$. $\Delta\phi_o$: change in rotation angle of the optical indicatrix around the b-axis.

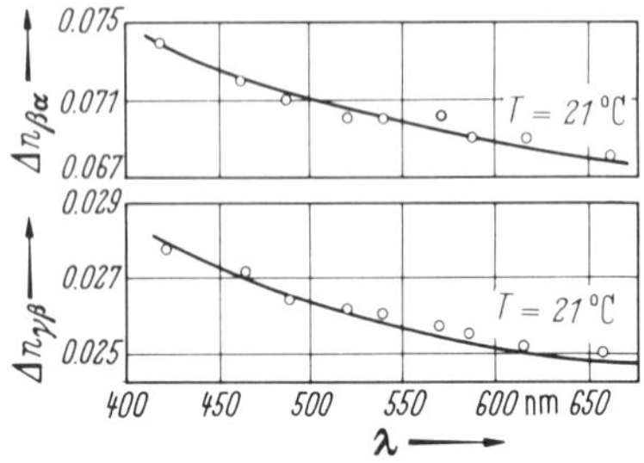

Fig. 1289. $(NH_2CH_2COOH)_3 \cdot H_2SO_4$ (TGS). $\Delta n_{\gamma\beta}$, $\Delta n_{\beta\alpha}$ vs. λ [66 I 6]. $\Delta n_{\gamma\beta} = n_\gamma - n_\beta$, $\Delta n_{\beta\alpha} = n_\beta - n_\alpha$.

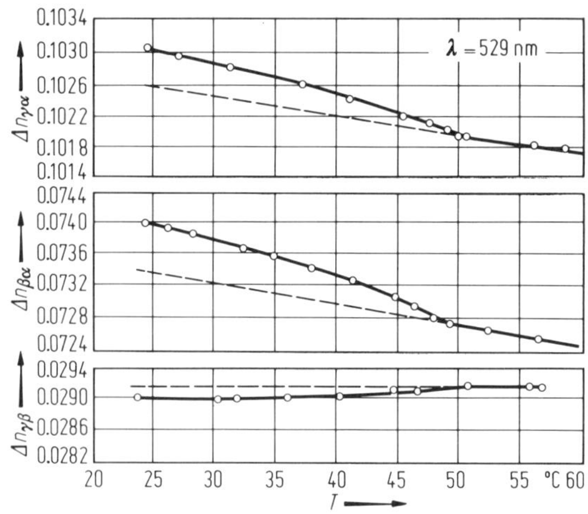

Fig. 1290. $(NH_2CH_2COOH)_3 \cdot H_2SO_4$ (TGS). $\Delta n_{\gamma\beta}$, $\Delta n_{\beta\alpha}$, $\Delta n_{\gamma\alpha}$ vs. T [68 L 3]. $\Delta n_{\lambda\mu} = n_\lambda - n_\mu$.

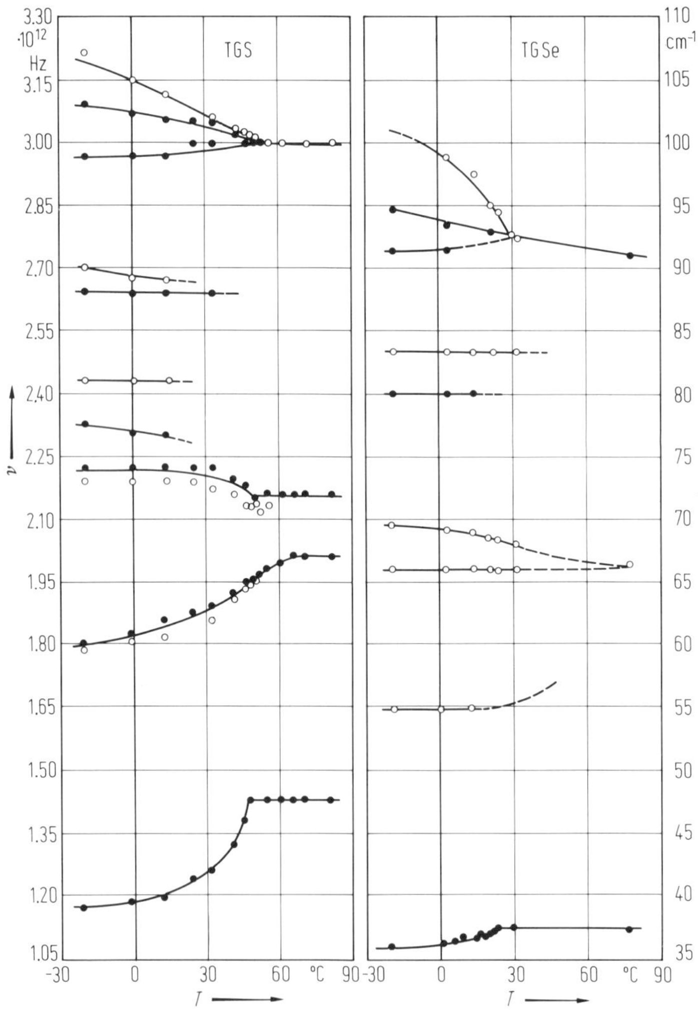

Fig. 1292. $(NH_2CH_2COOH)_3 \cdot H_2SO_4$ (TGS), $(NH_2CH_2COOH)_3 \cdot H_2SeO_4$ (TGSe). ν vs. T [77 W 7]. ν: frequency of the lattice vibration modes. Open circles: B-symmetry mode by IR spectra. Full circles: A-symmetry modes by Raman spectra.

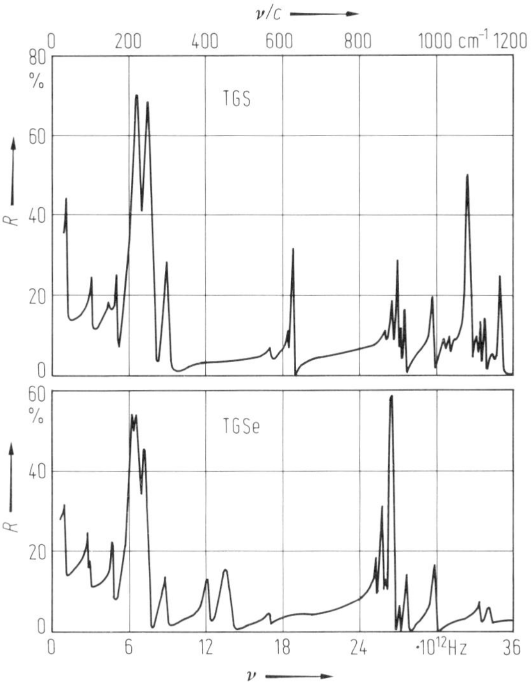

Fig. 1291. $(NH_2CH_2COOH)_3 \cdot H_2SO_4$ (TGS), $(NH_2CH_2COOH)_3 \cdot H_2SeO_4$ (TGSe). R vs. ν [77W7]. R: infrared reflectivity at 80 K. Polarization of the incident radiation is parallel to the b-axis.

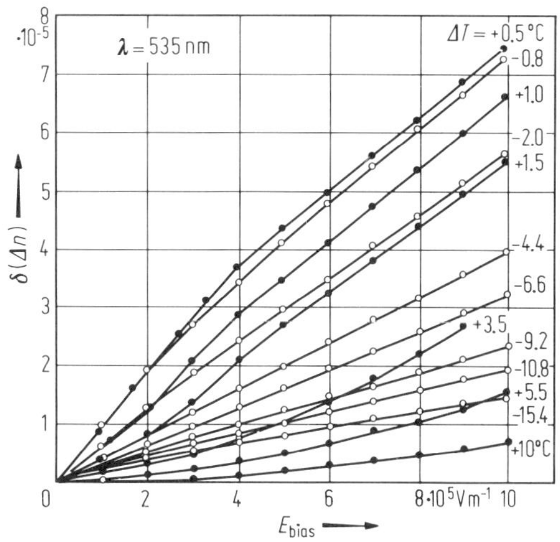

Fig. 1293. $(NH_2CH_2COOH)_3 \cdot H_2SO_4$ (TGS). $\delta(\Delta n)$ vs. E_{bias} [67S21]. $\delta(\Delta n) = \delta(n_\gamma - n_\alpha)$. $E_{bias} \| c$. Parameter: $\Delta T = T - \Theta_{II-I}$.

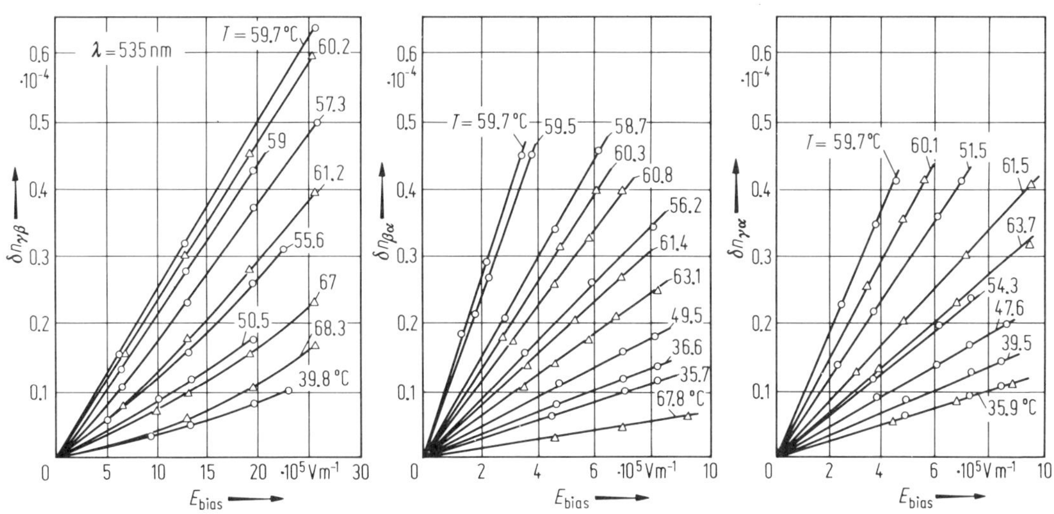

Fig. 1294. Deuterated triglycine sulfate. $\delta n_{\gamma\beta}$, $\delta n_{\beta\alpha}$, $\delta n_{\gamma\alpha}$ vs. E_{bias} [67S20]. $\delta n_{\lambda\mu} = \delta(n_\lambda - n_\mu)$. $E_{bias} \| b$.

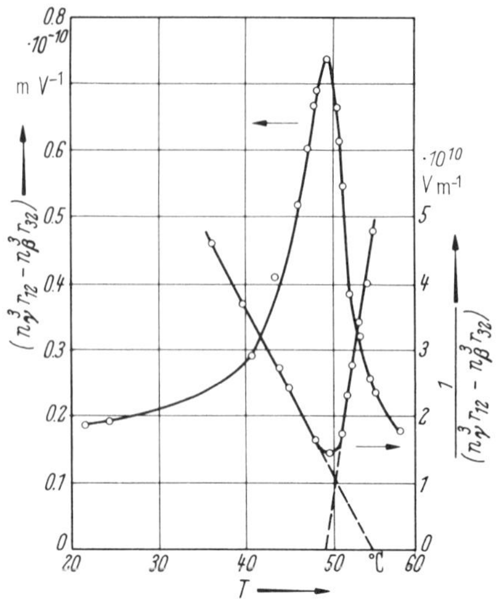

Fig. 1295. $(NH_2CH_2COOH)_3 \cdot H_2SO_4$ (TGS). $n_\gamma^3 r_{12} - n_\beta^3 r_{32}$ vs. T [66V4].

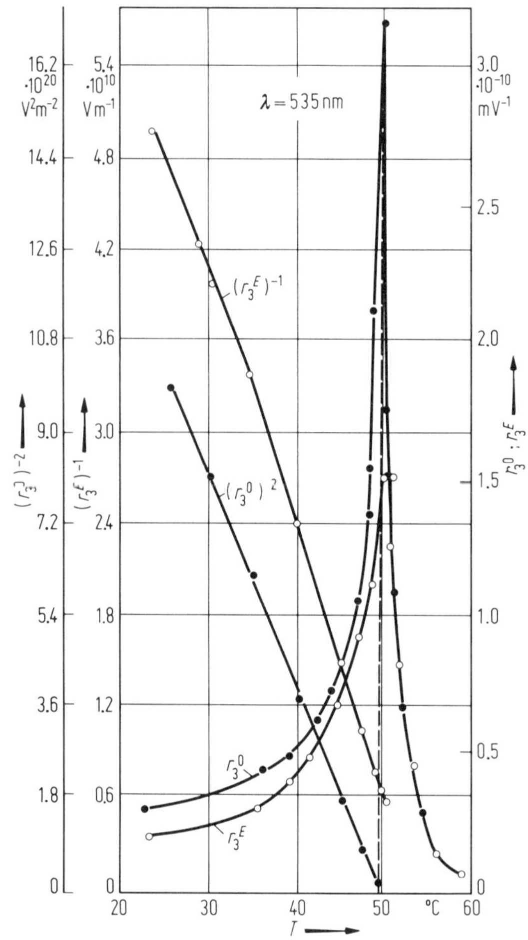

Fig. 1296. $(NH_2CH_2COOH)_3 \cdot H_2SO_4$ (TGS). r_3^0, r_3^E, $(r_3^0)^{-2}$, $(r_3^E)^{-1}$ vs. T [67S21]. $r_3 = n_\gamma^3 r_{12} - n_\alpha^3 r_{22}$. r_3^0: measured by weak fields. r_3^E: measured by strong fields.

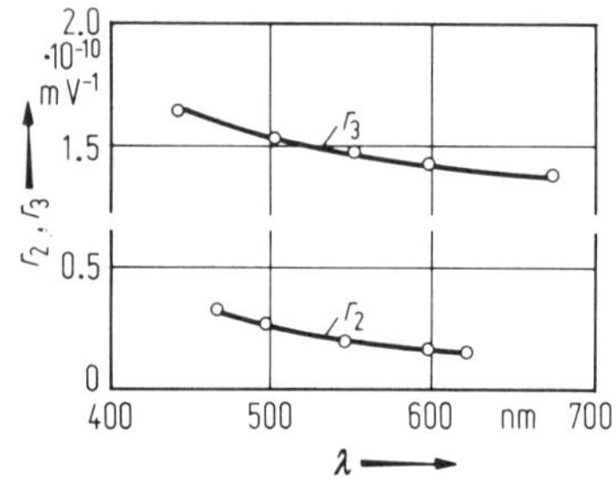

Fig. 1297. Deuterated triglycine sulfate. r_2, r_3 vs. λ [67S20]. $r_2 = n_\gamma^3 r_{12} - n_\beta^3 r_{32}$, $r_3 = n_\gamma^3 r_{12} - n_\alpha^3 r_{22}$.

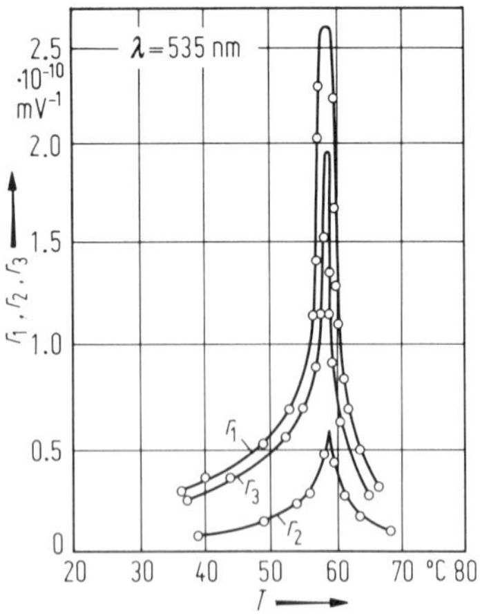

Fig. 1298. Deuterated triglycine sulfate. r_1, r_2, r_3 vs. T [67S20]. $r_1 = n_\beta^3 r_{32} - n_\alpha^3 r_{22}$, $r_2 = n_\gamma^3 r_{12} - n_\beta^3 r_{32}$, $r_3 = n_\gamma^3 r_{12} - n_\alpha^3 r_{22}$.

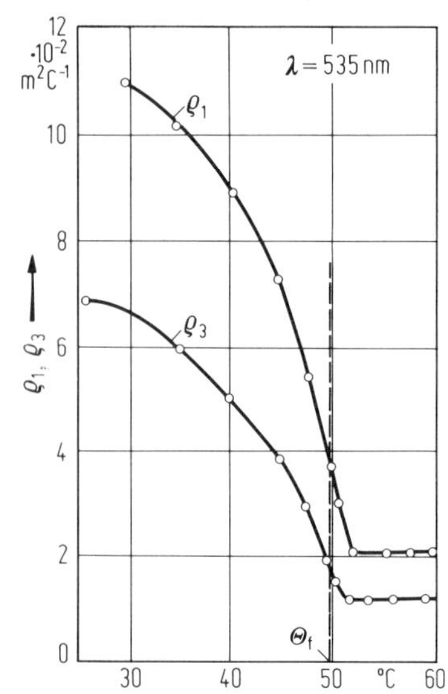

Fig. 1299. $(NH_2CH_2COOH)_3 \cdot H_2SO_4$ (TGS). ϱ_1, ϱ_3 vs. T [67S21]. $\varrho_1 = n_\beta^3 \varrho_{32} - n_\alpha^3 \varrho_{22}$, $\varrho_3 = n_\gamma^3 \varrho_{12} - n_\alpha^3 \varrho_{22}$.

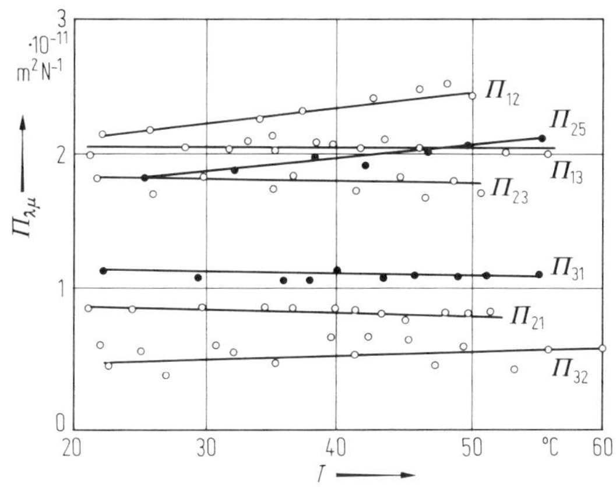

Fig. 1300. $(NH_2CH_2COOH)_3 \cdot H_2SO_4$ (TGS). $\Pi_{\lambda\mu}$ vs. T [67V2]. $\lambda = 535$ nm.

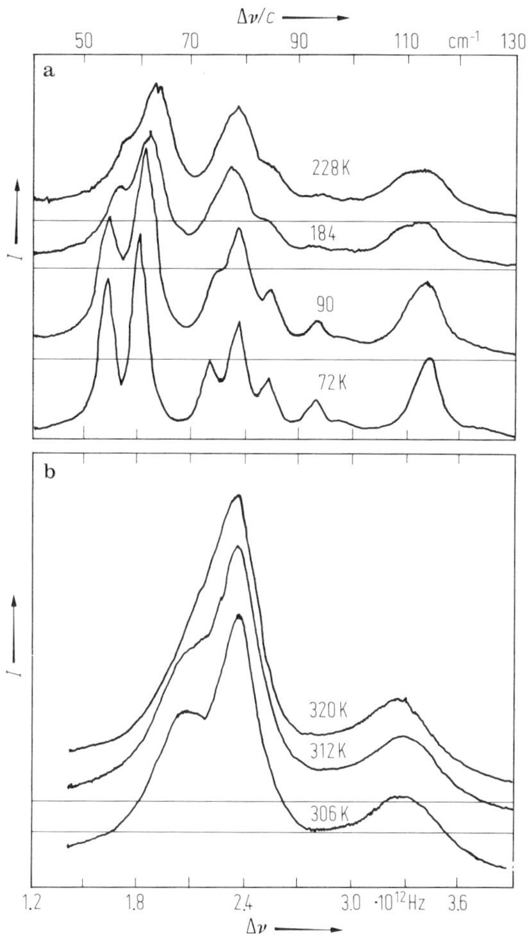

Fig. 1301. $(NH_2CH_2COOH)_3 \cdot H_2SO_4$ (TGS). I vs. $\Delta\nu$ [76C2]. I: Raman scattering intensity for the scattering geometry $X(Z,X)Z$. Parameter: T. Fig. a: $T <$ RT. Fig. b: $T >$ RT.

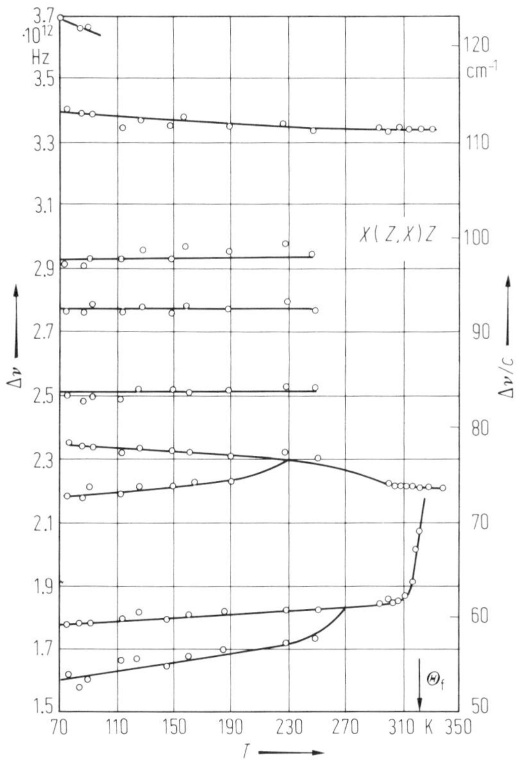

Fig. 1302. $(NH_2CH_2COOH)_3 \cdot H_2SO_4$ (TGS). $\Delta\nu$ vs. T [76C2]. $\Delta\nu$: Raman frequency shift in the scattering geometry of $X(Z,X)Z$.

Figs. 1303, 1304, see p. 622

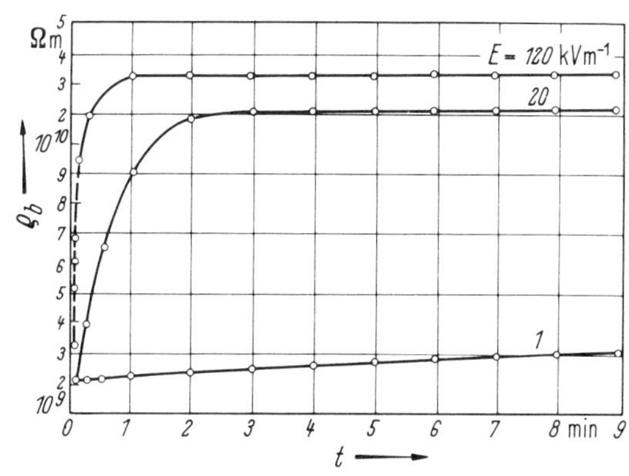

Fig. 1305. $(NH_2CH_2COOH)_3 \cdot H_2SO_4$ (TGS). ϱ_b vs. t [60G5]. Parameter: E. Crystal size $10 \times 10 \times 0.5$ mm^3. The low initial resistances are caused by changes in the domain structure.

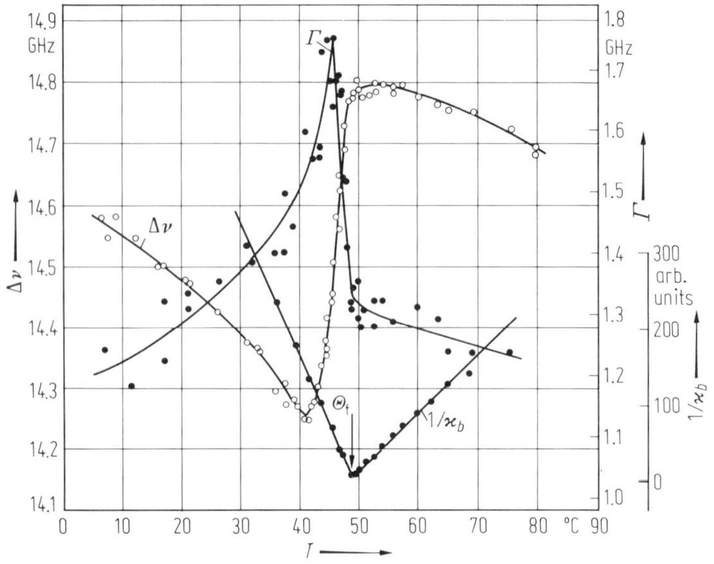

Fig. 1303. $(NH_2CH_2COOH)_3 \cdot H_2SO_4$ (TGS). $\Delta\nu$ vs. T, Γ vs. T, $\varkappa_b^{-1}$ vs. T [76Y1]. $\Delta\nu$: frequency shift of Brillouin scattering of c_{33} mode. Γ: full linewidth at half maximum.

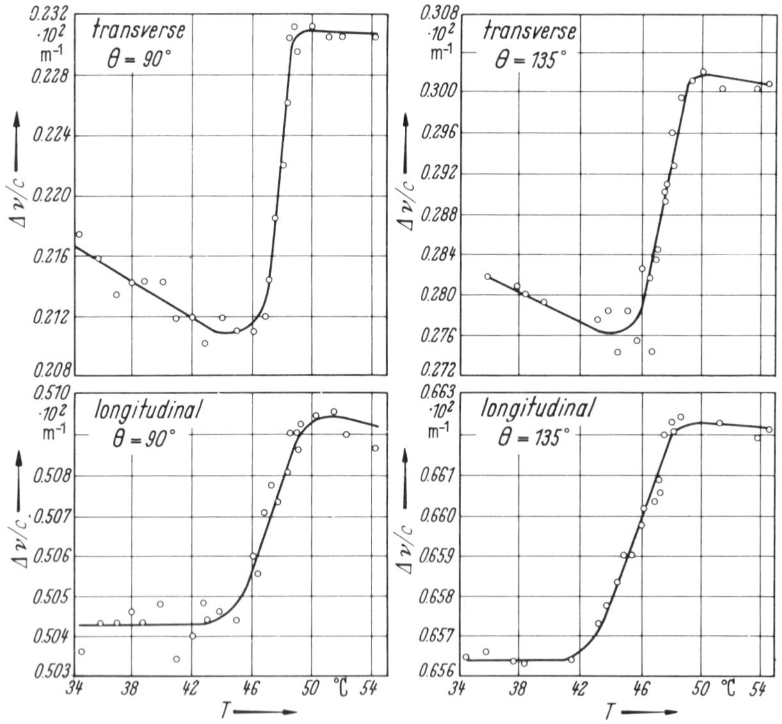

Fig. 1304. $(NH_2CH_2COOH)_3 \cdot H_2SO_4$ (TGS). $\Delta\nu/c$ vs. T [66G1]. $\Delta\nu/c$: Brillouin shift from visible He–Ne laser frequency. θ: scattering angle.

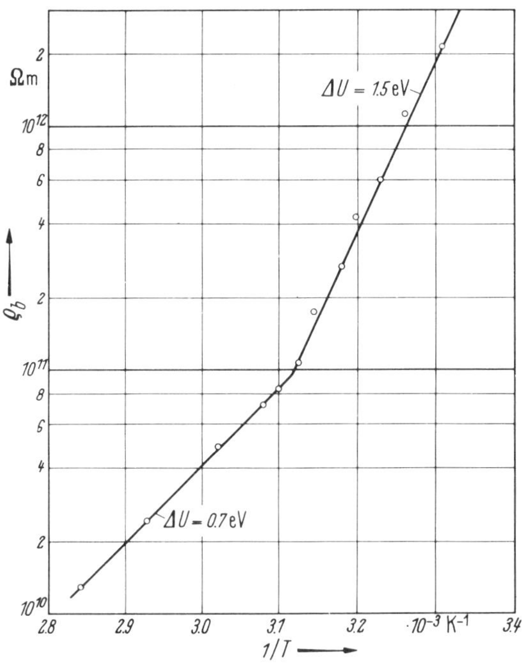

Fig. 1306. $(NH_2CH_2COOH)_3 \cdot H_2SO_4$ (TGS). ϱ_b vs. $1/T$ [61T2].

Fig. 1309, see p. 624

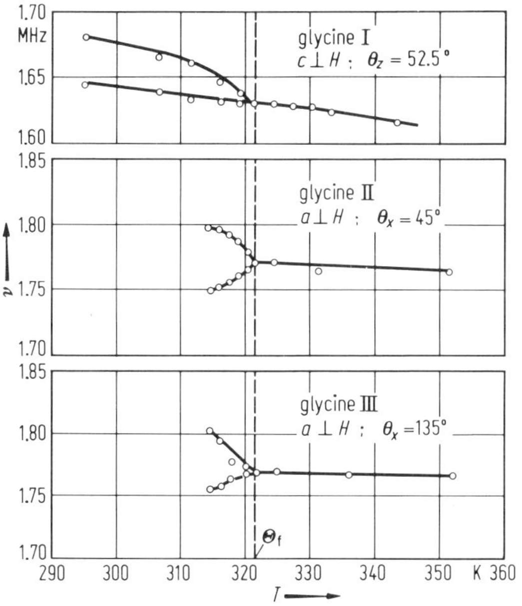

Fig. 1308. $(NH_2CH_2COOH)_3 \cdot H_2SO_4$ (TGS). ν vs. T [71B17]. ν: double NMR $^1H-^{14}N$, ^{14}N, transition frequency. θ_x: ***H*** direction measured from the c'-axis in the bc' plane. θ_z: ***H*** direction measured from the b-axis in the a^*b plane. $\boldsymbol{a}^* = (a' \sin\beta', 0, 0)$.

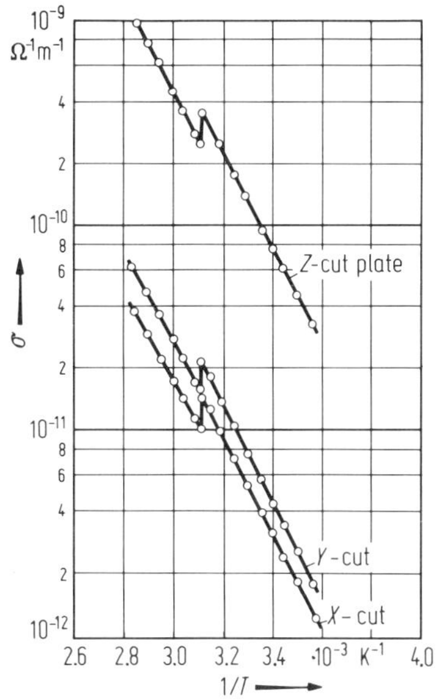

Fig. 1307. $(NH_2CH_2COOH)_3 \cdot H_2SO_4$ (TGS). σ vs. $1/T$ for X-cut, Y-cut and Z-cut single crystals [62G3]. Parameter: electric field direction. Three-electrode system with shielding electrode was adopted. $E = 120$ kV m^{-1}. Thickness of samples were 1 mm.

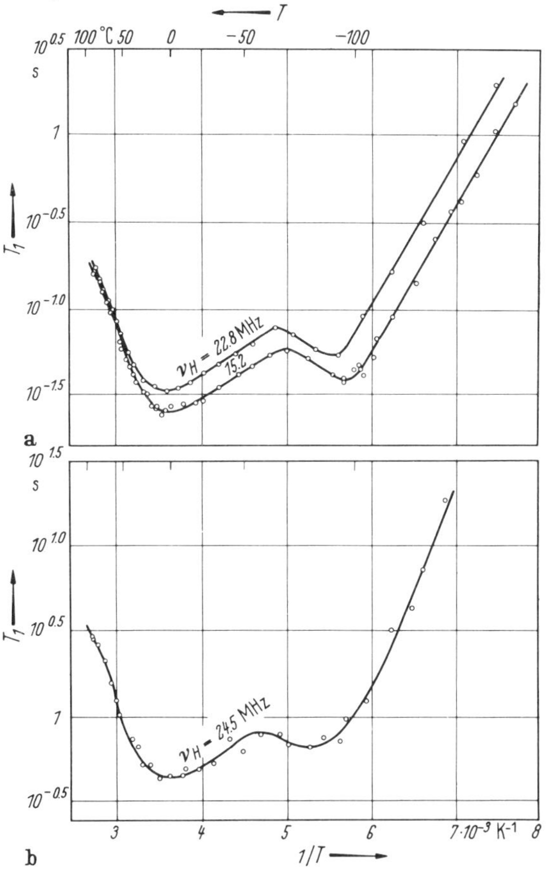

Fig. 1310. $(NH_2CH_2COOH)_3 \cdot H_2SO_4$ (TGS). T_1 vs. $1/T$ [66B5]. T_1: proton spin-lattice relaxation time. Fig. a: powdered TGS. Fig. b: partially deuterated TGS. ν_H: Larmor frequency of proton.

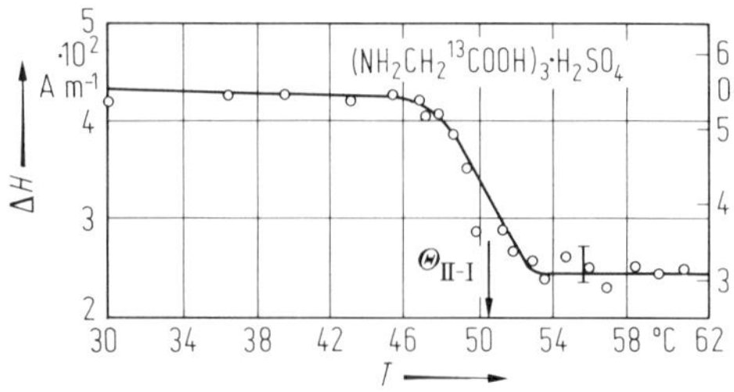

Fig. 1309. $(NH_2CH_2COOH)_3 \cdot H_2SO_4$ (TGS). ΔH vs. T [76M22]. ΔH: NMR line width of ^{13}C.

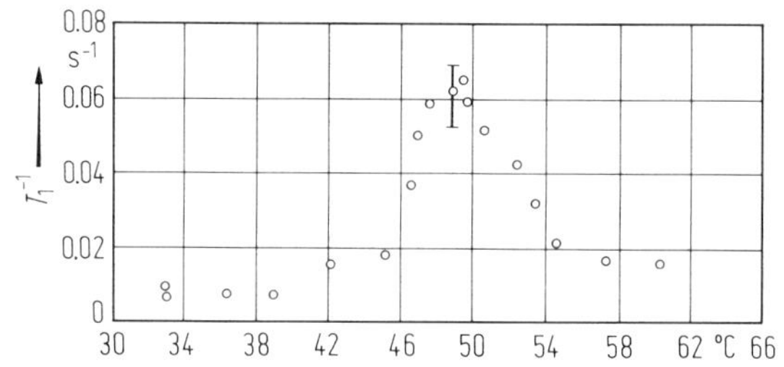

Fig. 1311. $(NH_2CH_2COOH)_3 \cdot H_2SO_4$ (TGS). T_1^{-1} vs. T [76M21]. T_1: spin lattice relaxation time of ^{13}C at 22.63 MHz.

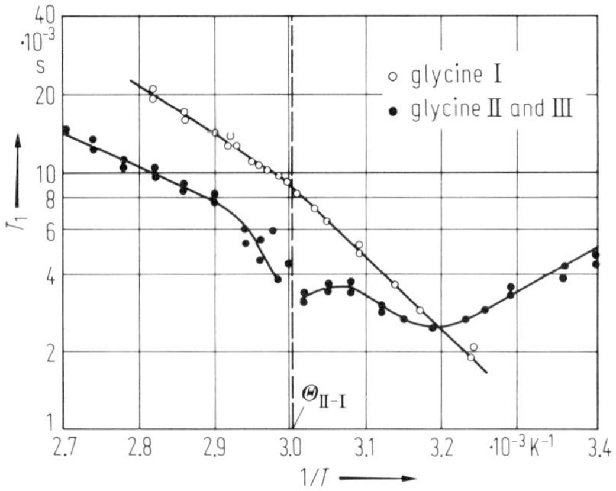

Fig. 1312. Deuterated triglycine sulfate. T_1 vs. T^{-1} [75S31]. T_1: ND_3 deuteron spin-lattice relaxation time.

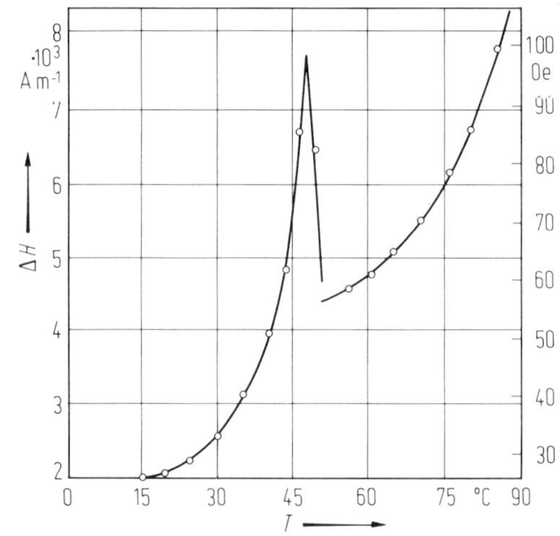

Fig. 1313. $(NH_2CH_2COOH)_3 \cdot H_2SO_4{:}Cr^{3+}$. ΔH vs. T [78O8]. ΔH: linewidth of the resonance due to the $M_s = 1/2$ to $M_s = 3/2$ transition. dc magnetic field is parallel to the z-axis of the spin Hamiltonian tensor.

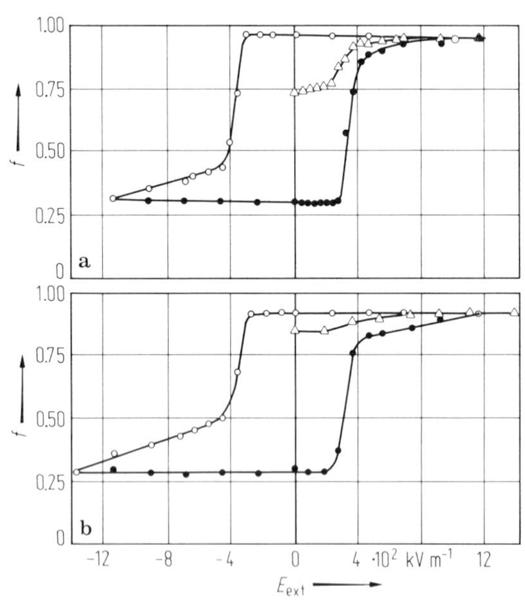

Fig. 1314. $(NH_2CH_2COOH)_3 \cdot H_2SO_4{:}Cr^{3+}$. f vs. E_{ext} [74W2]. $f = I_1/(I_1 + I_2)$. I_1, I_2: normalized intensities of the $M = -1/2 \leftrightarrow -3/2$ transition, which correspond to the opposite polarized domains, respectively. $E_{ext} \| [010]$, $\boldsymbol{H} \perp c' \sin\beta'$. Fig. a: -30 °C; Fig. b: -60 °C.

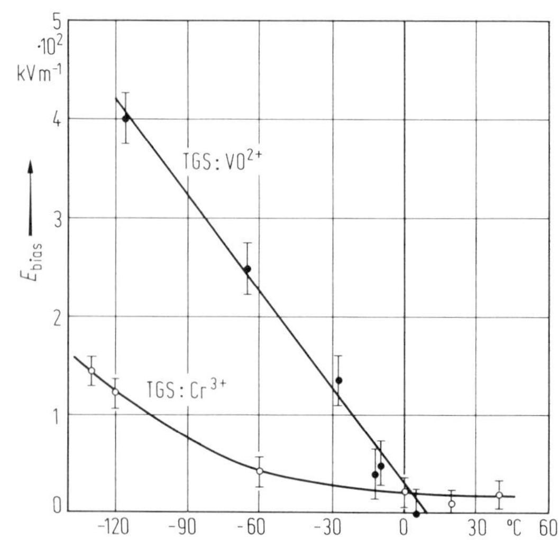

Fig. 1315. $(NH_2CH_2COOH)_3 \cdot H_2SO_4{:}VO^{2+}$; $:Cr^{3+}$. E_{bias} vs. T [78W4]. E_{bias}: bias field of ESR intensity hysteresis loop vs. temperature.

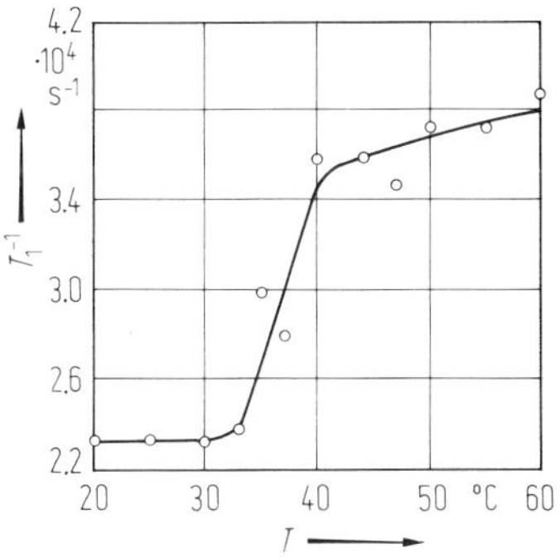

Fig. 1316. $(NH_2CH_2COOH)_3 \cdot H_2SO_4$ (TGS). T_1^{-1} vs. T of X-irradiated crystal [74V2]. T_1: electron spin-lattice relaxation time of radical produced by X-irradiation.

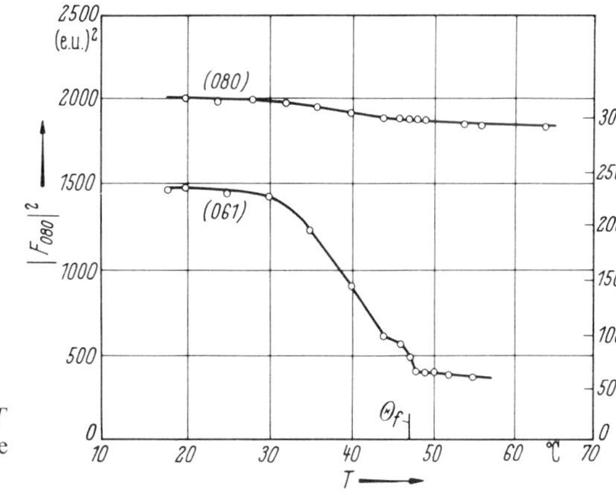

Fig. 1317. $(NH_2CH_2COOH)_3 \cdot H_2SO_4$(TGS). $|F_{080}|^2$, $|F_{061}|^2$ vs. T [61S7]. The Miller indices should follow the crystallographic axes chosen by Hoshino et al., see 3a. F_{hkl}: structure factors in e.u.: electron unit.

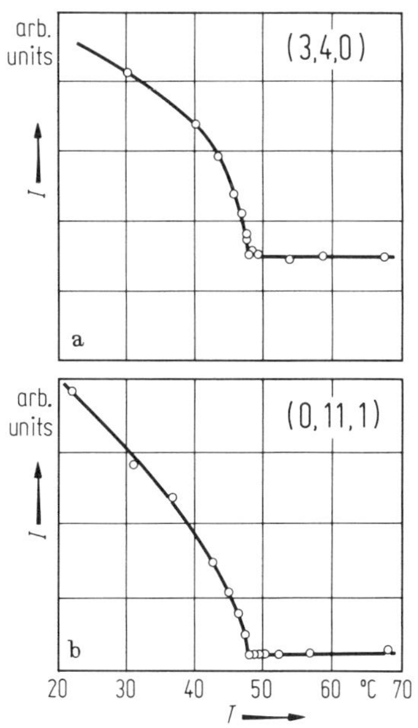

Fig. 1318. $(NH_2CH_2COOH)_3 \cdot H_2SO_4$ (TGS). I vs. T [71F5]. Fig. a: (3,4,0) reflection. Fig. b: (0,11,1) reflection. I: integrated intensity of Bragg reflection. The Miller indices should follow the crystallographic axes chosen by Hoshino et al., see 3a.

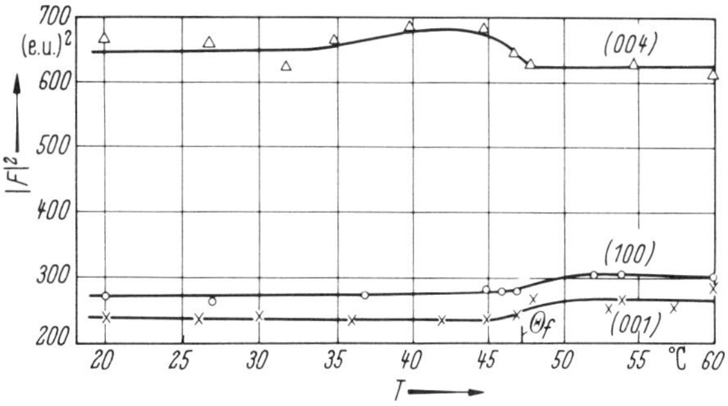

Fig. 1319. $(NH_2CH_2COOH)_3 \cdot H_2SO_4$(TGS). $|F_{004}|^2$, $|F_{100}|^2$, $|F_{001}|^2$ vs. T [61S7]. The Miller indices should follow the crystallographic axes chosen by Hoshino et al., see 3a. (see Fig. 1220). e.u.: electron unit.

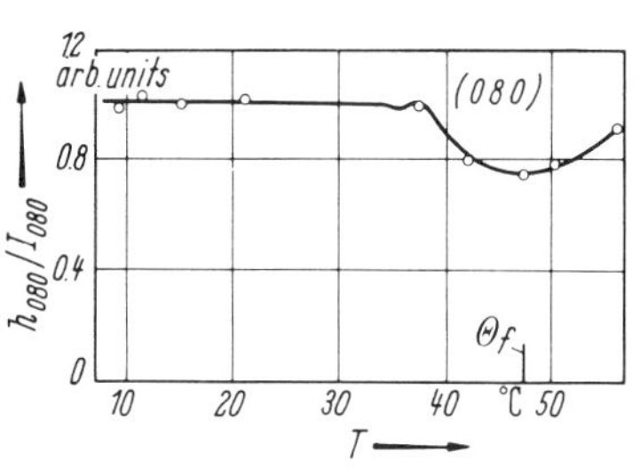

Fig. 1320. $(NH_2CH_2COOH)_3 \cdot H_2SO_4$ (TGS). h_{080}/I_{080} vs. T [61S7]. h_{080}: peak height of (080) Bragg reflection, I: integral intensity.

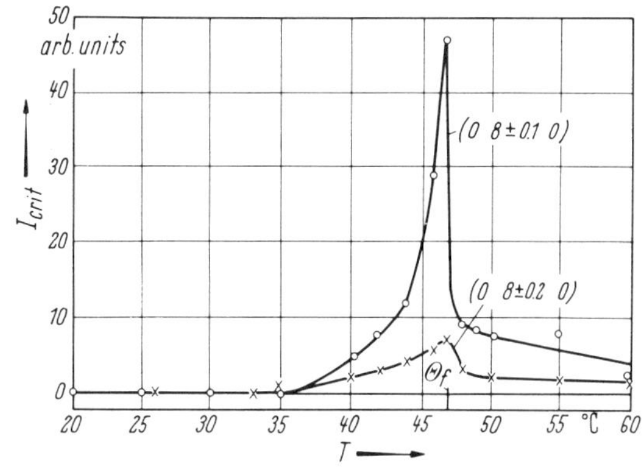

Fig. 1321. $(NH_2CH_2COOH)_3 \cdot H_2SO_4$ (TGS). I_{crit} at (0, 8 ± 0.1, 0) and (0, 8 ± 0.2, 0) vs. T [61S7]. I_{crit}: critical X-ray scattering intensity.

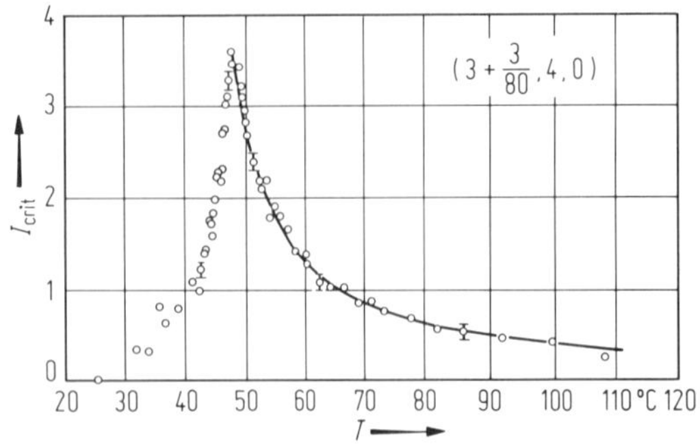

Fig. 1322. $(NH_2CH_2COOH)_3 \cdot H_2SO_4$ (TGS). I_{crit} at (3+3/80, 4, 0) vs. T [71F5]. I_{crit}: critical X-ray scattering intensity. The Miller indices should follow the crystallographic axes chosen by Hoshino et al., see 3a.

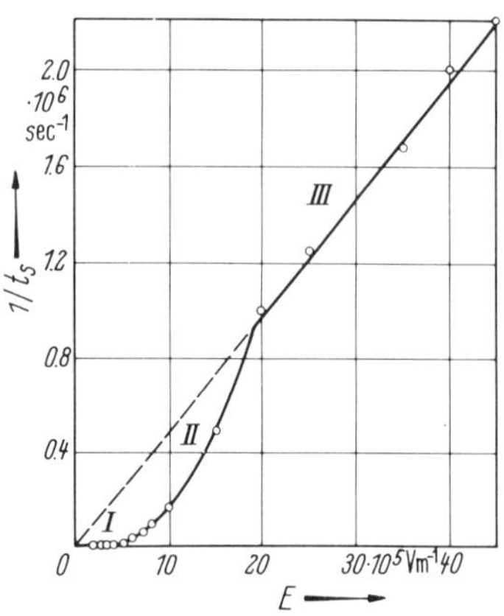

Fig. 1324. $(NH_2CH_2COOH)_3 \cdot H_2SO_4$ (TGS). $1/t_s$ vs. E [59F2]. I, II, and III merely distinguishes three parts of the curve of different behaviour.

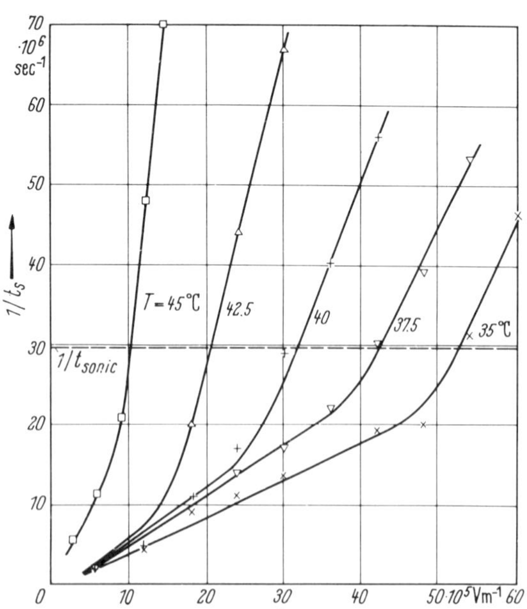

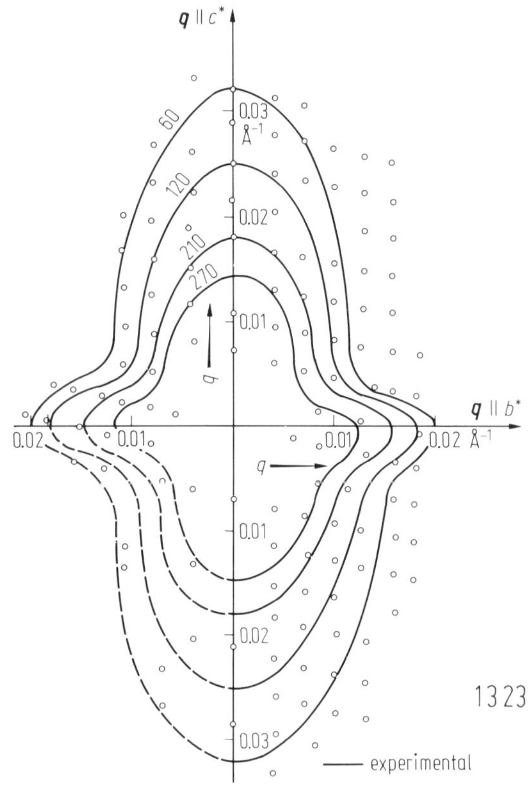

Fig. 1323. $(NH_2CH_2COOH)_3 \cdot H_2SO_4$ (TGS). X-ray diffuse scattering at 48.4 °C [73P15]. Constant intensity curves in the reciprocal lattice space around (040). The dots denote points where intensities were measured. The numbers at the curve give counts for 5 min. Transition temperature of the sample is 48 °C.

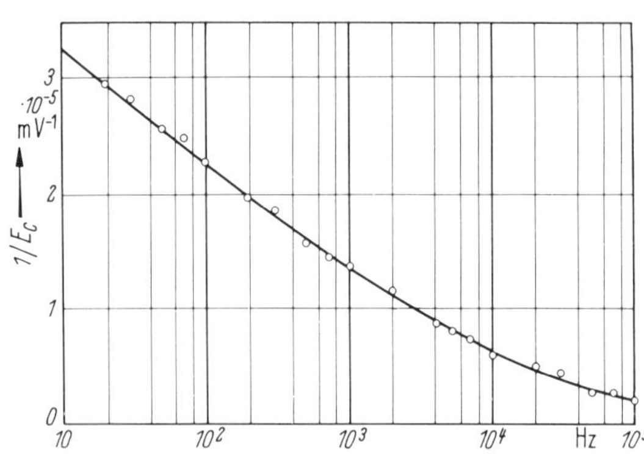

Fig. 1326. $(NH_2CH_2COOH)_3 \cdot H_2SO_4$ (TGS). E_c^{-1} vs. f [58P8].

◄

Fig. 1325. $(NH_2CH_2COOH)_3 \cdot H_2SO_4$ (TGS). $1/t_s$ vs. E [65B5]. Parameter: T. t_s: switching time. t_{sonic}: transit time of a longitudinal sound wave. Transit time: time necessary for growing of domain walls from one electrode to the opposite electrode.

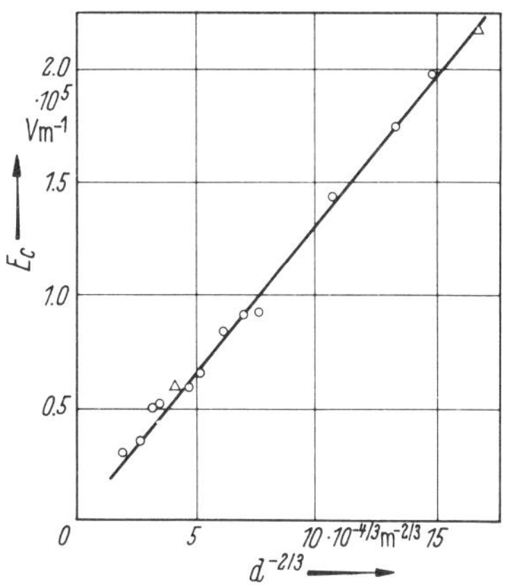

Fig. 1327. $(NH_2CH_2COOH)_3 \cdot H_2SO_4$ (TGS). E_c vs. $d^{-2/3}$ [62K5]. d: crystal thickness.

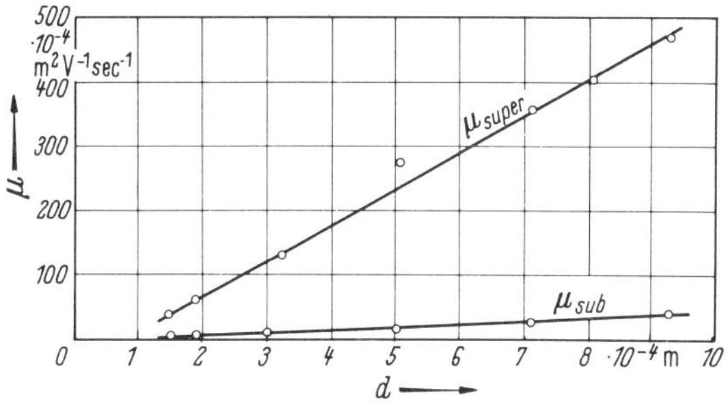

Fig. 1328. $(NH_2CH_2COOH)_3 \cdot H_2SO_4$ (TGS). μ vs. d [65B5]. μ_{sub}: subsonic wall mobility, μ_{super}: supersonic wall mobility. d: crystal thickness.

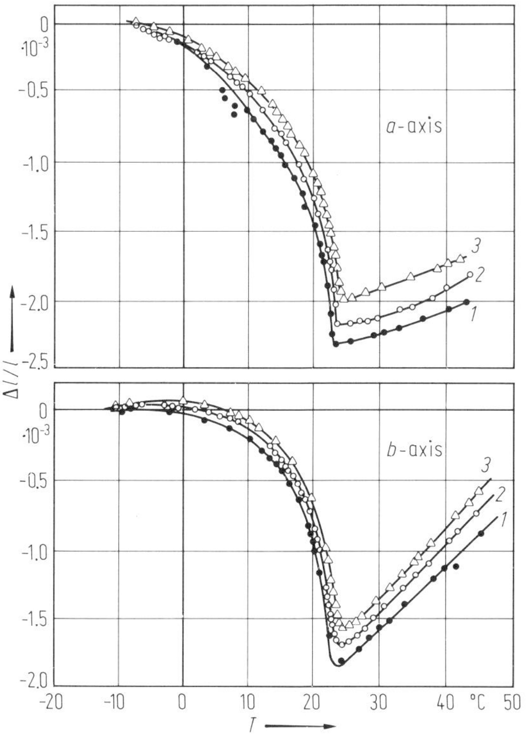

Fig. 1329. $(NH_2CH_2COOH)_3 \cdot H_2SeO_4$ (TGSe). $\Delta l/l$ vs. T [73V1]. $\Delta l/l$: linear thermal expansion at various bias fields E. Curve *1*: $E=0$; *2*: $E=50$ kV m^{-1}; *3*: $E=200$ kV m^{-1}.

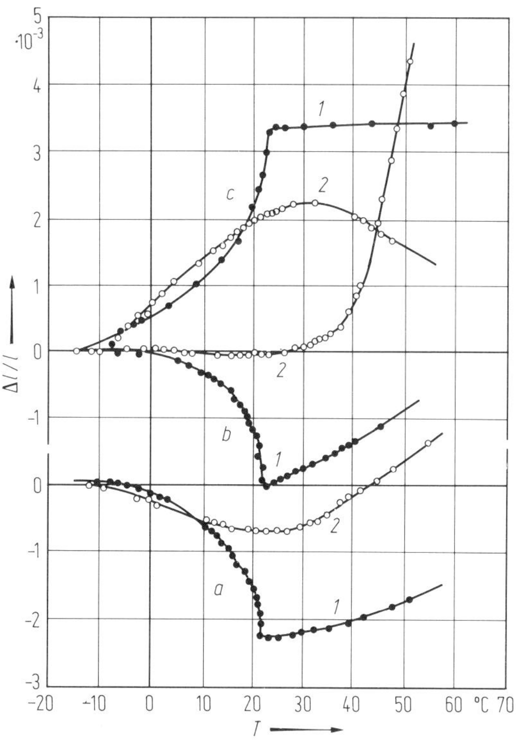

Fig. 1330. $(NH_2CH_2COOH)_3 \cdot H_2SeO_4$ (TGSe). $\Delta l/l$ vs. T [73V1]. $\Delta l/l$: linear thermal expansion along the a, b, c axes. Curves *1*: before γ-irradiation. Curves *2*: after γ-irradiation of 500 mR.

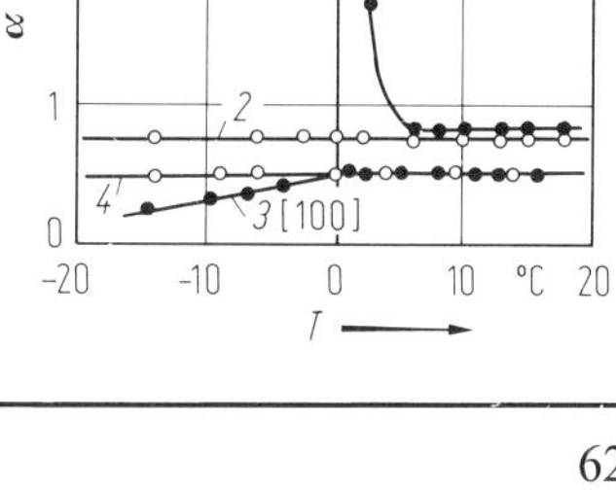

Fig. 1331. $(NH_2CH_2COOH)_3 \cdot H_2SeO_4$ (TGSe). α vs. T [73V1]. α: linear thermal expansion coefficients along the [100] and [001] directions. Curves *1*, *3*: before γ-irradiation. Curves *2*, *4*: after γ-irradiation with the doses 500 mR.

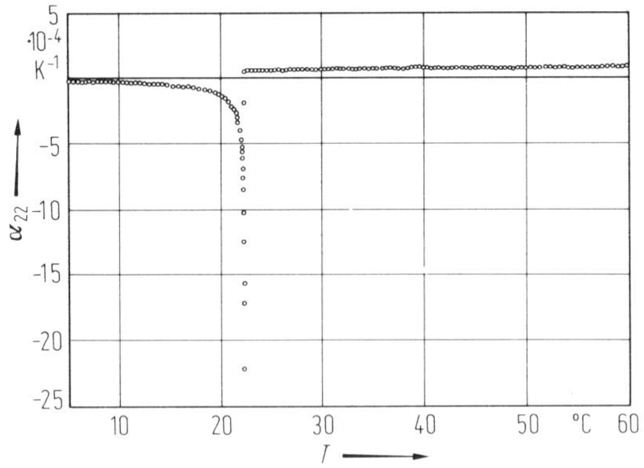

Fig. 1332. $(NH_2CH_2COOH)_3 \cdot H_2SeO_4$ (TGSe). α_{22} vs. T [79S4]. α_{22}: linear thermal expansion coefficient along the b-axis.

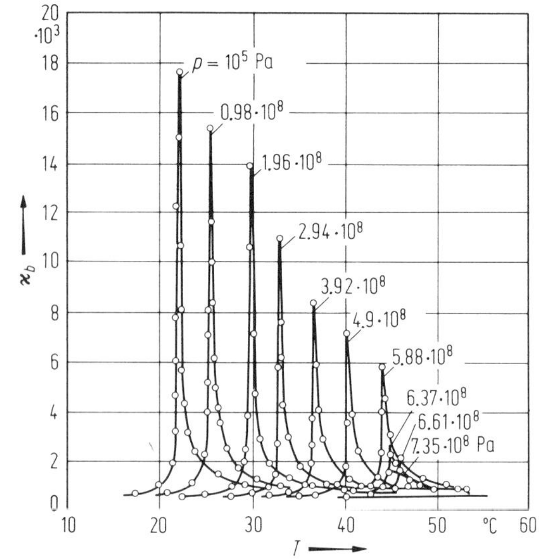

Fig. 1333. $(NH_2CH_2COOH)_3 \cdot H_2SeO_4$ (TGSe). $\varkappa_b$ vs. T [68P3]. Parameter: p.

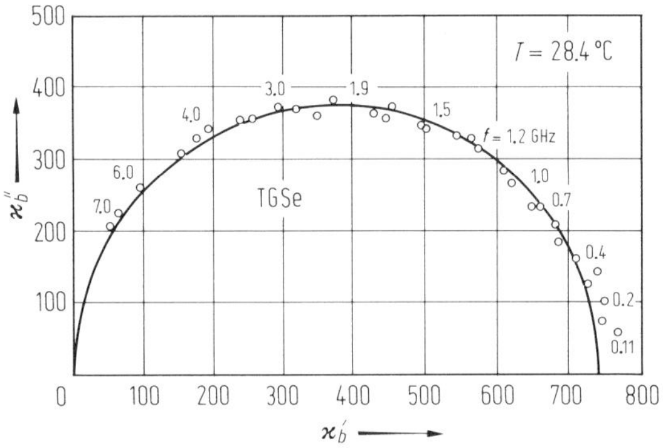

Fig. 1334. $(NH_2CH_2COOH)_3 \cdot H_2SeO_4$ (TGSe). Cole-Cole plot of the complex dielectric constant [74L14].

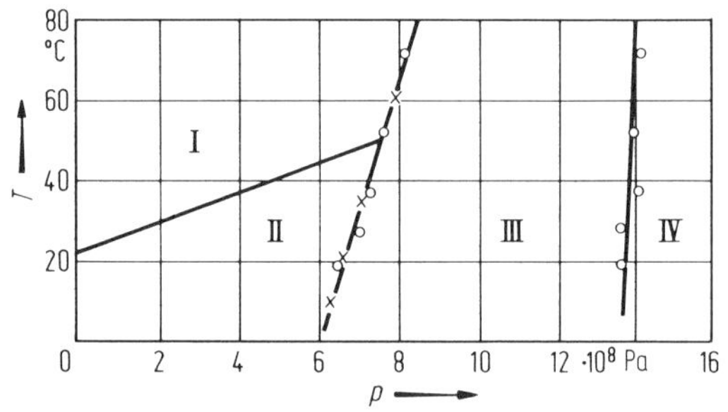

Fig. 1336. $(NH_2CH_2COOH)_3 \cdot H_2SeO_4$ (TGSe). Phase diagram [70M12].

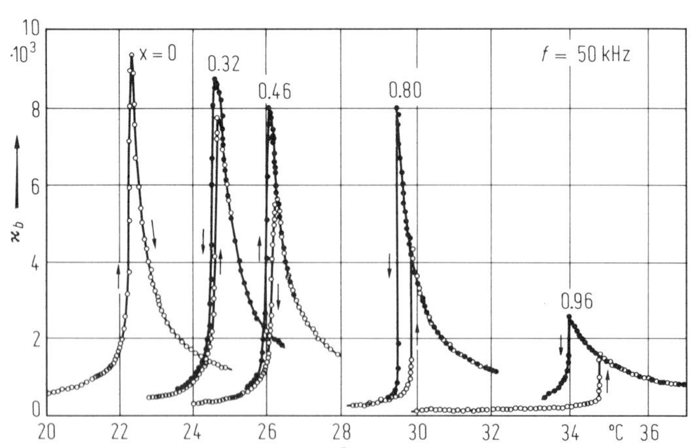

Fig. 1335. $[(NH_2CH_2COOH)_3 \cdot H_2SeO_4]_{1-x} \cdot [(ND_2CH_2COOD)_3 \cdot D_2SO_4]_x$. $\varkappa_b$ vs. T [76G7]. Parameter: x.

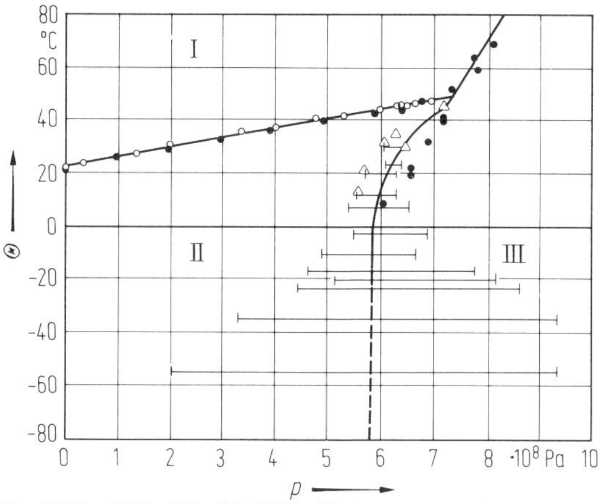

Fig. 1337. $(NH_2CH_2COOH)_3 \cdot H_2SeO_4$ (TGSe). Θ vs. T. Full circles, triangles [68 P 3]; open circles, bars [74 G 6].

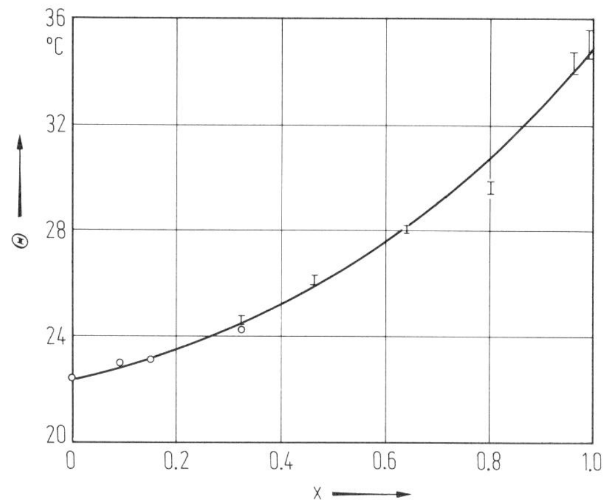

Fig. 1338. $[(NH_2CH_2COOH)_3 \cdot H_2SeO_4]_{1-x} \cdot [(ND_2CH_2COOD)_3 \cdot D_2SeO_3]_x$. Θ vs. x [76 G 7]. Vertical bars show thermal hysteresis of the ferroelectric transition.

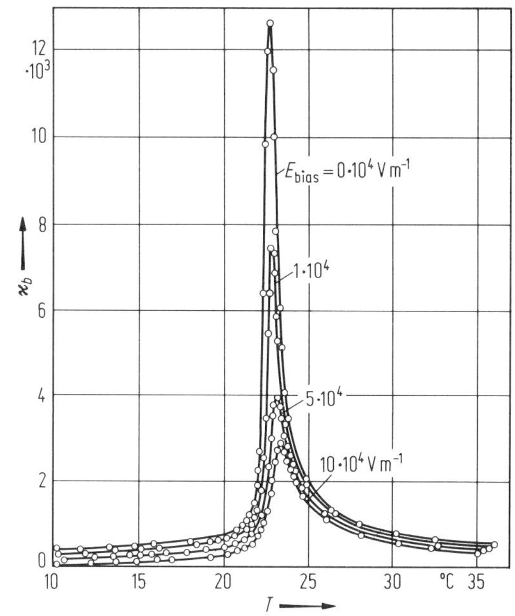

Fig. 1339. $(NH_2CH_2COOH)_3 \cdot H_2SeO_4$ (TGSe). $\varkappa_b$ vs. T [71 Z 2]. Parameter: E_{bias}.

Fig. 1340, see p. 630

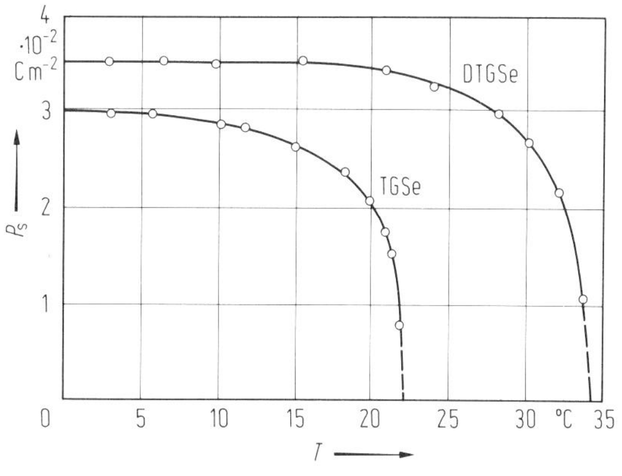

Fig. 1341. $(NH_2CH_2COOH)_3 \cdot H_2SeO_4$ (TGSe). Deuterated triglycine selenate. P_s vs. T [68 M 1].

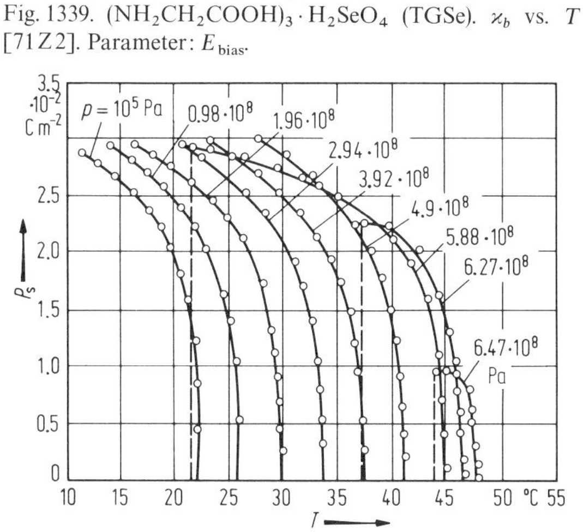

Fig. 1342. $(NH_2CH_2COOH)_3 \cdot H_2SeO_4$ (TGSe). P_s vs. T [68 P 1]. Parameter: p.

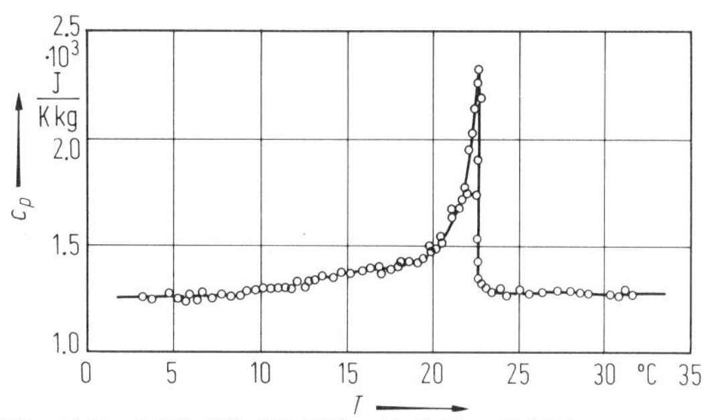

Fig. 1343. $(NH_2CH_2COOH)_3 \cdot H_2SeO_4$ (TGSe). c_p vs. T [68 S 22]. c_p: specific heat capacity at constant pressure.

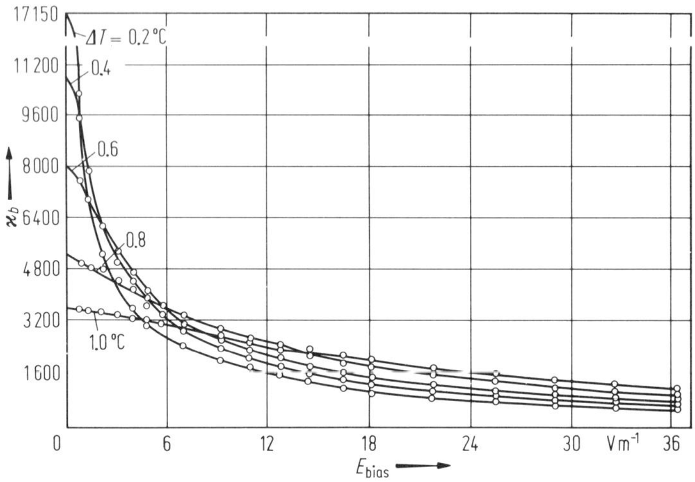

Fig. 1340. $(NH_2CH_2COOH)_3 \cdot H_2SeO_4$ (TGSe). $\varkappa_b$ vs. E_{bias} [71 Z 2]. Parameter: $\Delta T = T - \Theta_f$.

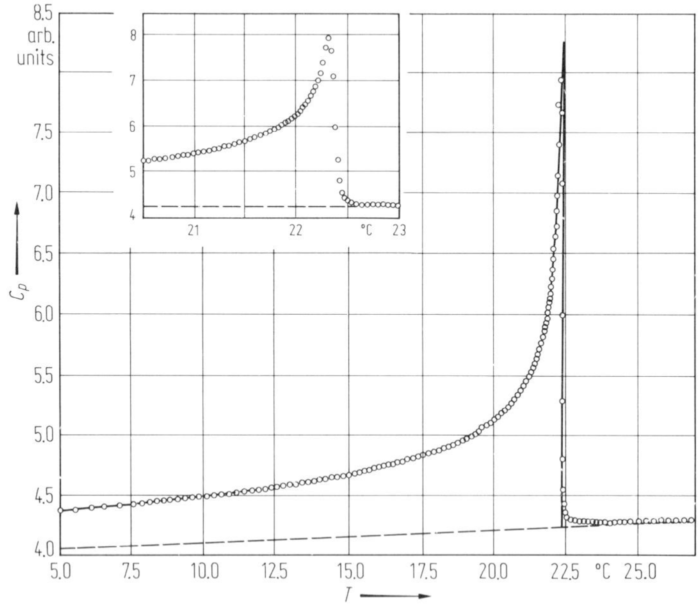

Fig. 1344. $(NH_2CH_2COOH)_3 \cdot H_2SeO_4$ (TGSe). C_p vs. T [77 E 2].

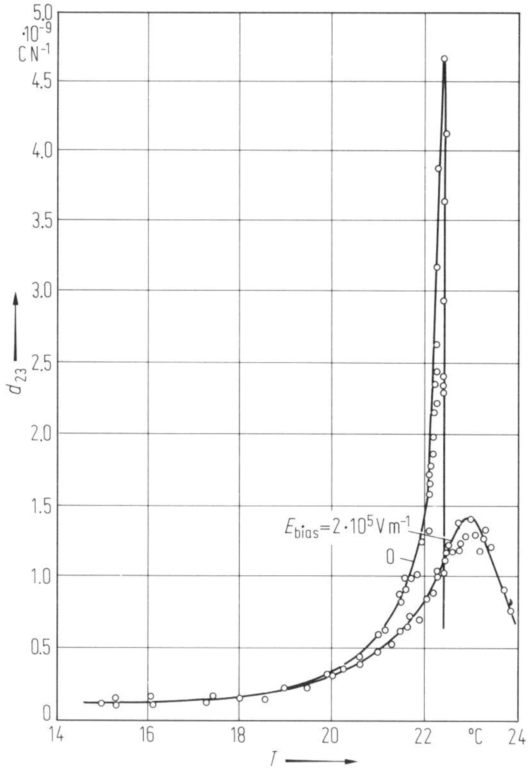

Fig. 1345. $(NH_2CH_2COOH)_3 \cdot H_2SeO_4$ (TGSe). d_{23} vs. T [74V1].

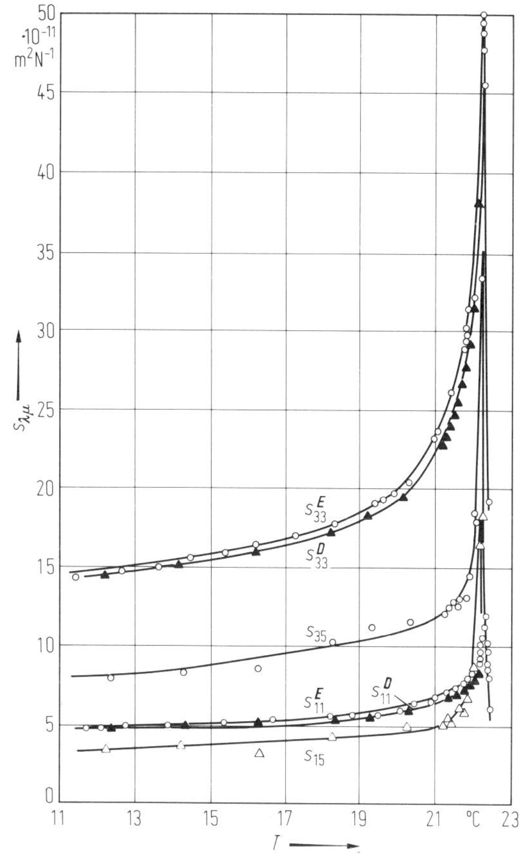

Fig. 1346. $(NH_2CH_2COOH)_3 \cdot H_2SeO_4$ (TGSe). $s_{\lambda\mu}$ vs. T [74V1].

For Fig. 1348, see p. 632

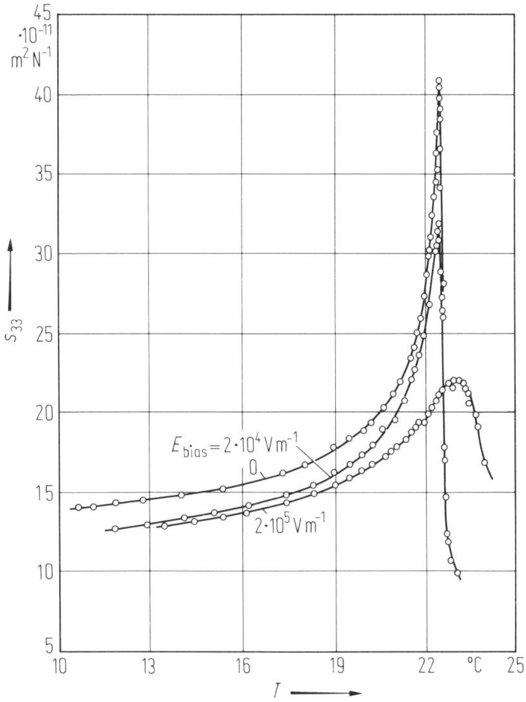

Fig. 1347. $(NH_2CH_2COOH)_3 \cdot H_2SeO_4$ (TGSe). s_{33} vs. T [74V1]. Parameter: E_{bias}.

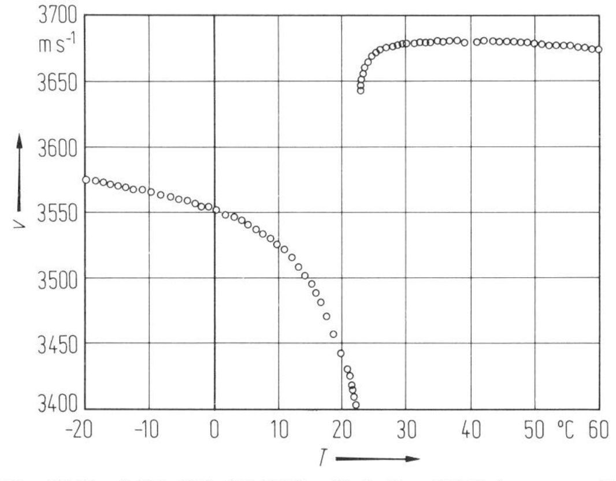

Fig. 1349. $(NH_2CH_2COOH)_3 \cdot H_2SeO_4$ (TGSe). v vs. T [75T12]. v: velocity of the longitudinal sound waves propagating along the c-axis. $f = 10$ MHz.

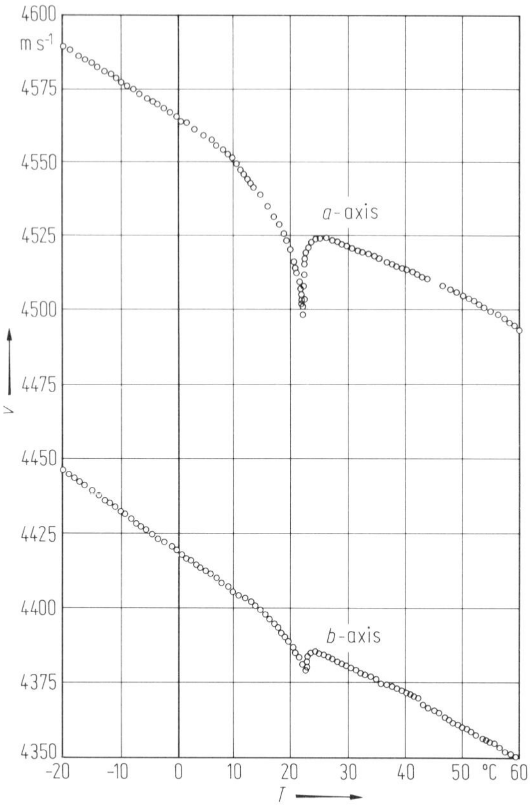

Fig. 1348. $(NH_2CH_2COOH)_3 \cdot H_2SeO_4$ (TGSe). v vs. T [75T12]. v: velocity of the longitudinal sound waves propagating along the a- and b-axes. $f = 10$ MHz.

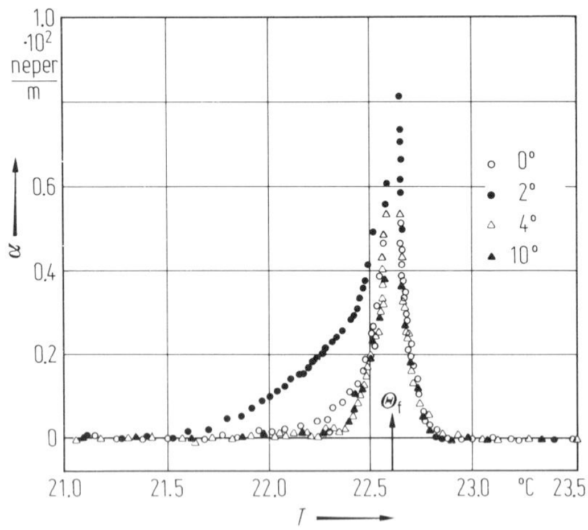

Fig. 1351. $(NH_2CH_2COOH)_3 \cdot H_2SeO_4$ (TGSe). α vs. T [74T7]. α: attenuation coefficients of the quasi-longitudinal ultrasound propagating in the bc plane with angles 0, 2, 4, and 10° to the c-axis. $f = 10$ MHz.

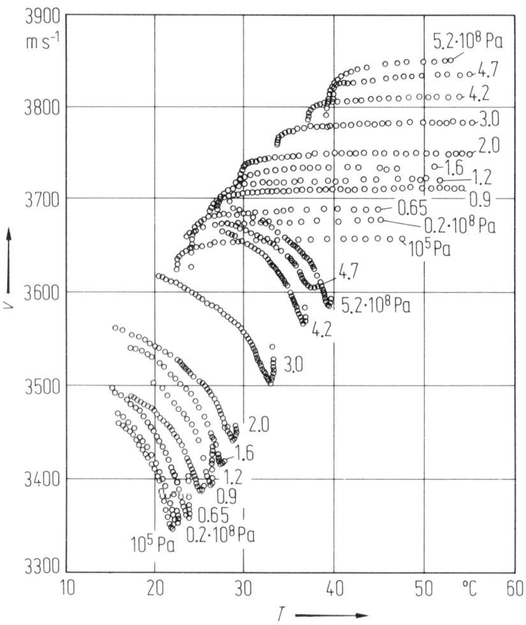

Fig. 1350. $(NH_2CH_2COOH)_3 \cdot H_2SeO_4$ (TGSe). v vs. T [78Y2]. v: longitudinal sound velocity propagating along the c-axis at various hydrostatic pressures. $f = 10$ MHz.

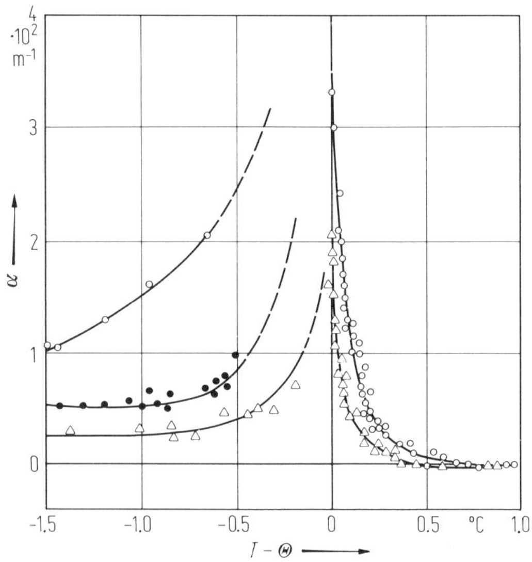

Fig. 1352. $(NH_2CH_2COOH)_3 \cdot H_2SeO_4$ (TGSe). α vs. $T - \Theta$ [67M6]. α: absorption coefficients of the longitudinal ultrasound propagating along the c-axis. Triangles: 5 MHz; full circles: 10 MHz; open circles: 15 MHz.

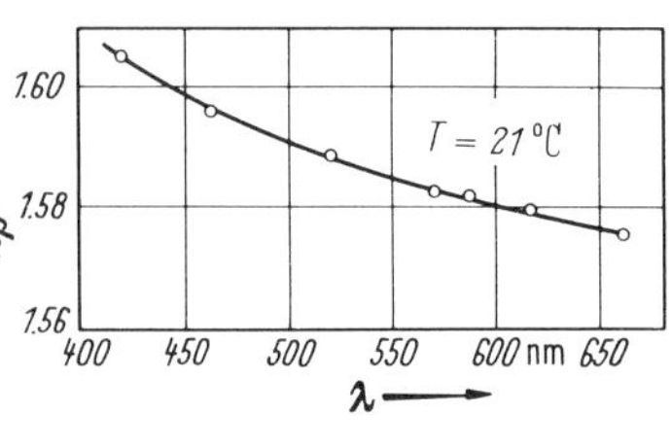

Fig. 1353. $(NH_2CH_2COOH)_3 \cdot H_2SeO_4$ (TGSe). n_β vs. λ [66I6].

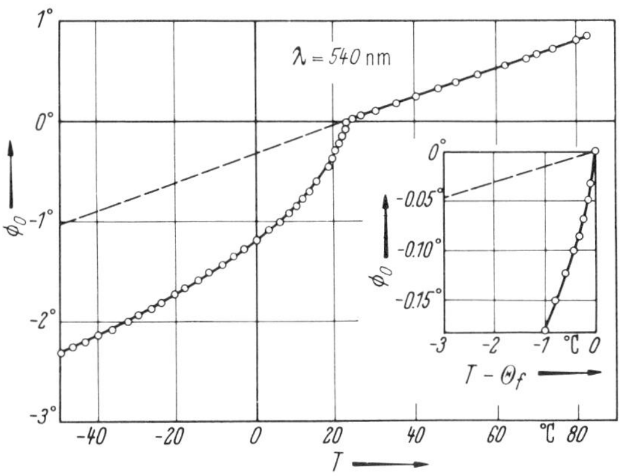

Fig. 1354. $(NH_2CH_2COOH)_3 \cdot H_2SeO_4$ (TGSe). ϕ_o vs. T [6616]. ϕ_o: rotation angle of the optical indicatrix around the b-axis.

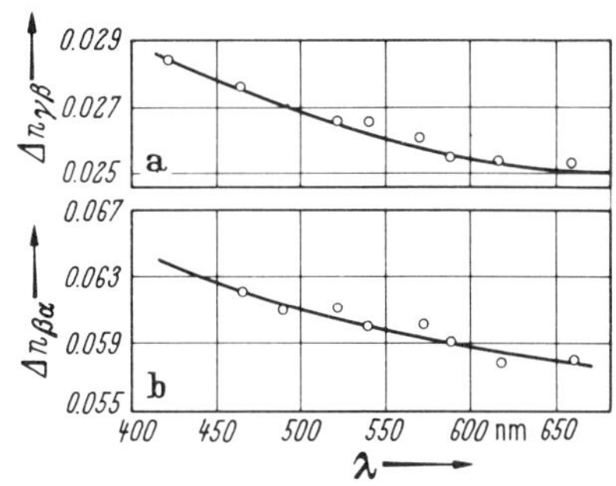

Fig. 1355. $(NH_2CH_2COOH)_3 \cdot H_2SeO_4$ (TGSe). $\Delta n_{\gamma\beta}$, $\Delta n_{\beta\alpha}$ vs. λ [6616]. a: $\Delta n_{\gamma\beta} = n_\gamma - n_\beta$. b: $\Delta n_{\beta\alpha} = n_\beta - n_\alpha$. $T = 21\ ^\circ C$.

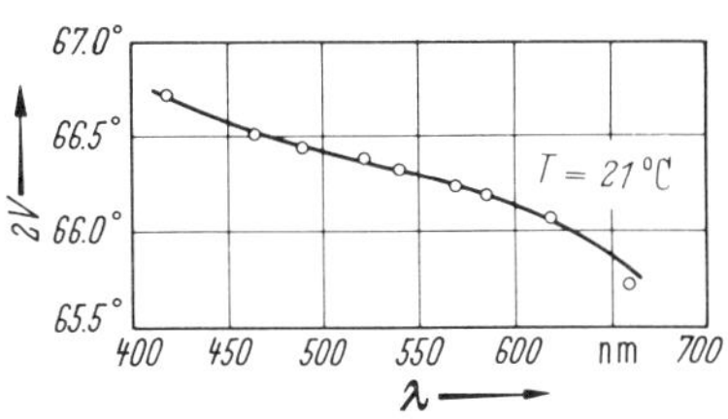

Fig. 1356. $(NH_2CH_2COOH)_3 \cdot H_2SeO_4$ (TGSe). $2V$ vs. λ [6616]. $2V$: optical axial angle.

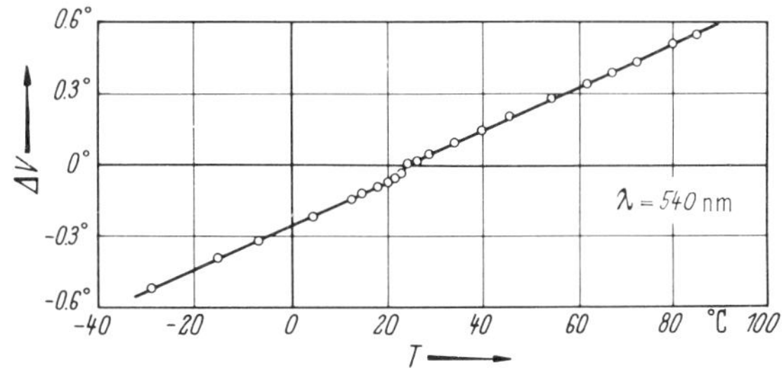

Fig. 1357. $(NH_2CH_2COOH)_3 \cdot H_2SeO_4$ (TGSe). ΔV vs. T [6615]. $\Delta V = V(T) - V(\Theta_{\text{II}-\text{I}})$. $2V(T)$: optical axial angle at T(°C).

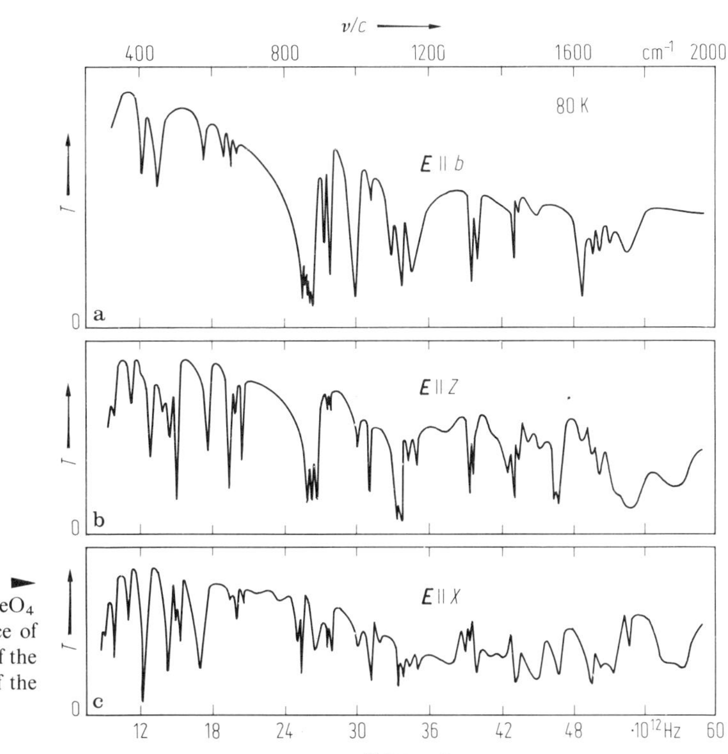

Fig. 1358. $(NH_2CH_2COOH)_3 \cdot H_2SeO_4$ (TGSe). T vs. ν [77G2]. T: transmittance of infrared radiation at 80 K. Polarization of the light: (a) ∥ b, (b) ∥ Z, (c) ∥ X. Thickness of the sample is less than 1 μm.

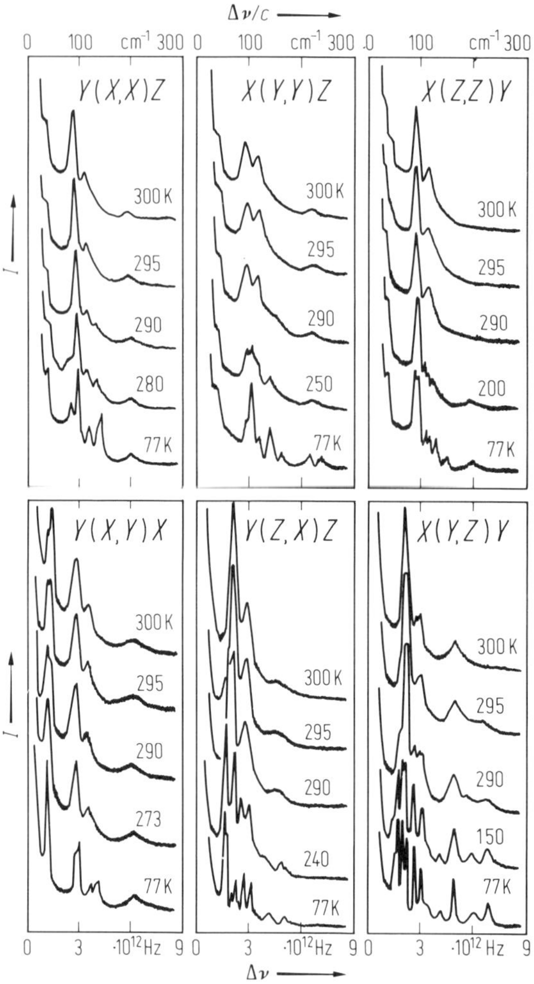

Fig. 1359. $(NH_2CH_2COOH)_3 \cdot H_2SeO_4$ (TGSe). I vs. $\Delta\nu$ [77K3]. I: Raman scattering intensity. Parameter: T.

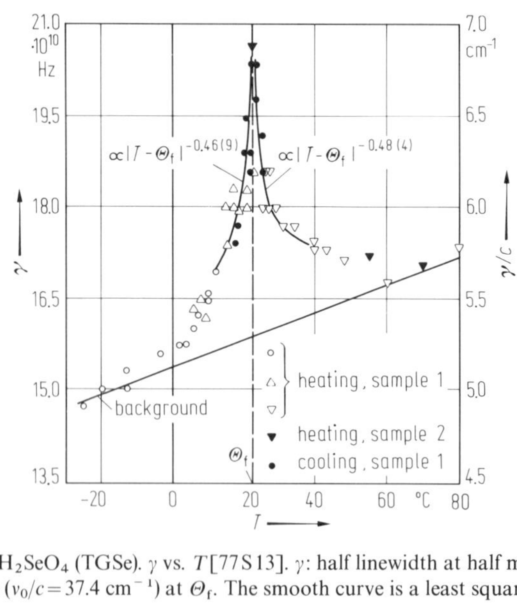

Fig. 1360. $(NH_2CH_2COOH)_3 \cdot H_2SeO_4$ (TGSe). γ vs. T [77S13]. γ: half linewidth at half maximum of the Raman spectrum of the lowest TO-phonon mode ($\nu_0/c = 37.4\ cm^{-1}$) at Θ_f. The smooth curve is a least square fit to the data.

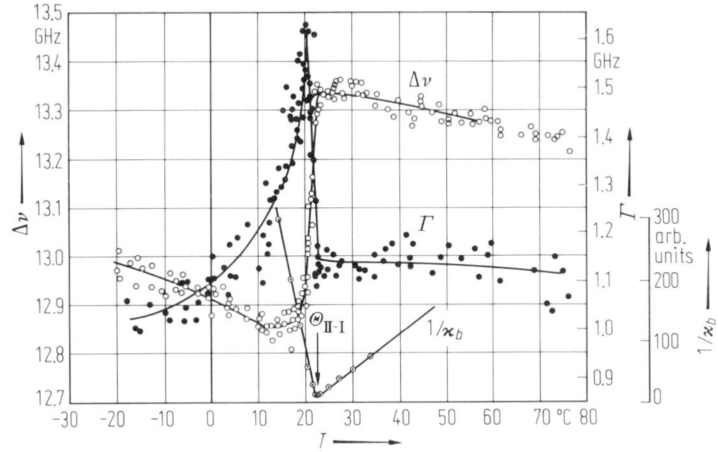

Fig. 1361. $(NH_2CH_2COOH)_3 \cdot H_2SeO_4$ (TGSe). $\Delta\nu$, Γ, $\varkappa_b^{-1}$ vs. T [76 Y 1]. $\Delta\nu$: frequency shift of Brillouin scattering; c_{33}-mode. Γ: full linewidth at half maximum.

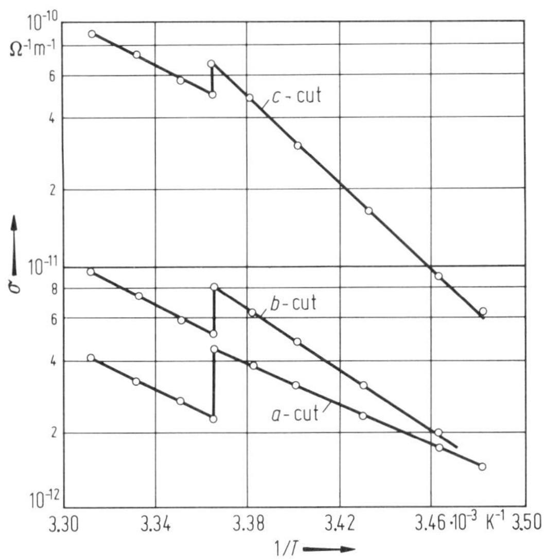

Fig. 1362. $(NH_2CH_2COOH)_3 \cdot H_2SeO_4$ (TGSe). σ vs. $1/T$ [69 S 18]. σ was measured by two terminal method. Applied field $E = 120\ \mathrm{kV\ m^{-1}}$.

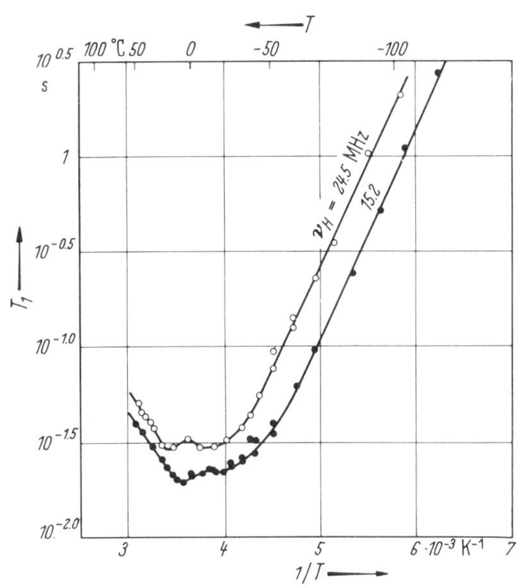

Fig. 1363. $(NH_2CH_2COOH)_3 \cdot H_2SeO_4$ (TGSe). T_1 vs. $1/T$ [66 B 5]. T_1: spin lattice relaxation time for proton. ν_H: Larmor frequency of proton.

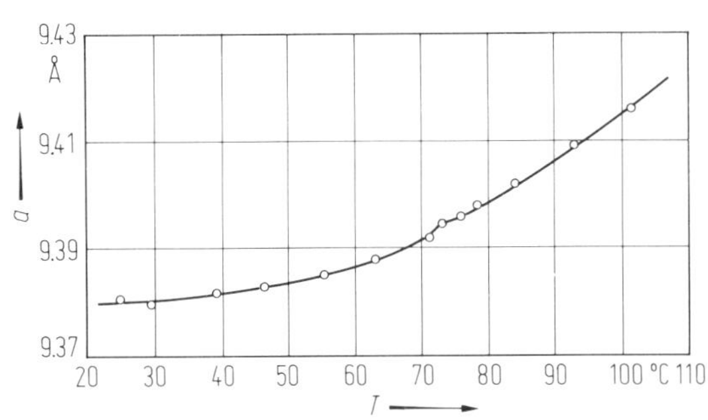

Fig. 1364. $(NH_2CH_2COOH)_3 \cdot H_2BeF_4$ (TGFB). a vs. T [77 W 5]. a: unit cell parameter.

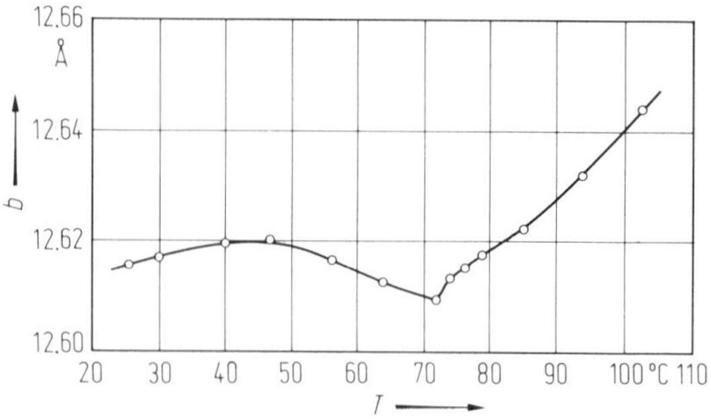

Fig. 1365. $(NH_2CH_2COOH)_3 \cdot H_2BeF_4$ (TGFB). b vs. T [77 W 5]. b: unit cell parameter.

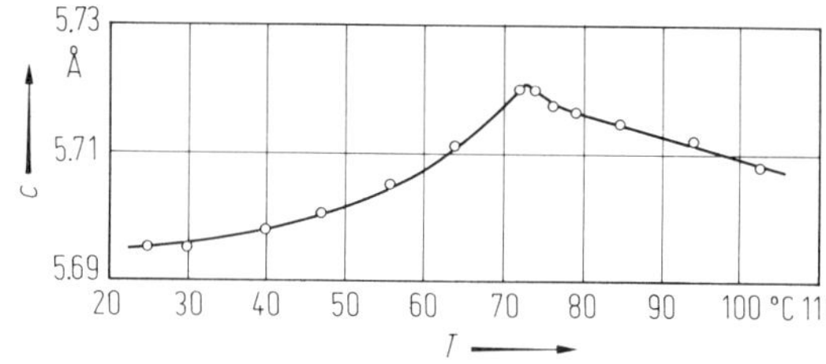

Fig. 1366. $(NH_2CH_2COOH)_3 \cdot H_2BeF_4$ (TGFB). c vs. T [77 W 5]. c: unit cell parameter.

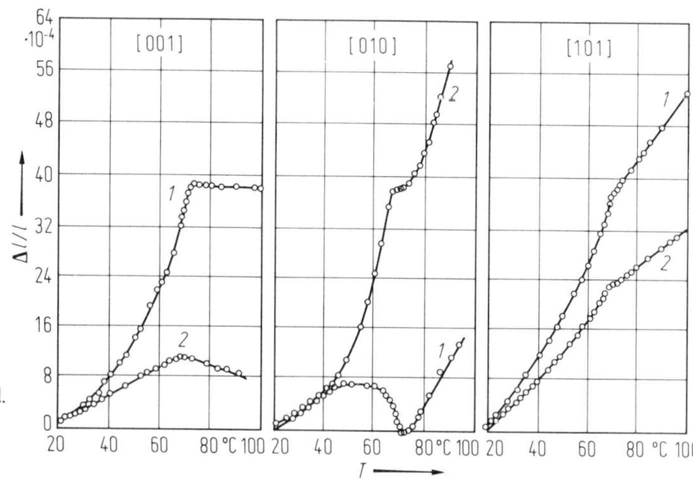

Fig. 1367. $(NH_2CH_2COOH)_3 \cdot H_2BeF_4$ (TGFB). $\Delta l/l$ vs. T [72 V 1]. $\Delta l/l$: linear thermal expansion. Curves *1*: unirradiated. Curves *2*: after γ-irradiation of 500 mR.

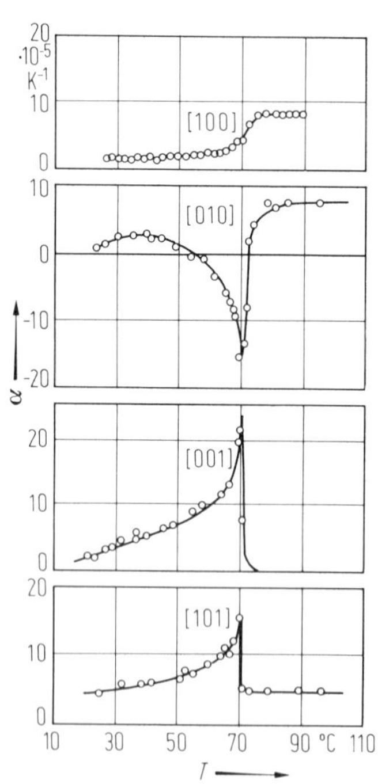

Fig. 1368. $(NH_2CH_2COOH)_3 \cdot H_2BeF_4$ (TGFB). α vs. T [72 Z 1]. α: linear thermal expansion coefficients along the [100], [010], [001], and [101] directions.

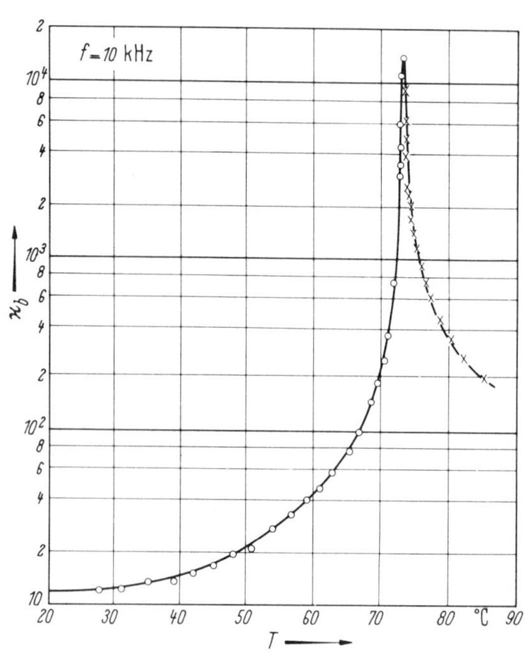

Fig. 1369. $(NH_2CH_2COOH)_3 \cdot H_2BeF_4$ (TGFB). $\varkappa_b$ vs. T [66 W 7].

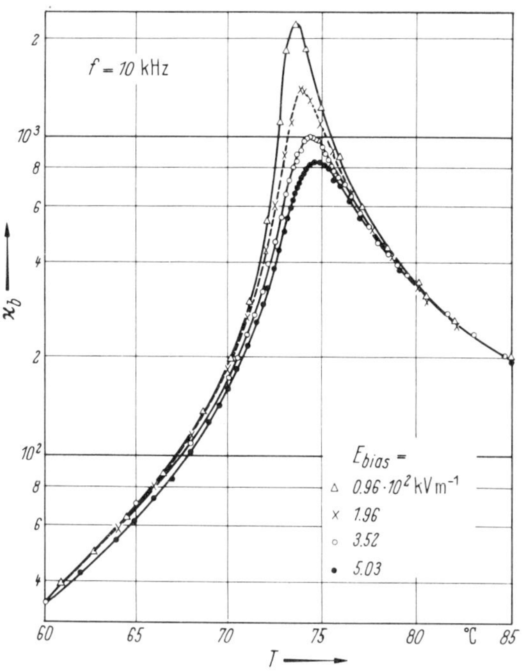

Fig. 1371. $(NH_2CH_2COOH)_3 \cdot H_2BeF_4$ (TGFB). $\varkappa_b$ vs. T [66 W 7]. Parameter: E_{bias}.

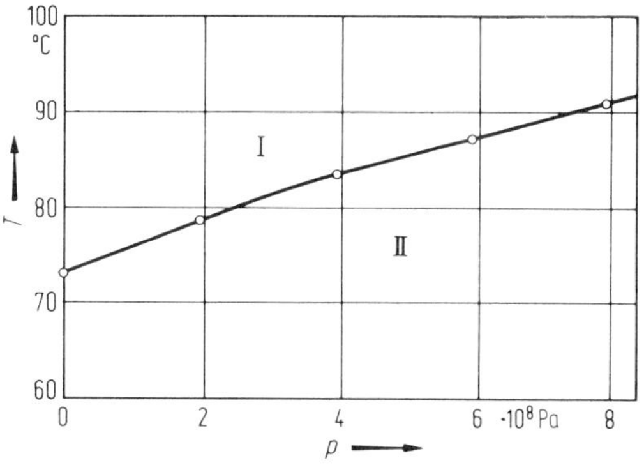

Fig. 1370. $(NH_2CH_2COOH)_3 \cdot H_2BeF_4$ (TGFB). Phase diagram [67 M 9].

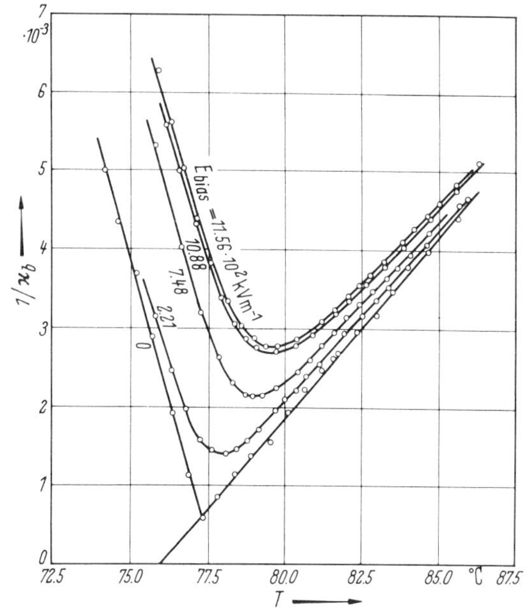

Fig. 1372. $(ND_2CH_2COOD)_3 \cdot D_2BeF_4$. $1/\varkappa_b$ vs. T [66 B 8]. Parameter: E_{bias}.

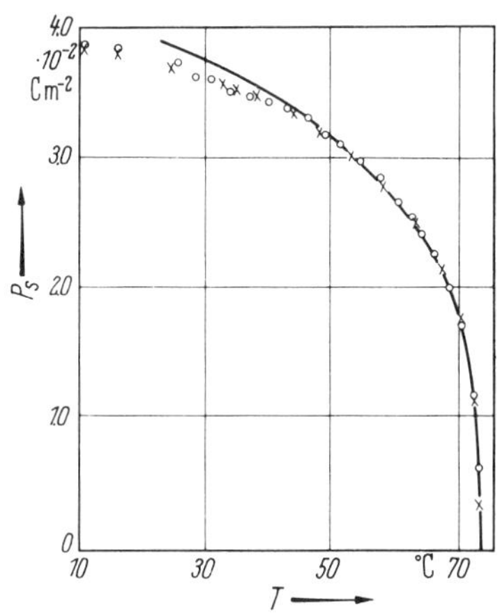

Fig. 1373. $(NH_2CH_2COOH)_3 \cdot H_2BeF_4$ (TGFB). P_s vs. T [66 W 7].

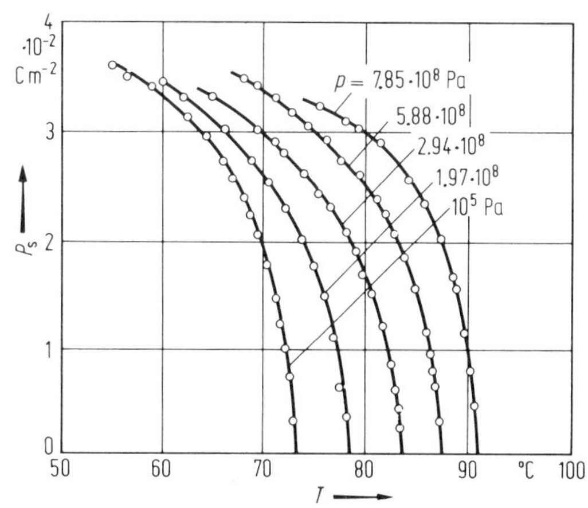

Fig. 1374. $(NH_2CH_2COOH)_3 \cdot H_2BeF_4$ (TGFB). P_s vs. T [67 M 9]. Parameter: p.

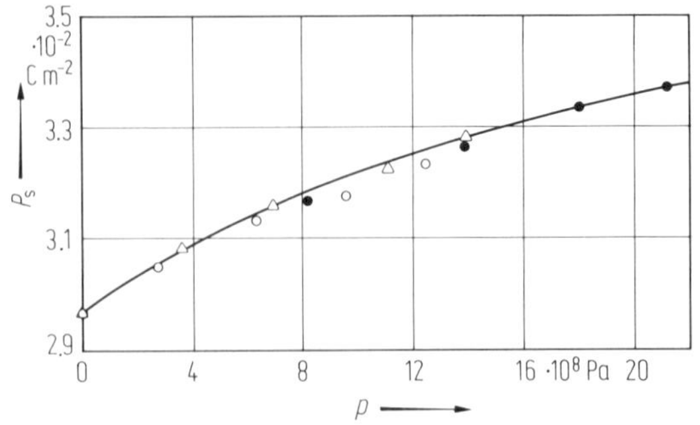

Fig. 1375. $(NH_2CH_2COOH)_3 \cdot H_2BeF_4$ (TGFB). P_s vs. p at RT [74S32].

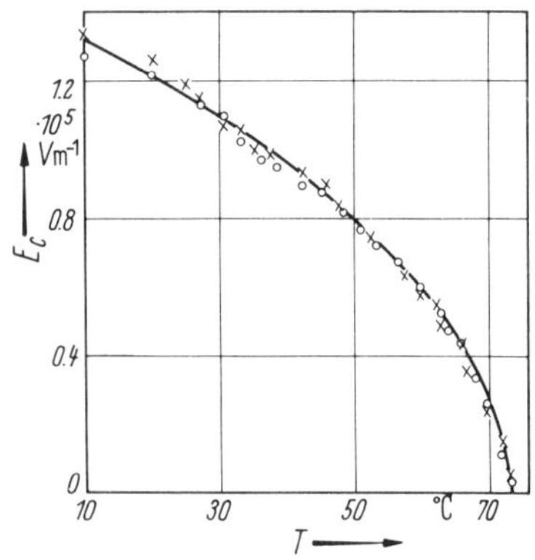

Fig. 1376. $(NH_2CH_2COOH)_3 \cdot H_2BeF_4$ (TGFB). E_c vs. T [66W7].

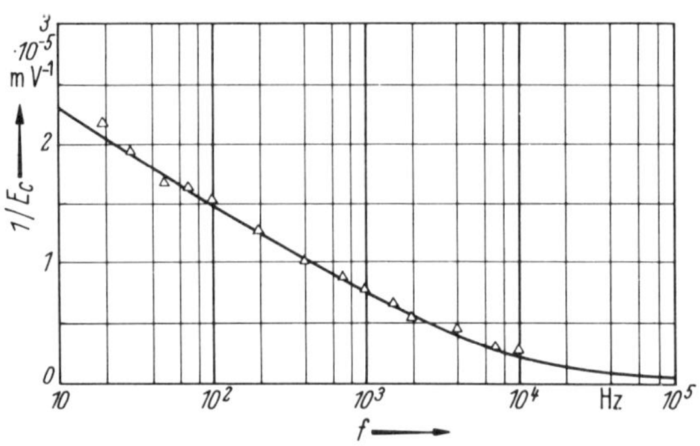

Fig. 1377. $(NH_2CH_2COOH)_3 \cdot H_2BeF_4$ (TGFB). $1/E_c$ vs. f [58P6].

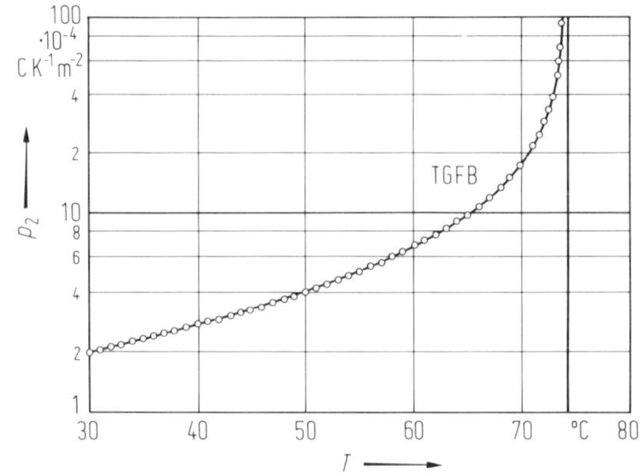

Fig. 1378. $(NH_2CH_2COOH)_3 \cdot H_2BeF_4$ (TGFB). p_2 vs. T [78F1]. p_2: pyroelectric coefficient along the [010] axis.

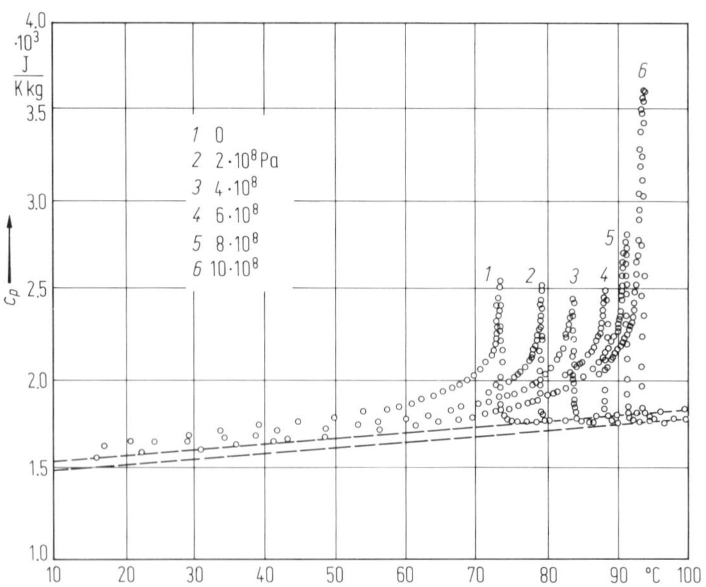

Fig. 1379. $(NH_2CH_2COOH)_3 \cdot H_2BeF_4$ (TGFB). c_p vs. T [79C1]. Parameter: p.

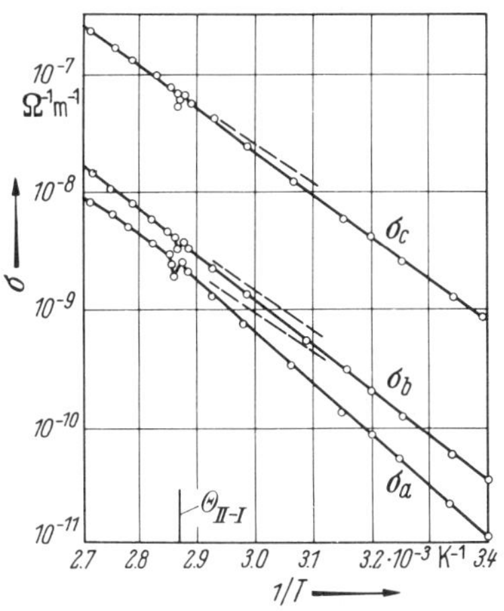

Fig. 1381. $(NH_2CH_2COOH)_3 \cdot H_2BeF_4$ (TGFB). σ_a, σ_b, σ_c vs. $1/T$ [61 G 5].

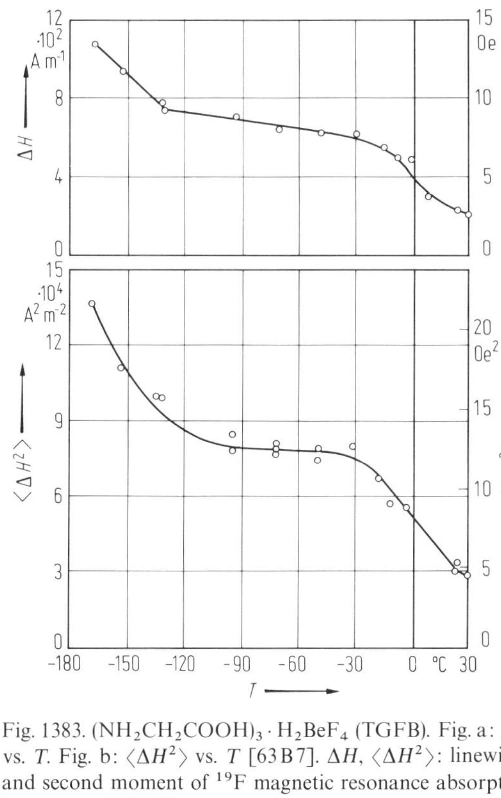

Fig. 1383. $(NH_2CH_2COOH)_3 \cdot H_2BeF_4$ (TGFB). Fig. a: ΔH vs. T. Fig. b: $\langle \Delta H^2 \rangle$ vs. T [63 B 7]. ΔH, $\langle \Delta H^2 \rangle$: linewidth and second moment of ^{19}F magnetic resonance absorption in powdered TGFB, respectively.

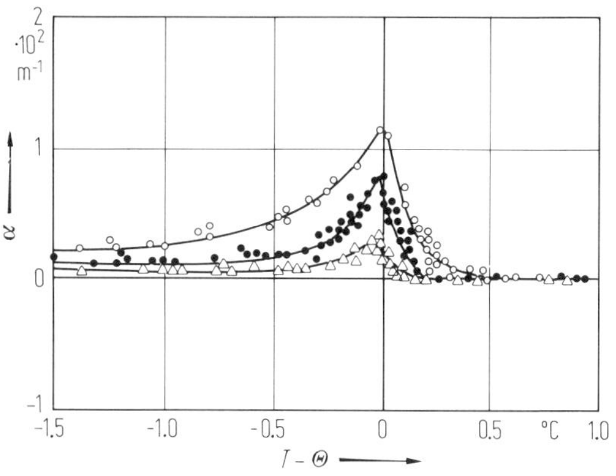

Fig. 1380. $(NH_2CH_2COOH)_3 \cdot H_2BeF_4$ (TGFB). α vs. $T-\Theta$ [67 M 6]. α: absorption coefficients of the longitudinal ultrasound propagating along the c-axis. Triangles: 5 MHz; full circles: 10 MHz; open circles: 15 MHz.

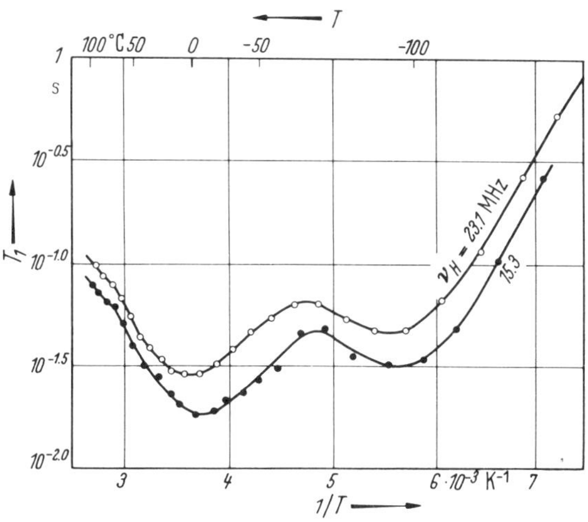

Fig. 1382. $(NH_2CH_2COOH)_3 \cdot H_2BeF_4$ (TGFB). T_1 vs. $1/T$ [66 B 5]. T_1: spin-lattice relaxation time for proton. ν_H: Larmor frequency of proton.

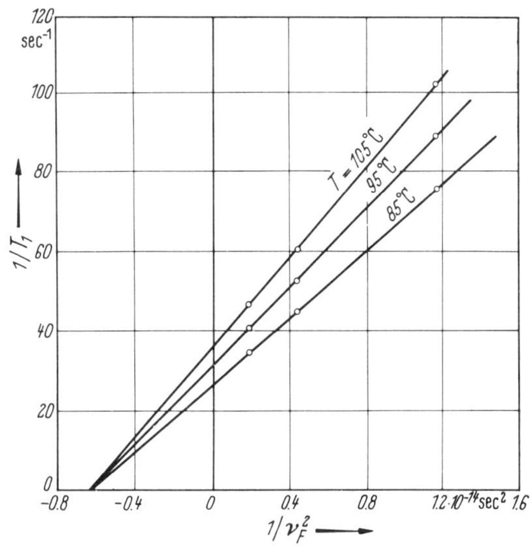

Fig. 1384. $(NH_2CH_2COOH)_3 \cdot H_2BeF_4$ (TGFB). $1/T_1$ of ^{19}F vs. $1/\nu_F^2$ [66 B 5]. Parameter: T. ν_F: Larmor frequency of ^{19}F.

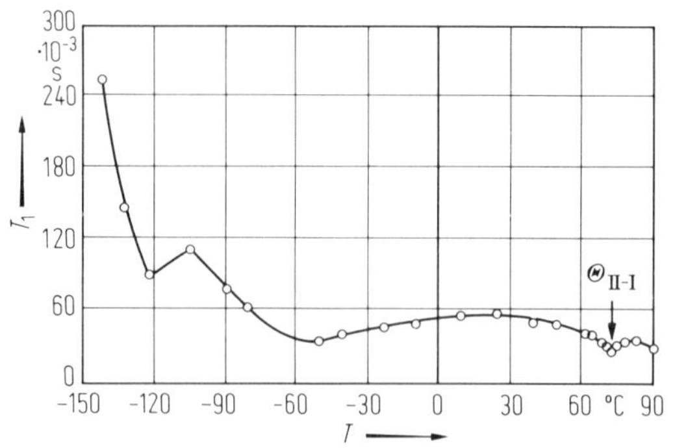

Fig. 1385. $(NH_2CH_2COOH)_3 \cdot H_2BeF_4$ (TGFB). T_1 vs. T [76M25]. T_1: spin-lattice relaxation time of ^{19}F at 27 MHz.

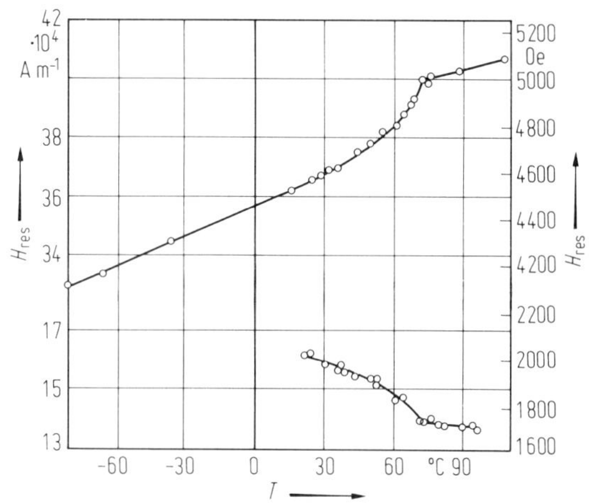

Fig. 1386. $(NH_2CH_2COOH)_3$ $H_2BeF_4{:}Cu^{2+}$. H_{res} vs. T [67S24]. H_{res}: ESR magnetic field strength due to Cu^{2+} ion pair.

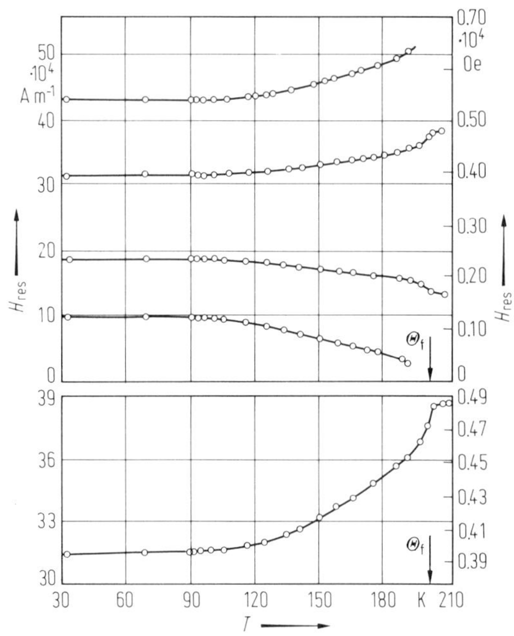

Fig. 1387. $(NH_2CH_2COOH)_3 \cdot H_2BeF_4{:}Cu^{2+}$. H_{res} vs. T [72S26]. H_{res}: resonant field strength due to four-copper complex.

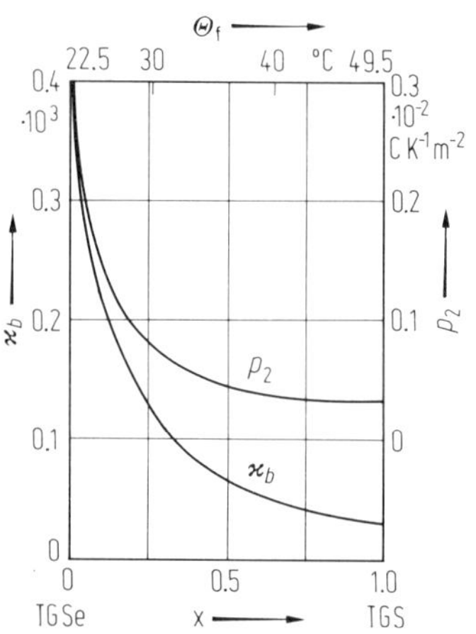

Fig. 1389. $\{(NH_2CH_2COOH)_3 \cdot H_2SO_4\}_x \cdot \{(NH_2CH_2COOH)_3 \cdot H_2SeO_4\}_{1-x}$ (TGS-TGSe). $\varkappa_b$ and p_2 vs. x at 21 °C [76B18]. $f = 1.6$ kHz. Θ_f of the composition is given at the top as an alternative abscissa.

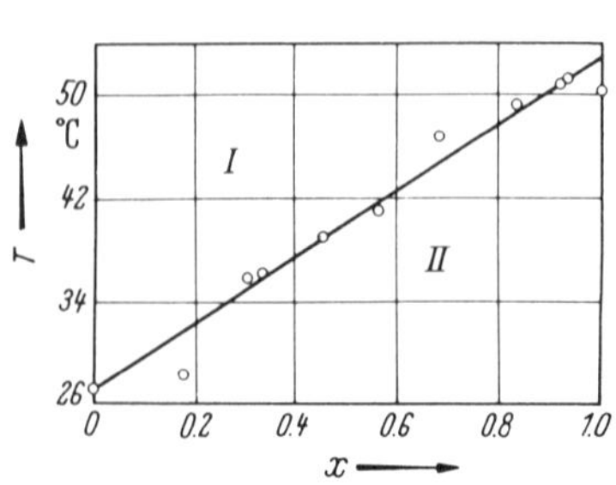

◄

Fig. 1388. $\{(NH_2CH_2COOH)_3 \cdot H_2SO_4\}_x \cdot \{(NH_2CH_2COOH)_3 \cdot H_2SeO_4\}_{1-x}$ (TGS-TGSe). Phase diagram [59F3].

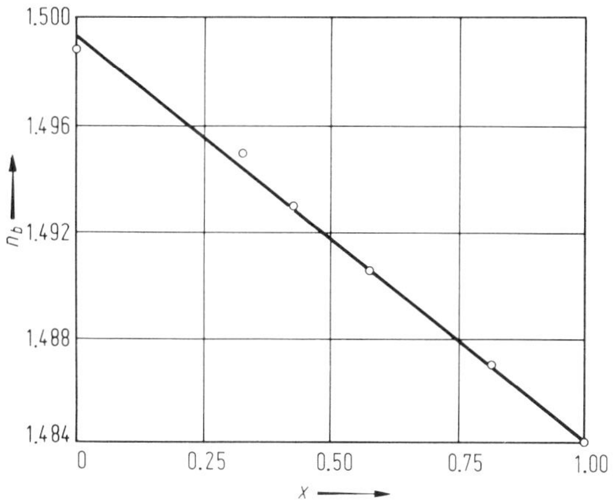

Fig. 1390α. $\{(NH_2CH_2COOH)_3 \cdot H_2SO_4\}_x \cdot \{(NH_2CH_2COOH)_3 \cdot H_2SeO_4\}_{1-x}$ (TGS-TGSe). n_b vs. x [72H15].

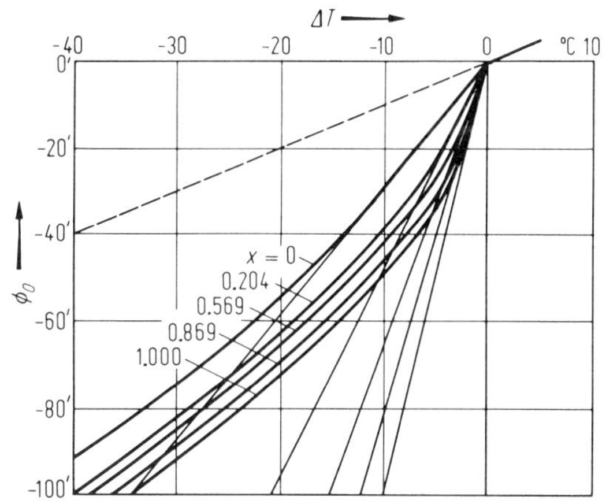

Fig. 1390β. $\{(NH_2CH_2COOH)_3 \cdot H_2SO_4\}_x \cdot \{(NH_2CH_2COOH)_3 \cdot H_2SeO_4\}_{1-x}$ (TGS-TGSe). ϕ_o vs. ΔT [68B12]. Parameter: x. ϕ_o: rotation angle of the optical indicatrix around the *b*-axis. $\Delta T = T - \Theta_f$.

37 $NH_2CH_2COOH \cdot AgNO_3$

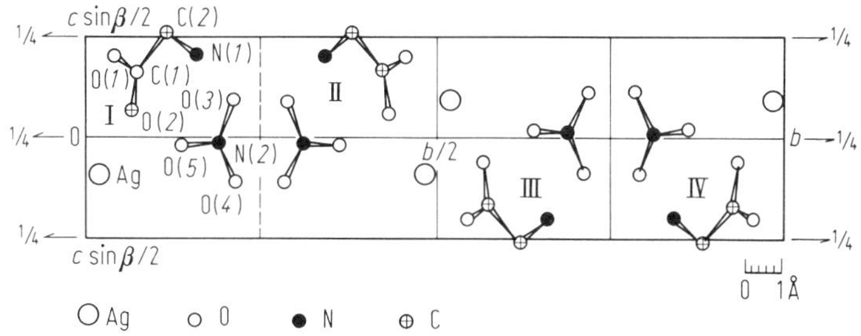

Fig. 1392. $NH_2CH_2COOH \cdot AgNO_3$. Structure of phase I [72M16]. View along the *a*-axis. The Roman numerals indicate the equivalent positions, and the relevant symmetry operations are as follows: I: *x, y, z*; II: $1/2+x$, $1/2-y$, *z*; III: $1/2-x$, $1/2+y$, $-z$; IV: $-x$, $-y+1$, $-z$.

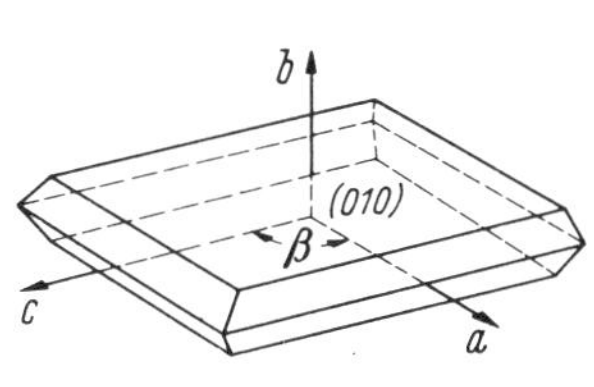

Fig. 1391. $NH_2CH_2COOH \cdot AgNO_3$. Crystal form [64M4].

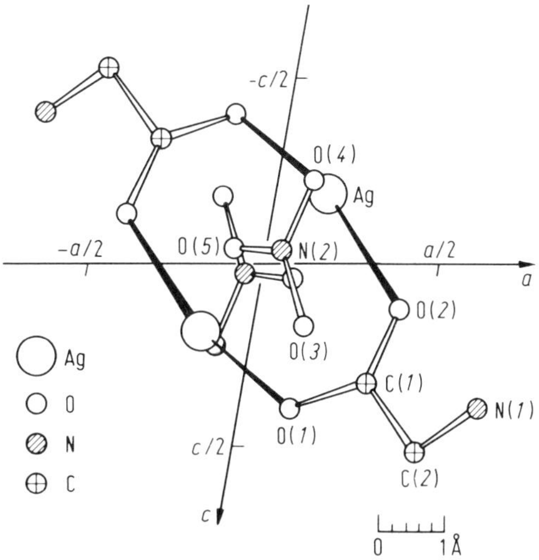

Fig. 1393. $NH_2CH_2COOH \cdot AgNO_3$. Structure of phase I [72M16]. View along the *b*-axis, showing the existence of glycine dimers bridged by silver atoms.

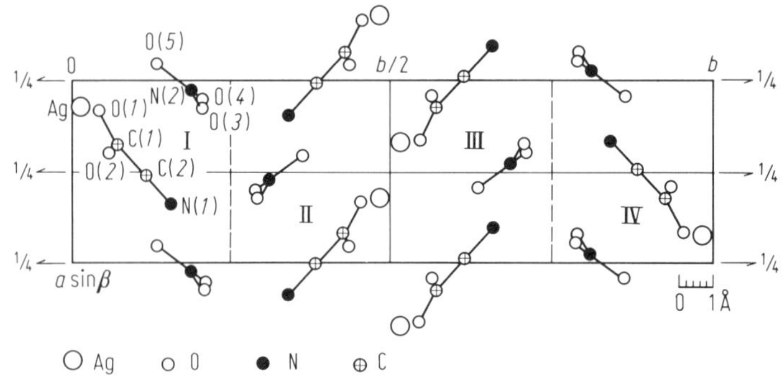

Fig. 1394. $NH_2CH_2COOH \cdot AgNO_3$. Structure of phase I [72M16]. View along the c-axis. See Fig. 1391 for the Roman numerals.

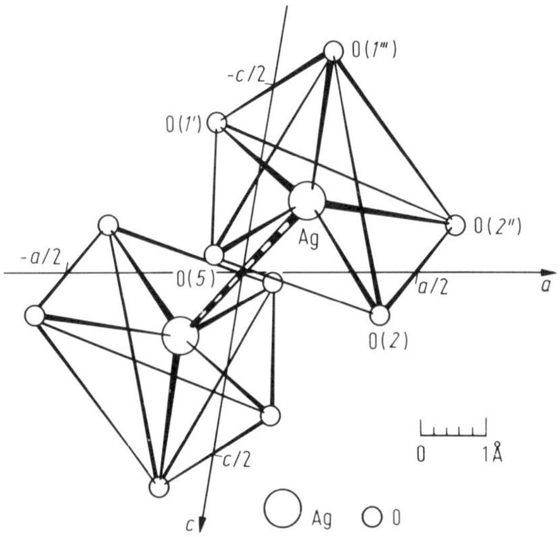

Fig. 1395. $NH_2CH_2COOH \cdot AgNO_3$. Structure of phase I [72M16]. Surroundings of the inversion center at (0, 0, 0). The Ag-Ag bond is represented by thick, partially shaded lines. The labelled atoms have the following coordinates: O(2), O(5), Ag: (x, y, z); O(1'): ($-x$, $-y$, $-z$); O(2″): ($-x+1$, $-y$, $-z$); O(1‴): (x, y, $z-1$).

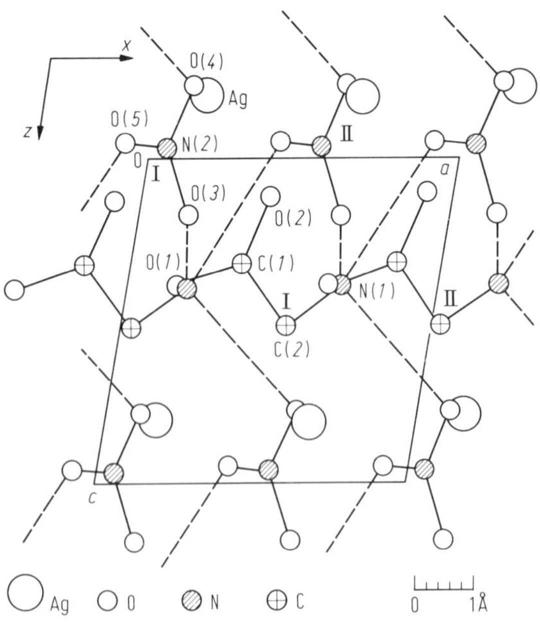

Fig. 1396. $NH_2CH_2COOH \cdot AgNO_3$. Structure of phase I [72M16]. Hydrogen bondings (broken lines) viewed along the b-axis. See Fig. 1392 for I and II.

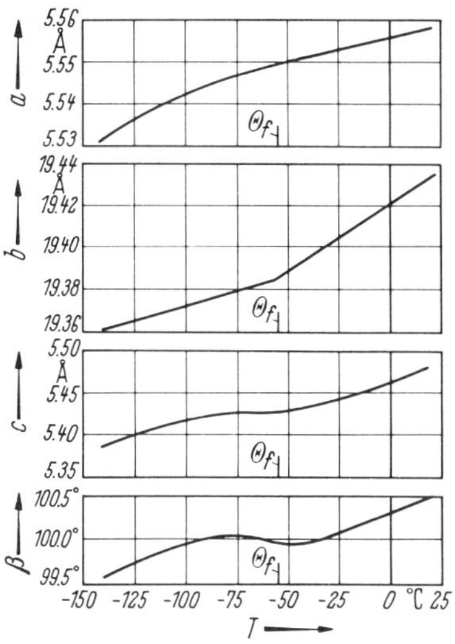

Fig. 1397. $NH_2CH_2COOH \cdot AgNO_3$. Lattice parameters vs. T [64M4].

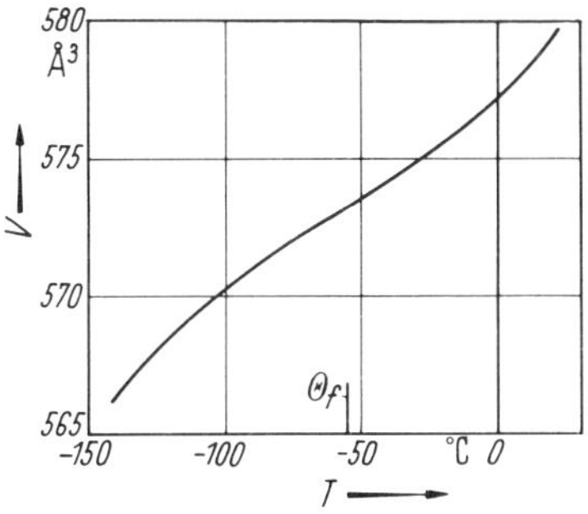

Fig. 1398. $NH_2CH_2COOH \cdot AgNO_3$. V vs. T [64M4]. V: unit cell volume.

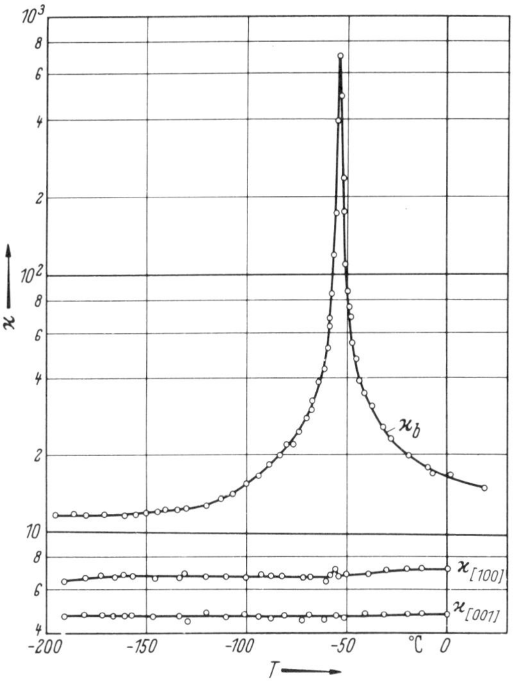

Fig. 1399. $NH_2CH_2COOH \cdot AgNO_3$. $\varkappa_{[100]}$, $\varkappa_b$, $\varkappa_{[001]}$ vs. T [64M4]. $f = 10$ kHz.

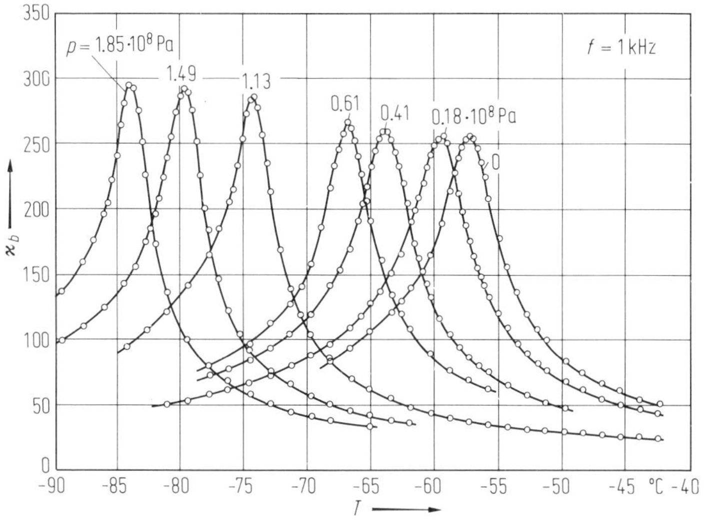

Fig. 1400. $NH_2CH_2COOH \cdot AgNO_3$. $\varkappa_b$ vs. T [77G6]. Parameter: p.

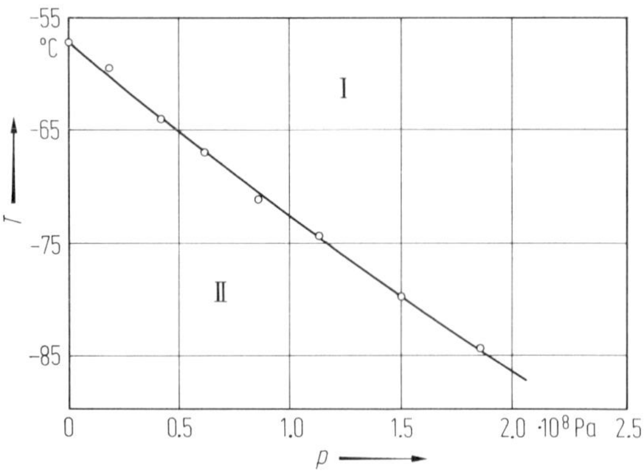

Fig. 1402. $NH_2CH_2COOH \cdot AgNO_3$. Phase diagram [77G6].

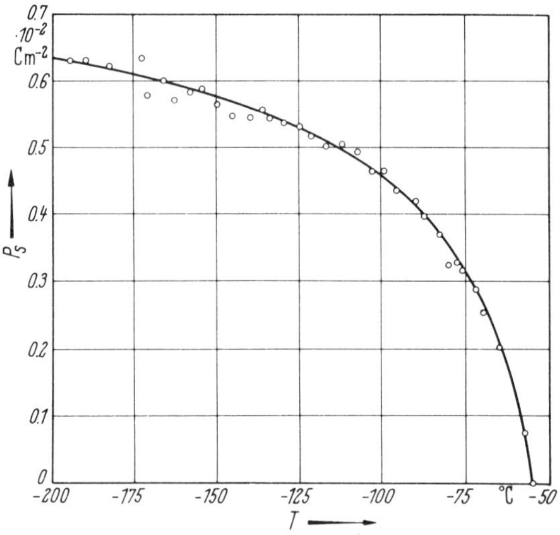

Fig. 1403. $NH_2CH_2COOH \cdot AgNO_3$. P_s vs. T [64M4].

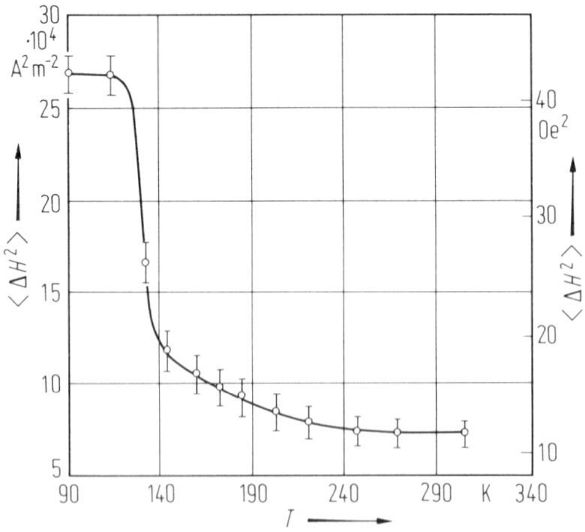

Fig. 1405. $NH_2CH_2COOH \cdot AgNO_3$. $\langle \Delta H^2 \rangle$ of proton vs. T [66E1]. $\langle \Delta H^2 \rangle$: second moment of magnetic resonance curve.

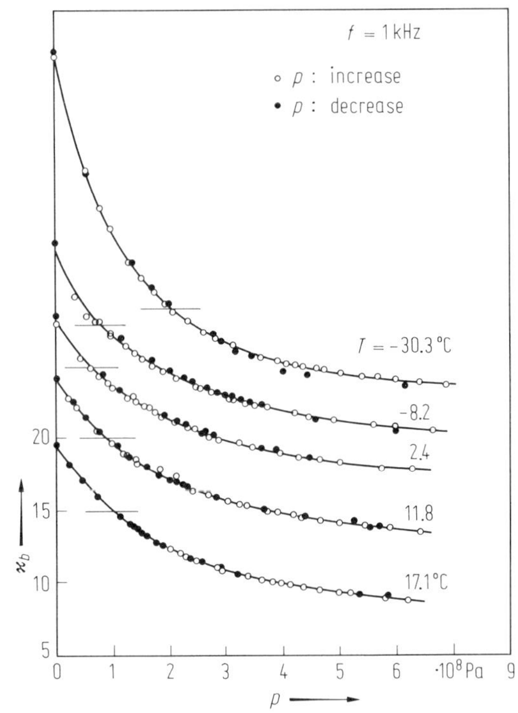

Fig. 1401. $NH_2CH_2COOH \cdot AgNO_3$. $\varkappa_b$ vs. p [77G6]. Parameter: T. Thin horizontal lines show the level of $\varkappa_b = 15$.

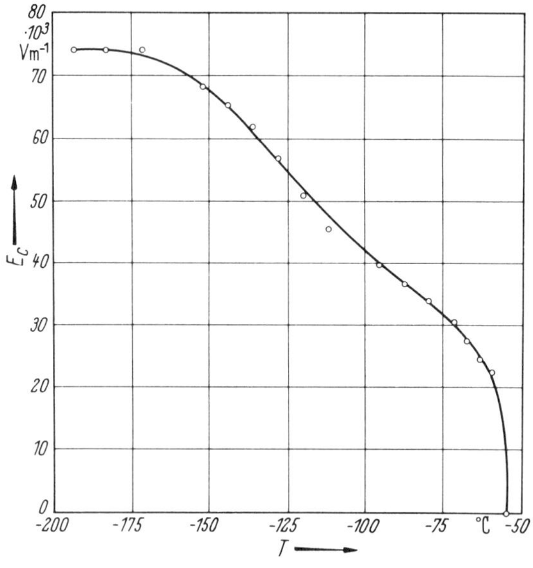

Fig. 1404. $NH_2CH_2COOH \cdot AgNO_3$. E_c vs. T [57P2].

38 $(NH_2CH_2COOH)_2 \cdot HNO_3$

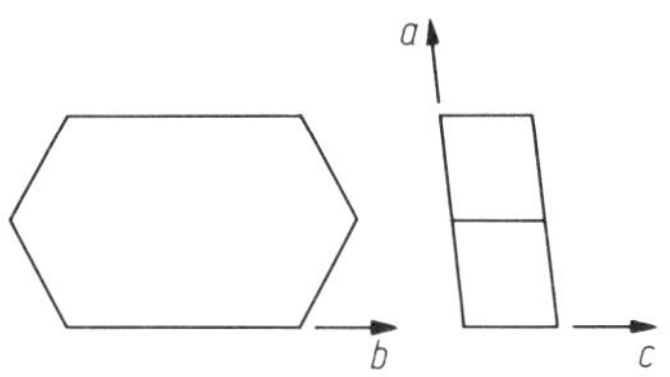

Fig. 1406. $(NH_2CH_2COOH)_2 \cdot HNO_3$. Crystal form [68S7].

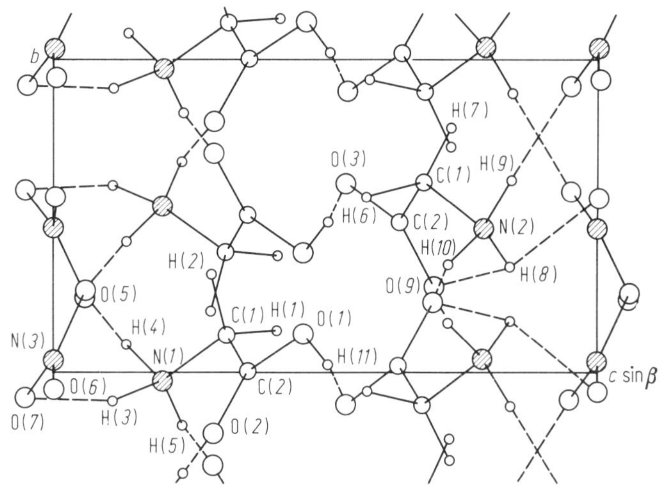

Fig. 1408. $(NH_2CH_2COOH)_2 \cdot HNO_3$. Structure of phase II [68S7]. Projection along the a-axis.

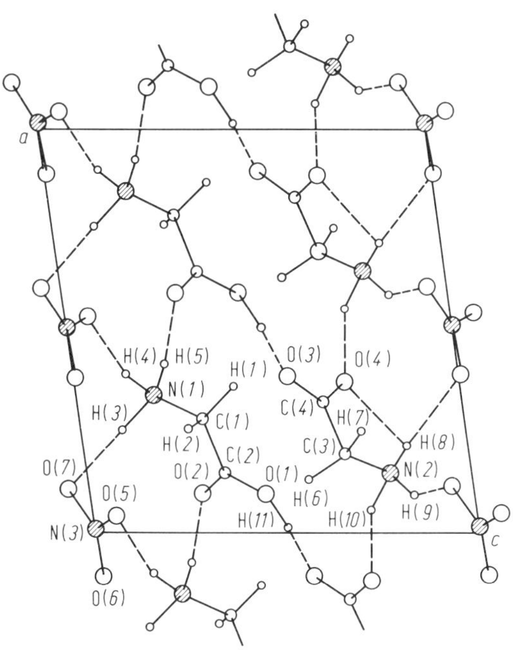

Fig. 1409. $(NH_2CH_2COOH)_2 \cdot HNO_3$. Structure of phase II [68S7]. Projection along the b-axis.

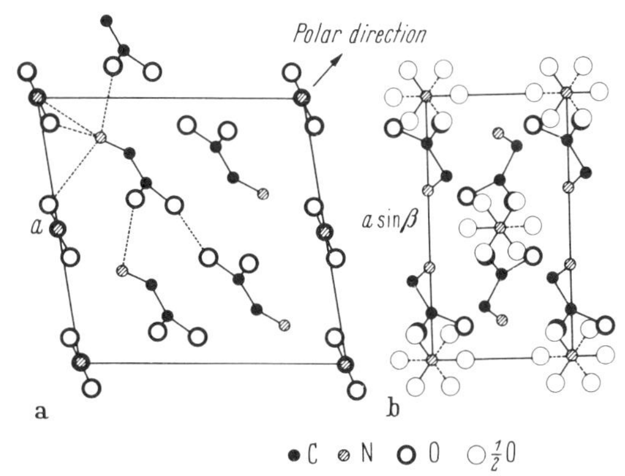

Fig. 1407. $(NH_2CH_2COOH)_2 \cdot HNO_3$. Structure of phase I [63H7]. Fig. a: Projection along the b-axis. Fig. b: projection along the c-axis.

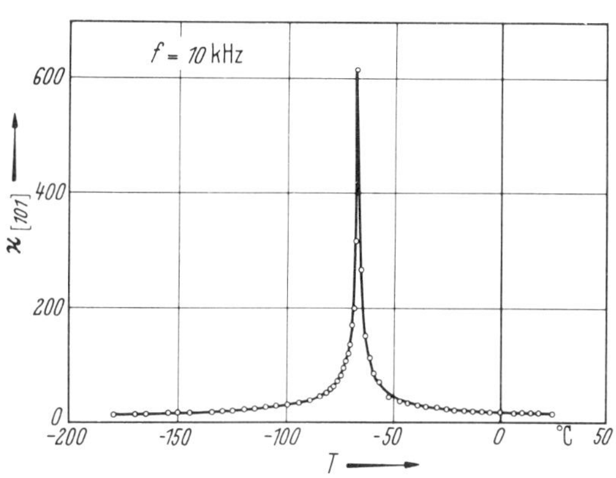

Fig. 1410α. $(NH_2CH_2COOH)_2 \cdot HNO_3$. $\varkappa_{[101]}$ vs. T [58P1]. $\varkappa_{[101]}$: dielectric constant along the [101] direction.

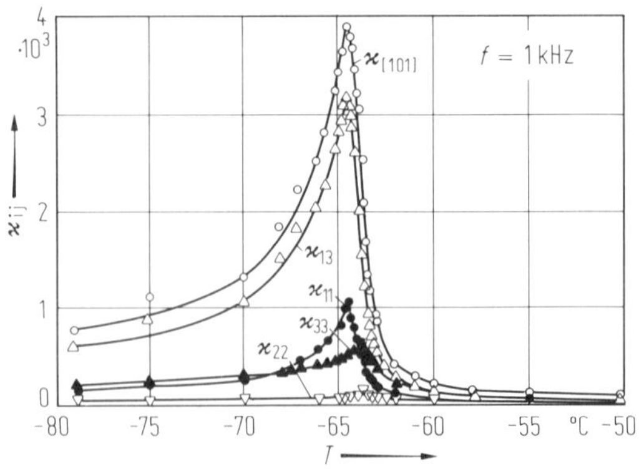

Fig. 1410 β. $(NH_2CH_2COOH)_2 \cdot HNO_3$. $\varkappa_{ij}$, $\varkappa_{[101]}$ vs. T in the vicinity of Θ_f [79 V 1]. $\varkappa_{[101]}$: dielectric constant along the [101] direction.

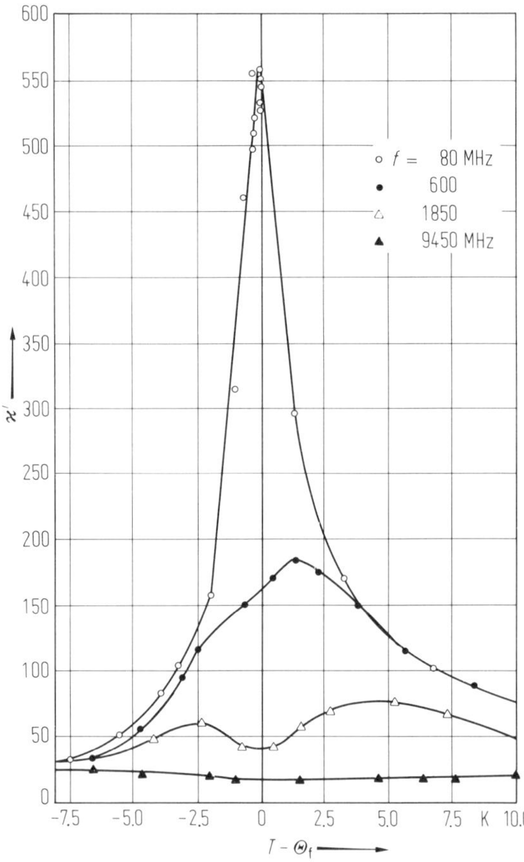

Fig. 1411 α. $(NH_2CH_2COOH)_2 \cdot HNO_3$. $\varkappa'$ vs. $T-\Theta_f$ [78 K 16]. Parameter: f.

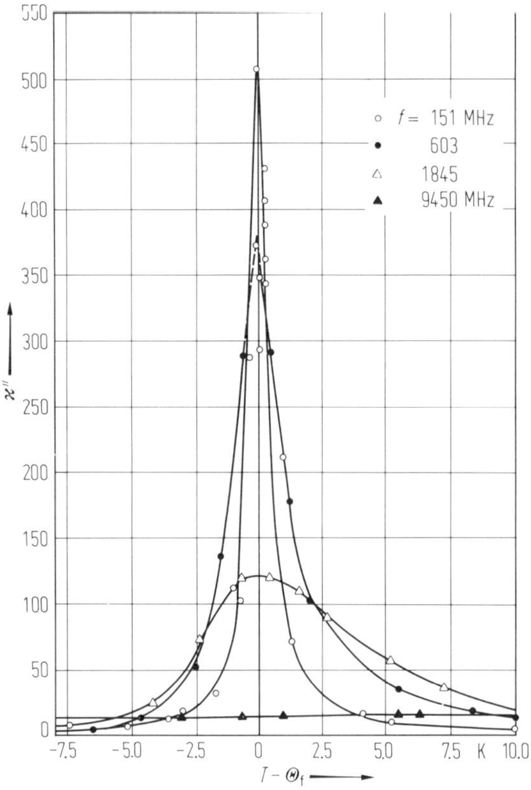

Fig. 1411 β. $(NH_2CH_2COOH)_2 \cdot HNO_3$. $\varkappa''$ vs. $T-\Theta_f$ [78 K 16]. Parameter: f.

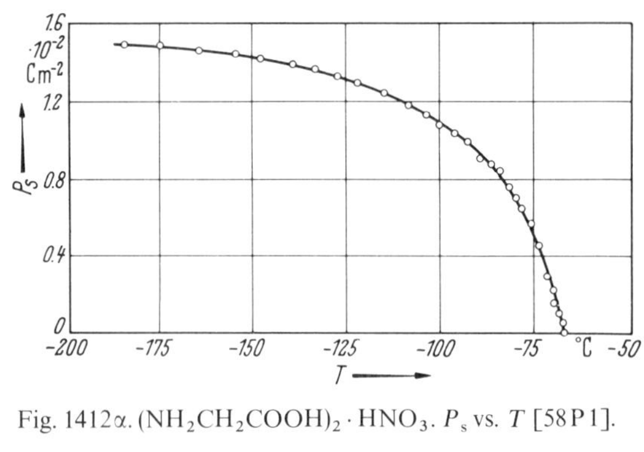

Fig. 1412 α. $(NH_2CH_2COOH)_2 \cdot HNO_3$. P_s vs. T [58 P 1].

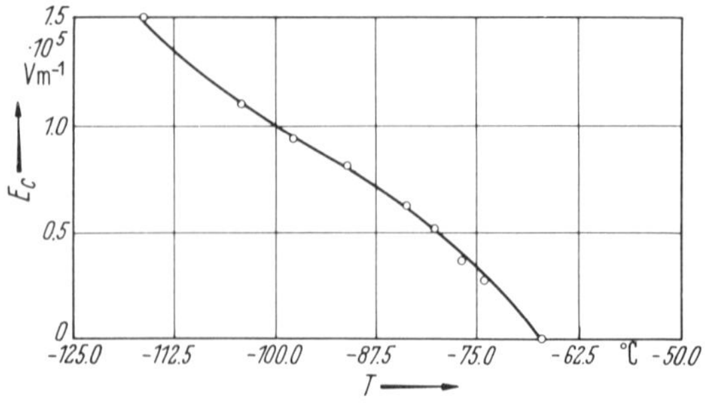

Fig. 1412 β. $(NH_2CH_2COOH)_2 \cdot HNO_3$. E_c vs. T [58 P 1].

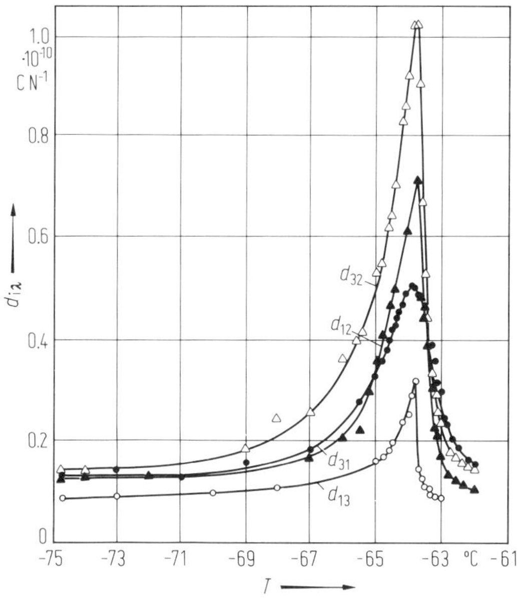

Fig. 1413. $(NH_2CH_2COOH)_2 \cdot HNO_3$. $d_{i\lambda}$ vs. T [79V1].

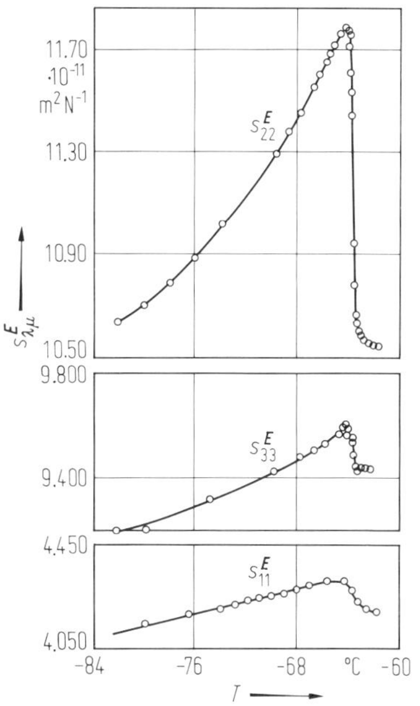

Fig. 1414. $(NH_2CH_2COOH)_2 \cdot HNO_3$. $s^E_{\lambda\mu}$ vs. T [79V1].

Fig. 1415. $(NH_2CH_2COO)_2 \cdot HNO_3$. $\langle \Delta H^2 \rangle$ vs. T [70B9]. $\langle \Delta H^2 \rangle$: second moment of proton.

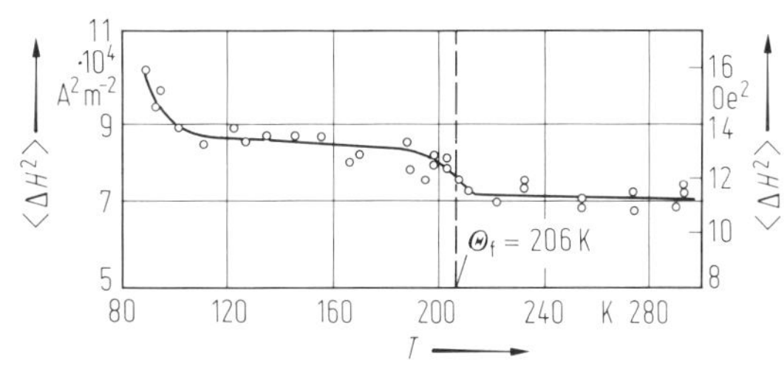

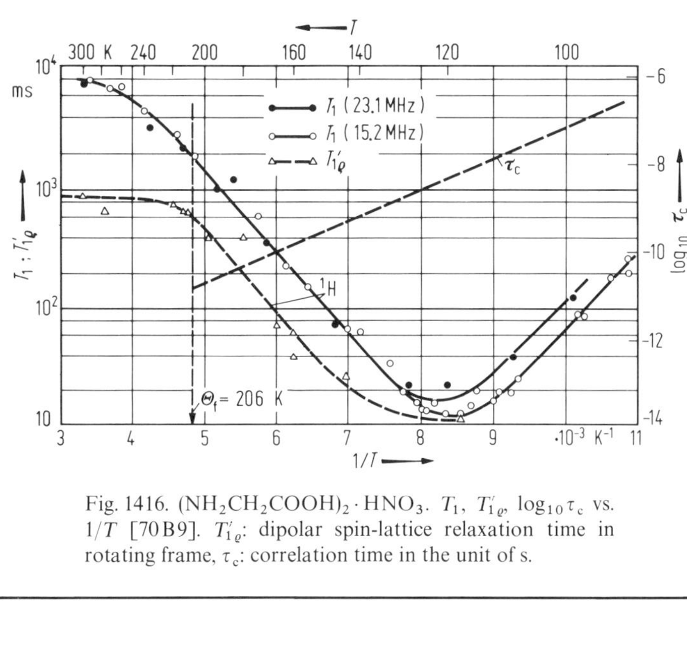

Fig. 1416. $(NH_2CH_2COOH)_2 \cdot HNO_3$. T_1, $T'_{1\varrho}$, $\log_{10}\tau_c$ vs. $1/T$ [70B9]. $T'_{1\varrho}$: dipolar spin-lattice relaxation time in rotating frame, τ_c: correlation time in the unit of s.

39 $(NH_2CH_2COOH)_2 \cdot MnCl_2 \cdot 2H_2O$

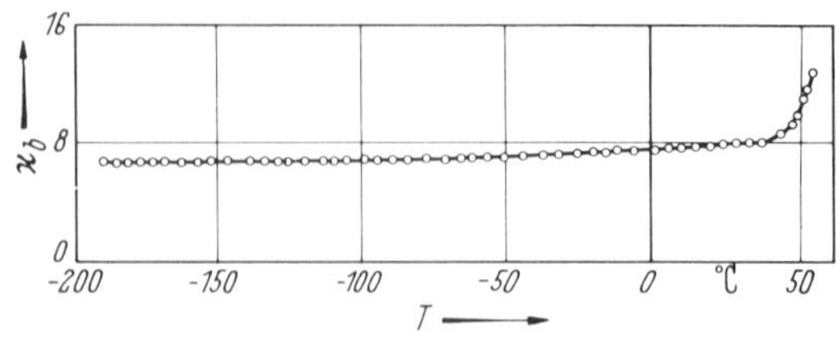

Fig. 1417. $(NH_2CH_2COOH)_2 \cdot MnCl_2 \cdot 2H_2O$. $\varkappa_b$ vs. T [58P3].

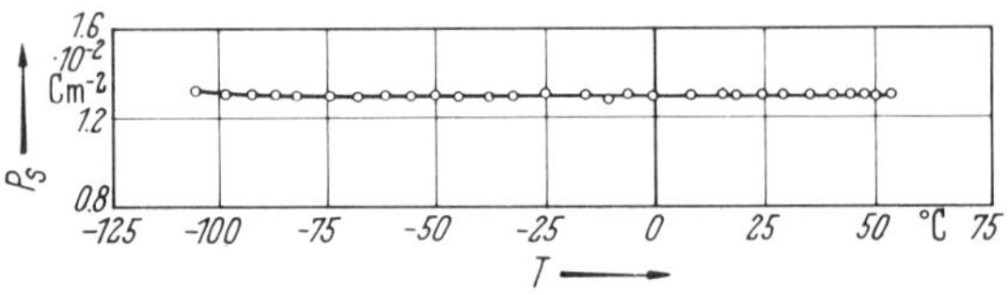

Fig. 1418. $(NH_2CH_2COOH)_2 \cdot MnCl_2 \cdot 2H_2O$. P_s vs. T [58P3].

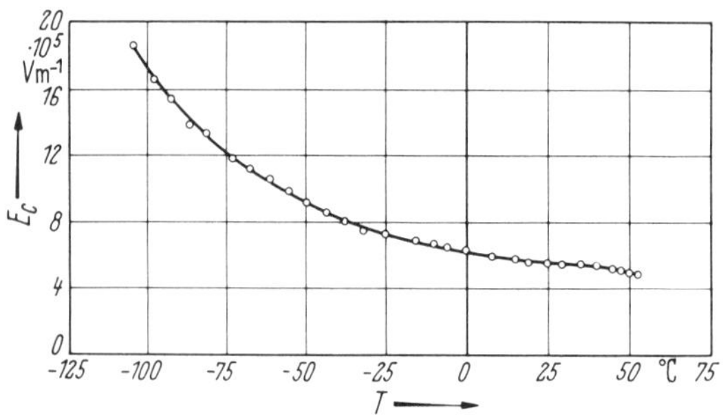

Fig. 1419. $(NH_2CH_2COOH)_2 \cdot MnCl_2 \cdot 2H_2O$. E_c vs. T [58P3].

40 $(CH_3NHCH_2COOH)_3 \cdot CaCl_2$

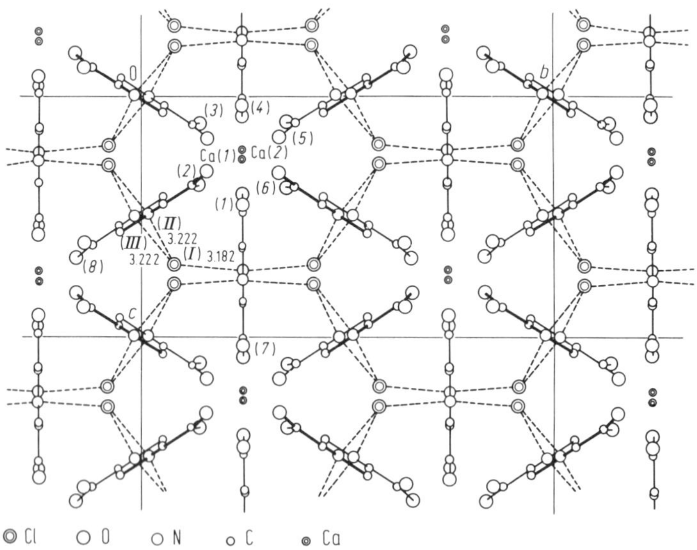

Fig. 1420. $(CH_3NHCH_2COOH)_3 \cdot CaCl_2$. Crystal structure of phase I [72A17]. Projection on (100). The sarcosine molecules are designated by Arabic numerals, the hydrogen bonds (dashed lines) by Roman numerals, whose notation is the same as that in Fig. 1421. Distances in Å.

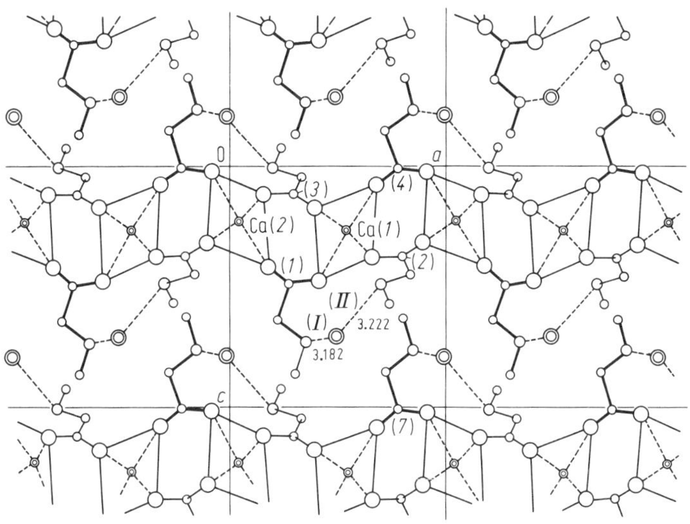

Fig. 1421. $(CH_3NHCH_2COOH)_3 \cdot CaCl_2$. Crystal structure of phase I [72A17]. Projection on (010). See notation and caption of Fig. 1420.

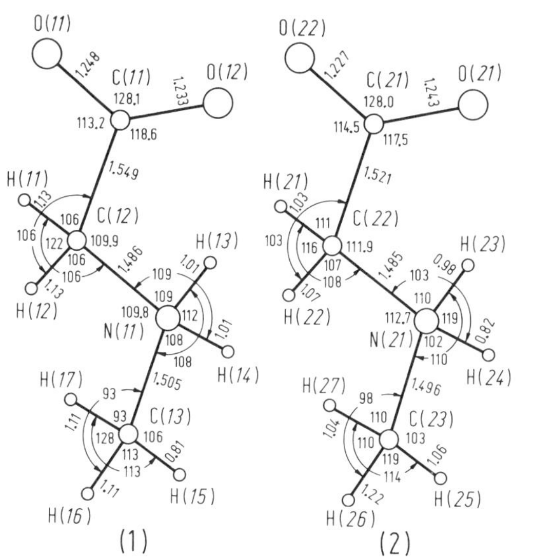

Fig. 1422. $(CH_3NHCH_2COOH)_3 \cdot CaCl_2$. Crystal structure of phase I [72A17]. Interatomic distances (in Å) and bond angles (in deg) in sarcosine molecules. The molecule on the left hand side, (1), is in the mirror plane, while the one on the right, (2), is in the general position.

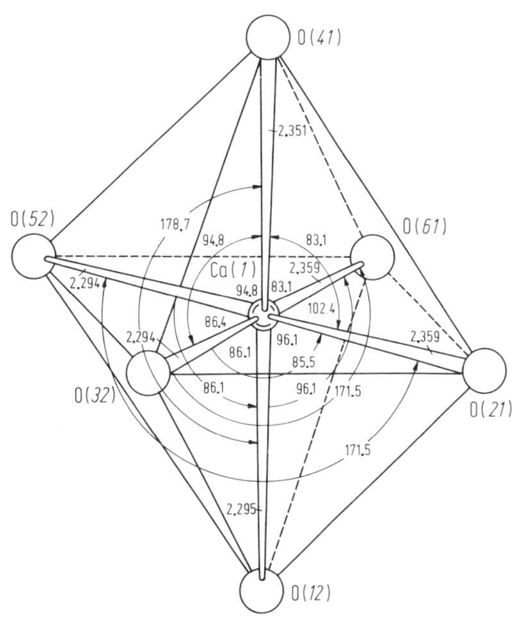

Fig. 1423. $(CH_3NHCH_2COOH)_3 \cdot CaCl_2$. Crystal structure of phase I [72A17]. Geometry of the octahedron around the Ca ion. Interatomic distances are in Å, and bond angles are in degrees.

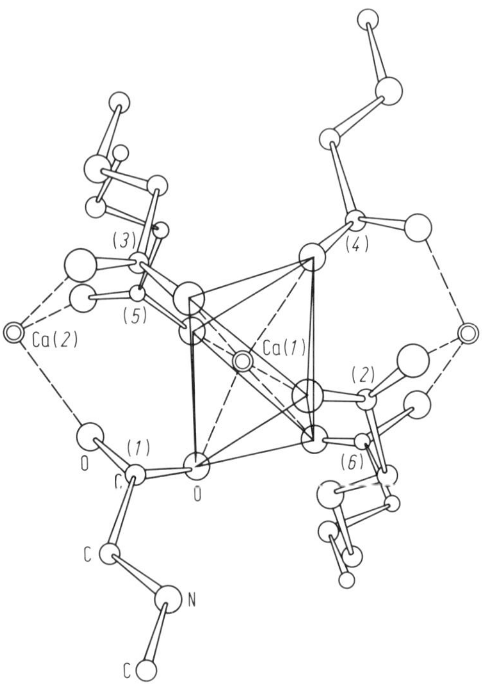

Fig. 1424. $(CH_3NHCH_2COOH)_3 \cdot CaCl_2$. Crystal structure of phase I [72A17]. A sketch of the neighbours of the Ca ion. The molecules (1) and (4) and the Ca ion are in the mirror plane, while others are in general position. The molecule (2) is the mirror image of the molecule (6), and (3) is the mirror image of (5).

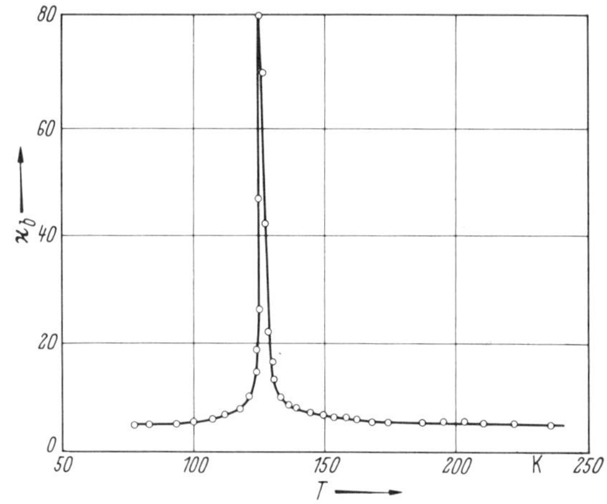

Fig. 1425α. $(CH_3NHCH_2COOH)_3 \cdot CaCl_2$. $\varkappa_b$ vs. T [65M3]. $f = 10$ kHz.

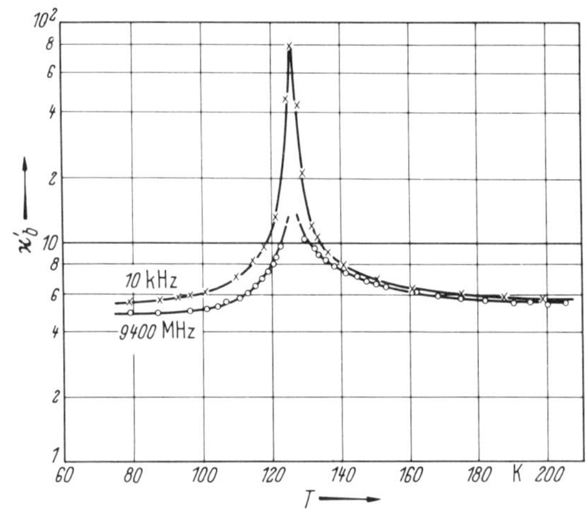

Fig. 1425β. $(CH_3NHCH_2COOH)_3 \cdot CaCl_2$. $\varkappa_b'$ vs. T [65M3]. Parameter: f.

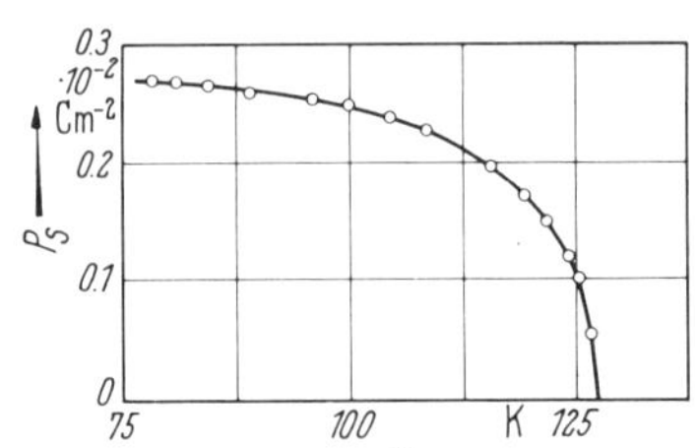

Fig. 1426α. $(CH_3NHCH_2COOH)_3 \cdot CaCl_2$. P_s vs. T [65M3].

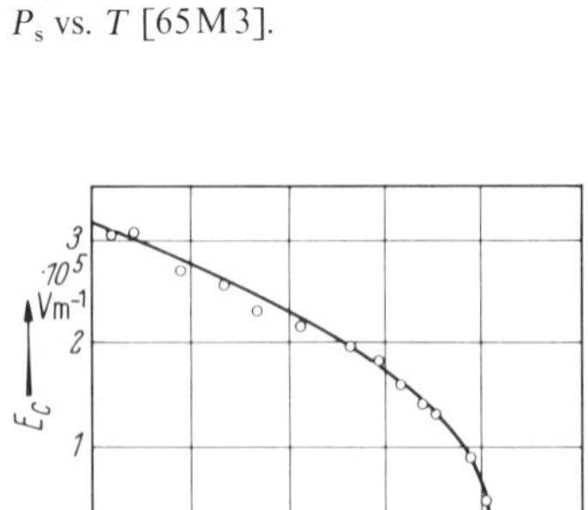

Fig. 1426β. $(CH_3NHCH_2COOH)_3 \cdot CaCl_2$. E_c vs. T [65M3].

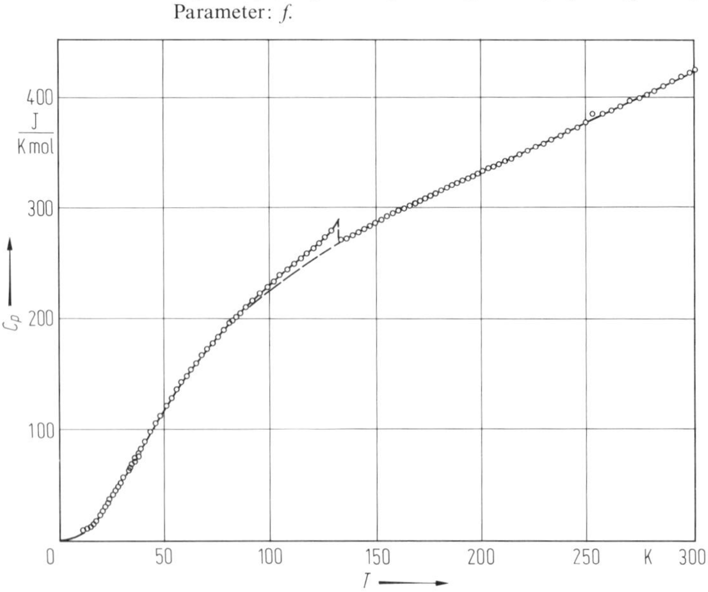

Fig. 1427. $(CH_3NHCH_2COOH)_3 \cdot CaCl_2$. C_p vs. T [79M4]. C_p: molar heat capacity at constant pressure.

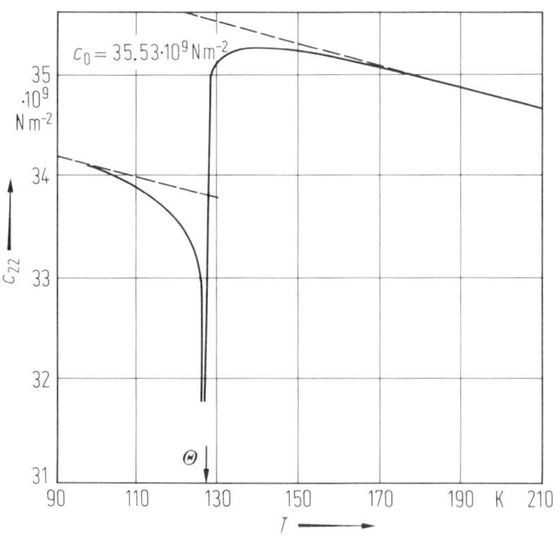

Fig. 1428. $(CH_3NHCH_2COOH)_3 \cdot CaCl_2$. c_{22} vs. T [78S19]. c_{22}: elastic stiffness. c_0 shows the c_{22} value at Θ obtained by extrapolation of the curve shown by the dashed line.

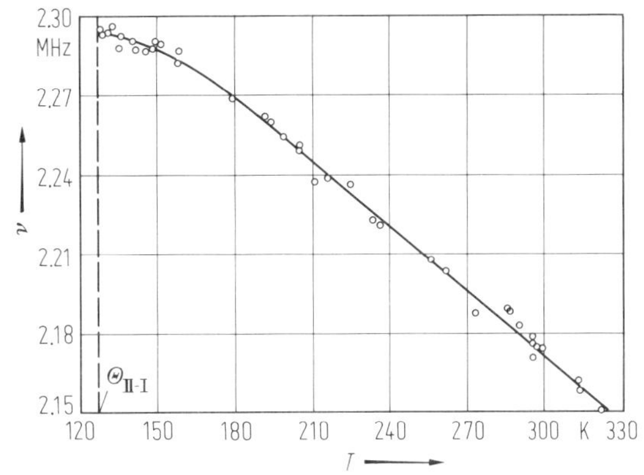

Fig. 1431. $(CH_3NHCH_2COOH)_3 \cdot CaCl_2$. ν vs. T [75K2]. ν: ^{35}Cl quadrupole resonance frequency.

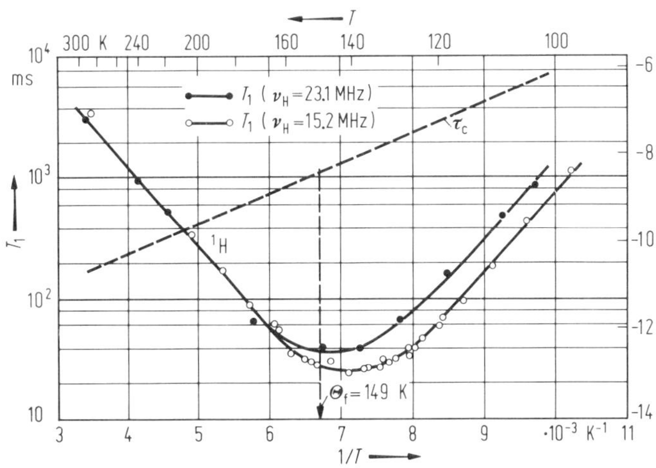

Fig. 1429. $(CH_3NHCH_2COOH)_3 \cdot CaCl_2$. T_1, $\log_{10} \tau_c$ vs. $1/T$ [70B9]. T_1: proton spin-lattice relaxation time. τ_c: correlation time in the unit of seconds. ν_4: Larmor frequency of proton.

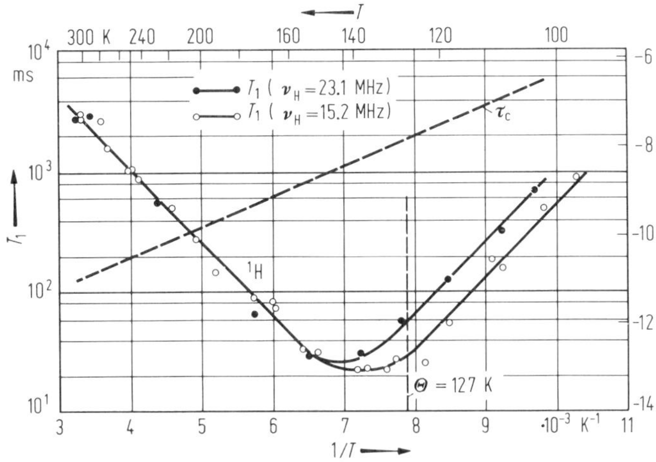

Fig. 1430. $(CH_3NDCH_2COOD)_3 \cdot CaCl_2$ (partially deuterated). T_1, $\log_{10} \tau_c$ vs. $1/T$ [70B9] T_1: proton spin-lattice relaxation time. τ_c: correlation time in the unit of s.

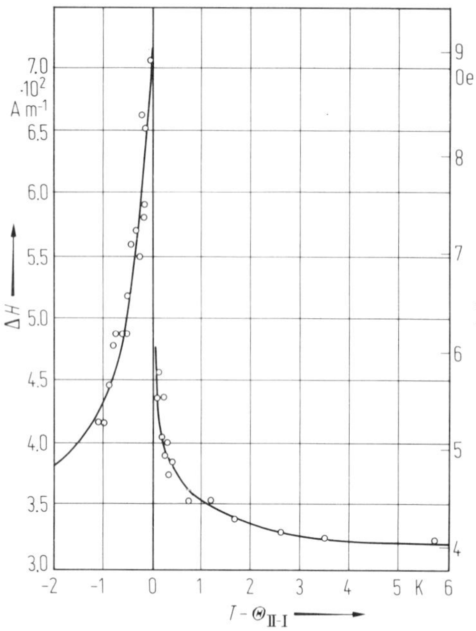

Fig. 1432. $(CH_3NHCH_2COOH)_3 \cdot CaCl_2$: Mn^{2+}. ΔH vs. $T - \Theta_{II-I}$ [76 L 5]. ΔH: ESR linewidth of Mn^{2+}.

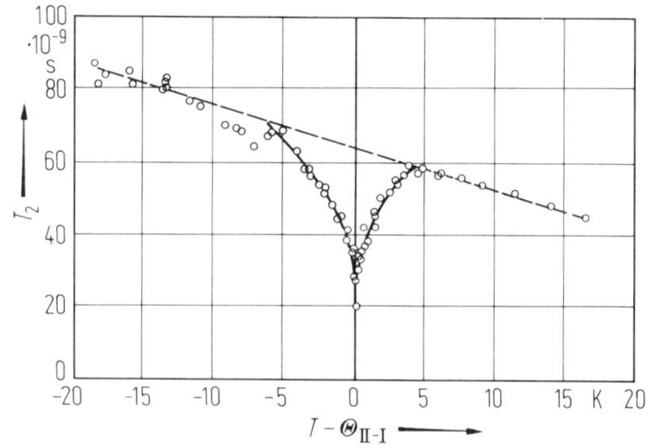

Fig. 1433. $(CH_3NHCH_2COOH)_3 \cdot CaCl_2$: Mn^{2+}. T_2 vs. $T - \Theta_{II-I}$ [76 W 4]. T_2: transverse relaxation time of Mn^{2+}.

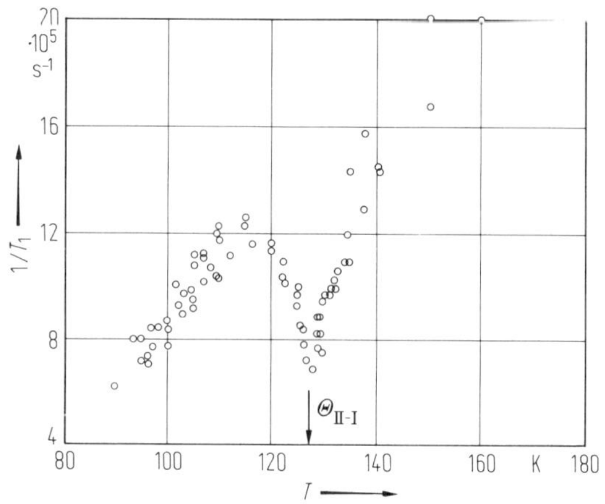

Fig. 1434. $(CH_3NHCH_2COOH)_3 \cdot CaCl_2$. T_1^{-1} vs. T [78 B 10]. T_1^{-1}: inverse of Mn^{2+} spin-lattice relaxation time.

41 Rochelle salt ($NaKC_4H_4O_6 \cdot 4H_2O$) family

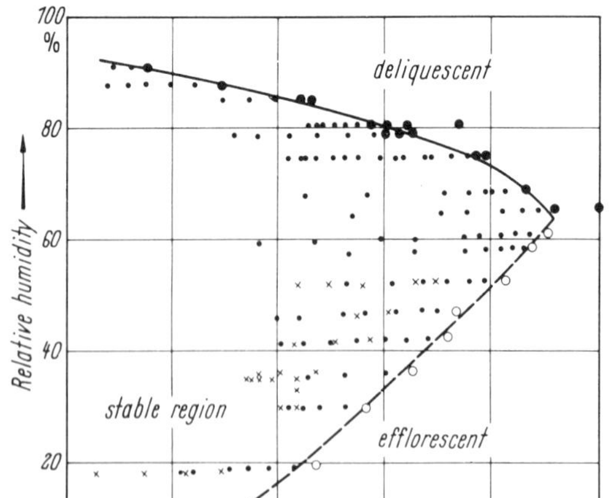

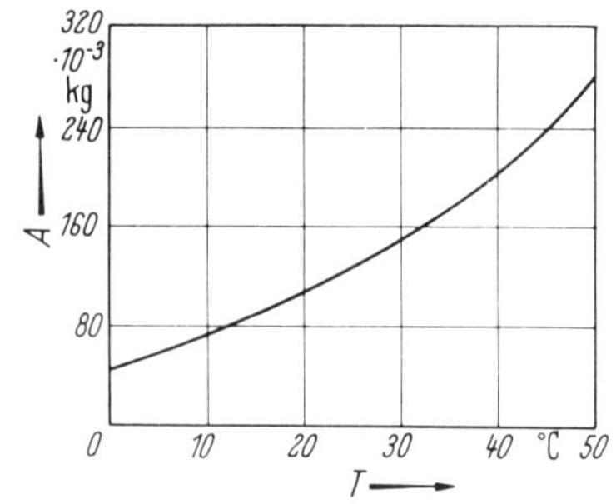

Fig. 1436. $NaKC_4H_4O_6 \cdot 4H_2O$ (RS). A vs. T [49 m 1]. A: weight of RS crystal soluble in 100 cm^3 (= 10^{-4} m^3) H_2O.

Fig. 1435. $NaKC_4H_4O_6 \cdot 4H_2O$ (RS). Stable region for deliquescence and efflorescence [43 K 1]. ○: a virgin sample which has undergone efflorescence or deliquescence; • and ×: a sample which has not undergone any change, •: a virgin sample, ×: a sample which has once undergone efflorescence.

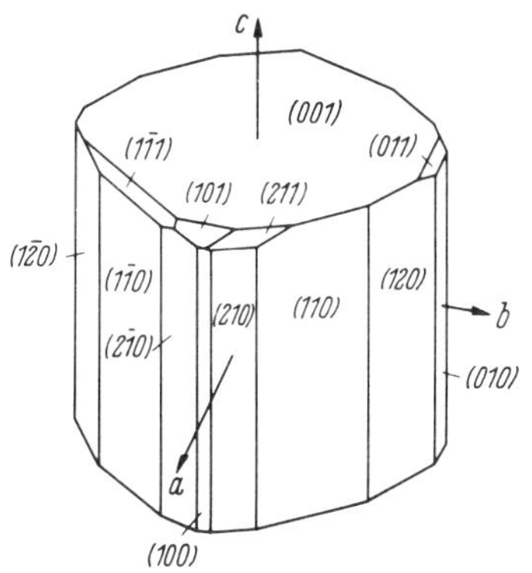

Fig. 1437. $NaKC_4H_4O_6 \cdot 4H_2O$ (RS). Crystal form of *d*-crystal. $X \| a$, $Y \| b$, $Z \| c$ [68M6].

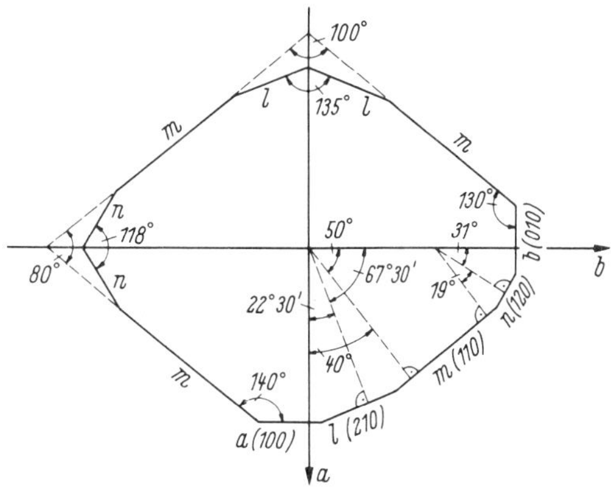

Fig. 1438. $NaKC_4H_4O_6 \cdot 4H_2O$ (RS). Crystal form [62j1]. The drawing represents the cross-section perpendicular to [001].

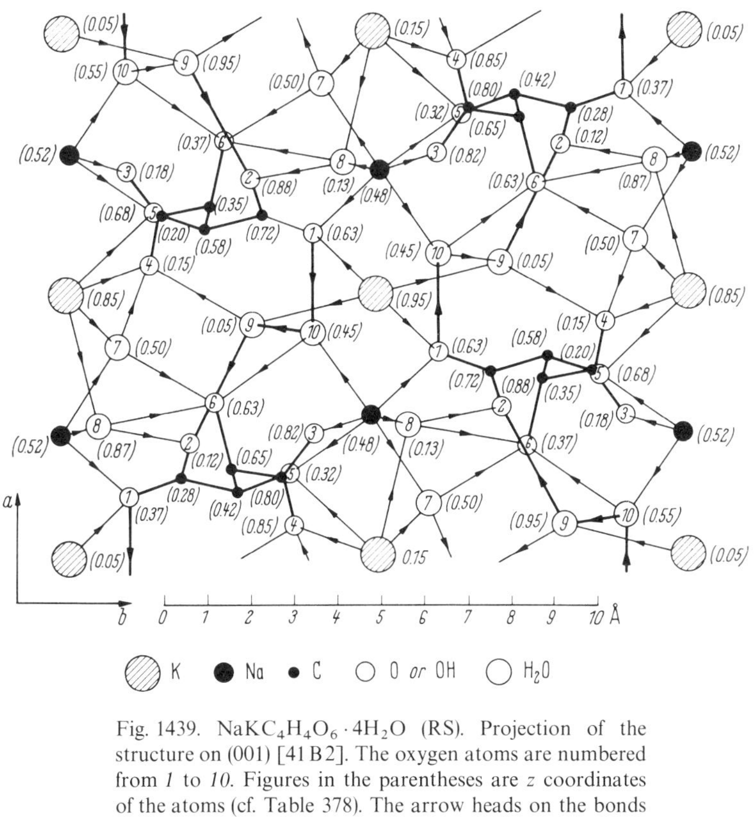

Fig. 1439. $NaKC_4H_4O_6 \cdot 4H_2O$ (RS). Projection of the structure on (001) [41B2]. The oxygen atoms are numbered from *1* to *10*. Figures in the parentheses are *z* coordinates of the atoms (cf. Table 378). The arrow heads on the bonds are drawn in a direction from positive atoms to negative atoms. The structure determination was made at RT, but the resolution was not good enough to show the effect of the monoclinic symmetry. For a refinement of the structure, see [57M3].

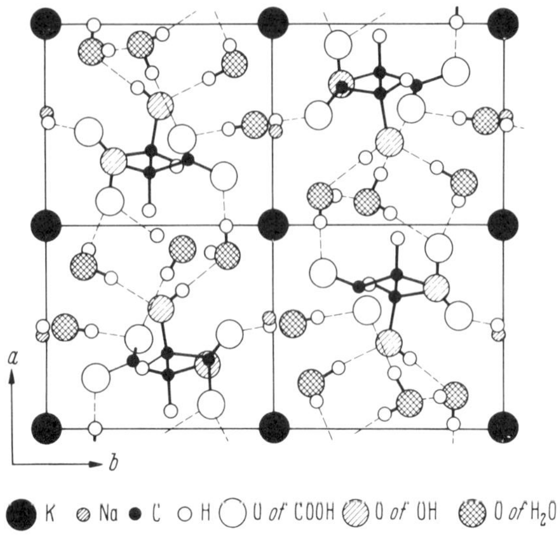

Fig. 1440. $NaKC_4H_4O_6 \cdot 4H_2O$ (RS). Projection of the structure on (001) [55 S 2]. The hydrogen bond system.

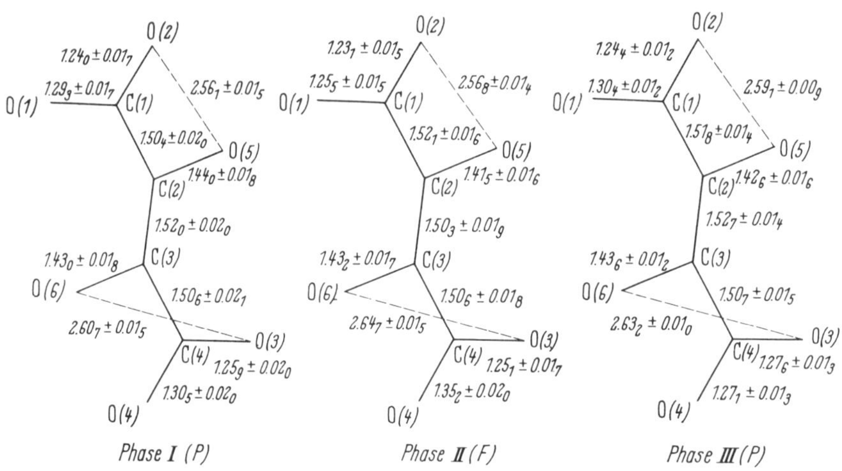

Fig. 1441. $NaKC_4H_4O_6 \cdot 4H_2O$ (RS). Interatomic distances in the tartrate ion in the phases I, II, and III [60 K 5].

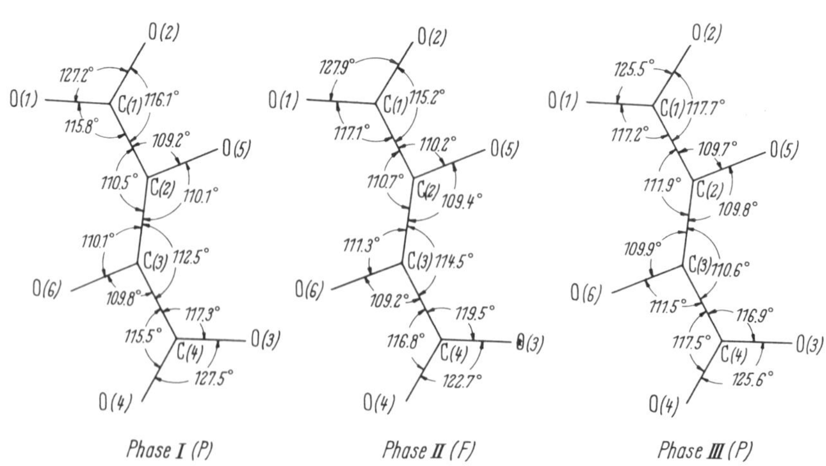

Fig. 1442. $NaKC_4H_4O_6 \cdot 4H_2O$ (RS). Bond angles in the tartrate ion in the phases I, II, and III [60 K 5].

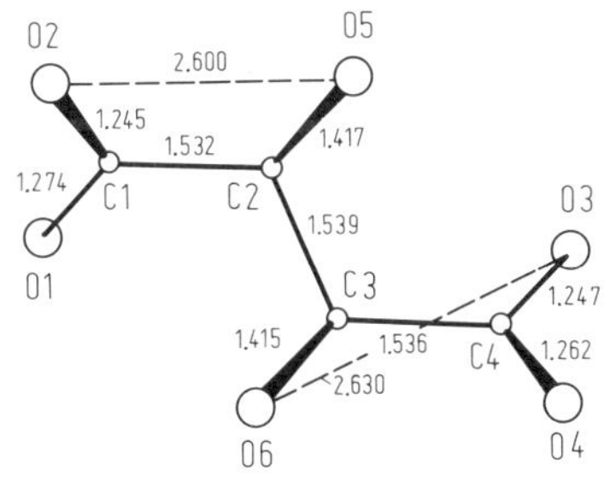

Fig. 1443. $NaKC_4H_4O_6 \cdot 4H_2O$ (RS). Interatomic distances (in Å) in the tartrate ion in the phase II [72S20].

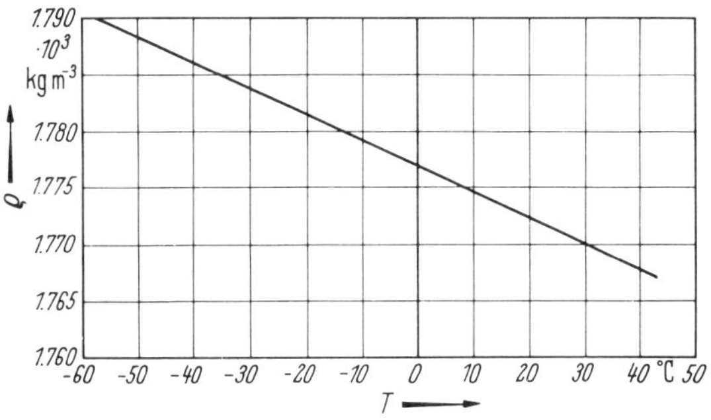

Fig. 1446. $NaKC_4H_4O_6 \cdot 4H_2O$ (RS). ϱ vs. T [50J1]. ϱ: density.

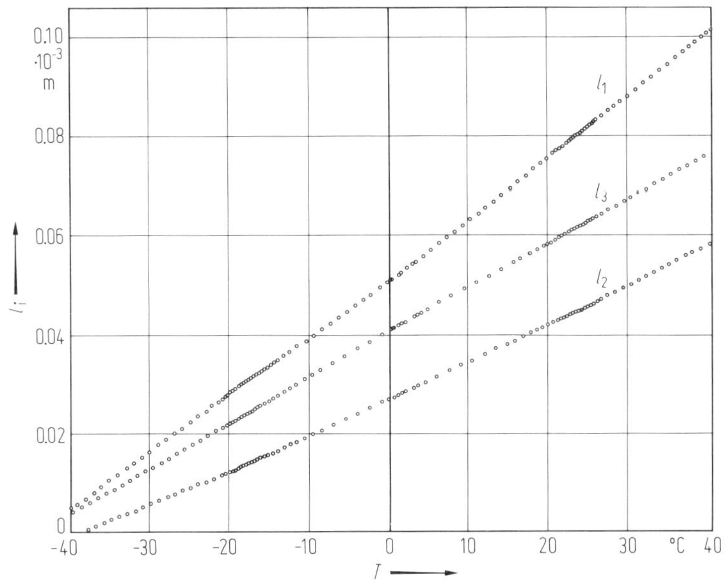

Fig. 1444. $NaKC_4H_4O_6 \cdot 4H_2O$ (RS). l_1, l_2, l_3 vs. T [76I2]. l_1, l_2, l_3: thermal dilatation along the a, b, c axes.

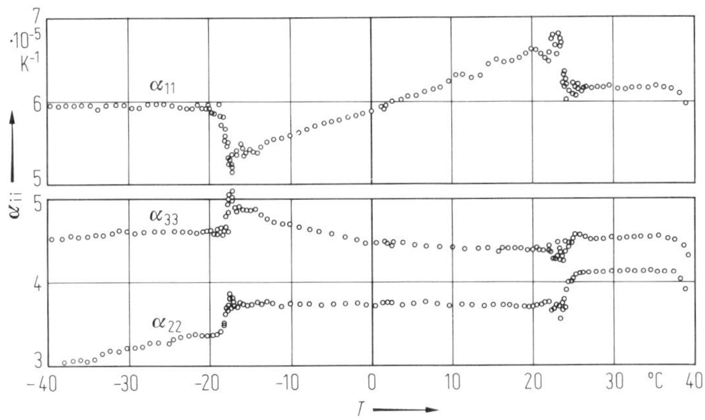

Fig. 1445. $NaKC_4H_4O_6 \cdot 4H_2O$ (RS). α_{11}, α_{22}, α_{33} vs. T [76I2]. α: linear thermal expansion coefficient.

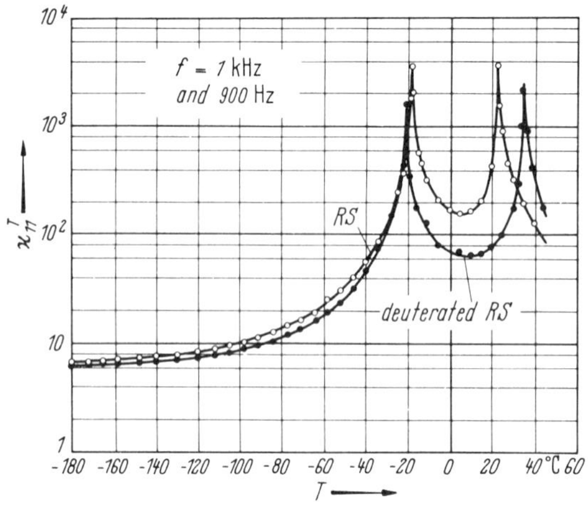

Fig. 1447. $NaKC_4H_4O_6 \cdot 4H_2O$ (RS), $NaKC_4H_2D_2O_6 \cdot 4D_2O$ (deuterated RS). $\varkappa_{11}^T$ vs. T [39H1].

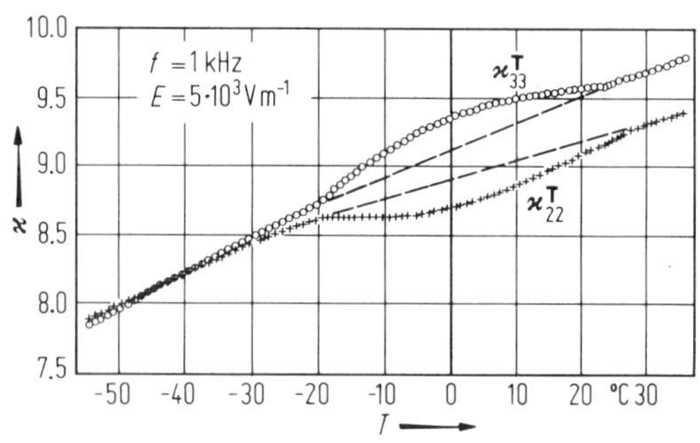

Fig. 1448. $NaKC_4H_4O_6 \cdot 4H_2O$ (RS). $\varkappa_{22}^T$, $\varkappa_{33}^T$ vs. T [68F4]. Relative humidity: 12%.

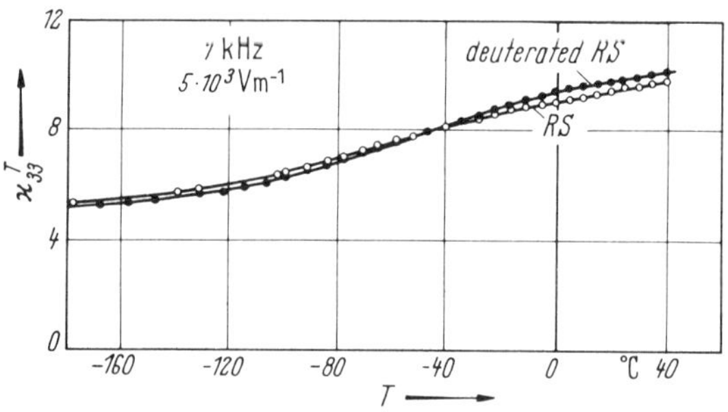

Fig. 1450. $NaKC_4H_4O_6 \cdot 4H_2O$ (RS), $NaKC_4H_2D_2O_6 \cdot 4D_2O$ (deuterated RS). $\varkappa_{33}^T$ vs. T [39H1].

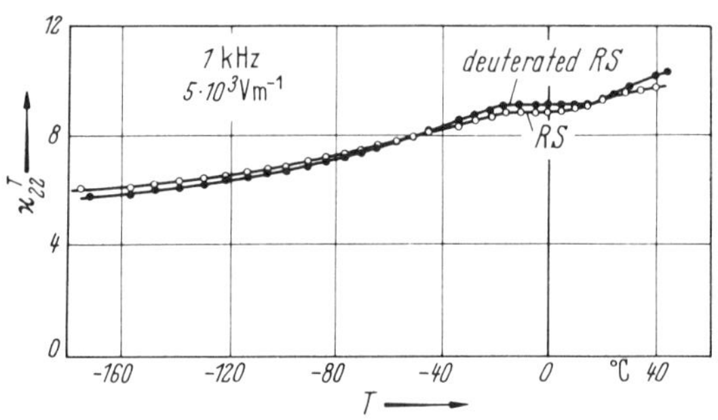

Fig. 1449. $NaKC_4H_4O_6 \cdot 4H_2O$ (RS), $NaK_4C_4H_2D_2O_6 \cdot 4D_2O$ (deuterated RS). $\varkappa_{22}^T$ vs. T [39H1].

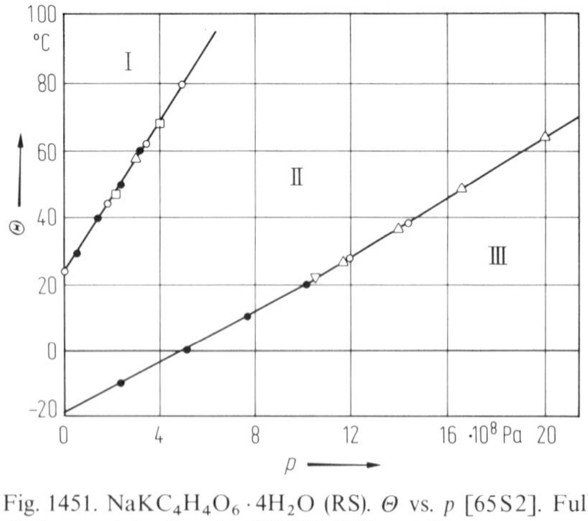

Fig. 1451. $NaKC_4H_4O_6 \cdot 4H_2O$ (RS). Θ vs. p [65S2]. Full circles are data given by [38B1]. Different figures represent results of different runs.

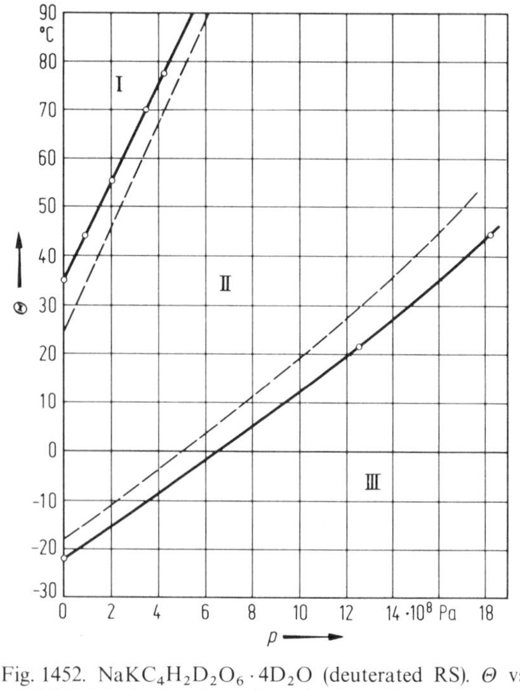

Fig. 1452. $NaKC_4H_2D_2O_6 \cdot 4D_2O$ (deuterated RS). Θ vs. p [68S3]. The broken lines show the data of normal salt.

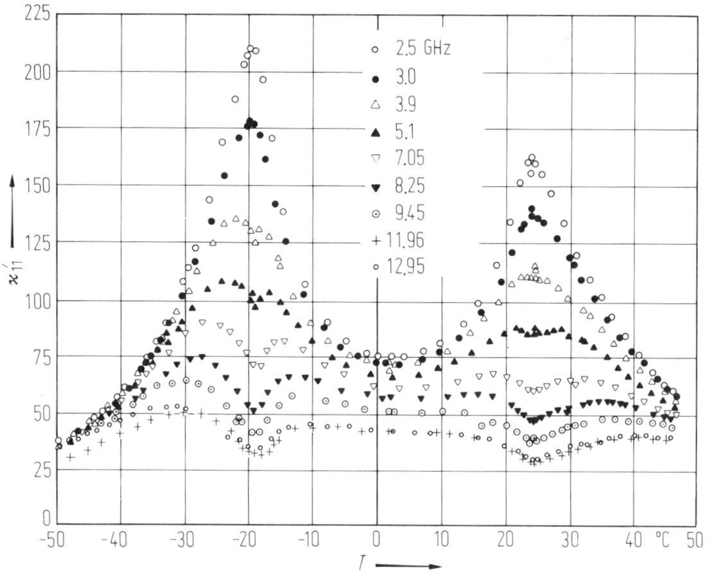

Fig. 1453. $NaKC_4H_4O_6 \cdot 4H_2O$ (RS). $\varkappa'_{11}$ vs. T [68S6]. Parameter: f.

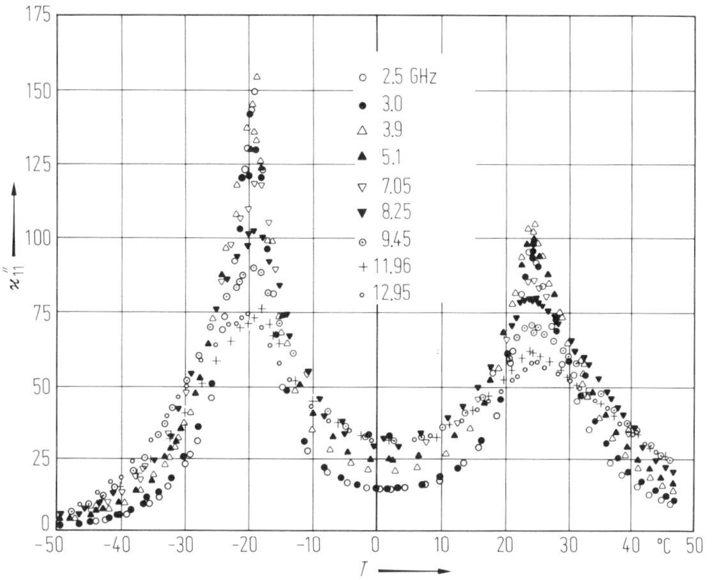

Fig. 1454. $NaKC_4H_4O_6 \cdot 4H_2O$ (RS). $\varkappa''_{11}$ vs. T [68S6]. Parameter: f.

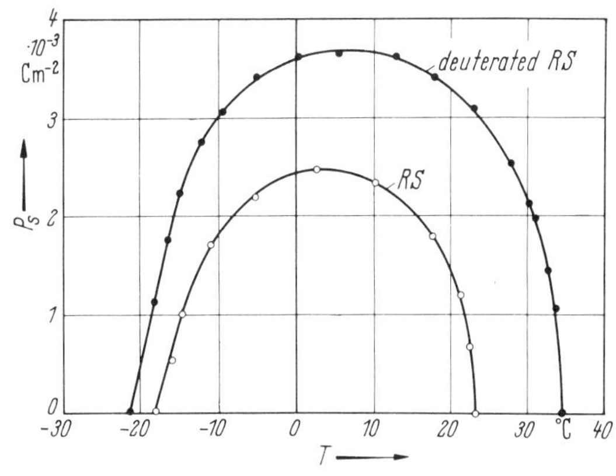

Fig. 1455. $NaKC_4H_4O_6 \cdot 4H_2O$ (RS), $NaKC_4H_2D_2O_6 \cdot 4D_2O$ (deuterated RS). P_s vs. T [39H1].

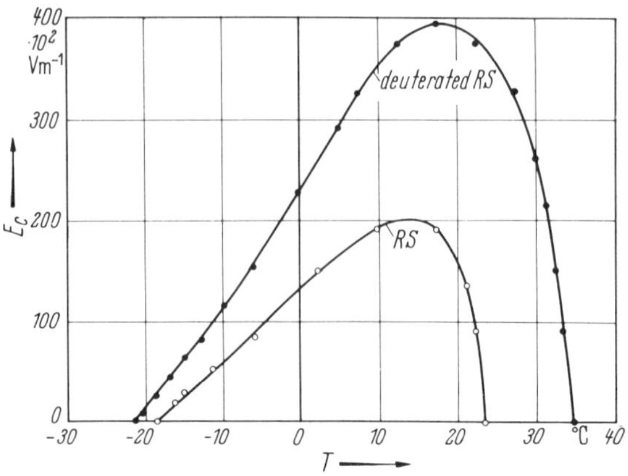

Fig. 1456. $NaKC_4H_4O_6 \cdot 4H_2O$ (RS), $NaKC_4H_2D_2O_6 \cdot 4D_2O$ (deuterated RS). E_c vs. T [39 H 1].

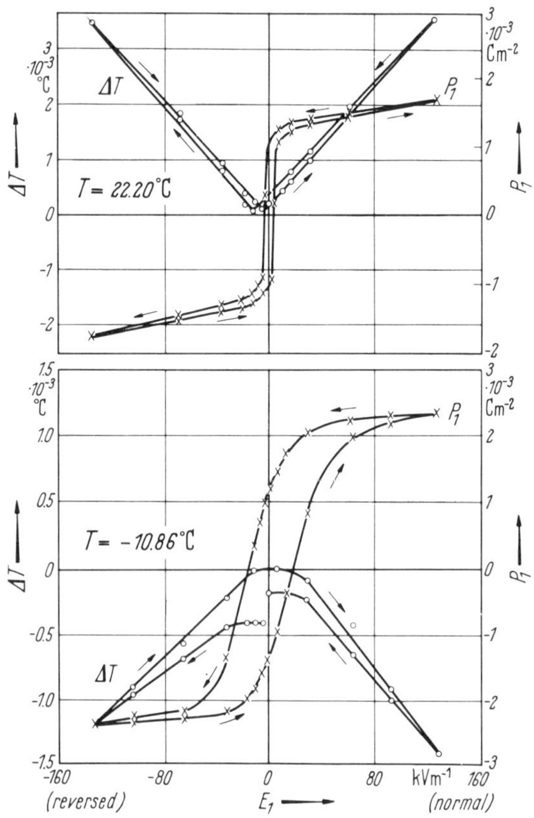

Fig. 1457. $NaKC_4H_4O_6 \cdot 4H_2O$ (RS). ΔT, P_1 vs. E_1 [63 W 7]. ΔT: electrocaloric temperature change; P_1, E_1: X component of $\boldsymbol{P}$, $\boldsymbol{E}$.

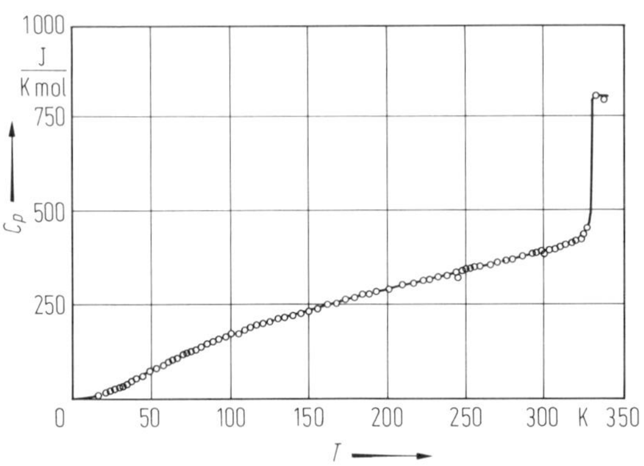

Fig. 1458. $NaKC_4H_4O_6 \cdot 4H_2O$ (RS). C_p vs. T [38 H 1].

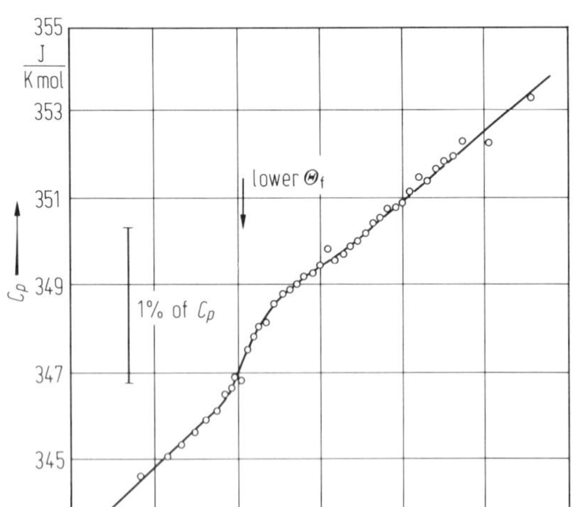

Fig. 1459. $NaKC_4H_4O_6 \cdot 4H_2O$ (RS). C_p vs. T [78 T 4].

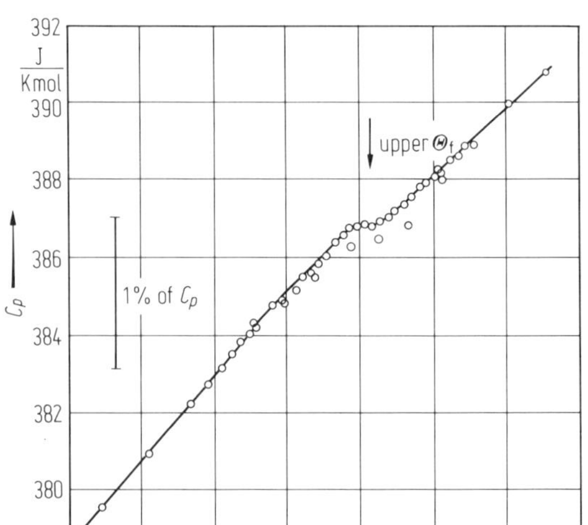

Fig. 1460. $NaKC_4H_4O_6 \cdot 4H_2O$ (RS). C_p vs. T [78 T 4].

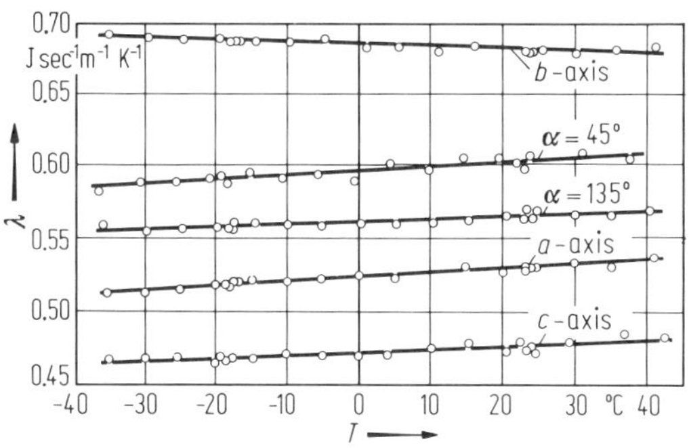

Fig. 1461. $NaKC_4H_4O_6 \cdot 4H_2O$ (RS). λ vs. T [72S8]. λ: thermal conductivity. $\alpha=45°$ and $\alpha=135°$ give two directions perpendicular to the a axis.

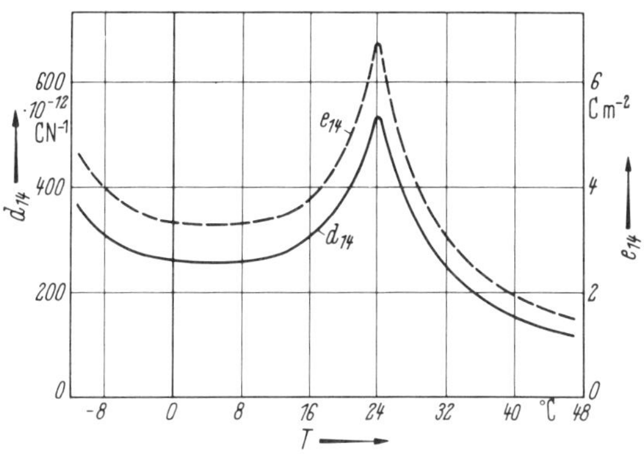

Fig. 1463. $NaKC_4H_4O_6 \cdot 4H_2O$ (RS). d_{14}, e_{14} vs. T [39M1].

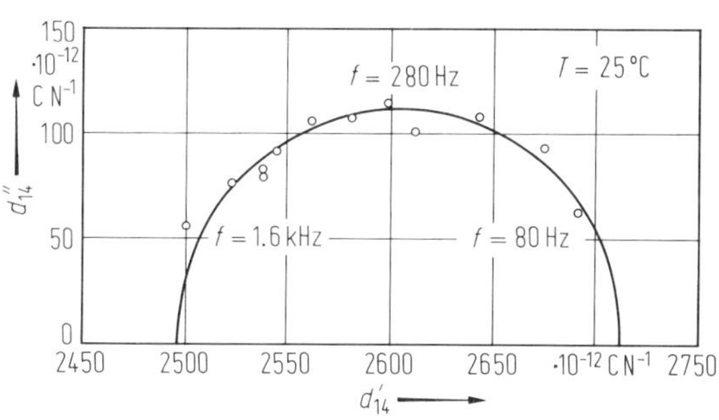

Fig. 1465. $NaKC_4H_4O_6 \cdot 4H_2O$ (RS). Debye circle for d_{14} in the complex plane ($d_{14}=d'_{14}-jd''_{14}$) [72M20].

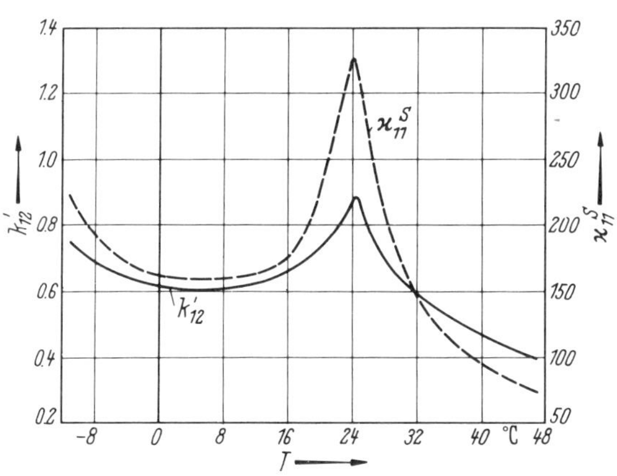

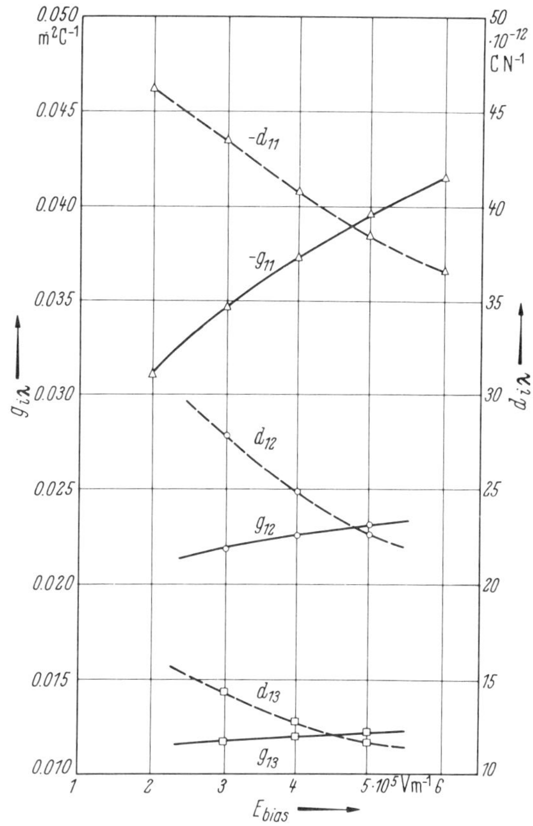

Fig. 1462. $NaKC_4H_4O_6 \cdot 4H_2O$ (RS). $g_{i\lambda}$, $d_{i\lambda}$ vs. E_{bias} at 17 °C [61S5]. See also [60F6].

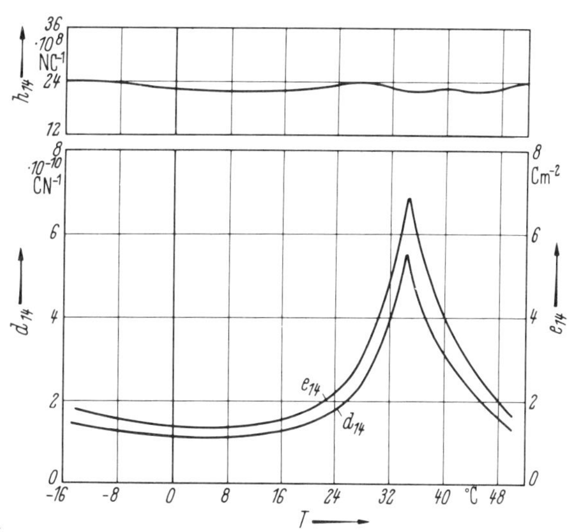

Fig. 1464. $NaKC_4H_2D_2O_6 \cdot 4D_2O$ (deuterated RS). d_{14}, e_{14}, h_{14} vs. T [40H1].

◄

Fig. 1466. $NaKC_4H_4O_6 \cdot 4H_2O$ (RS). $\varkappa^S_{11}$, k'_{12} vs. T [39M1]. k'_{12}: electromechanical coupling factor of a 45° X-cut bar (∥ [011]), namely $(k'_{12})^2=(d'_{12})^2/\varepsilon^T_{11}(s'_{22})^E$, $d'_{12}=d_{14}/2$, $(s'_{22})^E=(s_{22}+s_{33}+2s_{23}+s^E_{44})/4$.

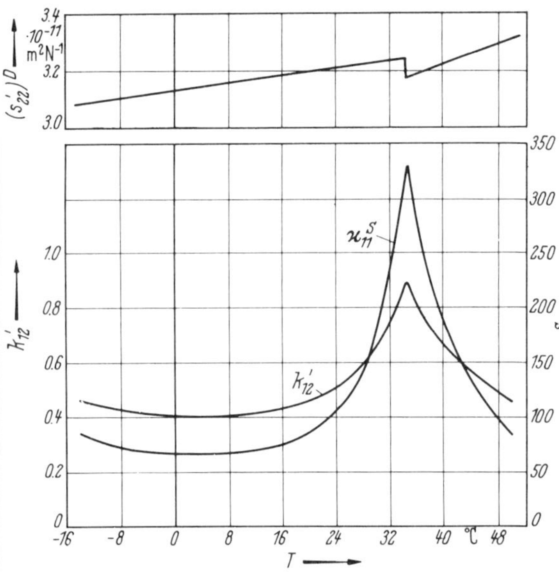

Fig. 1467. $NaKC_4H_2D_2O_6 \cdot 4D_2O$ (deuterated RS). $(s'_{22})^D$, k'_{12}, $\varkappa^S_{11}$ vs. T [40H1]. See the caption of Fig. 1466.

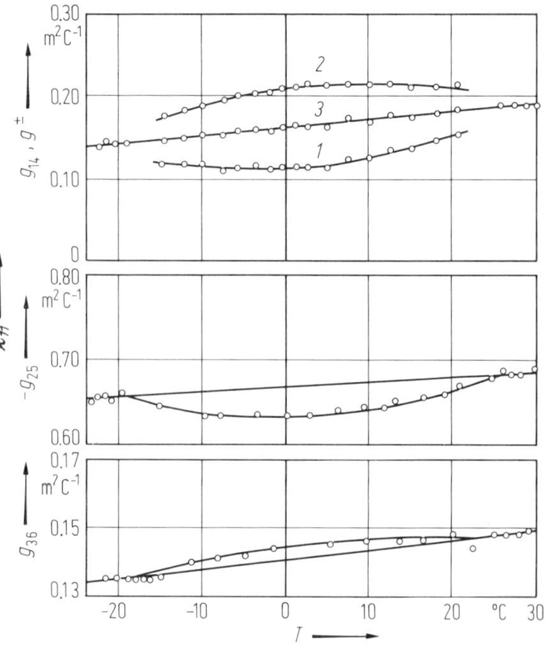

Fig. 1468. $NaKC_4H_4O_6 \cdot 4H_2O$ (RS). $g_{i\lambda}$ vs. T [76S9]. Strain is measured by capacitive dilatometric method. The curves *1*, *2* show $g^{\pm} = g_{14} \pm (g_{12} + g_{13})$, *3* shows $g_{14} = g^{+} = (g^{+} + g^{-})/2$. See also [67U3, 56M5].

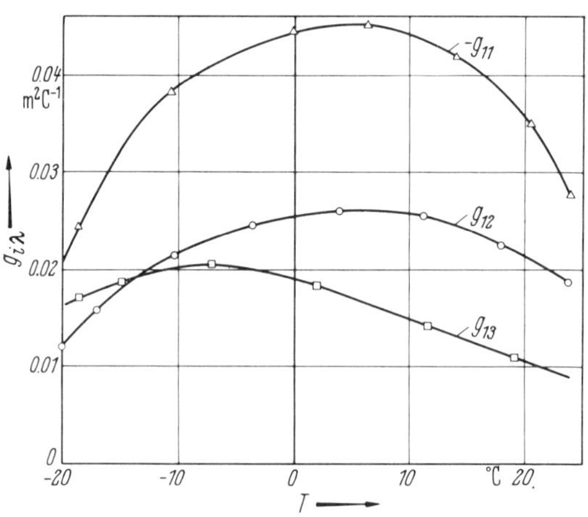

Fig. 1470. $NaKC_4H_4O_6 \cdot 4H_2O$ (RS). $-g_{11}$, g_{12}, g_{13} vs. T [61S5]. $E_{bias} = 5 \cdot 10^5 V\, m^{-1}$.

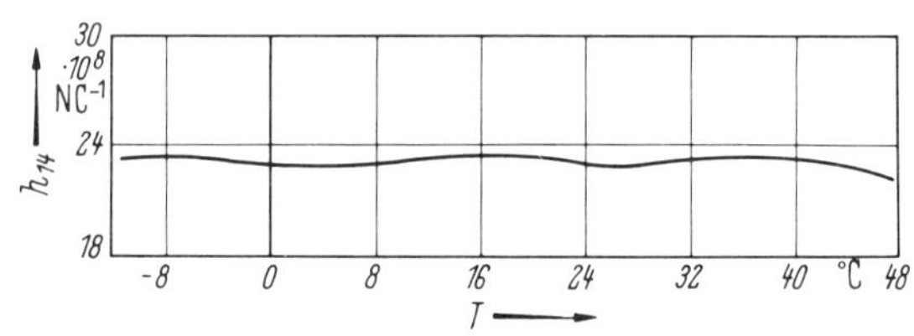

Fig. 1469. $NaKC_4H_4O_6 \cdot 4H_2O$ (RS). h_{14} vs. T [39M1].

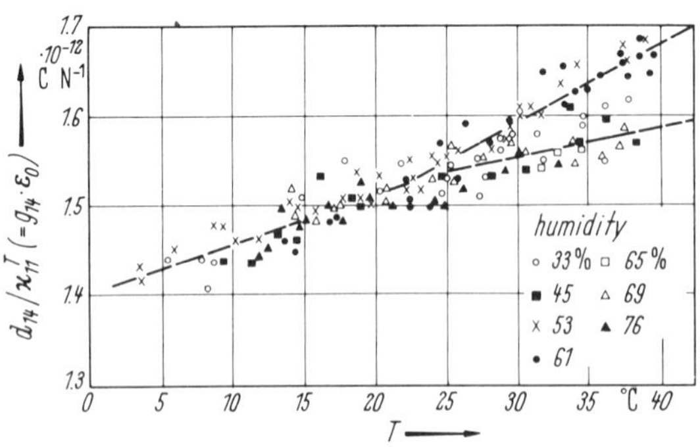

Fig. 1471. $NaKC_4H_4O_6 \cdot 4H_2O$ (RS). $d_{14}/\varkappa^T_{11}$ $(= g_{14}\varepsilon_0)$ vs. T [47K2]. Parameter: humidity. See also [49K1].

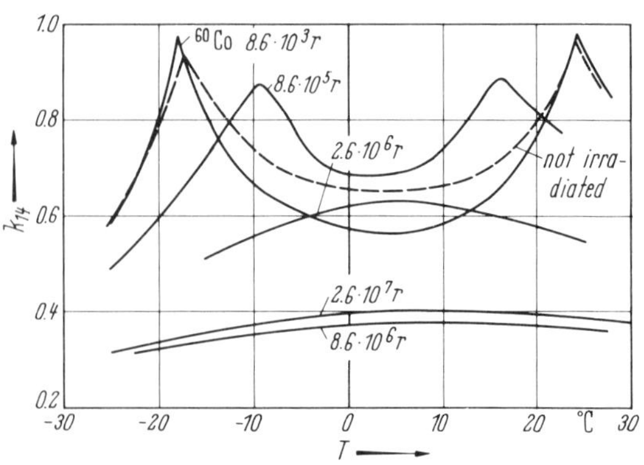

Fig. 1472. $NaKC_4H_4O_6 \cdot 4H_2O$ (RS). k_{14} vs. T [63K7]. Parameter: ^{60}Co irradiation level. $1r = 2.58 \cdot 10^{-4} C\, kg^{-1}$.

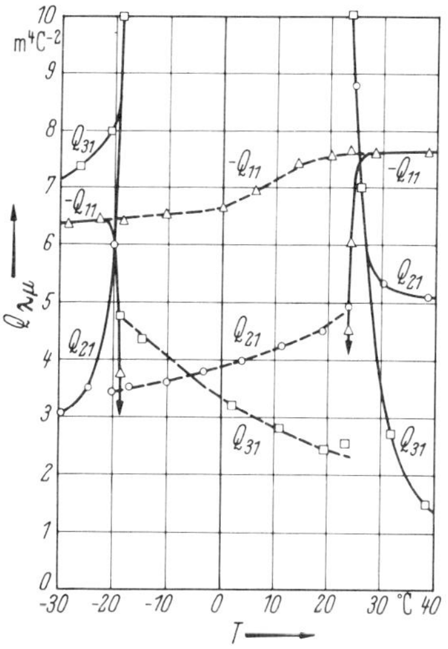

Fig. 1473. $NaKC_4H_4O_6 \cdot 4H_2O$ (RS). Q_{11}, Q_{21}, Q_{31} vs. T [61S5]. Q_{11}, Q_{21}, Q_{31}: electrostrictive constant. See also [67U3].

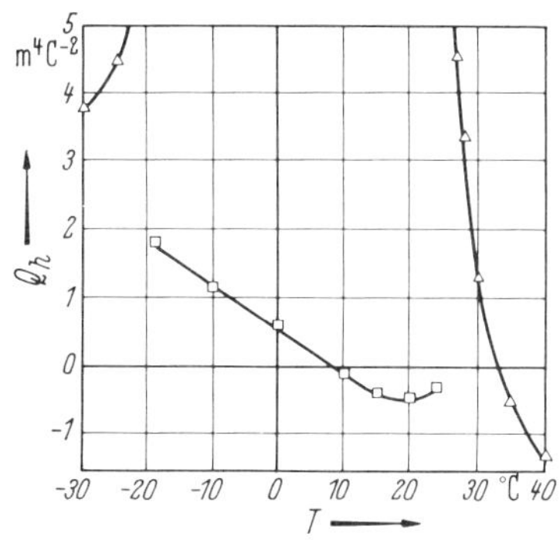

Fig. 1474. $NaKC_4H_4O_6 \cdot 4H_2O$ (RS). Q_h vs. T [61S5]. Q_h: volume electrostrictive constant.

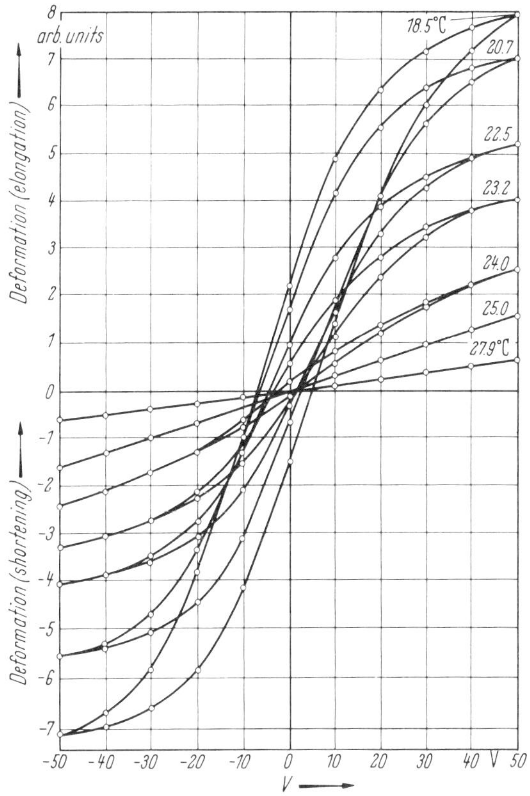

Fig. 1475. $NaKC_4H_4O_6 \cdot 4H_2O$ (RS). Deformation vs. V [39H2]. Parameter: T. 1 division of the deformation corresponds to an elongation of $3.6 \cdot 10^{-5}$ m m^{-1} along [011], and 1 V corresponds to $E = 5.7 \cdot 10^2$ V m^{-1} along [100].

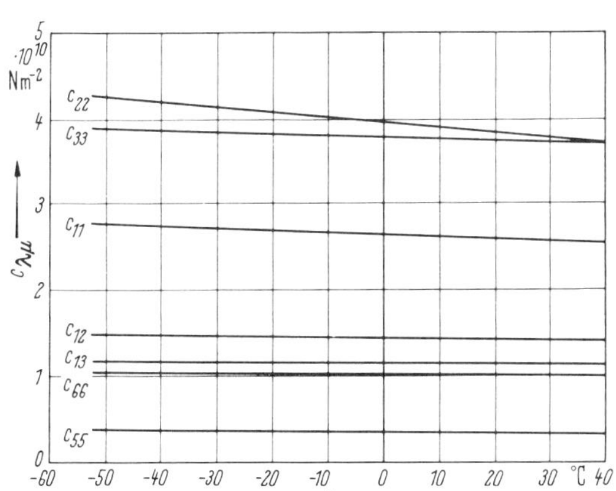

Fig. 1476. $NaKC_4H_4O_6 \cdot 4H_2O$ (RS). $c_{\lambda\mu}$ vs. T [50J1].

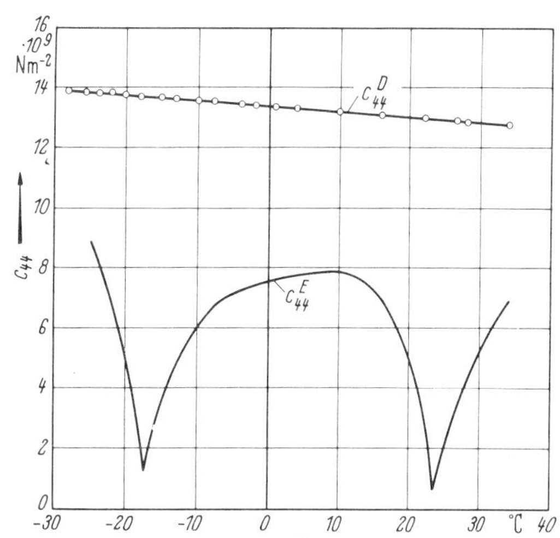

Fig. 1477. $NaKC_4H_4O_6 \cdot 4H_2O$ (RS). c^D_{44}, c^E_{44} vs. T [64B2].

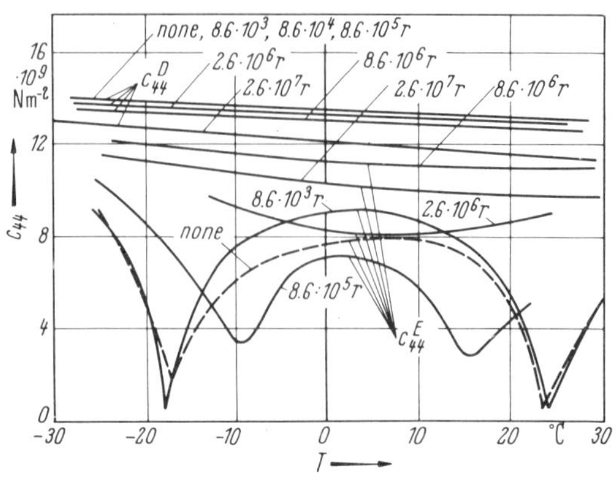

Fig. 1478. $NaKC_4H_4O_6 \cdot 4H_2O$ (RS). $c_{44}^{\boldsymbol{D}}$, $c_{44}^{\boldsymbol{E}}$ vs. T [63K7]. Parameter: ^{60}Co irradiation level. See also [61T1]. $1r = 2.58 \cdot 10^{-4} C\,kg^{-1}$.

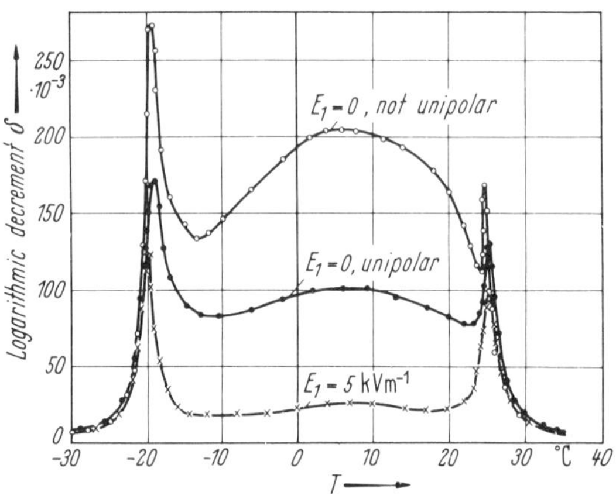

Fig. 1480. $NaKC_4H_4O_6 \cdot 4H_2O$ (RS). δ vs. T [63S7]. Parameter: bias field E_1 along [100]. δ: logarithmic decrement of a 45° X-cut bar.

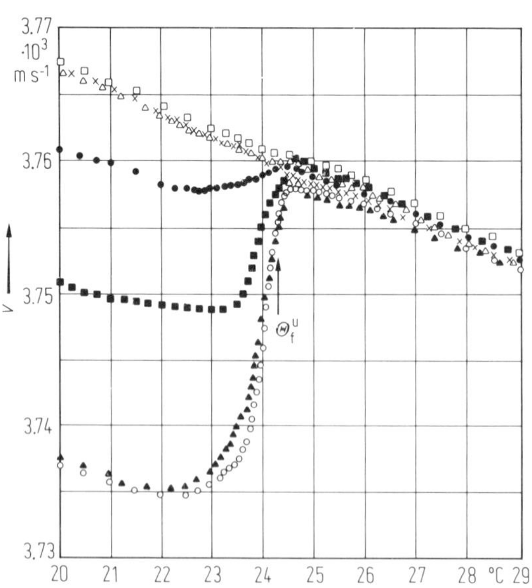

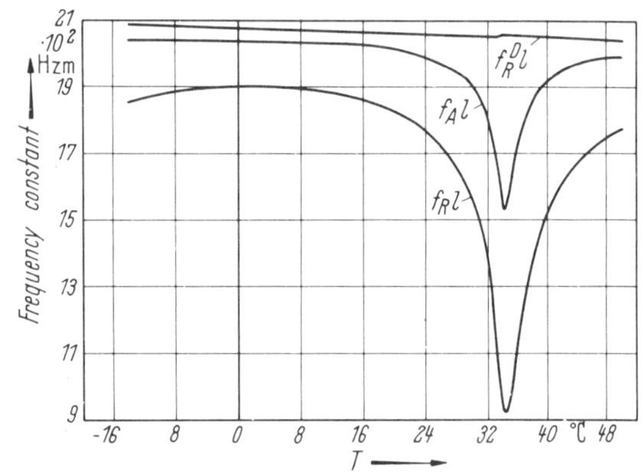

Fig. 1479. $NaKC_4H_2D_2O_6 \cdot 4D_2O$ (deuterated RS). $f_R l$, $f_A l$, $f_R^{\boldsymbol{D}} l$ vs. T [40H1]. $f_R l$, $f_A l$, $f_R^{\boldsymbol{D}} l$: frequency constant of a 45° X-cut bar. f_R, f_A: resonant, antiresonant frequency of a plated crystal, $f_R l - \{\varrho(s_{22}+s_{33}+2s_{23}+s_{44}^{\boldsymbol{E}})/4\}^{1/2}/2$. $f_R^{\boldsymbol{D}}$ resonant frequency of an unplated crystal, $f_R^{\boldsymbol{D}} l = \{\varrho(s_{22}+s_{33}+2s_{23}+s_{44}^{\boldsymbol{D}})/4\}^{-1/2}/2$. l: length, ϱ: density.

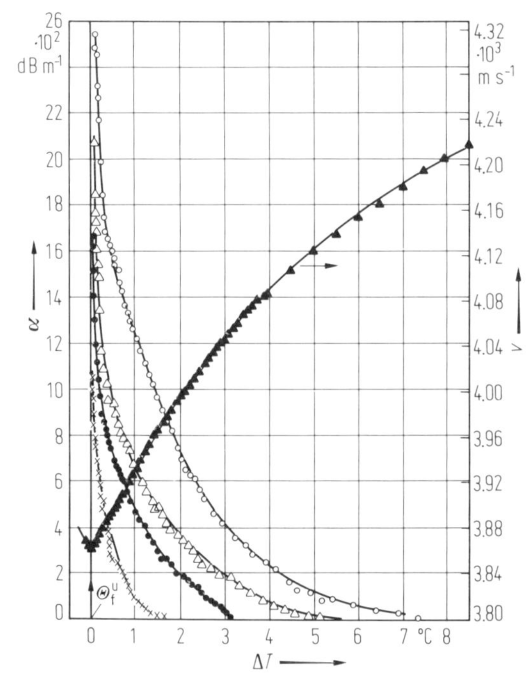

Fig. 1482. $NaKC_4H_4O_6 \cdot 4H_2O$ (RS). α, v vs. ΔT [77K8]. α: anomalous part of the attenuation of ultrasonic quasi-longitudinal wave propagated along the 45° direction in a 45° X-cut crystal, for 7 MHz (crosses), 21 MHz (full circles), 35 MHz (open triangles), 49 MHz (open circles). v: velocity, for 21 MHz (full triangles). $\Delta T = T - \Theta_f^u$.

◄ Fig. 1481. $NaKC_4H_4O_6 \cdot 4H_2O$ (RS). v vs. T [76K4]. v: ultrasonic velocity of longitudinal wave propagating along the a axis. $f = 15$ MHz. Open circles: warming run at $E_{bias} = 0$; full triangles: cooling run at $E_{bias} = 0$; open triangles: warming run at $18 \cdot 10^3\,V\,m^{-1}$; full circles: cooling run at $18 \cdot 10^3\,V\,m^{-1}$; crosses: warming run at $1 \cdot 10^5\,V\,m^{-1}$; open squares: cooling run at $1 \cdot 10^5\,V\,m^{-1}$; full squares: warming run for the poled crystal.

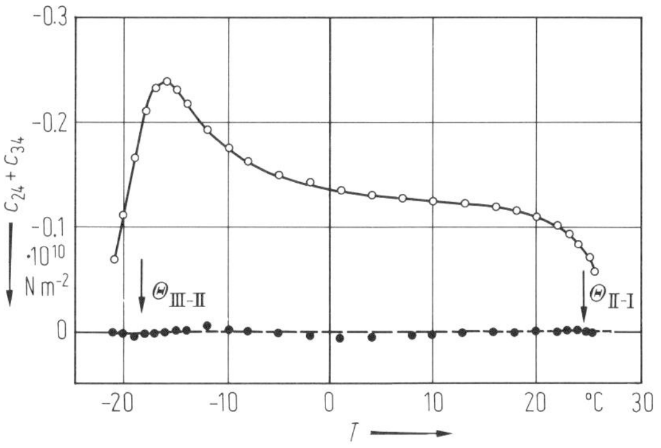

Fig. 1483. $NaKC_4H_4O_6 \cdot 4H_2O$ (RS). $c_{24}+c_{34}$ vs. T [76S2]. Open circles: $c_{24}^E+c_{34}^E$; full circles: $c_{24}^D+c_{34}^D$. See also [49P1].

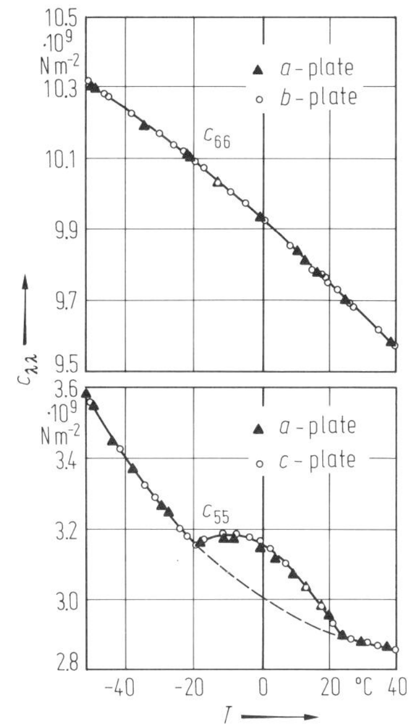

Fig. 1484. $NaKC_4H_4O_6 \cdot 4H_2O$ (RS). $c_{\lambda\lambda}$ vs. T [74S1]. Values are obtained by measuring frequencies of piezoelectric lines in NQR.

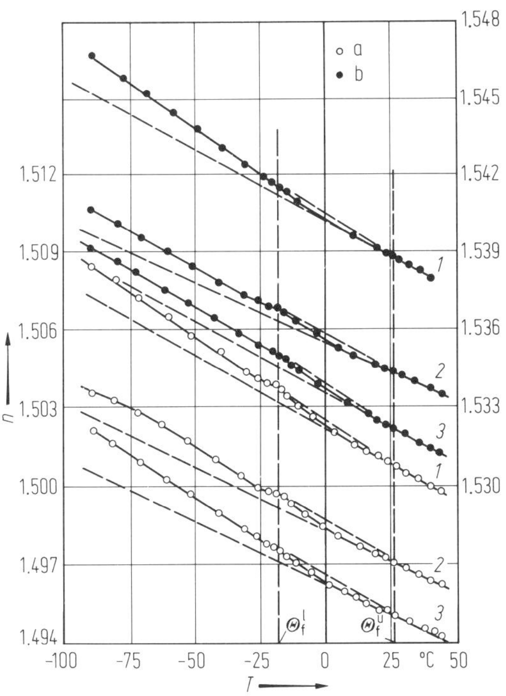

Fig. 1485. $NaKC_4H_4O_6 \cdot 4H_2O$ (RS). n vs. T [76R7]. a: $\lambda = 500$ nm (left-hand scale), b: $\lambda = 300$ nm (right-hand scale). 1: n_a, 2: n_b, 3: n_c.

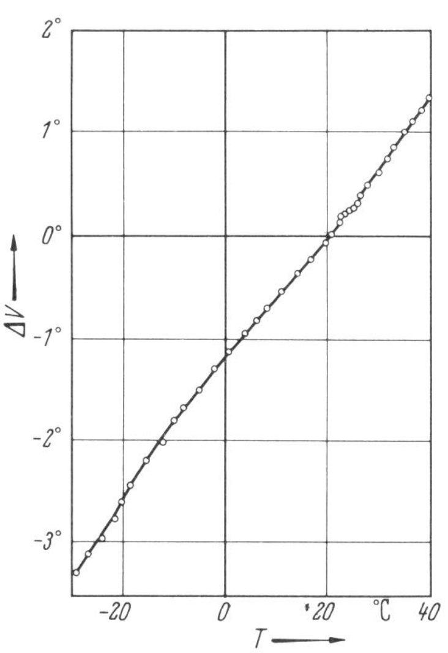

Fig. 1486. $NaKC_4H_4O_6 \cdot 4H_2O$ (RS). ΔV vs. T [66I6]. ΔV: change in V; $2V$: optical axial angle. $\lambda = 558$ nm. See also [67B13].

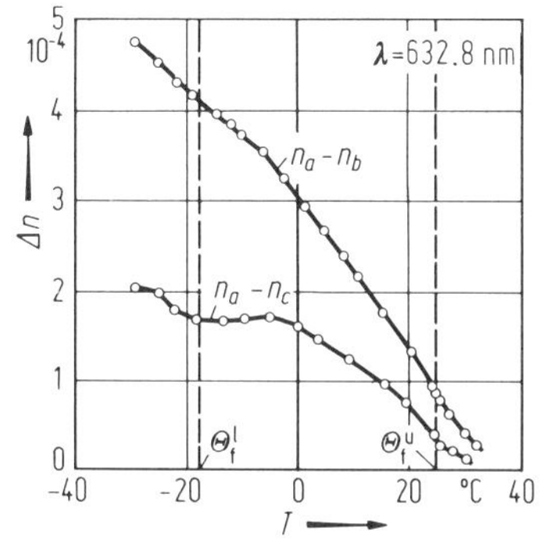

Fig. 1487. $NaKC_4H_4O_6 \cdot 4H_2O$ (RS). Δn vs. T [70L11].

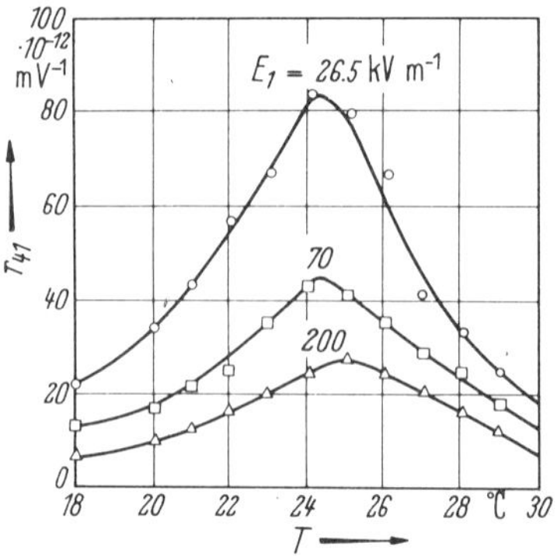

Fig. 1488. $NaKC_4H_4O_6 \cdot 4H_2O$ (RS). r_{41} vs. T [66A5]. Parameter: applied field E_1 along [100]. $\lambda = 546$ nm.

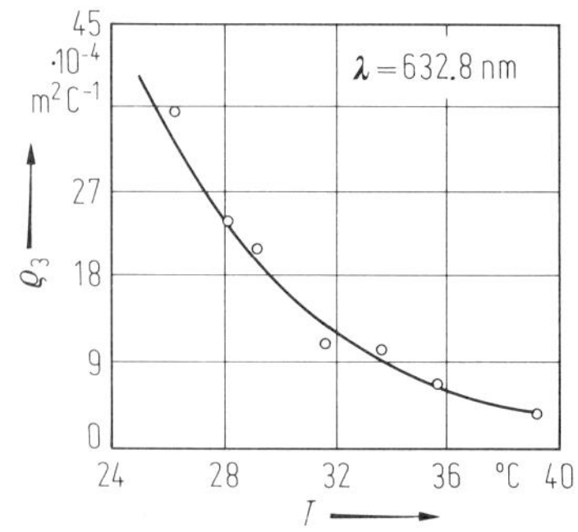

Fig. 1489. $NaKC_4H_4O_6 \cdot 4H_2O$ (RS). ϱ_3 vs. T [67A5]. $\varrho_3 = \varrho_{11} - (n_c/n_a)^3\, \varrho_{31}$.

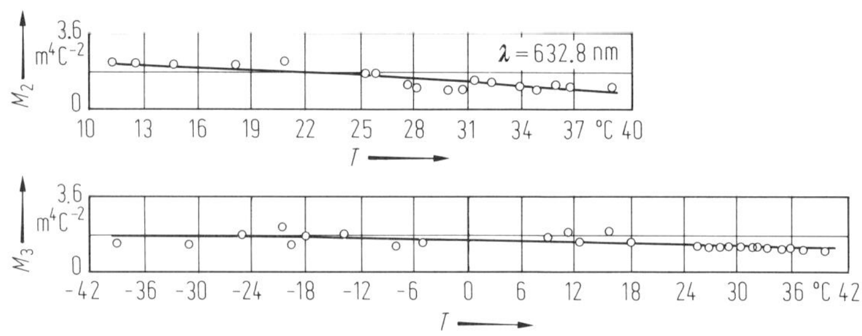

Fig. 1490. $NaKC_4H_4O_6 \cdot 4H_2O$ (RS). M_2, M_3 vs. T [67A5]. $M_2 = M_{11} - (n_b/n_a)^3 M_{21}$, $M_3 = M_{11} - (n_c/n_a)^3\, M_{31}$.

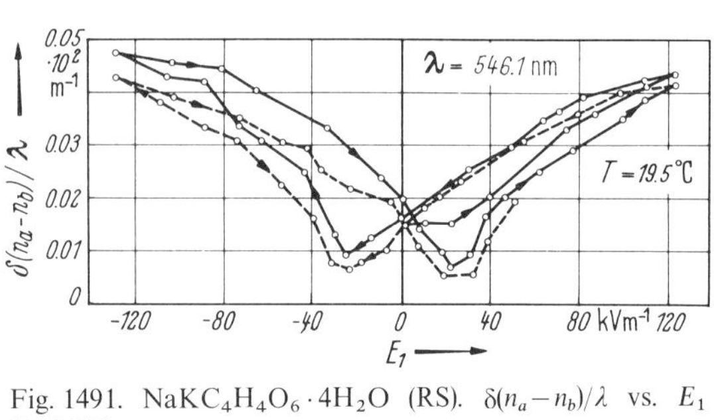

Fig. 1491. $NaKC_4H_4O_6 \cdot 4H_2O$ (RS). $\delta(n_a - n_b)/\lambda$ vs. E_1 [35M1].

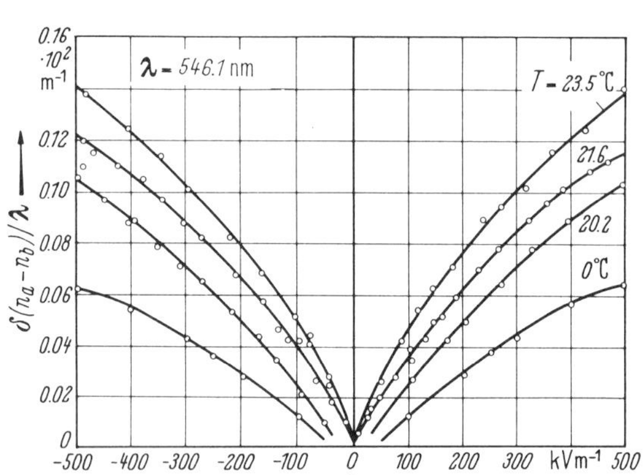

Fig. 1492. $NaKC_4H_4O_6 \cdot 4H_2O$ (RS). $\delta(n_a - n_b)/\lambda$ vs. E_1 [35M1]. Parameter: T.

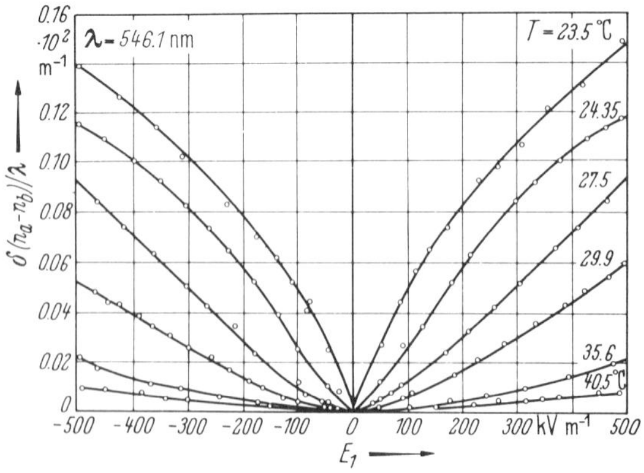

Fig. 1493. $NaKC_4H_4O_6 \cdot 4H_2O$ (RS). $\delta(n_a - n_b)/\lambda$ vs. E_1 [35M1]. Parameter: T.

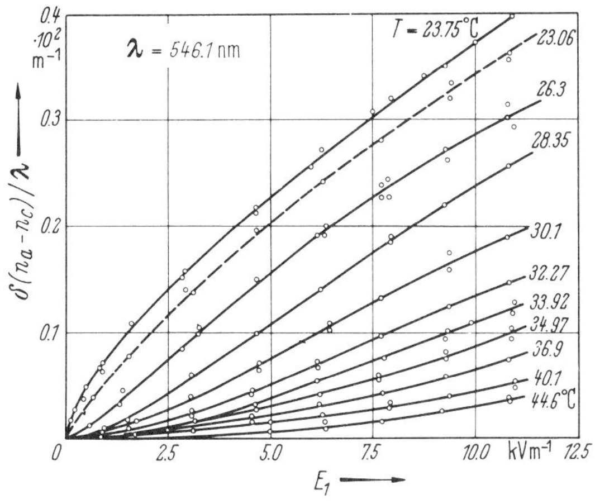

Fig. 1494. $NaKC_4H_4O_6 \cdot 4H_2O$ (RS). $\delta(n_a - n_c)/\lambda$ vs. E_1 [35M1]. Parameter: T.

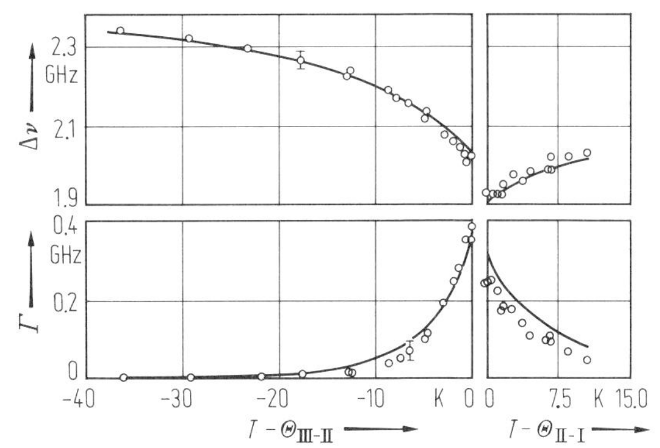

Fig. 1495. $NaKC_4H_4O_6 \cdot 4H_2O$ (RS). $\Delta\nu$, Γ vs. $T - \Theta_{III-II}$ and $T - \Theta_{II-I}$ [75S2]. $\Delta\nu$: frequency shift of Brillouin scattering. Γ: full width at half maximum of spectral line. Quasi-longitudinal [011] phonon. Scattering angle $\theta = 10.2°$.

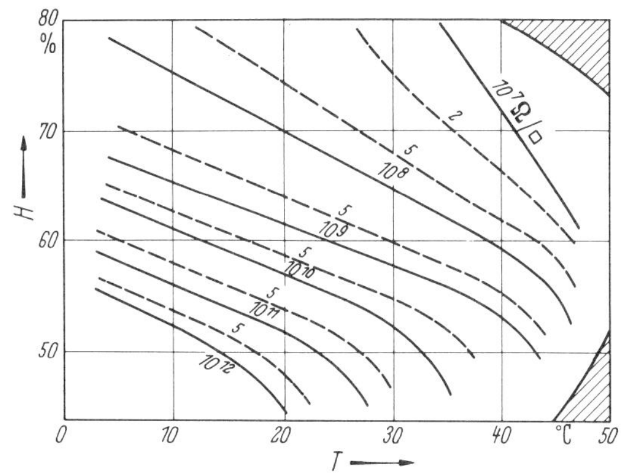

Fig. 1497. $NaKC_4H_4O_6 \cdot 4H_2O$ (RS). H vs. T [47K1]. Parameter: surface resistivity of (001) along [100]. H: relative humidity. Applied field: ≈700 V m^{-1}.

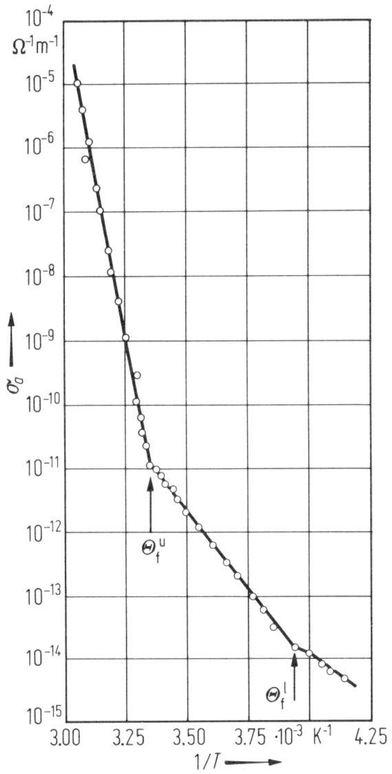

Fig. 1496. $NaKC_4H_4O_6 \cdot 4H_2O$ (RS). σ_a vs. $1/T$ [62G3]. Three electrode method with shielding electrode. Applied field: 60 kV m^{-1}.

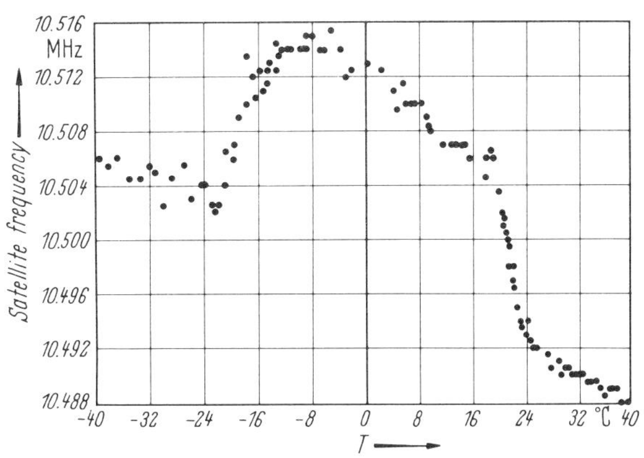

Fig. 1498. $NaKC_4H_4O_6 \cdot 4H_2O$ (RS). Satellite frequency vs. T [66M2]. Satellite frequency of one ^{23}Na.

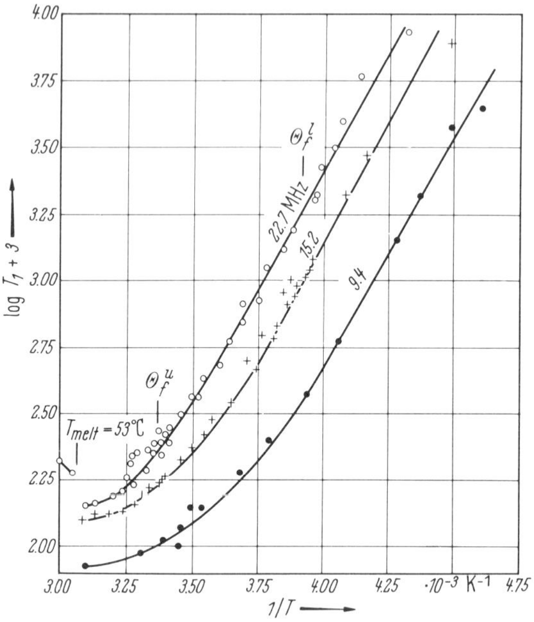

Fig. 1499. $NaKC_4H_4O_6 \cdot 4H_2O$ (RS). (log $T_1 + 3$) vs. $1/T$ [66B6]. Parameter: f. T_1: spin lattice relaxation time of proton, in s.

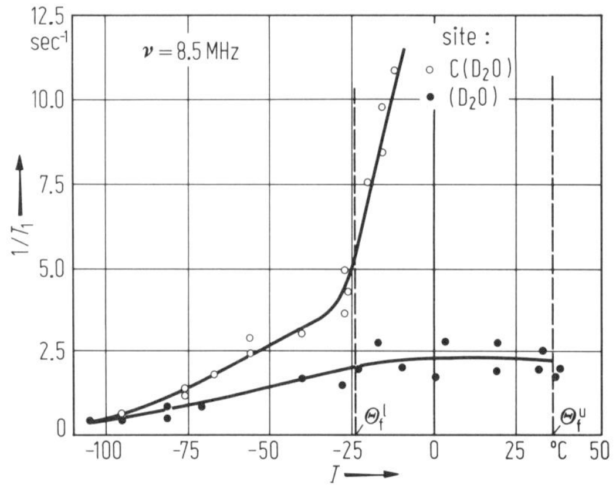

Fig. 1501. $NaKC_4H_2D_2O_6 \cdot 4D_2O$ (deuterated RS). $1/T_1$ vs. T [69B8]. T_1: deuteron spin-lattice relaxation time. Open circle: D_2O on C site, full circle: D_2O on A and B sites.

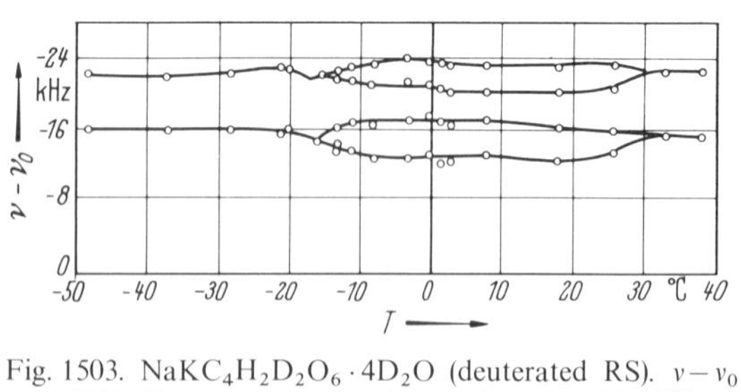

Fig. 1503. $NaKC_4H_2D_2O_6 \cdot 4D_2O$ (deuterated RS). $\nu - \nu_0$ vs. T[64B6]. $\nu - \nu_0$: second order quadrupole shift of ^{23}Na. Direction of magnetic field H: $c \perp \boldsymbol{H}$; ∢ $(b, \boldsymbol{H}) = 278°$. ν_0: Larmor frequency.

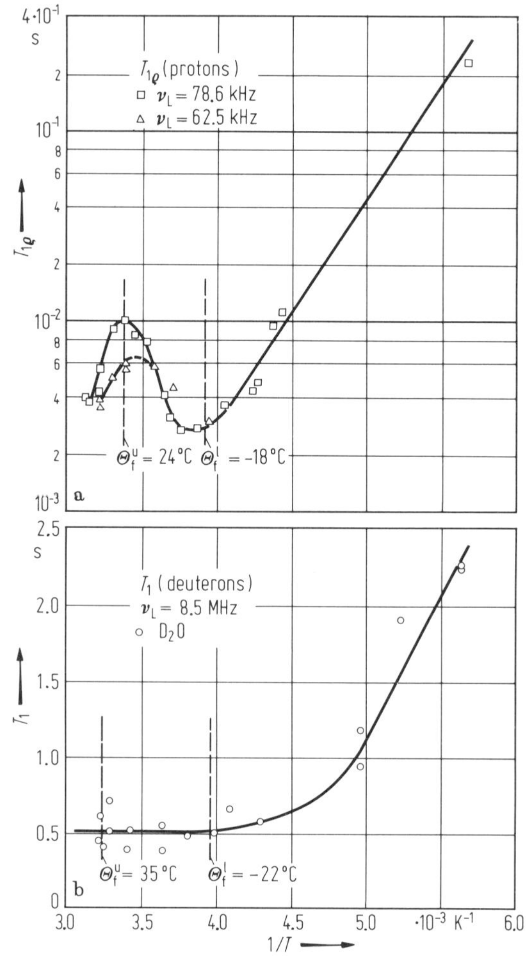

Fig. 1500. $NaKC_4H_4O_6 \cdot 4H_2O$ (RS) (Fig. a), $NaKC_4H_2D_2O_6 \cdot 4D_2O$ (deuterated RS) (Fig. b). $T_{1\varrho}$, T_1 vs. $1/T$ [69B11]. ν_L: Larmor frequency.

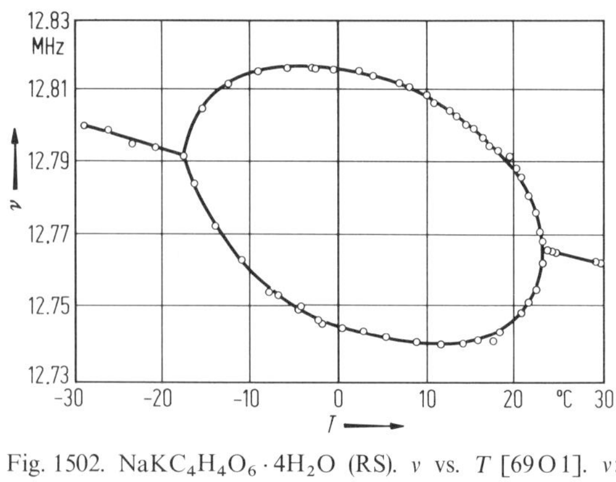

Fig. 1502. $NaKC_4H_4O_6 \cdot 4H_2O$ (RS). ν vs. T [69O1]. ν: frequency of satellite line. See also [70F2].

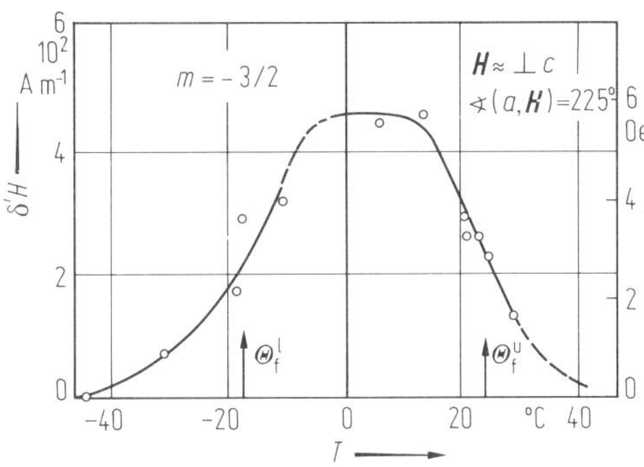

Fig. 1504. $NaKC_4H_4O_6 \cdot 4H_2O$ (RS). $\delta' H$ vs. T [71V4]. $\delta' H$: shift of resonance field under $E = 10^5\,V\,m^{-1}$ along [100].

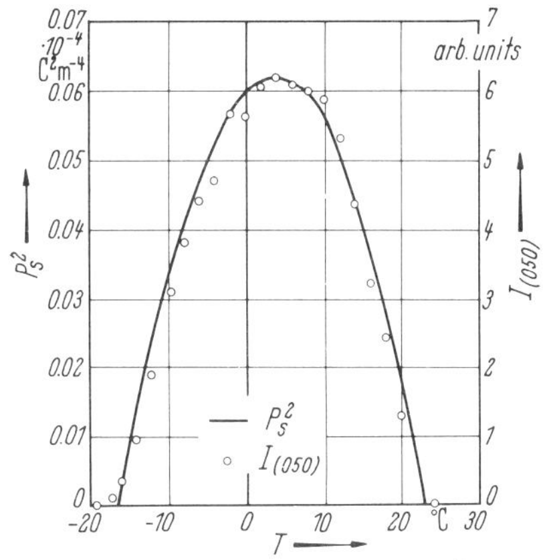

Fig. 1505. $NaKC_4H_4O_6 \cdot 4H_2O$ (RS). $I_{(050)}$, P_s^2 vs. T [67S12]. $I_{(050)}$: (050) integrated intensity.

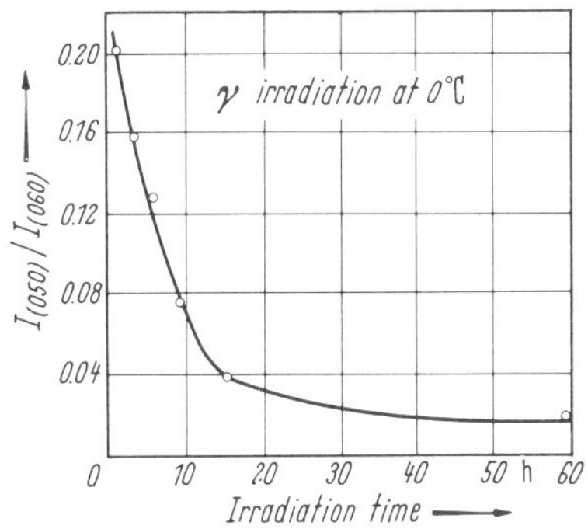

Fig. 1506. $NaKC_4H_2D_2O_6 \cdot 4D_2O$ (deuterated RS). $I_{(050)}/I_{(060)}$ vs. irradiation time at RT [63B10]. $I_{(050)}$, $I_{(060)}$: integrated intensity.

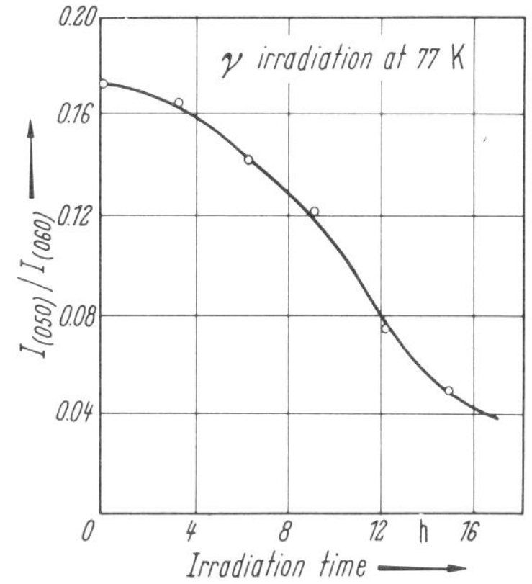

Fig. 1507. $NaKC_4H_2D_2O_6 \cdot 4D_2O$ (deuterated RS). $I_{(050)}/I_{(060)}$ vs. irradiation time [63B10]. See Fig. 1506.

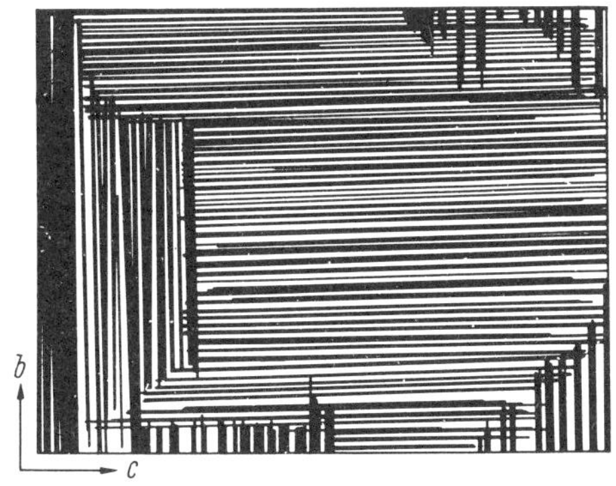

Fig. 1508. $NaKC_4H_4O_6 \cdot 4H_2O$ (RS). Example of complicated domain structure [53M2]. Observed with polarization microscope.

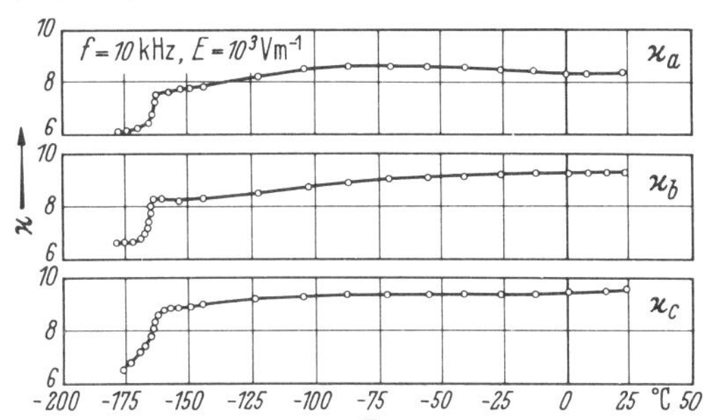

Fig. 1509. $NaNH_4C_4H_4O_6 \cdot 4H_2O$. $\varkappa_a$, $\varkappa_b$, $\varkappa_c$ vs. T [53J1].

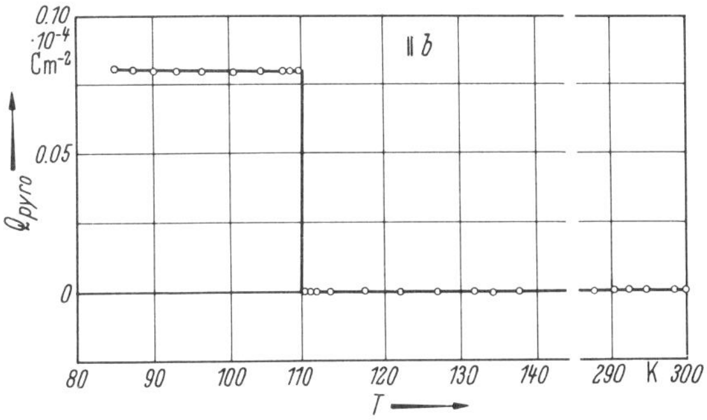

Fig. 1510. $NaNH_4C_4H_4O_6 \cdot 4H_2O$. Q_{pyro} vs. T [58T1]. Q_{pyro}: pyroelectric charge along b-axis.

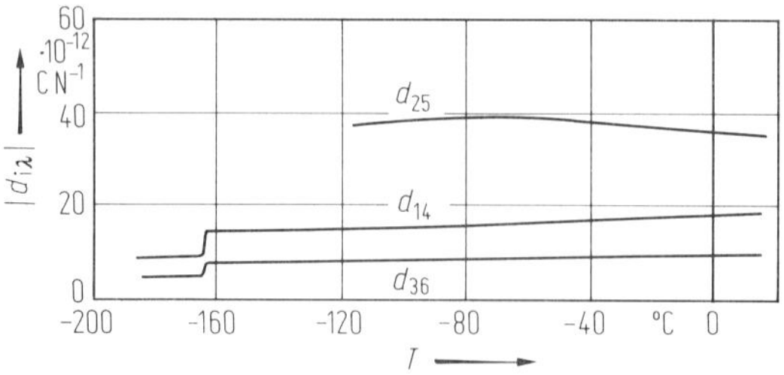

Fig. 1511. $NaNH_4C_4H_4O_6 \cdot 4H_2O$. $|d_{i\lambda}|$ vs. T [59 M 1].

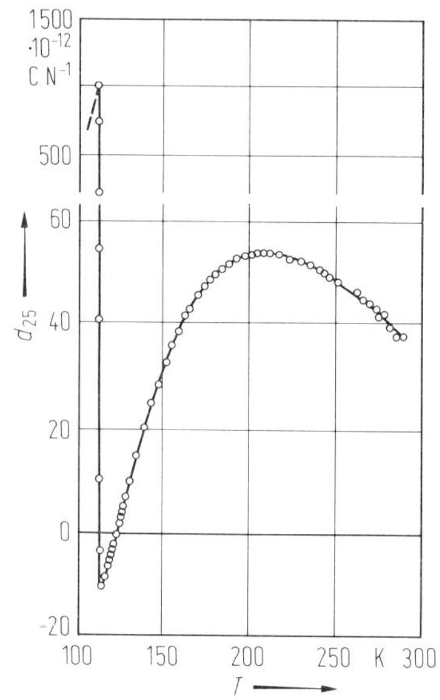

Fig. 1512. $NaNH_4C_4H_4O_6 \cdot 4H_2O$. d_{25} vs. T [77 G 10].

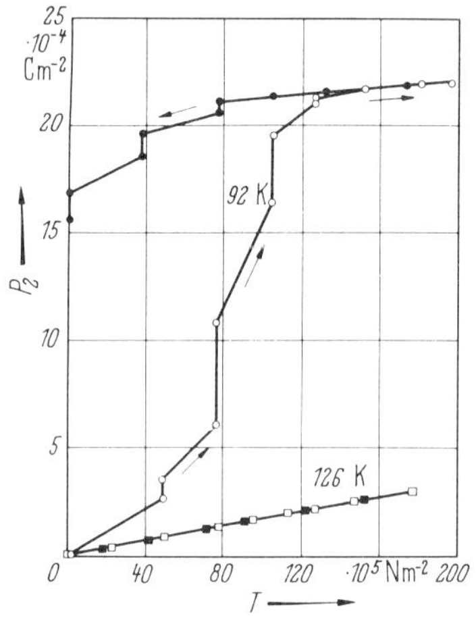

Fig. 1513. $NaNH_4C_4H_4O_6 \cdot 4H_2O$. P_2 vs. T [58 T 1]. P_2: polarization along [010], T: uniaxial pressure along [101].

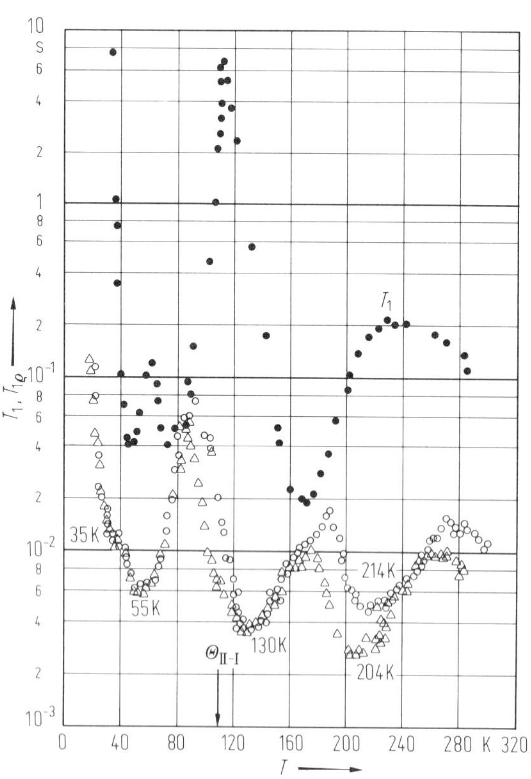

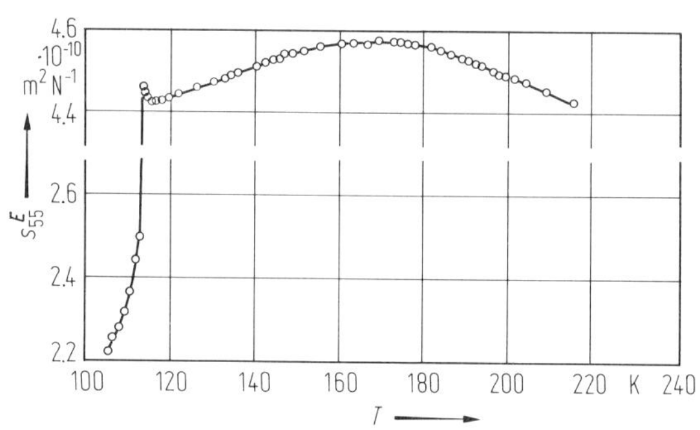

Fig. 1514. $NaNH_4C_4H_4O_6 \cdot 4H_2O$. s^E_{55} vs. T [77 G 10].

◄

Fig. 1515. $NaNH_4C_4H_4O_6 \cdot 4H_2O$. T_1, $T_{1\varrho}$ vs. T [76 M 19]. Full circles: T_1 at 10 MHz; open circles and triangles: $T_{1\varrho}$ at $H_1 \simeq 1.4 \cdot 10^3$ A m^{-1} and $0.72 \cdot 10^3$ A m^{-1}. T_1: proton spin-lattice relaxation time, $T_{1\varrho}$: proton spin-lattice relaxation time in rotating frame, H_1: amplitude of the radio frequency magnetic field.

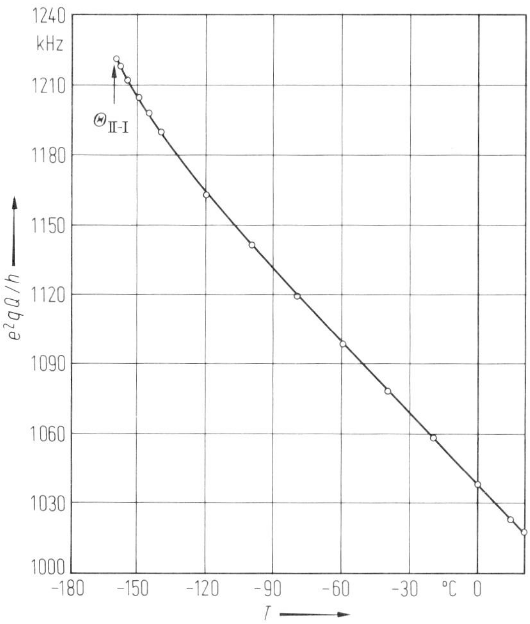

Fig. 1516. $NaNH_4C_4H_4O_6 \cdot 4H_2O$. e^2qQ/h vs. T [75T5]. e^2qQ/h: for ^{23}Na.

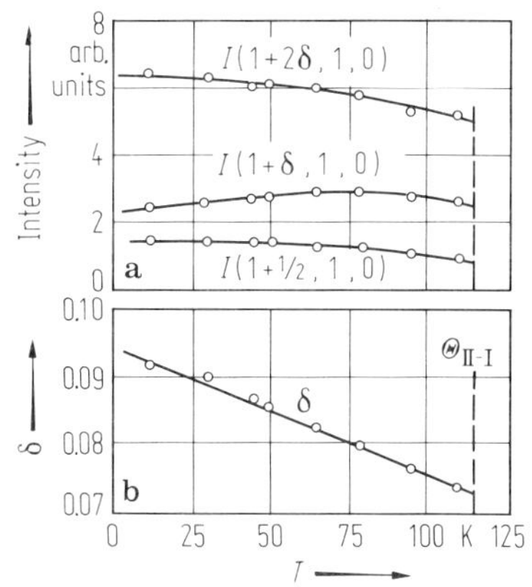

Fig. 1517. $NaND_4C_4H_2D_2O_6 \cdot 4D_2O$. Neutron diffraction in phase II [78I2]. (a) $I(1+2\delta, 1, 0)$, $I(1+\delta, 1, 0)$, $I(1+1/2, 1, 0)$ vs. T; (b) δ vs. T. δ in units of $2\pi/a$.

Fig. 1519. $NaNH_4C_4H_4O_6 \cdot 4H_2O$–$NaKC_4H_4O_6 \cdot 4H_2O$. Relation between the compositions of the crystal and the solution [58M1].

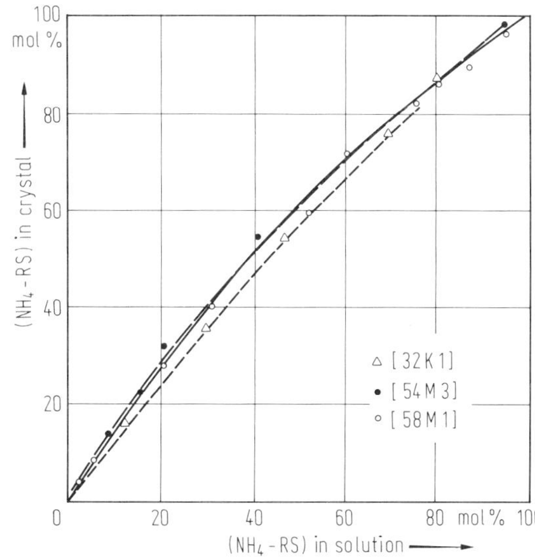

For Fig. 1520, see p. 670

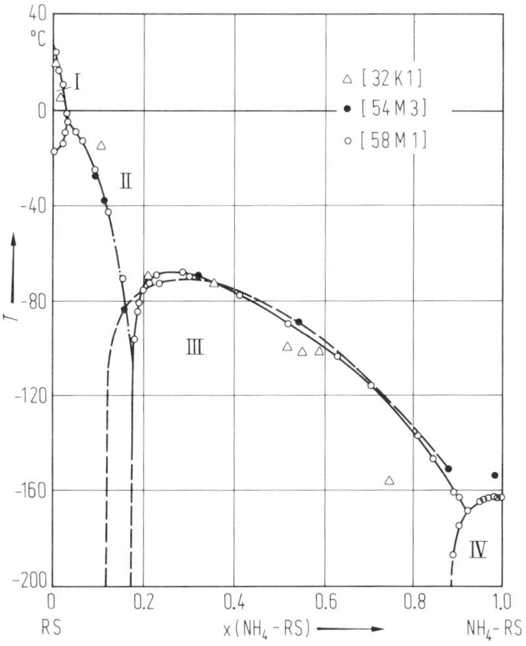

Fig. 1518. $NaK_{1-x}(NH_4)_xC_4H_4O_6 \cdot 4H_2O$. T vs. x [58M1]. The dashed-dotted line shows the temperature where the broad dielectric peak appears.

Fig. 1521. $NaK_{1-x}(NH_4)_xC_4H_4O_6 \cdot 4H_2O$. $\varkappa_a$ vs. T [58M1]. Parameter: x.

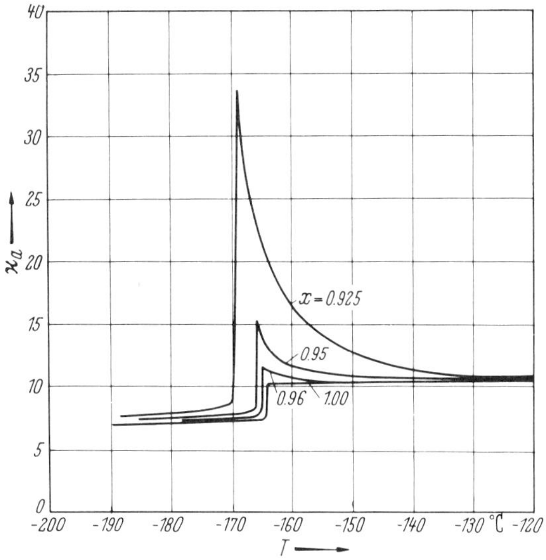

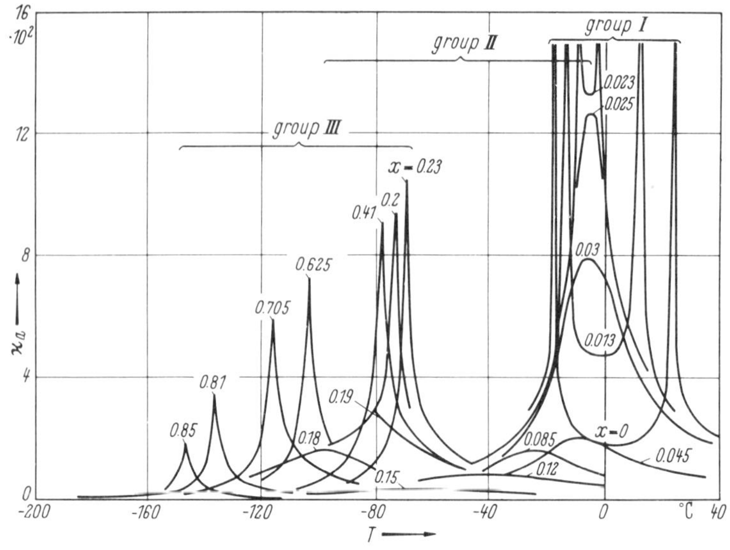

Fig. 1520. $NaK_{1-x}(NH_4)_xC_4H_4O_6 \cdot 4H_2O$. $\varkappa_a$ vs. T [58 M 1]. Parameter: x. For group I, II, III; see Fig. 1518.

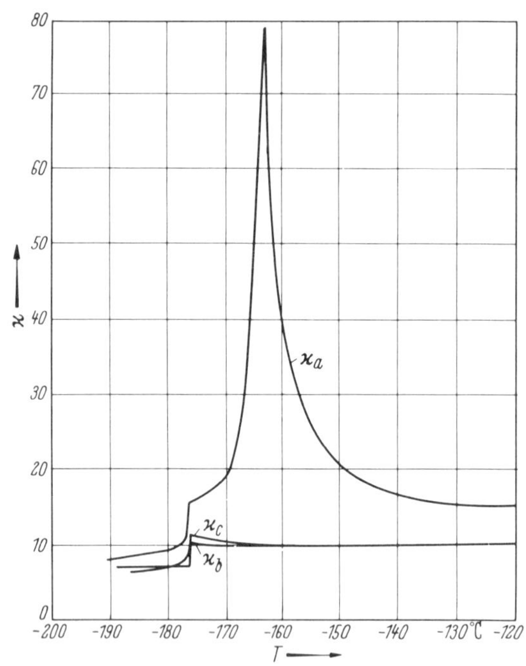

Fig. 1522. $NaK_{0.095}(NH_4)_{0.905}C_4H_4O_6 \cdot 4H_2O$. $\varkappa_a$, $\varkappa_b$, $\varkappa_c$ vs. T [58 M 1].

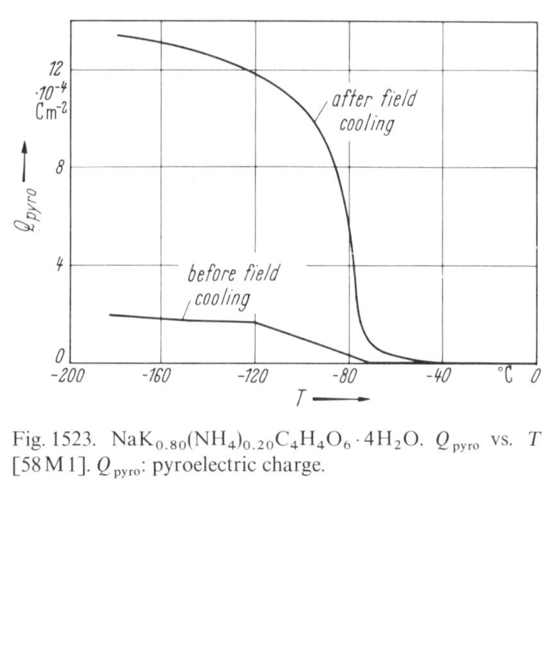

Fig. 1523. $NaK_{0.80}(NH_4)_{0.20}C_4H_4O_6 \cdot 4H_2O$. Q_{pyro} vs. T [58 M 1]. Q_{pyro}: pyroelectric charge.

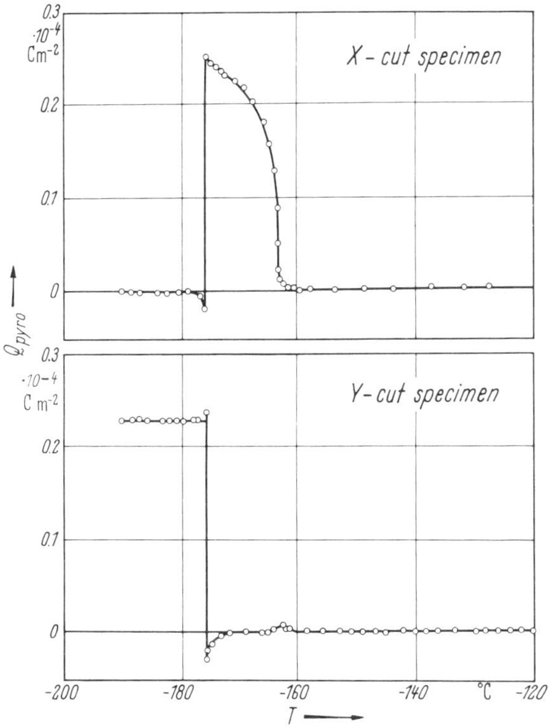

Fig. 1524. $NaK_{0.095}(NH_4)_{0.905}C_4H_4O_6 \cdot 4H_2O$. Q_{pyro} vs. T [58 M 1]. Q_{pyro}: pyroelectric charge.

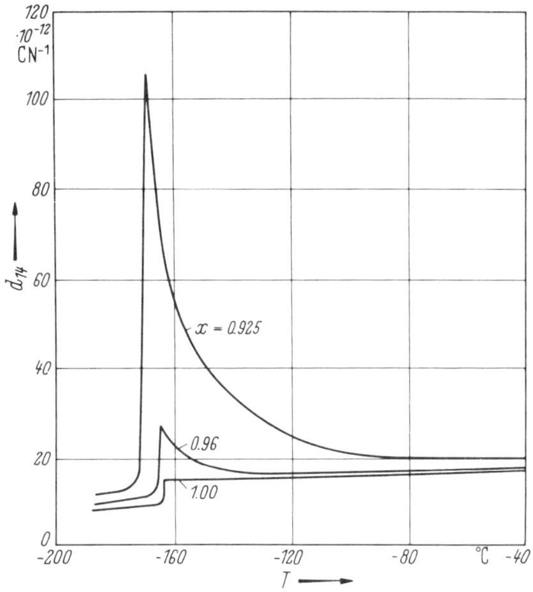

Fig. 1525. $NaK_{1-x}(NH_4)_xC_4H_4O_6 \cdot 4H_2O$. d_{14} vs. T [59 M 1]. Parameter: x.

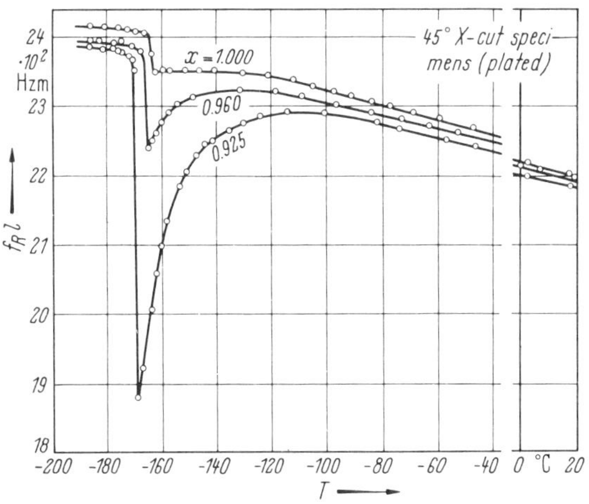

Fig. 1526. $NaK_{1-x}(NH_4)_xC_4H_4O_6 \cdot 4H_2O$. f_Rl vs. T [59 M 1]. Parameter: x. f_Rl: frequency constant of the 45° X-cut bar (see the caption of Fig. 1479).

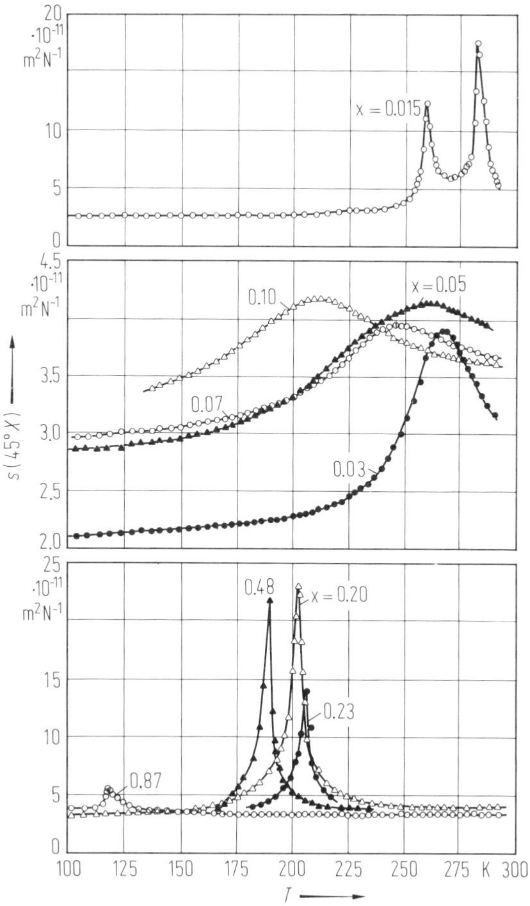

Fig. 1527. $NaK_{1-x}(NH_4)_xC_4H_4O_6 \cdot 4H_2O$. $s(45° X)$ vs. T [78 M 1]. Parameter: x. $s(45° X) = (1/4)(s_{22} + s_{33} + 2s_{23} + s_{44}^E)$.

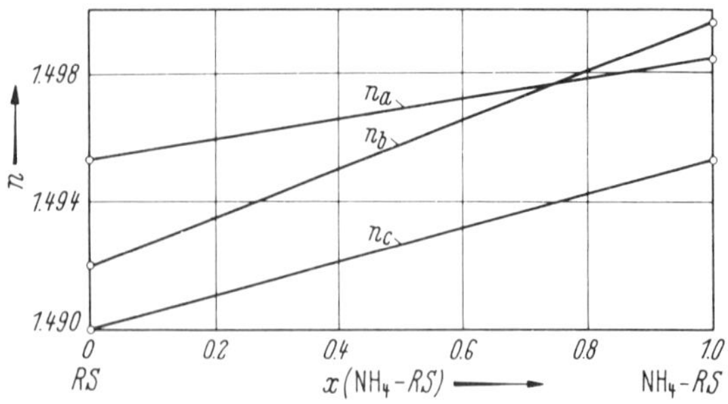

Fig. 1528. $NaK_{1-x}(NH_4)_xC_4H_4O_6 \cdot 4H_2O$. n vs. x [58M1].

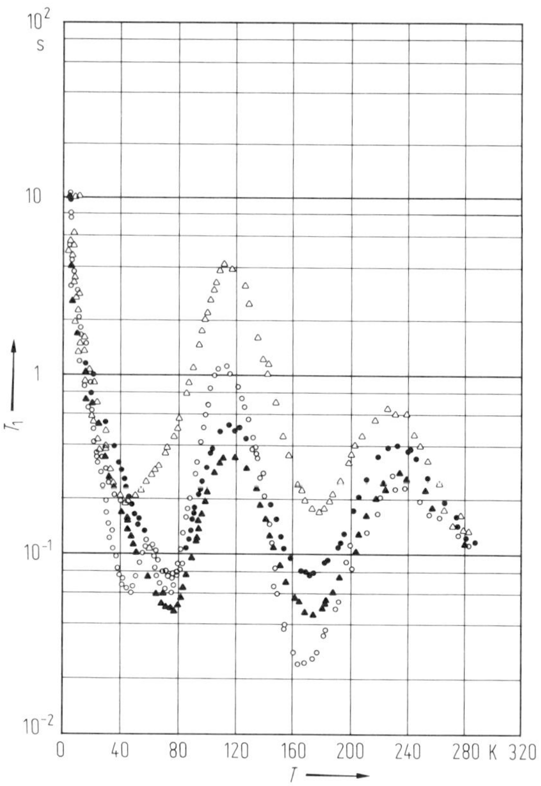

Fig. 1529. $NaK_{1-x}(NH_4)_xC_4H_4O_6 \cdot 4H_2O$. T_1 vs. T [74M15]. T_1: proton spin-lattice relaxation time at 10 MHz. Parameter: x. (open triangles: 0.20; full circles: 0.40; full triangles: 0.60; open circles: 0.87).

42 $LiNH_4C_4H_4O_6 \cdot H_2O$ family

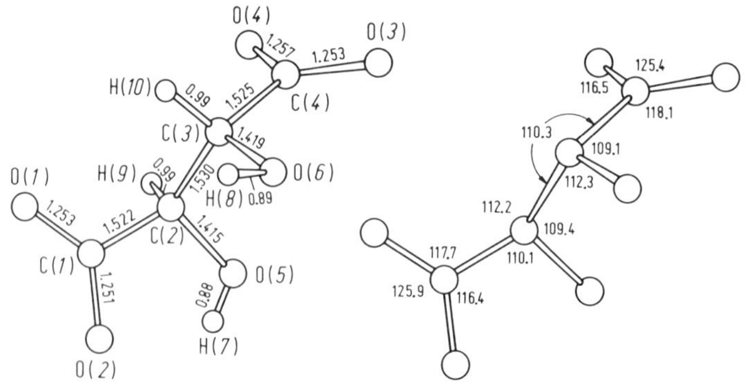

Fig. 1530. $LiNH_4C_4H_4O_6 \cdot H_2O$ (LAT). Structure of phase I [72H10]. Schematic view showing tartrate molecule. Bond lengths (in Å) and angles (in deg) are shown in the left and right figures, respectively.

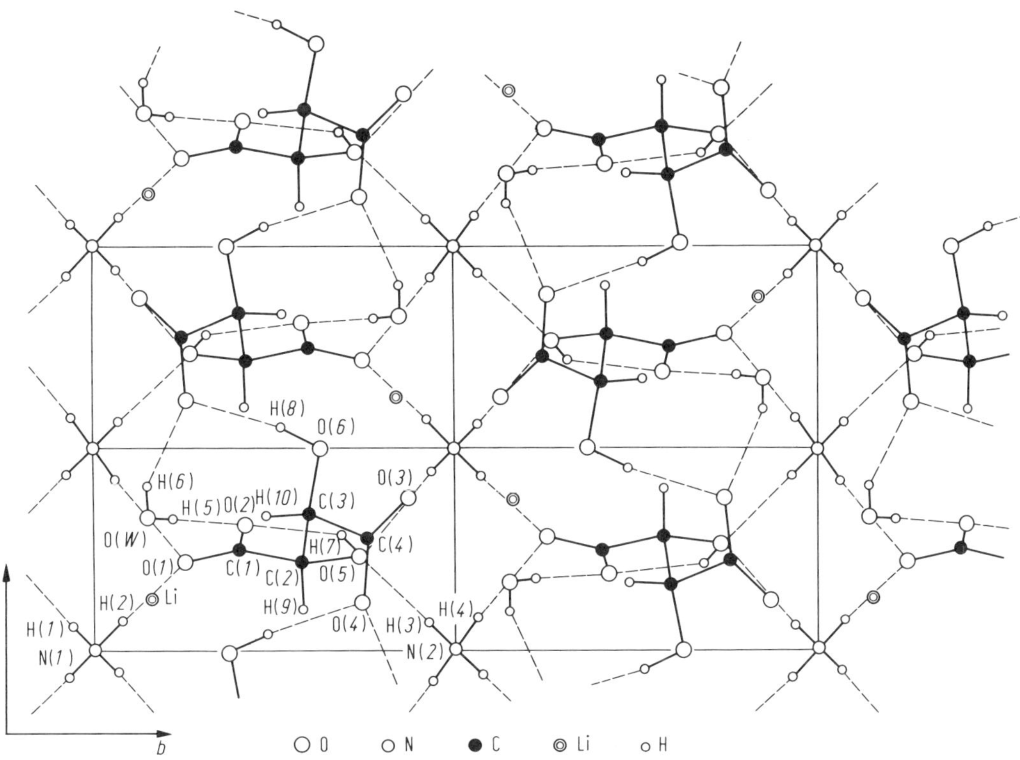

Fig. 1531. $LiNH_4C_4H_4O_6 \cdot H_2O$ (LAT). Structure of phase I [72H10]. Projection onto (001). The hydrogen bonds are shown by broken lines.

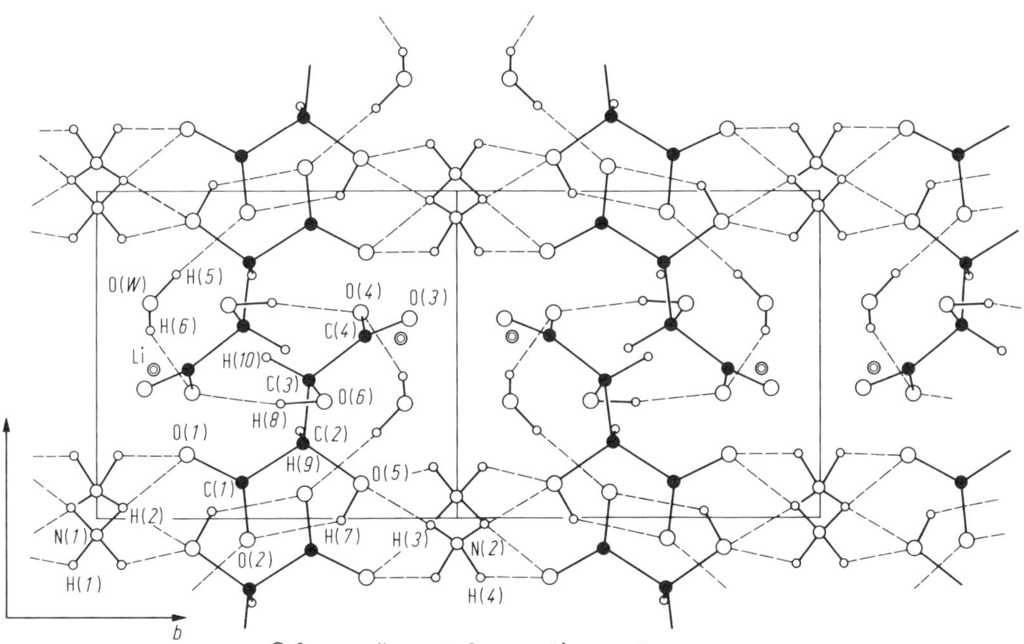

Fig. 1532. $LiNH_4C_4H_4O_6 \cdot H_2O$ (LAT). Structure of phase I [72H10]. Projection onto (100). The broken lines represent hydrogen bonds.

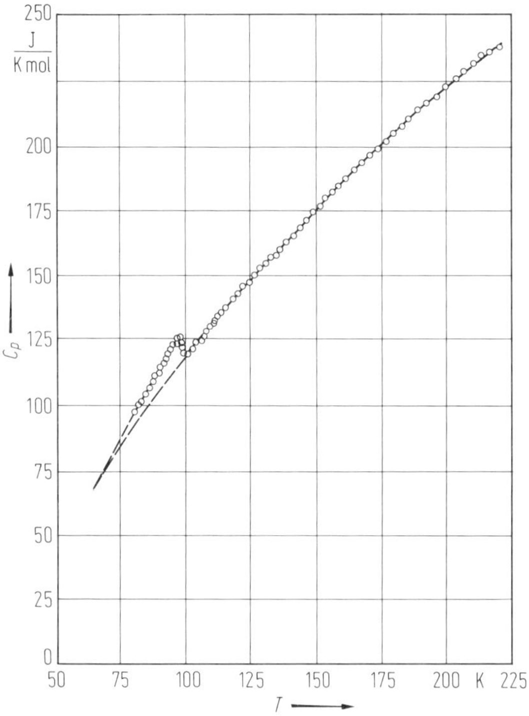

Fig. 1535. $LiNH_4C_4H_4O_6 \cdot H_2O$ (LAT). C_p vs. T [73 M 3].

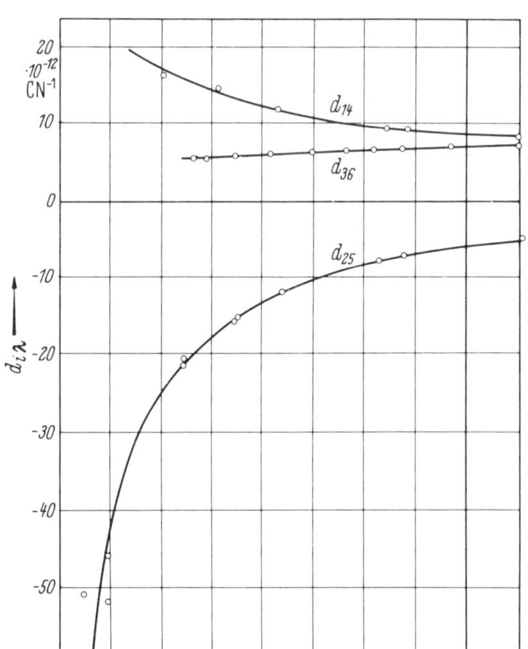

Fig. 1536. $LiNH_4C_4H_4O_6 \cdot H_2O$ (LAT). $d_{i\lambda}$ vs. T [66 b 1].

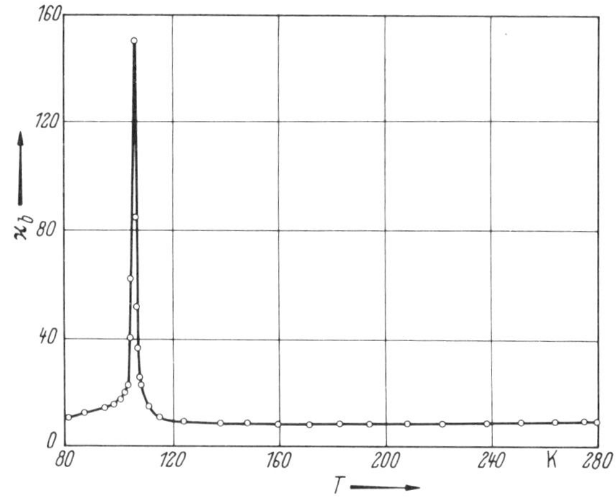

Fig. 1533. $LiNH_4C_4H_4O_6 \cdot H_2O$ (LAT). $\varkappa_b$ vs. T [51 M 3].

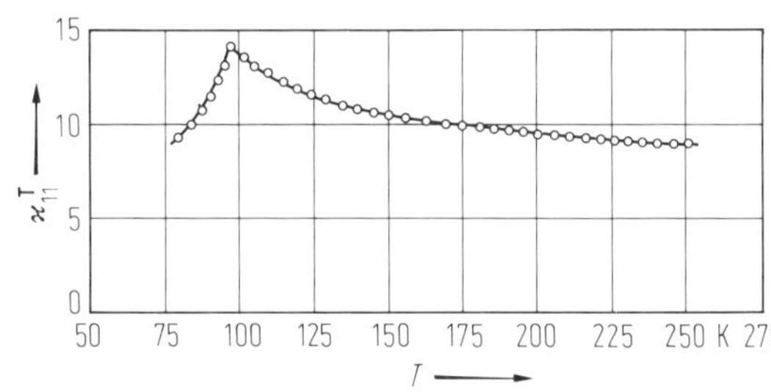

Fig. 1534. $LiNH_4C_4H_4O_6 \cdot H_2O$ (LAT). $\varkappa_{11}^T$ vs. T at 1 kHz [77 M 3].

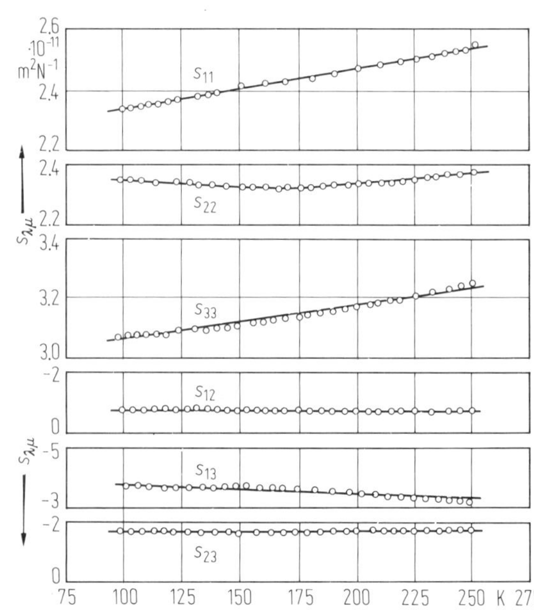

Fig. 1537. $LiNH_4C_4H_4O_6 \cdot H_2O$ (LAT). $s_{\lambda\mu}$ vs. T [77 M 3].

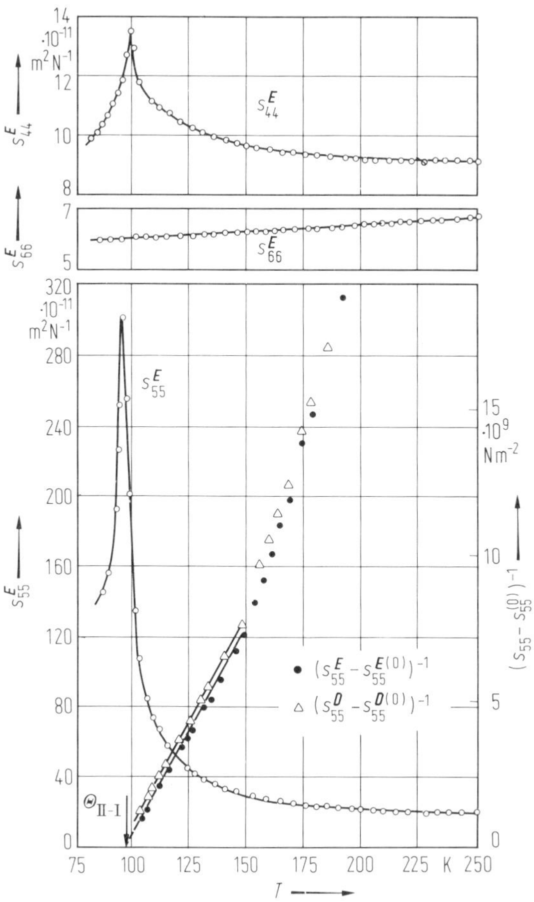

Fig. 1538. $LiNH_4C_4H_4O_6 \cdot H_2O$ (LAT). Elastic compliances (shear components) vs. T [77 M 3]. Two straight lines show $(s^E_{55} - s^{E(0)}_{55})^{-1}$ and $(s^D_{55} - s^{D(0)}_{55})^{-1}$, where the values for $s^{E(0)}_{55}$ and $s^{D(0)}_{55}$ are taken as 18.0 and $17.0 \cdot 10^{-11}\,m^2\,N^{-1}$, respectively.

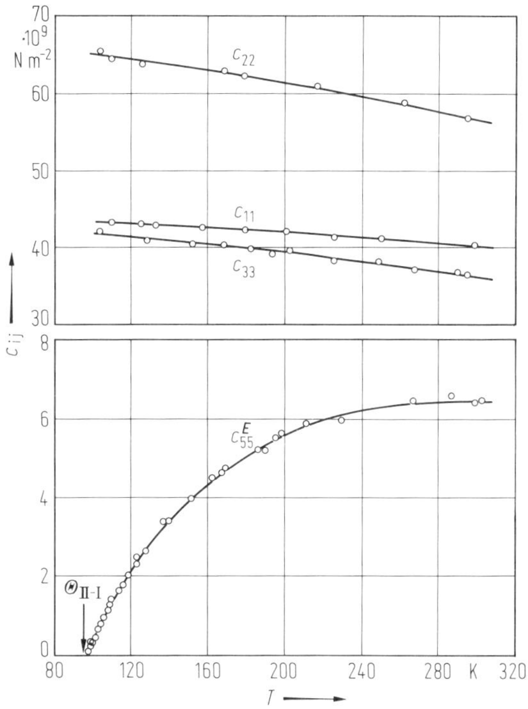

Fig. 1539. $LiNH_4C_4H_4O_6 \cdot H_2O$ (LAT). $c_{\lambda\mu}$ vs. T [78 U 2]. Brillouin scattering.

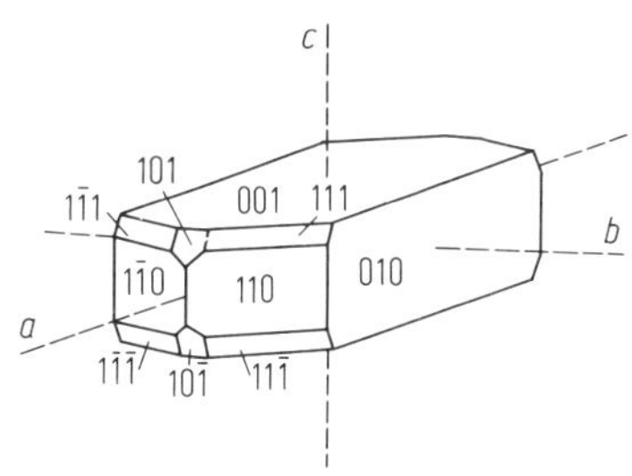

Fig. 1540. $LiTlC_4H_4O_6 \cdot H_2O$ (LTT). Crystal form [70 B 16].

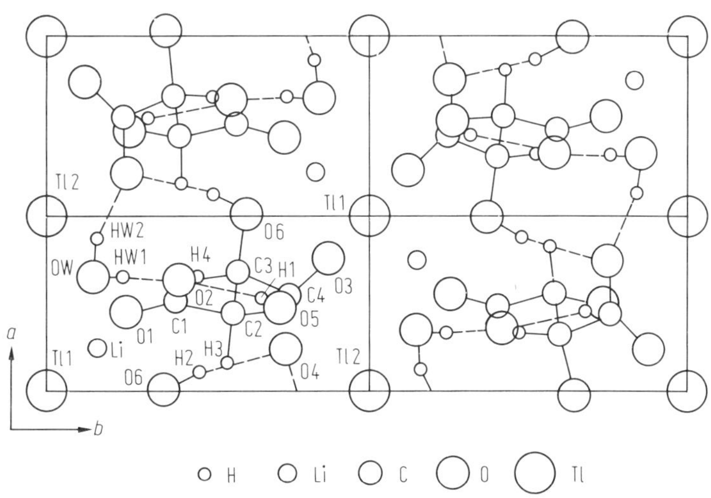

Fig. 1541. $LiTlC_4H_4O_6 \cdot H_2O$ (LTT). Structure of phase I [78 K 9]. Projection on (001).

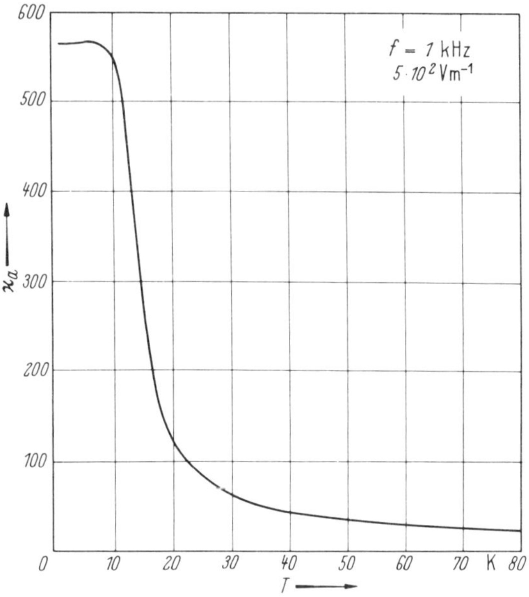

Fig. 1542. $LiTlC_4H_4O_6 \cdot H_2O$ (LTT). $\varkappa_a$ vs. T [51 M 1].

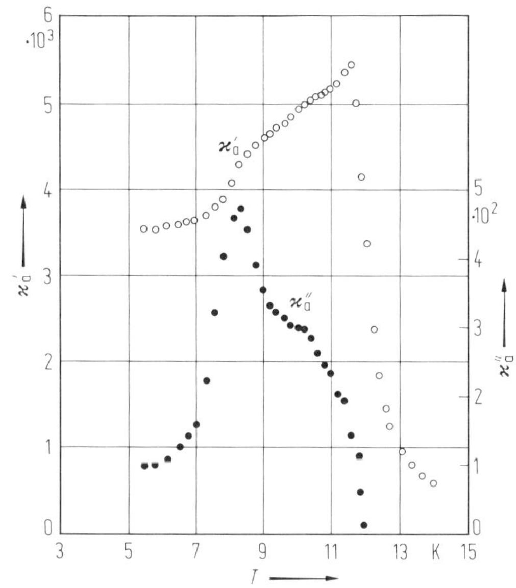

Fig. 1543. $LiTlC_4H_4O_6 \cdot H_2O$ (LTT). $\varkappa_a'$, $\varkappa_a''$ vs. T [74 A 1]. $f = 1$ kHz.

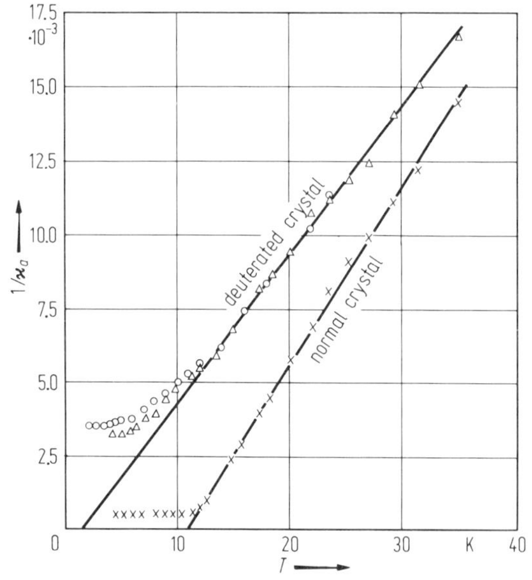

Fig. 1544. $LiTlC_4H_4O_6 \cdot H_2O$ (LTT). $1/\varkappa_a$ vs. T for normal crystal and deuterated crystal (more than 80% of hydrogen is replaced by D) [70 B 16]. $f = 1$ kHz, 200 V m^{-1}.

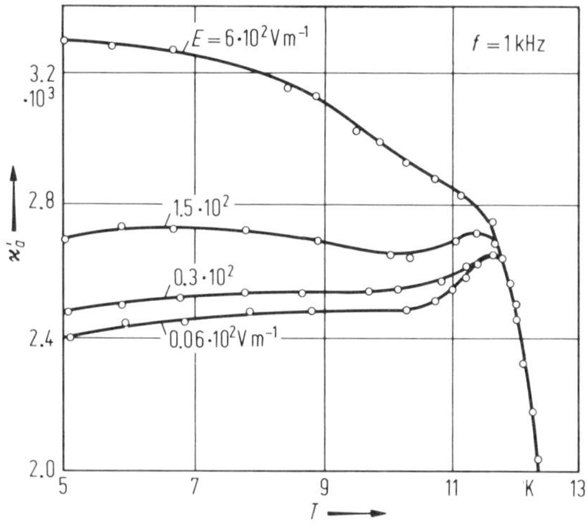

Fig. 1545. $LiTlC_4H_4O_6 \cdot H_2O$ (LTT). $\varkappa_a'$ vs. T [70 F 4]. Parameter: E, the measuring ac field at 1 kHz.

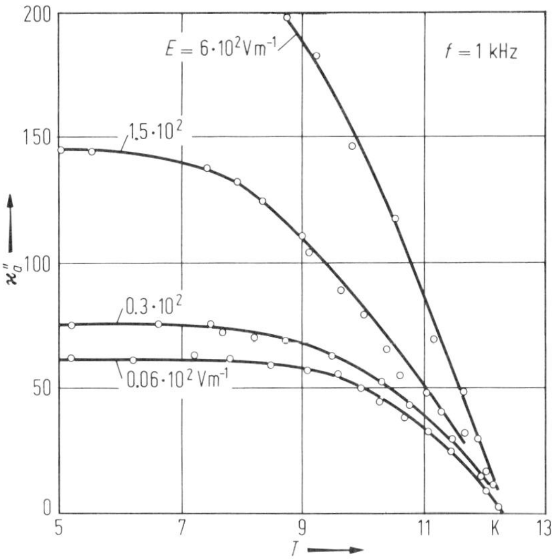

Fig. 1546. $LiTlC_4H_4O_6 \cdot H_2O$ (LTT). $\varkappa_a''$ vs. T [70F4]. Parameter: E, the measuring ac field at 1 kHz.

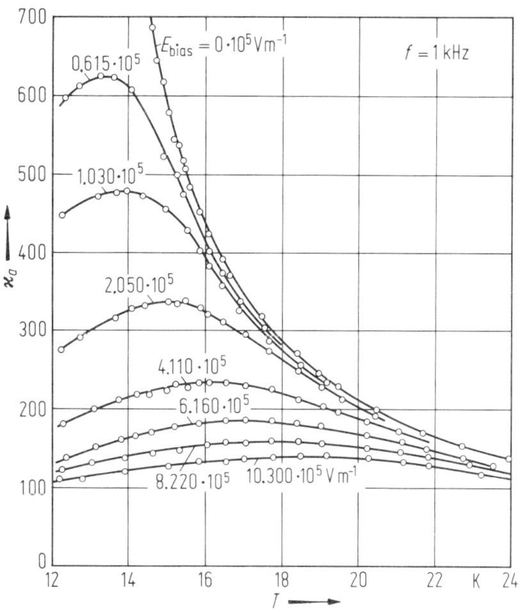

Fig. 1547. $LiTlC_4H_4O_6 \cdot H_2O$ (LTT). $\varkappa_a$ vs. T [70F4]. Parameter: E_{bias}.

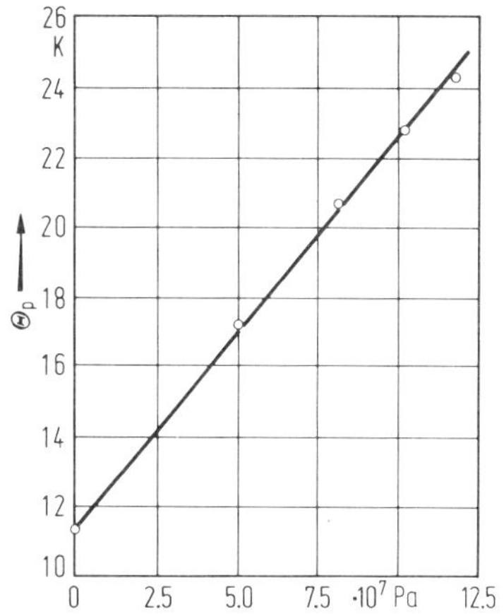

Fig. 1549. $LiTlC_4H_4O_6 \cdot H_2O$ (LTT). Θ_p vs. hydrostatic pressure p

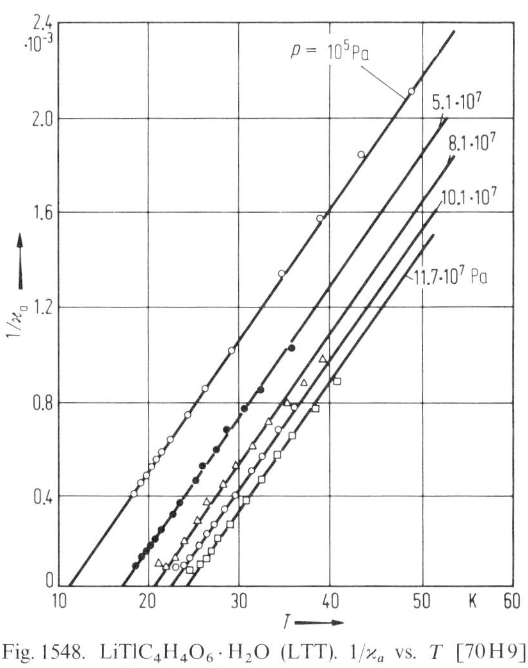

Fig. 1548. $LiTlC_4H_4O_6 \cdot H_2O$ (LTT). $1/\varkappa_a$ vs. T [70H9]. Parameter: p, hydrostatic pressure.

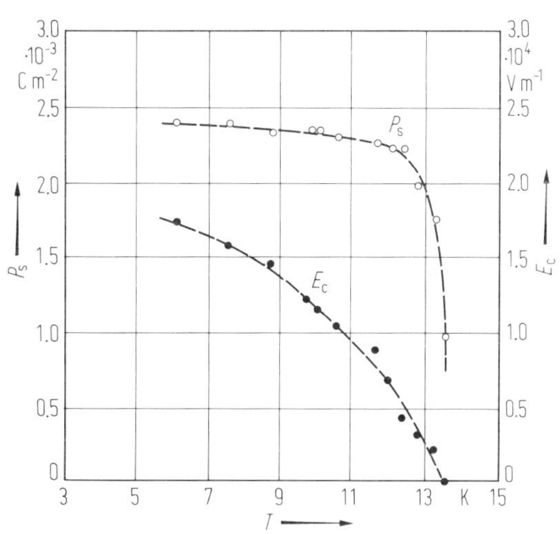

Fig. 1550. $LiTlC_4H_4O_6 \cdot H_2O$ (LTT). P_s, E_c vs. T [74A1]. At $f = 60$ Hz and field amplitude of 120 kVm^{-1}.

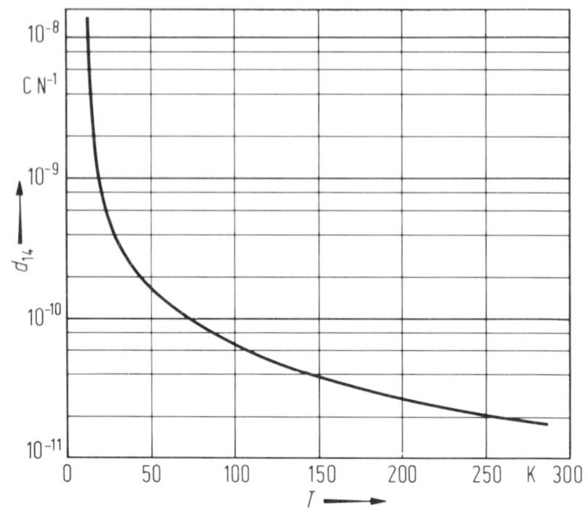

Fig. 1551. $LiTlC_4H_4O_6 \cdot H_2O$ (LTT). Piezoelectric strain constant d_{14} vs. T [71 S8].

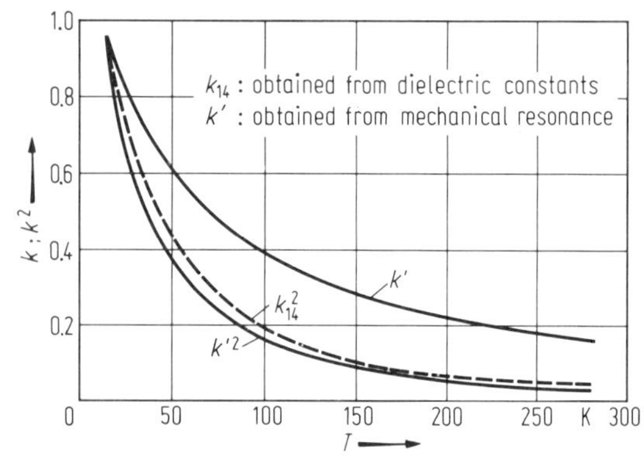

Fig. 1552. $LiTlC_4H_4O_6 \cdot H_2O$ (LTT). k, k^2 vs. T [71 S8]. k': electromechanical coupling factor of a 45° X-cut bar for the length extensional mode of vibration; k_{14}: the coupling factor determined from $k_{14} = (1 - \varkappa_a^S/\varkappa_a^T)^{1/2}$.

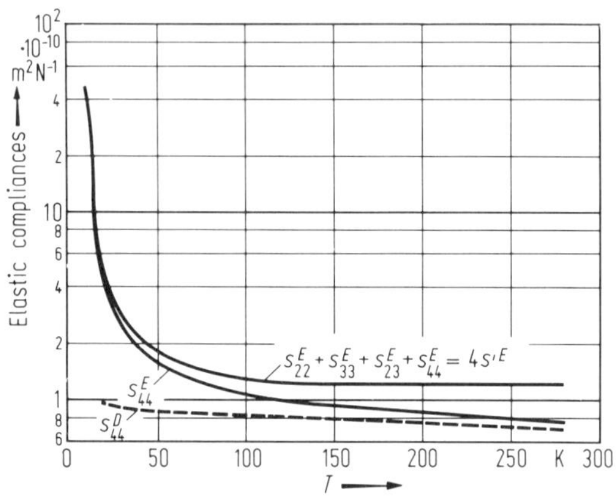

◄ Fig. 1553. $LiTlC_4H_4O_6 \cdot H_2O$ (LTT). Elastic compliances vs. T [71 S8]. s'^E: compliance for the length extensional strain of the 45° X-cut bar at constant electric field.

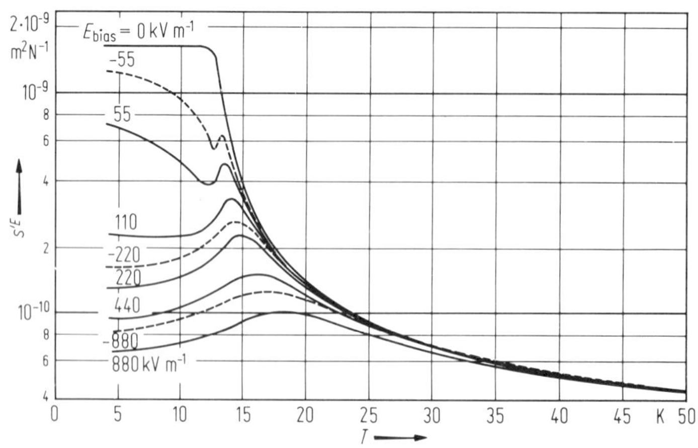

Fig. 1554. $LiTlC_4H_4O_6 \cdot H_2O$ (LTT). s'^E vs. T [71 S7]. Parameter: E_{bias}, the dc bias field along a. $s'^E = (s_{22}^E + s_{33}^E + 2s_{23}^E + s_{44}^E)/4$.

43 Miscellaneous crystals (including oxides) and polymers

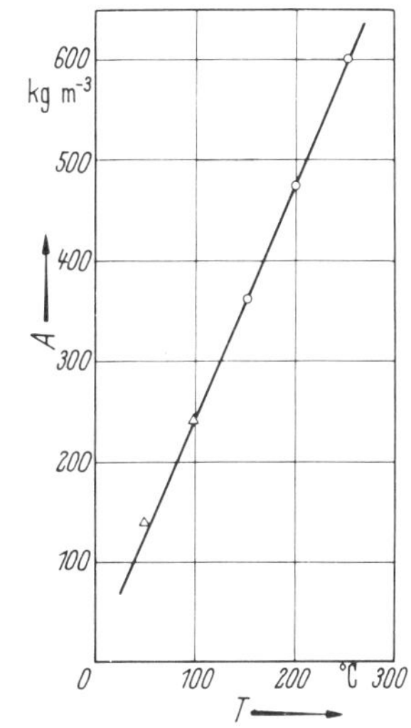

Fig. 1555. KIO_3. *A* vs. *T* [61 H 3]. *A*: solubility in water (solute/solvent).

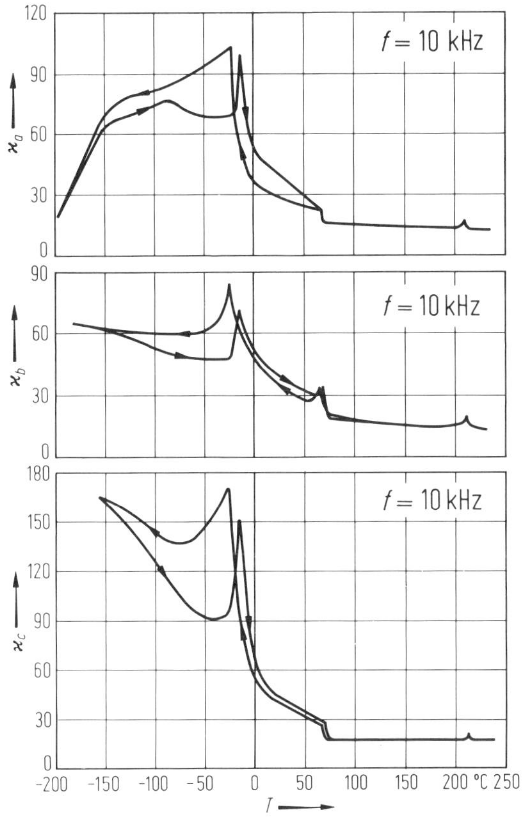

Fig. 1556. KIO_3. $\varkappa_a$, $\varkappa_b$, $\varkappa_c$ vs. *T* [70 H 10]. The axes *a*, *b*, *c* refer to the pseudocubic cell.

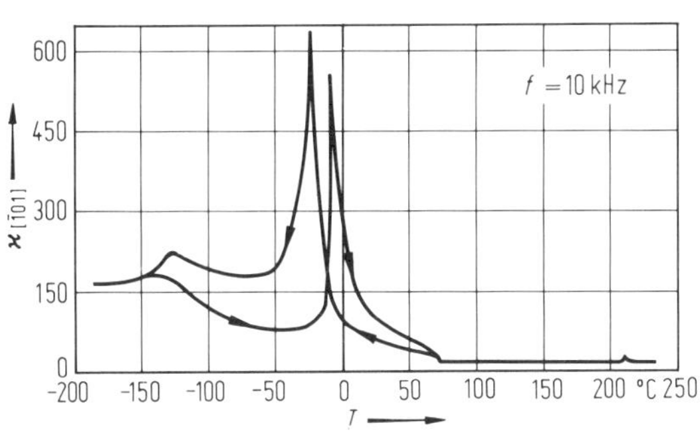

Fig. 1557. KIO_3. $\varkappa_{[\bar{1}01]}$ vs. *T* [70 H 10]. [$\bar{1}$01] refers to the pseudocubic cell.

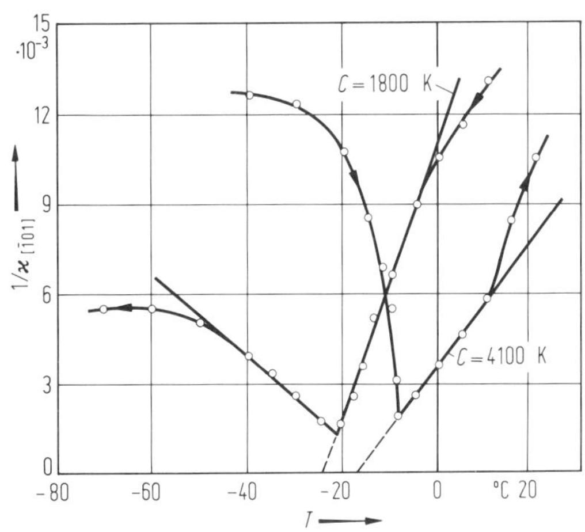

Fig. 1558. KIO_3. $1/\varkappa_{[\bar{1}01]}$ vs. *T* [70 H 10]. [$\bar{1}$01] refers to the pseudocubic cell. The two curves correspond to different treatment of specimens. *C*: Curie-Weiss constant.

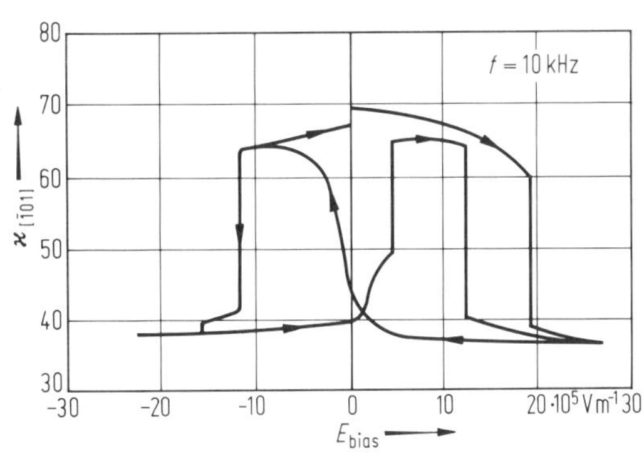

Fig. 1559. KIO_3. $\varkappa_{[\bar{1}01]}$ vs. E_{bias} [70 H 10]. $T = 22$ °C. [$\bar{1}$01] refers to the pseudocubic cell.

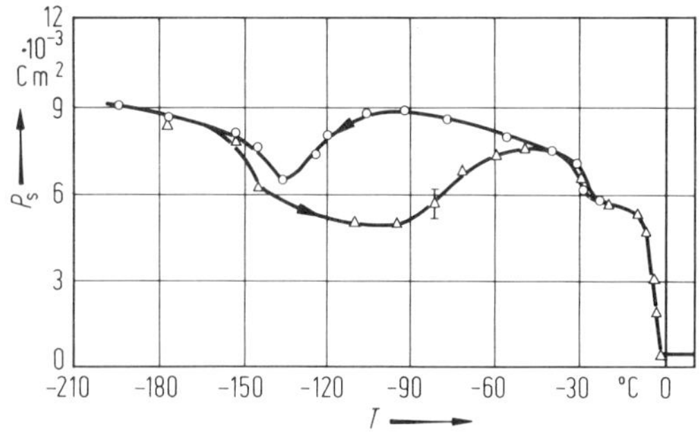

Fig. 1560. KIO_3. P_s vs. T [70H10]. $\boldsymbol{E}\parallel[\bar{1}01]$. $[\bar{1}01]$ refers to the pseudocubic cell.

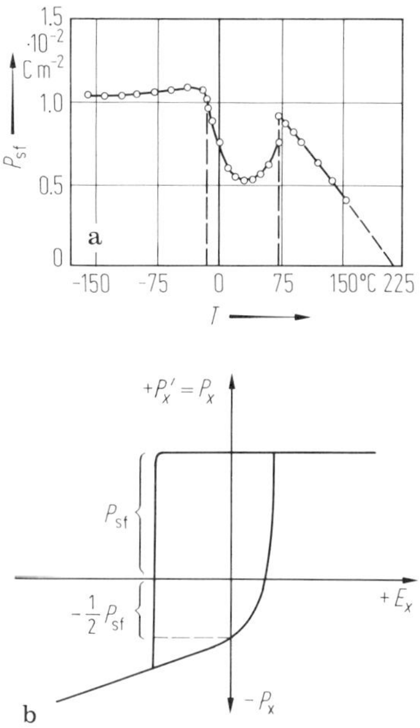

Fig. 1561. KIO_3. Fig. a: P_{sf} vs. T [73S11]. Fig. b: Hysteresis loop observed with an x-cut specimen, and definition of P_{sf}. Dashed lines in Fig. (a) indicate Θ_{IV-III} and Θ_{III-II}; see subsection 1b of Nr. 43-2.

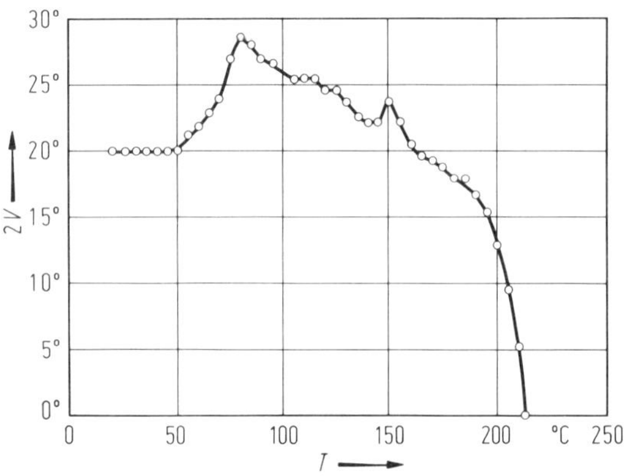

Fig. 1562. KIO_3. $2V$ vs. T [72C7]. $2V$: optical axial angle.

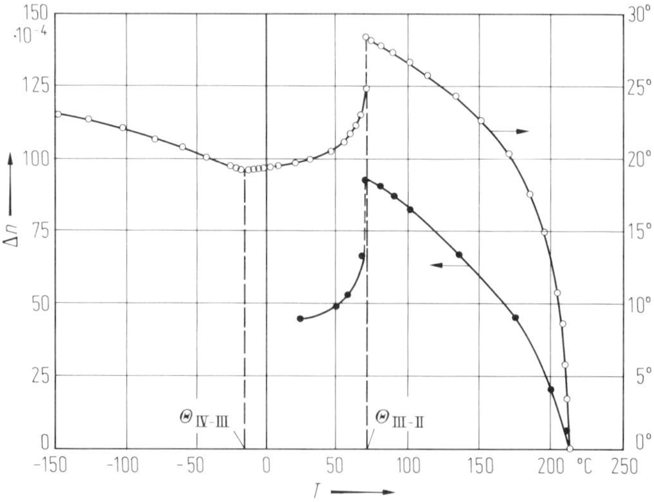

Fig. 1563. KIO_3. Δn, $2V$ vs. T [73S11]. $\Delta n = n_y - n_x$. $2V$: optical axial angle. n_X, n_Y, n_Z: refractive indices for $\boldsymbol{E}$ of light $\parallel X$, Y, Z axes, where Z is parallel to the c axis in the trigonal R3m phase above 212 °C, X lies in the mirror plane and Y is perpendicular to X and Z.

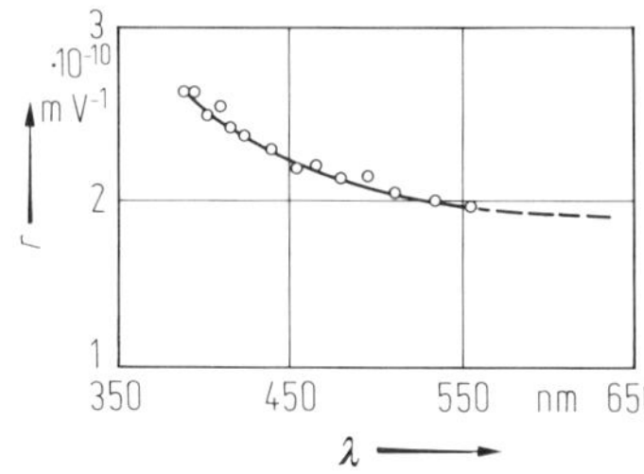

Fig. 1564. KIO_3. r vs. λ [73S11]. $r = r_{21} - (n_x/n_y)^3 r_{11}$.
$T = 20$ °C. See also Fig. 1563.

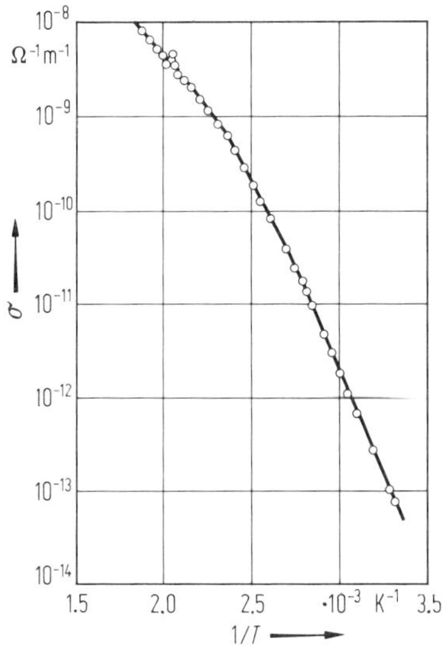

Fig. 1565. KIO_3. σ vs. $1/T$ [62G3]. σ was measured in dry air by the two-terminal method with vacuum evaporated Ag electrodes. Applied field $E = 120$ kV m^{-1}.

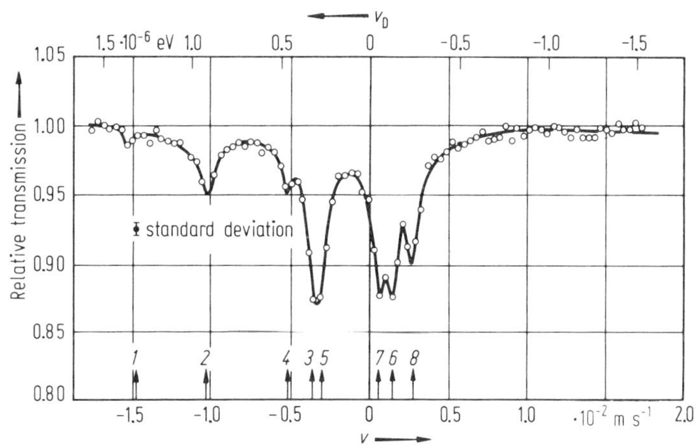

Fig. 1566. KIO_3. Relative transmission vs. v, v_D [63D3]. v: surface velocity. v_D: Doppler velocity. The source is $^{66}Zn^{129}Te$. The arrows indicate the calculated positions of the hyperfine components.

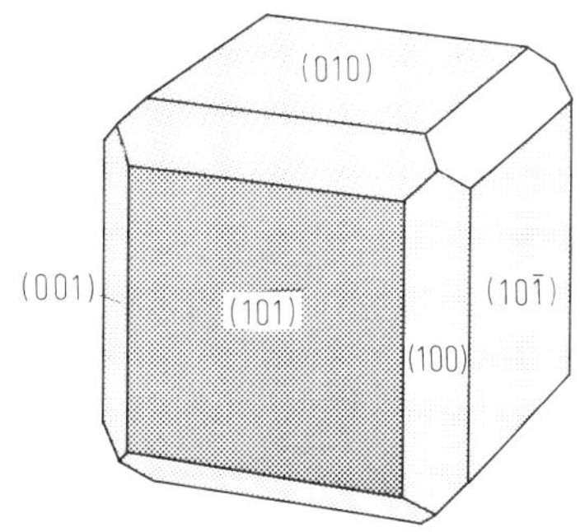

Fig. 1568. NH_4IO_3. Crystal form [69C6].

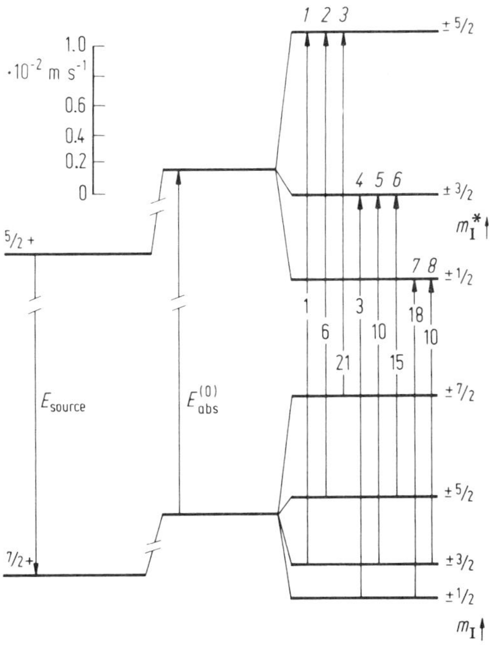

Fig. 1567. KIO_3. Nuclear energy levels for ground and first excited states of ^{129}I (source) and KIO_3 (absorber) [63D3]. m, m^*: magnetic quantum number of ground and excited states.

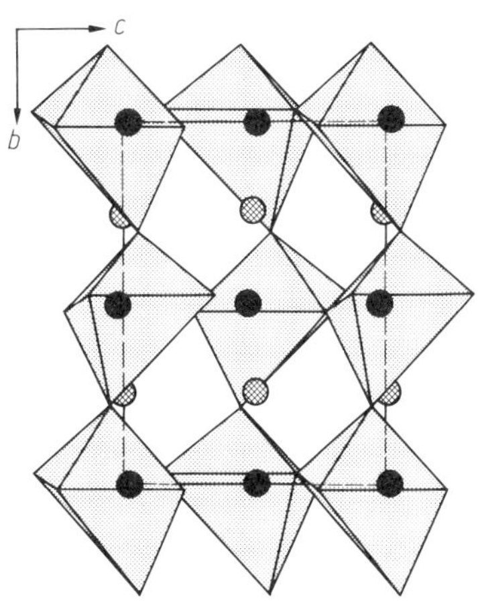

Fig. 1569. NH_4IO_3. Structure of phase III [71K3]. Projection along the a-axis. Black and hatched circles represent iodine atoms and NH_4-ions, respectively. Distorted octahedra show the oxygen-octahedra surrounding iodine atoms.

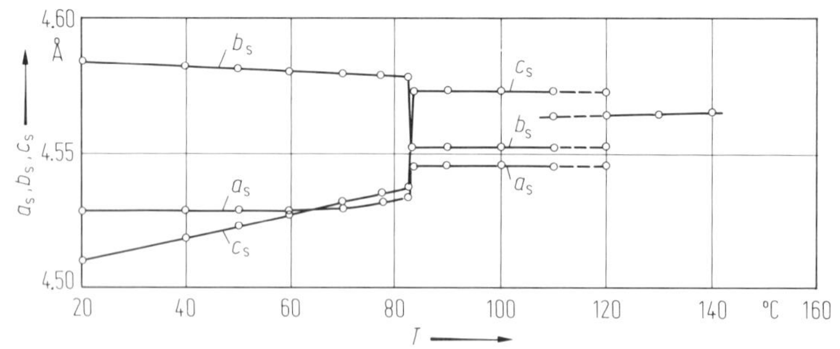

Fig. 1570. NH_4IO_3. a_s, b_s, c_s vs. T [75V7]. a_s, b_s, c_s: unit cell parameters for hypothetical pseudo-cubic subcell. In phases III and II the relation to the orthorhombic unit cell parameters a, b, c are; $a_s = a/\sqrt{2}$, $b_s = b/2$, $c_s = c/\sqrt{2}$.

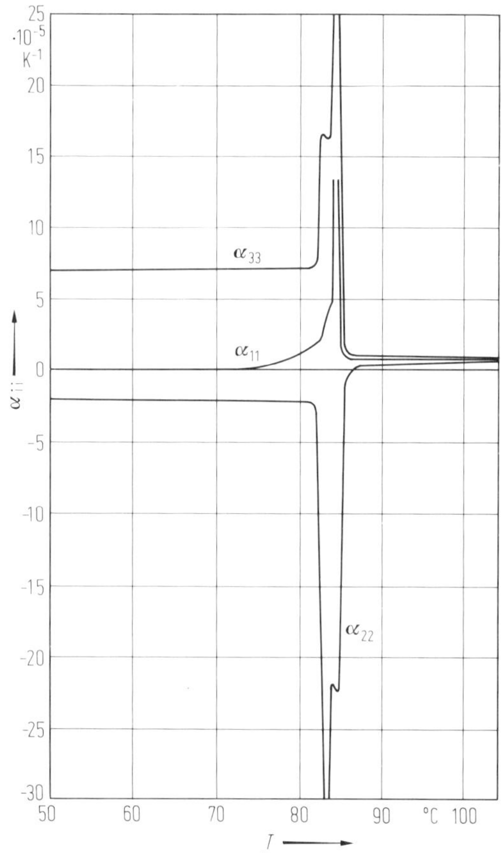

Fig. 1571. NH_4IO_3. α_{ii} vs. T [75V7]. α_{ii}: linear thermal expansion coefficient.

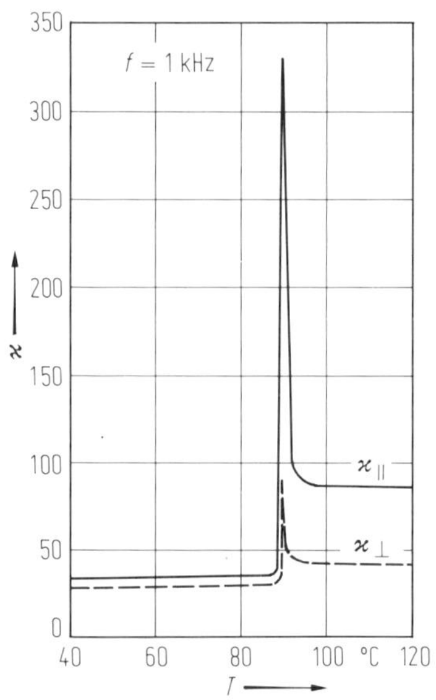

Fig. 1572. NH_4IO_3. $\varkappa_{\parallel}$, $\varkappa_{\perp}$ vs. T [69C6]. $\varkappa_{\parallel}$, $\varkappa_{\perp}$: dielectric constants along and perpendicular to the b-axis, respectively.

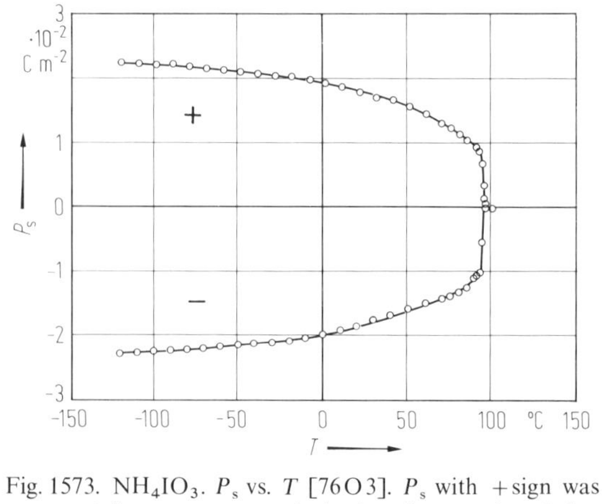

Fig. 1573. NH_4IO_3. P_s vs. T [76O3]. P_s with +sign was measured for an as-grown single domain crystal and that with −sign was measured after the polarity had been reversed by a dc field of $25 \cdot 10^5$ V m^{-1} for 10 min at 75 °C.

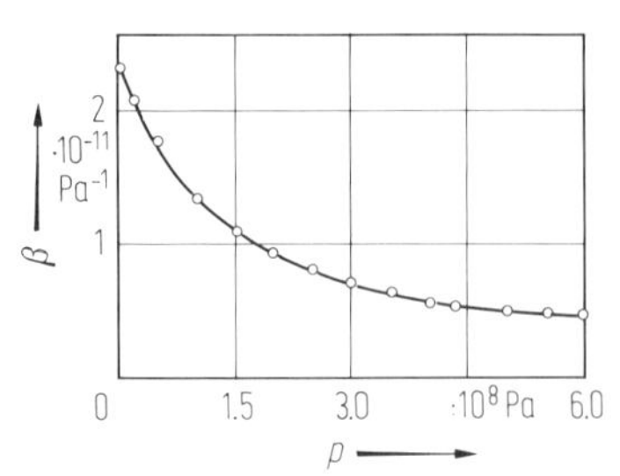

Fig. 1574. NH_4IO_3. β vs. p [75V7]. T=RT. β: volume compressibility.

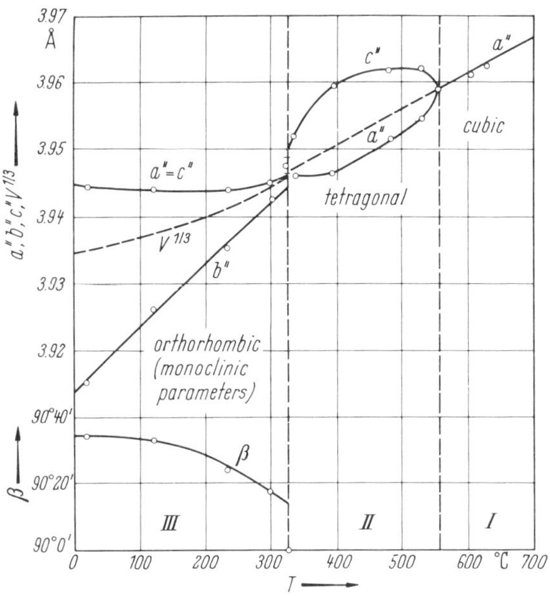

Fig. 1575. $AgNbO_3$. Lattice parameters of pseudocubic cell vs. T [58 F 1]. $a'' = \frac{1}{2}a'$, $b'' = \frac{1}{4}b'$ and $c'' = \frac{1}{2}c'$.

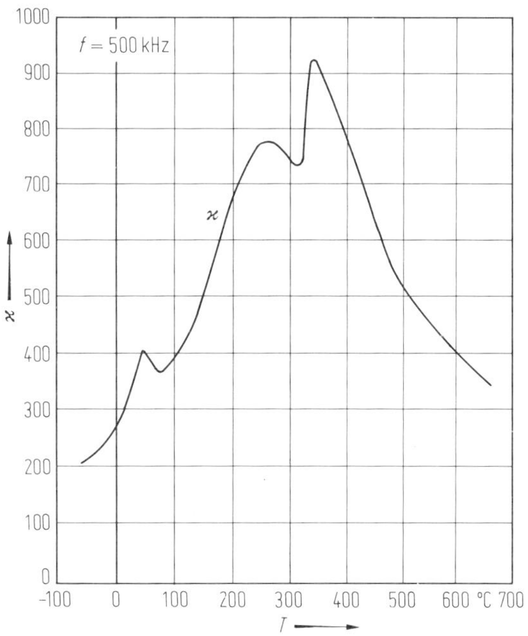

Fig. 1576. $AgNbO_3$. $\varkappa$ vs. T [58 F 1].

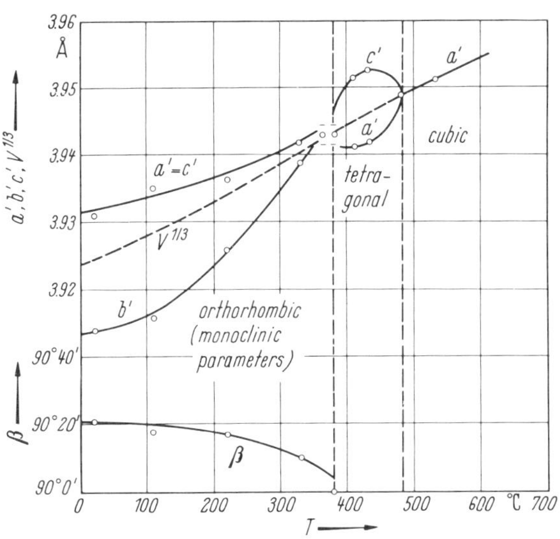

Fig. 1577. $AgTaO_3$. Lattice parameters of pseudocubic cell vs. T [58 F 1].

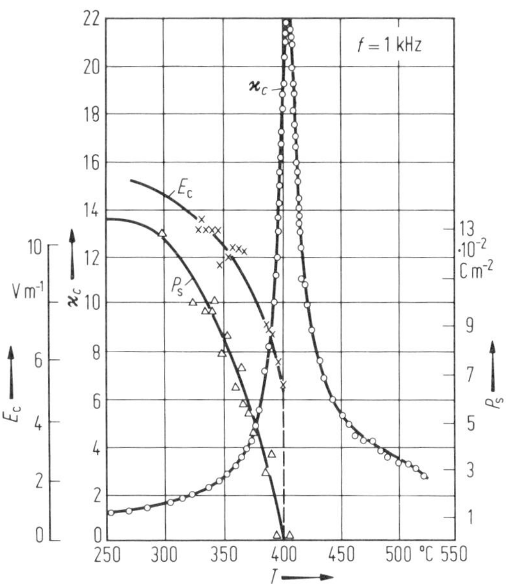

Fig. 1578. Stibiotantalite [Sb(Ta, Nb)O_4]. $\varkappa_c$, P_s, E_c vs. T [70 G 1]. See also [71 N 13, 72 K 3].

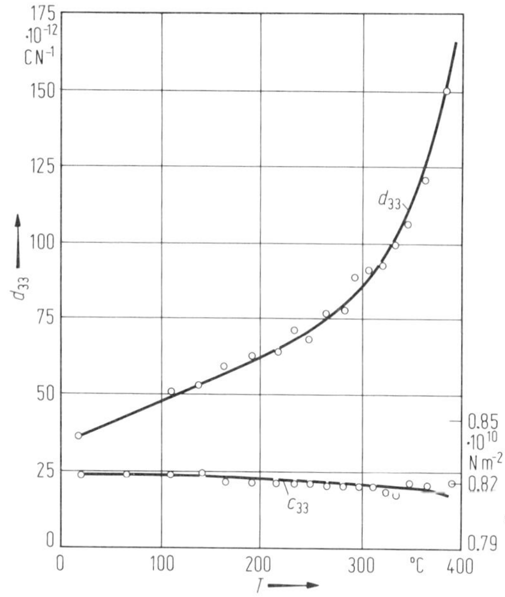

Fig. 1579. Stibiotantalite [Sb(Ta, Nb)O_4]. d_{33}, c_{33} vs. T [71N13].

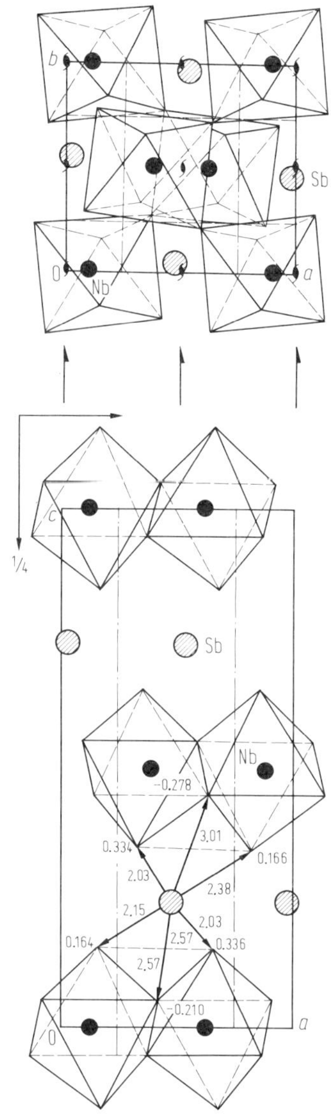

Fig. 1580. $SbNbO_4$. Crystal structure. (001) and (010) projections [76R1]. Interatomic distances (in Å) and fractional coordinates are included. See also Table 415.

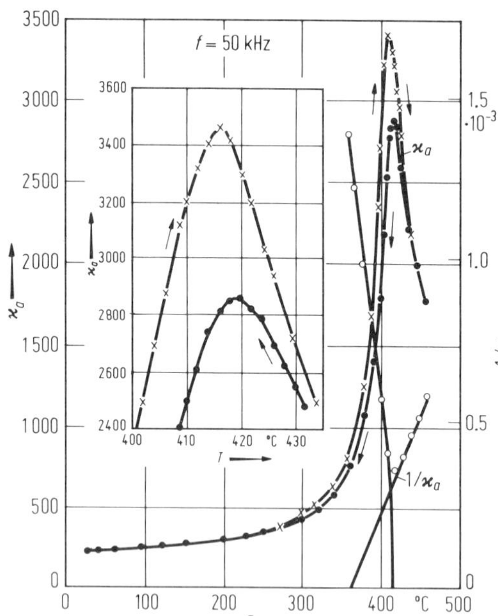

Fig. 1581. $SbNbO_4$. $\varkappa_a$, $1/\varkappa_a$ vs. T [72L7].

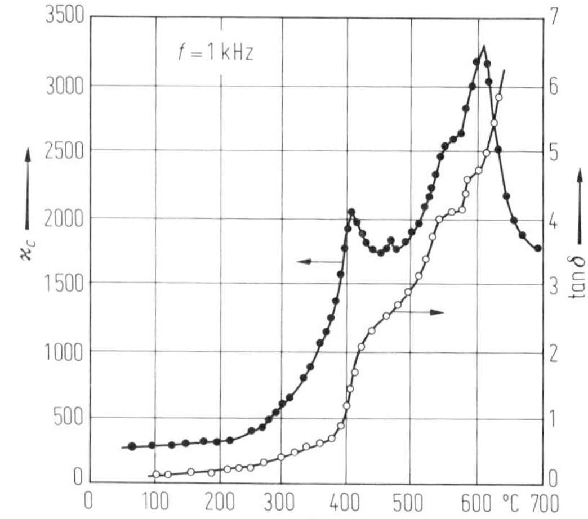

Fig. 1582. $SbNbO_4$. $\varkappa_c$, tan δ vs. T [74I16].

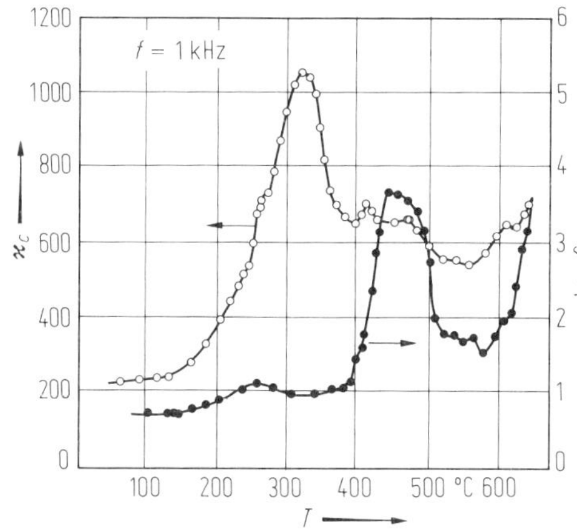

Fig. 1583. $SbTaO_4$. $\varkappa_c$, $\tan\delta$ vs. T [74I16].

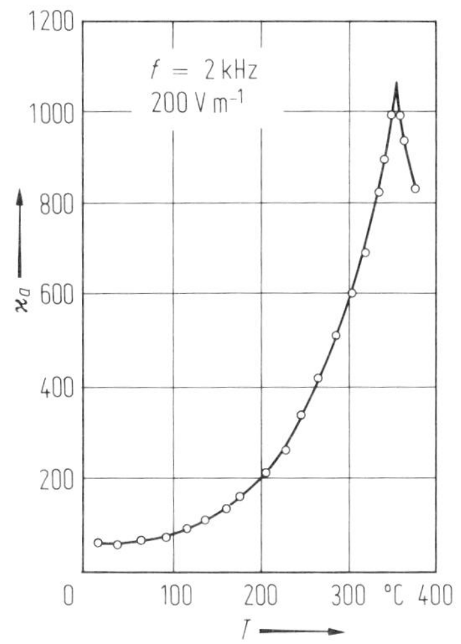

Fig. 1584. $BiNbO_4$. $\varkappa_a$ vs. T [75S33].

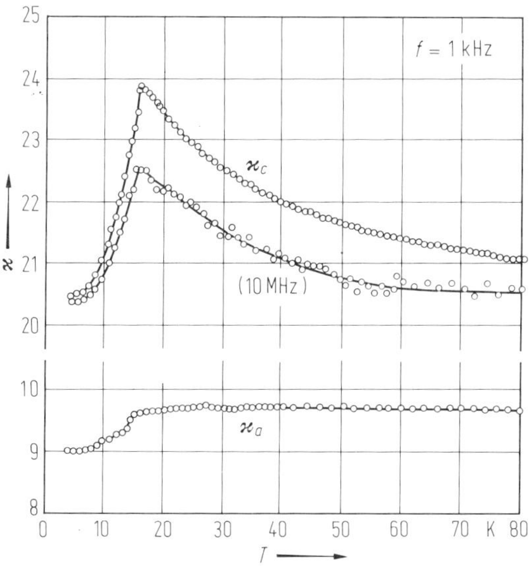

Fig. 1585. $DyVO_4$. $\varkappa_a$, $\varkappa_c$ vs. T [77U3].

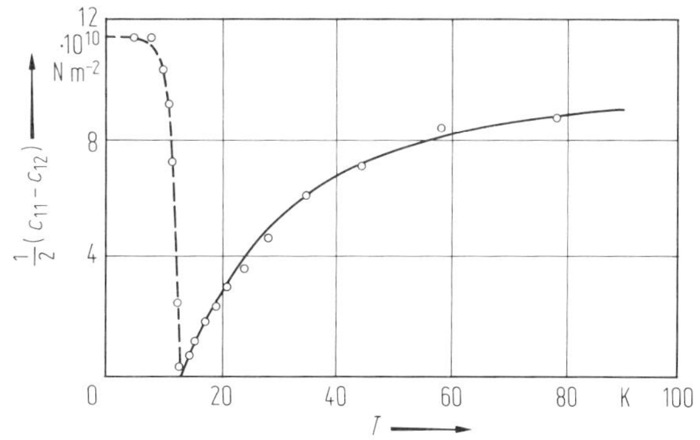

Fig. 1586. $DyVO_4$. $\frac{1}{2}(c_{11}-c_{12})$ vs. T [72S2]. Brillouin scattering method.

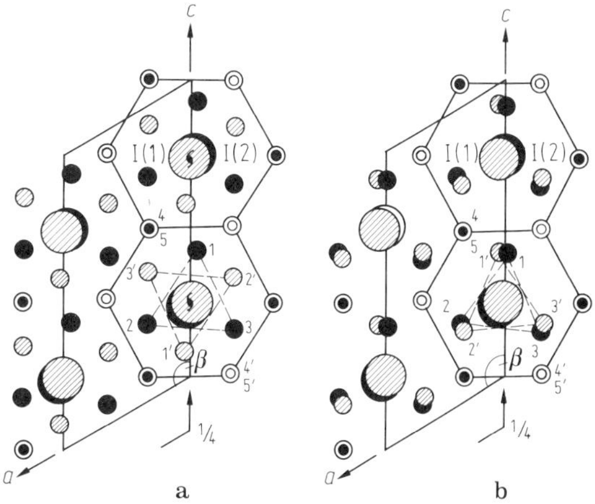

Fig. 1587. Sb_5O_7I. Sb–I framework of polytypes [77N6]. (a) 2MC and (b) 2MA. Viewed in b direction showing the typical Sb–I polyhedra. Approximate height of atoms: I(1): 0.0, I(2): 0.50, Sb(1) ~ Sb(3): 0.25, Sb(1′) ~ Sb(3′): 0.75, Sb(4): 0.05, Sb(5): 0.45, Sb(4′): 0.55, Sb(5′): 0.95.

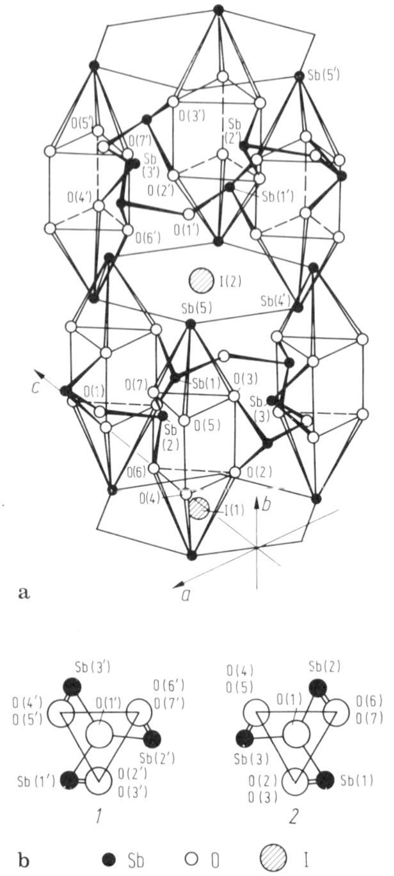

Fig. 1588. Sb_5O_7I (polytype 2MC). Fig. a Perspective view of SB – O slabs and their stacking [77 N 6]. Fig. b Stacking units of $[Sb_3O_7]^{5-}$ [78 K 18]. Polytype 2MC contains only one of these types, for example, type ②. In polytype 2MA, succeeding layers along *b*-axis are alternately built up from units of type 1 and 2. Therefore, positions of Sb(1′), Sb(3′) are different from those shown in the Fig. a. See also Fig. 1587.

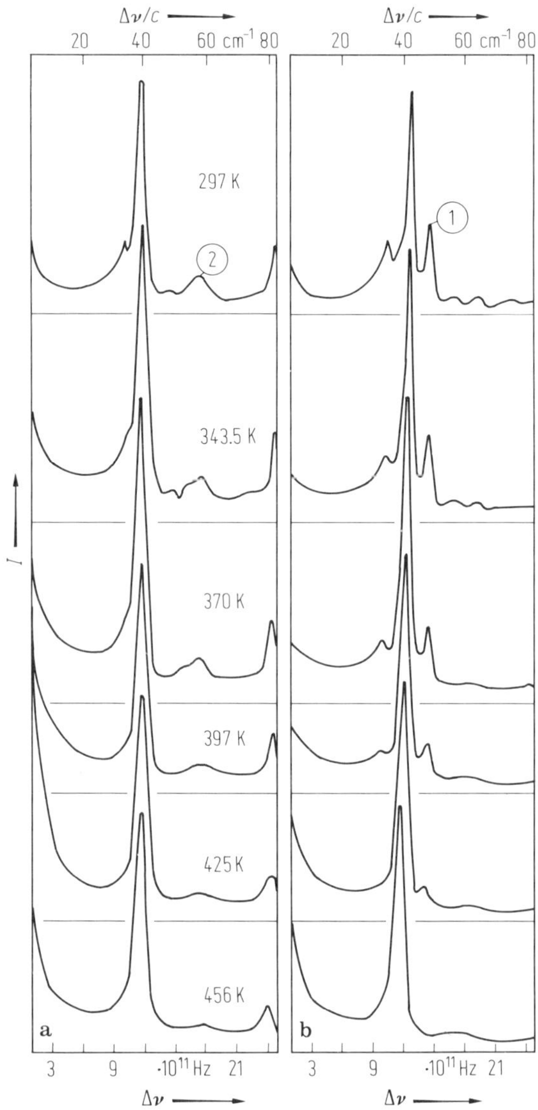

Fig. 1589. Sb_5O_7I (polytype 2MA). *I* vs. Δν [79 P 1]. *I*: Intensity of Raman scattering at low frequency region. Parameter: *T*. $\Theta = 438$ K. A′ modes measured in scattering geometries of (Fig. a) $Z(X, X)Y$ and (Fig. b) $Z(Y, X)Y$. See also Fig. 1590.

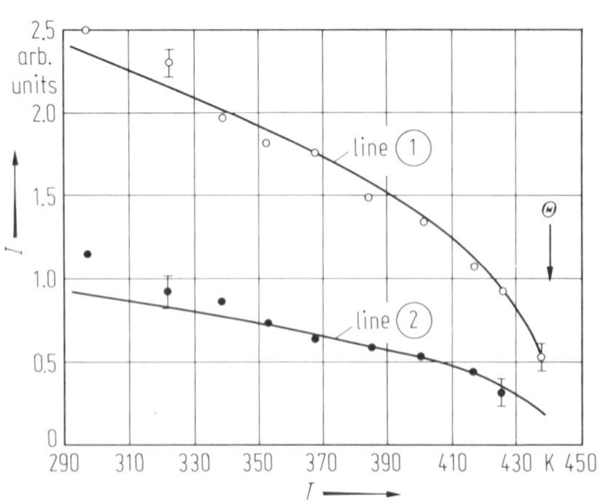

Fig. 1590. Sb_5O_7I (polytype 2MA). *I* vs. *T* [79 P 1]. *I*: Raman scattering intensity of the mode ① and ② in Fig. 1589.

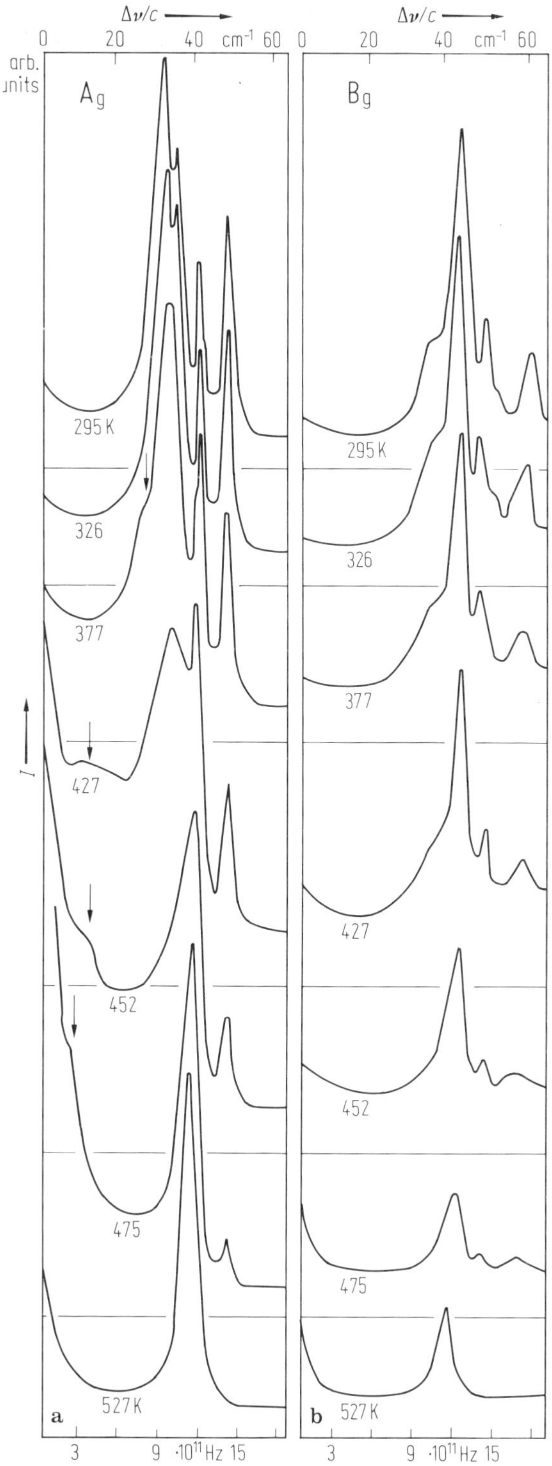

Fig. 1591. Sb_5O_7I (polytype 2MA). I vs. $\Delta\nu$ [76P18]. I: intensity of Raman scattering at low frequency region. Parameter: T. $\Theta = 481$ K. (a): A_g modes measured in a scattering geometry of $Z(X, X)Y$. (b): B_g modes measured in $Z(X, Z)Y$. The strongly damped soft mode is indicated by arrows.

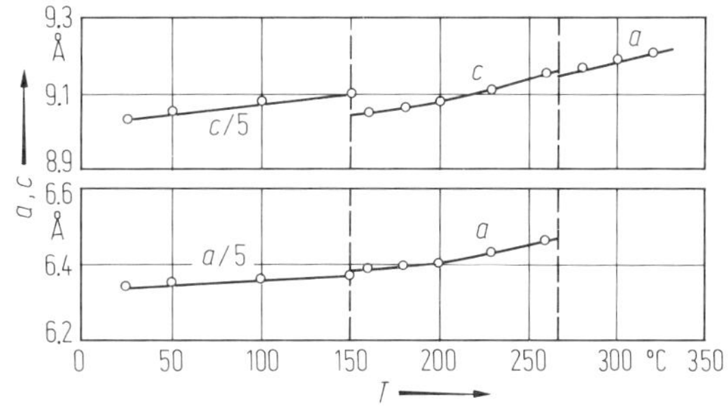

Fig. 1592. $Rb_3MoO_3F_3$. Lattice parameters vs. T [77P1].

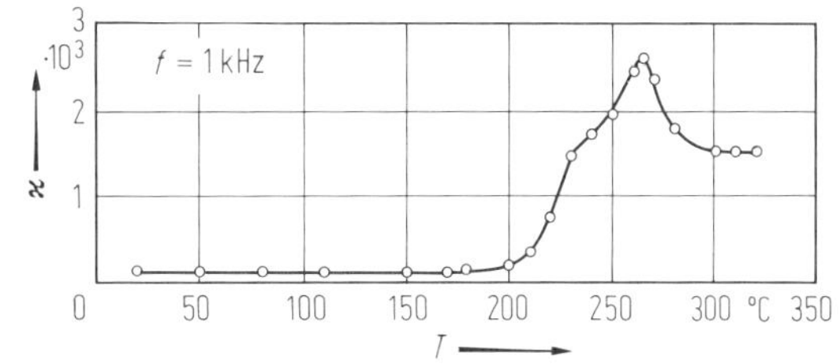

Fig. 1593. $Rb_3MoO_3F_3$ (polycrystalline). $\varkappa$ vs. T [77P1].

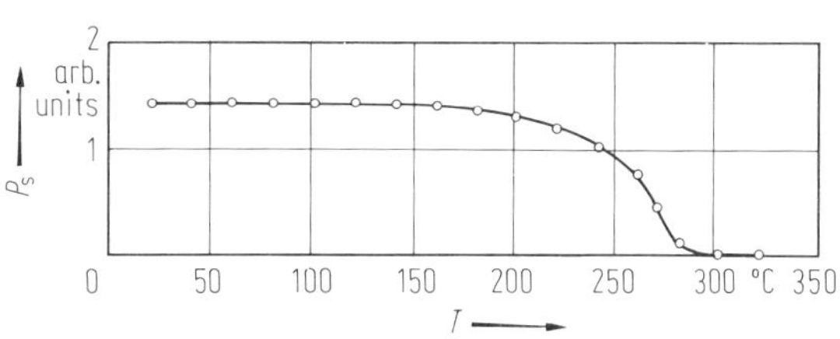

Fig. 1594. $Rb_3MoO_3F_3$ (polycrystalline). P_s vs. T [77P1].

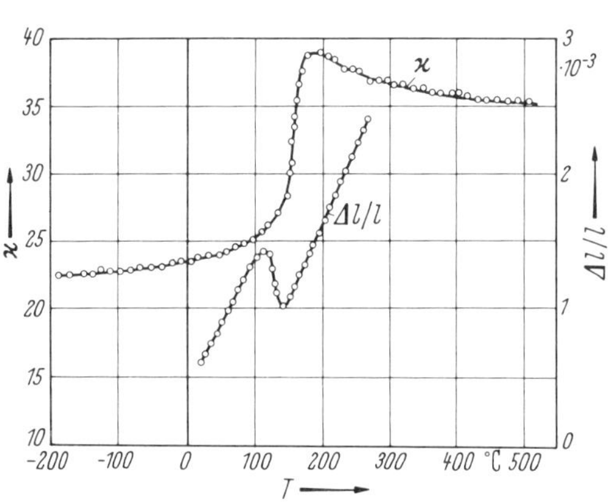

Fig. 1595. Pb_4SiO_6 (ceramics). $\varkappa$, $\Delta l/l$ vs. T [65I5]. $\varkappa$ at $f = 500$ kHz.

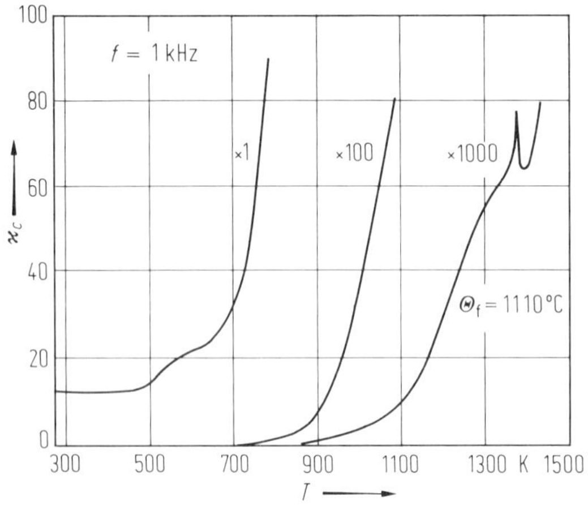

Fig. 1596. $Ca_3(VO_4)_2$. $\varkappa_c$ vs. T [78 G 8]. $f = 1$ kHz.

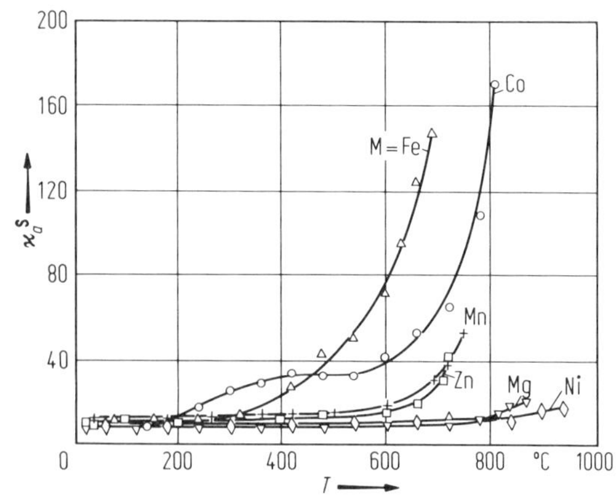

Fig. 1597. $BaMF_4$ (M = Mg, Zn, Mn, Fe, Co, Ni). $\varkappa_a^S$ vs. T measured at 100 MHz [69 D 2].

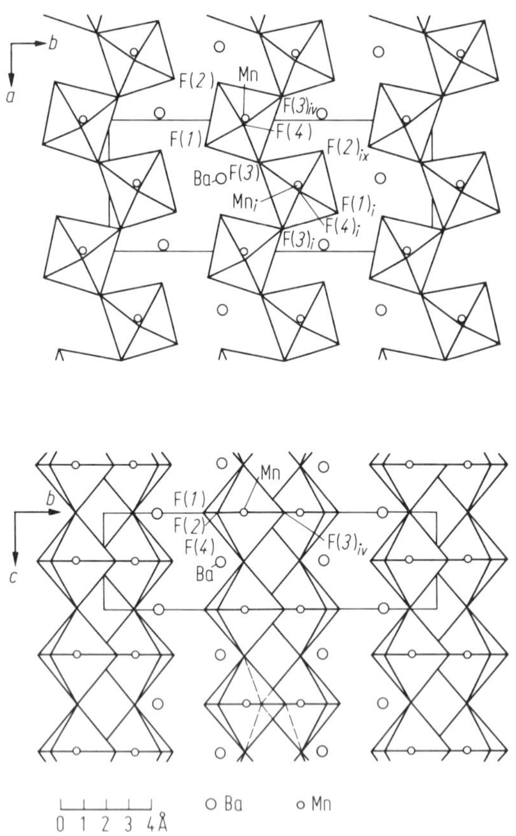

Fig. 1598. $BaMnF_4$. Crystal structure [69 K 3]. Schematic view showing MnF_6 octahedra in $BaMnF_4$ structure along the c axis (upper) and along the a axis (lower). Subscripts attached to names of atoms denote symmetry operations, see Table 430 about the nomenclature.

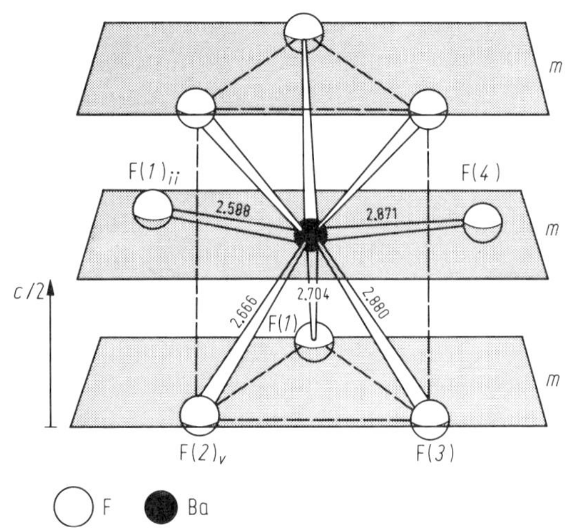

Fig. 1599. $BaMnF_4$. Crystal structure [69 K 3]. Coordination polyhedron of fluorine atoms around Ba. Distances are in Å units and subscripts attached to names of atoms denote symmetry operations; for the explanation of the nomenclature, see Table 430.

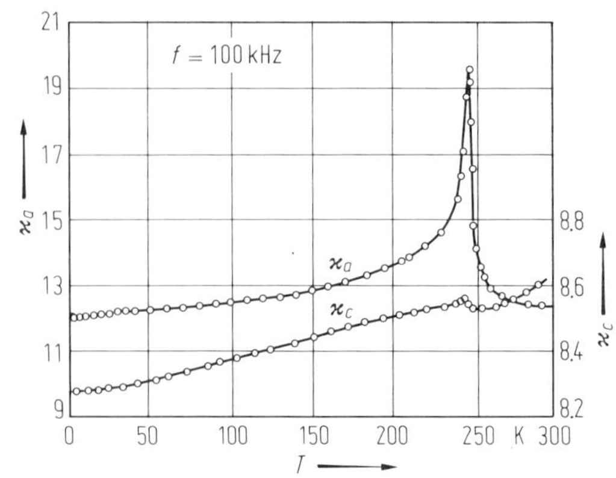

Fig. 1600. $BaMnF_4$. $\varkappa_a$, $\varkappa_c$ vs. T [76 S 4].

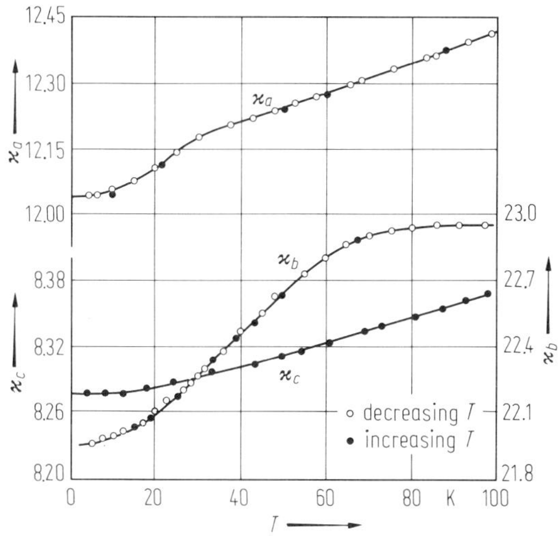

Fig. 1601. $BaMnF_4$. $\varkappa_a, \varkappa_b, \varkappa_c$ vs. T in low temperature region [77S4]. $f = 10 \cdots 100$ kHz.

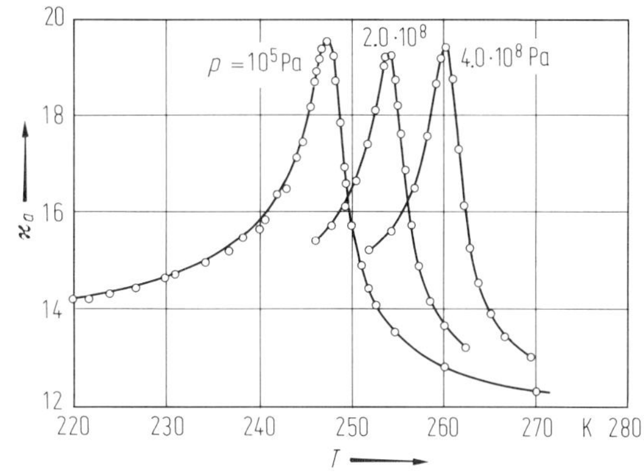

Fig. 1602. $BaMnF_4$. $\varkappa_a$ vs. T [76S4]. Parameter: p, hydrostatic pressure. $f = 1 \cdots 100$ kHz.

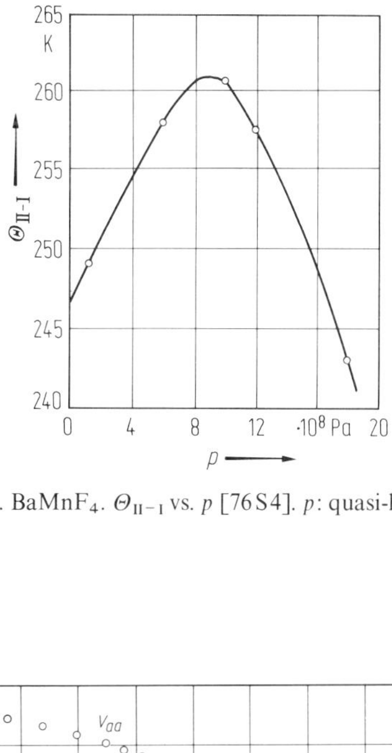

Fig. 1603. $BaMnF_4$. $\Theta_{\text{II-I}}$ vs. p [76S4]. p: quasi-hydrostatic pressure.

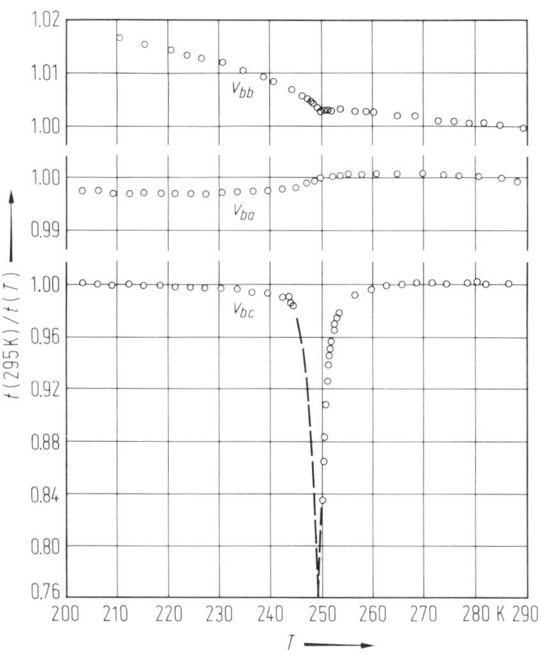

Fig. 1605. $BaMnF_4$. t(295 K)/$t(T)$ vs. T [75F8]. Normalized inverse transit time for modes propagating along the b-axis. Circles: for $f = 30$ MHz, and dashed line: for $f = 4$ MHz.

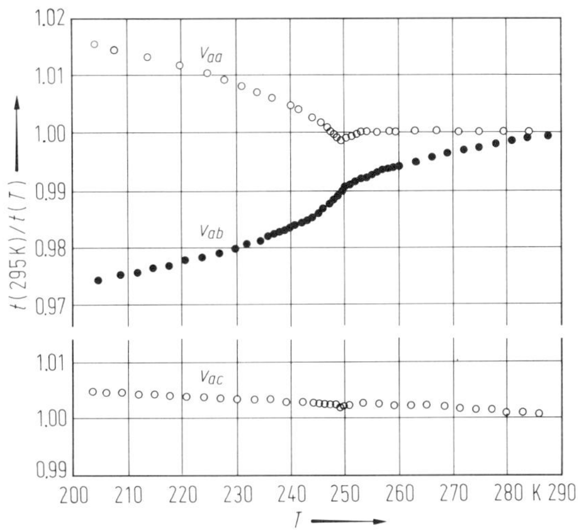

Fig. 1604. $BaMnF_4$. t(295 K)/$t(T)$ vs. T [75F8]. t: transit time for modes propagating along the a axis, $f = 30$ MHz. t is inversely proportional to the sound velocity. v_{aa}: longitudinal wave velocity. v_{ab}: velocity of a sound wave polarized along the b axis. v_{ac}: polarized along the c axis.

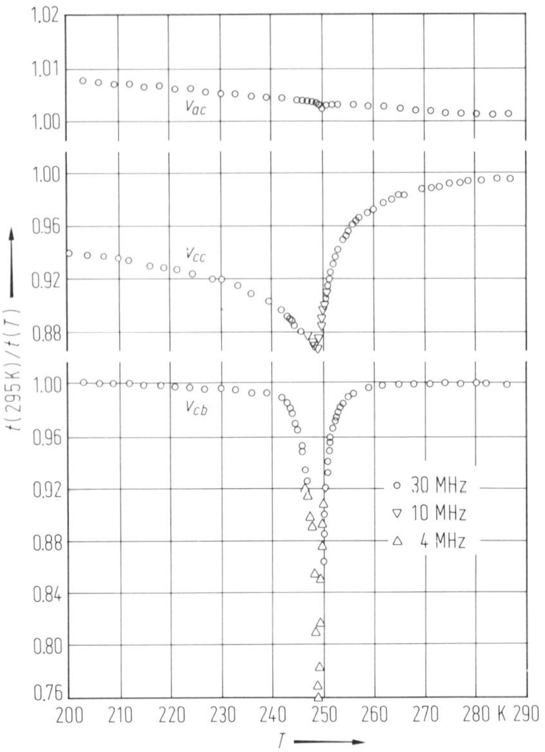

Fig. 1606. $BaMnF_4$. t (295 K)/t(T) vs. T [75F8]. Normalized inverse transit time for modes propagating along the c axis.

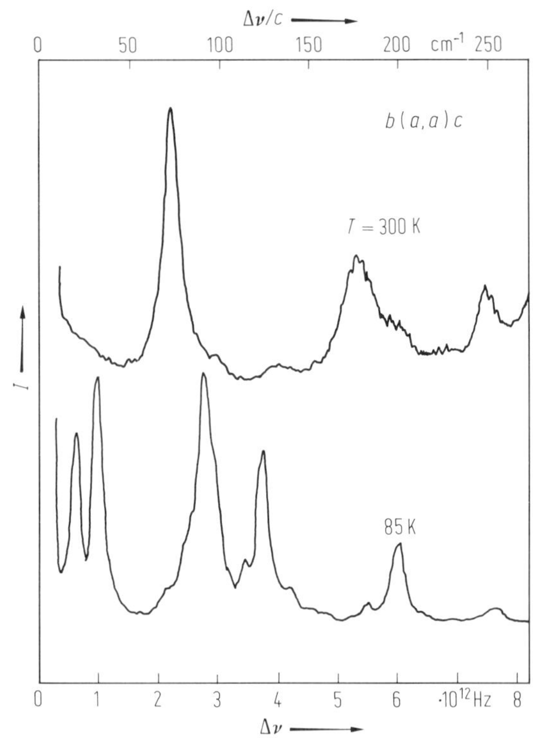

Fig. 1607. $BaMnF_4$. I vs. $\Delta\nu$ [76R10]. I: Raman scattering intensity for the scattering geometry of $b(a,a)$c in the pyroelectric phase at 300 K and the antiferroelectric phase at 80 K.

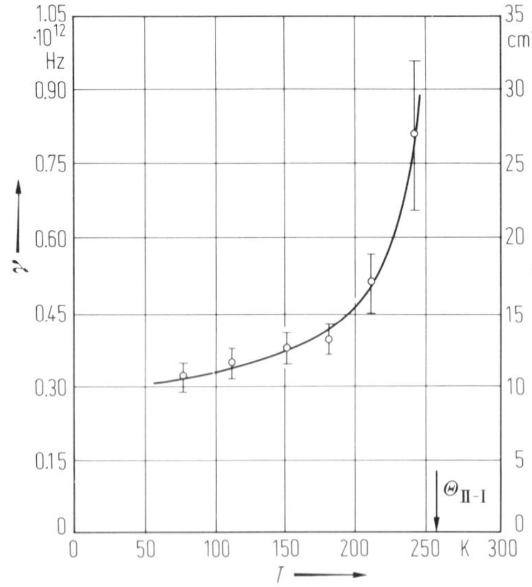

Fig. 1608. $BaMnF_4$. γ vs. T [76R10]. γ: damping constant of the soft mode obtained as the full width at half-maximum of the Raman spectrum.

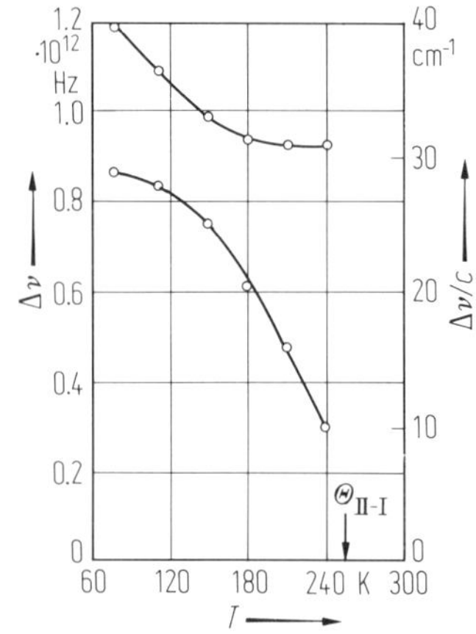

Fig. 1609. $BaMnF_4$. $\Delta\nu$ vs. T [76R10]. $\Delta\nu$: Raman frequency shift of the soft mode below $\Theta_{II-I}=255$ K. For the temperature range of 210 K $< T < \Theta_{II-I}$, the mode is overdamped. See also [74R7].

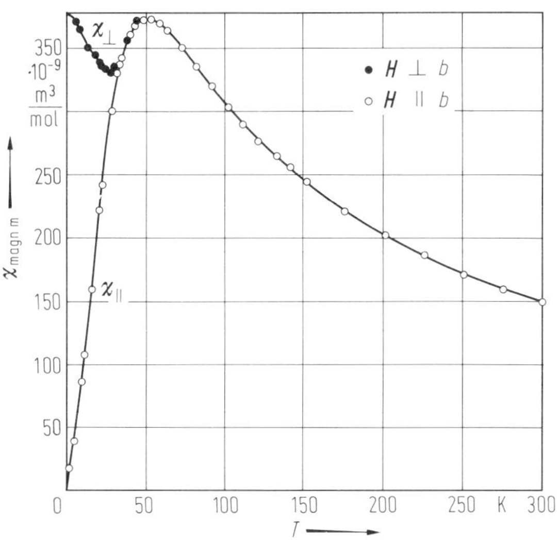

Fig. 1610. $BaMnF_4$. $\chi_{magn\,m}$ vs. T [69H8]. $\chi_\perp$ and $\chi_\parallel$ are the susceptibilities measured perpendicular to and parallel to the b axis, respectively.

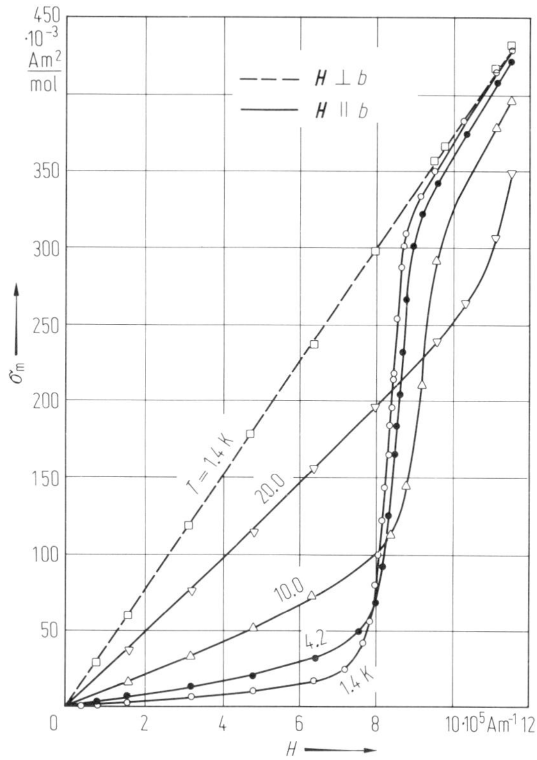

For Fig. 1612, see p. 692.

Fig. 1611. $BaMnF_4$. σ_m vs. H [69H8]. Parameter: T and field directions, the field is along the b axis except the case (dashed curve) where the field is perpendicular to b.

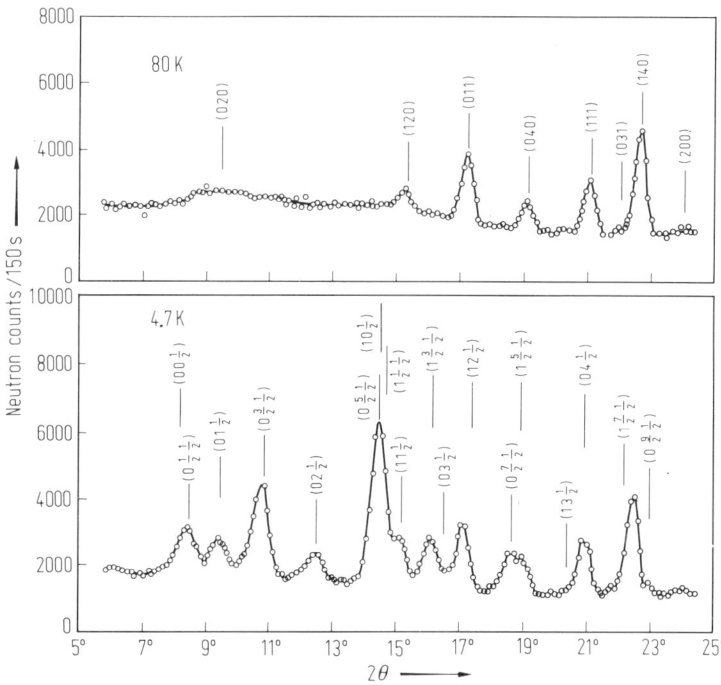

Fig. 1613. $BaCoF_4$. Neutron diffraction patterns from a powder [72E1]. All peaks are indexed in terms of the chemical cell.

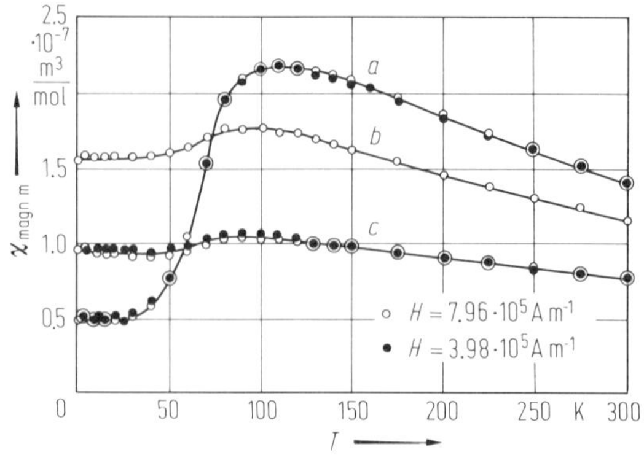

Fig. 1612. $BaCoF_4$. $\chi_{magn\ m}$ vs. T [72E1].

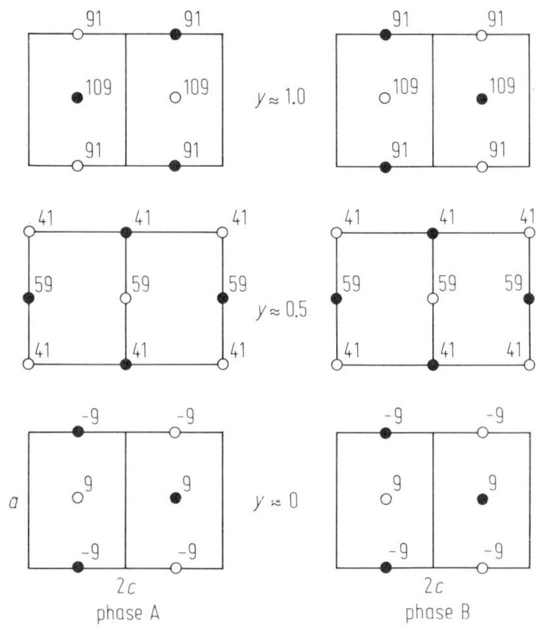

Fig. 1614. $BaCoF_4$. Magnetic structure [72E1]. Open and closed circles represent oppositely directed moments. Spin axis is parallel to [100] direction. Only Co^{2+} ions are shown. Phases A, B are two magnetic phases. Small numerals denote height in units of 0.01b.

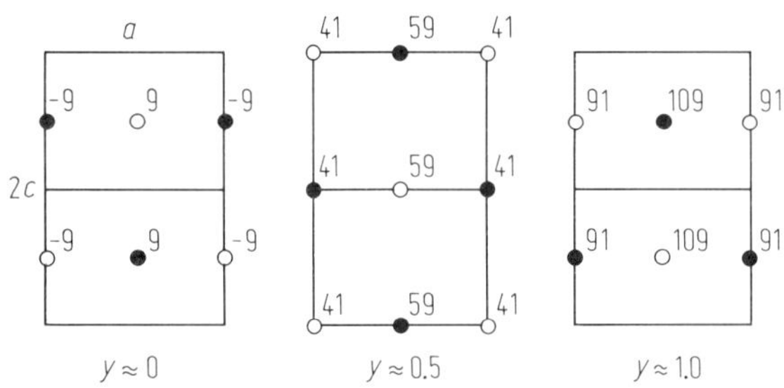

Fig. 1616. $BaNiF_4$. Magnetic structure [70C8]. Open and closed circles depict oppositely directed moments along the b axis. Small numerals denote height in units of 0.01 b. Only Ni ions are shown.

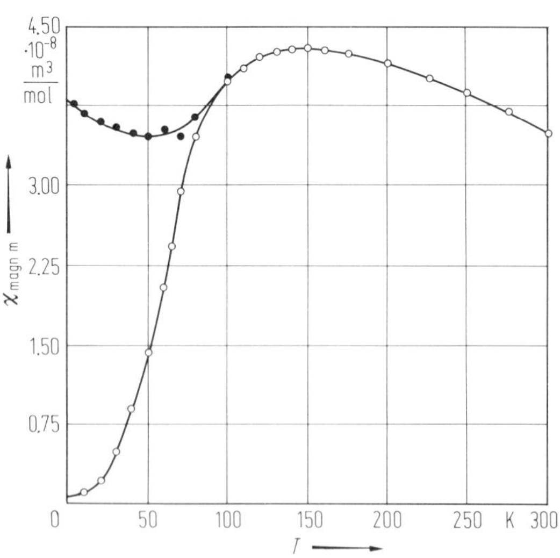

Fig. 1615. $BaNiF_4$. $\chi_{magn\ m}$ vs. T [70C8]. Open circles: $\boldsymbol{H} \| b$; full circles: $\boldsymbol{H} \perp b$.

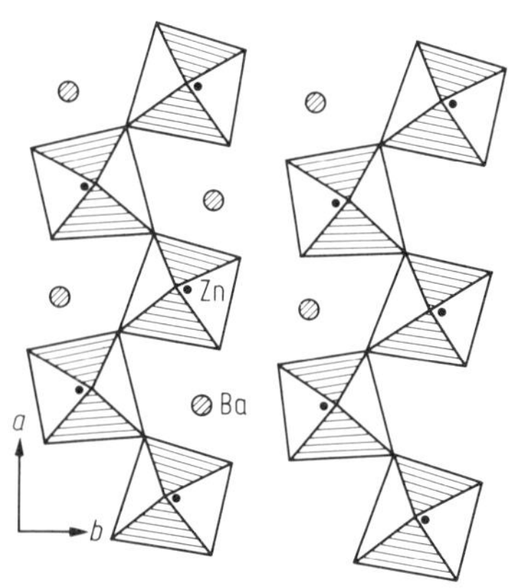

Fig. 1617. $BaZnF_4$. Crystal structure [69K3]. Schematic view along c. The polar axis is a.

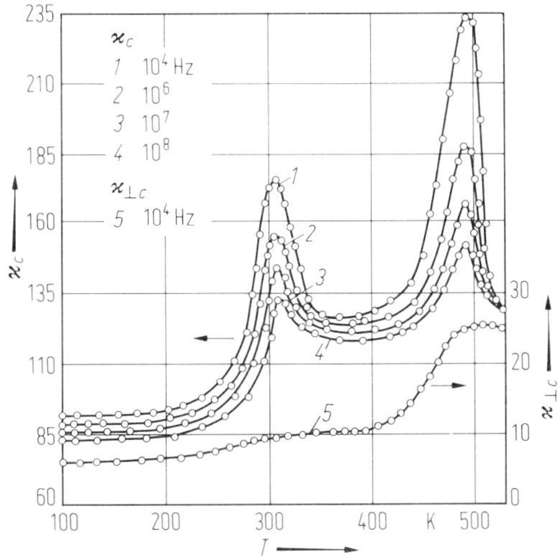

Fig. 1618. Sb_2S_3. $\varkappa_c$, $\varkappa_{\perp c}$ vs. T [74O4]. Parameter: f.

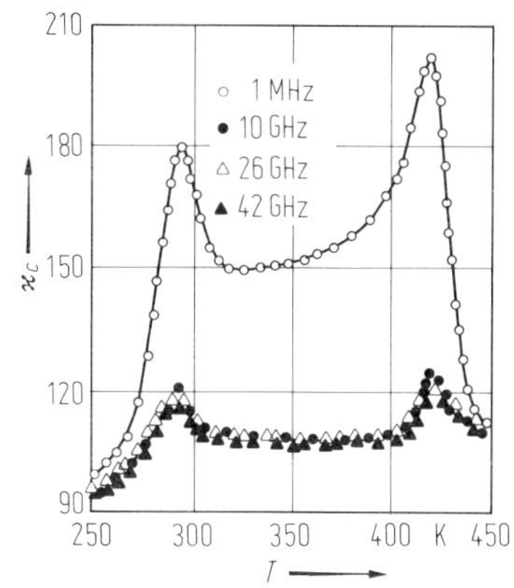

Fig. 1619. Sb_2S_3. $\varkappa_c$ vs. T in microwave region [76G16].

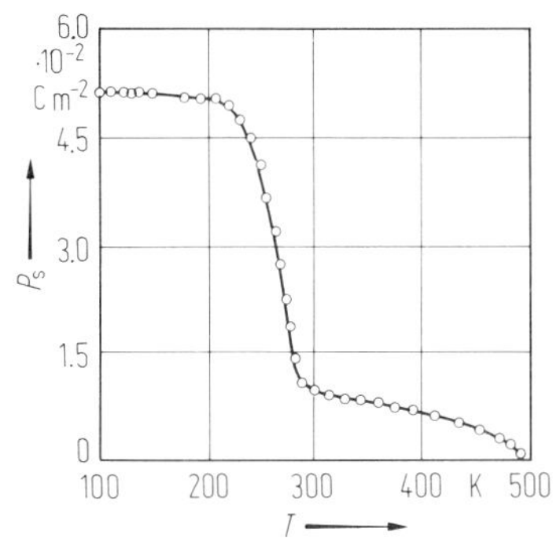

Fig. 1620. Sb_2S_3. P_s vs. T [74O4].

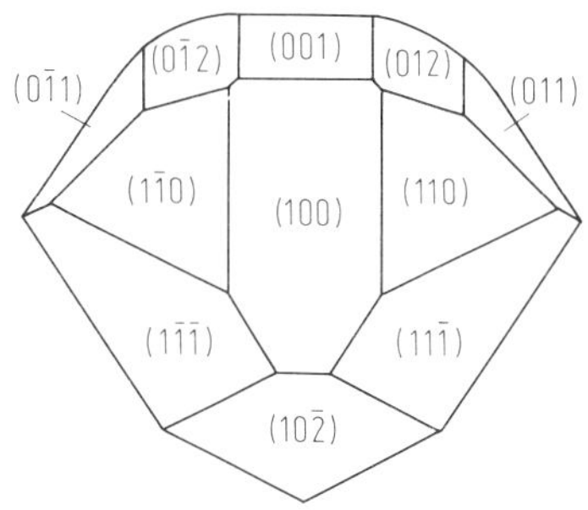

Fig. 1621. $Sn_2P_2S_6$. Crystal form [74C2].

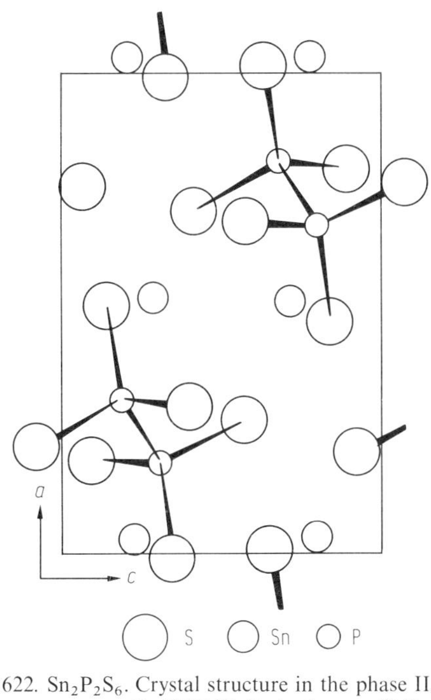

Fig. 1622. $Sn_2P_2S_6$. Crystal structure in the phase II, based on the space group Pn [74D8]. Projection on (010).

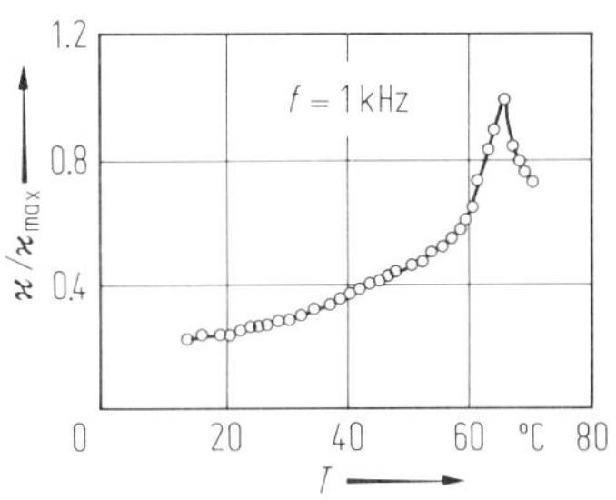

Fig. 1623. $Sn_2P_2S_6$. $\varkappa/\varkappa_{max}$ vs. T [77B19]. $\varkappa_{max}=7410$; maximum dielectric constant at Θ_f. Measurements were made along the [101] direction.

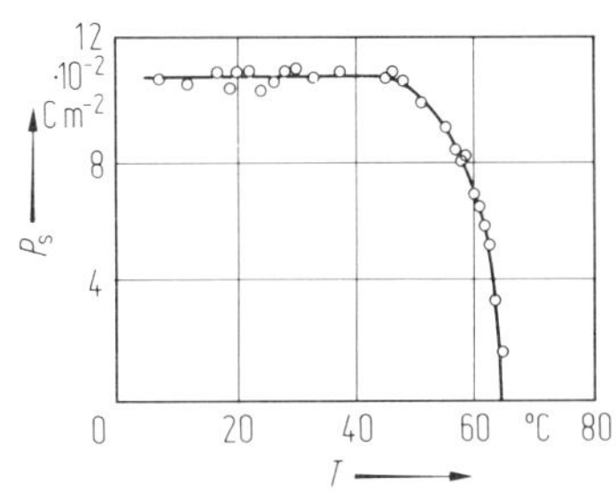

Fig. 1624. $Sn_2P_2S_6$. P_s vs. T [77B19]. Measurements were made along the [101] direction.

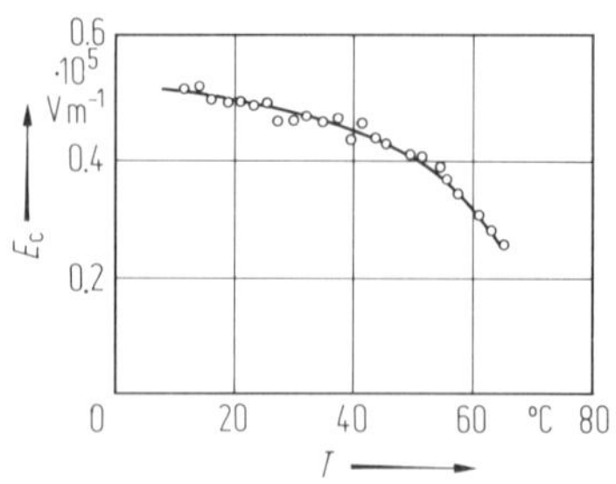

Fig. 1625. $Sn_2P_2S_6$. E_c vs. T [77B19]. Measurements were made along the [101] direction.

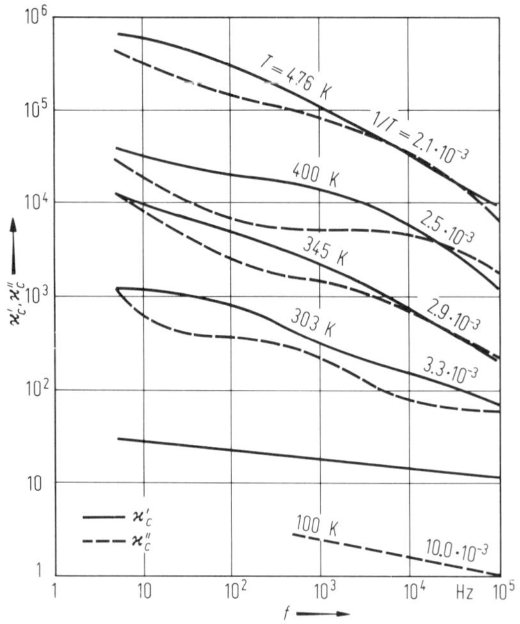

Fig. 1627. $Li(N_2H_5)SO_4$. $\varkappa'_c$, $\varkappa''_c$ vs. f [71S11]. Parameter: T.

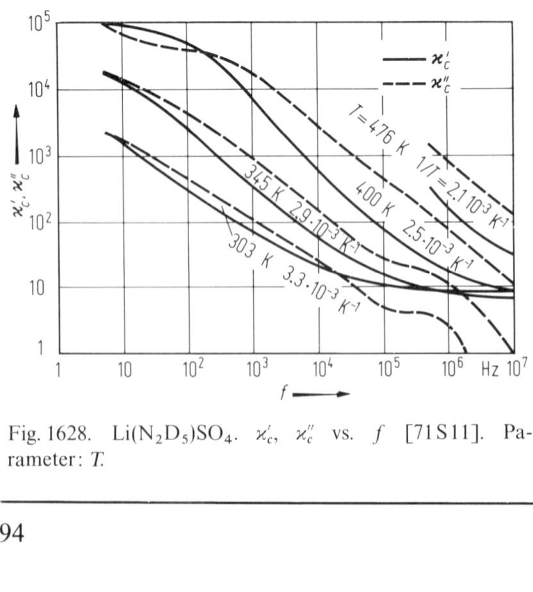

Fig. 1628. $Li(N_2D_5)SO_4$. $\varkappa'_c$, $\varkappa''_c$ vs. f [71S11]. Parameter: T.

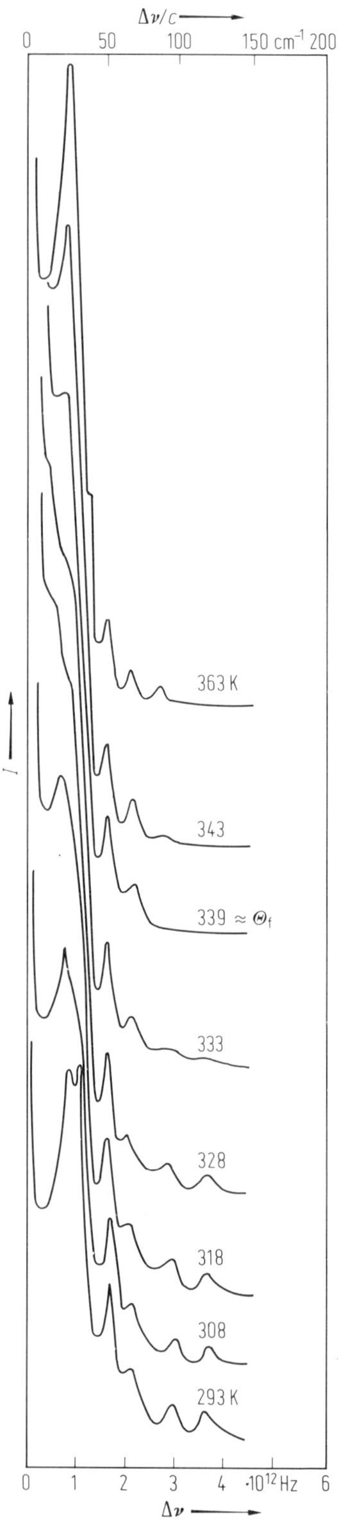

Fig. 1626. $Sn_2P_2S_6$. I vs. $\Delta\nu$ [78V2]. Parameter: T. I: Raman scattering intensity in a scattering geometry of $Z(Y, Y)X$. The frequency of an A' mode at about 30 cm^{-1} at RT decreases as the temperature approaches Θ_f.

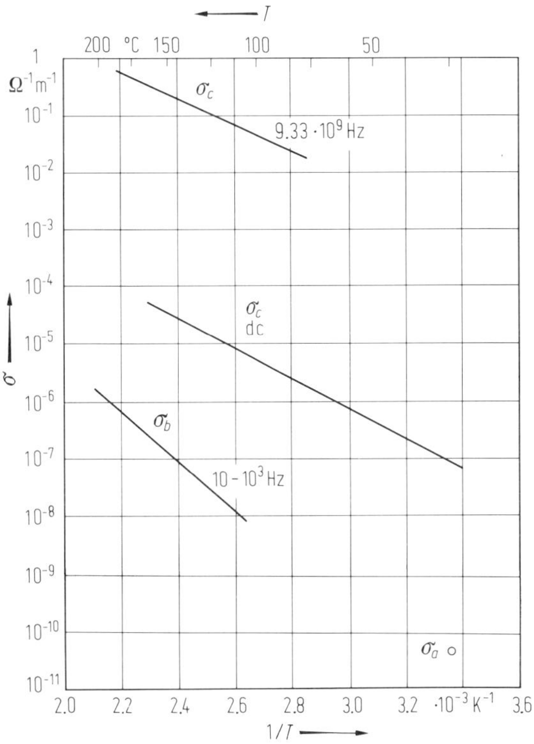

Fig. 1629. $Li(N_2H_5)SO_4$. σ_a, σ_b, σ_c vs. $1/T$ [71 S 11]. Parameter: f.

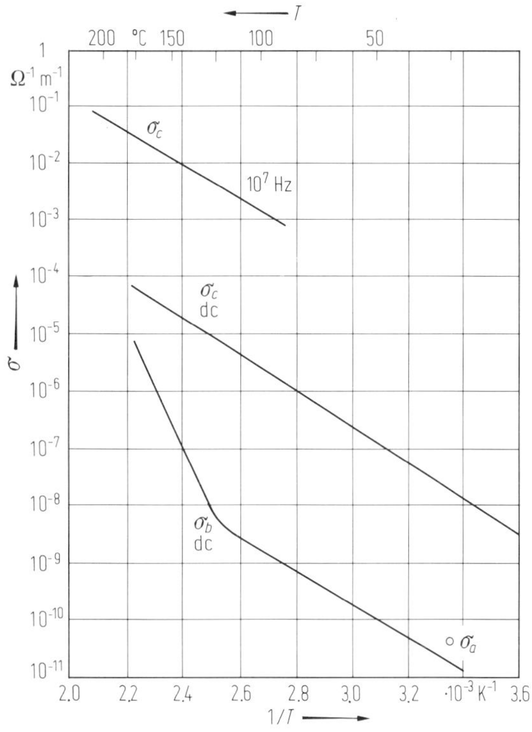

Fig. 1630. $Li(N_2D_5)SO_4$. σ_a, σ_b, σ_c vs. $1/T$ [71 S 11]. Parameter: f.

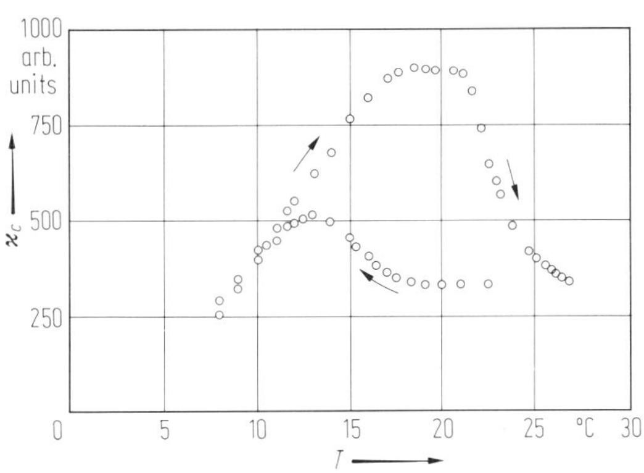

Fig. 1632. $H_2(UO_2)_2(AsO_4)_2 \cdot 8H_2O$. $\varkappa_c$ vs. T [75 D 1].

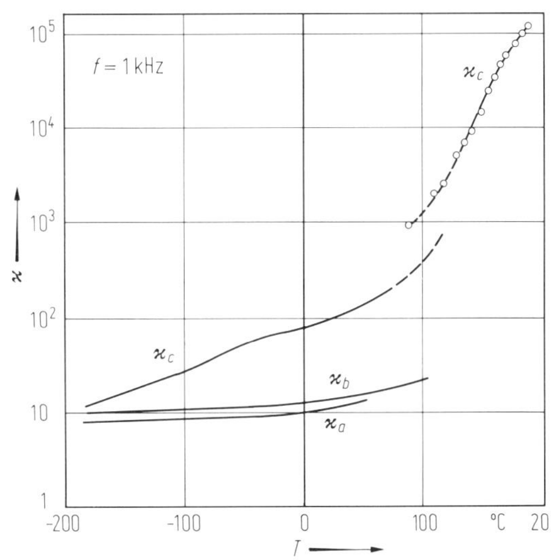

Fig. 1631. $Li(N_2H_5)BeF_4$. $\varkappa_a$, $\varkappa_b$, $\varkappa_c$ vs. T [71 P 1]. The value of $\varkappa_c$ strongly depends on quality and thermal hysteresis of the specimen.

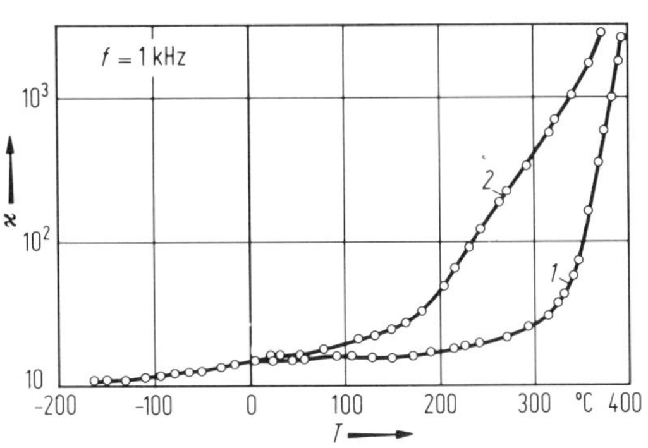

◄ Fig. 1633. $NaTh_2(PO_4)_3$. $\varkappa$ vs. T [72 T 12]. Curve *1*: first run, *2*: second run after 24 h. When measurements on the same specimen were repeated, after a short pause an increase of the permittivity was detected as shown by the curve *2*. After a pause of 10 days approximately the same results as for the first time were obtained indicating a recovery effect.

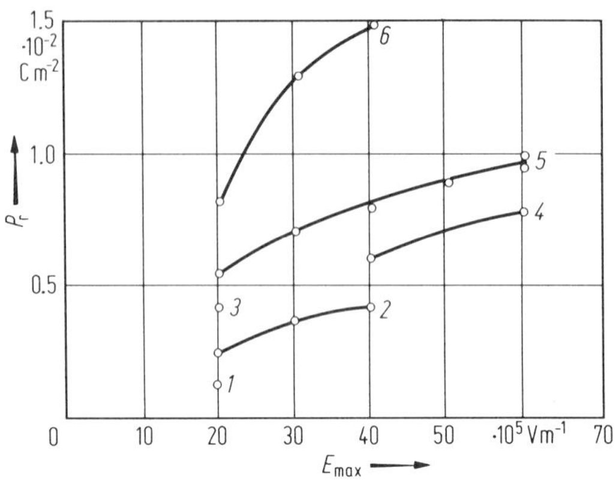

Fig. 1634. $NaTh_2(PO_4)_3$. P_r vs. E_{max} [69T3]. T=RT. P_r: remanent polarization, E_{max}: maximum applied field. Numbers *1*⋯*6* correspond to six different samples. Electric fields are applied in the ⌊100⌋ direction.

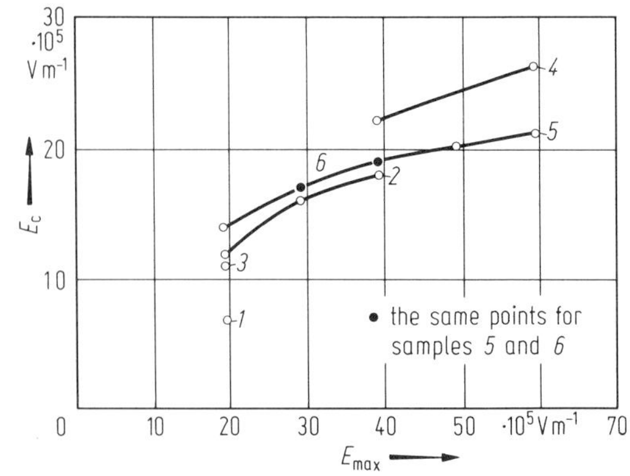

Fig. 1635. $NaTh_2(PO_4)_3$. E_c vs. E_{max} [69T3]. T=RT. E_{max}: maximum applied field. Numbers *1*⋯*6* correspond to six different samples. Electric fields are applied in the ⌊100⌋ directions.

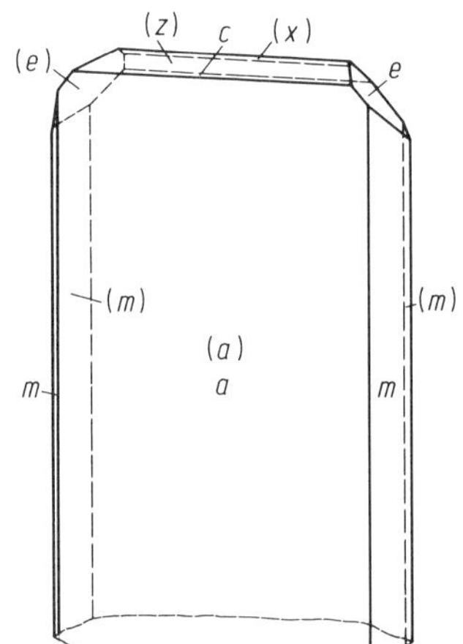

Fig. 1636. $AgTh_2(PO_4)_3$. Crystal habit [70T10]. a: {100}, c: {001}, m: {110}, e: {$\bar{1}$11}, x: {$\bar{1}$01}, z: {$\bar{2}$01}.

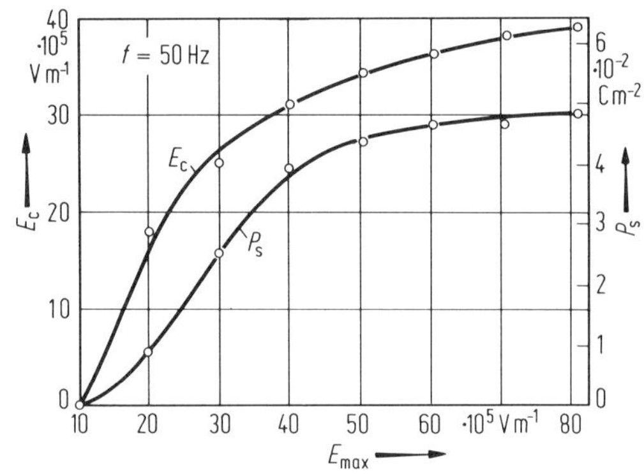

Fig. 1637. $AgTh_2(PO_4)_3$. P_s, E_c vs. E_{max} [70T10]. T=RT, f=50 Hz. E_{max}: maximum applied electric field.

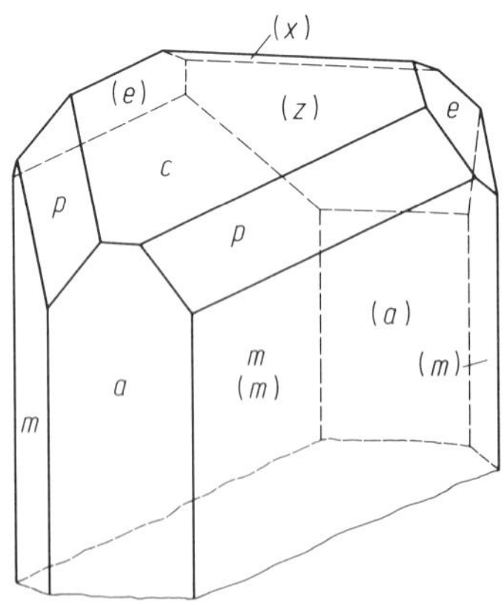

Fig. 1638. $NaU_2(PO_4)_3$. Crystal habit [68M2]. c: {001}, a: {100}, m: {110}, x: {$\bar{1}0\bar{1}$}, z: {$\bar{2}$01}, p: {111}, e: {$\bar{1}$11}.

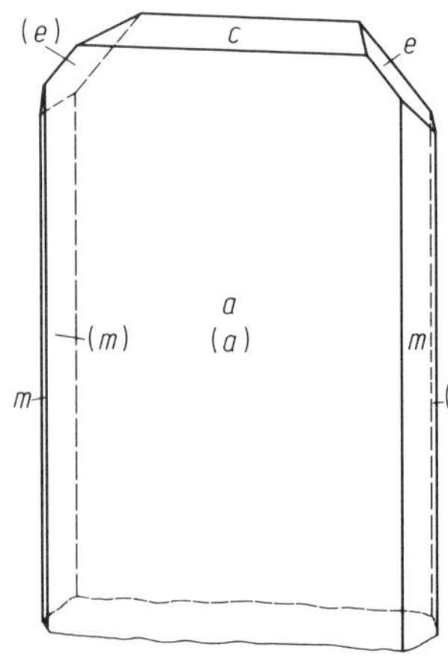

Fig. 1639. $NaU_2(PO_4)_3$. Crystal habit [68M2]. c: {001}, a: {100}, m: {110}, e: {$\bar{1}$11}.

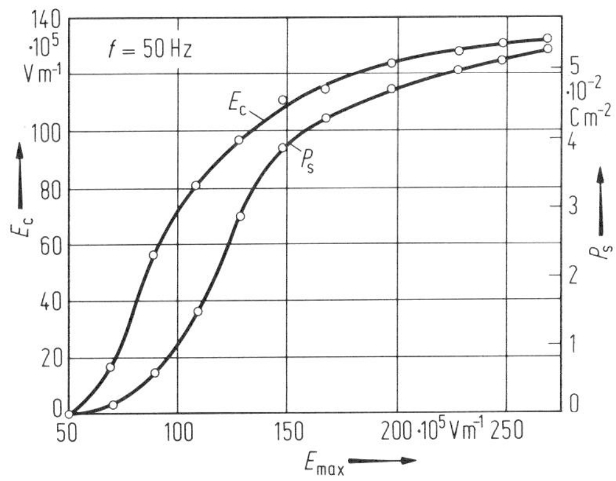

Fig. 1640. $NaU_2(PO_4)_3$. P_s, E_c vs. E_{max} [69T2]. T=RT. E_{max}: maximum applied electric field. Electric field is applied normal to (100).

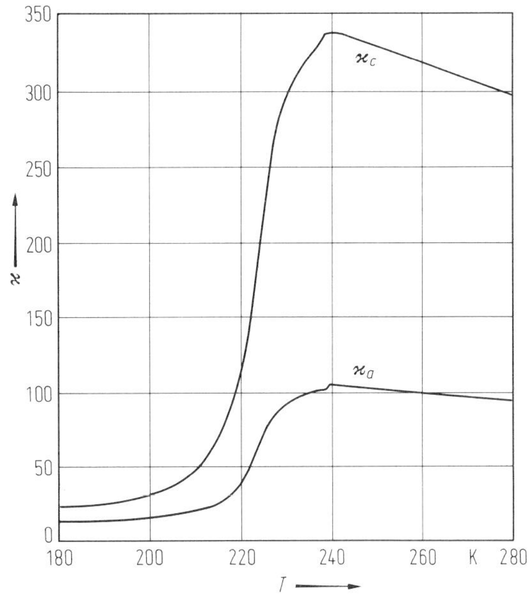

Fig. 1641. $Ag_2H_3IO_6$. $\varkappa_a$, $\varkappa_c$ vs. T [76R8]. $\varkappa_a$, $\varkappa_c$: static dielectric constants.

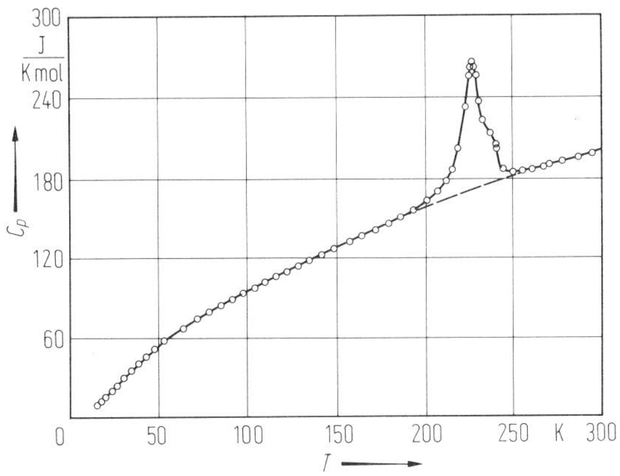

◄ Fig. 1642. $Ag_2H_3IO_6$. C_p vs. T [44S2].

For Fig. 1643, see p. 698.

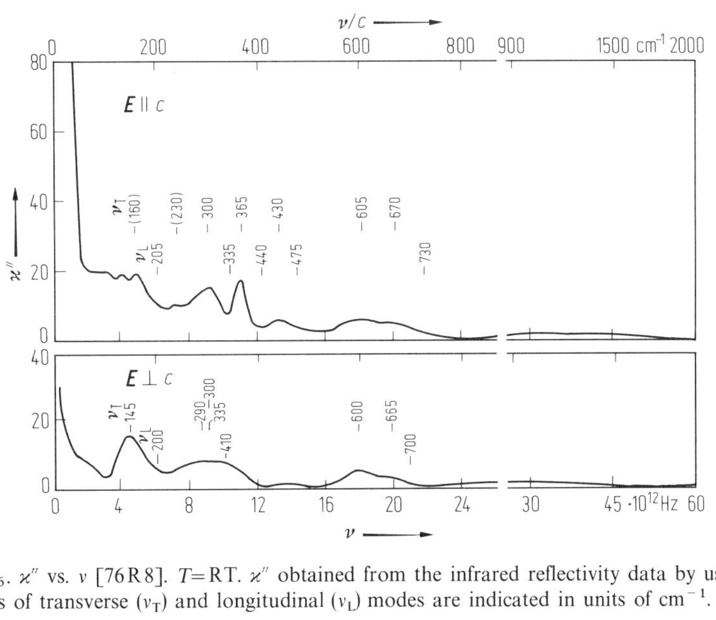

Fig. 1644. $Ag_2H_3IO_6$. $\varkappa''$ vs. ν [76R8]. T=RT. $\varkappa''$ obtained from the infrared reflectivity data by using Kramers-Kronig analysis. Frequencies of transverse (ν_T) and longitudinal (ν_L) modes are indicated in units of cm^{-1}. 1 cm^{-1} = $3 \cdot 10^{10}$ Hz.

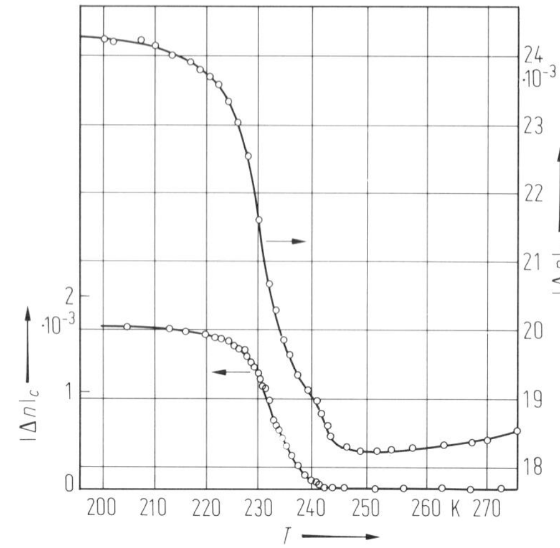

Fig. 1643. $Ag_2H_3IO_6$. $|\Delta n|_a$, $|\Delta n|_c$ vs. T [76R8]. $|\Delta n|_a$: $\boldsymbol{E}\|c$, $|\Delta n|_c$: $\boldsymbol{E}\perp c$. $\lambda=588$ nm.

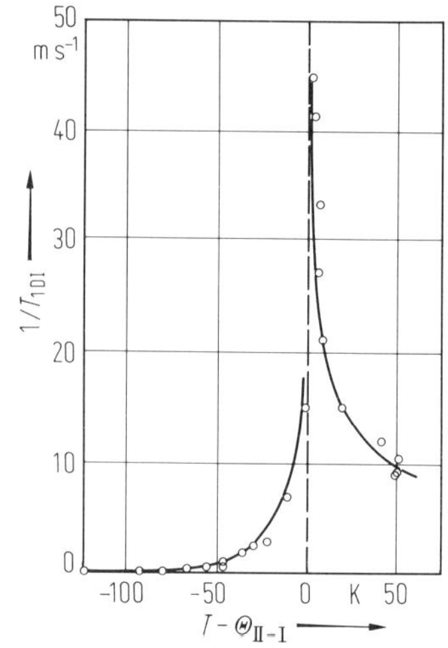

Fig. 1645. $Ag_2H_3IO_6$. $T_{1\,DI}^{-1}$ vs. $T-\Theta_{II-I}$ [75B16]. $T_{1\,DI}$: dipolar-frame spin-lattice relaxation time for iodine.

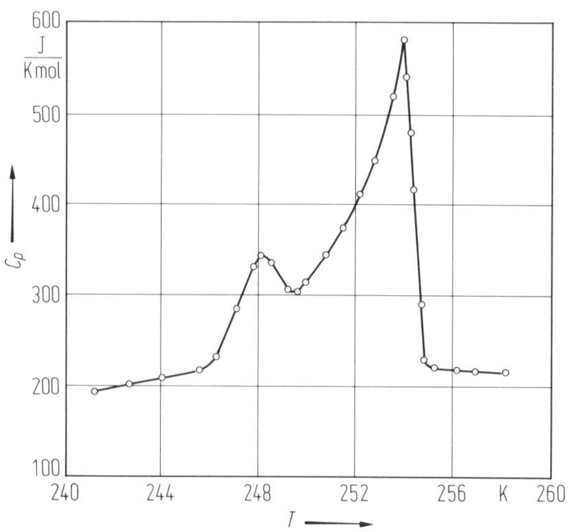

Fig. 1647. $(NH_4)_2H_3IO_6$. C_p vs. T [45B1].

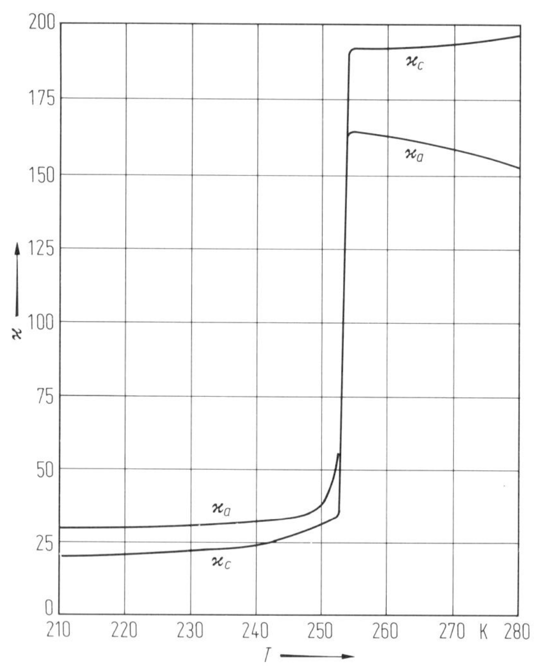

Fig. 1646. $(NH_4)_2H_3IO_6$. $\varkappa_a$, $\varkappa_c$ vs. T [76R8]. $\varkappa_a$, $\varkappa_c$: static dielectric constants.

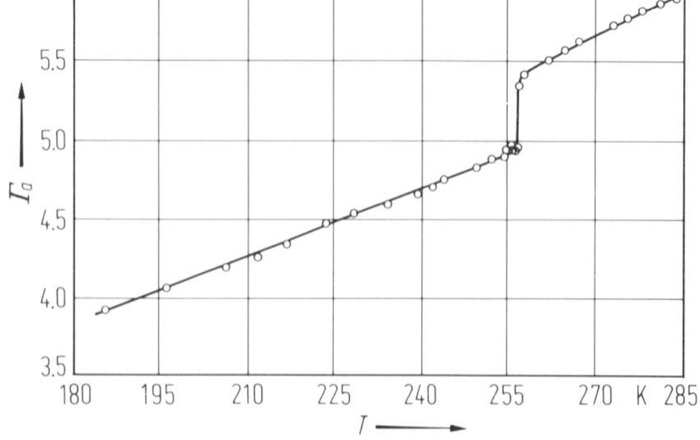

◄ Fig. 1648. $(NH_4)_2H_3IO_6$. Γ_a vs. T [76R8]. Γ_a: birefringence with polarization $\boldsymbol{E}\|c$ which is expressed as path difference in units of wavelength used ($\lambda=588$ nm).

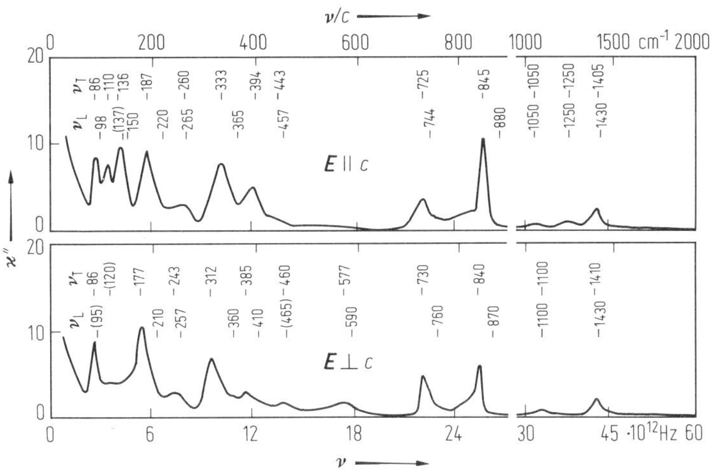

Fig. 1649. $(NH_4)_2H_3IO_6$. $\varkappa''$ vs. ν [76R8]. T = RT. $\varkappa''$ obtained from infrared reflectivity data by using Kramers-Kronig analysis. Frequencies of transverse (ν_T) and longitudinal (ν_L) modes are indicated in units of cm^{-1}. $1\ cm^{-1} = 3 \cdot 10^{10}$ Hz.

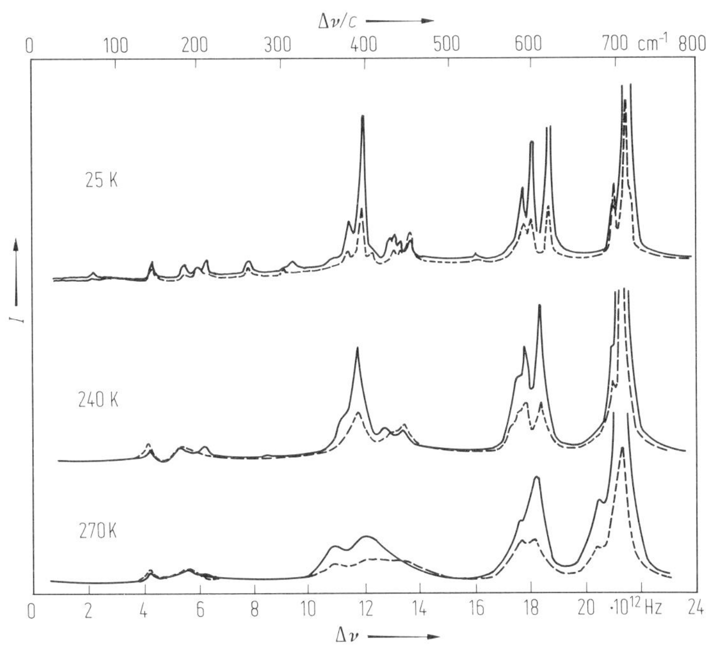

Fig. 1650. $(NH_4)_2H_3IO_6$. I vs. $\Delta\nu$ [76P16]. I: Raman scattering intensity. Solid lines: A-modes observed in the scattering geometry $X(Z,Z)Y$. Dotted lines: E-modes observed in the scattering geometry $X(Y,X)Y$ or $X(Y,Z)Y$. Parameter: T.

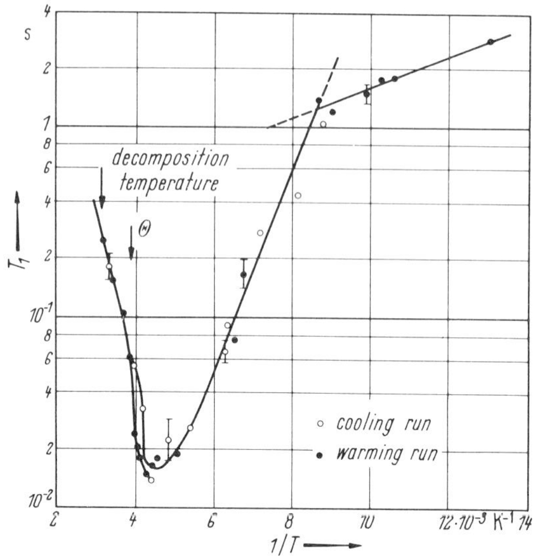

Fig. 1651. $(NH_4)_2H_3IO_6$. T_1 vs. $1/T$ [62M5]. T_1: Spin-lattice relaxation time for proton, $f = 30$ MHz.

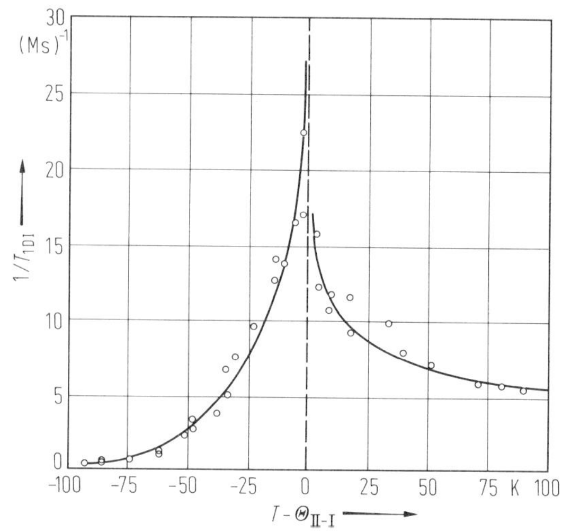

Fig. 1652. $(NH_4)_2H_3IO_6$.$T_{1\,DI}^{-1}$ vs. $T - \Theta_{II-I}$ [73B7]. $T_{1\,DI}$: dipolar-frame spin-lattice relaxation time for iodine.

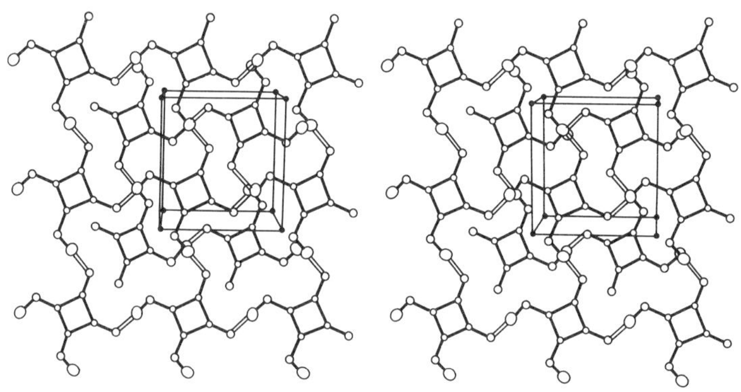

Fig. 1653. $C_4H_2O_4$. A stereoscopic view of the crystal packing at RT [77S18]. See Fig. 1654.

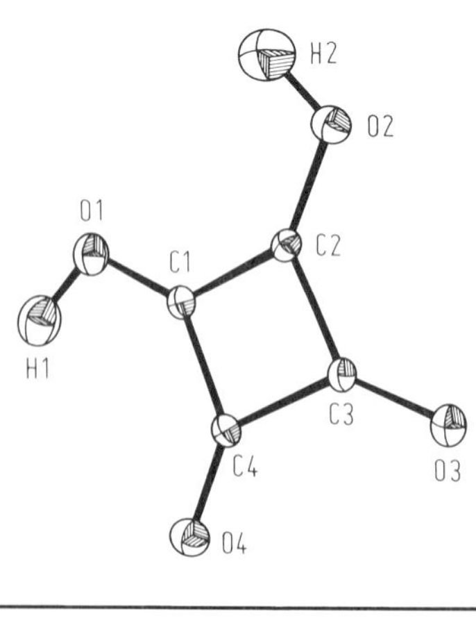

Fig. 1654. $C_4H_2O_4$. Illustration of the thermal ellipsoids in the molecule at RT [77S18]. The ellipsoids are drawn at the 50% level. See Fig. 1653.

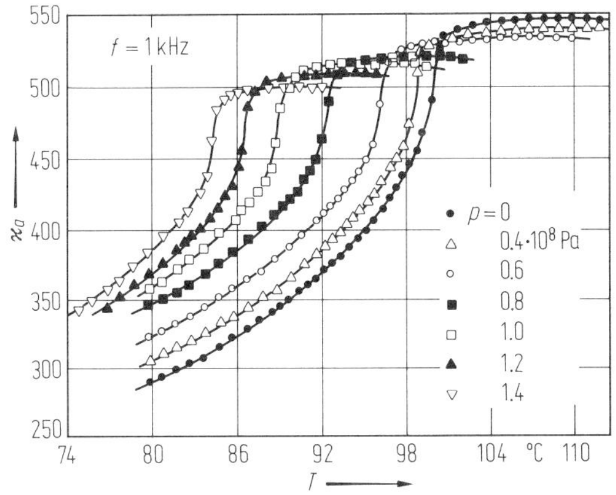

Fig. 1655α. $C_4H_2O_4$. $\varkappa_a$ vs. T [78Y6]. Parameter: p.

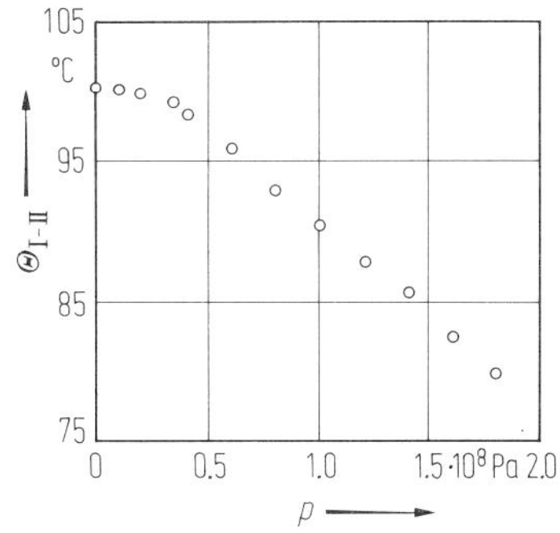

Fig. 1655β. $C_4H_2O_4$. Θ_{I-II} vs. p [78Y6].

For Fig. 1657, see p. 702.

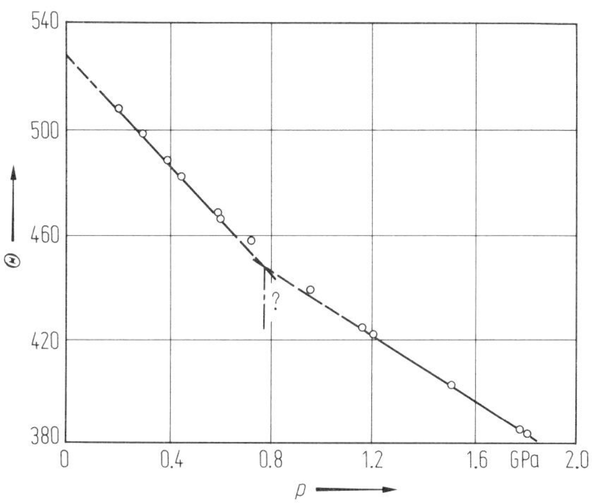

Fig. 1655γ. $C_4D_2O_4$. Θ_{I-II} vs. p [79S9]. Θ in K.

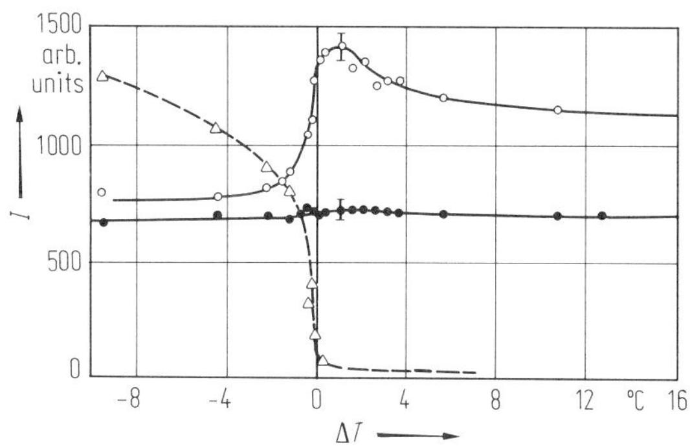

Fig. 1658. $C_4D_2O_4$. Temperature dependence of superlattice neutron reflection intensity and diffuse neutron scattering intensity at (410) in the vicinity of Θ_{I-II} [77S5]. Triangles: total peak intensity, open circles: diffuse peak intensity, full circles: background intensity, $\Delta T = T - \Theta_{I-II}$.

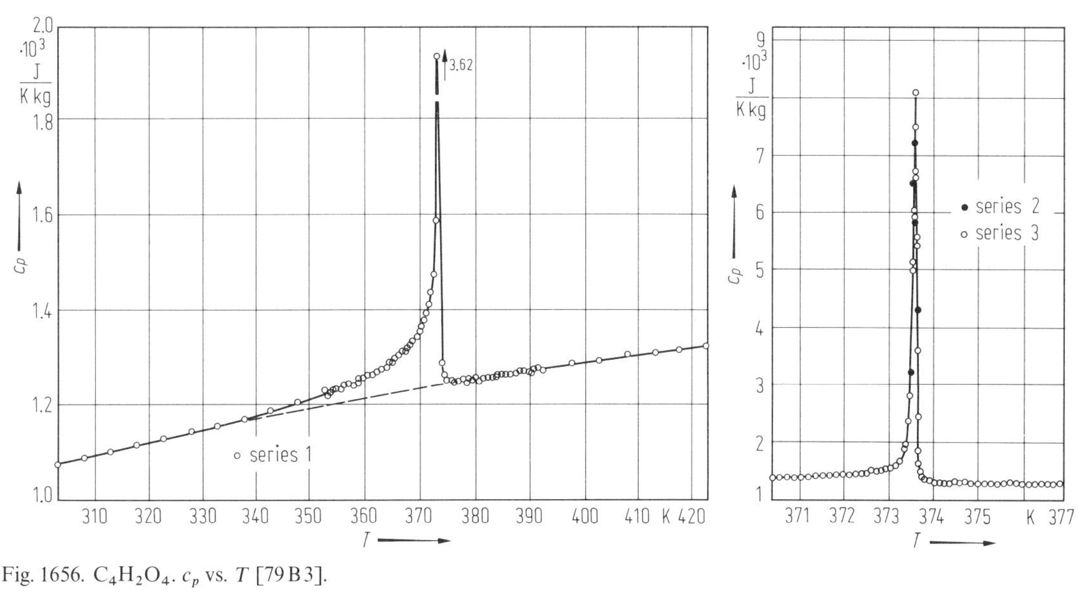

Fig. 1656. $C_4H_2O_4$. c_p vs. T [79B3].

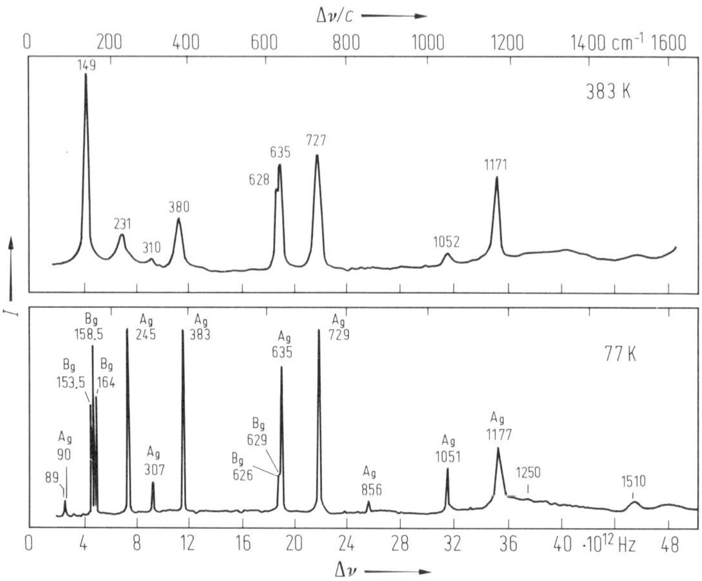

Fig. 1657. $C_4H_2O_4$. I vs. $\Delta\nu$ [76 N 3]. I: Raman scattering intensity below and above the transition temperature. Polarization of light is not specified. Frequency and the symmetry of each peak are given in the figure in units of cm^{-1}.

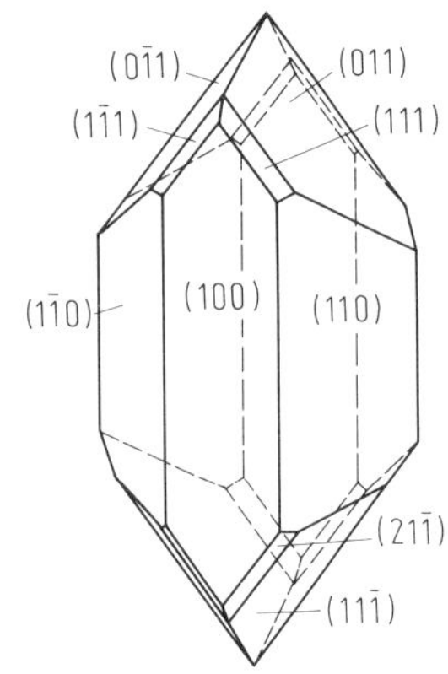

Fig. 1659. $SnCl_2 \cdot 2H_2O$. Crystal form [06 g 1].

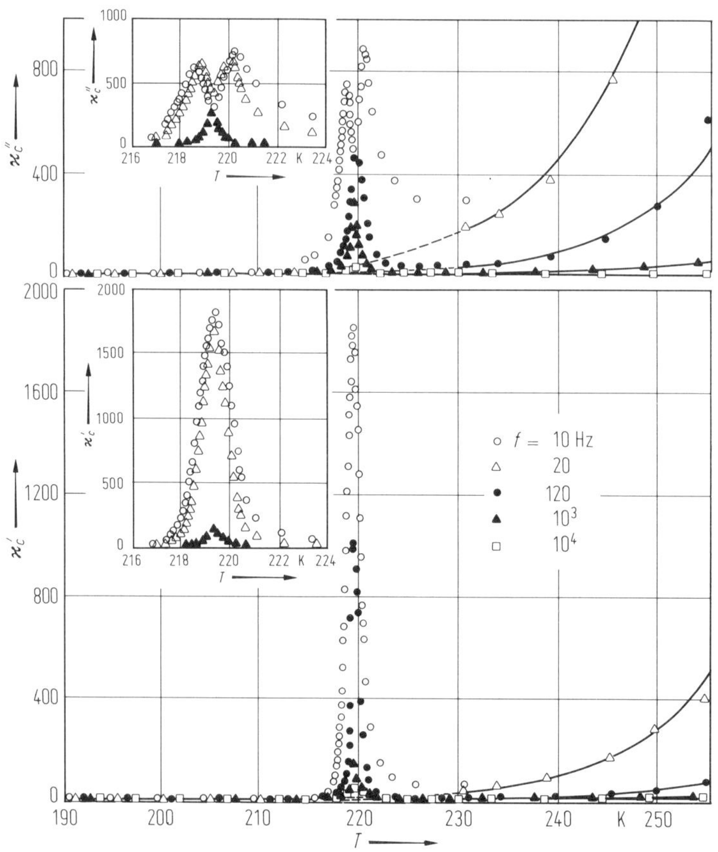

Fig. 1660. $SnCl_2 \cdot 2H_2O$. $\varkappa_c'$, $\varkappa_c''$ vs. T [76 M 18]. Parameter: f.

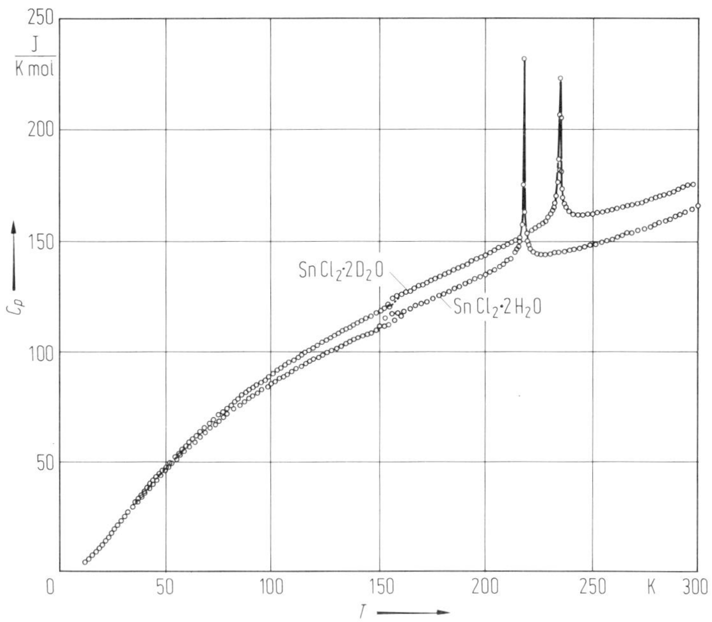

Fig. 1661. $SnCl_2 \cdot 2H_2O$, $SnCl_2 \cdot 2D_2O$. C_p vs. T [74M9].

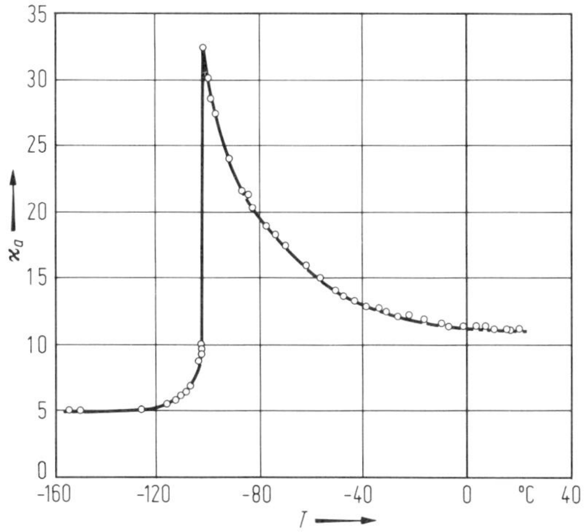

Fig. 1663. $NH_4PF_6 \cdot NH_4F$. $\varkappa_a$ vs. T[69T1]. ($f = 1$ kHz).

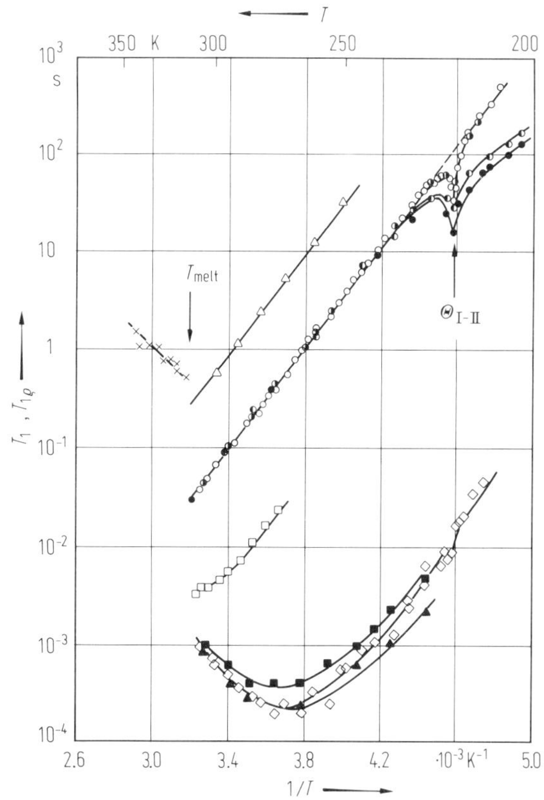

Fig. 1662. $SnCl_2 \cdot 2H_2O$. T_1, $T_{1\varrho}$ vs. 1/T [76K7]. T_1: proton spin-lattice relaxation time, $T_{1\varrho}$: proton spin-lattice relaxation time in rotary frame. T_1: △ 60 MHz, $H \perp (100)$, ○ 20 MHz, $H \perp (100)$, □ 4.5 MHz, $H \perp (100)$. ◑, ◐, ● 20 MHz, powders, × 60 MHz, melt, ◆ 20 MHz, melt. $T_{1\varrho}$: ■ $H_1 = 1.19 \cdot 10^3$ A m^{-1}, powder, ◇ $H_1 = 0.796 \cdot 10^3$ A m^{-1}, $H \perp (100)$, ▲ $H_1 = 0.637 \cdot 10^3$ A m^{-1}, powder. H_1: amplitude of rf field.

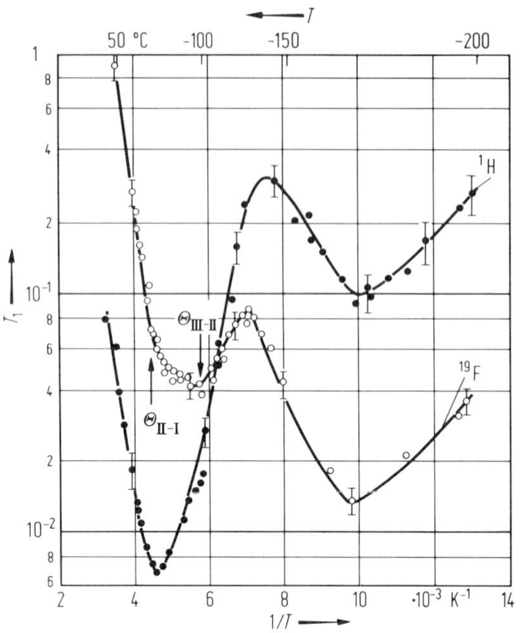

Fig. 1664. $NH_4PF_6 \cdot NH_4F$. T_1 vs. $1/T$ [69T1]. T_1: spin-lattice relaxation time in s.

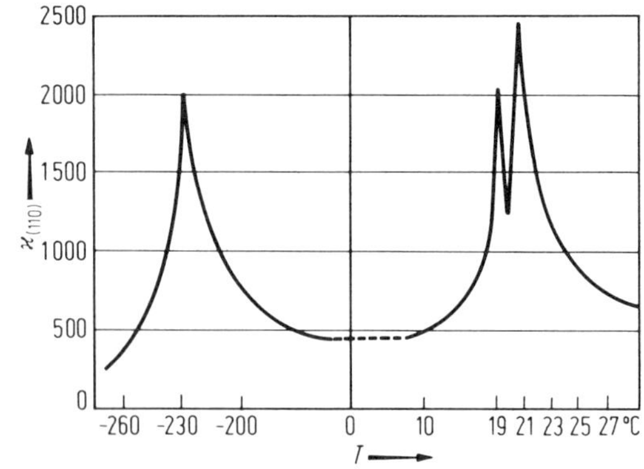

Fig. 1666. $H_2NCONHNH_2 \cdot HCl$. $\varkappa_{(110)}$ vs. T [72R5].

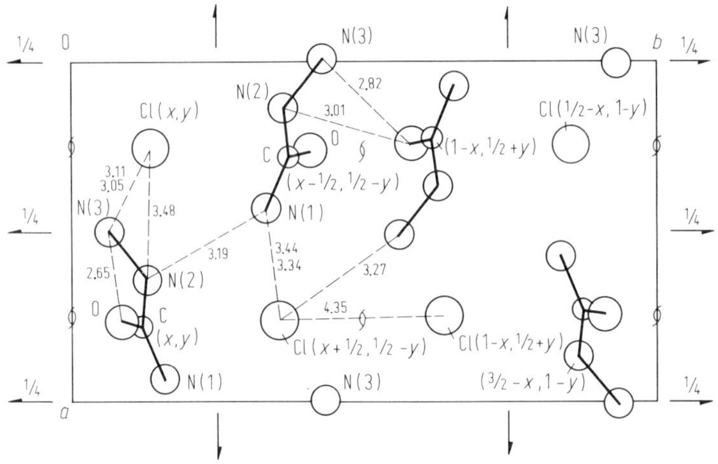

Fig. 1665. $H_2NCONHNH_2 \cdot HCl$. Crystal structure [65N3]. Projection on (001). Interatomic distances in Å.

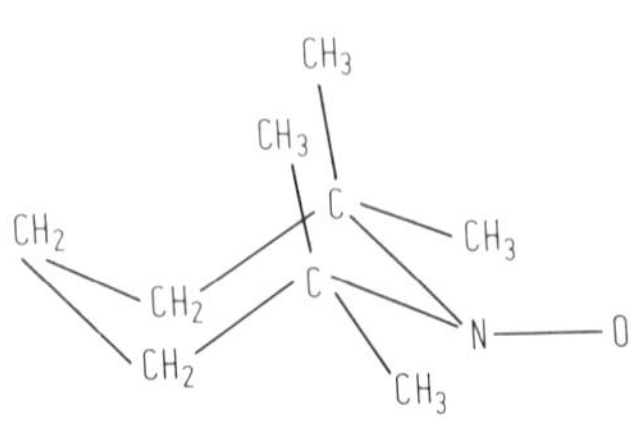

Fig. 1668. Tanane. Form of a molecule of tanane $C_9H_{18}NO$.

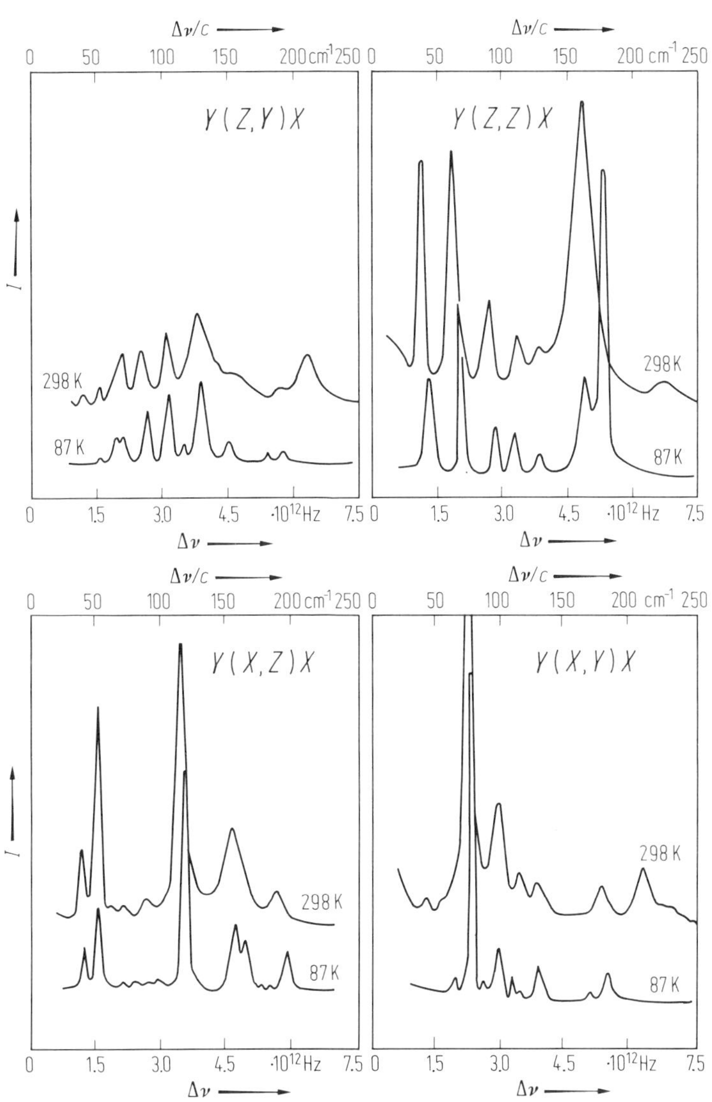

Fig. 1667. $H_2NCONHNH_2 \cdot HCl$. I vs. $\Delta\nu$ [76F1]. I: Raman scattering intensities for various scattering geometries. $T = 298$ and 87 K.

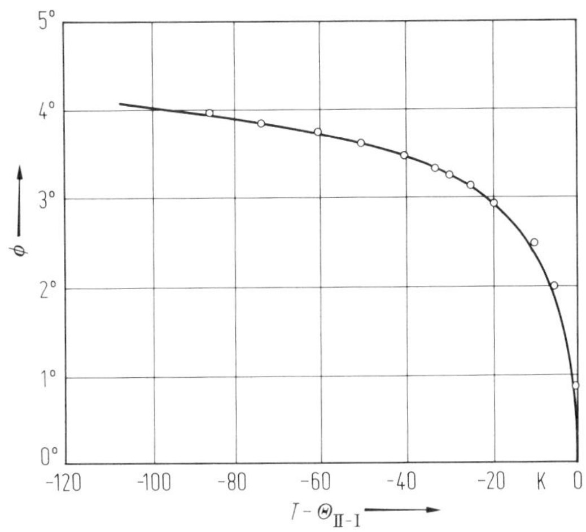

Fig. 1669. Tanane. ϕ vs. $T - \Theta_{II-I}$ [74B15]. ϕ: angle between a tetragonal axis in phase I and a pseudo-tetragonal axis in orthorhombic phase II.

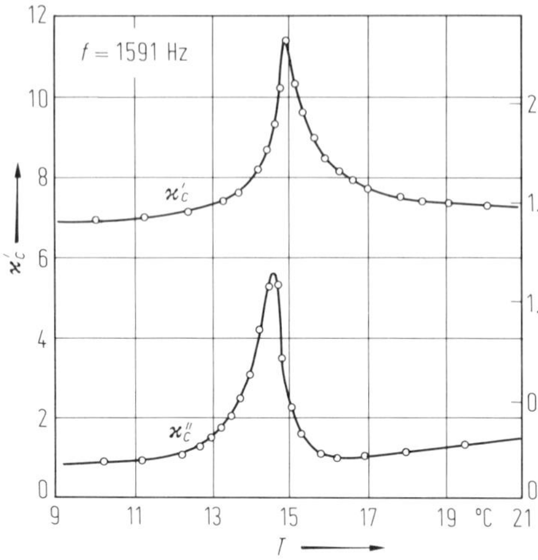

Fig. 1670. Tanane. $\varkappa_c'$, $\varkappa_c''$ vs. T [74 A 4]. f = 1 kHz.

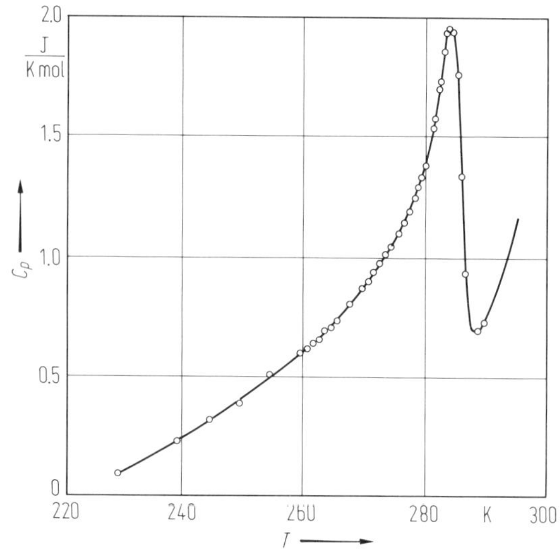

Fig. 1671. Tanane. C_p vs. T [73 B 8].

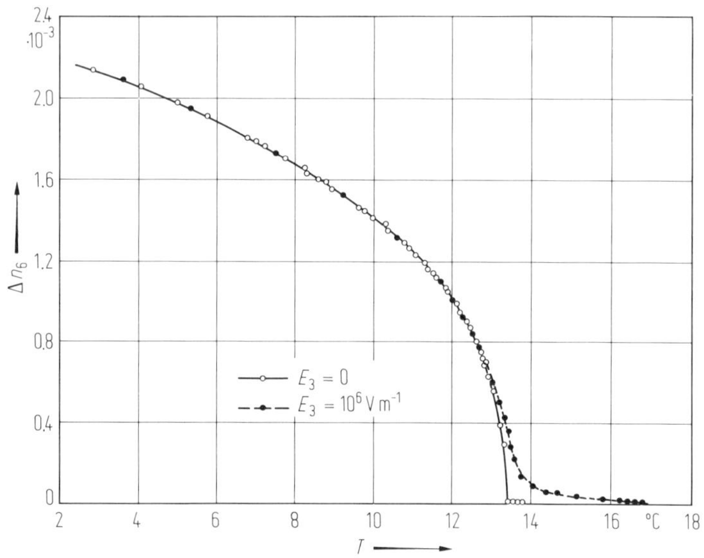

Fig. 1672. Tanane. Δn_6 vs. T [78 L 1]. Full circles: spontaneous birefringence on a short-circuited sample, open circles: birefringence induced by a dc electric field.

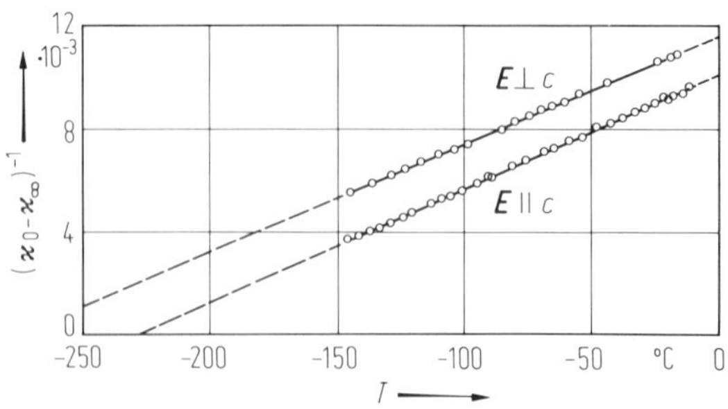

Fig. 1675. H_2O (Ice). $(\varkappa_0 - \varkappa_\infty)^{-1}$ vs. T of ice Ih [78 K 5]. $\varkappa_0$, $\varkappa_\infty$: static and high frequency dielectric constant, respectively.

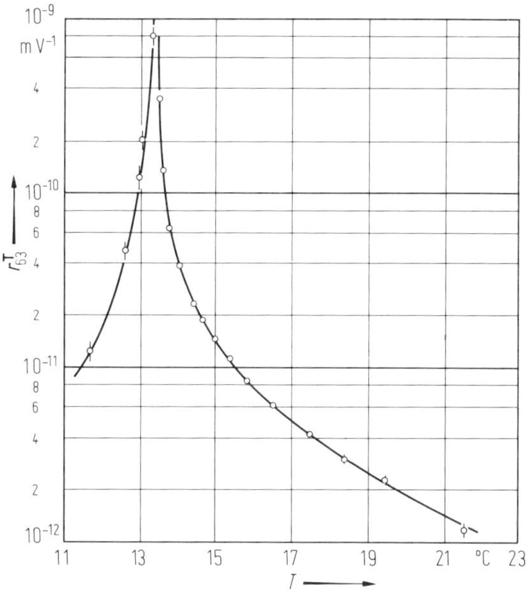

Fig. 1673. Tanane. r_{63}^{T} vs. T [78L1]. r_{63}^{T}: electrooptic constants at $\lambda=618$ nm.

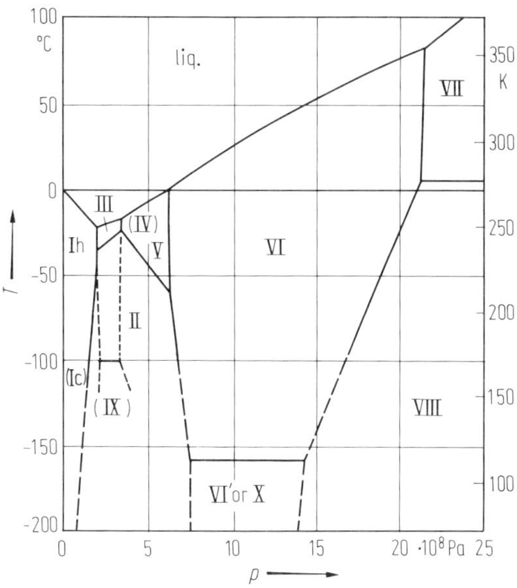

Fig. 1674. H_2O (Ice). Phase diagram [79M1]. Solid line: phase boundary, dashed line: phase boundary estimated by extrapolation, dotted line: boundary for metastable state obtained by extrapolation.

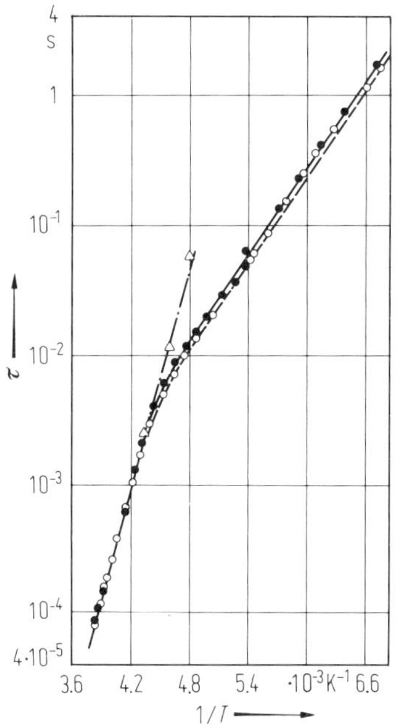

Fig. 1676. H_2O (Ice). τ vs. T^{-1} of ice Ih [78K5]. Open circles: $\boldsymbol{E}\|\boldsymbol{c}$; full circles: $\boldsymbol{E}\perp\boldsymbol{c}$; triangles: data by [52A2].

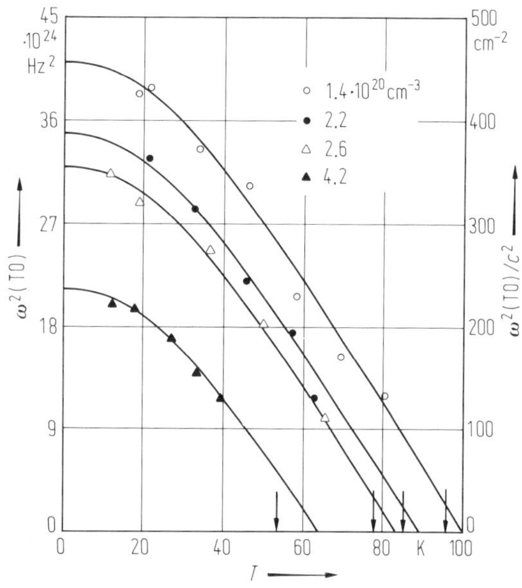

Fig. 1677. SnTe. ω^2(TO) vs. T for specimens with different hole concentrations [77S38]. Parameter: concentration of holes. ω(TO), frequency of TO-phonon observed by Raman scattering. The arrows show the temperatures of resistivity anomaly.

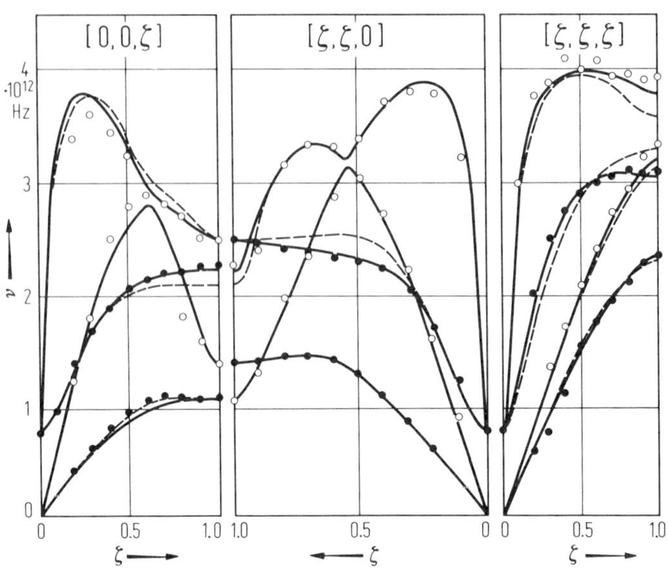

Fig. 1678. SnTe. ν vs. ζ at 100 K [69C4]. ν, phonon frequency and ζ, reduced wave vector coordinates. Solid and broken lines are theoretical results based on two different models.

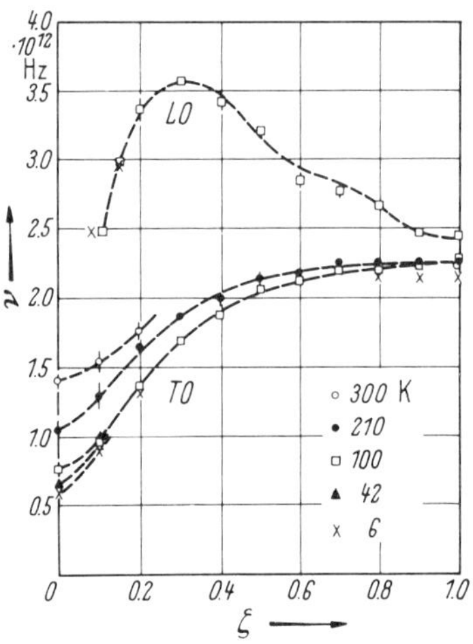

Fig. 1679. SnTe. ν vs. ζ at various temperatures [66P1]. ν, phonon frequency, and, ζ reduced wave vector coordinate, wave vector $\boldsymbol{q} \| [001]$. ($\zeta = aq/2\pi$).

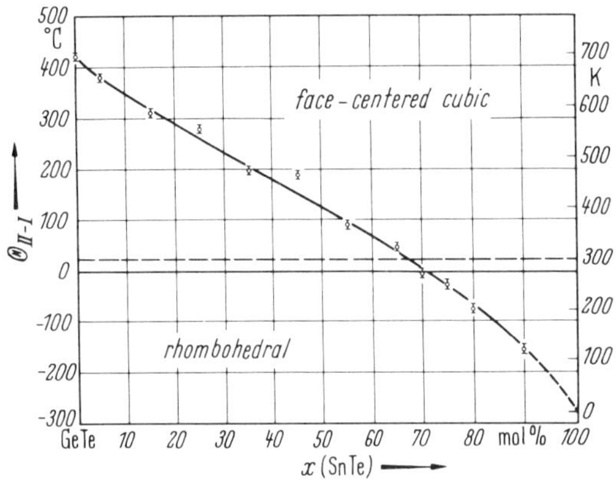

Fig. 1680. GeTe – SnTe. Θ_{II-I} vs. x [63B5]. x: mol% of SnTe.

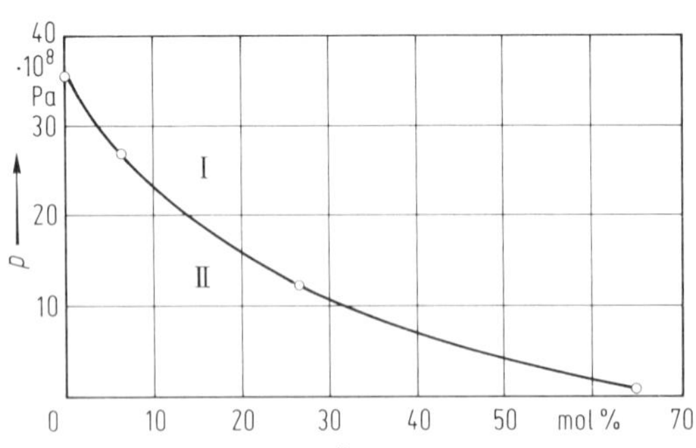

Fig. 1681. GeTe – SnTe. Phase diagram at RT [67K1]. p: pressure. x: concentration of SnTe.

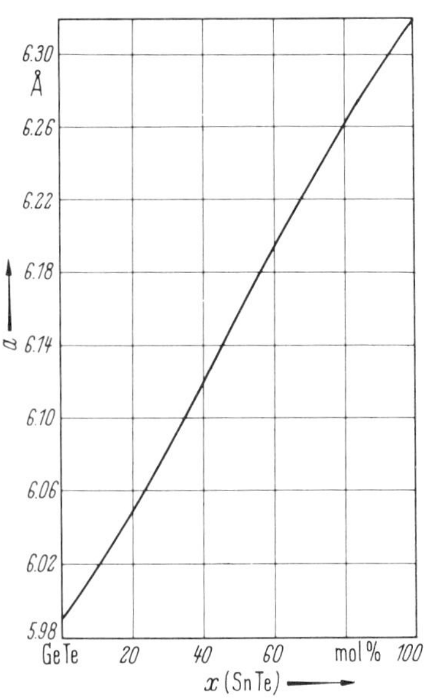

Fig. 1682. GeTe – SnTe. a vs. x at 300 K [63B5]. x: mol% of SnTe.

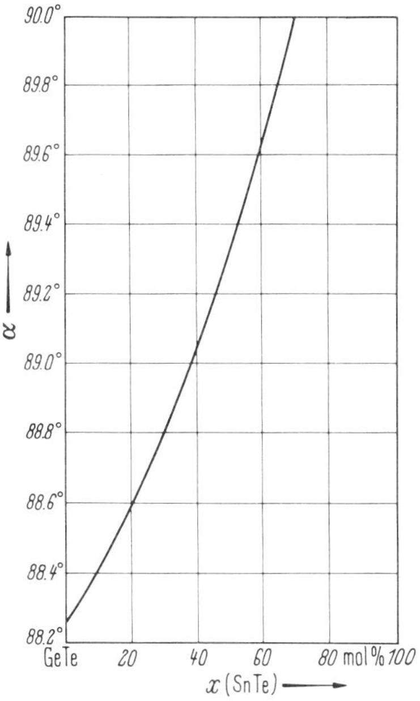

Fig. 1683. GeTe – SnTe. α vs. x at RT [63 B 5]. α: rhombohedral angle. x: mol% of SnTe.

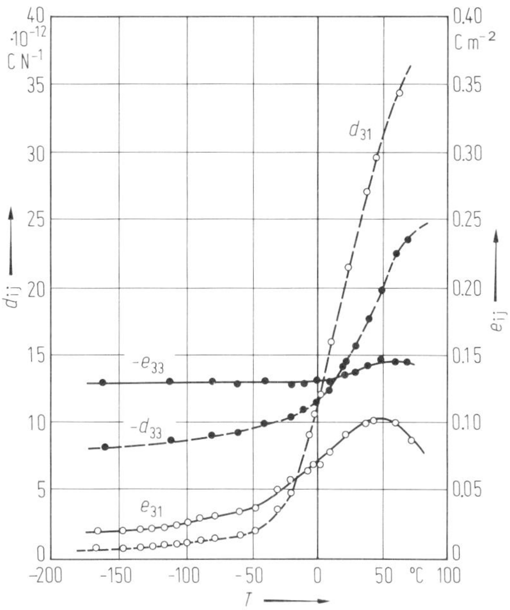

Fig. 1684. $(CH_2CF_2)_n(PVF_2)$ · $d_{31}, d_{33}, e_{31}, e_{33}$ vs. T [76 O 2]. UO film.

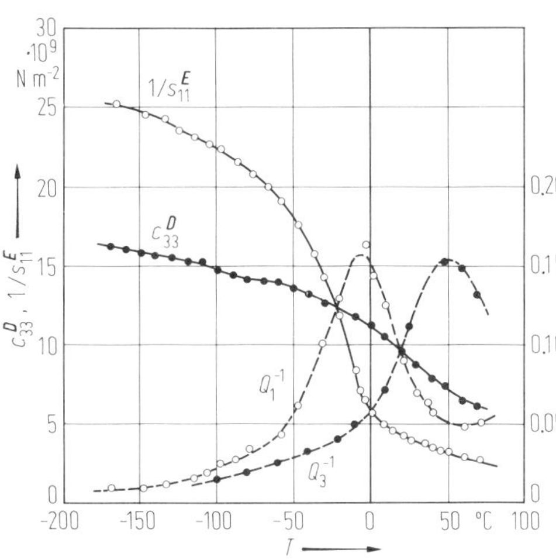

Fig. 1685. $(CH_2CF_2)_n\,(PVF_2)$. $(s^E_{11})^{-1}$, c^D_{33}, Q_1^{-1}, Q_3^{-1} vs. T [76 O 2]. Q_1, Q_3: mechanical quality factors for s^E_{11}, c^D_{33}. UO film.

44 Liquid crystals and related liquids

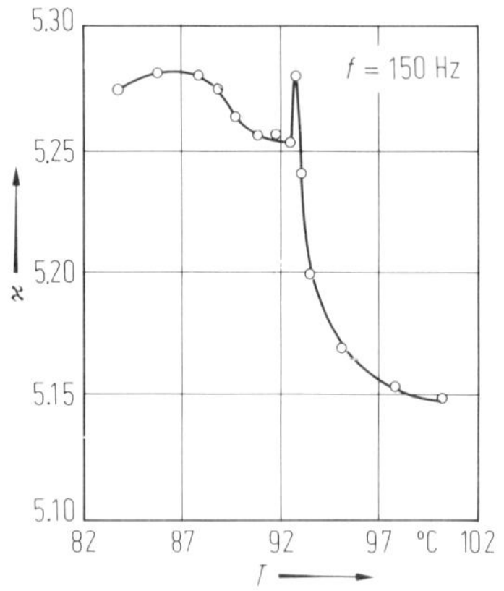

Fig. 1686. $C_{31}H_{43}O_3N$ (DOBAMBC). $\varkappa$ vs. T at 150 Hz [77Y8].

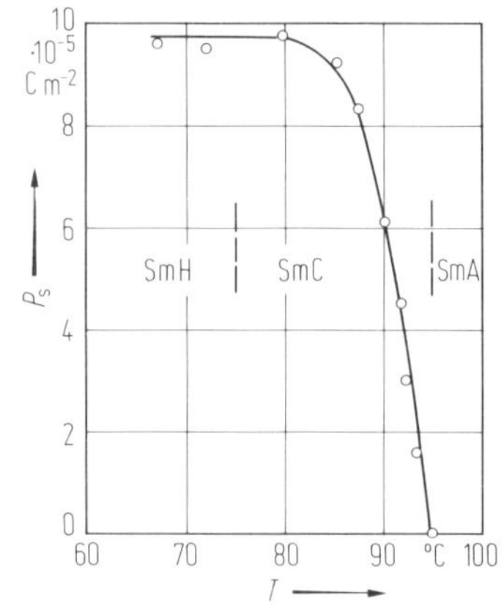

Fig. 1687. $C_{31}H_{43}O_3N$ (DOBAMBC). P_s vs. T [77Y8]. Sm: smectic.

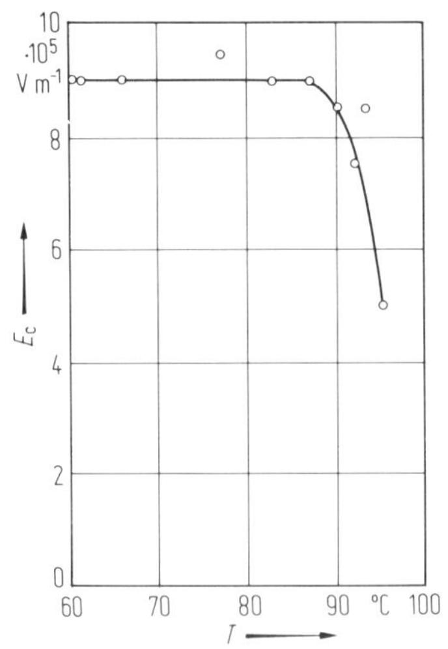

Fig. 1688. $C_{31}H_{43}O_3N$ (DOBAMBC). E_c vs. T [77Y8].

IV References

A Textbooks and review articles

06 g 1 Groth, P.: Chemische Kristallographie, Teil 1, Leipzig: Wilhelm Engelman 1906.

10 g 1 Groth, P.: Chemische Kristallographie, Teil 3, Leipzig: Wilhelm Engelman 1910.

30 a 1 Astbury, W. T.: p. 353 in [30 w 1].

30 m 1 Merwin, H. E.: p. 27 in [30 w 1].

30 w 1 Washburn, E. W. (Editor): International Critical Tables, Vol. 7, New York: McGraw-Hill Book Co., Inc. 1930.

31 w 1 Winchell, A. N., Emmos, R. C.: The Microscopic Characters of Artificial Inorganic Solid Substances or Artificial Minerals, New York: John Wiley & Sons 1931.

33 k 1 Kurchatov, I. V.: Segnetoelektriki, Moskva: Gosdarstvennoe Tekhniko-Teoreticheskoe Izdatel'stvo 1933 (in Russian); Le Champ Moléculaire dans les Diélectriques (le Sel de Seignette), Paris: Hermann & Cie 1936 (Abbreviated French version).

37 b 1 Bragg, W. L.: Atomic Structure of Minerals, New York: Cornell University Press 1937.

41 d 1 Dana, J. D., Dana, E. S.: The System of Mineralogy, Vol. 1, New York: John Wiley & Sons Inc. 1941.

44 d 1 Dana, J. D.: The System of Mineralogy, 7-th ed., Vol. 1, New York: John Wiley & Sons, Inc. 1944.

46 c 1 Cady, W. G.: Piezoelectricity, New York: McGraw-Hill Book Co., Inc. 1946; Piezoelectricity (in two volumes), New York: Dover Publications, Inc. 1964 (Revised and enlarged version).

49 m 1 Miyake, S.: Rochelle Salt, Tokyo: Koyama-Shoten 1949 (in Japanese).

50 m 1 Mason, W. P.: Piezoelectric Crystals and Their Applications to Ultrasonics, Princeton: D. Van Nostrand Co. 1950.

50 v 1 von Hippel, A.: Ferroelectricity, Domain Structure, and Phase Transitions of Barium Titanate, Rev. Mod. Phys. **22** (1950) 221.

53 b 1 Bleaney, B., Stevens, K. W. H.: Paramagnetic Resonance, Rep. Prog. Phys. **16** (1953) 108.

54 d 1 Devonshire, A. F.: Theory of Ferroelectrics, Advances in Phys. **3** (1954) 85.

54 v 1 von Hippel, A. R. (Editor): Dielectric Materials and Applications, New York: John Wiley & Sons, Inc. 1954.

55 b 1 Bowers, K. D., Owen, J.: Paramagnetic Resonance II, Rep. Prog. Phys. **18** (1955) 304.

55 t 1 Tanaka, T.: Barium Titanate Ceramics and Their Applications, Tokyo: The OHMsha Ltd. 1955 (in Japanese).

56 f 1 Forsbergh Jr., P. W.: Piezoelectricity, Electrostriction and Ferroelectricity, pp. 264–392 in Handbuch der Physik, Bd. XVII, Dielektrika (edited by Flugge, S.), Berlin, Göttingen, Heidelberg: Springer 1956.

56 s 1 Sachse, H.: Ferroelektrika, Berlin: Springer-Verlag 1956; Ferroelectrics, Washington, D. C.: Atomic Energy Commission 1956 (English transl.); Les Ferroélectriques, Paris: Dunod 1958 (French transl.).

57 k 1 Känzig, W.: Ferroelectrics and Antiferroelectrics, pp. 5–197 in Solid State Physics, Vol. 4 (edited by Seitz, F., Turnbull, D.), New York: Academic Press 1957.

57 k 2 Koster, G. F.: Space Groups and Their Representations, New York: Academic Press 1957; and pp. 173···256 in Solid State Physics, Vol. 5 (edited by Seitz, F., Turnbul, D.), New York: Academic Press 1957.

57 m 1 Megaw, H. D.: Ferroelectricity in Crystals, London: Methuen & Co., Ltd. 1957.

57 n 1 Nye, J. E.: Physical Properties of Crystals, Oxford: Clarendon Press 1957.

58 h 1 Hansen, M., Anderko, K.: Constitution of Binary Alloys, New York: McGraw-Hill Book Co., Inc. 1958.

59 g 1 Grant, F. A.: Rev. Mod. Phys. **31** (1959) 646.

59 s 1 Shubnikov, A. V., Sheftal', N. N. (Editors): Rost Kristallov, Tom 2, Moscow, 1959 (in Russian); Growth of Crystals, Vol. 2, New York: Consultants Bureau 1959 (English transl.).

60 c 1 Cochran, W.: Crystal Stability and the Theory of Ferroelectricity, Advances in Phys. **9** (1960) 387.

60 l 1 Low, W.: Paramagnetic Resonance in Solids, Solid State Phys., Suppl. 2, New York: Academic Press 1960 (edited by Seitz, F., Turnbul, D.).

60 r 1 Rowland, J. F., Bright, N. F. H., Jongejan, A.: Advances in X-Ray Analysis Vol. 2 (1960), New York: Plenum Press.

61 c 1 Cochran, W.: Crystal Stability and the Theory of Ferroelectricity, Part II. Piezoelectric Crystals, Advances in Phys. **10** (1961) 401.

61 g 1	Goldsmith, A., Waterman, T. E., Hirschborn, H. J.: Handbook of Thermophysical Properties of Solid Materials, Vol. III, Ceramics, New York: Macmillan 1961.
61 o 1	Ohara, G.: Synthetic Crystals and Their Application, Tokyo: Kyoritsu Shuppan Co. Ltd. 1961 (in Japanese).
61 r 1	Ramachandran, G. N., Ramaseshan, S.: Crystal Optics in Handbuch der Physik, Bd. XXV/1 (edited by Flugge, S.), Berlin, Göttingen, Heidelberg: Springer 1961.
61 s 1	Shubnikov, A. V., Sheftal', N. N. (Editors): Rost Kristallov, Tom 3, Moscow, 1961 (in Russian); Growth of Crystals, Vol. 3, New York: Consultants Bureau 1962.
62 j 1	Jona, F., Shirane, G.: Ferroelectric Crystals, Oxford: Pergamon Press 1962.
62 k 1	Kafarov, V. V.: Spravochnic po Rastvorimosti, Tom I, Moscow, USSR Academy of Science, 1962; Stephen, H., Stephen, T.: Solubilities of Inorganic and Organic Compounds, Vol. 1, London: Pergamon Press 1963.
62 k 2	Koritnig, S.: Landolt-Börnstein, 6. Aufl., Band II/8: Berlin, Göttingen, Heidelberg: Springer 1962, pp. 2–55.
62 w 1	Wooster, W. A.: Diffuse X-Ray Reflections from Crystals, Oxford: Clarendon Press 1962.
63 f 1	Fox, D., Labes, M. M., Weissenberger, A.: Physics and Chemistry of Organic Solid State, Vol. 1, New York/London: Interscience Publishers, a division of John Wiley & Sons 1963.
63 f 2	Frauenfelder, H.: The Mössbauer effect, W. A. Benjamin, Inc. 1963.
63 g 1	Gray, D. E. (Editor): Am. Inst. Phys. Handb., 2nd ed., New York: McGraw Hill, Inc. 1963.
63 l 1	Low, W. (Editor): Paramagnetic Resonance, Vol. I, Proceedings of the 1 st International Conference held at Jerusalem, New York: Academic Press 1963.
63 l 2	Low, W. (Editor): Paramagnetic Resonance, Vol. II, Proceedings of the 1 st International Conference held at Jerusalem, New York: Academic Press 1963.
63 m 1	Mason, W. P., Properties of Transducer Materials, pp. 3–98 in [63 g 1].
64 b 1	Berlincourt, D. A., Curran, D. R., Jaffe, H.: Piezoelectric and Piezomagnetic Materials and Their Function in Transducers, pp. 169–270 in Physical Acoustics, Vol. 1, Part A (edited by Mason, W. P.), New York: Academic Press 1964.
64 g 1	Guinier, A.:Théorie et Technique de la Radiocristallographie, Paris:Dunod 1964, Chap. 13.
64 h 1	Heine, V.: Group Theory in Quantum Mechanics, Oxford/London/New York/Paris: Pergamon Press 1964.
64 h 2	Hutchings, M. T.: Point-Charge Calculations of Energy Levels of Magnetic Ions in Crystalline Electric Fields, Solid State Physics, Vol. 16 (edited by Seitz, F., Turnbull, D.), New York: Academic Press 1964.
64 l 1	Levin, M. L., Robbino, C. R., McMurdie, H.: Phase Diagrams for Ceramists, Columbus, Ohio: American Ceramic Society, 1964.
64 m 1	Martin, H.-J.:Die Ferroelektrika, Leipzig: Akademische Verlagsgesellschaft 1964.
64 m 2	Mason, P. W. (Editor): Physical Acoustics, Vol. I, Part A, New York: Academic Press 1964.
64 w 1	Winchel, A. N., Winchel, H.: The Microscopical Characters of Artificial Inorganic Solid Substances, New York: Academic Press 1964.
65 b 1	Burgess, R. E. (Editor): Fluctuation Phenomena in Solids, New York: Academic Press 1965.
65 e 1	Egelstaff, P. A. (Editor): Thermal Neutron Scattering, London and New York: Academic Press 1965.
65 l 1	Lonsdale, K. (General Editor): International Tables for X-Ray Crystallography, Vol. I (1965), Vol. II (1967), Vol. III (1968) , Vol. IV (1974), Birmingham: Kynoch.
65 w 1	Wyckoff, R. W. G.: Crystal Structures, Vol. 3, New York: John Wiley & Sons, 1965.
66 b 1	Bechmann, R., Hearmon, R. F. S.: Landolt-Börnstein, Numerical Data and Functional Relationships in Science and Technology, New Series, Group III: Crystal and Solid State Physics, Vol. 1, Elastic, Piezoelectric, Piezooptic and Electrooptic Constants of Crystals, Berlin, Heidelberg, New York: Springer 1966.
66 b 2	Brockhouse, B. N.: Neutron Scattering by Phonons, in Phonons in Perfect Lattices and in Lattices with Point Imperfection, edited by Stevenson, R. W. H., Edinburgh/London: Oliver & Boyd 1966, pp 110–152.
66 d 1	Dvorak, V., Fouskova, A., Glogar, P. (Editors): Proceedings of the International Meeting on Ferroelectricity, held at Prague (1966) (in two volumes), Prague: Institute of Physics of the Czechoslovak Academy of Sciences 1966.
66 m 1	Mason, W. P.: Crystal Physics of Interaction Processes, New York: Academic Press 1966.
67 b 1	Blinc, R. (Editor): Magnetic Resonance and Relaxation, Proceedings of the XIX th Colloque Ampere, held at Ljubljana (1966), Amsterdam: North-Holland Publishing Co. 1967.

67b2	Blinc, R.: Magnetic Resonance in Hydrogen-Bonded Ferroelectrics, pp. 141–204 in Advances in Magnetic Resonance, Vol. 3 (edited by Waugh, J. S.), New York: Academic Press 1967.
67b3	Burfoot, J. C.: Ferroelectrics, An Introduction to the Physical Principles, London: D. Van Nostrand Co., Ltd. 1967.
67e1	Eyraud, L.: Diélectriques Solides Anisotropes et Ferroélectricité, Paris: Gauthier-Villars 1967.
67f1	Fatuzzo, E., Merz, W. J.: Ferroelectricity, Amsterdam: North-Holland Publishing Co. 1967.
67j1	James, R. W.: The Optical Principles of the Diffraction of X-Ray, London: Bell and Sons 1945, Chap. V.
67w1	Weller, E. F. (Editor): Ferroelectricity, Proceedings of the Symposium, held at Warren (1966), Amsterdam: Elsevier Publishing Co. 1967.
68c1	Campbell, J. J., Jones, W. R.: A Method for Estimating Optimal Crystal Cuts and Propagation Directions for Excitation of Piezoelectric Surface Waves, IEEE Trans. Sonics Ultrasonics, SU **15** (1968) 209.
68g1	Goldanskii, V. I., Herber, R. H.: Chemical Applications of Mössbauer Spectroscopy, New York/London: Academic Press 1968.
68s1	Smolenskii, G. A., Bokov, V. A., Isupov, V. A., Krainik, N. N., Nedlin, G. H.: Magnetically Ordered Ferroelectrics, Helv. Phys. Acta **41** (1968) 1187.
68s2	Smolenskii, G. A., Krainik, N. N.: Segnetoelektriki i Antisegnetoelektriki, Moskva: Izdatel'stvo "Nauka" 1968 (in Russian); Ferroelektrika und Antiferroelektrika, Leipzig: BSB B. G. Teubner Verlagsgesellschaft 1972 (German transl.).
68t1	Takagi, Y., Sawada, S.: Materials Engineering of Magnetic and Dielectric Substances, Tokyo: The OHMsha Ltd. 1968 (in Japanese).
68z1	Zheludev, I. S.: Fizika Kristallicheskikh Dielektrikov, Moskva: Izdatel'stvo "Nauka" 1968; Physics of Crystalline Dielectrics (in two volumes), New York: Plenum Press 1971 (English transl.).
69a1	Averbach, P. (Editor): Magnetic Resonance and Radiofrequency Spectroscopy, Proceedings of the XVth Colloque Ampere, held at Grenoble (1968), Amsterdam: North-Holland Publishing Co. 1969.
69b1	Bechmann, R., Hearmon, R. F. S., Kurtz, S. K.: Landolt-Börnstein, Numerical Data and Functional Relationships in Science and Technology, New Series, Group III: Crystal and Solid State Physics, Vol. 2, Elastic, Piezoelectric, Piezooptic, Electrooptic Constants, and Nonlinear Dielectric Susceptibilities of Crystals (Supplement and Extension to Vol. 1), Berlin, Heidelberg, New York: Springer 1969.
69c1	Cochran, W.: Dynamical Scattering and Dielectric Properties of Ferroelectric Crystals, Advances in Phys. **18** (1969) 157.
69g1	Glasso, F. S.: Structure, Properties and Preparation of Perovskite-Type Compounds, Oxford: Pergamon Press 1969.
69g2	Gurevich, V. M.: Elektroprovodnost' Segnetoelektrikov, Moskva: Izdatel'stvo Komiteta Standartov, Mer i Izmeritel'nykh Priboroy pri Sovete Ministrov SSSR 1969; Electric Conductivity of Ferroelectrics, Jerusalem: Israel Program for Scientific Translations 1971 (English transl.).
69h1	Holland, R., Eer Nisse, E. P.: Design of Resonant Piezoelectric Devices, Cambridge, Mass.: The M. I. T. Press 1969.
69m1	Mitsui, T., Abe, R., Furuhata, Y., Gesi, K., Ikeda, T., Kawabe, K., Makita, Y., Marutake, M., Nakamura, E., Nomura, S., Sawaguchi, E., Shiozaki, Y., Tatsuzaki, I., Toyoda, K.: Landolt-Börnstein, Numerical Data and Functional Relationships in Science and Technology, New Series, Group III: Crystal and Solid State Physics, Vol. 3, Ferro- and Antiferroelectric Substances, Berlin, Heidelberg, New York: Springer 1969.
69m2	Mitsui, T., Tatsuzaki, I., Nakamura, E.: Ferroelectrics, Tokyo: Maki-Shoten, Inc. 1969 (in Japanese); An Introduction to the Physics of Ferroelectrics, London: Gordon and Breach Science Publishers 1976 (English version).
69o1	Okazaki, K.: Ceramic Engineering for Dielectrics, Tokyo: Gakken-sha Publishing Co., Ltd. 1969 (in Japanese).
69s1	Samara, G. A.: The Effect of Hydrostatic Pressure on Ferroelectric Properties, pp. 155–239 in Advances in High Pressure Research, Vol. 3 (edited by Bradley, R. S.), London: Academic Press 1969.
69s2	Stadler, H. L. (Editor): Proceedings of the Symposium on Applications of Ferroelectricity, held in Washington, D. C. (1968), IEEE Trans. Electron Devices, ED **16** (1969) No. 6.
69t1	Tiersten, H. F.: Linear Piezoelectric Plate Vibrations, New York: Plenum Press 1969.

69w1	Wright, G. B. (Editor): Light Scattering Spectra of Solids, Proceedings of the International Conference, held in New York (1968), New York: Springer-Verlag 1969.
70a1	Abragam, A., Bleaney, B.: Electron Paramagnetic Resonance of Transition Ions, Oxford: Clarendon Press 1970.
70a2	Ascher, E.: The Interaction between Magnetization and Polarization: Phenomenological Symmetry Consideration on Boracites; J. Phys. Soc. Jpn. **28** (1970) Suppl. 7.
70b1	Barrett, H. H.: Acoustic Properties of Materials of the Perovskite Structure, pp. 65–105 in Physical Acoustics, Vol. 6 (edited by Mason, W. P., Thurston, R. N.), New York: Academic Press 1970.
70d1	Damon, R. W., Maloney, W. T., McMahon, D. H.: Interaction of Light with Ultrasound: Phenomena and Applications, pp. 273–366 in Physical Acoustics, Vol. 7 (edited by Mason, W. P., Thurston, R. N.), New York: Academic Press 1970.
70f1	Farnell, G. W.: Properties of Elastic Surface Waves, pp. 109–166 in Physical Acoustics, Vol. 6 (edited by Mason, W. P., Thurston, R. N.), New York: Academic Press 1970.
70f2	Fleury, P. A.: Light Scattering as a Probe of Phonons and Other Excitations, pp. 1–64 in Physical Acoustics, Vol. 6 (edited by Mason, W. P., Thurston, R. N.), New York: Academic Press 1970.
70g1	Garland, C. W.: Ultrasonic Investigation of Phase Transitions and Critical Points, pp. 52–148 in Physical Acoustics, Vol. 7 (edited by Mason, W. P., Thurston, R. N.), New York: Academic Press 1970.
70g2	Goodenough, J. B., Graper, W., Holtzberg, F., Huber, D. L., Lefever, R. A., Longo, J. M., McGuire, T. R., Methfessel, S.: Landolt-Börnstein, Numerical Data and Functional Relationships in Science and Technology, New Series, Group III: Crystal and Solid State Physics, Vol. 4, Magnetic and Other Properties of Oxides and Related Compounds, Part a, Berlin: Springer-Verlag 1970.
70g3	Goodenough, J. B., Longo, J. M.: Crystallographic and Magnetic Properties of Perovskite and Perovskite-Related Compounds, pp. 126–314 in [70g2].
70g4	Grindlay, J.: An Introduction to the Phenomenological Theory of Ferroelectricity, Oxford: Pergamon Press 1970.
70m1	Müser, H. E., Petersson, J. (Editors): Proceedings of the European Meeting on Ferroelectricity, held at Saarbrücken (1969), Stuttgart: Wissenschaftliche Verlagsgesellschaft mbH 1970.
70n1	Nettleton, R. E.: Ferroelectric Phase Transitions: A Review of Theory and Experiment, Ferroelectrics Vol. 1 (1970) 3, 87, 93, 111, 121, 127, 207, 221, Vol. 2 (1971) 5, 77, 93.
70o1	O'Dell, T. H.: The Electrodynamics of Magneto-electric Media, Amsterdam: North-Holland Publishing Co. 1970.
70p1	Putley, E. H.: The Pyroelectric Detector, pp. 259–285 Semiconductors and Semimetals, Vol. 5: Infrared Detectors (edited by Willardson, R. K., Beer, A. C.), New York: Academic Press 1970.
70s1	Skinner, S. M.: Magnetically Ordered Ferroelectric Materials, IEEE Trans. Parts Materials Packaging, PMP **6** (1970) 68.
70s2	Sonin, A. S., Strukov, B. A.: Vvedenie v Segnetoelektrichestvo, Moskva: Izdatel'stvo "Vysshaya Shkola" 1970 (in Russian):Einführung in die Ferroelektrizität, Berlin: Akademie-Verlag 1974 (German transl.).
70t1	Takagi, Y. (Editor): Proceedings of the Second International Meeting on Ferroelectricity, held in Kyoto (1969), Tokyo: Physical Society of Japan 1970; J. Phys. Soc. Japan **28** (1970) Suppl.
70t2	Tokunaga, M.: Models of the Phase Transition in KDP Type Ferroelectrics, Ferroelectrics **1** (1970) 195.
70w1	White, R. M.: Surface Elastic Waves, Proc. IEEE **58** (1970) 1238.
71b1	Balkanski, M. (Editor): Proceedings of the Second International Conference on Light Scattering in Solids, held in Paris (1971), Paris: Flammarion Sciences 1971.
71b2	Berlincourt, D.: Piezoelectric Crystals and Ceramics, pp. 63–124 in Ultrasonic Transducer Materials (edited by Mattiat, O. E.), New York: Plenum Press 1971.
71f1	Fesenko, E. G.: Semeistvo Perovskita i Segnetoelektrichestvo, Moskva: Atomizdat 1972 (in Russian).
71h1	Heywang, W.: Semiconducting Barium Titanate, J. Materials Sci. **6** (1971) 1214.
71j1	Jaffe, B., Cook Jr., W. R., Jaffe, H.: Piezoelectric Ceramics, London: Academic Press 1971.
71k1	Kawabe, K.: Ferroelectrics, Tokyo: Kyoritsu Shuppan Co. 1971 (in Japanese).
71k2	Kreher, K.: Ferroelektrische Halbleiter, Wiss. Z. Karl-Marx-Univ. Leipzig, Math.-Naturwiss. Reihe **20** (1971) 287.

71 m 2 Mason, W. P.: Fifty Years of Ferroelectricity, J. Acoust. Soc. Am. **50** (1971) 1281.

71 m 3 Moses, A. J.: Optical Materials Properties in Handbook of Electronic Materials, Vol. 1, New York: IFI/Plenum 1971.

71 p 1 Perry, C. H.: Dielectric Properties and Optical Phonons in Para- and Ferro-electric Perovskites, pp. 557–591 in Far-Infrared Spectroscopy (edited by Moller, K. D., Rothschild, W. G.), New York: Wiley-Interscience 1971.

71 p 2 Pressley, R. J. (Editor): Handbook of Lasers with Selected Data on Optical Technology, Cleveland: The Chemical Rubber Co. 1971.

71 p 3 Pell, E. M. (Editor): Proceedings of the Third International Conference on Photoconductivity, held at Stanford (1969), J. Phys. Chem. Solids **32** (1971) Suppl.

71 s 1 Samuelsen, E., Andersen, E., Feder, J. (Editors): Structural Phase Transitions and Soft Modes, Proceedings of the NATO Advanced Study Institute, held at Geilo (1971), Oslo: Universitetsforlaget 1971.

71 s 2 Singh, S.: Nonlinear Optical Materials in Handbook of Lasers with Selected Data on Optical Technology, edited by Presley, R. J., Cleveland: The Chemical Rubber Co. 1971.

71 s 3 Smolenskii, G. A., Bokov, V. A., Isupov, V. A., Krainik, N. N., Pasynkov, R. E., Shur, M. S.: Segnetoelektriki i Antisegnetoelektriki, Leningrad: Izdatel'stvo "Nauka" 1971 (in Russian).

71 v 1 Valasck, J.: Thc Early History of Ferroelectricity, Ferroelectrics **2** (1971) 239.

71 z 1 Zheludev, I. S.: Ferroelectricity and Symmetry, pp. 429–464 in Solid State Physics, Vol. 26 (edited by Ehrenreich, H., Seitz, F., Turnbull, D.), New York: Academic Press 1971.

72 a 1 Arecchi, F. T., Schulz-Dubois, E. O. (Editors): Laser Handbook, Vol. 1, Amsterdam: North-Holland Publishing Co. 1972.

72 a 2 Axe, J. D.: Dynamical Effects in Solid State Phase Transformations, pp. 749–773 in Physics of Electronic Ceramics, Part B (edited by Hench, L. L., Dove, D. B.), New York: Marcel Dekker, Inc. 1972.

72 b 1 Blinc, R., Zeks, B.: Dynamics of Order-Disorder-Type Ferroelectrics and Anti-ferroelectrics, Advances in Phys. **21** (1972) 693.

72 b 2 Bradley, C. J., Cracknell, A. P.: The Mathematical Theory of Symmetry in Solids, Oxford: Clarendon Press, 1972.

72 c 1 Cross, L. E., Kurtz, S. K. (Editors): Proceedings of the 1971 IEEE Symposium on Applications of Ferroelectrics, held at Yorktown Heights (1971), Ferroelectrics Vol. 3 (1972) Nos. 2, 3, 4; IEEE Trans. Sonics Ultrasonics, SU 19 (1972) No. 2.

72 c 2 Cross, L. E.: Thermodynamic Phenomenology of Ferroelectricity in Single Crystal and Ceramic Systems, pp. 707–747 in Physics of Electronic Ceramics, Part B (edited by Hench, L. L., Dove, D. B.), New York: Marcel Dekker, Inc. 1972.

72 f 1 Farnell, G. W., Adler, E. L.: Elastic Wave Propagation in Thin Layers, pp. 35–127 in Physical Acoustics, Vol. 9 (edited by Mason, W. P., Thurston, R. N.), New York: Academic Press 1972.

72 g 1 Godefroy, L., Godefroy, G. (Réducteurs): 2ème Congrès Européen de Ferroélectricité, Exposés et Communications présentés à Dijon (1971), J. Phys. (Paris) **33** (1972), Suppl. Colloque No. C-2.

72 h 1 Hoshino, S., Kohra, K., Sadanaga, R., Takeuchi, Y. (Editors): Collected Abstracts of the Ninth International Congress of Cristallography, held in Kyoto (1972), International Union of Crystallography, 1972; and Acta Crystallogr. A **28** (1972) Suppl.

72 k 1 Kielich, S.: Nonlinear Optical and Electro-optical Properties of Dielectrics and Ferroelectrics, Ferroelectrics **4** (1972) 257.

72 k 2 Kurtz, S. K.: Nonlinear Optical Materials, pp. 923–974 in [72 a 1].

72 l 1 Laudiese, R. A., Mullin, J. B., Mutaftschiev, B. (Editors): Crystal Growth 1971, Proceedings of the 3rd International Conference, held at Marseille (1971), Amsterdam: North-Holland Publishing Co. 1971; J. Cryst. Growth **13** (1972).

72 p 1 Pepinsky, R.: Some Structural Mechanisms in Ferroelectricity, pp. 567–706 in Physics of Electronic Ceramics, Part B (edited by Hench, L. L., Dove, D. B.), New York: Marcel Dekker, Inc. 1972.

72 p 2 Pinnow, D. A.: Elasto-Optic Materials, pp. 995–1008 in [72 a 1].

72 w 1 Wemple, S. H.: Electro-Optic Materials, pp. 975–994 in [72 a 1].

72 w 2 Wemple, S. H., DiDomenico Jr., M.: Electrooptical and Nonlinear Optical Properties of Crystals, pp. 263–383 in Applied Solid State Science, Vol. 3 (edited by Wolfe, R.), New York: Academic Press 1972.

73 b 1 Bhide, V. G.: Mössbauer Effect and Its Applications, New Delhi: Tate McGraw-Hill Publishing Co. 1973.

73 b 2 Bhide, V. G.: Ferroelectrics and Mössbauer Effect, pp. 130–165 in [73 b 1].

73m1	Mathieu, J. P. (Editor): Advances in Raman Spectroscopy, Vol. 1; Proceedings of the 3rd International Conference on Raman Spectroscopy, held at Reims (1972), London: Heyden & Son. Ltd. 1973.
73m2	Megaw, H. D.: Crystal Structures; A Working Approach, Philadelphia: W. B. Saunders Co. 1973.
73m3	Mitsui, T., Nakamura, E., Tokunaga, M.: Critical Phenomena in Ferroelectric Phase Transitions. I. Experimental Observations and Qualitative Discussion, Ferroelectrics **5** (1973) 185.
73r1	Rehwald, W.: The Study of Structural Phase Transitions by Means of Ultrasonic Experiments, Advances in Phys. **22** (1973) 721.
73t1	Tanaka, T., Okazaki, T., Ichinose, N. (Editors): Piezoelectric Ceramic Materials, Tokyo: Gakkensha Publishing Co. 1973 (in Japanese).
73v1	Vaks, V. G.: Vvedenie v Mikroskopicheskuyu Teoriya Segnetoelektrikov, Moskva: Izdatel'stvo "Nauka" 1973 (in Russian).
73z1	Zheludev, I. S.: Osnovy Segnetoelektrichestva, Moskva: Atomizdat 1973 (in Russian).
74a1	Abe, R., Ohtsuka, T., Tachikawa, K.: Ferroelectrics and Superconductors, Tokyo: Maruzen Co., Ltd. 1974 (in Japanese).
74b1	Bjorkstam, J. L.: NMR Studies of Collective Atomic Motion near Ferroelectric Phase Transitions, pp. 1–70 in Advances in Magnetic Resonance, Vol. 7 (edited by Waugh, J. S.), New York: Academic Press 1974.
74b2	Blinc, R., Zeks, B.: Soft Modes in Ferroelectrics and Antiferroelectrics, Amsterdam: North-Holland Publishing Co. 1974.
74c1	Cochran, W. (Chairman): Proceedings of the Third International Meeting on Ferroelectricity, held in Edinburgh (1974), Ferroelectrics Vol. 7 (1974) Nos. 1, 2, 3, 4 and Vol. 8 (1974) nos. 1,2.
74d1	Dvorak, V.: Improper Ferroelectrics, Ferroelectrics **7** (1974) 1.
74e1	Electronic Materials Manufacturers Association (Editor): Piezoelectric Ceramics and Their Applications, Tokyo: Dempa Shinbun-sha 1974 (in Japanese).
74g1	Garn, L. N., Sharp, E. J.: Pyroelectric Vidicon Target Materials, IEEE Trans. Parts Hybrid Packaging, PHP **10** (1974) 208.
74k1	Kaminow, I. P.: An Introduction to Electrooptic Devices, New York: Academic Press 1974.
74l1	Land, C. E., Thacher, P. D., Haertling, G. H.: Electrooptic Ceramics, pp. 137–233 in Applied Solid State Science, Vol. 4 (edited by Wolfe, R.), New York: Academic Press 1974.
74l2	Lang, S. B.: Sourcebook of Pyroelectricity, London: Gordon and Breach Science Publishers 1974.
74l3	Levanyuk, A. P., Sannikov, D. G.: Improper Ferroelectrics, Uspekhi Fiz. Nauk **112** (1974) 561 (in Russian); Soviet Phys. Uspekhi (English transl.) **17** (1974) 199.
74m1	Marie, G., Donjon, J., Hazan, J.-P.: Pockels-Effect Imaging Devices and Their Applications, pp. 225–302 in Advances in Image Pickup and Display, Vol. 1 (edited by Kazan, B.), New York: Academic Press 1974.
74n1	Nelmes, R. J.: Structural Studies of Boracites. A Review of the Properties of Boracites, J. Phys. C (Solid State Phys.) **7** (1974) 3840.
74r1	Riste, R. (Editor): Anharmonic Lattices, Structural Transitions and Melting, Proceedings of the NATO Advanced Study Institute, held at Geilo (1973), Leiden: Noordhoff International Publishing 1974.
74s1	Scott, J. F.: Soft-Mode Spectroscopy: Experimental Studies of Structural Phase Transitions, Rev. Mod. Phys. **46** (1974) 83.
74s2	Shirane, G.: Neutron Scattering Studies of Structural Phase Transitions at Brookhaven, Rev. Mod. Phys. **46** (1974) 437.
74v1	Vogt, H.: Study of Structural Phase Transitions by Techniques of Nonlinear Optics, Appl. Phys. **5** (1974) 85.
75b1	Broadhurst, M. G. (Editor): Proceedings of Piezoelectric and Pyroelectric Symposium-Workshop, held at Gaithersburg (1975), Springfield: National Technical Information Service 1975; NBSIR 75–760.
75f1	Freeman, A. J., Schmid, H. (Editors): Magnetoelectric Interaction Phenomena in Crystals, Proceedings of the Symposium, held in Seattle (1973), London: Gordon and Breach Science Publishers 1975.
75k1	Kuz'minov, Yu. S.: Niobat i Tantalat Litiya; Materialy dlya Nelineinoi Optiki, Moskva: Izdatel'stvo "Nauka" 1975 (in Russian).
75m1	Maldonado, J. R., Fraser, D. B., Meitzler, A. H.: Display Applications of PLZT Ceramics, pp. 65–168 in Advances in Image Pickup and Display, Vol. 2 (edited by Kazan, B.), New York: Academic Press 1975.

75m2	Mitsui, T., Marutake, M., Sawaguchi, E., Gesi, K., Ikeda, T., Kobayashi, J., Makita, Y., Nakamura, E., Niizeki, N., Nomura, S., Sakudo, T., Shiozaki, Y., Tatsuzaki, I., Toyoda, K.: Landolt-Börnstein, Numerical Data and Functional Relationships in Science and Technology, New Series, Group III: Crystal and Solid State Physics, Vol. 9, Ferro- and Antiferroelectric Substances (Supplement and Extension to Vol. 3), Berlin, Heidelberg, New York: Springer 1975.
75n1	Nakamura, T.: Light Scattering Studies on Soft Phonon Phase Transitions, Ferroelectrics **2** (1975) 159.
75n2	Novakovic, L.: The Pseudo-Spin Method in Magnetism and Ferroelectricity, Oxford: Pergamon Press 1975.
75r1	Riste, R. (Editor): Fluctuations, Instabilities, and Phase Transitions, Proceedings of the NATO Advanced Study Institute, held at Geilo (1975), New York: Plenum Press 1975.
75v1	von der Linde, D., Glass, A. M.: Photorefractive Effects for Reversible Holographic Storage of Information, Appl. Phys. **8** (1975) 85.
75w1	Waku, S.: Ferroelectric Materials, Elect. Commun. Lab. Tech. J. (1975) Suppl. No. 28 (in Japanese).
76a1	Axe, J. D.: Incommensurate Structural Phase Transformations, pp. 353–378 in Proceedings of the Conference on Neutron Scattering, held at Gatlinburg (1976), Vol. I (edited by Moon, R. M.), Springfield: National Technical Information Service 1976.
76b1	Balkanski, M., Leite, R. C. C., Porto, S. P. S. (Editors): Proceedings of the Third International Conference on Light Scattering in Solids, held in Campinas (1975), Paris: Flammarion Sciences 1976.
76b2	Bauer, A., Buhling, D., Gesemann, H.-J., Helke, G., Schreckenbach, W.:Technologie und Anwendungen von Ferroelektrika, Leipzig: Akademische Verlagsgesellschaft 1976.
76b3	Blinc, R.: Magnetic Resonance Studies of Hydrogen Bonding in Solids, pp. 831–887 in The Hydrogen Bond, II, Structure and Spectroscopy (edited by Schuster, P., Zundel, G., Sandorfy, G.), Amsterdam: North-Holland Publishing Co. 1976.
76d1	Daniels, J., Hardtl, K. H., Hennings, D., Wernicke, R.: Defect Chemistry and Electrical Conductivity of Doped Barium Titanate Ceramics, Philips Res. Rept. **31** (1976) No. 6.
76i1	Ishibashi, Y., Takagi, Y.: Improper Ferroelectric Phase Transitions, Japan. J. Appl. Phys. **15** (1976) 1621.
76l1	Land, C. E., Glass, A. M. (Editors): Proceedings of the 1975 IEEE Symposium on Applications of Ferroelectrics, held in Albuquerque (1975), Ferroelectrics **10** (1976) Nos. 1, 2, 3, 4, **11** (1976) Nos. 1, 2.
76m1	Müller, K. A., Merz, W. J. (Editors): Proceedings of the Third European Meeting on Ferroelectricity, held in Zürich (1975), Ferroelectrics **12** (1976) Nos. 1, 2, 3, 4, **13** (1976) Nos. 1, 2, 3, 4, **14** (1976) Nos. 1, 2.
76m2	Müller, K. A., Rigamonti, A. (Editors): Local Properties at Phase Transitions, Proceedings of the International School of Physics "Enrico Fermi", Course LIX, held at Varenna on Lake Como (1973), Amsterdam: North-Holland Publishing Co. 1976.
76n1	Niizeki, N.: Growth and Characterization of Ferroelectric Crystals, pp. 97–105 in Crystal Growth and Characterization, Proceedings of the Second International Spring School on Crystal Growth, held at Lake Kawaguchi (1974) (edited by Ueda, R., Mullin, J. B.), Amsterdam: North-Holland Publishing Co. 1976.
76s1	Schmidt, V. H.: Ferroelectric Hydrogen Bonded Systems, pp. 1109–1167 in The Hydrogen Bond, III, Dynamics, Thermodynamics and Special Systems (edited by Schuster, P., Zundel, G., Sandorfy, G.), Amsterdam: North-Holland Publishing Co. 1976.
76s2	Spiridonov, F. M., Perova, E. B., Belova, I. D., Ivanova, L. A., Roginskaya, Yu. E., Venevtsev, Yu. N.: Sintez, Struktura i Svoistva Soedinenii Semeistva Pirokhlora, in Nauchno-Tekhnicheskie Prognozy v Oblasti Sinteza Segnetomaterialov, Moskva: Nauchno-Issledovatel'skii Institut Tekhniko-Ekonomicheskikh Issledovanii 1976 (in Russian).
76s3	Stiller, H.: Coherent Neutron Scattering for Observations on Collective Proton Motions in Systems of Hydrogen Bonds, pp. 1169–1195 in The Hydrogen Bond, III, Dynamics, Thermodynamics and Special Systems (edited by Schuster, P., Zundel, G., Sandorfy, G.), Amsterdam: North-Holland Publishing Co. 1976.
76s4	Schmidt, E., Brandmuller, J., Kiefer, W., Schrader, B., Schrotter, H. W.: Proceedings of the Fifth International Conference on Raman Spectroscopy, held in Freiburg (1976), Freiburg/Breisgau: Hans Ferdinand Schultz Verlag 1976.
76t1	Tokunaga, M., Mitsui, T.: Critical Phenomena in Ferroelectric Phase Transitions. II. Discussion on the Width of the Critical Region, Ferroelectrics **11** (1976) 451.

76v1	Voronov, V. V., Kuz'minov, Yu. S., Osiko, V. V.: Optically Induced Changes in the Refractive Index of Ferroelectric Crystals and Their Use in Reusable Holographic Memories (Review), Kvantovaya Elektronika **3** (1976) 2101 (in Russian); Soviet J. Quantum Electronics **6** (1976) 1143.
76w1	Wada, Y., Hayakawa, R.: Piezoelectricity and Pyroelectricity of Polymers, Jpn. J. Appl. Phys. **15** (1976) 2041.
76w2	Whalley, E.: The Hydrogen Bond in Ice, pp. 1425–1470 in The Hydrogen Bond, III, Dynamics, Thermodynamics and Special Systems (edited by Schuster, P., Zundel, G., Sandorfy, G.), Amsterdam: North-Holland Publishing Co. 1976.
77a1	Aleksandrov, K. S., Kobayashi, J. (Chairmen): Proceedings of the First Soviet-Japanese Symposium on Ferroelectrics, held at Novosibirsk (1976), Izv. Akad. Nauk SSSR, Ser. Fiz. **41** (1977) Nos. 3, 4 (in Russian); Bull. Acad. Sci. USSR, Phys. Ser. (English transl.) **41** (1977) Nos. 3, 4.
77d1	Dorner, B., Comes, R.: Phonons and Structural Phase Transformations, pp. 127–196 in Dynamics of Solids and Liquids by Neutron Scattering, Topics in Current Physics, Vol. 3 (edited by Lovesey, S. W., Springer, T.), Berlin, Heidelberg, New York: Springer 1977.
77g1	Gomez, M., Gonzalo, J. A., Kay, M. I. (Editors): Proceedings of the International Conference on Low Lying Lattice Vibrational Modes and Their Relationship to Superconductivity and Ferroelectricity, held in San Juan and Mayaguez (1975), Ferroelectrics **16** (1977) Nos. 1, 2, 3, 4, **17** (1977) Nos. 1, 2.
77h1	Hashiguti, R. R., Mikoshiba, N.: Internal Friction and Ultrasonic Attenuation in Solids, Proceedings of the 6th International Conference, held in Tokyo (1977), Tokyo: Univ. of Tokyo Press 1977.
77l1	Lines, M. E., Glass, A. M.: Principles and Applications of Ferroelectrics and Related Materials, Oxford: Clarendon Press 1977.
77m1	Marshall, W., Lovesey, W. S.: Theory of Thermal Neutron Scattering, Oxford: Clarendon Press 1977.
77m2	Meyer, R. B.: Ferroelectric Liquid Crystals; A Review, Molec. Cryst. Liquid Cryst. **40** (1977) 33.
77p1	Publishing Committee of the Electronic Ceramics (Editor): Semiconducting Barium Titanates, Tokyo: Giken Co., Ltd. 1977 (in Japanese).
77p2	Putley, E. H.: The Pyroelectric Detector-An Update, pp. 441–449 in Semiconductors and Semimetals, Vol. 12: Infrared Detectors II (edited by Willardson, R. K., Beer, A. C.), New York: Academic Press 1977.
77p3	Pislewski, N. (Editor): Radio and Microwave Spectroscopy, Proceedings of the Conference, held in Poznan (1977), Poznan: Institute of Molecular Physics, Polish Academy of Science 1977.
77r1	Riste, T. (Editor): Electron-Phonon Interactions and Phase Transitions, Proceedings of the NATO Advanced Study Institute, held at Geilo (1977), New York: Plenum Press 1977.
77s1	Sawada, S., Hamano, K.: Ferroelectricity, pp. 437–533 in Jikken Butsurigaku Koza, Vol. 16: Denki Bussei (edited by Okada, T.), Tokyo: Kyoritsu Shuppan Co., Ltd. 1977 (in Japanese).
77s2	Singer, B.: Theory and Performance Characteristics of Pyroelectric Imaging Tubes, pp. 1–82 in Advances in Image Pickup and Display, Vol. 3 (edited by Kazan, B.), New York: Academic Press 1977.
77s3	Staebler, D. L.: Ferroelectric Crystals, pp. 101–132 in Holographic Recording Materials, Topics in Applied Physics, Vol. 20 (edited by Smith, H. M.), Berlin, Heidelberg, New York: Springer 1977.
77t1	Tanaka, T. (Chairman): Proceedings of the First Meeting on Ferroelectric Materials and Their Applications, held in Kyoto (1977), Kyoto: Department of Electronics, Kyoto University 1977.
77t2	Toyoda, K.: Bibliography on Barium Titanate Semiconductors, pp. 135–162 in [77p1].
78b1	Balkanski, M. (Editor): Proceedings of the International Conference on Lattice Dynamics, held in Paris (1977), Paris: Flammarion Sciences 1978.
78e1	Enke, K., Fleischhauer, J., Grusser, W., Hansen, P., Nomura, S., Tolksdorf, W., Winkler, G., Wolfmeier, U.: Landolt-Börnstein, Numerical Data and Functional Relationships in Science and Technology, New Series, Group III: Crystal and Solid State Physics, Vol. 12, Magnetic and Other Properties of Oxides and Related Compounds, Part a, Garnets and Perovskites (Supplement and Extension to Vol. 4), Berlin, Heidelberg, New York: Springer 1978.
78h1	Hayes, W., Loudon, R.: Scattering of Light by Crystals, New York: John Wiley & Sons, 1978.
78i1	IUPAP: Symbols, Units and Nomenclature in Physics, Physica, A **93** (1978) 1.
78k1	Kepler, R. G.: Piezoelectricity, Pyroelectricity, and Ferroelectricity in Organic Materials, pp. 497–518 in Annual Review of Physical Chemistry, Vol. 29 (edited by Ravinovich, B. S., Schurr, J. M., Strauss, H. L.), Palo Alto: Annual Reviews, Inc. 1978.
78l1	Liu, S. T., Long, D.: Pyroelectric Detectors and Materials, Proc. IEEE **66** (1978) 14.

78n1 Nomura, S.: Crystallographic and Magnetic Properties of Perovskite and Perovskite-Related Compounds, pp. 368–520 in [78e1].

78o1 Oliner, A. A. (Editor): Acoustic Surface Waves, Berlin, Heidelberg, New York: Springer 1978.

78r1 Rauber, A.: Chemistry and Physics of Lithium Niobate, pp. 481–601 in Current Topics in Materials Science, Vol. 1 (edited by Kaldis, E.), Amsterdam: North-Holland Publishing Co. 1978.

78s1 Samara, G. A.: High Pressure Studies of Soft Mode Transitions in Solids, pp. 255–277 in High-Pressure and Low-Temperature Physics, Proceedings of the International Conference, held at Cleveland (1977) (edited by Chu, C. W., Woolam, J. A.), New York: Plenum Press 1978.

78s2 Shuvalov, L. A. (Editor): Proceedings of the Fourth International Meeting on Ferroelectricity, held in Leningrad (1977), Ferroelectrics **20** (1978) Nos. 1, 2, 3, 4, **21** (1978) Nos. 1, 2, 3, 4, **22** (1978) Nos. 1, 2.

78s3 Schmid, E. D., Krishnan, R. S., Kiefer, W., Schrotter, H. W. (Editors): Proceedings of the Sixth International Conference on Raman Spectroscopy, held at Bangalore (1978), London: Heyden, 1978.

79c1 Choy, M. M., Cook, W. R., Hearmon, R. F. S., Jaffe, H., Jerphagnon, J., Kurtz, S. K., Liu, S. T., Nelson, D. F.: Landolt-Börnstein, Numerical Data and Functional Relationships in Science and Technology, New Series, Group III: Crystal and Solid State Physics, Vol. II, Elastic, Piezoelectric, Pyroelectric, Piezooptic, Electrooptic Constants, and Nonlinear Dielectric Susceptibilities of Crystals (Revised and Extended Edition of Vol. III/1 and III/2), Berlin, Heidelberg, New York: Springer 1979.

79k1 Kurtz, S. K., Jerphagnon, J., Choy, M. M.: Nonlinear Dielectric Susceptibilities, pp. 671–743 in [79c1].

79t1 Toyoda, K.: Bibliography on Ferroelectrics, stored in computer systems at Murata Manufacturing Co., Ltd., Kyoto and at Hokkaido University, Sapporo.

B Research papers

1860S1 Schrauf: Wien. Ber. **41** (1860) 769.
1874T1 Topsoe, H., Christiansen, C.: Ann. Phys. Chem. **6** (1874) 499.
1882F1 Fitz, A., Sanson, F.: Z. Kristallogr. **6** (1882) 67.
1894L1 Lavinir, A.: Bull. Soc. Franc. Miner. Crist. **17** (1894) 153.
1894P1 Pockels, F.: Abh. Ges. Wiss. Göttingen **39** (1894) 1.
1895D1 Dufet, H.: C. R. Acad. Sci (Paris) **120** (1895) 377.
03T1 Tutton, A. E. H.: J. Chem. Soc. **83** (1903) 1049.
04G1 Gossner, B.: Z. Kristallogr. **38** (1904) 110.
16B1 Bridgman, P. W.: Proc. Am. Acad. Arts Sci. **52** (1916) 91.
16O1 Ogg, A., Hopwood, F. L.: Philos. Mag. (6) **32** (1916) 518.
20E1 Eppendahl, R.: Ann. Physik **61** (1920) 591.
21V1 Valasek, J.: Phys. Rev. **17** (1921) 475.
24E1 Eucken, A., Karwat, E.: Z. Phys. Chem. (Leipzig) **112** (1924) 467.
24V1 Valasek, J.: Phys. Rev. **24** (1924) 560.
25B1 Bradley, A. J.: Philos. Mag. (6) **49** (1925) 523.
27M1 Mandell, W.: Proc. Roy. Soc. London, Ser. A **116** (1927) 623.
27S1 Spencer, M. A.: Mineral. Mag. **21** (1927) 149.
27V1 Valasek, J.: Science **65** (1927) 235.
28D1 Demeny, L., Nitta, I.: Bull. Chem. Soc. Jpn. **3** (1928) 128.
28G1 Giauque, W. F., Wiebe, R.: J. Am. Chem. Soc. **50** (1928) 101.
28M1 Mandell, W.: Proc. Roy. Soc. London, Ser. A **121** (1928) 130.
28O1 Ogg, A.: Philos. Mag. (7) **5** (1928) 354.
28T1 Taylor, W., Boyer, T.: Mem. Proc. Manchester Lit. Philos. Soc. **72** (1928) 125.
30K1 Kobeko, P., Kurtschatov, J.: Z. Physik **66** (1930) 192.
30K2 Kracek, F. C.: J. Phys. Chem. **34** (1930) 225.
30O1 Ogg, A.: Philos. Mag. (7) **9** (1930) 665.
30W1 West, J.: Z. Kristallogr. **74** (1930) 306.
31E1 Edwards, D. A.: Z. Kristallogr. **80** (1931) 154.
31Z1 Ziegler, C. E.: Phys. Rev. **38** (1931) 1040.

35N1	Nitta, I., Watanabe, T.: Sci. Pap. Inst. Phys. Chem. Res. **26** (1935) 164.
35V1	Vigness, I.: Phys. Rev. **48** (1935) 198.
36S1	Seidl, F.: Z. Physik **101** (1936) 234.
37B1	Bridgman, P. W.: Proc. Am. Acad. Arts Sci. **72** (1937) 45.
37B2	Busch, G.: Helv. Phys. Acta **10** (1937) 261.
37H1	Helmholz, L.: J. Am. Chem. Soc. **59** (1937) 2036.
31Z2	Ziegler, C. E.: Phys. Rev. **37** (1931) 1693.
32B1	Benrath, A., Thiemann, W.: Z. Anorg. Allg. Chem. **208** (1932) 177.
32K1	Kurtshatow, B., Eremejew, M.: Phys. Z. Sowjetunion **1** (1932) 140.
32W1	Wyckoff, R. W. G., Corey, R. B.: Z. Kristallogr. **81** (1932) 386.
33B1	Bloomenthal, S.: Physics **4** (1933) 172.
33K1	Krishnan, K. S., Dasgupta, A. C.: Indian J. Phys. **8** (1933) 49.
34S1	Staub, H.: Phys. Z. **35** (1934) 720.
35B1	Biefeld, L. P., Harris, P. M.: J. Am. Chem. Soc. **57** (1935) 396.
35B2	Busch, G., Scherrer, P.: Naturwissenschaften **23** (1935) 737.
35H1	Hablutzel, J.: Helv. Phys. Acta **8** (1935) 498.
35M1	Mueller, H.: Phys. Rev. **47** (1935) 175.
37J1	Jaffe, H. V. R.: Phys. Rev. **51** (1937) 43.
38B1	Bancroft, D.: Phys. Rev. **53** (1938) 587.
38B2	Bridgman, P. W.: Proc. Am. Acad. Arts Sci. **72** (1938) 226.
38B3	Busch, G.: Helv. Phys. Acta **11** (1938) 269.
38D1	Dihlstrom, K.: Z. Anorg. Allg. Chem. **239** (1938) 57.
38H1	Hicks Jr., J. F. G., Hooley, J. G.: J. Am. Chem. Soc. **60** (1938) 2994.
38S1	Seidl, F.: Phys. Z. **39** (1938) 714.
38S2	Spitzer, F.: Dissertation, Göttingen (1938).
39B1	Barth, T. F. W.: Z. Phys. Chem., B **43** (1939) 448.
39H1	Hablutzel, J.: Helv. Phys. Acta **12** (1939) 489.
39H2	Hinz, H.: Z. Physik **111** (1939) 617.
39M1	Mason, W. P.: Phys. Rev. **55** (1939) 1141.
39M2	Mason, W. P.: Phys. Rev. **55** (1939) 775.
39U1	Ubbelohde, A. R.: Proc. Roy. Soc. London, Ser. A **173** (1939) 417.
40H1	Holden, A. N., Mason, W. P.: Phys. Rev. **57** (1940) 54.
40M1	Mueller, H.: Phys. Rev. **58** (1940) 565.
40M2	Mueller, H.: Ann. N. Y. Acad. Sci. **40** (1940) 321.
41B1	Bantle, W., Scherrer, P.: Helv. Phys. Acta **14** (1941) 146.
41B2	Beevers, A. C., Hughes, S.: Proc. Roy. Soc. London, Ser. A **177** (1941) 251.
41M1	Miekk-Oja, H.: Ann. Acad. Sci. Fenn. AI, Math. Phys. (1941) No. 7.
42B1	Bantle, W.: Helv. Phys. Acta **15** (1942) 373.
42U1	Ubbelohde, A. R., Woodward, I.: Proc. Roy. Soc. (London), Ser. A **179** (1942) 39.
43B1	Baertschi, P.: Helv. Phys. Acta **16** (1943) 218.
43B2	Bantle, W., Caflisch, Ch.: Helv. Phys. Acta **16** (1943) 235.
43K1	Kawai, H.: J. Phys.-Math. Soc. Jpn. **17** (1943) 270.
43S1	Strijk, B., MacGillavry, C. H.: Rec. Trav. Chim. **62** (1943) 705.
44M1	Mukherjee, P. L.: Indian J. Phys. **18** (1944) 148.
44S1	Stephenson, C. C., Adams, H. E.: J. Am. Chem. Soc. **66** (1944) 1409.
44S2	Stephenson, C. C., Adams, H. E.: J. Am. Chem. Soc. **66** (1944) 1412.
44S3	Stephenson, C. C., Hooly, G. J.: J. Am. Chem. Soc. **66** (1944) 1397.
44S4	Stephenson, C. C., Zettlemoyer, A. C.: J. Am. Chem. Soc. **66** (1944) 1402.
44S5	Stephenson, C. C., Zettlemoyer, A. C.: J. Am. Chem. Soc. **66** (1944) 1405.
44V1	von Arx, A., Bantle, W.: Helv. Phys. Acta **17** (1944) 298.
44Z1	Zwicker, B., Scherrer, P.: Helv. Phys. Acta **17** (1944) 346.
45B1	Baertschi, P.: Helv. Phys. Acta **18** (1945) 267.
45B2	Baertschi, P., Matthias, B., Merz, W., Scherrer, P.: Helv. Phys. Acta **18** (1945) 240.
45B3	Bantle, W., Matthias, B., Scherrer, P.: Helv. Phys. Acta **18** (1945) 389.
45B4	Bridgman, P. W.: Proc. Am. Acad. Arts Sci. **76** (1945) 1.
45J1	Jaffe, H.: Meeting of Am. Phys. Soc., New York, **1945**.
45K1	Klug, H. P., Sears Jr., G. W.: J. Am. Chem. Soc. **67** (1945) 878.

45 S 1	Shomate, C. H.: J. Am. Chem. Soc. **67** (1945) 1096.
46 C 1	Clusius, K.: Z. Naturforsch. a **1** (1946) 142.
46 D 1	Davisson, J. W.: Phys. Rev. **70** (1946) 685.
46 M 1	Mason, W. P.: Phys. Rev. **69** (1946) 173.
46 U 1	Ubbelohde, A. R., Woodward, I.: Proc. Roy. Soc. (London) Ser. A **185** (1946) 448.
46 Z 1	Zwicker, B.: Helv. Phys. Acta **19** (1946) 523.
47 C 1	Clusius, K., Wolf, G.: Z. Naturforsch. a **2** (1947) 495.
47 H 1	Huntington, H. B.: Phys. Rev. **72** (1947) 321.
47 K 1	Kawai, H.: Butsuri **2** (1947) 193.
47 K 2	Kawai, H.: J. Phys. Soc. Jpn. **2** (1947) 113.
47 M 1	Matthias, B., Merz, W., Scherrer, P.: Helv. Phys. Acta **20** (1947) 273.
47 P 1	Pospelov, V. A., Zhdanov, G. S.: p. 421 in Structure Report, Vol. XI, Utrecht: Oosthoek's **1947**.
47 T 1	Tahvonen, P. E.: Ann. Acad. Sci. Fenn., AI, Math. Phys. (1947) No. 44.
47 U 1	Ubbelohde, A. R., Woodward, I.: Proc. Roy. Soc. London, Ser. A **188** (1947) 358.
48 F 1	Frazer, B. C.: Thesis, Alabama Polytechnic Institute, USA (1948).
48 J 1	Jaffe, H.: The Brush Development Co., Cleveland, Ohio, US Signal Corps. Contract No. W-28-003-sc-1583, Final report April 1, **1948**.
48 K 1	Kawai, H.: J. Phys. Soc. Jpn. **3** (1948) 105.
48 K 2	Klassen-Nekliudova, M. A. Chernysheva, Shternberg, A. A.: Dokl. Akad. Nauk SSSR **63** (1948) 527.
48 O 1	Officer, Ch. B.: Thesis, Wesleyan University, Middletown, Conn. (1948).
48 U 1	Ueda, R.: J. Phys. Soc. Jpn. **3** (1948) 328.
48 W 1	West, C. D., Makas, A. M.: J. Chem. Phys. **16** (1948) 427.
49 A 1	Alpert, N. L.: Phys. Rev. **75** (1949) 398.
49 B 1	Bleaney, B.: Phys. Rev. **75** (1949) 1962.
49 J 1	Jaffe, H.: USA Pat. No. 2463109, March **1** (1949); Brit. Pat. No. 634775.
49 K 1	Kawai, H., Marutake, M.: J. Phys. Soc. Jpn. **4** (1949) 91.
49 N 1	Niemiec, T.: Phys. Rev. **75** (1949) 215.
49 P 1	Price, W. J.: Phys. Rev. **75** (1949) 946.
49 S 1	Sundara Rao, R. V. G.: Proc. Indian Acad. Sci. A **30** (1949) 173.
49 T 1	Takahashi, H., Hara, S.: J. Phys. Soc. Jpn. **4** (1949) 261.
50 B 1	Bagguley, D. M. S., Criffiths, H. E.: Proc. Roy. Soc. (London) Ser. A **204** (1950) 188.
50 B 2	Baumgartner, H.: Helv. Phys. Acta **23** (1950) 651.
50 B 3	Beck, M., Gränicher, H.: Helv. Phys. Acta **23** (1950) 522.
50 B 4	Bleaney, B.: Proc. Roy. Soc. (London) Ser. A **204** (1950) 203.
50 C 1	Carpenter, R. O'B.: J. Opt. Soc. Am. **40** (1950) 225.
50 C 3	Chernysheva, M. A.: Dokl. Akad. Nauk SSSR **74** (1950) 247.
50 D 1	Dickson, D. H. W., Ubbelohde, A. R.: Acta Crystallogr. **3** (1950) 6.
50 D 2	Donges, E.: Z. Anorg. Allg. Chem. **263** (1950) 112.
50 D 3	Donges, E.: Z. Anorg. Allg. Chem. **263** (1950) 280.
50 F 1	Furuichi, J., Mitsui, T.: Phys. Rev. **80** (1950) 93.
50 J 1	Jona, F.: Helv. Phys. Acta **23** (1950) 795.
50 N 1	Newman, R.: J. Chem. Phys. **18** (1950) 669.
50 P 1	Price, W. J., Huntington, H. B.: J. Acoust. Soc. Am. **22** (1950) 32.
50 S 2	Seidl, F.: Tschermak's Mineral. Petrogr. Mitt. **1** (1950) 432.
50 V 1	Van Dyke, K. S., Gordon, G. D.: US Signal Corps Contract No. W-28-003-sc-1556, Tenth Report (1950), Wesleyan University, Middleton, Conn.
50 W 1	Walker, A. C.: J. Franklin Inst. **250** (1950) 481.
50 W 2	West, C. D., Makas, A. S.: Am. Mineralogist **35** (1950) 130.
50 W 3	Willard, G. W.: Unpublished work (1950) cited in [50 C 1].
51 C 1	Carpenter, R. O'B.: Thesis, Harvard University, Cambridge, Massachusetts (1951).
51 D 1	Donges, E.: Z. Anorg. Allg. Chem. **265** (1951) 56.
51 K 1	Korhonen, U.: Ann. Acad. Sci. Fenn., AI **102** (1951) 37.
51 M 1	Matthias, B. T., Hulm, J. K.: Phys. Rev. **82** (1951) 108.
51 M 3	Merz, W. J.: Phys. Rev. **82** (1951) 562; erratum **83** (1951) 656.
51 N 1	Nitta, I., Kiriyama, R., Haisa, M.: Sci. Pap. Osaka Univ. (1951) No. 30.
51 S 3	Schubert, K., Fricke, H.: Z. Naturforsch. a **6** (1951) 781.
52 A 2	Auty, R. P., Cole, R. H.: J. Chem. Phys. **20** (1952) 1309.

52F1 Fellner-Feldegg, H.: Tschermaks Mineral. Petrogr. Mitt. **3** (1952) 37.
52G1 Griffiths, J. H. E., Powell, J. A.: Proc. Phys. Soc. (London) Ser. A **65** (1952) 289.
52M1 Mason, W. P., Matthias, B. T.: Phys. Rev. **88** (1952) 477.
52N1 Nagamiya, T.: Progr. Theor. Phys. **7** (1952) 275.
53A1 Annaka, S., Ouchi, Y.: Kagaku **23** (1953) 538.
53B1 Bacon, G. E., Pease, R. S.: Proc. Roy. Soc. (London) Ser. A **220** (1953) 397.
53B2 Brown, N. L., Cole, R. H.: J. Chem. Phys. **21** (1953) 1920.
53B3 Busch, G., Känzig, W., Meier, W. M.: Helv. Phys. Acta **26** (1953) 385.
53C1 Cavalca, L., Nardelli, M., Braibanti, A.: Gazz. Chim. Ital. **83** (1953) 476.
53C2 Carpenter, R. O'B.: J. Acoust. Soc. Am. **25** (1953) 1145.
53F1 Frazer, B. C., Pepinsky, R.: Phys. Rev. **91** (1953) 212.
53G2 Grenier, G. E.: Bachelor Thesis, Massachusetts Institute of Technology (1953).
53J1 Jona, F., Pepinsky, R.: Phys. Rev. **92** (1953) 1577.
53K1 Korhonen, U.: Ann. Acad. Sci. Fenn. AI **150** (1953) 16.
53M2 Mitsui, T., Furuichi, J.: Phys. Rev. **90** (1953) 193.
53S2 Schubert, K., Fricke, H.: Z. Metallkunde **44** (1953) 457.
53S6 Stephenson, C. C., Corbella, J. M., Russell, L. A.: J. Chem. Phys. **21** (1953) 1110.
53V1 Vernon, R. C., Pepinsky, R.: Unpublished work (1953) cited in [62B11].
54B1 Bleaney, B., Trenan, R. S.: Proc. Roy. Soc. (London) Ser. A **223** (1954) 1.
54C1 Carpenter, R. O'B.: J. Opt. Soc. Am. **44** (1954) 425.
54G2 Gränicher, H., Meier, W. M., Petter, W.: Helv. Phys. Acta **27** (1954) 216.
54K2 Kiriyama, R., Ibamoto, H., Matsuo, K.: Acta Crystallogr. **7** (1954) 482.
54M3 Melmed, A., Jona, F., Pepinsky, R.: Tech. Rep. No. 20, Wright Air Development Center, Contract AF33 (616)-2133 (1954).
54S7 Swenson, R. W., Cole, R. H.: J. Chem. Phys. **22** (1954) 284.
54T1 Truter, M. R.: Acta Crystallogr. **7** (1954) 73.
55A1 Aboav, D., Gränicher, H., Petter, W.: Helv. Phys. Acta **28** (1955) 299.
55B1 Bacon, G. E., Pease, R. S.: Proc. Roy. Soc. (London) Ser. A **230** (1955) 359.
55B2 Baker, J. M., Bleaney, B.: Unpublished work (1955) cited in [55b1].
55B3 Bechmann, R.: Unpublished work (1955) cited in [66b1].
55C2 Carpenter, G. B.: Acta Crystallogr. **8** (1955) 852.
55D3 Deviot, B.: J. Phys. Radium **16** (1955) 162.
55F1 Frenzel, G.: Neues Jahrb. Mineral. Monatsh. (1955) 241.
55H1 Havailiak, S., Cole, R. H.: J. Chem. Phys. **23** (1955) 2455.
55H2 Holden, A. N., Matthias, B. T., Merz, W. J., Remeika, J. P.: Phys. Rev. **98** (1955) 546.
55H4 Hornig, D. F., Osberg, W. E.: J. Chem. Phys. **23** (1955) 662.
55I1 Ibuki, S., Yoshimatsu, S.: J. Phys. Soc. Jpn. **10** (1955) 549.
55K2 Keeling Jr., R. O., Pepinsky, R.: Z. Kristallogr. **106** (1955) 236.
55L1 Losche, V. A.: Exp. Tech. Phys. **3** (1955) 18.
55M1 Marutake, M.: Bull. Kobayashi Inst. Phys. Res. **5** (1955) 225.
55M4 Menary, J. W.: Acta Crystallogr. **8** (1955) 840.
55S1 Seki, S., Momotani, M., Nakatsu, K., Oshima, T.: Bull. Chem. Soc. Jpn. **28** (1955) 411.
55S2 Shirane, G., Jona, F., Pepinsky, R.: Proc. I.R.E. **43** (1955) 38.
56A1 Abe, R.: J. Phys. Soc. Jpn. **11** (1956) 104.
56B1 Baker, J. M.: Proc. Phys. Soc. (London) Ser. A **69** (1956) 633.
56B2 Bechmann, R.: J. Acoust. Soc. Am. **28** (1956) 347.
56C3 Couture, L., LeMontagner, S., LeBot, J., LeTraon, A.: C. R. Acad. Sci. (Paris) **242** (1956) 1804.
56G1 Goldsmith, G. J.: Bull. Am. Phys. Soc. (2) **1** (1956) 322.
56G3 Gränicher, H.: Unpublished work (1956) cited in [57k1].
56H1 Holden, A. N., Merz, W. J., Remeika, J. P., Matthias, B. T.: Phys. Rev. **101** (1956) 962.
56J1 Jona, F., Pepinsky, R.: Phys. Rev. **103** (1956) 1126.
56J2 Jona, F., Vidulich, S. M., Pepinsky, R.: Unpublished work (1956).
56L1 LeMontagner, S., LeBot, J., Hagene, M., Lasbleis, F., LePage, M.: C. R. Acad. Sci. (Paris) **242** (1956) 475.
56L3 Ludwig, G. W.: J. Chem. Phys. **25** (1956) 159.
56M1 Matthias, B. T., Miller, C. E., Remeika, J.: Phys. Rev. **104** (1956) 849.
56M2 Matthias, B. T., Remeika, J. P.: Phys. Rev. **103** (1956) 262.

56M3	Merz, W. J.: Phys. Rev. **103** (1956) 565.
56M5	Muser, H. E.: Z. Physik **145** (1956) 621.
56P2	Pepinsky, R., Jona, F., Shirane, G.: Phys. Rev. **102** (1956) 1181.
56P3	Plyuschev, V. E., Markina, I. B., Shklover, L. P.: Dokl. Akad. Nauk SSSR **108** (1956) 645.
56R2	Remeika, J. P., Merz, W. J.: Phys. Rev. **102** (1956) 295.
56R3	Rez, I. S., Varfolomeeva, L. A.: Article on p. 126 (in Russian) or p. 93 in [59s1].
56S4	Shimada, J., Abe, H., Ono, K.: J. Phys. Soc. Jpn. **11** (1956) 137.
56S6	Solomon, A. L.: Phys. Rev. **104** (1956) 1191.
56S7	Swanson, H. E., Gilferich, N. T., Cook, M. I.: Natl. Bur. Std. Circ. No. 539 (1956) Tl, 6, 1.
56W1	Wood, E. A.: Acta Crystallogr. **9** (1956) 618.
57B1	Baker, N., Webber, D. S.: J. Chem. Phys. **27** (1957) 689.
57B3	Bechmann, R., Taylor, R.: Selected Engineering Report, Post Office Research Station, London, Her Majesty's Stationary Office (1957).
57B5	Bogle, G. S., Gabriel, J. R., Bottomley, G. A.: Trans. Faraday Soc. **53** (1957) 1058.
57C1	Cole, R. H., Havailiak, S.: Discussions Faraday Soc. **23** (1957) 31.
57E1	Eastman, D. R. P.: Master Thesis, The Pennsylvania State Univ. (1957).
57H2	Hoshino, S., Mitsui, T., Jona, F., Pepinsky, R.: Phys. Rev. **107** (1957) 1255.
57H3	Hurlbut, C. S.: Am. Mineralogist **42** (1957) 178.
57J2	Jona, F., Mitsui, T., Pepinsky, R.: Unpublished work (1957) cited in [62j1].
57J5	Jona, F., Vedam, K., Mitsui, T., Pepinsky, R.: Unpublished work (1957) cited in [62j1].
57L2	LeMontagner, S., LeBot, J., Pasquet, M.: Arch. Sci. (Genève) **10** (1957) Special No. 18.
57M2	Matthias, B. T., Remeika, J. P.: Phys. Rev. **107** (1957) 1727.
57M3	Mazzi, F., Jona, F., Pepinsky, R.: Z. Kristallogr. **108** (1957) 359.
57O1	Okaya, Y., Ahmed, M. S., Pepinsky, R., Vand, V.: Z. Kristallogr. **109** (1957) 367.
57P1	Pepinsky, R., Jona, F.: Phys. Rev. **105** (1957) 344.
57P2	Pepinsky, R., Okaya, Y. Eastman, D. P., Mitsui, T.: Phys. Rev. **107** (1957) 1538.
57P3	Pepinsky, R., Okaya, Y., Jona, F.: Bull. Am. Phys. Soc. (2) **2** (1957) 220.
57P4	Pepinsky, R., Okaya, Y., Mitsui, T.: Acta Crystallogr. **10** (1957) 600.
57P5	Pepinsky, R., Vedam, K., Jona, F.: Unpublished work (1957) cited in [62j1].
57W1	Wieder, H. H.: Proc. I. R. E. **45** (1957) 1094.
57W3	Wood, E. A., Holden, A. N.: Acta Crystallogr. **10** (1957) 145.
58A1	Abrikosov, N., Vasserman, A. M., Poretskaya, L. V.: Dokl. Akad. Nauk SSSR **123** (1958) 279.
58C1	Christ, C. L., Clark, J. R., Evans Jr., H. T.: Acta Crystallogr. **11** (1958) 761.
58D1	Daniels, J. M., Wesemeyer, H.: Can. J. Phys. **36** (1958) 144.
58D2	Delain, C.: C. R. Acad. Sci. (Paris) **247** (1958) 1451.
58D3	Domanski, S.: Proc. Phys. Soc. London **72** (1958) 306.
58E1	Ezhkova, Z. I., Zhdanov, G. S., Umanskii, M. M.: Kristallografiya **3** (1958) 231; Sov. Phys. Crystallogr. (English Transl.) **3** (1958) 230.
58F1	Francombe, M. H., Lewis, B.: Acta Crystallogr. **11** (1958) 175.
58H2	Holuj, F., Petch, H. E.: Can. J. Phys. **36** (1958) 145.
58H3	Hoshino, S., Vedam, K., Okaya, Y., Pepinsky, R.: Phys. Rev. **112** (1958) 405.
58K5	Kuncher, N. R., Truter, M. R.: J. Chem. Soc. **517** (1958) 2551.
58M1	Makita, Y., Takagi, Y.: J. Phys. Soc. Jpn. **13** (1958) 367.
58M5	Mooser, E., Pearson, W. B.: J. Phys. Chem. Solids **7** (1958) 65.
58N1	Nitsche, R.: Helv. Phys. Acta **31** (1958) 306.
58O1	Ogawa, T.: Bull. Kobayashi Inst. Phys. Res. **8** (1958) 111.
58O2	Okaya, Y., Vedam, K., Pepinsky, R.: Acta Crystallogr. **11** (1958) 307.
58P1	Pepinsky, R., Vedam, K., Hoshino, S., Okaya, Y.: Phys. Rev. **111** (1958) 430.
58P2	Pepinsky, R., Vedam, K., Hoshino, S., Okaya, Y.: Phys. Rev. **111** (1958) 1508.
58P3	Pepinsky, R., Vedam, K., Okaya, Y.: Phys. Rev. **110** (1958) 1309.
58P4	Pepinsky, R., Vedam, K., Okaya, Y., Hoshino, S.: Phys. Rev. **111** (1958) 1467.
58P6	Poindexter, E., Giardini, A. A.: Phys. Rev. **110** (1958) 1069.
58P8	Pulvari, C. F., Kuebler, W.: J. Appl. Phys. **29** (1958) 1742.
58S1	Sawada, S., Nomura, S., Fujii, S.: J. Phys. Soc. Jpn. **13** (1958) 1549.
58S2	Sawada, S., Nomura, S., Fujii, S., Yoshida, I.: Phys. Rev. Letters **1** (1958) 320.
58T1	Takagi, Y., Makita, Y.: J. Phys. Soc. Jpn. **13** (1958) 272.
58T3	Tenzer, L., Frazer, B. C., Pepinsky, R.: Acta Crystallogr. **11** (1958) 505.

58T4	Triebwasser, S.: IBM J. Res. Dev. **2** (1958) 212.
58V2	von Gattow, G., Zemann, J.: Z. Anorg. Allg. Chem. **293** (1958) 16.
58W1	Wieder, H. H.: Phys. Rev. **110** (1958) 29.
59B1	Bjorkstam, J. L., Uehling, E. A.: Phys. Rev. **114** (1959) 961.
59B2	Blinc, R.: J. Chem. Phys. **31** (1959) 849.
59D4	Dunne, T.: Unpublished work (1959) cited in [62j1].
59F2	Fatuzzo, E., Merz, W. J.: Phys. Rev. **116** (1959) 61.
59F3	Fatuzzo, E., Nitsche, R.: Z. Elektrochem. **63** (1959) 970.
59F4	Ferroni, E., Orioli, P.: Z. Kristallogr. **111** (1959) 362.
59G1	Geller, S., Booth, D. P.: Z. Kristallogr. **111** (1959) 117.
59G2	Goldsmith, G. J., White, J. G.: J. Chem. Phys. **31** (1959) 1175.
59H2	Haussühl, S.: Z. Kristallogr. **111** (1959) 321.
59H3	Haussühl, S., Trost, F.: Z. Naturforsch., a **14** (1959) 437.
59H6	Hoshino, S., Okaya, Y., Pepinsky, R.: Phys. Rev. **115** (1959) 323.
59J1	Jaffe, H.: Private communication (1959) cited in [66b1].
59K2	Konstantinova, V. P., Sil'vestrova, I. M., Aleksandrov, K. S.: Kristallografiya **4** (1959) 70; Sov. Phys. Crystallogr. **4** (1959) 63.
59M1	Makita, Y.: Unpublished work (1959).
59M6	Morosin, B., Lingafelter, E. C.: Acta Crystallogr. **12** (1959) 611.
59N1	Nicolini, L.: Nuovo Cimento, Ser. 10, **13** (1959) 257; erratum **14** (1959) 905.
59P1	Pearson, G. L., Feldman, W. L.: J. Phys. Chem. Solids 9 (1959) 28.
59P2	Pepinsky, R., Vedam, K.: Unpublished work (1959) cited in [59H6].
59P3	Pepinsky, R., Vedam, K.: Phys. Rev. **114** (1959) 1217.
59P4	Pepinsky, R., Vedam, K., Okaya, Y., Unterleitner, F.: Bull. Am. Phys. Soc. (2) **4** (1959) 63.
59S5	Smith, P. L.: Unpublished work (1959) cited in [66b1].
59V1	Vedam, K., Pepinsky, R., Lajzerowicz, J., Okaya, Y., Stemple, N.: Bull. Am. Phys. Soc. (2) **4** (1959) 63.
59W1	Waku, S., Hirabayashi, H., Toyoda, H., Iwasaki, H., Kiriyama, R.: J. Phys. Soc. Jpn. **14** (1959) 973.
59W3	Wieder, H. H.: J. Appl. Phys. **30** (1959) 1010.
59Y1	Yamanaka, C., Yoshimune, Y., Yamamura, Y.: Denki Gakkai Zasshi (J. Inst. Electr. Eng. Jpn.) **79** (1959) 867.
60B4	Blinc, R., Levstek, I.: J. Phys. Chem. Solids **12** (1960) 295.
60C1	Chynoweth, A. G.: Phys. Rev. **117** (1960) 1235.
60D2	Dvoryankin, V. F., Vainshtein, B. K.: Kristallografiya **5** (1960) 589.
60E1	Eisner, I. Ya.: Izv. Akad. Nauk SSSR, Ser. Fiz. **24** (1960) 1326; Bull. Acad. Sci. USSR, Phys. Ser. (English Transl.) **24** (1960) 1327.
60F3	Fatuzzo, E.: Proc. Phys. Soc. London **76** (1960) 797.
60F4	Fatuzzo, E., Nitsche, R.: Phys. Rev. **117** (1960) 936.
60F6	Fotchenkov, A. A.: Kristallografiya **5** (1960) 415; Sov. Phys. Crystallogr. (English Transl.) **5** (1960) 390.
60G5	Gurevich, V. M., Zheludev, I. S., Rez, I. S.: Kristallografiya **5** (1960) 142; Sov. Phys. Crystallogr. (English Transl.) **5** (1960) 133.
60J1	Jackle, W.: Z. Angew. Phys. **12** (1960) 148.
60J2	Johnson, Q. C.: U. S. Atomic Energy Commission UCRL-9350 (1960), cited in [71B22].
60J3	Jona, F., Shirane, G.: Phys. Rev. **117** (1960) 139.
60K1	Konstantinova, V. P., Sil'vestrova, I. M., Shuvalov, L. A., Yurin, V. A.: Izv. Akad. Nauk SSSR, Ser. Fiz. **24** (1960) 1203; Bull. Acad. Sci. USSR, Phys. (English Transl.) **24** (1960) 1206.
60K5	Krstanovic, I., Okaya, Y., Pepinsky, R.: Unpublished work (1960).
60L1	Lurio, A., Sterm, E.: J. Appl. Phys. **31** (1960) 1125.
60N1	Nitsche, R., Merz, W. J.: J. Phys. Chem. Solids **13** (1960) 154.
60P1	Pepinsky, R., Vedam, K.: Phys. Rev. **117** (1960) 1502.
60R1	Rowland, J. F., Bright, N. F. H., Jongejan, A.: Adv. in X-Ray Anal., Vol. 2, New York: Plenum Press 1960.
60S3	Scavicar, S.: Z. Kristallogr. **114** (1960) 85.
60S4	Schinke, H., Sauerwald, F.: Z. Anorg. Allg. Chem. **304** (1960) 25.
60S12	Swanson, H. E., Cook, M. I., Isaacs, T., Evans, E. H.: Natl. Bur. Std. Circ. No. 539 (1960).
60T3	Toyoda, H., Niizeki, N., Waku, S.: J. Phys. Soc. Jpn. **15** (1960) 1831.

60 V 1	Vedam, K., Okaya, Y., Pepinsky, R.: Phys. Rev. **119** (1960) 1252.
60 W 1	Waku, S., Masuno, K., Tanaka, T.: J. Phys. Soc. Jpn. **15** (1960) 1698.
60 W 2	Waku, S., Masuno, K., Tanaka, T., Iwasaki, H.: J. Phys. Soc. Jpn. **15** (1960) 1185.
60 W 3	Weir, C. E., Van Valkenburg, A.: J. Res. Natl. Bur. Std., A **64** (1960) 103.
60 W 4	Wieder, H. H., Collins, D. A.: Phys. Rev. **120** (1960) 725.
60 W 5	Wieder, H. H.: J. Appl. Phys. **31** (1960) 180.
60 Y 2	Yoshida, I., Sawada, S.: J. Phys. Soc. Jpn. **15** (1960) 199.
61 A 1	Achyuthan, K., Breazeale, M. A.: J. Opt. Soc. Am. **51** (1961) 914.
61 B 1	Blinc, R., Brenman, M., Waugh, J. S.: J. Chem. Phys. **35** (1961) 1770.
61 B 2	Blinc, R., Detoni, S., Levstek, I., Pintar, M., Poberaj, S., Schara, M.: J. Phys. Chem. Solids **20** (1961) 187.
61 B 3	Blinc, R., Pintar, M.: J. Chem. Phys. **35** (1961) 1140.
61 B 4	Bogle, G. S., Symmons, H. F.: Proc. Phys. Soc. London **78** (1961) 812.
61 B 7	Burns, G.: Phys. Rev. **123** (1961) 1634.
61 D 1	Dunne, T. G., Wang, M. L.: J. Electrochem. Soc. **108** (1961) 111.
61 E 1	Emsley, J. W., Smith, J. A. S.: Trans. Faraday Soc. **57** (1961) 893.
61 E 2	Emsley, J. W., Smith, J. A. S.: Trans. Faraday Soc. **57** (1961) 1233.
61 G 2	Grosvald, G., Hagenmuller, P., LeBot, J.: Bull. Soc. Chim. Fr. (1961) 1955.
61 G 4	Gurevich, V. M., Zheludev, I. S.: Kristallografiya **6** (1961) 135; Sov. Phys. Crystallogr. (English Transl.) **6** (1961) 110.
61 G 5	Gurevich, V. M., Zheludev, I. S.: Kristallografiya **6** (1961) 778; Sov. Phys. Crystallogr. (English Transl.) **6** (1962) 624.
61 H 3	Herlach, F.: Helv. Phys. Acta **34** (1961) 305.
61 H 6	Hoshino, S., Shibuya, I.: J. Phys. Soc. Jpn. **16** (1961) 1254.
61 J 1	Jaffe, H., Smith, C. S.: Phys. Rev. **121** (1961) 1604.
61 K 1	Kaczmarek, F.: MIT Lab. Insul. Res. Tech. Rep. **163** (1961).
61 K 3	Kay, M. I., Frazer, B. C.: Acta Crystallogr. **14** (1961) 56.
61 K 4	Kobayashi, J., Yamada, N.: Bull. Sci. Eng. Res. Lab., Waseda Univ. **18** (1961) 35.
61 K 5	Krebs, H., Grun, K., Kallen, D.: Z. Anorg. Allg. Chem. **312** (1961) 307.
61 L 1	Lundin, A. G., Mikhailov, G. M., Gabuda, S. P.: Zh. Eksp. Teor. Fiz. **40** (1961) 1282; Sov. Phys. JETP (English Transl.) **13** (1961) 903.
61 L 2	Lundin, A. G., Mikhailov, G. M., Gabuda, S. P.: Dokl. Akad. Nauk SSSR **136** (1961) 864.
61 M 1	Mikhalov, G. M., Lundin, A. G., Gabuda, S. P.: Zh. Eksp. Teor. Fiz. **41** (1961) 1370; Sov. Phys. JETP (English Transl.) **14** (1962) 977.
61 M 3	Moulton, G. C., Moulton, W. G.: J. Chem. Phys. **35** (1961) 208.
61 N 1	Nakamura, T., Ohi, K.: J. Phys. Soc. Jpn. **16** (1961) 209.
61 N 2	Namba, S.: Rev. Sci. Instr. **32** (1961) 595.
61 N 3	Nomura, S.: J. Phys. Soc. Jpn. **16** (1961) 2440.
61 N 4	Nomura, S., Asao, Y., Sawada, S.: J. Phys. Soc. Jpn. **16** (1961) 917.
61 O 1	Okada, K.: J. Phys. Soc. Jpn. **16** (1961) 414.
61 O 2	Ovenall, D. W., Muller, K. A.: Helv. Phys. Acta **34** (1961) 786.
61 R 1	Rushton, E.: Brit. J. App. Phys. **12** (1961) 417.
61 S 1	Saburi, O.: J. Am. Ceram. Soc. **44** (1961) 54.
61 S 2	Sawada, S., Nomura, S., Asao, Y.: J. Phys. Soc. Jpn. **16** (1961) 2207.
61 S 3	Sawada, S., Nomura, S., Asao, Y.: J. Phys. Soc. Jpn. **16** (1961) 2486.
61 S 5	Schmidt, G.: Z. Physik **161** (1961) 579.
61 S 6	Shibuya, I.: J. Phys. Soc. Jpn. **16** (1961) 490.
61 S 7	Shibuya, I., Mitsui, T.: J. Phys. Soc. Jpn. **16** (1961) 479.
61 S 8	Shuvalov, L. A., Pluzhnikov, K. A.: Kristallografiya **6** (1961) 692; Sov. Phys. Crystallogr. (English Transl.) **6** (1962) 555.
61 S 9	Sil'vestrova, I. M.: Kristallografiya **6** (1961) 582; Sov. Phys. Crystallogr. (English Transl.) **6** (1962) 466.
61 S 13	Stemple, N.: Unpublished work (1961) cited in [61 B 7].
61 T 1	Toyoda, K., Kawabata, A., Tanaka, T.: Bull. Inst. Chem. Res., Kyoto Univ. **39** (1961) 189.
61 T 2	Toyoda, H., Tanaka, Y., Shiokawa, W.: Rev. Electr. Commun. Lab. **9** (1961) 485.
61 V 1	Vainshtein, B. K., Dvoryankin, V. F.: Dokl. Akad. Nauk SSSR **140** (1961) 111; Sov. Phys. Dokl. (English Transl.) **6** (1962) 747.

61W1	Waku, S., Masuno, K., Tanaka, T.: J. Phys. Soc. Jpn. **16** (1961) 1158.
61Y1	Yoshida, I., Sawada, S.: J. Phys. Soc. Jpn. **16** (1961) 2467.
61Z1	Zeldes, H., Livingston, R.: J. Chem. Phys. **35** (1961) 563.
62A1	Alexandrov, K. S., Ryabinskin, L. N.: Dokl. Akad. Nauk SSSR **142** (1962) 1298; Sov. Phys. Dokl. (English Transl.) **7** (1962) 99.
62A2	Anderson, A., Walmsley, S. H., Gebbie, H. A.: Philos. Mag. (8) **7** (1962) 1243.
62A3	Ando, R.: J. Phys. Soc. Jpn. **17** (1962) 937.
62A4	Asao, Y., Yoshida, I., Ando, R., Sawada, S.: J. Phys. Soc. Jpn. **17** (1962) 442.
62A5	Asch, P. G., Fontanel, M., Mesnard, G.: J. Phys. Radium **23** (1962) 1006.
62B4	Berlincourt, D. A., Curran, D. R.: Symp. Phys. and Nondestructive Testing, San Antonis, Texas (1962), Clevite Corporation, Elec. Res. Div., Cleveland, Ohio.
62B5	Bjorkstam, J. L., Jones, E. D., Silsbee, H. B., Uehling, E. A.: J. Chem. Phys. **37** (1962) 469.
62B8	Brock, E. G., Stripe, D., Hormats, E. I.: J. Chem. Phys. **37** (1962) 2735.
62B9	Brown, R. N., McLaren, A. C.: Acta Crystallogr. **15** (1962) 974.
62B11	Burns, G.: Phys. Rev. **127** (1962) 1193.
62C2	Cunningman, J., McMillan, J. A., Smaller, B., Yasaitis, E.: J. Phys. Chem. Solids **23** (1962) 167.
62D1	Davis, B. L., Adams, L. H.: Z. Kristallogr. **117** (1962) 399.
62F2	Fatuzzo, E.: J. Appl. Phys. **33** (1962) 2588.
62F3	Fatuzzo, E., Harbeke, G., Merz, W. J., Nitsche, R., Roetschi, H., Ruppel, W.: Phys. Rev. **127** (1962) 2036.
62F4	Fatuzzo, E., Nitsche, R., Roetschi, H., Zingg, S.: Phys. Rev. **125** (1962) 514.
62F8	Futama, H.: J. Phys. Soc. Jpn. **17** (1962) 434.
62F9	Futama, H., Pepinsky, R.: J. Phys. Soc. Jpn. **17** (1962) 725.
62G1	Garland, C. W., Park, K. C.: J. Appl. Phys. **33** (1962) 759.
62G2	Geller, S., Katz, H.: Bell Syst. Tech. J. **41** (1962) 425.
62G3	Gurevich, V. M.: Thesis of Candidate of Phys.-Math. Sciences, Moskva (1962) cited in [69g2].
62I2	Ikeda, T., Tanaka, Y., Toyoda, H.: Jpn. J. Appl. Phys. **1** (1962) 13.
62I5	Itoh, J., Kamiya, Y.: J. Phys. Soc. Jpn. **17** (1962) Suppl. B-I, 512.
62J1	Jaeger, R. E., Egerton, L.: J. Am. Ceram. Soc. **45** (1962) 209.
62K2	Kabalkina, S. A., Vereshchagin, L. F., Shulenin, B. M.: Dokl. Akad. Nauk SSSR **144** (1962) 1019; Sov. Phys. Dokl. (English Transl.) **7** (1962) 527.
62K3	Karpus, A., Mikalkevicius, M.: Liet. Fiz. Rinkinys **2** (1962) 151 (in Russian).
62K4	Kasatkin, A. P., Petrov, T. G., Treivus, E. B.: Kristallografiya **7** (1962) 952; Sov. Phys. Crystallogr. (English Transl.) **7** (1963) 770.
62K5	Kay, H. F., Dunn, J. W.: Philos. Mag. (8) **7** (1962) 2027.
62K6	Kay, M. I., Frazer, B. C., Ueda, R.: Acta Crystallogr. **15** (1962) 506.
62K8	Keller, C.: Z. Anorg. Allg. Chem. **318** (1962) 89.
62K9	Kern, R.: J. Phys. Chem. Solids **23** (1962) 249.
62K10	Kiriyama, H.: Bull. Chem. Soc. Jpn. **35** (1962) 1199.
62K11	Kiriyama, H.: Bull. Chem. Soc. Jpn. **35** (1962) 1205.
62L1	Leonidova, G. G., Polandov, I. N., Golentovskaya, I. P.: Fiz. Tverd. Tela **4** (1962) 3337; Sov. Phys. Solid State (English Transl.) **4** (1963) 2443.
62M1	Mac Chesney, J. B., Sauer, H. A.: J. Am. Ceram. Soc. **45** (1962) 416.
62M2	Makishima, S., Hasegawa, K., Shionoya, S.: J. Phys. Chem. Solids **23** (1962) 749.
62M3	Makita, Y., Pepinsky, R.: Bull. Am. Phys. Soc. (2) **7** (1962) 177.
62M5	Miller, S. R., Blinc, R., Brenman, M., Waugh, J. S.: Phys. Rev. **126** (1962) 528.
62M7	Minaeva, K. A.: Kristallografiya **7** (1962) 425; Sov. Phys. Crystallogr. **7** (1962) 335.
62M9	Myasnikova, T. P., Yatsenko, A. F.: Fiz. Tverd. Tela **4** (1962) 653; Sov. Phys. Solid State (English Transl.) **4** (1962) 475.
62P1	Pepinsky, R., Makita, Y.: Bull. Am. Phys. Soc. (2) **7** (1962) 241.
62P2	Pepinsky, R., Vedam, K., Lajzerowicz, J., Okaya, Y.: AD-285512, p. 1 (1962).
62S1	Sager, A., Miller, R. C.: Proc. Intern. Conf. Phys. Semiconductors, Exeter Eng. (1962) 653.
62S2	Savage, A., Miller, R. C.: Appl. Optics **1** (1962) 661.
62S4	Schmidt, V. H., Uehling, E. A.: Phys. Rev. **126** (1962) 447.
62S6	Shibuya, I., Hoshino, S.: Jpn. J. Appl. Phys. **1** (1962) 249.
62S7	Shinnaka, Y.: J. Phys. Soc. Jpn. **17** (1962) 820.
62S9	Singh, B.: Doctoral Dissertation, The Pennsylvania State University, **1962**.

62S11	Skalyo, J., Friedberg, S. A.: U. S. Dep. Commun. Office Tech. Serv. AD 285, 553, **1962**.
62S14	Stephenson, C. C., Lundell, O. R.: J. Phys. Chem. **66** (1962) 787.
62S15	Stewart, J. W.: J. Chem. Phys. **36** (1962) 400.
62W1	Weiss, A., Biedenkapp, D.: Z. Naturforsch. a **17** (1962) 794.
62W2	Wieder, H. H., Clawson, A. R., Parkerson, C. R.: J. Appl. Phys. **33** (1962) 1720.
62Z2	Zeldes, H.: Article on p. 764 in [6312].
62Z3	Zheludev, I. S., Lelekov, V. S.: Kristallografiya **7** (1962) 463; Sov. Phys. Crystallogr. (English Transl.) **7** (1962) 369.
63A1	Anderson, A., Gebbie, H. A., Walmsley, S. H.: Molec. Phys. **7** (1963) 401.
63B2	Berlincourt, D., Cook Jr., W. R., Rander, M. E.: Acta Crystallogr. **16** (1963) 163.
63B5	Bierly, J. N., Muldawer, L., Beckman, O.: Acta Metall. **11** (1963) 447.
63B7	Blinc, R., Zupančič, I.: J. Phys. Chem. Solids **24** (1963) 1379.
63B10	Boutin, H., Frazer, B. C., Jona, F.: J. Phys. Chem. Solids **24** (1963) 1341.
63C1	Chang, E., Westrum Jr., E. F.: Unpublished data cited in [63f1].
63C5	Cuthbert, J. D., Petch, H. E.: Can. J. Phys. **41** (1963) 1629.
63C6	Cleaver, B., Rhodes, E., Ubbelohde, A. R.: Proc. Roy. Soc. (London) Ser. A **276** (1963) 437.
63D1	Dantsiger, A. Ya., Fesenko, E. G.: Kristallografiya **8** (1963) 894; Sov. Phys. Crystallogr. (English Transl.) **8** (1964) 717.
63D2	Deshpande, V. T., Khan, A. A.: Acta Crystallogr. **16** (1963) 936.
63D3	De Waard, H., De Pasquali, G., Hafemeister, D.: Phys. Lett. **5** (1963) 217.
63G4	Gutiérrez, M., Canut, M. L., Amorós, J. L.: Acta Crystallogr. **16** (1963) Suppl. A166.
63H1	Hamano, K., Negishi, K., Marutake, M., Nomura, S.: Jpn. J. Appl. Phys. **2** (1963) 83.
63H2	Harbeke, G.: J. Phys. Chem. Solids **24** (1963) 957.
63H3	Hatta, I., Sawada, S., Asao, Y., Yanagi, T.: J. Phys. Soc. Jpn. **18** (1963) 1229.
63H5	Hill, R. M., Ichiki, S. K.: Phys. Rev. **130** (1963) 150.
63H6	Hill, R. M., Ichiki, S. K.: Phys. Rev. **132** (1963) 1603.
63H7	Hoshino, S., Sato, S., Toyoda, K.: Jpn. J. Appl. Phys. **2** (1963) 519.
63I2	Ikeda, T., Tanaka, Y., Toyoda, H.: Jpn. J. Appl. Phys. **2** (1963) 199.
63K5	Kobayashi, H., Haseda, T.: J. Phys. Soc. Jpn. **18** (1963) 541.
63K6	Kobayashi, J., Yamada, N.: J. Phys. Soc. Jpn. **18** (1963) 324.
63K7	Krueger, H. H. A., Cook Jr., W. R., Sartain, C. C., Yockey, H. P.: J. Appl. Phys. **34** (1963) 218.
63L1	LeDonche, L., LeMontagner, S.: C. R. Acad. Sci. (Paris) **256** (1963) 4406.
63L2	Leonidova, G. G.: Fiz. Tverd. Tela **5** (1963) 3430; Sov. Phys. Solid State (English Transl.) **5** (1964) 2519.
63M5	Myers, R. A.: Office of Naval Research, Contract No. 1866 (16); Tech. Rep. **433** (1963), Cruft Laboratory, Harvard University, Cambridge, Massachusetts.
63N1	Nakagawa, T., Sawada, S., Kawakubo, T., Nomura, S.: J. Phys. Soc. Jpn. **18** (1963) 1227.
63N2	Nakamura, E., Mitsui, T., Furuichi, J.: J. Phys. Soc. Jpn. **18** (1963) 1477.
63N3	Nakamura, T., Ohki, K.: J. Phys. Soc. Jpn. **18** (1963) 985.
63R2	Roth, R. S., Waring, J. L.: Am. Mineralogist **48** (1963) 1348.
63S4	Schmidt, V. H.: J. Chem. Phys. **38** (1963) 2783.
63S6	Schmidt, G., Pfannschmidt, P.: Phys. Status Solidi **3** (1963) 2215.
63S7	Shirokov, A. M., Shuvalov, L. A.: Kristallografiya **8** (1963) 733; Sov. Phys. Crystallogr. (English Transl.) **8** (1964) 586.
63S8	Sliker, T. R., Burlage, S. R.: J. Appl. Phys. **34** (1963) 1837.
63S10	Sonin, A. S., Zheludev, I. S.: Kristallografiya **8** (1963) 285; Sov. Phys. Crystallogr. (English Transl.) **8** (1963) 219.
63S14	Strukov, B. A., Danilycheva, M. N.: Fiz. Tverd. Tela **5** (1963) 1724; Sov. Phys. Solid State (English Transl.) **5** (1963) 1253.
63T1	Tanisaki, S.: J. Phys. Soc. Jpn. **18** (1963) 1181.
63T2	Toshev, S. D.: Kristallografiya **8** (1963) 680; Sov. Phys. Crystallogr. (English Transl.) **8** (1964) 541.
63U1	Udalova, V. V., Pinsker, Z. G.: Kristallografiya **8** (1963) 538; Sov. Phys. Crystallogr. (English Transl.) **8** (1964) 433.
63W1	Wachtman Jr., J. B., Wheat, M. L., Marzullo, S.: J. Res. Natl. Bur. Std., Sect. A **67** (1963) 193.
63W2	Wakuta, Y.: J. Phys. Soc. Jpn. **18** (1963) 671.
63W5	White, J. G.: Acta Crystallogr. **16** (1963) 397.
63W7	Wiseman, G. G., Kuebler, J. K.: Phys. Rev. **131** (1963) 2023.

63Y2	Yamada, Y., Shibuya, I., Hoshino, S.: J. Phys. Soc. Jpn. **18** (1963) 1594.
64A1	Abe, R.: Jpn. J. Appl. Phys. **3** (1964) 243.
64A2	Anderson, A., Gebbie, H. A., Walmsley, S. H.: Mol. Phys. **7** (1964) 401.
64A3	Arndt, R., Niggli, A.: Naturwissenschaften **51** (1964) 158.
64B2	Berlincourt, D., Curran, D. R., Jaffe, H.: Article on p. 169 of [64m2].
64B4	Berlincourt, D. A., Jaffe, H., Merz, W. J., Nitsche, R.: Appl. Phys. Lett. **4** (1964) 61.
64B5	Bis, R. F.: Solid State Commun. **2** (1964) 161.
64B6	Blinc, R., Petkovsek, J., Zupančič, I.: Phys. Rev. **136** (1964) A1684.
64B8	Brown, I. D.: Acta Crystallogr. **17** (1964) 654.
64B10	Buhrer, C. F., Ho, L.: Appl. Optics **3** (1964) 314.
64C1	Canut, M., Hoseman, R.: Acta Crystallogr. **17** (1964) 973.
64C2	Chiba, T.: J. Chem. Phys. **41** (1964) 1352.
64D2	Dengel, O., Eckener, U., Plitz, H., Riehl, N.: Phys. Lett. **9** (1964) 291.
64D3	Dork, R. A., Schubring, N. W., Nolta, J. P.: J. Appl. Phys. **35** (1964) 1984.
64F5	Flecher, R. O., Steeple, H.: Acta Crystallogr. **17** (1964) 290.
64F8	Futama, H., Tatsuzaki, I., Yokozawa, Y., Ito, K.: Proc. Meeting Phys. Soc. Jpn., Paper 9P-R-13, Nagoya **1964**.
64G2	Gcsi, K., Takagi, Y.: J. Phys. Soc. Jpn. **19** (1964) 632.
64G4	Gonzalo, J. A., Lopez-Alonso, J. R.: J. Phys. Chem. Solids **25** (1964) 303.
64H2	Hamano, K.: J. Phys. Soc. Jpn. **19** (1964) 945.
64H3	Haussühl, S.: Z. Kristallogr. **120** (1964) 401.
64H6	Hein, R. A., Gibson, J. W., Mazelsky, R., Miller, R. C., Hulm, J. K.: Phys. Rev. Lett. **12** (1964) 320.
64K1	Kafalas, J. K., Mariano, A. N.: Science **143** (1964) 952.
64K4	Kay, H. F., Dunn, J. W.: Philos. Mag. (8) **7** (1962) 2027.
64K5	Kiriyama, R., Kiriyama, H., Wada, T., Niizeki, N., Hirabayashi, H.: J. Phys. Soc. Jpn. **19** (1964) 540.
64K10	Kanzig, W., Hart Jr., H. R., Roberts, S.: Phys. Rev. Lett. **13** (1964) 543.
64L3	Livingston, R., Zeldes, H.: J. Chem. Phys. **41** (1964) 4011.
64M1	Makita, Y.: J. Phys. Soc. Jpn. **19** (1964) 576.
64M3	Miranda, S., Cases, J.: Rev. Chim. Miner. **1** (1964) 69.
64M4	Mitani, S.: J. Phys. Soc. Jpn. **19** (1964) 481.
64M5	Mori, T., Tamura, H.: J. Phys. Soc. Jpn. **19** (1964) 1247.
64M7	Murphy, E. J.: J. Appl. Phys. **35** (1964) 2609.
64N2	Niizeki, N., Koizumi, H.: J. Phys. Soc. Jpn. **19** (1964) 132.
64N3	Nitsche, R., Roetschi, H., Wild, P.: Appl. Phys. Lett. **4** (1964) 210.
64O1	O'Brien, E. J., Litovitz, T. A.: J. Appl. Phys. **35** (1964) 180.
64O2	Ott, J. H., Sliker, T. R.: J. Opt. Soc. Am. **54** (1964) 1442.
64P2	Peterson, E. R.: Doctor Thesis, Washington State Univ. (1964).
64S4	Santoro, R. P., Newnham, R. E.: J. Am. Ceram. Soc. **47** (1964) 491.
64S5	Sasaki, Y.: Jpn. J. Appl. Phys. **3** (1964) 558.
64S7	Sliker, T. R.: Engineering Memorandum 64-10 (1964), Clevite Corporation, Cleveland, Ohio.
64S8	Sliker, T. R.: J. Opt. Soc. Am. **54** (1964) 1348.
64S9	Smith, D. H., Cotts, R. M.: J. Chem. Phys. **41** (1964) 2403.
64S12	Strukov, B. A.: Fiz. Tverd. Tela **6** (1964) 2862; Sov. Phys. Solid State (English Transl.) **6** (1965) 2278.
64T3	Tateno, J., Gesi, K.: J. Chem. Phys. **40** (1964) 1317.
64V1	Vanderkooy, J., Cuthbert, J. D., Petch, H. E.: Can. J. Phys. **42** (1964) 1871.
64V2	Van der Ziel, J. P.: Appl. Phys. Lett. **5** (1964) 27.
64V3	Van der Ziel, J. P., Bloembergen, N.: Phys. Rev. **135** (1964) A1662.
64W2	Winchel, A. N., Winchel, H.: The Microscopic Character of Artificial Inorganic Solid Substances, 3rd Ed., New York, Academic Press, **1964**.
64Z1	Zernike Jr., F.: J. Opt. Soc. Am. **54** (1964) 1215.
65B2	Ballman, A. A.: J. Am. Ceram. Soc. **48** (1965) 112.
65B5	Binggeli, B., Fatuzzo, E.: J. Appl. Phys. **36** (1965) 1431.
65B6	Blinc, R., Jovanovic, A., Levstik, A., Prelesnik, A.: J. Phys. Chem. Solids **26** (1965) 1359.
65B9	Boutin, H., Safford, G. J.: Inelastic Neutron Scattering, International Atomic Energy Agency, Viena (1965); Proceedings of Symposium of Neutrons, 15–19 December 1964, Vol. II, p. 393.
65B11	Burke Jr., J. R., Allgaier, R. S., Houston Jr., B. B., Babiskin, J., Siebenmann, P. G.: Phys. Rev. Lett. **14** (1965) 360.

65B13	Bylander, E. G., Dixon, J. R., Riedl, H. R., Schoolar, R. B.: Phys. Rev. **138** (1965) A864.
65C2	Cardona, M.: Phys. Rev. **140** (1965) A651.
65C4	Chiba, T.: Bull. Chem. Soc. Jpn. **38** (1965) 490.
65D1	Dantsiger, A. Ya.: Fiz. Tverd. Tela **7** (1965) 2279; Sov. Phys. Solid State (English Transl.) **7** (1966) 1845.
65D2	Dantsiger, A. Ya., Freizon, I. A.: Fiz. Tverd. Tela **7** (1965) 1517; Sov. Phys. Solid State (English Transl.) **7** (1965) 1218.
65D3	Dantsiger, A. Ya., Freizon, I. A.: Fiz. Tverd. Tela **7** (1965) 2190; Sov. Phys. Solid State (English Transl.) **7** (1966) 1762.
65D4	Deshpande, V. T., Khan, A. A.: Acta Crystallogr. **18** (1965) 977.
65D7	Dantsiger, A. Ya.: Izv. Akad. Nauk SSSR, Ser. Fiz. **29** (1965) 1042; Bull. Acad. Sci. USSR, Phys. Ser. **29** (1965) 1043.
65E1	Eisuer, I. Ya.: Izv. Akad. Nauk SSSR, Ser. Fiz. **29** (1965) 2014; Bull. Acad. Sci. USSR, Phys. Ser (English Transl.) **29** (1965) 1848.
65G2	Gesi, K.: Jpn. J. Appl. Phys. **4** (1965) 818.
65G3	Gesi, K., Ozawa, K., Takagi, Y.: J. Phys. Soc. Jpn. **20** (1965) 1773.
65G6	Gulyamov, K., Lyakhovitakaya, V. A., Tikhomirova, N. A., Fridkin, V. M.: Dokl. Akad. Nauk SSSR **161** (1965) 1060; Sov. Phys. Dokl (English Transl.) **10** (1965) 331.
65H1	Hainsworth, F. N., Petch, H. E.: Solid State Commun. **3** (1965) 315.
65H2	Hamano, K., Nakamura, T., Ishibashi, Y., Ooyane, T.: J. Phys. Soc. Jpn. **20** (1965) 1886.
65H4	Hatta, I., Sawada, S.: Jpn. J. Appl. Phys. **4** (1965) 389.
65H5	Haussühl, V. S.: Acta Crystallogr. **18** (1965) 841.
65H9	Horak, J., Cermak, K.: Czech J. Phys. B **15** (1965) 536.
65I5	Isupov, V. A.: Fiz. Tverd. Tela **7** (1965) 2221; Sov. Phys. Solid State (English Transl.) **7** (1966) 1789.
65I6	Iwasaki, H.: J. Phys. Soc. Jpn. **20** (1965) 1783.
65K2	Kaminow, I. P.: Phys. Rev. **138** (1965) A1539.
65K3	Kapustin, A. P., Vistin', L. K.: Kristallografiya **10** (1965) 118; Sov. Phys. Crystallogr. (English Transl.) **10** (1965) 95.
65K4	Kawabe, U., Yanagi, T., Sawada, S.: J. Phys. Soc. Jpn. **20** (1965) 2059.
65K5	Khanna, R. K., Decius, J. C., Lippincott, E. R.: J. Chem. Phys. **43** (1965) 2974.
65K6	Krishnan, R. S., Krishnan, K.: Proc. Indian Acad. Sci., A **61** (1965) 122.
65K7	Kobayashi, J.: Butsuri (Proc. Phys. Soc. Jpn.) **20** (1965) 203.
65L1	LeMontagner, S., LeDonche, L., Gallier, J.: Bull. Soc. Sci. Bretagne **38** (1965) 159.
65L2	Losche, A., Windsch, W.: Phys. Status Solidi **11** (1965) K55.
65M2	Makita, Y.: J. Phys. Soc. Jpn. **20** (1965) 1567.
65M3	Makita, Y.: J. Phys. Soc. Jpn. **20** (1965) 2073.
65M4	Makita, Y., Sekido, T.: J. Phys. Soc. Jpn. **20** (1965) 954.
65M5	Maruyama, N., Sawada, S.: J. Phys. Soc. Jpn. **20** (1965) 811.
65M6	Minaeva, K. A., Levanyuk, A. P.: Izv. Akad. Nauk SSSR, Ser. Fiz. **29** (1965) 978; Bull. Acad. Sci. USSR, Phys. Ser. (English Transl.) **29** (1965) 978.
65M8	Mori, T., Tamura, H., Sawaguchi, E.: J. Phys. Soc. Jpn. **20** (1965) 281.
65N2	Nakamura, N., Suga, H., Chihara, H., Seki, S.: Bull. Chem. Soc. Jpn. **38** (1965) 1779.
65N3	Nardelli, M., Fava, G., Giraldi, G.: Acta Crystallogr. **19** (1965) 1038.
65O1	Obenshain, F. E., Roberts, L. D., Coleman, C. F., Forester, D. W.: Phys. Rev. Lett. **14** (1965) 365.
65O2	Okada, K.: Phys. Rev. Lett. **15** (1965) 252.
65O3	O'Reilly, D. E., Schacher, G. E.: J. Chem. Phys. **43** (1965) 4222.
65O5	Owens, B. B.: J. Chem. Phys. **42** (1965) 2259.
65P1	Pakhomov, V. I., Gerken, V. A.: Izv. Akad. Nauk SSSR, Ser. Fiz. **29** (1965) 901; Bull. Acad. Sci. USSR, Phys. Ser. (English Transl.) **29** (1965) 907.
65R1	Rapoport, E., Kennedy, G. C.: J. Phys. Chem. Solids **26** (1965) 1995.
65R2	Riedl, H. R., Dixon, J. R., Schoolar, R. B.: Solid State Commun. **3** (1965) 323.
65S1	Sakiyama, M., Kimoto, A., Seki, S.: J. Phys. Soc. Jpn. **20** (1965) 2180.
65S2	Samara, G. A.: J. Phys. Chem. Solids **26** (1965) 121.
65S5	Sasaki, Y.: Jpn. J. Appl. Phys. **4** (1965) 228.
65S6	Sasaki, Y.: Jpn. J. Appl. Phys. **4** (1965) 614.
65S7	Sawada, S.: Butsuri (Proc. Phys. Soc. Jpn.) **20** (1965) 188.
65S16	Skapsi, A. C., Rogers, D.: Chem. Commun. **1965,** 611.

65T5	Tomashpol'skii, Yu. Ya., Venevtsev, Yu. N.: Fiz. Tverd. Tela **7** (1965) 3126; Sov. Phys. Solid State (English Transl.) **7** (1966) 2529.
65T7	Tsang, T., O'Reilly, D. E.: J. Chem.Phys. **43** (1965) 4234.
65U3	Unruh, H. G.: Phys. Lett. **17** (1965) 8.
65V1	Veselc, J., Sroubek, Z., Zdansky, K.: Phys. Status Solidi **10** (1965) K99.
65V2	Vishnevskii, V. N., Romanyuk, N. A., Stefanskii, I. V.: Opt. Spektrosk. **18** (1965) 838; Opt. Spectrosc. (English Transl.) **18** (1965) 468.
65W1	Wang, C. C., Racette, G. W.: J. Appl. Phys. **36** (1965) 3281.
65Y1	Yanagi, T.: J. Phys. Soc. Jpn. **20** (1965) 1351.
65Z1	Zheludev, I. S., Ludupov, Ts.-Zh.: Kristallografiya **10** (1965) 764; Sov. Phys. Crystallogr. (English Transl.) **10** (1966) 645.
66A3	Alekseeva, V. G., Landsberg, E. G.: Fiz. Tverd. Tela **8** (1966) 3138; Sov. Phys. Solid State (English Transl.) **8** (1967) 2518.
66A5	Anistratov, A. T.: Kristallografiya **11** (1966) 823; Sov. Phys. Crystallogr. (English Transl.) **11** (1967) 700.
66B5	Blinc, R., Lahajnar, G., Pintar, M., Zupančič, I.: J. Chem. Phys. **44** (1966) 1784.
66B6	Blinc, R., Uehling, A.: Phys. Lett. **20** (1966) 337.
66B8	Brezina, B., Smutny, F.: Article on p. 182 in Vol. I of [66d1].
66C4	Craig, P. P.: Phys. Lett. **20** (1966) 140.
66D1	Devanarayanan, S., Easwaran, K. R. K.: Proc. Indian Acad. Sci., A **64** (1966) 173.
66D2	Deslattes, R. D., Torgesen, J. L., Paretzkin, B., Horton, A. T.: J. Appl. Phys. **37** (1966) 541.
66E1	Easwaran, K. R. K.: J. Phys. Soc. Jpn. **21** (1966) 1614.
66E2	Easwaran, K. R. K.: J. Chem. Phys. **45** (1966) 403.
66F1	Fort Worth Division of General Dynamics: Res. Summary, Appl. Res. Program, Jan.–June 1966, APR-10 (1966) 84, cited in [69B1].
66G1	Gammon, R., Cummins, H. Z.: Phys. Rev. Lett. **17** (1966) 193.
66G2	Goldak, J., Barrett, C. B., Innes, D., Youdelis, W.: J. Chem. Phys. **44** (1966) 3323.
66G3	Gonzalo, J. A.: Phys. Rev. **144** (1966) 662.
66G5	Guyon, P., Lajzerowicz: Phys. Status Solidi **16** (1966) 525.
66H1	Hainsworth, F. N., Petch, H. E.: Can. J. Phys. **44** (1966) 3083.
66H2	Hamano, K.: Jpn. J. Appl. Phys. **5** (1966) 865.
66H3	Hampton, M., Herring, F. G., Lin, W. C., McDowell, C. A.: Mol. Phys. **10** (1966) 565.
66H5	Harris, L. B., Vella, G. J.: J. Appl. Phys. **37** (1966) 4294.
66H9	Havraneova, V., Horak, J.: Collect. Czech. Chem. Commun. **31** (1966) 1256.
66H11	Hein, R. A.: Phys. Lett. **23** (1966) 435.
66H12	Hein, R. A., Gibson, J. W., Falge Jr., R. L., Mazelsky, R., Miller, R. C., Hulm, J. K.: J. Phys. Soc. Jpn. **21** (1966) Suppl. 643.
66H13	Hirotsu, S.: Master Thesis, Tokyo Institute of Technology (1966).
66I1	Ichikawa, M.: Master Thesis, Hokkaido University (1966).
66I2	Ichikawa, M., Mitsui, T.: Phys. Rev. **152** (1966) 495.
66I4	Imai, K., Kawada, S., Ida, M.: J. Phys. Soc. Jpn. **21** (1966) 1855.
66I5	Ivanov, N. R., Shuvalov, L. A.: Kristallografiya **11** (1966) 614; Sov. Phys. Crystallogr. (English Transl.) **11** (1967) 534.
66I6	Ivanov, N. R., Zotov, V. F.: Kristallografiya **11** (1966) 924; Sov. Phys. Crystallogr. (English Transl.) **11** (1967) 781.
66I7	Iwata, Y., Fukui, S., Koyano, N., Shibuya, I.: J. Phys. Soc. Jpn. **21** (1966) 1846.
66K4	Khanna, R. K., Horak, M., Lippincott, E. R.: J. Chem. Phys. **45** (1966) 982.
66K6	Kohin, R. P., Nadeau, P. G.: J. Chem. Phys. **44** (1966) 691.
66K7	Krainik, N. N., Popov, S. N., Myl'nikova, I. E.: Fiz. Tverd. Tela **8** (1966) 3664; Sov. Phys. Solid State (English Transl.) **8** (1967) 2933.
66K8	Krasnikova, A. Ya., Polandov, I. N., Mylov, V. P.: Fiz. Tverd. Tela **8** (1966) 3320; Sov. Phys. Solid State (English Transl.) **8** (1967) 2653.
66K10	Koldobskaya, M. F., Meleshina, V. A., Rez, I. S., Mironova, Z. A., Gavrilova, I. V., Safronov, G. M.: Kristallografiya **11** (1966) 785; Sov. Phys. Crystallogr. (English Transl.) **11** (1967) 665.
66L3	Look, D. C., Lowe, I. J., Northby, J. A.: J. Chem. Phys. **44** (1966) 3441.
66M2	Miller, N. C., Casabella, P. A.: Phys. Rev. **152** (1966) 228.
66M4	Mnatsakanyan, A. V., Shuvalov, L. A., Zheludev, I. S., Gavrilova, I. V.: Kristallografiya **11** (1966) 464; Sov. Phys. Crystallogr. (English Transl.) **11** (1966) 412.

66N4	Nosov, V. N., Fridkin, V. M.: Fiz. Tverd. Tela **8** (1966) 148; Sov. Phys. Solid State (English Transl.) **8** (1966) 113.
66O1	Ohshima, H., Nakamura, E.: J. Phys. Chem. Solids **27** (1966) 481.
66O2	Okada, K., Kay, M. I., Cromer, D. T., Almodovar, I.: J. Chem. Phys. **44** (1966) 1648.
66O3	Ostroff, E. D., Waugh, J. S.: Phys. Rev. Lett. **16** (1966) 1097.
66P1	Pawley, G. S., Cochran, W., Cowley, R. A., Dolling, G.: Phys. Rev. Lett. **17** (1966) 753.
66P3	Philips, R. A.: J. Opt. Soc. Am. **56** (1966) 629.
66R1	Rapoport, E.: J. Phys. Chem. Solids **27** (1966) 1349.
66R2	Rapoport, E.: J. Chem. Phys. **45** (1966) 2721.
66R7	Rush, J. J., Leung, P. S., Taylor, T. I.: J. Chem. Phys. **45** (1966) 1312.
66S1	Safrankova, M., Fousek, J., Kaczer, J.: Article on p. 51 in Vol. II of [66d1].
66S5	Samara, G. A., Anderson, D. H.: Solid State Commun. **4** (1966) 653.
66S6	Sato, Y.: J. Chem. Phys. **45** (1966) 275.
66S9	Sawaguchi, E., Mori, T.: J. Phys. Soc. Jpn. **21** (1966) 2077.
66S13	Schlemper, E. O., Hamilton, W. C.: J. Chem. Phys. **44** (1966) 4498.
66S17	Shuvalov, L. A., Mnatsakanyan, A. V.: Kristallografiya **11** (1966) 222; Sov. Phys. Crystallogr. (English Transl.) **11** (1966) 210.
66S22	Stankowski, J.: Article on p. 364 in Vol. II of [66d1].
66S23	Strukov, B. A.: Kristallografiya **11** (1966) 892; Sov. Phys. Crystallogr. (English Transl.) **11** (1967) 757.
66S25	Strukov, B. A., Taraskin, S. A., Koptsik, V. A.: Zh. Eksp. Teor. Fiz. **51** (1966) 1037; Sov. Phys. JETP (English Transl.) **24** (1967) 692.
66S26	Suzuki, S., Takagi, M.: J. Phys. Soc. Jpn. **21** (1966) 554.
66T1	Takagi, Y., Gesi, K.: Jpn. J. Appl. Phys. **5** (1966) 1118.
66T5	Tatsuzaki, I., Itoh, K., Ueda, S., Shindo, Y.: Phys. Rev. Lett. **17** (1966) 198.
66V2	Vasilevskaya, A. S.: Kristallografiya **11** (1966) 755; Sov. Phys. Crystallogr. (English Transl.) **11** (1967) 644.
66V3	Vasilevskaya, A. S., Sonin, A. S.: Fiz. Tverd. Tela **8** (1966) 3434; Sov. Phys. Solid State (English Transl.) **8** (1967) 2756.
66V4	Vasilevskaya, A. S., Sonin, A. S., Slepkov, I. A.: Kristallografiya **11** (1966) 749; Sov. Phys. Crystallogr. (English Transl.) **11** (1967) 640.
66V5	Vishnevskii, V. N., Stefanskii, I. V.: Opt. Spektrosk. **20** (1966) 357; Opt. Spectrosc. (English Transl.) **20** (1966) 195.
66W3	Warrier, A. V. R., Narayanan, P. S.: Proc. Indian Acad. Sci., A **64** (1966) 254.
66W4	Wartewig, S., Windsch, W.: Phys. Status Solidi **17** (1966) 163.
66W7	Wieder, H. H., Parkerson, C. R.: J. Phys. Chem. Solids **27** (1966) 247.
66W9	Williams, R., Heilmeier, G.: J. Chem. Phys. **44** (1966) 638.
66Y1	Yamazaki, M., Ogawa, T.: J. Opt. Soc. Am. **56** (1966) 1407.
67A4	Aleksandrova, I. P., Zherebtsova, L. I., Aleksandrov, K. S.: Kristallografiya **12** (1967) 894; Sov. Phys. Crystallogr. (English Transl.) **12** (1968) 780.
67A5	Anistratov, A. T., Aleksandrov, K. S.: Kristallografiya **12** (1967) 459; Sov. Phys. Crystallogr. (English Transl.) **12** (1967) 392.
67A6	Arnold, G. M., Heastie, R.: Chem. Phys. Lett. **1** (1967) 51.
67A7	Asch, G., Michel, R.: Jpn. J. Appl. Phys. **6** (1967) 1008.
67B9	Bjorkstam, J. L.: Phys. Rev. **153** (1967) 599.
67B10	Bjorkstam, J. L., Willmorth, J. H.: Article on p. 728 in [67b1].
67B11	Blinc, R., Cevc, P., Schara, M.: Phys. Rev. **159** (1967) 411.
67B12	Blinc, R., Pintar, M., Zupančič, I.: J. Phys. Chem. Solids **28** (1967) 405.
67B13	Blokh, O. G., Lutsiv-Shumskii, L. F.: Kristallografiya **12** (1967) 455; Sov. Phys. Crystallogr. (English Transl.) **12** (1967) 389.
67C2	Cevc, P., Modesto, M., Schara, M.: J. Cryst. Growth **1** (1967) 160.
67C3	Chen, A., Chernow, F.: Phys. Rev. **154** (1967) 493.
67C5	Cook Jr., W. R.: J. Appl. Phys. **38** (1967) 1637.
67C6	Cook Jr., W. R.: J. Inorg. Nucl. Chem. **29** (1967) 844.
67C7	Corazza, E., Sabelli, C., Giuseppetti, G.: Acta Crystallogr. **22** (1967) 683.
67C8	Cousseins, J. C., Samouel, M.: C. R. Acad. Sci. (Paris) C **265** (1967) 1121.
67C9	Cubiotti, G., Geracitano, R.: Phys. Lett. A **24** (1967) 179.

67C11	Chisler, E. V., Shur, M. S.: Fiz. Tverd. Tela **9** (1967) 1015; Sov. Phys. Solid State (English Transl.) **9** (1967) 796.
67D2	Dixon, R. W.: J. Appl. Phys. **38** (1967) 5149.
67E1	Ellison, P. R., Troup, G. J.: Proc. Phys. Soc. London **92** (1967) 1040.
67F3	Frova, A., Boddy, J., Chen, Y. S.: Phys. Rev. **157** (1967) 700.
67F4	Fukunishi, S., Kittaka, S., Makita, Y., Waku, S.: J. Phys. Soc. Jpn. **23** (1967) 1181.
67F7	Futama, H., Shiozaki, Y., Chiba, A., Tanaka, E., Mitsui, T., Furuichi, J.: Phys. Lett. A **25** (1967) 8.
67F8	Fermor, J. H., Kjekshus, A.: Acta Chem. Scand. **21** (1967) 1265.
67G2	Gavrilova-Poldol'skaya, G. V., Gabuda, S. P., Lundin, A. G.: Fiz. Tverd. Tela **9** (1967) 1167; Sov. Phys. Solid State (English Transl.) **9** (1967) 911.
67G3	Gerken, V. A., Pakhomov, V. I., Sonin, A. S., Marchenko, E. I., Nepomnyashchaya, V. N.: Kristallografiya **12** (1967) 607; Sov. Phys. Crystallogr. (English Transl.) **12** (1968) 531.
67G5	Gesi, K.: Jpn. J. Appl. Phys. **6** (1967) 781.
67G6	Gesi, K.: Doctor Thesis, Tokyo Institute of Technology (1967).
67G9	Giess, E. A., Scott, B. A., Burns, G., O'Kane, D. F., Segmuller, A.: J. Am. Ceram. Soc. **52** (1969) 276.
67G10	Gigas, B. P.: Fiz. Tverd. Tela **9** (1967) 2430; Sov. Phys. Solid State (English Transl.) **9** (1968) 1903.
67G12	Groenewegen, P. P. M., Cole, R. H.: J. Chem. Phys. **46** (1967) 1069.
67H1	Haas, W., Johannes, R.: Appl. Opt. **6** (1967) 2007.
67H8	Holmberg, R. W., Livingston, R.: J. Chem. Phys. **47** (1967) 2552.
67H10	Hoshino, S., Motegi, H.: Jpn. J. Appl. Phys. **6** (1967) 708.
67H11	Hoshino, S., Shimaoka, K., Niimura, N.: Phys. Rev. Lett **19** (1967) 1286.
67I1	Ismailzade, I. G., Nesterenko, V. I.: Kristallografiya **12** (1967) 717; Sov. Phys. Crystallogr. (English Transl.) **12** (1968) 625.
67J1	Johannes, R., Haas, W.: Appl. Opt. **6** (1967) 1059.
67K1	Kabalkina, S. S., Serebryanaya, N. R., Vereshchagin, L. F.: Fiz. Tverd. Tela **9** (1967) 3208; Sov. Phys. Solid State (English Transl.) **9** (1968) 2527.
67K4	Kikuchi, A., Oka, Y., Sawaguchi, E.: J. Phys. Soc. Jpn. 23 (1967) 337.
67K5	Knispel, R. R., Schmidt, V. H.: Bull. Am. Phys. Soc. (2) **12** (1967) 290.
67K6	Kremenchugskii, L. S., Samoilov, V. B.: Kristallografiya **12** (1967) 1077; Sov. Phys. Crystallogr. (English Transl.) **12** (1968) 940.
67K8	Kunitomo, M., Terao, T., Tsutsumi, Y., Hashi, T.: J. Phys. Soc. Jpn. **22** (1967) 945.
67L1	Labbe, J.: Phys. Rev. **158** (1967) 647.
67M1	Makita, Y., Tsuchiya, T., Yamagata, K.: J. Phys. Soc. Jpn. **22** (1967) 938.
67M4	Maruyama, H., Tomiie, Y., Mizutani, I., Yamazaki, Y., Kobayashi, J.: J. Phys. Soc. Jpn. **23** (1967) 899.
67M6	Minaeva, K. A., Levanyuk, A. P., Strukov, B. A., Koptsik, V. A.: Fiz. Tverd. Tela **9** (1967) 1220; Sov. Phys. Solid State (English Transl.) **9** (1967) 950.
67M7	Mizutani, I., Yamazaki, Y., Uesu, Y., Yamada, N., Kobayashi, J., Maruyama, H., Tomiie, Y.: J. Phys. Soc. Jpn. **23** (1967) 900.
67M8	Mochizuki, J., Futama, H.: J. Phys. Soc. Jpn. **23** (1967) 127.
67M9	Mylov, V. P., Polandov, I. N., Strukov, B. A.: Fiz. Tverd. Tela **9** (1967) 3012; Sov. Phys. Solid State (English Transl.) **9** (1968) 2375.
67N1	Nakamura, E., Hosoya, M.: J. Phys. Soc. Jpn. **23** (1967) 844.
67O1	Ohm, E. A.: Appl. Opt. **6** (1967) 1233.
67O2	Okada, K.: Phys. Rev. **164** (1967) 683.
67O3	Okada, K., Gonzalo, J. A., Rivera, J. M.: J. Phys. Chem. Solids **28** (1967) 689.
67O4	O'Keeffe, M., Perrino, C. T.: J. Phys. Chem. Solids **28** (1967) 211.
67O5	O'Keeffe, M., Perrino, C. T.: J. Phys. Chem. Solids **28** (1967) 1086.
67O6	O'Reilly, D. E., Tsang, T.: J. Chem. Phys. **46** (1967) 1291.
67O7	O'Reilly, D. E., Tsang, T.: J. Chem. Phys. **47** (1967) 4072.
67R1	Reese, W., May, L. F.: Phys. Rev. **162** (1967) 510.
67R3	Rinkyavichyus, V. S., Mikalkevichyus, M. P.: Fiz. Tverd. Tela **9** (1967) 2997; Sov. Phys. Solid State (English Transl.) **9** (1968) 2360.
67R5	Rosner, R. D., Turner, E. H., Kaminow, I. P.: Appl. Opt. **6** (1967) 778.
67S2	Sándor, E., Farrow, R. F. C.: Nature **213** (1967) 171.
67S3	Sándor, E., Farrow, R. F. C.: Nature **215** (1967) 1265.

67S4	Sartene, R., Parodi, O.: J. Phys. (Paris) **28** (1967) 301.
67S5	Sawada, A., Abe, R.: Jpn. J. Appl. Phys. **6** (1967) 659.
67S6	Schacher, G. E.: J. Chem. Phys. **46** (1967) 3565.
67S7	Schein, B. J. B., Lingafelter, E. C., Stewart, J. M.: J. Chem. Phys. **47** (1967) 5183.
67S8	Schmid, H., Trooster, J. M.: Solid State Commun. **5** (1967) 31.
67S10	Sekido, T., Mitsui, T.: J. Phys. Chem. Solids **28** (1967) 967.
67S11	Shapiro, S. M., Gammon, R. W., Cummins, H. Z.: Appl. Phys. Lett. **10** (1967) 113.
67S12	Shiozaki, Y.: J. Phys. Chem. Solids **28** (1967) 2381.
67S14	Shuvalov, L. A., Ivanov, N. R., Sitnik, T. K.: Kristallografiya **12** (1967) 366; Sov. Phys. Crystallogr. (English Transl.) **12** (1967) 315.
67S15	Shuvalov, L. A., Ivanov, N. R.: Phys. Status Solidi **22** (1967) 279.
67S17	Sljukic, M., Matkovic, B., Prodic, B., Scavnicar, S.: Croat. Chem. Acta **39** (1967) 145.
67S19	Sonin, A. S., Suvarov, V. S.: Fiz. Tverd. Tela **9** (1967) 1829; Sov. Phys. Solid State (English Transl.) **9** (1967) 1437.
67S20	Sonin, A. S., Vasilevskaya, A. S.: Kristallografiya **12** (1967) 451; Sov. Phys. Crystallogr. (English Transl.) **12** (1967) 386.
67S21	Sonin, A. S., Vasilevskaya, A. S.: Fiz. Tverd. Tela **9** (1967) 1951; Sov. Phys. Solid State (English Transl.) **9** (1968) 1536.
67S24	Stankowski, J.: Phys. Status Solidi **24** (1967) 451.
67S26	Strukov, B. A., Amin, M., Sonin, A. S.: Fiz. Tverd. Tela **9** (1967) 2421; Sov. Phys. Solid State (English Transl.) **9** (1968) 1894.
67S27	Şuemune, Y.: J. Phys. Soc. Jpn. **22** (1967) 735.
67S28	Sugai, S., Mochizuki, J., Futama, H.: Unpublished work (1967).
67S29	Suzuki, I., Abe, R.: J. Phys. Soc. Jpn. **23** (1967) 1049.
67T1	Takagi, Y., Gesi, K.: J. Phys. Soc. Jpn. **22** (1967) 979.
67T2	Takahashi, H., Schrader, B., Meier, W., Gottlieb, K.: J. Chem. Phys. **47** (1967) 3842.
67T3	Takama, T., Mitsui, T.: J. Phys. Soc. Jpn. **23** (1967) 331.
67T4	Takeda, T., Watanabe, A.: J. Phys. Soc. Jpn. **23** (1967) 469.
67T6	Truter, M. R.: Acta Crystallogr. **22** (1967) 556.
67T7	Tsu, R., Howard, W. E., Esaki, L.: Solid State Commun. **5** (1967) 167.
67T8	Tuberfield, K. C.: Solid State Commun. **5** (1967) 587.
67U1	Ueda, R., Ono, M., Tokugawa, F.: J. Phys. Chem. Solids Suppl. **1** (1967) 381 (Proceedings Intern. Conf. on Crystal Growth, held at Boston, 1966).
67U3	Unruh, H.-G., Muser, H. E.: Ann. Phys. (Leipzig) (7) **19** (1967) 28.
67V2	Vasilevskaya, A. S., Perfilova, V. E., Sonin, A. S.: Izv. Akad. Nauk SSSR, Ser. Fiz. **31** (1967) 1132; Bull. Acad. Sci. USSR, Phys. Ser. (English Transl.) **31** (1967) 1149.
67V3	Vasilevskaya, A. S., Sonin, A. S., Rez, I. S.: Fiz. Tverd. Tela **9** (1967) 1260; Sov. Phys. Solid State (Engl. Transl.) **9** (1967) 986.
67V4	Vasilevskaya, A. S., Volkova, E. N., Koptsik, V. A., Rshkovich, L. N., Regul'skaya, T. A., Rez, I. S., Sonin, A. S., Suvorov, V. S.: Kristallografiya **12** (1967) 518; Sov. Phys. Crystallogr. (English Transl.) **12** (1967) 446.
67V5	Verkhovskaya, K. A., Sonin, A. S.: Zh. Eksp. Teor. Fiz. **52** (1967) 383; Sov. Phys. JETP (English Transl.) **25** (1967) 249.
67W5	Weirauch, D. F., Tennery, V. J.: Am. Ceram. Soc. **50** (1967) 671.
67W6	Wiesner, J. R., Srivastava, R. C., Kennard, C. H. L., DiVaira, M., Lingafelter, E. C.: Acta Crystallogr. **23** (1967) 565.
67W7	Windsch, W., Welter, M.: Z. Naturforsch. **22**a (1967) 1.
67W8	Worlock, J. M., Fleury, P. A.: Phys. Rev. Lett. **19** (1967) 1176.
68A1	Adhav, R. S.: J. Acoust. Soc. Am. **43** (1968) 835.
68A2	Adhav, R. S.: J. Appl. Phys. **39** (1968) 4095.
68A3	Amin, M., Strukov, B. A.: Fiz. Tverd. Tela **10** (1968) 3158; Sov. Phys. Solid State (English Transl.) **10** (1969) 2498.
68A4	Ashworth, T., Steeple, H.: Cryogenics **8** (1968) 225.
68A6	Axe, J. D.: Phys. Rev. **167** (1968) 573.
68A7	Azoulay, J., Grinberg, Y., Pelah, I., Wiener, E.: J. Phys. Chem. Solids **29** (1968) 843.
68B1	Balkanski, M., Teng, M. K., Nusimovici, M.: Phys. Rev. **176** (1968) 1098.
68B4	Beezhold, W., Uehling, E. A.: Phys. Rev. **175** (1968) 624.

68 B 8	Blinc, R., Dimic, V., Kolar, D., Lahajnar, G., Stepišnik, J., Žumer, S., Vene, N., Hadzi, D.: J. Chem. Phys. **49** (1968) 4996.
68 B 9	Blinc, R., Jamsek, M., Stepišnik, J., Trontelj, Z., Bjorkstam, J. L.: Solid State Commun. **6** (1968) 821.
68 B 10	Blinc, R., Stepišnik, J., Zupančič, I.: Phys. Rev. **176** (1968) 732.
68 B 11	Blinc, R., Žumer, S.: Phys. Rev. Lett. **21** (1968) 1004.
68 B 12	Brezina, B., Galanov, E. K., Ivanov, N. R., Kislovskii, L. D., Shuvalov, L. A.: Kristallografiya **13** (1968) 821; Sov. Phys. Crystallogr. (English Transl.) **13** (1969) 710.
68 B 13	Brezina, B., Smutny, F.: Czech. J. Phys. B **18** (1968) 393.
68 B 14	Brody, E. M., Cummins, H. Z.: Phys. Rev. Lett. **21** (1968) 1263.
68 B 16	Bukshpan, S.: Solid State Commun. **6** (1968) 477.
68 B 20	Barnoski, M. K., Ballantyne, J. M.: Phys. Rev. **174** (1968) 946.
68 C 2	Chaddha, G.: J. Phys. Soc. Jpn. **24** (1968) 976.
68 C 3	Chincholkar, V. S., Unruh, H. G.: Phys. Status Solidi **29** (1968) 669.
68 D 2	Davis, T. A., Vedam, K.: J. Opt. Soc. Am. **58** (1968) 1446.
68 E 1	Eibschutz, M., Guggenheim, H. J.: Solid State Commun. **6** (1968) 737.
68 E 2	Elcombe, M. M., Taylor, J. C.: Acta Crystallogr. A **24** (1968) 410.
68 E 3	Emmenegger, F., Nitsche, R., Miller, A.: J. Appl. Phys. **39** (1968) 3039.
68 F 1	Fermor, J. H., Kjekshus, A.: Acta Chem. Scand. **22** (1968) 836.
68 F 4	Forsch, K., Muser, H. E.: Z. Naturforsch. a **23** (1968) 1231.
68 G 1	Genin, D. J., O'Reilly, D. E., Peterson, E. M., Tsang, T.: J. Chem. Phys. **48** (1968) 4525; erratum **50** (1969) 2786.
68 G 2	Genin, D. J., O'Reilly, D. E., Tsang, T.: Phys. Rev. **167** (1968) 445.
68 G 3	Gerzanich, E. I., Bryzgalov, I. A., Rakcheev, A. D., Lyakhovitskaya, V. A.: Kristallografiya **13** (1968) 898; Sov. Phys. Crystallogr. (English Transl.) **13** (1969) 776.
68 G 4	Gerzanich, E. I., Fridkin, V. M.: Fiz. Tverd. Tela **10** (1968) 3111; Sov. Phys. Solid State (English Transl.) **10** (1969) 2452.
68 G 6	Gill, D.: Phys. Lett. A **28** (1968) 234.
68 G 8	Gränicher, H., Kind, R., Meier, W. M., Petter, W.: Helv. Phys. Acta **41** (1968) 843.
68 H 1	Hatta, I.: J. Phys. Soc. Jpn. **24** (1968) 1043.
68 H 2	Horak, J., Turjanica, I. D., Klazar, J., Kozakova, M.: Krist. Tech. **3** (1968) 241.
68 I 2	Iwasaki, H., Toyoda, H., Ohtsuka, M.: Jpn. J. Appl. Phys. **7** (1968) 787.
68 K 1	Kaminov, I. P., Damen, T. C.: Phys. Rev. Lett. **20** (1968) 1105.
68 K 2	Kawada, S.: J. Phys. Soc. Jpn. **25** (1968) 919.
68 L 2	Ligenza, S., Kisyńska, K., Kopcewicz, M.: Nukleonika **13** (1968) 1133.
68 L 3	Lomov, L. G., Sonin, A. S., Regul'skaya, T. A.: Kristallografiya **13** (1968) 90; Sov. Phys. Crystallogr. (English Transl.) **13** (1968) 68.
68 L 4	Ludupov, Ts. Zh., Fiala, I., Zheludev, I. S.: Kristallografiya **13** (1968) 442; Sov. Phys. Crystallogr. (English Transl.) **13** (1968) 359.
68 L 5	Ledsham, A. H. C., Steeple, H.: Acta Crystallogr. B **24** (1968) 320.
68 M 1	Malek, Z., Strajblova, J., Novotny, J., Marecek, V.: Czech. J. Phys. B **18** (1968) 1224.
68 M 2	Matkovic, B., Prodic, B., Sljukic, M.: Bull. Soc. Chim. France (1968) Special No., 1777.
68 M 4	Mihaitov, M. P.: C. R. Acad. Bulg. Sci. **21** (1968) 1029.
68 M 5	Minaeva, K. A., Strukov, B. A., Varnstoff, K.: Fiz. Tverd. Tela **10** (1968) 2125; Sov. Phys. Solid State (English Transl.) **10** (1969) 1665.
68 M 6	Marutake, M.: Unpublished work (1968).
68 N 1	Nakamura, N., Suga, H., Chihara, H., Seki, S.: Bull. Chem. Soc. Jpn. **41** (1968) 291.
68 N 4	Norris, M. O., Strange, J. H., Powles, J. G., Rhodes, M., Marsden, K., Krynicki, K.: J. Phys. C **1** (1968) 422.
68 O 2	Ōkuma, H., Nakamura, N., Chihara, H.: J. Phys. Soc. Jpn. **24** (1968) 452.
68 O 4	Ohi, K.: J. Phys. Soc. Jpn. **25** (1968) 1369.
68 P 1	Peterson, G. E., Bridenbaugh, P. M.: J. Chem. Phys. **48** (1968) 3402.
68 P 2	Pikka, T. A., Fridkin, V. M.: Fiz Tverd. Tela **10** (1968) 3378; Sov. Phys. Solid State (English Transl.) **10** (1969) 2668.
68 P 3	Polandov, I. N., Mylov, V. P., Strukov, B. A., Varikash, V. M.: Fiz. Tverd. Tela **10** (1968) 1377; Sov. Phys. Solid State (English Transl.) **10** (1968) 1092.
68 P 4	Polandov, I. N., Mylov, V. P., Strukov, B. A.: Fiz. Tverd. Tela **10** (1968) 2232; Sov. Phys. Solid State (English Transl.) **10** (1969) 1754.

68 S 1	Salhotra, P. P., Subbarao, E. C., Venkateswarlu, P.: Phys. Status Solidi **29** (1968) 859.
68 S 2	Samara, G. A.: Phys. Rev. **173** (1968) 605.
68 S 3	Samara, G. A.: J. Phys. Chem. Solids **29** (1968) 870.
68 S 4	Samara, G. A.: Phys. Lett. A **27** (1968) 232.
68 S 5	Sandor, E., Johnson, M. W.: Nature **217** (1968) 541.
68 S 6	Sandy, F., Jones, R. V.: Phys. Rev. **168** (1968) 481.
68 S 7	Sato, S.: J. Phys. Soc. Jpn. **25** (1968) 185.
68 S 8	Schnering, H. G., Bleckmann, P.: Naturwissenschaften **55** (1968) 342.
68 S 10	Seehra, M. S., Castner Jr., T. G.: Phys. Kondens. Mater. **7** (1968) 185.
68 S 11	Seo, I., Kittaka, S., Makita, Y.: Jpn. J. Appl. Phys. **7** (1968) 780.
68 S 13	Silvidi, A. A.: J. Chem. Phys. **48** (1968) 1402.
68 S 17	Sonin, A. S., Filimonov, A. A., Suvorov, V. S.: Fiz. Tverd. Tela **10** (1968) 1883; Sov. Phys. Solid State (English Transl.) **10** (1968) 1481.
68 S 18	Sonin, A. S., Perfilova, V. E.: Fiz. Tverd. Tela **10** (1968) 305; Sov. Phys. Solid State (English Transl.) **10** (1968) 242.
68 S 21	Strukov, B. A., Amin, M., Kopchik, V. A.: Phys. Status Solidi **27** (1968) 741; erratum **28** (1968) 849.
68 S 22	Strukov, B. A., Taraskin, S. A., Koptsik, V. A., Varikash, V. M.: Kristallografiya **13** (1968) 541; Sov. Phys. Crystallogr. (English Transl.) **13** (1968) 447.
68 S 23	Stugrnuer, E. F., Merg, W. J.: Helv. Phys. Acta **41** (1968) 1206.
68 S 24	Suzuki, I., Abe, R.: J. Phys. Soc. Jpn. **25** (1968) 783.
68 T 2	Taylor, G. W., Lechner, B. J.: J. Appl. Phys. **39** (1968) 2372.
68 T 7	Tsutsumi, Y., Kunimoto, M., Terao, T., Hashi, T.: J. Phys. Soc. Jpn. **24** (1968) 429.
68 T 8	Tsuboi, M., Terada, M., Kajiura, T.: Bull. Chem. Soc. Jpn. **41** (1968) 2545.
68 V 2	Vijayan, M.: Acta Crystallogr. B **24** (1968) 1237.
68 Y 3	Yamamoto, T., Nakamura, N., Chihara, H.: J. Phys. Soc. Jpn. **25** (1968) 291.
68 Y 4	Yanagi, T.: Jpn. J. Appl. Phys. **7** (1968) 1056.
68 Z 1	Zherebtsova, L. I., Rostuntsova, A. I.: Kristallografiya **13** (1968) 538; Sov. Phys. Crystallogr. (English Transl.) **13** (1968) 444.
69 A 1	Adhav, R. S.: J. Phys. D **2** (1969) 171.
69 A 2	Adhav, R. S.: J. Opt. Soc. Am. **59** (1969) 414.
69 A 3	Adhav, R. S.: J. Appl. Phys. **40** (1969) 2725.
69 A 4	Aiki, K., Hududa, K., Matumura, O.: J. Phys. Soc. Jpn. **26** (1969) 1064.
69 A 7	Aleksandrov, K. S., Aleksandrova, I. P., Zherebtsova, L. I., Rostuntseva, A. I., Leibovich, T. A., Zaitseva, M. P.: Fiz. Tverd. Tela **11** (1969) 2027; Sov. Phys. Solid State (English Transl.) **11** (1970) 1639.
69 A 8	Aleksandrova, I. P., Shcherbakov, V. N.: Kristallografiya **14** (1969) 711; Sov. Phys. Crystallogr. (English Transl.) **14** (1979) 608.
69 A 9	Aleksandrov, K. S., Zaitseva, M. P., Shabanova, L. A.: Fiz. Tverd. Tela **11** (1969) 155; Sov. Phys. Solid State (English Transl.) **11** (1969) 114.
69 A 10	Allen, P. B., Cohen, M. L.: Phys. Rev. **177** (1969) 704.
69 A 12	Apkar'yants, P. A., Sonin, A. S.: Fiz. Tverd. Tela **11** (1969) 196; Sov. Phys. Solid State (English Transl.) **11** (1969) 148.
69 A 13	Apkar'yants, P. A., Lomova, L. G., Sonin, A. S.: Izv. Akad. Nauk Armyan. SSR, Fiz. **4** (1969) 25.
69 B 1	Barnett, J. D., Pack, J., Hall, H. T.: Trans. Am. Crystallogr. Assoc. **5** (1969) 113.
69 B 3	Baughman, R. J., Lefever, R. A.: Mater. Res. Bull. **4** (1969) 721.
69 B 4	Beattie, A. G.: J. Appl. Phys. **40** (1969) 4818.
69 B 6	Blinc, R., Bjorkstam, J. L.: Phys. Rev. Lett. **23** (1969) 589.
69 B 7	Blinc, R., Ferraro, J. R., Postmus, C.: J. Chem. Phys. **51** (1969) 732.
69 B 8	Blinc, R., Jamsek, M., Stepisnik, J., Trontelj, Z., Bjorkstam, J. L.: Solid State Commun. **7** (1969) 95.
69 B 9	Blinc, R., Mali, M.: Phys. Rev. **179** (1969) 552.
69 B 10	Blinc, R., O'Reilly, D. E., Peterson, E. M., Williams, J. M.: J. Chem. Phys. **50** (1969) 5408.
69 B 11	Blinc, R., Peterson, E. M., O'Reilly, D. E.: Phys. Lett. A **28** (1969) 624.
69 B 12	Bogomolov, A. A., Ivanov, V. V., Rudyak, V. M.: Kristallografiya **14** (1969) 1033; Sov. Phys. Crystallogr. (English Transl.) **14** (1969) 894.
69 B 14	Brody, E. M., Cummins, H. Z.: Phys. Rev. Lett. **23** (1969) 1039.
69 B 15	Burke, Jr., J. R., Riedl, H. R.: Phys. Rev. **184** (1969) 830.

69 B 17	Burns, G., Giess, E. A., O'Kane, D. F., Scott, B. A., Smith, A. W.: J. Appl. Phys. **40** (1969) 901.
69 B 19	Bury, P. C., McLaren, A. C.: Phys. Status Solidi **31** (1969) 799.
69 C 2	Clark, J. B.: High Temp. High Pressures **1** (1969) 553.
69 C 4	Cowley, E. R., Darby, J. K., Pawley, G. S.: J. Phys. C **2** (1969) 1916.
69 C 5	Cox, D. E., Frazer, B. C., Newnham, R. E., Santoro, R. P.: J. Appl. Phys. **40** (1969) 1124.
69 C 6	Crane, G. R., Bergman Jr., J. G., Glass, A. M.: J. Am. Ceram. Soc. **52** (1969) 655.
69 D 2	DiDomenico Jr., M., Eibschutz, M., Guggenheim, H. J., Camlibel, I.: Solid State Commun. **7** (1969) 1119.
69 D 3	Dollase, W. A.: Acta Crystallogr. B **25** (1969) 2298.
69 E 1	Eibschütz, M., Guggenheim, H. J., Wemple, S. H., Camlibel, I., DiDomenico Jr., M.: Phys. Lett. A **29** (1969) 409.
69 E 2	El Saffar, Z. M.: J. Chem. Phys. **50** (1969) 4520.
69 F 3	Fiory, A. T.: Bull. Am. Phys. Soc. (2) **14** (1969) 46.
69 G 1	Gelbwachs, J., Pantell, R. H., Puthoff, H. E., Yarborough, J. M.: Appl. Phys. Lett. **14** (1969) 258.
69 G 2	Genin, D. J., O'Reilly, D. E.: J. Chem. Phys. **50** (1969) 2842.
69 G 3	Gerzanich, E. I., Fridkin, V. M.: Zh. Eksp. Teor. Fiz. **56** (1969) 780; Sov. Phys. JETP (English Transl.) **29** (1969) 424.
69 G 4	Gerzanich, E. I., Fridkin, V. M.: Kristallografiya **14** (1969) 298; Sov. Phys. Crystallogr. (English Transl.) **14** (1969) 234.
69 G 5	Gesi, K.: J. Phys. Soc. Jpn. **26** (1969) 107.
69 G 6	Gesi, K.: J. Phys. Soc. Jpn. **26** (1969) 953.
69 G 7	Gesi, K.: J. Phys. Soc. Jpn. **26** (1969) 1554.
69 G 12	Grecu, N. M., Darabont, A.: Rev. Roumaine Phys. **14** (1969) 313.
69 G 13	Gupta, L. C.: J. Phys. Soc. Jpn. **27** (1969) 1229.
69 H 1	Hahn, Th., Lohre, G., Chung, S. J.: Naturwissenschaften **56** (1969) 459.
69 H 2	Hansen, F., Hazell, R. G., Rasmussen, S. E.: Acta Chem. Scand. **23** (1969) 2561.
69 H 4	Hartland, A.: J. Phys. C **2** (1969) 264.
69 H 5	Hatta, I., Ishiguro, T., Mikoshiba, N.: Phys. Lett. A **29** (1969) 421.
69 H 6	Hein, R. A., Meijer, P. H. E.: Phys. Rev. **179** (1969) 497.
69 H 8	Holmes, L., Eibschütz, M., Guggenheim, H. J.: Solid State Commun. **7** (1969) 973.
69 H 9	Howell, F. L., Schmidt, V. H.: J. Chem. Phys. **51** (1969) 1983.
69 I 3	Ito, M., Suzuki, M., Yokoyama, T.: J. Chem. Phys. **50** (1969) 2949.
69 J 2	Johnston, A. R., Nakamura, T.: J. Appl. Phys. **40** (1969) 3656.
69 K 1	Kato, T., Abe, R.: J. Phys. Soc. Jpn. **26** (1969) 948.
69 K 2	Kawai, H.: Jpn. J. Appl. Phys. **8** (1969) 975.
69 K 3	Keve, E. T., Abrahams, S. C., Bernstein, J. L.: J. Chem. Phys. **51** (1969) 4928.
69 K 6	Kobayashi, T., Ida, M., Kawada, S.: J. Phys. Soc. Jpn. **27** (1969) 1365.
69 K 8	Kozhin, V. M., Zaitseva, M. P.: Kristallografiya **14** (1969) 344; Sov. Phys. Crystallogr. (English Transl.) **14** (1969) 277.
69 K 9	Krasnikova, A. Ya., Polandov, I. N.: Fiz. Tverd. Tela **11** (1969) 1753; Sov. Phys. Solid State (English Transl.) **11** (1970) 1421.
69 K 10	Kravchenko, E. A., Dembovskii, S. A., Chernov, A. P., Semin, G. K.: Izv. Akad. Nauk SSSR, Ser. Fiz. **33** (1969) 279; Bull. Acad. Sci. USSR, Phys. Ser. (English Transl.) **33** (1969) 254.
69 K 12	Kudzin, A. Yu., Sukhinskii, A. N., Volnyanskii, M. D.: Fiz. Tverd. Tela **11** (1969) 2679; Sov. Phys. Solid State (English Transl.) **11** (1970) 2168.
69 K 14	Kunigelis, V., Samulionis, V., Garshka, E. P.: Fiz. Tverd. Tela **11** (1969) 2367; Sov. Phys. Solid State (English Transl.) **11** (1969) 1909.
69 K 15	Kydon, D. W., Petch, H. E., Pintar, M.: J. Chem. Phys. **51** (1969) 487.
69 L 5	Lundin, A. G., Zeer, E. P., Habuda, S. P.: Izv. Akad. Nauk SSSR, Ser. Fiz. **33** (1969) 261; Bull. Acad. Sci. USSR, Phys. Ser. (English Transl.) **33** (1969) 237.
69 L 6	Luther, G., Müser, H. E.: Z. Naturforsch. a **24** (1969) 389.
69 L 7	Luz, Z., Reuveni, A., Holmberg, R. W., Silver, B. L.: J. Chem. Phys. **51** (1969) 4017.
69 M 1	Makita, Y., Seo, I.: J. Chem. Phys. **51** (1969) 3058.
69 M 2	Masuda, Y., Sakata, K., Hasegawa, S., Ohara, G., Wada, M., Nishizawa, M.: Jpn. J. Appl. Phys. **8** (1969) 692.
69 M 3	Meister, H., Skalyo Jr., J., Frazer, B. C., Shirane, G.: Phys. Rev. **184** (1969) 550.
69 M 7	Milsch, B., Windsch, W.: Phys. Status Solidi **31** (1969) K79.

69 M 8	Minaeva, K. A., Levanyuk, A. P., Strukov, B. A.: Izv. Akad. Nauk SSSR, Ser. Fiz. **33** (1969) 328; Bull. Acad. Sci. USSR, Phys. Ser. (English Transl.) **33** (1969) 304.
69 N 1	Nakamura, E., Nagai, T., Hashimoto, S., Hosoya, M.: J. Phys. Soc. Jpn. **26** (1969) 979.
69 N 2	Narasimhamurty, T. S.: Phys. Rev. **186** (1969) 945.
69 N 4	Neumann, H., Kreher, K., Schmitz, W.: Phys. Lett. A **28** (1969) 743.
69 O 1	Oja, T., Casabella, P. A.: Phys. Rev. **177** (1969) 830.
69 O 3	Otani, A., Makishima, S.: J. Phys. Soc. Jpn. **26** (1969) 85.
69 P 1	Pandey, R. K.: J. Phys. Soc. Jpn. **27** (1969) 633.
69 P 3	Perfilova, V. E., Sirotin, Yu. I., Sonin, A. S.: Kristallografiya **14** (1969) 157; Sov. Phys. Crystallogr. (English Transl.) **14** (1969) 136.
69 P 5	Petzelt, J.: Phys. Status Solidi **36** (1969) 321.
69 P 7	Popolitov, V. I.: Kristallografiya **14** (1969) 375; Sov. Phys. Crystallogr. (English Transl.) **14** (1969) 312.
69 P 8	Popov, S. N., Krainik, N. N., Myl'nikova, I. E.: Izv. Akad. Nauk SSSR, Ser. Fiz. **33** (1969) 271; Bull. Acad. Sci. USSR, Phys. Ser. (English Transl.) **33** (1969) 247.
69 R 1	Rabenau, A., Rau, H.: Philips Tech. Rev. **30** (1969) 89.
69 R 3	Riede, V.: Phys. Lett. A **29** (1969) 715.
69 R 5	Robcrtson, D. S.: J. Phys. D **2** (1969) 1781.
69 R 6	Rudyak, V. M., Bogomolov, A. A., Ivanov, V. V.: Fiz. Tverd. Tela **11** (1969) 1700; Sov. Phys. Solid State (English Transl.) **11** (1969) 1377.
69 S 2	Salhotra, P. P., Subbarao, E. C., Venkateswarlu, P.: J. Phys. Soc. Jpn. **27** (1969) 621.
69 S 3	Samulionis, V. I., Kunigelis, V. F.: Fiz. Tverd. Tela **11** (1969) 1933; Sov. Phys. Solid State (English Transl.) **11** (1970) 1605.
69 S 4	Schmid, H., Kliegel, G., Kobayashi, J.: Helv. Phys. Acta **42** (1969) 599.
69 S 7	Shuvalov, L. A., Ivanov, N. R., Gordeeva, N. V., Kirpichnikova, L. F.: Kristallografiya **14** (1969) 658; Sov. Phys. Crystallogr. (English Transl.) **14** (1970) 554.
69 S 8	Shuvalov, L. A., Ivanov, N. R., Kirprichnikova, L. F., Schagina, N. M.: J. Phys. Chem. Solids **30** (1969) 57.
69 S 10	Silvidi, A. A., Falzone, A. J., Roupe, J. L.: Solid State Commun. **7** (1969) 359.
69 S 14	Soda, G., Chiba, T.: J. Phys. Soc. Jpn. **26** (1969) 249.
69 S 15	Soda, G., Chiba, T.: J. Phys. Soc. Jpn. **26** (1969) 717.
69 S 16	Soda, G., Chiba, T.: J. Phys. Soc. Jpn. **26** (1969) 723.
69 S 18	Sonin, A. S., Plotinskii, V. N.: Fig. 167 in [69 g 2].
69 S 19	Sonin, A. S., Vasilevskaya, A. S., Kalitina, M. P., Marchenko, E. I.: Kristallografiya **14** (1969) 343; Sov. Phys. Crystallogr. (English Transl.) **14** (1969) 275.
69 S 20	Suzuki, K., Sawada, S., Sugawara, F., Nakamura, T.: J. Phys. Soc. Jpn. **26** (1969) 1199.
69 S 21	Sandor, E., Farrow, R. F. C.: Disc. Faraday Soc. **48** (1969) 78.
69 T 1	Todo, I., Tatsuzaki, I.: Phys. Status Solidi **32** (1969) 263.
69 T 2	Topic, M., Prodic, B.: J. Appl. Crystallogr. **2** (1969) 230.
69 T 3	Topic, M., Prodic, B., Sljukic, M.: Czech. J. Phys. B **19** (1969) 1295.
69 T 4	Toyoda, K., Ishikawa, K., Kameyama, H.: J. Phys. Soc. Jpn. **27** (1969) 261.
69 T 6	Tsutsumi, Y., Kunitomo, M., Terao, T., Hashi, T.: J. Phys. Soc. Jpn. **26** (1969) 16.
69 T 8	Travina, T. S., Golik, L. L., Elnison, A. J., Isnatkin, V. J., Lyakhovitskaya, V. A.: Sov. Phys. Solid State (English Transl.) **11** (1969), 500; Fiz. Tverd. Tela **11** (1969) 622.
69 V 5	Vasilevskaya, A. S., Sonin, A. S.: Kristallografiya **14** (1969) 713; Sov. Phys. Crystallogr. (English Transl.) **14** (1970) 611.
69 V 7	Vinogradova, I. S., Falaleeva, L. G.: Izv. Akad. Nauk SSSR, Ser. Fiz. **33** (1969) 254; Bull. Acad. Sci. USSR, Phys. Ser. (English Transl.) **33** (1969) 231.
69 W 2	Weir, C. E., Piermarini, G. J., Block, S.: J. Chem. Phys. **50** (1969) 2089.
69 Y 1	Yagi, T., Tatsuzaki, I.: J. Phys. Soc. Jpn. **26** (1969) 865.
69 Z 1	Zavarizina, N. I., Gabuda, S. P.: Izv. Akad. Nauk SSSR, Ser. Fiz. **33** (1969) 258; Bull. Acad. Sci. USSR, Phys. Ser. (English Transl.) **33** (1969) 235.
69 Z 2	Zhukov, A. P., Rez, I. S., Pakhomov, V. I., Semin, G. K.: Izv. Akad. Nauk SSSR, Ser. Fiz. **33** (1969) 274; Bull. Acad. Sci. USSR, Phys. Ser. (English Transl.) **33** (1969) 250.
70 A 2	Aiki, K.: J. Phys. Soc. Jpn. **29** (1970) 379.
70 A 3	Aiki, K., Hukuda, K., Koga, H., Kobayashi, T.: J. Phys. Soc. Jpn. **28** (1970) 389.
70 A 5	Amin, M., Strukov, B. A.: Fiz. Tverd. Tela **12** (1970) 2035; Sov. Phys. Solid State (English Transl.) **12** (1971) 1616.

70 A 7	Anistratov, A. T., Martynov, V. G.: Kristallografiya **15** (1970) 308; Sov. Phys. Crystallogr. (English Transl.) **15** (1970) 256.
70 A 8	Apkar'yants, P. A., Izrailenko, A. N., Sonin, A. S.: Fiz. Tverd. Tela **12** (1970) 3193; Sov. Phys. Solid State (English Transl.) **12** (1971) 2580.
70 A 9	Ashmore, J. P., Petch, H. E.: Bull. Am. Phys. Soc. (2) **15** (1970) 775.
70 A 10	Audzionis, A. I., Grigas, I. P., Karpus, A. S.: Fiz. Tverd. Tela **12** (1970) 146; Sov. Phys. Solid State (English Transl.) **12** (1970) 115.
70 A 13	Aizu, K.: J. Phys. Soc. Jpn. **28** (1970) 706.
70 B 3	Belson, H. S., Houston, B.: J. Appl. Phys. **41** (1970) 422.
70 B 4	Belyaev, A. D., Gromashevskii, V. L.: Fiz. Tverd. Tela **12** (1970) 2806; Sov. Phys. Solid State (English Transl.) **12** (1971) 2266.
70 B 5	Belyaev, L. M., Verkhovskaya, K. A., Volk, T. R., Ogadzhanova, V. V., Soboleva, L. V., Tikhomirova, N. A., Fridkin, V. M.: Izv. Akad. Nauk SSSR, Ser. Fiz. **34** (1970) 2550; Bull. Acad. Sci. USSR, Phys. Ser. (English Transl.) **34** (1970) 2270.
70 B 7	Bezdetnyi, N. M., Zeinally, A. Kh., Lebedeva, N. N., Sheinkman, M. K.: Fiz. Tverd. Tela **12** (1970) 2480; Sov. Phys. Solid State (English Transl.) **12** (1971) 1990.
70 B 8	Bis, R. F., Dixon, J. R.: Phys. Rev. B **4** (1970) 1004.
70 B 9	Blinc, R., Jamšek-Vilfan, M., Lahajnar, G., Hajdukovič, G.: J. Chem. Phys. **52** (1970) 6407.
70 B 10	Blinc, R., Levstik, A.: Phys. Status Solidi (a) **2** (1970) K131.
70 B 11	Blinc, R., O'Reilly, D. E., Peterson, E. M.: Phys. Rev. B **1** (1970) 1953.
70 B 12	Boiko, A. A., Golovnin, V. A.: Kristallografiya **15** (1970) 186; Sov. Phys. Crystallogr. (English Transl.) **15** (1970) 153.
70 B 13	Bonera, G., Borsa, F., Rigamonti, A.: Phys. Rev. B **2** (1970) 2784.
70 B 16	Brezina, B., Janousek, V., Marecek, V., Smutny, F.: Article on p. 369 in [70 m 1].
70 B 17	Brunstein, M., Grinberg, J.: Solid State Commun. **8** (1970) 1211.
70 B 18	Bunting, J. G., Steeple, H., Ashworth, T.: Phys. Lett. A **33** (1970) 37.
70 B 22	Becker, W.-J., Will, G.: Z. Kristallogr. **131** (1970) 139.
70 C 7	Cooke, A. H., Ellis, C. J., Gehring, K. A., Leask, M. J. M., Martin, D. M., Wanklyn, B. M., Wells, M. R., White, R. L.: Solid State Commun. **8** (1970) 689.
70 C 8	Cox, D. E., Eibschütz, M., Guggenheim, H. J., Holmes, L.: J. Appl. Phys. **41** (1970) 943.
70 C 10	Chisler, E. V., Savatinova, I. T., Fridkin, V. M.: Sov. Phys. Solid State (English Transl.) **12** (1971) 2327; Fiz. Tverd. Tela **12** (1970) 2882.
70 F 2	Fitzgerald, M. E., Casabella, P. A.: Phys. Rev. B **2** (1970) 1350.
70 F 4	Fousek, J., Cross, L. E., Seely, K.: Ferroelectrics **1** (1970) 63.
70 F 6	Frenzel, C., Pietrass, B., Hegenbarth, E.: Phys. Status Solidi (a) **2** (1970) 273.
70 F 7	Fridkin, V. M., Belyaev, L. M., Grekov, A. A., Rodin, A. I.: J. Phys. Soc. Jpn. **28** (1970) Suppl. 448.
70 G 1	Gavrilova, N. D., Karyakina, N. F., Koptsik, V. A., Novik, V. K.: Dokl. Akad. Nauk SSSR **195** (1970) 823; Sov. Phys. Dokl. (English Transl.) **15** (1971) 1075.
70 G 3	Gesi, K.: J. Phys. Soc. Jpn. **28** (1970) 395.
70 G 4	Gesi, K.: J. Phys. Soc. Jpn. **28** (1970) 1365.
70 G 5	Gesi, K.: J. Phys. Soc. Jpn. **28** (1970) 1377.
70 G 8	Golik, L. L., Elinson, M. I.: Fiz. Tverd. Tela **12** (1970) 2895; Sov. Phys. Solid State (English Transl.) **12** (1971) 2338.
70 G 9	Golovei, M. P., Kalinkina, I. N., Kosourov, G. I.: Opt. Spektrosk. **28** (1970) 991; Opt. Spectrosc. (English Transl.) **28** (1970) 535.
70 G 10	Grekov, A. A., Rogach, E. D., Sukiyazov, A. G.: Fiz. Tverd. Tela **12** (1970) 3559; Sov. Phys. Solid State (English Transl.) **12** (1971) 2890.
70 H 2	Hafele, H. G., Wacherning, H., Irslinger, C., Grisar, R., Nitsche, R.: Phys. Status Solidi **42** (1970) 531.
70 H 6	Harbeke, G., Steigmeier, E. F., Wehner, R. K.: Solid State Commun. **8** (1970) 1765.
70 H 7	Hatta, I., Ishiguro, T., Mikoshiba, N.: Article on p. 211 in [70 t 1].
70 H 8	Hauret, G., Gharbi, A.: C. R. Acad. Sci. (Paris) B 271 (1970) 1072.
70 H 9	Hegenbarth, E. M.: Kristallografiya **15** (1970) 322; Sov. Phys. Crystallogr. (English Transl.) **15** (1970) 268.
70 H 10	Helg, U.: Z. Kristallogr. **131** (1970) 241.
70 H 12	Horak, J.: J. Phys. (Paris) **31** (1970) 121.
70 H 13	Hosoya, M., Nakamura, E.: Jpn. J. Appl. Phys. **9** (1970) 552.

70I4	Ito, M.: Chem. Phys. Lett. **7** (1970) 439.
70I6	Ivanov, N. R., Konstantinova, A. F.: Kristallografiya **15** (1970) 490; Sov. Phys. Crystallogr. (English Transl.) **15** (1970) 416.
70I7	Ivanov, N. R., Shuvalov, L. A.: Article on p. 72 in [70t1].
70I8	Ivanov, N. R., Tukhtasunov, I. T., Shuvalov, L. A.: Kristallografiya **15** (1970) 752; Sov. Phys. Crystallogr. (English Transl.) **15** (1971) 647.
70J1	Jacobi, H.: Z. Kristallogr. **132** (1970) 434.
70K1	Kadaba, P. K., O'Reilly, D. E.: J. Chem. Phys. **32** (1970) 2403.
70K2	Kalman, A., Stephens, J. S., Cruickshank, D. W. J.: Acta Crystallogr. B **26** (1970) 1451.
70K3	Kaplan, S. F., Kay, M. I., Morosin, B.: Ferroelectrics **1** (1970) 31.
70K4	Kasahara, M., Tatsuzaki, I.: J. Phys. Soc. Jpn. **29** (1970) 1392.
70K5	Kato, T., Abe, R.: J. Phys. Soc. Jpn. **29** (1970) 389.
70K6	Kato, T., Mizuno, O., Abe, R.: J. Phys. Soc. Jpn. **29** (1970) 393.
70K7	Kawano, T., Niimori, K., Hukuda, K., Fujita, N.: J. Phys. Soc. Jpn. **29** (1970) 633.
70K10	Kind, R., Gränicher, H.: Article on p. 109 in [70t1].
70K11	Kiriyama, H., Kiriyama, R.: Article on p. 114 in [70t1].
70K12	Kliya, M. O., Lyachovitskaya, V. A.: Article on p. 114 in [70t1].
70K13	Kliya, M. O., Lyachovitskaya, V. A.: Kristallografiya **15** (1970) 75; Sov. Phys. Crystallogr. (English Transl.) **15** (1970) 59.
70K14	Knauer, R. C.: Phys. Lett. A **33** (1970) 313.
70K18	Kobayashi, J., Uesu, Y., Mizutani, I., Enomoto, Y.: Phys. Status Solidi (a) **3** (1970) 63.
70K20	Kudzin, A. Yu., Sukhinskii, A. N., Butsychenko, V. A.: Fiz. Tverd. Tela **12** (1970) 2180; Sov. Phys. Solid State (English Transl.) **12** (1970) 1740.
70K21	Kunimoto, M., Terao, T., Hashi, T.: Phys. Lett. A **31** (1970) 14.
70L2	Lefkowitz, I., Shields, M., Dolling, G.: J. Cryst. Growth **6** (1970) 143.
70L6	Ledsham, A. H. C., Steeple, H., Hughes, W.: Acta Crystallogr. B **26** (1970) 1240.
70L7	Lewis, J. E.: Phys. Status Solidi **42** (1970) K97.
70L8	Lin-Chung, P. J.: J. Phys. Chem. Solids **31** (1970) 2199.
70L9	Litov, E., Garland, C. W.: Phys. Rev. B **2** (1970) 4597.
70L10	Litov, E., Uehling, E. A.: Phys. Rev. B **1** (1970) 3713.
70L11	Lomova, L. G., Sonin, A. S.: Fiz. Tverd. Tela **12** (1970) 3331; Sov. Phys. Solid State (English Transl.) **12** (1971) 2699.
70M1	Matsumura, S.: Oyo Buturi **39** (1970) 604 (in Japanese).
70M3	McKenzie, D. R., Ham, N. S., Whitfield, H. J.: Solid State Commun. **8** (1970) 2059.
70M4	Midorikawa, M., Takagi, Y., Ishibashi, Y.: J. Phys. Soc. Jpn. **28** (1970) 1001.
70M5	Miki, H., Makita, Y.: J. Phys. Soc. Jpn. **29** (1970) 143.
70M7	Minaeva, K. A., Strukov, B. A., Ihu, H. C.: Fiz. Tverd. Tela **12** (1970) 1584; Sov. Phys. Solid State (English Transl.) **12** (1970) 1256.
70M12	Mylov, V. P., Chuagulov, B. R., Leonidova, G. G.: Fiz. Tverd. Tela **12** (1970) 1288; Sov. Phys. Solid State (English Transl.) **12** (1970) 1012.
70M13	Mori, T., Tamura, H., Sawaguchi, E.: Article on p. 445 in [70t 1].
70N1	Narayanamurti, V., Pohl, R. O.: Rev. Mod. Phys. **42** (1970) 201.
70N2	Nassau, K., Shiever, J. W., Kowalchik, M.: J. Cryst. Growth **7** (1970) 237.
70N3	Novikova, S. I., Shelimova, A. E., Abrikosov, N. Kh., Galyutin, V. I., Evseev, B. A.: Fiz. Tverd. Tela **12** (1970) 3623; Sov. Phys. Solid State (English Transl.) **12** (1971) 2945.
70O2	Ogawa, K., Yamada, Y., Taguchi, I., Osaki, K., Watanabe, T., Nitta, I.: Sci. Rept. Coll. Gen. Educ., Osaka Univ. **19** (1970) 1.
70O3	Ohi, K.: Article on p. 84 in [70t1].
70O4	Ota, K., Ishibashi, Y., Takagi, Y.: J. Phys. Soc. Jpn. **29** (1970) 1545.
70P1	Pannetier, J., Lucas, J.: Mater. Res. Bull. **5** (1970) 797.
70P4	Petzelt, J.: Article on p. 287 in [70m1].
70P6	Popolitov, V. I.: Kristallografiya **15** (1970) 1270; Sov. Phys. Crystallogr. (English Transl.) **15** (1971) 1118.
70P7	Popolitov, V. I., Litvin, B. N., Lobachev, A. N.: Phys. Status Solidi (a) **3** (1970) K1.
70P8	Popov, S. N., Krainik, N. N.: Fiz. Tverd. Tela **12** (1970) 3022; Sov. Phys. Solid State (English Transl.) **12** (1971) 2444.
70P9	Popov, S. N., Krainik, N. N., Mylnikova, I. E.: Article on p. 120 in [70t1].

70 P 10	Perry, C. H., Agrawal, D. K.: Solid State Commun. **8** (1970) 225.
70 P 11	Popolitov, V. J., Litvin, R. N.: Sov. Phys. Cryst. (English Transl.) **15** (1971) 1113; Kristallogr. **15** (1970) 1267.
70 R 3	Rapoport, E.: J. Chem. Phys. **53** (1970) 311.
70 R 7	Riede, V.: Phys. Status Solidi **2** (1970) K193.
70 R 8	Rigamonti, A., Petrini, G.: Phys. Status Solidi **41** (1970) 591.
70 S 1	Sagar, A.: J. Appl. Phys. **41** (1970) 811.
70 S 3	Sakurai, J., Cowley, R. A., Dolling, G.: J. Phys. Soc. Jpn. **28** (1970) 1426.
70 S 5	Sandercock, J. R.: Opt. Commun. **2** (1970) 73.
70 S 9	Schmidt, V. H., Howell, F. L.: Article on p. 106 in [70 t 1].
70 S 12	Shimshoni, M., Harnik, E.: Phys. Lett. A **32** (1970) 321.
70 S 15	Simon, A.: Z. Naturforsch. **25b** (1970) 1489.
70 S 18	Skalyo Jr., J., Frazer, B. C., Shirane, G.: Phys. Rev. B **1** (1970) 278.
70 S 21	Steigmeier, E. F., Harbeke, G.: Solid State Commun. **8** (1970) 1275.
70 S 22	Sugawara, F., Nakamura, T.: J. Phys. Soc. Jpn. **28** (1970) 158.
70 S 23	Sugawara, F., Nakamura, T.: J. Phys. Soc. Jpn. **29** (1970) 162.
70 S 24	Sugawara, F., Nakamura, T.: Article on p. 221 in [70 t 1].
70 S 26	Spencer, E. G., Guggenheim, H. J., Kominak, G. J.: Appl. Phys. Lett. **17** (1979) 300.
70 T 2	Taraskin, S. A., Strukov, B. A., Meleshina, V. A.: Fiz. Tverd. Tela **12** (1970) 1386; Sov. Phys. Solid State (English Transl.) **12** (1970) 1089.
70 T 4	Taylor, J. C., Mueller, M. H., Hitterman, R. L.: Acta Crystallogr. A **26** (1970) 559.
70 T 6	Tello, M. J., Gonzalo, J. A.: Article on p. 199 in [70 t 1].
70 T 10	Topic, M., Kojic-Prodic, B., Popovic, S.: Czech. J. Phys. B **20** (1970) 1003.
70 T 12	Toyoda, K., Ishikawa, K.: Article on p. 451 in [70 t 1].
70 U 4	Unruh, H.-G.: Solid State Commun. **8** (1970) 1951.
70 V 3	Vogt, H., Happ, H., Häfele, H. G.: Phys. Status Solidi (a) **1** (1970) 439.
70 W 2	Wang, C. H., Fleury, P. A.: J. Chem. Phys. **53** (1970) 2243.
70 W 5	Wiener-Avnear, E., Levin, S., Pelah, I.: J. Chem. Phys. **52** (1970) 2881.
70 W 7	Wilson, C. M.: Doctor Thesis, The Johns Hopkins University (1970).
70 Y 1	Yagi, T., Tatsuzaki, I., Tōdō, I.: J. Phys. Soc. Jpn. **28** (1970) 321.
70 Y 3	Yanagi , T., Iio, K., Hanadate, H., Sawada, S.: Article on p. 78 in [70 t 1].
70 Y 5	Yoganarasimhan, S. R., Sood, R. K.: Phil. Mag. (8) **22** (1970) 1075.
70 Z 1	Zaporozhets, O. I., Lyakhovitskaya, V. A., Pekar, S. I., Polotskii, I. G., Sil'vestrova, I. M.: Fiz. Tverd. Tela **12** (1970) 671; Sov. Phys. Solid State (English Transl.) **12** (1970) 523.
70 Z 2	Zav'yalova, A. M., Zaks, P. L., Syrkin, L. N.: Fiz. Tverd. Tela **12** (1970) 1580; Sov. Phys. Solid State (English Transl.) **12** (1970) 1252.
71 A 6	Adriaenssens, G. J., Bjorkstam, J. L.: J. Chem. Phys. **55** (1971) 1137.
71 A 7	Agrawal, D. K., Perry, C. H.: Phys. Rev. B **4** (1971) 1893.
71 A 8	Aleksandrov, K. S., Aleksandrova, I. P., Zherebtsova, L. I., Zaitseva, M. P., Anistratov, A. T.: Ferroelectrics **2** (1971) 1.
71 A 12	Arend, H., Perison, J.: Mater. Res. Bull. **6** (1971) 1205.
71 B 1	Bajak, I. L.: Fyz. Casopis **21** (1971) 90.
71 B 4	Beliackas, R., Grigas, J., Orliukas, A., Shugurov, V.: Liet. Fiz. Rinkinys **11** (1971) 1029.
71 B 6	Bender, V. P., Fridkin, V. M.: Fiz. Tverd. Tela **13** (1971) 614; Sov. Phys. Solid State (English Transl.) **13** (1971) 501.
71 B 7	Benepe, J. W., Reese, W.: Phys. Rev. B **3** (1971) 3032.
71 B 9	Bergman Jr., J. G., McFee, J. H., Crane, G. R.: Appl. Phys. Lett. **18** (1971) 203.
71 B 14	Bird, M. J., Lomer, T. R.: Acta Crystallogr. B **27** (1971) 859.
71 B 15	Blank, S. L.: Unpublished work (1971) cited in [71 K 3].
71 B 17	Blinc, R., Mali, M., Osredkar, R., Prelesnik, A., Zupančič, I., Ehrenberg, L.: J. Chem. Phys. **55** (1971) 4843.
71 B 18	Blinc, R., Pirs, J.: J. Chem. Phys. **54** (1971) 1535.
71 B 19	Blinc, R., Slak, J., Stepišnik, J.: J. Chem. Phys. **55** (1971) 4848.
71 B 20	Blinc, R., Stepišnik, J., Jamšek-Vilfan, M., Žumer, S.: J. Chem. Phys. **54** (1971) 187.
71 B 22	Boldrini, P.: Acta Crystallogr. B**27** (1971) 860.
71 B 23	Boyer, L., Vacher, R.: Phys. Status Solidi (a) **6** (1971) K105.
71 B 24	Březina, B., Havránková, M.: Mater. Res. Bull. **6** (1971) 537.

71 B 29	Brillson, L., Burstein, E.: Phys. Rev. Lett. **27** (1971) 808.
71 C 2	Carlson, R. E., Friedlich, H. B.: J. Chem. Phys. **54** (1971) 2794.
71 F 2	Fiory, A. T.: Phys. Rev. B **4** (1971) 614.
71 F 5	Fujii, Y., Yamada, Y.: J. Phys. Soc. Jpn. **30** (1971) 1676.
71 G 4	Gladkii, V. V., Sidnenko, E. V.: Fiz. Tverd. Tela **13** (1971) 1642; Sov. Phys. Solid State (English Transl.) **13** (1971) 1642.
71 G 5	Gladkii, V. V., Sidnenko, E. V.: Fiz. Tverd. Tela **13** (1971) 3092; Sov. Phys. Solid State (English Transl.) **13** (1972) 2592.
71 G 6	Glass, A. M., McFee, J. H., Bergman Jr., J. G.: J. Appl. Phys. **42** (1971) 5219.
71 G 8	Gorbatyi, L. V., Ponomarev, V. I., Kheiker, D. M.: Kristallografiya **16** (1971) 899; Sov. Phys. Crystallogr. (English Transl.) **16** (1972) 781.
71 H 3	Hammer, H.: Article on p. 425 in [71 b 1].
71 H 5	Hatta, I., Ikushima, A.: Phys. Lett. A **37** (1971) 207.
71 H 6	Hauret, G., Taurel, L., Chapelle, J. P.: C. R. Acad. Sci. (Paris) B **273** (1971) 627.
71 H 7	Helwig, J., Albers, J.: Phys. Status Solidi (a) **7** (1971) 151.
71 H 9	Hill, J. C., Mohan, P. V.: Ferroelectrics **2** (1971) 201.
71 H 10	Hochli, U. T.: Phys. Rev. B **6** (1971) 1814.
71 H 12	Hartwig, C. M., Wiener-Avnear, E., Smit, J., Porto, S. P. S.: Phys. Rev. B **3** (1971) 2078.
71 J 1	Jeffery, J. W.: J. Appl. Cryst. **4** (1971) 334.
71 K 2	Katiyar, R. S., Ryan, R. S., Scott, J. F.: Article on p. 436 in [71 b 1].
71 K 3	Keve, E. T., Abrahams, S. C., Bernstein, J. L.: J. Chem. Phys. **54** (1971) 2556.
71 K 6	Khimich, T. A., Belov, V. F., Zhukov, O. K., Yurin, V. A., Korablin, L. N., Shipko, M. N., Lobachev, A. N., Popolitov, V. I.: Fiz. Tverd. Tela **13** (1971) 1507; Sov. Phys. Solid State (English Transl.) **13** (1971) 1265.
71 K 7	Kind, R.: Phys. Kondens. Mater. **13** (1971) 217.
71 K 9	Knispel, R. R., Petch, H. E.: Can. J. Phys. **49** (1971) 870.
71 K 10	Kobayashi, J., Bouillot, J., Kinoshita, K.: Phys. Status Solidi (b) **47** (1971) 619.
71 K 11	Kobayashi, J., Uesu, Y., Enomoto, Y.: Phys. Status Solidi (b) **45** (1971) 293.
71 K 14	Khimich, T. A., Belov, V. F., Zhukov, O. K., Yurin, V. A., Korablin, L. N., Shipko, M. N., Lobachev, A. N., Popolitov, V. I.: Fiz. Tverd. Tela **13** (1971) 1507; Sov. Phys. Solid State (English Transl.) **13** (1971) 1265.
71 K 15	Krasnikova, A. Ya., Toshev, S. D., Koptsik, V. A.: Izv. Akad. Nauk SSSR, Ser. Fiz. **35** (1971) 1882; Bull. Acad. Sci. USSR, Phys. Ser. (English Transl.) **35** (1971) 1709.
71 L 1	Lavrenčič, B.: Phys. Status Solidi (a) **5** (1971) K133.
71 L 5	Leong, J. T., Emrick, R. M.: J. Phys. Chem. Solids **32** (1971) 2593.
71 M 1	Makita, Y., Sumita, M.: J. Phys. Soc. Jpn. **31** (1971) 792.
71 M 3	McCarthy, G. J., Schlege, L. H., Sawaguchi, E.: J. Appl. Crystallogr. **4** (1971) 180.
71 M 4	Michard, F., Plicque, F.: C. R. Acad. Sci. (Paris) **272** (1971) 848.
71 M 5	Midorikawa, M., Ishibashi, Y., Takagi, Y.: J. Phys. Soc. Jpn. **30** (1971) 449.
71 M 6	Miki, H., Makita, Y., Gesi, K.: J. Phys. Soc. Jpn. **30** (1971) 1512.
71 M 10	Montano, P. A., Shechter, H., Biran, A.: Solid State Commun. **9** (1971) 2029.
71 M 11	Moreau, J. M., Michel, C., Gerson, R., James, W. J.: J. Phys. Chem. Solids **32** (1971) 1315.
71 N 3	Nakamura, K., Wada, Y.: J. Polymer Sci., Part A-2, **9** (1971) 161.
71 N 7	Neels, H., Schmitz, W., Hoffmann, H., Rossner, R., Topp, W.: Kristall Tech. **6** (1971) 225.
71 N 8	Nelmes, R. J.: Acta Crystallogr. B **27** (1971) 272.
71 N 9	Nesterenko, P. S., Savchenko, E. A., Tatarenko, L. N.: Fiz. Tverd. Tela **13** (1971) 2492; Sov. Phys. Solid State (English Transl.) **13** (1972) 2094.
71 N 11	Nikiforov, I. Ya., Khasabov, A. G.: Fiz. Tverd. Tela **13** (1971) 3589; Sov. Phys. Solid State (English Transl.) **13** (1971) 3030.
71 N 13	Novik, V. K., Karyakina, N. F., Bochkov, B. G., Koptsik, V. A., Gavrilova, N. D.: Izv. Akad. Nauk SSSR, Ser. Fiz. **35** (1971) 1874; Bull. Acad. Sci. USSR, Phys. Ser. (English Transl.) **35** (1971) 1703.
71 N 14	Novikova, S. I., Shelimova, L. E., Abrikosov, N. Kh., Evseev. B. A.: Fiz. Tverd. Tela **13** (1971) 2764; Sov. Phys. Solid State (English Transl.) **13** (1972) 2310.
71 O 1	Okada, K., Sugie, H.: Phys. Lett. A **37** (1971) 337.
71 O 2	O'Reilly, D. E., Peterson, E. M., El Saffar, Z. M.: J. Chem. Phys. **54** (1971) 1304.
71 P 1	Palau, J.-M., Lassabatere, L.: C. R. Acad. Sci. (Paris) **273** B (1971) 714.

71 P 4	Pierre, C., Dufour, J. P., Remoissenet, M.: Solid State Commun. **9** (1971) 1493; erratum **9** (1971) No. 23, v.
71 S 4	Samara, G. A.: Phys. Rev. Lett. **27** (1971) 103.
71 S 6	Sawada, A., Takagi, Y.: J. Phys. Soc. Jpn. **31** (1971) 952.
71 S 7	Sawaguchi, E., Cross, L. E.: Appl. Phys. Lett. **18** (1971) 1.
71 S 8	Sawaguchi, E., Cross, L. E.: Ferroelectrics **2** (1971) 37.
71 S 9	Samulionis, V. I., Kunigelis, V. F.: Fiz. Tverd. Tela **13** (1971) 740; Sov. Phys. Solid State (English Transl.) **13** (1971) 611.
71 S 10	Samulionis, V. I., Kunigelis, V. F., Girshovichus, M. N.: Zh. Eksp. Teor. Fiz. **61** (1971) 1941; Sov. Phys. JETP (English Transl.) **34** (1972) 1033.
71 S 11	Schmidt, V. H., Drumheller, J. E., Howell, F. L.: Phys. Rev. B **4** (1971) 4582.
71 S 12	Schutte, C. J. H., Van Rensburg, D. J. J.: J. Mol. Struct. **9** (1971) 77.
71 S 13	Seregin, P. P., Savin, E. P.: Fiz. Tverd. Tela **13** (1971) 2790; Sov. Phys. Solid State (English Transl.) **13** (1972) 2336.
71 S 16	Shigenari, T., Takagi, Y.: J. Phys. Soc. Jpn. **31** (1971) 312.
71 S 17	Shiozaki, Y.: Ferroelectrics **2** (1971) 245.
71 S 18	Shiraki, H., Tatsuzaki, I., Yagi, T.: Phys. Status Solidi (a) **7** (1971) 227.
71 S 20	Shirokov, A. M., Baranov, A. I., Shuvalov, L. A.: Izv. Akad. Nauk SSSR, Ser. Fiz. **35** (1971) 1903.
71 S 22	Silvidi, A. A., Workman, D. T.: J. Chem. Phys. **55** (1971) 4672.
71 S 23	Simon, A.: J. Appl. Crystallogr. **4** (1971) 138.
71 S 27	Sobolev, V. V.: Phys. Status Solidi (b) **45** (1971) K149.
71 S 28	Soeya, T., Yahata, Y., Nozaki, K.: Japan. J. Appl. Phys. **10** (1971) 279.
71 S 29	Someya, T., Kobayashi, J.: Phys. Status Solidi (a) **4** (1971) K161.
71 S 30	Steigmeier, E. F., Harbeke, G., Wehner, R. K.: Article on p. 396 in [71 b 1].
71 S 31	Strukov, B. A., Baddur, A., Velichko, I. A.: Fiz. Tverd. Tela **13** (1971) 2484; Sov. Phys. Solid State (English Transl.) **13** (1972) 2085.
71 S 33	Suzuki, I., Abe, R.: J. Phys. Soc. Jpn. **30** (1971) 1210.
71 S 34	Suzuki, I., Abe, R.: J. Phys. Soc. Jpn. **31** (1971) 179.
71 S 35	Suzuki, I., Abe, R.: J. Phys. Soc. Jpn. **31** (1971) 951.
71 S 36	Suzuki, S., Takagi, M.: J. Phys. Soc. Jpn. **30** (1971) 188.
71 S 37	Spitsyna, V. D., Lyakhovitskaya, V. A., Popovkin, B. A.: Kristallografiya **16** (1971) 218; Sov. Phys. Crystallogr. (English Transl.) **16** (1971) 175.
71 T 1	Tédenac, J. C., Vilminot, S., Cot, L., Norbert, A., Maurin, M.: Mater. Res. Bull. **6** (1971) 183.
71 T 2	Tivari, H. W., Narayanan, P. S.: Kristallografiya **16** (1971) 952; Sov. Phys. Crystallogr. (English Transl.) **16** (1972) 824.
71 T 3	Todo, I., Tatsuzaki, I.: J. Phys. Soc. Jpn. **31** (1971) 1479.
71 V 1	Vasilevskaya, A. S., Sonin, A. S.: Fiz. Tverd. Tela **13** (1971) 1550; Sov. Phys. Solid State (English Transl.) **13** (1971) 1299.
71 V 3	Vlokh, O. G., Lutsiv-Shumskii, V. F., Pylypyshin, B. P.: Kristallografiya **16** (1971) 828; Sov. Phys. Crystallogr. (English Transl.) **16** (1972) 717.
71 V 4	Volkel, G., Windsch, W.: Phys. Status Solidi (b) **43** (1971) 263.
71 V 5	Volkova, E. N., Berezhnoi, B. M., Izrailenko, A. N., Mishchenko, A. V., Rashkovich, L. N.: Izv. Akad. Nauk SSSR, Ser. Fiz. **35** (1971) 1858; Bull. Acad. Sci. USSR, Phys. Ser. (English Transl.) **35** (1971) 1690.
71 W 1	Will, G., Schafer, W.: J. Phys. C **4** (1971) 811.
71 Y 2	Yamagata, K., Hayama, M., Odaka, T.: J. Phys. Soc. Jpn. **31** (1971) 1279.
71 Y 6	Yuzvak, V. I., Aleksandrova, I. P., Shcherbakov, V. N.: Fiz. Tverd. Tela **13** (1971) 1932; Sov. Phys. Solid State (English Transl.) **13** (1972) 1620.
71 Z 1	Zaitseva, M. P., Zherebtsova, L. I., Shabanova, L. A.: Izv. Akad. Nauk SSSR, Ser. Fiz. **35** (1971) 1890; Bull. Acad. Sci. USSR, Phys. Ser. (English Transl.) **35** (1971) 1715.
71 Z 2	Zarembovskaya, T. A., Varikash, V. M.: Fiz. Tverd. Tela **13** (1971) 3537; Sov. Phys. Solid State (English Transl.) **13** (1972) 2989.
72 A 1	Adbulsabirov, R. Y., Greznev, Y. S., Zaripov, M. M.: Fiz. Tverd. Tela **14** (1972) 3430; Sov. Phys. Solid State (English Transl.) **14** (1973) 2894.
72 A 2	Abe, Y., Ohneda, Y., Abe, S., Kojima, S.: J. Phys. Soc. Jpn. **33** (1972) 864.
72 A 6	Aleksandrov, A. Yu., Baltrunas, D. I., Belyaev, L. M., Lyubutin, I. S., Lyakhovitskaya, V. A.: Kristallografiya **17** (1972) 332; Sov. Phys. Crystallogr. (English Transl.) **17** (1972) 281.

72A 7	Aleksandrov, K. S., Khlyustov, V. G., Flerov, I. A., Beznosikova, N. V.: Fiz. Tverd. Tela **14** (1972) 3374; Sov. Phys. Solid State (English Transl.) **14** (1973) 2853.
72A 8	Allgaier, R. S., Houston, B.: Phys. Rev. B **5** (1972) 2186.
72A 9	Al'shin, B. I., Astrov, D. N., Zorin, R. V.: Zh. Eksp. Teor. Fiz. **63** (1972) 2198; Sov. Phys. JETP (English Transl.) **36** (1973) 1161.
72A 10	Amov, I. G., Kirov, K. I., Toshev, S. D.: Kristallografiya **17** (1972) 1257; Sov. Phys. Crystallogr. (English Transl.) **17** (1973) 1101.
72A 11	Anderson, M. R., Brown, I. D.: Acta Crystallogr. A **28** (1972) 663.
72A 12	Anistratov, A. T., Mel'nikova, S. V.: Kristallografiya **17** (1972) 149; Sov. Phys. Crystallogr. (English Transl.) **17** (1972) 119.
72A 13	Anistratov, A. T., Zaitseva, M. P., Iskornev, I. M.: Kristallografiya **17** (1972) 974; Sov. Phys. Crystallogr. (English Transl.) **17** (1973) 858.
72A 14	Arend, H., Huber, W.: J. Cryst. Growth **12** (1972) 179.
72A 16	Artobolevskaya, E. S., Chenskii, E. V., Gvozdover, R. S., Petrov, V. I.: Fiz. Tverd. Tela **14** (1972) 2236; Sov. Phys. Solid State (English Transl.) **14** (1973) 1935.
72A 17	Ashida, T., Bando, S., Kakudo, M.: Acta Crystallogr. B **28** (1972) 1560.
72B 1	Bandy, A., Cessac, G. L., Lippincott, E. R.: Spectrochim. Acta A **28** (1972) 1807.
72B 3	Bartuch, H., Windsch, W.: Phys. Status Solidi (a) **14** (1972) K51.
72B 4	Bayliss, P., Nowacki, W.: Z. Kristallogr. **135** (1972) 308.
72B 10	Blinc, R., Mali, M., Osredkar, R., Prelesnik, A., Seliger, J., Zupančič, I.: Chem. Phys. Lett. **14** (1972) 49.
72B 11	Blinc, R., Mali, M., Slak, J., Stepišnik, J., Žumer, S.: J. Chem. Phys. **56** (1972) 3566.
72B 13	Boksha, S. S.: J. Cryst. Growth **12** (1972) 113.
72B 15	Bornarel, J., Lajzerowicz, J.: Ferroelectrics **4** (1972) 177.
72B 16	Brezina, B., Glogarova, M.: Phys. Status Solidi (a) **11** (1972) K39.
72B 18	Broberg, T. W., She, C. Y., Wall, L. S.: Phys. Rev. B **6** (1972) 3332.
72B 20	Brosowski, G., Luther, G., Müser, H. E.: Phys. Status Solidi (a) **14** (1972) K15.
72B 28	Bercha, D. M., Zayachkovskii, M. P.: Sov. Phys. Solid State (English Transl.) **14** (1972) 766; Fiz. Tverd. Tela. **14** (1972) 766.
72C 1	Capiomont, A., Bordeau, D., Lajzerowicż, J.: C. R. Acad. Sci. (Paris) C **275** (1972) 317.
72C 3	Chern, M.-J., Phillips, R. A.: J. Appl. Phys. **43** (1972) 496.
72C 4	Chihara, H., Inaba, A., Nakamura, N., Okuma, H., Soda, G., Yamamoto, T.: Acta Crystallogr. A **28** (1972) S182.
72C 7	Crane, G. R.: J. Appl. Crystallogr. **5** (1972) 360.
72D 1	Dalal, N. S., Dickinson, J. R., McDowell, C. A.: J. Chem. Phys. **57** (1972) 4254.
72D 2	Dalal, N. S., McDowell, C. A.: Phys. Rev. B **5** (1972) 1074.
72D 5	Dominquez, E., Jimenez, B., Mendiola, J., Vivas, E.: J. Mater. Sci. **7** (1972) 363.
72E 1	Eibschütz, M., Holmes, L., Guggenheim, H. J., Cox, D. E.: Phys. Rev. B **6** (1972) 2677.
72E 2	El Saffar, Z. M., O'Reilly, D. E., Peterson, E. M., Flick, C.: J. Chem. Phys. **57** (1972) 2372.
72F 2	Fermor, J. H., Kjekshus, A.: Acta Chem. Scand. **26** (1972) 2645.
72F 6	Fok, J., Hanscomb, J. R.: J. Appl. Phys. **43** (1972) 4824.
72G 1	Gaultier, M., Pannetier, G.: Rev. Chim. Miner. **9** (1972) 271.
72G 4	Gesi, K.: J. Phys. Soc. Jpn. **33** (1972) 108.
72G 5	Gesi, K., Ozawa, K.: J. Phys. Soc. Jpn. **33** (1972) 569.
72G 6	Gibby, M. G., Griffin, R. G., Pines, A., Waugh, J. S.: Chem. Phys. Lett. **17** (1972) 80.
72G 9	Glogarova, M., Fousek, J., Brezina, B.: Article on p. 75 in [72g 1].
72G 10	Glogarova, M., Frenzel, C., Hegenbarth, E.: Phys. Status Solidi (b) **53** (1972) 369.
72G 11	Gorbatyi, L. V., Ponomarev, V. I., Kheiker, D. M.: Kristallografiya **17** (1972) 1131; Sov. Phys. Crystallogr. (English Transl.) **17** (1973) 1002.
72G 12	Grassie, A. D. C., Benyon, A.: Phys. Lett. A **39** (1972) 199.
72G 13	Grigas, I. P., Kunigelis, V. F., Orlyukas, A. S., Samulionis, V. I.: Fiz. Tverd. Tela **14** (1972) 928; Sov. Phys. Solid State (English Transl.) **14** (1972) 796.
72G 14	Grunberg, J., Levin, S., Pelah, I., Gerlich, D.: Phys. Status Solidi (b) **49** (1972) 857.
72G 15	Gesi, K.: Ferroelectrics **4** (1972) 245.
72H 1	Hamakawa, Y., Yoshida, M., Yamanaka, K.: J. Non-Cryst. Solids **8, 9, 10** (1972) 868.
72H 2	Hamano, K., Shinmi, T.: J. Phys. Soc. Jpn. **33** (1972) 118.
72H 3	Hartwig, C. M., Wiener-Avnear, E., Porto, S. P. S.: Phys. Rev. B **5** (1972) 79.

72 H 4	Hasegawa, R., Takahashi, Y., Chatani, Y., Tadokoro, H.: Polymer J. **3** (1972) 600.
72 H 5	Hastings, R. N., Oja, T.: J. Chem. Phys. **57** (1972) 2139.
72 H 6	Hauret, G., Chapelle, J. P., Taurel, L.: Phys. Status Solidi (a) **11** (1972) 255.
72 H 7	Hauret, G., Taurel, L., Chapelle, J. P.: Solid State Commun. **10** (1972) 727.
72 H 10	Hinazumi, H., Mitsui, T.: Acta Crystallogr. B **28** (1972) 3299.
72 H 11	Hirakawa, K., Hayashi, H., Miike, H.: J. Phys. Soc. Jpn. **32** (1972) 1667.
72 H 15	Honeyman, W. N., Lee, M. K.: J. Phys. D **5** (1972) 188.
72 I 1	Ichikawa, M.: Acta Crystallogr. B **28** (1972) 755.
72 I 3	Inoue, K., Ishidate, T.: Rept. Fac. Sci., Shizuoka Univ. **7** (1972) 7.
72 I 4	Ishida, A., Mikami, O., Miyazawa, S., Sumi, M.: Appl. Phys. Lett. **21** (1972) 192.
72 I 5	Ishida, K., Mitsui, T.: Acta Crystallogr. A **28** (1972) S184.
72 I 6	Ishikawa, K., Tanaka, R., Toyoda, K.: Phys. Lett. A **42** (1972) 289.
72 J 1	Jacobi, H.: Z. Kristallogr. **135** (1972) 467.
72 K 1	Kachalov, O. V., Shpil'ko, I. A.: Zh. Eksp. Teor. Fiz. **62** (1972) 1840; Sov. Phys. JETP (English Transl.) **35** (1972) 957.
72 K 2	Kameyama, H., Ishibashi, Y., Takagi, Y.: J. Phys. Soc. Jpn. **33** (1972) 861.
72 K 3	Karyakina, N. F., Bochkov, B. G., Gavrilova, N. D., Novik, V. K., Koptsik, V. A.: Kristallografiya **17** (1972) 871; Sov. Phys. Crystallogr. (English Transl.) **17** (1973) 766.
72 K 4	Kato, T., Abe, R.: J. Phys. Soc. Jpn. **32** (1972) 717.
72 K 6	Kawano, T.: J. Phys. Soc. Jpn. **33** (1972) 1492.
72 K 7	Kay, M. I.: Ferroelectrics **4** (1972) 235.
72 K 8	Kay, M. I., Kleinberg, R.: Ferroelectrics **4** (1972) 147.
72 K 10	Kisynska, K., Kopcewicz, M., Kotlicki, A.: Phys. Status Solidi (b) **49** (1972) 85.
72 K 13	Kobayashi, T., Yakabe, M., Hukuda, K.: J. Phys. Soc. Jpn. **32** (1972) 578.
72 K 14	Krainik, N. N., Popov, S. N., Mylnikova, I. E.: Article on p. 179 in [72 g 1].
72 K 16	Kubarev, Y. G., Moskvich, Y. N., Zeer, E. P.: Phys. Status Solidi (b) **53** (1972) 41.
72 L 2	Laulicht, I., Pellach, E., Brith, M.: J. Chem. Phys. **57** (1972) 2857.
72 L 5	Levine, B. F., Bethea, C. G.: Appl. Phys. Lett. **20** (1972) 272.
72 L 7	Lobachev, A. N., Peskin, V. F., Popolitov, V. I., Syrkin, L. N., Feoktistova, N. N.: Fiz. Tverd. Tela **14** (1972) 604; Sov. Phys. Solid State (English Transl.) **14** (1972) 509.
72 L 9	Luspin, Y., Hauret, G.: C. R. Acad. Sci. (Paris) B **274** (1972) 995.
72 M 1	Mackowiak, M., Kurzynski, M.: Phys. Status Solidi (b) **51** (1972) 841.
72 M 10	McFee, J. H., Bergman, Jr., J. G., Crane, G. R.: Ferroelectrics **3** (1972) 305.
72 M 13	Melcher, R. L., Scott, B. A.: Phys. Rev. Lett. **28** (1972) 607.
72 M 14	Mielke, Z., Ratajczak, H.: Bull. Acad. Polon. Sci., Ser. Sci., Chim. **20** (1972) 255.
72 M 16	Mohana Rao, J. K., Viswamitra, M. A.: Acta Crystallogr. B **28** (1972) 1484.
72 M 17	Mori, K., Hayashi, M.: J. Phys. Soc. Jpn. **33** (1972) 1396.
72 M 18	Morosin, B., Samara, G. A.: Ferroelectrics **3** (1972) 49.
72 M 20	Muser, H. E., Schmitt, H.: Article on p. 103 in [72 g 1].
72 M 21	Mylov, V. P., Leonidova, G. G., Churagulov, B. R.: Fiz. Tverd. Tela **14** (1972) 2184; Sov. Phys. Solid State (English Transl.) **14** (1973) 1893.
72 N 3	Nelmes, R. J.: Phys. Status Solidi (b) **52** (1972) K89.
72 N 4	Nelmes, R. J.: Ferroelectrics **4** (1972) 133.
72 N 5	Newkirk, H. W., Quadflieg, P., Liebertz, J., Kockel, A.: Ferroelectrics **4** (1972) 51.
72 N 8	Niimura, N., Shimaoka, K., Motegi, H., Hoshino, S.: J. Phys. Soc. Jpn. **32** (1972) 1019.
72 O 1	Oddon, Y., Caranoni, C., Tranquard, A.: C. R. Acad. Sci. (Paris) C **274** (1972) 1671.
72 P 1	Parker, R. S., Schmidt, V. H.: J. Magn. Reson. **6** (1972) 507.
72 P 2	Petrov, S. V., Popov, M. A., Prozorova, L. A.: Zh. Eksp. Teor. Fiz. **62** (1972) 1884; Sov. Phys. JETP (English Transl.) **35** (1972) 981.
72 P 4	Polandov, I. N., Levina, M. E., Mylov, V. P.: Zh. Fiz. Khim. **46** (1972) 491; Russ. J. Phys. Chem. (English Transl.) **46** (1972) 281.
72 P 5	Prelesnik, B., Herak, R., Manojlovic-Muir, Lj., Muir, K. W.: Acta Crystallogr. B **28** (1972) 3104.
72 R 1	Ramaswamy, V.: Appl. Phys. Lett. **21** (1972) 183.
72 R 5	Rocaries, J.-C., Boldrini, P.: Appl. Phys. Lett. **20** (1972) 49.
72 R 7	Royal, G., Morlon, B., Godefroy, G.: C. R. Acad. Sci. (Paris) B **275** (1972) 353.
72 R 8	Ryan, J. K., Katiyar, R. J., Taylor, W.: Article on p. 49 in [72gl].
72 S 1	Salje, E.: Z. Kristallogr. **136** (1972) 135.

72S2	Sandercoock, J. R., Palmer, S. B., Elliott, R. J., Hayes, W., Smith, S. R. P., Young, A. P.: J. Phys. C **5** (1972) 3126.
72S4	Sastry, M. D.: Solid State Commun. **11** (1972) 1671.
72S5	Sato, S.: J. Phys. Soc. Jpn. **32** (1972) 1670.
72S6	Savage, H. T., Houston, B., Burke Jr., J. R.: Phys. Rev. B **6** (1972) 2292.
72S7	Sawaguchi, E., Cross, L. E.: Article on p. 327 in [72c1].
72S8	Schafer, H., Albers, J., Helwig, J.: Phys. Lett. A **39** (1972) 159.
72S9	Scheiding, C., Schmidt, G.: Phys. Status Solidi (a) **9** (1972) K77.
72S11	Schmidt, V. H., Parker, R. S.: Article on p. 109 in [72g1].
72S15	Scott, J. F., Wilson, C. M.: Solid State Commun. **10** (1972) 597.
72S16	Serdobol'skaya, O. Yu., Tam, K. T.: Fiz. Tverd. Tela **14** (1972) 2443; Sov. Phys. Solid State (English Transl.) **14** (1972) 2114.
72S17	She, C. Y., Broberg, T. W., Wall, L. S.: Phys. Rev. B **6** (1972) 1847.
72S19	Shimaoka, K., Tsuda, N., Yoshimura, Y.: Acta Crystallogr. A **28** (1972) S187.
72S20	Shiozaki, Y., Mitsui, T.: Acta Crystallogr. A **28** (1972) S185.
72S21	Shuvalov, L. A., Anisimova, V. N., Ivanov, N. R., Velichko, I. A.: Kristallografiya **17** (1972) 1256; Sov. Phys. Crystallogr. (English Transl.) **17** (1973) 1096.
72S22	Shuvalov, L. A., Ivanov, N. R., Kirpichnikova, L. F., Shchagina, N. M.: Kristallografiya **17** (1972) 966; Sov. Phys. Crystallogr. (English Transl.) **17** (1973) 851.
72S23	Sidnenko, E. V., Gladkii, V. V.: Kristallografiya **17** (1972) 978; Sov. Phys. Crystallogr. (English Transl.) **17** (1973) 861.
72S25	Smutny, F., Albers, J.: Phys. Status Solidi (b) **49** (1972) K159.
72S26	Stankowski, J., Mackowiak, M.: Phys. Status Solidi (b) **51** (1972) 449.
72S28	Strukov, B. A., Baddur, A., Koptsik, V. A., Velichko, I. A.: Fiz. Tverd. Tela **14** (1972) 1034; Sov. Phys. Solid State (English Transl.) **14** (1972) 885.
72S29	Suemune, Y.: J. Phys. Soc. Jpn. **20** (1965) 174.
72S31	Sugawara, F., Nakamura, T.: J. Phys. Chem. Solids **33** (1972) 1665.
72S33	Sun, T. S., Anderson, A.: Chem. Phys. Lett. **17** (1972) 104.
72S34	Suzuki, I., Maeda, M., Abe, R.: J. Phys. Soc. Jpn. **33** (1972) 860.
72S35	Semin, G. K.: Phys. Status Solidi **11** (1972) K61.
72S37	Scott, B. A., Burns, G.: J. Am. Ceram. Soc. **55** (1972) 331.
72T1	Takahashi, M., Manamatsu, S., Kimura, M.: Article on p. 681 in [72l1].
72T2	Taraskin, S. A., Lyakhovitskaya, V. A., Ivanov-Shits, A. K.: Kristallografiya **17** (1972) 681; Sov. Phys. Crystallogr. (English Transl.) **17** (1972) 597.
72T3	Tatsuzaki, I., Sakata, K., Todo, I., Tokunaga, M.: J. Phys. Soc. Jpn. **33** (1972) 438.
72T4	Tellgren, R., Ahmad, D., Liminga, R.: Chem. Scripta **2** (1972) 215.
72T5	Tellgren, R., Liminga, R.: J. Solid State Chem. **4** (1972) 255.
72T6	Temme, F. P., Waddington, T. C.: J. Chem. Soc., Faraday Trans. II **68** (1972) 350.
72T7	Teng, M. K., Balkanski, M., Massot, M.: Phys. Rev. B **5** (1972) 1031.
72T8	Terao, T., Hashi, T.: J. Magn. Reson. **7** (1972) 238.
72T12	Topic, M., Napijalo, M., Popovic, S., Zeljic, Z.: Phys. Status Solidi (a) **11** (1972) 787.
72T14	Torrie, B. H., Lin, C. C., Binbrek, O. S., Anderson, A.: J. Phys. Chem. Solids **33** (1972) 697.
72T16	Trontelj, Z., Rebic, M.: Solid State Commun. **11** (1972) 1337.
72T17	Tsunekawa, S., Ishibashi, Y., Takagi, Y.: J. Phys. Soc. Jpn. **33** (1972) 862.
72U3	Upreti, G. C.: Phys. Status Solidi (b) **54** (1972) 387.
72V1	Varikash, V. M., Zarembovskaya, T. A., Lagutina, Zh. P.: Kristallografiya **17** (1972) 473; Sov. Phys. Crystallogr. (English Transl.) **17** (1973) 768.
72V3	Vinogradova, I. S.: Kristallografiya **17** (1972) 410; Sov. Phys. Crystallogr. (English Transl.) **17** (1972) 350.
72V4	Volnyanskii, M. D., Kudzin, A. Yu., Sukhinskii, A. N.: Kristallografiya **17** (1972) 421; Sov. Phys. Crystallogr. (English Transl.) **17** (1972) 364.
72W2	Ward, R. W., Timusk, T.: Phys. Rev. B **5** (1972) 2351.
72W3	Whipps, P. W., Cosier, R. S., Bye, K. L.: J. Mater. Sci. **7** (1972) 1476.
72W4	Wiedemeier, H., Irene, E. A., Chaudhuri, A. K.: J. Cryst. Growth **13, 14** (1972) 393.
72Z1	Zarembovskaya, T. A., Varikash, V. M., Pupkevich, P. A.: Izv. VUZ Fiz. (1972) 153.
72Z2	Zorin, R. V., Al'shin, B. I., Astrov, D. N.: Zh. Eksp. Teor. Fiz. **62** (1972) 1201; Sov. Phys. JETP (English Transl.) **35** (1972) 634.

73 A 2	Acharya, P. K., Narayanan, P. S.: Indian J. Pure Appl. Phys. **11** (1973) 519.
73 A 3	Acharya, P. K., Narayanan, P. S.: Spectrochim. Acta A **29** (1973) 925.
73 A 4	Adriaenssens, G. J., Bjorkstam, J. L.: Phys. Status Solidi (a) **18** (1973) 129.
73 A 6	Aleksandrov, K. S., Anistratov, A. T., Ivanov, N. R., Mel'nikova, S. V., Shuvalov, L. A.: Fiz. Tverd. Tela **15** (1973) 456; Sov. Phys. Solid State (English Transl.) **15** (1973) 324.
73 A 7	Anderson, M. R., Brown, I. D., Vilminot, S.: Acta Crystallogr. B **29** (1973) 2625.
73 A 8	Andrade, P. da R., Prasad Rao, A. D., Katiyar, R. S., Porto, S. P. S.: Solid State Commun. **12** (1973) 847.
73 A 10	Arthur, J. W., Lockwood, D. J., Taylor, W.: Article on p. 144 in [73 m 1].
73 B 5	Blinc, R., Mali, M., Osredkar, R., Parker, R., Seliger, J., Žumer, S.: J. Chem. Phys. **59** (1973) 2947.
73 B 6	Blinc, R., Mali, M., Pirš, J., Žumer, S.: J. Chem. Phys. **58** (1973) 2262.
73 B 7	Blinc, R., Pirš, J., Žumer, S.: Phys. Rev. B **8** (1973) 15.
73 B 8	Bordeaux, D., Bornarel, J., Capiomont, A., Lajzerowicz-Bonneteau, J., Lajzerowicz, J., Legrand, J. F.: Phys. Rev. Lett. **31** (1973) 314.
73 B 9	Brooker, M. H.: J. Chem. Phys. **59** (1973) 5828.
73 B 10	Brosowski, G., Luther, G., Petersson, J.: Z. Naturforsch. a **28** (1973) 1814.
73 B 12	Buchman, P.: Ferroelectrics **5** (1973) 39.
73 B 16	Blinc, R., Lavrenčič, B., Levstek, I., Smolej, V., Žekš, B.: Phys. Status Solidi (b) **60** (1973) 255.
73 C 1	Canterford, R. P., Ninio, F.: J. Phys. C **6** (1973) 575.
73 C 3	Chihara, H., Higashigaki, Y.: Unpublished work (1973).
73 C 4	Chihara, H., Inaba, A., Nakamura, N., Yamamoto, T.: J. Phys. Soc. Jpn. **35** (1973) 1480.
73 D 4	Dietrich, W., Drissler, F., Schmid, D., Wolf, H. C.: Z. Naturforsch. **28a** (1973) 284.
73 F 1	Fairall, C. W., Reese, W.: Phys. Rev. B **8** (1973) 3475.
73 F 4	Fitzgerald, M. E., Casabella, P. A.: Phys. Rev. B **8** (1973) 3076.
73 F 5	Fleming, J. W., Siapkas, D., Lewis, J., Wilkinson, G. R.: High-Freq. Dielectr. Meas., Proc. Tutorial Conf. (1972), Guildford: IPC Sci. Technol. Press 1973.
73 F 8	Franke, V., Glogarova, M., Hegenbarth, E.: Phys. Status Solidi (b) **58** (1973) K69.
73 F 11	Fridkin, V. M., Grekov, A. A., Rodin, A. I., Savchenko, E. A., Volk, T. R.: Ferroelectrics **6** (1973) 71.
73 F 12	Fujimoto, S., Barnett, J. D.: Trans. Inst. Elect. Engrs. Jpn. A **93** (1973) 8 (in Japanese).
73 F 13	Furman, E., Brafman, O., Makovsky, J.: Phys. Rev. B **8** (1973) 2341.
73 G 1	Gesi, K., Ozawa, K.: Jpn. J. Appl. Phys. **12** (1973) 951.
73 G 2	Gesi, K., Ozawa, K.: J. Phys. Soc. Jpn. **35** (1973) 199.
73 G 3	Gesi, K., Ozawa, K.: J. Phys. Soc. Jpn. **35** (1973) 943.
73 G 4	Gesi, K., Ozawa, K.: J. Phys. Soc. Jpn. **35** (1973) 1562.
73 G 5	Gesi, K., Ozawa, K., Makita, Y.: Jpn. J. Appl. Phys. **12** (1973) 1962.
73 G 6	Gesi, K., Ozawa, K., Makita, Y.: Jpn. J. Appl. Phys. **12** (1973) 1963.
73 G 8	Gladkii, V. V., Magataev, V. K., Sidnenko, E. V.: Ferroelectrics **5** (1973) 107.
73 G 9	Glogarova, M., Fousek, J.: Phys. Status Solidi (a) **15** (1973) 579.
73 G 10	Golenishchev-Kutuzov, V. A., Bogdanova, H. G., Vladimirzev, Ju. V., Kopvillem, U. Ch.: Phys. Lett. A **43** (1973) 15.
73 G 11	Grosescu, R.: Chem. Phys. Lett. **21** (1973) 80.
73 G 13	Gopal, R., Calvo, C.: Z. Kristallogr. **137** (1973) 67.
73 H 1	Hamano, K.: J. Phys. Soc. Jpn. **35** (1973) 157.
73 H 2	Hamid, S. A.: Z. Kristallogr. **137** (1973) 412.
73 H 3	Hammer, H.: Article on p. 167 in [73 m 1].
73 H 4	Hanafusa, H., Kukuda, K., Furukawa, K.: J. Phys. Soc. Jpn. **35** (1973) 307.
73 H 5	Harris, L. B., Vella, G. J.: J. Chem. Phys. **58** (1973) 4550.
73 H 7	Hasse-Wessel, W.: Naturwissenschaften **60** (1973) 474.
73 H 8	Hatta, I., Ikushima, A.: J. Phys. Chem. Solids **34** (1973) 57.
73 H 10	Hewat, A. W.: Nature **246** (1973) 90.
73 H 11	Hikita, T., Sakata, K., Tatsuzaki, I.: J. Phys. Soc. Jpn. **34** (1973) 1248.
73 I 3	Ikeda, T., Fujibayashi, K., Nagai, T., Kobayashi, J.: Phys. Status Solidi (a) **16** (1973) 279.
73 I 5	Irie, K.: J. Phys. Soc. Jpn. **34** (1973) 1530.
73 I 7	Ishida, K., Mitsui, T., Shiozaki, Y.: Unpublished work (1973).
73 I 10	Itoh, K., Mitsui, T.: Ferroelectrics **5** (1973) 235.
73 J 2	Jain, Y. S., Bist, H. D., Upreti, G. C.: Chem. Phys. Lett. **22** (1973) 572.

73 J 3	Jaque, F., Herreros, J. M., Sanchez, C.: Solid State Commun. **12** (1973) 63.
73 K 2	Kameyama, H., Ishibashi, Y., Takagi, Y.: J. Phys. Soc. Jpn. **35** (1973) 1450.
73 K 7	Kato, T., Abe, R.: J. Phys. Soc. Jpn. **35** (1973) 1643.
73 K 8	Kawamura, T., Mitsuishi, A.: Technol. Rept., Osaka Univ. **23** (1973) 365.
73 K 9	Kawano, T., Furukata, K.: J. Phys. Soc. Jpn. **35** (1973) 1791.
73 K 10	Kay, M. I., Kleinberg, R.: Ferroelectrics **5** (1973) 45.
73 K 11	Keve, E. T., Skapski, A. C.: J. Solid State Chem. **8** (1973) 159.
73 K 12	Khan, A. A., Baur, W. H.: Acta Crystallogr. B **29** (1973) 2721.
73 K 13	Kiriyama, R., Kiriyama, H., Kitahama, K., Nakamura, O.: Chem. Lett. (1973) 1105.
73 K 15	Klyucharev, V. A., Bozhko, V. P., Zakurkin, V. V., Gaidamaka, A. P.: Fiz. Tverd. Tela **15** (1973) 587; Sov. Phys. Solid State (English Transl.) **15** (1973) 409.
73 K 17	Kramer, V., Schuhmacher, M., Nitsche, R.: Mater. Res. Bull. **8** (1973) 65.
73 K 18	Kruglik, A. I., Simonov, V. I., Yuzvak, V. I.: Kristallografiya **18** (1973) 287; Sov. Phys. Crystallogr. (English Transl.) **18** (1973) 177.
73 L 5	Luther, G.: Phys. Status Solidi (a) **20** (1973) 227.
73 M 1	Maciolek, R. B., Liu, S. T.: J. Electron. Mater. **2** (1973) 191.
73 M 2	Makita, Y., Suzuki, S.: J. Phys. Soc. Jpn. **34** (1973) 278.
73 M 3	Makita, Y., Wada, K.: J. Phys. Soc. Jpn. **34** (1973) 1111.
73 M 6	Martynov, V. G., Aleksandrov, K. S., Anistratov, A. T.: Fiz. Tverd. Tela **15** (1973) 2922; Sov. Phys. Solid State (English Transl.) **15** (1974) 1950.
73 M 7	Matsuo, T., Oguni, M., Suga, H., Seki, S.: Proc. Jpn. Acad. **49** (1973) 196.
73 M 8	Matsuo, T., Tatsumi, M., Suga, H., Seki, S.: Solid State Commun. **13** (1973) 1829.
73 M 10	McKenzie, D. R., Dryden, J. S.: J. Phys. C **6** (1973) 767.
73 M 12	Mercado, A., Gonzalo, J. A.: Ferroelectrics **5** (1973) 229.
73 M 14	Miki, H.: J. Phys. Soc. Jpn. **34** (1973) 1314.
73 N 3	Nakano, J.: Master Thesis, Hokkaido University (1973).
73 N 4	Nakano, J., Shiozaki, Y., Nakamura, E.: J. Phys. Soc. Jpn. **34** (1973) 1423.
73 N 5	Nakatani, N.: Jpn. J. Appl. Phys. **12** (1973) 1723.
73 N 6	Nako, K., Balkanski, M.: Phys. Rev. B **8** (1973) 5759.
73 N 7	Narasimhamurty, T. S., Veerabhadra Rao, K., Pettersen, H. E.: J. Mater. Sci. **8** (1973) 577.
73 N 10	Niimura, N., Fujii, Y., Motegi, H., Hoshino, S.: J. Phys. Soc. Jpn. **35** (1973) 842.
73 N 11	Nimmo, J. K., Lucas, B. W.: J. Phys. C **6** (1973) 201.
73 O 1	Ohashi, S., Abe, R., Kawamura, A.: J. Phys. Soc. Jpn. **34** (1973) 1694.
73 O 3	Oja, T.: J. Chem. Phys. **59** (1973) 2668.
73 P 3	Peercy, P. S.: Phys. Rev. B **8** (1973) 2033.
73 P 5	Petzelt, J.: Ferroelectrics **5** (1973) 219.
73 P 6	Petzelt, J., Grigas, J.: Ferroelectrics **5** (1973) 59.
73 P 7	Petzelt, J., Kislovskii, D., Shuvalov, L. A.: Kristallografiya **18** (1973) 560; Sov. Phys. Crystallogr. (English Transl.) **18** (1973) 351.
73 P 10	Poplavko, Yu. M., Meriakri, V. V., Pereverzeva, L. P., Aleshechkin, V. N., Molchanov, V. I.: Fiz. Tverd. Tela **15** (1973) 2515; Sov. Phys. Solid State (English Transl.) **15** (1974) 1672.
73 P 11	Poplavko, Yu. M., Pereverzeva, L. P., Meriakri, V. V., Ushatkin, E. F., Ogurtsov, S. V., Yashchikin, P. I.: Fiz. Tverd. Tela **15** (1973) 1250; Sov. Phys. Solid State (English Transl.) **15** (1973) 842.
73 P 13	Popolitov, V. I., Lobachev, A. N., Peskin, V. F., Syrkin, L. N., Feoktistova, N. N.: Kristallografiya **18** (1973) 411; Sov. Phys. Crystallogr. (English Transl.) **18** (1973) 258.
73 P 14	Pouga, G. D., Pouga, P. P., Maksimetz, V. V., Boretz, A. N., Hroshik, I. I., Bercha, D. M., Chepur, D. V.: Ferroelectrics **6** (1973) 111.
73 P 15	Pura, B., Przedmojski, J.: Phys. Lett. A **43** (1973) 217.
73 R 1	Ramani, K., Srinivasan, R.: Indian J. Pure Appl. Phys. **11** (1973) 223.
73 R 5	Reese, R. L., Fritz, I. J., Cummins, H. Z.: Phys. Rev. B **7** (1973) 4165.
73 S 1	Samara, G. A.: Ferroelectrics **5** (1973) 25.
73 S 5	Semmingsen, D.: Acta Chem. Scand. **27** (1973) 3961.
73 S 8	Shimizu, H., Umeno, M., Wakita, K., Kameyama, H., Ishibashi, Y.: J. Phys. Soc. Jpn. **34** (1973) 983.
73 S 10	Shustin, O. A., Velichkina, T. S., Mikheeva, L. F., Yakovlev, I. A.: Zh. Eksp. Teor. Fiz. Pis'ma **18** (1973) 632; JETP Lett. (English Transl.) **18** (1973) 370.
73 S 11	Shuvalov, L. A., Ivanov, N. R., Chikhladze, O. A., Izrailenko, A. N.: Kristallografiya **18** (1973) 1207; Sov. Phys. Crystallogr. (English Transl.) **18** (1974) 758.

73 S 15	Stefanovich, S. Yu., Venevtsev, Yu. N.: Phys. Status Solidi (a) **20** (1973) K49.
73 S 18	Strukov, B. A., Baddur, A., Zinenko, V. I., Mikailov, V. K., Koptsik, V. A.: Fiz. Tverd. Tela **15** (1973) 2018; Sov. Phys. Solid State (English Transl.) **15** (1974) 1347.
73 S 20	Strukov, B. A., Skomorokhova, T. L., Koptsik, V. A., Boiko, A. A., Izrailenko, A. N.: Kristallografiya **18** (1973) 143; Sov. Phys. Crystallogr. (English Transl.) **18** (1973) 86.
73 T 1	Tellgren, R., Ahmad, D., Liminga, R.: J. Solid State Chem. **6** (1973) 250.
73 T 2	Temme, F. P., Smith, J. A. S., Waddington, T. C.: J. Chem. Soc., Faraday Trans. II **69** (1973) 1.
73 T 3	Torrie, B. H., Knipsel, R. R.: Ferroelectrics **5** (1973) 53.
73 T 4	Tsuchida, K., Abe, R., Naito, M.: J. Phys. Soc. Jpn. **35** (1973) 806.
73 T 5	Tsunekawa, S., Ishibashi, Y., Takagi, Y.: J. Phys. Soc. Jpn. **34** (1973) 470.
73 T 6	Tumolillo, T.A.: Phys. Status Solidi (a) **17** (1973) 315.
73 V 1	Varikash, V. M., Zarembovskaya, T. A.: Izv. VUZ Fiz. (1973) **141**.
73 V 2	Volkova, E. N., Velichko, I. A.: Kristallografiya **18** (1973) 409; Sov. Phys. Crystallogr. (English Transl.) **18** (1973) 256.
73 W 1	Warkusz, F., Lukaszewicz, K.: Bull. Acad. Polon. Sci., Ser. Sci., Chim. **21** (1973) 271.
73 W 2	Warkusz, F., Lukaszewicz, K.: Bull. Acad. Polon. Sci., Ser. Sci., Chim. **21** (1973) 275.
73 W 3	Watton, A., Petch, H. E., Pintar, M. M.: Can. J. Phys. **51** (1973) 1005.
73 W 4	Wegener, H. H. F., Braunecker, B., Ritter, G., Seyboth, D.: Z. Phys. **262** (1973) 149.
73 Y 3	Yoshida, M., Yamanaka, K., Hamakawa, Y.: Jpn. J. Appl. Phys. **12** (1973) 1699.
74 A 1	Abe, R., Kamiya, N., Matsuda, M.: Ferroelectrics **8** (1974) 557.
74 A 2	Adriaenssens, G. J.: Ferroelectrics **7** (1974) 123.
74 A 4	Anderson, A., Bastie, P., Bornarel, J., Lajzerowicz, J., Legrand, J. F.: Ferroelectrics **8** (1974) 403.
74 A 5	Anderson, M. R., Brown, I. D.: Acta Crystallogr. B **30** (1974) 831.
74 A 6	Anistratov, A. T., Mel'nikova, S. V.: Kristallografiya **19** (1974) 815; Sov. Phys. Crystallogr. (English Transl.) **19** (1975) 504.
74 A 8	Azoulay, J., Gerlich, D., Wiener, E., Pelah, I.: Ferroelectrics **8** (1974) 599.
74 B 1	Badr, Ya. A.-Kh., Karpov, S. V., Shultin, A. A.: Fiz. Tverd. Tela **16** (1974) 2327; Sov. Phys. Solid State (English Transl.) **16** (1975) 1515.
74 B 4	Benoit, J. P., Chapelle, J. P.: Solid State Commun. **14** (1974) 883.
74 B 8	Bjorkstam, J. L., Wei, C. H.: Ferroelectrics **7** (1974) 127.
74 B 9	Blinc, R., Lahajnar, G., Levstik, A., Osredkar, R., Shuvalov, L. A.: Phys. Status Solidi (b) **65** (1974) 397.
74 B 10	Blinc, R., Seliger, J., Osredkar, R., Mali, M.: Phys. Lett. A **47** (1974) 131.
74 B 11	Blinc, R., Slak, J., Zupančič, I.: J. Chem. Phys. **61** (1974) 988.
74 B 14	Bonera, G., Massara, C. I., Mognaschi, E. R., Rigamonti, A.: Solid State Commun. **15** (1974) 19.
74 B 15	Bordeaux, P. D., Capiomont, A., Lajzerowicz-Bonneteau, J., Jouve, M., Thomas, M.: Acta Crystallogr. B **30** (1974) 2156.
74 B 16	Bornarel, J., Lajzerowicz, J., Legrand. J. F.: Ferroelectrics **7** (1974) 313.
74 B 17	Brezina, B., Havrenkova, M.: J. Cryst. Growth **21** (1974) 77.
74 B 18	Brillson, L. J., Burstein, E., Muldawer, L.: Phys. Rev. B **9** (1974) 1547.
74 B 19	Brody, E. M., Cummins, H. Z.: Phys. Rev. B **9** (1974) 179.
74 B 20	Brosowski, G., Buchheit, W., Muller, D., Petersson, J.: Phys. Status Solidi (b) **62** (1974) 93.
74 B 21	Burghoff, U., Rosenberger, H., Zeiss, R., Muller, R., Rashkovich, L. N.: Phys. Status Solidi (a) **26** (1974) K171.
74 B 22	Burghoff, U., Scheller, G., Muller, R.: Phys. Status Solidi (a) **25** (1974) K31.
74 B 23	Burkov, V. I., Ivanova, I. N., Kizel', V. A., Pakhomov, V. I.: Kristallografiya **19** (1974) 1093; Sov. Phys. Crystallogr. (English Transl.) **19** (1975) 676.
74 C 2	Carpentier, C. D., Nitsche, R.: Mater. Res. Bull. **9** (1974) 401.
74 C 3	Carpentier, C. D., Nitsche, R.: Mater. Res. Bull. **9** (1974) 1097.
74 C 4	Chabin, M., Gilletta, F.: Ferroelectrics **8** (1974) 563.
74 C 6	Chemla, D. S., Batifal, E.: Opt. Commun. **11** (1974) 57.
74 C 7	Chihara, H., Nakamura, N., Okuma, H.: J. Phys. Soc. Jpn. **37** (1974) 1385.
74 D 1	Dalal, N. S., McDowell, C. A., Srinivasan, R.: J. Chem. Phys. **60** (1974) 3787.
74 D 3	Day, G. W., Hamilton, C. A., Peterson, R. L., Phelan Jr., R. J., Mullen, L. O.: Appl. Phys. Lett. **24** (1974) 456.
74 D 6	Deutsch, M., Litov, E.: Ferroelectrics **7** (1974) 209.
74 D 7	Dietrich, O. W., Cowley, R. A., Shapiro, S. M.: J. Phys. C **7** (1974) L239.
74 D 8	Dittmar, G., Schäfer, H.: Z. Naturforsch. b **29** (1974) 312.

74E3	Eiriksson, V. R., Rouse, K. D., Nelmes, R. J.: Ferroelectrics **8** (1974) 485.
74F2	Fong, C. Y., Petroff, Y., Kohn, S., Shen, Y. R.: Solid State Commun. **14** (1974) 681.
74F5	Fujimoto, M., Dressel, L. A.: Ferroelectrics **8** (1974) 611.
74F7	Fano, V., Ortalli, I.: J. Chem. Phys. **61** (1974) 5017.
74G2	Gaillard, J., Gloux, P., Lamatte, B.: Solid State Commun. **14** (1974) 417.
74G3	Gaillard, J., Gloux, P., Lamotte, B.: Mol. Phys. **27** (1974) 1441.
74G5	Gesi, K., Ozawa, K.: J. Phys. Soc. Jpn. **36** (1974) 1496.
74G6	Gesi, K., Ozawa, K.: J. Phys. Soc. Jpn. **37** (1974) 1053.
74G7	Gillespie, S., Schurmann, H. K., Gunton, J. D., Mihalisin, T.: Solid State Commun. **15** (1974) 1753.
74G8	Glogarova, M.: Phys. Status Solidi (a) **22** (1974) K69.
74G10	Grosescu, R.: Rev. Roumaine Phys. **19** (1974) 565.
74G11	Grossmann, J., Müller, D., Petersson, J.: Z. Naturforsch. a **29** (1974) 1055.
74H2	Hamano, K., Yamaguchi, T.: Ferroelectrics **7** (1974) 241.
74H3	Hamid, S. A.: J. Cryst. Growth **22** (1974) 331.
74H5	Hatanaka, H., Terao, T., Morimoto, K., Hashi, T.: J. Phys. Soc. Jpn. **36** (1974) 1207.
74I5	Imai, K.: J. Phys. Soc. Jpn. **36** (1974) 1069.
74I6	Inoue, K., Ishidate, T.: Ferroelectrics **7** (1974) 105.
74I7	Irie, K., Ohi, K.: Ferroelectrics **8** (1974) 555.
74I9	Ishida, K., Mitsui, T.: Ferroelectrics **8** (1974) 475.
74I10	Ishidate, T., Inoue, K., Kameyama, H., Ishibashi, I.: J. Phys. Soc. Jpn. **37** (1974) 1176.
74I11	Isupov, V. A., Golovshchikova, G. I., Myl'nikova, I. E.: Ferroelectrics **8** (1974) 507.
74I14	Itoh, K., Ogusu, K., Shiozaki, Y., Toyoda, K.: Ferroelectrics **7** (1974) 79.
74I15	Ivanov, N. R., Koniak, C.: Kristallografiya **19** (1974) 1216; Sov. Phys Crystallogr. (English Transl.) **19** (1975) 755.
74I16	Ivanova, L. A., Popolitov, V. I., Stefanovich, S. Yu., Lobachev, A. N., Venevtsev, Yu. N.: Kristallografiya **19** (1974) 573; Sov. Phys. Crystallogr. (English Transl.) **19** (1974) 356.
74I18	Ishikawa, K., Shibata, Y., Toyoda, K.: Phys. Status Solidi (a) **25** (1974) K187.
74J1	Jeitschko, W., Sleight, A. W.: Acta Crystallogr. B **30** (1974) 2088.
74J2	Joshi, M. S., Paul, B. K.: J. Cryst. Growth **22** (1974) 321.
74K1	Kameyama, H., Ishibashi, Y., Takagi, Y.: J. Phys. Soc. Jpn. **36** (1974) 614.
74K2	Kamysheva, L. N., Zolototrubov, Yu. S., Grindnev, S. A.: Ferroelectrics **8** (1974) 559.
74K3	Kasahara, M., Tatsuzaki, I.: J. Phys. Soc. Jpn. **36** (1974) 786.
74K4	Kasahara, M., Tatsuzaki, I.: J. Phys. Soc. Jpn. **37** (1974) 167.
74K6	Kato, K.: IEEE J. Quantum Electron., QE **10** (1974) 616.
74K7	Kato, K.: IEEE J. Quantum Electron., QE **10** (1974) 622.
74K8	Kato, T., Abe, R.: J. Phys. Soc. Jpn. **36** (1974) 1065.
74K9	Kato, K., Nakao, S.: Jpn. J. Appl. Phys. **13** (1974) 1681.
74K10	Kawamura, T., Mitsuishi, A., Furuya, N., Shimomura, O.: Opt. Commun. **10** (1974) 337.
74K11	Kawano, T.: J. Phys. Soc. Jpn. **37** (1974) 848.
74K12	Keester, K. L., Jacobs, J. T.: Ferroelectrics **8** (1974) 657.
74K13	Keune, W.: Phys. Rev. B **10** (1974) 5057.
74K14	Khairetdinov, E. F., Boldyrev, V. V., Burshtein, A. I.: J. Solid State Chem. **10** (1974) 288.
74K17	Konak, C., Fousek, J., Ivanov, N. R.: Ferroelectrics **6** (1974) 235.
74K19	Kondo, T., Ishibashi, Y., Takagi, Y.: J. Phys. Soc. Jpn. **37** (1974) 1708.
74K21	Krämer, V., Nitsche, R., Schuhmacher, M.: J. Cryst. Growth **24, 25** (1974) 179.
74K23	Kuru, Y., Ikeda, T.: Unpublished work (1974).
74L1	Lagakos, N., Cummins, H. Z.: Phys. Rev. B **10** (1974) 1063.
74L2	Lahajnar, G., Blinc, R., Žumer, S.: Phys. Condens. Matter **18** (1974) 301.
74L4	Lewis, J. E.: Phys. Lett. A **47** (1974) 404.
74L9	Loiacono, G. M., Balascio, J. F., Osborne, W.: Appl. Phys. Lett. **24** (1974) 455.
74L10	Lowndes, R. P., Tornberg, N. E., Leung, R. C.: Phys. Rev. B **10** (1974) 911.
74L14	Luther, G., Birkenheier, C., Brosowski, G., Müser, H. E., Petersson, J.: Ferroelectrics **8** (1974) 569.
74M2	Makita, Y., Suzuki, S.: J. Phys. Soc. Jpn. **36** (1974) 1215.
74M3	Makita, Y., Yamauchi, Y.: J. Phys. Soc. Jpn. **37** (1974) 1470.
74M4	Manoogian, A., Leclerc, A.: Phys. Rev. B **10** (1974) 1052.
74M7	Massara, C. I., Mognaschi, E. R.: Ferroelectrics **7** (1974) 115.
74M9	Matsuo, T., Oguni, M., Suga, H., Seki, S., Nagle, J. F.: Bull. Chem. Soc. Jpn. **47** (1974) 57.
74M13	Miller, R. C., Nordland Jr., W. A.: J. Appl. Phys. **45** (1974) 898.

74M14	Mitani, S., Fukui, S., Shibuya, I., Shiozaki, Y., Toyoda, K., Pepinsky, R.: Ferroelectrics **8** (1974) 477.
74M15	Morimoto, K.: J. Phys. Soc. Jpn. **37** (1974) 1574.
74N1	Nakano, J., Shiozaki, Y., Nakamura, E.: Ferroelectrics **8** (1974) 483.
74N5	Negran, T. J., Glass, A. M., Brickenkamp, C. S., Rosenstein, R. D., Osterheld, R. K., Susott, R.: Ferroelectrics **6** (1974) 179.
74N7	Nelmes, R. J., Rouse, K. D.: Ferroelectrics **8** (1974) 487.
74N12	Nozik, Yu. Z.: Geokhimiya **1974**, 228.
74N13	Nelmes, R. J.: J. Phys. C **7** (1974) 3840.
74O1	Ohama, N.: Mater. Res. Bull. **9** (1974) 283.
74O3	Orlyukas, A., Grigas, J.: Liet. Fiz. Rinkinys **14** (1974) 313.
74O4	Orlyukas, A. S., Grigas, I. P.: Kristallografiya **19** (1974) 880; Sov. Phys. Crystallogr. (English Transl.) **19** (1975) 547.
74P1	Petzelt, J., Grigas, J., Mayerova, I.: Ferroelectrics. **6** (1974) 225.
74P2	Phelan Jr., R. J., Peterson, R. L., Hamilton, C. A., Day, G. W.: Ferroelectrics **7** (1974) 375.
74P5	Poplavko, Yu. M., Pereverzeva, L. P., Yashchishin, P. I., Mariakri, V. V., Ushatkin, E. F., Ogurtsov, S. V.: Ukr. Fiz. Zh (Russ. Ed.) **19** (1974) 1688.
74R2	Recker, K., Wallrafen, F., Haussuhl, S.: J. Cryst. Growth **26** (1974) 97.
74R5	Roos, J., Kind, R.: Ferroelectrics **8** (1974) 553.
74R6	Rucci, A., Serpi, A., Serpi-Macciotta, P.: Phys. Status Solidi (a) **25** (1974) K191.
74R7	Ryan, J. F., Scott, J. F.: Solid State Commun. **14** (1974) 5.
74S1	Sakami, J.: Phys. Lett. A **50** (1974) 109.
74S3	Salje, E.: Acta Crystallogr. A **30** (1974) 518.
74S4	Salje, E.: Z. Kristallogr. **139** (1974) 317.
74S6	Samara, G. A.: Ferroelectrics **7** (1974) 221.
74S7	Samulionis, V. I., Orlyukas, A. S., Grigas, I. P.: Fiz. Tverd. Tela **16** (1974) 206; Sov. Phys. Solid State (English Transl.) **16** (1974) 127.
74S8	Sasaki, Y.: J. Phys. Soc. Jpn. **37** (1974) 1570.
74S9	Sato, K., Sawada, A., Takagi, Y., Ishibashi, Y.: J. Phys. Soc. Jpn. **36** (1974) 616.
74S12	Schmidt, G., Warnstorff, K.: Phys. Status Solidi (a) **24** (1974) K137.
74S13	Sedlakova, L., Petrzilka, V., Kawalec, M., Cech, J., Michalec, R., Mikula, P., Chalupa, B., Zelenka, J.: Phys. Status Solidi (a) **22** (1974) K223.
74S14	Semmingsen, D., Feder, J.: Solid State Commun. **15** (1974) 1369.
74S15	Seryshev, S. A., Vinogradova, I. S., Buznik, V. M.: Fiz. Tverd. Tela **16** (1974) 881; Sov. Phys. Solid State (English Transl.) **16** (1974) 565.
74S17	Shimizu, H., Ishibashi, Y., Tsukamoto, M., Umeno, M.: Ferroelectrics **8** (1974) 521.
74S18	Shimizu, H., Tsukamoto, M., Ishibashi, Y., Umeno, M.: J. Phys. Soc. Jpn. **36** (1974) 498.
74S20	Silvidi, A. A., Workman, D. T.: Ferroelectrics **6** (1974) 183.
74S21	Singh, S., Singh, K.: J. Phys. Soc. Jpn. **36** (1974) 1558.
74S23	Sirota, N. N., Korina, R. V.: Kristall Tech. **9** (1974) 55.
74S31	Sorge, G., Frenzel, Ch.: Phys. Status Solidi (a) **23** (1974) K67.
74S32	Stankowski, J., Galezewski, A., Waplak, S., Gruszczynaka, U., Gierszal, H.: Ferroelectrics **6** (1974) 209.
74S33	Stankowski, J., Waplak, S., Yurin, V. A.: Phys. Status Solidi (a) **22** (1974) K41.
74S34	Stankowski, J., Wieckowski, A., Hedewy, S.: J. Magn. Reson. **15** (1974) 498.
74S36	Stepien-Damm, J. A., Suski, T., Meysner, L., Hilczer, B., Lukaszewicz, K.: Bull. Acad. Polon. Sci., Ser. Sci., Chim. **22** (1974) 685.
74S39	Suzuki, I.: J. Phys. Soc. Jpn. **37** (1974) 1379.
74T2	Tamura, M., Ogasawara, K., Ono, N., Hagiwara, S.: J. Appl. Phys. **45** (1974) 3768.
74T3	Tellgren, R., Liminga, R.: Acta Crystallogr. B **30** (1974) 2497.
74T4	Tellgren, R., Liminga, R.: Ferroelectrics **8** (1974) 629.
74T5	Terao, K., Hashi, T.: J. Phys. Soc. Jpn. **36** (1974) 989.
74T6	Todo, I., Tatsuzaki, I.: Phys. Status Solidi (a) **23** (1974) 591.
74T7	Todo, I., Tatsuzaki, I.: J. Phys. Soc. Jpn. **37** (1974) 1477.
74T10	Tomas, A., Benoit, J.-L., Herpin, P., Mercier, D.: C. R. Acad. Sci. (Paris) B **278** (1974) 779.
74T11	Trevino, S. F., Prask, H., Casella, R. C.: Phys. Rev. B **10** (1974) 739.
74U1	Uchida, K., Takahashi, H., Higasi, K.: Bull. Chem. Soc. Jpn. **47** (1974) 1545.
74U3	Uesu, Y., Tanaka, T., Kobayashi, J.: Ferroelectrics **7** (1974) 247.

74V1	Varikash, V. M., Sarembovskaja, T. A.: Kristall Tech. **9** (1974) 685.
74V2	Volkel, G., Bartuch, V., Brunner, W., Windsch, W.: Phys. Status Solidi (a) **25** (1974) 591.
74W2	Windsch, W., Kunze, R., Volkel, G.: Phys. Status Solidi (a) **26** (1974) K175.
74W3	Windsch, W., Welter, M., Driesel, W.: Ferroelectrics **8** (1974) 551.
74Y1	Yacoby, Y., Cerdeira, F., Schmidt, M., Holzapfel, W. B.: Solid State Commun. **14** (1974) 1325.
74Y2	Yagi, T., Tatsuzaki, I.: J. Phys. Soc. Jpn. **37** (1974) 1038.
74Y3	Yano, S., Tadano, K., Aoki, K.: J. Polymer Sci., Polymer Phys. Ed. **12** (1974) 1875.
74Y5	Yuzvak, V. I., Zherebtsova, L. I., Shkuryaeva, V. B., Aleksandrova, I. P.: Kristallografiya **19** (1974) 773; Sov. Phys. Crystallogr. (English Transl.) **19** (1975) 480.
74Z1	Zaitov, M. M., Zaripov, M. M., Shakirov, R. K.: Fiz. Tverd. Tela **16** (1974) 1579; Sov. Phys. Solid State (English Transl.) **16** (1974) 1033.
74Z2	Zaitseva, M. P., Anistratov, A. T., Krumin', A. E., Shabanova, L. A., Iskornev, I. M.: Kristallografiya **19** (1974) 1020; Sov. Phys. Crystallogr. (English Transl.) **19** (1975) 632.
74Z3	Zerem, J. Z., Halperin, A.: Ferroelectrics **7** (1974) 205.
74Z4	Zhdanova, V. V., Sergeev, V. P., Strukov, B. A., Izrailenko, A. N.: Fiz. Tverd. Tela **16** (1974) 1490; Sov. Phys. Solid State (English Transl.) **16** (1974) 958.
74Z5	Zheludev, I. S., Gladkii, V. V., Sidnenko, E. V., Magataev, V. K.: Ferroelectrics **8** (1974) 567.
74Z7	Žumer, S., Pirs, J.: Ferroelectrics **7** (1974) 119.
75A1	Adonin, A. A., Grekov, A. A.: Fiz. Tverd. Tela **17** (1975) 2410; Sov. Phys. Solid State (English Transl.) **17** (1976) 1591.
75A2	Adriaenssens, G. J.: Phys. Rev. B **12** (1975) 5116.
75A4	Agaronov, B. S., Levedeva, N. N., Sheinkman, M. K.: Fiz. Tverd. Tela **17** (1975) 2188; Sov. Phys. Solid State (English Transl.) **17** (1976) 1450.
75A6	Aleksandrov, K. S., Aleksandrova, I. P., Zherebtsova, L. I., Kruglik, A. I., Krupnyi, A. I., Mel'nikova, S. V., Shneider, V. E.: Izv. Akad. Nauk SSSR, Ser. Fiz. **39** (1975) 943; Bull. Acad. Sci. USSR, Phys. Ser. (English Transl.) **39**, No. 5 (1975) 44.
75A7	Aleksandrov, K. S., Zherebtsova, L. I., Zaitseva, M. P., Haussuhl, S.: Izv. Akad. Nauk SSSR, Ser. Fiz. **39** (1975) 939; Bull. Acad. Sci. USSR, Phys. Ser. (English Transl.) **39**, No. 5 (1975) 40.
75A8	Aleksandrova, I. P., Repeta, V. S., Konstantinova, V. P.: Izv. Akad. Nauk SSSR, Ser. Fiz. **39** (1975) 922; Bull. Acad. Sci. USSR, Phys. Ser. (English Transl.) **39**, No. 5 (1975) 27.
75A9	Anderson, M. R., Brown, I. D.: Acta Crystallogr. B **31** (1975) 1500.
75A11	Anistratov, A. T., Mel'nikova, S. V.: Izv. Akad. Nauk SSSR, Ser. Fiz. **39** (1975) 808; Bull. Acad. Sci. USSR, Phys. Ser. (English Transl.) **39**, No. 4 (1975) 144.
75A12	Anistratov, A. T., Mel'nikova, S. V., Komogortsev, S. V.: Kristallografiya **20** (1975) 848; Sov. Phys. Crystallogr. (English Transl.) **20** (1976) 522.
75A13	Ashmore, J. P., Petch, H. E.: Can. J. Phys. **53** (1975) 2694.
75A14	Avogadro, A., Bonera, G., Rigamonti, A.: J. Magn. Reson. **20** (1975) 399.
75B1	Bailey, C. E., Cuesta-Barro, R. C.: J. Chem. Phys. **63** (1975) 4120.
75B2	Baisa, L. F., Barabash, A. I., Belokon, A. T., Puchkovskaya, G. A., Semin, G. K., Frolkov, Yu. A.: Izv. Akad. Nauk SSSR, Ser. Fiz. **39** (1975) 702; Bull. Acad. Sci. USSR, Phys. Ser. (English Transl.) **39**, No. 4 (1975) 51.
75B3	Baisa, D. F., Barabash, A. I., Moskalev, A. K., Semin, G. K.: Phys. Status Solidi (a) **27** (1975) 315.
75B7	Beerman, H. P.: Infrared Phys. **15** (1975) 225.
75B10	Bercha, D. M., Zayachkovskii, M. P., Zayachkovskaya, N. F., Maslyuk, V. T.: Fiz. Tverd. Tela **17** (1975) 2122; Sov. Phys. Solid State (English Transl.) **17** (1976) 1393.
75B11	Bergman, J. G., Crane, G. R., Guggenheim, H.: J. Appl. Phys. **46** (1975) 4645.
75B13	Blinc, R., Burgar, M., Čižikov, S., Levstik, A., Kadaba, P., Shuvalov, L. A.: Phys. Status Solidi (b) **67** (1975) 689.
75B14	Blinc, R., Mali, M., Osredkar, R., Seliger, J.: J. Chem. Phys. **63** (1975) 35.
75B15	Blinc, R., Mali, M., Žumer, S.: J. Chem. Phys. **63** (1975) 2898.
75B16	Blinc, R., Žumer, S.: J. Chem. Phys. **62** (1975) 3118.
75B19	Březina, B., Smutný, F., Fousek, J.: Czech. J. Phys. B **25** (1975) 1411.
75C1	Camnasio, A. J., Gonzalo, J. A.: J. Phys. Soc. Jpn. **39** (1975) 451.
75C3	Cevc, P., Schara, M.: Phys. Status Solidi (a) **30** (1975) 357.
75C4	Chabin, M., Gilletta, F.: Ferroelectrics **9** (1975) 155.
75C5	Chihara, H.: Private communication (1975) cited in [75G7].
75C8	Crane, G. R.: Z. Kristallogr. **141** (1975) 312.
75C9	Czapla, Z., Kolodziej, H. A., Sobczyk, L.: J. Chem. Soc., Faraday Trans. II **71** (1975) 763.

75C10	Czapla, Z., Kolodziej, K. A., Sobczyk, L.: J. Chem. Soc., Faraday Trans. II **71** (1975) 767.
75D1	de Benyacar, M. A. R., de Dussel, H. L.: Ferroelectrics **9** (1975) 241.
75D3	Deigen, M. F., Gromovoi, Y. S., Zheru, I. I., Popovich, G. A.: Fiz. Tverd. Tela **17** (1975) 3075; Sov. Phys. Solid State (English Transl.) **17** (1976) 2038.
75D4	Delahaigue, A., Khelifa, B., Jouve, P.: Phys. Status Solidi (b) **72** (1975) 585.
75D6	Dimir, V., Osredkar, M.: Phys. Status Solidi (a) **30** (1975) K63.
75D7	Donaldson, J. D., Kjekshus, A., Nicholson, D. G., Southern, J.T.: Acta Chem. Scand. A **29** (1975) 220.
75D9	Dvorak, V.: Phys. Status Solidi (b) **71** (1975) 269.
75D11	Same as [75D7].
75E2	Ema, K., Hamano, K., Hatta, I.: J. Phys. Soc. Jpn. **39** (1975) 726.
75F2	Fawcett, V., Long, D. A., Sankaranarayanan, V. N.: J. Raman Spectrosc. **3** (1975) 177.
75F3	Fawcett, V., Long, D. A., Sankaranarayanan, V. N.: J. Raman Spectrosc. **3** (1975) 197.
75F4	Figuiere, P., Ghelfenstein, M., Szwarc, H.: Chem. Phys. Lett. **33** (1975) 99.
75F7	Franke, V., Hegenbarth, E., Brezina, B.: Phys. Status Solidi (a) **28** (1975) K77.
75F8	Fritz, I. J.: Phys. Lett. A **51** (1975) 219.
75F9	Fujimoto, M., Dressel, L. A.: Ferroelectrics **9** (1975) 221.
75G1	Gaillard, J., Gloux, P.: J. Phys. (Paris) **36** (1975) 683.
75G5	Gauss, K. E., Happ, H., Rother, G.: Phys. Status Solidi (b) **72** (1975) 623.
75G7	Gesi, K., Ozawa, K.: J. Phys. Soc. Jpn. **38** (1975) 459.
75G8	Gesi, K., Ozawa, K.: J. Phys. Soc. Jpn. **38** (1975) 467.
75G9	Gesi, K., Ozawa, K.: J. Phys. Soc. Jpn. **39** (1975) 1026.
75G12	Gupta, L. C., Vijaryaraghavan, R., Ripmeester, J. A., Garg, S. K.: J. Phys. Soc. Jpn. **38** (1975) 1784.
75H2	Hatta, I.: J. Phys. Soc. Jpn. **38** (1975) 1430.
75H3	Havlin, S., Litov, E., Sompolinsky, H.: Phys. Lett. A **51** (1975) 33.
75H4	Herlinger, A. W.: Spectrosc. Lett. **8** (1975) 787.
75H5	Hikita, T., Tatsuzaki, I.: J. Phys. Soc. Jpn. **38** (1975) 1083.
75H6	Hildmann, B. O., Hahn, Th., Cross, L. E., Newnham, R. E.: Appl. Phys. Lett. **27** (1975) 103.
75H7	Hok, N. N., Calvo, C.: Can. J. Chem. **53** (1975) 1449.
75H8	Holden, J. R., Dickinson, C. W.: J. Phys. Chem. **79** (1975) 249.
75I2	Ikeda, T., Yasuda, G.: Jpn. J. Appl. Phys. **14** (1975) 1287.
75I6	Itoh, K., Matsubayashi, T., Nakamura, E., Motegi, H.: J. Phys. Soc. Jpn. **39** (1975) 843.
75I7	Ivanov, N. R., Shuvalov, L. A., Schmidt, H., Stolp, E.: Izv. Akad. Nauk SSSR, Ser. Fiz. **39** (1975) 933; Bull. Acad. Sci. USSR, Phys. Ser. (English Transl.) **39**, No. 5 (1975) 35.
75I8	Ivanova, L. A., Popolitov, B. N., Belous, A. G., Venevtsev, Yu. N.: Izv. Akad. Nauk SSSR, Ser. Fiz. **39** (1975) 1036; Bull. Acad. Sci. USSR, Phys. Ser. (English Transl.) **39**, No. 5 (1975) 118.
75I9	Iizumi, M., Hamaguchi, Y., Komatsubara, K. F., Kato, Y.: J. Phys. Soc. Jpn. **38** (1975) 443.
75K1	Kachalov, N. P., Orlyukas, A., Polandov, I. N., Grigas, I.: Fiz. Tverd. Tela **17** (1975) 1790; Sov. Phys. Solid State (English Transl.) **17** (1975) 1164.
75K2	Kadaba, P. K., Pirnat, J., Trontelj, Z.: Chem. Phys. Lett. **32** (1975) 382.
75K3	Kameyama, H., Ishibashi, Y., Takagi, Y.: J. Phys. Soc. Jpn. **38** (1975) 1703.
75K5	Kanashiro, T., Ohno, T., Satoh, M.: J. Phys. Soc. Jpn. **38** (1975) 1293.
75K6	Kasahara, M., Tatsuzaki, I.: J. Phys. Soc. Jpn. **38** (1975) 1389.
75K7	Kasahara, M., Sasakawa, K., Tatsuzaki, I.: J. Phys. Soc. Jpn. **39** (1975) 1022.
75K8	Kasturi, S. R., Moran, P. R.: Phys. Rev. B **12** (1975) 1874.
75K9	Kay, M. I.: Ferroelectrics **9** (1975) 171.
75K10	Kay, M. I., Gonzalo, J. A., Maglic, R.: Ferroelectrics **9** (1975) 179.
75K13	Klimowski, J., Wanarski, W., Ozgo, D.: Acta Phys. Polon. A **48** (1975) 307.
75K15	Kobayashi, T., Shimizu, S.: Phys. Lett. A **54** (1975) 311.
75K18	Kopcewicz, M., Kotlicki, A., Szefer, M.: Phys. Status Solidi (a) **29** (1975) 635.
75K21	Krämer, V.: Acta Crystallogr. B **31** (1975) 234.
75K24	Krupnyi, A. I., Shabanova, L. A., Aleksandrov, K. S.: Kristallografiya **20** (1975) 355; Sov. Phys. Crystallogr. (English Transl.) **20** (1975) 218.
75K26	Kobayashi, K. L. I., Kato, Y., Katayama, Y., Komatsubara, K. F.: Solid State Commun. **17** (1975) 875.
75L1	Lagakos, N., Cummins, H. Z.: Phys. Rev. Lett. **34** (1975) 883.
75L2	Laulicht, I., Ofek, R.: J. Raman Spectrosc. **4** (1975) 41.

75L5	Levstik, A., Blinc, R., Kadaba, P., Čižikov, S., Levstik, I., Filipic, C.: Solid State Commun. **16** (1975) 1339.
75L7	Liu, Y. S., Shultz, A. R.: Appl. Phys. Lett. **27** (1975) 585.
75L8	Loiacono, G. M.: Acta Electron. **18** (1975) 241.
75L9	Lowndes, R. P., Leung, R. C., Spillman, W. B., Tornberg, N. E.: Proceedings of the Fourth International Conference on High Pressure, held in Kyoto (1974), edited by Osugi, J. (The Physico-Chemical Society of Japan, Kyoto, 1975) p. 260.
75M2	Maeda, M., Suzuki, I., Abe, R.: J. Phys. Soc. Jpn. **38** (1975) 592.
75M3	Maeda, M., Suzuki, I., Abe, R.: J. Phys. Soc. Jpn. **39** (1975) 1319.
75M4	Magataev, V. K., Gladkii, V. V., Zheludev, I. S.: Izv. Akad. Nauk SSSR, Ser. Fiz. **39** (1975) 778; Bull. Acad. Sci. USSR, Phys. Ser. (English Transl.) **39**, No. 4 (1975) 118.
75M5	Mariani, E., Dikant, J., Jónová, A., Dobrzhanskii, G. F.: Phys. Status Solidi (a) **31** (1975) 749.
75M8	McDowell, C. A., Raghunathan, P., Srinivasan, R.: Mol. Phys. **29** (1975) 815.
75M9	McKenzie, D. R.: J. Phys. C **8** (1975) 1607.
75M10	McKenzie, D. R.: J. Phys. C **8** (1975) 2003.
75M13	Meyer, R. B., Liebert, L., Strzelecki, L., Keller, P.: J. Phys. Lett. (Paris) **36** (1975) L69.
75M14	Milsch, B., Windsch, W.: Phys. Status Solidi (b) **71** (1975) K17.
75M15	Minakata, M., Noda, J., Uchida, N.: Appl. Phys. Lett. **26** (1975) 395.
75M16	Mitsui, T., Oka, T., Shiroishi, Y., Takashige, M., Ito, K., Sawada, S.: J. Phys. Soc. Jpn. **39** (1975) 845.
75N3	Navalgund, R., Gupta, L. C.: J. Phys. Soc. Jpn. **39** (1975) 880.
75N6	Nozik, Yu. Z.: Kristallografiya **20** (1975) 169; Sov. Phys. Crystallogr. (English Transl.) **20** (1975) 98.
75O1	Oguchi, M., Matsuo, T., Suga, H., Seki, S.: Bull. Chem. Soc. Jpn. **48** (1975) 379.
75O3	Okai, B., Yoshimoto, J.: J. Phys. Soc. Jpn. **39** (1975) 162.
75O4	Orliukas, A., Grigas, J., Mozgova, N.: Liet. Fiz. Rinkinys **15** (1975) 637.
75O5	Osaka, T., Makita, Y., Gesi, K.: J. Phys. Soc. Jpn. **38** (1975) 292.
75O6	Owens, F. J.: J. Magn. Reson. **20** (1975) 130.
75P1	Pandey, S. D., Venkateswarld, P.: J. Magn. Reson. **17** (1975) 137.
75P4	Peercy, P. S.: Phys. Rev. B **12** (1975) 2725.
75P5	Peercy, P. S.: Phys. Rev. Lett. **35** (1975) 1581.
75P7	Pieranski, P., Guyon, E., Keller, P.: J. Phys. (Paris) **36** (1975) 1005.
75P8	Piślewski, N., Voigtsberger, B., Hilczer, B., Szczpańska, L.: Phys. Status Solidi (a) **31** (1975) K21.
75P11	Pollina, R. J., Garland, C. W.: Phys. Rev. B **12** (1975) 362.
75P12	Popov, S. N., Krainik, N. N., Kamenstev, Y. S., My'lnikova, I. E.: Fiz. Tverd. Tela **17** (1975) 417.; Sov. Phys. Solid State (English Transl.) **17** (1975) 262.
75P14	Press, W., Cox, D. E., Axe, J. D., Shirane, G., Harada, J., Fujii, Y., Hoshino, S.: Solid State Commun. **16** (1975) 1365.
75P15	Prettl, W., Rieder, K. H., Nitsche, R.: Z. Phys. B **22** (1975) 49.
75R4	Rehwald, W., Lang, G. K.: J. Phys. C (Solid State Phys.) **8** (1975) 3287.
75S1	Sailer, E., Konak, C., Unruh, H. G., Fouskova, A.: Phys. Status Solidi (a) **29** (1975) K73.
75S2	Sailer, E., Unruh, H. G.: Solid State Commun. **16** (1975) 615.
75S6	Samara, G. A.: Ferroelectrics **9** (1975) 209.
75S8	Sawada, A., Ohya, S., Ishibashi, Y., Takagi, Y.: J. Phys. Soc. Jpn. **38** (1975) 1408.
75S14	Schmidt, G., Sorge, G., Titz, H.: Phys. Status Solidi (a) **29** (1975) K65.
75S16	Sergeev, N. A., Falaleev, O. V.: Izv. Akad. Nauk SSSR, Ser. Fiz. **39** (1975) 714; Bull. Acad. Sci. USSR, Phys. Ser. (English Transl.) **39**, No. 4 (1975) 63.
75S17	Seryshev, S. A., Vinogradova, I. S., Lundin, A. G.: Fiz. Tverd. Tela **17** (1975) 449; Sov. Phys. Solid State (English Transl.) **17** (1975) 281.
75S18	Shabanov, V. F., Fedotov, A. P.: Fiz. Tverd. Tela **17** (1975) 553; Sov. Phys. Solid State (English Transl.) **17** (1975) 347.
75S19	She, C. Y., Pan, C. L.: Solid State Commun. **17** (1975) 529.
75S21	Shimizu, H., Tsukamoto, M., Ishibashi, Y., Takagi, Y.: J. Phys. Soc. Jpn. **38** (1975) 195.
75S23	Shirokov, A. M., Aleksandrova, I. P., Yuzvak, V. I., Baranov, A. I., Serebryannikov, V. L.: Izv. Akad. Nauk SSSR, Ser. Fiz. **39** (1975) 947; Bull. Acad. Sci. USSR, Phys. Ser. (English Transl.) **39**, No. 5 (1975) 47.
75S25	Sleight, A. W., Bierlein, J. D., Bierstedt, P. E.: J. Chem. Phys. **62** (1975) 2826.
75S26	Sleight, A. W., Jones, G. A.: Acta Crystallogr. B **31** (1975) 2748.

75S29	Spitsyna, V. D., Lyakhovitskaya, V. A., Belyaev, L. M.: Kristallografiya **20** (1975) 840; Sov. Phys. Crystallogr. (English Transl.) **20** (1976) 515.
75S30	Steigmeier, E. F., Auderset, H., Harbeke, G.: Phys. Status Solidi (b) **70** (1975) 705.
75S31	Stepisnik, J.: J. Chem. Phys. **62** (1975) 34.
75S33	Sych, A. M., Popolitov, V. I., Eremenko, L. A.: Zh. Neorg. Khim. **20** (1975) 2605.
75S34	Strukov, B. A., Minaeva, K. A., Teleshevskii, V. I., Shirina, N. G., Kkhanna, S. K.: Izv. Akad. Nauk SSSR, Ser. Fiz. **39** (1975) 758; Bull. Acad. Sci. USSR, Phys. Ser. (English Transl.) **39**, No. 4 (1975) 101.
75S35	Seddon, T., Farley, J. M., Saunders, G. A.: Solide State Commun. **17** (1975) 55.
75T1	Takagi, Y., Shigenari, T.: J. Phys. Soc. Jpn. **39** (1975) 440.
75T3	Takashige, M., Iwamura, K., Hirotsu, S., Sawada, S.: J. Phys. Soc. Jpn. **38** (1975) 1217.
75T4	Takoshima, T., Tatsuzaki, I.: J. Phys. Soc. Jpn. **39** (1975) 427.
75T5	Takoshima, T., Tatsuzaki, I.: Solid State Commun. **16** (1975) 537.
75T8	Tello, M. J., Socias, C., Bocanegra, E. H.: An. Fis. **71** (1975) 113.
75T9	Terauchi, H., Takenaka, H., Shimaoka, K.: J. Phys. Soc. Jpn. **39** (1975) 435.
75T10	Thornley, F. R., Nelmes, R. J., Rouse, K. D.: Chem. Phys. Lett. **34** (1975) 175.
75T12	Todo, I.: J. Phys. Soc. Jpn. **39** (1975) 1538.
75T13	Tsuchida, K., Abe, R.: J. Phys. Soc. Jpn. **38** (1975) 1687.
75V2	Varikash, V. M., Drozhdin, S. N., Zarembovskaya, T. A., Novik, V. K.: Kristallografiya **20** (1975) 435; Sov. Phys. Crystallogr. (English Transl.) **20** (1975) 267.
75V4	Veerabhadra Rao, K., Narasimhamurty, T. S.: J. Mater. Sci. **10** (1975) 1019.
75V5	Vinogradova, I. S., Seryshev, S. A.: Fiz. Tverd. Tela **17** (1975) 3020; Sov. Phys. Solid State (English Transl.) (1976) 2003.
75V7	Viswanathan, K., Salje, E.: Acta Crystallogr. A **31** (1975) 810.
75V8	Vlokh, O. G., Krochuk, A. S., Lutsiv-Shumskii, L. F., Smishko, E. V.: Izv. Akad. Nauk SSSR, Ser. Fiz. **39** (1975) 983; Bull. Acad. Sci. USSR, Phys. Ser. (English Transl.) **39**, No. 5 (1975) 75.
75V9	Vlokh, O. G., Laz'ke, L. A., Zheludev, I. S.: Kristallografiya **20** (1975) 1056; Sov. Phys. Crystallogr. (English Transl.) **20** (1976) 646.
75V11	Völkel, G., Windsch, W., Urbanowitschius, W.: J. Magn. Reson. **18** (1975) 57.
75W2	Weber, K. E., Flanagan, T. B.: Solid State Commun. **16** (1975) 23.
75W3	Windsch, W.: Izv. Akad. Nauk SSSR, Ser. Fiz. **39** (1975) 914; Bull. Acad. Sci. USSR, Phys. Ser. (English Transl.) **39**, No. 5 (1975) 20.
75W4	Windsch, W., Lippe, R., Volkel, G.: Solid State Commun. **17** (1975) 1375.
75W5	Windsch, W., Volkel, G.: Ferroelectrics **9** (1975) 187.
75Y3	Yamamoto, Y., Matsuura, M., Haseda, T.: J. Phys. Soc. Jpn. **38** (1975) 1776.
75Z1	Zaitseva, M. P., Krupnii, A. I., Kokopin, Yu. I., Krasikov, V. S.: Izv. Akad. Nauk SSSR, Ser. Fiz. **39** (1975) 954; Bull. Acad. Sci. USSR, Phys. Ser. (English Transl.) **39**, No. 5 (1975) 52.
75Z2	Zaitseva, M. P., Rozhnova, G. P.: Kristallografiya **20** (1975) 178; Sov. Phys. Crystallogr. (English Transl.) **20** (1975) 105.
75Z3	Zaporozhets, O. I.: Fiz. Tverd. Tela **17** (1975) 3672; Sov. Phys. Solid State (English Transl.) **17** (1975) 2387.
75Z5	Zherebtsova, L. I., Shtain, A. V.: Kristallografiya **20** (1975) 176; Sov. Phys. Crystallogr. (English Transl.) **20** (1975) 103.
75Z6	Zhizhin, G. N., Myasnikova, T. P., Rogovoi, V. N.: Fiz. Tverd. Tela **17** (1975) 1270; Sov. Phys. Solid State (English Transl.) **17** (1975) 821.
76A1	Adriaenssens, G. J.: Ferroelectrics **12** (1976) 269.
76A4	Aleksandrov, K. S., Anistratov, A. T., Blat, D. X., Zherebtsova, L. I., Zinenko, V. I., Iskornev, I. M., Melnikova, S. V., Flerov, I. N.: Ferroelectrics **12** (1976) 191.
76A5	Aleksandrov, K. S., Anistratov, A. T., Krupnyi, A. I., Martynov, V. G., Popkov, Y. A., Fomin, V. I.: Kristallografiya **21** (1976) 534; Sov. Phys. Crystallogr. (English Transl.) **21** (1976) 296.
76A6	Aleksandrova, I. P., Repeta, V. S., Chekmasova, T. I., Konstantinova, V. P.: Fiz. Tverd. Tela **18** (1976) 1311; Sov. Phys. Solid State (English Transl.) **18** (1976) 754.
76A7	Aleksandrova, I. P., Yuzvak, V. I., Shuvalov, L. A.: Kristallografiya **21** (1976) 1136; Sov. Phys. Crystallogr. (English Transl.) **21** (1976) 658.
76A10	Ambe, F., Ambe, S.: J. Phys. (Paris) **37** (1976) Suppl. C6, 923.
76A11	Amer, N., Yu. P. Y., Petroff, Y., Shen, Y. R.: Article on p. 49 in [76b1].
76A13	Arend, H., Blinc, R., Kandusar, A.: Ferroelectrics **13** (1976) 511.
76B1	Bastie, P., Bornarel, J. , Lajzerowicz, J., Legrand, J. F.: Ferroelectrics **13** (1976) 319.

76B2	Bastie, P., Bornarel, J., Lajzerowicz, J., Schneider, J. R.: Ferroelectrics **14** (1976) 587.
76B3	Belouet, C., Monnier, M., Dunia, E., Petroff, J. F.: Mater. Res. Bull. **11** (1976) 903.
76B5	Bevolo, A. J., Shanks, H. R., Eckels, D. E.: Phys. Rev. B **13** (1976) 3523.
76B6	Blinc, R.: Article on p. 377 in [76m2].
76B7	Blinc, R., Arend, H., Kandušer, A.: Phys. Status Solidi (b) **74** (1976) 425.
76B8	Blinc, R., Pirc, R., Žekš, B.: Phys. Rev. B **13** (1976) 2943.
76B10	Brehat, F., Claudel, J., Strimer, P., Hadni, A.: J. Phys. Lett. **37** (1976) L229.
76B11	Březina, B., Havránková, M., Dušek, K.: J. Cryst. Growth **34** (1976) 248.
76B12	Buchheim, A., Grande, S.: Ann. Physik (7) **33** (1976) 374.
76B18	Bye, K. L., Whipps, P. W., Keve, E. T.: Ferroelectrics **11** (1976) 525.
76C1	Castellucci, E., Firnstein, L. A., Porto, S. P. S.: Article on p. 872 in [76b1].
76C2	Cervenka, P. O., Prasad Rao, A. D., Porto, S. P. S.: Article on p. 953 in [76b1].
76C3	Cevc, G., Cevc, P., Blinc, R.: Ferroelectrics **12** (1976) 275.
76C4	Chihara, H., Inaba, A.: J. Phys. Soc. Jpn. **40** (1976) 597.
76C5	Chihara, H., Inaba, A.: J. Phys. Soc. Jpn. **40** (1976) 1383.
76C6	Chu, K.: J. Magn. Reson. **21** (1976) 151.
76D3	Day, G. W., Hamilton, C. A., Gruzensky, P. M., Phelan Jr., R. J.: Ferroelectrics **10** (1976) 99.
76D4	Deguchi, K., Nakamura, E.: J. Phys. Soc. Jpn. **40** (1976) 478.
76D7	Dietrich, W., Schmid, D.: Phys. Status Solidi (b) **74** (1976) 609.
76D9	Durvasula, L. N., Gammon, R. W.: Article on p. 775 in [76b1].
76E1	Edlund, O., Lund, A.: J. Magn. Reson. **21** (1976) 197.
76E4	El Saffar, Z. M.: J. Chem. Phys. **64** (1976) 3282.
76F1	Fawcett, V., Long, D. A.: J. Chem. Soc., Faraday Trans. II **72** (1976) 313.
76F2	Fawcett, V., Long, D. A.: Article on p. 622 in [76s4].
76F7	Flerov, I. N., Iskornev, I. M.: Fiz. Tverd. Tela **18** (1976) 3666; Sov. Phys. Solid State (English Transl.) **18** (1976) 2135.
76F8	Fletcher, S. R., Keve, E. T., Skapski, A. C.: Ferroelectrics **14** (1976) 775.
76F9	Fletcher, S. R., Keve, E. T., Skapski, A. C.: Ferroelectrics **14** (1976) 789.
76F10	Fornges, D.: Arch Elektron. Übertragungstech. **30** (1976) 83.
76F11	Fousek, J., Konak, C.: Ferroelectrics **12** (1976) 185.
76F12	Fritz, I. J.: Phys. Rev. B **13** (1976) 705.
76F15	Fujimoto, M., Dressel, L. A.: Ferroelectrics **13** (1976) 449.
76F17	Furman, E., Brafman, D., Makovsky, J.: Phys. Rev. B **13** (1976) 1703.
76G2	Gaillard, J., Gloux, P.: J. Phys. (Paris) **37** (1976) 407.
76G3	Gaillard, J., Gloux, P.: J. Phys. (Paris) **37** (1976) 1307.
76G4	Gauss, K. E., Happ, H.: Phys. Status Solidi (b) **78** (1976) 133.
76G6	Gesi, K.: Phys. Status Solidi (a) **33** (1976) 479.
76G7	Gesi, K.: J. Phys. Soc. Jpn. **41** (1976) 565.
76G8	Gesi, K.: J. Phys. Soc. Jpn. **41** (1976) 1437.
76G9	Gladkii, V. V., Magataev, V. K., Shuvalov, L. A., Fedosyuk, R. M.: Phys. Lett. A **59** (1976) 391.
76G10	Gladkii, V. V., Magataev, V. K., Zheludev, I. S.: Ferroelectrics **13** (1976) 343.
76G13	Gorbatyi, L. V., Ponomarev, V. I., Kheiker, D. M.: Kristallografiya **21** (1976) 307; Sov. Phys. Crystallogr. (English Transl.) **21** (1976) 165.
76G16	Grigas, J., Meshkauskas, J., Orliukas, A.: Phys. Status Solidi (a) **37** (1976) K39.
76G17	Grigas, J., Zadrozhnaya, L. A., Lyakhovitskaya, V. A., Orliukas, A.: Liet. Fiz. Rinkinys **16** (1976) 833.
76G19	Grossmann, J., Müller, D., Petersson, J., Schneider, E.: Z. Naturforsch. a **31** (1976) 1089.
76H2	Harrison, J. P., Hessler, J. P., Taylor, D. R.: Phys. Rev. B **14** (1976) 2979.
76H4	Havlin, S., Litov, E., Sompolinsky, H.: Phys. Rev. B **13** (1976) 4999.
76H5	Hay, W. J., Nelmes, R. J.: Ferroelectrics **14** (1976) 599.
76H7	Hikita, T., Kudo, T., Chubachi, Y., Ikeda, T.: J. Phys. Soc. Jpn. **41** (1976) 349.
76H8	Holman, R. L.: Ferroelectrics **14** (1976) 675.
76H9	Holste, J. C., Lawless, W. N., Samara, G. A.: Ferroelectrics **11** (1976) 337.
76H10	Huong, P. V., Couzi, M., Vignalou, J. R., Tranquard, A.: Article on p. 845 in [76b1].
76I2	Imai, K.: J. Phys. Soc. Jpn. **41** (1976) 2005.
76I3	Iqbal, Z., Christoe, C. W.: Solid State Commun. **18** (1976) 269.
76I4	Iqbal, Z., Christoe, C. W.: Chem. Phys. Lett. **37** (1976) 460.
76I7	Itoh, K., Matsunaga, H., Nakamura, E.: J. Phys. Soc. Jpn. **41** (1976) 1679.

76I10	Iwata, Y., Koyano, N., Shibuya, I.: Annual Rept. Res. Reactor Inst., Kyoto Univ. **9** (1976) 9.
76J1	Jain, R. K., Gustafson, T. K.: IEEE J. Quantum Electron., QE **12** (1976) 555.
76J2	Jasinski, A., Sulek, Z.: Phys. Status Solidi (b) **74** (1976) K5.
76J4	Johari, G. P., Whalley, E.: J. Chem. Phys. **64** (1976) 4484.
76K4	Kawashima, R., Tatsuzaki, I.: Phys. Lett. A **56** (1976) 43.
76K6	Kennedy, N. S. J., Nelmes, R. J., Thornley, F. R.: Ferroelectrics **14** (1976) 591.
76K7	Kiriyama, H., Nakamura, O., Kiriyama, R.: Chem. Lett. **1976**, 689.
76K11	Kirikov, V. A., Gladkii, V. V., Magataev, V. K.: Kristallografiya **21** (1976) 1212; Sov. Phys. Crystallogr. (English Transl.) **21** (1976) 701.
76K12	Klimowski, J., Wanarski, W., Ozgo, D.: Phys. Status Solidi (a) **34** (1976) 697.
76K14	Konak, C., Matras, J.: Czech. J. Phys. B **26** (1976) 577.
76K15	Koralewski, M., Surma, M.: Acta Phys. Polon. A **49** (1976) 803.
76K16	Kotake, Y., Miyagawa, I.: J. Chem. Phys. **64** (1976) 3169.
76K18	Kruglik, A. I.: Dokl. Akad. Nauk SSSR **229** (1976) 853; Sov. Phys. Dokl. (English Transl.) **21** (1976) 420.
76K19	Kucharczyk, D., Pietraszko, A., Łukaszewicz, K.: Phys. Status Solidi (a) **37** (1976) 287.
76K20	Kobayashi, K. L. I., Kato, Y., Katayama, Y., Komatsubara, K. F.: Phys. Rev. Lett. **37** (1976) 772.
76L1	Lavrenčič, B. B., Levstek, I., Čopic, M., Trost, S.: Article on p. 928 in [76b1].
76L2	Lawless, W. N.: Phys. Rev. B **14** (1976) 134.
76L5	Lippe, R., Windsch, W., Volkel, G., Schulga, W.: Solid State Commun. **19** (1976) 587.
76L9	Luther, G.: Ferroelectrics **12** (1976) 243.
76M1	Mali, M., Osredkar, R., Zumer, S.: Ferroelectrics **12** (1976) 279.
76M3	Mansingh, A., Rai, G.: Can. J. Phys. **54** (1976) 2050.
76M4	Mansingh, A., SRJ, E. S.: J. Phys. D **9** (1976) 1379.
76M6	Matsuo, T., Kume, Y., Suga, H., Seki, S.: J. Phys. Chem. Solids **37** (1976) 499.
76M7	Matsuo, T., Tatsumi, M., Suga, H., Seki, S.: Unpublished work (1976).
76M8	Matyas, M., Horak, J.: Phys. Status Solidi (a) **36** (1976) K137.
76M9	Mazzacurati, V., Signorelli, G., Sampoli, M.: J. Raman Spectrosc. **4** (1976) 339.
76M10	McBrierty, V. J., Douglass, D. C., Weber, T. A.: J. Polymer Sci., Polymer Phys. Ed. **14** (1976) 1271.
76M13	Meshkauskas, I., Orlyukas, A., Grigas, I.: Fiz. Tverd. Tela **18** (1976) 3106; Sov. Phys. Solid State (English Transl.) **18** (1976) 1811.
76M15	Meyer, G. M., Dietrich, O. W.: Ferroelectrics **14** (1976) 595.
76M16	Meyer, G. M., Dietrich, O. W., Nelmes, R. J., Hay, W. J., Cowley, R. A.: J. Phys. C **9** (1976) L83.
76M18	Mognaschi, E. R., Rigamonti, A., Menafra, L.: Phys. Rev. B **14** (1976) 2005.
76M19	Morimoto, K.: J. Phys. Soc. Jpn. **40** (1976) 804.
76M20	Morrison, I. E. G., Rees, L. V. C., Silver, J., White, E. A. D.: J. Chem. Soc. Dalton (1976) 1103.
76M21	Müller, D.: Ferroelectrics **13** (1976) 431.
76M22	Müller, D.: Solid State Commun. **19** (1976) 1067.
76M23	Müller, K. A., Berlinger, W.: Phys. Rev. Lett. **37** (1976) 916.
76M24	Müller, K. A., Dalal, N. S., Berlinger, W.: Phys. Rev. Lett. **36** (1976) 1504.
76M25	Müller, D., Petersson, J.: Phys. Status Solidi (b) **78** (1976) 191.
76N1	Nagae, Y., Ishibashi, Y., Takagi, Y., Kameyama, H.: J. Phys. Soc. Jpn. **41** (1976) 1300.
76N2	Nagae, Y., Wada, M., Ishibashi, Y., Takagi, Y.: J. Phys. Soc. Jpn. **41** (1976) 1659.
76N3	Nakashima, S., Balkanski, M.: Solid State Commun. **19** (1976) 1225.
76N4	Navalgund, R., Gupta, L. C.: Solid State Commun. **19** (1976) 1205.
76N6	Nelmes, R. J., Thornley, F. R.: Ferroelectrics **13** (1976) 355.
76N7	Nimmo, J. K., Lucas, B. W.: Acta Crystallogr. B **32** (1976) 1968.
76N8	Navalgund, R., Gupta, L. C.: Ferroelectrics **14** (1976) 767.
76O2	Ohigashi, H.: J. Appl. Phys. **47** (1976) 949.
76O3	Oka, T., Mitsui, T., Shiroishi, Y., Sawada, S.: J. Phys. Soc. Jpn. **40** (1976) 913.
76O5	Onaka, R., Ito, H.: J. Phys. Soc. Jpn. **41** (1976) 1303.
76O6	Oshiki, M., Fukada, E.: Jpn. J. Appl. Phys. **15** (1976) 43.
76P2	Paul, B., Albers, J., Müser, H. E.: Ferroelectrics **14** (1976) 707.
76P3	Peercy, P. S.: Article on p. 782 in [76b1].
76P4	Peercy, P. S.: Phys. Rev. B **13** (1976) 3945.
76P8	Pereverzeva, L. P., Poplavko, Yu. M., Rez, I. S., Kuznetsova, L. I.: Kristallografiya **21** (1976) 981.; Sov. Phys. Crystallogr. (English Transl.) **21** (1976) 562.
76P10	Perrino, C. T., Bush Jr., R. C.: J. Solid State Chem. **17** (1976) 121.

76P13	Petersen, G., Bray, P. J.: J. Chem. Phys. **64** (1976) 522.
76P14	Pistorius, C. W. F. T.: Article in "Progress in Solid State Chemistry" (Editors: McCaldin, J., Somorjai, G.), Vol. 11, Part 1, Pergamon Press, New York (1976).
76P15	Placido, F.: J. Phys. (Paris) **37** (1976) Suppl. C6, 31.
76P16	Poulet, H., Mathieu, J. P.: J. Raman Spectrosc. **4** (1976) 231.
76P17	Prasad Rao, A. D., Porto, S. P. S.: Article on p. 877 in [76b1].
76P18	Prettl, W., Rieder, K. H.: Phys. Rev. B **14** (1976) 2171.
76R1	Rannev, N. V., Shchedrin, B. M., Venevtsev, Yu. N.: Ferroelectrics **13** (1976) 523.
76R7	Romanyuk, N. A., Kostetokii, A. M.: Fiz. Tverd. Tela **18** (1976) 1489; Sov. Phys. Solid State (English Transl.) **18** (1976) 867.
76R8	Roos, J., Kind, R., Petzelt, J.: Z. Phys. B **24** (1976) 99.
76R10	Ryan, J. F., Scott, J. F.: Article on p. 761 in [76b1].
76S1	Sader, E., Harnik, E.: Phys. Lett. A **58** (1976) 199.
76S2	Sailer, E., Unruh, H. G.: Ferroelectrics **12** (1976) 285.
76S4	Samara, G. A., Richards, P. M.: Phys. Rev. B **14** (1976) 5076.
76S5	Sastry, M. D.: Phys. Rev. B **13** (1976) 1156.
76S6	Savinkov, A. I., Steingart, L. M., Markovskii, V. Yu., Shaskol'skaya, M. P.: Opt. Mekh. Promst. **43**, No. 10 (1976) 58; Sov. J. Opt. Technol. (English Transl.) **43** (1976) 631.
76S7	Sawada, A., Makita, Y., Takagi, Y.: J. Phys. Soc. Jpn. **41** (1976) 174.
76S8	Scarparo, M. A. F., Srivastava, R., Katiyar, R. S.: Article on p. 839 in [76b1].
76S9	Schmitt, H.: Ferroelectrics **14** (1976) 575.
76S12	Shibata, N., Abe, R., Suzuki, I.: J. Phys. Soc. Jpn. **41** (1976) 2011.
76S18	Singh, G. P., Basu, B. K.: Phys. Lett. A **58** (1976) 39.
76S19	Slak, J., Blinc, R., Rutar, V., Žumer, S.: Ferroelectrics **12** (1976) 281.
76S26	Shiroishi, Y., Nakata, A., Sawada, S.: J. Phys. Soc. Jpn. **40** (1976) 911.
76T3	Takashige, M., Iwamura, H., Hirotsu, S., Sawada, S.: Ferroelectrics **11** (1976) 431.
76T4	Tamaki, A., Fujimura, T., Kamiyoshi, K.: J. Chem. Phys. **65** (1976) 4867.
76T6	Teng, M. K., Massot, M., Balkanski, M., Ziolkiewicz, M. K.: Article on p. 806 in [76b1].
76T8	Theveneau, H., Papon, P.: Phys. Rev. B **14** (1976) 3810.
76T11	Ti, S. S., Kettle, S. F. A., Ra, O.: J. Raman Spectrosc. **5** (1976) 325.
76T13	Toepfer, K. D., Helberg, H. W.: Phys. Status Solidi (a) **35** (1976) 131.
76T17	Tsuchida, K., Abe, R.: J. Phys. Soc. Jpn. **40** (1976) 204.
76U2	Uesu, Y., Kobayashi, J.: Phys. Status Solidi (a) **34** (1976) 475.
76V1	Veerabhadra Rao, K., Narasimhamurty, T. S.: Opt. Quantum Electron. **8** (1976) 255.
76V3	Verkhovskaya, K. A., Lobachev, A. N., Popov, B. N., Popolitov, V. I., Peskin, V. F., Fridkin, V. M.: Zh. Eksp. Teor. Fiz. Pis'ma **23** (1976) 522; JETP Lett. (English Transl.) **23** (1976) 476.
76V4	Vignalov, J. R., Tranquard, A., Couzi, M., Huong, P. V.: Article on p. 594 in [76s4].
76V5	Volkova, E. N., Faerman, Sh. L.: Kvantovaya Elektron. **3** (1976) 2508; Sov. J. Quantum Electron. (English Transl.) **6** (1976) 1380.
76W2	Weber, H.-J., Haussühl, S.: Acta Crystallogr. A **32** (1976) 892.
76W3	Welch, G. J., Miller, R. L.: J. Polymer Sci., Polymer Phys. Ed. **14** (1976) 1683.
76W4	Windsch, W.: Ferroelectrics **12** (1976) 63.
76W6	Wurfel, P., Batra, I. P.: Ferroelectrics **12** (1976) 55.
76Y1	Yagi, T., Tokunaga, M., Tatsuzaki, I.: J. Phys. Soc. Jpn. **40** (1976) 1659.
76Y2	Yamaguchi, T., Hamano, K.: J. Phys. Soc. Jpn. **41** (1976) 1663.
76Y4	Yu, L. J., Lee, H., Bak, C. S., Labes, M. M.: Phys. Rev. Lett. **36** (1976) 388.
76Y5	Yurin, V. A., Waplak, S., Stankowski, J., Ankudinov, M. A., Zheludev, I. S.: Kristallografiya **21** (1976) 327; Sov. Phys. Crystallogr. (English Transl.) **21** (1976) 177.
76Z2	Zeyen, C. M. E., Meister, H.: Ferroelectrics **14** (1976) 731.
77A1	Abdullin, R. S., Pen'kov, I. N., Nizamutdinov, N. M., Grigas, I., Safin, I. A.: Fiz. Tverd. Tela **19** (1977) 2632; Sov. Phys. Solid State (English Transl.) **19** (1977) 1542.
77A2	Abraham, S. C., Bernstein, J. L.: J. Chem. Phys. **67** (1977) 2146.
77A5	Aktsipetrov, O. A., Kitaeva, G. Kh., Penin, A. N.: Fiz. Tverd. Tela **19** (1977) 1001; Sov. Phys. Solid State (English Transl.) **19** (1977) 582.
77A6	Aleksandrov, K. S., Aleksandrova, I. P., Anistratov, A. T., Shabanov, V. F.: Izv. Akad. Nauk SSSR, Ser. Fiz. **41** (1977) 599; Bull. Acad. Sci. USSR, Phys. Ser. (English Transl.) **41**, No. 3 (1977) 114.
77A7	Aleksandrov, K. S., Anistratov, A. T., Zamkov, A. V., Rez, I. S.: Fiz. Tverd. Tela **19** (1977) 1863; Sov. Phys. Solid State (English Transl.) **19** (1977) 1090.

77A8	Aleksandrova, I. P.: Article on p. 49 in [77p3].
77A10	Ambrosetti, R., Angelone, R., Colligiani, A.: Phys. Rev. B **15** (1977) 4318.
77A11	Andreev, R. B., Volosov, V. D., Krylov, V. N.: Zh. Tekh. Fiz. **47** (1977) 1977; Sov. Phys. Tech. Phys. (English Transl.) **22** (1977) 1146.
77A12	Azoulay, J., Gerlich, D., Wiener-Avnear, E., Pelak, I.: Phys. Status Solidi (b) **81** (1977) 295.
77B9	Bespalov, V. I., Batyreva, I. A., Dmitrenko, L. A., Korolikhin, V. V.: Kvantovaya Elektron. **4** (1977) 1563; Soviet J. Quantum Electron. (English Transl.) **7** (1977) 885.
77B10	Billman, A., Levelnt, A.: J. Phys. (Paris) **38** (1977) 1307.
77B12	Blinc, R., Burgar, M., Rutar, V., Seliger, J., Zupančič, I.: Phys. Rev. Lett. **38** (1977) 92.
77B13	Blinc, R., Slak, J., Luzar, M.: Phys. Status Solidi (a) **40** (1977) K121.
77B17	Bottcher, R., Heinhold, D., Windsch, W.: Chem. Phys. Lett. **49** (1977) 148.
77B18	Budík, L., Koňák, Č., Szczepańska, L.: Czech. J. Phys. B **27** (1977) 597.
77B19	Buturlakin, A. P., Gurzan, M. I., Slivka, V. Yu.: Fiz. Tverd. Tela **19** (1977) 1990; Sov. Phys. Solid State (English Transl.) **19** (1977) 1165.
77C1	Chabin, M., Gilletta, F.: Ferroelectrics **15** (1977) 149.
77C2	Chekmasova, T. I., Kabanov, I. S., Yuzvak, V. I.: Phys. Status Solidi (a) **44** (1977) K155.
77C3	Choe, H. M., Judy, J. H., van der Ziel, A.: Ferroelectrics **15** (1977) 181.
77C6	Czapla, Z., Sobczyk, L.: Acta Phys. Polon. A **51** (1977) 189.
77D1	Dalal, N. S., Hebden, J. A., Kennedy, D. E., McDowell, C. A.: J. Chem. Phys. **66** (1977) 4425.
77D2	Deb, S. K.: Phys. Rev. B **16** (1977) 1020.
77D4	Dikant, J., Mariani, E., Jónová, A., Dobržanskij, G. F.: Acta Phys. Slovaca **27** (1977) 266.
77D7	Durvasula, L. N., Gammon, R. W.: Ferroelectrics **16** (1977) 199.
77D8	Durvasula, L. N., Gammon, R. W.: Phys. Rev. Lett. **38** (1977) 1081.
77E1	Eichler, H. J., Frey, H., Knof, J., Eichler, J.: Z. Phys. B **28** (1977) 297.
77E2	Ema, K., Hamano, K., Kurihara, K., Hatta, I.: J. Phys. Soc. Jpn. **43** (1977) 1954.
77F1	Fano, V.: Solid State Commun. **22** (1977) 467.
77F2	Fedotov, A. P., Shabanov, V. F., Shuvalov, L. A., Shchagina, N. M.: Kristallografiya **22** (1977) 1224; Sov. Phys. Crystallogr. (English Transl.) **22** (1977) 695.
77F4	Ferguson, A. I., Dunn, M. H.: Opt. Commun. **23** (1977) 177.
77F5	Fujimoto, M.: Article on p. 293 in [77p3].
77F6	Fujimoto, S., Yasuda, N.: Jpn. J. Appl. Phys. **16** (1977) 49.
77F7	Fujimoto, S., Yasuda, N., Shimizu, H., Tsuboi, S., Kawabe, K., Takagi, Y., Midorikawa, M.: J. Phys. Soc. Jpn. **42** (1977) 911.
77G1	Gaillard, J., Gloux, P., Müller, K. A.: Phys. Rev. Lett. **38** (1977) 1216.
77G2	Gerbaux, X., Hadni, A., Waldschmidt, J. M.: Opt. Commun. **21** (1977) 425.
77G3	Gesi, K.: J. Phys. Soc. Jpn. **42** (1977) 1785.
77G4	Gesi, K.: J. Phys. Soc. Jpn. **43** (1977) 1941.
77G5	Gesi, K.: J. Phys. Soc. Jpn. **43** (1977) 1949.
77G6	Gesi, K., Ozawa, K.: J. Phys. Soc. Jpn. **42** (1977) 923.
77G7	Gesi, K., Ozawa, K.: J. Phys. Soc. Jpn. **43** (1977) 563.
77G8	Gesi, K., Ozawa, K.: Jpn. J. Appl. Phys. **16** (1977) 1457.
77G9	Gladkii, V. V., Kirikov, V. A., Magataev, V. K., Shuvalov, L. A.: Fiz. Tverd. Tela **19** (1977) 291; Sov. Phys. Solid State (English Transl.) **19** (1977) 167.
77G10	Gladkii, V. V., Magamaev, V. K., Kirikov, V. A.: Fiz. Tverd. Tela **19** (1977) 1102; Sov. Phys. Solid State (English Transl.) **19** (1977) 641.
77G11	Glass, A. M., Lines, M. E., Eibschutz, M., Hsu, F. S. L., Guggenheim, H. J.: Commun. Phys. **2** (1977) 103.
77G14	Goffman, V. G., Ivchenkova, M. D., Mishchenko, A. V., Shaimerdinov, B. U.: Kristallografiya **22** (1977) 1107; Sov. Phys. Crystallogr. (English Transl.) **22** (1977) 631.
77G15	Gordon, J. E., Boal, D. G.: Ferroelectrics **17** (1977) 411.
77G19	Gurzan, M. I., Buturlakin, A. P., Gerasimenko, V. S., Korda, N. F., Slivka, V. Yu.: Fiz. Tverd. Tela **19** (1977) 3068; Sov. Phys. Solid State (English Transl.) **19** (1977) 1794.
77G22	Gesi, K., Ozawa, K.: J. Phys. Soc. Jpn. **43** (1977) 570.
77H2	Happ, H., Rother, G.: Phys. Status Solidi (b) **79** (1977) 473.
77H3	Hasebe, K., Tanisaki, S.: J. Phys. Soc. Jpn. **42** (1977) 568.
77H4	Hatta, I., Kobayashi, K. L. I.: Solid State Commun. **22** (1977) 775.
77H5	Hatta, I., Rehwald, W.: J. Phys. C **10** (1977) 2075.
77H8	Helwig, J., Petersson, J., Schneider, E.: Z. Naturforsch. **32a** (1977) 787.

77H9	Helwig, J., Petersson, J., Schneider, E.: Z. Phys. B **28** (1977) 87.
77H11	Hikita, T., Ikeda, T.: J. Phys. Soc. Jpn. **42** (1977) 351.
77H12	Hikita, T., Kudo, S., Ikeda, T.: Article on p. 203 in [77h1].
77H13	Hikita, T., Sato, S., Sekiguchi, H., Ikeda, T.: J. Phys. Soc. Jpn. **42** (1977) 1656.
77H14	Hikita, T., Sekiguchi, H., Ikeda, T.: J. Phys. Soc. Jpn. **43** (1977) 1327.
77H17	Hoffmann, S. K., Szczepaniak, L. S., Stankowski, J.: Article on p. 352 in [77p3].
77I2	Iizumi, M., Axe, J. D., Shirane, G., Shimaoka, K.: Phys. Rev. B **15** (1977) 4392.
77I3	Iizumi, M., Gesi, K.: Solid State Commun. **22** (1977) 37.
77I4	Iizumi, M., Gesi, K.: Unpublished work (1977).
77I6	Ikeda, T., Saeki, Y.: Jpn. J. Appl. Phys. **16** (1977) 1793.
77I7	Iskornev, I. M., Flerov, I. N.: Fiz. Tverd. Tela **19** (1977) 1040; Sov. Phys. Solid State (English Transl.) **19** (1977) 605.
77I10	Izumi, T., Inada, K., Nakajima, H., Kohra, K.: Jpn. J. Appl. Phys. **16** (1977) 1063.
77J1	Joshi, M. S., Antony, A. V.: J. Phys. D **10** (1977) 2393.
77K1	Kajikawa, H., Ozaki, T., Nakamura, E.: J. Phys. Soc. Jpn. **43** (1977) 937.
77K2	Kanda, E., Fujimura, T.: J. Phys. Soc. Jpn. **43** (1977) 1813.
77K3	Kaneko, N., Takahashi, H., Higasi, K.: J. Phys. Chem. Solids **38** (1977) 849.
77K4	Katiyar, R. S., Scarparo, A. F., Srivastava, R., Porto, S. P. S.: Ferroelectrics **16** (1977) 213.
77K5	Kawamori, A., Hashimoto, E., Mikami, Y.: J. Phys. Soc. Jpn. **43** (1977) 1809.
77K6	Kawano, T., Furukawa, K.: Rept. Fac. Sci., Kagoshima Univ. (Math. Phys. Chem.) **10** (1977) 27.
77K8	Kawashima, R., Tatsuzaki, I.: J. Phys. Soc. Jpn. **42** (1977) 564.
77K9	Kawashima, R., Tokunaga, M., Tatsuzaki, I.: J. Phys. Soc. Jpn. **42** (1977) 903.
77K10	Khelifa, B., Delahaigue, A., Jouve, P.: Phys. Status Solidi (b) **83** (1977) 139.
77K11	Kitahama, K., Kiriyama, H.: Bull. Chem. Soc. Jpn. **50** (1977) 3167.
77K12	Koňák, Č., Smutný, F., Březina, B., Budik, L.: Czech. J. Phys. B **27** (1977) 1313.
77L1	Lagakos, N., Cummins, H. Z.: Ferroelectrics **17** (1977) 433.
77L2	Lavrenčič , B. B., Petzelt, J.: J. Chem. Phys. **67** (1977) 3890.
77L4	Ledsham, D. A., Chambers, W. G., Parker, T. J.: Infrared Phys. **17** (1977) 165.
77L5	Leung, R. C., Tornberg, N. E., Lowndes, R. P.: J. Phys. C **10** (1977) 4855.
77L6	Loiacono, G. M., Osborne, W. N., Delfine, M.: J. Cryst. Growth **41** (1977) 45.
77L7	Lowe, W., Dernier, P. D., McWhan, D. B.: J. Phys. Chem. Solids **38** (1977) 1212.
77L8	Lu Hanh, S., Nomura, S.: Article on p. 315 in [77t1].
77L9	Luspin, Y., Hauret, G.: Ferroelectrics **15** (1977) 43.
77M1	Mackowiak, M., Brookeman, J. R.: Article on p. 293 in [77p3].
77M2	Madvaliev, U. M., Sandler, Yu. M., Serikov, V. I., Serdobol'skaya, O. Yu.: Fiz. Tverd. Tela **19** (1977) 900; Sov. Phys. Solid State (English Transl.) **19** (1977) 524.
77M3	Maeda, M., Ikeda, T.: J. Phys. Soc. Jpn. **42** (1977) 1931.
77M5	Makita, Y., Sakurai, F., Osaka, T., Tatsuzaki, I.: J. Phys. Soc. Jpn. **42** (1977) 518.
77M6	Makita, Y., Yamauchi, Y., Suzuki, S.: J. Phys. Soc. Jpn. **43** (1977) 181.
77M7	Malecka, A., Stankowski, J., Shuvalov, L. A.: Ferroelectrics **15** (1977) 1.
77M9	Mamedov, A. M.: Fiz. Tverd. Tela **19** (1977) 845; Sov. Phys. Solid State (English Transl.) **19** (1977) 488.
77M10	Manjunath, C. V., Srinivasan, R.: Phys. Status Solidi (a) **44** (1977) 687.
77M11	Mansingh, A., Srivastava, K. N., Singh, B.: J. Phys. D **10** (1977) 2117.
77M14	Matsuo, T., Suga, H.: Solid State Commun. **21** (1977) 923.
77M15	Mermelstein, M. D., Cummins, H. Z.: Phys. Rev. B **16** (1977) 2177.
77M16	Milla, F., Blinc, R.: J. Chem. Phys. **66** (1977) 1703.
77M17	Milsch, B., Windsch, W.: Phys. Status Solidi (a) **41** (1977) K141.
77M18	Misra, S. K., Giguere, P., Sharp, G. R.: J. Chem. Phys. **66** (1977) 1758.
77M21	Murphy, K. E., Flanagan, B.: J. Chem. Soc., Faraday Trans. I **73** (1977) 1188.
77N2	Nassau, K., Shiever, J. W.: J. Cryst. Growth **42** (1977) 588.
77N4	Nikogosyan, D. N.: Kvantovaya Elektron. **4** (1977) 5; Sov. J. Quantum Electron. (English Transl.) **7** (1977) 1.
77N6	Nitsche, R., Krämer, V., Schuhmacher, M., Bussmann, A.: J. Cryst. Growth **42** (1977) 549.
77N7	Novák, L., Smutný, F., Fousek, J.: Czech. J. Phys. B **27** (1977) 477.
77N8	Nozik, Yu. Z., Fykin, L. E., Muradyan, L. A., Sarin, N. A.: Kristallografiya **22** (1977) 69; Sov. Phys. Crystallogr. (English Transl.) **22** (1977) 37.
77O1	Okada, K., Sugie, H.: Ferroelectrics **17** (1977) 325.

77O2	Olejnik, S., Lukaszewicz, K.: Acta Univ. Wratislav. No. 341 (1977) 43.
77O3	Onodera, A., Shiozaki, Y.: J. Phys. Soc. Jpn. **42** (1977) 1425.
77O4	Osaka, T., Makita, Y., Gesi, K.: J. Phys. Soc. Jpn. **43** (1977) 933.
77O5	Owens, F. J.: Ferroelectrics **16** (1977) 123.
77P1	Péraudeau, G., Ravez, J., Tressaud, A., Hegenmuller, P., Arend, H., Chanussot, G.: Solid State Commun. **23** (1977) 543.
77P2	Perrino, C. T., Van de Ven, J.: J. Solid State Chem. **20** (1977) 201.
77P3	Palau, J. M., Lassabatere, L.: J. Phys. (Paris) **38** (1977) 259.
77P6	Poulet, H., Mathieu, J. P.: Solid State Commun. **21** (1977) 421.
77R4	Rashkovich, L. H., Meteva, K. B., Shevchik, Ya. E., Hoffman, V. G., Mischenko, A. V.: Kristallografiya **22** (1977) 1075; Sov. Phys. Crystallogr. (English Transl.) **22** (1977) 613.
77R6	Reintjes, J., Eckardt, R. C.: IEEE J. Quantum Electron. QE **13** (1977) 791.
77S1	Sader, E., Harnik, E.: J. Appl. Phys. **48** (1977) 3607.
77S2	Salje, E.: Acta Crystallogr. B **33** (1977) 574.
77S3	Salje, E., Bismayer, U.: Opt. Commun. **20** (1977) 303.
77S4	Samara, G. A., Scott, J. F.: Solid State Commun. **21** (1977) 167.
77S5	Samuelsen, E. J., Semmingsen, D.: J. Phys. Chem. Solids **38** (1977) 1275.
77S7	Sato, T.: J. Appl. Phys. **48** (1977) 3120.
77S8	Savatinova, I., Anachtova, E.: Phys. Status Solidi (b) **82** (1977) 677.
77S9	Sawada, A., Ishibashi, Y., Takagi, Y.: J. Phys. Soc. Jpn. **43** (1977) 195.
77S10	Sawada, S., Shiroishi, Y., Yamamoto, A., Takashige, M., Matsuo, M.: J. Phys. Soc. Jpn. **43** (1977) 2099.
77S11	Sawada, S., Shiroishi, Y., Yamamoto, A., Takashige, M., Matsuo, M.: J. Phys. Soc. Jpn. **43** (1977) 2101.
77S13	Schaack, G., Winterfeldt, V.: Ferroelectrics **15** (1977) 35.
77S17	Semmingsen, D., Ellenson, W. D., Frazer, B. C., Shirane, G.: Phys. Rev. Lett. **38** (1977) 1299.
77S18	Semmingsen, D., Hollonder, F. J., Koetzle, T. F.: J. Chem. Phys. **66** (1977) 4405.
77S20	Sharon, M., Kalia, A. K.: J. Solid State Chem. **21** (1977) 171.
77S21	Shevchik, Ya. E., Momtaz, R. Sh., Rashkovich, L. N.: Kristallografiya **22** (1977) 421; Sov. Phys. Crystallogr. (English Transl.) **22** (1977) 240.
77S22	Shigenari, T., Takagi, Y.: J. Phys. Soc. Jpn. **42** (1977) 1650.
77S24	Shimizu, H., Oguri, A., Yasuda, N., Fujimoto, S.: J. Phys. Soc. Jpn. **43** (1977) 349.
77S27	Shiozaki, Y., Koh, S., Sawaguchi, E.: J. Phys. Soc. Jpn. **43** (1977) 721.
77S28	Shiozaki, S., Sawada, A., Ishibashi, Y., Takagi, Y.: J. Phys. Soc. Jpn. **42** (1977) 353.
77S30	Shuldin, Yu. V., Semenchinskii, S. G.: Fiz. Tverd. Tela **19** (1977) 1213; Sov. Phys. Solid State (English Transl.) **19** (1977) 709.
77S31	Smith, W. L.: Appl. Opt. **16** (1977) 1798.
77S34	Spillman, W. B., Leung, R. C., Tornberg, N. E., Lowndes, R. P.: Ferroelectrics **17** (1977) 383.
77S36	Strel'nikova, R. V., Rashkovich, L. N.: Kristallografiya **22** (1977) 844; Sov. Phys. Crystallogr. (English Transl.) **22** (1977) 483.
77S38	Sugai, S., Murase, K., Katayama, S., Takaoka, S., Nishi, S., Kawamura, H.: Solid State Commun. **24** (1977) 407.
77S39	Sugai, S., Murase, K., Kawamura, H.: Solid State Commun. **23** (1977) 127.
77S40	Suski, T.: Acta Univ. Wratislav. No. 341 (1977) 54.
77S41	Suzuki, K., Inoue, K., Shibuya, M.: J. Phys. Soc. Jpn. **43** (1977) 1457.k
77S42	Savatinova, I., Anachkova, E.: Phys. Status Solidi (b) **84** (1977) 401.
77T2	Takenaka, H., Ohnishi, R., Terauchi, H., Sakai, T.: J. Phys. Soc. Jpn. **42** (1977) 929.
77T3	Takoshima, T., Tatsuzaki, I.: J. Phys. Soc. Jpn. **42** (1977) 916.
77T4	Tamura, M., Hagiwara, S., Matsumoto, S., Ono, N.: J. Appl. Phys. **48** (1977) 513.
77T5	Taraskin, S. A., Strukov, B. A., Fedorikhin, V. A., Belugina, N. V., Meleshina, V. A.: Fiz. Tverd. Tela **19** (1977) 2936; Sov. Phys. Solid State (English Transl.) **19** (1977) 1721.
77T7	Tellgren, R., Liminga, R.: Ferroelectrics **15** (1977) 15.
77T9	Tilley, R. J. D.: Nature **269** (1977) 229.
77T11	Tornberg, N. E., Lowndes, R. P.: J. Phys. C **10** (1977) L549.
77T15	Tsukamoto, M., Nakamura, E., Ozaki, T.: J. Phys. Soc. Jpn. **42** (1977) 190.
77U3	Unoki, H., Sakudo, T.: Phys. Rev. Lett. **38** (1977) 137.
77V2	Vazquez, F., Gonzalo, J. A.: Ferroelectrics **16** (1977) 223.
77V3	Vesel, J. E., Torrie, B. H.: Can. J. Phys. **55** (1977) 592.

77W1	Wada, M., Sawada, A., Ishibashi, Y., Takagi, Y.: J. Phys. Soc. Jpn. **42** (1977) 1229.
77W2	Wada, M., Sawada, A., Ishibashi, Y., Takagi, Y.: J. Phys. Soc. Jpn. **43** (1977) 950.
77W3	Wada, M., Uwe, H., Sawada, A., Ishibashi, Y., Takagi, Y., Sakudo, T.: J. Phys. Soc. Jpn. **43** (1977) 544.
77W4	Wang, C. H., Tatsumi, M., Matsuo, T., Suga, H.: J. Chem. Phys. **67** (1977) 3097.
77W5	Was'kowska, A.: Acta Univ. Wratislav. No. 341 (1977) 47.
77W7	Winterfeldt, V., Schaack, G., Klopperpieper, A.: Ferroelectrics **15** (1977) 21.
77Y1	Yagi, T., Tanaka, K., Tatsuzaki, I.: Phys. Rev. Lett. **38** (1977) 609.
77Y2	Yagi, T., Tatsuzaki, I., Tanaka, H.: Article on p. 209 in [77h1].
77Y3	Yamada, N., Kawano, S.: J. Phys. Soc. Jpn. **43** (1977) 1016.
77Y5	Yamamoto, S., Suematsu, Y., Shinnaka, Y.: J. Phys. Soc. Jpn. **43** (1977) 1962.
77Y7	Yoshino, K., Uemoto, T., Balakrishnan, K. G., Yanagida, S., Inuishi, Y.: Article on p. 115 in [70t1].
77Y8	Yoshino, K., Uemoto, T., Inuishi, Y.: Jpn. J. Appl. Phys. **16** (1977) 571.
77Y9	Yuzvak, V. J., Chekmasova, T. J., Aleksandrova, J. P.: Article on p. 293 in [77p3].
77Z4	Zyss, J., Chemla, D. S., Vilminot, S., Cot, L., Maurin, M.: Rev. Phys. Appl. **12** (1977) 1767.
78A1	Abrahams, S. C., Lissalde, F., Bernstein, J. L.: J. Chem. Phys. **68** (1978) 1926.
78A5	Al-Karaghouli, A. R., Abdul-Wahab, B., Ajaj, E., Sequeira, A.: Acta Crystallogr. B **34** (1978) 1040.
78B1	Barb, D., Grecu, N. M., Grecu, V. V., Popescu, E. F.: Chem. Phys. Lett. **56** (1978) 355.
78B2	Bastie, P., Vallade, M., Vettier, C., Zeyen, C. M. E.: Phys. Rev. Lett. **40** (1978) 337.
78B4	Bechthold, P. S., Liebertz, J., Deserno, U.: Opt. Commun. **27** (1978) 393.
78B6	Berdowski, J., Opilski, A.: J. Cryst. Growth **43** (1978) 381.
78B8	Bezdetny, N. M., Zeinally, A. K., Silvestrov, V. G.: Ferroelectrics **18** (1978) 117.
78B9	Brezina, B., Fouskova, A.: Kristal. Tech. **13** (1978) 623.
78B10	Brunner, W., Volkel, G., Windsch, W., Kurkin, I. N., Shlenkin, V. I.: Solid State Commun. **26** (1978) 853.
78B11	Böhm, H.: Z. Kristallogr. **148** (1978) 207.
78B12	Bechtle, D. W., Scott, J. F., Lockwood, D. J.: Phys. Rev. B **18** (1978) 6213.
78C1	Chomnilpan, S., Tellgren, R., Liminga, R.: Acta Crystallogr. B **34** (1978) 373.
78C4	Courtens, E.: Phys. Rev. Lett. **41** (1978) 1171.
78D1	Das, A. N., Ghosh, B.: J. Phys. C **11** (1978) 2483.
78D2	deDussel, H. L., Dewainer, L. S., Debenyacar, M. A. R.: Ferroelectrics **17** (1978) 487.
78D3	de Pater, C. J.: Phys. Status Solidi (a) **48** (1978) 503.
78D4	de Pater, C. J., van Dijk, C.: Phys. Rev. B **18** (1978) 1281.
78D7	Diemer, H. K., Bijvank, E. J., den Hartog, H. W.: Phys. Status Solidi (b) **87** (1978) 697.
78F1	Felix, P., Gamot, P., Lacheau, P., Raverdy, Y.: Ferroelectrics **17** (1978) 543.
78F4	Fujii, Y., Hoshino, S., Sakuma, T.: J. Phys. Soc. Jpn. **45** (1978) 175.
78G1	Galzerani, J. C., Katiyar, R. S.: Article on p. 398 in [78s3].
78G4	Gesi, K.: J. Phys. Soc. Jpn. **45** (1978) 1431.
78G5	Gesi, K., Iizumi, M.: J. Phys. Soc. Jpn. **45** (1978) 1777.
78G6	Gesi, K., Ozawa, K.: J. Phys. Soc. Jpn. **44** (1978) 689.
78G7	Gesi, K., Ozawa, K.: Japan. J. Appl. Phys. **17** (1978) 435.
78G8	Glass, A. M., Abrahams, S. C., Ballman, A. A., Loiacono, G.: Ferroelectrics **17** (1978) 579.
78G11	Grande, S., Mecke, H. D., Idziak, S.: Phys. Status Solidi (a) **45** (1978) K29.
78G12	Gridnev, S. A., Darinskii, B. M.: Phys. Status Solidi (a) **47** (1978) 379.
78G13	Grimm, H., Fitzgerald, W. J.: Acta Crystallogr. A **34** (1978) 268.
78G14	Grimm, H., Fitzgerald, W. J.: J. Phys. C **11** (1978) 829.
78G15	Grossmann, J., Müller, D., Petersson, J.: Phys. Status Solidi (b) **88** (1978) 99.
78G16	Grossmann, J., Müller, D., Petersson, J., Schneider, E.: Z. Phys. B **31** (1978) 187.
78G18	Gridnev, S. A., Postnikov, V. S., Prasolov, B. N., Shuvalov, L. A., Fedosyuk, R. M.: Ferroelectrics **21** (1978) 597.
78H1	Hamano, K., Ema, K., Iwane, Y.: J. Phys. Soc. Jpn. **44** (1978) 933.
78H4	Hatta, I., Shimizu, Y., Hamano, K.: J. Phys. Soc. Jpn. **44** (1978) 1887.
78H5	Haussühl, S., Lieberz, J.: Z. Kristallogr. **148** (1978) 87.
78H10	Hikita, T., Chubachi, Y., Ikeda, T.: J. Phys. Soc. Jpn. **44** (1978) 525.
78H11	Hirotsu, S.: Private communication (1978).
78H12	Hoffman, J., Kuczynski, W., Malecki, J.: Mol. Cryst. Liquid Cryst. **14** (1978) 287.
78H13	Hukuda, K., Nakagawa, Y.: J. Phys. Soc. Jpn. **44** (1978) 1588.
78H14	Hosokawa, T., Kobayashi, J., Uesu, Y., Miyazaki, H.: Ferroelectrics **20** (1978) 201.

78H15	Horvath, J.: Acta Crystallogr. A **34** (1978) Suppl. S. 34.
78I2	Iizumi, M., Gesi, K.: J. Phys. Soc. Jpn. **45** (1978) 711.
78I4	Irie, K.: Ferroelectrics **18** (1978) 231.
78I5	Ivanov, N. R.: Ferroelectrics **21** (1978) 591.
78J1	Jain, V. K.: J. Chem. Phys. **69** (1978) 500.
78J2	Jain, A. K., Upreti, G. C.: Chem. Phys. Lett. **55** (1978) 506.
78J3	Johnson, B. C., Guinn, K. R., Martin, W. E., Fountain, Wm. D.: J. Appl. Phys. **49** (1978) 75.
78J4	Joshi, M. S., Antony, A. V.: J. Mater. Sci. **13** (1978) 939.
78K2	Kasahara, M.: J. Phys. Soc. Jpn. **44** (1978) 537.
78K3	Kasahara, M., Tatsuzaki, I.: J. Phys. Soc. Jpn. **45** (1978) 1296.
78K5	Kawada, S.: J. Phys. Soc. Jpn. **44** (1978) 1881.
78K6	Kawashima, R., Tatsuzaki, I.: Phys. Lett. A **66** (1978) 407.
78K7	Kawashima, R., Tatsuzaki, I.: J. Phys. Soc. Jpn. **44** (1978) 509.
78K8	Kay, M. I.: Ferroelectrics **19** (1978) 151.
78K9	Kay, M. I.: Ferroelectrics **19** (1978) 159.
78K10	Kobayashi, J., Takahashi, T., Hosokawa, T., Uesu, Y.: J. Appl. Phys. **42** (1978) 809.
78K11	Kobayashi, J., Uesu, Y., Miyazaki, H., Hosokawa, T.: Ferroelectrics **21** (1978) 515.
78K12	Kohin, R. P., Žakelj, J., Jain, V. K.: J. Chem. Phys. **68** (1978) 778.
78K16	Kołodziej, H., A., Narewski, E., Sobczyk, L.: Acta Phys. Polon. A **53** (1978) 79.
78K17	Kozlov, G. V., Lebedev, S. P., Volkov, A. A.: Private communication (1978).
78K18	Krämer, V.: Acta Crystallogr. B **34** (1978) 2695.
78K19	Koňák, Č., Smutný, F., Březina, B., Budík, L.: Ferroelectrics **21** (1978) 361.
78K20	Kroupa, J., Petzelt, J., Kozlov, G. V., Volkov, A. A.: Ferroelectrics **21** (1978) 387.
78K21	Kobayashi, J., Uesu, Y., Sandow, T., Sorimachi, H., Shuvalov, L. A., Ivanov, N. R.: Ferroelectrics **20** (1978) 303.
78K22	Krasnikov, V. S., Serebrennikov, V. L.: Kristallografiya **23** (1978) 863; Sov. Phys. Cryst. (English Transl.) **23** (1978) 487.
78L1	Lajzerowicz, J., Legrand, J. F.: Phys. Rev. B **17** (1978) 1438.
78L2	Lettieri, T. R., Brody, E. M., Bassett, W. A.: Solid State Commun. **26** (1978) 235.
78L6	Lavrenčič, B. B., Čopič, M., Zgonik, M., Petzelt, J.: Ferroelectrics **21** (1978) 325.
78L7	Luther, G., Muser, H. E., Levstik, A.: Phys. Status Solidi (a) **50** (1978) 459.
78M1	Maeda, M., Ikeda, T.: J. Phys. Soc. Jpn. **45** (1978) 162.
78M2	Makita, Y., Osaka, T., Miyazaki, A.: J. Phys. Soc. Jpn. **44** (1978) 225.
78M3	Manjunath, C. V., Strinivasan, R.: Phys. Status Solidi (b) **87** (1978) 395.
78M4	Matsuki, K., Ohno, K., Sohma, J.: Jpn. J. Appl. Phys. **17** (1978) 1707.
78M6	McMillan, J. A., Clemens, J. M.: J. Chem. Phys. **68** (1978) 3627.
78M7	Metcalfe, B., Clark, J. B.: Thermochim. Acta **24** (1978) 149.
78M8	Müller, K. A., Berlinger, W.: Z. Phys. B **31** (1978) 151.
78M9	Mullen, D., Heger, G., Treutmann, W.: Z. Kristallogr. **148** (1978) 95.
78M10	Mullen, D., Hellner, E.: Acta Crystallogr. B **34** (1978) 2789.
78M12	Moskalev, A. K., Belobrova, I. A., Aleksandrova, I. P.: Fiz. Tverd. Tela **20** (1978) 3288; Sov. Phys. Solid State (English Transl.) **20** (1978) 1896.
78N1	Nakamura, E., Kajikawa, H.: J. Phys. Soc. Jpn. **44** (1978) 519.
78N2	Nelmes, R. J., Choudhary, R. N. P.: Solid State Commun. **26** (1978) 823.
78N3	Newnham, R. E., Kramer, J. J., Schulze, W. A., Cross, L. E.: J. Appl. Phys. **49** (1978) 6008.
78N4	Nelmes, R. J., Choudhary, R. N. P.: Ferroelectrics **21** (1978) 467.
78N5	Nozik, Yu. Z., Fykin, L. E., Dudarev, V. Ya., Muradyan, L. A., Rostuntseva, A. I.: Kristallografiya **23** (1978) 617; Sov. Phys. Cryst. (English Transl.) **23** (1978) 343.
78O2	Ohi, K., Osaka, J., Uno, H.: J. Phys. Soc. Jpn. **44** (1978) 529.
78O3	Oki, S., Futagami, K., Akashi, Y.: Jpn. J. Appl. Phys. **17** (1978) 23.
78O4	Onodera, A., Fujishita, H., Shiozaki, Y.: Solid State Commun. **27** (1978) 463.
78O5	Onodera, A., Shiozaki, Y.: Ferroelectrics **19** (1978) 23.
78O6	Osaka, T.: J. Phys. Soc. Jpn. **45** (1978) 571.
78O7	Osaka, T., Makita, Y., Gesi, K.: Unpublished work (1978).
78O8	Owens, F. J.: J. Phys. Chem. Solids **39** (1978) 291.
78P4	Pietraszko, A., Shuvalov, L. A.: Ferroelectrics **21** (1978) 573.
78R1	Ramanathan, K. V., Srinivasan, R.: Chem. Phys. Lett. **56** (1978) 359.
78R2	Rao, K. S., Bhanumathi, A.: J. Phys. Soc. Jpn. **44** (1978) 505.

78 R 3	Rapoport, E., Clark, J. B., Richter, P. W.: J. Solid State Chem. **24** (1978) 423.
78 R 4	Ratajczak, H., Baran, J.: J. Raman Spectrosc. **7** (1978) 54.
78 S 1	Sallivan, S., Thomas, E. L.: Opt. Commun. **25** (1978) 125.
78 S 2	Samara, G. A.: Phys. Rev. B **17** (1978) 3020.
78 S 3	Samskog, P. O., Nilsson, G., Lund, A.: J. Chem. Phys. **68** (1978) 4986.
78 S 4	Sathyanarayan, S. G., Narayana, M., Sastry, G. S.: J. Chem. Phys. **68** (1978) 3122.
78 S 5	Sawada, S., Shiroishi, Y., Yamamoto, A., Takashige, M., Matsuo, M.: Phys. Lett. A **67** (1978) 56.
78 S 6	Sawada, S., Shiroishi, Y., Yamamoto, A., Takashige, M., Matsuo, M.: J. Phys. Soc. Jpn. **44** (1978) 687.
78 S 8	Shapira, Y., Levin, S., Gerlich, D., Szapiro, S.: Ferroelectrics **17** (1978) 459.
78 S 9	Singh, G. P., Basu, B. K.: Phys. Lett. A **64** (1978) 425.
78 S 10	Slak, J., Lozar, B., Blinc, R., Shuvalov, L. A.: Phys. Lett. A **64** (1978) 423.
78 S 12	Srinivasan, M. R., Bhat, H. L., Rama Rao, A. H., Narayanan, P. S.: Indian J. Phys. **52** A (1978) 373.
78 S 13	Staples, E. J., Neurgaonkar, R. R., Lim, T. C.: Appl. Phys. Lett. **32** (1978) 197.
78 S 14	Suda, F., Hatano, J., Futama, H.: J. Phys. Soc. Jpn. **44** (1978) 244.
78 S 15	Suzuki, S., Makita, Y.: Acta Crystallogr. B **34** (1978) 732.
78 S 16	Suzuki, S., Oshino, Y., Gesi, K., Makita, Y.: Unpublished work (1978).
78 S 17	Smutny, F., Fousek, J.: Ferroelectrics **21** (1978) 385.
78 S 18	Sandler, Yu. M., Zaitseva, M. P., Sysoev, A. M., Kokorin, Yu. I.: Ferroelectrics **21** (1978) 531.
78 S 19	Sorge, G., Straube, U.: Ferroelectrics **21** (1978) 533.
78 T 1	Takahashi, K., Takagi, M.: J. Phys. Soc. Jpn. **44** (1978) 1266.
78 T 2	Tanaka, H., Yagi, T., Tatsuzaki, I.: J. Phys. Soc. Jpn. **44** (1978) 1257.
78 T 3	Tanaka, H., Yagi, T., Tatsuzaki, I.: J. Phys. Soc. Jpn. **44** (1978) 2009.
78 T 4	Tatsumi, M., Matsuo, T., Suga, H., Seki, S.: J. Phys. Chem. Solids **39** (1978) 427.
78 T 5	Teng, M. K., Massot, M., Balkanski, M.: Phys. Rev. B **17** (1978) 3695.
78 T 6	Tominaga, Y., Nakamura, T.: Solid State Commun. **27** (1978) 1375.
78 T 7	Takashige, M., Hirotsu, S., Sawada, S., Hamano, K.: J. Phys. Soc. Jpn. **45** (1978) 558.
78 T 8	Trontelj, Z., Pirnat, J., Lužnik, J.: Ferroelectrics **20** (1978) 213.
78 U 2	Udagawa, M., Kohn, K., Nakamura, T.: J. Phys. Soc. Jpn. **44** (1978) 1873.
78 U 5	Unruh, H. G., Sailer, E., Hussinger, H., Ayere, O.: Solid State Commun. **25** (1978) 871.
78 V 1	Veerabhadra Rao, K., Narasimhamurty, T. S.: J. Phys. C **11** (1978) 2343.
78 V 2	Vysochanskii, Yu. M., Slivka, V. Yu., Buturlakin, A. P., Gurzan, M. I., Chepur, D. V.: Fiz. Tverd. Tela **20** (1978) 90; Sov. Phys. Solid State (English Transl.) **20** (1978) 49.
78 W 1	Wada, M., Sawada, A., Ishibashi, Y.: J. Phys. Soc. Jpn. **45** (1978) 1429.
78 W 2	Watton, A., Reynhardt, E. C., Petch, H. E.: J. Chem. Phys. **69** (1978) 1263.
78 W 3	Western, A. B., Baker, A. G., Bacon, C. R., Schmidt, V. H.: Phys. Rev. B **17** (1978) 4461.
78 W 4	Windsch, W., Volkel, G.: Ferroelectrics **17** (1978) 491.
78 W 8	Waskowska, A., Olejnik, S., Lukaszewicz, K, Glowiak, T.: Acta Crystallogr. B **34** (1978) 3344.
78 Y 1	Yamada, N., Chubachi, Y., Ikeda, T.: J. Phys. Soc. Jpn. **45** (1978) 1638.
78 Y 2	Yamashita, H., Todo, I., Tatsuzaki, I.: J. Phys. Soc. Jpn. **44** (1978) 1261.
78 Y 4	Yasuda, N., Okamoto, M., Shimizu, H., Fujimoto, S.: Phys. Rev. Lett. **41** (1978) 1311.
78 Y 5	Yasuda, N., Okamoto, M., Shimizu, H., Fujimoto, S., Yoshino, K., Inuishi, Y.: Jpn. J. Appl. Phys. **17** (1978) 437.
78 Y 6	Yasuda, N., Sumi, K., Shimizu, H., Fujimoto, S., Okada, K., Suzuki, I., Sugie, H., Yoshino, K., Inuishi, Y.: J. Phys. C **11** (1978) L299.
78 Y 7	Yoshihara, A., Fujimura, T., Kamiyoshi, K.: J. Phys. Soc. Jpn. **44** (1978) 1241.
78 Y 8	Yoshino, K., Balakrishnan, K. G., Uemoto, T., Iwasaki, Y., Inuishi, Y.: Jpn. J. Appl. Phys. **17** (1978) 597.
78 Z 1	Zametin, V. I., Rabkin, L. M.: Ferroelectrics **18** (1978) 175.
78 Z 2	Zapart, W., Waplak, S., Stankowski, J.: J. Phys. Soc. Jpn. **44** (1978) 1600.
78 Z 3	Zherebtsova, L. I., Koval', T. A., Zaitseva, M. P.: Kristallografiya **23** (1978) 419; Sov. Phys. Crystallogr. (English Transl.) **23** (1978) 234.
79 B 1	Barnes, J. C., Sesay, L. J.: J. Less Common Metals **63** (1979) 301.
79 B2	Blinc, R., Burgar, M., Slak, J., Rutar, V., Milia, F.: Phys. Status Solidi (a) **56** (1979) K65.
79 B 3	Barth, E., Helwig, J., Maier, H.-D., Muser, H. E., Patersson, J.: Z. Phys. B **34** (1979) 393.
79 C 1	Chernenko, V. A., Polanov, I. N.: Fiz. Tverd. Tela **21** (1979) 1846; Sov. Phys. Solid State (English Transl.) **21** (1979) 1059.

79C2	Czapla, Z., Lis, T., Sobczyk, L.: Phys. Status Solidi (a) **51** (1979) 609.
79C3	Chomnilpan, S., Liminga, R., Tellgren, R.: Acta Crystallogr. B **35** (1979) 2692.
79C4	Cox, D. E., Shapiro, S. M., Cowley, R. A.: Phys. Rev. B **19** (1979) 5754.
79D1	Deguchi, K.: J. Phys. Soc. Jpn. **47** (1979) 153.
79D2	de Pater, C. J., Axe, J. D., Currat, R.: Phys. Rev. B **19** (1979) 4684.
79E1	Ema, K., Hamano, K., Ikeda, Y.: J. Phys. Soc. Jpn. **46** (1979) 345.
79E2	Ema, K., Katayama, M., Ikeda, Y., Hamano, K.: J. Phys. Soc. Jpn. **46** (1979) 347.
79F3	Fleury, P. A., Chiang, S., Lyons, K. B.: Solid State Commun. **31** (1979) 279.
79F4	Fernandes, J. R., Ganguly, S., Rao, C. N. R.: Spectrochim. Acta A **35** (1979) 1013.
79G1	Gesi, K., Iizumi, M.: J. Phys. Soc. Jpn. **46** (1979) 697.
79H1	Hangyo, M., Kato, R.: J. Phys. Soc. Jpn. **46** (1979) 349.
79H2	Hikita, T., Yuzuki, Y.: To be published (1979).
79H3	Hirotsu, S., Toyota, K., Hamano, K.: J. Phys. Soc. Jpn. **46** (1979) 1389.
79H4	Hoshizaki, H., Sawada, A., Ishibashi, Y.: J. Phys. Soc. Jpn. **47** (1979) 341.
79H5	Haider, G., Happ, H.: J. Phys. C. (Solid State Phys.) **12** (1979) 1011.
79H6	Hazen, R. M., Finger, L. W.: J. Appl. Phys. **50** (1979) 6826.
79I1	Iwata, Y., Koyano, N., Shibuya, I., Tokunaga, M.: J. Phys. Soc. Jpn. **47** (1979) 922.
79J1	Jain, A. K.: Solid State Commun. **31** (1979) 237.
79J2	Jumeau, D.: J. Phys. Chem. Solids **40** (1979) 669.
79J3	Jain, V. K., Venkateswarlu, P. V.: Chem. Phys. Lett. **62** (1979) 320.
79J4	Jakubas, R., Kolodziej, H. A., Sobezyk, L.: J. Phys. C **12** (1979) L49.
79K1	Kameyama, H., Ishibashi, Y., Takagi, Y.: J. Phys. Soc. Jpn. **47** (1979) 559.
79K2	Kameyama, H., Ishibashi, Y., Takagi, Y.: J. Phys. Soc. Jpn. **46** (1979) 566.
79K3	Krasikov, V. S., Kruglik, A. I.: Fiz. Tverd. Tela **21** (1979) 2834; Sov. Phys. Solid State **21** (1979) 1634.
79K4	Kalliomäki, M. S., Meisalo, V. P. J.: Acta Crystallogr. B **35** (1979) 2829.
79K5	Kumakura, H., Hinode, K., Tanigawa, S., Doyama, M.: Appl. Phys. **19** (1979) 363.
79M1	Mishima, O.: Thesis, Osaka University (1979).
79M2	Massa, N. E., Ullman, F. G., Hardy, J. R.: Solid State Commun. **32** (1979) 1005.
79M3	Milia, F.: Phys. Lett. A **70** (1979) 218.
79M4	Matsuo, T., Mansson, M., Sunner, S.: Acta Chem. Scand. A **33** (1979) 781.
79M5	Mehring, M., Suwelack, D.: Phys. Rev. Lett. **42** (1979) 317.
79N4	Niimori, K., Nakagawa, Y., Hukuda, K.: J. Phys. Soc. Jpn. **46** (1979) 1563.
79O1	Onodera, A., Shiozaki, Y.: J. Phys. Soc. Jpn. **46** (1979) 157.
79O2	Onodera, A., Shiozaki, Y.: To be published (1979).
79O3	Osaka, T., Makita, Y.: Unpublished work (1979).
79O4	Osaka, T., Makita, Y., Gesi, K.: J. Phys. Soc. Jpn. **46** (1979) 577.
79O5	Owens, F. J.: Chem. Phys. Lett. **64** (1979) 116.
79P1	Prettl, W., Rieder, K. H., Nitsche, R. N.: Z. Phys. B **32** (1979) 215.
79P2	Pietraszko, A., Waskowska, A., Olejnik, S., Lukaszewicz, K.: Phase Transitions **1** (1979) 99.
79P3	Poprawski, R., Mroz, J., Czapla, Z., Sobczyk, L.: Acta Phys. Polon. A **55** (1979) 641.
79P4	Pouget, J. P., Shaprio, S. M., Nassau, K.: J. Phys. Chem. Solids **40** (1979) 267.
79S1	Sawada, S.: Kotai Butsuri 14 (1979) **51**.
79S2	Sawada , A., Kikugawa, T., Ishibashi, Y.: J. Phys. Soc. Jpn. **46** (1979) 871.
79S3	Shimizu, H., Yamada, H., Yasuda, N., Fujimoto, S.: Phys. Status Solidi (a) **51** (1979) K121.
79S4	Sumi, T.: Master Thesis, Hiroshima University (1979).
79S5	Shimizu, H., Oguri, A., Abe, N., Yasuda, N., Fujimoto, S., Sawada, S., Shiroishi, Y., Takashige, M.: Solid State Commun. **29** (1979) 125.
79S6	Suzuki, S., Osaka, T., Makita, Y.: J. Phys. Soc. Jpn. **47** (1979) 1741.
79S7	Suzuki, S., Oshino, Y., Gesi, K., Makita, Y.: J. Phys. Soc. Jpn. **47** (1979) 874.
79S8	Sakai, A., Kasahara, M., Tatsuzaki, I.: J. Phys. Soc. Jpn. **47** (1979) 161.
79S9	Samara, G. A.: J. Chem. Phys. **71** (1979) 1401.
79S10	Samuelsen, E. J., Fjaer, E., Semmingsen, D.: J. Phys. C **12** (1979) 2007.
79S11	Suzuki, S.: J. Phys. Soc. Jpn. **47** (1979) 1205.
79T1	Takagi, Y., Shigenari, T.: J. Phys. Soc. Jpn. **47** (1979) 576.
79T2	Takenaka, H., Terauchi, H., Kawamori, A.: J. Phys. Soc. Jpn. **46** (1979) 914.
79T3	Tatsumi, M., Matsuo, T., Suga, H., Seki, S.: Bull. Chem. Soc. Jpn. **52** (1979) 716.
79T4	Tatsumi, M., Matsuo, T., Suga, H., Seki, S.: Bull. Chem. Soc. Jpn. **52** (1979) 728.
79T5	Takoshima, T., Kasahara, M., Tatsuzaki, I.: J. Phys. Soc. Jpn. **46** (1979) 1804.

79V1	Varikash, V. M., Rodin, S. V.: Kristallografiya **24** (1979) 383; Sov. Phys. Cryst. (English Transl.) **24** (1979) 221.
79V2	Vinogradova, I. S., Roslyakov, A. I.: Kristallografiya **24** (1979) **24** (1979) 284; Sov. Phys. Crystallogr. (English Transl.) **24** (1979) 161.
79W1	Wada, M., Sawada, A., Ishibashi, Y.: J. Phys. Soc. Jpn. **47** (1979) 1185.
79Y1	Yamada, M.: Master Thesis, Hokkaido University (1979).
79Y2	Yamada, N.: J. Phys. Soc. Jpn. **46** (1979) 561.
79Y3	Yasuda, N., Fujimoto, S., Asano, T., Shimizu, H., Yoshino, K., Inuishi, Y.: Jpn. J. Appl. Phys. **18** (1979) 1607.
79Y4	Yagi, T., Cho, M., Hidaka, Y.: J. Phys. Soc. Jpn. **46** (1979) 1957.
79Y5	Yamaguchi, T.: Doctor Thesis, Tokyo Institute of Technology (1979).
79Y6	Yasuda, N., Sumi, K., Shimizu, H., Fujimoto, S., Inuishi, Y.: Jpn. J. Appl. Phys. **18** (1979) 1485.
80A1	Athee, M., Hewat, A. W.: Phys. Status Solidi (a) **58** (1980) 525.
80B1	Blinc, R., Juznic, S., Rutar, V., Seliger, J., Zumer, S.: Phys. Rev. Lett. **44** (1980) 609.
80B2	Blinc, R., Rutar, V., Seliger, J., Zumer, S., Rasing, Th., Aleksandrova, I. P.: Solid State Commun. **34** (1980) 895.
80C1	Czapla, Z., Lis, T., Sobczyk, L.: Ferroelectrics **26** (1980) 771.
80F1	Francke, E., Postollec, M. L., Mathieu, J. P., Poulet, H.: Solid State Commun. **33** (1980) 155.
80F2	Fox, D. L., Scott, J. F., Guggenheim, H. J.: Ferroelectrics **25** (1980) 483.
80F3	Fox, D. L., Tilley, D. R., Scott, J. F.: Phys. Rev. B **21** (1980) 2926.
80F4	Fukui, M., Chikushi, S., Abe, R.: J. Phys. Soc. Jpn. **48** (1980) 1196.
80G1	Gesi, K.: J. Phys. Soc. Jpn. **48** (1980) 1399.
80G2	Gesi, K., Tizumi, M.: J. Phys. Soc. Jpn. **48** (1980) 337.
80G3	Gesi, K.: Jpn. J. Appl. Phys. **19** (1980) 1051.
80G4	Gesi, K., Ozawa, K., Osaka, T., Makita, Y.: J. Phys. Soc. Jpn. **49** (1980) 1083.
80H1	Horioka, M., Sawada, A., Abe, R.: Jpn. J. Appl. Phys. **19** (1980) L145.
80H2	Hangyo, M., Kato, R.: J. Phys. Soc. Jpn. **48** (1980) 1209.
80I1	Itoh, K., Kataoka, T., Matsunaga, H., Nakamura, E.: J. Phys. Soc. Jpn. **48** (1980) 1039.
80K1	Kroupa, J., Czapla, Z.: Ferroelectrics **25** (1980) 597.
80K2	Kudo, S., Ikeda, T.: Jpn. J. Appl. Phys. **19** (1980) L 45.
80L1	Lyons, K. B., Negran, T. J., Guggenheim, H. J.: J. Phys. C (Solid State Phys.) **13** (1980) L415.
80M1	Mashiyama, H., Tanisaki, S.: Phys. Lett. A **76** (1980) 347.
80M2	Matsuda, T., Hatta, I.: J. Phys. Soc. Jpn. **48** (1980) 157.
80M3	Milia, F., Voudouris, M.: Phys. Lett. A **76** (1980) 350.
80N1	Nakamura, T., Kasahara, M., Tatsuzaki, I.: J. Phys. Soc. Jpn. **49** (1980) 1429.
80O1	Osaka, T., Makita, Y., Gesi, K.: J. Phys. Soc. Jpn. **49** (1980) 593.
80P1	Poprawski, R., Mroz, J., Czapla, Z.: Acta Phys. Polon. A **57** (1980) 429.
80R1	Rasing, Th., Stoelinga, J. H. M., Wyder, P., Janner, A., Janssen, T.: Ferroelectrics **26** (1980) 707.
80T1	Takashige, M., Nakamura, T., Udagawa, M., Kojima, S., Hirotsu, S., Sawada, S.: J. Phys. Soc. Jpn. **48** (1980) 150.
80Y1	Yamada, N.: Private commun. (1980).
81K1	Kudo, S., Ikeda, T.: Private communication (1981).
81M1	Midorikawa, M.: Private communication (1981).
81O1	Onodera, A., Shiozaki, Y.: Ferroelectrics **31** (1981) 27.

V Combined index of substances for subvolumes a and b

This index lists all substances contained in subvolumes a and b of Vol. III/16. It consists of two parts: part A is for pure compounds [including complex compounds such as $Pb(Co_{1/2}Nb_{1/2})O_3$ or $Na_{0.5}Bi_{4.5}Ti_4O_{15}$] and part B is for solid solutions.

V A: In the 1st column the substances are ordered according to the alphabetically arranged gross formulae. Whenever their names (e.g. Rochelle salt) and/or their abbreviated names (e.g. TGS) are widely used, they are also listed in this column. In the gross formulae the numbers of the elements are determined by simple addition.

Compounds containing water of crystallization are listed twice: (1) H and O of the water molecules are included in the gross formula, (2) the water molecules are attached separately to the gross formula. For instance, Rochelle salt is listed in the 1st column as $C_4H_4KNaO_6 \cdot 4H_2O$, $C_4H_{12}KNaO_{10}$ and also as Rochelle salt, RS, Seignette salt.

Crystals in which H is replaced by D (e.g. KD_2PO_4) are listed separately, though data on their properties are presented in the same section as the data on the non-deuterated crystals (e.g. KH_2PO_4).

V B: In the 1st column the gross formula of each end material of a solid solution (e.g. $BaTiO_3$ or $PbTiO_3$ for the solid solution $BaTiO_3$–$PbTiO_3$) is listed in the same manner as in the case of pure compounds. Thus the solid solution $BaTiO_3$–$PbTiO_3$ can be found in the 2nd column by looking first for either BaO_3Ti or O_3PbTi in the 1st column of part B. If the solid solution is expressed by the formula such as $(Sb_xBi_{1-x})SI$, the end material can be obtained by putting x equal to either 0 or 1. Whenever solid solutions have abbreviated names (e.g. PLZT), these are also listed in the 1st column.

The 3rd column gives the section numbers, and the 4th column gives the first page of each section in subvolumes a and b, respectively.

V A Pure compounds

Gross formula	Chemical formula	Nr.	p. in subvol. a	p. in subvol. b
$BaCo_{1/2}O_3Te_{1/2}$	$Ba(Co_{1/2}Te_{1/2})O_3$	1B2–v	89	
$Ba_2DyNa_3Nb_{10}O_{30}$	$Ba_2Na_3DyNb_{10}O_{30}$	6B–35	188	
$Ba_2EuNa_3Nb_{10}O_{30}$	$Ba_2Na_3EuNb_{10}O_{30}$	6B–35	188	
BaF_4Fe	$BaFeF_4$	43–13–iii		287
BaF_4Mg	$BaMgF_4$	43–13–i		283
BaF_4Mn	$BaMnF_4$	43–13–ii		284
$BaFNa_2Nb_5O_{14}$	$BaNa_2Nb_5O_{14}F$	6B–82	196	
BaF_4Ni	$BaNiF_4$	43–13–v		288
BaF_4Zn	$BaZnF_4$	43–13–vi		288
$Ba_4Fe_2Gd_2Nb_8O_{30}$	$Ba_4Gd_2Fe_2Nb_8O_{30}$	6B–50	190	
$Ba_2Fe_3Nb_7Nd_4O_{30}$	$Ba_2Nd_4Fe_3Nb_7O_{30}$	6B–67	194	
$Ba_4Fe_2Nb_8Nd_2O_{30}$	$Ba_4Nd_2Fe_2Nb_8O_{30}$	6B–50	190	
$Ba_6FeNb_9O_{30}$	$Ba_6FeNb_9O_{30}$	6B–40	189	
$Ba_2Fe_3Nb_7O_{30}Sm_4$	$Ba_2Sm_4Fe_3Nb_7O_{30}$	6B–67	194	
$Ba_4Fe_2Nb_8O_{30}Sm_2$	$Ba_4Sm_2Fe_2Nb_8O_{30}$	6B–50	190	
$Ba_2GdNa_3Nb_{10}O_{30}$	$Ba_2Na_3GdNb_{10}O_{30}$	6B–35	188	
$Ba_3GdNaNb_{10}O_{30}$	$Ba_3NaGdNb_{10}O_{30}$	6B–34	187	
$Ba_2KNb_5O_{15}$	$Ba_2KNb_5O_{15}$	6B–13	183	
$Ba_5KNb_9O_{30}Ti$	$Ba_5KTiNb_9O_{30}$	6B–45	189	
$BaK_5Nb_7O_{30}W_3$	$BaK_5W_3Nb_7O_{30}$	6B–73	195	
$Ba_2KO_{15}Ta_5$	$Ba_2KTa_5O_{15}$	6B–29	186	
$BaLa_2Na_2Nb_{10}O_{30}$	$BaNa_2La_2Nb_{10}O_{30}$	6B–36	188	
$Ba_2LaNa_3Nb_{10}O_{30}$	$Ba_2Na_3LaNb_{10}O_{30}$	6B–35	188	
$Ba_3LaNaNb_{10}O_{30}$	$Ba_3NaLaNb_{10}O_{30}$	6B–34	187	
$Ba_3La_3Nb_5O_{30}Ti_5$	$Ba_3La_3Ti_5Nb_5O_{30}$	6B–80	196	
$Ba_4La_2Nb_6O_{30}Ti_4$	$Ba_4La_2Ti_4Nb_6O_{30}$	6B–77	195	
$Ba_5LaNb_7O_{30}Ti_3$	$Ba_5LaTi_3Nb_7O_{30}$	6B–71	195	
$Ba_4Li_4Na_2Nb_6O_{30}Ti_4$	$Ba_4Li_4Na_2Ti_4Nb_6O_{30}$	6B–79	196	
$Ba_2LiNb_5O_{15}$	$Ba_2LiNb_5O_{15}$	6B–11	179	
$Ba_5LiNb_9O_{30}Ti$	$Ba_5LiTiNb_9O_{30}$	6B–43	189	
$Ba_2LiO_{15}Ta_5$	$Ba_2LiTa_5O_{15}$	6B–27	186	
$Ba_9MgNb_{14}O_{45}$	$Ba_9MgNb_{14}O_{45}$	6B–38	188	
$BaMg_{1/2}O_3Te_{1/2}$	$Ba(Mg_{1/2}Te_{1/2})O_3$	1B2–ii	88	
$BaMn_{1/2}O_3Te_{1/2}$	$Ba(Mn_{1/2}Te_{1/2})O_3$	1B2–iv	89	
$Ba_3MoNa_3Nb_9O_{30}$	$Ba_3Na_3MoNb_9O_{30}$	6B–46	189	
$Ba_2NaNb_5O_{15}$	$Ba_2NaNb_5O_{15}$	6B–12	179	
$Ba_5NaNb_9O_{30}Ti$	$Ba_5NaTiNb_9O_{30}$	6B–44	189	
$Ba_2Na_4Nb_8O_{30}W_2$	$Ba_2Na_4W_2Nb_8O_{30}$	6B–65	194	
$Ba_3Na_3Nb_9O_{30}W$	$Ba_3Na_3WNb_9O_{30}$	6B–48	190	
$Ba_2Na_3Nb_{10}O_{30}Y$	$Ba_2Na_3YNb_{10}O_{30}$	6B–35	188	
$Ba_3NaNb_{10}O_{30}Y$	$Ba_3NaYNb_{10}O_{30}$	6B–34	187	
$Ba_2NaO_{15}Ta_5$	$Ba_2NaTa_5O_{15}$	6B–28	186	
$Ba_2Nb_8Nd_4Ni_2O_{30}$	$Ba_2Nd_4Ni_2Nb_8O_{30}$	6B–53	190	
$Ba_2Nb_8Ni_2O_{30}Sm_4$	$Ba_2Sm_4Ni_2Nb_8O_{30}$	6B–53	190	
$Ba_3Nb_{10}O_{28}$	$Ba_3Nb_{10}O_{28}$	6B–1	173	
$Ba_4Nb_8O_{30}Pb_2Ti_2$	$Ba_4Pb_2Ti_2Nb_8O_{30}$	6B–59	193	
$Ba_2Nb_5O_{15}Rb$	$Ba_2RbNb_5O_{15}$	6B–14	183	
$Ba_4Nb_8O_{30}Sr_2Ti_2$	$Sr_2Ba_4Ti_2Nb_8O_{30}$	6B–58	193	
$BaNb_5O_{15}SrTl$	$SrBaTlNb_5O_{15}$	6B–22	185	
$Ba_6Nb_8O_{30}Ti_2$	$Ba_6Ti_2Nb_8O_{30}$	6B–55	191	
$Ba_6Nb_8O_{30}Zr_2$	$Ba_6Zr_2Nb_8O_{30}$	6B–60	193	
$Ba_6O_{30}Ta_8Ti_2$	$Ba_6Ti_2Ta_8O_{30}$	6B–66	194	
BaO_3Ti	$BaTiO_3$	1A–8	66	
BaO_3Zr	$BaZrO_3$	1A–12	82	
$BeC_6F_4H_{17}N_3O_6$	$(NH_2CH_2COOH)_3 \cdot H_2BeF_4$	36A–3		241
$BeCr_2O_4$	Cr_2BeO_4	43–8		277
$BeF_4H_5LiN_2$	$Li(N_2H_5)BeF_4$	43–17–ii		293

Gross formula	Chemical formula	Nr.	p. in subvol. a	b
$C_4H_{11}N_3O_7$	$(NH_2CH_2COOH)_2 \cdot HNO_3$	38A–1		248
$C_{14}H_{14}N_2O_3$	$C_{14}H_{14}N_2O_3$, *p*-Azoxyanisole	44–1		306
$C_{16}H_{18}N_2O_3$	$C_{16}H_{18}N_2O_3$, *p*-Azoxyphenetole	44–2		307
$C_{31}H_{43}NO_3$	$C_{31}H_{43}O_3N$, DOBAMBC	44–7		307
$C_{33}H_{47}NO_3$	$C_{33}H_{47}O_3N$, DDOBAMBC	44–8		308
$C_{35}H_{51}NO_3$	$C_{35}H_{51}O_3N$, TDOBAMBC	44–9		308
$C_4H_{12}N_2O_8S \cdot H_2O$	$(NH_2CH_2COOH)_2 \cdot H_2SO_4 \cdot H_2O$	43–19		294
$C_4H_{14}N_2O_9S$	$(NH_2CH_2COOH)_2 \cdot H_2SO_4 \cdot H_2O$	43–19		294
$C_6H_{17}N_3O_{10}S$	$(NH_2CH_2COOH)_3 \cdot H_2SO_4$	36A–1		223
$C_2H_{12}N_6O_{10}S_2U \cdot 3H_2O$	$[C(NH_2)_3]_2UO_2(SO_4)_2 \cdot 3H_2O$	43–18		293
$C_2H_{18}N_6O_{13}S_2U$	$[C(NH_2)_3]_2UO_2(SO_4)_2 \cdot 3H_2O$	43–18		293
$CH_6NO_8S_2V \cdot 12H_2O$	$CH_3NH_3V(SO_4)_2 \cdot 12H_2O$	26A–5		161
$CH_6N_3O_8S_2V \cdot 6H_2O$	$C(NH_2)_3V(SO_4)_2 \cdot 6H_2O$	27A–2		169
$CH_{18}N_3O_{14}S_2V$	$C(NH_2)_3V(SO_4)_2 \cdot 6H_2O$	27A–2		169
$CH_{30}NO_{20}S_2V$	$CH_3NH_3V(SO_4)_2 \cdot 12H_2O$	26A–5		161
$C_6H_{17}N_3O_{10}Se$	$(NH_2CH_2COOH)_3 \cdot H_2SeO_4$	36A–2		236
CH_4N_2S	$SC(NH_2)_2$	31A–1		203
$C_4H_4NaO_6Rb \cdot 4H_2O$	$NaRbC_4H_4O_6 \cdot 4H_2O$	41A–3		262
$C_4H_{12}NaO_{10}Rb$	$NaRbC_4H_4O_6 \cdot 4H_2O$	41A–3		262
$C_4H_2O_4$	$C_4H_2O_4$	43–23		297
$C_{10}H_{10}O_3$	$C_{10}H_{10}O_3$, *p*-Methoxycinnamic acid	44–4		307
$C_{11}H_{14}O_3$	$C_{11}H_{14}O_3$, *p*-Butoxybenzoic acid	44–3		307
$C_6K_4MnN_6 \cdot 3H_2O$	$K_4Mn(CN)_6 \cdot 3H_2O$	30A–1		199
$C_6K_4N_6Os \cdot 3H_2O$	$K_4Os(CN)_6 \cdot 3H_2O$	30A–4		202
$C_6K_4N_6Ru \cdot 3H_2O$	$K_4Ru(CN)_6 \cdot 3H_2O$	30A–3		202
$Ca_{1/2}CdO_3Te_{1/2}$	$Cd(Ca_{1/2}Te_{1/2})O_3$	1B2–i	88	
$CaCd_{1/2}O_3Te_{1/2}$	$Ca(Cd_{1/2}Te_{1/2})O_3$	1B2–i	88	
$CaCo_{1/2}O_3Te_{1/2}$	$Ca(Co_{1/2}Te_{1/2})O_3$	1B2–i	88	
$Ca_2H_8N_2O_{12}S_3$	$(NH_4)_2Ca_2(SO_4)_3$	24A–5		142
$CaKNb_5O_{15}Sr$	$CaSrKNb_5O_{15}$	6B–18	185	
$Ca_2K_2O_{12}S_3$	$K_2Ca_2(SO_4)_3$	24A–3		142
$Ca_2KO_{15}Ta_5$	$Ca_2KTa_5O_{15}$	6B–24	186	
$CaMg_{1/2}O_3Te_{1/2}$	$Ca(Mg_{1/2}Te_{1/2})O_3$	1B2–i	88	
$Ca_{1/2}Mn_{1/2}O_6Pb_2Ta$	$Pb_2Ca_{1/2}Mn_{1/2}TaO_6$	7B–4	218	
$CaMn_{1/2}O_3Te_{1/2}$	$Ca(Mn_{1/2}Te_{1/2})O_3$	1B2–i	88	
$Ca_2Nb_2O_7$	$Ca_2Nb_2O_7$	8A–3	225	
$Ca_2Nb_8O_{30}Sr_4Ti_2$	$Ca_2Sr_4Ti_2Nb_8O_{30}$	6B–56	193	
$Ca_{1/2}O_3PbTe_{1/2}$	$Pb(Ca_{1/2}Te_{1/2})O_3$	1B2–viii	89	
$Ca_2O_{12}Rb_2S_3$	$Rb_2Ca_2(SO_4)_3$	24A–4		142
$Ca_{1/2}O_3SrTe_{1/2}$	$Sr(Ca_{1/2}Te_{1/2})O_3$	1B2–i	88	
$Ca_{3/2}O_3Te_{1/2}$	$Ca(Ca_{1/2}Te_{1/2})O_3$	1B2–i	88	
CaO_3Ti	$CaTiO_3$	1A–5	58	
$Ca_3O_8V_2$	$Ca_3(VO_4)_2$	43–12		282
CaO_3Zr	$CaZrO_3$	1A–10	81	
$CdCo_{1/3}Nb_{2/3}O_3$	$Cd(Co_{1/3}Nb_{2/3})O_3$	1B4–ii	100	
$CdCo_{1/2}Nb_{1/2}O_3$	$Cd(Co_{1/2}Nb_{1/2})O_3$	1B3–iv	94	
$CdCo_{1/2}O_3Te_{1/2}$	$Cd(Co_{1/2}Te_{1/2})O_3$	1B2–i	88	
$CdCr_{1/2}Nb_{1/2}O_3$	$Cd(Cr_{1/2}Nb_{1/2})O_3$	1B3–ii	93	
Cd_2CrNbO_6	Cd_2CrNbO_6	7B–2	217	
Cd_2CrO_6Ta	Cd_2CrTaO_6	7B–2	217	
$CdFe_{1/2}Nb_{1/2}O_3$	$Cd(Fe_{1/2}Nb_{1/2})O_3$	1B3–iii	93	
Cd_2FeNbO_6	Cd_2FeNbO_6	7B–2	217	
$CdFe_{1/2}O_3Ta_{1/2}$	$Cd(Fe_{1/2}Ta_{1/2})O_3$	1B3–vi	94	
Cd_2FeO_6Ta	Cd_2FeTaO_6	7B–2	217	
$Cd_2H_8N_2O_{12}S_3$	$(NH_4)_2Cd_2(SO_4)_3$	24A–13		148
$CdHfO_3$	$CdHfO_3$	1A–15	85	
$Cd_2K_2O_{12}S_3$	$K_2Cd_2(SO_4)_3$	24A–10		144

Gross formula	Chemical formula	Nr.	p. in subvol. a	b
$CdMg_{1/3}Nb_{2/3}O_3$	$Cd(Mg_{1/3}Nb_{2/3})O_3$	1B4–i	100	
$CdMg_{1/2}O_3Te_{1/2}$	$Cd(Mg_{1/2}Te_{1/2})O_3$	1B2–i	88	
$Cd_{1/4}Mn_{1/4}Nb_{1/2}O_3Pb$	$Pb(Cd_{1/4}Mn_{1/4}Nb_{1/2})O_3$	1B3–xxiv	100	
$Cd_{1/2}Mn_{1/2}NbO_6Pb_2$	$Pb_2Cd_{1/2}Mn_{1/2}NbO_6$	7B–4	218	
$Cd_{1/2}Mn_{1/2}O_6Pb_2Ta$	$Pb_2Cd_{1/2}Mn_{1/2}TaO_6$	7B–4	218	
$Cd_{1/4}Mn_{1/4}O_3PbW_{1/2}$	$Pb(Cd_{1/4}Mn_{1/4}W_{1/2})O_3$	1B3–xxiv	100	
Cd_2MnO_6Ta	Cd_2MnTaO_6	7B–2	217	
$CdMn_{1/2}O_3Te_{1/2}$	$Cd(Mn_{1/2}Te_{1/2})O_3$	1B2–i	88	
$CdNb_{2/3}Ni_{1/3}O_3$	$Cd(Ni_{1/3}Nb_{2/3})O_3$	1B4–iii	101	
$Cd_2Nb_2O_7$	$Cd_2Nb_2O_7$	7A–1	214	
$Cd_{1/3}Nb_{2/3}O_3Pb$	$Pb(Cd_{1/3}Nb_{2/3})O_3$	1B4–vi	103	
$Cd_{2/3}Nb_{4/3}O_6Pb_2$	$Pb_2(Cd_{2/3}Nb_{4/3})O_6$	7B–6	218	
$Cd_{4/9}Nb_{2/9}O_3PbW_{1/3}$	$Pb(Cd_{4/9}Nb_{2/9}W_{1/3})O_3$	1B6–i	106	
$Cd_2Nb_2O_6S$	$Cd_2Nb_2O_6S$	7A–2	215	
Cd_2NbO_6Sb	Cd_2SbNbO_6	7B–2	217	
$CdNb_{1/2}O_3Sc_{1/2}$	$Cd(Sc_{1/2}Nb_{1/2})O_3$	1B3–i	93	
$Cd_{1/2}O_6Pb_2SnW_{1/2}$	$Pb_2Cd_{1/2}W_{1/2}SnO_6$	7B–4	218	
$Cd_{1/2}O_6Pb_2TaTi_{1/2}$	$Pb_2Cd_{1/2}Ti_{1/2}TaO_6$	7B–4	218	
$Cd_{1/2}O_3PbTe_{1/2}$	$Pb(Cd_{1/2}Te_{1/2})O_3$	1B2–xiii	90	
$Cd_{1/2}O_3PbW_{1/2}$	$Pb(Cd_{1/2}W_{1/2})O_3$	1B2–xv	91	
$Cd_2O_{12}Rb_2S_3$	$Rb_2Cd_2(SO_4)_3$	24A–11		147
$Cd_2O_{12}S_3Tl_2$	$Tl_2Cd_2(SO_4)_3$	24A–12		148
$CdO_3Sc_{1/2}Ta_{1/2}$	$Cd(Sc_{1/2}Ta_{1/2})O_3$	1B3–v	94	
Cd_2O_6ScTa	Cd_2ScTaO_6	7B–2	217	
$Cd_{1/2}O_3SrTe_{1/2}$	$Sr(Cd_{1/2}Te_{1/2})O_3$	1B2–i	88	
$Cd_2O_7Ta_2$	$Cd_2Ta_2O_7$	7A–4	215	
$Cd_{3/2}O_3Te_{1/2}$	$Cd(Cd_{1/2}Te_{1/2})O_3$	1B2–i	88	
$CdO_3Te_{1/2}Zn_{1/2}$	$Cd(Zn_{1/2}Te_{1/2})O_3$	1B2–i	88	
CdO_3Ti	$CdTiO_3$	1A–7	66	
$CeK_2Nb_5O_{15}$	$K_2CeNb_5O_{15}$	6B–5	177	
ClH	HCl	15A–1		47
$Cl_2H_4O_2Sn$	$SnCl_2 \cdot 2H_2O$	43–24		299
ClK	Alkali halides with additives	43–29		302
Cl_4K_2Zn	K_2ZnCl_4	20A–6		123
$Cl_2Sn \cdot 2H_2O$	$SnCl_2 \cdot 2H_2O$	43–24		299
Cl_3Pr	$PrCl_3$	43–14		288
Cl_4Rb_2Zn	Rb_2ZnCl_4	20A–7		123
$Co_2K_2O_{12}S_3$	$K_2Co_2(SO_4)_3$	24A–8		143
Colemanite	$Ca_2B_6O_{11} \cdot 5H_2O$	29A–1		196
$Co_{1/4}Mn_{1/4}Nb_{1/2}O_3Pb$	$Pb(Co_{1/4}Mn_{1/4}Nb_{1/2})O_3$	1B3–xxiv	100	
$Co_{1/4}Mn_{1/4}O_3PbW_{1/2}$	$Pb(Co_{1/4}Mn_{1/4}W_{1/2})O_3$	1B3–xxiv	100	
$Co_{1/3}Nb_{2/3}O_3Pb$	$Pb(Co_{1/3}Nb_{2/3})O_3$	1B4–vii	103	
$Co_{1/2}Nb_{1/2}O_3Pb$	$Pb(Co_{1/2}Nb_{1/2})O_3$	1B3–x	96	
$Co_{1/3}O_3PbTa_{2/3}$	$Pb(Co_{1/3}Ta_{2/3})O_3$	1B4–x	104	
$Co_{1/2}O_3PbTa_{1/2}$	$Pb(Co_{1/2}Ta_{1/2})O_3$	1B3–xix	99	
$Co_{1/2}O_3PbTe_{1/2}$	$Pb(Co_{1/2}Te_{1/2})O_3$	1B2–x	90	
$Co_{1/2}O_3PbW_{1/2}$	$Pb(Co_{1/2}W_{1/2})O_3$	1B2–xvii	92	
$Co_{1/2}O_3SrTe_{1/2}$	$Sr(Co_{1/2}Te_{1/2})O_3$	1B2–i	88	
$Cr_{1/4}Nb_{1/2}O_3PbSc_{1/4}$	$Pb(Sc_{1/4}Cr_{1/4}Nb_{1/2})O_3$	1B3–xxv	100	
CrO_6Pb_2Ta	Pb_2CrTaO_6	7B–3	218	
CsD_2O_4P	CsD_2PO_4	19A–3		86
CsH_2O_4P	CsH_2PO_4	19A–3		86
$CsH_3O_6Se_2$	$CsH_3(SeO_3)_2$	28A–5		190
$CsNO_3$	$CsNO_3$	17A–4		65
$Cu_{2/3}Nb_{4/3}O_6Pb_2$	$Pb_2(Cu_{2/3}Nb_{4/3})O_6$	7B–6	218	

V B Solid solutions

Gross formula	Chemical formula	Nr.	p. in subvol.	
			a	b
$CdMgNb_2O_7$	$CdMgNb_2O_7$-$Cd_2Nb_2O_7$	7C–5	219	
$CdMnNb_2O_7$	$CdMnNb_2O_7$-$Cd_2Nb_2O_7$	7C–5	219	
$CdMnNb_2O_6S$	$CdMnNb_2O_6S$-$Cd_2Nb_2O_6S$	7C–15	220	
$CdNb_2NiO_7$	$CdNiNb_2O_7$-$Cd_2Nb_2O_7$	7C–5	219	
$CdNb_2NiO_6S$	$CdNiNb_2O_6S$-$Cd_2Nb_2O_6S$	7C–15	220	
$CdNb_2O_6$	$CdNb_2O_6$-$NaNbO_3$	1C–a14	111	
$CdNb_2O_6$	$CdNb_2O_6$-$PbNb_2O_6$	6C–7	198	
$CdNb_2O_6$	$CdNb_2O_6$-$NaNbO_3$-$SrNb_2O_6$	1C–c7	138	
$Cd_2Nb_2O_7$	$Cd_2Nb_2O_7$-$Ca_2Nb_2O_7$	7C–2	219	
$Cd_2Nb_2O_7$	$Cd_2Nb_2O_7$-$Cd_{4/3}BiNb_{5/3}O_7$	7C–9	220	
$Cd_2Nb_2O_7$	$Cd_2Nb_2O_7$-$CdBiNbTiO_7$	7C–7	219	
$Cd_2Nb_2O_7$	$Cd_2Nb_2O_7$-$CdBiNb_{5/3}Zn_{1/3}O_7$	7C–8	219	
$Cd_2Nb_2O_7$	$Cd_2Nb_2O_7$-$CdCoNb_2O_7$	7C–5	219	
$Cd_2Nb_2O_7$	$Cd_2Nb_2O_7$-$CdCuNb_2O_7$	7C–5	219	
$Cd_2Nb_2O_7$	$Cd_2Nb_2O_7$-$CdFeNb_2O_7$	7C–5	219	
$Cd_2Nb_2O_7$	$Cd_2Nb_2O_7$-$CdGdTi_2O_6F$	7C–11	220	
$Cd_2Nb_2O_7$	$Cd_2Nb_2O_7$-$CdMgNb_2O_7$	7C–5	219	
$Cd_2Nb_2O_7$	$Cd_2Nb_2O_7$-$CdMnNb_2O_7$	7C–5	219	
$Cd_2Nb_2O_7$	$Cd_2Nb_2O_7$-$Cd_2Nb_2O_6S$	7C–13	220	
$Cd_2Nb_2O_7$	$Cd_2Nb_2O_7$-Cd_2NbSbO_7	7C–6	219	
$Cd_2Nb_2O_7$	$Cd_2Nb_2O_7$-$CdNiNb_2O_7$	7C–5	219	
$Cd_2Nb_2O_7$	$Cd_2Nb_2O_7$-$Cd_2SnO_5F_2$	7C–12	220	
$Cd_2Nb_2O_7$	$Cd_2Nb_2O_7$-$Cd_2Ta_2O_7$	7C–4	219	
$Cd_2Nb_2O_7$	$Cd_2Nb_2O_7$-$CdZnNb_2O_7$	7C-5	219	
$Cd_2Nb_2O_7$	$Cd_2Nb_2O_7$-$CdZnNb_2O_6S$	7C–14	220	
$Cd_2Nb_2O_7$	$Cd_2Nb_2O_7$-$Mg_2Nb_2O_7$	7C–1	219	
$Cd_2Nb_2O_7$	$Cd_2Nb_2O_7$-$NaGdNbTiO_6F$	7C–10	220	
$Cd_2Nb_2O_7$	$Cd_2Nb_2O_7$-$Pb_2Nb_2O_7$	7C–3	219	
$Cd_2Nb_2O_7$	$Cd_2Nb_2O_7$-Bi_3NbO_7-$Zn_2Nb_2O_7$	7C–17	220	
$Cd_{1/3}Nb_{2/3}O_3Pb$	$Pb(Cd_{1/3}Nb_{2/3})O_3$-$PbZrO_3$-$PbTiO_3$	1C–c28	141	
$Cd_2Nb_2O_6S$	$Cd_2Nb_2O_6S$-$CdCoNb_2O_6S$	7C–15	220	
$Cd_2Nb_2O_6S$	$Cd_2Nb_2O_6S$-$CdCuNb_2O_6S$	7C–15	220	
$Cd_2Nb_2O_6S$	$Cd_2Nb_2O_6S$-$CdFeNb_2O_6S$	7C–15	220	
$Cd_2Nb_2O_6S$	$Cd_2Nb_2O_6S$-$CdMnNb_2O_6S$	7C–15	220	
$Cd_2Nb_2O_6S$	$Cd_2Nb_2O_6S$-$Cd_2Nb_2O_7$	7C–13	220	
$Cd_2Nb_2O_6S$	$Cd_2Nb_2O_6S$-$CdNiNb_2O_6S$	7C–15	220	
$Cd_2Nb_2O_6S$	$Cd_2Nb_2O_6S$-$CdZnNb_2O_6S$	7C–15	220	
Cd_2NbO_7Sb	Cd_2NbSbO_7-$Cd_2Nb_2O_7$	7C–6	219	
$CdNb_2O_6SZn$	$CdZnNb_2O_6S$-$Cd_2Nb_2O_7$	7C–14	220	
$CdNb_2O_6SZn$	$CdZnNb_2O_6S$-$Cd_2Nb_2O_6S$	7C–15	220	
$CdNb_2O_7Zn$	$CdZnNb_2O_7$-$Cd_2Nb_2O_7$	7C–5	219	
$Cd_{1/2}O_3PbW_{1/2}$	$Pb(Cd_{1/2}W_{1/2})O_3$-$PbTiO_3$-$PbZrO_3$	1C–c23	140	
$Cd_2O_{12}S_3Tl_2$	$Tl_2Cd_2(SO_4)_3$-$(NH_4)_2Cd_2(SO_4)_3$	24B–1		149
$Cd_2O_7Ta_2$	$Cd_2Ta_2O_7$-$Cd_2Nb_2O_7$	7C–4	219	
CdO_3Ti	$CdTiO_3$-$BaTiO_3$	1C–a32	116	
CdO_3Ti	$CdTiO_3$-$BiFeO_3$	1C–a33	116	
CdO_3Ti	$CdTiO_3$-$LiNbO_3$	1C–a34	116	
CdO_3Ti	$CdTiO_3$-$LiTaO_3$	1C–a35	116	
CdO_3Ti	$CdTiO_3$-$NaTaO_3$	1C–a21	114	
$Ce_{1/2}Li_{1/2}O_3Ti$	$(Li_{1/2}Ce_{1/2})TiO_3$-$PbTiO_3$	1C–b9	131	
$CoLaO_3$	$LaCoO_3$-$PbTiO_3$-$LaMnO_3$	1C–c34	143	
$Co_2Nb_2O_7$	$Co_2Nb_2O_7$-$BaTiO_3$	1C–a49	118	
$Co_{1/3}Nb_{2/3}O_3Pb$	$Pb(Co_{1/3}Nb_{2/3})O_3$-$PbTiO_3$-$PbZrO_3$	1C–c25	141	
$Co_{1/2}O_3PbW_{1/2}$	$Pb(Co_{1/2}W_{1/2})O_3$-$BaTiO_3$	1C–b3	130	
$Co_{1/2}O_3PbW_{1/2}$	$Pb(Co_{1/2}W_{1/2})O_3$-$Pb(Fe_{2/3}W_{1/3})O_3$	1C–b44	136	
$Co_{1/2}O_3PbW_{1/2}$	$Pb(Co_{1/2}W_{1/2})O_3$-$PbTiO_3$	1C–b15	132	
$Co_{1/2}O_3PbW_{1/2}$	$Pb(Co_{1/2}W_{1/2})O_3$-$Pb(Yb_{1/2}Nb_{1/2})O_3$	1C–b43	136	

Gross formula	Chemical formula	Nr.	p. in subvol.	
			a	b
$Nb_{1/2}O_3PbSb_{1/2}$	$Pb(Sb_{1/2}Nb_{1/2})O_3$-$PbTiO_3$-$PbZrO_3$	1C–c19	140	
$Nb_{1/2}O_3PbSc_{1/2}$	$Pb(Sc_{1/2}Nb_{1/2})O_3$-$PbHfO_3$	1C–b33	134	
$Nb_{1/2}O_3PbSc_{1/2}$	$Pb(Sc_{1/2}Nb_{1/2})O_3$-$PbTiO_3$	1C–b16	132	
$Nb_{1/2}O_3PbSc_{1/2}$	$Pb(Sc_{1/2}Nb_{1/2})O_3$-$PbZrO_3$	1C–b29	134	
$Nb_{1/2}O_3PbYb_{1/2}$	$Pb(Yb_{1/2}Nb_{1/2})O_3$-$Pb(Co_{1/2}W_{1/2})O_3$	1C–b43	136	
$Nb_{1/2}O_3PbYb_{1/2}$	$Pb(Yb_{1/2}Nb_{1/2})O_3$-$Pb(Fe_{2/3}W_{1/3})O_3$	1C–b45	136	
$Nb_{2/3}O_3PbZn_{1/3}$	$Pb(Zn_{1/3}Nb_{2/3})O_3$-$Ba(Zn_{1/3}Nb_{2/3})O_3$	1C–b47	136	
$Nb_{2/3}O_3PbZn_{1/3}$	$Pb(Zn_{1/3}Nb_{2/3})O_3$-$PbTiO_3$	1C–b23	133	
$Nb_{2/3}O_3PbZn_{1/3}$	$Pb(Zn_{1/3}Nb_{2/3})O_3$-$Ba(Zn_{1/3}Nb_{2/3})O_3$-$PbTiO_3$	1C–c33	142	
$Nb_{2/3}O_3PbZn_{1/3}$	$Pb(Zn_{1/3}Nb_{2/3})O_3$-$PbTiO_3$-$PbZrO_3$	1C–c27	141	
NbO_3Rb	$RbNbO_3$-$PbNb_2O_6$	6C–4	198	
NbO_3Rb	$RbNbO_3$-$BaNb_2O_6$-$NaNbO_3$	6C–30	206	
NbO_4Sb	$SbNbO_4$-$SbTaO_4$	43–6–i		274
Nb_3O_9Sm	$SmNb_3O_9$-$PbNb_2O_6$	6C–9	200	
Nb_2O_6Sr	$SrNb_2O_6$-$BaNb_2O_6$ (SBN)	6C–26	202	
Nb_2O_6Sr	$SrNb_2O_6$-$NaNbO_3$	1C–a13	111	
Nb_2O_6Sr	$SrNb_2O_6$-$PbNb_2O_6$	6C–6	198	
Nb_2O_6Sr	$SrNb_2O_6$-$SrTiO_3$	1C–a30	116	
Nb_2O_6Sr	$SrNb_2O_6$-$BaNb_2O_6$-$CaNb_2O_6$	6C–25	202	
Nb_2O_6Sr	$SrNb_2O_6$-$BaNb_2O_6$-$PbNb_2O_6$	6C–11	200	
Nb_2O_6Sr	$SrNb_2O_6$-$KNbO_3$-$LaNb_3O_9$	6C–27	205	
Nb_2O_6Sr	$SrNb_2O_6$-$NaNbO_3$-$CdNb_2O_6$	1C–c7	138	
Nb_2O_6Sr	$(Sr, Ba)Nb_2O_6$-$(Sr, Ba)_5Li_2Ti_2Nb_8O_{30}$	6D–30	214	
Nb_2O_6Sr	$SrNb_2O_6$-$SrTa_2O_6$-$BaNb_2O_6$-$BaTa_2O_6$-$PbNb_2O_6$-$PbTa_2O_6$	6C–20	201	
$Nb_2O_7Sr_2$	$Sr_2Nb_2O_7$-$Ba_2Nb_2O_7$	8B–3	229	
$Nb_2O_7Sr_2$	$Sr_2Nb_2O_7$-$Pb_2Nb_2O_7$	8B–4	229	
$Nb_2O_7Sr_2$	$Sr_2Nb_2O_7$-$Sr_2Ta_2O_7$	8B–6	229	
$Nb_2O_7Zn_2$	$Zn_2Nb_2O_7$-$Cd_2Nb_2O_7$-Bi_3NbO_7	7C–17	220	
Nb_3O_9Y	YNb_3O_9-$PbNb_2O_6$	6C–9	200	
$Ni_2O_7Ta_2$	$Ni_2Ta_2O_7$-$BaTiO_3$	1C–a49	118	
NiO_3Ti	$NiTiO_3$-$BaTiO_3$	1C–a45	118	
$O_{11}Pb_5Si_3$	$Pb_5Si_3O_{11}$-$Pb_5Ge_3O_{11}$	11B–2	249	
O_3PbSn	$PbO \cdot SnO_2$-$BaSnO_3$	1C–a79	129	
O_3PbSn	$PbO \cdot SnO_2$-$BaTiO_3$	1C–a40	117	
O_3PbSn	$PbO \cdot SnO_2$-$PbNb_2O_6$	6C–16	201	
O_3PbSn	$PbO \cdot SnO_2$-$PbTiO_3$	1C–a58	126	
O_3PbSn	$PbO \cdot SnO_2$-$PbZrO_3$	1C–a71	128	
O_3PbSn	$PbO \cdot SnO_2$-$PbHfO_3$-$PbTiO_3$	1C–c37	146	
O_3PbSn	$PbO \cdot SnO_2$-$PbTiO_3$-$PbZrO_3$	1C–c16	139	
O_3PbSn	(Pb–Ba) (Sn–Ti)O_3	1C–c6	138	
O_3PbSn	$PbO \cdot SnO_2$-$PbNb_2O_6$-$PbTiO_3$-$PbZrO_3$	1C–c17	140	
O_6PbTa_2	$PbTa_2O_6$-$KTaO_3$	6C–22	201	
O_6PbTa_2	$PbTa_2O_6$-$NaTaO_3$	6C–21	201	
O_6PbTa_2	$PbTa_2O_6$-$PbNb_2O_6$	6C–17	201	
O_6PbTa_2	$PbTa_2O_6$-$PbZrO_3$	1C–a75	129	
O_6PbTa_2	$PbTa_2O_6$-$SrTa_2O_6$	6C–23	201	
O_6PbTa_2	$PbTa_2O_6$-$BaNb_2O_6$-$BaTa_2O_6$-$PbNb_2O_6$	6C–18	201	
O_6PbTa_2	$PbTa_2O_6$-$CaNb_2O_6$-$CaTa_2O_6$-$BaNb_2O_6$-$BaTa_2O_6$-$PbNb_2O_6$	6C–19	201	
O_6PbTa_2	$PbTa_2O_6$-$SrNb_2O_6$-$SrTa_2O_6$-$BaNb_2O_6$-$BaTa_2O_6$-$PbNb_2O_6$	6C–20	201	
O_3PbTi	$PbTiO_3$-$BaNb_{4/5}O_3$	1C–a66	127	
O_3PbTi	$PbTiO_3$-$BaSnO_3$	1C–a57	126	
O_3PbTi	$PbTiO_3$-$BaTiO_3$	1C–a36	116	
O_3PbTi	$PbTiO_3$-$Ba(Zn_{1/3}Nb_{2/3})O_3$	1C–b22	133	

Gross formula	Chemical formula	Nr.	p. in subvol. a	b
$O_3TlW_{1/2}Zr_{1/2}$	$Tl(Zr_{1/2}W_{1/2})O_3$-$PbZrO_3$	1C–b28	134	
O_3W	WO_3-K_2O-Nb_2O_5	6C–35	208	
PLHT	$(Pb, La)_{1-\alpha}(Hf, Ti)_{1-\beta}O_3$	1C–c38	146	
PLZT	$(Pb, La)_{1-\alpha}(Zr, Ti)_{1-\beta}O_3$	1C–c35	143	
PLZT (Solid solution related to PLZT)		1C–c36	146	
PZT	$Pb(Zr_xTi_{1-x})O_3$ ($x = 0.5 \cdots 0.6$)	1C–a54	122	
SBN	$SrNb_2O_6$-$BaNb_2O_6$	6C–26	202	
Stibiotantalite	$Sb(Ta,Nb)O_4$	43–6–i		274
SnTe	SnTe-GeTe	43–30–iii		305